Diagram Symbols

The entries in this table are given roughly in order of appearance in the textbook, but consistency has been maintained in later chapters as well.

Mechanics

Symbol	Meaning
	displacement vector, component displacement
	velocity vector, component velocity
	acceleration vector
	force vector, component force
	momentum vector
	angle marking
	rotational motion

Thermodynamics

Symbol	Meaning
	energy transferred as heat
	energy transferred as work
	cycle or process

Waves and Electromagnetism

Symbol	Meaning
	ray (light or sound)
+	positive charge
−	negative charge
	electric field lines
	electric field vector
	electric current
	magnetic field lines
	magnetic field vector (into page, out of page)

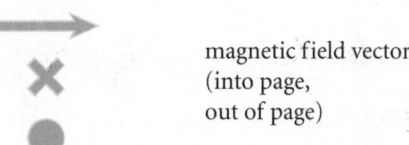

ANNOTATED TEACHER'S EDITION

HOLT
PHYSICS

TEXAS EDITION

SERWAY • FAUGHN

Track Your Students' Mastery
of the TEKS with On-page
Correlations!

Build Student Success
with Tutor CD-ROMs!

See page T20.

Keep Up-to-Date with
Physics Web Resources!

See page T21.

Make Organizing Resources
a Breeze with the New
One-Stop Planner CD-ROM!

See pages T22–T24.

HOLT, RINEHART AND WINSTON

A Harcourt Classroom Education Company

Austin • New York • Orlando • Atlanta • San Francisco • Boston • Dallas • Toronto • London

Authors

Raymond A. Serway
Professor Emeritus
James Madison University

Jerry S. Faughn
Professor Emeritus
Eastern Kentucky University

Cover Photo: © Lawrence Manning/CORBIS
Cover Design: Jason Wilson

Copyright © 2002 by Holt, Rinehart and Winston

All rights reserved. No part of this publication may be reproduced or transmitted in any form or by any means, electronic or mechanical, including photocopy, recording, or any information storage and retrieval system, without permission in writing from the publisher.

Requests for permission to make copies of any part of the work should be mailed to the following address: Permissions Department, Holt, Rinehart and Winston, 10801 North MoPac Expressway, Austin, Texas 78759.

For permission to reprint copyrighted material, grateful acknowledgment is made to the following sources:

sciLINKS is owned and provided by the National Science Teacher's Association. All rights reserved.

The name of the **Smithsonian Institution** and the sunburst logo are registered trademarks of the Smithsonian Institution. The copyright in the Smithsonian website and Smithsonian website pages are owned by the Smithsonian Institution. All other material owned and provided by Holt, Rinehart and Winston under copyright appearing above.

Copyright © 2000 **CNN** and **CNNfyi.com** are trademarks of Cable News Network LP, LLLP, a Time Warner Company. All rights reserved. Copyright © 2000 Turner Learning logos are trademarks of Turner Learning, Inc., a Time Warner Company. All rights reserved.

CBL, Calculator-Based Laboratory, and TI-GRAPH LINK are trademarks of Texas Instruments.

Printed in the United States of America
ISBN 0-03-064922-6
2 3 4 5 6 048 03 02 01

Acknowledgments

Contributing Writers

Robert W. Avakian
Instructor
Trinity School
Midland, TX

Phillip G. Bunce
James Bowie High School
Austin, TX

Robert Davisson
Science Writer
Delaware, OH

Judith Edgington, Ph.D.
Physics/Science Education
Consultant and Curriculum Designer
Austin, TX

John Jewett Jr., Ph.D.
Professor of Physics
Physics Department
California State Polytechnic University
Pomona, CA

Seth Madej
Associate Producer
Pulse of the Planet radio series
Jim Metzner Productions, Inc.
Yorktown Heights, NY

Jim Metzner
Pulse of the Planet radio series
Yorktown Heights, NY

John M. Stokes
Science Writer
El Paso, TX

Salvatore Tocci
Science Writer
East Hampton, NY

Lab Reviewers

Christopher Barnett
Richard DeCoster
Elizabeth Ramsayer
Joseph Serpico
Niles West High School
Niles, IL

Gregory Puskar
Laboratory Manager
Physics Department
West Virginia University
Morgantown, WV

Richard Sorensen
Vernier Software & Technology
Beaverton, OR

Lee Senholtz
Martin Taylor
Sargent-Welch/VWR
Buffalo Grove, IL

Text Reviewers

John Adamowski
Chairperson
Science Department
Fenton High School
Bensenville, IL

John Ahlquist, M.S.
Chemistry/Physics teacher
Science Department
Anoka High School
Anoka, MN

Stephen D. Baker
Physics Department
Rice University
Houston, TX

Maurice Belanger
Science Department Head
Nashua High School
Nashua, NH

Larry G. Brown
Science and Computer Science
  Instructor
Science Department
Morgan Park Academy
Chicago, IL

David S. Coco, Ph. D.
Senior Research Physicist
Applied Research Laboratories
University of Texas at Austin
Austin, TX

William K. Conway, Ph.D.
Advanced Placement Physics Teacher
Science Department
Lake Forest High School
Lake Forest, IL

Jack Cooper
Physics Teacher
Ennis High School
Ennis, TX

Brad de Young
Department of Physics & Physical
  Oceanography
Memorial University
St. John's, Newfoundland, Canada

Bill Deutschmann, Ph.D.
President
Oregon Laser Consultants
Klamath Falls, OR

Fred E. Ellis, Ph.D.
Physics & Geology
University of Texas-Pan American
Edinburg, TX

William D. Ellis
Advanced Placement Physics
  Instructor
Chairman: Science Department
Butler Senior High School
Butler, PA

Diego Enciso
Troy, MI

Bruce Esser
Physics Teacher
Marian High School
Omaha, NE

Arthur A. Few
Professor of Space Physics and
  Environmental Science
Rice University
Houston, TX

Scott Fricke, Ph.D.
Schlumberger Oilfield Services
Sugarland, TX

David J. Hamilton, Ed.D.
Physics Teacher
Benjamin Franklin High School
Portland, OR

Roy W. Hann, Jr., Ph.D.
Professor of Civil Engineering
Texas A & M University
College Station, TX

Sally Hicks, Ph.D.
Department of Physics
University of Dallas
Irving, TX

J. Philip Holden, Ph.D.
Physics Education Consultant
Michigan Dept. of Education
Lansing, MI

John L. Hubisz, Ph.D.
Physics Department
North Carolina State University
Raleigh, NC

Robert C. Hudson
Associate Professor Emeritus
Physics Department
Roanoke College
Salem, VA

William H. Ingham, Ph.D.
Professor
Department of Physics
James Madison University
Harrisonburg, VA

Douglas C. Jenkins
Chairman
Science Department
Warren Central High School
Bowling Green, KY

David S. Jones
Physics Teacher
Miami Sunset Senior High School
Miami, FL

Roger Kassebaum
Physics Department
Millard North High School
Omaha, NE

Mervin W. Koehlinger, M.S.
Physics Instructor
Concordia Lutheran High School
Fort Wayne, IN

Karen B. Kwitter, Ph.D.
Professor of Astronomy
Williams College
Williamstown, MA

Phillip LaRoe
Professor of Physics
Helena College of Technology
Helena, MT

William Lash
Westwood High School
Round Rock, TX

Norman A. Mankins
Science Curriculum Specialist
Canton City Schools
Canton, OH

Joseph McClure, Ph.D.
Associate Professor Emeritus
Department of Physics
Georgetown University
Washington, DC

John McGehee
Physics Teacher
Palos Verdes Peninsula High School
Rolling Hills Estates, CA

Ralph McGrew
Engineering Science Department
Broome Community College
Binghamton, NY

Clement J. Moses, Ph. D.
Associate Professor of Physics
Physics Department
Utica College
Utica, NY

Richard L. Pitter, Ph.D.

Alvin M. Saperstein, Ph.D.
Professor of Physics and Fellow of
  Center for Peace and
  Conflict Studies
Department of Physics
  and Astronomy
Wayne State University
Detroit, MI

Debra Schell
Physics Teacher
Austintown Fitch High School
Austintown, OH

Edward Schweber
Physics Teacher
Solomon Schechter Day School
West Orange, NJ

Joseph A. Taylor
Physics Instructor
Middletown Area High School
Middletown, PA

Leonard L. Thompson
Physics and Meteorology Teacher
Science Department
North Allegheny Sr. High School
Wexford, PA

Thomas H. Troland, Ph.D.
Physics Department
University of Kentucky
Lexington, KY

John T. Vieira
Science Department Head
B.M.C. Durfee High School
Fall River, MA

Mary L. White
Coastal Ecology Institute
Louisiana State University
Baton Rouge, LA

Jerome Williams M.S.
Professor Emeritus
Oceanography Department
US Naval Academy
Annapolis, MD

Tim Wright
Physics Instructor
Stevens Point Area Senior High
  School
Stevens Point, WI

Carol J. Zimmerman, Ph.D.
Exxon Exploration Company
Houston, TX

G. Patrick Zober
Science Curriculum Coordinator
Yough Senior High School
Herminie, PA

Patricia J. Zober
APC Physics Teacher
Ringgold High School
Monongahela, PA

Table of Contents In Brief

Chapters

Contents

5120 N

1520 N   950 N

4050 N

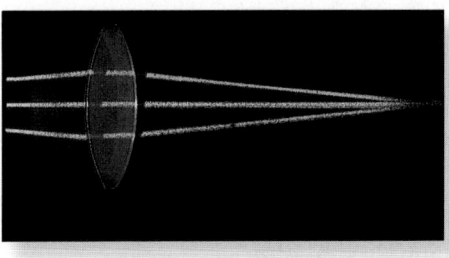

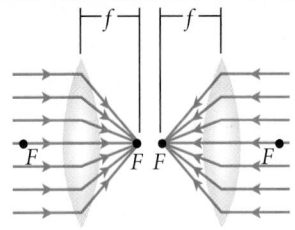

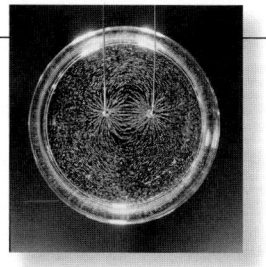
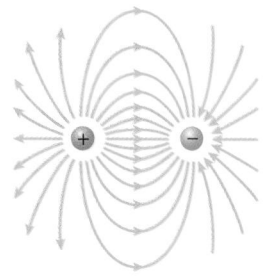

Science, Technology and Society: Electromagnetic
Fields: Can They Affect Your Health?   790

22 Induction and Alternating Current   792

23 Atomic Physics   828

Timelines—Physics and Its World: 1890–1950   862

24 Modern Electronics   864

Science, Technology, and Society: What Can We Do
With Nuclear Waste?   892

Potential difference   Atomic gas   Slit   Prism   Viewing screen

25 Subatomic Physics 894

Diagram symbols

Positive charge	$\oplus$ $+q$
Negative charge	$\ominus$ $-q$
Electric field vector	E
Electric field lines	

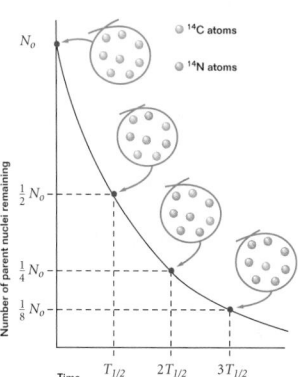

Reference Section 934

---

internet connect

This textbook contains the following on-line resources to help you make the most of your science experience.

Visit **go.hrw.com** for extra help and study aids matched to your textbook. Just type in the keyword HF2 HOME.

Visit **www.scilinks.org** to find resources specific to topics in your textbook. Keywords appear throughout your book to take you further.

Smithsonian Institution
Internet Connections
Visit **www.si.edu/hrw** for specifically chosen on-line materials from one of our nation's premier science museums.

Visit **www.cnnfyi.com** for late-breaking news and current events stories selected just for you.

Features

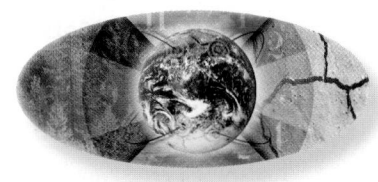

Laboratory Experiments

- Planning information and equipment notes for each lab are in the Teacher's Interleaves at the beginning of each chapter.

- Teacher background about the procedures and answers are in the extended margin of the Teacher's Edition of the lab pages.

- Lab safety information for the teacher is provided on p. T45.

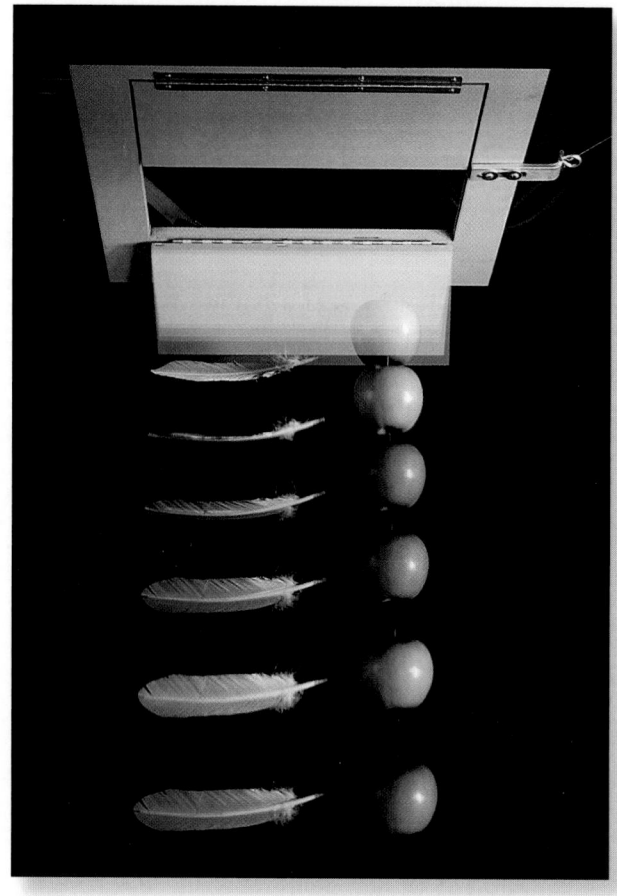

Quick Labs

HOLT PHYSICS by Serway and Faughn

The Best of Both Worlds

Until now, you've had to choose between a conceptual approach with little quantitative problem-solving, or a mathematically-based presentation with little development and assessment of students' understanding of the concepts behind the physics. But either approach by itself is not enough to build the true physics understanding necessary for success. Only *Holt Physics* delivers both.

Why we wrote this book:

As a high school teacher, you face challenges in preparing your students to understand the world around them. You also want to make your class as inviting, interesting, and inclusive as possible. We wanted to write the book that was both "user friendly" and one that would help you and your students achieve these goals.

Get the Physics Right First and foremost, we wanted to give you a book that was technically correct and one that provided good preparation for college. Our previous experience writing *College Physics* gave us the background we needed to write an authoritative, accurate, and up-to-date text that is appropriate for today's students.

Link Concepts and Problem-Solving Students need clear conceptual development and plenty of practice working with both fundamental physical concepts and problem-solving skills. We wanted this book to help students with both.

Focus on the Diagram Learning how to prepare an accurate and informative diagram for a situation is a crucial step that identifies the connection between the concrete world and the world of physics. We wanted to provide an abundance of support in preparing and interpreting such diagrams to sharpen students' skills.

Relate to the Student The best way to ensure learning that lasts is through practical applications and concrete examples that students can relate to and appreciate. Therefore, we wanted a book filled with examples from the text presentation to questions, problems, and other features.

Without a doubt, the most important elements in any learning environment are you, the instructor, and effective communication between you and your students. If you are excited, knowledgeable, and interested in what you teach, and convey this effectively, you will be very successful in the classroom. We applaud your contributions to the world and to our future, and we wish you and your students much success.

Regards,

Ray Serway   *Jerry Faughn*

What's New for This Edition

INTERNET RESOURCES (see p. T21)

Visit **www.scilinks.org** for Web resources specific to the topics in your textbook.

Smithsonian Institution®

Visit **www.si.edu/hrw** for specially chosen on-line materials from the Smithsonian Institution.

Visit **go.hrw.com** for extra help and study aids matched to your textbook.

Visit **www.cnnfyi.com** for late-breaking news and current-events stories selected just for you.

MEDIA RESOURCES (see p. T20–T21)

Holt Physics Interactive Tutor CD-ROM, the only interactive learning environment devoted to teaching problem-solving skills and strategies

CNN Presents: Physical Science, a videotape of news stories about physics topics

EXPANDED PROBLEM-SOLVING STRAND (see p. T34)

- There are more problems (2800) and a better selection than any other program offers.

- Practice sets start with items simply following a model and gradually increase in difficulty.

- Practice Guides in the ATE margin categorize all available problems in the textbook, *Problem Workbook,* and *One-Stop Planner CD-ROM* by the variable sought.

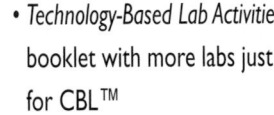

LAB ACTIVITIES FOR CBL™ TECHNOLOGY (see p. T28)

- Many alternative Calculator-Based Laboratory™ procedures given on the student textbook page

- *Technology-Based Lab Activities* booklet with more labs just for CBL™

ONE-STOP PLANNER CD-ROM (see p. T22)

Planning lessons, customizing lesson plans and tests, and organizing worksheets, labs, and other material is now a snap!

PHYSICS Program Resources

Complete and Easy to Use

All the core resources you need for effective teaching

Laboratory Experiments

contains 37 additional labs, including Discovery Labs for basic concept development, and Invention Labs for student-designed inquiry experiences. The companion Teacher's Edition includes all pages from this booklet as well as teacher's notes about equipment, safety, evaluating student work, clean-up, and disposal. **See pages T30 and T31.**

Technology-Based Lab Activities

provides lab experiments that integrate Calculator-Based Laboratory™ (CBL™) technology. The companion Teacher's Edition includes all pages from this booklet as well as complete teacher notes and resources. **See pages T28 and T29.**

Problem Workbook

includes an additional worked-out sample solving for another variable for each of the 97 problem types in the textbook, and practice problems different from those in the text.

Section Review Worksheets

target key skills and concepts from each section of the book. Also includes a 2-page mixed review worksheet for every chapter.

 **Interactive Tutor CD-ROMs**

explore concepts and problem-solving skills as students practice interactively with to 20 key physics topics. Printable worksheets and problem sets linked to each topic reinforce understanding. **See page T20.**

Teaching Transparencies

help you use the text's key illustrations and diagrams as visual teaching tools. The package includes 128 full-color transparencies from the textbook, along with 128 blackline transparency masters which include tables, diagrams, and related information.

CNN Presents Science in the News: Physical Science

videotape portrays the relevance of physics with current events stories and improves critical-thinking and viewing skills with accompanying work sheets. **See page T21.**

Assessment Item Listing

contains a hard copy print-out of all of the items in the test generator.

One-Stop Planner CD-ROM

This CD-ROM, which can be used by either Macintosh® or Windows® users, includes all of the program resources in printable format, as well as customizable lesson plans, and an easy-to-use test generator with more than 2000 items. (**See pages T22–T24.**)

HOLT SCIENCE
SKILLS WORKSHOP: READING IN THE CONTENT AREA

(**not shown**) trains your students in the specialized skills that make reading science different than reading a short story. (See page T33.)

HOLT SCIENCE
LABORATORY MANAGER'S PROFESSIONAL REFERENCE

(**not shown**) gives you ready-to-use checklists and information on the basics of risk management, typical hazards, and pertinent standards and regulations.

 PHYSICS CLASSIC

The Mechanical Universe High School Adaptation

Some sections in Holt Physics are correlated in the planning charts to The Mechanical Universe/High School Adaptation. This NSF-sponsored project provides smaller, more affordable videotaped modules of the original award-winning 52-program series, specifically selected for use in high school classrooms.

For information, contact:

INTELECOM
150 East Colorado Blvd., #300
Pasadena, CA 91105-1937
telephone: 626-796-7300
fax: 626-577-4282
e-mail: customerservice@intelecom.org

HOLT PHYSICS Interactive Tutor

One-on-One Instruction

Integrating physics concepts with interactive problem-solving practice

Only the *Holt Physics Interactive Tutor* provides an interactive environment in which students can explore concepts and practice problem-solving techniques. As students practice, their errors are gently pointed out and explained immediately—just as you would if you could work individually with every student.

Each module recaps a concept. Then, the Interactive Tutor demonstrates a step for solving a specific type of problem and explains its conceptual basis. Finally, the student is challenged to interact to do the same for a different problem situation. The cycle is repeated until the student achieves success. Then the student can print and complete worksheets and a problem set so you can assess their understanding.

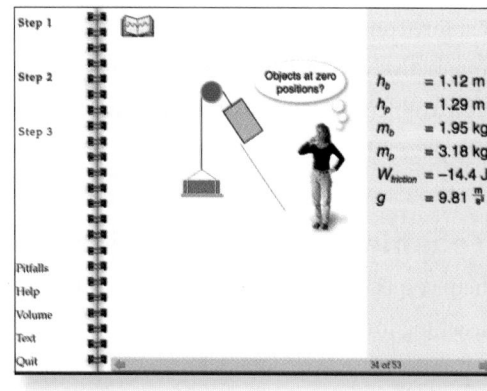

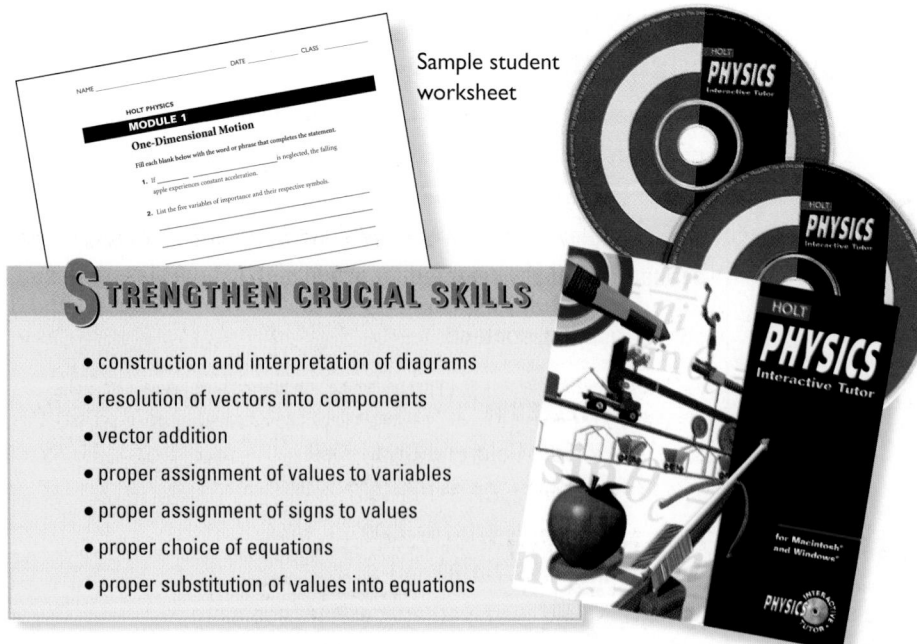

Sample student worksheet

STRENGTHEN CRUCIAL SKILLS

- construction and interpretation of diagrams
- resolution of vectors into components
- vector addition
- proper assignment of values to variables
- proper assignment of signs to values
- proper choice of equations
- proper substitution of values into equations

CD 1 Modules

1 **One-Dimensional Motion**
2 **Vectors**
3 **Two-Dimensional Motion**
4 **Net Force**
5 **Work**
6 **Work–Kinetic Energy Theorem**
7 **Conservation of Momentum**
8 **Angular Kinematics**
9 **Torque**
10 **Rotational Inertia**

CD 2 Modules

11 **Hooke's Law**
12 **Frequency and Wavelength**
13 **Doppler Effect**
14 **Reflection**
15 **Refraction**
16 **Force Between Charges**
17 **Electrical Circuits**
18 **Magnetic Field of a Wire**
19 **Magnetic Force on a Wire**
20 **Induction and Transformers**

HOLT PHYSICS and the Internet

Finding Value in the Vastness of Cyberspace

On-line resources do the legwork to provide up-to-date information for you and your students.

internet connect

SCiLINKS NSTA

TOPIC: Archimedes
GO TO: www.scilinks.org
sciLINKS CODE: HF2091

Instant access to web content selected specifically for *Holt Physics*

• All *sci*LINKS topics shown in icons throughout the text are managed and monitored by staff from the National Science Teachers Association.

• Sites are selected by teachers for appropriate content and grade level to match your text.

• Sites are continuously added and deleted from the system.

• Students don't end up at dead ends, dark sites, or sites under construction.

• *sci*LINKS pulls sites from all over the world to give a global view of the dynamic nature of science communication.

Downloadable resources for *Holt Physics* from Holt, Rinehart and Winston

Visit **www.go.hrw.com** to stay up to date as additional resources are developed for *Holt Physics*. Look for content updates to keep you current, additional articles, activities, teaching suggestions, lab tips, and other resources.

> It includes a link to a Web-based homework grading system that contains thousands of *Holt Physics* practice problems with randomized values. It also saves you time by customizing student assignments and grading the work for you!

Web materials from the Smithsonian Institution

The Smithsonian Institution maintains special Web sites for use with *Holt Physics*. Visit **www.si.edu/hrw** for a complete list of these resources. You can find interactive exhibits, classroom activities, interviews with scientists, and a variety of application and extension topics.

Smithsonian Institution
Internet Connections

Web and video resources from CNN show relevance

CNN fyi.com

CNN
Turner Le@rning
A Time Warner Company

Visit **www.cnnfyi.com** for late-breaking and archived news and current-events stories selected just for you. Content is updated frequently, and a special teacher's section provides detailed lesson plans that show you how to integrate news reports from CNN and other resources from the Web site into your classroom.

The *CNN Presents Science in the News: Physical Science* video package includes more than 20 short video clips that show the newsworthy nature of physics. The clips make great warm-up activities to start lessons. Teacher's notes and Critical Thinking Worksheets for the student reinforce the physics in each story.

Video News Features include:

Amusement Park Physics	Trebuchet Catapult
Crash-Test Dummies	Energy-Saving House
Egg-Drop Contest	Color-Deficiency Lenses

Holt PHYSICS

One-Stop Planner CD-ROM

Saves You Time

One easy-to-carry CD-ROM makes managing lessons and resources as easy as point and click!

Printable Teaching Resources

• The entire package is designed to be simple to navigate. Just point the mouse to the chapter and section you're covering, and click! Easy-to-understand menus put you just one more click away from previews of any resource you might need. **One CD-ROM is all you need when planning lessons, whether at home or at school!**

• You can print the previews you choose directly from the CD-ROM and photocopy them for classroom use. **You will no longer fumble with stacks of booklets and lists of page numbers when making photocopies for your classroom!**

• Resources can also be viewed by category, such as labs or worksheets. **You will have no need to carry the entire booklet home to know what's coming up later in the year!**

• You also get additional resources not found in printed format, such as **nearly 1,000 additional practice problems** and detailed solutions for all practice problems. **You won't have to spend all your free time making up more problems or writing solutions!**

Editable Lesson Plans

• All program resources are integrated into the suggested lesson plans with hyperlinks that show you previews. **You'll know exactly how to organize and when to use all the resources!**

• Lesson plans are arranged to show both traditional and block schedules. **You can hit the ground running, whatever your situation!**

• Suggested lesson plans are available as PDF files and as editable word processing documents. **You can easily customize lesson plans to match your calendar and your school district's requirements!**

Easy-to-Use Test Generator

• The Test Generator can be used to create customized tests from a bank of over 2,000 items. Each item is correlated to the section objectives from the textbook, and you can easily add your own questions. **You get a great foundation to build on that you can improve year after year!**

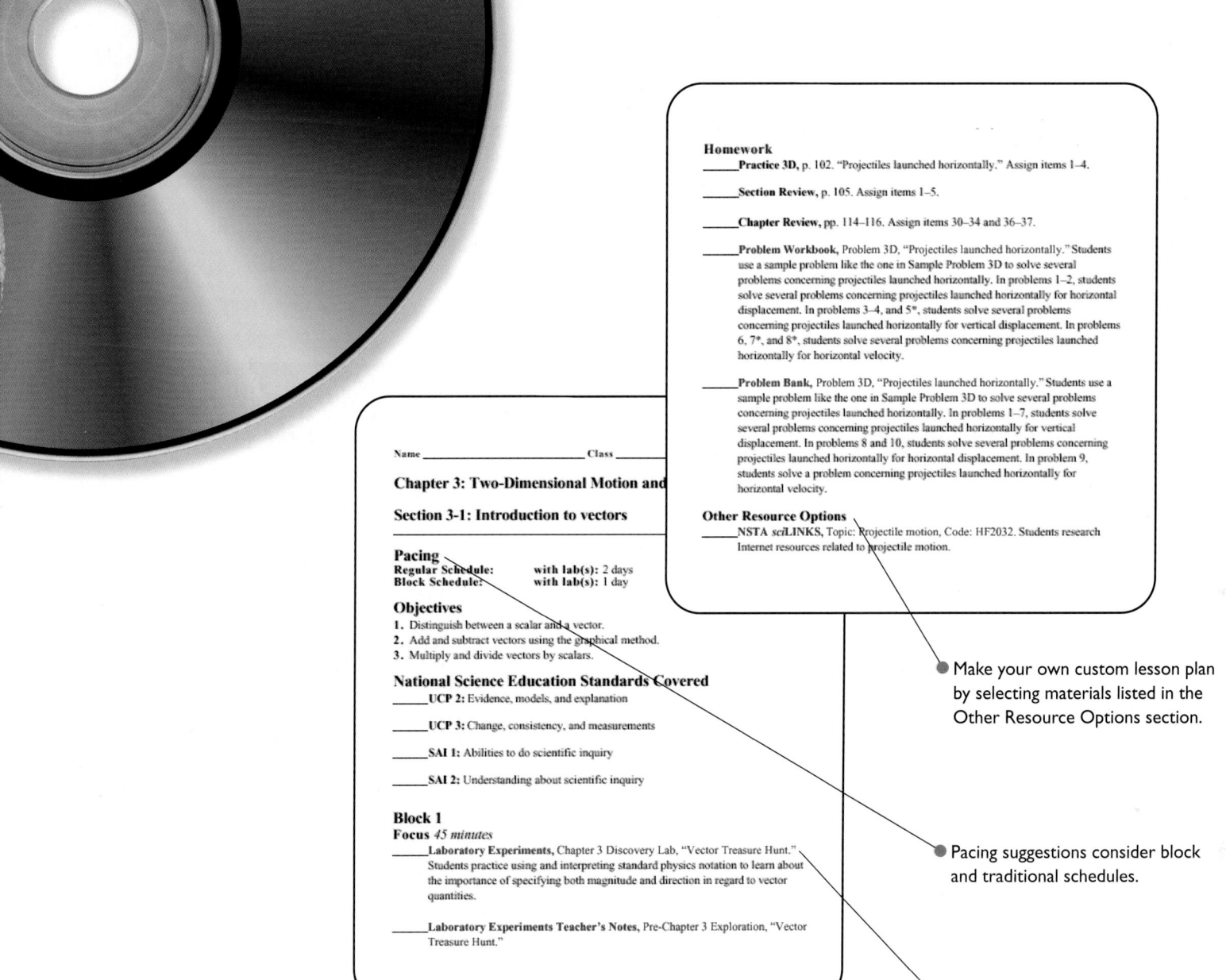

Homework

_____**Practice 3D,** p. 102. "Projectiles launched horizontally." Assign items 1–4.

_____**Section Review,** p. 105. Assign items 1–5.

_____**Chapter Review,** pp. 114–116. Assign items 30–34 and 36–37.

_____**Problem Workbook,** Problem 3D, "Projectiles launched horizontally." Students use a sample problem like the one in Sample Problem 3D to solve several problems concerning projectiles launched horizontally. In problems 1–2, students solve several problems concerning projectiles launched horizontally for horizontal displacement. In problems 3–4, and 5*, students solve several problems concerning projectiles launched horizontally for vertical displacement. In problems 6, 7*, and 8*, students solve several problems concerning projectiles launched horizontally for horizontal velocity.

_____**Problem Bank,** Problem 3D, "Projectiles launched horizontally." Students use a sample problem like the one in Sample Problem 3D to solve several problems concerning projectiles launched horizontally. In problems 1–7, students solve several problems concerning projectiles launched horizontally for vertical displacement. In problems 8 and 10, students solve several problems concerning projectiles launched horizontally for horizontal displacement. In problem 9, students solve a problem concerning projectiles launched horizontally for horizontal velocity.

Other Resource Options

_____**NSTA *sci*LINKS,** Topic: Projectile motion, Code: HF2032. Students research Internet resources related to projectile motion.

Name _____ Class _____

Chapter 3: Two-Dimensional Motion and

Section 3-1: Introduction to vectors

Pacing
Regular Schedule: **with lab(s):** 2 days
Block Schedule: **with lab(s):** 1 day

Objectives
1. Distinguish between a scalar and a vector.
2. Add and subtract vectors using the graphical method.
3. Multiply and divide vectors by scalars.

National Science Education Standards Covered
_____**UCP 2:** Evidence, models, and explanation

_____**UCP 3:** Change, consistency, and measurements

_____**SAI 1:** Abilities to do scientific inquiry

_____**SAI 2:** Understanding about scientific inquiry

Block 1
Focus *45 minutes*
_____**Laboratory Experiments,** Chapter 3 Discovery Lab, "Vector Treasure Hunt." Students practice using and interpreting standard physics notation to learn about the importance of specifying both magnitude and direction in regard to vector quantities.

_____**Laboratory Experiments Teacher's Notes,** Pre-Chapter 3 Exploration, "Vector Treasure Hunt."

● Make your own custom lesson plan by selecting materials listed in the Other Resource Options section.

● Pacing suggestions consider block and traditional schedules.

● Titles that appear in blue are hot-linked to the actual resource. Click to view any resource or to print any worksheet.

LESSON PLAN FORMATS

Microsoft Word
Microsoft Works
WordPerfect (Windows only)
ClarisWorks
PDF

Resources available on the *One-Stop Planner CD-ROM* include:

Teaching Resources

Lesson Plans
Teacher's Solution Manual and Answer Key*
Graphing Calculator Keystroking Guide* (see page T25)
Correlation to the National Science Education Standards

Laboratory Resources

Laboratory Experiments Booklet
Laboratory Experiments Teacher's Edition
Data Tables for Laboratory Experiments*
Technology-Based Lab Activities Booklet
Technology-Based Lab Activities Teacher's Edition

Practice and Reinforcement Resources

Section Reviews Worksheets
Problem Workbook
Problem Bank*

Assessment Resources

Chapter Tests and Answer Key
Test Generator*
Alternative Assessments from the NSF-Sponsored C³P Project*
Assessment Rubrics*

Media and Enrichment Resources

Teaching Transparencies and Black line Masters
Interactive Tutor Worksheets and Answer Key*
CNN Presents Science in the News: Physical Science Critical Thinking Worksheets
CNN Presents Science in the News: Physical Science User's Guide
Preprogrammed hyperlinks to the Web sites for *sciLINKS*, the
Smithsonian Institution, go.hrw.com, and cnnfyi.com.*

* Found only on the Holt Physics One-Stop Planner CD-ROM.

ONLY AVAILABLE ON THE ONE-STOP PLANNER

- **Problem Bank**
  contains another set of 97 worked-out samples and nearly 1,000 additional practice problems for the problem types featured in the student text.

- **Teacher's Solution Manual and Answer Key**
  contains answer keys for resources and detailed solutions for all of the practice problems in the textbook, *Problem Workbook,* and *Problem Bank.*

- **Data Tables for Laboratory Experiments**
  save time by providing blank, ready-to-fill data tables for your students and tables containing sample data for you.

Holt Physics Teacher's Edition

PHYSICS and the Texas One-Stop Planner

Complete alignment with the TEKS

TEKS-based lesson plans keep your course aligned with Texas requirements

The lesson plans on the Texas version of the *One-Stop Planner* for *Holt Physics* have all of the features described in the preceding pages. In addition, they indicate which of the Texas Essential Knowledge and Skills (TEKS) are being fulfilled in each lesson. You, your science supervisor, and your students' parents can have complete peace of mind, knowing that the instruction in *Holt Physics* matches the state's requirements.

TEKS-based assessment makes sure students are held accountable for the right concepts

On the *Texas One-Stop Planner* for *Holt Physics*, the Test Generator's more than 2000 items are coded not just by the objective in the textbook, but also by which of the TEKS they are fulfilling. You can even search the bank of items and sort it by the TEKS. After you select items for a test and save them as a document, you'll always have a record of which TEKS students were tested upon.

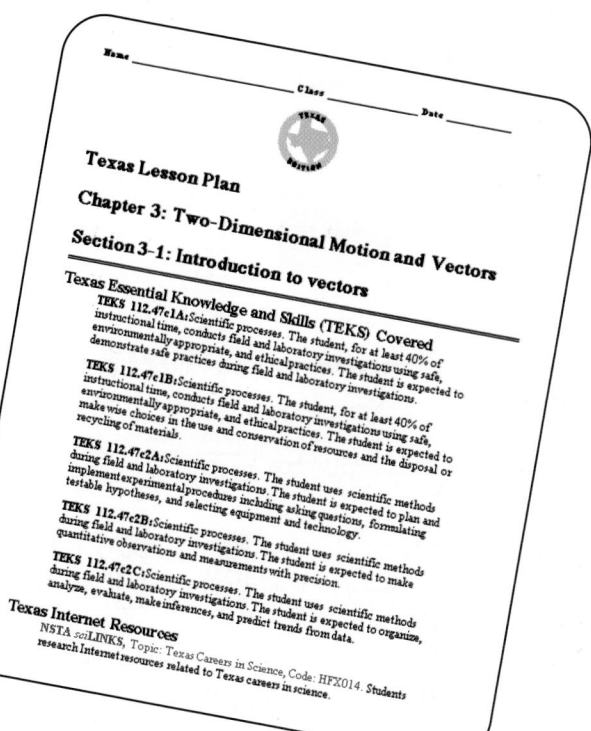

PHYSICS Texas Teacher's Edition

HOLT

Point-of-use TEKS icons

Clearly show you exactly what TEKS-related content is in the text

The detailed correlations and icons throughout the Teacher's Edition highlight not only the essential science content knowledge, but also the scientific skills that are expected of physics students in Texas. For a detailed analysis of the entire textbook, see pages TX6 and following.

TEKS Correlations

provided in the chapter interleaf Planning Guides give you a quick overview of the coverage in each section.

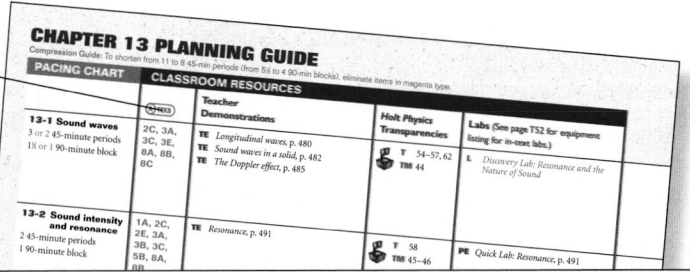

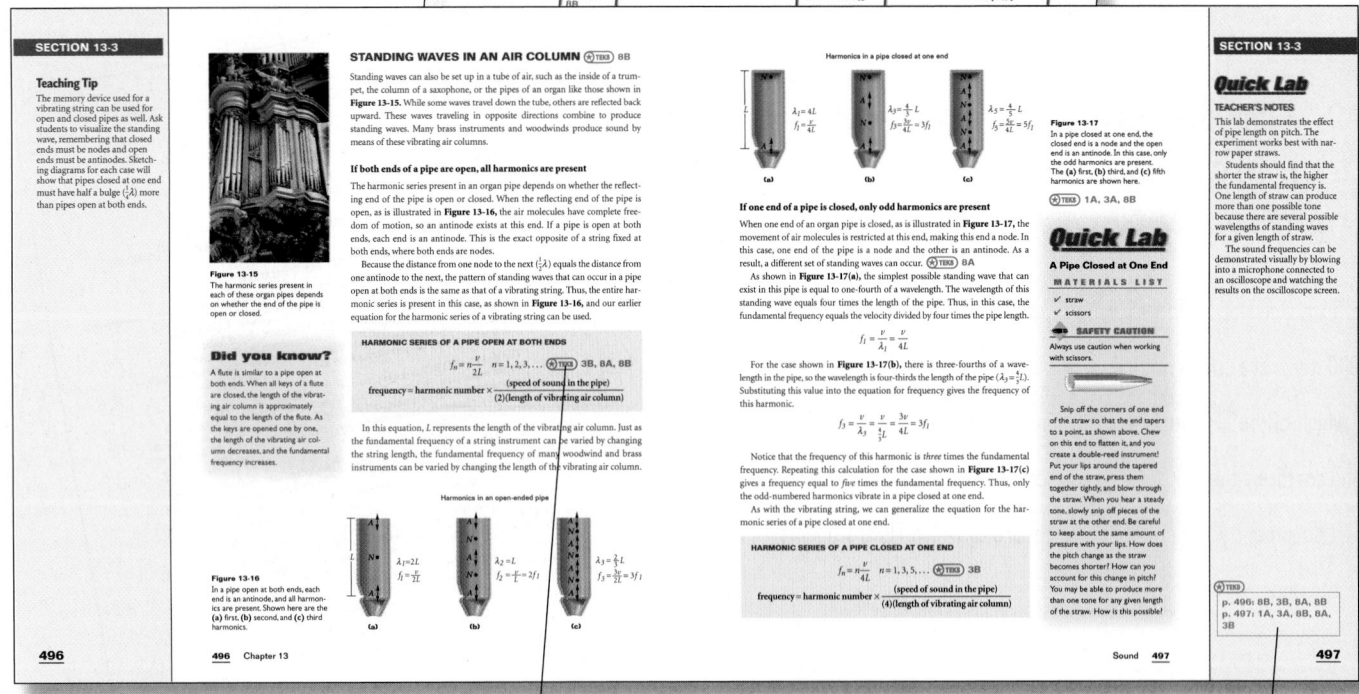

Individual point-of-use TEKS icons

indicate exactly **which** elements on each page are TEKS-related.

Spread-by-spread TEKS summary boxes

itemize the coverage found on each of the pages

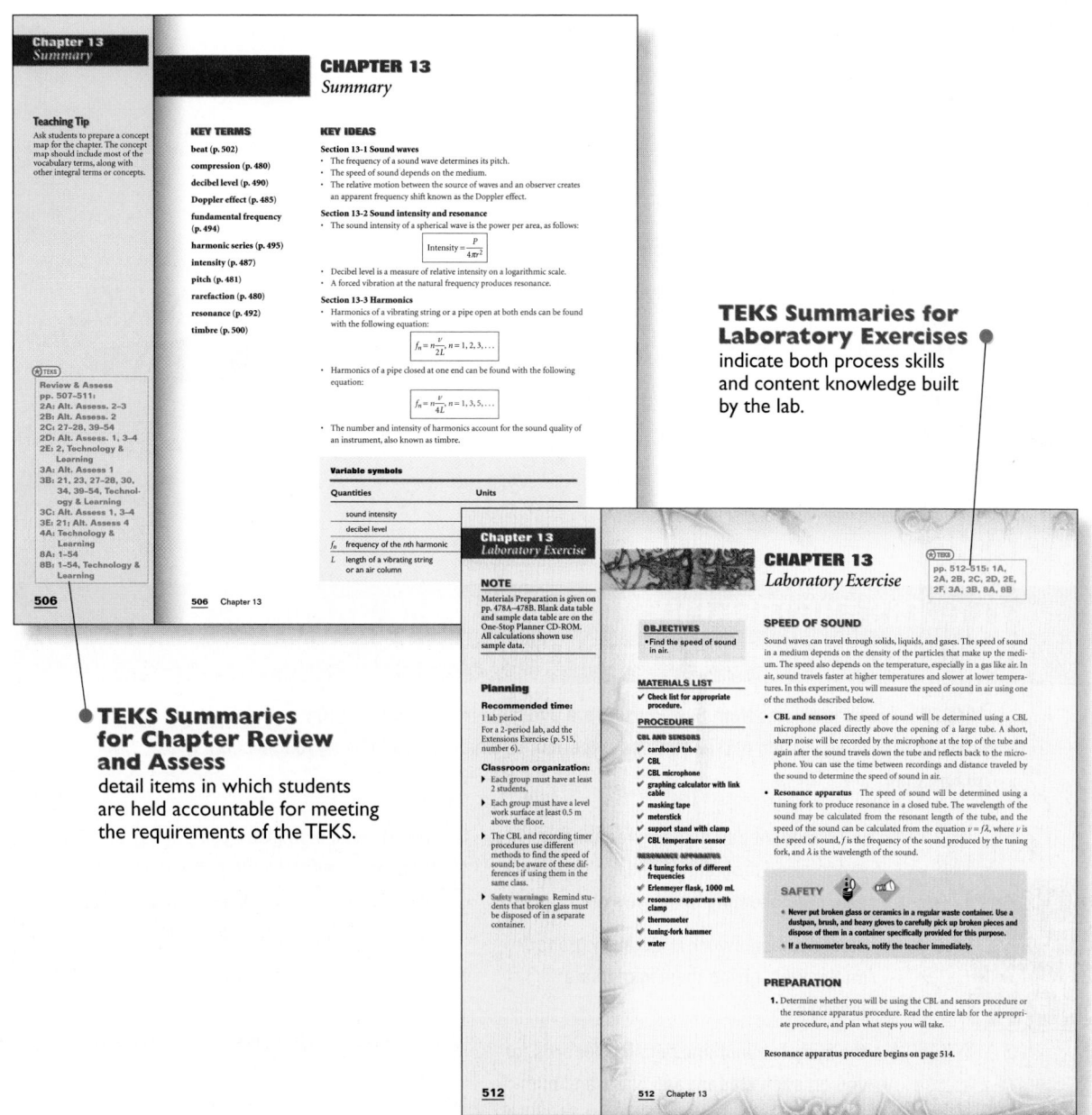

TEKS Summaries for Laboratory Exercises
indicate both process skills and content knowledge built by the lab.

TEKS Summaries for Chapter Review and Assess
detail items in which students are held accountable for meeting the requirements of the TEKS.

HOLT PHYSICS and Texas Internet Resources

Material customized for your needs

Connects you to pre-screened content

Icons throughout the margins of the Teacher's Edition give you *sci*LINKS codes that can take you to resources about Texas' connection to the sciences. A complete list of topics and codes is given on the next page. Just go to **www.scilinks.org** and type in the *sci*LINKS code given.

Like the *sci*LINKS codes found in the student book, the sites that are shown have been thoroughly scrutinized for validity, usefulness, and appropriateness by teachers and staff of the National Science Teacher's Association (NSTA).

ACCESS THE FREE PHYSICS HOMEWORK GRADING SYSTEM AT THE UNIVERSITY OF TEXAS

All 2800 practice problems from *Holt Physics* are available in randomized form on the web homework grading system of UT's Physics Department, which is available to you through **go.hrw.com**.

- each student can be assigned unique randomized problem values, helping you eliminate cheating on homework

- students' homework is instantly graded on the web, giving them immediate feedback

- the system maintains detailed records for you on each student and class and highlights problem areas that may need reteaching

- the UT bank also gives you access to thousands more problems from a variety of college texts, all labeled according to problem type and difficulty level

- the randomization process effectively gives you an INFINITE number of practice problems—no two assignments will ever be alike

Teaching Tip

Superconductors are introduced here because of their relationship to resistance. Superconductivity is discussed in greater detail in Chapter 24.

Visual Strategy

Figure 19-10
Point out to students that this type of levitation has major advantages (reducing friction) but that it has disadvantages as well.

Q Why are superconductors like this not used widely for trains or floating cars?

A *Answers may vary but could include high construction costs, high cooling costs, and temperature control problems, which could lead to crashes.*

internet connect

*sci*LINKS.
NSTA
TOPIC: Superconductor Research in Texas
GO TO: www.scilinks.org
*sci*LINKS CODE: HFX008

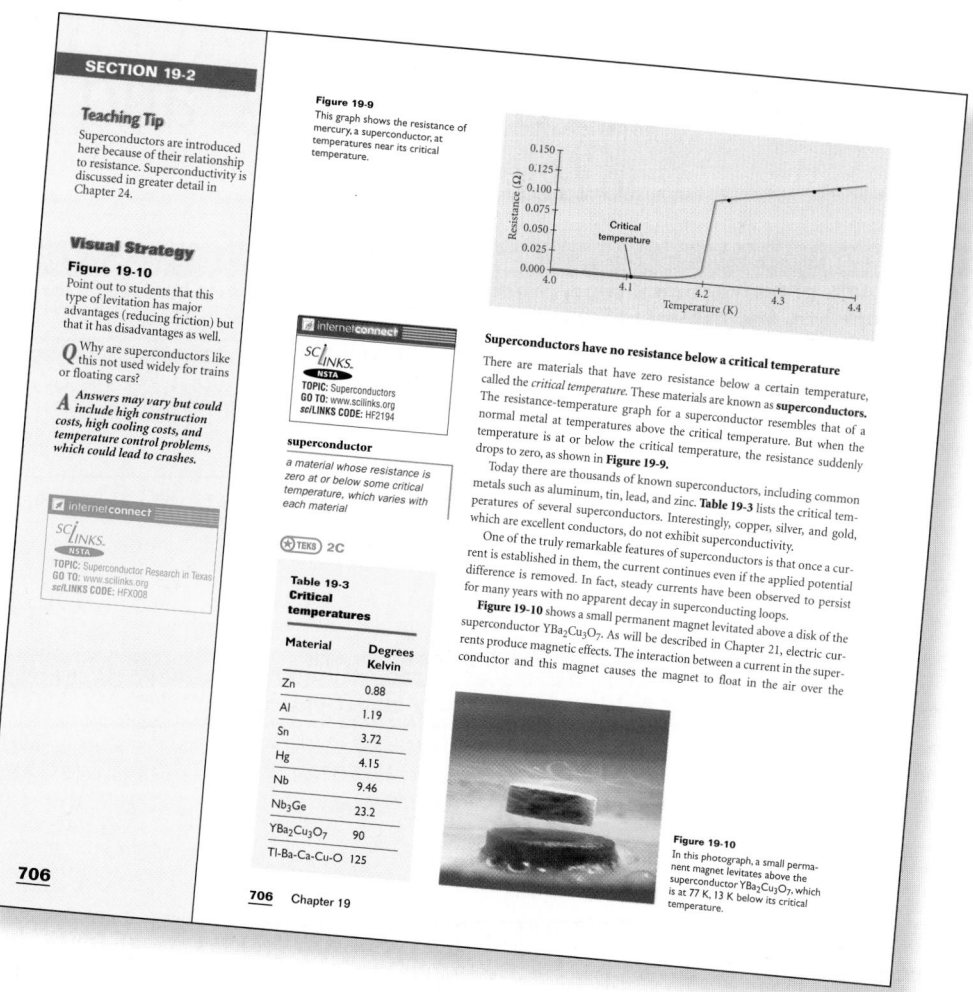

Figure 19-9
This graph shows the resistance of mercury, a superconductor, at temperatures near its critical temperature.

internet connect

*sci*LINKS.
NSTA
TOPIC: Superconductors
GO TO: www.scilinks.org
*sci*LINKS CODE: HF2194

superconductor

a material whose resistance is zero at or below some critical temperature, which varies with each material

⊛ TEKS 2C

Table 19-3
Critical temperatures

Material	Degrees Kelvin
Zn	0.88
Al	1.19
Sn	3.72
Hg	4.15
Nb	9.46
Nb_3Ge	23.2
$YBa_2Cu_3O_7$	90
Tl-Ba-Ca-Cu-O	125

Superconductors have no resistance below a critical temperature

There are materials that have zero resistance below a certain temperature, called the *critical temperature*. These materials are known as **superconductors.** The resistance-temperature graph for a superconductor resembles that of a normal metal at temperatures above the critical temperature. But when the temperature is at or below the critical temperature, the resistance suddenly drops to zero, as shown in **Figure 19-9.**

Today there are thousands of known superconductors, including common metals such as aluminum, tin, lead, and zinc. **Table 19-3** lists the critical temperatures of several superconductors. Interestingly, copper, silver, and gold, which are excellent conductors, do not exhibit superconductivity.

One of the truly remarkable features of superconductors is that once a current is established in them, the current continues even if the applied potential difference is removed. In fact, steady currents have been observed to persist for many years with no apparent decay in superconducting loops.

Figure 19-10 shows a small permanent magnet levitated above a disk of the superconductor $YBa_2Cu_3O_7$. As will be described in Chapter 21, electric currents produce magnetic effects. The interaction between a current in the superconductor and this magnet causes the magnet to float in the air over the

Figure 19-10
In this photograph, a small permanent magnet levitates above the superconductor $YBa_2Cu_3O_7$, which is at 77 K, 13 K below its critical temperature.

internet connect

*sci*LINKS.
NSTA
TOPIC: Research on the Atom in Texas
GO TO: www.scilinks.org
*sci*LINKS CODE: HFX005

Detailed Correlation of the
Texas Essential Knowledge and Skills
to *Holt Physics*

Code	Scientific Processes	Page Correlation
1. The student, for at least 40% of instructional time, conducts field and laboratory investigations using safe, environmentally appropriate, and ethical practices.		
1.A	The student is expected to demonstrate safe practices during field and laboratory investigations. **TAKS Obj. 1: The student will demonstrate an understanding of the nature of science.**	12; 32–37; 62; 76–81; 100; 120–121; 126; 134; 158–163; 199 #1; 200–205; 238–241; 245; 274–275; 279; 284; 313–314; 335; 339; 349 #3; 350–353; 358; 368; 392–397; 444; 474–475; 491; 497; 547; 711; 843
1.B	The student is expected to make wise choices in the use and conservation of resources and the disposal or recycling of materials. **TAKS Obj. 1: The student will demonstrate an understanding of the nature of science.**	15; 18; 31 #4; 435 #2; 514; 558; 559; 625; 663; 674; 690–691; 893; 933

Code	Scientific Processes	Page Correlation
2. The student uses scientific methods during field and laboratory investigations.		
2.A	The student is expected to plan and implement experimental procedures including asking questions, formulating testable hypotheses, and selecting equipment and technology. **TAKS Obj. 1: The student will demonstrate an understanding of the nature of science.**	31 #2; 81; 119 #2; 157 #2; 163 #11; 199 #3, #5; 237 #1; 273 #1; 312 #2, #3; 349 #1, #3; 391 #1, #3; 397; 473 #1, #2, #4; 515 #5, #6; 554 #1, #3, #6; 591 #3; 659 #1; 622 #1, #2, #3
2.B	The student is expected to make quantitative observations and measurements with precision. **TAKS Obj. 1: The student will demonstrate an understanding of the nature of science.**	12; 16–17; 31 #2, #4; 32–37; 46; 62; 75 #3; 76–81; 119 #2; 120–121; 157 #1, #2, #4; 158–163; 199 #1, #2, #3, #5; 200–205; 237 #1; 238–241; 245; 273 #1, #2; 274–275; 312 #4; 313–314; 660–663
2.C	The student is expected to organize, analyze, evaluate, make inferences, and predict trends from data. **TAKS Obj. 1: The student will demonstrate an understanding of the nature of science.**	10–14; 16–20; 22–23; 25; 28–37; 39; 41–49; 53; 55; 57–59; 62–65; 69–75; 71 #31; 76–81; 86–87; 90–91; 93–94; 97; 101–105; 109; 113–119; 120–121; 133; 135; 138; 140; 162–163; 173; 194 #15; 195 #28; 241 #6; 288 #1; 289 #1; 323–324; 353 #5; 360; 391 #3; 396 #5; 405 #2; 408 #4; 413 #2; 419 #4; 426; 431 #10; 432 #21; 435 #40
2.D	The student is expected to communicate valid conclusions. **TAKS Obj. 1: The student will demonstrate an understanding of the nature of science.**	31 #1, #5; 157 #2, #3, #4, #5; 237 #2, #3, #5; 273 #2, #3; 312 #4, #5; 349 #1, #4; 391 #1, #2, #4, #5, #6; 473 #3, #4, #6; 510 #1, #3, #4; 199 #1, #2, #3, #4, #5, #6; 397; 435 #1, #2, #3, #4, #5, #6; 515 #7; 554 #1
2.E	The student is expected to graph data to observe and identify relationships between variables.	20; 30; 34–37; 81 #3, #9; 861 #2, #3, #4; 113–119; 133; 140; 149; 151–157; 199 #2, #3; 236; 238–241; 273 #1; 274–275; 311; 350–353; 390; 391 #3; 458 #3; 474–475; 511; 559 #2, #4, #5; 595 #5, #6; 658; 686; 687 #4; 858; 861 #2, #3, #4; 933 #3
2.F	The student is expected to read the scale on scientific instruments with precision. **TAKS Obj. 1: The student will demonstrate an understanding of the nature of science.**	31 #2, #4; 32–37; 75 #2, #4; 76–81; 119 #2; 120–121; 157 #1, #2, #4; 158–163; 199 #1, #3, #5; 200–205; 237 #1; 238–241; 245; 274–275; 349 #3; 350–353; 392–397; 444; 473 #1, #2, #4; 474–475

Code	Scientific Processes	Page Correlation
3. The student uses critical thinking and scientific problem solving to make informed decisions.		
3.A	The student is expected to analyze, review, and critique scientific explanations, including hypotheses and theories, as to their strengths and weaknesses using scientific evidence and information. **TAKS Obj. 1: The student will demonstrate an understanding of the nature of science.**	8; 67; 75 #3; 106; 111; 125; 140; 141; 143; 144; 157 #1; 168; 187; 199 #6; 200–205; 208; 210; 213–214; 216; 227; 237 #5; 245; 250; 253; 262; 262–263; 266; 283–284; 286; 31 #1; 289 #6, #7; 349 #2, #4; 382 #2, #3; 486 #5, #6, #7; 312 #1; 320–321; 322; 328; 329; 334; 335; 339; 349 #5; 350–353; 358; 367; 379; 383; 391 #2; 405; 407; 420; 422 #1, #2; 428 #1, #2; 429 #3; 435 #3, #6; 438; 439; 444; 449; 473 #5; 474–475; 483; 491–492; 497; 515 #4; 545; 554 #2; 591 #4; 607; 622 #3; 648; 851; 871 #1; 888 #1; 893; 927 #1
3.B	The student is expected to express laws symbolically and employ mathematical procedures including vector addition and right-triangle geometry to solve physical problems.	21–22; 60; 84–86; 89–95; 97; 111; 114–119; 133; 135; 137–138; 140; 144; 147; 149; 151–157; 210; 216; 220; 223; 227; 245; 259; 260; 263; 280; 290; 320; 326; 329; 335; 338; 367; 411; 439; 466; 564; 566; 634; 638-639; 646–647; 700
3.C	The student is expected to evaluate the impact of research on scientific thought, society, and the environment.	6; 9; 110; 119 #5; 129; 134; 191; 212–213; 217; 237 #3; 300; 321; 333–334; 349 #3, #4, #5; 375; 384–385; 388 #26, #27, #28; 391 #5; 414; 416–417; 421; 442; 464; 467; 482; 484; 516–517; 659 #3; 690–691; 832; 837; 840–843; 846; 851; 853–843; 892; 905; 915; 916 #3; 923–925; 931 #2, #6
3.D	The student is expected to describe the connection between physics and future careers.	312 #6; 591 #1; 687 #2; 759 #2–#4; 785 #2; 888 #4; 931 #3
3.E	The student is expected to research and describe the history of physics and contributions of scientists.	8; 20; 31 #3; 56; 60; 66; 75 # 5, #6; 124; 129–130; 157 #3; 190; 200–205; 210; 233 #21; 237 #4; 264; 267; 273 #3; 302; 320; 326; 349 #4; 354; 391, #4, #5; 435 #1; 439; 447; 466–467; 485; 505; 523; 554 #2, #4, #5; 591 #5; 622 #4; 630; 634; 642; 659 #2, #5; 677; 687 #5; 700; 707; 714–715; 721 #2, #3, #5; 728; 785 #5; 796; 825 #4, #5; 831–832; 834–835; 837–838; 840; 843; 849–850; 852–853; 862–863; 878; 882; 909; 915; 931 #4

Code	Scientific Concepts	Page Correlation
4. The student knows the laws governing motion.		
4.A	The student is expected to generate and interpret graphs describing motion including the use of real-time technology.	21; 45–47; 50; 52; 57; 63; 65; 69 #1, #3, #10; 70 #17; 71 #30; 72 #42; 75 #1; 76; 80–81; 86–88; 92–95; 97–98; 101–105; 108–109; 113–119; 132; 149; 151–163; 272; 274–275; 443; 453–455; 472; 474–475
4.B	The student is expected to analyze examples of uniform and accelerated motion including linear, projectile, and circular. **TAKS Obj. 5: The student will demonstrate an understanding of motion, forces, and energy.**	8; 21; 40–65; 69 #1–#11; 70 #12–#17; 71 #18–#30; 72 #31–#42; 73 #43–#51; 74 #52–#57; 75 #58–#60; 76–81; 98–105; 107–109; 109 #68, #69; 113–121; 137; 151–157; 244; 246–248; 252–255; 259; 261; 266; 269–273
4.C	The student is expected to demonstrate the effects of forces on the motion of objects.	126; 128; 134; 151–163; 260–262; 270–275; 279–282; 286–291; 300–301; 305–314; 440–441; 443; 445
4.D	The student is expected to develop and interpret a free-body diagram for force analysis.	126–128; 132–133; 139–140; 146–147; 149; 151–157
4.E	The student is expected to identify and describe motion relative to different frames of reference.	40; 43; 67; 106–111; 116 #45–#52

Code	Scientific Concepts	Page Correlation
5. The student knows that changes occur within a physical system and recognizes that energy and momentum are conserved.		
5.A	The student is expected to interpret evidence for the work-energy theorem. **TAKS Obj. 5: The student will demonstrate an understanding of motion, forces, and energy.**	174–176; 194 #21, #22
5.B	The student is expected to observe and describe examples of kinetic and potential energy and their transformations. **TAKS Obj. 5: The student will demonstrate an understanding of motion, forces, and energy.**	172–173; 175–186; 193–205; 222; 224–230; 233–237; 295; 310 #64, #66, #70; 331; 335; 359–360; 364 #1, #5; 365; 366; 368; 370; 377; 388 #17; 389 #31, #42, 402; 409; 417; 427; 441; 444; 657 #58; 666–669; 675; 679–681; 683 #1–#6, #10, #11; 684 #47; 688–689; 690–691; 697–698; 708–709; 717 #8; 839 #2, #5; 856 #4; 859 #53; 860–861; 929 #45
5.C	The student is expected to calculate the mechanical energy and momentum in a physical system such as billiards, cars, and trains. **TAKS Obj. 5: The student will demonstrate an understanding of motion, forces, and energy.**	184–186; 195 #26–#32; 200–205; 209; 211–214; 232 #1, #2, #12, #13; 235 #43
5.D	The student is expected to demonstrate the conservation of energy and momentum. **TAKS Obj. 5: The student will demonstrate an understanding of motion, forces, and energy.**	183–187; 195 #26–#34; 200–205; 215; 218–219; 221; 223–224; 226–230; 233 #17–#26, #28; 234 #29–#40; 235 #44–#53; 236 #54–#55; 237 #56–#58; 238–241; 292–294; 296–297; 301; 303; 307 #35; 308 #36–#37, #38, #50; 309 #51, #52, #53, #58, #59; 310 #60, #62, #65, #66; 311 #71; 368–370; 372–374; 388 #20; 389 #31; 392–397; 432 #15; 689 #3; 735

Code	Scientific Concepts	Page Correlation
6. The student knows forces in nature.		
6.A	The student is expected to identify the influence of mass and distance on gravitational forces. **TAKS Obj. 5: The student will demonstrate an understanding of motion, forces, and energy.**	141; 177; 263–265; 270 #29, #31; 271 #39, #40; 272 #49; 635; 637
6.B	The student is expected to research and describe the historical development of the concepts of gravitational, electrical, and magnetic force.	264; 273; 273 #3; 630; 642; 659 #5; 687 #3, #5; 721 #5; 785 #5; 798; 814; 825 #4
6.C	The student is expected to identify and analyze the influences of charge and distance on electric forces. **TAKS Obj. 5: The student will demonstrate an understanding of motion, forces, and energy.**	628–629; 631–644; 646–649; 652; 654 #2, #3, #6, #9–#17; 655 #18–#24, #29; 656 #38–#40, #43–#45, #48; 657 #50, #52, #53, #56, #59, #60; 658 #61; 659 #1; 660–663; 671–673; 675; 683 #8, #12, #13; 684 #14, #21, #22, #24, #29; 685 #31–#34, #38–#41, 686 #42; 687 #48, #49
6.D	The student is expected to demonstrate the relationship between electricity and magnetism.	771; 785 #4; 786–789; 794; 796; 799–800; 802; 804–806; 813; 821 #1–#11; 822 #12, #19–#24; 823 #32, #37–#39; 825 #1; 826–827
6.E	The student is expected to design and analyze electric circuits. **TAKS Obj. 5: The student will demonstrate an understanding of motion, forces, and energy.**	695–696; 702–703; 710; 717 #15, #18; 718 #28, #29, #39; 719 #40, #48, #50, #52; 720 #54, #57; 721 #58, #59; 722–725; 730; 732–735; 736–745; 746–752; 754–759; 754 #1–#14; 755 #15–#23; 756 #24–#34; 757 #35–#43; 758 #44–#46; 759 #47–#49; 759 #1, #5; 760–763; 786–789; 876–877; 879; 889–891
6.F	The student is expected to identify examples of electrical and magnetic forces in everyday life.	645; 659 #2, #3; 676; 679; 687 #5; 690–691; 774; 777; 785 #4; 801; 812; 815–816; 818

Code	Scientific Concepts	Page Correlation

7. The student knows the laws of thermodynamics.

Code	Scientific Concepts	Page Correlation
7.A	The student is expected to analyze and explain everyday examples that illustrate the laws of thermodynamics. **TAKS Obj. 5: The student will demonstrate an understanding of motion, forces, and energy.**	412–413; 415–416; 418; 419 #2–#8; 421–424; 428; 432–435 #14, #18, #19, #25, #40; 435 #3, #5, #6
7.B	The student is expected to evaluate different methods of heat energy transfer that result in an increasng amount of disorder. **TAKS Obj. 5: The student will demonstrate an understanding of motion, forces, and energy.**	371; 376; 383; 388 #17; 389 #40; 403; 428; 429 #4; 433 #31, #32,#35; 435 #4

Code	Scientific Concepts	Page Correlation

8. The student knows the characteristics and behavior of waves.

Code	Scientific Concepts	Page Correlation
8.A	The student is expected to examine and describe a variety of waves propagated in various types of media and describe wave characteristics such as velocity, frequency, amplitude, and behaviors such as reflection, refraction, and interference. **TAKS Obj. 5: The student will demonstrate an understanding of motion, forces, and energy.**	452–458; 459–465; 467; 470–473; 470 #23–#35; 471–473; 494–504; 507–510; 521–525; 526–529; 531–542; 543; 545; 547–548; 550 #3–#6, #8–#11, #14; 551–554; 556–558; 562–567; 568–579; 580–585; 599–603; 604–612; 613–618; 620–622; 620–622; 622 #1–#5; 623; 714; 834; 838; 848–851; 854; 857–858; 857 #35–#37, #39–#42; 858 #43, #44
8.B	The student is expected to identify the characteristics and behaviors of sound and electromagnetic waves. **TAKS Obj. 5: The student will demonstrate an understanding of motion, forces, and energy.**	456–458; 459–460; 464; 470 #32; 472 #50, #51; 480; 481; 482–486; 487–493; 494–503; 504; 507–510; 511; 512–515; 520–525; 526–527; 536; 540; 542; 543–548; 550 #1, #2, #7, #12, #13; 556–559; 613; 621 #15; 645; 704; 718 #39; 777; 831; 833; 838; 843; 848; 854 #1; 856–859; 856 #1–#3, #9, #14–#18; 857 #29, #31; 858 #45; 859 #47, #49, #50
8.C	The student is expected to interpret the role of wave characteristics and behaviors found in medicinal and industrial applications. **TAKS Obj. 5: The student will demonstrate an understanding of motion, forces, and energy.**	464; 467; 542 #3, #5; 547; 554 #4; 618; 622 #5; 859 #3 884 #4

Code	Scientific Concepts	Page Correlation

9. The student knows simple examples of quantum physics.

Code	Scientific Concepts	Page Correlation
9.A	The student is expected to describe the photoelectric effect.	834–837; 839; 856; 858–859; 856 #5–#7, #10–#12; 858 #43, #46; 859 #48, #51, #52; 860–861
9.B	The student is expected to explain the line spectra from different gas-discharge tubes.	845; 847; 857 #21–#22, #26–#28

Detailed Correlation of *Holt Physics*
to the Texas Essential Knowledge and Skills for Physics

Chapter 1	The Science of Physics		
Page	**TEKS covered**	**Page**	**TEKS covered**
3–5	3.C	23	2.C
6	3.B	24	3.A
7–8	2.E, 3.A., 3.E, 4.B	25	2.C
9	3.B	28	2.C, 4.C
10–11	2.C	29	2.C
12	1.A, 2.A, 2.B	30	2.C, 2.E
13–14	2.C	31 #1	2.D, 3.A
15	1.B	31 #2	2.A, 2.B, 2.F
16–17	2.B, 2.C	31 #3	3.E
18	1.B	31 #4	1.B, 2.A, 2.B, 2.F
19	2.C	31 #5	2.D
20	2.C, 2.E, 3.E	32–33	1.A, 2.B, 2.C, 2.F
21	3.A, 4.A, 4.B	34–37	1.A, 2.B, 2.C, 2.E, 2.F
22	2.C, 3.A		

Chapter 2	Motion in One Dimension		
Page	**TEKS covered**	**Page**	**TEKS covered**
39	2.C	50	4.A, 4.B
40	4.B, 4.D	51	4.B
41–42	2.C, 4.B	52	4.A, 4.B
43	2.C, 4.D	53	2.C, 4.B
44–45	2.C, 4.A	54	4.B
46	2.C, 2.B, 4.A, 4.B	55	2.C, 4.B
47	2.C, 4.A, 4.B	56	3.E, 4.B
48–49	2.C, 4.B	57	2.C, 4.A, 4.B

Chapter 2   Motion in One Dimension (cont.)

Page	TEKS covered	Page	TEKS covered
58	2.C, 4.B	71	2.C, 4.A, 4.B
59	2.C, 4.A, 4.B	72–75	2.C, 4.B
60	3.A, 3.E, 4.B	75 #2	2.D
61	4.B	75 #3	2.B, 3.A
62	1.A, 2.B, 2.C, 4.A, 4.B	75 #4	2.D
63	2.C, 4.A, 4.B	75 #5	3.E
64	2.C, 4.B	75 #6	2.F, 3.E
65	2.C, 4.A, 4.B	76	1.A, 2.B, 2.C, 2.F, 4.A, 4.B
66	3.E	77–79	1.A, 2.B, 2.C, 2.F, 4.B
67	3.A, 4.D	80	1.A, 2.B, 2.C, 2.F, 4.A, 4.B
69	2.C, 4.A, 4.B	81	1.A, 2.A, 2.B, 2.C, 2.F, 4.A, 4.B
70	2.C, 4.B	81	2.E

Chapter 3   Two-Dimensional Motion and Vectors

Page	TEKS covered	Page	TEKS covered
84	3.A	99	4.B
85	3.B	100	1.A, 4.B
86	2.C, 3.B, 4.A	101–103	2.C, 4.A, 4.B
87	2.C	104–105	2.C, 4.A
88	4.A	106	3.A, 4.D
89	3.B	107	4.B, 4.D, 4.E
90–91	2.C, 3.B	108	4.A, 4.B, 4.D
91	2.C, 3.B	109	2.C, 4.B, 4.E
92	3.B, 4.A	110	3.B, 4.D
93	2.C, 3.B, 4.A	111	3.A, 4.D
94	2.C, 3.B	113	2.C
95	3.B, 4.A	114–115	2.C, 2.E, 3.B, 4.B
97	2.C, 3.B	116	2.C, 2.E, 3.B, 4.B, 4.E
98	4.A, 4.B	117	2.C, 2.E, 3.B, 4.B

Chapter 3   Two-Dimensional Motion and Vectors (cont.)

Page	TEKS covered	Page	TEKS covered
118	2.C, 2.E, 3.B, 4.A, 4.B	119 #3–#4	2.D
119	2.C, 2.E, 3.B., 4.B	119 #5	3.C
119 #1	2.D	120–121	1.A, 2.B, 2.C, 2.F, 4.B
119 #2	2.A, 2.B, 2.D, 2.F		

Chapter 4   Forces and the Laws of Motion

Page	TEKS covered	Page	TEKS covered
124	3.E	140	2.C, 2.E, 3.A, 3.B, 4.C
125	3.A	141	3.A, 5.D
126	1.A, 4.B, 4.D	142–144	3.A
127	4.D	146	4.A, 4.D
129	3.C, 3.E, 4.B	147	3.B, 4.C
130	3.E	149	2.E, 3.B, 4.C
131	3.A	151–157	2.E, 3.B, 4.A, 4.B, 4.C, 4.D
132	4.A, 4.D	157 #1	2.B, 2.F, 3.A
133	2.C, 2.E, 3.A, 3.B, 4.C	157 #2	2.A, 2.B, 2.D, 2.F
134	1.A, 3.C, 4.B	157 #3	2.D, 3.E
135	2.C, 3.A, 3.B, 4.A	157 #4	2.B, 2.D, 2.F
136	3.A	157 #5	2.D
137	3.A, 4.B	158–161	1.A, 2.B, 2.F, 4.A, 4.B
138	2.C, 3.A, 3.C	162–163	1.A, 2.B, 2.C, 2.F, 4.A, 4.B
139	3.A, 4.D	163 #11	2.A

Chapter 5   Work and Energy

Page	TEKS covered	Page	TEKS covered
168	3.A	191	3.C
172	5.A	193–199	5.A, 5.B
173	2.C, 5.B	194 #15	2.C
174	4.E	194 #21, #22	4.E
175	4.E, 5.A	195 #26–#27	5.C
176	4.E, 5.B	195 #28	2.C, 5.C
177	5.B, 6.A	195 #29–#34	5.C
178–181	5.B	199 #1	2.A, 2.B, 2.D, 2.F
182	5.B, 5.C	199 #2	2.B, 2.D, 2.E
183	5.A, 5.C	199 #3	2.A, 2.B, 2.D, 2.E, 2.F
184	5.A, 5.B, 5.C	199 #4	2.D
185–186	5.B, 5.C	199 #5	2.A, 2.B, 2.D, 2.F
187	3.A	199 #6	2.D, 3.A
190	3.E	200–205	1.A, 2.B, 2.F, 3.A, 3.E, 5.A, 5.B, 5.C

Chapter 6   Momentum and Collisions

Page	TEKS covered	Page	TEKS covered
208	3.A	215	5.D
209	5.C	216	3.A
210	3.A, 3.E	217	3.C
211	5.C	218–219	5.D
212	3.C, 5.C	220	3.A
213	3.A, 3.C, 5.C	221	5.D
214	3.A, 5.C	222	5.B

Chapter 6   Momentum and Collisions (cont.)

Page	TEKS covered	Page	TEKS covered
223	3.A, 5.D	236–237	2.E, 5.B, 5.C, 5.D
224	5.B, 5.C	237 #1	2.A, 2.B, 2.F
225	5.B	237 #2	2.D
226	5.B, 5.D	237 #3	2.D, 3.C
227	3.A, 5.B, 5.D	237 #4	3.E
228–230	5.B, 5.D	237 #5	2.D, 3.A
232	5.C	238–241	1.A, 2.B, 2.E, 2.F, 5.C, 5.D
233	3.E, 5.B, 5.C, 5.D	241 #6	2.C
234–235	5.B, 5.C, 5.D		

Chapter 7   Rotational Motion and the Law of Gravity

Page	TEKS covered	Page	TEKS covered
244	4.B	264	3.E, 5.D, 6.A
245	1.A, 2.B, 2.F, 3.A	265	5.D
246–249	4.B	266	3.A, 4.B
250	3.A, 4.B	267	3.E
251–252	4.B	269	4.B
253	3.A, 4.B	270–271	4.B, 5.D
254–255	4.B	272	4.A, 4.B, 5.D
256–258	4.B	273	4.B, 5.D
259	3.A, 3.B, 4.B	273 #1	2.A, 2.B, 2.E
260	3.A, 4.B	273 #2	2.B, 2.D
261	4.B	273 #3	2.D, 3.E
262	3.A, 4.B	274–275	1.A, 2.B, 2.E, 2.F, 4.A, 4.B
263	3.A, 5.D		

Chapter 8   Rotational Equilibrium and Dynamics

Page	TEKS covered	Page	TEKS covered
279	1.A, 2.A	300	3.C, 4.B
280	3.A	301	4.B, 5.C
283	3.A	302	3.E
284	1.A, 3.A	303	5.D
285	4.B	305–306	4.B
286	3.A, 4.B	307–311	4.B, 5.C, 5.D
287	4.B	310 #64, #66, #70	5.B
288–289	4.B	311	2.E
288 #1	2.C	312	4.B
289 #1	2.C	312 #1	3.A
289 #6, #7	3.A	312 #2	2.A
290	3.A, 4.B	312 #3	2.A
291	4.B	312 #4	2.B, 2.D
292–294	5.D	312 #5	2.D
295	5.A	312 #6	3.C, 3.D
296–297	5.C	313–314	1.A, 2.B, 4.B

Chapter 9   Fluid Mechanics

Page	TEKS covered	Page	TEKS covered
320	3.A, 3.C, 3.E	338	1.A, 3.A
321–322	3.A	339	3.A
323–324	2.C	349 #1	2.A, 2.D
326	3.A, 3.E	349 #2	3.A
328	3.A, 3.C	349 #3	2.A, 2.F, 3.C
329	3.A	349 #4	2.D, 3.A, 3.C, 3.E
331	5.A	353 #5	2.C, 3.A, 3.C
333	3.C, 5.B	350–353	1.A, 2.E, 2.F, 3.A
334	3.A, 3.C	354	3.E
335	1.A, 3.A, 5.B		

Chapter 10   Heat

Page	TEKS covered	Page	TEKS covered
358	1.A, 3.A	384–385	3.C
359	5.AB	388 #17	7.A
360	2.C, 5.A	388 #20	5.C
364 #1, #5	5.B	388 #26–28	3.C
365	5.B	389 #31	5.B, 5.C
366	5.A	389 #40	7.A
367	3.A	389 #42	5.B
368	1.A, 5.B, 5.C	390	2.E
369	5.C	391 #1	2.A, 2.D
370	5.A, 5.C	391 #2	3.A, 2.D
371	7.A	391 #3	2.A, 2.C, 2.E
372	5.C	391 #4	2.D, 3.E
373–374	5.C	391 #5	2.D, 3.C, 3.E
375	3.C, 7.A	391 #6	2.D
377–378	5.B	392–396	1.A, 2.F, 5.C
379	3.A	396 #5	2.C
382 #2, #3	3.A	397	1.A, 2.A, 2.D, 2.F, 5.C
383	3.A, 7.A		

Chapter 11   Thermodynamics

Page	TEKS covered	Page	TEKS covered
402	5.A	413	2.C, 7.A
403	7.A	414	3.C
405	2.C, 3.A	415	6.F
407	3.A	416	3.C, 6.F
408 #4	2.C	417	3.C, 5.B
409	5.B	418	6.F
411	3.A	419 #2–#3	7.A
412	6.F	419 #4	2.C, 7.A

Chapter 11 Thermodynamics (cont.)

Page	TEKS covered	Page	TEKS covered
419 #5–#8	7.A	432 #14	7.A
420	3.A	432 #15	5.C
421	3.C, 6.F	432 #19	7.A
422	3.A, 7.A	432 #21	2.C
423	6.F	433 #25, #31, #32, #35	7.A
424	7.A	435 #40	2.C, 7.A
426	2.C	435 #1	2.D, 3.E
427	5.A	435 #2	1.B, 2.D
428	3.A, 6.F, 7.A	435 #3	2.D, 3.A, 7.A
429	3.A, 7.A	435 #4–#5	2.D, 7.A
431 #10	2.C	435 #6	2.D, 3.A, 7.A

Chapter 12 Vibrations and Waves

Page	TEKS covered	Page	TEKS covered
438	3.A	459–460	7.B, 8.A, 8.B
439	3.A, 3.E	461–463	7.B, 8.A
440	4.B	464	3.C, 7.B, 8.A, 8.B, 8.C
441	4.B, 5.A, 5.B	465	7.B, 8.A
442	3.C	466	3.A, 3.E
443	4.A, 4.B	467	3.B, 3.C, 3.E, 7.B, 8.A, 8.C
444	1.A, 2.F, 3.A, 5.A, 5.B	470–471	7.B, 8.A, 8.B
445	4.B	472–473	4.A, 7.B, 8.A, 8.B
447	3.E	473 #1, #2	2.A, 2.F
449	3.A	473 #3	2.D
452	7.B	473 #4	2.D, 2.F
453	4.A, 7.B	473 #5	3.A
454–455	4.A, 7.B, 8.A	473 #6	2.D
456–457	7.B, 8.A, 8.B	474–475	1.A, 2.E, 2.F, 3.A
458	2.E, 7.B, 8.A, 8.B		

Chapter 13 Sound

Page	TEKS covered	Page	TEKS covered
480	8.B	499	7.B, 8.A, 8.B
481	8.A	500–503	7.B, 8.B
482	3.C, 8.B	504	7.B
483	3.A, 8.B	505	3.E
484	3.C, 8.B	507–510	8.A, 8.B
485	3.E, 8.B	510 #1	2.D
486	3.A, 8.B	510 #2	2.A
487	5.A, 8.B	510 #3, #4	2.D
488–490	8.B	511	2.E
491	1.A, 3.A, 8.B	512–515	8.A, 8.B
492	3.A	515 #4	3.A
493	8.B	515 #5, #6	2.A
494–496	7.B, 8.B	515 #7	2.D
497	1.A, 3.A, 7.B, 8.B	516–517	3.C, 8.B
498	7.B, 8.B		

Chapter 14 Light and Reflection

Page	TEKS covered	Page	TEKS covered
520–521	7.B, 8.B	534–535	7.B
522	8.A, 8.B	536	8.A, 8.B
523	3.E, 7.B, 8.A, 8.B	537	7.B
524	7.B	539	7.B
525	7.B, 8.A, 8.B	540	8.A, 8.B
526–528	7.B	541	7.B
529	8.A	542	8.A, 8.B, 8.C
531–532	8.A	543	7.B

Chapter 14 Light and Reflection (cont.)

Page	TEKS covered	Page	TEKS covered
545	3.A	554 #3,	2.A
546	8.B	554 #4	3.E, 8.C
547	7.B, 8.A, 8.C	554 #5	3.E
548	8.A	554 #6	2.A
550–554	8.A, 8.B	556–558	7.B, 8.A, 8.B
554 #1	2.A, 2.D	559 #2, #4, #5	2.E
554 #2	3.A, 3.E	559 #7	2.A

Chapter 15 Refraction

Page	TEKS covered	Page	TEKS covered
562–563	7.B	580	7.B
564	3.A, 7.B	581 (QL)	7.B, 8.A
565	7.B	582	7.B, 8.A
566	3.A, 7.B	583–584	7.B
567	8.A	585 #1	7.B, 8.A
568–569	7.B	587–591	8.A
570 (QL)	7.B, 8.A	591 #1	3.C, 3.D
571–574	7.B	591 #3	2.A
575	7.B, 8.A	591 #4	3.A
577 (QL)	8.A	591 #5	3.E
578	7.B	592–594	8.A
579	7.B, 8.A	595 #6	2.E, 8.A

Chapter 16 Interference and Diffraction

Page	TEKS covered	Page	TEKS covered
599–602	7.B	614–616	7.B
603	8.AB	617	8.C
604–606	7.B	618	8.A, 8.C
607	3.A, 7.B	620–623	8.A
608–609	7.B	621 #5	8.B, 8.C
610	8.A	621 #15	8.B
611	7.B	622 #1, #2	2.A
612	8.A	622 #3	2.A, 3.A
613	7.B, 8.B	622 #4	3.E

Chapter 17 Electric Forces and Fields

Page	TEKS covered	Page	TEKS covered
628–629	6.B	648	3.A, 6.C
630	3.E, 6.B	649	6.C
631–642	6.C	652	6.C
634	3.A, 3.E	654–659	6.C
635	6.A	657 #58	5.B
637	6.A	658	2.E
638–639	3.B	659 #1	2.A
642	3.E, 6.B	659 #2	3.E, 6.E
643	6.B	659 #3	3.C, 6.E
644	6.B, 6.C	659 #5	3.E, 6.B
645	6.E, 8.B	660–663	2.B, 6.C
646–647	3.B		

Chapter 18   Electrical Energy and Capacitance

Page	TEKS covered	Page	TEKS covered
666–670	5.B	684 #14, #23, #29	6.C
671	5.B, 5.C	685 #30–#34, #38, #40–42	6.C
672–673	6.C	685 #35, #36	5.B
674	1.B	686	2.E
675	5.B, 6.C	686 #44	5.B
676	5.B, 6.E	687 #2	3.C, 3.D
677	5.B, 3.E	689 #3	5.C, 6.B
679	5.B, 6.E	687 #4	2.E
680–681	5.B	687 #5	3.E, 6.B, 6.E
683 #1	5.B	687–689	5.B
683 #12–#13	6.C	690–691	1.B, 3.C, 6.E

Chapter 19   Current and Resistance

Page	TEKS covered	Page	TEKS covered
695–696	6.E	715	3.E
698–699	5.B	717, #8	5.B
700	3.A, 3.E	717–725	6.E
702–703	6.E	718 #39	8.B
704	8.B	721 #1	2.A
707	3.E	721 #2	2.A, 3.E
708–709	5.B	721 #3	3.E
710	6.E	721 #5	2.A, 3.E, 6.B
714	3.E, 7.B		

Chapter 20   Circuits and Circuit Elements

Page	TEKS covered	Page	TEKS covered
728	3.E	742–752	6.E
730	6.E	754–763	6.E
732–733	6.E	758 #45	6.D
734	2.A, 5.B, 6.E	759 #1	2.A, 6.D
735	5.B, 5.C, 6.E	759 #2, #3, #4	3.C, 3.D
736–740	6.E	760–763	6.D
741	2.A, 6.E		

Chapter 21   Magnetism

Page	TEKS covered	Page	TEKS covered
771	6.C	785 #2	2.A, 3.C, 3.D
774	6.F	785 #4	6.C, 6.F
777	6.F, 8.B	785 #5	3.E, 6.B
785 #1	2.A	786–789	6.C, 6.E

Chapter 22   Induction and Alternating Current

Page	TEKS covered	Page	TEKS covered
794	5.B, 6.C	813	6.C
796	3.E, 6.C	814	6.B
797	3.A	815–816	6.F
798	6.B	818	6.F
799–800	6.C	821–827	6.C
801	6.F	825 #2	2.A
802	6.C	825 #4	3.E, 6.B
804–806	6.C	825 #5	3.E
812	6.F		

Chapter 23 Atomic Physics

Page	TEKS covered	Page	TEKS covered
831	3.E, 8.B	849	3.E, 7.B, 8.A
832	3.B, 3.E	850	3.E, 7.B
833	8.B	851	3.A, 3.B
834	3.E, 5.A, 8.A, 8.C	851 #1, #2, #3	8.A
835	3.E, 5.A, 8.C	852	3.E
836	5.A, 8.C	853	3.B, 3.E
837	3.B, 3.E, 5.A, 8.C	854	3.B, 8.A
838	3.E, 8.A, 8.B	854 #1	8.B
839	8.C	856	8.B, 8.C
839 #2, #5	5.B	857	8.A, 8.B, 9.A
840	3.B, 3.E	858	2.E, 8.A, 8.B, 8.C
842	3.B, 9.B	859	8.B, 8.C
843	3.B, 3.E, 8.B, 9.B	859 #53	5.B
844	3.B, 9.B	860–861	8.C
845	3.B, 9.A, 9.B	861 #1	5.B
846	3.B	861 #2, #3, #4	2.E
847	9.A	862–863	3.E
848	8.A, 8.B		

Chapter 24 Modern Electronics

Page	TEKS covered	Page	TEKS covered
871 #1	3.A	888 #1	3.A
875	5.B	888 #2	2.A
876–877	6.E	888 #4	3.C, 3.D
878	3.E	889–891	6.E
879	6.E	892	3.C
882	3.E	893	1.B, 3.A
884	8.B, 8.C		

Chapter 25 Subatomic Physics

Page	TEKS covered	Page	TEKS covered
905	3.B	929 #45	5.B
909	3.E	931 #2	3.C
915	3.C, 3.E	931 #3	3.C, 3.D
916 #3	3.C	931 #4	3.E
923–925	3.B	931 #6	3.C
927 #1	3.A	933 #3	2.E

HOLT
PHYSICS and Technology Skills

Graphing Calculators

Using technology to improve learning and critical-thinking skills

You won't have to worry about how you'll integrate practice using technology into the *Holt Physics* program, because it's already been done!

• alternative procedures for the CBL™ and Graphing Calculator for 11 of the in-text chapter labs (see p. T26)

• all-new *Technology-Based Lab Activities* booklet containing even more CBL™ labs (see p. T28)

• "Technology and Learning" exercises in the chapter review give students practical exercise in using the graphing calculator to explore an aspect of the chapter's topics using NEW Flash™ technology for Texas Instruments TI-83 Plus and TI-82 Plus calculators.

Downloading Programs for TI-83 Plus or TI-82 Plus Calculators with Flash™ Technology

Once, using the graphing calculator required painstaking keystroking. Now, we've done the work for you. Just download the programs from the **go.hrw.com** Web site, and port them from your computer to your graphing calculator using the GRAPH-LINK cable. (Full instructions are given in Appendix B.) Then, your students can concentrate on interpreting graphs and explaining their significance instead of keystroking!

For users of calculators without Flash™ technology, such as the TI-82 and TI-83, keystroking guides are available on the *One-Stop Planner CD-ROM*. These guides provide an alternative exercise for the graphing calculator and show all of the keystroking and programming in detail.

PHYSICS in the Laboratory
HOLT

A Wide Variety of Options

Joining new technologies and new strategies with classic physics lab experiences

The Holt Physics textbook includes 25 full-length labs that appear at the end of the chapters. In addition, 33 more labs are included in the Holt Physics Laboratory Experiments.

(see pages T30 and T31).

CBL™ AND EQUIPMENT OPTIONS

Many in-text labs include two procedures: one for typical equipment, and the other for use with graphing calculators, CBL™ (Calculator-Based Laboratory), and sensors. The graphing calculators control the CBL™ using free software from Vernier Software and Technology. In addition, the *Technology-Based Lab Activities* booklet includes 17 more experiments for CBL™. For more-detailed information on this software, see page T28.

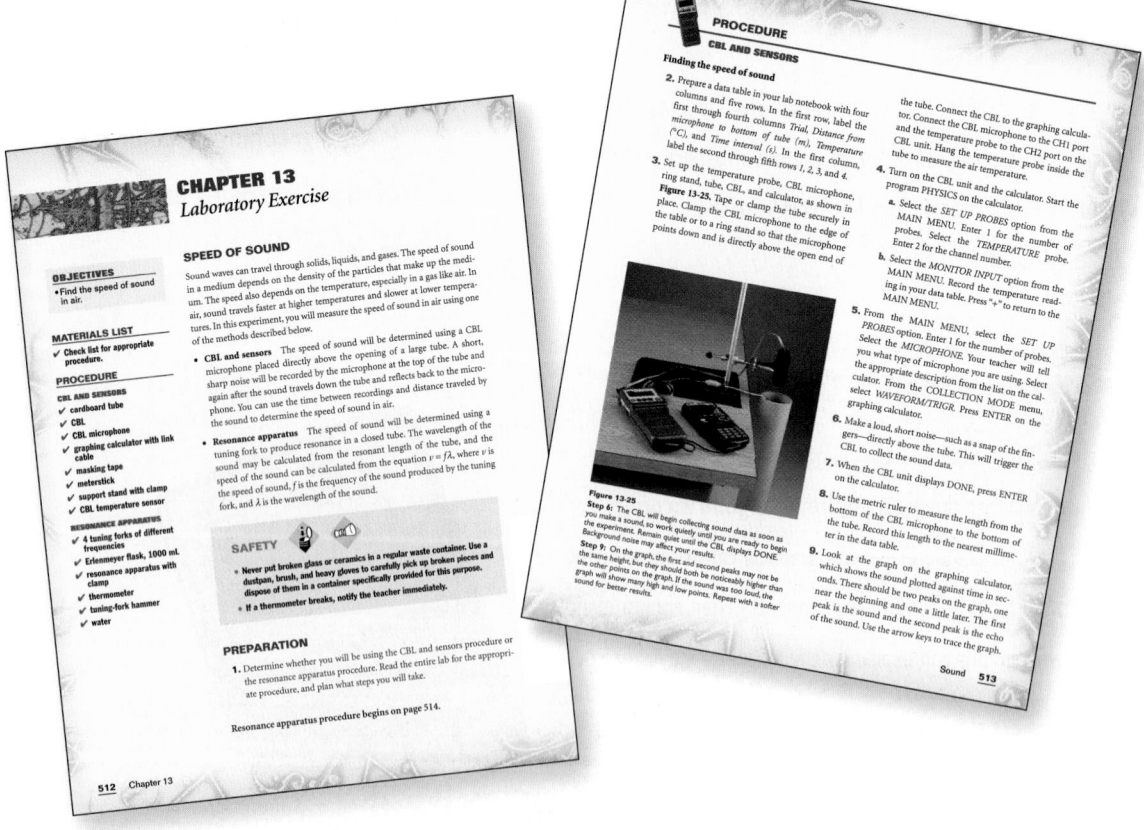

CHAPTER 13
Laboratory Exercise

SPEED OF SOUND

OBJECTIVES
• Find the speed of sound in air.

MATERIALS LIST
✔ Check list for appropriate procedure.

PROCEDURE

CBL AND SENSORS
✔ cardboard tube
✔ CBL
✔ CBL microphone
✔ graphing calculator with link cable
✔ masking tape
✔ meterstick
✔ support stand with clamp
✔ CBL temperature sensor

RESONANCE APPARATUS
✔ 4 tuning forks of different frequencies
✔ Erlenmeyer flask, 1000 mL
✔ resonance apparatus with clamp
✔ thermometer
✔ tuning-fork hammer
✔ water

Sound waves can travel through solids, liquids, and gases. The speed of sound in a medium depends on the density of the particles that make up the medium. The speed also depends on the temperature, especially in a gas like air. In air, sound travels faster at higher temperatures and slower at lower temperatures. In this experiment, you will measure the speed of sound in air using one of the methods described below.

• **CBL and sensors** The speed of sound will be determined using a CBL microphone placed directly above the opening of a large tube. A short, sharp noise will be recorded by the microphone at the top of the tube and again after the sound travels down the tube and reflects back to the microphone. You can use the time between recordings and distance traveled by the sound to determine the speed of sound in air.

• **Resonance apparatus** The speed of sound will be determined using a tuning fork to produce resonance in a closed tube. The wavelength of the sound may be calculated from the resonant length of the tube, and the speed of the sound can be calculated from the equation $v = f\lambda$, where v is the speed of the sound, f is the frequency of the sound produced by the tuning fork, and λ is the wavelength of the sound.

SAFETY

• Never put broken glass or ceramics in a regular waste container. Use a dustpan, brush, and heavy gloves to carefully pick up broken pieces and dispose of them in a container specifically provided for this purpose.

• If a thermometer breaks, notify the teacher immediately.

PREPARATION

1. Determine whether you will be using the CBL and sensors procedure or the resonance apparatus procedure. Read the entire lab for the appropriate procedure, and plan what steps you will take.

Resonance apparatus procedure begins on page 514.

512 Chapter 13

PROCEDURE
CBL AND SENSORS

Finding the speed of sound

2. Prepare a data table in your lab notebook with four columns and five rows. In the first row, label the first through fourth columns *Trial*, *Distance from microphone to bottom of tube (m)*, *Temperature (°C)*, and *Time interval (s)*. In the first column, label the second through fifth rows *1, 2, 3,* and *4*.

3. Set up the temperature probe, CBL microphone, ring stand, tube, CBL, and calculator, as shown in **Figure 13-25**. Tape or clamp the tube securely in place. Clamp the CBL microphone to the edge of the table or to a ring stand so that the microphone points down and is directly above the open end of the tube. Connect the CBL to the graphing calculator. Connect the CBL microphone to the CH1 port and the temperature probe to the CH2 port on the CBL unit. Hang the temperature probe inside the tube to measure the air temperature.

4. Turn on the CBL unit and the calculator. Start the program PHYSICS on the calculator.
   a. Select the *SET UP PROBES* option from the MAIN MENU. Enter 1 for the number of probes. Select the *TEMPERATURE* probe. Enter 2 for the channel number.
   b. Select the *MONITOR INPUT* option from the MAIN MENU. Record the temperature reading in your data table. Press "+" to return to the MAIN MENU.

5. From the MAIN MENU, select the *SET UP PROBES* option. Enter 1 for the number of probes. Select the *MICROPHONE*. Your teacher will tell you what type of microphone you are using. Select the appropriate description from the list on the calculator. From the *COLLECTION MODE* menu, select *WAVEFORM/TRIGR*. Press ENTER on the graphing calculator.

6. Make a loud, short noise—such as a snap of the fingers—directly above the tube. This will trigger the CBL to collect the sound data.

7. When the CBL unit displays DONE, press ENTER on the calculator.

8. Use the metric ruler to measure the length from the bottom of the CBL microphone to the bottom of the tube. Record this length to the nearest millimeter in the data table.

9. Look at the graph on the graphing calculator, which shows the sound plotted against time in seconds. There should be two peaks on the graph, one near the beginning and one a little later. The first peak is the sound and the second peak is the echo of the sound. Use the arrow keys to trace the graph.

Figure 13-25
Step 6: The CBL will begin collecting sound data as soon as you make a sound, so work quietly until you are ready to begin the experiment. Remain quiet until the CBL displays DONE. Background noise may affect your results.
Step 9: On the graph, the first and second peaks may not be the same height, but they should both be noticeably higher than the other points on the graph. If the sound was too loud, the graph will show many high and low points. Repeat with a softer sound for better results.

Sound **513**

Improved Analysis and Interpretation Items

Each of the in-text labs ends with a series of items designed to lead students step by step through the thinking and analysis necessary to reach the conclusions of the lab, just as the steps in the procedure lead students through the techniques of the lab.

Quick Labs

For teachers needing flexibility due to scheduling or equipment concerns, the text also includes 32 Quick Labs, brief thought-provoking demonstrations that students can perform in class or on their own at home. Quick Labs use easily found materials and involve little or no preparation time or clean-up afterward.

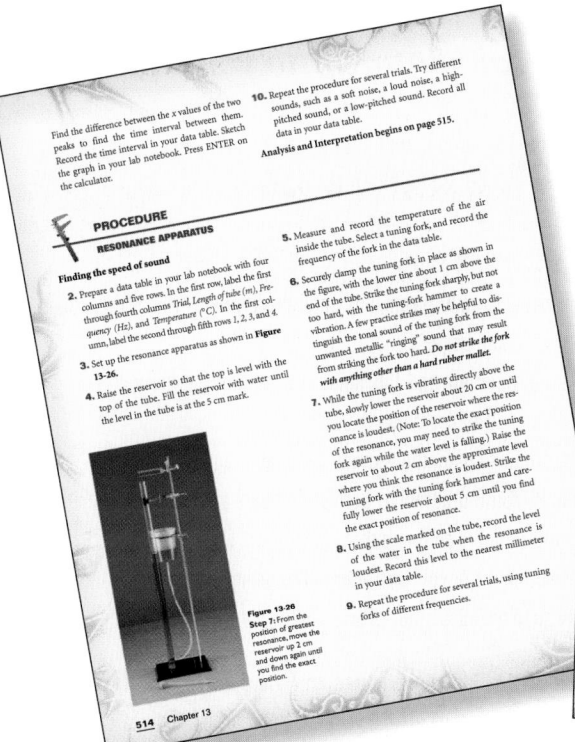

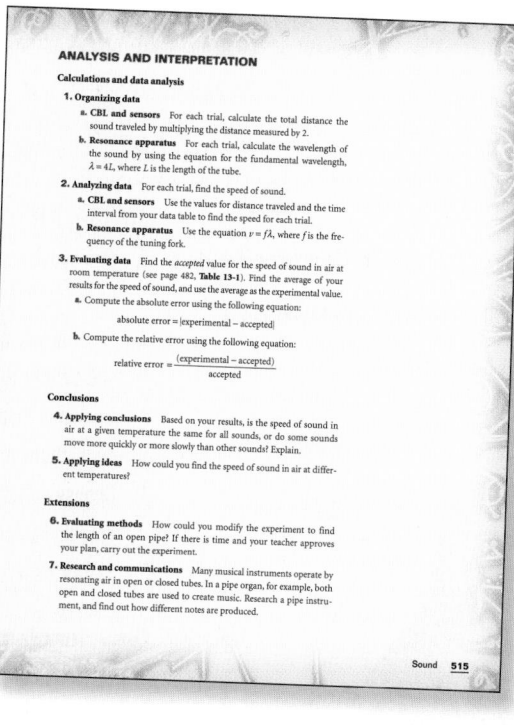

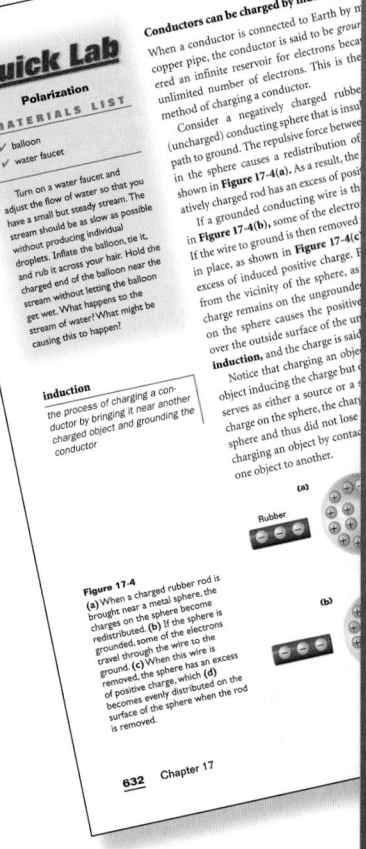

SAVING YOU TIME AND MONEY

✓ Tested Procedures

Physics teachers and the staff of Sargent-Welch/CENCO Scientific, Inc. rigorously bench-tested all procedures for the full-length labs, as well as the labs in the *Laboratory Experiments* booklet, for ease of use, safety, and practicality. As a result, many of the labs also include options for using the latest in cost-saving equipment.

✓ Sargent-Welch/CENCO Materials Ordering Software

With Sargent-Welch/CENCO's exclusive software-ordering system, specifically designed for use with *Holt Physics*, you can order your materials and supplies quickly and easily. The system lists all required and supplemental materials—per lab group or class size—needed for the full-length labs in the textbook and the labs in the Laboratory Experiments booklet. For the printed version of the Master Materials List for *Holt Physics*, see pages T48 through T53.

PHYSICS HOLT
Technology-Based Lab Activities

Using the Calculator-Based Laboratory (CBL™)
Simplifying the integration of CBL, sensors, and graphing calculators into your laboratory

FREE Physics Lab Software on the Web

Many of the laboratory experiments featured in this textbook contain procedures for the Texas Instruments Calculator-Based Laboratory (CBL) System, and all 17 of the labs found in the *Technology-Based Lab Activities* booklet do too.

WHAT TO DO

Finding PHYSICS Lab Software

1. Point your Web browser to **go.hrw.com,** and enter the keyword "HF2 HOME" to go to the *Holt Physics* home page.
2. Find the link to Vernier Software and Technology's Web site. This link will take you directly to where you can download the software.
3. Open and read (or print and read) the text file. It contains more-detailed instructions on the PHYSICS program by Vernier Software and Technology.
4. Choose only the version of the software that matches the CBL-compatible TI graphing calculator that you or your students will be using.
5. Follow the directions to download the software onto your hard drive.

Downloading PHYSICS Lab Software to Graphing Calculators

6. To download the PHYSICS group file from your computer to your calculator, follow the directions in Appendix B of the textbook. Because of the size of the PHYSICS program group, it is recommended that all other programs be removed before loading PHYSICS onto a calculator.
7. When the transfer is complete, a set of programs will appear on your graphing calculator. The set will include the program titled PHYSICS and a list of related programs that are to be used by the main PHYSICS program.
8. Use the link-to-link cable to copy your programs to each student's calculator.

Using PHYSICS Lab Software

9. To use the PHYSICS program, simply follow the instructions given in the lab procedure. The PHYSICS program can calibrate sensors, collect data, and perform data analysis. These features enable you to use the CBL™ to its full potential with many different Vernier probes and without writing your own programs or using many different interfaces.

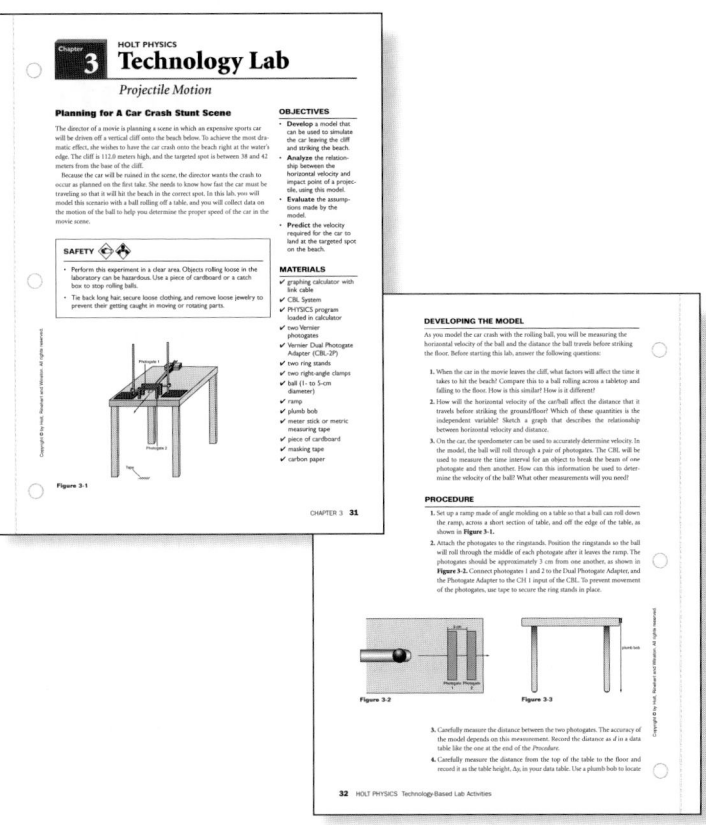

Features of CBL™ Procedures in Text and *Technology-Based Lab Activities*

- Step-by-step instructions refer to menus that appear on your calculator screen.

- All lab procedures were bench-tested with TI-82 and TI-83 calculators and the CBL™ system.

- Other TI calculators and the CBL2™ system can be used with minor adaptations.

Additional Features of *Technology-Based Lab Activities*

- A scenario-based approach allows students to make the connection between the physics principles studied in the lab and real-world situations.

- The procedures model complicated, large-scale phenomena using simplified situations.

- The data and conclusions are scaled back up to estimate the effects on the scenario.

- The accuracy of the models and the assumptions made are always examined and evaluated.

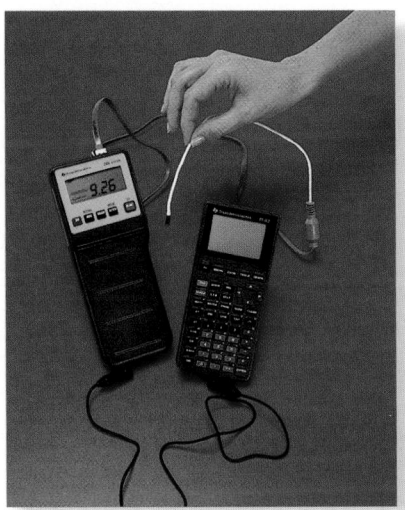

$\mathbb{S}$YSTEM REQUIREMENTS

You will need TI-GRAPH LINK™ software and a TI-GRAPH LINK computer-interface cable to get started using these programs with your TI graphing calculator and CBL™. These are available from Sargent-Welch/CENCO, from Vernier, or from educational- and office-supply companies.

Contact TI for Information and Technical Support
email: ti-cares@ti.com
address: Texas Instruments
P.O. Box 650311, MS 3962
Dallas, TX 75265

Customer Support Line
phone: 1-800-TI-CARES

Programming Assistance Technical Support
phone: (972) 917-8324

For more information on Vernier products
email: info@vernier.com
address: Vernier Software and Technology
13979 Millikan Way
Beaverton, OR 97005
phone: (503) 277-2299

HOLT PHYSICS Laboratory Experiments

An Innovative Approach

New strategies help students take responsibility for their own learning

Discovery Labs help students develop their own conceptual understanding of physics

Discovery Labs allow students to explore and experience a physics phenomenon before they study a chapter. The

procedures require no prior knowledge of the topic. Later, as students read the chapter, they will be able to draw upon

the concrete examples of the physics concepts they observed in lab to deepen their understanding.

● **Flexibility**

The lab procedure is broken down into small self-contained parts to maximize flexibility. If you have time, students can perform the whole lab. Alternatively, do each part immediately before students study the corresponding topics in the chapter. The lab can even be performed as a confirmatory exercise after the topic has been studied.

● **Integrated Procedure and Questions**

Each part of the procedure is followed by a brief set of items that are designed to solidify the students' observations and lead them to build their own understanding of the operative physics concepts.

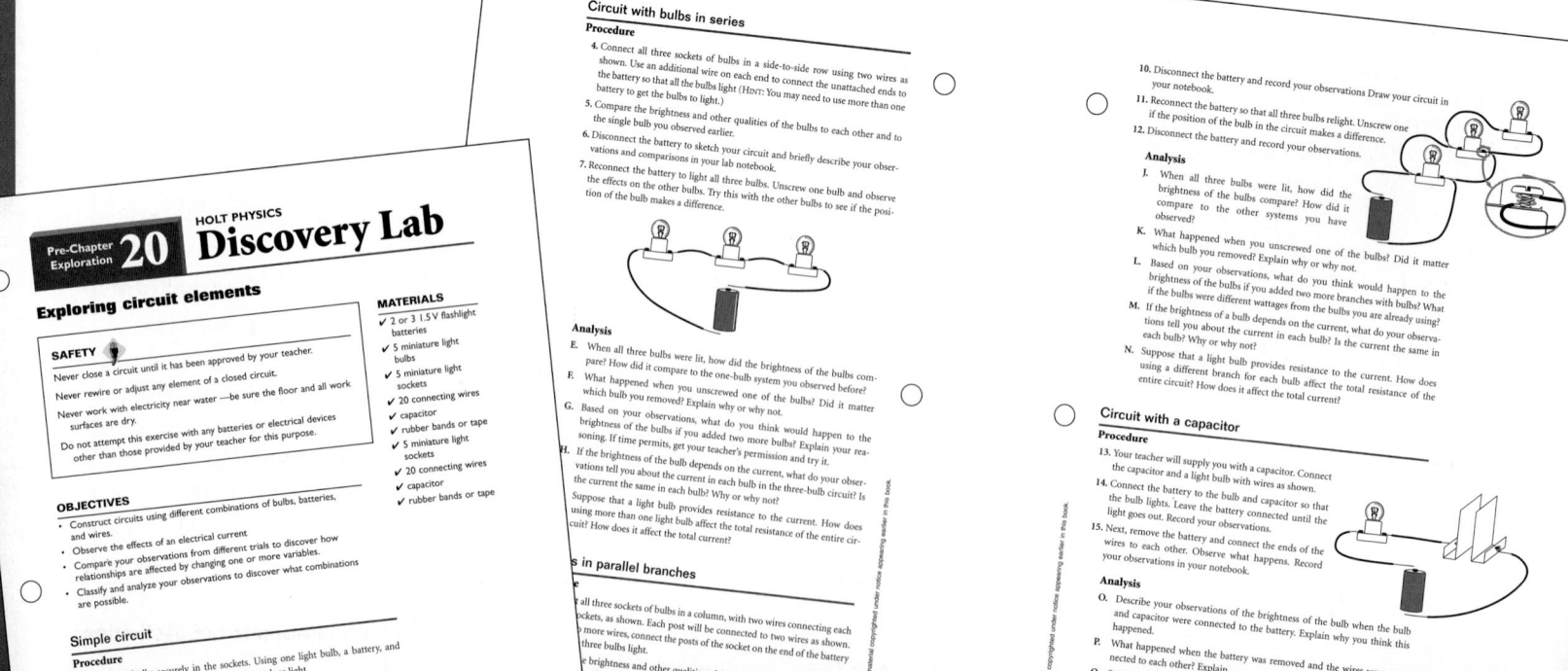

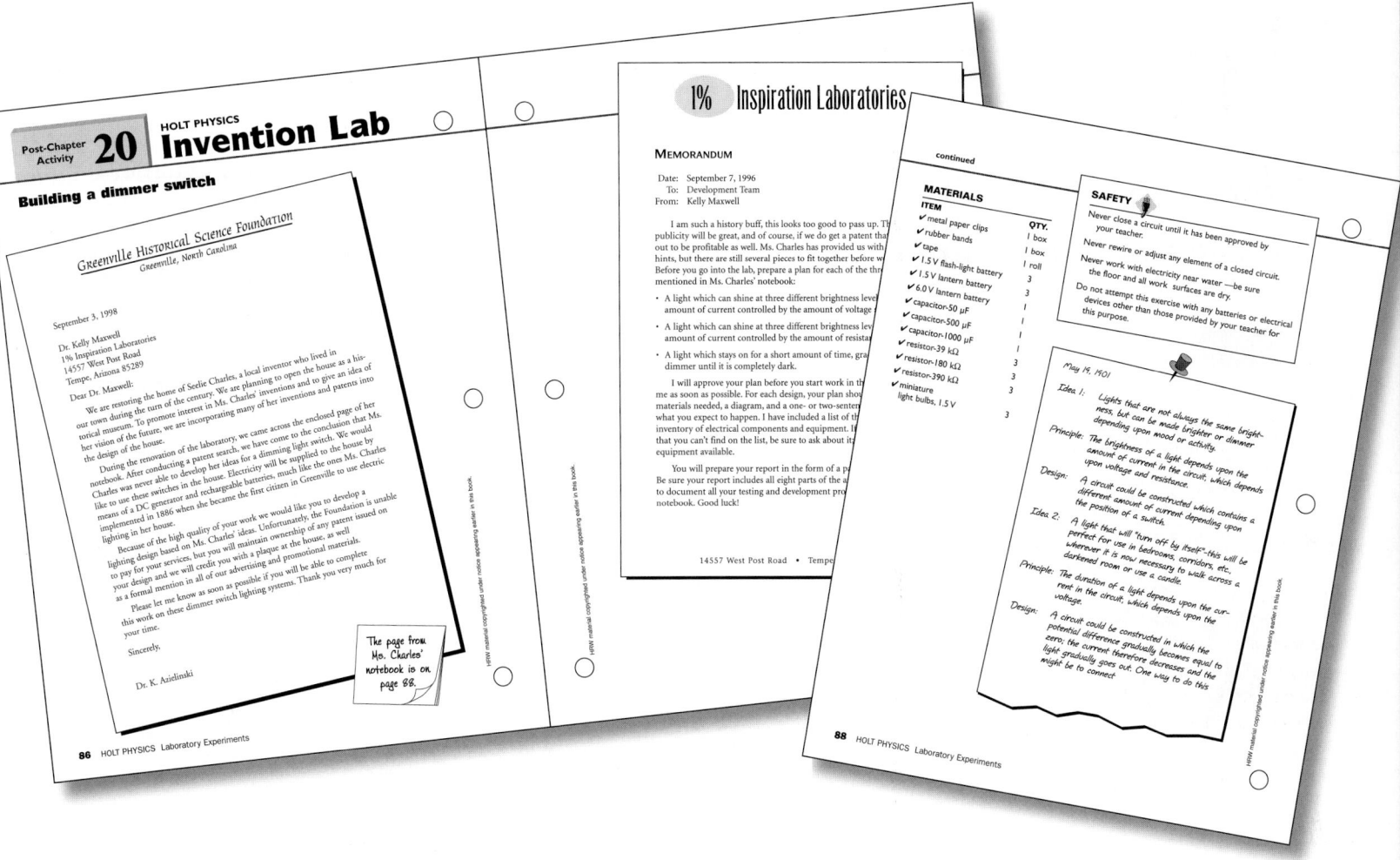

Invention Labs let students apply the concepts they've learned to a specific challenge

Invention labs provide an opportunity for open-ended inquiry labs.

All the invention labs are set in a working-world scenario in which students are employees of a research and development lab. Students are challenged to invent a device or test materials to solve a specific problem. Instead of being given a procedure, they receive only a list of some equipment that could be useful and a few hints. This way, they can create their own unique solution by applying their knowledge of physics.

● Initial Plan
is required from every student before they enter the lab. This helps focus student efforts so they can complete their work in the lab period. It also helps you identify students who may need more guidance from you before they proceed.

● Final Patent Application
is submitted by students instead of a traditional lab report. Students must make a sketch of their invention, delineate all of the parts, describe how it works, and include a discussion of the physics concepts underlying the operation of their invention.

HOLT
PHYSICS Lessons

Order, Simplicity, and Consistency

Sections are clearly organized with support for student understanding

15-1
Refraction

15-1 SECTION OBJECTIVES

- Recognize situations in which refraction will occur.
- Identify which direction light will bend when it passes from one medium to another.
- Solve problems using Snell's law.

refraction

the bending of a wave disturbance as it passes at an angle from one medium into another

REFRACTION OF LIGHT

Look at the tiny image of the flower that appears in the water droplet in **Figure 15-1**. The flower can be seen in the background of the photo. Why does the flower look different when viewed through the droplet? This phenomenon occurs because light is bent at the boundary between the water and the air around it. The bending of light as it travels from one medium to another is called **refraction**.

If light travels from one transparent medium to another at any angle other than straight on (parallel to the line normal to the surface), the light ray changes direction when it meets the boundary. As in the case of reflection, the angles of the incoming and refracted rays are measured with respect to the normal. For studying refraction, the normal line is extended into the refracting medium, as shown in **Figure 15-2**. The angle between the refracted ray and the normal is called the *angle of refraction, θ_r*. For refraction, the angle of incidence is designated as θ_i.

Refraction occurs when light's velocity changes

Glass, water, ice, diamonds, and quartz are all examples of transparent media through which light can pass. The speed of light in each of these materials is different. The speed of light in water, for instance, is less than the speed of light in air. And the speed of light in glass is less than the speed of light in water.

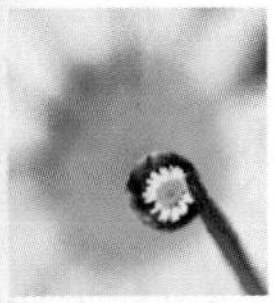

Figure 15-1
The flower looks small when viewed through the water droplet because the light from the flower is bent as it passes through the water.

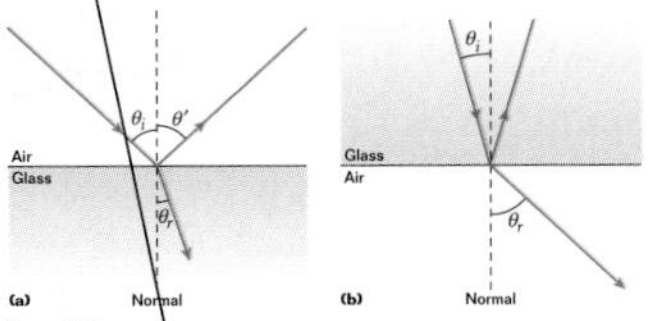

Figure 15-2
(a) When the light ray moves from air into glass, its path is bent toward the normal, (b) whereas the path of the light ray moving from glass into air is bent away from the normal.

562 Chapter 15

index of refraction

the ratio of the speed of light in a vacuum to its speed in a given transparent medium

THE LAW OF REFRACTION

An important property of transparent substances is the **index of refraction**. The index of refraction for a substance is the ratio of the speed of light in a vacuum to the speed of light in that substance.

INDEX OF REFRACTION

$$n = \frac{c}{v}$$

$$\text{index of refraction} = \frac{\text{speed of light in vacuum}}{\text{speed of light in medium}}$$

From this definition, we see that the index of refraction is a dimensionless number that is always greater than 1 because light always travels slower in a substance than in a vacuum. **Table 15-1** lists the indices of refraction for some representative substances. Note that the larger the index of refraction is, the slower light travels in that substance and the more a light ray will bend when it passes from a vacuum into that material.

Imagine, as an example, light passing between air and water. When light begins in the air (high speed of light and low index of refraction) and travels into the water (lower speed of light and higher index of refraction), the light rays are bent toward the normal. Conversely, when light passes from the water to the air, the light rays are bent away from the normal.

Note that the value for the index of refraction of air is nearly that of a vacuum. For simplicity, *use the value n = 1.00 for air when solving problems.*

Did you know?

The index of refraction of any medium can also be expressed as the ratio of the wavelength of light in a vacuum, λ_0, to the wavelength of light in that medium, λ_m, as shown in the following relation.

$$n = \frac{\lambda_0}{\lambda_m}$$

Table 15-1   Indices of refraction for various substances*

Solids at 20°C	n	Liquids at 20°C	n
Cubic zirconia	2.20	Benzene	1.501
Diamond	2.419	Carbon disulfide	1.628
Fluorite	1.434	Carbon tetrachloride	1.461
Fused quartz	1.458	Ethyl alcohol	1.361
Glass, crown	1.52	Glycerine	1.473
Glass, flint	1.66	Water	1.333
Ice (at 0°C)	1.309		
Polystyrene	1.49	**Gases at 0°C, I atm**	**n**
Sodium chloride	1.544	Air	1.000 293
Zircon	1.923	Carbon dioxide	1.000 450

*measured with light of vacuum wavelength = 589 nm

564 Chapter 15

Key Equations
are highlighted in a gold-colored box so that they are easy to find. To emphasize the concepts behind the equation, each of these key equations is titled and shown in words as well as symbols.

Headline Organization
includes both topic heads and summary heads that break the section's contents into manageable pieces. Summary heads also make locating specific content easier and can be used to facilitate review before a test.

Definitions
are placed in the margin and are located at point of use, so that they don't interrupt the narrative flow in the text.

Section Objectives
make it clear to students what they are expected to learn.

Did You Know?
cultivates student interest by including interesting information about the topic being studied.

ADDITIONAL CONTENT AREA READING SUPPORT

Reading science requires some different skills than those required to read a novel or short story. **Holt Science Skills Workshop: Reading in the Content Area** gives your students the strategic help they need to read in science class. Student worksheets develop skills such as comparison context clues, sequence context clues, probability context clues, and the use of cause and effect markers. Teaching strategies and a set of transparencies give you the tools to teach these skills, so students can make the most of their use of the textbook.

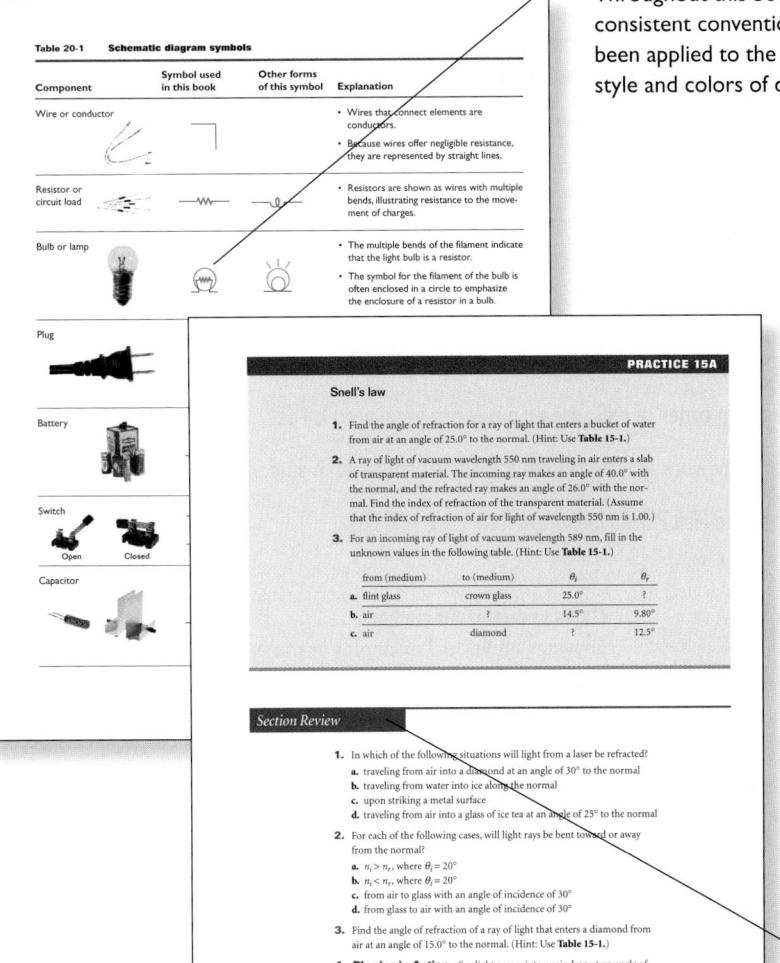

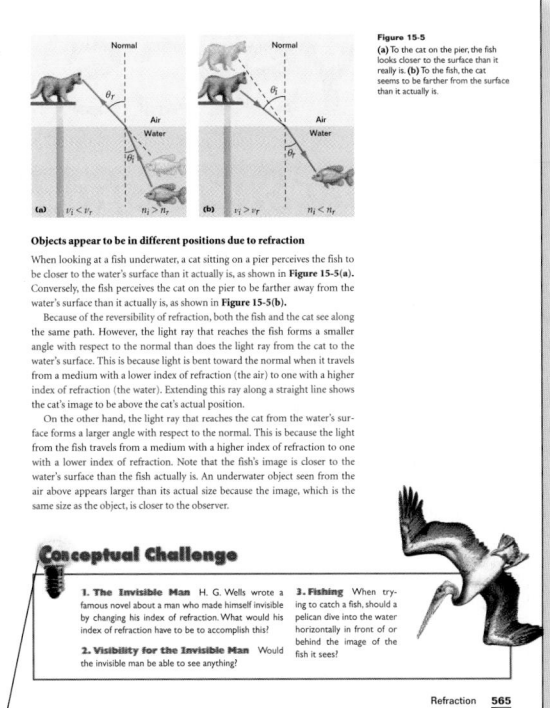

Physics Diagrams

Throughout this book, consistent conventions have been applied to the visual style and colors of diagrams.

Conceptual Challenge

provides a point-of-use check for students to ensure that they truly understand the conceptual basis of what they are studying in a new context.

Section Review

contains items that correspond to the Section Objectives, so you can accurately and immediately assess student achievement of the objectives. The items labeled "Physics in Action" require students to apply their new-found knowledge to some aspect of the application developed in the chapter's opening story.

SIGNIFICANT FIGURES

Holt Physics provides a clear but rigorous approach to significant figures in Chapter 1. This approach is maintained consistently throughout the rest of the program.

There are several valid approaches to significant figures that are used by different physics teachers. The discussion of significant figures in Holt Physics makes it clear that students should consult you, the teacher, for confirmation of which rules you believe are appropriate for your class.

Two people solving a problem with the same rules for significant figures can reach slightly different results depending on the order of operations they apply to the problem. As a result, you should be prepared to accept some small variances between the students' answers and the answers provided in the Teacher's Edition.

HOLT
PHYSICS Problem Solving

Modeling Successful Problem Solving

More practice problems than any other program

In the *Holt Physics* program, you get 291 worked-out sample problems showing three different rearrangements for each of 97 equations or problem types and more than 2800 practice problems. Practice problems are in order of difficulty, gradually progressing to harder problems. All practice problems are available in randomized form on the web homework grading system at go.hrw.com, so each student can be assigned unique problem values. Problem-solving skills are reinforced by the *Holt Physics Interactive Tutor* **(see p. T20).**

● Classroom Practice
provides additional examples in the extended margin of the Teacher's Edition to be used as team-work exercises or for demonstration in class.

Four Step Method ●
used for most problems, provides the structural framework students need to be able to connect the skills of problem solving with the conceptual reasons why the methods work.

● Sample Problem
provides the in-depth explanations students need and models the reasoning processes that can be used to solve similar problems.

**Alternative Problem- ●
Solving Approach**
in the extended margin of the Teacher's Edition highlights a different approach that can be used to solve the Sample Problem. Students often appreciate knowing that there is more than one path to an answer. In addition, this approach can be used to verify answers.

● Calculator Solution
contains information about special calculator functions and details the rounding necessary to maintain a consistent approach to significant figures.

**Interactive Problem- ●
Solving Tutor Icons**
in the margins alert you to problem types that are covered on the Interactive Tutor CD-ROMs.

SECTION 12-1

Classroom Practice

The following may be used as teamwork exercises or for demonstration at the chalkboard or on an overhead projector.

PROBLEM
Hooke's law

A 76 N crate is attached to a spring ($k = 450$ N/m). How much displacement is caused by the weight of this crate?

Answer
−0.17 m

A spring of $k = 1962$ N/m loses its elasticity if stretched more than 50.0 cm. What is the mass of the heaviest object the spring can support without being damaged?

Answer
1.00×10^2 kg

**Alternative Problem-
Solving Approach**
The weight of the object pulls downward.
$F_g = -mg = -5.4$ N
The spring stretches until its restoring force ($F_{elastic} = -kx$) balances the −5.4 N. This occurs when $x = -0.020$ m. Thus, 5.4 N = −k(−0.020 m) and $k = 270$ N/m.

PHYSICS Interactive Problem-Solving Tutor

See Module 11
"Hooke's Law" provides additional development of problem-solving skills for spring problems.

440

SAMPLE PROBLEM 12A

Hooke's law

PROBLEM

If a mass of 0.55 kg attached to a vertical spring stretches the spring 2.0 cm from its original equilibrium position, as shown in Figure 12-2, what is the spring constant?

$x = -2.0$ cm

Figure 12-2

SOLUTION

1. DEFINE **Given:** $m = 0.55$ kg $x = -2.0$ cm $= -0.020$ m
$g = 9.81$ m/s^2

Unknown: $k = ?$

Diagram:

$F_{elastic}$

F_g

2. PLAN **Choose an equation(s) or situation:** When the mass is attached to the spring, the equilibrium position changes. At the new equilibrium position, the net force acting on the mass is zero. So the spring force (given by Hooke's law) must be equal and opposite to the weight of the mass.

$$F_{net} = 0 = F_{elastic} + F_g$$
$$F_{elastic} = -kx$$
$$F_g = -mg$$
$$-kx - mg = 0$$

Rearrange the equation(s) to isolate the unknown(s):

$$kx = -mg$$
$$k = \frac{-mg}{x}$$

3. CALCULATE **Substitute the values into the equation(s) and solve:**

$$k = \frac{-(0.55 \text{ kg})(9.81 \text{ m/s}^2)}{-0.020 \text{ m}}$$

$$\boxed{k = 270 \text{ N/m}}$$

CALCULATOR SOLUTION
The calculator answer for k is 269.775. This answer is rounded to two significant figures, 270 N/m.

4. EVALUATE The value of k implies that about 300 N of force is required to displace the spring 1 m.

440 Chapter 12

Holt Physics Teacher's Edition

Practice problems

36. Microwaves travel at the speed of light, 3.00×10^8 m/s. When the frequency of microwaves is 9.00×10^9 Hz, what is their wavelength?
(See Sample Problem 12D.)

WAVE INTERACTIONS

Review questions

37. Using the superposition principle, draw the resultant waves for each of the examples in **Figure 12-29.**

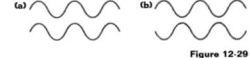

(a) **(b)**

Figure 12-29

38. What is the difference between constructive interference and destructive interference?

39. Which waveform of those shown in **Figure 12-30** is the resultant waveform?

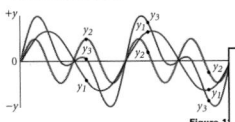

Figure 12-

40. Anthony sends a series of pulses of amplitude 24 down a string that is attached to a post at one Assuming the pulses are reflected with no lo amplitude, what is the amplitude at a point on string where the two pulses are crossing if
a. the string is rigidly attached to the post?
b. the end at which reflection occurs is fr slide up and down?

41. A wave of amplitude 0.75 m interferes with a ond wave of amplitude 0.53 m.
a. Find the maximum possible amplitude of resultant wave if the interference is construc
b. Find the maximum possible amplitude of resultant wave if the interference is destruc

Conceptual questions

42. Can more than two waves interfere in a given medium?

43. What is the resultant displacement at a position where destructive interference is complete?

44. When two waves interfere, can the resultant wave be larger than either of the two original waves? If so, under what conditions?

45. Which of the following wavelengths will produce standing waves on a string that is 3.5 m long?
a. 1.75 m
b. 3.5 m
c. 5.0 m
d. 7.0 m

MIXED REVIEW

46. In an arcade game, a 0.12 kg disk is shot across a frictionless horizontal surface by being compressed against a spring and then released. If the spring has a spring constant of 230 N/m and is compressed from its equilibrium position by 6.0 cm, what is the magnitude of the spring force on the disk at the moment it is released?

12 REVIEW & ASSESS

33. neither, because the speed of sound is constant in air
34. They can transport large amounts of energy rapidly.
35. a. 9.0 cm
b. 20.0 cm
c. 0.0400 s
d. 5.00 m/s
36. 0.0333 m
37. a. a sine wave with twice the amplitude
b. a straight line (the waves cancel each other completely)
38. In constructive interference, individual displacements are on the same side of equilibrium. In destructive interference, the individual displacements are on opposite sides of equilibrium.
39. y_3
40. a. 0 cm
b. 48 cm
41. a. 1.28 m

PRACTICE 12A

Hooke's law

1. Suppose the spring in Sample Problem 12A is replaced with a spring that stretches 36 cm from its equilibrium position.
a. What is the spring constant in this case?
b. Is this spring stiffer or less stiff than the one in Sample Problem 12A?

2. A load of 45 N attached to a spring that is hanging vertically stretches the spring 0.14 m. What is the spring constant?

3. A slingshot consists of a light leather cup attached between two rubber bands. If it takes a force of 32 N to stretch the bands 1.2 cm, what is the equivalent spring constant of the rubber bands?

4. How much force is required to pull the cup of the slingshot in problem 3 3.0 cm from its equilibrium position?

A stretched or compressed spring has elastic potential energy

As you saw in Chapter 5, a stretched or compressed spring stores elastic potential energy. To see how mechanical energy is conserved in an ideal mass-spring system, consider an archer shooting an arrow from a bow, as shown in **Figure 12-3.** Bending the bow by pulling back the bowstring is analogous to stretching a spring. To simplify this situation, we will disregard friction and internal energy.

Once the bowstring has been pulled back, the bow stores elastic potential energy. Because the bow, arrow, and bowstring (the system) are now at rest, the kinetic energy of the system is zero, and the mechanical energy of the system is solely elastic potential energy.

When the bowstring is released, the bow's elastic potential energy is converted to the kinetic energy of the arrow. At the moment the arrow leaves the bowstring, it gains most of the elastic potential energy originally stored in the bow. (The rest of the elastic potential energy is converted to the kinetic energy of the bow and the bowstring.) Thus, once the arrow has been released, the mechanical energy of the bow-and-arrow system is solely kinetic. Because mechanical energy must be conserved, the kinetic energy of the bow, arrow, and bowstring is equal to the elastic potential energy originally stored in the bow.

Module 11
"Hooke's Law" provides an interactive lesson with guided problem-solving practice to teach you about springs and spring constants.

Figure 12-3
The elastic potential energy stored in this stretched bow is converted into the kinetic energy of the arrow.

Vibrations and Waves **441**

SECTION 12-1

PRACTICE GUIDE 12A

Solving for:

k	📖 **PE** Sample, 1–3; Ch. Rvw. 8–9	
	💻 PW 3, 4*, 5*	
	PB 5–7	
F	📖 **PE** 4; Ch. Rvw. 46–47, 55	
	💻 PW 6*, 7, 8*	
	PB Sample, 1–4	
x	💻 PW Sample, 1–5	
	PB 8–10	

ANSWERS TO

Practice 12A
Hooke's law
1. a. 15 N/m
b. less stiff
2. 3.2×10^2 N/m
3. 2.7 ×
4. 81

NAME _____ DATE _____ CLASS _____

Holt Physics
Problem 12A
HOOKE'S LAW

PROBLEM
The pygmy shrew has an average mass of 2.0 g. How many shrews of average mass must be placed on a spring scale with a spring constant of 24 N/m to produce a 4.0 cm compression?

SOLUTION
1. DEFINE Given:
m = mass of one shrew = 2.0 g = 2.0×10^{-3} kg
g = 9.81 m/s^2
x = −4.0 cm = -4.0×10^{-2} m
k = 24 N/m

Unknown: n = total number of shrews = ?

2. PLAN Choose an equation(s) or situation: When the shrews are attached to the spring, the equilibrium position changes. At the new equilibrium position, the net force acting on the shrews is zero. So the spring force (given by Hooke's law) must be equal to and opposite the weight of the shrews.

$F_{net} = 0 = F_{elastic} + F_g$
$F_{elastic} = -kx$
$F_g = -m_{total} g = -nmg$
$-kx - nmg = 0$

Rearrange the equation(s) to isolate the unknown(s):
$$n = \frac{-kx}{mg}$$

3. CALCULATE Substitute the values into the equation(s) and solve:
$$n = \frac{-(24 \text{ N/m})(-4.0 \times 10^{-2} \text{ m})}{mg}$$

$$= \boxed{49}$$

4. EVALUATE Forty-nine shrews of 2.0 g each provide a total mass of about 0.1 kg, or a weight of just under 1 N. From the value of the spring constant, a force of 1 N should displace the spring by 1/24 of a meter, or about 4 cm. This indicates that the final result is consistent with the rest of the data.

ADDITIONAL PRACTICE

1. The largest meteorite of lunar origin reportedly has a mass of 19 g. If the meteorite produces a compression of 2.24 mm when placed on a spring scale, what is the spring constant of the spring?

Problem 12A **103**

• More Practice Problems

in the end-of-chapter Review and Assess are correlated to a specific Sample Problem in the chapter where students can go for help. Answers to selected problems are found in the back of the Pupil's Edition.

• Practice Guides

show you which problems solve for which variables so you can easily control the difficulty of the assignment.

Practice Problems •

follow each Sample Problem. Answers for selected problems are in the back of the Pupil's Edition. All answers are contained in the extended margin of the Teacher's Edition. Detailed solutions are in the Teacher's Solution Manual and Answer Key.

Problem Workbook • and Problem Bank

provide even more samples and practice.

Problem Solving in Holt Physics

T35

HOLT
PHYSICS Assessment

Measuring Conceptual Understanding and Problem-Solving Skills

Comprehensive assessment targets several levels of understanding

Each Review and Assess is divided into topics that include review items that focus on effective recall of key physics facts and relationships; conceptual questions that build in-depth understanding of the concepts behind those facts; and practice problems that gauge the ability to apply those facts, relationships, and concepts to solve problems.

Review Questions focus on basic comprehension of key physics facts and relationships.

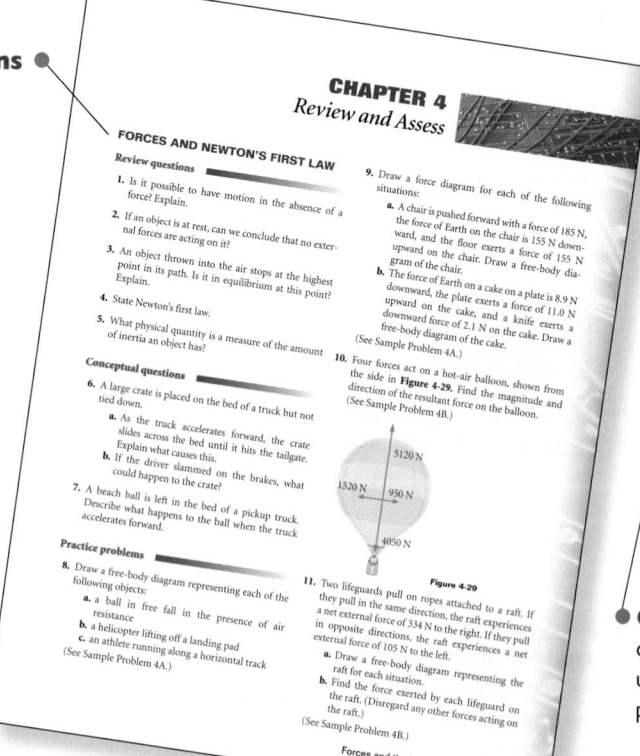

Practice Problems provide additional practice with the techniques presented in the Sample Problems. Each Practice Problem is correlated with a Sample Problem in the chapter.

Conceptual Questions challenge students to apply their understanding of the concepts of physics to a different situation.

Mixed Review

presents a series of additional problems of varying levels of difficulty that often require students to draw on their knowledge of other chapters, without indicating which sample problems model the problem.

Technology and Learning

gives students practical exercise in using the graphing and programming functions of a graphing calculator to explore an aspect of the chapter's topic using new Flash™ Technology in TI-82 Plus and TI-83 Plus graphing calculators.

Alternative Assessment

contains ideas that can be used to extend student understanding of the implications of a physics topic and to assess other student abilities. For guidelines to use in analyzing students' performance items and portfolio projects, see below.

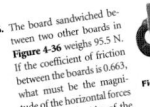

ASSESSMENT ON THE HOLT PHYSICS ONE-STOP PLANNER CD-ROM

Alternative Assessment: C³P Project's Assessments

More than 200 additional alternative assessment ideas are included, taken from the bank of items created by the Comprehensive Conceptual Curriculum for Physics (C³P) project, an NSF-funded effort to bring more hands-on experiences and less lecturing to physics teaching. Rubrics for these assessments and for lab activities are also found on the *One-Stop Planner CD-ROM*.

Traditional Assessment: Test Generator

The *Holt Physics* Test Generator with ExamView Software contains more than 2000 items, each of which is keyed to a specific chapter and section objective. Other codes detail the level of difficulty of the problem. This coding allows you to use the software to build customized chapter tests that focus on the areas you emphasize within each chapter. You can also add your own items to the bank of test questions.

One-Stop Planner
CD-ROM

HOLT
PHYSICS Features

Application of Concepts

Connecting the physics of the classroom to the real world and to the world of the working physicist

● **Tomorrow's Technology**
identifies ways in which basic physics concepts are being incorporated into new inventions and technologies that could change how we live. These features are adapted from the acclaimed *Pulse of the Planet* Public Radio Series.

● **Consumer Focus**
links the physics concepts discussed in the classroom to common experiences outside the classroom, such as driving cars, using microwave ovens, and choosing batteries.

● **Science, Technology, and Society**
explores some of the ways that science and technological innovation can impact societal issues. By connecting the knowledge of the classroom with real-world challenges, these features demonstrate how crucial science is to being an informed citizen.

Consumer Focus — Shock Absorbers and Damped Oscillation

Bumps in the road are certainly a nuisance, but without strategic use of damping devices, they could also prove deadly. To control a car going 110 km/h (70 mi/h), a driver needs all the wheels on the ground. Bumps in the road lift the wheels off the ground and rob the driver of control. A good solution is to fit the car with springs at each wheel. The springs absorb energy as the wheels rise over the bumps and push the wheels back to the pavement to keep the wheels on the road. However, once set in motion, springs tend to continue to go up and down in simple harmonic motion. This affects the driver's control of the car and also can be very uncomfortable.

One way to cut down on unwanted vibrations is to use stiff springs that compress only a few centimeters under thousands of newtons of force. Such springs have very high spring constants and thus do not vibrate as freely as softer springs with lower constants. However, this solution reduces the driver's ability to keep the car's wheels on the road.

To completely solve the problem, energy-absorbing devices known as shock absorbers are placed parallel to the springs in some automobiles, as shown in **Figure 12-4(a).** Shock absorbers are fluid-filled tubes that turn the simple harmonic motion of the springs into damped harmonic motion. In damped harmonic motion, each cycle of stretch and compression of the spring is much smaller than the previous cycle. Modern auto suspensions are set up so that all of a spring's energy is absorbed by the shock absorbers, eliminating vibrations in just one up-and-down cycle. This keeps the car from continually bouncing without sacrificing the spring's ability to keep the wheels on the road.

Different spring constants and shock absorber damping are combined to give a wide variety of road responses. For example, larger vehicles have heavy-duty leaf springs made of stacks of steel strips, which have a larger spring constant than coil springs do. In this type of suspension system, the shock absorber is perpendicular to the spring, as shown in **Figure 12-4(b).** The stiffness of the spring can affect steering response time, traction, and the general feel of the car.

As a result of the variety of combinations that are possible, your driving experiences can range from the luxurious floating of a limousine to the bone-rattling road feel of a sports car.

(a) Shock absorber / Coil spring
(b) Shock absorber / Leaf spring

Figure 12-4

Tomorrow's Technology — Deep-Sea Air Conditioning

Deep beneath the ocean, about half a mile down, sunlight barely penetrates the still waters. Scientists at Makai Ocean Engineering in Hawaii are now tapping into that pitch-dark region as a resource for air conditioning.

In tropical locations where buildings are cooled year-round, air-conditioning systems operate with cold water. Compressors cool the water, and pumps circulate it throughout the walls of a building, where the water absorbs heat from the rooms. Unfortunately, powering these compressors is neither cheap nor efficient.

Instead of using compressors to cool the water in their operating system, the systems designed by Makai use frigid water from the ocean's depths. First, engineers install a pipeline that reaches deep into the ocean, where the water is nearly freezing. Then, powerful pumps on the shoreline move the water directly into a building's air-conditioning system. There, a system of heat exchangers uses the sea water to cool the fresh water in the air-conditioning system.

One complicating factor is that the water must also be returned to the ocean in a manner that will not disrupt the local ecosystem. It must be either piped to a depth of a few hundred feet, where its temperature is close to that of the ocean at that level, or poured into onshore pits, where it eventually seeps through the land and comes to an acceptable temperature by the time it reaches the ocean.

"This deep-sea air conditioning benefits the environment by operating with a renewable resource instead of freon," said Dr. Van Ryzin, the president of Makai. "Because the system eliminates the need for compressors, it uses only about 10 percent of the electricity of current methods, saving fossil fuels and a lot of money." However, deep-sea air conditioning technology only works for buildings within a few kilometers of the shore and carries a hefty installation cost of several million dollars. For this reason, Dr. Van Ryzin thinks this type of system is most appropriate for large central air-conditioning systems, such as those necessary to cool resorts or large manufacturing plants, where the electricity savings can eventually make up for the installation costs. Under the right circumstances, air conditioning with sea water can be provided at one-third to one-half the cost of conventional air conditioning.

Circulation pump for fresh water / Warm fresh water / Building / Cool fresh water / Heat exchanger / Discharge of sea water into ground / Warm sea water / Cool sea water / Intake of sea water from ocean / Seawater pump

Science • Technology • Society

Climatic Warming

Scientists typically devise solutions to problems and then test the solution to determine if it indeed solves the problem. But sometimes the problem is only suggested by the evidence, and there are no chances to test the solutions. A current example of such a problem is climatic warming.

Data recorded from various locations around the world over the past century indicate that the average atmospheric temperature is 0.5°C higher now than it was 100 years ago. Although this sounds like a small amount, such an increase can have pronounced effects. Increased temperatures may eventually cause the ice in polar regions to melt, causing ocean levels to increase, which in turn may flood some coastal areas.

Small changes in temperature can also affect living organisms. Most trees can tolerate only about a 1°C increase in average temperature. If a tree does not reproduce often or easily enough to "migrate" through successive generations to a cooler location, it can become extinct in that region. Any organisms dependent on that type of tree also will suffer.

But such disasters depend on whether global temperatures continue to increase. Historical studies indicate that some short-term fluctuations in climate are natural, like the "little ice age" of the seventeenth century. If the current warming trend is part of a natural cycle, the dire predictions may be overstated or wrong.

Even if the warming is continuous, climatic systems are very complex, and involve many unexpected factors. For example, if polar ice melts, a sudden increase in humidity may result in snow in polar areas. This could counter the melting, thus causing ocean levels to remain stable.

Greenhouse Gases

Most of the current attention and concern about climatic warming has been focused on the increase in the amount of "greenhouse gases," primarily carbon dioxide and methane, in the atmosphere. Molecules of these gases absorb incoming solar infrared radiation and reradiate it in random directions, causing the radiation to be trapped and to warm the atmosphere, similar to the way the glass panes of a greenhouse trap heat and warm the air within.

While carbon dioxide [is one of the] components of the air, th[...] rapidly during the last hu[...] determined by analyzing[...] Greenland. Deeper sectio[...] earlier times. During the [...] 185 ppm of carbon dioxid[...] concentration from 130 ye[...] 300 ppm. Today, the levels [...] that can be accounted for b[...] combustion reactions, prim[...] petroleum burning, and by [...] consuming trees through de[...]

But does the well-docum[...] greenhouse gas concentratio[...] predictions? Atmospheric p[...] improved their models in re[...] able to correctly predict past [...] the heat-absorbing qualities o[...] models remain oversimplifie[...] of detailed long-term data. In [...] many variables, such as fluctu[...] output and volcanic processes[...] and cannot be factored into pr[...] factors into account would req[...] models and more-sophisticated[...] are currently available. As a res[...] whether meaningful decisions a[...]

Research[...]

1. Carbon dioxide levels in t[...] varied during Earth's history. F[...] volcanoes, plants, and limeston[...] determine whether these proce[...] on the current increase in CO_2 [...] you think of any practical mean[...] processes to reduce CO_2 concer[...] would be the advantages and dis[...]

HOLT PHYSICS Teacher's Edition

Convenience and Flexibility

Detailed planning guides help you organize all of the resources of Holt Physics

For each of the book's sections, the chapter interleaf pages show you at a glance what classroom and assignment options you have, and what instructional technology you can use to enliven the chapter's topics.

● Assignment Resources

- Content Mastery items gauge students' knowledge of the section's content
- Conceptual Understanding items assess how deeply students understand key physics principles.
- Problem-Solving Practice items build the critical thinking skills necessary to apply physics knowledge to solve problems.

● Classroom Resources

- Keep on target with detailed breakdown of science standards
- Attention-grabbing teacher demonstrations
- Appropriate transparencies and transparency masters
- All pertinent laboratory activities, including those in the *Laboratory Experiments* and *Technology-Based Lab Activities* booklets

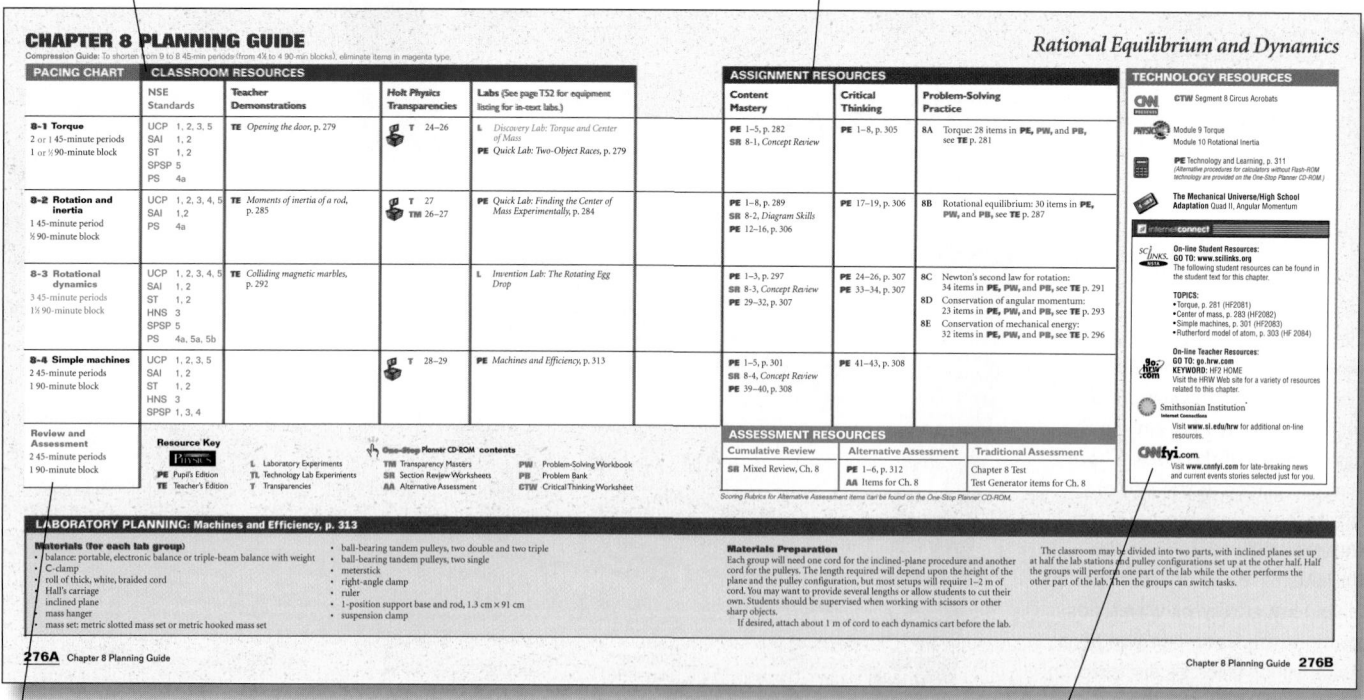

● Pacing Chart

- Indicates proposed timing for each section
- Includes compression guide: eliminate options in magenta type to shorten the course

Technology Resources ●

This part of the planning guide helps you identify specific places where you can integrate technology into your classroom, including internetconnect resources.

side of each bulb is connected to the positive terminal of the battery, and the right side of each bulb is connected to the negative terminal. Because the sides of each bulb are connected to common points, the potential difference across each bulb is the same. If the common points are the battery's terminals, as they are in the figure, the potential difference across each resistor is also equal to the terminal voltage of the battery. The current in each bulb, however, is not always the same.

The sum of currents in parallel resistors equals the total current

In **Figure 20-15**, when a certain amount of charge leaves the positive terminal and reaches the branch on the left side of the circuit, some of the charge moves through the top bulb and some moves through the bottom bulb. If one of the bulbs has less resistance, more charge moves through that bulb because the bulb offers less opposition to the flow of charges.

Because charge is conserved, the sum of the currents in each bulb equals the current delivered by the battery. This is true for all resistors in parallel.

$$I = I_1 + I_2 + I_3 \ldots$$

The parallel circuit shown in **Figure 20-15** can be simplified to an equivalent resistance with a method similar to the one used for series circuits. To do this, first show the relationship among the currents.

$$I = I_1 + I_2$$

Then substitute the equivalents for current according to $\Delta V = IR$.

$$\frac{\Delta V}{R_{eq}} = \frac{\Delta V_1}{R_1} + \frac{\Delta V_2}{R_2}$$

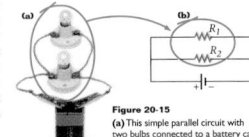

Figure 20-15
(a) This simple parallel circuit with two bulbs connected to a battery can be represented by **(b)** the schematic diagram shown on the right.

Quick Lab

Series and Parallel Circuits

MATERIALS LIST

- ✔ 4 regular drinking straws
- ✔ 4 stirring straws or coffee stirrers
- ✔ tape

Cut the regular drinking straws and thin stirring straws into equal lengths. Tape them end to end in long tubes to form series combinations. Form parallel combinations by taping the straws together side by side.

Try several combinations of like and unlike straws. Blow through each combination of tubes, holding your fingers in front of the opening(s) to compare the air flow (or current) that you achieve with each combination.

Straws in series

Straws in parallel

Rank the combinations according to how much resistance they offer. Classify them according to the amount of current created in each.

Circuits and Circuit Elements **741**

Visual Strategy

supplies ideas using diagrams and illustrations in the book to clarify the concepts covered. Each Visual Strategy includes a question for you to ask students, along with the answer.

Demonstrations

help you add interest and concrete experiences to your classroom presentation.

Quick Lab Teacher's Notes

detail how to use these mini labs effectively.

ADDITIONAL TEACHER'S EDITION FEATURES

● Practice Guides
concisely summarize the options for problem-solving practice available in the book and on the *One-Stop Planner*. (See pages T34–T35.)

● Interactive Problem-Solving Tutor Icons
remind you of where the tutor can be integrated into your lesson. For more on the Holt Physics Interactive Tutor, see page T20.

● Classroom Practice
appears on the same page as worked-out sample problems and provides additional problems for students to solve in groups or for classroom demonstration.

● Chapter Overview
quickly summarizes the content of the chapter's sections.

● Alternative Problem-Solving Approach
indicates other valid methods of solving a Sample Problem besides the one shown on the student page.

● Teaching Tips
alert you to information that may help you be more effective in communicating the key concepts to your students.

● Tapping Prior Knowledge
identifies the assumptions made about student understanding of key concepts, including questions that can help pinpoint prior misconceptions.

● The Language of Physics
provides an explanation of the wording or symbolic conventions used in the chapter for you to share with students.

Answers

for all problems and questions are found in the margin at point of use. Answers for selected problems are also found in the back of the student textbook. Completely worked-out solutions for quantitative problems are found on the One-Stop Planner CD-ROM.

Misconception Alert

describes common misconceptions students may have about the topic covered at that point in the text and provides tips on confronting and dispelling these misconceptions.

Key Models and Analogies

suggest alternative explanations that can help students more fully comprehend difficult concepts.

National Science Education Standards Correlation

Unifying concepts and processes	Science as inquiry	Science and technology	History and nature of science	Science in personal and social perspectives
Systems, order, and organization **UCP 1**	Abilities to do scientific inquiry **SAI 1**	Abilities of technological design **ST 1**	Science as a human endeavor **HNS 1**	Personal health **SPSP 1**
Evidence, models, and explanation **UCP 2**	Understanding about scientific inquiry **SAI 2**	Understanding about science and technology **ST 2**	Nature of science **HNS 2**	Populations, resources, and environments **SPSP 2**
Change, consistency, and measurements **UCP 3**			History of science **HNS 3**	Natural hazards **SPSP 3**
Evolution and equilibrium **UCP 4**				Risks and benefits **SPSP 4**
Form and function **UCP 5**				Science and technology in society **SPSP 5**

The following list shows the chapter correlation of *Holt Physics* with the National Science Education Standards (grades 9-12) for Physical Science content. For further detail, see the interleaf pages before each chapter.

Physical Science Content Standards	Code	Chapter Correlation
Structure of Atoms		
Matter is made of minute particles called atoms, and atoms are composed of even smaller components. These components have measurable properties, such as mass and electrical charge. Each atom has a positively charged nucleus surrounded by negatively charged electrons. The electric force between the nucleus and electrons holds the atom together.	**PS 1a**	Chapter 23
The atom's nucleus is composed of protons and neutrons, which are much more massive than electrons. When an element has atoms that differ in the number of neutrons, these atoms are called different isotopes of the element.	**PS 1b**	Chapter 25
The nuclear forces that hold the nucleus of an atom together, at nuclear distances, are usually stronger than the electric forces that would make it fly apart. Nuclear reactions convert a fraction of the mass of interacting particles into energy, and they can release much greater amounts of energy than atomic interactions. Fission is the splitting of a large nucleus into smaller pieces. Fusion is the joining of two nuclei at extremely high temperature and pressure, and is the process responsible for the energy of the sun and other stars.	**PS 1c**	Chapter 25
Radioactive isotopes are unstable and undergo spontaneous nuclear reactions, emitting particles and/or wavelike radiation. The decay of any one nucleus cannot be predicted, but a large group of identical nuclei decay at a predictable rate. This predictability can be used to estimate the age of materials that contain radioactive isotopes.	**PS 1d**	Chapter 25

Physical Science Content Standards	Code	Chapter Correlation	
Motion and Forces			
Objects change their motion only when a net force is applied. Laws of motion are used to calculate precisely the effects of forces on the motion of objects. The magnitude of the change in motion can be calculated using the relationship $\mathbf{F} = m\mathbf{a}$, which is independent of the nature of the force. Whenever one object exerts force on another, a force equal in magnitude and opposite in direction is exerted on the first object.	**PS 4a**	Chapter 4 Chapter 7 Chapter 8	Chapter 12 Chapter 21
Gravitation is a universal force that each mass exerts on any other mass. The strength of the gravitational attractive force between two masses is proportional to the masses and inversely proportional to the square of the distance between them.	**PS 4b**	Chapter 4 Chapter 7	
The electric force is a universal force that exists between any two charged objects. Opposite charges attract while like charges repel. The strength of the force is proportional to the charges, and, as with gravitation, inversely proportional to the square of the distance between them.	**PS 4c**	Chapter 21	
Between any two charged particles, electric force is vastly greater than the gravitational force. Most observable forces such as those exerted by a coiled spring or friction may be traced to electric forces acting between atoms and molecules.	**PS 4d**	Chapter 17	
Electricity and magnetism are two aspects of a single electromagnetic force. Moving electric charges produce magnetic forces, and moving magnets produce electric forces. These effects help students to understand electric motors and generators.	**PS 4e**	Chapter 21	
Conservation of Energy and the Increase in Disorder			
The total energy of the universe is constant. Energy can be transferred by collisions in chemical and nuclear reactions, by light waves and other radiations, and in many other ways. However, it can never be destroyed. As these transfers occur, the matter involved becomes steadily less ordered.	**PS 5a**	Chapter 5 Chapter 6 Chapter 8	Chapter 10 Chapter 11
All energy can be considered to be either kinetic energy, which is the energy of motion; potential energy, which depends on relative position; or energy contained by a field, such as electromagnetic waves.	**PS 5b**	Chapter 5 Chapter 8 Chapter 10	Chapter 11 Chapter 21
Heat consists of random motion and the vibrations of atoms, molecules, and ions. The higher the temperature, the greater the atomic or molecular motion.	**PS 5c**	Chapter 10 Chapter 11	
Everything tends to become less organized and less orderly over time. Thus, in all energy transfers, the overall effect is that the energy is spread out uniformly. Examples are the transfer of energy from hotter to cooler objects by conduction, radiation, or convection and the warming of our surroundings when we burn fuels.	**PS 5d**	Chapter 11	
Interactions of Energy and Matter			
Waves, including sound and seismic waves, waves on water, and light waves, have energy and can transfer energy when they interact with matter.	**PS 6a**	Chapter 12 Chapter 13	
Electromagnetic waves result when a charged object is accelerated or decelerated. Electromagnetic waves include radio waves (the longest wavelength), microwaves, infrared radiation (radiant heat), visible light, ultraviolet radiation, x-rays, and gamma rays. The energy of electromagnetic waves is carried in packets whose magnitude is inversely proportional to the wavelength.	**PS 6b**	Chapter 12 Chapter 14	
Each kind of atom or molecule can gain or lose energy only in particular discrete amounts and thus can absorb and emit light only at wavelengths corresponding to these amounts. These wavelengths can be used to identify the substance.	**PS 6c**	Chapter 23	
In some materials, such as metals, electrons flow easily, whereas in insulating materials such as glass they can hardly flow at all. Semiconducting materials have intermediate behavior. At low temperatures some materials become superconductors and offer no resistance to the flow of electrons.	**PS 6d**	Chapter 17 Chapter 19 Chapter 24	

Safety in the Physics Laboratory

Risk Assessment

Making your laboratory a safe place to work and learn

Concern for safety must begin before any activity in the classroom and before students enter the lab. A careful review of the facilities should be a basic part of preparation for each school term. You should investigate the physical environment and identify any safety risks, and inspect your work areas for compliance with safety regulations.

The review of the lab should be thorough, and all safety issues must be addressed immediately. Keep a file of your review, and add to the list each year; this will allow you to continue to raise the standard of safety in your classroom.

In physics, many of the experiments, demonstrations, and other activities are classics that have been used for years. This familiarity may lead to a comfort that can obscure inherent safety concerns. Review all experiments, demonstrations, and activities for safety concerns before presenting them to the class. Identify and eliminate potential safety hazards.

NOW AVAILABLE!

- **Laboratory Manager's Professional Reference** contains more detailed safety tips, ready-to-use checklists, and excerpts from relevant standards and regulations.

1. Identify the risks

Before introducing any activity to the class, analyze it and consider what could possibly go wrong. Carefully review the list of materials to make sure they are safe. Inspect the apparatus in your classroom to make sure it is in good working order. Read the procedures to make sure they are safe. Record any hazards or concerns you identify.

2. Evaluate the risks

Minimize the risks you identified in the last step without sacrificing learning. Remember that no activity you can perform in the lab is worth risking injury. Thus, extremely hazardous activities, or those that violate your school's policies, must be eliminated. For activities that present smaller risks, analyze each risk carefully to determine its likelihood. If the pedagogical value of the activity does not outweigh the risks, the activity must be eliminated.

3. Select controls to address risks

Even low-risk activities require controls to eliminate or minimize the risks. Make sure that in devising controls you do not substitute an equally or more hazardous alternative. Some control methods include the following:

- Explicit verbal and written warnings may be added or posted.
- Apparatus may be rebuilt or relocated, have parts replaced, or be replaced entirely by safer alternatives.
- Risky procedures may be eliminated.
- Activities may be changed from student activities to teacher demonstrations.

4. Implement and review selected controls

Controls do not help if they are forgotten or not enforced. The implementation and review of controls should be as systematic and thorough as the initial analysis of safety concerns in the lab and laboratory activities.

The following list describes several possible safety hazards and controls that can be implemented to resolve them. This list is not complete, but it can be used as a starting point to identify hazards in your laboratory.

Identified safety risk	Preventative control
Facilities and equipment	
Lab tables are in disrepair, room is poorly lighted, faucets and electrical outlets do not work or are difficult to use because of their locations.	Work surfaces should be level and stable. There should be adequate lighting. Water supplies, drains, and electrical outlets should be in good working order. Any equipment in a dangerous location should not be used; it should be relocated or rendered inoperable.
Wiring, plumbing, and air circulation systems do not work or do not meet current specifications.	Specifications should be kept on file. Conduct a periodic review of all equipment, and document compliance. Damaged fixtures must be labeled as such and repaired as soon as possible.
Labs are conducted in multipurpose rooms, and equipment from other courses remains accessible.	Only the items necessary for a given activity should be available to students. All equipment should be locked away when not in use.
Students are permitted to enter or work in the lab without teacher supervision.	Lock all laboratory rooms whenever a teacher is not present. Supervising teachers must be trained in lab safety and emergency procedures.
Safety equipment and emergency procedures	
Fire and other emergency drills are infrequent, and no records or measurements are made of the results of the drills.	Always carry out critical reviews of fire or other emergency drills. Be sure that plans include alternate routes. Don't wait until an emergency to find the flaws in your plans.
Emergency evacuation plans do not include instructions for securing the lab in the event of an evacuation during a lab activity.	Plan actions in case of an emergency: establish what devices should be turned off, which escape route to use, where to meet outside the building.
Fire extinguishers are in out-of-the-way locations, not on the escape route.	Place fire extinguishers near escape routes so that they will be of use during an emergency.
Fire extinguishers are not maintained. Teachers are not trained to use them.	Document regular maintenance of extinguishers. Train supervisory personnel in the proper use of extinguishers. Instruct students not to use an extinguisher but to call for a teacher.

Identified safety risk

Preventative control

Safety equipment and emergency procedures, *continued*

Teachers in labs and neighboring classrooms are not trained in CPR or first aid.

Teachers should receive training from the local chapter of the American Red Cross. Certifications should be kept current with frequent refresher courses.

Teachers are not aware of their legal responsibilities in case of an injury or accident.

Review your faculty handbook for your responsibilities regarding safety in the classroom and laboratory. Contact the legal counsel for your school district to find out the extent of their support and any rules, regulations, or procedures you must follow.

Emergency procedures are not posted. Emergency numbers are kept only at the switchboard. Instructions are given verbally only at the beginning of each school year.

Emergency procedures should be posted at all exits and near all safety equipment. Emergency numbers should be posted at all phones, and a script should be provided for the caller to use. Emergency procedures must be reviewed periodically, and students should be reminded of them at the beginning of each activity.

Spills are handled on a case-by-case basis and are cleaned up with whatever materials happen to be on hand.

Have the appropriate equipment and materials available for cleaning up; replace them before expiration dates. Make sure students know to alert you to spilled chemicals, blood, and broken glass.

Work habits and environment

Safety wear is only used for activities involving chemicals or hot plates.

Aprons and goggles should be worn in the lab at all times. Tie back long hair, secure loose clothing, and remove loose jewelry.

There is no dress code established for the laboratory; students are allowed to wear sandals or open-toed shoes.

Open-toed shoes should never be worn in the laboratory. Do not allow any footwear in the lab that does not cover feet completely.

Students are required to wear safety gear but teachers and visitors are not.

Keep extra safety gear on hand for visitors.

Safety is emphasized at the beginning of the term but is not mentioned later in the year.

Safety must be the first priority in all lab work. Students should be warned of risks and instructed in emergency procedures for each activity.

There is no assessment of students' knowledge and attitudes regarding safety.

Conduct frequent safety quizzes. Only students with perfect scores should be allowed to work in the lab.

Safety inspections are conducted irregularly and are not documented. Teachers and administrators are unaware of what documentation will be necessary in case of a lawsuit.

Safety reviews should be frequent and regular. All reviews should be documented, and improvements must be implemented immediately. Contact legal counsel for your district to make sure your procedures will protect you in case of a lawsuit.

Electrical Safety

Caustic Substances

Chemical Safety

Clothing Protection

Eye Protection

Glassware Safety

Hand Safety

Heating Safety

Sharp Object

Waste Disposal

Master Materials List

This materials list was prepared by Sargent-Welch/CENCO, the exclusive science supplier for Holt Physics. This list includes all equipment and supplies needed to perform the in-text Laboratory Exercises and Quick Labs.

Items are listed alphabetically. In many cases, equipment is available in several versions or styles. The recommended equipment is listed first, followed by alternatives. In most cases, the recommended equipment is the most economical, but in some cases it is a more expensive version that has proven to be the most reliable.

The descriptions of the equipment are followed by the quantity required for each student lab group, the CENCO catalog number, and the number of the chapter in which the materials are needed. This latter section is subdivided into three columns that indicate whether the equipment is required for the Quick Labs within the chapter, or the standard or CBL lab procedures for the labs after the chapters. Optional equipment is indicated next to the chapter number.

Items listed as "Local" should be obtained locally. Many of these items are readily available in most classrooms or schools. In most cases, classroom and school supplies such as paper, scissors, drawing compasses, pencils, and water, are not included in the materials list. Typical building equipment, such as gym lights and swing sets, also is not included in the list. The Planning Guide on the interleaf pages before each chapter gives the specific requirements and a detailed materials list for the in-text Laboratory Exercise.

DESCRIPTION	Quantity per group	CENCO Item number	Quick Lab	Standard procedure	CBL procedure
Alligator clip adapters, pair	4	CP30488-01		20	
Aperture tube for bulb or detector shield, blackened	2	Local		14	14
Aperture tube for probe, blackened	1	Local			14
Balance, electronic portable, 2000 g cap., or Balance with weights, triple-beam	1 1	WLS2646-59 WLS3455-02		throughout	throughout
Ball set, drilled	1	WLS4481-70	Q-134, Q-180, Q-274, Q-444	12	
Ball, drilled steel, 25 mm diameter	2	WLS4480-20E	Q-100	3	
Balloon, round, 130 × 75 mm	1	WLS71824-05	Q-632		
Batteries, 1.5 V, D cell	2	WLS30834	Q-734, Q-771		
Battery eliminator with alligator clips, 6 V/0.5 A or Battery eliminator, 1.5 V/3 V	1 1	CP31389-00 CP31388-00		14, 15, 16, 20, 22	
Beaker, low-form, 600 mL	2	WLS4678-ME		10	10
Black card, sheet	1	Local		16	
Black construction paper	1	Local		14, 22	

DESCRIPTION	Quantity per group	CENCO Item number	Laboratory exercise		
			Quick Lab	Standard procedure	CBL procedure
Bulb, 6.2 V/0.5 A, replacement	1	WLS44285-45E		14, 15	
or Bulb, 6.3 V/0.3 A, replacement	1	CP33002-00			
Calorimeter, low calorie	1	CP33734-00		18	
Calorimeter, student double-wall	1	CP32979-00			
or Calorimeter, CENCO double-wall	1	CP31024-01		10	10
Capacitor, 1 farad	1	CP32248-50		18	
Carbon disks, replacement	1	WL0850Z-20		2, 4, 6	
Carbon paper	4 sheets	WL0852C		3	
Card screen with millimeter scale	1	WL3612		15	
Cardboard box	1	Local		3	
Cardboard tube, 7.5 cm × 98 cm	1	Local			13
CBL current and voltage probe set	1	CP33672-00			20, 21
CBL DIN adapter	2	WLS13270-X			9, 20
CBL magnetic probe	1	CP33668-00			21
CBL microphone	1	WLS13270-E			13
CBL pressure-measuring module	1	WLS13270-C			9
CBL student force sensor	1	CP33663-00			4, 5
CBL system	1	WLS13270			throughout
CBL temperature probe (additional)	1	CP33785-00			10
CBL ultrasonic motion detector	1	WLS13270-A			1-6
Centripetal-force apparatus	1	WL0932A			
Clamp, C-type, 10 cm opening × 7 cm depth	1	WLS19405-B			1
Clamp, C-type, 7.5 cm opening × 5 cm depth	1	WLS19405-A		2, 4, 6	
Clamp, double 45°	1	WLS19320		13	
Clamp, pendulum	1	WL0828		12	
Clamp, right-angle, for rods up to 1.3 cm diameter	1	CP12241-01		3-8, 21	
or Clamp, right-angle, for rods up to 1.9 cm diameter		CP12241-02			
Clamp, round-jaw symmetrical, with holder	1	WLS18975		14, 17, 25	13
Clamp, suspension	1	WLS19501		8	
Clamp, table	1	CP75665-50		3, 5	4
or Clamp, C-type table	1	CP75671-50			
Clamp, V-jaw buret	1	WL4901			21
Clamp, V-jaw symmetrical, with holder	1	WLS18975			1, 2, 4, 6
Cloth towels or washcloths	several	Local		3	1-6

DESCRIPTION	Quantity per group	CENCO Item number	Quick Lab	Standard procedure	CBL procedure
Coils, primary and secondary, economy	1	CP32989-00		22	
or Coils, primary and secondary, CENCO	1	CP79750-00			
Compass, 16 mm	1	CP78430-02 (pkg. of 12)	Q-768, Q-771	21	
Copier paper, white	4 sheets	Local		3	
Cord, fine black nylon, 1 m	1	CP88066-00		3, 17	
Cord, pendulum suspension, 1 m	1	CP88066-00	Q-444	12	
Cord, white, thick, braided, 2 m	1	WLS22990-10 (9 m roll)	Q-284	4, 8	4
Culture dish, disposable, small	1	WLS83605-C		25	
Diode, dual-color light-emitting (forward voltages, red 2.0 V, green 2.1 V)	1	CP33693-30		24 (opt)	
Diode, germanium (1N34A)	1	Local		24 (opt)	
Diode, green-light-emitting (forward voltage 2.1 V)	1	CP33693-20		24 (opt)	
Diode, red-light-emitting (forward voltage 2.0 V)	1	CP33693-00		24 (opt)	
Diode, silicon (1N4001)	1	Local		24	
Diode, yellow-light-emitting (forward voltage 2.1 V)	1	CP33693-10		24 (opt)	
Dynamics carts set, student	1	CP30757-00	Q-134	4, 6	4, 6
or Dynamics carts		WL0849F			
Elasticity-of-gases apparatus	1	WL1077		9	
Electroscope, aluminum-leaf	1	WL1963A		17	
Erlenmeyer flask, 1000 mL	1	WLS34107-KE		13	
Erlenmeyer flask, 250 mL	1	WLS34105-F			10
Food color, set of four 8 mL bottles	1	WLC3799Y		13 (opt)	
Friction rod, borosilicate glass (hollow)	1	WL1925		17	
Friction rod, flint glass (solid)	1	WL1926		17	
Friction rod, hard rubber (solid)	1	WL1929		17	
Friction rod, polystyrene (solid)	1	WL1930		17	
Galvanometer kit, tangent	1	CP30167-00		21	21
Galvanometer, student, $-500\ \mu A$ to $+500\ \mu A$	1	WLS30663-32		22	
or Galvanometer, economical	1	CP82101-00			
or Galvanometer, CENCO 6 in 1	1	CP82120-00			
or Potentiometer, slide wire	1	CP83191-00			
Graphing calculator, TI-83	1	WLS13283-P			throughout

DESCRIPTION	Quantity per group	CENCO Item number	Laboratory exercise		
			Quick Lab	Standard procedure	CBL procedure
Grating, diffraction	1	CP86252-00	Q-843		
Grating, mounted film-transmission	1	CP86252-00		16	
Hall's carriage, student version	1	WL0813A		8	
or Hall's carriage	1	CP32491-00			
Hammer, soft rubber	1	WL3252		13	
Hooke's law apparatus, student	1	CP32483-00		5	
or Hooke's law apparatus, CENCO	1	CP73960-00			
Hooke's law apparatus, student (spring only)	1	CP32483-00		5	
or Spring, harmonic motion	1	CP75490-00			
Hot plate, student	1	WLS41056-50		10	10
or Hot plate, Cimarec		WLS41058-70			
or Bunsen burner, Burner gas		WLS11765			
connector, tripod ring,		WLS13121-A			
wire gauze, and lighter		WLS82505-A			
		WLS85325-A			
		WLS13110-10			
Inclined plane, plastic	1	CP32893-00	Q-100,	3, 8	
or inclined plane, wood	1	CP75845-00	Q-279		
Isogenerator set	1	CP32861-00		25	
Lamp and base, miniature	1	WL3571	Q-734	14-16	
Lamp base, with connecting leads	1	CP30083-10		23	
Lattice rod, 1.3 cm diameter, 30 cm long	2	WLS78454-D		5, 6	4, 5, 21
Lens, double-convex, 38 mm diameter, $f = 10$ cm	1	WL3400		15	
Lens, double-convex, 38 mm diameter, $f = 15$ cm	1	WL3404		15 (opt)	
Lens, double-convex, 38 mm diameter, $f = 20$ cm	1	WL3408		15 (opt)	
Light meter, digital (foot-candle/lux)	1	WLS44361-10A		14	
Magnet, Alnico V cylindrical	1	CP78291-30		22	
Magnifier, CENCO dual	1	WL8068	Q-570	10	
Mass hanger, 500 g capacity	1	CP09611-00		4, 8	4, 5
Mass set, metric hooked, 10 g to 1000 g	1	WLS4322-11		2-8	4, 5, 6
Mass set, metric, slotted, 1 g to 500 g	2	WLS4320-16		4-8	4, 5, 6
Mass, heavy, slotted, 1 kg	1	WLS4321-20B		9	
Mass, heavy, slotted, 2 kg	2	WLS4321-20C		9	
Metal shot, 100 g tin	1	WLC4893R		10	10
Metal shot, 500 g aluminum	1	WLC3082V		10	10
Metal shot, 500 g copper	1	WLC3604T		10	10
Metal shot, 500 g zinc	1	WLC4966T		10	10

DESCRIPTION	Quantity per group	CENCO Item number	Laboratory exercise		
			Quick Lab	Standard procedure	CBL procedure
Meterstick optical bench	1	WL3600G		15 (opt), 16 (opt)	
Meterstick, wood, 0.5 m long	1	WLS44685-20		8	
Meterstick, wood, 1 m long	1	WLS44685		throughout	throughout
Multimeter, basic digital,	2	WLS30712-53		18-24	
or Multimeter, student	2	CP32703-00			
or dc Ammeter AND dc Voltmeter	1	WLS30663-12			
(other meters are also acceptable)	1	WLS30665-12			
Nail, 9 cm long	1	Local	Q-771		
Nuclear lab station, basic	1	CP71299-53		25	
Pad, animal fur	1	WL1939		17	
Pad, silk	1	WL1935		17	
Pad, wool felt	1	WL1937		17	
Paper clips	6	WLS65380	Q-696	16	
Paper tape, 13 mm, replacement	1	WL0850Z-11		2, 4, 6	
Patch cord, 30 cm, white, with banana plugs	1	WLS31120-M		21	21
Patch cord, 60 cm, black, with banana plugs	1	WLS31120-P		20, 23	19, 20
Patch cord, 60 cm, red, with banana plugs	1	WLS31120-N		20, 23	19, 20
Patch cord, black, insulated alligator clip/stacking banana plug	2	Local		18-24	19, 20, 21
Patch cord, red, insulated alligator clip/stacking banana plug	2	Local		18-24	19, 20, 21
Photoelectric-effect device with amplifier and filters (red, green, blue)	1	CP32842-00		23	
Poster board, 25 cm × 25 cm piece	2	Local			4, 6
Power supply, CENCO	1	CP33031-00		18-24	19, 20, 21
6 V ac/dc low-voltage	1	CP32787-00			
or Power supply, 4-output dc regulated					
(other power supplies are also acceptable)					
Power supply, spectrum tube	1	WL2393D		16 (opt)	
Prism, right-angle	2		Q-581		
Pulley, adjustable table clamp	1	CP75665-03		4	
Pulley, ball-bearing tandem, double	2	CP75636-00		8	
Pulley, ball-bearing tandem, single	2	CP75635-00		8 (opt)	
Pulley, ball-bearing tandem, triple	2	CP75637-00		8	
Pulley, low-friction rod-mounted	1	WL0776E			4
Pulley, tandem, double	2	CP75644-00		8	
Pulley, tandem, triple	2	CP75646-00		8	
Resistance box, decade	1	CP82824-00		24	

DESCRIPTION	Quantity per group	CENCO Item number	Laboratory exercise		
			Quick Lab	Standard procedure	CBL procedure
Resistance coils, mounted	1	WL2821P		19	19
or Resistance coils, mounted	1	WL2821P			
Resistor, 1 kΩ, 0.5 W	1	CP32995-07		24	
Resistor, 1 kΩ, measuring	1	CP53614-00		24	
Resistor, 1 Ω, 10 W	1	CP32995-20		21	21
Resistor, 1 Ω, measuring	1	CP53610-00			19
Resistor, 10 Ω, 4-10 W	1	CP32995-50		18	
Resistor, 150 Ω, 0.5 W	1	CP32995-04		20	20
Resistor, 5 Ω, measuring	1	CP53611-00			19 (opt)
Resistor, 68 Ω, 0.5 W	1	CP32995-02		20	20
Resonance apparatus, student	1	CP32915-00		13	
or Resonance apparatus, CENCO	1	CP84930-02			
Rheostat, 10 Ω	1	CP83012-00		22	
Rubberbands	1	WLS73105	Q-368	5, 6	
Ruler, English/metric, 12 in./30 cm	1	WLS44625-50		8	
Ruler, metric, 15 cm	1	WLS44675		1, 15	
Safety mitt	1	WLS40266		throughout	throughout
Safety glasses	1	WLS40380-01A		throughout	throughout
Scale and slit	1	CP86260-01		16	
Screen support riders	2	WL3610B		15, 16	
Screen, illuminated object	1	WL3611G		15	
Silicone lubricant, tube	1	Local		9	
Spring scale	1	WLS3679-D	Q-180		
Static electricity tube, PVC	1	CP78610-00		17	
Steam generator	1	CP77936-01		10	10
Stopwatch, student electronic	1	WLA5615		throughout	
Straw, plastic drinking	2	WL5847K	Q-497, Q-741		
Support base and rod, 1-position, 1.3 cm × 91 cm	1	WLS78306-D		throughout	throughout
or Support base and rod, 1-position, 1.0 cm × 51 cm (other stand rods are also acceptable)	1	WLS78306-B			
Support base with rod, 51 cm long × 1 cm diameter	1	WLS78306-B			21
Support block for thermometer (used with low-calorie calorimeter)	1	Local		18	
Support, lens or mirror, for 4 cm lenses	1	WL3604A		14, 15, 23	
Support, lens, 7.5 cm	1	WL3606A		23	
Supports, metal meterstick	1	WL3602A		14, 15, 16, 23	14

DESCRIPTION	Quantity per group	CENCO Item number	Laboratory exercise		
			Quick Lab	Standard procedure	CBL procedure
Suspension stirrup (used with friction rods)	1	WL1820		17	
Switch, contact key	1	CP33638-00		19-24	19-24
or Switch, single-pole, single-throw	1	WLS31155-A			
Switch, double-pole, double-throw	1	WLS31155-D		18	21
Tape, black plastic electrical		WLS79500 (19 mm × 6 m roll)		18	
Tape, masking		WLS44182-80 (19 mm × 60 m roll)		3, 7	21
Tape, self-adhesive packing		Local (roll)		3	
Tape, self-adhesive, 1/2 in. wide		WLS-44178		throughout	throughout
Thermometer with fractional divisions of 0.1°C, -1 to 101°C	1	WLS80210-B		18 (opt)	
Thermometer with fractional divisions of 0.1°C, -1 to 51°C	1	WLS80210-A		10, 18	
Thermometer, red, liquid, -20 to 110°C	1	WLS80035-10B		10, 13	
TI-GRAPH LINK	1	WLS13280			throughout
Timer, acceleration	1	WL0850-ZA		2, 4, 6	
or Timer, compact spark (other timers are also acceptable)	1	WL0850L			
Timer, tabletop acceleration	1	CP30002-00		2 (opt)	
Toy car	1	Local	Q-126, Q-444		
Tuning fork, physical pitch, 1024 Hz	1	WL3227A		13 (opt)	
Tuning forks, physical pitch, set of 4	1	WL3229		13	
Washer, steel, 26 mm diameter	1	Local	Q-284	3	
Wire leads with alligator clips, 30 cm	1	Local		21	21
Wire leads with alligator clips, 60 cm	1	WLS31121-25D (pkg. of 4)		18-24	18-24
Wire stripper and cutter	1	WL0195	Q-245		
Wire, bare copper, 30 AWG	1	WLS85135-J (roll)	Q-245, Q-696	21	21
Wire, insulated copper, 18 AWG	1	WLS85135-30E (14 m roll)	Q-734, Q-771	15	
Wood, blocks, 2 in. × 4 in., set of 4 different lengths 10–20 cm	1	Local		1, 23	1, 2
Xenon light bulb (4.8 V, 850 mA)	1	Local		23	

CHAPTER 1 PLANNING GUIDE

Compression Guide: To shorten from 10 to 7 45-min periods (from 5 to 3½ 90-min blocks), eliminate items in magenta type.

PACING CHART	CLASSROOM RESOURCES			
	⭐TEKS	**Teacher Demonstrations**	**Holt Physics Transparencies**	**Labs** (See page T52 for equipment listing for in-text labs.)
1-1 What is physics? 3 or 2 45-minute periods 1½ or 1 90-minute block	3A, 3C, 3E, 4B	**TE** *Galileo's Hypothesis, p. 8*	**T** 1, 2, 3 **TM** 1, 2	**L** *Discovery Lab: Circumference-Diameter of a Circle*
1-2 Measurements in experiments 3 or 2 45-minute periods 1½ or 1 90-minute block	1A, 1B, 2A, 2B, 2C	**TE** *Measurements, p. 10* **TE** *Standard Units, p. 11* **TE** *Accuracy and Precision, p. 15* **TE** *Significant Figures, p. 15*	**T** 4 **TM** 1, 2	**PE** *Quick Lab: Metric Prefixes, p. 12* **PE** *Time and Measurement, p. 34* **PE** *Physics and Measurement, p. 32*
1-3 The language of physics 2 or 1 45-minute periods 1 or ½ 90-minute block	2C, 2E, 3A, 3B, 3E, 4A, 4B		**T** 8, 9, 10	**L** *Invention Lab: Bubble Solutions* **TL** *Graph Matching*

Review and Assessment
2 45-minute periods
1 90-minute block

Resource Key

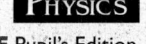
PE Pupil's Edition
TE Teacher's Edition

L Laboratory Experiments
TL Technology Lab Experiments
T Transparencies

🖐 **One-Stop Planner CD-ROM contents**

TM Transparency Masters
SR Section Review Worksheets
AA Alternative Assessment

PW Problem-Solving Workbook
PB Problem Bank
CTW Critical Thinking Worksheet

LABORATORY PLANNING: 1A Physics and Measurement, p. 32; 1B Measuring Time and Motion, p. 34

Materials (for each lab group)

Lab 1A
- balance: portable, electronic balance or triple-beam balance with weight
- meterstick
- metric ruler, 15 cm long

Materials (for each lab group)

Lab 1B
- wood block

Additional Equipment

CBL and Sensors Procedure
- 1-position support base and rod, 1.3 cm × 91 cm
- C-clamp
- CBL
- CBL ultrasonic motion detector
- graphing calculator
- TI Graph Link (recommended for downloading programs)
- V-jaw symmetrical clamp with holder

Stopwatch Procedure
- 12- or 24-hour alarm stopwatch

ASSIGNMENT RESOURCES

Content Mastery	Critical Thinking	Problem-Solving Practice
PE 1–5, p. 9 **SR** 1-1, *Concept Review* **PE** 1–4, p. 27		
PE 1–4, p. 19 **SR** 1-2, *Math Skills* **PE** 5–7, 15–22, pp. 27–28	**PE** 8–10, p. 27	**1A** Metric prefixes: 29 items in **PE, PW,** and **PB,** see **TE** p. 14
PE 1–5, p. 25 **SR** 1-3, *Math Skills* **PE** 23–27, p. 29	**PE** 28–36, p. 29	**PE** 37–42, 44–45, pp. 30–31

ASSESSMENT RESOURCES

Cumulative Review	Alternative Assessment	Traditional Assessment
SR Mixed Review, Ch. 1	**PE** 1–5, p. 31 **AA** Items for Ch. 1	Chapter 1 Test Test Generator items for Ch. 1

Scoring Rubrics for Alternative Assessment items can be found on the One-Stop Planner CD-ROM.

TECHNOLOGY RESOURCES

 CTW Segment 1 Amusement Park Physics

 PE Technology and Learning, p. 30
(Alternative procedures for calculators without Flash-ROM technology are provided on the One-Stop Planner CD-ROM.)

 internet**connect**

On-line Student Resources:
GO TO: www.scilinks.org
The following *sci*LINKS Internet resources can be found in the student text for this chapter.

TOPICS:
• Models in physics, p. 7 (HF2011)
• SI units, p. 11 (HF2012)
• Graphing, p. 56 (HF2013)
• Orders of magnitude, p. 23 (HF2014)

 On-line Teacher Resources:
GO TO: go.hrw.com
KEYWORD: HF2 HOME
Visit the HRW Web site for a variety of resources related to this chapter.

 Smithsonian Institution
Internet Connections
Visit **www.si.edu/hrw** for additional on-line resources.

 Visit **www.cnnfyi.com** for late-breaking news and current events stories selected just for you.

Required Precautions

Wear eye protection and other required safety equipment when cutting wood blocks. Follow all instructions and safety guidelines for the equipment. Do not work with powertools without other people present.

Materials Preparation

The wood blocks for the CBL and sensors procedure should be cut from standard 2 × 4 lumber. For future labs, each lab group should have 4 blocks, cut into lengths of approximately 10 cm, 15 cm, 20 cm, and 25 cm.

Section 1-1 describes the nature of physics and its related fields and activities, introduces the scientific method of inquiry, and discusses the role of models in science.

Section 1-2 introduces the basic SI units, precision versus accuracy, scientific notation, and significant digits.

Section 1-3 presents various ways of summarizing data, including tables, graphs, and equations; uses dimensional analysis to check the validity of expressions; and introduces estimation procedures.

About the Illustration

The photo shows a team heading upwind during a yachting competition near Key West, Florida. The principles of physics can be applied to several aspects of building a sailboat. Buoyancy, balance, and structural integrity are direct applications of concepts in physics, and the concepts of physics can be used to analyze the motion and forces involved in sailing.

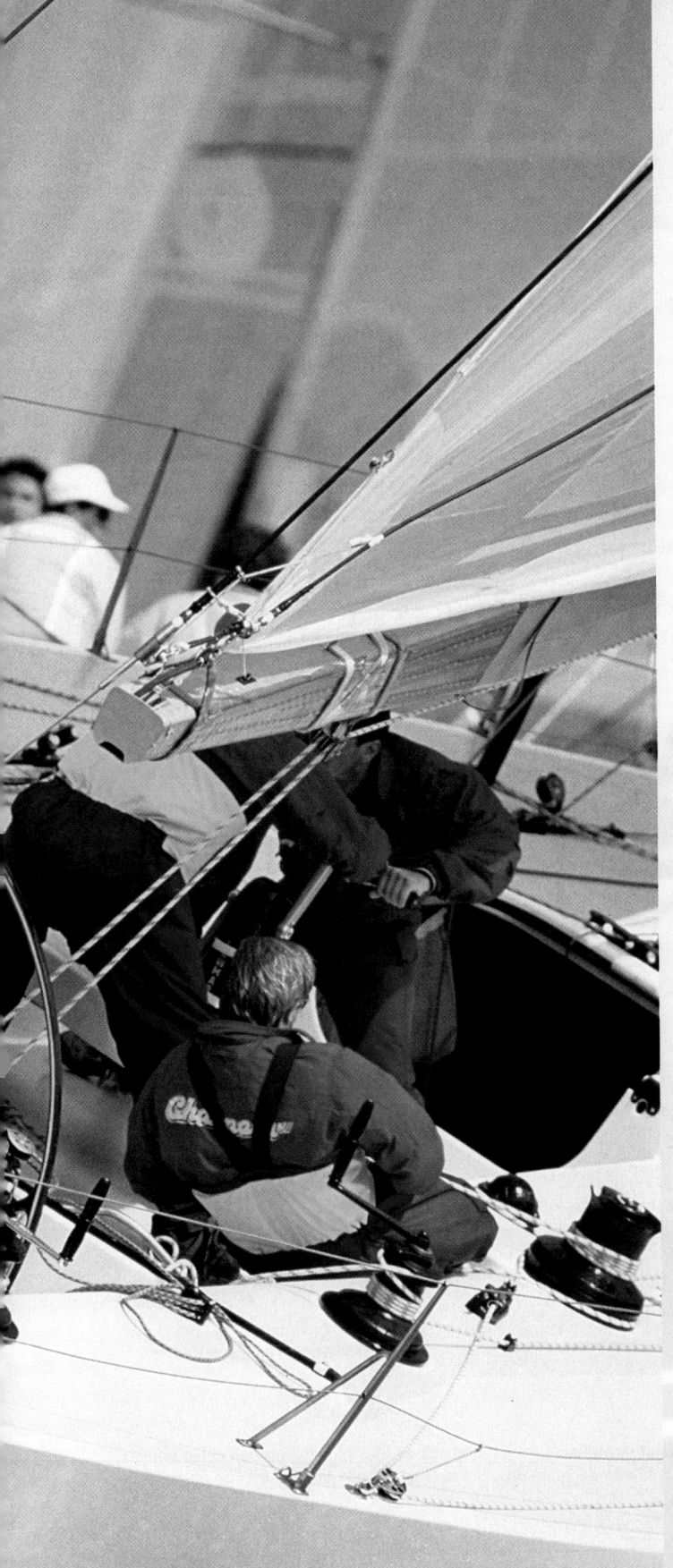

CHAPTER 1

The Science of Physics

PHYSICS IN ACTION

The sails of this boat billow out round and full. As the boat begins moving forward faster and faster, a ball on the deck rolls toward the stern. The boat pitches back and forth gently as it slices through the waves. Water slaps against the hull, occasionally spraying the crew and keeping them cool in the hot sun.

At first, this scene seems to have little to do with physics. Many people associate physics with complicated concepts studied by white-coated scientists working in laboratories with intricate machinery.

But physics can be used to explain anything in the physical world—from why the ball moves to the back as the boat speeds up to why different parts of the sail have different colors.

- *How are the principles of physics applied to sailboat design?*

- *How can the principles of physics be used to predict how a sailboat will move under various conditions?*

Tapping Prior Knowledge

Knowledge to Expect

✔ "Students learn that although there is no fixed set of steps that scientists follow, scientific investigations usually involve the collection of relevant evidence, the use of logical reasoning, and the application of imagination in devising hypotheses. If more than one variable changes at the same time in an experiment, the outcome of the experiment may not be clearly attributable to any one of the variables. Mathematical statements can be used to describe how one quantity changes when another changes. Graphs can show a variety of possible relationships between two variables. Different models can be used to represent the same thing." (AAAS's *Benchmarks for Science Literacy*, grades 6–8)

✔ "Students develop the ability to design and conduct a scientific investigation, use appropriate tools to gather, interpret, and analyze data, develop descriptions, explanations, predictions, and models using evidence, think critically and logically to make relationships between evidence and explanations, and use mathematics in all aspects of scientific inquiry." (NRC's *National Science Education Standards*, grades 5–8)

Items to Probe

✔ Ability to work with decimal numbers and powers of ten, and to draw, read, and interpret graphs: Ask students to calculate $(3 \times 10^{-5})(5 \times 10^{8})$.

1-1
What is physics?

Teaching Tip

Exposure to science, both formal and informal, shapes students' current views of physics. Students should be encouraged to express their perceptions and attitudes in a brainstorming session. Students need to realize that applied science, which is aimed at creating technology, interacts with basic science, which strives to develop fundamental ideas to explain natural phenomena. Point out that physics is a very broad field of study that presents an organized way of modeling and interpreting nature. The different areas of basic science specialize in interpreting different aspects of nature (living things, materials, celestial objects, and so forth). Sometimes, the boundaries between these areas are not clear, but the fundamental ideas of physics underlie all basic and applied sciences.

1-1 SECTION OBJECTIVES

- **Identify activities and fields that involve the major areas within physics.**

- **Describe the processes of the scientific method.**

- **Describe the role of models and diagrams in physics.**

THE TOPICS OF PHYSICS

Many people consider physics to be a difficult science that is far removed from their lives. This may be because many of the world's most famous physicists study topics such as the structure of the universe or the incredibly small particles within an atom, often using complicated tools to observe and measure what they are studying.

But physics is simply the study of the physical world. Everything around you can be described using the tools of physics. The goal of physics is to use a small number of basic concepts, equations, and assumptions to describe the physical world. Once the physical world has been described this way, the physics principles involved can be used to make predictions about a broad range of phenomena. For example, the same physics principles that are used to describe the interaction between two planets can also be used to describe the motion of a satellite orbiting the Earth.

Many of the inventions, appliances, tools, and buildings we live with today are made possible by the application of physics principles. Every time you take a step, catch a ball, open a door, whisper, or check your image in a mirror, you are unconsciously using your knowledge of physics. **Figure 1-1** indicates how the areas of physics apply to building and operating a car. (★)TEKS) **3C**

Figure 1-1
Without knowledge of many of the areas of physics, making cars would be impossible.

Thermodynamics Efficient engines, use of coolants

Electromagnetism
Battery, starter, headlights

Optics Headlights, rearview mirrors

Vibrations and mechanical waves
Shock absorbers, radio speakers, sound insulation

Mechanics Spinning motion of the wheels, tires that provide enough friction for traction

Physics is everywhere

We are surrounded by principles of physics in our everyday lives. In fact, most people know much more about physics than they realize. When you buy a carton of ice cream at the store and put it in the freezer at home, you do it because intuitively you know enough about the laws of physics to know that the ice cream will melt if you leave it on the counter.

Any problem that deals with temperature, size, motion, position, shape, or color involves physics. Physicists categorize the topics they study in a number of different ways. **Table 1-1** shows some of the major areas of physics that will be described in this book.

People who design, build, and operate sailboats need a working knowledge of the principles of physics. Designers figure out the best shape for the boat's hull so that it remains stable and floating yet quick-moving and maneuverable. This design requires knowledge of the physics of fluids. Determining the most efficient shapes for the sails and how to arrange them requires an understanding of the science of motion and its causes. Balancing loads in the construction of a sailboat requires knowledge of mechanics. Some of the same physics principles can also explain how the keel keeps the boat moving in one direction even when the wind is from a slightly different direction.

Table 1-1	Areas within physics ⭐TEKS 3C	
Name	**Subjects**	**Examples**
Mechanics	motion and its causes	falling objects, friction, weight, spinning objects
Thermodynamics	heat and temperature	melting and freezing processes, engines, refrigerators
Vibrations and wave phenomena	specific types of repetitive motions	springs, pendulums, sound
Optics	light	mirrors, lenses, color, astronomy
Electromagnetism	electricity, magnetism, and light	electrical charge, circuitry, permanent magnets, electromagnets
Relativity	particles moving at any speed, including very high speeds	particle collisions, particle accelerators, nuclear energy
Quantum mechanics	behavior of submicroscopic particles	the atom and its parts

Teaching Tip

Ask students to list the careers that they are interested in pursuing. Point out how a physics background may contribute to their effectiveness in those careers. Above all, a good understanding of science makes it possible to critically examine scientific theories and make educated decisions about science-related issues facing our society.

⭐TEKS

p. 4: 3C
p. 5: 3C

Key Models and Analogies

A police investigation provides a good analogy to help students see the scientific method at work. Have students link the procedures of a police investigation of a car accident with the stages of the scientific method.

- Observation/collection of data: The investigator examines the crime scene and fills out a report.

- Hypotheses: The investigator imagines several likely scenarios that might have caused the accident. Maybe the driver was intoxicated, fell asleep, or was speeding; maybe the car had mechanical failure; or maybe weather conditions affected the car's traction or the driver's ability to see the road well.

- Experiments/tests: The investigator might order a blood-alcohol-level test, check the car parts, test-drive the car in different weather conditions, or try to reproduce the skid marks left by the car.

- Interpret/revise hypothesis: The investigator must reexamine evidence and possibly revise his hypothesis. The evidence may be inconclusive.

- Conclusions: The investigator goes to court, reexamines the evidence, and defends his theory of how the accident occurred.

Make observations and collect data that lead to a question.

↓

Formulate and objectively test hypotheses by experiments.

↓

Interpret results, and revise the hypothesis if necessary.

↓

State conclusions in a form that can be evaluated by others.

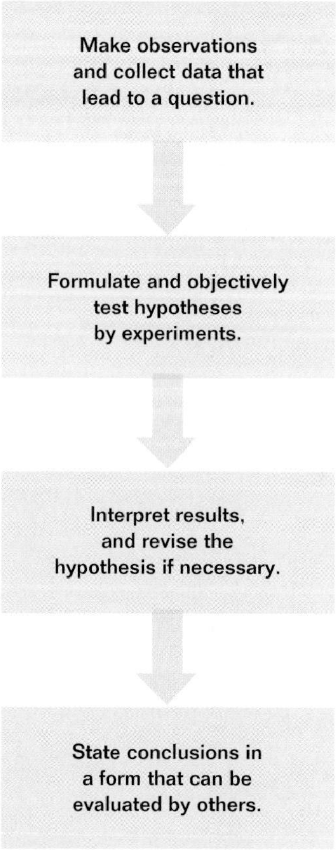

Figure 1-2
Physics, like all other sciences, is based on the scientific method.

model

a replica or description designed to show the structure or workings of an object, system, or concept

---

THE SCIENTIFIC METHOD

When scientists look at the world, they see a network of rules and relationships that determine what will happen in a given situation. Everything you will study in this course was learned because someone looked out at the world and asked questions about how things work.

There is no single procedure that scientists follow in their work. However, there are certain steps common to all good scientific investigations. These steps, called the *scientific method,* are summarized in **Figure 1-2.** This simple chart is easy to understand; but, in reality, most scientific work is not so easily separated. Sometimes, exploratory experiments are performed as a part of the first step in order to generate observations that can lead to a focused question. A revised hypothesis may require more experiments. ★TEKS 3C

Physics uses models that describe only part of reality

How is physics distinct from chemistry, biology, or the other sciences? One difference is the scope of the subject matter, as briefly referred to earlier in this section. Another difference is that although the physical world is very complex, physicists often use simple **models** to explain the most fundamental features of various phenomena. They use this approach because it is usually impossible to describe all aspects of a phenomenon at the same time. A common technique that physicists use to analyze an event or observation is to break it down into different parts. Then physicists decide which parts are important to what they want to study and which parts can be disregarded.

For example, let's say you wish to study the motion of the ball shown in **Figure 1-3.** There are many observations that can be made about the

Figure 1-3
This basketball game involves great complexity.

situation, including the ball's surroundings, size, spin, weight, color, time in the air, speed, and sound when hitting the ground. The first step toward simplifying this complicated situation is to decide what to study, the **system.** Typically, a single object and the items that immediately affect it are the focus of attention. Once you decide that the ball and its motion are what you want to study, you can eliminate all information about the surroundings of the ball except information that affects its motion, as indicated in **Figure 1-4(a).**

system

a set of items or interactions considered a distinct physical entity for the purpose of study

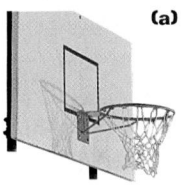

(a)

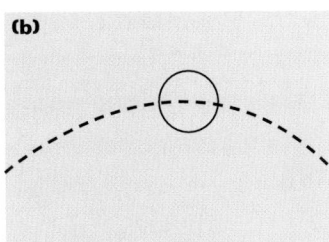

(b)

Figure 1-4
To analyze the motion of the basketball, **(a)** isolate the objects that will affect its motion. Then, **(b)** draw a diagram that includes only the motion.

You can disregard characteristics of the ball that have little or no effect on its motion, such as the ball's color and its sound when bouncing against the floor. In some studies of motion, even the ball's spin and size are disregarded, and the change in the ball's position will be the only quantity investigated, as shown in **Figure 1-4(b).** ⊛TEKS 4B

In effect, the physicist studies the motion of a ball by first creating a simple model of the ball and its motion. Unlike the real ball, the model object is isolated; it has no color, spin, or size, and it makes no noise on impact. Frequently, a model can be summarized with a diagram, like the one in **Figure 1-4(b).** Another way to summarize these models is to build a computer simulation or small-scale replica of the situation.

Without models to simplify matters, situations such as building a car or sailing a boat would be too complex to study. For instance, analyzing the motion of the sailboat is made easier by imagining that the push on the boat from the wind is steady and consistent. The boat is also treated as an object with a certain mass being pushed through the water. In other words, the color of the boat, the model of the boat, and the details of its shape are left out of the analysis. Furthermore, the water the boat moves through is treated as if it were a perfectly smooth-flowing liquid with no internal friction. In spite of these simplifications, the analysis can still make useful predictions of how the sailboat will move.

internet connect

sciLINKS
NSTA

TOPIC: Models in physics
GO TO: www.scilinks.org
*sci*LINKS **CODE:** HF2011

Visual Strategy

Figure 1-4
Ask students to compare the photograph in **Figure 1-4(a)** with the photograph in **Figure 1-3.** Point out that the details relevant to the study of the ball's motion are the direction that the ball was thrown; the forces on the ball; the ball's spin; the ball's location at every instant; the ball's shape, size, and mass; and the air surrounding the ball.

Q Which of these details are ignored in the diagram of the ball's motion in **Figure 1-4(b),** for the purpose of simplification?

A *Size, spin, and air resistance normally affect the ball's flight. However, as a first approximation, the study focuses on the trajectory of a model ball and assumes that the effects of air resistance and spin are negligible.*

⊛TEKS

p. 6: 3C
p. 7: 4B

Galileo's hypothesis

Purpose Illustrate the arguments for and against Galileo's thought experiment.

Materials pennies, tape, coffee filters, deep and clear container, water

Procedure Stack 10 pennies and tape them together. Hold one penny in one hand, and hold the stack of pennies in the other hand; drop them simultaneously. Tell students to listen carefully and decide which object hits the ground first. (It should be difficult to detect any difference in the falling time.)

Hold one coffee filter in one hand and a stack of 10 filters in the other hand; drop them simultaneously. This time, it should be easy to see that 10 filters fall faster than one. How could Galileo defend his hypothesis in the face of such evidence? *(He identified the role of air resistance.)*

Fill the clear container with water, and drop the single penny and the 10-penny stack into the water simultaneously. This time, the single penny will land on the bottom last. Have students use these experiments to defend or criticize the argument that heavy bodies fall faster than lighter ones.

★TEKS

p. 8: 3E, 3A, 4B
p. 9: 3C

Models can help build hypotheses

A scientific hypothesis is a reasonable explanation for observations—one that can be tested with additional experiments. The process of simplifying and modeling a situation can help you identify the relevant variables and identify a hypothesis for testing.

Consider the example of Galileo's "thought experiment," in which he modeled the behavior of falling objects in order to develop a hypothesis about how objects fell. At the time Galileo published his work on falling objects, in 1638, scientists believed that a heavy object would fall faster than a lighter object.

In Galileo's thought experiment, he imagined two identical objects being released at the same time from the same height. They should fall with the same speed. Galileo then imagined tying the two objects together while they were falling. Both scenarios are represented in **Figure 1-5.** ★TEKS **3E**

If common belief were correct, the two objects tied together would suddenly fall faster than they had fallen before because they would be one heavy object instead of two lighter ones. But Galileo believed that tying the two objects together should not cause such a sudden change. As a result of this reasoning, Galileo hypothesized that all objects fall at the same rate in the absence of air resistance, no matter what size they are. ★TEKS **3A**

Figure 1-5
Galileo used the thought experiment represented by this diagram as a way to organize his thoughts about falling bodies. Heavy objects must fall as fast as lighter ones **(a),** or else two bricks tied together would fall faster than they would if kept separate **(b).**

(a)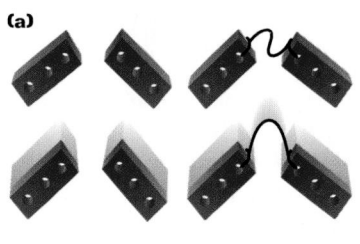

(b)

What does happen: Heavy objects fall as fast as lighter ones.

What does not happen: Heavy objects do *not* fall faster than lighter ones.

Models help guide experimental design

Galileo performed many experiments to test his hypothesis. To be certain he was observing differences due to weight, he kept all other variables the same: the objects he tested had the same size (but different weights) and were measured falling from the same point.

The measuring devices at that time were not precise enough to measure the motion of objects falling in air, and there was no way to eliminate air resistance. So Galileo used the motion of a ball rolling down a series of smooth ramps as a model of the motion of a falling ball. The steeper the ramp, the closer the model came to representing a falling object. These ramp experiments provided data that matched the predictions Galileo made in his hypothesis. ★TEKS **4B**

Like Galileo's hypothesis, any hypothesis must be tested in a **controlled experiment.** In an experiment to test a hypothesis, you must change one variable at a time to determine what influences the phenomenon you are observing. Galileo performed a series of experiments using balls of different weights on one ramp before determining the time they took to roll down a steeper ramp.

The best physics hypotheses can make predictions in new situations

Until the invention of the air pump, it was not possible to perform direct tests of Galileo's hypothesis by observing objects falling in the absence of air resistance. But even though it was not completely testable, Galileo's hypothesis was used to make reasonably accurate predictions about the motion of many objects, from raindrops to boulders (even though they all experience air resistance).

Even if some experiments produce results that support a certain hypothesis, at any time another experiment may produce results that do not support the hypothesis. When this occurs, scientists repeat the experiment until they are sure that the results are not in error. If the unexpected results are confirmed, the hypothesis must be abandoned or revised. That is why the last step of the scientific method is so important. A conclusion is valid only if it can be verified by other people. (★)TEKS 3C

controlled experiment

experiment involving manipulation of a single variable or factor

Section Review

1. Name the areas of physics.

2. Identify the area of physics that is most relevant to each of the following situations. Explain your reasoning.
   a. a high school football game
   b. food preparation for the prom
   c. playing in the school band
   d. lightning in a thunderstorm
   e. wearing a pair of sunglasses outside in the sun

3. What are the activities involved in the scientific method?

4. Give two examples of ways that physicists model the physical world.

5. Physics in Action   Identify the area of physics involved in each of the following tests of a lightweight metal alloy proposed for use in sailboat hulls:
   a. testing the effects of a collision on the alloy
   b. testing the effects of extreme heat and cold on the alloy
   c. testing whether the alloy can affect a magnetic compass needle

 Misconception Alert

Point out that even when the results of an inquiry appear convincing, the conclusions must be formulated carefully by specifying the circumstances under which the experiment was performed and the reasoning that led to accepting or refuting the evidence. Ask students which variable was manipulated when vacuum pumps made it possible to test Galileo's hypothesis (*air resistance*).

Section Review
ANSWERS

1. mechanics, thermodynamics, vibrations and wave phenomena, optics, electromagnetism, relativity, and quantum mechanics

2. **a.** mechanics; because the ball is a moving, spinning object
   b. thermodynamics; because cooking concerns increasing temperature
   c. vibrations and wave phenomena; because music is a type of sound
   d. electromagnetism; because lightning is a form of electricity
   e. optics; because sunglasses have lenses to decrease the light that reaches your eye

3. observing and collecting data, formulating and testing hypotheses, interpreting results and revising the hypothesis, stating conclusions

4. drawing diagrams, using a similar but slightly different situation

5. **a.** mechanics
   b. thermodynamics
   c. electromagnetism

1-2
Measurements in experiments

1-2 SECTION OBJECTIVES

- List basic SI units and the quantities they describe.

- Convert measurements into scientific notation.

- Distinguish between *accuracy* and *precision.*

- Use significant figures in measurements and calculations.

NUMBERS AS MEASUREMENTS

Physicists perform experiments to test hypotheses about how changing one variable in a situation affects another variable. An accurate analysis of such experiments requires numerical measurements.

Numerical measurements are different from the numbers used in a mathematics class. In mathematics, a number like 7 can stand alone and be used in equations. In science, measurements are more than just a number. For example, a measurement reported as 7 leads to several questions. What physical quantity is being measured—length, mass, time, or something else? If it is length that is being measured, what units were used for the measurement—meters, feet, inches, miles, or light-years?

The description of *what kind* of physical quantity is represented by a certain measurement is called *dimension.* In the next several chapters, you will encounter three basic dimensions: length, mass, and time. Many other measurements can be expressed in terms of these three dimensions. For example, physical quantities, such as force, velocity, energy, volume, and acceleration, can all be described as combinations of length, mass, and time. In later chapters, we will need to add two other dimensions to our list, for temperature and for electric current.

The description of *how much* of a physical quantity is represented by a certain numerical measurement depends on the *units* with which the quantity is measured. For example, small distances are more easily measured in millimeters than in kilometers or light-years. ⊛ TEKS 2C

SI is the standard measurement system for science

When scientists do research, they must communicate the results of their experiments with each other and agree on a system of units for their measurements. In 1960, an international committee agreed on a system of standards, such as the standard shown in **Figure 1-6,** and designations for the fundamental quantities needed for measurements. This system of units is called the Système International (SI). In SI, there are only seven base units, each describing a single dimension, such as length, mass, or time. The units of length, mass, and

Figure 1-6
The official standard kilogram mass is a platinum-iridium cylinder kept in a sealed container at the International Bureau of Weights and Measures at Sèvres, France.

Table 1-2   SI standards ⭐TEKS 2C

Unit	Original standard	Current standard
meter (length)	$\frac{1}{10\,000\,000}$ distance from equator to North Pole	the distance traveled by light in a vacuum in $3.33564095 \times 10^{-9}$ s
kilogram (mass)	mass of 0.001 cubic meters of water	the mass of a specific platinum-iridium alloy cylinder
second (time)	$\left(\frac{1}{60}\right)\left(\frac{1}{60}\right)\left(\frac{1}{24}\right) =$ 0.000 01574 average solar days	9 192 631 770 times the period of a radio wave emitted from a cesium-133 atom

internet connect

*SCI*LINKS

NSTA

TOPIC: SI units
GO TO: www.scilinks.org
*sci*LINKS CODE: HF2012

time are the meter, kilogram, and second, respectively. In most measurements, these units will be abbreviated as m, kg, and s, respectively.

These units are defined by the standards described in **Table 1-2** and are reproduced so that every meterstick, kilogram mass, and clock in the world is calibrated to give consistent results. We will use SI units throughout this book because they are almost universally accepted in science and industry.

Not every observation can be described using one of these units, but the units can be combined to form derived units. Derived units are formed by combining the seven base units with multiplication or division. For example, speeds are typically expressed in units of meters per second (m/s), with one unit divided by another unit.

In other cases, it may appear that a new unit that is not one of the base units is being introduced, but often these new units merely serve as shorthand ways to refer to combinations of units. For example, forces and weights are typically measured in units of newtons (N), but a newton is defined as being exactly equivalent to one kilogram multiplied by meters per second squared ($1\,\text{kg} \cdot \text{m/s}^2$). Derived units, such as newtons, will be explained throughout this book as they are introduced.

SI uses prefixes to accommodate extremes ⭐TEKS 2C

Physics is a science that describes a broad range of topics and requires a wide range of measurements, from very large to very small. For example, distance measurements can range from the distances between stars (about 100 000 000 000 000 000 m) to the distances between atoms in a solid (0.000 000 001 m). Because these numbers can be extremely difficult to read and write, they are often expressed in powers of 10, such as 1×10^{17} m or 1×10^{-9} m.

Another approach commonly used in SI is to combine the units with prefixes that symbolize certain powers of 10, as shown in **Figure 1-7.** The most

Did you know?

NIST-7, an atomic clock at the National Institute of Standards and Technology in Colorado, is one of the most accurate timing devices in the world. As a public service, the Institute sends out radio transmissions 24 hours a day in order to broadcast the time given by the atomic clock.

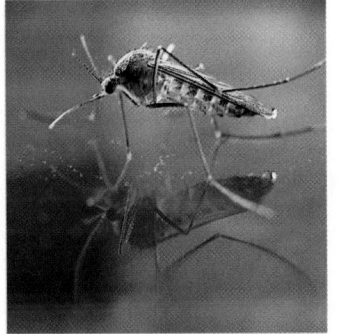

Figure 1-7
The mass of this mosquito can be expressed several different ways: 1×10^{-5} kg, 0.01 g, or 10 mg.

SECTION 1-2

Demonstration 3

Standard Units

Purpose Show that standard units of measurement are based on arbitrary convention.

Materials ribbon strips about 1 cm wide but of unequal lengths

Procedure Explain that the class will define its own standard unit of length with a ribbon strip. Have each group measure the length of the classroom in "ribbon units." Each group will have different results (10.5 ribbons, 9.75 ribbons, etc.). Have the groups compare their ribbons and discover that the lengths were not exactly equal. Let students decide which ribbon should be the standard. Ask them to compare their reasons for selecting a particular ribbon. There is no particular reason to prefer any one. Mark two points about 80 cm apart on a ribbon. Tape the ribbon on the chalkboard, and tell the class to calibrate their ribbon units against this new standard. Have them repeat their measurements. Results should be about the same number of standard ribbon units. Point out that inches, meters, and miles were defined by similar conventions. Scientific standards of measure are universally established through a similar but more rigorous process.

The Language of Physics

Derived units describe quantities, such as density, that are calculated from measurements in base units.

Quick Lab

Teaching Tip

In order to make the terminology relevant to students, have them propose examples of quantities they know to be measured in *milli-, micro-, kilo-*, etc. For example, the frequency of their favorite radio station could be 94.7 MHz, or the dosage of some vitamins may be 300 mg. Ask students to restate these quantities in powers of 10.

Quick Lab

TEACHER'S NOTES

This activity is intended to have students explore the use of prefixes for powers of 10 with metric units.

This lab works best when the papers are folded to fit on the pan of the balance. Students should record their measurements as given by the scale and try to convert them to smaller or larger units using different prefixes listed in **Table 1-3.** Let them realize that in this case the most convenient way to represent the measurements is to use grams and decimal numbers.

Teaching Tip

The prefix *kilo* stands for 1000, *mega* stands for 1 million, and *giga* stands for 1 billion. Ask students to write their computer's RAM and hard-drive memory in bytes. Ask how many watts a 20-megawatt power station supplies (*20 million*). How many watts does a 300 milliwatt bulb use? (*0.3 W*)

Table 1-3 ⭐TEKS 2C
Some prefixes for powers of 10 used with metric units

Power	Prefix	Abbreviation	Power	Prefix	Abbreviation
10^{-18}	atto-	a	10^{-1}	deci-	d
10^{-15}	femto-	f	10^{1}	deka-	da
10^{-12}	pico-	p	10^{3}	kilo-	k
10^{-9}	nano-	n	10^{6}	mega-	M
10^{-6}	micro-	μ (Greek letter *mu*)	10^{9}	giga-	G
			10^{12}	tera-	T
10^{-3}	milli-	m	10^{15}	peta-	P
10^{-2}	centi-	c	10^{18}	exa-	E

Quick Lab

Metric Prefixes

MATERIALS LIST

✔ balance (0.01 g precision or better)

✔ 50 sheets of loose-leaf paper

Record the following measurements (with appropriate units and metric prefixes):

• the mass of a single sheet of paper

• the mass of exactly 10 sheets of paper

• the mass of exactly 50 sheets of paper

Use each of these measurements to determine the mass of a single sheet of paper. How many different ways can you express each of these measurements? Use your results to estimate the mass of one ream (500 sheets) of paper. How many ways can you express this mass? Which is the most practical approach? Give reasons for your answer.

⭐TEKS 1A, 2A, 2B

---

common prefixes and their symbols are shown in **Table 1-3.** For example, the length of a housefly, 5×10^{-3} m, is equivalent to 5 millimeters (mm), and the distance of a satellite 8.25×10^{5} m from Earth's surface can be expressed as 825 kilometers (km). A year, which is 3.2×10^{7} s, can also be expressed as 32 megaseconds (Ms).

Converting a measurement from its prefix form is easy to do. You can build conversion factors from any equivalent relationship, including those in **Table 1-3,** such as 1.609 km = 1 mi and 3600 s = 1 h. Just put the quantity on one side of the equation in the numerator and the quantity on the other side in the denominator, as shown below for the case of the conversion 1 mm = 1×10^{-3} m. Because these two quantities are equal, the following equations are also true:

$$\frac{1 \text{ mm}}{10^{-3} \text{ m}} = 1 \quad \text{and} \quad \frac{10^{-3} \text{ m}}{1 \text{ mm}} = 1$$

Thus, any measurement multiplied by either one of these fractions will be multiplied by 1. The number and the unit will change, but the quantity described by the measurement will stay the same.

To convert measurements, use the conversion factor that will cancel with the units you are given to provide the units you need, as shown in the example below. Typically, the units to which you are converting should be placed in the numerator. It is useful to cross out units that cancel to help keep track of them.

$$\text{Units } \textit{don't } \text{cancel: } 37.2 \text{ mm} \times \frac{1 \text{ mm}}{10^{-3} \text{ m}} = 3.72 \times 10^{4} \frac{\text{mm}^2}{\text{m}}$$

$$\text{Units } \textit{do } \text{cancel: } 37.2 \text{ mm} \times \frac{10^{-3} \text{ m}}{1 \text{ mm}} = 3.72 \times 10^{-2} \text{ m}$$

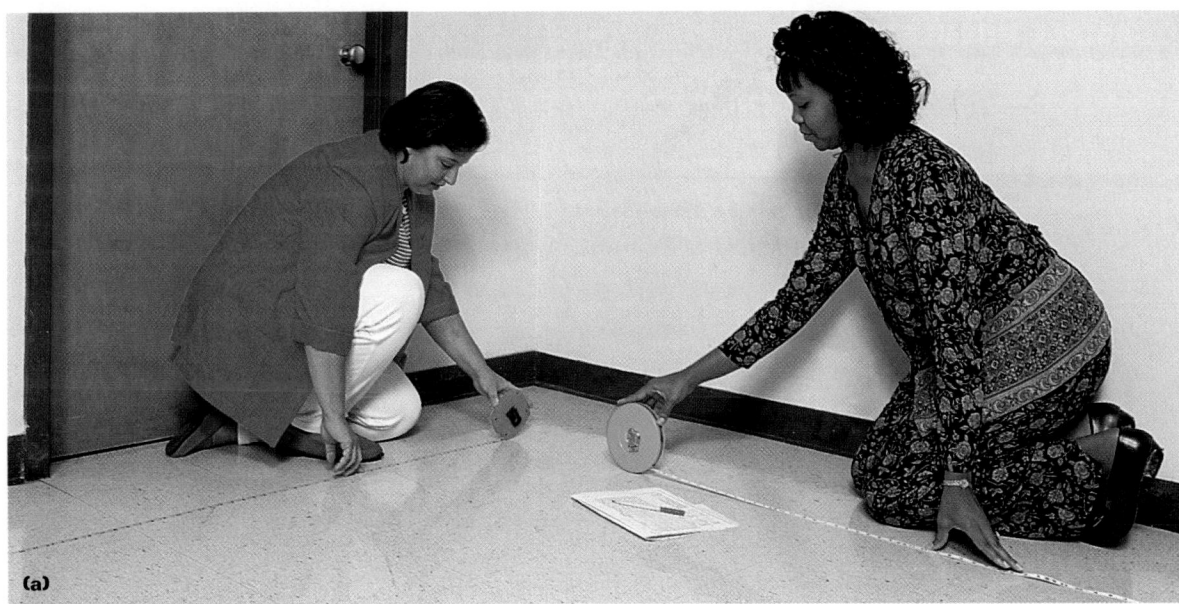

Figure 1-8
When determining area by multiplying measurements of
length and width, be sure the measurements are expressed in
the same units.

(b)

2035 cm
× 12.5 m
1017.5
4070
2035
25437.5
about ? ?
2.54 × 10⁴ cm•m

(c)

20.35 m
× 12.5 m
10.175
40.70
203.5
254.375
about ✓
2.54 × 10² m²

Both dimension and units must agree

Measurements of physical quantities must be expressed in units that match
the dimensions of that quantity. For example, measurements of length cannot
be expressed in units of kilograms because units of kilograms describe the
dimension of mass. It is very important to be certain that a measurement is
expressed in units that refer to the correct dimension. One good technique for
avoiding errors in physics is to check the units in an answer to be certain they
are appropriate for the dimension of the physical quantity that is being sought
in a problem or calculation.

If several people make independent measurements of the same physical
quantity, they may each use different units. As an example, consider **Figure
1-8(a)**, which shows two people measuring a room to determine the area of
carpet necessary to cover the floor. It is possible for one person to measure the
length of the room in meters and for the other person to measure the width of
the room in centimeters. When the numbers are multiplied to find the area,
they will give a difficult-to-interpret answer in units of cm•m, as shown in
Figure 1-8(b). On the other hand, if both measurements are made using the
same units, the calculated area is much easier to intepret because it is expressed
in units of m², as shown in **Figure 1-8(c).** Suppose that the measurements were
made in different units, as in the example above. Because centimeters and
meters are both units of length, one unit can be easily converted to the other. In
order to avoid confusion, it is better to make the conversion to the same units
before doing any more arithmetic. ⊛TEKS **2C**

Visual Strategy

Figure 1-8
Point out that all measurements
could have been converted to cm.

Q What would be the area
in cm²?

A (2035 cm)(1250 cm)=
2 540 000 cm² = 2.54 × 10⁶ cm²

Q This number is much greater
than the number of square
meters. Does it indicate an error?

A No. The same-sized area contains a small number of large
units (large tiles) or a large number of small units (small tiles),
just as 20 one-dollar bills have
the same value as 2000 pennies.

⊛TEKS
p. 12: 2C, 1A, 2A, 2B,
2C
p. 13: 2C

Classroom Practice

The following may be used as teamwork exercises or for demonstration at the board or on an overhead projector.

PROBLEM

Metric prefixes

The mass of an average woman is 60 000 000 mg. Express this in grams and kilograms.

Answer

60 000 g, 60 kg

PRACTICE GUIDE 1A

Converting:

SI units	📖 **PE** Sample, 1–5; Ch. Rvw. 11–14
	💿 **PW** Sample, 1–8
	PB Sample, 1–10

ANSWERS TO

Practice 1A
Metric prefixes
1. 5×10^{-5} m
2. 1×10^{-6} s
3. a. 1×10^{-8} m
   b. 1×10^{-5} mm
   c. 1×10^{-2} μm
4. 0.15 Tm, 1.5×10^{8} km
5. 1.440×10^{3} kg

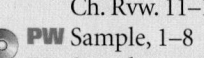

★TEKS

p. 14: 2C
p. 15: 1B

Metric prefixes ⭐TEKS 2C

PROBLEM

A typical bacterium has a mass of about 2.0 fg. Express this measurement in terms of grams and kilograms.

SOLUTION

Given: mass = 2.0 fg

Unknown: mass = ? g   mass = ? kg

Build conversion factors from the relationships given in **Table 1-3.** Two possibilities are shown below.

$$\frac{1 \times 10^{-15} \text{ g}}{1 \text{ fg}} \text{ and } \frac{1 \text{ fg}}{1 \times 10^{-15} \text{ g}}$$

Only the first one will cancel the units of femtograms to give units of grams.

$$(2.0 \text{ fg})\left(\frac{1 \times 10^{-15} \text{ g}}{1 \text{ fg}}\right) = \boxed{2.0 \times 10^{-15} \text{ g}}$$

Then, take this answer and use a similar process to cancel the units of grams to give units of kilograms.

$$(2.0 \times 10^{-15} \text{ g})\left(\frac{1 \text{ kg}}{1 \times 10^{3} \text{ g}}\right) = \boxed{2.0 \times 10^{-18} \text{ kg}}$$

PRACTICE 1A

Metric prefixes

1. A human hair is approximately 50 μm in diameter. Express this diameter in meters.

2. A typical radio wave has a period of 1 μs. Express this period in seconds.

3. A hydrogen atom has a diameter of about 10 nm.
   a. Express this diameter in meters.
   b. Express this diameter in millimeters.
   c. Express this diameter in micrometers.

4. The distance between the sun and the Earth is about 1.5×10^{11} m. Express this distance with an SI prefix and in kilometers.

5. The average mass of an automobile in the United States is about 1.440×10^{6} g. Express this mass in kilograms.

ACCURACY AND PRECISION

Because theories are based on observation and experiment, careful measurements are very important in physics. But in reality, no measurement is perfect. In describing the imperfection, there are two factors to consider: a measurement's **accuracy** and a measurement's **precision.** Although these terms are often used interchangeably in everyday speech, they have specific meanings in a scientific discussion.

Problems with accuracy are due to error

Experimental work is never free of error, but it is important to minimize error in order to obtain accurate results. Human error can occur, for example, if a mistake is made in reading an instrument or recording the results. One way to avoid human error is to take repeated measurements to be certain they are consistent.

If some measurements are taken using one method and some are taken using a different method, another type of error, called method error, will result. Method error can be greatly reduced by standardizing the method of taking measurements. For example, when measuring a length with a meterstick, choose a line of sight directly over what is being measured, as shown in **Figure 1-9(a).** If you are too far to one side, you are likely to overestimate or underestimate the measurement, as shown in **Figure 1-9(b)** and **(c).** This problem is due to the phenomenon known as *parallax.* Another example of parallax is the fact that the speedometer reading reported by a car's driver is more accurate than the speedometer reading as seen from the passenger seat in an automobile.

Another type of error is instrument error. If a meterstick or balance is not in good working order, this will introduce error into any measurements made with the device. For this reason, it is important to be careful with lab equipment. Rough handling can damage balances. If a wooden meterstick gets wet, it can warp, making accurate measurements difficult. (★TEKS) **1B**

accuracy

describes how close a measured value is to the true value of the quantity measured

precision

refers to the degree of exactness with which a measurement is made and stated

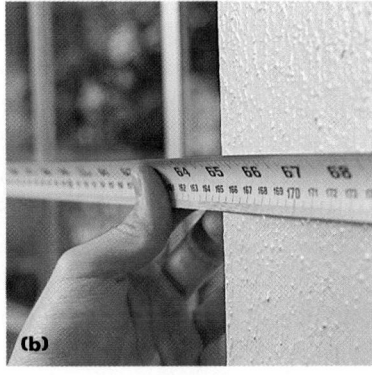

Figure 1-9
If you measure this window by keeping your line of sight directly over the measurement **(a),** you will find that it is 165.2 cm long. If you do not keep your eye directly above the mark, **(b)** and **(c),** you may report a measurement with some error.

Demonstration 4

Accuracy and precision

Purpose Show the difference between accuracy and precision in measuring length.

Materials shoe, metersticks

Procedure Have three students successively measure the length of a long table with a shoe. Ask students to compare the accuracy of each measurement and the precision of the measuring tool. *(The measurement has little accuracy; i.e., it is difficult to estimate a fractional part of a shoe. The tool lacks precision; i.e., repeated measurements produce little agreement.)* Have three students repeat the measurement with metersticks and compare their results. Are there still differences even though the meterstick is more precise?

Demonstration 5

Significant figures

Purpose Clarify the meaning of significant figures.

Materials pan balance, small paper cup, salt or sugar

Procedure Pour salt into the cup on one pan until it is balanced with 183 g. Slowly add small amounts of salt, and have students note that the reading remains 183 g until the balance tilts. Try to restore balance by adding 1 g masses (they will probably be too heavy or too light). Ask students to guess the weight of the salt. *(The answer is actually between 183 g and 184 g.)*

Teaching Tip

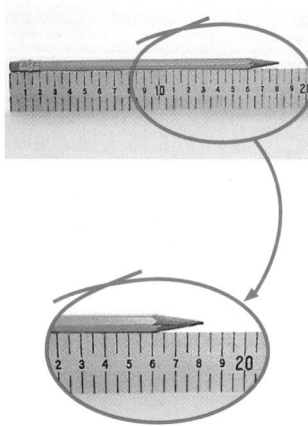

Figure 1-10
Even though this ruler is marked in only centimeters and half-centimeters, if you estimate, you can use it to report measurements to a precision of a millimeter.

significant figures

those digits in a measurement that are known with certainty plus the first digit that is uncertain

Because the ends of a meterstick can be easily damaged or worn, it is best to minimize instrument error by making measurements with a portion of the scale that is in the middle of the meterstick. Instead of measuring from the end (0 cm), try measuring from the 10 cm line.

Precision describes the limitations of the measuring instrument

Poor accuracy involves errors that can often be corrected. On the other hand, precision describes how exact a measurement can possibly be. For example, a measurement of 1.325 m is more precise than a measurement of 1.3 m. A lack of precision is typically due to limitations of the measuring instrument and is not the result of human error or lack of calibration. For example, if a meterstick is divided only into centimeters, it will be difficult to measure something only a few millimeters thick with it. ⭐TEKS **2B**

In many situations, you can improve the precision of a measurement. This can be done by making a reasonable estimation of where the mark on the instrument would have been. Suppose that in a laboratory experiment you are asked to measure the length of a pencil with a meterstick marked in centimeters, as shown in **Figure 1-10.** The end of the pencil lies somewhere between 18 cm and 18.5 cm. The length you have actually measured is slightly more than 18 cm. You can make a reasonable estimation of how far between the two marks the end of the pencil is and add a digit to the end of the actual measurement. In this case, the end of the pencil seems to be less than half way between the two marks, so you would report the measurement as 18.2 cm.

Significant figures help keep track of imprecision

It is important to record the precision of your measurements so that other people can understand and interpret your results. A common convention used in science to indicate precision is known as **significant figures.**

In the case of the measurement of the pencil as about 18.2 cm, the measurement has three significant figures. The significant figures of a measurement include all the digits that are actually measured (18 cm), plus one *estimated* digit. Note that the number of significant figures is determined by the precision of the markings on the measuring scale.

The last digit is reported as a 0.2 (for the estimated 0.2 cm past the 18 cm mark). Because this digit is an estimate, the true value for the measurement is actually somewhere between 18.15 cm and 18.25 cm.

When the last digit in a recorded measurement is a zero, it is difficult to tell whether the zero is there as a place holder or as a significant digit. For example, if a length is recorded as 230 mm, it is impossible to tell whether this number has two or three significant digits. In other words, it can be difficult to know whether the measurement of 230 mm means the measurement is known to be between 225 mm and 235 mm or is known more precisely to be between 229.5 mm and 230.5 mm. ⭐TEKS **2C**

One way to solve such problems is to report all values using scientific notation. In scientific notation, the measurement is recorded to a power of 10, and all of the figures given are significant. For example, if the length of 230 cm has two significant figures, it would be recorded in scientific notation as 2.3×10^2 cm. If it has three significant figures, it would be recorded as 2.30×10^2 cm.

Scientific notation is also helpful when the zero in a recorded measurement appears in front of the measured digits. For example, a measurement such as 0.000 15 cm should be expressed in scientific notation as 1.5×10^{-4} cm if it has two significant figures. The three zeros between the decimal point and the digit 1 are not counted as significant figures because they are present only to locate the decimal point and to indicate the order of magnitude. The rules for determining how many significant figures are in a measurement that includes zeros are shown in **Table 1-4.** (★)TEKS 2B

Significant figures in calculations require special rules

In calculations, the number of significant figures in your result depends on the number of significant figures in each measurement. For example, if someone reports that the height of a mountaintop, like the one shown in **Figure 1-11,** is 1710 m, that implies that its actual height is between 1705 and 1715 m. If another person builds a pile of rocks 0.20 m high on top of the mountain, that would not suddenly make the mountain's new height known accurately enough to be measured as 1710.20 m. The final answer cannot be more precise than the least precise measurement used to find the answer. Therefore, the answer should be rounded off to 1710 m even if the pile of rocks is included.

Figure 1-11
If a mountain's height is known with an uncertainty of 5 m, the addition of 0.20 m of rocks will not appreciably change the height.

Table 1-4	Rules for determining whether zeros are significant figures (★)TEKS 2C
Rule	**Examples**
1. Zeros between other nonzero digits are significant.	**a.** 50.3 m has three significant figures. **b.** 3.0025 s has five significant figures.
2. Zeros in front of nonzero digits are not significant.	**a.** 0.892 kg has three significant figures. **b.** 0.0008 ms has one significant figure.
3. Zeros that are at the end of a number and also to the right of the decimal are significant.	**a.** 57.00 g has four significant figures. **b.** 2.000 000 kg has seven significant figures.
4. Zeros at the end of a number but to the left of a decimal are significant if they have been measured or are the first estimated digit; otherwise, they are *not* significant. In this book, they will be treated as *not* significant.	**a.** 1000 m may contain from one to four significant figures, depending on the precision of the measurement, but in this book it will be assumed that measurements like this have one significant figure. **b.** 20 m may contain one or two significant figures, but in this book it will be assumed to have one significant figure.

SECTION 1-2

Teaching Tip

Have students assume that the precision of the time measurement in the 1500 m race from the previous Teaching Tip is 0.01 s. Therefore, if the possible 60 cm variation in the length is to be ignored between any two competitions, a swimmer would have to swim this difference in less than 0.01 s. Show that at a rate of 60 cm/0.01 s, a swimmer would have to swim 60 m (more than 30 body lengths) in 1 s. Thus, if it takes longer than 0.01 s to swim 60 cm, then there can be no valid comparisons of time that differ by less than 0.01 s from one 1500 m race to the next. The possible difference in the times may not be due to the ability of the individual swimmers; it may be due to a difference in the lengths of the two pools. Why do Olympic judges measure swimming times to the 0.1 s, 0.01 s, or even 0.001 s? Students can either explain this or have something new to think about as they watch or engage in sports activities.

Visual Strategy

Table 1-4

Students may need to practice with examples in applying the rules shown in **Table 1-4.**

Q Write each of the examples in scientific notation. Check the number of significant figures.

A 5.03×10^1 m; 3.0025 s; 8.92×10^{-1} kg; 8×10^{-4} ms; 5.700×10^1 g; 2.000 000 kg

(★)TEKS
> p. 16: 2C, 3B, 4A
> p. 17: 2B, 2C

The Language of Physics

In mathematics, we use 97.3, 97.30, 97.300, etc., to represent the same number with a single value (e.g., if $x = 97.3$ and $y = 97.300$, then $x = y$). When numbers represent the results of measurements, they are not pure mathematical quantities: they contain additional information about precision. Measured values actually stand for a range of possible values. This range is defined by the significant digits. Here, x could be 97.25 cm, 97.26 cm, or 97.34 cm. But y could be 97.295 cm, 97.296 cm, or 97.304 cm. There may be a difference between x and y, although it may not be measurable.

 Misconception Alert

Sometimes, students are reluctant to round off a product so that it has no decimal digits. They might even think they misapplied a rule. An example may familiarize them with such counterintuitive results. Ask them to multiply 53.5824 s by 2.14 m/s and round off. The result, 114.666336 m, needs to be reduced to three significant digits, that is, 115 m.

Similar rules apply to multiplication. Suppose that you calculate the area of a room by multiplying the width and length. If the room's dimensions are 4.6 m by 6.7 m, the product of these values would be 30.82 m². However, this answer contains four significant figures, which implies that it is more precise than the measurements of the length and width. Because the room could be as small as 4.55 m by 6.65 m or as large as 4.65 m by 6.75 m, the area of the room is known only to be between 30.26 m² and 31.39 m². The area of the room can have only two significant figures because each measurement has only two. So it must be rounded off to 31 m². **Table 1-5** summarizes the two basic rules for determining significant figures when you are performing calculations.

Table 1-5   Rules for calculating with significant figures ⭐TEKS 2C

Type of calculation	Rule	Example
addition or subtraction	The final answer should have the same number of digits to the right of the decimal as the measurement with the *smallest* number of digits *to the right of the decimal.*	97.3 + 5.85 —— 103.15 $\xrightarrow{\text{round off}}$ 103.2
multiplication or division	The final answer has the same number of significant figures as the measurement having the *smallest* number of *significant figures.*	123 × 5.35 —— 658.05 $\xrightarrow{\text{round off}}$ 658

Calculators do not pay attention to significant figures

When you use a calculator to analyze problems or measurements, you may be able to save time because the calculator can do the math more quickly than you. However, the calculator does not keep track of the significant figures in your measurements.

Calculators often exaggerate the precision of your final results by returning answers with as many digits as the display can show. To reinforce the correct approach, the answers to the sample problems in this book will always show only the number of significant figures that the measurements justify.

In order to provide answers with the correct number of significant figures, it will sometimes be necessary to round the results of a calculation. The rules described in **Table 1-6** will be used. In this book, the results of a calculation will be rounded after each type of mathematical operation. For example, the result of a series of multiplications should be rounded using the multiplication/division rule before it is added to another number. Similarly, the sum of several numbers should be rounded according to the addition/subtraction rule before the sum is multiplied by another number. You should consult your teacher to find out whether to round this way or to delay rounding until the end of all calculations. ⭐TEKS 1B

Table 1-6	**Rules for rounding**  **2C**	
What to do	**When to do it**	**Examples**
round down	• whenever the digit following the last significant figure is a 0, 1, 2, 3, or 4	30.24 becomes 30.2
	• if the last significant figure is an even number and the next digit is a 5, with no other nonzero digits	32.25 becomes 32.2 32.65000 becomes 32.6
round up	• whenever the digit following the last significant figure is a 6, 7, 8, or 9	22.49 becomes 22.5
	• if the digit following the last significant figure is a 5 followed by a nonzero digit	54.7511 becomes 54.8
	• if the last significant figure is an odd number and the next digit is a 5, with no other nonzero digits	54.75 becomes 54.8 79.3500 becomes 79.4

Section Review

1. Which SI units would you use for the following measurements?
   a. the length of a swimming pool
   b. the mass of the water in the pool
   c. the time it takes a swimmer to swim a lap

2. Express the following measurements as indicated.
   a. 6.20 mg in kilograms
   b. 3×10^{-9} s in milliseconds
   c. 88.0 km in meters

3. The following students measure the density of a piece of lead three times. The density of lead is actually 11.34 g/cm^3. Considering all of the results, which person's results were accurate? Which were precise? Were any both accurate and precise?
   a. Rachel: 11.32 g/cm^3, 11.35 g/cm^3, 11.33 g/cm^3
   b. Daniel: 11.43 g/cm^3, 11.44 g/cm^3, 11.42 g/cm^3
   c. Leah: 11.55 g/cm^3, 11.34 g/cm^3, 11.04 g/cm^3

4. Perform these calculations, following the rules for significant figures.
   a. $26 \times 0.02584 = ?$
   b. $15.3 \div 1.1 = ?$
   c. $782.45 - 3.5328 = ?$
   d. $63.258 + 734.2 = ?$

 Misconception Alert

The *symmetric rounding rules* used when the last digit is a five may appear odd. Students may not be clear about their reasons. Explain that when the last digit is five, by convention we round to the *nearest even digit*. Ask them to apply this to rounding 12.05, 12.15, 12.25, 12.35, 12.45 and so on until 12.95. (*12.0, 12.2, 12.2, 12.4, 12.4, etc.*) Students will realize that they alternatively round up and down. Point out that this convention allows them to reduce the average error because the numbers are sometimes increased and sometimes decreased.

Section Review
ANSWERS

1. a. meters
   b. kilograms
   c. seconds
2. a. 6.20×10^{-6} kg
   b. 3×10^{-6} ms
   c. 8.80×10^4 m
3. a. accurate and precise
   b. precise
   c. neither
4. a. 0.67
   b. 14
   c. 778.92
   d. 797.5

★TEKS

p. 18: 2C, 1B
p. 19: 2C

1-3
The language of physics

1-3 SECTION OBJECTIVES

- **Interpret data in tables and graphs, and recognize equations that summarize data.**

- **Distinguish between conventions for abbreviating units and quantities.**

- **Use dimensional analysis to check the validity of expressions.**

- **Perform order-of-magnitude calculations.**

(★TEKS) **2C, 3E**

(★TEKS) **2E**

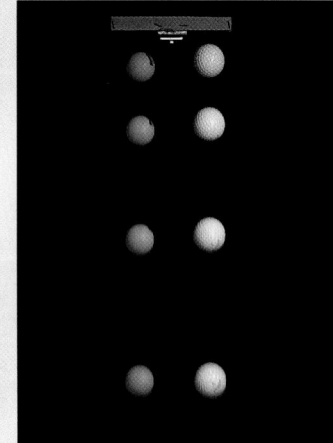

Figure 1-12
This experiment tests Galileo's hypothesis by having two balls with different masses dropped simultaneously.

MATHEMATICS AND PHYSICS

Just as physicists create simplified models to better understand the real world, they use the tools of mathematics to analyze and summarize their observations. Then they can use the mathematical relationships among physical quantities to help predict what will happen in new situations.

Tables, graphs, and equations can make data easier to understand

There are many ways to organize data. Consider the experiment shown in **Figure 1-12,** which tests Galileo's hypothesis by dropping a table-tennis ball and a golf ball in a vacuum and measuring how far each ball falls during a certain time interval. The results are recorded as a set of numbers corresponding to the times of the fall and the distance each ball falls. A convenient way to analyze the data is to form a table like **Table 1-7.** A clear trend can be seen in the data. The more time that passes after each ball is dropped, the farther the ball falls.

Table 1-7	Data from dropped-ball experiment	
Time (s)	Distance golf ball falls (cm)	Distance table-tennis ball falls (cm)
0.067	2.20	2.20
0.133	8.67	8.67
0.200	19.60	19.59
0.267	34.93	34.92
0.333	54.34	54.33
0.400	78.40	78.39

A better method for analyzing the data is to construct a graph of the distance the balls fall in each time interval, as shown in **Figure 1-13** on the next page. Using the graph, you can reconstruct the table by noting the values along the distance and time axes for each of the points shown.

In addition, because the graph shows an obvious pattern, we can draw a smooth curve through the data points to make estimations for times when we have no data, such as 0.225 s. The shape of the graph also provides information about the relationship between time and distance.

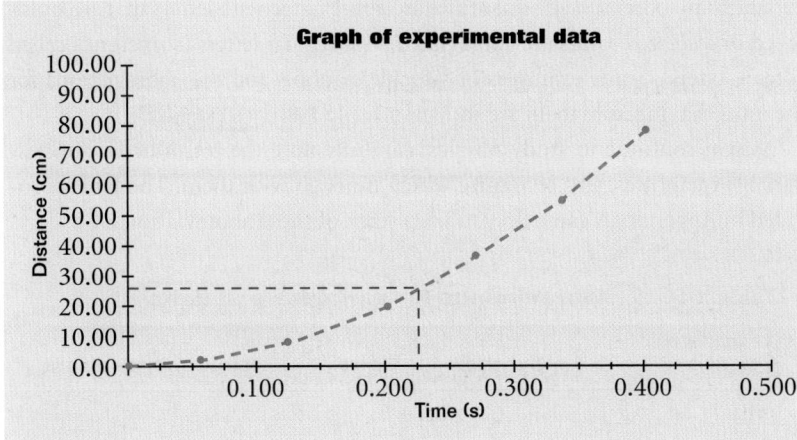

Graph of experimental data

We can also use the following equation to describe the relationship between the variables in the experiment:

(change in position in meters) = 4.9 × (time of fall in seconds)²

This equation allows you to reproduce the graph and make predictions about the change in position for any arbitrary time during the fall.

Physics equations indicate relationships

While mathematicians use equations to describe relationships between variables, physicists use the tools of mathematics to describe measured or predicted relationships between physical quantities in a situation. For example, one or more variables may affect the outcome of an experiment. In the case of a prediction, the physical equation is a compact statement based on a model of the situation. It shows how two or more variables are thought to be related. Many of the most important equations in physics do not contain numbers. Instead, they represent a simple description of the relationship between physical quantities.

To make expressions as simple as possible, physicists often use letters to describe specific quantities in an equation. For example, the letter v is used to denote speed. Sometimes, Greek letters are used to describe mathematical operations. For example, the Greek letter Δ (delta) is often used to mean "difference or change in," and the Greek letter Σ (sigma) is used to mean "sum" or "total."

With these conventions, the word equation above can be written as follows:

$$\Delta y = 4.9(\Delta t)^2$$

The abbreviation Δy indicates the change in a ball's position from its starting point, and Δt indicates the time elapsed. ⭐TEKS **3B, 4B**

As you saw in Section 1-2, the units in which these quantities are measured are also often abbreviated with symbols consisting of a letter or two. Most physics books provide some clues to help you keep track of which letters refer to quantities and variables and which letters are used to indicate units. Typically,

Figure 1-13
The graph of these data provides a convenient way to summarize the data and indicate the relationship between the time an object has been falling and the distance it has fallen.

⭐TEKS **4A**

internetconnect

SC INKS.

NSTA

TOPIC: Graphing
GO TO: www.scilinks.org
*sci*LINKS CODE: HF2013

Visual Strategy

Figure 1-13
Have students verify that the points on the graph correspond to the numbers listed in **Table 1-7.**

Q Was the point (0, 0) listed in the data table? How would you justify it?

A *No, we assume that the clock started exactly at the instant the balls were dropped and that distances were measured from the dropping point.*

Q At what time had the balls fallen 50 cm?

A *0.32 s*

The Language of Physics
In the equation, the term *change in position* stands for the vertical distance from the edge of the table. Likewise, *time of fall* specifies the number of seconds since the balls were dropped.

We use the symbols Δy and Δt to represent the vertical distance between two points and the time interval, respectively.

⭐TEKS

p. 20: 2C, 3E, 2E
p. 21: 4A, 3B, 4B

Misconception Alert

Students may think that algebraic symbols representing variable quantities must always be the same letters shown in the table. Point out that physicists tend to *choose* different symbols for each dimension in different contexts and that there are few fixed rules. For instance, as shown in **Table 1-8,** the symbol for change in position is Δx or Δy; change in position can also by symbolized by d, s, ℓ, or h.

Teaching Tip

Point out that dimensional analysis is a good tool for checking whether the equation you are using gives the kind of quantity you want to calculate. For example, when you want to find volume (V) based on density (ρ) and mass (m), you may find that you erroneously set $V = \dfrac{\rho}{m}$ before doing the calculations. Simply replace each variable with its dimensions and see whether the equation balances.

$$V = \frac{\rho}{m}$$

$$cm^3 = \frac{\dfrac{g}{cm^3}}{g} = \frac{\cancel{g}}{cm^3\,\cancel{g}} = \frac{1}{cm^3}$$

CONCEPT PREVIEW ➤

The differences between quantities denoted with boldfaced symbols and those denoted with italicized symbols will be further described in Chapter 3.

Figure 1-14
Dimensional analysis can be a useful check for many types of problems, including those involving the length of time it would take for this car to travel 725 km if it moves with a speed of 88 km/h.

variables and other specific quantities are abbreviated with letters that are **bold-faced** or *italicized*. Units are abbreviated with regular letters (sometimes called roman letters). Some examples of variable symbols and the abbreviations for the units that measure them are shown in **Table 1-8.** ⭐TEKS **3B**

As you continue to study physics, carefully note the introduction of new variable quantities, and recognize which units go with them. The tables provided in Appendix A can help you keep track of these abbreviations.

Table 1-8	Abbreviations for variables and units		
Quantity	**Symbol**	**Units**	**Unit abbreviations**
change in position	Δx, Δy	meters	m
time interval	Δt	seconds	s
mass	m	kilograms	kg

EVALUATING PHYSICS EXPRESSIONS

Like most models physicists build to describe the world around them, physics equations are valid only if they can be used to make predictions about situations. Although an experiment is the best way to test the validity of a physics expression, several techniques can be used to evaluate whether expressions are likely to be valid.

Dimensional analysis can weed out invalid equations

Suppose a car, such as the one in **Figure 1-14,** is moving at a speed of 88 km/h and you want to know how long it will take it to travel 725 km. How can you decide a good way to solve the problem?

You can use a powerful procedure called *dimensional analysis.* Dimensional analysis makes use of the fact that *dimensions can be treated as algebraic quantities.* For example, quantities can be added or subtracted only if they have the same dimensions, and the two sides of any given equation must have the same dimensions. ⭐TEKS **2C**

Let us apply this technique to the problem of the car moving at a speed of 88 km/h. This measurement is given in dimensions of length over time. The total distance traveled has the dimension of length. Multiplying these numbers together gives the dimensions indicated below. Clearly, the result of this calculation does not have the dimensions of time, which is what you are trying to calculate. That is,

$$\frac{length}{time} \times length = \frac{length^2}{time} \quad \text{or} \quad \frac{88\ km}{1.0\ h} \times 725\ km = \frac{6.4 \times 10^4\ km^2}{1.0\ h}$$

To calculate an answer that will have the dimension of time, you should take the distance and *divide* it by the speed of the car, as follows:

$$\text{length} \div \frac{\text{length}}{\text{time}} = \frac{\text{length} \times \text{time}}{\text{length}} = \text{time} \qquad \frac{725 \text{ km} \times 1.0 \text{ h}}{88 \text{ km}} = 8.2 \text{ h}$$

In a very simple example like this one, you might be able to solve the problem without dimensional analysis. But in more-complicated situations, dimensional analysis is a wise first step that can often save you a great deal of time.

Order-of-magnitude estimations check answers

Because the scope of physics is so wide and the numbers may be astronomically large or subatomically small, it is often useful to estimate an answer to a problem before trying to solve the problem exactly. This kind of estimate is called an *order-of-magnitude* calculation, which means determining the power of 10 that is closest to the actual numerical value of the quantity. Once you have done this, you will be in a position to judge whether the answer you get from a more exacting procedure is correct. (★)TEKS) 2C

For example, consider the car trip described in the discussion of dimensional analysis. We must divide the distance by the speed to find the time. The distance, 725 km, is closer to 10^3 km (or 1000 km) than to 10^2 km (or 100 km), so we use 10^3 km. The speed, 88 km/h, is about 10^2 km/h (or 100 km/h).

$$\frac{10^3 \text{ km}}{10^2 \text{ km/h}} = 10 \text{ h}$$

This estimate indicates that the answer should be closer to 10 than to 1 or to 100 (or 10^2). If you perform the calculation, you will find that the correct answer is 8.2 h, which certainly fits this range.

Order-of-magnitude estimates can also be used to estimate numbers in situations in which little information is given. For example, how could you estimate how many gallons of gasoline are used annually by all of the cars in the United States?

First, consider that the United States has about 250 million people. Assuming that each family of about five people has a car, an estimate of the number of cars in the country is 50 million.

Next, decide the order of magnitude of the average distance each car travels every year. Some cars travel as few as 1000 mi per year, while others travel more than 100 000 mi per year. The appropriate order of magnitude to include in the estimate is 10 000 mi, or 10^4 mi, per year.

If we assume that cars average 20 mi for every gallon of gas, each car needs about 500 gal per year.

$$\left(\frac{10\ 000 \text{ mi}}{1 \text{ year}} \right)\left(\frac{1 \text{ gal}}{20 \text{ mi}} \right) = 500 \text{ gal/year for each car}$$

Did you know?

The physicist Enrico Fermi made the first nuclear reactor at the University of Chicago in 1942. Fermi was also well known for his ability to make quick order-of-magnitude calculations, such as estimating the number of piano tuners in New York City.

internet**connect**

SC*i*LINKS

NSTA

TOPIC: Orders of magnitude
GO TO: www.scilinks.org
*sci*LINKS CODE: HF2014

(★)TEKS)

p. 22: 3B, 2C
p. 23: 2C

Consumer Focus

BACKGROUND

This feature gives a real-world example of the concepts introduced in this section. Dimensional analysis and order-of-magnitude estimation are used to approximate the number of M&M's eaten over time. The estimation is then checked through further dimensional analysis and estimation to find the number of candies eaten by each consumer per month.

EXTENSION

Ask students to estimate the average thickness of a page in their book by guessing the thickness of the book then dividing this measurement by the number of pages in the book.

You have seen the signs before. "Billions of Hamburgers Sold." "Over a Million Satisfied Customers." These are pretty large numbers. Who counted all of those burgers and customers? Can these claims be trusted?

For most large numbers, such as the number of stars in the universe, the exact quantity is not known. Nobody really needs the exact number of burgers, customers, or stars, so no one is paid to make an exact count. Typically, these numbers are estimates from other data, such as the amount of raw materials consumed, the sales income of a company, or the number of stars in a very small patch of sky.

To see how this works, consider M&M's™ chocolate candies. How many of these candies have been eaten over all of time? According to the head office of Mars, Inc., the manufacturer of M&M's candies, "well over 100 million candies a day have been consumed" for the last 10 years. To keep up with this rate of consumption, 10 years (3650 days) would require the production of the following number of candies:

$$\frac{10^8 \text{ candies}}{\text{day}} \times (3.6 \times 10^3 \text{ days}) = 3.6 \times 10^{11} \text{ candies}$$

M&M's candies have been produced in large quantities since 1942. Assuming that the average production rate for each decade is at least half the amount for the last 10 years, we can say that more than 1.3×10^{12}, or 1 trillion, M&M's candies have been eaten over time.

But if no one is counting, can you be sure these claims are reasonable? You can test the validity of these claims by making an order-of-magnitude esti-

mate of what this would mean for a single person and deciding whether it seems reasonable.

Is it reasonable to assume that over 100 million M&M's candies are eaten every day? First consider how many people are likely to be consuming these candies. The population of the United States is about 250 million people. Not all people in the United States eat M&M's candies, but these candies are also sold in many other countries, so a number like 200 million potential consumers seems about right.

Next, examine these numbers on a per-person (or *per capita*) basis by dividing the number eaten per day by the number of people who do the eating.

$$\frac{10^8 \text{ candies/day}}{2 \times 10^8 \text{ people}} = 0.5 \text{ candies per day per person}$$

Is it likely that 200 million people would carefully cut their M&M's candies in half? No, this number is an average over the period of one day. Instead of considering the average per day, it may be more appropriate to consider a longer period of time, like a month. Certainly it is possible to imagine a typical consumer eating 15 candies during a month. This would mean that the average consumer would consume a regular-sized packet every three months or so.

Thus, the claim that 100 million M&M's candies are eaten daily seems reasonable based on the estimates. Breaking large numbers like this down in terms of *per capita* consumption can help identify whether they are reasonable. ⭐TEKS **3A**

Multiplying this by the estimate of the total number of cars in the United States gives an annual consumption of more than 2×10^{10} gallons.

$$(5 \times 10^7 \text{ cars})\left(\frac{500 \text{ gal}}{1 \text{ car}}\right) = 2.5 \times 10^{10} \text{ gal}$$

This corresponds to a yearly consumer expenditure of more than $20 billion! Even so, this estimate may be low because we haven't accounted for commercial consumption and two-car families.

Section Review

⭐ TEKS 2C

1. Which of the following graphs best matches the data shown below?

Volume of air (m^3)	Mass of air (kg)
0.50	0.644
1.50	1.936
2.25	2.899
4.00	5.159
5.50	7.096

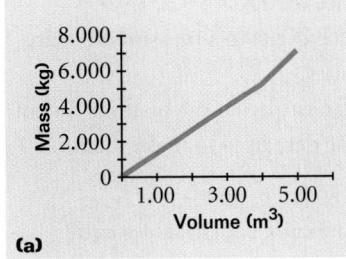
(a)

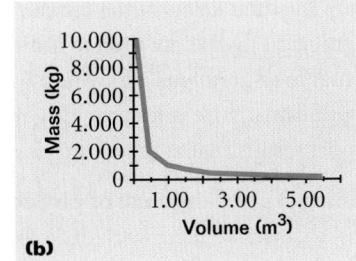
(b)

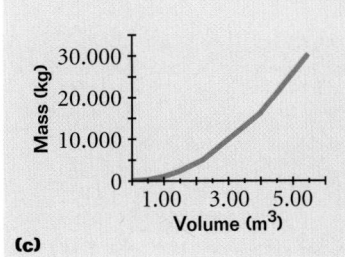
(c)

Figure 1-15

2. Which of the following equations best matches the data from item 1?
   a. $(\text{mass})^2 = 1.29 \,(\text{volume})$ **b.** $(\text{mass})(\text{volume}) = 1.29$
   c. $\text{mass} = 1.29 \,(\text{volume})$ **d.** $\text{mass} = 1.29 \,(\text{volume})^2$

3. Indicate which of the following physics symbols denote units and which denote variables or quantities.
   a. C   **b.** c **c.** C **d.** t **e.** T   **f.** T

4. Determine the units of the quantity described by the following combinations of units:
   a. kg (m/s) (1/s)   **b.** (kg/s) (m/s^2)
   c. (kg/s) $(\text{m/s})^2$ **d.** (kg/s) (m/s)

5. Which of the following is the best order of magnitude estimate in meters of the height of a mountain?
   a. 1 m   **b.** 10 m   **c.** 100 m   **d.** 1000 m

Section Review
ANSWERS

1. a
2. c
3. units: a, e
   variables/quantities: b, c, d, f
4. **a.** $\text{kg} \cdot \text{m/s}^2$
   b. $\text{kg} \cdot \text{m/s}^3$
   c. $\text{kg} \cdot \text{m}^2/\text{s}^3$
   d. $\text{kg} \cdot \text{m/s}^2$
5. d

⭐ TEKS

p. 24: 3A
p. 25: 2C

CHAPTER 1
Summary

KEY TERMS

accuracy (p. 15)

controlled experiment (p. 9)

model (p. 6)

precision (p. 15)

significant figures (p. 16)

system (p. 7)

KEY IDEAS

Section 1-1 What is physics?

- Physics is the study of the physical world, from motion and energy to light and electricity.
- Physics uses the scientific method to discover general laws that can be used to make predictions about a variety of situations.
- A common technique in physics for analyzing a complex situation is to disregard irrelevant factors and create a model that describes the essence of a system or situation.

Section 1-2 Measurements in experiments

- Physics measurements are typically made and expressed in SI, a system that uses a set of base units and prefixes to describe measurements of physical quantities.
- *Accuracy* describes how close a measurement is to reality. *Precision* results from the limitations of the measuring device used.
- Significant figures are used to indicate which digits in a measurement are actual measurements and which are estimates.
- Significant-figure rules provide a means to ensure that calculations do not report results that are more precise than the data used to make them.

Section 1-3 The language of physics

- Physicists make their work easier by summarizing data in tables and graphs and by abbreviating quantities in equations.
- Dimensional analysis can help identify whether a physics expression is a valid one.
- Order-of-magnitude calculations provide a quick way to evaluate the appropriateness of an answer.

Variable symbols

Quantities		Units	
$\Delta x, \Delta y$	change in position	m	meters
Δt	time interval	s	seconds
m	mass	kg	kilograms

CHAPTER 1
Review and Assess

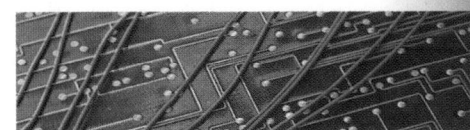

THE SCIENCE OF PHYSICS

Review questions

1. Refer to **Table 1-1** on page 5 to identify at least two areas of physics involved in the following:
   a. building a louder stereo system in your car
   b. bungee jumping
   c. judging how hot a stove burner is by looking at it
   d. cooling off on a hot day by diving into a swimming pool

2. Which of the following scenarios fit the approach of the scientific method?
   a. An auto mechanic listens to how a car runs and comes up with an idea of what might be wrong. The mechanic tests the idea by adjusting the idle speed. Then the mechanic decides his idea was wrong based on this evidence. Finally, the mechanic decides the only other problem could be the fuel pump, and he consults with the shop's other mechanics about his conclusion.
   b. Because of a difference of opinions about where to take the class trip, the class president holds an election. The majority of the students decide to go to the amusement park instead of to the shore.
   c. Your school's basketball team has advanced to the regional playoffs. A friend from another school says their team will win because their players want to win more than your school's team does.
   d. A water fountain does not squirt high enough. The handle on the fountain seems loose, so you try to push the handle in as you turn it. When you do this, the water squirts high enough that you can get a drink. You make sure to tell all your friends how you did it.

3. You have decided to select a new car by using the scientific method. How might you proceed?

4. Consider the phrase, "The quick brown fox jumped over the lazy dog." Which details of this situation would a physicist who is modeling the path of a fox ignore?

SI UNITS

Review questions

5. List an appropriate SI base unit (with a prefix as needed) for measuring the following:
   a. the time it takes to play a CD in your stereo
   b. the mass of a sports car
   c. the length of a soccer field
   d. the diameter of a large pizza
   e. the mass of a single slice of pepperoni
   f. a semester at your school
   g. the distance from your home to your school
   h. your mass
   i. the length of your physics lab room
   j. your height

6. If you square the speed expressed in meters per second, in what units will the answer be expressed?

7. If you divide a force measured in newtons (1 newton = 1 kg•m/s^2) by a speed expressed in meters per second, in what units will the answer be expressed?

Conceptual questions

8. The height of a horse is sometimes given in units of "hands." Why was this a poor standard of length before it was redefined to refer to exactly 4 in.?

9. Explain the advantages in having the meter officially defined in terms of the distance light travels in a given time rather than as the length of a specific metal bar.

10. Einstein's famous equation indicates that $E = mc^2$, where c is the speed of light and m is the object's mass. Given this, what is the SI unit for E?

11. a. 2×10^2 mm
    b. 7.8×10^3 s
    c. 1.6×10^7 μg
    d. 7.5×10^4 cm
    e. 6.75×10^{-4} g
    f. 4.62×10^{-2} cm
    g. 9.7 m/s
12. a. 1 dekaration
    b. 2 kilomockingbirds
    c. 1 microphone
    d. 1 nanogoat
    e. 1 examiner
13. 1.08×10^9 km
14. 11 people
15. yes; measurements can all be close to the same value but not close to the true value.
16. a. 1
    b. 5
    c. 4
    d. 4
    e. 6
17. No, the metric conversions have more significant digits than the original measurements.
18. a. 3.00×10^8 m/s
    b. 2.9979×10^8 m/s
    c. $2.997\ 925 \times 10^8$ m/s
19. a. 3
    b. 4
    c. 3
    d. 2
20. a. 797 g
    b. 0.90 m/s
    c. 17.8 mm
    d. 23.7 s
21. 228.8 cm
22. 115.9 m

Practice problems

11. Express each of the following as indicated:
    a. 2 dm expressed in millimeters
    b. 2 h 10 min expressed in seconds
    c. 16 g expressed in micrograms
    d. 0.75 km expressed in centimeters
    e. 0.675 mg expressed in grams
    f. 462 μm expressed in centimeters
    g. 35 km/h expressed in meters per second
    (See Sample Problem 1A.)

12. Use the SI prefixes in **Table 1-3** on page 12 to convert these *hypothetical* units of measure into appropriate quantities:
    a. 10 rations
    b. 2000 mockingbirds
    c. 10^{-6} phones
    d. 10^{-9} goats
    e. 10^{18} miners
    (See Sample Problem 1A.)

13. Use the fact that the speed of light in a vacuum is about 3.00×10^8 m/s to determine how many kilometers a pulse from a laser beam travels in exactly one hour.
    (See Sample Problem 1A.)

14. If a metric ton is 1.000×10^3 kg, how many 85 kg people can safely occupy an elevator that can hold a maximum mass of exactly 1 metric ton?
    (See Sample Problem 1A.)

ACCURACY, PRECISION, AND SIGNIFICANT FIGURES

Review questions

15. Can a set of measurements be precise but not accurate? Explain.

16. How many significant figures are in the following measurements?
    a. 300 000 000 m/s
    b. 25.030°C
    c. 0.006 070°C
    d. 1.004 J
    e. 1.305 20 MHz

17. **Figure 1-16** shows photographs of unit conversions on the labels of some grocery-store items. Check the accuracy of these conversions. Are the manufacturers using significant figures correctly?

(a) 　(b)

(c) 　(d)

Figure 1-16

18. The value of the speed of light is now known to be $2.997\ 924\ 58 \times 10^8$ m/s. Express the speed of light in the following ways:
    a. with three significant figures
    b. with five significant figures
    c. with seven significant figures

19. How many significant figures are there in the following measurements?
    a. 78.9 ± 0.2 m
    b. 3.788×10^9 s
    c. 2.46×10^6 kg
    d. 0.0032 mm

20. Carry out the following arithmetic operations:
    a. find the sum of the measurements 756 g, 37.2 g, 0.83 g, and 2.5 g
    b. find the quotient of 3.2 m/3.563 s
    c. find the product of 5.67 mm $\times$ π
    d. find the difference of 27.54 s and 3.8 s

21. A fisherman catches two sturgeons. The smaller of the two has a measured length of 93.46 cm (two decimal places and four significant figures), and the larger fish has a measured length of 135.3 cm (one decimal place and four significant figures). What is the total length of the two fish?

22. A farmer measures the distance around a rectangular field. The length of each long side of the rectangle is found to be 38.44 m, and the length of each short side is found to be 19.5 m. What is the total distance around the field?

DIMENSIONAL ANALYSIS AND ORDER-OF-MAGNITUDE ESTIMATES

Review questions

Note: In developing answers to order-of-magnitude calculations, you should state your important assumptions, including the numerical values assigned to parameters used in the solution. Since only order-of-magnitude results are expected, do not be surprised if your results differ from those of other students.

23. Suppose that two quantities, *A* and *B*, have different dimensions. Which of the following arithmetic operations *could* be physically meaningful?

    a. $A + B$
    b. A/B
    c. $A \times B$
    d. $A - B$

24. Estimate the order of magnitude of the length in meters of each of the following:

    a. a ladybug
    b. your leg
    c. your school building
    d. a giraffe
    e. a city block

25. If an equation is dimensionally correct, does this mean that the equation is true?

26. The radius of a circle inscribed in any triangle whose sides are *a*, *b*, and *c* is given by the following equation, in which *s* is an abbreviation for $(a + b + c) \div 2$. Check this formula for dimensional consistency.

$$r = \sqrt{\frac{(s-a)(s-b)(s-c)}{s}}$$

27. The period of a simple pendulum, defined as the time necessary for one complete oscillation, is measured in time units and is given by the equation

$$T = 2\pi \sqrt{\frac{L}{g}}$$

where *L* is the length of the pendulum and *g* is the acceleration due to gravity, which has units of length divided by time squared. Check this equation for dimensional consistency.

Conceptual questions

28. In a desperate attempt to come up with an equation to solve a problem during an examination, a student tries the following: (velocity in m/s)^2 = (acceleration in m/s^2) × (time in s). Use dimensional analysis to determine whether this equation might be valid.

29. Estimate the number of breaths taken during 70 years, the average life span of a person.

30. Estimate the number of times your heart beats in an average day.

31. Estimate the magnitude of your age, as measured in units of seconds.

32. An automobile tire is rated to last for 50 000 mi. Estimate the number of revolutions the tire will make in its lifetime.

33. Imagine that you are the equipment manager of a professional baseball team. One of your jobs is to keep a supply of baseballs for games in your home ballpark. Balls are sometimes lost when players hit them into the stands as either home runs or foul balls. Estimate how many baseballs you have to buy per season in order to make up for such losses. Assume your team plays an 81-game home schedule in a season.

34. A chain of hamburger restaurants advertises that it has sold more than 50 billion hamburgers over the years. Estimate how many pounds of hamburger meat must have been used by the restaurant chain to make 50 billion hamburgers and how many head of cattle were required to furnish the meat for these hamburgers.

35. Estimate the number of piano tuners living in New York City (The population of New York City is approximately 8 million). This problem was first proposed by the physicist Enrico Fermi, who was well known for his ability to quickly make order-of-magnitude calculations.

36. Estimate the number of table-tennis balls that would fit (without being crushed) into a room that is 4 m long, 4 m wide, and 3 m high. Assume that the diameter of a ball is 3.8 cm.

23. b, c
24. **a.** 10^{-2} m
    b. 10^0 m
    c. 10^1 m to 10^2 m
    d. 10^0 m
    e. 10^2 m
25. No, the values must also be correct.
26. The dimensions are consistent.
27. The dimensions are consistent.
28. The equation is not valid.

(Note: Because the following are estimates, student answers may vary from those shown here.)
29. 4×10^8 breaths
30. 9×10^4 beats
31. 5.4×10^8 s
32. 4×10^7 revolutions
33. 2×10^3 balls
34. 1×10^{10} lbs, 2×10^7 head of cattle
35. 8×10^2 tuners
36. 8×10^5 balls

37. a. 22 cm; 38 cm^2
b. 29.2 cm; 67.9 cm^2

38. Take the $5000 because it would take you 272 years to count out the $5 billion in single dollar bills.

39. 9.818×10^{-2} m

40. 1.79×10^{-9} m

ANSWERS TO

Technology & Learning

a. grams
b. 0.6 g; 0.2 g; 0.7 g
c. 9.2 g; 2.8 g; 11 g
d. 13 g; 3.9 g; 15 g
e. larger density

MIXED REVIEW PROBLEMS

37. Calculate the circumference and area for the following circles. (Use the following formulas: circumference $= 2\pi r$ and area $= \pi r^2$.)

   a. a circle of radius 3.5 cm

   b. a circle of radius 4.65 cm

38. A billionaire offers to give you $5 billion if you will count out the amount in $1 bills or a lump sum of $5000. Which offer should you accept? Explain your answer. (Assume that you can count at an average rate of one bill per second, and be sure to allow for the fact that you need about 10 hours a day for sleeping and eating. Your answer does not need to be limited to one significant figure.)

39. Exactly 1 quart of ice cream is to be made in the form of a cube. What should be the length of one side in meters for the container to have the appropriate volume? (Use the following conversion: $4 \text{ qt} = 3.786 \times 10^{-3} \text{ m}^3$)

40. You can obtain a rough estimate of the size of a molecule with the following simple experiment: Let a droplet of oil spread out on a fairly large but smooth water surface. The resulting "oil slick" that forms on the surface of the water will be approximately one molecule thick. Given an oil droplet with a mass of 9.00×10^{-7} kg and a density of 918 kg/m^3 that spreads out to form a circle with a radius of 41.8 cm on the water surface, what is the approximate diameter of an oil molecule?

Technology & Learning

Graphing calculators

Refer to Appendix B for instructions on downloading programs for your calculator. The program "Chap1" allows you to analyze the relationship between the mass and length of three wires, each made of a different substance.

All three wires have a diameter of 0.50 cm. Because the wires have the same diameter, their cross-sectional areas are the same. As for any circle, this area is equal to πr^2. Using this area, the wires can be described by the following equations:

$$Y_1 = 8.96X^* \pi (0.25)^2$$
$$Y_2 = 2.70X^* \pi (0.25)^2$$
$$Y_3 = 10.49X^* \pi (0.25)^2$$

In these equations, X represents the length of the wire in centimeters. Note that X is multiplied by a different factor in each equation. This factor signifies the mass per unit volume, or *density*, of the substance.

   a. Assuming the density is in units of grams per centimeter, use dimensional analysis to determine the units of Y.

Press PRGM, and scroll down to "Chap1" by pressing ▼. Press ENTER to execute the program.

Press ENTER twice to begin graphing. The calculator will display three lines. Each line represents one type of wire. The mass of the wire in grams is plotted on the *y*-axis, and the length of the wire in centimeters is plotted on the *x*-axis.

The calculator is already in TRACE mode; a blinking cursor should be visible on the central line. Press ◄ to move along this line. Press ► to move from one line to another. The equation for the line being traced appears in the top left corner of the screen. The values corresponding to the placement of the cursor appear at the bottom left corner. Use these values to complete the following exercises.

Find the approximate masses of each kind of wire at the following lengths:

   b. 0.3 cm

   c. 5.2 cm

   d. 7.3 cm

   e. Assuming equal size, does a larger or smaller density correspond to a larger mass? Press ENTER and CLEAR to end.

41. An ancient unit of length called the cubit was equal to approximately 50 centimeters, which is, of course, approximately .50 meters. It has been said that Noah's ark was 300 cubits long, 50 cubits wide, and 30 cubits high. Estimate the volume of the ark in cubic meters. Also estimate the volume of a typical home, and compare it with the ark's volume.

42. If one micrometeorite (a sphere with a diameter of 1.0×10^{-6} m) struck each square meter of the moon each second, it would take many years to cover the moon with micrometeorites to a depth of 1.0 m. Consider a cubic box, 1.0 m on a side, on the moon. Find how long it would take to completely fill the box with micrometeorites.

43. One cubic centimeter (1.0 cm^3) of water has a mass of 1.0×10^{-3} kg at 25°C. Determine the mass of 1.0 m^3 of water at 25°C.

44. Assuming biological substances are 90 percent water and the density of water is $1.0 \times 10^3 \text{ kg/m}^3$, estimate the masses (density multiplied by volume) of the following:

 a. a spherical cell with a diameter of 1.0 μm (volume $= \frac{4}{3}\pi r^3$)

 b. a fly, which can be approximated by a cylinder 4.0 mm long and 2.0 mm in diameter (volume $= \ell\pi r^2$)

45. The radius of the planet Saturn is 5.85×10^7 m, and its mass is 5.68×10^{26} kg.

 a. Find the density of Saturn (its mass divided by its volume) in grams per cubic centimeter. (The volume of a sphere is given by $\frac{4}{3}\pi r^3$.)

 b. Find the surface area of Saturn in square meters. (The surface area of a sphere is given by $4\pi r^2$.)

Alternative Assessment

Performance assessment

1. Imagine that you are a member of your state's highway board. In order to comply with a bill passed in the state legislature, all of your state's highway signs must show distances in miles and kilometers. Two plans are before you. One plan suggests adding metric equivalents to all highway signs as follows: Dallas 300 mi (483 km). Proponents of the other plan say that the first plan makes the metric system seem more cumbersome, so they propose replacing the old signs with new signs every 50 km as follows: Dallas 300 km (186 mi). Participate in a class debate about which plan should be followed.

2. Can you measure the mass of a five-cent coin with a bathroom scale? Record the mass in grams displayed by your scales as you place coins on the scales, one at a time. Then divide each measurement by the number of coins to determine the approximate mass of a single five-cent coin, but remember to follow the rules for significant figures in calculations. Which estimate do you think is the most accurate? Which is the most precise?

Portfolio projects

3. Find out who were the Nobel laureates for physics last year, and research their work. Alternatively, explore the history of the Nobel Prizes. Who founded the awards? Why? Who delivers the award? Where? For either topic, summarize your findings in a brochure, poster, or computer presentation.

4. You have a clock with a second hand, a ruler marked in millimeters, a graduated cylinder marked in milliliters, and scales sensitive to 1 mg. How would you measure the mass of a drop of water? How would you measure the period of a swing? How would you measure the volume of a paper clip? How can you improve the accuracy of your measurements? Write the procedures clearly so that a partner can follow them and obtain reasonable results.

5. Create a poster or other presentation depicting the possible ranges of measurement for a dimension, such as distance, time, temperature, speed, or mass. Depict examples ranging from the very large to the very small. Include several examples that are typical of your own experiences.

41. The ark ($5 \times 10^4 \text{ m}^3$) was about 10 to 100 times as large as a house (6×10^2 to 10^3 m^3).

42. 3.2×10^{10} years

43. 1.0×10^3 kg

44. **a.** 5×10^{-16} kg
 b. 1×10^{-5} kg

45. **a.** 0.677 g/cm^3
 b. $4.30 \times 10^{16} \text{ m}^2$

Alternative Assessment
ANSWERS

Performance assessment

1. Students should recognize that preferences for one scheme or the other have nothing to do with the accuracy of the measurements.

2. Students should recognize that their approximation of the mass of a single coin becomes both more accurate and more precise as they increase the number of coins.

 Portfolio projects

3. Students' presentations will vary depending on which scientists they choose. Alfred Nobel (1833–1896), inventor of dynamite, left a fund for the establishment of annual prizes in several fields.

4. Students' procedures should be safe and thoroughly explained. In some cases, they will need to measure several objects together.

5. Students' presentations will vary but should include references for sources of information. Check that all examples are within a reasonable order of magnitude.

NOTE

Materials Preparation is given on pp. 2A–2B. Blank data table and sample data table are on the One-Stop Planner CD-ROM. All calculations shown use sample data.

Planning

Recommended time:

1 lab period

Classroom organization:

▶ This lab should be performed by students alone. If students work in groups, they should alternate duties so that each student performs all steps.

Measurement Tips

◆ Remind students to record all measured digits plus one estimated digit.

◆ Students may repeat all measurements using the 15 cm ruler.

Techniques to Demonstrate

Show students the proper way to hold a meterstick when taking a measurement.

Show students how to adjust the balances to zero. If you are using triple-beam balances, demonstrate handling the balances and moving the masses to take a measurement.

✔ Checkpoints

Step 3: Make sure all students perform all measurements. Students should be able to explain

CHAPTER 1
Laboratory Exercise A

★ TEKS

pp. 32–33: 1A, 2B, 2C, 2E, 2F

OBJECTIVES

• Use typical laboratory equipment to make accurate measurements.

• Measure length and mass in SI units.

• Determine the appropriate number of significant figures for various measurements and calculations.

• Examine the relationships between measured physical quantities using graphs and data analysis.

MATERIALS LIST

✔ 2 rectangular wooden blocks
✔ 15 cm metric ruler
✔ balance
✔ meterstick

PHYSICS AND MEASUREMENT

In this laboratory exercise, you will gain experience making measurements as a physicist does. All measurements will be made using units to the precision allowed by your instruments.

SAFETY

• **Review lab safety guidelines. Always follow correct procedures in the lab.**

PREPARATION

1. Read the entire lab procedure, and plan the steps you will take.

2. Prepare a data table in your lab notebook with seven columns and five rows, as shown below. In the first row, label the second through seventh columns *Trial 1, Trial 2, Trial 3, Trial 4, Trial 5,* and *Trial 6.* In the first column, label the second through fifth rows *Length (cm), Width (cm), Thickness (cm),* and *Mass (kg).*

	Trial 1	Trial 2	Trial 3	Trial 4	Trial 5	Trial 6
Length (cm)						
Width (cm)						
Thickness (cm)						
Mass (kg)						

PROCEDURE

Measuring length, width, thickness, and mass

3. Use a meterstick to measure the length of the wood block. Record all measured digits plus one estimated digit.

4. Follow the same procedure to measure the width and thickness of the block. Repeat all measurements two more times. Record your data.

5. Carefully adjust the balance to obtain an average zero reading when there is no mass on it. Your teacher will show you how to adjust the balances in your classroom to obtain an average zero reading. Use the balance to find the mass of the block, as shown in **Figure 1-17.** Record the measurement in your data table.

6. Repeat the mass measurement three more times, and record the values in your data table. Each time, place the block on a different side.

7. For trials 4–6, repeat steps 3 through 6 with the second wood block.

ANALYSIS AND INTERPRETATION

Calculations and data analysis

1. **Organizing data** Using your data, calculate the volume of the wood block for each trial. The equation for the volume of a rectangular block is $volume = length \times width \times thickness$.

2. **Analyzing data** Use your measurements from different trials to answer the following questions.

   a. For each block, what is the difference between the smallest length measurement and the largest length measurement?

   b. For each block, what is the difference between the smallest calculated volume and the largest calculated volume?

   c. Based on your answers to (a) and (b), how does multiplying several length measurements together to find the volume affect the precision of the result?

Conclusions

3. **Interpreting results** For each trial, find the ratio between the mass and the volume. Based on your data, what is the relationship between the mass and volume?

4. **Evaluating methods** For each type of measurement you made, explain how error could have affected your results. Consider method error and instrument error. How could you find out whether error had a significant effect on your results for each part of the lab?

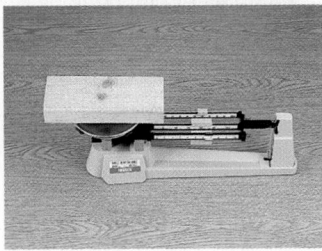

Figure 1-17

Step 3: Always record measurements to the precision allowed by your instruments.

Step 5: Make sure you know how to use the balances in your classroom. The balance should read zero when there is no mass on it. The number of significant figures in your measurement will be determined by your instrument, the object being measured, and the purpose of your measurement.

how they assign significant figures to their measurements.

Step 5: Students may need help using the balance for the first time. For some balances, instrument drift may prevent a continuous zero reading, so an average zero reading will be the goal.

ANSWERS TO

Analysis and Interpretation

CALCULATIONS AND DATA ANALYSIS

1. Answers will vary, depending on the blocks used. Make sure answers have the right number of significant figures. For sample data, values range from 160 cm^3 to 172 cm^3.

2. Answers will vary.

   a. For the sample data, the answer is 8.15 cm.

   b. For the sample data, the answer is 12 cm^3.

   c. Students should recognize that the answer becomes less precise as several values are multiplied and that the difference between the highest and lowest answers becomes greater.

CONCLUSIONS

3. Answers should state that both wood blocks have the same mass-to-volume ratio, even though the mass and volume are not the same. Some students may realize that this value is the density of the wood.

4. Answers will vary but should include an analysis of error in the laboratory.

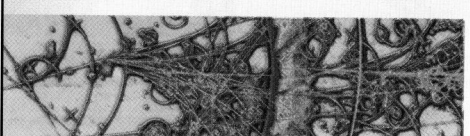

NOTE

Materials Preparation is given on pp. 2A–2B. Blank data table and sample data table are on the One-Stop Planner CD-ROM. All calculations shown use sample data.

Planning

Recommended time:

1 lab period

Classroom organization:

▶ Each lab group should have two students.

▶ Each group needs a large, clear area to work in. If possible, consider conducting this lab in a large, open space, such as outdoors or in a gymnasium.

▶ The CBL and sensors procedure and the stopwatch procedure may be used in the same class. If there is time, have students perform both procedures.

▶ **Safety warnings:** Remind students to be aware of other groups' activities. Falling objects can cause serious injury.

OBJECTIVES

- Use typical laboratory equipment to measure the distance and time of an observed motion.
- Measure distance and time in SI units.
- Determine the appropriate number of significant figures for various measurements.
- Use graphs and data analysis to examine the relationships between measured physical quantities.

MATERIALS LIST

✔ **meterstick**

✔ **rectangular wooden block**

PROCEDURE

CBL AND SENSORS

✔ **C-clamp**

✔ **CBL**

✔ **CBL motion detector**

✔ **graphing calculator with link cable**

✔ **support stand and clamp**

✔ **thin foam pad**

STOPWATCH

✔ **stopwatch**

CHAPTER 1
Laboratory Exercise B

TIME AND MEASUREMENT

Many fields of physics require experimenters to study events that take place over time. In this laboratory exercise, you will become familiar with the kinds of equipment used to make these measurements, such as metersticks and stopwatches; or motion detectors, CBL, and graphing calculators. All measurements will be made using SI units to the precision allowed by your instruments.

SAFETY

- **Perform this lab in a clear area. Falling or dropped masses can cause serious injury.**

PREPARATION

1. Determine whether you will be using the CBL and sensors or the stopwatch. Read the entire lab for the appropriate procedure, and plan the steps you will take.

2. Prepare a data table in your lab notebook with three columns and seven rows. In the first row, label the columns *Trial*, *Distance (m)*, and *Time (s)*. Label the second through seventh rows *1, 2, 3, 4, 5*, and *6*.

Trial	Distance (m)	Time (s)
1		
2		
3		
4		
5		
6		

Stopwatch procedure begins on page 36.

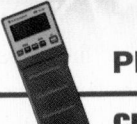

PROCEDURE

CBL AND SENSORS

Measuring distance and time

3. This exercise should be performed with a partner. Perform this in a clear area away from other groups. Connect the CBL to the calculator with the unit-to-unit link cable using the ports located on each unit. Connect the motion detector to the SONIC port.

4. Set up the apparatus as shown in **Figure 1-18.** Securely clamp the motion detector to the support stand so that it faces downward, over the edge of the table. Make sure the motion detector is far enough away from the edge of the table that the signal will not hit the tabletop, clamp, or table leg.

5. Use a meterstick to measure a distance 0.5 m below the motion detector, and mark the point with tape on the table or stand. This is the starting position from which the blocks will be dropped from rest. Measure the height of the tape mark above the floor. Record this distance in your data table.

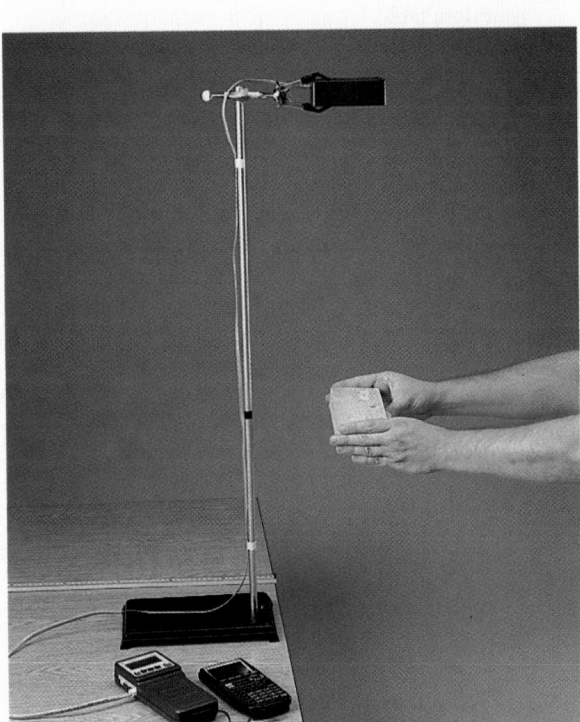

Figure 1-18

Step 4: The motion detector should be clamped securely to the stand, and the base of the stand should be clamped to the table if possible. Tape the cord to the stand to keep it out of the way.

Step 7: With the CBL in MONITOR INPUT mode, move the wooden block up and down below the motion detector to check the readings.

6. Start the program PHYSICS on your graphing calculator. Select option *SET UP PROBES* from the MAIN MENU. Enter 1 for the number of probes. Select *MOTION DETECTOR* from the list.

7. Select the *MONITOR INPUT* option from the DATA COLLECTION menu. Test to be sure the motion detector is positioned properly.

   a. Read the CBL measurement for the distance between the motion detector and the floor. Measure the distance with a meterstick to confirm the CBL value. If the CBL reading is too low, adjust the motion detector to make sure the signal is not hitting the table instead of the floor.

   b. Cover the floor under the motion detector with a foam pad to reduce feedback.

   c. Hold the wooden block directly beneath the motion detector, move the block up and down, and read the CBL measurements. Make sure the motion detector is not detecting other objects, such as the stand base, the tabletop, or the table leg. When the probe is functioning correctly, press + on the calculator to return to the MAIN MENU.

8. Select the *COLLECT DATA* option. Enter 0.02 for the time between samples. Enter 99 for the number of samples.

   a. Check the values you entered, and press ENTER. If the values are correct, select *USE TIME SETUP* to continue. If you made a mistake entering the time values, select *MODIFY SETUP*, reenter the values, and continue.

CBL and Sensors Tips

◆ Students should have the program PHYSICS on their graphing calculators. Refer to Appendix B for instructions.

◆ Better results will be obtained with blocks that have similar dimensions (length, width, and thickness), such that rotation does not greatly change measurements. The higher the drop, the better the results are likely to be.

Techniques to Demonstrate

Show students how to determine whether the motion detector can "see" the wood block by moving the block up and down under the detector.

Demonstrate how to hold the block between your hands and release the block by pulling your hands straight out to the sides. Show how this method prevents the block from turning, while other methods cause the block to turn.

✔ Checkpoints

Step 4: Make sure the CBL, calculator, and motion detector are connected properly and that the motion detector is clamped securely in position.

Step 7: Students should be able to demonstrate that the detector is working properly. They should be able to explain why the distance measurement increases as the block moves away and decreases as the block moves toward the detector.

Step 8: With the TI-83 calculator, the selection of *NON-LIVE DISPLAY* is automatic at this sample rate. Students using the TI-83 should ignore part (b).

Step 13: Show the students how to use the arrow keys to trace the graph. Explain how to choose the points and how to find the difference between the values of the points.

Stopwatch Tips

◆ Students should practice timing the falling block before recording data.

◆ Better results that are easier to reproduce will be obtained with a greater falling distance.

Techniques to Demonstrate

Demonstrate how to hold the block between your hands and release the block by pulling your hands straight out to the sides. Show how this method prevents the block from turning, while other methods cause the block to turn.

Make sure students know how to operate and read a stopwatch.

✔ Checkpoints

Step 4: Students should hold the block at about shoulder height for each trial. In step 7, the second student should hold the block at his or her own shoulder height.

Step 5: Before recording data, students should practice until they can demonstrate that the timer starts at the moment the block is released and stops when the block hits the floor.

b. If you are given a choice on the TIME GRAPH menu, select *NON-LIVE DISPLAY.* Otherwise, continue to the next step.

9. One student should hold the block horizontally between flat hands, as shown in **Figure 1-19** on the next page. Position the block directly below the motion detector and level with the 0.5 m mark.

10. Turn on the CBL and the graphing calculator. When the area is clear of people and objects, one student should press ENTER on the graphing calculator. As soon as the motion detector begins to click, the student holding the block should release the block by pulling both hands out to the side. Releasing the block this way will prevent the block from twisting as it falls, which could affect the results of this experiment.

11. When the motion detector has stopped clicking and the CBL displays DONE, press ENTER on the graphing calculator to get to the SELECT CHANNELS menu. Select the *SONIC* option, and then select *DISTANCE* to plot a graph of distance in meters against time in seconds. The graph should have a smooth shape. If it has spikes or black lines, repeat the trial to obtain a smooth graph, and continue on to the next step.

12. Examine the graph to find the section of the curve that represents the block's motion. On the far left and far right, the curve represents the position of the block before and after its motion. The middle section of the curve represents the motion of the falling block. Sketch the graph in your lab notebook.

13. Use the arrow keys to trace the graph. The x- and y-coordinates will be displayed as the cursor moves along the graph. Select a point from the beginning of the block's motion and another point from the end. Find the time interval between the two points by finding the difference between their x-values. Record this in your data table as the time in seconds. Find the distance moved by the block during that time by finding the difference between the y-values of the two points. Record this in your data table as the distance. Press ENTER on the graphing calculator.

14. Repeat for two more trials, recording all data in your data table. Try to drop the block from exactly the same height each time.

15. Switch roles so that the student who dropped the block is now operating the CBL, and repeat the experiment. Perform three trials. Record all data in your data table.

Analysis and Interpretation begins on page 37.

PROCEDURE

STOPWATCH

Measuring distance and time

3. Perform this exercise with a partner. One partner will drop the wooden block from a measured height, and the other partner will measure the time it takes the block to fall to the floor. Perform this in a clear area away from other groups.

4. One student should hold the wooden block held straight out in front of him or her at shoulder height. Hold the block between your hands, as shown in **Figure 1-19** on the next page. Use the

meterstick to measure the height of the wood block. Record this distance in your data table.

5. Use the stopwatch to time the fall of the block. Make sure the area is clear, and inform nearby groups that you are about to begin. The student holding the block should release it by pulling both hands straight out to the sides. The student with the stopwatch should begin timing the instant the block is released and stop timing as soon as the block hits the floor. In your data table, record the time required for the block to fall.

6. Repeat for two more trials, recording all data in your data table. Try to drop the block from exactly the same height each time.

7. Switch roles, and repeat steps 4 through 6. Perform three trials. Record all data in your data table.

Figure 1-19
Step 4: Hold the block between your hands.
Step 5: Release the block by pulling both hands straight out to the sides. It may take some practice to release the block so that it falls straight down without turning.

ANALYSIS AND INTERPRETATION

Calculations and data analysis

1. Organizing data   Did the block always fall from the same height in the same amount of time? Explain how you found the answer to this question.

2. Graphing data   Using the data from all trials, make a scatter plot of the distance versus the time of the block's fall. Use a graphing calculator, computer, or graph paper.

Conclusions

3. Evaluating methods   For each type of measurement you made, explain how error could have affected your results. Consider method error and instrument error. How could you find out whether error had a significant effect on your results for each part of the lab?

Extensions

4. Evaluating data   If there is time and your teacher approves, conduct the following experiment. Have one student drop the wooden block from shoulder height while all other class members time the fall. Perform three trials. Compare results each time. What does this exercise suggest about accuracy and precision in the laboratory?

ANSWERS TO

Analysis and Interpretation

Note: In this procedure, students should discover how human reaction time places a limitation on their measurements.

CALCULATIONS AND DATA ANALYSIS

1. Student answers should state that the block always falls from the same height in the same amount of time. If students answer the question by finding the ratio between distance and time, they will find different values for different heights.

2. For the stopwatch procedure, the graph should show one point for each height from which the block fell. For the CBL procedure, the graph should show points that can be connected to make a parabolic curve.

3. Student answers will vary but should include an analysis of error in the laboratory.

EXTENSIONS

4. Typically, a range of values is reported as different class members measure the time of the same event. This should lead students to realize that for short times, error can be a significant factor.

CHAPTER 2 PLANNING GUIDE

Compression Guide: To shorten from 12 to 9 45-min periods (from 6 to 4½ 90-min blocks), eliminate items in magenta type.

PACING CHART	CLASSROOM RESOURCES			
	⭐ TEKS	Teacher Demonstrations	*Holt Physics* Transparencies	Labs (See page T52 for equipment listing for in-text labs.)
2-1 Displacement and Velocity 3 or 2 45-minute periods 1½ or 1 90-minute block	2B, 2C, 4A, 4B, 4E	**TE** *Displacement, p. 42*	**T** 5 **TM** 11	**L** *Discovery Lab: Motion*
2-2 Acceleration 4 or 2 45-minute periods 2 or 1 90-minute block	2C, 3E, 4A, 4B	**TE** *Acceleration, p. 48* **TE** *Constant Acceleration, p. 51*	**TM** 12, 13	**L** *Invention Lab: Race Car Construction* **TL** *Acceleration*
2-3 Falling objects 3 or 2 45-minute periods 1½ or 1 90-minute block	1A, 2B, 2C, 3A, 3B, 3E, 4A, 4B, 4E			**PE** *Quick Lab: Time Interval of Free Fall, p. 62* **PE** *Measuring Time and Motion, p. 76* **TL** *Free Fall*
Review and Assessment 2 45-minute periods 1 90-minute block				

Resource Key

PHYSICS HOLT

PE Pupil's Edition
TE Teacher's Edition

L Laboratory Experiments
TL Technology Lab Experiments
T Transparencies

One-Stop Planner CD-ROM **contents**

TM Transparency Masters
SR Section Review Worksheets
AA Alternative Assessment

PW Problem-Solving Workbook
PB Problem Bank
CTW Critical Thinking Worksheet

LABORATORY PLANNING: Measuring Time and Motion, p. 76

Materials (for each lab group)
- balance: portable, electronic balance or triple-beam balance with weight
- C-clamp
- meterstick
- 1-position support base and rod, 1.3 cm × 91 cm

Additional Equipment
CBL and Sensors Procedure
- CBL
- graphing calculator
- CBL ultrasonic motion detector
- V-jaw symmetrical clamp with holder
- 4 wood blocks of different sizes
- TI Graph Link (recommended for downloading programs)

ASSIGNMENT RESOURCES

Content Mastery	Critical Thinking	Problem-Solving Practice
PE 1–6, p. 47 **SR** 2-1, *Graph Skills* **PE** 1–5, p. 69	**PE** 1–2, p. 41 **PE** 1–2, p. 45 **PE** 6–7, p. 69	**2A** Average velocity and displacement: 36 items in **PE, PW,** and **PB,** see **TE** p. 44
PE 1–6, p. 59 **SR** 2-2, *Math Skills* **PE** 16–17, p. 70	**PE** 1–3, p. 50	**2B** Average acceleration: 31 items in **PE, PW,** and **PB,** see **TE** p. 49 **2C** Displacement with constant acceleration: 30 items in **PE, PW,** and **PB,** see **TE** p. 53 **2D** Velocity and displacement with constant acceleration: 39 items in **PE, PW,** and **PB,** see **TE** p. 55 **2E** Final velocity after any displacement: 27 items, see **TE** p. 57
PE 1–6, p. 65 **SR** 2-3, *Math Skills*	**PE** 34–37, p. 72	**2F** Falling object: 37 items in **PE, PW,** and **PB,** see **TE** pp. 63–64

ASSESSMENT RESOURCES

Cumulative Review	Alternative Assessment	Traditional Assessment
SR Mixed Review, Ch. 2	**PE** 1–6, p. 75 **AA** Items for Ch. 2	Chapter 2 Test Test Generator items for Ch. 2

Scoring Rubrics for Alternative Assessment items can be found on the One-Stop Planner CD-ROM.

TECHNOLOGY RESOURCES

 **CTW** Segment 2 Land Speed Record

 Module 1 One-Dimensional Motion

 PE Technology and Learning, p. 74
(Alternative procedures for calculators without Flash-ROM technology are provided on the One-Stop Planner CD-ROM.)

 The Mechanical Universe/High School Adaptation Quad III, The Law of Falling Bodies

internetconnect

 On-line Student Resources:
GO TO: www.scilinks.org
The following *sci*LINKS Internet resources can be found in the student text for this chapter.

TOPICS:
- Motion, p. 43 (HF2021)
- Acceleration, p. 48 (HF2022)
- Galileo, p. 56 (HF2023)
- Free fall, p. 61 (HF2024)
- Relativity of Time, p. 67 (HF2025)

 On-line Teacher Resources:
GO TO: go.hrw.com
KEYWORD: HF2 HOME
Visit the HRW Website for a variety of resources related to this chapter.

Smithsonian Institution
Internet Connections
Visit **www.si.edu/hrw** for additional on-line resources.

CNNfyi.com
Visit **www.cnnfyi.com** for late-breaking news and current events stories selected just for you.

Recording Timer Procedure
- Recording timer: acceleration timer, tabletop acceleration timer, or compact spark timer
- metric hooked mass set
- 12- or 24-hour alarm stopwatch
- replacement paper tape, 13 mm
- replacement carbon disks

Required Precautions
Wear eye protection and other required safety equipment when cutting wood blocks. Follow all instructions and safety guidelines for the equipment. Do not work with powertools without other people present.

Materials Preparation
The wood blocks for the CBL and sensors procedure should be cut from standard 2 × 4 lumber. Each lab group should have 4 blocks cut into lengths of approximately 10 cm, 15 cm, 20 cm, and 25 cm.

Section 2-1 introduces the concepts and relationships between displacement, time, speed, and velocity.

Section 2-2 describes the difference between accelerated motion and nonaccelerated motion and introduces the kinematic equations for constant acceleration.

Section 2-3 explores freely falling bodies as examples of motion with constant acceleration.

About the Illustration

Japan's bullet trains are known for their speed and reliability. Up to 130 million passengers use the Shinkansen system every year. The combination of the trains' speed and the large number of trains on each line (nearly 300 trains per day on the Tokyo-Osaka route) make the trains known for their frequency as well.

The bullet train is also known for its safety record. Unlike the French *Train a Grande Vitesse* (TGV), which moves mostly through the country along a straight track, the Japanese bullet train must constantly accelerate and decelerate to accommodate the many turns and high-population areas. The bullet train has yet to have a derailment or collision.

Interactive Problem-Solving Tutor

See Module 1
"One-Dimensional Motion" provides additional development of problem-solving skills for this chapter.

CHAPTER 2

Motion in One Dimension

PHYSICS IN ACTION

The electric bullet trains in Japan have been running at speeds of 200 km/h since October 1964. Newer trains can achieve speeds of over 270 km/h. Unlike traditional locomotives, each car on the Japanese trains is powered by a motor that also acts as a brake. The individually powered cars can achieve much higher speeds than a long chain of cars pulled by a locomotive engine, and the bullet trains can accelerate and decelerate much more quickly. This allows the quiet bullet trains to be used for travel across Japan and within the cities.

- *When the speed of the trains increased by 70 km/h, how much less time did it take to travel 97 km from Tokyo to Mount Fuji?*

- *How far before a station must the train begin to slow in order to stop directly in front of the platform?*

CONCEPT REVIEW

Significant figures (Section 1-2)

SI measurements (Section 1-2)

Graphs and tables (Section 1-3)

Knowledge to Expect

✔ "Students can determine what units an answer should be expressed in from the units of the inputs to the calculations." (AAAS's *Benchmarks for Science Literacy*, grades 6–8)

✔ "Students can convert compound units (such as miles per hour into feet per second)." (AAAS's *Benchmarks for Science Literacy*, grades 6–8)

✔ "Students can decide what degree of precision is adequate and round off the result of calculator operations to enough significant figures to reasonably reflect those of the inputs." (AAAS's *Benchmarks for Science Literacy*, grades 6–8)

Knowledge to Review

✔ The number of significant figures in a number equals the number of reliably known digits. (Section 1-2)

✔ SI measurements are based on a set of seven worldwide standards: m, kg, s, K, A, mol, cd. (Section 1-2)

✔ Graphs and tables provide a way to organize data and recognize relationships. (Section 1-3)

Items to Probe

✔ Unit conversions: Ask the students to convert from m/s to km/h.

✔ Significant figures: Ask the students to identify the number of significant digits in a set of numbers on the board. Make sure that they understand that 50 m/s only has one significant digit.

2-1
Displacement and velocity

Section 2-1

The Language of Physics

Although this chapter discusses displacement, velocity, and acceleration, the concept of vectors is not introduced until Chapter 3. For the purposes of this chapter, it is sufficient to describe the direction of a quantity with a positive or negative sign because the focus is on motion in one dimension. The transition to two-dimensional motion and vectors is made in Section 3-1.

2-1 SECTION OBJECTIVES

- **Describe motion in terms of frame of reference, displacement, time, and velocity.**

- **Calculate the displacement of an object traveling at a known velocity for a specific time interval.**

- **Construct and interpret graphs of position versus time.**

frame of reference

a coordinate system for specifying the precise location of objects in space

MOTION

Motion happens all around us. Every day, we see objects such as cars, people, and soccer balls move in different directions with different speeds. We are so familiar with the idea of motion that it requires a special effort to analyze motion as a physicist does.

One-dimensional motion is the simplest form of motion

One way to simplify the concept of motion is to consider only the kinds of motion that take place in one direction. An example of this one-dimensional motion is the motion of a commuter train on a straight track, as in **Figure 2-1.**

In this one-dimensional motion, the train can move either forward or backward along the tracks. It cannot move left and right or up and down. This chapter deals only with one-dimensional motion. In later chapters, you will learn how to describe more complicated motions by breaking them down into examples of one-dimensional motion. ⭐TEKS **4B**

Motion takes place over time and depends upon the frame of reference

It seems simple to describe the motion of the train. As the train in **Figure 2-1** begins its route, it is at the first station. Later, it will be at another station farther down the tracks.

But what about all of the motions around the train? The Earth is spinning on its axis, so the train, stations, and the tracks are also moving around the axis. At the same time, the Earth is moving around the sun. The sun and the rest of the solar system are moving through our galaxy. This galaxy is traveling through space as well. ⭐TEKS **4E**

Whenever faced with a complex situation like this, physicists break it down into simpler parts. One key approach is to choose a **frame of reference** against which you can measure changes in posi-

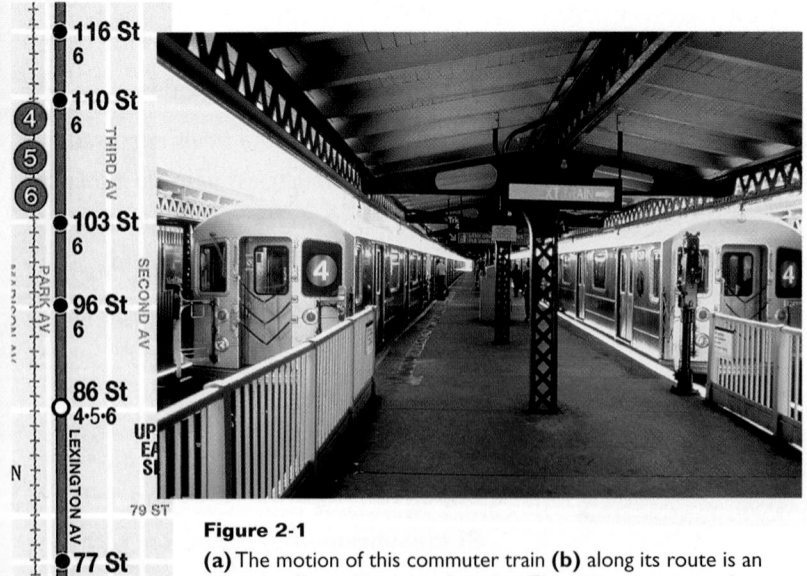

Figure 2-1

(a) The motion of this commuter train **(b)** along its route is an example of one-dimensional motion. The train can move only forward and backward along the track.

⭐TEKS

p. 40: 4B, 4E
p. 41: 2C, 4B

Δx

| 0 | 10 | 20 | 30 | 40 | 50 | 60 | 70 | 80 | 90 |

x_i x_f

Figure 2-2

A gecko moving along the x-axis from x_i to x_f undergoes a displacement of $\Delta x = x_f - x_i$.

tion. In the case of the train, the stations along its route are convenient frames of reference.

If an object is at rest (not moving), its position does not change with respect to a frame of reference. For example, the benches on the platform of one subway station never move down the tracks to another station.

In physics, any frame of reference can be chosen as long as it is used consistently. If you are consistent, you will get the same results, no matter which frame of reference you choose. But some frames of reference can make explaining things easier than other frames of reference.

For example, when considering the motion of the gecko in **Figure 2-2,** it is useful to imagine a stick marked in centimeters placed under the gecko's feet to define the frame of reference. The measuring stick serves as an x-axis. You can use it to identify the gecko's initial position and its final position.

DISPLACEMENT

As any object moves from one position to another, the length of the straight line drawn from its initial position to the object's final position is called the **displacement** of the object.

Displacement is a change in position

The gecko in **Figure 2-2** moves from left to right along the x-axis from an initial position, x_i, to a final position, x_f. The gecko's displacement is the difference between its final and initial coordinates, or $x_f - x_i$. In this case, the displacement is about 63 cm (85 cm − 22 cm). ⭐TEKS **2C, 4B**

DISPLACEMENT

$$\Delta x = x_f - x_i$$

displacement = change in position = final position − initial position

---

The Greek letter delta (Δ) before the x denotes a *change* in the position of an object.

displacement

the change in position of an object

Conceptual Challenge

1. Space shuttle

A space shuttle takes off from Florida and circles Earth several times, finally landing in California. While the shuttle is in flight, a photographer flies from Florida to California to take pictures of the astronauts when they step off the shuttle. Who undergoes the greater displacement, the photographer or the astronauts?

2. Roundtrip

What is the difference between the displacement of the photographer flying from Florida to California and the displacement of the astronauts flying from California back to Florida?

Visual Strategy

Figure 2-2

Point out to students that the choice of a reference point for the coordinate system is arbitrary, but once chosen, the same point must be used throughout the problem.

Q What would the displacement of the gecko be if the zero end of the meterstick had been lined up with the gecko's first position?

A *the same; where we put the meterstick does not affect how far the gecko moved.*

The Language of Physics

In this book, Δx refers to a change in position along the x-axis of whatever coordinate system is chosen and Δy refers to a change in position along the y-axis. In Chapter 3, these two variables refer to the x and y components of a displacement vector.

🛑 **Misconception Alert**

Many students may have difficulty understanding that displacement is the length of the straight-line path between two points rather than the distance traveled. Point out that although the odometer on a car shows that it has been driven 5 mi, the displacement may have been 0 mi.

ANSWERS TO

Conceptual Challenge

1. neither; the displacements are the same.
2. The difference between these two displacements is in their direction—one is positive and the other negative.

Displacement

Purpose Demonstrate the importance of direction in reference to displacement.

Materials one meterstick, 3 pieces of modeling clay, one toothpick or paper clip, one toy car

Procedure Place the meterstick on edge so that the 0 mark is to the students' left and the students can see the numbers. Put the toothpick in one of the pieces of modeling clay to represent the initial position.

For positive displacement, place the initial position marker and car somewhere between 0 and 10 cm. Roll the car down the meterstick to some point past the 50 cm mark, and place the final position marker (the second piece of modeling clay) at the car's new position. Ask the students to calculate the displacement of the car. *(It should be a positive number.)*

For negative displacement, move the initial position marker to the car, and roll the car back toward the zero end of the meterstick. Stop the car and place the third position marker. Ask the students to calculate the displacement for the second leg of the trip. *(It should be a negative number.)*

Now have them calculate the car's total displacement.

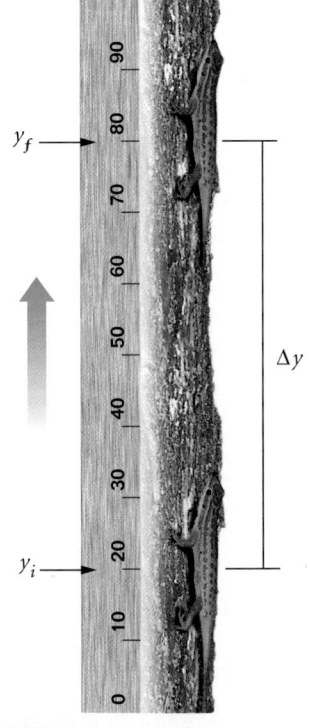

Figure 2-3
When the gecko is climbing a tree, the displacement is measured on the y-axis. Again, the gecko's position is determined by the position of the same point on its body.

Now suppose the gecko runs up a tree, as shown in **Figure 2-3.** In this case, we place the measuring stick parallel to the tree. The measuring stick can serve as the y-axis of our coordinate system. The gecko's initial and final positions are indicated by y_i and y_f, respectively, and the gecko's displacement is denoted as Δy.

Displacement is not always equal to the distance traveled

Displacement does not always tell you the distance an object has moved. For example, what if the gecko in **Figure 2-3** runs up the tree from the 20 cm marker (its initial position) to the 80 cm marker. After that, it retreats down the tree to the 50 cm marker (its final position). It has traveled a total distance of 90 cm. However, its displacement is only 30 cm ($y_f - y_i = 50$ cm $- 20$ cm $= 30$ cm). If the gecko was to return to its starting point, its displacement would be zero because its initial position and final position would be the same. ⭐TEKS 2C, 4B

Displacement can be positive or negative

Displacement also includes a description of the direction of motion. In one-dimensional motion, there are only two directions in which an object can move, and these directions can be described as positive or negative.

In this book, unless otherwise stated, the right will be considered the positive direction and the left will be considered the negative direction. Similarly, upward will be considered positive and downward will be considered negative. **Table 2-1** gives examples of determining displacements for a variety of situations. ⭐TEKS 2C, 4B

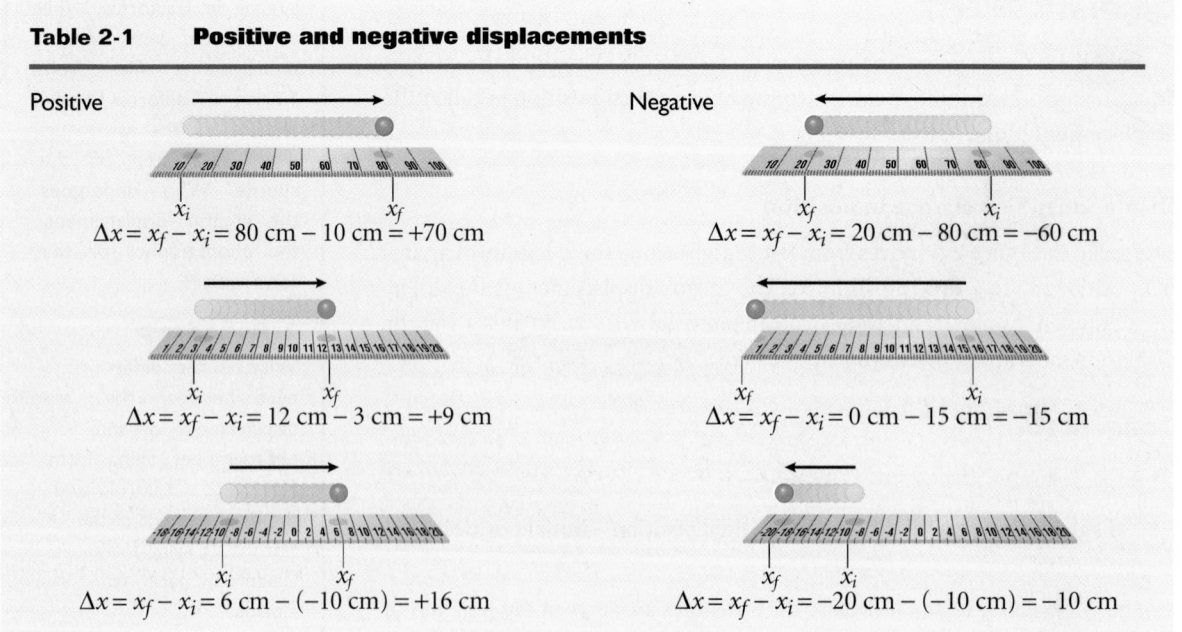

Table 2-1   Positive and negative displacements

Positive

$\Delta x = x_f - x_i = 80$ cm $- 10$ cm $= +70$ cm

$\Delta x = x_f - x_i = 12$ cm $- 3$ cm $= +9$ cm

$\Delta x = x_f - x_i = 6$ cm $- (-10$ cm$) = +16$ cm

Negative

$\Delta x = x_f - x_i = 20$ cm $- 80$ cm $= -60$ cm

$\Delta x = x_f - x_i = 0$ cm $- 15$ cm $= -15$ cm

$\Delta x = x_f - x_i = -20$ cm $- (-10$ cm$) = -10$ cm

VELOCITY

Where an object started and where it stopped does not completely describe the motion of the object. The ground nearby may move 8.0 cm to the left. This motion could take a full year, and be a sign of the normal slow movement of Earth's tectonic plates. If this motion takes place in just a second, however, you may be experiencing an earthquake or a landslide. Knowing the speed is important when evaluating motion. (★)TEKS 4E

Average velocity is displacement divided by the time interval

Consider the car in **Figure 2-4**. The car is moving along a highway in a straight line (the x-axis). Suppose that the positions of the car are x_i at time t_i and x_f at time t_f. In the time interval $\Delta t = t_f - t_i$, the displacement of the car is $\Delta x = x_f - x_i$. The **average velocity,** v_{avg}, is defined as the displacement divided by the time interval during which the displacement occurred. In SI, the unit of velocity is meters per second, abbreviated as m/s.

AVERAGE VELOCITY

$$v_{avg} = \frac{\Delta x}{\Delta t} = \frac{x_f - x_i}{t_f - t_i}$$

$$\text{average velocity} = \frac{\text{change in position}}{\text{change in time}} = \frac{\text{displacement}}{\text{time interval}}$$

The average velocity of an object can be positive or negative, depending on the sign of the displacement. (The time interval is always positive.) As an example, consider a car trip to a friend's house 370 km to the west (the negative direction) along a straight highway. If you left your house at 10 A.M. and arrived at your friend's house at 3 P.M., your average velocity would be as follows:

$$v_{avg} = \frac{\Delta x}{\Delta t} = \frac{-370 \text{ km}}{5.0 \text{ h}} = -74 \text{ km/h}$$

This value is an average. You probably did not travel exactly 74 km/h at every moment. You may have stopped to buy gas or have lunch. At other times, you may have traveled more slowly due to heavy traffic. To make up for such delays, when you were traveling slower than 74 km/h, there must also have been other times when you traveled faster than 74 km/h.

The average velocity is equal to the constant velocity needed to cover the given displacement in a given time interval. In the example above, if you left your house and maintained a velocity of 74 km/h to the west at every moment, it would take you 5.0 h to travel 370 km. (★)TEKS 2C, 4B

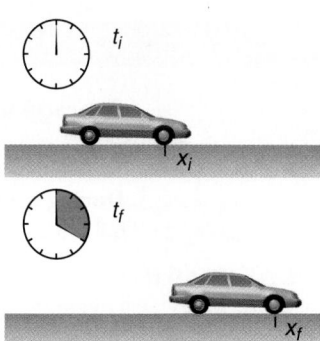

Figure 2-4
The average velocity of this car tells you how fast and in which direction it is moving.

average velocity

the total displacement divided by the time interval during which the displacement occurred

internetconnect

SCiLINKS
NSTA
TOPIC: Motion
GO TO: www.scilinks.org
*sci*LINKS CODE: HF2021

Did you know?

The branch of physics concerned with the study of motion is called *dynamics*. The part of dynamics that describes motion without regard to its causes is called *kinematics*.

Misconception Alert

Many students believe that the average velocity is always the average of the starting and ending velocities. Use counterexamples to address this misconception.

Example: A car travels from city A to city B (100 km). If the first half of the distance is driven at 50 km/h and the second half is driven at 100 km/h, the average speed is given by the following relation.

$$\frac{100 \text{ km}}{\dfrac{50 \text{ km}}{50 \text{ km/h}} + \dfrac{50 \text{ km}}{100 \text{ km/h}}} = 67 \text{ km/h}$$

The average velocity would be 75 km/h if the car spent equal *time* at 50 km/h and 100 km/h.

(★)TEKS
p. 42: 2C, 4B, 2C, 4B
p. 43: 4E, 2C, 4B

Classroom Practice

PROBLEM

A doctor travels to the east from city A to city B (75 km) in 1.0 h. What is the doctor's average velocity?

Answer

75 km/h to the east

PRACTICE GUIDE 2A		
Solving for:		
Δx	📖 **PE**	Sample, 1–3; Ch. Rvw. 8–14, 15*, 47b
	💿 **PW**	3b
	PB	8, 10
Δt	📖 **PE**	4–6a; Ch. Rvw. 43, 44*, 47a
	💿 **PW**	Sample, 1–3a, 4a, 8b
	PB	7, 9
v_{avg}	📖 **PE**	6b; Ch. Rvw. 12c–d, 13a, 14, 60*
	💿 **PW**	4b, 5–8b
	PB	Sample, 1–6

ANSWERS TO

Practice 2A
Average velocity and displacement

1. 2.0 km to the east
2. 3.1 km to the south
3. 680 m to the north
4. 3.00 h
5. 0.43 h
6. **a.** 6.4 h
   b. 77 km/h to the south

Average velocity and displacement ⭐TEKS 2C, 4B

PROBLEM

During a race on level ground, Andra runs with an average velocity of 6.02 m/s to the east. What distance does Andra cover in 137 s?

SOLUTION

Given:
$$v_{avg} = 6.02 \text{ m/s}$$
$$\Delta t = 137 \text{ s}$$

Unknown: $\Delta x = ?$

Rearrange the average velocity equation to solve for displacement.

$$v_{avg} = \frac{\Delta x}{\Delta t}$$

$$\Delta x = v_{avg}\Delta t$$

$$\Delta x = v_{avg}\Delta t = (6.02 \text{ m/s})(137 \text{ s}) = \boxed{825 \text{ m to the east}}$$

CALCULATOR SOLUTION

The calculator answer is 824.74 m, but both the values for velocity and time have three significant figures, so the displacement must be reported as 825 m.

PRACTICE 2A

Average velocity and displacement

1. Heather and Matthew walk eastward with a speed of 0.98 m/s. If it takes them 34 min to walk to the store, how far have they walked?

2. If Joe rides south on his bicycle in a straight line for 15 min with an average speed of 12.5 km/h, how far has he ridden?

3. It takes you 9.5 min to walk with an average velocity of 1.2 m/s to the north from the bus stop to the museum entrance. What is your displacement?

4. Simpson drives his car with an average velocity of 48.0 km/h to the east. How long will it take him to drive 144 km on a straight highway?

5. Look back at item 4. How much time would Simpson save by increasing his average velocity to 56.0 km/h to the east?

6. A bus travels 280 km south along a straight path with an average velocity of 88 km/h to the south. The bus stops for 24 min, then it travels 210 km south with an average velocity of 75 km/h to the south.
   a. How long does the total trip last?
   b. What is the average velocity for the total trip?

Velocity is not the same as speed

In everyday language, the terms *speed* and *velocity* are used interchangeably. In physics, however, there is an important distinction between these two terms. As we have seen, velocity describes motion with both a direction and a numerical value (a magnitude) indicating how fast something moves. However, speed has no direction, only magnitude. An object's average speed is equal to the distance traveled divided by the time interval for the motion.

$$\text{average speed} = \frac{\text{distance traveled}}{\text{time of travel}}$$

Velocity can be interpreted graphically

The velocity of an object can be determined if its position is known at specific times along its path. One way to determine this is to make a graph of the motion. **Figure 2-5** represents such a graph. Notice that time is plotted on the horizontal axis and position is plotted on the vertical axis. ⭐TEKS **4A**

The object moves 4.0 m in the time interval between $t = 0$ and $t = 4.0$ s. Likewise, the object moves 4.0 m in the time interval between $t = 4.0$ s and $t = 8.0$ s. From this data we see that the average velocity for each of these time intervals is 1.0 m/s (because $v_{avg} = \Delta x/\Delta t = 4.0$ m/4.0 s). Because the average velocity does not change, the object is moving with a constant velocity of 1.0 m/s, and its motion is represented by a straight line on the position-time graph.

For any position-time graph, we can also determine the average velocity by drawing a straight line between any two points on the graph. The slope of this line indicates the average velocity between the positions and times represented by these points. To better understand this concept, compare the equation for the slope of the line with the equation for the average velocity.

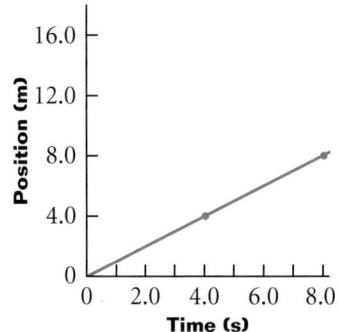

Figure 2-5
The motion of an object moving with constant velocity will provide a straight-line graph of position versus time. The slope of this graph indicates the average velocity.

⭐TEKS **2C, 4B**

Slope of a Line	**Average Velocity**
$\text{slope} = \dfrac{\text{rise}}{\text{run}} = \dfrac{\text{change in vertical coordinates}}{\text{change in horizontal coordinates}}$	$v_{avg} = \dfrac{\Delta x}{\Delta t} = \dfrac{x_f - x_i}{t_f - t_i}$

Conceptual Challenge ⭐TEKS 2C, 4B

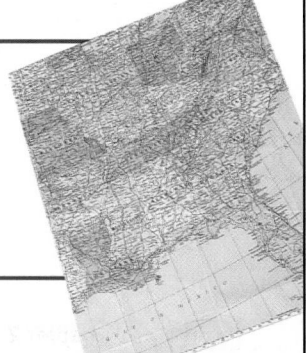

1. Book on a table A book is moved once around the edge of a tabletop with dimensions 1.75 m × 2.25 m. If the book ends up at its initial position, what is its displacement? If it completes its motion in 23 s, what is its average velocity? What is its average speed?

2. Travel Car A travels from New York to Miami at a speed of 25 m/s. Car B travels from New York to Chicago, also at a speed of 25 m/s. Are the velocities of the cars equal? Explain.

SECTION 2-1

Key Models and Analogies

The idea of a car race, such as the Indianapolis 500, may help students understand the difference between speed and velocity. Cars have *speedometers*, not "velocitometers." Indy cars spend hours going very fast (high speed) but getting nowhere (zero velocity). In order to have a nonzero velocity, the car must finish at some point other than the starting point.

Teaching Tip

Probe your students' understanding of the concept of *slope*. Many students may remember the concept from math class. Students will most likely remember slope from algebra as described by the phrase *rise over run*.

ANSWERS TO

Conceptual Challenge

1. The book's displacement is zero, its average velocity is zero, and its average speed is 0.35 m/s.
2. The velocity of car A does not equal the velocity of car B even though their speeds are the same because they are traveling in different directions.

⭐TEKS

p. 44: 2C, 4B
p. 45: 4A, 2C, 4A, 4B

Misconception Alert

Some students will use raw data points to calculate the slope of a line. Students should be shown how to draw a "best-fit" line and should be cautioned to use points on the best-fit line, not the raw data points, which may or may not be on the line.

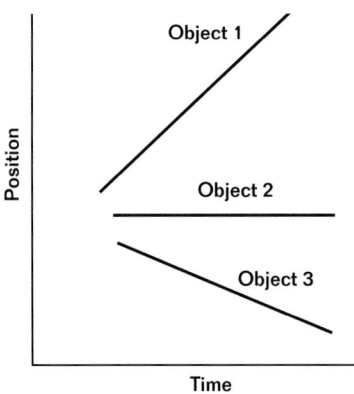

Figure 2-6

These position versus time graphs show that object 1 moves with a constant positive velocity. Object 2 is at rest. Object 3 moves with a constant negative velocity.

instantaneous velocity

the velocity of an object at some instant (or specific point in its path)

Did you know?

When you use a car's speedometer to check your speed, the speedometer tells you the *instantaneous speed* of the car. The speedometer's reading is speed, the magnitude of velocity. In order to give a reading of velocity, the speedometer would have to indicate not only how fast you are driving but also in what direction.

Table 2-2
Velocity-Time Data

t (s)	v (m/s)
0.0	0.0
1.0	4.0
2.0	8.0
3.0	12.0
4.0	16.0

Figure 2-6 represents straight-line graphs of position versus time for three different objects. Object 1 has a constant positive velocity because its position increases uniformly with time. Object 2 has zero velocity (the object is at rest) because its position remains constant with time. Object 3 has constant negative velocity because its position decreases with time. ⭐TEKS **4A**

Instantaneous velocity may not be the same as average velocity

Now consider an object whose position versus time graph is not a straight line, but a curve, as in **Figure 2-7**. The object moves through larger and larger displacements as each second passes. Thus, its velocity increases with time.

For example, between $t = 0$ s and $t = 2.0$ s, the object moves 8.0 m, and its average velocity in this time interval is 4.0 m/s (because $v_{avg} = 8.0$ m/2.0 s). However, between $t = 0$ s and $t = 4.0$ s, it moves 32 m, so its average velocity in this time interval is 8.0 m/s (because $v_{avg} = 32$ m/4.0 s). We obtain different average velocities, depending on the time interval we choose. But how can we find the velocity at an instant of time?

To determine the velocity at some instant, such as $t = 3.0$ s, we study a small time interval near that instant. As the intervals become smaller and smaller, the average velocity over that interval approaches the exact velocity at $t = 3.0$ s. This is called the **instantaneous velocity.** ⭐TEKS **2C, 4B**

One way to determine the instantaneous velocity is to construct a straight line that is *tangent* to the position versus time graph at that instant. The slope of this tangent line is equal to the value of the instantaneous velocity at that point. For example, the instantaneous velocity of the object in **Figure 2-7** at $t = 3.0$ s is 12 m/s. **Table 2-2** lists the instantaneous velocities of the object described by the graph in **Figure 2-7**. You can verify some of these values by carefully measuring the slope of the curve. ⭐TEKS **2B**

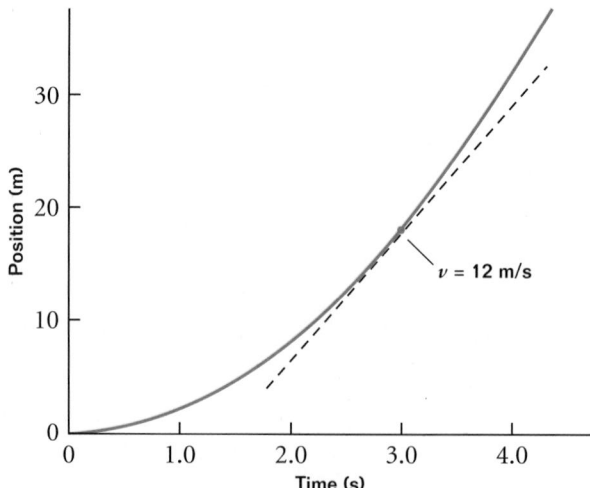

Figure 2-7

The instantaneous velocity at a given time can be determined by measuring the slope of the line that is tangent to that point on the position versus time graph.

Section Review

(★) TEKS 2C, 4A, 4B

1. Does knowing the distance between two objects give you enough information to locate the objects? Explain.

2. What is the shortest possible time in which a bacterium could drift at a constant speed of 3.5 mm/s across a petri dish with a diameter of 8.4 cm?

3. **Figure 2-8** shows position-time graphs of the straight-line movement of two brown bears in a wildlife preserve. Which bear has the greater average velocity over the entire period? Which bear has the greater velocity at $t = 8.0$ min? Is the velocity of bear A always positive? Is the velocity of bear B ever negative?

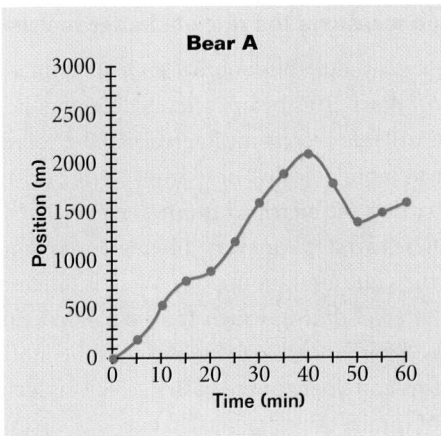
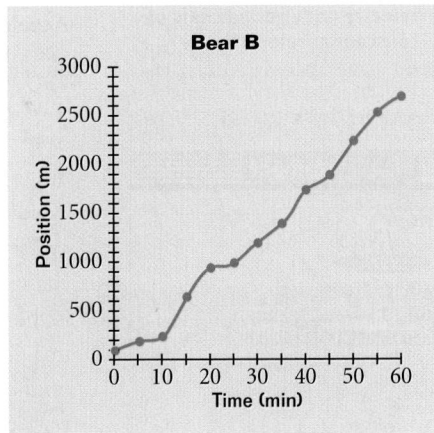

Figure 2-8

4. A child is pushing a shopping cart at a speed of 1.5 m/s. How long will it take this child to push the cart down an aisle with a length of 9.3 m?

5. An athlete swims from the north end to the south end of a 50.0 m pool in 20.0 s and makes the return trip to the starting position in 22.0 s.
   a. What is the average velocity for the first half of the swim?
   b. What is the average velocity for the second half of the swim?
   c. What is the average velocity for the roundtrip?

6. Two students walk in the same direction along a straight path, at a constant speed—one at 0.90 m/s and the other at 1.90 m/s.
   a. Assuming that they start at the same point and the same time, how much sooner does the faster student arrive at a destination 780 m away?
   b. How far would the students have to walk so that the faster student arrives 5.50 min before the slower student?

1. No, because a single distance could correspond to a variety of different positions of the objects.

2. 24 s

3. bear B; bear A; no; no

4. 6.2 s

5. a. 2.50 m/s to the south
   b. 2.27 m/s to the north
   c. 0.0 m/s

6. a. 460 s
   b. 570 m

(★) TEKS

p. 46: 4A, 2C, 4B, 2B
p. 47: 2C, 4A, 4B

Acceleration

Purpose Visually demonstrate acceleration and supply data for the students to calculate acceleration.

Materials marble, inclined plane, tape, stopwatch, meterstick, protractor, metal or glass cup

Procedure Ask students to observe the motion of the marble as it rolls down the plane. Release the marble from rest at the top of the plane. When the marble reaches the end, ask students to describe the motion. Ask students to identify the locations of slowest speed (*top*) and highest speed (*bottom*). Repeat the demonstration at a different angle, and have students compare the two trials.

Add quantitative values by timing the runs. Place a tape marker near the top of the plane to serve as a start line. Use a metal or glass cup at the end of the plane as a sound cue for stopping the stopwatch. Measure the length of the plane. Have students record the time and angle measurements for several trials and then calculate the acceleration for each trial. Each trial should have a different angle and therefore a different acceleration. With enough trials, a graph of acceleration versus angle can be generated.

⭐ TEKS

p. 48: 2C, 4B 4B
p. 49: 2C, 4B

48

2-2 SECTION OBJECTIVES

- **Describe motion in terms of changing velocity.**

- **Compare graphical representations of accelerated and nonaccelerated motions.**

- **Apply kinematic equations to calculate distance, time, or velocity under conditions of constant acceleration.**

internet connect

SCI LINKS

NSTA

TOPIC: Acceleration
GO TO: www.scilinks.org
*sci*LINKS CODE: HF2022

acceleration

the rate of change of velocity

2-2
Acceleration

CHANGES IN VELOCITY

The new bullet trains have a top speed of about 300 km/h. Because a train stops to load and unload passengers, it spends very little time traveling at that top speed. Most of the time the train is in motion, its velocity is either increasing or decreasing. It loses speed as it slows down to stop and gains speed as it pulls away and heads for the next station. ⭐ TEKS 2C, 4B

Acceleration measures the rate of change in velocity

Similarly, when a shuttle bus approaches a stop, the driver begins to apply the brakes to slow down 5.0 s before actually reaching the stop. The speed changes from 9.0 m/s to 0 m/s over a time interval of 5.0 s. Sometimes, however, the shuttle stops much more quickly. For example, if the driver slams on the brakes to avoid hitting a dog, the bus slows from 9.0 m/s to 0 m/s in just 1.5 s.

Clearly these two stops are very different, even though the shuttle's velocity changes by the same amount in both. What is different in these two examples is the time interval during which the change in velocity occurs. As you can imagine, this difference has a great effect on the motion of the bus, as well as on the comfort and safety of the passengers. A sudden change in velocity feels very different from a slow, gradual change.

The quantity that describes the rate of change of velocity in a given time interval is called **acceleration.** The magnitude of the acceleration is calculated by dividing the total change in an object's velocity by the time interval in which the change occurs. ⭐ TEKS 4B

AVERAGE ACCELERATION

$$a_{avg} = \frac{\Delta v}{\Delta t} = \frac{v_f - v_i}{t_f - t_i}$$

$$\text{average acceleration} = \frac{\text{change in velocity}}{\text{time required for change}}$$

Acceleration has dimensions of length divided by time squared. The units of acceleration in SI are meters per second per second, which is written as meters per second squared, as shown below. When measured in these units, acceleration describes how much the velocity changes in each second.

$$\frac{(m/s)}{s} = \frac{m}{s} \times \frac{1}{s} = \frac{m}{s^2}$$

Average acceleration (★)TEKS 2C, 4B

PROBLEM

A shuttle bus slows to a stop with an average acceleration of -1.8 m/s^2.
How long does it take the bus to slow from 9.0 m/s to 0.0 m/s?

SOLUTION

Given: $v_i = 9.0$ m/s $v_f = 0$ m/s
$a_{avg} = -1.8$ m/s^2

Unknown: $\Delta t = ?$

Rearrange the average acceleration equation to solve for the time interval.

$$a_{avg} = \frac{\Delta v}{\Delta t}$$

$$\Delta t = \frac{\Delta v}{a_{avg}}$$

$$\Delta v = v_f - v_i = 0 \text{ m/s} - 9.0 \text{ m/s} = -9.0 \text{ m/s}$$

$$\Delta t = \frac{-9.0 \text{ m/s}}{-1.8 \text{ m/s}^2}$$

$$\boxed{\Delta t = 5.0 \text{ s}}$$

Average acceleration

1. As the shuttle bus comes to a sudden stop to avoid hitting a dog, it accelerates uniformly at -4.1 m/s^2 as it slows from 9.0 m/s to 0 m/s. Find the time interval of acceleration for the bus.

2. A car traveling at 7.0 m/s accelerates uniformly at 2.5 m/s^2 to reach a speed of 12.0 m/s. How long does it take for this acceleration to occur?

3. With an average acceleration of -0.50 m/s^2, how long will it take a cyclist to bring a bicycle with an initial speed of 13.5 m/s to a complete stop?

4. Turner's treadmill runs with a velocity of -1.2 m/s and speeds up at regular intervals during a half-hour workout. After 25 min, the treadmill has a velocity of -6.5 m/s. What is the average acceleration of the treadmill during this period?

5. Suppose a treadmill has an average acceleration of 4.7×10^{-3} m/s^2.
   a. How much does its speed change after 5.0 min?
   b. If the treadmill's initial speed is 1.7 m/s, what will its final speed be?

SECTION 2-2

Classroom Practice

The following may be used as teamwork exercises or for demonstration at the chalkboard or on an overhead projector.

PROBLEM

Average acceleration

Find the acceleration of an amusement park ride that falls from rest to a speed of 28 m/s in 3.0 s.

Answer
9.3 m/s^2

PRACTICE GUIDE 2B	
Solving for:	
Δt	📖 **PE** Sample, 1–3; Ch. Rvw. 20, 31a
	💿 **PW** 7a, 8–9
	PB 6, 8, 9
Δv	📖 **PE** 5; Ch. Rvw. 23b
	💿 **PW** Sample, 1–4a, 5a, 8b
	PB 7, 10
a_{avg}	📖 **PE** 4; Ch. Rvw. 23a, 30, 54
	💿 **PW** 4b, 5b, 6, 7b, 10
	PB Sample, 1–5

ANSWERS TO

Practice 2B
Average acceleration
1. 2.2 s
2. 2.0 s
3. 27 s
4. -3.5×10^{-3} m/s^2
5. a. 1.4 m/s
   b. 3.1 m/s

Figure 2-9
High-speed trains like this can travel at speeds of about 300 km/h (186 mi/h).

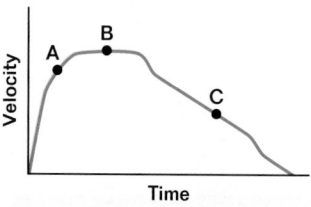

Figure 2-10
When the velocity in the positive direction is increasing, the acceleration is positive, as at point A. When the velocity is constant, there is no acceleration, as at point B. When the velocity in the positive direction is decreasing, the acceleration is negative, as at point C.

Acceleration has direction and magnitude

Figure 2-9 shows a high-speed train leaving a station. Imagine that the train is moving to the right, so that the displacement and the velocity are positive. The velocity increases in magnitude as the train picks up speed. Therefore, the final velocity will be greater than the initial velocity, and Δv will be positive. When Δv is positive, the acceleration is positive.

On long trips with no stops, the train may travel for a while at a constant velocity. In this situation, because the velocity is not changing, $\Delta v = 0$ m/s. When the velocity is constant, the acceleration is equal to zero.

Imagine that the train, still traveling in the positive direction, slows down as it approaches the next station. In this case, the velocity is still positive, but the acceleration is negative. This is because the initial velocity is larger than the final velocity, so Δv will be negative. ⭐TEKS **4B**

The slope and shape of the graph describe the object's motion

As with all motion graphs, the slope and shape of the velocity-time graph in **Figure 2-10** allow a detailed analysis of the train's motion over time. When the train leaves the station, its speed is increasing over time. The line on the graph plotting this motion slopes up and to the right, as at point **A** on the graph.

When the train moves with a constant velocity, the line on the graph continues to the right but it is horizontal, with a slope equal to zero. This indicates that the train's velocity is constant, as at point **B** on the graph.

Finally, as the train approaches the station, its velocity decreases over time. The graph segment representing this motion slopes down to the right, as at point **C** on the graph. This downward slope indicates that the velocity is decreasing over time.

A negative value for the acceleration does not always indicate a deceleration. For example, if the train were moving in the negative direction, the acceleration would be negative when the train gained speed to leave a station and positive when the train lost speed to enter a station. ⭐TEKS **4A**

Conceptual Challenge ⭐TEKS **4B**

1. Fly ball If a baseball has zero velocity at some instant, is the acceleration of the baseball necessarily zero at that instant? Explain, and give examples.

2. Runaway train If a passenger train is traveling on a straight track with a negative velocity and a positive acceleration, is it speeding up or slowing down?

3. Hike-and-bike trail
When Jennifer is out for a ride, she slows down on her bike as she approaches a group of hikers on a trail. Explain how her acceleration can be positive even though her speed is decreasing.

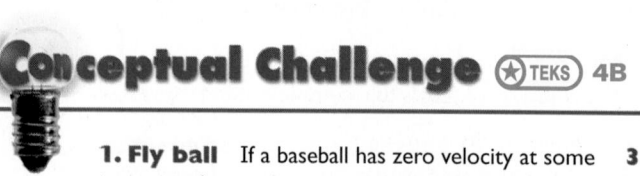

Table 2-3 shows how the signs of the velocity and acceleration can be combined to give a description of an object's motion. From this table, you can see that a negative acceleration can describe an object that is speeding up (when the velocity is negative) and an object that is slowing down (when the velocity is positive). Use this table to check your answers to problems involving acceleration.

For example, in **Figure 2-10** the initial velocity v_i of the train is positive. At point **A** on the graph the train's velocity is still increasing, so its acceleration is positive as well. The first entry in **Table 2-3** shows that in this situation, the train is speeding up. At point **C,** the velocity is still positive, but it is decreasing, so the train's acceleration is negative. **Table 2-3** tells you that in this case, the train is slowing down.

Table 2-3	Velocity and acceleration ⊛TEKS 4B	
v_i	a	Motion
+	+	speeding up
–	–	speeding up
+	–	slowing down
–	+	slowing down
– or +	0	constant velocity
0	– or +	speeding up from rest
0	0	remaining at rest

MOTION WITH CONSTANT ACCELERATION

Figure 2-11 is a strobe photograph of a ball moving in a straight line with constant acceleration. While the ball was moving, its image was captured ten times in one second, so the time interval between successive images is 0.10 s. As the ball's velocity increases, the ball travels a greater distance during each time interval. Because the acceleration is constant, the velocity increases by exactly the same amount during each time interval. Because the velocity increases by the same amount in each time interval, the displacement for each time interval increases by the same amount. In other words, the distance that the ball travels in each time interval is equal to the distance it traveled in the previous time interval, plus a constant distance. You can see this in the photograph by noting that the distance between images increases while the time interval between images remains constant. The relationships between displacement, velocity, and constant acceleration are expressed by equations that apply to any object moving with constant acceleration. ⊛TEKS 4B

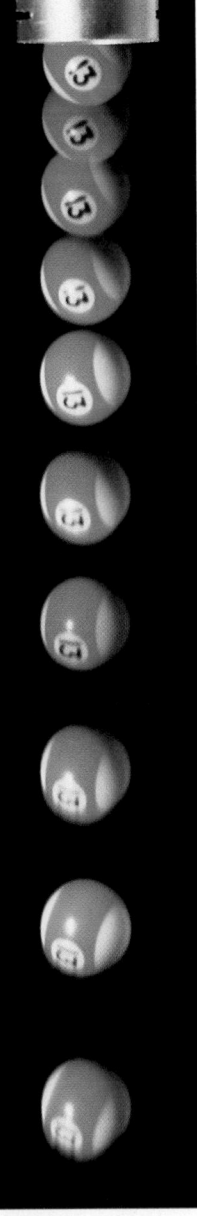

Figure 2-11
The motion in this picture took place in about 1.00 s. In this short time interval, your eyes could only detect a blur. This photo shows what really happens within that time.

Demonstration 3

Constant acceleration

Purpose Give several visual examples of constant acceleration.

Materials metronome (optional), tile floor (or tape)

Procedure Tell students you are going to demonstrate constant velocity and then constant acceleration. Use a metronome (or have students clap at regular intervals) to show time intervals. For the first part (constant velocity), walk in a straight line at a rate of one tile per time interval. (If you do not have a tile floor, use the tape to mark regular intervals on the floor.)

Now show a constant acceleration of one tile per interval. Walk a distance of one tile in the first interval, two tiles in the second interval, three in the third, and so on. Be sure to use a large enough time interval to make this possible.

Finally, show students a constant negative acceleration. Start walking at a rate of four or five tiles per interval, decreasing by one tile per interval with each interval. When you get to zero tiles per interval, you may want to continue by walking backward one tile per interval, then two tiles per interval, and so on. Explain that the acceleration was still present at the velocity of zero tiles per interval, so your velocity continued to change.

⊛TEKS

p. 50: 4B, 4A, 4B
p. 51: 4B, 4B

Teaching Tip

Show students that the area under the curve in a graph of velocity versus time equals the displacement during that time interval. Use the simplest case in **Figure 2-12,** where v_i equals zero, to illustrate this point. Choose a point on the graph, and draw a vertical line from the x-axis to the point and a horizontal line from the y-axis to the point, as shown below.

Use the corresponding velocity and time interval values to find the area of the rectangle ($A = v \times t$). Point out that the line in the graph bisects the box; thus, the area under the line equals $\frac{1}{2}v_f t$, which is the equation for the displacement of a constantly accelerated object that begins at rest.

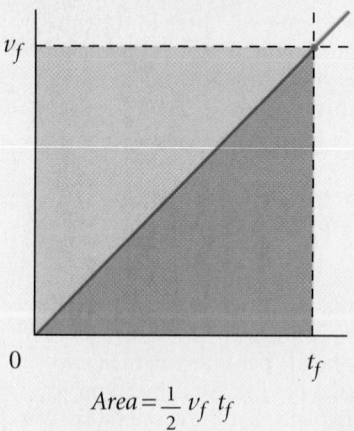

$$Area = \frac{1}{2} v_f \, t_f$$

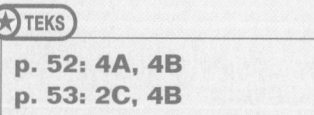
Figure 2-12

If a ball moved for the same time with a constant velocity equal to v_{avg}, it would have the same displacement as the ball in **Figure 2-11** moving with constant acceleration.

★ TEKS **4A**

Did you know?

Decreases in speed are sometimes called decelerations. Despite the sound of the name, decelerations are really a special case of acceleration in which the magnitude of the velocity—or the speed—decreases with time.

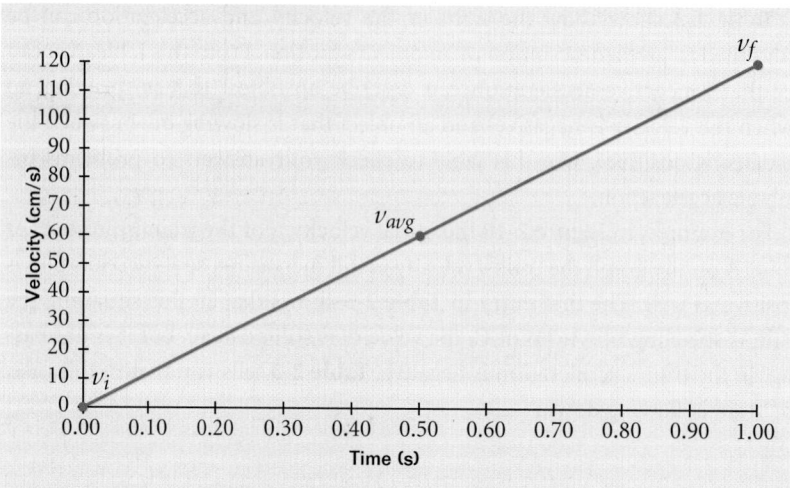

Displacement depends on acceleration, initial velocity, and time

Figure 2-12 is a graph of the ball's velocity plotted against time. The initial, final, and average velocities are marked on the graph. We know that the average velocity is equal to displacement divided by the time interval.

$$v_{avg} = \frac{\Delta x}{\Delta t}$$

For an object moving with constant acceleration, the average velocity is equal to the average of the initial velocity and the final velocity.

$$v_{avg} = \frac{v_i + v_f}{2} \qquad \text{average velocity} = \frac{\text{initial velocity} + \text{final velocity}}{2}$$

To find an expression for the displacement in terms of the initial and final velocity, we can set the expressions for average velocity equal to each other.

$$\frac{\Delta x}{\Delta t} = v_{avg} = \frac{v_i + v_f}{2}$$

$$\frac{\text{displacement}}{\text{time interval}} = \frac{\text{initial velocity} + \text{final velocity}}{2}$$

Multiplying both sides of the equation by Δt gives us an expression for the displacement as a function of time. This equation can be used to find the displacement of any object moving with constant acceleration. ★ TEKS **4B**

DISPLACEMENT WITH CONSTANT UNIFORM ACCELERATION

$$\Delta x = \frac{1}{2}(v_i + v_f)\Delta t$$

displacement $= \frac{1}{2}$(**initial velocity** + **final velocity**)(**time interval**)

Displacement with uniform acceleration ⭐TEKS 2C, 4B

PROBLEM

A racing car reaches a speed of 42 m/s. It then begins a uniform negative acceleration, using its parachute and braking system, and comes to rest 5.5 s later. Find how far the car moves while stopping.

SOLUTION

Given: $v_i = 42$ m/s $v_f = 0$ m/s
$\Delta t = 5.5$ s

Unknown: $\Delta x = ?$

Use the equation for displacement from page 52.

$$\Delta x = \tfrac{1}{2}(v_i + v_f)\Delta t$$

$$\Delta x = \tfrac{1}{2}(42 \text{ m/s} + 0 \text{ m/s})\,(5.5\text{ s})$$

$$= \tfrac{1}{2}(42 \text{ m/s})\,(5.5\text{ s})$$

$$\Delta x = (21 \text{ m/s})\,(5.5\text{ s})$$

$$\boxed{\Delta x = 120 \text{ m}}$$

CALCULATOR SOLUTION

The calculator answer is 115.5. However, the velocity and time values have only two significant figures each, so the answer must be reported as 120 m.

Displacement with constant uniform acceleration

1. A car accelerates uniformly from rest to a speed of 23.7 km/h in 6.5 s. Find the distance the car travels during this time.

2. When Maggie applies the brakes of her car, the car slows uniformly from 15.0 m/s to 0 m/s in 2.50 s. How many meters before a stop sign must she apply her brakes in order to stop at the sign?

3. A jet plane lands with a speed of 100 m/s and can accelerate uniformly at a maximum rate of −5.0 m/s^2 as it comes to rest. Can this plane land at an airport where the runway is 0.80 km long?

4. A driver in a car traveling at a speed of 78 km/h sees a cat 101 m away on the road. How long will it take for the car to accelerate uniformly to a stop in exactly 99 m?

5. A car enters the freeway with a speed of 6.4 m/s and accelerates uniformly for 3.2 km in 3.5 min. How fast is the car moving after this time?

Classroom Practice

The following may be used as teamwork exercises or for demonstration at the chalkboard or on an overhead projector.

PROBLEM

Displacement with uniform acceleration

A bicyclist accelerates from 5.0 m/s to a velocity of 16 m/s in 8 s. Assuming uniform acceleration, what distance does the bicyclist travel during this time interval?

Answer
84 m

PRACTICE GUIDE 2C		
Solving for:		
Δx	📖 **PE**	Sample, 1–3; Ch. Rvw. 23c, 25, 27
	💿 **PW**	4–6, 9
	PB	8
Δt	📖 **PE**	4; Ch. Rvw. 50, 52*
	💿 **PW**	Sample, 1–3
	PB	7, 9–10
Δv	📖 **PE**	5; Ch. Rvw. 23b
	💿 **PW**	7–8
	PB	Sample, 1–6

ANSWERS TO

Practice 2C
Displacement with uniform acceleration
1. 21 m
2. 18.8 m
3. no; the plane needs 1 km to land.
4. 9.1 s
5. 24 m/s

The Language of Physics

Other texts often use the term v_0 to represent the initial velocity (v_i) of the object and the term v to represent the final velocity (v_f) of the object.

Teaching Tip

You may want to present an interesting geometrical interpretation of this equation:

$$\Delta x = v_i \Delta t + \frac{1}{2} a (\Delta t)^2$$

Draw a velocity versus time graph representing the equation $v_f = v_i + a\Delta t$ as below.

The area beneath this curve has two parts. The lower part is a rectangle of area $v_i \Delta t$. The upper part is a triangle of area $\frac{1}{2} a (\Delta t)^2$. Therefore, the total area under the straight-line graph is equal to the displacement, Δx, of the object.

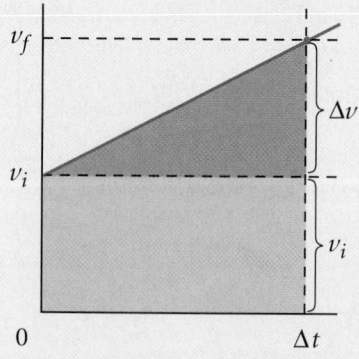

Final velocity depends on initial velocity, acceleration, and time

What if the final velocity of the ball is not known but we still want to calculate the displacement? If we know the initial velocity, the uniform acceleration, and the elapsed time, we can find the final velocity. We can then use this value for the final velocity to find the total displacement of the ball.

By rearranging the equation for acceleration, we can find a value for the final velocity. (★)**TEKS** **4B**

$$a = \frac{v_f - v_i}{t_f - t_i} = \frac{v_f - v_i}{\Delta t}$$

$$a\Delta t = v_f - v_i$$

By adding the initial velocity to both sides of the equation, we get an equation for the final velocity of the ball.

$$a\Delta t + v_i = v_f$$

VELOCITY WITH CONSTANT UNIFORM ACCELERATION

$$v_f = v_i + a\Delta t$$

final velocity = initial velocity + (acceleration × time interval)

You can use this equation to find the final velocity of an object moving with uniform acceleration after it has accelerated at a constant rate for any time interval, whether the time interval is a minute or half an hour.

If you want to know the displacement of an object moving with uniform acceleration over some certain time interval, you can obtain another useful expression for displacement by substituting the expression for v_f into the expression for Δx. (★)**TEKS** **4B**

$$\Delta x = \frac{1}{2}(v_i + v_f)\Delta t$$

$$\Delta x = \frac{1}{2}(v_i + v_i + a\Delta t)\Delta t$$

$$\Delta x = \frac{1}{2}[2v_i \Delta t + a(\Delta t)^2]$$

DISPLACEMENT WITH CONSTANT UNIFORM ACCELERATION

$$\Delta x = v_i \Delta t + \frac{1}{2} a (\Delta t)^2$$

displacement = (initial velocity × time interval) +
$$\frac{1}{2}\ \textbf{acceleration} × (\textbf{time interval})^2$$

This equation is useful not only for finding the displacement of an object moving with uniform acceleration but also for finding the displacement required for an object to reach a certain speed or to come to a stop.

CONCEPT PREVIEW

Galileo's achievements in the science of mechanics paved the way for Newton's development of the laws of motion, which we will study in Chapter 4.

Velocity and displacement with uniform acceleration ⓧTEKS 2C, 4B

PROBLEM

A plane starting at rest at one end of a runway undergoes a uniform acceleration of 4.8 m/s² for 15 s before takeoff. What is its speed at takeoff? How long must the runway be for the plane to be able to take off?

SOLUTION

Given: $v_i = 0$ m/s $a = 4.8$ m/s² $\Delta t = 15$ s

Unknowns: $v_f = ?$ $\Delta x = ?$

Use the equation for the velocity of a uniformly accelerated object from page 54.

$$v_f = v_i + a\Delta t$$
$$v_f = 0 \text{ m/s} + (4.8 \text{ m/s}^2)(15 \text{ s})$$

$$\boxed{v_f = 72 \text{ m/s}}$$

Use the equation for the displacement from page 54.

$$\Delta x = v_i\Delta t + \tfrac{1}{2}a(\Delta t)^2$$
$$\Delta x = (0 \text{ m/s})(15 \text{ s}) + \tfrac{1}{2}(4.8 \text{ m/s}^2)(15 \text{ s})^2$$

$$\boxed{\Delta x = 540 \text{ m}}$$

Velocity and displacement with uniform acceleration

1. A car with an initial speed of 23.7 km/h accelerates at a uniform rate of 0.92 m/s² for 3.6 s. Find the final speed and the displacement of the car during this time.

2. An automobile with an initial speed of 4.30 m/s accelerates uniformly at the rate of 3.0 m/s². Find the final speed and the displacement after 5.0 s.

3. A car starts from rest and travels for 5.0 s with a uniform acceleration of −1.5 m/s². What is the final velocity of the car? How far does the car travel in this time interval?

4. A driver of a car traveling at 15.0 m/s applies the brakes, causing a uniform acceleration of −2.0 m/s². How long does it take the car to accelerate to a final speed of 10.0 m/s? How far has the car moved during the braking period?

SECTION 2-2

PRACTICE GUIDE 2D

Solving for:

Δv	📖 **PE**	Sample, 1–3; Ch. Rvw. 21–22a, 26a, 28*, 58c*
	💿 **PW**	10–12
	PB	8
Δx	📖 **PE**	Sample, 1–4; Ch. Rvw. 24, 26b, 28, 29, 31b, 51b*, 58b*
	💿 **PW**	6*, 7–9
	PB	9b, 10b
Δt	📖 **PE**	4; Ch. Rvw. 22b, 31a, 51a*, 58a*
	💿 **PW**	Sample, 1–3, 4*, 5a*, 9
	PB	9a, 10a
a	💿 **PW**	5b, 7, 13–15
	PB	Sample, 1–7

ANSWERS TO

Practice 2D
Velocity and displacement with uniform acceleration

1. 36 km/h (9.9 m/s); +0.030 km
2. 19 m/s; 6.0×10^1 m
3. −7.5 m/s; 19 m
4. 2.5 s; 32 m

ⓧTEKS

p. 54: 4B, 4B
p. 55: 2C, 4B

(★) TEKS 3E

Did you know?

The word *physics* comes from the Ancient Greek word for "nature." According to Aristotle, who assigned the name, physics is the study of natural events. Aristotle believed that the study of motion was the basis of physics. However, Aristotle did not believe that mathematics could apply to this study. Galileo disagreed and developed the foundations for the modern study of motion using mathematics. In 1632, Galileo published the first mathematical treatment of motion.

internet**connect**

SC*i*LINKS.

NSTA

TOPIC: Galileo
GO TO: www.scilinks.org
*sci*LINKS CODE: HF2023

(★) TEKS

p. 56: 4B, 3E, 4B
p. 57: 2C, 4A, 4B

Time can be found from displacement and velocities (★) TEKS 4B

So far, all of the equations for motion under uniform acceleration have required knowing the time interval. We can also obtain an expression that relates displacement, velocity, and acceleration without using the time interval. This method involves rearranging one equation to solve for Δt and substituting that expression in another equation, making it possible to find the final velocity of a uniformly accelerated object without knowing how long it has been accelerating. Start with the equation for displacement from page 52.

$$\Delta x = \tfrac{1}{2}(v_i + v_f)\Delta t \quad \text{Now, multiply both sides by 2.}$$

$$2\Delta x = (v_i + v_f)\Delta t \quad \text{Next, divide both sides by } (v_i + v_f)$$
$$\text{to solve for } \Delta t.$$

$$\left(\frac{2\Delta x}{v_i + v_f}\right) = \Delta t$$

Now that we have an expression for Δt, we can substitute this expression into the equation for the final velocity.

$$v_f = v_i + a(\Delta t)$$

$$v_f = v_i + a\left(\frac{2\Delta x}{v_i + v_f}\right)$$

In its present form, this equation is not very helpful because v_f appears on both sides. To solve for v_f, first subtract v_i from both sides of the equation.

$$v_f - v_i = a\left(\frac{2\Delta x}{v_i + v_f}\right)$$

Next, multiply both sides by $(v_i + v_f)$ to get all the velocities on the same side of the equation.

$$(v_f - v_i)(v_f + v_i) = 2a\Delta x = v_f^2 - v_i^2$$

Add v_i^2 to both sides to solve for v_f^2.

FINAL VELOCITY AFTER ANY DISPLACEMENT

$$v_f^2 = v_i^2 + 2a\Delta x$$

$$(\text{final velocity})^2 = (\text{initial velocity})^2 + 2(\text{acceleration})(\text{displacement})$$

When using this equation, you must take the square root of the right side of the equation to find the final velocity. Remember that the square root may be either positive or negative. If you have been consistent in your use of the sign convention, you will be able to determine which value is the right answer by reasoning based on the direction of the motion. (★) TEKS 4B

Final velocity after any displacement ★TEKS 2C, 4A, 4B

PROBLEM

A person pushing a stroller starts from rest, uniformly accelerating at a rate of 0.500 m/s². What is the velocity of the stroller after it has traveled 4.75 m?

SOLUTION

1. DEFINE

Given: $v_i = 0$ m/s $a = 0.500$ m/s²
$\Delta x = 4.75$ m

Unknown: $v_f = ?$

Diagram:

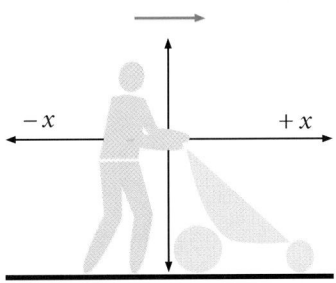

Choose a coordinate system. The most convenient one has an origin at the initial location of the stroller. The positive direction is to the right.

2. PLAN

Choose an equation or situation: Because the initial velocity, acceleration, and displacement are known, the final velocity can be found using the equation from page 56.

$$v_f^2 = v_i^2 + 2a\Delta x$$

3. CALCULATE

Substitute the values into the equation: Remember to take the square root in the final step.

$$v_f^2 = (0 \text{ m/s})^2 + 2(0.500 \text{ m/s}^2)(4.75 \text{ m})$$
$$v_f^2 = 4.75 \text{ m}^2/\text{s}^2$$
$$v_f = \sqrt{4.75 \text{ m}^2/\text{s}^2} = \pm 2.18 \text{ m/s}$$

$$\boxed{v_f = +2.18 \text{ m/s}}$$

4. EVALUATE

Because the stroller is moving to the right from rest with a positive acceleration, we expect the final velocity to be positive. **Table 2-3** shows that an object with a positive acceleration and a positive velocity is speeding up. However, a negative velocity and a positive acceleration indicates that the object is slowing down. Because the stroller starts from rest, only a positive velocity makes sense as the answer.

Classroom Practice

The following may be used as teamwork exercises or for demonstration at the chalkboard or on an overhead projector.

PROBLEM

Final velocity after any displacement

An aircraft has a landing speed of 302 km/h. The landing area of an aircraft carrier is 195 m long. What is the minimum uniform acceleration required for a safe landing?

Answer
-18.0 m/s²

An electron is accelerated uniformly from rest in an accelerator at 4.5×10^7 m/s² over a distance of 95 km. Assuming constant acceleration, what is the final velocity of the electron?

Answer
2.9×10^6 m/s

PRACTICE GUIDE 2E		
Solving for:		
Δv	📖 PE	Sample, 1, 2a–2c, 3a; Ch. Rvw. 32a, 33, 38, 48
	💿 PW	4–6, 7a, 7b
	PB	7, 10
Δx	📖 PE	4, 6; Ch. Rvw. 48, 53a*
	💿 PW	Sample, 1–3
	PB	6, 8–9
a	📖 PE	5
	💿 PW	8, 9
	PB	Sample, 1–5

Alternative Problem-Solving Approach

Because acceleration is uniform and $v_i = 0$, we know that $v_f = 2v_{avg}$.

$$v_{avg} = \frac{\Delta x}{\Delta t}$$

$$v_f = 2\frac{\Delta x}{\Delta t}$$

Because $\Delta x = \frac{1}{2}a(\Delta t)^2$,

$$\Delta t = \sqrt{\frac{2\Delta x}{a}}$$

Substitute this expression for Δt into the expression for v_f.

$$v_f = \sqrt{2a\Delta x} = +2.18 \text{ m/s}$$

ANSWERS TO

Practice 2E
Final velocity after any displacement

1. +2.51 m/s
2. a. +21 m/s
   b. +16 m/s
   c. +13 m/s
3. a. 16 m/s
   b. 7.0 s
4. 87 m
5. +2.3 m/s^2
6. 7.4 m

⭐ TEKS

p. 58: 2C, 4B, 4B
p. 59: 2C, 4A, 4B

Final velocity after any displacement ⭐TEKS 2C, 4B

1. Find the velocity after the stroller has traveled 6.32 m.

2. A car traveling initially at +7.0 m/s accelerates uniformly at the rate of +0.80 m/s^2 for a distance of 245 m.
   a. What is its velocity at the end of the acceleration?
   b. What is its velocity after it accelerates for 125 m?
   c. What is its velocity after it accelerates for 67 m?

3. A car accelerates uniformly in a straight line from rest at the rate of 2.3 m/s^2.
   a. What is the speed of the car after it has traveled 55 m?
   b. How long does it take the car to travel 55 m?

4. A certain car is capable of accelerating at a uniform rate of 0.85 m/s^2. What is the magnitude of the car's displacement as it accelerates uniformly from a speed of 83 km/h to one of 94 km/h?

5. An aircraft has a liftoff speed of 120 km/h. What minimum uniform acceleration does this require if the aircraft is to be airborne after a take-off run of 240 m?

6. A motorboat accelerates uniformly from a velocity of 6.5 m/s to the west to a velocity of 1.5 m/s to the west. If its acceleration was 2.7 m/s^2 to the east, how far did it travel during the acceleration?

With the four equations presented in this section, it is possible to solve any problem involving one-dimensional motion with uniform acceleration. For your convenience, the equations that are used most often are listed in **Table 2-4.**

The first column of the table gives the equations in their standard form. The second column gives the form to use for an object that starts at rest and then undergoes uniformly accelerated motion. For an object initially at rest, $v_i = 0$. Using this value for v_i in the equations in the first column will result in the equations in the second column. Referring back to the sample problems in this chapter will guide you through using these equations to solve many problems.

Table 2-4 ⭐TEKS 4B
Equations for uniformly accelerated straight-line motion

Form to use when accelerating object has an initial velocity	Form to use when accelerating object starts from rest
$\Delta x = \frac{1}{2}(v_i + v_f)\Delta t$	$\Delta x = \frac{1}{2}(v_f)\Delta t$
$v_f = v_i + a(\Delta t)$	$v_f = a(\Delta t)$
$\Delta x = v_i(\Delta t) + \frac{1}{2}a(\Delta t)^2$	$\Delta x = \frac{1}{2}a(\Delta t)^2$
$v_f^2 = v_i^2 + 2a\Delta x$	$v_f^2 = 2a\Delta x$

Section Review

TEKS 2C, 4A, 4B

1. Marissa's car accelerates uniformly at a rate of +2.60 m/s². How long does it take for Marissa's car to accelerate from a speed of 88.5 km/h to a speed of 96.5 km/h?

2. A bowling ball with a negative initial velocity slows down as it rolls down the lane toward the pins. Is the bowling ball's acceleration positive or negative as it rolls toward the pins?

3. Nathan accelerates his skateboard uniformly along a straight path from rest to 12.5 m/s in 2.5 s.
   a. What is Nathan's acceleration?
   b. What is Nathan's displacement during this time interval?
   c. What is Nathan's average velocity during this time interval?

4. Two cars are moving in the same direction in parallel lanes along a highway. At some instant, the instantaneous velocity of car A exceeds the instantaneous velocity of car B. Does this mean that car A's acceleration is greater than car B's? Explain, and use examples.

5. The velocity-versus-time graph for a shuttle bus moving along a straight path is shown in **Figure 2-13.**
   a. Identify the time intervals during which the velocity of the shuttle bus is constant.
   b. Identify the time intervals during which the acceleration of the shuttle bus is constant.
   c. Find the value for the average velocity of the shuttle bus during each time interval identified in **b.**
   d. Find the acceleration of the shuttle bus during each time interval identified in **b.**
   e. Identify the times at which the velocity of the shuttle bus is zero.
   f. Identify the times at which the acceleration of the shuttle bus is zero.
   g. Explain what the shape of the graph reveals about the acceleration in each time interval.

6. Is the shuttle bus in item 5 always moving in the same direction? Explain, and refer to the time intervals shown on the graph.

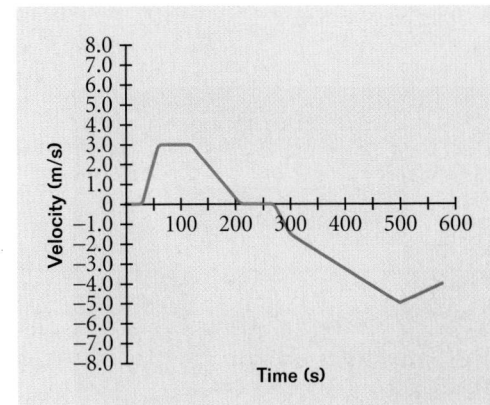

Figure 2-13

Section Review
ANSWERS

1. 0.85 s
2. positive
3. a. +5.0 m/s²
   b. +16 m
   c. +6.4 m/s
4. No, car A's acceleration is not necessarily greater than car B's acceleration. If the two cars are moving in the positive direction, car A could be slowing down (negative acceleration) while car B is speeding up (positive acceleration), even though car A's velocity is greater than car B's velocity.
5. a. 0 s to 30 s; 60 s to 125 s; 210 s to 275 s
   b. 0 s to 30 s; 30 s to 60 s; 60 s to 125 s; 125 s to 210 s; 210 s to 275 s; 275 s to 300 s; 300 s to 520 s; 520 s to 580 s
   c. 0 m/s; 1.5 m/s; 0 m/s; 1.5 m/s; 0 m/s; −0.75 m/s; −3.25 m/s; −4.5 m/s
   d. 0 m/s²; 0.1 m/s²; 0 m/s²; −0.04 m/s²; 0 m/s²; −0.06 m/s²; −0.02 m/s²; 0.02 m/s²
   e. 0 to 30 s; 210 to 275 s
   f. 0 s to 30s; 60 s to 125 s; 210 s to 275 s
   g. When the graph slopes upward, acceleration is positive. When it slopes downward, acceleration is negative.
6. No; the bus is moving in the positive direction from 30 s to 210 s (when velocity is positive) and in the negative direction from 275 s to 600 s (when velocity is negative).

2-3
Falling objects

Visual Strategy

Figure 2-14

Point out that the feather and apple will fall at the same rate regardless of the value of *g*.

Q If the feather and apple experiment were performed on the moon, where free-fall acceleration is approximately one-sixth the value of free-fall acceleration on Earth, how would the picture compare with **Figure 2-14**?

A *The feather and apple would still fall with equal accelerations (1/6 g); since the acceleration is less, the displacement in each time interval would be less.*

2-3 SECTION OBJECTIVES

- **Relate the motion of a freely falling body to motion with constant acceleration.**

- **Calculate displacement, velocity, and time at various points in the motion of a freely falling object.**

- **Compare the motions of different objects in free fall.**

free fall

motion of an object falling with a constant acceleration

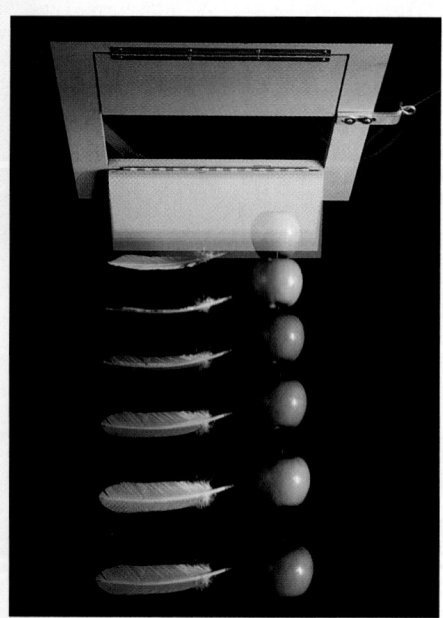

Figure 2-14
When there is no air resistance, all objects fall with the same acceleration regardless of their masses.

FREE FALL ⭐TEKS 3E

On August 2, 1971, a demonstration was conducted on the moon by astronaut David Scott. He simultaneously released a hammer and a feather from the same height above the moon's surface. The hammer and the feather both fell straight down and landed on the lunar surface at exactly the same moment. Although the hammer is more massive than the feather, both objects fell at the same rate. That is, they traveled the same displacement in the same amount of time.

Freely falling bodies undergo constant acceleration

In **Figure 2-14,** a feather and an apple are released from rest in a vacuum chamber. The trapdoor that released the two objects was opened with an electronic switch at the same instant the camera shutter was opened. The two objects fell at exactly the same rate, as indicated by the horizontal alignment of the multiple images.

It is now well known that in the absence of air resistance all objects dropped near the surface of a planet fall with the same constant acceleration. Such motion is referred to as **free fall.** ⭐TEKS 4B

The amount of time that passed between the first and second images is equal to the amount of time that passed between the fifth and sixth images. The picture, however, shows that the displacement in each time interval did not remain constant. Therefore, the velocity was not constant. The apple and the feather were accelerating.

Compare the displacement between the first and second images to the displacement between the second and third images. As you can see, within each time interval the displacement of the feather increased by the same amount as the displacement of the apple. Because the time intervals are the same, we know that the velocity of each object is increasing by the same amount in each time interval. In other words, the apple and the feather are falling with the same constant acceleration. ⭐TEKS 3B

The free-fall acceleration is denoted with the symbol *g*. At the surface of Earth the magnitude of *g* is approximately 9.81 m/s^2, or 981 cm/s^2, or 32 ft/s^2. Unless stated otherwise, this book will use the value 9.81 m/s^2 for calculations. This acceleration is directed downward, toward the center of the Earth. In our usual choice of coordinates, the downward direction is negative. Thus, the acceleration of objects in free fall near the surface of the Earth is $a = -g = -9.81$ m/s^2.

What goes up must come down ★TEKS 4B

Figure 2-15 is a strobe photograph of a ball thrown up into the air with an initial upward velocity of +10.5 m/s. The photo on the left shows the ball moving up from its release to the top of its path, and the photo on the right shows the ball moving down from the top of its path. Everyday experience shows that when we throw an object up in the air, it will continue to move upward for some time, stop momentarily at the peak, and then change direction and begin to fall. Because the object changes direction, it may seem like the velocity and acceleration are both changing. Actually, objects thrown into the air have a downward acceleration as soon as they are released.

In the photograph on the left, the upward displacement of the ball between each successive image is smaller and smaller until the ball stops and finally begins to move with an increasing downward velocity, as shown on the right. As soon as the ball is released with an initial upward velocity of +10.5 m/s, it has an acceleration of −9.81 m/s². After 1.0 s ($\Delta t = 1.0$ s) the ball's velocity will change by −9.81 m/s to 0.69 m/s upward. After 2.0 s ($\Delta t = 2.0$ s) the ball's velocity will again change by −9.81 m/s, to −9.12 m/s.

The graph in **Figure 2-16** shows the velocity of the ball plotted against time. As you can see, there is an instant when the velocity of the ball is equal to 0 m/s. This happens at the instant when the ball reaches the peak of its upward motion and is about to begin moving downward. Although the velocity is zero at the instant the ball reaches the peak, the acceleration is equal to −9.81 m/s² at every instant regardless of the magnitude or direction of the velocity. It is important to note that the acceleration is −9.81 m/s² even at the peak where the velocity is zero. The straight-line slope of the graph indicates that the acceleration is constant at every moment.

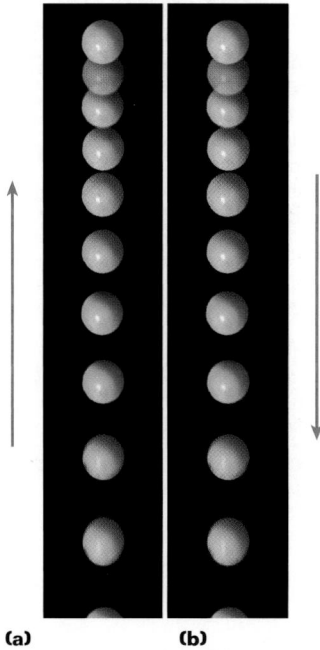

(a)   **(b)**

Figure 2-15
At the very top of its path, the ball's velocity is zero, but the ball's acceleration is −9.81 m/s² at every point—both when it is moving up **(a)** and when it is moving down **(b)**.

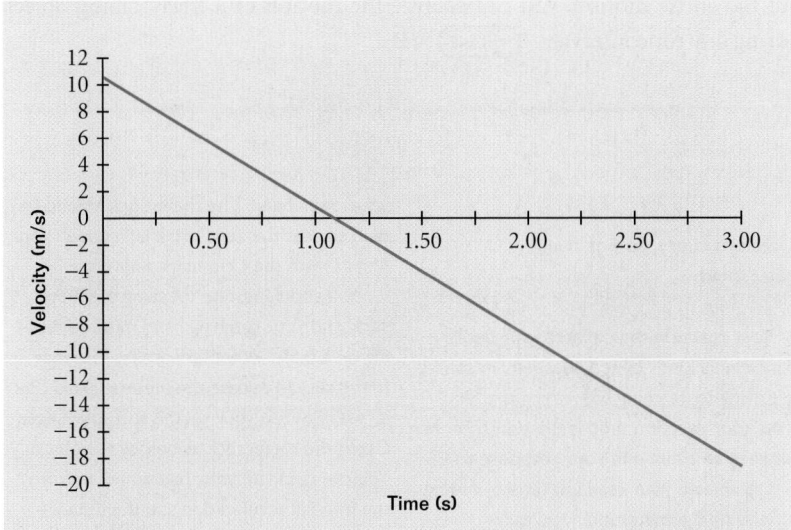

★TEKS 4A

Figure 2-16
On this velocity-time graph, the slope of the line, which is equal to the ball's acceleration, is constant from the moment the ball is released ($t = 0.00$) and throughout its motion.

☑ internetconnect

SCLINKS.

NSTA

TOPIC: Free fall
GO TO: www.scilinks.org
*sci*LINKS CODE: HF2024

SECTION 2-3

Visual Strategy

Figure 2-15
Because the ball undergoes constant acceleration, the ball's speed is the same going up as it is going down at any point in the object's flight.

Q If the initial velocity of the ball is 10.5 m/s when it is thrown from the ground, what is the final velocity of the ball at the end of its flight?

A *−10.5 m/s*

★TEKS

p. 60: 3E, 4B, 3B
p. 61: 4B, 4A

Misconception Alert

Many students find it very difficult to grasp the idea that an object can have a nonzero acceleration at the top of its flight when the velocity in the y direction is zero. Point out that the acceleration causes a change in the direction of the velocity at the instant the velocity is zero.

Quick Lab

TEACHER'S NOTES

In order to reduce error caused by anticipating the drop, have students look at the end of the meterstick between their fingers. If time allows, have each student test several times and use the mean distance in the calculation of reaction time.

The students releasing the meterstick should hold it between their thumb and index finger to reduce the tendency to rotate the meterstick when it is released.

⭐ TEKS

p. 62: 4B, 4B, 1A, 2B, 2C
p. 63: 2C, 4A, 4B

Module 1
"One-Dimensional Motion" provides an interactive lesson with guided problem-solving practice to teach you about all kinds of one-dimensional motion, including free fall.

CONCEPT PREVIEW

In Chapter 7 we will study how to deal with variations in g with altitude.

Freely falling objects always have the same downward acceleration

It may seem a little confusing to think of something that is moving upward, like the ball in the example, as having a downward acceleration. Thinking of this motion as motion with a positive velocity and a negative acceleration may help. The downward acceleration is the same when an object is moving up, when it is at rest at the top of its path, and when it is moving down. The only things changing are the position and the magnitude and direction of the velocity.

When an object is thrown up in the air, it has a positive velocity and a negative acceleration. From **Table 2-3,** we see that this means the object is slowing down. From the example of the ball and from everyday experience, we know that this makes sense. The object continues to move upward but with a smaller and smaller speed. In the photograph of the ball, this decrease in speed is shown by the smaller and smaller displacements as the ball moves up to the top of its path. ⭐TEKS **4B**

At the top of its path, the object's velocity has decreased until it is zero. Although it is impossible to see this because it happens so quickly, the object is actually at rest for an instant. Even though the velocity is zero at this instant, the acceleration is still -9.81 m/s^2.

When the object begins moving down, it has a negative velocity and its acceleration is still negative. From **Table 2-3,** we see that a negative acceleration and a negative velocity indicate an object that is speeding up. In fact, this is what happens when objects undergo free-fall acceleration. Objects that are falling toward Earth move faster and faster as they fall. In the photograph of the ball in **Figure 2-15,** this increase in speed is shown by the greater and greater displacements between the images as the ball falls.

Knowing the free-fall acceleration makes it easy to calculate the velocity, time, and displacement of many different motions using the equations for constantly accelerated motion. Because the acceleration is the same throughout the entire motion, you can analyze the motion of a freely-falling object during any time interval. ⭐TEKS **4B**

Quick Lab

Time Interval of Free Fall

MATERIALS LIST

✔ 1 meterstick or ruler

⭐TEKS **1A, 2B, 2C**

 SAFETY CAUTION

Avoid eye injury; do not swing metersticks.

Your reaction time affects your performance in all kinds of activities—from sports to driving to catching something that you drop. Your reaction time is the time interval between an event and your response to it.

Determine your reaction time by having a friend hold a meterstick vertically between the thumb and index finger of your open hand. The meterstick should be held so that the zero mark is between your fingers with the 1 cm mark above it.

You should not be touching the meterstick, and your catching hand must be resting on a table. Without warning you, your friend should release the meterstick so that it falls between your thumb and your finger. Catch the meterstick as quickly as you can. You can calculate your reaction time from the free-fall acceleration and the distance the meterstick has fallen through your grasp.

Falling object ★TEKS 2C, 4A, 4B

PROBLEM

Jason hits a volleyball so that it moves with an initial velocity of 6.0 m/s straight upward. If the volleyball starts from 2.0 m above the floor, how long will it be in the air before it strikes the floor? Assume that Jason is the last player to touch the ball before it hits the floor.

SOLUTION

1. DEFINE　**Given:**　　$v_i = +6.0 \text{ m/s}$　$a = -9.81 \text{ m/s}^2$

　　　　　　　　　$\Delta y = -2.0 \text{ m}$

Unknown:　$\Delta t = ?$

Diagram:　Choose a coordinate system. Place the
　　　　　　origin at the starting point of the ball.
　　　　　　$(y_i = 0 \text{ at } t_i = 0)$.

2. PLAN　**Choose the equation(s) or situation:** Both the time interval
and the final velocity are unknown. Therefore, first solve for
v_f using the equation that does not require time.

$$v_f^2 = v_i^2 + 2a\Delta y$$

Then the equation for v_f that does involve time can be used to solve for Δt.

$$v_f = v_i + a\Delta t$$

Rearrange the equation(s) to isolate the unknown(s): No rearrangement is
necessary to calculate v_f, but the second equation must be rearranged to solve
for Δt.

$$\Delta t = \frac{v_f - v_i}{a}$$

3. CALCULATE　**Substitute the values into the equation:** First find the velocity of the ball at
the moment that it hits the floor.

$$v_f^2 = v_i^2 + 2a\Delta y = (6.0 \text{ m/s})^2 + 2(-9.81 \text{ m/s}^2)(-2.0 \text{ m})$$

$$v_f^2 = 36 \text{ m}^2/\text{s}^2 + 39 \text{ m}^2/\text{s}^2 = 75 \text{ m}^2/\text{s}^2$$

When you take the square root to find v_f, the answer will be either positive or
negative. In this case, select the negative answer because the ball will be moving toward the floor, in the negative direction.

$$v_f = \sqrt{75 \text{ m}^2/\text{s}^2} = \pm 8.7 \text{ m/s}$$

$$v_f = -8.7 \text{ m/s}$$

continued on
next page

Classroom Practice

*The following may be used
as teamwork exercises or for
demonstration at the chalkboard
or on an overhead projector.*

PROBLEM

Falling object

A ball is thrown straight up into
the air at an initial velocity of
25.0 m/s. Create a table showing
the ball's position, velocity, and
acceleration each second for the
first 5.00 s of its motion.

Answer

t (s)	y (m)	v (m/s)	a (m/s^2)
1.00	20.1	+15.2	−9.81
2.00	30.4	+5.4	−9.81
3.00	30.9	−4.4	−9.81
4.00	21.6	−14.2	−9.81
5.00	2.50	−24.0	−9.81

Find the ball's time, position,
velocity, and acceleration at the
top of its flight.

Answer

$\Delta t = 2.55 \text{ s}; y = 31.9 \text{ m};$
$v = 0.00 \text{ m/s}; a = -9.81 \text{ m/s}^2$

**Interactive
Problem-
Solving
Tutor**

See Module 1
"One-Dimensional Motion" provides additional development of
problem-solving skills for this
chapter.

ANSWERS TO

Practice 2F
Falling object

1. **a.** −42 s
   b. 11 m/s
2. **a.** 22.1 m/s
   b. 2.25 s
3. **a.** 8.0 m/s
   b. 1.63 s
4. **a.** 3.7 m
   b. 0.76 s
5. **a.** No, the apple will reach only 1.6 m.
   b. 0.82 s
6. 1.8 m

Next, use this value of v_f in the second equation to solve for Δt.

$$\Delta t = \frac{v_f - v_i}{a} = \frac{-8.7 \text{ m/s} - 6.0 \text{ m/s}}{-9.81 \text{ m/s}^2} = \frac{-14.7 \text{ m/s}}{-9.81 \text{ m/s}^2}$$

$$\boxed{\Delta t = 1.50 \text{ s}}$$

4. EVALUATE Because the velocity will decrease by 9.81 m/s in 1 s and because v_i is only 6.0 m/s, it will take a little less than 1 s for the ball to reach its maximum height. Once the ball is at its maximum height, it will take less than 1 s to fall to its original position and a little additional time to fall the final 2.0 m to the floor. Therefore, a total time of between 1.0 s and 2.0 s is reasonable.

PRACTICE 2F

Falling object ⭐TEKS 2C, 4B

1. A robot probe drops a camera off the rim of a 239 m high cliff on Mars, where the free-fall acceleration is −3.7 m/s^2.
   a. Find the velocity with which the camera hits the ground.
   b. Find the time required for it to hit the ground.

2. A flowerpot falls from a windowsill 25.0 m above the sidewalk.
   a. How fast is the flowerpot moving when it strikes the ground?
   b. How much time does a passerby on the sidewalk below have to move out of the way before the flowerpot hits the ground?

3. A tennis ball is thrown vertically upward with an initial velocity of +8.0 m/s.
   a. What will the ball's speed be when it returns to its starting point?
   b. How long will the ball take to reach its starting point?

4. Stephanie hits a volleyball from a height of 0.80 m and gives it an initial velocity of +7.5 m/s straight up.
   a. How high will the volleyball go?
   b. How long will it take the ball to reach its maximum height? (Hint: At maximum height, $v = 0$ m/s.)

5. Maria throws an apple vertically upward from a height of 1.3 m with an initial velocity of +2.4 m/s.
   a. Will the apple reach a friend in a treehouse 5.3 m above the ground?
   b. If the apple is not caught, how long will the apple be in the air before it hits the ground?

6. Calculate the displacement of the volleyball in Sample Problem 2F when the volleyball's final velocity is 1.1 m/s upward.

Section Review

★TEKS 2C, 4A, 4B

1. A coin is tossed vertically upward.
   a. What happens to its velocity while it is in the air?
   b. Does its acceleration increase, decrease, or remain constant while it is in the air?

2. A pebble is dropped down a well and hits the water 1.5 s later. Using the equations for motion with constant acceleration, determine the distance from the edge of the well to the water's surface.

3. A ball is thrown vertically upward. What are its velocity and acceleration when it reaches its maximum altitude? What is its acceleration just before it hits the ground?

4. Two children are bouncing small rubber balls. One child simply drops a ball. At the same time, the second child throws a ball downward so that it has an initial speed of 10 m/s. What is the acceleration of each ball while in motion?

5. A gymnast practices two dismounts from the high bar on the uneven parallel bars. During one dismount, she swings up off the bar with an initial upward velocity of + 4.0 m/s. In the second, she releases from the same height but with an initial downward velocity of −3.0 m/s. How do the final velocities of the gymnast as she reaches the ground differ? What is her acceleration in each case?

6. **Figure 2-17** is a position-time graph of the motion of a basketball thrown straight up. Use the graph to sketch the path of the basketball and to sketch a velocity-time graph of the basketball's motion.
   a. Is the velocity of the basketball constant?
   b. Is the acceleration of the basketball constant?
   c. What is the initial velocity of the basketball?

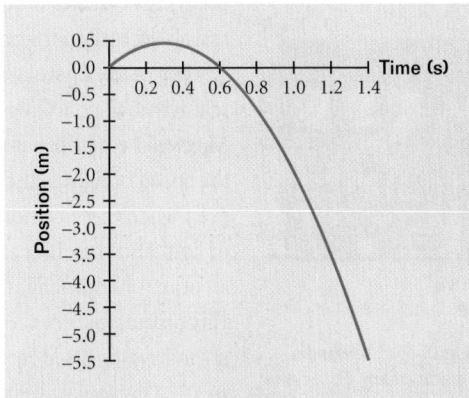

Figure 2-17

Section Review
ANSWERS

1. a. The coin's velocity decreases, becomes zero at its maximum height, and then increases in the negative direction until it hits the ground.
   b. The coin's acceleration remains constant.

2. 11 m

3. At maximum altitude, $v = 0$ and $a = -9.81$ m/s^2. Just before the ball hits the ground, $a = -9.81$ m/s^2.

4. −9.81 m/s^2 for each ball

5. The gymnast's final velocities will be determined by the equation $v_f^2 = v_i^2 + 2a\Delta y$. The acceleration and displacement are the same, so the final velocity is greater for the larger initial velocity, +4.0 m/s, and her acceleration (-9.81 m/s^2) will also be the same in each case.

6. The ball goes up 0.5 m, returns to its original position, and then falls another 5.5 m. The time axis of students' graphs should follow that in **Figure 2-17,** and the velocity coordinates should roughly correspond to $\frac{y}{x}$ in **Figure 2-17** at each point.
   a. no
   b. yes
   c. 3 m/s

★TEKS

p. 64: 2C, 4B
p. 65: 2C, 4A, 4B

PHYSICS ON THE EDGE

BACKGROUND

This feature discusses how Einstein's special theory of relativity modifies the notion of time as discussed in the rest of this chapter. Although the concept of time dilation is challenging, the basic aspects can be understood conceptually at this level, and the equation for time dilation can be used to see how the theories merge at small speeds but diverge at speeds that are large relative to the speed of light.

Before Einstein published the special theory of relativity in 1905, physicists were struggling with an apparent contradiction in physics. The principle of relativity, which states that the laws of physics are the same in any reference frame, did not allow for the fact that the speed of light is independent of the reference frame from which it is measured. (This contradiction is not immediately obvious to students; a class discussion of the subject will help to bring out the contradiction.)

While most physicists assumed that one of these postulates must be abandoned, Einstein had the insight that both postulates could be true if our ideas of space and time are modified. Time dilation is one of the necessary modifications.

Time Dilation

Earlier in this chapter, you worked with equations that describe motion in terms of a time interval (Δt). Before Einstein developed the special theory of relativity, everyone assumed that Δt must be the same for any observer, whether that observer is at rest or in motion with respect to the event being measured. This idea is often expressed by the statement that time is *absolute*.

The relativity of time ★ TEKS 3E

In 1905, Einstein challenged the assumption that time is absolute in a paper titled "The Electrodynamics of Moving Bodies," which contained his special theory of relativity. One of the consequences of this theory is that Δt *does* depend on the observer's motion. Consider a passenger in a train that is moving uniformly with respect to an observer standing beside the track, as shown in **Figure 2-18.** The passenger on the train shines a pulse of light toward a mirror directly above him and measures the amount of time it takes for the pulse to return. Because the passenger is moving along with the train, he sees the pulse of light travel directly up and then directly back down, as in **Figure 2-18(a).** The observer beside the track, however, sees the pulse hit the mirror at an angle, as in **Figure 2-18(b),** because the train is moving with respect to the track. Thus, the distance the light travels according to the observer is *greater* than the distance the light travels according to the passenger.

One of the postulates of Einstein's theory of relativity, which follows from James Clerk Maxwell's equations about light waves, is that the speed of light is the same for *any* observer, even when there is motion between the source of light and the observer. Light is different from all other phenomena in this respect. Although this postulate seems counterintuitive, it was strongly supported by an experiment performed in 1851 by Armand Fizeau. But if the speed of light is the same for both the passenger on the

Mirror

Passenger's perspective
(a)

(b)   **Observer's perspective**

Figure 2-18
(a) A passenger on a train sends a pulse of light toward a mirror directly above. **(b)** Relative to a stationary observer beside the track, the distance the light travels is greater than that measured by the passenger.

train and the observer beside the track while the distances traveled are different, the time intervals observed by each person must also be different. Thus, the observer beside the track measures a longer time interval than the passenger does. This effect is known as *time dilation*.

Calculating time dilation

Time dilation is given by the following equation, where Δt represents the time interval relative to the person beside the track (stationary system); and $\Delta t'$ represents the time interval relative to the person on the train (moving system): **(★)TEKS) 4E**

$$\Delta t = \frac{\Delta t'}{\sqrt{1 - \dfrac{v^2}{c^2}}}$$

In this equation, v represents the speed of the train relative to the person beside the track, and c is the speed of light in a vacuum, 3.00×10^8 m/s. At speeds with which we are familiar, where v is much smaller than c, $\dfrac{v^2}{c^2}$ is such a small fraction that Δt is almost equal to $\Delta t'$. That is why we do not observe the effects of time dilation in our typical experiences. But when speeds are closer to the speed of light, time dilation becomes more noticeable. As seen by this equation, time dilation becomes infinite as v approaches the speed of light.

According to Einstein, the motion between the train and the track is *relative*; that is, either system can be considered to be in motion with respect to the other. For the passenger, the train is stationary and the observer beside the track is in motion. If the light experiment is repeated by the observer beside the track, then the passenger would see the light travel a greater distance than the observer would. So, according to the passenger, it is the observer beside the track whose clock runs more slowly. Observers see their clocks running as if they were not moving. Any clocks in motion relative to the observers will seem to the observers to run slowly. Similarly, by comparing the differences between the time intervals of their own clocks and clocks moving relative to theirs, observers can determine how fast the other clocks are moving with respect to their own.

Experimental verification (★)TEKS) 3A

The effects we have been considering hold true for all physical processes, including chemical and biological reactions. Scientists have demonstrated time dilation by comparing the lifetime of muons (a type of unstable elementary particle) traveling at $0.9994c$ with the lifetime of stationary muons. In another experiment, atomic clocks on jet planes flying around the world were compared with identical clocks at the U.S. Naval Observatory. In both cases, time dilations were observed that matched the predictions of Einstein's theory within the limits of experimental error.

internet connect

SC*i*LINKS.
NSTA
TOPIC: Relativity of time
GO TO: www.scilinks.org
***sci*LINKS CODE:** HF2025

EXTENSION

- Have students research length contraction, another consequence of the theory of special relativity. Length contraction is given by the equation:

$$L = L_0 \sqrt{1 - \frac{v^2}{c^2}}$$

  Ask students to try different values of speed in the equations for time dilation and length contraction to determine at what speeds the effects of special relativity become significant.

- Have students research experiments that support the theory of special relativity, including the Michelson-Morley experiment and Armand Fizeau's measurement of the speed of light in water. In their report, they should include a discussion of how these results support the theory of relativity.

Teaching Tip

The theory of special relativity is valid only when the motion between the two reference frames is uniform. When one reference frame is accelerating in relation to the other, the general theory of relativity must be used.

(STOP) **Misconception Alert**

Because students have no direct experience with time dilation, many students initially think that time dilation is a theoretical idea rather than a physical effect that can be observed. Use the experiments discussed at the end of this feature to stress the fact that time dilation is a real physical effect.

CHAPTER 2
Summary

Teaching Tip

Many parents measure their child's height every year by marking notches on a door post. Point out that in this situation the child's growth each year is the displacement. The child's growth rate (velocity) is that displacement divided by the time (1 year), and the change in growth rate per unit of time (acceleration) is the difference in the growth rate from one year to the next.

★ TEKS

Review & Assess
pp. 69–75:
2B: 3
2C: 1–2, 8–16, 20–33,
　　38–60
2D: 2, 4
2F: 6
3A: 3
3E: 5–6
4A: 1, 3–5, 10, 30, 42,
　　Alt. Assess. 1,
　　Technology &
　　Learning
4B: 1–60

KEY TERMS

acceleration (p. 48)

average velocity (p. 43)

displacement (p. 41)

frame of reference (p. 40)

free fall (p. 60)

**instantaneous velocity
(p. 46)**

Key Symbols

Quantities		Units	
x	position	m	meters
y	position	m	meters
Δx	displacement	m	meters
Δy	displacement	m	meters
v	velocity	m/s	meters per second
a	acceleration	m/s^2	meters per second^2
g	free-fall acceleration	m/s^2	meters per second^2

KEY IDEAS

Section 2-1 Displacement and velocity

- Displacement is a change of position in a certain direction, not the total distance traveled.
- The average velocity of an object during some time interval is equal to the displacement of the object divided by the time interval. Like displacement, velocity indicates both speed and direction.
- The average velocity of an object is given by the following equation:

$$v_{avg} = \frac{\Delta x}{\Delta t} = \frac{x_f - x_i}{t_f - t_i}$$

- The average velocity is equal to the slope of the straight line connecting the initial and final points on a graph of the position of the object versus time.

Section 2-2 Acceleration

- The average acceleration of an object during a certain time interval is equal to the change in the object's velocity divided by the time interval. Acceleration has both magnitude and direction.
- The direction of the acceleration is not always the same as the direction of the velocity. The direction of the acceleration depends on the direction of the motion and on whether the velocity is increasing or decreasing.
- The average acceleration of an object is given by the following equation:

$$a_{avg} = \frac{\Delta v}{\Delta t} = \frac{v_f - v_i}{t_f - t_i}$$

- The average acceleration is equal to the slope of the straight line connecting the initial and final points on the graph of the velocity of the object versus time.
- The equations on page 58 are valid whenever acceleration is constant.

Section 2-3 Falling objects

- An object thrown or dropped in the presence of Earth's gravity experiences a constant acceleration directed toward the center of Earth. This acceleration is called the free-fall acceleration, or the acceleration due to gravity.
- Free-fall acceleration is the same for all objects, regardless of mass.
- The value for free-fall acceleration used in this book is $a = -g = -9.81$ m/s^2.
- In this book, the direction of the free-fall acceleration is considered to be negative because the object accelerates toward Earth.

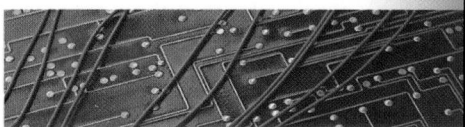
DISPLACEMENT AND VELOCITY

Review questions

1. On the graph in **Figure 2-19,** what is the total distance traveled during the recorded time interval? What is the displacement?

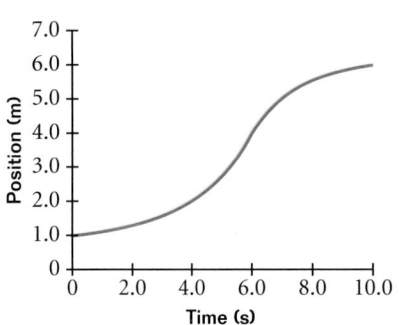

Figure 2-19

2. On a position-time graph such as **Figure 2-19,** what represents the instantaneous velocity?

3. Sketch a position-time graph for each of the following situations:
   a. an object at rest
   b. an object with constant positive velocity
   c. an object with constant negative velocity

4. The position-time graph for a bug crawling along a line is shown in **Figure 2-20.** Determine whether the velocity is positive, negative, or zero at each of the times marked on the graph.

5. Use the position-time graph in **Figure 2-20** to answer the following questions:
   a. During which time interval(s) does the velocity decrease?
   b. During which time interval(s) does the velocity increase?

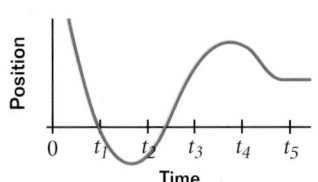

Figure 2-20

Conceptual questions

6. If the average velocity of a duck is zero in a given time interval, what can you say about the displacement of the duck for that interval?

7. Velocity can be either positive or negative, depending on the direction of the displacement. The time interval, Δt, is always positive. Why?

Practice problems

8. A bus travels from El Paso, Texas, to Chihuahua, Mexico, in 5.2 h with an average velocity of 73 km/h to the south. What is the bus's displacement? (See Sample Problem 2A.)

9. A school bus takes 0.530 h to reach the school from your house. If the average velocity of the bus is 19.0 km/h to the east, what is the displacement? (See Sample Problem 2A.)

10. Figure 2-21 is the position-time graph for a squirrel running along a clothesline.

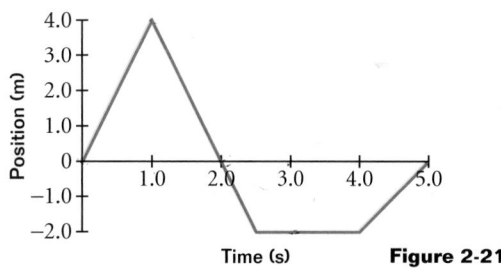

Figure 2-21

   a. What is the squirrel's displacement at the time $t = 3.0$ s?
   b. What is the squirrel's average velocity during the time interval between 0.0 s and 3.0 s?
   (See Sample Problem 2A.)

11. The Olympic record for the marathon is 2 h, 9 min, 21 s. If the average speed of a runner achieving this record is 5.436 m/s, what is the marathon distance? (See Sample Problem 2A.)

ANSWERS TO

Chapter 2
Review and Assess

1. 5.0 m; +5.0 m
2. the slope of the line at each point
3. a. slope must be zero
   b. slope must be positive and constant
   c. slope must be negative and constant
4. t_1: negative; t_2: positive; t_3: positive; t_4: negative; t_5: zero
5. a. 0 to t_1; t_1 to t_2; t_3 to t_4; t_4 to t_5
   b. t_1 to t_2; t_2 to t_3; t_3 to t_4
6. The duck's displacement must also be zero.
7. The time interval is always positive because time can only move in one direction (forward).
8. 380 km
9. 10.1 km
10. a. -2.0 m
   b. -0.67 m/s
11. 4.22×10^1 km

12. a. +70.0 m
   b. +140.0 m
   c. +14 m/s
   d. +28 m/s
13. a. 53.5 km/h
   b. 91.0 km
14. a. $\Delta x_1 = 2400$ m; $v_1 = 4.0$ m/s
   b. $\Delta x_2 = 1500$ m; $v_2 = 2.5$ m/s
   c. $\Delta x_3 = 900$ m; $v_3 = 1.5$ m/s
   d. $\Delta x_{tot} = 4800$ m; $v_{avg} = 2.7$ m/s
15. 0.2 km west of the flagpole
16. 0.00 m/s^2
17. a. slope is zero
   b. slope is positive
   c. slope is negative
   d. slope is negative
   e. slope is positive
18. a. left half of top photo,
   bottom photo
   b. right half of top photo
   c. middle photo

---

12. Two cars are traveling on a desert road, as shown in **Figure 2-22**. After 5.0 s, they are side by side at the next telephone pole. The distance between the poles is 70.0 m. Identify the following quantities:

   a. the displacement of car A after 5.0 s
   b. the displacement of car B after 5.0 s
   c. the average velocity of car A during 5.0 s
   d. the average velocity of car B during 5.0 s

(See Sample Problem 2A.)

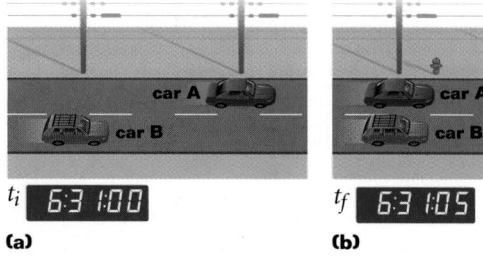

(a)   **(b)**

Figure 2-22

13. Sally travels by car from one city to another. She drives for 30.0 min at 80.0 km/h, 12.0 min at 105 km/h, and 45.0 min at 40.0 km/h, and she spends 15.0 min eating lunch and buying gas.

   a. Determine the average speed for the trip.
   b. Determine the total distance traveled.

(See Sample Problem 2A.)

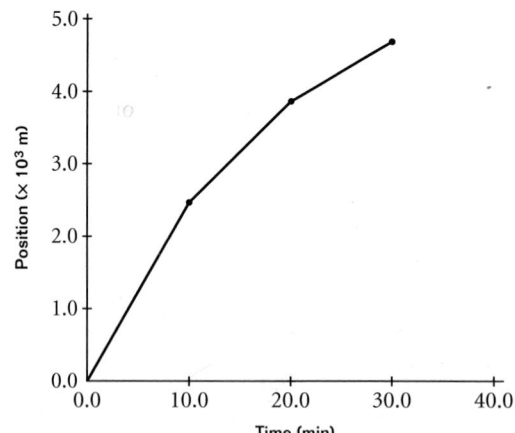

Figure 2-23

14. Figure 2-23 shows the position of a runner at different times during a run.

   a. For the time interval between $t = 0.0$ min and $t = 10.0$ min, what is the runner's displacement and average velocity?

   b. For the time interval between $t = 10.0$ min and $t = 20.0$ min, what is the runner's displacement? What is the runner's average velocity?

   c. For the time interval between $t = 20.0$ min and $t = 30.0$ min, what is the runner's displacement? For the same time period, what is the runner's average velocity?

   d. What is the runner's displacement for the total run? What is the runner's average velocity for the total run?

15. Runner A is initially 6.0 km west of a flagpole and is running with a constant velocity of 9.0 km/h due east. Runner B is initially 5.0 km east of the flagpole and is running with a constant velocity of 8.0 km/h due west. What will be the distance of the two runners from the flagpole when their paths cross? (It is not necessary to convert your answer from kilometers to meters for this problem. You may leave it in kilometers.)

(See Sample Problem 2A.)

VELOCITY AND ACCELERATION

Review questions

16. What would be the acceleration of a turtle that is moving with a constant velocity of 0.25 m/s to the right?

17. Sketch the velocity-time graphs for the following motions.

   a. a city bus that is moving with a constant velocity
   b. a wheelbarrow that is speeding up at a uniform rate of acceleration while moving in the positive direction
   c. a tiger that is speeding up at a uniform rate of acceleration while moving in the negative direction
   d. an iguana that is slowing down at a uniform rate of acceleration while moving in the positive direction
   e. a camel that is slowing down at a uniform rate of acceleration while moving in the negative direction

Conceptual questions

18. The strobe photographs in **Figure 2-24** show a disk moving from left to right under different conditions. The time interval between images is constant. Assuming that the direction to the right is positive, identify the following types of motion in each photograph.

    a. the acceleration is positive
    b. the acceleration is negative
    c. the velocity is constant

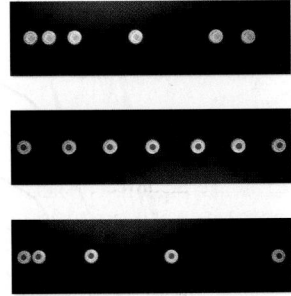

Figure 2-24

19. If a car is traveling eastward, can its acceleration be westward? Explain, and use examples.

Practice problems

20. A car traveling in a straight line has a velocity of +5.0 m/s. After an acceleration of 0.75 m/s^2, the car's velocity is +8.0 m/s. In what time interval did the acceleration occur?
(See Sample Problem 2B.)

21. A car traveling at +7.0 m/s accelerates at the rate of +0.80 m/s^2 for an interval of 2.0 s. Find v_f.
(See Sample Problem 2D.)

22. A snowmobile has an initial velocity of +3.0 m/s.
    a. If it accelerates at the rate of +0.50 m/s^2 for 7.0 s, what is the final velocity?
    b. If it accelerates at the rate of −0.60 m/s^2, how long will it take to reach a complete stop?
(See Sample Problems 2B and 2D.)

23. A car moving westward along a straight, level road increases its velocity uniformly from +16 m/s to +32 m/s in 10.0 s.
    a. What is the car's acceleration?
    b. What is its average velocity?

c. How far did it move while accelerating?
(See Sample Problems 2B and 2C.)

24. A ball initially at rest rolls down a hill with an acceleration of 3.3 m/s^2. If it accelerates for 7.5 s, how far will it move?
(See Sample Problem 2D.)

25. A bus slows down uniformly from 75.0 km/h (21 m/s) to 0 km/h in 21 s. How far does it travel before stopping?
(See Sample Problem 2C.)

26. A car accelerates from rest at −3.00 m/s^2.
    a. What is the velocity at the end of 5.0 s?
    b. What is the displacement after 5.0 s?
(See Sample Problem 2D.)

27. A car accelerates uniformly from rest to a speed of 65 km/h (18 m/s) in 12 s. Find the distance the car travels during this time.
(See Sample Problem 2C.)

28. A car starts from rest and travels for 5.0 s with a uniform acceleration of +1.5 m/s^2. The driver then applies the brakes, causing a uniform acceleration of −2.1 m/s^2. If the brakes are applied for 3.0 s, how fast is the car going at the end of the braking period, and how far has it gone from its start?
(See Sample Problem 2D.)

29. A boy sledding down a hill accelerates at 1.40 m/s^2. If he started from rest, in what distance would he reach a speed of 7.00 m/s?
(See Sample Problem 2D.)

30. The velocity-time graph for an object moving along a straight path is shown in **Figure 2-25**. Find the average accelerations during the time intervals 0.0 s to 5.0 s, 5.0 s to 15.0 s, and 0.0 s to 20.0 s.
(See Sample Problem 2B.)

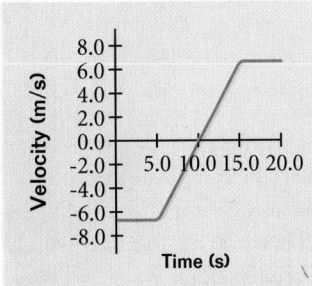

Figure 2-25

18. a. left half of top photo, bottom photo
    b. right half of top photo
    c. middle photo

19. Yes; a car traveling eastward and slowing down has a westward acceleration.

20. 4.0 s

21. +8.6 m/s

22. a. +6.5 m/s
    b. 5.0 s

23. a. +1.6 m/s^2
    b. +24 m/s
    c. 240 m

24. 93 m

25. 2.2×10^2 m

26. a. −15 m/s
    b. −38 m

27. 110 m

28. +1.5 m/s; +32 m

29. 17.5 m

30. 0.0 m/s^2; +1.4 m/s^2; +0.70 m/s^2

31. a. 2.0×10^1 s
  b. No; the plane needs at least 1.2 km to land.
32. a. 11 m/s
  b. 52 s
33. 0.99 m/s
34. a. The ball's velocity decreases, becomes zero at its maximum altitude, then increases in the negative direction.
  b. At maximum altitude, the ball's velocity is zero.
  c. -9.81 m/s^2
  d. -9.81 m/s^2
  e. The ball's acceleration remains constant.
35. For each ball, the velocity can be analyzed as the distance between images and the acceleration can be seen by comparing the distances. Both balls are accelerating; they start with approximately the same acceleration but the smaller ball's acceleration gradually decreases due to air resistance.
36. The two pins have the same acceleration (-9.81 m/s^2).
37. a. yes
  b. yes
  c. yes
38. -39.6 m/s
39. 3.94 s
40. 0.60 s
41. a. 2.55 s
  b. 2.63 s
42. a. 0.20 s
  b. 0.06 s and 0.34 s
  c. Answers may vary slightly but should be close to the following:
  $t = 0.05$ s, $v = 1.5$ m/s
  $t = 0.10$ s, $v = 1.0$ m/s
  $t = 0.15$ s, $v = 0.50$ m/s
  $t = 0.20$ s, $v = 0.00$ m/s
  d. Answers will vary but should be near -10 m/s^2.

31. A plane lands with a velocity of +120 m/s and accelerates at a maximum rate of -6.0 m/s^2.
  a. From the instant the plane touches the runway, what is the minimum time needed before it can come to rest?
  b. Can this plane land on a naval aircraft carrier where the runway is 0.80 km long?
  (See Sample Problems 2B and 2D.)

32. A sailboat starts from rest and accelerates at a rate of 0.21 m/s^2 over a distance of 280 m.
  a. Find the magnitude of the boat's final velocity.
  b. Find the time it takes the boat to travel this distance.
  (See Sample Problem 2E.)

33. An elevator is moving upward 1.20 m/s when it experiences an acceleration of 0.31 m/s^2 downward, over a distance of 0.75 m. What will its final velocity be?
  (See Sample Problem 2E.)

FALLING OBJECTS

Conceptual questions

34. A ball is thrown vertically upward.
  a. What happens to the ball's velocity while the ball is in the air?
  b. What is its velocity when it reaches its maximum altitude?
  c. What is its acceleration when it reaches its maximum altitude?
  d. What is its acceleration just before it hits the ground?
  e. Does its acceleration increase, decrease, or remain constant?

35. **Figure 2-26** is a strobe image of two falling balls released simultaneously. The ball on the left side is solid, and the ball on the right side is a hollow table-tennis ball. Analyze the motion of both balls in terms of velocity and acceleration.

36. A juggler throws a bowling pin into the air with an initial velocity v_i. Another juggler drops a pin at the same instant. Compare the accelerations of the two pins while they are in the air.

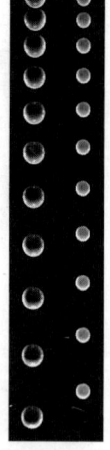

Figure 2-26

37. A bouquet is thrown upward.
  a. Will the value for the bouquet's displacement be the same no matter where you place the origin of the coordinate system?
  b. Will the value for the bouquet's velocity be the same?
  c. Will the value for the bouquet's acceleration be the same?

Practice problems

38. A worker drops a wrench from the top of a tower 80.0 m tall. What is the velocity when the wrench strikes the ground?
  (See Sample Problem 2F.)

39. A peregrine falcon dives at a pigeon. The falcon starts downward from rest with free-fall acceleration. If the pigeon is 76.0 m below the initial position of the falcon, how long does the falcon take to reach the pigeon? Assume that the pigeon remains at rest.
  (See Sample Problem 2F.)

40. A ball is thrown upward from the ground with an initial speed of 25 m/s; at the same instant, a ball is dropped from rest from a building 15 m high. After how long will the balls be at the same height?
  (See Sample Problem 2F.)

41. A ball is thrown vertically upward with a speed of 25.0 m/s from a height of 2.0 m.
  a. How long does it take to reach its highest point?
  b. How long does the ball take to hit the ground after it reaches its highest point?
  (See Sample Problem 2F.)

42. A ball is thrown directly upward into the air. **Figure 2-27** shows the vertical position of the ball with respect to time.
  a. How much time does the ball take to reach its maximum height?
  b. How much time does the ball take to reach one-half its maximum height?
  c. Estimate the slope of $\Delta y / \Delta t$ at $t = 0.05$ s, $t = 0.10$ s, $t = 0.15$ s, $t = 0.20$ s. On your paper, draw a coordinate system with velocity (v) on the y-axis and time (t) on the x-axis. Plot your velocity estimates against time.

d. From your graph, determine what the acceleration on the ball is.

(See Sample Problem 2F.)

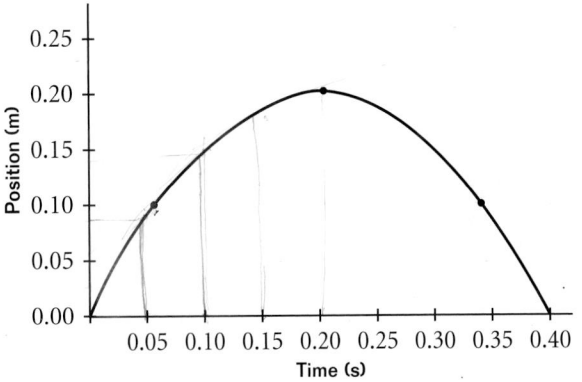

Figure 2-27

MIXED REVIEW

43. If the average speed of an orbiting space shuttle is 27 800 km/h, determine the time required for it to circle Earth. Make sure you consider the fact that the shuttle is orbiting about 320.0 km above Earth's surface. Assume that Earth's radius is 6380 km.

44. A train travels between stations 1 and 2, as shown in **Figure 2-28.** The engineer of the train is instructed to start from rest at station 1 and accelerate uniformly between points A and B, then coast with a uniform velocity between points B and C, and finally accelerate uniformly between points C and D until the train stops at station 2. The distances AB, BC, and CD are all equal, and it takes 5.00 min to travel between the two stations. Assume that the uniform accelerations have the same magnitude, even when they are opposite in direction.

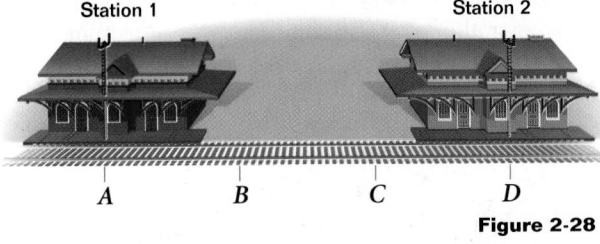

Figure 2-28

a. How much of this 5.00 min period does the train spend between points A and B?

b. How much of this 5.00 min period does the train spend between points B and C?

c. How much of this 5.00 min period does the train spend between points C and D?

45. Two students are on a balcony 19.6 m above the street. One student throws a ball vertically downward at 14.7 m/s. At the same instant, the other student throws a ball vertically upward at the same speed. The second ball just misses the balcony on the way down.

a. What is the difference in the time the balls spend in the air?

b. What is the velocity of each ball as it strikes the ground?

c. How far apart are the balls 0.800 s after they are thrown?

46. A rocket moves upward, starting from rest with an acceleration of +29.4 m/s^2 for 3.98 s. It runs out of fuel at the end of the 3.98 s but does not stop. How high does it rise above the ground?

47. Two cars travel westward along a straight highway, one at a constant velocity of 85 km/h, and the other at a constant velocity of 115 km/h.

a. Assuming that both cars start at the same point, how much sooner does the faster car arrive at a destination 16 km away?

b. How far must the cars travel for the faster car to arrive 15 min before the slower car?

48. A small first-aid kit is dropped by a rock climber who is descending steadily at 1.3 m/s. After 2.5 s, what is the velocity of the first-aid kit, and how far is the kit below the climber?

49. A small fish is dropped by a pelican that is rising steadily at 0.50 m/s.

a. After 2.5 s, what is the velocity of the fish?

b. How far below the pelican is the fish after 2.5 s?

50. A ranger in a national park is driving at 56 km/h when a deer jumps onto the road 65 m ahead of the vehicle. After a reaction time of t s, the ranger applies the brakes to produce an acceleration of −3.0 m/s^2. What is the maximum reaction time allowed if the ranger is to avoid hitting the deer?

51. A speeder passes a parked police car at 30.0 m/s. The police car starts from rest with a uniform acceleration of 2.44 m/s^2.

a. How much time passes before the speeder is overtaken by the police car?

43. 1.51 h
44. a. 2.00 min
    b. 1.00 min
    c. 2.00 min
45. a. 3.0 s
    b. −24.5 m/s for each
    c. 23.6 m
46. 931 m
47. a. 0.05 h
    b. 82 km
48. −26 m/s; 31 m
49. a. −24 m/s
    b. 33 m
50. 1.6 s
51. a. 24. 6 s
    b. 738 m

52. 5 s; 85 s; +60 m/s

53. a. 6100 m

    b. 9 s

54. -1.5×10^3 m/s^2

55. a. 2.33 s

    b. −32.9 m/s

56. a. 3.40 s

    b. −9.2 m/s

    c. −31 m/s; −33 m/s

57. a. 310 m

    b. 8.5 s

    c. 16.4 s

ANSWER TO

Technology & Learning

a. m/s^2

b. 41 m; 28 m/s

c. 91 m; 42 m/s

d. 250 m; 71 m/s

e. no

74

---

b. How far does the speeder get before being overtaken by the police car?

52. An ice sled powered by a rocket engine starts from rest on a large frozen lake and accelerates at +13.0 m/s^2. At t_1 the rocket engine is shut down and the sled moves with constant velocity v until t_2. The total distance traveled by the sled is 5.30×10^3 m and the total time is 90.0 s. Find t_1, t_2, and v.
(See Appendix A for hints on solving quadratic equations.)

53. At the 5800 m mark, the sled in the previous question begins to accelerate at −7.0 m/s^2. Use your answers from item 52 to answer the following questions.

    a. What is the final position of the sled when it comes to rest?

    b. How long does it take for the sled to come to rest?

54. A tennis ball with a velocity of +10.0 m/s to the right is thrown perpendicularly at a wall. After striking the wall, the ball rebounds in the opposite direction with a velocity of −8.0 m/s to the left. If the ball is in contact with the wall for 0.012 s, what is the average acceleration of the ball while it is in contact with the wall?

55. A parachutist descending at a speed of 10.0 m/s loses a shoe at an altitude of 50.0 m.

    a. When does the shoe reach the ground?

    b. What is the velocity of the shoe just before it hits the ground?

56. A mountain climber stands at the top of a 50.0 m cliff hanging over a calm pool of water. The climber throws two stones vertically 1.0 s apart and observes that they cause a single splash when they hit the water. The first stone has an initial velocity of +2.0 m/s.

    a. How long after release of the first stone will the two stones hit the water?

    b. What is the initial velocity of the second stone when it is thrown?

    c. What will the velocity of each stone be at the instant both stones hit the water?

57. A model rocket is launched straight upward with an initial speed of 50.0 m/s. It accelerates with a con-

Technology & Learning

Graphing calculators

Refer to Appendix B for instructions on downloading programs for your calculator. The program "Chap2" allows you to analyze graphs of displacement and speed for a falling hailstone.

    Your calculator will display two graphs: one for displacement versus time (Y$_1$) and the other for speed versus time (Y$_2$). These two graphs correspond to the following equations:

$$Y_1 = 4.9X^2$$
$$Y_2 = 9.8X$$

You should be able to correlate these equations with those shown in **Table 2-4.**

    a. Assuming time is measured in units of seconds and displacement is measured in meters, what are the units for the hailstone's acceleration?

Execute "Chap2" on the PRGM menu and press ENTER twice to begin graphing. The calculator will display two graphs. For the upper graph, the x value corresponds to the time in seconds and the y value corresponds to the displacement in meters. For the other graph, the x value corresponds to the time in seconds and the y value corresponds to the speed in meters per second.

    Identify the displacement and speed of the hailstone after the following time intervals (assuming that the hailstone starts from rest and that there is no air resistance):

    b. 2.9 s

    c. 4.3 s

    d. 7.2 s

    e. Does the hailstone's displacement increase at a constant rate?

Press ENTER and CLEAR to end.

stant upward acceleration of 2.00 m/s² until its engines stop at an altitude of 150 m.

    a. What is the maximum height reached by the rocket?

    b. When does the rocket reach maximum height?

    c. How long is the rocket in the air?

58. A professional race-car driver buys a car that can accelerate at +5.9 m/s². The racer decides to race against another driver in a souped-up stock car. Both start from rest, but the stock-car driver leaves 1.0 s before the driver of the sports car. The stock car moves with a constant acceleration of +3.6 m/s².

    a. Find the time it takes the sports-car driver to overtake the stock-car driver.

    b. Find the distance the two drivers travel before they are side by side.

    c. Find the velocities of both cars at the instant they are side by side.

(See Appendix A for hints on solving quadratic equations.)

59. Two cars are traveling along a straight line in the same direction, the lead car at 25 m/s and the other car at 35 m/s. At the moment the cars are 45 m apart, the lead driver applies the brakes, causing the car to have an acceleration of –2.0 m/s².

    a. How long does it take for the lead car to stop?

    b. Assume that the driver of the chasing car applies the brakes at the same time as the driver of the lead car. What must the chasing car's minimum negative acceleration be to avoid hitting the lead car?

    c. How long does it take the chasing car to stop?

60. One swimmer in a relay race has a 0.50 s lead and is swimming at a constant speed of 4.00 m/s. The swimmer has 50.0 m to swim before reaching the end of the pool. A second swimmer moves in the same direction as the leader. What constant speed must the second swimmer have in order to catch up to the leader at the end of the pool?

Alternative Assessment

Performance assessment

1. Can a boat moving eastward accelerate to the west? What happens to the boat's velocity? Name other examples of objects accelerating in the direction opposite their motion, including one with numerical values. Create diagrams and graphs.

2. The next time you are a passenger in a car, record the numbers displayed on the clock, the odometer, and the speedometer as frequently as you can for about 10 min. Create different representations of the car's motion, including maps, charts, and graphs. Exchange your representations with someone who made a different trip, and attempt to reconstruct that trip based on his or her report.

3. Two stones are thrown from a cliff at the same time with the same speed, one upward and one downward. Which stone, if either, hits the ground first? Which, if either, hits with the higher speed? In a group discussion, make your best argument for each possible prediction. Set up numerical examples and solve them to test your prediction.

Portfolio projects

4. Research typical values for velocities and acceleration of various objects. Include many examples, such as different animals, means of transportation, sports, continental drift, light, subatomic particles, and planets. Organize your findings for display on a poster or some other form.

5. Research Galileo's work on falling bodies. What did he want to demonstrate? What opinions or theories was he trying to refute? What arguments did he use to persuade others that he was right? Did he depend on experiments and observations, logic, or other approaches?

6. The study of various motions in nature requires devices for measuring periods of time. Prepare a presentation on a specific type of clock, such as water clocks, sand clocks, pendulum clocks, wind-up clocks, atomic clocks, or biological clocks. Who invented or discovered the clock? What scale of time does it measure? What are the principles or phenomena behind each clock? Can they be calibrated?

58. a. 4.6 s after stock car starts
    b. 38 m
    c. +17 m/s (stock car),
       +21 m/s (sports car)

59. a. 13 s
    b. –2.9 m/s²
    c. 12 s

60. 4.17 m/s

Alternative Assessment
ANSWERS

Performance assessment

1. All examples should involve decreasing speeds. The graphs should show velocity decreasing.

2. Students should be able to reconstruct a trip based on information about direction on a map and about speed from the speedometer.

3. The stone thrown downward hits the ground first, but both have the same speed when they hit the ground.

4. Students should record their values with appropriate units. Check the order of magnitude of their examples.

5. Galileo used both logic and detailed experiments to refute the Aristotelian view that bodies naturally tend to move downward and heavy bodies fall faster.

6. Student answers could include Galileo's use of his pulse to measure equal amounts of time and Huygens' clock, the first clock that could measure with the precision of 1 s.

NOTE

Materials Preparation is given on pp. 38A–38B. Blank data table and sample data table are on the One-Stop Planner CD-ROM. All calculations shown use sample data.

Planning

Recommended time:

1 lab period

Calibrating the recording timer may be done separately, or it may be skipped entirely if the period of the timer is known.

Classroom organization:

▶ Each group needs a level work surface at least 0.5 m above the floor with an edge to clamp the stand base onto. Each group needs clear floor space of a least 1.0 m².

▶ For the recording timer calibration, each group needs enough open space so the student can walk away from the timer in straight line for 3.0 s.

▶ Each lab group should have 2 students.

▶ The recording timer and CBL procedures may be used in the same class.

▶ **CBL and sensors:** If lab groups share balances, avoid traffic jams; some groups should mass all their blocks before setting up.

▶ **Safety warnings:** Remind students to attach masses to the paper tape securely and make sure the area is clear before dropping blocks or masses.

OBJECTIVES

- Measure motion in terms of the change in distance during a period of the timer or motion detector.
- Compare the speed and acceleration of different falling masses at different stages of free fall.
- Compare the experimental value for the average acceleration to the accepted value for free-fall acceleration.

MATERIALS LIST

✔ C-clamp
✔ meterstick
✔ masking tape

PROCEDURE

CBL AND SENSORS

✔ balance
✔ CBL motion detector
✔ CBL
✔ thin foam pad
✔ graphing calculator with link cable
✔ support stand with V-clamp
✔ wood blocks, 3 different masses

RECORDING TIMER

✔ cellophane tape
✔ recording timer and paper tape
✔ set of hooked masses
✔ stopwatch

MEASURING TIME AND MOTION

In this experiment, you will measure time intervals and displacements and determine the velocity and acceleration of free-falling bodies, using either a CBL™ unit with a motion detector or a recording timer.

- **CBL with motion detector**   The motion detector emits short bursts of ultrasonic waves and then "listens" for the echo of these waves as they return to it. Then the CBL uses the speed of the ultrasonic waves and the time it took for the waves to return to the motion detector to determine the distance to the object.

- **Recording timer**   A recording timer measures the time it takes an object to move a short distance by making marks at regular time intervals on a strip of paper attached to the moving object. As the paper tape is pulled through the timer, the distance between two dots on the tape is the distance the tape moved during one back-and-forth vibration of the clapper, the *period* of the timer.

SAFETY

- Tie back long hair, secure loose clothing, and remove loose jewelry to prevent their getting caught in moving or rotating parts.
- Attach masses securely. Falling or dropped masses can cause serious injury.

PREPARATION

1. Determine whether you will be using the CBL and sensors procedure or the recording timer procedure. Read the entire lab for the appropriate procedure, and plan what measurements you will take.

2. Prepare a data table with six columns and five rows in your lab notebook. In the first row, label the first two columns *Trial* and *Mass (kg)*. The space for the third through sixth columns should be labeled *Distance (m)*. Under this common label, columns 3–6 should be labeled *A–B, C–D, E–F,* and *G–H*. In the first column, label the second through fifth rows *1, 2, 3,* and *4*.

Recording timer procedure begins on page 79.

PROCEDURE

CBL AND SENSORS

Setting up the motion detector

3. Connect the CBL to the calculator with the unit-to-unit link cable using the link ports located on each unit. Connect the motion detector to the SONIC port.

4. Set up the apparatus as shown in **Figure 2-29**. Place the ring stand near the edge of the lab table. Use the C-clamp to clamp the base of the ring stand securely to the table. Position the clamp so that it protrudes as little as possible.

5. Using the V-clamp, securely clamp the motion detector to the ring stand so that it faces down, over the edge of the table. Make sure the motion detector is far enough away from the edge of the table that the signal will not hit the table top, clamp, or table leg.

6. Use a meterstick to measure a distance 0.5 m below the motion detector and mark the point with tape on the ring stand. This is the starting position from which the blocks will be dropped from rest.

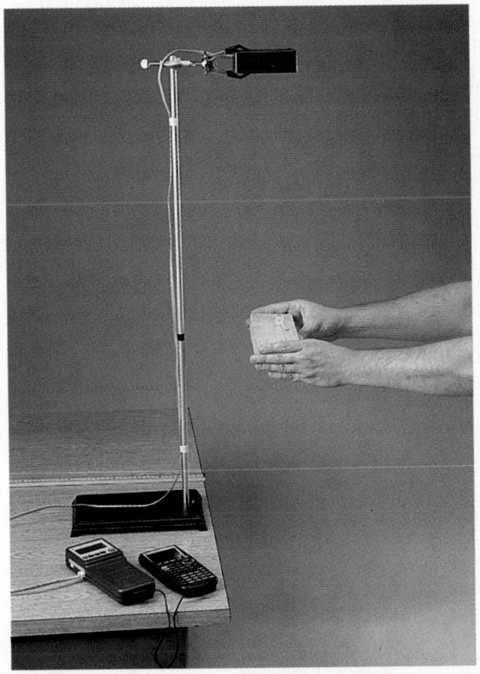

Figure 2-29

Step 9: With the CBL in MONITOR INPUT mode, move the wooden block up and down below the motion detector to check the readings.

Step 11: Hold the block flat and parallel to the tape mark.

Step 12: Release the block by pulling hands straight out to the sides. It may take some practice to release the block so that it falls straight down without turning. If the block turns while falling, the motion detector will measure the distance to the closest part of the block and introduce error into your results.

7. Start the program PHYSICS on your graphing calculator.

8. Select option *SET UP PROBES* from the MAIN MENU. Enter 1 for the number of probes. Select the motion detector from the list.

9. Select the *MONITOR INPUT* option from the DATA COLLECTION menu. Test to be sure the motion detector is positioned properly.

   a. Read the CBL measurement for the distance between the motion detector and the floor. Make sure the reading is a reasonable distance, or measure the distance with a meterstick to make sure the CBL gives the same value. If the CBL reading is too low, adjust the motion detector to make sure the signal is not hitting the table.

   b. Cover the floor under the motion detector with a foam pad to reduce unwanted feedback.

   c. Hold the wooden block directly beneath the motion detector, and move it up and down, reading the CBL measurements. Make sure the motion detector is not "seeing" other objects, such as the stand base, the table top, or the table leg. When the probe is functioning correctly, press "+" on the calculator to return to the MAIN MENU.

CBL and Sensors Tips

◆ Students should have the program PHYSICS on their graphing calculators. Refer to page Appendix B for instructions.

◆ The motion detector must be clamped tightly so that it can't move during the experiment.

◆ Students should practice dropping the blocks so they fall straight down without rotation. Remind them to keep their hands out of the way of the motion detector's signal.

Techniques to Demonstrate

Show the students how to use the arrow keys to trace the graph in step 15.

Explain how to choose the points and how to find the difference between the *y* values of the points in step 15.

Review what the graphs of velocity and acceleration in steps 16 and 17 reveal about the motion of the block.

Remind students to make sure the CBL and calculator are turned on before they perform each trial.

✔ Checkpoints

Step 6: Make sure students realize that the motion detector is the reference point for all distance measurements. The tape mark must be measured below the detector, not above the ground.

Step 9: Students should be able to demonstrate that the detector is working properly. They should be able to explain why the distance measurement increases as the block moves away and decreases as the block moves toward the detector.

Step 10: With the TI-83 calculator, the selection of NON-LIVE DISPLAY is automatic at this sample rate. Students using the TI-83 should ignore step (b).

Step 12: Make sure students can drop the block without letting it turn or getting their hands in the way.

Step 13: The graph should be smooth; a lot of black lines or spikes signify "drop-outs," times when the detector could not see the block. Check the setup and have students practice dropping the block before repeating the trial.

Step 14: Students should identify the portion of the curve that represents the block's motion. Guide students to choose four pairs of consecutive points on the graph.

Step 15: Students should be able to interpret the velocity graph on the calculator. They should be able to explain that the velocity of the block increased as the block fell.

Step 16: Students should be able to use the acceleration graph to explain whether the motion of the block meets their expectations. If there are any problems with the setup, students should be alerted by the graphs.

10. Select the *COLLECT DATA* option. Enter 0.02 for the time between samples. Enter 99 for the number of samples.

   a. Check the values you entered and press ENTER. If the values are correct, select *USE TIME SETUP* to continue. If you made a mistake entering the time values, select *MODIFY SETUP*, reenter the values, and continue.

   b. If you are given a choice on the TIME GRAPH menu, select *NON-LIVE DISPLAY*. Otherwise, continue to step 11.

Speed and acceleration of a falling object

11. Find the mass of the first wooden block, and label the block with tape. Record the mass in your data table. Hold the block horizontally between your hands, with your hands flat. Position the block directly below the motion detector and level with the 0.5 m mark.

12. Turn on the CBL and the graphing calculator. When the area is clear of people and objects, one student should press ENTER on the graphing calculator. As soon as the motion detector begins to click, the student holding the block should release the block by pulling both hands out to the side. Releasing the block this way will prevent the block from twisting as it falls, which may affect the results of this experiment.

13. When the motion detector has stopped clicking and the CBL displays DONE, press ENTER on the graphing calculator to get to the SELECT CHANNELS menu. Select the *SONIC* option, and then select *DISTANCE* to plot a graph of the distance in meters against time in seconds. (Note: The graph should have a smooth shape. If it has spikes or black lines, repeat the trial to obtain a smooth graph, then continue on to the next step.)

14. Use the arrow keys to trace along the curve. On the far left and the far right, the curve represents the position of the block before and after its motion. The middle section of the curve represents the

motion of the falling block. Sketch this graph in your lab notebook.

   a. Choose a point on the curve at the beginning of this middle section. Press the right arrow key once to select the next point. The difference between the x values for these two readings will be 0.02 s, the time interval between successive readings. Find the difference between the y values of these two points, and record it as $A–B$ for *Trial 1* in your data table.

   b. Press the right arrow key three times to move to another point on the curve. Press the right arrow key once to select the next point. Find the difference between the y values of these two points and record that as $C–D$ for *Trial 1* in your data table.

   c. Choose two more pairs of points along the curve. Select points at even intervals along the curve so that the first pair is at the beginning of the block's motion, and the last pair is at the end. Find the difference between the y values of each pair of points. Record them as $E–F$ and $G–H$ for *Trial 1*. Press ENTER to return to the SELECT CHANNELS menu.

15. Select the *SONIC* option again, and select *VELOCITY* to plot a graph of the velocity in m/s against time in seconds. Examine how the velocity of the block changed during the experiment. Sketch this graph in your lab notebook. Press ENTER.

16. Select the *SONIC* option again, and select *ACCELERATION* to plot a graph of the acceleration in m/s^2 against time in seconds. Examine how the acceleration of the block changed during the experiment. Sketch this graph in your lab notebook. Press ENTER.

17. Repeat this procedure using different wood blocks. Drop each block from the same level in each trial. Record all data for each trial.

18. Clean up your work area. Put equipment away safely so that it is ready to be used again.

Analysis and Interpretation begins on page 80.

PROCEDURE

RECORDING TIMER

Calibrating the timer

3. Prepare a second data table with three columns and five rows in your lab notebook. Label this table *Calibration*. In the first row, label the columns *Trial, Time (s),* and *No. of dots*. Fill in the first column by labeling the second through fifth rows *1, 2, 3,* and *4* for the number of trials.

4. Clamp the recording timer to the table to hold the timer in place. Choose a location that will allow you to pull a long section of paper tape through the timer in a straight line without hitting any obstacles. **Do not plug in the timer until your teacher approves your setup.**

5. Insert a strip of paper tape about 2.0 m long into the timer so that the paper can move freely and will be marked as it moves. Lay the tape flat behind the timer. One student should hold the end of the tape in front of the timer.

6. When your teacher approves your setup, plug the timer into the wall socket.

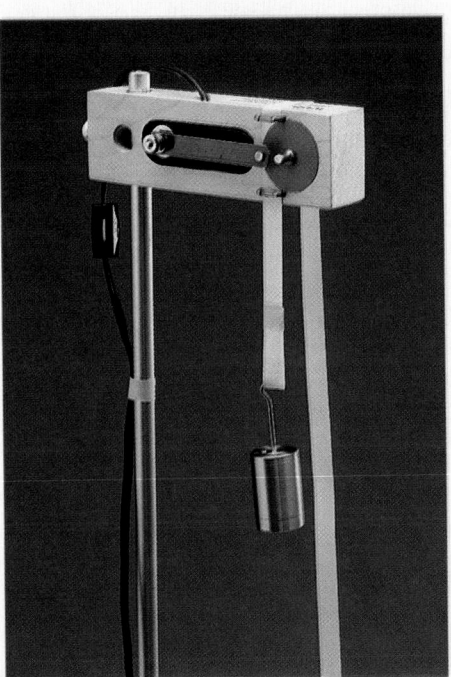

7. One student should start the timer and the stopwatch at the same time that the student holding the free end of the tape begins pulling the tape through the timer at a steady pace by walking away from the timer.

   a. After exactly 3.0 s, the first student should turn off the timer and stop the watch, just as the student with the tape stops walking. Mark the first and last dots on the tape. Tear or cut the dotted strip of tape from the roll and label it with the trial number and the time interval as measured by the stopwatch.

   b. Repeat this procedure three more times. Label all tapes.

8. Count the number of dots for each trial, starting with the second dot. Record this in your data table.

   a. Compute the period of the timer for each trial by dividing the 3.0 s time interval by the number of dots recorded in the table.

   b. Find the average value for the period of the recording timer. Use this value for all your calculations.

Speed and acceleration of a falling object

9. Set up the apparatus as shown in **Figure 2-30.** If possible, clamp the recording timer to a stand as shown. If the timer cannot be mounted on the stand, clamp the timer to the edge of the table as before.

Figure 2-30

Step 4: Clamp the timer to the edge of the table so the tape can be pulled through parallel to the floor for the calibration step. If your timer will not mount on a ring stand as shown, leave it clamped to the table throughout the experiment.

Step 5: Thread the tape through the timer and make sure the paper tape is *under* the carbon disk.

Step 9: When the timer is mounted on the ring stand, tape the cord to the stand to keep it out of the way.

Step 11: Use a piece of tape to mark the stand at the starting point, and start the mass from the same point for all trials.

Recording Timer Tips

◆ If 110 V recording timers are used, calibration may be omitted to save time; the average period of these timers is 0.017s.

◆ If possible, timers should be mounted on support rods for the second part of the lab, as shown.

Techniques to Demonstrate

Show students how to thread the paper tape through the recording timer guides under the carbon disk.

✔ Checkpoints

Step 6: Check each setup before the timer is plugged in. Make sure the tape is inserted properly, all clamps are tight and positioned where they will not protrude and cause injury.

Step 8: Students should use the spacing of the dots on the tape to show that they walked at a steady pace. They should explain how they find the period of the timer.

Step 9: All clamps must be tight and the recording timer must be positioned so the masses can fall without hitting any obstacles.

Step 11: Mass should be dropped from the same level for all trials. Students should hold the timer tape to suspend the mass in the starting position.

Step 12: Students should be able to explain how the dots on the tape represent the motion of the falling mass. They should be able to explain what order the dots were made in.

10. Cut a length of paper tape that is at least 20 cm longer than the distance between the timer and the floor. Thread the end of the tape through the timer.

11. Fold the end of the paper tape and fasten it with cellophane tape to make a loop. Hook a 200 g mass through the looped end of the paper tape, as shown.

12. Position the mass at a convenient level near the timer, as shown. Hold the mass in place by holding the tape behind the timer. Make sure the area is clear of people and objects. Simultaneously start the timer and release the tape so the mass falls to the floor. Stop the timer when the mass hits the floor.

13. Label the tape with the mass used. Label the second and third dots *A* and *B*, respectively. Count four

dots from *B* and label the seventh and eighth dots *C* and *D*, respectively. Label the twelfth and thirteenth dots *E* and *F*, and label the seventeenth and eighteenth dots *G* and *H*.

14. Repeat this procedure using different, larger masses, such as 300 g and 400 g masses. Drop each mass from the same level in each trial. Label all tapes, and record all data.

15. On each tape, measure the distance between *A* and *B*, between *C* and *D*, and so on. Record the distance in meters in your data table.

16. Clean up your work area. Put equipment away safely so that it is ready to be used again.

ANALYSIS AND INTERPRETATION

Calculations and data analysis

1. **Organizing data**   For each trial with the falling mass, find the magnitude of the average velocity, v_{avg}.

   a. **CBL and probeware**   Divide the distance *A–B* by 0.02 s, the time interval between distance readings. Repeat this calculation for the other distances you recorded for each trial.

   b. **Recording timer**   Divide the distance *A–B* by the average period of the timer. Repeat this calculation for the other marked distances for each trial.

2. **Analyzing results**   Using the results from item 1, calculate the average acceleration.

   a. **CBL and probeware**   Find the change of speed between the distance *A–B* and the distance *C–D*, between the distance *C–D* and the distance *E–F*, and so on. (Hint: Remember to use the total time interval—*A–D*, *C–F*, *E–H*—for each calculation.)

   b. **Recording timer**   Find the change of speed between the distance *A–B* and the distance *C–D*, between the distance *C–D* and the distance *E–F*, and so on. (Hint: Remember to use the total time interval for each calculation. For example, for the first calculation, use the time interval from *A* to *D*.)

ANSWERS TO

Analysis and Interpretation

CALCULATIONS AND DATA ANALYSIS

1. Students answers will vary. Answers should show that the velocity increases throughout the motion. Typical values will range from 0.210 m/s to 4.250 m/s.

2. Student answers will vary. Make sure students use the relationship $a_{avg} = \dfrac{\Delta v}{\Delta t}$.
   For sample data, calculated values range from 5.00 m/s^2 to 16.50 m/s^2.

3. **Graphing data** Use your data to plot the following graphs for each trial. On each graph, label the axes and indicate the trial number. Use a graphing calculator, computer, or graph paper.

    a. position versus time

    b. velocity versus time

    c. acceleration versus time

4. **Evaluating data** Use the values for the average acceleration for all four trials to find the average value.

5. **Evaluating data** Use the accepted value for the free-fall acceleration given in the text and the average of your results from item 4.

    a. Determine the absolute error of your results using the following equation:

$$\text{absolute error} = |\text{experimental} - \text{accepted}|$$

    b. Determine the relative error of your results using the following equation:

$$\text{relative error} = \frac{(\text{experimental} - \text{accepted})}{\text{accepted}}$$

Conclusions

6. **Relating ideas** Based on your results, how long would it take a 1000 kg mass to reach the floor if it were dropped from the same height as the masses in this experiment?

7. **Interpreting graphs** Calculate the slope of each velocity-time graph from item 3b above.

8. **Evaluating data** Find the average value for the slope of the velocity-time graphs. What is the relationship between this value and the values you found for the average accelerations of the masses?

Extensions

9. Devise a plan to perform this experiment to study the motion of an object thrown straight up into the air. Make sure you take into account any special safety requirements or equipment you might need to use. If there is time and your teacher approves your plan, perform the experiment. Use your data to plot graphs of the position, velocity, and acceleration versus time.

3. a. Student graphs should show a parabolic curve pointing up and to the right.

    b. Student graphs should show a straight line pointing up and to the right.

    c. Student graphs should show a straight line parallel to the x-axis.

4. CBL: all trials: $a_{avg} = 9.667$ m/s^2; **Recording timer:** all trials: $a_{avg} = 8.640$ m/s^2

5. Accepted value = 9.81 m/s^2; **CBL:** absolute error = 0.143 m/s^2, relative error = −0.015; **Recording timer:** absolute error = 1.17 m/s^2, relative error = −0.119

CONCLUSIONS

6. All blocks will fall with the same acceleration; therefore the 1000 kg block will fall in the same time as all the other blocks.

7. Answers should be similar to acceleration values.

8. They are the same or very close.

EXTENSIONS

9. Student plans should be safe and complete, including a list of equipment, measurements, and calculations required. Plans should recognize that the lab is essentially the same while also providing for the safety hazards and technical difficulties of studying an object that changes direction.

CHAPTER 3 PLANNING GUIDE

Compression Guide: To shorten from 10 to 7 45-min periods (from 5 to 3½ 90-min blocks), eliminate items in magenta type.

PACING CHART	CLASSROOM RESOURCES			
	(★)TEKS	**Teacher Demonstrations**	*Holt Physics* **Transparencies**	**Labs** (See page T52 for equipment listing for in-text labs.)
3-1 Introduction to vectors 2 or 1 45-minute periods 1 or ½ 90-minute block	2C, 3B, 4A	**TE** *Vector Addition*, p. 85	**T** 6 **TM** 15	**L** *Discovery Lab: Vector Treasure Hunt*
3-2 Vector operations 2 45-minute periods 1 90-minute block	2C, 3B, 4A		**TM** 16, 17, 18	
3-3 Projectile motion 3 or 2 45-minute periods 1½ or 1 90-minute block	1A, 2C, 4A, 4B	**TE** *Air Resistance*, p. 99 **TE** *Two-Dimensional Motion*, p. 100	**TM** 7	**PE** *Quick Lab: Projectile Motion*, p. 100 **PE** *Velocity of a Projectile*, p. 120 **L** *Invention Lab: The Path of a Human Cannonball* **TL** *Projectile Motion*
3-4 Relative motion 1 45-minute period ½ 90-minute block	2C, 3A, 3B, 3C, 4A, 4B, 4E		**TM** 8	

Review and Assessment
2 45-minute periods
1 90-minute block

Resource Key

PHYSICS

PE Pupil's Edition
TE Teacher's Edition

L Laboratory Experiments
TL Technology Lab Experiments
T Transparencies

One-Stop Planner CD-ROM **contents**

TM Transparency Masters
SR Section Review Worksheets
AA Alternative Assessment

PW Problem-Solving Workbook
PB Problem Bank
CTW Critical Thinking Worksheet

LABORATORY PLANNING: Velocity of a Projectile, p. 120

Materials (for each lab group)
- inclined plane
- 1-position support base and rod, 1.3 cm × 91 cm
- stand rod
- right-angle clamp for 1.3 cm rods
- C-clamp
- meterstick
- roll of black nylon cord
- steel washer, 26 mm diameter

- solid steel ball, 25 mm diameter
- carbon paper, 4 sheets
- white copier paper, 4 sheets
- roll of masking tape
- roll of adhesive tape, 0.5 in. wide
- 1 aluminum sheet, 12.5 cm × 25 cm, 0.0010 in. thick
- roll of adhesive packing tape
- cardboard box
- towel or washcloths

Two-Dimensional Motion and Vectors

ASSIGNMENT RESOURCES

Content Mastery	Critical Thinking	Problem-Solving Practice
PE 1–5, p. 87 **SR** 3-1, *Diagram Skills* **PE** 1–9, p. 113	**PE** 10–13, p. 113	
PE 1–4, p. 97 **SR** 3-2, *Diagram Skills* **PE** 14–18, p. 114	**PE** 19–21, p. 114	**3A** Finding resultant magnitude and direction: 27 items in **PE, PW,** and **PB,** see **TE** pp. 90–91 **3B** Resolving vectors: 30 items in **PE, PW,** and **PB,** see **TE** pp. 93–94 **3C** Adding vectors algebraically: 23 items in **PE, PW,** and **PB,** see **TE** pp. 95–96
PE 1–7, p. 105 **SR** 3-3, *Math Skills* **PE** 30–33, p. 115		**3D** Projectiles launched horizontally: 28 items in **PE, PW,** and **PB,** see **TE** p. 101 **3E** Projectiles launched at an angle: 40 items in **PE, PW,** and **PB,** see **TE** pp. 103–104
PE 1–3, p. 109 **SR** 3-4, *Diagram Skills* **PE** 42–46, p. 116	**PE** 1–2, p. 107	**3F** Relative velocity: 34 items in **PE, PW,** and **PB,** see **TE** pp. 108–109

ASSESSMENT RESOURCES

Cumulative Review	Alternative Assessment	Traditional Assessment
SR Mixed Review, Ch. 3	**PE** 1–6, p. 119 **AA** Items for Ch. 3	Chapter 3 Test Test Generator items for Ch. 3

Scoring Rubrics for Alternative Assessment items can be found on the One-Stop Planner CD-ROM.

TECHNOLOGY RESOURCES

 CTW Segment 3 Trebuchet Catapult

 Module 2 Vector Addition and Resolution
Module 3 Two-Dimensional Motion

 PE Technology and Learning, p. 118
(Alternative procedures for calculators without Flash-ROM technology are provided on the One-Stop Planner CD-ROM.)

 internet**connect**

 On-line Student Resources:
GO TO: www.scilinks.org
The following *sci*LINKS Internet resources can be found in the student text for this chapter.

TOPICS:
• Vectors, p. 86 (HF2031)
• Projectile motion, p. 99 (HF2032)
• Speed of light, p. 111 (HF2033)

 On-line Teacher Resources:
GO TO: go.hrw.com
KEYWORD: HF2 HOME
Visit the HRW Web site for a variety of resources related to this chapter.

 Smithsonian Institution
Internet Connections
Visit **www.si.edu/hrw** for additional on-line resources.

CNNfyi.com
Visit **www.cnnfyi.com** for late-breaking news and current events stories selected just for you.

Section 3-1 discusses scalar and vector quantities and graphical vector addition.

Section 3-2 explains the use of the Pythagorean theorem and trigonometric functions to find resultant vectors and vector components.

Section 3-3 explores projectile motion, neglecting air resistance.

Section 3-4 describes relative motion in terms of vector operations.

About the Illustration

The 20-acre Place de la Concorde, near the famous shopping district on the Champs-Elysées, was the site of several guillotine executions during the French Revolution's Reign of Terror (1793–1794). The square is now the site of the towering Obelisk of Luxor, seen in the background. Egypt gave France the obelisk as a gift in 1831. The fountain is one of two fountains that adorn the square.

Interactive Problem-Solving Tutor

See Module 2
"Vector Addition and Resolution" provides additional development of problem-solving skills for this chapter.

See Module 3
"Two-Dimensional Motion" provides additional development of problem-solving skills for this chapter.

CHAPTER 3

Two-Dimensional Motion and Vectors

PHYSICS IN ACTION

Although the streams of water of the Place de la Concorde fountain in Paris, France, seem to flow through transparent pipes, they are actually fired in highly concentrated solitary streams. Each of the streams follows a parabolic path, just like the path a baseball follows when it is thrown through the air. In this chapter you will analyze two-dimensional motion and solve problems involving objects projected into the air.

- *Why does all of the water in a stream follow the same path?*

- *How does the angle at which the water is fired affect its path?*

CONCEPT REVIEW

Displacement (Section 2-1)

Velocity (Section 2-1)

Acceleration (Section 2-2)

Free fall (Section 2-3)

Tapping Prior Knowledge

Knowledge to Expect

✔ "The motion of an object can be described by its position, direction of motion, and speed." (NRC's *National Science Education Standards*, grades 5–8)

✔ "The motion of an object is always judged with respect to some other object or point, and so the idea of absolute motion or rest is misleading." (AAAS's *Benchmarks for Science Literacy*, grades 6–8)

✔ "Unbalanced forces will cause changes in the speed or direction of an object's motion." (NRC's *National Science Education Standards*, grades 5–8)

Knowledge to Review

✔ Displacement is a change in location relative to a reference point. (Section 2-1)

✔ Velocity includes speed and direction. (Section 2-1)

✔ Acceleration is the change in velocity per unit time. It has both magnitude and direction. (Section 2-2)

Items to Probe

✔ Displacement versus distance: Have students decide displacement and distance values for scenarios listed on the board, such as round trips versus one-way trips.

✔ Acceleration and velocity: Ask students to describe the effects of the four possible combinations of acceleration and velocity in one dimension: $a > 0$, $v > 0$; $a < 0$, $v > 0$; $a > 0$, $v < 0$; $a < 0$, $v < 0$.

3-1
Introduction to vectors

Section 3-1

Teaching Tip

Carry an arrow with you to class, and use it often to describe the direction of the vector.

The Language of Physics

Establish a convention for distinguishing vectors from scalars for use in your classroom. Using an arrow above the variable is a good substitute for bold type, which would be hard to reproduce on a chalkboard. Make sure to be explicit when using this convention. For instance, if you use an arrow above the symbol, be sure to signify the *direction* of the vector as well as its magnitude and units.

Visual Strategy

Figure 3-1

Tell students that the arrows in the figure represent velocity, that is, both speed and direction.

Q Why do we consider velocity instead of speed in order to decide which player reaches the ball first?

A *Direction is important because a player may run very fast but not toward the ball.*

3-1 SECTION OBJECTIVES

- **Distinguish between a scalar and a vector.**
- **Add and subtract vectors using the graphical method.**
- **Multiply and divide vectors by scalars.**

scalar

a physical quantity that has only a magnitude but no direction

vector

a physical quantity that has both a magnitude and a direction

Figure 3-1
The lengths of the vector arrows represent the magnitudes of these two soccer players' velocities.

SCALARS AND VECTORS

In Chapter 2 our discussion of motion was limited to two directions, forward and backward. Mathematically, we described these directions of motion with a positive or negative sign. This chapter explains a method of describing the motion of objects that do not travel along a straight line.

Vectors indicate direction; scalars do not

Each of the physical quantities we will encounter in this book can be categorized as either a scalar or a vector quantity. A **scalar** is a quantity that can be completely specified by its magnitude with appropriate units; that is, a scalar has magnitude but no direction. Examples of scalar quantities are speed, volume, and the number of pages in this textbook. A **vector** is a physical quantity that has both direction and magnitude.

As we look back to Chapter 2, we can see that displacement is an example of a vector quantity. An airline pilot planning a trip must know exactly how far and which way to fly. Velocity is also a vector quantity. If we wish to describe the velocity of a bird, we must specify both its speed (say, 3.5 m/s) and the direction in which the bird is flying (say, northeast). Another example of a vector quantity is acceleration, which was also discussed in Chapter 2.

Vectors are represented by symbols

In physics, quantities are often represented by symbols, such as t for time. To help you keep track of which symbols represent vector quantities and which are used to indicate scalar quantities, this book will use **boldface** type to indicate vector quantities. Scalar quantities are designated by the use of *italics*. Thus, to describe the speed of a bird without giving its direction, you would write $v = 3.5$ m/s. But a velocity, which includes a direction, is written as **v** = 3.5 m/s to the northeast. Handwritten, a vector can be symbolized by showing an arrow drawn above the abbreviation for a quantity, such as $\vec{v} = 3.5$ m/s to the northeast.

One way to keep track of vectors and their directions is to use diagrams. In diagrams, vectors are shown as arrows that point in the direction of the vector. The length of a vector arrow in a diagram is related to the vector's magnitude. For example, in **Figure 3-1** the arrows represent the velocities of the two soccer players running toward the soccer ball. ⭐TEKS **3B**

Because the player on the right of the soccer ball is moving faster, the arrow representing that velocity is drawn longer than the arrow representing the velocity of the player on the left.

Vectors can be added graphically

When adding vectors, you must make certain that they have the same units and describe similar quantities. For example, it would be meaningless to add a velocity vector to a displacement vector because they describe different physical quantities. Similarly, it would be meaningless, as well as incorrect, to add two displacement vectors that are not expressed in the same units (in meters, in feet, and so on).

Section 2-1 covered vector addition and subtraction in one dimension. Think back to the example of the gecko that ran up a tree from a 20 cm marker to an 80 cm marker. Then the gecko reversed direction and ran back to the 50 cm marker. Because the two parts of this displacement are opposite, they can be added together to give a total displacement of 30 cm. The answer found by adding two vectors in this way is called the **resultant.** ⭐TEKS **3B**

Consider a student walking to school. The student walks 1600 m to a friend's house, then 1600 m to the school, as shown in **Figure 3-2.** The student's total displacement during his walk to school is in a direction from his house to the school, as shown by the dotted line. This direct path is the *vector sum* of the student's displacement from his house to his friend's house and his displacement from the friend's house to school. How can this resultant displacement be found?

One way to find the magnitude and direction of the student's total displacement is to draw the situation to scale on paper. Use a reasonable scale, such as 50 m on land equals 1 cm on paper. First draw the vector representing the student's displacement from his house to his friend's house, giving the proper direction and scaled magnitude. Then draw the vector representing his walk to the school, starting with the tail at the head of the first vector. Again give its scaled magnitude and the right direction. The magnitude of the resultant vector can then be determined by using a ruler to measure the length of the vector pointing from the tail of the first vector to the head of the second vector. The length of that vector can then be multiplied by 50 (or whatever scale you have chosen) to get the actual magnitude of the student's total displacement in meters.

The direction of the resultant vector may be determined by using a protractor to measure the angle between the first vector and the resultant.

resultant

a vector representing the sum of two or more vectors

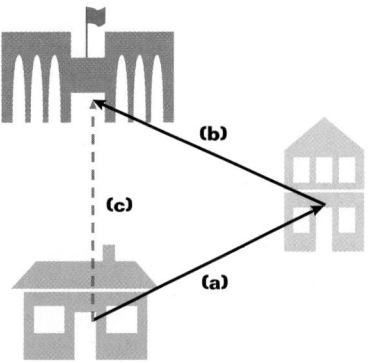

Figure 3-2
A student walks from his house to his friend's house **(a)**, then from his friend's house to the school **(b)**. The student's resultant displacement **(c)** can be found using a ruler and a protractor.

Demonstration 1

Vector addition

Purpose Preview force as a vector quantity to demonstrate vector addition.

Materials cart or table on wheels

Procedure Ask for a student volunteer. Explain to the class that you will push the table in one direction while the student will push the table in a second, perpendicular direction at the same time. Ask students to predict the motion of the table.

Have the student volunteer take a practice push alone, and emphasize that the table moves in the direction of the push. Take a practice push by yourself on the table, and again emphasize that the table moves in the direction of the push. Return the table to its original location, and both you and the student push the table perpendicular to each other at the count of three. Ask a student to explain why the table moved along the diagonal. *(The table moved in a direction between the directions of the two pushes.)*

Show the vector addition for the demonstration qualitatively on the chalkboard. This should lay the groundwork for a discussion of why vector addition is a valuable activity.

⭐TEKS

p. 84: 3B
p. 85: 3B

Misconception Alert

Students may have difficulty visualizing the movement of the toy car on the walkway. Illustrate the situation on the board, and emphasize that the motion represented in **Figure 3-3** is drawn as viewed by a stationary observer above the walkway. Frame of reference will be covered in Section 3-4.

The Language of Physics

The triangle method of vector addition is also called the head-to-tail method or the tip-to-tail method.

Teaching Tip

Students will better understand the commutative property of vector addition if they work several examples on paper with a ruler and protractor.

Visual Strategy

Figure 3-4

Reinforce students' understanding of the difference between distance and displacement by pointing out that the runner's displacement will be the same regardless of the path the runner chooses to follow.

Q How can the runner increase distance but begin and end at the same points?

A *Whatever distance the runner covers in one direction must be covered in the opposite direction so that the two extra displacements cancel.*

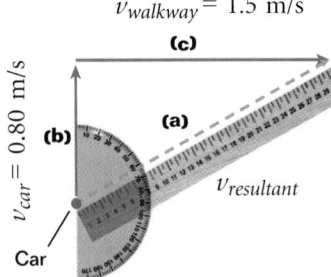

Figure 3-3

The resultant velocity **(a)** of a toy car moving at a velocity of 0.80 m/s **(b)** across a moving walkway with a velocity of 1.5 m/s **(c)** can be found using a ruler and a protractor.

internetconnect

SCiLINKS

NSTA

TOPIC: Vectors
GO TO: www.scilinks.org
*sci*LINKS CODE: HF2031

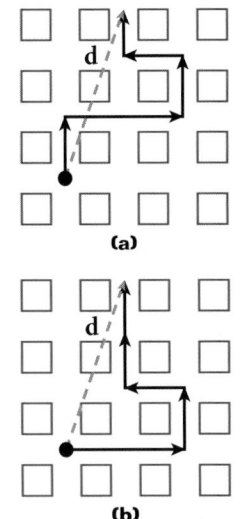

Figure 3-4

A marathon runner's displacement, **d,** will be the same regardless of whether the runner takes path **(a)** or **(b).**

PROPERTIES OF VECTORS

Now consider a case in which two or more vectors act at the same point. When this occurs, it is possible to find a resultant vector that has the same net effect as the combination of the individual vectors. Imagine looking down from the second level of an airport at a toy car moving at 0.80 m/s across a moving walkway that moves at 1.5 m/s, as graphically represented in **Figure 3-3.** How can you determine what the car's resultant velocity will look like from your vantage point? ⓣ TEKS **4A**

Vectors can be moved parallel to themselves in a diagram

Note that the car's resultant velocity while moving from one side of the walkway to the other will be the combination of two independent velocities. Thus, the car traveling for t seconds can be thought of as traveling first at 0.80 m/s across the walkway for t seconds, then down the walkway at 1.5 m/s for t seconds. In this way, we can draw a given vector anywhere in the diagram as long as the vector is parallel to its previous alignment and still points in the same direction. Thus, you can draw one vector with its tail starting at the tip of the other as long as the size and direction of each vector do not change.

Determining a resultant vector by drawing a vector from the tail of the first vector to the tip of the last vector is known as the *triangle method of addition.*

Again, the magnitude of the resultant vector can be measured using a ruler, and the angle can be measured with a protractor. In the next section, we will develop a technique for adding vectors that is less time-consuming because it involves a calculator instead of a ruler and protractor. ⓣ TEKS **3B**

Vectors can be added in any order

When two or more vectors are added, the sum is independent of the order of the addition. This idea is demonstrated by a runner practicing for a marathon along city streets, as represented in **Figure 3-4.** In **(a)** the runner takes one path during a run, and in **(b)** the runner takes another. Regardless of which path the runner takes, the runner will have the same total displacement, expressed as **d.** Similarly, the vector sum of two or more vectors is the same regardless of the order in which the vectors are added, provided that the magnitude and direction of each vector remain the same. ⓣ TEKS **3B**

To subtract a vector, add its opposite

Vector subtraction makes use of the definition of the negative of a vector. The negative of a vector is defined as a vector with the same magnitude as the original vector but opposite in direction. For instance, the negative of the velocity of a car traveling 30 m/s to the west is −30 m/s to the west, or 30 m/s to the east. Thus, as shown below, adding a vector to its negative vector gives zero.

$$v + (-v) = 30 \text{ m/s} + (-30 \text{ m/s}) = 30 \text{ m/s} - 30 \text{ m/s} = 0 \text{ m/s}$$

When adding vectors in two dimensions, you can add a negative vector to a positive vector that does not point along the same line by using the triangle method of addition. ⭐TEKS **2C**

Multiplying or dividing vectors by scalars results in vectors

There are mathematical operations in which vectors can multiply other vectors, but they are not necessary for the scope of this book. This book does, however, make use of vectors multiplied by scalars, with a vector as the result. For example, if a cab driver obeys a customer who tells him to go twice as fast, that cab's original velocity vector, $\mathbf{v_{cab}}$, is multiplied by the scalar number 2. The result, written $2\mathbf{v_{cab}}$, is a vector with a magnitude twice that of the original vector and pointing in the same direction.

On the other hand, if another cab driver is told to go twice as fast in the opposite direction, this is the same as multiplying by the scalar number −2. The result is a vector with a magnitude two times the initial velocity but pointing in the opposite direction, written as $-2\mathbf{v_{cab}}$. ⭐TEKS **2C**

Section Review

1. Which of the following quantities are scalars, and which are vectors?
   a. the acceleration of a plane as it takes off
   b. the number of passengers on the plane
   c. the duration of the flight
   d. the displacement of the flight
   e. the amount of fuel required for the flight ⭐TEKS **2C, 4A**

2. A roller coaster moves 85 m horizontally, then travels 45 m at an angle of 30.0° above the horizontal. What is its displacement from its starting point? Use graphical techniques.

3. A novice pilot sets a plane's controls, thinking the plane will fly at 2.50×10^2 km/h to the north. If the wind blows at 75 km/h toward the southeast, what is the plane's resultant velocity? Use graphical techniques.

4. While flying over the Grand Canyon, the pilot slows the plane's engines down to one-half the velocity in item 3. If the wind's velocity is still 75 km/h toward the southeast, what will the plane's new resultant velocity be? Use graphical techniques.

5. **Physics in Action** The water used in many fountains is recycled. For instance, a single water particle in a fountain travels through an 85 m system and then returns to the same point. What is the displacement of a water particle during one cycle?

SECTION 3-1

Teaching Tip
Some students may need further explanation and visual examples of the negative of a vector. Make sure you always specify which direction is positive and which is negative. Students will benefit most if the same conventions are used consistently throughout the course.

Section Review
ANSWERS

1. a. vector
   b. scalar
   c. scalar
   d. vector
   e. scalar
2. 126 m at $(1.0 \times 10^1)°$ above the horizontal
3. 204 km/h at 75° north of east
4. 89 km/h at 54° north of east
5. zero

⭐TEKS

p. 86: 4A, 3B, 3B
p. 87: 2C, 2C, 2C, 4A

3-2
Vector operations

3-2 SECTION OBJECTIVES

- **Identify appropriate coordinate systems for solving problems with vectors.**

- **Apply the Pythagorean theorem and tangent function to calculate the magnitude and direction of a resultant vector.**

- **Resolve vectors into components using the sine and cosine functions.**

- **Add vectors that are not perpendicular.**

Figure 3-5

A gecko's displacement while climbing a tree can be represented by an arrow pointing along the y-axis.

COORDINATE SYSTEMS IN TWO DIMENSIONS

In Chapter 2, the motion of a gecko climbing a tree was described as motion along the *y*-axis. The direction of the displacement of the gecko along the axis was denoted by a positive or negative sign. The displacement of the gecko can now be described by an arrow pointing along the *y*-axis, as shown in **Figure 3-5.** A more versatile system for diagraming the motion of an object, however, employs vectors and the use of both the *x*- and *y*-axes simultaneously.

The addition of another axis not only helps describe motion in two dimensions but also simplifies analysis of motion in one dimension. For example, two methods can be used to describe the motion of a jet moving at 300 m/s to the northeast. In one approach, the coordinate system can be turned so that the plane is depicted as moving along the *y*-axis, as in **Figure 3-6(a).** The jet's motion also can be depicted on a two-dimensional coordinate system whose axes point south to north and west to east, as shown in **Figure 3-6(b).**

The problem with the orientation in the first case is that the axis must be turned again if the direction of the plane changes. Also, it will become difficult to describe the direction of another plane that is not traveling exactly northeast. Thus, axes are often designated with the positive *y*-axis pointing north and the positive *x*-axis pointing east, as shown in **Figure 3-6(b).**

Similarly, when analyzing the motion of objects thrown into the air, orienting the *y*-axis perpendicular to the ground, and therefore parallel to the direction of free-fall acceleration, greatly simplifies things. ⭐TEKS **4A**

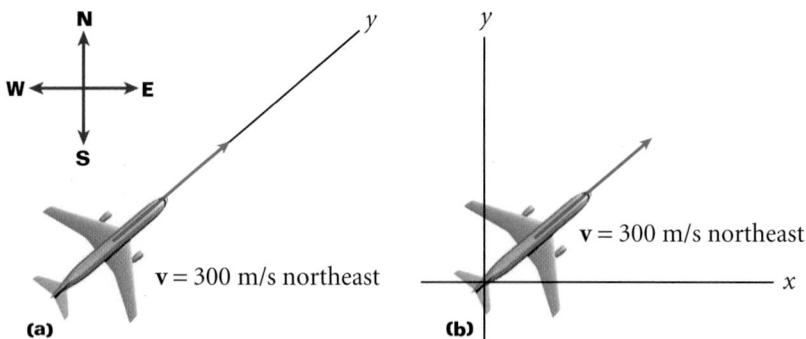

Figure 3-6

A plane traveling northeast at a velocity of 300 m/s can be represented as either **(a)** moving along a y-axis chosen to point to the northeast or **(b)** moving at an angle of 45° to both the x- and y-axes, which line up with west-east and north-south, respectively.

There are no firm rules for applying coordinate systems to situations involving vectors. As long as you are consistent, the final answer will be correct regardless of the system you choose. Perhaps the best choice for orienting axes is the approach that makes solving the problem easiest.

DETERMINING RESULTANT MAGNITUDE AND DIRECTION

In Section 3-1, the magnitude and direction of a resultant were found graphically by making a drawing. There are, however, drawbacks to this approach: it is time-consuming, and the accuracy of the answer depends on how carefully the diagram is drawn and measured. There is a better, simpler method using the Pythagorean theorem and the tangent function.

Use the Pythagorean theorem to find the magnitude of the resultant

Imagine a tourist climbing a pyramid in Egypt. The tourist knows the height and width of the pyramid and would like to know the distance covered in a climb from the bottom to the top of the pyramid.

As can be seen in **Figure 3-7,** the magnitude of the tourist's vertical displacement, Δy, is the height of the pyramid, and the magnitude of the horizontal displacement, Δx, equals the distance from one edge of the pyramid to the middle, or half the pyramid's width. Notice that these two vectors are perpendicular and form a right triangle with the displacement, **d.** **TEKS** **3B**

As shown in **Figure 3-8(a),** the Pythagorean theorem states that for any right triangle, the square of the hypotenuse—the side opposite the right angle—equals the sum of the squares of the other two sides, or legs.

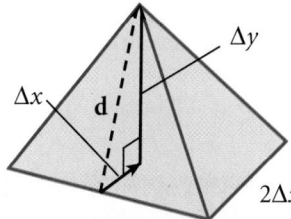

Figure 3-7
Because the base and height of a pyramid are perpendicular, we can find a tourist's total displacement, **d,** if we know the height, Δy, and width, $2\Delta x$, of the pyramid.

> **PYTHAGOREAN THEOREM FOR RIGHT TRIANGLES**
>
> $$c^2 = a^2 + b^2$$
>
> **(length of hypotenuse)^2 = (length of one leg)^2 + (length of other leg)^2**

In **Figure 3-8(b),** the Pythagorean theorem is applied to find the tourist's displacement. The square of the displacement is equal to the sum of the square of the horizontal displacement and the square of the vertical displacement. In this way, you can find out the magnitude of the displacement, d.

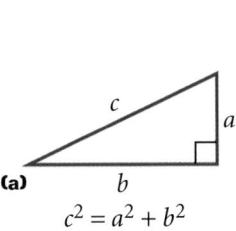

(a)
$$c^2 = a^2 + b^2$$

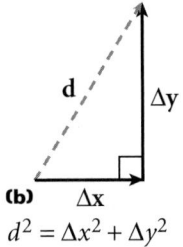

(b)
$$d^2 = \Delta x^2 + \Delta y^2$$

Figure 3-8
(a) The Pythagorean theorem can be applied to any right triangle. **(b)** It can also be applied to find the magnitude of a resultant displacement.

TEKS

p. 88: 4A
p. 89: 3B

> ⬛ **Misconception Alert**
>
> Students often try to apply the Pythagorean theorem to triangles that do not contain a right angle. Point out that the Pythagorean theorem can be used only when applied to a right triangle.

Teaching Tip
Point out that finding the resultant for the pyramid is fairly simple because the height, width, and hypotenuse form a right triangle. You may find it worthwhile to mention at this point that right triangles will also allow students to find the x and y components that are important for vector addition. Some students may know the Law of Cosines, which applies to all triangles, including right triangles. This equation and other trigonometry functions can be found in Appendix A.

The following may be used as teamwork exercises or for demonstration at the chalkboard or on an overhead projector.

PROBLEM

Finding resultant magnitude and direction

A plane travels from Houston, Texas, to Washington, D.C., which is 1540 km east and 1160 km north of Houston. What is the total displacement of the plane?

Answer

1930 km at 37.0° north of east

A camper travels 4.5 km northeast and 4.5 km northwest. What is the camper's total displacement?

Answer

6.4 km north

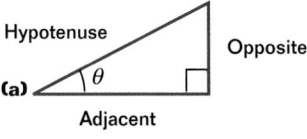

$$\tan \theta = \frac{\text{opp}}{\text{adj}}$$

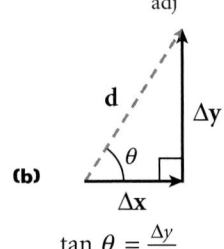

$$\tan \theta = \frac{\Delta y}{\Delta x}$$

$$\theta = \tan^{-1}\left(\frac{\Delta y}{\Delta x}\right)$$

Figure 3-9
(a) The tangent function can be applied to any right triangle, and **(b)** it can also be used to find the direction of a resultant displacement.

Use the tangent function to find the direction of the resultant

In order to completely describe the tourist's displacement, you must also know the direction of the tourist's motion. Because $\Delta\mathbf{x}$, $\Delta\mathbf{y}$, and **d** form a right triangle, as shown in **Figure 3-9(b),** the tangent function can be used to find the angle θ, which denotes the direction of the tourist's displacement.

For any right triangle, the tangent of an angle is defined as the ratio of the opposite and adjacent legs with respect to a specified acute angle of a right triangle, as shown in **Figure 3-9(a).**

As shown below, the quantity of the opposite leg divided by the magnitude of the adjacent leg equals the tangent of the angle. (★)TEKS **3B**

DEFINITION OF THE TANGENT FUNCTION FOR RIGHT TRIANGLES

$$\tan \theta = \frac{\text{opp}}{\text{adj}} \qquad\qquad \text{tangent of angle} = \frac{\text{opposite leg}}{\text{adjacent leg}}$$

The inverse of the tangent function, which is shown below, indicates the angle.

$$\theta = \tan^{-1}\left(\frac{\text{opp}}{\text{adj}}\right)$$

SAMPLE PROBLEM 3A

Finding resultant magnitude and direction

PROBLEM

An archaeologist climbs the Great Pyramid in Giza, Egypt. If the pyramid's height is 136 m and its width is 2.30×10^2 m, what is the magnitude and the direction of the archaeologist's displacement while climbing from the bottom of the pyramid to the top?

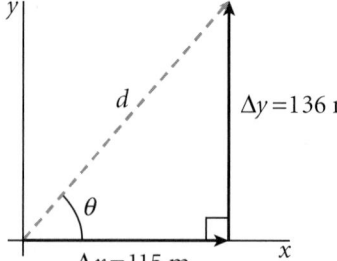

SOLUTION

1. DEFINE

Given: $\Delta y = 136 \text{ m}$ $\Delta x = \frac{1}{2}(\text{width}) = 115 \text{ m}$

Unknown: $d = ?$ $\theta = ?$

Diagram: Choose the archaeologist's starting position as the origin of the coordinate system.

2. PLAN

Choose an equation(s) or situation: The Pythagorean theorem can be used to find the magnitude of the archaeologist's displacement. The direction of the displacement can be found by using the tangent function. (★)TEKS **2C, 3B**

$$d^2 = \Delta x^2 + \Delta y^2$$

$$\tan \theta = \frac{\Delta y}{\Delta x}$$

Rearrange the equation(s) to isolate the unknown(s):

$$d = \sqrt{\Delta x^2 + \Delta y^2}$$

$$\theta = \tan^{-1}\left(\frac{\Delta y}{\Delta x}\right)$$

3. CALCULATE **Substitute the values into the equation(s) and solve:**

$$d = \sqrt{(115 \text{ m})^2 + (136 \text{ m})^2}$$

$$\boxed{d = 178 \text{ m}}$$

$$\theta = \tan^{-1}\left(\frac{136 \text{ m}}{115 \text{ m}}\right)$$

$$\boxed{\theta = 49.8°}$$

CALCULATOR SOLUTION

Be sure your calculator is set to calculate angles measured in degrees. Most calculators have a button labeled "DRG" that, when pressed, toggles between degrees, radians, and grads.

4. EVALUATE Because *d* is the hypotenuse, the archaeologist's displacement should be less than the sum of the height and half of the width. The angle is expected to be more than 45° because the height is greater than half of the width.

PRACTICE 3A

Finding resultant magnitude and direction

1. A truck driver attempting to deliver some furniture travels 8 km east, turns around and travels 3 km west, and then travels 12 km east to his destination.

    a. What distance has the driver traveled?

    b. What is the driver's total displacement?

2. While following the directions on a treasure map, a pirate walks 45.0 m north, then turns and walks 7.5 m east. What single straight-line displacement could the pirate have taken to reach the treasure?

3. Emily passes a soccer ball 6.0 m directly across the field to Kara, who then kicks the ball 14.5 m directly down the field to Luisa. What is the ball's total displacement as it travels between Emily and Luisa?

4. A hummingbird flies 1.2 m along a straight path at a height of 3.4 m above the ground. Upon spotting a flower below, the hummingbird drops directly downward 1.4 m to hover in front of the flower. What is the hummingbird's total displacement? ⭐TEKS **2C, 3B**

Alternative Problem-Solving Approach

The angle may be calculated using any trigonometric function, such as the following.

$$\theta = \sin^{-1}\left(\frac{\Delta y}{d}\right) = 49.8°$$

PRACTICE GUIDE 3A

Solving for:		
re-sultant	📖 **PE**	Sample, 1–4; Ch. Rvw. 22–24, 25*
	💿 **PW**	2, 4–5, 7*
	PB	Sample, 1–5
com-ponent	💿 **PW**	Sample, 1, 3*, 6*
	PB	6–10

ANSWERS TO

Practice 3A
Finding resultant magnitude and direction

1.   **a.** 23 km
     b. 17 km to the east
2. 45.6 m at 9.5° east of north
3. 15.7 m at 22° to the side of downfield
4. 1.8 m at 49° below the horizontal

⭐TEKS

p. 90: 3B, 2C, 3B
p. 91: 2C, 3B

components of a vector

the projections of a vector along the axes of a coordinate system

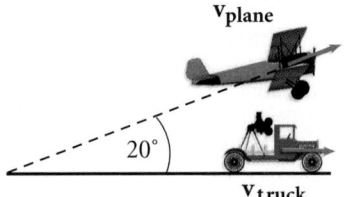

Figure 3-10
A truck carrying a film crew must be driven at the correct velocity to enable the crew to film the underside of a biplane flying at an angle of 20° to the ground at a speed of 95 km/h.

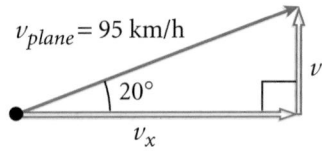

Figure 3-11
To stay beneath the biplane, the truck must be driven with a velocity equal to the *x* component (v_x) of the biplane's velocity.

RESOLVING VECTORS INTO COMPONENTS

In the pyramid example, the horizontal and vertical parts that add up to give the tourist's actual displacement are called **components.** The *x* component is parallel to the *x*-axis. The *y* component is parallel to the *y*-axis. These components can be either positive or negative numbers with units.

Any vector can be completely described by a set of perpendicular components. When a vector points along a single axis, as do the quantities in Chapter 2, the second component of the vector is equal to zero.

By breaking a single vector into two components, or *resolving* it into its components, an object's motion can sometimes be described more conveniently in terms of directions, such as north to south or east to west.

To illustrate this point, let's examine a scene on the set of a new action movie. For this scene a biplane travels at 95 km/h at an angle of 20° relative to the ground. Attempting to film the plane from below, a camera team travels in a truck, keeping the truck beneath the plane at all times, as shown in **Figure 3-10.** How fast must the truck travel to remain directly below the plane?

To find out the velocity that the truck must maintain to stay beneath the plane, we must know the horizontal component of the plane's velocity. Once more, the key to solving the problem is to recognize that a right triangle can be drawn using the plane's velocity and its *x* and *y* components. The situation can then be analyzed using trigonometry.

The sine and cosine functions are defined in terms of the lengths of the sides of such right triangles. The sine of an angle is the ratio of the leg opposite that angle to the hypotenuse. ★ TEKS **3B**

DEFINITION OF THE SINE FUNCTION FOR RIGHT TRIANGLES

$$\sin \theta = \frac{\text{opp}}{\text{hyp}} \qquad \text{sine of an angle} = \frac{\text{opposite leg}}{\text{hypotenuse}}$$

In **Figure 3-11,** the leg opposite the 20° angle represents the *y* component, v_y, which describes the vertical speed of the airplane. The hypotenuse, **v**$_{plane}$, is the resultant vector that describes the airplane's total motion.

The cosine of an angle is the ratio between the leg adjacent to that angle and the hypotenuse. ★ TEKS **3B**

DEFINITION OF THE COSINE FUNCTION FOR RIGHT TRIANGLES

$$\cos \theta = \frac{\text{adj}}{\text{hyp}} \qquad \text{cosine of an angle} = \frac{\text{adjacent leg}}{\text{hypotenuse}}$$

In **Figure 3-11,** the leg adjacent to the 20° angle represents the *x* component, v_x, which describes the horizontal speed of the airplane. This *x* component equals the speed that the truck must maintain to stay beneath the plane.

Resolving vectors

PROBLEM

Find the component velocities of a helicopter traveling 95 km/h at an angle of 35° to the ground. ⭐TEKS **2C, 3B, 4A**

SOLUTION

1. DEFINE

Given: $v = 95$ km/h $\theta = 35°$

Unknown: $v_x = ?$ $v_y = ?$

Diagram: The most convenient coordinate system is one with the x-axis directed along the ground and the y-axis directed vertically.

2. PLAN

Choose an equation(s) or situation: Because the axes are perpendicular, the sine and cosine functions can be used to find the components.

$$\sin \theta = \frac{v_y}{v}$$

$$\cos \theta = \frac{v_x}{v}$$

Rearrange the equation(s) to isolate the unknown(s):

$$v_y = v(\sin \theta)$$
$$v_x = v(\cos \theta)$$

3. CALCULATE

Substitute the values into the equation(s) and solve:

$$v_y = (95 \text{ km/h})(\sin 35°)$$

$$\boxed{v_y = 54 \text{ km/h}}$$

$$v_x = (95 \text{ km/h})(\cos 35°)$$

$$\boxed{v_x = 78 \text{ km/h}}$$

4. EVALUATE

Because the component velocities form a right triangle with the helicopter's actual velocity, the Pythagorean theorem can be used to check whether the magnitudes of the components are correct.

$$v^2 = v_x^2 + v_y^2$$
$$(95)^2 \approx (78)^2 + (54)^2$$
$$9025 \approx 9000$$

The slight difference is due to rounding.

CALCULATOR SOLUTION

When using your calculator to solve a problem, perform trigonometric functions such as *sin, cos,* and *tan* first, before multiplication. This approach will help you maintain the proper number of significant digits.

Classroom Practice

The following may be used as teamwork exercises or for demonstration at the chalkboard or on an overhead projector.

PROBLEM

Resolving vectors

An arrow is shot from a bow at an angle of 25° above the horizontal with an initial speed of 45 m/s. Find the horizontal and vertical components of the arrow's initial velocity.

Answer
   41 m/s, 19 m/s

The arrow strikes the target with a speed of 45 m/s at an angle of −25° with respect to the horizontal. Calculate the horizontal and vertical components of the arrow's final velocity.

Answer
   41 m/s, −19 m/s

⭐TEKS
   p. 92: 3B, 3B, 4A
   p. 93: 2C, 3B, 4A

ANSWERS TO

Practice 3B
Resolving vectors

1. 95 km/h
2. 44 km/h
3. 21 m/s, 5.7 m/s
4. 0 m, 5 m
5. 1.1×10^2 m, −53 m
6. 19.8 m, −11.7 m
7. 2.4 m/s^2, −0.77 m/s^2

Teaching Tip

Problems involving vectors that are not perpendicular use both vector addition and vector resolution. Because they act as a nice summary to the concepts of this section, you may want to do several examples involving this type of problem. These problems require a methodical approach to problem-solving, which should prove helpful to students while studying more-difficult subjects such as inclined-plane problems (Chapter 4) and equilibrium problems (Chapter 8).

Resolving vectors ⭐TEKS 2C, 3B, 4A

1. How fast must a truck travel to stay beneath an airplane that is moving 105 km/h at an angle of 25° to the ground?

2. What is the magnitude of the vertical component of the velocity of the plane in item 1?

3. A truck drives up a hill with a 15° incline. If the truck has a constant speed of 22 m/s, what are the horizontal and vertical components of the truck's velocity?

4. What are the horizontal and vertical components of a cat's displacement when it has climbed 5 m directly up a tree?

5. Find the horizontal and vertical components of the 125 m displacement of a superhero who flies down from the top of a tall building at an angle of 25° below the horizontal.

6. A child rides a toboggan down a hill that descends at an angle of 30.5° to the horizontal. If the hill is 23.0 m long, what are the horizontal and vertical components of the child's displacement?

7. A skier squats low and races down an 18° ski slope. During a 5 s interval, the skier accelerates at 2.5 m/s^2. What are the horizontal (perpendicular to the direction of free-fall acceleration) and vertical components of the skier's acceleration during this time interval?

ADDING VECTORS THAT ARE NOT PERPENDICULAR

Until this point, the vector-addition problems concerned vectors that are perpendicular to one another. However, many objects move in one direction, and then turn at an acute angle before continuing their motion.

Suppose that a plane initially travels 50 km at an angle of 35° to the ground, then climbs at only 10° to the ground for 220 km. How can you determine the magnitude and direction for the vector denoting the total displacement of the plane?

Because the original displacement vectors do not form a right triangle, it is not possible to directly apply the tangent function or the Pythagorean theorem when adding the original two vectors.

Determining the magnitude and the direction of the resultant can be achieved by resolving each of the plane's displacement vectors into their x and

y components. Then the components along each axis can be added together. As shown in **Figure 3-12,** these vector sums will be the two perpendicular components of the resultant, **d.** The magnitude of the resultant can then be found using the Pythagorean theorem, and its direction can be found using the tangent function. ⊛TEKS **3B**

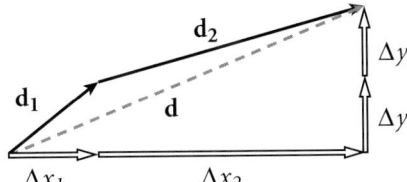

Figure 3-12
Add the components of the original displacement vectors to find two components that form a right triangle with the resultant vector.

Module 2 **"Vector Addition and Resolution"** provides an interactive lesson with guided problem-solving practice to teach you how to add different vectors, especially those that are not at right angles.

Visual Strategy

Figure 3-12
Point out that each type of vector is represented by a different type of arrow: displacement arrows are solid black, displacement components are outlined in black, and the resultant is dashed.

Q What does the dashed line, labeled **d,** represent?

A *the resultant of the two displacements*

SAMPLE PROBLEM 3C

Adding vectors algebraically

PROBLEM

A hiker walks 25.5 km from her base camp at 35° south of east. On the second day, she walks 41.0 km in a direction 65° north of east, at which point she discovers a forest ranger's tower. Determine the magnitude and direction of her resultant displacement between the base camp and the ranger's tower. ⊛TEKS **2C, 3B, 4A**

SOLUTION

1. **Select a coordinate system, draw a sketch of the vectors to be added, and label each vector.**

   Figure 3-13 depicts the situation drawn on a coordinate system. The positive y-axis points north and the positive x-axis points east. The origin of the axes is the base camp. In the chosen coordinate system, the hiker's direction θ_1 during the first day is signified by a negative angle because clockwise movement from the positive x-axis is conventionally considered to be a negative angle.

   Given: $\theta_1 = -35°$ $\theta_2 = 65°$ $d_1 = 25.5$ km $d_2 = 41.0$ km

   Unknown: $d = ?$ $\theta = ?$

2. **Find the x and y components of all vectors.**

   Make a separate sketch of the displacements for each day. The values for each of the displacement components can be determined by using the sine and cosine functions. Because the hiker's angle on the first day is negative, the y component of her displacement during that day is negative.

   $$\sin \theta = \frac{\Delta y}{d}$$

   $$\cos \theta = \frac{\Delta x}{d}$$

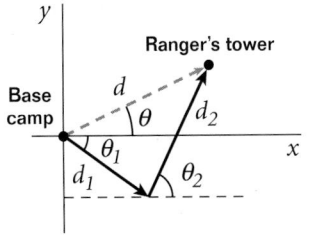

Figure 3-13

Classroom Practice

The following may be used as teamwork exercises or for demonstration at the chalkboard or on an overhead projector.

PROBLEM

Adding vectors algebraically

A camper walks 4.5 km northeast then 4.5 km south. Find the camper's total displacement (magnitude and direction).

Answer
3.5 km at 22° south of east

A plane flies 118 km at 15.0° south of east, then flies 118 km at 35.0° west of north. Find the magnitude and direction of the total displacement of the plane.

Answer
81 km at 55° north of east

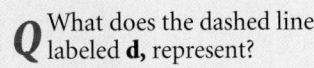

⊛TEKS

p. 94: 2C, 3B, 4A
p. 95: 3B, 2C, 3B, 4A

continued on
next page

Alternative Problem-Solving Approach

Any of the trigonometric functions may be used to solve for the angle.

$$\theta = \sin^{-1}\left(\frac{\Delta y}{d}\right) = (3.0 \times 10^1)°$$

$$\theta = \cos^{-1}\left(\frac{\Delta x}{d}\right) = (3.0 \times 10^1)°$$

You may want to review how each of these or other trigonometric functions can be used to find the answer.

Interactive Problem-Solving Tutor

See Module 2

"Vector Addition and Resolution" provides additional development of problem-solving skills for this chapter.

For day 1:
(Figure 3-14)

$$\Delta x_1 = d_1 (\cos \theta_1) = (25.5 \text{ km}) [\cos (-35°)]$$

$$\Delta x_1 = 21 \text{ km}$$

$$\Delta y_1 = d_1 (\sin \theta_1) = (25.5 \text{ km}) [\sin (-35°)]$$

$$\Delta y_1 = -15 \text{ km}$$

For day 2:
(Figure 3-15)

$$\Delta x_2 = d_2 (\cos \theta_2) = (41.0 \text{ km}) (\cos 65°)$$

$$\Delta x_2 = 17 \text{ km}$$

$$\Delta y_2 = d_2 (\sin \theta_2) = (41.0 \text{ km}) (\sin 65°)$$

$$\Delta y_2 = 37 \text{ km}$$

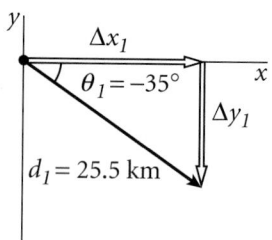

Figure 3-14

3. **Find the x and y components of the total displacement.**

First add together the x components to find the total displacement in the x direction. Then perform the same operation for the y direction.

$$\Delta x_{tot} = \Delta x_1 + \Delta x_2 = 21 \text{ km} + 17 \text{ km} = 38 \text{ km}$$

$$\Delta y_{tot} = \Delta y_1 + \Delta y_2 = -15 \text{ km} + 37 \text{ km} = 22 \text{ km}$$

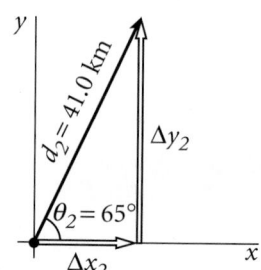

Figure 3-15

4. **Use the Pythagorean theorem to find the magnitude of the resultant vector.**

Because the components Δx_{tot} and Δy_{tot} are perpendicular, the Pythagorean theorem can be used to find the magnitude of the resultant vector.

$$d^2 = (\Delta x_{tot})^2 + (\Delta y_{tot})^2$$

$$d = \sqrt{(\Delta x_{tot})^2 + (\Delta y_{tot})^2} = \sqrt{(38 \text{ km})^2 + (22 \text{ km})^2}$$

$$\boxed{d = 44 \text{ km}}$$

5. **Use a suitable trigonometric function to find the angle the resultant vector makes with the x-axis.**

The direction of the resultant can be found using the tangent function.

$$\theta = \tan^{-1}\left(\frac{\Delta y}{\Delta x}\right)$$

$$\theta = \tan^{-1}\left(\frac{\Delta y_{tot}}{\Delta x_{tot}}\right) = \tan^{-1}\left(\frac{22 \text{ km}}{38 \text{ km}}\right)$$

$$\boxed{\theta = (3.0 \times 10^1)° \text{ north of east}}$$

6. **Evaluate your answer.**

If the diagram is drawn to scale, compare the algebraic results with the drawing. The calculated magnitude seems reasonable because the distance from the base camp to the ranger's tower is longer than the distance hiked during the first day and slightly longer than the distance hiked during the second day. The calculated direction of the resultant seems reasonable because the angle in **Figure 3-13** looks to be about 30°.

Adding vectors algebraically ⭐TEKS 2C, 3B, 4A

1. A football player runs directly down the field for 35 m before turning to the right at an angle of 25° from his original direction and running an additional 15 m before getting tackled. What is the magnitude and direction of the runner's total displacement?

2. A plane travels 2.5 km at an angle of 35° to the ground, then changes direction and travels 5.2 km at an angle of 22° to the ground. What is the magnitude and direction of the plane's total displacement?

3. During a rodeo, a clown runs 8.0 m north, turns 35° east of north, and runs 3.5 m. Then, after waiting for the bull to come near, the clown turns due east and runs 5.0 m to exit the arena. What is the clown's total displacement?

4. An airplane flying parallel to the ground undergoes two consecutive displacements. The first is 75 km 30.0° west of north, and the second is 155 km 60.0° east of north. What is the total displacement of the airplane?

Section Review

1. Identify a convenient coordinate system for analyzing each of the following situations:
   a. a dog walking along a sidewalk
   b. an acrobat walking along a high wire
   c. a submarine submerging at an angle of 30° to the horizontal

2. Find the magnitude and direction of the resultant velocity vector for the following perpendicular velocities:
   a. a fish swimming at 3.0 m/s relative to the water across a river that moves at 5.0 m/s
   b. a surfer traveling at 1.0 m/s relative to the water across a wave that is traveling at 6.0 m/s

3. Find the component vectors along the directions noted in parentheses.
   a. a car displaced northeast by 10.0 km (north and east)
   b. a duck accelerating away from a hunter at 2.0 m/s² at an angle of 35° to the ground (horizontal and vertical)

4. Find the resultant displacement of a fox that heads 55° north of west for 10.0 m, then turns and heads west for 5.0 m.

ANSWERS TO

Practice 3C
Adding vectors algebraically

1. 49 m at 7.3° to the right of down field
2. 7.5 km at 26° above the horizontal
3. 13.0 m at 33° east of north
4. 171 km at 34° east of north

Section Review
ANSWERS

1. a. x-axis: forward and backward on sidewalk
      y-axis: left and right on sidewalk
   b. x-axis: forward and backward on rope
      y-axis: up and down
   c. x-axis: horizontal at water level
      y-axis: up and down

2. a. 5.8 m/s at 59° downriver from its intended path
   b. 6.1 m/s at 9.5° from the direction the wave is traveling

3. a. 7.07 km north, 7.07 km east
   b. 1.6 m/s² horizontal, 1.1 m/s² vertical

4. 13.5 m at 37° north of west

⭐TEKS
p. 97: 2C, 3B, 4A

3-3
Projectile motion

Teaching Tip

On the chalkboard, show examples of vector components and the kinematic equations. Show the simplification of the *x*-direction equations when the *x* component of acceleration is zero.

Visual Strategy

Figure 3-17

Tell students that the long jumper builds up speed in the *x* direction and jumps, so there is also a component of speed in the *y* direction.

Q Does the angle of take-off matter to the jumper? Hint: Consider the difference between a very small angle (near 0°) and a very large angle (near 90°).

A *The angle matters because it affects how long the jumper stays off the ground and how far he goes while in the air.*

3-3 SECTION OBJECTIVES

- **Recognize examples of projectile motion.**

- **Describe the path of a projectile as a parabola.**

- **Resolve vectors into their components and apply the kinematic equations to solve problems involving projectile motion.**

Figure 3-16

When the long jumper is in the air, his velocity has both a horizontal and a vertical component.

Figure 3-17

(a) A long jumper's velocity while sprinting along the runway can be represented by a single vector.
(b) Once the jumper is airborne, the jumper's velocity at any instant can be described by the components of the velocity.

TWO-DIMENSIONAL MOTION

In the last section, quantities such as displacement and velocity were shown to be vectors that can be resolved into components. In this section, these components will be used to understand and predict the motion of objects thrown into the air.

Use of components avoids vector multiplication

How can you know the displacement, velocity, and acceleration of a ball at any point in time during its flight? All of the kinematic equations from Chapter 2 could be rewritten in terms of vector quantities. However, when an object is propelled into the air in a direction other than straight up or down, the velocity, acceleration, and displacement of the object do not all point in the same direction. This makes the vector forms of the equations difficult to solve.

One way to deal with these situations is to avoid using the complicated vector forms of the equations altogether. Instead, apply the technique of resolving vectors into components. Then you can apply the simpler one-dimensional forms of the equations for each component. Finally, you can recombine the components to determine the resultant. ⭐TEKS **4B**

Components simplify projectile motion

When a long jumper approaches his jump, he runs along a straight line, which can be called the *x*-axis. When he jumps, as shown in **Figure 3-16,** his velocity has both horizontal and vertical components. Movement in this plane can be depicted using both the *x*- and *y*-axes.

Note that in **Figure 3-17(b)** the jumper's velocity vector is resolved into its two component vectors. This way, the jumper's motion can be analyzed using the kinematic equations applied to one direction at a time. ⭐TEKS **4A**

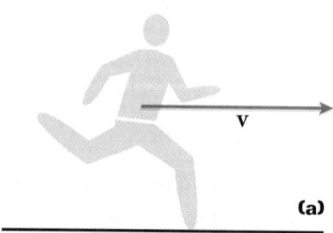

$\mathbf{v}$

(a)

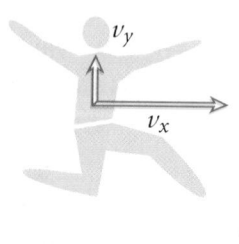

v_y

v_x

(b)

In this section, we will focus on the form of two-dimensional motion called **projectile motion.** Objects that are thrown or launched into the air and are subject to gravity are called *projectiles.* Some examples of projectiles are softballs, footballs, and arrows when they are projected through the air. Even a long jumper can be considered a projectile. (★)TEKS **4B**

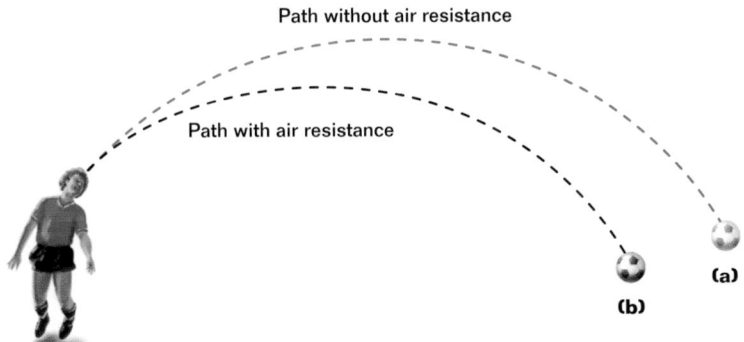

Path without air resistance

Path with air resistance

(a)

(b)

Projectiles follow parabolic trajectories

The path of a projectile is a curve called a parabola, as shown in **Figure 3-18(a).** Many people mistakenly believe that projectiles eventually fall straight down, much like a cartoon character does after running off a cliff. However, if an object has an initial horizontal velocity in any given time interval, there will be horizontal motion throughout the flight of the projectile. Note that for the purposes of samples and exercises in this book, the horizontal velocity of the projectile will be considered constant. This velocity would not be constant if we accounted for air resistance. With air resistance, a projectile slows down as it collides with air particles. Hence, as shown in **Figure 3-18(b),** the true path of a projectile traveling through Earth's atmosphere is not a parabola.

Projectile motion is free fall with an initial horizontal velocity

To understand the motion a projectile undergoes, first examine **Figure 3-19.** The red ball was dropped at the same instant the yellow ball was launched horizontally. If air resistance is disregarded, both balls hit the ground at the same time.

By examining each ball's position in relation to the horizontal lines and to one another, we see that the two balls fall at the same rate. This may seem impossible because one is given an initial velocity and the other begins from rest. But if the motion is analyzed one component at a time, it makes sense.

First, consider the red ball that falls straight down. It has no motion in the horizontal direction. In the vertical direction, it starts from rest ($v_{y,i} = 0$ m/s) and proceeds in free fall. Thus, the kinematic equations from Chapter 2 can be applied to analyze the vertical motion of the falling ball. Note that the acceleration, a, can be rewritten as $-g$ because the only vertical component of acceleration is free-fall acceleration. Note also that Δy is negative. (★)TEKS **4B**

projectile motion
_____

free-fall with an initial horizontal velocity

Figure 3-18
(a) Without air resistance, a soccer ball being headed into the air would be represented as traveling along a parabola. **(b)** With air resistance, the soccer ball would travel along a shorter path, which would not be a parabola.

internet**connect**

SC*i*LINKS.
NSTA
TOPIC: Projectile motion
GO TO: www.scilinks.org
*sci*LINKS CODE: HF2032

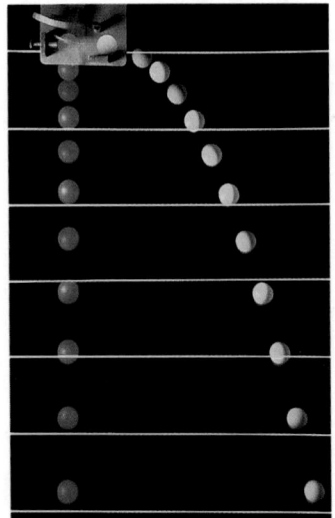

Figure 3-19
This is a strobe photograph of two table-tennis balls released at the same time. Even though the yellow ball is given an initial horizontal velocity and the red ball is simply dropped, both balls fall at the same rate.

SECTION 3-3

Demonstration 2

Air resistance

Purpose Show the effects of air resistance on the flight of a projectile.
Materials rubber stopper, table-tennis ball
Procedure Toss the stopper at an angle of 45°. Have students note the trajectory. Sketch the parabolic path on the chalkboard. Throw the table-tennis ball at an angle of 45°. Have students note the trajectory. Sketch its path on the chalkboard. Have students compare the two trajectories. Students should note that the path of the table-tennis ball is not symmetrical but has a steeper descent, demonstrating the effect of air resistance.

Teaching Tip

Review the definition of a parabola. A parabola is a curve in which every point is the same distance from a fixed point, called the *focus,* as it is from a fixed line, called the *directrix.* Illustrate this on the chalkboard for clarity.

(★)TEKS

p. 98: 4B, 4A
p. 99: 4B, 4B

Demonstration 3

Two-dimensional motion

Purpose Demonstrate that projectiles have a vertical acceleration equal to free-fall acceleration.

Materials two quarters

Procedure Place one quarter so that it extends halfway over the edge of a table. Tell students that you will flick the second quarter across the tabletop so that it will graze the first quarter. Explain that the first quarter will fly off the table horizontally while the second quarter will fall almost vertically. Ask them to predict which quarter will strike the ground first. Have students gather around the desk so that they can view the falling coins. Have them remain quiet so that they can hear the coins striking the floor. Repeat the demonstration several times. Ask students for their conclusions about the time it takes each coin to fall.

Quick Lab

TEACHER'S NOTES

Dropping the second ball as the first leaves the table is tricky. Students should try this several times in order to get the timing right. Holding the second ball just past the edge of the table and near the path of the first ball works well. You may have them try a glancing collision, as in Demonstration 3. You should point out the limitations of this Quick Lab because of human reaction time.

Did you know?

The greatest distance a regulation-size baseball has ever been thrown is 135.9 m, by Glen Gorbous in 1957.

VERTICAL MOTION OF A PROJECTILE THAT FALLS FROM REST

$$v_{y,f} = -g\Delta t$$

$$v_{y,f}^2 = -2g\Delta y$$

$$\Delta y = -\frac{1}{2}g(\Delta t)^2$$

Now consider the components of motion of the yellow ball that is launched in **Figure 3-19.** This ball undergoes the same horizontal displacement during each time interval. This means that the ball's horizontal velocity remains constant (if air resistance is assumed to be negligible). Thus, when using the kinematic equations from Chapter 2 to analyze the horizontal motion of a projectile, the initial horizontal velocity is equal to the horizontal velocity throughout the projectile's flight. A projectile's horizontal motion is described by the following equation. (★)TEKS **4B**

HORIZONTAL MOTION OF A PROJECTILE

$$v_x = v_{x,i} = \text{constant}$$

$$\Delta x = v_x \Delta t$$

Next consider the initial motion of the launched yellow ball in **Figure 3-19.** Despite having an initial horizontal velocity, the launched ball has no initial velocity in the vertical direction. Just like the red ball that falls straight down, the launched yellow ball is in free fall. Its vertical motion is described by the same free-fall equations. In any time interval, the launched ball undergoes the same vertical displacement as the ball that falls straight down. This is why both balls reach the ground at the same time.

To find the velocity of a projectile at any point during its flight, find the vector sum of the components of the velocity at that point. Use the Pythagorean theorem to find the magnitude of the velocity, and use the tangent function to find the direction of the velocity. (★)TEKS **4B**

Quick Lab

Projectile Motion

MATERIALS LIST

✔ 2 identical balls

✔ slope or ramp

Roll a ball off a table. At the instant the rolling ball leaves the table, drop a second ball from the same height above the floor. Do the two balls hit the floor at the same time? Try varying the speed at which you roll the first ball off the table. Does varying the speed affect whether the two balls strike the ground at the same time? Next roll one of the balls down a slope. Drop the other ball from the base of the slope at the instant the first ball leaves the slope. Which of the balls hits the ground first in this situation?

(★)TEKS **1A**

Projectiles launched horizontally

PROBLEM

The Royal Gorge Bridge in Colorado rises 321 m above the Arkansas River. Suppose you kick a little rock horizontally off the bridge. The rock hits the water such that the magnitude of its horizontal displacement is 45.0 m. Find the speed at which the rock was kicked. ⊛TEKS 2C, 4A, 4B

SOLUTION

1. DEFINE

Given: $\Delta y = -321$ m $\qquad \Delta x = 45.0$ m $\qquad a_y = g = 9.81$ m/s^2

Unknown: $v_i = ?$

Diagram: The initial velocity vector of the rock has only a horizontal component. Choose the coordinate system oriented so that the positive y direction points upward and the positive x direction points to the right.

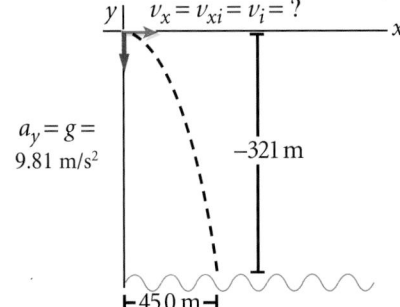

2. PLAN

Choose the equation(s) or situation:
Because air resistance can be neglected, the rock's horizontal velocity remains constant.

$$\Delta x = v_x \Delta t$$

Because there is no initial vertical velocity, the following equation applies.

$$\Delta y = -\tfrac{1}{2}g(\Delta t)^2$$

Rearrange the equation(s) to isolate the unknown(s):
Note that the time interval is the same for the vertical and horizontal displacements, so the second equation can be rearranged to solve for Δt.

$$\Delta t = \sqrt{\frac{2\Delta y}{-g}} \text{ where } \Delta y \text{ is negative}$$

$$v_x = \frac{\Delta x}{\Delta t} = \left(\sqrt{\frac{-g}{2\Delta y}}\right)\Delta x \; = \; V_x$$

3. CALCULATE

Substitute the values into the equation(s) and solve:
The value for v_x can be either positive or negative because of the square root. Because the direction was not asked for, use the positive root for v.

$$v_x = \sqrt{\frac{-9.81 \text{ m/s}^2}{(2)(-321 \text{ m})}}(45.0 \text{ m}) = \boxed{5.56 \text{ m/s}}$$

4. EVALUATE

To check your work, estimate the value of the time interval for Δx and solve for Δy. If v_x is about 5.5 m/s and $\Delta x = 45$ m, $\Delta t \approx 8$ s. If you use an approximate value of 10 m/s^2 for g, $\Delta y \approx -320$ m, almost identical to the given value.

Classroom Practice

The following may be used as teamwork exercises or for demonstration at the chalkboard or on an overhead projector.

PROBLEM

Projectiles launched horizontally

People in movies often jump from buildings into pools. If a person jumps from the 10th floor (30.0 m) to a pool that is 5.0 m away from the building, with what initial horizontal velocity must the person jump?

Answer
  2.0 m/s

PRACTICE GUIDE 3D		
Solving for:		
v_x	📖 PE	Sample, 1–3; Ch. Rvw. 34, 36, 58a
	💿 PW	6, 7*, 8*
	PB	9
Δx	📖 PE	4; Ch. Rvw. 37
	💿 PW	Sample, 1–2
	PB	8, 10
Δy	📖 PE	Ch. Rvw. 58b
	💿 PW	3–4, 5*
	PB	Sample, 1–7

⊛TEKS

p. 100: 4B, 4B, 1A
p. 101: 2C, 4A, 4B

Visual Strategy

Figure 3-20
Point out to students that from
the time immediately after firing
until it hits the ground, the projectile follows a parabolic path.

Q What is the acceleration of a
projectile just before it hits
the ground?

A *9.81 m/s², the same as at any
time during the flight*

Teaching Tip

Students often memorize equations without taking the time to
understand them. You may want
to carefully derive each of the
equations shown on this page
using the kinematic equations
from Chapter 2. Doing this will
also solidify students' understanding of projectile motion.

102

PRACTICE 3D

Projectiles launched horizontally ⭐TEKS 2C, 4A, 4B

1. An autographed baseball rolls off of a 0.70 m high desk and strikes the
    floor 0.25 m away from the base of the desk. How fast was it rolling?

2. A cat chases a mouse across a 1.0 m high table. The mouse steps out of
    the way, and the cat slides off the table and strikes the floor 2.2 m from
    the edge of the table. What was the cat's speed when it slid off the table?

3. A pelican flying along a horizontal path drops a fish from a height of
    5.4 m. The fish travels 8.0 m horizontally before it hits the water below.
    What is the pelican's initial speed?

4. If the pelican in item 3 was traveling at the same speed but was only
    2.7 m above the water, how far would the fish travel horizontally before
    hitting the water below?

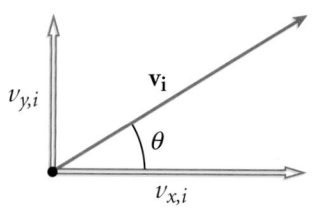

Figure 3-20
An object is projected with an initial
velocity, v_i, at an angle of θ. By
resolving the initial velocity into its
x and y components, the kinematic
equations can be applied to
describe the motion of the projectile throughout its flight.

⭐TEKS 4A

Module 3
"Two-Dimensional Motion"
provides an interactive lesson
with guided problem-solving
practice to teach you about
analyzing motion at angles,
including projectile motion.

Use components to analyze objects launched at an angle

Let us examine a case in which a projectile is launched at an angle to the horizontal, as shown in **Figure 3-20.** The projectile has an initial vertical component of velocity as well as a horizontal component of velocity.

Suppose the initial velocity vector makes an angle θ with the horizontal.
Again, to analyze the motion of such a projectile, the object's motion must be
resolved into its components. The sine and cosine functions can be used to
find the horizontal and vertical components of the initial velocity.

$$v_{x,i} = v_i (\cos \theta) \quad \text{and} \quad v_{y,i} = v_i (\sin \theta)$$

We can substitute these values for $v_{x,i}$ and $v_{y,i}$ into the kinematic equations
from Chapter 2 to obtain a set of equations that can be used to analyze the
motion of a projectile launched at an angle. ⭐TEKS 4B

PROJECTILES LAUNCHED AT AN ANGLE

$$v_x = v_i (\cos \theta) = \text{constant}$$
$$\Delta x = v_i (\cos \theta) \Delta t$$
$$v_{y,f} = v_i (\sin \theta) - g\Delta t$$
$$v_{y,f}^2 = v_i^2 (\sin \theta)^2 - 2g\Delta y$$
$$\Delta y = v_i (\sin \theta) \Delta t - \frac{1}{2}g(\Delta t)^2$$

As we have seen, the velocity of a projectile launched at an angle to the
ground has both horizontal and vertical components. The vertical motion is
similar to that of an object that is thrown straight up with an initial velocity.

Projectiles launched at an angle ⭐TEKS 2C, 4A, 4B

PROBLEM

A zookeeper finds an escaped monkey hanging from a light pole. Aiming her tranquilizer gun at the monkey, the zookeeper kneels 10.0 m from the light pole, which is 5.00 m high. The tip of her gun is 1.00 m above the ground. The monkey tries to trick the zookeeper by dropping a banana, then continues to hold onto the light pole. At the moment the monkey releases the banana, the zookeeper shoots. If the tranquilizer dart travels at 50.0 m/s, will the dart hit the monkey, the banana, or neither one?

SOLUTION

1. **Select a coordinate system.**

   As shown in **Figure 3-21,** the positive y-axis points up along the tip of the gun, and the positive x-axis points along the ground toward the light pole. Because the dart leaves the gun at a height of 1.00 m, the vertical distance to the monkey (and to the banana) is 4.00 m rather than 5.00 m.

   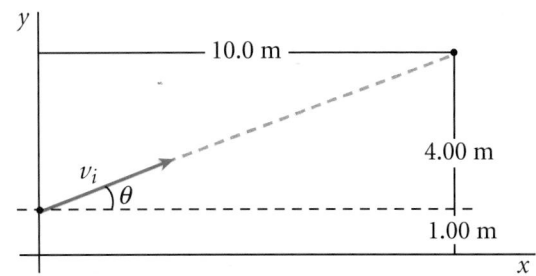

   Figure 3-21

2. **Use the tangent function to find the initial angle that the initial velocity makes with the x-axis.**

$$\theta = \tan^{-1}\left(\frac{\Delta y}{\Delta x}\right) = \tan^{-1}\left(\frac{4.00 \text{ m}}{10.0 \text{ m}}\right)$$

$$\theta = 21.8°$$

3. **Choose a kinematic equation to solve for time.**

   First rearrange the equation for motion along the x-axis to isolate the unknown, Δt, which is the time it takes the bullet to travel the horizontal distance from the tip of the gun to the pole.

$$\Delta x = v_i(\cos \theta)\Delta t$$

$$\Delta t = \frac{\Delta x}{v_i \cos \theta} = \frac{10.0 \text{ m}}{(50.0 \text{ m/s})(\cos 21.8°)}$$

$$\Delta t = 0.215 \text{ s}$$

4. **Using your knowledge of free fall, find out how far each object will fall during this time.**

   For the banana:

   This is a free-fall problem, where $v_i = 0$.

$$\Delta y = -\tfrac{1}{2}g(\Delta t)^2 = -\tfrac{1}{2}(9.81 \text{ m/s}^2)(0.215 \text{ s})^2$$

$$\Delta y = -0.227 \text{ m}$$

   Thus, the banana will be 0.227 m below its starting point.

continued on next page

Classroom Practice

The following may be used as teamwork exercises or for demonstration at the chalkboard or on an overhead projector.

PROBLEM

Projectiles launched at an angle

A golfer practices driving balls off a cliff and into water below. The cliff is 15 m from the water. If the golf ball is launched at 51 m/s at an angle of 15°, how far does the ball travel horizontally before hitting the water? (See Appendix A for hints on solving quadratic equations.)

Answer

1.7×10^2 m

 **Interactive Problem-Solving Tutor**

See Module 3
"Two-Dimensional Motion" provides additional development of problem-solving skills for this chapter.

⭐TEKS

p. 102: 2C, 4A, 4B, 4A, 4B

p. 103: 2C, 4A, 4B

Solving for:		
$\Delta y/\Delta x$	📖 PE	Sample, 1–3, 5*; Ch. Rvw. 35a, 38–40, 56*, 61a, 63b, 67–68, 70a, 70c
	💿 PW	4b, 5, 7–8
	PB	6
v_i	📖 PE	4–5*; Ch. Rvw. 53a*, 54a, 55*, 56*, 63a, 69
	💿 PW	Sample, 1–3, 4a
	PB	8, 10
Δt	📖 PE	Ch. Rvw. 35b, 53b*, 61b, 70b
	PB	7, 9
v_f	📖 PE	Ch. Rvw. 53c*, 54b
θ	💿 PW	6
	PB	Sample, 1–5

ANSWERS TO

Practice 3E
Projectiles launched at an angle

1. yes, $\Delta y = -2.3$ m
2. 70.3 m
3. 2.0 s; 4.8 m
4. 6.2 m/s
5. 17.7 m/s; 6.60 m

The dart has an initial vertical component of velocity equal to $v_i(\sin\theta)$, so:

$$\Delta y = v_i(\sin\theta)\Delta t - \tfrac{1}{2}g(\Delta t)^2$$
$$\Delta y = (50.0 \text{ m/s})(\sin 21.8°)(0.215 \text{ s}) - \tfrac{1}{2}(9.81 \text{ m/s}^2)(0.215 \text{ s})^2$$
$$\Delta y = 3.99 \text{ m} - 0.227 \text{ m} = 3.76 \text{ m}$$

The dart will be 3.76 m above its starting point.

5. **Analyze the results.**

By rearranging our equation for displacement, we can find the final height of both the banana and the dart.

$$y_f = y_i + \Delta y$$
$$y_{banana, f} = 5.00 \text{ m} + (-0.227 \text{ m}) = \boxed{4.77 \text{ m above the ground}}$$

$$y_{dart, f} = 1.00 \text{ m} + 3.76 \text{ m} = \boxed{4.76 \text{ m above the ground}}$$

The dart hits the banana. The slight difference is due to rounding.

PRACTICE 3E

Projectiles launched at an angle ⭐TEKS **2C, 4A, 4B**

1. In a scene in an action movie, a stuntman jumps from the top of one building to the top of another building 4.0 m away. After a running start, he leaps at an angle of 15° with respect to the flat roof while traveling at a speed of 5.0 m/s. Will he make it to the other roof, which is 2.5 m shorter than the building he jumps from?

2. A golfer can hit a golf ball a horizontal distance of over 300 m on a good drive. What maximum height will a 301.5 m drive reach if it is launched at an angle of 25.0° to the ground? (Hint: At the top of its flight, the ball's vertical velocity component will be zero.)

3. A baseball is thrown at an angle of 25° relative to the ground at a speed of 23.0 m/s. If the ball was caught 42.0 m from the thrower, how long was it in the air? How high was the tallest spot in the ball's path?

4. Salmon often jump waterfalls to reach their breeding grounds. Starting 2.00 m from a waterfall 0.55 m in height, at what minimum speed must a salmon jumping at an angle of 32.0° leave the water to continue upstream?

5. A quarterback throws the football to a stationary receiver who is 31.5 m down the field. If the football is thrown at an initial angle of 40.0° to the ground, at what initial speed must the quarterback throw the ball for it to reach the receiver? What is the ball's highest point during its flight?

Section Review

⭐TEKS 2C, 4A, 4B

1. Which of the following are examples of projectile motion?
   a. an airplane taking off
   b. a tennis ball lobbed over a net
   c. a plastic disk sailing across a lawn
   d. a hawk diving to catch a mouse
   e. a parachutist drifting to Earth
   f. a frog jumping from land into the water

2. Which of the following exhibit parabolic motion?
   a. a flat rock skipping across the surface of a lake
   b. a three-point shot in basketball
   c. the space shuttle while orbiting Earth
   d. a ball bouncing across a room
   e. a cliff diver
   f. a life preserver dropped from a stationary helicopter
   g. a person skipping

3. An Alaskan rescue plane drops a package of emergency rations to a stranded party of explorers, as illustrated in **Figure 3-22.** The plane is traveling horizontally at 100.0 m/s at a height of 50.0 m above the ground. What horizontal distance does the package travel before striking the ground?

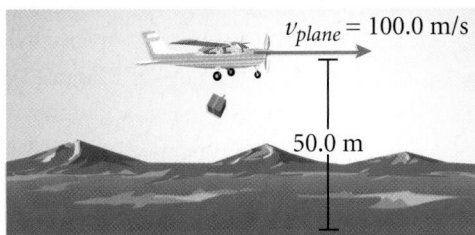

$v_{plane} = 100.0$ m/s

50.0 m

Figure 3-22

4. Find the velocity (magnitude and direction) of the package in item 3 just before it hits the ground.

5. During a thunderstorm, a tornado lifts a car to a height of 125 m above the ground. Increasing in strength, the tornado flings the car horizontally with an initial speed of 90.0 m/s. How long does the car take to reach the ground? How far horizontally does the car travel before hitting the ground?

6. **Physics in Action** Streams of water in a fountain shoot from one level to the next. A particle of water in a stream takes 0.50 s to travel between the first and second level. The receptacle on the second level is a horizontal distance of 1.5 m away from the spout on the first level. If the water is projected at an angle of 33°, what is the initial speed of the particle?

7. **Physics in Action** If a water particle in a stream of water in a fountain takes 0.35 s to travel from spout to receptacle when shot at an angle of 67° and an initial speed of 5.0 m/s, what is the vertical distance between the levels of the fountain?

Section Review
ANSWERS

1. b, f
2. a, b, d, e, g
3. 319 m
4. 104.8 m/s at 17.4° below the horizontal
5. 5.05 s; 454 m
6. 3.6 m/s
7. 1.0 m

⭐TEKS

p. 104: 2C, 4A, 4B
p. 105: 2C, 4A, 4B

3-4
Relative motion

Visual Strategy

Figure 3-23

Tell students that this diagram shows what would happen if there were no air resistance.

Q How would the two diagrams change if air resistance were included?

A *In the first diagram, the object would appear to fall down and backward as viewed by the pilot. In the second diagram, the object would follow a shortened path as viewed by the observer on the ground—it would become steeper as the object loses velocity in the x direction.*

3-4 SECTION OBJECTIVES

- **Describe situations in terms of frame of reference.**
- **Solve problems involving relative velocity.**

FRAMES OF REFERENCE

If you are moving at 80 km/h north and a car passes you going 90 km/h, to you the faster car seems to be moving north at 10 km/h. Someone standing on the side of the road would measure the velocity of the faster car as 90 km/h toward the north. This simple example demonstrates that velocity measurements depend on the frame of reference of the observer. (★)TEKS **4E**

Velocity measurements differ in different frames of reference

Observers using different frames of reference may measure different displacements or velocities for an object in motion. That is, two observers moving with respect to each other would generally not agree on some features of the motion.

Let us return to the example of the stunt dummy that is dropped from an airplane flying horizontally over Earth with a constant velocity. As shown in **Figure 3-23(a),** a passenger on the airplane would describe the motion of the dummy as a straight line toward Earth, whereas an observer on the ground would view the trajectory of the dummy as that of a projectile, as shown in **Figure 3-23(b).** Relative to the ground, the dummy would have a vertical component of velocity (resulting from free-fall acceleration and equal to the velocity measured by the observer in the airplane) *and* a horizontal component of velocity given to it by the airplane's motion. If the airplane continued to move horizontally with the same velocity, the dummy would enter the swimming pool directly beneath the airplane (assuming negligible air resistance). (★)TEKS **3A**

(a)

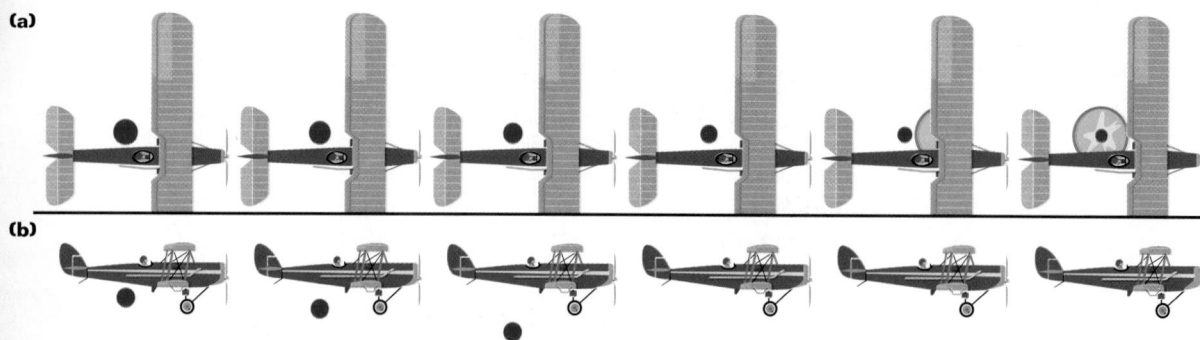

(b)

Figure 3-23 (★)TEKS **4E**
When viewed from the plane **(a),** the stunt dummy (represented by the maroon dot) falls straight down. When viewed from a stationary position on the ground **(b),** the stunt dummy follows a parabolic projectile path.

RELATIVE VELOCITY

The case of the faster car overtaking your car was easy to solve with a minimum of thought and effort, but you will encounter many situations in which a more systematic method of solving such problems is beneficial. To develop this method, write down all the information that is given and that you want to know in the form of velocities with subscripts appended.

$\mathbf{v_{se}} = +80$ km/h north (Here the subscript *se* means the velocity of the *slower* car with respect to *Earth.*)

$\mathbf{v_{fe}} = +90$ km/h north (The subscript *fe* means the velocity of the *fast* car with respect to *Earth.*)

We want to know $\mathbf{v_{fs}}$, which is the velocity of the fast car with respect to the slower car. To find this, we write an equation for $\mathbf{v_{fs}}$ in terms of the other velocities, so on the right side of the equation the subscripts start with *f* and eventually end with *s*. Also, each velocity subscript starts with the letter that ended the preceding velocity subscript.

$$\mathbf{v_{fs}} = \mathbf{v_{fe}} + \mathbf{v_{es}}$$

The boldface notation indicates that velocity is a vector quantity. This approach to adding and monitoring subscripts is similar to vector addition, in which vector arrows are placed head to tail to find a resultant.

If we take north to be the positive direction, we know that $\mathbf{v_{es}} = -\mathbf{v_{se}}$ because an observer in the slow car perceives Earth as moving south at a velocity of 80 km/h while a stationary observer on the ground (Earth) views the car as moving north at a velocity of 80 km/h. Thus, this problem can be solved as follows:

$$\mathbf{v_{fs}} = +90 \text{ km/h} - 80 \text{ km/h} = +10 \text{ km/h}$$

The positive sign means that the fast car appears (to the occupants of the slower car) to be moving north at 10 km/h.

There is no general equation to work relative velocity problems; instead, you should develop the necessary equations on your own by following the above technique for writing subscripts. ⊛TEKS **2C, 4B, 4E**

Did you know?

Like velocity, displacement and acceleration depend on the frame in which they are measured. In some cases, it is instructive to visualize gravity as the ground accelerating toward a projectile rather than the projectile accelerating toward the ground.

SECTION 3-4

Teaching Tip

Relative velocity can also be shown as the difference of two vectors.

$$\mathbf{v_{fs}} = \mathbf{v_{fe}} - \mathbf{v_{se}}$$

Another way of stating this is that the relative velocity of one moving object to another is the difference between their velocities relative to some common reference point.

You may want to demonstrate on the board that this works for noncollinear velocities, as in Sample Problem 3F, on the next page.

ANSWERS TO

Conceptual Challenge

1. greater than, because the elevator is accelerating upward toward the ball as it falls at free fall

2. because the plane's velocity is slower relative to the moving carrier and would have a better chance of stopping before reaching the far end of the carrier

Conceptual Challenge ⊛TEKS 4E

1. Elevator acceleration A boy bounces a small rubber ball in an elevator that is going down. If the boy drops the ball as the elevator is slowing down, is the ball's acceleration relative to the elevator less than or greater than its acceleration relative to the ground?

2. Aircraft carrier Why does a plane landing on an aircraft carrier approach the carrier from the rear instead of from the front?

⊛TEKS

p. 106: 4E, 3A, 4E
p. 107: 2C, 4B, 4E, 4E

PROBLEM

Relative velocity

A plane flies northeast at an airspeed of 563.0 km/h. A 48.0 km/h wind is blowing to the southeast. What is the plane's velocity relative to the ground?

Answer

565.0 km/h at 40.1° north of east

The wind shifts to blow 63.0 km/h towards the southwest. What is the plane's velocity relative to the ground?

Answer

500.0 km/h northeast

⭐TEKS

p. 108: 2C, 4A, 4B, 4E
p. 109: 2C, 4A, 4B, 4E

SAMPLE PROBLEM 3F

Relative velocity ⭐TEKS 2C, 4A, 4B, 4E

PROBLEM

A boat heading north crosses a wide river with a velocity of 10.00 km/h relative to the water. The river has a uniform velocity of 5.00 km/h due east. Determine the boat's velocity with respect to an observer on shore.

SOLUTION

1. DEFINE **Given:** $\mathbf{v_{br}}$ = 10.00 km/h due north (velocity of the boat, *b*, with respect to the *river, r*)

$\mathbf{v_{re}}$ = 5.00 km/h due east (velocity of the *river, r*, with respect to *Earth, e*)

Unknown: $\mathbf{v_{be}}$ = ?

Diagram:

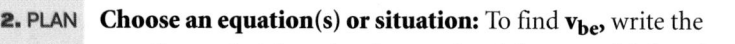

2. PLAN **Choose an equation(s) or situation:** To find $\mathbf{v_{be}}$, write the equation so that the subscripts on the right start with *b* and end with *e*.

$$\mathbf{v_{be}} = \mathbf{v_{br}} + \mathbf{v_{re}}$$

As in Section 3-2, we use the Pythagorean theorem to calculate the magnitude of the resultant velocity and the tangent function to find the direction.

$$(v_{be})^2 = (v_{br})^2 + (v_{re})^2$$

$$\tan \theta = \frac{v_{re}}{v_{br}}$$

Rearrange the equation(s) to isolate the unknown(s):

$$v_{be} = \sqrt{(v_{br})^2 + (v_{re})^2}$$

$$\theta = \tan^{-1}\left(\frac{v_{re}}{v_{br}}\right)$$

3. CALCULATE **Substitute the known values into the equation(s) and solve:**

$$v_{be} = \sqrt{(10.00 \text{ km/h})^2 + (5.00 \text{ km/h})^2}$$

$$\boxed{v_{be} = 11.18 \text{ km/h}}$$

$$\theta = \tan^{-1}\left(\frac{5.00}{10.00}\right)$$

$$\boxed{\theta = 26.6°}$$

4. EVALUATE The boat travels at a speed of 11.18 km/h in the direction 26.6° east of north with respect to Earth.

Relative velocity ⭐TEKS 2C, 4A, 4B, 4E

1. A passenger at the rear of a train traveling at 15 m/s relative to Earth throws a baseball with a speed of 15 m/s in the direction opposite the motion of the train. What is the velocity of the baseball relative to Earth as it leaves the thrower's hand? Show your work.

2. A spy runs from the front to the back of an aircraft carrier at a velocity of 3.5 m/s. If the aircraft carrier is moving forward at 18.0 m/s, how fast does the spy appear to be running when viewed by an observer on a nearby stationary submarine? Show your work.

3. A ferry is crossing a river. If the ferry is headed due north with a speed of 2.5 m/s relative to the water and the river's velocity is 3.0 m/s to the east, what will the boat's velocity be relative to Earth? (Hint: Remember to include the direction in describing the velocity.)

4. A pet-store supply truck moves at 25.0 m/s north along a highway. Inside, a dog moves at 1.75 m/s at an angle of 35.0° east of north. What is the velocity of the dog relative to the road?

Section Review

1. Describe the motion of the following objects if they are observed from the stated frames of reference:
   a. a person standing on a platform viewed from a train traveling north
   b. a train traveling north viewed by a person standing on a platform
   c. a ball dropped by a boy walking at a speed of 1 m/s viewed by the boy
   d. a ball dropped by a boy walking 1 m/s as seen by a nearby viewer who is stationary

2. A woman on a 10-speed bicycle travels at 9 m/s relative to the ground as she passes a little boy on a tricycle going in the opposite direction. If the boy is traveling at 1 m/s relative to the ground, how fast does the boy appear to be moving relative to the woman? Show your work.

3. A girl at an airport rolls a ball north on a moving walkway that moves east. If the ball's speed with respect to the walkway is 0.15 m/s and the walkway moves at a speed of 1.50 m/s, what is the velocity of the ball relative to the ground?

PRACTICE GUIDE 3F
Solving for:

v, θ		PE	Sample, 1–4; Ch. Rvw. 47, 48a, 49a–b, 51a–b, 52a, 64a, 64c, 65a–b
		PW	Sample, 1
		PB	7, 10
Δt		PE	Ch. Rvw. 50, 52b, 57a–b, 59, 62, 66
		PW	2, 4, 6
		PB	Sample, 1–5
$\Delta y/\Delta x$		PE	Ch. Rvw. 48b, 64b
		PW	5
		PB	6, 8–9

ANSWERS TO

Practice 3F
Relative velocity

1. 0 m/s
2. 14.5 m/s in the direction that the aircraft carrier is moving
3. 3.90 m/s at $(4.0 \times 10^{1})°$ north of east
4. 26.4 m/s at 2.17° east of north

Section Review
ANSWERS

1. a. south with a speed equal to the train's speed
   b. moves north
   c. appears to fall straight down
   d. moves in a parabola
2. 10 m/s away in the opposite direction
3. 1.51 m/s at 5.7° north of east

BACKGROUND

This feature builds on the Time Dilation feature in Chapter 2. In Chapter 2, the assumption that the speed of light is the same for all observers was used to explain why time measurements depend on an observer's frame of reference. In this feature, this assumption and its consequences are discussed in greater detail.

The feature begins by comparing the behavior of light with a typical case to show how the two differ. Next, the need to revise the classical addition of velocities is discussed. Finally, Einstein's relativistic addition of velocities is introduced, along with a discussion of how all cases are covered with this equation.

Today there is overwhelming physical evidence that the speed of light is absolute. Experiments in particle accelerators, in which particles reach speeds very close to c, support the relativistic rather than the classical addition of velocities.

Teaching Tip

The constancy of the speed of light is difficult for some students to grasp because we have no direct experience of this phenomenon. Use examples to familiarize your students with Einstein's theory of the constancy of the speed of light. For example, ask students to compare the speed of sound and light waves as viewed by two different observers, one at rest and one moving toward the source of the waves.

PHYSICS ON THE EDGE

RELATIVISTIC ADDITION OF VELOCITIES

In Section 3-4, you learned that velocity measurements are not absolute; every velocity measurement depends on the frame of reference of the observer with respect to the moving object. For example, imagine that someone riding a bike toward you at 25 m/s (v) throws a softball toward you. If the bicyclist measures the softball's speed (u') to be 15 m/s, you would perceive the ball to be moving toward you at 40 m/s (u) because you have a different frame of reference than the bicyclist does. This is expressed mathematically by the equation $u = v + u'$, which is also known as the classical addition of velocities. (★)TEKS **4E**

The speed of light

As stated in the "Time dilation" feature in Chapter 2, according to Einstein's special theory of relativity, the speed of light is absolute, or independent of all frames of reference. If, instead of a softball, the bicyclist were to shine a beam of light toward you, both you and the bicyclist would measure the light's speed as 3.0×10^8 m/s. This would remain true even if the bicyclist were moving toward you at 99 percent of the speed of light. Thus, Einstein's theory requires a different approach to the addition of velocities. Einstein's modification of the classical formula, which he derived in his 1905 paper on special relativity, covers both the case of the softball and the case of the light beam.

$$u = \frac{v + u'}{1 + (vu'/c^2)}$$

In the equation, u is the velocity of an object in a reference frame, u' is the velocity of the same object in another reference frame, v is the velocity of one reference frame relative to another, and c is the speed of light. (★)TEKS **3C**

The universality of Einstein's equation

How does Einstein's equation cover both cases? First we shall consider the bicyclist throwing a softball. Because c^2 is such a large number, the vu'/c^2 term in the denominator is very small for velocities typical of our everyday experience. As a result, the denominator of the equation is nearly equal to 1. Hence, for speeds that are small compared with c, the two theories give nearly the same result, $u = v + u'$, and the classical addition of velocities can be used.

However, when speeds approach the speed of light, vu'/c^2 increases, and the denominator becomes greater than 1 and less than 2. When this occurs, the difference between the two theories becomes significant. For example, if a bicyclist moving toward you at 80 percent of the speed of light were to throw a ball to you at 70 percent of the speed of light, you would observe the ball moving toward you at about 96 percent of the speed of light rather than the 150 percent of the speed of light predicted by classical theory. In this case, the difference between the velocities predicted by each theory cannot be ignored, and the relativistic addition of velocities must be used.

In this last example, it is significant that classical addition predicts a speed greater than the speed of light ($1.5c$), while the relativistic addition predicts a speed less than the speed of light ($0.96c$). In fact, no matter how close the speeds involved are to the speed of light, the relativistic equation yields a result less than the speed of light, as seen in **Table 3-1.**

How does Einstein's equation cover the second case, in which the bicyclist shines a beam of light toward you? Einstein's equation predicts that any object traveling at the speed of light ($u' = c$) will appear to travel at the speed of light ($u = c$) for an observer in any reference frame: ⭐TEKS **3B**

$$u = \frac{v + u'}{1 + (vu'/c^2)} = \frac{v + c}{1 + (vc/c^2)} = \frac{v + c}{1 + (v/c)} = \frac{v + c}{(c + v)/c} = c$$

This corresponds with our earlier statement that the bicyclist measures the beam of light traveling at the same speed that you do, 3.0×10^8 m/s, even though you have a different reference frame than the bicyclist does. This occurs regardless of how fast the bicycle is moving because v (the bicycle's speed) cancels from the equation. Thus, Einstein's relativistic equation successfully covers both cases. So, Einstein's equation is a more general case of the classical equation, which is simply the limiting case. ⭐TEKS **3A**

Figure 3-24
According to Einstein's relativistic equation for the addition of velocities, material particles can never reach the speed of light.

internet**connect**

SC*L*INKS.
NSTA

TOPIC: Speed of light
GO TO: www.scilinks.org
*sci*LINKS **CODE:** HF2033

SECTION 3-4

EXTENSION

- Have students research particle-accelerator experiments involving particles traveling at speeds close to *c*. Their reports should include a discussion of how observations support the relativistic addition of velocities.
- The *ether* was originally conceived as the medium through which light waves traveled. Have students investigate the concept of the ether and the Michelson-Morley experiment, which was intended to detect the ether. Then have a class discussion about why the concept of the ether was originally believed to be necessary and how the Michelson-Morley experiment and the special theory of relativity affected the theory of the ether.

Table 3-1		Classical and relativistic addition of velocities	
$c = 2.997\ 925\ 84 \times 10^8$ m/s		**Classical addition**	**Relativistic addition**
Speed between frames (v)	Speed in A (u')	Speed in B (u)	Speed in B (u)
25 m/s	15 m/s	40 m/s	40 m/s
100 000 m/s	100 000 m/s	200 000 m/s	200 000 m/s
50% of c	50% of c	299 792 584 m/s	239 834 067 m/s
90% of c	90% of c	539 626 651 m/s	298 136 271 m/s
99.99% of c	99.99% of c	599 525 210 m/s	299 792 582 m/s

⭐TEKS

p. 110: 4E, 3C
p. 111: 3B, 3A

CHAPTER 3
Summary

Teaching Tip

Ask students to prepare a concept map for the chapter. The concept map should include most of the vocabulary terms, along with other integral terms or concepts.

★ TEKS

Review & Assess
pp. 113–119:
2A: 2
2B: 2
2C: all
2D: 1, 2, 4, 5
2E: all
2F: 2
3B: all
3C: 5
4A: all
4B: all

KEY TERMS

components of a vector (p. 92)

projectile motion (p. 99)

resultant (p. 85)

scalar (p. 84)

vector (p. 84)

Diagram symbols

⟶ displacement vector

⟶ velocity vector

⟹ acceleration vector

- - - → resultant vector

⟹ component

KEY IDEAS

Section 3-1 Introduction to vectors

- A scalar is a quantity completely specified by only a number with appropriate units, whereas a vector is a quantity that has magnitude and direction.
- Vectors can be added graphically using the triangle method of addition, in which the tail of one vector is placed at the head of the other. The resultant is the vector drawn from the tail of the first vector to the head of the last vector.

Section 3-2 Vector operations

- The Pythagorean theorem and the tangent function can be used to find the magnitude and direction of a resultant vector.
- Any vector can be resolved into its component vectors using the sine and cosine functions.

Section 3-3 Projectile motion

- Neglecting air resistance, a projectile has a constant horizontal velocity and a constant downward free-fall acceleration.
- In the absence of air resistance, projectiles follow a parabolic path.

Section 3-4 Relative motion

- If the frame of reference is denoted with subscripts ($\mathbf{v_{ab}}$ is the velocity of a with respect to b), then the velocity of an object with respect to a different frame of reference can be found by adding the known velocities so that the subscript starts with the letter that ends the preceding velocity subscript, $\mathbf{v_{ab}} = \mathbf{v_{ac}} + \mathbf{v_{cb}}$.
- If the order of the subscripts is reversed, there is a change in sign, for example, $\mathbf{v_{cd}} = -\mathbf{v_{dc}}$.

Variable symbols

Quantities		Units	
$\mathbf{d}$ (vector)	displacement	m	meters
$\mathbf{v}$ (vector)	velocity	m/s	meters/second
$\mathbf{a}$ (vector)	acceleration	m/s^2	meters/second^2
Δx (scalar)	horizontal component	m	meters
Δy (scalar)	vertical component	m	meters

VECTORS AND THE GRAPHICAL METHOD

Review questions

1. The magnitude of a vector is a scalar. Explain this statement.

2. If two vectors have unequal magnitudes, can their sum be zero? Explain.

3. What is the relationship between instantaneous speed and instantaneous velocity?

4. What is another way of saying −30 m/s west?

5. Is it possible to add a vector quantity to a scalar quantity? Explain.

6. Vector **A** is 3.00 units in length and points along the positive *x*-axis. Vector **B** is 4.00 units in length and points along the negative *y*-axis. Use graphical methods to find the magnitude and direction of the following vectors:
   a. **A** + **B**
   b. **A** − **B**
   c. **A** + 2**B**
   d. **B** − **A**

7. Each of the displacement vectors **A** and **B** shown in **Figure 3-25** has a magnitude of 3.00 m. Graphically find the following:
   a. **A** + **B**
   b. **A** − **B**
   c. **B** − **A**
   d. **A** − 2**B**

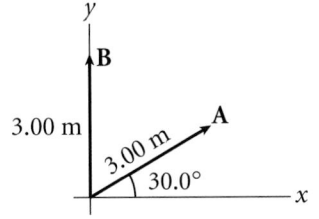

Figure 3-25

8. A dog searching for a bone walks 3.50 m south, then 8.20 m at an angle of 30.0° north of east, and finally 15.0 m west. Use graphical techniques to find the dog's resultant displacement vector.

9. A man lost in a maze makes three consecutive displacements so that at the end of the walk he is back where he started, as shown in **Figure 3-26.** The first displacement is 8.00 m westward, and the second is 13.0 m northward. Use the graphical method to find the third displacement.

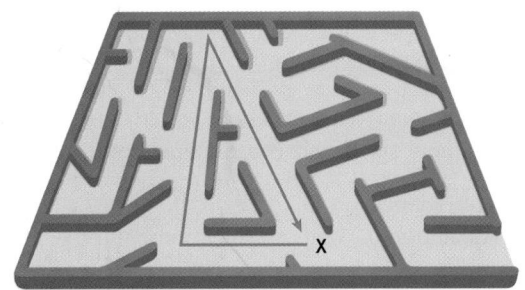

Figure 3-26

Conceptual questions

10. If **B** is added to **A,** under what conditions does the resultant have the magnitude equal to *A* + *B*?

11. Give an example of a moving object that has a velocity vector and an acceleration vector in the same direction and an example of one that has velocity and acceleration vectors in opposite directions.

12. A student accurately uses the method for combining vectors. The two vectors she combines have magnitudes of 55 and 25 units. The answer that she gets is either 85, 20, or 55. Pick the correct answer, and explain why it is the only one of the three that can be correct.

13. If a set of vectors laid head to tail forms a closed polygon, the resultant is zero. Is this statement true? Explain your reasoning.

14. Yes, the second component could be nonzero.

15. No, the hypotenuse is always greater than the legs.

16. Vector addition is combining vectors to find a resultant vector. Vector resolution is breaking a vector into its component vectors.

17. They are perpendicular.

18. Resolve both vectors into their vector components, add the parallel components together, and then find the vector sum of the perpendicular vectors.

19. They are equal and opposite.

20. oriented at 45° from the axis

21. They form a closed triangle when laid head to tail.

22. **a.** 5 blocks at 53° north of east
    b. 13 blocks

23. 8.07 m at 42.0° south of east

24. 42.7 yards

25. 61.8 m at 76.0° S of E (or S of W), 25.0 m at 53.1° S of E (or S of W)

26. 108 m, −19.1 m

27. 2.81 km east, 1.31 km north

28. 31.5 m, 26.4 m

29. 240.0 m at 57.23° south of west

30. Both hit at the same time.

31. yes, neglecting air resistance

VECTOR OPERATIONS

Review questions

14. Can a vector have a component equal to zero and still have a nonzero magnitude?

15. Can a vector have a component greater than its magnitude?

16. Explain the difference between vector addition and vector resolution.

17. If the component of one vector along the direction of another is zero, what can you conclude about these two vectors?

18. How would you add two vectors that are not perpendicular or parallel?

Conceptual questions

19. If **A** + **B** = 0, what can you say about the components of the two vectors?

20. Under what circumstances would a vector have components that are equal in magnitude?

21. The vector sum of three vectors gives a resultant equal to zero. What can you say about the vectors?

Practice problems

22. A girl delivering newspapers travels three blocks west, four blocks north, then six blocks east.
    a. What is her resultant displacement?
    b. What is the total distance she travels?
    (See Sample Problem 3A.)

23. A golfer takes two putts to sink his ball in the hole once he is on the green. The first putt displaces the ball 6.00 m east, and the second putt displaces it 5.40 m south. What displacement would put the ball in the hole in one putt? (See Sample Problem 3A.)

24. A quarterback takes the ball from the line of scrimmage, runs backward for 10.0 yards, then runs sideways parallel to the line of scrimmage for 15.0 yards. At this point, he throws a 50.0-yard forward pass straight down the field. What is the magnitude of the football's resultant displacement? (See Sample Problem 3A.)

25. A shopper pushing a cart through a store moves 40.0 m south down one aisle, then makes a 90.0° turn and moves 15.0 m. He then makes another 90.0° turn and moves 20.0 m. Find the shopper's total displacement. Note that you are not given the direction moved after any of the 90.0° turns, so there could be more than one answer. (See Sample Problem 3A.)

26. A submarine dives 110.0 m at an angle of 10.0° below the horizontal. What are the horizontal and vertical components of the submarine's displacement? (See Sample Problem 3B.)

27. A person walks 25.0° north of east for 3.10 km. How far would another person walk due north and due east to arrive at the same location? (See Sample Problem 3B.)

28. A roller coaster travels 41.1 m at an angle of 40.0° above the horizontal. How far does it move horizontally and vertically? (See Sample Problem 3B.)

29. A person walks the path shown in **Figure 3-27**. The total trip consists of four straight-line paths. At the end of the walk, what is the person's resultant displacement measured from the starting point? (See Sample Problem 3C.)

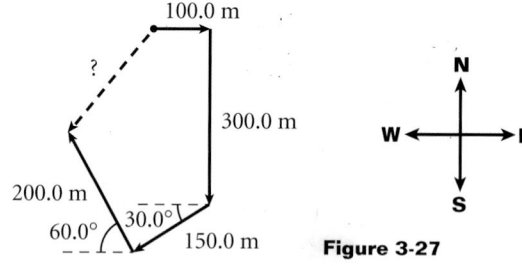

Figure 3-27

PROJECTILE MOTION

Review questions

30. A bullet is fired horizontally from a pistol, and another bullet is dropped simultaneously from the same height. If air resistance is neglected, which bullet hits the ground first?

31. If a rock is dropped from the top of a sailboat's mast, will it hit the deck at the same point whether the boat is at rest or in motion at constant velocity?

32. Does a ball dropped out of the window of a moving car take longer to reach the ground than one dropped at the same height from a car at rest?

33. A rock is dropped at the same instant that a ball at the same elevation is thrown horizontally. Which will have the greater velocity when it reaches ground level?

Practice problems

34. The fastest recorded pitch in Major League Baseball was thrown by Nolan Ryan in 1974. If this pitch were thrown horizontally, the ball would fall 0.809 m (2.65 ft) by the time it reached home plate, 18.3 m (60 ft) away. How fast was Ryan's pitch? (See Sample Problem 3D.)

35. A shell is fired from the ground with an initial speed of 1.70×10^3 m/s (approximately five times the speed of sound) at an initial angle of $55.0°$ to the horizontal. Neglecting air resistance, find
  a. the shell's horizontal range
  b. the amount of time the shell is in motion
(See Sample Problem 3E.)

36. A person standing at the edge of a seaside cliff kicks a stone over the edge with a speed of 18 m/s. The cliff is 52 m above the water's surface, as shown in **Figure 3-28.** How long does it take for the stone to fall to the water? With what speed does it strike the water? (See Sample Problem 3D.)

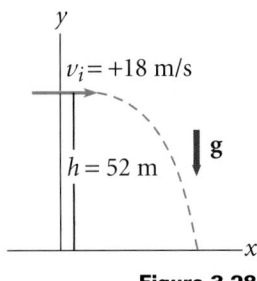

Figure 3-28

37. A spy in a speed boat is being chased down a river by government officials in a faster craft. Just as the officials' boat pulls up next to the spy's boat, both boats reach the edge of a 5.0 m waterfall. If the spy's speed is 15 m/s and the officials' speed is 26 m/s, how far apart will the two vessels be when they land below the waterfall?

38. A place kicker must kick a football from a point 36.0 m (about 40.0 yd) from the goal. As a result of the kick, the ball must clear the crossbar, which is 3.05 m high. When kicked, the ball leaves the ground with a speed of 20.0 m/s at an angle of $53°$ to the horizontal.
  a. By how much does the ball clear or fall short of clearing the crossbar?
  b. Does the ball approach the crossbar while still rising or while falling?
(See Sample Problem 3E.)

39. A daredevil is shot out of a cannon at $45.0°$ to the horizontal with an initial speed of 25.0 m/s. A net is positioned at a horizontal distance of 50.0 m from the cannon from which it is shot. At what height above the cannon should the net be placed in order to catch the daredevil? (See Sample Problem 3E.)

40. When a water gun is fired while being held horizontally at a height of 1.00 m above ground level, the water travels a horizontal distance of 5.00 m. A child, who is holding the same gun in a horizontal position, is also sliding down a $45.0°$ incline at a constant speed of 2.00 m/s. If the child fires the gun when it is 1.00 m above the ground and the water takes 0.329 s to reach the ground, how far will the water travel horizontally? (See Sample Problem 3E.)

41. A ship maneuvers to within 2.50×10^3 m of an island's 1.80×10^3 m high mountain peak and fires a projectile at an enemy ship 6.10×10^2 m on the other side of the peak, as illustrated in **Figure 3-29.** If the ship shoots the projectile with an initial velocity of 2.50×10^2 m/s at an angle of $75.0°$, how close to the enemy ship does the projectile land? How close (vertically) does the projectile come to the peak? (See Sample Problem 3E.)

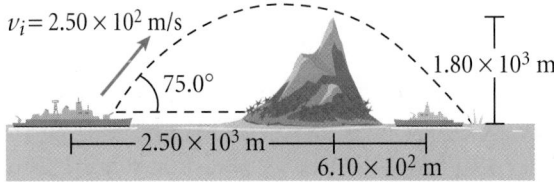

Figure 3-29

32. no, neglecting air resistance
33. Both vertical velocities will be the same, but the thrown ball will have a greater net velocity.
34. 45.1 m/s
35. a. 2.77×10^5 m
  b. 284 s
36. 3.3 s; 36 m/s
37. 11 m
38. a. clears the goal by 1 m
  b. falling
39. 10.8 m
40. 4.11 m
41. 80 m; 210 m

42. Displacement, velocity, and acceleration depend on the frame of reference in which they are measured.

43. the coordinate system used to describe the motion

44. Earth

45. a. 50 m/s, 20 m/s
b. 70 m/s east
c. 20 m/s

46. a. To the passenger, the ball appears to move in a straight line. To an outside observer, the ball moves along a parabolic trajectory.
b. The passenger would see the ball move backward, while the stationary observer would see no change from part (a).

47. 20 km/h south

48. a. 10.1 m/s at 8.53° east of north
b. 48.8 m

49. a. 14.1° north of west
b. 1.99 km/h

50. 7.5 min

51. a. 23.2° upstream from straight across
b. 8.72 m/s across the river

52. a. 12.5 m/s at 16° south of east
b. 113 s

53. a. 41.7 m/s
b. 3.81 s
c. $v_{y,f} = -13.5$ m/s,
$v_{x,f} = 34.2$ m/s,
$v_f = 36.7$ m/s

54. a. 15 m/s
b. 15 m/s

55. 10.5 m/s

RELATIVE MOTION

Review questions

42. Explain the statement, "All motion is relative."

43. What is a frame of reference?

44. When we describe motion, what is a common frame of reference?

45. A small airplane is flying at 50 m/s toward the east. A wind of 20 m/s toward the east suddenly begins to blow, giving the plane a velocity of 70 m/s east.

    a. Which of these are component vectors?
    b. Which is the resultant?
    c. What is the magnitude of the wind velocity?

46. A ball is thrown upward in the air by a passenger on a train that is moving with constant velocity.

    a. Describe the path of the ball as seen by the passenger. Describe the path as seen by a stationary observer outside the train.
    b. How would these observations change if the train were accelerating along the track?

Practice problems

47. The pilot of a plane measures an air velocity of 165 km/h south. An observer on the ground sees the plane pass overhead at a velocity of 145 km/h toward the north. What is the velocity of the wind that is affecting the plane?
(See Sample Problem 3F.)

48. A river flows due east at 1.50 m/s. A boat crosses the river from the south shore to the north shore by maintaining a constant velocity of 10.0 m/s due north relative to the water.

    a. What is the velocity of the boat as viewed by an observer on shore?
    b. If the river is 325 m wide, how far downstream is the boat when it reaches the north shore?
(See Sample Problem 3F.)

49. The pilot of an aircraft wishes to fly due west in a 50.0 km/h wind blowing toward the south. The speed of the aircraft in the absence of a wind is 205 km/h.

    a. In what direction should the aircraft head?
    b. What should its speed be relative to the ground?
(See Sample Problem 3F.)

50. A hunter wishes to cross a river that is 1.5 km wide and that flows with a speed of 5.0 km/h. The hunter uses a small powerboat that moves at a maximum speed of 12 km/h with respect to the water. What is the minimum time necessary for crossing?
(See Sample Problem 3F.)

51. A swimmer can swim in still water at a speed of 9.50 m/s. He intends to swim directly across a river that has a downstream current of 3.75 m/s.

    a. What must the swimmer's direction be?
    b. What is his velocity relative to the bank?
(See Sample Problem 3F.)

MIXED REVIEW PROBLEMS

52. A motorboat heads due east at 12.0 m/s across a river that flows toward the south at a speed of 3.5 m/s.

    a. What is the resultant velocity relative to an observer on the shore?
    b. If the river is 1360 m wide, how long does it take the boat to cross?

53. A ball player hits a home run, and the baseball just clears a wall 21.0 m high located 130.0 m from home plate. The ball is hit at an angle of 35.0° to the horizontal, and air resistance is negligible. Assume the ball is hit at a height of 1.0 m above the ground.

    a. What is the initial speed of the ball?
    b. How much time does it take for the ball to reach the wall?
    c. Find the velocity components and the speed of the ball when it reaches the wall.

54. A daredevil jumps a canyon 12 m wide. To do so, he drives a car up a 15° incline.

    a. What minimum speed must he achieve to clear the canyon?
    b. If the daredevil jumps at this minimum speed, what will his speed be when he reaches the other side?

55. A 2.00 m tall basketball player attempts a goal 10.00 m from the basket (3.05 m high). If he shoots the ball at a 45.0° angle, at what initial speed must he throw the basketball so that it goes through the hoop without striking the backboard?

56. A ball is thrown straight upward and returns to the thrower's hand after 3.00 s in the air. A second ball is thrown at an angle of 30.0° with the horizontal. At what speed must the second ball be thrown so that it reaches the same height as the one thrown vertically?

57. An escalator is 20.0 m long. If a person stands on the escalator, it takes 50.0 s to ride from the bottom to the top.

   a. If a person walks up the moving escalator with a speed of 0.500 m/s relative to the escalator, how long does it take the person to get to the top?

   b. If a person walks down the "up" escalator with the same relative speed as in item (a), how long does it take to reach the bottom?

58. A ball is projected horizontally from the edge of a table that is 1.00 m high, and it strikes the floor at a point 1.20 m from the base of the table.

   a. What is the initial speed of the ball?

   b. How high is the ball above the floor when its velocity vector makes a 45.0° angle with the horizontal?

59. How long does it take an automobile traveling 60.0 km/h to become even with a car that is traveling in another lane at 40.0 km/h if the cars' front bumpers are initially 125 m apart?

60. The eye of a hurricane passes over Grand Bahama Island. It is moving in a direction 60.0° north of west with a speed of 41.0 km/h. Exactly three hours later, the course of the hurricane shifts due north, and its speed slows to 25.0 km/h, as shown in **Figure 3-30.** How far from Grand Bahama is the hurricane 4.50 h after it passes over the island?

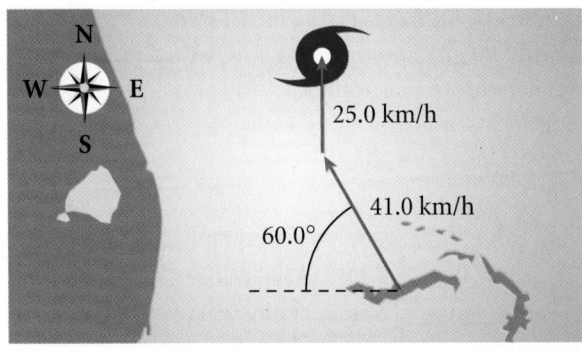

Figure 3-30

61. A car is parked on a cliff overlooking the ocean on an incline that makes an angle of 24.0° below the horizontal. The negligent driver leaves the car in neutral, and the emergency brakes are defective. The car rolls from rest down the incline with a constant acceleration of 4.00 m/s² and travels 50.0 m to the edge of the cliff. The cliff is 30.0 m above the ocean.

   a. What is the car's position relative to the base of the cliff when the car lands in the ocean?

   b. How long is the car in the air?

62. A boat moves through a river at 7.5 m/s relative to the water, regardless of the boat's direction. If the water in the river is flowing at 1.5 m/s, how long does it take the boat to make a round trip consisting of a 250 m displacement downstream followed by a 250 m displacement upstream?

63. A golf ball with an initial angle of 34° lands exactly 240 m down the range on a level course.

   a. Neglecting air friction, what initial speed would achieve this result?

   b. Using the speed determined in item (a), find the maximum height reached by the ball.

64. A water spider maintains an average position on the surface of a stream by darting upstream (against the current), then drifting downstream (with the current) to its original position. The current in the stream is 0.500 m/s relative to the shore, and the water spider darts upstream 0.560 m (relative to a spot on shore) in 0.800 s during the first part of its motion. Use upstream as the positive direction.

   a. Determine both the velocity of the water spider relative to the water during its dash upstream and its velocity during its drift downstream.

   b. How far upstream relative to the water does the water spider move during one cycle of this upstream and downstream motion?

   c. What is the average velocity of the water spider relative to the water for one complete cycle?

65. A car travels due east with a speed of 50.0 km/h. Rain is falling vertically with respect to Earth. The traces of the rain on the side windows of the car make an angle of 60.0° with the vertical. Find the velocity of the rain with respect to the following:

   a. the car

   b. Earth

56. 29.4 m/s
57. a. 22.2 s
   b. 2.00×10^2 s
58. a. 2.66 m/s
   b. 0.64 m
59. 22.5 s
60. 157 km
61. a. 32.5 m
   b. 1.78 s
62. 7.0×10^1 s
63. a. 5.0×10^1 m/s
   b. 4.0×10^1 m
64. a. 1.20 m/s, 0.00 m/s
   b. 0.960 m
   c. 0.500 m/s
65. a. 57.7 km/h at 60.0° west of the vertical
   b. 28.8 km/h straight down

66. 12.0 s

67. 18 m, 7.9 m

68. 15.3 m

69. 6.19 m/s downfield

66. A shopper in a department store can walk up a stationary (stalled) escalator in 30.0 s. If the normally functioning escalator can carry the standing shopper to the next floor in 20.0 s, how long would it take the shopper to walk up the moving escalator? Assume the same walking effort for the shopper whether the escalator is stalled or moving.

67. If a person can jump a horizontal distance of 3.0 m on Earth, how far could the person jump on the moon, where the free-fall acceleration is $g/6$ and $g = 9.81$ m/s^2? How far could the person jump on Mars, where the acceleration due to gravity is $0.38g$?

68. A science student riding on a flatcar of a train moving at a constant speed of 10.0 m/s throws a ball toward the caboose along a path that the student judges as making an initial angle of 60.0° with the horizontal. The teacher, who is standing on the ground nearby, observes the ball rising vertically. How high does the ball rise?

69. A football is thrown toward a receiver with an initial speed of 18.0 m/s at an angle of 35.0° above the horizontal. At that instant, the receiver is 18.0 m from the quarterback. In what direction and with what constant speed should the receiver run to catch the football at the level at which it was thrown?

Technology & Learning

Graphing calculators

Refer to Appendix B for instructions on downloading programs for your calculator. The program "Chap3" allows you to analyze a graph of height versus time for a baseball thrown straight up.

Recall the following equation from your studies of projectiles launched at an angle.

$$\Delta y = v_i(\sin \theta)\Delta t - \frac{1}{2}g(\Delta t)^2$$

The program "Chap3" stored on your graphing calculator makes use of the projectile motion equation. Given the initial velocity, your graphing calculator will use the following equation to graph the height (Y1) of the baseball versus the time interval (X) that the ball remains in the air. Note that the relationships in this equation are the same as those in the projectile motion equation shown above.

$$Y1 = VX - 4.9X^2$$

a. The two equations above differ in that the latter does not include the factor $\sin \theta$. Why has this factor been disregarded in the second equation?

Execute "Chap3" on the [PRGM] menu, and press [ENTER] to begin the program. Enter the value for the initial velocity (shown below), and press [ENTER] to begin graphing.

The calculator will provide the graph of the displacement function versus time. The x value corresponds to the time interval in seconds, and the y value corresponds to the height in meters. If the graph is not visible, press [WINDOW] and change the settings for the graph window.

Press [TRACE], and use the arrow keys to trace along the curve to the highest point of the graph. The y value there is the greatest height that the ball reaches. Trace the curve to the right, where the y value is zero. The x value is the duration of the ball's flight.

For each of the following initial velocities, identify the maximum height and flight time of a baseball thrown vertically.

b. 25 m/s

c. 35 m/s

d. 75 m/s

e. Does the appearance of the graph represent the actual trajectory of the ball? Explain.

Press [2nd] [QUIT] to stop graphing. Press [ENTER] to input a new value or [CLEAR] to end the program.

ANSWERS TO

Technology & Learning

a. because $\sin (90°) = 1$

b. 32 m, 5.1 s

c. 62 m, 7.2 s

d. 290 m, 15 s

e. No, the graph represents vertical displacement versus time.

70. A rocket is launched at an angle of 53° above the horizontal with an initial speed of 75 m/s, as shown in **Figure 3-31.** It moves for 25 s along its initial line of motion with an overall acceleration of 25 m/s². At this time its engines fail and the rocket proceeds to move as a free body.

  a. What is the rocket's maximum altitude?
  b. What is the rocket's total time of flight?
  c. What is the rocket's horizontal range?

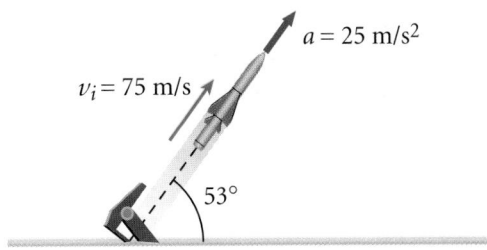

Figure 3-31

3 REVIEW & ASSESS

70. a. 2.4×10^4 m
  b. 152 s
  c. 7.8×10^4 m

Alternative Assessment

Performance assessment

1. Work in cooperative groups to analyze a game of chess in terms of displacement vectors. Make a model chessboard, and draw arrows showing all the possible moves for each piece as vectors made of horizontal and vertical components. Then have two members of your group play the game while the others keep track of each piece's moves. Be prepared to demonstrate how vector addition can be used to explain where a piece would be after several moves.

2. Use a garden hose to investigate the laws of projectile motion. Design experiments to investigate how the angle of the hose affects the range of the water stream. (Assume that the initial speed of water is constant and is determined by the pressure indicated by the faucet's setting.) What quantities will you measure, and how will you measure them? What variables do you need to control? What is the shape of the water stream? How can you reach the maximum range? How can you reach the highest point? Present your results to the rest of the class and discuss the conclusions.

3. How would a physics expert respond to the following suggestions made by three airline executives? Write a script of the expert's response for performance in front of the class.

Airline Executive A: Since the Earth rotates from west to east, we could operate "static flights"—helicopters that begin by hovering above New York City could begin their landing four hours later, when San Francisco arrives below.

Airline Executive B: This could work for one-way flights, but the return trip would take 20 hours.

Airline Executive C: That will never work. It's like when you throw a ball up in the air; it comes back to the same point.

Airline Executive A: That's only because the Earth's motion is not significant during that short a time.

Portfolio projects

4. You are helping NASA engineers design a basketball court for a colony on the moon. How do you anticipate the ball's motion compared with its motion on Earth? What changes will there be for the players—how they move and how they throw the ball? What changes would you recommend for the size of the court, the basket height, and other regulations in order to adapt the sport to the moon's low gravity? Create a presentation or a report presenting your suggestions, and include the physics concepts behind your recommendations.

5. There is conflicting testimony in a court case. A police officer claims that his radar monitor indicated that a car was traveling at 176 km/h (110 mi/h). The driver argues that the radar must have recorded the relative velocity because he was only going 88 km/h (55 mi/h). Is it possible that both are telling the truth? Could one be lying? Prepare scripts for expert witnesses, for both the prosecution and the defense, that use physics to justify their positions before the jury. Create visual aids to be used as evidence to support the different arguments.

Alternative Assessment
ANSWERS

Performance assessment

1. Chess pieces can move in ways that require more than one component. Student reports should show how several moves can be reported as a vector sum.

2. Student plans should be safe and include measurements of angle and range. They should find that 45° is the best angle for maximum range and 90° is the best angle for maximum height.

3. Student discussion should include the concept of relative velocity—when a helicopter takes off from Earth, it is already moving with Earth's velocity.

Portfolio projects

4. Students should recognize that balls will stay in the air longer. Players will need to shoot lower and more gently, or the court should be longer and the basket higher.

5. Students should recognize that if the cars were driving toward each other, both could be telling the truth.

NOTE

Materials Preparation is given on pp. 82A–82B. Blank data table and sample data table are on the One-Stop Planner CD-ROM. All calculations shown use sample data.

Planning

Recommended time:

1 lab period

Classroom organization:

▶ Each group must have at least 2 students.

▶ Each group needs a level surface at least 0.5 m above the floor, with at least 2.0 m of space in front of the table.

▶ **Safety warnings:** Remind students to be careful; balls on the floor represent a safety hazard.

Projectile motion tips

▶ With correct alignment, the box should catch the ball every time. Copy-paper boxes work well.

▶ Carbon paper may be taped down along one side. Handle carbon paper carefully and do not allow it to touch clothing.

Techniques to Demonstrate

Show students how to use the washer on a cord to find the baseline from which to measure the distance.

Some students may need help determining where to measure the height and length of the

OBJECTIVES

• Measure the velocity of projectiles in terms of the horizontal displacement during free fall.

• Compare the velocity and acceleration of projectiles accelerated down different inclined planes.

MATERIALS LIST

✔ aluminum sheet, edges covered with heavy tape
✔ C-clamp
✔ cardboard box
✔ cord
✔ inclined plane
✔ several sheets carbon paper
✔ several large sheets white paper
✔ masking tape
✔ meterstick
✔ packing tape
✔ small metal ball
✔ small metal washer
✔ support stand and clamp
✔ towel or cloth

CHAPTER 3
Laboratory Exercise

⭐ TEKS

pp. 120–121: 1A, 1B, 2B, 2C, 2F, 4B

VELOCITY OF A PROJECTILE

In this experiment, you will determine the horizontal velocity of a projectile and compare the effects of different inclined planes on this projectile motion.

SAFETY

• **Tie back long hair, secure loose clothing, and remove loose jewelry to prevent their getting caught in moving or rotating parts.**

• **Perform this experiment in a clear area. Falling or dropped masses can cause serious injury.**

PREPARATION

1. Read the entire lab, and plan what measurements you will take.

2. Prepare a data table in your lab notebook with five columns and nine rows. In the first row, label the columns *Trial, Height of ramp (m), Length of ramp (m), Displacement Δx (m), Displacement Δy (m)*. In the first column, label the second through ninth rows *1, 2, 3, 4, 5, 6, 7,* and *8*.

3. Set up the inclined plane at any angle, as shown in **Figure 3-32(a).** Tape the aluminum to the end of the plane and to the table. Leave at least 5 cm between the bottom end of the inclined plane and the edge of the table.

4. In front of the table, place the box to catch the ball after it bounces. Perform a practice trial to find the correct placement of the box. Use masking tape to secure the box to the floor. Cover the floor with white paper. Cover the white paper with the carbon paper, carbon side down. Tape the paper down.

5. Use a piece of cord and tape to hang a washer from the edge of the table so that the washer hangs a few centimeters above the floor. Mark the floor directly beneath the washer with tape. Move the washer to another point on the table edge, and repeat. Connect the two marks on the floor with masking tape.

PROCEDURE

6. Use tape to mark a starting line near the top of the inclined plane. Measure the height of the ramp from the tabletop to the tape mark. Measure the length along the ramp from the tape mark to the tabletop. Measure

120

the distance from the top of the tabletop to the floor. Enter these values in the data table for *Trial 1* as *Height of ramp (m)*, *Length of ramp (m)*, and *Displacement Δy*.

7. Place the metal ball on the inclined plane at the tape mark. Keep the area around the table clear of people and obstructions. Release the ball from rest so that it rolls down the inclined plane, off the table onto the carbon paper, and bounces into the box.

8. Lift the carbon paper. There should be a heavy carbon mark on the white paper where the ball landed. Label the mark with the trial number.

9. Replace the carbon paper and repeat this procedure as *Trial 2*.

10. With the inclined plane in the same position, place another tape mark about halfway down the inclined plane. Measure and record the height and length of the inclined plane from this mark. Use this mark as the starting point for *Trial 3* and *Trial 4*. Record all data.

11. Raise or lower the inclined plane and repeat the procedure. Perform two trials for each tape mark as *Trials 5, 6, 7*, and *8*. For each trial, measure the distance from the carbon mark to the tape line. Record this distance as *Displacement Δx*.

12. Clean up your work area. Put equipment away safely so it is ready to be used again. Recycle or dispose of used lab materials as directed by your teacher.

ANALYSIS AND INTERPRETATION

Calculations and data analysis

1. **Organizing data** Find the time interval for the ball's motion from the edge of the table to the floor using the equation for the vertical motion of a projectile from page 100, where Δy is the *Displacement Δy* recorded in your data table. The result is the time interval for each trial.

2. **Organizing data** Using the time interval from item 1 and the value for *Displacement Δx*, calculate the average horizontal velocity for each trial during the ball's motion from the edge of the table to the floor.

Conclusions

3. **Inferring conclusions** What is the relationship between the height of the inclined plane and the horizontal velocity of the ball? Explain.

4. **Inferring conclusions** What is the relationship between the length of the inclined plane and the horizontal velocity of the ball? Explain.

inclined plane.

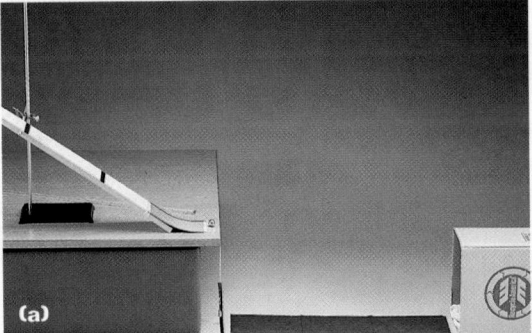

(a)

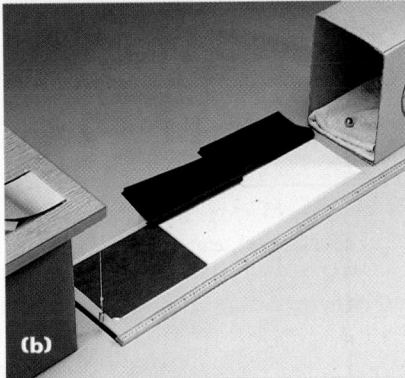

(b)

Figure 3-32

Step 3: Use tape to cover the sharp edges of the aluminum sheet before taping it to the end of the plane. The aluminum keeps the ball from bouncing as it rolls onto the table.

Step 4: Place a soft cloth in the box as shown. Throughout the lab, be careful not to have too much contact with the carbon paper.

Step 5: This tape line will serve as the baseline for measuring the horizontal distance traveled by the projectile.

✔ Checkpoints

Step 3: Make sure the plane is secure. Make sure all edges on the aluminum sheet are taped.

Step 4: Remind students to watch out for balls on the floor. The box should be taped to the floor and filled with a folded towel or cloth.

Step 5: Students should be able to explain why the distance must be measured from the tape line.

Step 10: For *Trials 3* and *4*, start the ball from the lower tape mark without moving the plane.

Step 11: *Trials 5* and *6*: release the ball from the higher tape mark; *Trials 7* and *8*: release from the lower mark. Students must measure the distances before lifting the paper from the floor.

CHAPTER 4 PLANNING GUIDE

Compression Guide: To shorten from 12 to 8 45-min periods (from 6 to 4 90-min blocks), eliminate items in magenta type.

PACING CHART	CLASSROOM RESOURCES			
	⊛ TEKS	Teacher Demonstrations	*Holt Physics* Transparencies	Labs (See page T52 for equipment listing for in-text labs.)
4-1 Changes in Motion 2 or 1 45-minute periods 1 or ½ 90-minute block	1A, 3A, 3C, 3E, 4B, 4C, 4D	**TE** *Contact vs. field forces*, p. 125 **TE** *Component forces*, p. 126	 **T** 9–10	**L** *Discovery Lab: Discovering Newton's Laws* **PE** *Quick Lab: Forces and Changes in Motion*, p. 126
4-2 Newton's first law 1 45-minute period ½ 90-minute block	2C, 2E, 3A, 3B, 3C, 3E, 4A, 4B, 4C, 4D	**TE** *Inertia*, p. 130 **TE** *Calculating net force*, p. 131	**T** 11–12	**PE** *Quick Lab: Inertia*, p. 134
4-3 Newton's second and third laws 3 45-minute periods 1½ 90-minute block	2C, 2E, 3A, 3B, 4B, 4C, 4D	**TE** *Newton's second law*, p. 136		**PE** *Force and Acceleration*, p. 158
4-4 Everyday forces 4 or 1 45-minute periods 2 or ½ 90-minute blocks	2C, 2E, 3A, 3B, 3C, 4A, 4B, 4C, 4D, 6A	**TE** *Static vs. kinetic friction*, p. 143 **TE** *Friction of different surfaces*, p. 143 **TE** *Friction and surface area*, p. 144	**T** 13–14 **TM** 19	**L** *Invention Lab: Friction Testing Materials* **TL** *Static and Kinetic Friction* **TL** *Air Resistance*

Review and Assessment
2 45-minute periods
1 90-minute block

Resource Key

PHYSICS

PE Pupil's Edition
TE Teacher's Edition

L Laboratory Experiments
TL Technology Lab Experiments
T Transparencies

👆 **One-Stop** Planner CD-ROM contents

TM Transparency Masters
SR Section Review Worksheets
AA Alternative Assessment

PW Problem-Solving Workbook
PB Problem Bank
CTW Critical Thinking Worksheet

LABORATORY PLANNING: Force and Acceleration, p. 158

Materials (for each lab group)
- 1 dynamics cart
- meterstick
- metric slotted mass set, 1 g–500 g
- metric hooked mass, 1000 g
- roll of thick, white, braided cord
- balance: portable, electronic balance or triple-beam balance with weight
- roll of adhesive tape, 0.5 in. wide

Additional Equipment CBL and Sensors Procedure
- CBL
- graphing calculator
- CBL ultrasonic motion detector
- CBL student force sensor
- V-jaw symmetrical clamp with holder
- 1-position support base and rod, 1.3 cm × 91 cm
- table clamp
- lattice rod, 1.3 cm × 30 cm

ASSIGNMENT RESOURCES

Content Mastery	Critical Thinking	Problem-Solving Practice	
PE 1–6, p. 128 **SR** 4-1, *Diagram Skills*			
PE 1–5, p. 135 **SR** 4-2, *Diagram Skills* **PE** 1–4, p. 151	**PE** 5–9, p. 151	**4A**	Net external force: 26 items in **PE, PW,** and **PB,** see **TE** pp. 132–133
PE 1–5, p. 140 **SR** 4-3, *Diagram Skills* **PE** 13–16, p. 152	**PE** 1–2, p. 138 **PE** 17–19, p. 152	**4B**	Newton's second law: 51 items in **PE, PW,** and **PB,** see **TE** pp. 137–138
PE 1–6, p. 149 **SR** 4-4, *Concept Review* **PE** 26–31, p. 153	**PE** 32–36, p. 153	**4C** **4D**	Coefficients of friction: 30 items in **PE, PW,** and **PB,** see **TE** p. 145 Overcoming friction: 34 items in **PE, PW,** and **PB,** see **TE** pp. 146–147

ASSESSMENT RESOURCES

Cumulative Review	Alternative Assessment	Traditional Assessment
SR Mixed Review, Ch. 4	**PE** 1–5, p. 157 **AA** Items for Ch. 4	Chapter 4 Test Test Generator items for Ch. 4

Scoring Rubrics for Alternative Assessment items can be found on the One-Stop Planner CD-ROM.

TECHNOLOGY RESOURCES

 **CTW** Segment 4 Crash-Test Dummies

 Module 4 Force and Inertia

 PE Technology and Learning, p. 156
(Alternative procedures for calculators without Flash-ROM technology are provided on the One-Stop Planner CD-ROM.)

 The Mechanical Universe/High School Adaptation Quad I, Newton's Laws Quad III, Inertia

internet connect

 On-line Student Resources:
GO TO: www.scilinks.org
The following student resources can be found in the student text for this chapter.

TOPICS:
• Forces, p. 126 (HF2041)
• Newton's laws, p. 139 (HF2042)
• Friction, p. 148 (HF2044)

 On-line Teacher Resources:
GO TO: go.hrw.com
KEYWORD: HF2 HOME
Visit the HRW Web site for a variety of resources related to this chapter.

Smithsonian Institution
Internet Connections
Visit **www.si.edu/hrw** for additional on-line resources.

 Visit **www.cnnfyi.com** for late-breaking news and current events stories selected just for you.

• right-angle clamp for 1.3 cm rod
• low-friction rod-mounted pulley
• piece of poster board, 25 cm × 25 cm
• TI Graph Link (recommended for downloading programs)

• 1-position support base and rod, 1.3 cm × 91 cm, or **C**-clamp
• adjustable table-clamp pulley
• replacement paper tape, 13 cm
• replacement carbon disks

Recording Timer Procedure
• recording timer: acceleration timer, table top acceleration timer, or compact spark timer

Materials Preparation
Set up a sample apparatus in the laboratory for students to refer to as they set up their equipment.

Section 4-1 defines *force* and introduces free-body diagrams.

Section 4-2 discusses Newton's first law and the relationship between mass and inertia.

Section 4-3 introduces the relationships between net external force, mass, and acceleration, and discusses action-reaction pairs.

Section 4-4 examines the familiar forces of weight, normal force, and friction.

About the Illustration

Crash-test dummies are equipped with up to 48 sensors: accelerometers and force meters are placed at different positions and depths to record the amount of force applied to the head, the bones, the organs, and the skin. Dummies are designed to resemble people of different shapes and sizes, from a six-month-old baby to a pregnant woman to a 223 lb, 7 ft tall man.

PHYSICS INTERACTIVE TUTOR

Interactive Problem-Solving Tutor

See Module 4

"Force and Inertia" provides additional development of problem-solving skills for this chapter.

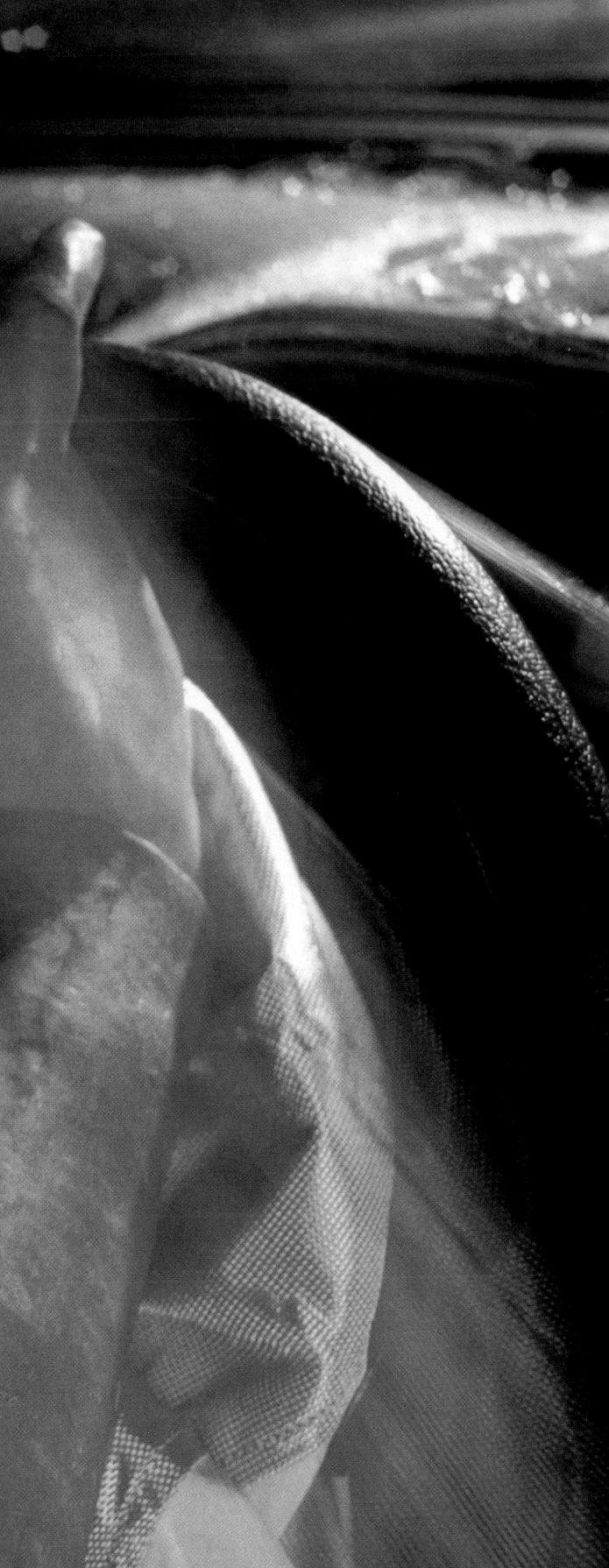

CHAPTER 4

Forces and the Laws of Motion

PHYSICS IN ACTION

At General Motors' Milford Proving Grounds, in Michigan, technicians place a crash-test dummy behind the wheel of a new car. While the car accelerates and slams into a concrete barrier, the scientists watch computer screens that display data from sensors implanted in the dummy. These sensors report the acceleration and forces acting on the dummy during the collision. The technicians study the data from the crash to design safer cars and more-effective restraint systems.

In this chapter you will study forces and learn to analyze motion in terms of force.

- *What are the forces acting on a dummy during a car crash?*
- *Why are seat belts necessary to keep dummies in their seats during a crash?*

CONCEPT REVIEW

Acceleration (Section 2-2)

Kinematic equations (Section 2-2)

Vector diagrams (Section 3-1)

Tapping Prior Knowledge

Knowledge to Expect
✔ "An object that is not being subjected to a force will continue to move at a constant speed in a straight line." (NRC's *National Science Education Standards,* Grades 5–8)

✔ "If more than one force acts on an object along a straight line, then the forces will reinforce or cancel one another, depending on their direction and magnitude. Unbalanced forces will cause changes in speed or direction of an object's motion." (NRC's *National Science Education Standards,* Grades 5–8)

Knowledge to Review
✔ Acceleration is the change in velocity (magnitude or direction) per unit time interval. (Section 2-2)

✔ Kinematic equations describe the motion of an object with constant acceleration. (Section 2-2)

✔ Vectors are quantities that have both magnitude and direction; they can be represented by arrows drawn in the appropriate direction, and the length of the arrows represents magnitude. (Section 3-1)

Items to Probe
✔ Vector addition: Have students practice resolving vectors into components, adding the components, and finding the resultant of the vector addition.

124

TEKS

p. 124: 3E
p. 125: 3A

Visual Strategy

Figure 4-1

Point out to students that the ball is experiencing force in all three pictures.

Q How can you tell that the ball experiences at least one force in each picture?

A by changes in the ball's speed or direction

Teaching Tip

Now that students have studied motion as complex as projectile motion, explore their understanding of force. Ask them what mechanism causes motion and why some objects accelerate at higher rates than others do. Point out that *force* describes any mechanism that causes or may cause a change in an object's velocity.

4-1
Changes in motion

4-1 SECTION OBJECTIVES

- **Explain how force affects the motion of an object.**
- **Distinguish between contact forces and field forces.**
- **Interpret and construct free-body diagrams.**

force

the cause of an acceleration, or the change in an object's velocity

FORCE

When we think of **force**, we usually imagine a *push* or a *pull* exerted on some object. For instance, you exert a force on a ball when you throw or kick it, and you exert a force on a chair when you sit in it. Force represents the interaction of an object with its environment.

Forces cause changes in velocity

In many situations, a force exerted on an object can change the object's velocity. Some examples of this are shown in **Figure 4-1.** A force can cause a stationary object to move, as when you throw a ball. Force also causes moving objects to stop, as when brakes stop your car or when you catch a ball. A force can also cause a moving object to change its direction, such as when a baseball collides with a bat and flies off in another direction. Notice that each of these cases involves a change in velocity—an acceleration.

(a) (b) (c)

Figure 4-1
Force can cause objects to
(a) start moving, **(b)** stop moving,
or **(c)** change direction.

The SI unit of force is the newton ⭐ TEKS 3E

The SI unit of force is the newton, named after Sir Isaac Newton (1642–1727), whose work contributed much to the modern understanding of force and motion. The newton (N) is defined as the amount of force that, when acting on a 1 kg mass, produces an acceleration of 1 m/s^2. Therefore, 1 N = 1 kg × 1 m/s^2.

The weight of an object is a measure of the magnitude of the gravitational force exerted on the object. It is the result of the interaction of an object's mass with the gravitational field of another object, such as Earth. Many of the terms

Table 4-1   **Units of mass, acceleration, and force**

System	Mass	Acceleration	Force
SI	kg	m/s^2	$N = kg \cdot m/s^2$
cgs	g	cm/s^2	$dyne = g \cdot cm/s^2$
Avoirdupois	slug	ft/s^2	$lb = slug \cdot ft/s^2$

and units you use every day to talk about weight are really units of force that can be converted to the SI unit of newtons. For example, a $\frac{1}{4}$ lb stick of margarine has a weight equivalent to a force of about 1 N, as shown in the following conversions:

$$1 \text{ lb} = 4.448 \text{ N}$$
$$1 \text{ N} = 0.225 \text{ lb}$$

Forces can act through contact or at a distance

If you pull on a spring, the spring stretches. If you pull on a wagon, the wagon moves. When a football is caught, its motion is stopped. These pushes and pulls are examples of **contact forces,** so named because they result from physical contact between two objects. This kind of force is usually easy to identify when you analyze a situation.

Another class of forces—called **field forces**—does not involve physical contact between two objects. One example of this kind of force is the force of gravity. Whenever an object falls to Earth, it is accelerated by Earth's gravity. In other words, Earth exerts a force on the object, even when it is not in immediate physical contact with the object.

Another common example of a field force is the attraction or repulsion between electrical charges. You can observe this force by rubbing a balloon against your hair and then observing how little pieces of paper appear to jump up and cling to the balloon's surface, as shown in **Figure 4-2.** The paper is pulled by the balloon's electric field.

The theory of fields was developed as a tool to explain how objects could exert force on each other without touching. According to this theory, the presence of an object affects the space around it so that a force is exerted on any other object placed within that space. This region of influence is called a field. Objects exert forces on one another when their fields interact. For example, an object falls to Earth because of the interaction between the object and the gravitational field of Earth. (★TEKS) **3A**

Field forces are especially important in the study of particle physics. The known fundamental forces in nature, which act on elementary particles, are all field forces. This means that elementary particles have no actual contact—all their interactions are the results of field forces.

contact force

force that arises from the physical contact of two objects

field force

force that can exist between objects, even in the absence of physical contact between the objects

Figure 4-2
The electric field around the balloon exerts an attractive force on the pieces of paper.

force diagram

a diagram of the objects involved in a situation and the forces exerted on the objects

★ TEKS **1A, 4C, 4D**

Quick Lab

Force and Changes in Motion

MATERIALS LIST

✔ 1 toy car
✔ 1 book

Use a toy car and a book to model a car colliding with a brick wall. Observe the motion of the car before and after the crash. Identify as many changes in its motion, such as a change in speed or in direction, as you can. Make a list of all the changes, and try to identify the forces that caused them. Make a force diagram of the collision.

FORCE DIAGRAMS

If you give a toy car a light push, it does not move as fast as it does when you give a harder push. The effect of a force depends on its magnitude, and the magnitude of the force of the second push is greater. The effect of a force on an object's motion also depends on the direction of the force. For example, if you push the toy car from the front, the car will move in a different direction than if you push it from behind.

Force is a vector

Because the effect of a force depends on its magnitude and direction, force is a vector quantity. Diagrams that show force vectors as arrows, like **Figure 4-3(a),** are called **force diagrams.** In this book, the arrows used to represent forces are blue. The tail of the arrow is attached to the object on which the force is acting. These diagrams can be used as tools in analyzing collisions and other situations.

At this point, we will disregard the size and shape of objects and assume that all forces act on a point at the center of an object. In force diagrams, all forces are drawn as if they act on that point, no matter where the force is applied.

A free-body diagram helps analyze a situation ★ TEKS **4D**

Once engineers analyzing a test-car crash have identified all the forces involved, they isolate the car from the other objects in its environment. One of their goals is to determine which forces affect the car and its passengers. **Figure 4-3(b)** is a free-body diagram. It represents the same collision as the force diagram **(a),** but it shows only the car and the forces acting on it. The forces exerted *by* the car on other objects are not included in the free-body diagram because they do not affect the motion of the car.

A free-body diagram is used to analyze only the forces affecting the motion of a single object. Free-body diagrams are constructed and analyzed just like other vector diagrams. In Section 4-2, you will learn to use free-body diagrams to find component and resultant forces. In this section, you will learn to draw free-body diagrams for situations described in this book.

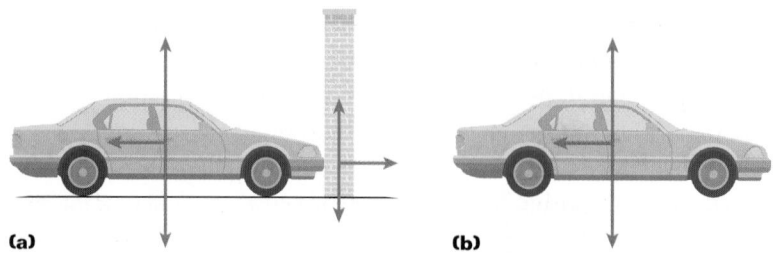

(a)   **(b)**

Figure 4-3
(a) In a force diagram, vector arrows represent all the forces acting in a situation. **(b)** A free-body diagram shows only the forces acting on the object of interest—in this case, the car.

The photograph in **Figure 4-4(a)** shows a car being pulled by a tow truck. In a situation like this, there are many forces acting on the car. The tow truck exerts a force on the car in the direction of the cable, the road exerts forces on the car, and the car is also acted on by a gravitational force. The relationships between these forces help describe the motion of the car as it is being towed. A free-body diagram of the car will show all the forces acting on the car as if the forces are acting on the center of the car. To draw the free-body diagram, you must first isolate and identify all the forces acting on the car.

 Misconception Alert

It is important to emphasize early and consistently that a free-body diagram shows only the forces acting *on* the object. A separate free-body diagram for the tow truck in **Figure 4-4** can be used to emphasize this point as well as to introduce Newton's third law.

Teaching Tip

A good understanding of free-body diagrams is essential to strong physics problem-solving skills. Take this time to make sure students can properly dissect a situation involving several forces. You may wish to do several examples on the board to further emphasize the importance of the diagram step of problem solving.

(a)

Free-body diagrams isolate an object and the forces acting on it

The figures shown in **Figure 4-4(b)** through **(f)** show all the steps required to draw a free-body diagram of the car. Following these general steps will allow you to draw a free-body diagram to isolate the forces acting on a single object for any situation. (★TEKS) **4D**

Before you begin identifying any of the forces acting in the situation, draw a diagram to represent the isolated object under consideration, as shown in step **(b).** It is often helpful to draw a very simple shape with some distinguishing characteristics that will help you visualize the object. In this case, the car is shown with a simple shape that is positioned in the same way as the car in the photograph. Positioning the car the same way in the diagram will help you draw the arrows representing the forces in the correct positions. Free-body diagrams are often drawn using simple squares, circles, or even points to represent the object. In most cases in this book, free-body diagrams will use shapes that help you visualize the specific situations.

The next step is to draw and label vector arrows representing all external forces acting on the object. The diagram in **(c)** shows the force exerted by the towing cable attached to the car. The arrow is shown originating at the center of the car because in this example all forces are assumed to act on a single point at the center of the object. When you draw the arrow representing the force, it is important to label the arrow either with the size of the force or a name that will distinguish it from the other forces acting on the car. In this case, the force the tow truck exerts on the car is 5800 N. The arrow is shown pointing in the same direction as the force of the cable on the car.

The gravitational force acting on the car is 14 700 N, directed toward the center of Earth. This force is shown in part **(d).** The road exerts an upward

Figure 4-4
The steps for drawing a free-body diagram to analyze the forces acting on this car **(a)** are shown in **(b)** through **(f).**

(b)

(c) 5800 N

(d) 5800 N

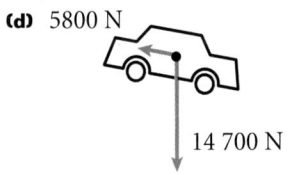

14 700 N

(e–f) on next page

(★TEKS)

p. 126: 1A, 4C, 4D, 4D
p. 127: 4D

1. **a.** accelerating a car, lifting a chess piece off the board, throwing a ball
   b. braking a car, catching a ball, pushing against a moving object
   c. turning a car, hitting a baseball with a bat, passing a soccer ball

2. gravity and electric force; hitting a baseball and catching a baseball (answers will vary); These are forces because they cause a change in motion.

3. Two force vectors should point from the center of the football: **F$_g$** points down and **F$_{kicker}$** points in the direction of the kick.

4. All force vectors should originate at the center of the object on which they act, with forward to the right, backward to the left, and downward pointing to the bottom of the page; because this is not a free-body diagram, the forces on both the car and the dummy can be drawn in one diagram.

5. The diagram should include the car and all the forces that act on the car. The dummy and the forces exerted on it should not be shown.

6. The diagram should include the dummy and all the forces that act on the dummy. The car and the forces exerted on it should not be shown.

128

(e)

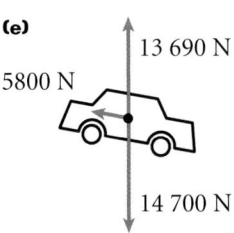

13 690 N
5800 N
14 700 N

(f)

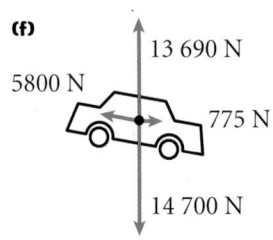

13 690 N
5800 N
775 N
14 700 N

Figure 4-4(e–f)

force on the car equal to 13 690 N, as shown in part **(e)**. Because of the interaction between the road and the car's tires, the road also exerts a backward force of friction on the car equal to 775 N, as shown in part **(f)**.

Make sure that only the forces acting on the car are included in your free-body diagram. Even though the car exerts forces on other objects, the only forces that should be included in the free-body diagram are the forces acting on the car.

The figure in part **(f)** is the completed free-body diagram of the car being towed. Compare the free-body diagram to the car in the photograph. Although the two images of the car are very different, there is a strong resemblance between them. As you learn more about physics, the free-body diagram may begin to look more and more like the real image of the car because it reveals more information about the car in physics terms than the photograph does.

A free-body diagram can be used to find the net external force acting on an object, using the rules for vector analysis from Chapter 3.

Section Review   ⭐TEKS **4B, 4C, 4D**

1. List three examples of each of the following:
   a. a force causing an object to start moving
   b. a force causing an object to stop moving
   c. a force causing an object to change direction

2. Give two examples of field forces described in this section and two examples of contact forces you observe in everyday life. Explain how you know that these are forces.

3. Draw a free-body diagram of a football being kicked. Assume that the only forces acting on the ball are the force of gravity and the force exerted by the kicker.

4. **Physics in Action** Draw a force diagram of a crash-test dummy in a car at the moment of collision. For this problem, assume that the forces acting on the car are 19 600 N downward, 17 800 N forward, and 25 000 N backward. The forces acting on the dummy are 585 N downward, 175 N backward, and 585 N upward.

5. **Physics in Action** Use the information given above to draw a free-body diagram showing only the forces acting on the car in item 4. Label all forces.

6. **Physics in Action** Use the information given above to draw a free-body diagram showing only the forces acting on the dummy in item 4. Label all forces.

Indestructible Alloy ⭐TEKS 3C, 3D, 3E

Impervium? Eternium? What do you call a nearly unbreakable metal? Dr. Richard Waterstrat has been pondering that question ever since he invented an alloy that is nearly impervious to wear.

Several years ago at the National Institute for Standards and Technology, Dr. Waterstrat was developing a metal for use in artificial hips and knees. Because the alloy would be implanted into the human body, it needed to be both nontoxic and crack-resistant.

He made one alloy by mixing three metals—zirconium, palladium, and ruthenium. The alloy seemed promising, so Dr. Waterstrat sent it down to the machine shop to prepare a sample for testing. Soon after, Dr. Waterstrat received a call from a worker in the shop who reported that he was unable to cut the alloy in a lathe, using conventional methods. At first, Dr. Waterstrat thought that the metal might simply be too hard to cut. But closer examination revealed that a very thin fibrous layer had formed wherever force had been applied to the metal. The crystalline structure of the alloy's surface had changed in response to the force to prevent new damage. The metal had actually "healed itself."

"The new crystals are harder and stronger than the original crystals," Dr. Waterstrat explained, "and that reinforces or 'heals' the defects that form as a result of the applied stress, making the material, in fact, stronger than it was to begin with. So the unusual wear-resistance is due to the fact that the metal is continually forming crystals under stress to resist further wear."

Dr. Waterstrat's alloy not only was resistant to cracking but also was found to be nearly impervious to wear. He submitted the alloy to a test to measure its wear-resistance, its ability to withstand intense wear over a long period of time. After a pin was rubbed against the metal for 5 million cycles, the alloy showed practically no wear. Artificial joints are constantly subjected to wear, so it seemed that Dr. Waterstrat had finally found his ideal metal. While the alloy is still being perfected for use in joint replacements, there are other applications for which it might be used. Any piece of metal that is subject to extreme wear, such as drill bits, bearings in machinery, or needles in sewing machines, could be coated with the wonder metal. The alloy also seems to be corrosion-resistant, suggesting that it could be used as the metal contact in an electric circuit, a part that is constantly subjected to high wear and corrosion. The indestructible alloy is here; now all it needs is a name.

BACKGROUND

Dr. Waterstrat's impressive alloy is resistant to very large forces, even forces applied over a very small area. It is important to recognize that the human body must be able to withstand many forces to function as well as it does.

Dr. Waterstrat's new alloy has possible applications not only in orthopedics but also in dentistry.

EXTENSION

Tell students that this feature relates not only to biotechnology, but also to material sciences, specifically metallurgy. Have students research careers in one of these two fields.

To summarize their findings, students should create a chart of information about the field, including salary, work hours, work environment, and benefits packages.

⭐TEKS
p. 128: 4B, 4C, 4D
p. 129: 3C, 3D, 3E

4-2
Newton's first law

4-2 SECTION OBJECTIVES

- **Explain the relationship between the motion of an object and the net external force acting on it.**

- **Determine the net external force on an object.**

- **Calculate the force required to bring an object into equilibrium.**

Figure 4-5
A hovercraft floats on a cushion of air above the water. Air provides less resistance to motion than water does.

inertia

the tendency of an object to maintain its state of motion

INERTIA

A hovercraft, such as the one in **Figure 4-5,** glides along the surface of the water on a cushion of air. Physicists study examples like this to determine exactly what effect forces have on the motion of objects. ⭐TEKS **4C**

The first step is to examine the motion of the object before forces are applied. A common misconception is that an object with no force acting on it will always be at rest. To test whether this is true, consider the example of sliding a block on different surfaces. First imagine a block on a deep, thick carpet. If you apply a force by pushing the block, it will begin sliding, but soon after you remove the force, the block will come to rest. Next imagine pushing the same block across a smooth waxed floor. When you push with the same force, the block will slide much farther before coming to rest.

In the 1630s, Galileo realized that a block sliding on a perfectly smooth surface would slide forever in the absence of an applied force. He concluded correctly that it is not the nature of an object to *stop* once it is set in motion; rather, it is an object's nature to *maintain its state of motion.* ⭐TEKS **3E**

This concept was further developed by Newton in 1687 and has come to be known as **Newton's first law of motion.**

> **NEWTON'S FIRST LAW**
>
> **An object at rest remains at rest, and an object in motion continues in motion with constant velocity (that is, constant speed in a straight line) unless the object experiences a net external force.**

The tendency of an object not to accelerate is called **inertia.** Newton's first law is often referred to as the *law of inertia* because it states that in the absence of forces, a body will preserve its state of motion. In other words, Newton's first law says that *when the net external force on an object is zero, its acceleration (or the change in its velocity) is zero.*

Acceleration is determined by net external force  **TEKS** 4B, 4C

Consider a car traveling at a constant velocity. Newton's first law tells us that the net external force on the car must be equal to zero. However, **Figure 4-6** shows that there are many forces acting on a car in motion. The vector $\mathbf{F_{forward}}$ represents the forward force of the road on the tires. The vector $\mathbf{F_{resistance}}$, which acts in the opposite direction, is due partly to friction between the road surface and tires and partly to air resistance. The vector $\mathbf{F_{gravity}}$ represents the downward force of gravity on the car, and the vector $\mathbf{F_{ground\text{-}on\text{-}car}}$ represents the upward force the road exerts on the car.

To understand how a car can maintain a constant velocity under the influence of so many forces, it is necessary to understand the distinction between external force and net external force. An *external force* is a single force that acts on an object as a result of the interaction between the object and its environment. All four forces in **Figure 4-6** are external forces acting on the car. The **net external force** is the vector sum of all the forces acting on an object.

When all external forces acting on an object are known, the net external force can be found using the methods for finding resultant vectors. That means an object's acceleration is determined by the combination of all the forces acting on it. The net external force is equivalent to the one force that would produce the same effect on the object as all the external forces combined. In other words, the object's acceleration is the same as if the net external force were the only force acting on the object.

Consider **Figure 4-7,** in which two teams are playing tug-of-war. If both teams pull on the rope with equal but opposite force **(a),** the net external force on the rope is equal to zero. The knot in the center of the rope will be at rest, even though forces are pulling the rope in opposite directions. When one team increases the force of its pull **(b),** the knot will experience a net external force equal to the difference in the forces, and it will accelerate in the direction of the greater pull.

Examples like this show how an object can be in equilibrium even when many forces act on it. For example, although there are four forces acting on the car in **Figure 4-6,** the car will maintain its constant velocity as long as the vector sum of these forces is equal to zero.

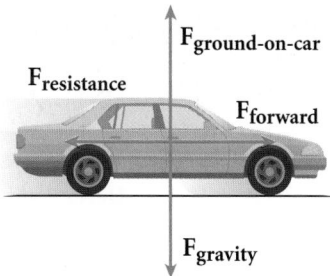

Figure 4-6
Although there are several forces acting on this car, the vector sum of the forces is zero, so the car moves at a constant velocity.

net external force

the total force resulting from a combination of external forces on an object; sometimes called the resultant force

Figure 4-7
(a) When there is no net force, the knot is at rest. **(b)** When there is a net force to the right, the knot moves to the right.

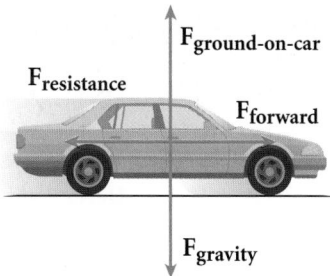

Demonstration 4

Calculating net force

Purpose Demonstrate that force applied at an angle to the direction of motion yields less of an effect on the object's motion than force applied in a straight line with the motion.

Materials two tables

Procedure Ask for a strong student volunteer. Have the student stand between two narrowly separated tables. Ask the student to place a hand on each table and to lift himself or herself from the floor. Have the student repeat the demonstration after you have moved the tables farther apart. Have the student repeat the demonstration after moving the tables still farther apart. Have the student report to the class how difficult this action is in each case.

Have students write in their notebooks why it became more difficult for the student to lift himself or herself as the tables were moved farther apart. Remind students that their explanation should utilize the concept of vector components. Use vector diagrams to show that as the tables were moved farther apart, the student had to exert greater forces along the arms in order to keep the vertical components of the lift equal to the student's weight.

TEKS

p. 130: 4C, 3E
p. 131: 4B, 4C

Classroom Practice

The following may be used as teamwork exercises or for demonstration at the chalkboard or on an overhead projector.

PROBLEM

Net external force

An agriculture student is designing a support to keep a tree upright. Two wires have been attached to the tree at right angles to each other. One wire exerts a force of 30.0 N on the tree; the other wire exerts a 40.0 N force. Determine the placement and force in the wire for a third wire so that the tree will have zero net force from the three wires.

Answer

50.0 N at 143° from the 40.0 N force and 127° from the 30.0 N force

A flying, stationary kite is acted on by a force of 9.8 N downward. The wind exerts a force of 45 N at an angle of 50.0° above the horizontal. Find the angle and force that the string exerts on the kite.

Answer

38 N, 40° below the horizontal

Net external force ⭐TEKS 2C, 3B, 4A, 4C, 4D

PROBLEM

Derek leaves his physics book on top of a drafting table that is inclined at a 35° angle. The free-body diagram in Figure 4-8 shows the forces acting on the book. Find the net external force acting on the book, and determine whether the book will remain at rest in this position.

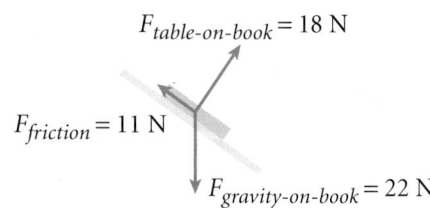

Figure 4-8

SOLUTION

1. Define the problem and identify the variables.

Given:
$$F_{gravity\text{-}on\text{-}book} = 22 \text{ N}$$
$$F_{friction} = 11 \text{ N}$$
$$F_{table\text{-}on\text{-}book} = 18 \text{ N}$$

Unknown: $F_{net} = ?$

2. Select a coordinate system, and apply it to the free-body diagram.

Choose the *x*-axis parallel to and the *y*-axis perpendicular to the incline of the table, as shown in **Figure 4-9.** This is the most convenient coordinate system because only the force of gravity on the book needs to be resolved into *x* and *y* components. All other forces are either along the *x*-axis or the *y*-axis.

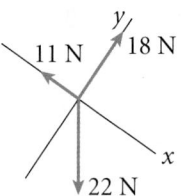

Figure 4-9

3. Find the *x* and *y* components of all vectors.

Draw a sketch as shown in **Figure 4-10** to help find the *x* component and *y* component of the vector **F**$_{gravity\text{-}on\text{-}book}$. Gravity acts at a 90° angle to the surface of Earth, and the *x*-axis is at a 35° angle to Earth's surface. Therefore, θ, the angle between the gravity vector and the *x*-axis, is the third angle in this right triangle and is equal to 55°.

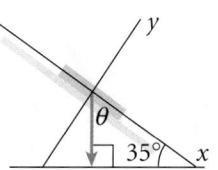

Figure 4-10

$$\cos \theta = \frac{F_{g,x}}{F_g}$$

$$F_{g,x} = F_g \cos \theta = (22 \text{ N})(\cos 55°) = 13 \text{ N}$$

$$\sin \theta = \frac{F_{g,y}}{F_g}$$

$$F_{g,y} = F_g \sin \theta = (22 \text{ N})(\sin 55°) = 18 \text{ N}$$

Now those components can be added to the free-body diagram, as shown in **Figure 4-11.**

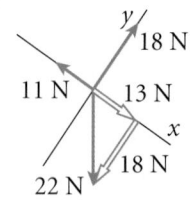

Figure 4-11

4. Find the net external force in both the *x* and *y* directions.

Figure 4-12 shows another free-body diagram of the book, now with forces acting only along the *x*- and *y*-axes.

For the *x* direction: $\Sigma F_x = F_{g,x} - F_{friction}$

$$\Sigma F_x = 13 \text{ N} - 11 \text{ N} = 2 \text{ N}$$

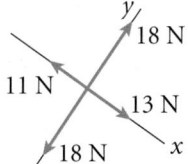

Figure 4-12

For the y direction: $\Sigma F_y = F_{table\text{-}on\text{-}book} - F_{g,y}$

$$\Sigma F_y = 18 \text{ N} - 18 \text{ N} = 0 \text{ N}$$

5. **Find the net external force.**

The net force in the y direction is equal to zero, so the net external force is equal to the net force in the positive x direction, 2 N positive.

$$\boxed{F_{net} = 2 \text{ N}}$$

6. **Evaluate your answer.**

The net external force acts on the book in the downhill direction. Therefore, the book will experience an acceleration in the downhill direction, and it will slide off the table, as shown **Figure 4-13**.

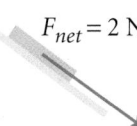

$F_{net} = 2$ N

Figure 4-13

PRACTICE 4A

Net external force ⭐TEKS 2C, 3B, 4B, 4C, 4D

1. A man is pulling on his dog with a force of 70.0 N directed at an angle of +30.0° to the horizontal. Find the x and y components of this force.

2. A crate is pulled to the right with a force of 82.0 N, to the left with a force of 115 N, upward with a force of 565 N, and downward with a force of 236 N.

  a. Find the net external force in the x direction.
  b. Find the net external force in the y direction.
  c. Find the magnitude and direction of the net external force on the crate.

3. A gust of wind blows an apple from a tree. As the apple falls, the force of gravity on the apple is 9.25 N downward, and the force of the wind on the apple is 1.05 N to the right. Find the magnitude and direction of the net external force on the apple.

4. The wind exerts a force of 452 N north on a sailboat, while the water exerts a force of 325 N west on the sailboat. Find the magnitude and direction of the net external force on the sailboat.

🛑 **Misconception Alert**

Students may think that if two objects are at rest, they have the same amount of inertia. After checking for this misconception, point to a pencil and a book, and ask which has more inertia.

Mass is a measurement of inertia ⭐TEKS 4B, 4C

Imagine a basketball and a golf ball at rest side by side on the ground. Newton's first law states that both remain at rest as long as no net external force acts on them. Now imagine supplying a net force by striking each ball with a golf club. If the two are struck with equal force, the golf ball will accelerate much more than the basketball. The basketball experiences a smaller acceleration because it has more inertia than the golf ball.

Quick Lab

TEACHER'S NOTES

Students must either hold the ball in place when pushing the skateboard up to speed or accelerate the skateboard slowly so that friction holds the ball in place.

Include trials that use large and small accelerations.

Visual Strategy

Figure 4-14

Many students have difficulty visualizing how mechanical devices operate. You may want to further describe how the seat belt mechanism works so that students fully benefit from this illustration.

Q Which way will the rod turn (clockwise or counterclockwise) if the car comes to an abrupt stop?

A *clockwise*

134

As the example of the golf ball and the basketball shows, the inertia of an object is proportional to its mass. The greater the mass of a body, the less the body accelerates under an applied force. Similarly, a light object will undergo a larger acceleration than a heavy object under the same force. Therefore, mass, which is a measure of the amount of matter in an object, is also a measure of the inertia of an object.

Objects in motion tend to stay in motion ★ TEKS 3C, 4C

Imagine a crash-test dummy not wearing a seat belt at the moment of a collision. Even after the car stops, the dummy will continue to move forward with the same velocity until acted on by some force. In this case, the force stopping the dummy would be the force exerted on the dummy by the dashboard of the car.

The purpose of a seat belt is to prevent serious injury by holding a passenger firmly in place in the event of a collision. A seat belt may also lock when a car rapidly slows down or turns a corner. While inertia causes passengers in a car to continue moving forward as the car slows down, inertia also causes seat belts to lock into place.

Figure 4-14 illustrates how one type of shoulder harness operates. Under normal conditions, the ratchet turns freely to allow the harness to wind or unwind along the pulley. In a collision, the car undergoes a large acceleration and rapidly comes to rest. The large mass under the seat, because of its inertia, continues to slide forward along the tracks. The pin connection between the mass and the rod causes the rod to pivot and lock the ratchet wheel in place. At this point the harness no longer unwinds, and the seat belt holds the passenger firmly in place.

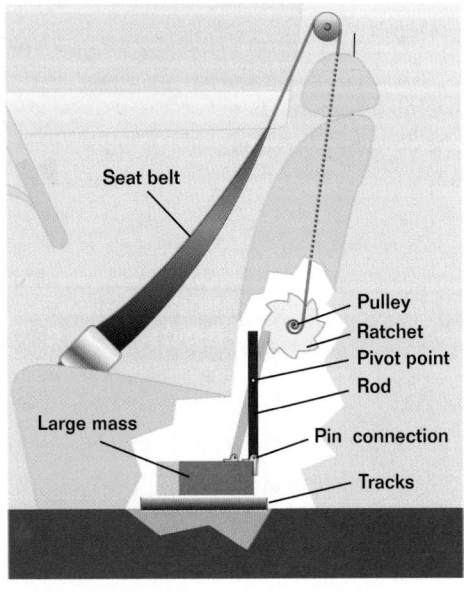

Seat belt

Pulley
Ratchet
Pivot point
Rod

Large mass

Pin connection

Tracks

Figure 4-14
When the car suddenly slows down, inertia causes the large mass under the seat to continue moving, which activates the lock on the safety belt.

EQUILIBRIUM

equilibrium

the state in which there is no change in a body's motion

★ TEKS 1A, 4B, 4C

Objects that are either at rest or moving with constant velocity are said to be in **equilibrium.** Newton's first law describes objects in equilibrium, whether they are at rest or moving with a constant velocity. Newton's first law states one condition that must be true for equilibrium: the net external force acting on a body in equilibrium must be equal to zero.

Quick Lab

Inertia

MATERIALS LIST

✔ skateboard or cart
✔ several toy balls with different masses

 **SAFETY CAUTION**

Perform this experiment away from walls and furniture that can be damaged.

Place a small ball near the back of a skateboard or cart. Push the skateboard across the floor and into a wall or other solid barrier. Observe what happens to the ball when the skateboard hits the wall. Perform several trials, placing the ball in different positions, such as in the middle of the skateboard and near the front of the skateboard. Repeat all trials using balls with different masses, and compare the results. Perform this experiment at different speeds, and compare the results.

The bob on the fishing line in **Figure 4-15(a)** is at rest. Therefore, we know that the net external force on the fishing bob is equal to zero. Imagine that a fish bites the bait, as shown in **Figure 4-15(b).** The fish exerts a force on the line. Because a net external force is acting on the line, the bob accelerates downward and disappears under the water. To return the bob to equilibrium, the person fishing must apply a force to the fishing line.

The force that brings an accelerating object into equilibrium must be equal and opposite to the force causing the object to accelerate. Therefore, if the person pulls on the line and applies a force to the bob that is equal and opposite to the force exerted by the fish, the net external force on the bob is equal to zero. This is shown in **Figure 4-15(c).**

An object is in equilibrium when the vector sum of the forces acting on it is equal to zero. Usually, the easiest way to determine whether a body is in equilibrium is to assign a coordinate system and resolve all forces into their x and y components. When the sum of all forces in the x direction is equal to zero (that is, $\Sigma F_x = 0$) and the sum of all forces in the y direction is equal to zero (that is, $\Sigma F_y = 0$), then the vector sum of *all* forces acting on the body is equal to zero and the body is in equilibrium. ⊛TEKS **3B, 4D**

(a)

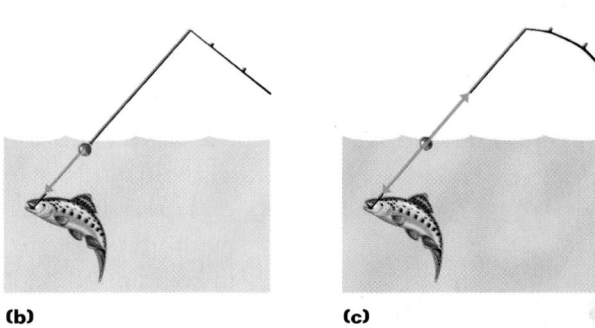

(b)　　　　　　　　**(c)**

Figure 4-15
(a) The bob on this fishing line is at rest. **(b)** When it is acted on by a net external force, it accelerates until **(c)** an equal and opposite external force is applied and the net external force is zero.

Visual Strategy

Figure 4-15
Point out that in order for the bob to be in equilibrium, all the forces must cancel. You may want to diagram this situation on the board and include the force of the water on the bob (buoyant force).

Q Other than the forces applied by the person and the fish, do any other forces act on the bob?

A *yes, the upward (buoyant) force of the water on the bob and the force of gravity*

Section Review

⊛TEKS **2C, 3B, 4B, 4C, 4D**

1. Can an object be in equilibrium if only one force acts on it?

2. If a car is traveling westward with a constant velocity of 20 m/s, what is the net external force acting on it?

3. If a car is accelerating downhill under a net force of 3674 N, what force must the brakes exert to cause the car to have a constant velocity?

4. **Physics in Action** The sensor in the torso of a crash-test dummy records the magnitude of the net external force acting on it. If the dummy is thrown forward with a force of 130.0 N while simultaneously being hit from the side with a force of 4500.0 N, what force will the sensor report?

5. **Physics in Action** What force will the seat belt have to exert on the dummy in item 4 to hold the dummy in the seat?

Section Review
ANSWERS

1. No, either zero force or two or more forces are required for equilibrium.
2. zero
3. −3674 N
4. 4502 N at 1.655° forward of the side
5. the same magnitude as the net force in item 4 but in the opposite direction

Demonstration 5

Newton's second law

Purpose Show students that force applied to an object to provide a given acceleration is proportional to the object's mass.

Materials tape, meterstick, metronome or stopwatch, physics book, balance

Procedure It is possible to redo Demonstration 3 on page 130 and quantify the results.

Standardize the distance of oscillation by placing two strips of tape 30 cm apart. Use a metronome or stopwatch to mark 1 s intervals. Assuming uniform acceleration over each half of the interval, you can use the kinematic equations to calculate the acceleration (60 cm/s^2, in this case). Measure the mass of the book, and calculate the force. Show students that the acceleration is independent of mass in this case (fixed distance and time), so the force required is proportional to the mass of the oscillating object.

⭐**TEKS**

p. 136: 4B, 4C
p. 137: 3B, 2C, 3B, 4B, 4C

4-3
Newton's second and third laws

4-3 SECTION OBJECTIVES

- **Describe the acceleration of an object in terms of its mass and the net external force acting on it.**

- **Predict the direction and magnitude of the acceleration caused by a known net external force.**

- **Identify action-reaction pairs.**

- **Explain why action-reaction pairs do not result in equilibrium.**

NEWTON'S SECOND LAW

From Newton's first law, we know that an object with no net external force acting on it is in a state of equilibrium. We also know that an object experiencing a net external force undergoes a change in its velocity. But exactly how much does a known force affect the motion of an object?

Force is proportional to mass and acceleration ⭐TEKS 4B, 4C

Imagine pushing a stalled car through a level intersection, as shown in **Figure 4-16.** Because a net force causes an object to accelerate, the speed of the car will increase. When you push the car by yourself, however, the acceleration will be so small that it will take a long time for you to notice an increase in the car's speed. If you get several friends to help you, the net force on the car is much greater, and the car will soon be moving so fast that you will have to run to keep up with it. This happens because the acceleration of an object is directly proportional to the net external force acting on it.

Experience pushing objects reveals that the mass of an object also affects its acceleration. If you push a bowling ball and a tennis ball with the same force, the tennis ball will accelerate quickly, while the bowling ball will accelerate slowly. Similarly, a lightweight car accelerates faster than a heavy truck if the same force is applied to both. It requires much less force to accelerate a low-mass object than it does to accelerate a high-mass object at the same rate. This is because an object with smaller mass has less inertia, or tendency to maintain its state of motion, than an object with greater mass.

Figure 4-16

(a) A small force on an object causes a small acceleration, but **(b)** a larger force causes a larger acceleration.

(a)

(b)

Newton's second law relates force, mass, and acceleration

The relationships between mass, force, and acceleration are quantified in **Newton's second law.**

NEWTON'S SECOND LAW

The acceleration of an object is directly proportional to the net external force acting on the object and inversely proportional to the object's mass.

According to Newton's second law, if equal forces are applied to two objects of different masses, the object with greater mass will experience a smaller acceleration, and the object with less mass will experience a greater acceleration.

In equation form, we can state Newton's law as

$$\Sigma\mathbf{F} = m\mathbf{a}$$

net external force = mass × acceleration

where **a** is the acceleration of the object, m is its mass, and $\Sigma\mathbf{F}$ represents the *vector sum of all external forces acting on the object*. This relationship makes it possible, when the mass of the object is known, to determine what effect a given force will have on an object's motion. (★)TEKS **3B**

Module 4
"Force and Inertia"
provides an interactive lesson with guided problem-solving practice to teach you about forces and Newton's laws.

SAMPLE PROBLEM 4B

Newton's second law (★)TEKS **2C, 3B, 4B, 4C**

PROBLEM

Roberto and Laura are studying across from each other at a wide table. Laura slides a 2.2 kg book toward Roberto. If the net external force acting on the book is 2.6 N to the right, what is the book's acceleration?

SOLUTION

Given: $m = 2.2$ kg
$\mathbf{F_{net}} = \Sigma\mathbf{F} = 2.6$ N, to the right

Unknown: $a = ?$

Use Newton's second law, and solve for a.

$$a = \frac{\Sigma\mathbf{F}}{m}$$

$$a = \frac{2.6\text{ N}}{2.2\text{ kg}}$$

$$\boxed{a = 1.2 \text{ m/s}^2 \text{ to the right}}$$

CALCULATOR SOLUTION

Your calculator will give you an answer of 1.1818 . . . for the acceleration. Because all values in the problem have only two significant digits, this value must be rounded up to 1.2.

Classroom Practice

The following may be used as teamwork exercises or for demonstration at the chalkboard or on an overhead projector.

PROBLEM

Newton's second law

A 7.5 kg bowling ball initially at rest is dropped from the top of an 11 m building. It hits the ground 1.5 s later. Find the net external force on the falling ball.

Answer
74 N

Space-shuttle astronauts experience accelerations of about 35 m/s^2 during takeoff. What force does a 75 kg astronaut experience during an acceleration of this magnitude?

Answer
2600 N

Interactive Problem-Solving Tutor

See Module 4
"Force and Inertia" provides additional development of problem-solving skills for this chapter.

ANSWERS TO

Practice 4B
Newton's second law

1. 2.2 m/s^2 forward
2. 1.4 m/s^2 north
3. 4.5 m/s^2 to the east
4. 14 N
5. 2.1 kg

ANSWERS TO

Conceptual Challenge

1. A greater force acts on the heavier rock, but the heavier rock also has greater mass, so the acceleration is the same. Free-fall acceleration is independent of mass.

2. The acceleration will increase as the mass decreases.

PRACTICE 4B

Newton's second law ⭐TEKS 2C, 3B, 4B, 4C

1. The net external force on the propeller of a 3.2 kg model airplane is 7.0 N forward. What is the acceleration of the airplane?

2. The net external force on a golf cart is 390 N north. If the cart has a total mass of 270 kg, what are the magnitude and direction of its acceleration?

3. A car has a mass of 1.50×10^3 kg. If the force acting on the car is 6.75×10^3 N to the east, what is the car's acceleration?

4. A 2.0 kg otter starts from rest at the top of a muddy incline 85 cm long and slides down to the bottom in 0.50 s. What net external force acts on the otter along the incline?

5. A soccer ball kicked with a force of 13.5 N accelerates at 6.5 m/s^2 to the right. What is the mass of the ball?

Conceptual Challenge

1. Gravity and rocks

The force of gravity is twice as great on a 2 kg rock as it is on a 1 kg rock. Why doesn't the 2 kg rock have a greater free-fall acceleration?

2. Leaking truck

A truck loaded with sand accelerates at 0.5 m/s^2 on the highway. If the driving force on the truck remains constant, what happens to the truck's acceleration if sand leaks at a constant rate from a hole in the truck bed?

In solving problems, it is often easier to break the equation for Newton's second law into components. The sum of forces acting in the *x* direction equals the mass times the acceleration in the *x* direction ($\Sigma F_x = ma_x$), and the sum of forces in the *y* direction equals the mass times the acceleration in the *y* direction ($\Sigma F_y = ma_y$). If the net external force is zero, then **a** = 0, which corresponds to the equilibrium situation where **v** is either constant or zero.

NEWTON'S THIRD LAW

A force is exerted on an object when that object interacts with some other object in its environment. Consider a moving car colliding with a concrete barrier. The car exerts a force on the barrier at the moment of collision, just as you apply force to a door when you push it or to a ball when you kick it. Furthermore, the barrier exerts a force on the car so that the car rapidly slows down after coming into contact with the barrier.

Forces always exist in pairs

From examples like this, Newton recognized that a single isolated force cannot exist. Instead, *forces always exist in pairs*. The car exerts a force on the barrier, and at the same time, the barrier exerts a force on the car. Newton described this type of situation with his **third law of motion.**

NEWTON'S THIRD LAW

If two objects interact, the magnitude of the force exerted on object 1 by object 2 is equal to the magnitude of the force simultaneously exerted on object 2 by object 1, and these two forces are opposite in direction.

internet connect

SCiLINKS
NSTA
TOPIC: Newton's laws
GO TO: www.scilinks.org
*sci*LINKS CODE: HF2042

An alternative statement of this law is that *for every action there is an equal and opposite reaction.* When two objects interact with one another, the forces they mutually exert on each other are called an **action-reaction pair.** The force that object 1 exerts on object 2 is sometimes called the *action force*, while the force that object 2 exerts on object 1 is called the *reaction force*. The action force is equal in magnitude and opposite in direction to the reaction force. The terms *action* and *reaction* sometimes cause confusion because they are used a little differently in physics than they are in everyday speech. In everyday speech, the word *reaction* is used to refer to something that happens *after* and *in response to* an event. In physics, however, the reaction force occurs at exactly the same time as the action force.

Because the forces coexist, either force can be called the action or the reaction. For example, you could call the force the car exerts on the barrier the action and the force the barrier exerts on the car the reaction. Likewise, you could choose to call the force the barrier exerts on the car the action and to call the force the car exerts on the barrier the reaction.

action-reaction pair

a pair of simultaneous equal but opposite forces resulting from the interaction of two objects

Action and reaction forces each act on different objects

The most important thing to remember about action-reaction pairs is that each force acts on a different object. Consider the task of driving a nail into a block of wood, as illustrated in **Figure 4-17.** To accelerate the nail and drive it into the block, the hammer exerts a force on the nail. According to Newton's third law, the nail exerts a force on the hammer that is equal to the magnitude of the force the hammer exerts on the nail.

The concept of action-reaction pairs is a common source of confusion because some people assume incorrectly that the equal and opposite forces balance one another and make any change in the state of motion impossible. If the nail exerts a force on the hammer that is equal to the force the hammer exerts on the nail, why doesn't the nail remain at rest? **(★)TEKS) 4C, 4D**

Because they act on different objects, action-reaction pairs do not result in equilibrium. The motion of the nail is affected only by the forces acting on the nail. The force the nail exerts on the hammer affects only the hammer, not the nail. To determine whether the nail will accelerate, draw a free-body diagram to isolate the forces acting on the nail alone, as shown in **Figure 4-18.** The force of the nail on the hammer is not included in the diagram because it does not affect the motion of the nail. According to the diagram, the nail will be driven into the wood because there is a net external force acting on the nail.

Figure 4-17
The nail exerts a force on the hammer that is equal and opposite to the force the hammer exerts on the nail.

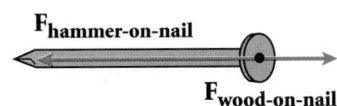

$$F_{\text{hammer-on-nail}} \qquad F_{\text{wood-on-nail}}$$

Figure 4-18
The net external force acting on the nail drives it to the left into the wood.

Misconception Alert

It is important to clear up any misconception that action and reaction forces cancel each other. One way to reinforce the true nature of Newton's third law is to use free-body diagrams. On the board, draw separate free-body diagrams for two or more interacting objects, such as a book on a table. Identify the third-law pairs, and point out that the force arrows are on separate bodies. The motion of the book is affected only by forces on the book. The motion of the table is affected only by forces on the table. Have students practice drawing free-body diagrams for multiple objects, building up levels of complexity with each new diagram (for example, a book on an inclined plane on a table on Earth).

(★)TEKS

p. 138: 2C, 3B, 4B, 4C
p. 139: 4C, 4D

Field forces also exist in pairs

Newton's third law also applies to field forces. For example, consider the gravitational force exerted by Earth on an object. During calibration at the crash-test site, engineers calibrate the sensors in the heads of crash-test dummies by removing the heads and dropping them from a known height.

The force Earth exerts on a dummy's head is $\mathbf{F_g}$. Let's call this force the action. What is the reaction? Because $\mathbf{F_g}$ is the force exerted on the falling head by Earth, the reaction to $\mathbf{F_g}$ is the force exerted on Earth by the falling head.

According to Newton's third law, the force of the dummy on Earth is equal to the force of Earth on the dummy. Thus, as a falling object accelerates toward Earth, Earth also accelerates toward the object.

The thought that Earth accelerates toward the dummy's head may seem to contradict our experience. One way to make sense of this is to refer to Newton's second law. The mass of Earth is much greater than that of the dummy's head. Therefore, while the dummy's head undergoes a large acceleration due to the force of Earth, the acceleration of Earth due to this reaction force is negligibly small because of its enormous mass. (★)TEKS **3A, 4C**

CONCEPT PREVIEW

Gravitational force is covered in Chapter 7, Section 7-3.

Section Review
ANSWERS

1. **a.** 12 N
   b. 3.0 m/s^2
2. The reaction force acts on the child, not on the wagon itself, so there is still a net force on the wagon.
3. **a.** person pushes on ground; ground pushes on person
   b. snowball exerts force on back; back exerts force on snowball
   c. ball exerts force on glove; glove exerts force on ball
   d. wind exerts force on window; window exerts force on wind
4. 1.6 m/s^2 at an angle of 65° north of east
5. each impact force has the same magnitude; the sports car experiences the larger acceleration because it has a smaller mass and acceleration is inversely proportional to mass

Section Review (★)TEKS **2C, 3B, 4B, 4D**

1. A 6.0 kg object undergoes an acceleration of 2.0 m/s^2.
   a. What is the magnitude of the net external force acting on it?
   b. If this same force is applied to a 4.0 kg object, what acceleration is produced?

2. A child pulls a wagon with a horizontal force, causing it to accelerate. Newton's third law says that the wagon exerts an equal and opposite force on the child. How can the wagon accelerate? (*Hint:* Draw a free-body diagram for each object to help you answer this question.)

3. Identify the action-reaction pairs in the following situations:
   a. a person takes a step
   b. a snowball hits someone in the back
   c. a baseball player catches a ball
   d. a gust of wind strikes a window

4. The forces acting on a sailboat are 390 N north and 180 N east. If the boat (including crew) has a mass of 270 kg, what are the magnitude and direction of its acceleration?

5. **Physics in Action**   If a small sports car collides head-on with a massive truck, which vehicle experiences the greater impact force? Which vehicle experiences the greater acceleration? Explain.

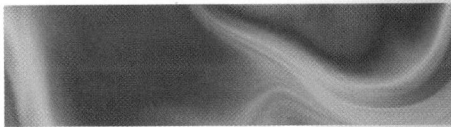

4-4
Everyday forces

WEIGHT ⭐TEKS 3A, 6A

You already know that a bowling ball weighs more than a tennis ball. But how do you know? If you imagine holding one ball in each hand, you can imagine the downward forces acting on your hands. Because the bowling ball has more mass than the tennis ball, the force of gravity pulls more strongly on the bowling ball, and it pushes your hand down with a stronger force than does the tennis ball.

The force of gravity exerted on the ball by Earth, **F$_g$**, is a vector quantity, directed toward the center of Earth. The magnitude of this force, F_g, is a scalar quantity called **weight**. When the mass and the acceleration due to gravity are known, the weight of an object can be calculated using the equation $F_g = mg$, where g is the magnitude of the acceleration due to gravity, or free-fall acceleration. In this book, $g = 9.81 \text{ m/s}^2$ unless otherwise specified.

Weight, unlike mass, is not an inherent property of an object. Because weight is dependent on the force of gravity, weight depends on location. For example, if the astronaut in **Figure 4-19** weighs 800 N (180 lb) on Earth, he would weigh only about 130 N (30 lb) on the moon. That is because the acceleration due to the moon's gravity is much smaller than 9.81 m/s^2.

Even on Earth, an object's weight may vary with location. Objects weigh less at higher altitudes than at sea level because the value of g decreases as distance from the center of Earth increases. Earth bulges out slightly at the equator because of the planet's rotation, so points on the equator are slightly farther from the center of Earth than points nearer the poles are. As a result, g also decreases slightly as latitude decreases.

THE NORMAL FORCE

Imagine a television set at rest on a table. We know that the force of gravity is acting on the television. How can we use Newton's laws to explain why the television does not continue to fall toward the center of Earth?

An analysis of the forces acting on the television will reveal the forces that cause equilibrium. First, we know that the force acting on a freely falling object is the downward force of gravity, **F$_g$**. Because the television is in equilibrium, we know that another force, equal in magnitude to **F$_g$** but in the opposite direction, must be acting on it. This force is the force exerted on the television by the table. This force is called the **normal force, F$_n$**.

4-4 SECTION OBJECTIVES

- **Explain the difference between mass and weight.**
- **Find the direction and magnitude of normal forces.**
- **Describe air resistance as a form of friction.**
- **Use coefficients of friction to calculate frictional force.**

weight

the magnitude of the force of gravity acting on an object

Figure 4-19
On the moon, astronauts weigh much less than they do on Earth because the acceleration due to the moon's gravity is so much less than Earth's.

normal force

a force exerted by one object on another in a direction perpendicular to the surface of contact

Visual Strategy

Figure 4-19
Point out that it is easier to lift massive objects on the moon because they weigh less on the moon, even though their mass remains the same. Also point out that an object's inertia is the same regardless of the magnitude of free-fall acceleration.

Q Disregarding air resistance, will a bullet shot from a gun go farther horizontally on Earth or on the moon?

A *The bullet will travel farther on the moon. Because the bullet is accelerated downward more slowly than on Earth, it is in the air for a longer time on the moon. The horizontal velocity will be the same in each case.*

Teaching Tip

For most practical purposes, the gravitational field near the surface of Earth is constant. For example, a person who weighs 180 lb at sea level would weigh 179.5 lb at an altitude of 9 km above sea level, a difference in weight of only 0.3 percent.

⭐TEKS

p. 140: 3A, 4C, 2C, 3B, 4B, 4D
p. 141: 3A, 6A

Visual Strategy

Figure 4-20

Tell students that the television is in equilibrium, so the normal force from the table must be equal in magnitude and opposite in direction to the weight of the television.

Have the students draw free-body diagrams for the television, table, and Earth and identify the third law pairs.

Q Do the forces in **Figure 4-20(a)** constitute an action-reaction pair (Newton's third law)?

A *No, both forces act on the television and therefore cannot be an action-reaction pair.*

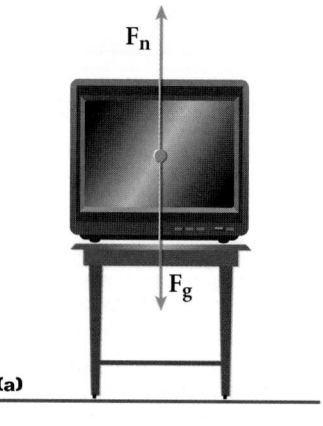

(a)

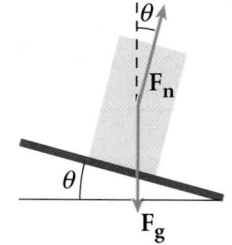

(b)

Figure 4-20

The normal force, $\mathbf{F_n}$, is always perpendicular to the surface but is not necessarily opposite the force of gravity.

static friction

the resistive force that opposes the relative motion of two contacting surfaces that are at rest with respect to one another

The word *normal* is used because the direction of the contact force is perpendicular to the table surface, and one meaning of the word *normal* is "perpendicular." **Figure 4-20(a)** shows the forces acting on the television.

The normal force is always perpendicular to the contact surface but is not always opposite in direction to the force of gravity. **Figure 4-20(b)** shows a free-body diagram of a refrigerator on a loading ramp. The normal force is perpendicular to the ramp, not directly opposite the force of gravity. In the absence of other forces, the normal force, $\mathbf{F_n}$, is equal and opposite to the component of $\mathbf{F_g}$ that is perpendicular to the contact surface. The magnitude of normal force can be calculated as $F_n = mg\cos\theta$. The angle θ is the angle between the normal force and a vertical line, and also the angle between the contact surface and a horizontal line. **★TEKS 3B, 4D**

THE FORCE OF FRICTION

Consider a jug of juice at rest on a table, as in **Figure 4-21(a)**. Because the jug is at rest, it is in equilibrium, and we know from Newton's first law that the net external force acting on it is zero. Newton's second law tells us that any additional force applied to the jug will cause it to accelerate and to remain in motion unless acted on by another force. However, experience tells us that the jug will not move at all if we apply a very small horizontal force. Even when we apply a force large enough to move the jug, it will stop moving almost as soon as we remove this applied force.

Friction opposes the applied force **★TEKS 4B, 4C, 4D**

When the jug is at rest, the only forces acting on it are the force of gravity and the normal force exerted by the table. These forces are equal and opposite, so the jug is in equilibrium. When you push the jug with a small horizontal force, as shown in **Figure 4-21(b)**, the table exerts an equal force in the opposite direction. As a result, the jug remains in equilibrium and therefore also remains at rest. The resistive force that keeps the jug from moving is called the force of **static friction** ($\mathbf{F_s}$).

Figure 4-21

(a)

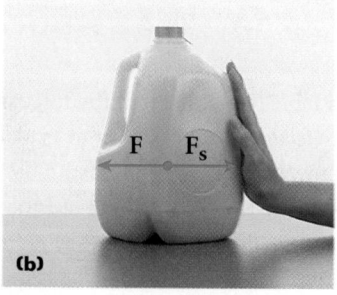

(b)

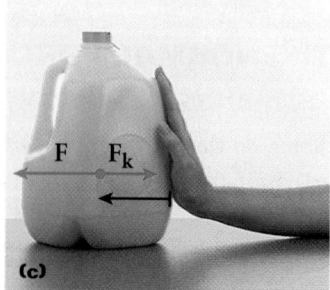

(c)

(a) Because this jug of juice is in equilibrium, any horizontal force applied to it should cause the jug to accelerate.

(b) When a small force is applied, the jug remains in equilibrium because the friction force is equal but opposite to the applied force.

(c) When a larger force is applied, the jug begins to move as soon as the applied force exceeds the opposing static friction force.

As long as the jug does not move, the force of static friction is always equal to and opposite in direction to the component of the applied force that is parallel to the surface ($\mathbf{F_s} = -\mathbf{F_{applied}}$). As the applied force increases, the force of static friction also increases; if the applied force decreases, the force of static friction also decreases. When the applied force is as great as it can be without causing the jug to move, the force of static friction reaches its maximum value, $\mathbf{F_{s,max}}$.

Kinetic friction is less than static friction ★TEKS 4C, 4D

When the applied force on the jug exceeds $\mathbf{F_{s,max}}$, the jug begins to move with an acceleration to the left, as shown in **Figure 4-21(c).** There is still a frictional force acting on the jug as it moves, but that force is actually less than $\mathbf{F_{s,max}}$. The retarding frictional force on an object in motion is called the force of **kinetic friction ($\mathbf{F_k}$).** The net external force acting on the object is equal to the difference between the applied force and the force of kinetic friction ($\mathbf{F} - \mathbf{F_k}$).

Frictional forces arise from complex interactions at the microscopic level between contacting surfaces. Most surfaces, even those that seem very smooth to the touch, are actually quite rough at the microscopic level, as illustrated in **Figure 4-22.** Notice that the surfaces are in contact at only a few points. When two surfaces are stationary with respect to each other, the surfaces stick together somewhat at the contact points. This *adhesion* is caused by electrostatic forces between molecules of the two surfaces. Because of adhesion, the force required to cause a stationary object to begin moving is usually greater than the force necessary to keep it moving at constant speed.

The force of friction is proportional to the normal force

It is easier to push a chair across the floor at a constant speed than to push a heavy desk across the floor at the same speed. In both cases, the amount of applied force is equal to the opposing force of friction, but the kinetic friction between the desk and the floor is greater than the kinetic friction between the chair and the floor.

Experimental observations show that the magnitude of the force of friction is proportional to the magnitude of the normal force exerted on an object by a surface. Because the desk is heavier than the chair, the desk also experiences a greater normal force and therefore greater friction.

Friction depends on the surfaces in contact

Keep in mind that the force of friction is really a macroscopic effect caused by a complex combination of forces at a microscopic level. The direct relationship between normal force and the force of friction is a good approximation of the friction between dry, flat surfaces that are at rest or sliding past one another. Under different conditions, the frictional force may depend on different factors. ★TEKS 3A

It is also easier to push a desk across a tile floor than across a floor covered with thick carpet. Although the normal force on the desk is the same in both

kinetic friction

the resistive force that opposes the relative motion of two contacting surfaces that are moving past one another

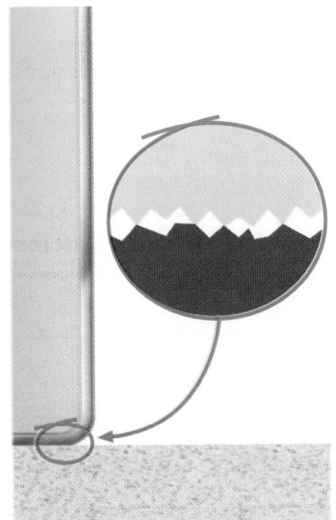

Figure 4-22
On the microscopic level, even very smooth surfaces make contact at only a few points.

Demonstration 6

Static vs. kinetic friction

Purpose Show that kinetic friction is less than static friction.

Materials rectangular block, hook, spring scale

Procedure Use the spring scale to measure the force required to start the rectangular block moving. Then use the spring scale to measure the frictional force for constant velocity. Perform several trials. Have students record all data and find the average for each. Point out that the normal force and the surfaces remain the same, so the only difference in the two average values is due to motion.

Demonstration 7

Friction of different surfaces

Purpose Show students that the force of friction depends on the surface.

Materials large cube with different materials (such as glass, carpeting, and sandpaper) covering each of four sides, with two sides left uncovered; hook; spring scale

Procedure Attach the hook to one of the two uncovered sides of the block. Pull the block across the table with the spring scale. Repeat the demonstration with a new surface of the cube exposed to the table. Repeat the demonstration for the two remaining covered sides. Have students summarize the results and reach a conclusion concerning the nature of the surfaces in contact and the frictional force.

Demonstration 8

Friction and surface area

Purpose Show the relation between surface area and frictional forces.

Materials rectangular block, hook, spring scale

Procedure Attach the hook to the block. Pull the block across the table with the spring scale. Have students note the force required to pull the block at a constant velocity. Repeat the demonstration for another surface area in contact with the table, and have students note the force. Have students summarize the results and reach a conclusion concerning the areas in contact and frictional forces.

Visual Strategy

Figure 4-23
Remind students that frictional force depends on the coefficient of friction and the normal force.

Q What changes in environment might cause a change in the frictional force experienced by the skier on the way down the hill?

A *Answers will vary but could include the following: surface conditions (wet or dry snow; ice; dirt), whether the skier is moving or not moving, the angle of the ski slope, and the change in g.*

144

---

coefficient of friction

the ratio of the force of friction to the normal force acting between two objects

Figure 4-23
The coefficients of friction between waxed wood and snow affect the acceleration of this skier on the slopes.

---

cases, the force of friction between the desk and the carpet is higher than that between the desk and the tile.

In addition to the normal force, the force of friction also depends on the composition and qualities of the surfaces in contact. The quantity that expresses the dependence of frictional forces on the particular surfaces in contact is called the **coefficient of friction.** The coefficient of friction is represented by the symbol μ, the lowercase Greek letter *mu*.

The coefficient of friction is a ratio of forces ⭐TEKS 3B

The coefficient of friction is defined as the ratio between the normal force and the force of friction between two surfaces. The *coefficient of kinetic friction* is the ratio of the force of kinetic friction to the normal force.

$$\mu_k = \frac{F_k}{F_n}$$

The *coefficient of static friction* is the ratio of the maximum value of the force of static friction to the normal force.

$$\mu_s = \frac{F_{s,max}}{F_n}$$

If the value of μ and the normal force on the object are known, then the magnitude of the force of friction can be calculated directly.

$$F_f = \mu F_n$$

Table 4-2 shows some experimental values of μ_s and μ_k for different materials. For example, the coefficients of friction between the skis and the wet snow in **Figure 4-23** range from $\mu_s = 0.14$ to $\mu_k = 0.1$. Because kinetic friction is less than or equal to the maximum static friction, the coefficient of kinetic friction is always less than or equal to the coefficient of static friction.

Table 4-2   Coefficients of friction (approximate values) ⭐TEKS 2C

	μ_s	μ_k		μ_s	μ_k
steel on steel	0.74	0.57	waxed wood on wet snow	0.14	0.1
aluminum on steel	0.61	0.47	waxed wood on dry snow	—	0.04
rubber on dry concrete	1.0	0.8	metal on metal (lubricated)	0.15	0.06
rubber on wet concrete	—	0.5	ice on ice	0.1	0.03
wood on wood	0.4	0.2	Teflon on Teflon	0.04	0.04
glass on glass	0.9	0.4	synovial joints in humans	0.01	0.003

Coefficients of friction ⭐TEKS 2C, 3B, 4B, 4C

PROBLEM

A 24 kg crate initially at rest on a horizontal floor requires a 75 N horizontal force to set it in motion. Find the coefficient of static friction between the crate and the floor.

SOLUTION

Given: $F_{s,max} = 75$ N $m = 24$ kg

Unknown: $\mu_s = ?$

Use the equation for the coefficient of static friction.

$$\mu_s = \frac{F_{s,max}}{F_n}$$

$$\mu_s = \frac{F_{s,max}}{mg} = \frac{75 \text{ N}}{24 \text{ kg} \times 9.81 \text{ m/s}^2}$$

$$\boxed{\mu_s = 0.32}$$

> **CALCULATOR SOLUTION**
>
> The answer your calculator will return for the coefficient of friction is 0.3185524975. The value for the mass has only two significant digits, so the answer must be rounded up to 0.32.

Coefficients of friction ⭐TEKS 2C, 3B, 4B, 4C

1. Once the crate in Sample Problem 4C is in motion, a horizontal force of 53 N keeps the crate moving with a constant velocity. Find μ_k, the coefficient of kinetic friction, between the crate and the floor.

2. A 25 kg chair initially at rest on a horizontal floor requires a 365 N horizontal force to set it in motion. Once the chair is in motion, a 327 N horizontal force keeps it moving at a constant velocity.
   a. Find the coefficient of static friction between the chair and the floor.
   b. Find the coefficient of kinetic friction between the chair and the floor.

3. A museum curator moves artifacts into place on many different display surfaces. Use the values in **Table 4-2** to find $F_{s,max}$ and F_k for the following situations:
   a. moving a 145 kg aluminum sculpture across a horizontal steel platform
   b. pulling a 15 kg steel sword across a horizontal steel shield
   c. pushing a 250 kg wood bed on a wood floor
   d. sliding a 0.55 kg glass amulet on a glass display case

Classroom Practice

The following may be used as a teamwork exercise or for demonstration at the chalkboard or on an overhead projector.

PROBLEM

Coefficients of friction

A 91 kg refrigerator is placed on a ramp. The refrigerator begins to slide when the ramp is raised to an angle of 34°. What is the coefficient of static friction?

Answer 0.67

PRACTICE GUIDE 4C

Solving for:		
μ	📖 **PE**	Sample, 1–2; Ch. Rvw. 37, 38*, 39*, 65b*, 67b, 69b*
	💿 **PW**	4–7, 10*
	PB	8–10
F_f	📖 **PE**	3; Ch. Rvw. 65a*, 69a*
	💿 **PW**	Sample, 1–3, 7, 10*
	PB	5–7
F_n, m	📖 **PE**	Ch. Rvw. 62*
	💿 **PW**	8–9
	PB	Sample, 1–4

ANSWERS TO

Practice 4C
1. 0.23
2. a. 1.5
   b. 1.3
3. a. 8.7×10^2 N, 6.7×10^2 N
   b. 1.1×10^2 N, 84N
   c. 1×10^3 N, 5×10^2 N
   d. 5 N, 2 N

Classroom Practice

The following may be used as teamwork exercises or for demonstration at the chalkboard or on an overhead projector.

PROBLEM

Overcoming friction

Two students are sliding a 225 kg sofa at constant speed across a wood floor. One student pulls with a force of 225 N at an angle of 13° above the horizontal. The other student pushes with a force of 250 N at an angle of 23° below the horizontal. What is the coefficient of kinetic friction between the sofa and the floor?

Answer
0.20

If the students carried the sofa, each would exert a force equal to half the weight of the sofa, which would be an improvement over the first scenario. What other changes could be made to make moving the sofa easier without lifting it off the floor?

Answer
Change the angles, or put the sofa on rollers.

★ TEKS

p. 146: 2C, 3B, 4B, 4C, 4D

p. 147: 2C, 3B, 4B, 4C, 4D

Overcoming friction ★ TEKS 2C, 3B, 4B, 4C, 4D

PROBLEM

A student moves a box of books by attaching a rope to the box and pulling with a force of 90.0 N at an angle of 30.0°, as shown in Figure 4-24. The box of books has a mass of 20.0 kg, and the coefficient of kinetic friction between the bottom of the box and the sidewalk is 0.50. Find the acceleration of the box.

Figure 4-24

SOLUTION

1. DEFINE Given:

$$m = 20.0 \text{ kg} \quad \mu_k = 0.50$$
$$\mathbf{F_{applied}} = 90.0 \text{ N at } \theta = 30.0°$$

Unknown: $\mathbf{a} = ?$

Diagram:

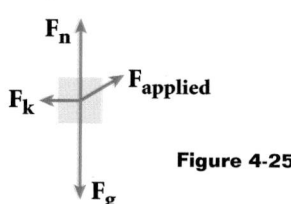

Figure 4-25

2. PLAN **Choose the equation(s) or situation:**
First find the normal force $\mathbf{F_n}$ by applying the first condition of equilibrium in the vertical direction.

$$\Sigma F_y = 0$$

Calculate the force of kinetic friction on the box.

$$F_k = \mu_k F_n$$

Apply Newton's second law along the horizontal direction to find the acceleration of the box.

$$\Sigma F_x = m a_x$$

3. CALCULATE **Choose a convenient coordinate system, and find the x and y components of all forces.**
The diagram in **Figure 4-26** shows the most convenient coordinate system. Find the y component of $\mathbf{F_{applied}}$.

$$F_{applied,y} = (90.0 \text{ N})(\sin 30.0°) = 45.0 \text{ N}$$

Find the x component of $\mathbf{F_{applied}}$.

$$F_{applied,x} = (90.0 \text{ N})(\cos 30.0°) = 77.9 \text{ N}$$

Use the mass to find the force of gravity acting on the box.

$$\mathbf{F_g} = (20.0 \text{ kg})(9.81 \text{ m/s}^2) = 196 \text{ N, downward}$$

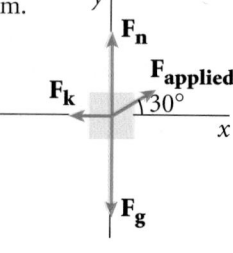

Figure 4-26

To find the normal force, find the sum of all forces in the y direction, set them equal to zero, and solve for F_n.

$$\Sigma F_y = F_n + F_{applied,\ y} - F_g = 0$$

$$F_n + 45.0\ N - 196\ N = 0$$

$$F_n = -45.0\ N + 196\ N = 151\ N$$

Use the normal force to find the force of kinetic friction.

$$F_k = \mu_k F_n = (0.50)(151\ N) = 75.5\ N \text{ to the left}$$

Determine the horizontal acceleration using Newton's second law.

$$\Sigma F_x = F_{applied,\ x} - F_k = ma_x$$

$$\Sigma F_x = 77.9\ N - 75.5\ N = (20.0\ kg)(a_x)$$

$$a_x = \frac{77.9\ N - 75.5\ N}{20.0\ kg} = \frac{2.4\ N}{20.0\ kg} = \frac{2.4\ kg \cdot m/s^2}{20.0\ kg}$$

$$\boxed{a = 0.12\ m/s^2 \text{ to the right}}$$

4. EVALUATE The normal force is not equal in magnitude to the weight because the y component of the student's pull on the rope helps support the box.

PRACTICE 4D

Overcoming friction ⭐TEKS 2C, 3B, 4B, 4C, 4D

1. A student moves a box of books down the hall by pulling on a rope attached to the box. The student pulls with a force of 185 N at an angle of 25.0° above the horizontal. The box has a mass of 35.0 kg, and μ_k between the box and the floor is 0.27. Find the acceleration of the box.

2. The student in item 1 moves the box up a ramp inclined at 12° with the horizontal. If the box starts from rest at the bottom of the ramp and is pulled at an angle of 25.0° with respect to the incline and with the same 185 N force, what is the acceleration up the ramp? Assume that $\mu_k = 0.27$.

3. A 75 kg box slides down a 25.0° ramp with an acceleration of 3.60 m/s^2.
   a. Find μ_k between the box and the ramp.
   b. What acceleration would a 175 kg box have on this ramp?

4. A box of books weighing 325 N moves with a constant velocity across the floor when it is pushed with a force of 425 N exerted downward at an angle of 35.2° below the horizontal. Find μ_k between the box and the floor.

PRACTICE GUIDE 4D

Solving for:		
$F_f,\ a$	📖 **PE**	Sample, 1–3; Ch. Rvw. 41*, 42*, 44b, 54a–b*, 55c*, 63a, 67a*
	💿 **PW**	5–7
	PB	4–7
$F_n,\ m$	📖 **PE**	Ch. Rvw. 40, 43, 45, 47, 57, 59*
	💿 **PW**	Sample, 1–3
	PB	8–10
μ	📖 **PE**	3, 4; Ch. Rvw. 38, 39, 55b*
	💿 **PW**	4
	PB	Sample, 1–3

ANSWERS TO

Practice 4D
Overcoming friction
1. 2.7 m/s^2 in the positive x direction
2. 0.77 m/s^2 up the ramp
3. **a.** 0.061
   b. 3.61 m/s^2 down the ramp
4. 0.609

Key Models and Analogies

Objects moving in space do not experience air resistance. This is why Earth continually orbits the sun without slowing down. (Earth's velocity is actually decreasing because of frequent collisions with small masses such as meteoroids, but this effect is minor.)

Consumer Focus

BACKGROUND

The coefficient of friction between the ground and the tires of a car is smaller when there is either rain or snow on the ground. Snow and rain tires are excellent examples of ways that we have adapted tires to regain some of the necessary frictional forces.

Point out to students that the friction between a tire and pavement is more complex than the simple sliding friction between dry surfaces that they have been studying. The force of friction on a car tire is not necessarily simply proportional to the normal force.

This is due in part to the fact that the tires are rolling, so they peel vertically away from the surface rather than continuously sliding across it. Also, when the road is covered with water or snow, other factors such as viscosity come into play.

TOPIC: Friction
GO TO: www.scilinks.org
*sci*LINKS CODE: HF2044

Air resistance is a form of friction (★)TEKS) 4B, 4C

Another type of friction, the retarding force produced by air resistance, is important in the analysis of motion. Whenever an object moves through a fluid medium, such as air or water, the fluid provides a resistance to the object's motion.

For example, the force of air resistance, $\mathbf{F_R}$, on a moving car acts in the direction opposite the direction of the car's motion. At low speeds, the magnitude of $\mathbf{F_R}$ is roughly proportional to the car's speed. At higher speeds, $\mathbf{F_R}$ is roughly proportional to the square of the car's speed. When the magnitude of $\mathbf{F_R}$ equals the magnitude of the force moving the car forward, the net force is zero and the car moves at a constant speed.

A similar situation occurs when an object falls through air. As a free-falling body accelerates, its velocity increases. As the velocity increases, the resistance of the air to the object's motion also constantly increases. When the upward force of air resistance balances the downward force of gravity, the net force on the object is zero and the object continues to move downward with a constant maximum speed, called the *terminal speed.*

Consumer Focus *Driving and Friction* (★)TEKS) 3C, 4B, 4C

*A*ccelerating a car seems simple to the driver. It is just a matter of pressing on a pedal or turning a wheel. But what are the forces involved?

The reason a car moves is that as its wheels turn, they push back against the road. It is actually the reaction force of the road pushing on the car that causes the car to accelerate. Without the friction between the tires and the road, the wheels would not be able to exert this force, and the car would not experience a reaction force. Thus, acceleration (whether speeding up, slowing down, or changing direction) requires this friction. Water and snow provide less friction and therefore reduce the amount of control the driver has over the direction and speed of the car.

As a car moves slowly over an area of water on the road, the water is squeezed out from under the tires. If the car moves too quickly, there is not enough time for the weight of the car to squeeze the water out from under the tires. The water trapped between the tires and the road will actually lift the tires and car off the road, a phenomenon called *hydroplaning.* When this occurs, there is very little friction between the tires and the water, and the car becomes difficult to control. To prevent hydroplaning, rain tires, such as the ones shown in **Figure 4-27,** keep water from accumulating between the tire and the road. Deep channels down the center of the tire provide a place for the water to accumulate, and curved grooves in the tread channel the water outward.

Because snow moves even less easily than water, snow tires are designed differently than rain tires. Snow tires have several deep grooves in their tread, enabling the tire to cut through the snow and make contact with the pavement. In the most extreme conditions, these deep grooves push against the snow and, like the paddle blades of a riverboat, use the snow's inertia to provide resistance.

Figure 4-27

1. Draw a free-body diagram for each of the following objects:

   a. a projectile in motion in the presence of air resistance

   b. a rocket leaving the launch pad with its engines operating

   c. a heavy crate being pushed across a rough surface

2. A bag of sugar has a mass of 2.26 kg.

   a. What is its weight in newtons on the moon, where the acceleration due to gravity is one-sixth that on Earth?

   b. What about on Jupiter, where the acceleration due to gravity is 2.64 times that on Earth?

3. A block of mass $m = 2.0$ kg is held in equilibrium on an incline of angle $\theta = 60.0°$ by a horizontal force.

   a. Determine the magnitude of this horizontal force. (Disregard friction.)

   b. Determine the magnitude of the normal force on the block.

4. A 55 kg ice skater, initially at rest on a flat skating rink, requires a 198 N horizontal force to be set in motion. However, after the skater is in motion, a horizontal force of 175 N keeps her moving at a constant velocity. Find the coefficients of static and kinetic friction between the skates and the ice.

5. The force of air resistance acting on a certain falling object is roughly proportional to the square of the object's velocity and is directed upward. If the object falls fast enough, will the force of air resistance eventually exceed the weight of the object and cause it to move upward? Explain your answer.

6. The graph in **Figure 4-28** shows the relationship between the applied force and the force of friction.

   a. What is the relationship between the forces at point A?

   b. What is the relationship between the forces at point B?

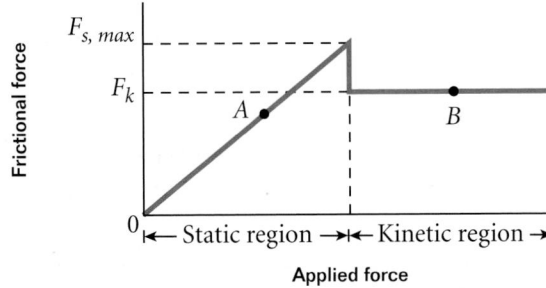

Figure 4-28

Section Review
ANSWERS

1. **a.** An arrow labeled F_g should point down, and an arrow labeled F_{air} should point opposite the direction of motion.

   b. $F_{engines}$ points up, F_g points down, and $F_{engines} > F_g$.

   c. F_g points down, an arrow representing F_n points up, $F_{applied}$ is horizontal, and $F_{friction}$ points in the opposite direction.

2. **a.** 3.70 N

   b. 58.5 N

3. **a.** 34 N

   b. 39 N

4. 0.37, 0.32

5. no; Once at equilibrium, the velocity will not increase, so the force of air resistance will not increase.

6. **a.** $F_s = F_{applied}$

   b. $F_k < F_{applied}$

⭐TEKS

p. 148: 4B, 4C, 3C, 4B, 4C

p. 149: 2C, 3B, 4A, 4B, 4C, 4D

Teaching Tip

Ask students to prepare a concept map for the chapter. The concept map should include most of the vocabulary terms, along with other integral terms or concepts.

TEKS

Review & Assess pp. 151–157:
2A: Alt. Assess. 2
2B: Alt. Assess. 1–2, 4
2C: 10–12, 20–25, 27–29, 37–69, Technology & Learning
2D: Alt. Assess. 2–5
2E: 60
2F: Alt. Assess. 1–2, 4
3A: Alt. Assess. 1
3B: 10–12, 20–25, 28–30, 37–69, Technology & Learning
3E: Alt. Assess. 3
4A: 60, Technology & Learning
4B: 1–3, 5–8, 10–12, 19–25, 27, 33, 35–69
4C: 5–6, 10–13, 15–25, 27, 33, 35–69, Technology & Learning
4D: 7–11, 16, 22–25, 28–30, 32–33, 37–59, 61–69

CHAPTER 4
Summary

KEY TERMS

action-reaction pair (p. 139)

coefficient of friction (p. 144)

contact force (p. 125)

equilibrium (p. 134)

field force (p. 125)

force (p. 124)

force diagram (p. 126)

inertia (p. 130)

kinetic friction (p. 143)

net external force (p. 131)

normal force (p. 141)

static friction (p. 143)

weight (p. 141)

KEY IDEAS

Section 4-1 Changes in motion

- Force is a vector quantity that causes changes in motion.
- Force can act either through the physical contact of two objects (contact force) or at a distance (field force).
- A free-body diagram shows only the forces that act on one object. These are the only forces that affect the motion of that object.

Section 4-2 Newton's first law

- The tendency of an object to maintain its state of motion is called inertia. Mass is the physical quantity used to measure inertia.
- The net external force acting on an object is the vector sum of all the forces acting on it. An object is in a state of equilibrium when the net external force acting on it is zero.

Section 4-3 Newton's second and third laws

- The net external force acting on an object is equal to the product of the mass of the object and its acceleration.
- When two bodies exert force on each other, the forces are equal in magnitude and opposite in direction. These forces are called action-reaction pairs. Forces always exist in pairs.

Section 4-4 Everyday forces

- The weight of an object is the force on the object due to gravity and is equal to the mass of the object times the acceleration due to gravity.
- A normal force is a force exerted by one surface on another in a direction perpendicular to the surface of contact.
- Friction is a resistive force that acts in a direction opposite to the direction of the relative motion of two contacting surfaces. The force of friction between two surfaces is proportional to the normal force.

Variable symbols

Quantities	Units	Conversions
F (vector) force	N newtons	$= kg\bullet m/s^2$
F (scalar)		
μ coefficient of friction	(no units)	

CHAPTER 4
Review and Assess

FORCES AND NEWTON'S FIRST LAW

Review questions

1. Is it possible for an object to move if no net force is acting on it? Explain.

2. If an object is at rest, can we conclude that no external forces are acting on it?

3. An object thrown into the air stops at the highest point in its path. Is it in equilibrium at this point? Explain.

4. What physical quantity is a measure of the amount of inertia an object has?

Conceptual questions

5. A beach ball is left in the bed of a pickup truck. Describe what happens to the ball when the truck accelerates forward.

6. A large crate is placed on the bed of a truck but is not tied down.
   a. As the truck accelerates forward, the crate slides across the bed until it hits the tailgate. Explain what causes this.
   b. If the driver slammed on the brakes, what could happen to the crate?

7. Draw a free-body diagram representing each of the following objects:
   a. a ball falling in the presence of air resistance
   b. a helicopter lifting off a landing pad
   c. an athlete running along a horizontal track

8. A chair is pushed forward with a force of 185 N. The gravitational force of Earth on the chair is 155 N downward, and the floor exerts a force of 155 N upward on the chair. Draw a free-body diagram showing the forces acting on the chair.

9. The gravitational force of Earth on a cake on a plate is 8.9 N downward. The plate exerts a force of 11.0 N upward on the cake, and a knife exerts a downward force of 2.1 N on the cake. Draw a free-body diagram of the cake.

Practice problems

10. Four forces act on a hot-air balloon, shown from the side in **Figure 4-29.** Find the magnitude and direction of the resultant force on the balloon. (See Sample Problem 4A.)

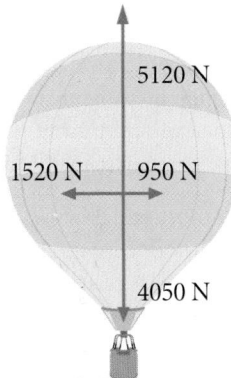

5120 N

1520 N   950 N

4050 N

Figure 4-29

11. Two lifeguards pull on ropes attached to a raft. If they pull in the same direction, the raft experiences a net external force of 334 N to the right. If they pull in opposite directions, the raft experiences a net external force of 106 N to the left.
   a. Draw a free-body diagram representing the raft for each situation.
   b. Find the force exerted by each lifeguard on the raft. (Disregard any other forces acting on the raft.)
   (See Sample Problem 4A.)

12. A dog pulls on a pillow with a force of 5 N at an angle of 37° above the horizontal. Find the x and y components of this force. (See Sample Problem 4A.)

13. because Earth has a very large mass

14. An object with greater mass requires a larger force for a given acceleration because it has more inertia than an object with less mass.

15. One-sixth of the force needed to lift an object on Earth is needed on the moon.

16. on the horse: the force of the cart, F_g down, F_n up, a reaction force of the ground on the hooves; on the cart: the force of the horse, F_g down, F_n up, kinetic friction

17. push it gently; With a smaller force, the astronaut will experience a smaller reaction force.

18. As the climber exerts a force downward, the rope supplies a reaction force that is directed upward. When this reaction force is greater than the climber's weight, the climber accelerates upward.

19. a. zero, as long as the speed is constant
    b. zero

20. 3.52 m/s^2

21. 4.7×10^2 s

22. a. 770 N at 8.13° to the right of forward
    b. 0.241 m/s^2 at 8.13° to the right of forward

23. a. 6.00 s
    b. 72.0 m
    c. 63.6 m/s

24. a. 4.78 m/s
    b. 7.80 m/s

25. a. 2.3 m
    b. 1.8 m

NEWTON'S SECOND AND THIRD LAWS

Review questions

13. Earth is attracted to an object with a force equal to and opposite the force Earth exerts on the object. Explain why Earth's acceleration is not equal to and opposite that of the object.

14. Explain Newton's second law in terms of inertia.

15. An astronaut on the moon has a 110 kg crate and a 230 kg crate. How do the forces required to lift the crates straight up on the moon compare with the forces required to lift them on Earth?

16. Draw a force diagram to identify all the action-reaction pairs that exist for a horse pulling a cart.

Conceptual questions

17. A space explorer is moving through space far from any planet or star and notices a large rock, taken as a specimen from an alien planet, floating around the cabin of the ship. Should the explorer push it gently or kick it toward the storage compartment? Why?

18. Explain why a rope climber must pull downward on the rope in order to move upward. Discuss the force exerted by the climber's arms in relation to the weight of the climber during the various stages of each "step" up the rope.

19. An 1850 kg car is moving to the right at a constant speed of 1.44 m/s.
    a. What is the net force on the car?
    b. What would be the net force on the car if it were moving to the left?

Practice problems

20. What acceleration will you give to a 24.3 kg box if you push it with a force of 85.5 N?
    (See Sample Problem 4B.)

21. A freight train has a mass of 1.5×10^7 kg. If the locomotive can exert a constant pull of 7.5×10^5 N, how long would it take to increase the speed of the train from rest to 85 km/h? (Disregard friction.)
    (See Sample Problem 4B.)

22. Two forces are applied to a car in an effort to accelerate it, as shown in **Figure 4-30**.
    a. What is the resultant of these two forces?
    b. If the car has a mass of 3200 kg, what acceleration does it have? (Disregard friction.)
    (See Sample Problem 4B.)

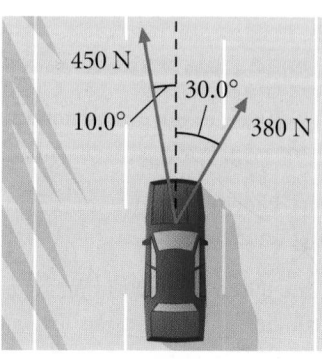

450 N

30.0°

10.0°

380 N

Figure 4-30

23. A 3.00 kg ball is dropped from the roof of a building 176.4 m high. While the ball is falling to Earth, a horizontal wind exerts a constant force of 12.0 N on the ball.
    a. How long does it take to hit the ground?
    b. How far from the building does the ball hit the ground?
    c. What is its speed when it hits the ground?
    (See Sample Problem 4B.)

24. A 40.0 kg wagon is towed up a hill inclined at 18.5° with respect to the horizontal. The tow rope is parallel to the incline and exerts a force of 140 N on the wagon. Assume that the wagon starts from rest at the bottom of the hill, and disregard friction.
    a. How fast is the wagon going after moving 30.0 m up the hill?
    b. How fast is the wagon going after moving 80.0 m up the hill?
    (See Sample Problem 4B.)

25. A shopper in a supermarket pushes a loaded 32 kg cart with a horizontal force of 12 N.
    a. How far will the cart move in 3.5 s, starting from rest? (Disregard friction.)
    b. How far will the cart move in 3.5 s if the shopper places an 85 N child in the cart before pushing?
    (See Sample Problem 4B.)

WEIGHT, FRICTION, AND NORMAL FORCE

Review questions

26. Explain the relationship between mass and weight.

27. A 0.150 kg baseball is thrown upward with an initial speed of 20.0 m/s.

    a. What is the force on the ball when it reaches half of its maximum height? (Disregard air resistance.)

    b. What is the force on the ball when it reaches its peak?

28. Draw free-body diagrams showing the weight and normal forces on a laundry basket in each of the following situations:

    a. at rest on a horizontal surface

    b. at rest on a ramp inclined 12° above the horizontal

    c. at rest on a ramp inclined 25° above the horizontal

    d. at rest on a ramp inclined 45° above the horizontal

29. If the basket in item 28 has a mass of 5.5 kg, find the magnitude of the normal force for the situations described in (a) through (d).

30. A teapot is initially at rest on a horizontal tabletop, then one end of the table is lifted slightly. Does the normal force increase or decrease? Does the force of static friction increase or decrease?

31. Which is usually greater, the maximum force of static friction or the force of kinetic friction?

Conceptual questions

32. Imagine an astronaut in space at the midpoint between two stars of equal mass. If all other objects are infinitely far away, how much does the astronaut weigh? Explain your answer.

33. A ball is held in a person's hand.

    a. Identify all the external forces acting on the ball and the reaction force to each.

    b. If the ball is dropped, what force is exerted on it while it is falling? Identify the reaction force in this case. (Disregard air resistance.)

34. Explain why pushing downward on a book as you push it across a table increases the force of friction between the table and the book.

35. Analyze the motion of a rock dropped in water in terms of its speed and acceleration. Assume that a resistive force acting on the rock increases as the speed increases.

36. A sky diver falls through the air. As the speed of the sky diver increases, what happens to the sky diver's acceleration? What is the acceleration when the sky diver reaches terminal speed?

Practice problems

37. A 95 kg clock initially at rest on a horizontal floor requires a 650 N horizontal force to set it in motion. After the clock is in motion, a horizontal force of 560 N keeps it moving with a constant velocity. Find μ_s and μ_k between the clock and the floor. (See Sample Problem 4C.)

38. A box slides down a 30.0° ramp with an acceleration of 1.20 m/s^2. Determine the coefficient of kinetic friction between the box and the ramp. (See Sample Problem 4C.)

39. A 4.00 kg block is pushed along the ceiling with a constant applied force of 85.0 N that acts at an angle of 55.0° with the horizontal, as in **Figure 4-31**. The block accelerates to the right at 6.00 m/s^2. Determine the coefficient of kinetic friction between the block and the ceiling. (See Sample Problem 4C.)

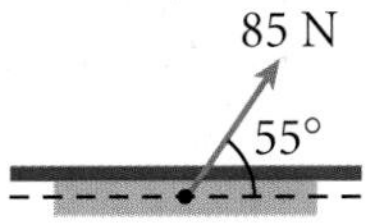

Figure 4-31

40. A 5.4 kg bag of groceries is in equilibrium on an incline of angle $\theta = 15°$. Find the magnitude of the normal force on the bag. (See Sample Problem 4D.)

41. A clerk moves a box of cans down an aisle by pulling on a strap attached to the box. The clerk pulls with a force of 185.0 N at an angle of 25.0° with the horizontal. The box has a mass of 35.0 kg, and the coefficient of kinetic friction between box and floor is 0.450. Find the acceleration of the box. (See Sample Problem 4D.)

26. Mass is the inertial property of matter. Weight is the force acting on an object due to gravity. Weight is equal to mass times the free-fall acceleration.

27. a. −1.47 N
    b. −1.47 N

28. a. $\mathbf{F_g}$ points down, and $\mathbf{F_n}$ points up.
    b. $\mathbf{F_g}$ points down, and $\mathbf{F_n}$ points up perpendicular to the surface of the ramp.
    c. same as (b)
    d. same as (b)

29. a. 54 N
    b. 53 N
    c. 49 N
    d. 38 N

30. The normal force decreases; The force of static friction increases to counteract the component of the weight along the table.

31. $F_{s,max}$

32. 0 N; The forces exerted by each star cancel.

33. a. the weight of the ball and an equal reaction force of the ball on Earth; the force of the person's hand on the ball and an equal reaction force of the ball on the hand

    b. F_g; the force of the ball on Earth

34. Pushing down on the book increases the normal force and therefore also increases the friction.

35. The rock will accelerate until the magnitude of the resistive force equals the weight of the object underwater. Then the rock's speed will be constant.

36. As the sky diver's speed increases, the acceleration decreases because the resistive force increases with increasing speed; zero

37. 0.70, 0.60

38. 0.436

39. 0.816

40. 51 N

41. 1.4 m/s² down the aisle

42. 1.0 m/s²

43. 15.9 N

44. a. If $F_{s,max} > ma$, then the box stays in place, but if $F_{s,max} < ma$, then the box will slide to the back of the truck bed.
b. 2.9 m/s² forward

45. 55 N to the right

46. 28 N; 56 N

47. 13 N down the incline

48. a. 9.81 m/s² downward
b. 22.2 N

49. a. 2.00 N
b. 6.04 N

50. 64 N upward

51. a. zero
b. 33.9 N

52. a. 0.250 m/s² forward
b. 18 m
c. 3.0 m/s

53. 5.0×10^1 m

54. a. 2.0 s
b. The box will never move. The force exerted is not enough to overcome friction.

55. a. 1.78 m/s²
b. 0.370
c. 9.40 N
d. 2.67 m/s

42. A 925 N crate is being pushed across a level floor by a force **F** of 325 N at an angle of 25° above the horizontal. The coefficient of kinetic friction between the crate and floor is 0.25. Find the magnitude of the acceleration of the crate.
(See Sample Problem 4D.)

43. A 2.00 kg block is in equilibrium on an incline of angle $\theta = 36.0°$. Find F_n of the incline on the block.
(See Sample Problem 4D.)

44. A 35 kg box rests on the back of a truck. The coefficient of static friction between the box and the truck bed is 0.300.
a. When the truck accelerates forward, what happens to the box?
b. Find the maximum acceleration the truck can have before the box slides backward.
(See Sample Problem 4D.)

MIXED REVIEW

45. What net external force is required to give a 25 kg suitcase an acceleration of 2.2 m/s² to the right?

46. A block with a mass of 5.0 kg is held in equilibrium on an incline of angle $\theta = 30.0°$ by the horizontal force, **F**, as shown in **Figure 4-32**. Find the magnitudes of **F** and of the normal force on the block. (Disregard friction.)

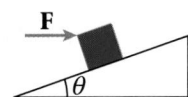

Figure 4-32

47. A 2.0 kg mass starts from rest and slides down an inclined plane 8.0×10^{-1} m long in 0.50 s. What net force is acting on the mass along the incline?

48. A 2.26 kg book is dropped from a height of 1.5 m.
a. What is its acceleration?
b. What is its weight in newtons?

49. A loaf of bread weighs 5.30 N on Earth.
a. What would it weigh, in newtons, on Mars, where the acceleration due to gravity is 0.378 times that on Earth?
b. What would it weigh on Neptune, where g is 1.14 times that on Earth?

50. A 5.0 kg bucket of water is raised from a well by a rope. If the upward acceleration of the bucket is 3.0 m/s², find the force exerted by the rope on the bucket of water.

51. A 3.46 kg briefcase is sitting at rest on a level floor.
a. What is its acceleration?
b. What is its weight in newtons?

52. A boat moves through the water with two forces acting on it. One is a 2.10×10^3 N forward push by the motor, and the other is a 1.80×10^3 N resistive force due to the water.
a. What is the acceleration of the 1200 kg boat?
b. If it starts from rest, how far will it move in 12 s?
c. What will its speed be at the end of this time interval?

53. A girl coasts down a hill on a sled, reaching level ground at the bottom with a speed of 7.0 m/s. The coefficient of kinetic friction between the sled's runners and the hard, icy snow is 0.050, and the girl and sled together weigh 645 N. How far does the sled travel on the level ground before coming to rest?

54. A box of books weighing 319 N is shoved across the floor by a force of 485 N exerted downward at an angle of 35° below the horizontal.
a. If μ_k between the box and the floor is 0.57, how long does it take to move the box 4.00 m, starting from rest?
b. If μ_k between the box and the floor is 0.75, how long does it take to move the box 4.00 m, starting from rest?

55. A 3.00 kg block starts from rest at the top of a 30.0° incline and accelerates uniformly down the incline, moving 2.00 m in 1.50 s.
a. Find the magnitude of the acceleration of the block.
b. Find the coefficient of kinetic friction between the block and the incline.
c. Find the magnitude of the frictional force acting on the block.
d. Find the speed of the block after it has slid a distance of 2.00 m.

56. A 75 kg person escapes from a burning building by jumping from a window 25 m above a catching net. Assuming that air resistance is simply a constant 95 N force on the person during the fall, determine the person's velocity just before hitting the net.

57. The parachute on a race car that weighs 8820 N opens at the end of a quarter-mile run when the car is traveling 35 m/s. What net retarding force must be supplied by the parachute to stop the car in a distance of 1100 m?

58. A 1250 kg car is pulling a 325 kg trailer. Together, the car and trailer have an acceleration of 2.15 m/s^2 directly forward.

   a. Determine the net force on the car.

   b. Determine the net force on the trailer.

59. The coefficient of static friction between the 3.00 kg crate and the 35.0° incline of **Figure 4-33** is 0.300. What is the magnitude of the minimum force, *F*, that must be applied to the crate perpendicular to the incline to prevent the crate from sliding down the incline?

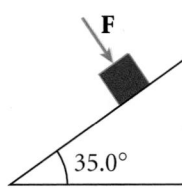

Figure 4-33

60. **Figure 4-34** shows a plot of the speed of a person's body during a chin-up. All motion is vertical and the mass of the person (excluding the arms) is 64.0 kg. Find the magnitude of the force exerted on the body by the arms at 0.50 s intervals.

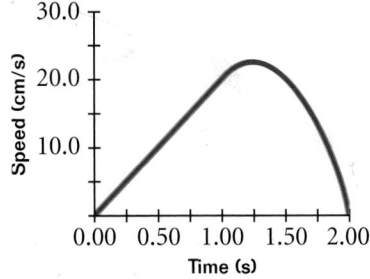

Figure 4-34

61. A machine in an ice factory is capable of exerting 3.00×10^2 N of force to pull large blocks of ice up a slope. The blocks each weigh 1.22×10^4 N. Assuming there is no friction, what is the maximum angle that the slope can make with the horizontal if the machine is to be able to complete the task?

62. The board sandwiched between two other boards in **Figure 4-35** weighs 95.5 N. If the coefficient of friction between the boards is 0.663, what must be the magnitude of the horizontal forces acting on both sides of the center board to keep it from slipping downward?

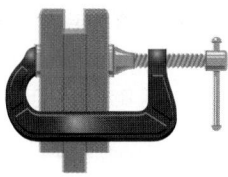

Figure 4-35

63. Three blocks are in contact with each other on a frictionless horizontal surface, as in **Figure 4-36**. A horizontal force, **F**, is applied to m_1. For this problem, $m_1 = 2.00$ kg, $m_2 = 3.00$ kg, $m_3 = 4.00$ kg, and **F** = 180 N to the right.

   a. Find the acceleration of the blocks.

   b. Find the resultant force on each block.

   c. Find the magnitude of the contact forces between the blocks.

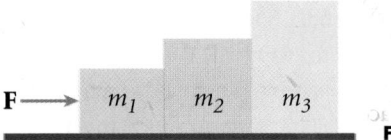

Figure 4-36

64. A car is traveling at 50.0 km/h on a flat highway.

   a. If the coefficient of kinetic friction between the road and the tires on a rainy day is 0.100, what is the minimum distance needed for the car to stop?

   b. What is the stopping distance when the surface is dry and the coefficient of kinetic friction is 0.600?

65. Two blocks with masses of 45.0 kg and 23.5 kg are stacked on a table with the heavier block on top. The coefficient of static friction is 0.60 between the two blocks and 0.30 between the bottom block and the table. A horizontal force is slowly applied to the top block until one of the blocks moves.

   a. Where does slippage occur first, between the two blocks or between the bottom block and the table? Explain.

   b. What value of the coefficient of static friction between the bottom block and the table would change the answer to part (a)?

56. 21 m/s downward
57. -5.0×10^2 N
58. a. 2690 N forward
b. 699 N forward
59. 32.2 N
60. 13 N, 13 N, 0 N, −26 N
61. 0.144°
62. 72.0 N
63. a. 2.0×10^1 m/s^2 to the right
b. $F_1 = 4.0 \times 10^1$ N,
$F_2 = 6.0 \times 10^1$ N,
$F_3 = 8.0 \times 10^1$ N, all to the right
c. m_3 by m_2: 8.0×10^1 N
m_2 by m_3: -8.0×10^1 N
m_2 by m_1: 1.40×10^2 N
m_1 by m_2: -1.40×10^2 N
64. a. 98.3 m
b. 16.4 m
65. a. between the bottom block and the table
b. 0.39

66. a. Δx
   b. $\frac{1}{4}(\Delta x)$

67. a. -1.2 m/s^2
   b. 0.12
   c. 45 m

ANSWERS TO
Technology & Learning

a. 10 N
b. 1.0×10^1 N
c. 130 N
d. 19 N
e. 240 N
f. $\mu_s = 1$

156

66. A truck driver slams on the brakes and skids to a stop through a displacement Δx.

   a. If the truck had double the mass, find the truck's "skidding distance" in terms of Δx. (Hint: Increasing the mass increases the normal force.)

   b. If the initial velocity of the truck were halved, what would be the truck's "skidding distance" in terms of Δx?

67. A hockey puck is hit on a frozen lake and starts moving with a speed of 12.0 m/s. Exactly 5.0 s later, its speed is 6.0 m/s.

   a. What is its average acceleration?
   b. What is the coefficient of kinetic friction between the puck and the ice?
   c. How far does the puck travel during this 5.0 s interval?

Technology & Learning

Graphing calculators

Refer to Appendix B for instructions on downloading programs for your calculator. The program "Chap4" allows you to analyze a graph of the maximum force of static friction versus the normal force in various situations.

Recall from your studies of static friction earlier in this chapter that the maximum force of static friction can be found using the following equation:

$$F_{s,max} = \mu_s F_n$$

The program "Chap4" stored on your graphing calculator makes use of the equation for static friction. Once the "Chap4" program is executed, your calculator will ask for MU, the coefficient of static friction. Given that value, your graphing calculator will use the following equation to graph the maximum force of static friction (Y_1) versus the normal force (X) for an object resting on that surface. Note that the relationships in this equation are the same as those in the static-friction equation above.

$$Y_1 = SX$$

a. If the Y_1 value for a 40 N object is equal to 10 N, how much force must be applied to this object to move it?

Execute "Chap4" on the PRGM menu, and press ENTER to begin the program. Enter the value for the coefficient of static friction (shown below), and press ENTER to begin graphing.

The calculator will provide the graph of the force of friction versus the normal force. (If the graph is not visible, press WINDOW and change the settings for the graph window, then press GRAPH.)

Press TRACE, and use the arrow keys to trace along the curve. The x-value corresponds to the normal force in newtons, and the y-value corresponds to the maximum force of static friction in newtons.

Find the maximum force of static friction for each of the following situations (b–e):

b. a 25 N wooden box resting on a wooden tabletop, with $\mu_s = 0.40$

c. a 325 N wooden box on the same tabletop

d. a 25 N steel box resting on a steel counter top, with $\mu_s = 0.74$

e. a 325 N steel box on the same counter top

f. At what value of μ_s is the force of static friction equal to the normal force at all points on the curve?

Press 2nd QUIT to stop graphing. Press ENTER to input a new value or CLEAR to end the program.

68. Find the magnitude of the force exerted by each cable to support the 625 N punching bag in **Figure 4-37.**

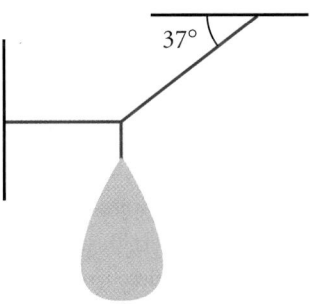

37°

Figure 4-37

69. Refer to the diagram in **Figure 4-38** to answer the following questions:

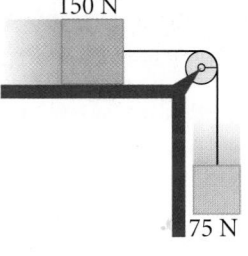

150 N

75 N

Figure 4-38

a. What is the magnitude of the minimum force of static friction on the 150 N block required to hold both blocks at rest?

b. What would the coefficient of static friction between the 150 N block and the table have to be to ensure that both blocks would be held at rest?

68. 1.0×10^3 N, 8.3×10^2 N
69. a. 75 N
     b. 0.50

Alternative Assessment

Performance assessment

1. Predict what will happen in the following test of the laws of motion. You and a partner face each other, each holding a bathroom scale. Place the scales back to back and slowly begin pushing on them. Record the measurements of both scales at the same time. Perform the experiment. Which of Newton's laws have you verified?

2. You can keep a 3 kg book from dropping by pushing it horizontally against a wall. Draw force diagrams and identify all the forces involved. How do they combine to result in a zero net force? Will the force you must supply to hold the book up be different for different types of walls? Design a series of experiments to test your answer. Identify exactly which measurements will be necessary and what equipment you will need. If your teacher approves your plan, perform the experiments and present your analysis and conclusions to the class.

Portfolio projects

3. Research how the work of scientists Antoine Lavoisier, Isaac Newton, and Albert Einstein related to the study of mass. Which might have said the following?

a. The mass of a body is a measure of the quantity of matter in it.

b. The mass of a body is its resistance to a change in motion.

c. The mass of a body depends on its velocity.

To what extent are these statements compatible or contradictory? Create a presentation, such as a poster, video, or computer presentation, of these scientists' conceptions of the meaning of mass.

4. Imagine an airplane with a series of special instruments anchored to its walls: a pendulum, a 100 kg mass on a spring balance, and a sealed half-full aquarium. What will happen to each instrument when the plane takes off, makes turns, slows down, lands, etc.? If possible, test your responses by simulating the experience in elevators, car rides, and other situations with similar instruments and with your body sensations. Write a report comparing your predictions with your experiences.

5. Research the effects of snow, ice, and water on friction and gliding in the context of skiing, ice skating, hockey, mountain climbing, and other activities. Explain how athletes in these sports need and use friction. Explain how friction—or its absence—can diminish these athletes' performance. How do your findings relate to your personal experience? Prepare a report or presentation and use specific examples.

Performance assessment

1. Scales will have identical readings because of Newton's third law.

2. Student plans should be safe and should involve measuring forces such as weight, force applied, normal force, and frictional force.

 **Portfolio projects**
3. Lavoisier: (a), based on his study of chemistry; Newton: (a) and (b), based on inertia and forces; Einstein: (a), (b), and (c), including effects at very high speeds

4. Students' explanations should rely on inertia. Spring-scale readings increase at takeoff.

5. In many of these contexts, the coefficient of friction is not a constant, and the force of friction is not simply proportional to the normal force. Items can glide over a low-friction surface without losing much speed, but controlling motion is difficult.

CHAPTER 4
Laboratory Exercise

TEKS
pp. 158–163: 1A,
2A, 2B, 2C, 2D, 2E,
2F, 3B, 4A, 4B, 4C

NOTE

Preparation is given on pp. 122A–122B. Blank data table and sample data table are on the One-Stop Planner CD-ROM. All calculations shown use sample data.

Planning

Recommended time:

1 lab period

For a 2-period lab, add the Extensions exercise (p. 163, number 11).

Classroom organization:

▶ Each group needs a level work surface that is at least 2.0 m long and 0.5 m above the floor, with an end to clamp the pulley to. To avoid crowding, lab groups may be larger than usual.

▶ Each group must have at least two students and no more than four. With larger groups, make sure all students are involved.

▶ The CBL and recording-timer procedures may be used in the same class.

▶ If lab groups share balances, avoid traffic jams by having some of the groups mass the cart before completing the setup.

▶ **Safety warnings:** Remind students to fasten masses securely, keep the area clear during the experiment, and prevent the cart from falling off the table.

OBJECTIVES

• Compare the accelerations of a mass acted on by different forces.

• Compare the accelerations of different masses acted on by the same force.

• Examine the relationships between mass, force, acceleration, and Newton's laws of motion.

MATERIALS LIST

✔ balance
✔ calibrated masses and holder
✔ cord
✔ dynamics cart
✔ mass hanger
✔ meterstick
✔ pulley with table clamp

PROCEDURE

CBL AND SENSORS

✔ CBL
✔ CBL force sensor with CBL-DIN adapter
✔ CBL motion detector
✔ graphing calculator with link cable
✔ support stand with V-jaw clamp
✔ rod and parallel clamp
✔ tape
✔ 25 cm × 25 cm square of poster board

RECORDING TIMER

✔ C-clamp
✔ recording timer and tape
✔ stopwatch

FORCE AND ACCELERATION

Newton's second law states that any net external force applied to a mass causes the mass to accelerate according to the equation $\mathbf{F} = m\mathbf{a}$. Because of frictional forces, experience does not always seem to support this. For example, when you are driving a car you must apply a constant force to keep the car moving with a constant velocity. In the absence of friction, the car would continue to move with a constant velocity after the force was removed. The continued application of force would cause the car to accelerate.

In this lab, you will study the motion of a dynamics cart pulled by the weight of masses falling from a table to the floor. The cart is set up so that any applied force will cause it to move with a constant velocity. In the first part of the experiment, the total mass will remain constant while the force acting on the cart will be different for each trial. In the second part, the force acting on the cart will remain constant, but the total mass will change for each trial.

SAFETY

• **Tie back long hair, secure loose clothing, and remove loose jewelry to prevent its getting caught in moving or rotating parts.**

• **Attach masses securely. Falling or dropped masses can cause serious injury.**

PREPARATION

1. Determine whether you will be using the CBL and sensors or the recording timer to take your measurements. Read the entire lab procedure for the appropriate method. Plan the steps you will take.

2. Prepare a data table in your lab notebook with six columns and six rows. In the first row, label the first through sixth columns *Trial*, *Total mass (kg)*, *Accelerating mass (kg)*, *Accelerating force (N)*, *Time interval (s)*, and *Distance (m)*. In the first column, label the second through sixth rows *1, 2, 3, 4,* and *5*.

3. Choose a location where the cart will be able to move a considerable distance without any obstacles and where you will be able to clamp the pulley to a table edge.

Recording-timer procedure begins on page 161.

PROCEDURE

CBL AND SENSORS

Apparatus setup

4. Connect the CBL to the calculator with the unit-to-unit link cable using the link ports located on each unit. Connect the force sensor to the CH1 port on the CBL and connect the motion detector to the SONIC port. Turn on the CBL and the graphing calculator.

5. Start the program PHYSICS on your graphing calculator.

   a. Select option *SET UP PROBES* from the MAIN MENU. Enter 2 for the number of probes. Select the motion detector. Next, select the appropriate force sensor from the list; your teacher will tell you what kind of force sensor you are using. Enter 1 for the channel number. Select *USED STORED* from the CALIBRATION menu.

   b. Select the *COLLECT DATA* option from the MAIN MENU. Select the *TIME GRAPH* option from the DATA COLLECTION menu. Enter 0.02 for the time between samples. Enter 99 for the number of samples. Check the values you entered and press ENTER. Press ENTER to continue. If you made a mistake entering the time values, select MODIFY SETUP, reenter the values, and continue.

6. Securely tape the force sensor to the dynamics cart, as shown in **Figure 4-39**. Tape the poster board to the opposite end of the dynamics cart to make a flat vertical surface.

7. Set up the apparatus as shown in **Figure 4-39**. Clamp the pulley to the table edge using a table clamp, rod, and parallel clamp so that it is level with the force sensor hook. Position the motion detector so that the cart will move away from it in a straight line. To keep the motion detector from moving during the experiment, securely clamp it to the ring stand. Place a piece of tape 0.5 m in front of the motion detector to serve as a starting line for the cart.

Constant mass with varying force

8. Carefully measure the mass of the cart assembly on the platform balance, making sure that the cart does not roll or fall off the balance. Then load it with masses equal to 0.60 kg. Lightly tape the masses to the cart to hold them in place. Add these masses to the mass of the cart.

9. Attach one end of the cord to a small mass hanger and the other end of the cord to the force sensor. Pass the cord over the pulley and fasten a small mass to the end to offset the frictional force on the cart.

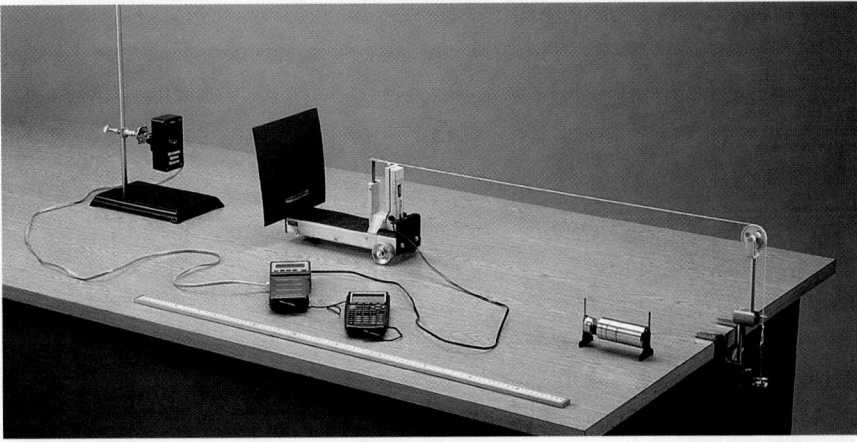

Figure 4-39

Step 7: Make sure the motion detector has a clear view of the cart. For all trials, start the cart from the same position.

Step 14: Using the CBL with the graphing calculator will allow you to view graphs of the motion after each trial.

CBL and Sensors Tips

◆ Students should have the program PHYSICS on their calculators. Refer to Appendix B for instructions.

◆ The motion detector must be stabilized so it can't move during the experiment.

Techniques to Demonstrate

Show the students how to use the arrow keys to trace the graph in steps 14 and 15.

Explain how to choose the points in step 15 and how to find the difference between the *x* and *y* values in step 16.

Remind students to make sure the CBL is on whenever they perform a trial.

✔ Checkpoints

Step 7: Make sure all clamps are placed so that they protrude as little as possible.

Step 9: Make sure masses are securely attached. Students should be able to demonstrate that the counterweight allows the cart to move at a constant velocity when given a small push.

Step 11: The force sensor cord must be kept out of the path of the cart; it must also be prevented from dragging behind the cart and slowing it down.

Step 13: It may help to clamp a wood block to the edge of the table to serve as a bumper to keep the cart from rolling off the table.

Step 14: The graph will not be a perfectly straight line. Students may be confused by the values above and below the line. Guide students to select the most frequently appearing value.

Step 15: Students should identify the part of the curve that shows the motion of the cart. If the curve does not show the cart moving away from the detector, make sure the detector points at the cart with no interference. If there is interference from the tabletop, tilt the detector slightly upward and clamp it in place. Cover the tabletop and any objects on the table, such as gas jets or faucets, with soft cloths.

Step 16: Students should be able to interpret the velocity and acceleration graphs on the graphing calculator. Students should be able to use the graphs to explain whether the motion of the cart meets their expectations. If there are any problems with the setup, the students should be alerted by the graphs.

Step 19: Students should recognize the difference between these two trials and the previous trials. They must add mass to the cart each time without changing the mass on the string.

The mass is correct when the car moves forward with a constant velocity when you give it a push. *This counterweight should stay on the string throughout the entire experiment.* Add the mass of the counterweight to the mass of the cart and masses, and record the sum as *Total mass* in your data table.

10. For the first trial, remove a 0.10 kg mass from the cart, and securely fasten it to the end of the cord along with the counterweight. Record 0.10 kg as the *Accelerating mass* in the data table.

11. Place the cart at the piece of tape 0.5 m in front of the motion detector. Place the cart so the poster board end is closest to the motion detector and lined up with the tape line. Keep the force sensor cord clear so that the cart will be able to move freely.

12. Make sure the CBL and the graphing calculator are turned on. Make sure the area under the falling mass is clear of obstacles. Press ENTER on the calculator to begin collecting data, and release the cart simultaneously. The motion detector will begin to click as it collects data.

13. Carefully stop the cart when the 0.10 kg mass hits the floor. **Do not let the cart fall off the table.**

14. When the motion detector stops clicking and the CBL displays *DONE*, press ENTER on the graphing calculator. Select the *ANALOG* option from the SELECT CHANNELS menu. This will plot a graph of the force sensor reading against time. Use the arrow keys to trace along the curve. The *y* value is the force in newtons; it should be fairly constant. Record this value as the *Accelerating force* in the data table. Press ENTER.

15. Select the *SONIC* option, and then select the *DISTANCE* option to plot a graph of the distance in meters against time. Use the arrow keys to trace along the curve. On the far left and the far right, the flat portion of the curve represents the positions of the cart before and after its motion. The middle section of the curve represents the motion of the cart. Choose a point on the curve near the begin-

ning of this middle section, and choose another point near the end.

16. Find the difference between the *y* values of these two points, and record it as the *Distance* for *Trial 1* in your data table. Find the difference between the *x* values for these two readings to find the time elapsed between measurements. Record this as the *Time interval* for *Trial 1* in your data table. Press ENTER on the calculator.

- Select the *SONIC* option again, and select the *VELOCITY* option to plot a graph of the velocity in m/s against time. Examine how the velocity of the cart changed during the experiment. Press ENTER.

- Select the *SONIC* option again, and select the *ACCELERATION* option to plot a graph of the acceleration in m/s^2 against time. Examine how the acceleration of the cart changed during the experiment. Press ENTER.

17. Replace the 0.10 kg mass in the cart. Remove 0.20 kg from the cart, and attach it securely to the end of the cord. Repeat the procedure.

18. Leave the 0.20 kg on the end of the cord, and attach the 0.10 kg mass from the cart securely to the end of the cord. Repeat the procedure.

Constant force with varying mass

19. For the two trials in this part of the experiment, keep 0.30 kg and the counterweight on the string. Be sure to include this mass when recording the total mass for these three trials.

20. Add 0.50 kg to the cart. Tape the mass to the cart to keep it in place. Run the experiment and record the total mass, accelerating mass, accelerating force, distance, and time under *Trial 4* in your data table.

21. Tape 1.00 kg to the cart, and repeat the procedure. Record the data under *Trial 5* in your data table.

22. Clean up your work area. Put equipment away safely so that it is ready to be used again.

Analysis and Interpretation begins on page 162.

PROCEDURE

RECORDING TIMER

Apparatus setup

4. Set up the apparatus as shown in **Figure 4-40.** Clamp the pulley to the edge of the table so that it is level with the top of the cart. Clamp the recording timer to a ring stand or to the edge of the table to hold it in place. If the timer is clamped to the table, leave 0.5 m between the timer and the initial position of the cart. Insert the carbon disk, and thread tape through the guides under the disk. When your teacher approves your setup, plug the timer into a wall outlet.

5. If you have not used the recording timer before, refer to the lab for Chapter 2 for instructions. Calibrate the recording timer with the stopwatch or use the previously determined value for the timer's period.

6. Record the value for the timer's period at the top of the data table.

7. Fasten the timing tape to one end of the cart.

Constant mass with varying force

8. Carefully measure the mass of the cart assembly on the platform balance, making sure that the cart does not roll or fall off the balance. Then load it with masses equal to 0.60 kg. Lightly tape the masses to the cart to hold them in place. Add these masses to the mass of the cart.

9. Attach one end of the cord to a small mass hanger and the other end of the cord to the cart. Pass the cord over the pulley and fasten a small mass to the end to offset the frictional force on the cart. The mass is correct when the car moves forward with a constant velocity when you give it a push. *This counterweight should stay on the string throughout the entire experiment.* Add the mass of the counterweight to the mass of the cart and masses, and record the sum as *Total mass* in your data table.

10. For the first trial, remove a 0.10 kg mass from the cart, and securely fasten it to the end of the string along with the counterweight. Record 0.10 kg as the *Accelerating mass* in the data table.

11. Hold the cart by holding the tape behind the timer. Make sure the area under the falling mass is clear of obstacles. Start the timer and release the tape simultaneously.

12. Carefully stop the cart when the 0.10 kg mass hits the floor, and then stop the timer. ***Do not let the cart fall off the table.***

13. Remove the tape and label it.

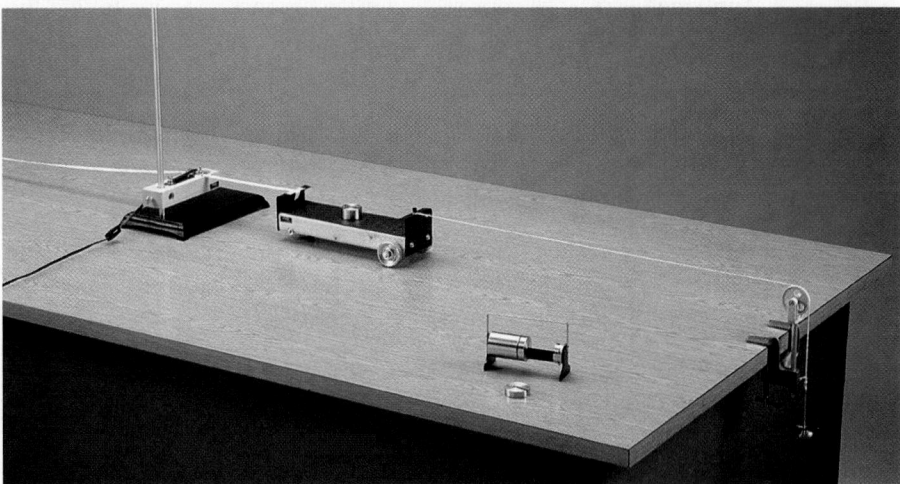

Figure 4-40
Step 4: Make sure the clamp protrudes as little as possible from the edge of the table.

Step 7: Fold the end of the recording tape over the edge of the rail on the cart and tape it down.

Step 10: Always hold the cart when you are removing and adding masses. When you are ready, release the cart from the same position each time.

Recording Timer Tips

◆ If 110V ac timers are used, calibration may be omitted: the average period of these timers is 0.017 s ($\frac{1}{60}$ s).

◆ If possible, mount timers on stand rods to adjust the height of the timer. The tape should be level from the cart to the timer. If the timer must be table-mounted, leave at least 0.5 m between the timer and the cart to prevent the difference in height from affecting the results.

Techniques to Demonstrate

Show students how to thread the paper tape through the recording timer guides under the carbon disk.

Students may need practice releasing the cart and starting the timer at the same time. The best method is to hold the cart by holding the tape straight out behind the timer.

The timing tape should be fastened to the cart by folding the end of the tape over, hooking the fold over the flat rail on the cart, and securing paper tape with cellophane tape.

✔ Checkpoints

Step 4: Check each setup before the timer is plugged in. Make sure the tape is level and inserted properly and that all clamps are tight and positioned where they will not protrude and cause injury.

Step 9: Make sure masses are securely attached. Students should be able to demonstrate that the counterweight allows the cart to move at a constant velocity when given a small push.

Step 15: Students should be able to explain how the dots on the tape represent the motion of the cart. They should be able to explain what order the dots were made in.

ANSWERS TO

Analysis and Interpretation

CALCULATIONS AND DATA ANALYSIS

1. Trial 1: $F = 0.981$ N, Trial 2: $F = 1.962$ N, Trial 3: $F = 2.943$ N, Trial 4: $F = 2.943$ N, Trial 5: $F = 2.943$ N

2. Trial 1: $a = 0.635$ m/s^2, Trial 2: $a = 1.250$ m/s^2, Trial 3: $a = 1.890$ m/s^2, Trial 4: $a = 1.458$ m/s^2, Trial 5: $a = 0.927$ m/s^2

3. Student graphs should show a straight line beginning at the origin and pointing up and to the right.

4. There is a direct relationship between the acceleration and the force.

14. Use a meterstick to measure the distance the weights fell. Record the *Distance* in your data table.

15. On the tape, measure this distance starting from the first clear dot. Mark the end of this distance. Count the number of dots between the first dot and this mark.

16. Calculate and record the *Time interval* represented by the number of dots. Fasten a new timing tape to the end of the cart.

17. Replace the 0.10 kg mass in the cart. Remove 0.20 kg from the cart and attach it securely to the end of the cord. Repeat the procedure, and record the results in your data table as *Trial 2*.

18. Leave the 0.20 kg on the end of the cord and attach the 0.10 kg mass from the cart securely to the end

of the cord. Repeat the procedure, and record the results in your data table as *Trial 3*.

Constant force with varying mass

19. For the two trials in this part of the experiment, keep 0.30 kg and the counterweight on the string. Be sure to include this mass when recording the total mass for these three trials.

20. Add 0.50 kg to the cart. Tape the mass to the cart to keep it in place. Run the experiment and record the total mass, accelerating mass, accelerating force, distance, and time under *Trial 4* in your data table.

21. Tape 1.00 kg to the cart and repeat the procedure. Record the data under *Trial 5* in your data table.

22. Clean up your work area. Put equipment away safely so that it is ready to be used again.

ANALYSIS AND INTERPRETATION

Calculations and data analysis

1. Analyzing data Using the Newton's second law equation, $\mathbf{F} = m\mathbf{a}$, where $a = g$, calculate the force acting on the mass in each trial.

    a. CBL and sensors Compare these values to the force values in your data table. Are they the same? If not, why not?

    b. Recording timer Enter these values in your data table as the *Accelerating force* for each trial.

2. Applying ideas Use your values for the distance and time to find the acceleration of the cart for each trial using the equation $\Delta x = \frac{1}{2}a\Delta t^2$ for constantly accelerated motion.

3. Graphing data Using the data from *Trials 1–3*, plot a graph of the accelerations of the cart versus the accelerating forces. Use a graphing calculator, computer, or graph paper.

4. Interpreting graphs Based on your graph from item 3, what is the relationship between the acceleration of the cart and the accelerating force? Explain how your graph supports your answer.

5. Graphing data Using the data from *Trials 3–5*, plot a graph of the total masses versus the accelerations. Use a graphing calculator, computer, or graph paper.

6. Interpreting graphs Based on your graph from item 5, what is the relationship between the total mass and the acceleration? Explain how your graph supports your answer.

Conclusions

7. Evaluating methods Why does the mass in *Trials 1–3* remain constant even though masses are removed from the cart during the trials?

8. Evaluating methods Do the carts move with the same velocity and acceleration as the accelerating masses that are dropped? If not, why not?

9. Applying conclusions Do your data support Newton's second law? Use your data and your analysis of your graphs to support your conclusions.

10. Applying ideas A team of automobile safety engineers developed a new type of car and performed some test crashes to find out whether the car is safe. The engineers tested the new car by involving it in a series of different types of accidents. For each test, the engineers applied a known force to the car and measured the acceleration of the car after the crash. The graph in **Figure 4-41** shows the acceleration of the car plotted against the applied force. Compare this with the data you collected and the graphs you made for this experiment to answer the following questions.

a. Based on the graph, what is the relationship between the acceleration of the new car and the force of the collision?

b. Does this graph support Newton's second law? Use your analysis of this graph to support your conclusions.

c. Does the data from the crash tests meet your expectations based on this lab? Explain what you think may have happened to affect the results. If you were on the engineering team, how would you find out whether your results were in error?

Extensions

11. Critical thinking How would your results be affected if you used the mass of the cart and its contents instead of the total mass? Predict what would happen if you performed *Trials 1–3* again keeping the mass of the cart and its contents constant while varying the accelerating mass. If there is time and your teacher approves your plan, go into the lab and try it. Plot your data using a graphing calculator, computer, or graph paper.

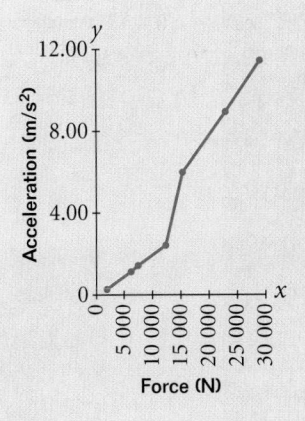

Figure 4-41

5. Students' graphs should show a straight line pointing down and to the right.

6. There is an indirect relationship between the acceleration and the mass.

CONCLUSIONS

7. The masses are moved from the cart to the string; all the mass in the system is part of the accelerated mass.

8. The velocity and acceleration are not the same. The dropped masses accelerate at 9.81 m/s², but the carts are accelerated by the force of the falling mass.

9. Data should support Newton's second law, a direct relationship between a and F, and an indirect relationship between m and a.

10. a. As F increases, a increases. The proportion between F and a is different above and below 15 000 N.

b. The graph does not support Newton's law because the relationship is not constant.

c. Student answers should reflect that the data are not what should be expected, possibly due to a change in mass or an error in measurement. To verify, the experiment should be repeated several times.

EXTENSIONS

11. Student answers should reflect an understanding that the acceleration of the cart depends on the force and the mass of the cart. Plans should be safe and complete.

Physics and Its World *Timeline 1540–1690*

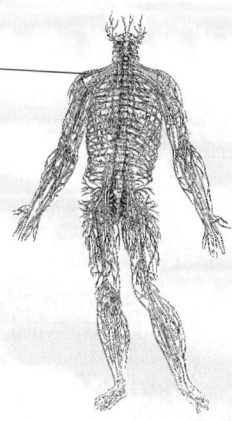

1556 – **Akbar** becomes ruler of
the Moghul Empire in North India,
Pakistan, and Afghanistan. By ensuring
religious tolerance, he establishes
greater unity in India, making it one of
the world's great powers.

1543 **Nicholas Copernicus'**
*On the Revolutions of the
Heavenly Bodies* is
published. It is the first
work on astronomy to provide
an analytical basis for the motion
of the planets, including Earth,
around the sun.

1543 – **Andries van Wesel,**
better known as Andreas
Vesalius, completes his *Seven
Books on the Structure of the
Human Body.* It is the first work
on anatomy to be based on the
dissection of human bodies.

1564 – English writers
Christopher Marlowe and
William Shakespeare are born.

1586 – Kabuki
theater is developed
in Japan.

1588 – **Queen Elizabeth I** of England
sends the English fleet to repel the invasion
by the Spanish Armada. The success of the
English navy marks the beginning of Great
Britain's status as a major naval power.

1592
$$\Delta x = v_i \Delta t + \frac{1}{2} a (\Delta t)^2$$

Galileo Galilei is appointed professor of mathematics
at the University of Padua. While there, he performs
experiments on the motions of bodies.

1605 – The first part of **Miguel de
Cervantes's** *Don Quixote* is published.

1540
1550
1560
1570
1580
1590
1600
1610

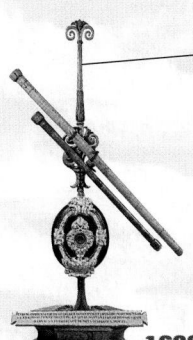

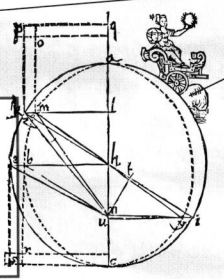

1609

$$T^2 \propto a^3$$

New Astronomy, by **Johannes Kepler**, is published. In it, Kepler demonstrates that the orbit of Mars is elliptical rather than circular.

1608 – The first telescopes are constructed in the Netherlands. Using these instruments as models, **Galileo** constructs his first telescope the following year.

1637 – **René Descartes's** *Discourse on Method* is published. According to Descartes's philosophy of rationalism, the laws of nature can be deduced by reason.

1644 – The Ch'ing, or Manchu, Dynasty is established in China. China becomes the most prosperous nation in the world, then declines until the Ch'ing Dynasty is replaced by the Chinese Republic in 1911.

1656 – The first paintings of Dutch artist **Jan Vermeer** are produced around this time. Vermeer's paintings portray middle- and working-class people in everyday situations.

1669 – Danish geologist **Niclaus Steno** correctly determines the structure of crystals and identifies fossils as organic remains.

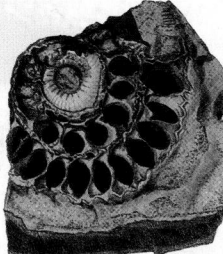

1678

$$c = f\lambda$$

Christian Huygens completes the bulk of his *Treatise on Light,* in which he presents his model of secondary wavelets, known today as Huygens' principle. The completed book is published 12 years later.

1687

$$F = ma$$

Issac Newton's masterpiece, the *Mathematical Principles of Nature,* is published. In this extensive work, Newton systematically presents a unified model of mechanics.

1610
1620
1630
1640
1650
1660
1670
1680
1690

CHAPTER 5 PLANNING GUIDE

Compression Guide: To shorten from 11 to 8 45-min periods (from 5½ to 4 90-min blocks), eliminate items in magenta type.

PACING CHART / CLASSROOM RESOURCES

PACING CHART	★ TEKS	Teacher Demonstrations	Holt Physics Transparencies	Labs (See page T52 for equipment listing for in-text labs.)	
5-1 Work 2 or 1 45-minute periods 1 or ½ 90-minute block	2C, 3A, 5B	**TE** *Work,* p. 168 **TE** *Quantifying work,* p. 170	**T** 15–16	**L** *Discovery Lab: Exploring Work and Energy*	
5-2 Energy 2 45-minute periods 1 90-minute block	2C, 5A, 5B, 6A	**TE** *Potential energy,* p. 178	**T** 17–18		
5-3 Conservation of energy 4 or 2 45-minute periods 2 or 1 90-minute blocks	1A, 2B, 2C, 2F, 3A, 3E, 5B, 5C, 5D	**TE** *Mechanical energy,* p. 181 **TE** *Conservation of energy,* p. 182	**T** 19 **TM** 20–21	**PE** *Quick Lab: Mechanical Energy,* p. 183 **PE** *Conservation of Mechanical Energy,* p. 200 **L** *Invention Lab: Bungee Jumping* **TL** *Loss of Mechanical Energy*	
5-4 Power 1 45-minute period ½ 90-minute block	2C, 3A, 3C, 3E		**TM** 22		

Review and Assessment
2 45-minute periods
1 90-minute block

Resource Key

PHYSICS (HOLT)

PE Pupil's Edition
TE Teacher's Edition

L Laboratory Experiments
TL Technology Lab Experiments
T Transparencies

 **One-Stop** Planner CD-ROM **contents**

TM Transparency Masters
SR Section Review Worksheets
AA Alternative Assessment

PW Problem-Solving Workbook
PB Problem Bank
CTW Critical Thinking Worksheet

LABORATORY PLANNING: Conservation of Mechanical Energy, p. 200

Materials (for each lab group)
- support base and rod: 1-position support base and rod, 1.3 cm × 91 cm, or tripod base and support rod
- lattice rod, 1.3 cm × 30 cm
- meterstick
- table clamp
- right-angle clamp for 1.3 cm rod
- metric slotted mass set, 1 g–500 g

Additional Equipment CBL and Sensors Procedure
- CBL
- graphing calculator
- CBL ultrasonic motion detector
- CBL student force sensor
- roll of adhesive tape, 0.5 in wide
- spring
- TI Graph Link (recommended for downloading programs)

ASSIGNMENT RESOURCES

Content Mastery	Critical Thinking	Problem-Solving Practice	
PE 1–6, p. 171 **SR** 5-1, *Math Skills* **PE** 1–3, p. 193	**PE** 4–6, p. 193	**5A**	Work: 36 items in **PE, PW,** and **PB,** see **TE** p. 169
PE 1–5, p. 180 **SR** 5-2, *Diagram Skills* **PE** 11–14, pp. 193–194	**PE** 15–18, p. 194	**5B** **5C** **5D**	Kinetic energy: 34 items in **PE, PW,** and **PB,** see **TE** p. 173 Work-kinetic energy theorem: 40 items in **PE, PW,** and **PB,** see **TE** p. 175 Potential energy: 29 items in **PE, PW,** and **PB,** see **TE** p. 179
PE 1–4, p. 186 **SR** 5-3, *Diagram Skills* **PE** 26–28, p. 195	**PE** 29–32, p. 195	**5E**	Conservation of mechanical energy: 42 items in **PE, PW,** and **PB,** see **TE** p. 184
PE 1–3, p. 189 **SR** 5-4, *Concept Review*	**PE** 1–2, p. 187	**5F**	Power: 22 items in **PE, PW,** and **PB,** see **TE** pp. 188–189

ASSESSMENT RESOURCES

Cumulative Review	Alternative Assessment	Traditional Assessment
SR Mixed Review, Ch. 5	**PE** 1–6, p. 199 **AA** Items for Ch. 5	Chapter 5 Test Test Generator items for Ch. 5

Scoring Rubrics for Alternative Assessment items can be found on the One-Stop Planner CD-ROM.

TECHNOLOGY RESOURCES

 PHYSICS TUTOR Module 5 Work

Module 6 Work-Kinetic Energy Theorem

 PE Technology and Learning, p. 198
(Alternative procedures for calculators without Flash-ROM technology are provided on the One-Stop Planner CD-ROM.)

 The Mechanical Universe/High School Adaptation Quad II, Conservation of Energy

 internet**connect**

 **SCILINKS NSTA**

On-line Student Resources:
GO TO: www.scilinks.org
The following student resources can be found in the student text for this chapter.

TOPICS:
• Work, p. 171 (HF2051)
• Potential and kinetic energy, p. 178 (HF2052)
• Conservation of energy, p. 185 (HF2053)
• Nuclear reactions, p. 191 (HF2054)

On-line Teacher Resources:
GO TO: go.hrw.com
KEYWORD: HF2 HOME
Visit the HRW Web site for a variety of resources related to this chapter.

 Smithsonian Institution
Internet Connections

Visit www.si.edu/hrw for additional on-line resources.

CNNfyi.com

Visit www.cnnfyi.com for late-breaking news and current events stories selected just for you.

Hooke's Law Apparatus Procedure
• Hooke's law apparatus with rubber bands

Section 5-1 introduces work and shows calculations of the work done in a variety of situations.

Section 5-2 identifies and shows calculations using kinetic energy, the work–kinetic energy theorem, and different types of potential energy.

Section 5-3 explores the conditions necessary for conservation of mechanical energy and applies this principle to problem solving.

Section 5-4 introduces the relationships between work, time, power, force, and speed.

About the Illustration

This audiokinetic sculpture was created by George Rhoads, whose sculptures can be seen at the Boston Museum of Science, at the Port Authority Bus Terminal in New York City, and in various shopping centers. After completing the chapter, have students return to this photograph and apply the concepts of work and the conservation of energy to describe which balls probably have mostly potential energy and which have mostly kinetic energy.

Interactive Problem-Solving Tutor

See Module 5

"Work" provides additional practice calculating net work.

See Module 6

"Work–Kinetic Energy Theorem" promotes additional development of problem-solving skills involving work.

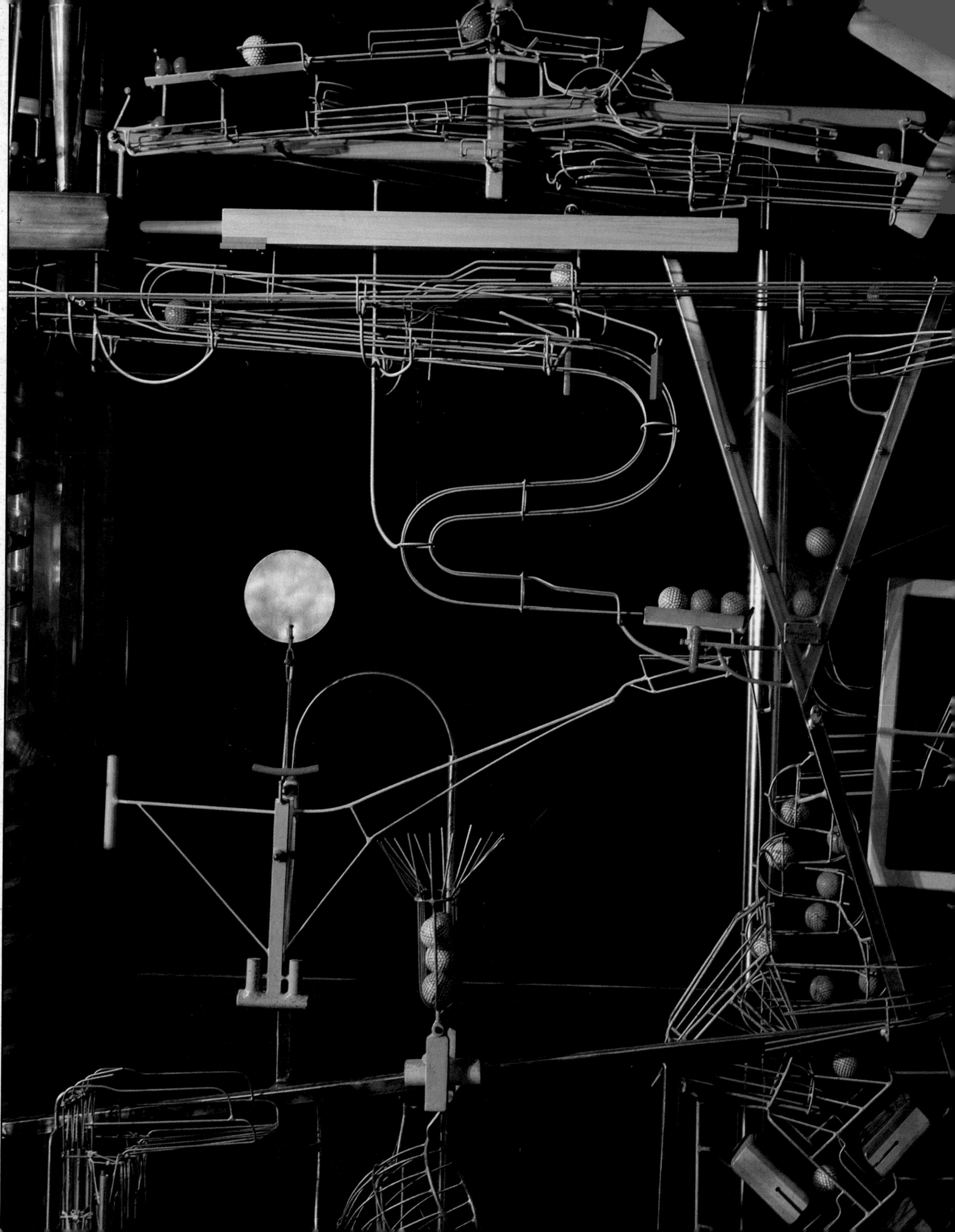

CHAPTER 5

Work and Energy

PHYSICS IN ACTION

This whimsical piece of art is a type of audio-kinetic sculpture. Balls are raised to a high point on the curved blue track. As the balls roll down the track, they turn levers, spin rotors, and bounce off elastic membranes.

As each ball travels along the track, the total energy of the system remains unchanged. The types of energy that are involved—whether associated with the ball's motion, the ball's position above the ground, or friction—vary in ways that keep the total energy of the system constant.

In this chapter, you will learn about work and the different types of energy that are relevant to mechanics.

- *How many different kinds of energy are used in this sculpture?*

- *How are work, energy, and power related in the functioning of this sculpture?*

CONCEPT REVIEW

Kinematics (Section 2-2)

Newton's second law (Section 4-3)

Force of friction (Section 4-4)

✔ "Students learn that energy cannot be created or destroyed, but only changed from one form to another." (AAAS's *Benchmarks for Science Literacy*, grades 6–8)

✔ "Energy is associated with heat, light, electricity, mechanical motion, sound, and the nature of a chemical. Energy is transferred in many ways." (NRC's *National Science Education Standards*, grades 5–8)

✔ "Students tend to (a) associate energy with living things; (b) believe that energy is a fuel-like quantity; (c) think that energy transformations involve only one form of energy at a time." (AAAS's *Benchmarks for Science Literacy*, The Research Base)

Knowledge to Review
✔ Review the kinematic equations. (Section 2-2)

✔ Newton's second law states that force = mass × acceleration ($\mathbf{F} = m\mathbf{a}$). (Section 4-3)

✔ Kinetic friction is a resistive force exerted on a moving body by its environment. (Section 4-4)

Items to Probe
✔ Familiarity with phenomena of energy transformation: Ask students to describe the action of jumping up and down on a trampoline in terms of energy.

✔ Preconceptions about energy dissipation: Ask students if energy is ever lost in a process.

5-1
Work

Work

Purpose Determine whether work is done in various situations.

Materials teacher's text, spring scale, string

Procedure Hang the textbook from the scale with the string. Hold the book stationary, have the students note the scale reading, and record the weight of the book *(mg)*. Ask the students if the spring is exerting a force on the book. *(Yes, the spring exerts a force on the book that is equal to and opposite the book's weight.)* Ask students if work is being done on the book. *(No, because the displacement of the book is zero.)*

Now lift the book about 1.5 m at a constant velocity and have the students note the scale reading. (After an initial increase, the reading will remain constant.) Again ask the students if work is being done on the book. *(Yes, the force of the lifting, equal in magnitude to the weight of the book, is upward, and the displacement is upward.)* Have students calculate the amount of work *(m × g × h)*.

Hold the book at shoulder height and carry it across the room at a constant speed. Ask students if work is being done on the book. *(No, because the upward force is perpendicular to the horizontal displacement.)*

⭐ TEKS

p. 168: 3A
p. 169: 2C

5-1 SECTION OBJECTIVES

- **Recognize the difference between the scientific and ordinary definitions of** *work.*

- **Define** *work* **by relating it to force and displacement.**

- **Identify where work is being performed in a variety of situations.**

- **Calculate the net work done when many forces are applied to an object.**

work

the product of the magnitudes of the component of a force along the direction of displacement and the displacement

Figure 5-1
This person exerts a constant force on the car and displaces it to the left. The work done on the car by the person is equal to the force times the displacement of the car.

DEFINITION OF WORK

Many of the terms you have encountered so far in this book have meanings in physics that are similar to their meanings in everyday life. In its everyday sense, the term *work* means to do something that takes physical or mental effort. But in physics, work has a distinctly different meaning. Consider the following situations:

- A student holds a heavy chair at arm's length for several minutes.
- A student carries a bucket of water along a horizontal path while walking at constant velocity.

It might surprise you to know that under the scientific definition of work, there is no work done on the chair or the bucket, even though effort is required in both cases. We will return to these examples later.

A force that causes a displacement of an object does work on the object

Imagine that your car, like the car shown in **Figure 5-1,** has run out of gas and you have to push it down the road to the gas station. If you push the car with a constant force, the **work** you do on the car is equal to the magnitude of the force, *F*, times the magnitude of the displacement of the car. Using the symbol *d* instead of Δx for displacement, we can define work as follows:

$$W = Fd$$

Work is not done on an object unless the object is moved because of the action of a force. The application of a force alone does not constitute work. For this reason, no work is done on the chair when a student holds the chair at arm's length. Even though the student exerts a force to support the chair, the chair does not move. The student's tired arms suggest that work is being done, which is indeed true. The quivering muscles in the student's arms go through many small displacements and do work within the student's body. However, work is not done on the chair. ⭐TEKS **3A**

Work is done only when components of a force are parallel to a displacement

When the force on an object and the object's displacement are in different directions, only the component of the force that is in the direction of the object's displacement does work. Components of the force perpendicular to a displacement do not do work.

For example, imagine pushing a crate along the ground. If the force you exert is horizontal, all of your effort moves the crate. If your force is other than horizontal, only the horizontal component of your applied force causes a displacement and does work. If the angle between the force and the direction of the displacement is θ, as in **Figure 5-2**, work can be written as follows:

$$W = Fd(\cos \theta)$$

If $\theta = 0°$, then $\cos 0° = 1$ and $W = Fd$, which is the definition of work given earlier. If $\theta = 90°$, however, then $\cos 90° = 0$ and $W = 0$. So, no work is done on a bucket of water being carried by a student walking horizontally. The upward force exerted to support the bucket is perpendicular to the displacement of the bucket, which results in no work done on the bucket. ⭐TEKS **2C**

Finally, if many constant forces are acting on an object, you can find the *net* work done by first finding the net force.

NET WORK DONE BY A CONSTANT NET FORCE

$$W_{net} = F_{net}d(\cos \theta)$$

net work = net force × displacement × cosine of the angle between them

---

Work has dimensions of force times length. In the SI system, work has a unit of newtons times meters (N•m), or joules (J). The work done in lifting an apple from your waist to the top of your head is about 1 J. Three push-ups require about 1000 J.

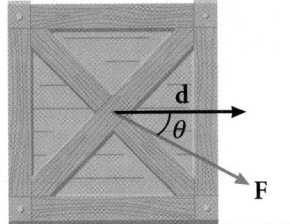

$$W = Fd(\cos \theta)$$

Figure 5-2
The work done on this crate is equal to the force times the displacement times the cosine of the angle between them.

Did you know?

The joule is named for the British physicist James Prescott Joule (1818–1889). Joule made major contributions to the understanding of energy, heat, and electricity.

SAMPLE PROBLEM 5A

Work ⭐TEKS 2C

PROBLEM

How much work is done on a vacuum cleaner pulled 3.0 m by a force of 50.0 N at an angle of 30.0° above the horizontal?

SOLUTION

Given: $F = 50.0$ N $\theta = 30.0°$ $d = 3.0$ m

Unknown: $W = ?$

Use the equation for net work done by a constant force:

$$W = Fd(\cos \theta)$$

Only the horizontal component of the applied force is doing work on the vacuum cleaner.

$$W = (50.0 \text{ N})(3.0 \text{ m})(\cos 30.0°)$$

$$\boxed{W = 130 \text{ J}}$$

Teaching Tip
Point out that although the crate in **Figure 5-2** is a large object, we think of its mass as reduced to a point at its center to simplify the situation.

Classroom Practice

The following may be used as teamwork exercises or for demonstration at the chalkboard or on an overhead projector.

PROBLEM

Work

A 20.0 kg suitcase is raised 3.0 m above a platform by a conveyor belt. How much work is done on the suitcase?

Answer
5.9×10^2 J

PRACTICE GUIDE 5A

Solving for:		
W	📖 **PE**	Sample, 1–3; Ch. Rvw. 7–10, 46, 48, 50, 60
	⊙ **PW**	8–11
	PB	5–7
F	📖 **PE**	Ch. Rvw. 60
	⊙ **PW**	5–7
	PB	Sample, 1–4
d	📖 **PE**	4; Ch. Rvw. 47
	⊙ **PW**	Sample, 1–4
	PB	8–10

Interactive Problem-Solving Tutor

See Module 5
"Work" provides additional practice calculating net work.

Demonstration 2

Quantifying work

Purpose Show how the angle between force and displacement affects work.

Materials wood block, spring scale, ruler

Procedure Attach the spring scale to the block. Hold the scale horizontally and pull the block 0.50 m across the table at a constant speed. Repeat the demonstration while holding the scale at angles of 30°, 45°, and 60° from the horizontal and pulling with the same force so that the scale shows the same reading. Have a student draw a data table on the chalkboard and note the readings of the scale (F), the angle (θ), and the displacement (d) of the block in each case. Add columns labeled "$\cos\theta$" and "W" to the data table on the chalkboard. Have students fill in the data table. Point out that although the force was the same in each case, the work done was different because $F(\cos\theta)$ decreased as the angle increased.

PRACTICE 5A

Work (★)TEKS 2C

1. A tugboat pulls a ship with a constant net horizontal force of 5.00×10^3 N and causes the ship to move through a harbor. How much work is done on the ship if it moves a distance of 3.00 km?

2. A weight lifter lifts a set of weights a vertical distance of 2.00 m. If a constant net force of 350 N is exerted on the weights, what is the net work done on the weights?

3. A shopper in a supermarket pushes a cart with a force of 35 N directed at an angle of 25° downward from the horizontal. Find the work done by the shopper on the cart as the shopper moves along a 50.0 m length of aisle.

4. If 2.0 J of work is done in raising a 180 g apple, how far is it lifted?

**Module 5
"Work"**
provides an interactive lesson with guided problem-solving practice to teach you about calculating net work.

The sign of work is important

Work is a scalar quantity and can be positive or negative, as shown in **Figure 5-3.** Work is positive when the component of force is in the same direction as the displacement. For example, when you lift a box, the work done by the force you exert on the box is positive because that force is upward, in the same

Figure 5-3
Depending on the angle, an applied force can either cause a moving car to slow down (left), which results in negative work done on the car, or speed up (right), which results in positive work done on the car.

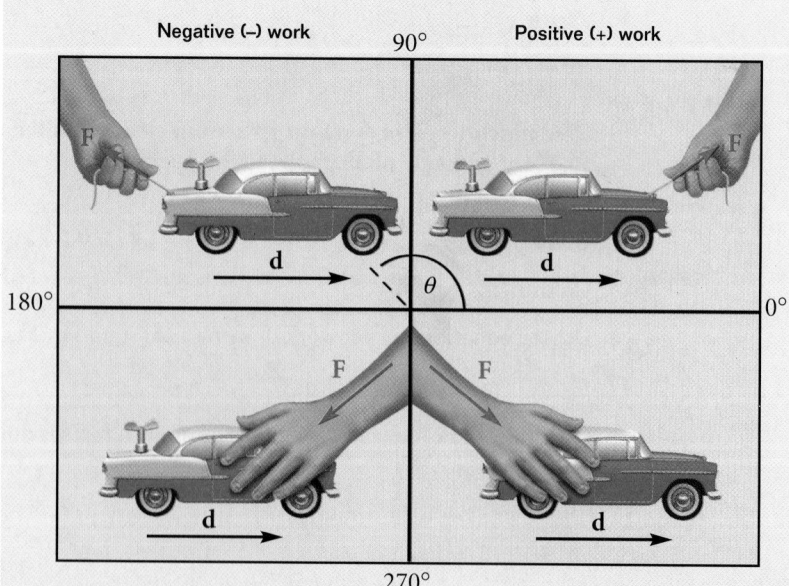

direction as the displacement. Work is negative when the force is in the direction opposite the displacement. For example, the force of kinetic friction between a sliding box and the floor is opposite to the displacement of the box, so the work done by the force on the box is negative. If you are very careful in applying the equation for work, your answer will have the correct sign: $\cos\theta$ is negative for angles greater than 90° but less than 270°.

If the work done on an object results only in a change in the object's speed, the sign of the net work on the object tells you whether the object's speed is increasing or decreasing. If the net work is positive, the object speeds up and the net force does work *on* the object. If the net work is negative, the object slows down and work is done *by* the object on another object.

internet **connect**

SC*LINKS*
NSTA
TOPIC: Work
GO TO: www.scilinks.org
*sci***LINKS CODE:** HF2051

Section Review

 TEKS 2C

1. For each of the following statements, identify whether the everyday or the scientific meaning of work is intended.
   a. Jack had to work against time as the deadline neared.
   b. Jill had to work on her homework before she went to bed.
   c. Jack did work carrying the pail of water up the hill.

2. If a neighbor pushes a lawnmower four times as far as you do but exerts only half the force, which one of you does more work and by how much?

3. For each of the following cases, indicate whether the work done on the second object in each example will have a positive or a negative value.
   a. The road exerts a friction force on a speeding car skidding to a stop.
   b. A rope exerts a force on a bucket as the bucket is raised up a well.
   c. Air exerts a force on a parachute as the parachutist slowly falls to Earth.

4. Determine whether work is being done in each of the following examples:
   a. a train engine pulling a loaded boxcar initially at rest
   b. a tug of war that is evenly matched
   c. a crane lifting a car

5. A worker pushes a 1.50×10^3 N crate with a horizontal force of 345 N a distance of 24.0 m. Assume the coefficient of kinetic friction between the crate and the floor is 0.220.
   a. How much work is done by the worker on the crate?
   b. How much work is done by the floor on the crate?
   c. What is the net work done on the crate?

6. **Physics in Action** A 0.075 kg ball in a kinetic sculpture is raised 1.32 m above the ground by a motorized vertical conveyor belt. A constant frictional force of 0.350 N acts in the direction opposite the conveyor belt's motion. What is the net work done on the ball?

Teaching Tip

Write the following table on the board to help students remember the various situations that affect the sign of work.

Force is in direction of motion.	positive work
Force opposes motion.	negative work
Force is 90° to motion.	no work
Object is not in motion.	no work

Section Review
ANSWERS

1. a. everyday sense
   b. everyday sense
   c. scientific sense
2. the neighbor; twice as much
3. a. negative
   b. positive
   c. negative
4. a. yes
   b. no
   c. yes
5. a. 8.28×10^3 J
   b. -7.92×10^3 J
   c. 3.6×10^2 J
6. 0.519 J

TEKS

p. 170: 2C
p. 171: 2C

Misconception Alert

Students may think that kinetic energy depends on the direction of motion. Ask them to compare the kinetic energy of identical cars traveling at the same speed in each of the following situations: one driving north, one driving south, one driving uphill, and one driving downhill. *(The kinetic energy is the same in each case because kinetic energy depends only on mass and speed, which are the same in each case.)*

The Language of Physics

The symbol for kinetic energy, *KE*, may look like the product of two variables (*K* and *E*) to some students. Point out that the two letters together designate kinetic energy. This symbol for kinetic energy is not universal. Some books use the letter *K* alone; others use *E* alone and specify the kind of energy in context.

⭐ TEKS

p. 173: 5B, 2C, 5B

5-2
Energy

5-2 SECTION OBJECTIVES

- **Identify several forms of energy.**

- **Calculate kinetic energy for an object.**

- **Apply the work–kinetic energy theorem to solve problems.**

- **Distinguish between kinetic and potential energy.**

- **Classify different types of potential energy.**

- **Calculate the potential energy associated with an object's position.**

kinetic energy

the energy of an object due to its motion

KINETIC ENERGY

Kinetic energy is energy associated with an object in motion. **Figure 5-4** shows a cart of mass m moving to the right on a frictionless air track under the action of a constant net force, **F**. Because the force is constant, we know from Newton's second law that the particle moves with a constant acceleration, **a**. While the force is applied, the cart accelerates from an initial velocity v_i to a final velocity v_f. If the particle is displaced a distance of Δx, the work done by **F** during this displacement is

$$W_{net} = F\Delta x = (ma)\Delta x$$

However, in Chapter 2 we found that the following relationship holds when an object undergoes constant acceleration:

$$v_f^2 = v_i^2 + 2a\Delta x$$

$$a\Delta x = \frac{v_f^2 - v_i^2}{2}$$

Substituting this result into the equation $W_{net} = (ma)\Delta x$ gives

$$W_{net} = m\left(\frac{v_f^2 - v_i^2}{2}\right)$$

$$W_{net} = \frac{1}{2}mv_f^2 - \frac{1}{2}mv_i^2$$

Kinetic energy depends on speed and mass

The quantity $\frac{1}{2}mv^2$ has a special name in physics: **kinetic energy.** The kinetic energy of an object with mass m and speed v, when treated as a particle, is given by the expression shown on the next page.

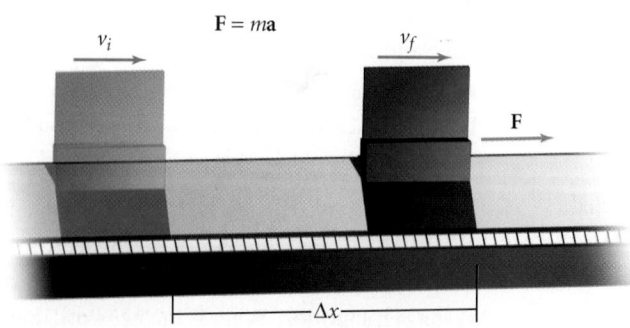

Figure 5-4

The work done on an object by a constant force equals the object's mass times its acceleration times its displacement.

KINETIC ENERGY

$$KE = \frac{1}{2}mv^2$$

$$\text{kinetic energy} = \frac{1}{2} \times \text{mass} \times (\text{speed})^2$$

Kinetic energy is a scalar quantity, and the SI unit for kinetic energy (and all other forms of energy) is the joule. Recall that a joule is also used as the basic unit for work.

Kinetic energy depends on both an object's speed and its mass. If a bowling ball and a volleyball are traveling at the same speed, which do you think has more kinetic energy? You may think that because they are moving with identical speeds they have exactly the same kinetic energy. However, the bowling ball has more kinetic energy than the volleyball traveling at the same speed because the bowling ball has more mass than the volleyball. ⭐TEKS **5B**

Module 6
"Work–Kinetic Energy Theorem"
provides an interactive lesson with guided problem-solving practice.

SAMPLE PROBLEM 5B

Kinetic energy ⭐TEKS **2C, 5B**

PROBLEM

A 7.00 kg bowling ball moves at 3.00 m/s. How much kinetic energy does the bowling ball have? How fast must a 2.45 g table-tennis ball move in order to have the same kinetic energy as the bowling ball? Is this speed reasonable for a table-tennis ball?

SOLUTION

Given: The subscripts b and t indicate the bowling ball and the table-tennis ball, respectively.

$$m_b = 7.00 \text{ kg} \qquad m_t = 2.45 \text{ g} \qquad v_b = 3.00 \text{ m/s}$$

Unknown: $KE_b = ? \quad v_t = ?$

Use the kinetic energy equation:

$$KE_b = \frac{1}{2}m_b v_b^2 = \frac{1}{2}(7.00 \text{ kg})(3.00 \text{ m/s})^2 = 31.5 \text{ J}$$

$$KE_t = \frac{1}{2}m_t v_t^2 = KE_b = 31.5 \text{ J}$$

$$v_t = \sqrt{\frac{2KE_b}{m_t}} = \sqrt{\frac{(2)(31.5 \text{ J})}{2.45 \times 10^{-3} \text{ kg}}}$$

$$\boxed{v_t = 1.60 \times 10^2 \text{ m/s}}$$

This speed is much too high to be reasonable for a table-tennis ball.

Kinetic energy

1. Calculate the speed of an 8.0×10^4 kg airliner with a kinetic energy of 1.1×10^9 J.

2. What is the speed of a 0.145 kg baseball if its kinetic energy is 109 J?

3. Two bullets have masses of 3.0 g and 6.0 g, respectively. Both are fired with a speed of 40.0 m/s. Which bullet has more kinetic energy? What is the ratio of their kinetic energies?

4. Two 3.0 g bullets are fired with speeds of 40.0 m/s and 80.0 m/s, respectively. What are their kinetic energies? Which bullet has more kinetic energy? What is the ratio of their kinetic energies?

5. A car has a kinetic energy of 4.32×10^5 J when traveling at a speed of 23 m/s. What is its mass?

ANSWERS TO

Practice 5B
Kinetic energy

1. 1.7×10^2 m/s

2. 38.8 m/s

3. the bullet with the greater mass; 2 to 1

4. 2.4 J, 9.6 J; the bullet with the greater speed; 1 to 4

5. 1.6×10^3 kg

The Language of Physics

The symbol Δ (the Greek letter *delta*) is used to denote change. Students should be familiar with this symbol from Chapter 2. Point out that although the context is different, the symbol means the same thing, namely, a difference between two quantities. The subscripts *i* and *f* used with *ME* stand for initial and final amounts, respectively, of mechanical energy. Thus, Δ*ME* is the difference between ME_f and ME_i, or $\Delta ME = ME_f - ME_i$.

---

work–kinetic energy theorem

the net work done on an object is equal to the change in the kinetic energy of the object

The equation $W_{net} = \frac{1}{2}mv_f^2 - \frac{1}{2}mv_i^2$ derived at the beginning of this section says that the net work done by a *net* force acting on an object is equal to the *change* in the kinetic energy of the object. This important relationship, known as the **work–kinetic energy theorem,** is often written as follows: ⭐TEKS **5A**

> **WORK–KINETIC ENERGY THEOREM**
>
> $$W_{net} = \Delta KE$$
>
> **net work = change in kinetic energy**

It is important to note that when we use this theorem, we must include all the forces that do work on the object in calculating the net work done. From this theorem, we see that the speed of the object increases if the net work done on it is positive, because the final kinetic energy is greater than the initial kinetic energy. The object's speed decreases if the net work is negative, because the final kinetic energy is less than the initial kinetic energy.

The work–kinetic energy theorem allows us to think of kinetic energy as the work an object can do as it comes to rest, or the amount of energy stored in the object. For example, the moving hammer on the verge of striking a nail in **Figure 5-5** has kinetic energy and can therefore do work on the nail. Part of this energy is used to drive the nail into the wall, and part goes into warming the hammer and nail upon impact.

Figure 5-5
A moving hammer has kinetic energy and so can do work on a nail, driving it into the wall.

Work–kinetic energy theorem ⭐TEKS 2C, 5A, 5B

PROBLEM

On a frozen pond, a person kicks a 10.0 kg sled, giving it an initial speed of 2.2 m/s. How far does the sled move if the coefficient of kinetic friction between the sled and the ice is 0.10?

SOLUTION

1. DEFINE

Given: $m = 10.0$ kg $v_i = 2.2$ m/s $v_f = 0$ m/s $\mu_k = 0.10$

Unknown: $d = ?$

Diagram:

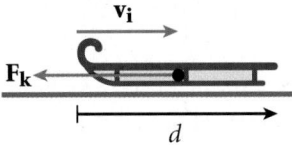

2. PLAN

Choose an equation(s) or situation: This problem can be solved using the definition of work and the work–kinetic energy theorem.

$$W_{net} = F_{net}d(\cos\theta)$$
$$W_{net} = \Delta KE$$

The initial kinetic energy is given to the sled by the person.

$$KE_i = \frac{1}{2}mv_i^2$$

Because the sled comes to rest, the final kinetic energy is zero.

$$KE_f = 0$$
$$\Delta KE = KE_f - KE_i = -\frac{1}{2}mv_i^2$$

The net work done on the sled is provided by the force of kinetic friction.

$$W_{net} = F_{net}d(\cos\theta) = \mu_k mgd(\cos\theta)$$

The force of kinetic friction is in the direction opposite d.

$$\theta = 180°$$

3. CALCULATE

Substitute values into the equations:

$$W_{net} = (0.10)(10.0\ \text{kg})(9.81\ \text{m/s}^2)\ d\ (\cos 180°)$$
$$W_{net} = (-9.8\ \text{N})d$$
$$\Delta KE = -KE_i = -(\tfrac{1}{2})(10.0\ \text{kg})(2.2\ \text{m/s})^2 = -24\ \text{J}$$

Use the work–kinetic energy theorem to solve for d.

$$W_{net} = \Delta KE$$
$$(-9.8\ \text{N})d = -24\ \text{J}$$

$$\boxed{d = 2.4\ \text{m}}$$

CALCULATOR SOLUTION

Your calculator should give an answer of 2.44898, but because the answer is limited to two significant figures, this number should be rounded to 2.4.

continued on next page

PRACTICE GUIDE 5C

Solving for:		
d	📖 **PE**	Sample, 1–3; Ch. Rvw. 21–22, 54
	💿 **PW**	7
	PB	5
F	📖 **PE**	4; Ch. Rvw. 49, 51
	💿 **PW**	3
	PB	3–4
W	📖 **PE**	5; Ch. Rvw. 45, 50, 58
	💿 **PW**	5
	PB	6
KE	📖 **PE**	5; Ch. Rvw. 40, 45, 58
	💿 **PW**	Sample, 1–2
	PB	7–8
v	📖 **PE**	5; Ch. Rvw. 38, 41, 45, 51
	💿 **PW**	4
	PB	9–10
m	💿 **PW**	6
	PB	Sample, 1–2

Interactive Problem-Solving Tutor

See Module 6
"Work–Kinetic Energy Theorem" promotes additional development of problem-solving skills.

⭐ TEKS
p. 174: 5A
p. 175: 2C, 5A, 5B

Alternative Problem-Solving Approach

First solve the problem with variables, and then substitute the numerical values given:

$$F_{net} = \mu_k mg$$
$$W_{net} = F_{net}d(\cos\theta) = \mu_k mgd(\cos\theta)$$
$$\text{Work done} = \Delta KE = KE_f - KE_i$$

$$\mu_k mgd(\cos\theta) = \tfrac{1}{2}mv_f^2 - \tfrac{1}{2}mv_i^2$$

$$d = \frac{mv_f^2 - mv_i^2}{2\mu_k mg(\cos\theta)} = \frac{v_f^2 - v_i^2}{2\mu_k g(\cos\theta)}$$

$$d = \frac{0 - (2.2 \text{ m/s})^2}{2(0.10)(9.81)(-1)}$$

$$d = 2.5 \text{ m}$$

(The difference between the two answers is due to rounding.)

ANSWERS TO

Practice 5C
Work–kinetic energy theorem

1. 7.8 m
2. 21 m
3. 5.1 m
4. 3.0×10^2 N
5. a. -1.9×10^2 J
   b. -2.8×10^2 J
   c. 7.5×10^2 J
   d. 2.8×10^2 J
   e. 7.6 m/s

★ TEKS

p. 176: 2C, 5A, 5B
p. 177: 5B, 5B, 6A

4. EVALUATE Note that because the direction of the force of kinetic friction is opposite the displacement, the net work done is negative. Also, according to Newton's second law, the acceleration of the sled is about -1 m/s^2 and the time it takes the sled to stop is about 2 s. Thus, the distance the sled traveled in the given amount of time should be less than the distance it would have traveled in the absence of friction.

$$2.4 \text{ m} < (2.2 \text{ m/s})(2 \text{ s}) = 4.4 \text{ m}$$

PRACTICE 5C

Work–kinetic energy theorem ★ TEKS 2C, 5A, 5B

1. A student wearing frictionless in-line skates on a horizontal surface is pushed by a friend with a constant force of 45 N. How far must the student be pushed, starting from rest, so that her final kinetic energy is 352 J?

2. A 2.0×10^3 kg car accelerates from rest under the actions of two forces. One is a forward force of 1140 N provided by traction between the wheels and the road. The other is a 950 N resistive force due to various frictional forces. Use the work–kinetic energy theorem to determine how far the car must travel for its speed to reach 2.0 m/s.

3. A 2.1×10^3 kg car starts from rest at the top of a driveway that is sloped at an angle of 20.0° with the horizontal. An average friction force of 4.0×10^3 N impedes the car's motion so that the car's speed at the bottom of the driveway is 3.8 m/s. What is the length of the driveway?

4. A 75 kg bobsled is pushed along a horizontal surface by two athletes. After the bobsled is pushed a distance of 4.5 m starting from rest, its speed is 6.0 m/s. Find the magnitude of the net force on the bobsled.

5. A 10.0 kg crate is pulled up a rough incline with an initial speed of 1.5 m/s. The pulling force is 100.0 N parallel to the incline, which makes an angle of 15.0° with the horizontal. Assuming the coefficient of kinetic friction is 0.40 and the crate is pulled a distance of 7.5 m, find the following:

   a. the work done by the Earth's gravity on the crate
   b. the work done by the force of friction on the crate
   c. the work done by the puller on the crate
   d. the change in kinetic energy of the crate
   e. the speed of the crate after it is pulled 7.5 m

POTENTIAL ENERGY

Consider the balanced boulder shown in **Figure 5-6.** While the boulder remains balanced, it has no kinetic energy. If it becomes unbalanced, it will fall vertically to the desert floor and will gain kinetic energy as it falls. A similar example is an arrow ready to be released on a bent bow. Once the arrow is in flight, it will have kinetic energy.

Potential energy is stored energy

As we have seen, an object in motion has kinetic energy. But a system can have other forms of energy. The examples above describe a form of energy that is due to the position of an object in relation to other objects or to a reference point. **Potential energy** is present in an object that has the potential to move because of its position relative to some other location. Unlike kinetic energy, potential energy depends not only on the properties of an object but also on the object's interaction with its environment. ⓉEKS **5B**

Gravitational potential energy depends on height from a zero level

In Chapter 3 we saw how gravitational force influences the motion of a projectile. If an object is thrown up in the air, the force of gravity will cause the object to eventually fall back down, provided the object was not thrown too hard. Similarly, the force of gravity will cause the unbalanced boulder in the previous example to fall. The energy associated with an object due to the object's position relative to a gravitational source is called **gravitational potential energy.**

Imagine an egg falling off a table. As it falls, it gains kinetic energy. But where does the egg's kinetic energy come from? It comes from the gravitational potential energy that is associated with the egg's initial position on the table relative to the floor. Gravitational potential energy can be determined using the following equation: ⓉEKS **5B**

> **GRAVITATIONAL POTENTIAL ENERGY**
>
> $$PE_g = mgh$$
>
> **gravitational potential energy = mass × free-fall acceleration × height**

The SI unit for gravitational potential energy, like for kinetic energy, is the joule. Note that the definition for gravitational potential energy in this chapter is valid only when the free-fall acceleration is constant over the entire height, such as at any point near the Earth's surface. Furthermore, gravitational potential energy depends on both the height and the free-fall acceleration, neither of which is a property of an object. ⓉEKS **6A**

Figure 5-6
Energy is present in this example, but it is not kinetic energy because there is no motion. What kind of energy is it?

potential energy

the energy associated with an object due to the position of the object

gravitational potential energy

the potential energy associated with an object due to the position of the object relative to the Earth or some other gravitational source

Did you know?

Another commonly used unit for energy besides the joule is the kilowatt-hour (kW•h). It is equal to 3.6×10^6 J. Electrical energy is often measured in kilowatt-hours.

The Language of Physics

In the symbol PE_g, PE stands for potential energy, and the subscript g specifies that the source of this potential energy is gravity. Some texts use U rather than PE to represent potential energy.

🛑 Misconception Alert

Some students do not realize that the potential energy of an object is relative. Point out that the zero-level for measuring height is arbitrarily defined in each problem. The potential energy is calculated *relative* to that level. Ask students how they would calculate the potential energy of a book on their desk relative to the desk, to the classroom floor, and to the roof.

Teaching Tip

Ask students whether it is possible to have a negative potential energy. (*Yes, a negative potential energy means that work needs to be done in order to bring an object to the zero-level.*) Then ask whether an object can have a positive potential energy relative to one reference point and a negative potential energy relative to another reference point. Have students give examples to support their answer. (*Yes, it is possible. For example, a book that is 0.5 m below a table and 0.5 m above the ground has a positive potential energy relative to the ground but a negative potential energy relative to the table.*)

Demonstration 3

Potential energy

Purpose Show that potential energy is stored energy.

Materials a racquetball cut in half

Caution *Do not face the area where you drop the ball, because it may rise up high enough to hit you.*

Procedure Pop the hollow hemisphere of the ball inside out and hold it hollow-side up. Before dropping it from a low height, ask students to predict whether it will bounce back and, if so, approximately how high.

Release the ball. The half ball will pop out on impact with the surface and will bounce up to a greater height with its hollow side facing down. Ask students where the additional gravitational potential energy came from. *(Elastic potential energy was stored in the half ball when it was inverted inside out.)*

Visual Strategy

Figure 5-8

Point out that the spring's potential energy depends on the difference between the spring's relaxed length and its compressed (or stretched) length.

⭐ **TEKS**

p. 178: **5B**
p. 179: **2C, 5B**

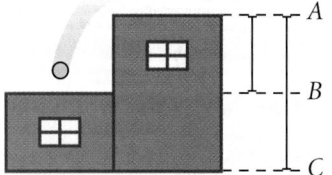

Figure 5-7
If *B* is the zero level, then all the gravitational potential energy is converted to kinetic energy as the ball falls from *A* to *B*. If *C* is the zero level, then only part of the total gravitational potential energy is converted to kinetic energy during the fall from *A* to *B*.

elastic potential energy

the potential energy in a stretched or compressed elastic object

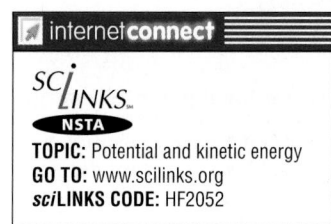

📶 internet**connect**

SC**i**LINKS
🔻 NSTA

TOPIC: Potential and kinetic energy
GO TO: www.scilinks.org
*sci***LINKS CODE:** HF2052

spring constant

a parameter that expresses how resistant a spring is to being compressed or stretched

Figure 5-8
The distance to use in the equation for elastic potential energy is the distance the spring is compressed or stretched from its relaxed length.

Suppose you drop a volleyball from a second-floor roof and it lands on the first-floor roof of an adjacent building (see **Figure 5-7**). If the height is measured from the ground, the gravitational potential energy is not zero because the ball is still above the ground. But if the height is measured from the first-floor roof, the potential energy is zero when the ball lands on the roof.

Gravitational potential energy is a result of an object's position, so it must be measured relative to some *zero level*. The zero level is the vertical coordinate at which gravitational potential energy is defined to be zero. This zero level is arbitrary, but it is chosen to make a specific problem easier to solve. In many cases, the statement of the problem suggests what to use as a zero level.

Elastic potential energy depends on distance compressed or stretched

Imagine you are playing with a spring on a tabletop. You push a block into the spring, compressing the spring, and then release the block. The block slides across the tabletop. The kinetic energy of the block came from the stored energy in the stretched or compressed spring. This potential energy is called **elastic potential energy.** Elastic potential energy is stored in any compressed or stretched object, such as a spring or the stretched strings of a tennis racket or guitar.

The length of a spring when no external forces are acting on it is called the *relaxed length* of the spring. When an external force compresses or stretches the spring, elastic potential energy is stored in the spring. The amount of energy depends on the distance the spring is compressed or stretched from its relaxed length, as shown in **Figure 5-8**. Elastic potential energy can be determined using the following equation: ⭐ TEKS **5B**

ELASTIC POTENTIAL ENERGY

$$PE_{elastic} = \frac{1}{2}kx^2$$

$$\text{elastic potential energy} = \frac{1}{2} \times \text{spring constant} \times \left(\begin{array}{c}\text{distance compressed} \\ \text{or stretched}\end{array}\right)^2$$

The symbol *k* is called the **spring constant,** or force constant. For a flexible spring, the spring constant is small, whereas for a stiff spring, the spring constant is large. Spring constants have units of newtons divided by meters (N/m).

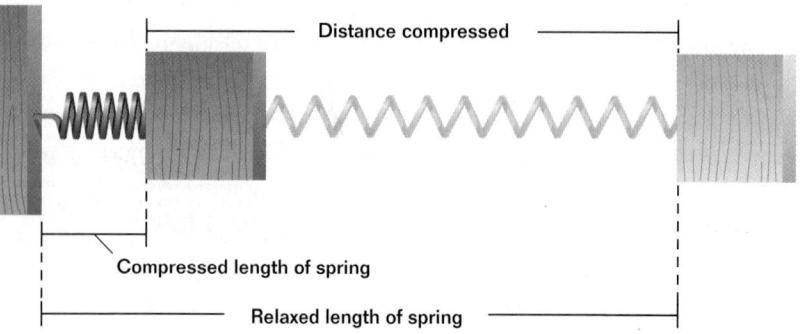

Distance compressed

Compressed length of spring

Relaxed length of spring

Potential energy ⭐TEKS 2C, 5B

PROBLEM

A 70.0 kg stuntman is attached to a bungee cord with an unstretched length of 15.0 m. He jumps off a bridge spanning a river from a height of 50.0 m. When he finally stops, the cord has a stretched length of 44.0 m. Treat the stuntman as a point mass, and disregard the weight of the bungee cord. Assuming the spring constant of the bungee cord is 71.8 N/m, what is the total potential energy relative to the water when the man stops falling?

SOLUTION

1. DEFINE **Given:** $m = 70.0$ kg $k = 71.8$ N/m $g = 9.81$ m/s^2

$h = 50.0$ m $- 44.0$ m $= 6.0$ m

$x = 44.0$ m $- 15.0$ m $= 29.0$ m

$PE = 0$ J at river level

Unknown: $PE_{tot} = ?$

Diagram:

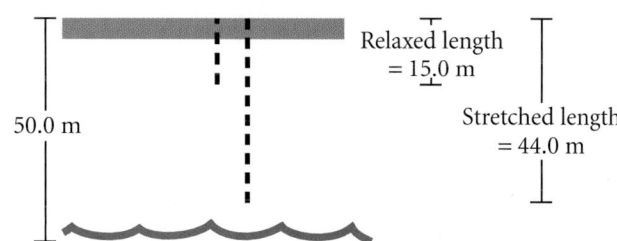

Relaxed length
= 15.0 m

50.0 m

Stretched length
= 44.0 m

2. PLAN **Choose an equation(s) or situation:** The zero level for gravitational potential energy is chosen to be at the surface of the water. The total potential energy is the sum of the gravitational and elastic potential energy.

$$PE_{tot} = PE_g + PE_{elastic}$$

$$PE_g = mgh$$

$$PE_{elastic} = \frac{1}{2}kx^2$$

3. CALCULATE **Substitute values into the equations:**

$$PE_g = (70.0 \text{ kg})(9.81 \text{ m/s}^2)(6.0 \text{ m}) = 4.1 \times 10^3 \text{ J}$$

$$PE_{elastic} = \frac{1}{2}(71.8 \text{ N/m})(29.0 \text{ m})^2 = 3.02 \times 10^4 \text{ J}$$

$$PE_{tot} = 4.1 \times 10^3 \text{ J} + 3.02 \times 10^4 \text{ J}$$

$$\boxed{PE_{tot} = 3.43 \times 10^4 \text{ J}}$$

4. EVALUATE One way to evaluate the answer is to make an order-of-magnitude estimate. The gravitational potential energy is on the order of 10^2 kg $\times 10$ m/s$^2 \times 10$ m $= 10^4$ J. The elastic potential energy is on the order of 1×10^2 N/m $\times 10^2$ m$^2 = 10^4$ J. Thus, the total potential energy should be on the order of 2×10^4 J. This number is close to the actual answer.

PROBLEM

Potential energy

When a 2.00 kg mass is attached to a vertical spring, the spring is stretched 10.0 cm such that the mass is 50.0 cm above the table.

a. What is the gravitational potential energy associated with this mass relative to the table?

b. What is the spring's elastic potential energy if the spring constant is 400.0 N/m?

c. What is the total potential energy of this system?

Answers
   a. 9.81 J
   b. 2.00 J
   c. 11.81 J

PRACTICE GUIDE 5D			
Solving for:			
PE	📖	**PE**	Sample, 1–3; Ch. Rvw. 23–25, 37
	💿	**PW**	7–9
		PB	9–10
k	💿	**PW**	10
		PB	Sample, 1–3
h or *d*	💿	**PW**	4–6, 10
		PB	4–6
m	💿	**PW**	Sample, 1–3
		PB	7–8

ANSWERS TO

Practice 5D
Potential energy
1. 3.3 J
2. 3.1×10^{-2} J
3. a. 785 J
   b. 105 J
   c. 0.00 J

Section Review
ANSWERS

1. a. kinetic energy
   b. nonmechanical energy
   c. kinetic energy, gravita-
      tional potential energy
   d. elastic potential energy
2. The heated water is an
   instance of nonmechanical
   energy, because its mass is
   not displaced with a velocity
   or with respect to a zero position, as would be the case for
   the various types of mechanical energy. The bicycle and
   football both have masses in
   motion, so they have kinetic
   energy. The wound spring
   has been displaced from its
   relaxed position and so has
   elastic potential energy, while
   the football is above the
   ground and therefore has a
   gravitational potential energy
   associated with it.
3. 4.4×10^{-3} J
4. 2.8 m/s
5. 6.18×10^{-2} J

PRACTICE 5D

Potential energy TEKS 2C, 5B

1. A spring with a force constant of 5.2 N/m has a relaxed length of 2.45 m. When a mass is attached to the end of the spring and allowed to come to rest, the vertical length of the spring is 3.57 m. Calculate the elastic potential energy stored in the spring.

2. The staples inside a stapler are kept in place by a spring with a relaxed length of 0.115 m. If the spring constant is 51.0 N/m, how much elastic potential energy is stored in the spring when its length is 0.150 m?

3. A 40.0 kg child is in a swing that is attached to ropes 2.00 m long. Find the gravitational potential energy associated with the child relative to the child's lowest position under the following conditions:
   a. when the ropes are horizontal
   b. when the ropes make a 30.0° angle with the vertical
   c. at the bottom of the circular arc

Section Review

1. What forms of energy are involved in the following situations?
   a. a bicycle coasting along a level road
   b. heating water
   c. throwing a football
   d. winding the mainspring of a clock

2. How do the forms of energy in item 1 differ from one another? Be sure to discuss mechanical versus nonmechanical energy, kinetic versus potential energy, and gravitational versus elastic potential energy.

3. A pinball bangs against a bumper, giving the ball a speed of 42 cm/s. If the ball has a mass of 50.0 g, what is the ball's kinetic energy in joules?

4. A student slides a 0.75 kg textbook across a table, and it comes to rest after traveling 1.2 m. Given that the coefficient of kinetic friction between the book and the table is 0.34, use the work–kinetic energy theorem to find the book's initial speed.

5. A spoon is raised 21.0 cm above a table. If the spoon and its contents have a mass of 30.0 g, what is the gravitational potential energy associated with the spoon at that height relative to the surface of the table?

Conservation of energy

CONSERVED QUANTITIES

When we say that something is *conserved*, we mean that it remains constant. If we have a certain amount of a conserved quantity at some instant of time, we will have the same amount of that quantity at a later time. This does not mean that the quantity cannot change form during that time, but if we consider all the forms that the quantity can take, we will find that we always have the same amount.

For example, the amount of money you now have is not a conserved quantity because it is likely to change over time. For the moment, however, let us assume that you do not spend the money you have, so your money is conserved. This means that if you have a dollar in your pocket, you will always have that same amount, although it may change form. One day it may be in the form of a bill. The next day you may have a hundred pennies, and the next day you may have an assortment of dimes and nickels. But when you total the change, you always have the equivalent of a dollar. It would be nice if money were like this, but of course it isn't. Because money is often acquired and spent, it is not a conserved quantity.

An example of a conserved quantity that you are already familiar with is mass. For instance, imagine that a light bulb is dropped on the floor and shatters into many pieces, as shown in **Figure 5-9.** No matter how the bulb shatters, the total mass of all of the pieces together is the same as the mass of the intact light bulb because mass is conserved.

MECHANICAL ENERGY

We have seen examples of objects that have either kinetic or potential energy. The description of the motion of many objects, however, often involves a combination of kinetic and potential energy as well as different forms of potential energy. Situations involving a combination of these different forms of energy can often be analyzed simply. For example, consider the motion of the different parts of a pendulum clock. The pendulum swings back and forth. At the highest point of its swing, there is only gravitational potential energy associated with its position. At other points in its swing, the pendulum is in motion, so it has kinetic energy as well. Elastic potential energy is also present in

5-3 SECTION OBJECTIVES

- **Identify situations in which conservation of mechanical energy is valid.**

- **Recognize the forms that conserved energy can take.**

- **Solve problems using conservation of mechanical energy.**

Figure 5-9
The mass of the light bulb, whether whole or in pieces, is constant and thus conserved.

The Language of Physics

The symbol ΣPE stands for "sum of the potential energies." Just as the Greek letter Δ (*delta*) is used to denote difference, the Greek letter Σ (*sigma*) is used to denote sum.

Demonstration 4

Mechanical energy

Purpose Show two kinds of energy in a mechanical system.

Materials pendulum attached to a ring stand

Procedure As the pendulum swings to and fro, have students describe the motion in terms of gravitational potential energy and kinetic energy when the bob is at different positions along its path. (*At maximum displacement, the gravitational potential energy is maximum and the bob's kinetic energy is zero. The potential energy is gradually converted into kinetic energy. At the equilibrium position, the kinetic energy is maximum and the gravitational potential energy is zero.*)

⭐ TEKS
p. 180: 2C, 5B

Demonstration 5

Conservation of energy

Purpose Demonstrate the conservation of mechanical energy.

Materials steel ball, spring balance, meterstick

Procedure Measure the weight of the steel ball, and record this value on the chalkboard. Drop the ball from shoulder height, and ask the students to describe the motion in terms of potential energy and kinetic energy when the ball is at different positions along its path.

Now measure the initial and final heights of the ball, and record these values on the chalkboard. Have students calculate the corresponding potential energy. Ask them to estimate how much kinetic energy the ball should have at its lowest point *(same as its initial potential energy)* and at its midway point, disregarding friction *(half of its initial potential energy)*. Explain that if friction can be disregarded, the ball's potential energy is converted into kinetic energy, while the total amount of mechanical energy remains constant.

Figure 5-10
Total potential energy and kinetic energy must be taken into account in order to describe the total energy of the pendulum in a clock.

mechanical energy

the sum of kinetic energy and all forms of potential energy

the many springs that are part of the inner workings of the clock. The motion of the pendulum in a clock is shown in **Figure 5-10.** (★)TEKS 5B

Analyzing situations involving kinetic, gravitational potential, and elastic potential energy is relatively simple. Unfortunately, analyzing situations involving other forms of energy—such as chemical potential energy—is not as easy.

We can ignore these other forms of energy if their influence is negligible or if they are not relevant to the situation being analyzed. In most situations that we are concerned with, these forms of energy are not involved in the motion of objects. In ignoring these other forms of energy, we will find it useful to define a quantity called **mechanical energy.** The mechanical energy is the sum of kinetic energy and all forms of potential energy associated with an object or group of objects.

$$ME = KE + \Sigma PE$$

All energy, such as nuclear, chemical, internal, and electrical, that is not mechanical energy is classified as *nonmechanical energy.* Do not be confused by the term *mechanical energy.* It is not a unique form of energy. It is merely a way of classifying energy, as shown in **Figure 5-11.** As you learn about new forms of energy in this book, you will be able to add them to this chart.

Mechanical energy is often conserved

Imagine a 75 g egg located on a countertop 1.0 m above the ground. The egg is knocked off the edge and falls to the ground. Because the acceleration of the egg is constant as it falls, you can use the kinematic formulas from Chapter 2 to determine the speed of the egg and the distance the egg has fallen at any subsequent time. The distance fallen can then be subtracted from the initial height to find the height of the egg above the ground at any subsequent time. For example, after 0.10 s, the egg has a speed of 0.98 m/s and has fallen a distance of 0.05 m, corresponding to a height above the ground of 0.95 m. Once the egg's speed and its height above the ground are known as a function of time, you can use what you have learned in this chapter to calculate both the kinetic energy of the egg and the gravitational potential energy associated with the position of the egg at any subsequent time. Adding the kinetic and potential energy gives the total mechanical energy at each position. (★)TEKS 2C

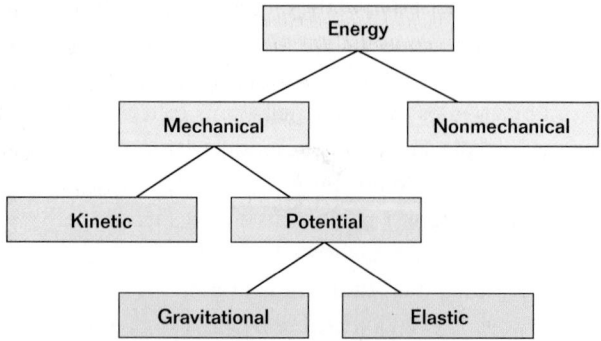

Figure 5-11
Energy can be classified in a number of ways.

Table 5-1   Energy of a falling egg TEKS 2C, 5B, 5D

Time (s)	Height (m)	Speed (m/s)	PE_g (J)	KE (J)	ME (J)
0.00	1.0	0.00	0.74	0.00	0.74
0.10	0.95	0.98	0.70	0.036	0.74
0.20	0.80	2.0	0.59	0.15	0.74
0.30	0.56	2.9	0.41	0.33	0.74
0.40	0.22	3.9	0.16	0.58	0.74

In the absence of friction, the total mechanical energy remains the same. This principle is called *conservation of mechanical energy*. Although the amount of mechanical energy is constant, mechanical energy itself can change form. For instance, consider the forms of energy for the falling egg, as shown in **Table 5-1.** As the egg falls, the potential energy is continuously converted into kinetic energy. If the egg were thrown up in the air, kinetic energy would be converted into gravitational potential energy. In either case, mechanical energy is conserved. The conservation of mechanical energy can be written symbolically as follows: ⭐TEKS **5B, 5D**

CONSERVATION OF MECHANICAL ENERGY

$$ME_i = ME_f$$

initial mechanical energy = final mechanical energy
(in the absence of friction)

The mathematical expression for the conservation of mechanical energy depends on the forms of potential energy in a given problem. For instance, if the only force acting on an object is the force of gravity, as in the egg example, the conservation law can be written as follows:

$$\tfrac{1}{2}mv_i^2 + mgh_i = \tfrac{1}{2}mv_f^2 + mgh_f$$

If other forces (except friction) are present, simply add the appropriate potential energy terms associated with each force. For instance, if the egg happened to compress or stretch a spring as it fell, the conservation law would also include an elastic potential energy term on each side of the equation.

In situations in which frictional forces are present, the principle of mechanical energy conservation no longer holds because kinetic energy is not simply converted to a form of potential energy. This special situation will be discussed more thoroughly on page 186.

Quick Lab

Mechanical Energy

MATERIALS LIST

- ✔ medium-sized spring (spring balance)
- ✔ assortment of small balls, each having a different mass
- ✔ ruler
- ✔ tape
- ✔ scale or balance

👓 **SAFETY CAUTION**

Students should wear goggles to perform this lab.

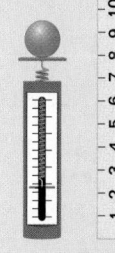

First determine the mass of each of the balls. Then tape the ruler to the side of a tabletop so that the ruler is vertical. Place the spring vertically on the tabletop near the ruler, and compress the spring with one of the balls. Release the ball, and measure the maximum height it achieves in the air. Repeat this process five times, and average the results. From the data, can you predict how high each of the other balls will rise? Test your predictions. (Hint: Assume mechanical energy is conserved.)

⭐TEKS **1A, 2B, 2C, 5B, 5D**

Quick Lab

TEACHER'S NOTES

This activity is meant to demonstrate energy transfer (from the spring to the ball) and the conservation of mechanical energy.

The lab is most effective when the balls have significantly different masses and when the spring is compressed the same amount in each case.

Because all of the balls have the same ME_i ($\tfrac{1}{2}kx_i^2$), which is converted into $ME_f = mgh_f$, balls with a larger mass will achieve a lower height.

Point out that if the measurements are reliable, they can be used to determine the spring constant.

⭐TEKS

p. 182: 5B, 2C
p. 183: 2C, 5B, 5D, 5B, 5D, 1A, 2B, 2C, 5B, 5D

Classroom Practice

The following may be used as teamwork exercises or for demonstration at the chalkboard or on an overhead projector.

PROBLEM

Conservation of mechanical energy

A small 10.0 g ball is held to a slingshot that is stretched 6.0 cm. The spring constant is 2.0×10^2 N/m.

a. What is the elastic potential energy of the slingshot before it is released?

b. What is the kinetic energy of the ball just after the slingshot is released?

c. What is the ball's speed at that instant?

d. How high does the ball rise if it is shot directly upward?

Answers

a. 0.36 J

b. 0.36 J

c. 8.5 m/s

d. 3.7 m

PRACTICE GUIDE 5E	
Solving for:	
v	📖 **PE** Sample, 1–3; Ch. Rvw. 33–34, 52–53, 57, 61–62
	💿 **PW** 4–5
	PB 8–10
h	📖 **PE** 4–5; Ch. Rvw. 39, 42–43, 52–53, 62
	💿 **PW** Sample, 1–3
	PB 5–7
E	📖 **PE** Ch. Rvw. 37, 46–47, 53, 55, 61
	💿 **PW** Sample, 6–7
	PB Sample, 1–4

SAMPLE PROBLEM 5E

Conservation of mechanical energy **TEKS** 2C, 5B, 5C, 5D

PROBLEM

Starting from rest, a child zooms down a frictionless slide from an initial height of 3.00 m. What is her speed at the bottom of the slide? Assume she has a mass of 25.0 kg.

SOLUTION

1. DEFINE

Given: $h = h_i = 3.00$ m $m = 25.0$ kg $v_i = 0.0$ m/s
$h_f = 0$ m

Unknown: $v_f = ?$

2. PLAN **Choose an equation(s) or situation:** The slide is frictionless, so mechanical energy is conserved. Kinetic energy and gravitational potential energy are the only forms of energy present.

$$KE = \tfrac{1}{2}mv^2 \qquad PE = mgh$$

The zero level chosen for gravitational potential energy is the bottom of the slide. Because the child ends at the zero level, the final gravitational potential energy is zero.

$$PE_{g,f} = 0$$

The initial gravitational potential energy at the top of the slide is

$$PE_{g,i} = mgh_i = mgh$$

Because the child starts at rest, the initial kinetic energy at the top is zero.

$$KE_i = 0$$

Therefore, the final kinetic energy is as follows:

$$KE_f = \tfrac{1}{2}mv_f^2$$

3. CALCULATE **Substitute values into the equations:**

$$PE_{g,i} = (25.0 \text{ kg})(9.81 \text{ m/s}^2)(3.00 \text{ m}) = 736 \text{ J}$$
$$KE_f = (\tfrac{1}{2})(25.0 \text{ kg})v_f^2$$

Now use the calculated quantities to evaluate the final velocity.

$$ME_i = ME_f$$
$$PE_i + KE_i = PE_f + KE_f$$
$$736 \text{ J} + 0 \text{ J} = 0 \text{ J} + (0.500)(25.0 \text{ kg})v_f^2$$

$$\boxed{v_f = 7.67 \text{ m/s}}$$

CALCULATOR SOLUTION

Your calculator should give an answer of 7.67333, but because the answer is limited to three significant figures, it should be rounded to 7.67.

4. EVALUATE The expression for the square of the final speed can be written as follows:

$$v_f^2 = \frac{2mgh}{m} = 2gh$$

Notice that the masses cancel, so the final speed does not depend on the mass of the child. This result makes sense because the acceleration of an object due to gravity does not depend on the mass of the object.

PRACTICE 5E

Conservation of mechanical energy ⓉTEKS 2C, 5B, 5C, 5D

1. A bird is flying with a speed of 18.0 m/s over water when it accidentally drops a 2.00 kg fish. If the altitude of the bird is 5.40 m and friction is disregarded, what is the speed of the fish when it hits the water?

2. A 755 N diver drops from a board 10.0 m above the water's surface. Find the diver's speed 5.00 m above the water's surface. Then find the diver's speed just before striking the water.

3. If the diver in item 2 leaves the board with an initial upward speed of 2.00 m/s, find the diver's speed when striking the water.

4. An Olympic runner leaps over a hurdle. If the runner's initial vertical speed is 2.2 m/s, how much will the runner's center of mass be raised during the jump?

5. A pendulum bob is released from some initial height such that the speed of the bob at the bottom of the swing is 1.9 m/s. What is the initial height of the bob?

Energy conservation occurs even when acceleration varies

If the slope of the slide in Sample Problem 5E was constant, the acceleration along the slide would also be constant and the kinematic formulas from Chapter 2 could have been used to solve the problem. However, you do not know the shape of the slide. Thus, the acceleration may not be constant, and the kinematic formulas could not be used.

But now we can apply a new method to solve such a problem. Because the slide is frictionless, mechanical energy is conserved. We simply equate the initial mechanical energy to the final mechanical energy and ignore all the details in the middle. The shape of the slide is not a contributing factor to the system's mechanical energy as long as friction can be ignored.

internet **connect**

SCI**LINKS**
NSTA
TOPIC: Conservation of energy
GO TO: www.scilinks.org
sci*LINKS* CODE: HF2053

Alternative Problem-Solving Approach

The process shown in Sample Problem 5E can be reversed. Rather than calculating each type of energy separately, begin with the conservation of mechanical energy:

$$ME_i = ME_f$$

Next determine what types of energy are involved and substitute the formulas for each type of energy into the equation.

In this case,

$$PE_i = KE_f$$
$$mgh_i = \tfrac{1}{2}mv_f^2$$

Solve for v in terms of the other variables, and then substitute the given values into this equation.

$$v_f^2 = 2gh_i$$
$$v_f = \sqrt{2gh_i}$$
$$v_f = \sqrt{2(9.81 \text{ m/s}^2)(3.00 \text{ m})}$$
$$v_f = 7.67 \text{ m/s}$$

ANSWERS TO

Practice 5E
Conservation of mechanical energy

1. 20.7 m/s
2. 9.9 m/s; 14.0 m/s
3. 14.1 m/s
4. 0.25 m
5. 0.18 m

ⓉTEKS

p. 184: 2C, 5B, 5C, 5D
p. 185: 2C, 5B, 5C, 5D

Some students may confuse the conservation of mechanical energy with the general energy conservation law. Point out that although mechanical energy is not always conserved, the total energy is always conserved. For example, as the sanding block's kinetic energy decreases, energy is transferred to the rough surface in the form of internal energy (this topic will be discussed in Chapter 10). As a result, the temperatures of the block and surface increase slightly. The total energy in the system remains constant, although the mechanical energy decreases.

Section Review
ANSWERS

1. 2.93 m/s

2. No, the roller coaster will not reach the top of the second hill. If the total mechanical energy is constant, the roller coaster will reach its initial height and then begin rolling back down the hill.

3. **a.** yes
   b. no
   c. yes, if air resistance is disregarded

4. Answers may vary. The downward-sloping track converts potential energy to kinetic energy. Levers employ kinetic energy to increase potential energy. Springs and elastic membranes convert kinetic energy to elastic potential energy and back again. Mechanical energy is not conserved; some energy is lost because of kinetic friction.

(a)

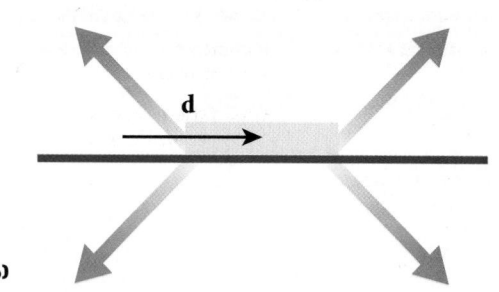

(b)

Figure 5-12
(a) As the block slides, its kinetic energy tends to decrease because of friction. The force from the hand keeps it moving.
(b) Kinetic energy is dissipated into the block and surface.

Mechanical energy is not conserved in the presence of friction

If you have ever used a sanding block to sand a rough surface, such as in **Figure 5-12,** you may have noticed that you had to keep applying a force to keep the block moving. The reason is that kinetic friction between the moving block and the surface causes the kinetic energy of the block to be converted into a nonmechanical form of energy. As you continue to exert a force on the block, you are replacing the kinetic energy that is lost because of kinetic friction. The observable result of this energy dissipation is that the sanding block and the tabletop become warmer.

In the presence of kinetic friction, nonmechanical energy is no longer negligible and mechanical energy is no longer conserved. This does not mean that energy in general is not conserved—total energy is *always* conserved. However, the mechanical energy is converted into forms of energy that are much more difficult to account for, and the mechanical energy is therefore considered to be "lost."

Section Review

⭐TEKS 2C, 5B, 5C, 5D

1. If the spring of a jack-in-the-box is compressed a distance of 8.00 cm from its relaxed length and then released, what is the speed of the toy head when the spring returns to its natural length? Assume the mass of the toy head is 50.0 g, the spring constant is 80.0 N/m, and the toy head moves only in the vertical direction. Also disregard the mass of the spring. (Hint: Remember that there are two forms of potential energy in the problem.)

2. You are designing a roller coaster in which a car will be pulled to the top of a hill of height *h* and then, starting from a momentary rest, will be released to roll freely down the hill and toward the peak of the next hill, which is 1.1 times as high. Will your design be successful? Explain your answer.

3. Is conservation of mechanical energy likely to hold in these situations?
   a. a hockey puck sliding on a frictionless surface of ice
   b. a toy car rolling on a carpeted floor
   c. a baseball being thrown into the air

4. **Physics in Action**   What parts of the kinetic sculpture on pp. 166 and 167 involve the conversion of one form of energy to another? Is mechanical energy conserved in these processes?

5-4
Power

RATE OF ENERGY TRANSFER

The rate at which work is done is called **power.** More generally, power is the rate of energy transfer by any method. Like the concepts of energy and work, power has a specific meaning in science that differs from its everyday meaning.

Imagine you are producing a play and you need to raise and lower the curtain between scenes in a specific amount of time. You decide to use a motor that will pull on a rope connected to the top of the curtain rod. Your assistant finds three motors but doesn't know which one to use. One way to decide is to consider the power output of each motor.

If the work done on an object is W in a time interval Δt, then the power delivered to the object over this time interval is written as follows:

POWER

$$P = \frac{W}{\Delta t}$$

power = work ÷ time

It is sometimes useful to rewrite this equation in an alternative form by substituting the definition of work into the definition of power.

$$W = Fd$$
$$P = \frac{W}{\Delta t} = F\frac{d}{\Delta t}$$

The distance moved per unit time is just the speed of the object.

5-4 SECTION OBJECTIVES

- **Relate the concepts of energy, time, and power.**
- **Calculate power in two different ways.**
- **Explain the effect of machines on work and power.**

power

the rate at which energy is transferred

Conceptual Challenge

1. Mountain roads   Many mountain roads are built so that they zigzag up the mountain rather than go straight up toward the peak. Discuss the advantages of such a design from the viewpoint of energy conservation and power. **3A**

2. Light bulbs   A light bulb is described as *having* 60 watts. What's wrong with this statement?

The following may be used as teamwork exercises or for demonstration at the chalkboard or on an overhead projector.

PROBLEM

Power

Two horses pull a cart. Each exerts a force of 250.0 N at a speed of 2.0 m/s for 10.0 min.

a. Calculate the power delivered by the horses.

b. How much work is done by the two horses?

Answers

    a. 1.0×10^3 W

    b. 6.0×10^5 J

Alternative Problem-Solving Approach

Calculate the time it would take each motor to do the same work:

$W = Fd = mgd = 14 \times 10^3$ J

$\Delta t = W/P$

$\Delta t_1 = 14 \times 10^3$ J$/1.0 \times 10^3$ W $= 14$ s

$\Delta t_2 = 14 \times 10^3$ J$/3.5 \times 10^3$ W $= 4.0$ s

$\Delta t_3 = 14 \times 10^3$ J$/5.5 \times 10^3$ W $= 2.5$ s

This approach shows that the second motor comes closest to 5.0 s.

Figure 5-13
The wattage of each of these bulbs tells you the rate at which energy is converted by the bulb. The bulbs in this photo require wattages ranging from 0.7 W to 200 W.

CONCEPT PREVIEW

Machines will be discussed further in Chapter 8.

POWER (ALTERNATIVE FORM)

$$P = Fv$$

power = force × speed

The SI unit of power is the *watt*, W, which is defined to be one joule per second. The *horsepower*, hp, is another unit of power that is sometimes used. One horsepower is equal to 746 watts.

The watt is perhaps most familiar to you from your everyday experience with light bulbs (see **Figure 5-13**). A dim light bulb uses about 40 W of power, while a bright bulb can use up to 500 W. Decorative lights use about 0.7 W each for indoor lights and 7.0 W each for outdoor lights.

Machines with different power ratings do the same work in different time intervals

In Sample Problem 5F, the three motors would lift the curtain at different rates because the power output for each motor is different. So each motor would do work on the curtain at different rates and would thus transfer energy to the curtain at different rates.

In a given amount of time, each motor would do different amounts of work on the curtain. The 5.5 kW motor would do the most amount of work in a given time, and the 1.0 kW motor would do the least amount of work in the same time. Yet all three motors would perform the same total amount of work in lifting the curtain. The important difference is that the more powerful motor could do the work in a shorter time interval.

SAMPLE PROBLEM 5F

Power ⭐TEKS 2C

PROBLEM

A 193 kg curtain needs to be raised 7.5 m, at constant speed, in as close to 5.0 s as possible. The power ratings for three motors are listed as 1.0 kW, 3.5 kW, and 5.5 kW. Which motor is best for the job?

SOLUTION

Given: $m = 193$ kg $\Delta t = 5.0$ s $d = 7.5$ m

Unknown: $P = ?$

Use the power equation from page 187.

$$P = \frac{W}{\Delta t} = \frac{Fd}{\Delta t} = \frac{mgd}{\Delta t}$$

$$= \frac{(193 \text{ kg})(9.81 \text{ m/s}^2)(7.5 \text{ m})}{5.0 \text{ s}}$$

$$P = 2.8 \times 10^3 \text{ W} = 2.8 \text{ kW}$$

The best motor to use is the 3.5 kW motor. The 1.0 kW motor will not lift the curtain fast enough, and the 5.5 kW motor will lift the curtain too fast.

Power ⭐TEKS 2C

1. A 1.0×10^3 kg elevator carries a maximum load of 800.0 kg. A constant frictional force of 4.0×10^3 N retards the elevator's motion upward. What minimum power, in kilowatts, must the motor deliver to lift the fully loaded elevator at a constant speed of 3.00 m/s?

2. A car with a mass of 1.50×10^3 kg starts from rest and accelerates to a speed of 18.0 m/s in 12.0 s. Assume that the force of resistance remains constant at 400.0 N during this time. What is the average power developed by the car's engine?

3. A rain cloud contains 2.66×10^7 kg of water vapor. How long would it take for a 2.00 kW pump to raise the same amount of water to the cloud's altitude, 2.00 km?

4. How long does it take a 19 kW steam engine to do 6.8×10^7 J of work?

5. A 1.50×10^3 kg car accelerates uniformly from rest to 10.0 m/s in 3.00 s.
   a. What is the work done on the car in·this time interval?
   b. What is the power delivered by the engine in this time interval?

Section Review

1. How are energy, time, and power related?

2. A 50.0 kg student climbs 5.00 m up a rope at a constant speed. If the student's power output is 200.0 W, how long does it take the student to climb the rope? How much work does the student do?

3. A motor-driven winch pulls the student in item 2 5.00 m up the rope at a constant speed of 1.25 m/s. How much power does the motor use in raising the student? How much work does the motor do on the student?

PRACTICE GUIDE 5F

Solving for:		
P	📖 **PE**	Sample, 1–2, 5; Ch. Rvw. 36
	💿 **PW**	5–6
	PB	8–10
Δt	📖 **PE**	3–4; Ch. Rvw. 35
	💿 **PW**	3–4
	PB	Sample, 1–3
W	📖 **PE**	5
	💿 **PW**	Sample, 1–2
	PB	4–7

ANSWERS TO

Practice 5F
Power
1. 66 kW
2. 2.38×10^4 W (23.8 kW)
3. 2.61×10^8 s (8.27 years)
4. 3.6×10^3 s (1.0 h)
5. a. 7.50×10^4 J
   b. 2.50×10^4 W

Section Review
ANSWERS

1. Power equals energy transferred divided by time of transfer.
2. 12.3 s; 2.45×10^3 J
3. 613 W; 2.45×10^3 J

BACKGROUND

The equivalence between mass and energy is a consequence of Einstein's special theory of relativity. Einstein introduced the subject in 1905 in a paper titled "Does the inertia of a body depend on its energy-content?"

Although the derivation of relativistic kinetic energy is beyond the scope of this book, the mass-energy equivalence can be understood conceptually at this level and will enhance students' understanding of energy.

The equivalence between mass and energy in terms of binding energy is explored in Chapter 25. Fission and fusion are also discussed in greater detail in Chapter 25.

EXTENSION

- Have students research fission and fusion as energy sources. In their reports, they should include a discussion of the following questions: Why are fusion reactors considered to be safer than fission reactors? What are the difficulties associated with developing a fusion reactor? What are the advantages and disadvantages of using nuclear reactors as sources of energy?

- Einstein was deeply concerned about the possibility of nuclear weapons suggested by his theory. Have students research Einstein's opinions on the subject of nuclear weapons to prepare for a classroom debate on the social responsibility of scientists.

PHYSICS ON THE EDGE

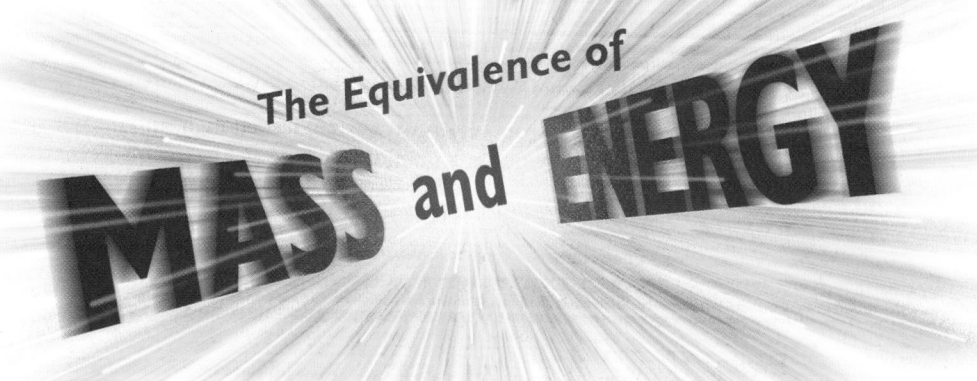

The Equivalence of MASS and ENERGY

Einstein's $E_R = mc^2$ is one of the most famous equations of the twentieth century. This equation was a surprise to Einstein, who discovered it through his work with relative velocity and kinetic energy.

Relativistic kinetic energy

In the "Relativistic addition of velocities" feature in Chapter 3, you learned how Einstein's special theory of relativity modifies the classical addition of velocities. The classical equation for kinetic energy ($KE = \frac{1}{2}mv^2$) must also be modified for relativity. In 1905, Einstein derived a new equation for kinetic energy based on the principles of special relativity: (★)TEKS 3E

$$KE = \frac{mc^2}{\sqrt{1 - \left(\frac{v^2}{c^2}\right)}} - mc^2$$

In this equation, m is the mass of the object, v is the velocity of the object, and c is the speed of light. (Some texts distinguish between a rest mass, m_0, and a relativistic mass, m; in this book, relativistic mass is not used, and m always refers to rest mass.) Although it isn't immediately obvious, this equation reduces to the classical equation $KE = \frac{1}{2}mv^2$ for speeds that are small relative to the speed of light, as shown in **Figure 5-14.**

Einstein's relativistic expression for kinetic energy has been confirmed by experiments in which electrons are accelerated to extremely high speeds in particle accelerators. In all cases, the experimental data correspond to Einstein's equation rather than to the classical equation. Nonetheless, the difference between the two theories at low speeds (relative to c) is so minimal that the classical equation can be used in all such cases when the speed is much less than c.

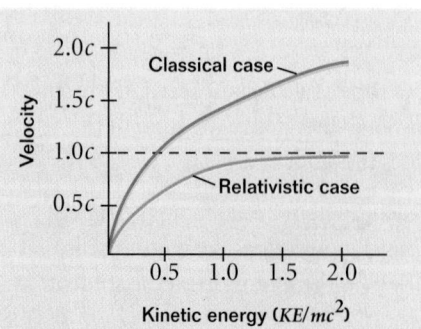

Figure 5-14

This graph of kinetic energy versus velocity for both the classical and relativistic equations shows that the two theories are in agreement when v is much less than c. Note that v is always less than c in the relativistic case.

Rest energy

The second term of Einstein's equation for kinetic energy, $-mc^2$, is required so that $KE = 0$ when $v = 0$. Note that this term is independent of velocity. This suggests that the *total* energy of an object equals its kinetic energy plus some additional form of energy equal to mc^2. The mathematical expression of this additional energy is the familiar Einstein equation:

$$E_R = mc^2$$

This equation shows that an object has a certain amount of energy (E_R), known as *rest energy,* simply by virtue of its mass. The rest energy of a body is equal to its mass, m, multiplied by the speed of light squared, c^2. Thus, the mass of a body is a measure of its rest energy. This equation is significant because rest energy is an aspect of special relativity that was not predicted by classical physics.

Experimental verification

The magnitude of the conversion factor between mass and rest energy $(c^2 = 9 \times 10^{16}\ \text{m}^2/\text{s}^2)$ is so great that even a very small mass has a huge amount of rest energy. Nuclear reactions utilize this relationship by converting mass (rest energy) into other forms of energy. In nuclear fission, which is the energy source of nuclear power plants, the nucleus of an atom is split into two or more nuclei. Taken together, the mass of these nuclei is slightly less than the mass of the original nucleus, and a very large amount of energy is released. In typical nuclear reactions, about one-thousandth of the initial mass is converted from rest energy into other forms of energy. This change in mass, although very small, can be detected experimentally. ⊛ TEKS **3C**

Another type of nuclear reaction that converts mass into energy is fusion, which is the source of energy for our sun and other stars. About 4.5 million tons of the sun's mass is converted into other forms of energy every second. Fortunately, the sun has enough mass to last approximately 5 billion more years.

Most of the energy changes encountered in your typical experiences are much smaller than the energy changes that occur in nuclear reactions. As a result, the change in mass is even less than that observed in nuclear reactions. Such changes are far too small to be detected experimentally. Thus, for typical cases, the classical equation still holds, and mass and energy can be thought of as separate.

Before Einstein's theory of relativity, conservation of energy and conservation of mass were regarded as two separate laws. The equivalence between mass and energy reveals that in fact these two laws are one. In the words of Einstein, "Prerelativity physics contains two conservation laws of fundamental importance. . . . Through relativity theory, they melt together into *one* principle."

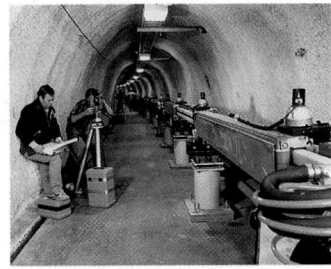

Figure 5-15
Electrons in the Stanford Linear Accelerator in California (SLAC) reach 99.999999967 percent of the speed of light. At such great speeds, the difference between classical and relativistic theories becomes significant.

TOPIC: Nuclear reactions
GO TO: www.scilinks.org
*sci*LINKS CODE: HF2054

Figure 5-16
Our sun uses a nuclear reaction called fusion to convert mass to energy. About 90 percent of the stars, including our sun, fuse hydrogen, while some older stars fuse helium.

The Language of Physics

The notion of mass in relativity has been undergoing a transformation in recent years. In earlier treatments of relativity, which you will still see in some current textbooks, the notion of *relativistic mass* is used. In these treatments, the symbol m represents the relativistic mass, which increases as the speed of the object increases. As a result, Einstein's equation is the commonly seen $E = mc^2$, and the energy is the relativistic *total energy*. In these treatments, the notion of *rest mass* is used to represent the mass of an object when its speed is zero.

In modern treatments of relativity, the symbol m represents simply the mass of an object, which remains constant. There is no notion of a change of mass with speed. Einstein's equation is $E_R = mc^2$, in which the energy is the *rest energy* of the object. The total energy of the object is then expressed as follows:

$$E = \frac{mc^2}{\sqrt{1 - \dfrac{v^2}{c^2}}}$$

There is no use of the term *rest mass.*

⊛ TEKS
p. 190: 3E
p. 191: 3C

CHAPTER 5
Summary

⭐ TEKS

Review & Assess
pp. 193–199:
2A: Alt. Assess. 1, 3, 5
2B: Alt. Assess. 1–3, 5
2C: Technology & Learning
2D: Alt. Assess. 1–6
2E: Alt. Assess. 2–3
2F: Alt. Assess. 1, 3, 5
3A: Alt. Assess. 6
5A: 15, 18, 21–22
5B: 11–14, 16, 19–20, 27, 30
5C: 26–32, 37, 40, 46, 47, 53, 55c, 61c
5D: 26–34

KEY TERMS

elastic potential energy (p. 178)

gravitational potential energy (p. 177)

kinetic energy (p. 172)

mechanical energy (p. 182)

potential energy (p. 177)

power (p. 187)

spring constant (p. 178)

work (p. 168)

work–kinetic energy theorem (p. 174)

KEY IDEAS

Section 5-1 Work

- Work is done on an object only when a net force acts on the object to displace it in the direction of a component of the net force.
- The amount of work done on an object is given by the following equation, where θ is the angle between the applied force, F, and the displacement of the object, d:

$$W = Fd(\cos \theta)$$

Section 5-2 Energy

- Objects in motion have kinetic energy because of their mass and speed.
- The net work done on or by an object is equal to the change in the kinetic energy of the object.
- Potential energy is energy associated with an object's position. Two forms of potential energy discussed in this chapter are gravitational potential energy and elastic potential energy.

Section 5-3 Conservation of energy

- Energy can change form but can never be created or destroyed.
- Mechanical energy is the total kinetic and potential energy present in a given situation.
- In the absence of friction, mechanical energy is conserved, so the amount of mechanical energy remains constant.

Section 5-4 Power

- Power is the rate at which work is done or the rate of energy transfer.
- Machines with different power ratings do the same amount of work in different time intervals.

Variable symbols			
Quantities		**Units**	**Conversions**
W	work	J joule	$= N \bullet m$ $= kg \bullet m^2/s^2$
KE	kinetic energy	J joule	
PE_g	gravitational potential energy	J joule	
$PE_{elastic}$	elastic potential energy	J joule	
P	power	W watt	$= J/s$

CHAPTER 5
Review and Assess

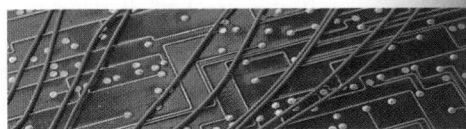

WORK

Review questions

1. Can the speed of an object change if the net work done on it is zero?

2. Discuss whether any work is being done by each of the following agents and, if so, whether the work is positive or negative.
   - **a.** a chicken scratching the ground
   - **b.** a person studying
   - **c.** a crane lifting a bucket of concrete
   - **d.** the force of gravity on the bucket in **(c)**

3. Furniture movers wish to load a truck using a ramp from the ground to the rear of the truck. One of the movers claims that less work would be required if the ramp's length were increased, reducing its angle with the horizontal. Is this claim valid? Explain.

Conceptual questions

4. A pendulum swings back and forth, as shown in **Figure 5-17.** Does the tension force in the string do work on the pendulum bob? Does the force of gravity do work on the bob? Explain your answers.

5. The drivers of two identical cars heading toward each other apply the brakes at the same instant. The skid marks of one of the cars are twice as long as the skid marks of the other vehicle. Assuming that the brakes of both cars apply the same force, what conclusions can you draw about the motion of the cars?

6. When a punter kicks a football, is he doing work on the ball while his toe is in contact with it? Is he doing work on the ball after the ball loses contact with his toe? Are any forces doing work on the ball while the ball is in flight?

Figure 5-17

Practice problems

7. A person lifts a 4.5 kg cement block a vertical distance of 1.2 m and then carries the block horizontally a distance of 7.3 m. Determine the work done by the person and by the force of gravity in this process. (See Sample Problem 5A.)

8. A plane designed for vertical takeoff has a mass of 8.0×10^3 kg. Find the net work done by all forces on the plane as it accelerates upward at 1.0 m/s^2 through a distance of 30.0 m after starting from rest. (See Sample Problem 5A.)

9. A catcher "gives" with a baseball when catching it. If the baseball exerts a force of 475 N on the glove such that the glove is displaced 10.0 cm, how much work is done by the ball? (See Sample Problem 5A.)

10. A flight attendant pulls her 70.0 N flight bag a distance of 253 m along a level airport floor at a constant velocity. The force she exerts is 40.0 N at an angle of 52.0° above the horizontal. Find the following:
    - **a.** the work she does on the flight bag
    - **b.** the work done by the force of friction on the flight bag
    - **c.** the coefficient of kinetic friction between the flight bag and the floor

    (See Sample Problem 5A.)

ENERGY

Review questions

11. A person drops a ball from the top of a building while another person on the ground observes the ball's motion. Will these two people always agree on the following?
    - **a.** the ball's potential energy
    - **b.** the ball's change in potential energy
    - **c.** the ball's kinetic energy

1. No, a change in speed corresponds to a change in kinetic energy, which cannot occur without work (either positive or negative) being done on the object.

2. **a.** yes, positive
   b. no
   c. yes, positive
   d. yes, negative

3. No, force would decrease, but distance would increase, which would keep work constant.

4. The tension is perpendicular to the bob's motion, so it does not do work on the bob. The component of the bob's weight that is perpendicular to the bob's motion does not do work on the bob, but the component that is in the direction of its motion does.

5. The car leaving longer skid marks was moving faster.

6. yes; no; yes, the ball's weight and air resistance

7. 53 J, −53 J

8. 2.4×10^5 J

9. 47.5 J

10. **a.** 6230 J
    b. −6230 J
    c. 0.~~352~~ .640

11. **a.** no
    b. yes
    c. yes

12. No, kinetic energy cannot be negative because mass is always positive and the speed term of the equation is squared.

13. yes, because potential energy depends on the distance to an arbitrary zero level, which can be above or below the object

14. 1 to 25

15. The gravitational force does not do work on the satellite because the satellite remains at the same height. Thus, the satellite's speed must be constant.

16. The work required to stop the car equals the car's initial kinetic energy. If speed is doubled, work is quadrupled. Thus, the car will travel 140 m. Its kinetic energy is changed into internal energy.

17. Work must be done against gravity in order to descend a staircase at a constant speed. Walking on a horizontal surface does not require work to be done against gravity.

18. The work done by friction equals the change in mechanical energy, so the particle's speed decreases.

19. 7.6×10^4 J

20. 1.7×10^4 m/s

21. 2.0×10^1 m

22. 1.4 m

23. a. 5400 J, 0 J; 5400 J
b. 0 J, −5400 J; 5400 J
c. 2700 J, −2700 J; 5400 J

24. a. −19.6 J
b. 39.2 J
c. 0 J

25. a. $(0.5)(500.0 \text{ N/m})$
$(4.00 \times 10^{-2} \text{ m})^2 = 0.400$ J
b. $(\frac{1}{2})(500.0 \text{ N/m})$
$(-3.00 \times 10^{-2} \text{ m})^2 = 0.225$ J
c. $(0.5)(500.0 \text{ N/m})(0 \text{ m})^2 = 0$ J

---

12. Can the kinetic energy of an object be negative? Explain your answer.

13. Can the gravitational potential energy of an object be negative? Explain your answer.

14. Two identical objects move with speeds of 5.0 m/s and 25.0 m/s. What is the ratio of their kinetic energies?

Conceptual questions

15. A satellite is in a circular orbit above Earth's surface. Why is the work done on the satellite by the gravitational force zero? What does the work–kinetic energy theorem predict about the satellite's speed?

16. A car traveling at 50.0 km/h skids a distance of 35 m after its brakes lock. Estimate how far it will skid if its brakes lock when its initial speed is 100.0 km/h. What happens to the car's kinetic energy as it comes to rest?

17. Explain why more energy is needed to walk down stairs than to walk horizontally at the same speed.

18. How can the work–kinetic energy theorem explain why the force of sliding friction reduces the kinetic energy of a particle?

Practice problems

19. What is the kinetic energy of an automobile with a mass of 1250 kg traveling at a speed of 11 m/s?
(See Sample Problem 5B.)

20. What speed would a fly with a mass of 0.55 g need in order to have the same kinetic energy as the automobile in item 19?
(See Sample Problem 5B.)

21. A 50.0 kg diver steps off a diving board and drops straight down into the water. The water provides an average net force of resistance of 1500 N to the diver's fall. If the diver comes to rest 5.0 m below the water's surface, what is the total distance between the diving board and the diver's stopping point underwater?
(See Sample Problem 5C.)

---

22. In a circus performance, a monkey on a sled is given an initial speed of 4.0 m/s up a 25° incline. The combined mass of the monkey and the sled is 20.0 kg, and the coefficient of kinetic friction between the sled and the incline is 0.20. How far up the incline does the sled move?
(See Sample Problem 5C.)

23. A 55 kg skier is at the top of a slope, as in **Figure 5-18**. At the initial point **A**, the skier is 10.0 m vertically above the final point **B**.

    a. Set the zero level for gravitational potential energy at **B**, and find the gravitational potential energy associated with the skier at **A** and at **B**. Then find the difference in potential energy between these two points.

    b. Repeat this problem with the zero level at point **A**.

    c. Repeat this problem with the zero level midway down the slope, at a height of 5.0 m.

(See Sample Problem 5D.)

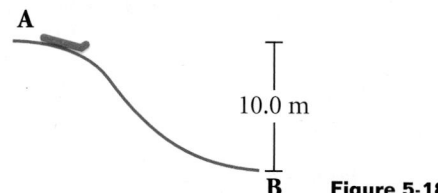

Figure 5-18

24. A 2.00 kg ball is attached to a ceiling by a 1.00 m long string. The height of the room is 3.00 m. What is the gravitational potential energy associated with the ball relative to each of the following?

    a. the ceiling
    b. the floor
    c. a point at the same elevation as the ball

(See Sample Problem 5D.)

25. A spring has a force constant of 500.0 N/m. Show that the potential energy stored in the spring is as follows:

    a. 0.400 J when the spring is stretched 4.00 cm from equilibrium

    b. 0.225 J when the spring is compressed 3.00 cm from equilibrium

    c. zero when the spring is unstretched

(See Sample Problem 5D.)

CONSERVATION OF MECHANICAL ENERGY

Review questions

26. Each of the following objects possesses energy. Which forms of energy are mechanical, which are nonmechanical, and which are a combination?
    a. glowing embers in a campfire
    b. a strong wind
    c. a swinging pendulum
    d. a person sitting on a mattress
    e. a rocket being launched into space

27. Discuss the energy transformations that occur during the pole-vault event shown in **Figure 5-19**. Disregard rotational motion and air resistance.

Figure 5-19

28. A bowling ball is suspended from the center of the ceiling of a lecture hall by a strong cord. The ball is drawn up to the tip of a lecturer's nose at the front of the room and then released. If the lecturer remains stationary, explain why the lecturer is not struck by the ball on its return swing. Would this person be safe if the ball were given a slight push from its starting position at the person's nose?

Conceptual questions

29. Discuss the work done and change in mechanical energy as an athlete does the following:
    a. lifts a weight
    b. holds the weight up in a fixed position
    c. lowers the weight slowly

30. A ball is thrown straight up. At what position is its kinetic energy at its maximum? At what position is gravitational potential energy at its maximum?

31. Advertisements for a toy ball once stated that it would rebound to a height greater than the height from which it was dropped. Is this possible?

32. A weight is connected to a spring that is suspended vertically from the ceiling. If the weight is displaced downward from its equilibrium position and released, it will oscillate up and down. How many forms of potential energy are involved? If air resistance and friction are disregarded, will the total mechanical energy be conserved? Explain.

Practice problems

33. A child and sled with a combined mass of 50.0 kg slide down a frictionless hill that is 7.34 m high. If the sled starts from rest, what is its speed at the bottom of the hill?
    (See Sample Problem 5E.)

34. Tarzan swings on a 30.0 m long vine initially inclined at an angle of 37.0° with the vertical. What is his speed at the bottom of the swing if he does the following?
    a. starts from rest
    b. pushes off with a speed of 4.00 m/s
    (See Sample Problem 5E.)

POWER

Practice problems

35. If an automobile engine delivers 50.0 hp of power, how much time will it take for the engine to do 6.40×10^5 J of work? (Hint: Note that one horsepower, 1 hp, is equal to 746 watts.)
    (See Sample Problem 5F.)

36. Water flows over a section of Niagara Falls at the rate of 1.2×10^6 kg/s and falls 50.0 m. How much power is generated by the falling water?
    (See Sample Problem 5F.)

26. a. nonmechanical
    b. mechanical
    c. mechanical
    d. mechanical
    e. both

27. As the athlete runs faster, KE increases. As he is lifted above the ground, KE decreases as PE_g and $PE_{elastic}$ increase ($PE_{elastic}$ comes from the bent pole). At the highest point, $KE = 0$ and PE_g is at its maximum value. As the athlete falls, KE increases and PE_g decreases. When the athlete lands, KE is at its maximum value and $PE_g = 0$.

28. The ball will not hit the lecturer because, according to the principle of energy conservation, it would need an input of energy to reach a height greater than its initial height. If the ball were given a push, the lecturer would be in danger.

29. a. athlete does work on the weight, PE_g increases
    b. no work done on the weight, PE_g is constant
    c. athlete does negative work on the weight, PE_g decreases

30. at the ball's lowest height; at its maximum height

31. no, because energy wouldn't be conserved

32. two, gravitational potential energy and elastic potential energy; yes, because total energy is conserved if there is no dissipation of energy

33. 12.0 m/s

34. a. 10.9 m/s
    b. 11.6 m/s

35. 17.2 s

36. 5.9×10^8 W

37. **a.** 0.633 J
    b. 0.633 J
    c. 2.43 m/s
    d. 0.422 J, 0.211 J
38. 0.265 m/s
39. 5.0 m
40. 1.2×10^3 J
41. 9.80 m/s
42. 2.5 m
43. 10.2 m
44. Although the total distance traveled by each ball is different, the displacements are the same, so the change in potential energy for each ball is the same. Also, each ball has the same initial kinetic energy, so the final kinetic energy of each ball (and thus the speed of each) will be the same.
45. **a.** 1.2 J
    b. 5.0 m/s
    c. 6.3 J
46. **a.** 61 J
    b. −45 J
    c. 0 J
47. **a.** −560 J
    b. 1.2 m
48. **a.** 9.0×10^2 J
    b. 0.38
49. **a.** 2.25×10^4 N
    b. 1.33×10^{-4} s

MIXED REVIEW

37. A 215 g particle is released from rest at point **A** inside a smooth hemispherical bowl of radius 30.0 cm, as shown in **Figure 5-20.** Calculate the following:

    a. the gravitational potential energy at **A** relative to **B**
    b. the particle's kinetic energy at **B**
    c. the particle's speed at **B**
    d. the potential energy and kinetic energy at **C**

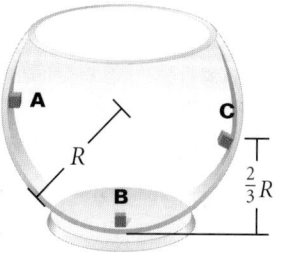

Figure 5-20

38. A person doing a chin-up weighs 700.0 N, disregarding the weight of the arms. During the first 25.0 cm of the lift, each arm exerts an upward force of 355 N on the torso. If the upward movement starts from rest, what is the person's speed at this point?

39. A 50.0 kg pole vaulter running at 10.0 m/s vaults over the bar. If the vaulter's horizontal component of velocity over the bar is 1.0 m/s and air resistance is disregarded, how high was the jump?

40. An 80.0 N box of clothes is pulled 20.0 m up a 30.0° ramp by a force of 115 N that points along the ramp. If the coefficient of kinetic friction between the box and ramp is 0.22, calculate the change in the box's kinetic energy.

41. A 98.0 N grocery cart is pushed 12.0 m along an aisle by a shopper who exerts a constant horizontal force of 40.0 N. If all frictional forces are neglected and the cart starts from rest, what is the grocery cart's final speed?

42. Tarzan and Jane, whose total mass is 130.0 kg, start their swing on a 5.0 m long vine when the vine is at an angle of 30.0° with the horizontal. At the bottom of the arc, Jane, whose mass is 50.0 kg, releases the vine. What is the maximum height at which Tarzan can land on a branch after his swing continues? (Hint: Treat Tarzan's and Jane's energies as separate quantities.)

43. A 0.250 kg block on a vertical spring with a spring constant of 5.00×10^3 N/m is pushed downward, compressing the spring 0.100 m. When released, the block leaves the spring and travels upward vertically. How high does it rise above the point of release?

44. Three identical balls, all with the same initial speed, are thrown by a juggling clown on a tightrope. The first ball is thrown horizontally, the second is thrown at some angle above the horizontal, and the third is thrown at some angle below the horizontal. Disregarding air resistance, describe the motions of the three balls, and compare the speeds of the balls as they reach the ground.

45. A 0.60 kg rubber ball has a speed of 2.0 m/s at point A and kinetic energy of 7.5 J at point B. Determine the following:

    a. the ball's kinetic energy at A
    b. the ball's speed at B
    c. the total work done on the ball as it moves from A to B

46. Starting from rest, a 5.0 kg block slides 2.5 m down a rough 30.0° incline in 2.0 s. Determine the following:

    a. the work done by the force of gravity
    b. the mechanical energy lost due to friction
    c. the work done by the normal force between the block and the incline

47. A 70.0 kg base runner begins his slide into second base while moving at a speed of 4.0 m/s. The coefficient of friction between his clothes and Earth is 0.70. He slides so that his speed is zero just as he reaches the base.

    a. How much mechanical energy is lost due to friction acting on the runner?
    b. How far does he slide?

48. A horizontal force of 150 N is used to push a 40.0 kg packing crate a distance of 6.00 m on a rough horizontal surface. If the crate moves with constant velocity, calculate the following:

    a. the work done by the force
    b. the coefficient of kinetic friction

49. A 5.00 g bullet moving at 600.0 m/s penetrates a tree trunk to a depth of 4.00 cm.

    a. Use work and energy considerations to find the magnitude of the force that stops the bullet.
    b. Assuming that the frictional force is constant, determine how much time elapses between the moment the bullet enters the tree and the moment the bullet stops moving.

50. A skier of mass 70.0 kg is pulled up a slope by a motor-driven cable. How much work is required to pull the skier 60.0 m up a 35° slope (assumed to be frictionless) at a constant speed of 2.0 m/s?

51. A 2.50×10^3 kg car requires 5.0 kJ of work to move from rest to some final speed. During this time, the car moves 25.0 m. Neglecting friction between the car and the road, find the following:
  a. the final speed
  b. the horizontal force exerted on the car

52. An acrobat on skis starts from rest 50.0 m above the ground on a frictionless track and flies off the track at a 45.0° angle above the horizontal and at a height of 10.0 m. Disregard air resistance.
  a. What is the skier's speed when leaving the track?
  b. What is the maximum height attained?

53. Figure 5-21 is a graph of the gravitational potential energy and kinetic energy of a 75 g yo-yo as it moves up and down on its string. Use the graph to answer the following questions:
  a. By what amount does the mechanical energy of the yo-yo change after 6.0 s?
  b. What is the speed of the yo-yo after 4.5 s?
  c. What is the maximum height of the yo-yo?

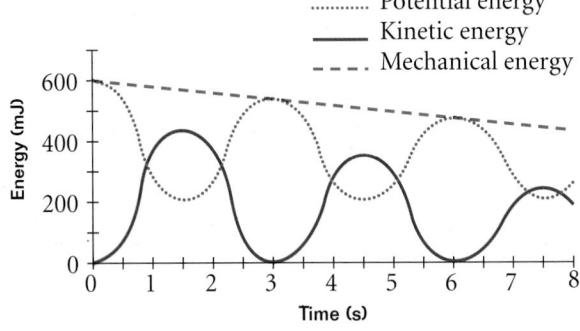

.... Potential energy
———— Kinetic energy
– – – Mechanical energy

Figure 5-21

54. A skier starts from rest at the top of a hill that is inclined at 10.5° with the horizontal. The hillside is 200.0 m long, and the coefficient of friction between the snow and the skis is 0.075. At the bottom of the hill, the snow is level and the coefficient of friction is unchanged. How far does the skier move along the horizontal portion of the snow before coming to rest?

55. Starting from rest, a 10.0 kg suitcase slides 3.00 m down a frictionless ramp inclined at 30.0° from the floor. The suitcase then slides an additional 5.00 m along the floor before coming to a stop. Determine the following:
  a. the speed of the suitcase at the bottom of the ramp
  b. the coefficient of kinetic friction between the suitcase and the floor
  c. the change in mechanical energy due to friction

56. An egg is dropped from a third-floor window and lands on a foam-rubber pad without breaking. If a 56.0 g egg falls 12.0 m from rest and the 5.00 cm thick foam pad stops it in 6.25 ms, by how much is the pad compressed? Assume constant upward acceleration as the egg compresses the foam-rubber pad. (Hint: Assume that the potential energy that the egg gains while the pad is being compressed is negligible.)

57. A 75 kg man jumps from a window 1.0 m above a sidewalk.
  a. What is his speed just before his feet strike the pavement?
  b. If the man jumps with his knees and ankles locked, the only cushion for his fall is approximately 0.50 cm in the pads of his feet. Calculate the magnitude of the average force exerted on him by the ground in this situation.

58. A projectile of mass 5.0 kg is shot horizontally with an initial speed of 17 m/s from a height of 25.0 m above a flat desert surface. For the instant before the projectile hits the surface, calculate each of the following quantities:
  a. the work done on the projectile by gravity
  b. the change in kinetic energy since the projectile was fired
  c. the final kinetic energy of the projectile

59. A light horizontal spring has a spring constant of 105 N/m. A 2.00 kg block is pressed against one end of the spring, compressing the spring 0.100 m. After the block is released, the block moves 0.250 m to the right before coming to rest. What is the coefficient of kinetic friction between the horizontal surface and the block?

50. 2.4×10^4 J
51. a. 2.0 m/s
  b. 2.0×10^2 N
52. a. 28.0 m/s
  b. 30.0 m above the ground
53. a. −100 mJ
  b. 3.1 m/s
  c. 0.82 m
54. 2.88×10^2 m
55. a. 5.42 m/s
  b. 0.300
  c. −147 J
56. 4.80 cm
57. a. 4.4 m/s
  b. 1.5×10^5 N
58. a. 1200 J
  b. 1200 J
  c. 1900 J
59. 0.107

ANSWERS TO

Technology & Learning

a. 955 N
b. 2.7×10^3 J
c. 4.0×10^3 J
d. 2.9×10^3 J
e. 4.3×10^3 J
f. d would have more KE

60. A 5.0 kg block is pushed 3.0 m at a constant velocity up a vertical wall by a constant force applied at an angle of 30.0° with the horizontal, as shown in **Figure 5-22.** If the coefficient of kinetic friction between the block and the wall is 0.30, determine the following:

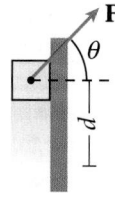

Figure 5-22

  a. the work done by the force on the block
  b. the work done by gravity on the block
  c. the magnitude of the normal force between the block and the wall

61. A 25 kg child on a 2.0 m long swing is released from rest when the swing supports make an angle of 30.0° with the vertical.

  a. What is the maximum potential energy associated with the child?
  b. Disregarding friction, find the child's speed at the lowest position.
  c. What is the child's total mechanical energy?
  d. If the speed of the child at the lowest position is 2.00 m/s, what is the change in mechanical energy due to friction?

Technology & Learning

Graphing calculators

Refer to Appendix B for instructions on downloading programs for your calculator. The program "Chap5" builds a table of work done for various displacements.

Work done, as you learned earlier in this chapter, is described by the following equation:

$$W_{net} = F_{net}d(\cos\theta)$$

The program "Chap5" stored on your graphing calculator makes use of the equation for work done. Once the "Chap5" program is executed, your calculator will ask for F, the net force acting on the object, and θ, the angle at which the force acts. The graphing calculator will use the following equation to create the table of work done (Y_1) for various displacements (X). Note that the relationships in this equation are the same as those in the work equation shown above.

$$Y_1 = FXCOS(\theta)$$

  a. An elephant applies a force of 2055 N against the front of a clown car. If the car pushes toward the elephant with a 3010 N force, what is the value of F in the equation above?

First, be certain the calculator is in degree mode by pressing [MODE] [▼] [▼] [▶] [ENTER].

Execute "Chap5" on the PRGM menu and press [ENTER] to begin the program. Enter the value for the net force applied (shown below) and press [ENTER]. Then enter the value for the angle at which the force is applied and press [ENTER].

The calculator will provide the table of the work done in joules for various displacements in meters. Press [▼] to scroll down through the table to find the displacement value you are looking for.

For each of the following situations, determine how much work is done on a sled by a person pulling on the sled on level ground.

  b. a constant force of 225 N at an angle of 35° for a distance of 15 m
  c. the same force at the same angle for a distance of 22 m
  d. a constant force of 215 N at an angle of 25° for a distance of 15 m
  e. the same force at the same angle for a distance of 22 m
  f. If the forces in b and d were applied over the same time interval, in which case would the sled have more kinetic energy?

Press [ENTER] to stop viewing the table. Press [ENTER] again to enter a new value or [CLEAR] to end the program.

62. A ball of mass 522 g starts at rest and slides down a frictionless track, as shown in **Figure 5-23.** It leaves the track horizontally, striking the ground.

    a. At what height above the ground does the ball start to move?

    b. What is the speed of the ball when it leaves the track?

    c. What is the speed of the ball when it hits the ground?

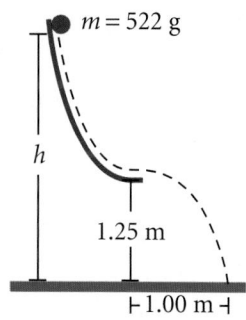

$m = 522$ g

h

1.25 m

⊢1.00 m⊣   **Figure 5-23**

Alternative Assessment

Performance assessment

1. Design experiments for measuring your power output when doing push-ups, running up a flight of stairs, pushing a car, loading boxes onto a truck, throwing a baseball, or performing other energy-transferring activities. What data do you need to measure or calculate? Form groups to present and discuss your plans. If your teacher approves your plans, perform the experiments.

2. Investigate the amount of kinetic energy involved when your car's speed is 60 km/h, 50 km/h, 40 km/h, 30 km/h, 20 km/h, and 10 km/h. (Hint: Find your car's mass in the owner's manual.) How much work does the brake system have to do to stop the car at each speed?

    If the owner's manual includes a table of braking distances at different speeds, determine the force the braking system must exert. Organize your findings in charts and graphs to study the questions and to present your conclusions.

3. Investigate the energy transformations of your body as you swing on a swingset. Working with a partner, measure the height of the swing at the high and low points of your motion. What points involve a maximum gravitational potential energy? What points involve a maximum kinetic energy? For three other points in the path of the swing, calculate the gravitational potential energy, the kinetic energy, and the velocity. Organize your findings in bar graphs.

Portfolio projects

4. In order to save fuel, an airline executive recommended the following changes in the airlines' largest jet flights:

    a. restrict the weight of personal luggage

    b. remove pillows, blankets, and magazines from the cabin

    c. lower flight altitudes by 5 percent

    d. reduce flying speeds by 5 percent

    Research the information necessary to calculate the approximate kinetic and potential energy of a large passenger aircraft. Which of the measures described above would result in significant savings? What might be their other consequences? Summarize your conclusions in a presentation or report.

5. Make a chart of the kinetic energies your body can have. Measure your mass and speed when walking, running, sprinting, riding a bicycle, and driving a car. Make a poster graphically comparing these findings.

6. You are trying to find a way to bring electricity to a remote village in order to run a water-purifying device. A donor is willing to provide battery chargers that connect to bicycles. Assuming the water-purification device requires 18.6 kW•h daily, how many bicycles would a village need if a person can average 100 W while riding a bicycle? Is this a useful way to help the village? How would you schedule the use of these chargers? Summarize your comments and suggestions in a letter to the donor.

Note

Materials Preparation is given on pp. 166A–166B. Blank data table and sample data table are on the One-Stop Planner CD-ROM. All calculations are performed using sample data.

Planning

Recommended time:

1 lab period

Classroom organization:

▸ Each group needs a level work surface. For the CBL and sensors procedure, students must be able to oscillate the mass hanger at least 0.5 m above the motion detector with the stand base clamped to the edge of the table.

▸ Each lab group should have two students.

▸ The Hooke's law apparatus and CBL procedures may be used in the same class.

▸ **Safety warning:** Remind students to attach masses securely and to make sure the area is clear before allowing masses to oscillate.

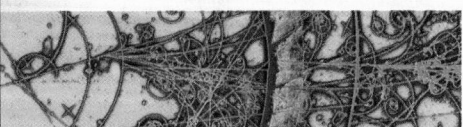

CHAPTER 5
Laboratory Exercise

★ TEKS

pp. 200–205: 1A, 2B, 2C, 2F, 3A, 3E, 5B, 5C, 5D

CONSERVATION OF MECHANICAL ENERGY

A mass on a spring will oscillate vertically when it is lifted to the length of the relaxed spring and released. The gravitational potential energy increases from a minimum at the lowest point to a maximum at the highest point. The elastic potential energy in the spring increases from a minimum at the highest point, where the spring is relaxed, to a maximum at the lowest point, where the spring is stretched. Because the mass is temporarily at rest, the kinetic energy of the mass is zero at the highest and lowest points. Thus, the total mechanical energy at those points is the sum of the elastic potential energy and the gravitational potential energy.

OBJECTIVES

- Determine the spring constant of a spring.
- Calculate elastic potential energy.
- Calculate gravitational potential energy.
- Determine whether mechanical energy is conserved in an oscillating spring.

MATERIALS LIST

✔ meterstick
✔ set of masses
✔ support stand and clamp

PROCEDURE

CBL AND SENSORS

✔ C-clamp
✔ CBL
✔ CBL motion detector
✔ force sensor with CBL-DIN adapter
✔ graphing calculator with link cable
✔ lattice rod and right angle clamp
✔ spring
✔ tape
✔ wire letter basket

HOOKE'S LAW APPARATUS

✔ Hooke's law apparatus
✔ rubber bands

SAFETY

- **Tie back long hair, secure loose clothing, and remove loose jewelry to prevent their getting caught in moving or rotating parts.**

- **Attach masses securely. Perform this experiment in a clear area. Swinging or dropped masses can cause serious injury.**

PREPARATION

1. Determine whether you will be using the CBL and sensors or the Hooke's law apparatus to perform this experiment. Read the entire lab procedure for the appropriate method. Plan the steps you will take.

2. Prepare a data table in your lab notebook with four columns and seven rows. In the first row, label the first through fourth columns *Trial, Mass (kg), Stretched spring (m),* and *Force (N)*. In the first column, label the second through seventh rows *1, 2, 3, 4, 5,* and *6*. Above or below the data table, make a space to enter the value for *Initial spring (m)*.

3. Prepare a second data table in your lab notebook with three columns and seven rows. In the first row, label the first through third columns *Trial, Highest point (m),* and *Lowest point (m)*. In the first column, label the second through seventh rows *1, 2, 3, 4, 5,* and *6*. Above or below the data table, make a space to enter the value for *Initial distance (m)*.

Hooke's law apparatus procedure begins on page 203.

PROCEDURE

CBL AND SENSORS

Spring constant

4. Set up the CBL, graphing calculator, force sensor, and motion detector as shown in **Figure 5-24.**

5. Connect the CBL unit to the calculator with the unit-to-unit link cable using the ports located on each unit. Connect the force sensor to the CH1 port on the CBL unit, and connect the motion detector to the SONIC port.

6. Place the ring stand near the edge of the lab table. Use the C-clamp to clamp the base of the ring stand securely to the table. Position the clamp so that it protrudes as little as possible from the edge of the table. Attach the force sensor to the ring stand with the lattice rod and clamp.

7. Hook one end of the spring securely onto the force sensor. Attach the mass hanger securely to the spring. Tape the motion detector onto a lab stool directly below the force sensor so that the motion detector faces up. The motion detector should be more than 0.5 m away from the mass hanger on the bottom of the spring. Place the wire letter basket upside down over the motion detector to protect the sensor in case the masses fall.

8. Measure the distance from the floor to the top of the motion detector. Record this as *Initial distance (m)* in your data table. This distance must remain constant throughout the lab.

9. Turn on the CBL and the graphing calculator. Start the program PHYSICS on the graphing calculator.

 a. Select option *SET UP PROBES* from the MAIN MENU. Enter 2 for the number of probes. Select the motion detector, and then select the force sensor from the list. Your teacher will tell you what kind of force sensor you are using. Enter 1 for the channel number. Select *USE STORED* from the CALIBRATION menu.

 b. Select the *COLLECT DATA* option from the MAIN MENU. Select the *TIME GRAPH* option from the DATA COLLECTION menu. Enter 0.02 for the time between samples. Enter 99 for the number of samples. Check the values you entered, and press ENTER. Press ENTER to continue. If you made a mistake entering the time values, select *MODIFY SETUP*, reenter the values, and continue.

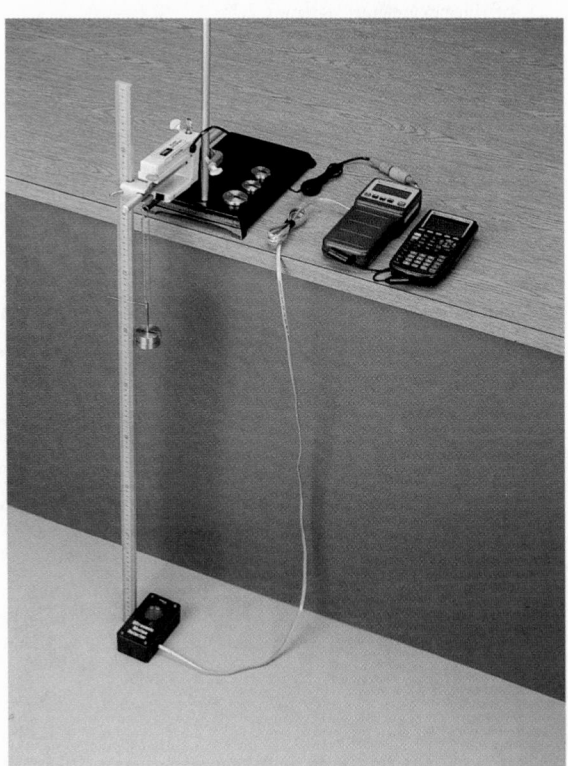

Figure 5-24

Step 6: Attach the force sensor securely to the ring stand. Tape the force sensor lead to the stand to keep it out of the way while you work.

Step 7: Make sure the motion detector is directly below the hanging mass. Make sure the force sensor is far enough over the edge of the table so that the motion detector will read the position of the mass without interference from the tabletop.

Step 9: In this part of the lab, you will collect data to find the spring constant of the spring.

Step 14: In this part of the lab, you will oscillate a mass on the spring to find out whether mechanical energy is conserved.

CBL and Sensors Tips

◆ Students should have the program PHYSICS on their graphing calculators. Refer to Appendix B for instructions.

◆ Tape the motion detector securely in place so that it can't move during the experiment.

◆ Remind students to keep their hands out of the way of the motion detector's signal.

◆ The motion detector may be placed directly on the floor instead of on a stool, and it should be at least 0.5 m away from the bottom of the mass hanger.

Techniques to Demonstrate

Show students how to place a meterstick as a guide, perpendicular to the motion detector and clamped or taped to the force sensor.

Demonstrate releasing the mass hanger from above.

Draw a diagram of the apparatus on the chalkboard, or refer to page 204. Label the distances students will be measuring in the lab: *Initial distance, Initial spring, Stretched spring, Highest point,* and *Lowest point.* Show students how to refer to the diagram to find the elongation of the spring and the height of the mass at each point.

✔ **Checkpoints**

Step 10: Students should be able to demonstrate that the motion detector reads the distance to the mass hanger.

Step 13: Students should be able to explain how these measurements will be used to find the spring constant.

Step 16: Students should be able to release the mass hanger so that it oscillates vertically without interfering with the motion detector.

Step 18: Make sure students can choose the appropriate points on the graph. Students should recognize that the motion detector is the reference point for all distance measurements.

10. Press ENTER on the graphing calculator to begin collecting data. The motion detector will begin to click as it collects data. When the motion detector stops clicking and the CBL shows DONE, press ENTER on the graphing calculator.

11. Select the *SONIC* option from the SELECT CHANNELS menu. Select *DISTANCE* to plot a graph of the distance in meters against the time in seconds. Press TRACE on the graphing calculator, and use the arrow keys to trace the graph. The distance (*y*) values should be fairly constant. Record this value as *Initial spring (m)* in your data table. Press ENTER on the graphing calculator. Select *RETURN* from the SELECT CHANNEL menu. Select *RETURN* again. Select *YES* from the REPEAT? menu.

12. Add enough mass to stretch the spring to about 1.25 times its original length. Record the mass in the first data table. Press ENTER on the graphing calculator. The motion detector will begin to click as it collects data. When the motion detector stops clicking and the CBL shows DONE, press ENTER on the graphing calculator.

    a. Select the *ANALOG* option from the SELECT CHANNELS menu to graph the force in newtons for each second. Use the arrow keys to trace the graph. The force (*y*) values should be fairly constant. Record this value in the first data table. Sketch the graph in your lab notebook. Press ENTER on the graphing calculator.

    b. Select the *SONIC* option from the SELECT CHANNELS menu. Select *DISTANCE* to plot a graph of the distance in meters against the time in seconds. Use the arrow keys to trace the graph. The distance (*y*) values should be fairly constant. Record this value as *Stretched spring (m)* in your data table. Press ENTER on the graphing calculator. Select *RETURN* from the SELECT CHANNEL menu. Select *RETURN* again. Select *YES* from the REPEAT? menu.

13. Perform several trials with increasing masses. Record the mass, force, and distance measurements in the first data table for each trial.

Conservation of mechanical energy

14. Starting with a small mass and gradually increasing the mass in small increments, place masses on the mass hanger until you find a mass that will stretch the spring to about twice its original length. Record the mass in the second data table. Leave the mass in place on the hanger so that the spring remains stretched.

15. Raise the mass hanger until the mass hanger is at the zero position, the position where you measured the *Initial spring* measurement.

16. Press ENTER on the graphing calculator and simultaneously release the hanger gently to let the hanger drop. The motion detector will begin to click as it collects data. It is best to release the hanger from above and pull your hand out quickly to the side. If your hand passes between the hanger and the motion detector, it will seriously affect your measurements.

17. When the motion detector stops clicking and the CBL shows DONE, press ENTER on the graphing calculator so that the calculator will receive the lists of data collected by the CBL.

18. Select *SONIC* from the SELECT CHANNELS menu. Select *DISTANCE* to plot a graph of the distance in meters against the time in seconds. Press TRACE on the graphing calculator, and use the arrow keys to trace the graph. The graph should move between high points and low points to reflect the oscillation of the mass on the spring. Record the distance (*y*) values of the *Highest point* and *Lowest point* in your data table. Sketch the graph in your lab notebook. Press ENTER on the graphing calculator.

19. Select *RETURN* from the SELECT CHANNEL menu. Select *RETURN* again. Select *YES* from the REPEAT? menu to continue to perform more trials.

20. Perform several more trials, using a different mass for each trial. Record all data in your data table.

21. Clean up your work area. Put equipment away safely so that it is ready to be used again.

Analysis and Interpretation begins on page 204.

PROCEDURE

HOOKE'S LAW APPARATUS

Spring constant

4. Set up the Hooke's law apparatus as shown in **Figure 5-25.**

5. Place a rubber band around the scale at the initial resting position of the pointer, or adjust the scale or pan to read 0.0 cm. Record this position of the pointer as *Initial spring (m)*. If you have set the scale at 0.0 cm, record 0.00 m as the initial spring position.

6. Measure the distance from the floor to the rubber band on the scale. Record this measurement in the second data table under *Initial distance (m)*. This distance must remain constant throughout the lab.

7. Find a mass that will stretch the spring so that the pointer moves approximately one-quarter of the way down the scale.

8. Record the value of the mass. Also record the position of the pointer under *Stretched spring* in the data table.

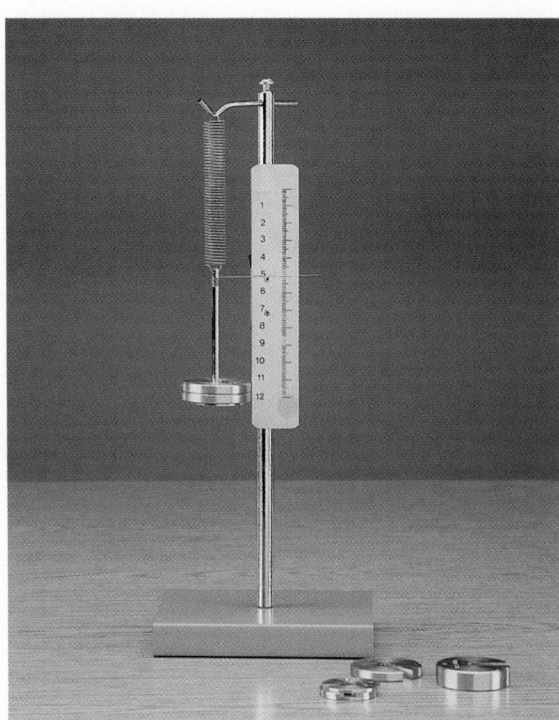

9. Perform several trials with increasing masses until the spring stretches to the bottom of the scale. Record the mass and the position of the pointer for each trial.

Conservation of mechanical energy

10. Find a mass that will stretch the spring to about twice its original length. Record the mass in the second data table. Leave the mass in place on the pan.

11. Raise the pan until the pointer is at the zero position, the position where you measured the *Initial spring* measurement.

12. Gently release the pan to let the pan drop. Watch closely to identify the high and low points of the oscillation.

13. Use a rubber band to mark the lowest position to which the pan falls, as indicated by the pointer. This point is the lowest point of the oscillation. Record the values as *Highest point* and *Lowest point* in your data table.

14. Perform several more trials, using a different mass for each trial. Record all data in your data table.

15. Clean up your work area. Put equipment away safely so that it is ready to be used again.

Figure 5-25

Step 5: If the scale is adjusted to read 0.0 cm, record 0.0 as the initial spring length in your data table.

Step 7: In this part of the lab, you will collect data to find the spring constant of the spring.

Step 10: In this part of the lab, you will oscillate a mass on the spring to find out whether mechanical energy is conserved.

Hooke's Law Apparatus Tips

◆ For best results, use weights of less than 1.0 N for steps 10–14.

Techniques to Demonstrate

Show students how to read the scale on the Hooke's law apparatus.

Demonstrate releasing the mass hanger so it will oscillate vertically without twisting.

Draw a diagram of the apparatus on the chalkboard, and label the distances students will be measuring in the lab: *Initial distance, Initial spring, Stretched spring, Highest point,* and *Lowest point.* Show students how to refer to the diagram to find the elongation of the spring and the height of the mass at each point.

✔ Checkpoints

Step 5: Students should adjust the scale to zero at the initial position of the spring if possible.

Step 6: Make sure students are measuring the vertical distance from the floor to the initial position of the spring.

Step 13: Students may need to practice a technique to identify the highest and lowest points while the mass is oscillating. Without disturbing the apparatus, they might use pencils as pointers to mark the place until they can place their rubber bands.

ANSWERS TO

Analysis and Interpretation

CALCULATIONS AND DATA ANALYSIS

1. a, b. Student answers will vary. Typical values range from 0.022 m to 0.118 m.

   c. For sample data, values range from $F = 0.050$ N to $F = 2.02$ N.

2. Student answers will vary. For sample data, values for k_{avg} range from 19.4 N/m to 25.5 N/m.

3. A stiffer spring would give greater values for the spring constant. The elastic potential energy would be greater, but the gravitational potential energy would not change.

4. Student answers will vary. Typical values range from 0.000 m to 0.010 m.

5. Typical values range from 0.041 m to 0.195 m.

6. For sample data, values range from 3.88×10^{-5} J to 1.28×10^{-3} J.

7. For sample data, values range from 2.16×10^{-3} J to 4.85×10^{-1} J.

8. $PE_{elastic}$ is greatest at the lowest point and least at the highest point because the elongation is greatest at the lowest point and because $PE_{elastic}$ depends on the elongation squared.

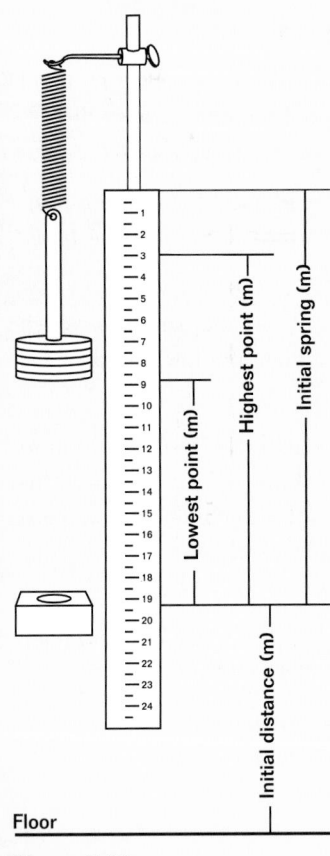

CBL and sensors

Highest point (m)

Initial spring (m)

Lowest point (m)

Initial distance (m)

Floor

Figure 5-26

ANALYSIS AND INTERPRETATION

Calculations and data analysis

1. Organizing data Use your data from the first data table to calculate the elongation of the spring.

   a. CBL and sensors Use the equation *elongation = initial spring − stretched spring.*

   b. Hooke's law apparatus Use the equation *elongation = stretched spring − initial spring.*

   c. Hooke's law apparatus Convert the masses used to measure the spring constant to their force equivalents.

2. Evaluating data For each trial, calculate the spring constant using the equation $k = \dfrac{force}{elongation}$. Take the average of all trials, and use this value as the spring constant.

3. Analyzing results How would using a stiffer spring affect the value for the spring constant? How would this change affect the values for the elastic and gravitational potential energies?

4. Organizing data Using your data from the second data table, calculate the elongation of the spring at the highest point of each trial.

   a. CBL and sensors Use the equation *elongation = initial spring − highest point.*

   b. Hooke's law apparatus Use the equation *elongation = highest point − initial spring.*

5. Organizing data Calculate the elongation of the spring at the lowest point of each trial.

   a. CBL and sensors Use the equation *elongation = initial spring − lowest point.*

   b. Hooke's law apparatus Use the equation *elongation = lowest point − initial spring.*

6. Analyzing information For each trial, calculate the elastic potential energy, $PE_{elastic} = \frac{1}{2}kx^2$, at the highest point of the oscillation.

7. Analyzing information For each trial, calculate the elastic potential energy at the lowest point of the oscillation.

8. Analyzing results Based on your calculations in items 6 and 7, where is the elastic potential energy greatest? Where is it the least? Explain these results in terms of the energy stored in the spring.

9. **Organizing data** Calculate the height of the mass at the highest point of each trial.

   a. **CBL and sensors** Use the equation *highest = initial distance + highest point*.

   b. **Hooke's law apparatus** Use the equation *highest = initial distance − elongation*.

10. **Organizing data** Calculate the height of the mass at the lowest point of each trial.

    a. **CBL and sensors** Use the equation *lowest = initial distance + lowest point*.

    b. **Hooke's law apparatus** Use the equation *lowest = initial distance − elongation*.

11. **Applying ideas** For each trial, calculate the gravitational potential energy, $PE_g = mgh$, at the highest point of the oscillation.

12. **Applying ideas** For each trial, calculate the gravitational potential energy at the lowest point of the oscillation.

13. **Analyzing results** According to your calculations in items 11 and 12, where is the gravitational potential energy the greatest? Where is it the least? Explain these results in terms of gravity and the height of the mass and the spring.

14. **Evaluating data** Find the total potential energy at the top of the oscillation and at the bottom of the oscillation.

Conclusions

15. **Drawing conclusions** Based on your data, is mechanical energy conserved in the oscillating mass on the spring? Explain how your data support your answers.

Extensions

16. **Extending ideas** Use your data to find the midpoint of the oscillation for each trial. Calculate the gravitational potential energy and the elastic potential energy at the midpoint. Use the principle of the conservation of mechanical energy to find the kinetic energy and the speed of the mass at the midpoint.

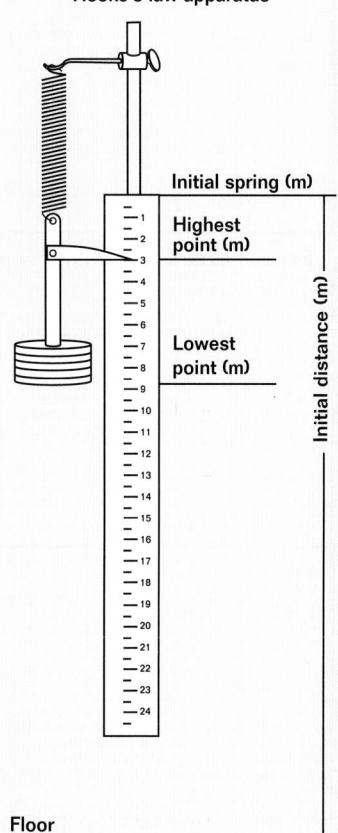

Hooke's law apparatus

Initial spring (m)

Highest point (m)

Lowest point (m)

Initial distance (m)

Floor

Figure 5-27

9. Student answers will vary. Typical values range from 0.235 m to 0.737 m.

10. Typical values range from 0.127 m to 0.680 m.

11. Typical values range from 1.2×10^{-1} J to 1.46 J.

12. Typical values range from 1.2×10^{-1} J to 1.1 J.

13. Gravitational *PE* is greatest at the highest point because it depends on the height of the mass.

14. Student answers will vary. Make sure students use the relationship $PE_{total} = PE_g + PE_{elastic}$. For sample data, values range from 1.2×10^{-1} J to 1.5 J at the high point and 1.6×10^{-1} J to 1.6 J at the low point.

CONCLUSIONS

15. Mechanical energy is conserved; the sum of the elastic and gravitational potential energies is the same at the top and bottom of the oscillation.

EXTENSIONS

16. Student answers will vary. For sample data, PE_g at the midpoint ranges from 1.6×10^{-3} J to 3.4×10^{-2} J, and $PE_{elastic}$ at the midpoint ranges from 0.07 J to 0.27 J. Values for *KE* at the midpoint range from 0.01 J to 0.10 J, and values for the speed at the midpoint range from 0.44 m/s to 0.89 m/s.

Compression Guide: To shorten from 9 to 6 45-min periods (from 4½ to 3 90-min blocks), eliminate items in magenta type.

PACING CHART	CLASSROOM RESOURCES			
	⭐TEKS	Teacher Demonstrations	*Holt Physics* Transparencies	Labs (See page T52 for equipment listing for in-text labs.)
6-1 Momentum and impulse 3 or 2 45-minute periods 1½ or 1 90-minute block	2C, 3A, 3B, 3C, 3E, 5C	**TE** *Impulse*, p. 210		**TL** *Impulse and Momentum*
6-2 Conservation of momentum 2 45-minute periods 1 90-minute block	2C, 3A, 3B, 3C, 5D	**TE** *Momentum and magnets*, p. 220		**PE** *Conservation of Momentum*, p. 238
6-3 Elastic and inelastic collisions 2 45-minute periods 1 90-minute block	2C, 3A, 3B, 5B, 5D	**TE** *Inelastic collisions*, p. 222	**T** 20	**PE** *Quick Lab: Elastic and Inelastic Collisions*, p. 227
Review and Assessment 2 45-minute periods 1 90-minute block				

Resource Key

PHYSICS

PE Pupil's Edition
TE Teacher's Edition

L Laboratory Experiments
TL Technology Lab Experiments
T Transparencies

🖱 **One-Stop** Planner CD-ROM **contents**

TM Transparency Masters
SR Section Review Worksheets
AA Alternative Assessment

PW Problem-Solving Workbook
PB Problem Bank
CTW Critical Thinking Worksheet

LABORATORY PLANNING: Conservation of Momentum, p. 238

Materials (for each lab group)
- balance: portable, electronic balance or triple-beam balance with weight
- C-clamp
- dynamics cart set with two carts
- meterstick
- metric slotted mass set, 1 g–500 5
- metric hooked mass, 1000 g
- 1-position support base and rod, 1.3 cm × 91 cm
- roll of adhesive tape, 0.5 in. wide

Additional Equipment
CBL and Sensors Procedure
- CBL
- graphing calculator
- CBL ultrasonic motion detector
- V-jaw symmetrical clamp with holder
- 2 pieces of poster board, 25 cm × 25 cm each
- TI Graph Link (recommended for downloading programs)

ASSIGNMENT RESOURCES

Content Mastery	Critical Thinking	Problem-Solving Practice	
PE 1–5, p. 214 **SR** 6-1, *Graph Skills* **PE** 1–4, p. 232	**PE** 5–11, p. 232	**6A**	Momentum: 24 items in **PE, PW,** and **PB,** see **TE** p. 209
		6B	Force and impulse: 29 items in **PE, PW,** and **PB,** see **TE** p. 211
		6C	Stopping distance: 25 items in **PE, PW,** and **PB,** see **TE** pp. 212–213
PE 1–4, p. 221 **SR** 6-2, *Concept Review* **PE** 17–19, p. 233	**PE** 1–2, p. 216 **PE** 20–23, p. 233	**6D**	Conservation of momentum: 30 items in **PE, PW,** and **PB,** see **TE** pp. 218–219
PE 1–5, p. 230 **SR** 6-3, *Diagram Skills*	**PE** 27–30, pp. 233–234	**6E**	Perfectly inelastic collisions: 34 items in **PE, PW,** and **PB,** see **TE** pp. 223–224
		6F	Kinetic energy in perfectly inelastic collisions: 25 items in **PE, PW,** and **PB,** see **TE** pp. 225–226
		6G	Elastic collisions: 30 items in **PE, PW,** and **PB,** see **TE** pp. 228–229

ASSESSMENT RESOURCES

Cumulative Review	Alternative Assessment	Traditional Assessment
SR Mixed Review, Ch. 6	**PE** 1–5, p. 237 **AA** Items for Ch. 6	Chapter 6 Test Test Generator items for Ch. 6

Scoring Rubrics for Alternative Assessment items can be found on the One-Stop Planner CD-ROM.

TECHNOLOGY RESOURCES

 CTW Segment 6 Egg Drop Contest

 Module 7 Conservation of Momentum

 PE Technology and Learning, p. 236
(Alternative procedures for calculators without Flash-ROM technology are provided on the One-Stop Planner CD-ROM.)

 The Mechanical Universe/High School Adaptation Quad II, Conservation of Momentum

internet connect

 On-line Student Resources:
GO TO: www.scilinks.org
The following *sci*LINKS Internet resources can be found in the student text for this chapter.

TOPICS:
- Momentum, p. 210 (HF2061)
- Rocketry, p. 216 (HF2062)
- Collisions, p. 226 (HF2064)

 On-line Teacher Resources:
GO TO: go.hrw.com
KEYWORD: HF2 HOME
Visit the HRW Web site for a variety of resources related to this chapter.

 Smithsonian Institution
Internet Connections
Visit **www.si.edu/hrw** for additional on-line resources.

Visit **www.cnnfyi.com** for late-breaking news and current events stories selected just for you.

Recording Timer Procedure
- recording timer: acceleration timer, tabletop acceleration timer, or compact spark timer
- 2 lattice rods, 1.3 cm × 30 cm
- 1 stand rod, 1.3 cm × 60 cm
- 1 right-angle clamp for 1.3 cm rods
- 2 rubber bands
- replacement paper tape, 13 mm
- replacement carbon disks

Materials Preparation
Set up a sample apparatus in the laboratory for students to refer to when setting up their equipment.

Section 6-1 defines momentum in terms of velocity, introduces the concept of impulse, and relates impulse and momentum.

Section 6-2 explores the law of conservation of momentum and uses this law to predict the final velocity of an object after a collision.

Section 6-3 distinguishes between elastic, perfectly inelastic, and inelastic collisions and discusses whether kinetic energy is conserved in each type of collision.

About the Illustration

Soccer is a good example to help students understand the concept of momentum and distinguish it from force, velocity, and kinetic energy. This photograph is a dramatic example of a player colliding with a ball and changing the momentum of the ball. Use this example to illustrate the vector nature of momentum; the photograph makes it clear that the direction as well as the magnitude of momentum is affected by the collision.

 **Interactive Problem-Solving Tutor**

See Module 7
"Conservation of Momentum" promotes additional development of problem-solving skills for this chapter.

CHAPTER 6

Momentum and Collisions

PHYSICS IN ACTION

Soccer players must consider an enormous amount of information every time they set the ball—or themselves—into motion. Once a player knows where the ball should go, the player has to decide how to get it there. The player also has to consider the ball's speed and direction in order to stop it or change its direction. The player in the photograph must determine how much force to exert on the ball and how much follow-through is needed. To do this, he must understand his own motion as well as the motion of the ball.

- *International regulations specify the mass of official soccer balls. How does the mass of a ball affect the way it behaves when kicked?*

- *How does the velocity of the player's foot affect the final velocity of the ball?*

CONCEPT REVIEW

Newton's laws of motion
  (Sections 4-2 and 4-3)

Kinetic energy (Section 5-2)

Conservation of energy
  (Section 5-3)

6-1
Momentum and impulse

6-1 SECTION OBJECTIVES

- **Compare the momentum of different moving objects.**
- **Compare the momentum of the same object moving with different velocities.**
- **Identify examples of change in the momentum of an object.**
- **Describe changes in momentum in terms of force and time.**

momentum

a vector quantity defined as the product of an object's mass and velocity

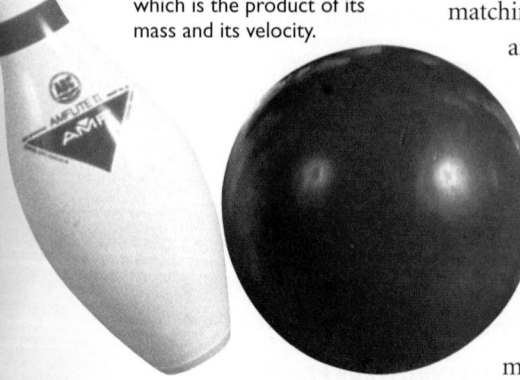

Figure 6-1
When the bowling ball strikes the pin, the acceleration of the pin depends on the ball's momentum, which is the product of its mass and its velocity.

LINEAR MOMENTUM

When a soccer player heads a moving ball during a game, the ball's velocity changes rapidly. The speed of the ball and the direction of the ball's motion change once it is struck so that the ball moves across the soccer field with a different speed than it had and in a different direction than it was traveling before the collision.

The quantities and kinematic equations from Chapter 2 can be used to describe the motion of the ball before and after the ball is struck. The concept of force and Newton's laws from Chapter 4 can be used to explain *why* the motion of the ball changes when it is struck. In this chapter, we will examine how the force and the duration of the collision between the ball and the soccer player affect the motion of the ball.

Momentum describes an object's motion

To address such questions, we need a new concept, **momentum.** *Momentum* is a word we use every day in a variety of situations. In physics, of course, this word has a specific meaning. The linear momentum of an object of mass *m* moving with a velocity **v** is defined as the product of the mass and the velocity. Momentum is represented by the symbol **p.**

MOMENTUM ⭐TEKS **3B**

$$\mathbf{p} = m\mathbf{v}$$

$$\text{momentum} = \text{mass} \times \text{velocity}$$

As its definition shows, momentum is a vector quantity, with its direction matching that of the velocity. Momentum has dimensions mass × length/time, and its SI units are kilogram-meters per second (kg•m/s).

If you think about some examples of the way the word *momentum* is used in everyday speech, you will see that the physics definition conveys a similar meaning. Imagine coasting down a hill of uniform slope on your bike without pedaling or using the brakes. Because of the force of gravity, you will accelerate at a constant rate so that your velocity will increase with time. This is often expressed by saying that you are "picking up speed" or "gathering momentum." The faster you move, the more momentum you have and the more difficult it is to come to a stop. ⭐TEKS **3A**

Imagine rolling a bowling ball down one lane at a bowling alley and rolling a playground ball down another lane at the same speed. The more-massive bowling ball, shown in **Figure 6-1,** will exert more force on the pins because the bowling ball has more momentum than the playground ball. When we think of a massive object moving at a high velocity, we often say that the object has a large momentum. A less massive object with the same velocity has a smaller momentum.

On the other hand, a small object moving with a very high velocity has a large momentum. One example of this is hailstones falling from very high clouds. By the time they reach Earth, they can have enough momentum to hurt you or cause serious damage to cars and buildings.

(★)TEKS) 3E

SAMPLE PROBLEM 6A

Momentum (★)TEKS) 2C, 3B, 5C

PROBLEM

A 2250 kg pickup truck has a velocity of 25 m/s to the east. What is the momentum of the truck?

SOLUTION

Given: $m = 2250$ kg $v = 25$ m/s to the east

Unknown: $p = ?$

Use the momentum equation from page 208.

$$\mathbf{p} = m\mathbf{v} = (2250 \text{ kg})(25 \text{ m/s})$$

$$\boxed{\mathbf{p} = 5.6 \times 10^4 \text{ kg} \cdot \text{m/s to the east}}$$

CALCULATOR SOLUTION

Your calculator will give you the answer 56250 for the momentum. The value for the velocity has only two significant figures, so the answer must be reported as 5.6×10^4.

PRACTICE 6A

Momentum (★)TEKS) 2C, 3B, 5C

1. An ostrich with a mass of 146 kg is running to the right with a velocity of 17 m/s. Find the momentum of the ostrich.

2. A 21 kg child is riding a 5.9 kg bike with a velocity of 4.5 m/s to the northwest.
   a. What is the total momentum of the child and the bike together?
   b. What is the momentum of the child?
   c. What is the momentum of the bike?

3. What velocity must a car with a mass of 1210 kg have in order to have the same momentum as the pickup truck in Sample Problem 6A?

PRACTICE GUIDE 6A

Solving for:

p	📖 **PE**	Sample, 1–2; Ch. Rvw. 12–13, 43*
	💿 **PW**	5–6
	PB	5–7
m	📖 **PE**	Ch. Rvw. 42*
	💿 **PW**	Sample, 1–2
	PB	8–10
v	📖 **PE**	3; Ch. Rvw. 41, 42*
	💿 **PW**	3–4
	PB	Sample, 1–4

ANSWERS TO

Practice 6A
Momentum
1. 2.5×10^3 kg•m/s to the right
2. a. 1.2×10^2 kg•m/s to the northwest
   b. 94 kg•m/s to the northwest
   c. 27 kg•m/s to the northwest
3. 46 m/s to the east

(★)TEKS)

p. 208: 3B, 3A
p. 209: 3E, 2C, 3B, 5C, 2C, 3B, 5C

Figure 6-2
When the ball is moving very fast, the player must exert a large force over a short time to change the ball's momentum and quickly bring the ball to a stop.

internet**connect**

SC*i*LINKS.
NSTA

TOPIC: Momentum
GO TO: www.scilinks.org
*sci*LINKS CODE: HF2061

impulse

for a constant external force, the product of the force and the time over which it acts on an object

A change in momentum takes force and time

Figure 6-2 shows a player stopping a moving soccer ball. In a given time interval, he must exert more force to stop a fast ball than to stop a ball that is moving more slowly. Now imagine a toy truck and a real dump truck starting from rest and rolling down the same hill at the same time. They would accelerate at the same rate, so their velocity at any instant would be the same, but it would take much more force to stop the massive dump truck than to stop the toy truck in the same time interval. You have probably also noticed that a ball moving very fast stings your hands when you catch it, while a slow-moving ball causes no discomfort when you catch it.

From examples like these, we see that momentum is closely related to force. In fact, when Newton first expressed his second law mathematically, he wrote it not as $\mathbf{F} = m\mathbf{a}$, but in the following form. ★ TEKS 3B, 3E

$$\mathbf{F} = \frac{\Delta \mathbf{p}}{\Delta t}$$

$$\text{force} = \frac{\text{change in momentum}}{\text{time interval}}$$

We can rearrange this equation to find the change in momentum in terms of the net external force and the time interval required to make this change.

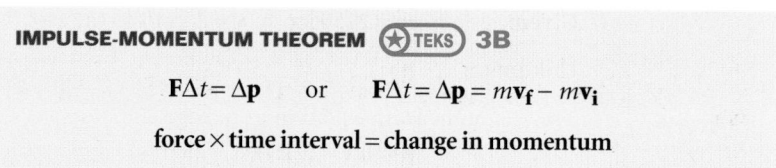

IMPULSE-MOMENTUM THEOREM ★ TEKS 3B

$$\mathbf{F}\Delta t = \Delta \mathbf{p} \quad \text{or} \quad \mathbf{F}\Delta t = \Delta \mathbf{p} = m\mathbf{v_f} - m\mathbf{v_i}$$

force × time interval = change in momentum

This equation states that a net external force, **F**, applied to an object for a certain time interval, Δt, will cause a change in the object's momentum equal to the product of the force and the time interval. In simple terms, a small force acting for a long time can produce the same change in momentum as a large force acting for a short time. In this book, all forces exerted on an object are assumed to be constant unless otherwise stated.

The expression $\mathbf{F}\Delta t = \Delta \mathbf{p}$ is called the impulse-momentum theorem. The term on the left side of the equation, $\mathbf{F}\Delta t$, is called the **impulse** of the force **F** for the time interval Δt.

The equation $\mathbf{F}\Delta t = \Delta \mathbf{p}$ explains why follow-through is important in so many sports, from karate and billiards to softball and croquet. For example, when a batter hits a ball, the ball will experience a greater change in momentum if the batter follows through and keeps the bat in contact with the ball for a longer time. Follow through is also important in many everyday activities, such as pushing a shopping cart or moving furniture. Extending the time interval over which a constant force is applied allows a smaller force to cause a greater change in momentum than would result if the force were applied for a very short time. ★ TEKS 3A, 3B

SAMPLE PROBLEM 6B

Force and impulse ⭐TEKS 2C, 3B, 5C

PROBLEM

A 1400 kg car moving westward with a velocity of 15 m/s collides with a utility pole and is brought to rest in 0.30 s. Find the magnitude of the force exerted on the car during the collision.

SOLUTION

Given: $m = 1400$ kg $v_i = 15$ m/s to the west $= -15$ m/s

$\Delta t = 0.30$ s $v_f = 0$ m/s

Unknown: $F = ?$

Use the impulse-momentum theorem.

$$\mathbf{F}\Delta t = \Delta \mathbf{p} = m\mathbf{v_f} - m\mathbf{v_i}$$

$$\mathbf{F} = \frac{m\mathbf{v_f} - m\mathbf{v_i}}{\Delta t}$$

$$\mathbf{F} = \frac{(1400 \text{ kg})(0 \text{ m/s}) - (1400 \text{ kg})(-15 \text{ m/s})}{0.30 \text{ s}} = \frac{21\,000 \text{ kg} \cdot \text{m/s}}{0.30 \text{ s}}$$

$$\boxed{\mathbf{F} = 7.0 \times 10^4 \text{ N to the east}}$$

PRACTICE 6B

Force and momentum ⭐TEKS 2C, 3B, 5C

1. A 0.50 kg football is thrown with a velocity of 15 m/s to the right. A stationary receiver catches the ball and brings it to rest in 0.020 s. What is the force exerted on the receiver?

2. An 82 kg man drops from rest on a diving board 3.0 m above the surface of the water and comes to rest 0.55 s after reaching the water. What force does the water exert on him?

3. A 0.40 kg soccer ball approaches a player horizontally with a velocity of 18 m/s to the north. The player strikes the ball and causes it to move in the opposite direction with a velocity of 22 m/s. What impulse was delivered to the ball by the player?

4. A 0.50 kg object is at rest. A 3.00 N force to the right acts on the object during a time interval of 1.50 s.
   a. What is the velocity of the object at the end of this interval?
   b. At the end of this interval, a constant force of 4.00 N to the left is applied for 3.00 s. What is the velocity at the end of the 3.00 s?

PRACTICE GUIDE 6B		
Solving for:		
F	📖	**PE** Sample, 1–2; Ch. Rvw. 14–15, 47, 56*
	💿	**PW** 7–9
		PB 5–7
Δt	📖	**PE** 3*
	💿	**PW** Sample, 1–3
		PB 8–10
$\Delta p, v$	📖	**PE** 4; Ch. Rvw. 55, 56*
	💿	**PW** 4–6
		PB Sample, 1–4

ANSWERS TO

Practice 6B
1. 3.8×10^2 N to the right
2. 1.1×10^3 N upward
3. 16.0 kg•m/s to the south
4. a. 9.0 m/s to the right
   b. 15 m/s to the left

Figure 6-3

Be sure students understand the relationship between stopping time and momentum.

Q Why is the loaded truck's stopping time twice as much as the empty truck's when acted on by the same force?

A *The loaded truck's momentum is twice as great, so its change in momentum is also twice as great. The applied forces are the same, so the time period must be twice as great because $\Delta p = F\Delta t$.*

Q How do the stopping distances of the trucks compare?

A *The loaded truck's time period is twice as great while its acceleration is half as much ($F = ma$). Because $x = v_i\Delta t + \frac{1}{2}a\Delta t^2$, the loaded truck's stopping distance is two times as great as the empty truck's.*

Classroom Practice

The following may be used as a teamwork exercise or for demonstration at the chalkboard or on an overhead projector.

PROBLEM

Stopping distance

If the maximum coefficient of kinetic friction between a 2300 kg car and a road is 0.50, what is the minimum stopping distance for a car moving at 29 m/s?

Answer

86 m

Stopping distances

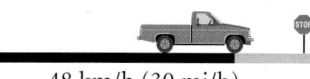

48 km/h (30 mi/h)

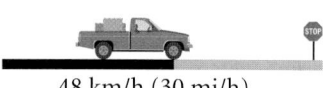

48 km/h (30 mi/h)

Figure 6-3
The loaded truck must undergo a greater change in momentum in order to stop than the truck without a load.

Stopping times and distances depend on the impulse-momentum theorem (★)TEKS 3C

Highway safety engineers use the impulse-momentum theorem to determine stopping distances and safe following distances for cars and trucks. For example, the truck hauling a load of bricks in **Figure 6-3** has twice the mass of the other truck, which has no load. Therefore, if both are traveling at 48 km/h, the loaded truck has twice as much momentum as the unloaded truck. If we assume that the brakes on each truck exert about the same force, we find that the stopping time is two times longer for the loaded truck than for the unloaded truck, and the stopping distance for the loaded truck is two times greater than the stopping distance for the truck without a load.

SAMPLE PROBLEM 6C

Stopping distance (★)TEKS 2C, 3B, 5C

PROBLEM

A 2250 kg car traveling to the west slows down uniformly from 20.0 m/s to 5.00 m/s. How long does it take the car to decelerate if the force on the car is 8450 N to the east? How far does the car travel during the deceleration?

SOLUTION

Given: $m = 2250$ kg $\mathbf{v_i} = 20.0$ m/s to the west $= -20.0$ m/s
$\mathbf{v_f} = 5.00$ m/s to the west $= -5.00$ m/s
$\mathbf{F} = 8450$ N to the east $= +8450$ N

Unknown: $\Delta t = ?$ $\Delta \mathbf{x} = ?$

Use the impulse-momentum theorem from page 210.

$$\mathbf{F}\Delta t = \Delta \mathbf{p}$$

$$\Delta t = \frac{\Delta \mathbf{p}}{\mathbf{F}} = \frac{m\mathbf{v_f} - m\mathbf{v_i}}{\mathbf{F}}$$

$$\Delta t = \frac{(2250 \text{ kg})(-5.00 \text{ m/s}) - (2250 \text{ kg})(-20.0 \text{ m/s})}{8450 \text{ kg} \cdot \text{m}^2/\text{s}^2}$$

$$\boxed{\Delta t = 4.00 \text{ s}}$$

$$\Delta \mathbf{x} = \frac{1}{2}(\mathbf{v_i} + \mathbf{v_f})\Delta t$$

$$\Delta \mathbf{x} = \frac{1}{2}(-20.0 \text{ m/s} - 5.00 \text{ m/s})(4.00 \text{ s})$$

$$\boxed{\Delta \mathbf{x} = -50.0 \text{ m} = 50.0 \text{ m to the west}}$$

Stopping distance ⭐TEKS 2C, 3B, 5C

1. How long would it take the car in Sample Problem 6C to come to a stop from 20.0 m/s to the west? How far would the car move before stopping? Assume a constant acceleration.

2. A 2500 kg car traveling to the north is slowed down uniformly from an initial velocity of 20.0 m/s by a 6250 N braking force acting opposite the car's motion. Use the impulse-momentum theorem to answer the following questions:

   a. What is the car's velocity after 2.50 s?
   b. How far does the car move during 2.50 s?
   c. How long does it take the car to come to a complete stop?

3. Assume that the car in Sample Problem 6C has a mass of 3250 kg.

   a. How much force would be required to cause the same acceleration as in item 1? Use the impulse-momentum theorem.
   b. How far would the car move before stopping?

A change in momentum over a longer time requires less force

The impulse-momentum theorem is used to design safety equipment that reduces the force exerted on the human body during collisions. Examples of this are the nets and giant air mattresses firefighters use to catch people who must jump out of tall burning buildings. The relationship is also used to design sports equipment and games. ⭐TEKS 3C

Figure 6-4 shows an Inupiat family playing a traditional game. Common sense tells us that it is much better for the girl to fall onto the outstretched blanket than onto the hard ground. In both cases, however, the change in momentum of the falling girl is exactly the same. The difference is that the blanket "gives way" and extends the time of collision so that the change in the girl's momentum occurs over a longer time interval. A longer time interval requires a smaller force to achieve the same change in the girl's momentum. Therefore, the force exerted on the girl when she lands on the outstretched blanket is less than the force would be if she were to land on the ground. ⭐TEKS 3A

Figure 6-4
In this game, the girl is protected from injury because the blanket reduces the force of the collision by allowing it to take place over a longer time interval.

SECTION 6-1

PRACTICE GUIDE 6C		
Solving for:		
Δx	📖 **PE**	Sample, 1–3; Ch. Rvw. 16
	💿 **PW**	7–9
	PB	5–7
Δt	📖 **PE**	Sample, 1–2; Ch. Rvw. 16
	💿 **PW**	Sample, 1–3
	PB	8–10
Δp	💿 **PW**	4–6
	PB	Sample, 1–4
F	📖 **PE**	3; Ch. Rvw. 47
	💿 **PW**	Sample, 1–3
	PB	5–7

ANSWERS TO

Practice 6C
Stopping distance
1. 5.33 s; 53.3 m to the west
2. a. 14 m/s to the north
   b. 42 m to the north
   c. 8.0 s
3. a. 1.22×10^4 N to the east
   b. 53.3 m to the west

⭐TEKS

p. 212: 3C, 2C, 3B, 5C
p. 213: 2C, 3B, 5C, 3C, 3A

SECTION 6-1

Teaching Tip

Make sure students understand that the correspondence between the time interval and the force is a result of the impulse-momentum theorem. When the egg hits the plate, as in **Figure 6-5(a),** the time period is shorter and the force is greater. When the egg hits the pillow, as in **(b),** the time increases and the force decreases. As the time period increases, the force continues to decrease.

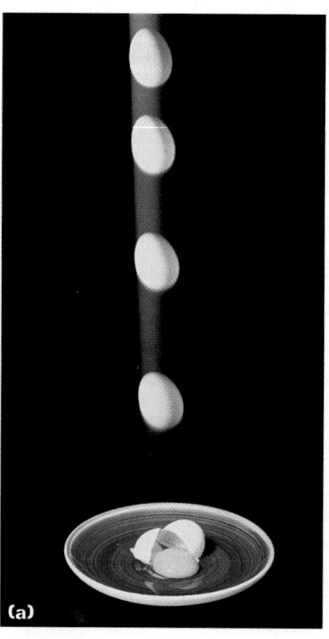

(a)   (b)

Figure 6-5
A large force exerted over a short time **(a)** causes the same change in the egg's momentum as a small force exerted over a longer time **(b).**

Now consider a falling egg. When the egg hits a hard surface, like the plate in **Figure 6-5(a),** the egg comes to rest in a very short time interval. The force the hard plate exerts on the egg due to the collision is large. When the egg hits a floor covered with a pillow, as in **Figure 6-5(b),** the egg undergoes the same change in momentum, but over a much longer time interval. In this case, the force required to accelerate the egg to rest is much smaller. By applying a small force to the egg over a longer time interval, the pillow causes the same change in the egg's momentum as the hard plate, which applies a large force over a short time interval. Because the force in the second situation is smaller, the egg can withstand it without breaking. ★TEKS **3A**

Section Review   ★TEKS **2C, 3B, 5C**

ANSWERS

1. **a.** momentum increases by a factor of two
   b. kinetic energy increases by a factor of four
2. **a.** 31.0 m/s
   b. the bullet
3. no; Because $\Delta\mathbf{p} = \mathbf{F}\Delta t$, it is possible for a large force applied over a very short time interval to change the momentum less than a smaller force applied over a longer time period.
4. Impulse is equivalent to change in momentum.
5. **a.** 2.6 kg•m/s downfield
   b. 1.3×10^2 N downfield

1. The speed of a particle is doubled.
   a. By what factor is its momentum changed?
   b. What happens to its kinetic energy?

2. A pitcher claims he can throw a 0.145 kg baseball with as much momentum as a speeding bullet. Assume that a 3.00 g bullet moves at a speed of 1.50×10^3 m/s.
   a. What must the baseball's speed be if the pitcher's claim is valid?
   b. Which has greater kinetic energy, the ball or the bullet?

3. When a force is exerted on an object, does a large force always produce a larger change in the object's momentum than a smaller force does? Explain.

4. What is the relationship between impulse and momentum?

5. **Physics in Action** A 0.42 kg soccer ball is moving downfield with a velocity of 12 m/s. A player kicks the ball so that it has a final velocity of 18 m/s downfield.
   a. What is the change in the ball's momentum?
   b. Find the constant force exerted by the player's foot on the ball if the two are in contact for 0.020 s.

6-2
Conservation of momentum

MOMENTUM IS CONSERVED

So far in this chapter, we have considered the momentum of only one object at a time. Now we will consider the momentum of two or more objects interacting with each other. **Figure 6-6** shows a stationary soccer ball set into motion by a collision with a moving soccer ball. Assume that both balls are on a smooth gym floor and that neither ball rotates before or after the collision. Before the collision, the momentum of ball B is equal to zero because the ball is stationary. During the collision, ball B gains momentum while ball A loses momentum. As it turns out, the momentum that ball A loses is exactly equal to the momentum that ball B gains.

6-2 SECTION OBJECTIVES

- **Describe the interaction between two objects in terms of the change in momentum of each object.**

- **Compare the total momentum of two objects before and after they interact.**

- **State the law of conservation of momentum.**

- **Predict the final velocities of objects after collisions, given the initial velocities.**

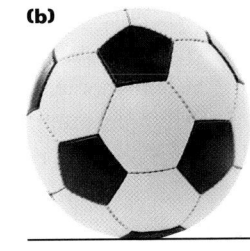

(a)   A   B   (b)   A   B

Table 6-1 shows the velocity and momentum of each soccer ball both before and after the collision. The momentum of each ball changes due to the collision, but the *total* momentum of the two balls together remains constant.

Figure 6-6
(a) Before the collision, ball A has momentum $\mathbf{p_A}$ and ball B has no momentum. (b) After the collision, ball B gains momentum $\mathbf{p_B}$.

Section 6-2

Visual Strategy

Figure 6-6
Point out to students that the two soccer balls interact by physically colliding.

Q How do the force exerted on ball A and the time interval over which it is exerted compare with the force exerted on ball B and its corresponding time interval?

A *The forces are equal in magnitude and opposite in direction (Newton's third law), and the time intervals are also equal.*

Q Using your answer to the previous question, determine the change in momentum of each ball.

A *The change in momentum of each ball must be equal in magnitude but opposite in direction because* $\Delta\mathbf{p} = \mathbf{F}\Delta t$.

Table 6-1	**Momentum in a collision** ⊛TEKS 2C, 5D					
	Ball A			**Ball B**		
	Mass	Velocity	Momentum	Mass	Velocity	Momentum
before collision	0.47 kg	0.84 m/s	0.40 kg•m/s	0.47 kg	0 m/s	0 kg•m/s
after collision	0.47 kg	0.04 m/s	0.02 kg•m/s	0.47 kg	0.80 m/s	0.38 kg•m/s

⊛TEKS
p. 214: 3A, 2C, 3B, 5C
p. 215: 2C, 5D

ANSWERS TO

Conceptual Challenge

1. No, the only possible way for the skaters' final total momentum to be zero is if the initial total momentum is also zero. This could happen only if both skaters initially have the same momentum but in opposite directions.

2. According to the principle of conservation of momentum, the momentum of the spacecraft and its fuel together before the rockets are fired must equal the momentum of the two after the rockets are fired. If both begin at rest, the total initial momentum is zero. When the rockets are fired, the combustion of the fuel gives the fuel momentum in a certain direction. The spacecraft will gain a momentum equal in magnitude but opposite in direction to keep the total momentum zero.

Teaching Tip

Explain to the students that the cumulative effects of frictional forces during the collision are very small if we consider the system immediately before and immediately after the collision. With this assumption, we can consider momentum to be conserved. If longer periods of time are considered, frictional forces do become significant.

★TEKS

p. 216: 3B, 3B, 3A, 5D, 5D
p. 217: 2C, 3C

internet**connect**

SCi**LINKS**
NSTA
TOPIC: Rocketry
GO TO: www.scilinks.org
*sci*LINKS CODE: HF2062

Conceptual Challenge

1. Ice skating

If a reckless ice skater collides with another skater who is standing on the ice, is it possible for both skaters to be at rest after the collision?

2. Space travel

A spacecraft undergoes a change of velocity when its rockets are fired. How does the spacecraft change velocity in empty space, where there is nothing for the gases emitted by the rockets to push against?

★TEKS 3A

In other words, the momentum of ball A plus the momentum of ball B before the collision is equal to the momentum of ball A plus the momentum of ball B after the collision.

$$\mathbf{p_{A,i}} + \mathbf{p_{B,i}} = \mathbf{p_{A,f}} + \mathbf{p_{B,f}}$$ ★TEKS 3B

This relationship is true for all interactions between isolated objects and is known as the *law of conservation of momentum*.

> **CONSERVATION OF MOMENTUM** ★TEKS 3B
>
> $$m_1\mathbf{v_{1,i}} + m_2\mathbf{v_{2,i}} = m_1\mathbf{v_{1,f}} + m_2\mathbf{v_{2,f}}$$
>
> **total initial momentum = total final momentum**

In its most general form, the law of conservation of momentum can be stated as follows:

The total momentum of all objects interacting with one another remains constant regardless of the nature of the forces between the objects.

Momentum is conserved in collisions

In the soccer-ball example, we found that the momentum of ball A does not remain constant and the momentum of ball B does not remain constant, but the total momentum of ball A and ball B does remain constant. In general, the total momentum remains constant for a system of objects that interact with one another. In this case, in which the floor is assumed to be frictionless, the soccer balls are the only two objects interacting. If a third object exerted a force on either ball A or ball B during the collision, the total momentum of ball A, ball B, and the third object would remain constant.

In this book, most conservation-of-momentum problems deal with only two isolated objects. However, when you use conservation of momentum to solve a problem or investigate a situation, it is important to include all objects that are involved in the interaction. Frictional forces—such as the frictional force between the soccer balls and the floor—will be disregarded in most conservation-of-momentum problems in this book. ★TEKS 5D

Momentum is conserved for objects pushing away from each other

Another example of conservation of momentum is when two or more interacting objects that initially have no momentum begin moving away from each other. Imagine that you initially stand at rest and then jump up, leaving the ground with a velocity **v.** Obviously, *your* momentum is not conserved; before the jump, it was zero, and it became *m***v** as you began to rise. However, the total momentum remains constant if you include Earth in your analysis. The total momentum for you and Earth remains constant. ★TEKS 5D

If your momentum after you jump is 60 kg•m/s upward, then Earth must have a corresponding momentum of 60 kg•m/s downward, because total momentum is conserved. However, because Earth has an enormous mass (6×10^{23} kg), that momentum corresponds to a tiny velocity (1×10^{-23} m/s).

Imagine two skaters pushing away from each other, as shown in **Figure 6-7.** The skaters are both initially at rest with a momentum of $\mathbf{p_{1,i}} = \mathbf{p_{2,i}} = 0$. When they push away from each other, they move in opposite directions with equal but opposite momentum so that the total final momentum is also zero ($\mathbf{p_{1,f}} + \mathbf{p_{2,f}} = 0$). ⭐TEKS 2C

(a)　　　　　　　　(b)

Figure 6-7

(a) When the skaters stand facing each other, both skaters have zero momentum, so the total momentum of both skaters is zero.

(b) When the skaters push away from each other, their momentum is equal but opposite, so the total momentum is still zero.

Consumer Focus *Surviving a Collision*

Pucks and carts collide in physics labs all the time with little damage. But when cars collide on a freeway, the resulting rapid change in speed can cause injury or death to the drivers and any passengers.

Many types of collisions are dangerous, but head-on collisions involve the greatest accelerations and thus the greatest forces. When two cars going 100 km/h (62 mi/h) collide head-on, each car dissipates the same amount of kinetic energy that it would dissipate if it hit the ground after being dropped from the roof of a 12-story building.

The key to many automobile-safety features is the concept of impulse. One way today's cars make use of the concept of impulse is by crumpling during impact. Pliable sheet metal and frame structures absorb energy until the force reaches the passenger compartment, which is built of rigid metal for protection. Because the crumpling slows the car gradually, it is an important factor in keeping the driver alive.

Even taking into account this built-in safety feature, the National Safety Council estimates that high-speed collisions involve accelerations of 20 times the free-fall acceleration. In other words, an 89 N (20 lb) infant could experience a force of 1780 N (400 lb) in a collision. If you are holding a baby in your lap during a collision, it is very likely that these large forces will break your hold on the baby. Because of inertia, the baby will continue at the car's original velocity and collide with the front windshield.

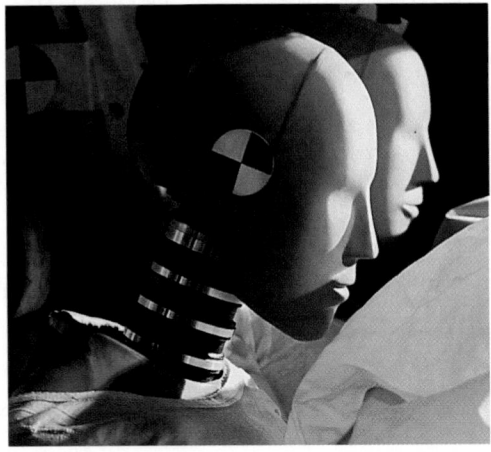

Seat belts are necessary to protect a body from forces of such large magnitudes. They stretch and extend the time it takes a passenger's body to stop, thereby reducing the force on the person. Seat belts also prevent passengers from hitting the inside frame of the car. During a collision, a person not wearing a seat belt is likely to hit the windshield, the steering wheel, or the dashboard—often with traumatic results. ⭐TEKS 3C

*The following may be used
as a teamwork exercise or for
demonstration at the chalkboard
or on an overhead projector.*

PROBLEM

Conservation of momentum

A 0.40 kg ball approaches a wall
perpendicularly at 15 m/s. It collides with the wall and rebounds
with an equal speed in the opposite direction. Calculate the
impulse exerted on the wall.

Answer

-12.0 kg•m/s

**Interactive
Problem-
Solving
Tutor**

See Module 7
"Conservation of Momentum"
promotes additional development of problem-solving skills
for this chapter.

SAMPLE PROBLEM 6D

Conservation of momentum ⭐TEKS 2C, 3B, 5D

PROBLEM

A 76 kg boater, initially at rest in a stationary 45 kg boat, steps out of the boat and onto the dock. If the boater moves out of the boat with a velocity of 2.5 m/s to the right, what is the final velocity of the boat?

SOLUTION

1. DEFINE   **Given:** $m_1 = 76$ kg $m_2 = 45$ kg $v_{1,i} = 0$

$v_{2,i} = 0$ $v_{1,f} = 2.5$ m/s to the right

Unknown: $v_{2,f} = ?$

Diagram: $m_1 = 76$ kg $v_{1,f} = 2.5$ m/s

$m_2 = 45$ kg

2. PLAN   **Choose an equation or situation:** Because the total momentum of an isolated system remains constant, the total initial momentum of the boater and
the boat will be equal to the total final momentum of the boater and the boat.

$$m_1 v_{1,i} + m_2 v_{2,i} = m_1 v_{1,f} + m_2 v_{2,f}$$

Because the boater and the boat are initially at rest, the total initial momentum of the system is equal to zero.

$$m_1 v_{1,i} + m_2 v_{2,i} = 0$$

Therefore, the final momentum of the system must also be equal to zero.

$$m_1 v_{1,f} + m_2 v_{2,f} = 0$$

3. CALCULATE   **Substitute the values into the equation(s) and solve:**

$$m_1 v_{1,f} + m_2 v_{2,f} = (76 \text{ kg} \times 2.5 \text{ m/s}) + (45 \text{ kg} \times v_{2,f})$$

$$190 \text{ kg•m/s} + 45 \text{ kg}(v_{2,f}) = 0$$

$$45 \text{ kg}(v_{2,f}) = -190 \text{ kg•m/s}$$

$$v_{2,f} = \frac{-190 \text{ kg•m/s}}{45 \text{ kg}}$$

$$\boxed{v_{2,f} = -4.2 \text{ m/s}}$$

4. EVALUATE   The negative sign for $v_{2,f}$ indicates that the boat is moving to the left, in the
direction *opposite* the motion of the boater.

$$\boxed{v_{2,f} = 4.2 \text{ m/s to the left}}$$

Conservation of momentum ⭐TEKS 2C, 3B, 5D

1. A 63.0 kg astronaut is on a spacewalk when the tether line to the shuttle breaks. The astronaut is able to throw a 10.0 kg oxygen tank in a direction away from the shuttle with a speed of 12.0 m/s, propelling the astronaut back to the shuttle. Assuming that the astronaut starts from rest, find the final speed of the astronaut after throwing the tank.

2. An 85.0 kg fisherman jumps from a dock into a 135.0 kg rowboat at rest on the west side of the dock. If the velocity of the fisherman is 4.30 m/s to the west as he leaves the dock, what is the final velocity of the fisherman and the boat?

3. Each croquet ball in a set has a mass of 0.50 kg. The green ball, traveling at 12.0 m/s, strikes the blue ball, which is at rest. Assuming that the balls slide on a frictionless surface and all collisions are head-on, find the final speed of the blue ball in each of the following situations:
   a. The green ball stops moving after it strikes the blue ball.
   b. The green ball continues moving after the collision at 2.4 m/s in the same direction.
   c. The green ball continues moving after the collision at 0.3 m/s in the same direction.

4. A boy on a 2.0 kg skateboard initially at rest tosses an 8.0 kg jug of water in the forward direction. If the jug has a speed of 3.0 m/s relative to the ground and the boy and skateboard move in the opposite direction at 0.60 m/s, find the boy's mass.

Newton's third law leads to conservation of momentum ⭐TEKS 5D

Consider two isolated bumper cars, m_1 and m_2, before and after they collide. Before the collision, the velocities of the two bumper cars are $\mathbf{v_{1,i}}$ and $\mathbf{v_{2,i}}$, respectively. After the collision, their velocities are $\mathbf{v_{1,f}}$ and $\mathbf{v_{2,f}}$, respectively. The impulse-momentum theorem, $\mathbf{F}\Delta t = \Delta\mathbf{p}$, describes the change in momentum of one of the bumper cars. Applied to m_1, the impulse-momentum theorem gives the following:

$$\mathbf{F_1}\Delta t = m_1\mathbf{v_{1,f}} - m_1\mathbf{v_{1,i}}$$

Likewise, for m_2 it gives the following:

$$\mathbf{F_2}\Delta t = m_2\mathbf{v_{2,f}} - m_2\mathbf{v_{2,i}}$$

Module 7
"Conservation of Momentum" provides an interactive lesson with guided problem-solving practice to teach you about momentum and momentum conservation.

PRACTICE GUIDE 6D

Solving for:		
v_f	📖 **PE**	Sample, 1–3; Ch. Rvw. 24–26, 46*, 52*, 53*, 57*, 59*
	💿 **PW**	5–7
	PB	5–7
v_i	💿 **PW**	3–4
	PB	Sample, 1–4
m	📖 **PE**	4
	💿 **PW**	Sample, 1–2
	PB	8–10

ANSWERS TO

Practice 6D
Conservation of momentum

1. 1.90 m/s
2. 1.66 m/s to the west
3. a. 12.0 m/s
   b. 9.6 m/s
   c. 11.7 m/s
4. 38 kg

Teaching Tip

A quick review of Newton's third law may help students better follow the derivation of the conservation of momentum in this section. Remind students that, according to Newton's third law, the force exerted by one body on another is equal in magnitude and opposite in direction to the force exerted on the first body by the second body.

⭐TEKS

p. 218: 2C, 3B, 5D
p. 219: 2C, 3B, 5D, 5D

Demonstration 2

Momentum and magnets

Purpose Demonstrate the principle of the conservation of momentum for objects acted on by field forces.

Materials 2 small horseshoe magnets, transparency graph paper, overhead projector

Procedure On the overhead projector, hold the two magnets centered on the origin of the axes so that like poles are together. Release the magnets, and have students observe the symmetrical pattern formed when the magnets stop. Point out that this pattern is due to equal and opposite momentums.

Teaching Tip

Remind students that conservation laws are valid only for a closed system. In the example of two bumper cars colliding, the system consists of the two cars. Most cases considered in this chapter involve just two objects in a collision, but a system can include any number of objects interacting with one another. All examples discussed in this chapter assume an isolated system unless stated otherwise.

⭐TEKS

p. 220: 3B, 3B
p. 221: 2C, 3B, 5D

Figure 6-8
Because of the collision, the force exerted on each bumper car causes a change in momentum for each car. The total momentum is the same before and after the collision.

$\mathbf{F_1}$ is the force that m_2 exerts on m_1 during the collision, and $\mathbf{F_2}$ is the force that m_1 exerts on m_2 during the collision, as shown in **Figure 6-8.** Because the only forces acting in the collision are the forces the two bumper cars exert on each other, Newton's third law tells us that the force on m_1 is equal to and opposite the force on m_2 ($\mathbf{F_1} = -\mathbf{F_2}$). Additionally, the two forces act over the same time interval, Δt. Therefore, the force m_2 exerts on m_1 multiplied by the time interval is equal to the force m_1 exerts on m_2 multiplied by the time interval, or $\mathbf{F_1}\Delta t = -\mathbf{F_2}\Delta t$. That is, the impulse on m_1 is equal to and opposite the impulse on m_2. This relationship is true in every collision or interaction between two isolated objects.

Before collision

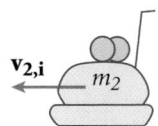

After collision

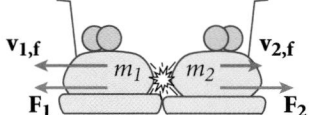

Because impulse is equal to the change in momentum, and the impulse on m_1 is equal to and opposite the impulse on m_2, the change in momentum of m_1 is equal to and opposite the change in momentum of m_2. This means that in every interaction between two isolated objects, the change in momentum of the first object is equal to and opposite the change in momentum of the second object. In equation form, this is expressed by the following equation.

$$m_1\mathbf{v_{1,f}} - m_1\mathbf{v_{1,i}} = -(m_2\mathbf{v_{2,f}} - m_2\mathbf{v_{2,i}})$$ ⭐TEKS **3B**

This equation means that if the momentum of one object increases after a collision, then the momentum of the other object in the situation must decrease by an equal amount. Rearranging this equation gives the following equation for the conservation of momentum.

$$m_1\mathbf{v_{1,i}} + m_2\mathbf{v_{2,i}} = m_1\mathbf{v_{1,f}} + m_2\mathbf{v_{2,f}}$$ ⭐TEKS **3B**

Forces in real collisions are not constant

As mentioned in Section 6-1, the forces involved in a collision are treated as though they are constant. In a real collision, however, the forces may vary in time in a complicated way. **Figure 6-9** shows the forces acting during the collision of the two bumper cars. At all times during the collision, the forces on the two cars are equal and opposite in direction. However, the magnitudes of the forces change throughout the collision—increasing, reaching a maximum, and then decreasing.

When solving impulse problems, you should use the average force during the collision as the value for force. In Chapter 2, you learned that the average velocity of an object undergoing a constant acceleration is equal to the constant velocity required for the object to travel the same displacement in the same time interval. Similarly, the average force during a collision is equal to the constant force required to cause the same change in momentum as the real, changing force.

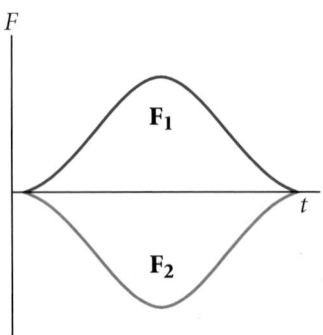

Figure 6-9
This graph shows the force on each bumper car during the collision. Although both forces vary with time, $\mathbf{F_1}$ and $\mathbf{F_2}$ are always equal in magnitude and opposite in direction.

Section Review

1. A 44 kg student on in-line skates is playing with a 22 kg exercise ball. Disregarding friction, explain what happens during the following situations.

   a. The student is holding the ball, and both are at rest. The student then throws the ball horizontally, causing the student to glide back at 3.5 m/s.

   b. Explain what happens to the ball in part (a) in terms of the momentum of the student and the momentum of the ball.

   c. The student is initially at rest. The student then catches the ball, which is initially moving to the right at 4.6 m/s.

   d. Explain what happens in part (c) in terms of the momentum of the student and the momentum of the ball.

2. A boy stands at one end of a floating raft that is stationary relative to the shore. He then walks in a straight line to the opposite end of the raft, away from the shore.

   a. Does the raft move? Explain.

   b. What is the total momentum of the boy and the raft before the boy walks across the raft?

   c. What is the total momentum of the boy and the raft after the boy walks across the raft?

3. High-speed stroboscopic photographs show the head of a 215 g golf club traveling at 55.0 m/s just before it strikes a 46 g golf ball at rest on a tee. After the collision, the club travels (in the same direction) at 42.0 m/s. Use the law of conservation of momentum to find the speed of the golf ball just after impact.

4. Two isolated objects have a head-on collision. For each of the following questions, explain your answer.

   a. If you know the change in momentum of one object, can you find the change in momentum of the other object?

   b. If you know the initial and final velocity of one object and the mass of the other object, do you have enough information to find the final velocity of the second object?

   c. If you know the masses of both objects and the final velocities of both objects, do you have enough information to find the initial velocities of both objects?

   d. If you know the masses and initial velocities of both objects and the final velocity of one object, do you have enough information to find the final velocity of the other object?

   e. If you know the change in momentum of one object and the initial and final velocities of the other object, do you have enough information to find the mass of either object?

Section Review ANSWERS

1. a. The ball will move away at 7.0 m/s.
   b. The momentum gained by the ball must be equal to and opposite the momentum gained by the student.
   c. The student and the ball will move to the right at 1.5 m/s.
   d. The student's initial momentum is zero. When the student catches the ball, some of the ball's momentum is transferred to the student.

2. a. yes; The total initial momentum is zero, so the boy and the raft must move in opposite directions to conserve momentum.
   b. zero
   c. zero

3. 61 m/s

4. a. Yes, the momentum lost by one object must equal the momentum gained by the other object.
   b. No, $v_{2,f}$ also depends on $v_{2,i}$ and m_1.
   c. No, using the conservation of momentum, you could only find a relationship between $v_{1,i}$ and $v_{2,i}$.
   d. Yes, using the conservation of momentum, you could substitute the given values and solve for v_f.
   e. Using the conservation of momentum, you could find m_1 if $v_{1,i}$ and $v_{1,f}$ are given, but you would need $v_{2,i}$ and $v_{2,f}$ to find m_2.

Inelastic collisions

Purpose Show the conservation of momentum in an inelastic collision.

Materials two balls with the same mass, string, tape, small piece of modeling clay, meterstick, paper or chalkboard

Procedure Tie a piece of string around each ball, using tape if necessary. Hold the two strings so that the balls hang at the same height in front of either the chalkboard or a length of paper taped to the wall. Place the clay on one of the balls so that the clay will hold the balls together when they collide. Hold up one of the balls, and have a student mark its displacement on the paper or chalkboard.

Release the ball. It should stick to the second ball; both balls should move together. Have a student mark the displacement of the two balls after the collision on the paper or chalkboard. Measure the two displacements with the meterstick. If momentum is conserved, the height of the two balls together will be $\frac{1}{4}$ the original height. Explain to the students that according to the conservation of momentum, $m_1\mathbf{v_{1,i}} + m_2\mathbf{v_{2,i}} = (m_1 + m_2)\mathbf{v_f}$ for a perfectly inelastic collision. Thus, since the second ball starts at rest, the final velocity of the two balls will be half the initial velocity of the first ball. Because the kinetic energy at the bottom of the swing equals the potential energy at the top $(mgh = \frac{1}{2}mv^2)$, the two balls should reach $\frac{1}{4}$ the initial height of the first ball.

6-3
Elastic and inelastic collisions

6-3 SECTION OBJECTIVES

- **Identify different types of collisions.**
- **Determine the changes in kinetic energy during perfectly inelastic collisions.**
- **Compare conservation of momentum and conservation of kinetic energy in perfectly inelastic and elastic collisions.**
- **Find the final velocity of an object in perfectly inelastic and elastic collisions.**

perfectly inelastic collision

a collision in which two objects stick together and move with a common velocity after colliding

COLLISIONS

As you go about your day-to-day activities, you probably witness many collisions without really thinking about them. In some collisions, two objects collide and stick together so that they travel together after the impact. An example of this is a collision between an arrow and a target, as shown in **Figure 6-10.** The arrow, sailing forward, collides with the target at rest. In an isolated system, the target and the arrow would both move together after the collision with a momentum equal to their combined momentum before the collision. In other collisions, like a collision between a tennis racquet and a tennis ball, two objects collide and bounce so that they move away with two different velocities.

The total momentum remains constant in any type of collision. However, the total kinetic energy is generally not conserved in a collision because some kinetic energy is converted to internal energy when the objects deform. In this section, we will examine different types of collisions and determine whether kinetic energy is conserved in each type. We will primarily explore two extreme types of collisions, elastic and perfectly inelastic. ⭐TEKS 5B

Perfectly inelastic collisions

When two objects collide and move together as one mass, like the arrow and the target, the collision is called **perfectly inelastic.** Likewise, if a meteorite collides head on with Earth, it becomes buried in Earth and the collision is nearly perfectly inelastic.

Figure 6-10
When an arrow pierces a target and remains stuck in the target, the arrow and target have undergone a perfectly inelastic collision (assuming no debris is thrown out).

Perfectly inelastic collisions are easy to analyze in terms of momentum because the objects become essentially one object after the collision. The final mass is equal to the combined mass of the two objects, and they move with the same velocity after colliding.

Consider two cars of masses m_1 and m_2 moving with initial velocities of $\mathbf{v_{1,i}}$ and $\mathbf{v_{2,i}}$ along a straight line, as shown in **Figure 6-11.** The two cars stick together and move with some common velocity, $\mathbf{v_f}$, along the same line of motion after the collision. The total momentum of the two cars before the collision is equal to the total momentum of the two cars after the collision.

(a)

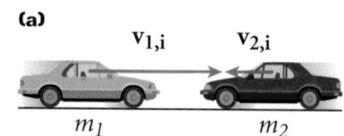

$\mathbf{v_{1,i}}$ $\mathbf{v_{2,i}}$

m_1 m_2

PERFECTLY INELASTIC COLLISION ⭐TEKS **3B**

$$m_1\mathbf{v_{1,i}} + m_2\mathbf{v_{2,i}} = (m_1 + m_2)\,\mathbf{v_f}$$

This simplified version of the equation for conservation of momentum is useful in analyzing perfectly inelastic collisions. When using this equation, it is important to pay attention to signs that indicate direction. In **Figure 6-11,** $\mathbf{v_{1,i}}$ has a positive value (m_1 moving to the right), while $\mathbf{v_{2,i}}$ has a negative value (m_2 moving to the left).

(b)

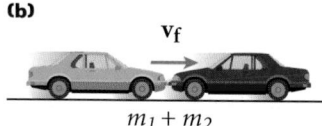

$\mathbf{v_f}$

$m_1 + m_2$

Figure 6-11
The total momentum of the two cars before the collision **(a)** is the same as the total momentum of the two cars after the inelastic collision **(b).**

SAMPLE PROBLEM 6E

Perfectly inelastic collisions ⭐TEKS 2C, 3B, 5D

PROBLEM

A 1850 kg luxury sedan stopped at a traffic light is struck from the rear by a compact car with a mass of 975 kg. The two cars become entangled as a result of the collision. If the compact car was moving at a velocity of 22.0 m/s to the north before the collision, what is the velocity of the entangled mass after the collision?

SOLUTION

Given: $m_1 = 1850$ kg $m_2 = 975$ kg $\mathbf{v_{1,i}} = 0$ m/s
             $\mathbf{v_{2,i}} = 22.0$ m/s to the north

Unknown: $\mathbf{v_f} = ?$

Use the equation for a perfectly inelastic collision.

$$m_1\mathbf{v_{1,i}} + m_2\mathbf{v_{2,i}} = (m_1 + m_2)\,\mathbf{v_f}$$

$$\mathbf{v_f} = \frac{m_1\mathbf{v_{1,i}} + m_2\mathbf{v_{2,i}}}{m_1 + m_2}$$

$$\mathbf{v_f} = \frac{(1850\text{ kg})(0\text{ m/s}) + (975\text{ kg})(22.0\text{ m/s})}{1850\text{ kg} + 975\text{ kg}} = \frac{2.14 \times 10^4\text{ kg}\cdot\text{m/s}}{2820\text{ kg}}$$

$$\boxed{\mathbf{v_f} = 7.59\text{ m/s to the north}}$$

Classroom Practice

The following may be used as teamwork exercises or for demonstration at the chalkboard or on an overhead projector.

PROBLEM

Perfectly inelastic collisions

An empty train car moving east at 21 m/s collides with a loaded train car initially at rest that has twice the mass of the empty car. The two cars stick together.

a. Find the velocity of the two cars after the collision.

b. Find the final speed if the loaded car moving at 17 m/s had hit the empty car initially at rest.

Answer
    a. 7.0 m/s to the east
    b. 11 m/s

An empty train car moving at 15 m/s collides with a loaded car of three times the mass moving at one-third the speed of the empty car. The cars stick together. Find the speed of the cars after the collision.

Answer
    7.5 m/s

⭐TEKS

p. 222: 5B
p. 223: 3B, 2C, 3B, 5D

ANSWERS TO

Practice 6E
Perfectly inelastic collisions

1. 3.8 m/s to the south
2. 1.8 m/s
3. 4.27 m/s to the north
4. 4.2 m/s to the right
5. **a.** 3.0 kg
   b. 5.32 m/s

Misconception Alert

Some students may confuse momentum (mv) and kinetic energy ($\frac{1}{2}mv^2$). Be sure students understand that in an inelastic collision *total* momentum and *total* energy are conserved but kinetic energy is *not* conserved (because some is converted to other forms of energy). Some students may also think that because the collision is perfectly inelastic, *all* of the initial kinetic energy is converted to other forms of energy. Use examples to show that this is not necessarily the case.

PRACTICE 6E

Perfectly inelastic collisions ⭐TEKS 2C, 3B, 5D

1. A 1500 kg car traveling at 15.0 m/s to the south collides with a 4500 kg truck that is initially at rest at a stoplight. The car and truck stick together and move together after the collision. What is the final velocity of the two-vehicle mass?

2. A grocery shopper tosses a 9.0 kg bag of rice into a stationary 18.0 kg grocery cart. The bag hits the cart with a horizontal speed of 5.5 m/s toward the front of the cart. What is the final speed of the cart and bag?

3. A 1.50×10^4 kg railroad car moving at 7.00 m/s to the north collides with and sticks to another railroad car of the same mass that is moving in the same direction at 1.50 m/s. What is the velocity of the joined cars after the collision?

4. A dry cleaner throws a 22 kg bag of laundry onto a stationary 9.0 kg cart. The cart and laundry bag begin moving at 3.0 m/s to the right. Find the velocity of the laundry bag before the collision.

5. A 47.4 kg student runs down the sidewalk and jumps with a horizontal speed of 4.20 m/s onto a stationary skateboard. The student and skateboard move down the sidewalk with a speed of 3.95 m/s. Find the following:
   a. the mass of the skateboard
   b. how fast the student would have to jump to have a final speed of 5.00 m/s

CONCEPT PREVIEW

Internal energy will be discussed in Chapter 10.

Kinetic energy is not constant in inelastic collisions ⭐TEKS 5B

In an inelastic collision, the total kinetic energy does not remain constant when the objects collide and stick together. Some of the kinetic energy is converted to sound energy and internal energy as the objects deform during the collision.

This phenomenon helps make sense of the special use of the words *elastic* and *inelastic* in physics. We normally think of *elastic* as referring to something that always returns to, or keeps, its original shape. In physics, the most important characteristic of an elastic collision is that the objects maintain their original shapes and are not deformed by the action of forces. Objects in an *inelastic* collision, on the other hand, are deformed during the collision and lose some kinetic energy.

The decrease in the total kinetic energy during an inelastic collision can be calculated using the formula for kinetic energy from Chapter 5, as shown in Sample Problem 6F. It is important to remember that not all of the initial kinetic energy is necessarily lost in a perfectly inelastic collision.

Kinetic energy in perfectly inelastic collisions ⭐TEKS 2C, 3B, 5B

PROBLEM

Two clay balls collide head-on in a perfectly inelastic collision. The first ball has a mass of 0.500 kg and an initial velocity of 4.00 m/s to the right. The mass of the second ball is 0.250 kg, and it has an initial velocity of 3.00 m/s to the left. What is the final velocity of the composite ball of clay after the collision? What is the decrease in kinetic energy during the collision?

SOLUTION

Given: $m_1 = 0.500$ kg $m_2 = 0.250$ kg $v_{1,i} = 4.00$ m/s to the right $= +4.00$ m/s
$v_{2,i} = 3.00$ m/s to the left $= -3.00$ m/s

Unknown: $v_f = ?$ $\Delta KE = ?$

Use the equation for perfectly inelastic collisions from page 223.

$$m_1 v_{1,i} + m_2 v_{2,i} = (m_1 + m_2)\, v_f$$

$$v_f = \frac{m_1 v_{1,i} + m_2 v_{2,i}}{m_1 + m_2}$$

$$v_f = \frac{(0.500 \text{ kg})(4.00 \text{ m/s}) + (0.250 \text{ kg})(-3.00 \text{ m/s})}{0.500 \text{ kg} + 0.250 \text{ kg}}$$

$$v_f = \frac{1.25 \text{ kg} \cdot \text{m/s}}{0.750 \text{ kg}}$$

$$\boxed{v_f = 1.67 \text{ m/s to the right}}$$

Use the equation for kinetic energy from Chapter 5.

Initial: $KE_i = KE_{1,i} + KE_{2,i}$

$KE_i = \frac{1}{2} m_1 v_{1,i}^2 + \frac{1}{2} m_2 v_{2,i}^2$

$KE_i = \frac{1}{2}(0.500 \text{ kg})(4.00 \text{ m/s})^2 + \frac{1}{2}(0.250 \text{ kg})(-3.00 \text{ m/s})^2$

$KE_i = 5.12$ J

Final: $KE_f = KE_{1,f} + KE_{2,f}$

$KE_f = \frac{1}{2}(m_1 + m_2) v_f^2$

$KE_f = \frac{1}{2}(0.750 \text{ kg})(1.67 \text{ m/s})^2$

$KE_f = 1.05$ J

$\Delta KE = KE_f - KE_i = 1.05 \text{ J} - 5.12 \text{ J}$

$$\boxed{\Delta KE = -4.07 \text{ J}}$$

Classroom Practice

The following may be used as teamwork exercises or for demonstration at the chalkboard or on an overhead projector.

PROBLEM

A clay ball with a mass of 0.35 kg hits another 0.35 kg ball at rest, and the two stick together. The first ball has an initial speed of 4.2 m/s.

a. What is the final speed of the balls?

b. Calculate the decrease in kinetic energy that occurs during the collision.

c. What percentage of the kinetic energy is converted to other forms of energy?

Answers

  a. 2.1 m/s

  b. 1.6 J

  c. 52 percent

A 0.75 kg ball moving at 3.8 m/s to the right strikes an identical ball moving at 3.8 m/s to the left. The balls stick together after the collision and stop. What percentage of the initial kinetic energy is converted to other forms?

Answer

100 percent

⭐TEKS

p. 224: 2C, 3B, 5D, 5B
p. 225: 2C, 3B, 5B

PRACTICE GUIDE 6F

Solving for:		
ΔKE	📖 **PE**	Sample, 1–3; Ch. Rvw. 34, 35, 49b*
	💿 **PW**	Sample, 1–7
	PB	Sample, 1–10

ANSWERS TO

Practice 6F

1. **a.** 0.43 m/s to the west
   b. 17 J
2. **a.** 6.2 m/s to the south
   b. 3 J
3. **a.** 4.6 m/s to the south
   b. 3.9×10^3 J

Key Models and Analogies

Just as friction is often disregarded to simplify situations, in a nearly elastic collision the decrease in kinetic energy can be disregarded so that the ideal case of an elastic collision can be used.

Teaching Tip

Discuss a variety of examples of collisions with students. For each example, ask whether the collision is closer to an elastic collision or to a perfectly inelastic collision. Also ask students where kinetic energy is converted to other forms of energy in each of the different examples.

PRACTICE 6F

Kinetic energy in perfectly inelastic collisions ⭐TEKS 2C, 3B, 5B, 5D

1. A 0.25 kg arrow with a velocity of 12 m/s to the west strikes and pierces the center of a 6.8 kg target.
   a. What is the final velocity of the combined mass?
   b. What is the decrease in kinetic energy during the collision?

2. During practice, a student kicks a 0.40 kg soccer ball with a velocity of 8.5 m/s to the south into a 0.15 kg bucket lying on its side. The bucket travels with the ball after the collision.
   a. What is the final velocity of the combined mass?
   b. What is the decrease in kinetic energy during the collision?

3. A 56 kg ice skater traveling at 4.0 m/s to the north suddenly grabs the hand of a 65 kg skater traveling at 12.0 m/s in the opposite direction as they pass. Without rotating, the two skaters continue skating together with joined hands.
   a. What is the final velocity of the two skaters?
   b. What is the decrease in kinetic energy during the collision?

ELASTIC COLLISIONS ⭐TEKS 5B

When a player kicks a soccer ball, the collision between the ball and the player's foot is much closer to elastic than the collisions we have studied so far. In this case, *elastic* means that the ball and the player's foot remain separate after the collision.

In an **elastic collision**, two objects collide and return to their original shapes with no change in total kinetic energy. After the collision, the two objects move separately. In an elastic collision, both the total momentum and the total kinetic energy remain constant.

elastic collision

a collision in which the total momentum and the total kinetic energy remain constant

internetconnect

SCI**LINKS**

NSTA

TOPIC: Collisions
GO TO: www.scilinks.org
sciLINKS CODE: HF2064

Most collisions are neither elastic nor perfectly inelastic ⭐TEKS 5B

In the everyday world, most collisions are not perfectly inelastic. That is, colliding objects do not usually stick together and continue to move as one object. However, most collisions are not elastic, either. Even nearly elastic collisions, such as those between billiard balls or between a football player's foot and the ball, result in some decrease in kinetic energy. For example, a football deforms when it is kicked. During this deformation, some of the kinetic energy is converted to internal elastic potential energy. In most collisions, some of the kinetic energy is also converted into sound, such as the click of billiard balls colliding. In fact, any collision that produces sound is not elastic; the sound represents a decrease in kinetic energy.

Elastic and perfectly inelastic collisions are limiting cases; most collisions actually fall into a category between these two extremes. In this third category of collisions, called *inelastic collisions,* the colliding objects bounce and move separately after the collision, but the total kinetic energy decreases in the collision. *For the problems in this book, we will consider all collisions in which the objects do not stick together to be elastic collisions.* This means that we will assume that the total momentum and the total kinetic energy remain constant in all collisions that are not perfectly inelastic.

Kinetic energy is conserved in elastic collisions ⓣTEKS 5B, 5D

Figure 6-12 shows an elastic head-on collision between two soccer balls of equal mass. Assume, as in earlier examples, that the balls are isolated on a frictionless surface and that they do not rotate. The first ball is moving to the right when it collides with the second ball, which is moving to the left. When considered as a whole, the entire system has momentum to the left.

After the elastic collision, the first ball moves to the left and the second ball moves to the right. The magnitude of the momentum of the first ball, which is now moving to the left, is greater than the magnitude of the momentum of the second ball, which is now moving to the right. When considered together, the entire system has momentum to the left, just as before the collision.

Another example of a nearly elastic collision is the collision between a golf ball and a club. After a golf club strikes a stationary golf ball, the golf ball moves at a very high speed in the same direction as the golf club. The golf club continues to move in the same direction, but its velocity decreases so that the momentum lost by the golf club is equal to and opposite the momentum gained by the golf ball. If a collision is perfectly elastic, *the total momentum and the total kinetic energy remain constant throughout the collision.*

MOMENTUM AND KINETIC ENERGY REMAIN CONSTANT ⓣTEKS 3B
IN AN ELASTIC COLLISION

$$m_1 \mathbf{v_{1,i}} + m_2 \mathbf{v_{2,i}} = m_1 \mathbf{v_{1,f}} + m_2 \mathbf{v_{2,f}}$$

$$\tfrac{1}{2} m_1 v_{1,i}^2 + \tfrac{1}{2} m_2 v_{2,i}^2 = \tfrac{1}{2} m_1 v_{1,f}^2 + \tfrac{1}{2} m_2 v_{2,f}^2$$

Remember that v is positive if an object moves to the right and negative if it moves to the left.

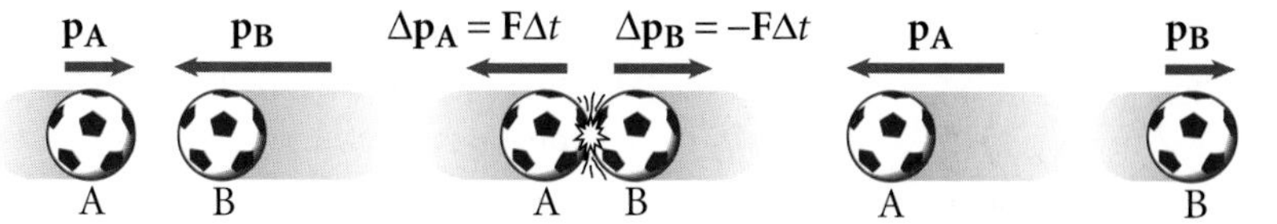

(a) Initial   **(b)** Impulse   **(c)** Final

p_A p_B $\Delta p_A = F \Delta t$ $\Delta p_B = -F \Delta t$ p_A p_B

A   B   A   B   A   B

Quick Lab

Elastic and Inelastic Collisions

MATERIALS LIST

✔ 2 or 3 small balls of different types

⚠ SAFETY CAUTION

Perform this lab in an open space, preferably outdoors, away from furniture and other people.

Drop one of the balls from shoulder height onto a hard-surfaced floor or sidewalk. Observe the motion of the ball before and after it collides with the ground. Next, throw the ball down from the same height. Perform several trials, giving the ball a different velocity each time. Repeat with the other balls.

During each trial, observe the height to which the ball bounces. Rate the collisions from most nearly elastic to most inelastic. Describe what evidence you have for or against conservation of kinetic energy and conservation of momentum for each collision.

ⓣTEKS 3B

Quick Lab

TEACHER'S NOTES

The purpose of this lab is to show that in any collision, the elasticity of the materials involved affects the changes in kinetic energy. Test the balls before the lab in order to ensure a noticeable difference in elasticity. An interesting contrast can be observed by comparing new tennis balls with older ones.

Teaching Tip

Point out to students that they should recognize the first equation in the box. This equation, which expresses the principle of conservation of momentum, holds for both types of collisions. The conservation of kinetic energy, on the other hand, which is expressed by the second equation in the box, is valid only for elastic collisions.

Figure 6-12
In an elastic collision like this one **(b)**, both objects return to their original shapes and move separately after the collision **(c)**.

ⓣTEKS

p. 226: 2C, 3B, 5B, 5D, 5B, 5B
p. 227: 5B, 5D, 3B

Classroom Practice

The following may be used as teamwork exercises or for demonstration at the chalkboard or on an overhead projector.

PROBLEM

Elastic collisions

Two billiard balls, each with a mass of 0.35 kg, strike each other head-on. One ball is initially moving left at 4.1 m/s and ends up moving right at 3.5 m/s. The second ball is initially moving to the right at 3.5 m/s. Assume that neither ball rotates before or after the collision and that both balls are moving on a frictionless surface. Find the final velocity of the second ball.

Answer

4.1 m/s to the left

Two nonrotating balls on a frictionless surface collide elastically head on. The first ball has a mass of 15 g and an initial velocity of 3.5 m/s to the right, while the second ball has a mass of 22 g and an initial velocity of 4.0 m/s to the left. The final velocity of the 15 g ball is 5.4 m/s to the left. What is the final velocity of the 22 g ball?

Answer

2.0 m/s to the right

Elastic collisions ⊛TEKS 2C, 3B, 5B, 5D

PROBLEM

A 0.015 kg marble moving to the right at 0.225 m/s makes an elastic head-on collision with a 0.030 kg shooter marble moving to the left at 0.180 m/s. After the collision, the smaller marble moves to the left at 0.315 m/s. Assume that neither marble rotates before or after the collision and that both marbles are moving on a frictionless surface. What is the velocity of the 0.030 kg marble after the collision?

SOLUTION

1. DEFINE   **Given:** $m_1 = 0.015$ kg $m_2 = 0.030$ kg

$\mathbf{v_{1,i}} = 0.225$ m/s to the right $= +0.225$ m/s

$\mathbf{v_{2,i}} = 0.180$ m/s to the left $= -0.180$ m/s

$\mathbf{v_{1,f}} = 0.315$ m/s to the left $= -0.315$ m/s

Unknown: $\mathbf{v_{2,f}} = ?$

Diagram:

0.225 m/s   −0.180 m/s

m_1
0.015 kg

m_2
0.030 kg

2. PLAN   **Choose an equation or situation:** Use the equation for the conservation of momentum to find the final velocity of m_2, the 0.030 kg marble.

$$m_1\mathbf{v_{1,i}} + m_2\mathbf{v_{2,i}} = m_1\mathbf{v_{1,f}} + m_2\mathbf{v_{2,f}}$$

Rearrange the equation(s) to solve for the unknown(s): Rearrange the equation to isolate the final velocity of m_2.

$$m_2\mathbf{v_{2,f}} = m_1\mathbf{v_{1,i}} + m_2\mathbf{v_{2,i}} - m_1\mathbf{v_{1,f}}$$

$$\mathbf{v_{2,f}} = \frac{m_1\mathbf{v_{1,i}} + m_2\mathbf{v_{2,i}} - m_1\mathbf{v_{1,f}}}{m_2}$$

3. CALCULATE   **Substitute the values into the equation(s) and solve:** The rearranged conservation-of-momentum equation will allow you to isolate and solve for the final velocity.

$$\mathbf{v_{2,f}} = \frac{(0.015 \text{ kg})(0.225 \text{ m/s}) + (0.030 \text{ kg})(-0.180 \text{ m/s}) - (0.015 \text{ kg})(-0.315 \text{ m/s})}{0.030 \text{ kg}}$$

$$\mathbf{v_{2,f}} = \frac{(3.4 \times 10^{-3} \text{ kg} \cdot \text{m/s}) + (-5.4 \times 10^{-3} \text{ kg} \cdot \text{m/s}) - (-4.7 \times 10^{-3} \text{ kg} \cdot \text{m/s})}{0.030 \text{ kg}}$$

$$\mathbf{v_{2,f}} = \frac{2.7 \times 10^{-3} \text{ kg} \cdot \text{m/s}}{3.0 \times 10^{-2} \text{ kg}}$$

$$\boxed{\mathbf{v_{2,f}} = 9.0 \times 10^{-2} \text{ m/s to the right}}$$

4. EVALUATE Confirm your answer by making sure kinetic energy is also conserved using these values.

Conservation of kinetic energy

$$\tfrac{1}{2}m_1v_{1,i}^2 + \tfrac{1}{2}m_2v_{2,i}^2 = \tfrac{1}{2}m_1v_{1,f}^2 + \tfrac{1}{2}m_2v_{2,f}^2$$

$$KE_i = \tfrac{1}{2}(0.015 \text{ kg})(0.225 \text{ m/s})^2 + \tfrac{1}{2}(0.030 \text{ kg})(-0.180 \text{ m/s})^2 =$$
$$8.7 \times 10^{-4} \text{ kg} \cdot \text{m}^2/\text{s}^2 = 8.7 \times 10^{-4} \text{ J}$$

$$KE_f = \tfrac{1}{2}(0.015 \text{ kg})(0.315 \text{ m/s})^2 + \tfrac{1}{2}(0.030 \text{ kg})(0.090 \text{ m/s})^2 =$$
$$8.7 \times 10^{-4} \text{ kg} \cdot \text{m}^2/\text{s}^2 = 8.7 \times 10^{-4} \text{ J}$$

Kinetic energy is conserved.

PRACTICE 6G

Elastic collisions ⊛TEKS 2C, 3B, 5B, 5D

1. A 0.015 kg marble sliding to the right at 22.5 cm/s on a frictionless surface makes an elastic head-on collision with a 0.015 kg marble moving to the left at 18.0 cm/s. After the collision, the first marble moves to the left at 18.0 cm/s.
   a. Find the velocity of the second marble after the collision.
   b. Verify your answer by calculating the total kinetic energy before and after the collision.

2. A 16.0 kg canoe moving to the left at 12 m/s makes an elastic head-on collision with a 4.0 kg raft moving to the right at 6.0 m/s. After the collision, the raft moves to the left at 22.7 m/s. Disregard any effects of the water.
   a. Find the velocity of the canoe after the collision.
   b. Verify your answer by calculating the total kinetic energy before and after the collision.

3. A 4.0 kg bowling ball sliding to the right at 8.0 m/s has an elastic head-on collision with another 4.0 kg bowling ball initially at rest. The first ball stops after the collision.
   a. Find the velocity of the second ball after the collision.
   b. Verify your answer by calculating the total kinetic energy before and after the collision.

4. A 25.0 kg bumper car moving to the right at 5.00 m/s overtakes and collides elastically with a 35.0 kg bumper car moving to the right. After the collision, the 25.0 kg bumper car slows to 1.50 m/s to the right, and the 35.0 kg car moves at 4.50 m/s to the right.
   a. Find the velocity of the 35 kg bumper car before the collision.
   b. Verify your answer by calculating the total kinetic energy before and after the collision.

PRACTICE GUIDE 6G

Solving for:		
v_f	📖 **PE**	Sample, 1–3; Ch. Rvw. 36–40, 50*, 55*
	💿 **PW**	Sample, 6–7
	PB	7–10
v_i	📖 **PE**	4
	💿 **PW**	Sample, 1–3
	PB	3–6
m	💿 **PW**	4–5
	PB	Sample, 1–2

ANSWERS TO

Practice 6G
Elastic collisions
1. a. 22.5 cm/s to the right
   b. $KE_i = 6.2 \times 10^{-4} \text{ J} = KE_f$
2. a. 4.7 m/s to the left
   b. $KE_i = 1.2 \times 10^3 \text{ J}$, $KE_f = 1.2 \times 10^3 \text{ J}$, so $KE_i = KE_f$
3. a. 8.0 m/s to the right
   b. $KE_i = 1.3 \times 10^2 \text{ J} = KE_f$
4. a. 2.0 m/s to the right
   b. $KE_i = 382 \text{ J} = KE_f$

⊛TEKS

p. 228: 2C, 3B, 5B, 5D
p. 229: 2C, 3B, 5B, 5D

Table 6-2

Point out that the third case (inelastic) contains elements of both ideal cases. Total *KE* is not conserved, as in perfectly inelastic collisions, but the two objects do separate from one another after the collision, as in perfectly elastic collisions.

Q What is common to all cases?

A *Momentum is conserved in each case.*

1. For elastic, answers may include billiard balls colliding, a soccer ball hitting a player's foot, or a tennis ball hitting a wall. For inelastic, answers may include a person catching a ball, a meteorite hitting Earth, or two clay balls colliding.

2. No, some *KE* is converted to sound energy and some is converted to internal elastic potential energy as the cars deform, so the collision cannot be elastic.

3. **a.** 0.92 m/s to the south
   b. 1.4×10^3 J

4. **a.** no; If the collision is perfectly elastic, total *KE* is conserved, but each object can gain or lose *KE*.
   b. no; Total **p** is conserved, but each object can gain or lose **p**.

5. **a.** 3.5 m/s
   b. 0 J
   c. 0 J

Table 6-2   Types of collisions

Type of collision	Diagram	What happens	Conserved quantity
perfectly inelastic		The two objects stick together after the collision so that their final velocities are the same.	momentum
elastic		The two objects bounce after the collision so that they move separately.	momentum kinetic energy
inelastic		The two objects deform during the collision so that the total kinetic energy decreases, but the objects move separately after the collision.	momentum

Section Review

(★TEKS) 2C, 3B, 5B, 5D

1. Give two examples of elastic collisions and two examples of perfectly inelastic collisions.

2. If two automobiles collide, they usually do not stick together. Does this mean the collision is elastic?

3. A 90.0 kg fullback moving south with a speed of 5.0 m/s has a perfectly inelastic collision with a 95.0 kg opponent running north at 3.0 m/s.
   a. Calculate the velocity of the players just after the tackle.
   b. Calculate the decrease in total kinetic energy as a result of the collision.

4. A rubber ball collides elastically with the sidewalk.
   a. Does each object have the same kinetic energy after the collision as it had before the collision? Explain.
   b. Does each object have the same momentum after the collision as it had before the collision? Explain.

5. **Physics in Action**   Two 0.40 kg soccer balls collide elastically in a head-on collision. The first ball starts at rest, and the second ball has a speed of 3.5 m/s. After the collision, the second ball is at rest.
   a. What is the final speed of the first ball?
   b. What is the kinetic energy of the first ball before the collision?
   c. What is the kinetic energy of the second ball after the collision?

CHAPTER 6
Summary

Chapter 6
Summary

KEY IDEAS

Section 6-1 Momentum and impulse

- Momentum is a vector quantity defined as the product of an object's mass and velocity, $\mathbf{p} = m\mathbf{v}$.
- A net external force applied constantly to an object for a certain time interval will cause a change in the object's momentum equal to the product of the force and the time interval, $\mathbf{F}\Delta t = \Delta\mathbf{p}$.
- The product of the constant applied force and the time interval during which the force is applied is called the impulse of the force for the time interval.

Section 6-2 Conservation of momentum

- In all interactions between isolated objects, momentum is conserved.
- In every interaction between two isolated objects, the change in momentum of the first object is equal to and opposite the change in momentum of the second object.

Section 6-3 Elastic and inelastic collisions

- In a perfectly inelastic collision, two objects stick together and move as one mass after the collision.
- Momentum is conserved but kinetic energy is not conserved in a perfectly inelastic collision.
- In an inelastic collision, kinetic energy is converted to internal elastic potential energy when the objects deform. Some kinetic energy is also converted to sound energy and internal energy.
- In an elastic collision, two objects return to their original shapes and move away from the collision separately.
- Both momentum and kinetic energy are conserved in an elastic collision.
- Few collisions are elastic or perfectly inelastic.

Variable symbols

Quantities		Units
$\mathbf{p}$	momentum	kg•m/s kilogram-meters per second
$\mathbf{F}\Delta t$	impulse	N•s Newton-seconds = kilogram-meters per second

KEY TERMS

elastic collision (p. 226)

impulse (p. 210)

momentum (p. 208)

perfectly inelastic collision (p. 222)

Teaching Tip

Ask students to prepare a concept map for the chapter. The concept map should include most of the vocabulary terms, along with other integral terms and concepts.

⭐ TEKS

p. 230: 2C, 3B, 5B, 5D
Review & Assess
pp. 232–237:
2A: Alt. Assess. 1
2B: Alt. Assess. 1
2C: 12–16, 24–26,
    31–59, Technology
    & Learning
2D: Alt. Assess. 2–3, 5
2E: Technology &
    Learning
2F: Alt. Assess. 1
3A: Alt. Assess. 5
3B: 12–16, 24–26,
    31–59, Technology
    & Learning
3C: Alt. Assess. 3
3E: 21; Alt. Assess. 4
4A: Technology &
    Learning
5B: 2–3, 27–29, 34–36,
    49; Alt. Assess. 2
5C: 1–2, 12–13,
    435D: 17–26,
    28–40, 44–58;
Alt. Assess. 4

CHAPTER 6
Review and Assess

MOMENTUM AND IMPULSE

Review questions

1. If an object is not moving, what is its momentum?

2. If a particle's kinetic energy is zero, what is its momentum?

3. If two particles have equal kinetic energies, do they have the same momentum? Explain.

4. Show that $\mathbf{F} = m\mathbf{a}$ and $\mathbf{F} = \dfrac{\Delta \mathbf{p}}{\Delta t}$ are equivalent.

Conceptual questions

5. A truck loaded with sand is moving down the highway in a straight path.
   a. What happens to the momentum of the truck if the truck's velocity is increasing?
   b. What happens to the momentum of the truck if sand leaks at a constant rate through a hole in the truck bed while the truck maintains a constant velocity?

6. Gymnasts always perform on padded mats. Use the impulse-momentum theorem to discuss how these mats protect the athletes.

7. When a car collision occurs, an air bag is inflated, protecting the passenger from serious injury. How does the air bag soften the blow? Discuss the physics involved in terms of momentum and impulse.

8. If you jump from a table onto the floor, are you more likely to be hurt if your legs are relaxed or if your legs are stiff and your knees are locked? Explain.

9. Consider a field of insects, all of which have essentially the same mass.
   a. If the total momentum of the insects is zero, what does this imply about their motion?
   b. If the total kinetic energy of the insects is zero, what does this imply about their motion?

10. Two students hold an open bed sheet loosely by its corners to form a "catching net." The instructor asks a third student to throw an egg into the middle of the sheet as hard as possible. Why doesn't the egg's shell break?

11. How do car bumpers that collapse on impact help protect a driver?

Practice problems

12. Calculate the linear momentum for each of the following cases:
    a. a proton with mass 1.67×10^{-27} kg moving with a velocity of 5.00×10^6 m/s straight up
    b. a 15.0 g bullet moving with a velocity of 325 m/s to the right
    c. a 75.0 kg sprinter running with a velocity of 10.0 m/s southwest
    d. Earth ($m = 5.98 \times 10^{24}$ kg) moving in its orbit with a velocity equal to 2.98×10^4 m/s forward
    (See Sample Problem 6A.)

13. What is the momentum of a 0.148 kg baseball thrown with a velocity of 35 m/s toward home plate? (See Sample Problem 6A.)

14. A 2.5 kg ball strikes a wall with a velocity of 8.5 m/s to the left. The ball bounces off with a velocity of 7.5 m/s to the right. If the ball is in contact with the wall for 0.25 s, what is the constant force exerted on the ball by the wall? (See Sample Problem 6B.)

15. A football punter accelerates a 0.55 kg football from rest to a speed of 8.0 m/s in 0.25 s. What constant force does the punter exert on the ball? (See Sample Problem 6B.)

16. A 0.15 kg baseball moving at +26 m/s is slowed to a stop by a catcher who exerts a constant force of −390 N. How long does it take this force to stop the ball? How far does the ball travel before stopping? (See Sample Problem 6C.)

CONSERVATION OF MOMENTUM

Review questions

17. Two skaters initially at rest push against each other so that they move in opposite directions. What is the total momentum of the two skaters when they begin moving? Explain.

18. In a collision between two soccer balls, momentum is conserved. Is momentum conserved for each soccer ball? Explain.

19. Explain how momentum is conserved when a ball bounces against a floor.

Conceptual questions

20. As a ball falls toward Earth, the momentum of the ball increases. How would you reconcile this observation with the law of conservation of momentum?

21. In the early 1900s, Robert Goddard proposed sending a rocket to the moon. Critics took the position that in a vacuum such as exists between Earth and the moon, the gases emitted by the rocket would have nothing to push against to propel the rocket. To settle the debate, Goddard placed a gun in a vacuum and fired a blank cartridge from it. (A blank cartridge fires only the hot gases of the burning gunpowder.) What happened when the gun was fired? Explain your answer.

22. An astronaut carrying a camera in space finds herself drifting away from a space shuttle after her tether becomes unfastened. If she has no propulsion device, what should she do to move back to the shuttle?

23. When a bullet is fired from a gun, what happens to the gun? Explain your answer using the principles of momentum discussed in this chapter.

Practice problems

24. A 65.0 kg ice skater moving to the right with a velocity of 2.50 m/s throws a 0.150 kg snowball to the right with a velocity of 32.0 m/s relative to the ground.
- **a.** What is the velocity of the ice skater after throwing the snowball? Disregard the friction between the skates and the ice.

- **b.** A second skater initially at rest with a mass of 60.0 kg catches the snowball. What is the velocity of the second skater after catching the snowball in a perfectly inelastic collision?

(See Sample Problem 6D.)

25. A tennis player places a 55 kg ball machine on a frictionless surface, as in **Figure 6-13**. The machine fires a 0.057 kg tennis ball horizontally with a velocity of 36 m/s toward the north. What is the final velocity of the machine?
(See Sample Problem 6D.)

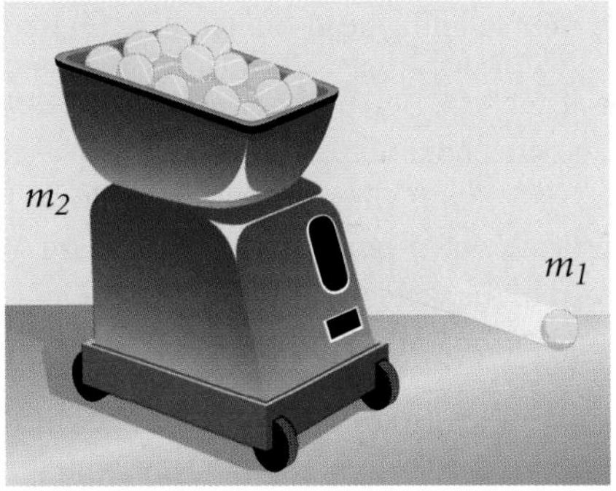

Figure 6-13

26. After being struck by a bowling ball, a 1.5 kg bowling pin sliding to the right at 3.0 m/s collides head-on with another 1.5 kg bowling pin initially at rest. Find the final velocity of the second pin in the following situations:
- **a.** the first pin moves to the right after the collision at 0.5 m/s
- **b.** the first pin stops moving when it hits the second pin

(See Sample Problem 6D.)

ELASTIC AND INELASTIC COLLISIONS

Review questions

27. Consider a perfectly inelastic head-on collision between a small car and a large truck traveling at the same speed. Which vehicle has a greater change in kinetic energy as a result of the collision?

28. Given the masses of two objects and their velocities before and after a head-on collision, how could you determine whether the collision was elastic, inelastic, or perfectly inelastic? Explain.

12. a. 8.35×10^{-21} kg•m/s upward
b. 4.88 kg•m/s to the right
c. 7.50×10^2 kg•m/s to the southwest
d. 1.78×10^{29} kg•m/s forward

13. 5.2 kg•m/s toward home plate

14. 160 N to the right

15. 18 N

16. 0.010 s; 0.13 m

17. Total momentum remains zero, so the two skaters have equal and opposite momentum.

18. no; Momentum can be transferred between balls.

19. Part of the ball's momentum is transferred to the ground; Earth's mass is so large that the resulting change in Earth's velocity is imperceptible.

20. As the ball accelerates toward Earth, Earth also accelerates toward the ball. Therefore, Earth is also gaining momentum in the direction opposite the ball's momentum.

21. The gun was pushed with a momentum equal in magnitude but opposite in direction to the momentum of the gases.

22. She should throw the camera in the direction away from the shuttle to cause her to move back toward the shuttle.

23. The gun recoils with a backward momentum equal to the forward momentum of the bullet. Because the gun's mass is so much greater than the bullet's, the gun's velocity will be smaller than the bullet's.

24. a. 2.43 m/s to the right
b. 7.97×10^{-2} m/s to the right

25. 0.037 m/s to the south

26. a. 2.5 m/s to the right
b. 3.0 m/s to the right

27. Because the initial velocities of the truck and the car are the same and the final velocity is the same, the change in *KE* depends only on the mass. The truck has a greater mass, so the change in its *KE* is greater.

28. by calculating the kinetic energy before and after the collision; If *KE* is conserved, the collision is elastic. If the collision is not elastic, use the final velocities and the conservation of momentum to determine whether it is perfectly inelastic or inelastic.

29. no; Total kinetic energy is conserved but kinetic energy can be transferred from one object to the other.

30. Both cannot be at rest after the collision because the total initial momentum was greater than zero; The object initially in motion can be at rest if its momentum is entirely transferred to the other object.

31. 1 m/s

32. 3.00 m/s

33. 4.2 m/s

34. a. 1.80 m/s
  b. 2.16×10^4 J

35. a. 0.81 m/s to the east
  b. 1.4×10^3 J

36. a. 12 cm/s to the right
  b. 1.1×10^{-4} J

37. 4.0 m/s

38. 17.2 cm/s to the right

39. 12.8 cm/s to the right

40. 5.00 m/s to the right

41. 42.0 m/s toward second base

42. 3.0 kg; 1.0×10^1 m/s

---

29. In an elastic collision between two objects, do both objects have the same kinetic energy after the collision as before? Explain.

30. If two objects collide and one is initially at rest, is it possible for both to be at rest after the collision? Is it possible for one to be at rest after the collision? Explain.

Practice problems

31. Two carts with masses of 4.0 kg and 3.0 kg move toward each other on a frictionless track with speeds of 5.0 m/s and 4.0 m/s respectively. The carts stick together after colliding head-on. Find the final speed. (See Sample Problem 6E.)

32. A 1.20 kg skateboard is coasting along the pavement at a speed of 5.00 m/s when a 0.800 kg cat drops from a tree vertically downward onto the skateboard. What is the speed of the skateboard-cat combination? (See Sample Problem 6E.)

33. Two carts with masses of 10.0 kg and 2.5 kg move in opposite directions on a frictionless horizontal track with speeds of 6.0 m/s and 3.0 m/s, respectively. The carts stick together after colliding head on. Find the final speed of the two carts. (See Sample Problem 6E.)

34. A railroad car with a mass of 2.00×10^4 kg moving at 3.00 m/s collides and joins with two railroad cars already joined together, each with the same mass as the single car and initially moving in the same direction at 1.20 m/s.
  a. What is the speed of the three joined cars after the collision?
  b. What is the decrease in kinetic energy during the collision?
  (See Sample Problem 6F.)

35. An 88 kg fullback moving east with a speed of 5.0 m/s is tackled by a 97 kg opponent running west at 3.0 m/s, and the collision is perfectly inelastic. Calculate the following:
  a. the velocity of the players just after the tackle
  b. the decrease in kinetic energy during the collision
  (See Sample Problem 6F.)

---

36. A 5.0 g coin sliding to the right at 25.0 cm/s makes an elastic head-on collision with a 15.0 g coin that is initially at rest. After the collision, the 5.0 g coin moves to the left at 12.5 cm/s.
  a. Find the final velocity of the other coin.
  b. Find the amount of kinetic energy transferred to the 15.0 g coin.
  (See Sample Problem 6G.)

37. A billiard ball traveling at 4.0 m/s has an elastic head-on collision with a billiard ball of equal mass that is initially at rest. The first ball is at rest after the collision. What is the speed of the second ball after the collision? (See Sample Problem 6G.)

38. A 25.0 g marble sliding to the right at 20.0 cm/s overtakes and collides elastically with a 10.0 g marble moving in the same direction at 15.0 cm/s. After the collision, the 10.0 g marble moves to the right at 22.1 cm/s. Find the velocity of the 25.0 g marble after the collision. (See Sample Problem 6G.)

39. A 15.0 g toy car moving to the right at 20.0 cm/s has an elastic head-on collision with a 20.0 g toy car moving in the opposite direction at 30.0 cm/s. After colliding, the 15.0 g car moves with a velocity of 37.1 cm/s to the left. Find the velocity of the 20.0 g car after the collision. (See Sample Problem 6G.)

40. Two shuffleboard disks of equal mass, one orange and the other yellow, are involved in an elastic collision. The yellow disk is initially at rest and is struck by the orange disk moving initially to the right at 5.00 m/s. After the collision, the orange disk is at rest. What is the velocity of the yellow disk after the collision? (See Sample Problem 6G.)

MIXED REVIEW

41. If a 0.147 kg baseball has a momentum of $\mathbf{p} = 6.17$ kg•m/s as it is thrown from home to second base, what is its velocity?

42. A moving object has a kinetic energy of 150 J and a momentum with a magnitude of 30.0 kg•m/s. Determine the mass and speed of the object.

43. A 0.10 kg ball of dough is thrown straight up into the air with an initial speed of 15 m/s.
   a. Find the momentum of the ball of dough at its maximum height.
   b. Find the momentum of the ball of dough halfway to its maximum height on the way up.

44. A 3.00 kg mud ball has a perfectly inelastic collision with a second mud ball that is initially at rest. The composite system moves with a speed equal to one-third the original speed of the 3.00 kg mud ball. What is the mass of the second mud ball?

45. A 5.5 g dart is fired into a block of wood with a mass of 22.6 g. The wood block is initially at rest on a 1.5 m tall post. After the collision, the wood block and dart land 2.5 m from the base of the post. Find the initial speed of the dart.

46. A 730 N student stands in the middle of a frozen pond having a radius of 5.0 m. He is unable to get to the other side because of a lack of friction between his shoes and the ice. To overcome this difficulty, he throws his 2.6 kg physics textbook horizontally toward the north shore at a speed of 5.0 m/s. How long does it take him to reach the south shore?

47. A 0.025 kg golf ball moving at 18.0 m/s crashes through the window of a house in 5.0×10^{-4} s. After the crash, the ball continues in the same direction with a speed of 10.0 m/s. Assuming the force exerted on the ball by the window was constant, what was the magnitude of this force?

48. A 1550 kg car moving south at 10.0 m/s collides with a 2550 kg car moving north. The cars stick together and move as a unit after the collision at a velocity of 5.22 m/s to the north. Find the velocity of the 2550 kg car before the collision.

49. A 2150 kg car moving east at 10.0 m/s collides with a 3250 kg car moving east. The cars stick together and move east as a unit after the collision at a velocity of 5.22 m/s.
   a. Find the velocity of the 3250 kg car before the collision.
   b. What is the decrease in kinetic energy during the collision?

50. A 0.400 kg bead slides on a straight frictionless wire with a velocity of 3.50 cm/s to the right, as shown in **Figure 6-14**. The bead collides elastically with a larger 0.600 kg bead initially at rest. After the collision, the smaller bead moves to the left with a velocity of 0.70 cm/s. Find the distance the larger bead moves along the wire in the first 5.0 s following the collision.

Figure 6-14

51. An 8.0 g bullet is fired into a 2.5 kg pendulum bob initially at rest and becomes embedded in it. If the pendulum rises a vertical distance of 6.0 cm, calculate the initial speed of the bullet.

52. The bird perched on the swing in **Figure 6-15** has a mass of 52.0 g, and the base of the swing has a mass of 153 g. The swing and bird are originally at rest, and then the bird takes off horizontally at 2.00 m/s. How high will the base of the swing rise above its original level? Disregard friction.

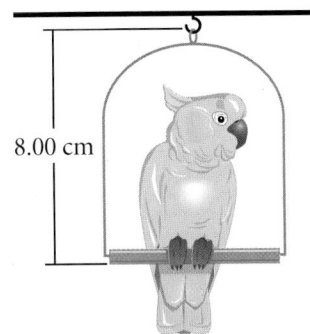

8.00 cm

Figure 6-15

53. An 85.0 kg astronaut is working on the engines of a spaceship that is drifting through space with a constant velocity. The astronaut turns away to look at Earth and several seconds later is 30.0 m behind the ship, at rest relative to the spaceship. The only way to return to the ship without a thruster is to throw a wrench directly away from the ship. If the wrench has a mass of 0.500 kg, and the astronaut throws the wrench with a speed of 20.0 m/s, how long does it take the astronaut to reach the ship?

43. a. 0.0 kg•m/s
   b. 1.0 kg•m/s upward
44. 6.00 kg
45. 23 m/s
46. 29 s
47. 4.0×10^2 N
48. 14.5 m/s to the north
49. a. 2.1 m/s to the east
   b. 4.1×10^4 J
50. 14 cm
51. 340 m/s
52. 2.36×10^{-2} m
53. 254 s

54. A 2250 kg car traveling at 10.0 m/s collides with a 2750 kg car that is initially at rest at a stoplight. The cars stick together and move 2.50 m before friction causes them to stop. Determine the coefficient of kinetic friction between the cars and the road, assuming that the negative acceleration is constant and that all wheels on both cars lock at the time of impact.

55. A constant force of 2.5 N to the right acts on a 1.5 kg mass for 0.50 s.

   a. Find the final velocity of the mass if it is initially at rest.

   b. Find the final velocity of the mass if it is initially moving along the x-axis with a velocity of 2.0 m/s to the left.

Technology & Learning

Graphing calculators

Refer to Appendix B for instructions on downloading programs for your calculator. The program "Chap6" allows you to analyze a graph of force versus time.

Force, as you learned earlier in this chapter, relates to momentum in the following way:

$$\mathbf{F} = \frac{\Delta \mathbf{p}}{\Delta t} \text{ where } \Delta \mathbf{p} = m\mathbf{v_f} - m\mathbf{v_i}$$

The program "Chap6" stored on your graphing calculator makes use of the equation that relates force and momentum. Once the "Chap6" program is executed, your calculator will ask for the mass, initial velocity, and final velocity. The graphing calculator will use the following equation to create a graph of the force (Y_1) versus the time interval (X). The relationships in this equation are the same as those in the force equation shown above. (Note that F in the equation below stands for "final," not force.)

$$Y_1 = M(F-I)/X$$

   a. The equation used by the calculator can also be derived from another equation that relates force and mass. What is this equation?

Execute "Chap6" on the [PRGM] menu, and press [ENTER] to begin the program. Enter the values for the mass, initial velocity, and final velocity (shown below), and press [ENTER] after each value.

The calculator will provide a graph of the force versus the time interval. (If the graph is not visible, press [WINDOW] and change the settings for the graph window, then press [GRAPH].)

Press [TRACE], and use the arrow keys to trace along the curve. The x-value corresponds to the time interval in seconds, and the y-value corresponds to the force in newtons. The force will be negative in cases where it opposes the ball's initial velocity.

Determine the force that must be exerted on a 0.43 kg soccer ball in the given time interval to cause the changes in momentum in the following situations (b–e). When entering negative values, make sure to use the [(-)] key, instead of the [-] key.

   b. the ball slows from 15 m/s to 0 m/s in 0.025 s

   c. the ball slows from 15 m/s to 0 m/s in 0.75 s

   d. the ball speeds up from 5.0 m/s in one direction to 22 m/s in the opposite direction in 0.045 s

   e. the ball speeds up from 5.0 m/s in one direction to 22 m/s in the opposite direction in 0.55 s

   f. In what quadrant would the graph appear if the ball accelerated from rest?

Press [2nd] [QUIT] to stop graphing. Press [ENTER] to input a new value or [CLEAR] to end the program.

56. A 55 kg pole-vaulter falls from rest from a height of 5.0 m onto a foam-rubber pad. The pole-vaulter comes to rest 0.30 s after landing on the pad.

    a. Calculate the athlete's velocity just before reaching the pad.

    b. Calculate the constant force exerted on the pole-vaulter due to the collision.

57. A 7.50 kg laundry bag is dropped from rest at an initial height of 3.00 m.

    a. What is the speed of Earth toward the bag just before the bag hits the ground? Use the value 5.98×10^{24} kg as the mass of Earth.

    b. Use your answer to part **(a)** to justify disregarding the motion of Earth when dealing with the motion of objects on Earth.

58. Two billiard balls with identical masses and sliding in opposite directions have an elastic head-on collision. Before the collision, each ball has a speed of 22 cm/s. Find the speed of each billiard ball immediately after the collision. (See Appendix A for hints on solving simultaneous equations.)

59. An unstable nucleus with a mass of 17.0×10^{-27} kg initially at rest disintegrates into three particles. One of the particles, of mass 5.0×10^{-27} kg, moves along the positive y-axis with a speed of 6.0×10^{6} m/s. Another particle, of mass 8.4×10^{-27} kg, moves along the positive x-axis with a speed of 4.0×10^{6} m/s. Determine the third particle's speed and direction of motion. (Assume that mass is conserved.)

Alternative Assessment

Performance assessment

1. Design an experiment that uses a dynamics cart with other easily found equipment to test whether it is safer to crash into a steel railing or into a container filled with sand. How can you measure the forces applied to the cart as it crashes into the barrier? If your teacher approves your plan, perform the experiment.

2. Obtain a videotape of one of your school's sports teams in action. Create a play-by-play description of a short segment of the videotape, explaining how momentum and kinetic energy change during impacts that take place in the segment.

3. An inventor has asked an Olympic biathlon team to test his new rifles during the target-shooting segment of the event. The new 0.75 kg guns shoot 25.0 g bullets at 615 m/s. The team's coach has hired you to advise him about how these guns could affect the biathletes' accuracy. Prepare figures to justify your answer. Be ready to defend your position.

Portfolio projects

4. Investigate the elastic collisions between atomic particles. What happens after an elastic collision between a hydrogen atom at rest and a helium atom moving at 150 m/s? Which direction will each particle move after the collision? Which particle will have a higher speed after the collision? What happens when a neutron moving at 150 m/s hits a hydrogen atom at rest? Research the masses of the particles involved. Draw diagrams of each collision.

5. An engineer working on a space mission claims that if momentum concerns are taken into account, a spaceship will need far less fuel for its return trip than it did for the first half of the mission. Prepare a detailed report on the validity of this hypothesis. Research the principles of rocket operations. Select specific examples of space missions, and study the nature of each mission and the amounts of fuel used. Your report should include diagrams and calculations.

Momentum and Collisions   **237**

NOTE

Materials Preparation is given on pp. 206A–206B. Blank data table and sample data table are on the One-Stop Planner CD-ROM. All calculations shown use sample data.

Planning

Recommended time:

1 lab period

For a 2-period lab, have students repeat the experiment to find the value of an unknown mass on one of the carts.

Classroom organization:

▸ Each group needs a level work surface with clear table space at least 2.0 m long.

▸ Each lab group should have at least 2 students.

▸ The recording timer and CBL procedures may be used in the same class.

▸ If lab groups share balances, some should mass their carts before setting up to avoid traffic jams.

▸ Safety warnings: Remind students to attach masses to carts securely and to make sure the carts do not fall off the table. Books or wooden blocks may be clamped to the ends of the table to serve as bumpers and keep the carts from falling.

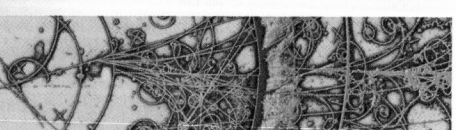

CHAPTER 6
Laboratory Exercise

⊛ **TEKS**

pp. 238–241: 1A, 2B, 2C, 2D, 2E, 2F, 3B, 4A, 5C, 5D

OBJECTIVES

• Measure the mass and velocity of two carts.

• Calculate the momentum of each cart.

• Verify the law of conservation of momentum.

MATERIALS LIST

✔ **two carts, one with a spring mechanism**

✔ **balance**

✔ **metric ruler**

PROCEDURE

CBL AND SENSORS

✔ **CBL**

✔ **CBL motion detector**

✔ **graphing calculator with link cable**

✔ **support stand with clamp**

✔ **set of masses**

✔ **tape**

✔ **two 25 × 25 cm squares of poster board**

RECORDING TIMER

✔ **recording timer**

✔ **paper tape**

✔ **stopwatch**

CONSERVATION OF MOMENTUM

The product of the mass of a moving object and its velocity is called its *momentum*. Studying momentum helps physicists understand the relationship between the motion of two interacting objects.

According to the impulse-momentum theorem, the change in an object's momentum is equal to the product of the force acting on the object and the time interval during which the force acts. Newton's third law states that every action is accompanied by an equal and opposite reaction. Thus, when a spring-loaded cart pushes off against another cart, the force on the first cart is accompanied by an equal and opposite force on the second cart. Both of these forces act for exactly the same time interval, so the change in momentum of the first cart is equal and opposite to the change in momentum of the second cart, in the absence of other forces.

In this experiment you will study the momentum of two carts with unequal masses. The carts will be placed together and will move apart when a compressed spring between them is released. You will find the mass and velocity of each cart in order to compare the momentum before and after the carts move apart.

SAFETY

• **Tie back long hair, secure loose clothing, and remove loose jewelry to prevent their getting caught in moving or rotating parts.**

PREPARATION

1. Determine whether you will be using the CBL and sensors or the recording timer to perform this experiment. Read the entire lab procedure for the appropriate method. Plan the steps you will take.

2. Prepare a data table in your lab notebook with seven columns and four rows. In the first row, label the first through seventh columns *Trial, m_1 (kg), m_2 (kg), Cart 1 Distance (m), Cart 2 Distance (m), Cart 1 Time interval (s)*, and *Cart 2 Time interval (s)*. In the first column, label the second through fourth rows *1, 2,* and *3*.

3. Choose a location where both carts will be able to move at least 1.0 m without any obstacles.

Recording timer procedure begins on page 240.

PROCEDURE

CBL AND SENSORS

Conservation of momentum

4. Connect the CBL to the calculator with the unit-to-unit link cable using the link ports located on each unit. Connect the motion detector to the SONIC port. Turn on the CBL and the graphing calculator.

5. Start the program PHYSICS on your graphing calculator.

    a. Select option *SET UP PROBES* from the MAIN MENU. Enter 1 for the number of probes. Select the motion detector from the list.

    b. Select the *COLLECT DATA* option from the MAIN MENU. Select the *TIME GRAPH* option from the DATA COLLECTION menu. Enter 0.02 for the time between samples. Enter 99 for the number of samples.

    c. Check the values you entered and press ENTER. If the values are correct, select *USE TIME SETUP* to continue. If you made a mistake entering the time values, select *MODIFY SETUP*, reenter the values, and continue.

    d. On the TIME GRAPH menu, select *NON-LIVE DISPLAY*.

6. Set up the apparatus, as shown in **Figure 6-16.** Position the motion detector so that it will be in a straight line with the motion of the carts. To keep the motion detector from moving during the experiment, clamp it to the support stand. Place a piece of tape 1.5 m in front of the motion detector to serve as a starting line for the cart.

7. Tape a square of poster board to the back of each cart to make a vertical surface. Label the carts *Cart 1* and *Cart 2*. Find the mass of cart 1, and record it in your data table. Securely tape a 1.0 kg mass to cart 2. Find the mass of cart 2 and the 1.0 kg mass, and record it in your data table.

8. Compress the spring mechanism and position the carts together, as shown in **Figure 6-16.** The poster board on cart 1 should be in a straight line with the motion detector, and it should be at least 1.5 m away from the motion detector. Use a piece of tape to mark the starting line on the table top. Cart 2 should be positioned to move away from the motion detector.

9. Press ENTER on the calculator. When the motion detector begins to click, quickly press the spring mechanism release on the carts. The carts will move away from each other. Let cart 1 move about 1.0 m, but catch the carts before they reach the edges of the table. ***Do not let the carts hit the motion detector or fall off the table.***

Figure 6-16

This photo shows all the equipment necessary to complete this lab, but the equipment is not placed at the appropriate distances for the experiment. Follow the written directions for correct placement.

Step 8: If the spring mechanism has more than one notch, choose the first notch. Make sure there is at least 1.5 m between the motion detector and the cart. Make sure the other cart is at least 1.0 m from the edge of the table.

Step 9: Let the carts move at least 1.0 m before you catch them, but do not let the carts hit the motion detector or fall off the table.

Step 13: Switch the positions of the carts to record data for the second cart.

CBL and Sensors Tips

◆ Students should have the program PHYSICS on their graphing calculators. Refer to Appendix B for instructions.

◆ The motion detector must be clamped tightly so that it can't move during the experiment.

◆ The cart should start 1.5 m away from the motion detector; this allows the cart to move 1.0 m after the spring is released.

◆ Make sure students understand how they will switch the apparatus to take readings for the second cart for each trial.

Techniques to Demonstrate

Remind students how to use the arrow keys to trace the graph, how to choose the points, and how to find the difference between the *y* values of the points.

Review what the graphs of velocity and acceleration in steps 12 and 13 reveal about the motion of the cart.

✔ Checkpoints

Step 6: Students should leave enough room for both carts to move 1.0 m. Remind students that they will have to catch both carts at the same time.

Step 12: The scale may be too large to easily see the value for the velocity. If so, students may quit the PHYSICS program, press WINDOW to set the Xmin and Xmax values to the initial and final times found in step 11, and press GRAPH to view the graph. They can then restart the PHYSICS program to view the acceleration graph and go on.

Recording-Timer Tips

◆ To attach the paper tape to the cart, create "sidearms" on the carts by securely attaching 30 cm lattice rods to the carts. The lattice rods must be included in the mass of the carts.

◆ Mount the timer on a support rod to level the tape path with the tops of the lattice rods.

Techniques to Demonstrate

Show students how to thread both tapes through the timer at the same time. The lower tape should pass under both carbon disks, and the upper tape should pass over both disks.

✔ Checkpoints

Step 9: Students should leave enough room for both carts to move about 1.0 m. Remind students that they will have to catch both carts at about the same time.

Step 11: Help students choose 3 dots on each tape where the velocity is fairly constant. Students should be able to explain how the dots represent the motion of the carts.

10. When the motion detector stops clicking and the CBL displays *DONE*, press ENTER on the graphing calculator. Select the *SONIC* option from the SELECT CHANNELS menu. Select *DISTANCE* to plot a graph of the distance between the cart and the motion detector in meters against time in seconds. Use the arrow keys to trace along the curve. On the far left and the far right, the curve represents the position of the cart before and after its motion. The middle section of the curve should be a straight line, and it represents the motion of the cart. Use the arrow keys to choose a point on the curve near the beginning of this middle section and another point near the end.

11. Find the difference between the *y*-values of these two points, and record it as the cart 1 distance for trial 1 in your data table. Find the difference between the *x*-values for these two readings to find the time elapsed between measurements. Record this as the cart 1 time interval for trial 1 in your data table. Press ENTER on the calculator.

12. Select the *SONIC* option again and select the *VELOCITY* option to plot a graph of the velocity against time. Examine how the velocity of the cart

PROCEDURE

RECORDING TIMER

Conservation of momentum

4. Set up the apparatus as shown in **Figure 6-17.** *Do not plug in the timer until your teacher approves your setup. Do not close the switch.*

5. If you have not used the recording timer before, refer to the Chapter 2 lab for instructions. Calibrate the recording timer with the stopwatch, or use the previously determined value for the timer's period.

6. Record the value for the timer's period on a line near the data table.

7. Measure the mass of one cart, and record it in the data table. Add a 1.0 kg mass to the second cart, and record the mass of the cart plus the 1.0 kg mass.

changed during the experiment. Make a sketch of the graph in your lab notebook. Press ENTER.

13. Select the *SONIC* option again and select the *ACCELERATION* option to plot a graph of the acceleration against time. Examine how the acceleration of the cart changed during the experiment. Make a sketch of the graph. Press ENTER.

14. To find the values for cart 2 distance and time interval, repeat the experiment by compressing the spring mechanism and repositioning the carts. Reverse the carts so that cart 2 is in a straight line with the motion detector while cart 1 moves away from the motion detector. Repeat step 8 through step 13. Record these values as the cart 2 distance and time interval in the data table for trial 1.

15. Using different masses, repeat the experiment two more times, for trials 2 and 3. For each trial, measure and record the values for cart 1 and for cart 2.

16. Clean up your work area. Put equipment away safely so that it is ready to be used again.

Analysis and Interpretation begins on page 241.

8. Fasten a timing tape to one end of each cart. Because both tapes pass through the same timer, place two carbon paper disks back to back between the paper tapes.

9. Compress the spring and position the carts. When your teacher approves your setup, plug the recording timer into the wall outlet. Start the timer and release the spring simultaneously.

10. Catch the carts before they reach the edge of the table and then stop the timer. *Do not let the carts fall off the table.* Remove the tapes. Label each tape so that it corresponds to the cart to which it was attached.

11. On each tape, find a portion where the distance between dots is fairly constant. Use the metric ruler to measure three distances between successive dots.

12. Find the average of these three values and record the average as the distance for that cart in your data table. Record the period of the timer as the time interval in your data table.

13. Using different masses, repeat the experiment two more times for trial 2 and trial 3.

14. Clean up your work area. Put equipment away safely so that it is ready to be used again.

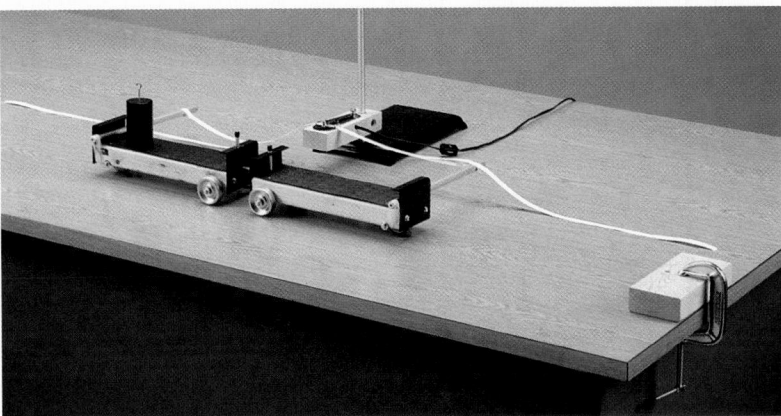

Figure 6-17

Step 7: The recording timer will mark the tapes for both carts at the same time. Place two carbon disks back to back with one tape above and one tape below.

Step 9: If the spring mechanism has more than one notch, choose the first notch. Press straight down to release the spring mechanism so that you do not affect the motion of the carts. Let the carts move at least 1.0 m before you catch them, but do not let the carts fall off the table.

ANALYSIS AND INTERPRETATION

Calculations and data analysis

1. **Organizing data**   For each trial, find the velocities v_1 and v_2. Because the carts are moving in opposite directions, assign one of the carts a negative velocity to indicate direction.

2. **Organizing data**   For each trial, calculate the momentum of each cart by multiplying its mass by its velocity.

3. **Organizing data**   For each trial, find the total momentum of the two carts.

4. **Applying ideas**   For each trial, what is the total momentum of the two carts before they start moving?

Conclusions

5. **Evaluating information**   Conservation of velocity would mean that the total velocity for both carts is the same after the spring mechanism is released as it was before the release. Is velocity conserved in this experiment? Support your answer with data from the experiment.

6. **Evaluating information**   On the basis of your data, is momentum conserved in this experiment? Support your answer with data from the experiment.

7. **Evaluating methods**   How would using two carts with identical masses affect your answers to items 5 and 6?

ANSWERS TO

Analysis and Interpretation

CALCULATIONS AND DATA ANALYSIS

1. Answers will vary. Make sure students use the relationship $v_{avg} = \dfrac{\Delta x}{\Delta t}$. Typical values will range from ± 0.250 m/s to ± 0.857 m/s.

2. Make sure students use the relationship $p = mv$. Typical values will range from ± 0.389 kg•m/s to ± 0.811 kg•m/s.

3. Make sure students use the relationship $p = p_1 + p_2$. Typical values will range from -0.004 kg•m/s to 0.004 kg•m/s.

4. For all trials, the total momentum of the two carts before they start moving is zero, because the carts have no velocity.

CONCLUSIONS

5. Velocity is not conserved in this experiment.

6. Momentum is conserved. The values for the total final momentum found in item 3 are very close to zero, the total initial momentum.

7. Momentum will always be conserved. If the carts have the same mass, velocity will also be conserved.

CHAPTER 7 PLANNING GUIDE

Compression Guide: To shorten from 9 to 8 45-min periods (from 4½ to 4 90-min blocks), eliminate items in magenta type.

PACING CHART	CLASSROOM RESOURCES			
	⭐TEKS	Teacher Demonstrations	Holt Physics Transparencies	Labs (See page T52 for equipment listing for in-text labs.)
7-1 Measuring rotational motion 3 or 2 45-minute periods 1½ or 1 90-minute block	1A, 2B, 2C, 2F, 3A, 3B, 4B	**TE** *Equal angular speed,* p. 250	**TM** 23–25	**L** *Discovery Lab: Circular Motion* **PE** *Quick Lab: Radians and Arc Length,* p. 245
7-2 Tangential and centripetal acceleration 1 45-minute period ½ 90-minute block	2C, 3A, 3B, 4B	**TE** *Tangential speed versus angular speed,* p. 253	**T** 21	**TL** *Centripetal Acceleration*
7-3 Causes of circular motion 3 45-minute periods 1½ 90-minute block	2C, 3A, 3B, 3E, 4B, 4C, 6A, 6B	**TE** *Force that maintains circular motion,* p. 260	**T** 22, 23	**PE** *Circular Motion,* p. 274

Review and Assessment
2 45-minute periods
1 90-minute block

Resource Key

PHYSICS
PE Pupil's Edition
TE Teacher's Edition

L Laboratory Experiments
TL Technology Lab Experiments
T Transparencies

 **One-Stop** Planner CD-ROM **contents**
TM Transparency Masters
SR Section Review Worksheets
AA Alternative Assessment

PW Problem-Solving Workbook
PB Problem Bank
CTW Critical Thinking Worksheet

LABORATORY PLANNING: Circular Motion, p. 274

Materials (for each lab group)
- centripetal force apparatus, contains
  - 2-hole rubber stopper
  - PVC tube
  - nylon cord
  - paper clips
  - set of 20 heavy washers
- meterstick

- metric hooked mass, 100 g
- metric hooked mass, 200 g
- balance: portable, electronic balance or triple-beam balance with weight
- roll of adhesive tape, 0.5 in. wide

Safety Equipment
- safety goggles or spectacles

ASSIGNMENT RESOURCES

Content Mastery	Critical Thinking	Problem-Solving Practice		
PE 1–5, p. 252 **SR** 7-1, *Concept Review* **PE** 1–4, p. 269		**7A**	Angular displacement: 23 items in **PE, PW,** and **PB,** see **TE** p. 246	
		7B	Angular speed: 23 items in **PE, PW,** and **PB,** see **TE** p. 248	
		7C	Angular acceleration: 22 items in **PE, PW,** and **PB,** see **TE** p. 249	
		7D	Angular kinematics: 34 items in **PE, PW,** and **PB,** see **TE** pp. 251–252	
PE 1–4, p. 259 **SR** 7-2, *Concept Review* **PE** 13–16, p. 269	**PE** 17–20, p. 270	**7E**	Tangential speed: 21 items in **PE, PW,** and **PB,** see **TE** pp. 254–255	
		7F	Tangential acceleration: 21 items in **PE, PW,** and **PB,** see **TE** p. 256	
		7G	Centripetal acceleration: 27 items in **PE, PW,** and **PB,** see **TE** p. 258	
PE 1–5, p. 265 **SR** 7-3, *Concept Review* **PE** 27–31, p. 270	**PE** 1–2, p. 262 **PE** 32–36, p. 270	**7H**	Force that maintains circular motion: 26 items in **PE, PW,** and **PB,** see **TE** p. 261	
		7I	Gravitational force: 22 items in **PE, PW,** and **PB,** see **TE** p. 264	

ASSESSMENT RESOURCES

Cumulative Review	Alternative Assessment	Traditional Assessment
SR Mixed Review, Ch. 7	**PE** 1–3, p. 273 **AA** Items for Ch. 7	Chapter 7 Test Test Generator items for Ch. 7

Scoring Rubrics for Alternative Assessment items can be found on the One-Stop Planner CD-ROM.

TECHNOLOGY RESOURCES

 CTW Segment 7 Zero-Gravity Plane

 Module 8 Angular Kinematics

 PE Technology and Learning, p. 272
(Alternative procedures for calculators without Flash-ROM technology are provided on the One-Stop Planner CD-ROM.)

 The Mechanical Universe/High School Adaptation Quad I, The Apple and the Moon Quad II, Moving in Circles

 internet**connect**

 On-line Student Resources:
GO TO: www.scilinks.org
The following *sci*LINKS Internet resources can be found in the student text for this chapter.

TOPICS:
• Rotational motion, p. 254 (HF2071)
• Circular motion, p. 260 (HF2072)
• Law of gravitation, p. 263 (HF2073)
• Black holes, p. 267 (HF2074)

 On-line Teacher Resources:
GO TO: go.hrw.com
KEYWORD: HF2 HOME
Visit the HRW Web site for a variety of resources related to this chapter.

Smithsonian Institution
Internet Connections
Visit **www.si.edu/hrw** for additional on-line resources.

 .com
Visit **www.cnnfyi.com** for late-breaking news and current events stories selected just for you.

Section 7-1 introduces the formulas and units for angular displacement, angular speed, and angular acceleration.

Section 7-2 explores tangential speed, tangential acceleration, and centripetal acceleration and discusses uniform circular motion.

Section 7-3 examines the causes of circular motion, including gravitation.

About the Illustration

Two forms of this ride can be found at amusement parks. In one case, as the angular speed of the ride increases, the floor of the ride drops so that riders are suspended against the wall in a vertical position. In the other case, the ride turns so that it spins in a vertical circular path. The ride pictured at right, called the Mexican Hat Ride, is an example of the latter case.

PHYSICS INTERACTIVE TUTOR

Interactive Problem-Solving Tutor

See Module 8
"Angular Kinematics" provides additional development of problem-solving skills for this chapter.

CHAPTER 7

Rotational Motion and the Law of Gravity

PHYSICS IN ACTION

When riding this spinning amusement-park ride, people feel as if a force is pressing them against the padding on the inside walls of the ride. However, it is actually inertia that causes their bodies to press against the padding. The inertia of their bodies tends to maintain motion in a straight-line path, while the walls of the ride exert a force on their bodies that makes them follow a circular path. This chapter will discuss the force that maintains circular motion and other rotational-motion quantities.

- *How can you determine the riders' average linear speed or acceleration during the ride?*

- *In what direction are the forces pushing or pulling the riders?*

CONCEPT REVIEW

Displacement (Section 2-1)

Velocity (Section 2-1)

Acceleration (Section 2-2)

Force (Section 4-1)

Tapping Prior Knowledge

Knowledge to Expect
✔ "Every object exerts gravitational force on every other object. The force depends on how much mass the objects have and on how far apart they are. The force is hard to detect unless at least one of the objects has a lot of mass." (AAAS's *Benchmarks for Science Literacy,* grades 6–8)

✔ "Gravity is the force that keeps planets in orbit around the sun and governs the motion of the rest of the solar system. Gravity alone holds us to the earth's surface and explains the phenomenon of the tides." (NRC's *National Science Education Standards,* grades 5–8)

Knowledge to Review
✔ Displacement is the change in position of an object. (Section 2-1)

✔ Velocity is speed and direction of travel. Average velocity is equal to displacement divided by time. (Section 2-1)

✔ Acceleration is the rate of change of velocity. (Section 2-2)

✔ Forces are the causes of accelerated motion. (Section 4-1)

Items to Probe
✔ **Net force:** Have students calculate the net force on an object experiencing multiple forces.

7-1
Measuring rotational motion

Note that rotational motion and circular motion are defined separately. The distinction between the two is that a solid object undergoes rotational motion whereas a point on a rotating object undergoes circular motion. A point mass moving about a central axis is also described by the term *circular motion*.

Visual Strategy

Figure 7-1

Point out that the angle through which the light bulb moves is related to the distance around the circle that the light bulb moves.

Q If the light bulb moves through an angle twice as large as the one shown in **(b)**, how would the new distance compare with the distance shown?

A *The new distance would be twice as large.*

7-1 SECTION OBJECTIVES

- **Relate radians to degrees.**
- **Calculate angular displacement using the arc length and the distance from the axis of rotation.**
- **Calculate angular speed or angular acceleration.**
- **Solve problems using the kinematic equations for rotational motion.**

---

rotational motion

motion of a body that spins about an axis

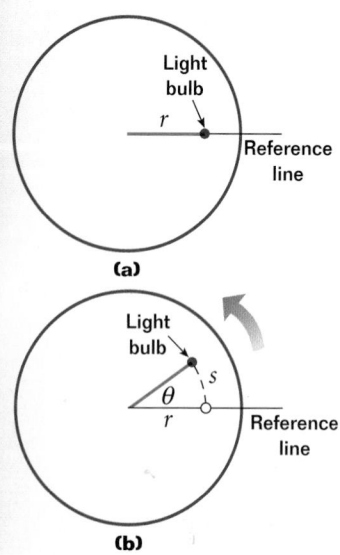

(a)

(b)

Figure 7-1
A light bulb on a rotating Ferris wheel **(a)** begins at a point along a reference line and **(b)** moves through an arc length *s*, and therefore through the angle θ.

ROTATIONAL QUANTITIES

When an object spins, it is said to undergo **rotational motion.** Consider a spinning Ferris wheel. The *axis of rotation* is the line about which the rotation occurs. In this case, it is a line perpendicular to the side of the Ferris wheel and passing through the wheel's center. How can we measure the distance traveled by an object on the edge of the Ferris wheel?

A point on an object that rotates about a single axis undergoes *circular motion* around that axis. In other words, regardless of the shape of the object, any single point on the object travels in a circle around the axis of rotation. It is difficult to describe the motion of a point moving in a circle using only the linear quantities introduced in Chapter 2 because the direction of motion in a circular path is constantly changing. For this reason, circular motion is described in terms of the angle through which the point on an object moves. When rotational motion is described using angles, all points on a rigid rotating object, except the points on the axis, move through the same angle during any time interval. **⊛ TEKS 4B**

In **Figure 7-1,** a light bulb at a distance *r* from the center of a Ferris wheel, like the one shown in **Figure 7-2,** moves about the axis in a circle of radius *r.* In fact, every point on the wheel undergoes circular motion about the center. To analyze such motion, it is convenient to set up a *fixed* reference line. Let us assume that at time $t = 0$, the bulb is on the reference line, as in **Figure 7-1(a),** and that a line is drawn from the center of the wheel to the bulb. After a time interval Δt, the bulb advances to a new position, as shown in **Figure 7-1(b).** In this time interval, the line from the center to the bulb (depicted with a red line in both diagrams) moved through the angle θ with respect to the reference line. Likewise, the bulb moved a distance *s*, measured along the circumference of the circle; *s* is the *arc length.*

Figure 7-2
Any point on a Ferris wheel that spins about a fixed axis undergoes circular motion.

Angles can be measured in radians

In the situations we have encountered so far, angles have been measured in degrees. However, in science, angles are often measured in **radians** (rad) rather than in degrees. Almost all of the equations used in this chapter and the next require that angles be measured in radians. In **Figure 7-1(b)**, when the arc length, s, is equal to the length of the radius, r, the angle θ swept by r is equal to 1 rad. In general, any angle θ measured in radians is defined by the following:

$$\theta = \frac{s}{r} \quad \text{(★) TEKS} \quad 3B$$

The radian is a pure number, with no dimensions. Because θ is the ratio of an arc length (a distance) to the length of the radius (also a distance), the units cancel and the abbreviation rad is substituted in their place.

When the bulb on the Ferris wheel moves through an angle of $360°$ (one revolution of the wheel), the arc length s is equal to the circumference of the circle, or $2\pi r$. Substituting this value for s in the above equation gives the corresponding angle in radians.

$$\theta = \frac{s}{r} = \frac{2\pi r}{r} = 2\pi \, \text{rad} \quad \text{(★) TEKS} \quad 2C$$

Thus, $360°$ equals 2π rad, or one complete revolution. In other words, one revolution corresponds to an angle of approximately $2(3.14) = 6.28$ rad. **Figure 7-3** depicts a circle marked with both radians and degrees.

It follows that any angle in degrees can be converted to an angle in radians by multiplying the angle measured in degrees by $2\pi/360°$. In this way, the degrees cancel out and the measurement is left in radians. The conversion relationship can be simplified as follows:

$$\theta(\text{rad}) = \frac{\pi}{180°}\theta(\text{deg})$$

radian

an angle whose arc length is equal to its radius, which is approximately equal to $57.3°$

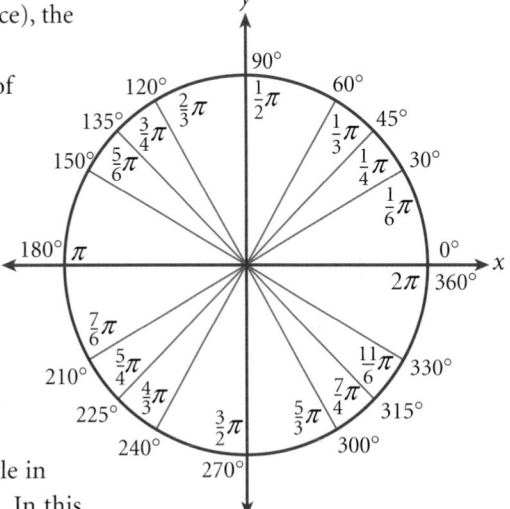

Figure 7-3
Angular motion is measured in units of radians. Because there are 2π radians in a full circle, radians are often expressed as a multiple of π.

SECTION 7-1

Visual Strategy

Figure 7-3
Strengthen students' understanding of radian measurement by having them use the values given in **Figure 7-3** to estimate the radian measures not shown. The students can then verify their answers by using the conversion equation shown on this page.

Q What is the radian measure equal to 75°?

A $\frac{5}{12}\pi$

Quick Lab

TEACHER'S NOTES

Students should use one of the full-length pieces of wire to draw the angle. The angle between lines drawn from the ends of the partial length of wire required to complete the circle will not equal 1 rad.

Quick Lab

Radians and Arc Length

MATERIALS

✔ drawing compass
✔ paper
✔ thin wire
✔ wire cutters or scissors

Use the compass to draw a circle on a sheet of paper, and mark the center point of the circle. Measure the distance from the center point to the outside of the circle. This is the radius of the circle. Using the wire cutters, cut several pieces of wire equal to the length of this radius. Bend the pieces of wire, and lay them along the circle you drew with your compass. Approximately how many pieces of wire do you use to go all the way around the circle? Draw lines from the center of the circle to each end of one of the wires. Note that the angle between these two lines equals 1 rad. How many of these angles are there in this circle? Draw a larger circle using your compass. How many pieces of wire (cut to the length of the radius) do you use to go all the way around this circle?

(★) TEKS 1A, 2B, 2C, 2F

(★) TEKS

p. 244: 4B
p. 245: 1A, 2B, 2C, 2F, 3B

The following may be used as teamwork exercises or for demonstration at the chalkboard or on an overhead projector.

PROBLEM

Angular displacement

Earth has an equatorial radius of approximately 6380 km and rotates 360° every 24 h.

a. What is the angular displacement (in degrees) of a person standing at the equator for 1.0 h?

b. Convert this angular displacement to radians.

c. What is the arc length traveled by this person?

Answers

   a. 15°

   b. 0.26 rad

   c. approximately 1700 km

Angular displacement describes how much an object has rotated

angular displacement

the angle through which a point, line, or body is rotated in a specified direction and about a specified axis

Just as an angle in radians is the ratio of the arc length to the radius, the **angular displacement** traveled by the bulb on the Ferris wheel is the change in the arc length, Δs, divided by the distance of the bulb from the axis of rotation. This relationship is depicted in **Figure 7-4**. ⭐TEKS **4B**

> **ANGULAR DISPLACEMENT**
>
> $$\Delta\theta = \frac{\Delta s}{r}$$
>
> $$\text{angular displacement (in radians)} = \frac{\text{change in arc length}}{\text{distance from axis}}$$

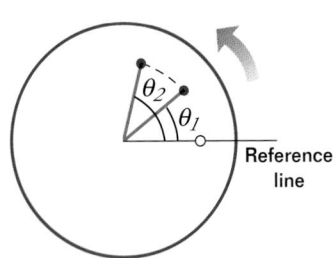

Figure 7-4
A light bulb on a rotating Ferris wheel rotates through an angular displacement of $\Delta\theta = \theta_2 - \theta_1$.

For the purposes of this textbook, when a rotating object is viewed from above, the arc length, s, is considered positive when the point rotates counterclockwise and negative when it rotates clockwise. In other words, $\Delta\theta$ is positive when the object rotates counterclockwise and negative when the object rotates clockwise.

SAMPLE PROBLEM 7A

Angular displacement ⭐TEKS **2C, 4B**

PROBLEM

While riding on a carousel that is rotating clockwise, a child travels through an arc length of 11.5 m. If the child's angular displacement is 165°, what is the radius of the carousel?

SOLUTION

Given: $\Delta\theta = -165°$ $\Delta s = -11.5$ m

Unknown: $r = ?$

First, convert the angular displacement to radians using the relationship on page 245.

$$\Delta\theta(\text{rad}) = \frac{\pi}{180°}\Delta\theta(\text{deg}) = \frac{\pi}{180°}(-165°)$$

$$\Delta\theta(\text{rad}) = -2.88 \text{ rad}$$

Use the angular displacement equation on this page. Rearrange to solve for r.

$$\Delta\theta = \frac{\Delta s}{r}$$

$$r = \frac{\Delta s}{\Delta\theta} = \frac{-11.5 \text{ m}}{-2.88 \text{ rad}}$$

$$\boxed{r = 3.99 \text{ m}}$$

> **CALCULATOR SOLUTION**
>
> Many calculators have a key labeled DEG▶ that converts from degrees to radians.

Angular displacement ⊛TEKS 2C, 4B

1. A girl sitting on a merry-go-round moves counterclockwise through an arc length of 2.50 m. If the girl's angular displacement is 1.67 rad, how far is she from the center of the merry-go-round?

2. A beetle sits at the top of a bicycle wheel and flies away just before it would be squashed. Assuming that the wheel turns clockwise, the beetle's angular displacement is π rad, which corresponds to an arc length of 1.2 m. What is the wheel's radius?

3. A car on a Ferris wheel has an angular displacement of $\frac{\pi}{4}$ rad, which corresponds to an arc length of 29.8 m. What is the Ferris wheel's radius?

4. Fill in the unknown quantities in the following table:

	$\Delta\theta$	Δs	r
a.	? rad	+0.25 m	0.10 m
b.	+0.75 rad	?	8.5 m
c.	? degrees	−4.2 m	0.75 m
d.	+135°	+2.6 m	?

Alternative Problem-Solving Approach

The angle traveled by the child, −165°, is equal to −45.8 percent of the circle. Therefore, the arc length −11.5 m also describes −45.8 percent of the circle. To find the circumference of the circle, calculate −11.5 ÷ −0.458 = 25.1. Because the circumference of a circle is given as $C = 2\pi r$, $C/2\pi = r = 3.99$ m.

ANSWERS TO

Practice 7A
Angular displacement
1. 1.50 m
2. 0.38 m
3. 37.9 m
4. a. 2.5 rad
   b. 6.4 m
   c. −320°
   d. 1.1 m

Angular speed describes rate of rotation

Linear speed describes the distance traveled in a specified interval of time. **Angular speed** is similarly defined. The average angular speed, ω_{avg} (ω is the Greek letter *omega*), of a rotating rigid object is the ratio of the angular displacement, $\Delta\theta$, to the time interval, Δt, that the object takes to undergo that displacement. Angular speed describes how quickly the rotation occurs. ⊛TEKS 4B

angular speed

the rate at which a body rotates about an axis, usually expressed in radians per second

ANGULAR SPEED

$$\omega_{avg} = \frac{\Delta\theta}{\Delta t}$$

$$\text{average angular speed} = \frac{\text{angular displacement}}{\text{time interval}}$$

Angular speed is given in units of radians per second (rad/s). Sometimes angular speeds are given in revolutions per unit time. Recall that 1 rev = 2π rad.

⊛TEKS

p. 246: 2C, 4B, 4B
p. 247: 4B, 2C, 4B

Classroom Practice

The following may be used as a teamwork exercise or for demonstration at the chalkboard or on an overhead projector.

PROBLEM
Angular speed

An Indy car can complete 120 laps in 1.5 h. Even though the track is an oval rather than a circle, you can still find the *average* angular speed. Calculate the average angular speed of the Indy car.

Answer
0.14 rad/s

PRACTICE GUIDE 7B

Solving for:		
Δt	📖 PE	Sample, 1–3, 4d; Ch. Rvw. 7
	💿 PW	5–6
	PB	4–6
$\Delta \theta$	📖 PE	4b
	💿 PW	Sample, 1–2
	PB	7–10
ω_{avg}	📖 PE	4a, 4c; Ch. Rvw. 41
	💿 PW	3–4
	PB	Sample, 1–3

ANSWERS TO

Practice 7B
Angular speed
1. 0.76 s
2. 1.5 s
3. 0.33 s
4. a. 0.23 rad/s
   b. 0.24 rad
   c. −6.3 rad/s
   d. 0.75 s

Angular speed ⭐TEKS 2C, 4B

PROBLEM

A child at an ice cream parlor spins on a stool. The child turns counterclockwise with an average angular speed of 4.0 rad/s. In what time interval will the child's feet have an angular displacement of 8.0π rad?

SOLUTION

Given: $\Delta\theta = 8.0\pi$ rad $\qquad \omega_{avg} = 4.0$ rad/s

Unknown: $\Delta t = ?$

Use the angular speed equation from page 247. Rearrange to solve for Δt.

$$\omega_{avg} = \frac{\Delta\theta}{\Delta t}$$

$$\Delta t = \frac{\Delta\theta}{\omega_{avg}}$$

$$\Delta t = \frac{8.0\pi \text{ rad}}{4.0 \text{ rad/s}} = 2.0\pi \text{ s}$$

$$\boxed{\Delta t = 6.3 \text{ s}}$$

PRACTICE 7B

Angular speed

1. A car tire rotates with an average angular speed of 29 rad/s. In what time interval will the tire rotate 3.5 times?

2. A girl ties a toy airplane to the end of a string and swings it around her head. The plane's average angular speed is 2.2 rad/s. In what time interval will the plane move through an angular displacement of 3.3 rad?

3. The average angular speed of a fly moving in a circle is 7.0 rad/s. How long does the fly take to move through 2.3 rad?

4. Fill in the unknown quantities in the following table:

	ω_{avg}	$\Delta\theta$	Δt
a.	?	+2.3 rad	10.0 s
b.	+0.75 rev/s	?	0.050 s
c.	?	−1.2 turns	1.2 s
d.	+2π rad/s	+1.5π rad	?

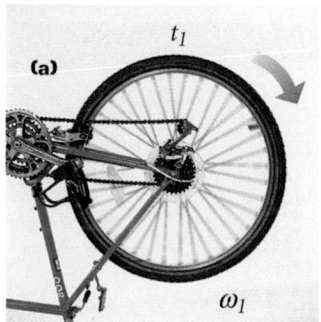

(a) t_1

ω_1

(b) t_2

ω_2

Figure 7-5
An accelerating bicycle wheel rotates with **(a)** an angular speed ω_1 at time t_1 and **(b)** an angular speed ω_2 at time t_2.

Angular acceleration occurs when angular speed changes

Figure 7-5 shows a bicycle turned upside down so that a repairperson can work on the rear wheel. The bicycle pedals are turned so that at time t_1 the wheel has angular speed ω_1, as shown in **Figure 7-5(a),** and at a later time, t_2, it has angular speed ω_2, as shown in **Figure 7-5(b).**

The average **angular acceleration,** α_{avg} (α is the Greek letter *alpha*), of an object is given by the relationship shown below. Angular acceleration has the units radians per second per second (rad/s^2). ⊛TEKS **4B**

angular acceleration

the time rate of change of angular speed, expressed in radians per second per second

---

ANGULAR ACCELERATION

$$\alpha_{avg} = \frac{\omega_2 - \omega_1}{t_2 - t_1} = \frac{\Delta\omega}{\Delta t}$$

$$\text{average angular acceleration} = \frac{\text{change in angular speed}}{\text{time interval}}$$

SAMPLE PROBLEM 7C

Angular acceleration ⊛TEKS 2C, 4B

PROBLEM

A car's tire rotates at an initial angular speed of 21.5 rad/s. The driver accelerates, and after 3.5 s the tire's angular speed is 28.0 rad/s. What is the tire's average angular acceleration during the 3.5 s time interval?

SOLUTION

Given: $\omega_1 = 21.5$ rad/s $\omega_2 = 28.0$ rad/s $\Delta t = 3.5$ s

Unknown: $\alpha_{avg} = ?$

Use the angular acceleration equation on this page.

$$\alpha_{avg} = \frac{\omega_2 - \omega_1}{\Delta t} = \frac{28.0 \text{ rad/s} - 21.5 \text{ rad/s}}{3.5 \text{ s}} = \frac{6.5 \text{ rad/s}}{3.5 \text{ s}}$$

$$\boxed{\alpha_{avg} = 1.9 \text{ rad/s}^2}$$

The following may be used as a teamwork exercise or for demonstration at the board or on an overhead projector.

PROBLEM

Angular acceleration

A top that is spinning at 15 rev/s spins for 55 s before coming to a stop. What is the average angular acceleration of the top while it is slowing?

Answer
    -1.7 rad/s^2

PRACTICE GUIDE 7C		
Solving for:		
α_{avg}	📖 **PE** Sample, 1–2, 3a, 3c; Ch. Rvw. 8–9	
	💿 **PW** 3–4	
	PB 7–10	
Δt	💿 **PW** Sample, 1–2	
	PB 4–6	
ω_2	💿 **PW** 5–6	
	PB Sample, 1–3	

⊛TEKS

p. 248: 2C, 4B
p. 249: 4B, 2C, 4B

Demonstration 1

Equal angular speed

Purpose Illustrate that angular speed is constant at any radius for a rigid extended object.

Materials record player/turntable; tape; colored markers

Procedure Use the tape and markers to make two brightly colored flags, and attach them to the turntable so that one flag is near the rim and the other is near, but not at, the center. Start the turntable at a moderately slow speed ($33\frac{1}{3}$ rpm) so that the flags are easily observed. Have students note the rotational speed of both flags. Point out that each flag makes a complete rotation in the same amount of time. Change speeds on the turntable, and repeat the observations.

250

PRACTICE 7C

Angular acceleration ⊛TEKS 2C, 4B

1. A figure skater begins spinning counterclockwise at an angular speed of 4.0π rad/s. During a 3.0 s interval, she slowly pulls her arms inward and finally spins at 8.0π rad/s. What is her average angular acceleration during this time interval?

2. What angular acceleration is necessary to increase the angular speed of a fan blade from 8.5 rad/s to 15.4 rad/s in 5.2 s?

3. Fill in the unknown quantities in the following table:

	α_{avg}	$\Delta\omega$	Δt
a.	?	+121.5 rad/s	7.0 s
b.	+0.75 rad/s^2	?	0.050 s
c.	?	−1.2 turns/s	1.2 s

All points on a rotating rigid object have the same angular acceleration and angular speed

If a point on the rim of a bicycle wheel had an angular speed greater than a point nearer the center, the shape of the wheel would be changing. Thus, for a rotating object to remain rigid, as does a bicycle wheel or a Ferris wheel, every portion of the object must have the same angular speed and the same angular acceleration. This fact is precisely what makes angular speed and angular acceleration so useful for describing rotational motion. ⊛TEKS 3A

COMPARING ANGULAR AND LINEAR QUANTITIES

Compare the equations we have found thus far for rotational motion with those we found for linear motion in Chapter 2. For example, compare the following defining equation for average angular speed with the defining equation for average linear speed: ⊛TEKS 4B

$$\omega_{avg} = \frac{\theta_f - \theta_i}{t_f - t_i} = \frac{\Delta\theta}{\Delta t} \qquad v_{avg} = \frac{x_f - x_i}{t_f - t_i} = \frac{\Delta x}{\Delta t}$$

The equations are similar, with θ replacing x and ω replacing v. Take careful note of such similarities as you study rotational motion because nearly every linear quantity we have encountered thus far has a corresponding twin in rotational motion, as shown in **Table 7-1.**

Table 7-1
Angular substitutes for linear quantities

Linear	Angular
x	θ
v	ω
a	α

Use kinematic equations for constant angular acceleration

In light of the similarities between variables in linear motion and those in rotational motion, it should be no surprise that the kinematic equations of rotational motion are similar to the linear kinematic equations in Chapter 2. The equations of rotational kinematics under constant angular acceleration, along with the corresponding equations for linear motion under constant acceleration, are summarized in **Table 7-2.** Note that the following rotational motion equations apply only for objects rotating about a fixed axis.

Table 7-2   Rotational and linear kinematic equations

Rotational motion with constant angular acceleration	Linear motion with constant acceleration
$\omega_f = \omega_i + \alpha\Delta t$	$v_f = v_i + a\Delta t$
$\Delta\theta = \omega_i\Delta t + \frac{1}{2}\alpha(\Delta t)^2$	$\Delta x = v_i\Delta t + \frac{1}{2}a(\Delta t)^2$
$\omega_f^2 = \omega_i^2 + 2\alpha(\Delta\theta)$	$v_f^2 = v_i^2 + 2a(\Delta x)$
$\Delta\theta = \frac{1}{2}(\omega_i + \omega_f)\Delta t$	$\Delta x = \frac{1}{2}(v_i + v_f)\Delta t$

Note the correlation between the rotational equations involving the angular variables θ, ω, and α and the equations of linear motion involving x, v, and a.

The quantity ω in these equations represents the *instantaneous angular speed* of the rotating object rather than the average angular speed.

Module 8
"Angular Kinematics"
provides an interactive lesson with guided problem-solving practice to teach you about different kinds of angular motion, including the types described here.

SAMPLE PROBLEM 7D

Angular kinematics ⭐TEKS 2C, 4B

PROBLEM

The wheel on an upside-down bicycle moves through 11.0 rad in 2.0 s. What is the wheel's angular acceleration if its initial angular speed is 2.0 rad/s?

SOLUTION

Given: $\Delta\theta = 11.0$ rad $\Delta t = 2.0$ s $\omega_i = 2.00$ rad/s

Unknown: $\alpha = ?$

Use the second angular kinematic equation from **Table 7-2** to solve for α.

$$\Delta\theta = \omega_i\Delta t + \frac{1}{2}\alpha(\Delta t)^2$$

$$\alpha = 2(\Delta\theta - \omega_i\Delta t)/(\Delta t)^2$$

$$\alpha = 2[11.0\ \text{rad} - (2.00\ \text{rad/s})(2.0\ \text{s})]/(2.0\ \text{s})^2$$

$$\boxed{\alpha = 3.5\ \text{rad/s}^2}$$

Classroom Practice

The following may be used as a teamwork exercise or for demonstration at the chalkboard or on an overhead projector.

PROBLEM

Angular kinematics

A barrel is given a downhill rolling start of 1.5 rad/s at the top of a hill. Assume a constant angular acceleration of 2.9 rad/s^2.

a. If the barrel takes 11.5 s to get to the bottom of the hill, what is the final angular speed of the barrel?

b. What angular displacement does the barrel experience during the 11.5 s ride?

Answers
   a. 35 rad/s
   b. 2.1×10^2 rad

Interactive Problem-Solving Tutor

See Module 8
"Angular Kinematics" provides additional development of problem-solving skills for this chapter.

⭐TEKS
   p. 250: 2C, 4B, 3A, 4B
   p. 251: 2C, 4B

ANSWERS TO

Practice 7D
Angular kinematics
1. 0.64 rad/s^2
2. 25 rad/s^2
3. 3.0 rad/s^2
4. 31.0 rad/s
5. 0.902 s

Section Review
ANSWERS

1. **a.** 0.44 rad   **c.** 2.23 rad
   b. 0.61 rad   **d.** 4.7 rad
2. −1.0 rad
3. 0.314 rad/s
4. 0.20 rad/s^2
5. 0.70 rad/s

Angular kinematics ⭐TEKS 2C, 4B

1. What is the angular acceleration of the upside-down bicycle wheel in Sample Problem 7D if it rotates through 18.0 rad in 5.00 s?

2. A diver performing a double somersault spins at an angular speed of 4.0π rad/s precisely 0.50 s after leaving the platform. Assuming the diver begins with zero initial angular speed and accelerates at a constant rate, what is the diver's angular acceleration during the double somersault?

3. A fish swimming behind an oil tanker gets caught in a whirlpool created by the ship's propellers. The fish has an angular speed of 1.0 rad/s. After 4.5 s, the fish's angular speed is 14.5 rad/s. If the water in the whirlpool accelerates at a constant rate, what is the angular acceleration?

4. A remote-controlled car's wheel accelerates at 22.4 rad/s^2. If the wheel begins with an angular speed of 10.8 rad/s, what is the wheel's angular speed after exactly three full turns?

5. How long does the wheel in item 4 take to make the three turns?

Section Review

1. Convert the following angles in degrees to radians:
   a. 25°
   b. 35°
   c. 128°
   d. 270°

2. A mosquito lands on a phonograph record 5.0 cm from the record's center. If the record turns clockwise so that the mosquito travels along an arc length of 5.0 cm, what is the mosquito's angular displacement?

3. A bicyclist rides along a circular track. If the bicyclist travels around exactly half the track in 10.0 s, what is his average angular speed?

4. **Physics in Action**   Find the angular acceleration of a spinning amusement-park ride that initially travels at 0.50 rad/s then accelerates to 0.60 rad/s during a 0.50 s time interval.

5. **Physics in Action**   What is the instantaneous angular speed of a spinning amusement-park ride that accelerates from 0.50 rad/s at a constant angular acceleration of 0.20 rad/s^2 for 1.0 s?

Tangential and centripetal acceleration

RELATIONSHIPS BETWEEN ANGULAR AND LINEAR QUANTITIES

As described at the beginning of Section 7-1, the motion of a point on a rotating object is most easily described in terms of an angle from a fixed reference line. In some cases, however, it is useful to understand how the angular speed and angular acceleration of a rotating object relate to the linear speed and linear acceleration of a point on the object.

Imagine a golfer swinging a golf club. The most effective method for hitting a golf ball a long distance involves swinging the club in an approximate circle around the body. If the club head undergoes a large angular acceleration, then the linear acceleration of the club head as it is swung will be large. This large linear acceleration causes the club head to strike the ball at a high speed and produce a significant force on the ball. This section will explore the relationships between angular and linear quantities. ⓧTEKS **3A**

Objects in circular motion have a tangential speed

Imagine an amusement-park carousel rotating about its center. Because a carousel is a rigid object, any two horses attached to the carousel have the same angular speed and angular acceleration regardless of their respective distances from the axis of rotation. However, if the two horses are different distances from the axis of rotation, they have different **tangential speeds.** The tangential speed of any point rotating about an axis is also called the instantaneous linear speed of that point. The tangential speed of a horse on the carousel is its speed along a line drawn tangent to its circular path. (Recall that the tangent to a circle is the line that touches the circle at one and only one point.) The tangential speeds of two horses at different distances from the center of a carousel are represented in **Figure 7-6.** ⓧTEKS **4B**

Note that the speed of the horse at point A is represented by a shorter arrow than the one that represents the speed of the horse at point B; this reflects the difference between the tangential speeds of the two horses. The horse on the outside must travel the same angular displacement during the same amount of time as the horse on the inside. To achieve this, the horse on the outside must travel a greater distance, Δs, than the horse on the inside. Thus, an object that is farther from the axis of a rigid rotating body, such as a carousel or a Ferris wheel, must travel at a higher tangential speed around the circular path, Δs, to travel the same angular displacement as would an object closer to the axis.

7-2 SECTION OBJECTIVES

- **Find the tangential speed of a point on a rigid rotating object using the angular speed and the radius.**
- **Solve problems involving tangential acceleration.**
- **Solve problems involving centripetal acceleration.**

tangential speed

the instantaneous linear speed of an object directed along the tangent to the object's circular path

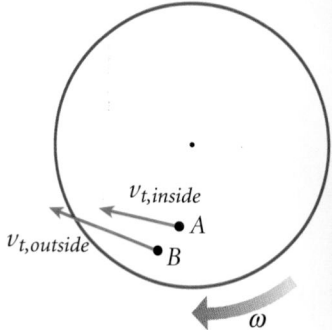

Figure 7-6

Horses on a carousel move at the same angular speed but different tangential speeds.

Demonstration 2

Tangential speed versus angular speed

Purpose Show that tangential speed depends on radius.

Materials two tennis balls attached to different lengths of string (approx. 1.0 m and 1.5 m)

Procedure Outside on an athletic field, hold the ends of both strings and whirl the tennis balls at constant angular speed over your head. Point out the equal angular speeds of the tennis balls. Ask students to predict the flights of the tennis balls when the strings are released.

Aiming away from students and any breakable items, release the strings and have students observe the flights. Discuss the longer horizontal displacement of the outer ball as a function of its tangential speed.

Visual Strategy

Figure 7-6

Q Which horse would travel farther before hitting the ground if the horses were released from the carousel?

A *The outer horse would travel farther because it has a higher tangential speed.*

ⓧTEKS

p. 252: 2C, 4B
p. 253: 3A, 4B

PROBLEM

Tangential speed

A golfer has a maximum angular speed of 6.3 rad/s for her swing. She can choose between two drivers, one placing the club head 1.9 m from her axis of rotation and the other placing it 1.7 m from the axis.

a. Find the tangential speed of the club head for each driver.

b. All other factors being equal, which driver is likely to hit the ball farther?

Answers

a. 12 m/s, 11 m/s

b. The longer driver will hit the ball farther because its club head has a higher tangential speed.

internet connect

*SCi*LINKS

NSTA

TOPIC: Rotational motion
GO TO: www.scilinks.org
*sci*LINKS CODE: HF2071

How can you find the tangential speed? Again consider the rotating carousel. If the carousel rotates through an angle $\Delta\theta$, a horse rotates through an arc length Δs in the interval Δt. The angular displacement of the horse is given by the equation for angular displacement.

$$\Delta\theta = \frac{\Delta s}{r}$$

To find the tangential speed of the horse, divide both sides of the equation by the time the horse takes to travel the distance Δs.

$$\frac{\Delta\theta}{\Delta t} = \frac{1}{r}\frac{\Delta s}{\Delta t}$$

From Section 7-1, you know that the left side of the equation equals ω_{avg}. Similarly, Δs is a linear distance, so Δs divided by Δt is a linear speed along an arc length. If Δt is very short, then Δs is so small that it is nearly tangent to the circle; therefore, the speed is the tangential speed. ⊛TEKS **4B**

> **TANGENTIAL SPEED**
>
> $$v_t = r\omega$$
>
> **tangential speed = distance from axis × angular speed**

Note that ω is the instantaneous angular speed, rather than the average angular speed, because the time interval is so short. This equation is valid only when ω is measured in radians per unit of time. Other measures of angular speed, such as degrees per second and revolutions per second, must not be used in this equation.

SAMPLE PROBLEM 7E

Tangential speed ⊛TEKS 2C, 4B

PROBLEM

The radius of a CD in a computer is 0.0600 m. If a microbe riding on the disc's rim has a tangential speed of 1.88 m/s, what is the disc's angular speed?

SOLUTION

Given: $r = 0.0600$ m $v_t = 1.88$ m/s

Unknown: $\omega = ?$

Use the tangential speed equation on this page to solve for angular speed.

$$v_t = r\omega$$

$$\omega = \frac{v_t}{r} = \frac{1.88 \text{ m/s}}{0.0600 \text{ m}}$$

$$\boxed{\omega = 31.3 \text{ rad/s}}$$

Tangential speed ⭐TEKS 2C, 4B

1. A woman passes through a revolving door with a tangential speed of 1.8 m/s. If she is 0.80 m from the center of the door, what is the door's angular speed?

2. A softball pitcher throws a ball with a tangential speed of 6.93 m/s. If the pitcher's arm is 0.660 m long, what is the angular speed of the ball before the pitcher releases it?

3. An athlete spins in a circle before releasing a discus with a tangential speed of 9.0 m/s. What is the angular speed of the spinning athlete? Assume the discus is 0.75 m from the athlete's axis of rotation.

4. Fill in the unknown quantities in the following table:

	v_t	ω	r
a.	?	121.5 rad/s	0.030 m
b.	0.75 m/s	?	0.050 m
c.	?	1.2 turns/s	3.8 m
d.	2.0π m/s	1.5π rad/s	?

SECTION 7-2

PRACTICE GUIDE 7E
Solving for:

ω	📖	**PE**	Sample, 1–3, 4b; Ch. Rvw. 21–22
	💿	**PW**	4
		PB	4–6
r	📖	**PE**	4d
	💿	**PW**	Sample, 1–3
		PB	7–10
v_t	📖	**PE**	4a, 4c
	💿	**PW**	5
		PB	Sample, 1–3

ANSWERS TO

Practice 7E
Tangential speed
1. 2.2 rad/s
2. 10.5 rad/s
3. 12 rad/s
4. a. 3.6 m/s
   b. 15 rad/s
   c. 29 m/s
   d. 1.3 m

Tangential acceleration is tangent to the circular path

If a carousel speeds up, the horses on it experience an angular acceleration. The linear acceleration related to this angular acceleration is tangent to the circular path and is called the **tangential acceleration.**

Imagine that an object rotating about a fixed axis changes its angular speed by $\Delta\omega$ in the interval Δt. At the end of this time, the speed of a point on the object has changed by the amount Δv_t. Using the equation for tangential velocity on page 254 gives the following:

tangential acceleration

the instantaneous linear acceleration of an object directed along the tangent to the object's circular path

$$\Delta v_t = r\Delta\omega$$

Dividing by Δt gives $\dfrac{\Delta v_t}{\Delta t} = r\dfrac{\Delta\omega}{\Delta t}$

If the time interval Δt is very small, then the left side of this relationship gives the tangential acceleration of the point. The angular speed divided by the time interval on the right side is the angular acceleration. Thus, the tangential acceleration of a point on a rotating object is given by the relationship on the next page. ⭐TEKS 4B

⭐TEKS

p. 254: 4B, 2C, 4B
p. 255: 2C, 4B, 4B

Classroom Practice

*The following may be used
as a teamwork exercise or for
demonstration at the chalkboard
or on an overhead projector.*

PROBLEM

Tangential acceleration

A yo-yo has a tangential accelera-tion of 0.98 m/s^2 when it is released. The string is wound around a central shaft of radius 0.35 cm. What is the angular acceleration of the yo-yo?

Answer

2.8×10^2 rad/s^2

PRACTICE GUIDE 7F		
Solving for:		
r	📖 **PE**	Sample, 1–2; Ch. Rvw. 23–24
	💿 **PW**	3–4
	PB	4–6
α	📖 **PE**	3
	💿 **PW**	Sample, 1–2
	PB	7–10
a_t	💿 **PW**	5–6
	PB	Sample, 1–3

ANSWERS TO

Practice 7F

Tangential acceleration

1. 1.5 m
2. 0.51 m
3. 0.63 rad/s^2

TANGENTIAL ACCELERATION

$$a_t = r\alpha$$

tangential acceleration = distance from axis $\times$ angular acceleration

Again, the angular acceleration in this equation refers to the instantaneous angular acceleration. This equation must use the unit radians to be valid. In SI, angular acceleration is expressed as radians per second per second.

SAMPLE PROBLEM 7F

• **Tangential acceleration** ⊛TEKS **2C, 4B**

PROBLEM

A spinning ride at a carnival has an angular acceleration of 0.50 rad/s^2. How far from the center is a rider who has a tangential acceleration of 3.3 m/s^2?

SOLUTION

Given: $\alpha = 0.50$ rad/s^2 $a_t = 3.3$ m/s^2

Unknown: $r = ?$

Use the tangential acceleration equation on this page. Rearrange to solve for r.

$$a_t = r\alpha$$

$$r = \frac{a_t}{\alpha} = \frac{3.3 \text{ m/s}^2}{0.50 \text{ rad/s}^2}$$

$$\boxed{r = 6.6 \text{ m}}$$

PRACTICE 7F

Tangential acceleration

1. A dog on a merry-go-round undergoes a 1.5 m/s^2 linear acceleration. If the merry-go-round's angular acceleration is 1.0 rad/s^2, how far is the dog from the axis of rotation?

2. A young boy swings a yo-yo horizontally above his head at an angular acceleration of 0.35 rad/s^2. If tangential acceleration of the yo-yo at the end of the string is 0.18 m/s^2, how long is the string?

3. What is a tire's angular acceleration if the tangential acceleration at a radius of 0.15 m is 9.4×10^{-2} m/s^2?

CENTRIPETAL ACCELERATION

Figure 7-7 shows a car moving in a circular path with a constant tangential speed of 30 km/h. Even though the car moves at a constant speed, it still has an acceleration. To see why this is, consider the defining equation for acceleration.

$$\mathbf{a} = \frac{\mathbf{v_f} - \mathbf{v_i}}{t_f - t_i}$$

Note that acceleration depends on a change in the velocity. Because velocity is a vector, there are two ways an acceleration can be produced: by a change in the *magnitude* of the velocity and by a change in the *direction* of the velocity. For a car moving in a circular path with constant speed, the acceleration is due to a change in direction. An acceleration of this nature is called a **centripetal** (center-seeking) **acceleration.** Its magnitude is given by the following equation:

$$a_c = \frac{v_t^2}{r}$$

Consider **Figure 7-8(a).** An object is seen first at point A, with tangential velocity $\mathbf{v_i}$ at time t_i, and then at point B, with tangential velocity $\mathbf{v_f}$ at a later time, t_f. Assume that $\mathbf{v_i}$ and $\mathbf{v_f}$ differ in direction only and their magnitudes are the same.

The change in velocity, $\Delta\mathbf{v} = \mathbf{v_f} - \mathbf{v_i}$, can be determined graphically, as shown by the vector triangle in **Figure 7-8(b).** Note that when Δt is very small (as Δt approaches zero), $\mathbf{v_f}$ will be almost parallel to $\mathbf{v_i}$ and the vector $\Delta\mathbf{v}$ will be approximately perpendicular to them, pointing toward the center of the circle. This means that the acceleration will also be directed toward the center of the circle because it is in the direction of $\Delta\mathbf{v}$.

Because the tangential speed is related to the angular speed through the relationship $v_t = r\omega$, the centripetal acceleration can be found using the angular speed as well. ⭐TEKS 4B

Figure 7-7
Although the car moves at a constant speed of 30 km/h, the car still has an acceleration because the direction of the velocity changes.

centripetal acceleration

> *acceleration directed toward the center of a circular path*

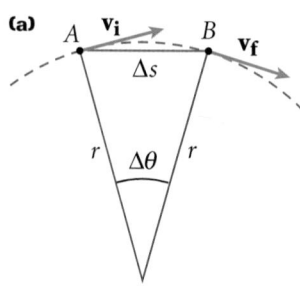

(a)

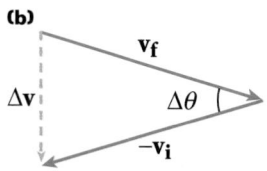

(b)

$$\Delta\mathbf{v} = \mathbf{v_f} - \mathbf{v_i} = \mathbf{v_f} + (-\mathbf{v_i})$$

Figure 7-8
(a) As the particle moves from A to B, the direction of the particle's velocity vector changes. **(b)** Vector addition is used to determine the direction of the change in velocity, $\Delta\mathbf{v}$, which for short time intervals is toward the center of the circle.

CENTRIPETAL ACCELERATION

$$a_c = \frac{v_t^2}{r}$$

$$a_c = r\omega^2$$

$$\text{centripetal acceleration} = \frac{(\text{tangential speed})^2}{\text{distance from axis}}$$

$$\text{centripetal acceleration} = \text{distance from axis} \times (\text{angular speed})^2$$

SECTION 7-2

🛑 **Misconception Alert**

Some students will have difficulty with terminology at this point because of previous familiarity with the term *centrifugal*. It is important to emphasize the distinction between *centripetal* (center-seeking) and *centrifugal* (center-fleeing). To avoid reinforcing this misconception, avoid using the term *centrifugal*.

🛑 **Misconception Alert**

Figure 7-8 is a complicated figure and may require some explanation and supplemental diagramming on the board. Make sure students understand that $\mathbf{v_i}$ in part **(b)** of the diagram is negative, and stress that part **(b)** represents the vector addition of $\mathbf{v_f}$ and $-\mathbf{v_i}$ for a *very short* time interval. Explain that the phrase "as Δt approaches zero" describes a situation in which the time interval is so small that it is *almost* equal to zero.

⭐TEKS

p. 256: 2C, 4B
p. 257: 4B

The following may be used as teamwork exercises or for demonstration at the board or on an overhead projector.

PROBLEM

Centripetal acceleration

A cylindrical space station with a 115 m radius rotates around its longitudinal axis at an angular speed of 0.292 rad/s. Calculate the centripetal acceleration:

a. halfway to the rim of the station

b. at the rim of the station

Answers
   a. 4.90 m/s^2
   b. 9.81 m/s^2

PRACTICE GUIDE 7G

Solving for:		
v_t	📖 **PE**	Sample, 1–3; Ch. Rvw. 25–26, 52b*
	💿 **PW**	3–4
	PB	4–6
r	📖 **PE**	4; Ch. Rvw. 43a
	💿 **PW**	Sample, 1–2
	PB	7–10
a_c	📖 **PE**	5; Ch. Rvw. 47a, 50
	💿 **PW**	5–6
	PB	Sample, 1–3

ANSWERS TO

Practice 7G
Centripetal acceleration

1. 2.5 m/s
2. 11 m/s
3. 1.5 m/s; 1.0 rad/s
4. 58.7 m
5. 84 m/s^2

Centripetal acceleration ⭐TEKS 2C, 4B

PROBLEM

A test car moves at a constant speed around a circular track. If the car is 48.2 m from the track's center and has a centripetal acceleration of 8.05 m/s^2, what is its tangential speed?

SOLUTION

Given: $r = 48.2 \text{ m}$ $a_c = 8.05 \text{ m/s}^2$

Unknown: $v_t = ?$

Use the first centripetal acceleration equation from page 257. Rearrange to solve for v_t.

$$a_c = \frac{v_t^2}{r}$$

$$v_t = \sqrt{a_c r} = \sqrt{(8.05 \text{ m/s}^2)(48.2 \text{ m})}$$

$$\boxed{v_t = 19.7 \text{ m/s}}$$

Centripetal acceleration

1. A girl sits on a tire that is attached to an overhanging tree limb by a rope. The girl's father pushes her so that her centripetal acceleration is 3.0 m/s^2. If the length of the rope is 2.1 m, what is the girl's tangential speed?

2. A young boy swings a yo-yo horizontally above his head so that the yo-yo has a centripetal acceleration of 250 m/s^2. If the yo-yo's string is 0.50 m long, what is the yo-yo's tangential speed?

3. A dog sits 1.5 m from the center of a merry-go-round. If the dog undergoes a 1.5 m/s^2 centripetal acceleration, what is the dog's linear speed? What is the angular speed of the merry-go-round?

4. A race car moves along a circular track at an angular speed of 0.512 rad/s. If the car's centripetal acceleration is 15.4 m/s^2, what is the distance between the car and the center of the track?

5. A piece of clay sits 0.20 m from the center of a potter's wheel. If the potter spins the wheel at an angular speed of 20.5 rad/s, what is the magnitude of the centripetal acceleration of the piece of clay on the wheel?

Tangential and centripetal accelerations are perpendicular

Centripetal and tangential acceleration are not the same. To understand why, consider a car moving around a circular track. Because the car is moving in a circular path, it always has a centripetal component of acceleration because its direction of travel, and hence the direction of its velocity, is continually changing. If the car's speed is increasing or decreasing, the car also has a tangential component of acceleration. To summarize, the tangential component of acceleration is due to changing speed; the centripetal component of acceleration is due to changing direction.

Find the total acceleration using the Pythagorean theorem

When both components of acceleration exist simultaneously, the tangential acceleration is tangent to the circular path and the centripetal acceleration points toward the center of the circular path. Because these components of acceleration are perpendicular to each other, the magnitude of the *total acceleration* can be found using the Pythagorean theorem, as follows:

$$a_{total} = \sqrt{a_t^2 + a_c^2}$$

The direction of the total acceleration, as shown in **Figure 7-9,** depends on the magnitude of each component of acceleration and can be found using the inverse of the tangent function. (★)TEKS **3B**

$$\theta = \tan^{-1}\frac{a_c}{a_t}$$

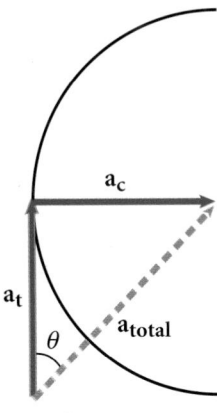

Figure 7-9
The direction of the total acceleration of a rotating object can be found using the tangent function.

SECTION 7-2

Teaching Tip

To simulate the large accelerations involved in spaceflight, *Mercury* astronauts rode in the U. S. Navy's centrifuge in Johnsville, Pennsylvania. The astronauts sat in a gondola at the end of a 15.2 m arm that spun around a central axis. During the spin, the astronauts experienced a combination of centripetal and tangential accelerations of the gondola that ranged from 8 to 10 times the acceleration of gravity.

Section Review   (★)TEKS **2C, 4B**

1. Find the tangential speed of a ball swung at a constant angular speed of 5.0 rad/s on a rope that is 5.0 m long.

2. If an object has a tangential acceleration of 10.0 m/s², the angular speed will do which of the following?

   a. decrease
   b. stay the same
   c. increase

3. **Physics in Action**   Find the tangential acceleration of a person standing 9.5 m from the center of a spinning amusement-park ride that has an angular acceleration of 0.15 rad/s².

4. **Physics in Action**   If a spinning amusement-park ride has an angular speed of 1.2 rad/s, what is the centripetal acceleration of a person standing 12 m from the center of the ride?

Section Review
ANSWERS

1. 25 m/s
2. c
3. 1.4 m/s²
4. 17 m/s²

(★)TEKS

p. 258: 2C, 4B
p. 259: 2C, 3B, 4B

7-3
Causes of circular motion

7-3 SECTION OBJECTIVES

- **Calculate the force that maintains circular motion.**

- **Explain how the apparent existence of an outward force in circular motion can be explained as inertia resisting the force that maintains circular motion.**

- **Apply Newton's universal law of gravitation to find the gravitational force between two masses.**

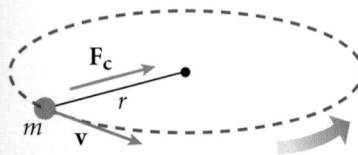

Figure 7-10
When a ball is whirled in a circle, a force directed toward the center of the ball's circular path acts on it.

FORCE THAT MAINTAINS CIRCULAR MOTION

Consider a ball of mass m tied to a string of length r that is being whirled in a horizontal circular path, as shown in **Figure 7-10.** Assume that the ball moves with constant speed. Because the velocity vector, **v,** changes direction continuously during the motion, the ball experiences a centripetal acceleration directed toward the center of motion, as described in Section 7-2, with magnitude given by the following equation: ⭐TEKS **4B**

$$a_c = \frac{v_t^2}{r}$$

The inertia of the ball tends to maintain the ball's motion in a straight-line path; however, the string counteracts this tendency by exerting a force on the ball that makes the ball follow a circular path. This force is directed along the length of the string toward the center of the circle, as shown in **Figure 7-10.** The magnitude of this force can be found by applying Newton's second law along the radial direction. ⭐TEKS **3B**

$$F_c = ma_c$$

The net force on an object directed toward the center of the object's circular path is the force that maintains the object's circular motion. ⭐TEKS **4C**

> **FORCE THAT MAINTAINS CIRCULAR MOTION**
>
> $$F_c = \frac{mv_t^2}{r}$$
>
> $$F_c = mr\omega^2$$
>
> $$\text{force that maintains circular motion} = \text{mass} \times \frac{(\text{tangential speed})^2}{\text{distance to axis}}$$
>
> $$\text{force that maintains circular motion} = \text{mass} \times \text{distance to axis} \times (\text{angular speed})^2$$

The force that maintains circular motion is measured in the SI unit of newtons. This force is no different from any of the other forces we have studied. For example, friction between a race car's tires and a circular racetrack provides the force that enables the car to travel in a circular path. As another example, the gravitational force exerted on the moon by Earth provides the force necessary to keep the moon in its orbit.

Force that maintains circular motion ⭐TEKS 2C, 4B, 4C

PROBLEM

A pilot is flying a small plane at 30.0 m/s in a circular path with a radius of 100.0 m. If a force of 635 N is needed to maintain the pilot's circular motion, what is the pilot's mass?

SOLUTION

Given: $v_t = 30.0$ m/s $r = 100.0$ m $F_c = 635$ N

Unknown: $m = ?$

Use the equation for force from page 260. Rearrange to solve for m.

$$F_c = m\frac{v_t^2}{r}$$

$$m = F_c\frac{r}{v_t^2} = 635 \text{ N}\frac{100.0 \text{ m}}{(30.0 \text{ m/s})^2}$$

$$\boxed{m = 70.6 \text{ kg}}$$

Force that maintains circular motion

1. A girl sits in a tire that is attached to an overhanging tree limb by a rope 2.10 m in length. The girl's father pushes her with a tangential speed of 2.50 m/s. If the magnitude of the force that maintains her circular motion is 88.0 N, what is the girl's mass?

2. A bicyclist is riding at a tangential speed of 13.2 m/s around a circular track with a radius of 40.0 m. If the magnitude of the force that maintains the bike's circular motion is 377 N, what is the combined mass of the bicycle and rider?

3. A dog sits 1.50 m from the center of a merry-go-round with an angular speed of 1.20 rad/s. If the magnitude of the force that maintains the dog's circular motion is 40.0 N, what is the dog's mass?

4. A 905 kg test car travels around a 3.25 km circular track. If the magnitude of the force that maintains the car's circular motion is 2140 N, what is the car's tangential speed?

SECTION 7-3

Classroom Practice

The following may be used as a teamwork exercise or for demonstration at the chalkboard or on an overhead projector.

PROBLEM

The moon (mass $= 7.36 \times 10^{22}$ kg) orbits Earth at a range of 3.84×10^5 km with a period of approximately 28 days. Determine the force that maintains the circular motion of the moon.

Answer
1.9×10^{20} N

PRACTICE GUIDE 7H

Solving for:		
m	📖 **PE**	Sample, 1–3; Ch. Rvw. 37a, 38*
	💿 **PW**	3
	PB	7–10
v_t	📖 **PE**	4; Ch. Rvw. 48*, 52b
	💿 **PW**	Sample, 1–2
	PB	4–5
r	💿 **PW**	4
	PB	6–7
F_c	📖 **PE**	Ch. Rvw. 37b, 43b, 47b*, 52a
	💿 **PW**	5
	PB	Sample, 1–3

ANSWERS TO

Practice 7H
1. 29.6 kg
2. 86.5 kg
3. 18.5 kg
4. 35.0 m/s

Teaching Tip

Explain how a washing machine removes excess water from clothes during the spin cycle.

An *erroneous* explanation from the reference frame of a stationary observer is that the rotating system creates some mysterious outward force on each drop of water and that this force causes the water to be hurled to the outer drum of the machine. The correct explanation is as follows: When the clothes are at rest in the machine, water is held to them by molecular forces between the water and the fabric. During the spin cycle, the clothes rotate and the molecular forces are not great enough to keep the water molecules moving in a circular path along with the clothes. Hence, the drops of water, because of their inertia, move in straight-line paths until they encounter the sides of the spinning drum.

ANSWERS TO

Conceptual Challenge

1. Each point on the crust (except the center) has an inertial tendency to move in a straight line. The cohesion of the dough keeps it from flying apart.

2. Inertia causes the swings and people to move outward. The chains provide the force that maintains the circular motion of the swings and people.

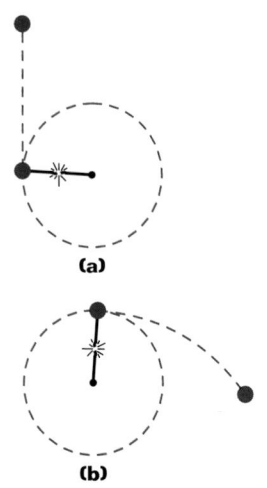

Figure 7-11
A ball is whirled in a vertical circular path on the end of a string. When the string breaks at the position shown in **(a)**, the ball moves vertically upward in free fall. **(b)** When the string breaks at the top of the ball's path, the ball moves along a parabolic path.

⭐ TEKS 3A, 4B, 4C

1. Pizza

Pizza makers traditionally form the crust by throwing the dough up in the air and spinning it. Why does this make the pizza crust bigger?

2. Swings

The amusement-park ride pictured below spins riders around on swings attached by cables from above. What causes the swings to move away from the center of the ride when the center column begins to turn?

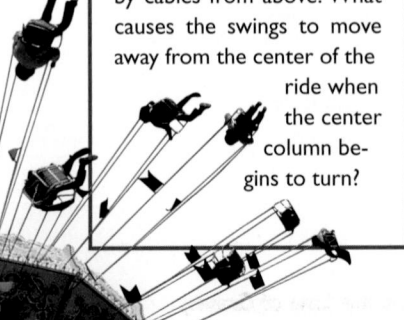

A force directed toward the center is necessary for circular motion

Because the force that maintains circular motion acts at right angles to the motion, it causes a change in the direction of the velocity. If this force vanishes, the object does not continue to move in its circular path. Instead, it moves along a straight-line path tangent to the circle. To see this point, consider a ball that is attached to a string and is being whirled in a vertical circle, as shown in **Figure 7-11.** If the string breaks when the ball is at the position shown in **Figure 7-11(a),** the force that maintains circular motion will vanish and the ball will move vertically upward. The motion of the ball will be that of a free-falling body. If the string breaks when the ball is at the top of its circular path, as shown in **Figure 7-11(b),** the ball will fly off horizontally in a direction tangent to the path, then move in the parabolic path of a projectile.

⭐ TEKS 3A, 4B

DESCRIBING THE MOTION OF A ROTATING SYSTEM

To better understand the motion of a rotating system, consider a car approaching a curved exit ramp to the left at high speed. As the driver makes the sharp left turn, the passenger slides to the right and hits the door. At that point, the force of the door keeps the passenger from being ejected from the car. What causes the passenger to move toward the door? A popular explanation is that there must be a force that pushes the passenger outward. This force is sometimes called the centrifugal force, but that term often creates confusion, so it is not used in this textbook.

Inertia is often misinterpreted as a force

The phenomenon is correctly explained as follows: Before the car enters the ramp, the passenger is moving in a straight-line path. As the car enters the ramp and travels along a curved path, the passenger, because of inertia, tends to move along the original straight-line path. This is in accordance with Newton's first law, which states that the natural tendency of a body is to continue moving in a straight line. However, if a sufficiently large force that maintains circular motion (toward the center of curvature) acts on the passenger, the person moves in a curved path, along with the car. The origin of the force that maintains the circular motion of the passenger is the force of friction between the passenger and the car seat. If this frictional force is not sufficient, the passenger slides across the seat as the car turns underneath. Because of inertia, the passenger continues to move in a straight-line path. Eventually, the passenger encounters the door, which provides a large enough force to enable the passenger to follow the same curved path as the car. The passenger slides toward the door not because of some mysterious outward force but because the force that maintains circular motion is not great enough to enable the passenger to travel along the circular path followed by the car.

⭐ TEKS 3A, 4B, 4C

NEWTON'S LAW OF UNIVERSAL GRAVITATION

Note that planets move in nearly circular orbits around the sun. As mentioned earlier, the force that keeps these planets from coasting off in a straight line is a **gravitational force.** The gravitational force is a field force that always exists between two masses, regardless of the medium that separates them. It exists not just between large masses like the sun, Earth, and moon but between any two masses, regardless of size or composition. For instance, desks in a classroom have a mutual attraction because of gravitational force. The force between the desks, however, is small relative to the force between the moon and Earth because the gravitational force is proportional to the product of the objects' masses. ⊛TEKS **6A**

Gravitational force acts such that objects are always attracted to one another. Examine the illustration of Earth and the moon in **Figure 7-12.** Note that the gravitational force between Earth and the moon is attractive, and recall that Newton's third law states that the force exerted on Earth by the moon, $\mathbf{F_{mE}}$, is equal in magnitude to and in the opposite direction of the force exerted on the moon by Earth, $\mathbf{F_{Em}}$. ⊛TEKS **3A**

gravitational force

the mutual force of attraction between particles of matter

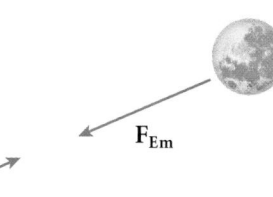

$\mathbf{F_{Em}}$

$\mathbf{F_{mE}}$

Figure 7-12

The gravitational force between Earth and the moon is attractive. According to Newton's third law, $F_{Em} = F_{mE}$.

Gravitational force depends on the distance between two masses

If masses m_1 and m_2 are separated by distance r, the magnitude of the gravitational force is given by the following equation: ⊛TEKS **3B, 6A**

NEWTON'S LAW OF UNIVERSAL GRAVITATION

$$F_g = G\frac{m_1 m_2}{r^2}$$

$$\text{gravitational force} = \text{constant} \times \frac{\text{mass 1} \times \text{mass 2}}{(\text{distance between center of masses})^2}$$

internet **connect**

$SC\overset{i}{L}INKS_{\text{\tiny .}}$

NSTA

TOPIC: Law of gravitation
GO TO: www.scilinks.org
***sci*LINKS CODE:** HF2073

G is a universal constant called the *constant of universal gravitation;* it can be used to calculate gravitational forces between any two particles and has been determined experimentally. ⊛TEKS **2C**

$$G = 6.673 \times 10^{-11}\frac{\text{N}\bullet\text{m}^2}{\text{kg}^2}$$

The law of universal gravitation is an example of an *inverse-square law,* because the force varies as the inverse square of the separation. That is, the force between two masses decreases as the masses move farther apart.

SECTION 7-3

Visual Strategy

Figure 7-12
Point out that in fact, the moon and Earth each orbit around the center of mass of the Earth-moon system. Because Earth has a much greater mass than the moon, the center of mass of the system is much closer to Earth—actually, it's inside Earth.

Q What might Earth's orbit around the moon look like from above the North Pole?

A *Because the axis of rotation for Earth's orbit is inside Earth, its orbital motion would look like a slight wobble.*

⊛TEKS

p. 262: 3A, 4B, 3A, 4B, 3A, 4B, 4C
p. 263: 6A, 3A, 3B, 6A, 2C

The following may be used as teamwork exercises or for demonstration at the chalkboard or on an overhead projector.

PROBLEM

Gravitational force

Find the gravitational force exerted on the moon (mass = 7.36×10^{22} kg) by Earth (mass = 5.98×10^{24} kg) when the distance between them is 3.84×10^8 m.

Answer

1.99×10^{20} N

Given that Earth's average distance from the sun (mass = 1.99×10^{30} kg) is 1.50×10^{11} m, find the net gravitational force exerted on the moon by the sun and Earth during a lunar eclipse (in which Earth is between the sun and the moon).

Answer

6.33×10^{20} N toward the sun

PRACTICE GUIDE 7I

Solving for:		
r	📖 **PE**	Sample, 1–2; Ch. Rvw. 39–40
	💿 **PW**	4–5
	PB	7–10
m	💿 **PW**	Sample, 1–3
	PB	4–6
F_g	📖 **PE**	3; Ch. Rvw. 49
	💿 **PW**	6–7
	PB	Sample, 1–3

Did you know?

Sir Isaac Newton knew from his first law that a net force had to be acting on the moon. Otherwise, the moon would move in a straight-line path rather than in an elliptical orbit. He reasoned that this force arises as a result of an attractive field force between the moon and Earth and that a force of the same origin causes an apple to fall from a tree to Earth.

⭐ **TEKS** 3E, 6B

Gravitational force is localized to the center of a spherical mass

The gravitational force exerted by a spherical mass on a particle outside the sphere is the same as it would be if the entire mass of the sphere were concentrated at its center. For example, the force on an object of mass m at Earth's surface has the following magnitude: ⭐ **TEKS** 6A

$$F_g = G\frac{M_E m}{R_E^2}$$

M_E is Earth's mass and R_E is its radius. This force is directed toward the center of Earth. Note that this force is in fact the weight of the mass, *mg*.

$$mg = G\frac{M_E m}{R_E^2}$$

By substituting the actual values for the mass and radius of Earth, we can find the value for *g* and compare it with the value of free-fall acceleration used throughout this book. ⭐ **TEKS** 2C

Because *m* occurs on both sides of the equation above, these masses cancel.

$$g = G\frac{M_E}{R_E^2} = \left(6.673 \times 10^{-11}\frac{\text{N} \cdot \text{m}^2}{\text{kg}^2}\right)\frac{5.98 \times 10^{24}\text{ kg}}{(6.37 \times 10^6\text{ m})^2} = 9.83\text{ m/s}^2$$

This value for *g* is approximately equal to the value used throughout this book. The difference is due to rounding the values for Earth's mass and radius.

SAMPLE PROBLEM 7I

Gravitational force ⭐ **TEKS** 2C, 6A

PROBLEM

Find the distance between a 0.300 kg billiard ball and a 0.400 kg billiard ball if the magnitude of the gravitational force is 8.92×10^{-11} N.

SOLUTION

Given: $m_1 = 0.300$ kg $m_2 = 0.400$ kg $F_g = 8.92 \times 10^{-11}$ N

Unknown: $r = ?$

Use the equation for Newton's Universal Law of Gravitation.

$$r^2 = \frac{G}{F_g}m_1 m_2 = \frac{6.673 \times 10^{-11}\frac{\text{N} \cdot \text{m}^2}{\text{kg}^2}}{8.92 \times 10^{-11}\text{ N}}(0.300\text{ kg})(0.400\text{ kg})$$

$$= 8.97 \times 10^{-2}\text{ m}^2$$

$$\boxed{r = \sqrt{8.97 \times 10^{-2}\text{ m}^2} = 3.00 \times 10^{-1}\text{ m}}$$

Gravitational force ⭐TEKS 2C, 6A

1. If the mass of each ball in Sample Problem 7I is 0.800 kg, at what distance between the balls will the gravitational force between the balls have the same magnitude as that in Sample Problem 7I?

2. Mars has a mass of about 6.4×10^{23} kg, and its moon Phobos has a mass of about 9.6×10^{15} kg. If the magnitude of the gravitational force between the two bodies is 4.6×10^{15} N, how far apart are Mars and Phobos?

3. Find the magnitude of the gravitational force a 67.5 kg person would experience while standing on the surface of each of the following planets:

Planet	m	r
a. Earth	5.98×10^{24} kg	6.37×10^{6} m
b. Mars	6.34×10^{23} kg	3.43×10^{6} m
c. Pluto	5×10^{23} kg	4×10^{5} m

Section Review

1. A roller coaster moves through a vertical loop at a constant speed and suspends its passengers upside down. In what direction is the force that causes the coaster and its passengers to move in a circle? What provides this force?

2. Identify the force that maintains the circular motion of the following:

   a. a *bicyclist* moving around a flat circular track
   b. a *bicycle* moving around a flat circular track
   c. a *bobsled* turning a corner on its track

3. A 90.0 kg person stands 1.00 m from a 60.0 kg person sitting on a bench nearby. What is the magnitude of the gravitational force between them?

4. **Physics in Action** A 90.0 kg person rides a spinning amusement-park ride that has an angular speed of 1.15 rad/s. If the radius of the ride is 11.5 m, what is the magnitude of the force that maintains the circular motion of the person?

5. **Physics in Action** Calculate the mass that a planet with the same radius as Earth would need in order to exert a gravitational force equal to the force on the person in item 4.

ANSWERS TO

Practice 7I
Gravitational force
1. 0.692 m
2. 9.4×10^{6} m (9.4×10^{3} km)
3. **a.** 664 N
   b. 243 N
   c. 1×10^{4} N

Section Review
ANSWERS

1. toward the center of the track; the track and gravity
2. **a.** the force of the bicycle seat on the bicyclist, friction between the bicyclist and the seat, and the force of the bicyclist's legs grasping the seat
   b. friction between the tires and the track
   c. the normal force from the curved side of the track
3. 3.60×10^{-7} N
4. 1.37×10^{3} N
5. 9.26×10^{24} kg

⭐TEKS

p. 264: 6A, 3E, 6B, 2C, 2C, 6A
p. 265: 2C, 6A

BACKGROUND

This feature builds on the discussion of gravitational force in this chapter. Stars collapse because of their own gravitational force after their source of fuel is depleted. It is theorized that when stars that are three to four times the size of our sun collapse, a black hole is formed.

The feature begins with a discussion of the two components of projectile motion, explaining that satellites are projectiles. Then the feature explains that a projectile's velocity must have a certain magnitude, calculated using the universal law of gravitation, to overcome the gravitational force exerted on the projectile by Earth. Next the discussion turns to the speed required to escape a very massive object with a small diameter, pointing out that the speed required to escape an object of such dimensions is greater than the speed of light. The feature concludes with a short history of black holes and current ideas related to them.

PHYSICS ON THE EDGE

Projectiles and satellites

As explained in Chapter 3, when a ball is thrown parallel to the ground, the motion of the ball has two components of motion: a horizontal velocity, which remains unchanged, and a vertical acceleration, which equals free-fall acceleration. Given this analysis, it may seem confusing to think of the moon and other satellites in orbit around Earth as projectiles. But in fact, satellites are projectiles. As shown in **Figure 7-13,** the larger the velocity parallel to Earth's surface, the farther the projectile moves before striking Earth. Note, however, that at some large velocity, the projectile returns to its point of origin without moving closer to Earth. In this case, the gravitational force between the projectile and Earth is just great enough to keep the projectile from moving along its inertial straight-line path. This is how satellites stay in orbit.

(★)TEKS **3A**

Escape speed

When the speed of an object, such as a rocket, is greater than the speed required to keep it in orbit, the object can escape the gravitational pull of Earth and soar off into space. The object soars off into space when its initial speed moves it out of the range in which the gravitational force is significant. Mathematically, the value of this *escape speed* (v_{esc}) is given by the following equation: (★)TEKS **4B**

$$v_{esc} = \sqrt{\frac{2MG}{R}}$$

Earth's radius, R, is about 6.37×10^6 m, and its mass is approximately 5.98×10^{24} kg. Thus, the escape speed of a projectile from Earth is 1.12×10^4 m/s. (Note that this value does not depend on the mass of the projectile in question.)

As the mass of a planet or other body increases and its radius decreases, the escape speed necessary for a projectile to escape the gravitational pull of that body increases, as shown in **Figure 7-14.** If the body has a very large mass and

Figure 7-13
When the speed of a projectile is large enough, the projectile orbits Earth as a satellite.

(★)TEKS **2C**

a small radius, the speed necessary for a projectile to escape the gravitational pull of that body reaches very high values. For example, an object with a mass three times that of the sun but with a diameter of about 10 km would require an escape speed equal to the speed of light. In other words, the force of gravity such an object exerts on a projectile is so great that even light does not move fast enough to escape it.

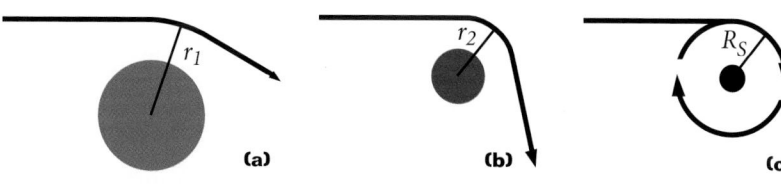

(a)   (b)   (c)

internet connect

SCI LINKS
NSTA

TOPIC: Black holes
GO TO: www.scilinks.org
***sci*LINKS CODE:** HF2074

Figure 7-14
(a) A projectile that passes near any massive object will be bent from its trajectory. **(b)** The more compact the object, the nearer the projectile can approach and the greater the escape speed it needs. **(c)** If the object is so small and massive that a projectile's escape speed at distance R_S is greater than the speed of light, then the object is classified as a black hole.

Black holes

The existence of such massive objects was first predicted in 1916 by Karl Schwarzschild, who used his solutions to Einstein's general-relativity equations to predict their properties. Thus, the distance from the center of the object to the circular orbit at which escape speed is equal to the speed of light is called the *Schwarzschild radius*. Because light cannot escape from any point within the sphere defined by the Schwarzschild radius, no information can be obtained about events that occur in this region. Hence, the edge of the sphere is called the *event horizon,* and the apparently lightless region within it is called a *black hole.* ⊛TEKS **3E**

Recent observations have provided strong evidence for the existence of black holes. Tremendous amounts of X rays and other radiation have been observed coming from regions that are near visible stars, although no stars appear to be the source of the radiation. If the visible star has a black hole as a companion, it could be losing some of its outer atmosphere to the black hole, and those atmospheric gases could be emitting radiation as they accelerate closer to the black hole. Candidates for black holes of these types include Scorpius X-1 and Cygnus X-1.

The large amount of energy that galactic centers produce suggests to many astrophysicists that the energy sources are supermassive black holes. The galaxy NGC 4261, shown in **Figure 7-15,** is likely to have a black hole at its center. Some astronomers believe the Milky Way, our own galaxy, contains a black hole about the size of our solar system.

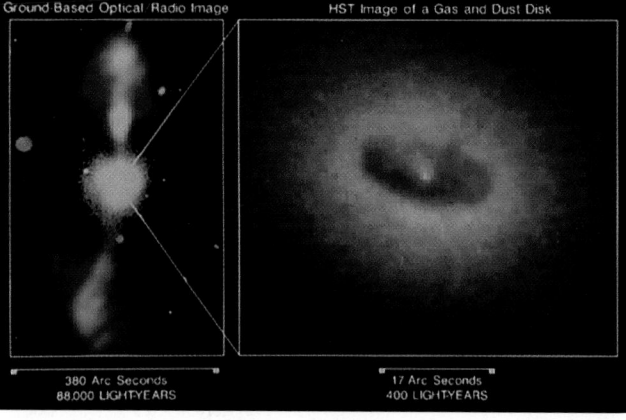

Ground-Based Optical Radio Image   HST Image of a Gas and Dust Disk

380 Arc Seconds   17 Arc Seconds
88,000 LIGHT-YEARS   400 LIGHT-YEARS

Figure 7-15
The gas jets and the central disk shown at right, created by combining an optical image with a radio telescope image of NGC 4261, show signs of a black hole in this galaxy's center.

SECTION 7-3

EXTENSION

- There is controversy over the existence of black holes. Have students research present speculations and experimental evidence about black holes, and have them create a list of convincing arguments for each side of the debate.

- After students have researched the subject of black holes, have them debate the topic in a class forum. Use the usual format for debate, allowing for an argument and a rebuttal. After the debate, ask students to write an essay about which side of the argument they find most compelling. Have them support their stance and disprove the opposition on each point.

⊛TEKS

p. 266: 3A, 4B, 2C
p. 267: 3E

CHAPTER 7
Summary

Chapter 7
Summary

Teaching Tip

This chapter contains several difficult concepts and equations. Students may find it helpful to compile a table of equations that includes short descriptions of the quantities involved. Students' tables may contain vocabulary terms or other integral terms or concepts. This table will prove useful to students solving problems on their own.

★ TEKS

**Review & Assess
pp. 269–273:**
2A: Alt. Assess. 1
2B: Alt. Assess. 1, 2
2C: 5–12, 21–26, 37–53
2D: Alt. Assess. 2, 3
2E: Alt. Assess. 1
3E: Alt. Assess. 3
4A: Technology & Learning
4B: 2–28, 30, 34, 36–38, 41–47, 50–53
4C: 27–29, 32–34, 36, 37, 47, 48, 51–53
6A: 29, 31, 39, 40, 49

KEY TERMS

angular acceleration (p. 249)

angular displacement (p. 246)

angular speed (p. 247)

centripetal acceleration (p. 257)

gravitational force (p. 263)

radian (p. 245)

rotational motion (p. 244)

tangential acceleration (p. 255)

tangential speed (p. 253)

Diagram symbols

Rotational motion

Angle marking

KEY IDEAS

Section 7-1 Measuring rotational motion

- The average angular speed, ω_{avg}, of a rigid, rotating object is defined as the ratio of the angular displacement, $\Delta\theta$, to the time interval, Δt.
- The average angular acceleration, α_{avg}, of a rigid, rotating object is defined as the ratio of the change in angular speed, $\Delta\omega$, to the time interval, Δt.

Section 7-2 Tangential and centripetal acceleration

- A point on an object rotating about a fixed axis has a tangential speed related to the object's angular speed. When the object's angular acceleration changes, the tangential acceleration of a point on the object changes.
- Uniform circular motion occurs when an acceleration of constant magnitude is perpendicular to the tangential velocity.

Section 7-3 Causes of circular motion

- Any object moving in a circular path must have a net force exerted on it that is directed toward the center of the circular path.
- Every particle in the universe attracts every other particle with a force that is directly proportional to the product of the particles' masses and inversely proportional to the square of the distance between the particles.

Variable symbols

Quantities		Units	
s	arc length	m	meters
$\Delta\theta$	angular displacement	rad	radians
ω	angular speed	rad/s	radians/second
α	angular acceleration	rad/s^2	$radians/second^2$
v_t	tangential speed	m/s	meters/second
a_t	tangential acceleration	m/s^2	$meters/second^2$
a_c	centripetal acceleration	m/s^2	$meters/second^2$
F_c	force that maintains circular motion	N	newtons
F_g	gravitational force	N	newtons
G	constant of universal gravitation	$\dfrac{N \cdot m^2}{kg^2}$	$\dfrac{newtons \cdot meters^2}{kilograms^2}$

CHAPTER 7
Review and Assess

RADIANS AND ANGULAR MOTION

Review questions

1. How many degrees equal π radians? How many revolutions equal π radians?

2. What units must be used for θ, ω, and α in the kinematic equations for rotational motion listed in **Table 7-2**?

3. Distinguish between linear speed and angular speed.

4. When a wheel rotates about a fixed axis, do all points on the wheel have the same angular speed?

Practice problems

5. A car on a Ferris wheel has an angular displacement of 0.34 rad. If the car moves through an arc length of 12 m, what is the radius of the Ferris wheel? (See Sample Problem 7A.)

6. When a wheel is rotated through an angle of 35°, a point on the circumference travels through an arc length of 2.5 m. When the wheel is rotated through angles of 35 rad and 35 rev, the same point travels through arc lengths of 143 m and 9.0×10^2 m, respectively. What is the radius of the wheel? (See Sample Problem 7A.)

7. How long does it take the second hand of a clock to move through 4.00 rad? (See Sample Problem 7B.)

8. A phonograph record has an initial angular speed of 33 rev/min. The record slows to 11 rev/min in 2.0 s. What is the record's average angular acceleration during this time interval? (See Sample Problem 7C.)

9. If a flywheel increases its average angular speed by 2.7 rad/s in 1.9 s, what is its angular acceleration? (See Sample Problem 7C.)

EQUATIONS FOR ANGULAR MOTION

Practice problems

10. A potter's wheel moves from rest to an angular speed of 0.20 rev/s in 30.0 s. Assuming constant angular acceleration, what is its angular acceleration in rad/s²? (See Sample Problem 7D.)

11. A drill starts from rest. After 3.20 s of constant angular acceleration, the drill turns at a rate of 2628 rad/s.
    a. Find the drill's angular acceleration.
    b. Determine the angle through which the drill rotates during this period.
    (See Sample Problem 7D.)

12. A tire placed on a balancing machine in a service station starts from rest and turns through 4.7 revs in 1.2 s before reaching its final angular speed. Assuming that the angular acceleration of the wheel is constant, calculate the wheel's angular acceleration. (See Sample Problem 7D.)

TANGENTIAL AND CENTRIPETAL ACCELERATION

Review questions

13. When a wheel rotates about a fixed axis, do all the points on the wheel have the same tangential speed?

14. Correct the following statement: The racing car rounds the turn at a constant velocity of 145 km/h.

15. Describe the path of a moving body whose acceleration is constant in magnitude at all times and is perpendicular to the velocity.

16. An object moves in a circular path with constant speed v.
    a. Is the object's velocity constant? Explain.
    b. Is its acceleration constant? Explain.

17. a car driving in a circle at a constant speed

18. No, a_c is necessary for circular motion.

19. yes; It changes **v** by changing the direction of movement.

20. a_c would equal g.

21. 1.5×10^2 rad/s

22. 1.8×10^{-3} rad/s, or 6.5 rad/h

23. 0.32 m

24. 0.023 m (2.3 cm)

25. 7.0 m/s

26. 2.7 m/s

27. yes; The object moves in a spiral path, increasing its distance outward, until the spring force ($F = -kx$) is great enough to keep the object at a constant radius.

28. The pail exerts a downward force that maintains circular motion. The water remains in the pail even when the pail is upside down because the pail and water have the same acceleration.

29. Gravitational forces are directly related to mass and inversely related to distance. In other words, the gravitational force between two objects increases as the mass of the objects increases. If the distance increases, the force decreases. The force varies as the inverse square of the distance.

30. a_c is directed toward the center of a circular path. α is a change in the angular speed in a given time interval.

31. Objects are so far apart that the gravitational forces are relatively small.

32. because the force holding Earth together is not large enough to resist the inertia of the points on the equator

Conceptual questions

17. Give an example of a situation in which an automobile driver can have a centripetal acceleration but no tangential acceleration.

18. Can a car move around a circular racetrack so that the car has a tangential acceleration but no centripetal acceleration?

19. The gas pedal and the brakes of a car accelerate and decelerate the car. Could a steering wheel perform either of these two actions? Explain.

20. It has been suggested that rotating cylinders about 16 km long and 8 km in diameter should be placed in space for future space colonies. The rotation would simulate gravity for the inhabitants of these colonies. Explain the concept behind this proposal.

Practice problems

21. A small pebble breaks loose from the treads of a tire with a radius of 32 cm. If the pebble's tangential speed is 49 m/s, what is the tire's angular speed? (See Sample Problem 7E.)

22. The Emerald Suite is a revolving restaurant at the top of the Space Needle in Seattle, Washington. If a customer sitting 12 m from the restaurant's center has a tangential speed of 2.18×10^{-2} m/s, what is the angular speed of the restaurant? (See Sample Problem 7E.)

23. A bicycle wheel has an angular acceleration of 1.5 rad/s^2. If a point on its rim has a tangential acceleration of 48 cm/s^2, what is the radius of the wheel? (See Sample Problem 7F.)

24. When the string is pulled in the correct direction on a window shade, a lever is released and the shaft that the shade is wound around spins. If the shaft's angular acceleration is 3.8 rad/s^2 and the shade accelerates upward at 0.086 m/s^2, what is the radius of the shaft? (See Sample Problem 7F.)

25. A building superintendent twirls a set of keys in a circle at the end of a cord. If the keys have a centripetal acceleration of 145 m/s^2 and the cord has a length of 0.34 m, what is the tangential speed of the keys? (See Sample Problem 7G.)

26. A sock stuck to the side of a clothes-dryer barrel has a centripetal acceleration of 28 m/s^2. If the dryer barrel has a radius of 27 cm, what is the tangential speed of the sock? (See Sample Problem 7G.)

CAUSES OF CIRCULAR MOTION

Review questions

27. Imagine that you attach a heavy object to one end of a spring and then, while holding the spring's other end, whirl the spring and object in a horizontal circle. Does the spring stretch? Why? Discuss your answer in terms of the force that maintains circular motion.

28. Why does the water remain in a pail that is whirled in a vertical path, as shown in **Figure 7-16**?

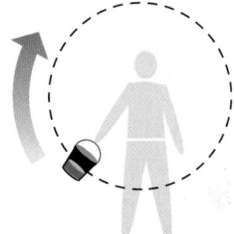

Figure 7-16

29. Identify the influence of mass and distance on gravitational forces.

30. Explain the difference between centripetal acceleration and angular acceleration.

31. Comment on the statement, "There is no gravity in outer space."

Conceptual questions

32. Explain why Earth is not spherical in shape and why it bulges at the equator.

33. Because of Earth's rotation, you would weigh slightly less at the equator than you would at the poles. Why?

34. Why does mud fly off a rapidly turning wheel?

35. Astronauts floating around inside the space shuttle are not actually in a zero-gravity environment. What is the real reason astronauts seem weightless?

36. A girl at a state fair swings a ball in a vertical circle at the end of a string. Is the force applied by the string greater than the weight of the ball at the bottom of the ball's path?

Practice problems

37. A roller-coaster car speeds down a hill past point A and then rolls up a hill past point B, as shown in **Figure 7-17**.
    a. The car has a speed of 20.0 m/s at point A. If the track exerts a force on the car of 2.06×10^4 N at this point, what is the mass of the car?
    b. What is the maximum speed the car can have at point B for the gravitational force to hold it on the track?
(See Sample Problem 7H.)

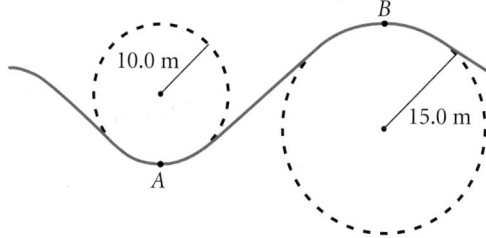

Figure 7-17

38. Tarzan tries to cross a river by swinging from one bank to the other on a vine that is 10.0 m long. His speed at the bottom of the swing, just as he clears the surface of the river, is 8.0 m/s. Tarzan does not know that the vine has a breaking strength of 1.0×10^3 N. What is the largest mass Tarzan can have and make it safely across the river?
(See Sample Problem 7H.)

39. The gravitational force of attraction between two students sitting at their desks in physics class is 3.20×10^{-8} N. If one student has a mass of 50.0 kg and the other has a mass of 60.0 kg, how far apart are the students sitting?
(See Sample Problem 7I.)

40. If the gravitational force between the electron (9.11×10^{-31} kg) and the proton (1.67×10^{-27} kg) in a hydrogen atom is 1.0×10^{-47} N, how far apart are the two particles?
(See Sample Problem 7I.)

MIXED REVIEW PROBLEMS

41. Find the average angular speed of Earth about the sun in radians per second.
(Hint: Earth orbits the sun once every 365.25 days.)

42. The tub within a washer goes into its spin cycle, starting from rest and reaching an angular speed of 11π rad/s in 8.0 s. At this point, the lid is opened, and a safety switch turns off the washer. The tub slows to rest in 12.0 s. Through how many revolutions does the tub turn? Assume constant angular acceleration while the machine is starting and stopping.

43. An airplane is flying in a horizontal circle at a speed of 105 m/s. The 80.0 kg pilot does not want the centripetal acceleration to exceed 7.00 times free-fall acceleration.
    a. Find the minimum radius of the plane's path.
    b. At this radius, what is the *net* force that maintains circular motion exerted on the pilot by the seat belts, the friction against the seat, and so forth?

44. A car traveling at 30.0 m/s undergoes a constant negative acceleration of magnitude 2.00 m/s² when the brakes are applied. How many revolutions does each tire make before the car comes to a stop, assuming that the car does not skid and that the tires have radii of 0.300 m?

45. A coin with a diameter of 2.40 cm is dropped onto a horizontal surface. The coin starts out with an initial angular speed of 18.0 rad/s and rolls in a straight line without slipping. If the rotation slows with an angular acceleration of magnitude 1.90 rad/s², how far does the coin roll before coming to rest?

46. A mass attached to a 50.0 cm string starts from rest and is rotated in a circular path exactly 40 times in 1.00 min before reaching a final angular speed. What is the angular speed of the mass after 1.00 min?

47. A 13 500 N car traveling at 50.0 km/h rounds a curve of radius 2.00×10^2 m. Find the following:
    a. the centripetal acceleration of the car
    b. the force that maintains centripetal acceleration
    c. the minimum coefficient of static friction between the tires and the road that will allow the car to round the curve safely

33. At the equator, part of the gravitational force with which Earth pulls on you is used to keep you traveling in a circle, so it is not available to pull you to the ground. (At the equator, $w = mg - mv^2/R$; at the poles, $w = mg$.)

34. The inertia of the pieces of mud exceeds the cohesive force that maintains the mud's circular motion.

35. The astronauts and space shuttle both have the same acceleration—that of free-fall acceleration. As a result, the astronauts see everything at rest and believe there is no force on them.

36. Yes, the string must exert a force equal to the ball's weight and the circular force that maintains circular motion.

37. a. 515 kg
    b. 12.1 m/s

38. 62 kg

39. 2.50 m

40. 1.0×10^{-10} m (0.10 nm)

41. 1.99×10^{-7} rad/s

42. 55.0 rev

43. a. 161 m
    b. 5.49×10^3 N

44. 119 rev

45. 1.02 m

46. 8.38 rad/s

47. a. 0.965 m/s²
    b. 1.33×10^3 N
    c. 0.0985

48. 12 m/s

49. a. 4.39×10^{20} N
    b. 1.99×10^{20} N
    c. 3.55×10^{22} N

50. a. 3.37×10^{-2} m/s^2
    b. 0 m/s^2

51. 8.3 s

ANSWERS TO

Technology & Learning

Answers may vary slightly, depending on viewing-window settings.

a. to convert revolutions to radians

b. 4.2 rad/s

c. 3.1 rad/s

d. 21 rad/s

e. 45 rad/s

f. The axes are always approaching but never meet the curve. For the curve to cross the *y*-axis, the object would have to complete its revolutions in a time interval of zero, which is impossible. For the curve to cross the *x*-axis, the object would have an angular speed of zero, in which case it would never complete the revolutions.

48. A 2.00×10^3 kg car rounds a circular turn of radius 20.0 m. If the road is flat and the coefficient of static friction between the tires and the road is 0.70, how fast can the car go without skidding?

49. During a solar eclipse, the moon, Earth, and sun lie on the same line, with the moon between Earth and the sun. What force is exerted on
    a. the moon by the sun?
    b. the moon by Earth?
    c. Earth by the sun?
(See the table in the appendix for data on the sun, moon, and Earth.)

50. Find the centripetal accelerations of the following:
    a. a point on the equator of Earth
    b. a point at the North Pole of Earth
(See the table in the appendix for data on Earth.)

51. A copper block rests 30.0 cm from the center of a steel turntable. The coefficient of static friction between the block and the surface is 0.53. The turntable starts from rest and rotates with a constant angular acceleration of 0.50 rad/s^2. After what time interval will the block start to slip on the turntable? (Hint: The normal force in this case equals the weight of the block.)

Technology & Learning

Graphing calculators

Refer to Appendix B for instructions on downloading programs for your calculator. The program "Chap7" allows you to analyze a graph of angular speed versus time interval.

Angular speed, as you learned earlier in this chapter, is described by the following equation:

$$\omega_{avg} = \frac{\Delta\theta}{\Delta t}$$

The program "Chap7" stored on your graphing calculator makes use of the equation for angular speed. Once the "Chap7" program is executed, your calculator will ask for the angular displacement in revolutions. The graphing calculator will use the following equation to create a graph of the angular speed (Y_1) versus the time interval (X). Note that the relationships in this equation are similar to those in the angular speed equation shown above.

$$Y_1 = (2\pi\,\theta)/X$$

a. Why is there a factor of 2π in the equation used by your graphing calculator?

First be certain your graphing calculator is in radian mode by pressing MODE ▼ ▼ ENTER.

Execute "Chap7" on the PRGM menu and press ENTER to begin the program. Enter the value for the angular displacement (shown below) and press ENTER.

The calculator will provide a graph of the angular speed versus the time interval. (If the graph is not visible, press WINDOW and change the settings for the graph window, then press GRAPH.)

Press TRACE and use the arrow keys to trace along the curve. The *x*-value corresponds to the time interval in seconds, and the *y*-value corresponds to the angular speed in radians per second.

Determine the angular speed in the following situations:

    b. a bowl on a mixer stand that turns 2.0 rev in 3.0 s
    c. the same bowl on a mixer stand that has slowed down to 2.0 rev in 4.0 s
    d. a bicycle wheel turning 2.5 rev in 0.75 s
    e. the same bicycle wheel turning 2.5 rev in 0.35 s
    f. The *x*- and *y*-axes are said to be asymptotic to the curve of angular speed versus time interval. What does this mean?

Press 2nd QUIT to stop graphing. Press ENTER to input a new value or CLEAR to end the program.

52. An air puck of mass 0.025 kg is tied to a string and allowed to revolve in a circle of radius 1.0 m on a frictionless horizontal surface. The other end of the string passes through a hole in the center of the surface, and a mass of 1.0 kg is tied to it, as shown in **Figure 7-19.** The suspended mass remains in equilibrium while the puck revolves on the surface.

    a. What is the magnitude of the force that maintains circular motion acting on the puck?

    b. What is the linear speed of the puck?

Figure 7-19

53. In a popular amusement-park ride, a cylinder of radius 3.00 m is set in rotation at an angular speed of 5.00 rad/s, as shown in **Figure 7-20.** The floor then drops away, leaving the riders suspended against the wall in a vertical position. What minimum coefficient of friction between a rider's clothing and the wall of the cylinder is needed to keep the rider from slipping? (Hint: Recall that $F_s = \mu_s F_n$, where the normal force is the force that maintains circular motion.)

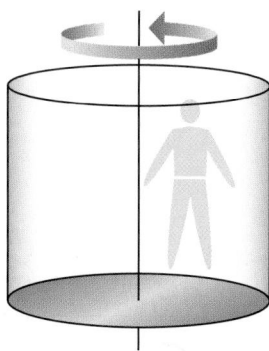

Figure 7-20

Alternative Assessment

Performance assessment

1. Turn a bicycle upside down. Make two marks on one spoke on the front wheel, one mark close to the rim and another mark closer to the axle. Then spin the front wheel. Which point seems to be moving fastest? Have partners count the rotations of one mark for 10 s or 20 s. Find the angular speed and the linear speed of each point. Reassign the observers to different points, and repeat the experiment. Make graphs to analyze the relationship between the linear and angular speeds.

2. When you ride a bicycle, the rotational motion you create on the pedals is transmitted to the back wheel through the primary sprocket wheel, the chain, and the secondary sprocket wheel. Study the connection between these components and measure how the angular and linear speeds change from one part of the bicycle to another. How does the velocity of the back wheel compare with that of the pedals on a bicycle? Demonstrate your findings in class.

Portfolio projects

3. Research the historical development of the concept of gravitational force. Find out how scientists' ideas about gravity have changed over time. Identify the contributions of different scientists, such as Galileo, Kepler, Newton, and Einstein. How did each scientist's work build on the work of earlier scientists? Analyze, review, and critique the different scientific explanations of gravity. Focus on each scientist's hypotheses and theories. What are their strengths? What are their weaknesses? What do scientists think about gravity now? Use scientific evidence and other information to support your answers. Write a report or prepare an oral presentation to share your conclusions.

52. a. 9.8 N

    b. 2.0×10^1 m/s

53. 0.131

Alternative Assessment
ANSWERS

Performance assessment

1. Student plans should be safe and should include measurements of the radius of rotation and either the angle or the distance each mark moves in a given time. A graph of v versus ω is a line with a slope equal to its radius, r.

2. Students should recognize that the gear on the wheel axle is so small that a small linear velocity from the chain causes a large angular speed in the axle and the tire, resulting in a large linear velocity of the bicycle.

Portfolio projects

3. Answers will vary. Newton built on the work of Kepler and Galileo to develop his law of universal gravitation. This law was considered to give a complete description of gravity until Einstein developed the general theory of relativity. Many of Einstein's predictions have been confirmed by observations, but experimenters are still trying to prove the existence of gravity waves and particles called gravitons.

Note

Materials Preparation is given on pp. 242A–242B. Blank data table and sample data table are on the One-Stop Planner CD-ROM. All calculations are performed using sample data.

Planning

Recommended time:

1 lab period

Classroom organization:

▶ Each lab group should have at least two students.

▶ Each group needs a large, clear area to work in. If possible, consider conducting this lab in a large open space, such as outdoors or in a gymnasium.

▶ **Safety warning:** Remind students to alert other groups before they begin each trial. Eye protection should be worn at all times.

Circular Motion Tips

◆ Have students practice timing the swing of the stopper before they begin taking data.

◆ Each trial should be performed twice to minimize error.

Techniques to Demonstrate

Show students how to make a loop in the cord to attach the masses. Remind them to tape the masses securely into the loop.

Demonstrate how to measure the radius using the masking tape as a reference.

OBJECTIVES

• Examine the relationship between the force that maintains circular motion, the radius, and the tangential speed of a whirling object.

MATERIALS LIST

✔ 1.5 m nylon cord
✔ 2-hole rubber stopper
✔ masking tape
✔ meterstick
✔ PVC tube, about 15 cm long and 1 cm in diameter
✔ set of masses
✔ stopwatch

CHAPTER 7
Laboratory Exercise

⭐ TEKS

pp. 274–275: 1A, 2B, 2C, 2E, 2F, 4A, 4B, 4C

CIRCULAR MOTION

In this experiment, you will construct a device for measuring the tangential speed of an object undergoing circular motion, and you will determine how the force and the radius affect the tangential speed of the object.

SAFETY

• **Tie back long hair, secure loose clothing, and remove loose jewelry to prevent their getting caught in moving or rotating parts.**

• **Wear eye protection and perform this experiment in a clear area. Swinging or dropped masses can cause serious injury.**

PREPARATION

1. Read the entire lab, and plan what measurements you will take.

2. Prepare a data table in your lab notebook with five columns and fifteen rows. In the first row, label the columns *Trial, Hanging mass (kg), Mass of stopper (kg), Total time (s), Radius (m)*. In the first column, label the second through fifteenth rows *1, 2, 3, 4, 5, 6, 7, 8, 9, 10, 11, 12, 13,* and *14.*

PROCEDURE

Constant radius with varying force

3. Measure the mass of the rubber stopper, and record it in your data table. Fasten one end of the nylon cord securely to the rubber stopper. Pass the other end of the cord through the PVC tube and securely fasten a 100 g mass to the other end, as shown in **Figure 7-21.** Leave approximately 0.75 m of cord between the top of the tube and the rubber stopper. Attach a piece of masking tape to the cord just below the bottom of the tube.

4. Make sure the area is clear of obstacles, and warn other students that you are beginning your experiment. Support the 100 g mass with one hand and hold the PVC tube in the other. Make the stopper at the end of the cord circle around the top of the tube by moving the tube in a circular motion.

5. Slowly release the 100 g mass, and adjust the speed of the stopper so that the masking tape stays just below the bottom of the tube. Make several practice runs before recording any data.

6. When you can keep the velocity of the stopper and the position of the masking tape relatively constant, measure the time required for 20 revolutions of the stopper. Record the time interval in your data table in the row labeled *Trial 1*. Repeat this trial and record the time interval in your data table as *Trial 2*.

7. Place the apparatus on the lab table. Extend the cord so that it is taut and the masking tape is in the same position it was in during the experiment. Measure the cord from the center of the top of the PVC tube to the center of the rubber stopper. Record this distance in the data table as *Radius* for *Trials 1* and *2*.

8. Repeat the procedure using three different masses for *Trials 3–8*. Keep the radius the same as in the first trial and use the same rubber stopper, but increase the mass at the end of the cord each time. Do not exceed 500 g. Attach all masses securely. Perform each trial two times, and record all data in your data table.

Constant force with varying radius

9. For *Trials 9–14*, use the same stopper and the 100 g mass, and try three different values for the radius in the range 0.50 m to 1.00 m. Make sure that you have a clear area of at least 2.5 m in diameter to work in. Record all data in your data table.

10. Clean up your work area. Put equipment away safely so that it is ready to be used again.

ANALYSIS AND INTERPRETATION

Calculations and data analysis

1. Organizing data   Calculate the weight of the hanging mass for each trial. This weight is the force that maintains circular motion, F_c.

2. Organizing data   For each trial, find the time necessary for one revolution of the stopper by dividing the total time required for 20 revolutions by 20.

3. Organizing data   Find the tangential speed for each trial.

a. Use the equation $v_t = \dfrac{2\pi r}{\Delta t}$.

b. Use the equation $v_t = \sqrt{\dfrac{F_c r}{m}}$, where r is the radius of revolution and m is the mass of the stopper.

4. Graphing data   Plot the following graphs:

a. Graph force versus tangential speed for *Trials 1–8*.

b. Graph tangential speed versus radius for *Trials 9–14*.

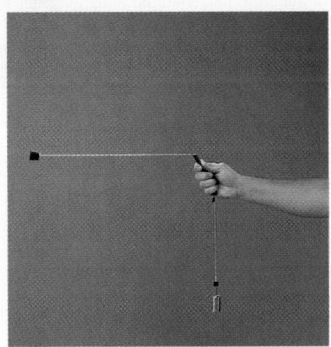

Figure 7-21

Step 3: To attach masses, make a loop in the cord, place the mass inside the loop, and secure the mass with masking tape.

Step 4: You will need a clear area larger than two times the radius.

Step 5: Spin the stopper so that the cord makes a 90° angle with the PVC tube. Release the mass slowly without changing the speed.

Step 7: Make sure the cord is held straight when you make your measurements.

✔ Checkpoints

Step 4: Make sure all groups have enough room to perform the lab and that they are aware of each other's activities.

Step 6: Students should be able to demonstrate that the stopper is spinning with constant speed. Remind them to pay attention to the position of the masking tape.

Step 9: Students should be able to demonstrate that they have changed the radius without changing any other conditions.

ANSWERS TO

Analysis and Interpretation

CALCULATIONS AND DATA ANALYSIS

1. Answers will vary. Make sure students use the relationship $F_g = mg$. Typical values will range from 0.981 N to 4.905 N.

2. Typical values will range from 0.330 s to 0.685 s.

3. a, b. Typical values will range from 6.49 m/s to 15.9 m/s.

4. a. The graph should be a parabolic curve.

b. The graph should be a straight line going up and to the right.

CHAPTER 8 PLANNING GUIDE

Compression Guide: To shorten from 9 to 8 45-min periods (from 4½ to 4 90-min blocks), eliminate items in magenta type.

PACING CHART	CLASSROOM RESOURCES			
	⭐TEKS	Teacher Demonstrations	*Holt Physics* Transparencies	Labs (See page T52 for equipment listing for in-text labs.)
8-1 Torque 2 or 1 45-minute periods 1 or ½ 90-minute block	1A, 2A, 2C, 3B, 4C	**TE** *Opening the door*, p. 279	**T** 24–26	**L** *Discovery Lab: Torque and Center of Mass* **PE** *Quick Lab: Two-Object Races*, p. 279
8-2 Rotation and inertia 1 45-minute period ½ 90-minute block	1A, 2C, 3A, 4B, 4C	**TE** *Moments of inertia of a rod*, p. 285	**T** 27 **TM** 26–27	**PE** *Quick Lab: Finding the Center of Mass Experimentally*, p. 284
8-3 Rotational dynamics 3 45-minute periods 1½ 90-minute block	2C, 3B, 4C, 5B, 5D	**TE** *Colliding magnetic marbles*, p. 292		**L** *Invention Lab: The Rotating Egg Drop*
8-4 Simple machines 2 45-minute periods 1 90-minute block	1A, 2B, 2C, 3C, 4C, 5D		**T** 28–29	**PE** *Machines and Efficiency*, p. 313
Review and Assessment 2 45-minute periods 1 90-minute block				

Resource Key

PHYSICS

PE Pupil's Edition
TE Teacher's Edition

L Laboratory Experiments
TL Technology Lab Experiments
T Transparencies

One-Stop Planner CD-ROM contents

TM Transparency Masters
SR Section Review Worksheets
AA Alternative Assessment

PW Problem-Solving Workbook
PB Problem Bank
CTW Critical Thinking Worksheet

LABORATORY PLANNING: Machines and Efficiency, p. 313

Materials (for each lab group)
- balance: portable, electronic balance or triple-beam balance with weight
- C-clamp
- roll of thick, white, braided cord
- Hall's carriage
- inclined plane
- mass hanger
- mass set: metric slotted mass set or metric hooked mass set
- ball-bearing tandem pulleys, two double and two triple
- ball-bearing tandem pulleys, two single
- meterstick
- right-angle clamp
- ruler
- 1-position support base and rod, 1.3 cm × 91 cm
- suspension clamp

ASSIGNMENT RESOURCES

Content Mastery	Critical Thinking	Problem-Solving Practice	
PE 1–5, p. 282 **SR** 8-1, *Concept Review*	**PE** 1–8, p. 305	**8A**	Torque: 28 items in **PE, PW,** and **PB,** see **TE** p. 281
PE 1–8, p. 289 **SR** 8-2, *Diagram Skills* **PE** 12–16, p. 306	**PE** 17–19, p. 306	**8B**	Rotational equilibrium: 30 items in **PE, PW,** and **PB,** see **TE** p. 287
PE 1–3, p. 297 **SR** 8-3, *Concept Review* **PE** 29–32, p. 307	**PE** 24–26, p. 307 **PE** 33–34, p. 307	**8C** **8D** **8E**	Newton's second law for rotation: 34 items in **PE, PW,** and **PB,** see **TE** p. 291 Conservation of angular momentum: 23 items in **PE, PW,** and **PB,** see **TE** p. 293 Conservation of mechanical energy: 32 items in **PE, PW,** and **PB,** see **TE** p. 296
PE 1–5, p. 301 **SR** 8-4, *Concept Review* **PE** 39–40, p. 308	**PE** 41–43, p. 308		

ASSESSMENT RESOURCES

Cumulative Review	Alternative Assessment	Traditional Assessment
SR Mixed Review, Ch. 8	**PE** 1–6, p. 312 **AA** Items for Ch. 8	Chapter 8 Test Test Generator items for Ch. 8

Scoring Rubrics for Alternative Assessment items can be found on the One-Stop Planner CD-ROM.

TECHNOLOGY RESOURCES

 CTW Segment 8 Circus Acrobats

 Module 9 Torque
Module 10 Rotational Inertia

 PE Technology and Learning, p. 311
(Alternative procedures for calculators without Flash-ROM technology are provided on the One-Stop Planner CD-ROM.)

 The Mechanical Universe/High School Adaptation Quad II, Angular Momentum

internet connect

 On-line Student Resources:
GO TO: www.scilinks.org
The following student resources can be found in the student text for this chapter.

TOPICS:
- Torque, p. 281 (HF2081)
- Center of mass, p. 283 (HF2082)
- Simple machines, p. 301 (HF2083)
- Rutherford model of atom, p. 303 (HF 2084)

On-line Teacher Resources:
GO TO: go.hrw.com
KEYWORD: HF2 HOME
Visit the HRW Web site for a variety of resources related to this chapter.

 Smithsonian Institution
Internet Connections
Visit **www.si.edu/hrw** for additional on-line resources.

Visit **www.cnnfyi.com** for late-breaking news and current events stories selected just for you.

Materials Preparation

Each group will need one cord for the inclined-plane procedure and another cord for the pulleys. The length required will depend upon the height of the plane and the pulley configuration, but most setups will require 1–2 m of cord. You may want to provide several lengths or allow students to cut their own. Students should be supervised when working with scissors or other sharp objects.

If desired, attach about 1 m of cord to each dynamics cart before the lab.

The classroom may be divided into two parts, with inclined planes set up at half the lab stations and pulley configurations set up at the other half. Half the groups will perform one part of the lab while the other performs the other part of the lab. Then the groups can switch tasks.

Section 8-1 introduces the concepts of torque, extended bodies, and lever arm and presents methods for calculating torque.

Section 8-2 describes center of mass and moment of inertia and uses the conditions of equilibrium for problem solving.

Section 8-3 introduces angular momentum, rotational kinetic energy, and Newton's second law for rotation. Rotational kinetic energy and conservation of angular momentum are used in problem-solving exercises.

Section 8-4 explores conservation of energy, including mechanical advantage and efficiency, in simple machines.

About the Illustration

Every February, the town of Evandale, Tasmania, holds a festival in which people dress up in costumes from the early 20th century. The main event is a penny-farthing race. This event receives international media attention. Spectators can rent a penny-farthing bicycle for recreational use.

Interactive Problem-Solving Tutor

See Module 9
"Torque" provides more opportunities for students to identify and calculate torque.

See Module 10
"Rotational Inertia" includes additional practice with moments of inertia and the rotational form of Newton's second law.

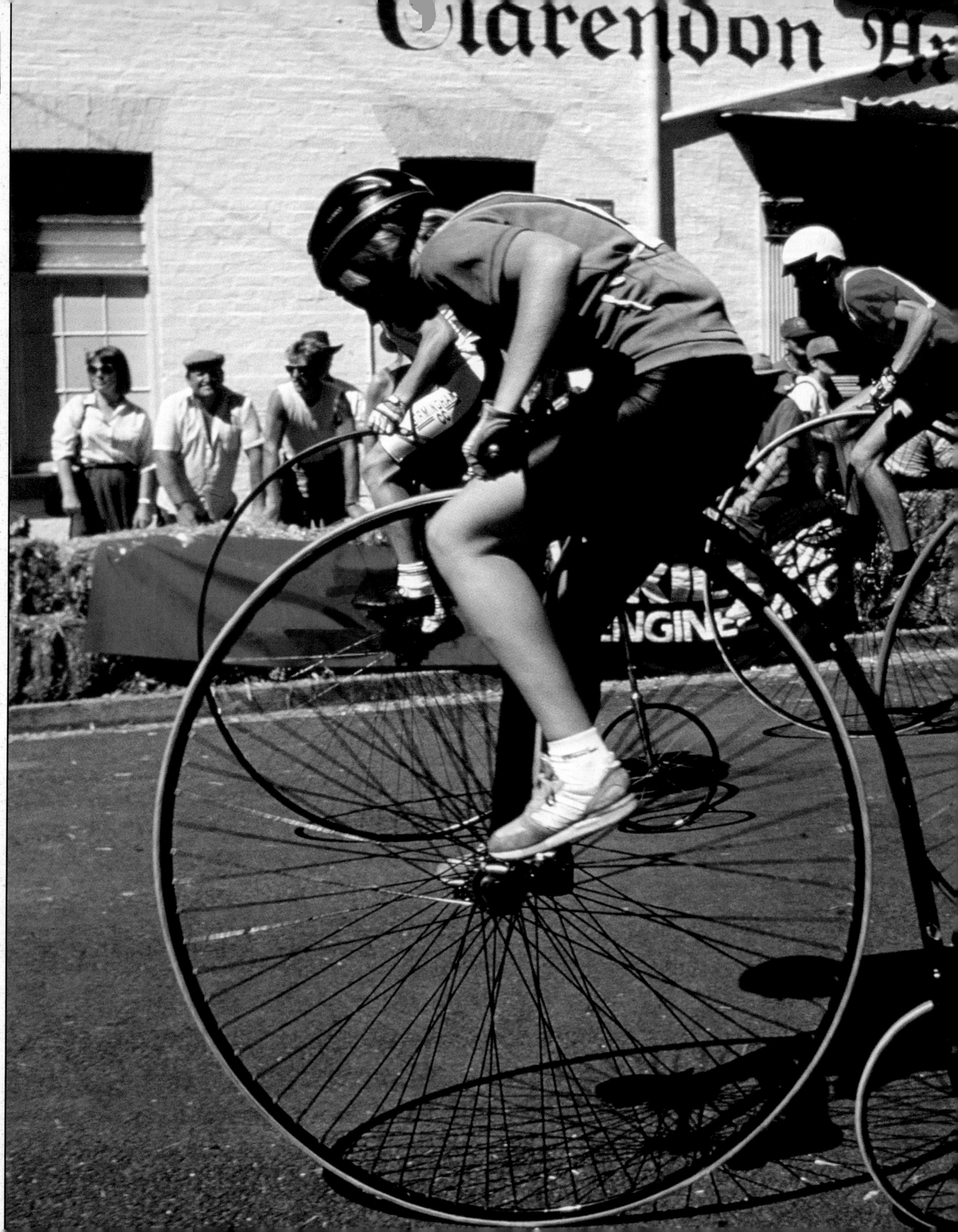

CHAPTER 8

Rotational Equilibrium and Dynamics

PHYSICS IN ACTION

One of the most popular early bicycles was the penny-farthing, first introduced in 1870. The bicycle was named for the relative sizes of its two wheels compared to the relative sizes of the penny and the farthing, two English coins. Early bicycles had no gears, just pedals attached directly to the wheel axle. This meant that the wheel turned once for every revolution of the pedals. For the penny-farthing, a gear was developed that allowed the wheel to turn twice for every turn of the pedals. More stable bicycles with gears and chains soon replaced the penny-farthing.

- *What makes a wheel difficult to rotate?*

- *How much does a wheel accelerate for a given applied force?*

CONCEPT REVIEW

Work   (Section 5-1)

Energy   (Section 5-2)

Momentum   (Section 6-1)

Angular speed and acceleration
   (Section 7-1)

Knowledge to Expect

✔ "Mechanical energy is in moving bodies; gravitational energy is in the separation of mutually attracting masses." (AAAS's *Benchmarks for Science Literacy*)

Knowledge to Review

✔ Work is done on an object to change the energy of the object ($W = Fd$). (Section 5-1)

✔ Conservation of energy states that energy is neither created nor destroyed. (Section 5-3)

✔ Momentum is a vector quantity that has a magnitude equal to the product of mass and velocity and that has the same direction as the velocity vector. (Section 6-1)

✔ Rotational measurements are expressed in radians. (Section 7-1)

Items to Probe

✔ Preconceptions about inertia and motion: Ask students to explain in their own words the relationship between inertia and motion. Students may mistakenly believe that inertia applies only to motionless objects.

8-1
Torque

Figure 8-1

Point out that the pins were initially at rest and were set in motion by the bowling ball.

Q Is the energy of the pins greater than, equal to, or less than the energy lost by the bowling ball?

A *The energy gained by the pins is equal to the energy lost by the bowling ball (except for a small loss of energy in the form of sound and a slight temperature increase in the pins).*

🛑 Misconception Alert

Many students may think that any force acting on an object will produce a torque. You may want to demonstrate that a force can be applied to an object without producing a torque by pushing an object without rotating it.

⭐ TEKS

p. 279: 4C, 1A, 2A

8-1 SECTION OBJECTIVES

- **Recognize the difference between a point mass and an extended object.**

- **Distinguish between torque and force.**

- **Calculate the magnitude of a torque on an object.**

- **Identify the lever arm associated with a torque on an object.**

THE MAGNITUDE OF A TORQUE

You have been chosen to judge a race involving three objects: a solid sphere, a solid cylinder, and a hollow cylinder. The spectators for the race demand that the race be fair, so you make sure that all of the objects have the same mass and radius and that they all start from rest. Then you let the three objects roll down a long ramp. Is there a way to predict which one will win and which one will lose?

If you really performed such a race (see the Quick Lab on the next page), you would discover that the sphere would come in first and that the hollow cylinder would come in last. This is a little surprising because in the absence of friction the acceleration due to gravity is the same for all objects near the Earth's surface. Yet the acceleration of each of these objects is different.

In earlier chapters, the motion of an object was described by assuming the object was a point mass. This description, however, does not account for the differences in the motion of the objects in the race. This is because these objects are *extended objects*. An extended object is an object that has a definite, finite size and shape. Although an extended object can be treated as a point mass to describe the motion of its center of mass, a more sophisticated model is required to describe its rotational motion.

Rotational and translational motion can be separated

Imagine that you roll a strike while bowling. What happens when the bowling ball strikes the pins, as shown in **Figure 8-1?** The pins fly backward, spinning in the air. The complicated motion of each pin can be separated into a translational motion and a rotational motion, each of which can be analyzed separately. For now, we will concentrate on an object's rotational motion. Then we will combine the rotational motion of an object with its translational motion.

Figure 8-1

In general, an extended object, such as one of these pins or this bowling ball, can exhibit rotational and translational motion.

Net torque produces rotation

Imagine a cat trying to leave a house by pushing perpendicularly on a cat-flap door. **Figure 8-2** shows a cat-flap door hinged at the top. In this configuration, the door is free to rotate around a line that passes through the hinge. This is the door's *axis of rotation*. When the cat pushes at the outer edge of the door with a force that is perpendicular to the door, the door opens. The ability of a force to rotate an object around some axis is measured by a quantity called **torque**.

Torque depends on a force and a lever arm

If a cat pushed on the door with the same force but at a point closer to the hinge, the door would be more difficult to rotate. How easily an object rotates depends not only on how much force is applied but also on where the force is applied. The farther the force is from the axis of rotation, the easier it is to rotate the object and the more torque is produced. The perpendicular distance from the axis of rotation to a line drawn along the direction of the force is called the **lever arm,** or moment arm. ⊛ TEKS 4C

Figure 8-3 shows a diagram of the force F applied by the pet perpendicular to the cat-flap door. If you examine the definition of *lever arm,* you will see that in this case the lever arm is the distance d shown in the figure, the distance from the pet's nose to the hinge. That is, d is the perpendicular distance from the axis of rotation to the line along which the applied force acts. If the pet pressed on the door at a higher point, the lever arm would be shorter. A smaller torque would be exerted for the shorter lever arm than for the one shown in **Figure 8-3.**

torque

a quantity that measures the ability of a force to rotate an object around some axis

lever arm

the perpendicular distance from the axis of rotation to a line drawn along the direction of the force

Figure 8-2
The cat-flap door rotates on a hinge, allowing pets to enter and leave a house at will.

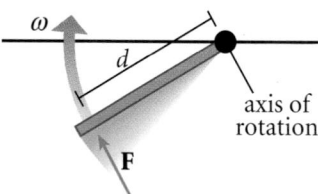

Figure 8-3
A force applied to an extended object can produce a torque. This torque, in turn, causes the object to rotate.

Quick Lab

Two-Object Races

MATERIALS LIST

✔ various solid cylinders, such as unopened soup cans

✔ various hollow cylinders, such as empty soup cans with the tops and bottoms removed or PVC pipes of different diameters and lengths

✔ various spheres, such as a golf ball, tennis ball, and baseball

✔ an incline about 1 m long

Place any two objects from the above list at the top of the incline, and release them simultaneously. Note which object reaches the bottom first. Repeat the race with various combinations of objects. See if you can discover a general rule to predict which object will win. (Hint: You may wish to consider factors such as mass, size, and shape.) ⊛ TEKS 1A, 2A

Demonstration 1

Opening the door

Purpose Review the relationship between torque, a lever arm, and the angle at which the force acts on an object.

Materials door

Procedure Open the door to the classroom with an applied force that is perpendicular to the door. Ask students whether a larger or smaller force is necessary if the applied force is at an angle other than 90°. Open the door using an applied force at a small (less than 45°) angle. Allow a student to try this comparison.

Then, using only perpendicular forces, open the door several times by applying force at different distances from the hinge. (You may have to tape the latch so that the door will open when you push without turning the knob.) Discuss with students the relative effort required to open a door when pushing near the middle compared with pushing near the hinged side of the door.

Quick Lab

TEACHER'S NOTES

Assemble and test the items before using them in class. To minimize internal friction, avoid soups that contain chunks of food. Freezing the can in advance is another way to diminish internal friction effects. Test several objects of each type to find examples that minimize rolling friction.

Students should recognize the general rule governing which types of objects have the lowest moment of inertia, but the results will be affected by rolling friction.

Interactive Problem-Solving Tutor

See Module 9

"Torque" provides more opportunities for students to identify and calculate torque.

Visual Strategy

Figure 8-5

Be certain students recognize how to identify the force, the distance to the axis, the angle between the force and the axis, and the lever arm.

Q Mechanics often use "cheater bars" to loosen stubborn bolts. A cheater bar is usually a length of pipe that fits over the handle of a wrench, making the handle longer. How do cheater bars help the mechanic?

A *Cheater bars increase the lever arm while keeping the force the same, thus increasing the torque applied to a stubborn bolt.*

Module 9
"Torque"

provides an interactive lesson with guided problem-solving practice to teach you about many aspects of rotational motion, including torque.

(★) TEKS

p. 280: 3B, 4C
p. 281: 2C, 4C

Figure 8-4
In each example, the cat is pushing on the same door at the same distance from the axis and with the same amount of force, but it is producing different amounts of torque.

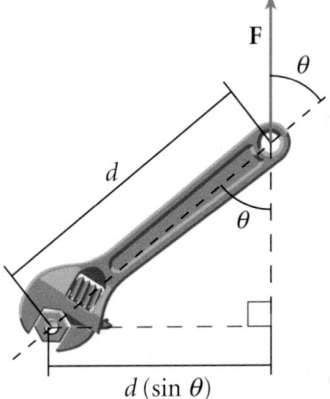

Figure 8-5
The direction of the lever arm is always perpendicular to the direction of the applied force.

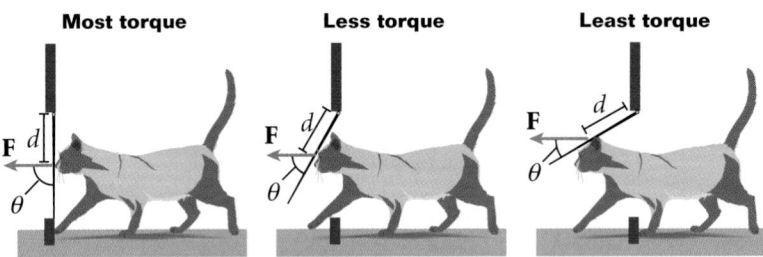

Torque also depends on the angle between a force and a lever arm

Forces do not have to be perpendicular to an object to cause the object to rotate. Imagine the cat-flap door again. What would happen if the cat pushed on the door at an angle to the door, rather than perpendicular to it as shown in **Figure 8-4**? The door would still rotate, but not as easily.

The symbol for torque is the Greek letter *tau* (τ), and the magnitude of the torque is given by the following equation: (★)TEKS **3B**

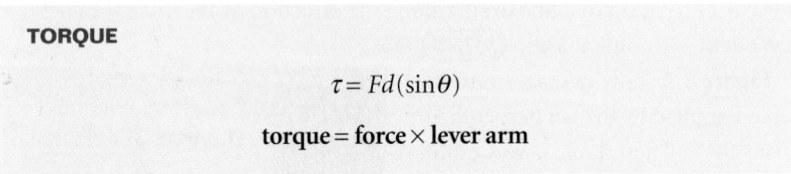

TORQUE

$$\tau = Fd(\sin\theta)$$

torque = force × lever arm

The SI unit of torque is the N•m. Notice that the inclusion of the angle θ, the angle between the force and the distance from the axis, in this equation takes into account the changes in torque shown in **Figure 8-4.**

Figure 8-5 shows a wrench pivoted around a bolt. In this case, the applied force acts at an angle to the wrench. The quantity d is the distance from the axis of rotation to the point where force is applied. The quantity $d(\sin\theta)$, however, is the *perpendicular* distance from the axis of rotation to a line drawn along the direction of the force, so it is the lever arm. (★)TEKS **4C**

THE SIGN OF A TORQUE

Torque, like displacement and force, is a vector quantity. However, for the purposes of this book we will primarily deal with torque as a scalar. Therefore, we will assign each torque a positive or negative sign, depending on the direction the force tends to rotate an object. We will use the convention that the sign of the torque resulting from a force is positive if the rotation is counterclockwise and negative if the rotation is clockwise. In calculations, remember to assign positive and negative values to forces and displacements according to the sign convention established in Chapter 2.

To determine the sign of a torque, imagine that it is the only torque acting on the object and that the object is free to rotate. Visualize the direction the object would rotate under these conditions. If more than one force is acting, then each force has a tendency to produce a rotation and should be treated separately. Be careful to associate the correct sign with each torque.

For example, imagine that you are pulling on a wishbone with a perpendicular force F_1 and that a friend is pulling in the opposite direction with a force F_2. If you pull the wishbone so that it would rotate counterclockwise, then you exert a positive torque of magnitude F_1d_1. Your friend, on the other hand, exerts a negative torque, $-F_2d_2$. To find the net torque acting on the wishbone, simply add up the individual torques.

$$\tau_{net} = \Sigma\tau = \tau_1 + \tau_2 = F_1d_1 + (-F_2d_2)$$

When you properly apply the sign convention, the sign of the net torque will tell you which way the object will rotate, if at all.

internet**connect**

SC*L*INKS.
NSTA
TOPIC: Torque
GO TO: www.scilinks.org
*sci*LINKS CODE: HF2081

SAMPLE PROBLEM 8A

Torque ⓉEKS 2C, 4C

PROBLEM

A basketball is being pushed by two players during tip-off. One player exerts a downward force of 11 N at a distance of 7.0 cm from the axis of rotation. The second player applies an upward force of 15 N at a perpendicular distance of 14 cm from the axis of rotation. Find the net torque acting on the ball.

SOLUTION

1. DEFINE   **Given:** $F_1 = 15$ N $F_2 = 11$ N
                              $d_1 = 0.14$ m $d_2 = 0.070$ m

   Unknown: $\tau_{net} = ?$

   Diagram:

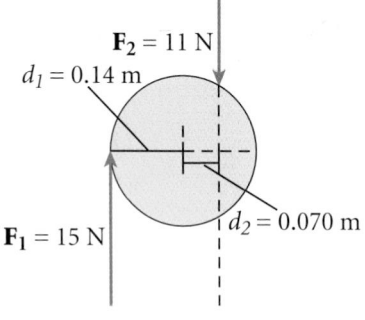

$F_2 = 11$ N
$d_1 = 0.14$ m
$F_1 = 15$ N
$d_2 = 0.070$ m

2. PLAN   **Choose an equation(s) or situation:** Apply the definition of torque to each force and add up the individual torques.

$$\tau = Fd$$
$$\tau_{net} = \tau_1 + \tau_2 = F_1d_1 + F_2d_2$$

3. CALCULATE   **Substitute the value(s) into the equation(s) and solve:** First, determine the torque produced by each force. Each force produces clockwise rotation, so both torques are negative.

$$\tau_1 = F_1d_1 = -(15 \text{ N})(0.14 \text{ m}) = -2.1 \text{ N}\bullet\text{m}$$
$$\tau_2 = F_2d_2 = -(11 \text{ N})(0.070 \text{ m}) = -0.77 \text{ N}\bullet\text{m}$$
$$\tau_{net} = -2.1 \text{ N}\bullet\text{m} - 0.77 \text{ N}\bullet\text{m}$$

$$\boxed{\tau_{net} = -2.9 \text{ N}\bullet\text{m}}$$

CALCULATOR SOLUTION

Your calculator will give the answer as 2.87. Because of the significant figure rule for addition, the answer should be rounded to 2.9.

4. EVALUATE   The net torque is negative, so the ball rotates in a clockwise direction.

Classroom Practice

The following may be used as teamwork exercises or for demonstration at the board or on an overhead projector.

PROBLEM

Torque

A student pushes with a minimum force of 50.0 N on the middle of a door to open it.

   a. What minimum force must be applied at the edge of the door in order for the door to open?

   b. What minimum force must be applied to the hinged side of the door in order for the door to open?

Answers

   a. 25.0 N
   b. The door cannot be opened by a force at the hinge location. It can be *broken* but not opened normally.

PRACTICE GUIDE 8A		
Solving for:		
τ	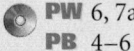 PE	Sample, 1–2; Ch. Rvw. 9–10, 11a, 47, 61*
	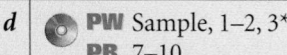 PW	6, 7a
	PB	4–6
d	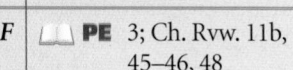 PW	Sample, 1–2, 3*
	PB	7–10
F	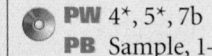 PE	3; Ch. Rvw. 11b, 45–46, 48
	PW	4*, 5*, 7b
	PB	Sample, 1–3

ANSWERS TO

Practice 8A
Torque
1. 0.75 N•m
2. **a.** 5.1 N•m
   b. 15 N•m
3. 133 N

Teaching Tip

Another way to calculate torque is to use the full distance from the pivot point to the applied force and to use only the perpendicular component of the applied force. The magnitude of this "effective force" is equal to $F(\sin\theta)$. The resulting torque will be equal to the result that would be obtained using the applied force and lever arm.

Section Review
ANSWERS

1. **a.** point mass
   b. extended object
   c. extended object
   d. point mass
2. torque; It is measured in units of N•m, not N. It is a measure of the ability of a force to accelerate an object around an axis; It depends on force, lever arm, and angle at which the force is applied.
3. $\tau_{30} = 0$ N•m, $\tau_{25} = 43$ N•m, $\tau_{10} = -16$ N•m;
   The bar will rotate counterclockwise.
4. Twice as much force would be needed to open the door.
5. The longer the pedal arm is, the greater the torque.

PRACTICE 8A

Torque ⊛TEKS 2C, 4C

1. Find the magnitude of the torque produced by a 3.0 N force applied to a door at a perpendicular distance of 0.25 m from the hinge.

2. A simple pendulum consists of a 3.0 kg point mass hanging at the end of a 2.0 m long light string that is connected to a pivot point.
   a. Calculate the magnitude of the torque (due to the force of gravity) around this pivot point when the string makes a 5.0° angle with the vertical.
   b. Repeat this calculation for an angle of 15.0°.

3. If the torque required to loosen a nut on the wheel of a car has a magnitude of 40.0 N•m, what *minimum* force must be exerted by a mechanic at the end of a 30.0 cm wrench to loosen the nut?

Section Review

1. In which of the following situations should the object(s) be treated as a point mass? In which should the object(s) be treated as an extended object?
   a. a baseball dropped from the roof of a house
   b. a baseball rolling toward third base
   c. a pinwheel in the wind
   d. Earth traveling around the sun

2. What is the rotational analog of a force? How does it differ from a force? On what quantities does it depend?

3. Calculate the torque for each force acting on the bar in **Figure 8-6.** Assume the axis is perpendicular to the page and passes through point *O*. In what direction will the object rotate?

4. How would the force needed to open a door change if you put the handle in the middle of the door?

5. **Physics in Action**   How does the length of the pedal arm on a penny-farthing bicycle affect the amount of torque applied to the front wheel?

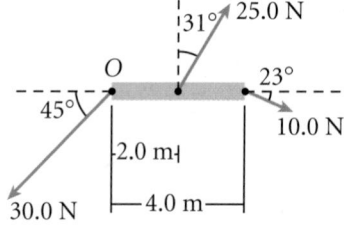

Figure 8-6

8-2
Rotation and inertia

CENTER OF MASS

Locating the axis of rotation for the cat-flap door is simple: it rotates on its hinges because the house applies a force that keeps the hinges in place. Now imagine you are playing fetch with your dog, and you throw a stick up into the air for the dog to retrieve. How can you determine the point around which the stick will rotate as it travels through the air? Unlike the cat-flap door, the stick is not attached to anything. There is a special point around which the stick rotates if gravity is the only force acting on the stick. This point is called the stick's **center of mass.**

Rotational and translational motion can be combined

The center of mass is also the point at which all the mass of the body can be considered to be concentrated. This means that the complete motion of the stick is a combination of both translational and rotational motion. The stick rotates in the air around its center of mass. The center of mass, in turn, moves as if the stick were a point mass, with all of its mass concentrated at that point for purposes of analyzing its translational motion.

Note that the hammer in **Figure 8-7** rotates about its center of mass as it moves through the air. As the rest of the hammer spins, the center of mass moves along the path of a projectile.

For regularly shaped objects, such as a sphere or a cube, the center of mass is at the geometric center of the object. For more complicated objects, calculating the location of the center of mass is more difficult and is beyond the scope of this book. While the center of mass is the position at which an extended object's mass can be treated as a point mass, the *center of gravity* is the position at which the gravitational force acts on the extended object as if it were a point mass. For most situations in this book, the center of mass and the center of gravity are equivalent.

8-2 SECTION OBJECTIVES

- **Identify the center of mass of an object.**
- **Distinguish between mass and moment of inertia.**
- **Define the second condition of equilibrium.**
- **Solve problems involving the first and second conditions of equilibrium.**

center of mass

the point at which all the mass of the body can be considered to be concentrated when analyzing translational motion

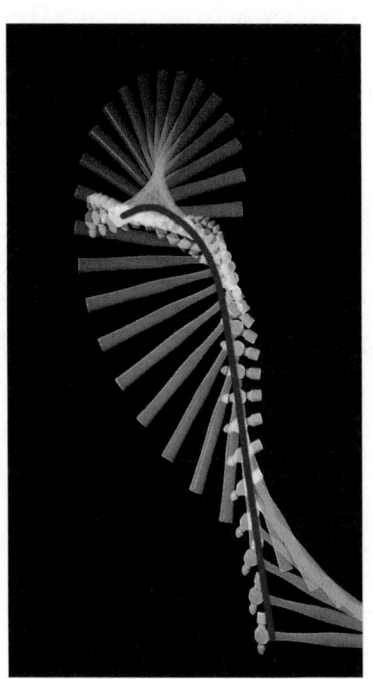

internet connect

SC*LINKS*

NSTA

TOPIC: Center of mass
GO TO: www.scilinks.org
sciLINKS CODE: HF2082

 **TEKS** 3A

Figure 8-7
The point around which this object rotates is the center of mass. The center of mass traces out a parabola.

 Misconception Alert

Point out that objects rotate around their center of mass *in the absence of other forces.* For example, a ruler thrown through the air will rotate around its own center of mass because air resistance is evenly distributed and produces zero net torque. However, a ruler with an index card taped to one end will not rotate around its own center of mass when it is thrown through the air because the card provides air resistance, which in turn produces a torque.

Teaching Tip

Let students examine objects that have a center of mass outside the object itself, such as a doughnut, a coat hanger, or a boomerang.

TEKS

p. 282: 2C, 4C
p. 283: 3A

Quick Lab

moment of inertia

the tendency of a body rotating about a fixed axis to resist a change in rotational motion

MOMENT OF INERTIA

Imagine you are rotating a baseball bat. There are many axes around which the bat can be rotated. But it is easier to rotate the bat around some axes than others, even though the bat's mass has not changed. The resistance of an object to changes in rotational motion is measured by a quantity called the **moment of inertia.** The term *moment* has a meaning in physics that is different from its everyday meaning. The moment of inertia is a measure of the object's resistance to a change in its rotational motion about some axis.

Moment of inertia is the rotational analog of mass

The moment of inertia is similar to mass because they are both forms of inertia. However, there is an important difference between the inertia that resists changes in translational motion (mass) and the inertia that resists changes in rotational motion (moment of inertia). Mass is an intrinsic property of an object, and the moment of inertia is not. It depends on the object's mass and the distribution of that mass around the axis of rotation. The farther the mass of an object is, on average, from the axis of rotation, the greater is the object's moment of inertia and the more difficult it is to rotate the object. This is why, in the race on page 278, the solid sphere came in first and the hollow cylinder came in last. The mass of the hollow cylinder is all concentrated around its rim (large moment of inertia), while the mass of the sphere is more evenly distributed throughout its volume (small moment of inertia). **(★)TEKS** **3A**

Calculating the moment of inertia

According to Newton's second law, when a net force acts on an object, the resulting acceleration of the object depends on the object's mass. Similarly, when a net torque acts on an object, the resulting change in the rotational motion of the object depends on the object's moment of inertia.

Finding the Center of Mass Experimentally

MATERIALS LIST

✔ cardboard
✔ scissors
✔ hole punch
✔ pushpin or nail
✔ corkboard or tackboard
✔ length of string, about 40 cm
✔ straightedge
✔ pencil or pen
✔ weight, such as a washer

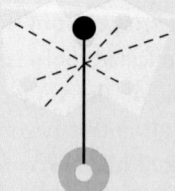

Cut out an irregular shape from the cardboard, and punch 3–5 holes around the edge of the shape. Put the pushpin through one of the holes, and tack the shape to a corkboard so that the shape can rotate freely. (You may also hang the shape from a nail in the wall.)

Attach the weight to the end of the string and hang the string from the push-pin or nail. When the string stops moving, trace a line on the cardboard that follows the string.

Repeat for each of the holes in the cardboard. The point where the lines intersect is the center of mass. **(★)TEKS** **1A**

The calculation of the moment of inertia is a straightforward but often tedious process. Fortunately, some simple formulas are available for common shapes. **Table 8-1** gives the moments of inertia for some common shapes. When the need arises, you can use this table to determine the moment of inertia of a body having one of the listed shapes.

The units for moment of inertia are kg•m^2. To get an idea of the size of this unit, note that bowling balls typically have moments of inertia about an axis through their centers ranging from about 0.7 kg•m^2 to 1.8 kg•m^2, depending on the mass and size of the ball. ⊛TEKS 2C

Notice that the moment of inertia for the solid sphere is indeed smaller than the moment of inertia for the thin hoop, as expected. In fact, the moment of inertia for the thin hoop about the symmetry axis through the center of mass is the largest moment of inertia that is possible for any shape.

Also notice that a point mass in a circular path, such as a ball on a string, has the same moment of inertia as the thin hoop if the distance of the point mass from its axis of rotation is equal to the hoop's radius. This shows that only the distance of a mass from the axis of rotation is important in determining the moment of inertia for a shape. At a given radius from an axis, it does not matter how the mass is distributed around the axis. ⊛TEKS 4B

Finally, recall the example of the rotating baseball bat that began this section. A bat can be modeled as a rotating thin rod. **Table 8-1** shows that the moment of

Table 8-1   The moment of inertia for a few shapes

Shape	Moment of inertia	Shape	Moment of inertia
thin hoop about symmetry axis	MR^2	thin rod about perpendicular axis through center	$\frac{1}{12}M\ell^2$
thin hoop about diameter	$\frac{1}{2}MR^2$	thin rod about perpendicular axis through end	$\frac{1}{3}M\ell^2$
point mass about axis	MR^2	solid sphere about diameter	$\frac{2}{5}MR^2$
disk or cylinder about symmetry axis	$\frac{1}{2}MR^2$	thin spherical shell about diameter	$\frac{2}{3}MR^2$

Demonstration 2

Moment of inertia of a rod

Purpose To give visual examples of the two cases of a thin rod and the case of a cylinder described in **Table 8-1.**

Materials broomstick or dowel

Procedure If possible, let student volunteers assist by trying the various demonstrations.

Thin rod about center: Hold the rod in the center with one hand. Rotate the rod back and forth through half rotations at regular time intervals. Note the force required to change the motion.

Thin rod about end: Hold rod at one end. Rotate the rod through half circles in the same time interval as previously used. Note the increased force required (corresponding to the larger moment of inertia).

Cylinder: Hold the rod vertically between your palms with your fingers extended. Rotate the cylinder by moving your palms back and forth in the same regular time interval as used previously. Note the much smaller force required in this case.

⊛TEKS
p. 284: 3A, 1A
p. 285: 2C, 4B

Students will likely confuse zero net torque with zero rotation. State explicitly that an object rotating at constant speed is experiencing zero net torque. This situation is analogous to linear motion: an object moving at a constant velocity has zero net force acting on it.

The Language of Physics

Another way to state the second equilibrium condition is to say that the sum of the clockwise torques must equal the sum of the counterclockwise torques.

inertia of a thin rod is larger if the rod is longer or more massive. When a bat is held at its end, its length is greatest with respect to the rotation axis, and so its moment of inertia is greatest. The moment of inertia decreases, and the bat is easier to swing if you hold the bat closer to the center. Baseball players sometimes do this either because a bat is too heavy (large M) or is too long (large ℓ). In both cases, the player decreases the bat's moment of inertia. ⭐TEKS **3A**

ROTATIONAL EQUILIBRIUM

Figure 8-8
The two forces exerted on this table are equal and opposite, yet the table moves. How is this possible?

Imagine that you and a friend are trying to move a piece of heavy furniture and that you are both a little confused. Instead of pushing from the same side, you push on opposite sides, as shown in **Figure 8-8.** The two forces acting on the furniture are equal in magnitude and opposite in direction. Your friend thinks the condition for equilibrium is satisfied because the two forces balance each other. He says the piece of furniture shouldn't move. But it does; it rotates in place.

Equilibrium requires zero net force and zero net torque

The piece of furniture can move even though the net force acting on it is zero because the net torque acting on it is not zero. If the net force on an object is zero, the object is in *translational equilibrium*. If the net torque on an object is zero, the object is in *rotational equilibrium*. For an object to be completely in equilibrium, both rotational and translational, there must be both zero net force and zero net torque, as summarized in **Table 8-2.** The dependence of equilibrium on the absence of net torque is called the *second condition for equilibrium*.

To apply the first condition for equilibrium to an object, it is necessary to add up all of the forces acting on the object (see Chapter 4). To apply the second condition for equilibrium to an object, it is also necessary to choose an axis of rotation around which to calculate the torque. Which axis should be chosen? The answer is that it does not matter. The resultant torque acting on an object in rotational equilibrium is independent of where the axis is placed. This fact is useful in solving rotational equilibrium problems because an unknown force that acts along a line passing through this axis of rotation will not produce any torque. Beginning a diagram by arbitrarily setting an axis where a force acts can eliminate an unknown in the problem.

Table 8-2	**Conditions for equilibrium** ⭐TEKS **4C**	
Type of equilibrium	**Symbolic equation**	**Meaning**
translational	$\sum F = 0$	The net force on an object must be zero.
rotational	$\sum \tau = 0$	The net torque on an object must be zero.

Rotational equilibrium ⊛TEKS 2C, 4C

PROBLEM

A uniform 5.00 m long horizontal beam that weighs 315 N is attached to a wall by a pin connection that allows the beam to rotate. Its far end is supported by a cable that makes an angle of 53° with the horizontal, and a 545 N person is standing 1.50 m from the pin. Find the force in the cable, F_T, and the force exerted on the beam by the wall, R, if the beam is in equilibrium.

SOLUTION

1. DEFINE

Given:

$L = 5.00$ m $F_{g,\,b} = 315$ N $\theta = 53°$

$F_{g,\,p} = 545$ N $d = 1.50$ m

Unknown: $F_T = ?$ $R = ?$

Diagram: The weight of a uniform extended object is assumed to be concentrated at the object's center of mass.

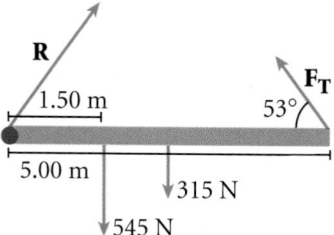

2. PLAN

Choose an equation(s) or situation: The unknowns are R_x, R_y, and F_T. The first condition of equilibrium for the x and y directions gives:

x component equation: $F_x = R_x - F_T(\cos\theta) = 0$

y component equation: $F_y = R_y + F_T(\sin\theta) - F_{g,p} - F_{g,b} = 0$

Because there are three unknowns and only two equations, we cannot find the solutions from only the first condition of equilibrium.

Choose a point for calculating the net torque: The pin connection is a convenient place to put the axis because the unknown force, R, will not contribute to the net torque about this point.

Apply the second condition of equilibrium: The necessary third equation can be found from the second condition of equilibrium.

$$\tau = F_T L(\sin\theta) - F_{g,b}\frac{L}{2} - F_{g,p}d = 0$$

3. CALCULATE

Substitute the values into the equation(s) and solve:

$\tau = F_T(\sin 53°)(5.00\ \text{m}) - (315\ \text{N})(2.50\ \text{m}) -$

$(545\ \text{N})(1.50\ \text{m}) = 0$

$\tau = F_T(4.0\ \text{m}) - 788\ \text{N}\bullet\text{m} - 818\ \text{N}\bullet\text{m} = 0$

$$F_T = \frac{1606\ \text{N}\bullet\text{m}}{4.0\ \text{m}}$$

$\boxed{F_T = 4.0 \times 10^2\ \text{N}}$

continued on next page

Classroom Practice

The following may be used as a teamwork exercise or for demonstration at the board or on an overhead projector.

PROBLEM

Rotational equilibrium

An 8.5 m long ladder weighs 350 N. The ladder leans against a frictionless vertical wall. If the ladder makes an angle of 60° with the ground, find the force exerted on the ladder by the wall and the force exerted on the ladder by the ground.

Answer

100 N out from the wall, 360 N at an angle of 74° from the ground

PRACTICE GUIDE 8B		
Solving for:		
F	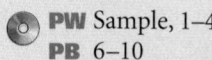 **PE**	Sample, 1–3; Ch. Rvw. 20–23, 55, 77a
	PW	Sample, 1–4
	PB	6–10
d	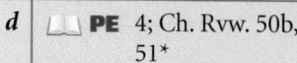 **PE**	4; Ch. Rvw. 50b, 51*
	PW	5–8
	PB	Sample, 1–5

⊛TEKS

p. 286: **3A, 4C**
p. 287: **2C, 4C**

Alternative Problem-Solving Approach

Since the object is in equilibrium, any choice for rotational axis will work. Choose the other end and F_T will be eliminated.

$$\tau = -R_y L + F_{g,p}(L-d) + F_{g,b}\left(\frac{L}{2}\right)$$

$$\tau = 0$$

$$R_y = \frac{1910 \text{ N}\bullet\text{m} + 788 \text{ N}\bullet\text{m}}{5.00 \text{ m}}$$

$$R_y = 5.40 \times 10^2 \text{ N}$$

By first condition of equilibrium for y-axis:

$$F_T = \frac{F_{g,p} + F_{g,b} - R_y}{\sin\theta}$$

$$F_T = 401 \text{ N}$$

By first condition of equilibrium for x-axis:

$$R_x = F_T(\cos\theta)$$
$$R_x = 241 \text{ N}$$

$$R = \sqrt{R_x^2 + R_y^2}$$

$$R = 592 \text{ N}$$

The slight differences are due to rounding.

ANSWERS TO

Practice 8B
Rotational equilibrium

1. 4.0×10^2 N, 5.9×10^2 N
2. 2.12×10^5 N, 2.08×10^5 N
3. 333 N, 567 N
4. **a.** 0.86 m from the 400.0 N child
   b. 0.49 m from the pivot point on the same side as the 300.0 N child

This value for the force in the wire is then substituted into the x and y equations to find R.

$$F_x = R_x - F_T(\cos 53°) = 0$$

$$R_x = (400 \text{ N})(\cos 53°)$$

$$R_x = 240 \text{ N}$$

$$F_y = R_y + F_T(\sin 53°) - 545 \text{ N} - 315 \text{ N} = 0$$

$$R_y = -3.2 \times 10^2 \text{ N} + 8.60 \times 10^2 \text{ N}$$

$$R_y = 540 \text{ N}$$

$$R = \sqrt{R_x^2 + R_y^2}$$

$$R = \sqrt{(240 \text{ N})^2 + (540 \text{ N})^2}$$

$$\boxed{R = 5.9 \times 10^2 \text{ N}}$$

4. EVALUATE The sum of the y components of the force in the wire and the force exerted by the wall must equal the weight of the beam and the person. Thus, the force in the wire and the force exerted by the wall must be greater than the sum of the two weights.

$$400 + 590 > 545 + 315$$

PRACTICE 8B

Rotational equilibrium (★)TEKS 2C, 4C

1. Rework the example problem above with the axis of rotation passing through the center of mass of the beam. Verify that the answers do not change even though the axis is different.

2. A uniform bridge 20.0 m long and weighing 4.00×10^5 N is supported by two pillars located 3.00 m from each end. If a 1.96×10^4 N car is parked 8.00 m from one end of the bridge, how much force does each pillar exert?

3. A 700.0 N window washer is standing on a uniform scaffold supported by a vertical rope at each end. The scaffold weighs 200.0 N and is 3.00 m long. What is the force in each rope when the window washer stands 1.00 m from one end?

4. A 400.0 N child and a 300.0 N child sit on either end of a 2.0 m long seesaw.
   a. Where along the seesaw should the pivot be placed to ensure rotational equilibrium? Disregard the mass of the seesaw.
   b. Suppose a 225 N child sits 0.200 m from the 400.0 N child. Where must a 325 N child sit to maintain rotational equilibrium?

1. At which of the seven positions indicated in **Figure 8-9** should the supporting pivot be located to produce the following?

50 N

a. a net positive torque

b. a net negative torque

c. no rotation

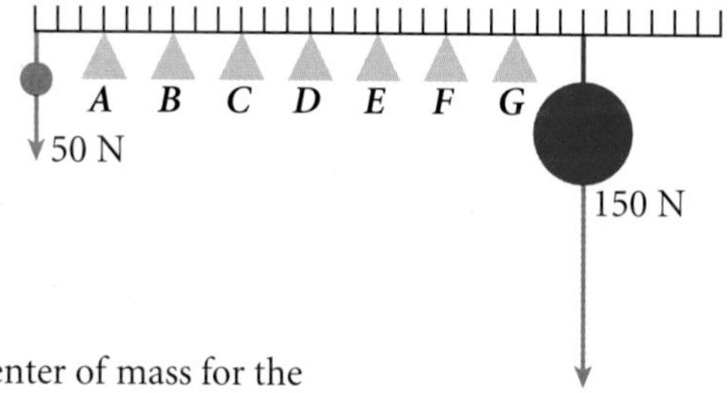

150 N

2. Describe the approximate location of the center of mass for the following objects:

Figure 8-9

a. a meterstick

b. a bowling ball

c. an ice cube

d. a doughnut

e. a banana

3. A student says that moment of inertia and mass are the same thing. Explain what is wrong with this reasoning.

4. Identify which, if any, conditions of equilibrium hold for the following situations:

a. a bicycle wheel rolling along a level highway at constant speed

b. a bicycle parked against a curb

c. the tires on a braking automobile that is still moving

d. a football traveling through the air

5. A uniform 40.0 N board supports two children, one weighing 510 N and the other weighing 350 N. The support is under the center of mass of the board, and the 510 N child is 1.50 m from the center.

a. Where should the 350 N child sit to balance the system?

b. How much force does the support exert on the board?

6. Physics in Action Why would it be beneficial for a bicycle to have a low center of mass when the rider rounds a turn?  TEKS **3A**

7. Physics in Action A bicycle designer recently modified a bicycle by adding cylindrical weights to the spokes of the wheels. He reasoned that this would make the mass of the wheel, on average, closer to the axle, in turn making the moment of inertia smaller and the wheel easier to rotate. Where did he go wrong?  TEKS **3A**

8. Physics in Action The front wheel of a penny-farthing bicycle is three times as large as the rear wheel. How much more massive must the rear wheel be to have the same moment of inertia as the front wheel?

SECTION 8-2

Section Review
ANSWERS

1. **a.** G
   b. A–E
   c. F

2. **a.** 50.0 cm mark
   b. near the center of the ball away from the fingerholes
   c. center of the cube
   d. center of the doughnut
   e. at the middle along the inside curve

3. Mass is inertia that resists changes in translational motion, while moment of inertia is inertia that resists changes in rotational motion.

4. **a.** rotational and translational
   b. rotational and translational
   c. neither
   d. rotational equilibrium

5. **a.** 2.2 m from center
   b. 9.0×10^2 N

6. The lever arm is shorter when the center of mass is lower. Thus, the torque on the bike is less likely to tip it over.

7. The designer increased the net mass of the wheel, increasing its moment of inertia.

8. nine times more massive

 TEKS

p. 288: 2C, 4C
p. 289: 2C, 4C, 3A, 3A

8-3
Rotational dynamics

Visual Strategy

Figure 8-10
Point out that the force of the falling water exerts a roughly constant torque on the water-wheel, making it spin.

Q If the torque due to the water is constant, why does the wheel spin at a constant angular velocity? That is, why doesn't the wheel accelerate without limit?

A *Friction between the axle and the wheel and friction due to air resistance exert a counter-torque so that at some speed the net torque is zero and the wheel turns at constant speed.*

Interactive Problem-Solving Tutor

See Module 10
"Rotational Inertia" includes additional practice with moments of inertia and the rotational form of Newton's second law.

⭐TEKS

p. 290: 3B, 4C
p. 291: 2C, 4C

8-3 SECTION OBJECTIVES

- **Describe Newton's second law for rotation.**

- **Calculate the angular momentum for various rotating objects.**

- **Solve problems involving rotational kinetic energy.**

Figure 8-10
The continuous flow of water exerts a torque on the waterwheel.

Module 10
"Rotational Inertia" provides an interactive lesson with guided problem-solving practice to teach you about rotational motion and Newton's second law for rotating objects.

NEWTON'S SECOND LAW FOR ROTATION

You learned in Section 8-2 that there is a relationship between the net torque on an object and the angular acceleration given to the object. This is analogous to Newton's second law, which relates the net force on an object to the translational acceleration given to the object. Newton's second law for rotating objects can be written as follows: ⭐TEKS **3B**

NEWTON'S SECOND LAW FOR ROTATING OBJECTS

$$\tau_{net} = I\alpha$$

net torque = moment of inertia × angular acceleration

Recall that a net positive torque causes an object to rotate counterclockwise. This means that the angular acceleration of the object is also counterclockwise. Similarly, a net negative torque will produce a clockwise angular acceleration. Thus, in calculating an object's angular acceleration, it is important to keep track of the signs of the torques acting on the object.

For example, consider **Figure 8-10,** which shows a continuous stream of water falling on a wheel. The falling water exerts a force on the rim of the wheel, producing a torque that causes the wheel to rotate. Other forces such as air resistance and friction between the axle and the wheel produce counteracting torques. When the net torque on the wheel is zero, the wheel rotates with constant angular velocity.

The relationship between these translational and rotational quantities is summarized in **Table 8-3.** ⭐TEKS **4C**

Table 8-3	Newton's second law for translational and rotational motion	
Translation	$F = ma$	force = mass × acceleration
Rotation	$\tau = I\alpha$	torque = moment of inertia × angular acceleration

Newton's second law for rotation  TEKS 2C, 4C

PROBLEM

A student tosses a dart using only the rotation of her forearm to accelerate the dart. The forearm rotates in a vertical plane about an axis at the elbow joint. The forearm and dart have a combined moment of inertia of 0.075 kg•m^2 about the axis, and the length of the forearm is 0.26 m. If the dart has a tangential acceleration of 45 m/s^2 just before it is released, what is the net torque on the arm and dart?

SOLUTION

Given: $I = 0.075$ kg•m^2 $a = 45$ m/s^2 $d = 0.26$ m

Unknown: $\tau = ?$

Use the equation for Newton's second law for rotating objects, given on page 290.

$$\tau = I\alpha \text{ where } \alpha = a/d$$

$$\tau = I(a/d)$$

$$\tau = (0.075 \text{ kg•m}^2)(45 \text{ m/s}^2)/0.26 \text{ m}$$

$$\boxed{\tau = 13 \text{ N•m}}$$

Newton's second law for rotation

1. A potter's wheel of radius 0.50 m and mass 100.0 kg is freely rotating at 50.0 rev/min. The potter can stop the wheel in 6.0 s by pressing a wet rag against the rim.
   a. What is the angular acceleration of the wheel?
   b. How much torque does the potter apply to the wheel?

2. A bicycle tire of radius 0.33 m and mass 1.5 kg is rotating at 98.7 rad/s. What torque is necessary to stop the tire in 2.0 s?

3. A light string 4.00 m long is wrapped around a solid cylindrical spool with a radius of 0.075 m and a mass of 0.500 kg. A 5.00 kg mass is then attached to the free end of the string, causing the string to unwind from the spool.
   a. What is the angular acceleration of the spool?
   b. How fast will the spool be rotating after all of the string has unwound?

Classroom Practice

The following may be used as a teamwork exercise or for demonstration at the board or on an overhead projector.

PROBLEM

Newton's 2nd law for rotation

A disk with a mass of 165.0 g and a radius of 13.5 cm that is spinning at 3.0×10^1 rad/s can be stopped by a hand in 0.10 s. If the mass of the disk is concentrated at the rim, what is the average torque on the disk by the hand?

Answer
−0.45 N•m

PRACTICE GUIDE 8C		
Solving for:		
τ	📖 **PE**	Sample, 1b, 2; Ch. Rvw. 27–28, 56a, 57b, 68b
	💿 **PW**	9–12
	PB	4–6
α	📖 **PE**	1a, 3; Ch. Rvw. 58, 60c, 68*, 69*
	💿 **PW**	Sample, 1–4
	PB	7–10
I	📖 **PE**	Ch. Rvw. 54b*, 57a
	💿 **PW**	5–8
	PB	Sample, 1–3

ANSWERS TO

Practice 8C
Newton's 2nd law for rotation
1. a. −0.87 rad/s^2
   b. −11 N•m
2. −8.1 N•m
3. a. 2600 rad/s^2
   b. 530 rad/s

Colliding magnetic marbles

Purpose This demonstration illustrates conservation of angular momentum and shows that objects moving linearly can also have angular momentum.

Materials 2 magnetic marbles, overhead projector

Procedure Place one magnetic marble in the center of the overhead projector. Send the second magnetic marble toward the first to cause a glancing collision. The marbles will stick together and spin about their center of mass. Point out to students that the linear momentum is **not** converted to angular momentum. Repeat the demonstration and have students observe the slight translation of the two marbles after the collision. This is the linear momentum, which is lost due to friction. The spinning is due to conservation of the angular momentum of the moving marble relative to the stationary marble.

MOMENTUM

Have you ever swung a sledgehammer or a similarly heavy object? You probably noticed that it took some effort to start the object rotating and that it also took an effort to stop it from rotating. This is because objects resist changes in their rotational motion as well as in their translational motion.

Rotating objects have angular momentum

angular momentum

the product of a rotating object's moment of inertia and angular speed about the same axis

Because a rotating object has inertia, it also possesses momentum associated with its rotation. This momentum is called **angular momentum.** Angular momentum is defined by the following equation:

ANGULAR MOMENTUM

$$L = I\omega$$

angular momentum = moment of inertia × angular speed

The unit of angular momentum is kg•m^2/s. To get an idea of how large this unit is, note that a 35 kg bowling ball rolling at an angular speed of 40 rad/s has an angular momentum of about 80 kg•m^2/s. The relationship between these translational and rotational quantities is summarized in **Table 8-4.** ⓧTEKS 2C

Table 8-4	Translational and angular momentum	
Translational	$p = mv$	momentum = mass × speed
Rotational	$L = I\omega$	rotational momentum = moment of inertia × angular speed

Figure 8-11
Angular momentum is conserved as the skater pulls his arms toward his body.

Angular momentum may be conserved

When the net external torque acting on an object or objects is zero, the angular momentum of the object(s) does not change. This is the law of *conservation of angular momentum.*

For example, assuming the friction between the skates and the ice is negligible, there is no torque acting on the skater in **Figure 8-11,** so his angular momentum is conserved. When he brings his hands and feet closer to his body, more of his mass, on average, is nearer his axis of rotation. As a result, the moment of inertia of his body decreases. Because his angular momentum is constant, his angular speed increases to compensate for his smaller moment of inertia. ⓧTEKS 5D

Conservation of angular momentum ⭐TEKS 2C, 5D

PROBLEM

A 65 kg student is spinning on a merry-go-round that has a mass of 5.25×10^2 kg and a radius of 2.00 m. She walks from the edge of the merry-go-round toward the center. If the angular speed of the merry-go-round is initially 0.20 rad/s, what is its angular speed when the student reaches a point 0.50 m from the center?

SOLUTION

1. DEFINE **Given:** $M = 5.25 \times 10^2$ kg $r_i = R = 2.00$ m

$r_f = 0.50$ m $m = 65$ kg $\omega_i = 0.20$ rad/s

Unknown: $\omega_f = ?$

Diagram:

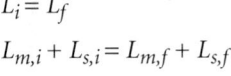

2. PLAN **Choose an equation(s) or situation:** Because there are no external torques, the angular momentum of the system (merry-go-round plus student) is conserved.

$$L_i = L_f$$

$$L_{m,i} + L_{s,i} = L_{m,f} + L_{s,f}$$

Determine the moments of inertia. Treat the merry-go-round as a solid disk, and treat the student as a point mass.

$$I_m = \tfrac{1}{2}MR^2$$

$$I_{s,i} = mR^2$$

$$I_{s,f} = mr_f^2$$

3. CALCULATE **Substitute the values into the equation(s) and solve:** Determine the initial moments of inertia, I_m and $I_{s,i}$, and the initial angular momentum, L_i.

$$I_m = \left(\tfrac{1}{2}\right)(5.25 \times 10^2 \text{ kg})(2.00 \text{ m})^2 = 1.05 \times 10^3 \text{ kg} \cdot \text{m}^2$$

$$I_{s,i} = (65 \text{ kg})(2.00 \text{ m})^2 = 260 \text{ kg} \cdot \text{m}^2$$

$$L_i = L_{m,i} + L_{s,i} = I_m\omega_i + I_{s,i}\omega_i$$

$$L_i = (1.05 \times 10^3 \text{ kg} \cdot \text{m}^2)(0.20 \text{ rad/s}) + (260 \text{ kg} \cdot \text{m}^2)(0.20 \text{ rad/s})$$

$$L_i = 260 \text{ kg} \cdot \text{m}^2/\text{s}$$

Determine the final moment of inertia, $I_{s,f}$, and the final angular momentum, L_f.

$$I_{s,f} = (65 \text{ kg})(0.50 \text{ m})^2 = 16 \text{ kg} \cdot \text{m}^2$$

$$L_f = L_{m,f} + L_{s,f} = I_m\omega_f + I_{s,f}\omega_f$$

$$L_f = (1.05 \times 10^3 \text{ kg} \cdot \text{m}^2 + 16 \text{ kg} \cdot \text{m}^2)\omega_f$$

$$L_f = (1.07 \times 10^3 \text{ kg} \cdot \text{m}^2)\omega_f$$

continued on
next page

The following may be used as a teamwork exercise or for demonstration at the board or on an overhead projector.

PROBLEM

Conservation of angular momentum

A 0.11 kg mouse rides on the edge of a Lazy Susan that has a mass of 1.3 kg and a radius of 0.25 m. If the Lazy Susan begins with an angular speed of 3.0 rad/s, what is its angular speed after the mouse walks from the edge to a point 0.15 m from the center?

Answer

3.2 rad/s

PRACTICE GUIDE 8D		
Solving for:		
ω	📖 **PE**	Sample, 1–3, 4*, 5; Ch. Rvw. 35–36, 71
	💿 **PW**	5, 6*
	PB	4–6
r	💿 **PW**	Sample, 1–2
	PB	7–10
I	💿 **PW**	3
	PB	Sample, 1–3

⭐TEKS

p. 292: 2C, 5D
p. 293: 2C, 5D

Alternative Problem-Solving Approach

Since the student is on the merry-go-round, we know that $\omega_s = \omega_m$ and we know that I_m does not change. Therefore,

$L_i = L_f$

$(L_s + L_m)_i = (L_s + L_m)_f$

$(I_{s,i} + I_m)\omega_i = (I_{s,f} + I_m)\omega_f$

$\omega_f = \dfrac{\omega_i(I_{s,i} + I_m)}{I_{s,f} + I_m}$

where $I_m = \frac{1}{2}MR^2$

$I_{s,i} = mR^2$

$I_{s,f} = mr^2, r = \frac{1}{4}R$

$I_{s,f} = \frac{1}{16}mR^2$

$\omega_f = \dfrac{\omega_i(m + \frac{1}{2}M)R^2}{\left(\frac{1}{16}m + \frac{1}{2}M\right)R^2}$

$\omega_f = 0.25$ rad/s

This answer is calculated without rounding numbers before the final step, and therefore differs slightly from the answer in step 3 of the sample problem.

ANSWERS TO

Practice 8D
Conservation of angular momentum

1. 0.35 rad/s
2. 24 rad/s
3. 6.73 rad/s
4. 9.1×10^2 m/s
5. 4.0×10^2 rad/s

Equate the initial and final angular momentum.

$$260 \text{ kg} \cdot \text{m}^2/\text{s} = (1.07 \times 10^3 \text{ kg} \cdot \text{m}^2)\,\omega_f$$

$$\boxed{\omega_f = 0.24 \text{ rad/s}}$$

4. EVALUATE Because the total moment of inertia decreases as the student moves toward the axis, the final angular speed should be greater than the initial angular speed.

$$0.24 \text{ rad/s} > 0.20 \text{ rad/s}$$

PRACTICE 8D

Conservation of angular momentum ⭐TEKS 2C, 5D

1. A merry-go-round rotates at the rate of 0.30 rad/s with an 80.0 kg man standing at a point 2.0 m from the axis of rotation. What is the new angular speed when the man walks to a point 1.0 m from the center? Assume that the merry-go-round is a solid 6.50×10^2 kg cylinder with a radius of 2.00 m.

2. A 2.0 kg bicycle wheel with a radius of 0.30 m turns at a constant angular speed of 25 rad/s when a 0.30 kg reflector is at a distance of 0.19 m from the axle. What is the angular speed of the wheel when the reflector slides to a distance of 0.25 m from the axle?

3. A solid, vertical cylinder with a mass of 10.0 kg and a radius of 1.00 m rotates with an angular speed of 7.00 rad/s about a fixed vertical axis through its center. A 0.250 kg piece of putty is dropped vertically at a point 0.900 m from the cylinder's center of rotation and sticks to the cylinder. Determine the final angular speed of the system.

4. As Halley's comet orbits the sun, its distance from the sun changes dramatically, from 8.8×10^{10} m to 5.2×10^{12} m. If the comet's speed at closest approach is 5.4×10^4 m/s, what is its speed when it is farthest from the sun if angular momentum is conserved?

5. The entrance of a science museum features a funnel into which marbles are rolled one at a time. The marbles circle around the wall of the funnel, eventually spiraling down into the neck of the funnel. The internal radius of the funnel at the top is 0.54 m. At the bottom, the funnel's neck narrows to an internal radius of 0.040 m. A 2.5×10^{-2} kg marble begins rolling in a large circular orbit around the funnel's rim at 0.35 rev/s. If it continues moving in a roughly circular path, what will the marble's angular speed be as it passes throught the neck of the funnel? (Consider only the effects of the conservation of angular momentum.)

KINETIC ENERGY

In Chapter 5 you learned that the mechanical energy of an object includes translational kinetic energy and potential energy, but that was for objects that can be modeled as point masses. In other words, this approach did not consider the possibility that objects could have rotational motion along with translational motion.

Rotating objects have rotational kinetic energy

Rotating objects possess kinetic energy associated with their angular speed. This form of energy is called **rotational kinetic energy** and is expressed by the following equation: ⭐TEKS 5B

CALCULATING ROTATIONAL KINETIC ENERGY

$$KE_{rot} = \frac{1}{2}I\omega^2$$

rotational kinetic energy $= \frac{1}{2} \times$ **moment of inertia** $\times$ **(angular speed)**^{2}

This is analogous to the translational kinetic energy of a particle, given by the expression $\frac{1}{2}mv^2$, where the moment of inertia replaces the mass and the angular speed replaces the translational speed. The unit of rotational kinetic energy is the joule, the SI unit for energy (see Chapter 5). The relationship between these translational and rotational quantities is summarized in **Table 8-5.**

Table 8-5	**Translational and rotational kinetic energy**	
Translational	$KE_{trans} = \frac{1}{2}mv^2$	translational kinetic energy $= \frac{1}{2}$mass $\times$ (speed)^2
Rotational	$KE_{rot} = \frac{1}{2}I\omega^2$	rotational kinetic energy $= \frac{1}{2}$moment of inertia $\times$ (angular speed)^2

Mechanical energy may be conserved

Recall the race between two objects on page 278. Because it is assumed that gravity is the only external force acting on the cylinders and spheres, the mechanical energy associated with each object is conserved. Unlike the examples of energy conservation in Chapter 5, however, the objects in this example are rotating. Recalling that mechanical energy is the sum of all types of kinetic and potential energy, we must include a rotational kinetic energy term in our formula for mechanical energy, as follows:

$$ME = KE_{trans} + KE_{rot} + PE_g$$
$$ME = \frac{1}{2}mv^2 + \frac{1}{2}I\omega^2 + mgh$$

Teaching Tip

Students may not be able to grasp the idea that rotational kinetic energy is indeed a mechanical energy that can be converted into other forms of energy. Have students think about what will happen if you hold a bicycle wheel vertically above the ground so that the wheel is spinning about its axis. If you slowly lower the spinning wheel so that it touches the ground and then release it, have students predict what will happen. Students may realize that the wheel will begin to move across the floor because some of the rotational kinetic energy will be converted to translational kinetic energy. If possible, demonstrate in a clear area using a bicycle wheel or pull-start toy car.

⭐TEKS

p. 294: 2C, 5D
p. 295: 5B

The following may be used as a teamwork exercise or for demonstration at the board or on an overhead projector.

PROBLEM

Conservation of ME

A moving volleyball has PE_g, KE_{trans}, and KE_{rot}. A ball with a mass of 350 g and a radius of 11.0 cm travels at 130 km/h, 3.0 m off the ground, and has a top spin rotation of 3.0 rev/s. (Treat the ball as a hollow sphere.)

a. If the ball started at rest 1.0 m off the ground, how much work was done on the ball?

b. What percentage of the energy of the ball is potential energy?

c. What percentage is translational kinetic energy?

d. What percentage is rotational kinetic energy?

Answers

a. 240 J
b. 4.2 percent
c. 96 percent
d. 0.011 percent

The total percentage is not 100 percent because of rounding.

Conservation of mechanical energy ★TEKS 2C, 5D

PROBLEM

A solid ball with a mass of 4.10 kg and a radius of 0.050 m starts from rest at a height of 2.00 m and rolls down a 30.0° slope, as shown in Figure 8-15. What is the translational speed of the ball when it leaves the incline?

SOLUTION

1. DEFINE   **Given:** $h = 2.00$ m $\theta = 30.0°$ $m = 4.10$ kg
   $r = 0.050$ m $v_i = 0.0$ m/s

Unknown: $v_f = ?$

Diagram:

2.00 m

30.0°

2. PLAN   **Choose an equation(s) or situation:** Apply the conservation of mechanical energy.

$$ME_i = ME_f$$

Initially, the system possesses only gravitational potential energy. When the ball reaches the bottom of the ramp, this potential energy has been converted to translational and rotational kinetic energy.

$$mgh = \tfrac{1}{2}mv_f^2 + \tfrac{1}{2}I\omega_f^2 \quad \text{where } \omega_f = \frac{v_f}{r}$$

The moment of inertia for a solid ball can be found in **Table 8-1,** on page 285.

$$I = \tfrac{2}{5}mr^2$$

Equate the initial and final mechanical energy.

$$mgh = \tfrac{1}{2}mv_f^2 + \tfrac{1}{2}\left(\tfrac{2}{5}mr^2\right)\left(\frac{v_f}{r}\right)^2 = \tfrac{1}{2}mv_f^2 + \tfrac{1}{5}mv_f^2 = \tfrac{7}{10}mv_f^2$$

Rearrange the equation(s) to isolate the unknown(s):

$$v_f^2 = \tfrac{10}{7}gh$$

3. CALCULATE   **Substitute the values into the equation(s) and solve:**

$$v_f^2 = \tfrac{10}{7}(9.81 \text{ m/s}^2)(2.00 \text{ m})$$

$$\boxed{v_f = 5.29 \text{ m/s}}$$

4. EVALUATE   This speed should be less than the speed of an object undergoing free fall from the same height because part of the energy goes into rotation.

$$v_f(\text{free fall}) = \sqrt{2gh} = 6.26 \text{ m/s}$$

5.29 m/s < 6.26 m/s

Conservation of mechanical energy (★)TEKS) 2C, 5D

1. Repeat Sample Problem 8E using a solid cylinder of the same mass and radius as the ball and releasing it from the same height. In a race between these two objects on an incline, which would win?

2. A 1.5 kg bicycle tire of radius 0.33 m starts from rest and rolls down from the top of a hill that is 14.8 m high. What is the translational speed of the tire when it reaches the bottom of the hill? (Assume that the tire is a hoop with $I = mr^2$.)

3. A regulation basketball has a 25 cm diameter and may be approximated as a thin spherical shell. How long will it take a basketball starting from rest to roll without slipping 4.0 m down an incline that makes an angle of 30.0° with the horizontal?

Section Review (★)TEKS) 2C, 5D

1. A student holds a 3.0 kg mass in each hand while sitting on a rotating stool. When his arms are extended horizontally, the masses are 1.0 m from the axis of rotation and he rotates with an angular speed of 0.75 rad/s. If the student pulls the masses horizontally to 0.30 m from the axis of rotation, what is his new angular speed? Assume the combined moment of inertia of the student and the stool together is 3.0 kg•m^2 and is constant.

2. A 4.0 kg mass is connected by a massless string over a massless and frictionless pulley to the center of an 8.0 kg wheel. Assume that the wheel has a radius of 0.50 m and a moment of inertia of 2.0 kg•m^2, as shown in **Figure 8-12**. The mass is released from rest at a height of 2.0 m above the ground. What will its speed be just before it strikes the ground? (Hint: Apply conservation of mechanical energy.)

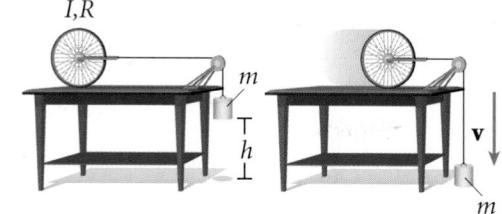

Figure 8-12

3. **Physics in Action**   A bicyclist exerts a constant force of 40.0 N on a pedal 0.15 m from the axis of rotation of a penny-farthing bicycle wheel with a radius of 50.0 cm. If his speed is 2.25 m/s 3.0 s after he starts from rest, what is the moment of inertia of the wheel? (Disregard friction and the moment of inertia of the small wheel.)

PRACTICE GUIDE 8E
Solving for:

v	📖 PE	Sample, 1–2, 3*; Ch. Rvw. 37, 38*, 65*
	💿 PW	2–3
	PB	6–7
h	📖 PE	Ch. Rvw. 52–53, 59, 62
	💿 PW	4–5, 6
	PB	8–10
I	📖 PE	Ch. Rvw. 49
	💿 PW	Sample, 1
	PB	4–5
KE	📖 PE	Ch. Rvw. 63–64, 67, 70a
	💿 PW	6–7
	PB	Sample, 1–3

ANSWERS TO

Practice 8E
Conservation of mechanical energy

1. 5.11 m/s, the ball wins the race
2. 12.0 m/s
3. 1.6 s

Section Review
ANSWERS

1. 1.9 rad/s
2. 2.8 m/s
3. 4.0 kg•m^2

8-4
Simple machines

8-4 SECTION OBJECTIVES

- **Identify the six types of simple machines.**

- **Explain how the operation of a simple machine alters the applied force and the distance moved.**

- **Calculate the mechanical advantage of a simple machine.**

TYPES OF SIMPLE MACHINES

What do you do when you need to drive a nail into a board? You probably hit the nail with a hammer. Similarly, you would probably use scissors to cut paper or a bottle opener to pry a cap off a bottle. All of these devices make your task easier. These devices are *machines*.

The term *machine* may bring to mind intricate systems with multicolored wires and complex gear-and-pulley systems. Compared with internal-combustion engines or airplanes, simple devices such as hammers, scissors, and bottle openers may not seem like machines, but they are.

A machine is any device that transmits or modifies force, usually by changing the force applied to an object. All machines are combinations or modifications of six fundamental types of machines, called *simple machines*. These six simple machines are the lever, pulley, inclined plane, wheel and axle, wedge, and screw, as shown in **Table 8-6.**

Table 8-6   Six simple machines

Lever (Fulcrum)	Inclined plane	Wheel / Axle
Wedge	Pulleys	Screw

USING SIMPLE MACHINES

Because the purpose of a simple machine is to change the direction or magnitude of an input force, a useful way of characterizing a simple machine is to compare how large the output force is relative to the input force. This ratio, called the machine's *mechanical advantage,* is written as follows:

$$MA = \frac{\text{output force}}{\text{input force}} = \frac{F_{out}}{F_{in}}$$

A good example of mechanical advantage is the claw hammer, which is a type of lever, as shown in **Figure 8-13.**

A person applies an input force to one end of the handle. The handle, in turn, exerts an output force on the head of a nail stuck in a board. Rotational equilibrium is maintained, so the input torque must balance the output torque. This can be written as follows:

$$\tau_{in} = \tau_{out}$$
$$F_{in}d_{in} = F_{out}d_{out}$$

Substituting this expression into the definition of mechanical advantage gives the following result:

$$MA = \frac{F_{out}}{F_{in}} = \frac{d_{in}}{d_{out}}$$

The longer the input lever arm is compared with the output lever arm, the greater the mechanical advantage is. This in turn indicates the factor by which the input force is amplified. If the force of the board on the nail is 99 N and if the mechanical advantage is 10, then only a force of 10 N is needed to pull out the nail. Without a machine, the nail could not be removed unless the force was greater than 100 N. (★)TEKS 2C

Machines can alter the force and the distance moved

You have learned that mechanical energy is conserved in the absence of friction. This law holds for machines as well. A machine can increase (or decrease) the force acting on an object at the expense (or gain) of the distance moved, but the product of the two—the work done on the object—is constant.

For example, imagine an incline. **Figure 8-14** shows two examples of a refrigerator being loaded onto a flatbed truck. In one example, the refrigerator is lifted directly onto the truck. In the other example, an incline is used.

In the first example, a force (F_1) of 1200 N is required to lift the refrigerator, which moves through a distance (d_1) of 1.5 m. This requires 1800 N•m of work. In the second example, a lesser force (F_2) of only 360 N is needed, but the refrigerator must be pushed a greater distance (d_2) of 5.0 m. This also requires 1800 N•m of work. As a result, the two methods require the same amount of energy. (★)TEKS 2C

Figure 8-13
A hammer makes it easier to pry a nail from a board by multiplying the input force. The hammer swivels around the point marked with a black dot.

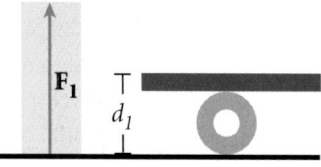

Small distance—Large force

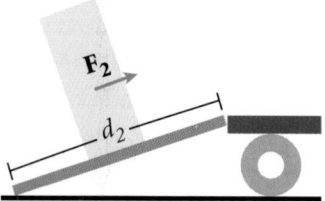

Large distance—Small force

Figure 8-14
Simple machines can alter both the force needed to perform a task and the distance through which the force acts.

SECTION 8-4

Visual Strategy

Figure 8-13
Point out that the reason the claw hammer is so useful is that the length of the handle is longer than the length of the claw. Thus, the amount of force needed to produce a large torque on the nail is small because the lever arm is long.

Q Would the claw hammer still provide a mechanical advantage if the length of the claw equaled the length of the handle?

A *No, if both the claw and the handle were the same length, the force input would equal the force output, and the mechanical advantage would equal 1.*

(STOP) **Misconception Alert**

Reinforce the idea that machines do not create something from nothing. Machines use the same amount of energy to achieve the goal. Use numerical examples to illustrate that the work done on the object is the same. Show the trade-off between force and distance.

(★)TEKS
p. 299: 2C, 2C

Tomorrow's Technology

BACKGROUND

Although Dr. Kazerooni's invention is not a simple machine, it does apply several of the basic principles that make simple machines useful.

The machine shown in the figure on this page is the top half of Dr. Kazerooni's *electric extender*. The bottom half is similar in construction and has large foot pads that enable the operator to move around with a minimum of effort; this is especially useful for carrying heavy loads. The commands of the operator are transferred to the extender via pressure-sensitive padding that senses the force applied by the operator, and the extender moves accordingly.

internet connect

SCLINKS.
NSTA

TOPIC: Texas careers in science
GO TO: www.scilinks.org
*sci*LINKS CODE: HFX014

(★)TEKS

p. 300: 4C, 3C
p. 301: 5D, 2C, 4C

Tomorrow's Technology

Human Extenders

A cyborg, as any science-fiction aficionado knows, is part human and part machine and is able to perform extraordinary tasks. Although cyborgs are still more fiction than science, Dr. Homayoon Kazerooni, of the University of California at Berkeley, has been inventing machines called *human extenders* that can give mere mortals superhuman strength.

"Human extenders are robotic systems worn by a human to move heavy objects," Dr. Kazerooni says. One of the first machines Dr. Kazerooni designed is a 1.5 m (5 ft) long steel arm that weighs thousands of newtons (several hundred pounds) and is attached to a pedestal on the floor. The operator inserts one arm into the device, and an attached computer senses the arm's movement and uses hydraulic pressure to move the extender in conjunction with the operator's arm. With the extender, a person can lift objects weighing as much as 890 N (200 lb) while exerting a force of only 89 N (20 lb). In this case, the extender yields a mechanical advantage of 10 (890 N/89 N = 10). (★)TEKS 4C

Dr. Kazerooni is developing a complete suit of human extenders that will be powered by electricity. Controlled completely by the movement of the user, the suit has two arms that sense and respond to both the force applied by the human and the weight of the object being lifted, taking most of the effort away from the operator. The machine's legs are able to balance the weight of the equipment, and they attach at the operator's feet to allow movement around the room. (★)TEKS 3C

Dr. Kazerooni envisions human extenders being used primarily as labor aids for factory workers. Approximately 30 percent of all workplace accidents in the United States are related to back injuries, and they are usually the result of a repeated lifting and moving of heavy objects. Human extenders could solve that problem. "The person who is wearing the machine," Dr. Kazerooni says, "will feel less force and less fatigue, and therefore the potential for back injuries or any kind of injury would be less."

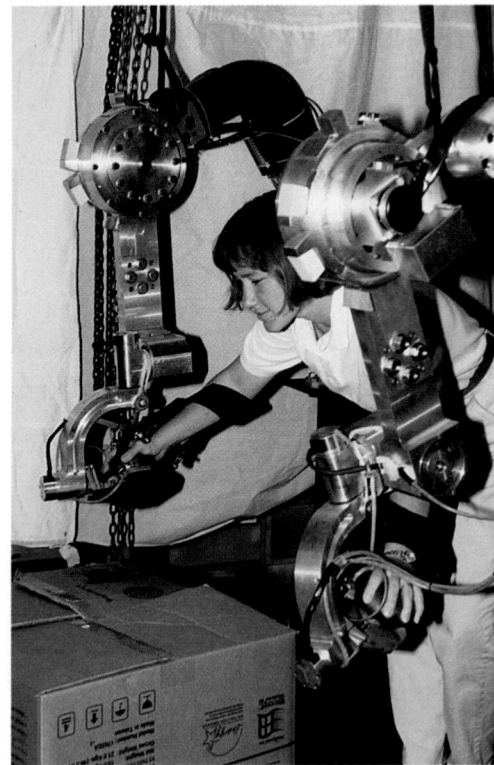

Figure 8-15

Efficiency is a measure of how well a machine works

The simple machines we have considered so far are ideal, frictionless machines. Real machines, however, are not frictionless. They dissipate energy. When the parts of a machine move and contact other objects, some of the input energy is dissipated as sound or heat. The *efficiency* of a machine is a measure of how much input energy is lost compared with how much energy is used to perform work on an object. It is defined by the following equation:

$$eff = \frac{W_{out}}{W_{in}}$$

If a machine is frictionless, then mechanical energy is conserved. This means that the work done on the machine (input work) is equal to the work done by the machine (output work) because work is a measure of energy transfer. Thus, the mechanical efficiency of an ideal machine is 1, or 100 percent. This is the best efficiency a machine can have. Because all real machines have at least a little friction, the efficiency of real machines is always less than 1. ⭐TEKS **5D**

internet**connect**

SC*LINKS*
NSTA
TOPIC: Simple machines
GO TO: www.scilinks.org
*sci*LINKS CODE: HF2083

Section Review

⭐TEKS **2C, 4C**

1. **Figure 8-16** shows an example of a Rube Goldberg machine. Identify two types of simple machines that are included in this compound machine.

2. The efficiency of a squeaky pulley system is 73 percent. The pulleys are used to raise a mass to a certain height. What force is exerted on the machine if a rope is pulled 18.0 m in order to raise a 58 kg mass a height of 3.0 m?

3. A person lifts a 950 N box by pushing it up an incline. If the person exerts a force of 350 N along the incline, what is the mechanical advantage of the incline?

4. You are attempting to move a large rock using a long lever. Will the work you do on the lever be greater than, the same as, or less than the work done by the lever on the rock? Explain.

5. **Physics in Action** A bicycle can be described as a combination of simple machines. Identify three types of simple machines that are used to propel a typical bicycle.

Figure 8-16

Section Review
ANSWERS

1. Possible answers include lever, wedge, and pulley.
2. 130 N
3. 2.7
4. Ideally, the amounts of work will be the same, but because no machine is perfectly efficient, the work you do will be greater.
5. the pedal arm (lever), the wheels (wheel and axle), and the gear system (pulley)

PHYSICS ON THE EDGE

BACKGROUND

This feature discusses the modern concept of angular momentum, in which there are two types of angular momentum associated with the electron: orbital angular momentum and electron spin. Unlike the classical examples studied earlier in this chapter, neither of these is a literal description of the electron's motion, and both are quantized rather than continuous.

The quantization of orbital angular momentum was first introduced by Niels Bohr as part of his revolutionary model of the hydrogen atom. Bohr, like Rutherford, considered the electron to be orbiting the nucleus. However, Bohr added the postulate that only certain orbits are stable; the electron is never found between these orbits. The stable orbits are those for which the electron's orbital angular momentum is an integral multiple of $h/2\pi$ (where h is Planck's constant, 6.63×10^{-34} J•s, and $n = 1, 2, 3 \ldots$), or $I\omega = nh/2\pi$.

Although more-recent models no longer view the electron as orbiting the nucleus, the electron is still considered to have an orbital angular momentum as described by Bohr. Electron spin likewise began as a literal description but is now considered to be a property of the electron independent of its location, similar to mass and charge.

Quantum Angular Momentum

Earlier in this chapter, we discussed angular momentum and its effects in the macroscopic world of your everyday experience. In the early twentieth century, scientists realized that they must modify their ideas about angular momentum when working with the microscopic world of atoms and subatomic particles.

Electron orbital angular momentum

In 1911, Ernest Rutherford proposed a model of the atom in which negatively charged particles called *electrons* orbit a positively charged nucleus containing particles called *protons,* much as the Earth orbits the sun, as shown in **Figure 8-17(a).** Because the electron orbits the nucleus, it has an orbital angular momentum. ⬥TEKS 3E

Further investigations into the microscopic realm revealed that the electron cannot be precisely located in space and therefore cannot be visualized as orbiting the proton. Instead, in modern theory the electron's location is depicted by an electron cloud, as shown in **Figure 8-17(b),** whose density varies throughout the cloud in proportion to the probability of finding the electron at a particular location in the cloud. Even though the electron does not orbit the nucleus in this model, the electron still has an orbital angular momentum that is very different from the angular momentum discussed earlier in this chapter.

For most of the history of science, it was assumed that angular momentum could have any possible value. But investigations at the atomic level have shown that this is not the case. The orbital angular momentum of the electron can have only certain possible values. Such a quantity is said to be discrete, and the angular momentum is said to be *quantized.* The branch of modern theory that deals with quanta is called *quantum mechanics.*

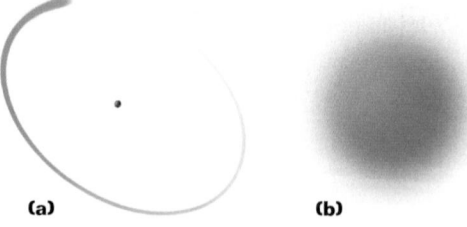

(a) **(b)**

Figure 8-17
(a) In Rutherford's model of the atom, electrons orbit the nucleus. **(b)** In quantum mechanics, an electron cloud is used to show the probability that the electron will be at different points. The densest regions of the cloud represent the most probable locations for the electron.

Electron spin

In addition to the orbital angular momentum of the electron, certain experimental evidence led scientists to postulate another type of angular momentum of the electron. This type of angular momentum is known as *spin* because the effects it explains are those that would result if the electron were to spin on its axis, much like the Earth spins on its axis of rotation. Scientists first imagined that the electron actually spins in this way, but it soon became clear that electron spin is not a literal description. Instead, electron spin is a property that is independent of the electron's motion in space. In this respect, electron spin is very different from the spin of the Earth.

Just as a wheel can turn either clockwise or counterclockwise, there are two possible types of electron spin, *spin up* and *spin down*, as shown in **Figure 8-18.** Thus, like orbital angular momentum, spin is quantized. Because the electron isn't really spinning in space, it should not be assumed that **Figure 8-18** is a physical description of the electron's motion.

Conservation of angular momentum

Although the quantum-mechanical concept of angular momentum is radically different from the classical concept of angular momentum, there is one fundamental similarity between the two models. Earlier in this chapter, you learned that angular momentum is always conserved. This principle still holds in quantum mechanics, where the total angular momentum, that is, the sum of the orbital angular momentum and spin, is always conserved.

As you have seen, the quantum-mechanical model of the atom cannot be visualized in the same way that previous atomic models could be. Although this may initially seem like a flaw of the modern theory, the accuracy of predictions based on quantum mechanics has convinced many scientists that physical models based on our experiences in the macroscopic realm cannot provide a complete picture of nature. Consequently, mathematical models must be used to describe the microscopic realm of atoms and subatomic particles.

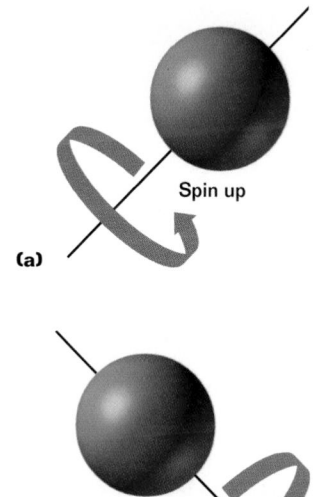

(a)

(b)

Figure 8-18
In the quantum-mechanical model of the atom, the electron has both an orbital angular momentum and an intrinsic angular momentum, known as spin. **(a)** Spin up and **(b)** spin down are the only possible values for electron spin.

Table 8-8	Angular momentum ⭐(TEKS) 5D	
Classical angular momentum	**Quantum angular momentum**	
corresponds to a literal rotation	does not correspond to a literal rotation	
can have any possible value	can have only certain discrete values	
total angular momentum (orbital + rotational) conserved	total angular momentum (orbital angular momentum + spin) conserved	

TOPIC: Rutherford model of atom
GO TO: www.scilinks.org
*sci*LINKS CODE: HF2084

CHAPTER 8
Summary

Teaching Tip

Have students create tables of comparison for the key terms and formulas of Chapter 8 and their translational counterparts. Have them include both similarities and differences.

★ TEKS

Review & Assess
pp. 269–273:
2A: 2–3
2B: 4
2C: 9–11, 20–23,
    35–38, 45–77
2D: Alt. Assess. 4–5
2E: Technology &
    Learning
3A: Alt. Assess. 1
3D: Alt. Assess. 6
4C: 1–3, 6–12, 16–17,
    19–28, 41, 45–49,
    54–57, 61, 68–69,
    72, 74, 77, Tech-
    nology & Learning
5B: 64, 67, 70
5D: 35–38, 50–53,
    58–60, 62, 65,
    66, 71

KEY TERMS

angular momentum (p. 292)

center of mass (p. 283)

lever arm (p. 279)

moment of inertia (p. 284)

**rotational kinetic energy
(p. 295)**

torque (p. 279)

KEY IDEAS

Section 8-1 Torque
- Torque is a measure of a force's ability to rotate an object.
- The torque on an object depends on the magnitude of the applied force and on the length of the lever arm, according to the following equation:

$$\tau = Fd(\sin\theta)$$

Section 8-2 Rotation and inertia
- The moment of inertia of an object is a measure of the resistance of the object to changes in rotational motion.
- For an extended object to be in complete equilibrium, it must be in both translational and rotational equilibrium.

Section 8-3 Rotational dynamics
- The rotational equation analogous to Newton's second law can be described as follows:

$$\tau = I\alpha$$

- A rotating object possesses angular momentum, which is conserved in the absence of any external forces on the object.
- A rotating object possesses rotational kinetic energy, which is conserved in the absence of any external forces on the object.

Section 8-4 Simple machines
- A simple machine can alter the force applied to an object or the distance an applied force moves an object.
- Simple machines can provide a mechanical advantage.

Key Symbols

Quantities		Units		Conversions
τ	torque	N•m	newton meter	$= \text{kg}\cdot\text{m}^2/\text{s}^2$
$d(\sin\theta)$	lever arm	m	meter	
I	moment of inertia	$\text{kg}\cdot\text{m}^2$	kilogram meter squared	
L	angular momentum	$\text{kg}\cdot\text{m}^2/\text{s}$	kilogram meter squared per second	
KE_{rot}	rotational kinetic energy	J	joule	$= \text{N}\cdot\text{m}$ $= \text{kg}\cdot\text{m}^2/\text{s}^2$

CHAPTER 8
Review and Assess

TORQUE AND MOMENT OF INERTIA

Conceptual questions

1. Explain how an orthodontist uses torque to straighten or realign teeth.

2. Which of the forces acting on the rod in **Figure 8-19** will produce a torque about the axis at the left end of the rod?

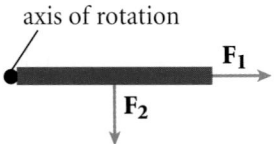

axis of rotation

F_1

F_2

Figure 8-19

3. Two children are rolling automobile tires down a hill. One child claims that the tire will roll faster if one of them curls up in the tire's center. The other child claims that will cause the tire to roll more slowly. Which child is correct?

4. The moment of inertia of Earth was recently measured to be $0.331MR^2$. What does this tell you about the distribution of mass inside Earth? (Hint: Compare this value with the moments of inertia in **Table 8-1.**)

5. The moment of inertia for a regular object can never be larger than MR^2, where M is the mass and R is the size of the object. Why is this so? (Hint: Which object has a moment of inertia of MR^2?)

6. Two forces of equal magnitude act on a wheel, as shown in **Figure 8-20.** Which force will produce the greater torque on the wheel?

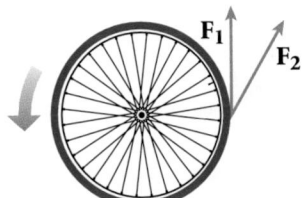

F_1

F_2

Figure 8-20

7. Two forces equal in magnitude but opposite in direction act at the same point on an object. Is it possible for there to be a net torque on the object? Explain.

8. It is more difficult to do a sit-up with your hands held behind your head than it is to do a sit-up with your arms stretched out in front of you. Explain why this statement is true.

Practice problems

9. A bucket filled with water has a mass of 54 kg and is hanging from a rope that is wound around a 0.050 m radius stationary cylinder. If the cylinder does not rotate and the bucket hangs straight down, what is the magnitude of the torque the bucket produces around the center of the cylinder?
(See Sample Problem 8A.)

10. A mechanic jacks up a car to an angle of 8.0° with the horizontal in order to change the front tires. The car is 3.05 m long and has a mass of 1130 kg. Its center of mass is located 1.12 m from the front end. The rear wheels are 0.40 m from the back end. Calculate the torque exerted by the car around the back wheels.
(See Sample Problem 8A.)

11. The arm of a crane at a construction site is 15.0 m long, and it makes an angle of 20.0° with the horizontal. Assume that the maximum load the crane can handle is limited by the amount of torque the load produces around the base of the arm.
   a. What is the magnitude of the maximum torque the crane can withstand if the maximum load the crane can handle is 450 N?
   b. What is the maximum load for this crane at an angle of 40.0° with the horizontal?
(See Sample Problem 8A.)

12. c

13. c

14. zero net force and zero net torque; The force of the see-saw support must equal the children's net weight, and the torque exerted by each child must be equal and opposite.

15. Its velocity (angular and linear) must be constant.

16. a. They spin about the baton's center of mass.
  b. parabolic motion

17. Neglecting air resistance, it moves along a parabolic path.

18. yes; Move the less massive object farther from the axis of rotation (increase the lever arm).

19. No, there could be a net translational force; yes, because with no torque there is no angular acceleration

20. 328 N, 552 N

21. a. 392 N
  b. $R_x = 339$ N, $R_y = 0$ N

22. 1500 N, 1400 N (x component), and 2600 N (y component)

CENTER OF MASS AND ROTATIONAL EQUILIBRIUM

Review questions

12. At a circus performance, a juggler is throwing two spinning clubs. One of the clubs is heavier than the other. Which of the following statements is true?

  a. The smaller club is likely to have a larger moment of inertia.
  b. The ends of each club will trace out parabolas as the club is thrown.
  c. The center of mass of each club will trace out a parabola as the club is thrown.

13. When the juggler in the previous problem stands up straight and holds each club at arm's length, his center of mass will probably be

  a. located at a point exactly in the middle of his body
  b. slightly to the side where he is holding the light club
  c. slightly to the side where he is holding the heavy club

14. What are the conditions for equilibrium? Explain how they apply to children attempting to balance a seesaw.

15. What must be true about the velocity of a moving object in equilibrium?

16. A twirler throws a baton in the air.

  a. Describe the motion of the ends of the baton as it moves through the air.
  b. Decribe the motion of the center of mass of the baton.

Conceptual questions

17. A projectile is fired into the air and suddenly explodes into several fragments. What can be said about the motion of the center of mass of the fragments after the explosion?

18. Is it possible to balance two objects that have different masses (and therefore weights) on a simple balance beam? Explain.

19. A particle moves in a straight line, and you are told that the torque acting on it is zero about some unspecified origin. Does this necessarily imply that the total force on the particle is zero? Can you conclude that the angular velocity of the particle is constant? Explain.

Practice problems

20. A window washer is standing on a scaffold supported by a vertical rope at each end. The scaffold weighs 205 N and is 3.00 m long. What is the force each rope exerts on the scaffold when the 675 N worker stands 1.00 m from one end of the scaffold? (See Sample Problem 8B.)

21. A floodlight with a mass of 20.0 kg is used to illuminate the parking lot in front of a library. The floodlight is supported at the end of a horizontal beam that is hinged to a vertical pole, as shown in **Figure 8-21**. A cable that makes an angle of 30.0° with the beam is attached to the pole to help support the floodlight. Find the following, assuming the mass of the beam is negligible when compared with the mass of the floodlight:

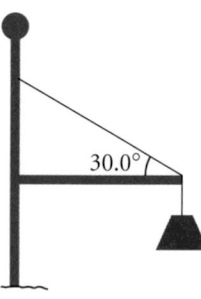

Figure 8-21

  a. the force provided by the cable
  b. the horizontal and vertical forces exerted on the beam by the pole

(See Sample Problem 8B.)

22. A 1200.0 N uniform boom is supported by a cable, as shown in **Figure 8-22.** The boom is pivoted at the bottom, and a 2000.0 N weight hangs from its top. Find the force applied by the supporting cable and the components of the reaction force on the bottom of the boom. (See Sample Problem 8B.)

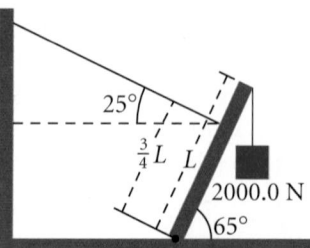

Figure 8-22

23. A uniform 10.0 N picture frame is supported as shown in **Figure 8-23.** Find the force in the cords and the magnitude of the horizontal force at P that are required to hold the frame in this position. (See Sample Problem 8B.)

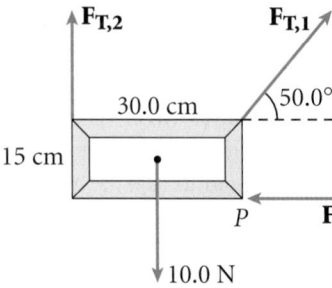

Figure 8-23

NEWTON'S SECOND LAW FOR ROTATION

Conceptual questions

24. An object rotates with a constant angular velocity. Can there be a net torque acting on the object? Explain your answer.

25. If an object is at rest, can you be certain that no external torques are acting on it?

26. Two uniformly solid disks of equal radii roll down an incline without slipping. The first disk has twice the mass of the second disk. How much torque was exerted on the first disk compared with the amount exerted on the second disk?

Practice problems

27. A 30.0 kg uniform solid cylinder has a radius of 0.180 m. If the cylinder accelerates at 2.30×10^{-2} rad/s^2 as it rotates about an axis through its center, how large is the torque acting on the cylinder? (See Sample Problem 8C.)

28. A 350 kg merry-go-round in the shape of a horizontal disk with a radius of 1.5 m is set in motion by wrapping a rope about the rim of the disk and pulling on the rope. How large a torque would have to be exerted to bring the merry-go-round from rest to an angular speed of 3.14 rad/s in 2.00 s? (See Sample Problem 8C.)

ANGULAR MOMENTUM AND ROTATIONAL KINETIC ENERGY

Review questions

29. Is angular momentum always conserved? Explain.

30. Is it possible for two objects with the same mass and the same rotational speeds to have different values of angular momentum? Explain.

31. A child on a merry-go-round moves from near the axis to the outer edge of the merry-go-round. What happens to the rotational speed of the merry-go-round? Explain.

32. Is it possible for an ice skater to change her rotational speed without any external torque? Explain.

Conceptual questions

33. Ice skaters use the conservation of angular momentum to produce high-speed spins when they bring their arms close to the rotation axis. Imagine that a skater moves her arms inward, cutting the moment of inertia in half and therefore doubling the angular speed. If we consider the rotational kinetic energy, we see that the energy is *doubled* in this situation. Thus, angular momentum is conserved, but kinetic energy is not. Where does this extra rotational kinetic energy come from?

34. A solid 2.0 kg ball with a radius of 0.50 m starts at a height of 3.0 m and rolls down a 20° slope. A solid disk and a ring start at the same time and the same height. Both the ring and the disk have the same mass and radius as the ball. Which of the three objects will win the race to the bottom if all roll without slipping?

Practice problems

35. A 15.0 kg turntable with a radius of 25 cm is covered with a uniform layer of dry ice that has a mass of 9.0 kg. The angular speed of the turntable and dry ice is initially 0.75 rad/s, but it increases as the dry ice evaporates. What is the angular speed of the turntable once all the dry ice has evaporated? (See Sample Problem 8D.)

23. 11 N, 1.6 N, 7.1 N

24. no; If there were a net torque, the object would accelerate angularly.

25. No, the torques may be in equilibrium.

26. twice as much torque

27. 1.12×10^{-2} N•m

28. 620 N•m

29. no; Only when there is no external torque on a system is the angular momentum conserved.

30. yes; Angular momentum depends on moment of inertia, not mass alone.

31. It slows because the moment of inertia has increased, requiring a slower rotational speed for angular momentum to be conserved.

32. yes; By pulling her arms in she can increase her rotational speed without an external torque.

33. As the skater pulls her arms in, she does work, which is then converted into rotational kinetic energy.

34. The ball wins because it has the smallest moment of inertia.

35. 1.2 rad/s

36. 0.13 rad/s counterclockwise
37. 7.0 m/s
38. 36 rad/s
39. The long-handled screwdriver provides a longer lever arm and greater mechanical advantage.
40. The advantage is in multiplying force applied or changing the direction of the applied force.
41. Placing the axis of rotation nearer to the rock will increase the mechanical advantage.
42. A perpetual motion machine would require 100 percent efficiency.
43. You would have to apply the force over a greater distance.
44. a
45. 220 N
46. 2.0×10^2 N
47. 1800 N•m
48. 885 N
49. 0.12 kg •m^2
50. a. 1.20 kg
    b. at the 0.595 m mark

36. A 65 kg woman stands at the rim of a horizontal turntable with a moment of inertia of 1.5×10^3 kg•m^2 and a radius of 2.0 m. The system is initially at rest, and the turntable is free to rotate about a frictionless vertical axle through its center. The woman then starts walking clockwise (when viewed from above) around the rim at a constant speed of 0.75 rad/s relative to Earth. In what direction and with what angular speed does the turntable rotate? (See Sample Problem 8D.)

37. A 35 kg bowling ball with a radius of 13 cm starts from rest at the top of an incline 3.5 m in height. Find the translational speed of the bowling ball after it has rolled to the bottom of the incline. (Assume the ball is a uniform solid sphere.) (See Sample Problem 8E.)

38. A solid 240 N ball with a radius of 0.20 m rolls 6.0 m down a ramp that is inclined at 37° with the horizontal. If the ball starts from rest at the top of the ramp, what is the angular speed of the ball at the bottom of the ramp? (See Sample Problem 8E.)

SIMPLE MACHINES

Review questions

39. Why is it easier to loosen the lid from the top of a paint can with a long-handled screwdriver than with a short-handled screwdriver?

40. If a machine cannot multiply the amount of work, what is the advantage of using such a machine?

Conceptual questions

41. You are attempting to move a large rock using a long lever. Is it more effective to place the lever's axis of rotation nearer to your hands or nearer to the rock? Explain.

42. A perpetual motion machine is a machine that, when set in motion, will never come to a halt. Why is such a machine not possible?

43. If you were to use a machine to increase the output force, what factor would have to be sacrificed? Give an example.

MIXED REVIEW PROBLEMS

44. Two spheres look identical and have the same mass. One is hollow, and the other is solid. Which method would determine which is which?
    a. roll them down an incline
    b. drop them from the same height
    c. weigh them on a scale

45. A wooden bucket filled with water has a mass of 75 kg and is attached to a rope that is wound around a cylinder with a radius of 0.075 m. A crank with a turning radius of 0.25 m is attached to the end of the cylinder. What minimum force directed perpendicularly to the crank handle is required to raise the bucket?

46. If the torque required to loosen a nut that holds a wheel on a car has a magnitude of 58 N•m, what force must be exerted at the end of a 0.35 m lug wrench to loosen the nut when the angle is 56°?

47. In a canyon between two mountains, a spherical boulder with a radius of 1.4 m is just set in motion by a force of 1600 N. The force is applied at an angle of 53.5° measured with respect to the radius of the boulder. What is the magnitude of the torque on the boulder?

48. A 23.0 cm screwdriver is used to pry open a can of paint. If the axis of rotation is 2.00 cm from the end of the screwdriver blade and a force of 84.3 N is exerted at the end of the screwdriver's handle, what force is applied to the lid?

49. The net work done in accelerating a propeller from rest to an angular speed of 220 rad/s is 3000.0 J. What is the moment of inertia of the propeller?

50. A 0.100 kg meterstick is supported at its 40.0 cm mark by a string attached to the ceiling. A 0.700 kg mass hangs vertically from the 5.00 cm mark. A mass is attached somewhere on the meterstick to keep it horizontal and in *both* rotational and translational equilibrium. If the force applied by the string attaching the meterstick to the ceiling is 19.6 N, determine the following:
    a. the value of the unknown mass
    b. the point where the mass attaches to the stick

51. A uniform ladder 8.00 m long and weighing 200.0 N rests against a smooth wall. The coefficient of static friction between the ladder and the ground is 0.600, and the ladder makes a 50.0° angle with the ground. How far up the ladder can an 800.0 N person climb before the ladder begins to slip?

52. A 0.0200 m diameter coin rolls up a 15.0° inclined plane. The coin starts with an initial angular speed of 45.0 rad/s and rolls in a straight line without slipping. How much vertical distance does it gain before it stops rolling?

53. In a circus performance, a large 4.0 kg hoop with a radius of 2.0 m rolls without slipping. If the hoop is given an angular speed of 6.0 rad/s while rolling on the horizontal and is allowed to roll up a ramp inclined at 15° with the horizontal, how far (measured along the incline) does the hoop roll?

54. A 12 kg mass is attached to a cord that is wrapped around a wheel with a radius of 10.0 cm, as shown in **Figure 8-24.** The acceleration of the mass down the frictionless incline is measured to be 2.0 m/s². Assuming the axle of the wheel to be frictionless, determine

   a. the force in the rope.
   b. the moment of inertia of the wheel.
   c. the angular speed of the wheel 2.0 s after it begins rotating, starting from rest.

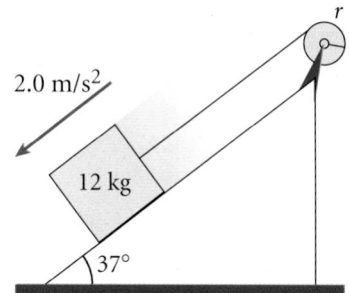

2.0 m/s²

12 kg

37°

Figure 8-24

55. A person is standing on tiptoe, and the person's total weight is supported by the force on the toe. A mechanical model for the situation is shown in **Figure 8-25,** where T is the force in the Achilles tendon and R is the force on the foot due to the tibia. Assume the total weight is 700.0 N, and find the values of T and R, if the angle labeled θ is 21.2°.

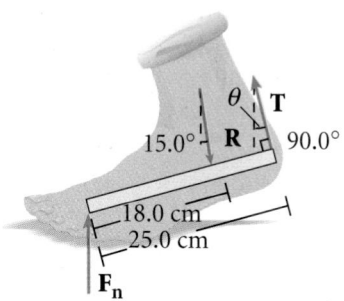

θ T

15.0° R 90.0°

18.0 cm
25.0 cm

F_n **Figure 8-25**

56. A cylindrical fishing reel has a mass of 0.85 kg and a radius of 4.0 cm. A friction clutch in the reel exerts a restraining torque of 1.3 N•m if a fish pulls on the line. The fisherman gets a bite, and the reel begins to spin with an angular acceleration of 66 rad/s². Find the following:

   a. the force of the fish on the line
   b. the amount of line that unwinds from the reel in 0.50 s

57. The combination of an applied force and a frictional force produces a constant torque of 36 N•m on a wheel rotating about a fixed axis. The applied force acts for 6.0 s, during which time the angular speed of the wheel increases from 0 to 12 rad/s. The applied force is then removed, and the wheel comes to rest in 65 s. Answer the following questions:

   a. What is the moment of inertia of the wheel?
   b. What is the frictional torque?
   c. How many revolutions does the wheel make during the entire 71 s time interval?

58. A cable passes over a pulley. Because of friction, the force in the cable is not the same on opposite sides of the pulley. The force on one side is 120.0 N, and the force on the other side is 100.0 N. Assuming that the pulley is a uniform disk with a mass of 2.1 kg and a radius of 0.81 m, determine the angular acceleration of the pulley.

59. As part of a kinetic sculpture, a 5.0 kg hoop with a radius of 3.0 m rolls without slipping. If the hoop is given an angular speed of 3.0 rad/s while rolling on the horizontal and then rolls up a ramp inclined at 20.0° with the horizontal, how far does the hoop roll along the incline?

51. 6.1 m
52. 1.55×10^{-2} m
53. 57 m
54. a. 47 N
   b. 0.24 kg•m²
   c. 4.0×10^{1} rad/s
55. $R = 2400$ N, $T = 1700$ N
56. a. 33 N
   b. 0.33 m
57. a. 18 kg•m²
   b. −3.3 N•m
   c. 68 rev
58. 24 rad/s²
59. 24 m

60. a. 5.35 m/s^2
   b. 42.8 m
   c. 8.92 rad/s^2

61. -2300 N

62. 0.124 m

63. (See *Teacher's Solution Manual and Answer Key.*)

64. 280 J

65. 149 rad/s

66. a. $5.84 \times 10^{33} \text{ kg} \cdot \text{m}^2/\text{s}$
   b. $2.66 \times 10^{40} \text{ kg} \cdot \text{m}^2/\text{s}$

67. a. $3.0 \times 10^3 \text{ J}, 3.5 \times 10^3 \text{ J}$
   b. increased
   c. work is done as arms are lowered

68. a. $1.1 \text{ m/s}^2, -1.1 \text{ m/s}^2$
   b. $22 \text{ N}, 43 \text{ N}$

69. a. 3.1 m/s^2
   b. $27 \text{ N}, 9.3 \text{ N}$

70. a. $5.48 \times 10^6 \text{ J}$
   b. 735 s

60. A cylindrical 5.00 kg pulley with a radius of 0.600 m is used to lower a 3.00 kg bucket into a well. The bucket starts from rest and falls for 4.00 s.

   a. What is the linear acceleration of the falling bucket?
   b. How far does it drop?
   c. What is the angular acceleration of the cylindrical pulley?

61. The hands of the clock in the famous Parliament Clock Tower in London are 2.7 m and 4.5 m long and have masses of 60.0 kg and 100.0 kg, respectively. Calculate the torque around the center of the clock due to the weight of these hands at 5:20. (Model the hands as thin rods.)

62. A coin with a diameter of 3.00 cm rolls up a 30.0° inclined plane. The coin starts with an initial angular speed of 60.0 rad/s and rolls in a straight line without slipping. How far does it roll up the inclined plane?

63. A solid sphere rolls along a horizontal, smooth surface at a constant linear speed without slipping. Show that the rotational kinetic energy about the center of the sphere is two-sevenths of its total kinetic energy.

64. A horizontal 800.0 N merry-go-round with a radius of 1.5 m is started from rest by a constant horizontal force of 50.0 N applied tangentially to the merry-go-round. Find the kinetic energy of the merry-go-round after 3.0 s. Assume it is a solid cylinder.

65. A top has a moment of inertia of $4.00 \times 10^{-4} \text{ kg} \cdot \text{m}^2$ and is initially at rest. It is free to rotate about a vertical stationary axis. A string around a peg along the axis of the top is pulled, maintaining a constant tension of 5.57 N in the string. If the string does not slip while it is wound around the peg, what is the angular speed of the top after 80.0 cm of string has been pulled off the peg? (Hint: Consider the work done.)

66. Calculate the following:

   a. the angular momentum of Earth that arises from its spinning motion on its axis
   b. the angular momentum of Earth that arises from its orbital motion about the sun

   (Hint: See item 4 on page 305 and Appendix E.)

67. A skater spins with an angular speed of 12.0 rad/s with his arms outstretched. He lowers his arms, decreasing his moment of inertia from $41 \text{ kg} \cdot \text{m}^2$ to $36 \text{ kg} \cdot \text{m}^2$.

   a. Calculate his initial and final rotational kinetic energy.
   b. Is the rotational kinetic energy increased or decreased?
   c. How do you account for this change in kinetic energy?

68. A pulley has a moment of inertia of $5.0 \text{ kg} \cdot \text{m}^2$ and a radius of 0.50 m. A cord is wrapped over the pulley and attached to a hanging object on either end. Assume the cord does not slip, the axle is frictionless, and the two hanging objects have masses of 2.0 kg and 5.0 kg.

   a. Find the acceleration of each mass.
   b. Find the force in the cord supporting each mass. (Note that they are different.)

69. A 4.0 kg mass is connected by a light cord to a 3.0 kg mass on a smooth surface as shown in **Figure 8-26.** The pulley rotates about a frictionless axle and has a moment of inertia of $0.50 \text{ kg} \cdot \text{m}^2$ and a radius of 0.30 m. Assuming that the cord does not slip on the pulley, answer the following questions:

   a. What is the acceleration of the two masses?
   b. What are the forces in the string F_1 and F_2?

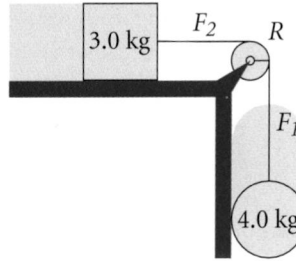

Figure 8-26

70. A car is designed to get its energy from a rotating flywheel with a radius of 2.00 m and a mass of 500.0 kg. Before a trip, the disk-shaped flywheel is attached to an electric motor, which brings the flywheel's rotational speed up to 1000.0 rev/min.

   a. Find the kinetic energy stored in the flywheel.
   b. If the flywheel is to supply as much energy to the car as a 7457 W motor would, find the length of time the car can run before the flywheel has to be brought back up to speed again.

71. Figure 8-27 shows a system of point masses that rotates at an angular speed of 2.0 rev/s. The masses are connected by light, flexible spokes that can be lengthened or shortened. What is the new angular speed if the spokes are shortened to 0.50 m? (An effect similar to this occurred in the early stages of the formation of our galaxy. As the massive cloud of gas and dust contracted, an initially small rotation increased with time.)

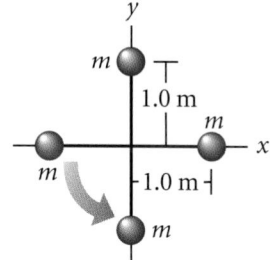

Figure 8-27

71. 8.0 rev/s

Technology  Learning

Graphing calculators

Refer to Appendix B for instructions on downloading programs for your calculator. The program "Chap8" allows you to analyze a graph of torque versus angle of applied force.

Torque, as you learned earlier in this chapter, is described by the following equation:

$$\tau = Fd(\sin\theta)$$

The program "Chap8" stored on your graphing calculator makes use of the equation for torque. Once the "Chap8" program is executed, your calculator will ask for the force and the distance from the axis of rotation. The graphing calculator will use the following equation to create a graph of the torque (Y1) versus the angle (X) at which the force is applied. The relationships in this equation are the same as those in the force equation shown above.

$$Y1 = Fd\sin(X)$$

Recall that the sine function is a periodic function that repeats every 360° and falls below the *x*-axis at 180°. Because the only values necessary for the torque calculation are less than 180°, the *x* values for the viewing window are preset. Xmin and Xmax values are set at 0 and 180, respectively.

a. What is a more straightforward way of saying, "The mechanic applied a force of −8 N at an angle of 200°"?

First, be certain that the calculator is in degree mode by pressing [MODE] [▼] [▼] [►] [ENTER].

Execute "Chap8" on the [PRGM] menu, and press [ENTER] to begin the program. Enter the values for the force and the distance from the axis of rotation (shown below), and press [ENTER] after each value.

The calculator will provide a graph of the torque versus the angle at which the force is applied. (If the graph is not visible, press [WINDOW] and change the *y*-value settings for the graph window, then press [GRAPH]. Adjusting the *x* values is not necessary.)

Press [TRACE] and use the arrow keys to trace along the curve. The *x* value corresponds to the angle in degrees, and the *y* value corresponds to the torque in newton•meters.

Determine the torque involved in each of the following situations:

b. a force of 15.0 N that is applied 0.45 m from a door's hinges makes an angle of 75° with the door

c. the same force makes an angle of 45° with the door

d. a force of 15.0 N that is applied 0.25 m from a door's hinges makes an angle of 45° with the door

e. the same force makes an angle of 25° with the door

f. At what *x* value do you find the largest torque?

Press [2nd] [QUIT] to stop graphing. Press [ENTER] to input a new value or [CLEAR] to end the program.

72. 2.0×10^2 N
73. 63%
74. 6.4 m
75. 72%
76. 0.289
77. **a.** 1320.0 N, 273 N
   b. 0.324

Alternative Assessment
ANSWERS

Performance assessment

1. The right arm is longer; Lengths are not required to be known; $xr = 5R$, and $xR = 15r$; $x^2 = 75$, or $x = 8.66$ g.

2. One solution uses two single beads to balance the 2-bead group, then hangs the middle of that set to balance the 4-bead group, and so on.

3. Gradually increasing the amount of overhang is the most successful approach; Five books give a one-book-length overhang; You can extend 52 cards to 2.5 times the length of one card.

 **Portfolio projects**
4. Student plans should be safe and should involve measuring forces and lever arms.

5. Student diagrams should clearly identify force input and force output.

6. Student research will vary; Most buildings must remain in equilibrium under a variety of conditions.

---

72. The efficiency of a pulley system is 64 percent. The pulleys are used to raise a mass of 78 kg to a height of 4.0 m. What force is exerted on the rope of the pulley system if the rope is pulled for 24 m in order to raise the mass to the required height?

73. A crate is pulled 2.0 m at constant velocity along a 15° incline. The coefficient of kinetic friction between the crate and the plane is 0.160. Calculate the efficiency of this procedure.

74. A pulley system has an efficiency of 87.5 percent. How much of the rope must be pulled in if a force of 648 N is needed to lift a 150 kg desk 2.46 m?

75. A pulley system is used to lift a piano 3.0 m. If a force of 2200 N is applied to the rope as the rope is pulled in 14 m, what is the efficiency of the machine? Assume the mass of the piano is 750 kg.

76. A uniform 6.0 m tall aluminum ladder is leaning against a frictionless vertical wall. The ladder has a weight of 250 N. The ladder slips when it makes a 60.0° angle with the horizontal floor. Determine the coefficient of friction between the ladder and the floor.

77. A ladder with a length of 15.0 m and a weight of 520.0 N rests against a frictionless wall, making an angle of 60.0° with the horizontal.
   a. Find the horizontal and vertical forces exerted on the base of the ladder by Earth when an 800.0 N firefighter is 4.00 m from the bottom of the ladder.
   b. If the ladder is just on the verge of slipping when the firefighter is 9.00 m up, what is the coefficient of static friction between the ladder and the ground?

Alternative Assessment

Performance assessment

1. Imagine a balance with unequal arms. An earring placed in the left basket was balanced by 5.00 g of standard masses on the right. When placed in the right basket, the same earring required 15.00 g on the left to balance. Which was the longer arm? Do you need to know the exact length of each arm to determine the mass of the earring? Explain.

2. You have 32 identical beads, 30 of which are assembled in clusters of 16, 8, 4, and 2. Design and build a mobile that will balance, using all the beads and clusters. Explain your design in terms of torque and rotational equilibrium.

3. A well-known problem in architecture is to stack bricks on top of one another in a way that provides a maximum offset. Design an experiment to determine how much offset you can have by stacking two, three, and four physics textbooks. How many books would be needed to prevent the center of mass of the top book from being directly over the bottom book? (Alternatively, try this with a deck of playing cards.)

Portfolio projects

4. Describe exactly which measurements you would need to make in order to identify the torques at work during a ride on a specific bicycle. (Your plans should include measurements you can make with equipment available to you.) If others in the class analyzed different bicycle models, compare the models for efficiency and mechanical advantage.

5. Prepare a poster or a series of models of simple machines, explaining their use and how they work. Include a schematic diagram next to each sample or picture to identify the fulcrum, lever arm, and resistance. Add your own examples to the following list: nail clipper, wheelbarrow, can opener, nutcracker, electric drill, screwdriver, tweezers, key in lock.

6. Research what architects do, and create a presentation on how they use physics in their work. What studies and training are necessary? What are some areas of specialization in architecture? What associations and professional groups keep architects informed about new developments?

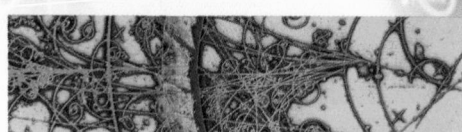

CHAPTER 8
Laboratory Exercise

MACHINES AND EFFICIENCY

In this experiment, you will raise objects using two different types of machines. You will find the work input and the work output for each machine. The ratio of the useful work output to the work input is called the *efficiency* of the machine. By calculating efficiency, you will be able to compare different machines for different jobs.

SAFETY

- **Tie back long hair, secure loose clothing, and remove loose jewelry to prevent their getting caught in moving parts and pulleys.**
- **Attach string to masses and objects securely. Falling or dropped masses can cause serious injury.**

PREPARATION

1. Read the entire lab, and plan what measurements you will take.

2. Prepare a data table with six columns and seven rows in your lab notebook. In the first row, label the first through sixth columns *Trial*, *Machine*, *mass$_1$ (kg)*, *Δh (m)*, *mass$_2$ (kg)*, and *Δd (m)*. In the first column, label the second through seventh rows *1, 2, 3, 4, 5,* and *6*.

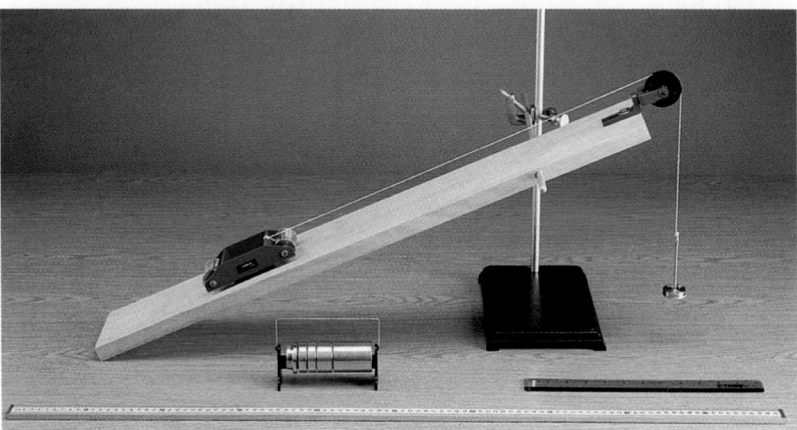

OBJECTIVES

- **Measure the work input and work output of several machines.**
- **Calculate the efficiency of each machine.**
- **Compare machines based on their efficiencies, and determine what factors affect efficiency.**

MATERIALS LIST

- ✔ **balance**
- ✔ **C-clamp**
- ✔ **cord**
- ✔ **dynamics cart**
- ✔ **inclined plane**
- ✔ **mass hanger**
- ✔ **pulleys, single and tandem**
- ✔ **meterstick**
- ✔ **set of hooked masses**
- ✔ **right-angle clamp**
- ✔ **support stand**
- ✔ **suspension clamp**

Figure 8-28

Step 3: Choose any angle, but make sure the top of the plane is at least 20 cm above the table.

Step 4: Make sure the string is long enough to help prevent the cart from falling off the top of the plane. Attach the mass hanger securely to the end of the string.

Step 7: Keep the angle the same for all trials.

NOTE

Preparation is given on pp. 276A–276B. Blank data table and sample data table are on the One-Stop Planner CD-ROM. All calculations shown use sample data.

Planning

Recommended time:

1 lab period
2-day option: plane/pulley

Classroom organization:

▶ This lab may be performed by 1 or more students.

▶ Half the class may work with pulleys while the other half works with the inclined plane.

▶ Safety warnings: Falling masses can cause injury. Goggles should be worn to shield eyes from hooked masses at eye level.

Inclined plane and pulley tips

▶ Place a wooden block or textbook at the bottom of the incline to stop the cart.

▶ Use single and tandem open-faced pulley sets; multiple-pulley sets are difficult to thread and confusing for students.

Techniques to Demonstrate

Demonstrate threading pulleys: Hang the upper set (A) from the stand, and hold the lower set (B). Connect B to A by threading the bottom A pulley to the top B pulley. Hang a mass on B to stabilize it, and release. Keep the cord tight and finish threading.

✔ Checkpoints

Step 3: Make sure the inclined plane is secure. The apparatus should be away from the edge of the table; if not, use a C-clamp to secure the base of the stand to the table. For best results the top of the plane should be at least 20 cm above the table.

Step 4: The cord should be attached through the hole in the body of the cart. Make sure the masses are attached securely. Remind students to include the mass hanger in the total suspended mass.

Step 5: Students should demonstrate that they are using the smallest mass that will allow the cart to move up the plane and that the cart has constant velocity. The hanging mass should hit the table before the cart reaches the top of the plane.

Step 8: Students will need help threading the pulley systems. See page 313 for hints.

Step 9: Measure the position from the bottom of each mass.

Step 10: Students should demonstrate that the added mass is the smallest that will cause the pulley set to move and that the pulley set has a constant velocity. Remind students to include the mass hanger in the total recorded mass.

PROCEDURE

Inclined plane

3. Set up the inclined plane as shown in **Figure 8-28** on page 313. Set the incline securely to any angle. Keep the angle constant during this part of the experiment. Place the inclined plane away from the edge of the table, or clamp its base to the edge of the table.

4. Measure the mass of the cart. Attach a piece of cord through the hole on the body of the cart. The cord should be long enough so that the other end of the cord reaches the table top before the cart reaches the top of the incline. Place the cart on the plane and run the cord over the pulley at the top of the plane. Attach a mass hanger to the free end of the cord.

5. Place a 200 g mass in the cart. Record the total mass of the cart and its contents as $mass_1$. Attach masses to the mass hanger until you find the lowest mass that will allow the cart to move up the plane with a constant velocity. Stop the cart before it reaches the top of the incline. Record the mass of the mass hanger plus the added mass as $mass_2$ in your data table.

6. Measure the distances, and record them. Δh is the *vertical* distance the cart moves, while the mass hanger on the cord moves the distance Δd.

7. Repeat steps 5 and 6 several times, increasing the mass in the cart by 100 g and finding the mass that will allow the cart to move with a constant velocity each time. Record all data for each trial in your data table.

Pulley

8. Set up a pulley system like the one shown in **Figure 8-29.** For the first trial, use five pulleys. Keep the area beneath the pulley system clear throughout the experiment. Measure the mass of the bottom set of pulleys before including them in the setup. Attach a 500 g mass to the bottom, as shown. Record the total mass of the 500 g mass plus the bottom set of pulleys as $mass_1$ in your data table.

9. Starting with 50 g, add enough mass to the mass hanger to prevent the pulleys from moving when released. Place the mass hanger just below the 500 g mass, and measure the initial positions of both masses by measuring the height of each mass above the base.

10. Add masses to the mass hanger until you find the mass that will make the 500 g mass move up with constant velocity once it has been started. Record the mass of the mass hanger plus the added mass as $mass_2$ in your data table.

11. Measure the final positions of both masses, and record the distances (final position − initial position) in your data table. Δh is the *vertical* dis-

tance through which the mass on the pulley is raised, while the mass on the mass hanger moves down through the distance Δd.

12. Using the same 500 g *mass₁*, perform two more trials using different pulley systems (four pulleys, six pulleys, and so on). Record all data. Be sure to include the mass of the bottom set of pulleys in the total mass that is raised in each trial.

13. Clean up your work area. Put equipment away safely so that it is ready to be used again.

ANALYSIS AND INTERPRETATION

Calculations and data analysis

1. Organizing data For each trial, make the following calculations:

    a. the weight of the mass being raised

    b. the weight of the mass on the string

    c. the work input and the work output

2. Analyzing results In which trial did a machine perform the most work? In which trial did a machine perform the least work?

3. Analyzing data Calculate the efficiency for each trial.

4. Evaluating data Is the machine that performed the most work also the most efficient? Is the machine that performed the least work also the least efficient? What is the relationship between work and efficiency?

5. Analyzing results Based on your calculations in item 4, which is more efficient, a pulley system or an inclined plane?

Conclusions

6. Evaluating methods Why is it important to calculate the work input and the work output from measurements made when the object is moving with constant velocity?

Extensions

7. Designing experiments Design an experiment to measure the efficiency of different lever setups. If there is time and your teacher approves, test your lever setups in the lab. How does the efficiency of a lever compare with the efficiency of the other types of machines you have studied?

8. Building models Compare the trial with the highest efficiency and the trial with the lowest efficiency. Based on their differences, design a more efficient machine than any you built in the lab. If there is time and your teacher approves, test the machine to test whether it is more efficient.

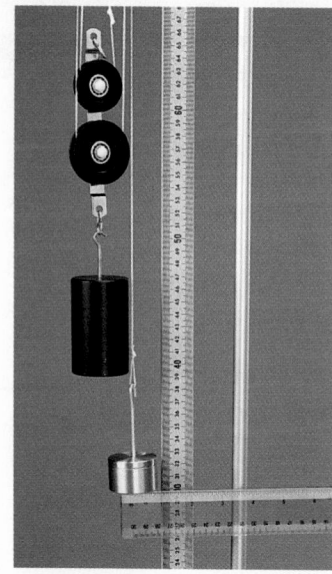

Figure 8-29

Step 8: Clamp a meterstick parallel to the stand to take measurements throughout the lab.

Step 9: Use another ruler as a straight edge to help you measure the positions.

Step 10: The pulleys should not begin moving when the mass is added, but they should move with a constant velocity after a gentle push.

Step 11: Measure and record the distance moved by the mass on the pulley as Δh and the mass hanger distance as Δd.

ANSWERS TO

Analysis and Interpretation

CALCULATIONS AND DATA ANALYSIS

1. a. Inclined Plane: T1: 2.49 N, T2: 4.91 N, T3: 6.87 N; **Pulley:** T1: 5.56 N, T2: 5.62 N, T3: 5.56 N

  b. Inclined Plane: T1: 1.12 N, T2: 2.18 N, T3: 3.05 N; **Pulley:** T1: 1.22 N, T2: 1.34 N, T3: 1.50 N

  c. Student answers will vary. Make sure students use the relationship $W = Fd$. Typical values will be within these ranges: 0.271 J–0.774 J (inclined plane), 0.256 J–0.351 J (pulley).

2. The most work was done by the Trial 3 plane. The least work was done by the Trial 3 pulleys.

3. Inclined Plane: T1: 91.3%, T2: 93.3%, T3: 93.2%; **Pulley:** T1: 90.4%, T2: 85.0%, T3: 90.7%

4. Efficiency depends on both the work input and the work output.

5. inclined plane

CONCLUSIONS

6. Measurements should be made when there is no net external force doing work.

EXTENSIONS

7, 8. Student plans should be safe and complete and should include calculations of work input and work output.

Compression Guide: To shorten from 10 to 8 45-min periods (from 5 to 4 90-min blocks), eliminate items in magenta type.

PACING CHART	CLASSROOM RESOURCES			
	⊛ TEKS	Teacher Demonstrations	*Holt Physics* Transparencies	Labs (See page T52 for equipment listing for in-text labs.)
9-1 Fluids and buoyant force 2 45-minute periods 1 90-minute block	2C, 3A, 3B, 3C, 3E, 4C	**TE** *Volume of liquids and gases*, p. 318 **TE** *Buoyant force*, p. 319 **TE** *Float an egg*, p. 321 **TE** *Lifting weights with a balloon*, p. 322	**T** 30–31 **TM** 38	
9-2 Fluid pressure and temperature 2 45-minute periods 1 90-minute block	2C, 3A, 3B, 3E, 4C, 5B	**TE** *Defining pressure*, p. 325 **TE** *Hydrostatic pressure*, p. 329	**T** 32	
9-3 Fluids in motion 2 45-minute periods 1 90-minute block	1A, 2C, 3A, 3B, 3C	**TE** *Fluid flow around a table-tennis ball*, p. 332 **TE** *Fluid flow through two cans*, p. 332 **TE** *Atomizer*, p. 335	**T** 33	**PE** *Quick Lab: Bernoulli's Principle*, p. 335
9-4 Properties of gases 2 45-minute periods 1 90-minute block	1A, 2C, 3A, 3B	**TE** *Temperature, pressure, and volume*, p. 338		**PE** *Quick Lab: Ideal Gas Law*, p. 339 **PE** *Boyle's Law*, p. 350
Review and Assessment 2 45-minute periods 1 90-minute block				

Resource Key

PHYSICS
HOLT

PE Pupil's Edition
TE Teacher's Edition

L Laboratory Experiments
TL Technology Lab Experiments
T Transparencies

One-Stop Planner CD-ROM contents

TM Transparency Masters
SR Section Review Worksheets
AA Alternative Assessment

PW Problem-Solving Workbook
PB Problem Bank
CTW Critical Thinking Worksheet

LABORATORY PLANNING: Boyle's Law, p. 350

Materials (for each lab group)
CBL and Sensors Procedure
- CBL
- graphing calculator
- pressure sensor module with syringe
- DIN adapter for CBL
- TI Graph Link (recommended for downloading programs)

Boyle's Law Apparatus Procedure
- elasticity of gases apparatus
- heavy slotted mass, 1 kg
- 2 heavy slotted masses, 2 kg
- tube of silicone lubricant

ASSIGNMENT RESOURCES

Content Mastery	Critical Thinking	Problem-Solving Practice
PE 1–5, p. 324 **SR** 9-1, *Concept Review*	**PE** 1–3, p. 322 **PE** 1–7, p. 343	**9A** Buoyant force: 38 items in **PE, PW,** and **PB,** see **TE** pp. 323–324
PE 1–5, p. 331 **SR** 9-2, *Concept Review*	**PE** 1–3, p. 328 **PE** 10–15, p. 343	**9B** Pressure: 33 items in **PE, PW,** and **PB,** see **TE** p. 327 **9C** Pressure as a function of depth: 26 items in **PE, PW,** and **PB,** see **TE** p. 330
PE 1–3, p. 337 **SR** 9-3, *Math Skills*	**PE** 20–22, p. 344	**9D** Bernoulli's equation: 31 items in **PE, PW,** and **PB,** see **TE** pp. 336–337
PE 1–5, p. 341 **SR** 9-4, *Concept Review*	**PE** 25–28, p. 344–345	**9E** The ideal gas law: 33 items in **PE, PW,** and **PB,** see **TE** pp. 340–341

ASSESSMENT RESOURCES

Cumulative Review	Alternative Assessment	Traditional Assessment
SR Mixed Review, Ch. 9	**PE** 1–5, p. 349 **AA** Items for Ch. 9	Chapter 9 Test Test Generator items for Ch. 9

Scoring Rubrics for Alternative Assessment items can be found on the One-Stop Planner CD-ROM.

TECHNOLOGY RESOURCES

 CTW Segment 9 Turbulent Flow
CTW Segment 10 Wet Design

 PE Technology and Learning, p. 348
(Alternative procedures for calculators without Flash-ROM technology are provided on the One-Stop Planner CD-ROM.)

 The Mechanical Universe/High School Adaptation Quad IV, Temperature and the Gas Laws

internet connect

 On-line Student Resources:
GO TO: www.scilinks.org
The following *sci*LINKS Internet resources can be found in the student text for this chapter.

TOPICS:
• Archimedes, p. 320 (HF2091)
• Buoyancy, p. 322 (HF2092)
• Atmospheric pressure, p. 329 (HF2093)
• Bernoulli's principle, p. 335 (HF2094)
• Gas laws, p. 339 (HF2095)

 On-line Teacher Resources:
GO TO: go.hrw.com
KEYWORD: HF2 HOME
Visit the HRW Web site for a variety of resources related to this chapter.

 Smithsonian Institution
Internet Connections
Visit www.si.edu/hrw for additional on-line resources.

CNNfyi.com
Visit www.cnnfyi.com for late-breaking news and current events stories selected just for you.

Safety Equipment
• safety goggles or spectacles

Materials Preparation
For the procedure using the Boyle's law apparatus, prepare syringes before class to save time. Remove the cap, pull the piston out, lubricate the cylinder wall very lightly with silicone lubricant, and reinsert the piston. This ensures low friction for the experiment.

Section 9-1 defines ideal fluids, calculates buoyant force, and explains why objects float or sink.

Section 9-2 calculates pressures transferred by a fluid in a hydraulic lift and explains how hydrostatic pressure varies with depth.

Section 9-3 introduces the equation of continuity and applies Bernoulli's equation to solve problems of fluids in motion.

Section 9-4 defines an ideal gas and uses the ideal gas law to predict the state of a gas under different conditions.

About the Illustration

This photograph, taken in December 1998, shows kayaker Grant Armal overlooking a set of rapids on the Rio Micos, located in the Eastern Sierras in the Mexican state of Potosi.

CHAPTER 9

Fluid Mechanics

PHYSICS IN ACTION

Whitewater rafters and kayakers know that a river does not flow at the same rate at all points in the river. Along some stretches of river, the water moves slowly and smoothly. In other parts, the water races and rolls in the turbulence of the rapids.

Rafters can plan for what's ahead in the river if they know how the speed of the river depends on local topography. The rate at which a river flows depends in large part on the cross-sectional area of the water at a given point along the river. Where the river is deep and wide, it moves slowly. Where the river is shallow or narrow, the water moves faster and may form turbulent rapids.

- *Why does a raft float on water?*

- *Why is water turbulent in the rapids and smooth in other places on a river?*

CONCEPT REVIEW

Force (Section 4-1)

Energy (Section 5-2)

Conservation laws (Section 5-3)

Knowledge to Expect

✔ "There are formulas for calculating the surface areas and volumes of regular shapes. When the linear size of a shape changes by some factor, its area changes in proportion to the square of the factor, and the volume in proportion to its cube." (AAAS's *Benchmarks for Science Literacy*, grades 9–12)

✔ "By the age of 14 (and depending on context), students typically understand conservation of length and amount, area, weight, and displaced volume." (AAAS's *Benchmarks for Science Literacy*, The Research Base)

Knowledge to Review

✔ Forces can cause changes in an object's motion or in its shape.
(Section 4-1)

✔ Energy can be kinetic energy or potential energy.
(Section 5-2)

✔ In the absence of friction, the total mechanical energy of a system is constant. The total mass of a closed system is constant.
(Section 5-3)

Items to Probe

✔ Operational understanding of the concepts of area and volume: Ask students to compare the volume of containers of different shapes (with approximately the same capacity).

✔ Ability to relate density, mass, and volume in a meaningful way: Ask students to calculate m, given ρ and V, using correct units.

Volume of liquids and gases

Purpose Demonstrate that, unlike gases, liquids have a definite volume and cannot be significantly compressed.

Materials large syringe, large eyedropper, or turkey baster

Procedure Pull the piston to about half the tube's length, letting air into the syringe. Hold your finger tightly over the opening and pull back as far as possible. Read the scale marking at the piston level. Now push back as far as possible. Have a student record the air volume on the chalkboard for each case. Point out that no air has been let into or out of the cavity, but the same amount of air occupies all the space that is available. Next, fill the syringe halfway with water and repeat. Let a student in class try to compress it.

Using the eyedropper, show that the volume of air that equals that of the bulb and the pipe can easily be reduced to the volume of the pipe by squeezing the bulb while holding the opening tightly sealed. Fill the eyedropper with water and record the water level. Holding the end sealed while squeezing will demonstrate that the air volume is reduced, while the water level remains the same.

9-1
Fluids and buoyant force

9-1 SECTION OBJECTIVES

- **Define a fluid.**
- **Distinguish a liquid from a gas.**
- **Determine the magnitude of the buoyant force exerted on a floating object or a submerged object.**
- **Explain why some objects float and some objects sink.**

fluid

> *a nonsolid state of matter in which the atoms or molecules are free to move past each other, as in a gas or a liquid*

DEFINING A FLUID

Matter is normally classified as being in one of three states—solid, liquid, or gaseous. Up to this point, this book's discussion of motion and its causes has dealt primarily with the behavior of solid objects. This chapter concerns the mechanics of liquids and gases.

Figure 9-1(a) is a photo of a liquid; **Figure 9-1(b)** shows an example of a gas. Pause a moment and see if you can identify a common trait between them. One property they have in common is the ability to flow and to alter their shape in the process. Materials that exhibit these properties are called **fluids.** Solid objects are not considered to be fluids because they cannot flow and therefore have a definite shape.

Liquids have a definite volume; gases do not

Even though both gases and liquids are fluids, there is a difference between them: one has a definite volume, and the other does not. Liquids, like solids, have a definite volume, but unlike solids, they do not have a definite shape. Imagine filling the tank of a lawn mower with gasoline. The gasoline, a liquid, changes its shape from that of its original container to that of the tank. If there is a gallon of gasoline in the container before you pour, there will be a gallon in the tank after you pour. Gases, on the other hand, have neither a definite volume nor a definite shape. When a gas is poured from one container into another, the gas not only changes its shape to fit the new container but also spreads out to fill the container.

Figure 9-1

Both **(a)** liquids and **(b)** gases are fluids because they can flow and change shape.

(a) (b)

DENSITY AND BUOYANT FORCE

Have you ever felt confined in a crowded elevator? You probably felt that way because there were too many people in the elevator for the amount of space available. In other words, the *density* of people was too high. In general, density is a measure of how much there is of a quantity in a given amount of space. The quantity can be anything from people or trees to mass or energy.

Mass density is mass per unit volume of a substance

When the word *density* is used to describe a fluid, what is really being measured is the fluid's **mass density.** Mass density is the mass per unit volume of a substance. It is often represented by the Greek letter ρ (*rho*).

MASS DENSITY ⭐TEKS **3B**

$$\rho = \frac{m}{V}$$

$$\text{mass density} = \frac{\text{mass}}{\text{volume}}$$

The SI unit of mass density is kilograms per cubic meter (kg/m^3). In this book we will follow the convention of using the word *density* to refer to *mass density*. **Table 9-1** lists the densities of some fluids and a few important solids.

Solids and liquids tend to be almost incompressible, meaning that their density changes very little with changes in pressure. Thus, the densities listed in **Table 9-1** for solids and liquids are approximately independent of pressure. Gases, on the other hand, are compressible and can have densities over a wide range of values. Thus, there is not a standard density for a gas, as there is for solids and liquids. The densities listed for gases in **Table 9-1** are the values of the density at a stated temperature and pressure. For deviations of temperature and pressure from these values, the density will vary significantly.

Buoyant forces can keep objects afloat

Have you ever wondered why things feel lighter underwater than they do in air? The reason is that a fluid exerts an upward force on objects that are partially or completely submerged in it. This upward force is called a **buoyant force.** If you have ever rested on an air mattress in a swimming pool, you have experienced a buoyant force. The buoyant force kept you and the mattress afloat.

Because the buoyant force acts in a direction opposite the force of gravity, objects submerged in a fluid such as water have a net force on them that is smaller than their weight. This means that they appear to weigh less in water than they do in air. The weight of an object immersed in a fluid is the object's *apparent weight.* In the case of a heavy object, such as a brick, its apparent weight is less in water than in air, but it may still sink in water because the buoyant force is not enough to keep it afloat. ⭐TEKS **4C**

---

mass density

the mass per unit volume of a substance

Table 9-1
Densities of some common substances*

Substance	ρ (kg/m^3)
hydrogen	0.0899
helium	0.179
steam (100°C)	0.598
air	1.29
oxygen	1.43
carbon dioxide	1.98
ethanol	0.806×10^3
ice	0.917×10^3
fresh water (4°C)	1.00×10^3
sea water (15°C)	1.025×10^3
iron	7.86×10^3
mercury	13.6×10^3
gold	19.3×10^3

*All densities are measured at 0°C and 1 atm unless otherwise noted.

---

buoyant force

a force that acts upward on an object submerged in a liquid or floating on the liquid's surface

Demonstration 2

Buoyant force

Purpose Show the relationship between buoyant force and submerged volume.

Materials large spring scale, several cylinders of the same shape but different materials (preferably low density), a clear container partially filled with water

Procedure Point out that all of the cylinders have the same volume. Measure this volume with a graduated cylinder or by the overflow method. Hang a cylinder on the scale, read the weight, and start lowering the cylinder into the water. Students should observe that the scale's reading drops continuously as more of the cylinder is submerged. Explain that the water is exerting an upward force on the cylinder. Have a student record the scale's reading before immersion, midway through immersion, and when completely immersed. Repeat with the other cylinders. Examining all of the data will show that for the same volume submerged, the buoyant force is the same, regardless of the cylinder's weight.

⭐TEKS

p. 319: 3B, 4C

Figure 9-2
(a) A brick is being lowered into a container of water. **(b)** The brick displaces water, causing the water to flow into a smaller container. **(c)** When the brick is completely submerged, the volume of the displaced water **(d)** is equal to the volume of the brick.

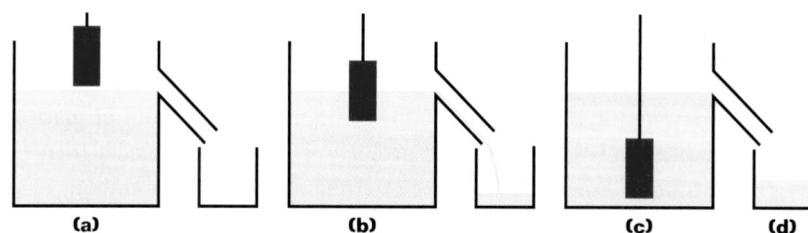

(a)   (b)   (c)   (d)

The Language of Physics

Students may need help interpreting all the symbols in the equations and relating them to prior knowledge. It may be helpful to remind students that g is free-fall acceleration, with a value of 9.81 m/s^2, which allows them to find an object's weight in newtons when its mass is known in kilograms.

Misconception Alert

Students may wonder why the buoyant force, F_B, is treated like weight (mg) even though it pushes upward. By carefully reading the statement of Archimedes' principle, they may realize that the *magnitude* of that force equals the weight of the fluid that would otherwise occupy the space taken up by the submerged object.

Teaching Tip

In order to help students understand the expression of F_{net} as a function of g and of the densities and volumes of the fluid and the object, ask them to write the units associated with each variable in the development of F_{net}.

Did you know?

Archimedes was a Greek mathematician who was born in Syracuse, a city on the island of Sicily. According to legend, the king of Syracuse suspected that a certain golden crown was not pure gold. While bathing, Archimedes figured out how to test the crown's authenticity when he discovered the buoyancy principle. He is reported to have then exclaimed, "Eureka!" meaning "I've found it!"

 TEKS **3E**

Archimedes' principle determines the amount of buoyancy

Imagine that you submerge a brick in a container of water, as shown in **Figure 9-2.** A spout on the side of the container at the water's surface allows water to flow out of the container. As the brick sinks, the water level rises and water flows through the spout into a smaller container. The total volume of water that collects in the smaller container is the *displaced volume* of water from the large container. The displaced volume of water is equal to the volume of the portion of the brick that is underwater.

The magnitude of the buoyant force acting on the brick at any given time can be calculated by using a rule known as *Archimedes' principle.* This principle can be stated as follows: *Any object completely or partially submerged in a fluid experiences an upward buoyant force equal in magnitude to the weight of the fluid displaced by the object.* Everyone has experienced Archimedes' principle. For example, recall that it is relatively easy to lift someone if you are both standing in a swimming pool, even if lifting that same person on dry land would be difficult. ⭐TEKS **3A**

Using m_f to represent the mass of the displaced fluid, Archimedes' principle can be written symbolically as follows:

BUOYANT FORCE ⭐TEKS **3B**

$$F_B = F_g \,(displaced\ fluid) = m_f g$$

magnitude of buoyant force = weight of fluid displaced

Whether an object will float or sink depends on the net force acting on it. This net force is the object's apparent weight and can be calculated as follows:

$$F_{net} = F_B - F_g \,(object)$$

Now we can apply Archimedes' principle, using m_o to represent the mass of the submerged object.

$$F_{net} = m_f g - m_o g$$

Remember that $m = \rho V$, so the expression can be rewritten as follows:

$$F_{net} = (\rho_f V_f - \rho_o V_o)g$$

Note that in this expression, the fluid quantities refer to the *displaced* fluid.

internet**connect**

SC*i*LINKS
NSTA

TOPIC: Archimedes
GO TO: www.scilinks.org
*sci*LINKS CODE: HF2091

For a floating object, the buoyant force equals the object's weight

Imagine a cargo-filled raft floating on a lake. There are two forces acting on the raft and its cargo: the downward force of gravity and the upward buoyant force of the water. Because the raft is floating in the water, the raft is in equilibrium and the two forces are balanced, as shown in **Figure 9-3.** For floating objects, the buoyant force and the weight of the object are equal in magnitude.

BUOYANT FORCE ON FLOATING OBJECTS

$$F_B = F_g \,(object) = m_o g$$

buoyant force = weight of floating object

Notice that Archimedes' principle is not required to find the buoyant force on a floating object if the weight of the object is known. ⊛TEKS **3A, 3B**

The density of an object determines the depth of submersion

When an object floats in a fluid, the net force on the object is zero. Relating the net force on the object to the buoyant force, using Archimedes' principle, gives the following result:

$$F_{net} = 0 = (\rho_f V_f - \rho_o V_o)g$$

This equation can then be rearranged to show two equal ratios:

$$\frac{\rho_f}{\rho_o} = \frac{V_o}{V_f}$$

Of course, the displaced volume of fluid can never be greater than the volume of the object itself. So for an object to float, the object's density can never be greater than the density of the fluid in which the object floats. Furthermore, the ratio of the total volume of a floating object, V_o, to the submerged volume of the object, V_f, is equal to the ratio of the two densities. If the densities are equal, the entire object is submerged, but the object does not sink.

Buoyancy can be changed by changing average density

The buoyancy of an object can be changed by changing the object's average density. For example, a fish can adjust its average density by inflating or deflating an organ called a swim bladder. A fish fills its swim bladder with gas either by gulping air at the surface or by secreting gas from its gas gland into the swim bladder.

The ballast tank of a submarine works in much the same way as the swim bladder of a fish. In submarines, compressed air is pumped into the ballast tanks (and water is pumped out) to make the submarine rise to the surface. When the submarine is ready to dive again, air in the tanks is replaced with water, which increases the overall average density of the ship. ⊛TEKS **3C**

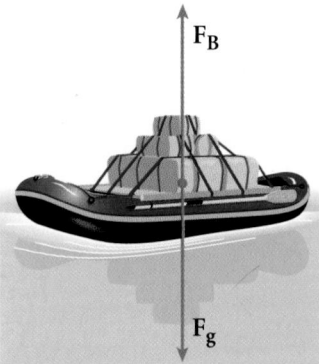

Figure 9-3
The raft and cargo are floating because their weight and the buoyant force are balanced.

Demonstration 3

Float an egg

Purpose Demonstrate the effect of liquid density on the buoyant force.

Materials raw egg, demonstration scale, glass jar, water, table salt, measuring spoon, stirring stick

Procedure Measure the weight of the egg and record it on the chalkboard. Fill the glass jar about two-thirds full with water. Demonstrate that the egg sinks in the water. Is a buoyant force acting on it? (*yes, but less than the egg's weight*) Explain that adding salt to the water will increase its density. Have the class watch the egg as you stir in one spoonful of salt at a time. The egg will start to float but will remain submerged. Ask students to estimate the buoyant force (*same as the egg's weight*). Keep adding salt until the egg floats partially above water. Again ask students to estimate the buoyant force (*same as the egg's weight*). Explain that the buoyant force balances the egg's weight in both cases, but as the density of the water increases, the volume of the egg that is immersed decreases.

⊛TEKS

p. 320: **3A, 3B, 3E**
p. 321: **3A, 3B, 3C**

Lifting weights with a balloon

Purpose Demonstrate Archimedes' principle in gases.

Materials large helium balloon, plastic-foam cup, small objects of known weight, string, meterstick

Procedure Hang the plastic-foam cup under the balloon and put weights in it to keep the cup on a table. Record the weight in the cup and compare it with the buoyant force exerted by the air on the balloon (*weight > buoyant force*). Measure the circumference of the balloon with a string and have students calculate its radius (divide by 2π) and its volume (assume it is approximately spherical). Ask students to estimate the buoyant force from the air ($\rho_{air}Vg$). Remove weight from the cup until it floats just above the table. On the chalkboard, record the weight remaining in the cup. Compare this weight with the calculated buoyant force.

ANSWERS TO

Conceptual Challenge

1. The buoyant force causes the net force on the astronaut to be close to zero. In space, the astronauts accelerate at the same rate as the craft, so they feel as if they have no net force acting on them.

2. disagree; because the buoyant force is also proportional to g

3. Helium is less dense than air and therefore floats in air.

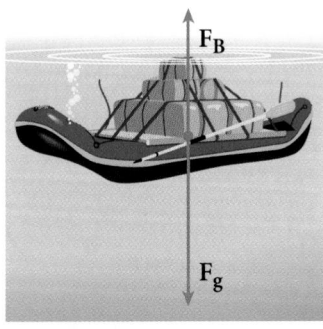

Figure 9-4
The raft and cargo sink because their density is greater than the density of water.

internet**connect**

SC*i*LINKS

NSTA

TOPIC: Buoyancy
GO TO: www.scilinks.org
***sci*LINKS CODE:** HF2092

The apparent weight of a submerged object depends on density

Imagine that a hole is accidentally punched in the raft shown in **Figure 9-3** and that the raft begins to sink. The cargo and raft eventually sink below the water's surface, as shown in **Figure 9-4.** The net force on the raft and cargo is the vector sum of the buoyant force and the weight of the raft and cargo. As the volume of the raft decreases, the volume of water displaced by the raft and cargo also decreases, as does the magnitude of the buoyant force. This can be written by using the expression for the net force: (★)TEKS **3B**

$$F_{net} = (\rho_f V_f - \rho_o V_o)g$$

Because the raft and cargo are completely submerged, the two volumes are equal:

$$F_{net} = (\rho_f - \rho_o)Vg$$

Notice that both the direction and the magnitude of the net force depend on the difference between the density of the object and the density of the fluid in which it is immersed. If the object's density is greater than the fluid density, the net force is negative (downward) and the object sinks. If the object's density is less than the fluid density, the net force is positive (upward) and the object rises to the surface and floats. If the densities are the same, the object floats suspended underwater.

A simple relationship between the weight of a submerged object and the buoyant force on the object can be found by considering their ratio as follows:

$$\frac{F_g(object)}{F_B} = \frac{\rho_o \cancel{Vg}}{\rho_f \cancel{Vg}}$$

$$\frac{F_g(object)}{F_B} = \frac{\rho_o}{\rho_f}$$

This last expression is often useful in solving buoyancy problems.

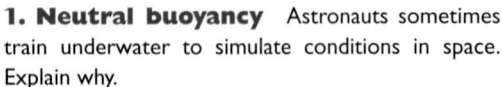

Conceptual Challenge (★)TEKS **3A**

1. Neutral buoyancy Astronauts sometimes train underwater to simulate conditions in space. Explain why.

2. More gravity A student claims that if the strength of Earth's gravity doubled, people would be unable to float on water. Do you agree or disagree with this statement? Why?

3. Ballooning Explain why balloonists use helium instead of pure oxygen in balloons.

Buoyant force ⭐TEKS 2C, 3B

PROBLEM

A bargain hunter purchases a "gold" crown, like the one shown in Figure 9-5, at a flea market. After she gets home, she hangs the crown from a scale and finds its weight to be 7.84 N. She then weighs the crown while it is immersed in water, and the scale reads 6.86 N. Is the crown made of pure gold? Explain.

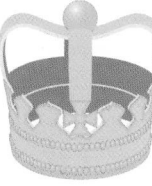

Figure 9-5

SOLUTION

1. DEFINE

Given: $F_g = 7.84$ N   apparent weight = 6.86 N
$\rho_f = \rho_{water} = 1.00 \times 10^3$ kg/m^3

Unknown: $\rho_o = ?$

Diagram:

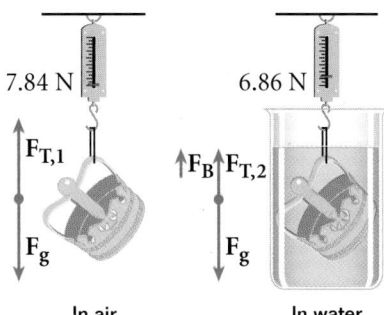

In air   In water

2. PLAN

Choose an equation(s) or situation: Because the object is completely submerged, consider the ratio of the weight to the buoyant force.

$$F_g - F_B = \text{apparent weight}$$

$$\frac{F_g}{F_B} = \frac{\rho_o}{\rho_f}$$

Rearrange the equation(s) to isolate the unknown(s):

$$F_B = F_g - (\text{apparent weight})$$

$$\rho_o = \frac{F_g}{F_B}\rho_f$$

3. CALCULATE

Substitute the values into the equation(s) and solve:

$$F_B = 7.84 \text{ N} - 6.86 \text{ N} = 0.98 \text{ N}$$

$$\rho_o = \frac{F_g}{F_B}\rho_f = \frac{7.84 \text{ N}}{0.98 \text{ N}}(1.00 \times 10^3 \text{ kg/m}^3)$$

$$\boxed{\rho_o = 8.0 \times 10^3 \text{ kg/m}^3}$$

4. EVALUATE   From **Table 9-1,** we know the density of gold is 19.3×10^3 kg/m^3. Because 8.0×10^3 kg/m^3 < 19.3×10^3 kg/m^3, the crown cannot be pure gold.

Alternative Problem-Solving Approach

$F_B = F_{actual} - F_{apparent}$
$F_B = 7.84$ N $- 6.86$ N $= 0.98$ N
$F_B = \rho_{fluid}Vg$
$V =$

$$\frac{0.98 \text{ N}}{(1.00 \times 10^3 \text{ kg/m}^3)(9.81 \text{ m/s}^2)}$$

$V = 0.00010$ m^3
This volume of gold should weigh 18.9 N (0.00010 m^3 $\times$ 9.81 m/s^2 $\times$ density of gold), which is much greater than the 7.84 N weight given in the problem.

Classroom Practice

The following may be used as teamwork exercises or for demonstration at the chalkboard or on an overhead projector.

PROBLEM

Buoyant force

Calculate the actual weight, the buoyant force, and the apparent weight of a 5.00×10^{-5} m^3 iron ball floating at rest in mercury.

Answer
3.86 N, 3.86 N, 0 N

How much of the ball's volume is immersed in mercury?

Answer
2.89×10^{-5} m^3

⭐TEKS

p. 322: 3B, 3A
p. 323: 2C, 3B

PRACTICE GUIDE 9A
Solving for:

ρ	📖 **PE**	Sample, 1; Ch. Rvw. 8–9, 54
	💿 **PW** 6–8	
	PB 5–7	
m	📖 **PE** 2, 3; Ch. Rvw. 36	
	💿 **PW** Sample, 1–3	
	PB 8–10	
F_B	📖 **PE** 4; Ch. Rvw. 41*, 43*, 44*, 52*, 53*, 59*, 63*, 64*, 65*, 68*, 69*	
	💿 **PW** 4–5	
	PB Sample, 1–4	

ANSWERS TO

Practice 9A
Buoyant force

1. **a.** 3.57×10^3 kg/m³
   b. 6.4×10^2 kg/m³
2. 97 kg
3. 9.4×10^3 N
4. **a.** 6.63 N
   b. 5.58 N

Section Review
ANSWERS

1. A solid has a definite shape, while a fluid does not; a liquid has a definite volume, while a gas does not.
2. b, c, and d
3. 4.7×10^3 m³
4. 9.92×10^2 kg/m³
5. The effective density of the kayak includes both the material composing the kayak and the air within the kayak below the level of the water.

PRACTICE 9A

Buoyant force ⭐TEKS 2C, 3B

1. A piece of metal weighs 50.0 N in air, 36.0 N in water, and 41.0 N in an unknown liquid. Find the densities of the following:
   a. the metal
   b. the unknown liquid

2. A 2.8 kg rectangular air mattress is 2.00 m long, 0.500 m wide, and 0.100 m thick. What mass can it support in water before sinking?

3. A ferry boat is 4.0 m wide and 6.0 m long. When a truck pulls onto it, the boat sinks 4.00 cm in the water. What is the combined weight of the truck and the ferry?

4. An empty rubber balloon has a mass of 0.0120 kg. The balloon is filled with helium at 0°C, 1 atm pressure, and a density of 0.179 kg/m³. The filled balloon has a radius of 0.500 m.
   a. What is the magnitude of the buoyant force acting on the balloon? (Hint: See **Table 9-1** for the density of air.)
   b. What is the magnitude of the net force acting on the balloon?

Section Review  ⭐TEKS 2C, 3B

1. What is the difference between a solid and a fluid? What is the difference between a gas and a liquid?

2. Which of the following objects will float in a tub of mercury?
   a. a solid gold bead
   b. an ice cube
   c. an iron bolt
   d. 5 mL of water

3. A 650 kg weather balloon is designed to lift a 4600 kg package. What volume should the balloon have after being inflated with helium at 0°C and 1 atm pressure to lift the total load? (Hint: Use the density values in **Table 9-1.**)

4. A submerged submarine alters its buoyancy so that it initially accelerates upward at 0.325 m/s². What is the submarine's average density at this time? (Hint: the density of sea water is 1.025×10^3 kg/m³.)

5. **Physics in Action** Many kayaks are made of plastics and other composite materials that are denser than water. How are such kayaks able to float in water?

9-2
Fluid pressure and temperature

PRESSURE

Deep-sea explorers wear atmospheric diving suits like the one shown in **Figure 9-6** to resist the forces exerted by water in the depths of the ocean. You experience similar forces on your ears when you dive to the bottom of a swimming pool, drive up a mountain, or ride in an airplane.

Pressure is force per unit area

In the examples above the fluids exert **pressure** on your eardrums. Pressure is a measure of how much force is applied over a given area. It can be written as follows:

PRESSURE ⭐TEKS 3B

$$P = \frac{F}{A}$$

$$\text{pressure} = \frac{\text{force}}{\text{area}}$$

The SI unit of pressure is the *pascal* (Pa), which is equal to 1 N/m^2. The pascal is a small unit of pressure. The pressure of the atmosphere at sea level is about 10^5 Pa. This amount of air pressure under normal conditions is the basis for another unit, the *atmosphere* (atm). When calculating pressure, 10^5 Pa is about the same as 1 atm. The absolute air pressure inside a typical automobile tire is about 3×10^5 Pa, or 3 atm. **Table 9-2** lists some additional pressures.

Table 9-2   Some pressures ⭐TEKS 2C

Location	P (Pa)
Center of the sun	2×10^{16}
Center of Earth	4×10^{11}
Bottom of the Pacific Ocean	6×10^7
Atmosphere at sea level	1.01×10^5
Atmosphere at 10 km above sea level	2.8×10^4
Best vacuum in a laboratory	1×10^{-12}

9-2 SECTION OBJECTIVES

- **Calculate the pressure exerted by a fluid.**

- **Calculate how pressure varies with depth in a fluid.**

- **Describe fluids in terms of temperature.**

pressure

the magnitude of the force on a surface per unit area

Figure 9-6
Atmospheric diving suits allow divers to withstand the pressure exerted by the fluid in the ocean at depths of up to 610 m.

Misconception Alert

Students may confuse the pressure *increase* in Pascal's principle with the pressure of the fluid itself. While an increase in pressure is transmitted equally throughout a fluid, the total pressure at different points in the fluid may vary, for example, with depth.

Did you know?

Each time you squeeze a tube of toothpaste, you experience Pascal's principle in action. The pressure you apply by squeezing the sides of the tube is transmitted throughout the toothpaste. The increased pressure near the open mouth of the tube forces the paste out and onto your toothbrush.

Applied pressure is transmitted equally throughout a fluid

When you pump a bicycle tire, you apply a force on the pump that in turn exerts a force on the air inside the tire. The air responds by pushing not only against the pump but also against the walls of the tire. As a result, the pressure increases by an equal amount throughout the tire.

In general, if the pressure in a fluid is increased at any point in a container (such as at the valve of the tire), the pressure increases at all points inside the container by exactly the same amount. Blaise Pascal (1623–1662) noted this fact in what is now called *Pascal's principle* (or *Pascal's law*): (★)TEKS **3E**

> **PASCAL'S PRINCIPLE**
>
> **Pressure applied to a fluid in a closed container is transmitted equally to every point of the fluid and to the walls of the container.**

A hydraulic lift, such as the one shown in **Figure 9-7,** makes use of Pascal's principle. A small force F_1 applied to a small piston of area A_1 causes a pressure increase in a fluid, such as oil. According to Pascal's law, this increase in pressure, P_{inc}, is transmitted to a larger piston of area A_2 and the fluid exerts a force F_2 on this piston. Applying Pascal's principle and the definition of pressure gives the following equation:

$$P_{inc} = \frac{F_1}{A_1} = \frac{F_2}{A_2}$$ (★)TEKS **3B**

Rearranging this equation to solve for F_2 produces the following:

$$F_2 = \frac{A_2}{A_1} F_1$$

This second equation shows that the output force, F_2, is larger than the input force, F_1, by a factor equal to the ratio of the areas of the two pistons.

Figure 9-7
Because the pressure is the same on both sides of the enclosed fluid in a hydraulic lift, a small force on the smaller piston (left) produces a much larger force on the larger piston (right).

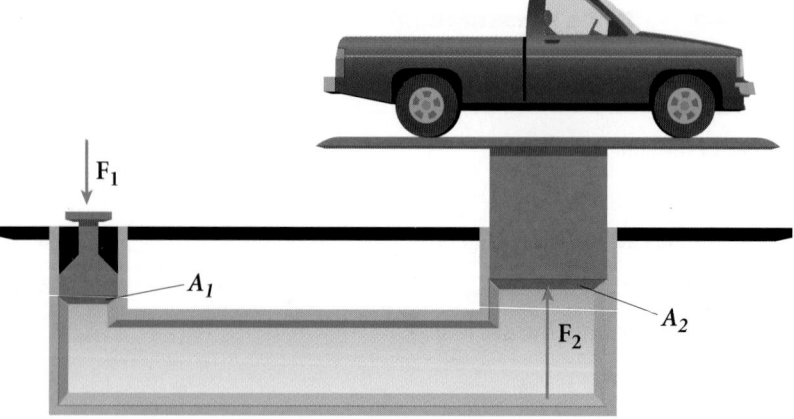

Pressure ⭐TEKS 2C, 3B

PROBLEM

The small piston of a hydraulic lift has an area of 0.20 m². A car weighing 1.20×10^4 N sits on a rack mounted on the large piston. The large piston has an area of 0.90 m². How large a force must be applied to the small piston to support the car?

SOLUTION

Given: $A_1 = 0.20 \text{ m}^2$ $A_2 = 0.90 \text{ m}^2$
     $F_2 = 1.20 \times 10^4 \text{ N}$

Unknown: $F_1 = ?$

Use the equation for pressure from page 325.

$$\frac{F_1}{A_1} = \frac{F_2}{A_2}$$

$$F_1 = \left(\frac{A_1}{A_2}\right)F_2 = \left(\frac{0.20 \text{ m}^2}{0.90 \text{ m}^2}\right)(1.20 \times 10^4 \text{ N})$$

$$\boxed{F_1 = 2.7 \times 10^3 \text{ N}}$$

PRACTICE 9B

Pressure ⭐TEKS 2C, 3B

1. In a car lift, compressed air exerts a force on a piston with a radius of 5.00 cm. This pressure is transmitted to a second piston with a radius of 15.0 cm.
   a. How large a force must the compressed air exert to lift a 1.33×10^4 N car?
   b. What pressure produces this force? Neglect the weight of the pistons.

2. A 1.5 m wide by 2.5 m long water bed weighs 1025 N. Find the pressure that the water bed exerts on the floor. Assume that the entire lower surface of the bed makes contact with the floor.

3. A person rides up a lift to a mountaintop, but the person's ears fail to "pop"—that is, the pressure of the inner ear does not equalize with the outside atmosphere. The radius of each eardrum is 0.40 cm. The pressure of the atmosphere drops from 1.010×10^5 Pa at the bottom of the lift to 0.998×10^5 Pa at the top.
   a. What is the pressure on the inner ear at the top of the mountain?
   b. What is the magnitude of the net force on each eardrum?

Classroom Practice

The following may be used as a teamwork exercise or for demonstration at the chalkboard or on an overhead projector.

PROBLEM

Pressure

In a hydraulic lift, a 620 N force is exerted on a 0.20 m² piston in order to support a weight that is placed on a 2.0 m² piston. How much pressure is exerted on the narrow piston? How much weight can the wide piston lift?

Answer
     3.1×10^3 Pa; 6.2×10^3 N

PRACTICE GUIDE 9B		
Solving for:		
F	**PE**	Sample, 1a, 3b; Ch. Rvw. 16–18, 19b*, 32, 50*
	PW	Sample, 1–2
	PB	5–7
P	**PE**	1b, 2, 3a; Ch. Rvw. 33, 37, 38b*, 47, 55*, 67*
	PW	6–7
	PB	8–10
A	**PW**	3–5
	PB	Sample, 1–4

ANSWERS TO

Practice 9B
Pressure
1. **a.** 1.48×10^3 N
   b. 1.88×10^5 Pa
2. 2.7×10^2 Pa
3. **a.** 1.2×10^3 Pa
   b. 6.0×10^{-2} N

Pressure varies with depth in a fluid

As a submarine dives deeper in the water, the pressure of the water against the hull of the submarine increases, and the resistance of the hull must be strong enough to withstand large pressures. Water pressure increases with depth because the water at a given depth must support the weight of the water above it.

Imagine a small area on the hull of a submarine. The weight of the entire column of water above that area exerts a force on the area. The column of water has a volume equal to Ah, where A is the cross-sectional area of the column and h is its height. Hence the mass of this column of water is $m = \rho V = \rho Ah$. Using the definitions of density and pressure, the pressure at this depth due to the weight of the column of water can be calculated as follows:

$$P = \frac{F}{A} = \frac{mg}{A} = \frac{\rho V g}{A} = \frac{\rho A h g}{A} = \rho h g$$

Note that this equation is valid only if the density is the same throughout the fluid.

The pressure in the equation above is referred to as *gauge pressure*. It is not the total pressure at this depth because the atmosphere itself also exerts a pressure at the surface. Thus, the gauge pressure is actually the total pressure minus the atmospheric pressure. By using the symbol P_0 for the atmospheric pressure at the surface, we can express the total pressure, or *absolute pressure*, at a given depth in a fluid of uniform density ρ as follows:

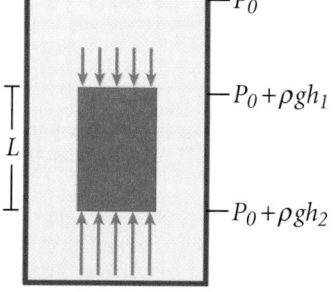

P_0

$P_0 + \rho g h_1$

L

$P_0 + \rho g h_2$

Figure 9-8
The fluid pressure at the bottom of the box is greater than the fluid pressure at the top of the box.

FLUID PRESSURE AS A FUNCTION OF DEPTH ⊛TEKS **3B**

$$P = P_0 + \rho g h$$

absolute pressure =
atmospheric pressure + (density × free-fall acceleration × depth)

This expression for pressure in a fluid can be used to help understand buoyant forces. Consider a rectangular box submerged in a container of water. The water pressure pushing down on the top of the box is $-(P_0 + \rho g h_1)$, and

Conceptual Challenge ⊛TEKS 3A, 4C

1. Atmospheric pressure Why doesn't the roof of a building collapse under the tremendous pressure exerted by our atmosphere?

2. Force and work In a hydraulic lift, which of the two pistons (large or small) moves through a longer distance while the hydraulic lift is lifting an object? (Hint: Remember that for an ideal machine, the input work equals the output work.)

3. Snowshoes A woman wearing snowshoes stands safely in the snow. If she removes her snowshoes, she quickly begins to sink. Explain what happens in terms of force and pressure.

the water pressure pushing up on the bottom of the box is $P_0 + \rho g h_2$. The net pressure on the box is the sum of these two pressures.

$$P_{net} = P_{bottom} + P_{top} = (P_0 + \rho g h_2) - (P_0 + \rho g h_1) = \rho g (h_2 - h_1) = \rho g L$$

From this result, we can find the net vertical force due to the pressure on the box as follows:

$$F_{net} = P_{net} A = \rho g L A = \rho g V = m_f g$$

Note that this is an expression of Archimedes' principle. In general, we can say that buoyant forces arise from the differences in fluid pressure between the top and the bottom of an immersed object. ⭐TEKS **3A, 3B, 4C**

Atmospheric pressure is pressure from above

The weight of the air in the upper portion of Earth's atmosphere exerts pressure on the layers of air below. This pressure is called *atmospheric pressure;* the force it exerts on our bodies (assuming a body area of 2 m²) is extremely large, on the order of 200 000 N (40 000 lb). How can we exist under such tremendous forces without our bodies collapsing? The answer is that our body cavities and tissues are permeated with fluids and gases that are pushing outward with a pressure equal to that of the atmosphere. Consequently, our bodies are in equilibrium—the force of the atmosphere pushing in equals the internal force pushing out. ⭐TEKS **3A**

An instrument that is commonly used to measure atmospheric pressure is the *mercury barometer.* **Figure 9-9** shows a very simple mercury barometer. A long tube that is open at one end and closed at the other is filled with mercury and then inverted into a dish of mercury. Once inverted, the mercury does not empty into the bowl; rather, the atmosphere exerts a pressure on the mercury in the bowl and pushes the mercury in the tube to some height above the bowl. In this way, the force exerted on the bowl of mercury by the atmosphere is equal to the weight of the column of mercury in the tube. Any change in the height of the column of mercury means that the atmosphere's pressure has changed.

Kinetic theory of gases can describe the origin of gas pressure

Many models of a gas have been developed over the years. Almost all of these models attempt to explain the macroscopic properties of a gas, such as pressure, in terms of events occurring in the gas on a microscopic scale. The most successful model by far is the *kinetic theory of gases.*

In kinetic theory, gas particles are likened to a collection of billiard balls that constantly collide with one another. This simple model is successful in explaining many of the macroscopic properties of a gas. For instance, as these particles strike a wall of a container, they transfer some of their momentum during the collision. The rate of transfer of momentum to the container wall is equal to the force exerted by the gas on the container wall (see Chapter 6). This force per unit area is the gas pressure.

internet connect

SCiLINKS
NSTA

TOPIC: Atmospheric pressure
GO TO: www.scilinks.org
*sci*LINKS CODE: HF2093

Figure 9-9
The height of the mercury in the tube of a barometer indicates the atmospheric pressure.

SECTION 9-2

Demonstration 6

Hydrostatic pressure

Purpose Demonstrate that pressure increases with depth.

Materials 32 oz plastic soda bottle, tape, water, bucket; optional: plastic straw or small pieces of glass or metal tubing and modeling clay or silicon

Procedure Drill three holes about 10 cm apart along the side of the bottle, with the lowest hole close to the bottom, and tape them. The holes should not be aligned vertically; they should be displaced horizontally about 1 cm so that the water streams do not collide. Put small pieces of tubing into the holes to improve the flow. Cover the holes with tape, fill the bottle with water, and place it at the edge of a table. Place it high enough above the bucket so that the effects of depth on the range of each stream will be easily observed. Ask students to predict how the water streams will compare. Quickly pull the tape away from all three holes, and have students observe the shape of each stream. Water shooting out from near the bottom of the bottle exits the bottle at a higher speed than water shooting out from near the top does. The reason is that the pressure in the bottle increases with depth.

⭐TEKS

p.328: 3B, 3A, 4C
p.329: 3A, 3B, 4C, 3A

Classroom Practice

The following may be used as a teamwork exercise or for demonstration at the chalkboard or on an overhead projector.

PROBLEM

Pressure as a function of depth

Find the atmospheric pressure at an altitude of 1.0×10^3 m if the air density is constant. Assume that the air density is uniformly 1.29 kg/m^3 and $P_0 = 1.01 \times 10^5$ Pa.

Answer

8.8×10^4 Pa

PRACTICE GUIDE 9C

Solving for:		
P	PE	Sample, 1, 2; Ch. Rvw. 19a, 34*, 35*, 46*, 57b*
	PW	5–6
	PB	5–7
h	PE	3, 4; Ch. Rvw. 57a*
	PW	3–4
	PB	Sample, 1–4
ρ	PW	Sample, 1–2
	PB	8–10

ANSWERS TO

Practice 9C

Pressure as a function of depth

1. 1.11×10^8 Pa
2. **a.** 1.03×10^5 Pa
   b. 1.05×10^5 Pa
3. 0.20 m
4. 20.1 m

SAMPLE PROBLEM 9C

Pressure as a function of depth ⭐TEKS 2C, 3B

PROBLEM

Calculate the absolute pressure at an ocean depth of 1.00×10^3 m. Assume that the density of the water is 1.025×10^3 kg/m^3 and that $P_0 = 1.01 \times 10^5$ Pa.

SOLUTION

Given:
$$h = 1.00 \times 10^3 \text{ m} \quad P_0 = 1.01 \times 10^5 \text{ Pa} \quad \rho = 1.025 \times 10^3 \text{ kg/m}^3$$
$$g = 9.81 \text{ m/s}^2$$

Unknown: $P = ?$

Use the equation for fluid pressure as a function of depth from page 328.

$$P = P_0 + \rho g h$$
$$P = P_0 + (1.025 \times 10^3 \text{ kg/m}^3)(9.81 \text{ m/s}^2)(1.00 \times 10^3 \text{ m})$$
$$P = 1.01 \times 10^5 \text{ Pa} + 1.01 \times 10^7 \text{ Pa}$$

$$\boxed{P = 1.02 \times 10^7 \text{ Pa}}$$

PRACTICE 9C

Pressure as a function of depth ⭐TEKS 2C, 3B

1. The Mariana Trench, in the Pacific Ocean, is about 11.0 km deep. If atmospheric pressure at sea level is 1.01×10^5 Pa, how much pressure would a submarine need to be able to withstand to reach this depth? (Use the value for the density of sea water given in **Table 9-1**.)

2. A container is filled with water to a depth of 20.0 cm. On top of the water floats a 30.0 cm thick layer of oil with a density of 0.70×10^3 kg/m^3.
   a. What is the pressure at the surface of the water?
   b. What is the absolute pressure at the bottom of the container?

3. A beaker containing mercury is placed inside a vacuum chamber in a laboratory. The pressure at the bottom of the beaker is 2.7×10^4 Pa. What is the height of the mercury in the beaker? (See **Table 9-1** for the density of mercury. Hint: Think carefully about what value to use for atmospheric pressure.)

4. Calculate the depth in the ocean at which the pressure is three times atmospheric pressure. (Use the value for the density of sea water given in **Table 9-1**.)

TEMPERATURE IN A GAS ⭐TEKS 5B

Density and pressure are not the only two quantities useful in describing a fluid. The **temperature** of a fluid is also important. We often associate the concept of temperature with how hot or cold an object feels when we touch it. Thus, our senses provide us with qualitative indications of temperature. But to understand what the temperature of a gas really measures, we must turn again to the kinetic theory of gases.

Like pressure, temperature in a gas can be understood on the basis of what is happening on the atomic scale. Kinetic theory predicts that temperature is proportional to the average kinetic energy of the particles in the gas. The higher the temperature of the gas, the faster the particles move. As the speed of the particles increases, the rate of collisions against the walls of the container increases. More momentum is transferred to the container walls in a given time interval, resulting in an increase in pressure. Thus, kinetic theory predicts that the pressure and temperature of a gas are related. As we will see later in this chapter, this is indeed the case.

The SI units for temperature are kelvins and degrees Celsius (written K and °C). To quickly convert from the Celsius scale to the Kelvin scale, add 273. Room temperature is about 293 K (20°C).

temperature

a measure of the average kinetic energy of the particles in a substance

CONCEPT PREVIEW

Temperature and temperature scales will be discussed further in Chapter 10.

Section Review ⭐TEKS 2C, 3B

1. Which of the following exerts the most pressure while resting on a floor?
   a. a 25 N box with 1.5 m sides
   b. a 15 N cylinder with a base radius of 1.0 m
   c. a 25 N box with 2.0 m sides
   d. a 25 N cylinder with a base radius of 1.0 m

2. Water is to be pumped to the top of the Empire State Building, which is 366 m high. What gauge pressure is needed in the water line at the base of the building to raise the water to this height? (Hint: See **Table 9-1** for the density of water.)

3. A room on the first floor of a hospital has a temperature of 20°C. A room on the top floor has a temperature of 22°C. In which of these two rooms is the average kinetic energy of the air particles greater?

4. The temperature of the air outside on a cool morning is 11°C. What is this temperature on the Kelvin scale?

5. When a submarine dives to a depth of 5.0×10^2 m, how much pressure, in Pa, must its hull be able to withstand? How many times larger is this pressure than the pressure at the surface? (Hint: See **Table 9-1** for the density of sea water.)

Section Review
ANSWERS

1. a
2. 3.59×10^6 Pa
3. room on the top floor
4. 284 K
5. 5.0×10^6 Pa; 5.0×10^1

⭐TEKS

p. 330: 2C, 3B, 2C, 3B
p. 331: 5B, 2C, 3B

332

Demonstration 7

Fluid flow around a table-tennis ball

Purpose Introduce students to some of the interesting effects of fluid flow.

Materials table-tennis ball glued to the end of a piece of light string, water faucet

Procedure Tell the class that this section covers the surprising phenomena of fluids in motion "sucking in" things around them.

Hold the end of the string so that the ball is a few centimeters below the faucet but away from the stream path. When you turn on the faucet, the ball is attracted to the water.

Demonstration 8

Fluid flow through two cans

Purpose Further introduce the effects of fluid flow.

Materials two soda cans, flat surface

Procedure Lay the soda cans on their sides approximately 2 cm apart on a table or other flat surface. Ask students what would happen if you were to blow air between the two cans. Ask a student volunteer to blow between the cans. The cans will move toward each other. Explain that the air moves faster between the cans, lowering the pressure between them and drawing them together.

9-3
Fluids in motion

9-3 SECTION OBJECTIVES

- Examine the motion of a fluid using the continuity equation.
- Apply Bernoulli's equation to solve fluid-flow problems.
- Recognize the effects of Bernoulli's principle on fluid motion.

ideal fluid

a fluid that has no internal friction or viscosity and is incompressible

FLUID FLOW

Have you ever gone canoeing or rafting down a river? If so, you may have noticed that part of the river flowed smoothly, allowing you to float calmly or to simply paddle along. At other places in the river, there may have been rocks or dramatic bends that created foamy whitewater rapids.

When a fluid, such as river water, is in motion, the flow can be characterized in one of two ways. The flow is said to be *laminar* if every particle that passes a particular point moves along the same smooth path traveled by the particles that passed that point earlier. This path is called a *streamline*. Different streamlines cannot cross each other, and the streamline at any point coincides with the direction of fluid velocity at that point. The smooth stretches of a river are regions of laminar flow.

In contrast, the flow of a fluid becomes irregular, or *turbulent,* above a certain velocity or under conditions that can cause abrupt changes in velocity, such as where there are obstacles or sharp turns in a river. Irregular motions of the fluid, called *eddy currents,* are characteristic of turbulent flow. Examples of turbulent flow are found in water in the wake of a ship or in the air currents of a severe thunderstorm.

Figure 9-10 shows a photograph of water flowing past a cylinder. Hydrogen bubbles were added to the water to make the streamlines and the eddy currents visible. Notice the dramatic difference in flow patterns between the laminar flow and the turbulent flow. Laminar flow is much easier to model because it is predictable. Turbulent flow is extremely chaotic and unpredictable.

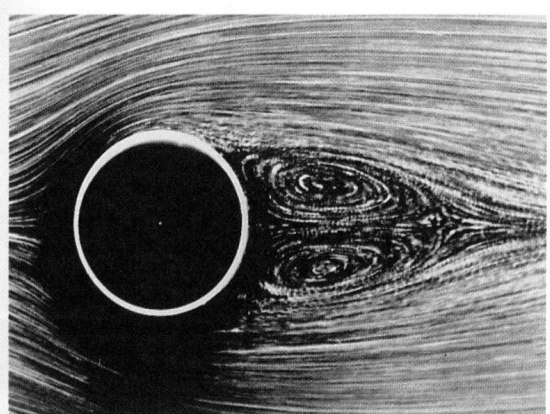

Figure 9-10
The water flowing around this cylinder exhibits laminar flow and turbulent flow.

The ideal fluid model simplifies fluid-flow analysis

Many features of fluid motion can be understood by considering the behavior of an **ideal fluid.** While discussing density and buoyancy, we assumed all of the fluids used in problems were practically incompressible. A fluid is incompressible if the density of the fluid always remains constant. Incompressibility is one of the characteristics of an ideal fluid.

The term *viscosity* refers to the amount of internal friction within a fluid. Internal friction can occur when one layer of fluid slides past another layer. A fluid with a high viscosity flows more slowly through a pipe than does a fluid with a low viscosity. As a viscous fluid flows, part of the kinetic energy of the fluid is transformed into internal energy because of the internal friction. Ideal fluids are considered *nonviscous*, so they lose no kinetic energy due to friction as they flow.

Ideal fluids are also characterized by a *steady flow*. In other words, the velocity, density, and pressure at each point in the fluid are constant. The flow of an ideal fluid is also *nonturbulent*, which means that there are no eddy currents in the moving fluid.

Although no real fluid has all the properties of an ideal fluid, the ideal fluid model does help explain many properties of real fluids, so the model is a useful tool for analysis. Unless otherwise stated, the fluids in the rest of our discussion of fluid flow will be treated as ideal fluids.

PRINCIPLES OF FLUID FLOW

Fluid behavior is often very complex. A detailed analysis of the forces acting on a fluid may be so complicated that even a supercomputer cannot create an accurate model. However, several general principles describing the flow of fluids can be derived relatively easily from basic physical laws.

The continuity equation results from mass conservation

Imagine that an ideal fluid flows into one end of a pipe and out the other end, as shown in **Figure 9-11.** The diameter of the pipe is different on each end. How does the speed of fluid flow change as the fluid passes through the pipe?

Because mass is conserved and because the fluid is incompressible, we know that the mass flowing into the bottom of the pipe, m_1, must equal the mass flowing out of the top of the pipe, m_2, during any given time interval:

$$m_1 = m_2$$

This simple equation can be expanded by recalling that $m = \rho V$ and by using the formula for the volume of a cylinder, $V = A\Delta x$.

$$\rho_1 V_1 = \rho_2 V_2$$
$$\rho_1 A_1 \Delta x_1 = \rho_2 A_2 \Delta x_2$$

The length of the cylinder, Δx, is also the distance the fluid travels, which is equal to the speed of flow multiplied by the time interval ($\Delta x = v\Delta t$).

$$\rho_1 A_1 v_1 \Delta t = \rho_2 A_2 v_2 \Delta t$$

For an ideal fluid, both the time interval and the density are the same on each side of the equation, so they cancel each other out. The resulting equation is called the continuity equation:

CONTINUITY EQUATION  **3B**

$$A_1 v_1 = A_2 v_2$$

area × speed in region 1 = area × speed in region 2

(★) TEKS 3C

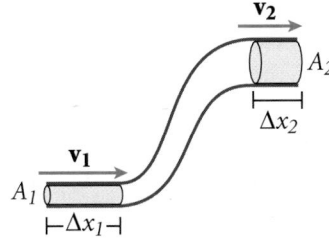

Figure 9-11
The mass flowing into the pipe must equal the mass flowing out of the pipe in the same time interval.

Teaching Tip

The tilt of an airplane wing also adds to the lift on the plane. The front of the wing is tilted upward so that air striking the bottom of the wing is deflected downward. At the same time, the deflected air produces an upward reaction force on the wing, contributing substantially to the lift.

Figure 9-12
Placing your thumb over the end of a garden hose reduces the area of the opening and increases the speed of the water.

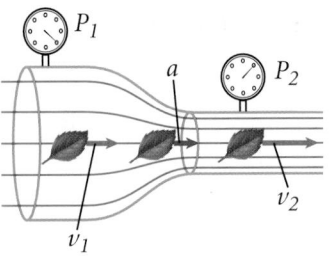

Figure 9-13
As a leaf passes into a constriction in a drainage pipe, the leaf speeds up. Pressure gauges show that the water pressure on the right is less than the pressure on the left.

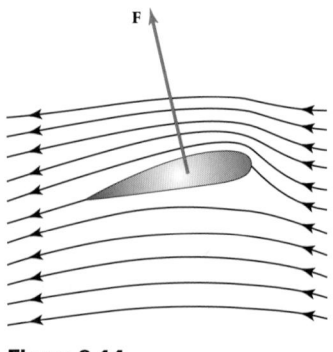

Figure 9-14
As air flows around an airplane wing, the air above the wing moves faster than the air below, producing lift.

The speed of fluid flow depends on cross-sectional area TEKS 3A

Note in the continuity equation that A_1 and A_2 can represent any two different cross-sectional areas of the pipe, not just the ends. This equation implies that the fluid speed is faster where the pipe is narrow and slower where the pipe is wide. The product Av, which has units of volume per unit time, is called the *flow rate*. The flow rate is constant throughout the pipe.

The continuity equation explains an effect you may have experienced when placing your thumb over the end of a garden hose, as shown in **Figure 9-12.** Because your thumb blocks some of the area through which the water can exit the hose, the water exits at a higher speed than it would otherwise. The continuity equation also explains why a river tends to flow more rapidly in places where the river is shallow or narrow than in places where the river is deep and wide.

The pressure in a fluid is related to the speed of flow TEKS 3A

Suppose there is a water-logged leaf carried along by the water in a drainage pipe, as shown in **Figure 9-13.** The continuity equation shows that the water moves faster through the narrow part of the tube than through the wider part of the tube. Therefore, as the water carries the leaf into the constriction, the leaf speeds up.

If the water and the leaf are accelerating as they enter the constriction, an unbalanced force must be causing the acceleration, according to Newton's second law. This unbalanced force is a result of the fact that the water pressure in front of the leaf is less than the water pressure behind the leaf. The pressure difference causes the leaf and the water around it to accelerate as it enters the narrow part of the tube. This behavior illustrates a general principle, known as *Bernoulli's principle*, which can be stated as follows:

> **BERNOULLI'S PRINCIPLE**
>
> **The pressure in a fluid decreases as the fluid's velocity increases.**

The lift on an airplane wing can be explained, in part, with Bernoulli's principle. As an airplane flies, air flows around the wings and body of the plane, as shown in **Figure 9-14.** Airplane wings are designed to direct the flow of air so that the air speed above the wing is greater than the air speed below the wing. Therefore, the air pressure above the wing is less than the pressure below, and there is a net upward force on the wing, called *lift*. TEKS 3C

Bernoulli's equation relates pressure to energy in a moving fluid

Imagine a fluid moving through a pipe of varying cross-section and elevation, as shown in **Figure 9-15.** As the fluid flows into regions of different cross-sectional area, the pressure and speed of the fluid along a given streamline in the pipe can change. However, if the speed of a fluid changes, then the fluid's kinetic energy also changes. The change in kinetic energy may be compensated for by a change in gravitational potential energy or by a change in pressure so energy is still conserved.

The expression for the conservation of energy in fluids is called *Bernoulli's equation,* and it can be expressed as follows:

BERNOULLI'S EQUATION ⭐TEKS 3B

$$P + \frac{1}{2}\rho v^2 + \rho g h = \text{constant}$$

**pressure + kinetic energy per unit volume +
gravitational potential energy per unit volume =
constant along a given streamline**

Note that Bernoulli's equation differs slightly from the law of conservation of energy given in Chapter 5. First of all, two of the terms on the left side of the equation look like the terms for kinetic energy and gravitational potential energy, but they contain density, ρ, instead of mass, m. That is because the conserved quantity in Bernoulli's equation is energy per unit volume—not just energy—and density is equivalent to mass per unit volume. This statement of the conservation of energy in fluids also includes an additional term: pressure, P. Note that the units of pressure are equivalent to the units for energy per unit volume.

If you wish to compare the energy in a given volume of fluid at two different points, Bernoulli's equation takes the following equivalent form:

$$P_1 + \frac{1}{2}\rho v_1^2 + \rho g h_1 = P_2 + \frac{1}{2}\rho v_2^2 + \rho g h_2$$

Bernoulli's principle is a special case of Bernoulli's equation

Two special cases of Bernoulli's equation are worth mentioning here. First, if the fluid is not moving, then both speeds are zero. This case is a static situation, such as a column of water in a cylinder. If the height at the top of the column, h_1, is defined as zero, and h_2 is the depth, then Bernoulli's equation reduces to the equation for pressure as a function of depth, introduced in **Section 9-2:**

$$P_1 = P_2 + \rho g h_2 \text{ (static fluid)}$$

Second, imagine again a fluid flowing through a horizontal pipe with a constriction. Because the height of the fluid is constant, the gravitational potential energy does not change. Bernoulli's equation then reduces to the following:

$$P_1 + \frac{1}{2}\rho v_1^2 = P_2 + \frac{1}{2}\rho v_2^2 \text{ (horizontal pipe)}$$

This equation suggests that if v_1 is greater than v_2 at two different points in the flow, then P_1 must be less than P_2. In other words, the pressure decreases as speed increases. This is Bernoulli's principle again, which we now can see as a special case of Bernoulli's equation. The conditions required for this case tell us that Bernoulli's principle is strictly true only when elevation is constant. ⭐TEKS 3A, 3B

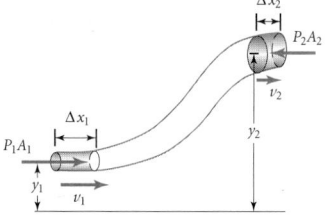

Figure 9-15
As a fluid flows through this pipe, it may change velocity, pressure, and elevation.

Quick Lab

Bernoulli's Principle

MATERIALS LIST

✔ single sheet of paper

You can see the effects of Bernoulli's principle on a sheet of paper by holding the edge of the sheet horizontally and blowing across the top surface. The sheet should rise as a result of the lower air pressure above the sheet.

⭐TEKS 1A, 3A, 5B

internet connect

SC*LINKS*
NSTA

TOPIC: Bernoulli's principle
GO TO: www.scilinks.org
*sci*LINKS CODE: HF2094

Teaching Tip

The equivalence of units in Bernoulli's equation can be seen more clearly if you multiply the units of pressure, N/m^2, by a factor equivalent to unity, m/m:

$$\frac{N}{m^2} \times \frac{m}{m} = \frac{N \cdot m}{m^3} = \frac{J}{m^3}$$

Quick Lab

TEACHER'S NOTES

For this lab to be effective, it is best to hold a piece of very light paper or soft plastic with both hands and blow over it through a straw.

Demonstration 9

Atomizer

Purpose Demonstrate an application of Bernoulli's principle.

Materials glass jar, water, plastic straw, bicycle-horn bulb attached to thin hollow tube

Procedure Place the straw in the water. Use the bulb and tube to blow air across the top of the straw. Water will come up the straw and spray out. The increased air flow above the straw results in a pressure decrease inside it.

⭐TEKS

p. 334: 3A, 3A, 3C
p. 335: 3B, 1A, 3A, 3A,
3B, 5B

The following may be used as teamwork exercises or for demonstration at the chalkboard or on an overhead projector.

PROBLEM

Bernoulli's equation

A camper creates a shower by attaching a tube to the bottom of a hanging bucket that is open to the atmosphere on top. If the water level in the tank is 3.15 m above the end of the tube (the shower head), then what is the speed of the water exiting the tube?

Answer
7.86 m/s

A pipe narrows from a cross section of 2.0 m^2 to 0.30 m^2. If the speed of the water flowing through the wider area of the pipe is 8.0 m/s, what is the speed of the water flowing through the narrow part?

Answer
53 m/s

Assuming incompressible flow, what is the change in pressure as the pipe narrows?

Answer
1.4 × 10^6 Pa

★ **TEKS**

p. 336: 2C, 3B
p. 337: 2C, 3B, 2C, 3B

336

SAMPLE PROBLEM 9D

Bernoulli's equation ★ TEKS 2C, 3B

PROBLEM

A water tank has a spigot near its bottom. If the top of the tank is open to the atmosphere, determine the speed at which the water leaves the spigot when the water level is 0.500 m above the spigot.

SOLUTION

1. DEFINE **Given:** $h_2 - h_1 = 0.500$ m

Unknown: $v_1 = ?$

Diagram:

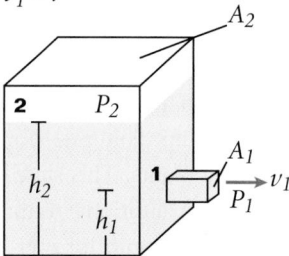

2. PLAN **Choose an equation(s) or situation:** Because this problem involves fluid flow and differences in height, it requires the application of Bernoulli's equation.

$$P_1 + \frac{1}{2}\rho v_1^2 + \rho g h_1 = P_2 + \frac{1}{2}\rho v_2^2 + \rho g h_2$$

Point 1 is at the hole, and point 2 is at the top of the tank. If we assume that the hole is small, then the water level drops very slowly, so we can assume that v_2 is approximately zero. Also, note that $P_1 = P_0$ and $P_2 = P_0$ because both the top of the tank and the spigot are open to the atmosphere.

$$P_0 + \frac{1}{2}\rho v_1^2 + \rho g h_1 = P_0 + \rho g h_2$$

Rearrange the equation(s) to isolate the unknown(s):

$$\frac{1}{2}\rho v_1^2 = \rho g h_2 - \rho g h_1$$

$$v_1^2 = 2g(h_2 - h_1)$$

$$v_1 = \sqrt{2g(h_2 - h_1)}$$

3. CALCULATE **Substitute the values into the equation(s) and solve:**

$$v_1 = \sqrt{2(9.81 \text{ m/s}^2)(0.500 \text{ m})}$$

$$\boxed{v_1 = 3.13 \text{ m/s}}$$

4. EVALUATE A quick estimate gives the following:

$$v_1 \approx \sqrt{2(10)(0.5)} \approx 3$$

PRACTICE 9D

Bernoulli's equation ⭐TEKS 2C, 3B

1. A large storage tank, open to the atmosphere at the top and filled with water, develops a small hole in its side at a point 16 m below the water level. If the rate of flow of water from the leak is 2.5×10^{-3} m³/min, determine the following:

   a. the speed at which the water leaves the hole
   b. the diameter of the hole

2. A liquid with a density of 1.65×10^3 kg/m³ flows through two horizontal sections of tubing joined end to end. In the first section, the cross-sectional area is 10.0 cm², the flow speed is 275 cm/s, and the pressure is 1.20×10^5 Pa. In the second section, the cross-sectional area is 2.50 cm². Calculate the following:

   a. the flow speed in the smaller section
   b. the pressure in the smaller section

3. When a person inhales, air moves down the windpipe at 15 cm/s. The average flow speed of the air doubles when passing through a constriction in the bronchus. Assuming incompressible flow, determine the pressure drop in the constriction.

Section Review   ⭐TEKS 2C, 3B

1. The time required to fill a bucket with water from a certain garden hose is 30.0 s. If you cover part of the hose's nozzle with your thumb so that the speed of the water leaving the nozzle doubles, how long does it take to fill the bucket?

2. Water at a pressure of 3.00×10^5 Pa flows through a horizontal pipe at a speed of 1.00 m/s. The pipe narrows to one-fourth its original diameter. Find the following:

   a. the flow speed in the narrow section
   b. the pressure in the narrow section

3. The water supply of a building is fed through a main entrance pipe that is 6.0 cm in diameter. A 2.0 cm diameter faucet tap positioned 2.00 m above the main pipe fills a 2.5×10^{-2} m³ container in 30.0 s.

   a. What is the speed at which the water leaves the faucet?
   b. What is the gauge pressure in the main pipe?

PRACTICE GUIDE 9D

Solving for:

v	📖 **PE**	Sample, 1, 2a; Ch. Rvw. 23–24
	💿 **PW**	5–7
	PB	5–7
P	📖 **PE**	2b, 3; Ch. Rvw. 38, 39*, 58
	💿 **PW**	Sample, 1–2
	PB	8–10
h	📖 **PE**	Ch. Rvw. 48*, 49*, 56*
	💿 **PW**	3–4
	PB	Sample, 1–4

ANSWERS TO

Practice 9D
Bernoulli's equation

1. **a.** 18 m/s
   b. 1.7×10^{-3} m
2. **a.** 11.0 m/s
   b. 2.7×10^4 Pa
3. -4.4×10^{-2} Pa

Section Review
ANSWERS

1. 30.0 s because the volume rate of flow is constant
2. **a.** 16.0 m/s
   b. 1.72×10^5 Pa
3. **a.** 2.7 m/s
   b. 2.32×10^4 Pa

9-4
Properties of gases

9-4 SECTION OBJECTIVES

- **Define the general properties of an ideal gas.**

- **Use the ideal gas law to predict the properties of an ideal gas under different conditions.**

Figure 9-16
The balloon is inflated because the volume and pressure of the air inside are both increasing.

GAS LAWS

When the density of a gas is sufficiently low, the pressure, volume, and temperature of the gas tend to be related to one another in a fairly simple way. In addition, the relationship is a good approximation for the behavior of many real gases over a wide range of temperatures and pressures. These observations have led scientists to develop the concept of an *ideal gas.*

Volume, pressure, and temperature are the three variables that completely describe the macroscopic state of an ideal gas. One of the most important equations in fluid mechanics relates these three quantities to each other.

The ideal gas law relates gas volume, pressure, and temperature

The *ideal gas law* is an expression that relates the volume, pressure, and temperature of a gas. This relationship can be written as follows:

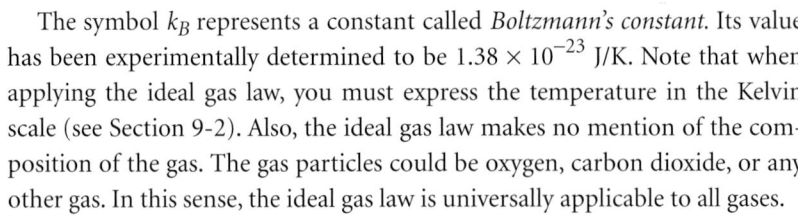

IDEAL GAS LAW ⭐TEKS **3B**

$$PV = Nk_BT$$

pressure × volume =
number of gas particles × Boltzmann's constant × temperature

The symbol k_B represents a constant called *Boltzmann's constant.* Its value has been experimentally determined to be 1.38×10^{-23} J/K. Note that when applying the ideal gas law, you must express the temperature in the Kelvin scale (see Section 9-2). Also, the ideal gas law makes no mention of the composition of the gas. The gas particles could be oxygen, carbon dioxide, or any other gas. In this sense, the ideal gas law is universally applicable to all gases.

If a gas undergoes a change in volume, pressure, or temperature (or any combination of these), the ideal gas law can be expressed in a particularly useful form. If the number of particles in the gas is constant, the initial and final states of the gas are related as follows:

$$N_1 = N_2$$
$$\frac{P_1V_1}{T_1} = \frac{P_2V_2}{T_2}$$

This relation is illustrated in the experiment shown in **Figure 9-16.** In this experiment, a flask filled with air (V_1 equals the volume of the flask) at room temperature (T_1) and atmospheric pressure ($P_1 = P_0$) is placed over a heat source, with a balloon placed over the opening of the flask. As the flask sits over the burner, the temperature of the air inside it increases from T_1 to T_2. According to the ideal gas law, when the temperature increases, either the pressure or the volume—or both—must also increase. Thus, the air inside the flask exerts a pressure (P_2) on the balloon that serves to inflate the balloon. Because the flask is not completely closed, the air expands to a larger volume (V_2) to fill the balloon. When the flask is taken off the burner, the pressure, volume, and temperature of the air inside will slowly return to their initial states.

Another alternative form of the ideal gas law indicates the law's dependence on mass density. Assuming each particle in the gas has a mass m, the total mass of the gas is $N \times m = M$. The ideal gas law can then be written as follows:

$$PV = Nk_BT = \frac{Mk_BT}{m}$$

$$P = \frac{Mk_BT}{mV} = \left(\frac{M}{V}\right)\frac{k_BT}{m} = \frac{\rho k_BT}{m}$$

A real gas can often be modeled as an ideal gas

An ideal gas is defined as a gas whose behavior is accurately described by the ideal gas law. Although no real gas obeys the ideal gas law exactly for all temperatures and pressures, the ideal gas law holds for a broad range of physical conditions for all gases. The behavior of real gases departs from the behavior of an ideal gas at high pressures or low temperatures, conditions under which the gas nearly liquefies. However, when a real gas has a relatively high temperature and a relatively low pressure, such as at room temperature and atmospheric pressure, its behavior approximates that of an ideal gas.

For problems involving the motion of fluids, we have assumed that all gases and liquids are ideal fluids. Recall that an ideal fluid is a liquid or gas that is assumed to be incompressible. This is usually a good assumption because it is difficult to compress a fluid—even a gas—when it is not confined to a container. A fluid will tend to flow under the action of a force, changing its shape while maintaining a constant volume, rather than compress.

(★) TEKS 1A, 3A

Quick Lab

Ideal Gas Law

MATERIALS

✔ 1 plastic 1 L bottle
✔ 1 quarter

Make sure the bottle is empty, and remove the cap. Place the bottle in the freezer for at least 10 min. Wet the quarter with water, and place the quarter over the bottle's opening as you take the bottle out of the freezer. Set the bottle on a nearby tabletop; then observe the bottle and quarter while the air in the bottle warms up. As the air inside the bottle

begins to return to room temperature, the quarter begins to jiggle around on top of the bottle. What does this movement tell you about the pressure and volume inside the bottle? What causes this increase in pressure and volume? Hypothesize as to why you need to wet the quarter before placing it on top of the bottle.

internet**connect**

SC*i*LINKS
NSTA

TOPIC: Gas laws
GO TO: www.scilinks.org
*sci*LINKS CODE: HF2095

Teaching Tip

Point out that the n in $PV = nRT$ and the N in $PV = Nk_BT$ are related through Avogadro's number (the same remark applies to R and k_B).

Key Models and Analogies

Graphs offer a convenient way to represent the relationship between temperature, volume, and pressure for the following special cases of the ideal gas law:

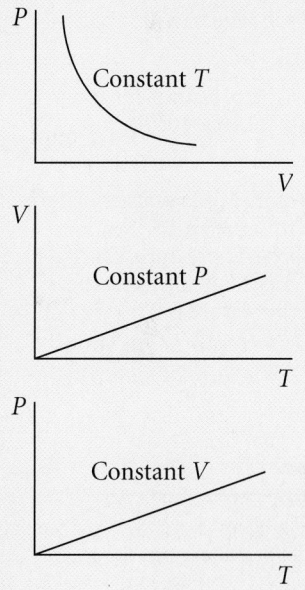

Quick Lab

TEACHER'S NOTES

This activity is meant to demonstrate that pressure increases with temperature when the volume of a gas is constant. Eventually, the pressure becomes great enough to overcome the weight of the quarter, and it jumps up slightly, allowing air to escape.

PROBLEM

The ideal gas law

A sealed tank with a volume of 0.10 m^3 contains air at $27°C$ under pressure 1.8×10^4 Pa. The valve can withstand pressures up to 5.0×10^4 Pa. Will the valve hold if the air inside the tank is heated to $227°C$?

Answer

Yes, the pressure will go up only to 3.0×10^4 Pa.

The volume of a weather balloon is 8.0 m^3 at 3.0×10^2 K under atmospheric pressure. The balloon rises to an altitude where the temperature is 2.0×10^2 K and the pressure is $\frac{1}{2}$ of atmospheric pressure. What is its new volume?

Answer

11 m^3

⭐TEKS

p. 340: 3A, 2C, 3B
p. 341: 2C, 3B, 2C, 3A, 3B

This section, however, considers confined gases whose pressure, volume, and temperature may change. For example, when a force is applied to a piston, the gas inside the cylinder below the piston is compressed. Even though an ideal gas behaves like an ideal fluid in many situations, it cannot be treated as incompressible when confined to a container. ⭐TEKS **3A**

SAMPLE PROBLEM 9E

The ideal gas law ⭐TEKS **2C, 3B**

Figure 9-17

PROBLEM

Pure helium gas is contained in a leakproof cylinder containing a movable piston, as shown in Figure 9-17. The initial volume, pressure, and temperature of the gas are 15 L, 2.0 atm, and 310 K, respectively. If the gas is rapidly compressed to 12 L and the pressure increases to 3.5 atm, find the final temperature of the gas.

SOLUTION

1. DEFINE   **Given:** $V_1 = 15 \text{ L}$ $P_1 = 2.0 \text{ atm}$ $T_1 = 310 \text{ K}$
                                  $V_2 = 12 \text{ L}$ $P_2 = 3.5 \text{ atm}$

Unknown: $T_2 = ?$

2. PLAN   **Choose an equation(s) or situation:** Because the gas undergoes a change and no gas particles are lost, the form of the ideal gas law relating the initial and final states should be used.

$$\frac{P_1 V_1}{T_1} = \frac{P_2 V_2}{T_2}$$

Rearrange the equation(s) to isolate the unknown(s):

$$T_2 P_1 V_1 = T_1 P_2 V_2$$

$$T_2 = T_1 \left(\frac{P_2 V_2}{P_1 V_1} \right)$$

3. CALCULATE   **Substitute the values into the equation(s) and solve:**

$$T_2 = (310 \text{ K}) \left[\frac{(3.5 \text{ atm})(12 \text{ L})}{(2.0 \text{ atm})(15 \text{ L})} \right]$$

$$\boxed{T_2 = 4.3 \times 10^2 \text{ K}}$$

> **CALCULATOR SOLUTION**
>
> Your calculator should give the answer as 434. However, because the values in the problem are known to only two significant figures, the answer should be rounded to 430 and expressed in scientific notation.

4. EVALUATE   Because the gas was compressed and the pressure increased, the gas temperature should have increased.

$$430 \text{ K} > 310 \text{ K}$$

The ideal gas law ⭐TEKS 2C, 3B

1. A cylinder with a movable piston contains gas at a temperature of 27°C, with a volume of 1.5 m^3 and a pressure of 0.20×10^5 Pa. What will be the final temperature of the gas if it is compressed to 0.70 m^3 and its pressure is increased to 0.80×10^5 Pa?

2. Gas is confined in a tank at a pressure of 1.0×10^8 Pa and a temperature of 15.0°C. If half the gas is withdrawn and the temperature is raised to 65.0°C, what is the new pressure in the tank in Pa?

3. A gas bubble with a volume of 0.10 cm^3 is formed at the bottom of a 10.0 cm deep container of mercury. If the temperature is 27°C at the bottom of the container and 37°C at the top of the container, what is the volume of the bubble just beneath the surface of the mercury? Assume that the surface is at atmospheric pressure. (Hint: Use the density of mercury from **Table 9-1**.)

Section Review  ⭐TEKS 2C, 3A, 3B

1. Name some conditions under which a real gas is likely to behave like an ideal gas.

2. What happens to the size of a helium balloon as it rises? Why?

3. Two identical cylinders at the same temperature contain the same kind of gas. If cylinder A contains three times as many gas particles as cylinder B, what can you say about the relative pressures in the cylinders?

4. The pressure on an ideal gas is cut in half, resulting in a decrease in temperature to three-fourths the original value. Calculate the ratio of the final volume to the original volume of the gas.

5. A container of oxygen gas in a chemistry lab room is at a pressure of 6.0 atm and a temperature of 27°C.
   a. If the gas is heated at constant volume until the pressure triples, what is the final temperature?
   b. If the gas is heated so that both the pressure and volume are doubled, what is the final temperature?

PRACTICE GUIDE 9E

Solving for:

T	📖 **PE**	Sample, 1; Ch. Rvw. 29–30, 45a*
	💿 **PW**	7–9
	PB	5–7
P	📖 **PE**	2; Ch. Rvw. 66
	💿 **PW**	Sample, 1–3
	PB	8–10
V	📖 **PE**	3; Ch. Rvw. 35*, 36*, 40*, 42*, 45b*, 46*
	💿 **PW**	4–6
	PB	Sample, 1–4

ANSWERS TO

Practice 9E
The ideal gas law
1. 5.6×10^2 K
2. 5.9×10^7 Pa
3. 1.2×10^{-7} m^3

Section Review
ANSWERS

1. at relatively high temperature and relatively low pressure
2. It expands; The air pressure decreases as altitude increases.
3. The pressure is three times as great in cylinder A.
4. 3:2
5. a. 9.0×10^2 K
   b. 1.2×10^3 K

CHAPTER 9
Summary

Teaching Tip

Ask students to prepare a concept map of the chapter. The concept map should include most of the vocabulary terms, along with other integral terms or concepts.

KEY TERMS

buoyant force (p. 319)

fluid (p. 318)

ideal fluid (p. 332)

mass density (p. 319)

pressure (p. 325)

temperature (p. 331)

KEY IDEAS

Section 9-1 Fluids and buoyant force

- A fluid is a material that can flow, and thus it has no definite shape. Both gases and liquids are fluids.
- Buoyant force is an upward force exerted by a fluid on an object floating on or submerged in the fluid.
- The magnitude of a buoyant force for a submerged object is determined by Archimedes' principle and is equal to the weight of the displaced fluid.

Section 9-2 Fluid pressure and temperature

- Pressure is a measure of how much force is exerted over a given area.
- The pressure in a fluid increases with depth.

Section 9-3 Fluids in motion

- Moving fluids can exhibit laminar (smooth) flow or turbulent flow.
- An ideal fluid is incompressible, nonviscous, and nonturbulent.
- According to the continuity equation, the amount of fluid leaving a pipe during some time interval equals the amount entering the pipe during that same time interval.
- According to Bernoulli's principle, swift-moving fluids exert less pressure than slower-moving fluids.

Section 9-4 Properties of gases

- An ideal gas obeys the ideal gas law. The ideal gas law relates the volume, pressure, and temperature of a gas confined to a container.

⭐TEKS

Review & Assess
pp. 342–349:
2A: Alt. Assess. 1, 3
2B: Alt. Assess. 1
2C: 8–9, 16–19, 23–24, 29–69, Technology & Learning
2D: Alt. Assess. 1, 4
2F: Alt. Assess. 3
3A: Alt. Assess. 2, 4, 5
3B: 8–9, 16–19, 23–24, 29–69, Technology & Learning
3C: Alt. Assess. 3, 4, 5
3E: Alt. Assess. 4

Variable symbols

Quantities		Units		Conversions
ρ	density	kg/m^3	kilogram per meter^3	$= 10^{-3} \, g/cm^3$
P	pressure	Pa	pascal	$= N/m^2$
				$= 10^{-5} \, atm$
T	temperature	K	kelvin	
		°C	degrees Celsius	$= K - 273$
k_B	Boltzmann's constant	J/K	joules per kelvin	

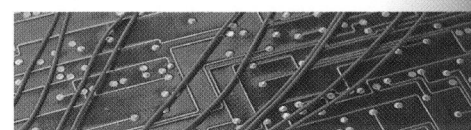

DENSITY AND BUOYANCY

Conceptual questions

1. If an inflated beach ball is placed beneath the surface of a pool of water and released, it shoots upward, out of the water. Use Archimedes' principle to explain why.

2. Will an ice cube float higher in water or in mercury? (Hint: See **Table 9-1,** on page 319.)

3. An ice cube is submerged in a glass of water. What happens to the level of the water as the ice melts?

4. Will a ship ride higher in an inland freshwater lake or in the ocean? Why?

5. Steel is much denser than water. How, then, do steel boats float?

6. A small piece of steel is tied to a block of wood. When the wood is placed in a tub of water with the steel on top, half the block is submerged. If the block is inverted so that the steel is underwater, will the amount of the wooden block that is submerged increase, decrease, or remain the same?

7. A fish rests on the bottom of a bucket of water while the bucket is being weighed. When the fish begins to swim around, does the reading on the scale change?

Practice problems

8. An object weighs 315 N in air. When tied to a string, connected to a balance, and immersed in water, it weighs 265 N. When it is immersed in oil, it weighs 269 N. Find the following:
 a. the density of the object
 b. the density of the oil
(See Sample Problem 9A.)

9. A sample of an unknown material weighs 300.0 N in air and 200.0 N when submerged in an alcohol solution with a density of 0.70×10^3 kg/m^3. What is the density of the material?
(See Sample Problem 9A.)

PRESSURE

Conceptual questions

10. After a long class, a physics teacher stretches out for a nap on a bed of nails. How is this possible?

11. If you lay a steel needle horizontally on water, it will float. If you place the needle vertically into the water, it will sink. Explain why.

12. A typical silo on a farm has many bands wrapped around its perimeter, as shown in **Figure 9-18.** Why is the spacing between successive bands smaller toward the bottom?

13. Which dam must be stronger, one that holds back 1.0×10^5 m^3 of water 10 m deep or one that holds back 1000 m^3 of water 20 m deep?

Figure 9-18

14. In terms of the kinetic theory of gases, explain why gases do the following:
 a. expand when heated
 b. exert pressure

15. When drinking through a straw, you reduce the pressure in your mouth and the atmosphere moves the liquid. Could you use a straw to drink on the moon?

ANSWERS TO

Chapter 9
Review and Assess

1. The weight of the water displaced is greater than the weight of the ball.

2. mercury, because $\rho_{mercury} > \rho_{water}$

3. The level falls because $\rho_{ice} < \rho_{water}$.

4. in the ocean; Because $\rho_{sea\ water} > \rho_{fresh\ water}$, F_B is greater in sea water.

5. because the average density of the boat, including air inside the hollow hull, is less than the density of the water

6. The amount of the block that is submerged will decrease. The volume of water displaced remains the same when the block is inverted. When the block is inverted, the steel occupies some of the displaced volume. Thus, the amount of wood displaced is less than half the block.

7. The weight may vary slightly because of impulses by the fish's tail, but the mass of the system will stay the same.

8. a. 6.3×10^3 kg/m^3
 b. 9.2×10^2 kg/m^3

9. 2.1×10^3 kg/m^3

10. The force opposing F_g is spread out over a large number of nails, so no single nail exerts very much pressure.

11. In the horizontal case, the weight of the needle is spread out along the length of the needle so that the surface tension is greater than the weight at each point.

12. because the pressure the grains exert increases with increasing depth

13. 20 m deep, because pressure increases with increasing depth

14. **a.** Molecules are moving faster, so the particles collide more often and, in the absence of any other force, push each other outward.

    b. As particles move about in a gas, they exert a force $(P = \frac{F}{A})$ on the walls of the container.

15. No, there would be no way to get the pressure in your mouth lower than the zero atmospheric pressure outside the liquid.

16. 1.9×10^4 N

17. 6.28 N

18. 14 N downward

19. **a.** 2.61×10^6 Pa

    b. 1.84×10^5 N

20. The air flow causes the pressure over the entrance in the mound to be lower than the pressure over the other entrance. Thus, air is pushed through the mound by the higher-pressure area.

21. The water on the first floor has only kinetic energy $(\frac{1}{2}\rho v_1^2)$, whereas the water on the second floor has both kinetic and potential energy $(\frac{1}{2}\rho v_2^2 + \rho gh)$. Thus, because energy is conserved, $v_1 > v_2$.

22. The moving air above the ball creates a low pressure area so that the air below the ball exerts a force that is equal and opposite F_g.

23. 2.4 m/s

24. 12.6 m/s

25. Pressure decreases, so volume increases.

Practice problems

16. The four tires of an automobile are inflated to an absolute pressure of 2.0×10^5 Pa. Each tire has an area of 0.024 m^2 in contact with the ground. Determine the weight of the automobile.
(See Sample Problem 9B.)

17. A pipe contains water at 5.00×10^5 Pa above atmospheric pressure. If you patch a 4.00 mm diameter hole in the pipe with a piece of bubble gum, how much force must the gum be able to withstand?
(See Sample Problem 9B.)

18. A piston, A, has a diameter of 0.64 cm, as in **Figure 9-19.** A second piston, B, has a diameter of 3.8 cm. In the absence of friction, determine the force, F, necessary to support the 500.0 N weight.
(See Sample Problem 9B.)

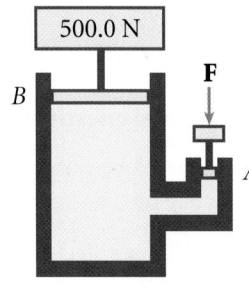

Figure 9-19

19. A submarine is at an ocean depth of 250 m.

    a. Calculate the absolute pressure at this depth. Assume that the density of water is 1.025×10^3 kg/m^3 and that atmospheric pressure is 1.01×10^5 Pa.
    (See Sample Problem 9C.)

    b. Calculate the magnitude of the total force exerted at this depth on a circular submarine window with a diameter of 30.0 cm.
    (See Sample Problem 9B.)

FLUID FLOW

Conceptual questions

20. Prairie dogs live in underground burrows with at least two entrances. They ventilate their burrows by building a mound around one entrance, which is open to a stream of air. A second entrance at ground level is open to almost stagnant air. Use Bernoulli's principle to explain how this construction creates air flow through the burrow.

21. Municipal water supplies are often provided by reservoirs built on high ground. Why does water from such a reservoir flow more rapidly out of a faucet on the ground floor of a building than out of an identical faucet on a higher floor?

22. If air from a hair dryer is blown over the top of a table-tennis ball, the ball can be suspended in air. Explain how this suspension is possible.

Practice problems

23. A dairy farmer notices that a circular water trough near the barn has become rusty and now has a hole near the base. The hole is 0.30 m below the level of the water that is in the tank. If the top of the trough is open to the atmosphere, what is the speed of the water as it leaves the hole?
(See Sample Problem 9D.)

24. The hypodermic syringe shown in **Figure 9-20** contains a medicine with the same density as water. The barrel of the syringe has a cross-sectional area of 2.50×10^{-5} m^2. The cross-sectional area of the needle is 1.00×10^{-8} m^2. In the absence of a force on the plunger, the pressure everywhere is atmospheric pressure. A 2.00 N force is exerted on the plunger, making medicine squirt from the needle. Determine the speed of the emerging fluid. Assume that the pressure in the needle remains at atmospheric pressure, that the syringe is horizontal, and that the speed of the emerging fluid is the same as the speed of the fluid in the needle.
(See Sample Problem 9D.)

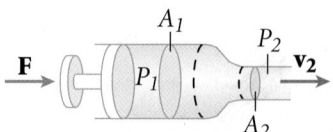

Figure 9-20

GASES AND THE IDEAL GAS LAW

Conceptual questions

25. Why do underwater bubbles grow as they rise?

26. What happens to a helium-filled balloon released into the air? Does it expand or contract? Does it stop rising at some height?

27. What increase of pressure is required to change the temperature of a sample of nitrogen by 1.00 percent? Assume the volume of the sample remains constant.

28. A balloon filled with air is compressed to half its initial volume. If the temperature inside the balloon remains constant, what happens to the pressure of the air inside the balloon?

Practice problems

29. An ideal gas is contained in a vessel of fixed volume at a temperature of 325 K and a pressure of 1.22×10^5 Pa. If the pressure is increased to 1.78×10^5 Pa, what is the final temperature of the gas?
(See Sample Problem 9E.)

30. The pressure in a constant-volume gas thermometer is 7.09×10^5 Pa at 100.0°C and 5.19×10^4 Pa at 0.0°C. What is the temperature when the pressure is 4.05×10^3 Pa?
(See Sample Problem 9E.)

MIXED REVIEW

31. An engineer weighs a sample of mercury ($\rho = 13.6 \times 10^3$ kg/m^3) and finds that the weight of the sample is 4.5 N. What is the sample's volume.

32. How much force does the atmosphere exert on 1.00 km^2 of land at sea level?

33. A 70.0 kg man sits in a 5.0 kg chair so that his weight is evenly distributed on the legs of the chair. Assume that each leg makes contact with the floor over a circular area with a radius of 1.0 cm. What is the pressure exerted on the floor by each leg?

34. A swimmer has 8.20×10^{-4} m^3 of air in his lungs when he dives into a lake. Assuming the pressure of the air is 95 percent of the external pressure at all times, what is the volume of the air at a depth of 10.0 m? Assume that the atmospheric pressure at the surface is 1.013×10^5 Pa, the density of the lake water is 1.00×10^3 kg/m^3, and the temperature is constant.

35. An air bubble has a volume of 1.50 cm^3 when it is released by a submarine 100.0 m below the surface of the sea. What is the volume of the bubble when it reaches the surface? Assume that the temperature of the air in the bubble remains constant during ascent.

36. A frog in a hemi-spherical bowl, as shown in **Figure 9-21,** just floats in a fluid with a density of 1.35×10^3 kg/m^3. If the bowl has a radius of 6.00 cm and negligible mass, what is the mass of the frog?

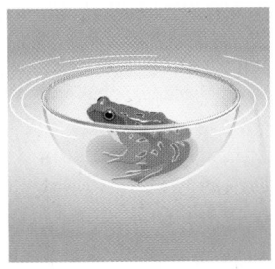

Figure 9-21

37. A circular swimming pool at sea level has a flat bottom and a 6.00 m diameter. It is filled with water to a depth of 1.50 m.
  a. What is the absolute pressure at the bottom?
  b. Two people with a combined mass of 150 kg float in the pool. What is the resulting increase in the average absolute pressure at the bottom?

38. The wind blows with a speed of 30.0 m/s over the roof of your house.
  a. Assuming the air inside the house is relatively stagnant, what is the pressure difference at the roof between the inside air and the outside air?
  b. What net force does this pressure difference produce on a roof having an area of 175 m^2?

39. A bag of blood with a density of 1050 kg/m^3 is raised about 1.00 m higher than the level of a patient's arm. How much greater is the blood pressure at the patient's arm than it would be if the bag were at the same height as the arm?

40. The density of helium gas at 0.0°C is 0.179 kg/m^3. The temperature is then raised to 100.0°C, but the pressure is kept constant. Assuming that helium is an ideal gas, calculate the new density of the gas.

41. When a load of 1.0×10^6 N is placed on a battleship, the ship sinks only 2.5 cm in the water. Estimate the cross-sectional area of the ship at water level. (Hint: See **Table 9-1** for the density of sea water.)

26. It expands as it rises and stops expanding when the density of the air equals the total average density of the helium and the balloon. Because of the mass of the balloon itself, the density of the helium will be slightly less than the density of the air at that height.
27. 1.00 percent
28. The pressure doubles.
29. 474 K
30. 21.3 K
31. 3.4×10^{-5} m^3
32. 1.01×10^{11} N
33. 5.9×10^5 Pa
34. 4.2×10^{-4} m^3
35. 16.5 cm^3
36. 6.11×10^{-1} kg
37. a. 1.16×10^5 Pa
      b. 52 Pa
38. a. 5.80×10^2 Pa
      b. 1.02×10^5 N upward
39. 1.03×10^4 Pa
40. 0.131 kg/m^3
41. 4.0×10^3 m^2

42. A weather balloon is designed to expand to a maximum radius of 20.0 m when the air pressure is 3.0×10^3 Pa and the temperature of the air surrounding it is 200.0 K. If the balloon is filled at a pressure of 1.01×10^5 Pa and 300.0 K, what is the radius of the balloon at the time of liftoff?

43. A 1.0 kg beaker containing 2.0 kg of oil with a density of 916 kg/m^3 rests on a scale. A 2.0 kg block of iron is suspended from a spring scale and completely submerged in the oil, as shown in **Figure 9-22.** Find the equilibrium readings of both scales. (Hint: See **Table 9-1** for the density of iron.)

Figure 9-22

44. A raft is constructed of wood having a density of 600.0 kg/m^3. The surface area of the bottom of the raft is 5.7 m^2, and the volume of the raft is 0.60 m^3. When the raft is placed in fresh water having a density of 1.0×10^3 kg/m^3, how deep is the bottom of the raft below water level?

45. Before beginning a long trip on a hot day, a driver inflates an automobile tire to a gauge pressure of 1.8 atm at 293 K. At the end of the trip, the gauge pressure in the tire has increased to 2.1 atm.

   a. Assuming the volume of the air inside the tire has remained constant, what is its temperature at the end of the trip?

   b. What volume of air (measured at atmospheric pressure) should be released from the tire so that the pressure returns to its initial value? Assume that the air is released during a short time interval during which the temperature remains at the value found in part **(a).** Write your answer in terms of the initial volume, V_i.

46. A cylindrical diving bell 3.0 m in diameter and 4.0 m tall with an open bottom is submerged to a depth of 220 m in the ocean. The temperature of the air at the surface is 25°C, and the air's temperature 220 m down is 5.0°C. The density of sea water is 1025 kg/m^3. How high does the sea water rise in the bell when the bell is submerged?

47. A physics book has a height of 26 cm, a width of 21 cm, and a thickness of 3.5 cm.

   a. What is the density of the physics book if it weighs 19 N?

   b. Find the pressure that the physics book exerts on a desktop when the book lies face up.

   c. Find the pressure that the physics book exerts on the surface of a desktop when the book is balanced on its spine.

48. A jet of water squirts horizontally from a hole near the bottom of the tank as shown in **Figure 9-23.** If the hole has a diameter of 3.50 mm and the top of the tank is open, what is the height of the water in the tank?

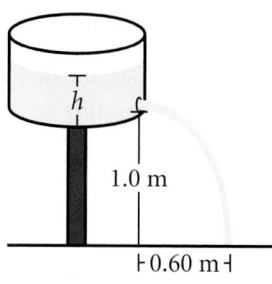

Figure 9-23

49. A water tank open to the atmosphere at the top has two holes punched in its side, one above the other. The holes are 5.00 cm and 12.0 cm above the ground. What is the height of the water in the tank if the two streams of water hit the ground at the same place?

50. A hydraulic brake system is shown in **Figure 9-24.** The area of the piston in the master cylinder is 6.40 cm^2, and the area of the piston in the brake cylinder is 1.75 cm^2. The coefficient of friction between the brake shoe and the wheel drum is 0.50. Determine the frictional force between the brake shoe and the wheel drum when a force of 44 N is exerted on the pedal.

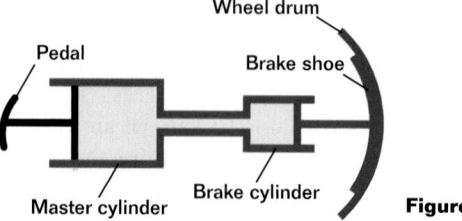

Figure 9-24

51. A natural-gas pipeline with a diameter of 0.250 m delivers 1.55 m^3 of gas per second. What is the flow speed of the gas?

52. A 2.0 cm thick bar of soap is floating in water, with 1.5 cm of the bar underwater. Bath oil with a density of 900.0 kg/m^3 is added and floats on top of the water. What is the depth of the oil layer when the top of the soap is just level with the upper surface of the oil?

53. Oil having a density of 930 kg/m^3 floats on water. A rectangular block of wood 4.00 cm high and with a density of 960 kg/m^3 floats partly in the oil and partly in the water. The oil completely covers the block. How far below the interface between the two liquids is the bottom of the block?

54. A block of wood weighs 50.0 N in air. A sinker is hanging from the block, and the weight of the wood-sinker combination is 200.0 N when the sinker alone is immersed in water. When the wood-sinker combination is completely immersed, the weight is 140.0 N. Find the density of the block.

55. In a time interval of 1.0 s, 5.0×10^{23} nitrogen molecules strike a wall area of 8.0 cm^2. The mass of one nitrogen molecule is 4.68×10^{-26} kg. If the molecules move at 300.0 m/s and strike the wall head-on in an elastic collision, what is the pressure exerted on the wall?

56. **Figure 9-25** shows a water tank with a valve at the bottom. If this valve is opened, what is the maximum height attained by the water stream coming out of the right side of the tank? Assume that $h = 10.0$ m, $L = 2.0$ m, and $\theta = 30.0°$ and that the cross-sectional area at A is very large compared with that at B.

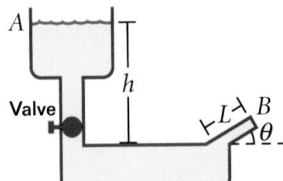

Figure 9-25

57. An air bubble originating from a deep-sea diver has a radius of 2.0 mm at the depth of the diver. When the bubble reaches the surface of the water, it has a radius of 3.0 mm. Assuming that the temperature of the air in the bubble remains constant, determine the following:
   a. the depth of the diver
   b. the absolute pressure at this depth

58. Water flows through a 0.30 m radius pipe at the rate of 0.20 m^3/s. The pressure in the pipe is atmospheric. The pipe slants downhill and feeds into a second pipe with a radius of 0.15 m, positioned 0.60 m lower. What is the gauge pressure in the lower pipe?

59. A light spring with a spring constant of 90.0 N/m rests vertically on a table, as shown in **Figure 9-26(a)**. A 2.00 g balloon is filled with helium (0°C and 1 atm pressure) to a volume of 5.00 m^3 and connected to the spring, causing the spring to stretch, as in **Figure 9-26(b)**. How much does the spring stretch when the system is in equilibrium? (Hint: See **Table 9-1** for the density of helium. The magnitude of the spring force equals $k\Delta x$.)

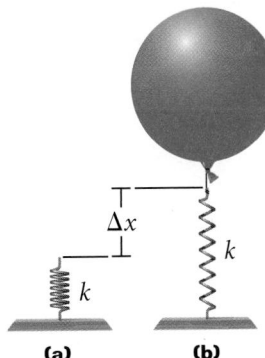

(a)   (b)   **Figure 9-26**

60. The aorta in an average adult has a cross-sectional area of 2.0 cm^2.
   a. Calculate the flow rate (in grams per second) of blood ($\rho = 1.0$ g/cm^3) in the aorta if the flow speed is 42 cm/s.
   b. Assume that the aorta branches to form a large number of capillaries with a combined cross-sectional area of 3.0×10^3 cm^2. What is the flow speed in the capillaries?

61. The approximate inside diameter of the aorta is 1.6 cm, and that of a capillary is 1.0×10^{-6} m. The average flow speed is about 1.0 m/s in the aorta and 1.0 cm/s in the capillaries. If all the blood in the aorta eventually flows through the capillaries, estimate the number of capillaries in the circulatory system.

62. A cowboy at a ranch fills a water trough that is 1.5 m long, 65 cm wide, and 45 cm deep. He uses a hose having a diameter of 2.0 cm, and the water emerges from the hose at 1.5 m/s. How long does it take the cowboy to fill the trough?

52. 5×10^{-2} m
53. 1.71×10^{-2} m
54. 833 kg/m^3
55. 1.8×10^4 Pa
56. 2.2 m above the spout opening
57. **a.** 24 m
   b. 3.4×10^5 Pa
58. 2.2×10^3 Pa
59. 0.605 m
60. **a.** 84 g/s
   b. 2.8×10^{-4} m/s
61. 2.6×10^{10} capillaries
62. 9.3×10^2 s

63. 60.9 m/s^2
64. 6.3 m
65. 15 m
66. $1.68 \times 10^5 \text{ Pa}$

63. A light balloon is filled with helium at 0.0°C and 1.0 atm and then released from the ground. Determine its initial acceleration. Disregard the air resistance on the balloon. (Hint: See **Table 9-1** for information on the densities of helium and air.)

64. A 1.0 kg hollow ball with a radius of 0.10 m is filled with air and is released from rest at the bottom of a 2.0 m deep pool of water. How high above the water does the ball rise? Disregard friction and the ball's motion when it is only partially submerged.

65. A small ball 0.60 times as dense as water is dropped from a height of 10.0 m above the surface of a smooth lake. Determine the maximum depth to which the ball will sink. Disregard any energy transferred to the water during impact and sinking.

66. A sealed glass bottle at 27°C contains air at a pressure of 1.01×10^5 Pa and has a volume of 30.0 cm^3. The bottle is tossed into an open fire. When the temperature of the air in the bottle reaches 225°C, what is the pressure inside the bottle? Assume the volume of the bottle is constant.

Technology & Learning

Graphing calculators

Refer to Appendix B for instructions on downloading programs for your calculator. The program "Chap9" builds a table of flow rates for various hose diameters and flow speeds.

Flow rate, as you learned earlier in this chapter, is described by the following equation:

$$\text{flow rate} = Av$$

The program "Chap9" stored on your graphing calculator makes use of the equation for flow rate. Once the "Chap9" program is executed, your calculator will ask for the flow speed. The graphing calculator will use the following equation to build a table of flow rates (Y$_1$) versus hose diameters (X).

$$Y_1 = \pi * V(X/2)^2$$

Note that the relationships in this equation are the same as those in the flow-rate equation shown above.

a. Using the variables used on the graphing calculator, write the expression for the cross-sectional area of the hose.

Execute "Chap9" on the PRGM menu, and press ENTER to begin the program. Enter the value for the flow speed of the liquid (shown in items b–g below), and press ENTER.

The calculator will provide a table of flow rates in cm^3/s versus hose diameters in cm. Scroll down the table to find the values you need. Press ENTER only when you are ready to exit the table.

Determine the flow rates in each of the following situations (b–f):

b. a 2.0 cm garden hose with water traveling through it at 25 cm/s
c. a 4.5 cm diameter fire hose with water traveling through it at 275 cm/s
d. a 2.5 cm diameter fire hose with water traveling through it at 275 cm/s
e. a 3.5 cm diameter fire hose with water traveling through it at 425 cm/s
f. a 5.5 cm diameter fire hose with water traveling through it at 425 cm/s
g. Hose A has a diameter that is twice as large as the diameter of hose B. How many times larger is the flow rate in A than the flow rate in B?

Press ENTER to exit the table. Press ENTER to enter a new value or CLEAR to end the program.

ANSWERS TO

Technology & Learning
a. $\pi(X/2)^2$
b. $79 \text{ cm}^3/\text{s}$
c. $4.4 \times 10^3 \text{ cm}^3/\text{s}$
d. $1.3 \times 10^3 \text{ cm}^3/\text{s}$
e. $4.1 \times 10^3 \text{ cm}^3/\text{s}$
f. $1.0 \times 10^4 \text{ cm}^3/\text{s}$
g. The flow rate of A is four times the flow rate of B.

67. In testing a new material for shielding spacecraft, 150 ball bearings each moving at a supersonic speed of 400.0 m/s collide head-on and elastically with the material during a 1.00 min interval. If the ball bearings each have a mass of 8.0 g and the area of the tested material is 0.75 m², what is the pressure exerted on the material?

68. A thin, rigid, spherical shell with a mass of 4.00 kg and diameter of 0.200 m is filled with helium at 0°C and 1 atm pressure. It is then released from rest on the bottom of a pool of water that is 4.00 m deep.

    a. Determine the upward acceleration of the shell.

    b. How long will it take for the top of the shell to reach the surface? Disregard frictional effects.

69. A light spring with a spring constant of 16.0 N/m rests vertically on the bottom of a large beaker of water, as shown in **Figure 9-27(a)**. A 5.00×10^{-3} kg block of wood with a density of 650.0 kg/m³ is connected to the spring, and the mass-spring system is allowed to come to static equilibrium, as shown in **Figure 9-27(b)**. How much does the spring stretch?

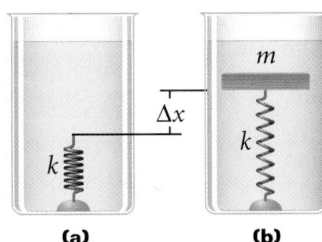

(a)   (b)   **Figure 9-27**

Alternative Assessment

Performance assessment

1. Build a hydrometer from a long test tube with some sand at the bottom and a stopper. Adjust the amount of sand as needed so that the tube floats in most liquids. Calibrate it, and place a label with markings on the tube. Measure the densities of the following liquid foods: skim milk, whole milk, vegetable oil, pancake syrup, and molasses. Summarize your findings in a chart or table.

2. Explain how you can use differences in pressure to measure changes in altitude, assuming air density is constant. How accurate would your barometer need to be to provide an answer accurate to one meter for a 55 m tall building, a 255 m tall building, and a 2200 m tall mountain?

3. The owner of a fleet of tractor-trailers has contacted you after a series of accidents involving tractor-trailers passing each other on the highway. The owner wants to know how drivers can minimize the pull exerted as one tractor-trailer passes another going in the same direction. Should the passing tractor-trailer try to pass as quickly as possible or as slowly as possible? Design experiments to deter-

mine the answer by using model motor boats in a swimming pool. Indicate exactly what you will measure and how. If your teacher approves your plan and you are able to locate the necessary equipment, perform the experiment.

Portfolio projects

4. Record any examples of pumps in the tools, machines, and appliances you encounter in one week, and briefly describe the appearance and function of each pump. Research how one of these pumps works, contrasting it with a water pump in a car and an air pump used by bicyclists. Share your findings in a group meeting and create a presentation, model, or diagram that summarizes the group's findings.

5. You have been hired as a consultant to help the instructors of a diving school. They want you to develop materials that explain the physics involved in the following diving-safety rules: diving tanks must be kept immersed in cold water while they are being filled, and divers must exhale while ascending to the surface. Use the gas laws to explain what dangers (if any) these rules are designed to prevent.

CHAPTER 9
Laboratory Exercise

⭐TEKS

pp. 350–353: 1A,
2B, 2C, 2D, 2E, 2F,
3A, 3B

NOTE

Materials Preparation is given on pp. 316A–316B. Blank data table and sample data table are on the One-Stop Planner CD-ROM. All calculations are performed using sample data.

Planning

Recommended time:

1 lab period

Classroom organization:

▶ This lab may be performed by students alone or in pairs.

▶ The CBL and sensors procedure and the Boyle's law apparatus may be used in the same class.

▶ **Safety warnings:** This is a low-risk laboratory experiment. Remind students to follow all standard laboratory rules and procedures.

OBJECTIVES

• Measure the volume and pressure of a gas at constant temperature.

• Explore the relationships between the volume and pressure of a gas.

MATERIALS LIST

✔ **Check list for appropriate procedure.**

PROCEDURE

CBL AND SENSORS

✔ **CBL**

✔ **graphing calculator with link cable**

✔ **CBL pressure sensor with CBL-DIN adapter and syringe**

✔ **airline tubing (10 cm)**

BOYLE'S LAW APPARATUS

✔ **Boyle's law apparatus**

✔ **set of five 1 kg masses**

BOYLE'S LAW

The ideal gas law states the relationship between the temperature, pressure, and volume of a confined ideal gas. At room temperature and atmospheric pressure, air behaves nearly like an ideal gas. In this lab, you will hold the temperature constant and explore the relationship between the volume and pressure of a fixed amount of air at a constant temperature. Because the air will be contained in an airtight syringe, the quantity of gas will be constant throughout the experiment.

You will perform this experiment using either a CBL with pressure sensor or the Boyle's law apparatus.

• **CBL and sensors** You will use the pressure sensor to measure the pressure of the air at different volumes, starting with an initial volume of 10 cm^3 and decreasing by 1 cm^3 increments. You will graph your data and analyze the graphs to find the relationship between pressure and volume for a gas.

• **Boyle's law apparatus** You will increase the pressure on a fixed quantity of air in a syringe by adding weight to the end of the plunger. As the pressure is increased, you will measure the change in volume using the markings on the syringe. You will graph your data and analyze the graphs to find the relationship between pressure and volume for a gas.

SAFETY

• Tie back long hair, secure loose clothing, and remove loose jewelry to prevent their getting caught in moving or rotating parts.

• Wear eye protection. Contents under pressure may become projectiles and cause serious injury.

PREPARATION

1. Determine whether you will be using the CBL and sensors procedure or the Boyle's law apparatus. Read the entire lab for the appropriate procedure, and plan what steps you will take.

Boyle's law apparatus procedure begins on page 352.

PROCEDURE

CBL AND SENSORS

Pressure and volume of a gas

2. Prepare a data table in your lab notebook with four columns and nine rows. In the first row, label the first four columns *Volume (cm³)*, *Trial 1 Pressure (atm)*, *Trial 2 Pressure (atm)*, and *Trial 3 Pressure (atm)*.

3. Set up the pressure sensor, CBL, and graphing calculator as shown in **Figure 9-28**. Connect the pressure sensor to the CH1 port on the CBL. Turn on the CBL and the graphing calculator.

4. Start the program PHYSICS on the calculator. Select option *SET UP PROBES* from the MAIN MENU. Enter 1 for the number of probes. Select the pressure sensor from the list. Enter 1 for the channel number. Select *USE STORED* from the CALIBRATION menu. Select *ATM* from the PRESSURE UNITS menu.

5. Select the *COLLECT DATA* option from the MAIN MENU. Select the *TRIGGER* option from the DATA COLLECTION menu.

6. Turn the stopcock so that the valve handle points downward. This will allow the syringe to fill with air. Pull the plunger on the syringe to the 10 cm³ mark. Turn the stopcock so that the valve handle points upward to set the volume of air in the syringe to 10 cm³. Attach one end of the tubing to the end of the syringe and attach the other end to the pressure sensor.

7. Slowly push in the plunger until the volume of air in the syringe is near 8 cm³. Press TRIGGER on the CBL to collect the pressure reading at this volume. Record the reading from the CBL under *Trial 1 Pressure* in your data table. Read the volume, and record it in your data table. Select *CONTINUE* from the TRIGGER menu on the graphing calculator.

8. Slowly push in the plunger until the volume of air in the syringe is near 7 cm³. Press TRIGGER on the CBL to collect the pressure reading. Record the reading under *Trial 1 Pressure* in your data table. Read the volume, and record it in your data table. Select *CONTINUE* on the graphing calculator.

9. Repeat several times, decreasing the volume each time by 1 cm³ until the volume is 2 cm³.

10. Repeat steps 4–9 twice and record the new values in the data table under *Trial 2* and *Trial 3*.

11. Clean up your work area. Put equipment away safely so that it is ready to be used again.

Analysis and Interpretation begins on page 353.

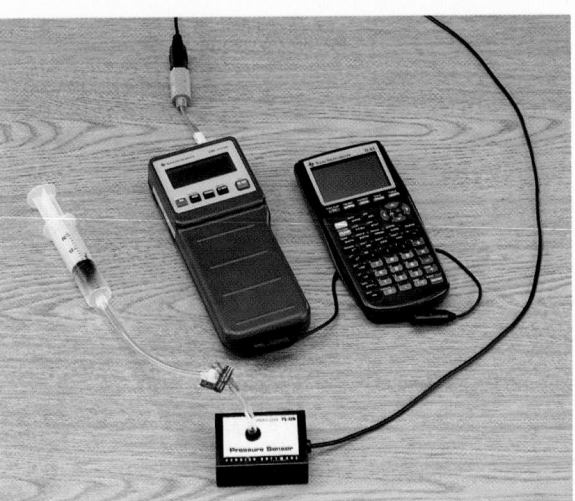

Figure 9-28
Step 6: Set the volume of air in the syringe and attach the tubing securely to the syringe and pressure sensor.
Step 7: Record each volume in your data table, and use the same volumes for each trial.
Step 9: Decrease the volume by 1 cm³ until you reach 2 cm³ or until the plunger will no longer move.

CBL and Sensors Tips

◆ Students should have the program PHYSICS on their graphing calculators. Refer to Appendix B for instructions.

◆ The CBL will measure the pressure in atm (atmospheres).

Techniques to Demonstrate

Demonstrate turning the stopcock to operate the valve on the syringe. Show students how the valve works, and remind them that it is important to open the valve so that the pressure of the initial volume of air is the same as the pressure of the air in the room. Demonstrate how closing the valve seals the syringe to keep the air in the syringe throughout the experiment.

✔ Checkpoints

Step 6: Make sure students attach the tubing securely.

Step 7: Remind students to record both the volume and the pressure readings. Make sure students read the volume reading on the syringe and do not simply record the value given in the procedure.

Step 9: Students may find that it is easy to push the plunger in beyond the 2 cm³ mark. If so, they may want to continue as far as they can. However, do not allow students to force the plunger when it becomes extremely difficult to push.

Boyle's Law Apparatus Tips

◆ The syringe has a removable plastic cap beneath the lower block. To ensure low friction, remove the cap, pull the piston out, *lightly* lubricate the cylinder wall with silicone grease, and reinsert the piston.

◆ The mass load on the piston should not exceed 5 kg; masses greater than 5 kg cause the piston to twist and become unstable.

Techniques to Demonstrate

Show students how to replace the cap while holding the piston firmly in place.

Demonstrate twisting the piston without pushing it down or pulling it up. Explain that this helps keep the friction between the piston and the cylinder from opposing the motion of the piston.

✔ Checkpoints

Step 4: Make sure students twist the piston without exerting force on the air in the syringe.

PROCEDURE

BOYLE'S LAW APPARATUS

Pressure and volume of a gas

2. Prepare a data table in your lab notebook with four columns and six rows. In the first row, label the first four columns *Number of weights*, *Trial 1 Volume (cm³)*, *Trial 2 Volume (cm³)*, and *Trial 3 Volume (cm³)*. In the first column, label the second through sixth rows *0, 1, 2, 3,* and *4*.

3. Remove the plastic cap, and adjust the piston head so that it reads between 30 cm³ and 35 cm³. See **Figure 9-29.**

4. While holding the piston in place, carefully replace the cap. Twist the piston several times to allow the head to overcome any frictional forces.

5. When the piston comes to rest, read the volume to the nearest 0.25 cm³. Record this value as the volume for zero weight in your data table.

6. Carefully place one 1 kg mass on the piston. Twist the piston several times.

7. When the piston comes to rest, read the volume and record it in your data table.

8. Carefully add another 1 kg mass to the piston so that the total mass on the piston is 2 kg. Twist the piston several times.

9. When the piston comes to rest, read the volume and record it in your data table.

10. Carefully add another 1 kg mass to the piston so that the total mass on the piston is 3 kg. Twist the piston several times.

11. When the piston comes to rest, read the volume and record it in your data table.

12. Add another 1 kg mass to the piston. Twist the piston several times. When the piston comes to rest, read the volume and record it in your data table.

13. Repeat steps 3–12 twice and record the new values in the data table under *Trial 2* and *Trial 3.*

14. Clean up your work area. Put equipment away safely so that it is ready to be used again.

Analysis and Interpretation begins on page 353.

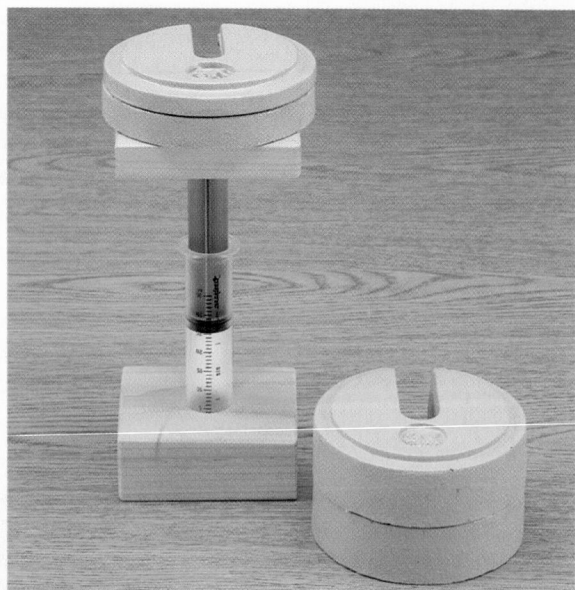

Figure 9-29

Step 3: Adjust the piston head to set the volume of air in the syringe between 30 cm³ and 35 cm³.

Step 5: Wait for the piston to come to rest before reading the volume.

ANALYSIS AND INTERPRETATION

Calculations and data analysis

1. **Organizing data**   Using data from all three trials, make the following calculations:

   a. **CBL and sensors**   Calculate the average pressure at each volume of air in the syringe.

   b. **Boyle's law apparatus**   Calculate the average volume for each level of weight on the piston.

2. **Organizing data**   Calculate the inverse of each of the pressures or volumes from item 1.

3. **Graphing data**   Plot a graph using the calculated averages from item 1. Use a graphing calculator, computer, or graph paper.

   a. **CBL and sensors**   Graph *Average Pressure* versus *Volume*.

   b. **Boyle's law apparatus**   Graph *Average Volume* versus *Number of Weights*.

4. **Graphing data**   Using a graphing calculator, computer, or graph paper, make a second graph using the inverse averages from item 2.

   a. **CBL and sensors**   Plot a graph of the *Inverse Average Pressure* versus *Volume*.

   b. **Boyle's law apparatus**   Plot a graph of the *Inverse Average Volume* versus *Number of Weights*.

Conclusions

5. **Analyzing graphs**   Based on your graphs, what is the relationship between the volume and the pressure of a gas held at a constant temperature? Explain how your graphs support your answer.

6. **Evaluating methods**

   a. **CBL and sensors**   Because the pressure sensor was not calibrated to your altitude, the pressure readings do not accurately reflect the pressure of the gas in the syringe. Explain why calibrating the probe is unnecessary for finding the relationship between the pressure readings and the volume of the gas.

   b. **Boyle's law apparatus**   In plotting the graphs, the number of weights was used instead of the amount of pressure. Explain how the weight on the piston serves as a measure of the pressure of the air inside the syringe.

ANSWERS TO

Analysis and Interpretation

CALCULATIONS AND DATA ANALYSIS

1. a. For sample data, values range from 1.14 atm to 2.19 atm.

   b. For sample data, values range from 15.17 cm^3 to 29.00 cm^3.

2. a. For sample data, values range from 0.457 atm^{-1} to 0.877 atm^{-1}.

   b. For sample data, values range from 0.0345 cm^{-3} to 0.0659 cm^{-3}.

3. a, b. Student graphs should show a straight line pointing down and to the right.

4. a, b. Student graphs should show a straight line pointing up and to the right.

CONCLUSIONS

5. At a constant temperature, the volume of gas decreases as the pressure increases.

6. a. **CBL and sensors** The pressure sensor is able to measure the difference between one pressure reading and another.

   b. **Boyle's law apparatus** Students should recognize that because equal masses were used, the weight applied to the gas is directly proportional to the number of masses. The pressure is a direct result of the weight.

1690

Physics and Its World *Timeline 1690–1785*

1695 – The Ashanti, the last of the major African kingdoms, emerges in what is now Ghana. The Ashanti's strong centralized government and effective bureaucracy enable them to control the region for nearly two centuries.

1700

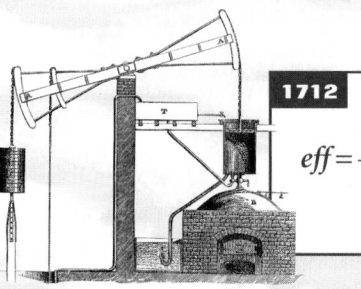

1712

$$eff = \frac{W_{net}}{Q_h}$$

Thomas Newcomen invents the first practical steam engine. Over 60 years later, **James Watt** makes significant improvements to the Newcomen engine.

1715 – Chinese writer **Ts'ao Chen** is born. His book *The Dream of the Red Chamber* is widely regarded today as the greatest Chinese novel.

1710

1721 – **Johann Sebastian Bach** composes the six *Brandenburg Concertos.*

1720

1738 – Under the leadership of **Nadir Shah,** the Persian Empire expands into India as the Moghul Empire enters a stage of decline.

1735 – **John Harrison** constructs the first of four chronometers that will allow navigators to accurately determine a ship's longitude.

1730

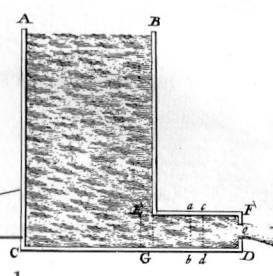

1738

$$P + \frac{1}{2}\rho v^2 + \rho gh = \text{constant}$$

Daniel Bernoulli's *Hydrodynamics,* which includes his research on the mechanical behavior of fluids, is published.

1740

354

1752

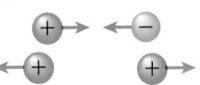

Benjamin Franklin builds on the first studies of electricity performed earlier in the century by describing electricity as having positive and negative charge. He also performs the dangerous "kite experiment," in which he demonstrates that lightning consists of electric charge.

1744 – Contrary to the favored idea that heat is a fluid, Russian chemist **Mikhail V. Lomonosov** suggests that heat is the result of motion. Four years later, Lomonosov formulates conservation laws for mass and energy.

1740

1750

1756 – The Seven Years War begins.

1757 – German musician **William Herschel** emigrates to England to avoid fighting in the Seven Years War. Over the next 60 years he pursues astronomy, constructing the largest reflecting telescopes of the era and discovering new objects such as binary stars and the planet Uranus.

1760

1770 – **Antoine Laurent Lavoisier** begins his research on chemical reactions, notably oxidation and combustion.

1770

1782 – **Caroline Herschel**, sister of astronomer **William Herschel**, joins her brother in England. She compiles the most comprehensive star catalog of the era, and discovers several nebulous objects that are eventually recognized as galaxies.

1775 – The American Revolution begins.

1780

1785

$$F_{electric} = k_c\left(\frac{q_1 q_2}{r^2}\right)$$

Charles Augustin Coulomb begins a series of experiments that will systematically and conclusively prove the inverse-square law for electric force. The law has been suggested for over 30 years by scientists such as **Daniel Bernoulli, Joseph Priestly,** and **Henry Cavendish.**

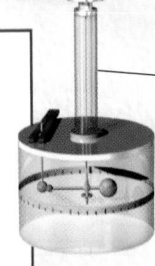

1790

CHAPTER 10 PLANNING GUIDE

Compression Guide: To shorten from 12 to 8 45-min periods (from 5½ to 4 90-min blocks), eliminate items in magenta type.

PACING CHART	CLASSROOM RESOURCES			
	ⓉⒺⒿⓀ**TEKS**	**Teacher Demonstrations**	**Holt Physics Transparencies**	**Labs** (See page T52 for equipment listing for in-text labs.)
10-1 Temperature and thermal equilibrium 2 or 1 45-minute periods 1 or 1½ 90-minute block	**1A, 2C, 2D, 3A, 3B, 5B**	**TE** *Temperature and internal energy*, p. 359	**T** 34 **TM** 29–31	**L** *Discovery Lab: Temperature and Internal Energy* **PE** *Quick Lab: Sensing Temperature*, p. 358
10-2 Defining heat 2 45-minute periods 1 90-minute block	**1A, 2C, 3A, 3B, 5B, 5D**	**TE** *Internal energy*, p. 368	**T** 35 **TM** 32	**PE** *Quick Lab: Work and Heat*, p. 368
10-3 Changes in temperature and phase 5 or 3 45-minute periods 2½ or 1½ 90-minute blocks	**2C, 3A, 3C, 5B, 5D, 7B**	**TE** *Lowering boiling points*, p. 379	**T** 36–38 **TM** 33–34	**PE** *Specific Heat Capacity*, p. 392 **L** *Invention Lab: Thermal Conduction* **TL** *Newton's Law of Cooling*
10-4 Controlling heat 1 45-minute period ½ 90-minute block	**3A, 3C, 7B**	**TE** *Thermal conduction*, p. 383		
Review and Assessment 2 45-minute periods 1 90-minute block				

Resource Key

PHYSICS *(Holt)*

PE Pupil's Edition
TE Teacher's Edition

L Laboratory Experiments
TL Technology Lab Experiments
T Transparencies

One-Stop Planner CD-ROM contents

TM Transparency Masters
SR Section Review Worksheets
AA Alternative Assessment

PW Problem-Solving Workbook
PB Problem Bank
CTW Critical Thinking Worksheet

LABORATORY PLANNING: Specific Heat Capacity, p. 392

Materials (for each lab group)
- balance: portable, electronic balance, 2000 g capacity or triple-beam balance with weight
- 2 low-form beakers, 600 mL
- hot plate
- metal shot:
  - 100 g tin
  - 500 g copper
  - 500 g zinc
  - 500 g aluminum
- double-wall metal calorimeter and stirring rod
- steam generator (metal heating vessel, including metal heating dipper)
- small plastic dish
- ice
- water

Safety Equipment
- safety hot mitt
- safety goggles or spectacles

ASSIGNMENT RESOURCES

Content Mastery	Critical Thinking	Problem-Solving Practice
PE 1–6, p. 364 **SR** 10-1, *Math Skills* **PE** 1–4, p. 387	**PE** 1–2, p. 360 **PE** 5–8, p. 387	**10A** Temperature conversion: 42 items in **PE**, **PW**, and **PB**, see **TE** p. 363
PE 1–4, p. 370 **SR** 10-2, *Concept Review* **PE** 11–13, p. 387	**PE** 14–18, pp. 387–388	**10B** Conservation of energy: 24 items in **PE**, **PW**, and **PB**, see **TE** p. 369
PE 1–6, p. 382 **SR** 10-3, *Graph Skills* **PE** 21–23, p. 388	**PE** 24–29, p. 388	**10C** Calorimetry: 27 items in **PE**, **PW**, and **PB**, see **TE** pp. 373–374 **10D** Heat of phase change: 26 items in **PE**, **PW**, and **PB**, see **TE** pp. 380–381
PE 1–3, p. 385 **SR** 10-4, *Concept Review* **PE** 34–35, p. 389	**PE** 36–40, p. 389	

ASSESSMENT RESOURCES

Cumulative Review	Alternative Assessment	Traditional Assessment
SR Mixed Review, Ch. 10	**PE** 1–6, p. 391 **AA** Items for Ch. 10	Chapter 10 Test Test Generator items for Ch. 10

Scoring Rubrics for Alternative Assessment items can be found on the One-Stop Planner CD-ROM.

TECHNOLOGY RESOURCES

 CTW Segment 11 Energy-Saving House
CTW Segment 12 Urban Heat Islands

 PE Technology and Learning, p. 390
(Alternative procedures for calculators without Flash-ROM technology are provided on the One-Stop Planner CD-ROM.)

 The Mechanical Universe/High School Adaptation Quad IV, Temperature and the Gas Laws

internetconnect

 On-line Student Resources:
GO TO: www.scilinks.org
The following *sci*LINKS Internet resources can be found in the student text for this chapter.

TOPICS:
•Temperature scales, p. 361 (HF2101)
•James Prescott Joule, p. 366 (HF2102)
•Specific heat, p. 372 (HF2103)
•Heat pumps, p. 376 (HF2104)
•Conduction and convection, p. 384 (HF2105)

 On-line Teacher Resources:
GO TO: go.hrw.com
KEYWORD: HF2 HOME
Visit the HRW Web site for a variety of resources related to this chapter.

 Smithsonian Institution
Internet Connections
Visit **www.si.edu/hrw** for additional on-line resources.

Visit **www.cnnfyi.com** for late-breaking news and current events stories selected just for you.

Additional Equipment (for each lab group)

CBL and Sensors Procedure
- CBL
- graphing calculator with link cable
- 2 temperature probes (If using only one probe, see special instructions in Teachers Notes.)
- TI Graph Link (recommended for downloading programs)

Thermometer Procedure
- dual magnifier
- red liquid thermometer, −20°C to 110°C
- fractional thermometer, −5°C to 50°C

Required Precautions
Use a hot mitt and wear safety goggles or spectacles when drying the samples in a lab oven.

Materials Preparation
Metal shot should be clean and dry. Between classes, dry the metal shot by placing it in a laboratory oven at low heat. Cool samples before class begins. Each type of metal should be labeled with a number, not the name of the metal. Record the sample numbers and their corresponding metals to check students' results. Metal samples should be placed in large beakers with spatulas for students to measure out. For best results, use crushed ice or clean snow.

Section 10-1 introduces the concepts of temperature, internal energy, and thermal equilibrium and identifies the Fahrenheit, Celsius, and Kelvin temperature scales.

Section 10-2 relates heat and temperature change to molecular motion and internal energy and demonstrates how the conservation of energy can be used to calculate internal energy.

Section 10-3 introduces specific heat capacity and latent heat and describes the relationship between the amount of energy transferred as heat and either temperature change or phase change.

Section 10-4 distinguishes between thermal conduction and thermal insulation and discusses ways to control energy transfer.

About the Illustration

Cooking popcorn is a familiar example of the transfer of energy as heat as well as a dramatic example of what happens when water rapidly undergoes a phase change to become steam. The high temperature of the steam and the increase in volume that water undergoes during vaporization cause the popcorn kernels to suddenly explode.

Note that the trajectories shown in this time exposure are parabolic, as predicted in Chapter 3.

CHAPTER 10

Heat

PHYSICS IN ACTION

Whether you pop corn by putting the kernels in a pan of hot oil or in a microwave oven, the hard kernels will absorb energy until, at a high temperature, they rupture. At this point, superheated water suddenly turns to steam and rushes outward, and the kernels burst open to form the fluffy, edible puffs of starch. But what actually happens when water turns into steam, and what do we mean when we talk about heat and temperature?

In this chapter you will study what distinguishes temperature and heat and how different substances behave when energy is added to or removed from them, causing a change in their temperature or phase.

- *Why do the kernels require steam, not just superheated water, to produce popcorn?*

- *What role does oil play in the preparation of popcorn?*

CONCEPT REVIEW

Work   (Section 5-1)

Energy   (Section 5-2)

Conservation of energy
   (Section 5-3)

Knowledge to Expect

✔ "Students learn that a warmer object can warm a cooler one by contact or at a distance and some materials conduct heat much better than others." (AAAS's *Benchmarks for Science Literacy*, grades 3–5)

✔ "Students learn that heat can be transferred through materials by the collisions of atoms." (AAAS's *Benchmarks for Science Literacy*, grades 6–8)

✔ "Heat moves in predictable ways, flowing from warmer objects to cooler ones, until both reach the same temperature." (NRC's *Science Education Standards*, grades 5–8)

Knowledge to Review

✔ Work is the product of force and displacement. (Section 5-1)

✔ Kinetic energy is the energy of an object due to its motion. Potential energy is the energy of an object due to its position. (Section 5-2)

✔ Conservation of mechanical energy states that in the absence of friction, the total mechanical energy of a system remains the same. (Section 5-3)

Items to Probe

✔ Familiarity with heat and friction: Ask what two metallic objects feel like after they have been rubbed together.

✔ Preconceptions about heat and temperature: Ask students what happens when warm things are put with cooler ones.

358

 Misconception Alert

Some students may confuse their perceptions of hot and cold with the temperature of an object; these students think that objects that feel hot have high temperatures and objects that feel cool have low temperatures. The discussion and Quick Lab on this page address this misconception.

Quick Lab

TEACHER'S NOTES

This experiment is meant to demonstrate that whether an object feels hot or cold is not a reliable indicator of the object's temperature. After the experiment, ask students to explain why this is the case (*perceived temperature depends on the temperature difference between the water and your hand*).

The experiment works best if the ice cubes have just melted and the mixture is stirred to ensure that all parts of the water have the same temperature.

10-1
Temperature and thermal equilibrium

10-1 SECTION OBJECTIVES

- **Relate temperature to the kinetic energy of atoms and molecules.**

- **Describe the changes in the temperatures of two objects reaching thermal equilibrium.**

- **Identify the various temperature scales, and be able to convert from one scale to another.**

DEFINING TEMPERATURE

When you hold a glass of lemonade with ice, like that shown in **Figure 10-1,** you feel a sharp sensation in your hand that we describe as "cold." Likewise, you experience a "hot" feeling when you touch a cup of hot chocolate. We often associate temperature with how hot or cold an object feels when we touch it. Our sense of touch serves as a qualitative indicator of temperature. However, this sensation of hot or cold also depends on the temperature of the skin and therefore is misleading. The same object may feel warm or cool, depending on the properties of the object and on the conditions of your body.

Determining an object's temperature with precision requires a standard definition of temperature and a procedure for making measurements that establish how "hot" or "cold" objects are. ⭐TEKS **3A**

Figure 10-1
Objects at low temperatures feel cold to the touch, while objects at high temperatures feel hot. However, the sensation of hot and cold can be misleading.

Adding or removing energy usually changes temperature

Consider what happens when you use an electric range to cook food. By turning the dial that controls the electric current delivered to the heating element, you can adjust the element's temperature. As the current is increased, the temperature of the element increases. Similarly, as the current is reduced, the temperature of the element decreases. In general, energy must be either added to or removed from a substance to change its temperature.

Quick Lab

Sensing Temperature

MATERIALS LIST

✔ 3 identical basins
✔ hot and cold tap water
✔ ice

 SAFETY CAUTION

Use only hot tap water. The temperature of the hot water must not exceed 50°C (122°F).

Fill one basin with hot tap water. Fill another with cold tap water, and add ice until about one-third of the mixture is ice.

Fill the third basin with an equal mixture of hot and cold tap water.

Place your left hand in the hot water and your right hand in the cold water for 15 s. Then place both hands in the basin of lukewarm water for 15 s. Describe whether the water feels hot or cold to either of your hands. ⭐TEKS **1A**

Temperature is proportional to the kinetic energy of atoms and molecules

In Section 9-2 you learned that temperature is proportional to the average kinetic energy of particles in a substance. A substance's temperature increases as a direct result of added energy being distributed among the particles of the substance, as shown in **Figure 10-2.**

For a monatomic gas, temperature can be understood in terms of the translational kinetic energy of the atoms in the gas. For other kinds of substances, molecules can rotate or vibrate, so rotational kinetic energy or vibrational kinetic and potential energies also exist (see **Table 10-1**). ⭐TEKS **5B**

The energies associated with atomic motion are referred to as **internal energy,** which is proportional to the substance's temperature. For an ideal gas, the internal energy depends only on the temperature of the gas. For gases with two or more atoms per molecule, as well as for liquids and solids, other properties besides temperature contribute to the internal energy. The symbol U stands for internal energy, and ΔU stands for a change in internal energy. ⭐TEKS **3B**

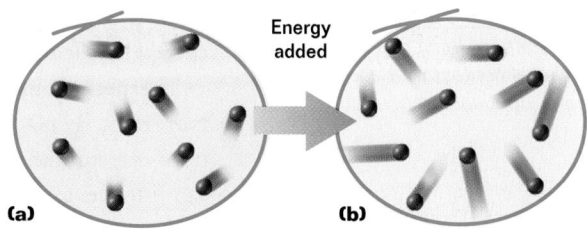

Figure 10-2
The low average kinetic energy of the particles **(a),** and thus the temperature of the gas, increases when energy is added to the gas **(b).**

internal energy

the energy of a substance due to the random motions of its component particles and equal to the total energy of those particles

Temperature is meaningful only when it is stable

Imagine a can of warm fruit juice immersed in a large beaker of cold water. After about 15 minutes, the can of fruit juice will be cooler and the water surrounding it will be slightly warmer. Eventually, both the can of fruit juice and

Table 10-1	Examples of different forms of energy		
Form of energy	**Macroscopic examples**	**Microscopic examples**	**Energy type**
Translational	airplane in flight, roller coaster at bottom of rise	CO_2 molecule in linear motion	kinetic energy
Rotational	spinning top	CO_2 molecule spinning about its center of mass	kinetic energy
Vibrational	plucked guitar string	bending and stretching of bonds between atoms in a CO_2 molecule	kinetic and potential energy

ANSWERS TO
Conceptual Challenge

1. b; If the final temperature were less than 50°C (a) or greater than 60°C (c), energy would not be conserved. For energy to be conserved, the equilibrium temperature must be between the initial temperatures of the substances.

2. The water in the swimming pool has more internal energy. The much larger volume, and therefore the much larger number of particles, more than makes up for the lower temperature; The hot tea has a higher average kinetic energy, because temperature is proportional to average kinetic energy.

360

---

thermal equilibrium

the state in which two bodies in physical contact with each other have identical temperatures

⊛ TEKS 2C, 5B

Conceptual Challenge

1. Hot chocolate

If two cups of hot chocolate, one at 50°C and the other at 60°C, are poured together in a large container, will the final temperature of the double batch be

a. less than 50°C?
b. between 50°C and 60°C?
c. greater than 60°C?

Explain your answer.

2. Hot and cold liquids

A cup of hot tea is poured from a teapot, and a swimming pool is filled with cold water. Which one has a higher total internal energy? Which has a higher average kinetic energy? Explain.

---

the water will be at the same temperature. That temperature will not change as long as conditions remain unchanged in the beaker. Another way of expressing this is to say that the water and can of juice are in **thermal equilibrium** with each other.

Thermal equilibrium is the basis for measuring temperature with thermometers. By placing a thermometer in contact with an object and waiting until the column of liquid in the thermometer stops rising or falling, you can find the temperature of the object. This is because the thermometer is at the same temperature as, or is in thermal equilibrium with, the object. Just as in the case of the can of fruit juice in the cold water, the temperature of any two objects at thermal equilibrium always lies between their initial temperatures.

Matter expands as its temperature increases

You have learned that increasing the temperature of a gas may cause the volume of the gas to increase. This occurs not only for gases, but also for liquids and solids. In general, if the temperature of a substance increases, so does its volume. This phenomenon is known as *thermal expansion*.

You may have noticed that the concrete roadway segments of a bridge are separated by gaps several centimeters wide. This is necessary because concrete expands with increasing temperature. Without these gaps, the force from the thermal expansion would cause the segments to push against each other, and they would eventually buckle and break apart.

Different substances undergo different amounts of expansion for a given temperature change. The thermal expansion characteristics of a material are indicated by a quantity called the *coefficient of volume expansion*. Gases have the largest values for this coefficient. Liquids have much smaller values.

In general, the volume of a liquid tends to increase with increasing temperature. However, the volume of water increases with decreasing temperature in the range between 0°C and 4°C. This explains why ice floats in liquid water. It also explains why a pond freezes from the top down instead of from the bottom up. If this did not happen, fish would likely not survive in freezing temperatures.

Solids typically have the smallest coefficient of volume expansion values. For this reason, liquids in solid containers expand more than the container. This property allows some liquids to be used to measure changes in temperature.

MEASURING TEMPERATURE

In order for a device to be used as a thermometer, it must make use of a change in some physical property that corresponds to changing temperature, such as the volume of a gas or liquid, or the pressure of a gas at constant volume. The most common thermometers use a glass tube containing a thin column of mercury, colored alcohol, or colored mineral spirits. When the thermometer is heated, the volume of the liquid expands. Because the cross-sectional area of the tube remains nearly constant during temperature changes, the change in length of the liquid column is proportional to the temperature change (see **Figure 10-3**).

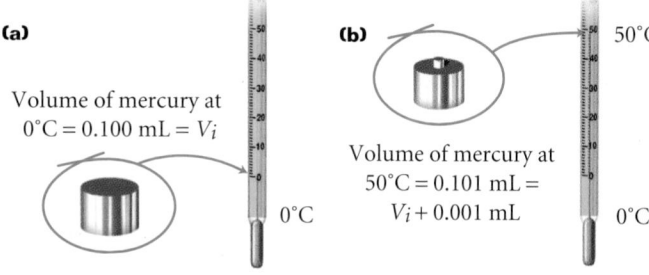

(a) Volume of mercury at $0°C = 0.100$ mL $= V_i$ $0°C$

(b) Volume of mercury at $50°C = 0.101$ mL $= V_i + 0.001$ mL $50°C$ $0°C$

Figure 10-3
The change in the mercury's volume from a temperature of 0°C **(a)** to a temperature of 50°C **(b)** is small, but because the mercury is limited to expansion in only one direction, the linear change is large.

Calibrating thermometers requires fixed temperatures

A thermometer must be more than an unmarked, thin glass tube of liquid. For a thermometer to measure temperature in a variety of situations, the length of the liquid column at different temperatures must be known. One reference point is etched on the tube and refers to when the thermometer is in thermal equilibrium with a mixture of water and ice at one atmosphere of pressure. This temperature is called the *ice point* of water and is defined as zero degrees Celsius, or 0°C. A second reference mark is made at the point when the thermometer is in thermal equilibrium with a mixture of steam and water at one atmosphere of pressure. This temperature is called the *steam point* of water and is defined as 100°C.

A temperature scale can be made by dividing the distance between the reference marks into equally spaced units, called *degrees*. The scale assumes the expansion of the mercury is linear. ⭐TEKS **2C**

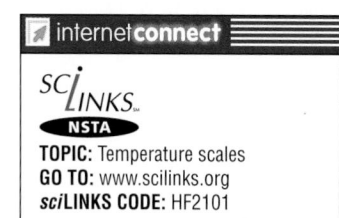

internet**connect**

SCI**LINKS**

NSTA

TOPIC: Temperature scales
GO TO: www.scilinks.org
*sci***LINKS CODE:** HF2101

Temperature units depend on the scale used

The temperature scales most widely used today are the Fahrenheit, Celsius, and Kelvin (or absolute) scales. The Fahrenheit scale is commonly used in the United States. The Celsius scale is used in countries that have adopted the metric system and by the scientific community worldwide.

Celsius and Fahrenheit temperature measurements can be converted to each other using this equation.

CELSIUS-FAHRENHEIT TEMPERATURE CONVERSION

$$T_F = \frac{9}{5}T_C + 32.0$$

Fahrenheit temperature $= \left(\frac{9}{5} \times \textbf{Celsius temperature}\right) + 32.0$

The number 32.0 in the equation indicates the difference between the ice point value in each scale. The point at which water freezes is 0.0 degrees in the Celsius scale and 32.0 degrees in the Fahrenheit scale. ⭐TEKS **2C**

Temperature values in the Celsius and Fahrenheit scales can have positive, negative, or zero values. But because the kinetic energy of the atoms in a substance is positive, the absolute temperature that is proportional to that energy should be positive also. A temperature scale with only positive values is suggested

Did you know?

When a thermometer reaches thermal equilibrium with an object, the object's temperature changes slightly. In most cases the object is so massive compared with the thermometer that the object's temperature change is insignificant.

SECTION 10-1

Visual Strategy

Figure 10-3
Point out that the tube containing mercury is sealed.

Q Is there more mercury in the tube at 50°C than there is at 0°C?

A No, because of thermal expansion, the same amount of mercury occupies a larger volume at 50°C.

Teaching Tip
Point out that the first half of the Celsius-Fahrenheit conversion equation ($\frac{9}{5}T_C$) accounts for the difference in the size of a degree in each scale and that the second half of the equation (+32.0) adjusts the zero point of the Fahrenheit scale so that the total number of degrees is measured from the same reference point (in this case, the freezing point of water, 0°C and 32°F). To convert a range of degrees rather than a specific temperature, only the first half of the equation should be used.

⭐TEKS

p. 360: 2C, 5B
p. 361: 2C, 2C

Figure 10-4
Point out that, unlike the ideal gas represented in this graph, real gases turn to liquids at low temperatures. Make sure that students interpret the labels and graph properly.

Q According to the graph, what is the volume of this amount of ideal gas at 0°C?

A *approximately 600 mL*

Teaching Tip

Compare the Celsius-Kelvin conversion equation with the Celsius-Fahrenheit conversion equation on the previous page. Ask students why T_C is multiplied by $\frac{9}{5}$ in one equation but not in the other *(the size of a degree differs between the Celsius and Fahrenheit scales but not between the Celsius and Kelvin scales)* and why one equation has 32.0 while the other has 273.15 *(the scales have different zero points).*

The Language of Physics

In common speech, we speak of *hot temperatures* and *cold temperatures*. In physics, we say that objects are *hot* or *cold* relative to our body or senses, but their temperature is *high* or *low* relative to a temperature scale.

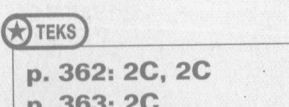
TEKS
p. 362: 2C, 2C
p. 363: 2C

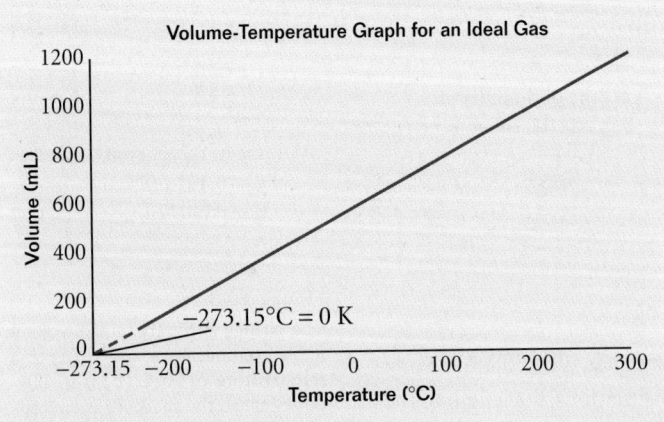

Volume-Temperature Graph for an Ideal Gas

−273.15°C = 0 K

Figure 10-4
If an ideal gas could be compressed to zero volume, its temperature would be −273.15°C, or 0 K.

in the graph of volume versus temperature for an ideal gas, shown in **Figure 10-4.** As the temperature of the gas decreases, so does its volume. If it were possible to compress the matter in a gas to zero volume, the gas temperature would equal −273.15°C. This temperature is designated in the Kelvin scale as 0.00 K, where K is the symbol for the temperature unit called the kelvin. Temperatures in this scale are indicated by the symbol T. ★TEKS **2C**

A temperature difference of one degree is the same on the Celsius and Kelvin scales. The two scales differ only in the choice of zero point. Thus, the ice point (0.00°C) equals 273.15 K, and the steam point (100.00°C) equals 373.15 K (see **Table 10-2**). The Celsius temperature can therefore be converted to the Kelvin temperature by adding 273.15. ★TEKS **2C**

CELSIUS-KELVIN TEMPERATURE CONVERSION

$$T = T_C + 273.15$$

Kelvin temperature = Celsius temperature + 273.15

Kelvin temperatures for various physical processes can range from around 1 000 000 000 K (10^9 K), which is the temperature of the interiors of the most massive stars, to less than 1 K, which is slightly cooler than the boiling point of liquid helium. The temperature 0 K is often referred to as *absolute zero*. Absolute zero has never been reached, although laboratory experiments have reached temperatures of 0.000 001 K.

Table 10-2 **Temperature scales and their uses**

Scale	Ice point	Steam point	Applications
Fahrenheit	32°F	212°F	meteorology, medicine, and non-scientific uses (U.S.)
Celsius	0°C	100°C	meteorology, medicine, and non-scientific uses (outside U.S.); other sciences (international)
Kelvin (absolute)	273.15 K	373.15 K	physical chemistry, gas laws, astrophysics, thermodynamics, low-temperature physics

Temperature conversion  2C

PROBLEM

What are the equivalent Celsius and Kelvin temperatures of 50.0°F?

SOLUTION

Given: $T_F = 50.0°F$

Unknown: $T_C = ?$ $T = ?$

Use the Celsius-Fahrenheit equation from page 361.

$$T_F = \frac{9}{5}T_C + 32.0$$

$$T_C = \frac{5}{9}(T_F - 32.0)$$

$$T_C = \frac{5}{9}(50.0 - 32.0)°C = 10.0°C$$

Use the Celsius-Kelvin equation from page 362.

$$T = T_C + 273.15$$

$$T = (10.0 + 273.2)K = 283.2 \text{ K}$$

$$\boxed{\begin{array}{l} T_C = 10.0°C \\ T = 283.2 \text{ K} \end{array}}$$

Temperature conversion

1. The lowest outdoor temperature ever recorded on Earth is −128.6°F, recorded at Vostok Station, Antarctica, in 1983. What is this temperature on the Celsius and Kelvin scales?

2. The temperatures of one northeastern state range from 105°F in the summer to −25°F in winter. Express this temperature range in degrees Celsius and in kelvins.

3. The normal human body temperature is 98.6°F. A person with a fever may record 102°F. Express these temperatures in degrees Celsius.

4. A pan of water is heated from 23°C to 78°C. What is the change in its temperature on the Kelvin and Fahrenheit scales?

5. Liquid nitrogen is used to cool substances to very low temperatures. Express the boiling point of liquid nitrogen (77.34 K at 1 atm of pressure) in degrees Fahrenheit and in degrees Celsius.

Classroom Practice

The following may be used as teamwork exercises or for demonstration at the chalkboard or on an overhead projector.

PROBLEM

Temperature conversion

One day it was −40°C at the top of Mont Blanc and −40°F at the top of Mount Whitney. Which place was colder?

Answer
  neither (−40°C = −40°F)

PRACTICE GUIDE 10A

Solving for:		
T_C	PE	Sample, 1–3, 5; Ch. Rvw. 9–10, 45
	PW	4–5
	PB	Sample, 1–3, 8–10
T_F	PE	4–5; Ch. Rvw. 43, 45, 48
	PW	Sample, 1–3, 5
	PB	Sample, 1–3
T	PE	Sample, 1–2, 4; Ch. Rvw. 9–10, 43, 48
	PW	Sample, 1–2, 6–7
	PB	4–7

ANSWERS TO

Practice 10A
Temperature conversion
1. −89.22°C, 183.93 K
2. 41°C to −32°C, 314 K to 241K
3. 37.0°C, 39°C
4. 55 K, 99°F
5. −320.5°F, −195.8°C

Section Review
ANSWERS

1. c, a, b; The blue particles in (c) have the greatest average kinetic energy and therefore have the highest temperature.

2. The pan's temperature decreases if the water's temperature increases; The water and pan have reached thermal equilibrium when their temperatures are the same.

3. −183.0°C, −297.4°F

4. **a.** 119.0°C
   b. 246.2°F, 832.3°F
   c. 392.2 K, 717.8 K

5. a, c

6. a, b (assuming the hot plate has been on for a long time and has supplied energy to the interior of the popper)

Section Review    2C, 2D, 5B

1. Two gases that are in physical contact with each other consist of particles of identical mass. In what order should the images shown in **Figure 10-5** be placed to correctly describe the changing distribution of kinetic energy among the gas particles? Which group of particles has the highest temperature at any time? Explain.

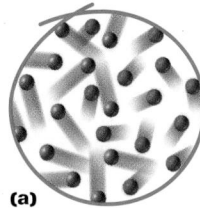 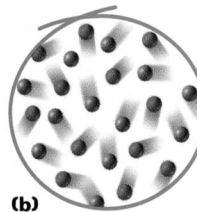 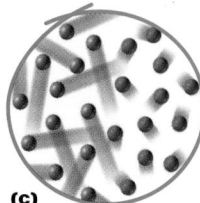

(a)   (b)   (c)

Figure 10-5

2. A hot copper pan is dropped into a tub of water. If the water's temperature rises, what happens to the temperature of the pan? How will you know when the water and copper pan reach thermal equilibrium?

3. Oxygen condenses into a liquid at approximately 90.2 K. To what temperature does this correspond on both the Celsius and Fahrenheit temperature scales?

4. The boiling point of sulfur is 444.6°C. Sulfur's melting point is 586.1°F lower than its boiling point.

   a. Determine the melting point of sulfur in degrees Celsius.
   b. Find the melting and boiling points in degrees Fahrenheit.
   c. Find the melting and boiling points in kelvins.

5. **Physics in Action**   Which of the following is true for the water molecules inside popcorn kernels during popping?

   a. Their temperature increases.
   b. They are destroyed.
   c. Their kinetic energy increases.
   d. Their mass changes.

6. **Physics in Action**   Referring to **Figure 10-6,** determine which of the following pairs represent objects that are in thermal equilibrium with each other.

   a. the hot plate and the glass pot
   b. the hot oil and the popcorn kernels
   c. the outside air and the hot plate

Figure 10-6

10-2
Defining heat

HEAT AND ENERGY

Thermal physics often appears mysterious at the macroscopic level. Hot objects become cool without any obvious cause. To understand thermal processes, it is helpful to shift attention to the behavior of atoms and molecules. Mechanics can be used to explain much of what is happening at the molecular, or microscopic, level. This in turn accounts for what you observe at the macroscopic level. Throughout this chapter, the focus will shift between these two viewpoints.

Recall the can of warm fruit juice immersed in the beaker of cold water (shown in **Figure 10-7**). The temperature of the can and the juice in it is lowered, and the water's temperature is slightly increased, until at thermal equilibrium both final temperatures are the same. Energy is transferred from the can of juice to the water because the two objects are at different temperatures. This energy that is transferred is defined as **heat.**

The word *heat* is sometimes used to refer to the *process* by which energy is transferred between objects because of a difference in their temperatures. This textbook will use *heat* to refer only to the energy itself.

Energy is transferred between substances as heat

From a macroscopic viewpoint, energy transferred as heat always moves from an object at higher temperature to an object at lower temperature. This is similar to the mechanical behavior of objects moving from a higher gravitational potential energy to a lower gravitational potential energy. Just as a pencil will drop from your desk to the floor but will not jump from the floor to your

10-2 SECTION OBJECTIVES

- **Explain heat as the energy transferred between substances that are at different temperatures.**

- **Relate heat and temperature change on the macroscopic level to particle motion on the microscopic level.**

- **Apply the principle of energy conservation to calculate changes in potential, kinetic, and internal energy.**

heat

the energy transferred between objects because of a difference in their temperatures

Figure 10-7
Energy is transferred as heat from objects with higher temperatures (the fruit juice and can) to those with lower temperatures (the cold water).

The Language of Physics

The *internal energy, U,* also called the *thermal energy* of an object, is the total energy of all of the molecules in an object, including both their kinetic energy and their potential energy. However, in this chapter we are primarily concerned with changes in the kinetic energy of the molecules.

Key Models and Analogies

The kinetic and potential energies of an object depend on its motion and position, respectively, as well as the object's mass, but not on the substance of which it is made. For example, a 6.0 kg crate moving at 5.0 m/s has 75 J of kinetic energy regardless of whether it is full of sand, bricks, ice, or gold. Similarly, every molecule has kinetic and potential energy due to its mass, motion, and position inside a substance regardless of what the substance is.

Misconception Alert

Many students do not clearly distinguish between temperature, heat, and internal energy. Temperature measures the average kinetic energy of the molecules in an object. Heat is the energy transferred from one object to another. That is, heat is energy in transit. Internal energy is the sum of the energies of the molecules. Drawing schematic representations of these statements using a specific example will help students understand these relationships.

Figure 10-8

Energy is transferred as heat from the higher-energy particles to lower-energy particles **(a).** The net energy transferred is zero when thermal equilibrium is reached **(b).**

internet connect

SCI LINKS
NSTA

TOPIC: James Prescott Joule
GO TO: www.scilinks.org
*sci*LINKS CODE: HF2102

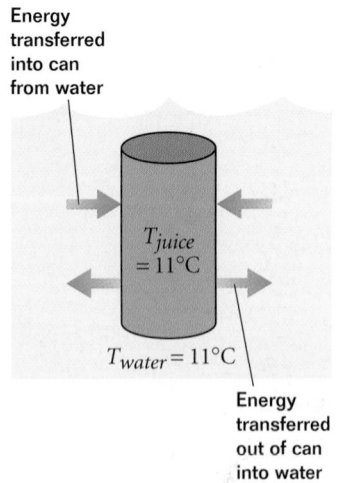

Energy transferred into can from water

T_{juice} = 11°C

T_{water} = 11°C

Energy transferred out of can into water

Figure 10-9

At thermal equilibrium, the net energy exchanged between two objects equals zero.

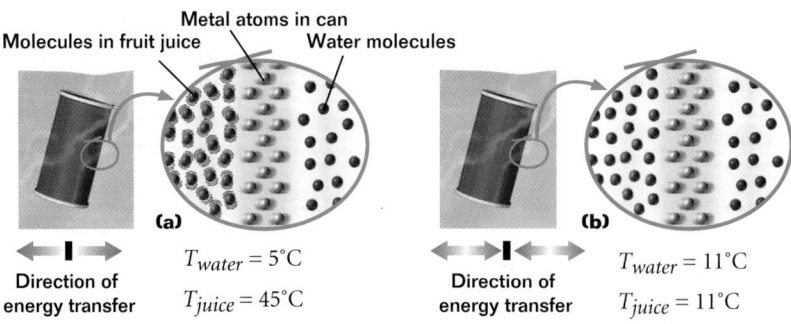

Molecules in fruit juice — Metal atoms in can — Water molecules

(a)

Direction of energy transfer

T_{water} = 5°C
T_{juice} = 45°C

(b)

Direction of energy transfer

T_{water} = 11°C
T_{juice} = 11°C

desk, so energy will travel spontaneously from an object at higher temperature to one at lower temperature and not the other way around. ⓉⒺⓀⓈ **5B**

The direction in which energy travels as heat can be explained at the atomic level. At first, the molecules in the fruit juice have higher average kinetic energies than do the water molecules that surround the can, as shown in **Figure 10-8.** This energy is transferred from the juice to the can by the molecules in the juice colliding with the metal atoms of the can. The atoms vibrate more because of their increased energy, and this energy is transferred to the surrounding water molecules.

As the energy of the water molecules gradually increases, the energy of the molecules in the fruit juice and the atoms of the can decreases until all of the particles have, on the average, equal kinetic energies. However, it is also possible for some of the energy to be transferred through collisions from the lower-energy water molecules to the higher-energy metal atoms and fruit-juice particles. Therefore, energy can move in both directions. Because the average kinetic energy of particles is higher in the body at higher temperature, more energy is transferred out of it as heat than is transferred into it. The net result is that energy is transferred as heat in only one direction.

The transfer of energy as heat alters an object's temperature

Thermal equilibrium may be understood in terms of energy exchange between two objects at equal temperature. When the can of fruit juice and the surrounding water are at the same temperature, as depicted in **Figure 10-9,** the quantity of energy transferred from the can of fruit juice to the water is the same as the energy transferred from the water to the can of juice. The net energy transferred between the two objects is zero.

This reveals the difference between temperature and heat. The atoms of all objects are in continuous motion, so all objects have some internal energy. Because temperature is a measure of that energy, all objects have some temperature. Heat, on the other hand, is the energy transferred from one object to another because of the temperature difference between them. When there is no temperature difference between a substance and its surroundings, no net energy is transferred as heat.

Energy transfer depends on the difference of the temperatures of the two objects. The greater the temperature difference between two objects, the greater the amount of energy that is transferred between them as heat.

For example, in winter, energy is transferred as heat from a car's surface at 30°C to a cold raindrop at 5°C. In the summer, energy is transferred as heat from a car's surface at 45°C to a warm raindrop at 20°C. In each case, the amount of energy transferred is the same, because the substances and the temperature difference (25°C) are the same (see **Figure 10-10**). ⭐TEKS **2C**

The concepts of heat and temperature help to explain why hands held in separate bowls containing hot and cold water subsequently sense the temperature of lukewarm water differently. The nerves in the outer skin of your hand detect energy passing through the skin from objects with temperatures different than your body temperature. If one hand is at thermal equilibrium with cold water, more energy is transferred from the outer layers of your hand than can be replaced by the blood, which has a temperature of about 37.0°C (98.6°F). When the hand is immediately placed in water that is at a higher temperature, energy is transferred from the water to the cooler hand. The energy transferred into the skin causes the water to feel warm. Likewise, the hand that has been in hot water temporarily gains energy from the water. The loss of this energy to the lukewarm water makes that water feel cool. ⭐TEKS **3A**

Heat has the units of energy

Before scientists arrived at the current model for heat, several different units for measuring heat had already been developed. These units are still widely used in many applications and therefore are listed in **Table 10-3**. Because heat, like work, is energy in transit, all heat units can be converted to joules, the SI unit for energy.

Just as other forms of energy have a symbol that identifies them (*PE* for potential energy, *KE* for kinetic energy, *U* for internal energy, *W* for work), heat is indicated by the symbol *Q*. ⭐TEKS **3B**

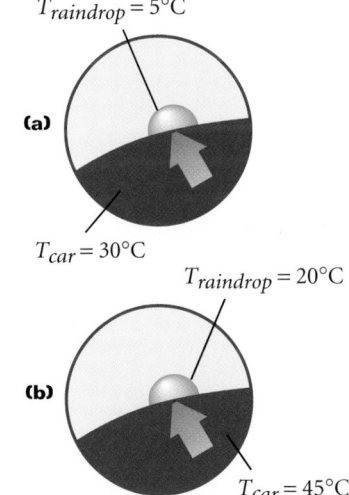

Figure 10-10
The energy transferred as heat from the car's surface to the raindrop is the same for low temperatures **(a)** as for high temperatures **(b)**, provided the temperature differences are the same.

Table 10-3	Thermal units and their values in joules	
Heat unit	**Equivalent value**	**Uses**
joule (J)	equal to $1 \text{ kg} \cdot \left(\dfrac{\text{m}^2}{\text{s}^2}\right)$	SI unit of energy
calorie (cal)	4.186 J	non-SI unit of heat; found especially in older works of physics and chemistry
kilocalorie (kcal)	4.186×10^3 J	non-SI unit of heat
Calorie, or dietary Calorie	4.186×10^3 J = 1 kcal	food and nutritional science
British thermal unit (Btu)	1.055×10^3 J	English unit of heat; used in engineering, air-conditioning, and refrigeration
therm	1.055×10^8 J	equal to 100 000 Btu; used to measure natural-gas usage

Key Models and Analogies

Just as the work, *W*, done on an object equals the change in kinetic energy of the object ($W = \Delta KE$ in the work–kinetic energy theorem), the heat, *Q*, transferred to or from an object equals the increase or decrease of the internal energy in the object through temperature change ($Q = \Delta U$).

The Language of Physics

Texts that use the term *heat energy* (rather than internal energy) and *heat flow* (rather than heat) likewise refer to *Q* as heat flow rather than heat.

Teaching Tip

Make sure students understand the meaning of the symbols in the table. Remind students that SI stands for *Système International d'Unités*, or International System of Units.

⭐TEKS

p. 366: 5B
p. 367: 2C, 3A, 3B

Internal energy

Purpose Show the conversion of work into internal energy.

Materials mixing bowl, cold water, thermometer, electric egg beater

Procedure Fill $\frac{1}{2}$ to $\frac{3}{4}$ of the mixing bowl with cold water. Measure and record the temperature. Run the electric beater in the bowl for about 10 min and record the final temperature. Have students describe the chain of energy transfers that took place (*work done by the beater converted to kinetic energy of the water molecules, or the internal energy of the water*). The experiment can be quantified by recording the temperature of the water at regular intervals of time throughout the experiment.

Quick Lab

TEACHER'S NOTES

This exercise is meant to demonstrate how work increases an object's internal energy.

Some students may assert that the increase in temperature is caused by the transfer of energy from their hand to the rubber band. This should be recognized as an astute observation. Address it by comparing results reported by students with warm hands with those reported by students with cold hands.

HEAT AND WORK

Hammer a nail into a block of wood. After several minutes, pry the nail loose from the block and touch the side of the nail. It feels warm to the touch, indicating that energy is being transferred from the nail to your hand. Work is done in pulling the nail out of the wood. The nail encounters friction with the wood, and most of the energy required to overcome this friction is transformed into internal energy. The increase in the internal energy of the nail raises the nail's temperature, and the temperature difference between the nail and your hand results in the transfer of energy to your hand as heat.

Friction is just one way of increasing a substance's internal energy. In the case of solids, internal energy can be increased by deforming their structure. Common examples of this are when a rubber band is stretched or a piece of metal is bent. ⭐TEKS 5B

Total energy is conserved

When the concept of mechanical energy was introduced in Chapter 5, you discovered that whenever friction between two objects exists, not all of the work done in overcoming friction appears as mechanical energy. Similarly, not all of the kinetic energy in inelastic collisions remains as kinetic energy. Some of this energy is absorbed by the objects as internal energy. This is why, in the case of the nail pulled from the wood, the nail (and if you could touch it, the wood inside the hole) feels warm. If changes in internal energy are taken into account along with changes in mechanical energy, the total energy is a universally conserved property. ⭐TEKS 3B, 5D

CONSERVATION OF ENERGY

$$\Delta PE + \Delta KE + \Delta U = 0$$

the change in potential energy + the change in kinetic energy + the change in internal energy = 0

⭐TEKS 1A

Work and Heat

MATERIALS LIST

✔ 1 large rubber band about 7–10 mm wide

 SAFETY CAUTION

To avoid breaking the rubber band, do not stretch it more than a few inches. Do not point a stretched rubber band at another person.

Hold the rubber band between your thumbs. Touch the middle section of the rubber band to your lip and note how it feels. Rapidly stretch the rubber band and keep it stretched. Touch the middle section of the rubber band to your lip again. Notice whether the rubber band's temperature has changed. (You may have to repeat this procedure several times before you can clearly distinguish the temperature difference.)

Conservation of energy ⭐TEKS 2C, 5D

PROBLEM

An arrangement similar to the one used to demonstrate energy conservation is shown at right. A vessel contains water. Paddles that are propelled by falling masses turn in the water. This agitation warms the water and increases its internal energy. The temperature of the water is then measured, giving an indication of the water's internal-energy increase. If a total mass of 11.5 kg falls 1.3 m and all of the mechanical energy is converted to internal energy, by how much will the internal energy of the water increase? (Assume no energy is transferred as heat out of the vessel to the surroundings or from the surroundings to the vessel's interior.)

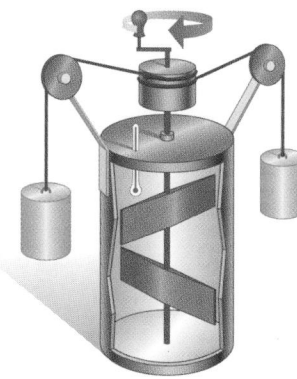

SOLUTION

1. DEFINE

Given: $m = 11.5 \text{ kg}$ $\qquad h = 1.3 \text{ m}$ $\qquad g = 9.81 \text{ m/s}^2$

Unknown: $\Delta PE = ?$ $\qquad \Delta KE = ?$ $\qquad \Delta U = ?$

2. PLAN

Choose an equation(s) or situation: The equation for conservation of energy can be expressed as the initial total energy equal to the final total energy. Because there is no kinetic energy in the apparatus when the mass is released or when it comes to rest, both KE_i and KE_f equal zero. Because all of the potential energy is assumed to be converted to internal energy, PE_i can be set equal to mgh if PE_f is set equal to zero.

$$\Delta PE + \Delta KE + \Delta U = 0$$
$$PE_i + KE_i + U_i = PE_f + KE_f + U_f$$
$$PE_i = mgh$$
$$PE_f = 0$$
$$KE_i = 0$$
$$KE_f = 0$$
$$mgh + 0 + U_i = 0 + 0 + U_f$$
$$\Delta U = U_f - U_i = mgh$$

3. CALCULATE

Substitute values into the equation(s) and solve:

$$\Delta U = (11.5 \text{ kg})(9.81 \text{ m/s}^2)(1.3 \text{ m})$$
$$= 1.5 \times 10^2 \text{ J}$$

$$\boxed{\Delta U = 1.5 \times 10^2 \text{ J}}$$

CALCULATOR SOLUTION

Because the minimum number of significant figures in the data is two, the calculator answer, 146.6595 J, should be rounded to two digits.

4. EVALUATE

The answer can be estimated using rounded values for m and g. If $m \approx 10$ kg and $g \approx 10 \text{ m/s}^2$, then $\Delta U \approx 130$ J, which is close to the actual value calculated.

PRACTICE GUIDE 10B

Solving for:

ΔU	📖 **PE**	Sample, 1–3; Ch. Rvw. 19–20, 42*
	💿 **PW**	5–6, 7*
	PB	8–10
m	📖 **PE**	4
	💿 **PW**	Sample, 1–2
	PB	5–7
PE	💿 **PW**	3–4
	PB	Sample, 1–4

Classroom Practice

The following may be used as teamwork exercises or for demonstration at the chalkboard or on an overhead projector.

PROBLEM

Conservation of energy

A 0.10 kg ball falls 10.0 m onto a hard floor and then bounces back up to 9.0 m. How much of its mechanical energy is transformed to the internal energy of the ball and the floor?

Answer
0.98 J

⭐**TEKS**

p. 368: 5B, 3B, 5D, 1A
p. 369: 2C, 5D

ANSWERS TO

Practice 10B
Conservation of energy

1. 755 J
2. 1.76×10^3 J
3. 0.96 J
4. 1.2×10^{-4} kg
5. 41 m/s

Section Review
ANSWERS

1. You would need to know the comparative temperature changes for each bottle of water (assuming the water doesn't freeze in either case).

2. In the first case, molecules in the exhaled air have a greater average kinetic energy than the air surrounding your cold hands. Energy is transferred to the hands, causing their temperature to increase. In the second case, the molecules in the soup have a greater average kinetic energy than the exhaled air passing over the soup's surface. Energy is therefore transferred from the soup to the relatively cooler air.

3. yes; Shaking the bottle adds kinetic energy to the system, and this kinetic energy is converted into the internal energy of the water molecules.

4. 10.1°C

PRACTICE 10B

Conservation of energy  TEKS 2C, 5D

1. In the arrangement described in Sample Problem 10B, how much would the water's internal energy increase if the mass fell 6.69 m?

2. A worker drives a 0.500 kg spike into a rail tie with a 2.50 kg sledgehammer. The hammer hits the spike with a speed of 65.0 m/s. If one-third of the hammer's kinetic energy is converted to the internal energy of the hammer and spike, how much does the total internal energy increase?

3. A 3.0×10^{-3} kg copper penny drops a distance of 50.0 m to the ground. If 65 percent of the initial potential energy goes into increasing the internal energy of the penny, determine the magnitude of that increase.

4. A 2.5 kg block of ice at a temperature of 0.0°C and an initial speed of 5.7 m/s slides across a level floor. If 3.3×10^5 J are required to melt 1.0 kg of ice, how much ice melts, assuming that the initial kinetic energy of the ice block is entirely converted to the ice's internal energy?

5. The amount of internal energy needed to raise the temperature of 0.25 kg of water by 0.2°C is 209.3 J. How fast must a 0.25 kg baseball travel in order for its kinetic energy to equal this internal energy?

Section Review TEKS 2C, 5B

1. A bottle of water at room temperature is placed in a freezer for a short time. An identical bottle of water that has been lying in the sunlight is placed in a refrigerator for the same amount of time. What must you know to determine which situation involves more energy transfer?

2. Use the microscopic interpretations of temperature and heat to explain how you can blow on your hands to warm them and also blow on a bowl of hot soup to cool it.

3. If a bottle of water is shaken vigorously, will the internal energy of the water change? Why or why not?

4. Water at the top of Niagara Falls has a temperature of 10.0°C. Assume that all of the potential energy goes into increasing the internal energy of the water and that it takes 4186 J/kg to increase the water's temperature by 1°C. If 505 kg of water falls a distance of 50.0 m, what will the temperature of the water be at the bottom of the falls?

10-3
Changes in temperature and phase

SPECIFIC HEAT CAPACITY

You have probably noticed on a hot day that the air around a swimming pool (like the one shown in **Figure 10-11**) is hot but the pool water is cool. This may seem odd, because both the air and water receive energy from sunlight. The water may be cooler than the air, in part because of evaporation, which is a cooling process. However, there is another property of all substances that causes their temperatures to vary by different amounts when equal amounts of energy are added to or removed from them. **⭐TEKS** **7B**

This property can be explained in terms of the motion of atoms and molecules in a substance, which in turn affects how much the substance's temperature changes for a given amount of energy that is added or removed. Each substance has a unique value for the energy required to change the temperature of 1 kg of that substance by 1°C. This value, known as the **specific heat capacity** (or sometimes just specific heat) of the substance, relates mass, temperature change, and energy transferred as heat.

The specific heat capacity is related to energy transferred, mass, and temperature change by the following equation:

SPECIFIC HEAT CAPACITY

$$c_p = \frac{Q}{m\Delta T}$$

$$\text{specific heat capacity} = \frac{\text{energy transferred as heat}}{\text{mass} \times \text{change in temperature}}$$

The subscript p indicates that the specific heat capacity is measured at constant pressure. Maintaining constant pressure is an important detail when determining certain thermal properties of gases, which are much more affected by changes in pressure than are solids or liquids. Note that a temperature change of 1°C is equal in magnitude to a temperature change of 1 K, so that ΔT gives the temperature change in either scale.

The equation for specific heat capacity applies to both substances that absorb energy from their surroundings and those that transfer energy to their surroundings. When the temperature increases, ΔT and Q are taken to be positive, which corresponds to energy transferred into the substance. Likewise, when the temperature decreases, ΔT and Q are negative and energy is

---

10-3 SECTION OBJECTIVES

- **Perform calculations with specific heat capacity.**

- **Perform calculations involving latent heat.**

- **Interpret the various sections of a heating curve.**

Figure 10-11
The air around the pool and the water in the pool receive energy from sunlight. However, the increase in temperature is greater for the air than for the water.

specific heat capacity

> *the quantity of energy needed to raise the temperature of 1 kg of a substance by 1°C at constant pressure*

372

Teaching Tip

Point out that the specific heat capacities of ice, water, and steam do not have the same value; in other words, the specific heat capacity of a substance depends on its phase.

Visual Strategy

Figure 10-12

Be sure students understand this schematic representation of a calorimeter.

Q Why is water used in a calorimeter?

A *Water's specific heat capacity has a well-known value, and thus the change in the water's temperature can be used to determine the specific heat capacity of the sample.*

internet**connect**

SC*LINKS*

NSTA

TOPIC: Specific heat
GO TO: www.scilinks.org
***sci*LINKS CODE:** HF2103

calorimetry

an experimental procedure used to measure the energy transferred from one substance to another as heat

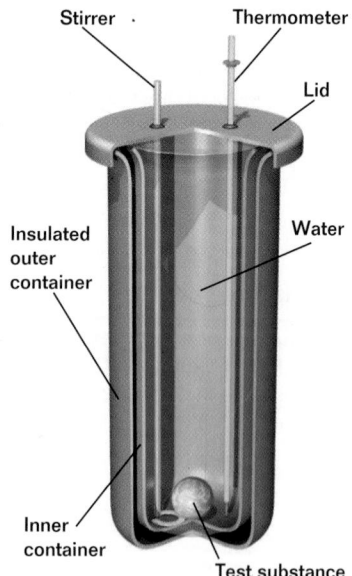

Stirrer · Thermometer

Lid

Insulated outer container

Water

Inner container

Test substance

Figure 10-12
A simple calorimeter allows the specific heat capacity of a substance to be determined.

Table 10-4	**Specific heat capacities**		
Substance	c_p **(J/kg•°C)**	**Substance**	c_p **(J/kg•°C)**
aluminum	8.99×10^2	lead	1.28×10^2
copper	3.87×10^2	mercury	1.38×10^2
glass	8.37×10^2	silver	2.34×10^2
gold	1.29×10^2	steam	2.01×10^3
ice	2.09×10^3	water	4.186×10^3
iron	4.48×10^2		

transferred from the substance. **Table 10-4** lists specific heat capacities that have been determined for several substances. ⊛TEKS **2C**

Determining specific heat capacity

To measure the specific heat capacity of a substance, it is necessary to measure mass, temperature change, and energy transferred as heat. Mass and temperature change are directly measurable, but the direct measurement of heat is difficult. However, the specific heat capacity of water (4.186 kJ/kg•°C) is well known, so the energy transferred as heat between an object of unknown specific heat capacity and a known quantity of water can be measured.

If a hot substance is placed in an insulated container of cool water, energy conservation requires that the energy the substance gives up must equal the energy absorbed by the water. Although some energy is transferred to the surrounding container, this effect is small and will be ignored in this discussion. Energy conservation can be used to calculate the specific heat capacity, $c_{p,x}$, of the substance (indicated by the subscript x). For simplicity, a subscript w will always stand for "water" in problems involving specific heat capacities. ⊛TEKS **5D**

energy absorbed by water = energy released by the substance

$$Q_w = Q_x$$

$$c_{p,w} m_w \Delta T_w = c_{p,x} m_x \Delta T_x$$

The energy gained by a substance is usually expressed as a positive quantity, and energy released usually has a negative value. The minus sign of the latter quantity can be eliminated if ΔT_x and ΔT_{water} are written as the larger temperature value minus the smaller one. Therefore, ΔT should always be written as a positive quantity for this equation.

This approach to determining a substance's specific heat capacity is called **calorimetry,** and devices that are used for making this measurement are called *calorimeters*. A calorimeter also contains a thermometer for measuring the final temperature when the substances are at thermal equilibrium and a stirrer to ensure the uniform mixture of energy throughout the water (see **Figure 10-12**).

Calorimetry (★TEKS) 2C, 5D

PROBLEM

A 0.050 kg metal bolt is heated to an unknown initial temperature. It is then dropped into a beaker containing 0.15 kg of water with an initial temperature of 21.0°C. The bolt and the water then reach a final temperature of 25.0°C. If the metal has a specific heat capacity of 899 J/kg•°C, find the initial temperature of the metal.

SOLUTION

1. DEFINE **Given:**

$m_{metal} = m_m = 0.050$ kg $c_{p,m} = 899$ J/kg•°C
$m_{water} = m_w = 0.15$ kg $c_{p,w} = 4186$ J/kg•°C
$T_{water} = T_w = 21.0$°C $T_{final} = T_f = 25.0$°C

Unknown: $T_{metal} = T_m = ?$

Diagram:

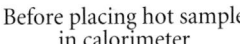

Before placing hot sample in calorimeter

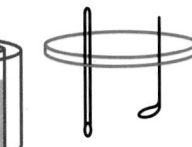

After thermal equilibrium has been reached

$m_m = 0.050$ kg $m_w = 0.15$ kg $T_f = 25.0$°C
$T_w = 21.0$°C

2. PLAN **Choose an equation(s) or situation:** Equate the energy removed from the bolt to the energy absorbed by the water.

energy removed from metal = energy absorbed by water

$$c_{p,m} m_m \Delta T_m = c_{p,w} m_w \Delta T_w$$

Rearrange the equation to isolate the unknown:

$$\Delta T_m = \frac{m_w c_{p,w} \Delta T_w}{m_m c_{p,m}}$$

3. CALCULATE **Substitute values into the equation(s) and solve:**

Note that ΔT_w has been made positive.

$$\Delta T_w = T_f - T_w = 25.0°C - 21.0°C = 4.0°C$$

$$\Delta T_m = \frac{(0.15 \text{ kg})\left(\dfrac{4186 \text{ J}}{\text{kg•°C}}\right)(4.0°C)}{(0.050 \text{ kg})\left(\dfrac{899 \text{ J}}{\text{kg•°C}}\right)}$$

continued on next page

Classroom Practice

The following may be used as teamwork exercises or for demonstration at the chalkboard or on an overhead projector.

PROBLEM

Calorimetry

You are preparing to take a bath. The cold-water faucet supplies water at 20.0°C, and the water from the hot-water faucet is 60.0°C. Each faucet has poured 25.0 kg of water into the tub. What is the temperature of the bath?

Answer
40.0°C

You prefer your bath at 30.0°C. The hot-water faucet has already poured 20.0 kg of water at 60.0°C into the tub. How much cold water (20.0°C) should you add?

Answer
60.0 kg

(★TEKS)

p. 372: 2C, 5D
p. 373: 2C, 5D

ANSWERS TO

Practice 10C
Calorimetry
1. 47°C
2. 18.0°C
3. 79°C
4. 2500 J/kg•°C
5. 390 J/kg•°C
6. 3.2 m^3
7. 135 g

$$\Delta T_m = 56°C$$

$$T_m = T_f + \Delta T_m$$

$$T_m = 25°C + 56°C = 81°C$$

$$\boxed{T_m = 81°C}$$

PRACTICE 10C

Calorimetry

1. What is the final temperature when a 3.0 kg gold bar at 99°C is dropped into 0.22 kg of water at 25°C?

2. A 0.225 kg sample of tin initially at 97.5°C is dropped into 0.115 kg of water initially at 10.0°C. If the specific heat capacity of tin is 230 J/kg•°C, what is the final equilibrium temperature of the tin-water mixture?

3. What is the final temperature when 0.032 kg of milk at 11°C is added to 0.16 kg of coffee at 91°C? Assume the specific heat capacities of the two liquids are the same as water, and disregard any energy transfer to the liquids' surroundings.

4. A cup is made of an experimental material that can hold hot liquids without significantly increasing its own temperature. The 0.75 kg cup has an initial temperature of 36.5°C when it is submerged in 1.25 kg of water with an initial temperature of 20.0°C. What is the cup's specific heat capacity if the final temperature is 24.4°C?

5. Brass is an alloy made from copper and zinc. A 0.59 kg brass sample at 98.0°C is dropped into 2.80 kg of water at 5.0°C. If the equilibrium temperature is 6.8°C, what is the specific heat capacity of brass?

6. The air temperature above coastal areas is profoundly influenced by the large specific heat capacity of water. How large of a volume of air can be cooled by 1.0°C if energy is transferred as heat from the air to the water, thus increasing the temperature of 1.0 kg of water by 1.0°C? The specific heat capacity of air is approximately 1000.0 J/kg•°C, and the density of air is approximately 1.29 kg/m^3.

7. A hot, just-minted copper coin is placed in 101 g of water to cool. The water temperature changes by 8.39°C and the temperature of the coin changes by 68.0°C. What is the mass of the coin? Disregard any energy transfer to the water's surroundings.

Tomorrow's Technology

Heating and Cooling from the Ground Up ⭐TEKS 3C

As the earliest cave dwellers knew, a good way to stay warm in the winter and cool in the summer is to go underground. Now scientists and engineers are using the same premise—and using existing technology in a new, more efficient way—to heat and cool aboveground homes for a fraction of the cost of conventional systems.

"At any given occasion, the earth temperature is the seasonal average temperature," said Gunnar Walmet, of the New York State Energy Research and Development Authority (NYSERDA). "In New York state, that's typically about 50°F all year long."

Although the average specific heat capacity of earth has a smaller value than the specific heat capacity of air, the earth has a greater density. That means there are more kilograms of earth than there are of air near a house and that a 1°C change in temperature involves transferring more energy to or from the ground than to or from the air. Thus, in the wintertime, the ground will probably have a higher temperature than the air above it, while in the summer, the ground will likely have a lower temperature than the air.

An earth-coupled heat pump enables homeowners to tap the earth's belowground temperature to heat their homes in the winter or cool them during the summer. The system includes a network of plastic pipes placed in trenches or inserted in holes drilled 2 to 3 m (6 to 10 ft) beneath the ground's surface. To heat a home, a fluid circulates through the pipe, absorbs energy from the surrounding earth, and transfers this energy to a heat pump inside the house.

The heat pump uses a compressor, tubing, and refrigerant to transfer the energy from the liquid to the air inside the house. A blower-and-duct system distributes the warm air through the home. According to NYSERDA, the system can deliver up to four times as much energy into the house as the electrical energy needed to drive it.

Like other heat pumps, the system is reversible. In the summer, it can transfer energy from the air in the house to the system of pipes belowground.

There are currently tens of thousands of earth-coupled heat pumps installed throughout the United States. Although the system can function anywhere on Earth's surface, it is most appropriate in severe climates, where dramatic temperature swings may not be ideal for air-based systems.

Tomorrow's Technology

BACKGROUND

The earth-coupled heat pump demonstrates an excellent use of the transfer of energy as heat.

The Department of Energy is now campaigning to promote this technology and increase installations from 40,000 per year to 400,000 per year. This has not happened yet because electricity prices have continually increased, but fossil fuel prices have remained nearly the same, keeping heating costs steady. There are approximately a few hundred thousand units currently installed.

EXTENSION

Have students brainstorm about a campaign scheme to increase the use of earth-coupled heat pumps. Sending these ideas in a letter to the U.S. Department of Energy will teach students to become more active participants in government.

⭐TEKS
p. 374: 2C, 5D
p. 375: 3C

Visual Strategy

Visual Strategy

Figure 10-13

Make sure students understand the meaning of symbols in the graph. Explain that the ⊸⋏⊸ indicates a break in the scale on the *x*-axis.

Q What do the horizontal segments of the graph (**B** and **D**) indicate?

A *They represent the times when the temperature was constant and the substance was undergoing a phase change. In B the ice was melting, and in D the water was turning into steam.*

Q Did energy transfer happen continually throughout the process? Did the temperature increase at the same rate throughout the process?

A *The water absorbed energy continuously, but the temperature increased only when all of the water was in one phase (segments A, C, and E); The rate of temperature increase varied for each phase because ice, water, and steam have different specific heat capacities.*

internet connect

SC*LINKS*
NSTA
TOPIC: Heat pumps
GO TO: www.scilinks.org
*sci*LINKS CODE: HF2104

LATENT HEAT

If you place an ice cube with a temperature of −25°C in a pan and then place the pan on a hot range element or burner, the temperature of the ice will increase until the ice begins to melt at 0°C. By knowing the mass and specific heat capacity of ice, you can calculate how much energy is being added to the ice from the element. However, this procedure only works as long as the ice remains ice and its temperature continues to rise as energy is transferred to it. ⊛TEKS **7B**

The graph in **Figure 10-13** and data in **Table 10-5** show how the temperature of 10.0 g of ice changes as energy is added. You can see that as the ice is heated there is a steady increase in temperature from −25°C to 0°C (segment **A** of the graph). ⊛TEKS **2C**

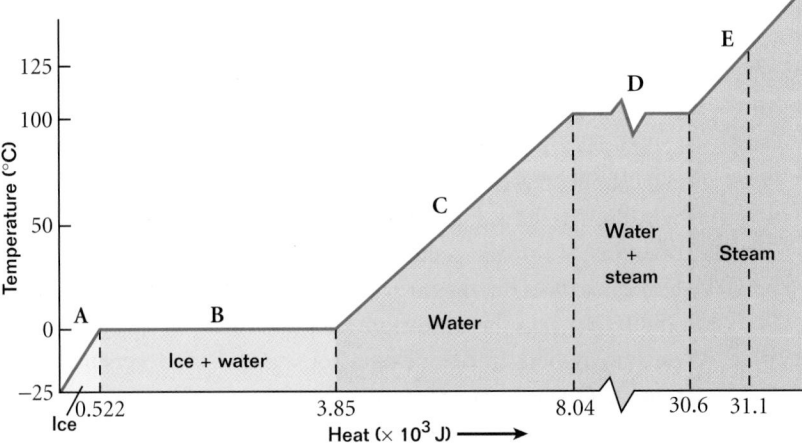

Figure 10-13

This idealized graph shows the temperature change of 10.0 g of ice as it is heated from −25°C in the ice phase to steam above 125°C at atmospheric pressure.

The situation at 0°C is very different. Despite the fact that energy is continuously being added, there is no change in temperature. Instead, the nature of the ice changes. The ice begins to melt and change into water at 0°C (segment **B**). The ice-and-water mixture remains at this temperature until all of the ice melts. From 0°C to 100°C, the water's temperature steadily increases (segment **C**). At 100°C, however, the temperature stops rising and the water turns into

Table 10-5	Changes occurring during the heating of 10.0 g of ice		
Segment of graph	**Type of change**	**Amount of energy transferred as heat**	**Temperature range of segment**
A	temperature of ice increases	522 J	−25°C to 0°C
B	ice melts; becomes water	3.33×10^3 J	0°C
C	temperature of water increases	4.19×10^3 J	0°C to 100°C
D	water boils; becomes steam	2.26×10^4 J	100°C
E	temperature of steam increases	502 J	100°C to 125°C

steam (segment **D**). Once the water has completely vaporized, the temperature of the steam increases (segment **E**). Steam whose temperature is greater than the boiling point of water is referred to as *superheated.*

When substances melt, freeze, boil, condense, or sublime (change from a solid to vapor or from vapor to a solid), the energy added or removed changes the internal energy of the substance without changing its temperature. These changes in matter are called **phase changes.**

The existence of phase changes requires that the definition of heat be expanded. *Heat is the energy that is exchanged between two objects at different temperatures or between two objects at the same temperature when one of them is undergoing a phase change.*

Phase changes involve potential energy between particles ★TEKS 5B

To understand the behavior of a substance undergoing a phase change, you will need to recall how energy is transferred in collisions. Potential energy is the energy an object has because of its position relative to another object. Examples of potential energy are a pencil about to fall from your desk or a rubber band held in a tightly stretched position. Potential energy is present among a collection of particles in a solid or a liquid in the form of attractive bonds. These bonds result from the charges within atoms and molecules. Potential energy is associated with the electric forces between these charges.

The equilibrium separation between atoms or molecules corresponds to a position at which there is a minimum potential energy. The potential energy increases with increasing atomic separation from the equilibrium position. This resembles the elastic potential energy of a spring, as discussed in Chapter 5. For this reason, a collection of individual atoms or molecules and the bonds between them are often modeled as masses at the ends of springs.

If the particles are far enough apart, the bonds between them can break. The work needed to increase potential energy and break a bond is provided by collisions with energetic atoms or molecules, as shown in **Figure 10-14.** Just as bonds can be broken, new bonds can be formed if atoms or molecules are brought close together. This involves the collection of particles going from a high potential energy (large average separation) to a lower potential energy (small average separation). This decrease in potential energy involves a release of energy in the form of increasing kinetic energy of nearby particles.

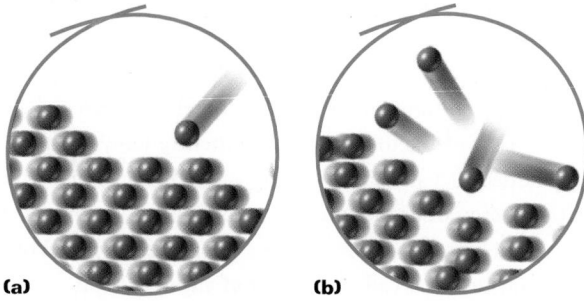

(a)   **(b)**

phase change

---

the physical change of a substance from one state (solid, liquid, or gas) to another at constant temperature and pressure

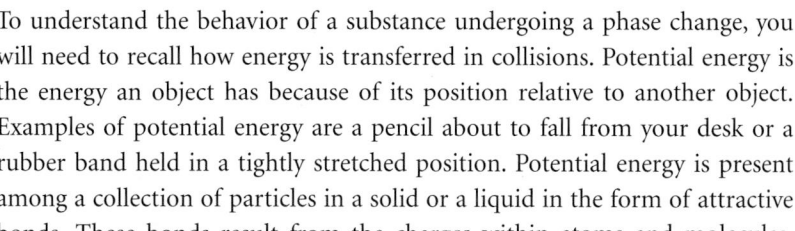

Figure 10-14
Energy added to a substance **(a)** can increase the vibrational kinetic energy of its particles or break the bonds between those particles **(b).**

 Misconception Alert

Students know from lessons in chemistry that heating and cooling can cause changes in the chemical properties of materials. To ensure they distinguish these phenomena from the physical process of phase change, ask if a substance chemically changes in the course of melting or boiling.

★TEKS

p. 376: 7B, 2C
p. 377: 5B

The Language of Physics

There is more than one use of the term *fusion* in physics. Some students may be familiar with fusion as a nuclear process. In this context, the term *fusion* refers not to a nuclear process but to the phase change from a liquid to a solid. Point out that the reverse process of fusion is melting and that it occurs at the same temperature, called the melting point (or freezing point). Likewise, the reverse of vaporization is condensation. Vaporization and condensation occur at the boiling point, which is different for each substance. Discuss whether the term *boiling* always means "high-temperature."

Teaching Tip

Be sure that students understand the relationship between heat of fusion, heat of vaporization, and latent heat. Latent heat can be either the heat of fusion or the heat of vaporization. The heat of fusion corresponds to melting or solidification, and the heat of vaporization corresponds to vaporization or condensation.

Figure 10-15
The heat of fusion is the difference between the energy needed to break bonds in a solid and the energy released when bonds form in a liquid.

heat of fusion

the energy per unit mass transferred in order to change a substance from solid to liquid or from liquid to solid at constant temperature and pressure

heat of vaporization

the energy per unit mass transferred in order to change a substance from liquid to vapor or from vapor to liquid at constant temperature and pressure

Energy required to melt a substance goes into rearranging the molecules

Phase changes result from a change in the potential energy between particles of a substance. When energy is added to or removed from a substance undergoing a phase change, the particles of the substance rearrange themselves to make up for their change of energy. This occurs without a change in the average kinetic energy of the particles.

For instance, if ice is melting, the absorbed energy is sufficient to break the weak bonds that hold the water molecules together as a well-ordered crystal. New but different bonds form between the liquid water molecules that have separated from the crystal, so some of the absorbed energy is released again. The difference between the potential energies of the broken bonds and the newly formed bonds is equal to the net energy added to the ice, as shown in **Figure 10-15.** As a result, the energy used to rearrange the molecules is not available to increase the molecules' kinetic energy, and therefore no increase in the temperature of the ice-and-water mixture occurs. ⭐TEKS 7B

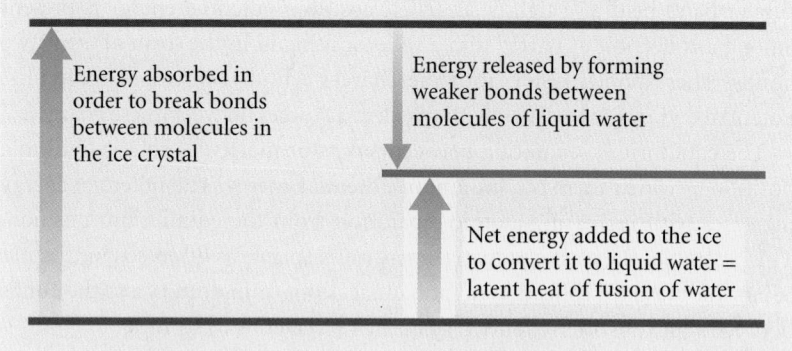

Energy absorbed in order to break bonds between molecules in the ice crystal

Energy released by forming weaker bonds between molecules of liquid water

Net energy added to the ice to convert it to liquid water = latent heat of fusion of water

For any substance, the energy added to the substance during melting equals the difference between the total potential energies for particles in the solid and the liquid phases. This energy per unit mass is called the **heat of fusion.**

Energy required to vaporize a substance mostly goes into separating the molecules

As in a solid, the molecules are close together in a liquid, and in liquid water they are even closer together than in ice. The forces between liquid water molecules are stronger than those that exist between the more widely separated water molecules in steam. Therefore, at 100°C, all the energy absorbed by water goes into overcoming the attractive forces between the liquid water molecules. None is available to increase the kinetic energy of the molecules.

The energy added to a substance during vaporization equals the difference in the potential energy of attraction between the particles of a liquid and the potential energy of attraction between the gas particles (see **Figure 10-16**). This energy per unit mass is called the **heat of vaporization.**

Because they have few close neighbors, the particles in the gas phase gain very little energy from weak bonding. Therefore, more energy is required to vaporize a given mass of substance than to melt it. As a result, the heat of vaporization is much greater than the heat of fusion. Both the heat of fusion and the heat of vaporization are classified as **latent heat.** ⭐TEKS 3A

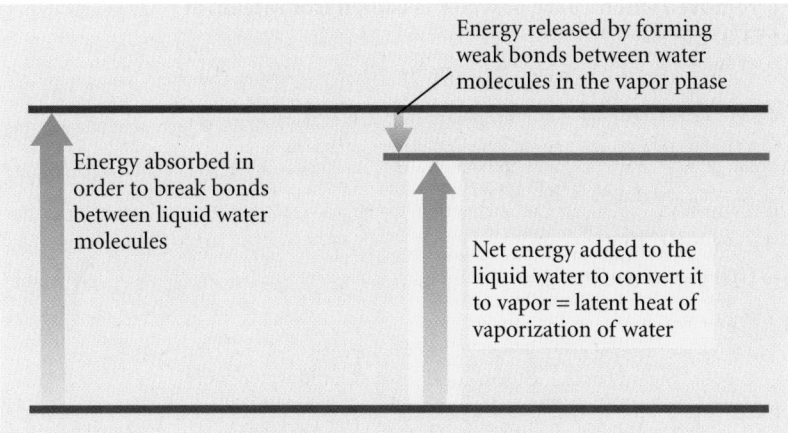

Energy released by forming weak bonds between water molecules in the vapor phase

Energy absorbed in order to break bonds between liquid water molecules

Net energy added to the liquid water to convert it to vapor = latent heat of vaporization of water

latent heat

the energy per unit mass that is transferred during a phase change of a substance

Figure 10-16
The heat of vaporization is mostly the energy required to separate molecules from the liquid phase.

LATENT HEAT

$$Q = mL$$

Energy transferred as heat during a phase change = mass × latent heat

For calculations involving melting or freezing, the latent heat of fusion is noted by the symbol L_f. Similarly, for calculations involving vaporizing or condensing, the symbol L_v is used for latent heat of vaporization. **Table 10-6** lists latent heats for a few substances. ⭐TEKS 2C

Table 10-6	Latent heats of fusion and vaporization at standard pressure			
Substance	Melting point (°C)	L_f (J/kg)	Boiling point (°C)	L_v (J/kg)
nitrogen	−209.97	2.55×10^4	−195.81	2.01×10^5
oxygen	−218.79	1.38×10^4	−182.97	2.13×10^5
ethyl alcohol	−114	1.04×10^5	78	8.54×10^5
water	0.00	3.33×10^5	100.00	2.26×10^6
lead	327.3	2.45×10^4	1745	8.70×10^5
aluminum	660.4	3.97×10^5	2467	1.14×10^7

SECTION 10-3

Misconception Alert

Notice that the values given in **Table 10-6** are valid for pure substances at standard pressure only. The boiling point can be raised or lowered by increasing or decreasing pressure. Furthermore, mixtures may not have the same melting point as pure substances. For example, salt lowers the melting point of ice.

Demonstration 3

Lowering boiling points

Purpose Show that the boiling point of a liquid depends on pressure.

Materials bell jar connected to a vacuum pump, glass of warm water

Procedure Place the glass of warm water in the bell jar, and begin to exhaust the air. Have the students observe the water as the air is evacuated from the bell jar. The water will soon boil. Ask the students what effect lowering air pressure has on the temperature at which water boils. (*Decreasing pressure lowers the boiling point.*) Ask students why it is necessary to boil a three-minute egg for more than 3 min at higher altitudes. (*Water boils at a lower temperature at higher altitudes because atmospheric pressure is less and the internal energy of the boiling water is therefore lower.*)

⭐TEKS
p. 378: 7B
p. 379: 3A, 2C

Classroom Practice

*The following may be used
as a teamwork exercise or for
demonstration at the chalkboard
or on an overhead projector.*

PROBLEM

Heat of phase change

How much energy is needed to
melt a 100.0 g sample of aluminum whose initial temperature is 20.0°C?

Answer

9.73×10^4 J

Teaching Tip

Point out to students that when
they are solving problems, it is
important to identify the needed
information before beginning the
problem. Sometimes this information is not given in the problem and must be found in other
reference sources. For example,
not all of the necessary data for
Sample Problem 10D are found
in the problem text. Ask students
to identify what data they would
need to find in this case and what
reference sources they could use.
(**Table 10-6** *gives the boiling point
and latent heat of vaporization for
water, and* **Table 10-4** *gives the
specific heat capacities of water
and steam.*)

⭐ TEKS
p. 380: 2C, 5B
p. 381: 2C, 5B

SAMPLE PROBLEM 10D

Heat of phase change ⭐TEKS 2C, 5B

PROBLEM

**How much energy is removed when 10.0 g of water is cooled from steam at
133.0°C to liquid at 53.0°C?**

SOLUTION

1. DEFINE   **Given:** $T_{steam} = T_s = 133.0$ °C $T_{water} = T_w = 53.0$°C

$c_{p,steam} = c_{p,s} = 2.01 \times 10^3$ J/kg•°C
$c_{p,water} = c_{p,w} = 4.186 \times 10^3$ J/kg•°C
$L_v = 2.26 \times 10^6$ J/kg
$m = 10.0$ g $= 10.0 \times 10^{-3}$ kg

Unknown: $Q_{total} = ?$

Diagram:

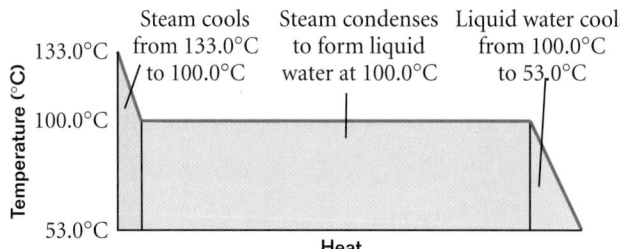

Steam cools | Steam condenses | Liquid water cools
from 133.0°C | to form liquid | from 100.0°C
to 100.0°C | water at 100.0°C | to 53.0°C

2. PLAN   **Choose an equation(s) or situation:** Heat is calculated by using $Q = mc_p\Delta T$
when no phase changes occur. When steam changes to liquid water, a phase
change occurs and the equation for the heat of vaporization, $Q = mL_v$, must
be used. Be sure that ΔT is positive for each step.

To cool the steam to 100.0°C: $Q_1 = mc_{p,s}\Delta T$

To change steam to water at 100.0°C: $Q_2 = mL_v$

To cool the water to 53.0°C: $Q_3 = mc_{p,w}\Delta T$

3. CALCULATE   **Substitute values into the equation(s) and solve:** Find ΔT for steam cooling
and water cooling. Calculate Q for both cooling steps and the phase change.

For the cooling steam:

$$\Delta T_s = 133.0°C - 100.0°C = 33.0°C$$

$$Q_1 = mc_{p,s}\Delta T = (10.0 \times 10^{-3} \text{ kg})\left(2.01 \times 10^3 \frac{J}{kg \cdot °C}\right)(33.0°C)$$

$$= 663 \text{ J}$$

For the steam condensing to water:

$$Q_2 = mL_v = (10.0 \times 10^{-3} \text{ kg})\left(2.26 \times 10^6 \frac{J}{kg}\right)$$

$$= 2.26 \times 10^4 \text{ J}$$

For the cooling water:

$$\Delta T_w = 100.0°C - 53.0°C = 47.0°C$$

$$Q_3 = mc_{p,w}\Delta T = (10.0 \times 10^{-3}\text{ kg})\left(4.186 \times 10^3\frac{\text{J}}{\text{kg}\cdot°C}\right)(47.0°C)$$

$$= 1.97 \times 10^3\text{ J}$$

$$Q_{total} = Q_1 + Q_2 + Q_3 =$$
$$663\text{ J} + (2.26 \times 10^4\text{ J})$$
$$+ (1.97 \times 10^3\text{ J}) = 2.52 \times 10^4\text{ J}$$

$$\boxed{Q_{total} = 2.52 \times 10^4\text{ J removed}}$$

CALCULATOR SOLUTION

Because of the significant figure rule for addition, the calculator answer, 25233, should be rounded to 2.52×10^4.

4. EVALUATE Most of the energy is added to or removed from a substance during phase changes. In this example, about 90 percent of the energy removed from the steam is accounted for by the heat of vaporization.

PRACTICE GUIDE 10D

Solving for:		
Q	📖 PE	Sample, 1–4; Ch. Rvw. 32–33
	💿 PW	6–7
	PB	9–10
ΔT	📖 PE	5–6
	💿 PW	Sample, 1–2
	PB	6–8
L	💿 PW	3, 5
	PB	Sample, 1–3
m	💿 PW	4, 6
	PB	4–5

PRACTICE 10D

Heat of phase change ⭐TEKS **2C, 5B**

1. How much energy is required to change a 42 g ice cube from ice at −11°C to steam at 111°C? (Hint: Refer to **Tables 10-4** and **10-6**.)

2. Liquid nitrogen, which has a boiling point of 77 K, is commonly used to cool substances to low temperatures. How much energy must be removed from 1.0 kg of gaseous nitrogen at 77 K for it to completely liquefy?

3. How much energy is needed to melt 0.225 kg of lead so that it can be used to make a lead sinker for fishing? The sample has an initial temperature of 27.3°C and is poured in the mold immediately after it has melted.

4. How much energy is needed to melt exactly 1000 aluminum cans, each with a mass of 14.0 g, for recycling? Assume an initial temperature of 26.4°C.

5. A 0.011 kg cube of ice at 0.0°C is added to 0.450 kg of soup at 80.0°C. Assuming that the soup has the same specific heat capacity as water, find the final temperature of the soup after the ice has melted. (Hint: There is a temperature change after the ice melts.)

6. At a foundry, 25 kg of molten aluminum with a temperature of 660.4°C is poured into a mold. If this is carried out in a room containing 130 kg of air at 25°C, what is the temperature of the air after the aluminum is completely solidified? Assume that the specific heat capacity of air is 1.0×10^3 J/kg·°C.

ANSWERS TO

Practice 10D
Heat of phase change
1. 1.29×10^5 J
2. 2.0×10^5 J
3. 1.415×10^4 J
4. 1.354×10^7 J
5. 76.2°C
6. 101°C

Section Review

ANSWERS

1. 6.8 g

2. When firewood is damp, a large amount of energy is used to increase the water's temperature and then to vaporize the water (because water has both a high specific heat capacity and a high latent heat of vaporization). After this has been accomplished, the remaining energy is used to burn the wood. Thus, much more energy is required when the wood is damp.

3. The steam has more internal energy in the form of latent heat and thus will transfer a great deal of energy at 100°C before its temperature decreases. The temperature of 100°C water will immediately begin decreasing as energy is transferred to the body.

4. (Estimated values may vary.)
   a. 2×10^3 J/kg•°C
   b. 4.7×10^5 J/kg
   c. 1×10^3 J/kg•°C
   d. 7×10^2 J/kg•°C
   e. 5.2×10^7 J/kg

5. 3340 J

6. 2.7×10^4 J

1. A jeweler working with a heated 47 g gold ring must lower the ring's temperature to make it safe to handle. If the ring is initially at 99°C, what mass of water at 25°C is needed to lower the ring's temperature to 38°C?

2. Using the concepts of latent heat and internal energy, explain why it is difficult to build a fire with damp wood.

3. Why does steam at 100°C cause more severe burns than does liquid water at 100°C?

4. From the heating curve for a 15 g sample, as shown in **Figure 10-17,** estimate the following properties of the substance.
   a. the specific heat capacity of the liquid
   b. the latent heat of fusion
   c. the specific heat capacity of the solid
   d. the specific heat capacity of the vapor
   e. the latent heat of vaporization

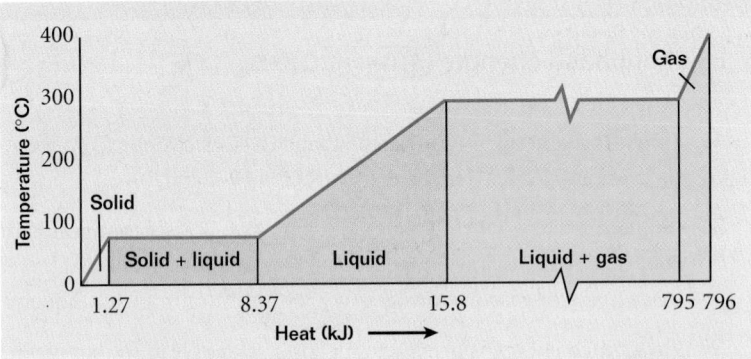

Figure 10-17

5. **Physics in Action** How much energy must be added to a bowl of 125 popcorn kernels in order for them to reach a popping temperature of 175°C? Assume that their initial temperature is 21°C, that the specific heat capacity of popcorn is 1650 J/kg•°C, and that each kernel has a mass of 0.105 g.

6. **Physics in Action** Because of the pressure inside a popcorn kernel, water does not vaporize at 100°C. Instead, it stays liquid until its temperature is about 175°C, at which point the kernel ruptures and the superheated water turns into steam. How much energy is needed to pop 95.0 g of corn if 14 percent of a kernel's mass consists of water? Assume that the latent heat of vaporization for water at 175°C is 0.90 times its value at 100°C and that the kernels have an initial temperature of 175°C.

10-4
Controlling heat

THERMAL CONDUCTION

When you first place an iron skillet on a range burner or element, the metal handle feels comfortable to the touch. But after a few minutes the handle becomes too hot to touch without a cooking mitt. During that time, energy is transferred as heat from the high-temperature burner to the skillet. This type of energy transfer is called **thermal conduction.**

Thermal conduction can be understood by the behavior of atoms in a metal. Before the skillet is placed on the heating element, the skillet's iron atoms have an energy proportional to the temperature of the room. As the skillet is heated, the atoms nearest the heating element vibrate with greater energy. These vibrating atoms jostle their less energetic neighbors and transfer some of their energy in the process. Gradually, iron atoms farther away from the element gain more energy. ⭐TEKS 7B

The rate of thermal conduction depends on the properties of the substance being heated. A metal ice tray and a package of frozen food removed from the freezer are at the same temperature. However, the metal tray feels colder because metal conducts energy more easily and more rapidly than cardboard at the place where it comes into contact with your hand. In contrast, a piece of ceramic conducts energy very slowly, as may be seen in **Figure 10-18.** The end of the ceramic piece that is embedded in ice is barely affected by the energy of the flame surrounding the other end. Substances that rapidly transfer energy as heat are called *thermal conductors,* while those that slowly transfer energy as heat are called *thermal insulators.* ⭐TEKS 3A

In general, metals are good thermal conductors. Materials such as asbestos, cork, ceramic, cardboard, and fiberglass are poor thermal conductors (and therefore good thermal insulators). Gases also are poor thermal conductors. The gas particles are so far apart with respect to their size that collisions between them are rare, and their kinetic energy is transferred slowly.

Although cooking oil is not any better as a thermal conductor than most nonmetals, it is useful for transferring energy uniformly around the surface of the food being cooked. When popping popcorn, for instance, coating the kernels with oil improves the energy transfer to each kernel, so that a higher percentage of them pop.

10-4 SECTION OBJECTIVES

- **Explain how energy is transferred as heat through the process of thermal conduction.**

- **Recognize how energy transfer can be controlled with clothing.**

thermal conduction

the process by which energy is transferred as heat through a material between two points at different temperatures

Figure 10-18
Ceramics are poor thermal conductors, as indicated in this photograph.

Teaching Tip

Point out that the electromagnetic radiation emitted by the human body cannot be seen by humans because the radiation's wavelengths are primarily in the infrared portion of the electromagnetic spectrum. However, this radiation can be seen by some snakes. By detecting infrared radiation, these snakes seek out their prey in what appears to humans to be complete darkness. A similar approach is used in some types of night-vision goggles.

Teaching Tip

Point out that the wavelengths emitted by any object depend on the object's temperature. As the temperature of an object increases, the emitted wavelengths shift toward shorter wavelengths in the electromagnetic spectrum. This accounts for the red glow of an electric burner and, at an even higher temperature, the white glow of the tungsten filament in a light bulb.

 TEKS

p. 384: 3C
p. 385: 3C

internet connect

SCiLINKS

NSTA

TOPIC: Conduction and convection
GO TO: www.scilinks.org
sciLINKS CODE: HF2105

CONCEPT PREVIEW →

Electromagnetic radiation will be discussed in more detail in Chapter 14.

Figure 10-19
The Inupiat parka, called an *atigi,* consists today of a canvas shell over sheepskin. The wool provides layers of insulating air between the wearer and the cold.

Convection and radiation also transfer energy

There are two other mechanisms for transferring energy between places or objects at different temperatures. *Convection* involves the displacement of cold matter by hot matter, such as when hot air over a flame rises upward. This mechanism does not involve heat alone. Instead, it uses the combined effects of pressure differences, conduction, and buoyancy. In the case of air over a flame, the air is heated through particle collisions (conduction), causing it to expand and its density to decrease. The warm air is then displaced by denser, colder air from above.

The other principal energy transfer mechanism is *electromagnetic radiation,* which includes visible light. Unlike convection, energy in this form does not involve the transfer of matter. Instead, objects reduce their internal energy by radiating electromagnetic radiation with particular wavelengths. This energy transfer often takes place from high temperature to low temperature, so that radiation is frequently associated with heat. The human body emits energy in the infrared portion of the electromagnetic spectrum.

CLOTHING AND CLIMATE

To remain healthy, the human body must maintain a temperature close to 37.0°C (98.6°F). This becomes increasingly difficult as the surrounding air becomes hotter or colder than body temperature.

Without proper insulation, the body's temperature will drop in its attempt to reach thermal equilibrium with very cold surroundings. If this situation is not corrected in time, the body will enter a state of hypothermia, which lowers pulse, blood pressure, and respiration. Once the body temperature reaches 32.2°C (90.0°F), the person can lose consciousness. When the body temperature reaches 25.6°C (78.0°F), hypothermia is almost always fatal.

Insulating materials retain energy for cold climates

To prevent hypothermia, the transfer of energy from the human body to the surrounding air must be hindered. This is done by surrounding the body with heat-insulating material. An extremely effective and common thermal insulator is air. Like most gases, air is a very poor thermal conductor, so even a thin layer of air near the skin provides a barrier to energy transfer.

The Inupiat Eskimo people of northern Alaska have designed clothing to protect them from the severe Arctic climate, where average air temperatures range from 10°C (50°F) to −37°C (−35°F). The Inupiat clothing is made from animal skins that make use of air's insulating properties. Until recently, the traditional parka *(atigi)* was made from caribou skins. Two separate parkas are worn in layers, with the fur lining the inside of the inner parka and the outside of the outer parka. Insulation is provided by air trapped between the short inner hairs and within the long, hollow hairs of the fur. Today, inner parkas are made from sheepskin (see **Figure 10-19**). **TEKS** 3C

Evaporation aids energy transfer in hot climates ⊛TEKS 3C

At the other extreme, the Bedouins of the Arabian Desert have developed clothing that permits them to survive another of the harshest environments on Earth. Bedouin garments cover most of the body, thus protecting the wearer from direct sunlight and preventing excessive loss of body water from evaporation. These clothes are also designed to cool the wearer. The Bedouins must keep their body temperatures from becoming too high in desert temperatures, which often are in excess of 38°C (100°F). Heat exhaustion or heatstroke will result if the body's temperature becomes too high.

Although there are a number of differences among the types of clothing worn by different tribes and by men and women within tribes, a few basic garments are common to all Bedouins. One of these is the loose-fitting, elongated linen shirt called a *dish-dash* or *dish-dasha*, depending on whether it is worn by men or women, respectively. This shirt is worn close to the body, usually over an undergarment.

The loose fit and flared cut of the *dish-dash* permits air to flow over the wearer's skin. This causes any perspiration that has collected on the skin's surface to evaporate. During evaporation, water molecules enter the vapor phase. Because of the high specific heat capacity and latent heat of vaporization for water, evaporation removes a good deal of energy from the skin and air, thus causing the skin to cool.

Another common article of clothing is the *kefiyah*, a headcloth worn by Bedouin men, as shown in **Figure 10-20.** A similar garment made of two separate cloths, which are called a *mandil* and a *hatta*, is worn by Bedouin women. Firmly wrapped around the head of the wearer, the cloth absorbs perspiration and cools the wearer during evaporation. This same garment is also useful during cold periods in the desert. Wound snugly around the head, the garment traps air within its folds, thus providing an insulating layer to keep the head warm.

Figure 10-20
The Bedouin headcloth, called a *kefiyah*, employs evaporation to remove energy from the air close to the head, thus cooling the wearer.

Section Review

1. Why do fluffy down comforters feel warmer than thin cloth blankets?

2. Explain how conduction causes water on the surface of a bridge to freeze sooner than water on the road surface on either side of the bridge.

3. On a camping trip, your friend tells you that fluffing up a down sleeping bag before you go to bed will keep you warmer than sleeping in the same bag when it is still crushed from being in its storage sack. Explain why this happens. (Hint: A large amount of air is present in an uncrushed sleeping bag.)

Section Review
ANSWERS

1. Down comforters feel warmer because they contain more air, which is a thermal insulator.

2. The bridge's surface is directly exposed to the cold air both above and below. The surface of the road is exposed to the air on only one side, so the transfer of energy as heat is less rapid on the road than it is on the bridge. In addition, the large, stable temperature of the ground below the road helps to keep the temperature of the road stable. Because the ground temperature is often above freezing, the temperature of the road pavement is also above freezing. The pavement on the bridge does not have this same source of energy, so its temperature rapidly approaches that of the cold air.

3. If the insulating layer of air is squeezed out of the sleeping bag, you will not feel as warm.

CHAPTER 10
Summary

KEY TERMS

calorimetry (p. 372)

heat (p. 365)

heat of fusion (p. 378)

heat of vaporization (p. 378)

internal energy (p. 359)

latent heat (p. 379)

phase change (p. 377)

specific heat capacity (p. 371)

thermal conduction (p. 383)

thermal equilibrium (p. 360)

KEY IDEAS

Section 10-1 Temperature and thermal equilibrium
- Temperature can be changed by transferring energy to or from a substance.
- Thermal equilibrium is the condition in which the temperature of two objects in physical contact with each other is the same.

Section 10-2 Defining heat
- Heat is energy that is transferred from objects at higher temperatures to objects at lower temperatures.
- Energy is conserved when mechanical energy and internal energy are taken into account.

Section 10-3 Changes in temperature and phase
- Specific heat capacity, which is a measure of the energy needed to change a substance's temperature, is described by the following formula:

$$c_p = \frac{Q}{m\Delta T}$$

- Latent heat, the energy required to change the phase of a substance, is described by the following formula:

$$L = \frac{Q}{m}$$

Section 10-4 Controlling heat
- Energy is transferred by thermal conduction through particle collisions.

Variable symbols

Quantities		Units	
T	temperature (Kelvin)	K	kelvins
T_C	temperature (Celsius)	°C	degrees Celsius
T_F	temperature (Fahrenheit)	°F	degrees Fahrenheit
ΔU	change in internal energy	J	joules
Q	heat	J	joules
c_p	specific heat capacity at constant pressure	$\dfrac{J}{kg \cdot °C}$	
L	latent heat	$\dfrac{J}{kg}$	

CHAPTER 10
Review and Assess

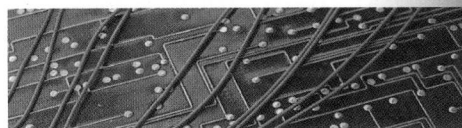

TEMPERATURE AND THERMAL EQUILIBRIUM

Review questions

1. What is the relationship between temperature and internal energy?

2. What property of two objects determines if the two are in a state of thermal equilibrium?

3. What are some physical properties that could be used in developing a temperature scale?

4. What property must a substance have in order to be used for calibrating a thermometer?

Conceptual questions

5. Which object in each of the following pairs has greater total internal energy, assuming that both objects in each pair are in thermal equilibrium? Explain your reasoning in each case.
   a. a metal knife in thermal equilibrium with a hot griddle
   b. a 1 kg block of ice at −25°C or seven 12 g ice cubes at −25°C

6. Assume that each pair of objects in item 5 has the same internal energy instead of the same temperature. Which item in each pair will have the higher temperature?

7. Why are the steam and ice points of water better fixed points for a thermometer than the temperature of a human body?

8. How does the temperature of a tub of hot water as measured by a thermometer differ from the water's temperature before the measurement is made? What property of a thermometer is necessary for the difference between these two temperatures to be minimized?

Practice problems

9. The highest recorded temperature on Earth was 136°F, at Azizia, Libya, in 1922. Express this temperature in degrees Celsius and in kelvins.
   (See Sample Problem 10A.)

10. The melting point of gold is 1947°F. Express this temperature in degrees Celsius and in kelvins.
    (See Sample Problem 10A.)

DEFINING HEAT

Review questions

11. Which drawing in **Figure 10-21** correctly shows the direction in which the net energy is transferred by heat between an ice cube and the freezer walls when the temperature of both is −10°C? Explain your answer.

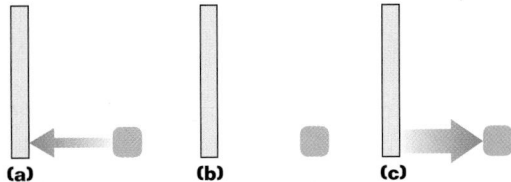

(a)   **(b)**   **(c)**

Figure 10-21

12. A glass of water has a temperature of 8°C. In which situation will more energy be transferred, when the air's temperature is 25°C or 35°C?

13. How much energy is transferred between a piece of toast and an oven when both are at a temperature of 55°C? Explain.

Conceptual questions

14. If water in a sealed, insulated container is stirred, is its temperature likely to increase slightly, decrease slightly, or stay the same? Explain your answer.

15. More liquid is in contact with the air, so internal energy is transferred from coffee to air.

16. Heat is energy in transit, not a substance. Temperature is proportional to internal energy, not to heat.

17. More energy is transferred from the higher-temperature object because the average KE of its particles is greater.

18. b; larger temperature difference

19. 7.9×10^4 J

20. a. 2.9 J
b. It dissipates into the air, the ground, and the hammer.

21. masses and initial values of T for sample and water, T_{eq}, $c_{p,w}$

22. conservation of energy; Energy removed from sample equals energy added to water.

23. Energy is used to break bonds between molecules in ice.

24. The air gives up energy to water, which changes phase, so the air's temperature decreases.

25. ethyl alcohol

26. A large mass of cool earth around the cellar does not undergo as large a temperature change during the year as does the outside air.

27. The water's large latent heat of fusion prevented it and the adjacent produce from being quickly frozen by the cold air.

28. Water on an orange's surface gives up energy before the orange's interior does, preventing the orange from freezing.

29. a. 1.1×10^3 J/kg•°C
b. 4.4×10^5 J/kg
c. 5.0×10^2 J/kg•°C
d. 4×10^2 J/kg•°C
e. 3.6×10^7 J/kg
(Estimates on the specific heat capacities may vary.)

15. Given your answer to item 14, why does stirring a hot cup of coffee cool it down?

16. Given any two bodies, the one with the higher temperature contains more heat. What is wrong with this statement?

17. Use the kinetic theory of atoms and molecules to explain why energy that is transferred as heat always goes from objects at higher temperatures to those at lower temperatures.

18. In which of the two situations described is more energy transferred? Explain your answer.
a. a cup of hot chocolate with a temperature of 40°C inside a freezer at −20°C
b. the same cup of hot chocolate at 90°C in a room at 25°C

Practice problems

19. A force of 315 N is applied horizontally to a wooden crate in order to displace it 35.0 m across a level floor at a constant velocity. As a result of this work the crate's internal energy is increased by an amount equal to 14 percent of the crate's initial internal energy. Calculate the initial internal energy of the crate. (See Sample Problem 10B.)

20. A 0.75 kg spike is hammered into a railroad tie. The initial speed of the spike is equal to 3.0 m/s.
a. If the tie and spike together absorb 85 percent of the spike's initial kinetic energy as internal energy, calculate the increase in internal energy of the tie and spike.
b. What happens to the remaining energy?
(See Sample Problem 10B.)

CHANGES IN TEMPERATURE AND PHASE

Review questions

21. What data are required in order to determine the specific heat capacity of an unknown substance by means of calorimetry?

22. What principle permits calorimetry to be used to determine the specific heat capacity of a substance? Explain.

23. Why does the temperature of melting ice not change even though energy is being transferred as heat to the ice?

Conceptual questions

24. Why does the evaporation of water cool the air near the water's surface?

25. Ethyl alcohol has about one-half the specific heat capacity of water. If equal masses of alcohol and water in separate beakers at the same temperature are supplied with the same amount of energy, which will have the higher final temperature?

26. Until refrigerators were invented, many people stored fruits and vegetables in underground cellars. Why was this more effective than keeping them in the open air?

27. During the winter, the people mentioned in item 26 would often place an open barrel of water in the cellar alongside their produce. Explain why this was done and why it would be effective.

28. During a cold spell, Florida orange growers often spray a mist of water over their trees during the night. What does this accomplish?

29. From the heating curve for a 23 g sample (see **Figure 10-22**), estimate the following properties of the substance.
a. the specific heat capacity of the liquid
b. the latent heat of fusion
c. the specific heat capacity of the solid
d. the specific heat capacity of the vapor
e. the latent heat of vaporization

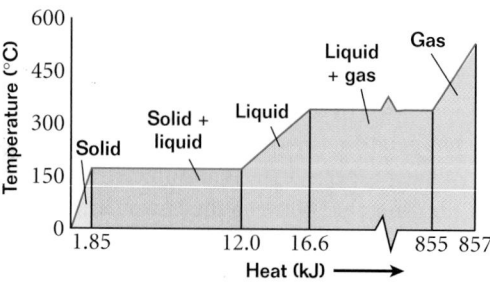

Figure 10-22

Practice problems

30. A 25.5 g silver ring ($c_p = 234$ J/kg•°C) is heated to a temperature of 84.0°C and then placed in a calorimeter containing 5.00×10^{-2} kg of water at 24.0°C. The calorimeter is not perfectly insulated, however, and 0.140 kJ of energy is transferred to the surroundings before a final temperature is reached. What is the final temperature?
(See Sample Problem 10C.)

31. When a driver brakes an automobile, friction between the brake disks and the brake pads converts part of the car's translational kinetic energy to internal energy. If a 1500 kg automobile traveling at 32 m/s comes to a halt after its brakes are applied, how much can the temperature rise in each of the four 3.5 kg steel brake disks? Assume the disks are made of iron ($c_p = 448$ J/kg•°C) and that all of the kinetic energy is distributed in equal parts to the internal energy of the brakes.
(See Sample Problem 10C.)

32. A plastic-foam container used as a picnic cooler contains a block of ice at 0°C. If 225 g of ice melts, how much heat passes through the walls of the container?
(See Sample Problem 10D.)

33. The largest of the Great Lakes, Lake Superior, contains about 1.20×10^{16} kg of water. If the lake had a temperature of 12.0°C, how much energy would have to be removed to freeze the whole lake at 0°C?
(See Sample Problem 10D.)

THERMAL CONDUCTION AND INSULATION

Review questions

34. How does a metal rod conduct energy from one end, which has been placed in a fire, to the other end, which is at room temperature?

35. How does air within winter clothing keep you warm on cold winter days?

Conceptual questions

36. A metal spoon is placed in one of two identical cups of hot coffee. Which cup will be cooler after a few minutes?

37. A tile floor may feel uncomfortably cold to your bare feet, but a carpeted floor in an adjoining room at the same temperature feels warm. Why?

38. Why is it recommended that several items of clothing be worn in layers on cold days?

39. Why does a fan make you feel cooler on a hot day?

40. A paper cup is filled with water and then placed over an open flame, as shown in **Figure 10-23**. Explain why the cup does not catch fire and burn.

MIXED REVIEW

Figure 10-23

41. Absolute zero on the Rankine temperature scale is $T_R = 0°R$, and the scale's unit is the same size as the Fahrenheit degree.
  a. Write a formula that relates the Rankine scale to the Fahrenheit scale.
  b. Write a formula that relates the Rankine scale to the Kelvin scale.

42. A 3.0 kg rock is initially at rest at the top of a cliff. Assuming the rock falls into the sea at the foot of the cliff and that its kinetic energy is transferred entirely to the water, how high is the cliff if the temperature of 1.0 kg of water is raised 0.10°C?

43. Convert the following temperatures to degrees Fahrenheit and kelvins.
  a. the boiling point of liquid hydrogen (−252.87°C)
  b. the temperature of a room at 20.5°C

44. The freezing and boiling points of water on the imaginary "Too Hot" temperature scale are selected to be exactly 50 and 200 degrees TH.
  a. Derive an equation relating the Too Hot scale to the Celsius scale. (Hint: Make a graph of one temperature scale versus the other, and solve for the equation of the line.)
  b. Calculate absolute zero in degrees TH.

30. 15.0°C

31. 120°C

32. 7.49×10^4 J

33. 4.60×10^{21} J

34. The atoms and free electrons in the metal bump into one another, thereby transferring energy.

35. Air is an insulator, so the internal energy of your body is not as easily transferred to the environment.

36. The coffee in the cup with the metal spoon is cooler because the metal absorbs energy more rapidly from the coffee.

37. The tile floor is a better thermal conductor than the carpet. Thus, the energy transfer from your feet to the tile happens faster.

38. The air between each of the layers acts as a thermal insulator.

39. Evaporation requires energy and thus cools the surrounding air. A fan continually supplies new, dry air, so perspiration is continually evaporated and the body is cooled.

40. The paper nearest to the flame maintains a temperature equal to that of the boiling water as long as the water has not entirely boiled away. The cup doesn't burn because the heat input at the bottom of the cup is rapidly conducted to the boiling water.

41. a. $T_R = T_F + 459.7$, or $T_F = T_R - 459.7$
  b. $T = \frac{5}{9} T_R$, or $T_R = \frac{9}{5} T$

42. 14 m

43. a. −423.2°F, 20.28 K
  b. 68.9°F, 293.6 K

44. a. $T_{TH} = \frac{3}{2} T_C + 50$, or $T_C = \frac{2}{3}(T_{TH} - 50)$
  b. −360° TH

ANSWERS TO

45. $T_F = \frac{9}{5}(-40°C) + 32.0 = -40°F$
46. 5.1×10^4 s, or 14 h
47. 330 g

Technology & Learning

Answers may vary slightly, depending on viewing window settings.
a. Y_1 = final temperature
   T = initial temperature
   X = energy absorbed
   M = mass
   C = specific heat capacity
b. 27°C
c. 3.0×10^1°C
d. 36°C
e. 43°C
f. same slope, y intercept at 10

---

45. Show that the temperature $-40°$ is unique in that it has the same numerical value on the Celsius and Fahrenheit scales.

46. A hot-water heater is operated by solar power. If the solar collector has an area of 6.0 m^2 and the power delivered by sunlight is 550 W/m^2, how long will it take to increase the temperature of 1.0 m^3 of water from 21°C to 61°C?

47. A student drops two metallic objects into a 120 g steel container holding 150 g of water at 25°C. One object is a 253 g cube of copper that is initially at 85°C, and the other is a chunk of aluminum that is initially at 5°C. To the surprise of the student, the water reaches a final temperature of 25°C, its initial temperature. What is the mass of the aluminum chunk?

Technology & Learning

Graphing calculators

Refer to Appendix B for instructions on downloading programs for your calculator. The program "Chap10" allows you to analyze a graph of temperature versus energy absorbed for a sample with a known mass and specific heat capacity.

Specific heat capacity, as you learned earlier in this chapter, is described by the following equation:

$$c_p = \frac{Q}{m\Delta t}$$

The program "Chap10" stored on your graphing calculator makes use of the equation for specific heat capacity. Once the "Chap10" program is executed, your calculator will ask for the initial temperature, mass, and specific heat capacity of the sample. The graphing calculator will use the following equation to create a graph of temperature (Y_1) versus the energy absorbed (X).

$$Y_1 = T + (X/(MC))$$

a. The graphing calculator equation is the same as the specific heat capacity equation shown above. Specify what each variable in the graphing calculator equation represents.

Execute "Chap10" on the [PRGM] menu and press [ENTER] to begin the program. Enter the values for the mass, specific heat capacity, and initial temperature (shown below), pressing [ENTER] after each one.

The calculator will provide a graph of the temperature versus the energy absorbed. (If the graph is not visible, press [WINDOW] and change the settings for the graph window so that Xmin is the lowest energy value required and Xmax is the highest value required, then press [ENTER].)

Press [TRACE], and use the arrow keys to trace along the curve. The x value corresponds to the absorbed energy in joules, and the y value corresponds to the temperature in degrees Celsius.

Determine the temperature of a 0.050 kg piece of aluminum foil (specific heat capacity equals 899 J/kg•°C) originally at 25°C that absorbs the following amounts of energy by heat:

b. 75 J
c. 225 J
d. 475 J
e. 825 J
f. If the initial temperature were 10°C instead of 25°C, how would the graph be different?

Press [2nd] [QUIT] to stop graphing. Press [ENTER] to input a new value or [CLEAR] to end the program.

48. At what Fahrenheit temperature are the Kelvin and Fahrenheit temperatures numerically equal? (See Sample Problem 10A.)

49. A 250 g aluminum cup holds and is in thermal equilibrium with 850 g of water at 83°C. The combination of cup and water is cooled uniformly so that the temperature decreases by 1.5°C per minute. At what rate is energy being removed?

50. A jar of tea is placed in sunlight until it reaches an equilibrium temperature of 32°C. In an attempt to cool the liquid, which has a mass of 180 g, 112 g of ice at 0°C is added. At the time at which the temperature of the tea is 15°C, determine the mass of the remaining ice in the jar. Assume the specific heat capacity of the tea to be that of pure liquid water.

48. 574.6 K = 574.6°F
49. 5.7×10^3 J/min = 95 J/s
50. 8.0×10^1 g

Alternative Assessment
ANSWERS

Performance assessment

1. Student plans should be safe and complete. They should include a list of equipment, measurements, and calculations. One technique is to measure the ring's mass and to then measure the temperature change of the ring and warm water when they are placed in a calorimeter.

2. Student answers should indicate that metal is a good conductor and that the faster way to transfer energy to the potato's interior is with the nail.

3. Student plans should be safe and complete and should include a list of equipment, measurements, and calculations. Graphs for insulators should show slow energy loss.

 **Portfolio projects** **4.** Student research will vary. Joule's work was accepted after Helmholtz confirmed its theoretical foundations.

5. Student analyses should include considerations of economics, environment, and convenience.

6. Answers will vary. For temperatures below mercury's freezing point, organic liquids are used in thermometers. Temperatures of distant objects are measured indirectly.

Alternative Assessment

Performance assessment

1. According to legend, Archimedes determined whether the king's crown was pure gold by comparing its water displacement with the displacement of a piece of pure gold of equal mass. But this procedure is difficult to apply to very small objects. Design a method for determining whether a ring is pure gold using the concept of specific heat capacity. Present your plan to the class, and ask others to suggest improvements to your design. Discuss each suggestion's advantages and disadvantages.

2. The host of a cooking show on television claims that you can greatly reduce the baking time for potatoes by inserting a nail through each potato. Explain whether this advice has a scientific basis. Would this approach be more efficient than wrapping the potatoes in aluminum foil? Discuss and list all arguments.

3. The graph of decreasing temperature versus time of a hot object is called its cooling curve. Design and perform an experiment to determine the cooling curve of water in containers of various materials and shapes. Draw cooling curves for each one. Which trends represent good insulation? Use your findings and graphs to design a lunch box that keeps food warm or cold.

Portfolio projects

4. Research the life and work of James Prescott Joule, who is best known for his apparatus demonstrating the equivalence of work and heat and the conservation of energy. Many scientists of the day initially did not accept Joule's conclusions. Research the reasoning behind their objections. Prepare a presentation for a class discussion either supporting the objections of Joule's critics or defending Joule's conclusion before England's Royal Academy of Sciences.

5. Get information on solar water heaters available where you live. How does each type work? Compare prices and operating expenses for solar water heaters versus gas water heaters. What are some of the other advantages and limitations of solar water heaters? Prepare an informative brochure for homeowners interested in this technology.

6. Research how scientists measure the temperature of the following: the sun, a flame, a volcano, outer space, liquid hydrogen, mice, and insects. Find out what instruments are used in each case and how they are calibrated to known temperatures. Using what you learn, prepare a chart or other presentation on the tools used to measure temperature and the limitations on their ranges.

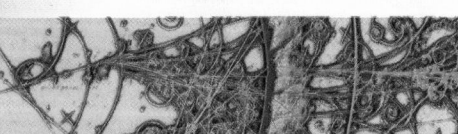

CHAPTER 10
Laboratory Exercise

★ TEKS

pp. 392–397: 1A, 2A, 2B, 2C, 2D, 2F, 3C, 5D

Planning

Recommended time:

1 lab period for 3 samples; additional samples may require more time.

Classroom organization:

▶ Each group must have at least two students. Groups must perform more than one step at a time, so groups may be larger than usual. With larger groups, make sure all students are involved and paying attention to safety.

▶ Bringing water to a boil takes 10–20 min; other tasks can be completed as the water heats.

▶ Each group must have a level work surface large enough so that students are not too close to the hot plates.

▶ The CBL and thermometer procedures may be used in the same class.

▶ Students should place wet metal shot in designated containers at the end of the lab to be dried and used again.

▶ **Safety warnings:** This lab presents several safety hazards. Remind students never to leave a hot plate unattended while it is on. Make sure all students wear the appropriate safety gear at all times.

OBJECTIVES

• Measure temperature.

• Apply the specific heat capacity equation for calorimetry to calculate the specific heat capacity of a metal.

• Identify unknown metals by comparing their specific heat capacities with accepted values for specific heat capacities.

MATERIALS LIST

✔ 2 beakers

✔ samples of various metals

✔ hot plate

✔ metal calorimeter and stirring rod

✔ ice

✔ balance

✔ metal heating vessel with metal heating dipper

✔ small plastic dish

PROCEDURE

CBL AND SENSORS

✔ CBL

✔ graphing calculator with link cable

✔ 2 temperature probes

THERMOMETER

✔ hand-held magnifying lens

✔ 2 thermometers

SPECIFIC HEAT CAPACITY

In this experiment, you will use calorimetry to identify various metals. In each trial, you will heat a sample of metal by placing it above a bath of water and bringing the water to a boil. When the sample is heated, you will place it in a calorimeter containing cold water. The water in the calorimeter will be warmed by the metal as the metal cools. According to the principle of energy conservation, the total amount of energy transferred out of the metal sample as it cools equals the energy transferred into the water and calorimeter as they are warmed. In this lab, you will use your measurements to determine the specific heat capacity and identity of each metal.

SAFETY

• When using a burner or hot plate, always wear goggles and an apron to protect your eyes and clothing. Tie back long hair, secure loose clothing, and remove loose jewelry. If your clothing catches on fire, walk to the emergency lab shower and use the shower to put out the fire.

• Never leave a hot plate unattended while it is turned on.

• If a thermometer breaks, notify the teacher immediately.

• Do not heat glassware that is broken, chipped, or cracked. Use tongs or a mitt to handle heated glassware and other equipment because it does not always look hot when it is hot. Allow all equipment to cool before storing it.

• Never put broken glass or ceramics in a regular waste container. Use a dustpan, brush, and heavy gloves to carefully pick up broken pieces and dispose of them in a container specifically provided for this purpose.

PREPARATION

1. Determine whether you will be using the CBL and sensors procedure or the thermometers. Read the entire lab for the appropriate procedure, and plan what steps you will take. Plan efficiently. Make sure you know which steps can be performed while you are waiting for the water to heat.

2. Prepare a data table with four columns and eight rows in your lab notebook. In the first row, label the second through fourth columns *Trial 1, Trial 2,* and *Trial 3.* In the first column, label the second through eighth rows *Sample number, Mass of metal, Mass of calorimeter cup and stirrer, Mass of water, Initial temperature of metal, Initial temperature of water and calorimeter,* and *Final temperature of metal, water, and calorimeter.*

3. In Appendix E, look up the specific heat capacity of the material the calorimeter is made of and record the information in the top left corner of your data table.

Thermometer procedure begins on page 395.

PROCEDURE

CBL AND SENSORS

Finding the specific heat capacity of a metal

4. Choose a location where you can set up the experiment away from the edge of the table and away from other groups. Make sure the hot plate is in the "off" position before you plug it in.

5. Fill a metal heating vessel with 200 mL of water and place it on the hot plate. Turn on the hot plate and adjust the heating controls to heat the water.

6. Set up the temperature probe, CBL, and graphing calculator as shown in **Figure 10-24.** Connect the CBL to the graphing calculator with the unit-to-unit link cable using the I/O ports located on each unit. Connect the first temperature probe to the CH1 port. Connect the second temperature probe to the CH2 port. Turn on the CBL and the graphing calculator. Start the program PHYSICS on the graphing calculator.

a. Select option *SET UP PROBES* from the MAIN MENU. Enter 2 for the number of probes. Select the temperature probe from the list. Enter 1 for the channel number. Select the temperature probe from the list again, and enter 2 for the channel number.

b. Select the *COLLECT DATA* option from the MAIN MENU. Select the *TRIGGER* option from the DATA COLLECTION menu.

7. Obtain about 100 g of the metal sample. First find the mass of the small plastic dish. Place the metal shot in the dish and determine the mass of the shot. Record the number and the mass of the sample in your data table. Place one temperature probe in the metal heating dipper, and carefully pour the sample into the metal heating dipper. Make sure the temperature probe is surrounded by the metal sample.

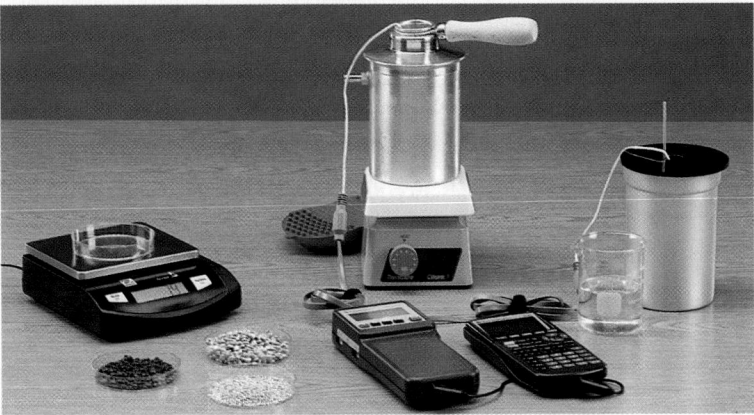

Figure 10-24

Step 5: Start heating the water before you set up the CBL and temperature probes. Never leave a hot plate unattended when it is turned on.

Step 7: Be very careful when pouring the metal sample in the dipper around the temperature probe.

Step 17: Begin taking temperature readings a few seconds before adding the sample to the calorimeter.

Step 19: Record the *highest* temperature reached by the water, sample, and calorimeter combination, not the final temperature.

CBL and Sensors Tips

◆ Students should have the program PHYSICS on their graphing calculators. Refer to Appendix B for instructions.

◆ If using one temperature probe per group, the probe must be cooled after the temperature of the metal sample is taken and before the calorimeter temperature is taken. While the probe cools, the sample cools; this introduces error. The heat capacity of the uncooled probe introduces errors. Using two probes will solve these problems.

◆ Students should begin collecting data for the calorimeter *before* adding the metal sample to the calorimeter.

◆ When adding the metal to the calorimeter, don't let it touch the sensor in the water; it will produce a false maximum value.

Techniques to Demonstrate

Demonstrate how to carefully pour the shot into the heating dipper so that it surrounds the temperature probe. The probe must not touch the dipper's sides.

✔ Checkpoints

Step 4: All hot plates and containers of liquids must be kept away from the edge of the table. Metal shot must be kept in a container at all times.

Step 5: The hot plate must be turned up to the highest level.

Step 6: The CBL, graphing calculator, and cords must be kept away from the hot plate.

Step 8: Students must exercise care when placing the heating dipper on top of the heating vessel.

Step 9: For the next few steps, make sure the hot plate is not left unattended.

Step 10: Only a small amount of ice is required; if too much is used, it should be removed before students measure the mass of the water.

Step 11: If students are not doing more trials, make sure all hot plates are turned off. If they remain on, make sure that heating vessels have enough water in them and that hot plates are attended to. The vessel, water, and hot plate are still hot; student groups should move away from the hot plate and exercise caution.

Step 13: If using only one temperature probe, cool the probe in a beaker of cool water, not in the calorimeter, until the probe is cool and the CBL unit displays a constant temperature. Dry the probe thoroughly before placing it in the calorimeter.

Step 18: Students should use mitts when handling the hot metal heating dipper. Remind students not to let the hot metal sample touch the temperature probe in the calorimeter.

Step 19: Students should be able to use the graph to find the temperature at specific times during the experiment.

Step 21: Because the water is still hot, any subsequent trials will take less time than the initial trial.

8. Place the dipper with metal contents into the top of the heating vessel, as shown in **Figure 10-24.** Make sure the temperature probe leads do not touch the hot plate or any heated surface.

9. While the sample is heating, find the mass of the empty inner cup of the calorimeter and the stirring rod. Record the mass in your data table. Do not leave the hot plate unattended.

10. For the water in the calorimeter, you will need about 100 g of water that is a little colder than room temperature. Put the water in a beaker. Use the second CBL temperature probe to find room temperature. Look at the temperatures on the CBL and read the temperature reading for the second probe. Place the second sensor in the water to check the water's temperature. (Do not use water colder than 5°C below room temperature. You may need to use ice to get the initial temperature low enough, but make sure all the ice has melted before pouring the water into the calorimeter.)

11. Place the calorimeter and stirrer on the balance and carefully add 100 g of the water. Record the mass of the water in your data table. Replace the cup in its insulating shell, and cover.

12. When the CBL displays a constant temperature for several readings, press TRIGGER on the CBL to collect the temperature reading of the metal sample. Record the temperature reading in your data table as the metal's initial temperature. Select *CONTINUE* from the TRIGGER menu on the graphing calculator.

13. Carefully remove the first temperature probe from the dipper. Set the probe aside to cool.

14. Use the stirring rod to stir the water in the calorimeter. Place the second temperature probe in the calorimeter. When the CBL displays a constant temperature for several readings for the calorimeter water, press TRIGGER on the CBL to collect the temperature readings. Record the initial temperature of the water and calorimeter. Select *STOP* from

the TRIGGER menu on the graphing calculator. Leave the probe in the calorimeter.

15. Select *COLLECT DATA* from the MAIN MENU on the graphing calculator. Select the *TIME GRAPH* option from the DATA COLLECTION menu. Enter 2.0 for the time between samples. Enter 99 for the number of samples. Check the values you entered, and then press ENTER. Press ENTER to continue. If you made a mistake entering the time values, select *MODIFY SETUP*, reenter the values, and continue.

16. From the TIME GRAPH menu, select *LIVE DISPLAY*. Enter 0 for *Ymin*, enter 100 for *Ymax*, and enter 5 for *Yscl*.

17. Press ENTER on the graphing calculator to begin collecting the temperature readings for the water in the calorimeter.

18. Quickly transfer the metal sample to the calorimeter of cold water and replace the cover. Use a mitt when handling the metal heating dipper. Use the stirring rod to gently agitate the sample and to stir the water in the calorimeter. If you are not doing any more trials, make sure the hot plate is turned off. Otherwise, make sure there is plenty of water in the heating vessel, and do not leave the hot plate unattended.

19. When the CBL displays DONE, use the arrow keys to trace the graph. Time in seconds is graphed on the *x*-axis, and the temperature readings are graphed on the *y*-axis. Record the highest temperature reading from the CBL in your data table.

20. Press ENTER on the graphing calculator. On the REPEAT? menu, select *NO*. If you are going to perform another trial, select the *COLLECT DATA* option from the MAIN MENU. Select *TRIGGER* from the DATA COLLECTION menu.

21. If time permits, make additional trials with other samples. Record data for all trials in your data table.

Analysis and Interpretation begins on page 396.

PROCEDURE

THERMOMETER

Finding the specific heat capacity of a metal

4. Choose a location where you can set up the experiment away from the edge of the table and from other groups. Make sure the hot plate is in the "off" position before you plug it in.

5. Fill a metal heating vessel with 200 mL of water and place it on the hot plate, as shown in **Figure 10-25.** Turn on the hot plate and adjust the heating control to heat the water.

6. Measure out about 100 g of the metal sample. Record the number of the metal sample in your data table. Hold the thermometer in the metal heating dipper, and very carefully pour the sample into the metal heating dipper. Make sure the bulb of the thermometer is surrounded by the metal. Place the dipper with metal contents into the heating vessel. Hold the thermometer while the sample is heating.

7. While the sample is heating, determine the mass of the stirring rod and empty inner cup of the calorimeter. Record the mass in your data table. Do not leave the hot plate unattended.

8. Use the second thermometer to measure room temperature. For the water in the calorimeter, you will need about 100 g of water that is a little colder than room temperature. Put the water in a beaker. Place the thermometer in the water to check the temperature of the water. (Do not use water colder than 5°C below room temperature. You may need to use ice to get the initial temperature low enough, but make sure all the ice has melted before pouring the water into the calorimeter.)

9. Place the calorimeter and stirrer on the balance, and carefully add 100 g of the water. Record the mass of the water in your data table. Replace the cup in its insulating shell, and cover.

10. Use the thermometer to measure the temperature of the sample when the water is boiling and the sample reaches a constant temperature. Record this temperature as the initial temperature of the metal sample. (Note: When making temperature readings, take care not to touch the hot plate and the water.) Use the hand-held magnifying lens to estimate to the nearest 0.5°C. Make sure that the

Figure 10-25

Step 5: Start heating the water before you begin the rest of the lab. Never leave a hot plate unattended when it is turned on.

Step 6: Be very careful when pouring the metal sample in the dipper around the thermometer. Make sure the thermometer bulb is surrounded by the metal sample.

Step 12: Begin taking temperature readings a few seconds before adding the sample to the calorimeter.

Step 15: Record the *highest* temperature reached by the water, sample, and calorimeter combination.

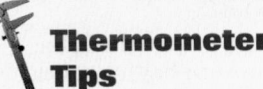

Thermometer Tips

◆ Using two thermometers improves the speed and accuracy of the lab: use a −20°C to 110°C thermometer to measure the temperature of the sample and a −5°C to 50°C thermometer to measure the temperature of the water in the calorimeter.

◆ Students must keep their hands out of the steam.

Techniques to Demonstrate

Demonstrate how to carefully pour the shot into the metal heating dipper around the thermometer. The bulb must be surrounded by shot. The thermometer must be removed very slowly and carefully, and it must *never* be inserted into the dipper full of shot.

✔ Checkpoints

Step 4: All hot plates and liquids must be kept away from the edge of the table. Keep metal shot in a container at all times.

Step 5: The hot plate must be turned up to the highest level.

Step 6: Students must be careful when pouring metal shot into the dipper. The thermometer must be held upright until it is removed from the heating dipper.

Step 7: For the next few steps, do not leave the hot plate unattended.

Step 8: Only a small amount of ice is required; it should be removed before students measure the mass of the water.

Step 10: If not doing more trials, make sure hot plates are turned off. If they remain on, make sure that heating vessels have enough water in them and that hot plates are attended to.

Step 13: Students should use mitts to handle the hot dipper.

Step 14: Stir gently; violent motion of the sample in the water could break the thermometer.

Step 16: Because the water is still hot, subsequent trials will take less time.

ANSWERS TO

Analysis and Interpretation

CALCULATIONS AND DATA ANALYSIS

1. Student answers will vary. Make sure students use the relationship $\Delta T = T_f - T_i$. Typical values range from 0.9°C to 7.58°C.

2. Student answers will vary. Make sure students use the relationship $c_p = \dfrac{Q}{m\Delta T}$.

   a. Typical values range from 43.7 J to 411 J.

   b. Typical values range from 558 J to 3380 J.

3. Make sure students use the relationship $Q_{total} = Q_1 + Q_2$. Typical values range from 601.3 J to 3793 J.

4. Make sure students use the relationship $c_p = \dfrac{Q}{m\Delta T}$. Typical values for the specific heat capacity range from 210.2 J/kg•°C to 952.8 J/kg•°C.

thermometer bulb is completely surrounded by the metal sample, and keep your line of sight at a right angle to the stem of the thermometer. Reading the thermometer at an angle will cause considerable errors in your measurements. Carefully remove the thermometer and set it aside in a secure place.

11. Use the stirring rod to gently stir the water in the calorimeter. **Do not use the thermometer to stir the water.**

12. Place the second thermometer in the covered calorimeter. Measure the temperature of the water in the calorimeter to the nearest 0.1°C. Record this temperature in your data table as the initial temperature of the water and calorimeter.

13. Quickly transfer the sample to the cold water in the calorimeter and replace the cover. Use a mitt when handling the metal heating dipper. If you are not doing any more trials, make sure the hot plate is turned off. Otherwise, make sure there is plenty of water in the heating vessel, and do not leave the hot plate unattended.

14. Use the stirring rod to gently agitate the sample and stir the water in the calorimeter. **Do not use the thermometer to stir the water.**

15. Take readings every 5.0 s until five consecutive readings are the same. Record the highest reading in your data table.

16. If time permits, make additional trials with other metals. Record the data for all trials in your data table.

17. Clean up your work area. Put equipment away safely so that it is ready to be used again.

ANALYSIS AND INTERPRETATION

Calculations and data analysis

1. **Organizing data** For each trial, calculate the temperature change of the water and calorimeter.

2. **Analyzing data** Use your data for each trial.

   a. Calculate the energy transferred to the calorimeter cup and stirring rod as heat, using the value for the specific heat capacity you found in step 3.

   b. Calculate the energy transferred to the water as heat.

3. **Applying ideas** Calculate the total energy transferred as heat into the water and the calorimeter.

4. **Analyzing results** For each trial, find the temperature change of the sample and calculate the specific heat capacity of the sample.

Conclusions

5. **Evaluating data** Use the accepted values for the specific heat capacities of various metals in **Table 10-4** on page 372, to determine what metal each sample is made of.

6. Evaluating results   Calculate the absolute and relative errors of the experimental values. Check with your teacher to see if you have correctly identified the metals.

   a. Use the following equation to compute the absolute error:

$$\text{absolute error} = |\,\text{experimental} - \text{accepted}\,|$$

   b. Use the following equation to compute the relative error:

$$\text{relative error} = \frac{(\text{experimental} - \text{accepted})}{\text{accepted}}$$

7. Evaluating methods   Explain why the energy transferred as heat into the calorimeter and the water is equal to the energy transferred as heat from the metal sample.

8. Evaluating methods   Explain why it is important to calculate the temperature change using the highest temperature as the final temperature, rather than the last temperature recorded.

9. Evaluating methods   Why should the water be a few degrees colder than room temperature when the initial temperature is taken?

10. Applying conclusions   How would your results be affected if the initial temperature of the water in the calorimeter were 50°C instead of slightly cooler than room temperature?

11. Relating ideas   How is the temperature change of the calorimeter and the water within the calorimeter affected by the specific heat capacity of the metal? Did a metal with a high specific heat capacity raise the temperature of the water and the calorimeter more or less than a metal with a low specific heat capacity?

12. Building models   An environmentally conscious engineering team wants to design tea kettles out of a metal that will allow the water to reach its boiling point using the least possible amount of energy from a range or other heating source. Using the values for specific heat capacity in **Table 10-4** on page 372, choose a material that would work well, considering only the implications of transfer of energy as heat. Explain how the specific heat capacity of water will affect the operation of the tea kettle.

Extensions

13. Evaluating methods   What is the purpose of the outer shell of the calorimeter and the insulating ring in this experiment?

14. Designing experiments   If there is time and your teacher approves, design an experiment to measure the specific heat capacity of the calorimeter. Compare this measured value with the accepted value from **Table 10-4** on page 372. Are they the same? If not, how would using the experimental value affect your results in this lab?

CONCLUSIONS

5. Student answers should correctly identify the metal used in each trial.

6. a. For sample data, values range from 2.5 J/kg•°C to 3.34 J/kg•°C.

   b. For sample data, values range from 0.007 to 0.151.

7. Energy is conserved in a closed system.

8. The highest temperature represents the total energy transferred as heat.

9. A greater temperature difference between the water and the sample allows more accurate results to be obtained; the smaller the difference, the higher the precision that is required.

10. The difference would be smaller, so error would be greater.

11. The higher specific heat capacities caused a greater change in the temperature of the water.

12. Student should choose a metal with a low specific heat capacity.

EXTENSIONS

13. The shell and insulated ring prevent energy transfer.

14. Plans should be safe and complete, and should include a list of equipment, measurements, and calculations required. One technique is to repeat the experiment using a known metal.

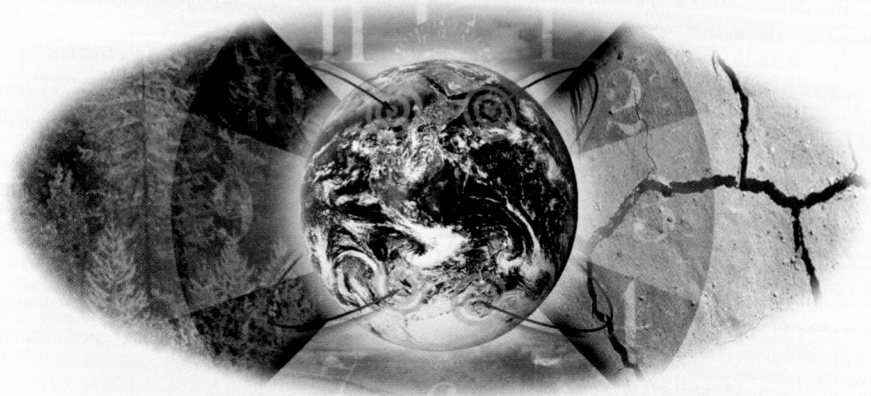

Climatic Warming

Scientists typically devise solutions to problems and then test the solution to determine if it indeed solves the problem. But sometimes the problem is only suggested by the evidence, and there are no chances to test the solutions. A current example of such a problem is climatic warming.

Data recorded from various locations around the world over the past century indicate that the average atmospheric temperature is 0.5°C higher now than it was 100 years ago. Although this sounds like a small amount, such an increase can have pronounced effects. Increased temperatures may eventually cause the ice in polar regions to melt, causing ocean levels to increase, which in turn may flood some coastal areas.

Small changes in temperature can also affect living organisms. Most trees can tolerate only about a 1°C increase in average temperature. If a tree does not reproduce often or easily enough to "migrate" through successive generations to a cooler location, it can become extinct in that region. Any organisms dependent on that type of tree also will suffer.

But such disasters depend on whether global temperatures continue to increase. Historical studies indicate that some short-term fluctuations in climate are natural, like the "little ice age" of the seventeenth

century. If the current warming trend is part of a natural cycle, the dire predictions may be overstated or wrong.

Even if the warming is continuous, climatic systems are very complex and involve many unexpected factors. For example, if polar ice melts, a sudden increase in humidity may result in snow in polar areas. This could counter the melting, thus causing ocean levels to remain stable.

Greenhouse Gases

Most of the current attention and concern about climatic warming has been focused on the increase in the amount of "greenhouse gases," primarily carbon dioxide and methane, in the atmosphere. Molecules of these gases absorb energy that is radiated from Earth's surface, causing their temperature to rise. These molecules then release energy as heat, causing the atmosphere to be warmer than it would be without these gases.

While carbon dioxide and methane are natural components of the air, their levels have increased rapidly during the last hundred years. This has been determined by analyzing air trapped in the ice layers of Greenland. Deeper sections of the ice contain air from earlier times. During the last ice age, there were about 185 ppm of carbon dioxide, CO_2, in the air, but the concentration from 130 years ago was slightly below 300 ppm. Today, the levels are 350 ppm, an increase that can be accounted for by the increase in combustion reactions, primarily from coal and petroleum burning, and by the decrease in CO_2-consuming trees through deforestation.

But does the well-documented increase in greenhouse gas concentrations enable detailed predictions? Atmospheric physicists have greatly improved their models in recent years, and they are able to correctly predict past ice ages and account for the energy-absorbing qualities of oceans. But such models remain oversimplified, partly because of a lack of detailed long-term data. In addition, the impact of many variables, such as fluctuations in solar energy output and volcanic processes, are poorly understood and cannot be factored into predictions. To take all factors into account would require more-complex models and more-sophisticated supercomputers than are currently available. As a result, many question whether meaningful decisions and planning can occur.

Risk of Action and Inaction

The evidence for climatic warming is suggestive but not conclusive. What should be done? Basically, there are two choices: either do something or do nothing.

The risks of doing nothing are that the situation may worsen. But it is also possible that waiting for better evidence will allow for a greater consensus among the world's nations about how to solve the problem efficiently. Convincing the world's population that action taken now will have the desired benefit decades from now will not be easy.

Acting now also involves risks. Gas and coal could be rationed or taxed to limit consumption. The development of existing energy-efficient technologies, such as low-power electric lights and more efficient motors and engines, could cut use of coal and gasoline in half. However, the economic effects could be as severe as those resulting from climatic warming.

But none of these options can guarantee results. Even if the trend toward climatic warming stops, it will be hard to prove whether this was due to human reduction in greenhouse gases, to natural cyclic patterns, or to other causes.

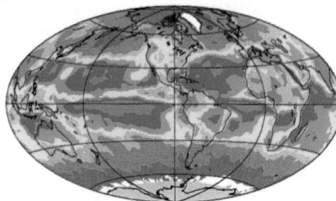

This map shows the reflecting properties of the Earth's surface. Regions colored in blue or green absorb much of the energy striking them.

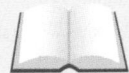 **Researching the Issue**

1. Carbon dioxide levels in the atmosphere have varied during Earth's history. Research the roles of volcanoes, plants, and limestone formation, and determine whether these processes have any bearing on the current increase in CO_2 concentrations. Can you think of any practical means of using these processes to reduce CO_2 concentrations? What would be the advantages and disadvantages?

2. Find out what technological developments have been suggested for slowing climatic warming. Can they be easily implemented? What are the drawbacks of these methods?

internet**connect**

SCi**LINKS**
NSTA

TOPIC: Greenhouse gases
GO TO: www.scilinks.org
***sci*LINKS CODE:** HF2106

CHAPTER 11 PLANNING GUIDE

Compression Guide: To shorten from 7 to 6 45-min periods (from 3½ to 3 90-min blocks), eliminate items in magenta type.

PACING CHART	CLASSROOM RESOURCES			
	⭐TEKS	Teacher Demonstrations	Holt Physics Transparencies	Labs (See page T52 for equipment listing for in-text labs.)
11-1 Relationships between heat and work 1 45-minute period ½ 90-minute block	2C, 3A, 5B, 7B	**TE** *Work from heat*, p. 402 **TE** *Adiabatic processes*, p. 407	**T** 39–40	
11-2 Thermodynamic processes 2 45-minute periods 1 90-minute block	2C, 3B, 3C, 5B, 7A, 7B		**T** 41–42 **TM** 35–38	
11-3 Efficiency of heat engines 1 45-minute period ½ 90-minute block	2C, 3A, 3C, 7A			
11-4 Entropy 1 45-minute period ½ 90-minute block	2C, 3A, 5B, 7A, 7B	**TE** *Order, disorder, and probability,* p. 425	**T** 43–44 **TM** 40–41	**PE** *Quick Lab: Entropy and Probability,* p. 426
Review and Assessment 2 45-minute periods 1 90-minute block				

Resource Key

HOLT PHYSICS

PE Pupil's Edition
TE Teacher's Edition

L Laboratory Experiments
TL Technology Lab Experiments
T Transparencies

👆 **One-Stop** Planner CD-ROM **contents**

TM Transparency Masters
SR Section Review Worksheets
AA Alternative Assessment

PW Problem-Solving Workbook
PB Problem Bank
CTW Critical Thinking Worksheet

ASSIGNMENT RESOURCES

Content Mastery	Critical Thinking	Problem-Solving Practice
PE 1–4, p. 408 **SR** 11-1, *Concept Review* **PE** 1–4, p. 431	**PE** 5–9, p. 431	**11A** Work done on or by a gas: 22 items in **PE, PW,** and **PB,** see **TE** pp. 404–405
PE 1–8, p. 419 **SR** 11-2, *Diagram Skills* **PE** 12–15, pp. 431–432	**PE** 16–18, p. 432	**11B** The first law of thermodynamics: 27 items in **PE, PW,** and **PB,** see **TE** p. 412
PE 1–3, p. 424 **SR** 11-3, *Concept Review* **PE** 22–24, pp. 432–433	**PE** 1–2, p. 422 **PE** 25–27, p. 433	**11C** Heat-engine efficiency: 25 items in **PE, PW,** and **PB,** see **TE** p. 423
PE 1–4, p. 429 **SR** 10-4, *Math Skills* **PE** 31–34, p. 433	**PE** 1–2, p. 428 **PE** 35–38, pp. 433–434	

ASSESSMENT RESOURCES

Cumulative Review	Alternative Assessment	Traditional Assessment
SR Mixed Review, Ch. 11	**PE** 1–6, p. 435 **AA** Items for Ch. 11	Chapter 11 Test Test Generator items for Ch. 11

Scoring Rubrics for Alternative Assessment items can be found on the One-Stop Planner CD-ROM.

TECHNOLOGY RESOURCES

 CTW Segment 13 Water-Cooled City

 PE Technology and Learning, p. 434
(Alternative procedures for calculators without Flash-ROM technology are provided on the One-Stop Planner CD-ROM.)

internetconnect

 On-line Student Resources:
GO TO: www.scilinks.org
The following *sci*LINKS Internet resources can be found in the student text for this chapter.

TOPICS:
• Energy transfer, p. 403 (HF2111)
• Thermodynamics, p. 410 (HF2112)
• Heat engines, p. 417 (HF2113)
• Stirling engines, p. 422 (HF2114)
• Entropy, p. 425 (HF2115)

 On-line Teacher Resources:
GO TO: go.hrw.com
KEYWORD: HF2 HOME
Visit the HRW Web site for a variety of resources related to this chapter.

 Smithsonian Institution*
Internet Connections
Visit **www.si.edu/hrw** for additional on-line resources.

CNNfyi.com
Visit **www.cnnfyi.com** for late-breaking news and current events stories selected just for you.

Section 11-1 explains that a system can absorb energy by heat or work and then transfer energy to its surroundings as work or heat, and it distinguishes between isovolumetric, isothermal, and adiabatic processes.

Section 11-2 introduces the first law of thermodynamics and the relationships between heat, work, and internal energy, and it applies the first law of thermodynamics to cyclic processes in refrigeration, heat engines, and combustion engines.

Section 11-3 introduces the second law of thermodynamics, discusses why two bodies must be at different temperatures in order for work to be done, and shows how to calculate the efficiency of heat engines.

Section 11-4 introduces entropy and its relationship to disorder and probability.

About the Illustration

For many students, the concepts of thermodynamics are difficult and abstract. Using examples that students are familiar with and that can be represented by fairly simple models, such as the balloon shown in this photograph, helps students see how thermodynamics applies to the world around them. Return to this example throughout the chapter as new concepts are introduced.

CHAPTER 11

Thermodynamics

PHYSICS IN ACTION

This balloon, which is used to lift scientific instruments into the upper atmosphere, can be modeled as a simple thermodynamic system. For instance, changes in temperature outside the balloon may cause temperature changes inside the balloon. These changes occur because energy is transferred between the gas in the balloon and the outside air. This energy can be transferred as either heat or work. In either case, the internal energy of the balloon's gas changes.

In this chapter, you will study how work and heat serve to change a system's internal energy and how machine efficiency is limited.

- *What roles do heat and work play in changing the volume and temperature of a balloon?*

- *Why must a heat engine operate between two temperatures?*

CONCEPT REVIEW

Work   (Section 5-1)

Internal energy   (Section 10-1)

Heat   (Section 10-2)

Conservation of energy
  (Section 10-2)

11-1
Relationships between heat and work

11-1 SECTION OBJECTIVES

- Recognize that a system can absorb or release energy as heat in order for work to be done on or by the system and that work done on or by a system can result in the transfer of energy as heat.

- Compute the amount of work done during a thermodynamic process.

- Distinguish between isovolumetric, isothermal, and adiabatic thermodynamic processes.

Figure 11-1
Work increases the nail's internal energy at the nail's surface. This energy is transferred away from the nail's surface as heat.

HEAT, WORK, AND INTERNAL ENERGY

Recall from Chapter 10 how pulling a nail from a piece of wood (as shown in **Figure 11-1**) causes the temperature of the nail to increase. The work done to overcome friction between the nail and the wood fibers transfers energy to the iron atoms in the nail and the molecules in the wood. This increase in the kinetic energy of these particles accounts for the increase in the internal energies of the nail and wood along their surface of contact. **(★)TEKS 5B**

As long as a substance does not change phase, its internal energy will increase as its temperature increases. The increase in internal energy of the nail's surface corresponds to an increase in the surface temperature, which is higher than the temperature of the nail's interior. As a result, energy is transferred as heat from the surface to the interior of the nail. When all portions of the nail reach thermal equilibrium, this energy transfer ceases.

As we have seen, the work done in removing a nail from the wood increases the nail's internal energy. Part of this energy is transferred to the wood as heat. The final equilibrium temperature of the nail and wood is greater than their initial temperatures. This increase in temperature indicates that the internal energy of both the wood and nail increased as the result of work done on them.

Note that this description requires a more complete definition of energy than the one found in Chapter 5. The mechanical energy added to the nail and wood as work does not appear to be conserved. The conservation of mechanical energy discussed in Chapter 5 did not take into consideration changes in the internal energy or energy transferred by heat (as discussed in Chapter 10). In some cases, the principle of energy conservation can be more useful if its scope is broadened to include internal energy and heat.

Heat can be used to do work

Work can transfer energy to a substance, which increases the internal energy of the substance. This internal energy can then decrease through the transfer of energy as heat. The reverse is also possible. Energy can be transferred to a substance as heat and from the substance as work.

Consider a flask of water. A balloon is placed over the mouth of the flask, and the flask is heated until the water boils. Energy transferred as heat from the flame of the gas burner to the water increases the internal energy of the water. When the water's temperature reaches the boiling point, the water changes phase and becomes steam. At this constant temperature, the volume of the steam increases. This expansion provides a force that pushes the balloon outward against the force exerted by the atmosphere, as shown in **Figure 11-2.** Thus, the steam does work on the balloon, and the steam's internal energy decreases as predicted by the principle of energy conservation. (★TEKS) **7B**

Heat and work are energy transferred to or from a system

On a microscopic scale, heat and work are very similar. Neither is a property of a substance; rather, both are energy that is transferred to or from a substance, thus changing the substance's internal energy. This change in internal energy is apparent by the change in the substance's temperature or phase.

So far, the internal energy of a substance or combination of substances has been treated as a single quantity to which energy is added or from which energy is taken away. Such a substance or combination of substances is called a **system.** All of the parts of a system are in thermal equilibrium with each other before and after a process adds or removes energy.

An example of a system would be the flask, balloon, water, and steam that were heated over the burner. As the burner transferred energy as heat to the system, that system's internal energy increased. When the expanding steam, a part of the system, did work on the balloon, the system's internal energy decreased. Some of the energy transferred to the system as heat was transferred out of the system as work done on the balloon.

Although a system is often treated as being isolated, in most cases it interacts with its surroundings. In the example above, a heat interaction occurs between the burner and the system, and work is done by the system on the surroundings (the balloon moves the outside air outward). Energy is also transferred as heat to the air surrounding the flask because of the temperature difference between the flask and the surrounding air. The surroundings with which the system interacts are referred to as its **environment.**

Work done on or by a gas is the pressure multiplied by the change in volume

As defined in Chapter 5, work is the product of a force applied to an object and the distance moved by that object in the direction of the force. By the simplest description of work in thermodynamic systems, work is the product of the change in the volume of a gas and the pressure applied on or by the gas. This chapter will use only this description of work. The expression in Chapter 5 can be derived from this equation for work using the definition for pressure (force divided by area) and the definition for the change of volume (area multiplied by the displacement).

Figure 11-2
Energy transferred as heat turns water into steam. Energy from the steam does work against the force exerted by air outside the balloon.

system

a collection of matter within a clearly defined boundary across which no matter passes

*internet***connect**

SCI*LINKS*

NSTA

TOPIC: Energy transfer
GO TO: www.scilinks.org
*sci***LINKS CODE:** HF2111

environment

everything outside a system that can affect or be affected by the system's behavior

Teaching Tip

Some of the first experiments demonstrating the equivalence between heat and work were performed by James Prescott Joule. In perhaps the most well known experiment, a paddle wheel was turned by falling weights. When the paddle wheel was placed in water, work was done on the water by the friction between the wheel and the water. As a result, the water's temperature increased. Joule found that the increase in temperature was proportional to the energy expended. In this experiment, the work done on a system was used to increase the system's internal energy by raising its temperature. In the example given on this page (in which steam is used to inflate a balloon), the reverse process occurs. The steam does work, and its internal energy decreases.

(STOP) **Misconception Alert**

The concepts in this chapter are often confused by students. Small group discussions will help students review and organize their concepts of heat, work, pressure, temperature, and volume of a gas. Ask how they could use different containers (soft or rigid, thermally isolated or not) to control these variables.

(★TEKS)

p. 402: 5B
p. 403: 7B

The following may be used as teamwork exercises or for demonstration at the chalkboard or on an overhead projector.

PROBLEM

Work done on or by a gas

The cross-sectional area of the piston in **Figure 11-3** is 0.20 m². A 400.0 N weight pushes the piston down 0.15 m and compresses the gas in the cylinder. How much pressure is exerted on the gas?

Answer

2.0×10^3 Pa

By how much did the gas volume decrease?

Answer

0.030 m³

PRACTICE GUIDE 11A

Solving for:	
W	📖 **PE** Sample, 1–3; Ch. Rvw. 10–11
	💿 **PW** 5–6
	PB 8–10
P	📖 **PE** 4
	💿 **PW** 3–4
	PB Sample, 1–3
ΔV	💿 **PW** Sample, 1–2
	PB 4–7

 TEKS

p. 404: 2C
p. 405: 2C, 3A

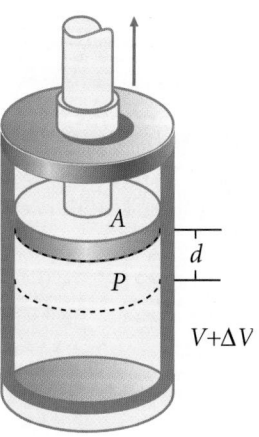

Figure 11-3
Work done on or by the gas is the product of the volume change (area *A* multiplied by the displacement *d*) and the pressure of the gas.

DEFINING WORK IN TERMS OF CHANGING VOLUME

$$W = Fd$$

$$\text{work} = \text{force} \times \text{distance moved}$$

$$W = Fd\left(\frac{A}{A}\right) = \left(\frac{F}{A}\right)(Ad) = P\Delta V$$

$$\text{work} = \text{pressure} \times \text{volume change}$$

When *P* is constant, work can be written as $P\Delta V$ rather than as $\Delta(PV)$. If the gas expands, as shown in **Figure 11-3,** ΔV is positive and the work done by the gas on the piston is positive. If the gas is compressed, ΔV is negative and the work done by the gas on the piston is negative (the piston does work on the gas). When the gas volume remains constant, there is no displacement and no work is done on or by the system.

Although the pressure can change during a process, work is done only if the volume changes. A situation in which pressure increases and volume remains constant is comparable to one in which a force does not displace a mass even as the force is increased. Work is not done in either situation.

SAMPLE PROBLEM 11A

Work done on or by a gas ⭐**TEKS** **2C**

PROBLEM

An engine cylinder has a cross-sectional area of 0.010 m². How much work can be done by a gas in the cylinder if the gas exerts a constant pressure of 7.5×10^5 Pa on the piston and moves the piston a distance of 0.040 m?

SOLUTION

Given: $A = 0.010 \text{ m}^2 \qquad d = 0.040 \text{ m}$

$P = 7.5 \times 10^5 \text{ Pa} = 7.5 \times 10^5 \text{ N/m}^2$

Unknown: $\Delta V = ? \qquad W = ?$

Use the area-times-distance formula for volume change. Use the definition of work in terms of changing volume on this page.

$$\Delta V = Ad$$

$$W = P\Delta V$$

$$\Delta V = (0.010 \text{ m}^2)(0.040 \text{ m}) = 4.0 \times 10^{-4} \text{ m}^3$$

$$W = (7.5 \times 10^5 \text{ N/m}^2)(4.0 \times 10^{-4} \text{ m}^3) = 3.0 \times 10^2 \text{ J}$$

$$\boxed{W \text{ done by the gas} = 3.0 \times 10^2 \text{ J}}$$

Work done on or by a gas (★)TEKS 2C

1. Gas in a container is at a pressure of 1.6×10^5 Pa and a volume of 4.0 m^3. What is the work done by the gas if

   a. it expands at constant pressure to twice its initial volume?
   b. it is compressed at constant pressure to one-quarter of its initial volume?

2. A gas is enclosed in a container fitted with a piston. The applied pressure is maintained at 599.5 kPa as the piston moves inward, which changes the volume of the gas from 5.317×10^{-4} m^3 to 2.523×10^{-4} m^3. How much work is done? Is the work done *on* or *by* the gas? Explain your answer.

3. A toy balloon is inflated with helium at a constant pressure that is 4.3×10^5 Pa in excess of atmospheric pressure. If the balloon inflates from a volume of 1.8×10^{-4} m^3 to 9.5×10^{-4} m^3, how much work is done on the balloon by the helium?

4. Steam moves into the cylinder of a steam engine at a constant pressure and does 0.84 J of work on a piston. The diameter of the piston is 1.6 cm, and the piston travels 2.1 cm in one stroke. What is the pressure of the steam?

THERMODYNAMIC PROCESSES

In this section three distinct quantities have been related to each other: internal energy (U), heat (Q), and work (W). However, they do not all appear in every thermodynamic process. Work can be performed in some processes with a change in internal energy and with no energy transferred as heat. In other cases, internal energy can change as energy is transferred as heat without work being done. In reality, processes that involve only work or only heat are rare. In most processes, energy is transferred to or from the system as both heat and work. However, a real process can often be approximated and described by one of the ideal processes.

No work is done in a constant-volume process

Suppose a car with closed windows and doors is parked inside a hot garage. The internal energy of the system (inside the car) increases as energy is transferred as heat into the car from the hot air in the garage. The car's heavy steel walls and sealed windows keep the system's volume nearly constant. As a result, no work is done by the system. All changes in the system's internal energy are due to the transfer of energy as heat. (★)TEKS 3A

ANSWERS TO

Practice 11A
Work done on or by a gas
1. **a.** 6.4×10^5 J
   b. -4.8×10^5 J
2. -167.5 J; Work is done on the gas because the volume change is negative.
3. 3.3×10^2 J
4. 2.0×10^5 Pa

Teaching Tip

Explain to students why the rest of the chapter deals essentially with thermodynamic processes in *gases*. Tell them that all objects (solids, liquids, and gases) have internal energy, which is the sum of the kinetic and potential energies of their molecules. However, monatomic gases present a simpler situation to study because all of their internal energy is kinetic (their molecules are too far apart to interact with each other significantly).

Teaching Tip

Point out that keeping pressure, volume, temperature, or internal energy constant simplifies the problem by reducing the number of variables.

(STOP) Misconception Alert

Make sure students realize that the closed car is only an approximation of a system. The seals on the windows and the air conditioning vents allow some air to be transferred between the inside and outside of the car.

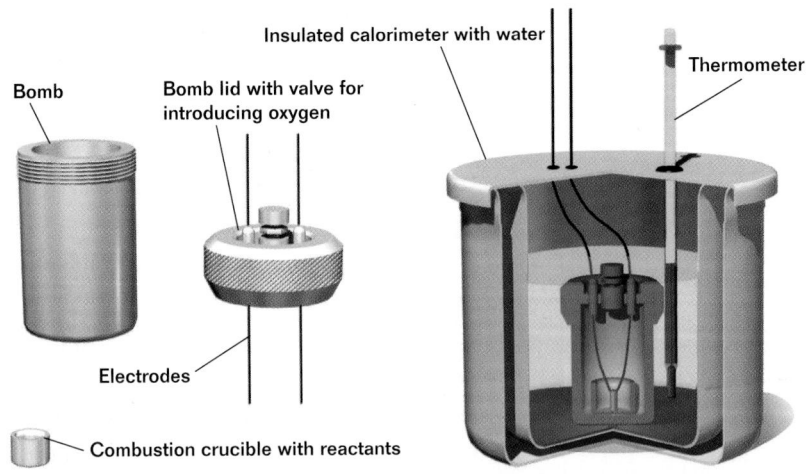

Figure 11-4
The volume inside the bomb calorimeter is nearly constant, so most of the energy is transferred to or from the calorimeter as heat.

isovolumetric process

a thermodynamic process that takes place at constant volume so that no work is done on or by the system

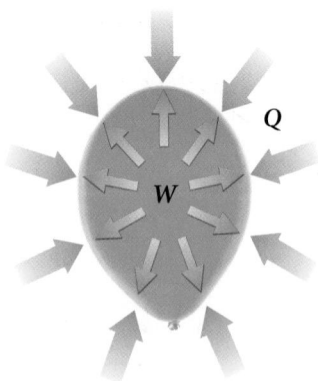

Figure 11-5
An isothermal process can be approximated if energy is slowly removed from a system as work while an equivalent amount of energy is added as heat.

In general, when a gas undergoes a change in temperature but no change in volume, no work is done on or by the system. Such a process is called a constant-volume, or **isovolumetric,** process.

Another example of an isovolumetric process takes place inside a *bomb calorimeter,* shown in **Figure 11-4.** This device is a thick container in which a small quantity of a substance undergoes a combustion reaction. The energy released by the reaction increases the pressure and temperature of the gaseous reaction products. Because the container's walls are thick, there is no change in the volume of the gas. The energy can be transferred from the container only as heat. As in the case of the simple calorimeter discussed in Chapter 10, the increase in the temperature of water surrounding the bomb calorimeter provides information for calculating the total amount of energy produced by the reaction.

Internal energy is constant in a constant-temperature process

Although you may think of a toy balloon that has been inflated and sealed as a static system, it is subject to continuous thermodynamic effects. Consider what happens to such a balloon during an approaching storm. During the few hours before the storm arrives, the barometric pressure of the atmosphere steadily decreases by about 2000 Pa. If you are indoors and the temperature of the building is controlled, any change in outside temperature will not take place indoors. But because no building is perfectly sealed, changes in the pressure of the air outside also take place inside.

As the atmospheric pressure inside the building slowly decreases, the balloon expands and slowly does work on the air outside the balloon. At the same time, energy is slowly transferred into the balloon as heat. The net result of these two processes is that the air inside the balloon is at the same temperature as the air outside the balloon. So the internal energy of the balloon's air does not change, and the energy transferred out of the balloon as work is matched by the energy transferred into the balloon as heat (see **Figure 11-5**).

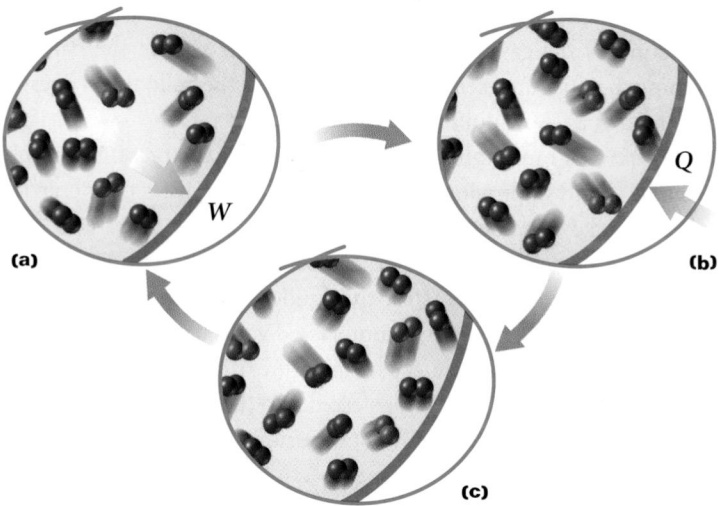

(a) W

(b) Q

(c)

Figure 11-6

In an isothermal process in a balloon, **(a)** small amounts of energy are removed as work as the balloon's surface is stretched by the expanding gas. **(b)** Energy is added to the gas within the balloon's interior as heat so that **(c)** thermal equilibrium is quickly restored.

This procedure is a close approximation of an **isothermal process.** In an isothermal process, the system's temperature remains constant and internal energy does not change when energy is transferred to or from the system as heat or work.

You may wonder how energy can be transferred as heat from the air outside the balloon to the air inside when both gases are at the same constant temperature. After all, if no temperature difference exists, then energy cannot be transferred as heat. However, the transfer of energy as heat can occur in an isothermal process if it is assumed that the process takes place as a large number of very gradual, very small changes, as shown in **Figure 11-6.**

When the air inside the balloon expands, its internal energy and temperature decrease slightly. But as soon as they decrease, energy is transferred as heat from the higher-temperature outside air to the air inside the balloon. As a result, the temperature and internal energy of the air inside the balloon rise to their original values. Because the decrease and increase occur more rapidly than the overall processes of transferring energy as work or heat, the internal energy of the balloon's air effectively remains constant. (★TEKS) **3A**

Energy is not transferred as heat in an adiabatic process

When a tank of compressed gas is opened to fill a toy balloon, the process of inflation occurs rapidly, unlike the previously described isothermal process, in which the inflation occurs gradually. The internal energy of the gas does not remain constant. Instead, as the pressure of the gas in the tank decreases, so does the gas's internal energy and temperature.

If the balloon and the tank are thermally insulated, no energy can be transferred from the expanding gas as heat because there is not enough time for the transfer of energy to occur. When energy is not transferred to or from a system as heat, the process is called **adiabatic.** The decrease in internal energy

isothermal process

a thermodynamic process that takes place at constant temperature and in which the internal energy of a system remains unchanged

Did you know?

The absence of a change in the internal energy of an isothermal process is true only for systems in which there is no phase change. During a phase change, such as when water becomes steam, the temperature remains constant but internal energy increases.

adiabatic process

a thermodynamic process during which work is done on or by the system but no energy is transferred to or from the system as heat

SECTION 11-1

Teaching Tip

An isothermal process can be closely achieved when the gas container is a good thermal conductor, is relatively small, and is submerged in a steady temperature environment.

Demonstration 2

Adiabatic processes

Purpose Show that quick expansion of a gas results in a temperature drop.

Materials one or more of the following: helium tank, nitrogen tank, CO_2 fire extinguisher (not the chemical kind), CO_2 canister used for inflating life jackets

CAUTION *Pressures are very high. Wear safety goggles.*

Procedure Have students touch the tanks and nozzles to verify that they are at room temperature. Open a tank for 2–3 seconds and then close it. Have students touch the nozzle again. (It should be very cold.) Repeat with other tanks and gases, if available. Explain that these examples were nearly adiabatic processes in which the gas's high pressure dropped suddenly. As a result, the gas expanded very quickly, did work on its environment, and lost internal energy faster than it could absorb energy from the environment. Thus, the gas's temperature decreased. Because the nozzle and tank wall are good thermal conductors, internal energy was immediately transferred from the nozzle and tank wall to the gas in order to restore the cold gas to thermal equilibrium.

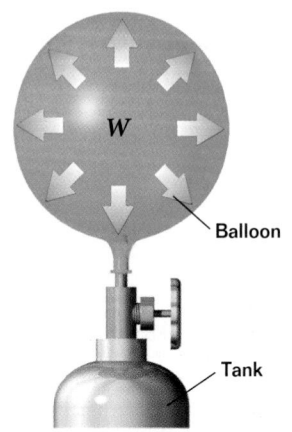

Figure 11-7
As the gas inside the tank and balloon rapidly expands, its internal energy decreases. This energy leaves the system by means of work done against the outside air.

must therefore be equal to the energy transferred from the gas as work. This work is done by the gas pushing against the inner wall of the balloon and overcoming the pressure exerted by the air outside the balloon. As a result, the balloon inflates, as shown in **Figure 11-7.**

Notice that this rapid inflation of the balloon in the present example is a process that is only approximately adiabatic. Some transfer of energy as heat actually does take place because neither the balloon nor the tank is perfectly insulated. The decrease in the internal energy and the temperature of the rapidly expanding gas accounts for the sudden drop in temperature of the outside surface of the tank when a compressed gas is being released. Once the adiabatic expansion is complete, the temperature of the gas gradually increases as energy from the outside air is transferred into the tank as heat.

The adiabatic expansion and compression of gases is found in many applications. As you will learn in the next section, both refrigerators and internal combustion engines require that gases be compressed or expanded rapidly. By assuming these processes to be adiabatic, one can make predictions about how a machine will operate.

Section Review  **2C**

1. In which of the situations listed below is energy being transferred as heat to the system in order for the system to do work? In which situation is work being done on the system in order for energy to be transferred from the system as heat?

   a. Two sticks are rubbed together to start a fire.
   b. A firecracker explodes.
   c. A red-hot iron bar is set aside to cool.

2. Identify the following processes as isothermal, isovolumetric, or adiabatic:

   a. a tire being rapidly inflated
   b. a tire expanding gradually with heating
   c. a tire being heated while in a rigid metal container

3. A mixture of gasoline vapor and air is placed in an engine cylinder. The piston has an area of 7.4×10^{-3} m^2 and is displaced inward by 7.2×10^{-2} m. If 9.5×10^5 Pa of pressure is placed on the piston, how much work is done during this process? Is work being done *on* or *by* the gas mixture?

4. **Physics in Action** A weather balloon slowly expands as energy is transferred as heat from the outside air. If the average net pressure is 1.5×10^3 Pa and the balloon's volume increases by 5.4×10^{-5} m^3, how much work is done by the expanding gas? What thermodynamic process does this situation resemble?

11-2
Thermodynamic processes

THE FIRST LAW OF THERMODYNAMICS

Imagine a roller coaster that operates without friction. The car is raised against the force of gravity by work. Once the car is freely moving, it will have a certain kinetic energy (*KE*) and a certain potential energy (*PE*). The coaster will move slower (have a small *KE*) at the top of the rise (have a large *PE*) and faster (have a larger *KE*) at low points in the track (have a smaller *PE*). However, the mechanical energy, *KE* + *PE*, remains constant throughout the ride's duration.

If friction is taken into account, mechanical energy is not conserved, as shown in **Figure 11-8.** A steady decrease in the car's total mechanical energy occurs because of work being done against the friction between the car's axles and its bearings and between the car's wheels and the coaster track. Mechanical energy is transferred to the atoms and molecules throughout the entire roller coaster (both the car and the track). Thus, the roller coaster's internal energy increases by an amount equal to the decrease in the mechanical energy. Most of this energy is then gradually dissipated to the air surrounding the roller coaster as heat and sound. If the internal energy for the roller coaster (the system) and the energy dissipated to the surrounding air (the environment) are taken into account, then the total energy will be constant. ⭐(TEKS) 5B

11-2 SECTION OBJECTIVES

- **Illustrate how the first law of thermodynamics is a statement of energy conservation.**

- **Calculate heat, work, and the change in internal energy by applying the first law of thermodynamics.**

- **Apply the first law of thermodynamics to describe cyclic processes.**

Visual Strategy

Figure 11-8
Point out that the *KE* bars represent the energy of the car alone, the *PE* bars represent the energy of the car-Earth system, and the *U* bars represent the sum of the internal energies of the car and the track.

Q How does the potential energy vary as the car rolls up and down the track? How is this reflected in the energy bars?

A *The potential energy depends only on the car's elevation. Accordingly, the PE bar is highest at (b), second highest at (e), second lowest at (c), and lowest at (d).*

Q Draw a fourth bar representing the *mechanical* energy, *ME*, at locations **(b), (c), (d),** and **(e).** How do these *ME* bars relate to the *U* bars?

A *The new ME bar should equal KE + PE in each case. Thus, it will get shorter from (b) to (e) as the U bar gets taller by the same amount.*

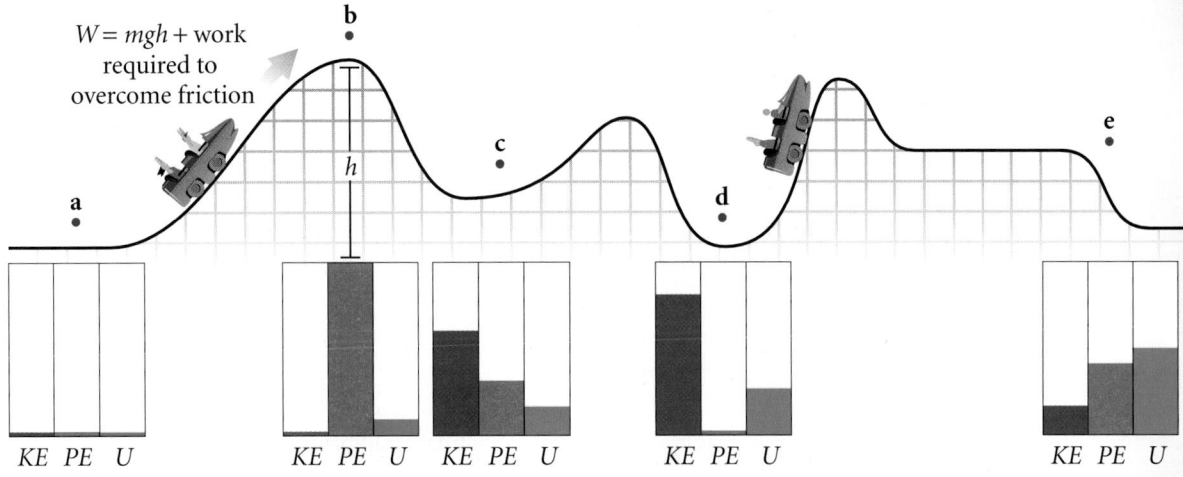

$W = mgh + \text{work}$
required to
overcome friction

Figure 11-8
In the presence of friction, the internal energy (*U*) of the roller coaster increases as *KE* + *PE* decreases.

⭐(TEKS)

p. 408: 2C
p. 409: 5B

Teaching Tip

To remember whether a system's internal energy increases or decreases, one may find it helpful to visualize the system as a circle. When work is done on the system or energy is transferred as heat into the system, an arrow points into the circle, which shows that internal energy increases. When work is done by the system or energy is transferred as heat out of the system, the arrow points out of the circle, which shows that internal energy decreases.

Key Models and Analogies

Show these simple examples to model the signs of Q and W in typical situations. Students can reduce problems to one case or to a combination of two cases.

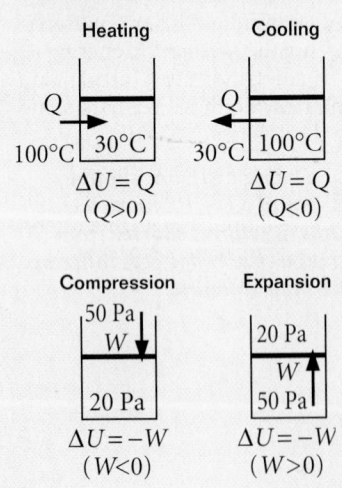

The principle of energy conservation that takes into account a system's internal energy as well as work and heat is called the *first law of thermodynamics.*

Imagine that the isothermally expanding toy balloon in the previous section is squeezed rapidly. The process is no longer isothermal. Instead, it is a combination of two processes. On the one hand, work (W) is done on the system. The balloon and the air inside it (the system) are compressed, so the air's internal energy and temperature increase. Work is being done on the system, so W is a negative quantity. The rapid squeezing of the balloon can be treated as an adiabatic process, so $Q = 0$ and, therefore, $\Delta U = -W$.

After the compression step, energy is transferred from the system as heat (Q). Some of the internal energy of the air inside the balloon is transferred to the air outside the balloon. During this step, the internal energy of the gas decreases, so ΔU has a negative value. Similarly, because energy is removed from the system, Q has a negative value. The change in internal energy for this step, in which the volume of the air inside the balloon stays constant, can be expressed as $-\Delta U = -Q$, or $\Delta U = Q$.

The signs for heat and work for a system are summarized in **Table 11-1.**

Table 11-1	Signs of Q and W for a system ⊛TEKS 2C	
$Q > 0$	energy added to system as heat	internal energy increases if $W = 0$
$Q < 0$	energy removed from system as heat	internal energy decreases if $W = 0$
$Q = 0$	no transfer of energy as heat	no internal energy change if $W = 0$
$W > 0$	work done by system (expansion of gas)	internal energy decreases if $Q = 0$
$W < 0$	work done on system (compression of gas)	internal energy increases if $Q = 0$
$W = 0$	no work done	no internal energy change if $Q = 0$

The first law of thermodynamics can be expressed mathematically

In all the thermodynamic processes described so far, energy has been conserved. To describe the overall change in the system's internal energy, one must account for the transfer of energy to or from the system as heat and work. The total change in the internal energy is the difference between the final internal energy value (U_f) and the initial internal energy value (U_i). That is, $\Delta U = U_f - U_i$. Energy conservation requires that the total change in internal energy from its initial to its final equilibrium conditions be equal to the transfer of energy as both heat and work. This statement of total energy conservation is the first law of thermodynamics.

THE FIRST LAW OF THERMODYNAMICS 3B

$$\Delta U = Q - W$$

Change in system's internal energy = energy transferred to or from system as heat – energy transferred to or from system as work

When this equation is used, all quantities must have the same energy units. Throughout this chapter, the SI unit for energy, the joule, will be used.

According to the first law of thermodynamics, a system's internal energy can be changed by transferring energy as either work, heat, or a combination of the two. The thermodynamic processes discussed on pp. 405–408 can therefore be expressed using the equation for the first law of thermodynamics, as shown in **Table 11-2.**

Table 11-2   **First law of thermodynamics for special processes** 2C

Process	Conditions	First law of thermodynamics	Interpretation
Isovolumetric	no work done	$\Delta V = 0$, so $P\Delta V = 0$ and $W = 0$; therefore, $\Delta U = Q$	Energy added to the system as heat ($Q > 0$) increases the system's internal energy.
			Energy removed from the system as heat ($Q < 0$) decreases the system's internal energy.
Isothermal	no change in temperature and internal energy	$\Delta T = 0$, so $\Delta U = 0$; therefore, $\Delta U = Q - W = 0$, or $Q = W$	Energy added to the system as heat is removed from the system as work done by the system.
			Energy added to the system by work done on it is removed from the system as heat.
Adiabatic	no energy transferred as heat	$Q = 0$, so $\Delta U = -W$	Work done on the system ($W < 0$) increases the system's internal energy.
			Work done by the system ($W > 0$) decreases the system's internal energy.
Isolated system	no heat or work interaction with surroundings	$Q = W = 0$, so $\Delta U = 0$ and $U_i = U_f$	There is no change in the system's internal energy.

 Misconception Alert

Students may not be aware that $\Delta U = U_f - U_i$ is simply a definition based on the conventional use of symbols.

The Language of Physics

The first law is a statement of conservation of energy. Because the Q term is positive, it represents the *energy added to* the system as heat. Because the W term is negative, it represents the *work done by* the system.

Teaching Tip

Ask students to hide all but the first column of **Table 11-2** and try to reconstruct as much information as possible from the initial conditions. For example, in an isovolumetric process, $\Delta V = 0$. Because $W = P\Delta V$, W must also equal zero. In this case, $\Delta U = Q - W$ becomes $\Delta U = Q$.

Ask students under which conditions work can be done on a gas without changing its internal energy (*in an isothermal process, where $\Delta U = 0$*).

⭐ TEKS

p. 410: 2C
p. 411: 3B, 2C

The first law of thermodynamics (★)TEKS 2C, 7A

Classroom Practice

The following may be used as teamwork exercises or for demonstration at the chalkboard or on an overhead projector.

PROBLEM

The first law of thermodynamics

A gas is trapped in a small metal cylinder with a movable piston and is submerged in a large amount of ice water so that the initial temperature of the gas is 0°C. A total of 1200 J of work is done by a force that slowly pushes the piston inward.

a. Is this process isothermal, adiabatic, or isovolumetric?

b. How much energy is transferred as heat between the gas and the ice water?

Answers

a. isothermal (the large amount of ice water and the slow process maintain the gas at 0°C)

b. Q from the gas to the ice water = 1200 J

PRACTICE GUIDE 11B

Solving for:		
Q	📖 **PE**	Sample, 1–3; Ch. Rvw. 19–21, 39
	💿 **PW**	3, 5
	PB	7–10
W	📖 **PE**	4; Ch. Rvw. 21
	💿 **PW**	2, 6–7
	PB	Sample, 1–3
ΔU	📖 **PE**	5; Ch. Rvw. 21, 39–40
	💿 **PW**	Sample, 1, 2–3, 4
	PB	4–6

PROBLEM

A total of 135 J of work is done on a gaseous refrigerant as it undergoes compression. If the internal energy of the gas increases by 114 J during the process, what is the total amount of energy transferred as heat? Has energy been added to or removed from the refrigerant as heat?

SOLUTION

1. DEFINE **Given:** $W = -135$ J $\Delta U = 114$ J

Work is done on the gas, so work (W) has a negative value. The internal energy increases during the process, so the change in internal energy (ΔU) has a positive value.

Unknown: $Q = ?$

Diagram:

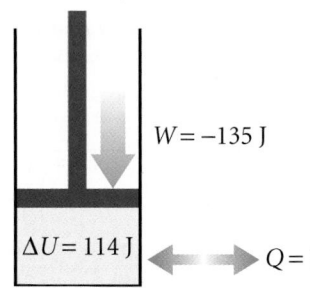

$W = -135$ J

$\Delta U = 114$ J ⬌ $Q = ?$

2. PLAN **Choose an equation(s) or situation:** Apply the first law of thermodynamics using the values for the change in internal energy (ΔU) and the work (W) done on the system in order to find the value for heat (Q).

$$\Delta U = Q - W$$

Rearrange the equation(s) to isolate the unknown(s):

$$Q = \Delta U + W$$

3. CALCULATE **Substitute the values into the equation(s) and solve:**

$$Q = 114 \text{ J} + (-135 \text{ J}) = -21 \text{ J}$$

$$\boxed{Q = -21 \text{ J}}$$

The sign for the value of Q is negative. From **Table 11-1,** $Q < 0$ indicates that energy is transferred as heat *from* the refrigerant.

4. EVALUATE Although the internal energy of the refrigerant increases under compression, more energy is added as work than can be accounted for by the increase in the internal energy. This energy is removed from the gas as heat, as indicated by the minus sign preceding the value for Q.

The first law of thermodynamics ⭐TEKS 2C, 7A

1. A system's initial internal energy is 27 J. Then heat is added to the system. If the final internal energy is 34 J and the system does 26 J of work, how much heat is added to the system?

2. The internal energy of the gas in a gasoline engine's cylinder decreases by 195 J. If 52.0 J of work is done by the gas, how much energy is transferred as heat? Is this energy added to or removed from the gas?

3. A 2.0 kg quantity of water is held at constant volume in a pressure cooker and heated by a range element. The system's internal energy increases by 8.0×10^3 J. However, the pressure cooker is not well insulated, and 2.0×10^3 J of energy is transferred to the surrounding air. How much energy is transferred from the range element to the pressure cooker as heat?

4. The internal energy of a gas decreases by 344 J. If the process is adiabatic, how much energy is transferred as heat? How much work is done on or by the gas?

5. A steam engine's boiler completely converts 155 kg of water to steam. This process involves the transfer of 3.50×10^8 J as heat. If steam escaping through a safety valve does 1.76×10^8 J of work expanding against the outside atmosphere, what is the net change in the internal energy of the water-steam system?

CYCLIC PROCESSES

A refrigerator performs work to create a temperature difference between its closed interior and its environment (the air in the room). A refrigerator can be represented schematically as a system that transfers energy from a body at low temperature, T_{cold} or T_c, to one at a high temperature, T_{hot} or T_h (see **Figure 11-9**). The refrigerator uses work performed by an electric motor to compress the refrigerant.

The process by which a refrigerator accomplishes this task consists of four basic steps. The system to and from which energy is transferred is defined here as the refrigerant contained within the inner surface of the tubing.

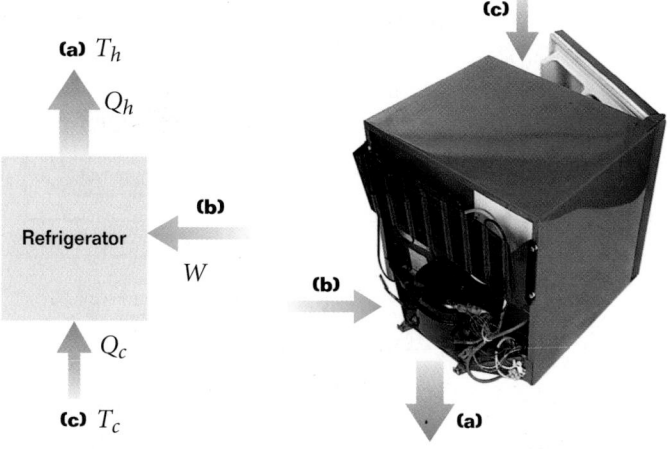

Figure 11-9
A refrigerator does work **(b)** in order to transfer energy as heat from a low-temperature region (the inside of the refrigerator) at T_c **(c)** to a high-temperature region (the air outside the refrigerator) at T_h **(a)**.

Thermodynamics **413**

SECTION 11-2

ANSWERS TO

Practice 11B
The first law of thermodynamics
1. 33 J
2. −143 J; removed as heat
3. 1.00×10^4 J
4. 0 J; 344 J done by gas
5. 1.74×10^8 J

Teaching Tip

Point out that the refrigerant as a system is in thermal contact with two environments (the inside of the refrigerator and the outside air) at two separate moments in time. Between these steps of the cycle, the refrigerant does work and work is done on it adiabatically. The end result of drawing energy as heat from inside (Q_c) to outside (Q_h) occurs when the electric motor in the refrigerator does work to compress the refrigerant.

⭐TEKS

p. 412: 2C, 7A
p. 413: 2C, 7A

Visual Strategy

Figure 11-10

Be sure that students recognize areas of low pressure (the refrigerant inside the refrigerator) and high pressure (the refrigerant outside the refrigerator) in the fluid shown in the diagram.

Q What happens to the refrigerant's volume, internal energy, and temperature when the refrigerant passes through the expansion valve?

A *During the adiabatic expansion of liquid refrigerant, the refrigerant's volume increases, which causes its internal energy to decrease ($-\Delta U < 0$). This internal energy decrease results in a lowering of the refrigerant's temperature. The energy is transferred from the refrigerant as it does work against the spring in the expansion valve.*

Q As the refrigerant goes from **(d)** to **(b)** through the coiled pipe **(a)**, which area will be at a lower temperature?

A *The lowest temperature zone will be near (d)—where the freezer compartment would be.*

The entire process is shown in **Figure 11-10.** Initially, liquid refrigerant is at a low temperature and pressure so that it is colder than the air inside the refrigerator. The refrigerant absorbs energy from inside the refrigerator and lowers the refrigerator's interior temperature. This transfer of energy as heat increases the temperature of the liquid refrigerant until it begins to boil, as shown in **(a).** The refrigerant continues to absorb energy at the constant boiling temperature until the refrigerant has completely vaporized. (★)**TEKS** **3C**

Once it is in the vapor phase, the refrigerant is passed through a *compressor.* The compressor does work on the gas by decreasing its volume without transferring energy as heat, as shown in **(b).** This adiabatic process increases the pressure and internal energy (temperature) of the gaseous refrigerant.

In the next step, the refrigerant is moved to the outer parts of the refrigerator, where thermal contact is made with the air in the room. The refrigerant gives up energy to the environment, which is at a lower temperature, as shown in **(c).** The gaseous refrigerant at high pressure then condenses at a constant temperature to a liquid.

The liquefied refrigerant is then brought back into the refrigerator. Just outside the low-temperature interior of the refrigerator, the refrigerant goes through an *expansion valve* and expands without absorbing energy as heat. The liquid then does work as it moves from a high-pressure region to a low-pressure region and its volume increases, as shown in **(d).** By doing work against the spring in the valve, the refrigerant reduces its internal energy.

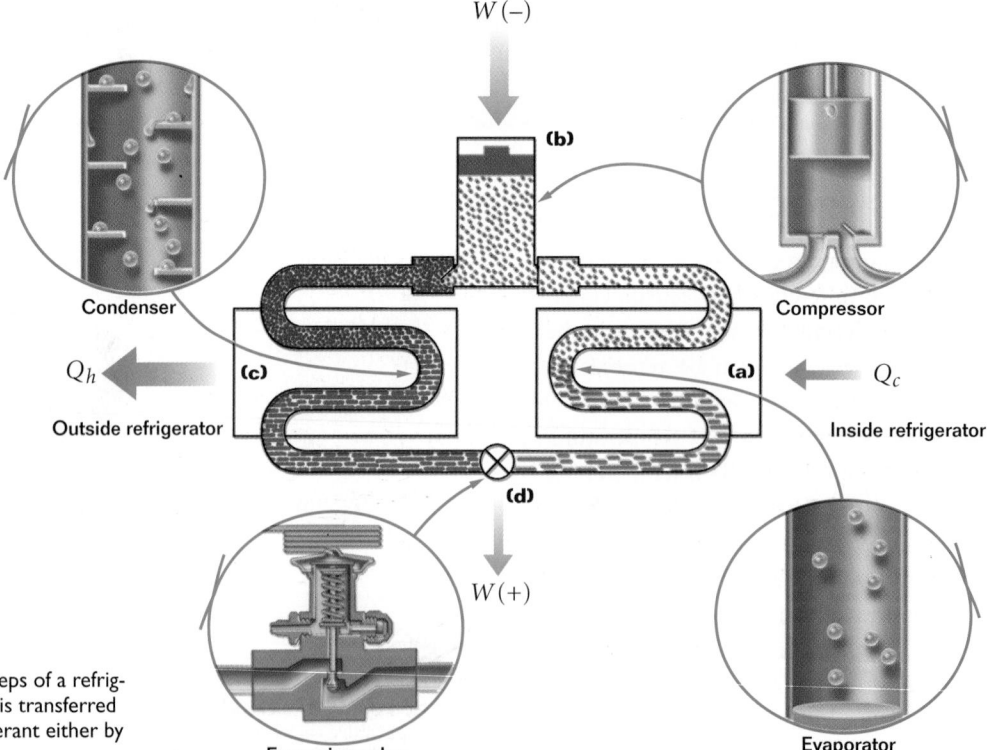

Figure 11-10

In each of the four steps of a refrigeration cycle, energy is transferred to or from the refrigerant either by heat or by work.

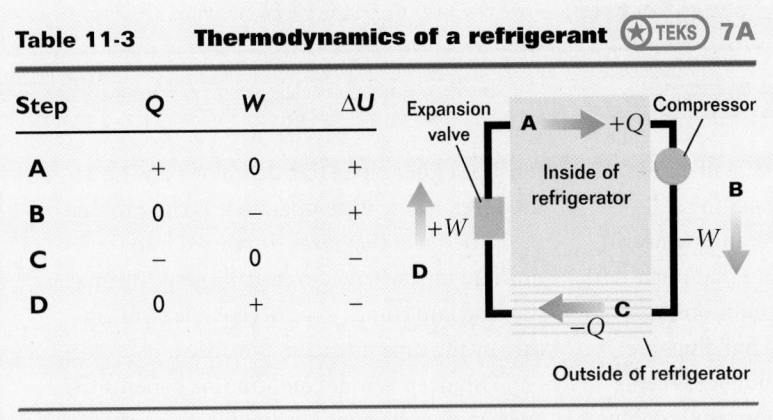

Table 11-3   **Thermodynamics of a refrigerant** (★)TEKS) 7A

Step	Q	W	ΔU
A	+	0	+
B	0	−	+
C	−	0	−
D	0	+	−

The refrigerant now has the same internal energy and phase as it did at the start of the process. If its temperature is still lower than the temperature of the air inside the refrigerator, the cycle will repeat.

The first law of thermodynamics can be used to describe the signs of each thermodynamic quantity in the four steps listed, as indicated in **Table 11-3.**

The change in internal energy of a system is zero in a cyclic process

The fact that the refrigerant's final internal energy is the same as its initial internal energy is extremely important, not only with refrigerators but with all machines that use heat to do work or that do work to create temperature differences. If a system's properties at the end of a process are identical to the system's properties before the process took place, the final and initial values of internal energy are the same and the change in internal energy is zero.

$$\Delta U_{net} = 0 \text{ and } Q_{net} = W_{net}$$

This process resembles an isothermal process in that all energy is transferred as work and heat. But now the process is repeated with no net change in the system's internal energy. Such a process is called a **cyclic process.**

In a refrigerator, the net amount of work done during each complete cycle must equal the net amount of energy transferred as heat to and from the refrigerant. Energy is transferred as heat from the cold interior of the refrigerator to the even colder evaporating refrigerant (Q_{cold}, or Q_c). Energy is also transferred as heat from the hot condensing refrigerant to the relatively cooler air outside the refrigerator (Q_{hot}, or Q_h). Therefore, the difference between Q_h and Q_c equals the net energy transferred as heat—and thus the net work done—during one cycle of the refrigeration process.

$$W_{net} = Q_h - Q_c, \text{ where } Q_h > Q_c$$

The colder you want the inside of a refrigerator to be, the greater the net energy transferred as heat ($Q_h - Q_c$) must be. The net energy transferred as heat can be increased only if the refrigerator does more work.

Did you know?

The process of changing phase can involve work. The volume of a liquid increases as the liquid vaporizes and does work on the gases above it. Similarly, work is done on a gas that condenses to a smaller-volume liquid. Because the system for a refrigerator is defined as the refrigerant within the tubing, energy is not transferred as work between the refrigerant and its surroundings outside the tubing during vaporization or condensation.

cyclic process

a thermodynamic process in which a system returns to the same conditions under which it started

SECTION 11-2

Visual Strategy

Table 11-3
Point out that the table refers to transfers of energy as heat and work to and from the refrigerant (the system). Students should be able to correctly identify the system-environment relationship in each step of the cycle.

Q Describe the source and nature of the energy transfers in each step of the process.

A *(a) Energy is transferred as heat from refrigeration compartment to refrigerant in the tubing inside the refrigerator.*

(b) Work is done by the compressor on the refrigerant in an approximately adiabatic process.

(c) Energy is transferred as heat from the refrigerant in the tubing outside the refrigerator to the air (the exterior of the refrigerator).

(d) Work is done by the refrigerant on the expansion valve in an approximately adiabatic process.

(★)TEKS)

p. 414: 3C
p. 415: 7A

Deep-Sea Air Conditioning ⭐TEKS 3C, 7A

Deep beneath the ocean, about half a mile down, sunlight barely penetrates the still waters. Scientists at Makai Ocean Engineering in Hawaii are now tapping into that pitch-dark region as a resource for air conditioning.

In tropical locations where buildings are cooled year-round, air-conditioning systems operate with cold water. Compressors cool the water, and pumps circulate it throughout the walls of a building, where the water absorbs heat from the rooms. Unfortunately, powering these compressors is neither cheap nor efficient.

Instead of using compressors to cool the water in their operating system, the systems designed by Makai use frigid water from the ocean's depths. First, engineers install a pipeline that reaches deep into the ocean, where the water is nearly freezing. Then, powerful pumps on the shoreline move the water directly into a building's air-conditioning system. There, a system of heat exchangers uses the sea water to cool the fresh water in the air-conditioning system.

One complicating factor is that the water must also be returned to the ocean in a manner that will not disrupt the local ecosystem. It must be either piped to a depth of a few hundred feet, where its temperature is close to that of the ocean at that level, or poured into onshore pits, where it eventually seeps through the land and comes to an acceptable temperature by the time it reaches the ocean.

"This deep-sea air conditioning benefits the environment by operating with a renewable resource instead of freon," said Dr. Van Ryzin, the president of Makai. "Because the system eliminates the need for compressors, it uses only about 10 percent of the electricity of current methods, saving fossil fuels and a lot of money." However, deep-sea air-conditioning technology works only for buildings within a few kilometers of the shore and carries a hefty installation cost of several million dollars. For this reason, Dr. Van Ryzin thinks this type of system is most appropriate for large central air-conditioning systems, such as those necessary to cool resorts or large manufacturing plants, where the electricity savings can eventually make up for the installation costs. Under the right circumstances, air conditioning with sea water can be provided at one-third to one-half the cost of conventional air conditioning.

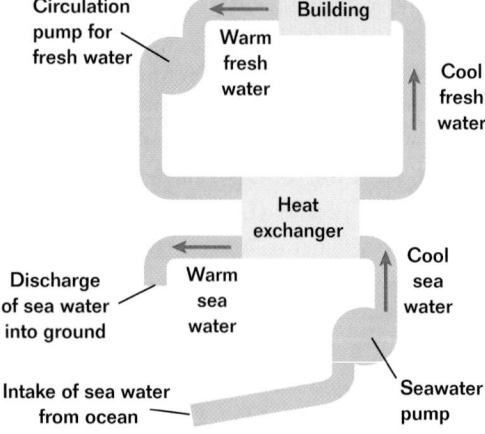

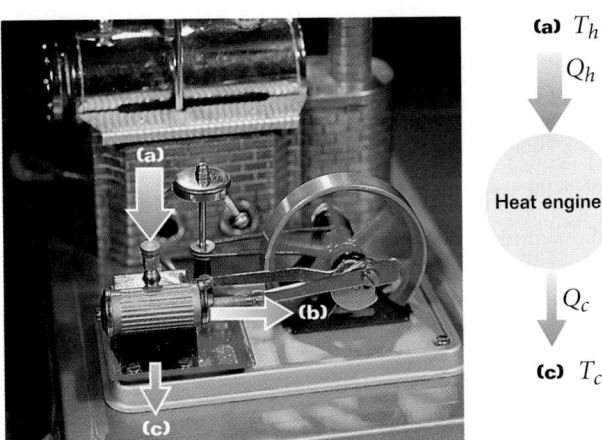

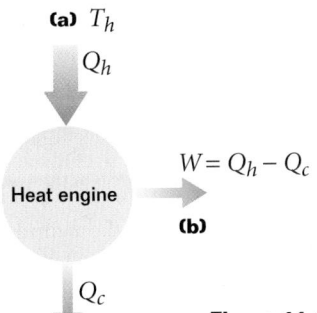

(a) T_h

Q_h

Heat engine

$W = Q_h - Q_c$

(b)

Q_c

(c) T_c

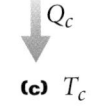

Figure 11-11

A heat engine is able to do work **(b)** by transferring energy from a high-temperature substance (the boiler) at T_h **(a)** to a substance at a lower temperature (the air surrounding the engine) at T_c **(c)**.

Heat engines use heat to do work

A refrigerator uses mechanical work to create a difference in temperature and thus transfers energy as heat. A heat engine is a device that does the opposite. It uses heat to do mechanical work.

One analogy for a heat engine is a water wheel. A water wheel uses the energy of water falling from one level above Earth's surface to another. The change in potential energy increases the water's kinetic energy so that the water can do work on one side of the wheel and thus turn it. The greater the difference between the initial and final values of the potential energy, the greater the amount of work that can be done. ⊛TEKS **5B**

Instead of using the difference in potential energy to do work, heat engines do work by transferring energy from a high-temperature substance to a lower-temperature substance, as indicated for the steam engine shown in **Figure 11-11.** For each complete cycle of the heat engine, the net work done will equal the difference between the energy transferred as heat from a high-temperature substance to the engine (Q_h) and the energy transferred as heat from the engine to a lower-temperature substance (Q_c).

$$W_{net} = Q_h - Q_c$$

The larger the difference between the amount of energy transferred as heat into the engine and out of the engine, the more work the engine can do.

The internal-combustion engine found in most vehicles is an example of a heat engine. Internal-combustion engines burn fuel within a closed chamber (the cylinder). The potential energy of the chemical bonds in the reactant gases is converted to kinetic energy of the particle products of the reaction. These gaseous products push against a piston and thus do work on the environment (in this case, a crankshaft that transforms the linear motion of the piston to the rotational motion of the axle and wheels). ⊛TEKS **3C, 7B**

Although the basic operation of any internal-combustion engine resembles that of an ideal heat engine, certain steps do not fit the model. When gas is taken in or removed from the cylinder, matter enters or leaves the system so

---

internetconnect

SC**LINKS**

NSTA

TOPIC: Heat engines
GO TO: www.scilinks.org
*sci***LINKS CODE:** HF2113

Key Models and Analogies

The model of a system interacting with its environment is useful when dealing with thermodynamic cycles. Point out that the steam in the cylinder is the thermodynamic system, which exchanges energy with its surroundings. Remind students that energy is transferred as heat when the temperature of the system and that of the environment are different. The air is the environment for the cooling phase, and the boiler is the environment in the heating phase.

Teaching Tip

Have students describe the four steps in the cycle in terms of energy transfer, work done, and changes in internal energy of the steam.

Heater-boiler: Q_{in} into steam
W_{out} done by steam on piston
Air: Q_{out} out of steam
W_{in} done by piston on steam

🛑 Misconception Alert

Students tend to believe that $W_{out} = W_{in}$ because volume increases and then decreases by the same amount as the piston returns to its original position. Remind them that $W = P\Delta V$. Ask if the pressure in the cylinder is the same during expansion of the steam at a high temperature as it is during compression at a low temperature. (*No, the pressure exerted by the steam at high temperature is greater than the pressure exerted by the steam at low temperature.*)

Teaching Tip

Point out that another difference between an ideal heat engine and the internal combustion engine is that in the heat engine, the energy source is external and energy is transferred to the engine as heat. In the internal combustion engine, the energy source is internal (because it comes from a chemical reaction in the cylinder), so energy is not transferred into the engine as heat.

Key Models and Analogies

A graph of pressure (P) versus volume (V) is often used for depicting the steps of thermodynamic cycles. Draw a simple graph as shown below on the chalkboard. Ask students which segment of the graph corresponds to heating (*ab*), cooling (*cd*), compression (*da*), and expansion (*bc*). Use this simple graph to tell students that the net work done in the cycle equals the area enclosed by the graph connecting points **a, b, c,** and **d.**

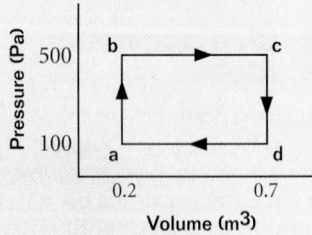

★ TEKS

p. 418: 7A
p. 419: 2C, 7A

that the matter in the system is not isolated. No heat engine operates perfectly. Only part of the available internal energy leaves the engine as work done on the environment; most of the energy is removed as heat.

In **Figure 11-12,** the steps in one cycle of operation for a gasoline engine (one type of internal combustion engine) are shown. During compression, shown in (**a**), work is done by the piston as it adiabatically compresses the fuel-and-air mixture in the cylinder. Once maximum compression of the gas is reached, combustion takes place. The chemical potential energy released during combustion increases the internal energy of the gas, as shown in (**b**). The hot, high-pressure gases from the combustion reaction expand in volume, which causes the piston to do work on the crankshaft, as shown in (**c**). Once all of the work is done by the piston, energy is transferred as heat through the walls of the cylinder and by the physical transfer of the hot exhaust gases from the cylinder, as shown in (**d**). A new fuel-air mixture is then drawn through the intake valve into the cylinder by the downward-moving piston, as shown in (**e**).

 **7A**

Figure 11-12
The steps below show one complete cycle of a gasoline engine.

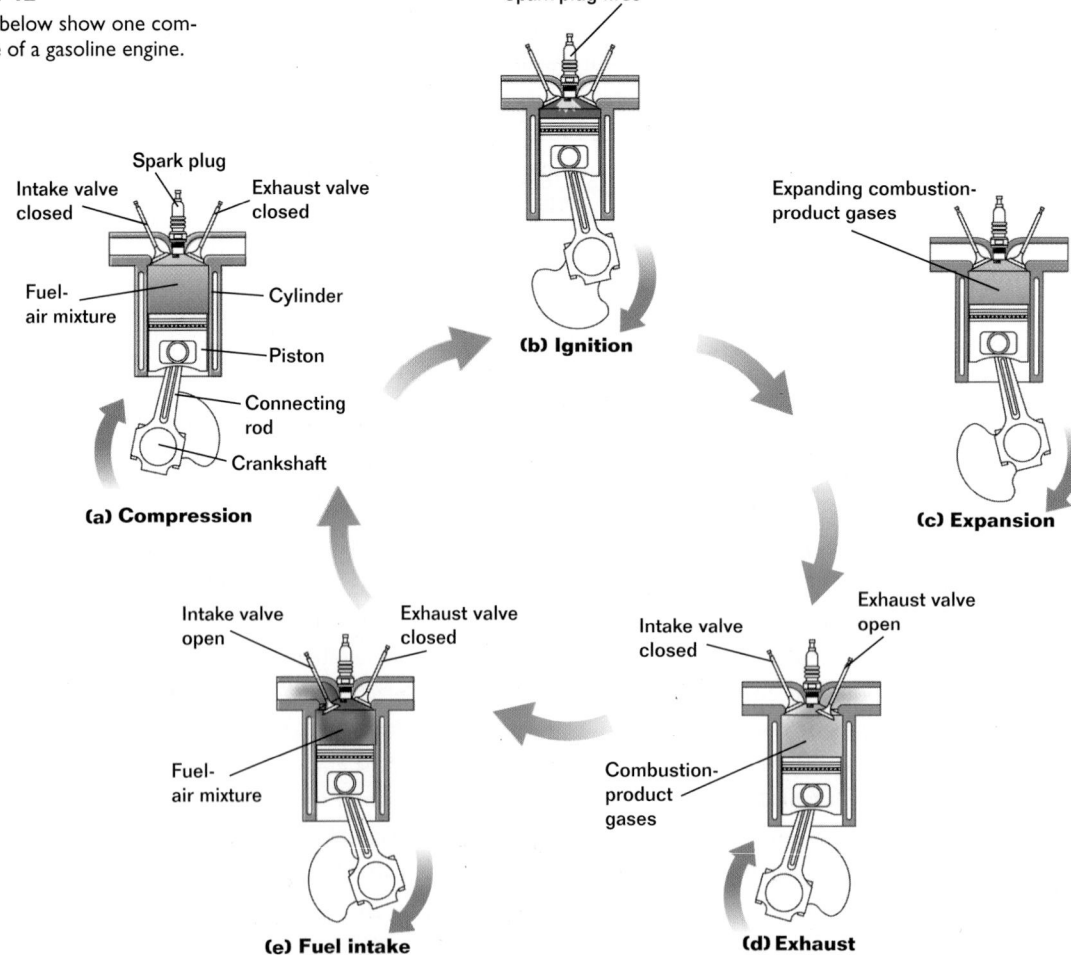

Section Review  ⭐TEKS 2C, 7A

1. Use the first law of thermodynamics to show that the internal energy of an isolated system is always conserved.

2. In the systems listed below, identify where energy is transferred as heat and work and where changes in internal energy occur. Is energy conserved in each case?
   a. the steam in a steam engine consisting of a boiler, a firebox, a cylinder, a piston, and a flywheel
   b. the drill bit of a power drill and a metal block into which a hole is being drilled

3. Does the system's overall internal energy increase, decrease, or remain unchanged in either of the situations in item 2? Explain your answer in terms of the first law of thermodynamics.

4. A compressor for a jackhammer expands the air in the hammer's cylinder at a constant pressure of 8.6×10^5 Pa. The increase in the hammer's volume is 4.05×10^{-4} m^3. During the process, 9.5 J of energy is transferred out of the cylinder as heat.
   a. What is the work done by the air?
   b. What is the change in the air's internal energy?
   c. What type of ideal thermodynamic process does this approximate?

5. A mixture of fuel and air is enclosed in an engine cylinder fitted with a piston. The gas pressure is maintained at 7.07×10^5 Pa as the piston moves slowly inward. If the gas volume decreases by 1.1×10^{-4} m^3 and the internal energy of the gas increases by 62 J, how much energy is added to or removed from the system as heat?

6. Using what you learned in this section, explain why opening the refrigerator door on a hot day does not cause your kitchen to become cooler.

7. Over several cycles, a refrigerator does 1.51×10^4 J of work on the refrigerant. The refrigerant in turn removes 7.55×10^4 J as heat from the air inside the refrigerator.
   a. How much energy is transferred as heat from the refrigerator's inner compartment to the outside air?
   b. What is the net change in the internal energy of the refrigerant?
   c. What is the amount of work done on the air inside the refrigerator?
   d. What is the net change in the internal energy of the air inside the refrigerator?

8. **Physics in Action** If a weather balloon in flight gives up 15 J of energy as heat and the gas within it does 13 J of work on the outside air, by how much does its internal energy change?

11-3
Efficiency of heat engines

THE SECOND LAW OF THERMODYNAMICS

In the previous section, you learned how a heat engine absorbs a quantity of energy from a high-temperature body as heat, does work on the environment, and then gives up energy to a low-temperature body as heat. The work derived from each cycle of a heat engine equals the difference between the heat input and heat output during the cycle ($W_{net} = Q_{net} = Q_h - Q_c$).

This equation, obtained from the first law of thermodynamics, indicates that all energy entering and leaving the system is accounted for and is thus conserved. The equation also suggests that more work is gained by taking more energy at a higher temperature and giving up less energy at a lower temperature. If no energy is given up at the lower temperature ($Q_c = 0$), then it seems that work could be obtained from energy transferred as heat from any body, such as the air around the engine. Such an engine would be able to do more work on hot days than on cold days, but it would always do work as long as the engine's temperature was less than the temperature of the surrounding air.

A heat engine cannot transfer all energy from heat to do work

Unfortunately, it is impossible to make such an engine. While the transfer of energy as heat from the high-temperature source to the engine would cause the engine to do work, it would not be a cyclic process. In order for the cycle to be completed, the engine would have to transfer energy away as heat. Because the only body to which this energy can be transferred is the high-temperature source, the engine must do work to transfer this energy. This is the same amount of work that was made available through the energy transferred as heat from the high-temperature body in the first place. Thus, no net work is obtained from this engine in a cyclic process. (★TEKS) **3A**

The requirement that a heat engine give up some energy at a lower temperature in order to do work does not follow from the first law of thermodynamics. This requirement is the basis of what is called the *second law of thermodynamics.* The second law of thermodynamics can be stated as follows: *No cyclic process that converts heat entirely into work is possible.*

According to the second law of thermodynamics, *W* can never be equal to Q_h in a cyclic process. In other words, some energy must always be transferred as heat to the system's surroundings ($Q_c > 0$).

Tomorrow's Technology

Solar Thermal Power Systems ⓣTEKS 3C, 7A

Because the fossil fuels used to run our generators are being rapidly depleted, we must find new methods of producing electricity. While water and wind power are already in use, the most promising source of electricity may be something Earth has more than enough of—sunlight.

At Sandia National Laboratories, in Albuquerque, New Mexico, engineers are working to harness the sun's energy to generate electricity. One of their projects involves the Stirling engine, a machine that was invented by Robert Stirling in 1816. A large, dish-shaped mirror is used to reflect sunlight onto an absorber, which collects the energy and uses it to increase the internal energy of helium inside the engine. At that point, the engine works much like an automobile engine. The heated helium gas is used to move a piston, but instead of spinning a set of wheels, this piston turns an electric generator.

The Stirling engine operates very efficiently; it holds a world record for converting solar energy into electricity. It is ideal for remote locations, where normal power lines cannot be run, or for power-specific devices, such as water pumps for agricultural purposes.

Sandia is also developing a solar power plant that uses the sun's energy to melt large quantities of salt. The energy transferred as heat from the salt is then used to generate steam, which can turn a turbine to make electricity. Also, the hot salt can be kept in insulated tanks, which enables the salt's high internal energy to be stored. Previous solar power systems simply heated water to the boiling point, but the water boiled only while the

sun was shining. Sandia's salt-heated device stores energy more efficiently than water and maintains the higher temperature long enough to produce electricity even at night.

Greg Kolb, an engineer at Sandia, envisions such a power source replacing the central power stations we have today. "Imagine a tower about the size of the Washington Monument surrounded by a field of mirrors on the ground approximately one square mile in area," Kolb says. "The mirrors are reflecting the sunlight to the top of the tower, where all the light is focused and the energy is absorbed in a large heat exchanger." The engineer estimates that about 10 000 such setups spread throughout the nation could provide as much energy as the United States consumes annually.

BACKGROUND

This feature describes two devices that convert solar energy into electricity. In the first, a Stirling engine converts some of the energy transferred as heat from the solar collectors into work, which is then used to generate electricity. The Stirling engine is another example of a heat engine. This engine has proven to be highly efficient at converting solar energy into electricity. The second energy source, also being developed by Sandia National Laboratories, uses salt as a high-temperature reservoir. The energy transferred from the salt as it cools is used to create steam that turns the turbine of an electrical generator.

EXTENSION

Students may show interest in learning more about solar power. Stage a solar-energy fair in which students invent devices (or reproduce devices) that use solar energy for some productive purpose. This type of competition is an excellent introduction to the field of engineering.

The Language of Physics

The term *efficiency* is used to compare the amount of work output to the amount of energy invested in a process. Efficiency is usually represented either by *eff* or simply by *e*. In this text, the abbreviation *eff* is used. Efficiency may be expressed as a decimal number or as a percentage.

Teaching Tip

Students may need to be reminded that W_{net} denotes the difference between work output and work input in a cycle. $W_{net} = Q_h - Q_c$ because the first law of thermodynamics requires that $W_{out} - W_{in} = Q_h - Q_c$ over a cycle. Working through a simple numerical example may help students understand the meaning of efficiency. For example, if $W_{out} = 1000$ J, $W_{in} = 600$ J, and $Q_h = 2000$ J, they can calculate $W_{net} = 400$ J and $eff = 0.2$.

ANSWERS TO

Conceptual Challenge

1. According to the second law of thermodynamics, some energy must be lost to the environment. For the engine to be efficient, it must be able to give up energy as heat to a substance at a much lower temperature. This is provided by passing water, coolant, or air around the engine's cylinders.

2. Exhaust energy must be removed as heat to a low-temperature substance, such as water. A river provides a continuously replenished source of cool water.

Table 11-4
Typical efficiencies for engines

Engine type	eff (calculated maximum values)
steam engine	0.29
steam turbine	0.40
gasoline engine	0.60
diesel engine	0.56

Engine type	eff (measured values)
steam engine	0.17
steam turbine	0.30
gasoline engine	0.25
diesel engine	0.35

internetconnect

SCI**LINKS**
NSTA
TOPIC: Stirling engines
GO TO: www.scilinks.org
*sci*LINKS CODE: HF2114

THERMODYNAMIC EFFICIENCY

A cyclic process cannot completely convert energy transferred as heat into work, nor can it transfer energy as heat from a low-temperature body to a high-temperature body without work being done in the process. However, a cyclic process can be made to approach these ideal situations. A measure of how well an engine operates is given by the engine's *efficiency* (*eff*). In general, efficiency is a measure of the useful energy taken out of a process relative to the total energy that is put into the process. Efficiencies for different types of engines are listed in **Table 11-4.**

Recall from the first law of thermodynamics that the work done on the environment by the engine is equal to the difference between the energy transferred to and from the system as heat. For a heat engine, the efficiency is the ratio of work done by the engine to the energy added to the system as heat during one cycle.

EQUATION FOR THE EFFICIENCY OF A HEAT ENGINE

$$eff = \frac{W_{net}}{Q_h} = \frac{Q_h - Q_c}{Q_h} = 1 - \frac{Q_c}{Q_h}$$

$$\text{efficiency} = \frac{\text{net work done by engine}}{\text{energy added to engine as heat}}$$

$$= \frac{\text{energy added as heat} - \text{energy removed as heat}}{\text{energy added as heat}}$$

$$= 1 - \frac{\text{energy removed as heat}}{\text{energy added as heat}}$$

Notice that efficiency is a unitless quantity that can be calculated using only the *magnitudes* for the energies added to and taken away from the engine.

This equation confirms that a heat engine has 100 percent efficiency (*eff* = 1) only if there is no energy transferred away from the engine as heat ($Q_c = 0$).

(★ TEKS) **3A, 7A**

Conceptual Challenge

1. Cooling engines Use the second law of thermodynamics to explain why an automobile engine requires a cooling system to operate.

2. Power plants Why are many coal-burning and nuclear power plants located near rivers?

Unfortunately, there can be no such heat engine, so the efficiencies of all engines are less than 1.0. The smaller the fraction of usable energy that an engine can provide, the lower its efficiency is.

The equation also provides some important information for increasing engine efficiency. If the amount of energy added to the system as heat is increased or the amount of energy given up by the system is reduced, the ratio of Q_c/Q_h becomes much smaller and the engine's efficiency comes closer to 1.0.

The efficiency equation gives only a maximum value for an engine's efficiency. Friction and thermal conduction in the engine hinder the engine's performance, and experimentally measured efficiencies are usually lower than the calculated efficiencies (see **Table 11-4**).

SAMPLE PROBLEM 11C

Heat-engine efficiency ⭐TEKS 2C, 7A

PROBLEM

Find the efficiency of a gasoline engine that, during one cycle, receives 204 J of energy from combustion and loses 153 J as heat to the exhaust.

SOLUTION

1. DEFINE   **Given:** $Q_h = 204$ J $Q_c = 153$ J

Unknown: $eff = ?$

Diagram:

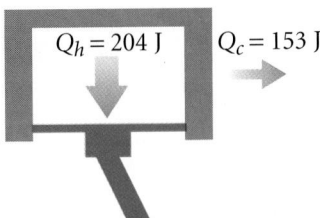

$Q_h = 204$ J $Q_c = 153$ J

2. PLAN   **Choose an equation(s) or situation:** The efficiency of a heat engine is the ratio of the work done by the engine to the energy transferred to it as heat.

$$eff = \frac{W_{net}}{Q_h} = 1 - \frac{Q_c}{Q_h}$$

3. CALCULATE   **Substitute the values into the equation(s) and solve:**

$$eff = 1 - \frac{153 \text{ J}}{204 \text{ J}} = 0.250$$

$$\boxed{eff = 0.250}$$

4. EVALUATE   Only 25 percent of the energy added as heat is used by the engine to do work. As expected, the efficiency is less than 1.0.

Thermodynamics   **423**

Classroom Practice

The following may be used as teamwork exercises or for demonstration at the chalkboard or on an overhead projector.

PROBLEM

Heat-engine efficiency

A steam engine takes in 198×10^3 J and exhausts 149×10^3 J as heat per cycle. What is its efficiency?

Answer
   0.247

A turbine takes in 67 500 J as heat and does 18 100 J of work during each cycle. Calculate its efficiency.

Answer
   0.268

PRACTICE GUIDE 11C		
Solving for:		
eff	📖 **PE**	Sample, 1–3; Ch. Rvw. 28–30
	💿 **PW**	4–6
	PB	8–10
W	📖 **PE**	4
	💿 **PW**	Sample, 1
	PB	Sample, 1–4
Q	📖 **PE**	5–6
	💿 **PW**	Sample, 1–4
	PB	5–7

⭐TEKS

p. 422: 3A, 7A
p. 423: 2C, 7A

ANSWERS TO

Practice 11C
Heat-engine efficiency

1. 0.1504
2. 0.59
3. **a.** 0.247
   b. 4.9×10^4 J
4. 210 J
5. 755 J
6. 8.7×10^2 J

Section Review
ANSWERS

1. no; In order for a heat engine to do work in a thermodynamic cycle, some energy must be transferred as heat to surroundings at a temperature lower than that of the engine.
2. This increases the amount of energy transferred to the engine as heat (Q_h) and thus raises the engine efficiency.
3. **a.** 4.0×10^4 J
   b. 0.53
   c. yes; because *eff* < 1
   d. no; Real engines dissipate energy and thus have efficiencies less than 0.50.

⭐TEKS

p. 424: 2C, 7A, 2C, 7A
p. 425: 2C

PRACTICE 11C

Heat-engine efficiency ⭐TEKS 2C, 7A

1. If a steam engine takes in 2.254×10^4 kJ from the boiler and gives up 1.915×10^4 kJ in exhaust during one cycle, what is the engine's efficiency?

2. A test model for an experimental gasoline engine does 45 J of work in one cycle and gives up 31 J as heat. What is the engine's efficiency?

3. A steam engine absorbs 1.98×10^5 J and expels 1.49×10^5 J in each cycle. Assume that all of the remaining energy is used to do work.
   a. What is the engine's efficiency?
   b. How much work is done in each cycle?

4. If a gasoline engine has an efficiency of 21 percent and loses 780 J to the cooling system and exhaust during each cycle, how much work is done by the engine?

5. A certain diesel engine performs 372 J of work in each cycle with an efficiency of 33.0 percent. How much energy is transferred from the engine to the exhaust and cooling system as heat?

6. If the energy removed from an engine as heat during one cycle is 6.0×10^2 J, how much energy must be added to the engine during one cycle in order for it to operate at 31 percent efficiency?

Section Review  ⭐TEKS 2C, 7A

1. Is it possible to construct a heat engine that doesn't transfer energy to its surroundings? Explain.

2. A steam-driven turbine is one major component of an electric power plant. Why is it advantageous to increase the steam's temperature as much as possible?

3. An engineer claims to have built an engine that takes in 75 000 J and expels 35 000 J.
   a. How much energy can the engine provide by doing work?
   b. What is the efficiency of the engine?
   c. Is this efficiency possible? Explain your answer.
   d. Compare the calculated efficiency for the proposed engine with the measured efficiencies for engines in **Table 11-4.** Is the value for the proposed engine's efficiency probable? Explain your answer.

11-4
Entropy

HEAT AND ENTROPY

A forest such as the one shown in the background of **Figure 11-13** appears to have random spacing between the trees. However, if you were to encounter a forest where all the trees appear to be equally spaced, such as the one shown in the foreground of **Figure 11-13,** you would conclude that someone had planted the trees.

In general, if the laws of nature are allowed to act without interference, it is believed that a disorderly arrangement is much more likely to result than an orderly one. For instance, when you shuffle a deck of cards, the cards are most likely to be all mixed up. It is highly improbable that shuffled cards would end up separated by suit and in numerical sequence. Such a highly ordered arrangement can be formed in only a few ways, but there are more than 8×10^{67} ways to arrange 52 cards.

Entropy is a measure of a system's disorder ⊛(TEKS) 2C

The assertion that a disordered arrangement has a vastly greater probability of forming than a highly ordered arrangement can be demonstrated with a smaller number of components. If two light blue marbles and two purple marbles are placed in two groups of two, four possible combinations can be formed. Two combinations have all of the marbles separated by color. Two combinations have two marbles of each color in each group. There is only one way to form each of the first two combinations, but there are two ways to form

11-4 SECTION OBJECTIVES

- **Relate the disorder of a system to its ability to do work or transfer energy as heat.**

- **Identify systems with high and low entropy.**

- **Distinguish between entropy changes within systems and the entropy change for the universe as a whole.**

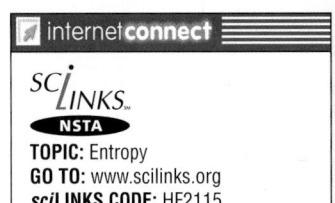

internet connect

SC*i*LINKS.
NSTA

TOPIC: Entropy
GO TO: www.scilinks.org
*sci*LINKS CODE: HF2115

Figure 11-13
An orchard represents a well-ordered system, whereas the randomly spaced trees in the forest make for a fairly disordered system.

Section 11-4

Demonstration 3

Order, disorder, and probability

Purpose Show that in large systems, ordered arrangements are less probable than disordered ones.

Materials pennies, cup

Procedure Tell students that they will be tossing pennies and predicting one of two possible outcomes: *mixed* or *same*. Show that the odds for either possibility are the same when playing with two pennies because there are two ways to get *same* (two heads, HH, and two tails, TT) and two ways to get *mixed* (HT, TH). Ask students to make predictions for playing with three pennies. Have one student toss three pennies while another student records the results on the chalkboard. Repeat several times. Ask students to describe all the possible outcomes of tossing three pennies (*HHH, HHT, HTH, HTT, THH, THT, TTH, TTT*). Which arrangements are more ordered? (*same:* $\frac{2}{8}$) Which are more likely to be formed? (*mixed:* $\frac{6}{8}$) Ask if the odds would change when playing with four pennies (*yes*). What is the probability of the outcome *same* in this case? ($\frac{2}{16}$) Repeat for 5 and 10 pennies ($\frac{2}{32}$, $\frac{2}{1024}$). Point out that as the number of pennies increases, the probability of the outcome *same* rapidly decreases.

Teaching Tip

Have students use their calculators to verify that $52! = 8 \times 10^{67}$.

425

Quick Lab

TEACHER'S NOTES

This experiment is intended to illustrate that a system can be in any one of many states but that some states are formed more frequently than others.

Students benefit more from this activity when they visualize all of the possibilities. They may use various representations to list all cases. One possibility for the case with two dice is to draw a 6 × 6 table and fill each cell with the sum of its row and column numbers, as follows:

	1	2	3	4	5	6
1	2	3	4	5	6	7
2	3	4	5	6	7	8
3	4	5	6	7	8	9
4	5	6	7	8	9	10
5	6	7	8	9	10	11
6	7	8	9	10	11	12

Then have students count all of the cells with sums equal to 2, 3, 4, and so on.

Their results should be:

2–1 way	8–5 ways
3–2 ways	9–4 ways
4–3 ways	10–3 ways
5–4 ways	11–2 ways
6–5 ways	12–1 way
7–6 ways	

Table 11-5 Ordered and disordered arrangements

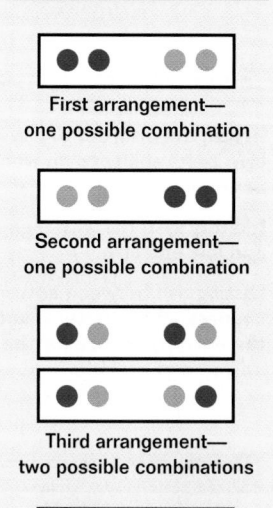

First arrangement—
one possible combination

Second arrangement—
one possible combination

Third arrangement—
two possible combinations

Fourth arrangement—
two possible combinations

entropy

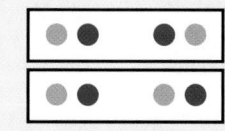
a measure of the disorder of a system

⭐ TEKS 2C

Quick Lab

Entropy and Probability

MATERIALS LIST

✔ 3 dice

✔ a sheet of paper

✔ a pencil

Take two dice from a board game. Record all the possible ways to obtain the numbers 2 through 12 on the sheet of paper. How many possible dice combinations can be rolled? How many combinations of both dice will produce the number 5? the number 8? the number 11? What number(s) from 2 through 12 are most probable? How many ways out of the total

number of ways can this number(s) be rolled? What number(s) from 2 through 12 are least probable? How many ways out of the total number of ways can this number(s) be rolled?

Repeat the experiment with three dice. Write down all of the possible combinations that will produce the numbers 3 through 18. What number is most probable?

each of the last two combinations. These different arrangements are illustrated in **Table 11-5.**

This set of marbles may be thought of as a simplified thermodynamic system. Think of the purple marbles as high-energy particles and the light blue marbles as low-energy particles. There are two states with ordered, unmixed energies. They are described by the first two combinations (two purple and two light blue, or two light blue and two purple). There are, by contrast, four relatively disordered, mixed-energy states. These are indicated by the last two combinations. Thus, there is a 2-to-1 probability that the particles will be in the mixed energy, low-order states.

In thermodynamics, a system left to itself tends to go from a state with a very ordered set of energies (one that has only a small probability of being randomly formed) to one in which there is less order (or that has a high probability of being randomly formed). The measure of a system's disorder is called the **entropy** of the system. The greater the entropy of a system is, the greater the system's disorder.

Systems with maximum disorder are favored

To understand how entropy applies to a natural process, imagine a box that contains four pits into which the marbles can fit. Two pits are on one side of the box and two are on the other. The purple marbles are placed on one side of the box and the light blue marbles are placed on the other. If the box is closed and then shaken, the marbles are twice as likely to be mixed than unmixed. If the box is shaken again, the marbles are still likely to be mixed.

The greater probability of a disordered arrangement indicates that an ordered system (whether the system is composed of separated marbles or particles of separate energies) is likely to become disordered. Put another way, the entropy of a system increases. This greater probability also reduces the chance that a disordered system will become ordered at random. Thus, once a system has reached a state of greatest disorder, it will tend to remain in that state and have *maximum entropy.*

Increasing disorder reduces the energy available for work

Entropy's role with respect to the second law of thermodynamics still requires explanation. You now know that a system will move toward a state of disorder on its own. But how does this tendency toward disorder affect the ability of a system to do work?

Consider a ball thrown toward a wall, as shown in **Figure 11-14.** The kinetic energy of the particles in the ball depends on their individual random motions and the kinetic energy of the ball as a whole. The wide range of kinetic energies that the particles have because of their random motions make up the ball's internal energy. These energies are in a disordered state. The translational kinetic energy of the particles due to the motion of the entire ball in one direction and with one velocity is, by contrast, well ordered.

When the ball hits the wall, the wall itself does not appear to move and the ball does not fall apart. But the particles within each object undergo collisions with some particles in the other object. These collisions increase the random individual motions of the particles. Thus, some particles within the wall increase their kinetic energy. Some particles in the ball will also increase their kinetic energy, but the ball as a whole still moves in one direction.

According to the first law of thermodynamics, this increase in internal energy must be balanced by a decrease of energy elsewhere. The kinetic energy of the entire ball is thus reduced. Before the collision, the ball could do a certain amount of work. With enough energy, it could knock over a bucket, for example. With the transformation of part of this ordered energy into disordered internal energy, this capability of doing work is reduced. The ball rebounds with less total kinetic energy than it originally had. ⭐TEKS 5B

Greater disorder means there is less energy to do work

In a similar way, heat engines are limited in that only some of the energy added as heat can be used to do work. Not all of the gas particles move in an orderly fashion toward the piston and give up all of their energy in collision with the piston, as shown on the left in **Figure 11-15.** Instead, they move in all available directions, as shown on the right in **Figure 11-15,** and transfer energy through collisions with the walls of the engine cylinder as well as with each other. Although energy is conserved, not all of it is available to do useful work. The motion of the particles of a system (in this case, the gas in the cylinder) is not well ordered and therefore is less useful for doing work.

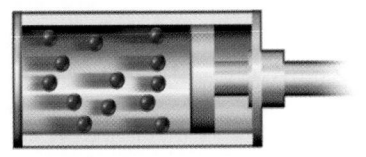

Well ordered; high efficiency and highly improbable distribution of velocities

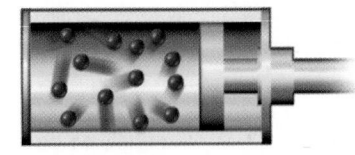

Highly disordered; average efficiency and highly probable distribution of velocities

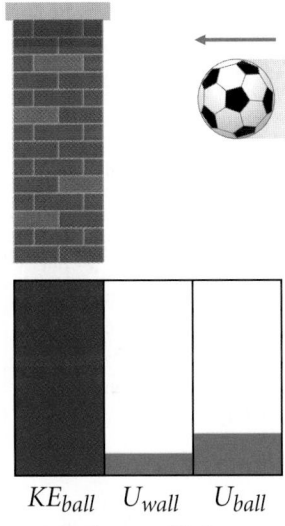

KE_{ball} U_{wall} U_{ball}
Before collision

KE_{ball} U_{wall} U_{ball}
After collision

Figure 11-14
The kinetic energies of the particles in the ball and the wall increase at the expense of the kinetic energy of the ball as a whole.

Figure 11-15
If all gas particles moved toward the piston, all of the internal energy could be used to do work. This extremely well ordered situation is highly improbable.

Visual Strategy

Figure 11-14
Point out that the U bars indicate that both the wall and the ball have internal energy before the collision because all objects have internal energy at all times. The initial level of the U bars is relative to some arbitrary zero level.

Q Write an equation relating the KE and U of each of the objects involved in this process.

A
$(KE_{ball} + U_{ball} + U_{wall})_{BEFORE}$
$= (KE_{ball} + U_{ball} + U_{wall})_{AFTER}$
or $\Delta U_{ball} + \Delta U_{wall}$
$= -\Delta KE_{ball}$

Q Which of the energy changes in the previous expression are an increase, and which are a decrease?

A increase ($\Delta U_{ball} > 0$)
increase ($\Delta U_{wall} > 0$)
decrease ($\Delta KE_{ball} < 0$)

The Language of Physics

The term *entropy* refers to the measure of the disorder in a system. This measure is related to the ability of a system to do useful work. It depends on the temperature and other system characteristics. The term *entropy* is also used in reference to disorder in other areas. In the field of information theory, well-organized information systems have a lower entropy than random ones. For example, when the key terms in this text are indexed, the information in the book is more highly organized. The entropy of the book (as an information system, not as a physical system) is lower. Similarly, a library without a catalog has a higher entropy of information.

ANSWERS TO

Conceptual Challenge

1. A perfectly ordered alternating pattern of heads and tails is very unlikely. The overall distribution will be half heads and half tails, but less ordered.

2. no change in the total energy (similar to no change in the total number of colored marbles); increase in entropy (lukewarm water has more disorder and higher entropy than the separated hot and cold water, just like mixed marbles have higher entropy than separated marbles)

Conceptual Challenge

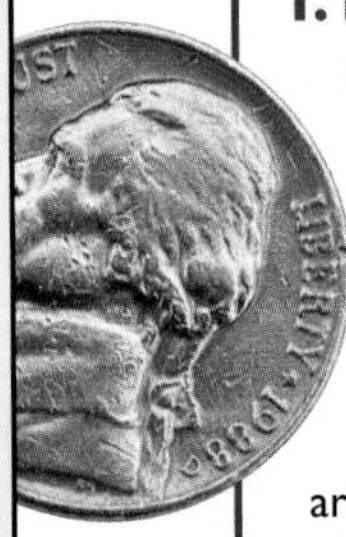

1. Flipping a coin

Explain why the repeated flipping of a coin does not produce a perfectly ordered, alternating pattern of heads and tails, even though the overall odds are 50:50.

2. Boiling water

A pan of boiling water is mixed with a pan of ice-cold water. The resulting lukewarm contents are then separated back into the two pans. Is there a change in the total energy of the system? Is there a change in entropy? Explain. (Hint: Compare this situation to the mixing of different-colored marbles.)

The second law of thermodynamics can be expressed in terms of entropy change

When an ice cube is surrounded by warm air, it melts at a constant temperature. The water molecules go from an ordered crystalline arrangement with a narrow range of molecular energies to a fairly disordered liquid with a wider range of energies. The entropy of the system increases because energy is transferred as heat to the system. Heat and entropy are therefore related.

When energy is added to or removed from a substance as heat, the temperature of the substance will usually change as well. The exception is when the substance changes phase. In this case, the temperature remains constant while the entropy of the system changes. For this reason, liquids solidify and become more ordered when energy is removed from them, and liquids evaporate and become less ordered when energy is transferred to them.

Because of the connection between a system's entropy, its ability to do work, and the direction of energy transfer, the second law of thermodynamics can also be expressed in terms of entropy change. This law applies to the entire universe, not only to a system that interacts with its environment. So the second law can be stated as follows: *The entropy of the universe increases in all natural processes.*

Entropy can increase or decrease within a system

Entropy can decrease for parts of systems, such as the water in the freezer shown in **Figure 11-16,** provided this decrease is offset by a greater increase in entropy elsewhere in the universe.

When work is done on a system, the system's entropy can be decreased. Just as it takes work to put a messy room in order or to put the hundreds of pieces

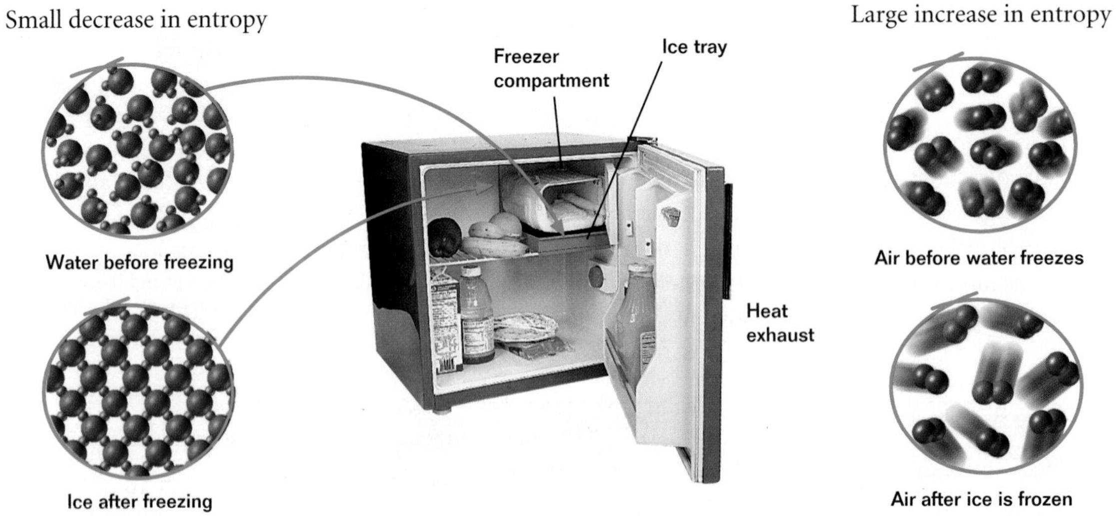

Small decrease in entropy

Large increase in entropy

Water before freezing

Freezer compartment

Ice tray

Air before water freezes

Heat exhaust

Ice after freezing

Air after ice is frozen

Figure 11-16

Because of the refrigerator's less-than-perfect efficiency, the entropy of the outside air molecules increases more than the entropy of the freezing water decreases. ★ TEKS 7A

of a jigsaw puzzle together to make one cohesive picture, work is done by the freezer to reduce the entropy of the water in the ice tray. In each of these examples, the environment's entropy increases to offset the decrease in the system's entropy. The surrounding air's entropy increases because of energy transferred to it as heat, either from the person cleaning the room or from the freezer freezing the water.

As a system's entropy decreases by some amount, the entropy of the system's environment increases by a greater amount. The water's entropy decreases as it becomes ice, but the entropy of the air in the room is increased by a greater amount as energy is transferred by heat from the refrigerator. The result is that the total entropy of the refrigerator and the room together has increased.

Ultimately, the entropy of the universe should reach a maximum value. At that time, the universe will be in a state of thermal equilibrium and its temperature will be uniform throughout. All of the energy that is present now will still be in the universe. However, because there will be no difference in temperature anywhere in the universe, the transfer of energy as heat—and thus work—will be impossible. All physical, chemical, and biological processes will cease because a state of perfect disorder implies an absence of energy that can be used to do work. This gloomy state of affairs is sometimes referred to as an ultimate "heat death" of the universe. Estimates based on present models of stellar formation and evolution place this time of universal thermal equilibrium about 100 trillion years in the future. (★)TEKS 3A

Did you know?

Entropy decreases in many systems on Earth. These appear to be spontaneous because we think of the Earth itself as a closed system. So much energy comes from the sun that the disorder in chemical and biological systems is reduced, while the total entropy of the Earth, sun, and intervening space increases.

Section Review   (★)TEKS 3A, 7B

1. Use the discussion on pp. 425–426 as a model to show that three purple marbles and three light blue marbles in two groups of three marbles each can be arranged in four combinations: two with only one possible arrangement each and two with nine possible arrangements each.

2. Which of the following systems have high entropy? Which systems have low entropy?
   a. papers scattered randomly across a desk
   b. a freshly opened pack of cards
   c. a cultivated field of cabbages
   d. a room after a party

3. Some molecules have been observed to form spontaneously, even though they are more ordered than their components. Interpret the second law of thermodynamics at the particle level to explain why this can occur.

4. Discuss three common examples of natural processes that involve decreases in entropy. Identify the corresponding entropy increases in the environments of these processes.

Section Review
ANSWERS

1. all purple marbles on left (PPP/BBB)—1 combination, all blue marbles on left (BBB/PPP)—1 combination, 2 purple marbles on left (PPB/BBP, PPB/BPB, PPB/PBB, PBP/BBP, PBP/BPB, PBP/PBB, BPP/BBP, BPP/BPB, BPP/PBB)—9 combinations, 1 purple marble on left (opposite of previous 9 cases)—9 combinations (there are 2 ordered states and 18 mixed states)

2. a. high entropy
   b. low entropy
   c. low entropy
   d. high entropy

3. Individual molecules can decrease entropy spontaneously, just as energy can be transferred from low-energy to high-energy particles. The conditions for this decrease require an exchange of energy between the molecule and its environment such that the entropy of the environment increases by more than the molecule's entropy decreases.

4. Answers may vary. Examples include water freezing in winter (energy transfer from water to air increases air's entropy), formation of limestone (water's evaporation increases surrounding air's entropy), and growing plants (plants draw energy from sunlight, take water and minerals from soil, and remove CO_2 from and add oxygen to the air, all of which increase the environment's entropy).

CHAPTER 11
Summary

Teaching Tip

Have students create concept maps of the following: compression, expansion, heating, cooling, work, heat, and internal energy. Let students add terms such as *volume, temperature,* or *entropy* to link related concepts. There are many ways to represent these concepts and their relationships. This exercise will help students synthesize the information presented in the chapter.

⭐ TEKS

Review & Assess
pp. 431–435:
1B: Alt. Assess. 2
2C: 10–11, 19–21,
     28–30, 39–40
2D: Alt. Assess. 1–6
3A: Alt. Assess. 3, 6
3E: Alt. Assess. 1
5D: 15
7A: 14, 18–19, 25, 40;
     Alt. Assess. 3, 5–6
7B: 31–32, 35; Alt.
     Assess. 4

KEY TERMS

adiabatic process (p. 407)

cyclic process (p. 415)

entropy (p. 426)

environment (p. 403)

isothermal process (p. 407)

isovolumetric process (p. 406)

system (p. 403)

Diagram symbols

Energy transferred as heat

Energy transferred as work

Thermodynamic cycle

KEY IDEAS

Section 11-1 Relationships between heat and work

- An object or grouping of objects that are in thermal equilibrium and contain an unchanging amount of matter make up a system, while their surroundings make up the system's environment.
- Energy can be transferred to or from a system as heat and/or work, changing the system's internal energy in the process.
- For gases at constant pressure, work is defined as the product of gas pressure and the change in the volume of the gas.

$$W = P\Delta V$$

Section 11-2 Thermodynamic processes

- Energy is conserved for any system and its environment and is described by the first law of thermodynamics.

$$\Delta U = Q - W$$

- A cyclic process returns a system to conditions identical to those it had before the process began, so its internal energy is unchanged.

Section 11-3 Efficiency of heat engines

- The second law of thermodynamics states that no machine can transfer all of its absorbed energy as work.
- The efficiency of a heat engine depends on the amount of energy transferred as heat to and from the engine.

$$eff = 1 - \frac{Q_c}{Q_h}$$

Section 11-4 Entropy

- Entropy is a measure of the disorder of a system. The more disordered a system is, the less energy that is available to do work.
- The entropy of a system can increase or decrease, but the total entropy of the universe is always increasing.

Variable symbols

Quantities		Units	
ΔU	change in internal energy	J	joules
Q	heat	J	joules
W	work	J	joules
eff	efficiency		(unitless)

CHAPTER 11
Review and Assess

HEAT, WORK, AND INTERNAL ENERGY

Review questions

1. Define a thermodynamic system and its environment.

2. In what two ways can the internal energy of a system be increased?

3. Which of the following expressions have units that are equivalent to the units of work?

   a. mg **d.** Fd
   b. $\frac{1}{2}mv^2$ **e.** $P\Delta V$
   c. mgh **f.** $V\Delta T$

4. For each of the following, which thermodynamic quantities have values equal to zero?

   a. an isothermal process
   b. an adiabatic process
   c. an isovolumetric process

Conceptual questions

5. Can energy be added to or removed from a substance without causing the substance's temperature to rise or its phase to change? Explain your answer.

6. When an ideal gas expands adiabatically, it does work on its surroundings. Describe the various transfers of energy that take place.

7. In each of the following cases, trace the chain of energy transfers (as heat or as work) as well as changes in internal energy.

   a. You rub your hands together to warm them on a cold day.
   b. A hole is drilled into a block of metal. When a small amount of water is placed in the drilled hole, steam rises from the hole.

8. Paint from an aerosol can is sprayed continuously for 30 s. The can was initially at room temperature, but now it feels cold to the touch. What type of thermodynamic process occurs for a small sample of gas as it leaves the high-pressure interior of the can and moves to the outside atmosphere?

9. The can of spray paint in item 8 is set aside for an hour. During this time the contents of the can return to room temperature. What type of thermodynamic process takes place in the can during the time the can is not in use?

Practice problems

10. How much work is done when a tire's volume increases from 35.25×10^{-3} m^3 to 39.47×10^{-3} m^3 at a pressure of 2.55×10^5 Pa in excess of atmospheric pressure? Is work done on or by the gas? (See Sample Problem 11A.)

11. Helium in a toy balloon does work on its surroundings as it expands with a constant pressure of 2.52×10^5 Pa in excess of atmospheric pressure. The balloon's initial volume is 1.1×10^{-4} m^3, and its final volume is 1.50×10^{-3} m^3. From this information, determine the amount of work done by the gas in the balloon. (See Sample Problem 11A.)

ENERGY CONSERVATION AND CYCLIC PROCESSES

Review questions

12. Write the equation for the first law of thermodynamics, and explain why it is an expression of energy conservation.

13. Rewrite the equation for the first law of thermodynamics for each of the following special thermodynamic processes:

    a. an isothermal process
    b. an adiabatic process
    c. an isovolumetric process

ANSWERS TO
Chapter 11
Review and Assess

1. see definitions on page 403
2. energy transfers to the system as heat or as work
3. b, c, d, e
4. **a.** ΔU
   b. Q
   c. $W, \Delta V$
5. yes; if the energy is quickly removed as work or heat, as in isothermal processes in which thermal equilibrium is maintained macroscopically
6. Work done by gas causes a decrease in U and T of gas and an increase in U of surroundings.
7. **a.** work done to overcome friction between hands, which raises hands' internal energy, energy transferred as heat from warmed hands to cold air
   b. work done by drill overcomes friction, which increases U of drill and block, with water, U of drill and block decreases as energy is transferred as heat to water, U of water increases until large enough for phase change
8. adiabatic
9. isovolumetric
10. 1.08×10^3 J; done by the gas
11. 3.50×10^2 J
12. $\Delta U = Q - W$; change in a system's internal energy equals energy transferred as heat or work to or from a system
13. **a.** $\Delta U = 0$, $Q = W$
    b. $\Delta U = -W$, $Q = 0$
    c. $\Delta U = Q$, $W = 0$

14. Answers should agree with the discussion on pp. 414–415.

15. The additional energy is provided as work done to cool the refrigerator's interior.

16. a. $\Delta U < 0$, $Q < 0$, $W = 0$
b. $\Delta U > 0$, $Q > 0$, $W = 0$

17. a. none (Q and W positive, $\Delta U = 0$ if process is treated as isothermal)
b. $\Delta U < 0$, $Q < 0$ for refrigerator interior ($W > 0$)
c. $\Delta U < 0$ ($Q = 0$, $W > 0$)

18.

	ΔU	Q	W
a.	+	+	0
b.	–	0	+
c.	–	–	0
d.	+	0	–

19. 2.14×10^9 J/s

20. 647 kJ

21. a. 1.7×10^6 J, to the rod
b. 3.3×10^2 J; by the rod
c. 1.7×10^6 J; it increases

22. Energy is always conserved (first law), but not all of the energy transferred into a system can be used to do work (second law). Energy that must be wasted, according to the second law, is still accounted for by the first law.

23. All energy transferred into the engine must be used to do work ($Q_c = W$). No energy is wasted ($Q_c = 0$). This condition cannot be met by real cyclic heat engines.

24. Efficiency increases as Q_h increases and Q_c decreases.

---

14. List the four thermodynamic processes that take place in one cycle of operation for a refrigerator. What is the difference between the system's internal energy before and after one cycle?

15. How is energy conserved if more energy is transferred as heat from a refrigerator to the outside air than is removed from the inside air of the refrigerator?

Conceptual questions

16. A bomb calorimeter is placed in a water bath and remains at constant volume while a mixture of fuel and oxygen is burned inside it. The temperature of the water is observed to rise during the combustion reaction. The calorimeter and the water remain at constant volume.

  a. If the reaction products are the system, which thermodynamic quantities—ΔU, Q, or W—are positive and which are negative?
  b. If the water bath is the system, which thermodynamic quantities—ΔU, Q, or W—are positive and which are negative?

17. Which of the thermodynamic values (ΔU, Q, or W) would be negative for the following systems?

  a. a steel rail (system) undergoing slow thermal expansion on a hot day displaces the spikes and ties that hold the rail in place
  b. a closed refrigerator (system) in a kitchen
  c. the helium in a thermally insulated weather balloon (system) expands during inflation

18. A system consists of steam within the confines of a steam engine (boiler, steam pipes, cylinder, and piston). Construct a table similar to **Table 11-3**, and determine the signs of the quantities ΔU, Q, and W in the following steps of the engine's cycle:

  a. Water in a rigid boiler is heated until it vaporizes.
  b. Steam is conveyed through steam pipes from the boiler to the cylinder of the engine, where the steam rapidly pushes the piston toward the outside of the cylinder.
  c. The steam condenses to liquid water at constant volume within the cylinder.
  d. The piston moves into the cylinder, pushing the liquid water out and back into the boiler.

Practice problems

19. A power plant has a power output of 1055 MW and operates with an efficiency of 33.0 percent. Excess energy is carried away as heat from the plant to a nearby river that has a flow rate of 1.1×10^6 kg/s. How much energy is transferred as heat to the river each second?
(See Sample Problem 11B.)

20. Heat is added to an open pan of water at 100.0°C, vaporizing the water. The expanding steam that results does 43.0 kJ of work, and the internal energy of the system increases by 604 kJ. How much energy is transferred to the system as heat?
(See Sample Problem 11B.)

21. A 150 kg steel rod in a building under construction supports a load of 6050 kg. During the day the rod's temperature increases from 22°C to 47°C, causing the rod to thermally expand and raise the load 5.5 mm.

  a. Find the energy transferred as heat to or from the rod. (Hint: Assume the specific heat capacity of steel is the same as for iron.)
  b. Find the work done in this process. Is work done on or by the rod?
  c. How great is the change in the rod's internal energy? Does the rod's internal energy increase or decrease?
(See Sample Problem 11B.)

EFFICIENCY OF HEAT ENGINES

Review questions

22. The first law of thermodynamics states that you cannot obtain more energy from a process than you originally put in. The second law states that you cannot obtain as much usable energy from a system as you put into it. Explain why these two statements do not contradict each other.

23. What conditions are necessary for a heat engine to have an efficiency of 1.0?

24. How do the temperature of combustion and the temperatures of coolant and exhaust affect the efficiency of automobile engines?

Conceptual questions

25. If a cup of very hot water is used as an energy source and a cup of cold water is used as an energy "sink," the cups can, in principle, be used to do work, as shown in **Figure 11-17.** If the contents are mixed together and the resulting lukewarm contents are separated into two cups, no work can be done. Use the second law of thermodynamics to explain this. Has the first law of thermodynamics been violated by mixing and separating the contents of the two cups?

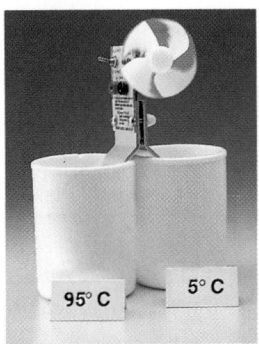 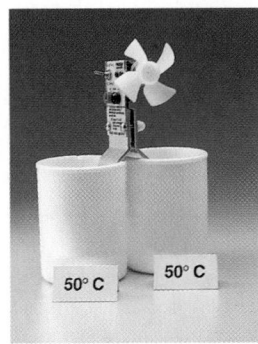

Figure 11-17

26. What happens to the temperature of a room if an air conditioner is left running on a table in the middle of the room? Take into account the energy transferred as heat from the motor that drives the air conditioner's compressor.

27. Suppose the waste heat at a power plant is exhausted to a pond of water. Could the efficiency of the plant be increased by refrigerating the water in the pond?

Practice problems

28. In one cycle, an engine burning a mixture of air and methanol (methyl alcohol) absorbs 525 J and expels 415 J. What is the engine's efficiency?
(See Sample Problem 11C.)

29. The energy provided each hour by heat to the turbine in an electric power plant is 9.5×10^{12} J. If 6.5×10^{12} J of energy is exhausted each hour from the engine as heat, what is the efficiency of this heat engine?
(See Sample Problem 11C.)

30. A heat engine absorbs 850 J of energy per cycle from a high-temperature source. The engine does 3.5×10^2 J of work during each cycle, expelling 5.0×10^2 J as heat. What is the engine's efficiency?
(See Sample Problem 11C.)

ENTROPY

Review questions

31. Explain how increasing entropy reduces the ability of energy to do work when you hammer a nail into a piece of wood.

32. In which of the following systems is entropy increasing?
    a. An egg is broken and scrambled.
    b. A cluttered room is cleaned and organized.
    c. A thin stick is placed in a glass of sugar-saturated water, and sugar crystals form on the stick.

33. A thermodynamic process occurs in which the entropy of a system decreases. From the second law of thermodynamics, what can you conclude about the entropy change of the environment?

34. Why is it not possible for all of the energy transferred as heat from a high-temperature source to be expelled from an engine by work?

Conceptual questions

35. A salt solution is placed in a bowl and set in sunlight. The salt crystals that remain after the water has evaporated are more highly ordered than the randomly dispersed sodium and chloride ions in the solution. Has the requirement that total entropy increase been violated? Explain your answer.

36. Use a discussion of internal energy and entropy to explain why the statement, "Energy is not conserved in an inelastic collision," is not true.

37. An ideal heat engine uses energy transferred as heat from a high-temperature body to do work. The engine then transfers energy as heat to a low-temperature body in order to complete the cycle. Find the sign of the entropy changes for the following:
    a. the high-temperature body
    b. the low-temperature body
    c. the engine
    d. the universe

25. Energy must be transferred as heat for work to be done. This cannot occur if the water in both cups has same temperature; No, the total energy in the water is unchanged, although usable energy has decreased.

26. The temperature rises. The energy removed from air is returned to air on the exhaust side and added to the energy from the electric motor.

27. The plant's efficiency would increase, but the advantage gained would be canceled by the use of energy needed to refrigerate the water.

28. 0.210

29. 0.32

30. 0.41

31. Some of the energy increases the internal energy of the nail and wood instead of doing work on the nail.

32. (a)

33. The environment's entropy must have increased.

34. If all the energy from a high-temperature source were used by the engine to do work, the engine would not be able to expel energy to a lower-temperature body. As a result, the engine could not return to the beginning of the cycle without doing work to replace the energy taken from the high-temperature source.

35. no; Entropy of water and air during the water's evaporation increases by more than the entropy of the sodium and chloride ions decreases.

At the top of the middle column: heat. What is the engine's efficiency? (See Sample Problem 11C.)

36. In inelastic collisions, some kinetic energy is converted to the internal energy of the colliding objects, so the system's entropy increases. Kinetic energy is not conserved, but total energy is conserved.

37. a. negative
b. positive
c. zero (for a full cycle with no wear on the engine)
d. positive

38. a. According to the second law, disorganized energy cannot be fully converted into organized energy that is used to do work.
b. Disorganized energy is removed from water to form ice, but a greater amount of organized energy must become disorganized in order to operate the freezer.
c. More organized energy from the body becomes disorganized than is used to do work on the objects in the room (increase in overall entropy exceeds decrease in the room's entropy).

ANSWERS TO

Technology & Learning

a. 0 K (absolute zero)
b. 0.322
c. 0.404
d. 0.169
e. increase T_h

38. One way to look at the transfer of energy between substances is to think of energy transferred by heat as a "disorganized" form of energy and energy transferred by work as an "organized" form. Use this interpretation to show the following:

a. A heat engine can never have an efficiency of 1.0.
b. The increased order obtained by freezing water is less than the total disorder that results from the freezer used to form the ice.
c. The increased order obtained from cleaning a room is less than the total disorder that results from the heat engine (you) used to do the work.

MIXED REVIEW

39. A gas expands when 606 J of energy is added to it by heat. The expanding gas does 418 J of work on its surroundings.

Technology & Learning

Graphing calculators

Refer to Appendix B for instructions on downloading programs for your calculator. The program "Chap11" allows you to predict the highest theoretical efficiency for a Carnot engine operating between two specified heat reservoirs.

A French engineer named Sadi Carnot (1796–1832) studied the efficiencies of heat engines. He described an ideal engine—now called the Carnot engine—that consists of an ideal gas inside a thermally nonconductive cylinder with a piston and a replaceable base. In the Carnot engine, the piston moves upward as the cylinder's conductive base is brought in contact with a heat reservoir, T_h. The piston then continues to rise when the base is replaced by a nonconductive base. The energy is then transferred to a cooler reservoir at a temperature, T_c, followed by further compression when the base is again replaced. Carnot discovered that the efficiency of such an engine can be determined by the following equation:

$$highest\ theoretical\ efficiency = 1 - \frac{T_c}{T_h}$$

a. What temperature must the cooler reservoir be for a Carnot engine to be 100 percent efficient?

The program "Chap11" stored on your graphing calculator makes use of Carnot's equation. Once the "Chap11" program is executed, your calculator will ask for the temperature of the hotter (T_h) reservoir and the cooler (T_c) reservoir. The graphing calculator will use this temperature range to find the highest theoretical efficiency.

Execute "Chap11" on the PRGM menu, and press ENTER to begin the program. Enter the value for the hotter temperature; then enter the value for the cooler temperature, and press ENTER after each value.

The calculator will provide the highest theoretical efficiency. Because of friction and other problems, the actual efficiency of a heat engine in these situations will be lower than the calculated efficiency.

Determine the highest theoretical efficiency for a Carnot engine operating within the following sets of temperatures:

b. 435 K and 295 K
c. 495 K and 295 K
d. 295 K and 245 K
e. How could you increase the efficiency if the cold reservoir cannot be cooled more than 245 K?

Press ENTER to input a new value or CLEAR to end the program.

a. What is the overall change in the internal energy of the gas?

b. If the work done by the gas equals 1212 J, how much energy must have been added as heat in order for the internal energy at the end of the process to equal the initial internal energy?

40. The lid of a pressure cooker forms a nearly airtight seal. Steam builds up pressure and increases temperature within the pressure cooker so that food cooks faster than it does in an ordinary pot. The system is defined as the pressure cooker and the water and steam within it.

a. If 2.0 g of water is sealed in a pressure cooker and then vaporized by heating, what happens to the water's internal energy?

b. Is energy transferred as heat to or from the system?

c. Is energy transferred as work to or from the system?

d. If 5175 J must be added as heat to completely vaporize the water, what is the change in the water's internal energy?

39. a. $\Delta U_i = 188$ J
  b. $Q_2 = 1.400 \times 10^3$ J
40. a. it increases $(\Delta U > 0)$
  b. to the system $(Q > 0)$
  c. no $(\Delta U = 0$; therefore, $W = 0)$
  d. 5175 J

Alternative Assessment

Performance assessment

1. Work in groups to create a classroom presentation on the life, times, and work of James Watt, inventor of the first commercially successful steam engine in the early nineteenth century. Include material about how this machine affected transportation and industry in the United States.

2. Talk to someone who works on air conditioners or refrigerators to find out what fluids are used in these systems. What properties should refrigerant fluids have? Research the use of freon and freon substitutes. Why is using freon forbidden by international treaty? What fluids are now used in refrigerators and car air conditioners? For what temperature ranges are these fluids appropriate? What are the advantages and disadvantages of each fluid? Summarize your research in the form of a presentation or report.

3. The law of entropy can also be called the law of increasing disorder, but this law seems to contradict the existence of living organisms that are able to organize chemicals into organic molecules. Prepare for a class debate on the validity of the following arguments:

a. Living things are not subject to the laws of thermodynamics.

b. The increase in the universe's entropy due to life processes is greater than the decrease in entropy within a living organism.

Portfolio projects

4. During one day, record examples of any processes involving energy transfers that you observe. Identify whether entropy increases in each case. Display your examples, including an explanation for each, on a poster for a collective class exhibit on entropy.

5. Research how an internal-combustion engine operates. Describe the four steps of a combustion cycle. What materials go in and out of the engine during each step? How many cylinders are involved in one cycle? What energy processes take place during each stroke? In which steps is work done? Summarize your findings with diagrams or in a report. Contact an expert auto mechanic, and ask the mechanic to review your report for accuracy.

6. Imagine that an inventor is asking you to invest your savings in the development of a new turbine that will produce cheap electricity. The turbine will take in 1000 J of energy from fuel to supply 650 J of work, which can then be used to power a generator. The energy removed as heat to a cooling system will raise the temperature of 0.10 kg of water by 1.2°C. Are these figures consistent with the first and second laws of thermodynamics? Would you consider investing in this project? Write a business letter to the inventor explaining how your analysis affected your decision.

Alternative Assessment
ANSWERS

Performance assessment

1. Student answers will vary. Watt's engine marked the start of the mechanization of the industrial revolution.

2. Students' answers will vary. Freon is banned because of harm it may do to the ozone layer. Factors in choosing coolants include latent heat, boiling point, cost, and safety.

3. Students should realize that **(b)** best describes the second law.

 **Portfolio projects**

4. Student displays will vary. Check that system and surroundings for each example are identified.

5. Air and fuel go in; exhaust goes out. The number of cylinders depends on the type of engine. Work is done as the gas expands and as the piston expels the exhaust.

6. The proposal is invalid because the energy provided by work is greater than the differences in the energy transferred as heat to and from the engine.

CHAPTER 12 PLANNING GUIDE

Compression Guide: To shorten from 13 to 10 45-min periods (from 6½ to 5 90-min blocks), eliminate items in magenta type.

PACING CHART	CLASSROOM RESOURCES			
	⭐TEKS	Teacher Demonstrations	Holt Physics Transparencies	Labs (See page T52 for equipment listing for in-text labs.)
12-1 Simple harmonic motion 3 or 2 45-minute periods 1½ or 1 90-minute block	1A, 2B, 2C, 2F, 3A, 3B, 3C, 3E, 4A, 4C, 5B	**TE** *A vibrating spring*, p. 438 **TE** *An oscillating pendulum*, p. 438 **TE** *Hooke's law*, p. 439	**T** 45–47 **TM** 42	**L** *Discovery Lab: Pendulum and Spring Waves* **PE** *Quick Lab: Energy of a Pendulum*, p. 444
12-2 Measuring simple harmonic motion 4 or 2 45-minute periods 2 or 1 90-minute blocks	2C, 3A, 3B, 3E	**TE** *Period and frequency*, p. 446 **TE** *Relationship between string length and the period of a pendulum*, p. 447	**TM** 43	**PE** *The Pendulum and Simple Harmonic Motion*, p. 474 **L** *Invention Lab: Tensile Strength and Hooke's Law* **TL** *Pendulum Periods*
12-3 Properties of waves 2 45-minute periods 1 90-minute block	2C, 2E, 3A, 3B, 4A, 8A, 8B	**TE** *Wave motion*, p. 452 **TE** *Transverse waves*, p. 453 **TE** *Longitudinal waves*, p. 455 **TE** *Amplitude, wavelength, and wave speed*, p. 456	**T** 48–49	
12-4 Wave interactions 2 45-minute periods 1 90-minute block	2C, 3A, 3B, 3C, 3E, 8A, 8B	**TE** *Wave superposition*, p. 459 **TE** *Waves passing each other*, p. 460 **TE** *Wave reflection*, p. 462	**T** 50–53	
Review and Assessment 2 45-minute periods 1 90-minute block				

Resource Key

PHYSICS

PE Pupil's Edition
TE Teacher's Edition

L Laboratory Experiments
TL Technology Lab Experiments
T Transparencies

🖐 **One-Stop Planner CD-ROM contents**

TM Transparency Masters
SR Section Review Worksheets
AA Alternative Assessment

PW Problem-Solving Workbook
PB Problem Bank
CTW Critical Thinking Worksheet

LABORATORY PLANNING: The Pendulum and Simple Harmonic Motion, p. 474

Materials (for each lab group)
- balance: portable, electronic balance or triple-beam balance with weight
- roll of pendulum suspension cord
- meterstick
- drilled-ball set (provides pendulum bobs for two workstations)

- pendulum clamp
- protractor
- 12- or 24-hour alarm stopwatch
- 1-position support base and rod, 1.3 cm × 91 cm

ASSIGNMENT RESOURCES

Content Mastery	Critical Thinking	Problem-Solving Practice
PE 1–4, p. 445 **SR** 12-1, *Concept Review* **PE** 1–5, p. 469	**PE** 1–2, p. 439 **PE** 6–7, p. 469	**12A** Hooke's law: 29 items in **PE, PW,** and **PB,** see **TE** pp. 440–441
PE 1–4, p. 451 **SR** 12-2, *Math Skills* **PE** 10–12, p. 469	**PE** 1–2, p. 449 **PE** 13–18, p. 469	**12B** Simple harmonic motion of a simple pendulum: 28 items in **PE, PW,** and **PB,** see **TE** pp. 448–449 **12C** Simple harmonic motion of a mass-spring system: 23 items in **PE, PW,** and **PB,** see **TE** pp. 450–451
PE 1–5, p. 458 **SR** 12-3, *Concept Review* **PE** 23–29, p. 470	**PE** 30–35, p. 470	**12D** Wave speed: 26 items in **PE, PW,** and **PB,** see **TE** p. 457
PE 1–5, p. 465 **SR** 12-4, *Graph Skills* **PE** 37–41, p. 471	**PE** 42–45, p. 471	

TECHNOLOGY RESOURCES

 PHYSICS Module 11 Hooke's Law
Module 12 Wave Frequency and Wavelength

 PE Technology and Learning, p. 472
(Alternative procedures for calculators without Flash-ROM technology are provided on the One-Stop Planner CD-ROM.)

 The Mechanical Universe/High School Adaptation Quad I, Harmonic Motion
Quad II, Introduction to Waves

internet connect

 SCI LINKS **On-line Student Resources:**
GO TO: www.scilinks.org
The following *sci*LINKS Internet resources can be found in the student text for this chapter.

TOPICS:
• Hooke's law, p. 439 (HF2121)
• Pendulums, p. 444 (HF2122)
• Wave motion, p. 453 (HF2123)
• Electron microscope, p. 467 (HF2124)

 On-line Teacher Resources:
GO TO: go.hrw.com
KEYWORD: HF2 HOME
Visit the HRW Web site for a variety of resources related to this chapter.

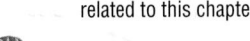 Smithsonian Institution
Internet Connections
Visit **www.si.edu/hrw** for additional on-line resources.

 CNNfyi.com
Visit **www.cnnfyi.com** for late-breaking news and current events stories selected just for you.

ASSESSMENT RESOURCES

Cumulative Review	Alternative Assessment	Traditional Assessment
SR Mixed Review, Ch. 12	**PE** 1–6, p. 473 **AA** Items for Ch. 12	Chapter 12 Test Test Generator items for Ch. 12

Scoring Rubrics for Alternative Assessment items can be found on the One-Stop Planner CD-ROM.

Section 12-1 introduces restoring force, the conditions of simple harmonic motion, Hooke's law, and the relationship between force, velocity, and acceleration in simple harmonic motion.

Section 12-2 identifies the variables affecting amplitude, period, and frequency in a simple pendulum and in a mass-spring system.

Section 12-3 introduces concepts of wave motion, including wave speed, frequency, wavelength, amplitude, and energy, and discusses their relationships.

Section 12-4 explores how to use the superposition principle to predict patterns of interference and to identify the conditions for standing waves.

About the Illustration

The mechanical metronome was invented by Dietrich Winkel (c. 1776–1826) but was patented by Johann N. Maelzel in 1816. Today digital metronomes, which typically include both a flashing light and a ticking sound, are often used.

PHYSICS **Interactive Problem-Solving Tutor**

See Module 11

"Hooke's Law" provides additional development of problem-solving skills for spring problems.

See Module 12

"Wave Frequency and Wavelength" provides additional practice with the wave speed equation.

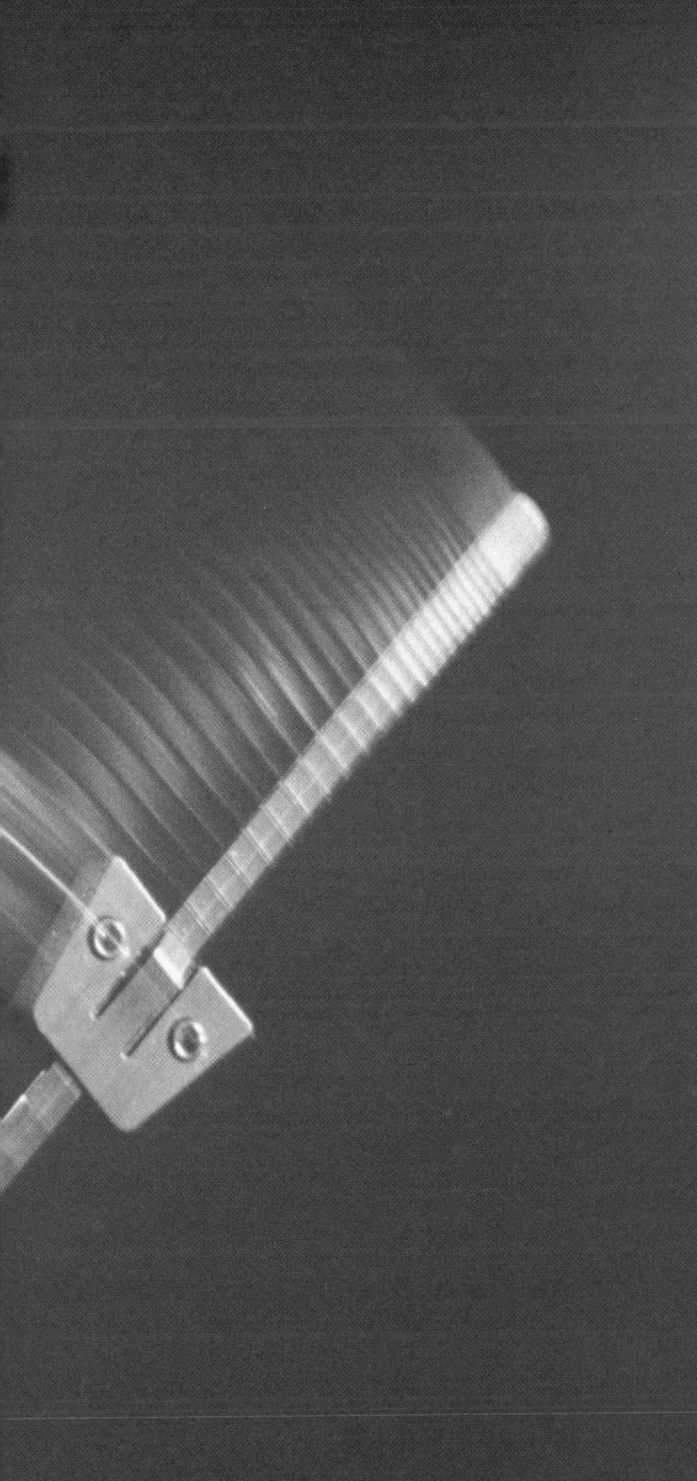

CHAPTER 12

Vibrations and Waves

PHYSICS IN ACTION

A mechanical metronome consists of a pendulum that vibrates around a pivot. Below the pivot is a counterweight. A sliding weight above the pivot can be positioned to change the rate of vibration. As the pendulum vibrates, the metronome produces a ticking sound, which musicians use to keep a steady tempo.

The vibrations of a metronome are an example of a periodic motion known as *simple harmonic motion.* In this chapter you will study simple harmonic motion and learn about the relationship between simple harmonic vibrations and wave motions.

- *What is the relationship between the period and the frequency of a metronome?*

- *Are all pendulums examples of simple harmonic motion?*

CONCEPT REVIEW

Elastic potential energy
(Section 5-2)

Gravitational potential energy
(Section 5-2)

Tapping Prior Knowledge

Knowledge to Expect
- ✔ "Students increase their inventory of examples of periodic motion and devise ways of measuring different rates of vibration." (AAAS's *Benchmarks for Science Literacy,* grades 3–5)

- ✔ "Students learn some of the properties of waves by using water tables, ropes, and springs." (AAAS's *Benchmarks for Science Literacy,* grades 6–8)

- ✔ "An object's motion can be described by tracing and measuring its position over time." (NRC's *National Science Education Standards,* grades K–4)

Knowledge to Review
- ✔ Elastic potential energy is the energy stored in a stretched or compressed elastic object. (Section 5-2)

- ✔ A spring constant is a parameter that expresses how resistant a spring is to being compressed or stretched. (Section 5-2)

- ✔ Gravitational potential energy is the energy associated with an object due to its position relative to Earth. (Section 5-2)

Items to Probe
- ✔ Familiarity with periodic motion: Ask students to describe the motion of objects that move in a closed path.

- ✔ Preconceptions about waves: Ask students to identify the source and the propagating medium in various examples of wave phenomena.

Simple harmonic motion

Demonstration 1

A vibrating spring

Purpose Show the changes in velocity and restoring force for a vibrating mass-spring system.

Materials spring, ring stand, weight holder, weight

Procedure Attach the spring to the ring stand, suspend a weight holder, and add a weight to the holder. Stretch the weight down to start a harmonic motion. Have the students sketch diagrams representing the position and direction of motion of the weight at various points of the cycle. Ask them whether the magnitude of the velocity increases or decreases, and have them draw the force vector that may cause this motion in each case. Point out that the force always pulls the weight toward its equilibrium position.

Demonstration 2

An oscillating pendulum

Purpose Show the changes in velocity and restoring force for an oscillating pendulum.

Materials pendulum bob, string, ring stand

Procedure Attach the bob to the string and suspend the string from the ring stand. Start the pendulum swinging, then repeat the process described in Demonstration 1. Point out that the two cases are analogous and that the force always pulls the pendulum bob toward its equilibrium position.

12-1 SECTION OBJECTIVES

- **Identify the conditions of simple harmonic motion.**

- **Explain how force, velocity, and acceleration change as an object vibrates with simple harmonic motion.**

- **Calculate the spring force using Hooke's law.**

HOOKE'S LAW

A repeated motion, such as that of an acrobat swinging on a trapeze, is called a periodic motion. Other periodic motions include those made by a child on a playground swing, a wrecking ball swaying to and fro, and the pendulum of a grandfather clock or a metronome. In each of these cases, the periodic motion is back and forth over the same path.

One of the simplest types of back-and-forth periodic motion is a mass attached to a spring, as shown in **Figure 12-1.** Let us assume that the mass moves on a frictionless horizontal surface. When the spring is stretched or compressed and then released, it vibrates back and forth around its unstretched position. We will begin by considering this example, and then we will apply our conclusions to the swinging motion of a trapeze acrobat.

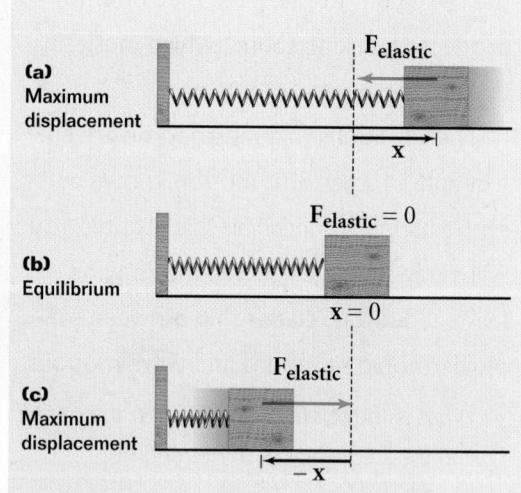

Figure 12-1

The direction of the force acting on the mass ($\mathbf{F}_{elastic}$) is always opposite the direction of the mass's displacement from equilibrium ($\mathbf{x} = 0$).
(a) When the spring is stretched to the right, the spring force pulls the mass to the left. **(b)** When the spring is unstretched, the spring force is zero.
(c) When the spring is compressed to the left, the spring force is directed to the right.

At the equilibrium position, velocity reaches a maximum

In **Figure 12-1(a),** the spring is stretched away from its unstretched, or equilibrium, position ($x = 0$). When released, the spring exerts a force on the mass toward the equilibrium position. This spring force decreases as the spring moves toward the equilibrium position, and it reaches zero at equilibrium, as illustrated in **Figure 12-1(b).** The mass's acceleration also becomes zero at equilibrium.

Though the spring force and acceleration decrease as the mass moves toward the equilibrium position, the velocity of the mass increases. At the equilibrium position, when acceleration reaches zero, the velocity reaches a maximum. At that point, although no net force is acting on the mass, the mass's momentum causes it to overshoot the equilibrium position and compress the spring. ⭐TEKS **3A**

At maximum displacement, spring force and acceleration reach a maximum

As the mass moves beyond equilibrium, the spring force and the acceleration increase. But the direction of the spring force and the acceleration (toward equilibrium) is opposite the mass's direction of motion (away from equilibrium), and the mass begins to slow down.

When the spring's compression is equal to the distance the spring was originally stretched away from the equilibrium position (x), as shown in **Figure 12-1(c),** the mass is at maximum displacement, and the spring force and acceleration of the mass reach a maximum. At this point, the velocity of the mass becomes zero. The spring force acting to the right causes the mass to change its direction, and the mass begins moving back toward the equilibrium position. Then the entire process begins again, and the mass continues to oscillate back and forth over the same path.

In an ideal system, the mass-spring system would oscillate indefinitely. But in the physical world, friction retards the motion of the vibrating mass, and the mass-spring system eventually comes to rest. This effect is called *damping*. In most cases, the effect of damping is minimal over a short period of time, so the ideal mass-spring system provides an approximation for the motion of a physical mass-spring system.

In simple harmonic motion, restoring force is proportional to displacement

As you have seen, the spring force always pushes or pulls the mass back toward its original equilibrium position. For this reason, it is sometimes called a *restoring force*. Measurements show that the restoring force is directly proportional to the displacement of the mass. Any periodic motion that is the result of a restoring force that is proportional to displacement is described by the term **simple harmonic motion.** Because simple harmonic motion involves a restoring force, every simple harmonic motion is a back-and-forth motion over the same path.

In 1678, Robert Hooke found that most mass-spring systems obey a simple relationship between force and displacement. For small displacements from equilibrium, the following equation describes the relationship: (★)TEKS **3E**

HOOKE'S LAW (★)TEKS **3B**

$$F_{elastic} = -kx$$

spring force = −(spring constant × displacement)

The negative sign in the equation signifies that the direction of the spring force is always opposite the direction of the mass's displacement from equilibrium. In other words, the negative sign shows that the spring force will tend to move the object back to its equilibrium position.

As mentioned in Chapter 5, the term k is a positive constant called the spring constant. The value of the spring constant is a measure of the stiffness of the spring. A greater value of k means a stiffer spring because a greater force is needed to stretch or compress that spring. The SI units of k are N/m. As a result, N is the unit of the spring force when the spring constant (N/m) is multiplied by the displacement (m).

internet **connect**

SC**LINKS**
NSTA

TOPIC: Hooke's law
GO TO: www.scilinks.org
*sci*LINKS CODE: HF2121

simple harmonic motion

vibration about an equilibrium position in which a restoring force is proportional to the displacement from equilibrium

Conceptual Challenge

1. Earth's orbit

The motion of Earth orbiting the sun is periodic. Is this motion simple harmonic? Why or why not?

2. Pinball

In pinball games, the force exerted by a compressed spring is used to release a ball. If the distance the spring is compressed is doubled, how will the force acting on the ball change? If the spring is replaced with one that is half as stiff, how will the force acting on the ball change?

(★)TEKS **3A**

Demonstration 3

Hooke's law

Purpose Verify Hooke's law experimentally.

Materials 2 springs with different spring constants, 2 ring stands, 2 weight holders, incremental weights, ruler

Procedure Attach the springs to the ring stands, and suspend a weight holder from each spring. Add incremental weights to the holders, measure the resulting displacements, and record these values on the chalkboard.

From the data for each spring, sketch a graph of force versus displacement on the chalkboard. Show that the relationship between force and displacement is linear, and calculate the slope (which equals the spring constant). Ask students why a negative sign appears in the equation. *(The elastic force on the weight is opposite the weight's displacement from equilibrium.)*

ANSWERS TO

Conceptual Challenge

1. no; because Earth does not oscillate about an equilibrium position

2. The force will double; The force will be half as large.

(★)TEKS

p. 438: 3A
p. 439: 3E, 3B, 3A

439

Classroom Practice

The following may be used as teamwork exercises or for demonstration at the chalkboard or on an overhead projector.

PROBLEM

Hooke's law

A 76 N crate is attached to a spring ($k = 450$ N/m). How much displacement is caused by the weight of this crate?

Answer
−0.17 m

A spring of $k = 1962$ N/m loses its elasticity if stretched more than 50.0 cm. What is the mass of the heaviest object the spring can support without being damaged?

Answer
1.00×10^2 kg

Alternative Problem-Solving Approach

The weight of the object pulls downward.
$$F_g = -mg = -5.4 \text{ N}$$
The spring stretches until its restoring force ($F_{elastic} = -kx$) balances the −5.4 N. This occurs when $x = -0.020$ m. Thus, 5.4 N = $-k(-0.020$ m) and $k = 270$ N/m.

Interactive Problem-Solving Tutor

See Module 11
"Hooke's Law" provides additional development of problem-solving skills for spring problems.

SAMPLE PROBLEM 12A

Hooke's law ⭐TEKS 2C, 3B, 4C

PROBLEM

If a mass of 0.55 kg attached to a vertical spring stretches the spring 2.0 cm from its original equilibrium position, as shown in Figure 12-2, what is the spring constant?

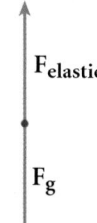

$x = -2.0$ cm

m

Figure 12-2

SOLUTION

1. DEFINE **Given:** $m = 0.55$ kg $x = -2.0$ cm $= -0.020$ m
$g = 9.81$ m/s^2

Unknown: $k = ?$

Diagram:

$F_{elastic}$

F_g

2. PLAN **Choose an equation(s) or situation:** When the mass is attached to the spring, the equilibrium position changes. At the new equilibrium position, the net force acting on the mass is zero. So the spring force (given by Hooke's law) must be equal and opposite to the weight of the mass.

$$\mathbf{F_{net}} = 0 = \mathbf{F_{elastic}} + \mathbf{F_g}$$

$$F_{elastic} = -kx$$

$$F_g = -mg$$

$$-kx - mg = 0$$

Rearrange the equation(s) to isolate the unknown(s):

$$kx = -mg$$

$$k = \frac{-mg}{x}$$

3. CALCULATE **Substitute the values into the equation(s) and solve:**

$$k = \frac{-(0.55 \text{ kg})(9.81 \text{ m/s}^2)}{-0.020 \text{ m}}$$

$$\boxed{k = 270 \text{ N/m}}$$

CALCULATOR SOLUTION

The calculator answer for *k* is 269.775. This answer is rounded to two significant figures, 270 N/m.

4. EVALUATE The value of *k* implies that about 300 N of force is required to displace the spring 1 m.

Hooke's law ⭐TEKS 2C, 3B, 4C

1. Suppose the spring in Sample Problem 12A is replaced with a spring that stretches 36 cm from its equilibrium position.
   a. What is the spring constant in this case?
   b. Is this spring stiffer or less stiff than the one in Sample Problem 12A?

2. A load of 45 N attached to a spring that is hanging vertically stretches the spring 0.14 m. What is the spring constant?

3. A slingshot consists of a light leather cup attached between two rubber bands. If it takes a force of 32 N to stretch the bands 1.2 cm, what is the equivalent spring constant of the rubber bands?

4. How much force is required to pull the cup of the slingshot in problem 3 3.0 cm from its equilibrium position?

A stretched or compressed spring has elastic potential energy

As you saw in Chapter 5, a stretched or compressed spring stores elastic potential energy. To see how mechanical energy is conserved in an ideal mass-spring system, consider an archer shooting an arrow from a bow, as shown in **Figure 12-3.** Bending the bow by pulling back the bowstring is analogous to stretching a spring. To simplify this situation, we will disregard friction and internal energy.

Once the bowstring has been pulled back, the bow stores elastic potential energy. Because the bow, arrow, and bowstring (the system) are now at rest, the kinetic energy of the system is zero, and the mechanical energy of the system is solely elastic potential energy.

When the bowstring is released, the bow's elastic potential energy is converted to the kinetic energy of the arrow. At the moment the arrow leaves the bowstring, it gains most of the elastic potential energy originally stored in the bow. (The rest of the elastic potential energy is converted to the kinetic energy of the bow and the bowstring.) Thus, once the arrow has been released, the mechanical energy of the bow-and-arrow system is solely kinetic. Because mechanical energy must be conserved, the kinetic energy of the bow, arrow, and bowstring is equal to the elastic potential energy originally stored in the bow. ⭐TEKS 5B

PHYSICS INTERACTIVE TUTOR

Module 11
"Hooke's Law" provides an interactive lesson with guided problem-solving practice to teach you about springs and spring constants.

Figure 12-3
The elastic potential energy stored in this stretched bow is converted into the kinetic energy of the arrow.

PRACTICE GUIDE 12A		
Solving for:		
k	📖 **PE**	Sample, 1–3; Ch. Rvw. 8–9
	💿 **PW**	3, 4*, 5*
	PB	5–7
F	📖 **PE**	4; Ch. Rvw. 46–47, 55
	💿 **PW**	6*, 7, 8*
	PB	Sample, 1–4
x	💿 **PW**	Sample, 1–5
	PB	8–10

ANSWERS TO

Practice 12A
Hooke's law

1. **a.** 15 N/m
   b. less stiff
2. 3.2×10^2 N/m
3. 2.7×10^3 N/m
4. 81 N

⭐TEKS

p. 440: 2C, 3B, 4C
p. 441: 2C, 3B, 4C, 5B

Consumer Focus   *Shock Absorbers and Damped Oscillation*

BACKGROUND

The spring–shock absorber system on modern cars is an excellent example of damped harmonic oscillation.

A shock absorber consists of a piston moving up and down in a chamber filled with oil. As the piston moves, the oil is squeezed through the channels between the piston and the tube, causing the piston to decelerate.

*B*umps in the road are certainly a nuisance, but without strategic use of damping devices, they could also prove deadly. To control a car going 110 km/h (70 mi/h), a driver needs all the wheels on the ground. Bumps in the road lift the wheels off the ground and rob the driver of control. A good solution is to fit the car with springs at each wheel. The springs absorb energy as the wheels rise over the bumps and push the wheels back to the pavement to keep the wheels on the road. However, once set in motion, springs tend to continue to go up and down in simple harmonic motion. This affects the driver's control of the car and can also be uncomfortable.

One way to cut down on unwanted vibrations is to use stiff springs that compress only a few centimeters under thousands of newtons of force. Such springs have very high spring constants and thus do not vibrate as freely as softer springs with lower constants. However, this solution reduces the driver's ability to keep the car's wheels on the road.

⭐ **TEKS** 3C

To completely solve the problem, energy-absorbing devices known as shock absorbers are placed parallel to the springs in some automobiles, as shown in **Figure 12-4(a)**. Shock absorbers are fluid-filled tubes that turn the simple harmonic motion of the springs into damped harmonic motion. In damped harmonic motion, each cycle of stretch and compression of the spring is much smaller than the previous cycle. Modern auto suspensions are set up so that all of a spring's energy is absorbed by the shock absorbers, eliminating vibrations in just one up-and-down cycle. This keeps the car from continually bouncing without sacrificing the spring's ability to keep the wheels on the road.

Different spring constants and shock absorber damping are combined to give a wide variety of road responses. For example, larger vehicles have heavy-duty leaf springs made of stacks of steel strips, which have a larger spring constant than coil springs do. In this type of suspension system, the shock absorber is perpendicular to the spring, as shown in **Figure 12-4(b).** The stiffness of the spring can affect steering response time, traction, and the general feel of the car.

As a result of the variety of combinations that are possible, your driving experiences can range from the luxurious floating of a limousine to the bone-rattling road feel of a sports car.

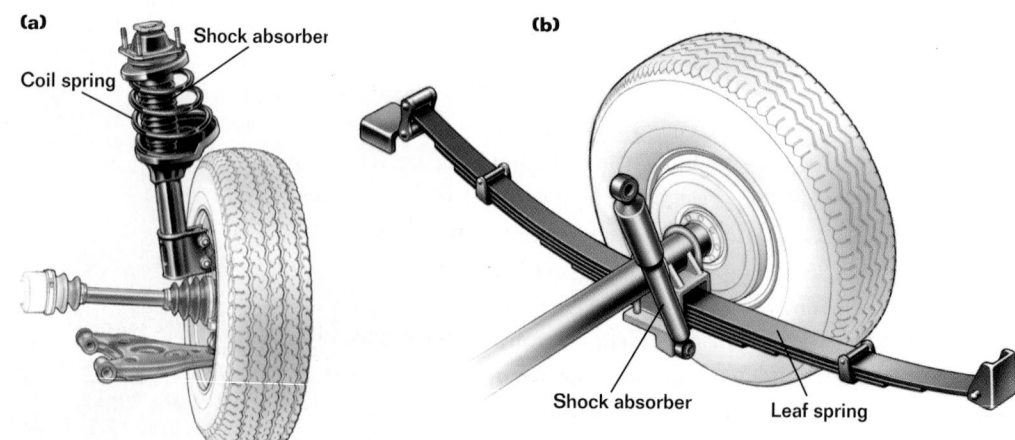

(a)

Shock absorber

Coil spring

(b)

Shock absorber   Leaf spring

Figure 12-4

THE SIMPLE PENDULUM

As you have seen, the periodic motion of a mass-spring system is one example of simple harmonic motion. Now consider the trapeze acrobats shown in **Figure 12-5(a).** Like the vibrating mass-spring system, the swinging motion of a trapeze acrobat is a periodic vibration. Is a trapeze acrobat's motion an example of simple harmonic motion?

To answer this question, we will use a simple pendulum as a model of the acrobat's motion, which is a physical pendulum. A simple pendulum consists of a mass called a *bob*, which is attached to a fixed string, as shown in **Figure 12-5(b).** When working with a simple pendulum, we assume that the mass of the bob is concentrated at a point and that the mass of the string is negligible. Furthermore, we disregard the effects of friction and air resistance. For a physical pendulum, on the other hand, the distribution of the mass must be considered, and friction and air resistance also must be taken into account. To simplify our analysis, we will disregard these complications and use a simple pendulum to approximate a physical pendulum in all of our examples.

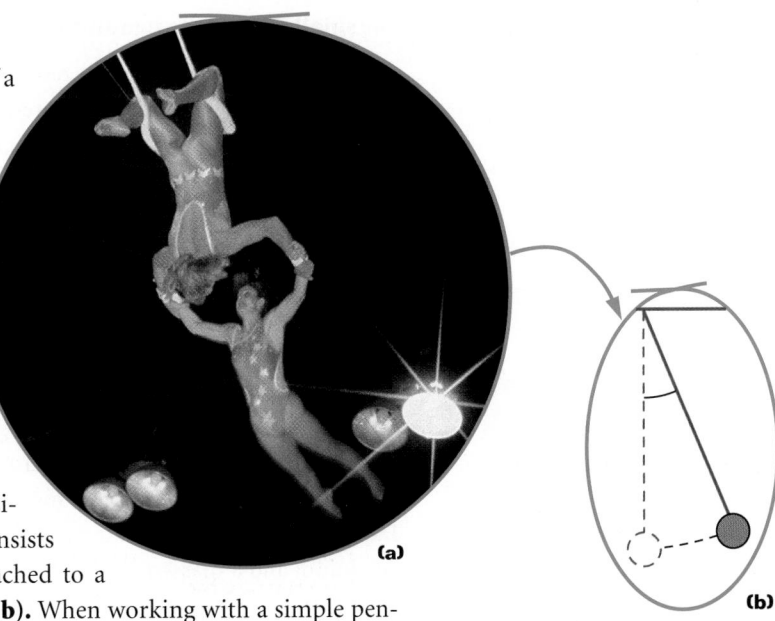
(a)
(b)

Figure 12-5
(a) The motion of these trapeze acrobats is modeled by **(b)** a simple pendulum.

The restoring force of a pendulum is a component of the bob's weight

To see whether the pendulum's motion is simple harmonic, we must first examine the forces exerted on the pendulum's bob to determine which force acts as the restoring force. If the restoring force is proportional to the displacement, then the pendulum's motion is simple harmonic. Let us select a coordinate system in which the *x*-axis is tangent to the direction of motion and the *y*-axis is perpendicular to the direction of motion. Because the bob is always changing its position, these axes will change at each point of the bob's motion.

The forces acting on the bob at any point include the force exerted by the string and the bob's weight. The force exerted by the string always acts along the *y*-axis, which is along the string. At any point other than the equilibrium position, the bob's weight can be resolved into two components along the chosen axes, as shown in **Figure 12-6.** Because both the force exerted by the string and the *y* component of the bob's weight are perpendicular to the bob's motion, the *x* component of the bob's weight is the net force acting on the bob in the direction of its motion. In this case, the *x* component of the bob's weight always pushes or pulls the bob toward its equilibrium position and hence is the restoring force. ⭐TEKS **4A, 4C, 4D**

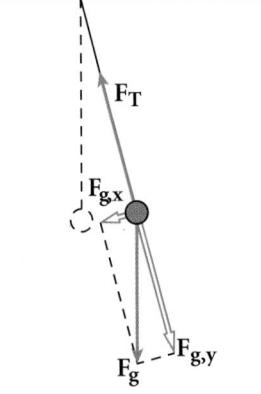

Figure 12-6
At any displacement from equilibrium, the weight of the bob **(F$_g$)** can be resolved into two components. The *x* component **(F$_{g,x}$)**, which is perpendicular to the string, is the only force acting on the bob in the direction of its motion.

Vibrations and Waves   443

The Language of Physics

Point out that the term *simple pendulum* is used because we have simplified our analysis by disregarding complications such as friction and air resistance. The term *ideal mass-spring system,* seen earlier in this section, was used to express the same concept for a mass-spring system.

Visual Strategy

Figure 12-6

Be certain students make a distinction between force diagrams and schematic diagrams representing physical objects. In particular, string *length* and string *force* should not be confused.

Q Draw the vectors representing the forces on the bob at the equilibrium position when the pendulum is at rest. Which component of **F$_g$** is 0? Which component of **F$_g$** is equal to **F$_g$**?

A *Students' vector diagrams should show that F$_g$ is equal and opposite to F$_T$ (both are vertical); F$_{g,x}$ = 0; F$_{g,y}$ = F$_g$.*

⭐TEKS

p. 442: 3C
p. 443: 4A, 4C, 4D

Quick Lab

TEACHER'S NOTES

This activity is meant to demonstrate that the kinetic energy of the pendulum at the equilibrium position increases as the pendulum's maximum displacement from equilibrium increases.

For this lab to be effective, it is best to arrange the pendulum so that it neither swings back nor moves on, but rather transfers all of its energy to the toy car. This is best achieved when the collision between the car and the bob is head-on and when the mass of the bob is nearly equal to the mass of the car.

Of course, the energy transferred to the car will be quickly dissipated, in part because of the inelasticity of the collision and in part because of friction on the wheels of the car. As a result, the displacement of the car is only a very rough indication of the energy of the pendulum.

Visual Strategy

Figure 12-7

Point out to students that the total mechanical energy of the system is represented by a horizontal line because the sum of the kinetic energy and the potential energy is always constant. This means that as one increases, the other decreases by the same amount, and vice versa.

Q Does this graph apply to a vibrating mass-spring system as well?

A *yes, because the energy changes in a mass-spring system are analogous to those in a simple pendulum*

444

internet**connect**

SCI**LINKS**

NSTA

TOPIC: Pendulums
GO TO: www.scilinks.org
sciLINKS CODE: HF2122

⭐TEKS **1A, 2B, 2C, 2F**

Quick Lab

Energy of a Pendulum

MATERIALS LIST

✔ pendulum bob and string
✔ tape
✔ toy car
✔ protractor
✔ meterstick or tape measure

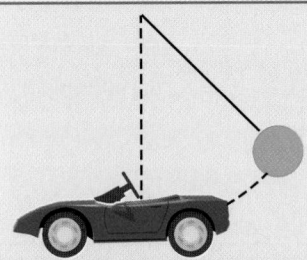

Tie one end of a piece of string around the pendulum bob, and use tape to secure it in place. Set the toy car on a smooth surface, and hold the string of the pendulum directly above the car so that the bob rests on the car. Use your other hand to pull back the bob of the pendulum, and have your partner measure the angle of the pendulum with the protractor.

Release the pendulum so that the bob strikes the car. Measure the displacement of the car. What happened to the pendulum's potential energy after you released the bob? Repeat the process using different angles. How can you account for your results?

For small angles, the pendulum's motion is simple harmonic

As with a mass-spring system, the restoring force of a simple pendulum is not constant. Instead, the magnitude of the restoring force varies with the bob's distance from the equilibrium position. The magnitude of the restoring force decreases as the bob moves toward the equilibrium position and becomes zero at the equilibrium position. When the angle of displacement is relatively small ($<15°$), the restoring force is proportional to the displacement. For such small angles of displacement, the pendulum's motion is simple harmonic. We will assume small angles of displacement unless otherwise noted. ⭐TEKS **3A**

Because a simple pendulum vibrates with simple harmonic motion, many of our earlier conclusions for a mass-spring system apply here. At maximum displacement, the restoring force and acceleration reach a maximum while the velocity becomes zero. Conversely, at equilibrium the restoring force and acceleration become zero and velocity reaches a maximum. **Table 12-1** on the following page illustrates the analogy between a simple pendulum and a mass-spring system.

Gravitational potential increases as a pendulum's displacement increases

As with the mass-spring system, the mechanical energy of a simple pendulum is conserved in an ideal (frictionless) system. However, the spring's potential energy is elastic, while the pendulum's potential energy is gravitational. We define the gravitational potential energy of a pendulum to be zero when it is at the lowest point of its swing. ⭐TEKS **5B**

Figure 12-7 illustrates how a pendulum's mechanical energy changes as the pendulum oscillates. At maximum displacement from equilibrium, a pendulum's energy is entirely gravitational potential energy. As the pendulum swings toward equilibrium, it gains kinetic energy and loses potential energy. At the equilibrium position, its energy becomes solely kinetic.

As the pendulum swings past its equilibrium position, the kinetic energy decreases while the gravitational potential energy increases. At maximum displacement from equilibrium, the pendulum's energy is once again entirely gravitational potential energy.

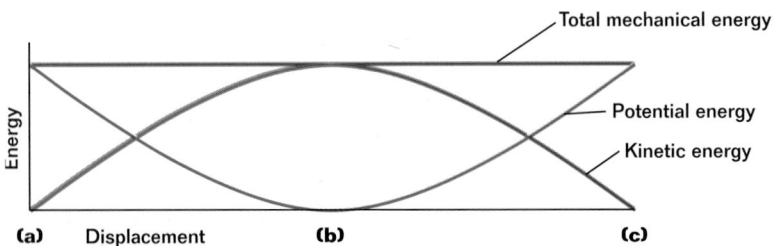

Figure 12-7
Whether at maximum displacement **(a)**, equilibrium **(b)**, or maximum displacement in the other direction **(c)**, the pendulum's total mechanical energy remains the same. However, as the graph shows, the pendulum's kinetic energy and potential energy are constantly changing.

Table 12-1   Simple harmonic motion

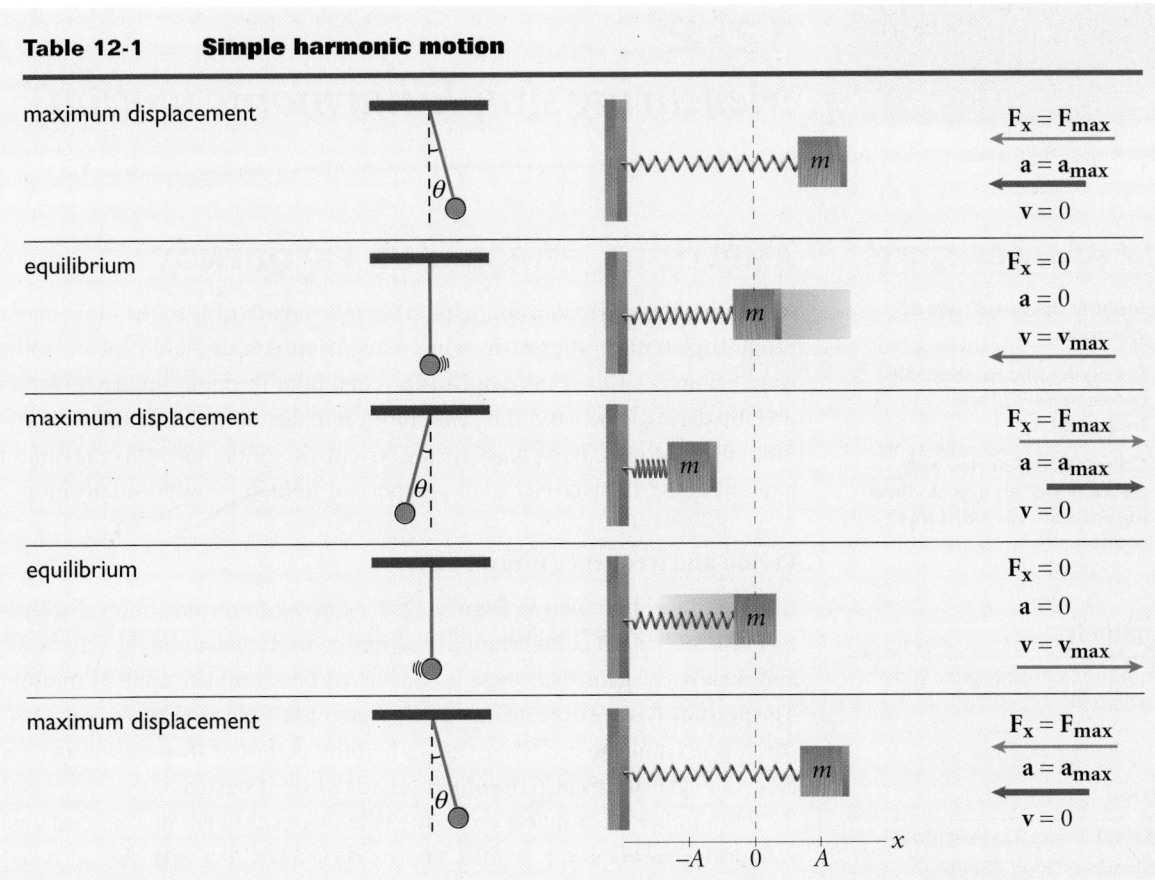

maximum displacement		$F_x = F_{max}$ $a = a_{max}$ $v = 0$
equilibrium		$F_x = 0$ $a = 0$ $v = v_{max}$
maximum displacement		$F_x = F_{max}$ $a = a_{max}$ $v = 0$
equilibrium		$F_x = 0$ $a = 0$ $v = v_{max}$
maximum displacement		$F_x = F_{max}$ $a = a_{max}$ $v = 0$

$-A \quad 0 \quad A$ x

Section Review

★ TEKS 2C, 3B, 4C

1. Which of these periodic motions are simple harmonic?
   a. a child swinging on a playground swing (at a small angle)
   b. a record rotating on a turntable
   c. an oscillating clock pendulum

2. A pinball machine uses a spring that is compressed 4.0 cm to launch a ball. If the spring constant is 13 N/m, what is the force on the ball at the moment the spring is released?

3. How does the restoring force acting on a pendulum bob change as the bob swings toward the equilibrium position? How do the bob's acceleration (along the direction of motion) and velocity change?

4. When an acrobat reaches the equilibrium position, the net force acting along the direction of motion is zero. Why does the acrobat swing past the equilibrium position?

Visual Strategy

Table 12-1

Table 12-1 can be used to review the concepts discussed in this section. Comparing these two different types of simple harmonic motion will help students grasp the essential aspects of simple harmonic motion.

Q Does the fourth column of the table refer to the simple pendulum or to the mass-spring system?

A *both; The displacements, velocities, and restoring forces in these two cases are analogous.*

Section Review
ANSWERS

1. a, c
2. 0.52 N
3. force decreases; acceleration decreases; velocity increases
4. because the acrobat's velocity is at a maximum from previous acceleration

★ TEKS

p. 444: 1A, 2B, 2C, 2F, 3A, 5B
p. 445: 2C, 3B, 4C

Period and frequency

Purpose Show period and frequency empirically and verify their inverse relationship.

Materials pendulum bob, string, ring stand, clock

Procedure Attach the pendulum bob to the string and suspend the string from the ring stand. Set the pendulum in motion. Have a student record the time required to complete 20 oscillations. Meanwhile, have another student record how many times the pendulum bob returns to the same place each second.

Have students use the first measurement to find the pendulum's period (T = number of seconds/20), and ask the students what the second measurement indicates (*frequency*). Compare the values for the period and frequency of the pendulum. (*The two should be inversely related.*)

The Language of Physics

Students may be confused by the transition from cycles/s as a measure of frequency to the SI unit s^{-1} (hertz). Point out that the term *cycle* is not part of the SI unit because this term refers to an event rather than a unit of measure. The same holds true for the period, which can be considered as s/cycle but whose SI unit is simply s. This should help clarify the inverse relationship between frequency (cycles/s or s^{-1}) and period (s/cycle or s).

12-2 SECTION OBJECTIVES

- **Identify the amplitude of vibration.**
- **Recognize the relationship between period and frequency.**
- **Calculate the period and frequency of an object vibrating with simple harmonic motion.**

amplitude

the maximum displacement from equilibrium

period

the time it takes to execute a complete cycle of motion

frequency

the number of cycles or vibrations per unit of time

12-2
Measuring simple harmonic motion

AMPLITUDE, PERIOD, AND FREQUENCY

In the absence of friction, a moving trapeze always returns to the same maximum displacement after each swing. This maximum displacement from the equilibrium position is the **amplitude.** A pendulum's amplitude can be measured by the angle between the pendulum's equilibrium position and its maximum displacement. For a mass-spring system, the amplitude is the maximum amount the spring is stretched or compressed from its equilibrium position.

Period and frequency measure time

Imagine the ride shown in **Figure 12-8** swinging from maximum displacement on one side of equilibrium to maximum displacement on the other side, and then back again. This cycle is considered one complete cycle of motion. The **period,** T, is the time it takes for this complete cycle of motion. For example, if one complete cycle takes 20 s, then the period of this motion is 20 s. Note that after the time T, the object is back where it started.

The number of complete cycles the ride swings through in a unit of time is the ride's **frequency,** f. If one complete cycle takes 20 s, then the ride's frequency is $\frac{1}{20}$ cycles/s, or 0.05 cycles/s. The SI unit of frequency is s^{-1}, known as hertz (Hz). In this case, the ride's frequency is 0.05 Hz. ⭐TEKS 2C

Period and frequency can be confusing because both are concepts involving time in simple harmonic motion. Notice that the period is the time per cycle and that the frequency is the number of cycles per unit time, so they are inversely related. ⭐TEKS 3B

$$f = \frac{1}{T} \text{ or } T = \frac{1}{f}$$

This relationship was used to determine the frequency of the ride.

$$f = \frac{1}{T} = \frac{1}{20 \text{ s}} = 0.05 \text{ Hz}$$

In any problem where you have a value for period or frequency, you can use this relationship to calculate the other value.

Figure 12-8
For any periodic motion—such as the motion of this amusement park ride in Helsinki, Finland—period and frequency are inversely related.

Table 12-2 Measures of simple harmonic motion

Term	Example	Definition	SI unit
amplitude	θ	maximum displacement from equilibrium	radian, rad meter, m
period, T		time it takes to execute a complete cycle of motion	second, s
frequency, f		number of cycles or vibrations per unit of time	hertz, Hz ($Hz = s^{-1}$)

The period of a simple pendulum depends on pendulum length and free-fall acceleration

Although both a simple pendulum and a mass-spring system vibrate with simple harmonic motion, calculating the period and frequency of each requires a separate equation. This is because in each, the period and frequency depend on different physical factors.

Consider an experimental setup of two pendulums of the same length but with bobs of different masses. The length of a pendulum is measured from the pivot point to the center of mass of the pendulum bob. If you were to pull each bob aside the same small distance and then release them at the same time, each pendulum would complete one vibration in the same amount of time. If you then changed the amplitude of one of the pendulums, you would find that they would still have the same period. Thus, for small amplitudes, the period of a pendulum does not depend on the amplitude.

However, changing the length of a pendulum *does* affect its period. A change in the free-fall acceleration also affects the period of a pendulum. The exact relationship between these variables can be derived mathematically or found experimentally. ⭐TEKS **3A**

PERIOD OF A SIMPLE PENDULUM IN SIMPLE HARMONIC MOTION

⭐TEKS **3B**

$$T = 2\pi\sqrt{\frac{L}{g}}$$

period = $2\pi \times$ square root of (length divided by free-fall acceleration)

Did you know?

Galileo is credited as the first person to notice that the motion of a pendulum depends on its length and is independent of its amplitude (for small angles). He supposedly observed this while attending church services at a cathedral in Pisa. The pendulum he studied was a swinging chandelier that was set in motion when someone bumped it while lighting the candles. Galileo is said to have measured its frequency, and hence its period, by timing the swings with his pulse.

⭐TEKS **3E**

Demonstration 5

Relationship between the length and the period of a pendulum

Purpose Verify the equation for a pendulum's period experimentally.

Materials pendulum bob, string, ring stand, clock, meterstick

Procedure Repeat Demonstration 4 with a variety of lengths. Record each length and its corresponding period. (Frequency does not need to be measured in this demonstration.) Verify that the results are consistent with the following equation:

$$T = 2\pi\sqrt{\frac{L}{g}}$$

Next ask the students to calculate the length required for a pendulum to have a period of 1.0 s. Have the students construct such a pendulum to test their prediction.

Misconception Alert

Remind students that, as seen in Section 12-1, a pendulum's amplitude must be less than about 15° in order for its motion to be simple harmonic. For greater amplitudes, the pendulum's amplitude *does* affect its period, and in those cases this equation for a pendulum's period would not apply.

⭐TEKS

p. 446: 2C, 3B
p. 447: 3A, 3E, 3B

The following may be used as teamwork exercises or for demonstration at the chalkboard or on an overhead projector.

PROBLEM

SHM of a simple pendulum

What is the period of a 3.98 m long pendulum? What is the period of a 99.4 cm long pendulum?

Answer
   4.00 s; 2.00 s

A desktop toy swings back and forth once every 1.0 s. How tall is this toy?

Answer
   0.25 m

What is the free-fall acceleration at a location where a 6.00 m long pendulum swings through exactly 100 cycles in 492 s?

Answer
   9.79 m/s^2

PRACTICE GUIDE 12B

Solving for:		
L	📖 **PE** Sample, 1–3; Ch. Rvw. 19–20	
	💿 **PW** 4, 5	
	PB 5–7	
T, f	📖 **PE** 4; Ch. Rvw. 21, 28a	
	💿 **PW** Sample, 1–3	
	PB 8–10	
g	💿 **PW** 6	
	PB Sample, 1–4	

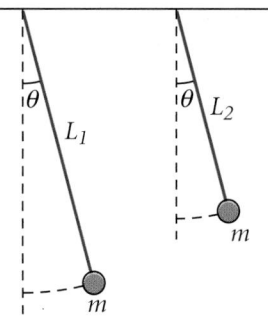

Figure 12-9
When the length of one pendulum is decreased, the distance that the pendulum travels to equilibrium is also decreased. Because the accelerations of the two pendulums are equal, the shorter pendulum will have a smaller period.

Why does the period of a pendulum depend on pendulum length and free-fall acceleration? When two pendulums have different lengths but the same amplitude, the shorter pendulum will have a smaller arc to travel through, as shown in **Figure 12-9.** Because the distance the mass travels from maximum displacement to equilibrium is less while the acceleration of both pendulums remains the same, the shorter pendulum will have a shorter period.

Why don't mass and amplitude affect the period of a pendulum? When the bobs of two pendulums differ in mass, the heavier mass provides a larger restoring force, but it also needs a larger force to achieve the same acceleration. This is similar to the situation for objects in free fall, which all have the same acceleration regardless of their mass. Because the acceleration of both pendulums is the same, the period for both is also the same. ⭐TEKS **3A**

For small angles (<15°), when the amplitude of a pendulum increases, the restoring force also increases proportionally. Because force is proportional to acceleration, the initial acceleration will be greater. However, the distance this pendulum must cover is also greater. For small angles, the effects of the two increasing quantities cancel and the pendulum's period remains the same.

SAMPLE PROBLEM 12B

Simple harmonic motion of a simple pendulum ⭐TEKS 2C, 3B

PROBLEM

You need to know the height of a tower, but darkness obscures the ceiling. You note that a pendulum extending from the ceiling almost touches the floor and that its period is 12 s. How tall is the tower?

SOLUTION

Given: $T = 12$ s $g = 9.81$ m/s^2

Unknown: $L = ?$

Use the equation given on page 447 to solve for L.

$$T = 2\pi \sqrt{\frac{L}{g}}$$

$$\frac{T\sqrt{g}}{2\pi} = \sqrt{L}$$

$$\frac{T^2 g}{4\pi^2} = L$$

$$L = \frac{(12 \text{ s})^2 (9.81 \text{ m/s}^2)}{4\pi^2}$$

$$\boxed{L = 36 \text{ m}}$$

PRACTICE 12B

Simple harmonic motion of a simple pendulum ★TEKS 2C, 3B

1. If the period of the pendulum in the preceding sample problem were 24 s, how tall would the tower be?

2. You are designing a pendulum clock to have a period of 1.0 s. How long should the pendulum be?

3. A trapeze artist swings in simple harmonic motion with a period of 3.8 s. Calculate the length of the cables supporting the trapeze.

4. Calculate the period and frequency of a 3.500 m long pendulum at the following locations:

   a. the North Pole, where $g = 9.832$ m/s^2

   b. Chicago, where $g = 9.803$ m/s^2

   c. Jakarta, Indonesia, where $g = 9.782$ m/s^2

Period of a mass-spring system depends on mass and spring constant

Now consider the period of a mass-spring system. In this case, according to Hooke's law, the restoring force acting on the mass is determined by the displacement of the mass and by the spring constant ($F_{elastic} = -kx$). The magnitude of the mass does not affect the restoring force. So unlike the case of the pendulum, where a heavier mass increased both the force on the bob and the bob's inertia, a heavier mass attached to a spring increases inertia without providing a compensating increase in force. Because of this increase in inertia, a heavy mass has a smaller acceleration than a light mass has. Thus, a heavy mass will take more time to complete one cycle of motion. In other words, the heavy mass has a greater period. Thus, as mass increases, the period of vibration likewise increases. ★TEKS 3A

Conceptual Challenge

1. Pendulum on the moon

The free-fall acceleration on the surface of the moon is approximately one-sixth of the free-fall acceleration on the surface of Earth. Compare the period of a pendulum on Earth with that of an identical pendulum set in motion on the moon. ★TEKS 3A

2. Pendulum clocks

Why is a pendulum a reliable time-keeping device, even if its oscillations gradually decrease in amplitude over time? ★TEKS 3A

The following may be used as a teamwork exercise or for demonstration at the chalkboard or on an overhead projector.

PROBLEM

SHM of a mass-spring system

A 1.0 kg mass attached to one end of a spring completes one oscillation every 2.0 s. Find the spring constant.

Answer

  9.9 N/m

Alternative Problem-Solving Approach

Think of the total mass (1275 kg + 153 kg = 1428 kg) as suspended to one spring with four times the strength of each spring:

$k = 4 \times (2.0 \times 10^4 \text{ N/m})$

$k = 8.0 \times 10^4 \text{ N/m}$

$T = 2\pi \sqrt{\dfrac{m}{k}}$

$T = 2\pi \sqrt{\dfrac{1428 \text{ kg}}{8.0 \times 10^4 \text{ N/m}}}$

$T = 0.84 \text{ s}$

★ TEKS

  p. 450: 3B, 3A, 2C, 3B
  p. 451: 2C, 3B, 2C, 3B

The greater the spring constant (k), the stiffer the spring; hence a greater force is required to stretch or compress the spring. When force is greater, acceleration is greater and the amount of time required for a single cycle should decrease (assuming that the amplitude remains constant). Thus, for a given amplitude, a stiffer spring will take less time to complete one cycle of motion than one that is less stiff.

As with the pendulum, the equation for the period of a mass-spring system can be derived mathematically or found experimentally.

PERIOD OF A MASS-SPRING SYSTEM IN SIMPLE HARMONIC MOTION

$$T = 2\pi \sqrt{\frac{m}{k}} \quad \text{★ TEKS} \quad \textbf{3B}$$

period = $2\pi \times$ square root of (mass divided by spring constant)

Note that, as with the pendulum, changing the amplitude of the vibration does not affect the period. This statement is true only for systems and circumstances in which the spring obeys Hooke's law. ★ TEKS **3A**

SAMPLE PROBLEM 12C

Simple harmonic motion of a mass-spring system ★ TEKS 2C, 3B

PROBLEM

The body of a 1275 kg car is supported on a frame by four springs. Two people riding in the car have a combined mass of 153 kg. When driven over a pothole in the road, the frame vibrates with a period of 0.840 s. For the first few seconds, the vibration approximates simple harmonic motion. Find the spring constant of a single spring.

SOLUTION

Given: $\quad m = \dfrac{(1275 \text{ kg} + 153 \text{ kg})}{4} = 357 \text{ kg} \qquad T = 0.840 \text{ s}$

Unknown: $\quad k = ?$

Use the equation for the period of a mass-spring system to solve for k:

$$T = 2\pi \sqrt{\frac{m}{k}}$$

$$T^2 = 4\pi^2 \left(\frac{m}{k}\right)$$

$$k = \frac{4\pi^2 m}{T^2} = \frac{4\pi^2 (357 \text{ kg})}{(0.84 \text{ s})^2}$$

$$\boxed{k = 2.00 \times 10^4 \text{ N/m}}$$

Simple harmonic motion of a mass-spring system ⭐TEKS 2C, 3B

1. A mass of 0.30 kg is attached to a spring and is set into vibration with a period of 0.24 s. What is the spring constant of the spring?

2. When a mass of 25 g is attached to a certain spring, it makes 20 complete vibrations in 4.0 s. What is the spring constant of the spring?

3. A 125 N object vibrates with a period of 3.56 s when hanging from a spring. What is the spring constant of the spring?

4. When two more people get into the car described in Sample Problem 12C, the total mass of all four occupants of the car becomes 255 kg. Now what is the period of vibration of the car when it is driven over a pothole in the road?

5. A spring of spring constant 30.0 N/m is attached to different masses, and the system is set in motion. Find the period and frequency of vibration for masses of the following magnitudes:
   a. 2.3 kg
   b. 15 g
   c. 1.9 kg

Section Review ⭐TEKS 2C, 3B

1. Two mass-spring systems vibrate with simple harmonic motion. If the spring constants of each system are equal and the mass of one is twice that of the other, which system has a greater period?

2. A child swings on a playground swing with a 2.5 m long chain.
   a. What is the period of the child's motion?
   b. What is the frequency of vibration?

3. A 0.75 kg mass attached to a vertical spring stretches the spring 0.30 m.
   a. What is the spring constant?
   b. The mass-spring system is now placed on a horizontal surface and set vibrating. What is the period of the vibration?

4. **Physics in Action** The reading on a metronome indicates the number of oscillations per minute. What are the period and frequency of the metronome's vibration when the metronome is set at 180?

PRACTICE GUIDE 12C

Solving for:		
k	📖 **PE** Sample, 1–3 💿 **PW** 4 **PB** 4–6	
T, f	📖 **PE** 4, 5; Ch. Rvw. 22 💿 **PW** Sample, 1, 2, 3* **PB** 7–10	
m	💿 **PW** 5, 6 **PB** Sample, 1–3	

ANSWERS TO

Practice 12C
SHM of a mass-spring system

1. 2.1×10^2 N/m
2. 25 N/m
3. 39.6 N/m
4. 0.869 s
5. a. 1.7 s, 0.59 Hz
   b. 0.14 s, 7.1 Hz
   c. 1.6 s, 0.62 Hz

Section Review
ANSWERS

1. The system with the larger mass has a greater period.
2. a. 3.2 s
   b. 0.31 Hz
3. a. 25 N/m
   b. 1.1 s
4. 0.33 s, 3.0 Hz

12-3
Properties of waves

Wave motion

Purpose Distinguish between wave motion and particle vibration.

Materials long, coiled spring

Procedure Stretch the spring about 5–10 m. Have a student hold one end of the spring securely. As you hold the other end, send a single pulse along the spring by quickly jerking the spring sideways and then back to its equilibrium position. This will produce a transverse pulse. Have students note that the wave pulse is moving from you toward the student and that no part of the spring is being carried from you to the student. Emphasize this by having a student attach a small piece of paper to one of the coils of the spring.

Then send another pulse along the spring and show that while the wave moves along the spring, the coil returns to its original position. Tell students that this is an example of a mechanical wave whose medium is the spring.

⭐ TEKS

p. 452: 8A
p. 453: 8A

12-3 SECTION OBJECTIVES

- **Distinguish local particle vibrations from overall wave motion.**
- **Differentiate between pulse waves and periodic waves.**
- **Interpret waveforms of transverse and longitudinal waves.**
- **Apply the relationship among wave speed, frequency, and wavelength to solve problems.**
- **Relate energy and amplitude.**

Figure 12-10
A pebble dropped into a pond creates ripple waves similar to those shown here.

medium

the material through which a disturbance travels

mechanical wave

a wave that propagates through a deformable, elastic medium

12-3
Properties of waves

WAVE MOTION

Consider what happens to the surface of a pond when you drop a pebble into the water. The disturbance created by the pebble generates water waves that travel away from the disturbance, as seen in **Figure 12-10.** If you examined the motion of a leaf floating near the disturbance, you would see that the leaf moves up and down and back and forth about its original position. However, the leaf does not undergo any net displacement from the motion of the waves.

The leaf's motion indicates the motion of the particles in the water. The water molecules move locally, like the leaf does, but they do not travel across the pond. That is, the water wave moves from one place to another, but the water itself is not carried with it. ⭐TEKS **8A**

A wave is the motion of a disturbance

Ripple waves in a pond start with a disturbance at some point in the water. This disturbance causes water on the surface near that point to move, which in turn causes points farther away to move. In this way, the waves travel outward in a circular pattern away from the original disturbance.

In this example, the water in the pond is the **medium** through which the disturbance travels. Particles in the medium—in this case, water molecules—vibrate up and down as waves pass. Note that the medium does not actually travel with the waves. After the waves have passed, the water returns to its original position.

Waves of almost every kind require a material medium in which to travel. Sound waves, for example, cannot travel through outer space, because space is very nearly a vacuum. In order for sound waves to travel, they must have a medium such as air or water. Waves that require a material medium are called **mechanical waves.**

Not all wave propagation requires a medium. Electromagnetic waves, such as visible light, radio waves, microwaves, and X rays, can travel through a vacuum. You will study electromagnetic waves in Chapters 14–16.

WAVE TYPES

One of the simplest ways to demonstrate wave motion is to flip one end of a taut rope whose opposite end is fixed, as shown in **Figure 12-11.** The flip of your wrist creates a pulse that travels to the fixed end with a definite speed. A wave that consists of a single traveling pulse is called a **pulse wave.**

Now imagine that you continue to generate pulses at one end of the rope. Together, these pulses form what is called a **periodic wave.** Whenever the source of a wave's motion is a periodic motion, such as the motion of your hand moving up and down repeatedly, a periodic wave is produced.

pulse wave

a single, nonperiodic disturbance

periodic wave

a wave whose source is some form of periodic motion

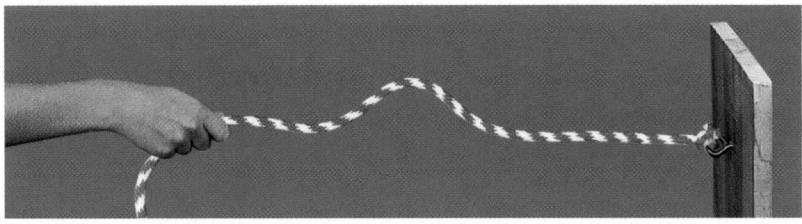

Figure 12-11
A single flip of a wrist on a taut rope creates a pulse wave.

Sine waves describe particles vibrating with simple harmonic motion

Figure 12-12 depicts a periodic wave on a string whose source is a blade vibrating with simple harmonic motion. As the wave travels to the right, any single point on the string vibrates up and down. Because the blade is vibrating with simple harmonic motion, the vibration of each point of the string is also simple harmonic. A wave whose source vibrates with simple harmonic motion is called a *sine wave.* Thus, a sine wave is a special case of a periodic wave in which the periodic motion is simple harmonic. The wave in **Figure 12-12** is called a sine wave because a graph of the trigonometric function $y = \sin x$ produces this curve when plotted. ⭐(TEKS) **8A**

Imagine that each particle of the string shown in **Figure 12-12** is a mass attached to a vibrating spring. As the wave travels to the right, each spring would vibrate around its equilibrium position with simple harmonic motion. This relationship between simple harmonic motion and wave motion enables us to use some of the terms and concepts from simple harmonic motion in our study of wave motion.

TOPIC: Wave motion
GO TO: www.scilinks.org
*sci*LINKS CODE: HF2123

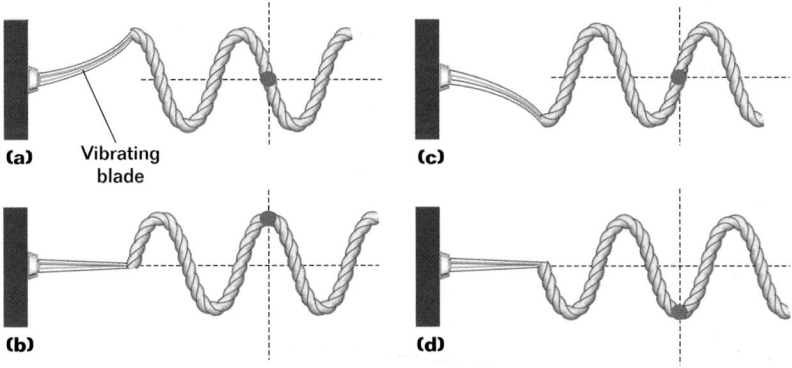

(a) Vibrating blade
(b)
(c)
(d)

Figure 12-12
As the sine wave created by this vibrating blade travels to the right, a single point on the string vibrates up and down with simple harmonic motion.

Demonstration 7

Transverse waves

Purpose Demonstrate that in a transverse pulse, particle vibration and wave motion are perpendicular to each other.

Materials long, coiled spring

Procedure Generate a pulse as you did in Demonstration 6. Have the students point in the direction in which the spring is displaced *(perpendicular to the spring).* Then, have them point in the direction in which the wave moves along the spring *(parallel to the spring).* The students should see that the disturbance of the spring is perpendicular to the direction of the motion of the disturbance along the spring. In other words, the medium is displaced perpendicular to the direction of the motion of the pulse. Tell students this is an example of a transverse pulse.

Visual Strategy

Figure 12-12

Ask the students to visualize a particle attached to the red point on the wave. As the wave travels, the particle would, like a mass on a spring, vibrate with simple harmonic motion.

Q For each component of the figure, determine in what part of its cycle the vibrating particle would be.

A (a) *equilibrium*
(b) *maximum displacement*
(c) *equilibrium*
(d) *maximum displacement*

Figure 12-13

Make sure students understand the meanings of displacement, amplitude, and wavelength.

Q Ask students to sketch graphs describing the waves that will be produced when the hand shaking the rope makes the following changes:

(a) makes motions that are twice as wide

(b) starts shaking the rope in the opposite direction.

Compare the displacements, amplitudes, and wavelengths of each graph with those in the original case (**Figure 12-13**).

A *(a) The wavelength doesn't change, the amplitude is twice the original, and the displacement is twice the original at each point. Thus, the crests of this graph are twice as high and the troughs are twice as low, but they occur at the same positions along the x-axis.*

(b) The amplitude and wavelength are the same as the original, but the displacements have opposite signs. This graph is a mirror image of the original one.

---

transverse wave

a wave whose particles vibrate perpendicularly to the direction of wave motion

crest

the highest point above the equilibrium position

trough

the lowest point below the equilibrium position

wavelength

the distance between two adjacent similar points of the wave, such as from crest to crest or from trough to trough

Figure 12-13
(a) A picture of a transverse wave at some instant t can be turned into **(b)** a graph. The x-axis represents the equilibrium position of the string. The curve shows the displacements of the string at time t.

Vibrations of a transverse wave are perpendicular to the wave motion

Figure 12-13(a) is a representation of the wave shown in **Figure 12-12** (on page 453) at a specific instant of time, *t*. This wave travels to the right as the particles of the rope vibrate up and down. Thus, the vibrations are perpendicular to the direction of the wave's motion. A wave such as this, in which the particles of the disturbed medium move perpendicularly to the wave motion, is called a **transverse wave.**

The wave shown in **Figure 12-13(a)** can be placed on a coordinate system, as shown in **Figure 12-13(b)**. A picture of a wave like the one in **Figure 12-13(b)** is sometimes called a *waveform*. A waveform can represent either the displacements of each point of the wave at a single moment in time or the displacements of a single particle as time passes. ⊛TEKS **4A**

In this case, the waveform depicts the displacements at a single instant. The *x*-axis represents the equilibrium position of the string, and the *y* coordinates of the curve represent the displacement of each point of the string at time *t*. For example, points where the curve crosses the *x*-axis (where $y = 0$) have zero displacement. Conversely, at the highest and lowest points of the curve, where displacement is greatest, the absolute values of *y* are greatest.

A wave is measured in terms of both its displacement from equilibrium and the distance between neighboring displacements. The highest point above the equilibrium position is called the wave **crest.** The lowest point below the equilibrium position is the **trough** of the wave. As in simple harmonic motion, amplitude is a measure of maximum displacement from equilibrium. Since the *x*-axis represents the equilibrium position of the string, the amplitude of a wave is the distance from the equilibrium position to a crest or to a trough, as shown in **Figure 12-13(b)**. ⊛TEKS **8A**

Notice that there is a series of crests and troughs in the waveform shown in **Figure 12-13(b)** and that the distance between adjacent crests or troughs is always the same. Thus, a wave can be thought of as a cyclical motion. A particle in the wave is displaced first in one direction, then in the other direction, finally returning to its original equilibrium position ready to repeat the cycle. The distance the wave travels during one cycle is called the **wavelength,** λ (the Greek letter *lambda*). This is also the distance between neighboring maximum positive (or negative) displacements. The most practical way to find the wavelength is to measure the distance between two adjacent similar points of the wave, such as from crest to crest or from trough to trough.

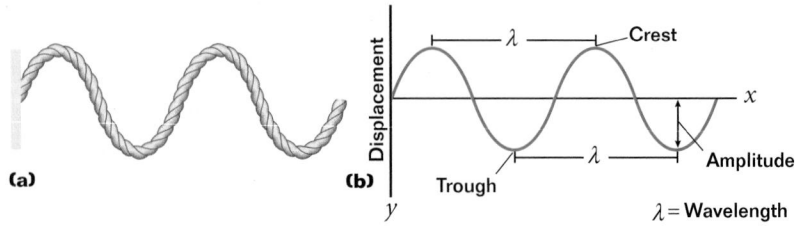

(a)

(b) λ = Wavelength

Vibrations of a longitudinal wave are parallel to the wave motion

You can create another type of wave with a spring. Suppose that one end of the spring is fixed and that the free end is pumped back and forth along the length of the spring, as shown in **Figure 12-14.** This action produces compressed and stretched regions of the coil that travel along the spring. The displacement of the coils is in the direction of wave motion. In other words, the vibrations are parallel to the motion of the wave. (★)TEKS **8A**

Compressed   Stretched   Compressed   Stretched

When the particles of the medium vibrate parallel to the direction of wave motion, the wave is called a **longitudinal wave.** Sound waves in the air are longitudinal waves because air particles vibrate back and forth in a direction parallel to the direction of wave motion.

A longitudinal wave can also be described by a sine curve. Consider a longitudinal wave traveling on a spring. **Figure 12-15(a)** is a snapshot of the longitudinal wave at some instant t, and **Figure 12-15(b)** shows the sine curve representing the wave. The compressed regions correspond to the crests of the waveform, and the stretched regions correspond to troughs. (★)TEKS **4A**

The type of wave represented by the curve in **Figure 12-15(b)** is often called a *density wave* or a *pressure wave* because the crests, where the spring coils are compressed, are regions of high density and pressure. Conversely, the troughs, where the coils are stretched, are regions of low density and pressure.

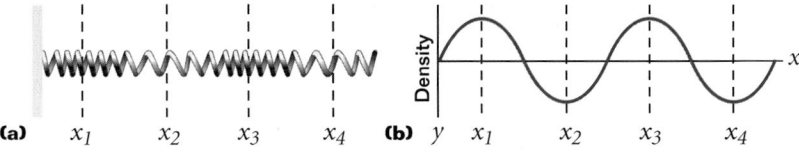

(a) x_1 x_2 x_3 x_4 (b) y x_1 x_2 x_3 x_4

PERIOD, FREQUENCY, AND WAVE SPEED

The up-and-down movement of your hand or a vibrating blade can generate waves on a string. Sound waves may begin, for example, with the vibrations of your vocal cords, a guitar string, or a taut drumhead. In each of these cases, the source of wave motion is a vibrating object. (★)TEKS **8A**

The vibrating object that causes a sine wave always has a characteristic frequency. Because this motion is transferred to the particles in the wave, the frequency of vibration of the particles is equal to the frequency of the source. When the vibrating particles of the medium complete one full cycle, one complete wavelength passes any given point. Thus, wave frequency describes the number of crests or troughs that pass a given point in a unit of time.

Figure 12-14
As this wave travels to the right, the coils of the spring are tighter in some regions and looser in others. The displacement of the coils is parallel to the direction of wave motion, so this wave is longitudinal.

longitudinal wave

a wave whose particles vibrate parallel to the direction of wave motion

Figure 12-15
(a) A longitudinal wave at some instant t can also be represented by **(b)** a graph. The crests of this waveform correspond to compressed regions, and the troughs correspond to stretched regions.

Demonstration 8

Longitudinal waves

Purpose Demonstrate that in a longitudinal pulse, particle vibration and wave motion are parallel.
Materials long, coiled spring
Procedure With the spring lying flat on the floor, compress approximately 10 cm of the coil. Instruct the students to observe the spring and to listen carefully as you release the pulse. Ask them to indicate the direction of the displacement of the spring and the direction in which the disturbance moved along the spring.

The students should see and hear that the displacement of the spring and the motion of the displacement along the spring are in the same direction. In this case, the direction in which the medium is disturbed is the same as the direction in which the disturbance moves through the medium. This is an example of a longitudinal pulse.

 Misconception Alert

Some students may confuse the graph of a transverse pulse, **Figure 12-13(b),** with the graph of a longitudinal pulse, **Figure 12-15(b).** Point out that in the first case, the *y*-axis represents *displacement*, while in the latter case, the *y*-axis represents *density*. Although the graphs look similar, this difference must be kept in mind when interpreting the two different kinds of graphs.

Demonstration 9

Amplitude, wavelength, and wave speed

Purpose Show that wave speed is independent of amplitude and wavelength.

Materials long, coiled spring and a clock or stopwatch

Procedure Generate a transverse pulse with a small amplitude, and have a student record the time it takes for the pulse to travel the length of the spring. Repeat this process for a pulse with a larger amplitude. Have students compare the two times *(they should be approximately equal)* and ask what conclusion they can draw from this observation *(that wave speed is independent of amplitude).* Repeat the process, but vary the wavelength rather than the amplitude by varying the frequency. Have students compare the times *(they should be approximately equal)*, and ask what conclusion they can draw from this observation *(that wave speed is independent of wavelength).*

Teaching Tip

Although this chapter primarily deals with mechanical waves, the wave-speed equation holds true for mechanical and electromagnetic waves, and both are included in practice problems.

Interactive Problem-Solving Tutor

See Module 12

"Wave Frequency and Wavelength" provides additional practice with the wave-speed equation.

Did you know?

The frequencies of sound waves audible to humans range from 20 Hz to 20 000 Hz. Electromagnetic waves, which include visible light, radio waves, and microwaves, have an even broader range of frequencies—from about 10 Hz to 10^{25} Hz and higher.

PHYSICS INTERACTIVE TUTOR

Module 12
"Wave Frequency and Wavelength" provides an interactive lesson with guided problem-solving practice to teach you about wave properties and the wave-speed equation.

The period of a wave is the amount of time required for one complete vibration of the particles of the medium. As the particles of the medium complete one full cycle of vibration at any point of the wave, one wavelength passes by that same point. Thus, the period of a wave describes the time it takes for a complete wavelength to pass a given point. The relationship between period and frequency seen earlier in this chapter holds true for waves as well; the period of a wave is inversely related to its frequency.

Wave speed equals frequency times wavelength ⭐TEKS 3B, 8A

We can now derive an expression for the speed of a wave in terms of its period or frequency. We know that speed is equal to displacement divided by the time it takes to undergo that displacement.

$$v = \frac{\Delta x}{\Delta t}$$

For waves, a displacement of one wavelength (λ) occurs in a time interval equal to one period of the vibration (T).

$$v = \frac{\lambda}{T}$$

As you saw earlier in this chapter, frequency and period are inversely related.

$$f = \frac{1}{T}$$

Substituting this frequency relationship into the previous equation for speed gives a new equation for the speed of a wave.

$$v = \frac{\lambda}{T} = f\lambda$$

SPEED OF A WAVE ⭐TEKS 3B, 8A

$$v = f\lambda$$

speed of a wave = frequency × wavelength

The speed of a mechanical wave is constant for any given medium. For example, at a concert, sound waves from different instruments reach your ears at the same moment, even when the frequencies of the sound waves are different. Thus, although the frequencies and wavelengths of the sounds produced by each instrument may be different, the product of the two is always the same at the same temperature. ⭐TEKS 3A, 8A, 8B

When a mechanical wave's frequency is increased, its wavelength must decrease in order for its speed to remain constant. The speed of a wave changes only when the wave moves from one medium to another or when certain properties of the medium are varied.

Wave speed (★ TEKS) 2C, 3B, 8A, 8B

PROBLEM

The piano string tuned to middle C vibrates with a frequency of 264 Hz. Assuming the speed of sound in air is 343 m/s, find the wavelength of the sound waves produced by the string.

SOLUTION

Given: $v = 343$ m/s $f = 264$ Hz

Unknown: $\lambda = ?$

Use the equation relating speed, wavelength, and frequency for a wave, given on page 456.

$$v = f\lambda$$

$$\lambda = \frac{v}{f} = \frac{343 \text{ m/s}}{264 \text{ Hz}} = \frac{343 \text{ m} \cdot \text{s}^{-1}}{264 \text{ s}^{-1}}$$

$$\boxed{\lambda = 1.30 \text{ m}}$$

Wave speed (★ TEKS) 2C, 3B, 8A, 8B

1. A piano emits frequencies that range from a low of about 28 Hz to a high of about 4200 Hz. Find the range of wavelengths in air attained by this instrument when the speed of sound in air is 340 m/s.

2. The speed of all electromagnetic waves in empty space is 3.00×10^8 m/s. Calculate the wavelength of electromagnetic waves emitted at the following frequencies:
   a. radio waves at 88.0 MHz
   b. visible light at 6.0×10^8 MHz
   c. X rays at 3.0×10^{12} MHz

3. The red light emitted by a He-Ne laser has a wavelength of 633 nm in air and travels at 3.00×10^8 m/s. Find the frequency of the laser light.

4. A tuning fork produces a sound with a frequency of 256 Hz and a wavelength in air of 1.35 m.
   a. What value does this give for the speed of sound in air?
   b. What would be the wavelength of the wave produced by this tuning fork in water in which sound travels at 1500 m/s?

SECTION 12-3

PRACTICE GUIDE 12D

Solving for:

λ	📖 **PE**	Sample, 1–2, 4b*; Ch. Rvw. 36, 51, 59*
	💿 **PW**	4
	PB	6, 7
f	📖 **PE**	3; Ch. Rvw. 48, 54
	💿 **PW**	Sample, 1–3
	PB	8–10
v	📖 **PE**	4a; Ch. Rvw. 57
	💿 **PW**	5, 6
	PB	Sample, 1–5

ANSWERS TO

Practice 12D
Wave speed
1. $0.081 \text{ m} \leq \lambda \leq 12 \text{ m}$
2. **a.** 3.41 m
   b. 5.0×10^{-7} m
   c. 1.0×10^{-10} m
3. 4.74×10^{14} Hz
4. **a.** 346 m/s
   b. 5.86 m

(★ TEKS)

p. 456: 3B, 8A, 3B, 8A, 3A, 8A, 8B
p. 457: 2C, 3B, 8A, 8B, 2C, 3B, 8A, 8B

Section Review
ANSWERS

1. The disturbance moves, not the medium.

2. **a.** One portion of the spring should have a single compressed region and a single stretched region.
   b. The spring should have several compressed regions and several stretched regions.
   c. The spring should contain a single hump either above or below its equilibrium position.
   d. The spring should contain several humps above and below its equilibrium position.

3. The graph for **(b)** should look like **Figure 12-13(b)** with the *y*-axis labeled *density*. The graph for **(d)** should resemble **Figure 12-13(b)**.

4. The energy will be 16 times as great.

5. 6.0×10^4 Hz

Waves transfer energy

When a pebble is dropped into a pond, the water wave that is produced carries a certain amount of energy. As the wave spreads to other parts of the pond, the energy likewise moves across the pond. Thus, the wave transfers energy from one place in the pond to another while the water remains in essentially the same place. In other words, waves transfer energy by transferring the *motion* of matter rather than by transferring matter itself. For this reason, waves are often able to transport energy efficiently. (★)**TEKS** **8A**

The rate at which a wave transfers energy depends on the amplitude at which the particles of the medium are vibrating. The greater the amplitude, the more energy a wave carries in a given time interval. For a mechanical wave, the energy transferred is proportional to the square of the wave's amplitude. When the amplitude of a mechanical wave is doubled, the energy it carries in a given time interval increases by a factor of four. Conversely, when the amplitude is halved, the energy decreases by a factor of four.

As with a mass-spring system or a simple pendulum, the amplitude of a wave gradually diminishes over time as its energy is dissipated. This effect, called *damping,* is usually minimal over relatively short distances. For simplicity, we have disregarded damping in our analysis of wave motions.

Section Review   (★)**TEKS** **2C, 2E, 3B, 8A, 8B**

1. As waves pass by a duck floating on a lake, the duck bobs up and down but remains in essentially one place. Explain why the duck is not carried along by the wave motion.

2. Sketch each of the following waves on a spring that is attached to a wall at one end:
   a. a pulse wave that is longitudinal
   b. a periodic wave that is longitudinal
   c. a pulse wave that is transverse
   d. a periodic wave that is transverse

3. Draw a graph for each of the waves described in items **(b)** and **(d)** above, and label the *y*-axis of each graph with the appropriate variable. Label the following on each graph: crest, trough, wavelength, and amplitude.

4. If the amplitude of a sound wave is increased by a factor of four, how does the energy carried by the sound wave in a given time interval change?

5. The smallest insects that a bat can detect are approximately the size of one wavelength of the sound the bat makes. What is the minimum frequency of sound waves required for the bat to detect an insect that is 0.57 cm long? (Assume the speed of sound is 340 m/s.)

12-4
Wave interactions

WAVE INTERFERENCE

When two bumper boats collide, as shown in **Figure 12-16,** each bounces back in another direction. The two bumper boats cannot occupy the same space, and so they are forced to change the direction of their motion. This is true not just of bumper boats but of all matter. Two different material objects can never occupy the same space at the same time.

When two waves come together, they do not bounce back as bumper boats do. If you listen carefully at a concert, you can distinguish the sounds of different instruments. Trumpet sounds are different from flute sounds, even when the two instruments are played at the same time. This means the sound waves of each instrument are unaffected by the other waves that are passing through the same space at the same moment. Because mechanical waves are not matter but rather are displacements of matter, two waves can occupy the same space at the same time. The combination of two overlapping waves is called *superposition*.

Figure 12-17 shows two sets of water waves in a ripple tank. As the waves move outward from their respective sources, they pass through one another. As they pass through one another, the waves interact to form an *interference pattern* of light and dark bands. Although this superposition of mechanical waves is fairly easy to observe, these are not the only kind of waves that can pass through the same space at the same time. Visible light and other forms of electromagnetic radiation also undergo superposition, and can interact to perform interference patterns. (★)TEKS **8A, 8B**

12-4 SECTION OBJECTIVES

- **Apply the superposition principle.**
- **Differentiate between constructive and destructive interference.**
- **Predict when a reflected wave will be inverted.**
- **Predict whether specific traveling waves will produce a standing wave.**
- **Identify nodes and antinodes of a standing wave.**

Figure 12-16
Two of these bumper boats cannot be in the same place at one time. Waves, on the other hand, can pass through one another.

Figure 12-17
This ripple tank demonstrates the interference of water waves.

**Waves passing
each other**

Purpose Show that wave pulses
are unaffected after they pass
through one another.

Materials long, coiled spring

Procedure Repeat Demonstration 10, but in this case the two
pulses should have opposite displacements. Have the students
observe the pulse that reaches
your hand after the two pulses
have passed through each other;
repeat this demonstration several
times. Next repeat the process for
two pulses of different amplitudes, and also repeat the process
for pulses with displacements on
the same side. Have the students
conclude that in all cases the two
pulses passed through each other
and were unaffected by the presence of the other pulse. Also have
students determine which examples were constructive *(displacements on the same side)* and
which were destructive *(displacements on opposite sides).*

Displacements in the same direction produce constructive interference

In **Figure 12-18(a),** two wave pulses are traveling toward each other on a
stretched rope. The larger pulse is moving to the right, while the smaller pulse
moves toward the left. At the moment the two wave pulses meet, a resultant
wave is formed, as shown in **Figure 12-18(b).**

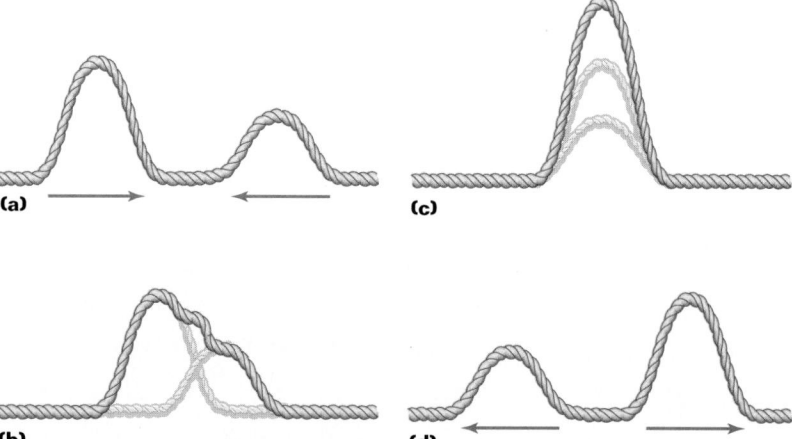

Figure 12-18
When these two wave pulses meet,
the displacements at each point
add up to form a resultant wave.
This is an example of constructive
interference.

At each point along the rope, the displacements due to the two pulses are
added together, and the result is the displacement of the resultant wave. For
example, when the two pulses exactly coincide, as they do in **Figure 12-18(c),**
the amplitude of the resultant wave is equal to the sum of the amplitudes of
each pulse. This method of summing the displacements of waves is known as
the *superposition principle.* According to this principle, when two or more
waves travel through a medium, the resultant wave is the sum of the displacements of the individual waves at each point. The superposition principle
holds true for all types of waves, both mechanical and electromagnetic. However, experiments show that the superposition principle is valid only when the
individual waves have small amplitudes of displacement—an assumption we
make in all our examples. ⭐TEKS **8A, 8B**

Notice that after the two pulses pass through each other, each pulse has the
same shape it had before the waves met and each is still traveling in the same
direction, as shown in **Figure 12-18(d).** This is true for sound waves at a concert, water waves in a pond, light waves, and other types of waves. Each wave
maintains its own characteristics after interference, just as the two pulses do in
our example above.

You have seen that when more than one wave travels through the same space,
the resultant wave is equal to the sum of the individual displacements. If the displacements are on the same side of equilibrium, as in **Figure 12-18,** they have
the same sign. When added together, the resultant wave is larger than the individual displacements. This is called **constructive interference.**

constructive interference

*interference in which individual
displacements on the same
side of the equilibrium position
are added together to form the
resultant wave*

Displacements in opposite directions produce destructive interference

What happens if the pulses are on opposite sides of the equilibrium position, as they are in **Figure 12-19(a)**? In this case, the displacements have different signs, one positive and one negative. When the positive and negative displacements are added, as shown in **Figure 12-19(b)** and **(c),** the resultant wave is the difference between the pulses. This is called **destructive interference.** After the pulses separate, their shapes are unchanged, as seen in **Figure 12-19(d).**

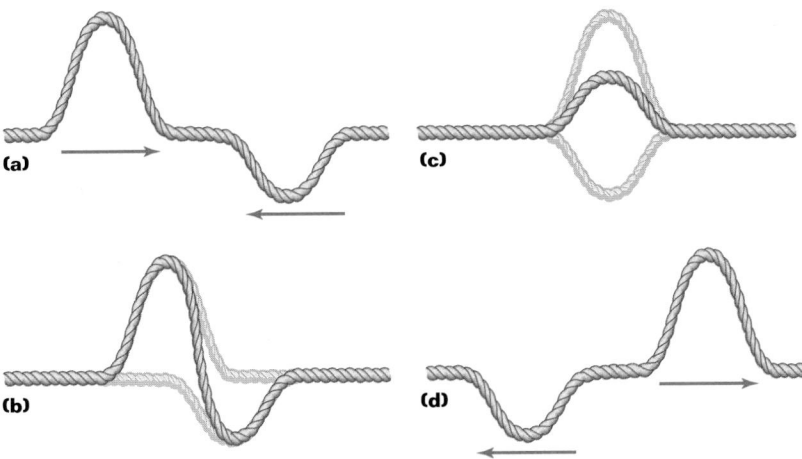

(a)

(c)

(b)

(d)

Figure 12-20 shows two pulses of equal amplitude but with displacements of opposite signs. When the two pulses coincide and the displacements are added, the resultant wave has a displacement of zero. In other words, the two pulses completely cancel each other; it is as if there were no disturbance at all at the instant the two pulses overlap. This situation is known as *complete destructive interference.*

If these waves were water waves coming together, one of the waves would be trying to pull an individual drop of water upward at the same instant and with the same force that another wave would be trying to pull it downward. The result would be no net force on the drop, and there would be no motion of the water at all at that moment.

Thus far, we have considered the interference produced by two transverse pulse waves. The superposition principle is valid for longitudinal waves as well. A *compression* involves a force on a particle in one direction, while a *rarefaction* involves a force on the same particle in the opposite direction. Hence, when a compression and a rarefaction interfere, there is destructive interference, and the net force on the particle is reduced.

In our examples, we have considered constructive and destructive interference separately, and we have dealt only with pulse waves. With periodic waves, complicated patterns arise that involve regions of constructive and destructive interference at different points, and the locations of these regions may vary with time as the individual waves travel. ⭐TEKS **8A**

destructive interference

interference in which individual displacements on opposite sides of the equilibrium position are added together to form the resultant wave

Figure 12-19
In this case, known as destructive interference, the displacement of one pulse is subtracted from the displacement of the other.

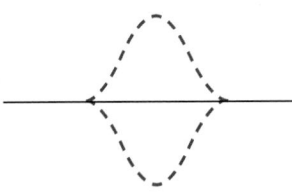

Figure 12-20
The resultant displacement at each point of the string is zero, so the two pulses cancel one another. This is complete destructive interference.

Misconception Alert

Students may believe that a new wave is created that replaces the original two wave pulses. Common sense and daily experience with collisions that *do* affect the objects involved support this preconception.

The *resultant wave* occurs only at the time and place that the waves meet each other. Each wave keeps its characteristics and continues with its original speed and wavelength after the encounter.

⭐TEKS
p. 460: 8A, 8B
p. 461: 8A

REFLECTION

In our discussion of waves so far, we have assumed that the waves being analyzed could travel indefinitely without striking anything that would stop them or otherwise change their motion. But what happens to the motion of a wave when it reaches a boundary?

At a free boundary, waves are reflected

Consider a pulse wave traveling on a stretched rope whose end forms a ring around a post, as shown in **Figure 12-21(a).** We will assume that the ring is free to slide along the post without friction.

As the pulse travels to the right, each point of the rope moves up once and then back down. When the pulse reaches the boundary, the rope is free to move up as usual, and it pulls the ring up with it. Then the ring falls back down to its previous level. The movement of the rope at the post is similar to the movement that would result if someone were to whip the rope upward to send a pulse to the left, which will cause a pulse to travel back along the rope to the left. This is called *reflection.* Note that the reflected pulse is identical to the incident pulse. ⭐TEKS **8A**

At a fixed boundary, waves are reflected and inverted

Now consider a pulse traveling on a stretched rope that is fixed at one end, as in **Figure 12-21(b).** When the pulse reaches the wall, the rope exerts an upward force on the wall, and the wall in turn exerts an equal and opposite reaction force on the rope. This downward force on the rope causes a displacement in the direction opposite the displacement of the original pulse. As a result, the pulse is inverted after reflection.

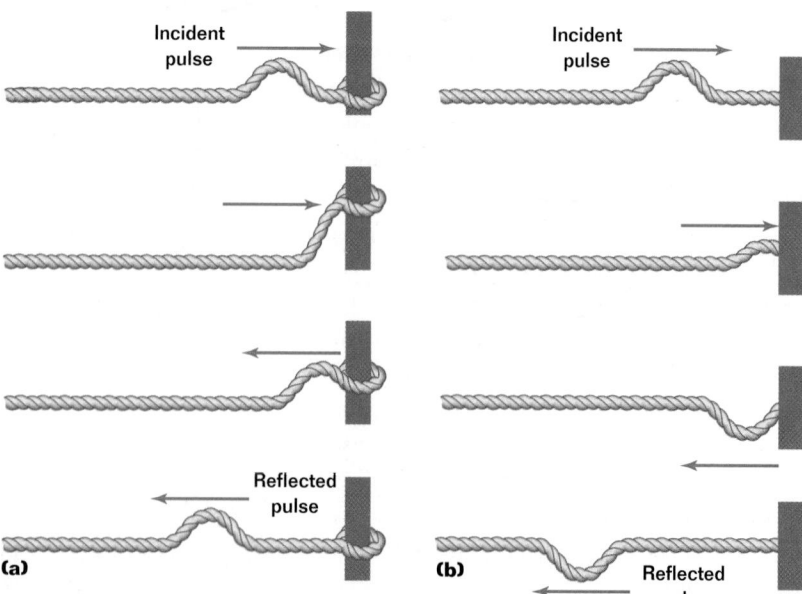

Figure 12-21

(a) When a pulse travels down a rope whose end is free to slide up the post, the pulse is reflected from the free end. **(b)** When a pulse travels down a rope that is fixed at one end, the reflected pulse is inverted.

STANDING WAVES

Consider a string that is attached on one end to a rigid support and that is shaken up and down in a regular motion at the other end. The regular motion produces waves of a certain frequency, wavelength, and amplitude traveling down the string. When the waves reach the other end, they are reflected back toward the oncoming waves. If the string is vibrated at exactly the right frequency, a **standing wave**—a resultant wave pattern that does not move along the string—is produced. The standing wave consists of alternating regions of constructive and destructive interference. ⭐TEKS **8A**

Standing waves have nodes and antinodes

Figure 12-22(a) shows four possible standing waves for a given string length. The points at which the two waves cancel are called **nodes.** There is no motion in the string at the nodes, but midway between two adjacent nodes, the string vibrates with the largest amplitude. These points are called **antinodes.**

Figure 12-22(b) shows the oscillation of the second case shown in **Figure 12-22(a)** during half a cycle. All points on the string oscillate vertically with the same frequency, except for the nodes, which are stationary. In this case, there are three nodes (N) and two antinodes (A), as illustrated in the figure. Notice that different points of the string have different amplitudes.

standing wave

> a wave pattern that results when two waves of the same frequency, wavelength, and amplitude travel in opposite directions and interfere

node

> a point in a standing wave that always undergoes complete destructive interference and therefore is stationary

antinode

> a point in a standing wave, halfway between two nodes, at which the largest amplitude occurs

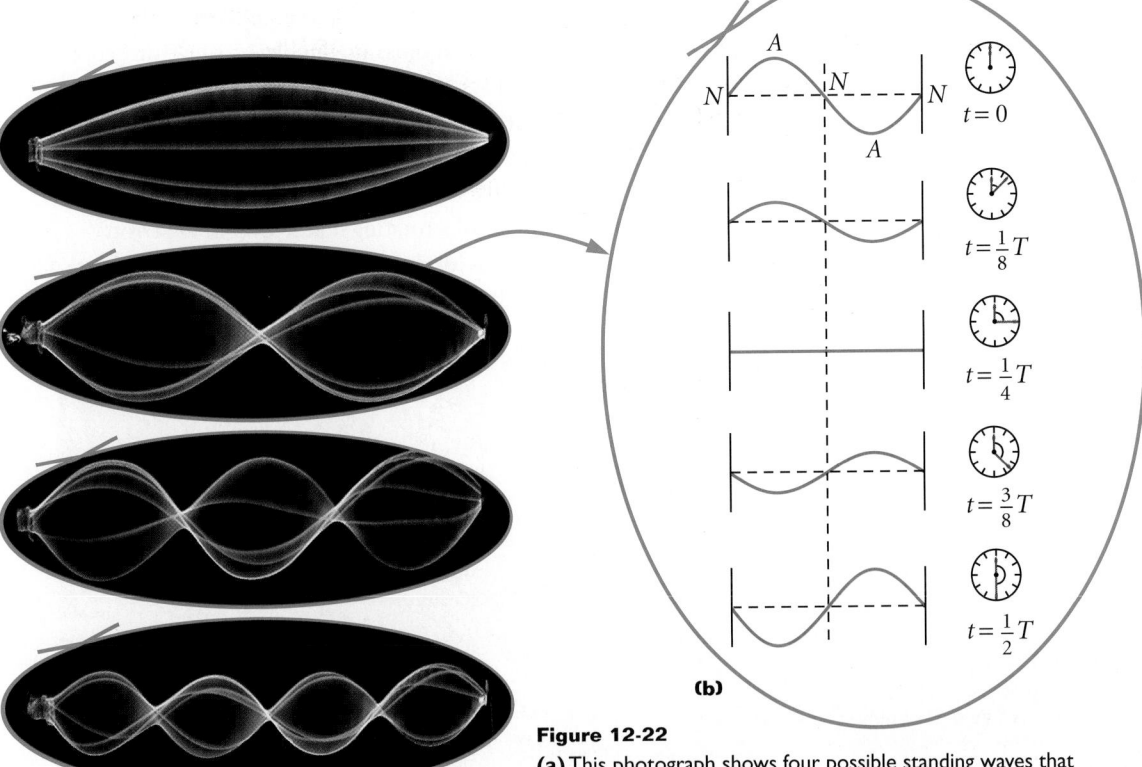

(b)

Figure 12-22

(a) This photograph shows four possible standing waves that can exist on a given string. **(b)** The diagram shows the progression of the second standing wave for one-half of a cycle.

(a)

The Language of Physics

The term *standing wave* may mislead students. Point out that the individual waves that compose standing waves are actually *traveling* waves. It is only the resultant wave, which is the superposition of various individual traveling waves, that appears to stand still.

Visual Strategy

Figure 12-22

Make sure students understand that the waves in **Figure 12-22(a)** are examples of different possible standing waves for a given string length, while the diagrams in **Figure 12-22(b)** represent the vibrations of just one of these standing waves. Specifically, the diagram corresponds to the standing wave shown in the second photograph from the top.

Q Have students draw a schematic diagram, like the one shown in **Figure 12-22(b),** for the wave shown in the top photograph, in which the wavelength equals twice the string length.

A *Students' diagrams should look like the left half (left of the dotted vertical line) of Figure 12-22(b).*

⭐TEKS

p. 462: 8A
p. 463: 8A

Tomorrow's Technology

BACKGROUND

The acoustic levitation device described here demonstrates that particles within a compressible fluid are displaced at all points on a wave except the pressure nodes. This device dramatically demonstrates that the movement of particles results in a force that not only allows us to hear sounds by pressing on our eardrums but also can be used to levitate matter.

Most of Jet Propulsion Labs' (JPL) work focused on levitating objects at the center of the chamber using the fundamental frequency of the chamber, where the wavelength of the sound waves equals the width of the device (one node). However, JPL has done work with other wavelengths, where the object was successfully levitated at a different node.

JPL is no longer doing this research on acoustic levitation. A company in Canada, however, is doing a lot of work, and they are developing a beam levitator that focuses 20 beams of sound on the sample to be levitated. Research is also being conducted on a silent levitation system that uses sound at 20 kHz, just outside the range of human hearing.

⭐TEKS

p. 464: 3C, 8A, 8B, 8C
p. 465: 2C, 3B, 8A

Tomorrow's Technology

Acoustic Levitation ⭐TEKS 3C, 8A, 8B, 8C

In the movies, levitating objects are the products of a hefty special-effects budget. For Dr. Martin Barmatz, a physicist at Jet Propulsion Laboratory, in Pasadena, California, all it takes is sound.

Barmatz studies the properties of a chamber in which lightweight objects can be levitated, rotated, and oscillated by using sound waves.

"It's a rectangular chamber," he said. "It has two equal sides and a slightly larger third side. High-powered speakers introduce sound energy through a hole in the center of each of the chamber's sides. Three standing waves, each one perpendicular to the others, are produced within the chamber."

"The horizontal standing waves center the object along the vertical axis of the chamber. Then, the force of the vertical standing wave levitates the object from below," he said. These waves have a frequency of about 2000 Hz, well within the range of human hearing. Because the level of sound produced is around 160 decibels—loud enough to cause extreme pain to the ears—acoustic-levitation chambers must be very well insulated against sound leakage.

Barmatz said that this technology was originally designed for NASA as part of a space-

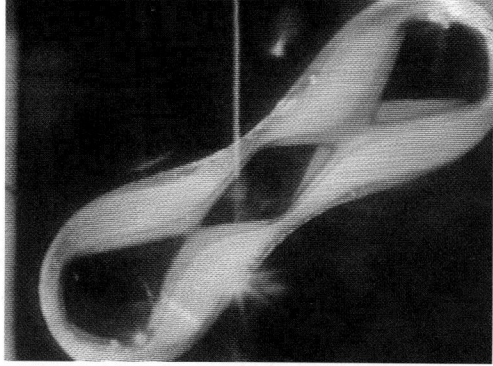

based furnace. In normal furnaces, substances being heated often become contaminated by their containers. By using these techniques in a space environment, however, the substances remain much purer because they are isolated from any container surfaces.

Acoustic levitation is used to study the properties of liquids and bubbles. A single drop of a liquid is isolated and levitated in the chamber. Then scientists study the path of bubbles in the liquid drop or the drop's behavior in a zero-gravity situation. The figure above shows an image of a rotating liquid drop in the Drop Physics Module, taken during the STS 50 Space Shuttle flight in June 1992 (shown below).

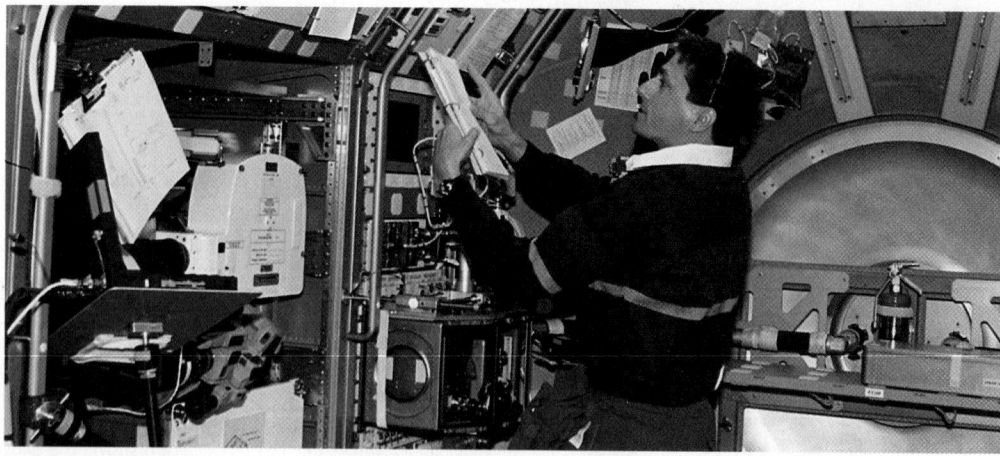

Only certain frequencies of vibration produce standing wave patterns. **Figure 12-23** shows different possible standing waves for a given string length. In each case, the curves represent the position of the string at different instants of time. If the string were vibrating rapidly, the several positions would blur together and give the appearance of loops, like those shown in the diagram. A single loop corresponds to either a crest or trough alone, while two loops correspond to a crest and a trough together, or one wavelength.

The ends of the string must be nodes because these points cannot vibrate. As you can see in **Figure 12-23,** standing waves can be produced for any wavelength that allows both ends of the string to be nodes. One possibility appears in **Figure 12-23(b).** In this case, each end is a node, and there are no nodes in between. Because a single loop corresponds to either a crest or trough alone, this standing wave corresponds to one-half of a wavelength. Thus, the wavelength in this case is equal to twice the string length ($2L$).

The next possible standing wave, shown in **Figure 12-23(c),** has three nodes: one at either end and one in the middle. In this case there are two loops, corresponding to a crest and a trough. Thus, this standing wave has a wavelength equal to the string length (L). The next case, shown in **Figure 12-23(d),** has a wavelength equal to two-thirds of the string length $\left(\frac{2}{3}L\right)$, and the pattern continues. Wavelengths between the values shown here do not produce standing waves because they do not allow both ends of the string to be nodes.

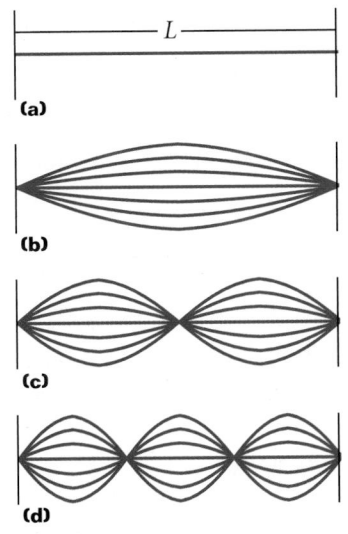

(a)
(b)
(c)
(d)

Figure 12-23

Only certain frequencies produce standing waves on this fixed string. The wavelength of these standing waves depends on the string length. Possible wavelengths include 2L **(b),** L **(c),** and $\frac{2}{3}$ L **(d).**

Section Review

⭐TEKS **2C, 3B, 8A**

1. A wave of amplitude 0.30 m interferes with a second wave of amplitude 0.20 m. What is the largest resultant displacement that may occur?

2. A string is rigidly attached to a post at one end. Several pulses of amplitude 0.15 m sent down the string are reflected at the post and travel back down the string without a loss of amplitude. What is the amplitude at a point on the string where the maximum displacement points of two pulses cross? What type of interference is this?

3. How would your answer to item 2 change if the same pulses were sent down a string whose end is free? What type of interference is this?

4. Look at the standing wave shown in **Figure 12-24.** How many nodes does this wave have? How many antinodes?

5. A stretched string fixed at both ends is 2.0 m long. What are three wavelengths that will produce standing waves on this string? Name at least one wavelength that would not produce a standing wave pattern, and explain your answer.

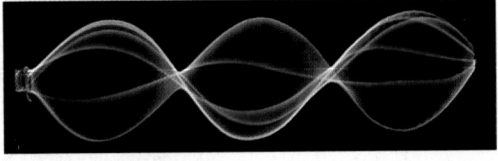

Figure 12-24

BACKGROUND

The last two sections of Chapter 12 primarily focused on wave motions that students are familiar with, such as waves on ropes and on springs. This feature extends wave concepts with a discussion of de Broglie waves, also known as matter waves. Matter waves are discussed further in Chapter 23, but the basic concepts can be understood at this time and are often very interesting to students.

Louis de Broglie, a French physicist, was awarded the Nobel Prize in 1929 for his prediction of the wave nature of material particles. De Broglie originally proposed this theory in his doctoral thesis in 1924.

The de Broglie family was an aristocratic French family, and Louis de Broglie held the title of Prince. De Broglie came late to the study of theoretical physics because he first studied history. Only after serving as a radio operator in World War I did he begin his study of physics.

De Broglie later found a connection between the wave nature of electrons and the Bohr model of hydrogen, in which only certain orbits of the electron are stable. De Broglie postulated that the stable orbits are those that contain an integral number of wavelengths; this is analogous to the standing waves that can exist on a vibrating string, as discussed in Section 12-4. This connection showed that the stable orbits allowed in Bohr's model are a result of the interference patterns of the electrons. (This is discussed in Chapter 23, where the Bohr model is introduced.)

De BROGLIE WAVES

Earlier in this chapter, we treated waves and particles as if there were a clear distinction between the two. For most of the history of science, this was believed to be the case. However, in the early twentieth century, scientists were confronted with experimental evidence suggesting that the distinction between waves and particles is not as clear-cut as everyone had assumed.

The dual nature of light ⭐TEKS 3E

This scientific revolution began in 1900, when Max Planck introduced the possibility that energy could come in discrete units. In 1905, Einstein extended Planck's theory, suggesting that all electromagnetic waves (such as light) sometimes behave like particles. According to this theory, light is both a wave and a particle; some experiments reveal its wave nature, and other experiments display its particle nature. Although this idea was initially greeted with skepticism, it explained certain phenomena that the wave theory of light could not account for and was soon confirmed empirically in a variety of experiments.

Matter waves

The idea that light has a dual nature led Louis de Broglie to propose in his doctoral thesis that perhaps all matter has wavelike characteristics. De Broglie's thesis, submitted in 1924, was based on his belief that there should not be two separate branches of physics, one for electromagnetic waves and another for matter. At that time, there was no experimental evidence to support his theory.

De Broglie's calculations suggested that matter waves had a wavelength, λ, often called the de Broglie wavelength, given by the following equation:

$$\lambda = \frac{h}{p} = \frac{h}{mv}$$

The variable h in this equation is called Planck's constant, which is approximately equal to 6.63×10^{-34} J•s, and p is the object's momentum, which is equivalent to its mass, m, times its velocity, v. Note that the dual nature of matter suggested by de Broglie is seen by this equation, which includes both a wave concept (λ) and a particle concept (mv). ⭐TEKS 3B

De Broglie's equation shows that the smaller the momentum of an object, the larger its de Broglie wavelength. But even when the momentum of an object is very small from our perspective, h is so small that the wavelength is still much too small for us to detect. In order to detect wavelengths this small, one must use an opening equal to or smaller than the wavelength because waves passing through such an opening will display patterns of constructive and destructive interference. When the opening is much larger than the wavelength, waves travel through it without being affected.

The de Broglie wavelength of a 0.15 kg baseball moving at 30 m/s is about 1.5×10^{-34} m. This is almost a trillion trillion times smaller than the diameter of a typical air molecule—much smaller than any possible opening through which we could observe interference effects. This explains why the de Broglie wavelength of objects cannot be observed in our everyday experience.

However, in the microscopic world, the wave effects of matter can be observed. The distance between atoms in a crystal is about 10^{-10} m. Electrons ($m = 9.109 \times 10^{-31}$ kg) accelerated to a speed of 1.4×10^{7} m/s have a de Broglie wavelength of approximately this size. Thus, the atoms in a crystal can act as a three-dimensional grating that should diffract electron waves. Such an experiment was performed three years after de Broglie's thesis by Clinton J. Davisson and Lester H. Germer, and the electrons did create patterns of constructive and destructive interference. This experiment gave confirmation of de Broglie's theory of the dual nature of matter.

The electron microscope   (★)TEKS 8C

A practical device that relies on the wave characteristics of matter is the electron microscope. In principle, the electron microscope is similar to an ordinary compound microscope. But while ordinary microscopes use lenses to bend rays of light that are reflected from a small object, electron microscopes use electric and magnetic fields to accelerate and focus a beam of electrons. Rather than examining the image through an eyepiece, as in an ordinary microscope, a magnetic lens forms an image on a fluorescent screen. Without the fluorescent screen, the image would not be visible.

Electron microscopes are able to distinguish details about 100 times smaller than optical microscopes. Because of their great resolving power, electron microscopes are widely used in many areas of scientific research.

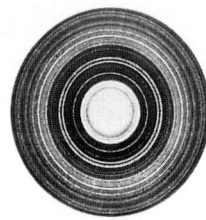

Figure 12-25

In this photograph, electron waves are diffracted by a crystal. Experiments such as this show the wave nature of electrons and thereby provide empirical evidence for de Broglie's theory of the dual nature of matter.

(★)TEKS **3C, 3E, 8A**

Figure 12-26

This image of cat hairs, produced by an electron microscope, is magnified 500 times.

internet connect

SCILINKS

TOPIC: Electron microscope
GO TO: www.scilinks.org
NSTA   *sci*LINKS CODE: HF2124

EXTENSION

Have students research the electron microscope and its uses. In their research, ask students to compare the electron microscope with a typical optical microscope. Then have students research the various fields in which electron microscopes are used and what kinds of images they produce. Have students share their results with the class.

Teaching Tip

The equation for the de Broglie wavelength of an electron is derived from the equation for the momentum of a particle of light, also known as a photon ($p = h/\lambda$). De Broglie postulated that this equation might apply to the electron as well ($\lambda = h/p = h/mv$).

(★)TEKS

p. 466: 3E, 3B
p. 467: 3C, 3E, 8A, 8C

467

CHAPTER 12
Summary

Teaching Tip

Ask students to prepare a concept map of the chapter. The concept map should include most of the vocabulary terms, along with other integral terms or concepts.

⊛ **TEKS**

Review & Assess
pp. 113–119:
2A: Alt. Assess 1–2, 4
2B: Alt. Assess 1–2, 4
2C: 8–9, 19–22, 35–36,
    46–59, Technology
    & Learning
2D: Alt. Assess 3–4, 6
2F: Alt. Assess 1–2, 4
3A: Alt. Assess 5
3B: 8–9, 19–22, 35–36,
    46–59, Technology
    & Learning
4A: Technology &
    Learning
8A: 23–59
8B: 32, 50, 51

KEY TERMS

amplitude (p. 446)

antinode (p. 463)

constructive interference (p. 460)

crest (p. 454)

destructive interference (p. 461)

frequency (p. 446)

longitudinal wave (p. 455)

mechanical wave (p. 452)

medium (p. 452)

node (p. 463)

period (p. 446)

periodic wave (p. 453)

pulse wave (p. 453)

simple harmonic motion (p. 439)

standing wave (p. 463)

transverse wave (p. 454)

trough (p. 454)

wavelength (p. 454)

KEY IDEAS

Section 12-1 Simple harmonic motion

- In simple harmonic motion, restoring force is proportional to displacement.
- A mass-spring system vibrates with simple harmonic motion, and the spring force is given by Hooke's law: $\boxed{F_{elastic} = -kx}$
- For small angles of displacement (<15°), a pendulum swings with simple harmonic motion.
- In simple harmonic motion, restoring force and acceleration are maximum at maximum displacement and velocity is maximum at equilibrium.

Section 12-2 Measuring simple harmonic motion

- The period of a mass-spring system depends only on the mass and the spring constant. The period of a simple pendulum depends only on the string length and the free-fall acceleration.
- Frequency is the inverse of period.

Section 12-3 Properties of waves

- Wave particles vibrate around an equilibrium position as the wave travels.
- In a transverse wave, vibrations are *perpendicular* to the direction of wave motion. In a longitudinal wave, vibrations are *parallel* to the direction of wave motion.
- Wave speed equals frequency times wavelength: $\boxed{v = f\lambda}$

Section 12-4 Wave interactions

- If two or more waves are moving through a medium, the resultant wave is found by adding the individual displacements together point by point.
- Standing waves are formed when two waves having the same frequency, amplitude, and wavelength travel in opposite directions through a medium and interfere.

Variable symbols			
Quantities		**Units**	
$F_{elastic}$	spring force	N	newtons
k	spring constant	N/m	newtons/meter
T	period	s	seconds
f	frequency	Hz	hertz = s^{-1}
λ	wavelength	m	meters

CHAPTER 12
Review and Assess

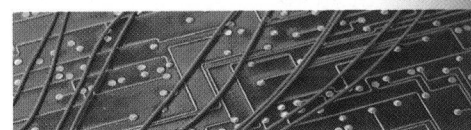

SIMPLE HARMONIC MOTION

Review questions

1. What characterizes an object's motion as simple harmonic?

2. List four examples of simple harmonic motion.

3. Does the acceleration of a simple harmonic oscillator remain constant during its motion? Is the acceleration ever zero? Explain.

4. A pendulum is released 40° from its resting position. Is its motion simple harmonic?

5. April is about to release the bob of a pendulum. Before she lets go, what sort of potential energy does the bob have? How does the energy of the bob change as it swings through one full cycle of motion?

Conceptual questions

6. An ideal mass-spring system vibrating with simple harmonic motion would oscillate indefinitely. Explain why.

7. In a simple pendulum, the weight of the bob can be divided into two components, one tangent to the bob's direction of motion and the other perpendicular to the bob's direction of motion. Which of these is the restoring force, and why?

Practice problems

8. Janet wants to find the spring constant of a given spring, so she hangs the spring vertically and attaches a 0.40 kg mass to the spring's other end. If the spring stretches 3.0 cm from its equilibrium position, what is the spring constant?
(See Sample Problem 12A.)

9. In preparing to shoot an arrow, an archer pulls a bow string back 0.40 m by exerting a force that increases uniformly from 0 to 230 N. What is the equivalent spring constant of the bow?
(See Sample Problem 12A.)

PERIOD AND FREQUENCY

Review questions

10. A child swings on a playground swing. How many times does the child swing through the swing's equilibrium position during the course of a single period of motion?

11. What is the total distance traveled by an object moving back and forth in simple harmonic motion in a time interval equal to its period when its amplitude is equal to A?

12. How is the period of a simple harmonic vibration related to its frequency?

Conceptual questions

13. What happens to the period of a simple pendulum when the pendulum's length is doubled? What happens when the suspended mass is doubled?

14. A pendulum bob is made with a ball filled with water. What would happen to the frequency of vibration of this pendulum if a hole in the ball allowed water to slowly leak out?

15. If a pendulum clock keeps perfect time at the base of a mountain, will it also keep perfect time when moved to the top of the mountain? Explain.

16. If a grandfather clock is running slow, how can you adjust the length of the pendulum to correct the time?

17. A simple pendulum can be used as an altimeter on a plane. How will the period of the pendulum vary as the plane rises from the ground to its cruising altitude of 1.00×10^4 m?

18. Will the period of a vibrating mass-spring system on Earth be different from the period of an identical mass-spring system on the moon? Why or why not?

ANSWERS TO

Chapter 12
Review and Assess

1. oscillation about an equilibrium position in which a restoring force is proportional to displacement

2. mass-spring system, trapeze artists, wrecking ball, pendulum of a grandfather clock, metronome, vibrating guitar string

3. No, acceleration changes throughout the oscillator's motion; It is zero at equilibrium and greatest at maximum displacement.

4. No, a pendulum's displacement is proportional to its restoring force only at angles smaller than 15°.

5. gravitational potential energy; When April lets go of the bob, $PE = $ max and $KE = 0$; at the bottom of its swing, $KE = $ max and $PE = 0$.

6. because frictional forces are not accounted for in an ideal mass-spring system

7. the tangent component; because it always pulls the bob toward the equilibrium position

8. 130 N/m

9. 580 N/m

10. twice

11. $4A$

12. They are inversely related.

13. period is $\sqrt{2}$ times as long; remains the same because mass does not affect period

14. The frequency would decrease as the distance from the pivot to the ball's center of mass increased, then increase when the water level reached the halfway point. When the ball became empty, the frequency would return to its original value.

15. no; g would change slightly, so T would also change.

16. Make the pendulum shorter to decrease the period.

17. The period will increase as the altitude increases.

18. They will be the same because the period is independent of free-fall acceleration.

19. 9.7 m

20. 22.4 m

21. **a.** 2.000 s
    b. 9.812 m/s^2
    c. 9.798 m/s^2

22. **a.** 0.57 s
    b. 1.8 Hz

23. movement of a disturbance

24. Transverse wave particles vibrate perpendicular to wave motion. Longitudinal wave particles vibrate parallel to wave motion.

25. **a.** perpendicular to wave motion
    b. transverse

26. longitudinal

27. one wavelength

28. 1/3 s; 3 Hz

29. sound waves, water waves, waves on a spring; Light waves do not need a medium to move through, but mechanical waves do.

30. up and down, no horizontal movement

31. It becomes half as long; It stays the same.

32. because sound waves are vibrations of air particles; Without particles, no propagation occurs.

Practice problems

19. Find the length of a pendulum that oscillates with a frequency of 0.16 Hz.
    (See Sample Problem 12B.)

20. A visitor to a lighthouse wishes to determine the height of the tower. The visitor ties a spool of thread to a small rock to make a simple pendulum, then hangs the pendulum down a spiral staircase in the center of the tower. The period of oscillation is 9.49 s. What is the height of the tower?
    (See Sample Problem 12B.)

21. A pendulum that moves through its equilibrium position once every 1.000 s is sometimes called a "seconds pendulum."
    a. What is the period of any seconds pendulum?
    b. In Cambridge, England, a seconds pendulum is 0.9942 m long. What is the free-fall acceleration in Cambridge?
    c. In Tokyo, Japan, a seconds pendulum is 0.9927 m long. What is the free-fall acceleration in Tokyo?
    (See Sample Problem 12B.)

22. A spring with a spring constant of 1.8×10^2 N/m is attached to a 1.5 kg mass and then set in motion.
    a. What is the period of the mass-spring system?
    b. What is the frequency of the vibration?
    (See Sample Problem 12C.)

PROPERTIES OF WAVES

Review questions

23. What is common to all waves?

24. How do transverse and longitudinal waves differ?

25. **Figure 12-27** depicts a pulse wave traveling on a spring.
    a. In which direction are the particles of the medium vibrating?
    b. Is this wave transverse or longitudinal?

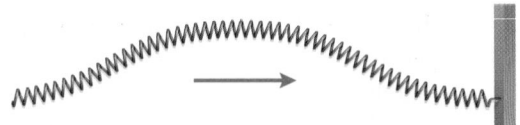

Figure 12-27

---

26. In a stretched spring, several coils are pinched together and others are spread farther apart than usual. What sort of wave is this?

27. How far does a wave travel in one period?

28. If you shook the end of a rope up and down three times each second, what would be the period of the waves set up in the rope? What would be the frequency?

29. Give three examples of mechanical waves. How are these different from electromagnetic waves, such as light waves?

Conceptual questions

30. How does a single point on a string move as a transverse wave passes by that point?

31. What happens to the wavelength of a wave on a string when the frequency is doubled? What happens to the speed of the wave?

32. Why do sound waves need a medium through which to travel?

33. Two tuning forks with frequencies of 256 Hz and 512 Hz are struck. Which of the sounds will move faster through the air?

34. What is one advantage of transferring energy by electromagnetic waves?

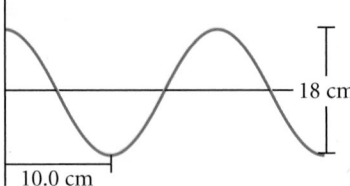

18 cm

10.0 cm

Figure 12-28

35. A wave traveling in the positive x direction with a frequency of 25.0 Hz is shown in **Figure 12-28** above. Find the following values for this wave:
    a. amplitude
    b. wavelength
    c. period
    d. speed

Practice problems

36. Microwaves travel at the speed of light, 3.00×10^8 m/s. When the frequency of microwaves is 9.00×10^9 Hz, what is their wavelength?
   (See Sample Problem 12D.)

WAVE INTERACTIONS

Review questions

37. Using the superposition principle, draw the resultant waves for each of the examples in **Figure 12-29.**

Figure 12-29

38. What is the difference between constructive interference and destructive interference?

39. Which waveform of those shown in **Figure 12-30** is the resultant waveform?

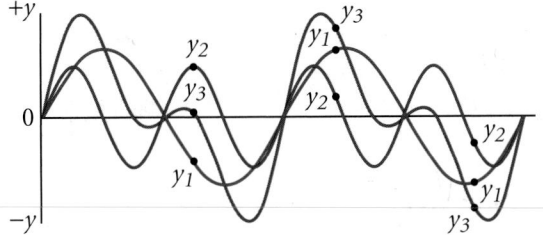

Figure 12-30

40. Anthony sends a series of pulses of amplitude 24 cm down a string that is attached to a post at one end. Assuming the pulses are reflected with no loss of amplitude, what is the amplitude at a point on the string where the two pulses are crossing if
   a. the string is rigidly attached to the post?
   b. the end at which reflection occurs is free to slide up and down?

41. A wave of amplitude 0.75 m interferes with a second wave of amplitude 0.53 m.
   a. Find the maximum possible amplitude of the resultant wave if the interference is constructive.
   b. Find the maximum possible amplitude of the resultant wave if the interference is destructive.

Conceptual questions

42. Can more than two waves interfere in a given medium?

43. What is the resultant displacement at a position where destructive interference is complete?

44. When two waves interfere, can the resultant wave be larger than either of the two original waves? If so, under what conditions?

45. Which of the following wavelengths will produce standing waves on a string that is 3.5 m long?
   a. 1.75 m
   b. 3.5 m
   c. 5.0 m
   d. 7.0 m

MIXED REVIEW

46. In an arcade game, a 0.12 kg disk is shot across a frictionless horizontal surface by being compressed against a spring and then released. If the spring has a spring constant of 230 N/m and is compressed from its equilibrium position by 6.0 cm, what is the magnitude of the spring force on the disk at the moment it is released?

47. A child's toy consists of a piece of plastic attached to a spring, as shown in **Figure 12-31.** The spring is compressed against the floor a distance of 2.0 cm and released. If the spring constant is 85 N/m, what is the magnitude of the spring force acting on the toy at the moment it is released?

Figure 12-31

48. Green light has a wavelength of 5.20×10^{-7} m and travels through the air at a speed of 3.00×10^8 m/s. Calculate the frequency and the period of green light waves with this wavelength.

49. You dip your finger into a pan of water twice each second, producing waves with crests that are separated by 0.15 m. Determine the frequency, period, and speed of these water waves.

33. neither, because the speed of sound is constant in air
34. They can transport large amounts of energy rapidly.
35. **a.** 9.0 cm
   b. 20.0 cm
   c. 0.0400 s
   d. 5.00 m/s
36. 0.0333 m
37. **a.** a sine wave with twice the amplitude
   b. a straight line (the waves cancel each other completely)
38. In constructive interference, individual displacements are on the same side of equilibrium. In destructive interference, the individual displacements are on opposite sides of equilibrium.
39. y_3
40. **a.** 0 cm
   b. 48 cm
41. **a.** 1.28 m
   b. 0.22 m
42. yes, because waves do not collide like other matter; They add to form a resultant wave.
43. zero
44. yes; when constructive interference occurs
45. a, b, d
46. 14 N
47. 1.7 N
48. 5.77×10^{14} Hz, 1.73×10^{-15} s
49. 2.0 Hz, 0.50 s, 0.30 m/s

50. A sound wave traveling at 343 m/s is emitted by the foghorn of a tugboat. An echo is heard 2.60 s later. How far away is the reflecting object?

51. The notes produced by a violin range in frequency from approximately 196 Hz to 2637 Hz. Find the possible range of wavelengths in air produced by this instrument when the speed of sound in air is 340 m/s.

52. What is the free-fall acceleration in a location where the period of a 0.850 m long pendulum is 1.86 s?

53. A mass-spring system oscillates with an amplitude of 3.5 cm. The spring constant is 250 N/m and the mass is 0.50 kg.

 a. Calculate the mechanical energy of the mass-spring system. Recall the potential energy of a mass-spring system: $PE_{elastic} = \frac{1}{2}kx^2$.
 b. Calculate the maximum acceleration of the mass-spring system.

54. A simple 2.00 m long pendulum oscillates in a location where $g = 9.80\ \text{m/s}^2$. How many complete oscillations does this pendulum make in 5.00 min?

Technology & Learning

Graphing calculators

Refer to Appendix B for instructions on downloading programs for your calculator. The program "Chap12" allows you to analyze a graph of period versus free-fall acceleration for a clock pendulum on various planets.

A pendulum's period, as you learned earlier in this chapter, is described by the following equation:

$$T = 2\pi\sqrt{\frac{L}{g}}$$

The program "Chap12" stored on your graphing calculator makes use of the equation for the period of a pendulum. Once the "Chap12" program is executed, your calculator will ask for the period of the pendulum on Earth. The graphing calculator will use the following equation to create a graph of the pendulum's period (Y_1) versus the free-fall acceleration (X).

$$Y_1 = 2\pi\sqrt{(L/X)} \text{ where } L = (9.81T^2)/(4\pi^2)$$

Note that the relationships in this equation are the same as those in the period of a pendulum equation shown above.

a. What do the variables Y_1 and T represent in the calculator equation?

Execute "Chap12" on the [PRGM] menu and press [ENTER] to begin the program. Enter the value for the pendulum's period on Earth (shown below) and press [ENTER].

The calculator will provide a graph of the pendulum's period versus the free-fall acceleration. (If the graph is not visible, press [WINDOW] and change the settings for the graph window, then press [GRAPH].)

Press [TRACE] and use the arrow keys to trace along the curve. The x value corresponds to the free-fall acceleration in meters per second squared, and the y value corresponds to the period in seconds.

Determine the period of a clock pendulum with a period of 2.0 s on Earth if the pendulum were moved to the following locations, which have the free-fall accelerations indicated:

b. the surface of Mars, $3.71\ \text{m/s}^2$

c. the surface of Venus, $8.78\ \text{m/s}^2$

d. the surface of Neptune, $11.8\ \text{m/s}^2$

e. Is the graph shifted up or down for a pendulum whose period is longer than 2.0 s on Earth?

Press [2nd] [QUIT] to stop graphing. Press [ENTER] to input a new value or [CLEAR] to end the program.

55. Yellow light travels through a certain glass block at a speed of 1.97×10^8 m/s. The wavelength of the light in this particular type of glass is 3.81×10^{-7} m (381 nm). What is the frequency of the yellow light in the glass block?

56. A 0.40 kg mass is attached to a spring with a spring constant of 160 N/m so that the mass is allowed to move on a horizontal frictionless surface. The mass is released from rest when the spring is compressed 0.15 m.
  a. Find the force on the mass at the instant the spring is released.
  b. Find the acceleration of the mass at the instant the spring is released.
  c. What are the maximum values of force and acceleration?

57. The distance between two successive crests of a certain transverse wave is 1.20 m. Eight crests pass a given point along the direction of travel every 12.0 s. Calculate the wave speed.

58. A certain pendulum clock that works perfectly on Earth is taken to the moon, where $g = 1.63$ m/s^2. If the clock is started at 12:00 A.M., what will it read after 24.0 h?

59. A harmonic wave is traveling along a rope. The oscillator that generates the wave completes 40.0 vibrations in 30.0 s. A given crest of the wave travels 425 cm along the rope in a time period of 10.0 s. What is the wavelength?

Alternative Assessment

Performance assessment

1. Design an experiment to compare the spring constant and period of oscillation of a system built with two (or more) springs connected in two ways: in series (attached end to end) and in parallel (one end of each spring anchored to a common point). If your teacher approves your plan, obtain the necessary equipment and perform the experiment.

2. The rule that the period of a pendulum is determined by its length is a good approximation for amplitudes below 15°. Design an experiment to investigate how amplitudes of oscillation greater than 15° affect the motion of a pendulum.

  List what equipment you would need, what measurements you would perform, what data you would record, and what you would calculate. If your teacher approves your plan, obtain the necessary equipment and perform the experiment.

3. Imagine you are sitting on a small boat on a calm lake. A stone thrown in the water makes a splashing noise and some ripples. Predict whether the boat will oscillate before, after, or at the same time that you hear the splash. Write a justification supporting your prediction. Identify any additional information that would improve your prediction.

Portfolio projects

4. Investigate the relationship between uniform circular motion and simple harmonic motion by studying the motion of a crank handle and its shadow on the ground. Determine the displacement and period for both the handle and its shadow at different points of a cycle. You may create geometrical drawings, build a model, or videotape your experiment.

5. Research earthquakes and different kinds of seismic waves. Create a presentation about earthquakes that includes answers to the following questions as well as additional information: Do earthquakes travel through oceans? What is transferred from place to place as seismic waves propagate? What determines their speed?

6. Identify examples of periodic motion in nature. Create a chart describing the objects involved, their path of motion, their periods, and the forces involved. Which of the periodic motions are harmonic and which are not?

55. 5.17×10^{14} Hz
56. a. 24 N
  b. 6.0×10^1 m/s^2
  c. 24 N, 6.0×10^1 m/s^2
57. 0.800 m/s
58. 9:48 A.M.
59. 0.319 m

Alternative Assessment
ANSWERS

Performance assessment

1. Student plans should be safe and complete, and should include a list of equipment, measurements, and calculations. For springs in series, $1/k = 1/k_1 + 1/k_2$; for springs in parallel, $k = k_1 + k_2$.

2. Student plans should be safe and complete, including a list of equipment, measurements, and calculations.

3. Students may need to research the speed of sound to recognize that water waves move slower than sound.

 **Portfolio projects**

4. The best results will be obtained if students perform calculations for at least 12 data points between 0° and 360°. Graphs should look like a sine or cosine wave.

5. Answers should indicate that wave speed depends on the medium. Earthquakes involve longitudinal P waves, transverse S waves, and Rayleigh waves (circular motion).

6. Examples should involve repetitive motion. Circular motions are typically not harmonic; vibrations often are.

NOTE

Materials Preparation is given on pp. 436A–436B. Blank data table and sample data table are on the One-Stop Planner CD-ROM. All calculations shown use sample data.

Planning

Recommended time:

1 lab period

Classroom organization:

▸ Each group must have at least 2 students.

▸ **Safety warnings:** Falling masses can cause injury. Students should wear goggles to shield eyes from clamps and swinging masses at eye level.

Pendulum Tips

◆ Use a tall support stand and pendulum clamp for best results and ease of adjusting the length of the cord.

✔ Checkpoints

Step 6: Make sure students hold the cord straight when lifting the bob. The angle should be 10° to 15°. If the swing traces a circle instead of an arc, the angle may be too large.

Step 8: To change the length of the cord, students should loosen the clamp, move the cord up or down, and replace the clamp securely.

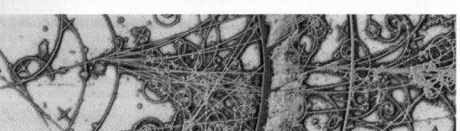

OBJECTIVES

- Construct simple pendulums, and find their periods.
- Calculate the value for *g*, the free-fall acceleration.
- Examine the relationships between length, mass, and period for different pendulums.

MATERIALS LIST

✔ balance
✔ cord
✔ meterstick
✔ pendulum bobs
✔ pendulum clamp
✔ protractor
✔ stopwatch
✔ support stand

CHAPTER 12
Laboratory Exercise

⭐TEKS

pp. 474–475: 1A, 2B, 2C, 2E, 2F, 3A, 4A

THE PENDULUM AND SIMPLE HARMONIC MOTION

In this experiment, you will construct models of a simple pendulum using different masses and lengths of cord. For each model, you will measure how long it takes the pendulum to complete 20 cycles. You can then use the period—the time required for the pendulum to complete one cycle—and the length of the cord to calculate *g* at your geographical location.

SAFETY

- **Tie back long hair, secure loose clothing, and remove loose jewelry to prevent their getting caught in moving parts or pulleys.**

- **Attach masses to the thread and the thread to clamps securely. Swing masses in areas free of people and obstacles. Swinging or dropped masses can cause serious injury.**

PREPARATION

1. Read the entire lab, and plan what measurements you will take.

2. Prepare a data table with four columns and seven rows in your lab notebook. In the first row, label the second through fourth columns *Mass (kg)*, *Length (m)*, and *Time (s)*. In the first column, label the first row *Trial* and label the second through seventh rows *1, 2, 3, 4, 5*, and *6*.

PROCEDURE

Constant mass with varying length

3. Measure the mass of the bob. Record it in your data table.

4. Set up a ring stand with a clamp to serve as the support. Place the ring stand away from the table's edge. Use cord to tie the bob to the clamp on the ring stand, securing it at the desired length (about 75 cm). Choose a location away from other groups, where the pendulum can swing freely.

5. Measure the length of the pendulum to the center of the bob.

6. Use the protractor to find the position of the pendulum where the amplitude is equal to 10°. Lift the bob to this position. Make sure that the path is free of obstructions, and release the bob so that it swings freely.

7. Measure and record the time required for 20 cycles of the pendulum bob. Keep the amplitude in all trials between 5° and 15°. Do not exceed 15°.

8. Perform two more trials using the same mass but different cord lengths.

Constant length with varying mass

9. Replace the bob with a different mass, and repeat the entire procedure. Record the mass and length.

10. Perform two more trials using the same length with different masses.

11. Clean up your work area. Put equipment away safely so that it is ready to be used again.

ANALYSIS AND INTERPRETATION

Calculations and data analysis

1. Applying ideas   For each trial, calculate the period of the pendulum.

2. Applying ideas   Using the equation for the period of a pendulum, calculate the value for the free-fall acceleration, *g*, for each trial.

3. Analyzing results   Use 9.81 m/s² as the accepted value for *g*.

a. Compute the absolute error for each trial using the following equation:

$$\text{absolute error} = |\text{experimental} - \text{accepted}|$$

b. Compute the relative error for each trial using the following equation:

$$\text{relative error} = \frac{(\text{experimental} - \text{accepted})}{\text{accepted}}$$

4. Graphing data   Plot the following graphs:

a. the period vs. the length for trials 1–3

b. the period vs. the mass of the bob for trials 4–6

c. the period vs. the square root of the length for trials 4–6

Conclusions

5. Evaluating results   Based on your data and your graphs, how does the mass of the pendulum bob affect its period of vibration?

6. Evaluating results   Based on your data and your graphs, how does the length of the pendulum affect its period of vibration?

Figure 12-32

Step 6: Hold the bob so that the cord is perfectly straight while you measure the angle.

Step 7: Release the bob gently so that it swings smoothly. Practice counting and timing cycles to get good results.

ANSWERS TO

Analysis and Interpretation

CALCULATIONS AND DATA ANALYSIS

1. Constant mass: Trial 1: 1.77 s; Trial 2: 1.41 s; Trial 3: 1.06 s. **Constant length:** Trial 4: 1.77 s; Trial 5: 1.77 s; Trial 6: 1.76 s.

2. Student answers will vary. Make sure students use the relationship $T = 2\pi \sqrt{\dfrac{L}{g}}$. Typical values will range from 9.50 m/s² to 9.82 m/s².

3. **a.** For sample data, values range from 0.01 m/s² to 0.29 m/s².

b. For sample data, values range from 0.001 to 0.030.

4. **a.** The graph should show a parabolic curve.

b. Graphs should show a straight line parallel to the *x*-axis.

c. Graphs should show a straight line pointing up and to the right.

CONCLUSIONS

5. The mass has no effect.

6. The longer the pendulum, the longer the period.

1780

1790

1800

1810

Physics and Its World *Timeline 1785–1830*

1789 – The storming of the Bastille marks the beginning of the French Revolution.

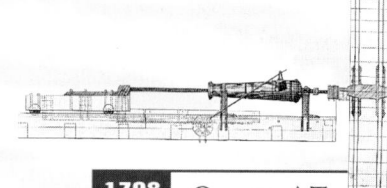

1798

$$Q = mc_p\Delta T$$

Benjamin Thompson (Count Rumford) demonstrates that energy transferred as heat results from mechanical processes, rather than the release of caloric, the heat fluid that has been widely believed to exist in all substances.

1796 – **Edward Jenner** develops the smallpox vaccine.

1800

$$\Delta V = \frac{\Delta PE_{electric}}{q}$$

Alessandro Volta develops the first current-electricity cell using alternating plates of silver and zinc.

1802 $m\lambda = d(\sin\theta)$

Thomas Young demonstrates that light rays interfere, providing the first substantial support for a wave theory of light.

1804 – Saint-Domingue, under the control of the French-African majority led by **Toussaint-Louverture,** becomes the independent Republic of Haiti. Over the next two decades most of Europe's western colonies become independent.

1804 – **Richard Trevithick** builds and tests the first steam locomotive. It pulls 10 tons along a distance of 15 km at a speed of 8 km/h.

1811 – Mathematician **Sophie Germain** writes the first of three papers on the mathematics of vibrating surfaces. She later addresses one of the most famous problems in mathematics—Fermat's last theorem—proving it to be true for a wide range of conditions.

1810 – **Kamehameha I** unites the Hawaiian islands under a monarchy.

1814

$$\sin\theta = \frac{m\lambda}{a}$$

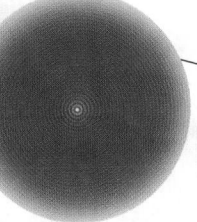

Augustin Fresnel begins his research in optics, the results of which will confirm and explain **Thomas Young's** discovery of interference and will firmly establish the wave model of light first suggested by **Christian Huygens** over a century earlier.

1818 – **Mary Shelley** writes *Frankenstein, or the Modern Prometheus.* Primarily thought of as a horror novel, the book's emphasis on science and its moral consequences also qualifies it as the first "science fiction" novel.

1820

$$F_{magnetic} = BI\ell$$

Hans Christian Oersted demonstrates that an electric current produces a magnetic field. (**Gian Dominico Romagnosi**, an amateur scientist, discovered the effect 18 years earlier, but at the time attracted no attention.) **André-Marie Ampere** repeats Oersted's experiment and formulates the law of electro-magnetism that today bears his name.

1826 – **Katsushika Hokusai** begins his series of prints *Thirty-Six Views of Mount Fuji.*

1830 – **Hector Berlioz** composes his *Symphonie Fantastique*, one of the first Romantic works for large orchestra that tells a story with music.

1800

1810

1820

1830

CHAPTER 13 PLANNING GUIDE

Compression Guide: To shorten from 11 to 8 45-min periods (from 5½ to 4 90-min blocks), eliminate items in magenta type.

PACING CHART	CLASSROOM RESOURCES			
	⭐TEKS	Teacher Demonstrations	Holt Physics Transparencies	Labs (See page T52 for equipment listing for in-text labs.)
13-1 Sound waves 3 45-minute periods 1½ 90-minute block	2C, 3A, 3C, 3E, 8A, 8B, 8C	**TE** *Longitudinal waves*, p. 480 **TE** *Sound waves in a solid*, p. 482 **TE** *The Doppler effect*, p. 485	**T** 54–57, 62 **TM** 44	**PE** *Speed of Sound*, p. 512
13-2 Sound intensity and resonance 2 or 1 45-minute periods 1 or ½ 90-minute block	1A, 2C, 2E, 3A, 3B, 3C, 5B, 8A, 8B	**TE** *Resonance*, p. 491	**T** 58 **TM** 45–46	**PE** *Quick Lab: Resonance*, p. 491 **L** *Discovery Lab: Resonance and the Nature of Sound*
13-3 Harmonics 4 or 2 45-minute periods 2 or 1 90-minute blocks	1A, 2C, 3A, 3B, 3C, 3E, 8A, 8B, 9B	**TE** *Seeing sounds*, p. 494	**T** 59–61 **TM** 47	**PE** *Quick Lab: A Pipe Closed at One End*, p. 497 **TL** *Sound Waves and Beats* **L** *Invention Lab: Building a Musical Instrument*
Review and Assessment 2 45-minute periods 1 90-minute block				

Resource Key

PHYSICS

PE Pupil's Edition
TE Teacher's Edition

L Laboratory Experiments
TL Technology Lab Experiments
T Transparencies

🖑 **One-Stop** **Planner CD-ROM** **contents**

TM Transparency Masters
SR Section Review Worksheets
AA Alternative Assessment

PW Problem-Solving Workbook
PB Problem Bank
CTW Critical Thinking Worksheet

LABORATORY PLANNING: Speed of Sound, p. 512

Materials (for each lab group):

CBL and Sensors Procedure
- CBL
- graphing calculator
- CBL microphone
- CBL temperature sensor
- 1-position support base and rod, 1.3 cm × 91 cm

- small symmetrical clamp with holder
- meterstick
- cardboard tube, 7.5 cm × 98 cm
- roll of adhesive tape, 0.5 in. wide
- TI Graph Link (recommended for downloading programs)

ASSIGNMENT RESOURCES

Content Mastery	Critical Thinking	Problem-Solving Practice
PE 1–7, p. 486 **SR** 13-1, *Concept Review* **PE** 1–9, p. 507	**PE** 1–2, p. 483 **PE** 10–15, p. 507	
PE 1–5, p. 493 **SR** 13-2, *Concept Review* **PE** 16–20, pp. 507–508	**PE** 1–4, p. 492 **PE** 21–26, p. 508	**13A** Intensity of sound waves: 29 items in **PE, PW,** and **PB,** see **TE** p. 488
PE 1–5, p. 503 **SR** 13-3, *Diagram Skills* **PE** 29–32, p. 508	**PE** 1–3, p. 502 **PE** 33–38, pp. 508–509	**13B** Harmonics: 30 items in **PE, PW,** and **PB,** see **TE** pp. 498–499

ASSESSMENT RESOURCES

Cumulative Review	Alternative Assessment	Traditional Assessment
SR Mixed Review, Ch. 13	**PE** 1–4, p. 510 **AA** Items for Ch. 13	Chapter 13 Test Test Generator items for Ch. 13

Scoring Rubrics for Alternative Assessment items can be found on the One-Stop Planner CD-ROM.

TECHNOLOGY RESOURCES

 CTW Segment 14 Virtual Practice Room

 Module 13 Doppler Effect

 PE Technology and Learning, p. 510
(Alternative procedures for calculators without Flash-ROM technology are provided on the One-Stop Planner CD-ROM.)

internet connect

 On-line Student Resources:
GO TO: www.scilinks.org
The following *sci*LINKS Internet resources can be found in the student text for this chapter.

TOPICS:
• Sound, p. 481 (HF2131)
• Doppler effect, p. 485 (HF2133)
• Resonance, p. 491 (HF2134)
• Harmonics, p. 495 (HF2135)
• Acoustics, p. 500 (HF2132)

 On-line Teacher Resources:
GO TO: go.hrw.com
KEYWORD: HF2 HOME
Visit the HRW Web site for a variety of resources related to this chapter.

 Smithsonian Institution®
Internet Connections
Visit **www.si.edu/hrw** for additional on-line resources.

CNNfyi.com
Visit **www.cnnfyi.com** for late-breaking news and current events stories selected just for you.

Resonance Apparatus Procedure

• resonance apparatus with double 45° clamp
• set of four tuning forks, physical pitch
• soft rubber hammer
• Erlenmeyer flask, 1000 mL
• food coloring (optional)
• red liquid thermometer, −20°C to 110°C

Materials Preparation

For better visibility, water for the resonance apparatus procedure may be tinted with food coloring. Mix a large quantity of colored water from which students may fill their apparatus, or set out the food coloring for students to use.

Section 13-1 explains how sound waves are produced, explores the basic characteristics of sound waves, and introduces the Doppler effect.

Section 13-2 explains how to calculate intensity; relates intensity, decibel level, and perceived loudness; and explores the phenomenon of resonance.

Section 13-3 introduces standing waves on a vibrating string and in open and closed pipes, calculates harmonics, relates harmonics and timbre, and discusses how beats occur.

About the Illustration

This photograph shows an Atlantic bottle-nosed dolphin off the coast of Belize. Dolphins use sounds for navigation, communication, and echolocation. A variety of other marine mammals and most bats also use sound waves to echolocate.

Interactive Problem-Solving Tutor

See Module 13

"Doppler Effect" provides a more detailed and quantitative treatment of the Doppler effect.

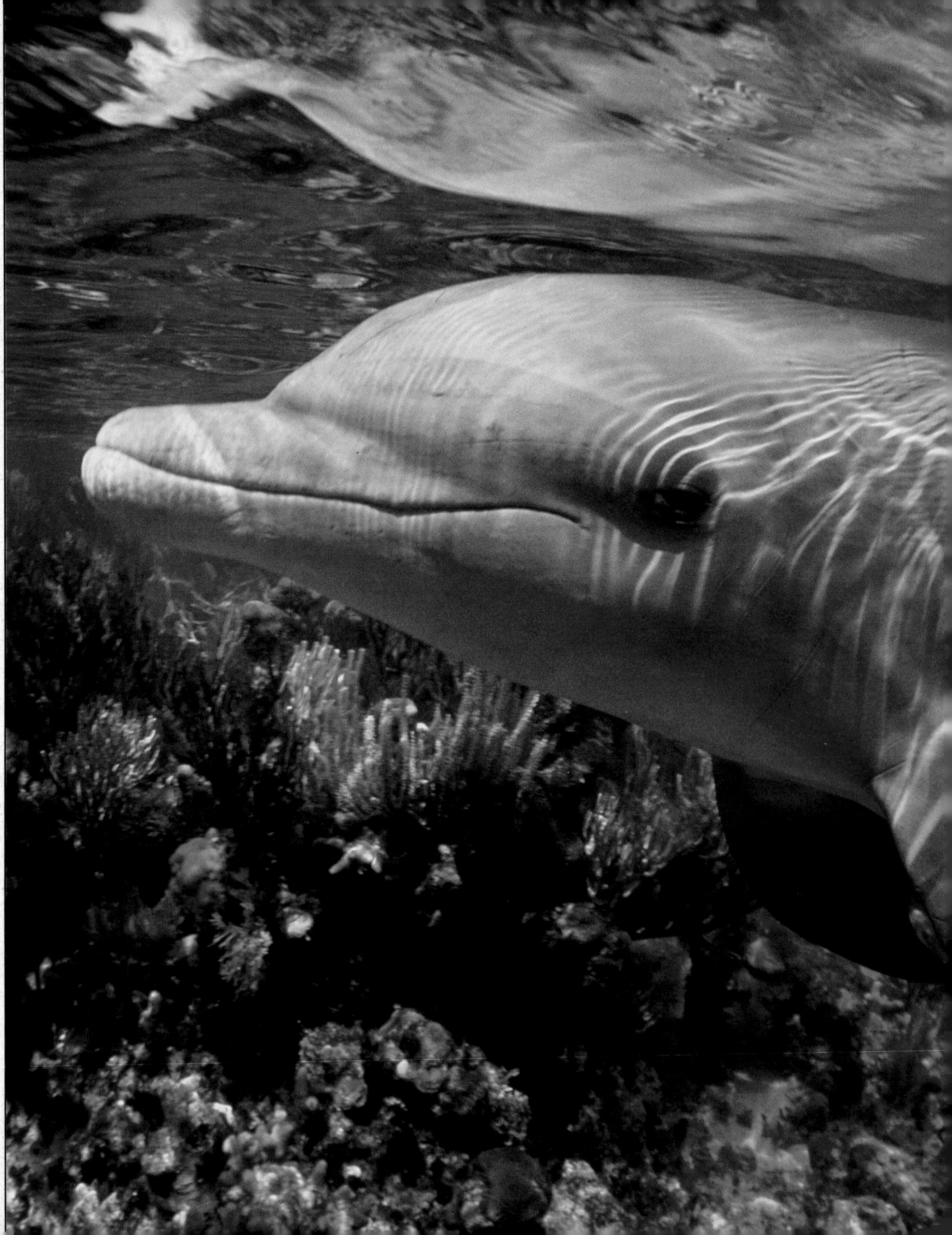

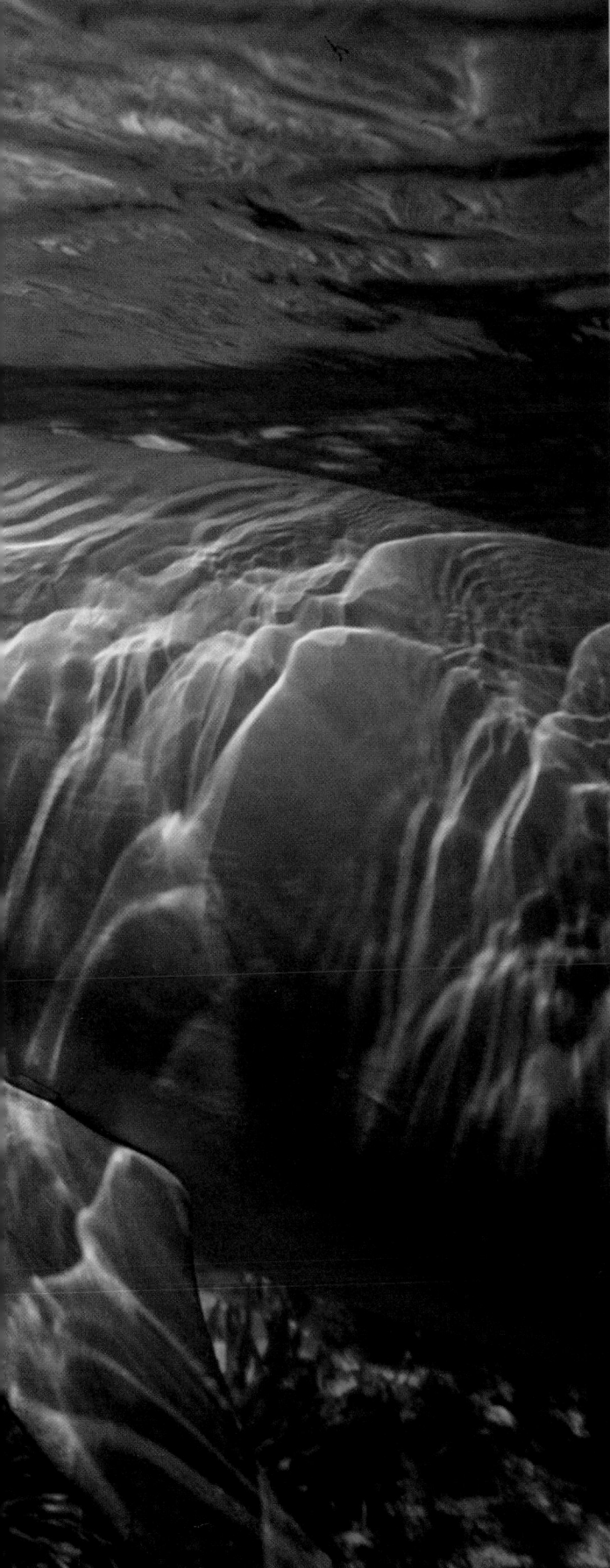

CHAPTER 13

Sound

PHYSICS IN ACTION

Some marine mammals, such as dolphins, use sound waves to locate distant objects. In this process, called *echolocation,* a dolphin produces a rapid train of short sound pulses that travel through the water, bounce off distant objects, and reflect back to the dolphin. From these echoes, dolphins can determine the size, shape, speed, and distance of their potential prey.

A dolphin's echolocation is extremely sophisticated. Experiments have shown that at a distance of 114 m, a blindfolded dolphin can locate a stainless-steel sphere with a diameter of 7.5 cm and can distinguish between a sheet of aluminum and a sheet of copper. In this chapter you will study sound waves and see how dolphins echolocate.

- *How does dolphin echolocation work?*
- *How does a dolphin determine the direction a fish is moving?*

CONCEPT REVIEW

Longitudinal waves (Section 12-3)

Wave speed (Section 12-3)

Standing waves (Section 12-4)

Knowledge to Expect

✔ "Sound is produced by vibrating objects. The pitch of the sound can be varied by changing the rate of vibration." (NRC's *National Science Education Standards,* grades K–4)

✔ "Vibrations in materials set up wavelike disturbances that spread away from the source, like sound and earthquake waves. Sound waves move at different speeds in different materials." (AAAS's *Benchmarks for Science Literacy,* grades 6–8)

Knowledge to Review

✔ Longitudinal waves are waves in which the particles vibrate parallel to the direction of wave motion. (Section 12-3)

✔ Wave speed is the product of wavelength and frequency. (Section 12-3)

✔ Standing waves are formed when two waves of the same frequency, amplitude, and wavelength travel in opposite directions through a medium and interfere with each other. (Section 12-4)

Items to Probe

✔ Familiarity with wave-related phenomena: Ask students to describe how sound travels between a speaker and a listener.

✔ Preconceptions about pitch and loudness: Ask students what determines the pitch and loudness associated with a sound.

13-1
Sound waves

13-1 SECTION OBJECTIVES

- **Explain how sound waves are produced.**

- **Relate frequency to pitch.**

- **Compare the speed of sound in various media.**

- **Relate plane waves to spherical waves.**

- **Recognize the Doppler effect, and determine the direction of a frequency shift when there is relative motion between a source and an observer.**

compression

the region of a longitudinal wave in which the density and pressure are greater than normal

rarefaction

the region of a longitudinal wave in which the density and pressure are less than normal

Figure 13-1
(a) The sound from a tuning fork is produced by **(b)** the vibrations of each of its prongs. **(c)** When a prong swings to the right, there is a region of high density and pressure. **(d)** When the prong swings back to the left, a region of lower density and pressure exists.

THE PRODUCTION OF SOUND WAVES ⊛TEKS 8A, 8B

Whether a sound wave conveys the shrill whine of a jet engine or the melodic whistling of a bird, it begins with a vibrating object. We will explore how sound waves are produced by considering a vibrating tuning fork, as shown in **Figure 13-1(a).**

The vibrating prong of a tuning fork, shown in **Figure 13-1(b),** sets the air molecules near it in motion. As the prong swings to the right, as shown in **Figure 13-1(c),** the air molecules in front of the movement are forced closer together. (This situation is exaggerated in the figure for clarity.) Such a region of high molecular density and high air pressure is called a **compression.** As the prong moves to the left, as in **Figure 13-1(d),** the molecules to the right spread apart, and the density and air pressure in this region become lower than normal. This region of lower density and pressure is called a **rarefaction.**

As the tuning fork continues to vibrate, a series of compressions and rarefactions form and spread away from each prong. These compressions and rarefactions expand and spread out in all directions, like ripple waves on a pond. When the tuning fork vibrates with simple harmonic motion, the air molecules also vibrate back and forth with simple harmonic motion.

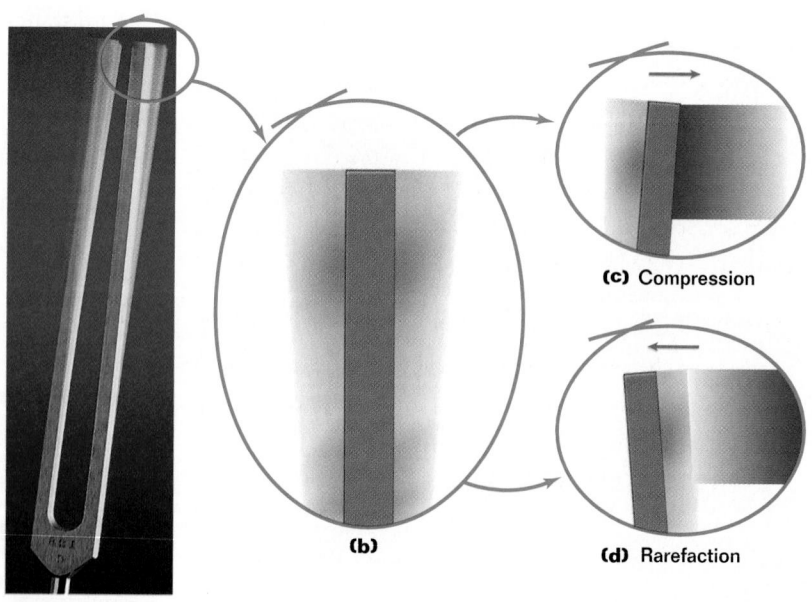

(c) Compression

(d) Rarefaction

(b)

(a)

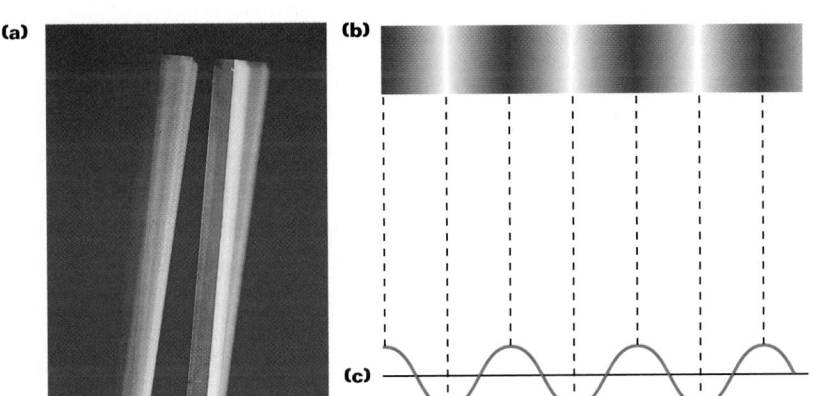

(a)

(b)

(c)

Figure 13-2
(a) As this tuning fork vibrates, **(b)** a series of compressions and rarefactions move away from each prong. **(c)** The crests of this sine wave correspond to compressions, and the troughs correspond to rarefactions.

Sound waves are longitudinal

In sound waves, the vibrations of air molecules are parallel to the direction of wave motion. Thus, sound waves are longitudinal. As you saw in Chapter 12, a longitudinal wave produced by a vibrating object can be represented by a sine curve. In **Figure 13-2** the crests of the sine curve correspond to compressions in the sound wave, and the troughs correspond to rarefactions. Because compressions are regions of higher pressure and rarefactions are regions of lower pressure, the sine curve represents the changes in air pressure due to the propagation of the sound waves. ⭐TEKS **8A, 8B**

CHARACTERISTICS OF SOUND WAVES

In Chapter 12, frequency was defined as the number of cycles per unit of time. Sound waves that the average human ear can hear, called *audible* sound waves, have frequencies between 20 and 20 000 Hz. (An individual's hearing depends on a variety of factors, including age and experiences with loud noises.) Sound waves with frequencies less than 20 Hz are called *infrasonic* waves, and those above 20 000 Hz are called *ultrasonic* waves.

It may seem confusing to use the term *sound waves* for infrasonic or ultrasonic waves since humans cannot hear these sounds, but ultrasonic and infrasonic waves consist of the same types of vibrations as the sounds that we can hear. The range of sound waves that are considered to be audible depends on the ability of the average human ear to detect their vibrations. Dogs can hear ultrasonic waves that humans cannot. ⭐TEKS **8A, 8B**

Frequency determines pitch

The frequency of an audible sound wave determines how high or low we perceive the sound to be, which is known as **pitch.** As the frequency of a sound wave increases, the pitch rises. The frequency of a wave is an objective quantity that can be measured, while pitch refers to how different frequencies are perceived by the human ear. ⭐TEKS **8A, 8B**

📶 internet**connect**

SC*i*/LINKS.

NSTA

TOPIC: Sound
GO TO: www.scilinks.org
*sci***LINKS CODE:** HF2131

Did you know?

Elephants use infrasonic sound waves to communicate with one another. Their large ears enable them to detect these low-frequency sound waves, which have relatively long wavelengths. Elephants can effectively communicate in this way, even when they are separated by many kilometers.

---

pitch

the perceived highness or lowness of a sound, depending on the frequency of the sound waves

---

Teaching Tip

Figure 13-2 uses a sine curve to represent the compressions and rarefactions of a longitudinal wave produced by a vibrating object. Compressions correspond to crests, and rarefactions correspond to troughs. Sometimes a sine curve is used to represent *displacement* rather than pressure and density. For any given longitudinal wave, the sine curve representing pressure and the sine curve representing displacement are 90° out of phase.

🛑 Misconception Alert

Point out that some individuals may be able to hear sounds slightly below 20 Hz or above 20 000 Hz because the range of frequencies defined as audible is based on the ability of the *average* human ear.

⭐TEKS

p. 480: 8A, 8B
p. 481: 8A, 8B, 8A, 8B, 8A, 8B

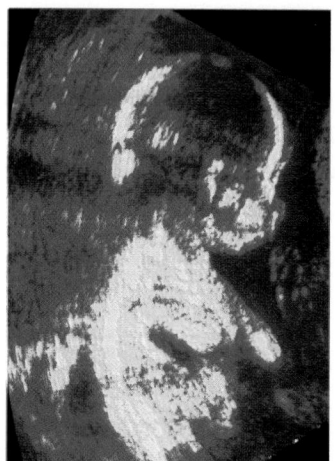

Figure 13-3
Ultrasound images, such as this one, are formed with reflected sound waves. This colorized image depicts a fetus after 21 weeks of development.

⭐TEKS 2C

Table 13-1 Speed of sound in various media	
Medium	v (m/s)
Gases	
air (0°C)	331
air (25°C)	346
air (100°C)	366
helium (0°C)	972
hydrogen (0°C)	1290
oxygen (0°C)	317
Liquids at 25°C	
methyl alcohol	1140
sea water	1530
water	1490
Solids	
aluminum	5100
copper	3560
iron	5130
lead	1320
vulcanized rubber	54

Ultrasonic waves can produce images ⭐TEKS 3C, 8A, 8B, 8C

As discussed in Chapter 12, wavelength decreases as frequency increases. Thus, infrasonic waves have longer wavelengths than audible sound waves, and ultrasonic waves have shorter wavelengths. Because of their short wavelengths, ultrasonic waves have widespread medical applications.

For example, ultrasonic waves can be used to produce images of objects inside the body. Such imaging is possible because sound waves are partially reflected when they reach a boundary between two materials of different densities. The images produced by ultrasonic waves are clearer and more detailed than those that could be produced by lower-frequency sound waves because the short wavelengths of ultrasonic waves are easily reflected off small objects. Audible and infrasonic sound waves are not as effective because their longer wavelengths pass around small objects.

In order for ultrasonic waves to "see" an object inside the body, the wavelength of the waves used must be about the same size or smaller than the object. A typical frequency used in an ultrasonic device is about 10 Mhz. The speed of an ultrasonic wave in human tissue is about 1500 m/s, so the wavelength of 10 Mhz waves is $\lambda = v/f = 1.5$ mm. This device will not detect objects smaller than this size.

Physicians commonly use ultrasonic waves to observe fetuses. In this process, a crystal emits ultrasonic pulses. The same crystal acts as a receiver and detects the reflected sound waves. These reflected sound waves are converted to an electric signal, which forms an image on a fluorescent screen, as in **Figure 13-3.** By repeating this process for different portions of the mother's abdomen, a physician can obtain a complete picture of the fetus. **Figure 13-3** shows the ultrasound image of a fetus in the womb after 21 weeks of development. In this profile view, the head is at the upper right of the image, and an outline of the spine and upper arm can also be seen. At this stage, the nose, lips, and chin are fully developed, and the fetus weighs about 500 g. These images allow doctors to detect some types of fetal abnormalities.

Dolphin echolocation works in a similar manner. A dolphin sends out pulses of sound, which return in the form of reflected sound waves. These reflected waves allow the dolphin to form an image of the object that reflected the waves. Dolphins use high-frequency waves for echolocation because shorter wavelengths are most effective for detecting smaller objects.

Speed of sound depends on the medium ⭐TEKS 8A, 8B

Sound waves can travel through solids, liquids, and gases. Because waves consist of particle vibrations, the speed of a wave depends on how quickly one particle can transfer its motion to another particle. For example, solid particles respond more rapidly to a disturbance than gas particles do because the molecules of a solid are closer together than those of a gas are. As a result, sound waves generally travel faster through solids than through gases. **Table 13-1** shows the speed of sound waves in various media.

The speed of sound also depends on the temperature of the medium. As temperature rises, the particles of a gas collide more frequently. Thus, in a gas, the disturbance can spread faster at higher temperatures than at lower temperatures. In liquids and solids, the particles are close enough together that the difference due to temperature changes is less noticeable.

Sound waves propagate in three dimensions ⓣ TEKS 8B

In Chapter 12, waves were shown as traveling in a single direction. But sound waves actually travel away from a vibrating source in all three dimensions. When a musician plays a saxophone in the middle of a room, the resulting sound can be heard throughout the room because the sound waves spread out in all directions. Such three-dimensional sound waves are approximately spherical. To simplify, we shall assume that sound waves are exactly spherical unless stated otherwise.

Spherical waves can be represented graphically in two dimensions with a series of circles surrounding the source, as shown in **Figure 13-4.** The circles represent the centers of compressions, called *wave fronts*. Because we are considering a three-dimensional phenomenon in two dimensions, each circle represents a spherical area.

Because each wave front corresponds to the center of a compression, the distance between adjacent wave fronts is equal to one wavelength, λ. The radial lines perpendicular to the wave fronts are called *rays*. Rays indicate the direction of the wave motion. The sine curve used in our previous representation of sound waves, also shown in **Figure 13-4,** corresponds to a single ray. Because crests of the sine curve represent compressions, each wave front crossed by this ray corresponds to a crest of the sine curve.

Now consider a small portion of a spherical wave front that is many wavelengths away from the source, as shown in **Figure 13-5.** In this case, the rays are nearly parallel lines, and the wave fronts are nearly parallel planes. Thus, at distances from the source that are great relative to the wavelength, we can approximate spherical wave fronts with parallel planes. Such waves are called *plane waves.* Any small portion of a spherical wave that is far from the source can be considered a plane wave. Plane waves can be treated as a series of identical one-dimensional waves, like those in Chapter 12, all traveling in the same direction.

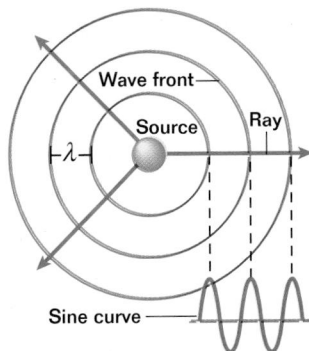

Figure 13-4
In this representation of a spherical wave, the wave fronts represent compressions and the rays show the direction of wave motion. Each wave front corresponds to a crest of the sine curve, which in turn corresponds to a single ray.

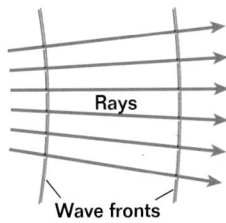

Figure 13-5
Spherical wave fronts that are a great distance from the source can be approximated with parallel planes known as plane waves.

Conceptual Challenge ⓣ TEKS 3A, 8A, 8B

1. Hydrogen and air Hydrogen atoms have a smaller mass than the primary components of air, so they are accelerated much more easily. How does this fact account for the great difference in the speed of sound waves traveling through air and hydrogen, as shown in **Table 13-1** on page 482?

2. Lightning and thunder Light waves travel nearly 1 million times faster than sound waves in air. With this in mind, explain how the distance to a lightning bolt can be determined by counting the seconds between the flash and the sound of the thunder.

ANSWERS TO

Conceptual Challenge

1. Because hydrogen molecules are accelerated more easily than air molecules are, hydrogen molecules spread vibrations much faster than air molecules (primarily nitrogen). As a result, sound waves travel faster in hydrogen than they do in air.

2. Because light travels so much faster than sound, the speed of light can be considered to be effectively infinite in this case. Thus, the time it takes the sound wave to reach the listener multiplied by the speed of sound in air gives the distance between the observer and the lightning bolt.

ⓣ TEKS

p. 482: 3C, 8A, 8B, 8C, 2C, 8A, 8B
p. 483: 8B, 3A, 8A, 8B

Tomorrow's Technology

Tomorrow's Technology

BACKGROUND

Because air is the most common medium for sound waves to travel through, students sometimes fail to recognize that sound can travel through solids and liquids as well. The fact that Prine's device listens for sounds inside a bridge demonstrates that sound waves can travel through a solid medium, such as steel. The device effectively filters out the lower-frequency droning of cars and other background sounds to concentrate on the high-frequency sounds that are more characteristic of a damaged bridge.

In addition to improving the safety of bridges, Prine's technology might help make bridges a little quieter. Steel bridges often give off sudden, loud noises that can disturb the surrounding neighborhood. These noises may be indicative of severe problems in the bridge. Until now, the sounds have been notoriously difficult to control.

⭐TEKS

p. 484: 3C, 8B, 8C
p. 485: 3E, 8B

Acoustic Bridge Inspection ⭐TEKS 3C, 8B, 8C

When we drive over a bridge, we usually take its structural integrity for granted. Unfortunately, drivers should not always make that assumption. Many of the almost 600 000 bridges in the United States were built over 80 years ago, and most of the rest are 35 to 40 years old. As a bridge ages, cracks form in the steel that can weaken the bridge or even lead to eventual collapse. Inspectors search for those cracks visually, but that method is generally unreliable. David Prine, a senior research scientist at Northwestern University, thinks that the most reliable way to find damage in bridges is to listen to them.

"When a piece of steel cracks, high-frequency sound is emitted," explained Prine. These noises often sound like banging noises that echo throughout the bridge.

Prine has developed a way to inspect bridges using a system of acoustic-emissions sensors—sensitive microphones attached to the steel in a bridge—to capture audio signals from the structure. The signals are put into a computer, which filters out traffic and other noise and searches for sounds around 150 kHz, the frequency range of cracking steel. The computer locates where on the bridge each sound is com-

ing from and then determines whether the recorded sounds match the pattern of a spreading crack. Damage can then be tracked down and repaired.

The device works best as a preventive measure. "The name of the game is to determine what the problems are at the earliest possible stage so you can do a minimal level of repair and not wait until it gets too bad," Prine said.

Still, the acoustic-emissions sensor comes in handy even after a bridge has been fixed. The sensor can monitor different repair methods after they have been tried to determine which methods best correct the problem. This is a much more efficient means of checking repairs than the old way—waiting until the bridge falls apart.

THE DOPPLER EFFECT

If you stand on the street while someone drives by honking a car horn, you will notice the pitch of the horn change. The pitch will be higher as the car approaches and will be lower as the car moves away. As you read earlier in this section, the pitch of a sound depends on its frequency. But in this case, the car horn is not changing its frequency. How can we account for this change in pitch?

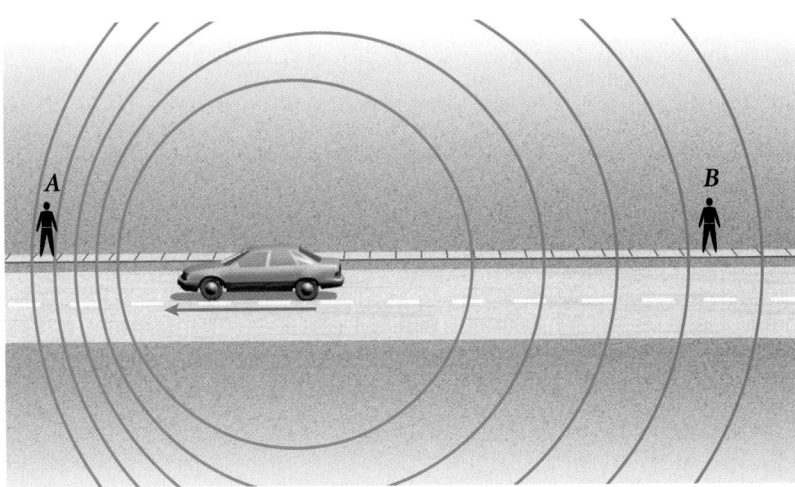

Relative motion creates a change in frequency ⭐TEKS 3E, 8B

In our earlier examples, we assumed that both the source of the sound waves and the listener were stationary. If a horn is honked in a parked car, an observer standing on the street hears the same pitch that the driver hears, as you would expect. For simplicity's sake, we will assume that the sound waves produced by the car horn are spherical.

When the car shown in **Figure 13-6** is moving, there is relative motion between the moving car and a stationary observer. This relative motion affects the way the wave fronts of the sound waves produced by the car's horn are perceived by an observer.

Although the frequency of the car horn (the source frequency) remains constant, the wave fronts reach an observer in front of the car, at point A, more often than they would if the car were stationary. This is because the source of the sound waves is moving toward the observer. Thus, the frequency heard by this observer is *greater* than the source frequency. (Note that the speed of the sound waves does not change.) For the same reason, the wave fronts reach an observer behind the car, at point B, less often than they would if the car were stationary. As a result, the frequency heard by this observer is *less* than the source frequency. This frequency shift is known as the **Doppler effect,** named for the Austrian physicist Christian Doppler (1803–1853), who first described it.

internet connect

SC*i*LINKS
NSTA

TOPIC: Doppler effect
GO TO: www.scilinks.org
*sci*LINKS CODE: HF2133

Figure 13-6
As this car moves to the left, an observer in front of the car, at point A, hears the car horn at a higher pitch than the driver, while an observer behind the car, at point B, hears a lower pitch.

Did you know?

The Doppler effect occurs with all types of waves. In the radar systems used by police to monitor car speeds, a computer compares the frequency of radar waves emitted with those reflected from a moving car and then uses this comparison to calculate the speed of the car.

Doppler effect

a frequency shift that is the result of relative motion between the source of waves and an observer

Demonstration 3

The Doppler effect

Purpose Show that the pitch of a sound depends on the relative motion between the source of the sound waves and the observer.

Materials battery-operated high-volume oscillator (available at most local electronics stores), appropriate batteries for the oscillator, foam ball large enough to hold the oscillator and batteries

Procedure Carefully cut into the foam ball and remove enough material so that the oscillator and the batteries will snugly fit inside the ball. Connect the batteries to the oscillator, and place the oscillator and batteries securely inside the ball. Allow the students to toss the ball about the classroom. Have the students note the differences in the pitch of the sound of the oscillator when the ball is traveling toward them, when the ball is traveling away from them, and when the ball is at rest.

Misconception Alert

Some students may think that the pitch *rises* as the source of sound approaches an observer and *decreases* as the source moves away. Stress the fact that the pitch is *higher* or *lower* and that it changes only when the source passes the observer. This is illustrated in **Figure 13-6,** which shows that the distance between wave fronts is constant for each observer.

Interactive Problem-Solving Tutor

See Module 13

"Doppler Effect" provides a more detailed and quantitative treatment of the Doppler effect.

Module 13
"Doppler Effect"
provides an interactive lesson with guided problem-solving practice to teach you more about the Doppler effect.

Because frequency determines pitch, the Doppler effect affects the pitch heard by each listener on the street. The observer in front of the car hears a higher pitch, while the observer behind the car hears a lower pitch.

We have considered a moving source with respect to a stationary observer, but the Doppler effect also occurs when the observer is moving with respect to a stationary source or when both are moving at different speeds. In other words, the Doppler effect occurs whenever there is *relative motion* between the source of waves and an observer. Although the Doppler effect is most commonly experienced with sound waves, it is a phenomenon common to all waves, including electromagnetic waves, such as visible light.

Section Review   ★TEKS 3A, 8B

1. If you hear a higher pitch from a trumpet than from a saxophone, how do the frequencies of the sound waves from the trumpet compare with those from the saxophone?

2. Could a small portion of the innermost wave front shown in **Figure 13-7** be approximated by a plane wave? Why or why not?

Figure 13-7

3. **Figure 13-8** is a diagram of the Doppler effect in a ripple tank. In which direction is the source of these ripple waves moving?

4. If the source of the waves in **Figure 13-8** is stationary, which way must the ripple tank be moving?

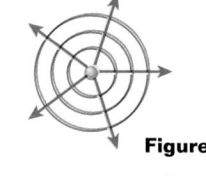

Figure 13-8

5. **Physics in Action**   Dolphins can produce sound waves with frequencies ranging from 0.25 kHz to 220 kHz, but only those at the upper end of this spectrum are used in echolocation. Explain why high-frequency waves work better than low-frequency waves.

6. **Physics in Action**   Sound pulses emitted by a dolphin travel through 20°C ocean water at a rate of 1450 m/s. In 20°C air, these pulses would travel 342.9 m/s. How can you account for this difference in speed?

7. **Physics in Action**   As a dolphin swims toward a fish, it sends out sound waves to determine the direction the fish is moving. If the frequency of the reflected waves is increased, is the dolphin catching up to the fish or falling behind?

SOUND INTENSITY

When a piano player strikes a piano key, a hammer inside the piano strikes a wire and causes it to vibrate, as shown in **Figure 13-9.** The wire's vibrations are then transferred to the piano's soundboard. As the soundboard vibrates, it exerts a force on air molecules around it, causing air molecules to move. Because this force is exerted through displacement of the soundboard, the soundboard does work on the air. Thus, as the soundboard vibrates back and forth, its kinetic energy is converted into sound waves. This is one reason that the vibration of the soundboard gradually dies out.

Intensity is the rate of energy flow through a given area

As described in Section 13-1, sound waves traveling in air are longitudinal waves. As the sound waves travel outward from the source, energy is transferred from one air molecule to the next. The rate at which this energy is transferred through a unit area of the plane wave is called the **intensity** of the wave. Because power, P, is defined as the rate of energy transfer, intensity can also be described in terms of power. ⭐TEKS **8A, 8B**

$$\text{intensity} = \frac{\Delta E/\Delta t}{\text{area}} = \frac{P}{\text{area}}$$

As seen in Chapter 5, the SI unit for power is the watt. Thus, intensity has units of watts per square meter (W/m^2). In a spherical wave, energy propagates equally in all directions; no one direction is preferred over any other. In this case, the power emitted by the source (P) is distributed over a spherical surface (area $= 4\pi r^2$), assuming that there is no absorption in the medium.

INTENSITY OF A SPHERICAL WAVE ⭐TEKS **3B**

$$\text{intensity} = \frac{P}{4\pi r^2}$$

$$\text{intensity} = \frac{(\text{power})}{(4\pi)(\text{distance from the source})^2}$$

This equation shows that the intensity of a sound wave decreases as the distance from the source (r) increases. This occurs because the same amount of energy is spread over a larger area.

13-2 SECTION OBJECTIVES

- **Calculate the intensity of sound waves.**

- **Relate intensity, decibel level, and perceived loudness.**

- **Explain why resonance occurs.**

---

intensity

the rate at which energy flows through a unit area perpendicular to the direction of wave motion

---

Figure 13-9
As this piano wire vibrates, it transfers energy to the piano's soundboard, which in turn transfers energy into the air in the form of sound.

🛑 Misconception Alert

Some students may have trouble distinguishing between energy and the *rate* of energy transfer, or power. Briefly review the idea of power (introduced in Chapter 5) to make sure this distinction is clear to students before intensity is introduced.

Key Models and Analogies

Ask students to imagine a balloon that is being inflated. The surface of the balloon is analogous to a spherical sound wave. Point out that the same amount of material is spread over a larger area as the balloon is inflated. In a similar way, the same amount of energy is spread over a larger area at greater distances from the source of the sound waves. Thus, intensity and distance are inversely related, as seen in the equation for the intensity of a spherical wave.

⭐TEKS

p. 486: 3A, 8B
p. 487: 8A, 8B, 3B

Classroom Practice

The following may be used as a teamwork exercise or for demonstration at the chalkboard or on an overhead projector.

PROBLEM

Intensity of sound waves

The intensity of the sound from an explosion is 0.10 W/m^2 at a distance of 1.0×10^3 m. Find the intensity of the explosion at distances of 5.0×10^2 m, 1.0×10^2 m, and 10.0 m.

Answer

0.41 W/m^2, 1.0×10^1 W/m^2, 1.0×10^3 W/m^2

PRACTICE GUIDE 13A	
Solving for:	
I	📖 **PE** Sample, 1–2; Ch. Rvw. 27–28 💿 **PW** 7–9 **PB** 5–7
P	📖 **PE** 3–4; Ch. Rvw. 42, 54* 💿 **PW** Sample, 1–3 **PB** 8–10
r	📖 **PE** 5; Ch. Rvw. 50* 💿 **PW** 4–6 **PB** Sample, 1–4

ANSWERS TO

Practice 13A
Intensity of sound waves
1. **a.** 8.0×10^{-4} W/m^2
   b. 1.6×10^{-3} W/m^2
   c. 6.4×10^{-3} W/m^2
2. 8.91×10^{-3} W/m^2
3. 2.3×10^{-5} W
4. 4.5 W
5. 4.8 m

Intensity of sound waves ⭐TEKS 2C, 3B, 8B

PROBLEM

What is the intensity of the sound waves produced by a trumpet at a distance of 3.2 m when the power output of the trumpet is 0.20 W? Assume that the sound waves are spherical.

SOLUTION

Given: $P = 0.20$ W $r = 3.2$ m

Unknown: Intensity $= ?$

Use the equation for the intensity of a spherical wave, given on page 487.

$$\text{Intensity} = \frac{P}{4\pi r^2}$$

$$\text{Intensity} = \frac{0.20 \text{ W}}{4\pi (3.2 \text{ m})^2}$$

$$\boxed{\text{Intensity} = 1.6 \times 10^{-3} \text{ W/m}^2}$$

CALCULATOR SOLUTION

The calculator answer for intensity is 0.0015542. This is rounded to 1.6×10^{-3} because each of the given quantities has two significant figures.

PRACTICE 13A

Intensity of sound waves ⭐TEKS 2C, 3B, 8B

1. Calculate the intensity of the sound waves from an electric guitar's amplifier at a distance of 5.0 m when its power output is equal to each of the following values:
   a. 0.25 W
   b. 0.50 W
   c. 2.0 W

2. At a maximum level of loudness, the power output of a 75-piece orchestra radiated as sound is 70.0 W. What is the intensity of these sound waves to a listener who is sitting 25.0 m from the orchestra?

3. If the intensity of a person's voice is 4.6×10^{-7} W/m^2 at a distance of 2.0 m, how much sound power does that person generate?

4. How much power is radiated as sound from a band whose intensity is 1.6×10^{-3} W/m^2 at a distance of 15 m?

5. The power output of a tuba is 0.35 W. At what distance is the sound intensity of the tuba 1.2×10^{-3} W/m^2?

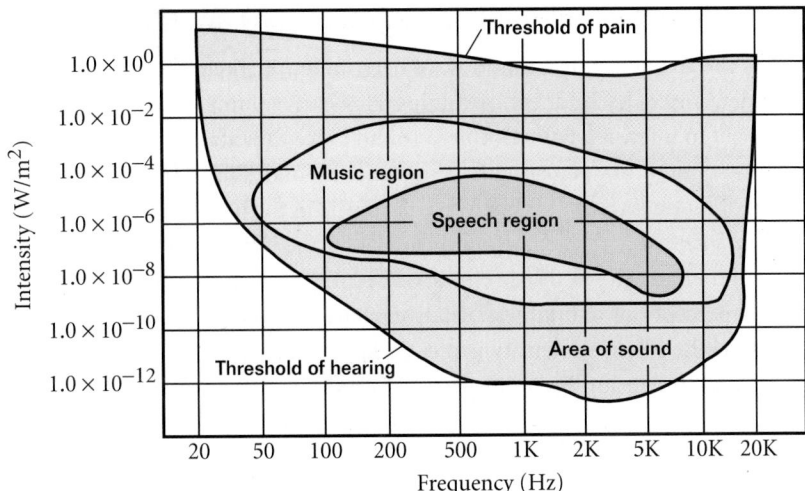

Figure 13-10
Human hearing depends on both
the frequency and the intensity of
sound waves. Sounds in the middle
of the spectrum of frequencies can
be heard more easily (at lower
intensities) than those at lower and
higher frequencies.

⊛ TEKS 2E, 8B

Intensity and frequency determine which sounds are audible

As you saw in Section 13-1, the frequency of sound waves heard by the average human ranges from 20 to 20 000 Hz. Intensity is also a factor in determining which sound waves are audible. **Figure 13-10** shows how the range of audibility of the average human ear depends on both frequency and intensity. As you can see in this graph, sounds at low frequencies (those below 50 Hz) or high frequencies (those above 12 000 Hz) must be relatively intense to be heard, whereas sounds in the middle of the spectrum are audible at lower intensities.

The softest sounds that can be heard by the average human ear occur at a frequency of about 1000 Hz and an intensity of 1.0×10^{-12} W/m^2. Such a sound is said to be at the *threshold of hearing*. (Note that some humans can hear slightly softer sounds, at a frequency of about 3300 Hz.) The threshold of hearing at each frequency is represented by the lowest curve in **Figure 13-10.** At the threshold of hearing, the changes in pressure due to compressions and rarefactions are about three ten-billionths of atmospheric pressure.

The maximum displacement of an air molecule at the threshold of hearing is approximately 1×10^{-11} m. Comparing this number to the diameter of a typical air molecule (about 1×10^{-10} m) reveals that the ear is an extremely sensitive detector of sound waves. ⊛ TEKS 2C

The loudest sounds that the human ear can tolerate have an intensity of about 1.0 W/m^2. This is known as the *threshold of pain* because sounds with greater intensities can produce pain in addition to hearing. The highest curve in **Figure 13-10** represents the threshold of pain at each frequency. Exposure to sounds above the threshold of pain can cause immediate damage to the ear, even if no pain is felt. Prolonged exposure to sounds of lower intensities can also damage the ear. For this reason, many rock musicians wear earplugs during their performances, and some rock stars must wear hearing aids. Note that the threshold of hearing and the threshold of pain merge at both high and low ends of the spectrum. ⊛ TEKS 3C

Did you know?

A 75-piece orchestra produces about 75 W at its loudest. This is comparable to the power required to keep one medium-sized electric light bulb burning. Speech has even less power. It would take the conversation of about 2 million people to provide the amount of power required to keep a 50 W light bulb burning.

Visual Strategy

Figure 13-10
Be certain that students understand the information contained in the different regions of the graph. Also point out that the scale of the *y*-axis of this graph is logarithmic. Thus, the intensity represented by each horizontal line is 100 times greater than the intensity represented by the line immediately below that line.

Q Are there musical sounds of 1000 Hz and 1.0×10^{-6} W/m^2?

A *yes, because the speech region is a subset of the music region*

Q Does this graph describe an individual's hearing exactly?

A *No, the graph is based on the average human ear. Each individual's hearing may vary.*

 Misconception Alert

The relationship between frequency, intensity, and audibility is complex and often confusing to students. Stress that neither frequency nor intensity alone can determine which sounds are audible; both factors must be taken into account.

⊛ TEKS

p. 488: 2C, 3B, 8B, 2C, 3B, 8B
p. 489: 2E, 8B, 2C, 3C

Misconception Alert

Some students may confuse decibel level with intensity. Point out that the ratio of one decibel level to another does *not* give the ratio between the intensities of these sounds because the decibel scale is logarithmic. Work with examples in **Table 13-2** to correct this misconception.

Key Models and Analogies

The decibel scale of sound loudness is designed similar to the Richter scale, which measures earthquake intensity, and the pH scale, which measures acidity levels. The values on these scales correspond to the order of magnitude (powers of 10) of the original quantity because their measurements span a very large range of values.

Visual Strategy

Table 13-2

Point out that zero on the decibel scale does not mean zero intensity or that there is no sound.

Q How does the sound intensity of a subway compare with that of a conversation?

A *At equal distances, noise from a subway brings to our ears 100 000 ($10^{-2}/10^{-7}$) times more energy than a conversation. (Some students may think the answer is twice as much, or the ratio of 100 to 50. Explain why this is not the case.)*

decibel level

relative intensity, determined by relating the intensity of a sound wave to the intensity at the threshold of hearing

Did you know?

The original unit of decibel level is the *bel,* named in honor of Alexander Graham Bell, the inventor of the telephone. The decibel is equivalent to 0.1 bel.

Relative intensity is measured in decibels ⊛TEKS 8A, 8B

Just as the frequency of a sound wave determines its pitch, the intensity of a wave determines its loudness, or volume. However, volume is not directly proportional to intensity. For example, a sound twice the intensity of the faintest audible sound is not perceived as being twice as loud. This is because the sensation of loudness is approximately logarithmic in the human ear.

Relative intensity, which is found by relating the intensity of a given sound wave to the intensity at the threshold of hearing, corresponds more closely to human perceptions of loudness. Relative intensity is also referred to as **decibel level** because relative intensity is measured in units called *decibels* (dB). The decibel is a dimensionless unit because it relates one intensity to another.

The conversion of intensity to decibel level is shown in **Table 13-2.** Notice in **Table 13-2** that when the intensity is multiplied by 10, 10 dB are added to the decibel level. A difference in 10 dB means the sound is approximately twice as loud. Although much more intensity (0.9 W/m^2) is added between 110 and 120 dB than between 10 and 20 dB (9×10^{-11} W/m^2), in each case the volume doubles. Because the volume doubles each time the decibel level increases by 10, sounds at the threshold of pain are 4096 times as loud as sounds at the threshold of hearing.

Table 13-2	Conversion of intensity to decibel level	
Intensity (W/m^2)	**Decibel level (dB)**	**Examples**
1.0×10^{-12}	0	threshold of hearing
1.0×10^{-11}	10	rustling leaves
1.0×10^{-10}	20	quiet whisper
1.0×10^{-9}	30	whisper
1.0×10^{-8}	40	mosquito buzzing
1.0×10^{-7}	50	normal conversation
1.0×10^{-6}	60	air conditioning at 6 m
1.0×10^{-5}	70	vacuum cleaner
1.0×10^{-4}	80	busy traffic, alarm clock
1.0×10^{-3}	90	lawn mower
1.0×10^{-2}	100	subway, power motor
1.0×10^{-1}	110	auto horn at 1 m
1.0×10^{0}	120	threshold of pain
1.0×10^{1}	130	thunderclap, machine gun
1.0×10^{3}	150	nearby jet airplane

FORCED VIBRATIONS AND RESONANCE

When an isolated guitar string is held taut and plucked, hardly any sound is heard. When the same string is placed on a guitar and plucked, the intensity of the sound increases dramatically. What is responsible for this difference? To find the answer to this question, consider a set of pendulums suspended from a beam and bound by a loose rubber band, as shown in **Figure 13-11.** If one of the pendulums is set in motion, its vibrations are transferred by the rubber band to the other pendulums, which will also begin vibrating. This is called a *forced vibration.* ★TEKS **3A, 8B**

The vibrating strings of a guitar force the bridge of the guitar to vibrate, and the bridge in turn transfers its vibrations to the guitar body. These forced vibrations are called *sympathetic vibrations.* The guitar body enables the strings' vibrations to be transferred to the air much more quickly because it has a larger area than the strings. As a result, the intensity of the sound is increased, and the strings' vibrations die out faster than they would if they were not attached to the body of the guitar. In other words, the guitar body allows the energy exchange between the strings and the air to happen more efficiently, thereby increasing the intensity of the sound produced.

In an electric guitar, string vibrations are translated into electrical impulses, which can be amplified as much as desired. An electric guitar can produce sounds that are much more intense than those of an unamplified acoustic guitar, which uses only the forced vibrations of the guitar's body to increase the intensity of the sound from the vibrating strings.

Vibration at the natural frequency produces resonance ★TEKS **8B**

As you saw in Chapter 12, the frequency of a pendulum depends on its string length. Thus, every pendulum will vibrate at a certain frequency, known as its *natural frequency.* In **Figure 13-11,** the two blue pendulums have the same natural frequency, while the red and green pendulums have different natural frequencies. When the first blue pendulum is set in motion, the red and green pendulums will vibrate only slightly, but the second blue pendulum will oscillate with a much larger amplitude because its natural frequency matches the

★TEKS **1A, 3A, 8B**

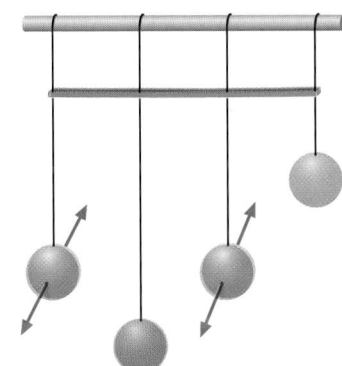

Figure 13-11
If one blue pendulum is set in motion, only the other blue pendulum, whose length is the same, will eventually oscillate with a large amplitude, or resonate.

internetconnect

SCI*LINKS*
NSTA

TOPIC: Resonance
GO TO: www.scilinks.org
*sci*LINKS CODE: HF2134

Resonance

MATERIALS LIST

✔ swing set

Go to a playground, and swing on one of the swings. Try pumping (or being pushed) at different rates—faster than, slower than, and equal to the natural frequency of the swing. Observe whether the rate at which you pump (or are pushed) affects how easily the amplitude of the vibration increases. Are some rates more effective at building your amplitude than others? You should find that the pushes are most effective when they match the swing's natural frequency. Explain how your results support the statement that resonance works best when the frequency of the applied force matches the system's natural frequency.

Demonstration 4

Resonance

Purpose Show resonance with tuning forks.

Materials several tuning forks (including two of the same frequency), two wooden crates, a rubber mallet

Procedure Place the wooden crates 15 cm apart with the open ends facing each other. Using the two forks that have the same frequency, mount a tuning fork on each crate. Strike one fork with the mallet. Students who are listening carefully should begin to hear a faint sound from the second tuning fork. Explain that because energy is transferred through air and wood, the second fork picked up these vibrations because its natural frequency matched that of the first fork. Repeat the experiment with forks of different frequencies to show that resonance does not occur in those cases.

Quick Lab

TEACHER'S NOTES

To extend this activity, have students compare two cases, starting with low frequencies. First have two partners stand on each side of the swing and push the swing so that it gets two pulses of energy per cycle. Then have the same partners stand on the same side and push with the same force once per cycle.

★TEKS

p. 490: 8A, 8B
p. 491: 3A, 8B, 8B, 1A, 3A, 8B

resonance

a condition that exists when the frequency of a force applied to a system matches the natural frequency of vibration of the system

frequency of the pendulum that was initially set in motion. This system is said to be in **resonance.** Since energy is transferred from one pendulum to the other, the amplitude of vibration of the first blue pendulum will decrease as the second blue pendulum's amplitude increases.

A striking example of structural resonance occurred in 1940, when the Tacoma Narrows bridge, in Washington, shown in **Figure 13-12,** was set in motion by the wind. High winds set up standing waves in the bridge, causing the bridge to oscillate at one of its natural frequencies. The amplitude of the vibrations increased until the bridge collapsed. A more recent example of structural resonance occurred during the Loma Prieta earthquake near Oakland, California, in 1989, when part of the upper deck of a freeway collapsed. The collapse of this particular section of roadway has been traced to the fact that the earthquake waves had a frequency of 1.5 Hz, very close to the natural frequency of that section of the roadway. ⊛TEKS 3C

Figure 13-12
On November 7, 1940, the Tacoma Narrows suspension bridge collapsed, just four months after it opened. Standing waves caused by strong winds set the bridge in motion and led to its collapse.

Conceptual Challenge ⊛TEKS 3A, 8B, 8C

1. Concert If a 15-person musical ensemble gains 15 new members, so that its size doubles, will a listener perceive the music created by the ensemble to be twice as loud? Why or why not?

2. A noisy factory Federal regulations require that no office or factory worker be exposed to noise levels that average above 90 dB over an 8 h day. Thus, a factory that currently averages 100 dB must reduce its noise level by 10 dB. Assuming that each piece of machinery produces the same amount of noise, what percentage of equipment must be removed? Explain your answer.

3. Broken crystal Opera singers have been known to set crystal goblets in vibration with their powerful voices. In fact, an amplified human voice can shatter the glass, but only at certain fundamental frequencies. Speculate about why only certain fundamental frequencies will break the glass.

4. Electric guitars Electric guitars, which use electric amplifiers to magnify their sound, can have a variety of shapes, but acoustic guitars must have an hourglass shape. Explain why.

The human ear transmits vibrations that cause nerve impulses

The human ear is divided into three sections—outer, middle, and inner—as shown in **Figure 13-13.** Sound waves from the atmosphere travel down the ear canal of the outer ear. The ear canal terminates at a thin, flat piece of tissue called the eardrum.

The eardrum vibrates with the sound waves and transfers these vibrations to the three small bones of the middle ear, known as the hammer, the anvil, and the stirrup. These bones in turn transmit the vibrations to the inner ear, which contains a snail-shaped tube about 2 cm long called the cochlea.

The cochlea is divided along its length by the basilar membrane, which consists of small hairs and nerve fibers. This membrane has different natural frequencies at different positions. Sound waves of varying frequencies resonate at different spots along the basilar membrane, creating impulses in different nerve fibers. These impulses are then sent to the brain, which interprets them as sounds of varying frequencies. ⓉEKS **3A**

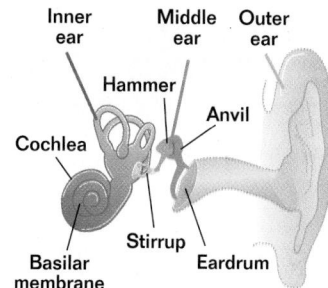

Figure 13-13
Sound waves travel through the three regions of the ear and are then transmitted to the brain as impulses through nerve endings on the basilar membrane.

Section Review
ⓉEKS **2C, 3B, 8B**

1. When the decibel level of traffic in the street goes from 40 to 60 dB, how much louder does the traffic noise seem? How much greater is the intensity?

2. If two flutists play their instruments together at the same intensity, is the sound twice as loud as that of either flutist playing alone at that intensity? Why or why not?

3. Which of the following factors change when a sound gets louder? Which change when a pitch gets higher?
   a. intensity
   b. speed of the sound waves
   c. frequency
   d. decibel level
   e. wavelength
   f. amplitude

4. A tuning fork consists of two metal prongs that vibrate at a single frequency when struck lightly. What will happen if a vibrating tuning fork is placed near another tuning fork of the same frequency? Explain.

5. **Physics in Action**   A certain microphone placed in the ocean is sensitive to sounds emitted by dolphins. To produce a usable signal, sound waves striking the microphone must have a decibel level of 10 dB. If dolphins emit sound waves with a power of 0.050 W, how far can a dolphin be from the microphone and still be heard? (Assume the sound waves propagate spherically, and disregard absorption of the sound waves.)

Section Review
ANSWERS

1. The traffic noise seems about four times as loud. The intensity has increased by a factor of 100 (10^2).

2. no; To double the volume, the intensity would need to increase by a factor of 10. In this case, the intensity has only doubled.

3. a, d, f; c, e

4. The second tuning fork will pick up the vibrations of the first tuning fork, and a faint sound will be heard from the second fork. This occurs because the two forks have the same natural frequency, which is the condition required for resonance.

5. 2.0×10^4 m

ⓉEKS
p. 492: 3C, 3A, 8B, 8C
p. 493: 3A, 2C, 3B, 8B

13-3
Harmonics

13-3 SECTION OBJECTIVES

- **Differentiate between the harmonic series of open and closed pipes.**

- **Calculate the harmonics of a vibrating string and of open and closed pipes.**

- **Relate harmonics and timbre.**

- **Relate the frequency difference between two waves to the number of beats heard per second.**

Figure 13-14
The vibrating strings of a violin produce standing waves whose frequencies depend on the string lengths.

fundamental frequency

the lowest frequency of vibration of a standing wave

STANDING WAVES ON A VIBRATING STRING

As discussed in Chapter 12, a variety of standing waves can occur when a string is fixed at one end and set into vibration at the other by a tuning fork or your moving hand. The vibrations on the string of a musical instrument, such as the violin in **Figure 13-14,** usually consist of many standing waves together at the same time, each of which has a different wavelength and frequency. So the sounds you hear from a stringed instrument, even those that sound like a single pitch, actually consist of multiple frequencies. ⊛TEKS **8A, 8B**

Table 13-3, on page 495, shows several possible vibrations on an idealized string. The ends of the string, which cannot vibrate, must always be nodes. The simplest vibration that can occur is shown in the first row of **Table 13-3.** In this case, the center of the string experiences the most displacement, and so it is an antinode. Because the distance from one node to the next is always half a wavelength, the string length must equal $\lambda_1/2$. Thus, the wavelength is twice the string length ($\lambda_1 = 2L$).

As described in Chapter 12, the speed of a wave equals the frequency times the wavelength, which can be rearranged as shown.

$$v = f\lambda, \text{ so } f = \frac{v}{\lambda}$$

By substituting the value for wavelength found above into this equation for frequency, we see that the frequency of this vibration is equal to the speed of the wave divided by twice the string length.

$$\text{fundamental frequency} = f_1 = \frac{v}{\lambda_1} = \frac{v}{2L} \quad ⊛\text{TEKS} \ \textbf{3B}$$

This frequency of vibration is called the **fundamental frequency** of the vibrating string. Because frequency is inversely proportional to wavelength and because we are considering the greatest possible wavelength, the fundamental frequency is the lowest possible frequency of a standing wave.

Harmonics are integral multiples of the fundamental frequency

The next possible standing wave for a string is shown in the second row of **Table 13-3.** In this case, there are three nodes instead of two, so the string length is equal to one wavelength. Because this wavelength is half the previous wavelength, the frequency of this wave is twice as much. ⊛TEKS **8A, 8B**

$$f_2 = 2f_1$$

Table 13-3   The harmonic series

$\lambda_1 = 2L$	f_1	fundamental frequency, or first harmonic
$\lambda_2 = L$	$f_2 = 2f_1$	second harmonic
$\lambda_3 = \frac{2}{3}L$	$f_3 = 3f_1$	third harmonic
$\lambda_4 = \frac{1}{2}L$	$f_4 = 4f_1$	fourth harmonic

This pattern continues, and the frequency of the standing wave shown in the third row of **Table 13-3** is three times the fundamental frequency. Hence, the frequencies of the standing wave patterns are all integral multiples of the fundamental frequency. These frequencies form what is called a **harmonic series.** The fundamental frequency (f_1) corresponds to the first harmonic, the next frequency (f_2) corresponds to the second harmonic, and so on.

Because each harmonic is an integral multiple of the fundamental frequency, the equation for the fundamental frequency can be generalized to include the entire harmonic series. Thus, $f_n = nf_1$, where f_1 is the fundamental frequency ($f_1 = \frac{v}{2L}$) and f_n is the frequency of the nth harmonic. Note that v is the speed of waves on the vibrating string and not the speed of the resultant sound waves in air.

harmonic series

a series of frequencies that includes the fundamental frequency and integral multiples of the fundamental frequency

HARMONIC SERIES OF STANDING WAVES ON A VIBRATING STRING

$$f_n = n\frac{v}{2L} \quad n = 1, 2, 3, \ldots \quad \bigstar \text{TEKS} \quad \textbf{3B, 8A, 8B}$$

$$\text{frequency} = \text{harmonic number} \times \frac{\text{(speed of waves on the string)}}{\text{(2)(length of vibrating string)}}$$

When a guitar player presses down on a guitar string at any point, that point becomes a node and only a portion of the string vibrates. As a result, a single string can be used to create a variety of fundamental frequencies. In the previous equation, L refers to the portion of the string that is vibrating.

Teaching Tip

As a memory aid, have students visualize the first harmonic. On a string fixed at each end, both ends must be nodes; hence, one bulge, which is $\frac{1}{2}\lambda$ (or a multiple of $\frac{1}{2}\lambda$ for other harmonics), must fit in one L.

Visual Strategy

Table 13-3

Be sure students understand that each bulge corresponds to half a wavelength.

Q Find the wavelength (λ_5) and frequency (f_5) for the next possible case in the harmonic series.

A $\lambda_5 = \frac{2}{5}L, f_5 = 5f_1$

Teaching Tip

Point out that frequency depends on both string length and wave speed, as shown by the equation for the harmonic series. Thus, two strings of the same length will not necessarily have the same fundamental frequency. The string's tension and mass per unit length affect the speed of waves on the string, so the fundamental frequency can also be changed by varying either of these factors.

⭐ TEKS

p. 494: 8A, 8B, 3B, 8A, 8B
p. 495: 3B, 8A, 8B

Teaching Tip

The memory device used for a vibrating string can be used for open and closed pipes as well. Ask students to visualize the standing wave, remembering that closed ends must be nodes and open ends must be antinodes. Sketching diagrams for each case will show that pipes closed at one end must have half a bulge ($\frac{1}{4}\lambda$) more than pipes open at both ends.

Figure 13-15
The harmonic series present in each of these organ pipes depends on whether the end of the pipe is open or closed.

Did you know?

A flute is similar to a pipe open at both ends. When all keys of a flute are closed, the length of the vibrating air column is approximately equal to the length of the flute. As the keys are opened one by one, the length of the vibrating air column decreases, and the fundamental frequency increases.

STANDING WAVES IN AN AIR COLUMN ⊛TEKS 8B

Standing waves can also be set up in a tube of air, such as the inside of a trumpet, the column of a saxophone, or the pipes of an organ like those shown in **Figure 13-15.** While some waves travel down the tube, others are reflected back upward. These waves traveling in opposite directions combine to produce standing waves. Many brass instruments and woodwinds produce sound by means of these vibrating air columns.

If both ends of a pipe are open, all harmonics are present

The harmonic series present in an organ pipe depends on whether the reflecting end of the pipe is open or closed. When the reflecting end of the pipe is open, as is illustrated in **Figure 13-16,** the air molecules have complete freedom of motion, so an antinode exists at this end. If a pipe is open at both ends, each end is an antinode. This is the exact opposite of a string fixed at both ends, where both ends are nodes.

Because the distance from one node to the next ($\frac{1}{2}\lambda$) equals the distance from one antinode to the next, the pattern of standing waves that can occur in a pipe open at both ends is the same as that of a vibrating string. Thus, the entire harmonic series is present in this case, as shown in **Figure 13-16,** and our earlier equation for the harmonic series of a vibrating string can be used.

HARMONIC SERIES OF A PIPE OPEN AT BOTH ENDS

$$f_n = n\frac{v}{2L} \quad n = 1, 2, 3, \ldots \quad \text{⊛TEKS 3B, 8A, 8B}$$

$$\text{frequency} = \text{harmonic number} \times \frac{\text{(speed of sound in the pipe)}}{\text{(2)(length of vibrating air column)}}$$

In this equation, L represents the length of the vibrating air column. Just as the fundamental frequency of a string instrument can be varied by changing the string length, the fundamental frequency of many woodwind and brass instruments can be varied by changing the length of the vibrating air column.

Harmonics in an open-ended pipe

Figure 13-16
In a pipe open at both ends, each end is an antinode, and all harmonics are present. Shown here are the **(a)** first, **(b)** second, and **(c)** third harmonics.

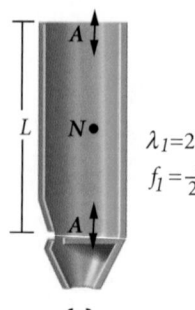

$\lambda_1 = 2L$
$f_1 = \frac{v}{2L}$

(a)

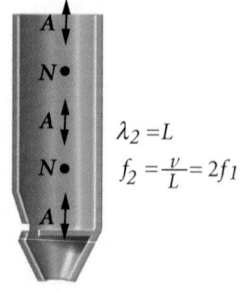

$\lambda_2 = L$
$f_2 = \frac{v}{L} = 2f_1$

(b)

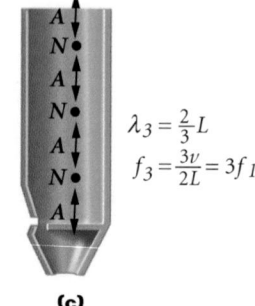

$\lambda_3 = \frac{2}{3}L$
$f_3 = \frac{3v}{2L} = 3f_1$

(c)

Harmonics in a pipe closed at one end

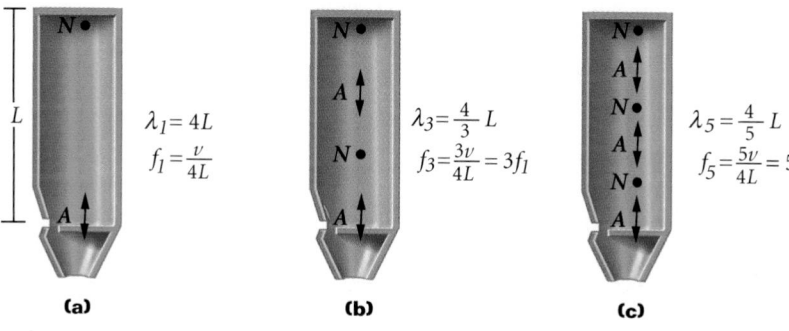

$\lambda_1 = 4L$

$f_1 = \dfrac{v}{4L}$

$\lambda_3 = \dfrac{4}{3}L$

$f_3 = \dfrac{3v}{4L} = 3f_1$

$\lambda_5 = \dfrac{4}{5}L$

$f_5 = \dfrac{5v}{4L} = 5f_1$

(a)　　　　(b)　　　　(c)

Figure 13-17

In a pipe closed at one end, the closed end is a node and the open end is an antinode. In this case, only the odd harmonics are present. The **(a)** first, **(b)** third, and **(c)** fifth harmonics are shown here.

⊛ TEKS　**1A, 3A, 8B**

If one end of a pipe is closed, only odd harmonics are present

When one end of an organ pipe is closed, as is illustrated in **Figure 13-17,** the movement of air molecules is restricted at this end, making this end a node. In this case, one end of the pipe is a node and the other is an antinode. As a result, a different set of standing waves can occur. ⊛ TEKS **8A**

As shown in **Figure 13-17(a),** the simplest possible standing wave that can exist in this pipe is equal to one-fourth of a wavelength. The wavelength of this standing wave equals four times the length of the pipe. Thus, in this case, the fundamental frequency equals the velocity divided by four times the pipe length.

$$f_1 = \dfrac{v}{\lambda_1} = \dfrac{v}{4L}$$

For the case shown in **Figure 13-17(b),** there is three-fourths of a wavelength in the pipe, so the wavelength is four-thirds the length of the pipe ($\lambda_3 = \frac{4}{3}L$). Substituting this value into the equation for frequency gives the frequency of this harmonic.

$$f_3 = \dfrac{v}{\lambda_3} = \dfrac{v}{\frac{4}{3}L} = \dfrac{3v}{4L} = 3f_1$$

Notice that the frequency of this harmonic is *three* times the fundamental frequency. Repeating this calculation for the case shown in **Figure 13-17(c)** gives a frequency equal to *five* times the fundamental frequency. Thus, only the odd-numbered harmonics vibrate in a pipe closed at one end.

As with the vibrating string, we can generalize the equation for the harmonic series of a pipe closed at one end.

HARMONIC SERIES OF A PIPE CLOSED AT ONE END

$$f_n = n\dfrac{v}{4L} \quad n = 1, 3, 5, \dots \quad ⊛\text{TEKS } \mathbf{3B}$$

$$\text{frequency} = \text{harmonic number} \times \dfrac{\text{(speed of sound in the pipe)}}{\text{(4)(length of vibrating air column)}}$$

Quick Lab

A Pipe Closed at One End

MATERIALS LIST

✔ straw

✔ scissors

⬛ SAFETY CAUTION

Always use caution when working with scissors.

Snip off the corners of one end of the straw so that the end tapers to a point, as shown above. Chew on this end to flatten it, and you create a double-reed instrument! Put your lips around the tapered end of the straw, press them together tightly, and blow through the straw. When you hear a steady tone, slowly snip off pieces of the straw at the other end. Be careful to keep about the same amount of pressure with your lips. How does the pitch change as the straw becomes shorter? How can you account for this change in pitch? You may be able to produce more than one tone for any given length of the straw. How is this possible?

Quick Lab

TEACHER'S NOTES

This lab demonstrates the effect of pipe length on pitch. The experiment works best with narrow paper straws.

Students should find that the shorter the straw is, the higher the fundamental frequency is. One length of straw can produce more than one possible tone because there are several possible wavelengths of standing waves for a given length of straw.

The sound frequencies can be demonstrated visually by blowing into a microphone connected to an oscilloscope and watching the results on the oscilloscope screen.

⊛ TEKS

p. 496: **8B, 3B, 8A, 8B**

p. 497: **1A, 3A, 8B, 8A, 3B**

The following may be used as a teamwork exercise or for demonstration at the chalkboard or on an overhead projector.

PROBLEM

Harmonics

One string on a toy guitar is 34.5 cm long.

a. What is the wavelength of its first harmonic?

b. The string is plucked, and the speed of waves on the string is 410 m/s. What are the first three harmonics?

Answer

    a. 69.0 cm

    b. 590 Hz, 1200 Hz, 1800 Hz

Alternative Problem-Solving Approach

For the open pipe, both open ends must be antinodes; thus $L = \frac{1}{2}\lambda_1$, or $\lambda_1 = 2L = 2(2.45 \text{ m}) = 4.90$ m. The first harmonic can now be found with the equation $f = \frac{v}{\lambda}$, as follows:

$$f_1 = \frac{v}{\lambda_1} = \frac{345 \text{ m/s}}{4.90 \text{ m}} = 70.4 \text{ Hz}$$

For the closed pipe, one end is a node and one is an antinode; hence, $L = \frac{1}{4}\lambda_1$, or $\lambda_1 = 4L = 4(2.45\text{m}) = 9.80$ m. Thus, the first harmonic is as follows:

$$f_1 = \frac{v}{\lambda_1} = \frac{345 \text{ m/s}}{9.80 \text{ m}} = 35.2 \text{ Hz}$$

The other harmonics can be found by multiplying the harmonic number by the fundamental frequency, as in Sample Problem 13B.

Trumpets, saxophones, and clarinets are similar to a pipe closed at one end. Although a trumpet has two open ends, the player's mouth effectively closes one end of the instrument. In a saxophone or a clarinet, the reed closes one end.

Despite the similarity between these instruments and a pipe closed at one end, our equation for the harmonic series of pipes does not directly apply to such instruments because any deviation from the cylindrical shape of a pipe affects the harmonic series of an instrument. For example, a clarinet is primarily cylindrical, but because the open end of the instrument is bell-shaped, there are some even harmonics in a clarinet's tone at relatively small intensities. The shape of a saxophone is such that the harmonic series in a saxophone is similar to a cylindrical pipe open at both ends even though only one end of the saxophone is open. These deviations are in part responsible for the variety of sounds that can be produced by different instruments. **(★)TEKS 8B**

SAMPLE PROBLEM 13B

Harmonics (★)TEKS 2C, 3B, 8A, 8B

PROBLEM

What are the first three harmonics in a 2.45 m long pipe that is open at both ends? What are the first three harmonics of this pipe when one end of the pipe is closed? Assume that the speed of sound in air is 345 m/s for both of these situations.

SOLUTION

1. DEFINE   **Given:** $L = 2.45$ m $v = 345$ m/s

              Unknown: Pipe open at both ends: f_1 f_2 f_3

                                 Pipe closed at one end: f_1 f_3 f_5

2. PLAN   **Choose an equation(s) or situation:**

When the pipe is open, all harmonics are present. Thus, the fundamental frequency can be found by using the equation for the entire harmonic series, given on page 496:

$$f_n = n\frac{v}{2L}, \; n = 1, 2, 3, \dots$$

When the pipe is closed at one end, only odd harmonics are present. In this case, the fundamental frequency is found by using the equation for the odd harmonic series, given on page 497:

$$f_n = n\frac{v}{4L}, \; n = 1, 3, 5, \dots$$

In both cases, the second two harmonics can be found by multiplying the harmonic numbers by the fundamental frequency.

3. CALCULATE For a pipe open at both ends:

$$f_1 = n\frac{v}{2L} = (1)\left(\frac{345 \text{ m/s}}{(2)(2.45 \text{ m})}\right) = \boxed{70.4 \text{ Hz}}$$

Because all harmonics are present in this case, the next two harmonics are the second and the third:

$$f_2 = 2f_1 = (2)(70.4 \text{ Hz}) = \boxed{141 \text{ Hz}}$$

$$f_3 = 3f_1 = (3)(70.4 \text{ Hz}) = \boxed{211 \text{ Hz}}$$

For a pipe closed at one end:

$$f_1 = n\frac{v}{4L} = (1)\left(\frac{345 \text{ m/s}}{(4)(2.45 \text{ m})}\right) = \boxed{35.2 \text{ Hz}}$$

Only the odd harmonics are present in this case, so the next possible harmonics are the third and the fifth:

$$f_3 = 3f_1 = (3)(35.2 \text{ Hz}) = \boxed{106 \text{ Hz}}$$

$$f_5 = 5f_1 = (5)(35.2 \text{ Hz}) = \boxed{176 \text{ Hz}}$$

4. EVALUATE In a pipe open at both ends, the first possible wavelength is $2L$; in a pipe closed at one end, the first possible wavelength is $4L$. Because frequency and wavelength are inversely proportional, the fundamental frequency of the open pipe should be twice that of the closed pipe, that is, $70.4 = (2)(35.2)$.

PRACTICE 13B

Harmonics ⭐TEKS 2C, 3B, 8A, 8B

1. What is the fundamental frequency of a 0.20 m long organ pipe that is closed at one end, when the speed of sound in the pipe is 352 m/s?

2. A flute is essentially a pipe open at both ends. The length of a flute is approximately 66.0 cm. What are the first three harmonics of a flute when all keys are closed, making the vibrating air column approximately equal to the length of the flute? The speed of sound in the flute is 340 m/s.

3. What is the fundamental frequency of a guitar string when the speed of waves on the string is 115 m/s and the effective string lengths are as follows:

   a. 70.0 cm **b.** 50.0 cm **c.** 40.0 cm

4. A violin string that is 50.0 cm long has a fundamental frequency of 440 Hz. What is the speed of the waves on this string?

PRACTICE GUIDE 13B

Solving for:		
f_n	📖 **PE**	Sample, 1–3; Ch. Rvw. 39–40, 41b–c*, 43a, 47
	💿 **PW**	7–9
	PB	5–7
v	📖 **PE**	4
	💿 **PW**	Sample, 1–3
	PB	8–10
L	📖 **PE**	Ch. Rvw. 41a, 43b*, 51*
	💿 **PW**	4–6
	PB	Sample, 1–4

ANSWERS TO

Practice 13B
Harmonics
1. 440 Hz
2. 260 Hz, 520 Hz, 780 Hz
3. **a.** 82.1 Hz
   b. 115 Hz
   c. 144 Hz
4. 440 m/s

⭐TEKS

p. 498: 8B, 2C, 3B, 8A, 8B
p. 499: 2C, 3B, 8A, 8B

Teaching Tip

Explain the distinction sometimes made between musical sounds and noise. When individual waveforms are regular, the resultant waveform, although complex, is composed of repeating patterns. On the other hand, when the individual waveforms are not periodic, the resultant waveform is nonrepeating. Sounds that are considered to be musical typically have waveforms with repeating patterns, while those that are considered to be noise do not.

Consumer Focus

BACKGROUND

The discussion of harmonics in this section considers the interference of the sound waves from a single instrument. This feature discusses the interference that occurs when echoes interfere with the original sound waves and with one another.

EXTENSION

Students may be surprised to learn that reverberation is an important consideration when buildings are being designed. This opportunity can be used to discuss acoustical engineering as a possible career choice for students who are interested in acoustics.

500

internet connect

SC_iLINKS.

NSTA

TOPIC: Acoustics
GO TO: www.scilinks.org
*sci*LINKS CODE: HF2132

timbre

the quality of a steady musical sound that is the result of a mixture of harmonics present at different intensities

Harmonics account for sound quality, or timbre ★TEKS 8A, 8B

Table 13-4, on page 501, shows the harmonics present in a tuning fork, a clarinet, and a viola when each sounds the musical note A-natural. Each instrument has its own characteristic mixture of harmonics at varying intensities.

The harmonics shown in the second column of **Table 13-4** add together according to the principle of superposition to give the resultant waveform shown in the third column. Since a tuning fork vibrates at only its fundamental frequency, its waveform is simply a sine wave. (Some tuning forks also vibrate at higher frequencies when they are struck hard enough.) The waveforms of the other instruments are more complex because they consist of many harmonics, each at different intensities. Each individual harmonic waveform is a sine wave, but the resultant wave is more complex than a sine wave because each individual waveform has a different frequency.

The different waveforms shown in the third column of **Table 13-4** explain why a clarinet sounds different from a viola, even when both instruments are sounding the same note at the same volume. In music, the mixture of harmonics that produces the characteristic sound of an instrument is referred to as the spectrum of the sound, which results in a response in the listener called sound quality, or **timbre.** The rich harmonics of most instruments provide a much fuller sound than that of a tuning fork.

Consumer Focus   *Reverberation* ★TEKS 3C, 8B, 8C

*A*uditoriums, churches, concert halls, libraries, and music rooms are designed with specific functions in mind. One auditorium may be made for rock concerts, while another is constructed for use as a lecture hall. Your school's auditorium, for instance, may allow you to hear a speaker well but make a band sound damped and muffled.

Rooms are often constructed so that sounds made by a speaker or a musical instrument bounce back and forth against the ceiling, walls, floor, and other surfaces. This repetitive echo is called *reverberation.* The reverberation time is the amount of time it takes for a sound's intensity to decrease by 60 dB.

For speech, the auditorium should be designed so that the reverberation time is relatively short. A repeated echo of each word could become confusing to listeners.

Music halls may also differ in construction depending on the type of music usually played there. For

example, rock music is generally less pleasing with a large amount of reverberation, but more reverberation is sometimes desired for orchestral and choral music.

For these reasons, you may notice a difference in the way ceilings, walls, and furnishings are designed in different rooms. Ceilings designed for a lot of reverberation are flat and hard. Ceilings in libraries and other quiet places are often made of soft or textured material to muffle sounds. Padded furnishings and plants can also be strategically arranged to absorb sound. All of these different factors are considered and combined to accomodate the auditory function of a room.

Table 13-4   Harmonics of a tuning fork, a clarinet, and a viola at the same pitch

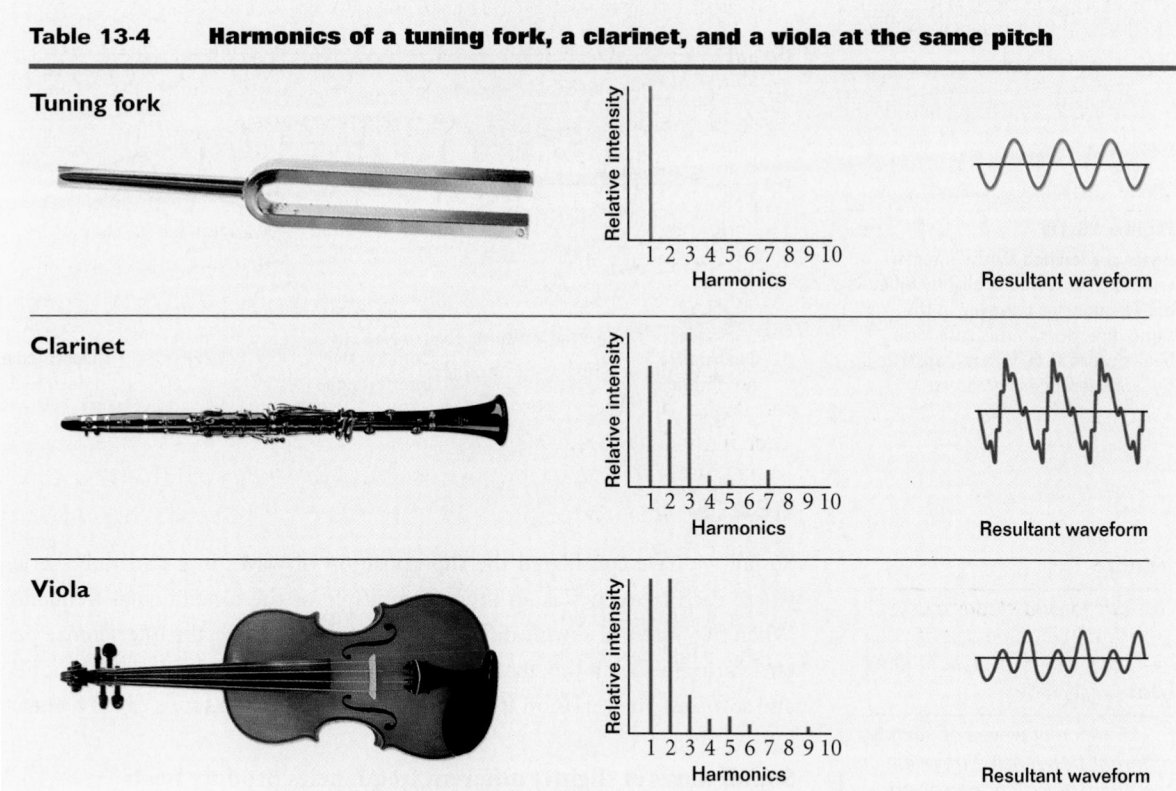

The intensity of each harmonic varies within a particular instrument, depending on frequency, amplitude of vibration, and a variety of other factors. With a violin, for example, the intensity of each harmonic depends on where the string is bowed, the speed of the bow on the string, and the force the bow exerts on the string. Because there are so many factors involved, most instruments can produce a wide variety of tones.

Even though the waveforms of a clarinet and a viola are more complex than those of a tuning fork, note that each consists of repeating patterns. Such waveforms are said to be *periodic*. These repeating patterns occur because each frequency is an integral multiple of the fundamental frequency.

Fundamental frequency determines pitch ⭐TEKS 8A, 8B

As you saw in Section 13-1, the frequency of a sound determines its pitch. In musical instruments, the fundamental frequency of a vibration typically determines pitch. Other harmonics are sometimes referred to as overtones. In the chromatic (half-step) musical scale, there are 12 notes, each of which has a characteristic frequency. The frequency of the thirteenth note is exactly twice that of the first note, and together the 13 notes constitute an *octave*. For stringed instruments and open-ended wind instruments, the frequency of the second harmonic of a note corresponds to the frequency of the octave above that note.

Visual Strategy

Table 13-4

Point out that the second column in **Table 13-4** represents the spectrum of wave amplitudes of harmonics produced by each instrument. Each line on that spectrum corresponds to a sound wave.

Q If the tuning fork sounds a fundamental A (220 Hz), which frequencies does the clarinet produce? Assume the clarinet is like an ideal pipe closed on one end.

A f_1 (220 Hz), f_2 (440 Hz), f_3 (660 Hz), f_4 (880 Hz), f_5 (1100 Hz), and f_7 (1540 Hz)

Q List the frequencies produced by the clarinet in the previous question in order of intensity, from greatest to least.

A 660 Hz, 220 Hz, 440 Hz, 1540 Hz, 880 Hz, 1100 Hz

⭐TEKS

p. 500: 8A, 8B, 3C, 8B, 8C

p. 501: 8A, 8B

Figure 13-18

Be sure students understand that this figure depicts two waves at a particular point in space as time passes (rather than an expanse of space at an instant of time, as in previous wave representations).

Q At what times are the two waves exactly out of phase? At what times are the two waves exactly in phase?

A t_1 and t_3; t_2

ANSWERS TO

Conceptual Challenge

1. The number of beats heard is the frequency difference between the tuning fork and the piano wire. By adjusting the wire until there are no beats heard, the tuner can match the wire's frequency to the tuning fork's frequency.

2. The fundamental frequencies are getting closer together because if the number of beats heard each second is decreasing, the two waves are closer to being completely in phase at all points.

3. The speed of the sound waves in air will not be the same as the speed of waves on the string. There is no frequency change because the vibrations are still occurring at the same rate, so the wavelength must change with the wave speed (because $v = \lambda f$).

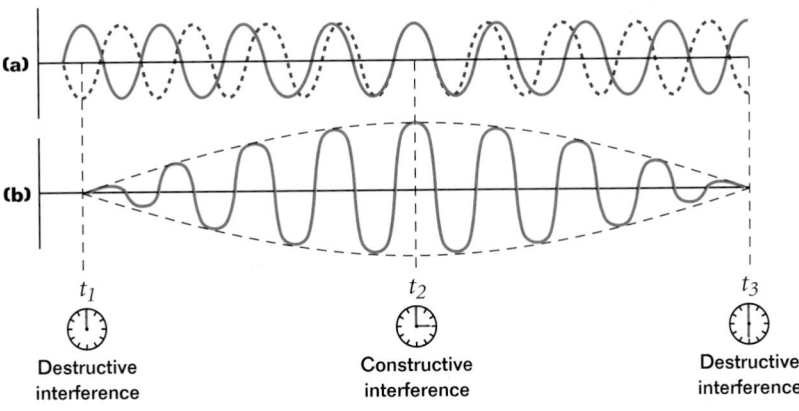

Figure 13-18

Beats are formed by the interference of two waves of slightly different frequencies traveling in the same direction. In this case, one beat occurs at t_2, where constructive interference is greatest.

(a)

(b)

t_1 — Destructive interference

t_2 — Constructive interference

t_3 — Destructive interference

beat

interference of waves of slightly different frequencies traveling in the same direction, perceived as a variation in loudness

BEATS

So far, we have considered the superposition of waves in a harmonic series, where each frequency is an integral multiple of the fundamental frequency. When two waves of *slightly* different frequencies interfere, the interference pattern varies in such a way that a listener hears an alternation between loudness and softness. The variation from soft to loud and back to soft is called a **beat**.

Sound waves at slightly different frequencies produce beats

Figure 13-18 shows how beats occur. In **Figure 13-18(a),** the waves produced by two tuning forks of different frequencies start exactly opposite one another. These waves combine according to the superposition principle, as shown in **Figure 13-18(b).** When the two waves are exactly opposite one another, they are said to be *out of phase*, and complete destructive interference occurs. For this reason, no sound is heard at t_1. (★ TEKS) **8A, 8B**

Because these waves have different frequencies, after a few more cycles, the crest of the blue wave matches up with the crest of the red wave, as at t_2. At this

Conceptual Challenge (★ TEKS) 3A, 8A, 8B

1. Piano tuning How does a piano tuner use a tuning fork to adjust a piano wire to a certain fundamental frequency?

2. Concert violins Before a performance, musicians tune their instruments to match their fundamental frequencies. If a conductor hears the number of beats decreasing as two violin players are tuning, are the fundamental frequencies of these violins becoming closer together or farther apart? Explain.

3. Sounds from a guitar Will the speed of waves on a vibrating guitar string be the same as the speed of the sound waves in the air that are generated by this vibration? How will the frequency and wavelength of the waves on the string compare with the frequency and wavelength of the sound waves in the air?

point, the waves are said to be *in phase*. Now constructive interference occurs, and the sound is louder. Because the blue wave has a higher frequency than the red wave, the waves are out of phase again at t_3, and no sound is heard.

As time passes, the waves continue to be in and out of phase, the interference constantly shifts between constructive interference and destructive interference, and the listener hears the sound getting softer and louder and then softer again. You may have noticed a similar phenomenon on a playground swing set. If two people are swinging next to one another at different frequencies, the two swings may alternate between being in phase and being out of phase.

The number of beats per second corresponds to the difference between frequencies ⓧTEKS 8A, 8B

In our previous example, there is one beat, which occurs at t_2. One beat corresponds to the blue wave gaining one entire cycle on the red wave. This is because to go from one destructive interference to the next, the red wave must lag one entire cycle behind the blue wave. If the time that lapses from t_1 to t_3 is one second, then the blue wave completes one more cycle per second than the red wave. In other words, its frequency is greater by 1 Hz. By generalizing this, you can see that the frequency difference between two sounds can be found by the number of beats heard per second.

Section Review   ⓧTEKS 2C, 3B, 8A, 8B

1. On a piano, the note middle C has a fundamental frequency of 264 Hz. What is the second harmonic of this note?

2. If the piano wire in item 1 is 66.0 cm long, what is the speed of waves on this wire?

3. A piano tuner using a 392 Hz tuning fork to tune the wire for G-natural hears four beats per second. What are the two possible frequencies of vibration of this piano wire?

4. In a clarinet, the reed end of the instrument acts as a node and the first open hole acts as an antinode. Because the shape of the clarinet is nearly cylindrical, its harmonic series approximately follows that of a pipe closed at one end. What harmonic series is predominant in a clarinet?

5. Which of the following must be different for a trumpet and a banjo when notes are being played by both at the same fundamental frequency?
   a. wavelength in air of the first harmonic
   b. number of harmonics present
   c. intensity of each harmonic
   d. speed of sound in air

BACKGROUND

This feature discusses the Doppler effect with light waves. It discusses how the idea of an expanding universe originated from observed red shifts and how this in turn suggests that at some point in the past the universe was infinitely small and dense. The current scientific model of the evolution of matter from the moment of the big bang to the present day is discussed at the end of Chapter 25.

Teaching Tip

The image behind the title of this feature represents small temperature fluctuations in the cosmic microwave background radiation detected by the United States satellite COBE (Cosmic Background Explorer); hot spots are red, and cold spots are blue. These findings, along with experiments that examine temperature fluctuations on a smaller scale, may reveal how clusters of matter such as stars and galaxies formed out of a uniform primordial soup of radiation and particles.

(★)TEKS

p. 504: 8A, 8B, 9B
p. 505: 3A, 3A, 3E

504

PHYSICS ON THE EDGE

The Doppler Effect and the BIG BANG

Earlier in this chapter, you learned that relative motion between the source of sound waves and an observer creates a frequency shift known as the Doppler effect. For visible light, the Doppler effect is observed as a change in color because the frequency of light waves determines color.

Frequency shifts (★)TEKS 8A, 8B, 9B

Of the colors of the visible spectrum, red light has the lowest frequency and violet light has the highest. When the source of electromagnetic waves is moving toward an observer, the frequency detected is higher than the source frequency. This corresponds to a shift toward the blue end of the spectrum, which is called a *blue shift*. When the source of electromagnetic waves is moving away from an observer, the observer detects a lower frequency, which corresponds to a shift toward the red end of the spectrum, called a *red shift*. These two types of frequency shifts are known as *blue shift* and *red shift*, respectively, even though this shift occurs for any type of radiation, not just visible light.

In astronomy, the light from distant stars or galaxies is analyzed by a process called *spectroscopy*. In this process, starlight is passed through a prism or diffraction grating to produce a spectrum. Dark lines appear in the spectrum at specific frequencies determined by the elements present in the atmospheres of stars. When these lines are shifted toward the red end of the spectrum, astronomers know the star is moving away from Earth; when the lines are shifted toward the blue end, the star is moving toward Earth.

Table 13-5   The Doppler effect for light

stationary source	$v = 0$	no shift
approaching source	v	blue shift
receding source	v	red shift

The expansion of the universe  **3A**

As scientists began to study other galaxies with spectroscopy, the results were astonishing: nearly all of the galaxies that were observed exhibited a red shift, which suggested that they were moving away from Earth. If all galaxies are moving away from Earth, the universe must be expanding. This does not suggest that Earth is at the center of the expansion; from any other point in the universe, the same phenomenon would be observed.

The expansion of the universe suggests that at some point in the past the universe must have been confined to a point of infinite density. The eruption of the universe is often referred to as the *big bang,* which is generally considered to have occurred between 10 billion and 20 billion years ago. Current models indicate that the big bang involved such great amounts of energy in such a small space that matter could not form clumps or even individual atoms. It took about 700 000 years for the universe to cool from around 10^{32} K to around 3000 K, a temperature cool enough for atoms to begin forming.

Experimental verification **TEKS** **3A, 3E**

In the 1960s, a group of scientists at Princeton predicted that the explosion of the big bang was so momentous that a small amount of radiation—the leftover glow from the big bang—should still be found in the universe. Around this time, Arno Penzias and Robert Wilson, of Bell Labs, noticed a faint background hiss interfering with satellite-communications experiments they were conducting. This signal, which was detected in equal amounts in all directions, remained despite all attempts to remove it. Penzias and Wilson learned of the Princeton group's work and realized that the interference they were experiencing matched the characteristics of the radiation expected from the big bang. Subsequent experiments have confirmed the existence of this radiation, known as *cosmic microwave background radiation.* This background radiation is considered to be the most conclusive evidence for the big bang theory.

The big bang theory is generally accepted by scientists today. Research now focuses on more-detailed issues. However, there are certain phenomena that the standard big bang model cannot account for, such as the uniform distribution of matter on a large scale and the large-scale clustering of galaxies. As a result, some scientists are currently working on modifications and refinements to the standard big bang theory.

In December 1995, the Hubble Space Telescope obtained an image that reveals galaxies so far away from Earth that their light must have left them 10 billion to 20 billion years ago. This image shows the galaxies as they existed 10 billion to 20 billion years in the past, when the universe was less than a billion years old. As technology improves, scientists can see galaxies even farther away and, hence, even farther back in time. Such observations may resolve many of the current questions regarding the origin of the universe.

Figure 13-19
Penzias and Wilson detected microwave background radiation, presumably left over from the big bang, with the horn antenna (in background) at Bell Telephone Laboratories, in New Jersey.

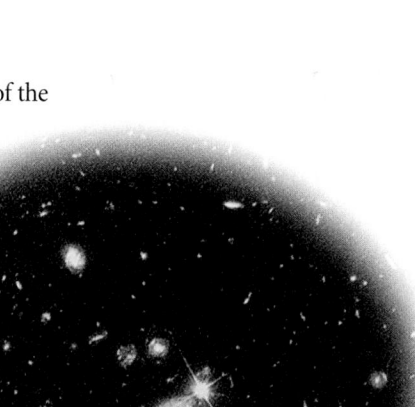

Figure 13-20
This image, called the Hubble Deep Field, is a composite of 342 separate exposures taken by NASA's Hubble Space Telescope during 10 consecutive days. Most of the objects are galaxies, each containing billions of stars. The full image contains over 1500 galaxies, some perhaps dating back to when galaxies first began forming.

Key Models and Analogies

Ask students to visualize a balloon with dots placed randomly on its surface. From the perspective of a single dot, as the balloon is inflated, all other dots appear to be moving away. This is analogous to the expansion of the universe; the fact that we observe all galaxies as moving away from Earth does not mean that Earth is the center of the expansion. The same phenomenon would be observed from any other point in the universe.

EXTENSION

- Have students research various theories regarding the age of the universe. Their reports should include a discussion of the following: How is the Hubble constant related to the age of the universe? What methods are used to measure the distances to stars? How do current theories on the age of the universe differ?

- Have students research Alan Guth's inflation theory, a modification of the standard big bang theory. What difficulties with the big bang theory does the inflation theory resolve?

- Spectroscopy is widely used in astronomy for a variety of purposes. Have students study its uses along with other practical applications of the Doppler effect.

- Have students explore various possibilities for the future of the universe. Their research should include an investigation into the question of how much dark matter the universe contains and how the answer to this question will affect the fate of the universe.

CHAPTER 13
Summary

KEY TERMS

beat (p. 502)

compression (p. 480)

decibel level (p. 490)

Doppler effect (p. 485)

**fundamental frequency
(p. 494)**

harmonic series (p. 495)

intensity (p. 487)

pitch (p. 481)

rarefaction (p. 480)

resonance (p. 492)

timbre (p. 500)

KEY IDEAS

Section 13-1 Sound waves
- The frequency of a sound wave determines its pitch.
- The speed of sound depends on the medium.
- The relative motion between the source of waves and an observer creates an apparent frequency shift known as the Doppler effect.

Section 13-2 Sound intensity and resonance
- The sound intensity of a spherical wave is the power per area, as follows:

$$\text{Intensity} = \frac{P}{4\pi r^2}$$

- Decibel level is a measure of relative intensity on a logarithmic scale.
- A forced vibration at the natural frequency produces resonance.

Section 13-3 Harmonics
- Harmonics of a vibrating string or a pipe open at both ends can be found with the following equation:

$$f_n = n\frac{v}{2L}, \, n = 1, 2, 3, \ldots$$

- Harmonics of a pipe closed at one end can be found with the following equation:

$$f_n = n\frac{v}{4L}, \, n = 1, 3, 5, \ldots$$

- The number and intensity of harmonics account for the sound quality of an instrument, also known as timbre.

Variable symbols

Quantities		Units	
	sound intensity	W/m^2	watts/meters squared
	decibel level	dB	decibels
f_n	frequency of the nth harmonic	Hz	Hertz = s^{-1}
L	length of a vibrating string or an air column	m	meters

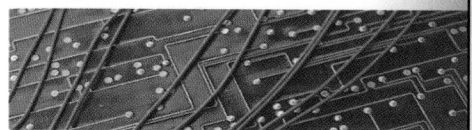
SOUND WAVES

Review questions

1. Why are sound waves in air characterized as longitudinal?

2. Draw the sine curve that corresponds to the sound wave depicted in **Figure 13-21.**

Figure 13-21

3. What is the difference between frequency and pitch?

4. Why can a dog hear a sound produced by a dog whistle, while his owner cannot?

5. What are the differences between infrasonic, audible, and ultrasonic sound waves?

6. Explain why the speed of sound depends on the temperature of the medium. Why is this temperature dependence more noticeable in a gas than in a solid or a liquid?

7. The Doppler effect occurs when
   a. a source of sound moves toward a listener.
   b. a listener moves toward a source of sound.
   c. a listener and a source of sound move away from each other.
   d. a listener and a source of sound move toward each other.
   e. All of the above

8. You are at a street corner and hear an ambulance siren. Without looking, how can you tell when the ambulance passes by?

9. Ultrasound waves are often used to produce images of objects inside the body. Why are ultrasound waves effective for this purpose?

Conceptual questions

10. If the wavelength of a sound source is reduced by a factor of 2, what happens to the wave's frequency? What happens to its speed?

11. As a result of a distant explosion, an observer first senses a ground tremor, then hears the explosion. What accounts for this time lag?

12. By listening to a band or an orchestra, how can you determine that the speed of sound is the same for all frequencies?

13. A sound wave travels in air at a frequency of 500 Hz. If part of the wave travels from air into water, does its frequency change? Does its wavelength change? Note that the speed of sound in air is about 340 m/s, whereas the speed of sound in water is about 1500 m/s.

14. A fire engine is moving at 40 m/s and sounding its horn. A car in front of the fire engine is moving at 30 m/s, and a van in front of the car is stationary. Which observer hears the fire engine's horn at a higher pitch, the driver of the car or the driver of the van?

15. A bat flying toward a wall emits a chirp at 40 kHz. Is the frequency of the echo received by the bat greater than, less than, or equal to 40 kHz?

SOUND INTENSITY AND RESONANCE

Review questions

16. If a sound seems to be getting louder, which of the following is probably increasing?
   a. intensity
   b. frequency
   c. speed of sound
   d. wavelength

17. Intensity is power per area; decibel level is a measure of *relative* intensity; volume doubles when intensity increases by a factor of 10.

18. 90 dB, 30 dB, 20 dB, 60 dB (Answers may vary slightly.)

19. because the threshold of hearing depends on both frequency and intensity

20. when a forced vibration is the same as the natural frequency of an object

21. Intensity decreases by a factor of 9.

22. because intensity decreases with distance and the sound has traveled from the source to a reflecting surface and back

23. The violin's sound intensity is $\frac{1}{100}$ that of the orchestra's, and its volume is $\frac{1}{4}$ that of the orchestra's.

24. 9 machines (for a total of 10)

25. The swing's amplitude is maximized when the pushes match the swing's natural frequency.

26. Vibrations could set the bridge in motion if they match one of the bridge's natural frequencies.

27. 70 dB

28. 7.96×10^{-2} W/m^2

29. the lowest possible frequency; They are integral multiples of the fundamental frequency.

30. **a.** 4.0 m
b. 2.0 m
c. 1.3 m
d. 1.0 m

31. because a closed end is a node, while an open end is an antinode

32. The instruments have different harmonics present at various intensity levels.

17. What is the difference between intensity, decibel level, and volume?

18. Using **Table 13-2** (page 490) as a guide, estimate the decibel levels of the following sounds: a cheering crowd at a football game, background noise in a church, the pages of this textbook being turned, and light traffic.

19. Why is the threshold of hearing represented as a curve in **Figure 13-10** (page 489) rather than as a single point?

20. Under what conditions does resonance occur?

Conceptual questions

21. If the distance from a point source of sound is tripled, by what factor does the sound intensity decrease? Assume there are no reflections from nearby objects to affect your results.

22. Why is the intensity of an echo less than that of the original sound?

23. The decibel level of an orchestra is 90 dB, and a single violin achieves a level of 70 dB. How do the intensity and volume of the sound of the full orchestra compare with those of the violin's sound?

24. A noisy machine in a factory produces a decibel rating of 80 dB. How many identical machines could you add to the factory without exceeding the 90 dB limit set by federal regulations?

25. Why are pushes given to a playground swing more effective if they are given at certain, regular intervals than if they are given at random positions in the swing's cycle?

26. Although soldiers are usually required to march together in step, they must break their march when crossing a bridge. Explain the possible danger of crossing a rickety bridge without taking this precaution.

Practice problems

27. A baseball coach shouts loudly at an umpire standing 5.0 meters away. If the sound power produced by the coach is 3.1×10^{-3} W, what is the decibel level of the sound when it reaches the umpire? (Hint: See Sample Problem 13A, then use **Table 13-2** on page 490.)

28. A stereo speaker represented by P in **Figure 13-22** emits sound waves with a power output of 100.0 W. What is the intensity of the sound waves at point x when $r = 10.0$ m? (See Sample Problem 13A.)

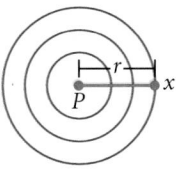

Figure 13-22

HARMONICS

Review questions

29. What is fundamental frequency? How are harmonics related to the fundamental frequency?

30. **Figure 13-23** shows a stretched string vibrating in several of its modes. If the length of the string is 2.0 m, what is the wavelength of the wave on the string in **(a)**, **(b)**, **(c)**, and **(d)**?

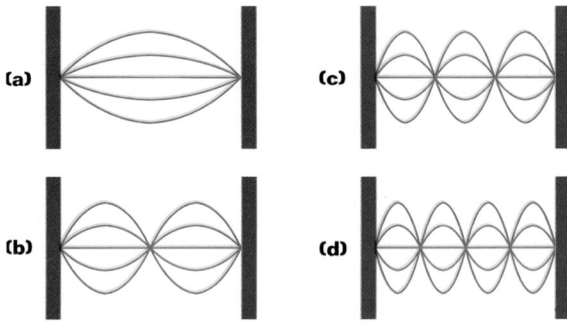

Figure 13-23

31. Why does a pipe closed at one end have a different harmonic series than an open pipe?

32. Explain why a saxophone sounds different from a clarinet, even when they sound the same fundamental frequency at the same decibel level.

Conceptual questions

33. Why does a vibrating guitar string sound louder when it is on the instrument than it does when it is stretched on a work bench?

34. Two violin players tuning their instruments together hear six beats in 2 s. What is the frequency difference between the two violins?

35. What is the purpose of the slide on a trombone and the valves on a trumpet?

36. A student records the first 10 harmonics for a pipe. Is it possible to determine whether the pipe is open or closed by comparing the difference in frequencies between the adjacent harmonics with the fundamental frequency? Explain.

37. A flute is similar to a pipe open at both ends, while a clarinet is similar to a pipe closed at one end. Explain why the fundamental frequency of a flute is about twice that of the clarinet, even though the length of these two instruments is approximately the same.

38. The fundamental frequency of any note produced by a flute will vary slightly with temperature changes in the air. For any given note, will an increase in temperature produce a slightly higher fundamental frequency or a slightly lower one?

Practice problems

39. What are the first three harmonics of a note produced on a 31.0 cm long violin string if waves on this string have a speed of 274.4 m/s? (See Sample Problem 13B.)

40. The human ear canal is about 2.8 cm long and can be regarded as a tube open at one end and closed at the eardrum. What is the fundamental frequency around which we would expect hearing to be best when the speed of sound in air is 340 m/s? (See Sample Problem 13B.)

MIXED REVIEW

41. A pipe that is open at both ends has a fundamental frequency of 320 Hz when the speed of sound in air is 331 m/s.
  a. What is the length of this pipe?
  b. What are the next two harmonics?
  c. What is the fundamental frequency of this pipe when the speed of sound in air is increased to 367 m/s due to a rise in the temperature of the air?

42. The area of a typical eardrum is approximately 5.0×10^{-5} m^2. Calculate the sound power (the energy per second) incident on the eardrum at
  a. the threshold of hearing.
  b. the threshold of pain.

43. The frequency of a tuning fork can be found by the method shown in **Figure 13-24**. A long tube open at both ends is submerged in a beaker of water, and the vibrating tuning fork is placed near the top of the tube. The length of the air column, L, is adjusted by moving the tube vertically. The sound waves generated by the fork are reinforced when the length of the air column corresponds to one of the resonant frequencies of the tube. The largest value for L for which a peak occurs in sound intensity is 9.00 cm. (Use 345 m/s as the speed of sound in air.)

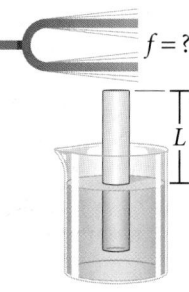

Figure 13-24

  a. What is the frequency of the tuning fork?
  b. What is the value of L for the next two harmonics?

44. When two tuning forks of 132 Hz and 137 Hz, respectively, are sounded simultaneously, how many beats per second are heard?

45. The range of human hearing extends from approximately 20 Hz to 20 000 Hz. Find the wavelengths of these extremes when the speed of sound in air is equal to 343 m/s.

46. A dolphin in 25°C sea water emits a sound directed toward the bottom of the ocean 150 m below. How much time passes before it hears an echo? (See **Table 13-1** on page 482 for the speed of the sound.)

47. An open organ pipe is 2.46 m long, and the speed of the air in the pipe is 345 m/s.
  a. What is the fundamental frequency of this pipe?
  b. How many harmonics are possible in the normal hearing range, 20 Hz to 20 000 Hz?

48. The greatest value ever achieved for the speed of sound in air is about 1.0×10^4 m/s, and the highest frequency ever produced is about 2.0×10^{10} Hz. Find the wavelength of this wave.

49. If you blow across the open end of a soda bottle and produce a tone of 250 Hz, what will be the frequency of the next harmonic heard if you blow much harder?

33. The guitar's body transfers the string's vibrations to the air more rapidly, thereby increasing the intensity of the sound.

34. 3 Hz

35. to change the length of the air column, thereby changing the fundamental frequency

36. yes; This difference will equal the fundamental frequency if the pipe is open at both ends but will equal twice the fundamental frequency if the pipe is closed at one end.

37. The first possible wavelength is $2L$ for the flute and $4L$ for the clarinet. Because the speed of sound is the same in each and $v = \lambda f$, the flute's fundamental frequency is twice the clarinet's.

38. As temperature increases, the speed of sound in air increases. Because f_1 is proportional to v, fundamental frequency likewise increases.

39. 443 Hz, 886 Hz, 1330 Hz

40. 3.0×10^3 Hz

41. a. 52 cm
  b. 640 Hz, 960 Hz
  c. 350 Hz

42. a. 5.0×10^{-17} W
  b. 5.0×10^{-5} W

43. a. 958 Hz
  b. 27.0 cm, 45.0 cm

44. 5 Hz

45. 20 m, 2×10^{-2} m

46. 0.20 s

47. a. 70.1 Hz
  b. 285

48. 5.0×10^{-7} m

49. 750 Hz

50. 3.2×10^3 m

51. $L_{closed} = 1.5 \, (L_{open})$

52. 10 mosquitoes

53. 1.9×10^{-2} m

54. a. 5.0×10^4 W

    b. 2.8×10^{-3} W

Alternative Assessment
ANSWERS

Performance assessment

1. Plans should consider the threshold of pain (120 dB). To achieve levels of 90 to 70 dB, the school must be moved 5 to 50 km. Soundproofing will require work on the school's buildings and landscaping.

2. When the bottle is tapped, the water vibrates. More water corresponds to a lower pitch. When air is blown over the bottle, the air vibrates. More water (less air) corresponds to a higher pitch.

3. Student reports should relate the physics of sound to the physiology of the ear and the workings of hearing aids and testing devices.

4. Soft irregular surfaces do not reflect sound waves, decreasing noise. Plans to reduce echo effects could include using gaps of about 8 cm or insulating materials.

510

---

50. A rock group is playing in a club. Sound emerging outdoors from an open door spreads uniformly in all directions. If the decibel level is 70 dB at a distance of 1.0 m from the door, at what distance is the music just barely audible to a person with a normal threshold of hearing? Disregard absorption.

51. The fundamental frequency of an open organ pipe corresponds to the note middle C ($f = 261.6$ Hz on the chromatic musical scale). The third harmonic (f_3) of another organ pipe that is closed at one end has the same frequency. Compare the lengths of these two pipes.

52. A typical decibel level for a buzzing mosquito is 40 dB, and normal conversation is approximately 50 dB. How many buzzing mosquitoes will produce a sound intensity equal to that of normal conversation?

53. Some studies indicate that the upper frequency limit of hearing is determined by the diameter of the eardrum. The wavelength of the sound wave and the diameter of the eardrum are approximately equal at this upper limit. If this is so, what is the diameter of the eardrum of a person capable of hearing 2.0×10^4 Hz? Assume 378 m/s is the speed of sound in the ear.

54. The decibel level of the noise from a jet aircraft is 130 dB when measured 20.0 m from the aircraft.

    a. How much sound power does the jet aircraft emit?

    b. How much sound power would strike the eardrum of an airport worker 20.0 m from the aircraft? (Use the diameter found in item 53 to calculate the area of the eardrum.)

Alternative Assessment

Performance assessment

1. A new airport is being built 750 m from your school. The noise level 50 m from planes that will land at the airport is 130 dB. In open spaces, such as the fields between the school and the airport, the level decreases by 20 dB each time the distance increases tenfold. Work in a cooperative group to research the options for keeping the noise level tolerable at the school. How far away would the school have to be moved to make the sound manageable? Research the cost of land near your school. What options are available for soundproofing the school's buildings? How expensive are these options? Have each member in the group present the advantages and disadvantages of such options.

2. Use soft-drink bottles and water to make a musical instrument. Adjust the amount of water in different bottles to create musical notes. Play them as percussion instruments (by tapping the bottles) or as wind instruments (by blowing over the mouths of individual bottles). What media are vibrating in each case? What affects the fundamental frequency? Use a microphone and an oscilloscope to analyze your performance and to demonstrate the effects of tuning your instrument.

Portfolio projects

3. Interview members of the medical profession to learn about human hearing. What are some types of hearing disabilities? How are hearing disabilities related to disease, age, and occupational or environmental hazards? What procedures and instruments are used to test hearing? How do hearing aids help? What are the limitations of hearing aids? Present your findings to the class.

4. Do research on the types of architectural acoustics that would affect a restaurant. What are some of the acoustics problems in places where many people gather? How do odd-shaped ceilings, decorative panels, draperies, and glass windows affect echo and noise? Find the shortest wavelengths of sounds that should be absorbed, considering that conversation sounds range from 500 to 5000 Hz. Prepare a plan or a model of your school cafeteria showing what approaches you would use to keep the level of noise to a minimum.

Technology & Learning

Graphing calculators

Refer to Appendix B for instructions on downloading programs for your calculator. The program "Chap13" allows you to analyze a graph of the frequency of a sound versus its apparent frequency to a stationary observer.

As you learned earlier in this chapter, a Doppler effect is experienced whenever there is relative motion between a source of sound and an observer. The frequencies heard by the observer can be described by the following two equations in which f' represents the apparent frequency and f represents the actual frequency.

$$f' = f\left(\frac{v_{sound}}{v_{sound} - v_{source}}\right)$$

$$f' = f\left(\frac{v_{sound}}{v_{sound} + v_{source}}\right)$$

The program "Chap13" stored on your graphing calculator makes use of the Doppler effect equations. Once the "Chap13" program is executed, your calculator will ask for the speed of sound and the speed of the source.

The graphing calculator will use the following equations to create two graphs: the apparent frequency (Y_1) versus the actual frequency (X) as the source approaches the observer, and the apparent frequency (Y_2) versus the actual frequency (X) as the source moves away from the observer. The relationships in these equations are the same as those in the Doppler effect equations shown above.

$$Y_1 = SX/(S-V)$$

$$Y_2 = SX/(S+V)$$

a. Which frequency is higher: Y_1 or Y_2?

Execute "Chap13" on the [PRGM] menu, and press [ENTER] to begin the program. Enter the magnitudes of the speed of sound and the speed of the source (shown below), pressing [ENTER] after each value.

The calculator will provide graphs of the actual frequency versus the apparent frequencies. (If the graphs are not visible, press [WINDOW] and change the settings for the graph window, then press [GRAPH].)

Press [TRACE], and use the arrow keys to trace along the curves. The x-value corresponds to the source's actual frequency in hertz. The y-value in the upper graph corresponds to the frequency of the source as heard by the observer as the source approaches the observer. The y-value in the lower graph corresponds to the frequency of the source as heard by the observer as the source moves away from the observer. Use the [▲] and [▼] keys to toggle between the two graphs.

Determine the apparent frequencies in the following cases (b–e) if the speed of sound is 346 m/s:

b. a car horn tuned to middle C (264 Hz) passing the listener at a speed of 25 m/s

c. a car horn tuned to G (392 Hz) passing the listener at a speed of 25 m/s

d. a trumpet player playing middle C (264 Hz) on a parade float that passes the listener at a speed of 5.0 m/s

e. a trumpet player playing G (392 Hz) on a parade float that passes the listener at a speed of 5.0 m/s

f. Two police cars are in pursuit of a criminal. Car 54 drives past you at 25 m/s, then car 42 passes you at 30 m/s. Both cars have the same siren set to play a constant frequency. Which car's siren will sound the most different when moving toward you versus moving away from you?

Press [2nd] [QUIT] to stop graphing. Press [ENTER] to input new values or [CLEAR] to end the program.

NOTE

NOTE

Materials Preparation is given on pp. 478A–478B. Blank data table and sample data table are on the One-Stop Planner CD-ROM. All calculations shown use sample data.

Planning

Recommended time:

1 lab period

For a 2-period lab, add the Extensions Exercise (p. 515, number 6).

Classroom organization:

▸ Each group must have at least 2 students.

▸ Each group must have a level work surface at least 0.5 m above the floor.

▸ The CBL and recording timer procedures use different methods to find the speed of sound; be aware of these differences if using them in the same class.

▸ **Safety warnings:** Remind students that broken glass must be disposed of in a separate container.

CHAPTER 13
Laboratory Exercise

⊛ TEKS
pp. 512–515: 1A, 1B, 2A, 2B, 2C, 2D, 2E, 2F, 3A, 3B, 8A, 8B

OBJECTIVES

• Find the speed of sound in air.

MATERIALS LIST

✔ Check list for appropriate procedure.

PROCEDURE

CBL AND SENSORS

✔ cardboard tube
✔ CBL
✔ CBL microphone
✔ graphing calculator with link cable
✔ masking tape
✔ meterstick
✔ support stand with clamp
✔ CBL temperature sensor

RESONANCE APPARATUS

✔ 4 tuning forks of different frequencies
✔ Erlenmeyer flask, 1000 mL
✔ resonance apparatus with clamp
✔ thermometer
✔ tuning-fork hammer
✔ water

SPEED OF SOUND

Sound waves can travel through solids, liquids, and gases. The speed of sound in a medium depends on the density of the particles that make up the medium. The speed also depends on the temperature, especially in a gas like air. In air, sound travels faster at higher temperatures and slower at lower temperatures. In this experiment, you will measure the speed of sound in air using one of the methods described below.

• **CBL and sensors** The speed of sound will be determined using a CBL microphone placed directly above the opening of a large tube. A short, sharp noise will be recorded by the microphone at the top of the tube and again after the sound travels down the tube and reflects back to the microphone. You can use the time between recordings and distance traveled by the sound to determine the speed of sound in air.

• **Resonance apparatus** The speed of sound will be determined using a tuning fork to produce resonance in a closed tube. The wavelength of the sound may be calculated from the resonant length of the tube, and the speed of the sound can be calculated from the equation $v = f\lambda$, where v is the speed of sound, f is the frequency of the sound produced by the tuning fork, and λ is the wavelength of the sound.

SAFETY

• **Never put broken glass or ceramics in a regular waste container. Use a dustpan, brush, and heavy gloves to carefully pick up broken pieces and dispose of them in a container specifically provided for this purpose.**

• **If a thermometer breaks, notify the teacher immediately.**

PREPARATION

1. Determine whether you will be using the CBL and sensors procedure or the resonance apparatus procedure. Read the entire lab for the appropriate procedure, and plan what steps you will take.

Resonance apparatus procedure begins on page 514.

PROCEDURE

CBL AND SENSORS

Finding the speed of sound

2. Prepare a data table in your lab notebook with four columns and five rows. In the first row, label the first through fourth columns *Trial, Distance from microphone to bottom of tube (m), Temperature (°C),* and *Time interval (s)*. In the first column, label the second through fifth rows *1, 2, 3,* and *4*.

3. Set up the temperature probe, CBL microphone, ring stand, tube, CBL, and calculator, as shown in **Figure 13-25**. Tape or clamp the tube securely in place. Clamp the CBL microphone to the edge of the table or to a ring stand so that the microphone points down and is directly above the open end of

the tube. Connect the CBL to the graphing calculator. Connect the CBL microphone to the CH1 port and the temperature probe to the CH2 port on the CBL unit. Hang the temperature probe inside the tube to measure the air temperature.

4. Turn on the CBL unit and the calculator. Start the program PHYSICS on the calculator.

   a. Select the *SET UP PROBES* option from the MAIN MENU. Enter 1 for the number of probes. Select the *TEMPERATURE* probe. Enter 2 for the channel number.

   b. Select the *MONITOR INPUT* option from the MAIN MENU. Record the temperature reading in your data table. Press "+" to return to the MAIN MENU.

5. From the MAIN MENU, select the *SET UP PROBES* option. Enter 1 for the number of probes. Select the *MICROPHONE*. Your teacher will tell you what type of microphone you are using. Select the appropriate description from the list on the calculator. From the COLLECTION MODE menu, select *WAVEFORM/TRIGR*. Press ENTER on the graphing calculator.

6. Make a loud, short noise—such as a snap of the fingers—directly above the tube. This will trigger the CBL to collect the sound data.

7. When the CBL unit displays DONE, press ENTER on the calculator.

8. Use the metric ruler to measure the length from the bottom of the CBL microphone to the bottom of the tube. Record this length to the nearest millimeter in the data table.

9. Look at the graph on the graphing calculator, which shows the sound plotted against time in seconds. There should be two peaks on the graph, one near the beginning and one a little later. The first peak is the sound and the second peak is the echo of the sound. Use the arrow keys to trace the graph.

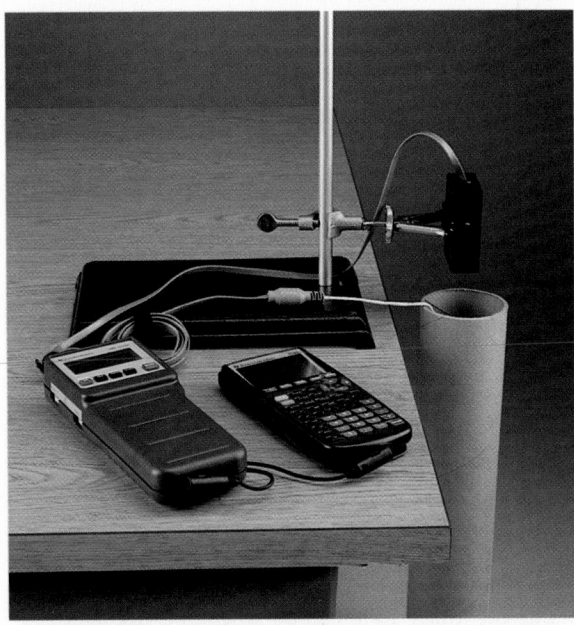

Figure 13-25

Step 6: The CBL will begin collecting sound data as soon as you make a sound, so work quietly until you are ready to begin the experiment. Remain quiet until the CBL displays DONE. Background noise may affect your results.

Step 9: On the graph, the first and second peaks may not be the same height, but they should both be noticeably higher than the other points on the graph. If the sound was too loud, the graph will show many high and low points. Repeat with a softer sound for better results.

CBL and Sensors Tips

◆ Students should have the program PHYSICS on their graphing calculators. Refer to Appendix B for instructions.

◆ The microphone cannot be used with other probes, so students must SET UP PROBES twice.

◆ The sound used should not be too loud (over 70 mV on the CBL); a loud sound causes the graph to have too many peaks.

◆ The graphs for different types of sounds will look very similar; the tube is dominated by its own low-resonance frequency for most sounds.

Techniques to Demonstrate

Show students how to identify the second peak on the graph and how to determine whether the sound was too loud.

✔ Checkpoints

Step 3: Make sure students wear goggles if there are clamps and support rods at eye level.

Step 5: Remind students to work quietly; random noises may affect other groups' results.

Step 6: Any quiet, sharp sound will work well. After making the initial sound, students must be silent until the CBL displays DONE.

Step 8: Students should realize that the sound traveled twice the length of the tube during the time interval.

Step 9: Guide students to find the time values for each peak.

Resonance Apparatus Tips

◆ Adding a few drops of food coloring to the water makes it easier for students to see the water levels in the tube.

◆ Students may need practice striking the tuning fork. They should also practice moving the reservoir up and down.

◆ Values for λ can be improved by using the end correction: effective tube length = $L + 0.4d$, where d is the internal diameter of the tube. This takes into account the air above the tube end that also vibrates. If you want to use this correction, have students measure the internal diameter of the tube.

Techniques to Demonstrate

Demonstrate the difference between resonance and other sounds students will hear during the lab.

✔ Checkpoints

Step 3: Make sure the apparatus is secure. The apparatus should be away from the edge of the table; otherwise use a C-clamp to secure the base of the support stand to the edge of the table. Make sure students wear goggles to protect themselves from clamps and support rods at eye level.

Step 6: The lower tine of the tuning fork should be 1 cm away from the top of the tube.

Step 7: Remind students to work quietly and to move the reservoir very slowly and steadily. They should demonstrate that they have found the greatest resonance.

10. Find the difference between the *x*-values of the two peaks to find the time interval between them. Record the time interval in your data table. Sketch the graph in your lab notebook. Press ENTER on the calculator.

11. Repeat the procedure for several trials. Try different sounds, such as a soft noise, a loud noise, a high-pitched sound, or a low-pitched sound. Record all data in your data table.

12. Clean up your work area. Put equipment away safely so that it is ready to be used again. Recycle or dispose of used materials as directed by your teacher.

Analysis and Interpretation begins on page 515.

PROCEDURE

RESONANCE APPARATUS

Finding the speed of sound

2. Prepare a data table in your lab notebook with four columns and five rows. In the first row, label the first through fourth columns *Trial, Length of tube (m), Frequency (Hz)*, and *Temperature (°C)*. In the first column, label the second through fifth rows *1, 2, 3*, and *4*.

3. Set up the resonance apparatus as shown in **Figure 13-26.**

4. Raise the reservoir so that the top is level with the top of the tube. Fill the reservoir with water until the level in the tube is at the 5 cm mark.

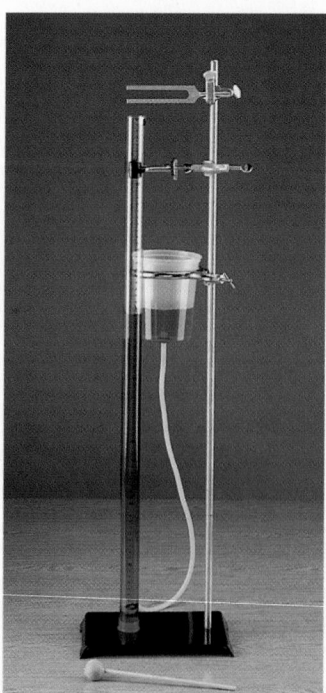

Figure 13-26
Step 7: From the position of greatest resonance, move the reservoir up 2 cm and down again until you find the exact position.

5. Measure and record the temperature of the air inside the tube. Select a tuning fork, and record the frequency of the fork in the data table.

6. Securely clamp the tuning fork in place as shown in the figure, with the lower tine about 1 cm above the end of the tube. Strike the tuning fork sharply, but not too hard, with the tuning-fork hammer to create a vibration. A few practice strikes may be helpful to distinguish the tonal sound of the tuning fork from the unwanted metallic "ringing" sound that may result from striking the fork too hard. ***Do not strike the fork with anything other than a hard rubber mallet.***

7. While the tuning fork is vibrating directly above the tube, slowly lower the reservoir about 20 cm or until you locate the position of the reservoir where the resonance is loudest. (Note: To locate the exact position of the resonance, you may need to strike the tuning fork again while the water level is falling.) Raise the reservoir to about 2 cm above the approximate level where you think the resonance is loudest. Strike the tuning fork with the tuning fork hammer and carefully lower the reservoir about 5 cm until you find the exact position of resonance.

8. Using the scale marked on the tube, record the level of the water in the tube when the resonance is loudest. Record this level to the nearest millimeter in your data table.

9. Repeat the procedure for several trials, using tuning forks of different frequencies.

10. Clean up your work area. Put equipment away safely so that it is ready to be used again. Recycle or dispose of used materials as directed by your teacher.

ANALYSIS AND INTERPRETATION

Calculations and data analysis

1. Organizing data

    a. CBL and sensors   For each trial, calculate the total distance the sound traveled by multiplying the distance measured by 2.

    b. Resonance apparatus   For each trial, calculate the wavelength of the sound by using the equation for the fundamental wavelength, $\lambda = 4L$, where L is the length of the tube.

2. Analyzing data   For each trial, find the speed of sound.

    a. CBL and sensors   Use the values for distance traveled and the time interval from your data table to find the speed for each trial.

    b. Resonance apparatus   Use the equation $v = f\lambda$, where f is the frequency of the tuning fork.

3. Evaluating data   Find the *accepted* value for the speed of sound in air at room temperature (see page 482, **Table 13-1**). Find the average of your results for the speed of sound, and use the average as the experimental value.

    a. Compute the absolute error using the following equation:

$$\text{absolute error} = |\text{experimental} - \text{accepted}|$$

    b. Compute the relative error using the following equation:

$$\text{relative error} = \frac{(\text{experimental} - \text{accepted})}{\text{accepted}}$$

Conclusions

4. Applying conclusions   Based on your results, is the speed of sound in air at a given temperature the same for all sounds, or do some sounds move more quickly or more slowly than other sounds? Explain.

5. Applying ideas   How could you find the speed of sound in air at different temperatures?

Extensions

6. Evaluating methods   How could you modify the experiment to find the length of an open pipe? If there is time and your teacher approves your plan, carry out the experiment.

7. Research and communications   Many musical instruments operate by resonating air in open or closed tubes. In a pipe organ, for example, both open and closed tubes are used to create music. Research a pipe instrument, and find out how different notes are produced.

ANSWERS TO

Analysis and Interpretation

CALCULATIONS AND DATA ANALYSIS

1. a. distance = 1.962 m

    b. Trial 1: $\lambda = 1.030$ m
        Trial 2: $\lambda = 0.848$ m
        Trial 3: $\lambda = 0.632$ m
        Trial 4: $\lambda = 0.292$ m

2. a. Trial 1: $v = 327.0$ m/s
        Trial 2: $v = 280.3$ m/s
        Trial 3: $v = 392.4$ m/s
        Trial 4: $v = 327.0$ m/s

    b. Trial 1: $v = 329.60$ m/s
        Trial 2: $v = 325.63$ m/s
        Trial 3: $v = 323.58$ m/s
        Trial 4: $v = 299.01$ m/s

3. Accepted value: 346 m/s

    a. CBL: 14.6 m/s; resonance apparatus: 25.8 m/s

    b. CBL: 0.04; resonance apparatus: −0.08

CONCLUSIONS

4. Students should realize that all sound waves travel at the same speed.

5. Answers will vary. Students could look up the values in a table or perform the experiment in different places or under different conditions.

EXTENSIONS

6. Student plans should be safe and complete and should involve measurements of wavelengths.

7. Student answers should reflect an understanding of how sound resonates in open and closed tubes.

Suppose you are spending some quiet time alone–reading, studying, or just daydreaming. Suddenly your peaceful mood is shattered by the sound of a lawn mower, loud music, or an airplane taking off. If this has happened to you, then you have experienced noise pollution.

Noise is defined as any loud, discordant, or disagreeable sound, so classifying sounds as noise is often a matter of personal opinion. When you are at a party, you might enjoy listening to loud music, but when you are at home trying to sleep, you may find the same music very disturbing.

There are two kinds of noise pollution, both of which can result in long-term hearing problems and even physical damage to the ear. Chapter 13 explains how we receive and interpret sound.

How can noise damage hearing?

The small bones and hairlike cells of the inner ear are delicate and very sensitive to the compression waves we interpret as sounds. The first type of noise pollution involves noises that are so loud they endanger the sensitive parts of the ear. Prolonged exposure to sounds of about 85 dB can begin to damage hearing irreversibly. Certain sounds above 120 dB can cause immediate damage. The sound level produced by a food blender or by diesel truck traffic is about 85 dB. A jet engine heard from a few meters away is about 140 dB.

Have you ever noticed the "headphones" worn by ground crew at an airport or by workers using chain saws or jackhammers? In most cases, these are ear protectors worn to prevent the hearing loss brought on by damage to the inner ear.

Whose noise annoys?

The second kind of noise pollution is more controversial because it involves noises that are considered annoyances. No one knows for sure how to measure levels of annoyance, but sometimes annoying noise becomes intolerable. Lack of

sleep due to noise causes people to have slow reaction times and poor judgment, which can result in mistakes at work or school and accidents on the job or on the road. Scientists have found that continuous, irritating noise often produces high blood pressure, which leads to other health problems.

A major debate involves noise made by aircraft. Airport traffic in the United States nearly doubled from 1980 to 1990 and continues to grow at a rapid pace. People who live near airports once found aircraft noise an occasional annoyance, but because of increased traffic and runways added to accommodate growth, they now suffer sleep disruptions and other health effects.

Many people have organized groups to oppose airport expansion. Their primary concerns are the increase in noise and the decrease in property values associated with airport expansion.

But, city governments argue that an airport benefits the entire community both socially and economically and that airports must expand to meet the needs of increased populations. Officials have also argued that people knew they were taking chances by building or buying near an airport and that the community cannot compensate for their losses. Airlines contend that attempts to reduce noise by using less power during takeoffs or by veering away from populated areas can pose a serious threat to passenger safety.

Other annoyances

Besides airports, people currently complain most about noise pollution from nearby construction sites, jet skis, loud stereos in homes and cars, all-terrain vehicles, snowmobiles, and power lawn equipment, such as mowers and leaf blowers. Many people want to control such noise by passing laws to limit the use of this equipment to certain times of the day or by requiring that sound-muffling devices be used.

Opponents to these measures argue that much of this activity takes place on private property and that, in the case of building sites and industries, noise limitation would increase costs. Some public officials would like to control annoying noise but point out that laws to do so fall under the category of nuisance laws, which are notoriously difficult to enforce.

Noise pollution is also a problem in areas where few or no people live. Unwanted noise in wilderness areas can affect animal behavior and reproduction. Sometimes animals are simply scared away from their habitats. For this reason, the government has taken action in some national parks to reduce sightseeing flights, get rid of noisy campers, and limit or eliminate certain noisy vehicles. Some parks have drastically limited the number of people who can be in a park at any one time.

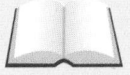

Researching the Issue

1. Obtain a sound-level meter, and measure the noise level at places where you and your friends might be during an average week. Also make some measurements at locations where sound is annoyingly loud. Be sure to hold the meter at head level and read the meter for 30 seconds to obtain an average. Present your findings to the class in a graphic display.

2. Measure the sound levels at increasing distances from two sources of steady, loud noise. Record all of your locations and measurements. Graph your data, and write an interpretation describing how sound level varies with distance from the source.

3. Is there a source of noise in your community that most people recognize to be a problem? If so, find out what causes the noise and what people want to do to relieve the problem. Hold a panel discussion to analyze the opinions of each side, and propose your own solution.

(★) TEKS
pp. 516–517: 3C, 8B

CHAPTER 14 PLANNING GUIDE

Compression Guide: To shorten from 11 to 8 45-min periods (from 5½ to 4 90-min blocks), eliminate items in magenta type.

PACING CHART	CLASSROOM RESOURCES			
	⭐TEKS	**Teacher** Demonstrations	*Holt Physics* Transparencies	**Labs** (See page T52 for equipment listing for in-text labs.)
14-1 Characteristics of light 2 45-minute periods 1 90-minute block	2C, 3A, 3B, 3C, 3E, 8A, 8B, 9B	TE *Infrared light*, p. 520 TE *Radio waves*, p. 521 TE *Light travels in straight lines*, p. 522	T 63 TM 48–49	PE *Brightness of Light*, p. 556
14-2 Flat mirrors 2 or 1 45-minute periods 1 or ½ 90-minute block	1A, 2C, 3A, 3C, 8A, 8B	TE *Diffuse reflection*, p. 526 TE *Specular reflection*, p. 527 TE *Flat mirror images*, p. 528	T 64	L *Discovery Lab: Light and Mirrors*
14-3 Curved mirrors 2 45-minute periods 1 90-minute block	1A, 2C, 3A, 3B, 3C, 8A, 8B, 8C	TE *Image formed by a concave mirror*, p. 530 TE *Focal point of a concave mirror*, p. 532 TE *Beams reflected from a concave mirror*, p. 533 TE *Convex mirror*, p. 537	T 65–68 TM 50–51	PE *Quick Lab: Curved Mirrors*, p. 532
14-4 Color and polarization 3 or 1 45-minute periods 1½ or ½ 90-minute block	1A, 3A, 3C, 8A, 8B, 8C	TE *Reflection and absorption of color*, p. 543 TE *Mixing light colors*, p. 544 TE *Polarization of waves*, p. 546 TE *Polarizing light by transmission*, p. 547 TE *Polarizing light by reflection*, p. 548	T 69, 70–71 TM 52	L *Invention Lab: Designing a Device to Trace Drawings* PE *Quick Lab: Polarization of Sunlight*, p. 547 TL *Polarized Light*
Review and Assessment 2 45-minute periods 1 90-minute block				

Resource Key

PHYSICS

PE Pupil's Edition
TE Teacher's Edition

L Laboratory Experiments
TL Technology Lab Experiments
T Transparencies

🖐 **One-Stop** Planner CD-ROM **contents**

TM Transparency Masters
SR Section Review Worksheets
AA Alternative Assessment

PW Problem-Solving Workbook
PB Problem Bank
CTW Critical Thinking Worksheet

LABORATORY PLANNING: Brightness of Light, p. 556

Materials (for each lab group):
- 1-position support base and rod, 1.3 cm × 91 cm
- battery eliminator with alligator clips, 6 V/0.5 A
- black construction paper
- blackened card tube for bulb shield
- meterstick, plain wood, 1 m long and metal supports
- miniature bulb and base
- replacement bulb, 6.3 V/0.3 A

- roll of adhesive tape, 0.5 in. wide
- round-jaw symmetrical clamp with holder

Additional Equipment
CBL and Sensors Procedure
- CBL
- graphing calculator
- CBL light sensor
- TI Graph Link (recommended for downloading programs)

ASSIGNMENT RESOURCES

Content Mastery	Critical Thinking	Problem-Solving Practice
PE 1–4, p. 525 **SR** 14-1, *Concept Review* **PE** 1–5, p. 550	**PE** 6–9, p. 550	**14A** Electromagnetic waves: 28 items in **PE, PW,** and **PB,** see **TE** p. 523
PE 1–6, p. 529 **SR** 14-2, *Diagram Skills* **PE** 14–16, pp. 550–551	**PE** 17–22, p. 551	
PE 1–5, p. 542 **SR** 14-3, *Diagram Skills* **PE** 23–27, p. 551	**PE** 28–33, p. 551	**14B** Concave mirrors: 32 items in **PE, PW,** and **PB,** see **TE** pp. 535–536 **14C** Convex mirrors: 33 items in **PE, PW,** and **PB,** see **TE** pp. 539–540
PE 1–3, 4, p. 548 **SR** 14-4, *Concept Review* **PE** 37–38, 39, p. 552	**PE** 1–2, p. 545 **PE** 40–42, 43–45, p. 552	

ASSESSMENT RESOURCES

Cumulative Review	Alternative Assessment	Traditional Assessment
SR Mixed Review, Ch. 14	**PE** 1–6, p. 554 **AA** Items for Ch. 14	Chapter 14 Test Test Generator items for Ch. 14

Scoring Rubrics for Alternative Assessment items can be found on the One-Stop Planner CD-ROM.

TECHNOLOGY RESOURCES

 CTW Segment 15 Japanese Telescope

 Module 14 Reflection and Mirrors

 PE Technology and Learning, p. 555
(Alternative procedures for calculators without Flash-ROM technology are provided on the One-Stop Planner CD-ROM.)

 The Mechanical Universe/High School Adaptation Quad VII, *The Wave Nature of Light*

 internet**connect**

 On-line Student Resources:
GO TO: www.scilinks.org
The following *sci*LINKS Internet resources can be found in the student text for this chapter.

TOPICS:
- Electromagnetic spectrum, p. 520 (HF2141)
- Light bulbs, p. 525 (HF2142)
- Mirrors, p. 528 (HF2143)
- Telescopes, p. 541 (HF2144)
- Color, p. 543 (HF2145)

 On-line Teacher Resources:
GO TO: go.hrw.com
KEYWORD: HF2 HOME
Visit the HRW Web site for a variety of resources related to this chapter.

 Smithsonian Institution
Internet Connections

Visit **www.si.edu/hrw** for additional on-line resources.

 CNN**fyi**.com

Visit **www.cnnfyi.com** for late-breaking news and current events stories selected just for you.

Light Meter Procedure
- blackened card tube for detector shield
- digital foot-candle/lux meter
- lens or mirror support, for 4 cm lenses

Required Precautions
The greatest danger in this activity is presented by exposed electrical connections. All bulb sockets should have enclosed contacts and insulated connectors should be used on all wire connectors.

Materials Preparation
Set up a sample apparatus in the laboratory for students to refer to as they set up their equipment. The blackened card tube to be used as a bulb shield may be constructed using a cardboard toilet paper tube with a hole at the level of the bulb filament. This hole should be adjusted to line up with the probe or meter. For the CBL light sensor, a shield can be constructed by rolling a piece of black construction paper into a tube. For the light meter, use another cardboard tube as a shield for the detector. Use the lens support to hold the tube in place during the lab.

Section 14-1 identifies the components of the electromagnetic spectrum, relates their frequency and wavelength to the speed of light, and introduces the relationship between brightness and distance for a light source.

Section 14-2 applies the laws of reflection to plane mirrors and uses ray diagrams to determine image location.

Section 14-3 shows how image location and magnification are calculated for concave and convex mirrors, uses ray diagrams to confirm calculated results, and explains spherical aberration.

Section 14-4 investigates additive and subtractive colors and explores the phenomenon of polarization.

About the Illustration

The radio telescope at Arecibo, Puerto Rico, is the largest single radio telescope in the world. The 304.8 m dish focuses radio signals from space onto receiving equipment at the focal point of the dish. The receiver converts the radio waves into electric signals, which are amplified and then recorded by a computer. The computer produces a radio "picture" of the source.

Interactive Problem-Solving Tutor

See Module 14

"Reflection and Mirrors" provides additional development of problem-solving skills for this chapter.

CHAPTER 14

Light and Reflection

PHYSICS IN ACTION

Although much larger and different in appearance, the radio telescope at Arecibo, Puerto Rico, operates on the same principles as an optical telescope. Just as the mirror of an optical telescope reflects waves of visible light to form an image at the eyepiece, the dish of a radio telescope reflects radio waves to form a radio image at a receiver poised above the dish.

The two types of instruments function similarly because both visible light and radio waves are forms of electromagnetic radiation and, as such, reflect from certain materials. In this chapter, the nature of electromagnetic waves and their behavior during reflection will be examined.

- Why are telescope mirrors always concave rather than flat or convex?

- How can the location and size of mirror images be predicted?

CONCEPT REVIEW

Properties of waves (Section 12-3)

Reflection of waves (Section 12-4)

Tapping Prior Knowledge

Knowledge to Expect

✔ "Students learn that light from the sun is made of a mixture of many different colors of light, even though to the eye the light looks almost white. Other things that give off or reflect light have a different mix of colors." (AAAS's *Benchmarks for Science Literacy,* grades 6–8)

✔ "Light interacts with matter by transmission, absorption, or scattering (including reflection). To see an object, light from that object—emitted by or scattered from it—must enter the eye." (NRC's *National Science Education Standards,* grades 5–8)

Knowledge to Review

✔ Waves transport energy. They can be transverse or longitudinal. Waves have amplitude, frequency, wavelength, and velocity. (Section 12-3)

Items to Probe

✔ The ability to describe spatial relationships in geometric terms: Make sure students understand terms such as *perpendicular* and *parallel* and can solve equations of the form $\left(\dfrac{1}{A}\right) + \left(\dfrac{1}{B}\right) = \left(\dfrac{1}{C}\right).$

✔ Preconceptions about light: Ask students to describe the path of sound waves when we hear something and the path of light rays when we see an object.

14-1
Characteristics of light

Infrared light

Purpose Demonstrate one form of invisible electromagnetic radiation.

Materials incandescent light source, black box, prism, white paper, thermometer or temperature sensor with probeware system

Procedure Make a slit in one side of the black box, and place the light source inside the box. Set the prism in the path of the beam of light emerging from the slit. Lower the classroom lights, and place the white paper on the other side of the prism so that the light beam's spectrum is cast on the paper. Tape the thermometer on the paper so that the red light of the spectrum shines on the thermometer bulb. Have a student record the initial temperature and the temperature after 5 min.

Repeat this procedure, placing the thermometer bulb in the dark region to the left of the red part of the spectrum. As an optional step, repeat the procedure for other colors in the spectrum and for the region just past the violet light. Explain that infrared (IR) radiation, which is one example of nonvisible electromagnetic radiation, increases a substance's temperature. Energy transferred away from the substance by heat can be photographically detected with infrared-sensitive film. (Note: If using a probeware system for this demonstration, place the temperature sensor in the same positions as the thermometer bulb.)

14-1 SECTION OBJECTIVES

- **Identify the components of the electromagnetic spectrum.**

- **Calculate the frequency or wavelength of electromagnetic radiation.**

- **Recognize that light has a finite speed.**

- **Describe how the brightness of a light source is affected by distance.**

electromagnetic wave

a transverse wave consisting of oscillating electric and magnetic fields at right angles to each other

internetconnect

SCiLINKS
NSTA

TOPIC: Electromagnetic spectrum
GO TO: www.scilinks.org
*sci*LINKS CODE: HF2141

ELECTROMAGNETIC WAVES

Nearly everyone has experienced light. When most people think of light, they think of light they can see. Some examples include the bright, white light that is produced by a light bulb or the sun. However, there is more to light than this.

When you hold a piece of green glass or plastic in front of a source of white light, you see green light pass through. This is also true for other colors. What your eyes recognize as "white" light is actually light that can be separated into six elementary colors of the visible *spectrum:* red, orange, yellow, green, blue, and violet.

If you examine a glass prism, like the one shown in **Figure 14-1,** or any thick, triangular-shaped piece of glass, you will find that sunlight passes through the glass and emerges as a rainbowlike band of colors.

The spectrum includes more than visible light ⭐TEKS 8B

Not all light is visible to the human eye. If you were to examine the light dispersed through a prism using certain types of photographic film, you would find that the film records a much wider spectrum than the one you see. A variety of forms of radiation—including X rays, microwaves, and radio waves—have many of the same properties as visible light. This is because they are all examples of **electromagnetic waves.**

Light has been described as a particle, as a wave, and even as a combination of the two. While the current model has incorporated aspects of both particle and wave theories, the wave model is best suited for an introductory discussion of light, and it is the one that will be used in this section.

Figure 14-1
A prism separates light into its component colors.

Electromagnetic waves vary depending on frequency and wavelength

In classical electromagnetic wave theory, light is a wave composed of oscillating electric and magnetic fields. These fields are perpendicular to the direction in which the wave moves, as shown in **Figure 14-2.** Therefore, electromagnetic waves are transverse waves. The electric and magnetic fields are also at right angles to each other.

Electromagnetic waves are distinguished by their different frequencies and wavelengths. In visible light, these differences in frequency and wavelength account for different colors. The difference in frequencies and wavelengths also distinguishes visible light from invisible electromagnetic radiation, such as X rays.

Types of electromagnetic waves are listed in **Table 14-1.** Note the wide range of wavelengths and frequencies. Although specific ranges are indicated in the table, the electromagnetic spectrum is, in reality, continuous. There is no sharp division between one kind of wave and the next. Some types of waves even have overlapping ranges.

★ TEKS **8A, 8B**

CONCEPT PREVIEW ➤

Electric and magnetic fields will be discussed in greater detail in Chapter 17 and Chapter 21.

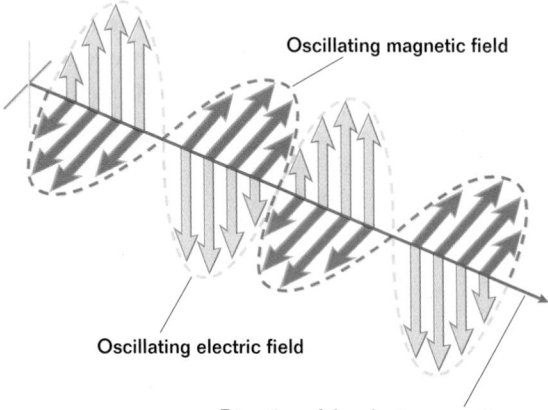

Oscillating magnetic field

Oscillating electric field

Direction of the electromagnetic wave

Figure 14-2
An electromagnetic wave consists of electric and magnetic field waves at right angles to each other.

Table 14-1   The electromagnetic spectrum ★ TEKS **2C, 3C**

Classification	Range	Applications
radio waves	$\lambda > 30$ cm $f < 1.0 \times 10^9$ Hz	AM and FM radio; television
microwaves	30 cm $> \lambda > 1$ mm 1.0×10^9 Hz $< f < 3.0 \times 10^{11}$ Hz	radar; atomic and molecular research; aircraft navigation; microwave ovens
infrared (IR) waves	1 mm $> \lambda > 700$ nm 3.0×10^{11} Hz $< f < 4.3 \times 10^{14}$ Hz	molecular vibrational spectra; infrared photography; physical therapy
visible light	700 nm (red) $> \lambda > 400$ nm (violet) 4.3×10^{14} Hz $< f < 7.5 \times 10^{14}$ Hz	visible-light photography; optical microscopy; optical astronomy
ultraviolet (UV) light	400 nm $> \lambda > 60$ nm 7.5×10^{14} Hz $< f < 5.0 \times 10^{15}$ Hz	sterilization of medical instruments; identification of fluorescent minerals
x rays	60 nm $> \lambda > 10^{-4}$ nm 5.0×10^{15} Hz $< f < 3.0 \times 10^{21}$ Hz	medical examination of bones, teeth, and vital organs; treatment for types of cancer
gamma rays	0.1 nm $> \lambda > 10^{-5}$ nm 3.0×10^{18} Hz $< f < 3.0 \times 10^{22}$ Hz	examination of thick materials for structural flaws; treatment of types of cancer; food irradiation

Demonstration 2

Radio waves

Purpose Demonstrate another example of radiation in the electromagnetic spectrum.

Materials pocket-size transistor radio, metal can with metal lid, glass jar, tin foil, paper, plastic wrap

Procedure Tell students that radio waves have a longer wavelength than infrared waves and that these waves can be detected with a radio. Turn on the radio, and tune it to a station. Have students notice changes in reception as you place the radio in the metal can and then in the glass jar. Wrap the radio in the paper, in the tin foil, and in the plastic wrap. Have students notice that paper is "transparent" to radio waves. Point out that different types of radiation can penetrate different materials.

Teaching Tip

Point out that all of the frequencies in **Table 14-1** are expressed in hertz but that the wavelengths are in centimeters, millimeters, and nanometers, which stand for 10^{-2}, 10^{-3}, and 10^{-9} m, respectively. Ask students if the wavelengths increase or decrease as the frequencies decrease (*increase*).

★ TEKS

p. 520: 8B
p. 521: 8A, 8B, 2C, 3C

Demonstration 3

Light travels in straight lines

Purpose Demonstrate that light waves can be approximated as rays.

Materials laser, index card, two dusty chalkboard erasers, plane mirror

CAUTION *Avoid pointing the laser beam near students' eyes; retinal damage may occur.*

Procedure Direct the laser beam across the room. Point out that in order for students to see the beam, it is necessary to have an object in the path of the beam that will reflect some light.

Place the index card in the path of the beam and slowly walk across the room while keeping the beam centered on the card. Students should see that the beam travels in a straight line.

Stand beside the beam and tap the erasers together above the beam. As the chalk dust falls from the erasers, the beam will become visible. Quickly walk along the length of the beam, tapping the erasers until the entire beam becomes visible.

All electromagnetic waves move at the speed of light ⊛ TEKS **8B**

All forms of electromagnetic radiation travel at a single high speed in a vacuum. Early experimental attempts to determine the speed of light failed because this speed is so great. As experimental techniques improved, especially during the nineteenth and early-twentieth centuries, the speed of light was determined with increasing accuracy and precision. By the mid-twentieth century, the experimental error was less than 0.001 percent. The currently accepted value for light traveling in a vacuum is $2.997\ 924\ 58 \times 10^8$ m/s. Light travels slightly slower in air, with a speed of $2.997\ 09 \times 10^8$ m/s. For calculations in this book, the value used for both situations will be 3.00×10^8 m/s.

The relationship between frequency, wavelength, and speed described in Chapter 12 also holds true for light waves.

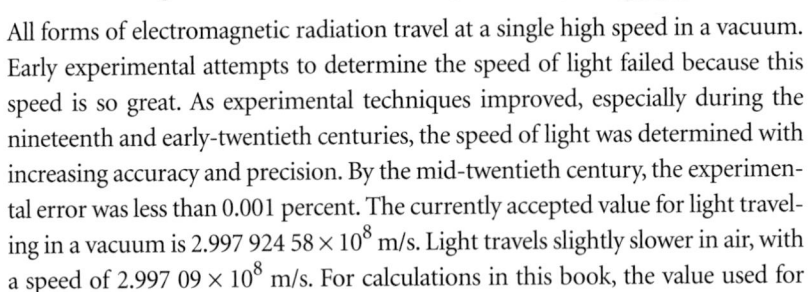

WAVE SPEED EQUATION ⊛ TEKS **3B**

$$c = f\lambda$$

speed of light = frequency × wavelength

SAMPLE PROBLEM 14A

Electromagnetic waves ⊛ TEKS **2C, 3B, 8A, 8B**

PROBLEM

The AM radio band extends from 5.4×10^5 Hz to 1.7×10^6 Hz. What are the longest and shortest wavelengths in this frequency range?

SOLUTION

Given: $f_1 = 5.4 \times 10^5$ Hz $\qquad f_2 = 1.7 \times 10^6$ Hz $\qquad c = 3.00 \times 10^8$ m/s

Unknown: $\lambda_1 = ?$ $\quad \lambda_2 = ?$

Use the wave speed equation on this page:

$$c = f\lambda \qquad \lambda = \frac{c}{f}$$

$$\lambda_1 = \frac{3.00 \times 10^8 \text{ m/s}}{5.4 \times 10^5 \text{ Hz}}$$

$$\boxed{\lambda_1 = 5.6 \times 10^2 \text{ m}}$$

$$\lambda_2 = \frac{3.00 \times 10^8 \text{ m/s}}{1.7 \times 10^6 \text{ Hz}}$$

$$\boxed{\lambda_2 = 1.8 \times 10^2 \text{ m}}$$

CALCULATOR SOLUTION

Although the calculator solutions are 555.5555556 m and 176.470588 m, both answers must be rounded to two digits because the frequencies have only two significant figures.

Electromagnetic waves ⊛TEKS 2C, 3B, 8A, 8B

1. Gamma-ray bursters are objects in the universe that emit pulses of gamma rays with high energies. The frequency of the most energetic bursts has been measured at around 3.0×10^{21} Hz. What is the wavelength of these gamma rays?

2. What is the wavelength range for the FM radio band (88 MHz–108 MHz)?

3. Shortwave radio is broadcast between 3.50 and 29.7 MHz. To what range of wavelengths does this correspond? Why do you suppose this part of the spectrum is called shortwave radio?

4. What is the frequency of an electromagnetic wave if it has a wavelength of 1.0 km?

5. The portion of the visible spectrum that appears brightest to the human eye is around 560 nm in wavelength, which corresponds to yellow-green. What is the frequency of 560 nm light?

6. What is the frequency of highly energetic ultraviolet radiation that has a wavelength of 125 nm?

Waves can be approximated as rays ⊛TEKS 8A, 3A, 3E

Consider an ocean wave coming toward the shore. The broad crest of the wave that is perpendicular to the wave's motion consists of a line of water particles. Similarly, another line of water particles forms a low-lying trough in the wave, and still another line of particles forms another crest. In any type of wave, these lines of particles are called *wave fronts*.

All the points on the wave front of a plane wave can be treated as point sources. A few of these points are shown on the initial wave front in **Figure 14-3**. Each of these point sources produces a circular or spherical secondary wave, or *wavelet*. The radii of these wavelets are indicated by the blue arrows in **Figure 14-3**. The line that is tangent to each of these wavelets at some later time determines the new position of the initial wave front (the new wave front in **Figure 14-3**). This approach to analyzing waves is called Huygens' principle, named for the physicist Christian Huygens, who developed it.

Huygens' principle can be used to derive the properties of any wave (including light) that interacts with matter, but the same results can be obtained by treating the propagating wave as a straight line perpendicular to the wave front. This line is called a *ray,* and this simplification is called the *ray approximation.*

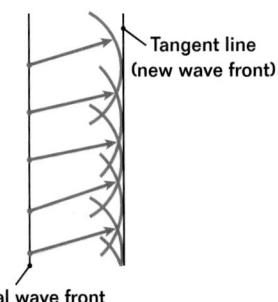

Figure 14-3
According to Huygens' principle, a wave front can be divided into point sources. The line tangent to the wavelets from these sources marks the wave front's new position.

PRACTICE GUIDE 14A	
Solving for:	
λ	📖 **PE** Sample, 1–3; Ch. Rvw. 10–13
	💿 **PW** 5–7
	PB 7–10
f	📖 **PE** 4–6
	💿 **PW** Sample, 1–4
	PB 3–6
c	💿 **PW** 8
	PB Sample, 1, 2

ANSWERS TO

Practice 14A
Electromagnetic waves
1. 1.0×10^{-13} m
2. 3.4 m–2.78 m
3. 85.7 m–10.1 m; The wavelengths are shorter than those of the AM radio band.
4. 3.0×10^{5} Hz
5. 5.4×10^{14} Hz
6. 2.40×10^{15} Hz

⊛TEKS

p. 522: 8B, 3B, 2C, 3B, 8A, 8B
p. 523: 2C, 3B, 8A, 8B, 8A, 3A, 3E

Tomorrow's Technology

BACKGROUND

The light created by sulfur bulbs spans a large range of frequencies, making the bulbs useful for a variety of applications.

With older, metal-halide bulbs, the color of the light emitted by the bulb changes as the amount of power changes because there are several different vaporized metals present in the gaseous discharge. Their relative amounts may vary with applied power. But sulfur bulbs use only sulfur, so the color of the light never changes.

Caution students about the hazards of constructing microwave-powered lamps as science projects. Microwaves can be dangerous if they are not correctly confined, and the high-voltage powered magnetrons—used to generate microwave radiation—can cause serious (even lethal) injury.

EXTENSION

Several new light sources have been introduced during the last decade. Have students research these sources and report on the advantages of each.

★ TEKS

p. 524: 3C, 8B, 9B
p. 525: 8B, 2C, 3B, 8A, 8B

Tomorrow's Technology

Sulfur Light Bulbs ★TEKS 3C, 8B, 9B

A new kind of lamp uses sulfur and microwaves to create light that is bright and energy-efficient and looks a lot like sunlight.

The sulfur bulb shown at the right of a metal-halide bulb in the photograph below is flooded with microwave energy. In a microwave oven, an electric field creates microwave radiation that causes water molecules to oscillate and the food's temperature to increase. In the lamp, the electric field accelerates electrons that strike sulfur molecules, causing them to become excited and emit light.

In the photo at right, light from two 3000 W microwave sulfur lamps is routed through light pipes to illuminate a region 85 m wide. Because of its tremendous intensity, the lamp is not practical for home use, says Michael Ury, vice-president of research and development at Fusion Lighting, in Rockville, Maryland. "It certainly will not be in your kitchen until we can make a lower-power version. So for the next few years, you'll find this being used in factories, and

perhaps in sports arenas, aircraft hangars, and shopping malls, where the ceilings are high," says Ury. The lamp cannot be used where the high temperature of overhead lighting is considered to be a negative factor, such as in hockey rinks. But, adds Ury, "this is one of the few light sources with which you can light plants to solar levels without cooking them."

Compared with other forms of outdoor lighting, the sulfur lamp makes it easier to discern a full spectrum of color. For instance, if you are looking for your blue-green car in a parking lot at night, you may not recognize your car because the high-pressure sodium lamps used in most parking lots are deficient in the blue and green frequency ranges, making your car look black. By contrast, says Ury, "The sulfur lamp spectrum contains all the colors. If we were to light up a parking lot with this light, you'd find your car; you'd recognize its color."

Other advantages of the sulfur lamp are its high efficiency and reliability. It operates at about 1350 W and produces nearly six times as much light per watt as a 100 W incandescent bulb. A sulfur light bulb is designed to last 60 000 hours, outlasting its microwave source, which usually has to be replaced between 15 000 and 20 000 hours.

Brightness decreases by the square of the distance from the source

You have probably noticed that it is easier to read a book when a lamp that is 1 m away from you has a 100 W bulb installed in it instead of a 25 W bulb. This experience suggests that the brightness of light depends on how much light is actually emitted from a source. However, the amount of light that you detect also depends on your distance from the source. (★)TEKS 8B

The farther you are from a light source, the less bright the light appears to be. This is because the light emitted by the source spreads outward in all directions. The farther the light is from the source, the more spread out the light becomes. Therefore, less light is available per unit area at a greater distance from the source than at a smaller one. The brightness you perceive from a light source is related to the amount of light that falls on a unit surface area at some distance from the source. As you may note from **Figure 14-4,** the apparent brightness is proportional to the actual brightness of the source divided by the square of the distance between the source and the observer. For example, if you move twice as far away from the light source, one-fourth as much light falls on the book.

internet connect
SC*LINKS
NSTA
TOPIC: Light bulbs
GO TO: www.scilinks.org
sciLINKS CODE: HF2142

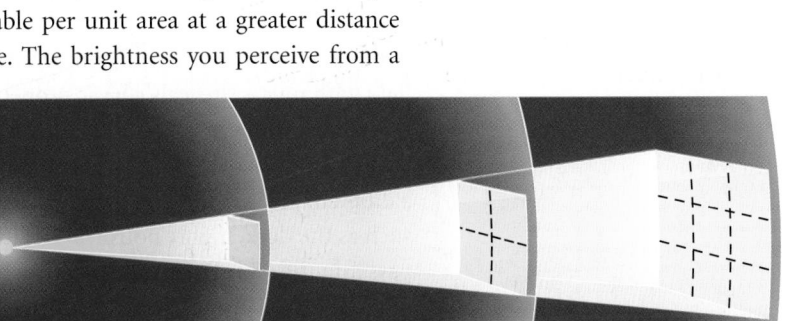

1 m   2 m   3 m

Figure 14-4
Less light falls on each unit square as the distance from the source increases.

Section Review

(★)TEKS 2C, 3B, 8A, 8B

1. Identify which portions of the electromagnetic spectrum are used in each of the devices listed.
   a. a microwave oven
   b. a television set
   c. a single-lens reflex camera

2. If an electromagnetic wave has a frequency of 7.57×10^{14} Hz, what is its wavelength? To what part of the spectrum does this wave belong?

3. Galileo performed an experiment to measure the speed of light by timing how long it took light to travel from a lamp he was holding to an assistant about 1.5 km away and back again. Why was Galileo unable to conclude that light had a finite speed?

4. How bright would the sun appear to an observer on Earth if the sun were four times farther from Earth than it actually is? Express your answer as a fraction of the sun's brightness on Earth's surface.

Teaching Tip

Point out that the brightness at any point is the power per unit of area at that point. Ask students to compare the surface areas of two spherical lampshades and the brightness of the light that hits their surface when identical light bulbs are placed at the center of each. Use values for their radii, such as 10 cm and 50 cm (surface area of a sphere = $4\pi r^2$).

Section Review
ANSWERS

1. a. microwave
   b. radio waves, visible light
   c. visible light
2. 3.96×10^{-7} m; near ultraviolet
3. The speed of light is too great to be measured over such a short distance. The time of travel for the light in Galileo's experiment was about 1.0×10^{-5} s.
4. $\frac{1}{16}$ of the sun's brightness on Earth's surface

14-2
Flat mirrors

Diffuse reflection

Purpose Demonstrate that light reflected from a rough textured surface is reflected in many directions.

Materials laser, index card

CAUTION *Avoid directing the laser beam near students' eyes.*

Procedure Tape the index card to a wall in the classroom. Direct the laser beam onto the card from across the room. Make sure the beam is not perpendicular to the card's surface. Explain to students that the light strikes the index card as a single beam but that because the card is a diffuse reflector, the beam undergoes reflection in all directions. Only a small part of the reflected light will go to each part of the room, enabling everyone to see the spot where the laser beam strikes the card.

Visual Strategy

Figure 14-6

Point out that the incident light rays are parallel in each part but that the reflected rays in **(a)** are not parallel.

Q Why are the reflected rays from the surface in **(a)** not parallel?

A *Because the surface in (a) has irregularities, it forms a different angle with the incident ray at each point.*

14-2 SECTION OBJECTIVES

- **Distinguish between specular and diffuse reflection of light.**
- **Apply the law of reflection for flat mirrors.**
- **Describe the nature of images formed by flat mirrors.**

reflection

the turning back of an electromagnetic wave at the surface of a substance

Figure 14-5

Mirrors reflect nearly all incoming light, so multiple images of an object between two mirrors are easily formed.

REFLECTION OF LIGHT

Suppose you have just had your hair cut and you want to know what the back of your head looks like. You can do this seemingly impossible task by using two mirrors to direct light from behind your head to your eyes. Redirecting light with mirrors reveals a basic property of light's interaction with matter.

Light traveling through a uniform substance, whether it is air, water, or a vacuum, always travels in a straight line. However, when the light encounters a different substance, its path will change. If a material is opaque to the light, such as the dark, highly polished surface of a wooden table, the light will not pass into the table more than a few wavelengths. Part of the light is absorbed, and the rest of it is deflected at the surface. This change in the direction of the light is called **reflection.** Most substances absorb at least some incoming light and reflect the rest. A good mirror can reflect about 90 percent of the incident light, but no surface is a perfect reflector. Notice in **Figure 14-5** that the images of the golf ball get successively darker. ⊛TEKS **8A, 8B**

The texture of a surface affects how it reflects light

The manner in which light is reflected from a surface depends on the surface's smoothness. Light that is reflected from a rough, textured surface, such as paper, cloth, or unpolished wood, is reflected in many different directions, as shown in **Figure 14-6(a).** This type of reflection is called *diffuse reflection.* Diffuse reflection will be discussed further in Section 14-4.

Light reflected from smooth, shiny surfaces, such as a mirror or water in a pond, is reflected in one direction only, as shown in **Figure 14-6(b).** This type of reflection is called *specular reflection.* A surface is considered smooth if its surface variations are small compared with the wavelength of the incoming light. For our discussion, reflection will be used to mean only specular reflection.

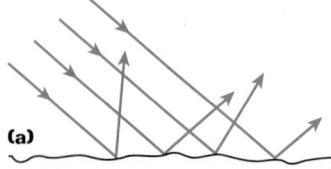

(a)

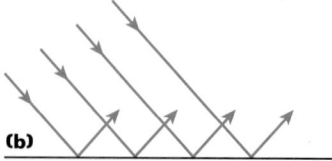

(b)

Figure 14-6

Diffusely reflected light is reflected in many directions **(a)**, whereas specularly reflected light is reflected in the same forward direction only **(b)**.

(a)

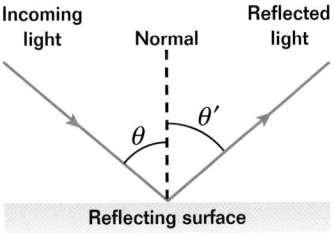

Incoming light **Normal** **Reflected light**

θ θ'

Reflecting surface

(b)

Figure 14-7
The symmetry of reflected light **(a)** is described by the law of reflection, which states that the angles of the incoming and reflected rays are equal **(b).**

Incoming and reflected angles are equal (★ TEKS) 8A, 8B

You probably have noticed that when incoming rays of light strike a smooth reflecting surface, such as a polished table or mirror, at an angle close to the surface, the reflected rays are also close to the surface. When the incoming rays are high above the reflecting surface, the reflected rays are also high above the surface. An example of this similarity between incoming and reflected rays is shown in **Figure 14-7(a).**

If a straight line is drawn perpendicular to the reflecting surface at the point where the incoming ray strikes the surface, the **angle of incidence** and the **angle of reflection** can be defined with respect to the line. Careful measurements of the incident and reflected angles θ and θ', respectively, reveal that the angles are equal, as illustrated in **Figure 14-7(b).**

$$\theta = \theta'$$
angle of incoming light ray = angle of reflected light ray

Note that the line perpendicular to the reflecting surface is referred to as the *normal* to the surface. It therefore follows that the angle between the incoming ray and the surface equals $90° - \theta$, and the angle between the reflected ray and the surface equals $90° - \theta'$.

angle of incidence

the angle between a ray that strikes a surface and the normal to that surface at the point of contact

angle of reflection

the angle formed by the line normal to a surface and the direction in which a reflected ray moves

FLAT MIRRORS

The simplest mirror is the *flat mirror.* If an object, such as a pencil, is placed at a distance in front of a flat mirror and light is bounced off the pencil, light rays will spread out from the pencil and reflect from the mirror's surface. To an observer looking at the mirror, these rays appear to come from a location on the other side of the mirror. As a convention, an object's image is said to be at this location behind the mirror. The relationship between the *object distance* from the mirror, which is represented as *p,* and the *image distance* (that is, the distance the image appears to be behind the mirror's surface), which is represented as *q,* is such that the object and image distances are equal. Similarly, the image of the object is the same size as the object.

SECTION 14-2

Demonstration 5

Specular reflection

Purpose Demonstrate that all parallel rays of light reflected from a smooth surface are reflected in the same direction.

Materials laser, flat mirror, dusty chalkboard erasers

CAUTION *Avoid directing the reflected beam from the laser toward the students.*

Procedure Tape the flat mirror to a wall in the classroom. Direct the beam of the laser onto the mirror from across the room. Make sure the beam is not perpendicular to the mirror's surface. Explain to students that the beam is reflected at the mirror's surface and that the angle of incidence is equal to the angle of reflection. This can be shown qualitatively by gently tapping the erasers in front of the mirror so that both the incoming and reflected beams become visible.

(★ TEKS)

p. 526: 8A, 8B
p. 527: 8A, 8B

Flat mirror images

Purpose Demonstrate that the image behind a mirror is virtual.

Materials sheet of high-quality plate glass (0.5 m × 0.5 m), two identical candles, black chalkboard or black surface

Procedure Place the sheet of glass vertically in front of the chalkboard. Place one candle about 30 cm in front of the glass, and place the other candle 30 cm behind it.

Have students who might be able to see the candle behind the glass without looking through the glass move to the rear of the classroom. Have students close their eyes while you light the front candle. Discreetly adjust the position of the candle behind the glass so that the image of the front candle as seen in the glass coincides with the position of the back candle. When properly positioned, the back candle will appear lit when it is viewed through the glass sheet.

Hold a match near the back candle and tell students to open their eyes. Raise the match and blow it out, giving the impression that you have just finished lighting the back candle. Ask how many candles are lit. Then lift the glass. Explain that the image of the flame appeared to be at the same distance behind the glass as the rear candle.

★ TEKS

p. 528: 8A, 3A
p. 529: 3A, 8A

---

virtual image

an image formed by light rays that only appear to intersect

internet connect

SCi*LINKS*

NSTA

TOPIC: Mirrors
GO TO: www.scilinks.org
*sci***LINKS CODE:** HF2143

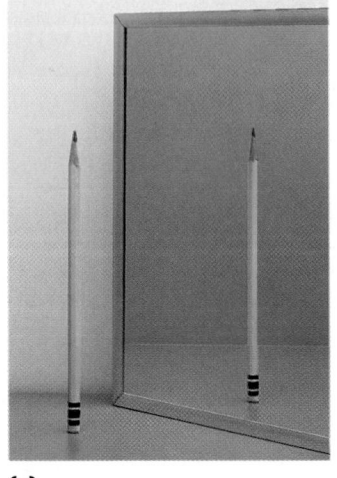

(a)

The image formed by rays that appear to come together at the image point behind the mirror—but never really do—is called a **virtual image.** As shown in **Figure 14-8(a),** a flat mirror always forms a virtual image, which can only be seen "behind" the surface of the mirror. For this reason, a virtual image can never be displayed on a physical surface.

Image location can be predicted with ray diagrams ★TEKS 8A

Ray diagrams, such as the one shown in **Figure 14-8(b),** are drawings that use simple geometry to locate an image formed by a mirror. Suppose you want to make a ray diagram for a pencil placed in front of a flat mirror. First, sketch the situation. Draw the location and arrangement of the mirror and the position of the pencil with respect to the mirror. Construct the drawing so that the object and the image distances (p and q, respectively) are proportional to their actual sizes. To simplify matters, we will consider only the tip of the pencil.

To pinpoint the location of the pencil tip's image, draw two rays on your diagram. Draw the first ray from the pencil tip perpendicular to the mirror's surface. Because this ray makes an angle of $0°$ with the normal to the mirror, the angle of reflection also equals $0°$, causing the ray to reflect back on itself. In **Figure 14-8(b),** this ray is denoted as **1** and is shown with arrows pointing in both directions because the incident ray reflects back on itself.

Draw the second ray from the tip of the pencil to the mirror, but this time place the ray at an angle that is not perpendicular to the surface of the mirror. Then draw the reflected ray, keeping in mind that it will reflect away from the surface of the mirror at an angle, θ', equal to the angle of incidence, θ. This ray is denoted in **Figure 14-8(b)** by the number **2.**

Next, trace both reflected rays back to the point from which they appear to have originated, that is, behind the mirror. Use dotted lines when drawing these rays that appear to emerge from behind the mirror to distinguish them from the actual rays of light (the solid lines) in front of the mirror. The point at which these dotted lines meet is the image point, which in this case is where the image of the pencil's tip forms.

By continuing this process for all of the other parts of the pencil, you can locate the complete virtual image of the pencil. Note that the pencil's image appears as far behind the mirror as the pencil is in front of the mirror ($p = q$). Likewise, the object height, h, equals the image height, h'. ★TEKS 3A

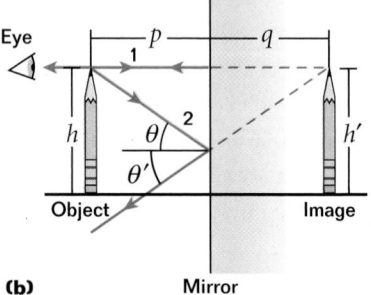

(b)   Mirror

Figure 14-8
The position and size of the virtual image that forms in a flat mirror **(a)** can be predicted by constructing a ray diagram **(b).**

This ray-tracing procedure will work for any object placed in front of a flat mirror. By selecting a single point on the object (usually its uppermost tip or edge), you can use ray tracing to locate the same point on the image. The rest of the image can be added once the image point and image distance have been determined.

The image formed by a flat mirror appears to have right-to-left reversal; that is, the right side of an object is the image's left side. You can see this by placing a watch in front of a mirror, as shown in **Figure 14-9**. In the mirror, the numbers on the watch are turned around and the hands point in the opposite direction.

Figure 14-9
The right side of an object becomes the left side of its image.

Section Review

⭐TEKS **3A, 8A**

1. Which of the following are examples of specular reflection and which are examples of diffuse reflection?

   a. reflection of light from the surface of a lake on a calm day
   b. reflection of light from a plastic trash bag
   c. reflection of light from the lens of eyeglasses
   d. reflection of light from a carpet

2. The photograph in **Figure 14-5** on page 526 shows multiple images that were created by multiple reflections between two flat mirrors. What conclusion can you make about the relative orientation of the mirrors? Explain your answer.

3. Suppose you are holding a flat mirror and standing at the center of a giant clock face built into the floor. Someone standing at 12 o'clock shines a beam of light toward you, and you want to use the mirror to reflect the beam toward an observer standing at 5 o'clock. What should the angle of incidence be to achieve this? What should the angle of reflection be?

4. Some department-store windows are slanted inward at the bottom. This is to decrease the glare from brightly illuminated buildings across the street, which would make it difficult for shoppers to see the display inside and near the bottom of the window. Sketch a light ray reflecting from such a window to show how this technique works.

5. If one wall of a room consists of a large flat mirror, how much larger will the room appear to be? Explain your answer.

6. Why does a flat mirror appear to reverse images left to right, but not up and down?

Teaching Tip
Point out that the object and its image are equidistant from the mirror's surface ($p = q$). Students can demonstrate this conclusion using simple geometry to show that the ray-tracing procedure produces congruent triangles.

Section Review
ANSWERS

1. **a.** specular
   b. diffuse
   c. specular
   d. diffuse

2. The light must be reflected from one mirror to the other in order to form images of images. Therefore, the mirrors must be exactly or very nearly parallel to each other, with their mirrored surfaces facing each other.

3. 75°; 75°

4. Diagrams will vary. Verify that the rays' angles of incidence equal the angles of reflection in the students' drawings.

5. It will seem twice as large, because the object distance (the distance between the mirror and any point in the room, including the back wall) equals the image distance.

6. An object facing the mirror produces an image that faces the object. Thus, the top of the image corresponds to the top of the object, while the image's left side corresponds to the object's right side.

530

Demonstration 7

Image formed by a concave mirror

Purpose Demonstrate that changing the curvature of a concave mirror produces different images.

Materials mylar sheet with reflective coating (flexible polyester material about 1 m × 0.5 m), poster board

Procedure Tape the mylar flat to the poster board, and place a can or a bottle with a large label on it on a table 20 to 50 cm away. Let students observe that the image in the mylar is like one they would expect to see in a flat mirror. Slowly start bending the sides of the mylar sheet to turn it into a concave mirror (until it has a cylindrical shape). Have students describe changes in the image as you roll the mylar. (*The virtual image grows wider and moves farther away, eventually disappearing when the radius of curvature becomes too small.*) Repeat if necessary.

Explain that by bending the mirror you changed its radius of curvature. Have a student hold the mirror at one constant curvature as you move the object slowly closer to the mirror and then farther away. Ask students to describe how changes in the object's distance affect the position and size of the object's image.

14-3
Curved mirrors

14-3 SECTION OBJECTIVES

- Calculate distances and focal lengths using the mirror equation for concave and convex spherical mirrors.

- Draw ray diagrams to find the image distance and magnification for concave and convex spherical mirrors.

- Distinguish between real and virtual images.

- Describe how parabolic mirrors differ from spherical mirrors.

concave spherical mirror

an inwardly curved, mirrored surface that is a portion of a sphere and that converges incoming light rays

CONCAVE SPHERICAL MIRRORS

You may have noticed small circular mirrors on dressing tables. Although such mirrors may appear at first glance to be the same as flat mirrors, the images they form differ from those formed by flat mirrors. For example, the images for objects close to the mirror are larger than the object, as shown in **Figure 14-10(a),** while the images of objects far from the mirror are smaller and upside down, as shown in **Figure 14-10(b).** Images such as these are characteristic of curved mirrors.

Concave mirrors focus light to form real images

One basic type of curved mirror is the spherical mirror. A spherical mirror, as its name implies, has the shape of part of a sphere's surface. A spherical mirror with light reflecting from its silvered, concave surface (that is, the inner surface of a sphere) is called a **concave spherical mirror.** Concave spherical mirrors are used whenever a magnified image of an object is needed, as in the case of the dressing-table mirror.

One factor that determines where the image will appear in a concave spherical mirror and how large that image will be is the amount by which the mirror is curved. This in turn depends on the radius of curvature, *R*, of the mirror. The radius of curvature is the same as the radius of the spherical shell of which the mirror is a small part; *R* is therefore the distance from the mirror's surface to the center of curvature, *C*.

(a)   **(b)**

Figure 14-10
Curved mirrors can be used to form images that are larger **(a)** or smaller **(b)** than the object.

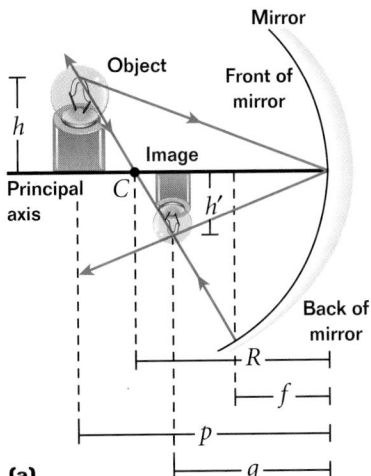

(a)

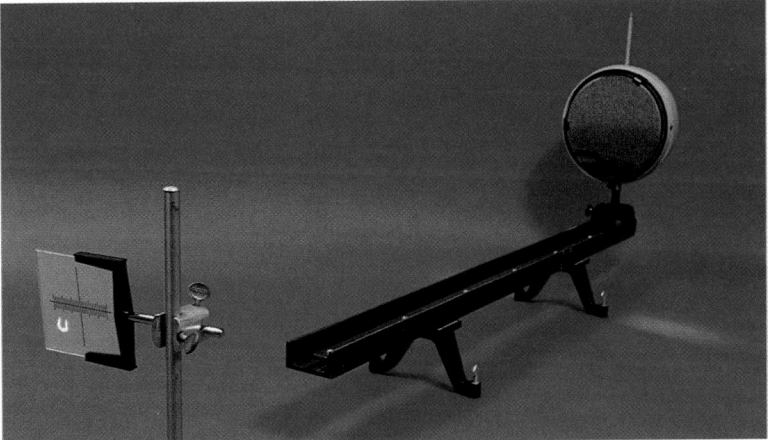

(b)

Figure 14-11
(a) The rays from an object, such as a light bulb, converge to form a real image in front of a concave mirror. **(b)** In this lab setup, the real image of a light-bulb filament appears on a glass plate in front of a concave mirror.

Imagine a light bulb placed upright at a distance *p* from a concave spherical mirror, as shown in **Figure 14-11(a).** The base of the bulb is along the mirror's principal axis, which is the line that extends infinitely from the center of the mirror's surface through the center of curvature, *C*. Light rays diverge from the light bulb, reflect from the mirror's surface, and converge at some distance, *q*, in front of the mirror. Because the light rays reflected by the mirror actually pass through the image point, which in this case is below the principal axis, the image forms in front of the mirror.

If you place a piece of paper at the image point, you will see on the paper a sharp and clear image of the light bulb. As you move the paper in either direction away from the image point, the rays diverge and the image becomes unfocused. An image of this type is called a **real image.** Unlike the virtual images that appear behind a flat mirror, real images can be displayed on a surface, like the images on a movie screen. **Figure 14-11(b)** shows a real image of a light-bulb filament on a glass plate in front of a concave mirror. This light bulb itself is outside the photograph, to the left. ⭐ TEKS **8A**

real image

an image formed when rays of light actually intersect at a single point

Image location can be predicted with the mirror equation

By looking at **Figure 14-11(a),** you can see that object distance, image distance, and radius of curvature are interdependent. If the object distance and radius of curvature of the mirror are known, you can predict where the image will appear. Alternatively, the radius of curvature of a mirror can be determined if you know where the image appears for a given object distance. The equation relating object distance, image distance, and the radius of curvature is called the mirror equation.

$$\frac{1}{p} + \frac{1}{q} = \frac{2}{R}$$

$$\frac{1}{\text{object distance}} + \frac{1}{\text{image distance}} = \frac{2}{\text{radius of curvature}}$$

Visual Strategy

Figure 14-11
This image shows a real image of a light bulb on a glass plate. The bulb itself is off to the left, too far away to fit in the photo frame. There is actually an image of the entire bulb on the plate, but only the image of the filament is bright enough to be seen in the photograph.

Have students follow the path of light rays starting from the object and arriving at the image. The real image appears at the point where many light rays come together after reflecting off different parts of the mirror.

Q What represents the size of the object? What represents its distance from the mirror? What represents the size of the image? What represents the image's distance from *C*? What does *f* refer to?

A *h; p; h'; R – q; distance from focal point to mirror*

Key Models and Analogies

To help students understand the reflections in **Figure 14-11,** point out that light striking the mirror is reflected according to the law of reflection, as if the curved mirror was made of many small plane mirrors positioned to form a circle. The ray through *C* would be normal to such a mirror, so it is reflected back in the same direction from which it came.

⭐ TEKS

p. 531: 8A

Quick Lab

TEACHER'S NOTES

This activity is intended to explore how an object's distance affects the object's image in concave and convex mirrors. This experiment works best with a very shiny spoon that has a large radius of curvature.

Demonstration 8

Focal point of a concave mirror

Purpose Demonstrate that rays parallel to the principal axis are reflected through the focal point, and show that $f = \dfrac{R}{2}$.

Materials light source, ray filter, concave mirror, white paper

Procedure Use the ray filter to produce five beams. Dim the lights, and hold the sheet of paper in front of the beams to let students observe that the incident rays are parallel. Place the concave mirror 20 to 30 cm from the light source, and let students observe the beams converging. Tell them that the point of convergence is called the focal point. Explain that past that point, the beams diverge. Draw the mirror's curve and the principal axis, and mark the focal point on the chalkboard. Ask students if that point could be the center of the circle from which the mirror was cut (*no*). Have students mark where the approximate center of the circle is. Measure *R*, and compare it with *f*.

532

TEKS 1A, 8A

Quick Lab

Curved Mirrors

MATERIALS LIST

✔ stainless-steel or silver spoon
✔ short pencil

Observe the pencil's reflection in the inner portion of the spoon. Slowly move the spoon closer to the pencil. Note any changes in the appearance of the pencil's reflection. Repeat these steps using the other side of the spoon as the mirror.

CONCEPT PREVIEW

When expressed in terms of focal length, the mirror equation becomes identical to the equation for a simple thin lens. The properties of lenses will be discussed in Chapter 15.

If the light bulb is placed very far from the mirror, the object distance, *p*, is great enough compared with *R* that $1/p$ is almost 0. In this case, *q* is almost *R*/2, so the image forms halfway between the center of curvature and the center of the mirror's surface. The image point, as shown in **Figure 14-12(a)** and **(b)**, is in this special case called the *focal point* of the mirror and is denoted by the capital letter *F*. Because the light rays are reversible, the reflected rays from a light source at the focal point will emerge parallel to each other.

When the image point is at the focal point, *F*, the image distance is called the *focal length*, and it is denoted by the lowercase letter *f*. For a spherical mirror, the focal length is equal to half the radius of curvature of the mirror. The mirror equation can therefore be expressed in terms of the focal length.

MIRROR EQUATION TEKS 3B, 8A

$$\frac{1}{p} + \frac{1}{q} = \frac{1}{f}$$

$$\frac{1}{\text{object distance}} + \frac{1}{\text{image distance}} = \frac{1}{\text{focal length}}$$

To use the mirror equation, a set of sign conventions for the three variables must be established. The region in which light rays reflect and form real images is called the front side of the mirror. The other side, where light rays do not exist—and where virtual images are formed—is called the back side of the mirror. The mirror is usually drawn so that the front side is to the left of the mirror's surface and the back side is to the right.

Object and image distances have a positive sign when measured from the center of the mirror to any point on the mirror's front side. Distances for images that form on the back side of the mirror always have a negative sign. Because the mirrored surface is on the front side of a concave mirror, its focal length always has a positive sign. The object and image heights are positive when both are above the principal axis and negative when either is below. **Table 14-4,** on page 538, lists these and other sign conventions for mirrors.

(a)

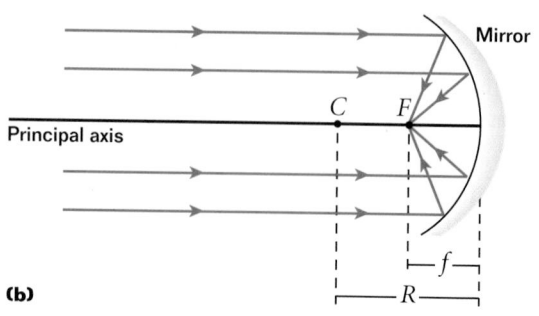

(b)

Figure 14-12

Light rays that are parallel converge at a single point **(a)**, which can be represented in a diagram **(b)**, when the rays are assumed to be from a distant object ($p \approx \infty$).

Magnification relates image and object sizes ⊛TEKS 8A

Unlike flat mirrors, curved mirrors form images that are not the same size as the object. The measure of how large or small the image is with respect to the original object's size is called the *magnification* of the image.

If you know where the light bulb's image will form for a given object distance, you can determine the magnification of the image. Magnification, *M*, is defined as the ratio of the height of the bulb's image to the bulb's actual height. *M* also equals the negative of the ratio of the image distance to the object distance. If an image is smaller than the object, the magnitude of its magnification is less than 1. If the image is larger than the object, the magnitude of its magnification is greater than 1. Magnification is a unitless quantity.

EQUATION FOR MAGNIFICATION ⊛TEKS 3B

$$M = \frac{h'}{h} = -\frac{q}{p}$$

$$\text{magnification} = \frac{\text{image height}}{\text{object height}} = -\frac{\text{image distance}}{\text{object distance}}$$

For an image in front of the mirror, *M* is negative and the image is upside down, or *inverted*, with respect to the object. When the image is behind the mirror, *M* is positive and the image is *upright* with respect to the object. The conventions for magnification are listed in **Table 14-2.**

Table 14-2   Sign conventions for magnification

Orientation of image with respect to object	Sign of *M*	Type of image this applies to
upright	+	virtual
inverted	−	real

Ray diagrams can be used for concave spherical mirrors

Ray diagrams are useful for checking values calculated from the mirror and magnification equations. The techniques for ray diagrams that were used to locate the image for an object in front of a flat mirror can also be used for concave spherical mirrors. When drawing ray diagrams for concave mirrors, follow the basic procedure for a flat mirror, but also measure all distances along the principal axis and mark the center of curvature, *C*, and the focal point, *F*. As with a flat mirror, draw the diagram to scale. For instance, if the object distance is 50 cm, you can draw the object distance as 5 cm.

For spherical mirrors, three reference rays are used to find the image point. The intersection of any *two* rays locates the image. The third ray should intersect

 Misconception Alert

The term *magnification* sometimes leads students to think that the image is larger than the object. Use a numerical example to show that this is not always the case. Ask students to calculate the image height of an 8 cm object when *p* = 12 cm and *q* = 3 cm (*−2 cm*).

Demonstration 9

Beams reflected from a concave mirror

Purpose Demonstrate the formation of a virtual image.

Materials light source, ray filter, concave mirror

Procedure Ask students if reflected beams always converge from concave mirrors. Lower the lights in the room so that the light can be easily seen. Place the mirror as far away as possible from the front of the light source, and move the mirror closer and closer to the light source until the reflected beams diverge. Tell students that the point behind the mirror from which the beam seems to come is a virtual image of the source.

⊛TEKS

p. 532: 1A, 8A, 3B, 8A
p. 533: 8A, 3B

Interactive Problem-Solving Tutor

See Module 14

"Reflection and Mirrors" provides additional development of problem-solving skills for this chapter.

Visual Strategy

Figure 14-13

Point out that from all the possible rays coming from the tip of the pencil, these three rays are selected because they are reflected according to the simple rules listed in **Table 14-3.** Make sure that students understand the application of these rules in the three diagrams.

Q Draw the path of two rays to form the image of a dot drawn at the middle of the pencil in **Figure 14-13(a).** Where do these rays intersect?

A *Ray 1$_{dot}$ would run parallel to 1 and hit the mirror at about half the height that 1 does. When returning through F, ray 1$_{dot}$ would pass through the middle of the image. Ray 2$_{dot}$ would be drawn by the rules of a ray coming to the mirror through the focal point. Accurate drawings would also show ray 2$_{dot}$ parallel to the principal axis and passing through the middle of the image after reflection.*

Module 14
"Reflection and Mirrors" provides an interactive lesson with guided problem-solving practice to teach you more about mirrors and images.

at the same point and can be used to check the diagram. These reference rays are described in **Table 14-3.**

Table 14-3	Rules for drawing reference rays	
Ray	**Line drawn from object to mirror**	**Line drawn from mirror to image after reflection**
1	parallel to principal axis	through focal point *F*
2	through focal point *F*	parallel to principal axis
3	through center of curvature *C*	back along itself through *C*

The image distance in the diagram should agree with the value for *q* calculated from the mirror equation. However, the image distance may differ because of inaccuracies that arise from drawing the ray diagrams at a reduced scale. Ray diagrams should therefore be used to obtain approximate values only; they should not be relied on for the best quantitative results.

When an object is moved toward a concave spherical mirror, its image changes, as shown in **Figure 14-13.** For object distances greater than the focal length, the image is real and inverted, as shown in **(a).** When the object is at the focal point, the image is infinitely far to the left and therefore is not seen, as is indicated in **(b).** When the object lies between the focal point and the mirror surface, the image forms again, only now it becomes virtual and upright, as shown in **(c).** ⭐TEKS **8A**

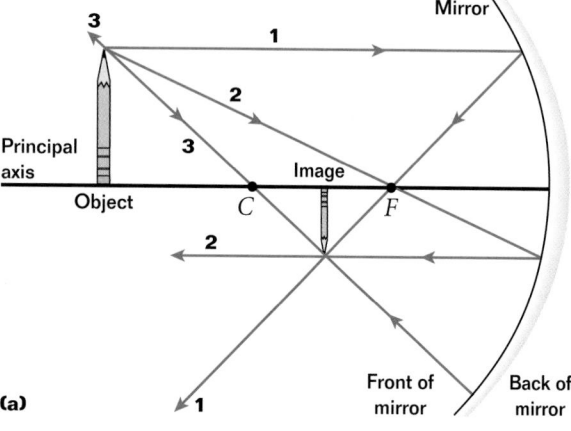

(a)

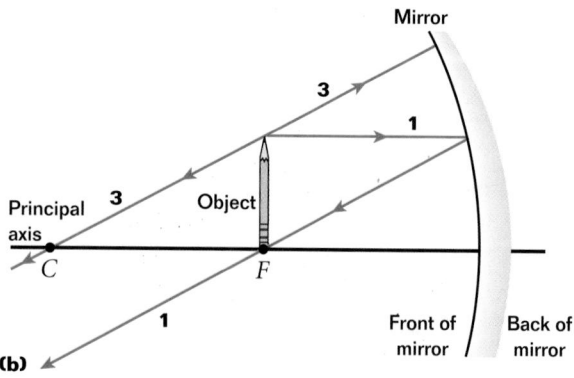

(b)

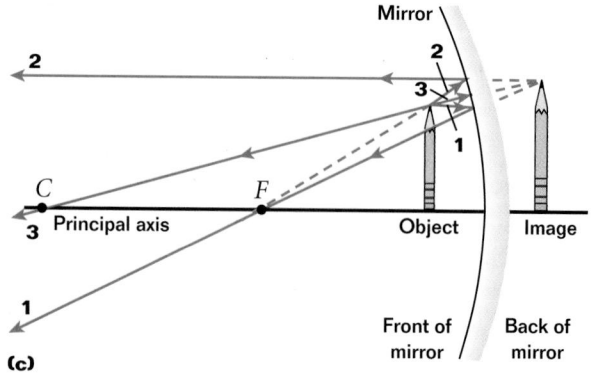

(c)

Figure 14-13
The image produced by a concave spherical mirror depends on whether the object distance is **(a)** greater than, **(b)** equal to, or **(c)** less than the focal length of the mirror. In each drawing, the rays are numbered according to their description in **Table 14-3.**

Concave mirrors ⭐TEKS 2C, 8A

PROBLEM

A concave spherical mirror has a focal length of 10.0 cm. Locate the image of a pencil that is placed upright 30.0 cm from the mirror. Find the magnification of the image. Draw a ray diagram to confirm your answer.

SOLUTION

1. DEFINE **Given:** $f = +10.0$ cm $p = +30.0$ cm

The mirror is concave, so f is positive. The object is in front of the mirror, so p is positive.

Unknown: $q = ?$ $M = ?$

Diagram:

Use the rules on page 534 to construct a ray diagram.

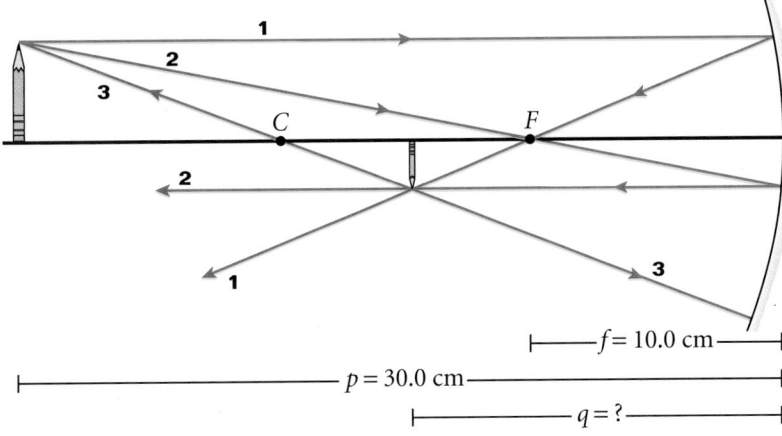

2. PLAN **Choose an equation(s) or situation:** Use the mirror equation that relates the object and image distances to focal length.

$$\frac{1}{p} + \frac{1}{q} = \frac{1}{f}$$

Use the magnification equation in terms of object and image distances.

$$M = -\frac{q}{p}$$

Rearrange the equation(s) to isolate the unknown(s): Subtract the reciprocal of the object distance from the reciprocal of the focal length to obtain an expression for the unknown image distance.

$$\frac{1}{q} = \frac{1}{f} - \frac{1}{p}$$

continued on next page

The following may be used as teamwork exercises or for demonstration at the chalkboard or on an overhead projector.

PROBLEM

Concave mirrors

When an object is placed 30.0 cm in front of a concave mirror, a real image is formed 60.0 cm from the mirror's surface. Find the focal length.

Answer
   20.0 cm

A square object is placed 15 cm in front of a concave mirror with a focal length of 25 cm. A round object is placed 45 cm in front of the same mirror. Find the image distance, magnification, and type of image formed for each object. Draw ray diagrams for each object to confirm your answers.

Answer
   $q_{square} = -37$ cm, $M_{square} = 2.5$, virtual and upright;
   $q_{round} = 56$ cm, $M_{round} = -1.2$, real and inverted

⭐TEKS

p. 534: 8A
p. 535: 2C, 8A

535

ANSWERS TO

Practice 14B
Concave mirrors

1. $p = 10.0$ cm: no image (infinite q); $p = 5.00$ cm: $q = -10.0$ cm, $M = 2.00$; virtual, upright image
2. $q = 53$ cm; $M = -0.57$, real, inverted image
3. $R = 1.00 \times 10^2$ cm; $M = 2.00$; virtual image
4. $f = 5.999$ cm; $M = -1.20$; $q = 7.710$ cm; $M = -0.286$; real image

3. CALCULATE **Substitute the values into the equation(s) and solve:** Substitute the values for f and p into the mirror equation and the magnification equation to find the image distance and magnification.

$$\frac{1}{q} = \frac{1}{10.0\ \text{cm}} - \frac{1}{30.0\ \text{cm}} = \frac{0.100}{1\ \text{cm}} - \frac{0.033}{1\ \text{cm}} = \frac{0.067}{1\ \text{cm}}$$

$$\boxed{q = 15\ \text{cm}}$$

$$\boxed{M = -\frac{q}{p} = -\frac{15\ \text{cm}}{30.0\ \text{cm}} = -0.50}$$

4. EVALUATE The image appears between the focal point (10.0 cm) and the center of curvature (20.0 cm), as confirmed by the ray diagram. The image is smaller than the object and inverted ($M < 0$), as is also confirmed by the ray diagram. The image is therefore real.

PRACTICE 14B

Concave mirrors ⭐TEKS 2C, 8A, 8B

1. Find the image distance and magnification of the mirror in the sample problem when the object distances are 10.0 cm and 5.00 cm. Are the images real or virtual? Are the images inverted or upright? Draw a ray diagram for each case to confirm your results.

2. A concave shaving mirror has a focal length of 33 cm. Calculate the image position of a cologne bottle placed in front of the mirror at a distance of 93 cm. Calculate the magnification of the image. Is the image real or virtual? Is the image inverted or upright? Draw a ray diagram to show where the image forms and how large it is with respect to the object.

3. A concave makeup mirror is designed so that a person 25.0 cm in front of it sees an upright image at a distance of 50.0 cm behind the mirror. What is the radius of curvature of the mirror? What is the magnification of the image? Is the image real or virtual?

4. A pen placed 11.0 cm from a concave spherical mirror produces a real image 13.2 cm from the mirror. What is the focal length of the mirror? What is the magnification of the image? If the pen is placed 27.0 cm from the mirror, what is the new position of the image? What is the magnification of the new image? Is the new image real or virtual? Draw ray diagrams to confirm your results.

CONVEX SPHERICAL MIRRORS ⭐TEKS 8A

On recent models of automobiles, there is a side-view mirror on the passenger's side of the car. Unlike the flat mirror on the driver's side, which produces unmagnified images, the passenger's mirror bulges outward at the center. Images in this mirror are distorted near the mirror's edges, and the image is smaller than the object. This type of mirror is called a **convex spherical mirror.**

A convex spherical mirror is a segment of a sphere that is silvered so that light is reflected from the sphere's outer, convex surface. This type of mirror is also called a diverging mirror because the incoming rays diverge after reflection as though they were coming from some point behind the mirror. The resulting image is therefore always virtual, and the image distance is always negative. Because the mirrored surface is on the side opposite the radius of curvature, a convex spherical mirror also has a negative focal length. The sign conventions for all mirrors are summarized in **Table 14-4** on page 538.

The technique for drawing ray diagrams for a convex mirror differs slightly from that for concave mirrors. The focal point and center of curvature are situated behind the mirror's surface. Dotted lines are extended along the reflected reference rays to points behind the mirror, as shown in **Figure 14-14(a).** A virtual, upright image forms where the three rays apparently intersect. Magnification for convex mirrors is always less than 1, as shown in **Figure 14-14(b).**

Convex spherical mirrors take the objects in a large field of view and produce a small image, so they are well suited for providing a fixed observer with a complete view of a large area. Convex mirrors are often placed in stores to help employees monitor customers and at the intersections of busy hallways so that people in both hallways can tell when others are approaching.

The side-view mirror on the passenger's side of a car is another application of the convex mirror. This mirror usually carries the warning, "objects are closer than they appear." Without this warning, a driver might think that he or she is looking into a flat mirror, which does not alter the size of the image. The driver could therefore be fooled into believing that a vehicle is farther away than it is. ⭐TEKS 3C, 8C

convex spherical mirror

an outwardly curved, mirrored surface that is a portion of a sphere and that diverges incoming light rays

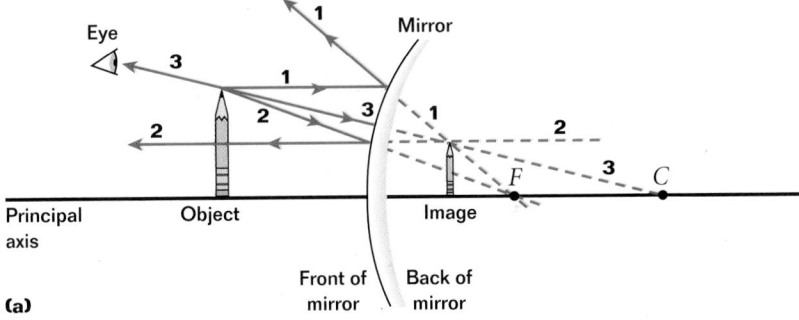

(a)

Figure 14-14

Light rays diverge upon reflection with a convex mirror **(a),** forming a virtual image that is always smaller than the object **(b).**

(b)

SECTION 14-3

Demonstration 10

Convex mirror

Purpose Demonstrate that parallel beams reflected by convex mirrors are diverging.

Materials light source, ray filter, convex mirror, white paper

Procedure Use the ray filter to produce five beams. Dim the lights. Place the paper in front of the beams and let students observe that the incident rays are parallel. Place the convex mirror as far as possible from the front of the light source, and let students observe the beams diverging. Move the mirror closer to the light source. Students will notice that the beam is always diverging. Explain that unlike concave mirrors, which can produce real and virtual images, convex mirrors produce only virtual images.

⭐TEKS

p. 536: 2C, 8A, 8B
p. 537: 8A, 3C, 8C

Visual Strategy

Table 14-4

Make sure that students properly interpret information related to all the cases listed in **Table 14-4.** Point out that as a general rule, distances in front of the mirror are assigned a positive sign and distances behind the mirror are assigned a negative sign.

Q The image formed by a concave mirror is upright and virtual. What would be the signs of R, f, q, and h'?

A +, +, −, +

Q The image formed by a convex mirror is also upright and virtual. What would be the signs of R, f, q, and h'?

A −, −, −, +

Did you know?

There are certain circumstances in which the object for one mirror is the image that appears behind another mirror. In these cases, the object is virtual and has a negative object distance. Because of the rarity of these situations, virtual object distance ($p < 0$) has not been listed in **Table 14-4.**

Table 14-4   Sign conventions for mirrors ⊛ TEKS **3B, 8A**

Symbol	Situation	Sign	
p	object is in front of the mirror (real object)	+	$p > 0$
q	image is in front of the mirror (real image)	+	$q > 0$
q	image is behind the mirror (virtual image)	−	$q < 0$
R, f	center of curvature is in front of the mirror (concave spherical mirror)	+	$R > 0$ $f > 0$
R, f	center of curvature is behind the mirror (convex spherical mirror)	−	$R < 0$ $f < 0$
R, f	mirror has no curvature (flat mirror)	∞	$-R, f \rightarrow \infty-$
h'	image is above the principal axis	+	$h, h' > 0$
h'	image is below the principal axis	−	$h > 0, h' < 0$

Convex mirrors ⭐TEKS 2C, 3B, 8A

PROBLEM

An upright pencil is placed in front of a convex spherical mirror with a focal length of 8.00 cm. An erect image 2.50 cm tall is formed 4.44 cm behind the mirror. Find the position of the object, the magnification of the image, and the height of the pencil.

SOLUTION

1. DEFINE   **Given:** $f = -8.00$ cm $q = -4.44$ cm $h' = 2.50$ cm

Because the mirror is convex, the focal length is negative.
The object is behind the mirror, so q is also negative.

Unknown: $p = ?$ $h = ?$

Diagram: Use the rules on page 534 to construct a ray diagram.

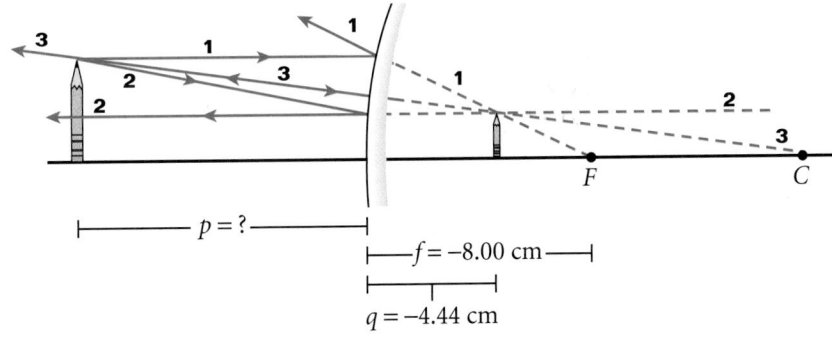

2. PLAN   **Choose an equation(s) or situation:** Use the mirror equation.

$$\frac{1}{p} + \frac{1}{q} = \frac{1}{f}$$

Use the magnification formula.

$$M = \frac{h'}{h} = -\frac{q}{p}$$

Rearrange the equation(s) to isolate the unknown(s):

$$\frac{1}{p} = \frac{1}{f} - \frac{1}{q} \quad \text{and} \quad h = -\frac{p}{q}h'$$

3. CALCULATE   **Substitute the values into the equation(s) and solve:**

$$\frac{1}{p} = \frac{1}{-8.00 \text{ cm}} - \frac{1}{-4.44 \text{ cm}}$$

$$\frac{1}{p} = \frac{-0.125}{1 \text{ cm}} - \frac{-0.225}{1 \text{ cm}} = \frac{0.100}{1 \text{ cm}}$$

$$\boxed{p = 10.0 \text{ cm}}$$

continued on next page

Classroom Practice

The following may be used as teamwork exercises or for demonstration at the chalkboard or on an overhead projector.

PROBLEM

Convex mirrors

The radius of curvature of a convex mirror is 12.0 cm. Where is the focal point located?

Answer
6.00 cm behind the mirror ($f = -6.00$ cm)

Find the position of the image for an object placed at the following distances from the mirror in the previous question: $p = 1.00$ cm, 2.00 cm, 3.00 cm, 6.00 cm, 12.0 cm, 30.0 cm, 50.0 cm

Answer
−0.855 cm, −1.50 cm, −2.00 cm, −2.99 cm, −4.00 cm, −5.00 cm, −5.35 cm

How does the position of the image vary as the object in the previous question moves farther away from the mirror?

Answer
The image is always behind the mirror and between the mirror and the focal point. It moves from $q = 0$ to $q = f$ as the object moves away from the mirror (from $p = 0$ to infinity).

⭐TEKS

p. 538: 3B, 8A
p. 539: 2C, 3B, 8A

PRACTICE GUIDE 14C

Solving for:

p	📖 **PE**	Sample, 1–3; Ch. Rvw. 36, 51*
	💿 **PW**	7–9
	PB	3–6
q	📖 **PE**	4–6
	💿 **PW**	Sample, 1–3
	PB	7–10
R, f	📖 **PE**	Ch. Rvw. 36, 48, 50, 55
	💿 **PW**	4–6
	PB	Sample, 1, 2
M	📖 **PE**	Sample 1–6; Ch. Rvw. 50
	💿 **PW**	Sample, 1–2, 4–5, 7–8
	PB	Sample, 1, 3–5, 7–8
h, h'	💿 **PW**	3, 6, 9
	PB	2, 6, 9

ANSWERS TO

Practice 14C
Convex mirrors

1. $p = 45.9$ cm; $M = 0.501$; virtual, upright image; $h = 3.39$ cm
2. $M = 0.048$; $p = 5.0$ m; $h = 1.7$ m; virtual, upright image
3. $p = 43$ cm; $h = 16$ cm; $M = 0.44$; virtual, upright image
4. $q = -0.25$ m; $M = 0.081$; virtual, upright image
5. $q = -1.31$ cm (behind the ornament's surface); $M = 0.125$; virtual, upright image
6. $q = -2.0 \times 10^1$ cm; $M = 0.41$; virtual, upright image

Substitute the values for p and q to find the magnification of the image.

$$M = -\frac{q}{p} = -\frac{-4.44 \text{ cm}}{10.0 \text{ cm}}$$

$$\boxed{M = 0.444}$$

Substitute the values for p, q, and h' to find the height of the object.

$$h = -\frac{p}{q}h' = -\frac{10.0 \text{ cm}}{-4.44 \text{ cm}} (2.50 \text{ cm})$$

$$\boxed{h = 5.63 \text{ cm}}$$

PRACTICE 14C

Convex mirrors ⭐TEKS 2C, 3B, 8A, 8B

1. The image of a crayon appears to be 23.0 cm behind the surface of a convex mirror and is 1.70 cm tall. If the mirror's focal length is 46.0 cm, how far in front of the mirror is the crayon positioned? What is the magnification of the image? Is the image virtual or real? Is the image inverted or upright? How tall is the actual crayon?

2. A convex mirror with a focal length of 0.25 m forms a 0.080 m tall image of an automobile at a distance of 0.24 m behind the mirror. What is the magnification of the image? Where is the car located, and what is its height? Is the image real or virtual? Is the image upright or inverted?

3. A convex mirror of focal length 33 cm forms an image of a soda bottle at a distance of 19 cm behind the mirror. If the height of the image is 7.0 cm, where is the object located, and how tall is it? What is the magnification of the image? Is the image virtual or real? Is the image inverted or upright? Draw a ray diagram to confirm your results.

4. A convex mirror with a radius of curvature of 0.550 m is placed above the aisles in a store. Determine the image distance and magnification of a customer lying on the floor 3.1 m below the mirror. Is the image virtual or real? Is the image inverted or upright?

5. A spherical glass ornament is 6.00 cm in diameter. If an object is placed 10.5 cm away from the ornament, where will its image form? What is the magnification? Is the image virtual or real? Is the image inverted or upright?

6. A candle is 49 cm in front of a convex spherical mirror that has a focal length of 35 cm. What are the image distance and magnification? Is the image virtual or real? Is the image inverted or upright? Draw a ray diagram to confirm your results.

PARABOLIC MIRRORS ⊛TEKS 8A

You have probably noticed that certain rays in ray diagrams do not intersect exactly at the image point. This occurs especially with rays that reflect at the mirror's surface far from the principal axis. The situation also occurs with real light rays and real spherical mirrors.

If light rays from an object are near the principal axis, all of the reflected rays pass through the image point. Rays that reflect at points on the mirror far from the principal axis converge at slightly different points on the principal axis, as shown in **Figure 14-15.** This produces a blurred image. This effect, called *spherical aberration*, is present to some extent in any spherical mirror.

Parabolic mirrors eliminate spherical aberration

A simple way to reduce the effect of spherical aberration is to use a mirror with a small diameter; that way, the rays are never far from the principal axis. If the mirror is large to begin with, shielding its outer portion will limit how much of the mirror is used and thus will accomplish the same effect. However, many concave mirrors, such as those used in astronomical telescopes, are made large so that they will collect a large amount of light. Therefore, it is not desirable to limit how much of the mirror is used in order to reduce spherical aberration. An alternative approach is to use a mirror that is not a segment of a sphere but still focuses light rays in a manner similar to a small spherical concave mirror. This is accomplished with a parabolic mirror.

Parabolic mirrors are segments of a paraboloid (a three-dimensional parabola) whose inner surface is reflecting. All rays parallel to the principal axis converge at the focal point regardless of where on the mirror's surface the rays reflect. Thus, a real image forms without spherical aberration, as illustrated in **Figure 14-16.** Similarly, light rays from an object at the focal point of a parabolic mirror will be reflected from the mirror in parallel rays. Parabolic reflectors are ideal for flashlights and automobile headlights.

Reflecting telescopes use parabolic mirrors ⊛TEKS 3C, 8C

A telescope permits you to view distant objects, whether they are buildings a few kilometers away or galaxies that are millions of light-years from Earth. Not all telescopes are intended for visible light. Because all electromagnetic radiation obeys the law of reflection, parabolic surfaces can be constructed to reflect and focus electromagnetic radiation of different wavelengths. For instance, a radio telescope consists of a large metal parabolic surface that reflects radio waves in order to receive radio signals from objects in space.

There are two types of telescopes that use visible light. One type, called a *refracting telescope,* uses a combination of lenses to form an image. It will be discussed in Chapter 15. The other kind uses a curved mirror and small lenses to form an image. This type of telescope is called a *reflecting telescope.*

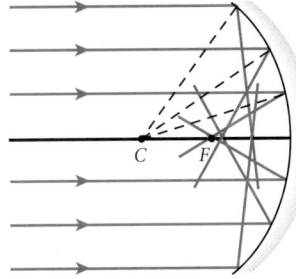

Figure 14-15
Spherical aberration occurs when parallel rays far from the principal axis converge away from the mirror's focal point.

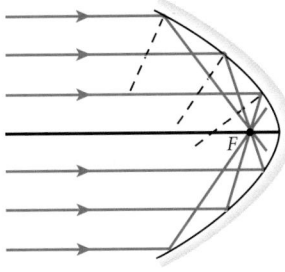

Figure 14-16
All parallel rays converge at a parabolic mirror's focal point. The curvature in this figure is much greater than it is in real parabolic mirrors.

internet connect

SC*LINKS*
NSTA

TOPIC: Telescopes
GO TO: www.scilinks.org
*sci*LINKS CODE: HF2144

SECTION 14-3

Visual Strategy

Figure 14-15
Remind students that for a spherical mirror, the normal to the surface at any point lies along the radius at that point. Point out that each pair of reflected rays at equal distances on opposite sides of the principal axis cross one another on the principal axis.

Q How does the angle of incidence vary when the incoming ray is farther away from the principal axis of a spherical mirror? Mark the points where each pair of reflected rays intersects. Where will the next pair intersect if the mirror's surface is extended?

A *The angle of incidence increases; The points should be marked on the principal axis, to the right of F; The points of intersection come closer to the mirror's surface, and the focal point becomes more ill-defined.*

 Misconception Alert

Students may be unclear about the difference between segments of a circle and of a parabolic curve. Ask them to draw the continuation of the circular and parabolic curves of the mirrors to see that the curves are similar for a short segment. Students may also explore the point at which the circle's approximation of a parabola starts to fail. They can do this by using their graphing calculators to compare the curves whose equations are $y = \sqrt{1 - x^2} - 1$ and $y = -x^2$.

⊛TEKS
p. 540: 2C, 3B, 8A, 8B
p. 541: 8A, 3C, 8C

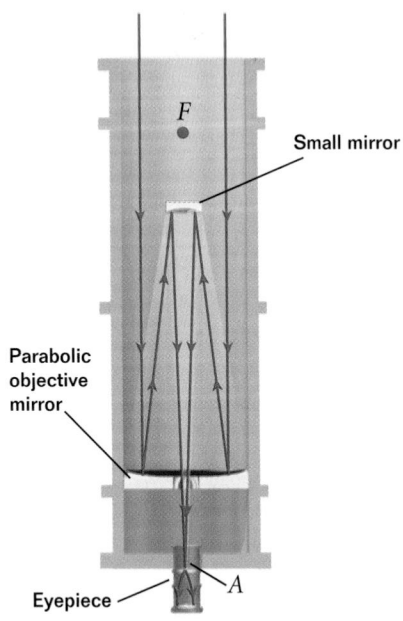

Figure 14-17
The parabolic objective mirror in a Cassegrain reflector focuses incoming light.

Reflecting telescopes employ a parabolic mirror (called an *objective mirror*) to focus light. One type of reflecting telescope, called a *Cassegrain reflector,* is shown in **Figure 14-17.** Parallel light rays pass down the barrel of the telescope and are reflected by the parabolic objective mirror at the telescope's base. These rays converge toward the objective mirror's focal point, *F,* where a real image would normally form. However, a small curved mirror that lies in the path of the light rays reflects the light back toward the center of the objective mirror. The light then passes through a small hole in the center of the objective mirror and comes to a focus at point *A.* An eyepiece near point *A* magnifies the image.

You may wonder how a hole can be placed in the objective mirror without affecting the final image formed by the telescope. Each part of the mirror's surface reflects light from distant objects, so a complete image is always formed. The presence of the hole merely reduces the amount of light that is reflected. Even that is not severely affected by the hole because the light-gathering capacity of an objective mirror is dependent on the mirror's area. For instance, a 1 m diameter hole in a mirror that is 4 m in diameter reduces the mirror's reflecting surface by only $\frac{1}{16}$, or 6.25 percent. (★)TEKS **3A**

Section Review   (★)TEKS **2C, 3B, 8A, 8B, 8C**

1. A steel ball bearing with a radius of 1.5 cm forms an image of an object that has been placed 1.1 cm away from the bearing's surface. Determine the image distance and magnification. Is the image virtual or real? Is the image inverted or upright? Draw a ray diagram to confirm your results.

2. Why is an image formed by a parabolic mirror brighter than the image of the same object formed by a concave spherical mirror?

3. A spherical mirror is to be used in a motion-picture projector to form an inverted, real image 95 times as tall as the picture in a single frame of film. The image is projected onto a screen 13 m from the mirror. What type of mirror is required, and how far should it be from the film?

4. Which of the following images are real and which are virtual?
   a. the image of a distant illuminated building projected onto a piece of heavy, white cardboard by a small reflecting telescope
   b. the image of an automobile in a flat rearview mirror
   c. the image of shop aisles in a convex observation mirror

5. **Physics in Action**   The reflector of the radio telescope at Arecibo has a radius of curvature of 265.0 m. How far above the reflector must the radio-detecting equipment be placed in order to obtain clear radio images?

Section Review
ANSWERS

1. $q = -0.45$ cm; $M = 0.41$; virtual, upright image

2. The rays reflected by a parabolic mirror all focus at one point, whereas the rays reflected by a concave spherical mirror reflect along a line that includes the mirror's focal point. The spherical mirror's image therefore has less light converging at the focal point, and therefore it is dimmer than the image formed by the parabolic mirror.

3. concave mirror; $p = 0.14$ m

4. a. real
   b. virtual
   c. virtual

5. 132.5 m, which is the telescope's focal length

COLOR

You have probably noticed that the color of an object can appear different under different lighting conditions. A plant that appears green in sunlight will appear black under a red light. These differences are due to differences in the reflecting and light-absorbing properties of the object being illuminated.

So far, we have assumed that objects are either like mirrors, which reflect almost all light uniformly, or like rough objects, which reflect light diffusely in several directions. But as mentioned in Section 14-1, objects absorb certain wavelengths from the light falling on them and reflect the rest. The color of an object depends on which wavelengths of light shine on the object and which wavelengths are reflected (see **Figure 14-18**).

If all wavelengths of incoming light are completely reflected by an object, that object appears the same color as the light illuminating it. This gives the object the same appearance as a white object illuminated by the light. An object of a particular color, such as the green leaf in **Figure 14-18,** absorbs light of all colors except the light whose color is the same as the object's color. By contrast, an object that reflects no light appears black. Green leaves appear black under red light because there is no green light present for them to reflect and the red light is absorbed. (★)TEKS **8A, 8B**

Additive primary colors produce white light when combined

Because white light can be dispersed into its elementary colors, it is reasonable to suppose that elementary colors can be combined to form white light. One way of doing this is to use a prism to recombine light that has been dispersed by another prism. Another way is to combine light that has been passed through red, green, and blue filters. These colors are called the *additive primary colors* because when they are added in varying proportions they can form all of the colors of the spectrum.

14-4 SECTION OBJECTIVES

- **Recognize how additive colors affect the color of light.**
- **Recognize how pigments affect the color of reflected light.**
- **Explain how linearly polarized light is formed and detected.**

---

internet connect

SCI LINKS

NSTA

TOPIC: Color
GO TO: www.scilinks.org
*sci***LINKS CODE:** HF2145

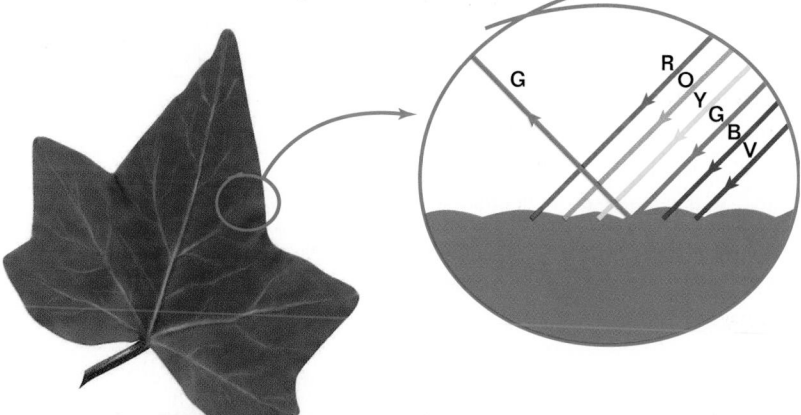

Figure 14-18
A leaf appears green under white light because the pigment in the leaf reflects only green light.

Mixing light colors

Purpose Demonstrate that the color we see when mixing light depends on the brightness of red, green, and blue.

Materials Macintosh® computer with system 7.5

Procedure Select Color in the Control Panel menu (under the apple). Choose Other from the Highlight Color menu box. A new box appears. Select the Apple RGB monitor icon (Note: You may need to click on the More Choices button to get to this option). Three meter bars of primary colors (red, green, and blue) appear (ignore the narrow bars above each meter bar). Each meter bar has a sliding cursor for brightness adjustment. The brightness can also be set by typing numbers in the percentage box to the right of each meter bar. The new rectangle displays the resulting color of the combined brightness settings. Set all of the bars to 0%, and have students note that the result is black. Now ask students to predict the result if each color meter is set at 100% (*white*). Set the blue meter at 0%. Explain that yellow is the complementary color of blue. Reset the blue meter at 100% and set the green meter at 0%. Then reset the green meter at 100% and set the red meter at 0%. Have students observe the color combinations. Suggest that students further explore this simulation later on their own.

⭐ TEKS

p. 544: 3C, 8C
p. 545: 8A, 8B, 3A

Figure 14-19
The combination of the additive primary colors in any two circles produces the complementary color of the third additive primary color.

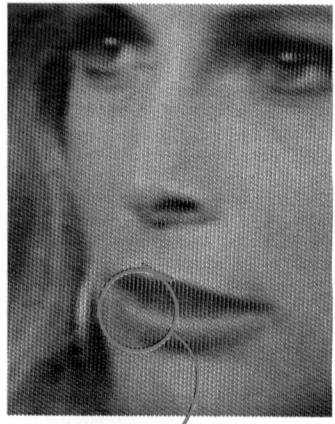

Figure 14-20
The brightness of the red, green, and blue pixels of a color television screen are adjusted to produce all of the colors in a single picture.

When light passed through a red filter is combined with green light produced with a green filter, a patch of yellow light appears. If this yellow light is combined with blue light, the resulting light will be colorless, or "white," as shown in **Figure 14-19.** Because yellow is the color added to the primary additive color blue to produce white light, yellow is called the *complementary* color of blue. Two primary colors combine to produce the complement of the third primary color, as indicated in **Table 14-5.**

One application of additive primary colors is the use of certain chemical compounds to give color to glass. Iron compounds give glass a green color. Manganese compounds give glass a magenta, or reddish blue, color. Green and magenta are complementary colors, so the right proportions of these compounds produces an equal combination of green and magenta light, and the resulting glass appears colorless. ⭐TEKS **3C, 8C**

Another example of additive colors is the image produced on a color television screen. A television screen consists of small, luminous dots, or *pixels*, that glow either red, green, or blue when they are struck by electrons (see **Figure 14-20**). By varying the brightness of different pixels in different parts of the picture, a picture with many colors present at one time is produced.

Humans can see in color because there are three kinds of color receptors in the eye. Each receptor, called a *cone cell*, is sensitive to either red, green, or blue light. Light of different wavelengths stimulates a combination of these receptors so that a wide range of colors can be perceived.

Table 14-5	Additive and subtractive primary colors	
Colors	**Additive (mixing light)**	**Subtractive (mixing pigments)**
red	primary	complementary to cyan
green	primary	complementary to magenta
blue	primary	complementary to yellow
cyan (blue green)	complementary to red	primary
magenta (red blue)	complementary to green	primary
yellow	complementary to blue	primary

Subtractive primary colors filter out all light when combined

When blue light and yellow light are mixed, white light results. However, if you mix a blue pigment (such as paint or the colored wax of a crayon) with a yellow pigment, the resulting color is green, not white. This difference is due to the fact that pigments rely on colors of light that are absorbed, or subtracted, from the incoming light.

For example, yellow pigment subtracts blue and violet colors from white light and reflects red, orange, yellow, and green light. Blue pigment subtracts red, orange, and yellow from the light and reflects green, blue, and violet. When yellow and blue pigments are combined, only green light is reflected. (★TEKS) 8A, 8B

When pigments are mixed, each one subtracts certain colors from white light, and the resulting color depends on the frequencies that are not absorbed. The primary pigments (or *primary subtractive colors,* as they are sometimes called) are cyan, magenta, and yellow. These are the same colors that are complementary to the additive primary colors (see **Table 14-5** on page 544). When any two primary subtractive colors are combined, they produce either red, green, or blue pigments. When the three primary pigments are mixed together in the proper proportions, all of the colors are subtracted from white light and the mixture is black, as shown in **Figure 14-21.**

Combining yellow pigment and its complementary color, blue, should produce a black pigment. Yet earlier, blue and yellow were combined to produce green. The difference between these two situations is explained by the broad use of color names. The "blue" pigment that is added to a "yellow" pigment to produce green is not a pure blue. If it were, only blue light would be reflected from it. Similarly, a pure yellow pigment will reflect only yellow light. Because most pigments found in paints and dyes are combinations of different substances, they reflect light from nearby parts of the visible spectrum. Without knowledge of the light-absorption characteristics of these pigments, it is hard to predict exactly what colors will result from different combinations.

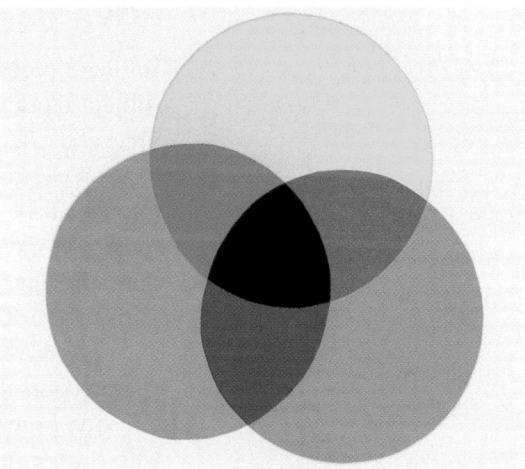

Figure 14-21
The combination of the subtractive primary colors by any two filters produces the complementary color of the third subtractive primary color.

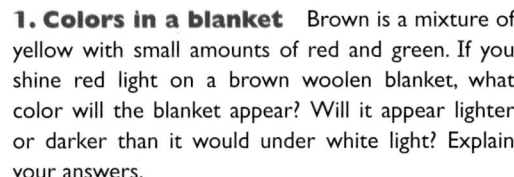
Conceptual Challenge (★TEKS) 3A

1. Colors in a blanket Brown is a mixture of yellow with small amounts of red and green. If you shine red light on a brown woolen blanket, what color will the blanket appear? Will it appear lighter or darker than it would under white light? Explain your answers.

2. Blueprints If a blueprint (a blue drawing on a white background) is viewed under blue light, will you still be able to perceive the drawing? What will the blueprint look like under yellow light?

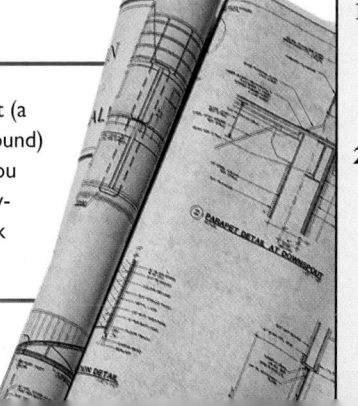

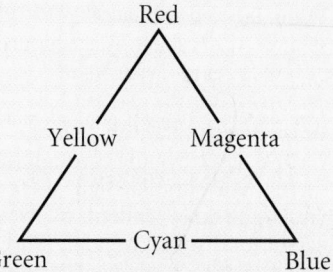

Polarization of waves

Purpose Simulate the polarization of light waves.

Materials 3 flexible coiled springs, thick sheet of cardboard (1.0 m × 2.5 m) with 3 slits (0.1 m × 0.6 m)

Procedure Explain that transverse waves traveling through the springs will simulate light waves. Designate three students as "wave generators," and have each hold one end of each spring. Let them generate pulses randomly (vertical, horizontal, and diagonal). Explain that randomly oscillating charges produce unpolarized light, that is, light waves that may vibrate in any direction.

Have students thread the springs through the slits in the cardboard, and have one student hold the cardboard "polarizer" so that the slits are *vertical*. Ask students what would happen if they tried to send vertical, horizontal, or random diagonal pulses. *(The vertical components would be transmitted, and the horizontal components would be reflected.)* Have students who are not holding springs stand on either side of the cardboard and take notes describing the incoming and the transmitted pulses in all of the experiments. Have students generate vertical, horizontal, and diagonal transverse pulses in the springs, and let them test their predictions.

POLARIZATION OF LIGHT WAVES 8A, 8B

You have probably seen sunglasses with Polaroid™ lenses that reduce glare without blocking the light entirely. There is a property of light that allows some of the light to be filtered by certain materials in the lenses.

In an electromagnetic wave, the electric field is at right angles to both the magnetic field and the direction of propagation. Light from a typical source consists of waves that have electric fields oscillating in random directions, as shown in **Figure 14-22.** Light of this sort is said to be *unpolarized.*

Electric-field oscillations of unpolarized light waves can be treated as combinations of vertical and horizontal electric-field oscillations. There are certain processes that separate waves with electric-field oscillations in the vertical direction from those in the horizontal direction, producing a beam of light with electric-field waves oriented in the same direction, as shown in **Figure 14-23.** These waves are said to have **linear polarization.**

Light can be linearly polarized through transmission

Certain transparent crystals cause unpolarized light that passes through them to become linearly polarized. The direction in which the electric fields are polarized is determined by the arrangement of the atoms or molecules in the

linear polarization

the alignment of electromagnetic waves in such a way that the vibrations of the electric fields in each of the waves are parallel to each other

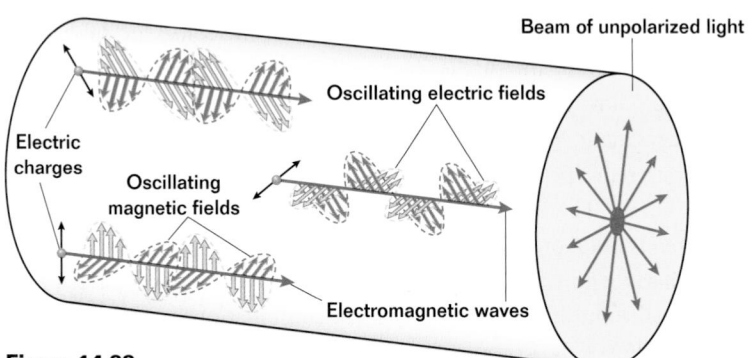

Figure 14-22
Randomly oscillating charges produce unpolarized light.

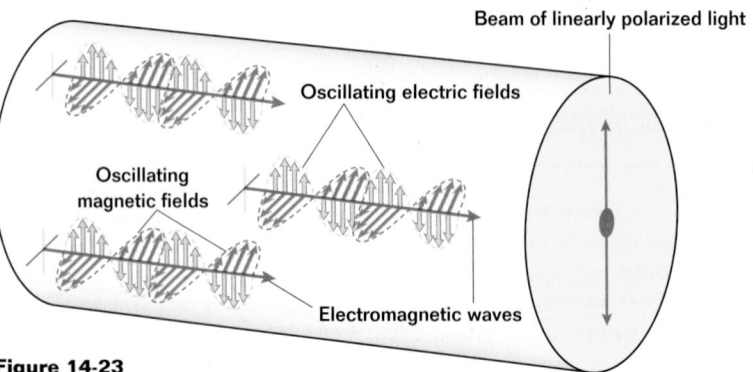

Figure 14-23
Light waves with aligned electric fields are linearly polarized.

crystal. The effect is similar to sending a transverse wave along a rope or spring through the slats of a picket fence, as shown in **Figure 14-24.** Only transverse waves in the up-and-down direction can pass through the fence.

For substances that polarize light by transmission, the line along which light is polarized is called the *transmission axis* of the substance. Only light waves that are linearly polarized with respect to the transmission axis of the polarizing substance can pass freely through the substance. All light that is polarized at an angle of 90° to the transmission axis does not pass through.

A polarizing substance can be used not only to linearly polarize light but also to determine if and how light is linearly polarized. By rotating a polarizing substance as a beam of polarized light passes through it, a change in the intensity of the light can be seen (see **Figure 14-25**). The light is brightest when its plane of polarization is parallel to the transmission axis. The larger the angle is between the electric-field waves and the transmission axis, the smaller the component of light that passes through the polarizer will be and the less bright the light will be. When the transmission axis is perpendicular to the plane of polarization for the light, no light passes through.

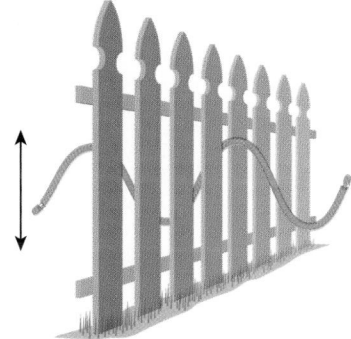

Figure 14-24
Only transverse waves that are aligned in the direction of the slats pass through the fence.

★ TEKS 8A

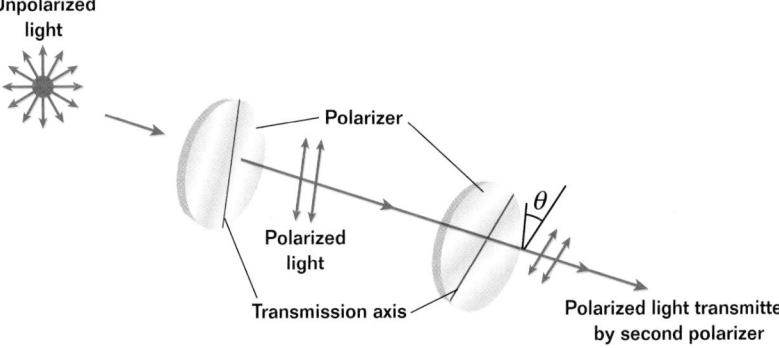

Figure 14-25
The brightness of the polarized light decreases as the angle, θ, increases between the transmission axis of the second polarizer and the plane of polarization of the light.

Light can be polarized by reflection and scattering ★ TEKS 8A, 8B

When light is reflected at a certain angle from a surface, the reflected light is completely polarized parallel to the reflecting surface. If the surface is parallel to the ground, the light is polarized horizontally. This is the case with glaring light that reflects at a low angle from roads, bodies of water, and car hoods.

★ TEKS 1A, 8A, 8B, 8C

Quick Lab

Polarization of Sunlight

MATERIALS LIST

✔ a sheet of Polaroid filter or sunglasses with polarizing lenses

 SAFETY CAUTION

Never look directly at the sun.

During mid-morning or mid-afternoon, when the sun is well above the horizon but not directly overhead, look directly up at the sky through the polarizing filter. Note how the light's intensity is reduced.

Rotate the polarizer. Take note of which orientations of the polarizer make the sky darker and thus best reduce the amount of transmitted light.

Repeat the test with light from other parts of the sky. Test light reflected off a table near a window. Compare the results of these various experiments.

Demonstration 14

Polarizing light by transmission

Purpose Demonstrate linear polarization of light.

Materials two polarizing sheets (approximately 20 cm × 25 cm), overhead projector, projection screen

Procedure Turn on the overhead projector, and have students observe the intensity of the light on the screen. Place one of the polarizing sheets on the overhead projector. Have students note the decreased intensity of the light. Explain to the class that the light from the projector is randomly oriented in all directions. Only the components of light parallel to the polarizer's transmission axis are transmitted, and therefore the intensity of light on the screen is reduced.

Hold the second polarizer in front of the top lens of the overhead projector. Have students note that the intensity of the light remains constant when the second polarizer's transmission axis is parallel to the transmission axis of the original polarizer. Rotate the second polarizer 90° so that its transmission axis is perpendicular to the transmission axis of the original polarizer. Have students note that the intensity of the light is almost zero.

★ TEKS

p. 546: 8A, 8B
p. 547: 8A, 8A, 8B, 1A, 8A, 8B, 8C

Polarizing light by reflection

Purpose Demonstrate that a reflected beam is polarized.

Materials focusable flashlight or projector, glass sheet (or the glass surface of an overhead projector), polarizing sheet, projection screen

Procedure Dim the classroom lights. Shine a flashlight beam directly on the screen. Place the polarizing sheet across the beam, and rotate the sheet. Have students observe that the initial intensity is reduced and that rotating the polarizing sheet does not make a difference. Explain that this is because the light is randomly polarized. Now shine the light beam onto the glass plate. Place the polarizing sheet across the reflected beam, and rotate the sheet. The brightness of the light on the screen will vary with the rotation. With a particular angle (about 34° between the beam and the glass plate) complete polarization may be obtained and the image on the screen will completely fade.

Section Review
ANSWERS

1. blue
2. magenta; red
3. yellow and cyan (more yellow than cyan); green and red (more green than red)
4. If you turn the polarizer and hardly any light gets through, then the reflected light is polarized.

Because the light that causes glare is in most cases horizontally polarized, it can be filtered out by a polarizing substance whose transmission axis is oriented vertically. This is the case with polarizing sunglasses. As shown in **Figure 14-26,** the angle between the polarized reflected light and the transmission axis of the polarizer is 90°. Thus, none of the polarized light passes through.

In addition to reflection and absorption, scattering can also polarize light. Scattering, or the absorption and reradiation of light by particles in the atmosphere, causes sunlight to be polarized, as shown in **Figure 14-27.** When an unpolarized beam of sunlight strikes air molecules, the electrons in the molecules begin vibrating with the electric field of the incoming wave. A horizontally polarized wave is emitted by the electrons as a result of their horizontal motion, and a vertically polarized wave is emitted parallel to Earth as a result of their vertical motion. Thus, an observer with his or her back to the sun will see polarized light when looking up toward the sky. ★TEKS **8A, 8B**

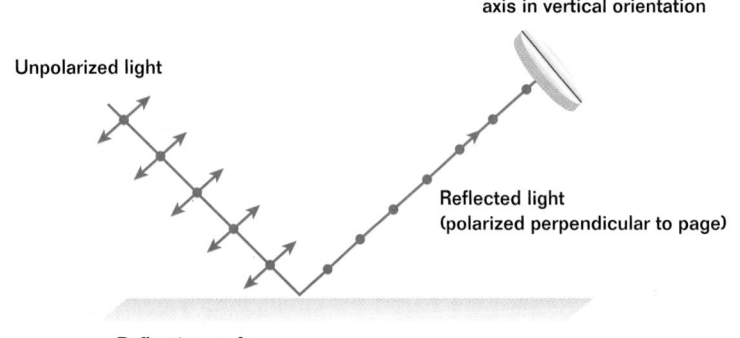

Figure 14-26
Reflected light is polarized horizontally. This light can be blocked by aligning the transmission axes of the sunglasses vertically.

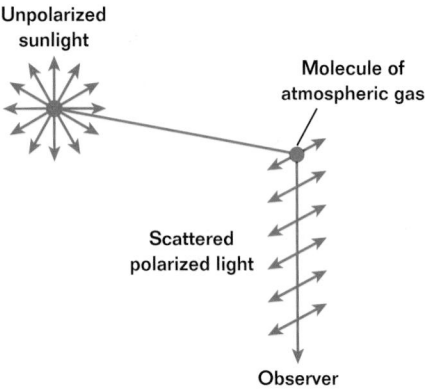

Figure 14-27
The sunlight scattered by air molecules is polarized for an observer on Earth's surface.

Section Review   ★TEKS **8A, 8B**

1. A lens for a spotlight is coated so that it does not transmit yellow light. If the light source is white, what color is the spotlight?

2. A house is painted with pigments that reflect red and blue light but absorb all other colors. What color does the house appear to be when it is illuminated by white light? What color does it appear to be under red light?

3. What primary pigments would an artist need to mix to obtain a pale yellow green color? What primary additive colors would a theater-lighting designer need to mix in order to produce the same color with light?

4. The light reflected from the surface of a pool of water is observed through a polarizer. How can you tell if the reflected light is polarized?

KEY IDEAS

Section 14-1 Characteristics of light

- Light is electromagnetic radiation that consists of oscillating electric and magnetic fields with different wavelengths.
- The relationship between the frequency, wavelength, and speed of electromagnetic radiation is given by the equation at right. $\boxed{c = f\lambda}$
- The brightness of light is inversely proportional to the square of the distance from the light source.

Section 14-2 Flat mirrors

- Light obeys the law of reflection, which states that the incident and reflected angles of light are equal.
- Flat mirrors form virtual images that are the same distance from the mirror's surface as the object is.

Section 14-3 Curved mirrors

- The mirror equation, at right, relates object distance, image distance, and focal length of a spherical mirror.

$$\frac{1}{p} + \frac{1}{q} = \frac{1}{f}$$

- The magnification equation, at right, relates image height or distance to object height or distance, respectively.

$$M = \frac{h'}{h} = -\frac{q}{p}$$

Section 14-4 Color and polarization

- Light of different colors can be produced by adding light consisting of the primary additive colors (red, green, and blue).
- Pigments can be produced by combining subtractive colors (magenta, yellow, and cyan).
- Light can be linearly polarized by transmission, reflection, or scattering.

Variable symbols

Quantities		Units	
p	object distance	m	meters
q	image distance	m	meters
R	radius of curvature	m	meters
f	focal length	m	meters
M	magnification		(unitless)

KEY TERMS

angle of incidence (p. 527)

angle of reflection (p. 527)

concave spherical mirror (p. 530)

convex spherical mirror (p. 537)

electromagnetic wave (p. 520)

linear polarization (p. 546)

real image (p. 531)

reflection (p. 526)

virtual image (p. 528)

Diagram symbols

Light rays (real)	⟶
Light rays (apparent)	– – – ⤍ – –
Normal lines	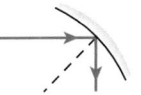
Flat mirror	
Concave mirror Convex mirror	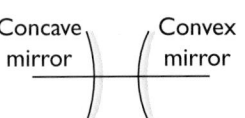

Teaching Tip

Explaining difficult concepts in written form helps to solidify students' understanding and enforces good communication skills. Have students summarize the differences between images formed by convex mirrors and images formed by concave mirrors. Their writings should include a thorough explanation of the mirror equation, sign conventions, and ray diagrams for each case. Be sure students explain concepts clearly and correctly and use good sentence structure.

(★) TEKS

p. 548: 8A, 8B, 8A, 8B
Review & Assess
pp. 550–555:
2A: Alt. Assess. 1, 3, 6
2C: 10–13, 20–21, 34–36, 46–60
2D: Alt. Assess. 1–2, 4–6
2E: Technology & Learning
3A: Alt. Assess. 2, 5
3B: 6, 10–13, 20–21, 34–36, 46–60, Technology & Learning
3C: Alt. Assess. 4
3E: Alt. Assess. 2, 4, 5
4A: Technology & Learning
8A: 3–6, 8–11, 14–60
8B: 1–2, 7, 12–13
8C: Alt. Assess. 4

CHAPTER 14
Review and Assess

ANSWERS TO

Chapter 14
Review and Assess

1. **a.** radio waves
   b. gamma rays
2. b
3. Its speed is accurately known. Measuring the time it takes light to travel a distance allows the distance to be determined. (Alternatively, if the source's brightness is known, its apparent brightness can be measured and its distance calculated.)
4. The wave front at *B* would be an arc of a large circle. The rays would point radially outward from *A* to *B*.
5. Apparent brightness equals the actual brightness divided by the square of the distance between observer and source.
6. $1999 + 2(95) = 2189$
7. 3.00×10^8 m/s
8. The light from galaxies was emitted millions of years ago.
9. no; Those stars may be closer, and so appear brighter.
10. 4.0×10^{-7} m, 3.0×10^{-7} m
11. 1×10^{-6} m
12. 3.02 m
13. 9.1×10^{-3} m (9.1 mm)
14. **a.** diffusely
    b. specularly
    c. specularly
    d. diffusely
    e. specularly

CHARACTERISTICS OF LIGHT

Review questions

1. Which band of the electromagnetic spectrum has
   a. the lowest frequency?
   b. the shortest wavelength?

2. Which of the following electromagnetic waves has the highest frequency?
   a. radio
   b. ultraviolet radiation
   c. blue light
   d. infrared radiation

3. Why can light be used to measure distances accurately? What must be known in order to make distance measurements?

4. For the diagram in **Figure 14-28**, use Huygens' principle to show what the wave front at point A will look like at point B. How would you represent this wave front in the ray approximation?

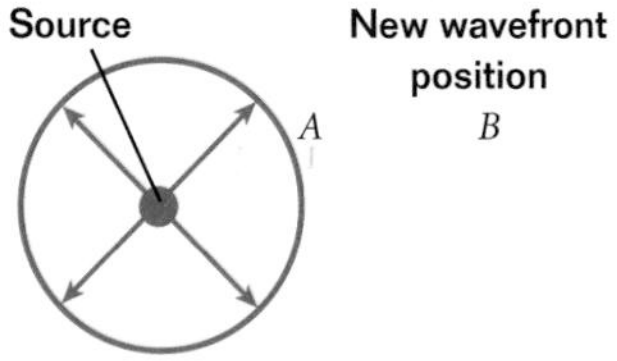

Source   New wavefront position
A   *B*

Figure 14-28

5. What is the relationship between the actual brightness of a light source and its apparent brightness from where you see it?

Conceptual questions

6. Suppose an intelligent society capable of receiving and transmitting radio signals lives on a planet orbiting Procyon, a star 95 light-years away from Earth. If a signal were sent toward Procyon in 1999, what is the earliest year that Earth could expect to receive a return message? (Hint: A light-year is the distance a ray of light travels in one year.)

7. How fast do X rays travel in a vacuum?

8. Why do astronomers observing distant galaxies talk about looking backward in time?

9. Do the brightest stars that you see in the night sky necessarily give off more light than dimmer stars? Explain your answer.

Practice problems

10. The compound eyes of bees and other insects are highly sensitive to light in the ultraviolet portion of the spectrum, particularly light with frequencies between 7.5×10^{14} Hz and 1.0×10^{15} Hz. To what wavelengths do these frequencies correspond? (See Sample Problem 14A.)

11. The brightest light detected from the star Antares has a frequency of about 3×10^{14} Hz. What is the wavelength of this light? (See Sample Problem 14A.)

12. What is the wavelength for an FM radio signal if the number on the dial reads 99.5 MHz? (See Sample Problem 14A.)

13. What is the wavelength of a radar signal that has a frequency of 33 GHz? (See Sample Problem 14A.)

FLAT MIRRORS

Review questions

14. For each of the objects listed below, identify whether light is reflected diffusely or specularly.
    a. a concrete driveway
    b. an undisturbed pond
    c. a polished silver tray
    d. a sheet of paper
    e. a mercury column in a thermometer

15. If you are stranded on an island, where would you align a mirror to use sunlight to signal a searching aircraft?

16. If you are standing 2 m in front of a flat mirror, how far behind the mirror is your image? What is the magnification of the image? Where is the image's right side with respect to your right side?

Conceptual questions

17. When you shine a flashlight across a room, you see the beam of light on the wall. Why do you not see the light in the air?

18. How can an object be a specular reflector for some electromagnetic waves yet be diffuse for others?

19. A flat mirror that is 0.85 m tall is nailed to a wall so that its upper edge is 1.7 m above the floor. Use the law of reflection and a ray diagram to determine if this mirror will show a person who is 1.7 m tall his or her complete reflection.

20. Two flat mirrors make an angle of 90.0° with each other, as diagrammed in **Figure 14-29**. An incoming ray makes an angle of 35° with the normal of mirror A. Use the law of reflection to determine the angle of reflection from mirror B. What is unusual about the incoming and reflected rays of light for this arrangement of mirrors?

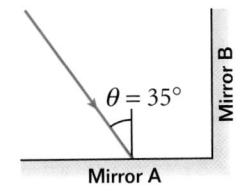

Figure 14-29

21. If you walk 1.2 m/s toward a flat mirror, how fast does your image move? In what direction does your image move with respect to you?

22. Why do the images produced by two opposing flat mirrors appear to be progressively smaller?

CURVED MIRRORS

Review questions

23. What type of mirror should be used to project movie images on a large screen?

24. If an object is placed outside the focal length of a concave mirror, what type of image will be formed? Will it appear in front of or behind the mirror?

25. Can you use a convex mirror to burn a hole in paper by focusing light rays at the mirror's focal point?

26. Can the image produced by a convex mirror ever be larger than the object?

27. Why are parabolic mirrors preferred over spherical concave mirrors for use in reflecting telescopes?

Conceptual questions

28. Where does a ray of light that is parallel to the principal axis of a concave mirror go after it is reflected at the mirror's surface?

29. What happens to the real image produced by a concave mirror if you move the original object to the location of the image?

30. Consider a concave spherical mirror and a real object. Is the image always inverted? Is the image always real? Give conditions for your answers.

31. Explain why magnified images seem dimmer than the original objects.

32. What test could you perform to determine if an image is real or virtual?

33. You've been given a concave mirror that may or may not be parabolic. What test could you perform to determine whether it is parabolic?

Practice problems

34. A concave shaving mirror has a radius of curvature of 25.0 cm. For each of the following cases, find the magnification and determine whether the image formed is real or virtual and upright or inverted.

    a. an upright pencil placed 45.0 cm from the mirror

    b. an upright pencil placed 25.0 cm from the mirror

    c. an upright pencil placed 5.00 cm from the mirror

(See Sample Problem 14B.)

15. point normal halfway between sun and aircraft

16. 2 m behind; $M = 1$; on the observer's left side

17. The gas molecules in air do not reflect the light.

18. Reflection is diffuse if λ is smaller than surface irregularities of reflector.

19. Diagram should show ray from feet reflected at bottom of mirror toward observer's eyes. Ray from top of head is reflected at top of mirror.

20. $\theta_2' = 55°$; Ray reflected from the second mirror is always parallel to the incoming ray.

21. 1.2 m/s; The image moves toward the mirror's surface.

22. Images serve as objects for more images. Each reflection doubles the apparent distance from "object" to mirror.

23. concave

24. real, inverted image; in front

25. No, rays always diverge from a convex mirror.

26. no, $h' < h$ for convex mirrors

27. no spherical aberration

28. through the focal point

29. A real image appears at the former object position.

30. no; no; image is upright and virtual when $p < f$

31. The light is spread out more in the larger image.

32. try to project image on paper

33. Produce rays parallel to and far from the principal axis. All rays focus at F for a parabolic mirror.

34. **a.** $M = -0.384$; real, inverted
    b. $M = -1.00$; real, inverted
    c. $M = 1.67$; virtual, upright

35. $p = 13$ cm; real, inverted; $M = -2.0$

36. $p = 53.5$ cm; $h = 5.76$ cm; $M = 0.295$; virtual, upright

37. red, green, blue; They make white light.

38. cyan, magenta, yellow; They make black pigment.

39. The polarized light from the first polarizer is blocked by the second polarizer when the component of the light that is parallel to the second polarizer's transmission axis equals zero; The light must be perpendicular (90°) to the second polarizer's transmission axis.

40. a. green pigment
b. white light
c. black pigment
d. yellow light
e. cyan pigment

41. a. magenta
b. red
c. blue
d. black
e. red

42. cyan; blue

43. Rotate the sunglasses while looking at the sky or sunlight reflecting off a horizontal surface. If brightness changes, the glasses have polarizing lenses.

44. Light reflected from a horizontal surface like an auto hood is polarized horizontally and is blocked by the lenses. Light reflected from tall narrow surfaces like the tank will be vertically polarized, and almost all of it will pass through the lenses.

45. yes; Light from the sky is polarized, but light from the clouds is not polarized.

46. $p = 4.1 \times 10^2$ cm; $f = 32$ cm; $R = 64$ cm; real, inverted image

35. A concave spherical mirror can be used to project an image onto a sheet of paper, allowing the magnified image of an illuminated real object to be accurately traced. If you have a concave mirror with a focal length of 8.5 cm, where would you place a sheet of paper so that the image projected onto it is twice as far from the mirror as the object is? Is the image upright or inverted, real or virtual? What would the magnification of the image be?
(See Sample Problem 14B.)

36. A convex mirror with a radius of curvature of 45.0 cm forms a 1.70 cm tall image of a pencil at a distance of 15.8 cm behind the mirror. Calculate the object distance for the pencil and its height. Is the image real or virtual? What is the magnification? Is the image inverted or upright?
(See Sample Problem 14C.)

COLOR AND POLARIZATION

Review questions

37. What are the three primary additive colors? What happens when you mix them?

38. What are the three primary subtractive colors (or primary pigments)? What happens when you mix them?

39. Explain why a polarizing disk used to analyze light can block light from a beam that has been passed through another polarizer. What is the relative orientation of the two polarizing disks?

Conceptual questions

40. Explain what could happen when you mix the following:
 a. cyan and yellow pigment
 b. blue and yellow light
 c. pure blue and pure yellow pigment
 d. green and red light
 e. pure green and pure blue pigment

41. What color would an opaque magenta shirt appear to be under the following colors of light?
 a. white   **d.** green
 b. red   **e.** yellow
 c. cyan

42. A substance is known to reflect green and blue light. What color would it appear to be when it is illuminated by white light? by blue light?

43. How can you tell if a pair of sunglasses has polarizing lenses?

44. Why would sunglasses with polarizing lenses remove the glare from your view of the hood of your car or a distant body of water but not from a tall metal tank used for storing liquids?

45. Is light from the sky polarized? Why do clouds seen through Polaroid™ glasses stand out in bold contrast to the sky?

MIXED REVIEW

46. The real image of a tree is magnified −0.085 times by a telescope's primary mirror. If the tree's image forms 35 cm in front of the mirror, what is the distance between the mirror and the tree? What is the focal length of the mirror? What is the value for the mirror's radius of curvature? Is the image virtual or real? Is the image inverted or upright?

47. A candlestick holder has a concave reflector behind the candle, as shown in **Figure 14-30.** The reflector magnifies a candle −0.75 times and forms an image 4.6 cm away from the reflector's surface. Is the image inverted or upright? What are the object distance and the reflector's focal length? Is the image virtual or real?

Figure 14-30

48. A child holds a candy bar 15.5 cm in front of the convex side-view mirror of an automobile. The image height is reduced by one-half. What is the radius of curvature of the mirror?

49. A glowing electric light bulb placed 15 cm from a concave spherical mirror produces a real image 8.5 cm from the mirror. If the light bulb is moved to a position 25 cm from the mirror, what is the position of the image? Is the final image real or virtual? What are the magnifications of the first and final images? Are the two images inverted or upright?

50. A convex mirror is placed on the ceiling at the intersection of two hallways. If a young man stands directly underneath the mirror, his shoe, which is a distance of 195 cm from the mirror, forms an image that appears 12.8 cm behind the mirror's surface. What is the mirror's focal length? What is the magnification of the image? Is the image real or virtual? Is the image upright or inverted?

51. The side-view mirror of an automobile has a radius of curvature of 11.3 cm. The mirror produces a virtual image one-third the size of the object. How far is the object from the mirror?

52. An object is placed 10.0 cm in front of a mirror. What type must the mirror be to form an image of the object on a wall 2.00 m away from the mirror? What is the magnification of the image? Is the image real or virtual? Is the image inverted or upright?

53. The reflecting surfaces of two intersecting flat mirrors are at an angle of θ ($0° < \theta < 90°$), as shown in **Figure 14-31**. A light ray strikes the horizontal mirror. Use the law of reflection to show that the emerging ray will intersect the incident ray at an angle of $\phi = 180° - 2\theta$.

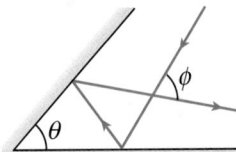

Figure 14-31

54. Show that if a flat mirror is assumed to have an "infinite" radius of curvature, the mirror equation reduces to $q = -p$.

55. A real object is placed at the zero end of a meterstick. A large concave mirror at the 100.0 cm end of the meterstick forms an image of the object at the 70.0 cm position. A small convex mirror placed at the 20.0 cm position forms a final image at the 10.0 cm point. What is the radius of curvature of the convex mirror?

56. A dedicated sports-car enthusiast polishes the inside and outside surfaces of a hubcap that is a section of a sphere. When he looks into one side of the hubcap, he sees an image of his face 30.0 cm behind the hubcap. He then turns the hubcap over and sees another image of his face 10.0 cm behind the hubcap.
   a. How far is his face from the hubcap?
   b. What is the radius of curvature of the hubcap?
   c. What is the magnification for each image?
   d. Are the images real or virtual?
   e. Are the images upright or inverted?

57. An object 2.70 cm tall is placed 12.0 cm in front of a mirror. What type of mirror and what radius of curvature are needed to create an upright image that is 5.40 cm in height? What is the magnification of the image? Is the image real or virtual?

58. A "floating coin" illusion consists of two parabolic mirrors, each with a focal length of 7.5 cm, facing each other so that their centers are 7.5 cm apart (see **Figure 14-32**). If a few coins are placed on the lower mirror, an image of the coins forms in the small opening at the center of the top mirror. Use the mirror equation and draw a ray diagram to show that the final image forms at that location. Show that the magnification is 1 and that the image is real and upright. (Note: A flashlight beam shone on these images has a very startling effect. Even at a glancing angle, the incoming light beam is seemingly reflected off the images of the coins. Do you understand why?)

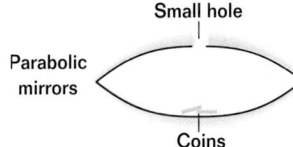

Figure 14-32

47. inverted; $p = 6.1$ cm; $f = 2.6$ cm; real

48. $R = -31.0$ cm

49. $q_2 = 6.7$ cm; real; $M_1 = -0.57$, $M_2 = -0.27$; inverted

50. $f = -13.7$ cm; $M = 0.0656$; virtual, upright

51. $p = 11.3$ cm

52. concave; $M = -20.0$; real; inverted

53. (See *Teacher's Solution Manual and Answer Key.*)

54. (See *Teacher's Solution Manual and Answer Key.*)

55. $R = -25.0$ cm

56. a. 15.0 cm
   b. 59.9 cm
   c. $M_{convex} = 2.00$, $M_{concave} = 0.667$
   d. virtual
   e. upright

57. concave, $R = 48.1$ cm; $M = 2.00$; virtual

58. (See *Teacher's Solution Manual and Answer Key.*)

59. (See *Teacher's Solution Manual and Answer Key*.)

60. (See *Teacher's Solution Manual and Answer Key*.)

Alternative Assessment
ANSWERS

Performance assessment

1. Students' answers will vary. Be sure plans are safe and experimental designs reflect a controlled experiment.

2. Alhazen (965–1038) studied light, pinhole cameras, and parabolic mirrors. When he failed to invent a machine to control the Nile's floods, the Caliph sentenced him to death. Alhazen escaped by pretending to be insane.

3. Students' discussions will vary. Students should recognize that convex mirrors will work best. Check that plans take into account the law of reflection.

Portfolio projects

4. Students' answers will vary. Be sure the class covers the entire spectrum. Students should provide source references.

5. Students' answers will vary, but should indicate that when the sun is on one side of Earth and the moon is on the other side, the moon will be bright at night.

6. Students' summaries will vary. Parallel mirrors produce an infinite number of images that are smaller and smaller. The angles for one, two, three, five, and seven images are 180°, 120°, 90°, 60°, and 45°, respectively.

59. Use the mirror equation and the equation for magnification to prove that the image of a real object formed by a convex mirror is always upright, virtual, and smaller than the object. Use the same equations to prove that the image of a real object placed in front of any spherical mirror is always virtual and upright when $p < |f|$.

60. Use trigonometry to derive the mirror and magnification equations from the ray diagram in **Figure 14-11** on page 531. (Hint: Note that the incoming ray between the light-bulb filament and the mirror forms the hypotenuse of a right triangle. The reflected ray between the image point and the mirror is also the hypotenuse of a right triangle.)

Alternative Assessment

Performance assessment

1. Suntan lotions include compounds that absorb the ultraviolet radiation in sunlight and therefore prevent the ultraviolet radiation from damaging skin cells. Design experiments to test the properties of varying grades (SPFs) of suntan lotions. Plan to use blueprint paper, film, plants, or other light-sensitive items. Write down the questions that will guide your inquiry, the materials you will need, the procedures you plan to follow, and the measurements you will take. If your teacher approves your plan, perform the experiments and report or demonstrate your findings in class.

2. The Egyptian scholar Alhazen studied lenses, mirrors, rainbows, and other light phenomena early in the Middle Ages. Research his scholarly work, his life, and his relationship with the Caliph al-Hakim. How advanced were Alhazen's inventions and theories? Summarize your findings and report them to the class.

3. Work in cooperative groups to explore the use of corner and ceiling mirrors as low-tech surveillance devices. Make a floor plan of an existing store, or devise a floor plan for an imaginary one. Determine how much of the store could be monitored by a clerk if flat mirrors were placed in the corners. If you could use curved mirrors in such a system, would you use concave or convex mirrors? Where would you place them? Identify which parts of the store could be observed with the curved mirrors in place. Note any disadvantages that your choice of mirrors may have.

Portfolio projects

4. Research the characteristics, effects, and applications of a specific type of electromagnetic wave in the spectrum. Find information about the range of wavelengths, frequencies, and energies; natural and artificial sources of the waves; and the methods used to detect them. Find out how they were discovered and how they affect matter. Learn about any dangers associated with them and about their uses in technology. Work together with others in the class who are researching other parts of the spectrum to build a group presentation, brochure, chart, or Web page that covers the entire spectrum.

5. The Chinese astronomer Chang Heng (A.D. 78–139) recognized that moonlight was a reflection of sunlight. He applied this theory to explain lunar eclipses. Make diagrams showing how Heng might have represented the moon's illumination and the path of light when the Earth, moon, and sun were in various positions on ordinary nights and on nights when there were lunar eclipses. Find out more about Heng's other scientific work, and report your findings to the class.

6. Explore how many images are produced when you stand between two flat mirrors whose reflecting surfaces face each other. What are the locations of the images? Are they identical? Investigate these questions with diagrams and calculations. Then test your calculated results with parallel mirrors, perpendicular mirrors, and mirrors at angles in between. Which angles produce one, two, three, five, and seven images? Summarize your results with a chart, diagram, or computer presentation.

Technology & Learning

Graphing calculators

Refer to Appendix B for instructions on downloading programs for your calculator. The program "Chap14" builds a table of image distance and magnification for various object distances for a curved mirror with a known focal length.

Image distance, as you learned earlier in this chapter, can be found using the mirror equation:

$$\frac{1}{p} + \frac{1}{q} = \frac{1}{f}$$

Meanwhile, the magnification of a mirror can be found using the equation for magnification, which makes use of the image distance and object distance.

$$M = -\frac{q}{p}$$

The program "Chap14" stored on your graphing calculator makes use of both the mirror equation and the equation for magnification. Once the "Chap14" program is executed, your calculator will ask for the focal length of the mirror. The graphing calculator will use the following equations to create a table of image distance (Y_1) and magnification (Y_2) for various object distances (X). Note that the relationships in these equations are the same as those in the mirror equation and magnification equation shown above.

$$Y_1 = (XF)/(X-F)$$
$$Y_2 = -Y_1/X$$

a. The mirror equation used by your calculator looks different than the mirror equation shown at the top of the page. The calculator equation has been solved for Y_1, the image distance. Rewrite the equation used by your calculator in the form shown at the top of the page.

Execute "Chap14" on the PRGM menu, and press ENTER to begin the program. Enter the value for the focal length (shown in b–g below), and press ENTER. Remember to use the (-) key, instead of the - key, for entering negative values.

The calculator will provide a table of image distances in meters (Y_1) versus object distances in meters (X). Press ▼ to scroll down through the table to find the image distance values you need. The column labeled Y_2 gives the value of the magnification. Recall that negative magnification values indicate that an image is real and inverted, while positive magnification values indicate that an image is virtual and upright. Magnification values greater than 1 or less than −1 indicate that the image is larger than the object.

Find the image distance and magnification in each of the following situations (b–g):

b. concave mirror of focal length +0.32 m, object distance of 0.20 m

c. concave mirror of focal length +0.32 m, object distance of 0.70 m

d. convex mirror of focal length −0.50 m, object distance of 0.40 m

e. convex mirror of focal length −0.50 m, object distance of 0.85 m

f. A boy holds an action figure at a distance of 0.25 m from a convex mirror that has a focal length equal to 0.45 m.

g. The boy in item f moves the action figure 0.15 m farther away from the mirror.

h. For items b through e, indicate the characteristics of the images, including whether the image is upright or inverted and whether it is a virtual or real image.

Press ENTER to stop viewing the table. Press ENTER again to enter a new value or CLEAR to end the program.

ANSWERS TO

Technology & Learning

a. $\frac{1}{X} + \frac{1}{Y_1} = \frac{1}{F}$

b. −0.53 m, 2.7

c. 0.59 m, −0.84

d. −0.22 m, 0.56

e. −0.31 m, 0.37

f. −0.56 m, 2.2

g. −3.6 m, 9.0

h. b) upright, virtual image;
c) inverted, real image;
d) upright, virtual image;
e) upright, virtual image

NOTE

Materials Preparation is given on pp. 518A–518B. Blank data table and sample data table are on the One-Stop Planner CD-ROM. All calculations shown use sample data.

Planning

Recommended time:

1 lab period

For a 2-period lab, add the Extensions Exercise (p. 559, item 7).

Classroom organization:

▶ Each lab group must have at least two students. Because of the complexity of the exercise, lab groups may have more than two students.

▶ Each lab group needs a level work surface that is near an electrical outlet and away from any sources of water. Each work area must be at least 1.5 m long.

▶ The CBL and sensors procedure and the light meter procedure may be used in the same class.

▶ **Safety warnings:** Remind students to report all breakage immediately. Students should be instructed not to look directly at a light source.

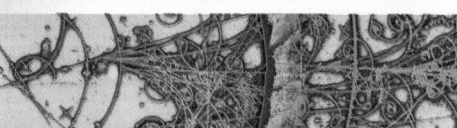

CHAPTER 14
Laboratory Exercise

⭐**TEKS**

pp. 556–559: 1A, 1B, 2B, 2C, 2D, 2E, 2F, 3A, 3B, 8A, 8B

OBJECTIVES

- Find the relationship between the intensity of the light emitted by a light source and the distance from the source.
- Explore the inverse square law in terms of the intensity of light.

MATERIALS LIST

✔ small, clear incandescent bulb
✔ meterstick-mounted bulb socket
✔ black tube to cover bulb and socket
✔ power supply
✔ meterstick
✔ meterstick supports
✔ support stand
✔ clamp for support stand

PROCEDURE

CBL AND SENSORS

✔ CBL
✔ CBL light probe
✔ graphing calculator with link cable
✔ black aperture tube for CBL light probe
✔ adhesive tape

LIGHT METER

✔ light meter
✔ black aperture tube for light meter
✔ black paper aperture stop for light meter

BRIGHTNESS OF LIGHT

The brightness, or intensity, of a light source may be measured with a light meter. In this lab, you will use a light meter to measure the intensity of light at different distances from the light source. The measured brightness of a light depends on the distance between the light meter and the light source. This relationship is an example of an inverse-square law. According to the inverse-square law, the brightness of light at a certain point is proportional to the square of the distance from the light source to the light meter. You will use your results from this experiment to investigate the relationship between the distance and the brightness of a light source and to examine the inverse-square law as it relates to the brightness of light. You will use your data to calculate the square of the distance, and you will analyze the relationship using graphs of your data.

SAFETY

- **Use a hot mitt to handle resistors, light sources, and other equipment that may be hot. Allow all equipment to cool before storing it.**

- **If a bulb breaks, notify your teacher immediately. Do not remove broken bulbs from sockets.**

- **Never put broken glass or ceramics in a regular waste container. Use a dustpan, brush, and heavy gloves to carefully pick up broken pieces, and dispose of them in a container specifically provided for this purpose.**

- **Avoid looking directly at a light source. Looking directly at a light source may cause permanent eye damage.**

PREPARATION

1. Determine whether you will be using the CBL and sensors procedure or the light meter. Read the entire lab for the appropriate procedure, and plan what steps you will take.

2. Prepare a data table in your lab notebook with two columns and nine rows. In the first row, label the columns *Distance (m)* and *Intensity.*

Light meter procedure begins on page 558.

PROCEDURE

CBL AND SENSORS

Brightness of light

3. In the first column of your data table, label the second through ninth rows *0.10*, *0.15*, *0.20*, *0.25*, *0.30*, *0.50*, *0.75*, and *1.00*. These values represent the distances at which you will take readings.

4. Set up the meterstick, meterstick supports, light source (bulb and socket), and power supply, as shown in **Figure 14-33.** Carefully screw the bulb into the lamp socket. Tape the meterstick and its supports to the lab table.

5. Set up the CBL light probe with aperture, CBL, and graphing calculator as shown in **Figure 14-33.** Set the CBL light probe directly above the 0.00 m mark on the meterstick, as shown. Connect the CBL to the graphing calculator with the unit-to-unit link cable using the ports located on each unit. Connect the light probe to the CH1 port. Turn on the CBL and the graphing calculator. Start the program PHYSICS on the graphing calculator.

   a. Select option *SET UP PROBES* from the MAIN MENU. Enter 1 for the number of probes. Select MORE PROBES, then select the light probe from the list. Enter 1 for the channel number.

   b. Select the *COLLECT DATA* option from the MAIN MENU. Select the *TRIGGER* option from the DATA COLLECTION menu.

   c. Draw a line above or below your data table, and label it *Background*. Press TRIGGER on the CBL to record the value for the background intensity. Record this value as *Background*. On the calculator, select *STOP*. From the MAIN MENU, select *TRIGGER/PROMPT*.

6. Set the bulb and socket 0.10 m from the end of the CBL sensor. Carefully align the sensor clamp and the aperture of the sensor so that the aperture is level, parallel to the meterstick, and at the same height as the hole in the tube that covers the bulb.

7. Set the power supply at 4.5 V, and connect it carefully with the wires from the light socket. **Do not plug in the power supply until your teacher has approved your setup.** When your teacher has approved your setup, carefully plug the power supply into the wall outlet to light the lamp.

8. Press TRIGGER on the CBL to collect data for the light intensity. The calculator will prompt you for another value. Enter 0.10 as the value of the distance for this trial.

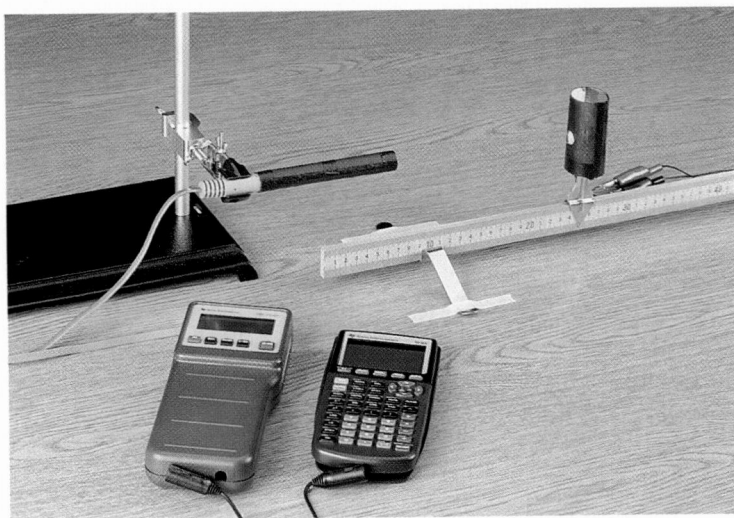

Figure 14-33

Step 4: Make sure the apparatus is set up securely. The meterstick should be taped to the table so that it cannot move during the experiment. Use a support stand and clamp to hold the light probe in place.

Step 6: Use the meterstick to place the light source at the correct position. Make sure the light probe and the light source are lined up.

CBL and Sensors Tips

◆ Students should have the program PHYSICS on their graphing calculators. Refer to Appendix B for instructions.

Techniques to Demonstrate

Show students how to set up the apparatus. Demonstrate how placing the light sensor at the 0.00 mark on the meterstick allows them to use the markings on the stick to measure the distances directly.

If the students are using a dc power supply, show them how to adjust the voltage. Power supplies should be set at around 5.0 V for this exercise.

✓ Checkpoints

Step 6: The bulb must be securely placed on the meterstick. If necessary, use electrical tape to hold the bulb in position. Students should be able to demonstrate that the light sensor is properly positioned.

Step 7: Students should be able to demonstrate that the power supply is connected correctly and set to the correct level before it is plugged in.

Step 8: Remind students that both the CBL and the calculator must be turned on before data can be collected. Students should enter the value for the distance in meters.

Light Meter Tips

◆ Most commercially available light meters measure in the metric unit of *lux*. Make sure you know what units are used by the meters in your class.

Techniques to Demonstrate

Show students how to adjust the light meter if the values for the intensity are fluctuating too much for them to read a constant value. Set the light meter to the "fast" mode, and use the "data hold" option to find the correct value.

Show students how to set up the apparatus. Demonstrate how placing the light sensor at the 0.00 mark on the meterstick allows students to use the markings on the stick to measure the distance directly.

If the students are using a dc power supply, show them how to adjust the voltage. Power supplies should be set at around 5.0 V for this exercise.

✔ Checkpoints

Step 4: The bulb must be securely placed on the meterstick. If necessary, use electrical tape to hold the bulb in position.

Step 5: Students should be able to demonstrate that the light meter is properly positioned.

Step 6: Students should be able to demonstrate that the power supply is connected correctly and set to the correct level before it is plugged in.

9. Select MORE DATA on the graphing calculator.

10. Carefully move the bulb to 0.15 m. Press TRIGGER on the CBL to collect data for the light intensity. When prompted, enter 0.15 for distance in this trial.

11. Repeat this procedure for all the distances in your data table. After each trial, select MORE DATA on the graphing calculator.

12. After the last trial, select QUIT on the graphing calculator. Carefully unplug the power supply from the wall outlet.

13. Record the data from all the trials in your data table. On the calculator, press STAT. Select EDIT. The column under L1 contains the distance values you entered for each trial. The column under L2 contains the light intensity values recorded by the CBL. Record the intensity values in the appropriate rows in your data table.

14. Clean up your work area. Put equipment away safely so that it is ready to be used again. Recycle or dispose of used materials as directed by your teacher.

Analysis and interpretation begins on page 559.

PROCEDURE

LIGHT METER

Brightness of light

3. In the first column of your data table, label the second through ninth rows *0.20, 0.25, 0.30, 0.35, 0.40, 0.50, 0.75,* and *1.00*. These values represent the distances at which you will take readings.

4. Set up the meterstick, meterstick supports, light source (bulb and socket), power supply, and light meter with aperture as shown in **Figure 14-34.** Carefully screw the bulb into the socket. Tape the meterstick and supports to the lab table. Set the 0.00 m mark on the meterstick directly below the face of the light meter as shown.

5. Set the bulb socket at a distance of 0.20 m from the light meter. Carefully align the clamp and the aperture of the detector so that the aperture is level, parallel to the meterstick, and at the same height as the hole in the tube covering the bulb. Adjust the bulb socket to the 0.20 m mark on the meterstick.

6. Set the power supply at 5.0 V, and connect it carefully with the wires from the light socket. **Do not plug in the power supply until your teacher has approved your setup.** When your teacher has approved your setup, carefully plug the power supply into the wall outlet to light the lamp.

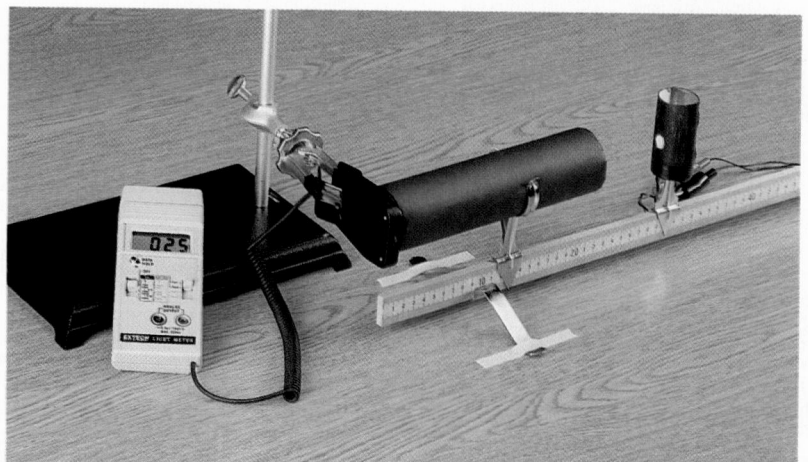

Figure 14-34

Step 4: Make sure the apparatus is set up securely. The meterstick should be taped to the table so that it cannot move during the experiment. Use a support stand and clamp to hold the light meter in place.

Step 5: Use the meterstick to place the light source at the correct position. Make sure the light source and light meter are lined up.

7. Use the light meter to read the intensity. Select the "fast" mode, and use the "data hold" option if the intensity values continually fluctuate. Record the intensity value for that distance in your data table.

8. Repeat this procedure for all other bulb distances recorded in your data table. Record all data in your data table.

9. After the last trial, carefully unplug the power supply from the wall outlet. Draw a line above or below your data table, and label it *Background*. Use this space to record the intensity reading of the light meter without the bulb illuminated.

10. Clean up your work area. Put equipment away safely so that it is ready to be used again. Recycle or dispose of used materials as directed by your teacher.

ANALYSIS AND INTERPRETATION

Calculations and data analysis

1. **Organizing data**   For each trial, find the real value of the measured light intensity by subtracting the background from the measured value.

2. **Graphing data**   Using your answers from item 1, make a graph of the intensity plotted against the distance. Use a graphing calculator, computer, or graph paper.

3. **Organizing data**   For each trial, calculate $1/(Distance^2)$. This value represents the inverse of the distance squared.

4. **Graphing data**   Using your answers from item 1 and item 3, make a graph of the intensity plotted against the inverse of the distance squared. Use a graphing calculator, computer, or graph paper.

Conclusions

5. **Interpreting graphs**   Based on your graphs, what is the relationship between the intensity of the light and the distance from the light source? Explain how your graphs support your answer.

6. **Evaluating methods**   What is the purpose of the shields used on the light bulb and the detector? Why are they important?

Extensions

7. **Designing experiments**   Devise a way to use this experiment to compare the intensities of light sources of different colors. If there is time and if your teacher approves your plan, perform the experiment, and record your results. Use your data to answer questions 1–5 above for each color. Write a brief report detailing your procedure and evaluating your results. Explain how color affects the intensity of a light source.

Step 7: Students may need assistance reading the light meter.

Step 8: Remind students to exercise care when handling the bulb, which may be hot.

ANSWERS TO

Analysis and Interpretation

CALCULATIONS AND DATA ANALYSIS

1. Answers will vary. Typical measured values will range from 7 lux to 125 lux for the light meter, and from 10 mW/cm^2 to 350 mW/cm^2 for the CBL.

2. The graph should have a parabolic shape and show that intensity decreases as distance increases.

3. Typical values will range from 1.00 m^{-2} to 100.00 m^{-2}.

4. The graph should have a straight line pointing up and to the right.

CONCLUSIONS

5. The intensity of a light varies directly with the inverse square of the distance from the light source.

6. The shields block out light from the room and direct all light to the meter; This gives higher values to improve the accuracy of the results.

EXTENSIONS

7. Student plans should be safe and complete, including a list of equipment, measurements, and calculations required.

Compression Guide: To shorten from 9 to 7 45-min periods (from 4½ to 3½ 90-min blocks), eliminate items in magenta type.

PACING CHART	CLASSROOM RESOURCES			
	⭐ TEKS	Teacher Demonstrations	Holt Physics Transparencies	Labs (See page T52 for equipment listing for in-text labs.)
15-1 Refraction 2 or 1 45-minute periods 1 or ½ 90-minute block	2C, 3A, 3B, 8A	**TE** *Refraction from air to water*, p. 562 **TE** *Refraction in various materials*, p. 564 **TE** *Underwater appearance*, p. 565	**T** 72–74 **TM** 53	**L** *Discovery Lab: Refraction and Lenses*
15-2 Thin lenses 3 45-minute periods 1½ 90-minute block	1A, 2B, 2C, 2E, 2F, 3A, 3C, 8A	**TE** *The effect of lenses on light beams*, p. 568 **TE** *Focal lengths of lenses*, p. 569 **TE** *Angular size of an object*, p. 572 **TE** *Microscope*, p. 578	**T** 75–78	**PE** *Quick Lab: Focal Length*, p. 570 **PE** *Quick Lab: Prescription Glasses*, p. 577 **PE** *Converging Lenses*, p. 593
15-3 Optical phenomena 2 or 1 45-minute periods 1 or ½ 90-minute block	1A, 2A, 2C, 3A, 3C, 8A	**TE** *Critical angle*, p. 580 **TE** *Fiber optics—bending light*, p. 582 **TE** *Dispersion*, p. 583 **TE** *Rainbow*, p. 584 **TE** *Chromatic aberration*, p. 585	**T** 79	**PE** *Quick Lab: Periscope*, p. 581 **L** *Invention Lab: Camera Design*

Review and Assessment
2 45-minute periods
1 90-minute block

Resource Key

PHYSICS

PE Pupil's Edition
TE Teacher's Edition

L Laboratory Experiments
TL Technology Lab Experiments
T Transparencies

🖑 **One-Stop** Planner CD-ROM **contents**

TM Transparency Masters
SR Section Review Worksheets
AA Alternative Assessment

PW Problem-Solving Workbook
PB Problem Bank
CTW Critical Thinking Worksheet

LABORATORY PLANNING: Converging Lenses, p. 593

Materials (for each lab group)
- battery eliminator with alligator clips, 6 V/0.5 A
- miniature bulb and base
- meterstick optical bench set includes:
    - card screen with mm scale (set of 5)
    - lens or mirror support, for 4 cm lenses
    - marker/object riders
    - meterstick, plain wood, 1 m long
    - pair of metal meterstick supports
    - screen support riders

- metric ruler, 15 cm long
- double convex lens
    - 38 mm diam. $f = 10$ cm, or
    - 38 mm diam. $f = 15$ cm, or
    - 38 mm diam. $f = 20$ cm
- illuminated object screen
- pkg. of 10 replacement lamps, 6.2 V/0.5 A
- roll of insulated copper wire, 18 awg, 30. m

ASSIGNMENT RESOURCES

Content Mastery	Critical Thinking	Problem-Solving Practice
PE 1–4, p. 567 **SR** 15-1, *Concept Review* **PE** 1–5, p. 587	**PE** 1–3, p. 565 **PE** 6–9, p. 687	**15A** Snell's law: 30 items in **PE, PW,** and **PB,** see **TE** pp. 566–567
PE 1–6, p. 579 **SR** 15-2, *Diagram Skills* **PE** 15–19, p. 588	**PE** 20–23, p. 588	**15B** Lenses: 42 items in **PE, PW,** and **PB,** see **TE** pp. 575–576
PE 1–4, p. 585 **SR** 15-3, *Concept Review* **PE** 27–31, p. 589	**PE** 32–35, p. 589	**15C** Critical angle: 29 items in **PE, PW,** and **PB,** see **TE** pp. 581–582

ASSESSMENT RESOURCES

Cumulative Review	Alternative Assessment	Traditional Assessment
SR Mixed Review, Ch. 15	**PE** 1–5, p. 591 **AA** Items for Ch. 15	Chapter 15 Test Test Generator items for Ch. 15

Scoring Rubrics for Alternative Assessment items can be found on the One-Stop Planner CD-ROM.

TECHNOLOGY RESOURCES

 CTW Segment 16 Color-Deficiency Lenses

 Module 15 Refraction and Lenses

 PE Technology and Learning, p. 592
(Alternative procedures for calculators without Flash-ROM technology are provided on the One-Stop Planner CD-ROM.)

internet connect

 On-line Student Resources:
GO TO: www.scilinks.org
The following *sci*LINKS Internet resources can be found in the student text for this chapter.

TOPICS:
• Snell's law, p. 566 (HF2151)
• Lenses, p. 570 (HF2152)
• Abnormalities of the eye, p. 577 (HF2153)
• Fiber optics, p. 583 (HF2154)
• Dispersion of light, p. 584 (HF2155)

 On-line Teacher Resources:
GO TO: go.hrw.com
KEYWORD: HF2 HOME
Visit the HRW Web site for a variety of resources related to this chapter.

 Smithsonian Institution
Internet Connections
Visit **www.si.edu/hrw** for additional on-line resources.

 Visit **www.cnnfyi.com** for late-breaking news and current events stories selected just for you.

Required Precautions

The greatest danger in this activity is presented by exposed electrical connections. All bulb sockets should have enclosed contacts and insulated connectors should be used on all wire connections.

Materials Preparation

Set up a sample apparatus in the laboratory for students to refer to as they set up their equipment.

If you are not using complete meterstick optical bench sets, you will need all the equipment listed under the *optical bench* heading for each lab group.

Section 15-1 investigates which direction light will bend when it enters another medium and uses Snell's law to solve problems.

Section 15-2 solves problems involving image formation by converging and diverging lenses using ray diagrams and the thin-lens equation, explores eye disorders and eyeglasses, and examines the positioning of lenses in microscopes and refracting telescopes.

Section 15-3 calculates critical angle; predicts when total internal reflection will occur; explains atmospheric phenomena, including mirages and rainbows; and briefly describes lens aberrations.

About the Illustration

Farmers near the Everglades in southern Florida use water trucks to irrigate their fields.

The formation of rainbows is explained in Section 15-3.

PHYSICS Interactive Problem-Solving Tutor

See Module 15

"Refraction and Lenses" provides additional development of problem-solving skills for this chapter.

CHAPTER 15

Refraction

PHYSICS IN ACTION

Most people associate rainbows with rain showers. However, rainbows can be observed any time an observer is between a source of light and water droplets in the air. Sunlight striking the droplets passes through their front surface and is partially reflected back toward the viewer from the back of the droplet. But why do we see the rainbow of colors? A rainbow occurs when sunlight is bent as it passes from air to water then from water to air. The degree to which the light bends depends on the frequency of the light. In this way, observers see only the frequency (the color) that is directed toward them from each droplet.

- *What causes the sunlight to bend as it passes through the droplets?*

- *Why do rainbows form in the shape of a semicircle?*

CONCEPT REVIEW

Wave speed   (Section 12-3)

Reflection   (Section 14-2)

Focal point   (Section 14-3)

Focal length   (Section 14-3)

Spherical aberration   (Section 14-3)

Tapping Prior Knowledge

Knowledge to Expect

✔ "Students learn that light from the sun is made up of a mixture of many different colors of light, even though to the eye the light looks almost white. Human eyes respond to only a narrow range of wavelengths of electromagnetic radiation—visible light. Differences in wavelength within the range are perceived as differences in color." (AAAS's *Benchmarks for Science Literacy*, grades 6–8)

Knowledge to Review

✔ Wave speed equals frequency times wavelength. (Section 12–3)

✔ Reflection is the turning back of an electromagnetic wave at the surface of a substance. (Section 14-2)

✔ The focal point is the point at which a beam parallel to the principal axis will converge after reflection from a concave mirror. (Section 14-3)

✔ The focal length is the distance from the focal point to the mirror. (Section 14-3)

✔ Spherical aberration is an effect in which the image produced by a curved mirror is blurred. It results from light rays converging at different points when the mirror is not parabolic. (Section 14-3)

Items to Probe

✔ Sine function: Ask what the highest value is that the sine of an angle may have and what the angle is at that value (*1, 90°*).

15-1
Refraction

Refraction from air to water

Purpose Demonstrate the phenomenon of refraction, and explore how the angle of incidence affects the angle of refraction.

Materials laser, dusty chalkboard erasers, aquarium filled with water and a few drops of whole milk

CAUTION *Avoid directing the laser beam toward the students.*

Procedure Adjust the position of the laser so that it shines into the water at an angle of about 30° from the normal. Gently tap the erasers together above the water. Have students observe the bending of the light ray as it enters the water. Now move the laser so that the beam strikes the water at various angles. Have students observe that the path of the light ray in water depends on the angle at which the light strikes the surface of the water.

Visual Strategy

Figure 15-2

Point out that a portion of the incident light is reflected. Make sure students realize that all angles are measured relative to the normal.

Q What is the angle between the normal line and the boundary between air and water?

A *90°*

15-1 SECTION OBJECTIVES

- **Recognize situations in which refraction will occur.**
- **Identify which direction light will bend when it passes from one medium to another.**
- **Solve problems using Snell's law.**

refraction

the bending of a wave disturbance as it passes at an angle from one medium into another

Figure 15-1
The flower looks small when viewed through the water droplet because the light from the flower is bent as it passes through the water.

REFRACTION OF LIGHT

Look at the tiny image of the flower that appears in the water droplet in **Figure 15-1.** The flower can be seen in the background of the photo. Why does the flower look different when viewed through the droplet? This phenomenon occurs because light is bent at the boundary between the water and the air around it. The bending of light as it travels from one medium to another is called **refraction.** ⊛TEKS **8A**

If light travels from one transparent medium to another at any angle other than straight on (parallel to the line normal to the surface), the light ray changes direction when it meets the boundary. As in the case of reflection, the angles of the incoming and refracted rays are measured with respect to the normal. For studying refraction, the normal line is extended into the refracting medium, as shown in **Figure 15-2.** The angle between the refracted ray and the normal is called the *angle of refraction, θ_r.* For refraction, the angle of incidence is designated as θ_i.

Refraction occurs when light's velocity changes

Glass, water, ice, diamonds, and quartz are all examples of transparent media through which light can pass. The speed of light in each of these materials is different. The speed of light in water, for instance, is less than the speed of light in air. And the speed of light in glass is less than the speed of light in water.

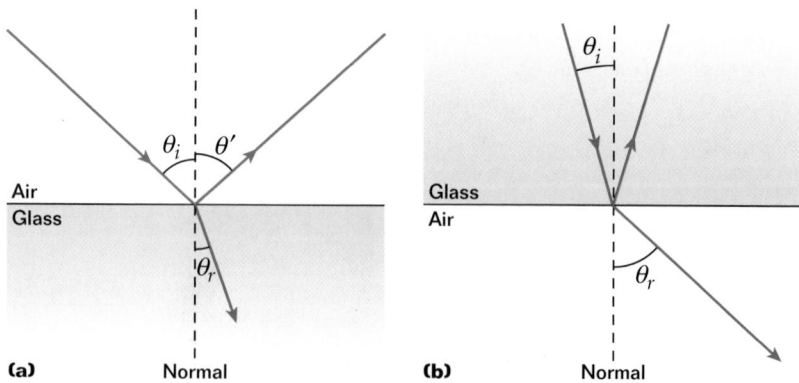

Figure 15-2
(a) When the light ray moves from air into glass, its path is bent toward the normal, **(b)** whereas the path of the light ray moving from glass into air is bent away from the normal.

When light moves from a material in which its speed is higher to a material in which its speed is lower, such as from air to glass, the ray is bent toward the normal, as shown in **Figure 15-2(a).** If the ray moves from a material in which its speed is lower to one in which its speed is higher, as in **Figure 15-2(b),** the ray is bent away from the normal. If the incident ray of light is parallel to the normal, then no refraction (bending) occurs in either case. ⭐TEKS **3A**

Note that the path of a light ray that crosses a boundary between two different media is reversible. If the ray in **Figure 15-2(a)** originated inside the glass block, it would follow the same path as shown in the figure, but the reflected ray would be inside the block.

Refraction can be explained in terms of the wave model of light

When a person rolls a barrel onto a lawn from a sidewalk, the barrel is harder to push on the lawn, just as a bicycle is harder to pedal across a lawn. If the person does not push harder on the lawn, the barrel will slow down. If the person pushes the barrel at an angle to the lawn without pushing harder once the barrel reaches the lawn, one end of the barrel will go slower, while the other end will continue at the same speed because it is still on the sidewalk. The slower end of the barrel will therefore act like a moving pivot, and the barrel will turn, as shown in **Figure 15-3.** ⭐TEKS **8A**

Light passing from one transparent medium into another can be thought of in a similar way. In **Figure 15-4,** wave fronts of a plane wave of light are traveling at an angle to the surface of a block of glass. As the light enters the glass, the wave fronts slow down, but the wave fronts that have not yet reached the surface of the glass continue traveling at the speed of light in air. During this time, the slower wave fronts travel a smaller distance than do the wave fronts in the air, so the entire plane wave changes directions.

Note the difference in wavelength (the space between the wave fronts) between the plane wave in air and the plane wave in the glass. Because the wave fronts inside the glass are traveling more slowly, in the same time interval they move through a shorter distance than the wave fronts that are still traveling in air. Thus, the wavelength of the light in the glass, λ_{glass}, is shorter than the wavelength of the incoming light, λ_{air}.

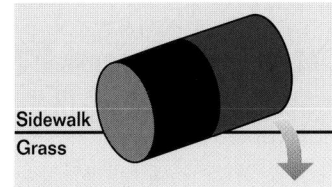

Figure 15-3
When this barrel is rolled at an angle from a sidewalk onto grass, the barrel turns because the gray end of the barrel, which is on the grass, travels slower than the maroon end, which is still on the sidewalk.

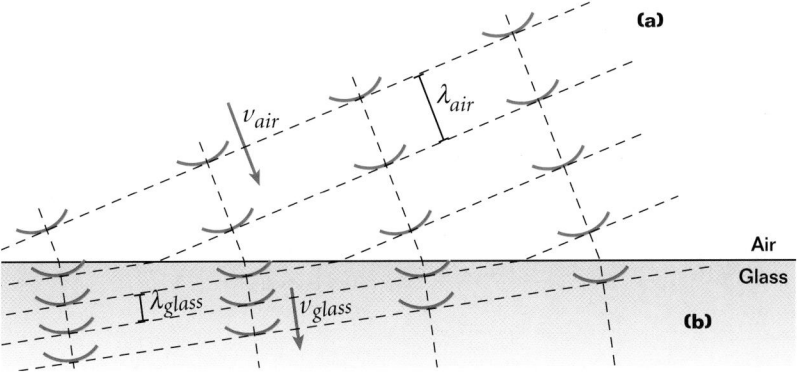

Figure 15-4
A plane wave traveling in air **(a)** has a wavelength of λ_{air} and velocity of v_{air}. Each wave front turns as it strikes the glass, just as the barrel turns when it is moving from the sidewalk onto the lawn. Because the speed of the wave fronts in the glass **(b),** v_{glass}, is slower, the wavelength of the light becomes shorter and the wave fronts change direction.

Key Models and Analogies

Some students may find the barrel analogy more helpful if they think about the barrel's line of motion as the path of a light ray (rather than a plane wave): the turning of the barrel corresponds to the bending of the light ray toward the normal.

🛑 Misconception Alert

Students may think that the frequency of light changes as light enters a different medium. Point out that the frequency cannot change. If this were not the case, the wave front would have two different frequencies at the interface between the two mediums. Because this is physically impossible, the frequency of the wave leaving one medium must be the same as the frequency of the wave that enters the second medium.

⭐TEKS
p. 562: 8A
p. 563: 3A, 8A

Refraction **563**

Refraction in various materials

Purpose Demonstrate that different materials have different refractive indices.

Materials laser, dusty chalkboard erasers, wide rectangular piece of plastic (Lucite), rectangular piece of glass or other material, aquarium filled with water and a few drops of milk

CAUTION *Avoid directing the laser beam toward the students.*

Procedure Adjust the position of the laser so that it shines into the side of the plastic, at an angle of about 30° from the normal (choose a point of incidence such that the light inside the plastic will exit the opposite parallel side of the block. Gently tap the erasers together to make the incoming laser beam visible. Have students observe and record the bending of the light rays as they enter and exit the material. Use chalk dust to see where the refracted beam goes.

Repeat this demonstration with the other materials. Point out that the angle of incidence is the same in every experiment. Ask students to compare the angles of refraction in the different materials. Ask in which material the speed of light is lowest (*the most refractive one*). Place the blocks inside the water and repeat. Have students observe that the bending is less dramatic.

★TEKS

p. 564: 2C, 3B, 8A
p. 565: 3A, 8A, 3A

index of refraction

the ratio of the speed of light in a vacuum to its speed in a given transparent medium

Did you know?

The index of refraction of any medium can also be expressed as the ratio of the wavelength of light in a vacuum, λ_0, to the wavelength of light in that medium, λ_n, as shown in the following relation.

$$n = \frac{\lambda_0}{\lambda_n}$$

THE LAW OF REFRACTION

An important property of transparent substances is the **index of refraction.** The index of refraction for a substance is the ratio of the speed of light in a vacuum to the speed of light in that substance. **★TEKS** 3B

INDEX OF REFRACTION

$$n = \frac{c}{v}$$

$$\text{index of refraction} = \frac{\text{speed of light in vacuum}}{\text{speed of light in medium}}$$

From this definition, we see that the index of refraction is a dimensionless number that is always greater than 1 because light always travels slower in a substance than in a vacuum. **Table 15-1** lists the indices of refraction for some representative substances. Note that the larger the index of refraction is, the slower light travels in that substance and the more a light ray will bend when it passes from a vacuum into that material. **★TEKS** 8A, 2C

Imagine, as an example, light passing between air and water. When light begins in the air (high speed of light and low index of refraction) and travels into the water (lower speed of light and higher index of refraction), the light rays are bent toward the normal. Conversely, when light passes from the water to the air, the light rays are bent away from the normal.

Note that the value for the index of refraction of air is nearly that of a vacuum. For simplicity, *use the value n = 1.00 for air when solving problems.*

Table 15-1	Indices of refraction for various substances*		
Solids at 20°C	**n**	**Liquids at 20°C**	**n**
Cubic zirconia	2.20	Benzene	1.501
Diamond	2.419	Carbon disulfide	1.628
Fluorite	1.434	Carbon tetrachloride	1.461
Fused quartz	1.458	Ethyl alcohol	1.361
Glass, crown	1.52	Glycerine	1.473
Glass, flint	1.66	Water	1.333
Ice (at 0°C)	1.309		
Polystyrene	1.49	**Gases at 0°C, 1 atm**	**n**
Sodium chloride	1.544	Air	1.000 293
Zircon	1.923	Carbon dioxide	1.000 450

*measured with light of vacuum wavelength = 589 nm

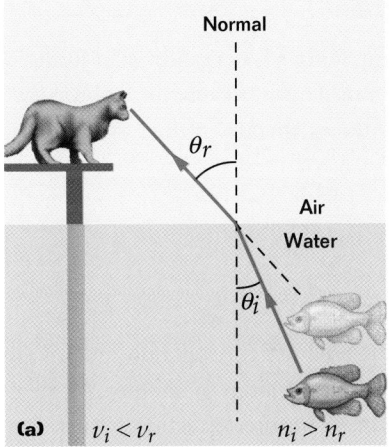

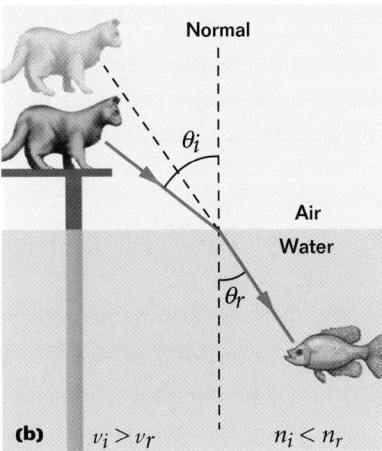

(a) $v_i < v_r$ $n_i > n_r$ (b) $v_i > v_r$ $n_i < n_r$

Figure 15-5
(a) To the cat on the pier, the fish looks closer to the surface than it really is. **(b)** To the fish, the cat seems to be farther from the surface than it actually is.

Objects appear to be in different positions due to refraction

When looking at a fish underwater, a cat sitting on a pier perceives the fish to be closer to the water's surface than it actually is, as shown in **Figure 15-5(a).** Conversely, the fish perceives the cat on the pier to be farther from the water's surface than it actually is, as shown in **Figure 15-5(b).** ⭐TEKS **3A, 8A**

Because of the reversibility of refraction, both the fish and the cat see along the same path. However, the light ray that reaches the fish forms a smaller angle with respect to the normal than does the light ray from the cat to the water's surface. This is because light is bent toward the normal when it travels from a medium with a lower index of refraction (the air) to one with a higher index of refraction (the water). Extending this ray along a straight line shows the cat's image to be above the cat's actual position.

On the other hand, the light ray that reaches the cat from the water's surface forms a larger angle with respect to the normal. This is because the light from the fish travels from a medium with a higher index of refraction to one with a lower index of refraction. Note that the fish's image is closer to the water's surface than the fish actually is. An underwater object seen from the air above appears larger than its actual size because the image, which is the same size as the object, is closer to the observer.

  ⭐TEKS **3A**

1. The Invisible Man H. G. Wells wrote a famous novel about a man who made himself invisible by changing his index of refraction. What would his index of refraction have to be to accomplish this?

2. Visibility for the Invisible Man Would the invisible man be able to see anything?

3. Fishing When trying to catch a fish, should a pelican dive into the water horizontally in front of or behind the image of the fish it sees?

Refraction   **565**

Point out that the spectrum of visible light is between 400 nm (violet) and 700 nm (red). Light with 589 nm wavelength is yellow.

Classroom Practice

The following may be used as teamwork exercises or for demonstration at the chalkboard or on an overhead projector.

PROBLEM

Snell's law

Find the angle of refraction of a light ray entering diamond from the following materials at an angle of 30.00°. (Hint: Use data from **Table 15-1.**)

a. water

b. cubic zirconia

Answer

  a. 15.99°

  b. 27.0°

A light ray (589 nm) traveling through air strikes an unknown substance at 60.00° and forms an angle of 41.42° with the normal inside. What material is it?

Answer

  ice ($n = 1.309$)

The Language of Physics

Sin^{-1} denotes the inverse function of sine, not $\dfrac{1}{\sin}$.
Just as $(\sin 30°) = 0.5$,
$\sin^{-1}(0.5) = 30°$.

Wavelength affects the index of refraction ⭐TEKS 8A

Note that the indices of refraction listed in **Table 15-1** are only valid for light that has a wavelength of 589 nm in a vacuum. This is because the amount that light bends when entering a different medium depends on the wavelength of the light as well as the speed. This is why a spectrum is produced when white light passes through a prism. Each color of light has a different wavelength. Therefore, each color of the spectrum is refracted by a different amount.

Snell's law determines the angle of refraction

The index of refraction of a material can be used to figure out how much a ray of light will be refracted as it passes from one medium to another. As mentioned, the greater the index of refraction, the more refraction occurs. But how can the angle of refraction be found?

In 1621, Willebrord Snell experimented with light passing through different media. He developed a relationship called Snell's law, which can be used to find the angle of refraction for light traveling between any two media. ⭐TEKS 3B

> **SNELL'S LAW**
>
> $$n_i(\sin \theta_i) = n_r(\sin \theta_r)$$
>
> **index of refraction of first medium × sine of the angle of incidence =
> index of refraction of second medium × sine of the angle of refraction**

internet connect

SC*i*LINKS
NSTA
TOPIC: Snell's law
GO TO: www.scilinks.org
*sci*LINKS CODE: HF2151

SAMPLE PROBLEM 15A

Snell's law ⭐TEKS 2C, 8A

PROBLEM

A light ray of wavelength 589 nm (produced by a sodium lamp) traveling through air strikes a smooth, flat slab of crown glass at an angle of 30.0° to the normal. Find the angle of refraction, θ_r.

SOLUTION

Given: $\theta_i = 30.0°$ $n_i = 1.00$ $n_r = 1.52$

Unknown: $\theta_r = ?$

Use the equation for Snell's law.

$$n_i(\sin \theta_i) = n_r(\sin \theta_r)$$

$$\theta_r = \sin^{-1}\left[\frac{n_i}{n_r}(\sin \theta_i)\right] = \sin^{-1}\left[\frac{1.00}{1.52}(\sin 30.0°)\right]$$

$$\boxed{\theta_r = 19.2°}$$

Snell's law ★TEKS 2C, 8A

1. Find the angle of refraction for a ray of light that enters a bucket of water from air at an angle of 25.0° to the normal. (Hint: Use **Table 15-1.**)

2. For an incoming ray of light of vacuum wavelength 589 nm, fill in the unknown values in the following table. (Hint: Use **Table 15-1.**)

from (medium)	to (medium)	θ_i	θ_r
a. flint glass	crown glass	25.0°	?
b. air	?	14.5°	9.80°
c. air	diamond	31.6°	?

3. A ray of light of vacuum wavelength 550 nm traveling in air enters a slab of transparent material. The incoming ray makes an angle of 40.0° with the normal, and the refracted ray makes an angle of 26.0° with the normal. Find the index of refraction of the transparent material. (Assume that the index of refraction of air for light of wavelength 550 nm is 1.00.)

Section Review ★TEKS 2C, 8A

1. In which of the following situations will light from a laser be refracted?
   a. traveling from air into a diamond at an angle of 30° to the normal
   b. traveling from water into ice along the normal
   c. upon striking a metal surface
   d. traveling from air into a glass of iced tea at an angle of 25° to the normal

2. For each of the following cases, will light rays be bent toward or away from the normal?
   a. $n_i > n_r$, where $\theta_i = 20°$
   b. $n_i < n_r$, where $\theta_i = 20°$
   c. from air to glass with an angle of incidence of 30°
   d. from glass to air with an angle of incidence of 30°

3. Find the angle of refraction of a ray of light that enters a diamond from air at an angle of 15.0° to the normal. (Hint: Use **Table 15-1.**)

4. **Physics in Action** Sunlight passes into a raindrop at an angle of 22.5° from the normal at one point on the droplet. What is the angle of refraction?

PRACTICE GUIDE 15A

Solving for:		
θ_r	📖 **PE**	Sample, 1–2; Ch. Rvw. 10–14, 42a, 49
	💿 **PW**	5
	PB	7–10
θ_i	📖 **PE**	Ch. Rvw. 40, 42b
	💿 **PW**	3
	PB	Sample, 1–3
n	📖 **PE**	3; Ch. Rvw. 39, 41
	💿 **PW**	Sample, 1–2, 4
	PB	4–6

ANSWERS TO

Practice 15A
Snell's law
1. 18.5°
2. a. 27.5°
   b. glycerine ($n = 1.47$)
   c. 12.5°
3. 1.47

Section Review
ANSWERS

1. a, d
2. a. away
   b. toward
   c. toward
   d. away
3. 6.14°
4. 16.7°

★TEKS

p. 566: 8A, 3B, 2C, 8A
p. 567: 2C, 8A, 2C, 8A

15-2
Thin lenses

15-2 SECTION OBJECTIVES

- **Use ray diagrams to find the position of an image produced by a converging or diverging lens, and identify the image as real or virtual.**

- **Solve problems using the thin-lens equation.**

- **Calculate the magnification of lenses.**

- **Describe the positioning of lenses in compound microscopes and refracting telescopes.**

lens

a transparent object that refracts light rays, causing them to converge or diverge to create an image

Figure 15-6

When rays of light pass through **(a)** a converging lens (thicker at the middle), they are bent inward. When they pass through **(b)** a diverging lens (thicker at the edge), they are bent outward.

TYPES OF LENSES

When light traveling in air enters a pane of glass, it is bent toward the normal. As the light exits the pane of glass, it is bent again. When the light exits, however, its speed increases as it enters the air, so the light bends away from the normal. Because the amount of refraction is the same regardless of whether light is entering or exiting a medium, the light rays are bent as much on exiting the pane of glass as they were on entering. Thus, the exiting ray of light is parallel to the ray that entered the pane, but it is displaced sideways by an amount that depends on the thickness of the pane, the index of refraction of the glass, and the angle of incidence of the light ray. **★TEKS 8A**

Curved surfaces change the direction of light

When the surfaces of a medium are curved, the alignment of the normal line differs for each spot on the surface of the medium. Thus, when light passes through a medium that has one or more curved surfaces, the change in the direction of the light rays varies from point to point. This principle is applied in objects called **lenses.** Like mirrors, lenses form images, but lenses do so by refraction rather than by reflection. The images formed can be either real or virtual, depending on the type of lens and on the placement of the object. Lenses are commonly used to form images in optical instruments, such as cameras, telescopes, and microscopes. In fact, the front of the human eyeball acts as a type of lens, converging light toward the light-sensitive retina at the back of the eyeball.

A typical lens consists of a piece of glass or plastic ground so that each of its two refracting surfaces is a segment of either a sphere or a plane. **Figure 15-6** shows examples of lenses. Notice that the lenses are shaped differently. The lens that is thicker at the middle than it is at the rim, shown in **Figure 15-6(a),** is an example of a *converging* lens. The lens that is thinner at the mid-

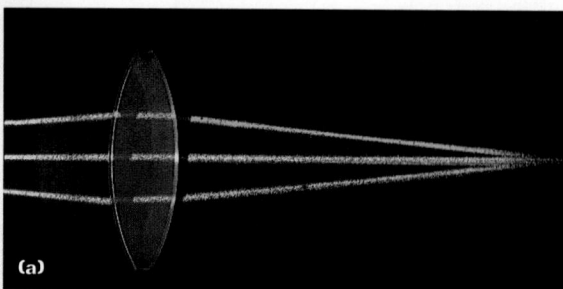

(a)

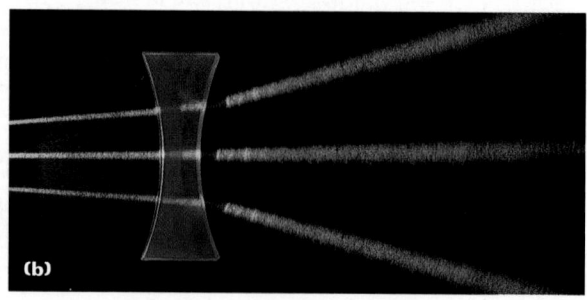

(b)

dle than it is at the rim, shown in **Figure 15-6(b),** is an example of a *diverging* lens. The light rays show why the names *converging* and *diverging* are applied to these lenses.

Focal length is the image distance for an infinite object distance

As with mirrors, it is convenient to define a point called the *focal point* for a lens. Note that light rays from an object far away are nearly parallel. The focal point of a converging lens is the location where the image of an object at an infinite distance from the lens is focused. For example, in **Figure 15-7(a)** a group of rays parallel to the principal axis passes through a focal point, *F,* after being bent inward by the lens. Unlike mirrors, every lens has a focal point on each side of the lens because light can pass through the lens from either side, as illustrated in **Figure 15-7.** The distance from the focal point to the center of the lens is called the *focal length, f.* The focal length is the image distance that corresponds to an infinite object distance. ⊛TEKS **8A**

Rays parallel to the principal axis diverge after passing through a diverging lens, as shown in **Figure 15-7(b).** In this case, the focal point is defined as the point from which the diverged rays appear to originate. Again, the focal length is defined as the distance from the center of the lens to the focal point.

Use ray diagrams to identify the image height and distance

Just as a set of standard rays and a ray diagram were used in Chapter 14 to predict the characteristics of images formed by spherical mirrors, a similar approach can be used for lenses. **Table 15-2** outlines the general rules for drawing these rays for lenses.

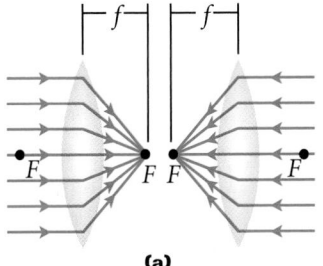

(a)

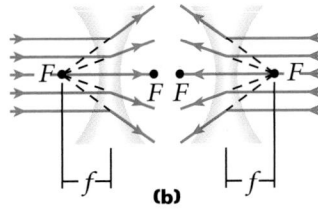

(b)

Figure 15-7
Both **(a)** converging lenses and **(b)** diverging lenses have two focal points but only one focal length.

Table 15-2	Rules for drawing reference rays	
Ray	**From object to lens**	**From *converging* lens to image**
Parallel ray	parallel to principal axis	passes through focal point, *F*
Central ray	to the center of the lens	from the center of the lens
Focal ray	passes through focal point, *F*	parallel to principal axis
Ray	**From object to lens**	**From *diverging* lens to image**
Parallel ray	parallel to principal axis	directed away from focal point, *F*
Central ray	to the center of the lens	from the center of the lens
Focal ray	proceeding toward back focal point, *F*	parallel to principal axis

Demonstration 5

Focal lengths of lenses

Purpose Locate the focal point of a converging lens.

Materials optical bench, light source, ray filter, converging lens, screen

Procedure Use the ray filter to produce five beams. Place the converging lens in front of the light source. Move the screen back and forth on the other side of the lens until the image is sharp. This is the focal point of the converging lens. Have students record the focal length.

⊛ TEKS
p. 568: 8A
p. 569: 8A

Key Models and Analogies

Converging lenses can be compared to concave mirrors, and diverging lenses can be compared to convex mirrors. Light rays pass through lenses but behave the same way as reflected rays do in mirrors. This analogy allows students to apply the rules for mirrors when studying lenses.

The equations and sign conventions for distances are also analogous. For example, remembering that the focal length in convex mirrors has a negative sign and that such mirrors form virtual images of real objects allows students to anticipate image formation in diverging lenses.

Quick Lab

TEACHER'S NOTES

This activity is meant to show a simple way to locate the focal point and to measure the focal length of a converging lens.

On a cloudy day, this lab can be done with a strong flashlight or penlight.

internetconnect

SC**LINKS**
NSTA

TOPIC: Lenses
GO TO: www.scilinks.org
*sci***LINKS CODE:** HF2152

Quick Lab

Focal Length

MATERIALS

✔ magnifying glass
✔ ruler

SAFETY CAUTION

Care should be taken not to focus the sunlight onto a flammable surface or any body parts, such as hands or arms. Also, do not look at the sun through the magnifying glass because serious eye injury can result.

On a sunny day, hold the magnifying glass, which is a converging lens, above a nonflammable surface, such as a sidewalk, so that a round spot of light is formed on the surface. Note where the spot formed by the lens is most distinct, or smallest. Use the ruler to measure the distance between the magnifying glass and the surface. This distance is the approximate focal length of the lens.

⭐ TEKS **1A, 2B, 8A**

The reasons why these rules work relate to concepts already covered in this textbook. From the definition of a focal point, we know that light traveling parallel to the principal axis (parallel ray) will be focused at the focal point. For a converging lens, this means that light will come together at the focal point in back of the lens. (In this book, the *front* of the lens is defined as the side of the lens that the light rays first encounter. The *back* of the lens refers to the side of the lens opposite where the light rays first encounter the lens.) But a similar ray passing through a diverging lens will exit the lens as if it originated from the focal point in front of the lens. Because refraction is reversible, a ray entering a converging lens from either focal point will be refracted so that it is parallel to the principal axis. ⭐ TEKS **3A**

For both lenses, a ray passing through the center of the lens will continue in a straight line with no net refraction. This occurs because both sides of a lens are parallel to one another along any path through the center of the lens. As with a pane of glass, the exiting ray will be parallel to the ray that entered the lens. For ray diagrams, however, the usual assumption is that the lens is negligibly thin, so it is assumed that the ray is not displaced sideways but instead continues in a straight line.

CHARACTERISTICS OF LENSES

Table 15-3 summarizes the possible relationships between object and image positions for converging lenses. The rules for drawing reference rays were used to create each of these diagrams. Note that applications are listed along with each ray diagram to show the varied uses of the different configurations.

Converging lenses can produce real or virtual images

Notice that an object infinitely far away from a converging lens will create a point image at the focal point, as shown in the first diagram in **Table 15-3**. This image is real, which means that it can be projected on a screen, whereas a virtual image cannot be projected onto a screen. ⭐ TEKS **8A**

As a distant object approaches the focal point, the image becomes larger and farther away, as shown in the second, third, and fourth diagrams in **Table 15-3**. When the object is at the focal point, as shown in the fifth diagram, the light rays from the object are refracted so that they exit the lens parallel to each other. (Note that because the object is at the focal point, it is impossible to draw a third ray that passes through that focal point, the lens, and the tip of the object.)

When the object is between a converging lens and its focal point, the light rays from the object diverge when they pass through the lens, as shown in the sixth diagram in **Table 15-3**. This image appears to an observer in back of the lens as being on the same side of the lens as the object. In other words, the brain interprets these diverging rays as coming from an object directly along the path of the rays that reach the eye. The ray diagram for this final case is less straightforward than those drawn for the other cases in the table. The first two

Table 15-3 **Images created by converging lenses** ⭐TEKS 8A

Ray diagrams

1.

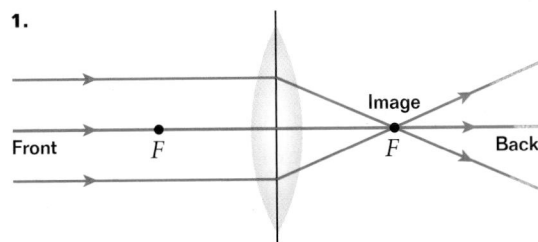

Configuration: object at infinity; point image at *F*

Applications: burning a hole with a magnifying glass

2.

Configuration: object outside 2*F*; real, smaller image between *F* and 2*F*

Applications: lens of a camera, human eyeball lens, and objective lens of a refracting telescope

3.

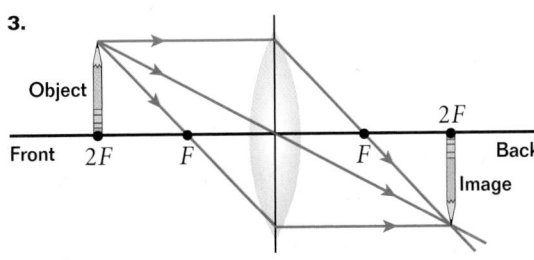

Configuration: object at 2*F*; real image at 2*F* same size as object

Applications: inverting lens of a field telescope

4.

Configuration: object between *F* and 2*F*; magnified real image outside 2*F*

Applications: motion-picture or slide projector and objective lens in a compound microscope

5.

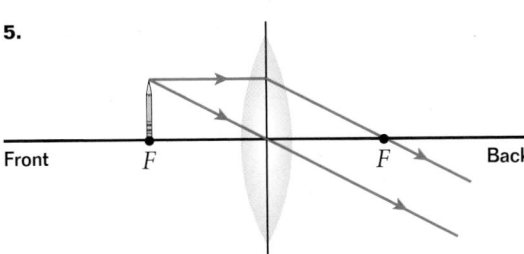

Configuration: object at *F*; image at infinity

Applications: lenses used in lighthouses and searchlights

6.

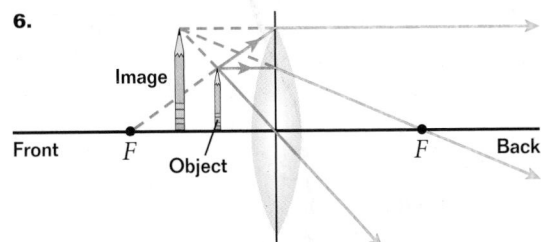

Configuration: object inside *F*; magnified virtual image on the same side of the lens as the object

Applications: magnifying with a magnifying glass; eyepiece lens of microscope, binoculars, and telescope

Visual Strategy

Table 15-3
Point out that in the first case, the object is so far away that the rays from all its points (top to bottom) converge at the focal point on the axis. In the other diagrams, rays are drawn from the object's top only, and their intersection—after passing through the lens—determines the image's top (the bottom is assumed to be on the axis at the same distance from the lens as the top).

Q Suppose that in each case the lens's focal length is 10 cm. What values (or range of values) will the object and image distances have for each case?

A *case 1: object at ∞, image at 10 cm*

case 2: ∞ > p > 20 cm, 20 cm > q > 10 cm

case 3: p = 20 cm, q = 20 cm

case 4: 20 cm > p > 10 cm, ∞ > q > 20 cm

case 5: p = 10 cm, image at infinity

case 6: p < 10 cm, negative image distance

⭐TEKS

p. 570: 3A, 1A, 2B, 8A
p. 571: 8A

Demonstration 6

Angular size of an object

Purpose Illustrate that angular size depends on the distance from the object.

Materials several metersticks, protractors, drinking straws, tape

Procedure Experiment in a long hallway. Divide students into pairs, and assign each pair one of the following distances: 3 m, 4 m, 5 m, or 6 m. Use a pin to fasten two drinking straws so that they pivot at the center of each protractor. In each pair, have one student hold the meterstick vertically and stand at the specified distance from his or her partner, who should be holding the protractor. Have students adjust the straws so that each straw lines up with one end of the meterstick and record the angle between the straws. Students should also describe how well they can see the marks or the numbers on the meterstick.

Back in the classroom, have each pair of students report their measurements in a data table on the chalkboard. Students should conclude that although all the metersticks were 1 m tall, they appeared smaller as their angular size decreased.

rays (parallel to the axis and through the center of the lens) are drawn in the usual fashion. The third ray, however, is drawn so that if it were extended, it would connect the focal point in front of the lens, the tip of the object, and the lens in a straight line. To determine where the image is, draw lines extending from the rays exiting the lens back to the point where they would appear to have originated to an observer on the back side of the lens (these lines are dashed in the sixth diagram in **Table 15-3**).

A magnifying glass increases the angular size of an object

A magnifying glass is a converging lens and an example of a *simple magnifier*. As the name implies, this device is used to increase the apparent size of an object. When an object is placed just inside the focal point of a lens, a magnified virtual image appears. The last diagram in **Table 15-3** shows why the image appears upright, but why does it appear larger? (★)TEKS **3A, 8A**

The size of the image of an object formed at the retina of the eye depends on the *angular size* of the object. As the object moves closer, its angular size increases; as it moves farther away, its angular size decreases. For instance, consider **Figure 15-8.** When a person is standing 100 m from a house, the numbers on the house will not be visible. However, as the viewer moves closer to the house, the numbers become visible. If the viewer moves even closer, a

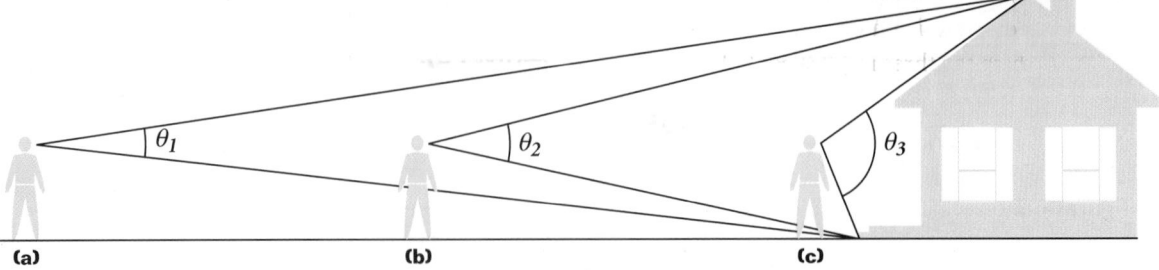

Figure 15-8
The angular size of the house is smaller at position **(a)** than it is when the viewer stands at position **(b).** At position **(c),** the angular size of the house is the largest and the most detail can be seen.

fly crawling across one of the numbers can be seen. Even more detail would be visible, such as the mouthparts of the fly, if the human eye could focus a very small distance from the object. What a simple magnifier does is enable the viewer to examine a small object by producing an enlarged image that fills a larger portion (angle) of the observer's field of vision.

Diverging lenses produce only a virtual image

As shown in **Figure 15-9,** diverging lenses can produce only one kind of image—virtual. Again, this image is upright, but diverging lenses *reduce* the size of the image. The image appears inside the focal point for any placement of the object.

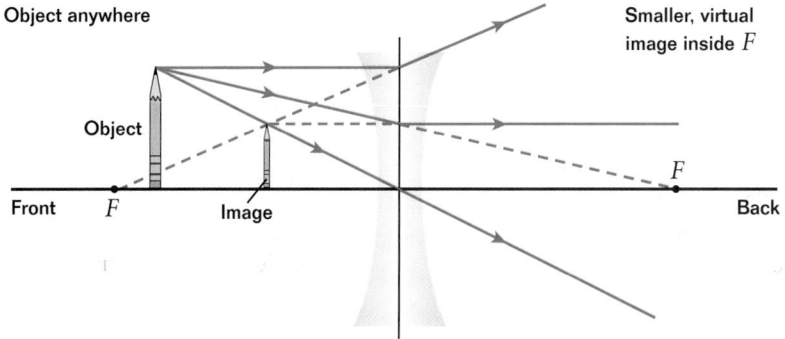

Figure 15-9
The image created by a diverging lens is always a virtual, smaller image.

The ray diagram shown in **Figure 15-9** for diverging lenses was created using the ray diagram rules for a diverging lens given in **Table 15-2.** The first ray, parallel to the axis, appears to come from the focal point on the same side of the lens as the object. This is indicated by the oblique dashed line. The second ray passes through the center of the lens and is not refracted. The third ray is drawn as if it were going to the focal point in back of the lens. As this ray passes through the lens, it is refracted parallel to the principal axis. Finally, this third ray must be extended backward, as shown by the dashed line. As with the sixth diagram in **Table 15-3,** the location of the tip of the image is the point at which the three rays appear to have originated. (★TEKS) **8A**

THE THIN-LENS EQUATION AND MAGNIFICATION

Ray diagrams for lenses give an estimate of image size and distance, but it is possible to calculate these values. The equation that relates object and image distances for a lens is identical to the mirror equation in Chapter 14. This equation can be used to calculate any one of the three distances involved in the equation if the other two are known.

Misconception Alert

Students might wonder about the need for diverging lenses because virtual images are also formed with converging lenses, as shown in the sixth case in **Table 15-3.** Ask them to compare the object's location when using each kind of lens for creating a virtual image. A diverging lens creates a virtual image of objects at any distance in front of it. The converging lens can do that only for objects inside *F,* and this virtual image is always magnified. The image produced by a diverging lens is always virtual and smaller than the object regardless of the object's location. This is useful for correcting nearsightedness with eyeglasses.

Did you know?

The lens of a camera forms an inverted image on the film in the back of the camera. To view this image before taking a picture, two methods are used. In one, a system of mirrors and prisms reflects the image to the viewfinder, uprighting the image in the process. In the other method, the viewfinder is a diverging lens that is separate from the main lens system. This lens forms an upright virtual image that resembles the image that will be projected onto the film.

(★)TEKS
p. 572: 3A, 8A
p. 573: 8A

Misconception Alert

Students may not be sure about all the cases shown in **Table 15-4,** particularly about the meaning of a negative distance for the object, which suggests that the object is in back of the lens. Some students may need to be reminded that in these conventions, light rays always travel from the front to the back of a lens. Ask them if an object could be virtual. Point out that the real image formed by a converging lens may become a virtual object for another lens if the lens is located so that it interrupts the rays before they converge.

 **Interactive Problem-Solving Tutor**

See Module 15
"Refraction and Lenses" provides additional development of problem-solving skills for this chapter.

 TEKS

p. 574: 2C, 8A
p. 575: 2C, 8A

574

Table 15-4
Sign conventions for lenses

	+	−
p	object in front of the lens	object in back of the lens
q	image in back of the lens	image in front of the lens
f	converging lens	diverging lens

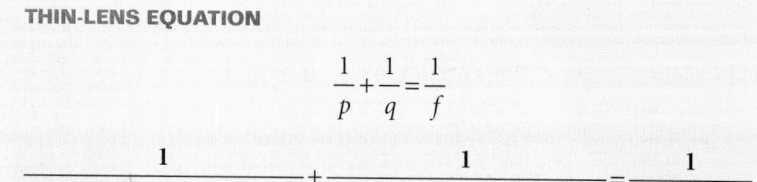 **THIN-LENS EQUATION**

$$\frac{1}{p} + \frac{1}{q} = \frac{1}{f}$$

$$\frac{1}{\text{distance from object to lens}} + \frac{1}{\text{distance from image to lens}} = \frac{1}{\text{focal length}}$$

This equation is called the *thin-lens equation* because it is derived using the assumption that the lens is very thin. In other words, this equation applies when the lens thickness is much smaller than its focal length. If this is true, it makes no difference whether we take the focal length to be the distance from the focal point to the surface of the lens or the distance from the focal point to the center of the lens. (★)TEKS 2C, 8A

The thin-lens equation can be applied to both converging and diverging lenses if we adhere to a set of sign conventions. **Table 15-4** gives the sign conventions for lenses. Under this convention, an image in back of the lens (that is, a real image) has a positive image distance, and an image in front of the lens, or a virtual image, has a negative image distance. A converging lens has a positive focal length and a diverging lens has a negative focal length. Because of this, converging lenses are sometimes called *positive lenses* and diverging lenses are sometimes called *negative lenses.*

Magnification of a lens depends on object and image distances

Recall that magnification (M) is defined as the ratio of image height to object height. The following equation can be used to calculate the magnification of both converging and diverging lenses. (★)TEKS 2C, 8A

MAGNIFICATION OF A LENS

$$M = \frac{h'}{h} = -\frac{q}{p}$$

$$\text{magnification} = \frac{\text{image height}}{\text{object height}} = -\frac{\text{distance from image to lens}}{\text{distance from object to lens}}$$

Note that the magnification equation for a lens is the same as the magnification equation for a mirror. If close attention is given to the sign conventions defined in **Table 15-4,** then the magnification will describe the image's size and orientation. When the magnitude of the magnification of an object is less than one, the image is smaller than the object. Conversely, when the magnitude of the magnification is greater than one, the image is larger than the object.

Additionally, a negative sign for the magnification indicates that the image is real and inverted. A positive magnification signifies that the image is upright and virtual.

Module 15
"Refraction and Lenses" provides an interactive lesson with guided problem-solving practice to teach you about the images produced with different types of lenses.

Lenses ⭐TEKS 2C, 8A

PROBLEM

An object is placed 30.0 cm in front of a converging lens and then 12.5 cm in front of a diverging lens. Both lenses have a focal length of 10.0 cm. For both cases, find the image distance and the magnification. Describe the images.

SOLUTION

1. DEFINE

Given:

$f_{converging} = 10.0$ cm $\qquad f_{diverging} = -10.0$ cm

$p_{converging} = 30.0$ cm $\qquad p_{diverging} = 12.5$ cm

Unknown: $q_{converging} = ?$ $\quad M = ?$ $\quad q_{diverging} = ?$ $\quad M = ?$

Diagrams:

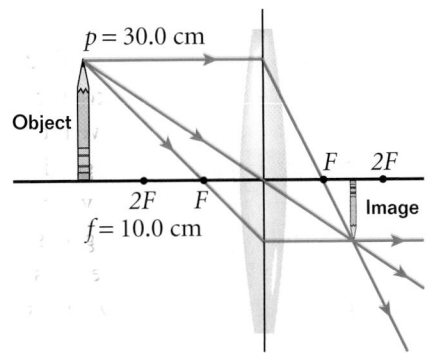

$p = 30.0$ cm

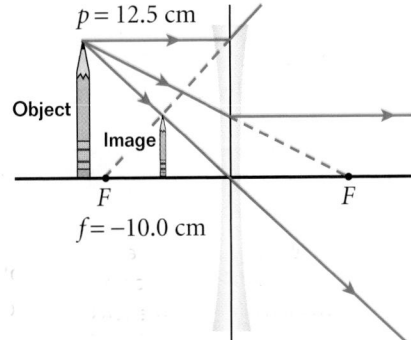

$p = 12.5$ cm

2. PLAN

Choose an equation(s) or situation:

The thin-lens equation can be used to find the image distance, while the equation for magnification will serve to describe the size and orientation of the image.

$$\frac{1}{p} + \frac{1}{q} = \frac{1}{f} \qquad M = -\frac{q}{p}$$

Rearrange the equation(s) to isolate the unknown(s):

$$\frac{1}{q} = \frac{1}{f} - \frac{1}{p}$$

3. CALCULATE

For the converging lens:

$$\frac{1}{q} = \frac{1}{f} - \frac{1}{p} = \frac{1}{10.0 \text{ cm}} - \frac{1}{30.0 \text{ cm}} = \frac{0.067}{1 \text{ cm}}$$

$$\boxed{q = 15 \text{ cm}}$$

$$M = -\frac{q}{p} = -\frac{15 \text{ cm}}{30.0 \text{ cm}}$$

$$\boxed{M = -0.50}$$

continued on next page

The following may be used as teamwork exercises or for demonstration at the board or on an overhead projector.

PROBLEM

Lenses

When an object is placed 3.00 cm in front of a converging lens, a real image is formed 6.00 cm in back of the lens. Find the focal distance.

Answer

$f = 2.00$ cm

Where would you place an object in order to produce a virtual image 15.0 cm in front of a converging lens with a focal length of 10.0 cm? How about a diverging lens with the same focal length?

Answer

converging lens: 5.98 cm; diverging lens: -3.0×10^1 cm

Alternative Problem-Solving Approach

Draw a ray diagram to scale on graph paper. Determine the answer using two rays, and draw the third one to confirm.

Another possibility is to apply this method using a computer art/paint/draw application. The number of pixels can be translated to scale. Working with a grid in the background may be helpful.

PRACTICE GUIDE 15B
Solving for:

q	📖 **PE**	Sample, 1–4; Ch. Rvw. 24, 26, 55*
	💿 **PW**	3, 6–9, 16
	PB	4–6
M	📖 **PE**	Sample, 1–4; Ch. Rvw. 24–26, 48
	💿 **PW**	1–2
	PB	4–6
p	📖 **PE**	4; Ch. Rvw. 43, 45, 47–48
	💿 **PW**	Sample, 1–7, 16
	PB	3–4, 7–10
h	💿 **PW**	Sample, 1–2, 4–5, 7–9, 14–15
	PB	Sample, 1–2
f	📖 **PE**	4; Ch. Rvw. 46, 64–65
	💿 **PW**	10–15
	PB	Sample, 1–2

ANSWERS TO

Practice 15B
Lenses

1. 2.0×10^1 cm, $M = -1.00$; real, inverted image
2. -3.0×10^1 cm, $M = 3.00$; virtual, upright image
3. -6.67 cm, $M = 0.333$; virtual, upright image
4. a. $p = 2.0$ cm, $M = 1.5$
    b. $p = 5.0$ cm, $M = -1.4$
    c. $q = -2.4$ cm, $M = 0.60$
    d. $f = -5.0$ cm, $q = -2.5$ cm

For the diverging lens:

$$\frac{1}{q} = \frac{1}{f} - \frac{1}{p} = \frac{1}{-10.0 \text{ cm}} - \frac{1}{12.5 \text{ cm}} = -\frac{0.180}{1 \text{ cm}}$$

$$\boxed{q = -5.56 \text{ cm}}$$

$$M = -\frac{q}{p} = -\frac{-5.56 \text{ cm}}{12.5 \text{ cm}}$$

$$\boxed{M = 0.445}$$

4. EVALUATE These values and signs for the converging lens indicate a real, inverted, smaller image. This is expected because the object distance is longer than twice the focal length of the converging lens. The values and signs for the diverging lens indicate a virtual, upright, smaller image formed inside the focal point. This is the only kind of image diverging lenses form.

PRACTICE 15B

Lenses ⭐TEKS 2C, 8A

1. An object is placed 20.0 cm in front of a converging lens of focal length 10.0 cm. Find the image distance and the magnification. Describe the image.

2. Sherlock Holmes examines a clue by holding his magnifying glass (with a focal length of 15.0 cm) 10.0 cm away from an object. Find the image distance and the magnification. Describe the image that he observes.

3. An object is placed 20.0 cm in front of a diverging lens of focal length 10.0 cm. Find the image distance and the magnification. Describe the image.

4. Fill in the missing values in the following table.

	f	p	q	M
		Converging lens		
a.	6.0 cm	?	−3.0 cm	?
b.	2.9 cm	?	7.0 cm	?
		Diverging lens		
c.	−6.0 cm	4.0 cm	?	?
d.	?	5.0 cm	?	0.50

EYEGLASSES AND CONTACT LENSES

The transparent front of the eyeball, called the *cornea*, acts like a lens, directing light rays toward the light-sensitive *retina* in the back of the eye. Although most of the refraction of light occurs at the cornea, the eye also contains a small lens, called the *crystalline lens,* that refracts light as well.

When the relaxed eye attempts to produce a focused image of a nearby object but the image position is behind the retina, the abnormality is known as *hyperopia* and the person is said to be *farsighted.* With this defect, distant objects are seen clearly but near objects are blurred. Either the hyperopic eyeball is too short or the ciliary muscle that adjusts the shape of the lens cannot adjust enough to properly focus the image. **Table 15-5** shows how hyperopia can be corrected with a converging lens. ⓣTEKS **3C**

Another condition, known as *myopia,* or *nearsightedness,* occurs either when the eyeball is longer than normal or when the maximum focal length of the lens is insufficient to produce a clear image on the retina. In this case, light from a distant object is focused in front of the retina. The distinguishing feature of this imperfection is that distant objects are not seen clearly. Nearsightedness can be corrected with a diverging lens, as shown in **Table 15-5.**

A contact lens is simply a lens worn directly over the cornea of the eye. The lens floats on a thin layer of tears.

Table 15-5   **Nearsighted and farsighted**

Farsighted

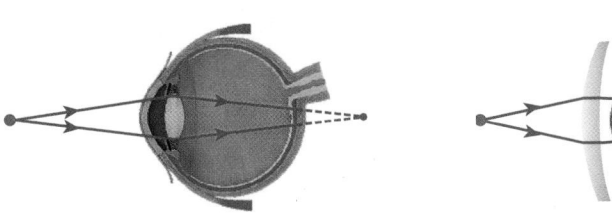

Hyperopia

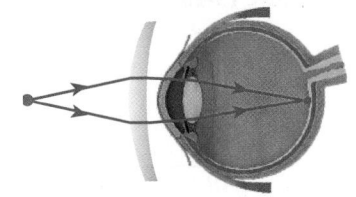

Corrected with a converging lens

Nearsighted

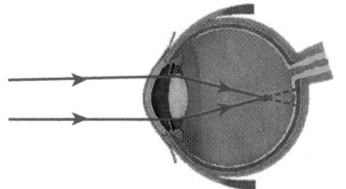

Myopia

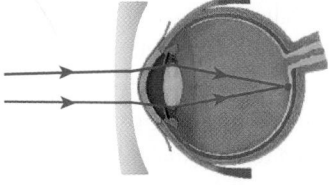

Corrected with a diverging lens

Quick Lab

Prescription Glasses

MATERIALS

✔ several pairs of prescription eyeglasses

Hold a pair of prescription glasses about 12 cm from your eye, and look at different objects through the lenses. Try this with different types of glasses, such as those for farsightedness and nearsightedness, and describe what effect the differences have on the image you see. If you have bifocals, how do the images produced by the top and bottom portions of the bifocal lens compare?

⭐TEKS **1A, 8A**

---

internetconnect

SC*i*LINKS
NSTA

TOPIC: Abnormalities of the eye
GO TO: www.scilinks.org
*sci*LINKS CODE: HF2153

Quick Lab

TEACHER'S NOTES

This activity is meant to let students experience how eyeglasses work to correct hyperopia and myopia.

This lab is more effective when using eyeglasses of 1.5 diopters or higher and when students make systematic observations of the same object at increasing distance or of objects of increasing size at a constant distance. The lines and numbers on a ruler make good objects for this experiment.

Key Models and Analogies

The optical functions of different parts of the eye may be compared to those of a camera. Ask: Which part of the eye corresponds to the film in a camera? *(retina)* Which components of the projector or camera might be compared to the eye's ability to accommodate? *(the ability to move the objective lens in a camera back and forth and to add lenses to it)*

⭐TEKS

p. 576: 2C, 8A
p. 577: 3C, 1A, 8A

Microscope

Purpose Explain the function of lenses in a microscope.

Materials optical bench, light source with small aperture, two converging lenses (one with a very short *f* and one with a long *f*), screen

Procedure Place the lens with the short focal length in front of the light source. Position the screen to show that the image is real and inverted. Tell students that this is what the objective does in a microscope. Point out that a screen is needed in order to see the image. Ask if it is magnified (*barely*).

Place the second lens in back of the screen and remove the screen. Tell students that this is the eyepiece. Ask students to try to locate the new image. (They may try to place the screen behind the second lens and move the second lens or the screen.) Have them look through the second lens toward the light source and move the second lens until they see the virtual image, upright and magnified.

Visual Strategy

Figure 15-10

Make sure that students transfer their understanding of the diagram to that of a real microscope.

Q Normally, a microscope is placed in a vertical position. Redraw the diagram so that it represents a microscope in a vertical position, and indicate where you would place the slide.

A *slide at O*

Figure 15-10

In a compound microscope, the real, inverted image produced by the objective lens is used as the object for the eyepiece lens.

COMBINATION OF THIN LENSES

If two lenses are used to form an image, the system can be treated in the following manner. First, the image of the first lens is calculated as though the second lens were not present. The light then approaches the second lens as if it had come from the image formed by the first lens. Hence, *the image formed by the first lens is treated as the object for the second lens.* The image formed by the second lens is the final image of the system. The overall magnification of a system of lenses is the product of the magnifications of the separate lenses. If the image formed by the first lens is in back of the second lens, then the image is treated as a virtual object for the second lens (that is, *p* is negative). The same procedure can be extended to a system of three or more lenses.

Compound microscopes use two converging lenses

A simple magnifier, such as a magnifying glass, provides only limited assistance when inspecting the minute details of an object. Greater magnification can be achieved by combining two lenses in a device called a compound microscope. It consists of two lenses: an objective lens (near the object) with a focal length of less than 1 cm and an eyepiece with a focal length of a few centimeters. As shown in **Figure 15-10,** the object placed just outside the focal point of the objective lens forms a real, inverted, and enlarged image that is at or just inside the focal point of the eyepiece. The eyepiece, which serves as a simple magnifier, uses this enlarged image as its object and produces an even more enlarged virtual image. The image viewed through a microscope is upside-down with respect to the actual orientation of the specimen, as shown in **Figure 15-10.** ⊛TEKS 8A

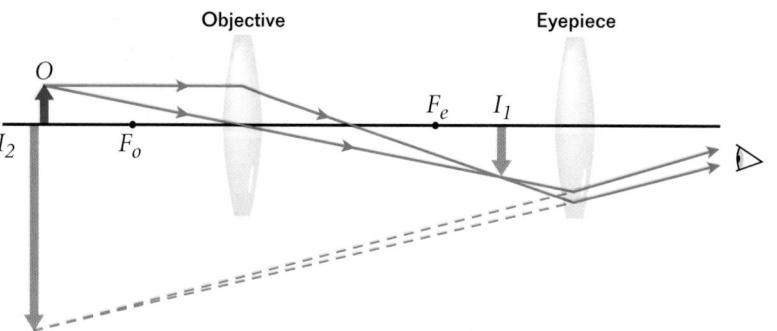

The microscope has extended our vision into the previously unknown realm of incredibly small objects. A question that is often asked about microscopes is, "With extreme patience and care, would it be possible to construct a microscope that would enable us to see an atom?" As long as visible light is used to illuminate the object, the answer is no. In order to be seen, the object under a microscope must be at least as large as a wavelength of light. An atom is many times smaller than a wavelength of visible light, so its mysteries must be probed through other techniques. ⊛TEKS 3C

Refracting telescopes also use two converging lenses

As mentioned in Chapter 14, there are two types of telescopes, reflecting and refracting. In a refracting telescope, an image is formed at the eye in much the same manner as is done with a microscope. A small, inverted image is formed at the focal point of the objective lens because the object is essentially at infinity. The eyepiece is positioned so that its focal point lies very close to the focal point of the objective lens, where the image is formed, as shown in **Figure 15-11**. Because the image is now just inside the focal point of the eyepiece, the eyepiece acts like a simple magnifier and allows the viewer to examine the object in detail. (★)TEKS **3C, 8A**

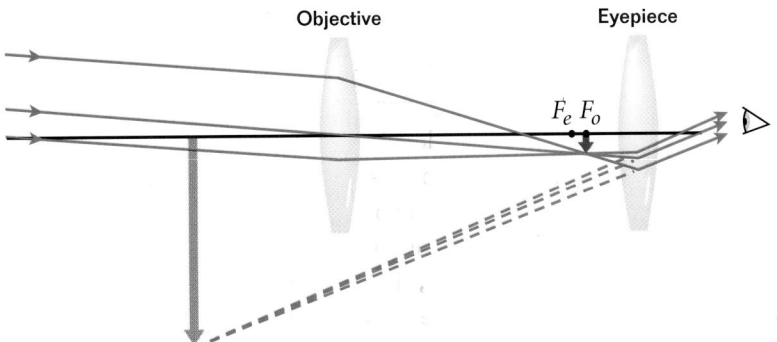

Figure 15-11
The image produced by the objective lens of a refracting telescope is a real, inverted image that is at its focal point. This inverted image, in turn, is the object from which the eyepiece creates a magnified, virtual image.

Section Review

(★)TEKS **2C, 8A**

1. Using a ray diagram, find the position and height of an image produced by a viewfinder in a camera with a focal length of 5.0 cm if the object is 1.0 cm tall and 10.0 cm in front of the lens. A camera viewfinder is a diverging lens.

2. What type of image, virtual or real, is produced in the following cases?
   a. an object inside the focal point of a camera lens
   b. an object outside the focal point of a refracting telescope's objective lens
   c. an object outside the focal point of a camera's viewfinder

3. Find the image position for an object placed 3.0 cm outside the focal point of a converging lens with a 4.0 cm focal length.

4. What is the magnification of the object from item 3?

5. Compare the length of a refracting telescope with the sum of the focal lengths of its two lenses.

6. What type of image is produced by the cornea and the lens on the retina?

SECTION 15-2

Teaching Tip

Viewing objects on land upside-down is troublesome. Discuss ways in which this problem can be solved.

Terrestrial (or field) telescopes contain an extra converging lens to make the images of objects on land appear upright. Galileo, who invented the first refracting telescope, used a diverging lens as the eyepiece to solve this inversion problem.

internet**connect**

SC*i*LINKS

NSTA

TOPIC: Telescopes in Texas
GO TO: www.scilinks.org
*sci*LINKS CODE: HFX010

Section Review
ANSWERS

1. $q = -3.3$ cm, $h' = 0.33$ cm
2. a. virtual
   b. real
   c. virtual
3. 9.1 cm
4. −1.3
5. The length of the telescope is slightly shorter than $f_0 + f_e$.
6. real, inverted

(★)TEKS

p. 578: 8A, 3C
p. 579: 3C, 8A, 2C, 8A

15-3
Optical phenomena

Critical angle

Purpose Demonstrate critical angle and total internal reflection.

Materials 90° prism, laser, chalk dust

CAUTION *Avoid directing the laser beam toward the students.*

Procedure Adjust the position of the laser so that the beam is in position 1, as shown below.

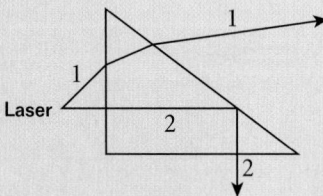

Gently tap the erasers together to make the incoming and the exiting laser beam visible. Let students observe the refracted beam inside and outside the prism.

Slowly rotate the beam from position 1 to position 2. When the exiting beam no longer passes through the hypotenuse of the prism, add chalk dust near the other side of the right angle (where beam 2 would exit). Have students observe the beam inside the prism and observe that it is totally internally reflected.

15-3 SECTION OBJECTIVES

- **Predict whether light will be refracted or undergo total internal reflection.**

- **Recognize atmospheric conditions that cause refraction.**

- **Explain dispersion and phenomena such as rainbows in terms of the relationship between the index of refraction and the wavelength.**

total internal reflection

the complete reflection of light at the boundary of two transparent media; this effect occurs when the angle of incidence exceeds the critical angle

critical angle

the minimum angle of incidence for which total internal reflection occurs

TOTAL INTERNAL REFLECTION

An interesting effect called **total internal reflection** can occur when light moves along a path from a medium with a *higher* index of refraction to one with a *lower* index of refraction. Consider light rays traveling from water into air, as shown in **Figure 15-12(a).** Four possible directions of the rays are shown in the figure. ⊛TEKS **8A**

At some particular angle of incidence, called the **critical angle,** the refracted ray moves parallel to the boundary, making the angle of refraction equal to 90°, as shown in **Figure 15-12(b).** For angles of incidence greater than the critical angle, the ray is entirely reflected at the boundary, as shown in **Figure 15-12.** This ray is reflected at the boundary as though it had struck a perfectly reflecting surface. Its path and the path of all rays like it can be predicted by the law of reflection; that is, the angle of incidence equals the angle of reflection.

In optical equipment, prisms are arranged so that light entering the prism is totally internally reflected off the back surface of the prism. Prisms are used in place of silvered or aluminized mirrors because they reflect light more efficiently and are more scratch resistant.

Snell's law can be used to find the critical angle. As mentioned above, when the angle of incidence, θ_i, equals the critical angle, θ_c, then the angle of refraction, θ_r, equals 90°. Substituting these values into Snell's law gives the following relation.

$$n_i(\sin \theta_c) = n_r(\sin 90°)$$

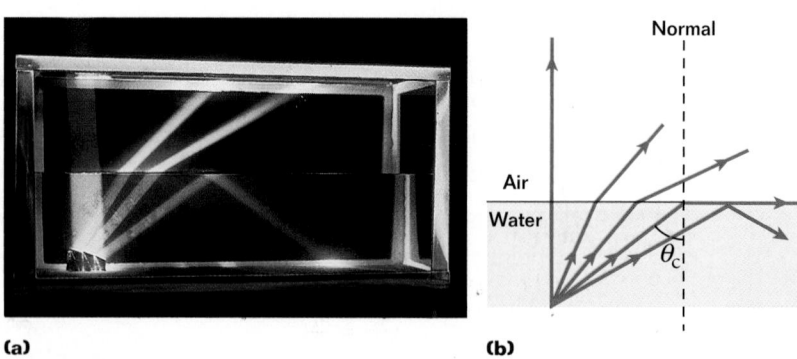

(a)　　　　　　　　　　**(b)**

Figure 15-12
(a) This photo demonstrates several different paths of light radiated from the bottom of an aquarium. **(b)** At the critical angle, θ_c, a light ray will travel parallel to the boundary. Any rays with an angle of incidence greater than θ_c will be totally internally reflected at the boundary.

Because the sine of 90° equals 1, the following relationship results.

CRITICAL ANGLE

$$\sin \theta_c = \frac{n_r}{n_i} \text{ for } n_i > n_r$$

$$\text{sine (critical angle)} = \frac{\text{index of refraction of second medium}}{\text{index of refraction of first medium}}$$

but only if index of refraction of first medium >
index of refraction of second medium

Note that this equation can be used only when n_i is greater than n_r. In other words, *total internal reflection occurs only when light moves along a path from a medium of high index of refraction to a medium of lower index of refraction.* If n_i were less than n_r, this equation would give $\sin \theta_c > 1$, which is an impossible result because by definition the sine of an angle can never be greater than one.

When the second substance is air, the critical angle is small for substances with large indices of refraction. Diamonds, which have an index of refraction of 2.419, have a critical angle of 24.4°. By comparison, the critical angle for crown glass, where $n = 1.52$, is 41.0°. Because diamonds have such a small critical angle, most of the light that enters a cut diamond is totally internally reflected. The reflected light eventually exits the diamond from the most visible faces of the diamond. Jewelers cut diamonds so that the maximum light entering the upper surface is reflected back to these faces. **(★)TEKS 3C**

SAMPLE PROBLEM 15C

Critical angle (★)TEKS 2C, 8A

PROBLEM

Find the critical angle for a water-air boundary if the index of refraction of water is 1.333.

SOLUTION

Given: $n_i = 1.333$ $\quad n_r = 1.00$

Unknown: $\theta_c = ?$

Use the equation for critical angle on this page.

$$\sin \theta_c = \frac{n_r}{n_i}$$

$$\theta_c = \sin^{-1}\left(\frac{n_r}{n_i}\right) = \sin^{-1}\left(\frac{1.00}{1.333}\right)$$

$$\boxed{\theta_c = 48.6°}$$

Periscope

MATERIALS

✔ two 90° prisms

Align the two prisms side by side as shown below.

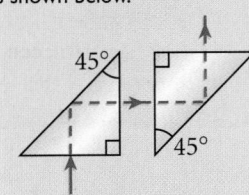

Note that this configuration can be used like a periscope to see an object above your line of sight if the configuration is oriented vertically and to see around a corner if it is oriented horizontally. How would you arrange the prisms to see behind you? Draw your design on paper and test it.

(★)TEKS 1A, 2A, 8A

Quick Lab

TEACHER'S NOTES

This activity is meant to demonstrate an application of total internal reflection. Point out that the critical angle for glass is about 41° and that the angle of incidence from glass to air in this arrangement is 45°.

This lab is more effective if students look through each prism separately to experience the path followed by the light rays and realize that the prisms could be replaced by mirrors.

Classroom Practice

The following may be used as teamwork exercises or for demonstration at the board or on an overhead projector.

PROBLEM

Critical angle

Calculate the critical angle of light traveling from the following substances into air.

a. quartz ($n = 1.46$)
b. acrylic resin ($n = 1.51$)
c. flint glass ($n = 1.66$)

Answer
    a. 43.2° **b.** 41.5° **c.** 37.0°

(★)**TEKS**

p. 580: 8A
p. 581: 1A, 2A, 8A, 3C,
2C, 8A

ANSWERS TO

Practice 15C
Critical angle

1. 42.8°
2. 64.82°
3. 49.8°
4. diamond (24.4°); cubic zirconia (27.0°)

⭐TEKS

p. 582: 2C, 8A, 8A, 3C, 3A
p. 583: 3A, 8A, 8A

PRACTICE 15C

Critical angle ⭐TEKS 2C, 8A

1. Find the critical angle for light traveling from glycerine ($n = 1.473$) into air.

2. Calculate the critical angle for light traveling from glycerine ($n = 1.473$) into water ($n = 1.333$).

3. Find the critical angle for light traveling from ice ($n = 1.309$) into air.

4. Which has a smaller critical angle in air, diamond ($n = 2.419$) or cubic zirconia ($n = 2.20$)? Show your work.

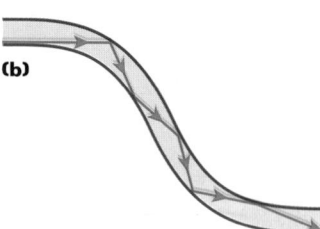

(a)

(b)

Figure 15-13
(a) A fiber optic cable consists of several fibers. **(b)** Light is guided along a fiber by multiple internal reflections.

Fiber optics can guide light over a long distance

Another interesting application of total internal reflection is the use of glass or transparent plastic rods, like the ones shown in **Figure 15-13(a),** to transfer light from one place to another. As indicated in **Figure 15-13(b),** light is confined to traveling within the rods, even around gentle curves, as a result of successive internal reflections. Such a *light pipe* can be flexible if thin fibers rather than thick rods are used. If a bundle of parallel fibers is used to construct an optical transmission line, images can be transferred from one point to another.

This technique is used in a technology known as *fiber optics.* Very little light intensity is lost in these fibers as a result of reflections on the sides. Any loss of intensity is due essentially to reflections from the two ends and absorption by the fiber material. Fiber-optic devices are particularly useful for viewing images produced at inaccessible locations. For example, a fiber-optic cable can be threaded through the esophagus and into the stomach to look for ulcers.

The field of fiber optics is also being increasingly used in telecommunications because the fibers can carry much higher volumes of telephone calls and computer signals than can electrical wires. ⭐TEKS 8A, 3C

ATMOSPHERIC REFRACTION

We see an example of refraction every day: the sun can be seen even after it has passed below the horizon. Rays of light from the sun strike Earth's atmosphere and are bent because the atmosphere has an index of refraction different from that of the near-vacuum of space. The bending in this situation is gradual and continuous because the light moves through layers of air that have a continuously changing index of refraction. When the rays reach the observer, the eye follows them back along the direction from which they appear to have come. ⭐TEKS 3A

Refracted light produces mirages

The *mirage* is another phenomenon of nature produced by refraction in the atmosphere. A mirage can be observed when the ground is so hot that the air directly above it is warmer than the air at higher elevations. The desert is, of course, a region in which such circumstances prevail, but mirages are also seen on heated roadways during the summer.

These layers of air at different heights above Earth have different densities and different refractive indices. The effect this can have is pictured in **Figure 15-14.** In this situation, the observer sees a tree in two different ways. One group of light rays reaches the observer by the straight-line path *A,* and the eye traces these rays back to see the tree in the normal fashion. In addition, a second group of rays travels along the curved path *B.* These rays are directed toward the ground and are then bent as a result of refraction. Consequently, the observer also sees an inverted image of the tree by tracing these rays back to the point at which they appear to have originated. ★TEKS **3A, 8A**

Because both an upright image and an inverted image are seen when the image of a tree is observed in a reflecting pool of water, the observer subconsciously calls upon this past experience and concludes that a pool of water must be in front of the tree.

Mirages also give the appearance of water even when no object is nearby. For instance, motorists driving along the road often see what appear to be wet spots on the road. In this case, light rays from the blue sky above are refracted by the warm air next to the dark, hot road.

DISPERSION

An important property of the index of refraction is that its value in anything but a vacuum depends on the wavelength of light. This phenomenon, described briefly in Chapter 14, is called **dispersion.** Since the index of refraction is a function of wavelength, Snell's law indicates that incoming light of different wavelengths is bent at different angles as it moves into a refracting material. As mentioned in Section 15-1, the index of refraction decreases with increasing wavelength. For instance, blue light ($\lambda \approx 470$ nm) bends more than red light ($\lambda \approx 650$ nm) when passing into a refracting material. ★TEKS **8A**

White light passed through a prism produces a visible spectrum

To understand how dispersion can affect light, consider what happens when light strikes a prism, as in **Figure 15-15.** Because of dispersion, the blue component of the incoming ray is bent more than the red component, and the rays that emerge from the second face of the prism fan out in a series of colors known as a *visible spectrum.* As mentioned in Chapter 14, these colors, in order of decreasing wavelength, are red, orange, yellow, green, blue, and violet.

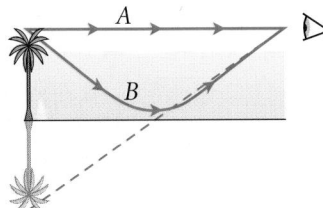

Figure 15-14

A mirage is produced by the bending of light rays in the atmosphere when there are large temperature differences between the ground and the air.

dispersion

the process of separating polychromatic light into its component wavelengths

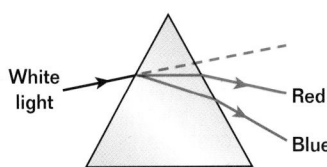

Figure 15-15

When white light enters a prism, the blue light is bent more than the red, and the prism disperses the white light into its various spectral components.

📡 internet**connect**

SCiLINKS

NSTA

TOPIC: Fiber optics
GO TO: www.scilinks.org
*sci*LINKS CODE: HF2154

Rainbow

Purpose Create a visual, rainbow-like display.

Materials overhead projector; large, clear plastic cup with steeply sloped sides filled with water

Procedure Place the plastic cup filled with water on the overhead projector. Upward light rays from the projector are at an angle of incidence with the steep sides; refraction separates the colors, causing them to form a rainbow on the ceiling. Point out that this demonstration does not model how a true rainbow is formed because the cup is not spherical.

internet connect

SC*i*LINKS.
NSTA
TOPIC: Dispersion of light
GO TO: www.scilinks.org
*sci*LINKS CODE: HF2155

Rainbows are created by dispersion of light in water droplets

The dispersion of light into a spectrum is demonstrated most vividly in nature by a rainbow, often seen by an observer positioned between the sun and a rain shower. When a ray of sunlight strikes a drop of water in the atmosphere, it is first refracted at the front surface of the drop, with the violet light refracting the most and the red light the least. Then at the back surface of the drop, the light is

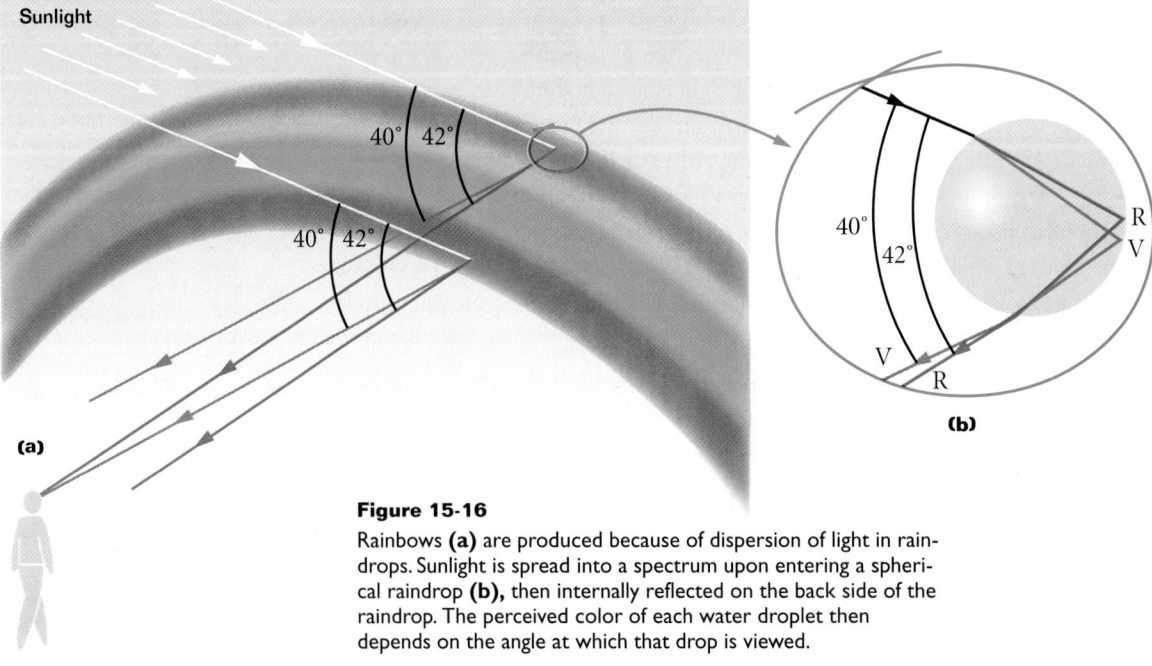

Figure 15-16
Rainbows **(a)** are produced because of dispersion of light in rain-drops. Sunlight is spread into a spectrum upon entering a spherical raindrop **(b)**, then internally reflected on the back side of the raindrop. The perceived color of each water droplet then depends on the angle at which that drop is viewed.

reflected and returns to the front surface, where it again undergoes refraction as it moves from water into air. The rays leave the drop so that the angle between the incident white light and the returning violet ray is 40° and the angle between the white light and the returning red ray is 42°, as shown in **Figure 15-16(b).** ⭐TEKS **3A, 8A**

Now consider **Figure 15-16(a).** When an observer views a raindrop high in the sky, the red light reaches the observer, but the violet light, like the other spectral colors, passes over the observer because it deviates from the path of the white light more than the red light does. Hence, the observer sees this drop as being red. Similarly, a drop lower in the sky would direct violet light toward the observer and appear to be violet. (The red light from this drop would strike the ground and not be seen.) The remaining colors of the spectrum would reach the observer from raindrops lying between these two extreme positions.

Note that rainbows are most commonly seen above the horizon, where the ends of the rainbow disappear into the ground. However, if an observer is at an elevated vantage point, such as on an airplane or at the rim of a canyon, a complete circular rainbow can be seen.

LENS ABERRATIONS

One of the basic problems of lenses and lens systems is the imperfect quality of the images. The simple theory of mirrors and lenses assumes that rays make small angles with the principal axis and that all rays reaching the lens or mirror from a point source are focused at a single point, producing a sharp image. Clearly, this is not always true in the real world. Where the approximations used in this theory do not hold, imperfect images are formed. ⊛TEKS 8A

As with spherical mirrors, *spherical aberration* occurs for lenses also. It results from the fact that the focal points of light rays far from the principal axis of a spherical lens are different from the focal points of rays with the same wavelength passing near the axis. Rays near the middle of the lens are focused farther from the lens than rays at the edges.

Another type of aberration, called **chromatic aberration,** arises from the wavelength dependence of refraction. Because the index of refraction of a material varies with wavelength, different wavelengths of light are focused at different focal points by a lens. For example, when white light passes through a lens, violet light is refracted more than red light, as shown in **Figure 15-17;** thus, the focal length for red light is greater than that for violet light. Other colors' wavelengths have intermediate focal points. Because a diverging lens has the opposite shape, the chromatic aberration for a diverging lens is opposite that for a converging lens. Chromatic aberration can be greatly reduced by the use of a combination of converging and diverging lenses made from two different types of glass.

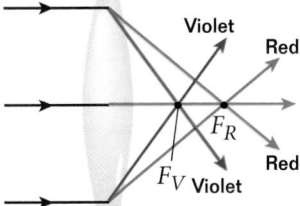

Figure 15-17
Because of dispersion, white light passing through a converging lens is focused at different focal points for each wavelength of light. (The angles in this figure are exaggerated for clarity.)

chromatic aberration

the focusing of different colors of light at different distances behind a lens

---

Section Review   ⊛TEKS 2C, 8A

1. Find the critical angle for light traveling from water ($n = 1.333$) into ice ($n = 1.309$).

2. Which of the following describe places where a mirage is likely to appear?
   a. above a warm lake on a warm day
   b. above an asphalt road on a hot day
   c. above a ski slope on a cold day
   d. above the sand on a beach on a hot day
   e. above a black car on a sunny day

3. When white light passes through a prism, which will be bent more, the red or green light?

4. **Physics in Action** After a storm, a man walks out onto his porch. Looking to the east, he sees a rainbow that has formed above his neighbor's house. What time of day is it, morning or evening?

CHAPTER 15
Summary

KEY TERMS

chromatic aberration (p. 585)

critical angle (p. 580)

dispersion (p. 583)

index of refraction (p. 564)

lens (p. 568)

refraction (p. 562)

total internal reflection (p. 580)

KEY IDEAS

Section 15-1 Refraction
- According to Snell's law, as a light ray travels from one medium into another medium where its speed is different, the light ray will change its direction unless it travels along the normal.
- When light passes from a medium with a smaller index of refraction to one with a larger index of refraction, the ray bends towards the normal. For the opposite situation, the ray bends away from the normal.

Section 15-2 Thin lenses
- The image produced by a converging lens is real and inverted when the object is outside the focal point and virtual and upright when the object is inside the focal point. Diverging lenses always produce upright, virtual images.
- The location of an image created by a lens can be found using either a ray diagram or the thin-lens equation.

Section 15-3 Optical phenomena
- Total internal reflection can occur when light attempts to move from a material with a higher index of refraction to one with a lower index of refraction. If the angle of incidence of a ray is greater than the critical angle, the ray is totally reflected at the boundary.
- Mirages and the visibility of the sun after it has physically set are natural phenomena that can be attributed to refraction of light in Earth's atmosphere.

Variable symbols

	Quantities		Units
θ_i	angle of incidence	°	degrees
θ_r	angle of refraction	°	degrees
n	index of refraction		
p	distance from object to lens	m	meters
q	distance from image to lens	m	meters
h'	image height	m	meters
h	object height	m	meters
θ_c	critical angle	°	degrees

CHAPTER 15
Review and Assess

REFRACTION AND SNELL'S LAW

Review questions

1. Does a light ray traveling from one medium into another always bend toward the normal?

2. As light travels from a vacuum ($n = 1$) to a medium such as glass ($n > 1$), does its wavelength change? Does its velocity change?

3. What is the relationship between the velocity of light and the index of refraction of a transparent substance?

4. Why does a clear stream always appear to be shallower than it actually is?

5. What are the three conditions that must be met for refraction to occur?

Conceptual questions

6. Two colors of light (X and Y) are sent through a glass prism, and X is bent more than Y. Which color travels more slowly in the prism?

7. Why does an oar appear to be bent when part of it is in the water?

8. A friend throws a coin into a pool. You close your eyes and dive toward the spot where you saw it from the edge of the pool. When you reach the bottom, will the coin be in front of you or behind you?

9. The level of water in a clear glass container is easily observed with the naked eye. The level of liquid helium in a clear glass container is extremely difficult to see with the naked eye. Explain why.

Practice problems

10. Light passes from air into water at an angle of incidence of 42.3°. Determine the angle of refraction in the water.
(See Sample Problem 15A.)

11. A ray of light enters the top of a glass of water at an angle of 36° with the vertical. What is the angle between the refracted ray and the vertical?
(See Sample Problem 15A.)

12. A narrow ray of yellow light from glowing sodium ($\lambda_0 = 589$ nm) traveling in air strikes a smooth surface of water at an angle of $\theta_i = 35.0°$. Determine the angle of refraction, θ_r.
(See Sample Problem 15A.)

13. A ray of light traveling in air strikes a flat 2.00 cm thick block of glass ($n = 1.50$) at an angle of 30.0° with the normal. Trace the light ray through the glass, and find the angles of incidence and refraction at each surface.
(See Sample Problem 15A.)

14. The light ray shown in **Figure 15-18** makes an angle of 20.0° with the normal line at the boundary of linseed oil and water. Determine the angles θ_1 and θ_2. Note that $n = 1.48$ for linseed oil.
(See Sample Problem 15A.)

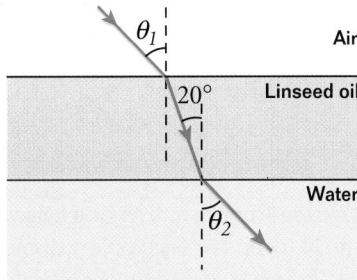

Figure 15-18

RAY DIAGRAMS AND THIN LENSES

Review questions

15. Which type of lens can focus the sun's rays?

16. Why is no image formed when an object is at the focal point of a converging lens?

ANSWERS TO

Chapter 15
Review and Assess

1. no, not when $n_i > n_r$

2. Yes, its wavelength gets shorter; Yes, its velocity gets slower.

3. $n = \dfrac{c}{v}$, where c is the speed of light in a vacuum

4. Light rays from the bottom bend away from the normal so that the image is closer to the observer.

5. $\theta_i \neq 0$, $n_i \neq n_r$, both media must be transparent

6. X

7. The image of the oar underwater is closer to the observer.

8. behind—when you are outside the water, the image of the coin appears farther away than the actual coin

9. n of liquid helium is approximately equal to n of air

10. 30.3°

11. 26°

12. 25.5°

13. 30.0°, 19.5°, 19.5°, 30.0°

14. $\theta_1 = 30.4°$, $\theta_2 = 22.3°$

15. converging

16. Rays are refracted parallel to one another.

17. a. object outside F
 b. object inside F
 c. object outside F
 d. object inside F
 e. object inside $2F$
 f. object outside $2F$

18. a. never **b.** always **c.** never
 d. always **e.** never **f.** always

19. Light from a point image at ∞ will enter the lens parallel to the principal axis and converge at F; Focus sunlight on the ground and measure f.

20. longer

21. converging lens; outside

22. The image produced by the objective lens is inside F_e. Therefore, the eyepiece produces a virtual image of a real, inverted image.

23. yes, because $n_{ice} > n_{air}$

24. a. -13.3 cm, $M = 0.332$; virtual, upright
 b. -10.0 cm, $M = 0.500$; virtual, upright
 c. -6.67 cm, $M = 0.667$; virtual, upright

25. 3.40; upright

26. a. 40.0 cm, $M = -1.00$; real, inverted
 b. -20.0 cm, $M = 2.0$; virtual, upright

27. no; $n_{air} < n_{water}$

28. The air next to the ground must be hotter than the air above it.

29. Light from the blue sky is refracted upward as it passes close to the ground.

30. Violet light is deviated more than red light as it passes through the drops of water.

31. a. chromatic
 b. spherical
 c. spherical
 d. spherical

17. Consider the image formed by a thin converging lens. Under what conditions will the image be
 a. inverted?
 b. upright?
 c. real?
 d. virtual?
 e. larger than the object?
 f. smaller than the object?

18. Repeat a–f of item 17 for a thin diverging lens.

19. Explain this statement: The focal point of a converging lens is the location of an image of a point object at infinity. Based on this statement, can you think of a quick method for determining the focal length of a positive lens?

Conceptual questions

20. If a glass converging lens is submerged in water, will its focal length be longer or shorter than when the lens is in air?

21. In order to get an upright image, slides must be placed upside down in a slide projector. What type of lens must the slide projector have? Is the slide inside or outside the focal point of the lens?

22. If there are two converging lenses in a compound microscope, why is the image still inverted?

23. In a Jules Verne novel, a piece of ice is shaped into the form of a magnifying lens to focus sunlight and thereby start a fire. Is this possible?

Practice problems

24. An object is placed in front of a diverging lens with a focal length of 20.0 cm. For each object distance, find the image distance and the magnification. Describe each image.
 a. 40.0 cm
 b. 20.0 cm
 c. 10.0 cm
 (See Sample Problem 15B.)

25. A person looks at a gem using a converging lens with a focal length of 12.5 cm. The lens forms a virtual image 30.0 cm from the lens. Determine the magnification. Is the image upright or inverted? (See Sample Problem 15B.)

26. An object is placed in front of a converging lens with a focal length of 20.0 cm. For each object distance, find the image distance and the magnification. Describe each image.
 a. 40.0 cm
 b. 10.0 cm
 (See Sample Problem 15B.)

TOTAL INTERNAL REFLECTION, ATMOSPHERIC REFRACTION, AND ABERRATIONS

Review questions

27. Is it possible to have total internal reflection for light incident from air on water? Explain.

28. What are the conditions necessary for the occurrence of a mirage?

29. On a hot day, what is it that we are seeing when we observe a "water on the road" mirage?

30. Why does the arc of a rainbow appear with red colors on top and violet colors on the bottom?

31. What type of aberration is involved in each of the following situations?
 a. The edges of the image appear reddish.
 b. The central portion of the image cannot be clearly focused.
 c. The outer portion of the image cannot be clearly focused.
 d. The central portion of the image is enlarged relative to the outer portions.

Conceptual questions

32. A laser beam passing through a nonhomogeneous sugar solution follows a curved path. Explain.

33. On a warm day, the image of a boat floating on cold water appears above the boat. Explain.

34. Explain why a mirror cannot give rise to chromatic aberration.

35. Why does a diamond show flashes of color when observed under ordinary white light?

Practice problems

36. Calculate the critical angle for light going from glycerine into air.
(See Sample Problem 15C.)

37. Assuming that $\lambda = 589$ nm, calculate the critical angles for the following materials when they are surrounded by air:
   a. zircon
   b. fluorite
   c. ice
(See Sample Problem 15C.)

38. Light traveling in air enters the flat side of a prism made of crown glass ($n = 1.52$), as shown in **Figure 15-19**. Will the light pass through the other side of the prism or will it be totally internally reflected? Show your work.
(See Sample Problem 15C.)

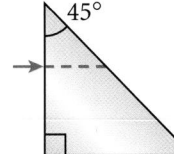

Figure 15-19

MIXED REVIEW

39. The angle of incidence and the angle of refraction for light going from air into a material with a higher index of refraction are 63.5° and 42.9°, respectively. What is the index of refraction of this material?

40. A person shines a light at a friend who is swimming underwater. If the ray in the water makes an angle of 36.2° with the normal, what is the angle of incidence?

41. What is the index of refraction of a material in which the speed of light is 1.85×10^8 m/s? Look at the indices of refraction in **Table 15-1** to identify this material.

42. Light moves from flint glass into water at an angle of incidence of 28.7°.
   a. What is the angle of refraction?
   b. At what angle would the light have to be incident to give an angle of refraction of 90.0°?

43. A magnifying glass has a converging lens of focal length 15.0 cm. At what distance from a nickel should you hold this lens to get an image with a magnification of +2.00?

Figure 15-20

44. The image of the United States postage stamps in **Figure 15-20** is 1.50 times the size of the actual stamps in front of the lens. Determine the focal length of the lens if the distance from the lens to the stamps is 2.84 cm.

45. Where must an object be placed to have a magnification of 2.00 in each of the following cases? Show your work.
   a. a converging lens of focal length 12.0 cm
   b. a diverging lens of focal length 12.0 cm

46. A diverging lens is used to form a virtual image of an object. The object is 80.0 cm in front of the lens, and the image is 40.0 cm in front of the lens. Determine the focal length of the lens.

47. A microscope slide is placed in front of a converging lens with a focal length of 2.44 cm. The lens forms an image of the slide 12.9 cm from the slide.
   a. How far is the lens from the slide if the image is real?
   b. How far is the lens from the slide if the image is virtual?

48. Where must an object be placed to form an image 30.0 cm from a diverging lens with a focal length of 40.0 cm? Determine the magnification of the image.

49. The index of refraction for red light in water is 1.331, and that for blue light is 1.340. If a ray of white light traveling in air enters the water at an angle of incidence of 83.0°, what are the angles of refraction for the red and blue components of the light?

50. A ray of light traveling in air strikes the surface of mineral oil at an angle of 23.1° with the normal to the surface. If the light travels at 2.17×10^8 m/s through the oil, what is the angle of refraction? (Hint: Remember the definition of the index of refraction.)

32. As ρ changes, the speed of the light through it changes. Thus, the light is continually refracted as n changes.

33. Rays initially moving upward are bent because T increases with height.

34. θ' does not depend on the wavelength of the light.

35. Light entering the diamond is dispersed. Each color is totally internally reflected until $\theta_i < \theta_c$.

36. 42.8°

37. a. 31.3°
   b. 44.2°
   c. 49.8°

38. It will be totally internally reflected because θ_i (45°) > θ_c (41.1°).

39. 1.31

40. 51.9°

41. 1.62; carbon disulfide

42. a. 36.7°
   b. 53.4°

43. 7.50 cm

44. 8.55 cm

45. a. 6.00 cm
   b. A diverging lens cannot form an image larger than the object.

46. −80.0 cm

47. a. 3.01 cm
   b. 2.05 cm

48. 120 cm; 0.25

49. blue: 47.8°, red: 48.2°

50. 16.5°

51. 48.8°

52. 67°

53. 4.54 m

54. 110.6°

55. $\dfrac{10}{9}f$

56. a. 40.8°
   b. 60.6°

57. a. 24.7°
   b. It will pass through the bottom surface because $\theta_i < \theta_c$ ($\theta_c = 41.8°$).

58. 1.3

59. 1.38

60. 36.6°

51. A ray of light traveling in air strikes the surface of a liquid. If the angle of incidence is 30.0° and the angle of refraction is 22.0°, find the critical angle for light traveling from the liquid back into the air.

52. The laws of refraction and reflection are the same for sound and for light. The speed of sound is 340 m/s in air and 1510 m/s in water. If a sound wave that is traveling in air approaches a flat water surface with an angle of incidence of 12.0°, what is the angle of refraction?

53. A jewel thief decides to hide a stolen diamond by placing it at the bottom of a crystal-clear fountain. He places a circular piece of wood on the surface of the water and anchors it directly above the diamond at the bottom of the fountain, as shown in **Figure 15-21.** If the fountain is 2.00 m deep, find the minimum diameter of the piece of wood that would prevent the diamond from being seen from outside the water.

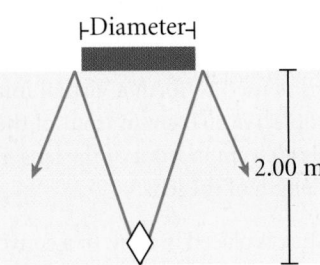

Figure 15-21

54. A ray of light traveling in air strikes the surface of a block of clear ice at an angle of 40.0° with the normal. Part of the light is reflected and part is refracted. Find the angle between the reflected and refracted light.

55. An object's distance from a converging lens is 10 times the focal length. How far is the image from the lens? Express the answer as a fraction of the focal length.

56. A fiber-optic cable used for telecommunications has an index of refraction of 1.53. For total internal reflection of light inside the cable, what is the minimum angle of incidence to the inside wall of the cable if the cable is in the following:
   a. air
   b. water

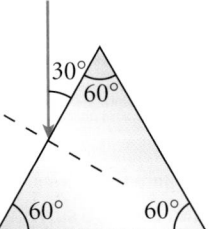

Figure 15-22

57. A ray of light traveling in air strikes the midpoint of one face of an equiangular glass prism ($n = 1.50$) at an angle of exactly 30°, as shown in **Figure 15-22.**
   a. Trace the path of the light ray through the glass and find the angle of incidence of the ray at the bottom of the prism.
   b. Will the ray pass through the bottom surface of the prism or will it be totally internally reflected?

58. Light strikes the surface of a prism, $n = 1.8$, as shown in **Figure 15-23.** If the prism is surrounded by a fluid, what is the maximum index of refraction of the fluid that will still cause total internal reflection within the prism?

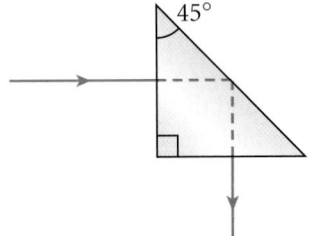

Figure 15-23

59. A fiber-optic rod consists of a central strand of material surrounded by an outer coating. The interior portion of the rod has an index of refraction of 1.60. If all rays striking the interior walls of the rod with incident angles greater than 59.5° are subject to total internal reflection, what is the index of refraction of the coating?

60. A flashlight on the bottom of a 4.00 m deep swimming pool sends a ray upward and at an angle so that the ray strikes the surface of the water 2.00 m from the point directly above the flashlight. What angle (in air) does the emerging ray make with the water's surface? (Hint: To determine the angle of incidence, consider the right triangle formed by the light ray, the pool bottom, and the imaginary line straight down from where the ray strikes the surface of the water.)

61. A submarine is 325 m horizontally out from the shore and 115 m beneath the surface of the water. A laser beam is sent from the submarine so that it strikes the surface of the water at a point 205 m from the shore. If the beam strikes the top of a building standing directly at the water's edge, find the height of the building. (Hint: To determine the angle of incidence, consider the right triangle formed by the light beam, the horizontal line drawn at the depth of the submarine, and the imaginary line straight down from where the beam strikes the surface of the water.)

62. A laser beam traveling in air strikes the midpoint of one end of a slab of material as shown in **Figure 15-24.** The index of refraction of the slab is 1.48. Determine the number of internal reflections of the laser beam before it finally emerges from the opposite end of the slab.

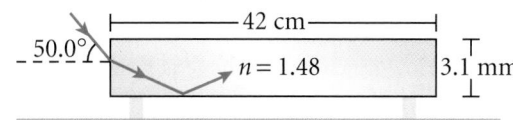

Figure 15-24

Alternative Assessment

Performance assessment

1. Interview an optometrist, optician, or ophthalmologist. Find out what equipment and tools they use. What kinds of eye problems are they able to correct? What training is necessary for each career?

2. Obtain permission to use a microscope and slides from your school's biology teacher. Identify the optical components (lenses, mirror, object, and light source) and knobs. Find out how they function at different magnifications and what adjustments must be made to obtain a clear image. Sketch a ray diagram for the microscope's image formation. Estimate the size of the images you see, and calculate the approximate size of the actual cells or microorganisms you observe. How closely do your estimates match the magnification indicated on the microscope?

3. Construct your own telescope with mailing tubes (one small enough to slide inside the other), two lenses, cardboard disks for mounting the lenses, glue, and masking tape. Test your instrument at night. Try to combine different lenses and explore ways to improve your telescope's performance. Keep records of your results to make a brochure documenting the development of your telescope.

Portfolio projects

4. Research how phone, television, and radio signals are transmitted over long distances through fiber-optic devices. Obtain information from companies that provide telephone or cable television service. What materials are fiber-optic cables made of? What are their most important properties? Are there limits on the kind of light that travels in these cables? What are the advantages of fiber-optic technology over broadcast transmission? Produce a brochure or informational video to explain this technology to consumers.

5. When the Indian physicist Venkata Raman first saw the Mediterranean Sea, he proposed that its blue color was due to the structure of water molecules rather than to the scattering of light from suspended particles. Later, he won the Nobel Prize for work relating to the implications of this hypothesis. Research Raman's life and work. Find out about his background and the challenges and opportunities he met on his way to becoming a physicist. Create a presentation about him in the form of a report, poster, short video, or computer presentation.

---

61. 58.0 m
62. 81 internal reflections

Alternative Assessment
ANSWERS

Performance assessment

1. Interview questions should demonstrate an understanding that refraction of light underlies the physiology of vision and corrective procedures.

2. Students' diagrams should indicate that microscopes have two lenses, an objective lens and an eyepiece. The magnification of the lenses multiplied together equals the net magnification.

3. Students should recognize that having one tube sliding within another will allow for easy adjustments to the distance between lenses when focusing.

 **4.** Students should recognize that one advantage of fiber optic transmission over broadcast technology is that absorption and dispersion of the signal are limited.

5. Raman (1888–1970) was the first Asian to win the Nobel Prize in one of the sciences. At that time in India, there were no opportunities for becoming a career physicist. Like Einstein, he took another job and studied science in his spare time. He later obtained a position at the University of Calcutta.

63. a. 4.83 cm

    b. The lens must be moved 0.12 cm.

64. 1.80 cm

65. 1.90 cm

63. A nature photographer is using a camera that has a lens with a focal length of 4.80 cm. The photographer is taking pictures of ancient trees in a forest and wants the lens to be focused on a very old tree that is 10.0 m away.

    a. How far must the lens be from the film in order for the resulting picture to be clearly focused?

    b. How much would the lens have to be moved to take a picture of another tree that is only 1.75 m away?

64. The distance from the front to the back of your eye is approximately 1.90 cm. If you can see a clear image of a book when it is 35.0 cm from your eye, what is the focal length of the lens/cornea system?

65. Suppose you look out the window and see your friend, who is standing 15.0 m away. To what focal length must your eye muscles adjust the lens of your eye so that you may see your friend clearly? Remember that the distance from the front to the back of your eye is about 1.90 cm.

Technology & Learning

Graphing calculators

Refer to Appendix B for instructions on downloading programs for your graphing calculator. The program "Chap15" allows you to analyze a graph of the angle of refraction versus the index of refraction for a light ray moving from air into a substance.

The relationship between the angle of refraction and the index of refraction, as you learned earlier in this chapter, is described by Snell's law:

$$n_i(\sin \theta_i) = n_r(\sin \theta_r)$$

The program "Chap15" stored on your graphing calculator makes use of Snell's law. Once the "Chap15" program is executed, your calculator will ask for the angle of incidence. The graphing calculator will use the following equation to create a graph of the index of refraction of the substance (Y_1) versus the angle of refraction (X). The relationships in this equation are the same as those in Snell's law shown above.

$$Y_1 = \sin(I)/\sin(X)$$

a. There is no mention of the index of refraction of the first medium in the equation used by your graphing calculator. Why has this factor been neglected?

Before executing the program, press MODE ▼ ▼ ▶ ENTER to set the calculator in degree mode.

Execute "Chap15" on the PRGM menu and press ENTER to begin the program. Enter the angle of incidence (shown below) and press ENTER.

The calculator will provide a graph of the angle of refraction versus the index of refraction. (There is no need to change the settings for the graph window; the window settings have been preset.)

Press TRACE and use the arrow keys to trace along the curve. The x value corresponds to the angle of refraction in degrees, and the y value corresponds to the index of refraction.

Determine the index of refraction for the following situations:

b. a light ray moving from air into a substance with an angle of incidence of 40° and an angle of refraction of 30°

c. a light ray moving from air into a substance with an angle of incidence of 40° and an angle of refraction of 25°

d. a light ray moving from air into a substance with an angle of incidence of 60° and an angle of refraction of 45°

e. a light ray moving from air into a substance with an angle of incidence of 60° and an angle of refraction of 30°

f. In items b–e, is the light bent toward or away from the normal?

Press 2nd QUIT to stop graphing. Press ENTER to input new values or CLEAR to end the program.

ANSWERS TO

Technology & Learning

a. $n_i = 1$ because the first medium is air

b. 1.30

c. 1.53

d. 1.22

e. 1.75

f. toward

CHAPTER 15
Laboratory Exercise

CONVERGING LENSES

Converging lenses can produce both real and virtual images, and they can produce images that are smaller, the same size, or larger than the object. In this experiment you will study image formation using a converging lens.

SAFETY

- **Use a hot mitt to handle resistors, light sources, and other equipment that may be hot. Allow all equipment to cool before storing it.**

- **Never put broken glass or ceramics in a regular waste container. Use a dustpan, brush, and heavy gloves to carefully pick up broken pieces, and dispose of them in a container specifically provided for this purpose.**

PREPARATION

1. Read the entire lab, and plan what measurements you will take.

2. Prepare a data table in your lab notebook with eight columns and six rows. Label this table *Focal Length*. In the first row, label the columns *Trial, Position of lens (cm), Position of object (cm), Position of image (cm), q (cm), p (cm), h_o (cm)*, and *h_i (cm)*. In the first column, label the second through sixth rows *1, 2, 3, 4,* and *5*. Above or below the data table, prepare a space to record the focal length, *f*, of the lens in centimeters.

PROCEDURE

Finding the focal length

3. Set up the meterstick, meterstick supports, image screen, and lens as shown in **Figure 15-25.** Make sure the image screen is securely held in the screen support rider to prevent its moving during the experiment. Locate and mark the point on the mounted screen where it intercepts the principal axis of the mounted lens.

4. Place the lens at the 50 cm mark and the image screen near the 70 cm mark on the meterstick. Through a laboratory window, choose an object, such as a building or tree, several hundred meters away, and point the mounted lens toward it. Stand with your back to the object, and carefully align the meterstick until the object's image is observed on the image screen.

OBJECTIVES

- Determine the focal length of a converging lens.

- Investigate the relationships between the positions of the lens and object, and the position and size of the image.

MATERIALS LIST

✔ 2 screen support riders

✔ cardboard image screen with metric scale

✔ converging lens

✔ dc power supply

✔ insulated copper wire, 2 lengths

✔ lens support rider

✔ meterstick and meterstick supports

✔ metric ruler

✔ miniature lamp and base on rider

✔ object screen

NOTE

Materials Preparation is given on pp. 560A–560B. Blank data table and sample data table are on the One-Stop Planner CD-ROM. All calculations are performed using sample data.

Planning

Recommended time:

1 lab period

Classroom organization:

▶ This lab can be performed by students working alone, but it is better with groups of two or more students.

▶ Each lab group needs a level work surface that is near an electrical outlet and away from any sources of water. Each work area must be at least 1.5 m long.

▶ **Safety warnings:** Remind students to report all breakage immediately. Students should be instructed not to look directly at a light source.

Converging Lenses Tips

◆ If power supplies are used that can supply more than the rated voltage of the bulb, make sure students do not burn out bulbs by using excessive voltage.

◆ In rooms with no serviceable windows, a lamp placed at least 8 m from the students' work areas can serve as the object for finding the focal length.

Techniques to Demonstrate

Show students the best way to find the image in the first part of the lab. Stand with your back to the object, and move the meterstick until the object's image can be seen on the screen.

✔ Checkpoints

Step 4: Make sure lenses are securely placed while students move the metersticks to locate the image.

Step 6: Students may think that the center of the screen must coincide exactly with the principal axis. Actually, as long as the principal axis is near the center of the screen, the procedure will work.

Step 7: Some students may be confused by the description of the distance in terms of the focal length of the lens. Help them find the appropriate position.

Step 10: With the lens in this position, students will not see an image on the screen.

Step 11: With the lens in this position, students will not see an image on the screen.

5. While viewing the image on the screen, move the screen along the meterstick until the image is as distinct as possible. Jot down the distance between the lens and the screen. Move the lens to the 60 cm position, and repeat the procedure using a different distant object. Average the two results for the distance between the lens and the screen. Record this average image distance as the *focal length* of the lens in your data table.

Figure 15-25

Step 4: Move the end of the meterstick until you find the position where the object, the lens, and the image screen are all in alignment.

Step 5: Find the position of the image screen where the image is most distinct, then move the lens and repeat. The average distance between the lens and the screen is the focal length.

Step 6: Use the illuminated object screen as the object for this part of the lab.

Formation of images

6. Place the illuminated object screen at one end of the meterstick. Make adjustments so that the center of the object screen coincides with the principal axis of the lens.

7. Place the lens far enough from the object screen to give an object distance greater than twice the focal length of the lens. Move the image screen along the meterstick until the image is as well defined as possible. Read and record in the data table the positions of the object, lens, and image on the meterstick. Also record the object distance, p, the image distance, q, the height of the object, h_o, and the height of the image, h_i.

8. For the second trial, place the lens at a position for an object distance exactly twice the focal length of the lens. Adjust the image screen for maximum image definition. Read and record in the data table the positions of the object, lens, and image on the meterstick. Also record the object distance, p, the image distance, q, and the height of the image, h_i. Record the height of the object, h_o, from step 7.

9. For the third trial, place the lens between one and two focal lengths from the object screen. Adjust the image screen for maximum image definition, and make all your measurements. Record all measurements in your data table.

10. For the fourth trial, place the lens exactly one focal length away from the object. Adjust the image screen for maximum image definition, and record all measurements or place X's in your data table.

11. For the fifth trial, place the lens less than one focal length away from the object. Adjust the image screen for maximum image definition, and record all measurements or place X's in your data table. Remove the image screen, and place your eye close to the lens. Look through the lens at the object. Record your observations in your lab notebook.

12. Clean up your work area. Put equipment away safely so that it is ready to be used again. Recycle or dispose of used materials as directed by your teacher.

ANALYSIS AND INTERPRETATION

Calculations and data analysis

1. Organizing data   For each trial recorded in the data table, perform the following calculations:

   a. Find the reciprocal of the object distance, p.

   b. Find the reciprocal of the image distance, q.

   c. Add the reciprocals found in **(a)** and **(b)**.

2. Analyzing data   Compare the sum of the reciprocals for each trial with the focal length of the lens. What is the relationship? Is this true for all trials? Explain.

3. Organizing data   For each trial, perform the following calculations:

   a. Find the ratio between q and p.

   b. Find the ratio between h_i and h_o.

Conclusions

4. Analyzing data   For each trial, compare the ratios found in item 3.

   a. Based on your results, what physical quantity is expressed by each of the ratios found in item 3?

   b. What is the relationship between the two ratios for each trial? Is this true for all trials? Explain.

5. Graphing data   Use a graphing calculator, computer, or graph paper. Plot a graph of the ratio between h_i and h_o (the magnification) versus p (the object distance).

6. Evaluating results   What does the shape of the graph reveal about the relationship between the magnification of the image and the distance of the object from the lens?

CHAPTER 16 PLANNING GUIDE

PACING CHART	CLASSROOM RESOURCES			
	(★) TEKS	Teacher Demonstrations	Holt Physics Transparencies	Labs (See page T52 for equipment listing for in-text labs.)
16-1 Interference 1 45-minute period ½ 90-minute block	2C, 3A, 8A	**TE** *Interference in sound waves*, p. 598 **TE** *Interference in a ripple tank*, p. 599 **TE** *How distance traveled affects interference*, p. 600 **TE** *Thin-film interference*, p. 601	**T** 80–81 **TM** 54–56	
16-2 Diffraction 3 45-minute periods 1½ 90-minute block	1A, 2B, 3A, 3C, 8A, 8B	**TE** *Waves bending around corners*, p. 604 **TE** *Diffraction and interference by a single slit*, p. 605 **TE** *Light diffraction by an obstacle: Poisson spot*, p. 606 **TE** *Effect of slit size on diffraction patterns*, p. 607 **TE** *Multiple-slit diffraction*, p. 608	**T** 82–85 **TM** 57	**PE** *Diffraction*, p. 624
16-3 Lasers 1 45-minute period ½ 90-minute block	3A, 3C, 8A, 8B, 8C	**TE** *Dancing light*, p. 613 **TE** *Interference in laser light*, p. 614 **TE** *How a CD works*, p. 617	**T** 86–87 **TM** 58	

Review and Assessment
2 45-minute periods
1 90-minute block

Resource Key

PE Pupil's Edition
TE Teacher's Edition

L Laboratory Experiments
TL Technology Lab Experiments
T Transparencies

One-Stop Planner CD-ROM contents

TM Transparency Masters
SR Section Review Worksheets
AA Alternative Assessment

PW Problem-Solving Workbook
PB Problem Bank
CTW Critical Thinking Worksheet

LABORATORY PLANNING: Diffraction, p. 624

Materials (for each lab group)
- battery eliminator with alligator clips, 6 V/0.5 A
- black card, 1 sheet
- meterstick optical bench set includes:
  - card screen with mm scale (set of 5)
  - lens or mirror support, for 4 cm lenses
  - marker/object riders
  - meterstick, plain wood, 1 m long

- pair of metal meterstick supports
- screen support riders
- miniature bulb and base
- mounted film transmission grating
- paperclips
- roll of self-adhesive tape, 0.5 in. wide
- scale and slit

ASSIGNMENT RESOURCES

Content Mastery	Critical Thinking	Problem-Solving Practice
PE 1–4, p. 603 **SR** 16-1, *Concept Review* **PE** 1–4, p. 620	**PE** 5–8, p. 620	**16A** Interference: 29 items in **PE**, **PW**, and **PB**, see **TE** pp. 602–603
PE 1–6, p. 612 **SR** 16-2, *Concept Review* **PE** 12–15, pp. 620–621	**PE** 1–2, p. 607 **PE** 16–18, p. 621	**16B** Diffraction gratings: 30 items in **PE**, **PW**, and **PB**, see **TE** pp. 609–610
PE 1–3, p. 618 **SR** 16-3, *Concept Review* **PE** 22–25, p. 621		

ASSESSMENT RESOURCES

Cumulative Review	Alternative Assessment	Traditional Assessment
SR Mixed Review, Ch. 16	**PE** 1–5, p. 622 **AA** Items for Ch. 16	Chapter 16 Test Test Generator items for Ch. 16

Scoring Rubrics for Alternative Assessment items can be found on the One-Stop Planner CD-ROM.

TECHNOLOGY RESOURCES

 CTW Segment 17 Holograms

 PE Technology and Learning, p. 623
(Alternative procedures for calculators without Flash-ROM technology are provided on the One-Stop Planner CD-ROM.)

 internet**connect**

 On-line Student Resources:
GO TO: www.scilinks.org
The following *sci*LINKS Internet resources can be found in the student text for this chapter.

TOPICS:
• Interference, p. 599 (HF2161)
• Diffraction, p. 606 (HF2162)
• Lasers, p. 614 (HF2163)
• Bar codes, p. 617 (HF2165)

 On-line Teacher Resources:
GO TO: go.hrw.com
KEYWORD: HF2 HOME
Visit the HRW Web site for a variety of resources related to this chapter.

 Smithsonian Institution
Internet Connections
Visit **www.si.edu/hrw** for additional on-line resources.

 CNN**fyi**.com
Visit **www.cnnfyi.com** for late-breaking news and current events stories selected just for you.

Materials Preparation

Set up a sample apparatus in the laboratory for students to refer to as they set up their equipment.

If you are not using complete meterstick optical bench sets, you will need all the equipment listed under the *optical bench* heading for each lab group.

Bent paperclips can be used as riders to mark the positions of the images on the meterstick scale. To make the riders, pull the small part of the paperclip forward. The bent paperclips can be hung over the edge of the meterstick at the position of the images.

Section 16-1 identifies the conditions required for interference to occur and shows how to calculate the location of bright and dark fringes in double-slit interference.

Section 16-2 describes how diffracted light waves interfere, shows how to calculate the position of fringes produced by a diffraction grating, and discusses the resolving power of optical instruments.

Section 16-3 describes how a laser produces coherent light and explores applications of lasers.

About the Illustration

The photograph shows compact laserdiscs during a stage in their manufacture. The colors visible on the discs' surfaces are characteristic of the light used in the photograph. Have students verify the ability of a compact disc to separate light into its particular spectrum by reflecting light from different sources off a compact disc's surface. For instance, sunlight is separated into all of the visible colors, while most fluorescent lighting produces the colors found in a mercury emission spectrum.

Interference and Diffraction

PHYSICS IN ACTION

The streaks of colored light you see coming from a compact disc resemble the colors that appear when white light passes through a prism. However, the compact disc does not separate light by means of refraction. Instead, the light waves undergo interference.

In interference, light waves combine to produce resultant waves that are either brighter or less bright than the component waves. For light coming from the compact disc, certain wavelengths are visible only at particular angles. Devices called diffraction gratings use the principle of interference to separate light into its component colors.

- *Why are light waves not easily observed bending around obstacles?*

- *What limits the separation of the colored light coming from a compact disc?*

CONCEPT REVIEW

Superposition principle
(Section 12-4)
Interference of waves (Section 12-4)

Tapping Prior Knowledge

Knowledge to Expect

✔ "Waves can superpose one another, bend around corners, reflect off surfaces, be absorbed and change direction when entering new materials. All these effects vary with wavelength." (AAAS's *Benchmarks for Science Literacy*, grades 9–12)

✔ "Each kind of atom or molecule can absorb and emit light only at certain wavelengths." (NRC's *National Science Education Standards*, grades 9–12)

Knowledge to Review

✔ The superposition principle: When two mechanical waves pass through the same space at the same time, their displacements at each point add. (Section 12-4)

✔ When two waves with the same frequency and amplitude overlap, the resulting wave has the same frequency as the individual waves. If the waves are in phase, the resultant wave has twice their amplitude. If they are 180° out of phase, the amplitudes cancel. (Section 12-4)

Items to Probe

✔ Preconceptions about waves: Ask students what happens when two waves (*A* and *B*) travel toward each other on a string and meet at a point where *A*'s displacement is 4 cm up and *B*'s displacement is 3 cm up. (*The displacement at that point is 7 cm up.*) What if *A* is 4 cm up and *B* is 3 cm down? (*The displacement is 1 cm up.*)

Interference in sound waves

Purpose Introduce students to interference patterns using sound waves from two coherent sources.

Materials sine-wave generator, amplifier, two speakers, tape measures, overhead projector with grid transparency

Procedure Connect the generator and the speakers to the amplifier. Place the speakers about 3 m apart so that both face the class. Have students stand in rows perpendicular to a line joining the two speakers. To reduce the effect of echoes from the walls, have students cover the ear that is opposite the speakers.

Set the generator to a frequency of about 440 Hz. Turn on the generator and amplifier, and adjust the two speakers to equal intensity. Have students slowly walk forward in their rows and listen to the intensity of the sound. Tell them to stand still when they find a location where the sound is at a minimum. Have students measure the distance between each point where sound intensity is at a minimum. Using these data, have students draw the *destructive interference* fringes on the transparency. When all of the fringe positions are recorded, turn the overhead projector on and have students note that they were standing in places where the superposition of waves resulted in destructive interference.

16-1 SECTION OBJECTIVES

- **Describe how light waves interfere with each other to produce bright and dark fringes.**
- **Identify the conditions required for interference to occur.**
- **Predict the location of interference fringes using the equation for double-slit interference.**

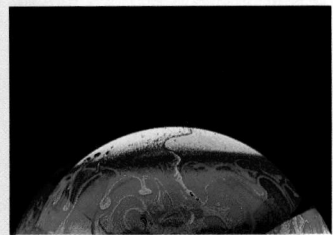

Figure 16-1
Light waves interfere to form bands of color on a soap bubble's surface.

Figure 16-2
Two waves can interfere **(a)** constructively or **(b)** destructively. In interference, energy is not lost but is instead redistributed.

16-1
Interference

LIGHT WAVES COMBINE WITH EACH OTHER

You have probably noticed the bands of color that form on the surface of a soap bubble, as shown in **Figure 16-1.** Unlike the colors that appear when light passes through a refracting substance, these colors are the result of light waves combining with each other.

Interference takes place only between waves with the same wavelength

To understand how light waves combine with each other, let us review how other kinds of waves combine. If two waves with identical wavelengths cross each other's path, they combine to form a resultant wave. This resultant wave has the same wavelength as the component waves, but according to the superposition principle, its displacement at any instant equals the sum of the displacements of the component waves. As you learned in Chapter 12, the resultant wave is the consequence of the *interference* between the two waves.

Figure 16-2 can be used to describe pairs of mechanical waves or electromagnetic waves with the same wavelength. Light waves that have the same wavelength are called *monochromatic,* which means single colored. In the case of *constructive interference,* the component waves combine to form a resultant wave with an amplitude that is greater than the amplitude of either of the individual component waves. For light, the result of constructive interference is light that is brighter than the light from the contributing waves. In the case of *destructive interference,* the resultant amplitude is less than the amplitude of the larger component wave. For light, the result of destructive interference is dimmer light or dark spots. (★)TEKS **8A, 8B**

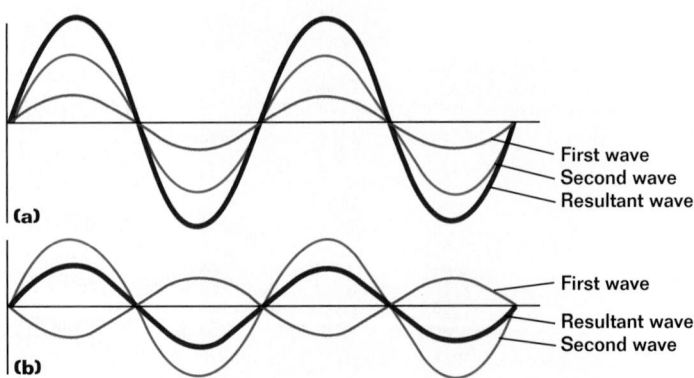

(a)
— First wave
— Second wave
— Resultant wave

(b)
— First wave
— Resultant wave
— Second wave

(a) 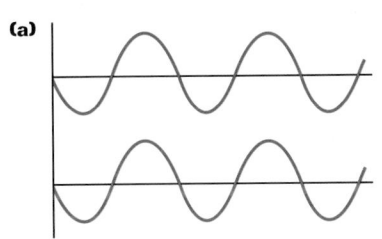 **(b)**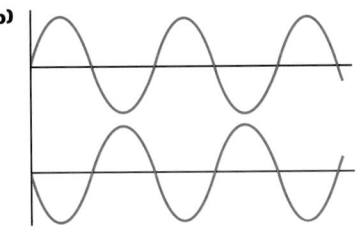

Figure 16-3
(a) The features of two waves in phase completely match, whereas **(b)** they are opposite each other in waves that are 180° out of phase.

internetconnect

SC/LINKS
NSTA
TOPIC: Interference
GO TO: www.scilinks.org
sciLINKS CODE: HF2161

Waves must have a constant phase difference for interference to be observed

For two waves to produce a stable interference pattern, the phases of the individual waves must remain unchanged relative to one another. If the crest of one wave overlaps the crest of another wave, as in **Figure 16-3(a),** the two have a phase difference of 0° and are said to be *in phase.* If the crest of one wave overlaps the trough of the other wave, as in **Figure 16-3(b),** the two waves have a phase difference of 180° and are said to be *out of phase.* ⭐**TEKS** **8A**

When the phase difference between two waves is constant and the waves do not shift relative to each other as time passes, the waves are said to have **coherence.** Sources of such waves are said to be *coherent.*

When two light bulbs are placed side by side, no interference is observed, even if the lights are the same color. The reason is that the light waves from one bulb are emitted independently of the waves from the other bulb. Changes occurring in the light from one bulb do not necessarily occur in the light from the other bulb. Thus, the phase difference between the light waves from the two bulbs is not constant. The light waves still interfere, but the conditions for the interference change with each phase change and, therefore, no single interference pattern is observed. Light sources of this type are said to be *incoherent.* ⭐**TEKS** **3A**

coherence

the property by which two waves with identical wavelengths maintain a constant phase relationship

DEMONSTRATING INTERFERENCE

Interference in light waves from two sources can be demonstrated in the following way. Light from a single source is passed through a narrow slit and then through two narrow parallel slits. The slits serve as a pair of coherent light sources because the waves emerging from them come from the same source. Any random change in the light emitted by the source will occur in the two separate beams at the same time.

If monochromatic light is used, the light from the two slits produces a series of bright and dark parallel bands, or *fringes,* on a distant viewing screen, as shown in **Figure 16-4.** When the light from the two slits arrives at a point on the viewing screen where constructive interference occurs, a bright fringe

Figure 16-4
An interference pattern consists of alternating light and dark fringes.

Demonstration 2

Interference in a ripple tank

Purpose Demonstrate interference patterns using two types of point sources.

Materials ripple tank, ripple generator, 2 pencils, overhead projector or light source, screen

Procedure Touch the water in the tank with a pencil point several times to generate a wave. Let students observe the wave pattern. Repeat the procedure with two pencils 20 cm apart so that the two waves vary in frequency and phase. Explain that the pencils act as point sources but that they generate waves of different phase or wavelength, or both. Use the ripple generator to create two coherent sources. Have students compare the interference pattern of this demonstration with the pattern that emerged in Demonstration 1. Point out that the speakers were also coherent sources. Explain that when the sources are coherent, they produce a stable interference pattern.

⭐**TEKS**
p. 598: 8A, 8B
p. 599: 8A, 3A

Demonstration 3

How distance traveled affects interference

Purpose Demonstrate that for constructive interference the difference between the distances traveled by two coherent wave fronts is equal to a whole number times the wavelength.

Materials sine-wave generator, amplifier, two speakers, tape measures

Procedure Set up the equipment as in Demonstration 1, and ask students to return to their positions in line. Set the generator to a frequency of about 440 Hz, and ask students to step forward and back until they find a place where the sound is at a maximum. Have them measure the distance from their location to each of the speakers and calculate the difference in distance (path difference). Given the sound's frequency and speed (330 m/s), ask students to calculate its wavelength (*0.75 m*), compare it to the path difference, and record the results on the chalkboard.

Increase the frequency to 660 Hz ($\lambda = 0.50$ m), and repeat the demonstration. Have students notice that they are now standing closer to each other.

Have students examine all the results and note that although each of them has detected a different wave crest, the path difference between the speakers and the wave crest equals a whole number times the wavelength in all cases.

★ TEKS

p. 600: 3A, 8A
p. 601: 2C, 8A, 2C, 8A

Figure 16-5
When waves of white light from two coherent sources interfere, the pattern is indistinct because different colors interfere constructively at different positions.

appears at that location. When the light from the two slits combines destructively at a point on the viewing screen, a dark fringe appears at that location.

When a white light source is used to observe interference, the situation becomes more complicated. The reason is that white light includes waves of many wavelengths. An example of a white-light interference pattern is shown in **Figure 16-5.** The interference pattern is stable or well defined at positions where there is constructive interference between light waves of the same wavelength. This explains the color bands on either side of the center band of white light. This effect also accounts for the bands of color seen on soap bubbles.

Figure 16-6 shows some of the ways that two coherent waves leaving the slits can combine at the viewing screen. When the waves arrive at the central point of the screen, as in **Figure 16-6(a),** they have traveled equal distances. Thus, they arrive in phase at the center of the screen, constructive interference occurs, and a bright fringe forms at that location. ⭐ TEKS **3A, 8A**

When the two light waves combine at a certain point off the center of the screen, as in **Figure 16-6(b),** the wave from the more distant slit must travel one wavelength farther than the wave from the nearer slit. Because the second wave has traveled exactly one wavelength farther than the first wave, the two waves are in phase when they combine at the screen. Constructive interference therefore occurs, and a second bright fringe appears on the screen.

If the waves meet midway between the locations of the two bright fringes, as in **Figure 16-6(c),** the first wave travels half a wavelength farther than the second wave. In this case, the trough of the first wave overlaps the crest of the second wave, giving rise to destructive interference. Consequently, a dark fringe appears on the viewing screen between the bright fringes.

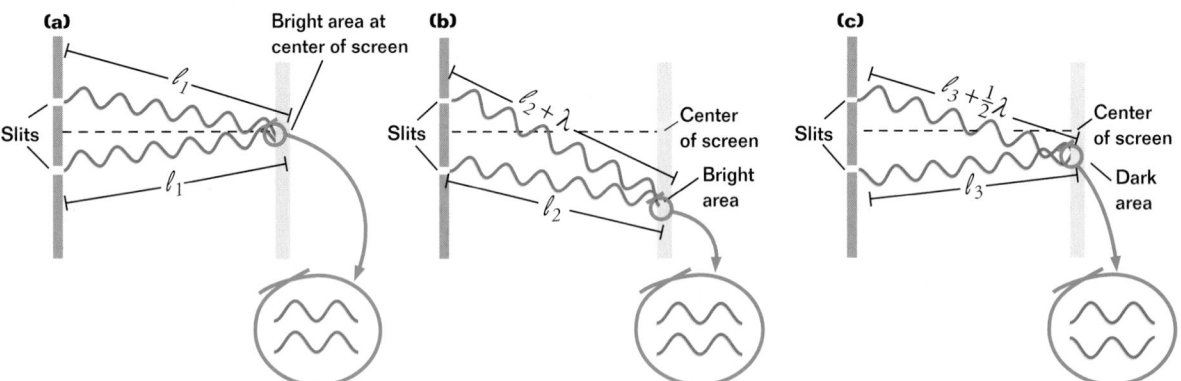

Figure 16-6
(a) When both waves of light travel the same distance (ℓ_1), they arrive at the screen in phase and interfere constructively. **(b)** If the difference between the distances traveled by the light from each source equals a whole wavelength, the waves still interfere constructively. **(c)** If the distances traveled by the light differ by a half wavelength, the waves interfere destructively.

Predicting the location of interference fringes

Consider two narrow slits that are separated by a distance d, as shown in **Figure 16-7,** and through which two coherent, monochromatic light waves, l_1 and l_2, pass. If the distance from the slits to the viewing screen is very large compared with the distance between the slits, then l_1 and l_2 are nearly parallel. As a result of this approximation, l_1 and l_2 make the same angle, θ, with the horizontal dotted lines that are perpendicular to the slit separation, d. The angle θ also indicates the position at which the waves combine with respect to the central point of the viewing screen.

The difference in the distance traveled by the two waves is called their **path difference.** Study the right triangle shown in **Figure 16-7,** and note that the path difference is equal to $d (\sin \theta)$. The value for the path difference varies with angle θ because the waves travel different distances for each fringe position on the distant viewing screen.

The value of the path difference determines whether the two waves are in or out of phase when they arrive at the viewing screen. If the path difference is either zero or some whole-number multiple of the wavelength, the two waves are in phase, and constructive interference results. The condition for bright fringes (constructive interference) is given by: (★)TEKS) **2C, 8A**

EQUATION FOR CONSTRUCTIVE INTERFERENCE

$$d (\sin \theta) = m\lambda \quad m = 0, \pm 1, \pm 2, \ldots$$

the path difference between two waves =
a whole-number multiple of the wavelength

In this equation, m is the **order number** of the fringe. The central bright fringe at $\theta = 0$ ($m = 0$) is called the *zeroth-order maximum,* or the *central maximum;* the first maximum on either side of the central maximum, which occurs when $m = \pm 1$, is called the *first-order maximum,* and so forth.

Similarly, when the path difference is an odd multiple of $\frac{1}{2}\lambda$, the two waves arriving at the screen are $180°$ out of phase, giving rise to destructive interference. The condition for dark fringes, or destructive interference, is given by the following equation: (★)TEKS) **2C, 8A**

EQUATION FOR DESTRUCTIVE INTERFERENCE

$$d (\sin \theta) = (m + \tfrac{1}{2})\lambda \quad m = 0, \pm 1, \pm 2, \ldots$$

the path difference between two waves =
an odd number of half wavelengths

If $m = 0$ in this equation, the path difference is $\frac{1}{2}\lambda$, which is the condition under which the first dark fringe forms on either side of the bright central maximum.

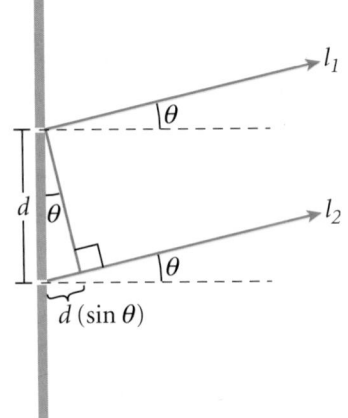

Figure 16-7
The path difference for two light waves equals $d (\sin \theta)$. In order to emphasize the path difference, the figure is not drawn to scale.

path difference

the difference in the distance traveled by two interfering light waves

order number

the number assigned to interference fringes with respect to the central bright fringe

SECTION 16-1

Visual Strategy

Figure 16-7
Point out that lines l_1 and l_2 should intersect for interference to occur. They are represented as parallel only as an approximation so that the path difference can be found mathematically.

Q If $d = 0.50$ mm and $\theta = 0.30°$, how large is the path difference? How many wavelengths does this equal for green light with a wavelength of 520 nm?

A $d (\sin \theta) = 2600$ nm, or 5 wavelengths of the light.

Demonstration 4

Thin-film interference

Purpose Demonstrate that wavelength affects the position of interference fringes.

Materials bottles of soap solution; overhead or slide projector; red, green, and blue cellophane

Procedure Explain that the path difference between reflected waves is so short that students will be able to observe interference. Have students blow bubbles and observe them in the light of the projector. Have students note how the bands of color swirl over the bubbles' surfaces. Cover the light source with one of the pieces of cellophane, and have students repeat the demonstration. This time, have students note the widths of the alternating monochromatic and black bands. After using all three cellophane covers, point out that the fringe width in each pattern varies with color (wavelength).

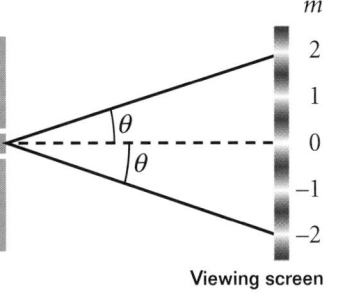

Figure 16-8
The higher-order ($m = \pm1, \pm2$) maxima appear on either side of the central maximum ($m = 0$).

Likewise, if $m = \pm1$, the path difference is $\frac{3}{2}\lambda$, which is the condition for the second dark fringe on each side of the central maximum, and so forth.

A representation of the interference pattern formed by double-slit interference is shown in **Figure 16-8.** The numbers indicate the two *maxima* (the plural of maximum) that form on either side of the central (zeroth-order) maximum. The darkest areas indicate the positions of the dark fringes, or *minima* (the plural of minimum), that also appear in the pattern.

Because the separation between interference fringes varies for light of different wavelengths, double-slit interference provides a method of measuring the wavelength of light. In fact, this technique was used to make the first measurement of the wavelength of light.

SAMPLE PROBLEM 16A

Interference ⭐TEKS 2C, 8A

PROBLEM

The distance between the two slits is 0.030 mm. The second-order bright fringe ($m = 2$) is measured on a viewing screen at an angle of 2.15° from the central maximum. Determine the wavelength of the light.

SOLUTION

1. DEFINE   **Given:** $d = 3.0 \times 10^{-5}$ m $m = 2$ $\theta = 2.15°$

Unknown: $\lambda = ?$

Diagram:

Second-order bright fringe ($m = 2$)
$d = 0.030$ mm *$\theta = 2.15°$* *Zeroth-order bright fringe ($m = 0$)*
Diagram not to scale

2. PLAN   **Choose an equation(s) or situation:** Use the equation for constructive interference.

$$d(\sin \theta) = m\lambda$$

Rearrange the equation(s) to isolate the unknown(s):

$$\lambda = \frac{d(\sin \theta)}{m}$$

3. CALCULATE   **Substitute the values into the equation(s) and solve:**

$$\lambda = \frac{(3.0 \times 10^{-5}\ \text{m})(\sin 2.15°)}{2}$$

$$\lambda = 5.6 \times 10^{-7}\ \text{m} = 5.6 \times 10^2\ \text{nm}$$

$$\boxed{\lambda = 5.6 \times 10^2\ \text{nm}}$$

CALCULATOR SOLUTION

Because the minimum number of significant figures for the data is two, the calculator answer 5.627366×10^{-7} should be rounded to two significant figures.

4. EVALUATE   As indicated by **Table 14-1** on page 521, this light is in the visible spectrum. The wavelength corresponds to light of a yellow-green color.

Interference ⭐TEKS 2C, 8A

1. A double-slit interference experiment is performed with blue-green light from an argon-gas laser (lasers will be discussed further in Section 16-3). The separation between the slits is 0.50 mm, and the first-order maximum of the interference pattern is at an angle of 0.059° from the center of the pattern. What is the wavelength of argon laser light?

2. Light falls on a double slit with slit separation of 2.02×10^{-6} m, and the first bright fringe is seen at an angle of 16.5° relative to the central maximum. Find the wavelength of the light.

3. A pair of narrow parallel slits separated by a distance of 0.250 mm are illuminated by the green component from a mercury vapor lamp ($\lambda = 546.1$ nm). Calculate the angle from the central maximum to the first bright fringe on either side of the central maximum.

4. Using the data from item 2, determine the angle between the central maximum and the second dark fringe in the interference pattern.

Section Review   ⭐TEKS 2C, 8A

1. What is the necessary condition for a path length difference between two waves that interfere constructively? destructively?

2. If white light is used instead of monochromatic light to demonstrate interference, how does the interference pattern change?

3. If the distance between two slits is 0.0550 mm, find the angle between the first-order and second-order bright fringes for yellow light with a wavelength of 605 nm.

4. Two radio antennas simultaneously transmit identical signals with a wavelength of 3.35 m, as shown in **Figure 16-9.** A radio several miles away in a car traveling parallel to the straight line between the antennas receives the signals. If the second maximum is located at an angle of 1.28° north of the central maximum for the interfering signals, what is the distance, d, between the two antennas?

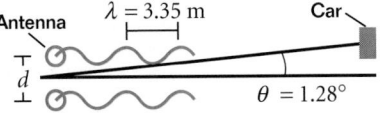

Figure 16-9

PRACTICE GUIDE 16A

Solving for:		
λ	📖 PE	Sample, 1–2; Ch. Rvw. 9–11, 29
	💿 PW	3–4
	PB	4–6
θ	📖 PE	3–4
	💿 PW	Sample, 1–2
	PB	7–10
d	📖 PE	Ch. Rvw. 28
	💿 PW	5–7
	PB	Sample, 1–3

ANSWERS TO

Practice 16A
Interference
1. 5.1×10^{-7} m $= 5.1 \times 10^2$ nm
2. 574 nm
3. 0.125°
4. 25.2°

Section Review
ANSWERS

1. a difference of an integral number of wavelengths; a difference of an odd integral number of half-wavelengths

2. It becomes blurred and the bright fringes are made up of narrow, colored bands.

3. 0.63°

4. 3.00×10^2 m

16-2
Diffraction

Section 16-2

Demonstration 5

Waves bending around corners

Purpose Demonstrate wave diffraction in a ripple tank.

Materials ripple tank, straight-wave generator, barrier, overhead projector or light source, screen

Procedure Place the barrier in the tank, and turn on the straight-wave generator. Let students examine the edges of the "shadow" of quiet water extending beyond the barrier. Explain that the waves that appear to start at the corners illustrate Huygens' principle. Ask students to sketch the patterns that they observe and to describe areas of light and shadow that would be formed if this were a light wave. Point out that the divergence of a wave from its initial path by an obstacle is called diffraction.

16-2 SECTION OBJECTIVES

- **Describe how light waves diffract around obstacles and produce bright and dark fringes.**

- **Calculate the positions of fringes for a diffraction grating.**

- **Describe how diffraction determines an optical instrument's ability to resolve images.**

diffraction

the spreading of waves into a region behind an obstruction

LIGHT WAVES BEND ⭐TEKS 8A

If you stand near the corner of a building, you can hear someone who is talking around the corner but you cannot see the person. This is because sound waves are able to bend around the corner. In a similar fashion, water waves bend around obstacles, such as the barrier shown in **Figure 16-10.** Light waves can also bend around obstacles, but because of their short wavelengths, the amount they bend is too small to be easily observed.

If light traveled only in straight lines, you would not be able to observe an interference pattern in the double-slit demonstration. Instead, you would see two thin strips of light where each slit and the source were lined up perfectly. The rest of the screen would be completely dark. The edges of the slits would appear on the screen as sharply defined shadows. But this does not happen. Some of the light bends to the right and to the left as it passes through each slit.

The bending of light as it passes through each of the two slits can be understood using Huygens' principle, which states that any point on a wave front can be treated as a point source of waves. Because each slit serves as a point source of light, the waves spread out from the slits. The result is that light deviates from a straight-line path and enters the region that would otherwise be shadowed. This divergence of light from its initial direction of travel is called **diffraction.**

In general, diffraction occurs when waves pass through small openings, around obstacles, or by sharp edges. When a wide slit (1 mm or more) is placed between a distant light source and a screen, the light produces a bright rectangle with clearly marked edges on the screen. But if the slit is narrowed,

Figure 16-10

A property of all waves is that they bend, or *diffract*, around objects.

the light begins to spread out and produce a *diffraction pattern,* like that shown in **Figure 16-11.** Like the interference fringes in the double-slit demonstration, this pattern of light and dark bands arises from the combination of light waves.

Wavelets in a wave front interfere with each other

Diffraction patterns resemble interference patterns because they also result from constructive and destructive interference. In the case of interference, it is assumed that the slits behave as point sources of light. For diffraction, the actual width of a single slit is considered.

According to Huygens' principle, each portion of a slit acts as a source of waves. Hence, light from one portion of the slit can interfere with light from another portion. The resultant intensity of the diffracted light on the screen depends on the angle, θ, that the light is diffracted.

To help you understand the single-slit diffraction pattern, consider **Figure 16-12(a),** which shows an incoming plane wave passing through a slit of width *a.* Each point within the slit is a source of Huygens wavelets. The figure is simplified by showing only five of these sources. As with double-slit interference, the viewing screen is assumed to be so far from the slit that the rays emerging from the slit are nearly parallel. At the viewing screen's midpoint, all rays from the slit travel the same distance, so a bright fringe appears.

The wavelets from the five sources can also interfere destructively when they arrive at the screen, as shown in **Figure 16-12(b).** The wave from source 1 travels the shortest distance to the screen, and the wave from source 5 travels the farthest. The first dark fringe occurs when the extra distance traveled by the wave from source 3 is one-half a wavelength. In this case, the wavelets from sources 1 and 3 are entirely out of phase and interfere destructively upon reaching the screen. In fact, at this angle, each wavelet from the upper half of the slit cancels a wavelet that travels one-half wavelength farther from the lower half, and a dark fringe results. ⊛TEKS **3A, 8A**

For angles other than those at which destructive interference completely occurs, some of the light waves remain uncanceled. At these angles light appears on the screen as part of a bright band. The brightest band appears in the pattern's center, while the bands to either side are much dimmer.

Slit width

| d | 0.8 d | 0.2 d |

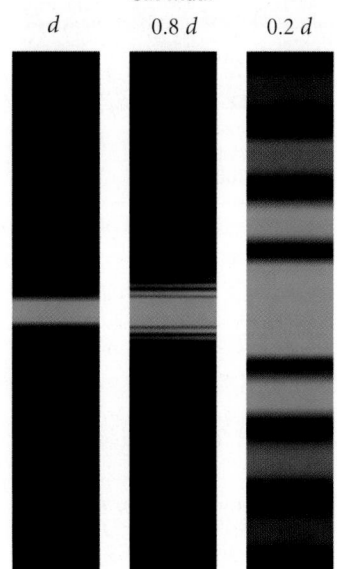

Figure 16-11
Diffraction becomes more evident as the width of the slit is narrowed. (Note: The wavelength of this light is 510 nm.)

⊛TEKS **8A**

Did you know?

Diffraction patterns are less spread out—and therefore less visible—when light passes through a wider slit. The reason is that the slit no longer resembles a point source. The light from broad sources is responsible for the sharp shadows produced by obstacles.

(a)

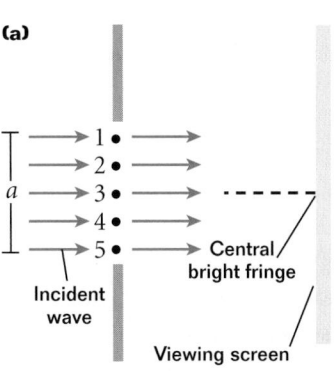

(b)

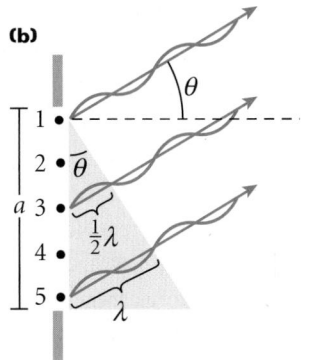

Figure 16-12
(a) By treating the light coming through the slit as a line of point sources along the slit's width, one can determine **(b)** the conditions at which destructive interference occurs between the waves in the upper half of the slit and the waves in the lower half.

Demonstration 6

Diffraction and interference by a single slit

Purpose Demonstrate diffraction and single-slit interference in a ripple tank.

Materials ripple tank, straight-wave generator, two straight barriers, overhead projector or light source, screen

Procedure Place two barriers about 10 cm from the straight-wave generator and leave a wide opening between them. Ask students to predict how these obstacles will affect the wave; then generate the wave to confirm their prediction. Ask what will change when the width of the opening decreases. Have students note that diffraction becomes more evident as the width of the slit narrows.

Tell students that you will increase the wavelength by decreasing the wave generator's frequency until interference patterns appear. (Increase the slit's width, if necessary.) Explain that each portion of the slit acts as a point source. Ask students under what conditions they might expect light to produce similar diffraction and interference patterns (*when the slit is wider—but not vastly wider—than the wavelength*).

⊛TEKS
p. 604: 8A
p. 605: 3A, 8A, 8A

Light diffraction by an obstacle: Poisson spot

Purpose Demonstrate the bright spot of light produced by interference of diffracted light around the edge of an obstacle.

Materials laser, pin with a round head, clay, screen

CAUTION *Direct the laser beam away from the students.*

Procedure Place the pin tip in the clay and place the pinhead in the path of the laser beam (far enough away that the beam is a little larger than the pinhead). Ask students to describe the pattern they expect to see on the screen. (Some students may expect a dark shadow of the pinhead; others may correctly expect an interference pattern in the shadow due to light bending around the edges of the head.) Have students note the bright spot at the center of the shadow. Explain that this experiment was crucial when the French physicist Simon Poisson challenged the wave theory of light. (Poisson argued that if the wave theory of light were true, a bright spot should appear at the center of an object's shadow. The supposed impossibility of this spot was believed to be sufficient to refute the wave theory of light.)

⭐ TEKS

p. 606: 3A, 8A
p. 607: 3A, 8A, 2C, 3A, 8A

internet**connect**

SC*i*LINKS
NSTA
TOPIC: Diffraction
GO TO: www.scilinks.org
*sci***LINKS CODE:** HF2162

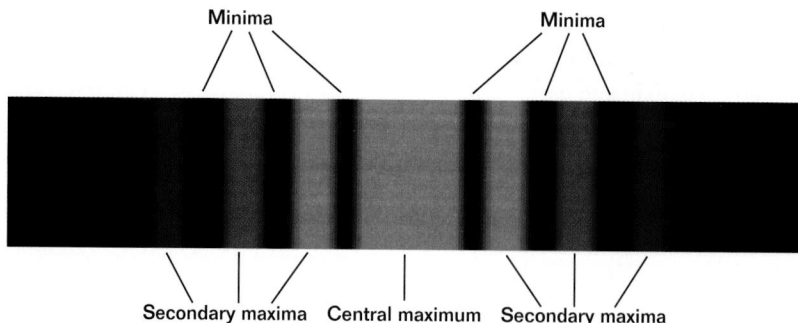

Minima   Minima

Secondary maxima   Central maximum   Secondary maxima

Figure 16-13
In a diffraction pattern, the central maximum is twice as wide as the secondary maxima.

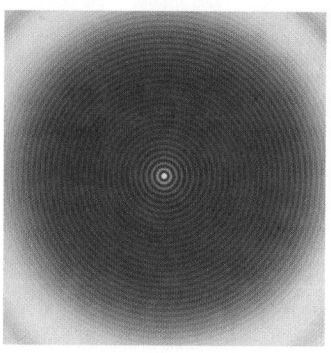

Figure 16-14
A diffraction pattern forms in the penny's shadow when light is diffracted at the penny's edge. Note the bright spot that is formed at the center of the shadow.

Figure 16-15
Compact discs disperse light into its component colors in a manner similar to that of a diffraction grating.

Light diffracted by an obstacle also produces a pattern

The diffraction pattern that results from monochromatic light passing through a single slit consists of a broad, intense central band—the *central maximum*—flanked by a series of narrower, less intense secondary bands (called *secondary maxima*) and a series of dark bands, or *minima*. An example of such a pattern is shown in **Figure 16-13.** The points at which maximum constructive interference occurs lie approximately halfway between the dark fringes. Note that the central bright fringe is twice as wide as the next brightest maxima and is about 22 times as bright. ⭐TEKS **8A**

Diffraction occurs around the edges of all objects. **Figure 16-14** shows the diffraction pattern that appears in the shadow of a penny. The pattern consists of the shadow, with a bright spot at its center, and a series of bright and dark bands of light that continue to the shadow's edge. The penny is large compared with the wavelength of the light, and a magnifying glass is required to observe the pattern.

DIFFRACTION GRATINGS

You have probably noticed that a compact disc disperses light in streaks of color. This is because the digital information (alternating pits and smooth reflecting surfaces) on the disc forms closely spaced rows. These rows of data do not reflect nearly as much light as the thin portions of the disc that separate them. These areas consist entirely of reflecting material, so light reflected from them undergoes constructive interference in certain directions. This constructive interference depends on the direction of the incoming light, the orientation of the disc, and the light's wavelength. When white light is reflected from the disc, as shown in **Figure 16-15,** each wavelength of light can be seen at a particular angle with respect to the disc's surface, causing you to see a "rainbow" of color. ⭐TEKS **3A**

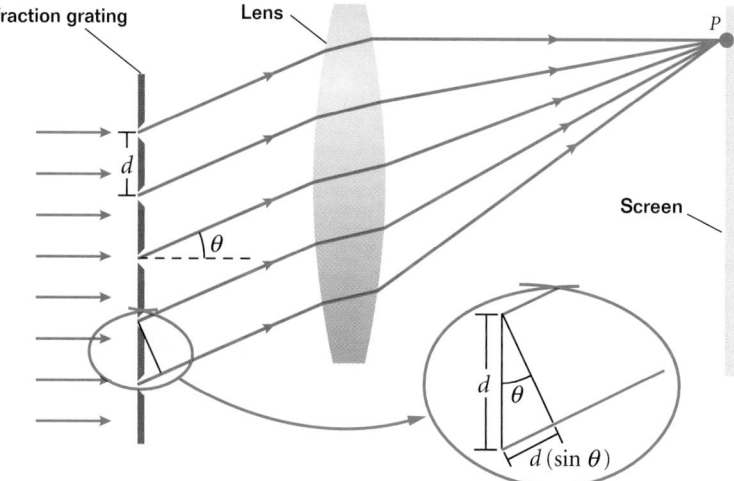

Diffraction grating
Lens
P
d
θ
Screen
d
θ
$d\,(\sin\,\theta)$

Figure 16-16

Light of a single wavelength passes through each of the slits of a diffraction grating to constructively interfere at a particular angle θ.

This phenomenon has been put to practical use in a device called a *diffraction grating*. A diffraction grating, which can be constructed to either transmit or reflect light, uses diffraction and interference to disperse light into its component colors with an effect similar to that of a glass prism. A transmission grating consists of many equally spaced parallel slits. Gratings are made by ruling equally spaced lines on a piece of glass using a diamond cutting point driven by an elaborate machine called a ruling engine. Replicas are then made by pouring liquid plastic on the grating and then peeling it off once it has set. This plastic grating is then fastened to a flat piece of glass or plastic for support.

Figure 16-16 shows a schematic diagram of a section of a diffraction grating. A monochromatic plane wave is incoming from the left, normal to the plane of the grating. The waves that emerge nearly parallel from the grating are brought together at a point *P* on the screen by the lens. The intensity of the pattern on the screen is the result of the combined effects of interference and diffraction. Each slit produces diffraction, and the diffracted beams in turn interfere with one another to produce the pattern.

For some arbitrary angle, θ, measured from the original direction of travel of the wave, the waves must travel *different* path lengths before reaching point *P* on the screen. Note that the path difference between waves from any two adjacent slits is $d\,(\sin\,\theta)$. If this path difference equals one wavelength or some integral multiple of a wavelength, waves from all slits will be in phase at *P*, and a bright line will be observed. The condition for bright line formation at angle θ is therefore given by the equation for constructive interference:

$$d\,(\sin\,\theta) = m\lambda \qquad m = 0, \pm1, \pm2, \dots$$

This equation can be used to calculate the wavelength of light if you know the grating spacing and the angle of deviation. The integer *m* is the order number for the bright lines of a given wavelength. If the incident radiation contains several wavelengths, each wavelength deviates by a specific angle, which can be determined from the equation. ⭐TEKS **2C, 3A, 8A**

⭐TEKS **3A, 8A**

Conceptual Challenge

1. Spiked stars

Photographs of stars always show spikes extending from the stars. Given that the aperture of a camera's rectangular shutter has straight edges, explain how diffraction accounts for the spikes.

2. Radio diffraction

Visible light waves are not observed diffracting around buildings or other obstacles. However, radio waves can be detected around buildings or mountains, even when the transmitter is not visible. Explain why diffraction is more evident for radio waves than for visible light.

The Language of Physics

As in double-slit interference, d represents the distance between two slits. Bright lines occur for special values of θ that exist when the path difference from adjacent slits is a whole number of wavelengths ($m = 0, \pm 1, \pm 2, \pm 3$, and so forth).

Demonstration 9

Multiple-slit diffraction

Purpose Demonstrate patterns formed by a diffraction grating and the effect of different grating line separations.

Materials laser, two optical gratings with different grating constants, screen

CAUTION *Avoid directing the laser beam toward the students.*

Procedure Shine the laser beam onto the screen, and place the optical grating directly in front of the beam. Have students note the bright central fringe (the zeroth-order maximum) and the first-order maxima, one on each side of the zeroth-order maximum.

Replace the first optical grating with the second. Ask students why there are differences in the separation between the zeroth-order and first-order maxima produced by each grating. *(Smaller line spacing produces greater separation between the zeroth-order and first-order maxima.)*

Remind students that diffraction is greatest when the size of the openings and the wavelength of waves are of the same order of magnitude.

Figure 16-17
Light is dispersed by a diffraction grating. The angle of deviation for the first-order maximum is smaller for blue light than for yellow light.

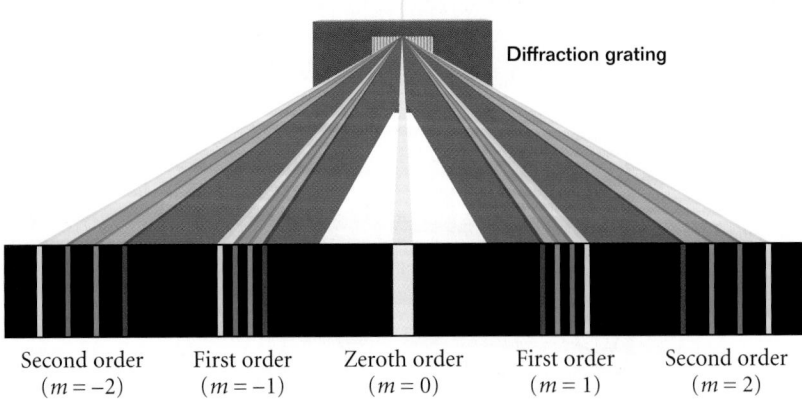

Second order ($m = -2$)	First order ($m = -1$)	Zeroth order ($m = 0$)	First order ($m = 1$)	Second order ($m = 2$)

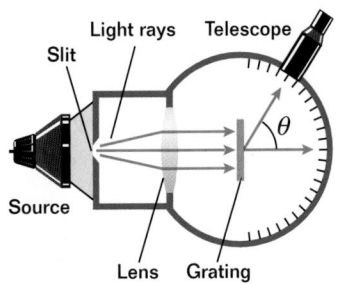

Figure 16-18
The spectrometer uses a grating to disperse the light from a source.

> **CONCEPT PREVIEW**

Spectrometers and the emission spectra they produce will be discussed in greater detail in Chapter 23.

Figure 16-19
The light from mercury vapor is passed through a diffraction grating, producing the spectrum shown.

Note in **Figure 16-17** that all wavelengths combine at $\theta = 0$, which corresponds to $m = 0$. This is called the *zeroth-order maximum.* The *first-order maximum*, corresponding to $m = \pm 1$, is observed at an angle that satisfies the relationship $\sin \theta = \lambda / d$. The *second-order maximum*, corresponding to $m = \pm 2$, is observed at an angle where $\sin \theta = 2\lambda / d$. ⊛TEKS 2C, 8A

The sharpness of the principal maxima and the broad range of the dark areas depend on the number of lines in a grating. The number of lines per unit length in a grating is the inverse of the line separation d. For example, a grating ruled with 5000 lines/cm has a slit spacing, d, equal to the inverse of this number; hence, $d = (1/5000)$ cm $= 2 \times 10^{-4}$ cm. The greater the number of lines per unit length in a grating, the less separation between the slits and the farther spread apart the individual wavelengths of light are.

Diffraction gratings are frequently used in devices called spectrometers, which separate the light from a source into its monochromatic components. A diagram of the basic components of a spectrometer is shown in **Figure 16-18.** The light to be analyzed passes through a slit and is formed into a parallel beam by a lens. The light then passes through the grating. The diffracted light leaves the grating at angles that satisfy the diffraction grating equation. A telescope with a calibrated scale is used to observe the first-order maxima and to measure the angles at which they appear. From these measurements, the wavelengths of the light can be determined and the chemical composition of the light source can be identified. An example of a spectrum produced by a spectrometer is shown in **Figure 16-19.** Spectrometers are used extensively in astronomy to study the chemical compositions and temperatures of stars, interstellar gas clouds, and galaxies. ⊛TEKS 3C, 8A

Diffraction gratings ⊛TEKS 2C, 8A

PROBLEM

Monochromatic light from a helium-neon laser ($\lambda = 632.8$ nm) shines at a right angle to the surface of a diffraction grating that contains 150 500 lines/m. Find the angles at which one would observe the first-order and second-order maxima.

SOLUTION

1. DEFINE   **Given:** $\lambda = 632.8$ nm $= 6.328 \times 10^{-7}$ m $m = 1$ and 2

$$d = \frac{1}{150\ 500\dfrac{\text{lines}}{\text{m}}} = \frac{1}{150\ 500}\,\text{m}$$

Unknown: $\theta_1 = ?$ $\theta_2 = ?$

Diagram:

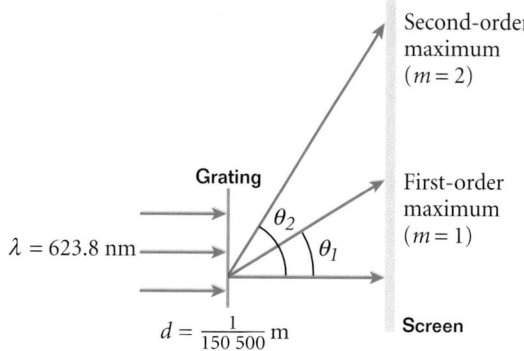

Second-order maximum ($m = 2$)

Grating

θ_2

First-order maximum ($m = 1$)

$\lambda = 623.8$ nm

θ_1

$d = \frac{1}{150\ 500}$ m   Screen

2. PLAN   **Choose an equation(s) or situation:** Use the equation for a diffraction grating.

$$d\,(\sin\theta) = m\lambda$$

Rearrange the equation(s) to isolate the unknown(s):

$$\theta = \sin^{-1}\left(\frac{m\lambda}{d}\right)$$

3. CALCULATE   **Substitute the values into the equation(s) and solve:**

For the first-order maximum, $m = 1$:

$$\theta_1 = \sin^{-1}\left(\frac{\lambda}{d}\right) = \sin^{-1}\left(\frac{6.328 \times 10^{-7}\ \text{m}}{\dfrac{1}{150\ 500}\ \text{m}}\right)$$

$$\boxed{\theta_1 = 5.465°}$$

continued on next page

PROBLEM

Diffraction gratings

Monochromatic light shines at the surface of a diffraction grating with 5.0×10^3 lines/cm. The first-order maximum is observed at a $15°$ angle. Find the wavelength.

Answer
  5.2×10^2 nm

Find the first-order and the second-order angles of diffraction observed through a 1.00×10^4 lines/cm diffraction grating with light of wavelengths 400.0 nm, 500.0 nm, and 600.0 nm.

Answers
  400.0 nm: $\theta_1 = 23.6°$,
  $\theta_2 = 53.1°$
  500.0 nm: $\theta_1 = 30.0°$,
  $\theta_2 = 90.0°$ (does not occur)
  600.0 nm: $\theta_1 = 36.9°$,
  $\sin\theta_2 = 1.2$ (does not occur)

Light of 400.0 nm wavelength is shined on a 5.0×10^3 lines/cm grating. How many diffraction lines can be observed?

Answer
  four (The fifth fringe cannot be seen because it is at an angle of 90°.)

⊛TEKS

p. 608: 2C, 8A, 3C, 8A
p. 609: 2C, 8A

First calculate sin θ explicitly. This allows you to evaluate your answer by dimensional analysis ($\sin\theta$ should have no units) and to assess whether diffraction lines of this order may exist. (The value of sin θ should be between 0 and 1.)

For $m = 2$:

$$\theta_2 = \sin^{-1}\left(\frac{2\lambda}{d}\right)$$

$$\theta_2 = \sin^{-1}\left(\frac{2(6.328 \times 10^{-7}\ \text{m})}{\dfrac{1}{150\ 500}\ \text{m}}\right)$$

$$\boxed{\theta_2 = 10.98°}$$

4. EVALUATE The second-order maximum is spread slightly more than twice as far from the center as the first-order maximum. This diffraction grating does not have high dispersion, and it can produce spectral lines up to the tenth-order maxima (where sin $\theta = 0.9524$).

PRACTICE GUIDE 16B

Solving for:		
θ	📖 **PE**	Sample, 1–2; Ch. Rvw. 19–21
	💿 **PW**	4–5
	PB	4–6
λ	💿 **PW**	Sample, 1–2
	PB	7–10
d	📖 **PE**	5*; Ch. Rvw. 28, 30
	💿 **PW**	3
	PB	3–4
m	📖 **PE**	3–4
	💿 **PW**	6–7
	PB	Sample, 1–2

PRACTICE 16B

Diffraction gratings ⊛TEKS 2C, 8A

1. A diffraction grating with 2.500×10^3 lines/cm is used to examine the sodium spectrum. Calculate the angular separation of the two closely spaced yellow lines of sodium (588.995 nm and 589.592 nm) in each of the first three orders.

2. A diffraction grating with 4525 lines/cm is illuminated by direct sunlight. The first-order solar spectrum is spread out on a white screen hanging on a wall opposite the grating.
   a. At what angle does the first-order maximum for blue light with a wavelength of 422 nm appear?
   b. At what angle does the first-order maximum for red light with a wavelength of 655 nm appear?

3. A grating with 1555 lines/cm is illuminated with light of wavelength 565 nm. What is the highest-order number that can be observed with this grating? (Hint: Remember that sin θ can never be greater than 1 for a diffraction grating.)

4. Repeat item 3 for a diffraction grating with 15 550 lines/cm that is illuminated with light of wavelength 565 nm.

5. A diffraction grating is calibrated by using the 546.1 nm line of mercury vapor. The first-order maximum is found at an angle of 21.2°. Calculate the number of lines per centimeter on this grating.

ANSWERS TO

Practice 16B
Diffraction gratings
1. 0.008°, 0.02°, 0.02°
2. a. 11.0°
   b. 17.2°
3. 11
4. 1
5. 6.62×10^3 lines/cm

DIFFRACTION AND INSTRUMENT RESOLUTION

The ability of an optical system, such as a microscope or a telescope, to distinguish between closely spaced objects is limited by the wave nature of light. To understand this limitation, consider **Figure 16-20,** which shows two light sources far from a narrow slit. The sources can be taken as two point sources that are not coherent. For example, they could be two distant stars that appear close to each other in the night sky.

If no diffraction occurred, you would observe two distinct bright spots (or images) on the screen at the far right. However, because of diffraction, each source is shown to have a bright central region flanked by weaker bright and dark rings. What is observed on the screen is the resultant from the superposition of two diffraction patterns, one from each source.

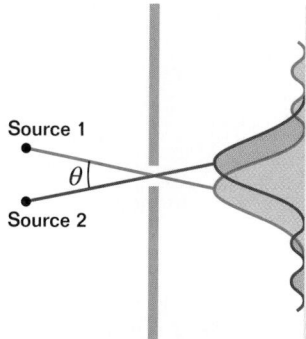

Figure 16-20
Each of two distant point sources produces a diffraction pattern.

Resolution depends on wavelength and aperture width

If the two sources are separated so that their central maxima do not overlap, as in **Figure 16-21,** their images can just be distinguished, and are said to be barely *resolved*. The angle between the resolved objects should be as small as possible for high resolution, or **resolving power.** The shorter the wavelength of the incoming light or the wider the opening, or *aperture*, through which the light passes, the smaller the angle of resolution, θ, will be and the greater the resolving power will be. For visible-light telescopes, the aperture width, D, is approximately equal to the diameter of the mirror or lens. The equation to determine the limiting angle of resolution in radians for an optical instrument with a circular aperture is as follows: (★)TEKS **2C, 3C, 8A**

$$\theta = 1.22\frac{\lambda}{D}$$

The constant 1.22 comes from the derivation of the equation for circular apertures and is absent for square apertures. The equation indicates that for light with a short wavelength, such as an X ray, a small aperture is sufficient for high resolution. On the other hand, if the wavelength of the light is long, as in the case of a radio wave, the aperture must be large in order to resolve distant objects. This explains why radio telescopes have large dishlike antennas.

Yet even with their large sizes, radio telescopes cannot resolve sources as easily as visible-light telescopes resolve visible-light sources. At the shortest radio

resolving power

the ability of an optical instrument to separate two images that are close together

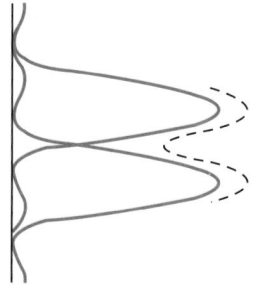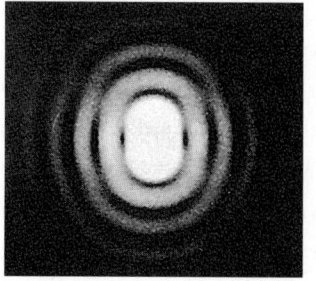

Figure 16-21
Two point sources are barely resolved if the central maxima of their diffraction patterns do not overlap.

Teaching Tip

Ask which color, blue or red, gives a better resolution with a microscope (*blue, because it has a shorter wavelength*). Point out that electron microscopes allow us to distinguish smaller details in cells than is possible with visible light because the effective wavelength of electrons used in electron microscopes typically is measured in thousandths of a nanometer. This is about 100 000 times shorter than the wavelength of red light.

Section Review
ANSWERS

1. The image of the point source casts a sharp shadow when the hole is much wider than the light's wavelength. As the hole narrows, the light waves bend more around the hole's edges, causing the image to widen. Light waves from the hole destructively interfere, producing a regular pattern of dark fringes on either side of the bright central fringe.

2. The width of the central maximum increases as the width of the slit decreases.

3. a human hair, because its size is closest to the size of the wavelength of visible light

4. orange light; Longer wavelengths are diffracted more.

5. **a.** 5.89×10^{-7} m (589 nm)
   b. 24.7°

6. yes; Ultraviolet light has a shorter wavelength than visible light, and resolving power is greater for short wavelengths.

Figure 16-22
The 27 antennas at the Very Large Array in New Mexico are used together to provide improved resolution for observing distant radio sources. The antennas can be arranged to have the resolving power of a 36 km wide radio telescope.

wavelength (1 mm), the largest single antenna for a radio telescope—the 305 m dish at Arecibo, Puerto Rico—has a resolution angle of 4×10^{-6} rad. The same resolution angle can be obtained for the longest visible light waves (700 nm) by an optical telescope with a 21 cm mirror. A Cassegrain reflector telescope with a mirror of this size can easily fit in the trunk of a car. By contrast, the radio dish at Arecibo is built into the top of a mountain.

To compensate for the poor resolution of radio waves, one can combine several radio telescopes so that they will function like a much larger telescope. An example of this is shown in **Figure 16-22.** If the radio antennas are arranged in a line and computers are used to process the signals that each antenna receives, the resolution of the radio "images" is the same as it would be if the radio telescope had a diameter of several kilometers. ⊛TEKS 3C

It should be noted that the resolving power for optical telescopes on Earth is limited by the constantly moving layers of air in the atmosphere, which blur the light from objects in space. The images from the Hubble Space Telescope are of superior quality largely because the telescope operates in the vacuum of space. Under these conditions, the actual resolving power of the telescope is close to the telescope's theoretical resolving power.

Section Review   ⊛TEKS 2C, 8A

1. A point source of light is inside a container that is opaque except for a single hole. Discuss what happens to the image of the point source projected onto a screen as the hole's width is reduced.

2. Describe the change in width of the central maximum of the single-slit diffraction pattern as the width of the slit is made smaller.

3. Which object would produce the most distinct diffraction pattern, an apple, a pencil lead, or a human hair? Explain your answer.

4. Would orange light or blue light produce a wider diffraction pattern? Explain why.

5. Light passes through a diffraction grating with 3550 lines/cm and forms a first-order maximum at an angle of 12.07°.
   a. What is the wavelength of the light?
   b. At what angle will the second maximum appear?

6. Would it be easier to resolve nearby objects if you detected them using ultraviolet radiation rather than visible light? Why?

16-3
Lasers

LASERS AND COHERENCE

At this point, you are familiar with electromagnetic radiation that is produced by glowing, or *incandescent*, light sources. This includes light from light bulbs, candle flames, or the sun. You are probably somewhat familiar with another form of light that is very different from the light produced by incandescent sources. The light produced by a **laser** has unique properties that make it very useful for many applications.

To understand how laser light is different from conventional light, consider the light produced by an incandescent light bulb, as shown in **Figure 16-23.** When electric charges move through the filament, electromagnetic waves are emitted in the form of visible light. In a typical light bulb, there are variations in the structure of the filament and in the way charges move through it. As a result, electromagnetic waves are emitted at different times from different parts of the filament. These waves have different intensities and move in different directions. The light also covers a wide range of the electromagnetic spectrum because it includes light of different wavelengths. Because so many different wavelengths exist, and because the light is changing almost constantly, the light produced is incoherent. That is, the component waves do not maintain a constant phase difference at all times. The wave fronts of incoherent light are like the wave fronts that result when rain falls on the surface of a pond. No two wave fronts are caused by the same event, and they therefore do not produce a stable interference pattern. (★)TEKS) **8A, 8B**

16-3 SECTION OBJECTIVES

- **Describe the properties of laser light.**

- **Explain how laser light has particular advantages in certain applications.**

laser

a device that produces an intense, nearly parallel beam of coherent light

Did you know?

The light from an ordinary electric lamp undergoes about 100 million (10^8) random changes every second.

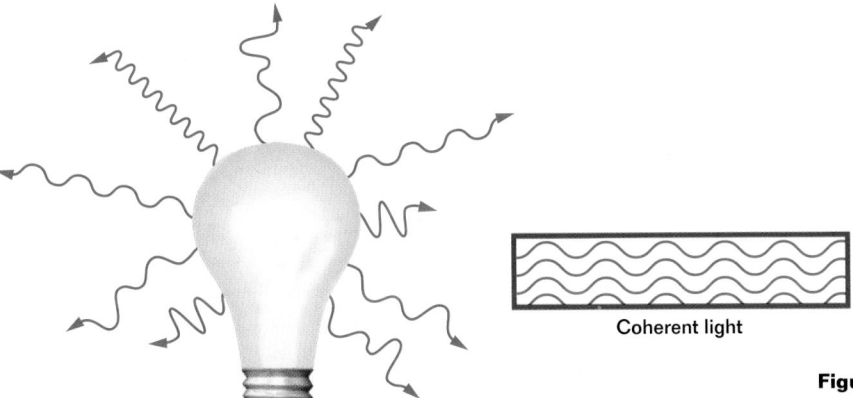

Coherent light

Noncoherent light

Figure 16-23
Unlike waves in coherent light, waves in incoherent light have changing phase relationships.

Demonstration 10

Dancing light

Purpose Demonstrate characteristics of a laser beam.

Materials cassette-tape or CD player with an open speaker, clear wrap, mirror (reflective mylar), laser, light pen

CAUTION *Avoid directing the laser beam toward students.*

Procedure Before students arrive, pull the clear wrap tightly around the speaker, and secure it with tape. Tape the mirror to the clear wrap in front of the speaker. In class, direct the laser beam so that it is reflected off the mirror attached to the speaker and onto the ceiling. Play the music, and darken the room. Replace the laser with a light pen, and have students compare the two beams. Have them note that laser light is emitted in one narrow beam in one direction and is almost monochromatic.

(★)TEKS)
p. 612: 3C, 2C, 8A
p. 613: 8A, 8B

Demonstration 11

Interference in laser light

Purpose Demonstrate interference of light waves with the same wavelength.

Materials laser, light pen, projection screen, opaque slide with two slits (to make one, spray a layer of paint on a glass plate and use a pin to etch two parallel slits about 1 mm apart through the paint)

CAUTION *Avoid directing the light from the laser at the students.*

Procedure Shine the laser through two slits and project the beam toward the screen. Point out that each slit acts as a point source of light. Have students examine the bands of brightness and darkness. Explain that the areas of darkness correspond to the nodal lines observed in the experiments with sound and water waves in Section 16-1, Demonstration 1 and Demonstration 2, respectively.

Repeat the experiment with a light pen. Explain that although the light from the two slits is nearly coherent, white light contains many wavelengths. Although these interfere with each other, the pattern is blurred.

internet**connect**

SCi**LINKS**
NSTA

TOPIC: Lasers
GO TO: www.scilinks.org
*sci***LINKS CODE:** HF2163

Did you know?

The word *laser* is an acronym (a word made from the first letters of several words) that stands for "light amplification by stimulated emission of radiation."

⭐ TEKS 8A

Figure 16-24
(a) Atoms or molecules in the active medium of a laser store energy from an external source.
(b) When a spontaneously emitted light wave interacts with an atom, it may cause the atom to emit an identical light wave. **(c)** Stimulated emission increases the amount of coherent light in the active medium, and the coherent waves behave as a single wave.

Lasers, on the other hand, typically produce a narrow beam of coherent light. The waves emitted by a laser are in phase, and they do not shift relative to each other as time progresses. Because all the waves are in phase, they interfere constructively at all points. The individual waves effectively behave like a single wave with a very large amplitude. In addition, the light produced by a laser is monochromatic, so all the waves have exactly the same wavelength. As a result of these properties, the intensity, or brightness, of laser light can be made much greater than that of incoherent light. For light, intensity is a measure of the energy transferred per unit time over a given area. Therefore, lasers are very high-energy light sources. ⭐TEKS **8A**

Lasers transform energy into coherent light

A laser is a device that converts light, electrical energy, or chemical energy into coherent light. There are a variety of different types of lasers, but they all have some common features. They all use a substance called the *active medium* to which energy is added to produce coherent light. The active medium can be a solid, liquid, or gas. The composition of the active medium determines the wavelength, or color, of the light produced by the laser.

The basic operation of a laser is shown in **Figure 16-24.** When high-energy light or electrical or chemical energy is added to the active medium, as in **Figure 16-24(a),** the atoms in the active medium absorb some of the energy.

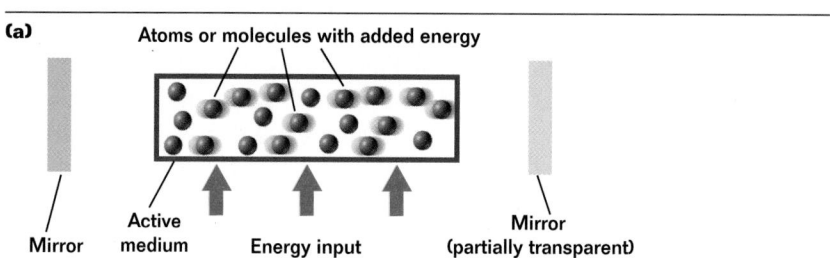

(a) Atoms or molecules with added energy

Mirror Active medium Energy input Mirror (partially transparent)

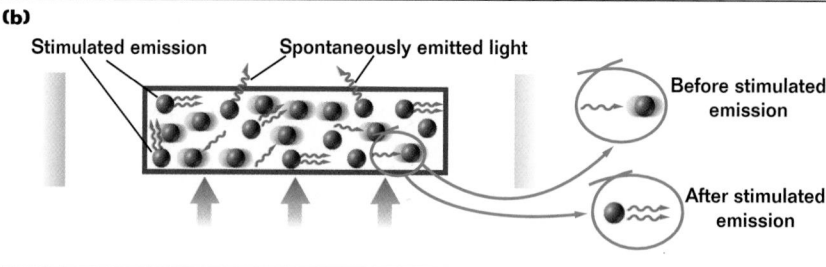

(b) Stimulated emission Spontaneously emitted light

Before stimulated emission

After stimulated emission

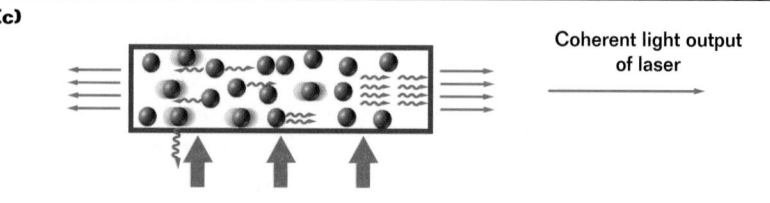

(c) Coherent light output of laser

Atoms exist at different energy levels. When energy is added to atoms at low energy levels, the atoms can be excited to higher energy levels. The atoms then release the energy in the form of light waves when they return to their original lower energy levels.

However, when light of a certain wavelength is applied to excited atoms, they can be stimulated to release light waves that have the same properties. After one atom spontaneously releases its energy in the form of a light wave, this initial wave can cause other energized atoms to release their excess energy as light waves with the same wavelength, phase, and direction as the initial wave, as shown in **Figure 16-24(b).** This process is called *stimulated emission.*

While some of the light produced by stimulated emission passes through the sides of the active medium, other light waves move along the length of the active medium. Mirrors on the ends of the material return these coherent light waves into the active medium, where they stimulate the emission of more coherent light waves, as shown in **Figure 16-24(c).** As the light passes back and forth through the active medium, it becomes more and more intense. One of the mirrors is slightly transparent, which allows some of the intense coherent light to be emitted by the laser.

APPLICATIONS OF LASERS

Of the properties of laser light, the one that is most evident is that it emerges from the laser as a narrow beam. Unlike the light from a light bulb or even the light that is focused by a parabolic reflector, the light from a laser undergoes very little spreading with distance. One reason is that all the light waves emitted by the laser have the same direction. As a result, a laser can be used to measure large distances, because it can be pointed at distant reflectors and detected again. Laser light is used by surveyors to measure distances and to guarantee that segments of bridges and tunnels are in line. TEKS **3C, 8A**

As shown in **Figure 16-25,** astronomers direct laser light at particular points on the moon's surface to determine the Earth-to-moon distance. A pulse of light is directed toward one of several 0.25 m^2 reflectors that were placed on the moon's surface by astronauts during the *Apollo* missions. Measuring the time the light takes to travel to the moon and back allows the Earth-to-moon distance (3.84×10^5 km) to be measured. Geologists use repeated measurements to record changes that occur in the height of Earth's crust because of geological processes. Lasers can be used for these measurements even when the height changes by only a few centimeters.

Lasers improve information storage and retrieval

A recent application of the laser is the compact-disc (CD) player. In a CD player, light from a laser passes through a glass plate and is then redirected by a mirror through a lens toward the compact disc, on which the music has been *digitally recorded.* The sound produced by the CD player is determined by the

Figure 16-25
A laser beam is fired at reflectors on the moon, which is more than 380 000 km away.

Key Models and Analogies

Have students think of a busy shopping mall, where people walk in all directions and with different strides. Tell students that this scene can represent non-coherent light emitted by a light bulb, with each person being analogous to a light wave with its own direction, wavelength, and phase. Then have students think of a marching band, with all members of the band lifting their feet at precisely the same time and marching in the same direction with steps that are the same size. Explain that this is analogous to laser light.

(STOP) Misconception Alert

Students may think that fluorescence and phosphorescence are related to the way lasers work. Make sure they realize that in fluorescent lamps, any atoms that become excited (by absorbing energy) *immediately* lose the energy by emitting light. In phosphorescent paint, the excited atoms are able to remain at a higher level of energy for a longer period of time, but they emit light *spontaneously.* In lasers, most of the atoms remain in an excited state *until* something triggers the excited atoms to emit light.

CONCEPT PREVIEW
The absorption and emission of light by atoms will be discussed in greater detail in Chapter 23.

(★)TEKS
p. 614: 8A
p. 615: 3C, 8A

Consumer Focus

BACKGROUND

Holography depends on the coherent nature of laser light to create an interference pattern that produces a three-dimensional image when illuminated by a light source.

Point out to students that many pairs of beams form an extremely complicated interference pattern on the film, one that can be produced only if the phase relationship of the two-wave pairs is constant throughout the exposure of the film. Tell them this condition is met through the use of light from a laser because of the coherence of laser light.

★ TEKS

p. 616: 3C, 8A
p. 617: 3C, 3A, 8A

616

Consumer Focus *Holograms* ★ TEKS 3C, 8A

*O*nce considered a modern miracle, holograms are now so commonplace that these three-dimensional images can be found practically everywhere, from credit cards to magazine covers to postage stamps. But how can such three-dimensional images be made from just a laser, some mirrors, and film?

Holography, the technique for making holograms, makes use of the fact that all the light leaving a laser is in phase. One way to make a hologram is to begin by splitting the laser beam with a mirror or prism. Half of the laser beam, called the reference beam, is directed by mirrors toward a light-sensitive film. The other half of the laser beam is reflected off the object toward the film.

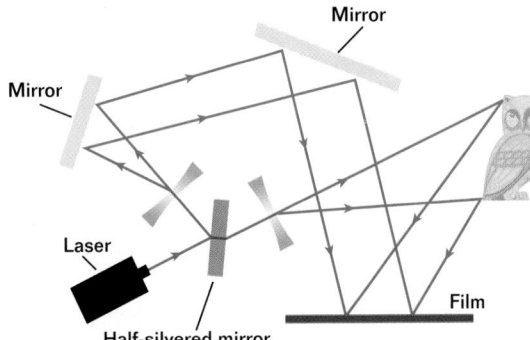

Because these two halves of the laser beam have not traveled the same distance, they are no longer in phase when they meet at the film. The degree to which they are out of phase at any spot on the film depends on the shape of the object. As a result of the phase difference, the two beams interfere with each other, producing what appears to be a pattern of dots and spirals on the film.

Once developed, the film acts like a diffraction grating. When a laser is shined onto the hologram from the same direction as the reference beam, the laser light interacts with the patterns on the film, changing the direction of the light waves in this beam. When viewed from the side of the film opposite the laser, the waves appear to come from the original illuminated object. The resulting three-dimensional image is a virtual image that appears to hang in the air on the same side of the film as the laser. A real image is also produced. This image appears on the same side as the observer and appears inverted, like the real image from a concave mirror.

Several varieties of holograms exist. The type just described, *transmission* holograms, are illuminated from behind. They can also be illuminated with white light, such as sunlight, but this often creates rainbow-like bands of color across the image. This occurs because each wavelength of light is diffracted a different amount by the film. *Reflection* holograms are lighted from the front and can be viewed with white light. *Embossed* holograms are similar to reflection holograms in that they are lighted from the front with white light; however, they are simply transmission holograms that are backed by a mirror. Embossed holograms are a low-cost alternative to reflection holograms. Finally, *pulsed* holograms use short bursts of light from high-powered lasers to make holograms of moving objects. With this technique, holograms can be made of people, scenes, and very large objects, and the short exposure time allows holographers to create a series of three-dimensional holograms of objects in motion.

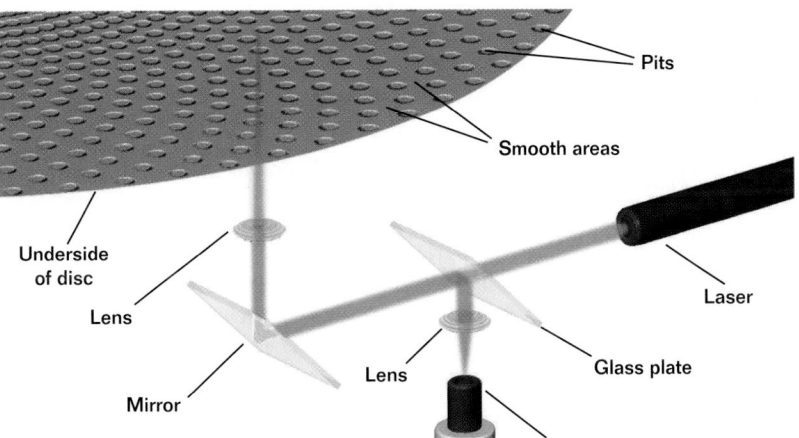

Pits

Smooth areas

Underside
of disc

Lens

Laser

Mirror

Lens

Glass plate

Detector

Figure 16-26
Light from a laser is directed toward the surface of the compact disc. Smooth parts of the disc reflect the light back to the detector.

way the laser light is reflected from the compact disc. The mechanism contained in a CD player is shown in **Figure 16-26.** ⊛TEKS 3C

In digital recording, a sound signal is sampled at regular intervals of time. Each sampling is converted to an electrical signal, which in turn is converted into a series of binary numbers. Binary numbers consist only of zeros and ones. The binary numbers are coded to contain information about the signal, including the frequencies and harmonics that are present, the volume for the left and right channels, and the speed of the motor that rotates the disc. Altogether, 44 000 samples of 16-digit (or 16-*bit*) binary numbers are taken each second. This sequence of numbers is recorded by a laser, which encodes the sequence by either burning a tiny pit into the disc or passing over a spot. ⊛TEKS 3A, 8A

When a compact disc is played, the laser light in the CD player is reflected sequentially from the string of binary bits etched on the disc. When the light is reflected from one of the untouched places on the disc, it reaches the detector. When the light strikes a pit on the rotating disc, the reflected light is out-of-phase with light reflected from the surrounding areas. Destructive interference prevents light from reaching the detector. Each "bright" area is interpreted as a binary 1. Each "dark" pit is interpreted as a binary 0. Electronic circuits connected to the detector translate the binary data to an electrical signal, which is then amplified so that it drives the stereo unit's loudspeakers. The resulting sound is a reconstruction of the original recorded sound.

A digital videodisc (DVD) player operates on the same principle. The laser in a DVD player is higher powered than the laser in a CD player. This higher-powered laser can read information that is coded in much smaller pits and smooth spaces. DVDs can hold more than seven times as much data as CDs, which allows them to hold more information or to produce higher-quality audio and video signals. DVD players can also be used to read standard CDs, but CD players cannot read DVDs.

Did you know?

The principle behind reading the information stored on a compact disc is also the basis for the reading of bar codes found on many products. When these products are scanned, laser light reflected from the bars and spaces of the bar code reproduces the binary codes that represent the product's inventory number. This information is transmitted to the store's computer system, which returns the product's name and price to the cash register.

internet **connect**

SCi **L** *INKS*

NSTA

TOPIC: Bar codes
GO TO: www.scilinks.org
*sci***LINKS CODE:** HF2165

Demonstration 12

How a CD works

Purpose Demonstrate how CDs transform light to sound.

Materials laser, turntable, mirror, transparency, solar cell connected to an amplifier/speaker

CAUTION *Avoid directing the laser beam toward students.*

Procedure Prepare a transparency with the pattern shown below. Place the mirror on the turntable, and tape the transparency to the top of the mirror's surface.

Shine the laser onto the mirror through the transparency, and place the solar cell so that it receives the reflected beam. When you turn on the turntable, the speaker will emit humming sounds that vary according to the patterns on the transparency. (The sound will be more pleasant if the bands are regularly spaced.)

Tell students that this simplified model simulates the microscopic absorbing and reflecting areas on a CD. These areas are encoded to generate sounds that are nearly identical to those of the original source.

Teaching Tip

Point out that there are many types of lasers, including semiconductor (diode) lasers, CO_2 lasers, ruby lasers, microwave lasers, X-ray lasers, and tunable lasers. Lasers now have medical, military, industrial, and scientific applications. They are used as bloodless scalpels, saws, and moonquake detectors. Astronomers have even found laser action in stars. In 1996, the Hubble Space Telescope discovered an ultraviolet laser star. Have students find more information about how lasers are used, and have them report on one of these topics.

Section Review
ANSWERS

1. Waves emitted by a laser do not shift relative to each other as time progresses (they are coherent and continuously in phase).

2. no; The second light wave is obtained from the energy that is added to an atom in the active medium by an external energy source.

3. Laser light is nearly monochromatic, so it does not spread out very much into different components with different wavelengths as it passes between the fiber and transmission and receiving equipment (that is, dispersion is nearly absent over short distances).

p. 618: 8A, 8C

Lasers are also used for many medical procedures. Many medical applications make use of the fact that specific body tissues absorb different wavelengths of lasers. For example, lasers can be used to lighten or remove scars and certain types of birthmarks without affecting surrounding tissues. The scar tissue responds to the wavelength of light used in the laser, but other body tissues are protected. ⊛TEKS 8C

Many medical applications of lasers take advantage of the fact that water can be vaporized by high-intensity infrared light produced by carbon dioxide lasers having a wavelength of 10 μm. Carbon dioxide lasers can cut through muscle tissue by heating and evaporating the water contained in the cells. One advantage of a laser is that the energy from the laser also coagulates blood in the newly opened blood vessels, thereby reducing blood loss and decreasing the risk of infection. A laser beam can also be trapped in an optical fiber endoscope, which can be inserted through an orifice and directed to internal body structures. As a result, surgeons can stop internal bleeding or remove tumors without performing massive surgery. ⊛TEKS 8C

Lasers can also be used to treat tissues that cannot be reached by conventional surgical methods. For example, lasers can pass through the outer structures of the eye, such as the cornea and lens, without damaging them. Therefore, lasers are effective at treating lesions of the inner eye. Lasers are used for other eye surgeries, including surgery to correct *glaucoma*, a condition in which the fluid pressure within the eye is too great. Left untreated, glaucoma can lead to damage of the optic nerve and eventual blindness. Focusing a laser at the clogged membrane allows a tiny hole to be burned in the membrane, which relieves the pressure. Lasers can also be used to correct nearsightedness by making cuts in the eye to slightly change its shape and focus.

Section Review   ⊛TEKS 8A

1. How does light from a laser differ from light whose waves all have the same wavelength but are not coherent?

2. The process of stimulated emission involves producing a second wave that is identical to the first. Does this gaining of a second wave violate the principle of energy conservation? Explain your answer.

3. Fiber-optics systems transmit light by means of internal reflection within thin strands of extremely pure glass. In these fiber-optics systems, laser light is used instead of white light to transmit the signal. Apply your knowledge of refraction to explain why.

CHAPTER 16
Summary

KEY IDEAS

Section 16-1 Interference

- Light waves with the same wavelength and constant phase differences interfere with each other to produce light and dark interference patterns.
- In double-slit interference, the position of a bright fringe requires that the path difference between two interfering point sources be equal to a whole number of wavelengths.

$$d\,(\sin\theta) = m\lambda \quad m = 0, \pm1, \pm2, \ldots$$

- In double-slit interference, the position of a dark fringe requires that the path difference between two interfering point sources be equal to an odd number of half-wavelengths.

$$d\,(\sin\theta) = \left(m + \tfrac{1}{2}\right)\lambda \quad m = 0, \pm1, \pm2, \ldots$$

Section 16-2 Diffraction

- Light waves form a diffraction pattern by passing around an obstacle or bending through a slit and interfering with each other.
- The position of a maximum in a pattern created by a diffraction grating depends on the separation of the slits in the grating, the order of the maximum, and the wavelength of the light.

$$d\,(\sin\theta) = m\lambda \quad m = 0, \pm1, \pm2, \ldots$$

Section 16-3 Coherence

- A laser is a device that transforms energy into a beam of coherent monochromatic light.

Variable symbols

Quantities		Units	
λ	wavelength	m	meters
θ	angle from the center of an interference pattern	°	degrees
d	slit separation	m	meters
m	order number		(unitless)

KEY TERMS

coherence (p. 599)

diffraction (p. 604)

laser (p. 613)

order number (p. 601)

path difference (p. 601)

resolving power (p. 611)

Teaching Tip

Because double-slit interference can be a difficult subject to understand, students may find it helpful to write an expository essay with diagrams, explaining how an interference pattern is created. Essays should include a discussion of diffraction as it relates to double-slit interference.

⊛ TEKS

Review & Assess pp. 620–622:
2A: Alt. Assess. 1–3
2B: Alt. Assess. 2
2C: 10–11, 19–21, 26–31, Technology & Learning
2D: Alt. Assess. 4–5
3A: Alt. Assess. 3
3C: Alt. Assess. 5
3E: Alt. Assess. 4
8A: 1–31, Alt. Assess. 1–5, Technology & Learning
8B: 15
8C: Alt. Assess. 5

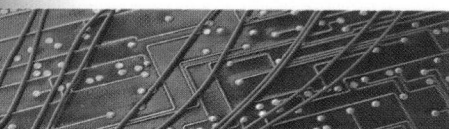

CHAPTER 16
Review and Assess

INTERFERENCE

Review questions

1. What happens if two light waves with the same amplitude interfere constructively? What happens if they interfere destructively?

2. Interference in sound is recognized by differences in volume; how is interference in light recognized?

3. A double-slit interference experiment is performed with red light and then again with blue light. In what ways do the two interference patterns differ? (Hint: Consider the difference in wavelength for the two colors of light.)

4. What data would you need to collect to correctly calculate the wavelength of light in a double-slit interference experiment?

Conceptual questions

5. If a double-slit experiment were performed underwater, how would the observed interference pattern be affected? (Hint: Consider how light changes in a medium with a higher index of refraction.)

6. Because of their great distance from us, stars are essentially point sources of light. If two stars were near each other in the sky, would the light from them produce an interference pattern? Explain your answer.

7. Assume that white light is provided by a single source in a double-slit experiment. Describe the interference pattern if one slit is covered with a red filter and the other slit is covered with a blue filter.

8. An interference pattern is formed by using green light and an apparatus in which the two slits can move. If the slits are moved farther apart, will the separation of the bright fringes in the pattern decrease, increase, or remain unchanged? Why?

Practice problems

9. Light falls on two slits spaced 0.33 mm apart. If the angle between the first dark fringe and the central maximum is 0.08°, what is the wavelength of the light?
(See Sample Problem 16A.)

10. A sodium-vapor street lamp produces light that is nearly monochromatic. If the light shines on a wooden door in which there are two straight, parallel cracks, an interference pattern will form on a distant wall behind the door. The slits have a separation of 0.3096 mm, and the second-order maximum occurs at an angle of 0.218° from the central maximum. Determine the following quantities:
    a. the wavelength of the light
    b. the angle of the third-order maximum
    c. the angle of the fourth-order minimum
(See Sample Problem 16A.)

11. All but two gaps within a set of venetian blinds have been blocked off to create a double-slit system. These gaps are separated by a distance of 3.2 cm. Infrared radiation is then passed through the two gaps in the blinds. If the angle between the central and the second-order maxima in the interference pattern is 0.56°, what is the wavelength of the radiation?
(See Sample Problem 16A.)

DIFFRACTION

Review questions

12. Why does light produce a pattern similar to an interference pattern when it passes through a single slit?

13. How does the width of the central region of a single-slit diffraction pattern change as the wavelength of the light increases?

14. Why is white light separated into a spectrum of colors when it is passed through a diffraction grating?

15. Why are orbiting telescopes designed for use in all of the electromagnetic spectrum except for radio?

Conceptual questions

16. Monochromatic light shines through two different diffraction gratings. The second grating produces a pattern in which the first-order and second-order maxima are more widely spread apart. Use this information to tell if there are more or fewer lines per centimeter in the second grating than in the first.

17. Why is the resolving power of your eye better at night than during the day?

18. Globular clusters, such as the one shown in **Figure 16-27,** are spherical groupings of stars that form a ring around the Milky Way galaxy. Because there can be millions of stars in a single cluster and because they are distant, resolving individual stars within the cluster is a challenge. Of the following conditions, which would make it easier to resolve the component stars? Which would make it more difficult?

Figure 16-27

  a. The number of stars per unit volume is half as great.
  b. The cluster is twice as far away.
  c. The cluster is observed in the ultraviolet portion instead of in the visible region of the electromagnetic spectrum.
  d. The telescope's mirror or lens is twice as wide.

Practice problems

19. Light with a wavelength of 707 nm is passed through a diffraction grating with 795 slits/cm. Find the angle at which one would observe the first-order maximum.
(See Sample Problem 16B.)

20. If light with a wavelength of 353 nm is passed through the same diffraction grating (with 795 slits/cm), find the angle at which one would observe the second-order maximum.
(See Sample Problem 16B.)

21. By attaching a diffraction-grating spectroscope to an astronomical telescope, one can measure the spectral lines from a star and determine the star's chemical composition. Assume the grating has 3661 slits/cm.

  a. If the wavelengths of the star's light are 478.5 nm, 647.4 nm, and 696.4 nm, what are the angles at which the first-order spectral lines occur?
  b. At what angles are these lines found in the second-order spectrum?
(See Sample Problem 16B.)

LASERS

Review questions

22. What properties does laser light have that are not found in the light used to light your home?

23. Laser light is commonly used to demonstrate double-slit interference. Explain why laser light is preferable to light from other sources for observing interference.

24. Give two examples in which the uniform direction of laser light is advantageous. Give two examples in which the high intensity of laser light is advantageous.

25. Laser light is often linearly polarized. How would you show that this statement is true?

MIXED REVIEW

26. The 546.1 nm line in mercury is measured at an angle of 81.0° in the third-order spectrum of a diffraction grating. Calculate the number of lines per millimeter for the grating.

27. Recall from Section 11-4 (pages 425–429) that the entropy of a system is a measure of that system's disorder. Why is it appropriate to describe a laser as an entropy-reducing device?

is to place in orbit. In order to have adequate resolution, an orbiting radio telescope would have to be very large.

16. The sin θ is inversely proportional to d, which is the reciprocal of the number of lines per centimeter. The grating that spreads the pattern the most has the most lines per centimeter.

17. The pupil is larger at night, so the angle of resolution is smaller and resolving power is greater.

18. easier: a, c, d
harder: b

19. 3.22°

20. 3.22°

21. a. 10.09°, 13.71°, 14.77°
 b. 20.51°, 28.30°, 30.66°

22. Laser light is coherent and monochromatic.

23. Light must be coherent for an interference pattern to form. An interference pattern is well defined with monochromatic light.

24. Answers may include distance measurements, compact disc players, and fiber optic communications; Answers may include laser surgery and fiber optic communications.

25. by rotating a polarizing sheet in front of the laser light—if the intensity of the transmitted light varies, the light is polarized

26. 603 lines/mm

27. Energy (light, for example) added to the active medium's atoms is not highly ordered. This light travels in all directions, is noncoherent, and is not monochromatic. The laser converts a portion of this energy into a beam of more coherent, monochromatic light.

28. 2.41×10^{-4} m

29. 432.0 nm

30. 8.000×10^{-7} m

31. 1.93×10^{-3} mm = 3 λ; a maximum

Alternative Assessment
ANSWERS

Performance assessment

1. To represent constant wavelength, the distances between lines should not exceed the lines' thickness. Students can create the same effects by overlapping drawings on computer screens.

2. One hole should produce concentric diffraction rings; two holes should produce a striped pattern. Suggested improvements will vary.

3. Students should use path difference and refraction to predict wavelengths.

 **Portfolio projects**

4. Presentations will vary but should indicate the source material used. Young (1773–1829) was a child prodigy. Fresnel (1788–1827) was eight years old before he could read.

5. Student presentations will vary. Findings should be explained clearly.

---

28. A double-slit interference experiment is performed using blue light from a hydrogen discharge tube (λ = 486 nm). The fifth-order bright fringe in the interference pattern is 0.578° from the central maximum. How far apart are the two slits separated?

29. A beam containing light of wavelengths λ_1 and λ_2 passes through a set of parallel slits. In the interference pattern, the fourth bright line of the λ_1 light occurs at the same position as the fifth bright line of the λ_2 light. If λ_1 is known to be 540.0 nm, what is the value of λ_2?

30. Visible light from an incandescent light bulb ranges from 400.0 nm to 700.0 nm. When this light is focused on a diffraction grating, the entire first-order spectrum is seen, but none of the second-order spectrum is seen. What is the maximum spacing between lines on this grating?

31. In an arrangement to demonstrate double-slit interference, λ = 643 nm, θ = 0.737°, and d = 0.150 mm. For light from the two slits interfering at this angle, what is the path difference both in millimeters and in terms of the number of wavelengths? Will the interference correspond to a maximum, a minimum, or an intermediate condition?

Alternative Assessment

Performance assessment

1. Design simulations of interference patterns. Use a computer to draw many concentric circles at regular distances to represent waves traveling from a point source. Photocopy the page onto two transparencies, and lay them on an overhead projector. Vary the distances between "source points," and observe how these variations affect interference patterns. Design transparencies with thicker lines with larger separations to explore the effect of wavelength on interference.

2. Investigate the effect of slit separation on interference patterns. Wrap a flashlight or a pen light tightly with tin foil and make pinholes in the foil. First record the pattern you see on a screen a few inches away with one hole, then do the same with two holes. How does the distance between the holes affect the distance between the bright parts of the pattern? Draw schematic diagrams of your observations, and compare them with the results of double-slit interference. How would you improve your equipment?

3. Soap bubbles exhibit different colors because light that is reflected from the outer layer of the soap film interferes with light that is refracted and then reflected from the inner layer of the soap film. Given a refractive index of n = 1.35 and thicknesses ranging from 600 nm to 1000 nm for a soap film, can you predict the colors of a bubble? Test your answer by making soap bubbles and observing the order in which the different colors appear. Can you tell the thickness of a soap bubble from its colors? Organize your findings into a chart, or create a computer program to predict the thicknesses of a bubble based on the wavelengths of light it appears to reflect.

Portfolio projects

4. Thomas Young's 1803 experiment provided crucial evidence for the wave nature of light, but it was met with strong opposition in England until Augustin Fresnel presented his wave theory of light to the French Academy of Sciences in 1819. Research the lives and careers of these two scientists. Create a presentation about one of them. The presentation can be in the form of a report, poster, short video, or computer presentation.

5. Research waves that surround you, including those used in commercial, medicinal, and industrial applications. Interpret how the waves' characteristics and behaviors make them useful. For example, investigate what kinds of waves are used in medical procedures such as MRI and ultrasound. What are their wavelengths? Research how lasers are used in medicine. How are they used in industry? Prepare a poster or chart describing your findings and present it to the class.

Technology & Learning

Graphing calculators

Refer to Appendix B for instructions on downloading programs for your calculator. The program "Chap16" builds a table of fringe angles versus order numbers for a double-slit interference pattern viewed on a screen.

Constructive interference in a double-slit interference experiment, as you learned earlier in this chapter, is described by the following equation:

$$d(\sin \theta) = m\lambda$$

The program "Chap16" stored on your graphing calculator makes use of the equation for constructive interference. Once the "Chap16" program is executed, your calculator will ask for the wavelength of the two waves and the slit separation. The graphing calculator will use the following equation to create a table of fringe angles (Y_1) versus order numbers (X). Note that the relationships in this equation are the same as those in the constructive interference equation above.

$$Y_1 = \sin^{-1}(XW/D)$$

Note that the values for wavelength and slit separation must be expressed in the same units so that the units cancel during calculation of the fringe angle. Thus, if the wavelength and slit separation are expressed in different units, the units of one must be converted to match the other.

a. The slit separation in a double-slit experiment is measured in millimeters. If the wavelength of the two waves is measured in nanometers, then what factor must you multiply the wavelength by to make its units match the slit separation?

Before executing the program, press MODE ▼ ▼ ► ENTER to be certain that your graphing calculator is in degree mode.

Execute "Chap16" on the PRGM menu and press ENTER to begin the program. Enter the value for the wavelength (shown below) and press ENTER. Then enter the value for the slit separation and press ENTER. Remember to use the (-) key, instead of the – key, for entering negative values. Also, use the exponent function key to enter powers of ten by pressing 2nd EE.

The calculator will provide a table of fringe angles in degrees (Y_1) versus order number (X). Press ▼ to scroll down through the table to find the fringe angle values you need. Remember that only the first few fringes will be bright enough to be visible.

Find the fringe angles on the viewing screen in a double-slit experiment for the first three bright fringes, given the following situations:

b. a yellow light of 589 nm that passes through two slits 2.00 mm apart and strikes a screen

c. a green light of 546 nm that passes through two slits 2.00 mm apart and strikes a screen

d. a green light of 546 nm that passes through two slits 3.50 mm apart and strikes a screen

e. a violet light of 437 nm that passes through two slits 3.50 mm apart and strikes a screen

f. a red light of 660 nm that passes through two slits 3.50 mm apart and strikes a screen

g. You want the constructive interference fringe that corresponds to order number one to be close to the center of the screen. Would you choose a long or short wavelength? Would you choose a large or small slit separation?

Press ENTER to stop viewing the table. Press ENTER again to enter a new value or CLEAR to end the program.

ANSWERS TO
Technology & Learning

a. 10^{-6}

b. 0.0169°, 0.0338°, 0.0506°

c. 0.0156°, 0.0313°, 0.0469°

d. 0.00894°, 0.0179°, 0.0268°

e. 0.00715°, 0.0143°, 0.0215°

f. 0.0108°, 0.0216°, 0.0324°

g. short wavelength, large slit separation

Planning

Recommended time:

1 lab period

Classroom organization:

▸ Each lab group needs a level work surface at least 1 m long.

▸ Each lab group should have two or more students.

▸ **Safety warnings:** Make sure all students know the proper procedure for cleaning up broken glass. Remind students not to attempt to remove broken bulbs from sockets.

Techniques to Demonstrate

Before the lab, demonstrate diffraction using an excited gas source. Set up the optical bench as described in step 3 of the lab, and place the light source about 5 cm away from the slit. Adjust the grating and the slit scale distance and angle so a full first-order and part of a second-order spectrum appear centered about the slit. Show students how to recognize the images and how to bend paper clips to form riders for marking each similar first-order image and each similar second-order image.

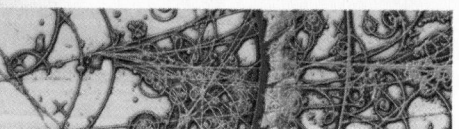

CHAPTER 16
Laboratory Exercise

(★)TEKS

pp. 624–625: 1A, 1B, 2B, 8A, 8B

OBJECTIVES

• Find wavelengths of diffracted light.

MATERIALS LIST

✔ bent paper-clip riders for meterstick
✔ black cardboard
✔ cellophane tape
✔ diffraction grating
✔ grating holder and support
✔ incandescent light source and power supply
✔ meterstick and 2 supports
✔ metric scale and slit

DIFFRACTION

In this experiment, you will pass white light through a diffraction grating and use your measurement to determine wavelengths of the light's components.

SAFETY

• **Use a hot mitt to handle resistors, light sources, and other equipment that may be hot. Allow all equipment to cool before storing it.**

• **If a bulb breaks, notify your teacher immediately. Do not remove broken bulbs from sockets.**

• **Never put broken glass or ceramics in a regular waste container. Use a dustpan, brush, and heavy gloves to carefully pick up broken pieces and dispose of them in a container specifically provided for this purpose.**

• **Avoid looking directly at a light source. Looking directly at a light source can cause permanent eye damage.**

PREPARATION

1. Read the entire lab, and plan the steps you will take.

2. Prepare a data table in your lab notebook with six columns and four rows. In the first row, label the columns *Light source, Image color, Order, Image 1 (m), Image 2 (m),* and *Slit (m)*. In the first column, label the second through fourth rows *White*. Above or below the data table, prepare a space to record the slit spacing, *d*, of the grating.

PROCEDURE

Wavelengths of white light

3. Set up the optical bench as shown in **Figure 16-29.** Mount the scale and slit on one end of the optical bench, and place a piece of tape over the slit. Place a cardboard shield around the light source to direct all the light through the slit. Illuminate the slit with white light. Mount the grating near the opposite end of the optical bench.

4. Adjust the apparatus so that the white-light source is centered on the slit and the slit scale is perpendicular to the optical bench. Tape the optical bench and the white-light source securely in place.

5. With your eye close to the grating, observe the first-order spectra. Move the grating forward or backward as required so that the entire spectrum appears on each side of the scale. Place a bent paper-clip rider on the scale at the point in each first-order spectrum where the yellow light is the purest. Adjust the grating and slit scale by rotating the grating around its vertical axis so that the two yellow points end up equidistant from the source slit. Reposition the riders if necessary.

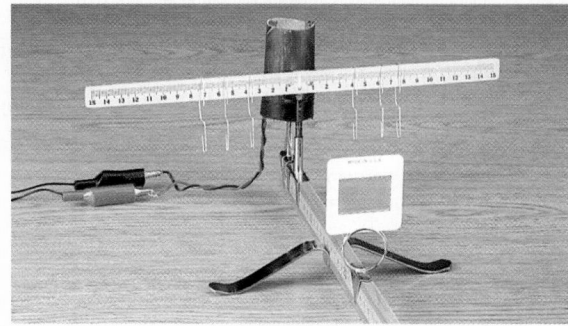

Figure 16-29

Step 3: Use the cardboard to make a shield around the light source. Make sure the light is directed through the slit.

Step 4: From above, the slit scale should form right angles with the meterstick. Measurements can be strongly affected if the equipment is moved even slightly during the procedure.

Step 5: Place a bent-paperclip rider to mark the position of the images on the scale.

6. Use the scale to measure the distance from the slit to each rider. Record these distances in your data table as *Image 1* and *Image 2*. Also measure the distance from the slit to the grating. Record this distance in your data table as the *Slit (m)*. Record the order number and the image color.

7. Next, adjust the grating and slit to find the clearest first-order continuous spectrum. Measure and record the distance from the slit to the grating. Place a rider on the scale at the point in each first-order spectrum where you see the extreme end of the violet spectrum. Measure and record the distance from the slit to each rider.

8. Repeat step 7 for the extreme red end of the spectrum. Record all data.

9. Clean up your work area. Put equipment away safely so that it is ready to be used again. Recycle or dispose of used materials as directed by your teacher.

ANALYSIS AND INTERPRETATION

Calculations and data analysis

1. **Organizing data** Use your data for each trial.

   a. For each trial, find the average image position.

   b. Use the average image position and the distance from the slit to the grating to find the distance from the grating to the image for each trial. (*Hint:* use the Pythagorean theorem.)

   c. To find sin θ for each trial, divide the average image position by the distance found in (b).

Conclusions

2. **Organizing information** For each trial, find the wavelength of the light using the equation $\lambda = \dfrac{d\,(\sin\theta)}{m}$, where λ is the wavelength of light (in meters, m), d is the diffraction-grating spacing (1/[number of lines/m]), and m is the order number of the spectrum containing the image.

✔ **Checkpoints**

Step 4: Make sure the apparatus is firmly attached to the table. Measurements can be greatly affected if the equipment is moved even slightly during the procedure.

Step 5: Students may need help finding the correct position of the grating. Students should be able to demonstrate that they have found the point in the spectrum where the color is purest.

Step 6: Students should be able to demonstrate how they used the scale to measure the distances.

ANSWERS TO

Analysis and Interpretation

CALCULATIONS AND DATA ANALYSIS

1. **a.** yellow: 0.096 m
   violet: 0.067 m
   red: 0.118 m

   b. Student answers will vary. For sample data, values range from 0.302 m to 0.318 m.

   c. For sample data, values range from 0.2199 to 0.3700.

CONCLUSIONS

2. yellow: $\lambda = 5.68 \times 10^{-7}$ m
   violet: $\lambda = 4.04 \times 10^{-7}$ m
   red: $\lambda = 6.80 \times 10^{-7}$ m

CHAPTER 17 PLANNING GUIDE

Compression Guide: To shorten from 10 to 8 45-min periods (from 5 to 4 90-min blocks), eliminate items in magenta type.

PACING CHART	CLASSROOM RESOURCES			
	⭐ TEKS	Teacher Demonstrations	*Holt Physics* Transparencies	Labs (See page T52 for equipment listing for in-text labs.)
17-1 Electric charge 3 or 2 45-minute periods 1½ or 1 90-minute block	1A, 2B, 2C, 3A, 3C, 3E, 6B, 6C	**TE** *Effects of charge*, p. 629 **TE** *Jumping spices*, p. 631 **TE** *Polarization*, p. 633	**T** 88–90	**L** *Discovery Lab: Charges and Electrostatics* **PE** *Quick Lab: Polarization*, p. 632 **PE** *Electrostatics*, p. 660
17-2 Electric force 2 45-minute periods 1 90-minute block	2C, 3A, 3B, 3E, 6A, 6B, 6C	**TE** *Electric force*, p. 634		
17-3 The electric field 3 or 2 45-minute periods 1½ or 1 90-minute block	2C, 3A, 3B, 3C, 6F, 6C, 8B	**TE** *Electric field strength*, p. 644 **TE** *Charge accumulation*, p. 651	**TM** 59–61 **T** 91	**L** *Invention Lab: Levitating Toys*
Review and Assessment 2 45-minute periods 1 90-minute block				

Resource Key

PHYSICS

PE Pupil's Edition
TE Teacher's Edition

L Laboratory Experiments
TL Technology Lab Experiments
T Transparencies

One-Stop Planner CD-ROM contents

TM Transparency Masters
SR Section Review Worksheets
AA Alternative Assessment

PW Problem-Solving Workbook
PB Problem Bank
CTW Critical Thinking Worksheet

LABORATORY PLANNING: Electrostatics, p. 660

Materials (for each lab group)
- 1-position support base and rod, 1.3 cm × 91 cm
- electroscope, aluminum leaf
- exciting pad, animal fur
- exciting pad, silk
- exciting pad, wool felt
- friction rod, borosilicate glass (hollow)
- friction rod, flint glass (solid)
- friction rod, hard rubber (solid)
- friction rod, polystyrene (solid)
- meterstick, wooden
- roll of fine black nylon cord, 9 m

ASSIGNMENT RESOURCES

Content Mastery	Critical Thinking	Problem-Solving Practice
PE 1–6, p. 633 **SR** 17-1, *Concept Review* **PE** 1–3, p. 654	**PE** 1–3, p. 631 **PE** 4–10, p. 654	
PE 1–5, p. 642 **SR** 14-2, *Math Skills* **PE** 11–15, p. 654	**PE** 1–3, p. 637 **PE** 16–17, p. 654	**17A** Coulomb's law: 28 items in **PE, PW,** and **PB,** see **TE** pp. 635–636 **17B** The superposition principle: 22 items in **PE, PW,** and **PB,** see **TE** p. 638–639 **17C** Equilibrium: 31 items in **PE, PW,** and **PB,** see **TE** p. 640–641
PE 1–5, p. 652 **SR** 17-3, *Concept Review* **PE** 25–32, p. 655	**PE** 33–37, pp. 655–656	**17D** Electric field strength: 28 items in **PE, PW,** and **PB,** see **TE** pp. 646–647

ASSESSMENT RESOURCES

Cumulative Review	Alternative Assessment	Traditional Assessment
SR Mixed Review, Ch. 17	**PE** 1–5, p. 659 **AA** Items for Ch. 17	Chapter 17 Test Test Generator items for Ch. 17

Scoring Rubrics for Alternative Assessment items can be found on the One-Stop Planner CD-ROM.

TECHNOLOGY RESOURCES

CTW Segment 16 Force Between Charges

PE Technology and Learning, p. 658
(Alternative procedures for calculators without Flash-ROM technology are provided on the One-Stop Planner CD-ROM.)

The Mechanical Universe/High School Adaptation Quad II, The Fundamental Forces
Quad III, The Millikan Experiment
Quad V, Electric Fields and Electric Forces

internet connect

On-line Student Resources:
GO TO: www.scilinks.org
The following *sci*LINKS Internet resources can be found in the student text for this chapter.

TOPICS:
• Electric charge, p. 630 (HF2171)
• Conductors and insulators, p. 633 (HF2172)
• Coulomb's law, p. 634 (HF2173)
• Microwaves, p. 644 (HF2174)
• Van de Graaff generator, p. 650 (HF2175)

On-line Teacher Resources:
GO TO: go.hrw.com
KEYWORD: HF2 HOME
Visit the HRW Web site for a variety of resources related to this chapter.

Smithsonian Institution
Internet Connections
Visit **www.si.edu/hrw** for additional on-line resources.

CNNfyi.com.
Visit **www.cnnfyi.com** for late-breaking news and current events stories selected just for you.

• roll of adhesive tape, 0.5 in. wide
• round-jaw symmetrical clamp, with holder
• static electricity tube, PVC
• suspension stirrup for rods

Materials Preparation
Electrostatic laboratories are difficult if undesired charge cannot be eliminated. A common ground consisting of a wire connected to a water pipe will allow students to eliminate this charge.

Section 17-1 introduces positive and negative electric charge, the conservation of charge, and the quantization of charge, and discusses conductors, insulators, and methods of charging.

Section 17-2 examines Coulomb's law and calculations of net electric forces using the superposition principle.

Section 17-3 introduces the electric field, electric field lines, and electric field strength; explores the electric fields around various charged objects; and discusses the properties of conductors in electrostatic equilibrium.

About the Illustration

Electrostatic spray painting is used in a variety of industries. Automobile bodies, furniture, toys, refrigerators, and various other mass-produced items are often painted electrostatically. Electrostatic spray painting can be used with metal and some types of wood. Plastic, rubber, and glass cannot be painted electrostatically.

PHYSICS Interactive
Problem-
Solving
Tutor

See Module 16
"Force Between Charges" promotes additional development of problem-solving skills for this chapter.

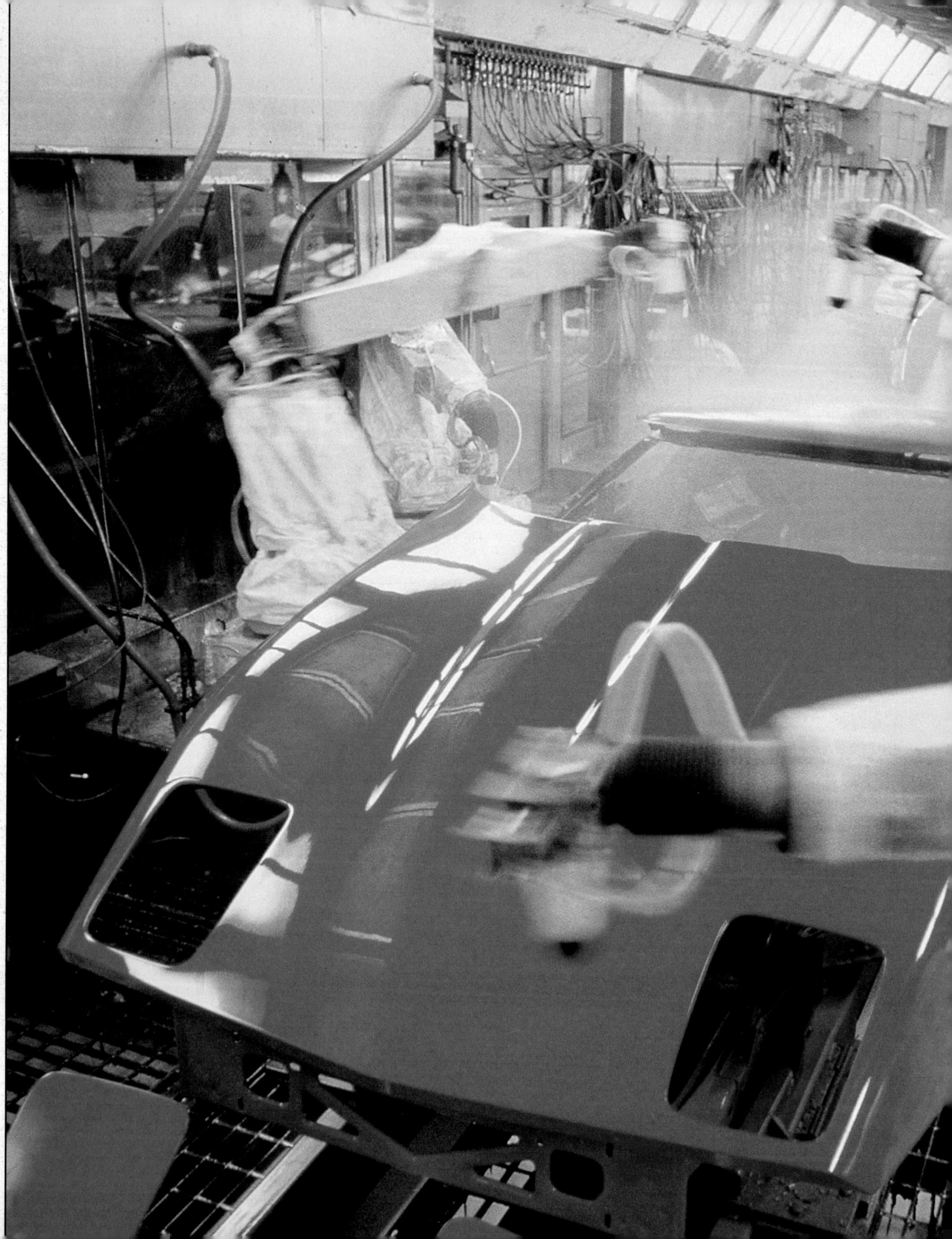

CHAPTER 17

Electric Forces and Fields

PHYSICS IN ACTION

In this factory in Bowling Green, Kentucky, a fresh coat of paint is being applied to an automobile by spray guns. With ordinary spray guns, any paint that does not happen to hit the body of the car is wasted. A special type of spray painting, known as *electrostatic spray painting*, utilizes electric force to minimize the amount of paint that is wasted. Electrostatic spray painting is used in a variety of industries. According to one estimate, electrostatic spray painting saves industries in the United States up to $50 million each year.

In this chapter, you will study electric force and its effects and see how this force is used in electrostatic spray painting.

- *How does an electrostatic spray gun work?*
- *Why are electrostatic spray guns more efficient than typical spray guns?*

CONCEPT REVIEW

Vector addition   (Section 3-2)

Field forces   (Section 4-1)

Tapping Prior Knowledge

Knowledge to Expect:

✔ "Without touching them, material that has been electrically charged pulls on all other materials and may either push or pull other charged matter." (AAAS's *Benchmarks for Science Literacy,* grades 3–5)

✔ "An unbalanced force acting on an object changes its speed or direction of motion, or both." (AAAS's *Benchmarks for Science Literacy,* grades 6–8)

Knowledge to Review:

✔ Vector addition gives a resultant vector that is equivalent to the added effects of each of the individual vectors. Vector addition may be accomplished graphically or mathematically. (Section 3-2)

✔ Field forces are forces that act on objects without physical contact. Field forces act at a distance. (Section 4-1)

Items to Probe:

✔ Vector addition and the superposition principle: Have students calculate the resultant force on an object experiencing two or three individual forces.

17-1
Electric charge

17-1 SECTION OBJECTIVES

- **Understand the basic properties of electric charge.**

- **Differentiate between conductors and insulators.**

- **Distinguish between charging by contact, charging by induction, and charging by polarization.**

CONCEPT PREVIEW

This chapter and Chapter 18 are concerned with stationary charges, known as static electricity or *electrostatics*. Chapters 19 and 20 discuss charges in motion, or *electrodynamics*.

PROPERTIES OF ELECTRIC CHARGE

You have probably noticed that after running a plastic comb through your hair on a dry day, the comb attracts strands of your hair or small pieces of paper. Another simple experiment is to rub an inflated balloon back and forth across your hair. You may find that the balloon is attracted to your hair, as shown in **Figure 17-1(a).** On a dry day, a rubbed balloon will stick to the wall of a room, often for hours. When materials behave this way, they are said to have become *electrically charged*. Experiments such as these work best on a dry day because excessive moisture can provide a pathway for charge to leak off a charged object.

You can give your body an electric charge by vigorously rubbing your shoes on a wool rug or by sliding across a car seat. You can then remove the charge on your body by lightly touching another person. Under the right conditions, you will see a spark when you touch, and both of you will feel a slight tingle.

Another way to observe static electricity is to rub two balloons across your hair and then hold them near one another, as shown in **Figure 17-1(b).** In this case, you will see the two balloons pushing each other apart. Why is a rubbed balloon attracted to your hair but repelled by another rubbed balloon?

There are two kinds of electric charge

The two balloons must have the same kind of charge because each became charged in the same way. Because the two charged balloons repel one another, we see that *like charges repel.* Conversely, a rubbed balloon and your hair, which do not have the same kind of charge, are attracted to one another. Thus, *unlike charges attract.* ⭐TEKS **6C**

Figure 17-1

(**a**) If you rub a balloon across your hair on a dry day, the balloon and your hair become charged and are attracted to one another. (**b**) Two charged balloons, on the other hand, repel one another.

(a)

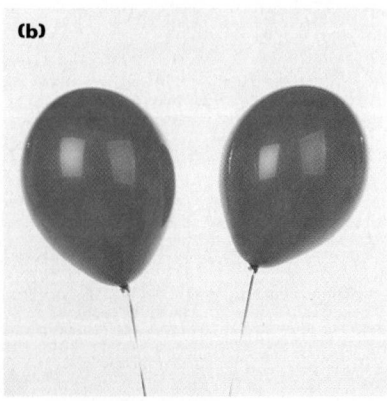
(b)

Benjamin Franklin (1706–1790) named the two different kinds of charge *positive* and *negative*. By convention, when you rub a balloon across your hair, the charge on your hair is *positive* and that on the balloon is *negative*, as shown in **Figure 17-2.** Positive and negative charges are said to be *opposite* because an object with an equal amount of positive and negative charge has no net charge.

Electrostatic spray painting utilizes the principle of attraction between unlike charges. Paint droplets are given a negative charge, and the object to be painted is given a positive charge. In ordinary spray painting, many paint droplets drift past the object being painted. But in electrostatic spray painting, the negatively charged paint droplets are attracted to the positively charged target object, so more of the paint droplets hit the object being painted and less paint is wasted. (★)TEKS **3C, 6C**

Electric charge is conserved

When you rub a balloon across your hair, how do the balloon and your hair become electrically charged? To answer this question, you'll need to know a little about the atoms that make up the matter you see around you. Every atom contains even smaller particles. Positively charged particles, called *protons,* and neutral particles, called *neutrons,* are located in the center of the atom, called the *nucleus.* Negatively charged particles, known as *electrons,* are located outside the nucleus and move around it. (You will study the structure of the atom and the particles within the atom in greater detail in Chapters 23 and 25, respectively.)

Protons and neutrons are relatively fixed in the nucleus of the atom, but electrons are easily transferred from one atom to another. When the electrons in an atom are balanced by an equal number of protons, the atom has no net charge. If an electron is transferred from one neutral atom to another, one of the atoms gains a negative charge and the other loses a negative charge, thereby becoming positive. Atoms that are positively or negatively charged are called *ions.*

Both a balloon and your hair contain a very large number of neutral atoms. Charge has a natural tendency to be transferred between unlike materials. Rubbing the two materials together serves to increase the area of contact and thus enhance the charge-transfer process. When a balloon is rubbed against your hair, some of your hair's electrons are transferred to the balloon. Thus, the balloon gains a certain amount of negative charge while your hair loses an equal amount of negative charge and hence is left with a positive charge. In this and similar experiments, only a small portion of the total available charge is transferred from one object to another. (★)TEKS **3A**

The positive charge on your hair is equal in magnitude to the negative charge on the balloon. Electric charge is conserved in this process; no charge is created or destroyed. This principle of conservation of charge is one of the fundamental laws of nature.

(a)

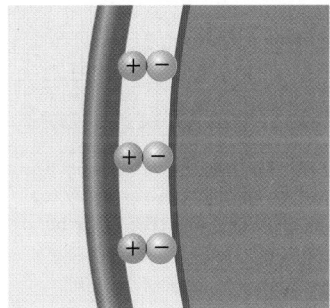

(b)

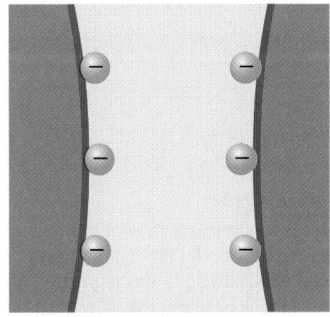

Figure 17-2

(a) This negatively charged balloon is attracted to positively charged hair because the two have opposite charges. **(b)** Two negatively charged balloons repel one another because they have the same charge.

(★)TEKS **3C, 6C**

Did you know?

Some cosmetic products contain an organic compound called *chitin,* which is found in crabs, lobsters, and butterflies and other insects. Chitin is positively charged, so it helps cosmetic products stick to human hair and skin, which are usually slightly negatively charged.

Demonstration 1

Effects of charge

Purpose Show two kinds of charge by use of a pith ball.

Materials pith ball suspended by thread, rubber rod and fur (or balloon and hair), glass rod and silk (or plastic wrap)

Procedure Charge the rubber rod with the fur (or the balloon with hair) by friction. Briefly touch the pith ball with the rubber rod and withdraw the rod. Have students observe the reaction of the pith ball, which is now negatively charged, as you bring the rubber rod near, and ask students to explain this reaction.

Now ground the pith ball with your finger and explain the grounding effect. *(Excess electrons leave the pith ball.)*

Repeat the above procedure with the glass rod, which becomes positively charged, and have students observe the same results. Without grounding the pith ball, bring a negatively charged rubber rod near the pith ball. Have students explain the resulting attraction. Explain that at least two kinds of charges are demonstrated here: one that repels the charged pith ball and one that attracts it.

(★)TEKS

p. 628: 6C
p. 629: 3C, 6C, 3A, 3C, 6C

Visual Strategy

Figure 17-3

Point out to students that Millikan used a positively charged metal plate to cause a negatively charged drop to rise.

Q What evidence do you see in the picture to indicate that electric forces are stronger than gravitational forces?

A *A small amount of electric charge provides a force that more than balances the gravitational attraction of the entire Earth.*

Q Will all of the oil drops rise toward the top plate when it is given a positive charge?

A *No, those that are positively charged will continue moving downward.*

Teaching Tip

Advanced students may want more information about how Millikan found that charge is quantized. Millikan measured the time intervals of a drop falling due to the force of gravity and rising due to electrical attraction and the corresponding distances. He then calculated the drop's velocities with this data. Through a relationship between the drop's upward velocity and the electric field strength ($v_{upward} \propto Eq - mg$), Millikan determined the charge on the drop. By repeating this process for thousands of drops, Millikan found that the charge on each drop was an integral multiple of a fundamental unit of charge. Encourage interested students to research these details after electric field strength has been introduced (Section 17-3) and present their findings to the class.

(★)TEKS 2C

Did you know?

In typical electrostatic experiments, in which an object is charged by rubbing, a net charge on the order of 10^{-6} C ($= 1\ \mu$C) is obtained. This is a very small fraction of the total amount of charge within each object.

Figure 17-3
This is a schematic view of the apparatus used by Millikan in his oil-drop experiment. In this experiment, Millikan found that there is a fundamental unit of charge.

CONCEPT PREVIEW ➤

Certain experiments have suggested the existence of fundamental particles, called *quarks,* that have charges of $\pm\frac{1}{3}e$ or $\pm\frac{2}{3}e$. You will learn about these particles and their properties in Chapter 25.

internet connect
SCiLINKS.
NSTA
TOPIC: Electric charge
GO TO: www.scilinks.org
*sci*LINKS CODE: HF2171

Electric charge is quantized

In 1909, Robert Millikan (1886–1953) performed an experiment at the University of Chicago in which he observed the motion of tiny oil droplets between two parallel metal plates, as shown in **Figure 17-3.** The oil droplets were charged by friction in an atomizer and allowed to pass through a hole in the top plate. Initially the droplets fell due to their weight. The top plate was given a positive charge as the droplets fell, and the droplets with a negative charge were attracted back upward toward the positively charged plate. By turning the charge on this plate on and off, Millikan was able to watch a single oil droplet for many hours as it alternately rose and fell. **(★)TEKS** 3E, 6B

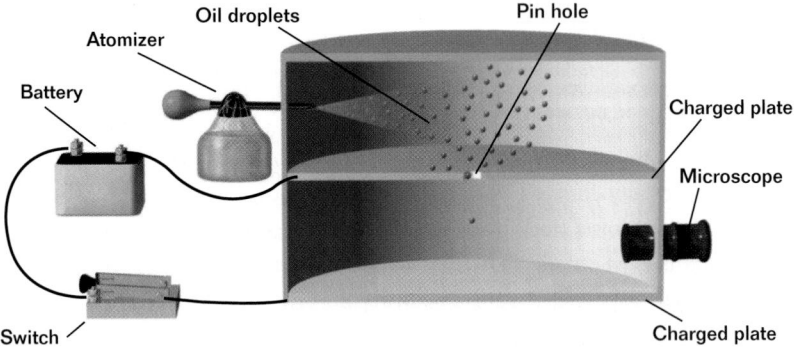

After repeating this process for thousands of drops, Millikan found that when an object is charged, its charge is always a multiple of a fundamental unit of charge, symbolized by the letter *e*. In modern terms, charge is said to be *quantized*. This means that charge occurs as discrete amounts in nature. Thus, an object may have a charge of $\pm e$, or $\pm 2e$, or $\pm 3e$, and so on.

Other experiments in Millikan's time demonstrated that the electron has a charge of $-e$ and the proton has an equal and opposite charge, $+e$. The value of *e* has since been determined to be $1.602\ 19 \times 10^{-19}$ C, where the coulomb (C) is the SI unit of electric charge. For calculations, this book will use the approximate value given in **Table 17-1.** A total charge of -1.0 C contains 6.2×10^{18} electrons ($1/e$). Comparing this with the number of free electrons in 1 cm³ of copper, which is on the order of 10^{23}, shows that 1.0 C is a substantial amount of charge. **(★)TEKS** 2C

Table 17-1	Charge and mass of atomic particles	
Particle	**Charge (C)**	**Mass (kg)**
electron	-1.60×10^{-19}	9.109×10^{-31}
proton	$+1.60 \times 10^{-19}$	1.673×10^{-27}
neutron	0	1.675×10^{-27}

TRANSFER OF ELECTRIC CHARGE

When a balloon and your hair are charged by rubbing, only the rubbed areas become charged, and there is no tendency for the charge to move into other regions of the material. In contrast, when materials such as copper, aluminum, and silver are charged in some small region, the charge readily distributes itself over the entire surface of the material. For this reason, it is convenient to classify substances in terms of their ability to transfer electric charge.

Materials in which electric charges move freely, such as copper and aluminum, are called **conductors.** Most metals are conductors. Materials in which electric charges do not move freely, such as glass, rubber, silk, and plastic, are called **insulators.**

Semiconductors are a third class of materials characterized by electrical properties that are somewhere between those of insulators and conductors. In their pure state, semiconductors are insulators. But the carefully controlled addition of specific atoms as impurities can dramatically increase a semiconductor's ability to conduct electric charge. Silicon and germanium are two well-known semiconductors that are used in a variety of electronic devices.

Certain metals belong to a fourth class of materials, called *superconductors.* Superconductors become perfect conductors when they are at or below a certain temperature. You will study semiconductors in greater detail in Chapter 24, and you will learn more about superconductors in Chapters 19 and 24.

Insulators and conductors can be charged by contact

In the experiments discussed above, a balloon and hair become charged because they are rubbed together. This process is known as *charging by contact.* Another example of charging by contact is a common experiment in which a glass rod is rubbed with silk and a rubber rod is rubbed with wool or fur. The two rods become oppositely charged and attract one another, like a balloon and your hair do. If two glass rods are charged, the rods have the same charge and repel each other, just as two charged balloons do. Likewise, two charged rubber rods repel one another. All of the materials used in these experiments—glass, rubber, silk, wool, and fur—are insulators. Can conductors also be charged by contact?

If you try a similar experiment with a copper rod, the rod does not attract or repel another charged rod. This might suggest that a metal cannot be charged by contact. However, if you hold the copper rod with an insulator and then rub it with wool or fur, the rod attracts a charged glass rod and repels a charged rubber rod.

In the first case, the electric charges produced by rubbing readily move from the copper through your body and finally to Earth because copper is a conductor. So the copper rod does become charged, but it soon becomes neutral again. In the second case, the insulating handle prevents the flow of charge to Earth, and the copper rod remains charged. Thus, both insulators and conductors can become charged by contact.

---

conductor

material that transfers charge easily

insulator

material that does not transfer charge easily

(★)TEKS 3A, 6C

Conceptual Challenge

1. Plastic wrap

Plastic wrap becomes electrically charged as it is pulled from its container, and, as a result, it is attracted to objects such as food containers. Explain why plastic is a good material for this purpose.

2. Charge transfer

If a glass rod is rubbed with silk, the glass becomes positive and the silk becomes negative. Compare the mass of the glass rod before and after it is charged.

3. Electrons

Many objects in the large-scale world have no net charge, even though they contain an extremely large amount of electrons. How is this possible?

Demonstration 2

Jumping spices

Purpose Show charging by contact and charging by induction.

Materials sheet of Plexiglas or plastic (plastic picture holder works well), cloth (wool or silk), dried spices (parsley, sage, etc.), overhead projector

Procedure *Note:* Try both wool and silk cloth ahead of time to see which creates a greater charge on the plastic.

Sprinkle spices on the projector. Use cloth to charge the plastic. Hold the charged plastic $\frac{1}{2}$ in. to 1 in. above the spices. Some pieces will stick to the plastic, but many will jump up and down repeatedly. Challenge students to explain the jumping of the initially neutral spices. This will work even in high humidity with vigorous rubbing for approx. 45 s.

ANSWERS TO

Conceptual Challenge

1. because plastic is an insulator, which holds electric charge
2. The glass rod's mass is slightly less after the rubbing because some of its electrons are transferred to the silk.
3. Each electron is neutralized by a positively charged proton.

(★)TEKS

p. 630: 2C, 3E, 6B, 2C
p. 631: 3A, 6C

Quick Lab

TEACHER'S NOTES

In this experiment, a polarized stream of water is deflected by a charged balloon. The lab works extremely well, even on humid days. Make sure the stream is small. Have students draw a schematic diagram representing the charges of the water and balloon.

By changing the height and horizontal distance of the balloon, students can drastically change the path of the stream. Challenge the students to explain these results in terms of Newton's laws and projectile motion.

Visual Strategy

Figure 17-4

Be sure students understand that the positive charge on the sphere does not move. In each case, the rearrangement of positive charge is due to the motion of negative charge.

Q Why do electrons leave the sphere in **(b)**?

A *The electrons are repelled by the negatively charged rod.*

⭐TEKS

p. 632: 1A, 6C, 3A
p. 633: 3A, 2C, 6C

Quick Lab

Polarization

MATERIALS LIST

✔ balloon

✔ water faucet

Turn on a water faucet and adjust the flow of water so that you have a small but steady stream. The stream should be as slow as possible without producing individual droplets. Inflate the balloon, tie it, and rub it across your hair. Hold the charged end of the balloon near the stream without letting the balloon get wet. What happens to the stream of water? What might be causing this to happen?

⭐TEKS **1A, 6C**

induction

the process of charging a conductor by bringing it near another charged object and grounding the conductor

Figure 17-4
(a) When a charged rubber rod is brought near a metal sphere, the charge on the sphere becomes redistributed. **(b)** If the sphere is grounded, some of the electrons travel through the wire to the ground. **(c)** When this wire is removed, the sphere has an excess of positive charge, which **(d)** becomes evenly distributed on the surface of the sphere when the rod is removed.

Conductors can be charged by induction

When a conductor is connected to Earth by means of a conducting wire or copper pipe, the conductor is said to be *grounded.* The Earth can be considered an infinite reservoir for electrons because it can accept or supply an unlimited number of electrons. This is the key to understanding another method of charging a conductor.

Consider a negatively charged rubber rod brought near a neutral (uncharged) conducting sphere that is insulated so that there is no conducting path to ground. The repulsive force between the electrons in the rod and those in the sphere causes a redistribution of negative charge on the sphere, as shown in **Figure 17-4(a).** As a result, the region of the sphere nearest the negatively charged rod has an excess of positive charge.

If a grounded conducting wire is then connected to the sphere, as shown in **Figure 17-4(b),** some of the electrons leave the sphere and travel to Earth. If the wire to ground is then removed while the negatively charged rod is held in place, as shown in **Figure 17-4(c),** the conducting sphere is left with an excess of induced positive charge. Finally, when the rubber rod is removed from the vicinity of the sphere, as in **Figure 17-4(d),** the induced positive charge remains on the ungrounded sphere. The motion of negative charges on the sphere causes the positive charge to become uniformly distributed over the outside surface of the ungrounded sphere. This process is known as **induction,** and the charge is said to be *induced* on the sphere. ⭐TEKS **3A**

Notice that charging an object by induction requires no contact with the object inducing the charge but does require contact with a third object, which serves as either a source or a sink of electrons. In the process of inducing a charge on the sphere, the charged rubber rod did not come in contact with the sphere and thus did not lose any of its negative charge. This is in contrast to charging an object by contact, in which charges are transferred directly from one object to another.

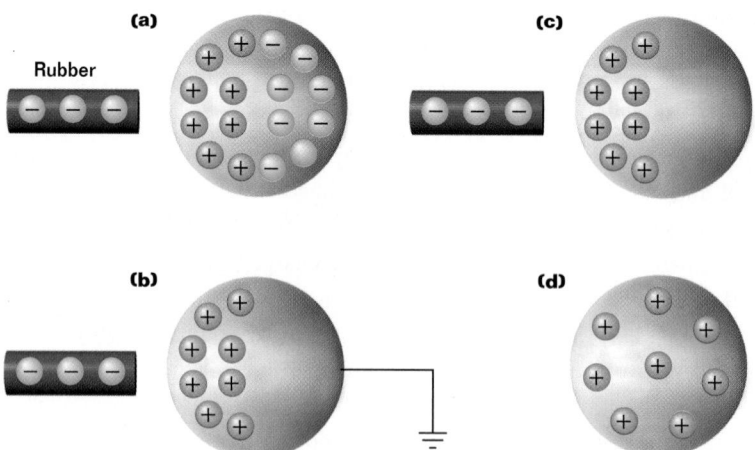

A surface charge can be induced on insulators by polarization

A process very similar to charging by induction in conductors takes place in insulators. In most neutral atoms or molecules, the center of positive charge coincides with the center of negative charge. In the presence of a charged object, these centers may shift slightly, resulting in more positive charge on one side of a molecule than on the other. This is known as *polarization*. (★)TEKS **3A**

This realignment of charge within individual molecules produces an induced charge on the surface of the insulator, as shown in **Figure 17-5(a).** When an object becomes polarized, it has no net charge but is still able to attract or repel objects due to this realignment of charge. This explains why a plastic comb can attract small pieces of paper that have no net charge, as shown in **Figure 17-5(b).** As with induction, in polarization one object induces a charge on the surface of another object with no physical contact.

internetconnect

SCI*LINKS*
NSTA

TOPIC: Conductors and insulators
GO TO: www.scilinks.org
*sci*LINKS CODE: HF2172

(a)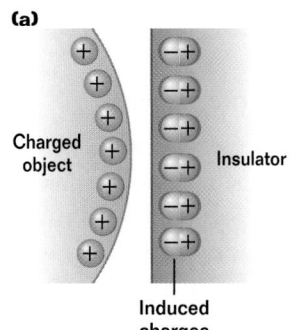

Charged object

Induced charges

Insulator

(b)

Figure 17-5
(a) The charged object on the left induces charges on the surface of an insulator, which is said to be *polarized.* **(b)** This charged comb induces a charge on the surface of small pieces of paper that have no net charge.

Section Review

 (★)TEKS **2C, 6C**

1. When a rubber rod is rubbed with wool, the rod becomes negatively charged. What can you conclude about the magnitude of the wool's charge after the rubbing process? Why?

2. What did Millikan's oil-drop experiment reveal about the nature of electric charge?

3. A typical lightning bolt has about 10.0 C of charge. How many excess electrons are in a typical lightning bolt?

4. If you stick a piece of transparent tape on your desk and then quickly pull it off, you will find that the tape is attracted to other areas of your desk that are not charged. Why does this happen?

5. Metals, such as copper and silver, can become charged by induction, while plastic materials cannot. Explain why.

6. **Physics in Action** Why is an electrostatic spray gun more efficient than an ordinary spray gun?

Demonstration 3

Polarization

Purpose Show the effects of polarization.

Materials rubber rod and fur, small piece of paper (approximately 5 mm × 5 mm), watch glass, meterstick

Procedure Charge the rubber rod with the fur, touch the rod to the paper, and lift the rod. The paper will be lifted up by the rod. Explain that the negatively charged rod induced a positive charge on the paper, and the two are attracted to one another.

Next balance the meterstick on the inverted watch glass. Charge the rod with the wool and bring it close to the end of the meterstick, which slowly begins to rotate. Point out that again there was a redistribution of electrons, producing areas of induced positive and negative charge on the end of the meterstick.

Section Review
ANSWERS

1. It is equal to the magnitude of the rod's charge; Charge is conserved.
2. Charge is quantized.
3. 6.25×10^{19} electrons
4. The tape induces a surface charge on the desk, so the two are attracted to one another.
5. because plastic, an insulator, does not easily conduct charge
6. More paint hits the object being painted because of an electrical attraction between the charged droplets and the oppositely charged object.

Demonstration 4

Electric force

Purpose Qualitatively illustrate the dependence of electric force on distance. (*Optional:* Qualitatively illustrate the dependence of electric force on charge.)

Materials balloon, water faucet, Leyden jar (optional)

Procedure Repeat the Quick Lab "Polarization" on page 632 a number of times, varying the distance between the balloon and the water stream in each trial. Have students observe the displacement of the water stream for different distances, and ask them to describe the relationship between distance and the force exerted on the stream.

Optional: Using a constant distance, show the effects of varying the charge of a Leyden jar on the water stream. Have students describe the relationship between the amount of charge and the force on the stream.

The Language of Physics

Some texts simply use k for the Coulomb constant. This book uses k_C instead so that students do not confuse the Coulomb constant with the spring constant (k).

⭐ TEKS

p. 634: 3B, 3E, 6C
p. 635: 2C, 6C, 2C, 6A, 6C

17-2 SECTION OBJECTIVES

- **Calculate electric force using Coulomb's law.**

- **Compare electric force with gravitational force.**

- **Apply the superposition principle to find the resultant force on a charge and to find the position at which the net force on a charge is zero.**

⬛ internet**connect**

SC*LINKS*
NSTA
TOPIC: Coulomb's law
GO TO: www.scilinks.org
*sci***LINKS CODE:** HF2173

COULOMB'S LAW

In Chapter 4, a force is defined as the cause of a change in motion. Because two charged objects near one another may experience motion either toward or away from each other, each object exerts a force on the other object. This force is called the *electric force.* The two balloon experiments described in Section 17-1 demonstrate that the electric force is attractive between opposite charges and repulsive between like charges. What determines how small or large the electric force will be?

The closer two charges are, the greater the force between them

If you rub a balloon against your hair, you will find that the closer the balloon is to your hair, the stronger the attraction. Likewise, the repulsion between two charged balloons becomes stronger as the distance between the balloons decreases. Thus, the distance between two objects affects the magnitude of the electric force between them. Further, it seems reasonable that the amount of charge on the objects will also affect the magnitude of the electric force. What is the precise relationship between distance, charge, and the electric force?

In the 1780s, Charles Coulomb conducted a variety of experiments in an attempt to determine the magnitude of the electric force between two charged objects. Coulomb found that the electric force between two charges is proportional to the product of the two charges. Hence, if one charge is doubled, the electric force likewise doubles, and if both charges are doubled, the electric force increases by a factor of four. Coulomb also found that the electric force is inversely proportional to the square of the distance between the charges. Thus, when the distance between two charges is halved, the force between them increases by a factor of four. The following equation, known as Coulomb's law, expresses these conclusions mathematically for two charges separated by a distance, r. ⭐ TEKS **3B, 3E, 6C**

COULOMB'S LAW

$$F_{electric} = k_C \left(\frac{q_1 q_2}{r^2} \right)$$

$$\text{electric force} = \text{Coulomb constant} \times \frac{(\text{charge 1})(\text{charge 2})}{(\text{distance})^2}$$

The symbol k_C, called the Coulomb constant, has SI units of $N \cdot m^2/C^2$ because this gives N as the unit of electric force. The value of k_C depends on the choice of units. Experiments have determined that in SI units, k_C has the value $8.9875 \times 10^9 \ N \cdot m^2/C^2$. This book will simplify calculations by using the approximate value $8.99 \times 10^9 \ N \cdot m^2/C^2$. (★)TEKS) 2C

When dealing with Coulomb's law, remember that force is a vector quantity and must be treated accordingly. The electric force between two objects always acts along the line between the objects. Also note that Coulomb's law applies exactly only to point charges or particles and to spherical distributions of charge. When applying Coulomb's law to spherical distributions of charge, use the distance between the centers of the spheres as r. (★)TEKS) 6C

Module 16
"Force Between Charges"
provides an interactive lesson with guided problem-solving practice to teach you about the electric forces between all kinds of objects, including point charges.

SAMPLE PROBLEM 17A

Coulomb's law (★)TEKS) 2C, 6A, 6C

PROBLEM

The electron and proton of a hydrogen atom are separated, on average, by a distance of about 5.3×10^{-11} m. Find the magnitudes of the electric force and the gravitational force that each particle exerts on the other.

SOLUTION

1. DEFINE   **Given:**
$$r = 5.3 \times 10^{-11} \ m$$
$$k_C = 8.99 \times 10^9 \ N \cdot m^2/C^2$$
$$m_e = 9.109 \times 10^{-31} \ kg$$
$$m_p = 1.673 \times 10^{-27} \ kg$$
$$q_e = -1.60 \times 10^{-19} \ C$$
$$q_p = +1.60 \times 10^{-19} \ C$$
$$G = 6.67 \times 10^{-11} \ N \cdot m^2/kg^2$$

Unknown: $F_{electric} = ?$ $F_g = ?$

2. PLAN   **Choose an equation(s) or situation:**
Find the magnitude of the electric force using Coulomb's law and the magnitude of the gravitational force using Newton's law of gravitation (introduced in Chapter 7).

$$F_{electric} = k_C \frac{q_1 q_2}{r^2} \qquad F_g = G \frac{m_e m_p}{r^2}$$

3. CALCULATE   **Substitute the values into the equations(s) and solve:**
Because we are finding the magnitude of the electric force, which is a scalar, we can disregard the sign of each charge in our calculation.

continued on next page

Classroom Practice

The following may be used as teamwork exercises or for demonstration at the board or on an overhead projector.

PROBLEM

Coulomb's law

Consider the forces $F_1 = F$ and $F_2 = -F$ acting on two charged particles separated by a distance, d. Explain the change in the forces exerted on each particle under the following conditions:

   a. the distance between the two particles doubles

   b. the charge on one particle doubles

   c. the charge on each particle doubles

   d. the charge on each particle and the distance between the two particles double

Answer
   a. $F_1 = \frac{1}{4}F, \ F_2 = -\frac{1}{4}F$
   b. $F_1 = 2F, \ F_2 = -2F$
   c. $F_1 = 4F, \ F_2 = -4F$
   d. $F_1 = F, \ F_2 = -F$

Interactive Problem-Solving Tutor

See Module 16
"Force Between Charges" promotes additional development of problem-solving skills for this chapter.

Teaching Tip

Relate the material in this chapter to previous discussions of force. According to Newton's third law, the force exerted by one body on another is equal in magnitude to the force exerted by the second body on the first. Thus, finding the magnitude of the electric force *between* two bodies is equivalent to finding the force exerted by one body on the other. To find the acceleration that each body experiences due to that force, the masses of each body must be considered ($F = ma$).

PRACTICE GUIDE 17A

Solving for:		
$F_{electric}$	📖 **PE**	Sample, 1–2, 3a; Ch. Rvw. 18–20
	💿 **PW**	5–7
	PB	4–6
q	💿 **PW**	Sample, 1–4
	PB	7–10
r	📖 **PE**	4
	💿 **PW**	8–10
	PB	Sample, 1–3

ANSWERS TO

Practice 17A
Coulomb's law

1. 230 N (attractive)
2. **a.** 2.2×10^{-5} N (attractive)
   b. 9.0×10^{-7} N (repulsive)
3. **a.** 16 N
   b. attractive
   c. 2.7×10^{13} electrons
   d. 3.8×10^{13} electrons
4. 39.3 cm

$$F_{electric} = k_C \frac{q_e q_p}{r^2} = \left(8.99 \times 10^9 \frac{\text{N} \cdot \text{m}^2}{\text{C}^2}\right)\left(\frac{(1.60 \times 10^{-19} \text{ C})^2}{(5.3 \times 10^{-11} \text{ m})^2}\right)$$

$$\boxed{F_{electric} = 8.2 \times 10^{-8} \text{ N}}$$

$$F_g = G \frac{m_e m_p}{r^2} =$$

$$\left(6.67 \times 10^{-11} \frac{\text{N} \cdot \text{m}^2}{\text{kg}^2}\right)\left(\frac{(9.109 \times 10^{-31} \text{ kg})(1.673 \times 10^{-27} \text{ kg})}{(5.3 \times 10^{-11} \text{ m})^2}\right)$$

$$\boxed{F_g = 3.6 \times 10^{-47} \text{ N}}$$

4. EVALUATE   The electron and the proton have opposite signs, so the electric force between the two particles is attractive. The ratio $F_{electric}/F_g \approx 2 \times 10^{39}$; hence, the gravitational force between the particles is negligible compared with the electric force between them. Because each force is inversely proportional to distance squared, this result is independent of the distance between the two particles.

PRACTICE 17A

Coulomb's law ⭐TEKS 2C, 6C

1. A balloon rubbed against denim gains a charge of -8.0 μC. What is the electric force between the balloon and the denim when the two are separated by a distance of 5.0 cm? (Assume that the charges are located at a point.)

2. Two identical conducting spheres are placed with their centers 0.30 m apart. One is given a charge of $+12 \times 10^{-9}$ C and the other is given a charge of -18×10^{-9} C.
   a. Find the electric force exerted on one sphere by the other.
   b. The spheres are connected by a conducting wire. After equilibrium has occurred, find the electric force between the two spheres.

3. A small cork with an excess charge of $+6.0$ μC (1 μC $= 10^{-6}$ C) is placed 0.12 m from another cork, which carries a charge of -4.3 μC.
   a. What is the magnitude of the electric force between the corks?
   b. Is this force attractive or repulsive?
   c. How many excess electrons are on the negative cork?
   d. How many electrons has the positive cork lost?

4. Two electrostatic point charges of $+60.0$ μC and $+50.0$ μC exert a repulsive force on each other of 175 N. What is the distance between the two charges?

Electric force is a field force

The Coulomb force is the second example we have studied of a force that is exerted by one object on another even though there is no physical contact between the two objects. As you learned in Chapter 4, such a force is known as a *field force*. Recall that another example of a field force is gravitational attraction. Notice that the mathematical form of the Coulomb force is very similar to that of the gravitational force. That is, both forces are inversely proportional to the square of the distance of separation. ⭐TEKS **6A, 6C**

However, there are some important differences between electric and gravitational forces. First of all, as you have seen, electric forces can be either attractive or repulsive. Gravitational forces, on the other hand, are always attractive. This is because objects can have either a positive or a negative charge, while mass is always positive.

Another difference between the gravitational force and the electric force is their relative strength. As shown in Sample Problem 17A, the electric force is significantly stronger than the gravitational force. As a result, the electric force between charged atomic particles is much stronger than their gravitational attraction to Earth.

In the large-scale world, the relative strength of these two forces can be seen by noting that the amount of charge required to overcome the gravitational force is relatively small. For example, if you rub a balloon against your hair and hold the balloon directly above your hair, your hair will stand on end because it is attracted toward the balloon. Although only a small amount of charge is transferred from your hair to the balloon, the electric force between the two is nonetheless stronger than the gravitational force that usually pulls your hair toward the ground.

Resultant force on a charge is the vector sum of the individual forces on that charge

Frequently, more than two charges are present, and it is necessary to find the net electric force on one of them. As demonstrated in Sample Problem 17A, Coulomb's law gives the electric force between any pair of charges. Coulomb's law applies even when more than two charges are present. Thus, the resultant force on any single charge equals the vector sum of the individual forces exerted on that charge by all of the other individual charges that are present. This is an example of the *principle of superposition*.

This method was introduced in Chapter 3 and was used to add forces together vectorially in Chapter 4. The process is essentially the same here; the only difference is that in this chapter we are considering electric forces, whereas in Chapter 4 we worked primarily with other kinds of forces. Once the magnitudes of the individual electric forces are found by Coulomb's law, they are added together exactly as forces were added together in Chapter 4. This process is demonstrated in Sample Problem 17B.

⭐TEKS **3A, 6C**

Conceptual Challenge

1. Electric force

The electric force is significantly stronger than the gravitational force. However, although we are attracted to Earth by gravity, we do not usually feel the effects of the electric force. Explain why.

2. Electrons in a coin

An ordinary nickel contains about 10^{24} electrons, all repelling one another. Why don't these electrons fly off the nickel?

3. Charged balloons

When the distance between two negatively charged balloons is doubled, by what factor does the repulsive force between them change?

🛑 Misconception Alert

Because the effects of the gravitational force are more common in our typical experiences than the effects of the electric force and because gravity acts on such a large scale, some students may think that the gravitational force is stronger than the electric force. Remind them that the electric force between a proton and an electron is much larger than the gravitational force between them (shown in Sample Problem 17A).

ANSWERS TO

Conceptual Challenge

1. Earth's gravitational effects are significant because Earth has such a large mass. Electric effects are not usually felt because most objects are electrically neutral, that is, they have the same number of electrons and protons.

2. Their attraction to the protons in the nickel overcomes their repulsion and holds them in the nickel.

3. The force decreases by a factor of four.

⭐TEKS

p. 636: 2C, 6C
p. 637: 6A, 6C, 3A 6C

Classroom Practice

The following may be used as teamwork exercises or for demonstration at the board or on an overhead projector.

PROBLEM

The superposition principle

Four equal charges of 1.5 μC are placed at the corners of a square with 5.0 cm sides. Find the net force on a fifth charge placed in the center of the square if the new charge is

a. −1.5 μC

b. +3.0 μC

Answer

   a. 0.0 N

   b. 0.0 N

Three charges are located on the *x*-axis. A 5.0 μC charge is located at *x* = 0.0 cm, a 1.5 μC charge is located at *x* = 3.0 cm, and a −3.0 μC charge is located at *x* = 5.0 cm. Find the resultant force on the 5.0 μC charge.

Answer

   21 N, along the negative *x*-axis

★ TEKS

p. 638: 2C, 3B, 6C
p. 639: 2C, 3B, 6C

The superposition principle   **★ TEKS**   2C, 3B, 6C

PROBLEM

Consider three point charges at the corners of a triangle, as shown in Figure 17-6, where $q_1 = 6.00 \times 10^{-9}$ C, $q_2 = -2.00 \times 10^{-9}$ C, and $q_3 = 5.00 \times 10^{-9}$ C. Find the magnitude and direction of the resultant force on q_3.

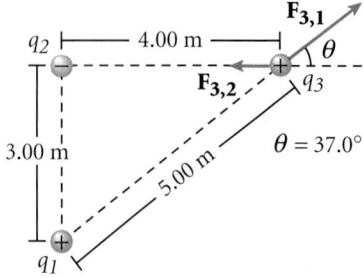

Figure 17-6

REASONING

According to the superposition principle, the resultant force on the charge q_3 is the vector sum of the forces exerted by q_1 and q_2 on q_3. First find the force exerted on q_3 by each, and then add these two forces together vectorially to get the resultant force on q_3.

SOLUTION

Given: $q_1 = +6.00 \times 10^{-9}$ C $r_{2,1} = 3.00$ m

               $q_2 = -2.00 \times 10^{-9}$ C $r_{3,2} = 4.00$ m

               $q_3 = +5.00 \times 10^{-9}$ C $r_{3,1} = 5.00$ m

               $k_C = 8.99 \times 10^9$ N•m²/C² $\theta = 37.0°$

Unknown: $\mathbf{F}_{3,\text{tot}} = ?$

Diagram:

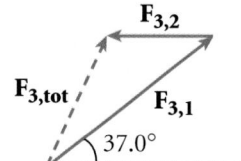

Figure 17-7

1. **Calculate the magnitude of the forces with Coulomb's law:**

$$F_{3,1} = k_C \frac{q_3 q_1}{(r_{3,1})^2} = (8.99 \times 10^9 \text{ N•m}^2/\text{C}^2)\left(\frac{(5.00 \times 10^{-9} \text{ C})(6.00 \times 10^{-9} \text{ C})}{(5.00 \text{ m})^2}\right)$$

$$F_{3,1} = 1.08 \times 10^{-8} \text{ N}$$

$$F_{3,2} = k_C \frac{q_3 q_2}{(r_{3,2})^2} = (8.99 \times 10^9 \text{ N•m}^2/\text{C}^2)\left(\frac{(5.00 \times 10^{-9} \text{ C})(2.00 \times 10^{-9} \text{ C})}{(4.00 \text{ m})^2}\right)$$

$$F_{3,2} = 5.62 \times 10^{-9} \text{ N}$$

2. **Determine the direction of the forces by analyzing the charges:**

The force $\mathbf{F}_{3,1}$ is repulsive because q_1 and q_3 have the same sign.

The force $\mathbf{F}_{3,2}$ is attractive because q_2 and q_3 have opposite signs.

3. **Find the *x* and *y* components of each force:**

At this point, the direction of each component must be taken into account.

For $\mathbf{F}_{3,1}$: $F_x = (F_{3,1})(\cos 37.0°) = (1.08 \times 10^{-8} \text{ N})(\cos 37.0°) = 8.63 \times 10^{-9} \text{ N}$

            $F_y = (F_{3,1})(\sin 37.0°) = (1.08 \times 10^{-8} \text{ N})(\sin 37.0°) = 6.50 \times 10^{-9} \text{ N}$

For $\mathbf{F_{3,2}}$: $F_x = F_{3,2} = -5.62 \times 10^{-9}$ N

$\qquad\qquad F_y = 0$ N

4. **Calculate the magnitude of the total force acting in both directions:**

$F_{x,tot} = 8.63 \times 10^{-9}$ N $- 5.62 \times 10^{-9}$ N $= 3.01 \times 10^{-9}$ N

$F_{y,tot} = 6.50 \times 10^{-9}$ N $+ 0$ N $= 6.50 \times 10^{-9}$ N

5. **Use the Pythagorean theorem to find the magnitude of the resultant force:**

$$F_{3,tot} = \sqrt{(F_{x,tot})^2 + (F_{y,tot})^2} = \sqrt{(3.01 \times 10^{-9}\ \text{N})^2 + (6.50 \times 10^{-9}\ \text{N})^2}$$

$$\boxed{F_{3,tot} = 7.16 \times 10^{-9}\ \text{N}}$$

6. **Use a suitable trigonometric function to find the direction of the resultant force:**

   In this case, you can use the inverse tangent function:

$$\tan \varphi = \frac{F_{y,tot}}{F_{x,tot}} = \frac{6.50 \times 10^{-9}\ \text{N}}{3.01 \times 10^{-9}\ \text{N}}$$

$$\boxed{\varphi = 65.2°}$$

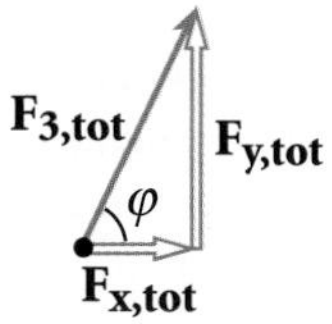

$\mathbf{F_{3,tot}}$ $\mathbf{F_{y,tot}}$ φ $\mathbf{F_{x,tot}}$

Figure 17-8

7. **Evaluate your answer:**

   Because $F_{3,1}$ is repulsive and $F_{3,2}$ is attractive, the magnitude of the resultant force ($F_{3,tot}$) should be between these two values:

1.08×10^{-8} N $(F_{3,1}) > 7.16 \times 10^{-9}$ N $(F_{3,tot}) > 5.62 \times 10^{-9}$ N $(F_{3,2})$.

PRACTICE 17B

The superposition principle ⭐(TEKS) 2C, 3B, 6C

1. Three point charges, q_1, q_2, and q_3, lie along the x-axis at $x = 0$, $x = 3.0$ cm, and $x = 5.0$ cm, respectively. Calculate the magnitude and direction of the electric force on each of the three point charges when $q_1 = +6.0\ \mu$C, $q_2 = +1.5\ \mu$C, and $q_3 = -2.0\ \mu$C.

2. Four charged particles are placed so that each particle is at the corner of a square. The sides of the square are 15 cm. The charge at the upper left corner is $+3.0\ \mu$C, the charge at the upper right corner is $-6.0\ \mu$C, the charge at the lower left corner is $-2.4\ \mu$C, and the charge at the lower right corner is $-9.0\ \mu$C.

   a. What is the net electric force on the $+3.0\ \mu$C charge?
   b. What is the net electric force on the $-6.0\ \mu$C charge?
   c. What is the net electric force on the $-9.0\ \mu$C charge?

Teaching Tip

Point out that in Sample Problem 17B, the angle between $\mathbf{F_{3,tot}}$ and the horizontal is the same in **Figure 17-7** (previous page) and **Figure 17-8** (this page). The difference between these two figures is that different components are depicted in each. **Figure 17-7** shows $\mathbf{F_{3,tot}}$ as the resultant force of the two individual forces acting on q_3. **Figure 17-8,** on the other hand, depicts the x and y components of this resultant force; they are used to find the magnitude of the angle between the resultant force and the horizontal.

PRACTICE GUIDE 17B		
Solving for:		
F	📖 PE	Sample, 1–2; Ch. Rvw. 21–22
	💿 PW	Sample, 1–6
	PB	Sample, 1–10

ANSWERS TO

Practice 17B

The superposition principle

1. 47 N, along the negative x-axis; 157 N, along the positive x-axis; 11.0×10^1 N, along the negative x-axis

2. a. 13.0 N, 31° below the positive x-axis
   b. 25 N, 78° above the negative x-axis
   c. 18 N, 75° below the positive x-axis

PROBLEM

Equilibrium

Two charges, q_1 and q_2, lie on the x-axis. The first charge is at the origin and the second is at $x = 1.0$ m. Determine the equilibrium position for a third charge, q_3, with respect to q_1 and q_2 for each of the following cases:

a. $q_1 = +10.0 \, \mu C$, $q_2 = +7.5 \, \mu C$

b. $q_1 = +3.7$ nC, $q_2 = +5.2$ nC

c. $q_1 = -3.7$ nC, $q_2 = -5.2$ nC

Answer

    a. 0.55 m from q_1

    b. 0.46 m from q_1

    c. 0.46 m from q_1

As seen in Chapter 4, objects that are at rest are said to be in equilibrium. According to Newton's first law, the net external force acting on a body in equilibrium must equal zero. In electrostatic situations, the equilibrium position of a charge is the location at which the net electric force on the charge is zero. To find this location, you must find the position at which the electric force from one charge is equal and opposite the electric force from another charge. This can be done by setting the forces (found by Coulomb's law) equal and then solving for the distance between either charge and the equilibrium position. This is demonstrated in Sample Problem 17C. ★TEKS **3A, 6C**

SAMPLE PROBLEM 17C

Equilibrium ★TEKS 2C, 6C

PROBLEM

Three charges lie along the x-axis. One positive charge, $q_1 = 15 \, \mu C$, is at $x = 2.0$ m, and another positive charge, $q_2 = 6.0 \, \mu C$, is at the origin. At what point on the x-axis must a negative charge, q_3, be placed so that the resultant force on it is zero?

SOLUTION

1. DEFINE

Given:

$q_1 = 15 \, \mu C \, (15 \times 10^{-6} \, C)$ $r_{3,1} = 2.0 \, m - P$

$q_2 = 6.0 \, \mu C \, (6.0 \times 10^{-6} \, C)$ $r_{3,2} = P$

Unknown: the distance (P) between the negative charge q_3 and the positive charge q_2 such that the resultant force on q_3 is zero

Diagram:

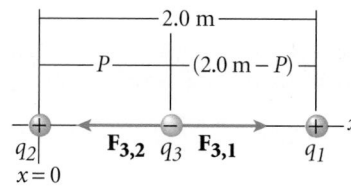

2. PLAN

Choose an equation(s) or situation:

The force exerted on q_3 by q_2 will be opposite the force exerted on q_3 by q_1 if q_3 lies on the x-axis between q_1 and q_2. Since we require that the resultant force on q_3 be zero, then $F_{3,1}$ and $F_{3,2}$, which can each be found by Coulomb's law, must equal one another.

$$F_{3,1} = k_C \left(\frac{q_3 q_1}{(r_{3,1})^2} \right) \text{ and } F_{3,2} = k_C \left(\frac{q_3 q_2}{(r_{3,2})^2} \right)$$

$$F_{3,1} = F_{3,2}$$

$$k_C \left(\frac{q_3 q_1}{(r_{3,1})^2} \right) = k_C \left(\frac{q_3 q_2}{(r_{3,2})^2} \right)$$

3. CALCULATE   Substitute the values into the equation(s) and solve:

$$k_C\left(\frac{q_3(15\times10^{-6}\text{ C})}{(2.0\text{ m}-P)^2}\right) = k_C\left(\frac{q_3(6.0\times10^{-6}\text{ C})}{P^2}\right)$$

Because k_C, 10^{-6}, and q_3 are common terms, these can be canceled from both sides of the equation. The unit C also cancels.

$$\frac{15}{(2.0\text{ m}-P)^2} = \frac{6.0}{P^2}$$

Now solve for P to find the location of q_3.

$$(P^2)(15) = (2.0\text{ m}-P)^2(6.0)$$

Take the square root of both sides, then isolate P.

$$P\sqrt{15} = (2.0\text{ m}-P)\sqrt{6.0}$$

$$P = (2.0\text{ m}-P)\sqrt{\frac{6.0}{15}}$$

$$P = 1.3\text{ m} - (0.63)(P)$$

$$(1.63)(P) = 1.3\text{ m}$$

$$\boxed{P = 0.80\text{ m from }q_2}$$

4. EVALUATE   Because q_1 is greater than q_2, we would expect the location of q_3 to be closer to q_2 in order for the forces to balance. Our answer seems reasonable because q_3 is 0.80 m from q_2 and 1.2 m (2.0 m − 0.80 m) from q_1.

PRACTICE 17C

Equilibrium ⭐TEKS 2C, 6C

1. A charge of $+2.00\times10^{-9}$ C is placed at the origin, and another charge of $+4.00\times10^{-9}$ C is placed at $x=1.5$ m. Find the point between these two charges where a charge of $+3.00\times10^{-9}$ C should be placed so that the net electric force on it is zero.

2. A charge q_1 of -5.00×10^{-9} C and a charge q_2 of -2.00×10^{-9} C are separated by a distance of 40.0 cm. Find the equilibrium position for a third charge of $+15.0\times10^{-9}$ C.

3. An electron is released above the Earth's surface. A second electron directly below it exerts just enough of an electric force on the first electron to cancel the gravitational force on it. Find the distance between the two electrons.

Teaching Tip

Point out that for equilibrium problems such as Sample Problem 17C, the location of the equilibrium position does not depend on the magnitude or sign of the third charge.

PRACTICE GUIDE 17C		
Solving for:		
P	📖 **PE**	Sample, 1–2; Ch. Rvw. 23–24
	💿 **PW**	7–10
	PB	4–6
q	📖 **PE**	Ch. Rvw. 47*, 48
	💿 **PW**	Sample, 1–6
	PB	7–10
F	📖 **PW**	11–12, 13*
	💿 **PB**	Sample, 1–3

ANSWERS TO

Practice 17C
Equilibrium

1. $x=0.64$ m
2. 24.5 cm from q_1 (15.5 cm from q_2)
3. 5.07 m

Coulomb quantified electric force with a torsion balance

Earlier in this chapter, you learned that Charles Coulomb was the first person to quantify the electric force and establish the inverse square law for electric charges. Coulomb measured electric forces between charged objects with a torsion balance, as shown in **Figure 17-9**. A torsion balance consists of two small spheres fixed to the ends of a light horizontal rod. The rod is made of an insulating material and is suspended by a silk thread. (★)TEKS 3E, 6C

In this experiment, one of the spheres is given a charge and another charged object is brought near the charged sphere. The attractive or repulsive force between the two causes the rod to rotate and to twist the suspension. The angle through which the rod rotates is measured by the deflection of a light beam reflected from a mirror attached to the suspension. The rod rotates through some angle against the restoring force of the twisted thread before reaching equilibrium. The value of the angle of rotation increases as the charge increases, thereby providing a quantitative measure of the electric force. With this experiment, Coulomb established the equation for electric force introduced at the beginning of this section. More-recent experiments have verified these results to within a very small uncertainty.

Charged object

Charged sphere

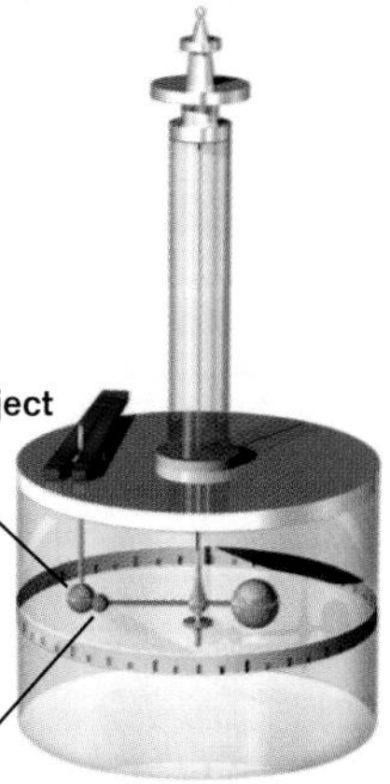

Figure 17-9
Coulomb's torsion balance was used to establish the inverse square law for the electric force between two charges.

Section Review

(★)TEKS 2C, 6B, 6C

1. A small glass ball rubbed with silk gains a charge of $+2.0 \ \mu C$. The glass ball is placed 12 cm from a small charged rubber ball that carries a charge of $-3.5 \ \mu C$.
    a. What is the magnitude of the electric force between the two balls?
    b. Is this force attractive or repulsive?
    c. How many electrons has the glass ball lost in the rubbing process?

2. What are some similarities between the electric force and the gravitational force? What are some differences between the two?

3. A $+2.2 \times 10^{-9}$ C charge is on the x-axis at $x = 1.5$ m, a $+5.4 \times 10^{-9}$ C charge is on the x-axis at $x = 2.0$ m, and a $+3.5 \times 10^{-9}$ C charge is at the origin. Find the net force on the charge at the origin.

4. A charge q_1 of -6.00×10^{-9} C and a charge q_2 of -3.00×10^{-9} C are separated by a distance of 60.0 cm. Where could a third charge be placed so that the net electric force on it is zero?

5. **Physics in Action** The electric force between a negatively charged paint droplet and a positively charged automobile body is increased by a factor of two, but the charges on each remain constant. How has the distance between the two changed? (Assume that the charge on the automobile is located at a single point.)

17-3
The electric field

ELECTRIC FIELD STRENGTH

As discussed earlier in this chapter, electric force, like gravitational force, is a field force. Unlike contact forces, which require physical contact between objects, field forces are capable of acting through space, producing an effect even when there is no physical contact between the objects involved. The concept of a field can help explain how two objects can exert forces on each other at a distance. For example, a charged object sets up an **electric field** in the space around it. When a second charged object enters this field, forces of an electrical nature arise. In other words, the second object interacts with the field of the first particle.

To define an electric field more precisely, consider **Figure 17-10(a),** which shows an object with a small positive charge, q_0, placed near a second object with a larger positive charge, Q. The strength of the electric field, E, at the location of q_0 is defined as the magnitude of the electric force acting on q_0 divided by the charge of q_0: **(★)TEKS) 6C**

$$E = \frac{F_{electric}}{q_0}$$

Note that this is the electric field at the location of q_0 produced by the charge Q, and *not* the field produced by q_0.

Because electric field strength is a ratio of force to charge, the SI units of E are newtons per coulomb (N/C). The electric field is a vector quantity. By convention, the direction of **E** at a point is defined as the direction of the electric force that would be exerted on a small *positive* charge (called a test charge) placed at that point. Thus, in **Figure 17-10(a),** the direction of the electric field is horizontal and to the right because a positive charge would be repelled by the positive sphere. In **Figure 17-10(b),** the direction of the electric field is to the left because a positive charge would be attracted toward the negatively charged sphere. In other words, the direction of **E** depends on the sign of the charge producing the field.

17-3 SECTION OBJECTIVES

- **Calculate electric field strength.**

- **Draw and interpret electric field lines.**

- **Identify the four properties associated with a conductor in electrostatic equilibrium.**

electric field

a region in space around a charged object in which a stationary charged object experiences an electric force because of its charge

(a)
(b)

Figure 17-10
(a) A small object with a positive charge placed in the field, **E,** of an object with a larger positive charge experiences an electric force directed to the right. **(b)** A small object with a positive charge placed in the field, **E,** of a negatively charged object experiences an electric force directed to the left.

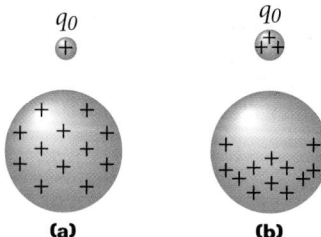

(a)   **(b)**

Figure 17-11
We must assume a small test charge, as in **(a)**, because a larger test charge, as in **(b)**, can cause a redistribution of the charge on the sphere, which changes the electric field strength.

CONCEPT PREVIEW

Just as gravitational potential energy depends on an object's position in a gravitational field, a type of potential energy known as electrical potential energy depends on a charge's position in an electric field. You will learn more about electrical potential energy in Chapter 18.

Now consider the positively charged conducting sphere in **Figure 17-11(a)**. The field in the region surrounding the sphere could be explored by placing a positive test charge, q_0, in a variety of places near the sphere. To find the electric field at each point, you would first find the electric force on this charge, then divide this force by the magnitude of the test charge.

However, when the magnitude of the test charge is great enough to influence the charge on the conducting sphere, a difficulty with our definition arises. According to Coulomb's law, a strong test charge will cause a rearrangement of the charges on the sphere, as shown in **Figure 17-11(b)**. As a result, the force exerted on the test charge is different from what the force would be if the movement of charge on the sphere had not taken place. Furthermore, the strength of the measured electric field is different from what it would be in the absence of the test charge. To eliminate this problem, we assume that the test charge is small enough to have a negligible effect on the location of the charges on the sphere. ⭐TEKS **6C**

Electric field strength depends on charge and distance

To reformulate our equation for electric field strength from a point charge, consider a charge, q, located a distance, r, from a small test charge, q_0. According to Coulomb's law, the magnitude of the force on the test charge is given by the following equation:

$$F_{electric} = k_C \frac{q q_0}{r^2}$$

We can find the magnitude of the electric field due to the point charge q at the position of q_0 by substituting this value into our previous equation for electric field strength.

$$E = \frac{F_{electric}}{q_0} = k_C \frac{q q_0}{r^2 q_0}$$

Notice that q_0 cancels, and we have a new equation for electric field strength from a point charge. ⭐TEKS **6C**

ELECTRIC FIELD STRENGTH FROM A POINT CHARGE

$$E = k_C \frac{q}{r^2}$$

electric field strength = Coulomb constant × $\dfrac{\text{charge producing the field}}{(\text{distance})^2}$

As stated above, electric field, **E,** is a vector. If q is positive, the field due to this charge is directed outward radially from q. If q is negative, the field is directed toward q. As with electric force, the electric field due to more than one charge is calculated by applying the principle of superposition. This is demonstrated in Sample Problem 17D on page 646.

Our new equation for electric field strength points out an important property of electric fields. As the equation indicates, an electric field at a given point depends only on the charge, q, of the object setting up the field and on the distance, r, from that object to a specific point in space. As a result, we can say that an electric field exists at any point near a charged body even when there is no test charge at that point. The examples in **Table 17-2** show the magnitudes of various electric fields. ⭐TEKS 2C

Table 17-2	Electric fields
Examples	**E, N/C**
in a fluorescent lighting tube	10
in the atmosphere during fair weather	100
under a thundercloud or in a lightning bolt	10 000
at the electron in a hydrogen atom	5.1×10^{11}

Consumer Focus *Microwave Ovens* ⭐TEKS 3C, 6F, 8B

*I*t would be hard to find a town in America that does not have a microwave oven. Homes, convenience stores, and restaurants all have this marvelous invention that somehow heats only the soft parts of the food and leaves the inorganic and hard materials, like ceramic and bone, at approximately the same temperature. A neat trick, indeed, but how is it done?

Microwave ovens take advantage of a property of water molecules called *bipolarity*. Water molecules are considered bipolar because each molecule has a positive and a negative end. In other words, more of the electrons in these molecules are at one end of the molecule than the other.

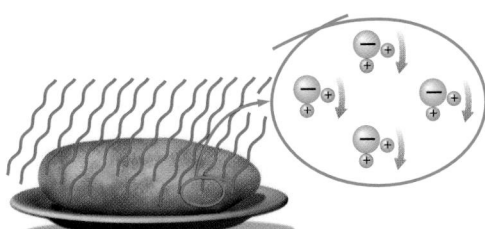

Figure 17-12

Because microwaves are a high-frequency form of electromagnetic radiation, they supply an electric field that changes polarity billions of times a second. As this electric field passes a bipolar molecule, the positive side of the particle experiences a force in one direction, and the negative side of the particle is pushed or pulled in the other direction. When the field changes polarity, the directions of these forces are reversed. Instead of tearing apart, the particles swing around and line up with the electric field.

As the bipolar molecules swing around, they rub against one another, producing friction. This friction in turn increases the internal energy of the food. Energy is transferred to the food by radiation (the microwaves) as opposed to conduction from hot air, as in a conventional oven.

Depending on the microwave oven's power and design, this rotational motion can generate up to about 3 J of internal energy each second in 1 g of water. At this rate, a top-power microwave oven can boil a cup (250 mL) of water in 2 min using about 0.033 kW•h of electricity.

Items such as bones, dry plates, and the air in the oven are unaffected by the fluctuating electric field because they are not polarized. Because energy is not wasted on heating these nonpolar items, the microwave oven cooks food faster and more efficiently than other ovens.

Teaching Tip

Students may ask how we can say there is an electric field without a test charge present. Point out that this conclusion is drawn from the fact that q_0 (the test charge) cancels from the equation for electric field strength. This conclusion cannot be demonstrated experimentally because the only way to observe a field is to observe its effects on an object. This could lead to a discussion of the fact that we do not *know* the field exists without a test charge, but since it is there whenever we test the field, we assume the field is there at all times.

Consumer Focus

BACKGROUND

Students should be familiar with the changing polarity of electric fields. With this background knowledge, this feature will serve as an introduction to the working principles behind microwave ovens.

EXTENSION

Controversy over the safety of microwave ovens in the home has surrounded this technology since its introduction. Have students research microwave-oven safety tests and present their results to the class.

⭐TEKS

p. 644: 6C, 6C
p. 645: 2C, 3C, 6F, 8B

Classroom Practice

The following may be used as teamwork exercises or for demonstration at the board or on an overhead projector.

PROBLEM

Electric field strength

An electric field around a charged object is 5.95×10^6 N/C at a distance of 10.0 cm. Find the charge on the object.

Answer

$6.62 \ \mu C$

A charge $q_1 = 4.50 \ \mu C$ experiences an attractive force of 1.35 N at a distance of 15.0 cm from a charged object, q_2.

a. Find the strength of the electric field due to q_2 at a distance of 15.0 cm.

b. Find the charge, q_2.

Answer

a. 3.00×10^5 N/C

b. $0.751 \ \mu C$

Electric field strength ⭐TEKS 2C, 3B, 6C

PROBLEM

A charge $q_1 = +7.00 \ \mu C$ is at the origin, and a charge $q_2 = -5.00 \ \mu C$ is on the x-axis 0.300 m from the origin, as shown in Figure 17-13. Find the electric field strength at point P, which is on the y-axis 0.400 m from the origin.

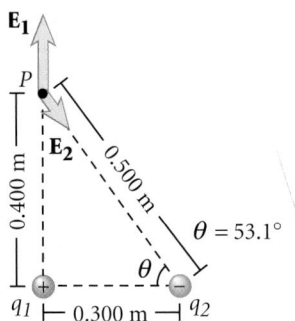

Figure 17-13

REASONING

The principle of superposition must be applied in order to calculate the electric field due to a group of point charges. You must first calculate the electric field produced by each charge individually at point P, then add these fields together as vectors.

SOLUTION

Given:

$q_1 = +7.00 \ \mu C \ (7.00 \times 10^{-6} \ C)$ $r_1 = 0.400$ m

$q_2 = -5.00 \ \mu C \ (-5.00 \times 10^{-6} \ C)$ $r_2 = 0.500$ m

$k_C = 8.99 \times 10^9 \ N \cdot m^2/C^2$ $\theta = 53.1°$

Unknown: **E** at $P \ (y = 0.400$ m$)$

1. **Calculate the magnitude of the electric field strength produced by each charge:**

Because we are finding the magnitude of the electric field, we can neglect the sign of each charge.

$$E_1 = k_C \frac{q_1}{r_1^2} = (8.99 \times 10^9 \ N \cdot m^2/C^2) \left(\frac{7.00 \times 10^{-6} \ C}{(0.400 \ m)^2} \right) = 3.93 \times 10^5 \ N/C$$

$$E_2 = k_C \frac{q_2}{r_2^2} = (8.99 \times 10^9 \ N \cdot m^2/C^2) \left(\frac{5.00 \times 10^{-6} \ C}{(0.500 \ m)^2} \right) = 1.80 \times 10^5 \ N/C$$

2. **Determine the direction of each electric field vector by analyzing the signs of the charges:**

The field vector $\mathbf{E_1}$ at P due to q_1 is directed vertically upward, as shown in **Figure 17-13,** because q_1 is positive. Likewise, the field vector $\mathbf{E_2}$ at P due to q_2 is directed toward q_2 because q_2 is negative.

3. **Find the *x* and *y* components of each electric field vector:**

At this point, the direction of each component must be taken into account.

For $\mathbf{E_1}$: $E_{x,1} = 0$ N/C

      $E_{y,1} = 3.93 \times 10^5$ N/C

For $\mathbf{E_2}$: $E_{x,2} = (E_2)(\cos 53.1°) = (1.80 \times 10^5 \ N/C)(\cos 53.1°) = 1.08 \times 10^5 \ N/C$

      $E_{y,2} = -(E_2)(\sin 53.1°) = (1.80 \times 10^5 \ N/C)(\sin 53.1°) = -1.44 \times 10^5 \ N/C$

4. Calculate the magnitude of the total electric field strength in both the x and y directions:

$E_{x,tot} = E_{x,1} + E_{x,2} = 0 \text{ N/C} + 1.08 \times 10^5 \text{ N/C} = 1.08 \times 10^5 \text{ N/C}$

$E_{y,tot} = E_{y,1} + E_{y,2} = 3.93 \times 10^5 \text{ N/C} - 1.44 \times 10^5 \text{ N/C} = 2.49 \times 10^5 \text{ N/C}$

5. Use the Pythagorean theorem to find the magnitude of the resultant electric field strength vector:

$E_{tot} = \sqrt{(E_{x,tot})^2 + (E_{y,tot})^2} = \sqrt{(1.08 \times 10^5 \text{ N/C})^2 + (2.49 \times 10^5 \text{ N/C})^2}$

$$\boxed{E_{tot} = 2.71 \times 10^5 \text{ N/C}}$$

6. Use a suitable trigonometric function to find the direction of the resultant electric field strength vector:

In this case, you can use the inverse tangent function:

$\tan \varphi = \dfrac{E_{y,tot}}{E_{x,tot}} = \dfrac{2.49 \times 10^5 \text{ N/C}}{1.08 \times 10^5 \text{ N/C}}$

$$\boxed{\varphi = 66.6°}$$

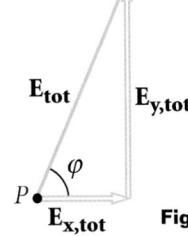

Figure 17-14

7. Evaluate your answer:

Because the electric field at point P is the result of a positive and a negative charge, E_{tot} should be between E_1 and E_2:

$3.93 \times 10^5 (E_1) > 2.71 \times 10^5 (E_{tot}) > 1.80 \times 10^5 (E_2)$

PRACTICE 17D

Electric field strength ⭐TEKS 2C, 3B, 6C

1. A charge, $q_1 = 5.00 \ \mu C$, is at the origin, and a second charge, $q_2 = -3.00 \ \mu C$, is on the x-axis 0.800 m from the origin. Find the electric field at a point on the y-axis 0.500 m from the origin.

2. A proton and an electron in a hydrogen atom are separated on the average by about 5.3×10^{-11} m. What is the magnitude and direction of the electric field set up by the proton at the position of the electron?

3. An electric field of 2.0×10^4 N/C is directed along the positive x-axis.

    a. What is the electric force on an electron in this field?

    b. What is the electric force on a proton in this field?

PRACTICE GUIDE 17D

Solving for:

E	📖 **PE**	Sample, 1–2; Ch. Rvw. 38–39, 41b, 43a, 44b, 50, 52, 56, 61, 67
	💿 **PW**	5–6
	PB	4–5
q	💿 **PW**	Sample, 1–2
	PB	6–7
F	📖 **PE**	3
	💿 **PB**	Sample, 1–3
r	💿 **PW**	3–4
	PB	8–10

ANSWERS TO

Practice 17D
Electric field strength
1. 1.66×10^5 N/C, 81.1° above the positive x-axis
2. 5.1×10^{11} N/C, away from the proton
3. **a.** 3.2×10^{-15} N, along the negative x-axis
    b. 3.2×10^{-15} N, along the positive x-axis

⭐TEKS

p. 646: 2C, 3B, 6C
p. 647: 2C, 3B, 6C

Misconception Alert

Some students may think that electric field lines are a physical phenomenon. Much like the magnetic field lines with which students may be familiar, electric field lines do not actually exist. Students should be told this explicitly. The lines are a visual representation of the field that would be experienced by a test charge. Explain to the students that an electric field is much like a gravitational field in that it cannot be directly observed. We can only observe the effects of the field.

Visual Strategy

Figure 17-15

Be sure students understand the similarities and differences between the two cases shown in this figure.

Q What can you conclude about the charges in **(a)** and **(b)** by comparing the electric field lines for each case?

A *Because the number of field lines leaving the charge in (a) is equal to the number of field lines approaching the charge in (b), the charges must be equal in magnitude. The field lines also show that the charge in (a) is positive (because the lines are pointing away from the charge) and that the charge in (b) is negative (because the lines are pointing toward the charge).*

electric field lines

lines that represent both the magnitude and the direction of the electric field

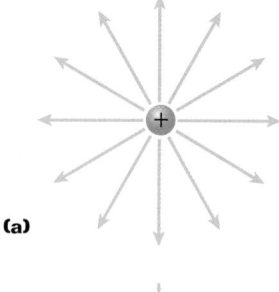

(a)

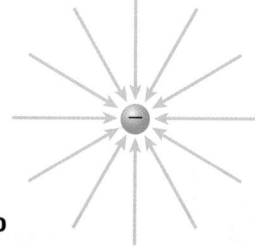

(b)

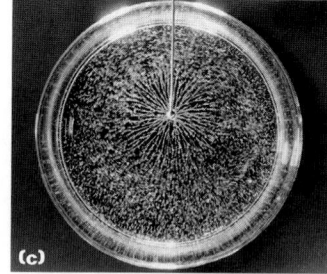

(c)

Figure 17-15
The diagrams **(a)** and **(b)** show some representative electric field lines for a positive and a negative point charge. In **(c)**, grass seeds align with a similar field produced by a charged body.

ELECTRIC FIELD LINES

A convenient aid for visualizing electric field patterns is to draw lines pointing in the direction of the electric field, called **electric field lines.** Although electric field lines do not really exist, they offer a useful means of analyzing fields by representing both the strength and the direction of the field at different points in space. This is useful because the field at each point is often the result of more than one charge, as seen in Sample Problem 17D on page 646. Field lines make it easier to visualize the net field at each point. ⭐TEKS **3A**

The number of field lines is proportional to the electric field strength

By convention, electric field lines are drawn so that the electric field vector, **E,** is tangent to the lines at each point. Further, the number of lines per unit area through a surface perpendicular to the lines is proportional to the strength of the electric field in a given region. Thus, E is stronger where the field lines are close together and weaker where they are far apart. ⭐TEKS **6C**

Figure 17-15(a) shows some representative electric field lines for a positive point charge. Note that this two-dimensional drawing contains only the field lines that lie in the plane containing the point charge. The lines are actually directed outward radially from the charge in all directions, somewhat like quills radiate from the body of a porcupine. Because a positive test charge placed in this field would be repelled by the positive charge q, the lines are directed away from the positive charge, extending to infinity. Similarly, the electric field lines for a single negative point charge, which begin at infinity, are directed inward toward the charge, as shown in **Figure 17-15(b).** Note that the lines are closer together as they get near the charge, indicating that the strength of the field is increasing. This is consistent with our equation for electric field strength, which is inversely proportional to distance squared. **Figure 17-15(c)** shows grass seeds in an insulating liquid. When a small charged conductor is placed in the center, these seeds align with the electric field produced by the charged body.

The rules for drawing electric field lines are summarized in **Table 17-3.** Note that no two field lines from the same field can cross one another. The reason is that at every point in space, the electric field vector points in a single direction and any field line at that point must also point in that direction.

Table 17-3	Rules for drawing electric field lines
The lines must begin on positive charges or at infinity and must terminate on negative charges or at infinity.
The number of lines drawn leaving a positive charge or approaching a negative charge is proportional to the magnitude of the charge.
No two field lines from the same field can cross each other.

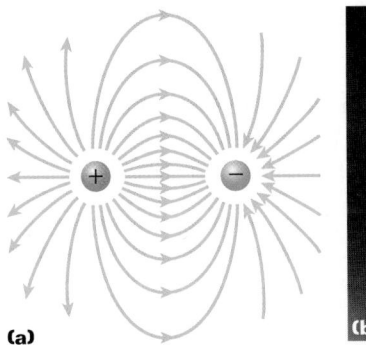

(a)

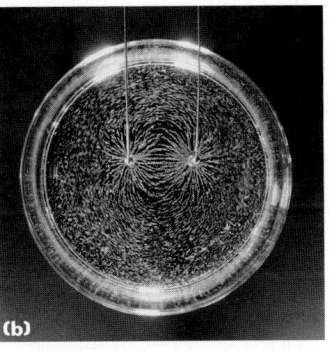

(b)

Figure 17-16 shows the electric field lines for two point charges of equal magnitudes but opposite signs. This charge configuration is called an *electric dipole*. In this case, the number of lines that begin at the positive charge must equal the number of lines that terminate on the negative charge. At points very near the charges, the lines are nearly radial. The high density of lines between the charges indicates a strong electric field in this region.

In electrostatic spray painting, field lines between a negatively charged spray gun and a positively charged target object are similar to those shown in **Figure 17-16.** As you can see, the field lines suggest that paint droplets that narrowly miss the target object still experience a force directed toward the object, sometimes causing them to wrap around from behind and hit it. This does happen and increases the efficiency of an electrostatic spray gun. (★)TEKS **3C, 6C**

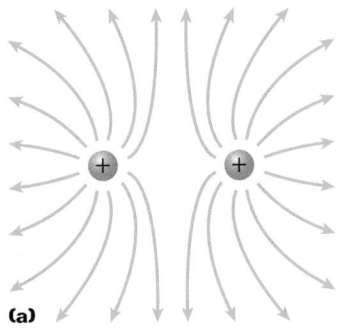

(a)

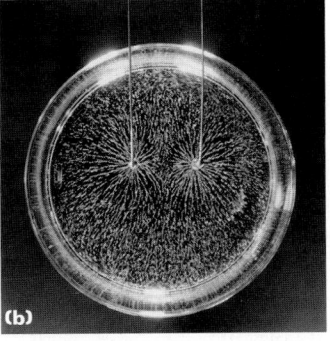

(b)

Figure 17-17 shows the electric field lines in the vicinity of two equal positive point charges. Again, close to either charge, the lines are nearly radial. The same number of lines emerges from each charge because the charges are equal in magnitude. At great distances from the charges, the field approximately equals that of a single point charge of magnitude $2q$.

Finally, **Figure 17-18** is a sketch of the electric field lines associated with a positive charge $+2q$ and a negative charge $-q$. In this case, the number of lines leaving the charge $+2q$ is twice the number terminating on the charge $-q$. Hence, only half the lines that leave the positive charge end at the negative charge. The remaining half terminate at infinity. At distances that are great compared with the separation between the charges, the pattern of electric field lines is equivalent to that of a single charge, $+q$.

Figure 17-16
(a) This diagram shows the electric field lines for two equal and opposite point charges. Note that the number of lines leaving the positive charge equals the number of lines terminating on the negative charge.
(b) In this photograph, grass seeds in an insulating liquid align with a similar electric field produced by two oppositely charged conductors.

Figure 17-17
(a) This diagram shows the electric field lines for two positive point charges. (b) The photograph shows the analogous case for grass seeds in an insulating liquid around two conductors with the same charge.

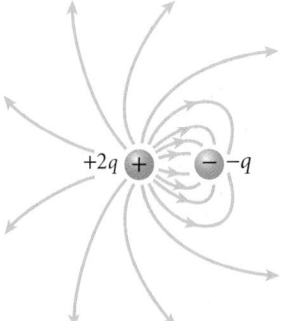

$+2q$ $-q$

Figure 17-18
In this case, only half the lines originating from the positive charge terminate on the negative charge because the positive charge is twice as great as the negative charge.

Visual Strategy

Figures 17-16 and 17-17
Have students examine these two figures and think about the spacing of the field lines.

Q How is the space between field lines related to the strength of the field?

A *The electric field is strongest where the lines are closest together and weakest where the lines are farthest apart.*

Q Where would an object in either of these fields experience the greatest force?

A *at the location where the field strength is greatest (where the lines are closest together)*

(★)TEKS

p. 648: 3A, 6C
p. 649: 3C, 6C

Electric Forces and Fields   **649**

Teaching Tip

The properties of conductors in electrostatic equilibrium are difficult for most students to understand initially. Present the students with diagrams showing the properties as a result of the electric forces in the conductor and the movement of charges. Also, remind students that when we derive each of these properties, we are assuming that charges are not moving. These properties do not hold true for situations involving moving charges.

Teaching Tip

Point out that the properties discussed for conductors in equilibrium do not hold true for insulators because charges do not move readily in insulators.

Visual Strategy

Figure 17-19

Be sure students understand the difference between the two cases.

Q Why is there a force vector directed to the left in (a) but not in (b)?

A *In (a), the vector E has a component parallel to the surface directed to the right. Because E represents the direction of force for a positive charge, negative charges experience a force directed to the left. There is no such force in (b) because E does not have a component parallel to the surface.*

internet connect

SC/LINKS
NSTA
TOPIC: Van de Graaff generator
GO TO: www.scilinks.org
sciLINKS CODE: HF2175

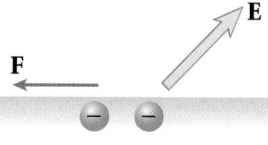

(a) Charges in motion

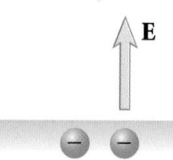

(b) Charges in equilibrium

Figure 17-19
(a) If **E** were not perpendicular to the surface of the charged object, the component of the field that is along the surface would cause charges to move. **(b)** But charges are not moving, so **E** must instead be perpendicular to the surface.

CONDUCTORS IN ELECTROSTATIC EQUILIBRIUM

A good electric conductor, such as copper, contains charges (electrons) that are not bound to any atoms and are free to move about within the material. When no net motion of charge is occurring within a conductor, the conductor is said to be in *electrostatic equilibrium*. As we shall see, such a conductor that is isolated has the four properties summarized in **Table 17-4.**

Table 17-4 Conductors in electrostatic equilibrium
The electric field is zero everywhere inside the conductor.
Any excess charge on an isolated conductor resides entirely on the conductor's outer surface.
The electric field just outside a charged conductor is perpendicular to the conductor's surface.
On an irregularly shaped conductor, charge tends to accumulate where the radius of curvature of the surface is smallest, that is, at sharp points.

The first property, which states that the electric field is zero inside a conductor in electrostatic equilibrium, can be understood by examining what would happen if this were not true. If there were an electric field inside a conductor, the free charges would move and a flow of charge, or current, would be created. However, if there were a net movement of charge, the conductor would no longer be in electrostatic equilibrium.

The fact that any excess charge resides on the outer surface of the conductor is a direct result of the repulsion between like charges described by Coulomb's law. If an excess of charge is placed inside a conductor, the repulsive forces arising between the charges force them as far apart as possible, causing them to quickly migrate to the surface. Although we will not demonstrate it in this text, this property is a result of the fact that the electric force obeys an inverse square law. With any other power law, an excess of charge would still exist on the surface, but there would also be a distribution of charge, either of the same or opposite sign, inside the conductor. ⊛TEKS 3A

We can understand why the electric field just outside a conductor must be perpendicular to the conductor's surface by considering what would happen if this were not true. If the electric field were not perpendicular to the surface, as in **Figure 17-19(a)**, the electric field would have a component along the surface. This would cause the free negative charges within the conductor to move to the left. But if the charges moved, a current would be created and there would no longer be electrostatic equilibrium. Hence, **E** must be perpendicular to the surface, as in **Figure 17-19(b).**

To see why charge tends to accumulate at sharp points, consider a conductor that is fairly flat at one end and relatively pointed at the other. Any excess charge placed on the object moves to its surface. **Figure 17-20** shows the forces between two charges at each end of such an object. At the flatter end, these forces are predominantly directed parallel to the surface. Thus, the charges move apart until repulsive forces from other nearby charges create a state of equilibrium.

At the sharp end, however, the forces of repulsion between two charges are directed predominantly perpendicular to the surface. As a result, there is less tendency for the charges to move apart along the surface and the amount of charge per unit area is greater than at the flat end. The cumulative effect of many such outward forces from nearby charges at the sharp end produces a large electric field directed away from the surface.

When the field becomes great enough, the air around the sharp end becomes ionized and a bluish glow called a *corona* is seen. Although air does not typically transfer electric charges easily, ionized air is a good conductor, and so charges leap from the surface into the surrounding air.

Electrostatic spray guns use this phenomenon to give paint droplets a negative charge. Electrostatic spray guns have a number of sharp points. When a spray gun is charged, the electric field around the sharp points can be strong enough to produce a corona around the gun. Air molecules in this area are ionized, and paint droplets pick up negative charges from these molecules as they move through the ionized air. ⓍTEKS **3C**

A Van de Graaff generator collects electric charge

In nuclear physics research, a Van de Graaff generator is often used to generate electric charge. Some generators are designed to accumulate positive charge. If protons are introduced into a tube attached to the dome of a Van de Graaff generator, the large electric field of the dome exerts a repulsive force on the protons, causing them to accelerate to energies high enough to initiate nuclear reactions between the protons and various target nuclei.

A simple Van de Graaff generator is shown in **Figure 17-21.** The Van de Graaff generator is designed to avoid ionizing the air, which would allow charge to "leak off" into the atmosphere.

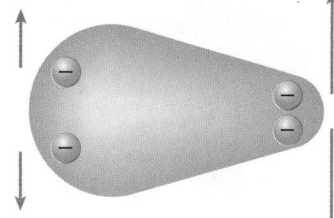

Figure 17-20
When one end of a conductor is sharper than the other, excess charge tends to accumulate at the sharper end, resulting in a larger charge per unit area and therefore a larger repulsive electric force between charges at this end.

Did you know?

The electric field intensity in the atmosphere is high when a thunderstorm is approaching. For this reason, sailors in the past sometimes observed a natural corona at the mast of their ship during bad weather. This type of corona was named "St. Elmo's fire" after the patron saint of Mediterranean sailors.

Figure 17-21
This person, insulated from the ground and touching the dome of a Van de Graaff generator, finds that her hair stands out on end because the strands become charged and repel one another.

Demonstration 6

Charge accumulation

Purpose Provide a dramatic visual demonstration of charge accumulation by a Van de Graaff generator.

Materials Van de Graaff generator, tape, aluminum pie pans, puffed rice (or small pieces of paper), glass beaker

Procedure Stack 10 to 12 pie pans upside down on top of the Van de Graaff generator. Turn the generator on. Have students explain the flying pans.

Next, tape one pie pan upright to the top of the generator, and pour in some of the puffed rice or paper pieces. Turn the generator on. Once again, have students explain the results.

Finally, tape the glass beaker upright to the top of the generator. Pour in some puffed rice. Turn the generator on. Have students explain the results.

ⓍTEKS

p. 650: 3A
p. 651: 3C

Electric Forces and Fields **651**

Section Review
ANSWERS

1. 8.0×10^3 N/C, directed toward the 40.0×10^{-9} C charge

2. **a.** $-\frac{3}{8}$
   b. q_1 is negative; q_2 is positive

3. **a.** All lines should point away from the two charges, and one charge should have four times as many lines as the other.
   b. All lines should point toward the two charges, and one charge should have four times as many lines as the other.

4. because charge accumulates at sharp points

5. They experience an electrical attraction that pulls them back toward the object.

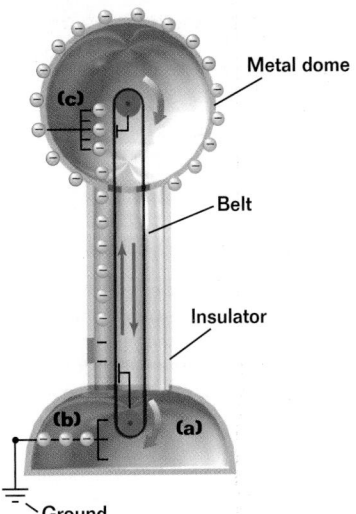

Metal dome

Belt

Insulator

(c)

(b) **(a)**

Ground

Figure 17-22
In a Van de Graaff generator, charge is transferred to the dome by a rotating belt that is kept in motion by a pulley **(a).** The charge is deposited on the belt at **(b)** and then transferred to the dome at **(c).**

Figure 17-22 shows the basic construction of a Van de Graaff generator. A motor-driven pulley **(a)** moves a non-conducting belt around a loop. The pulley and belt are made of different materials, and the friction of the pulley driving the belt causes a separation of charges. A strong positive charge builds up on the pulley, while the belt becomes negatively charged. The negative charge spreads over the length of the belt. Because the positive charge is concentrated on the comparatively small pulley while the corresponding negative charge is spread over the much larger belt, there is a net positive charge in the area around the lower pulley.

A comb of conducting needles **(b)** is located near where the belt passes over the pulley with its needles very close to, but not touching, the moving belt. The comb is connected by a wire to ground. The net positive charge in the area around the pulley attracts negative charge from ground onto the needles of the comb. The electrostatic attraction of the positive charge is great enough to cause the negative charges to leap off toward the pulley. The negative charges hit the belt, and the moving belt carries them up. The negative charges are drawn off the belt by another comb of needles **(c)** at the top. These charges distribute themselves over the outer surface of the dome. A large amount of negative charge can be accumulated on the dome in this manner. Because the electric field is greatest at sharp points, a sphere is the most effective shape for building up large amounts of charge without producing a corona.

Section Review ⭐ TEKS 2C, 6C

1. Find the electric field at a point midway between two charges of $+40.0 \times 10^{-9}$ C and $+60.0 \times 10^{-9}$ C separated by a distance of 30.0 cm.

2. **Figure 17-23** shows the electric field lines for two point charges separated by a small distance.
   a. Determine the ratio q_1/q_2.
   b. What are the signs of q_1 and q_2?

3. Two point charges are a small distance apart.
   a. Sketch the electric field lines for the two if one has a charge four times that of the other and if both charges are positive.
   b. Repeat **(a),** but assume both charges are negative.

4. Explain why you're more likely to get a shock from static electricity by touching a metal object with your finger instead of with your entire hand.

5. **Physics in Action**   In an electrostatic spray gun, paint droplets that miss the target object sometimes wrap around and hit the object from behind. Explain why.

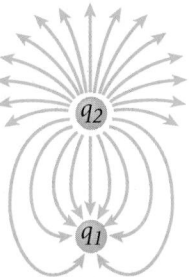

q_2

q_1

Figure 17-23

KEY IDEAS

Section 17-1 Electric charge

- There are two kinds of electric charge; likes repel, and unlikes attract.
- Electric charge is conserved.
- The fundamental unit of charge, e, is the charge of a single electron or proton.
- Conductors and insulators can be charged by contact. Conductors can also be charged by induction. A surface charge can be induced on an insulator by polarization.

Section 17-2 Electric force

- According to Coulomb's law, the electric force between two charges is proportional to the magnitude of the charges and inversely proportional to the square of the distance between them.

$$F_{electric} = k_C \frac{q_1 q_2}{r^2}$$

- The electric force is a field force.
- The resultant electric force on any charge is the vector sum of the individual electric forces on that charge.

Section 17-3 The electric field

- An electric field exists in the region around a charged object.
- Electric field strength depends on the magnitude of the charge producing the field and the distance between that charge and a point in the field.

$$E = \frac{F_{electric}}{q_0} = \frac{k_C q}{r^2}$$

- The direction of the electric field vector, **E**, is the direction in which an electric force would act on a positive test charge.
- Field lines are tangent to the electric field vector at any point, and the number of lines is proportional to the magnitude of the field strength.

KEY TERMS

conductor (p. 631)

electric field (p. 643)

electric field lines (p. 648)

induction (p. 632)

insulator (p. 631)

Diagram symbols

Positive charge	$\oplus$ $+q$
Negative charge	$\ominus$ $-q$
Electric field vector	→ **E**
Electric field lines	

Teaching Tip

Explaining concepts in written form helps to solidify students' understanding of difficult concepts and helps to enforce good communication skills. Have students summarize the main topics of this chapter in an essay, including methods of charging, the conservation of charge, Coulomb's law, and the electric field. Be sure students explain concepts clearly and correctly, and use good sentence structure.

★ TEKS

Review & Assess pp. 654–659:
2A: Alt. Assess. 1
2C: 3, 18–24, 38–62
2D: Alt. Assess. 2–4
3C: Alt. Assess. 3
3E: Alt. Assess. 2, 5
5B: 58
6B: Alt. Assess. 5
6C: 2–3, 6, 9–24, 29, 38–40, 43–45, 48, 50, 52–53, 56, 59–61, Alt. Assess. 1
6F: Alt. Assess. 2–3

Variable symbols

Quantities		Units		Conversions
$F_{electric}$	electric force	N	newtons	$= \text{kg} \cdot \text{m/s}^2$
q	charge	C	coulomb (SI unit of charge)	$= 6.3 \times 10^{18}$ e
		e	fundamental unit of charge	$= 1.60 \times 10^{-19}$ C
k_C	Coulomb constant	$\text{N} \cdot \frac{\text{m}^2}{\text{C}^2}$	newtons $\times \frac{\text{meters}^2}{\text{coulombs}^2}$	
E	electric field strength	N/C	newtons/coulomb	

CHAPTER 17
Review and Assess

ANSWERS TO

Chapter 17
Review and Assess

1. conductors transfer charge easily, insulators do not
2. opposite
3. 2.2×10^{13} electrons
4. c
5. no; *Positive* and *negative* are arbitrary designations.
6. No, if a charged object induces a surface charge on the suspended object, the two are attracted, but the suspended object has no net charge.
7. Protons are relatively fixed in the nucleus, whereas electrons can be transferred from one atom to another.
8. winter; because more charge can accumulate before electric discharge occurs
9. No, the balloon clings because its charge induces a surface charge on the wall.
10. repulsion; because attraction can be the result of an induced surface charge, but repulsion occurs only when two objects each have a net charge
11. the signs of the charges
12. to the left
13. mass is positive, charges are positive or negative
14. Each force exerted on an object is found, then the forces are added together vectorially.
15. Answers will vary but may include the force between hair and a comb or the force that acts when people receive a "shock" by touching an object.

ELECTRIC CHARGE

Review questions

1. How are conductors different from insulators?

2. When a conductor is charged by induction, is the induced surface charge on the conductor the same or opposite the charge of the object inducing the surface charge?

3. A negatively charged balloon has 3.5 μC of charge. How many excess electrons are on this balloon?

Conceptual questions

4. Which activity does not produce the same results as the other three?
   a. sliding over a plastic-covered automobile seat
   b. walking across a woolen carpet
   c. scraping food from a metal bowl with a metal spoon
   d. brushing dry hair with a plastic comb

5. Would life be different if the electron were positively charged and the proton were negatively charged? Explain your answer.

6. If a suspended object is attracted to another object that is charged, can you conclude that the suspended object is charged?

7. Explain from an atomic viewpoint why charge is usually transferred by electrons.

8. Because of a higher moisture content, air is a better conductor of charge in the summer than in the winter. Would you expect the shocks from static electricity to be more severe in summer or winter? Explain your answer.

9. A balloon is negatively charged by rubbing and then clings to a wall. Does this mean that the wall is positively charged?

10. Which effect proves more conclusively that an object is charged, attraction to or repulsion from another object? Explain.

ELECTRIC FORCE

Review questions

11. What determines the direction of the electric force between two charges?

12. In which direction will the electric force from the two equal positive charges pull the negative charge shown in **Figure 17-24**?

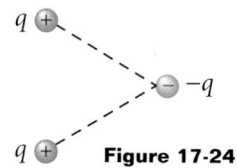

Figure 17-24

13. The gravitational force is always attractive, while the electric force is both attractive and repulsive. What accounts for this difference?

14. When more than one charged object is present in an area, how can the total electric force on one of the charged objects be found?

15. Identify examples of electric forces in everyday life.

Conceptual questions

16. How does the electric force between two charges change when the distance between them is doubled? How does it change when the distance is halved?

17. According to Newton's third law, every action has an equal and opposite reaction. When a comb is charged and held near small pieces of paper, the comb exerts an electric force on the paper pieces and pulls them toward it. Why don't you observe the comb moving toward the paper pieces as well?

Practice problems

18. At the point of fission, a nucleus of ^{235}U that has 92 protons is divided into two smaller spheres, each of which has 46 protons and a radius of 5.9×10^{-15} m. What is the repulsive force pushing these two spheres apart?
(See Sample Problem 17A.)

19. What is the electric force between a glass ball that has $+2.5 \ \mu C$ of charge and a rubber ball that has $-5.0 \ \mu C$ of charge when they are separated by a distance of 5.0 cm?
(See Sample Problem 17A.)

20. An alpha particle (charge $= +2.0e$) is sent at high speed toward a gold nucleus (charge $= +79e$). What is the electric force acting on the alpha particle when the alpha particle is 2.0×10^{-14} m from the gold nucleus?
(See Sample Problem 17A.)

21. Three positive point charges of 3.0 nC, 6.0 nC, and 2.0 nC, respectively, are arranged in a triangular pattern, as shown in **Figure 17-25.** Find the magnitude and direction of the electric force on the 6.0 nC charge.
(See Sample Problem 17B.)

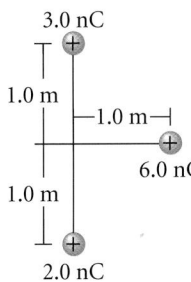

3.0 nC

1.0 m

—1.0 m—

6.0 nC

1.0 m

2.0 nC

Figure 17-25

22. Two positive point charges, each of which has a charge of 2.5×10^{-9} C, are located at $y = +0.50$ m and $y = -0.50$ m. Find the magnitude and direction of the resultant electrical force on a charge of 3.0×10^{-9} C located at $x = 0.70$ m.
(See Sample Problem 17B.)

23. Three point charges lie in a straight line along the y-axis. A charge of $q_1 = -9.0 \ \mu C$ is at $y = 6.0$ m, and a charge of $q_2 = -8.0 \ \mu C$ is at $y = -4.0$ m. The net electric force on the third point charge is zero. Where is this charge located?
(See Sample Problem 17C.)

24. A charge of $+3.5$ nC and a charge of $+5.0$ nC are separated by 40.0 cm. Find the equilibrium position for a -6.0 nC charge.
(See Sample Problem 17C.)

THE ELECTRIC FIELD

Review questions

25. What is an electric field?

26. Show that the definition of electric field strength ($E = F_{electric}/q_0$) is equivalent to the equation $E = k_C q/r^2$ for point charges.

27. In an irregularly shaped conductor, a corona forms around a sharp end sooner than around a smoother end. Explain why.

28. When electric field lines are being drawn, what determines the number of lines originating from a charge? What determines whether the lines originate from or terminate on a charge?

29. Draw some representative electric field lines for two charges of $+q$ and $-3q$ separated by a small distance.

30. Consider the electric field lines in **Figure 17-26.**
   a. Where is charge density the highest? Where is it the lowest?
   b. If an opposite charge were brought into the vicinity, where would charge on the pear-shaped object "leak off" most readily?

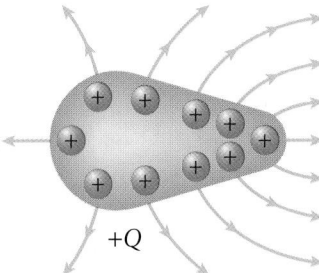

$+Q$

Figure 17-26

31. Do electric field lines actually exist?

32. Why does the dome of a Van de Graaff generator have a spherical surface?

Conceptual questions

33. When defining the electric field, why must the magnitude of the test charge be very small?

34. Why can't two field lines from the same field cross one another?

35. no; because there is no path for charge to escape; because the strands are charged

36. at infinity

37. Electric forces are equal and opposite; The proton's acceleration is less because it has a greater mass ($F = ma$).

38. 12.0×10^3 N/C, toward the 30.0×10^{-9} C charge

39. 5.7×10^3 N/C, $75°$ above the positive x-axis

40. 4.8×10^{-6} C

41. a. 5.7×10^{-27} N, in a direction opposite **E**
  b. 3.6×10^{-8} N/C

42. a. 2.75×10^{23} electrons
  b. 4.40×10^4 C

43. a. 2.0×10^7 N/C, along the positive x-axis
  b. 4.0×10^1 N

44. a. 1.3×10^{-5} N, $77°$ below the negative x-axis
  b. 2.6×10^3 N/C, $77°$ below the negative x-axis

45. 5.12×10^5 N

46. There are no lines inside the sphere. All lines outside the sphere are evenly spaced and are directed away from the sphere radially.

47. 5.72×10^{13} C

48. 7.2×10^{-9} C

---

35. A student stands on a piece of insulating material, places her hand on a Van de Graaff generator, and then turns on the generator (see **Figure 17-21** on page 651). Is she shocked? Why or why not? The student finds that her hair stands on end. Why does this occur?

36. In **Figure 17-18** (page 649), where do the extra lines leaving the $+2q$ charge end?

37. A "free" electron and "free" proton are placed in an identical electric field. Compare the electric force on each particle. How do their accelerations compare?

Practice problems

38. Find the electric field at a point midway between two charges of $+30.0 \times 10^{-9}$ C and $+60.0 \times 10^{-9}$ C separated by a distance of 30.0 cm. (See Sample Problem 17D.)

39. A $+5.7$ μC point charge is on the x-axis at $x = -3.0$ m, and a $+2.0$ μC point charge is on the x-axis at $x = +1.0$ m. Determine the net electric field (magnitude and direction) on the y-axis at $y = +2.0$ m. (See Sample Problem 17D.)

MIXED REVIEW

40. Calculate the net charge on a substance consisting of a combination of 7.0×10^{13} protons and 4.0×10^{13} electrons.

41. An electron moving through an electric field experiences an acceleration of 6.3×10^3 m/s^2.
  a. Find the electric force acting on the electron.
  b. What is the strength of the electric field?

42. One gram of copper has 9.48×10^{21} atoms, and each copper atom has 29 electrons.
  a. How many electrons are contained in 1.00 g of copper?
  b. What is the total charge of these electrons?

43. Consider three charges arranged as shown in **Figure 17-27**.
  a. What is the electric field strength at a point 1.0 cm to the left of the middle charge?

---

b. What is the magnitude of the force on a -2.0 μC charge placed at this point?

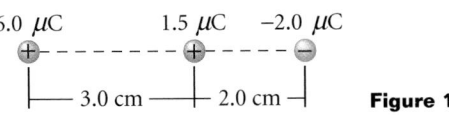

Figure 17-27

44. Consider three charges arranged in a triangle as shown in **Figure 17-28**.
  a. What is the net electric force on the charge at the origin?
  b. What is the net electric field at the position of the charge at the origin?

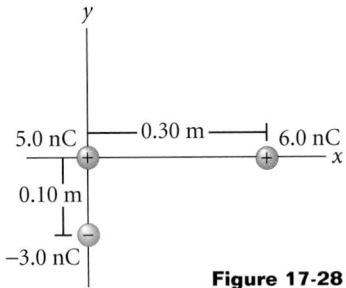

Figure 17-28

45. 1.00 g of hydrogen contains 6.02×10^{23} atoms, each with one electron and one proton. Suppose that 1.00 g of hydrogen is separated into protons and electrons, that the protons are placed at Earth's north pole, and that the electrons are placed at Earth's south pole. Find the magnitude of the resulting compressional force on Earth. (The radius of Earth is approximately 6.38×10^6 m.)

46. Sketch the electric field pattern set up by a positively charged hollow conducting sphere. Include regions both inside and outside the sphere.

47. The moon ($m = 7.36 \times 10^{22}$ kg) is bound to Earth ($m = 5.98 \times 10^{24}$ kg) by gravity. If, instead, the force of attraction were the result of each having a charge of the same magnitude but opposite in sign, find the quantity of charge that would have to be placed on each to produce the required force.

48. Two small metallic spheres, each with a mass of 0.20 g, are suspended as pendulums by light strings from a common point. They are given the same electric charge, and the two come to equilibrium when each string is at an angle of $5.0°$ with the vertical. If the string is 30.0 cm long, what is the magnitude of the charge on each sphere?

49. What are the magnitude and the direction of the electric field that will balance the weight of an electron? What are the magnitude and direction of the electric field that will balance the weight of a proton?

50. Three positive charges are arranged as shown in **Figure 17-29.** Find the electric field at the fourth corner of the rectangle.

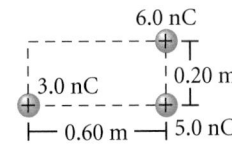

6.0 nC

3.0 nC   0.20 m

0.60 m   5.0 nC

Figure 17-29

51. An electron and a proton are each placed at rest in an external uniform electric field of 520 N/C. Calculate the speed of each particle after 48 ns.

52. The dome of a Van de Graaff generator receives a charge of 2.0×10^{-4} C. The radius of the dome is 1.0 m. Find the magnitude of the electric field strength at the following locations:

a. inside the dome
b. at the surface of the dome
c. 4.0 m from the center of the dome

53. Air becomes a conductor when the electric field strength exceeds 3.0×10^6 N/C. Determine the maximum amount of charge that can be carried by a metal sphere 2.0 m in radius.

54. A Van de Graaff generator is charged so that the electric field at its surface is 3.0×10^4 N/C.

a. What is the electric force on a proton released at the surface of the generator?
b. Find the proton's acceleration at this instant.

55. Thunderstorms can have an electric field of up to 3.4×10^5 N/C. What is the magnitude of the electric force on an electron in such a field?

56. Three identical charges ($q = +5.0 \ \mu C$) are along a circle with a radius of 2.0 m at angles of 30°, 150°, and 270°, as shown in **Figure 17-30.** What is the resultant electric field at the center?

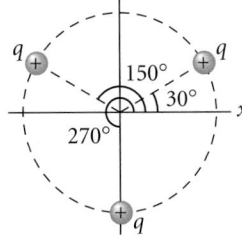

150°
30°
x
270°

Figure 17-30

57. An object with a net charge of 24 μC is placed in a uniform electric field of 610 N/C, directed vertically. What is the mass of this object if it floats in this electric field?

58. A proton accelerates from rest in a uniform electric field of 640 N/C. At some time later, its speed is 1.20×10^6 m/s.

a. What is the magnitude of the acceleration of the proton?
b. How long does it take the proton to reach this speed?
c. How far has it moved in this time interval?
d. What is its kinetic energy at the later time?

59. Three identical point charges, each of mass $m = 0.10$ kg, hang from three strings, as shown in **Figure 17-31.** If $L = 30.0$ cm and $\theta = 45°$, what is the value of q?

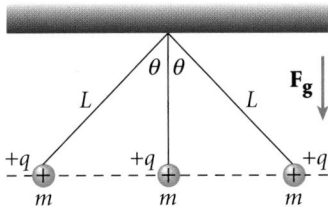

$\theta \ \theta$
F_g
L L

$+q$ $+q$ $+q$
m m m

Figure 17-31

60. A small 2.0 g plastic ball is suspended by a 20.0 cm string in a uniform electric field of 1.0×10^4 N/C, as shown in **Figure 17-32.**

a. Is the ball's charge positive or negative?
b. If the ball is in equilibrium when the string makes a 15° angle with the vertical as indicated, what is the net charge on the ball?

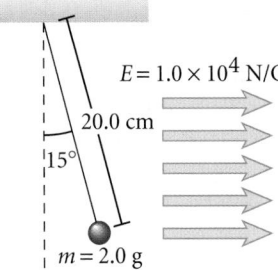

$E = 1.0 \times 10^4$ N/C

20.0 cm

15°

$m = 2.0$ g

Figure 17-32

61. 0 N/C

62. a. 5.3×10^{17} m/s^2
  b. 8.5×10^{-4} m
  c. 2.9×10^{14} m/s^2

ANSWERS TO

Technology & Learning

Answers may vary slightly, depending on viewing window settings.

a. decrease
b. 54 N
c. 8400 N
d. 6200 N
e. 440 N
f. curve shifts upward

---

61. In a laboratory experiment, five equal negative point charges are placed symmetrically around the circumference of a circle of radius r. Calculate the electric field at the center of the circle.

62. If the electric field strength is increased to about 3.0×10^6 N/C, air "breaks down" and loses its insulating quality. Under these conditions, sparking results.

  a. What acceleration does an electron experience when the electron is placed in such an electric field?

  b. If the electron starts from rest when it is placed in an electric field under these conditions, in what distance does it acquire a speed equal to 10.0 percent of the speed of light?

  c. What acceleration does a proton experience when the proton is placed in such an electric field?

Technology & Learning

Graphing calculators

Refer to Appendix B for instructions on downloading programs for your graphing calculator. The program "Chap17" allows you to analyze a graph of force versus distance for two positive charges.

Electric force, as you learned earlier in this chapter, is described by Coulomb's law:

$$F_{electric} = k_C \left(\frac{q_1 q_2}{r^2} \right)$$

The program "Chap17" stored on your graphing calculator makes use of Coulomb's law. Once the "Chap17" program is executed, your calculator will ask for the two charges. The graphing calculator will use the following equation to create a graph of the electric force (Y1) versus the distance (X) between the charges. The relationships in this equation are the same as those in the force equation shown above.

$$Y_1 = 8.99E9(AB/X^2)$$

  a. Using the graphing calculator equation above, predict whether the y values will increase or decrease as the x values increase.

Execute "Chap17" on the [PRGM] menu, and press [ENTER] to begin the program. Enter the magnitudes of the two charges (shown below), pressing [ENTER] after each value.

The calculator will provide a graph of the electric force versus the distance of separation. (If the graph is not visible, press [WINDOW] and change the settings for the graph window, then press [GRAPH].)

Press [TRACE], and use the arrow keys to trace along the curve. The x value corresponds to the distance in meters, and the y value corresponds to the electric force in newtons.

Determine the electric force involved in each of the following situations. Remember to use the [(-)] key, instead of the [-] key, for entering negative exponents. Also, use the exponent function key to enter powers of ten by pressing [2nd] [EE].

  b. 7.5×10^{-5} C and 3.2×10^{-6} C, 0.20 m apart
  c. 7.5×10^{-5} C and 3.2×10^{-6} C, 0.016 m apart
  d. 5.5×10^{-6} C and 3.2×10^{-5} C, 0.016 m apart
  e. 5.5×10^{-6} C and 3.2×10^{-5} C, 0.060 m apart
  f. Is the graph shifted up or down relative to the x-axis when the magnitude of the charges increases?

Press [2nd] [QUIT] to stop graphing. Press [ENTER] to input new values or [CLEAR] to end the program.

63. A DNA molecule (deoxyribonucleic acid) is 2.17 μm long. The ends of the molecule become singly ionized so that there is -1.60×10^{-19} C on one end and $+1.60 \times 10^{-19}$ C on the other. The helical molecule acts as a spring and compresses 1.00 percent upon becoming charged. Find the effective spring constant of the molecule.

64. An electron and a proton both start from rest and from the same point in a uniform electric field of 370.0 N/C. How far apart are they 1.00 μs after they are released? Ignore the attraction between the electron and the proton. (Hint: Imagine the experiment performed with the proton only, and then repeat with the electron only.)

65. An electron is accelerated by a constant electric field of 300.0 N/C.

    a. Find the acceleration of the electron.

    b. Find the electron's speed after 1.00×10^{-8} s, assuming it starts from rest.

66. A constant electric field directed along the positive x-axis has a strength of 2.0×10^3 N/C.

    a. Find the electric force exerted on a proton by the field.

    b. Find the acceleration of the proton.

    c. Find the time required for the proton to reach a speed of 1.00×10^6 m/s, assuming it starts from rest.

67. Consider an electron that is released from rest in a uniform electric field.

    a. If the electron is accelerated to 1.0 percent of the speed of light after traveling 2.0 mm, what is the strength of the electric field?

    b. What speed does the electron have after traveling 4.0 mm from rest?

68. Each of the protons in a particle beam has a kinetic energy of 3.25×10^{-15} J. What are the magnitude and direction of the electric field that will stop these protons in a distance of 1.25 m?

63. 2.25×10^{-9} N/m

64. 32.5 m

65. a. 5.27×10^{13} m/s^2
    b. 5.27×10^5 m/s

66. a. 3.2×10^{-16} N, along the positive x-axis
    b. 1.9×10^{11} m/s^2
    c. 5.3×10^{-6} s

67. a. 1.3×10^4 N/C
    b. 4.2×10^6 m/s

68. 1.62×10^4 N/C, opposite the proton's velocity

Alternative Assessment

Performance assessment

1. A metal can is placed on a wooden table. If a positively charged ball suspended by a thread is brought close to the can, the ball will swing toward the can, make contact, then move away. Explain why this happens and predict whether the ball is likely to make contact a second time. Sketch diagrams showing the charges on the ball and on the can at each phase. How can you test whether your explanation is correct? If your teacher approves of your plan, try testing your explanation.

2. The common copying machine was designed in the 1960s, after the American inventor Chester Carlson developed a practical device for attracting carbon-black to paper using localized electrostatic action. Research how this process works and determine why the last copy made when several hundred copies are made can be noticeably less sharp than the first copy. Create a report, poster, or brochure for office workers containing tips for using copiers.

Portfolio projects

3. Research how an electrostatic precipitator works to remove smoke and dust particles from the polluting emissions of fuel-burning industries. Find out what industries in your community use precipitators. What are their advantages and costs? What alternatives are available? Summarize your findings in a brochure, poster, or chart.

4. Imagine you are a member of a research team interested in lightning and you are preparing a grant proposal. Research information about the frequency, location, and effects of thunderstorms. Write a proposal that includes background information, research questions, a description of necessary equipment, and recommended locations for data collection.

5. Electric force is also known as the *Coulomb force* or *Coulomb interaction*. Research the historical development of the concept of electric force. Describe the work of Coulomb and other scientists such as Priestley, Cavendish, and Benjamin Franklin.

Performance Assessment

1. The ball induces an opposite charge, some of which is transferred upon contact, repelling the ball. Student plans should include methods for measuring charge polarity.

2. Repeated charging of the cylinder makes the pattern spread, causing blurry copies.

 Portfolio projects

3. Student presentations will vary. Advantages include pollution reduction. Disadvantages include costs.

4. Student plans will vary. Be certain plans correspond to the research questions asked.

5. Coulomb published the results of his investigations in 1785. His work built on the work of Priestley, Franklin, and others.

Planning

Recommended time:

1 lab period

Alternatively, each section of the lab could be performed for part of a period on different days.

Classroom organization:

▶ Each lab group needs a level work surface.

▶ Students may perform this lab alone or in groups of two or more students.

▶ **Safety warnings:** Remind students to make sure rods are secure when not in use. Glass rods may break, and all rods present a safety hazard if they roll onto the floor.

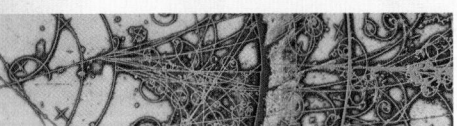

CHAPTER 17
Laboratory Exercise

★ TEKS

pp. 660–663: 1A, 1B, 2B, 2C, 6C

OBJECTIVES

• Investigate the use of an electroscope.

• Use an electroscope and other materials to analyze properties of static electricity.

• Determine the number of the kinds of electric charge.

MATERIALS LIST

✔ 2 polystyrene rods
✔ 2 PVC rods
✔ insulated copper wire
✔ demonstration capacitor
✔ electroscope
✔ flint glass rod
✔ insulated wire with 2 alligator clips
✔ metric ruler
✔ nylon cord
✔ roll of cellophane tape
✔ silk cloth
✔ silk thread
✔ support stand with clamp
✔ suspension support for rod
✔ wool pad

ELECTROSTATICS

When objects made of two different materials are rubbed together, electric charges accumulate on both objects. This phenomenon is known as static electricity. When an object has an electric charge, it attracts some things and repels others. The charges can also be transferred to some objects and not to others. In this experiment, you will produce charges on different objects and distinguish between the types of charges produced. You will also use an electroscope to examine the transfer of charges and the conductivity of different materials.

SAFETY

• **Never put broken glass or ceramics in a regular waste container. Use a dustpan, brush, and heavy gloves to carefully pick up broken pieces and dispose of them in a container specifically provided for this purpose.**

PREPARATION

1. Read the entire lab, and plan what steps you will take.

2. In your lab notebook, prepare an observation table with two wide columns. Label the columns *Experiment* and *Observation*. For each part of the lab, you will write a brief description of what you do in each step under *Experiment*. In the *Observation* column, record your observations of what happens.

PROCEDURE

Electric charge

3. Take four strips of cellophane tape. Each strip should be 20 cm long. Fold over a tab at the end of each tape. Tape strips to the lab table, and label the strips *A*, *B*, *C*, and *D* with a pencil.

4. Vigorously rub tapes A and B with a wool pad. Grasp the tabbed ends of A and B and carefully remove the tapes from the table. Slowly bring the tapes close together, but do not allow them to touch. Observe how they affect one another. Record your observations in your lab notebook. Carefully place tapes A and B back on the lab table.

5. Carefully remove tape C from the lab table. Tape it firmly down on top of tape D. Carefully remove tapes C and D together from the lab table. Quickly separate them, being careful not to tangle the tapes. Bring the tapes close together—but not touching—and observe how they affect one another. Record your observations in your lab notebook. Place tape D back on the lab table, and place tape C down on top of tape D.

6. Vigorously rub tape A with a wool pad again. Grasp tape A by the tab and carefully remove it from the table. Remove C and D together from the table. Quickly pull C and D apart. Bring C close to tape A, but do not let them touch. Observe how they affect each other. Move C away, and bring D close to tape A. Record your observations in your lab notebook. Throw the four tapes away.

7. Tape a meterstick flat on the surface of the table so that the end of the meterstick extends over the edge of the table. Take another 20 cm long piece of tape, fold a tab on one end, and tape it down on the table. Vigorously rub the tape with a wool pad. Grasp the tab, and carefully remove the tape. Attach the tape to the end of the meterstick so that it freely hangs straight down.

8. Rub a polystyrene rod with wool. Bring the rod near the end of the tape that is hanging down, and observe the effect on the tape. Record your observations in your lab notebook. Throw away the tape, and remove the meterstick from the tabletop.

9. Tie the suspension support securely to a string, and suspend the string from the support stand and clamp. Attach a polystyrene rod to the support, and rub the rod with a wool pad. Rub a second polystyrene rod with wool, and bring this rod near one end of the suspended rod. Observe what happens, and record your observations in your lab notebook.

10. Rub the PVC rod with wool, and bring the rod near one end of the suspended polystyrene rod. Observe what happens, and record your observations.

11. Suspend a glass rod on the support, and rub the rod with silk. Rub the PVC rod with wool, and bring the rod near one end of the glass rod. Observe what happens, and record your observations.

12. Suspend a PVC rod on the support, and rub the rod with wool. Rub another PVC rod with wool, and bring the rod near one end of the suspended PVC rod. Observe what happens, and record your observations.

Figure 17-33

Step 9: Suspend the rod securely so that it hangs freely.

Step 10: Charge the rod by rubbing it vigorously with the wool pad. Bring the charged rod near one end of the suspended rod.

Electrostatics Tips

Before the lab, remind students that they will need to determine how many types of charges there are, based on their observations.

Techniques to Demonstrate

Before beginning the lab, give a brief demonstration using two strips of adhesive tape. Prepare the tape as in Step 2 of the lab. Rub the strips with a wool pad, and pull them up. Hold them near each other, as in Step 4 of the lab. Touch one of the strips of tape with your hand to remove the charge. Remind students that touching items in this lab will affect their results.

✔ Checkpoints

Step 7: Make sure metersticks are firmly attached to the table-top and that they do not extend too far over the edge. Remind students to be careful and to keep their work areas clear of people and other obstacles.

Step 9: Make sure all setups are secure. Students should be starting to develop a hypothesis about the number of types of charge there are. They should be able to explain how their results will help them decide.

Step 14: Encourage students to determine what type of charge is on the electroscope.

Step 17: Students may discharge the electroscope by connecting the wire to any object at ground potential. A good choice is a water faucet or pipe.

Step 23: Based on their observation when the second capacitor plate is removed, students should begin to formulate an idea of how a capacitor works.

Charging an electroscope by conduction and induction

13. Charge a polystyrene rod, and touch it briefly against the knob of the electroscope. Record your observations.

14. Touch the knob of the electroscope with your hand. Observe what happens.

15. Bring a charged polystyrene rod near, but not touching, the electroscope knob. Observe what happens.

16. Continue holding the polystyrene rod near the electroscope knob. Briefly touch the knob of the electroscope with your finger. Observe what happens. Remove the rod, and observe what happens.

17. Discharge the electroscope by connecting a wire from it to a grounded object. Your teacher will tell you where to connect the wire.

18. Charge a glass rod, and repeat the procedure in steps 13–17 using the glass rod instead of the polystyrene rod.

Conductors, insulators, and capacitors

19. Set up the apparatus as shown in **Figure 17-34** but do not include the second capacitor plate until Step 22. The insulated wire has an alligator clip at each end. One alligator clip connects to the rod beneath the ball on the electroscope. Run the insulated copper wire from the other alligator clip to one of the plates of the demonstration capacitor.

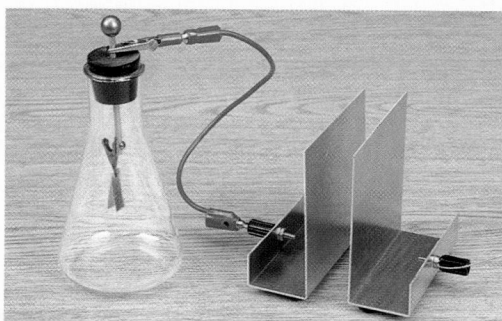

Figure 17-34
Step 20: Charge a polystyrene rod, and bring it near plate B of the capacitor.

Step 21: Repeat the procedure using thread instead of the copper wire.

Step 22: Use copper wire to connect the electroscope to one plate of the capacitor. Touch the charged rod to the plate, and then move the second plate close to the first.

20. Follow the procedure in steps 13–17, but this time bring the rod near the capacitor plate. Observe what happens to the leaves of the electroscope when the rod is brought near the capacitor plate. Record your observations.

21. Replace the copper wire with a piece of thread. Follow the procedure in steps 13–17, but this time bring the rod near the capacitor plate. Observe what happens to the leaves of the electroscope when the rod is brought near the capacitor plate. Record your observations.

22. Connect the rod beneath the knob of the electroscope to one plate of the demonstration capacitor with a short piece of copper wire. Touch the charged polystyrene rod to the plate, and observe what happens to the leaves of the electroscope. Bring the second plate of the capacitor near, but not touching, the first. Observe what happens, and record your observations in your notebook.

23. Remove the second capacitor plate. Observe what happens. Record your observations in your notebook.

24. Bring the second plate near the first again. Using both capacitor plates, try to cause the same result as you obtained using only one plate.

25. Clean up your work area. Put equipment away safely so that it is ready to be used again. Recycle or dispose of used materials as directed by your teacher.

ANALYSIS AND INTERPRETATION

Calculations and data analysis

1. Organizing data Use examples from your observations to explain your answers to the following questions. Assume that the polystyrene rod takes a negative charge when it is rubbed with wool. To answer these questions, assume that like charges repel one another and unlike charges attract one another.

   a. What type of charge is on tape A?

   b. What type of charge is on tape B?

   c. Are the charges on C and D the same?

   d. Are the charges on C or D the same as the charges on A or B?

   e. What type of charge is on the charged suspended glass rod? Is the charge on the suspended glass rod the same or different from the charge on tape A?

   f. What type of charge is on the charged suspended polystyrene rod?

   g. What type of charge is on the second charged polystyrene rod?

   h. What type of charge is on the charged suspended PVC rod?

2. Analyzing results Use your observations to answer the following:

   a. After you touch the knob of the electroscope with your hand, what type of charge is on the electroscope? Explain how your observations support this conclusion.

   b. When the charged polystyrene rod is used to charge an electroscope by induction, what type of charge is on the electroscope?

   c. What type of charge is on the electroscope when it is charged by induction using the charged glass rod?

3. Analyzing information Is copper a conductor or an insulator? Is silk a conductor or an insulator? Use your observations to support your answers.

Conclusions

4. Analyzing conclusions Based on your observations, how many types of charge are there? Explain how your observations support this conclusion.

5. Organizing ideas Use your results to explain what a capacitor does.

ANSWERS TO

Analysis and Interpretation

CALCULATIONS AND DATA ANALYSIS

   1. a. Tape A is negatively charged.

   b. Tape B is negatively charged.

   c. C and D do not have the same charge.

   d. Either C has the same charge as A and B or D has the same charge as A and B.

   e. The glass rod is positively charged; it does not have the same charge as A.

   f. The polystyrene rod is negatively charged.

   g. The second polystyrene rod is negatively charged.

   h. The PVC rod is negatively charged.

   2. a. no charge
   b. positive
   c. negative

   3. Copper is a conductor; silk is an insulator.

CONCLUSIONS

   4. Students should recognize that there are two types of charge.

   5. Capacitors store charge.

CHAPTER 18 PLANNING GUIDE

Compression Guide: To shorten from 7 to 6 45-min periods (from 3½ to 3 90-min blocks), eliminate items in magenta type.

PACING CHART	CLASSROOM RESOURCES			
	⭐TEKS	Teacher Demonstrations	*Holt Physics* Transparencies	**Labs** (See page T52 for equipment listing for in-text labs.)
18-1 Electrical potential energy 1 45-minute period ½ 90-minute block	**2C, 3A, 3B, 5B, 6C**	**TE** *Potential energy,* p. 666		
18-2 Potential difference 1 45-minute period ½ 90-minute block	**1B, 2C, 3B, 3C, 5B, 6C, 6F**			
18-3 Capacitance 3 or 2 45-minute periods 1½ or 1 90-minute block	**2C, 3A, 3B, 3E, 5B, 6B, 6C, 6F**	**TE** *Capacitor discharge,* p. 677 **TE** *Functions of a capacitor,* p. 679	**T** 92–94	**PE** *Capacitance and Electrical Energy,* p. 688 **TL** *Capacitors*
Review and Assessment 2 45-minute periods 1 90-minute block				

Resource Key

HOLT PHYSICS
PE Pupil's Edition
TE Teacher's Edition

L Laboratory Experiments
TL Technology Lab Experiments
T Transparencies

One-Stop Planner CD-ROM **contents**

TM Transparency Masters
SR Section Review Worksheets
AA Alternative Assessment

PW Problem-Solving Workbook
PB Problem Bank
CTW Critical Thinking Worksheet

LABORATORY PLANNING: Capacitance and Electrical Energy, p. 688

Materials (for each lab group):
- 10 Ω resistor, 4–10 W
- black plastic electrical tape, 19 mm × 21 mm
- knife switch (double pole, double throw)
- low calorie calorimeter, with support block for thermometer
- multimeter or voltmeter:
  - basic digital multimeter, or
  - student multimeter, or
  - CENCO standard movement, triple range, dc voltmeter (0–3 V, 10 V, 15 V)
- patch cord, black, insulated alligator clip/stacking banana plug

- patch cord, red, insulated alligator clip/stacking banana plug
- power supply:
  - CENCO 6 V ac/dc low-voltage power supply, or
  - CENCO universal power supply, or
  - 4-output dc regulated power supply
- student electronic stopwatch
- thermometer with fractional divisions:
  - Celsius 0.1°C, −1 to 101°C, or
  - Celsius 0.1°C, −1 to 51°C
- wire leads with alligator clips, 60 cm (pkg. of 10)

ASSIGNMENT RESOURCES

Content Mastery	Critical Thinking	Problem-Solving Practice
PE 1–6, p. 669 **SR** 18-1, *Concept Review* **PE** 1–3, p. 683		**18A** Electrical potential energy: 27 items in **PE**, **PW**, and **PB**, see **TE** pp. 668–669
PE 1–6, p. 675 **SR** 18-2, *Concept Review* **PE** 6–8, p. 683	**PE** 9–11, p. 683	**18B** Potential difference: 30 items in **PE**, **PW**, and **PB**, see **TE** pp. 672–673
PE 1–4, p. 681 **SR** 18-3, *Concept Review* **PE** 15–17, p. 684	**PE** 1–2, p. 678 **PE** 18–25, p. 684	**18C** Capacitance: 29 items in **PE**, **PW**, and **PB**, see **TE** pp. 680–681

ASSESSMENT RESOURCES

Cumulative Review	Alternative Assessment	Traditional Assessment
SR Mixed Review, Ch. 18	**PE** 1–5, p. 687 **AA** Items for Ch. 18	Chapter 18 Test Test Generator items for Ch. 18

Scoring Rubrics for Alternative Assessment items can be found on the One-Stop Planner CD-ROM.

TECHNOLOGY RESOURCES

PE Technology and Learning, p. 686
(Alternative procedures for calculators without Flash-ROM technology are provided on the One-Stop Planner CD-ROM.)

The Mechanical Universe/High School Adaptation Quad V, Equipotentials and Fields

  internet**connect**

On-line Student Resources:
GO TO: www.scilinks.org
The following *sci*LINKS Internet resources can be found in the student text for this chapter.

TOPICS:
- Electrical energy, p. 667 (HF2181)
- Batteries, p. 674 (HF2182)
- Michael Faraday, p. 677 (HF2184)
- Capacitance, p. 678 (HF2183)
- Electric vehicles, p. 691 (HF 2185)

On-line Teacher Resources:
GO TO: go.hrw.com
KEYWORD: HF2 HOME
Visit the HRW Web site for a variety of resources related to this chapter.

 Smithsonian Institution*
Internet Connections
Visit **www.si.edu/hrw** for additional on-line resources.

Visit **www.cnnfyi.com** for late-breaking news and current events stories selected just for you.

Required Precautions

The low-calorie calorimeter must be allowed to cool between trials.

Materials Preparation

The low-calorie calorimeter used in this lab is available from Sargent-Welch/CENCO. It consists of a plastic foam block surrounding a wire coil into which the thermometer can be inserted. When there is current in the wire, the wire heats up. Students can use the temperature change of the wire to find the energy transferred by heat. A calibration constant, in J/°C, is supplied with the calorimeter. Be sure to give students the value for the calibra-tion constant for the calorimeter; they will need it to complete their data analysis for their lab reports.

See pp. 692A–692B for instructions on using multimeters to measure voltage.

Chapter 18
Overview

Section 18-1 introduces electrical potential energy as a form of mechanical energy and calculates the electrical potential energy associated with a pair of charges.

Section 18-2 introduces electric potential, potential difference in a uniform field, and potential difference between an infinite distance and a location near a point charge. It applies the concepts of electric potential and potential difference to a battery in a circuit.

Section 18-3 relates capacitance to the storage of electrical potential energy in the form of separated charges, discusses the dependence of capacitance on the shape of the capacitor and on the material between its plates, and calculates the energy stored by a capacitor.

About the Illustration

This photograph of a lightning storm was taken in southern Arizona. Lightning, a familiar sight to all students, can be used to clarify the concepts of potential difference, capacitance, and electrical breakdown.

CHAPTER 18

Electrical Energy and Capacitance

PHYSICS IN ACTION

During a thunderstorm, particles having different charges accumulate in different parts of a cloud. This separation of charges creates an electric field between a cloud and the ground. Normally, the charges remain separate because air is a nearly perfect insulator. But as charges continue to accumulate, the strength of the electric field increases, creating a potential difference of as much as 100 million volts. As the potential difference increases, a crucial *breakdown voltage* is sometimes reached; that is, the air is ionized by the potential difference and becomes a conductor. Electric charge then flows between the cloud and the ground, an event we perceive as lightning in the sky.

- *What is the relationship between electric potential and the electric field?*

- *How is a thunderstorm like a large natural capacitor?*

CONCEPT REVIEW

- **Potential energy** **(Section 5-2)**

- **Electric field** **(Section 17-3)**

Knowledge to Expect

✔ "Without touching them, material that has been electrically charged pulls on all other materials and may either push or pull other charged matter." (AAAS's *Benchmarks for Science Literacy,* grades 3–5)

✔ "Electrical energy can be produced from a variety of energy sources and can be transformed into almost any other form of energy. Moreover, electricity is used to distribute energy quickly and conveniently to distant locations." (AAAS's *Benchmarks for Science Literacy,* grades 6–8)

✔ "Energy cannot be created or destroyed, but only changed from one form to another." (AAAS's *Benchmarks for Science Literacy,* grades 6–8)

Knowledge to Review

✔ Potential energy is the energy associated with an object due to its position relative to some arbitrary zero level.
(Section 5-2)

✔ An electric field permeates the space around a charged object. The strength of an electric field is the electric force per unit charge.
(Section 17-3)

Items to Probe

✔ Electric force versus electric field strength: Have students explain in their own words the relationship between electric force and electric field strength.

Potential energy

Purpose Show the effects of an electric field on the electrical potential energy by using an analogy with gravitational potential energy.

Materials lump of clay

Procedure Place the clay on the floor, and ask students to estimate the gravitational potential energy of the clay relative to the floor. (*The clay-Earth system's gravitational potential energy is zero because h = 0.*)

Next hold the clay about 1 m above the floor, and have students estimate the new gravitational potential energy (*mgh, h = 1 m*). Ask students how the clay-Earth system acquired gravitational potential energy (*work was done on the system as the clay was lifted in a direction opposite the direction of the field*). Explain that this is analogous to moving a negative charge in an electric field in the direction of **E** or moving a positive charge in an electric field in a direction opposite to **E**. Work is required to move the charge, just as work was required to lift the clay.

Next ask students to describe what happens if you let go of the clay. (*The gravitational field of the Earth exerts a force on the clay, which accelerates toward the Earth.*) Point out that the results would be the same in an electric field if the Earth and the clay were particles having opposite charge.

18-1
Electrical potential energy

electrical potential energy

potential energy associated with an object due to its position relative to a source of electric force

ELECTRICAL ENERGY AND ELECTRIC FORCE

When two charges interact, there is an electric force between them, as described in Chapter 17. As with the gravitational force associated with an object's position relative to Earth, there is a potential energy associated with this force. This kind of potential energy is called **electrical potential energy.** Unlike gravitational potential energy, electrical potential energy results from the interaction of two objects' charges, not their masses. (★)TEKS **5B, 6C**

Electrical potential energy is a form of mechanical energy

Mechanical energy is conserved as long as friction and radiation are not present. As with gravitational and elastic potential energy, electrical potential energy can be included in the expression for mechanical energy (see Chapter 5). If gravitational force, elastic force, and electric force are all acting on an object, the mechanical energy can be written as follows:

$$ME = KE + PE_{grav} + PE_{elastic} + PE_{electric}$$

To account for the forces (except friction) that may also be present in a problem, the appropriate potential-energy terms associated with each force are added to the expression for mechanical energy.

Any time a charge moves because of an electric force, whether from a uniform electric field or from another charge or group of charges, work is done on that charge. For example, in **Figure 18-1,** the electrical potential energy associated with each charge decreases as the charge moves.

Figure 18-1

As the charges in these sparks move, the electrical potential energy decreases, just as gravitational potential energy decreases as an object falls.

Electrical potential energy can be associated with a charge in a uniform field

Consider a positive charge in a uniform electric field. (A uniform field is a field that has the same value and direction at all points.) Assume the charge is displaced at a constant velocity *in the same direction as the electric field*, as shown in **Figure 18-2.**

There is a change in the electrical potential energy associated with the charge's new position in the electric field. The change in the electrical potential energy depends on the charge, q, as well as the strength of the electric field, E, and the displacement, d. It can be written as follows:

$$\Delta PE_{electric} = -qE\Delta d$$

The negative sign indicates that the electrical potential energy will increase if the charge is negative and decrease if the charge is positive.

As with other forms of potential energy, it is the *difference* in electrical potential energy that is physically important. If the displacement in the expression above is chosen so that it is the distance in the direction of the field from the reference point, or zero level, then the initial electrical potential energy is zero and the expression can be rewritten as shown below. As with other forms of energy, the SI unit for electrical potential energy is the joule (J).

> **ELECTRICAL POTENTIAL ENERGY IN A UNIFORM ELECTRIC FIELD**
>
> $$PE_{electric} = -qEd \quad \bigstar \text{TEKS} \quad \textbf{3B, 5B, 6C}$$
>
> **electrical potential energy =**
> **−(charge × electric field strength × displacement from the reference**
> **point in the direction of the field)**

This equation is valid only for a uniform electric field. As described in Chapter 17, the electric field lines for a point charge are farther apart as the distance from the charge increases. Thus, the electric field of a point charge is an example of a nonuniform field because the field strength decreases as distance from the charge increases.

When electrical potential energy is calculated, d is the magnitude of the displacement's component *in the direction of the electric field*. Any displacement of the charge in a direction perpendicular to the electric field does not change the electrical potential energy. This is similar to gravitational potential energy, in which only the vertical distance from the zero level is important.

Electrical potential energy can be associated with a pair of charges

Recall that a single point charge produces a nonuniform electric field. If a second charge is placed nearby, there will be an electrical potential energy associated with the two charges. Because the electric field is not uniform, the electrical potential energy of the system of two charges requires a different expression.

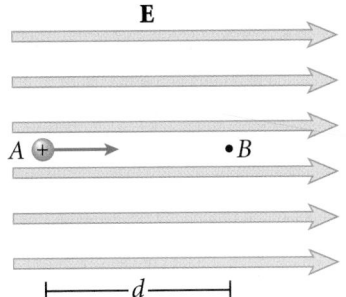

E

$A \oplus \longrightarrow \qquad \bullet B$

$\longmapsto\!\!-\!d\!-\!\!\longmapsto$

Figure 18-2
A positive charge moves from point A to point B in a uniform electric field, and the potential energy changes as a result.

internet **connect**

SC*i*LINKS
NSTA
TOPIC: Electrical energy
GO TO: www.scilinks.org
*sci***LINKS CODE:** HF2181

SECTION 18-1

Teaching Tip

Remind students that the vector for the electric field points in the direction in which a *positive* test charge experiences an electric force. The following table can be used as a memory aid:

	+ charge	− charge
Toward **E**	loses $PE_{electric}$	gains $PE_{electric}$
Opposite **E**	gains $PE_{electric}$	loses $PE_{electric}$

Key Models and Analogies

Compare an electric field to a gravitational field and a positive charge to a mass. If the field does work on the charge by moving it in the direction of **E** (just as Earth's gravitational field does work on a mass by moving it toward Earth), the final potential energy of the system is less than its initial potential energy.

A negative charge interacts in an opposite manner. Because a negative charge feels a force in the opposite direction, the final potential energy is less than the initial potential energy when the charge moves in the direction opposite **E.**

Point out that moving an object horizontally in a gravitational field is analogous to moving a charge in a direction perpendicular to **E.** No work is done, and potential energy remains constant.

$\bigstar$TEKS

p. 666: 5B, 6C
p. 667: 3B, 5B, 6C

Teaching Tip

Use dimensional analysis to show that the equation for the electrical potential energy associated with a pair of charges yields energy in joules when SI units are used:

$$PE_{electric} = k_C \frac{q_1 q_1}{r}$$

$$PE_{electric} = \left(\frac{N \bullet m^2}{C^2}\right)\left(\frac{C \bullet C}{m}\right)$$

$$PE_{electric} = N \bullet m = J$$

PRACTICE GUIDE 18A

Solving for:		
r	📖 **PE**	Sample, 1–3; Ch. Rvw. 4–5, 29*
	💿 **PW**	4–5, 8
	PB	4–6
q	📖 **PE**	4, Ch. Rvw. 29*
	💿 **PW**	Sample, 1–3, 8
	PB	7–10
$PE_{electric}$	📖 **PE**	Ch. Rvw. 31
	💿 **PW**	6–7
	PB	Sample, 1–3

ELECTRICAL POTENTIAL ENERGY FOR A PAIR OF CHARGES

$$PE_{electric} = k_C \frac{q_1 q_2}{r} \quad \text{(★ TEKS)} \quad \text{3B, 5B, 6C}$$

$$\text{electrical potential energy} = \text{Coulomb constant} \times \frac{\text{charge 1} \times \text{charge 2}}{\text{distance}}$$

The quantity k_C is the *Coulomb constant*, which has an approximate value of $8.99 \times 10^9 \, \text{N} \bullet \text{m}^2/\text{C}^2$.

There are several things to keep in mind regarding this expression. First, the reference point for the electrical potential energy is assumed to be at infinity. This can be verified by noting that the electrical potential energy goes to zero as the distance between the charges, r, goes to infinity. Second, because like charges repel, positive work must be done to bring them together, so the electrical potential energy is positive for like charges and negative for unlike charges.

This expression can be used to determine the electrical potential energy associated with more than two charges. In such cases, determine the electrical potential energy associated with *each pair* of charges, and then add the energies.

SAMPLE PROBLEM 18A

Electrical potential energy (★ TEKS) 2C, 3B, 5B, 6C

PROBLEM

The electrical potential energy associated with an electron and a proton is -4.35×10^{-18} J. What is the distance between these two charges?

SOLUTION

Given:

$$q_1 = -1.60 \times 10^{-19} \, \text{C} \qquad q_2 = +1.60 \times 10^{-19} \, \text{C}$$
$$PE_{electric} = -4.35 \times 10^{-18} \, \text{J}$$

Unknown: $r = ?$

Use the equation for the electrical potential energy associated with a pair of charges:

$$PE_{electric} = k_C \frac{q_1 q_2}{r}$$

Rearrange to solve for r:

$$r = k_C \frac{q_1 q_2}{PE_{electric}}$$

$$r = (8.99 \times 10^9 \, \text{N} \bullet \text{m}^2/\text{C}^2)\left(\frac{(-1.60 \times 10^{-19} \, \text{C})(1.60 \times 10^{-19} \, \text{C})}{(-4.35 \times 10^{-18} \, \text{J})}\right)$$

$$\boxed{r = 5.29 \times 10^{-11} \, \text{m}}$$

Electrical potential energy ⭐TEKS 2C, 3B, 5B, 6C

1. Two alpha particles (helium nuclei), each consisting of two protons and two neutrons, have an electrical potential energy of 6.32×10^{-19} J. What is the distance between these particles at this time?

2. Two charges are located along the x-axis. One has a charge of 6.4 μC, and the second has a charge of -3.2 μC. If the electrical potential energy associated with the pair of charges is -4.1×10^{-2} J, what is the distance between the charges?

3. In a charging process, 10^{13} electrons are removed from a metal sphere and placed on a second sphere that is initially uncharged. Then the electrical potential energy associated with the two spheres is found to be -7.2×10^{-2} J. What is the distance between the two spheres?

4. A charge moves a distance of 2.0 cm in the direction of a uniform electric field having a magnitude of 215 N/C. The electrical potential energy of the charge decreases by 6.9×10^{-19} J as it moves. Find the magnitude of the charge on the moving particle. (Hint: The electrical potential energy depends on the distance moved in the direction of the field.)

Section Review ⭐TEKS 2C, 3A, 3B, 5B, 6C

1. What is the difference between $\Delta PE_{electric}$ and $PE_{electric}$?

2. In a uniform electric field, what factors does the electrical potential energy depend on?

3. Describe the conditions that are necessary for mechanical energy to be a conserved quantity.

4. Is there a single correct reference point from which all electrical potential energy measurements must be taken?

5. A uniform electric field with a magnitude of 250 N/C is directed in the positive x direction. A 12 μC charge moves from the origin to the point (20.0 cm, 50.0 cm). What is the change in the electrical potential energy of the system as a result of the change in position of this charge?

6. **Physics in Action**   What is the change in the electrical potential energy in a lightning bolt if 35 C of charge travel to the ground from a cloud 2.0 km above the ground in the direction of the field? Assume the electric field is uniform and has a magnitude of 1.0×10^6 N/C.

ANSWERS TO

Practice 18A
Electrical potential energy

1. 1.46×10^{-9} m
2. 4.5 m
3. 0.32 m
4. 1.6×10^{-19} C

Section Review
ANSWERS

1. $\Delta PE_{electric} = -qE\Delta d = PE_f - PE_i$ is the change in electrical potential energy between two points. If the initial position is considered to be the zero level, $PE_i = 0$, so $PE_{electric} = PE_f = -qEd$.

2. charge, electric field strength, and position in the direction of the field

3. Mechanical energy is conserved in the absence of friction and radiation.

4. No, any reference point can be used, but the initial position is typically used as the zero level to simplify calculations.

5. -6.0×10^{-4} J

6. -7.0×10^{10} J

⭐TEKS

p. 668: 3B, 5B, 6C, 2C, 3B, 5B, 6C
p. 669: 2C, 3B, 5B, 6C, 2C, 3A, 3B, 5B, 6C

Key Models and Analogies

Drawing a parallel between gravitational and electric potential difference can be a useful tool. First, explain to students that gravitational potential is the gravitational potential energy per unit mass, while electric potential is the electrical potential energy per unit charge. Then ask students to imagine carrying a book up to the second floor of a building. The change in the gravitational potential energy of the book is mgh. Thus, the change in gravitational potential (gravitational potential difference) is $\frac{mgh}{m}$, or gh.

Point out that this gravitational potential is independent of the mass of the book; it depends only on the field and on the change in height. Likewise, in a uniform electric field, electric potential position depends only on the field and on the change in position, as seen by the equation for potential difference in a uniform electric field given on page 671 ($\Delta V = -E\Delta d$).

The Language of Physics

Make sure students distinguish between the symbols for electric potential (V) and potential difference (ΔV) with the abbreviation for the unit of potential difference, the volt (V). There is a relationship between the two. Because it is measured in volts, potential difference is sometimes referred to as voltage. This terminology led to the symbol V for electric potential and ΔV for a change in electric potential, or potential difference.

- **Distinguish between electrical potential energy, electric potential, and potential difference.**

- **Compute the electric potential for various charge distributions.**

electric potential

the electrical potential energy associated with a charged particle divided by the charge of the particle

potential difference

the change in electrical potential energy associated with a charged particle divided by the charge of the particle

Figure 18-3

For a typical car battery, there is a potential difference of 12 V between the negative (black) and the positive (red) terminals.

18-2
Potential difference

CHANGING ELECTRIC POTENTIAL

Electrical potential energy is useful in solving problems, particularly those involving charged particles. But at any point in an electric field, as the value of the charge increases, the value of the electrical potential energy increases. A more practical concept in the study of electricity is **electric potential.**

The electric potential at some point is defined as the electrical potential energy associated with a charged particle in an electric field divided by the charge of the particle.

$$V = \frac{PE_{electric}}{q} \quad \text{(★) TEKS} \quad \textbf{3B}$$

Although a greater charge will involve a greater amount of electrical potential energy, the ratio of that energy to the charge is the same as it would be if a smaller charge were at the same position in the field. In other words, the electric potential at a point *is independent of the charge at that point.*

Potential difference is a change in electric potential

Because the reference point for measuring electrical potential energy is arbitrary, the reference point for measuring electric potential is also arbitrary. Thus, only changes in electric potential are significant. The **potential difference** between two points can be expressed as follows:

POTENTIAL DIFFERENCE (★) TEKS **3B**

$$\Delta V = \frac{\Delta PE_{electric}}{q}$$

$$\text{potential difference} = \frac{\text{change in electrical potential energy}}{\text{electric charge}}$$

Potential difference is a measure of the change in the electrical potential energy divided by the charge. The SI unit for potential difference (and electric potential) is the *volt*, V, and is equivalent to one joule per coulomb. As a 1 C charge moves through a potential difference of 1 V, the charge gains (or loses) 1 J of energy. The potential difference between the two terminals of a battery, for instance, can range from about 1.5 V for a small battery to about 12 V for a car battery like the one shown in **Figure 18-3.** The potential difference between the two slots in a household electrical outlet is about 120 V. (★) TEKS **6F**

Remember that only electrical potential energy is a quantity of energy, with units in joules. Electric potential and potential difference are both measures of energy per unit charge (measured in units of volts), and potential difference describes a change in energy per unit charge.

The potential difference in a uniform field varies with the displacement from a reference point

The expression for potential difference can be combined with the expressions for electrical potential energy. The resulting equations are often simpler to apply in certain situations. For example, consider the electrical potential energy of a charge in a uniform electric field.

$$PE_{electric} = -qEd$$

This can be substituted into the equation for potential difference.

$$\Delta V = \frac{\Delta PE_{electric}}{q}$$

$$\Delta V = \frac{\Delta(-qEd)}{q}$$

As the charge moves in the electric field, the only quantity in the parentheses that changes is the displacement from the reference point. Thus, the potential difference in this case can be rewritten as follows:

POTENTIAL DIFFERENCE IN A UNIFORM ELECTRIC FIELD

$$\Delta V = -E\Delta d \quad \text{(★ TEKS)} \quad \textbf{3B, 6C}$$

potential difference =
−(magnitude of the electric field × displacement)

Again, keep in mind that the quantity Δd in this expression is the displacement moved in the direction of the field. Any displacement perpendicular to the field does not change the electrical potential energy.

The reference point for potential difference near a point charge is often at infinity

In Section 18-1, we were given the expression for the electrical potential energy associated with a pair of charges. To determine the potential difference between two points in the field of a point charge, we first calculate the electric potential associated with each point. Imagine a point charge q_2 in the electric field of a point charge q_1 at point B some distance away, as shown in **Figure 18-4.** The electric potential at point A due to q_1 can be expressed as follows:

$$V = \frac{PE_{electric}}{q_2} = k_C \frac{q_1 q_2}{r q_2} = k_C \frac{q_1}{r}$$

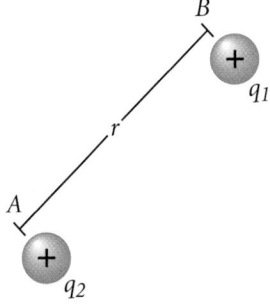

Figure 18-4
The electric potential at point A depends on the charge at point B and the distance r.

SECTION 18-2

Teaching Tip

Remind students that $F_{electric} = qE$, as seen in Chapter 17. Thus, the first equation on this page can be derived as follows:

$PE_{electric} = -W$
$PE_{electric} = -Fd$
$PE_{electric} = -(qE)d$

Visual Strategy

Figure 18-4

Point out to students that the potential energy of q_2 increases as the distance r decreases.

Q How would you modify the figure to make this case similar to gravitational potential?

A *Change q_2 to a negative charge. When q_2 is negative and q_1 is positive, the potential energy of q_2 increases as the distance r increases.*

★ TEKS

p. 670: 3B, 3B, 6F
p. 671: 3B, 6C

Misconception Alert

Some students may confuse electric potential ($V = \dfrac{k_C q_1}{r}$) and potential difference ($\Delta V = \dfrac{k_C q_1}{r}$) because of the similarity between their equations. Remind them that this similarity is the result of choosing zero potential at infinite distance as a reference point. To reinforce the individual nature of the two concepts, work out an example on the chalkboard or overhead projector using a different reference point.

Classroom Practice

The following may be used as a teamwork exercise or for demonstration at the chalkboard or on an overhead projector.

PROBLEM

Potential difference

Between a point some distance r_i from a point charge q_i and an infinite distance, there exists a potential difference of 1.0 V. Determine the resulting potential difference for the following cases:

a. $r = r_i$, $q = 2q_i$
b. $r = \frac{1}{2}r_i$, $q = q_i$
c. $r = 2r_i$, $q = q_i$
d. $r = 2r_i$, $q = 2q_i$

Answer
    a. $\Delta V = 2.0$ V
    b. $\Delta V = 2.0$ V
    c. $\Delta V = 0.50$ V
    d. $\Delta V = 1.0$ V

Did you know?

The volt is named after the Italian physicist Alessandro Volta (1745–1827), who developed the first practical electric battery, known as a voltaic pile. Because potential difference is measured in units of volts, it is sometimes referred to as *voltage*.

Do not confuse the two charges in this example. The charge q_1 is responsible for the electric potential at point *A*. Therefore, *an electric potential exists at some point in an electric field regardless of whether there is a charge at that point.* In this case, the electric potential at a point depends on only two quantities: the charge responsible for the electric potential (in this case q_1) and the distance *r* from this charge to the point in question.

To determine the potential difference between any two points near the point charge q_1, first note that the electric potential at each point depends only on the distance from each point to the charge q_1. If the two distances are r_1 and r_2, then the potential difference between these two points can be written as follows:

$$\Delta V = k_C \frac{q_1}{r_2} - k_C \frac{q_1}{r_1} = k_C q_1 \left(\frac{1}{r_2} - \frac{1}{r_1} \right)$$

If the distance r_1 between the point and q_1 is large enough, it is assumed to be infinitely far from the charge q_1. In that case, the quantity $1/r_1$ is zero. The expression then simplifies to the following (dropping the subscripts):

POTENTIAL DIFFERENCE BETWEEN A POINT AT INFINITY AND A POINT NEAR A POINT CHARGE

$$\Delta V = k_C \frac{q}{r} \quad \text{(★)TEKS} \quad \text{3B, 6C}$$

$$\text{potential difference} = \text{Coulomb constant} \times \frac{\text{value of the point charge}}{\text{distance to the point charge}}$$

This result for the potential difference associated with a point charge appears identical to the electric potential associated with a point charge. The two expressions look the same only because we have chosen a special reference point from which to measure the potential difference.

The superposition principle can be used to calculate the electric potential for a group of charges

The electric potential at a point near two or more charges is obtained by applying a rule called the *superposition principle*. This rule states that the total electric potential at some point near several point charges is the algebraic sum of the electric potentials resulting from each of the individual charges. While this is similar to the method used in Chapter 17 to find the resultant electric field at a point in space, here the summation is much easier to evaluate because the electric potentials are scalar quantities, not vector quantities. There are no components to worry about.

To evaluate the electric potential at a point near a group of point charges, you simply take the algebraic sum of the potentials resulting from all charges. Remember, you must keep track of signs. The electric potential at some point near a positive charge is positive, and the potential near a negative charge is negative.

Potential difference ⭐TEKS 2C, 3B, 6C

PROBLEM

A 5.0 μC point charge is at the origin, and a point charge of -2.0 μC is on the x-axis at (3.0 m, 0.0 m), as shown in Figure 18-5. Find the total potential difference resulting from these charges between point *P*, with coordinates (0.0 m, 4.0 m), and a point infinitely far away.

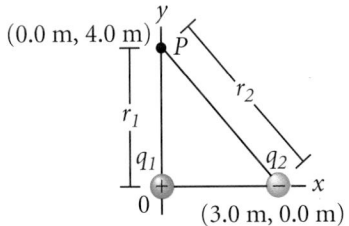

Figure 18-5

SOLUTION

Given: $q_1 = 5.0 \times 10^{-6}$ C $q_2 = -2.0 \times 10^{-6}$ C
$r_1 = 4.0$ m

Unknown: $\Delta V = ?$

Use the equation for the potential difference near a point charge.

$$\Delta V_1 = k_C \frac{q_1}{r_1} = \left(8.99 \times 10^9 \frac{\text{N} \cdot \text{m}^2}{\text{C}^2}\right)\left(\frac{5.0 \times 10^{-6} \text{ C}}{4.0 \text{ m}}\right) = 1.1 \times 10^4 \text{ V}$$

$$r_2 = \sqrt{(3.0 \text{ m})^2 + (4.0 \text{ m})^2} = \sqrt{25 \text{ m}^2} = 5.0 \text{ m}$$

$$\Delta V_2 = k_C \frac{q_2}{r_2} = \left(8.99 \times 10^9 \frac{\text{N} \cdot \text{m}^2}{\text{C}^2}\right)\left(\frac{-2.0 \times 10^{-6} \text{ C}}{5.0 \text{ m}}\right) = -0.36 \times 10^4 \text{ V}$$

$$\Delta V = \Delta V_1 + \Delta V_2 = (1.1 \times 10^4 \text{ V}) - (0.36 \times 10^4 \text{ V})$$

$$\boxed{\Delta V = 7000 \text{ V}}$$

PRACTICE GUIDE 18B

Solving for:		
ΔV	📖 PE	Sample, 1–3; Ch. Rvw. 12–13, 33
	💿 PW	4–5
	PB	5–7
r	📖 PE	Ch. Rvw. 40
	💿 PW	Sample, 1–2
	PB	3–4
q	📖 PE	Ch. Rvw. 42
	💿 PW	3–5
	PB	Sample, 1–2
V	📖 PE	Ch. Rvw. 14, 39
	💿 PW	5–8
	PB	8–10

ANSWERS TO

Practice 18B
Potential difference
1. 1.4×10^{-7} V
2. 110 V
3. 3.8×10^4 V

Potential difference ⭐TEKS 2C, 3B, 6C

1. Find the potential difference between a point infinitely far away and a point 1.0 cm from a proton.

2. Two point charges of magnitude 5.0 nC and -3.0 nC are separated by 35.0 cm. What is the potential difference between a point infinitely far away and a point midway between the charges?

3. Four particles with charges of 5.0 μC, 3.0 μC, 3.0 μC, and -5.0 μC are placed at the corners of a 2.0 m $\times$ 2.0 m square. Determine the potential difference between the center of the square and infinity.

⭐TEKS

p. 672: 3B, 6C
p. 673: 2C, 3B, 6C, 2C, 3B, 6C

Consumer Focus

BACKGROUND

This feature describes three types of household batteries. The feature highlights rechargeable batteries as the most economical and environmentally sound choice.

EXTENSION

Have students compare the prices of the three types of batteries discussed in this feature. Their comparisons should include a variety of brands and potential differences, and should also account for the fact that the cost of a rechargeable battery decreases with repeated uses. If projected lifetimes of the different batteries are available, this information should also be included. Have students present their findings in the form of a table.

internetconnect

SCI**LINKS**
NSTA

TOPIC: Batteries
GO TO: www.scilinks.org
*sci***LINKS CODE:** HF2182

A battery does work to move charges

One of the best applications of the concepts of electric potential and potential difference is in the operation of a battery in some electrical apparatus. The battery is useful when it is connected by conducting wires to devices such as light bulbs, radios, power windows, motors, and so forth. (⭐TEKS) **3C, 6F**

Recall that the reference point for determining the electric potential at some point is arbitrary. When a battery is connected by conductors to an electrical device, the reference point is often defined by *grounding* some point of the arrangement. (A point is said to be grounded when it is connected to an object having an electric potential of zero.) For example, imagine a typical 12 V automobile battery. Such a battery maintains a potential difference across its terminals, where the positive terminal is 12 V higher in potential than the negative terminal. If the negative plate of the battery is grounded, the positive plate would then have a potential of 12 V.

Consumer Focus *Finding the Right Battery*

*H*eavy duty," "long-lasting alkaline," and "environmentally friendly rechargeable" are some of the labels that manufacturers put on batteries. But how do you know which one is the best?

The answer depends on how you will use it. Some batteries are used continuously, but others are turned off and on frequently, as in a stereo. Still others must be able to hold a charge without being used, especially if they will be used in smoke detectors and flashlights.

In terms of price, "heavy duty" batteries typically cost the least, but they last only about 30 percent as long as alkaline batteries. This makes them prohibitively expensive for most uses and makes them an unnecessary source of landfill clutter.

Alkaline batteries are more expensive but have longer lives, lasting up to 6 h in continuous use and up to 18 h in intermittent use. They hold a full charge for years, making them good for use in flashlights and similar devices. They are now less of an environmental problem because manufacturers stopped using mercury in such products several years ago. However, because they are single-use batteries, they also end up in landfills very quickly.

Rechargeable cells are the most expensive to purchase initially. They can cost up to $8, but if recycled, they are the most economical in the long term and are the most environmentally sound choice. These cells, often called NiCads because they contain nickel (Ni) and cadmium (Cd) metals, can be recharged hundreds of times. Although NiCads last only about half as long on one charge as alkaline batteries, the electricity to recharge them costs pennies. NiCads lose about 1 percent of their stored energy each day they are not used and should therefore never be put in smoke detectors or flashlights. (⭐TEKS) **1B, 3C, 6F**

Now imagine a charge of 1 C moving around a battery connected to an electrical device. The charge moves inside the battery from the negative terminal (which is at an electrical potential of 0 V) to the positive terminal (which is at an electrical potential of 12 V). The electric field inside the battery does work on the charge to move it from the negative terminal to the positive terminal and to increase the electrical potential energy associated with the charge. The net result is an electrical potential increase of 12 V. This means that every coulomb of positive charge that leaves the positive terminal of the battery is associated with a total of 12 J of electrical potential energy.

As charge moves through the conductors and devices toward the negative battery terminal, it gives up its 12 J of electrical potential energy to the external devices. When the charge reaches the negative terminal, its electrical potential energy is zero. At this point, the battery provides another 12 J of energy to the charge as it is moved from the negative terminal to the positive terminal of the battery, allowing the charge to make another transit of the device and battery. ⭐TEKS **5B**

CONCEPT PREVIEW

The role of batteries in electric circuits will be discussed further in Chapters 19 and 20.

Section Review

⭐TEKS **2C, 3B, 5B, 6C**

1. The gap between electrodes in a spark plug is 0.060 cm. To produce an electric spark in a gasoline-air mixture, there must be an electric field of 3.0×10^6 V/m. What minimum potential difference must be supplied by the ignition circuit to start a car?

2. Given the electrical potential energy, how do you calculate electric potential?

3. Why is electric potential a more useful quantity for most calculations than electrical potential energy?

4. Explain how electric potential and potential difference are related. What units are used for each one?

5. A proton is released from rest in a uniform electric field with a magnitude of 8.0×10^4 V/m. The proton is displaced 0.50 m as a result.
   a. Find the potential difference between the proton's initial and final positions.
   b. Find the change in electrical potential energy of the proton as a result of this displacement.

6. **Physics in Action** In a thunderstorm, the air must be ionized by a high voltage before a conducting path for a lightning bolt can be created. An electric field of about 1.0×10^6 V/m is required to ionize dry air. What would the breakdown voltage in air be if a thundercloud were 1.60 km above ground? Assume that the electric field between the cloud and the ground is uniform.

Teaching Tip
Use the example of a battery to make distinctions between electrical potential energy, electric potential, and potential difference. The potential difference between the two terminals of a battery is fixed because the electric potential at each terminal is constant ($12 \text{ V} - 0 \text{ V} = 12 \text{ V}$). The electrical potential energy associated with the charge, on the other hand, changes as it moves through the circuit. The electrical potential energy associated with a charge increases as it moves within the battery, then decreases as it moves through external devices.

Section Review
ANSWERS

1. 1.8×10^3 V
2. $V = \dfrac{PE_{electric}}{q}$
3. Electrical potential energy at a point depends on the charge located at that point, while the electric potential at any point is independent of the charge at that point.
4. Potential difference is the change in electric potential. As a result, both have the same SI unit, called the volt (V).
5. a. -4.0×10^4 V
   b. -6.4×10^{-15} J
6. 1.6×10^9 V

⭐TEKS

p. 674: 3C, 6F, 1B, 3C, 6F
p. 675: 5B, 2C, 3B, 5B, 6C

18-3
Capacitance

- **Relate capacitance to the storage of electrical potential energy in the form of separated charges.**

- **Calculate the capacitance of various devices.**

- **Calculate the energy stored in a capacitor.**

Key Models and Analogies

Explain to students that a capacitor stores electrical potential energy much in the way a stretched or a compressed spring stores elastic potential energy. Work must be done to charge a capacitor, just as work must be done to stretch or compress a spring. In both cases, the potential energy acquired by doing work is stored and used at a later time.

Visual Strategy

Figure 18-6

Point out to the students that **Figure 18-6** represents the net charge accumulation on a charged capacitor.

Q Is the net charge of a charged capacitor greater than, less than, or equal to the net charge of the same capacitor when it is uncharged?

A **The net charges of the two capacitors are equal.**

capacitance

the ability of a conductor to store energy in the form of electrically separated charges

CAPACITORS AND CHARGE STORAGE

A capacitor is a device that is used in a variety of electric circuits to perform many functions. Uses include tuning the frequency of radio receivers, eliminating sparking in automobile ignition systems, and storing energy in electronic flash units. ⊛ TEKS 6F

A charged capacitor is useful because it acts as a storehouse of charge and energy that can be reclaimed when needed for a specific application. A typical design for a capacitor consists of two parallel metal plates separated by a small distance. This type of capacitor is called a *parallel-plate capacitor*.

When used in an electric circuit, the plates are connected to the two terminals of a battery or other potential difference, as shown in **Figure 18-6.** When this connection is made, charges are removed from one of the plates, leaving the plate with a net charge. An equal and opposite amount of charge accumulates on the other plate. Charge transfer between the plates stops when the potential difference between the plates is equal to the potential difference between the terminals of the battery. This charging process is illustrated in **Figure 18-6(b).**

Capacitance is the ratio of charge to potential difference

The ability of a conductor to store energy in the form of electrically separated charges is measured by the **capacitance** of the conductor. The capacitance is defined as the ratio of the net charge on each plate to the potential difference created by the separated charges.

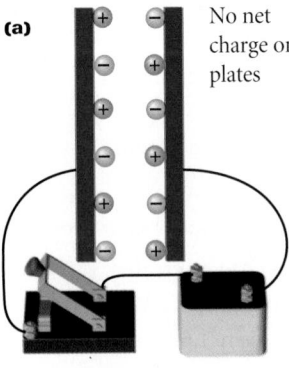
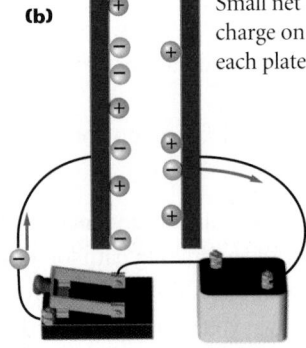
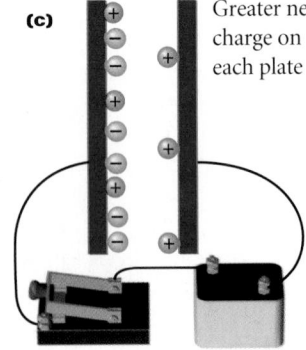

(a) No net charge on plates

(b) Small net charge on each plate

(c) Greater net charge on each plate

Before charging   During charging   After charging

Figure 18-6
When connected to a battery, the plates of a parallel-plate capacitor become oppositely charged.

CAPACITANCE

$$C = \frac{Q}{\Delta V} \quad \text{(★ TEKS) 3B}$$

$$\text{capacitance} = \frac{\text{magnitude of charge on each plate}}{\text{potential difference}}$$

The SI unit for capacitance is the *farad*, F, which is equivalent to a coulomb per volt (C/V). In practice, most typical capacitors have capacitances ranging from microfarads ($1\ \mu\text{F} = 1 \times 10^{-6}$ F) to picofarads ($1\ \text{pF} = 1 \times 10^{-12}$ F).

Capacitance depends on the size and shape of the capacitor

The capacitance of a parallel-plate capacitor with no material between its plates is given by the following expression:

CAPACITANCE FOR A PARALLEL-PLATE CAPACITOR IN A VACUUM

$$C = \varepsilon_0 \frac{A}{d} \quad \text{(★ TEKS) 3B, 6C}$$

$$\text{capacitance} = \text{permittivity of a vacuum} \times \frac{\text{area of one of the plates}}{\text{distance between the plates}}$$

In this expression, the Greek letter ε (epsilon) represents a constant called the *permittivity* of the medium. When it is followed by a subscripted zero, it refers to a vacuum. It has a magnitude of 8.85×10^{-12} $\text{C}^2/\text{N}\cdot\text{m}^2$.

Notice that the amount of charge a parallel-plate capacitor can store for a given potential difference increases as the plate area increases. A capacitor constructed from large plates has a larger capacitance than one having small plates if their plate separations are the same. For a given potential difference, the charge on the plates—and thus the capacitance—increases with decreasing plate separation (d).

Suppose an isolated conducting sphere has a radius R and a charge Q. The potential difference between the surface of the sphere and infinity is the same as it would be for an equal point charge at the center of the sphere.

$$\Delta V = k_C \frac{Q}{R}$$

Substituting this expression into the definition of capacitance results in the following expression:

$$C_{sphere} = \frac{Q}{\Delta V} = \frac{Q}{k_C \left(\dfrac{Q}{R}\right)} = \frac{R}{k_C}$$

(★ TEKS) **3E, 6B**

Did you know?

The farad is named after Michael Faraday (1791–1867), a prominent nineteenth-century English chemist and physicist. Faraday made many contributions to our understanding of electromagnetic phenomena.

internet connect

SciLINKS
NSTA

TOPIC: Michael Faraday
GO TO: www.scilinks.org
***sci*LINKS CODE:** HF2184

Demonstration 2

Capacitor discharge

Purpose Show a capacitor being charged and discharged.

Materials 9 V battery, 10 μF capacitor, two switches, ammeter, light bulb and socket, insulated wire

Procedure Construct the circuit so that the battery charges the capacitor when switch 1 is closed and the capacitor powers the bulb when switch 2 is closed. Explain to students that a circuit must be complete for charges to move. Charge the capacitor by closing switch 1 and opening switch 2. Ask what will happen if switch 1 is opened (*nothing*), and then open the switch.

Next ask what will happen if switch 1 is kept open and switch 2 is closed. Have the students observe the bulb as you close switch 2. To convince students that the battery is not affecting the bulb, remove the battery from the circuit, and repeat the demonstration. Repeat the demonstration a third time with an ammeter placed in series with the light bulb. Point out to students that a charged capacitor would work well in a flashlight if only a flash of light were needed. However, in most cases a continuous current is needed.

(★ TEKS)

p. 676: 6F
p. 677: 3B, 3E, 6B, 3B, 6C

Teaching Tip

When a dielectric completely fills the region between two plates of a capacitor, the capacitance is multiplied by a factor known as the dielectric constant, κ. The dielectric constant of air is very close to that of a vacuum, so calculations assuming a vacuum provide a good approximation for air. A few dielectric constants are given in the following table:

Material	Dielectric constant (κ)
Vacuum	1.000 00
Air	1.000 59
Nylon	3.4
Pyrex glass	5.6
Paper	3.7
Water	80

internet connect

SC/LINKS
NSTA

TOPIC: Capacitors
GO TO: www.scilinks.org
sciLINKS CODE: HF2183

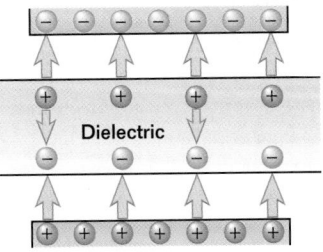

Figure 18-7
The effect of a dielectric is to reduce the strength of the electric field in a capacitor.

This equation indicates that the capacitance of a sphere increases as the size of the sphere increases. Because Earth is so large, it has an extremely large capacitance. This means that Earth can provide or accept a large amount of charge without its electric potential changing too much. This is why Earth is often used as a ground in electric circuits. The ground is the reference point from which the potential differences in a circuit are measured, so it is important that this reference point not change.

The material between the plates of a capacitor can change its capacitance

So far, we have assumed that the space between the plates of a parallel-plate capacitor is a vacuum. However, in many parallel-plate capacitors, the space is filled with a material called a *dielectric*. A dielectric is an insulating material, such as air, rubber, glass, or waxed paper. When a dielectric is inserted between the plates of a capacitor, the capacitance increases. This is because the molecules in a dielectric can align with the applied electric field, causing an excess negative charge near the surface of the dielectric at the positive plate and an excess positive charge near the surface of the dielectric at the negative plate. The surface charge on the dielectric effectively reduces the charge on the capacitor plates, as shown in **Figure 18-7.** Thus, the plates can store more charge for a given potential difference. According to the expression $Q = C\Delta V$, if the charge increases and the potential difference is constant, the capacitance must increase. A capacitor with a dielectric can store more charge and energy for a given potential difference than can the same capacitor without a dielectric. In this book, problems will assume that capacitors are in a vacuum, with no dielectrics.

Discharging capacitors rapidly release their charge

Once a capacitor is charged, the battery or other source of potential difference that charged it can be removed from the circuit. The two plates of the capacitor will remain charged unless they are connected with a material that conducts. Once the plates are connected, the capacitor will *discharge*. This process

ANSWERS TO

Conceptual Challenge

1. yes, because charge is conserved
2. electrical potential energy

Conceptual Challenge ⊛TEKS 3A, 5B

1. Charge on a capacitor plate A certain capacitor is designed so that one plate is large and the other is small. Do the plates have the same charge when connected to a battery?

2. Capacitor storage What does a capacitor store, given that the net charge in a parallel-plate capacitor is always zero?

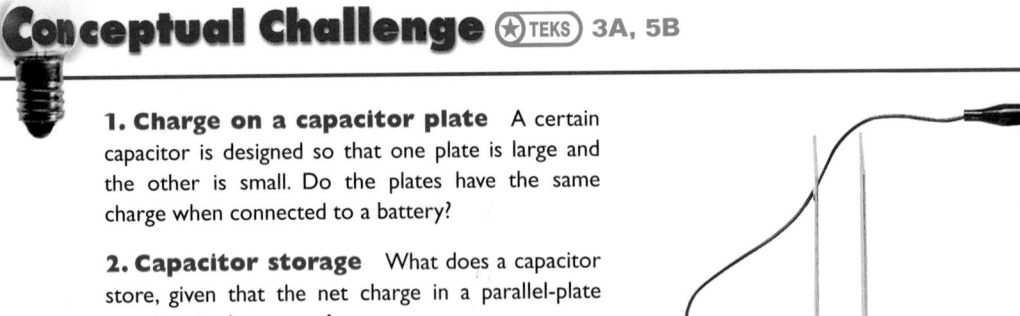

is the opposite of charging. The charges move back from one plate to another until both plates are uncharged again because this is the state of lowest potential energy.

One device that uses a capacitor is the flash attachment of a camera. A battery is used to charge the capacitor, and this stored charge is then released when the shutter-release button is pressed to take a picture. One advantage of using a discharging capacitor instead of a battery to power a flash is that with a capacitor, the stored charge can be delivered to a flash tube much faster, illuminating the subject at the instant more light is needed. ⊛TEKS 6F

Computers make use of capacitors in many ways. For example, one type of computer keyboard has capacitors at the base of its keys, as shown in **Figure 18-8.** Each key is connected to a movable plate, which represents one side of the capacitor. The fixed plate on the bottom of the keyboard represents the other side of the capacitor. When a key is pressed, the capacitor spacing decreases, causing an increase in capacitance. External electronic circuits recognize that a key has been pressed when its capacitance changes.

Because the area of the plates and the distance between the plates can be controlled, the capacitance, and thus the electric field strength, can also be easily controlled.

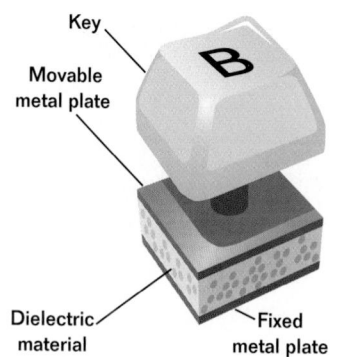

Figure 18-8
A parallel-plate capacitor is often used in keyboards.

ENERGY AND CAPACITORS ⊛TEKS 5B

A charged capacitor stores electrical potential energy because it requires work to move charges through a circuit to the opposite plates of a capacitor. The work done on these charges is a measure of the transfer of energy (see Chapter 5).

For example, if a capacitor is initially uncharged so that the plates are at the same electric potential, that is, if both plates are neutral, then almost no work is required to transfer a small amount of charge from one plate to the other. However, once a charge has been transferred, a small potential difference appears between the plates. As additional charge is transferred through this potential difference, the electrical potential energy of the system increases. This increase in energy is the result of work done on the charge. The electrical potential energy stored in a capacitor that is charged from zero to some charge, Q, is given by the following expression:

ELECTRICAL POTENTIAL ENERGY STORED IN A CHARGED CAPACITOR

$$PE_{electric} = \frac{1}{2}Q\Delta V \quad ⊛TEKS \; 3B, 5B$$

electrical potential energy =
$\frac{1}{2}$(**charge on one plate**)(**final potential difference**)

Teaching Tip

Living cells have characteristics similar to capacitors. Charged ions in a cell and in the fluid surrounding the cell set up a charge distribution across the membrane wall. The cell is equivalent to a small capacitor separated by a dielectric (the membrane wall). The potential difference across a cell wall is typically on the order of 100 mV.

Demonstration 3

Functions of a capacitor

Purpose Show the energy storage and delivery functions of a capacitor.

Materials hand-crank generator and large (1 F) capacitor

Procedure Before connecting the hand-crank generator to the capacitor, crank the generator and show the students how quickly the handle stops when it is released.

Connect the generator to the capacitor. Crank the generator, and release the handle. Have students observe the motor effect of the crank handle being driven by the stored energy in the capacitor.

⊛TEKS

p. 678: 3A, 5B
p. 679: 6F, 5B, 3B, 5B

Figure 18-9
The markings caused by electrical breakdown in this material look similar to the lightning bolts produced when air undergoes electrical breakdown.

By substituting the definition of capacitance ($C = Q/\Delta V$), these alternative forms can also be shown to be valid:

$$PE_{electric} = \tfrac{1}{2}C(\Delta V)^2$$

$$PE_{electric} = \frac{Q^2}{2C} \quad \text{⭐TEKS} \quad \textbf{3B, 5B}$$

These results apply to any capacitor. In practice, there is a limit to the maximum energy (or charge) that can be stored because electrical breakdown ultimately occurs between the plates of the capacitor for a sufficiently large potential difference. For this reason, capacitors are usually labeled with a maximum operating potential difference. Electrical breakdown in a capacitor is similar to a lightning discharge in the atmosphere. **Figure 18-9** shows a pattern created in a block of Plexiglass that has undergone electrical breakdown. This book's problems assume that all potential differences are below the maximum.

SAMPLE PROBLEM 18C

Capacitance ⭐TEKS 2C, 3B, 5B

PROBLEM

A 3.0 μF capacitor is connected to a 12 V battery. What is the magnitude of the charge on each plate of the capacitor, and how much electrical potential energy is stored in the capacitor?

SOLUTION

Given: $C = 3.0\ \mu F = 3.0 \times 10^{-6}\ F$ $\qquad \Delta V = 12\ V$

Unknown: $Q = ?$ $\qquad PE_{electric} = ?$

To determine the charge, use the equation for capacitance on page 677:

$$Q = C\Delta V$$
$$Q = (3.0 \times 10^{-6}\ F)(12\ V) = 36 \times 10^{-6}\ C$$

$$\boxed{Q = 36\ \mu C}$$

To determine the potential energy, use the alternative form of the equation for the potential energy of a charged capacitor shown on this page:

$$PE_{electric} = \tfrac{1}{2}C(\Delta V)^2$$
$$PE_{electric} = (0.5)(3.0 \times 10^{-6}\ F)(12\ V)^2$$

$$\boxed{PE_{electric} = 2.2 \times 10^{-4}\ J}$$

PRACTICE 18C

Capacitance ⭐TEKS 2C, 3B, 5B

1. A 4.00 μF capacitor is connected to a 12.0 V battery.
   a. What is the charge on each plate of the capacitor?
   b. If this same capacitor is connected to a 1.50 V battery, how much electrical potential energy is stored?

2. A parallel-plate capacitor has a charge of 6.0 μC when charged by a potential difference of 1.25 V.
   a. Find its capacitance.
   b. How much electrical potential energy is stored when this capacitor is connected to a 1.50 V battery?

3. A capacitor has a capacitance of 2.00 pF.
   a. What potential difference would be required to store 18.0 pC?
   b. How much charge is stored when the potential difference is 2.5 V?

4. You are asked to design a parallel-plate capacitor having a capacitance of 1.00 F and a plate separation of 1.00 mm. Calculate the required surface area of each plate. Is this a realistic size for a capacitor?

Section Review   ⭐TEKS 2C, 3B, 5B

1. Explain why two metal plates near each other will not become charged unless they are connected to a source of potential difference.

2. A parallel-plate capacitor has an area of 2.0 cm^2, and the plates are separated by 2.0 mm.
   a. What is the capacitance?
   b. How much charge does this capacitor store when connected to a 6.0 V battery?

3. A parallel-plate capacitor has a capacitance of 1.35 pF. If a 12.0 V battery is connected to this capacitor, how much electrical potential energy would it store?

4. **Physics in Action**   Assume Earth and a cloud layer 800.0 m above the Earth can be treated as plates of a parallel-plate capacitor.
   a. If the cloud layer has an area of 1.00×10^6 m^2, what is the capacitance?
   b. If an electric field strength of 2.0×10^6 N/C causes the air to conduct charge (lightning), what charge can the cloud hold?

Electrical Energy and Capacitance   **681**

ANSWERS TO

Practice 18C
Capacitance
1. a. 4.80×10^{-5} C
   b. 4.50×10^{-6} J
2. a. 4.8×10^{-6} F
   b. 5.4×10^{-6} J
3. a. 9.00 V
   b. 5.0×10^{-12} C
4. 1.13×10^8 m^2; no

Section Review
ANSWERS

1. If there is no potential difference, there is no electric force to set charges in motion.
2. a. 8.8×10^{-13} F
   b. 5.3×10^{-12} C
3. 9.72×10^{-11} J
4. a. 1.11×10^{-8} F
   b. ±18 C

⭐TEKS

p. 680: 3B, 5B, 2C, 3B, 5B
p. 681: 2C, 3B, 5B, 2C, 3B, 5B

CHAPTER 18
Summary

Teaching Tip

Written explanations help to solidify students' understanding of difficult concepts and to enforce good communication skills. Have students write essays in which they summarize the differences between electrical potential energy, electric potential, and potential difference. Essays should also include a thorough discussion of the factors that affect capacitance. Be sure students explain concepts clearly and correctly, and use good sentence structure.

★ TEKS

Review & Assess
pp. 683–687:
2C: 4–5, 12–14, 26–49
2D: Alt. Assess. 1, 3, 5
2E: Technology & Learning; Alt. Assess. 4
3B: 2–5, 12–14, 18, 21, 24, 26–49; Technology & Learning
3C: Alt. Assess. 5
3D: Alt. Assess. 2
3E: Alt. Assess. 5
5B: 1–6, 10–11, 27–28, 31, 35–38, 41–42, 44–47; Technology & Learning
6B: Alt Assess. 3, 5
6C: 8, 12–14, 21–22, 24, 29, 31–34, 38–42, 48–49
6F: Alt. Assess. 5

KEY TERMS

capacitance (p. 676)

electric potential (p. 670)

electrical potential energy (p. 666)

potential difference (p. 670)

KEY IDEAS

Section 18-1 Electrical potential energy

- Electrical potential energy is energy associated with a charged object due to its position relative to a source of electric force.
- Electrical potential energy is a form of mechanical energy.

Section 18-2 Potential difference

- Electric potential is electrical potential energy divided by charge.
- The electric potential at a given point in an electric field is independent of the charge at that point.
- Only differences in electric potential (potential differences) from one position to another are useful in calculations.

Section 18-3 Capacitance

- The capacitance, C, of an object is the amount of charge, Q, the object can store for a given potential difference, ΔV, as shown by the equation at right:

$$C = \frac{Q}{\Delta V}$$

- Capacitance depends on the shape of the capacitor, the distance between the plates, and the dielectric between the plates.
- A capacitor is a device that is used to store electrical potential energy.
- Capacitors will charge if a potential difference is applied. Once charged, a capacitor can discharge if its plates are connected by a conducting path.
- The potential energy stored in a charged capacitor depends on the charge and the final potential difference between the capacitor's two plates:

$$PE_{electric} = \frac{1}{2} Q \Delta V$$

Diagram symbols

Positive charge	
Negative charge	
Electric field vector	**E**

Variable symbols

Quantities		Units		Conversions
$PE_{electric}$	electrical potential energy	J	joule	$= N \cdot m$
				$= kg \cdot m^2/s^2$
V	electric potential	V	volt	$= J/C$
ΔV	potential difference	V	volt	
C	capacitance	F	farad	$= C/V$

ELECTRICAL POTENTIAL ENERGY

Review questions

1. Describe the motion and explain the energy conversions that are involved when a positive charge is placed in a uniform electric field. Be sure your discussion includes the following terms: *electrical potential energy, work,* and *kinetic energy.*

2. If a point charge is displaced perpendicular to a uniform electric field, which of the following expressions is likely to be equal to the change in electrical potential energy?
   a. $-qEd$
   b. 0
   c. $-k_c\left(\dfrac{q^2}{r^2}\right)$

3. Explain why the electrical potential energy is positive for a pair of like charges but negative for a pair of unlike charges.

Practice problems

4. A point charge of 9.00×10^{-9} C is located at the origin of a coordinate system. A positive charge of 3.00×10^{-9} C is brought in from infinity to a point such that the electrical potential energy associated with the two charges is 8.09×10^{-7} J. How far apart are the charges at this time?
   (See Sample Problem 18A.)

5. An electron that is initially 55 cm away from a proton is displaced to another point. If the change in the electrical potential energy as a result of this movement is 2.1×10^{-28} J, what is the final distance between the electron and the proton?
   (See Sample Problem 18A.)

ELECTRIC POTENTIAL AND POTENTIAL DIFFERENCE

Review questions

6. Differentiate between electrical potential energy and electric potential.

7. Differentiate between electric potential and potential difference.

8. At what location in relationship to a point charge is the electric potential considered by convention to be zero?

Conceptual questions

9. If the electric field in some region is zero, must the electric potential in that same region also be zero? Explain your answer.

10. If a proton is released from rest in a uniform electric field, does the corresponding electric potential at the proton's changing locations increase or decrease? What about the electrical potential energy?

11. If an electron is released from rest in a uniform electric field, does the corresponding electric potential at the electron's changing locations increase or decrease? What about the electrical potential energy?

Practice problems

12. The magnitude of a uniform electric field between the two plates is about 1.7×10^6 N/C. If the distance between these plates is 1.5 cm, find the potential difference between the plates.
    (See Sample Problem 18B.)

13. A force of 4.30×10^{-2} N is needed to move a charge of 56.0 μC a distance of 20.0 cm in the direction of a uniform electric field. What is the potential difference that will provide this force?
    (See Sample Problem 18B.)

ANSWERS TO

Chapter 18
Review and Assess

1. The positive charge moves in the direction of the electric field. As the field does work on the charge, it converts the electrical potential energy associated with the charge to kinetic energy.

2. b

3. The work required to bring two like charges together is positive, while the work required to bring two opposite charges together is negative.

4. 30.0 cm

5. 1.1 m

6. Electric potential is the electrical potential energy associated with a charge divided by the magnitude of the charge. Thus, electric potential is a characteristic of a point in space and is independent of the charge at that point.

7. Potential difference is the change in electric potential between two points in space.

8. at infinity

9. no; An electric field can be considered to be the rate of change of electric potential in space. Thus, if the electric field is zero, there is no *change* in the potential, but the potential can still have a constant nonzero value.

10. decrease; decrease

11. decrease; decrease

12. 2.6×10^4 V

13. -154 V

14. -4.2×10^5 V

15. The charge on each plate doubles.

16. Use a dielectric between the capacitor plates.

17. The Earth is large enough that it can accept or supply an unlimited number of charges without its electric potential changing significantly; Any object with this ability can act as a ground.

18. $PE_{electric}$ is 4 times as great.

19. yes; The plates' capacitance depends on the area of the plates and the distance between them.

20. The plates should be close together, have a large surface area, and have a dielectric between them.

21. C is 4 times as great.

22. $PE_{electric}$ is doubled when d is doubled ($PE_{electric} = \frac{1Q^2}{2C} = \frac{1}{2}Q^2\frac{d}{\varepsilon_0 A}$).

23. The capacitor stores electrical potential energy that could be discharged through the body, a conductor. This could be prevented by using an insulator.

24. a. 800 V
b. $\frac{1}{2}$

25. Because the potential difference is the same, the only difference between the two capacitors is the amount of charge stored. Thus, the shock must be caused by a large amount of charge passing through your body in a short time interval.

26. $\pm 7.2 \times 10^{-11}$ C

27. a. 1.3×10^{-3} C
b. 4.2 J

28. 0.22 J

29. 2.00×10^{-7} C, 3.000 m

30. 0.11 m

---

14. In **Figure 18-10,** find the electric potential at point P due to the grouping of charges at the other corners of the rectangle.
(See Sample Problem 18B.)

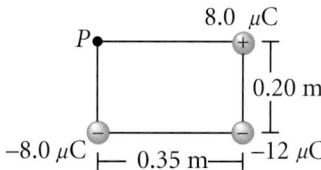

Figure 18-10

CAPACITANCE

Review questions

15. What happens to the charge on a parallel-plate capacitor if the potential difference doubles?

16. You want to increase the maximum potential difference of a parallel-plate capacitor. Describe how you can do this for a fixed plate separation.

17. Why is the Earth considered a "ground" in electric terms? Can any other object act as a ground?

Conceptual questions

18. If the potential difference across a capacitor is doubled, by what factor is the electrical potential energy stored in the capacitor multiplied?

19. Two parallel plates are uncharged. Does the set of plates have a capacitance? Explain.

20. If you were asked to design a small capacitor with high capacitance, what factors would be important in your design?

21. If the area of the plates of a parallel-plate capacitor is doubled while the spacing between the plates is halved, how is the capacitance affected?

22. A parallel-plate capacitor is charged and then disconnected from a battery. How much does the stored energy change when the plate separation is doubled?

23. Why is it dangerous to touch the terminals of a high-voltage capacitor even after the potential difference has been removed? What can be done to make the capacitor safe to handle?

---

24. The potential difference between a pair of oppositely charged parallel plates is 400 V.
a. If the spacing between the plates is doubled without altering the charge on the plates, what is the new potential difference between the plates?
b. If the plate spacing is doubled while the potential difference between the plates is kept constant, what is the ratio of the final charge on one of the plates to the original charge?

25. Two capacitors have the same potential difference across them, but one has a large capacitance and the other has a small capacitance. You would get a greater shock from touching the leads of the one with high capacitance than you would from touching the leads of the other one. From this information, explain what causes the sensation of shock.

Practice problems

26. A 12.0 V battery is connected to a 6.0 pF parallel-plate capacitor. What is the charge on each plate?
(See Sample Problem 18C.)

27. A parallel-plate capacitor has a capacitance of 0.20 μF and is to be operated at 6500 V.
a. Calculate the charge stored.
b. What is the electrical potential energy stored in the capacitor at the operating potential difference?
(See Sample Problem 18C.)

28. Two devices with capacitances of 25 μF and 5.0 μF are each charged with separate 120 V power supplies. Calculate the total energy stored in the two capacitors.
(See Sample Problem 18C.)

MIXED REVIEW

29. At some distance from a point charge, the electric potential is 600.0 V and the magnitude of the electric field is 200.0 N/C. Determine the charge and the distance from the charge.

30. A circular parallel-plate capacitor with a spacing of 3.0 mm is charged to produce a uniform electric field with a strength of 3.0×10^6 N/C. What plate radius is required if the stored charge is -1.0 μC?

31. Three charges are situated at three corners of a rectangle, as shown in **Figure 18-11.** How much electrical potential energy would be expended in moving the 8.0 μC charge to infinity?

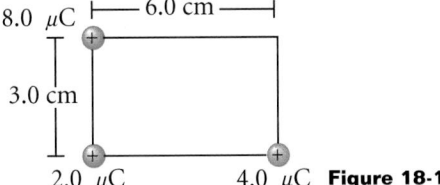
Figure 18-11

32. A 12 V battery is connected across two parallel metal plates separated by 0.30 cm. Find the magnitude of the electric field.

33. A parallel-plate capacitor has an area of 5.00 cm^2, and the plates are separated by 1.00 mm. The capacitor stores a charge of 400.0 pC.

    a. What is the potential difference across the plates of the capacitor?

    b. What is the magnitude of the uniform electric field in the region that is located between the plates?

34. A parallel-plate capacitor has a plate area of 175 cm^2 and a plate separation of 0.0400 mm. Determine the following:

    a. the capacitance

    b. the potential difference when the charge on the capacitor is 500.0 pC

35. A certain moving electron has a kinetic energy of 1.00×10^{-19} J.

    a. Calculate the speed necessary for the electron to have this energy.

    b. Repeat the calculation for a proton having a kinetic energy of 1.00×10^{-19} J.

36. A proton is accelerated from rest through a potential difference of 25 700 V.

    a. What is the kinetic energy of this proton in joules after this acceleration?

    b. What is the speed of the proton after this acceleration?

37. A proton is accelerated from rest through a potential difference of 120 V. Calculate the final speed of this proton.

38. A pair of oppositely charged parallel plates are separated by 5.33 mm. A potential difference of 600.0 V exists between the plates.

    a. What is the magnitude of the electric field strength in the region that is located between the plates?

    b. What is the magnitude of the force on an electron that is in the region between the plates at a point that is exactly 2.90 mm from the positive plate?

    c. The electron is moved to the negative plate from an initial position 2.90 mm from the positive plate. What is the change in electrical potential energy due to the movement of this electron?

39. The three charges shown in **Figure 18-12** are located at the vertices of an isosceles triangle. Calculate the electric potential at the midpoint of the base if each one of the charges at the corners has a magnitude of 5.0×10^{-9} C.

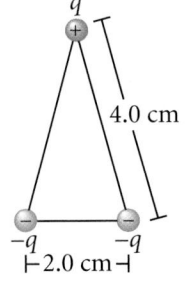
Figure 18-12

40. A charge of -3.00×10^{-9} C is at the origin of a coordinate system, and a charge of 8.00×10^{-9} C is on the x-axis at 2.00 m. At what two locations on the x-axis is the electric potential zero?

(Hint: One location is between the charges, and the other is to the left of the y-axis.)

41. A pair of oppositely charged parallel plates are separated by a distance of 5.0 cm with a potential difference of 550 V between the plates. A proton is released from rest at the positive plate at the same time that an electron is released from rest at the negative plate. Disregard any interaction between the proton and the electron.

    a. How long does it take for the paths of the proton and the electron to cross?

    b. How fast will each particle be going when their paths cross?

    c. How much time will elapse before the proton reaches the opposite plate?

(See Appendix A for hints on solving simultaneous equations.)

31. 9.1 J

32. 4.0×10^3 V/m

33. a. 90.4 V

    b. 9.04×10^4 V/m

34. a. 3.87×10^{-9} F

    b. 0.129 V

35. a. 4.69×10^5 m/s

    b. 1.09×10^4 m/s

36. a. 4.11×10^{-15} J

    b. 2.22×10^6 m/s

37. 1.5×10^5 m/s

38. a. 1.13×10^5 V/m

    b. 1.81×10^{-14} N

    c. 4.39×10^{-17} J

39. −7800 V

40. 0.545 m, −1.20 m

41. a. 7.2×10^{-9} s

    b. 1.4×10^7 m/s, 7.6×10^3 m/s

    c. 3.1×10^{-7} s

42. 3.20×10^{-19} C

43. 4.000×10^{-6} F

44. a. 7.2×10^{-13} J
  b. 2.9×10^{7} m/s

45. 2.1×10^{6} m/s

46. a. -9.22×10^{4} V
  b. 9.22×10^{4} V

ANSWERS TO

Technology & Learning

a. 2.5×10^{-6} or 2.5E–6
b. 8.28 V
c. 14.7 V
d. 6.12 V
e. 3.87 V
f. 9.30×10^{-5} C

42. An ion is displaced through a potential difference of 60.0 V and experiences an increase of electrical potential energy of 1.92×10^{-17} J. Calculate the charge on the ion.

43. A potential difference of 100.0 V exists across the plates of a capacitor when the charge on each plate is 400.0 μC. What is the capacitance?

44. A proton is accelerated through a potential difference of 4.5×10^{6} V.

  a. How much kinetic energy has the proton acquired?

  b. If the proton started at rest, how fast is it moving?

45. A positron (a particle with a charge of $+e$ and a mass equal to that of an electron) that is accelerated from rest between two points at a fixed potential difference acquires a speed of 9.0×10^{7} m/s. What speed is achieved by a *proton* accelerated from rest between the same two points? (Disregard relativistic effects.)

46. The speed of light is 3.00×10^{8} m/s.

  a. Through what potential difference would an electron starting from rest need to accelerate to achieve a speed of 60.0% of the speed of light? (Disregard relativistic effects.)

  b. Repeat this calculation for a positron.

Technology  Learning

Graphing calculators

Refer to Appendix B for instructions on downloading programs for your calculator. The program "Chap18" allows you to analyze a graph of potential difference versus electrical potential energy stored in a capacitor.

The electrical potential energy stored in a charged capacitor, as you learned earlier in this chapter, is described by the following equation:

$$PE_{electric} = \tfrac{1}{2}C(\Delta V)^2$$

The program "Chap18" stored on your graphing calculator makes use of this equation. Once the "Chap18" program is executed, your calculator will ask for the capacitance. The graphing calculator will use the following equation to create a graph of the electrical potential energy (Y_1) versus the potential difference (X). The relationships in this equation are the same as those in the equation shown above.

$$Y_1 = CX^2/2$$

  a. If the capacitance is 2.50 μF, what value should you key in on your graphing calculator?

Execute "Chap18" on the [PRGM] menu, and press [ENTER] to begin the program. Enter the value for the capacitance (shown below), and press [ENTER]. Remember to use the [(-)] key for entering negative values and [2nd] [EE] to enter exponents.

The calculator will provide a graph of the electrical potential energy versus the potential difference. (If the graph is not visible, press [WINDOW] and change the settings for the graph window, then press [GRAPH].)

Press [TRACE], and use the arrow keys to trace along the curve. The x-value corresponds to potential difference in volts, and the y-value corresponds to the potential energy in joules.

Determine the potential difference required in the following situations:

  b. a 3.50 μF capacitor storing 1.20×10^{-4} J of energy
  c. the same capacitor storing 3.80×10^{-4} J of energy
  d. a 24.0 μF capacitor storing 4.50×10^{-4} J of energy
  e. the same capacitor storing 1.80×10^{-4} J of energy
  f. What is the charge on one plate of the capacitor in item (e)?

Press [2nd] [QUIT] to stop graphing. Press [ENTER] to input a new value or [CLEAR] to end the program.

47. An electron moves from one plate of a capacitor to another, through a potential difference of 2200 V.

    a. Find the speed with which the electron strikes the positive plate.

    b. Repeat part **(a)** for a proton moving from the positive plate to the negative plate.

48. Each plate on a 3750 pF capacitor carries a charge with a magnitude of 1.75×10^{-8} C.

    a. What is the potential difference across the plates when the capacitor has been fully charged?

    b. If the plates are 6.50×10^{-4} m apart, what is the magnitude of the electric field between the two plates?

49. A parallel-plate capacitor is made of two circular plates, each with a diameter of 2.50×10^{-3} m. The plates of this capacitor are separated by a space of 1.40×10^{-4} m.

    a. Assuming that the capacitor is operating in a vacuum and that the permittivity of a vacuum can be used, determine the capacitance for this arrangement.

    b. How much charge will be stored on each plate of this capacitor when it is connected across a potential difference of 0.12 V?

    c. What is the electrical potential energy stored in this capacitor when it is fully charged by the potential difference of 0.12 V?

    d. What is the potential difference between a point midway between the plates and a point that is 1.10×10^{-4} m from one of the plates?

    e. If the potential difference of 0.12 V is removed from the circuit and if the circuit is allowed to discharge until the charge on the plates has decreased to 70.7 percent of its fully charged value, what will the potential difference across the capacitor be?

Alternative Assessment

Performance assessment

1. If the electric potential in a certain region of space is zero, can you infer that there is no electric charge in that area? Come up with examples of sets of charges that support your inference, and be prepared to defend your conclusion in a class discussion.

2. Imagine that you are assisting nuclear scientists who need to accelerate electrons between electrically charged plates. Design and sketch a piece of equipment that could accelerate electrons to 10^7 m/s. What should the potential difference be between the plates? How would protons move inside this device? What would you change in order to accelerate the electrons to 100 m/s?

Portfolio projects

3. Tantalum is an element widely used in electrolytic capacitors. Research tantalum and its properties. Where on Earth is it found? In what form is it found? How expensive is it? Present your findings to the class in the form of a report, poster, or computer presentation.

4. Investigate how the shape of equally charged objects affects the potential around them. Examine the case of a point charge of 1 C, a hollow conducting sphere with a charge of 1 C, and a 1 m × 1 m square plate with a charge of 1 C. Find the electric potential at increasing distances from 0.1 m to 10 m along a line through the center of each charge. Draw graphs of the electric potential versus the distance from the charged object, and compare them.

5. Research what types of capacitors are used in different electrical and electronic devices. If possible, obtain old radios, telephones, or other electronic instruments that you can take apart. Find the capacitors, and describe them. Complete this research by visiting electronics stores or by examining catalogs of electrical and electronic equipment. Do the capacitors that are the largest in size have the largest capacitance values? Can you find out what substances are used to make these capacitors? Compile the class findings in a poster or an exhibit about these capacitors, and indicate their relative sizes, structures, capacitances, and uses.

47. a. 2.8×10^7 m/s
   b. 6.5×10^5 m/s

48. a. 4.67 V
   b. 7180 V/m

49. a. 3.10×10^{-13} F
   b. 3.7×10^{-14} C
   c. 2.2×10^{-15} J
   d. 3.4×10^{-2} V
   e. 8.5×10^{-2} V

Alternative Assessment
ANSWERS

Performance assessment

1. Zero potential means that the total work done to bring a charge to that area is zero.

2. Student answers will vary but should include the need for potential differences of 284 V and 2.8×10^{-8} V to accelerate electrons to each speed.

3. Student answers will vary. Be sure students include sources. Thin films of Ta_2O_5 are used as dielectrics in some capacitors. Powder price for tantalum ranges from $5/g to $10/g.

4. For the point charge, students should draw a $\frac{1}{x}$ hyperbolic graph. For the sphere, the graph is similar for distances greater than its radius $(d > R)$. The graph for the plate should be a horizontal line that drops off far from the plate.

5. Students should recognize that size is not always a good predictor of capacitance. They can identify capacitors by their labels (μF, nF, or pF) or categorize them as fixed or variable, tubular or plate, and by their usage.

NOTE

Materials Preparation is given on pp. 664A–664B. Blank data table and sample data table are on the One-Stop Planner CD-ROM. All calculations shown use sample data.

Planning

Recommended time:

1 lab period

Classroom organization:

▶ Each lab group needs a level work surface that is near an electrical outlet and away from any sources of water.

▶ Each lab group should have two students.

▶ **Safety warnings:** Emphasize the dangers of working with electricity. Remind students to have you check their circuits before turning on the power supply or closing the switch.

Capacitance Tips

◆ Remind students that calorimetry allows them to measure the amount of energy transferred as heat, as in the Chapter 10 calorimetry lab.

Techniques to Demonstrate

Make sure students understand how to use the low-calorie calorimeter. Demonstrate wrapping tape around the bulb of the thermometer to ensure fit, and show the students how to use the support to hold the thermometer securely during the lab.

OBJECTIVES

- Explore the relationship between electrical energy and energy transferred by heat.
- Calculate the energy stored in a charged capacitor.

MATERIALS LIST

✔ 1-farad capacitor
✔ 10 Ω resistor
✔ insulated connecting wire
✔ low-calorie calorimeter with thermometer support
✔ momentary contact switch
✔ multimeter or dc voltmeter
✔ plastic electrical tape
✔ power supply
✔ stopwatch
✔ thermometer

CHAPTER 18
Laboratory Exercise

⭐ TEKS
pp. 688–689: 1A, 2B, 2C, 2D, 2F, 3B, 5B, 5D

CAPACITANCE AND ELECTRICAL ENERGY

When a capacitor is charged, it stores electrical potential energy, which can be measured by a change in temperature. In this experiment, you will charge a capacitor, then discharge it through a wire coil inside a special calorimeter. You will use the initial and final temperatures of the calorimeter to calculate the amount of energy transferred from the capacitor to the wire.

SAFETY

- **Use a hot mitt to handle resistors, light sources, and other equipment that may be hot. Allow all equipment to cool before storing it.**

- **Never put broken glass or ceramics in a regular waste container. Use a dustpan, brush, and heavy gloves to carefully pick up broken pieces, and dispose of them in a container specifically provided for this purpose.**

PREPARATION

1. Read the entire lab, and plan the steps you will take.

2. Prepare a data table in your lab notebook with six columns and five rows. In the first row, label the columns *Trial, Mass of calorimeter (kg), T_i (°C), T_{max} (°C), ΔV_c (V),* and *$C_{capacitor}$ (F)*. In the first column, label the second through fifth rows *1, 2, 3,* and *4*.

PROCEDURE

Charging the capacitor

3. Record the value of the capacitor in farads as $C_{capacitor}$.

4. Set up the apparatus as shown in **Figure 18-13.** Place the switch in front of you so that the switch moves from right to left. Using leads with alligator clips on both ends, connect the low-calorie calorimeter to the pins on the right side of the switch, and connect the capacitor to the center pins of the switch. Using leads with alligator clips on one end and banana terminals on the other, connect the resistor in series with the power supply and the pins on the left side of the switch. Connect the voltage meter to measure the potential difference across the capacitor. ***Do not close the switch until your teacher approves your circuit.***

5. Place the bulb end of the thermometer in the calorimeter. If necessary, carefully wrap one or two turns of electrical tape around the bulb to ensure a good fit. Place the other end of the thermometer in a support so that the thermometer is easy to read. Record the initial temperature of the calorimeter in your data table as T_i. **Do not close the switch.**

6. When your teacher has approved your circuit, turn on the power supply and close the switch to the left in order to charge the capacitor. While the capacitor is charging, watch the voltage meter. When the potential difference across the capacitor remains constant for 30 s, record the value.

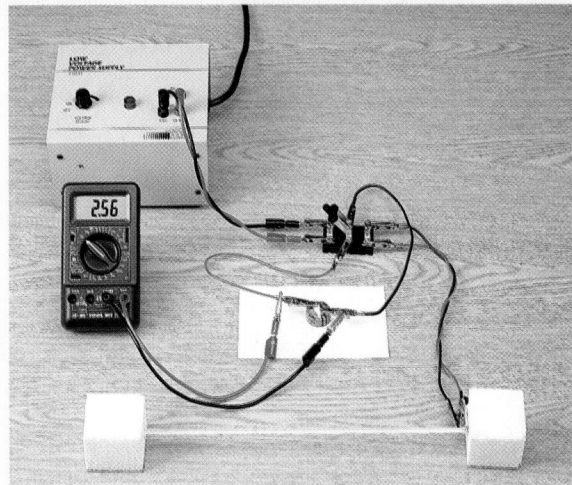

Discharging the capacitor

7. Close the switch to the right to discharge the capacitor so that the current will be in the wire inside the calorimeter. Read the temperature of the calorimeter at 5.0 s intervals until five consecutive readings are the same. Record the maximum temperature reached in your data table.

8. Repeat the procedure for three more trials. **Do not close the switch until your teacher approves your circuit.** Record all data for all trials.

9. Clean up your work area. Put equipment away safely so that it is ready to be used again.

Figure 18-13

Step 4: Place the switch in front of you so that it moves from right to left. Check all connections carefully.

Step 5: Carefully place the bulb end of the thermometer in the calorimeter. Support the other end of the thermometer by placing it in a support, as shown.

Step 7: While the capacitor is discharging, read and record the temperature of the calorimeter at 5.0 s intervals.

ANALYSIS AND INTERPRETATION

Calculations and data analysis

1. Organizing data   For each trial recorded in your data table, perform the following calculations.

   a. Calculate the temperature change of the calorimeter.

   b. Your teacher will give you the value for the calibration constant of the calorimeter in J/°C. With that value, calculate the energy transferred to the calorimeter as heat.

2. Organizing information   Use the equation $PE_{electric} = \frac{1}{2}C(\Delta V)^2$ to calculate the energy stored in the capacitor for each trial.

Conclusions

3. Evaluating results   For each trial, compare the energy transferred to the calorimeter as heat to the energy stored in the capacitor. Based on your observations, is energy conserved in this experiment? Explain.

If the students are using a dc power supply, show them how to adjust the voltage. Power supplies should be set at around 20 V for this exercise.

✔ Checkpoints

Step 4: Make sure the power supply, resistor, capacitor, and meter are connected properly.

Step 6: Students should be able to explain that the capacitor is fully charged when the voltage across it is constant.

Step 7: Remind students to begin taking readings as soon as they close the switch to the right.

ANSWERS TO

Analysis and Interpretation

CALCULATIONS AND DATA ANALYSIS

1. a. Student answers will vary. Make sure students use the relationship $\Delta T = T_f - T_i$. For sample data, values range from 2.8°C to 4.9°C.

   b. Student answers will vary. For sample data, values range from 8.7 J to 15 J.

2. Student answers will vary. For sample data, values range from 14 J to 16 J.

CONCLUSIONS

3. Student answers will vary. If energies are not the same, answers should include an analysis of possible causes of the discrepancy.

Are Electric Cars an Answer to Pollution?

In 1909, more than 50 percent of the automobiles in the United States were propelled by electric motors. These cars could go only a few miles before their batteries needed recharging, but they were more reliable than gasoline-powered cars, and starting them did not require hand cranking. Electric cars worked well for most in-city use, but as more roads were paved and as more people wanted to travel farther, electric cars were abandoned.

The return of electric cars

As industry spread and the number of cars on the road increased, the air in North America became more polluted and people searched for ways to reduce the pollution and its harmful effects on human health. In recent decades, federal and state laws have required industries and businesses–from steelmakers to dry cleaners–to limit polluting emissions. Although overall air quality has improved due to these regulations, pollution within many cities remains a serious problem, primarily because of emissions from motor-powered vehicles that use gasoline-burning internal combustion engines, or ICEs.

Gasoline-fueled cars emit nitrogen oxides, carbon monoxide, and unburned hydrocarbons—all of which, along with ozone, make up a major part of urban air pollution. In addition, ICEs give off large quantities of carbon dioxide, which contributes to Earth's greenhouse effect. So, because they emit no polluting exhaust fumes, people are interested once again in electric vehicles.

Modern electric vehicles

An electric vehicle, or EV, is powered by an electric motor with current supplied by an array of batteries. To recharge the batteries, the owner plugs the car into a recharging unit at home. The charging energy comes from the same electric-generating plants that produce household current. Besides producing no polluting exhaust, the main advantages of EVs are that they need no hydrocarbon fuel, they are quiet, and, because there are fewer moving parts to wear out, they are likely to last longer than cars with ICEs.

However, even EV proponents and manufacturers realize that EVs available now and in the near future have some major drawbacks. EVs can travel only about 65 km to 160 km before needing a recharge, which is not far enough for intercity travel or family trips. Also, EVs presently cost between $30,000 and $75,000, prices that are unacceptable to most consumers, especially when they consider the cars' limited range. Another problem is that the lead-acid batteries used in EVs, which cost about $3000, have to be replaced every two to three years.

The question of pollution

Many people wonder whether replacement of ICE vehicles by EVs will reduce air pollution and greenhouse gases or will increase the problem. After all, in charging a vehicle, the owner would be using electricity supplied by a power plant, which in its operation most likely causes air pollution. In some regions, such as southern California and the Pacific Northwest, abundant electrical energy is available from relatively nonpolluting hydroelectric, nuclear, and natural-gas-burning power plants. The need to charge large numbers of EVs is unlikely to increase pollution at these sites.

On the other hand, 55 percent of plants in the United States are powered by burning coal. Power plants will have to increase output by burning more coal, and new plants may have to be built. One scientist estimates that complete conversion to EVs would double the output of sulfur dioxide, a cause of acid rain, from coal-burning plants.

Supporters of EVs say that most people would recharge their vehicles overnight when plants are operating their capacity. EV proponents also have estimated that 20 million EVs could be supported in the United States without any new power plants. In addition,

they claim that burning coal to generate electric current to power an EV is up to twice as efficient as burning gasoline to power a car.

To settle this question, the U.S. Department of Energy has assigned Argonne National Laboratories to perform a Total Energy Cycle Assessment comparing the energy efficiency of EVs with that of vehicles that run on gasoline. Technicians will consider every step, from producing fuel to using the energy to move a car. In addition, they will keep track of all emissions, leftover material, and all activities involved in using and maintaining the vehicles. The results of this study should provide answers that will shape the destiny of electric vehicles.

internet connect

SCi LINKS

NSTA

TOPIC: Electric vehicles
GO TO: www.scilinks.org
sciLINKS CODE: HF2185

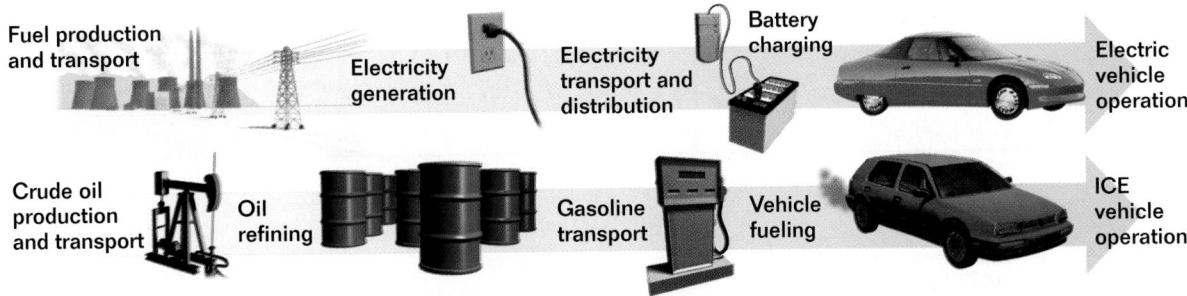

Fuel production and transport — Electricity generation — Electricity transport and distribution — Battery charging — Electric vehicle operation

Crude oil production and transport — Oil refining — Gasoline transport — Vehicle fueling — ICE vehicle operation

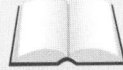 **Researching the Issue**

1. If the Argonne study reveals that gasoline-powered cars are more energy-efficient than EVs, do you think that EVs should be discontinued? Account for your opinion.

2. The limited range of EVs is one of their biggest drawbacks. Do research to find out about hybrid electric vehicles, or HEVs. Describe how they operate and explain the differences between series and parallel HEVs. How might these vehicles overcome the range limitations of EVs?

3. The state of California has mandated that by the year 2003 at least 10 percent of the vehicles for sale in the state must be zero-emission vehicles, and EVs are the only vehicles that meet that requirement. Other states have passed similar laws. Because EVs are so expensive, the federal and state governments are offering subsidies and tax credits to people who will buy them. Hold a discussion or debate on the question, "Should the government spend taxpayers' money to subsidize the purchase of nonpolluting vehicles that people might not otherwise buy?"

TEKS

pp. 690–691: 1B, 2C, 2D, 3C, 5B, 6F

CHAPTER 19 PLANNING GUIDE

Compression Guide: To shorten from 9 to 6 45-min periods (from 4½ to 3 90-min blocks), eliminate items in magenta type.

PACING CHART	CLASSROOM RESOURCES			
	★ TEKS	Teacher Demonstrations	Holt Physics Transparencies	Labs (See page T52 for equipment listing for in-text labs.)
19-1 Electric current 1 45-minute period ½ 90-minute block	1A, 2C, 3A, 3B, 3C, 3E, 5B, 6E, 6F	**TE** *Drift speed*, p. 697 **TE** *Potential difference as a source of current*, p. 698	 **T** 95	**PE** *Quick Lab: A Lemon Battery*, p. 696
19-2 Resistance 3 or 2 45-minute periods 1½ or 1 90-minute block	2C, 3A, 3B, 3C, 3E, 6E, 6F, 8B	**TE** *Non-ohmic resistance*, p. 701 **TE** *Factors that affect resistance*, p. 701	**TM** 62–63 **T** 96–97	**PE** *Current and Resistance*, p. 722 **L** *Discovery Lab: Resistors and Current*
19-3 Electric power 3 or 1 45-minute periods 1½ or ½ 90-minute block	1A, 2C, 3A, 3B, 3C, 3E, 5B, 6E, 6F, 8A			**PE** *Quick Lab: Energy Use in Home Appliances*, p. 711 **L** *Invention Lab: Battery-Operated Portable Heater* **TL** *Electrical Energy*

Review and Assessment
2 45-minute periods
1 90-minute block

Resource Key

PHYSICS

PE Pupil's Edition
TE Teacher's Edition

L Laboratory Experiments
TL Technology Lab Experiments
T Transparencies

One-Stop Planner CD-ROM contents

TM Transparency Masters
SR Section Review Worksheets
AA Alternative Assessment

PW Problem-Solving Workbook
PB Problem Bank
CTW Critical Thinking Worksheet

LABORATORY PLANNING: Current and Resistance, p. 722

Materials (for each lab group)
- mounted resistance coils
- patch cord, black, insulated alligator clip/stacking banana plug
- patch cord, red, insulated alligator clip/stacking banana plug
- power supply:
  - 4-output dc regulated power supply, or
  - CENCO 6 V ac/dc low-voltage power supply, or
  - CENCO universal power supply
- switch, contact key, or knife switch (single pole/single throw)
- wire leads with alligator clips (60 cm)

Additional Equipment
CBL and Sensors Procedure
- CBL
- CBL voltage probe
- graphing calculator with link cable
- measuring resistor, $1\ \Omega$
- measuring resistor, $5\ \Omega$ (optional)
- patch cord, 60 cm, black, with banana plugs
- patch cord, 60 cm, red, with banana plugs
- TI Graph Link (recommended for downloading programs)

ASSIGNMENT RESOURCES

Content Mastery	Critical Thinking	Problem-Solving Practice
PE 1–5, p. 699 **SR** 19-1, *Concept Review* **PE** 1–10, p. 717	**PE** 1–3, p. 697 **PE** 11–16, p. 717	**19A** Current: 28 items in **PE, PW,** and **PB,** see **TE** p. 695
PE 1–7, p. 707 **SR** 19-2, *Concept Review* **PE** 20–23, p. 718	**PE** 1–3, p. 704 **PE** 24–27, p. 718	**19B** Resistance: 29 items in **PE, PW,** and **PB,** see **TE** pp. 702–703
PE 1–5, p. 713 **SR** 19-3, *Concept Review* **PE** 31–35, p. 718	**PE** 1–3, p. 709 **PE** 36–39, p. 718	**19C** Electric power: 27 items in **PE, PW,** and **PB,** see **TE** p. 710 **19D** Cost of electrical energy: 27 items in **PE, PW,** and **PB,** see **TE** pp. 712–713

ASSESSMENT RESOURCES

Cumulative Review	Alternative Assessment	Traditional Assessment
SR Mixed Review, Ch. 19	**PE** 1–5, p. 721 **AA** Items for Ch. 19	Chapter 19 Test Test Generator items for Ch. 19

Scoring Rubrics for Alternative Assessment items can be found on the One-Stop Planner CD-ROM.

TECHNOLOGY RESOURCES

 CTW Segment 19 Edison's Lab

 PE Technology and Learning, p. 720
(Alternative procedures for calculators without Flash-ROM technology are provided on the One-Stop Planner CD-ROM.)

internet connect

 On-line Student Resources:
GO TO: www.scilinks.org
The following *sci*LINKS Internet resources can be found in the student text for this chapter.

TOPICS:
• Electric current, p. 696 (HF2191)
• Generators, p. 698 (HF2192)
• Ohm's law, p. 700 (HF2193)
• Superconductors, p. 706 (HF2194)

 On-line Teacher Resources:
GO TO: go.hrw.com
KEYWORD: HF2 HOME
Visit the HRW Web site for a variety of resources related to this chapter.

 Smithsonian Institution
Internet Connections
Visit www.si.edu/hrw for additional on-line resources.

Visit www.cnnfyi.com for late-breaking news and current events stories selected just for you.

Meters Procedure

One of the following:
2 basic digital multimeters; 2 student multimeters; 2 general purpose multimeters; or 1 CENCO standard movement, triple range dc ammeter (50 mA, 500 mA, 5 A) **and** 1 CENCO standard movement, triple range dc voltmeter (0–3, 10, 15 V)

Materials Preparation

To use multimeters to measure current and potential difference (voltage), follow this setup procedure.

Current: On the front of the multimeter, set the range switch pointer to *DC A* to read direct current in amperes. Set the pointer to the *200 mA* mark. Use a piece of tape to label this meter *A*. Make connections to the sockets labeled *mA* and *COM*.

Potential difference (voltage): On the front of the multimeter, set the range switch pointer to *DC V* to read direct current in volts. Set the pointer to the *20 V* mark. Use a piece of tape to label this meter *V*. Make connections to the sockets labeled *V*Ω and *COM*.

When checking students' circuits, always make sure voltage meters are connected in parallel and current meters are connected in series. Also make sure leads are connected to the correct terminals of the meters.

Section 19-1 introduces the concept of current as the rate of charge movement; discusses conventional current, drift velocity, and sources of current; and distinguishes between direct and alternating currents.

Section 19-2 explores resistance in terms of both Ohm's law and the factors that affect resistance, distinguishes between ohmic and non-ohmic materials, and introduces superconductors.

Section 19-3 introduces electric power as the rate at which electrical energy is transferred, solves problems involving electric power and the cost of electrical energy, and discusses why electrical energy is transported at high potential differences.

About the Illustration

Circuit boards that involve separate components joined by wires and attached to a base are called *conventional circuits*. In *integrated circuits,* all individual components are formed on a single chip that is made of silicon or another semiconducting material. Because integrated circuits are small, they allow signals to travel much faster between components. Integrated circuits are used in video games, computers, microwave ovens, and a variety of other electronic devices. In some cases, integrated circuits are used as components of conventional circuits.

CHAPTER 19

Current and Resistance

PHYSICS IN ACTION

Each device attached to a computer, such as a modem or a monitor, is typically controlled through a circuit board. These circuit boards are in turn connected to a main circuit board, called the *motherboard*. Each circuit board is studded with resistors, capacitors, and other tiny components that are involved with the movement and storage of electric charge. Copper "wires" printed onto the circuit boards during their manufacture conduct charges between components.

In this chapter, you will study the movement of electric charge and learn what factors affect the ease with which charges move through different materials.

- *What is the function of a resistor?*
- *What hinders the movement of charge?*

CONCEPT REVIEW

Power (Section 5-4)

Electrical potential energy
    (Section 18-1)

Potential difference (Section 18-2)

Tapping Prior Knowledge

Knowledge to Expect

✔ "Electrical energy can be produced from a variety of energy sources and can be transformed into almost any other form of energy. Moreover, electricity is used to distribute energy quickly and conveniently to distant locations." (AAAS's *Benchmarks for Science Literacy,* grades 6–8)

✔ "People try to conserve energy in order to slow down the depletion of energy resources and/or to save money." (AAAS's *Benchmarks for Science Literacy,* grades 3–5)

Knowledge to Review

✔ Power is the rate at which energy is transferred or work is done, as in the equation $P = W/\Delta t$. (Section 5-4)

✔ Electric field strength is the magnitude of the electric force per unit charge, as in the equation $E = F/q$. (Section 17-3)

✔ Electrical potential energy is the energy of a charge due to its position in an electric field. (Section 18-1)

✔ Potential difference is the electrical potential energy per unit charge. (Section 18-2)

Items to Probe

✔ Electrical potential energy: Have students explain the difference between potential energy and potential difference using mass and gravitational attraction instead of charges and electric fields.

694

Misconception Alert

Some students may think that electric current is the flow of electrical energy. Explain that current refers to the movement of matter, not the movement of energy. The flow of electrical energy corresponds to the movement of electric and magnetic fields, not to the movement of charge carriers.

Visual Strategy

Figure 19-1

Use this diagram to reinforce the definition of *current* as the rate of charge movement.

Q How does the current change if the number of charge carriers increases?

A *The current increases.*

Q How does the current change if the time interval during which a given number of charge carriers pass the cross-sectional area increases?

A *The current decreases.*

⭐ TEKS

p. 694: 6F, 3E, 3B
p. 695: 2C, 3B, 6E, 2C, 3B, 6E

19-1
Electric current

19-1 SECTION OBJECTIVES

- **Describe the basic properties of electric current.**

- **Solve problems relating current, charge, and time.**

- **Distinguish between the drift speed of a charge carrier and the average speed of the charge carrier between collisions.**

- **Differentiate between direct current and alternating current.**

Figure 19-1

The current in this wire is defined as the rate at which electric charges pass through a cross-sectional area of the wire.

current

the rate at which electric charges move through a given area

CURRENT AND CHARGE MOVEMENT ⭐TEKS 6F

Although many practical applications and devices are based on the principles of static electricity, electricity did not become an integral part of our daily lives until scientists learned to control the movement of electric charge, known as *current*. Electric currents power our lights, radios, television sets, air conditioners, and refrigerators. Currents also ignite the gasoline in automobile engines, travel through miniature components that make up the chips of computers, and perform countless other invaluable tasks.

Electric currents are even part of the human body. This connection between physics and biology was discovered by Luigi Galvani (1737–1798). While conducting electrical experiments near a frog he had recently dissected, Galvani noticed that electrical sparks caused the frog's legs to twitch and even convulse. After further research, Galvani concluded that electricity was present in the frog. Today, we know that electric currents are responsible for transmitting messages between body muscles and the brain. In fact, every action involving muscles is initiated by electrical activity. ⭐TEKS 3E

Current is the rate of charge movement

A current exists whenever there is a net movement of electric charge through a medium. To define *current* more precisely, suppose positive charges are moving through a wire, as shown in **Figure 19-1**. The **current** is the rate at which these charges move through the cross section of the wire. If ΔQ is the amount of charge that passes through this area in a time interval, Δt, then the current, I, is the ratio of the amount of charge to the time interval.

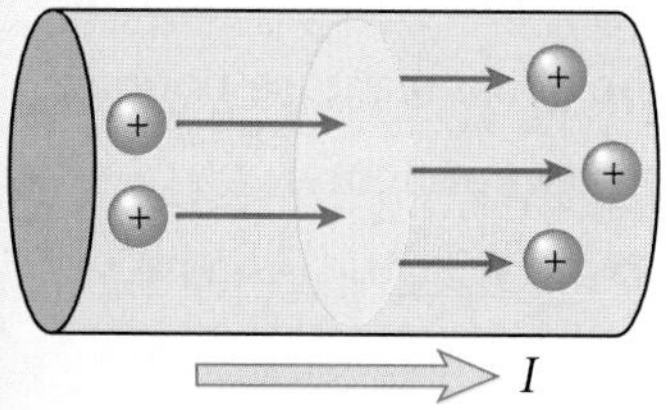

ELECTRIC CURRENT ⭐TEKS 3B

$$I = \frac{\Delta Q}{\Delta t}$$

$$\text{electric current} = \frac{\text{charge passing through a given area}}{\text{time interval}}$$

The SI unit for current is the *ampere*, A. One ampere is equivalent to one coulomb of charge passing through a cross-sectional area in a time interval of one second (1 A = 1 C/s).

Current ⭐TEKS 2C, 3B, 6E

PROBLEM

The current in a light bulb is 0.835 A. How long does it take for a total charge of 1.67 C to pass a point in the wire?

SOLUTION

Given: $\Delta Q = 1.67$ C $I = 0.835$ A

Unknown: $\Delta t = ?$

Use the equation for electric current given on page 694. Rearrange to solve for the time interval.

$$I = \frac{\Delta Q}{\Delta t}$$

$$\Delta t = \frac{\Delta Q}{I}$$

$$\Delta t = \frac{1.67 \text{ C}}{0.835 \text{ A}} = 2.00 \text{ s}$$

Current ⭐TEKS 2C, 3B, 6E

1. If the current in a wire of a CD player is 5.00 mA, how long would it take for 2.00 C of charge to pass a point in this wire?

2. In a particular television tube, the beam current is 60.0 μA. How long does it take for 3.75×10^{14} electrons to strike the screen?

3. If a metal wire carries a current of 80.0 mA, how long does it take for 3.00×10^{20} electrons to pass a given cross-sectional area of the wire?

4. The compressor on an air conditioner draws 40.0 A when it starts up. If the start-up time is 0.50 s, how much charge passes a cross-sectional area of the circuit in this time?

5. A total charge of 9.0 mC passes through a cross-sectional area of a nichrome wire in 3.5 s.
   a. What is the current in the wire?
   b. How many electrons pass through the cross-sectional area in 10.0 s?
   c. If the number of charges that pass through the cross-sectional area during the given time interval doubles, what is the resulting current?

Classroom Practice

The following may be used as a teamwork exercise or for demonstration at the chalkboard or on an overhead projector.

PROBLEM

Relating current and charge

A 100.0 W light bulb draws 0.83 A of current. How long does it take for 1.9×10^{22} electrons to pass a given cross-sectional area of the filament?

Answer
   1.0 h

PRACTICE GUIDE 19A

Solving for:		
Δt	📖 **PE**	Sample, 1–3; Ch. Rvw. 17–19
	💿 **PW**	6
	PB	4–6
ΔQ	📖 **PE**	4; Ch. Rvw. 46, 58a*
	💿 **PW**	Sample, 1–2, 3*
	PB	7–10
I	📖 **PE**	5; Ch. Rvw. 44, 58b
	💿 **PW**	4–5
	PB	Sample, 1–3

ANSWERS TO

Practice 19A
Current
1. 4.00×10^2 s
2. 1.00 s
3. 6.00×10^2 s
4. 2.0×10^1 C
5. a. 2.6×10^{-3} A
   b. 1.6×10^{17} electrons
   c. 5.1×10^{-3} A

Teaching Tip

Currents that consist of both positive and negative charge carriers in motion include those that exist in batteries, in the human body, in the ocean, and in the ground. Electric currents in the brain and nerves of the human body consist of moving sodium and potassium atoms that are charged.

Quick Lab

TEACHER'S NOTES

This lab is meant to give an example of charge movement through an electrolytic solution. Be sure that the ends of the copper wire are not sharp. Use sandpaper to smooth any sharp ends.

 Misconception Alert

Many students have the misconception that charge carriers move at the speed of light. When discussing the concept of drift velocity, address this misconception directly with students.

⭐ TEKS

p. 696: 6E, 1A, 6E, 6E, 6F
p. 697: 5B, 2C, 3A

696

TOPIC: Electric current
GO TO: www.scilinks.org
sciLINKS CODE: HF2191

⭐ TEKS **1A, 6E**

Quick Lab

A Lemon Battery

MATERIALS LIST

✔ lemon
✔ copper wire
✔ paper clip

Straighten the paper clip, and insert it and the copper wire into the lemon to construct a chemical cell. Touch the ends of both wires with your tongue. Because a potential difference exists across the two metals and because your saliva provides an electrolytic solution that conducts electric current, you should feel a slight tingling sensation on your tongue.

Conventional current is defined in terms of positive charge movement

The moving charges that make up a current can be positive, negative, or a combination of the two. In a common conductor, such as copper, current is due to the motion of negatively charged electrons. This is because the atomic structure of solid conductors allows the electrons to be transferred easily from one atom to the next, while the protons are relatively fixed inside the nucleus of the atom. In certain particle accelerators, a current exists when positively charged protons are set in motion. In some cases—in gases and dissolved salts, for example—current is the result of positive charges moving in one direction and negative charges moving in the opposite direction.

Positive and negative charges in motion are sometimes called *charge carriers*. *Conventional current* is defined as the current consisting of positive charge that would have the same effect as the actual motion of the charge carriers—regardless of whether the charge carriers are positive, negative, or a combination of the two. The three possible cases are shown in **Table 19-1.** We will use conventional current in this book unless stated otherwise. ⭐TEKS 6E

Table 19-1   Conventional current

	First case	Second case	Third case
motion of charge carriers	⟵ −	+ ⟶	+ ⟶ / ⟵ −
equivalent conventional current	+ ⟶	+ ⟶	+ ⟶ / + ⟶

As was explained in Chapter 18, an electric field in a material sets charges in motion. For a material to be a good conductor, charge carriers in the material must be able to move easily through the material. Many metals are good conductors because metals usually contain a large number of free electrons. Body fluids and salt water are able to conduct electric charge because they contain charged atoms called *ions*. Because dissolved ions can move through a solution easily, they can be charge carriers. A solute that consists of charge carriers is called an *electrolyte*.

DRIFT VELOCITY ⭐TEKS 6E, 6F

When you turn on a light switch, the light comes on almost immediately. For this reason, many people think that electrons flow very rapidly from the socket to the light bulb. However, this is not the case. When you turn on the switch, an electric field is established in the wire. This field, which sets electric charges in motion, travels through the wire at nearly the speed of light. The charges themselves, however, travel much more slowly.

Drift velocity is the net velocity of charge carriers

To see how the electrons move, consider a solid conductor in which the charge carriers are free electrons. When the conductor is in electrostatic equilibrium, the electrons move randomly, similar to the movement of molecules in a gas. When a potential difference is applied across the conductor, an electric field is set up inside the conductor. The force due to that field sets the electrons in motion, thereby creating a current.

These electrons do not move in straight lines along the conductor in a direction opposite the electric field. Instead, they undergo repeated collisions with the vibrating metal atoms of the conductor. If these collisions were charted, the result would be a complicated zigzag pattern like the one shown in **Figure 19-2.** The energy transferred from the electrons to the metal atoms during the collisions increases the vibrational energy of the atoms, and the conductor's temperature increases.

The average energy gained by the electrons as they are accelerated by the electric field is greater than the average loss in energy due to the collisions. Thus, despite the internal collisions, the individual electrons move slowly along the conductor in a direction opposite the electric field, **E,** with a velocity known as the **drift velocity, v_{drift}.** (★)TEKS 5B

Drift speeds are relatively small

The magnitudes of drift velocities, or drift speeds, are typically very small. In fact, the drift speed is much less than the average speed between collisions. For example, in a copper wire that has a current of 10.0 A, the drift speed of electrons is 2.46×10^{-4} m/s. These electrons would take about 68 min to travel 1 m! The electric field, on the other hand, reaches electrons throughout the wire at a speed approximately equal to the speed of light. (★)TEKS 2C

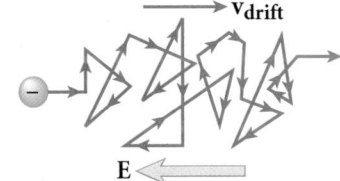

Figure 19-2
When an electron moves through a conductor, collisions with the vibrating metal atoms of the conductor force the electron to change its direction constantly.

drift velocity

the net velocity of a charge carrier moving in an electric field

Conceptual Challenge (★)TEKS 3A

1. Electric field inside a conductor We concluded in our study of electrostatics that the field inside a conductor is zero, yet we have seen that an electric field exists inside a conductor that carries a current. How is this electric field possible?

2. Turning on a light If charges travel very slowly through a metal (approximately 10^{-4} m/s), why doesn't it take several hours for a light to come on after you flip a switch?

3. Particle accelerator The positively charged dome of a Van de Graaff generator can be used to accelerate positively charged protons. A current exists due to the motion of these protons. In this case, how does the direction of conventional current compare with the direction in which the charge carriers move?

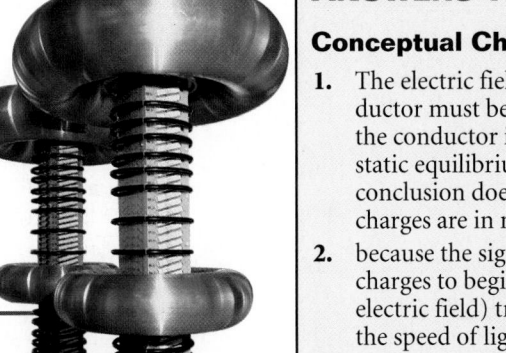

SECTION 19-1

Demonstration 1

Drift speed

Purpose Illustrate the difference between drift speed and the speed of a signal in a circuit.

Materials 30 cm plastic ruler with center groove, about 35 marbles

Procedure Place enough marbles in the center groove of the ruler to fill the ruler from end to end. Tell the students that the marbles represent the free electrons in a conductor. Add one more marble to the ruler by sliding it horizontally onto the end of the ruler. The marble at the other end of the ruler should roll off the ruler almost immediately.

Point out that each individual marble moved only a small distance (drift speed of about 10^{-4} m/s in a circuit) but the information traveled to the other end of the ruler at a high speed (nearly the speed of light in a circuit). Repeat the demonstration several times while students observe the motion of an individual marble.

ANSWERS TO

Conceptual Challenge

1. The electric field in a conductor must be zero when the conductor is in electrostatic equilibrium, but this conclusion does not apply if charges are in motion.
2. because the signal for charges to begin moving (the electric field) travels at nearly the speed of light
3. They are the same.

697

Demonstration 2

Potential difference as a source of current

Purpose Show the function of a battery or generator and the conservation of charge.

Materials inclined plane, marbles

Procedure Set up the inclined plane, and roll the marbles down the plane. As the marbles reach the bottom of the incline, return them to the top and let them roll down again. Have the students explain the parallels between this demonstration and a battery supplying energy to a circuit. (*The hand provides gravitational potential energy just as the battery provides electrical potential energy. The marbles lose energy on the ramp but are not "used up" in the process, just as charge carriers are not "used up" in a circuit.*)

Figure 19-3
Batteries maintain electric current by converting chemical energy into electrical energy.

```
internet connect
SCiLINKS
NSTA
TOPIC: Generators
GO TO: www.scilinks.org
sciLINKS CODE: HF2192
```

SOURCES AND TYPES OF CURRENT ⭐TEKS 5B

When you drop a ball, it falls to the ground, moving from a place of higher gravitational potential energy to one of lower gravitational potential energy. As discussed in Chapter 18, charges have similar behavior. For example, free electrons in a conductor move randomly when all points in the conductor are at the same potential. But when a potential difference is applied across the conductor, they will move slowly from a higher electric potential to a lower electric potential. Thus, a difference in potential maintains current in a circuit.

Batteries and generators supply energy to charge carriers

Both batteries and generators maintain a potential difference across their terminals by converting other forms of energy into electrical energy. **Figure 19-3** shows an assortment of batteries, which convert *chemical* energy to electrical potential energy.

As charge carriers collide with the atoms of a device, such as a light bulb or a heater, their electrical potential energy is converted into kinetic energy. Note that electrical energy, not charge, is "used up" in this process. The battery continues to supply electrical energy to the charge carriers until its chemical energy is depleted. At this point, the battery must be replaced or recharged.

Because batteries must often be replaced or recharged, generators are sometimes preferable. Generators convert *mechanical* energy into electrical energy. One type of generator, which is housed in dams like the one shown in **Figure 19-4,** converts the kinetic energy of falling water into electrical energy. Generators are the source of the potential difference across the two holes of a socket in a wall outlet in your home, which supplies the energy to operate your appliances. When you plug an appliance into an outlet, an average potential difference of 120 V is applied to the device. ⭐TEKS **3C, 5B**

Figure 19-4
In the electrical generators of this hydroelectric power plant, the mechanical energy of falling water is transformed into electrical energy.

Current can be direct or alternating  **TEKS** 6E

There are two different types of current: *direct current* (dc) and *alternating current* (ac). The difference between the two types of current is just what their names suggest. In direct current, charges move in only one direction. In alternating current, the motion of charges continuously changes in the forward and reverse directions.

Consider a light bulb connected to a battery. Because the positive terminal of the battery has a higher electric potential than the negative terminal has, charge carriers always move in one direction. Thus, the light bulb operates with a direct current. Because the potential difference between the terminals of a battery is fixed, batteries always generate a direct current.

In alternating current, the terminals of the source of potential difference are constantly changing sign. Hence, there is no net motion of the charge carriers in alternating current; they simply vibrate back and forth. If this vibration were slow enough, you would notice flickering in lights and similar effects in other appliances. To eliminate this problem, alternating current is made to change direction rapidly. In the United States, alternating current oscillates 60 times every second. Thus, its frequency is 60 Hz. The graphs in **Figure 19-5** compare direct and alternating current.

Unlike batteries, generators can produce either direct or alternating current, depending on their design. However, alternating current has advantages that make it more practical for use in transferring electrical energy. For this reason, the current supplied to your home by power companies is alternating current rather than direct current.

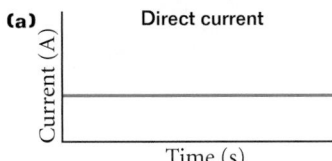

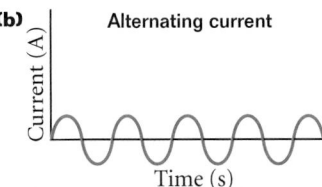

Figure 19-5
(a) The direction of direct current does not change, while **(b)** the direction of alternating current continually changes.

CONCEPT PREVIEW

Alternating current and generators will be discussed in greater detail in Chapter 22.

Section Review  **TEKS** 2C, 6E

1. Can the direction of conventional current ever be opposite the direction of charge movement? If so, when?

2. The charge that passes through the filament of a certain light bulb in 5.00 s is 3.0 C.
   a. What is the current in the light bulb?
   b. How many electrons pass through the filament of the light bulb in a time interval of 1.0 min?

3. In a conductor that carries a current, which is less, the drift speed of an electron or the average speed of the electron between collisions? Explain your answer.

4. What are the functions of batteries and generators?

5. In direct current, charge carriers have a drift velocity, but in alternating current, there is no net velocity of the charge carriers. Explain why.

Current and Resistance **699**

Teaching Tip

The decision to supply homes and businesses with alternating current rather than direct current came after a long debate between George Westinghouse's company (Westinghouse Electric Company), which supplied alternating current, and Thomas Edison's company (Edison Electric Light Company), which supplied direct current. Have students research this debate and the advantages and disadvantages of transporting each type of current.

Section Review
ANSWERS

1. yes; when charge carriers are negative

2. **a.** 0.60 A
   b. 2.2×10^{20} electrons

3. Although electrons undergo a force that moves them across the wire, collisions with atoms continually randomize their motion. As a result, the drift speed is much less than the average speed between collisions.

4. Batteries and generators provide a potential difference that maintains a current by converting chemical and mechanical energy into electrical energy.

5. In dc, the field always points in the same direction, so charge movement is in one direction. In ac, the direction of the field continually alternates, so there is no net motion of charge carriers and hence no drift velocity.

19-2
Resistance

 Misconception Alert

Some students may develop the idea that resistance is a variable that can change, like force or acceleration. Emphasize that the resistance of an object is more like mass (assuming temperature remains constant). Once a resistor is built, it usually has a fixed resistance. (Exceptions are non-ohmic resistors.) The resistance of a circuit can be changed by adding or changing resistors, just as adding mass to an object changes its total mass.

Teaching Tip

Many books refer to the equation $\Delta V = IR$ as Ohm's law. However, this equation can be used with both ohmic and non-ohmic materials. Ohm's law refers to the empirical observation that for many materials, the resistance is constant over a wide range of potential differences. Thus, the equation for resistance ($\Delta V = IR$) is an expression of Ohm's law only for cases in which R is independent of ΔV. To avoid misconceptions, this book introduces the equation for resistance first, then Ohm's law.

★ TEKS

p. 700: 6E, 6F, 3B, 3E, 3B
p. 701: 6E, 6E

19-2 SECTION OBJECTIVES

- **Calculate resistance, current, and potential difference using the definition of resistance.**

- **Distinguish between ohmic and non-ohmic materials.**

- **Know what factors affect resistance.**

- **Describe what is unique about superconductors.**

resistance

the opposition to the flow of current in a conductor

BEHAVIORS OF RESISTORS ★TEKS 6E, 6F

When a light bulb is connected to a battery, the current in the bulb depends on the potential difference across the battery. For example, a 9.0 V battery connected to a light bulb generates a greater current than a 6.0 V battery connected to the same bulb. But potential difference is not the only factor that determines the current in the light bulb. The materials that make up the connecting wires and the bulb's filament also affect the current in the bulb. Even though most materials can be classified as conductors or insulators, some conductors allow charges to move through them more easily than others. The opposition to the motion of charge through a conductor is the conductor's **resistance**. Quantitatively, resistance is defined as the ratio of potential difference to current, as follows:

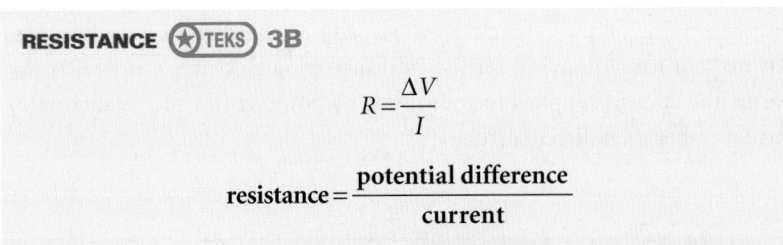

RESISTANCE ★TEKS 3B

$$R = \frac{\Delta V}{I}$$

$$\text{resistance} = \frac{\text{potential difference}}{\text{current}}$$

The SI unit for resistance, the *ohm*, is equal to volts per ampere and is represented by the Greek letter Ω (*omega*). If a potential difference of 1 V across a conductor produces a current of 1 A, the resistance of the conductor is 1 Ω.

Resistance is constant over a range of potential differences ★TEKS 3E

For many materials, including most metals, experiments show that *the resistance is constant over a wide range of applied potential differences.* This statement, known as Ohm's law, is named for Georg Simon Ohm (1789–1854), who was the first to conduct a systematic study of electrical resistance. Mathematically, Ohm's law is stated as follows:

$$\frac{\Delta V}{I} = \text{constant}$$

As can be seen by comparing the definition of resistance with Ohm's law, the constant of proportionality in the Ohm's law equation is resistance. It is common practice to express Ohm's law as $\Delta V = IR$, where R is understood to be independent of ΔV. ★TEKS 3B

Ohm's law does not hold for all materials  6E

Ohm's law is not a fundamental law of nature like the conservation of energy or the universal law of gravitation. Instead, it is a behavior that is valid only for certain materials. Materials that have a constant resistance over a wide range of potential differences are said to be *ohmic*. A graph of current versus potential difference for an ohmic material is linear, as shown in **Figure 19-6(a)**. This is because the slope of such a graph ($I/\Delta V$) is inversely proportional to resistance. When resistance is constant, the slope is constant and the resulting graph is a straight line.

Materials that do not function according to Ohm's law are said to be *non-ohmic*. **Figure 19-6(b)** shows a graph of current versus potential difference for a non-ohmic material. In this case, the slope is not constant because resistance varies. Hence, the resulting graph is nonlinear. One common semiconducting device that is non-ohmic is the *diode*. Its resistance is small for currents in one direction and large for currents in the reverse direction. Diodes are used in circuits to control the direction of current. This book assumes that all resistors function according to Ohm's law unless stated otherwise.

Resistance depends on length, cross-sectional area, material, and temperature  6E

In Section 19-1 we pointed out that electrons do not move in straight-line paths through a conductor. Instead, they undergo repeated collisions with the metal atoms. These collisions affect the motion of charges somewhat as a force of internal friction would. This is the origin of a material's resistance. Thus, any factors that affect the number of collisions will also affect a material's resistance. Some of these factors are shown in **Table 19-2.**

(a)

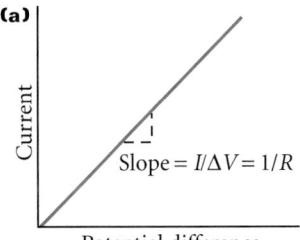

Slope $= I/\Delta V = 1/R$

(b)

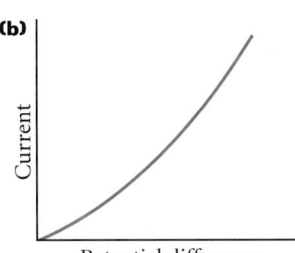

Figure 19-6
(a) The current–potential difference curve of an ohmic material is linear, and the slope is the inverse of the material's resistance. **(b)** The current–potential difference curve of a non-ohmic material is nonlinear.

Table 19-2   Factors that affect resistance

Factor	Less resistance	Greater resistance
length	L_1	L_2
cross-sectional area	A_1	A_2
material	Copper	Aluminum
temperature	T_1	T_2

Demonstration 3

Non-ohmic resistance

Purpose Show the non-ohmic response of a resistor.
Materials flashlight bulb, 12 V power supply, ammeter, ohmmeter
Procedure Measure the resistance of the bulb. Plot ΔV versus I on the chalkboard. Measure I in 2 V increments up to 10 V. Do not exceed 12 V, which will burn out the bulb. As you plot the curve, you will see that the curve is nonlinear. Why? (*The bulb is non-ohmic. Resistance in the filament varies as a function of temperature.*)

Demonstration 4

Factors that affect resistance

Purpose Show the dependence of resistance on cross-sectional area and length.
Materials two 50.0 Ω resistors, ohmmeter, connecting wires
Procedure Tell the students you have two equal resistors with the same length, cross-sectional area, and temperature. Ask students to predict whether the two resistors attached side by side will allow easier flow (lower R) or make the flow more difficult (higher R). Connect the two resistors side by side (in parallel), measure the resistance, and have students explain the results. (*Area is doubled, so resistance decreases.*) Repeat the demonstration for two resistors attached end to end (in series). (*Length is doubled, so resistance increases.*)

Teaching Tip

Teaching Tip

The first color band of a typical four-band resistor represents the first digit of the resistor's resistance, the second band represents the second digit, and the third band represents the power of ten by which these digits are multiplied. The fourth band, which is gold or silver, indicates the accuracy of the resistor. Color codes are as follows:

black	0
brown	1
red	2
orange	3
yellow	4
green	5
blue	6
violet	7
gray	8
white	9

Classroom Practice

The following may be used as a teamwork exercise or for demonstration at the chalkboard or on an overhead projector.

PROBLEM

Resistance

A potential difference ΔV_1 is applied across a resistance R_1, resulting in a current I_1. Determine the new current, I, when the following changes are made:

a. $\Delta V = 2\Delta V_1$, $R = R_1$

b. $\Delta V = \Delta V_1$, $R = 2R_1$

c. $\Delta V = 2\Delta V_1$, $R = 2R_1$

Answers

a. $I = 2I_1$

b. $I = \frac{1}{2}I_1$

c. $I = I_1$

Figure 19-7

Resistors, such as those shown here, are used to control current. The colors of the bands represent a code for the values of the resistances.

CONCEPT PREVIEW

You will work with different combinations of resistors in Chapter 20.

Resistors can be used to control the amount of current in a conductor

One way to change the current in a conductor is to change the potential difference across the ends of the conductor. But in many cases, such as in household circuits, the potential difference does not change. How can the current in a certain wire be changed if the potential difference remains constant?

According to the definition of resistance, if ΔV remains constant, current decreases when resistance increases. Thus, the current in a wire can be decreased by replacing the wire with one of higher resistance. The same effect can be accomplished by making the wire longer or by connecting a *resistor* to the wire. A resistor is a simple electrical element that provides a specified resistance. **Figure 19-7** shows a group of resistors in a circuit board. Resistors are sometimes used to control the current in an attached conductor because this is often more practical than changing the potential difference or the properties of the conductor.

SAMPLE PROBLEM 19B

Resistance ⭐TEKS 2C, 3B, 6E

PROBLEM

The resistance of a steam iron is 19.0 Ω. What is the current in the iron when it is connected across a potential difference of 120 V?

SOLUTION

Given: $R = 19.0\ \Omega \quad \Delta V = 120\ V$

Unknown: $I = ?$

Use the resistance equation given on page 700. Rearrange to solve for current.

$$R = \frac{\Delta V}{I}$$

$$I = \frac{\Delta V}{R} = \frac{120\ V}{19.0\ \Omega} = 6.32\ A$$

PRACTICE 19B

Resistance ⭐TEKS 2C, 3B, 6E

1. A 1.5 V battery is connected to a small light bulb with a resistance of 3.5 Ω. What is the current in the bulb?

2. A stereo with a resistance of 65 Ω is connected across a potential difference of 120 V. What is the current in this device?

3. Find the current in the following devices when they are connected across a potential difference of 120 V.

a. a hot plate with a resistance of 48 Ω
b. a microwave oven with a resistance of 20 Ω

4. The current in a microwave oven is 6.25 A. If the resistance of the oven's circuitry is 17.6 Ω, what is the potential difference across the oven?

5. A typical color television draws 2.5 A of current when connected across a potential difference of 115 V. What is the effective resistance of the television set?

6. The current in a certain resistor is 0.50 A when it is connected to a potential difference of 110 V. What is the current in this same resistor if

a. the operating potential difference is 90.0 V?
b. the operating potential difference is 130 V?

PRACTICE GUIDE 19B

Solving for:		
I	📖 PE	Sample, 1–3, 6*; Ch. Rvw. 28–30
	💿 PW	6–7
	PB	4–6
ΔV	📖 PE	4; Ch. Rvw. 47
	💿 PW	4–5
	PB	Sample, 1–3
R	📖 PE	5; Ch. Rvw. 45
	💿 PW	Sample, 1–3
	PB	7–10

ANSWERS TO

Practice 19B
Resistance

1. 0.43 A
2. 1.8 A
3. **a.** 2.5 A
   b. 6.0 A
4. 1.10×10^2 V
5. 46 Ω
6. **a.** 0.41 A
   b. 0.59 A

Salt water and perspiration lower the body's resistance ⭐TEKS 3A, 6F

The human body's resistance to current is on the order of 500 000 Ω when the skin is dry. However, the body's resistance decreases when the skin is wet. If the body is soaked with salt water, its resistance can be as low as 100 Ω. This is because ions in salt water readily conduct electric charge. Such low resistances can be dangerous if a large potential difference is applied between parts of the body because current increases as resistance decreases. Currents in the body that are less than 0.01 A either are imperceptible or generate a slight tingling feeling. Greater currents are painful and can disturb breathing, and currents above 0.15 A through the chest cavity can be fatal.

Perspiration also contains ions that conduct electric charge. In a *galvanic skin response* (GSR) test, commonly used as a stress test and as part of some lie detectors, a very small potential difference is set up across the body. Perspiration increases when a person is nervous or stressed, thereby decreasing the resistance of the body. In GSR tests, a state of low stress and high resistance, or "normal" state, is used as a control, and a state of higher stress is reflected as a decreased resistance compared with the normal state. ⭐TEKS 3C

Did you know?

In reality, very large changes in potential difference will affect the resistance of a conductor. However, this variation is negligible at the level of potential differences supplied to homes and used in other common applications.

⭐TEKS

p. 702: 2C, 3B, 6E
p. 703: 2C, 3B, 6E, 3A, 6F, 3C

Teaching Tip

Be sure students understand the relationship between the changes in the sound wave and the changes in the size of the resistance medium (the carbon granules). These changes in resistance allow the sound waves to be converted into electrical impulses, which can easily be transported over long distances.

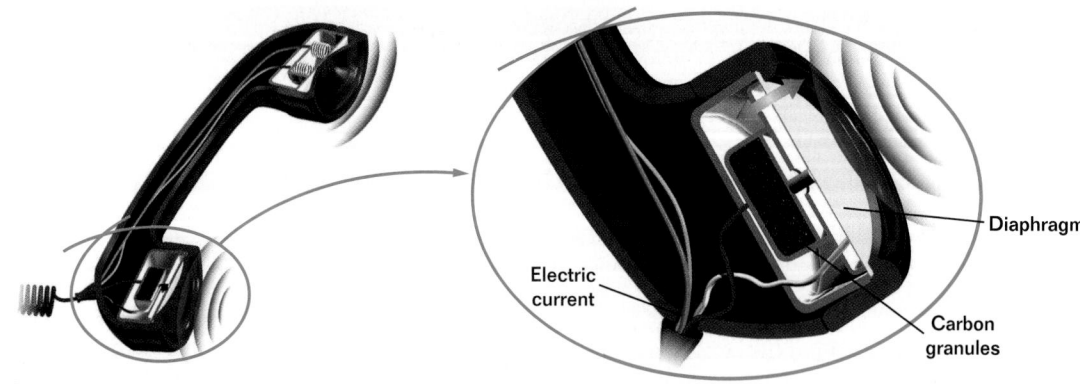

Figure 19-8

In the carbon microphone, used in some telephones, changes in sound waves affect the resistance medium (the carbon granules). The resulting changes in current are transmitted through the phone line and then converted back to sound waves in the listener's earpiece.

The carbon microphone uses varying resistance ✪TEKS 6F, 8B

Figure 19-8 illustrates the carbon microphone, commonly used in the mouthpiece of some telephones. The carbon microphone uses the inverse relationship between current and resistance to convert sound waves to electrical impulses. A flexible steel diaphragm is placed in contact with carbon granules inside a container. The carbon granules serve as the microphone's primary resistance medium. The microphone also contains a source of current.

The magnitude of the current in the microphone changes when the compressions and rarefactions of a sound wave strike the diaphragm. When a compression arrives at the microphone, the diaphragm flexes inward, causing the carbon granules to press together into a smaller-than-normal volume. This corresponds to a decrease in the length of the resistance medium, which results in a lower resistance and hence a greater current. When a rarefaction arrives, the reverse process occurs, resulting in a decrease in current. These variations in current, following the changes of the sound wave, are sent through the transformer to the telephone company's transmission line. A speaker in the listener's earpiece then converts the electric signals back to sound waves.

ANSWERS TO

Conceptual Challenge

1. According to $\Delta V = IR$, if the dryer is ohmic, the current in the dryer will increase if the potential difference across it increases. Thus, using a 120 V dryer in a 220 V outlet will generate a greater current than the dryer was designed for. This increased current could damage the dryer.

2. Current could be increased by increasing potential difference or by decreasing resistance. Resistance could be decreased by using shorter wires, wider wires, wires of a material with less resistance, or wires at a cooler temperature.

3. Water decreases the body's resistance. Thus, for a given potential difference, the current in the body will be greater if the body is wet.

Conceptual Challenge ✪TEKS 3A, 6E

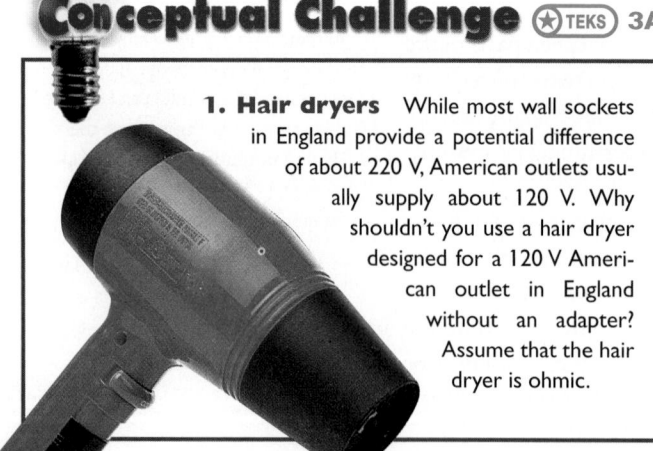

1. Hair dryers While most wall sockets in England provide a potential difference of about 220 V, American outlets usually supply about 120 V. Why shouldn't you use a hair dryer designed for a 120 V American outlet in England without an adapter? Assume that the hair dryer is ohmic.

2. Light bulbs While working in the laboratory, you need to increase the glow of a light bulb, so you wish to increase the current in the bulb. List all of the different factors you could adjust to increase the current in the bulb.

3. Faulty outlet If you touch a faulty electrical outlet, there is a potential difference across your body that generates a current in your body. This situation is always dangerous, but the risk increases greatly if your body is wet. Explain why.

Tomorrow's Technology ⊛TEKS 3C

Electroporation

Most people hate getting shots at the doctor's office. In the future, drugs may be delivered through a patient's skin without the use of needles—and without the pain they cause. Dr. Mark Prausnitz, of the Georgia Institute of Technology, has been working on a process that uses electricity to create tiny pores in the skin through which drug molecules can pass. This process is called electroporation.

Ordinarily, there are membranes in the outer part of the skin that prevent substances from entering the body. However, when high-voltage electrical pulses are applied to the skin, molecules are able to move through these membranes up to 10 000 times easier than under normal conditions.

"Charged molecules want to move in an electric field," explained Dr. Prausnitz. "But if there is a cell membrane in the way, suddenly they hit this membrane and they can't go anymore.

"As more and more of these charged molecules build up, you develop a voltage across this membrane. When the voltage gets high enough, the structure of the membrane itself changes to allow these molecules to cross the membrane, and this is the creation of an electropore."

This method could also be used to deliver drugs that would be destroyed by the stomach if taken in pill form. And it might add a level of convenience: Dr. Prausnitz envisions a device that could be worn like a watch that would deliver controlled doses of a drug over an entire day. Such a device would function like a nicotine patch except that it would also contain the electric equipment to carry out electroporation.

Although using electricity to deliver drugs through the skin is still in testing stages, electroporation is already being used to administer drugs to tumors in cancer patients. When the medication is injected into the patient's bloodstream, the cells of the tumor are electroporated, allowing more of the drug to enter the tumor and thus increasing the chance of successful treatment.

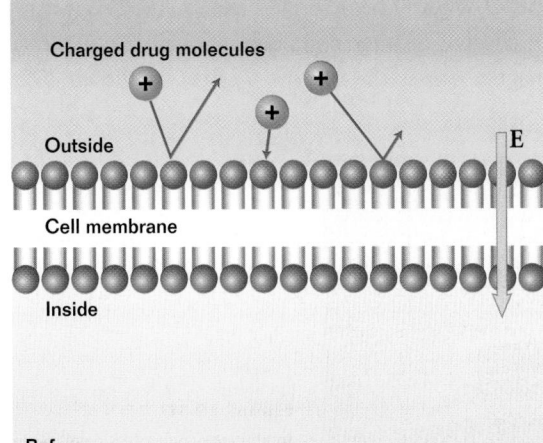

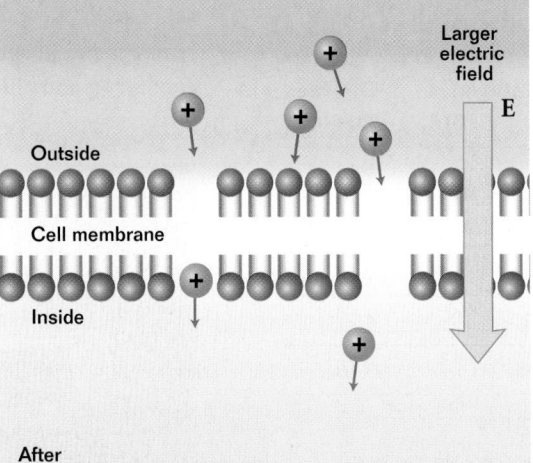

SECTION 19-2

Tomorrow's Technology

BACKGROUND

This feature discusses a new type of technology known as *electroporation*. In electroporation, electrical pulses are used to decrease the skin's resistance, allowing charged molecules to pass through cell membranes. This technology has now been used to inject drugs into tumors inside the body. This method, which enhances chemotherapy by concentrating drugs within the tumor, has been used on a number of patients.

Teaching Tip

Remind students that the term *voltage* is equivalent to the term *potential difference*, which was introduced in Chapter 18.

⊛TEKS
p. 704: 6F, 8B, 3A, 6E
p. 705: 3C

Figure 19-9
This graph shows the resistance of mercury, a superconductor, at temperatures near its critical temperature.

superconductor

a material whose resistance is zero at or below some critical temperature, which varies with each material

TEKS 2C

Table 19-3
Critical temperatures

Material	Degrees Kelvin
Zn	0.88
Al	1.19
Sn	3.72
Hg	4.15
Nb	9.46
Nb_3Ge	23.2
$YBa_2Cu_3O_7$	90
Tl-Ba-Ca-Cu-O	125

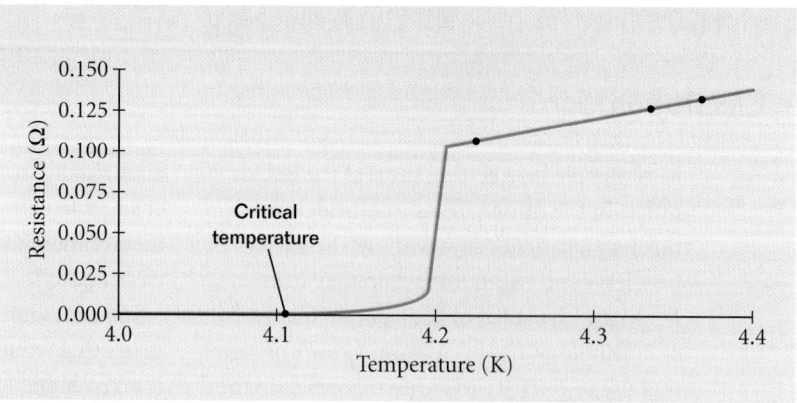

Superconductors have no resistance below a critical temperature

There are materials that have zero resistance below a certain temperature, called the *critical temperature*. These materials are known as **superconductors.** The resistance-temperature graph for a superconductor resembles that of a normal metal at temperatures above the critical temperature. But when the temperature is at or below the critical temperature, the resistance suddenly drops to zero, as shown in **Figure 19-9.**

Today there are thousands of known superconductors, including common metals such as aluminum, tin, lead, and zinc. **Table 19-3** lists the critical temperatures of several superconductors. Interestingly, copper, silver, and gold, which are excellent conductors, do not exhibit superconductivity.

One of the truly remarkable features of superconductors is that once a current is established in them, the current continues even if the applied potential difference is removed. In fact, steady currents have been observed to persist for many years with no apparent decay in superconducting loops.

Figure 19-10 shows a small permanent magnet levitated above a disk of the superconductor $YBa_2Cu_3O_7$. As will be described in Chapter 21, electric currents produce magnetic effects. The interaction between a current in the superconductor and this magnet causes the magnet to float in the air over the

Figure 19-10
In this photograph, a small permanent magnet levitates above the superconductor $YBa_2Cu_3O_7$, which is at 77 K, 13 K below its critical temperature.

superconductor. This is known as the *Meissner effect*. One application of the Meissner effect is the high-speed express train shown in **Figure 19-11,** which levitates a few inches above the track.

An important recent development in physics is the discovery of high-temperature superconductors. The excitement began with a 1986 publication by scientists at the IBM Zurich Research Laboratory in Switzerland. In this publication, scientists reported evidence for superconductivity at a temperature near 30 K. More recently, scientists have found superconductivity at temperatures as high as 150 K. The search continues for a material that has superconducting qualities at room temperature. It is an important search that has both scientific and practical applications.

One useful application of superconductivity is superconducting magnets. Such magnets are being considered for storing energy. The idea of using superconducting power lines to transmit power more efficiently is also being researched. Modern superconducting electronic devices that consist of two thin-film superconductors separated by a thin insulator have been constructed. They include magnetometers (magnetic-field measuring devices) and various microwave devices. ⊛TEKS 3C, 3E, 6F

Figure 19-11
This express train in Tokyo, Japan, which utilizes the Meissner effect, levitates above the track and is capable of speeds exceeding 225 km/h.

Section Review   ⊛TEKS 2C, 3B, 6E

1. How much current would a 10.2 Ω toaster oven draw when connected to a 120 V outlet?

2. An ammeter registers 2.5 A of current in a wire that is connected to a 9.0 V battery. What is the wire's resistance?

3. In a particular diode, the current triples when the applied potential difference is doubled. What can you conclude about the diode?

4. You have only one type of wire. If you are connecting a battery to a light bulb with this wire, how could you decrease the current in the wire?

5. How is the resistance of aluminum, which is a superconductor, different from that of gold, which does not exhibit superconductivity?

6. **Physics in Action**   What is the function of resistors in a circuit board? What is the function of diodes in a circuit board?

7. **Physics in Action**   Calculate the current in a 75 Ω resistor when a potential difference of 115 V is placed across it. What will the current be if the resistor is replaced with a 47 Ω resistor?

Section Review
ANSWERS

1. 12 A
2. 3.6 Ω
3. The diode is non-ohmic.
4. You could decrease current by making the wire as long as possible, thereby increasing its resistance.
5. Aluminum's resistance becomes zero at its critical temperature and remains zero at lower temperatures. Gold's resistance does not become zero at any temperature.
6. Resistors regulate current; Diodes regulate the direction of current.
7. 1.5 A; 2.4 A

⊛TEKS

p. 706: 2C
p. 707: 3C, 3E, 6F, 2C, 3B, 6E

Visual Strategy

Figure 19-12

Use an analogy with mechanical energy to help students understand the changes in electrical potential energy.

Q Compare the electrical energy changes shown in **Figure 19-12** to the mechanical energy changes involved in carrying tennis balls up a flight of stairs and then dropping the balls back down to the ground.

A *The tennis balls are analogous to charges, and the bulb uses energy just as dropping the balls does. Charges moving through the battery are analogous to the balls being carried up the stairs; the battery increases electrical potential energy just as raising the balls increases gravitational potential energy. The movement of charge through the wire is analogous to the balls being carried horizontally because the potential energies do not change in either case.*

⭐ **TEKS**

p. 708: 5B, 5B
p. 709: 3B, 5B, 3A, 6E

19-3 SECTION OBJECTIVES

- **Relate electric power to the rate at which electrical energy is converted to other forms of energy.**

- **Calculate electric power.**

- **Calculate the cost of running electrical appliances.**

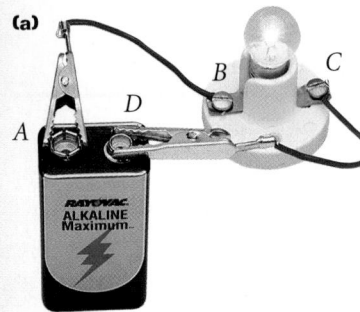
(a)

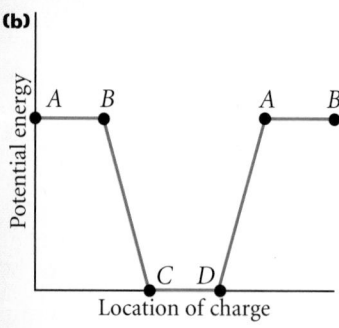
(b)

Figure 19-12
A charge leaves the battery at *A* with a certain amount of electrical potential energy. The charge loses this energy while moving from *B* to *C*, and then regains the energy as it moves from *D* to *A*.

ENERGY TRANSFER ⭐TEKS 5B

When a battery is used to maintain an electric current in a conductor, chemical energy stored in the battery is continuously converted to the electrical energy of the charge carriers. As the charge carriers move through the conductor, this electrical energy is converted to internal energy due to collisions between the charge carriers and other particles in the conductor.

For example, consider a light bulb connected to a battery, as shown in **Figure 19-12(a).** Imagine a charge *Q* moving from the battery's terminal to the light bulb and then back to the other terminal. The changes in electrical potential energy are shown in **Figure 19-12(b).** If we disregard the resistance of the connecting wire, no loss in energy occurs as the charge moves through the wire (*A* to *B*). But when the charge moves through the filament of the light bulb (*B* to *C*), which has a higher resistance than the wire has, it loses electrical potential energy due to collisions. This electrical energy is converted into internal energy, and the filament warms up.

When the charge first returns to the battery's terminal (*D*), its potential energy is zero, and the battery must do work on the charge. As the charge moves between the terminals of the battery (*D* to *A*), its electrical potential energy increases by $Q\Delta V$ (where ΔV is the potential difference across the two terminals). The battery's chemical energy decreases by the same amount. At this point, the process begins again. (Remember that this process happens very slowly compared with how quickly the bulb is illuminated.)

Electric power is the rate of conversion of electrical energy ⭐TEKS 5B

In Chapter 5, power was described as the rate at which work is done. *Electric power,* then, is the rate at which charge carriers do work. Put another way, electric power is the rate at which charge carriers convert electrical potential energy to nonelectrical forms of energy.

$$P = \frac{W}{\Delta t} = \frac{\Delta PE}{\Delta t}$$

As discussed in Chapter 18, potential difference is defined as the change in potential energy per unit of charge.

$$\Delta V = \frac{\Delta PE}{q}$$

This equation can be rewritten in terms of potential energy.

$$\Delta PE = q\Delta V$$

We can then substitute this expression for potential energy into the equation for power.

$$P = \frac{\Delta PE}{\Delta t} = \frac{q\Delta V}{\Delta t}$$

Because current, I, is defined as the rate of charge movement ($q/\Delta t$), we can express electric power as current multiplied by potential difference.

ELECTRIC POWER ⭐TEKS **3B**

$$P = I\Delta V$$

electric power = current × potential difference

This equation describes the rate at which charge carriers lose electrical potential energy. In other words, power is the rate of conversion of electrical energy. As described in Chapter 5, the SI unit of power is the *watt*, W. In terms of the dissipation of electrical energy, 1 W is equivalent to 1 J of electrical energy being converted to other forms of energy per second.

Most light bulbs are labeled with their power ratings. The amount of heat and light given off by a bulb is related to the power rating, also known as wattage, of the bulb.

Because $\Delta V = IR$, we can express the power dissipated by a resistor in the following alternative forms:

$$P = I\Delta V = I(IR) = I^2R$$

$$P = I\Delta V = \left(\frac{\Delta V}{R}\right)\Delta V = \frac{(\Delta V)^2}{R}$$

The conversion of electrical energy to internal energy in a resistant material is called *joule heating*, also often referred to as an I^2R loss. ⭐TEKS **5B**

Conceptual Challenge ⭐TEKS 3A, 6E

1. Power delivered to a light bulb Explain why the filament of a light bulb connected to a battery receives much more power than the wire connecting the bulb and the battery.

2. Power and resistance Compare the two alternative forms for the equation that expresses the power dissipated by a resistor. In the first equation ($P = I^2R$), power is proportional to resistance; in the second equation ($P = (\Delta V)^2/R$), power is *inversely* proportional to resistance. How can you reconcile this apparent discrepancy?

3. Different wattages Which has a greater resistance when connected to a 120 V outlet, a 40 W light bulb or a 100 W light bulb?

SECTION 19-3

Teaching Tip

Reviewing the concepts of electrical potential energy and potential difference (introduced in Chapter 18) will help students follow the derivation of the equation for electric power.

ANSWERS TO

Conceptual Challenge

1. The current in each is the same. Thus, according to $P = I^2R$, because the light bulb has a much higher resistance than the wire has, it receives much more power than the wire does.

2. The two equations refer to different cases. If current is constant, power is proportional to resistance ($P = I^2R$), while if potential difference is constant, power is inversely proportional to resistance ($P = (\Delta V)^2/R$).

3. the 40 W bulb

ANSWERS TO

Practice 19C
Electric power
1. $14\ \Omega$
2. $5.8 \times 10^4\ \Omega$
3. $22\ \Omega$
4. 6.25 A; 312 W

SAMPLE PROBLEM 19C

Electric power ⭐TEKS 2C, 3B, 6E

PROBLEM

An electric space heater is connected across a 120 V outlet. The heater dissipates 1320 W of power in the form of electromagnetic radiation and heat. Calculate the resistance of the heater.

SOLUTION

Given: $\Delta V = 120$ V $P = 1320$ W

Unknown: $R = ?$

Because power and potential difference are given but resistance is unknown, use the third form of the power equation on page 709, which includes these three variables.

$$P = \frac{(\Delta V)^2}{R}$$

$$R = \frac{(\Delta V)^2}{P} = \frac{(120\ \text{V})^2}{1320\ \text{W}} = \frac{(120)^2\ \text{J}^2/\text{C}^2}{1320\ \text{J/s}}$$

$$R = \frac{(120)^2\ \text{J/C}}{1320\ \text{C/s}} = 10.9\ \text{V/A}$$

$$\boxed{R = 10.9\ \Omega}$$

PRACTICE 19C

Electric power ⭐TEKS 2C, 3B, 6E

1. A 1050 W electric toaster operates on a household circuit of 120 V. What is the resistance of the wire that makes up the heating element of the toaster?

2. A small electronic device is rated at 0.25 W when connected to 120 V. What is the resistance of this device?

3. A calculator is rated at 0.10 W when connected to a 1.50 V battery. What is the resistance of this device?

4. An electric heater is operated by applying a potential difference of 50.0 V across a nichrome wire of total resistance 8.00 Ω. Find the current in the wire and the power rating of the heater.

Electric companies measure energy consumed in kilowatt-hours

Electric power, as discussed previously, is the rate of energy transfer. Power companies charge for energy, not power. However, the unit of energy used by electric companies to calculate consumption, the *kilowatt-hour,* is defined in terms of power. One kilowatt-hour (kW•h) is the energy delivered in 1 h at the constant rate of 1 kW. The following equation shows the relationship between the kilowatt-hour and the SI unit of energy, the joule:

$$1 \text{ kW•h} \times \frac{10^3 \text{ W}}{1 \text{ kW}} \times \frac{60 \text{ min}}{1 \text{ h}} \times \frac{60 \text{ s}}{1 \text{ min}} = 3.6 \times 10^6 \text{ W•s} = 3.6 \times 10^6 \text{ J}$$

On an electric bill, the electrical energy used in a given period is usually stated in multiples of kilowatt-hours, as shown in **Figure 19-13(a).** The cost of energy ranges from about 5 to 20 cents per kilowatt-hour, depending on where you live. An electric meter, like the one shown in **Figure 19-13(b),** is used by the electric company to determine how much energy is consumed over some period of time.

The electrical energy supplied by power companies is used to generate currents, which in turn are used to operate household appliances. As seen earlier in this section, as the charge carriers that make up a current encounter resistance, some of the electrical energy is converted to internal energy by collisions between moving electrons and atoms, and the conductor warms up. This effect is made useful in many common appliances, such as hair dryers, electric heaters, clothes dryers, toasters, and steam irons. ⊛ TEKS **3C, 6F**

Hair dryers contain a long, thin heating coil that becomes very hot when the hair dryer is turned on. A fan behind the heating coil blows air through the area that contains the coils and out of the hair dryer. In this case, the warm air is used to dry hair; the same principle is used in clothes dryers and electric heaters.

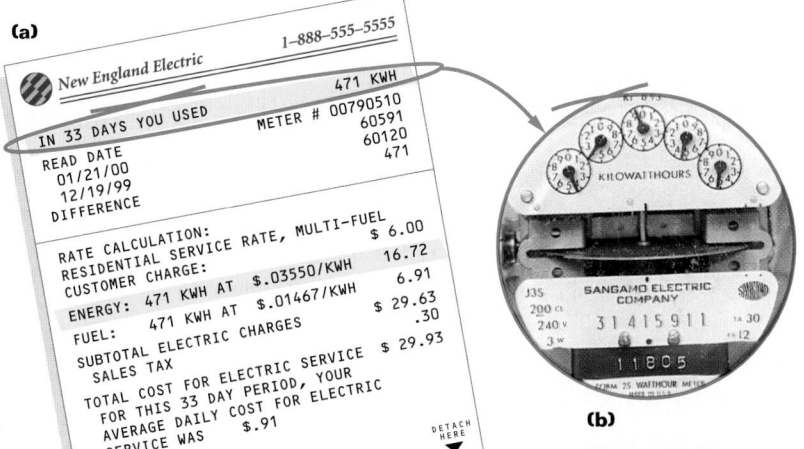

(a)

Figure 19-13
(a) Consumers are charged for the amount of energy they use in units of kilowatt-hours. **(b)** An electric meter, such as the one shown here, records the amount of energy consumed.

(b)

ANSWERS TO

Practice 19D

Cost of electrical energy

1. a. $0.14
   b. $4.40
   c. $0.42

2. a. 6.5×10^6 J
   b. 2.0×10^8 J
   c. 1.9×10^7 J

In electric crock pots, a heating coil located at the base of the pot warms food inside the pot. In a steam iron, a heating coil warms the bottom of the iron and also turns water to steam, which is sprayed from jets in the bottom of the iron. Electric toasters have heating elements around the edges and in the center of the toaster. When bread is loaded into the toaster, the heating coils turn on, and a timer determines the length of time the heating elements remain on before the bread pops out of the toaster. ⭐TEKS **3C, 6F**

SAMPLE PROBLEM 19D

Cost of electrical energy ⭐TEKS **2C, 3B, 6F**

PROBLEM

How much does it cost to operate a 100.0 W light bulb for 24 h if electrical energy costs $0.080 per kW•h?

SOLUTION

Given: Cost of energy = $0.080/kW•h
$P = 100.0$ W $= 0.1000$ kW $\Delta t = 24$ h

Unknown: Cost to operate the light bulb for 24 h

First calculate the energy used in units of kilowatt-hours by multiplying the power (in kW) by the time interval (in h). Then multiply the amount of energy by the cost per kilowatt-hour to find the total cost.

Energy $= P\Delta t = (0.1000$ kW$)(24$ h$) = 2.4$ kW•h
Cost $= (2.4$ kW•h$)($0.080/kW•h$)$

$$\boxed{\text{Cost} = \$0.19}$$

PRACTICE 19D

Cost of electrical energy ⭐TEKS **2C, 3B, 6F**

1. Assuming electrical energy costs $0.080 per kW•h, calculate the cost of running each of the following appliances for 24 h if 115 V is supplied to each:

   a. a 75.0 W stereo

   b. an electric oven that draws 20.0 A of current

   c. a television with a resistance of 60.0 Ω

2. Determine how many joules of energy are used by each appliance in item 1 in the 24 h period.

Electrical energy is transferred at high potential differences to minimize energy loss

When transporting electrical energy by power lines, such as those shown in **Figure 19-14,** power companies want to minimize the I^2R loss and maximize the energy delivered to a consumer. This can be done by decreasing either current or resistance. Although wires have little resistance, recall that resistance is proportional to length. Hence, resistance becomes a factor when power is transported over long distances. Even though power lines are designed to minimize resistance, some energy will be lost due to the length of the power lines.

As expressed by the equation $P = I^2R$, energy loss is proportional to the *square* of the current in the wire. For this reason, decreasing current is even more important than decreasing resistance. Because $P = I\Delta V$, the same amount of power can be transported either at high currents and low potential differences or at low currents and high potential differences. Thus, transferring electrical energy at low currents, thereby minimizing the I^2R loss, requires that electrical energy be transported at very high potential differences. Power plants transport electrical energy at potential differences of up to 765 000 V. In your city, this potential difference is reduced by a transformer to about 4000 V. At your home, this potential difference is reduced again to about 120 V by another transformer. ⊛TEKS **3C, 6F**

Figure 19-14
Power companies transfer electrical energy at high potential differences in order to minimize the I^2R loss.

Section Review ⊛TEKS **2C, 3B, 6E**

1. What does the power rating on a light bulb describe?

2. If the resistance of a light bulb is increased, how will the electrical energy used by the light bulb over the same time period change?

3. The potential difference across a resting neuron in the human body is about 70 mV, and the current in it is approximately 200 μA. How much power does the neuron release?

4. How much does it cost to watch an entire World Series (21 h) on a 90.0 W black-and-white television set? Assume that electrical energy costs $0.070/kW•h.

5. Explain why it is more efficient to transport electrical energy at high potential differences and low currents rather than at low potential differences and high currents.

BACKGROUND

This feature discusses some basic ideas from quantum mechanics and builds on these ideas to explore a phenomenon called *electron tunneling*. The feature concludes with a discussion of the scanning tunneling microscope, which utilizes electron tunneling to generate highly detailed images that depict the atomic structure on the surface of a material.

 Misconception Alert

Some students may think that the potential well shown in **Figure 19-15** represents a physical well or area. Be sure they understand that the height of the well corresponds to an amount of energy rather than to a physical height. Classically, the electron is confined to the well if its energy is less than *U*.

PHYSICS ON THE EDGE

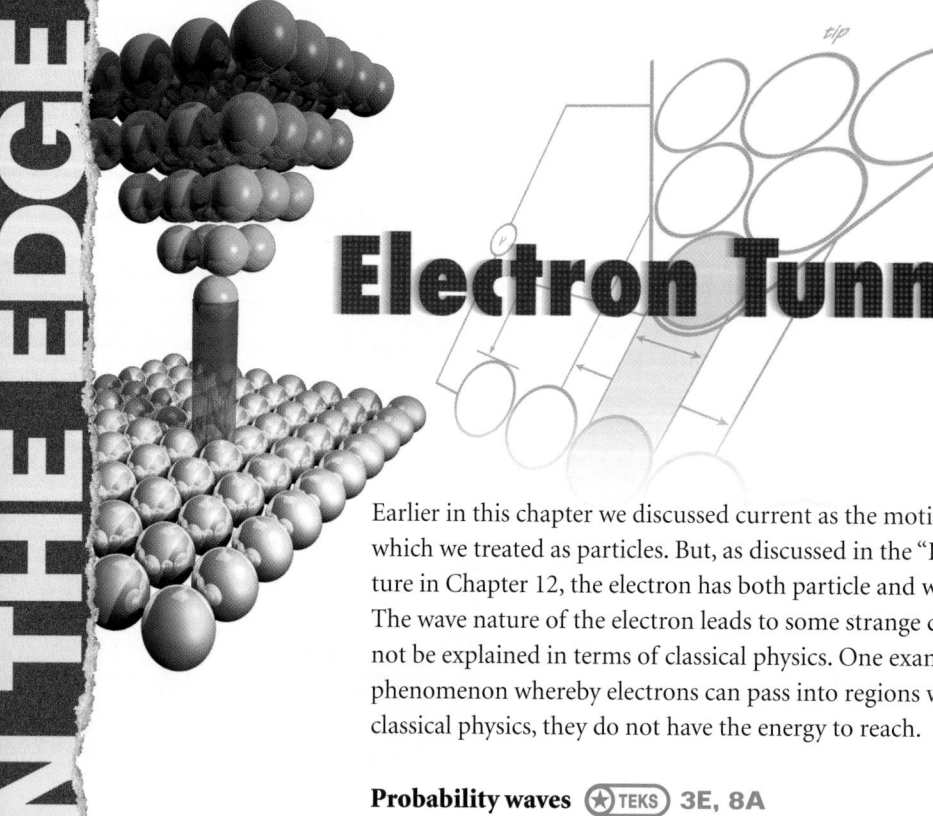

Electron Tunneling

Earlier in this chapter we discussed current as the motion of charge carriers, which we treated as particles. But, as discussed in the "De Broglie Waves" feature in Chapter 12, the electron has both particle and wave characteristics. The wave nature of the electron leads to some strange consequences that cannot be explained in terms of classical physics. One example is *tunneling*, a phenomenon whereby electrons can pass into regions which, according to classical physics, they do not have the energy to reach.

Probability waves ⊛TEKS 3E, 8A

To see how tunneling is possible, we must explore matter waves in greater detail. De Broglie's revolutionary idea that particles have a wave nature raised the question of how these matter waves behave. In 1926, Erwin Schrödinger proposed a wave equation that described the manner in which de Broglie matter waves change in space and time. Two years later, in an attempt to relate the wave and particle natures of matter, Max Born suggested that the square of the amplitude of a matter wave is proportional to the probability of finding the corresponding particle at that location.

Tunneling

Born's interpretation makes it possible for a particle to be found in a location that is not allowed by classical physics. Consider an electron with a potential energy of zero in the region between 0 and *L* (region II), which we call the *potential well*, and with a potential energy of some finite value *U* outside this

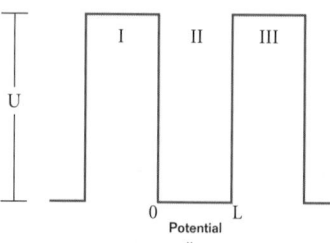

Figure 19-15

An electron has a potential energy of zero inside the well (Region II) and a potential energy of *U* outside the well. According to classical physics, if the electron's energy is less than *U*, it cannot escape the well without absorbing energy.

area (regions I and III), as shown in **Figure 19-15.** If the energy of the electron is less than *U*, then according to classical physics, the electron cannot escape the well without first acquiring additional energy.

The probability wave for this electron (in its lowest energy state) is shown in **Figure 19-16.** Between any two points of this curve, the area under the corresponding part of the curve is proportional to the probability of finding the electron in that region. The highest point of the curve corresponds to the most probable location of the electron, while the lower points correspond to less probable locations. Note that the curve never actually meets the *x*-axis. This means that the electron has some finite probability of being anywhere in space. Hence, there is a probability that the electron will actually be found outside the potential well. In other words, according to quantum mechanics, the electron is no longer confined to strict boundaries because of its energy. When the electron is found outside the boundaries established by classical physics, it is said to have *tunneled* to its new location.

The scanning tunneling microscope ⊛TEKS 3E, 3C

In 1981, Gerd Binnig and Heinrich Rohrer, at IBM Zurich, discovered a practical application of tunneling current: a powerful microscope called the *scanning tunneling microscope,* or *STM.* The STM can produce highly detailed images with resolution comparable to the size of a single atom. The image of the surface of nickel shown in **Figure 19-17** demonstrates the power of the STM. Note that individual nickel atoms are recognizable. The smallest detail that can be discerned is about 0.2 nm, or approximately the size of an atom's radius. A typical optical microscope has a resolution no better than 200 nm, or about half the wavelength of visible light, and so it could never show the detail shown in **Figure 19-17.**

In the STM, a conducting probe with a very sharp tip (about the width of an atom) is brought near the surface to be studied. According to classical physics, electrons cannot move between the surface and the tip because they lack the energy to escape either material. But according to quantum theory, electrons can tunnel across the barrier, provided the distance is small enough (about 1 nm). By applying a potential difference between the surface and the tip, the electrons can be made to tunnel preferentially from surface to tip. In this way, the tip samples the distribution of electrons just above the surface.

The STM works because the probability of tunneling decreases exponentially with distance. By monitoring changes in the tunneling current as the tip is scanned over the surface, scientists obtain a sensitive measure of the topography of the electron distribution on the surface. The result is used to make images like the one in **Figure 19-17.** The STM can measure the height of surface features to within 0.001 nm, approximately 1/100 of an atomic diameter.

Although the STM was originally designed for imaging atoms, other practical applications are being developed. Engineers have greatly reduced the size of the STM and hope to someday develop a computer in which every piece of data is held by a single atom or by small groups of atoms and then read by an STM.

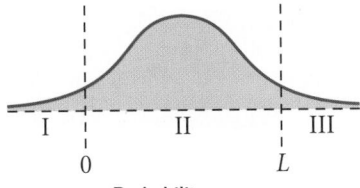

Probability wave

Figure 19-16
The probability curve for an electron in its lowest energy state shows that there is a certain probability of finding the electron outside the potential well.

Figure 19-17
A scanning tunneling microscope (STM) was used to produce this image of the surface of nickel. The contours represent the arrangement of individual nickel atoms on the surface. An STM enables scientists to see small details on surfaces with a lateral resolution of 0.2 nm and a vertical resolution of 0.001 nm.

SECTION 19-3

EXTENSION

Have students form small groups to brainstorm about possible uses of the STM. Then have them research uses of the STM that are currently being developed. Have each group share its ideas and research with the class.

Teaching Tip

Be sure students understand that regions I, II, and III of the potential well (**Figure 19-15**) correspond to regions I, II, and III of the probability wave (**Figure 19-16**). Because there is a finite probability that the electron will be found in regions I and III of **Figure 19-16,** there is a probability that the electron will be found outside the potential well.

 TEKS

p. 714: 3E, 8A
p. 715: 3E, 3C

CHAPTER 19
Summary

CHAPTER 19
Summary

KEY TERMS

current (p. 694)

drift velocity (p. 697)

resistance (p. 700)

superconductor (p. 706)

Diagram symbols

Current	I ⟶
Positive charge	+
Negative charge	−

KEY IDEAS

Section 19-1 Electric current

- Current is the rate of charge movement.
- Conventional current is defined in terms of *positive* charge movement.
- Drift velocity is the net velocity of charge carriers; its magnitude is much less than the average speed between collisions.
- Batteries and generators supply energy to charge carriers.
- In direct current, charges move in a single direction; in alternating current, the direction of charge movement continually alternates.

Section 19-2 Resistance

- According to the definition of resistance, potential difference equals current times resistance, as follows: $\Delta V = IR$
- Resistance depends on length, cross-sectional area, temperature, and material.
- Superconductors are materials that have resistances of zero below a critical temperature, which varies with each metal.

Section 19-3 Electric power

- Electric power is the rate of conversion of electrical energy: $P = I\Delta V$
- The power dissipated by a resistor can be calculated with the following equations: $P = I^2 R = \dfrac{(\Delta V)^2}{R}$
- Electric companies measure energy consumed in kilowatt-hours.

Variable symbols

Quantities		Units		Conversions	
ΔV	potential difference	V	volt	= J/C	= joules of energy/ coulomb of charge
I	current	A	ampere	= C/s	= coulombs of charge/second
R	resistance	Ω	ohm	= V/A	= volts of potential difference/ ampere of current
P	electric power	W	watt	= J/s	= joules of energy/second

ELECTRIC CURRENT

Review questions

1. What is electric current? What is the SI unit for electric current?

2. In a metal conductor, current is the result of moving electrons. Can charge carriers ever be positive?

3. What is meant by the term *conventional current*?

4. What is the difference between the drift speed of an electron in a metal wire and the average speed of the electron between collisions with the atoms of the metal wire?

5. There is a current in a metal wire due to the motion of electrons. Sketch a possible path for the motion of a single electron in this wire, the direction of the electric field vector, and the direction of conventional current.

6. What is an electrolyte?

7. What is the direction of conventional current in each case shown in **Figure 19-18**?

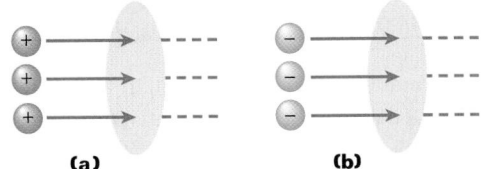

(a)　　　　　　　　**(b)**

Figure 19-18

8. Why must energy be continuously pumped into a circuit by a battery or a generator to maintain an electric current?

9. Name at least two differences between batteries and generators.

10. What is the difference between direct current and alternating current? Which type of current is supplied to the appliances in your home?

Conceptual questions

11. In an analogy between traffic flow and electric current, what would correspond to the charge, Q? What would correspond to the current, I?

12. Is current ever "used up"? Explain your answer.

13. Why do wires usually warm up when an electric current is in them?

14. A student in your class claims that batteries work by supplying the charges that move in a conductor, generating a current. What is wrong with this reasoning?

15. When a light bulb is connected to a battery, charges begin moving almost immediately, although each electron travels very slowly across the wire. Explain why the bulb lights up so quickly.

16. What is the net drift velocity of an electron in a wire that has alternating current in it?

Practice problems

17. How long does it take a total charge of 10.0 C to pass through a cross-sectional area of a copper wire that carries a current of 5.0 A?
(See Sample Problem 19A.)

18. A hair dryer draws a current of 9.1 A.
    a. How long does it take for 1.9×10^3 C of charge to pass through the hair dryer?
    b. How many electrons does this amount of charge represent?
(See Sample Problem 19A.)

19. How long does it take for 5.0 C of charge to pass through a cross-sectional area of a copper wire if $I = 5.0$ A?
(See Sample Problem 19A.)

20. length, cross-sectional area, temperature, material

21. c (greatest); a (least)

22. to regulate current

23. It becomes zero.

24. They are proportional.

25. They are inversely proportional.

26. At a higher temperature, atoms vibrate with greater amplitudes, which makes it more difficult for electrons to move through the material.

27. Answers will vary but could include increased efficiency of energy transportation and more efficient forms of transportation, such as the levitating train.

28. 0.20 A

29. 3.4 A

30. a. 1.8 A
    b. 4.5 A
    c. 0.45 A

31. Electric power is the rate at which electrical energy is converted. Mechanical power is the rate at which mechanical energy is converted. Electrical energy is a form of mechanical energy.

32. electrical energy; power

33. because resistance increases as length increases

34. 50.0 J

35. 3.6×10^6 J

36. the 75 W bulb

37. the conductor with the smaller resistance

38. 2.0×10^{16} J

39. The wires have much less resistance than the filament of the light bulb.

RESISTANCE

Review questions

20. What factors affect the resistance of a conductor?

21. Each of the wires shown in **Figure 19-19** is made of copper. Assuming each piece of wire is at the same temperature, which has the greatest resistance? Which has the least resistance?

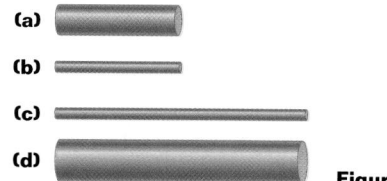

Figure 19-19

22. Why are resistors used in circuit boards?

23. The critical temperature of aluminum is 1.19 K. What happens to the resistance of aluminum at temperatures lower than 1.19 K?

Conceptual questions

24. For a constant resistance, how are potential difference and current related?

25. If the potential difference across a conductor is constant, how is current dependent on resistance?

26. Using the atomic theory of matter, explain why the resistance of a material should increase as its temperature increases.

27. Recent discoveries have led some scientists to hope that a material will be found that is superconducting at room temperature. Why would such a material be useful?

Practice problems

28. A nichrome wire with a resistance of 15 Ω is connected across the terminals of a 3.0 V flashlight battery. How much current is in the wire?
    (See Sample Problem 19B.)

29. How much current is drawn by a television with a resistance of 35 Ω that is connected across a potential difference of 120 V?
    (See Sample Problem 19B.)

30. Calculate the current that each resistor shown in **Figure 19-20** would draw when connected to a 9.0 V battery.
    (See Sample Problem 19B.)

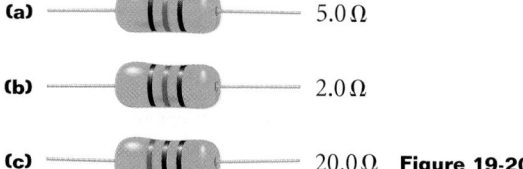

(a) 5.0 Ω

(b) 2.0 Ω

(c) 20.0 Ω **Figure 19-20**

ELECTRIC POWER

Review questions

31. Compare and contrast mechanical power with electric power.

32. What quantity is measured in kilowatt-hours? What quantity is measured in kilowatts?

33. If electrical energy is transmitted over long distances, the resistance of the wires becomes significant. Why?

34. How many joules of energy are dissipated by a 50.0 W light bulb in 1.00 s?

35. How many joules are in a kilowatt-hour?

Conceptual questions

36. A 60 W light bulb and a 75 W light bulb operate from 120 V. Which bulb has a greater current in it?

37. Two conductors of the same length and radius are connected across the same potential difference. One conductor has twice as much resistance as the other. Which conductor dissipates more power?

38. It is estimated that in the United States (population 250 million) there is one electric clock per person, with each clock using energy at a rate of 2.5 W. Using this estimate, how much energy is consumed by all of the electric clocks in the United States in a year?

39. When a small lamp is connected to a battery, the filament becomes hot enough to emit electromagnetic radiation in the form of visible light, while the wires do not. What does this tell you about their relative resistances of the filament and the wires?

Practice problems

40. A computer is connected across a 110 V power supply. The computer dissipates 130 W of power in the form of electromagnetic radiation and heat. Calculate the resistance of the computer.
(See Sample Problem 19C.)

41. The operating potential difference of a light bulb is 120 V. The power rating of the bulb is 75 W. Find the current in the bulb and the bulb's resistance.
(See Sample Problem 19C.)

42. How much would it cost to watch a football game for 3.0 h on a 325 W television described in item 43 if electrical energy costs $0.08/kW•h?
(See Sample Problem 19D.)

43. Calculate the cost of operating a 75 W light bulb continuously for a 30-day month when electrical energy costs $0.15/kW•h.
(See Sample Problem 19D.)

MIXED REVIEW

44. A net charge of 45 mC passes through the cross-sectional area of a wire in 15 s.
  a. What is the current in the wire?
  b. How many electrons pass the cross-sectional area in 1.0 min?

45. A potential difference of 12 V produces a current of 0.40 A in a piece of copper wire. What is the resistance of the wire?

46. The current in a lightning bolt is 2.0×10^5 A. How many coulombs of charge pass through a cross-sectional area of the lightning bolt in 0.50 s?

47. A person notices a mild shock if the current along a path through the thumb and index finger exceeds 80.0 μA. Determine the maximum allowable potential difference without shock across the thumb and index finger for the following:
  a. a dry-skin resistance of 4.0×10^5 Ω
  b. a wet-skin resistance of 2.0×10^3 Ω

48. How much power is needed to operate a radio that draws 7.0 A of current when a potential difference of 115 V is applied across it?

49. A color television has a power rating of 325 W. How much current does this set draw from a potential difference of 120 V?

50. An X-ray tube used for cancer therapy operates at 4.0 MV with a beam current of 25 mA striking a metal target. Calculate the power of this beam.

51. A steam iron draws 6.0 A when connected to a potential difference of 120 V.
  a. What is the power rating of this iron?
  b. How many joules of energy are produced in 20.0 min?
  c. How much does it cost to run the iron for 20.0 min at $0.010/kW•h?

52. An 11.0 W energy-efficient fluorescent lamp is designed to produce the same illumination as a conventional 40.0 W lamp.
  a. How much energy does this lamp save during 100.0 h of use?
  b. If electrical energy costs $0.080/kW•h, how much money is saved in 100.0 h?

53. Use the electric bill shown in **Figure 19-21** to answer the following questions:
  a. How many joules of energy were consumed in this billing cycle?
  b. What is the average amount of energy consumed per day in joules and kilowatt-hours?
  c. If the cost of energy were increased to $0.15/kW•h, how much more would energy cost in this billing cycle? (Assume that the price of fuel remains constant.)

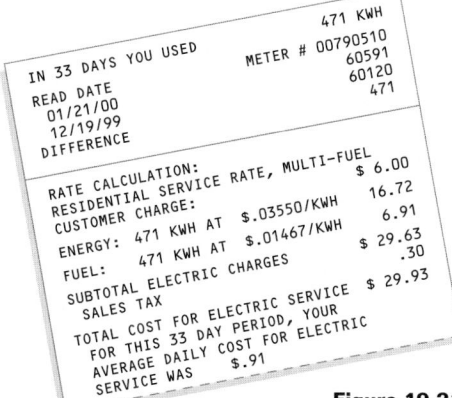

Figure 19-21

40. 93 Ω
41. 0.62 A; 190 Ω
42. $0.08
43. $8.10
44. a. 3.0×10^{-3} A
  b. 1.1×10^{18} electrons/min
45. 3.0×10^1 Ω
46. 1.0×10^5 C
47. a. 32 V
  b. 0.16 V
48. 8.0×10^2 W
49. 2.7 A
50. 1.0×10^5 W
51. a. 720 W
  b. 8.6×10^5 J
  c. $0.0024
52. a. 1.04×10^7 J
  b. $0.23
53. a. 1.70×10^9 J
  b. 5.15×10^7 J/day, 14.3 kW•h/day
  c. $54

54. 61.1 A

55. 3.2×10^5 J

56. 1.1×10^3 s (18 min)

57. 13.5 h

ANSWERS TO

Technology & Learning

a. $P = (Y_2)^2 Y_1$

b. 192 Ω, 0.625 A

c. 5.33 Ω, 3.75 A

d. 72.0 Ω, 1.67 A

e. 2.00 Ω, 10.0 A

f. 144 Ω, 0.833 A

g. bulb attached to the 120 V source

54. The mass of a gold atom is 3.27×10^{-25} kg. If 1.25 kg of gold is deposited on the negative electrode of an electrolytic cell in a period of 2.78 h, what is the current in the cell in this period? Assume that each gold ion carries one elementary unit of positive charge.

55. The power supplied to a typical black-and-white television is 90.0 W when the set is connected across a potential difference of 120 V. How much electrical energy does this set consume in 1.0 h?

56. A color television set draws about 2.5 A of current when connected to a potential difference of 120 V. How much time is required for it to consume the same energy that the black-and-white model described in item 57 consumes in 1.0 h?

57. The headlights on a car are rated at 80.0 W. If they are connected to a fully charged 90.0 A•h, 12.0 V battery, how long does it take the battery to completely discharge?

Technology & Learning

Graphing calculators

Refer to Appendix B for instructions on downloading programs for your calculator. The program "Chap19" builds a table of potential difference, resistance, and current, given the power dissipated by a resistor.

The power dissipated by a resistor, as you learned earlier in this chapter, is described by the following two equations:

$$P = \frac{(\Delta V)^2}{R} \text{ and } P = I\Delta V$$

The program "Chap19" stored on your graphing calculator makes use of these equations for the power dissipated by a resistor. Once the "Chap19" program is executed, your calculator will ask for the power dissipated by the resistor. The graphing calculator will use the following equations to create a table of resistance (Y1) and current (Y2) versus potential difference (X). Note that the relationships in these equations are the same as those in the power equations above; the variables have just been rearranged.

$$Y_1 = X^2/P \text{ and } Y_2 = P/X$$

a. The power dissipated by a resistor can also be expressed in terms of the variables Y1 and Y2 only. Write this expression.

Execute "Chap19" on the PRGM menu, and press ENTER to begin the program. Enter the value for the power dissipated (shown below), and press ENTER.

The calculator will provide a table of resistance in ohms (Y1) and current in amperes (Y2) versus potential difference in volts (X). Press ▼ to scroll down through the table to find the resistance and current values you need.

Determine the resistance of and current in the light bulbs in the following situations (b–f):

b. a 75.0 W bulb with a potential difference of 120.0 V across it

c. a 75.0 W bulb with a potential difference of 20.0 V across it

d. a 200.0 W bulb with a potential difference of 120.0 V across it

e. a 200.0 W bulb with a potential difference of 20.0 V across it

f. a 100.0 W bulb that you plug into the socket in your house where the source of potential difference has a magnitude of 120.0 V

g. Two light bulbs both dissipate the same amount of power. Which bulb has a higher resistance: a bulb attached to a 120 V source or a bulb attached to a 110 V source?

Press ENTER to stop viewing the table. Press ENTER again to enter a new value or CLEAR to end the program.

58. The current in a conductor varies over time as shown in **Figure 19-22.**

    a. How many coulombs of charge pass through a cross section of the conductor in the time interval $t = 0$ to $t = 5.0$ s?

    b. What constant current would transport the same total charge during the 5.0 s interval as does the actual current?

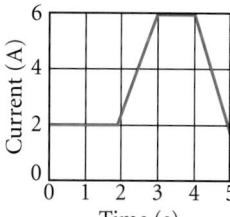

Time (s)   **Figure 19-22**

59. Birds resting on high-voltage power lines are a common sight. A certain copper power line carries a current of 50.0 A, and its resistance per unit length is 1.12×10^{-5} Ω/m. If a bird is standing on this line with its feet 4.0 cm apart, what is the potential difference across the bird's feet?

60. An electric car is designed to run on a bank of batteries with a total potential difference of 12 V and a total energy storage of 2.0×10^7 J.

    a. If the electric motor draws 8.0 kW, what is the current delivered to the motor?

    b. If the electric motor draws 8.0 kW as the car moves at a steady speed of 20.0 m/s, how far will the car travel before it is "out of juice"?

Alternative Assessment

Performance assessment

1. Design an experiment to investigate how the characteristics of a conducting wire affect its resistance. In particular, plan to explore the effects of length, shape, mass, thickness, and the nature of the material. If your teacher approves your plan, obtain the necessary equipment and perform the experiment. Share the results of your experiment with your class.

2. Construct a voltaic pile like the first battery made by Alessandro Volta (1745–1827). Make a stack of alternating copper and zinc disks, inserting cardboard moistened in salt water between the disks. (Copper pennies and dimes can be used as well.) How many layers do you need to make an LED light up? How many layers are required to light a flashlight bulb? How could you measure the relationship between stack size and potential difference? If your teacher approves your plan, carry out an experiment testing the relationships between the stack size and potential difference. Compare your method and results with those of other students in your class.

Portfolio projects

3. When Edison invented the electric light bulb in 1879, his bulb lasted only a week. In 1881, Lewis Howard Latimer received patents for bulbs that could operate for months. Research the life and accomplishments of Latimer, and prepare a presentation in the form of a report, poster, short video, or computer presentation.

4. Visit an electric parts or electronic parts store or consult a print or on-line catalog to learn about different kinds of resistors. Find out what the different resistors look like, what they are made of, what their resistance is, how they are labeled, and what they are used for. Summarize your findings in a poster or a brochure entitled *A Consumer's Guide to Resistors*.

5. The units of measurement you learned about in this chapter were named after three famous scientists: Andre-Marie Ampere, Georg Simon Ohm, and Alessandro Volta. Research their lives, works, discoveries, and contributions. Create a presentation about one of these scientists. The presentation can be in the form of a report, poster, short video, or computer presentation.

---

58. a. 18 C
   b. 3.6 A
59. 2.2×10^{-5} V
60. a. 670 A
   b. 5.0×10^4 m

Alternative Assessment
ANSWERS

Performance assessment

1. Student answers will vary. Students may measure resistance directly or measure potential difference across and current in the resistor.

2. Student answers will vary. Students should find that potential difference increases with more layers.

 **Portfolio projects**

3. Student answers will vary. Latimer (1848–1928) was a pioneer among African American inventors. He invented an inexpensive and durable light bulb cotton-thread filament. Later he worked closely with Alexander Graham Bell.

4. Student answers will vary. Presentations should describe various types and sizes of resistors. Students may include the color codes for resistors. Be sure students include references.

5. Ampere (1775–1836) founded the field of electrodynamics. Ohm (1787–1854) discovered the relationship between resistance and the length and cross section of a wire. Volta (1745–1827) invented the first battery, the "Voltaic pile."

NOTE

Materials Preparation is given on pp. 692A–692B. Blank data table and sample data table are on the One-Stop Planner CD-ROM. All calculations shown use sample data.

Planning

Recommended time:

1 lab period

For a 2-period lab, the procedure for meters may be repeated using a known resistor of a different value.

Classroom organization:

▶ Each lab group should have a level surface to work on that is near an electrical outlet and away from any sources of water.

▶ Each lab group should have two students.

▶ The meters and CBL procedures may be used in the same class.

▶ **Safety warnings:** Emphasize the dangers of working with electricity. For the safety of the students and of the equipment, remind students to have you check their circuits before turning on the power supply or closing the switch.

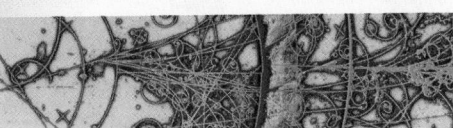

CHAPTER 19
Laboratory Exercise

★ TEKS
pp. 722–725: 1A, 2A, 2B, 2C, 2D, 2F, 3B, 6E

OBJECTIVES

- Determine the resistance of conductors using the definition of resistance.
- Explore the relationships between length, diameter, material, and the resistance of a conductor.

MATERIALS LIST

✔ insulated connecting wire
✔ momentary contact switch
✔ mounted resistance coils
✔ power supply

PROCEDURE

CBL AND SENSORS

✔ CBL
✔ CBL voltage probe
✔ graphing calculator with link cable
✔ resistor of known resistance

METERS

✔ 2 multimeters or 1 dc ammeter and 1 voltmeter

CURRENT AND RESISTANCE

Different substances offer different amounts of resistance to an electric current. Physicists have found that temperature, length, cross-sectional area, and the material the conductor is made of determine the resistance of a conductor. In this experiment, you will observe the effects of length, cross-sectional area, and material on the resistance of conductors.

You will use a set of mounted resistance coils, which will provide wire coils of different lengths, diameters, and metals. The resistance provided by each resistance coil will affect the current in the resistor. You will measure the potential difference across the resistance coil, and you will find the current in the conductor. Then you will use these values to calculate the resistance of each resistance coil using the definition of resistance.

SAFETY

- **Never close a circuit until it has been approved by your teacher. Never rewire or adjust any element of a closed circuit. Never work with electricity near water; be sure the floor and all work surfaces are dry.**

- **If the pointer on any kind of meter moves off scale, open the circuit immediately by opening the switch.**

- **Do not attempt this exercise with any batteries or electrical devices other than those provided by your teacher for this purpose.**

PREPARATION

1. Determine whether you will be using the CBL and sensors procedure or the meters. Read the entire lab for the appropriate procedure, and plan what steps you will take.

2. Prepare a data table in your lab notebook with seven columns and six rows. For the CBL and sensors procedure, label the first through seventh columns *Trial, Metal, Gauge number, Length (cm), Cross-sectional area* (cm^2), $\Delta V_C\ (V)$, and $\Delta V_R\ (V)$. Prepare a space near the data table to record the resistance of the known resistor. For the meters procedure, label the first through seventh columns *Trial, Metal, Gauge number, Length (cm), Cross-sectional area* (cm^2), $\Delta V_x\ (V)$, and $I\ (A)$. In the first column, label the second through sixth rows *1, 2, 3, 4,* and *5*.

Meters procedure begins on page 724.

PROCEDURE

CBL AND SENSORS

Potential difference and resistance

3. Set up the apparatus as shown in **Figure 19-23**. Construct a circuit that includes a power supply, a switch, a resistor of known resistance, and the mounted resistance coils. For each trial, you will measure the potential difference across one unknown resistance coil and then across the known resistor using the CBL voltage probe. ***Do not turn on the power supply. Do not close the switch until your teacher has approved your circuit.***

4. With the switch open, connect the voltage probe to measure the potential difference across the first unknown resistance coil. Connect the black lead of the voltage probe to the side of the coil that is connected to the black pin on the power supply. Connect the red lead to the other side of the coil. ***Do not close the switch.***

5. Connect the CBL and graphing calculator. Connect the voltage probe to CH1 on the CBL. Turn on the CBL and the graphing calculator, and start the program PHYSICS on the calculator.

6. Select option *SET UP PROBES* from the MAIN MENU. Enter 1 for the number of probes. Select *MORE PROBES* from the SELECT PROBE menu. Select the *VOLTAGE PROBE* from the list. Enter 1 for the channel number.

7. Select the *COLLECT DATA* option from the MAIN MENU. Select the *MONITOR INPUT* option from the DATA COLLECTION menu. The graphing calculator will begin to display values for the potential difference across the coil. ***Do not close the switch.***

8. When your teacher has approved your circuit, make sure the power supply dial is turned completely counterclockwise. Turn on the power supply, and slowly turn the dial clockwise. Periodically close the switch briefly and read the value for the potential difference on the CBL. When the potential difference is approximately 0.5 V, read and record the value as ΔV_C in your data table. Open the switch.

9. Remove the leads of the voltage probe from the resistance coil, and connect the probe to the known resistor. Connect the black lead to the side of the resistor that is connected to the black pin on the power supply, and connect the red lead to the other side of the resistor. Have your teacher approve your circuit.

10. When your teacher has approved your circuit, close the switch and read the value for the potential difference across the known resistor. Record this value as ΔV_R in your data table.

Figure 19-23

Step 3: The set of mounted resistance coils shown includes five different resistance coils. In this lab, you will measure the potential difference across each coil in turn.

Step 4: Always make sure the black lead on the probe is connected to the side of the coil that connects to the black pin on the power supply. Use your finger to trace the circuit from the black pin on the power supply through the circuit to the red pin on the power supply to check for proper connections.

Step 8: Close the switch only long enough to take readings. Open the switch as soon as you have taken the reading.

CBL and Sensors Tips

◆ Students should have the program PHYSICS on their graphing calculators. Refer to Appendix B for instructions.

Techniques to Demonstrate

Show students how to wire the circuit to include one of the resistance coils and how to move the leads to the next coil.

Demonstrate how to follow the circuit with your finger to make sure all the connections are correct.

✔ Checkpoints

Step 8: Check each circuit to make sure all connections are made properly and power supplies are set at appropriate levels. Students should be able to demonstrate that they can trace the circuit to check the connections.

Step 9: Check each circuit to make sure it is set up to measure the potential difference across the known resistor. Students should be able to demonstrate that they have made all connections properly and that the only change in the circuit is a change of the position of the voltage probe.

Step 12: For each trial, make sure students connect all circuits properly. At the beginning of each trial, the knob on the power supply should be turned completely counterclockwise.

◆ See pp. 692A–692B for instructions on setting up multimeters to measure potential difference and current.

◆ Make sure students understand how to wire current meters (in series) and voltage meters (in parallel) in a circuit.

Techniques to Demonstrate

Show students how to wire the circuit to include one of the resistance coils and how to move the meters to the next coil.

Demonstrate how to follow the circuit with your finger to make sure all the connections are correct.

✔ Checkpoints

Step 4: Check each circuit to make sure all connections are made properly and power supplies are set at appropriate levels. Students should be able to demonstrate that they can trace the circuit to check the connections.

Step 7: For each trial, make sure students connect all circuits properly. At the beginning of each trial, the knob on the power supply should be turned completely counterclockwise.

11. When your teacher has approved your circuit, close the switch and read the value for the potential difference across the unknown resistance coil. Record this value as ΔV_C in your data table. Open the switch.

12. For Trial 2, remove the voltage probe from the known resistor and connect it to the next coil of unknown resistance. Connect the black lead of the probe to the side of the coil that is connected to the black pin on the power supply, and connect the red lead to the other side. Repeat the procedure in steps 8–11. Repeat the procedure until the potential difference across all five coils has been recorded. Record the potential difference across the known resistor for each trial. Your teacher will supply the length and cross-sectional area for each unknown resistance coil. Record these in your data table.

13. Clean up your work area. Put equipment away safely so that it is ready to be used again.

Analysis and Interpretation begins on page 725.

PROCEDURE

METERS

Current at varied resistances

3. Set up the apparatus as shown in **Figure 19-24.** Construct a circuit that includes a power supply, a switch, a current meter, a voltage meter, and the mounted resistance coils. For each trial, you will measure the current and the potential difference for one of the coils to determine the value of the resistance. ***Do not turn on the power supply. Do not close the switch until your teacher has approved your circuit.***

4. With the switch open, connect the current meter in a straight line with the mounted resistance coils. Make sure the black lead on the meter is connected to the black pin on the power supply. Connect the black lead on the voltage meter to the side of the first resistance coil that is connected to the black pin on the power supply, and connect the red lead to the other side of the coil. ***Do not close the switch until your teacher approves your circuit.***

5. When your teacher has approved your circuit, make sure the power supply dial is turned completely counterclockwise. Turn on the power supply, and slowly turn the dial clockwise. Periodically close the switch briefly and read the current value on the current meter. Adjust the dial until the current is approximately 0.15 A.

6. Close the switch. Quickly record the current in and the potential difference across the resistance coil in your data table. Open the switch immediately. Turn off the power supply by turning the dial completely counterclockwise. Your teacher will supply the length and cross-sectional area of the wire on the coil. Record these in your data table.

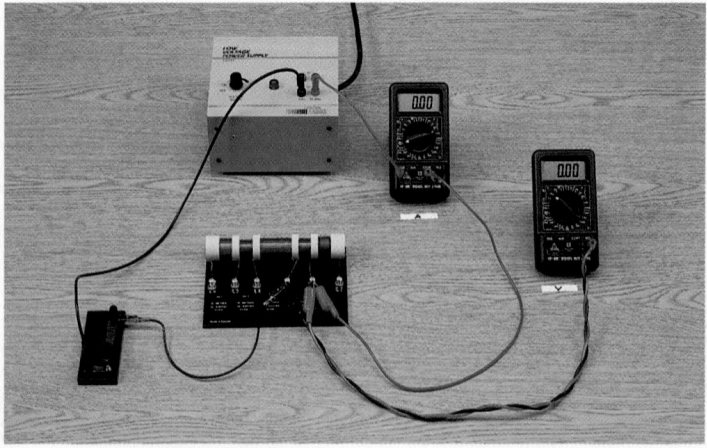

Figure 19-24

Step 3: The set of mounted resistance coils shown includes five different resistance coils. In this lab, you will measure the current and potential difference for each coil in turn.

Step 4: Use your finger to trace the circuit from the black pin on the power supply through the circuit to the red pin on the power supply to check for proper connections.

Step 6: Close the switch only long enough to take readings. Open the switch as soon as you have taken the reading.

7. Remove the leads of the voltmeter from the resistance coil, and connect them in parallel to the next adjacent coil. Repeat steps 5 and 6 until five coils have been studied. For each coil, adjust the current, and follow the same procedure as above. Record the current and potential difference readings for each coil. Your teacher will supply the length and cross-sectional area for each coil. Record these in your data table.

8. Clean up your work area. Put equipment away safely so that it is ready to be used again.

ANALYSIS AND INTERPRETATION

Calculations and data analysis

1. **Organizing data** Use the definition of resistance, $R = \dfrac{\Delta V}{I}$.

   a. **CBL and sensors** Use the value for the known resistance and the potential difference across the known resistor to calculate the current in the circuit for each trial. Use the value for the current to calculate the resistance, R_C, for each resistance coil you tested.

   b. **Meters** Use the measurements for current and potential difference to calculate the resistance, R_C, for each resistance coil you tested.

Conclusions

2. **Analyzing data** Rate the coils from lowest to highest resistance. Record your ratings.

   a. According to your results for this experiment, how does the length of the wire affect the resistance of the coil?

   b. According to your results for this experiment, how does the cross-sectional area affect the resistance of the coil?

3. **Evaluating results** Based on your results for the metals used in this experiment, which metal has the greatest resistance? Explain how you arrived at this conclusion.

4. **Evaluating results** Based on your results for the metals used in this experiment, which metal has the lowest resistance? Explain how you arrived at this conclusion.

Extension

5. Devise a method for identifying a resistance coil made of an unknown metal by placing the coil in a circuit and finding its resistance. Research and include in your plans a way to use the value for the resistance to identify the metal. If there is time and your teacher approves your plan, perform the experiment. Write a report detailing your procedure and results.

ANSWERS TO

Analysis and Interpretation

CALCULATIONS AND DATA ANALYSIS

1. Student answers will vary. Make sure students use the relationships $I = \Delta V_R / R_R$ and $R_C = \Delta V_C / I$. For sample data, values for the current range from 0.13 A to 1.24 A. Values for R_c range from 0.53 Ω to 16 Ω.

CONCLUSIONS

2. a. Longer wires provide more resistance.

   b. Thinner wires provide more resistance.

3. For resistance coils of Cu and Cu/Ni, Cu/Ni provides much greater resistance. Students should compare resistors of the same length and cross-sectional area.

4. For resistance coils of Cu and Cu/Ni, Cu provides much lower resistance. Students should compare resistors of the same length and cross-sectional area.

EXTENSION

5. Student plans should be safe and complete, including a list of equipment, measurements, and calculations required.

Physics and Its World *Timeline 1830–1890*

1831

$$\text{emf} = -N\Delta \frac{[AB(\cos\theta)]}{\Delta t}$$

Michael Faraday begins experiments demonstrating electromagnetic induction. Similar experiments are conducted around the same time by **Joseph Henry** in the United States, but he doesn't publish the results of his work at this time.

1831 – Charles Darwin sets sail on the H.M.S. *Beagle* to begin studies of life-forms in South America, New Zealand, and Australia. His discoveries form the foundation for the theory of natural selection.

1837 – Queen Victoria ascends the British throne at the age of 18. Her reign continues for 64 years, setting the tone for the Victorian era.

1839

Samuel Morse sends the first telegraph message from Washington, D. C. to Baltimore.

1843

$$\Delta U = Q - W$$

James Prescott Joule determines that mechanical energy is equivalent to energy transferred as heat, laying the foundation for the principle of energy conservation.

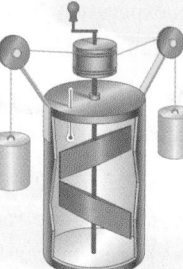

1843 – Richard Wagner's first major operatic success, *The Flying Dutchman,* premieres in Dresden, Germany.

1850 – Harriet Tubman, an ex-slave from Maryland, becomes a "conductor" on the Underground Railroad. Over the next decade, she helps more than 300 slaves escape to northern "free" states.

1850

$$W = Q_h - Q_c$$

Rudolph Clausius formulates the second law of thermodynamics, the first step in the transformation of thermo-dynamics into an exact science.

1820
1830
1840
1850
1860

1861 – **Benito Juárez** is elected president of Mexico. During his administration, the invasion by France is repelled and basic social reforms are implemented.

Scala/Art Resource, NY

1861 – The American Civil War begins at Fort Sumter in Charleston, South Carolina.

1873

$$c = \frac{1}{\sqrt{\mu_0 \varepsilon_0}}$$

James Clerk Maxwell completes his *Treatise on Electricity and Magnetism*. In this work, Maxwell gives **Michael Faraday's** discoveries a mathematical framework.

1874 – The first exhibition of impressionist paintings, including works by **Claude Monet, Camille Pissarro,** and **Pierre-Auguste Renoir,** takes place in Paris.

Giraudon/Art Resource, NY

1878 – The first commercial telephone exchange in the United States begins operation in New Haven, Connecticut.

1884 – *Huckleberry Finn,* by **Samuel L. Clemens** (better known as Mark Twain), is published.

1888

$$\lambda = \frac{c}{f}$$

Heinrich Hertz experimentally demonstrates the existence of electromagnetic waves, which were predicted by **James Clerk Maxwell. Oliver Lodge** makes the same discovery independently.

1860

1870

1880

1890

1900

Physics and Its World 1830–1890 **727**

CHAPTER 20 PLANNING GUIDE

Compression Guide: To shorten from 11 to 8 45-min periods (from 5½ to 4 90-min blocks), eliminate items in magenta type.

PACING CHART	CLASSROOM RESOURCES			
	⭐ TEKS	Teacher Demonstrations	Holt Physics Transparencies	Labs (See page T52 for equipment listing for in-text labs.)
20-1 Schematic diagrams and circuits 2 or 1 45-minute periods 1 or ½ 90-minute block	1A, 2A, 2C, 3A, 5B, 5D, 6E		🖥 **T** 98–99 **TM** 64	**L** *Discovery Lab: Exploring Circuit Elements* **PE** *Quick Lab: Simple Circuits*, p. 734
20-2 Resistors in series or in parallel 4 or 3 45-minute periods 2 or 1½ 90-minute blocks	1A, 2A, 2B, 2C, 2F, 3A, 3C, 6E	**TE** *Resistors in series*, p. 736 **TE** *Resistors in parallel*, p. 741	🖥 **TM** 65–66	**PE** *Quick Lab: Series and Parallel Circuits*, p. 741 **PE** *Resistors in Series and in Parallel*, p. 760 **TL** *Series and Parallel Circuits*
20-3 Complex resistor combinations 3 or 2 45-minute periods 1½ or 1 90-minute block	2C, 3C, 6E		🖥 **T** 100–101 **TM** 67	**L** *Invention Lab: Building a Dimmer Switch*
Review and Assessment 2 45-minute periods 1 90-minute block				

Resource Key

PHYSICS HOLT

PE Pupil's Edition
TE Teacher's Edition

L Laboratory Experiments
TL Technology Lab Experiments
T Transparencies

One-Stop Planner CD-ROM contents

TM Transparency Masters
SR Section Review Worksheets
AA Alternative Assessment

PW Problem-Solving Workbook
PB Problem Bank
CTW Critical Thinking Worksheet

LABORATORY PLANNING: Resistors in Series and in Parallel, p. 760

Materials (for each lab group)
- power supply: 1.5 V/3 V battery eliminator, or 20 V dc/500 mA regulated power supply
- switch: contact key or knife switch (single pole/single throw)
- 150 Ω resistor, 0.5 W
- 68 Ω resistor, 0.5 W
- pair of alligator clip adapters
- patch cord, 30 cm, black, with banana plugs
- patch cord, 30 cm, red, with banana plugs

Additional Equipment
CBL and Sensors Procedure
- CBL
- CBL current and voltage probe set with amplifier box
- 2 DIN adapters for CBL graphing calculator with link cable
- TI Graph Link (recommended for downloading programs)

ASSIGNMENT RESOURCES

Content Mastery	Critical Thinking	Problem-Solving Practice
PE 1–5, p. 735 **SR** 20-1, *Diagram Skills* **PE** 1–5, p. 754	**PE** 1–3, p. 732 **PE** 6–10, p. 754	
PE 1–6, p. 745 **SR** 20-2, *Concept Review* **PE** 11–12, p. 754	**PE** 1–2, p. 742 **PE** 13–15, pp. 754–755	**20A** Resistors in series: 26 items in **PE, PW,** and **PB,** see **TE** pp. 738–739 **20B** Resistors in parallel: 24 items in **PE, PW,** and **PB,** see **TE** pp. 743–744
PE 1–9, p. 752 **SR** 20-3, *Concept Review*	**PE** 20–22, p. 755	**20C** Equivalent resistance: 22 items in **PE, PW,** and **PB,** see **TE** pp. 747–748 **20D** Current in and potential difference across a resistor: 17 items in **PE, PW,** and **PB,** see **TE** pp. 749–751

ASSESSMENT RESOURCES

Cumulative Review	Alternative Assessment	Traditional Assessment
SR Mixed Review, Ch. 20	**PE** 1–5, p. 759 **AA** Items for Ch. 20	Chapter 20 Test Test Generator items for Ch. 20

Scoring Rubrics for Alternative Assessment items can be found on the One-Stop Planner CD-ROM.

TECHNOLOGY RESOURCES

 CTW Segment 20 Eagle Electrocution

 Module 17 Electric Circuits

 PE Technology and Learning, p. 758
(Alternative procedures for calculators without Flash-ROM technology are provided on the One-Stop Planner CD-ROM.)

 The Mechanical Universe/High School Adaptation Quad V, Simple dc Circuits

 internet**connect**

 On-line Student Resources:
GO TO: www.scilinks.org
The following *sci*LINKS Internet resources can be found in the student text for this chapter.

TOPICS:
• Electric circuit, p. 734 (HF2201)
• Resistors, p. 740 (HF2202)

 On-line Teacher Resources:
GO TO: go.hrw.com
KEYWORD: HF2 HOME
Visit the HRW Web site for a variety of resources related to this chapter.

Smithsonian Institution*
Internet Connections
Visit **www.si.edu/hrw** for additional on-line resources.

 .com
Visit **www.cnnfyi.com** for late-breaking news and current events stories selected just for you.

Meters Procedure
One of the following: 2 basic digital multimeters; 2 student multimeters; 2 general purpose multimeters; or 1 CENCO standard movement, triple range dc ammeter (50 mA, 500 mA, 5 A) and 1 CENCO standard movement, triple range dc voltmeter (0–3,10, 15 V)

Materials Preparation
See pages 692A–692B for instructions on using multimeters to measure current and voltage.

Section 20-1 introduces the concept of an electric circuit, distinguishes between open and closed circuits, and describes the concept of a *short circuit*.

Section 20-2 describes the relationships between equivalent resistance, current, and potential difference for series circuits and parallel circuits.

Section 20-3 explores complicated circuits containing portions in series and portions in parallel.

About the Illustration

The Riverwalk in San Antonio, Texas, is a downtown shopping and entertainment district built on the banks of the San Antonio River. The Riverwalk began as a Works Progress Administration project in the Great Depression of the 1930s. In the 1970s and 1980s, redevelopment and expansion of the Riverwalk sparked an economic revival of downtown San Antonio.

Interactive Problem-Solving Tutor

See Module 17

"Electric Circuits" provides additional development of problem-solving skills for this chapter.

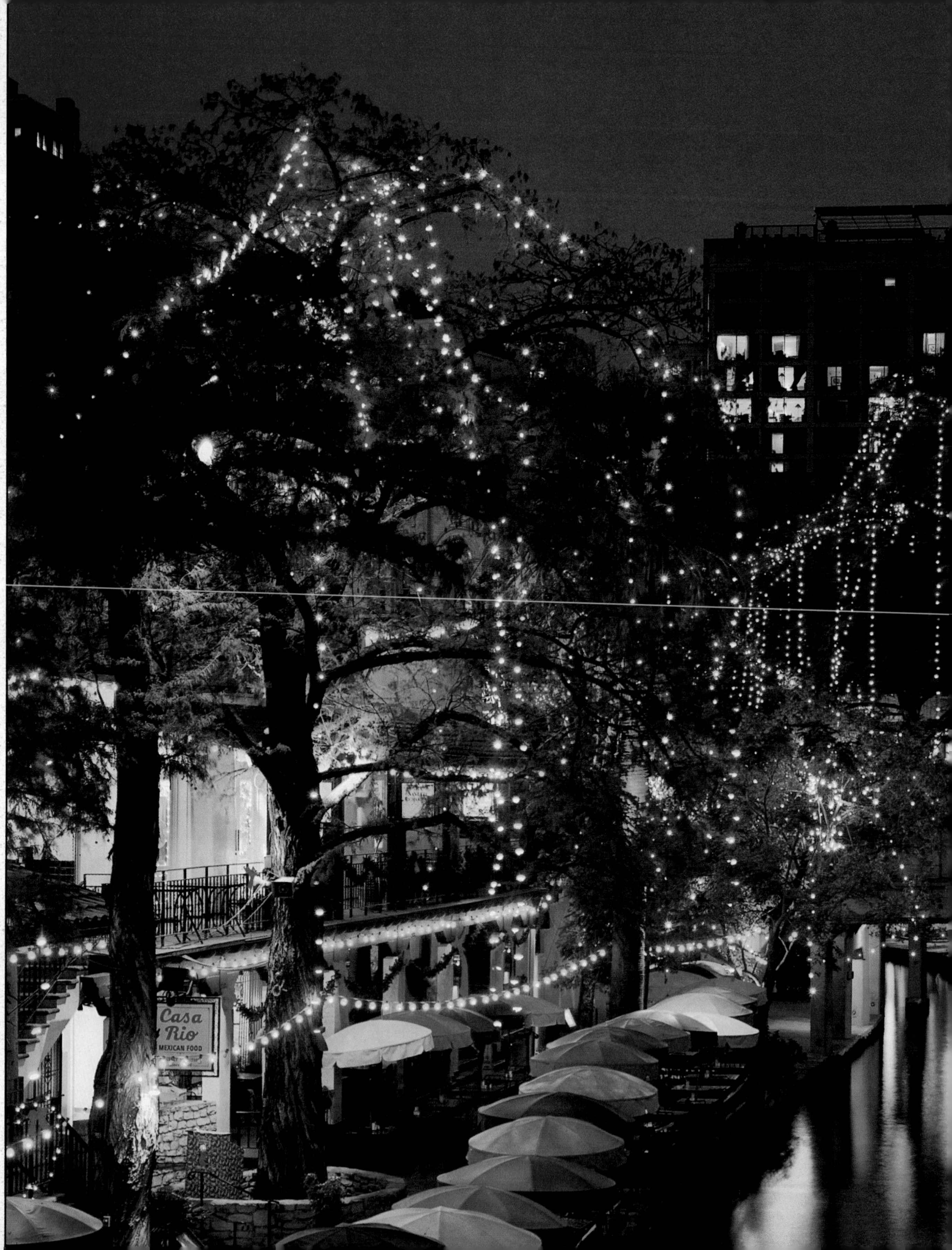

Circuits and Circuit Elements

PHYSICS IN ACTION

For strings of decorative lights—like these that illuminate the Riverwalk in San Antonio, Texas—two types of electric circuits can be used. The first type is a series circuit. You may have seen an older version of one, in which the entire string of lights goes dark when one bulb burns out. Newer series lights bypass the burnt-out bulb so that the others remain lit. The second type is a parallel circuit. In this type, the other bulbs remain lighted even when one or more bulbs burn out. In this chapter, you will explore the basic properties of these two types of circuits.

- *How does the wiring of a circuit alter the current in and potential difference across each element?*

- *What happens when both types of circuits are combined in a single circuit?*

CONCEPT REVIEW

Potential difference   (Section 18-2)

Current   (Section 19-1)

Resistance and Ohm's law   (Section 19-2)

Circuits and Circuit Elements   **729**

Tapping Prior Knowledge

Knowledge to Expect

✔ "Students can make safe electrical connections with plugs, sockets, and terminals." (AAAS's *Benchmarks for Science Literacy,* grades 3–5)

✔ "Students have built battery-driven electric circuits." (AAAS's *Benchmarks for Science Literacy,* grades 6–8)

✔ "Electric circuits provide a means of transferring electrical energy when heat, light, sound, or chemical changes are produced." (NRC's *National Science Education Standards,* grades 5–8)

Knowledge to Review

✔ Potential difference is the change in electric potential energy per unit charge from one point to another. (Section 18-2)

✔ Current is the rate at which electric charges move through a given area. (Section 19-1)

✔ $\Delta V = IR$ can be used to relate current and potential difference for specific electrical devices. (Section 19-2)

Items to Probe

✔ Preconceptions about wiring: Ask students to trace the path of charges moving in a string of decorative lights.

✔ Familiarity with schematic diagrams: Ask students to draw their own picture representing the wiring of decorative lights.

20-1
Schematic diagrams and circuits

Visual Strategy

Figure 20-1
Students should be encouraged to create alternative representations of the circuit shown in (**a**). Students should discuss what their symbols stand for, how convenient their symbols would be for others to use, and in what way each symbol reflects *relevant information.*

Q List information about the group of elements that is *not* relevant to its function and is unnecessary in a schematic.

A *The colors and sizes of the items shown and whether the wires are coiled, bent, or straight are irrelevant to the function of the group of elements.*

Visual Strategy

Figure 20-1
Students should recognize that the straight-line symbols connecting the battery symbol with the bulb symbol in (**b**) represent not only the wire but also all parts of the conducting connection between the bulb and battery.

Q List the parts of the photo symbolized by the black straight lines in the diagrams.

A *The black lines symbolize the conducting path provided by the wires, clips, and socket.*

20-1 SECTION OBJECTIVES

- **Interpret and construct circuit diagrams.**

- **Identify circuits as open or closed.**

- **Deduce the potential difference across the circuit load, given the potential difference across the battery's terminals.**

schematic diagram

a graphic representation of an electric circuit, with standardized symbols representing circuit components

SCHEMATIC DIAGRAMS

Take a few minutes to examine the battery and light bulb in **Figure 20-1(a),** then draw a diagram of each element in the photograph and its connection. How easily could your diagram be interpreted by someone else? Could the elements in your diagram be used to depict a string of decorative lights, such as those draped over the trees of the San Antonio Riverwalk? ⭐**TEKS** **6E**

A diagram that depicts the construction of an electrical apparatus is called a **schematic diagram.** The schematic diagram shown in **Figure 20-1(b)** uses symbols to represent the bulb, battery, and wire from **Figure 20-1(a).** Note that these same symbols can be used to describe these elements in any electrical apparatus. This way, schematic diagrams can be read by anyone familiar with the standard set of symbols.

Reading schematic diagrams is necessary to determine how the parts in an electrical device are arranged. In this chapter, you will see how the arrangement of resistors in an electrical device can affect the current in and potential difference across the other elements in the device. The ability to interpret schematic diagrams for complicated electrical equipment is an essential skill for solving problems involving electricity.

As shown in **Table 20-1,** each element used in a piece of electrical equipment is represented by a symbol in schematic diagrams that reflects the element's construction or function. For example, the schematic-diagram symbol that represents an open switch resembles the open knife switch that is shown in the corresponding photograph. Note that **Table 20-1** also includes other forms of schematic-diagram symbols; these alternative symbols will not be used in this book.

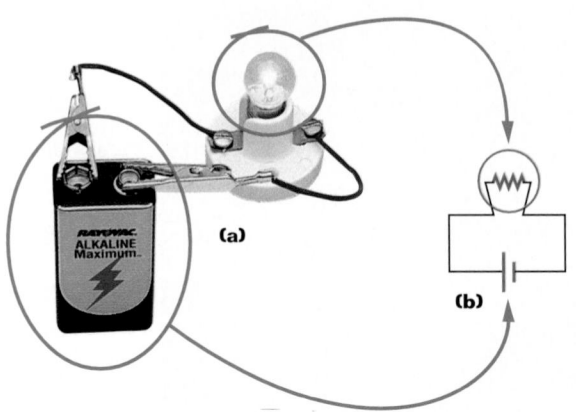

(a)

(b)

Figure 20-1
(**a**) When this battery is connected to a light bulb, the potential difference across the battery generates a current that illuminates the bulb.
(**b**) The connections between the light bulb and battery can be represented in a schematic diagram.

Table 20-1   Schematic diagram symbols

Component	Symbol used in this book	Other forms of this symbol	Explanation
Wire or conductor			• Wires that connect elements are conductors. • Because wires offer negligible resistance, they are represented by straight lines.
Resistor or circuit load			• Resistors are shown as wires with multiple bends, illustrating resistance to the movement of charges.
Bulb or lamp			• The multiple bends of the filament indicate that the light bulb is a resistor. • The symbol for the filament of the bulb is often enclosed in a circle to emphasize the enclosure of a resistor in a bulb.
Plug			• The plug symbol looks like a container for two prongs. • The potential difference between the two prongs of a plug is symbolized by lines of unequal height.
Battery		Multiple cells	• Differences in line height indicate a potential difference between positive and negative terminals of the battery. • The taller line represents the positive terminal of the battery.
Switch	Open Closed	Open Closed	• The small circles indicate the two places where the switch makes contact with the wires. Most switches work by breaking only one of the contacts, not both.
Capacitor			• The two parallel plates of a capacitor are symbolized by two parallel lines of equal height. • One curved line indicates that the capacitor can be used with only direct current sources with the polarity as shown.

Visual Strategy

Table 20-1
Be sure students recognize that the different symbols represent devices with different functions.

Q Challenge students to identify which devices have the following functions: storing energy, transforming energy, and conducting current.

A *batteries and capacitors store energy, resistors, bulbs, and batteries transform energy, wires, resistors, bulbs, plugs, closed switches, and batteries conduct current*

The Language of Physics

Although **Table 20-1** contains several schematic diagram symbols, several stylistic variations exist. For example, some other symbols for light bulbs are shown below.

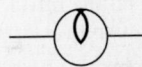

Because light bulbs are resistors, the symbols for resistors are often used for light bulbs.

⭐TEKS

p. 730: 6E

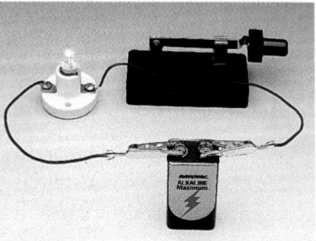

Figure 20-2
When all electrical components are connected, charges can move freely in a circuit. The movement of charges in a circuit can be halted by opening the switch.

electric circuit

a set of electrical components connected so that they provide one or more complete paths for the movement of charges

ELECTRIC CIRCUITS

Think about how you get the bulb in **Figure 20-2** to light up. Will the bulb stay lit if the switch is opened? Is there any way to light the bulb without connecting the wires to the battery?

The filament of the light bulb is a resistor. When a wire connects the terminals of the battery to the light bulb, as shown in **Figure 20-2,** charges built up on one terminal of the battery have a path to follow to reach the opposite charges on the other terminal. Because there are charges moving uniformly, a current exists. This current causes the filament to heat up and glow.

Together, the bulb, battery, switch, and wire form an **electric circuit.** An electric circuit is a path through which charges can be conducted. A schematic diagram for a circuit is sometimes called a circuit diagram. ⭐TEKS 6E

Any element or group of elements in a circuit that dissipates energy is called a *load.* Therefore, a simple circuit consists of a source of potential difference and electrical energy, such as a battery, and a load, such as a bulb or group of bulbs. Because the connecting wire and switch have negligible resistance, we will not consider these elements as part of the load.

In **Figure 20-2,** the path from one battery terminal to the other is complete, a potential difference exists, and electrons move from one terminal to the other. In other words, there is a closed-loop path for electrons to follow. This is called a *closed circuit.* The switch in the circuit in **Figure 20-2** must be closed in order for a steady current to exist.

Without a complete path, there is no charge flow and therefore no current. This situation is an *open circuit.* If the switch in **Figure 20-2** were open, as shown in **Table 20-1,** the circuit would be open, the current would be zero, and the bulb would not light up.

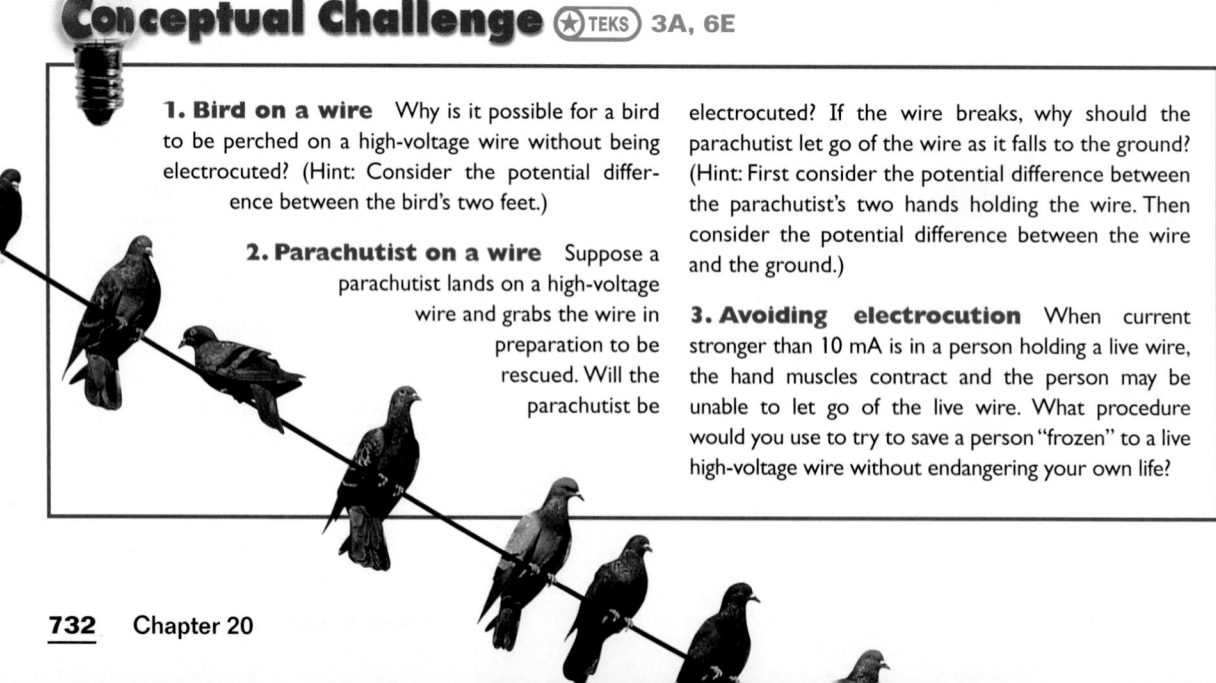

Conceptual Challenge ⭐TEKS 3A, 6E

1. Bird on a wire Why is it possible for a bird to be perched on a high-voltage wire without being electrocuted? (Hint: Consider the potential difference between the bird's two feet.)

2. Parachutist on a wire Suppose a parachutist lands on a high-voltage wire and grabs the wire in preparation to be rescued. Will the parachutist be electrocuted? If the wire breaks, why should the parachutist let go of the wire as it falls to the ground? (Hint: First consider the potential difference between the parachutist's two hands holding the wire. Then consider the potential difference between the wire and the ground.)

3. Avoiding electrocution When current stronger than 10 mA is in a person holding a live wire, the hand muscles contract and the person may be unable to let go of the live wire. What procedure would you use to try to save a person "frozen" to a live high-voltage wire without endangering your own life?

Light bulbs contain a complete conducting path

How does a light bulb contain a complete conducting path? When you look at a clear light bulb, you can see the twisted filament inside that provides a portion of the conducting path for the circuit. However, the bulb screws into a single socket; it seems to have only a single contact, the rounded part at the bulb's base.

Closer examination of the socket reveals that it has two contacts inside. One contact, in the bottom of the socket, is connected to the wire going to one side of the filament. The other contact is in the side of the socket, and it is connected to the wire going to the other side of the filament.

The placement of the contacts within the socket indicates how the bulb completes the circuit, as shown in **Figure 20-3.** Within the bulb, one side of the filament is connected with wires to the contact at the light bulb's base, **(a).** The other side of the filament is connected to the side of the metal base, **(c).** Insulating material between the side of the base and the contact on the bottom prevents the wires from being connected to each other with a conducting material. In this way, charges have only one path to follow when passing through a light bulb—through the filament, **(b).**

When a light bulb is screwed in, the contact on one side of the socket touches the threads on the side of the bulb's base. The contact on the bottom of the socket touches the contact on the bottom of the bulb's base. Charges then enter through the bulb's base, move through the bulb to the filament, and exit the bulb through the threads. For most light bulbs, the bulb will glow regardless of which direction the charges move. Thus, the positive terminal of a battery can be connected to either the base of the bulb or the threads of the bulb, as long as the negative terminal is connected to the threads or base, respectively. All that matters is that there is a complete conducting path for the charges to move through the circuit. (★)TEKS 6E

Short circuits can be hazardous

Without a load, such as a bulb or other resistor, the circuit contains little resistance to the movement of charges. This situation is called a short circuit. For example, a short circuit occurs when a wire is connected from one terminal of a battery to the other by a wire with little resistance. This commonly occurs when uninsulated wires connected to different terminals come into contact with each other. (★)TEKS 6E

When short circuits occur in the wiring of your home, the increase in current can become unsafe. Most wires cannot withstand the increased current, and they begin to overheat. The wire's insulation may even melt or cause a fire.

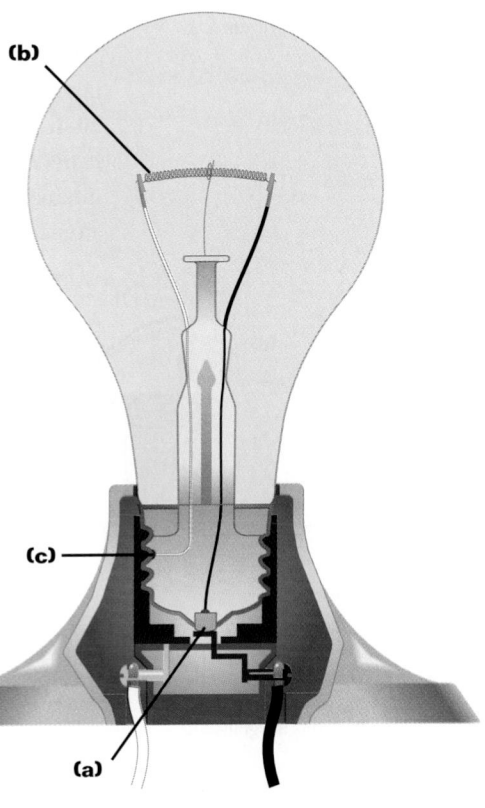

(b)

(c)

(a)

Figure 20-3
Light bulbs contain a complete conducting path. When a light bulb is screwed in, charges can enter through the base **(a),** move along the wire to the filament **(b),** and exit the bulb through the threads **(c).**

(STOP) Misconception Alert

Because batteries are said to *run down,* many students believe that electricity is consumed by a circuit.

To check for this misconception, ask students to draw arrows representing the current in a simple circuit.

Some may believe that current is used up in the resistor. Their diagrams will show charges moving only from the battery to the bulb.

Others may think that the current comes back to the battery but has decreased in magnitude. Arrows representing current in their diagrams may get smaller after the resistor.

Point out that the number of charges entering a part of the circuit in some time interval equals the number of charges leaving it in the same time interval.

(★)TEKS

p. 732: 6E, 3A, 6E
p. 733: 6E, 6E

The Language of Physics

The term *emf* originally stood for *electromotive force.* This term may be misleading because emf is not a force. Rather, it refers to a potential difference measured in volts. The voltage value on a battery label denotes its emf.

In this text, internal resistance will be disregarded unless specifically noted. The value of the terminal voltage, ΔV, can be found with the following equation:

$$\Delta V = \varepsilon - Ir$$

Quick Lab

TEACHER'S NOTES

To light the bulb, students should connect the bottom of the bulb to one terminal of the battery and the side of the bulb's base to the other terminal. The bulb can be lit with one wire by holding the base of the bulb to one of the battery's terminals and using the wire to connect the side of the bulb's base to the other terminal.

internet connect

*SCi*LINKS

NSTA

TOPIC: Electric circuits
GO TO: www.scilinks.org
*sci*LINKS CODE: HF2201

emf

the energy per unit charge supplied by a source of electric current

The source of potential difference and electrical energy is the circuit's emf

Is it possible to light a bulb using only wires and no battery? Without a potential difference, there is no charge flow and no current. A battery is necessary because the battery is the source of potential difference and electrical energy for the circuit. So the bulb must be connected to a battery to be lit. ★TEKS 5B, 6E

Any device that increases the potential energy of charges circulating in a circuit is a source of **emf.** The emf is the energy per unit charge supplied by a source of electric current. Think of such a source as a "charge pump" that forces electrons to move in a certain direction. Batteries and generators are examples of emf sources.

For conventional current, the terminal voltage is less than the emf

Look at the battery attached to the light bulb in the circuit shown in **Figure 20-4.** As shown in the inset, instead of behaving only like a source of emf, the battery behaves as if it contains both an emf source and a resistor. The battery's internal resistance to current is the result of moving charges colliding with atoms inside the battery while the charges are traveling from one terminal to the other. Thus, when charges move conventionally in a battery, the potential difference across the battery's terminals, the *terminal voltage,* is actually slightly less than the emf.

Unless otherwise stated, any reference in this book to the potential difference across a battery should be thought of as the potential difference measured across the battery's terminals rather than as the emf of the battery. In other words, all examples and end-of-chapter problems will disregard internal resistance.

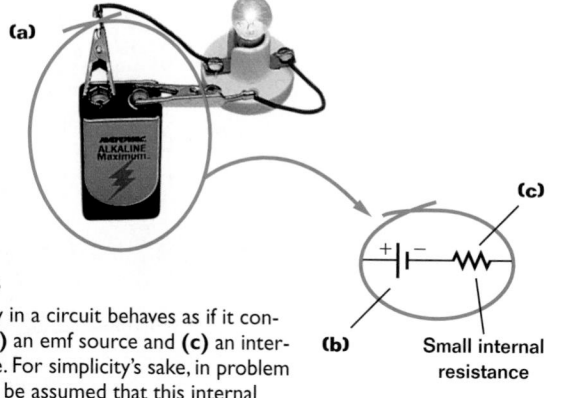

Figure 20-4
(a) A battery in a circuit behaves as if it contains both **(b)** an emf source and **(c)** an internal resistance. For simplicity's sake, in problem solving it will be assumed that this internal resistance is insignificant.

★TEKS 1A, 2A, 6E

Quick Lab

Simple Circuits

MATERIALS LIST

✔ 1 miniature light bulb
✔ 1 D-cell battery
✔ wires
✔ rubber band or tape

⚡ SAFETY CAUTION

Do not perform this lab with any batteries or electrical devices other than those listed here.

Never work with electricity near water. Be sure the floor and all work surfaces are dry.

Connect the bulb to the battery using two wires, using a rubber band or tape to

hold the wire to the battery. Once you have gotten the bulb to light, try different arrangements to see whether there is more than one way to get the bulb to light. Can you make the bulb light using just one wire? Diagram each arrangement that you try, and note whether it produces light.

Explain exactly which parts of the bulb, battery, and wire must be connected for the light bulb to produce light.

Potential difference across a load equals the terminal voltage

When a charge moves within a battery from one terminal to the other, the chemical energy of the battery is converted to the electrical potential energy of the charge. Then, as the charge leaves the battery terminal and moves through the circuit, its electrical potential energy is converted to other forms of energy. For instance, when the load is a resistor, the electrical potential energy of the charge is converted to the internal energy of the resistor and dissipated as thermal energy and light energy. **★TEKS** **5B, 5D**

Because of conservation of energy, the charge must gain as much energy as it loses in one complete trip around the circuit (starting and ending at the same place). Thus, the electrical potential energy gained in the battery must equal the energy dissipated by the load. Because the potential difference is the measurement of potential energy per amount of charge, the potential increase across the battery must equal the potential decrease across the load.

Section Review   **★TEKS** **2C, 6E**

1. Identify the types of elements in the schematic diagram illustrated in **Figure 20-5** and the number of each type.

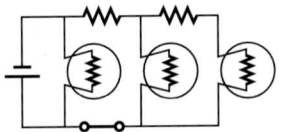

Figure 20-5

2. Using the symbols listed in **Table 20-1,** draw a schematic diagram that contains two resistors, an emf source, and a closed switch.

3. In which of the schematic diagrams shown below will there be no current?

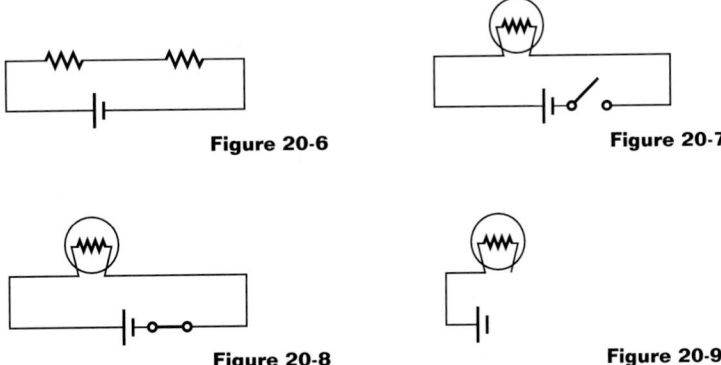

Figure 20-6   **Figure 20-7**

Figure 20-8   **Figure 20-9**

4. If the potential difference across the bulb in a certain flashlight is 3.0 V, what is the potential difference across the combination of batteries used to power it?

5. **Physics in Action**   In what form is the electrical energy that is supplied to a string of decorative lights dissipated?

Key Models and Analogies

From an energy-transformation perspective, think of batteries as electrical-energy-supply devices and of resistors and light bulbs as electrical-energy-consuming devices. The electric current conveys this energy from the battery to the resistor.

Section Review
ANSWERS

1. one battery, one closed switch, two resistors, and three bulbs

2. The circuit should include the circuit elements as they appear in **Table 20-1.**

3. **Figure 20-7** and **Figure 20-9** will have no current in them.

4. 3.0 V

5. It is converted to thermal energy and light energy.

★TEKS

p. 734: 5B, 6E, 1A, 2A, 6E
p. 735: 5B, 5D, 2C, 6E

Section 20-2

20-2
Resistors in series or in parallel

Resistors in series

Purpose Demonstrate that series circuits require all elements to conduct.

Materials 2 flashlight bulbs, bulb holders, battery, battery holder, 3 short pieces of wire

Procedure Wire the bulbs in series with the battery, and point out the lit bulbs. Trace the path for the movement of charges. Ask students to predict what will happen if you unscrew the second bulb. Unscrew it. Point out that the charges no longer have a complete path.

Visual Strategy

Figure 20-10

Point out that even though the resistors are different and must be labeled R_1 and R_2, there is only one value for current.

Q Is the current within the battery less than, equal to, or greater than the circuit current?

A *The current within the battery is the same as the circuit current.*

20-2 SECTION OBJECTIVES

- **Calculate the equivalent resistance for a circuit of resistors in series, and find the current in and potential difference across each resistor in the circuit.**

- **Calculate the equivalent resistance for a circuit of resistors in parallel, and find the current in and potential difference across each resistor in the circuit.**

series

describes a circuit or portion of a circuit that provides a single conducting path without junctions

RESISTORS IN SERIES

In a circuit that consists of a single bulb and a battery, the potential difference across the bulb equals the terminal voltage. The total current in the circuit can be found using the equation $\Delta V = IR$.

What happens when a second bulb is added to such a circuit, as shown in **Figure 20-10**? When moving through this circuit, charges that pass through one bulb must also move through the second bulb. Because all charges in the circuit must follow the same conducting path, these bulbs are said to be connected in **series**. ⍟TEKS **6E**

Resistors in series have the same current

Light-bulb filaments are resistors; thus, **Figure 20-10(b)** represents the two bulbs in **Figure 20-10(a)** as resistors. According to the conservation of charge, charges cannot build up or disappear at a point. For this reason, the amount of charge that enters one bulb in a given time interval equals the amount of charge that exits that bulb in the same amount of time. Because there is only one path for a charge to follow, the amount of charge entering and exiting the first bulb must equal the amount of charge that enters and exits the second bulb in the same time interval. ⍟TEKS **3A, 6E**

Because the current is the amount of charge moving past a point per unit of time, the current in the first bulb must equal the current in the second bulb. This is true for any number of resistors arranged in series. *When many resistors are connected in series, the current in each resistor is the same.*

The total current in a series circuit depends on how many resistors are present and on how much resistance each offers. Thus, to find the total current, first use the individual resistance values to find the total resistance of the circuit, called the *equivalent resistance*. Then the equivalent resistance can be used to find the current.

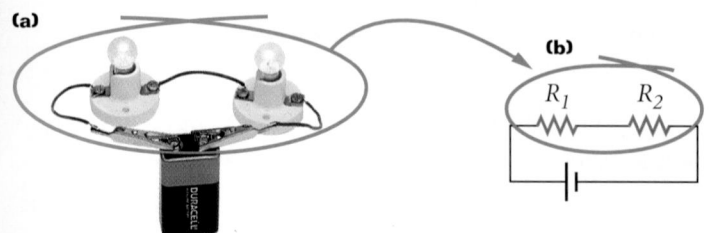

(a)

(b)

R_1 R_2

Figure 20-10
Because light-bulb filaments are resistors, **(a)** the two bulbs in this series circuit can be represented by **(b)** two resistors in the schematic diagram shown on the right.

The equivalent resistance in a series circuit is the sum of the circuit's resistances

As described in Section 20-1, the potential difference across the battery, ΔV, must equal the potential difference across the load, $\Delta V_1 + \Delta V_2$, where ΔV_1 is the potential difference across R_1 and ΔV_2 is the potential difference across R_2.

$$\Delta V = \Delta V_1 + \Delta V_2$$

According to $\Delta V = IR$, the potential difference across each resistor is equal to the current in that resistor multiplied by the resistance.

$$\Delta V = I_1 R_1 + I_2 R_2$$

Because the resistors are in series, the current in each is the same. For this reason, I_1 and I_2 can be replaced with a single variable for the current, I.

$$\Delta V = I(R_1 + R_2)$$

Finding a value for the equivalent resistance of the circuit is now possible. If you imagine the equivalent resistance replacing the original two resistors, as shown in **Figure 20-11,** you can treat the circuit as if it contains only one resistor and use $\Delta V = IR$ to relate the total potential difference, current, and equivalent resistance.

$$\Delta V = I(R_{eq})$$

Now set the last two equations for ΔV equal to each other, and divide by the current.

$$\Delta V = I(R_{eq}) = I(R_1 + R_2)$$

$$R_{eq} = R_1 + R_2$$

Thus, the equivalent resistance of the series combination is the sum of the individual resistances. An extension of this analysis shows that the equivalent resistance of two or more resistors connected in series can be calculated using the following equation. ⊛TEKS **6E**

RESISTORS IN SERIES

$$R_{eq} = R_1 + R_2 + R_3 \ldots$$

Equivalent resistance equals the total of individual resistances in series.

Because R_{eq} represents the sum of the individual resistances that have been connected in series, *the equivalent resistance of a series combination of resistors is always greater than any individual resistance.*

To find the total current in a series circuit, first simplify the circuit to a single equivalent resistance using the boxed equation above, then use $\Delta V = IR$ to calculate the current.

$$I = \frac{\Delta V}{R_{eq}}$$

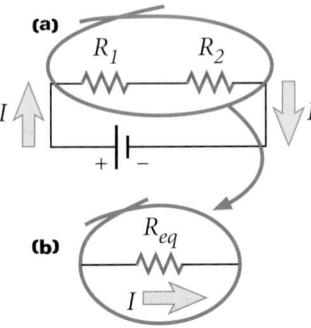

(a)

(b)

Figure 20-11
(a) The two resistors in the actual circuit have the same effect on the current in the circuit as **(b)** the equivalent resistor.

Visual Strategy

Figure 20-11
Be certain students understand what is meant by the idea that the resistor labeled R_{eq} can replace the other two resistors. The current in and potential difference across the equivalent resistor is the same as if the two resistors are taken together.

Q Explain why it was not necessary to label the current in **Figure 20-11(b)** as I_{eq}.

A *The current will be the same in this equivalent resistor as in the original circuit.*

⊛TEKS

p. 736: 6E, 3A, 6E
p. 737: 6E

The following may be used as a teamwork exercise or for demonstration at the board or on an overhead projector.

PROBLEM

Resistors in series

Calculate the equivalent resistance, the current in each resistor, and the potential difference across each resistor if a 24.0 V battery is connected in series to the following:

a. five 2.0 Ω resistors

b. 50 2.0 Ω resistors

c. 500 2.0 Ω resistors

Answers

a. 1.0×10^1 Ω, 2.4 A, 4.8 V
b. 1.0×10^2 Ω, 0.24 A, 0.48 V
c. 1.0×10^3 Ω, 0.024 A, 0.048 V

Alternative Problem-Solving Approach

Because the resistors are in series, the current is the same in each resistor. If I is the value of current, the problem can also be solved by first applying $\Delta V = IR$ to each resistor and then using the sum of potential differences to calculate I:

$$\Delta V_1 = R_1 I = 2.0I$$
$$\Delta V_2 = R_2 I = 4.0I$$
$$\Delta V_3 = R_3 I = 5.0I$$
$$\Delta V_4 = R_4 I = 7.0I$$
$$\Delta V_1 + \Delta V_2 + \Delta V_3 + \Delta V_4 = \Delta V$$
$$\Delta V = 18.0I$$

Now substitute the given value for ΔV:

$$18.0I = 9.0 \text{ V}$$
$$I = 0.50 \text{ A}$$

Because the current in each bulb is equal to the total current, you can also use $\Delta V = IR$ to calculate the potential difference across each resistor.

$$\Delta V_1 = IR_1 \text{ and } \Delta V_2 = IR_2$$

The method described above can be used to find the potential difference across resistors in a series circuit containing any number of resistors.

SAMPLE PROBLEM 20A

Resistors in series ⭐TEKS 2C, 6E

PROBLEM

A 9.0 V battery is connected to four light bulbs, as shown in Figure 20-12. Find the equivalent resistance for the circuit and the current in the circuit.

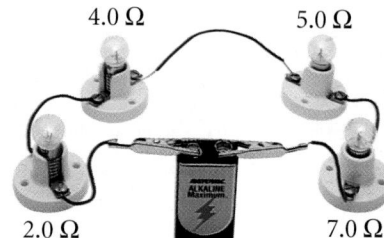

4.0 Ω   5.0 Ω

2.0 Ω   7.0 Ω

Figure 20-12

SOLUTION

1. DEFINE   **Given:** $\Delta V = 9.0$ V $R_1 = 2.0$ Ω
$R_2 = 4.0$ Ω $R_3 = 5.0$ Ω
$R_4 = 7.0$ Ω

Unknown: $R_{eq} = ?$ $I = ?$

Diagram:

4.0 Ω   5.0 Ω   7.0 Ω

2.0 Ω   9.0 V

2. PLAN   **Choose an equation(s) or situation:** Because the resistors are connected end to end, they are in series. Thus, the equivalent resistance can be calculated with the equation from page 737.

$$R_{eq} = R_1 + R_2 + R_3 \ldots$$

The following equation can be used to calculate the current.

$$\Delta V = IR_{eq}$$

Rearrange the equation(s) to isolate the unknown(s): No rearrangement is necessary to calculate R_{eq}, but $\Delta V = IR_{eq}$ must be rearranged to calculate current.

$$I = \frac{\Delta V}{R_{eq}}$$

3. CALCULATE   **Substitute the values into the equation(s) and solve:**

$$R_{eq} = 2.0 \text{ Ω} + 4.0 \text{ Ω} + 5.0 \text{ Ω} + 7.0 \text{ Ω}$$

$$\boxed{R_{eq} = 18.0 \text{ Ω}}$$

Substitute the equivalent resistance value into the equation for current.

$$I = \frac{\Delta V}{R_{eq}} = \frac{9.0 \text{ V}}{18.0 \text{ }\Omega}$$

$$\boxed{I = 0.50 \text{ A}}$$

4. EVALUATE For resistors connected in series, the equivalent resistance should be greater than the largest resistance in the circuit.

$$18.0 \text{ }\Omega > 7.0 \text{ }\Omega$$

PRACTICE 20A

Resistors in series ⭐TEKS 2C, 6E

1. A 12.0 V storage battery is connected to three resistors, 6.75 Ω, 15.3 Ω, and 21.6 Ω, respectively. The resistors are joined in series.

   a. Calculate the equivalent resistance.
   b. What is the current in the circuit?

2. A 4.0 Ω resistor, an 8.0 Ω resistor, and a 12.0 Ω resistor are connected in series with a 24.0 V battery.

   a. Calculate the equivalent resistance.
   b. Calculate the current in the circuit.
   c. What is the current in each resistor?

3. Because the current in the equivalent resistor of Sample Problem 20A is 0.50 A, it must also be the current in each resistor of the original circuit. Find the potential difference across each resistor.

4. A series combination of two resistors, 7.25 Ω and 4.03 Ω, is connected to a 9.00 V battery.

   a. Calculate the equivalent resistance of the circuit and the current.
   b. What is the potential difference across each resistor?

5. A 7.0 Ω resistor is connected in series with another resistor and a 4.5 V battery. The current in the circuit is 0.60 A. Calculate the value of the unknown resistance.

6. Several light bulbs are connected in series across a 115 V source of emf.

   a. What is the equivalent resistance if the current in the circuit is 1.70 A?
   b. If each light bulb has a resistance of 1.50 Ω, how many light bulbs are in the circuit?

PRACTICE GUIDE 20A

Solving for:

R_{eq}	📖 **PE** Sample, 1–2, 4, 6; Ch. Rvw. 16–17
	💿 **PW** Sample, 1–2, 5
	PB 4–6
I	📖 **PE** Sample, 1–2, 4; Ch. Rvw. 17
	💿 **PW** Sample, 4
	PB 7–10
R	📖 **PE** 5
	💿 **PW** 3, 6
	PB Sample, 1–3
ΔV	📖 **PE** 3, 4
	💿 **PW** 4
P	💿 **PW** 7

ANSWERS TO

Practice 20A
Resistors in series

1. **a.** 43.6 Ω
   b. 0.275 A
2. **a.** 24.0 Ω
   b. 1.00 A
   c. 1.00 A
3. 1.0 V, 2.0 V, 2.5 V, 3.5 V
4. **a.** 11.28 Ω, 0.798 A
   b. 5.79 V, 3.22 V
5. 0.5 Ω
6. **a.** 67.6 Ω
   b. 45 bulbs

⭐**TEKS**

p. 738: 2C, 6E
p. 739: 2C, 6E

Visual Strategy

Figure 20-13

Tell students that a bulb is said to *burn out* when its filament breaks. Remind students how a bulb works, referring them to **Figure 20-3** for clarification. Point out that when the filament is broken, charges no longer have a complete pathway from the base of the bulb to the threads.

Q Would the other bulbs light if a wire were attached from the base of the burnt-out bulb to its threads?

A *Yes, charges would have a complete path to follow.*

Visual Strategy

Figure 20-14

Have students trace each of the alternative pathways in their book. Point out that if any of these pathways remains intact, there will be current in the circuit.

Q What parts of the circuit would have current if all of the bulbs except the last one had broken filaments?

A *There would be current in the intact bulb and in the wires that connect the bulb across the potential difference.*

740

Figure 20-13

A burnt-out filament in a bulb has the same effect as an open switch. Because this series circuit is no longer complete, there is no current in the circuit.

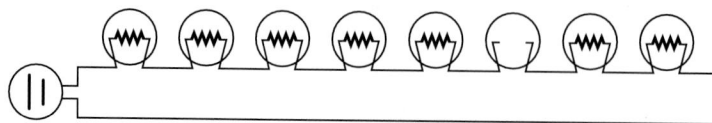

internet connect

SciLINKS

NSTA

TOPIC: Resistors
GO TO: www.scilinks.org
*sci*LINKS CODE: HF2202

parallel

describes two or more components in a circuit that are connected across common points or junctions, providing separate conducting paths for the current

Series circuits require all elements to conduct

What happens to a series circuit when a single bulb burns out? Consider what a circuit diagram for a string of lights with one broken filament would look like. As the schematic diagram in **Figure 20-13** shows, the broken filament means that there is a gap in the conducting pathway used to make up the circuit. Because the circuit is no longer closed, there is no current in it and all of the bulbs go dark. (★)TEKS 3A, 6E

Why, then, would anyone arrange resistors in series? Resistors can be placed in series with a device in order to regulate the current in that device. In the case of decorative lights, adding an additional bulb will decrease the current in each bulb. Thus, the filament of each bulb need not withstand such a high current. Another advantage to placing resistors in series is that several lesser resistances can be used to add up to a single greater resistance that is unavailable. Finally, in some cases, it is important to have a circuit that will have no current if any one of its component parts fails. This technique is used in a variety of contexts, including some burglar alarm systems. (★)TEKS 3C

RESISTORS IN PARALLEL

As discussed above, when a single bulb in a series light set burns out, the entire string of lights goes dark because the circuit is no longer closed. What would happen if there were alternative pathways for the movement of charge, as shown in **Figure 20-14**?

A wiring arrangement that provides alternative pathways for the movement of a charge is a **parallel** arrangement. The bulbs of the decorative light set shown in the schematic diagram in **Figure 20-14** are arranged in parallel with each other. (★)TEKS 6E

Resistors in parallel have the same potential differences across them

To explore the consequences of arranging resistors in parallel, consider the two bulbs connected to a battery in **Figure 20-15(a).** In this arrangement, the left

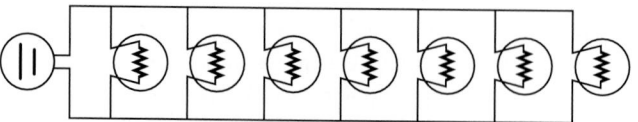

Figure 20-14

These decorative lights are wired in parallel. Notice that in a parallel arrangement there is more than one path for current.

side of each bulb is connected to the positive terminal of the battery, and the right side of each bulb is connected to the negative terminal. Because the sides of each bulb are connected to common points, the potential difference across each bulb is the same. If the common points are the battery's terminals, as they are in the figure, the potential difference across each resistor is also equal to the terminal voltage of the battery. The current in each bulb, however, is not always the same. ⭐TEKS **3A, 6E**

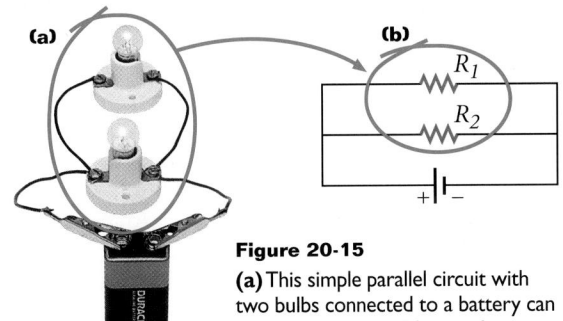

Figure 20-15
(a) This simple parallel circuit with two bulbs connected to a battery can be represented by **(b)** the schematic diagram shown on the right.

The sum of currents in parallel resistors equals the total current

In **Figure 20-15,** when a certain amount of charge leaves the positive terminal and reaches the branch on the left side of the circuit, some of the charge moves through the top bulb and some moves through the bottom bulb. If one of the bulbs has less resistance, more charge moves through that bulb because the bulb offers less opposition to the flow of charges.

Because charge is conserved, the sum of the currents in each bulb equals the current delivered by the battery. This is true for all resistors in parallel.

$$I = I_1 + I_2 + I_3 \dots$$

The parallel circuit shown in **Figure 20-15** can be simplified to an equivalent resistance with a method similar to the one used for series circuits. To do this, first show the relationship among the currents.

$$I = I_1 + I_2$$

Then substitute the equivalents for current according to $\Delta V = IR$.

$$\frac{\Delta V}{R_{eq}} = \frac{\Delta V_1}{R_1} + \frac{\Delta V_2}{R_2}$$

⭐TEKS **2A, 6E**

Quick Lab

Series and Parallel Circuits

MATERIALS LIST

✔ 4 regular drinking straws
✔ 4 stirring straws or coffee stirrers
✔ tape

Cut the regular drinking straws and thin stirring straws into equal lengths. Tape them end to end in long tubes to form series combinations. Form parallel combinations by taping the straws together side by side.

Try several combinations of like and unlike straws. Blow through each combination of tubes, holding your fingers in front of the opening(s) to compare the air flow (or current) that you achieve with each combination.

Straws in series

Straws in parallel

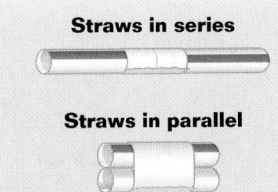

Rank the combinations according to how much resistance they offer. Classify them according to the amount of current created in each.

Visual Strategy

Figure 20-15
Working through a diagram like **Figure 20-15(b)** with numerical examples may help students understand the relationships for current in a parallel circuit.

Q Assume that I from the battery = 5 A and I_1 = 2 A. What must I_2 be?

A **3 A**

Demonstration 2

Resistors in parallel

Purpose Demonstrate that parallel circuits do not require all elements to conduct.

Materials two flashlight bulbs, bulb holders, battery, battery holder, four short pieces of wire

Procedure Connect the bulbs in series with the battery as shown in **Figure 20-15.** Trace the path for the movement of the charges. Ask students to predict what will happen if you unscrew the second bulb. Unscrew it. Point out that the charges still have a complete path in the other bulb.

Quick Lab

TEACHER'S NOTES

For this lab to be effective, it is very important that the straws be taped together. Crimping one end of a straw and stuffing it into another straw will not work well.

ANSWERS TO

Conceptual Challenge

1. Car headlights must be wired in parallel so that if one burns out, the other will stay lit. If they were wired in series, the brightness would diminish.

2. There are four possible circuits: all resistors in series, all resistors in parallel, one resistor in series with two others in parallel, and one resistor in parallel with two others in series.

Misconception Alert

Use a simple numerical example to demonstrate that mathematically adding the inverses is not the same as taking the inverse of the sum.

Correct: $\frac{1}{2} + \frac{1}{3} = \frac{5}{6}$, $R_{eq} = \frac{6}{5}$

Incorrect: $2 + 3 = 5$, $R_{eq} \neq \frac{1}{5}$

Key Models and Analogies

Resistors can be compared to construction areas that slow down traffic on a highway.

Having resistors in series is similar to having more than one construction area on the same stretch of highway. All cars will have to pass through all the construction areas.

Having resistors in parallel is like having more than one possible route, each route containing a construction area. Because the cars can go through different routes, the overall effect is less of a slowdown.

Conceptual Challenge

1. Car headlights

How can you tell that the headlights on a car are wired in parallel rather than in series? How would the brightness of the bulbs differ if they were wired in series across the same 12 V battery instead of in parallel?

2. Simple circuits

Sketch as many different circuits as you can using three light bulbs—each of which has the same resistance—and a battery.

★TEKS 3A, 6E

Because the potential difference across each bulb in a parallel arrangement equals the terminal voltage ($\Delta V = \Delta V_1 = \Delta V_2$), you can divide each side of the equation by ΔV to get the following equation.

$$\frac{1}{R_{eq}} = \frac{1}{R_1} + \frac{1}{R_2}$$

An extension of this analysis shows that the equivalent resistance of two or more resistors connected in parallel can be calculated using the following equation. **★TEKS** 6E

RESISTORS IN PARALLEL

$$\frac{1}{R_{eq}} = \frac{1}{R_1} + \frac{1}{R_2} + \frac{1}{R_3} \ldots$$

The equivalent resistance of resistors in parallel can be calculated using a reciprocal relationship.

Notice that this equation does not give the value of the equivalent resistance directly. You must take the reciprocal of your answer to obtain the value of the equivalent resistance.

Because of the reciprocal relationship, *the equivalent resistance for a parallel arrangement of resistors must always be less than the smallest resistance in the group of resistors.*

The conclusions made about both series and parallel circuits are summarized in **Table 20-2.**

Table 20-2   Resistors in series or in parallel

Circuit type	Series	Parallel
schematic diagram	●—ⱮⱮ—ⱮⱮ—●	●—[ⱮⱮ ⱮⱮ]—●
current	$I = I_1 = I_2 = I_3 \ldots$ = same for each resistor	$I = I_1 + I_2 + I_3 \ldots$ = sum of currents
potential difference	$\Delta V = \Delta V_1 + \Delta V_2 + \Delta V_3 \ldots$ = sum of potential differences	$\Delta V = \Delta V_1 = \Delta V_2 = \Delta V_3 \ldots$ = same for each resistor
equivalent resistance	$R_{eq} = R_1 + R_2 + R_3 \ldots$ = sum of individual resistances	$\frac{1}{R_{eq}} = \frac{1}{R_1} + \frac{1}{R_2} + \frac{1}{R_3} \ldots$ = reciprocal sum of resistances

Resistors in parallel ⭐TEKS 2C, 6E

PROBLEM

A 9.0 V battery is connected to four resistors, as shown in Figure 20-16. Find the equivalent resistance for the circuit and the total current in the circuit.

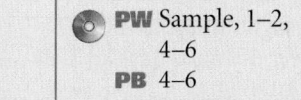
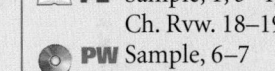

Figure 20-16

SOLUTION

1. DEFINE   **Given:** $\Delta V = 9.0 \text{ V}$ $R_1 = 2.0 \ \Omega$
$R_2 = 4.0 \ \Omega$ $R_3 = 5.0 \ \Omega$
$R_4 = 7.0 \ \Omega$

Unknown: $R_{eq} = ? \quad I = ?$

Diagram:

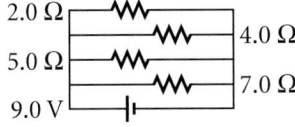

2. PLAN   **Choose an equation(s) or situation:** Because both sides of each resistor are connected to common points, they are in parallel. Thus, the equivalent resistance can be calculated with the equation from p. 742.

$$\frac{1}{R_{eq}} = \frac{1}{R_1} + \frac{1}{R_2} + \frac{1}{R_3} \ldots \text{ for parallel}$$

The following equation can be used to calculate the current.

$$\Delta V = IR_{eq}$$

Rearrange the equation(s) to isolate the unknown(s): No rearrangement is necessary to calculate R_{eq}; rearrange $\Delta V = IR_{eq}$ to calculate current.

$$I = \frac{\Delta V}{R_{eq}}$$

3. CALCULATE   **Substitute the values into the equation(s) and solve:** Remember to take the reciprocal in the final step.

$$\frac{1}{R_{eq}} = \frac{1}{2.0 \ \Omega} + \frac{1}{4.0 \ \Omega} + \frac{1}{5.0 \ \Omega} + \frac{1}{7.0 \ \Omega}$$

$$\frac{1}{R_{eq}} = \frac{0.50}{1 \ \Omega} + \frac{0.25}{1 \ \Omega} + \frac{0.20}{1 \ \Omega} + \frac{0.14}{1 \ \Omega} = \frac{1.09}{1 \ \Omega}$$

$$R_{eq} = \frac{1 \ \Omega}{1.09}$$

$$\boxed{R_{eq} = 0.917 \ \Omega}$$

continued on next page

Classroom Practice

The following may be used as a teamwork exercise or for demonstration at the board or on an overhead projector.

PROBLEM

Resistors in parallel

Find the equivalent resistance, the current in each resistor, and the current drawn by the circuit load for a 9.0 V battery connected in parallel to three 30.0 Ω resistors.

Answer
10.0 Ω, 0.30 A, 0.90 A

PRACTICE GUIDE 20B		
Solving for:		
R_{eq}	📖 PE	Sample, 2–4; Ch. Rvw. 18–19
	💿 PW	Sample, 1–2, 4–6
	PB	4–6
I	📖 PE	Sample, 1, 3–4; Ch. Rvw. 18–19
	💿 PW	Sample, 6–7
	PB	7–10
R	💿 PW	3
	PB	Sample, 1–3
ΔV	📖 PE	4b

⭐TEKS

p. 742: 6E, 3A, 6E
p. 743: 2C, 6E

Alternative Problem-Solving Approach

The problem can also be solved by applying $\Delta V = IR$ to each resistor to find its current, then adding these to get the total current. Finally, use $R_{eq} = \dfrac{\Delta V}{I_{tot}}$ to find R_{eq}.

$$I_1 = \frac{\Delta V}{R_1} = \frac{9.0 \text{ V}}{2.0 \text{ }\Omega} = 4.5 \text{ A}$$

$$I_2 = \frac{\Delta V}{R_2} = \frac{9.0 \text{ V}}{4.0 \text{ }\Omega} = 2.2 \text{ A}$$

$$I_3 = \frac{\Delta V}{R_3} = \frac{9.0 \text{ V}}{5.0 \text{ }\Omega} = 1.8 \text{ A}$$

$$I_4 = \frac{\Delta V}{R_4} = \frac{9.0 \text{ V}}{7.0 \text{ }\Omega} = 1.3 \text{ A}$$

$$I_{tot} = I_1 + I_2 + I_3 + I_4$$

$$I_{tot} = 9.8 \text{ A}$$

$$R_{eq} = \frac{9.0 \text{ V}}{9.8 \text{ A}} = 0.92 \text{ }\Omega$$

The slight difference in the answer obtained this way is due to rounding.

ANSWERS TO

Practice 20B
Resistors in parallel
1. 4.5 A, 2.2 A, 1.8 A, 1.3 A
2. 50.0 Ω
3. a. 2.2 Ω
   b. 6.0 A, 3.0 A, 2.00 A
4. a. 2.99 Ω
   b. 36.0 V
   c. 2.00 A, 6.00 A

Substitute that equivalent resistance value in the equation for current.

$$I = \frac{\Delta V_{tot}}{R_{eq}} = \frac{9.0 \text{ V}}{0.917 \text{ }\Omega}$$

$$\boxed{I = 9.8 \text{ A}}$$

4. EVALUATE For resistors connected in parallel, the equivalent resistance should be less than the smallest resistance.

$$0.917 \text{ }\Omega < 2.0 \text{ }\Omega$$

CALCULATOR SOLUTION

The calculator answer is 9.814612868, but because the potential difference, 9.0 V, has only two significant digits, the answer is reported as 9.8 A.

PRACTICE 20B

Resistors in parallel ⭐TEKS **2C, 6E**

1. The potential difference across the equivalent resistance in Sample Problem 20B equals the potential difference across each of the individual parallel resistors. Calculate the value for the current in each resistor.

2. A length of wire is cut into five equal pieces. The five pieces are then connected in parallel, with the resulting resistance being 2.00 Ω. What was the resistance of the original length of wire before it was cut up?

3. A 4.0 Ω resistor, an 8.0 Ω resistor, and a 12.0 Ω resistor are connected in parallel across a 24.0 V battery.
   a. What is the equivalent resistance of the circuit?
   b. What is the current in each resistor?

4. An 18.0 Ω, 9.00 Ω, and 6.00 Ω resistor are connected in parallel to an emf source. A current of 4.00 A is in the 9.00 Ω resistor.
   a. Calculate the equivalent resistance of the circuit.
   b. What is the potential difference across the source?
   c. Calculate the current in the other resistors.

Parallel circuits do not require all elements to conduct

What happens when a bulb burns out in a string of decorative lights that is wired in parallel? There is no current in that branch of the circuit, but each of the parallel branches provides a separate alternative pathway for current. Thus, the potential difference supplied to the other branches and the current in these branches remain the same, and the bulbs in these branches remain lit.

⭐TEKS **3A, 6E**

When resistors are wired in parallel with an emf source, the potential difference across each resistor always equals the potential difference across the source. Because household circuits are arranged in parallel, appliance manufacturers are able to standardize their design, producing devices that all operate at the same potential difference. (Because the potential difference provided by a wall outlet in a home in North America is not the same as the potential difference that is standard on other continents, appliances made in North America are not always compatible with wall outlets in homes on other continents.)

Because household devices operate at the same potential difference, manufacturers can choose the resistance of the device to ensure that the current in the device will be neither too high nor too low for the internal wiring and other components that make up the device. (★)TEKS 3C, 6E

Additionally, the equivalent resistance of several parallel resistors is less than the resistance of any of the individual resistors. Thus, a low equivalent resistance can be created with a group of resistors of higher resistances.

Section Review (★)TEKS 2C, 6E

1. Two resistors are wired in series. In another circuit, the same two resistors are wired in parallel. In which circuit is the equivalent resistance greater?

2. A 5 Ω, a 10 Ω, and a 15 Ω resistor are connected in series.
   a. Which resistor has the most current in it?
   b. Which resistor has the largest potential difference across it?

3. A 5 Ω, a 10 Ω, and a 15 Ω resistor are connected in parallel.
   a. Which resistor has the most current in it?
   b. Which resistor has the largest potential difference across it?

4. Find the current in and potential difference across each of the resistors in the following circuits:
   a. a 2.0 Ω and a 4.0 Ω resistor wired in series with a 12 V source
   b. a 2.0 Ω and a 4.0 Ω resistor wired in parallel with a 12 V source
   c. a 4.0 Ω and a 12.0 Ω resistor wired in series with a 4.0 V source
   d. a 4.0 Ω and a 12.0 Ω resistor wired in parallel with a 4.0 V source

5. Find the current in and potential difference across each of the resistors in the following circuits:
   a. a 150 Ω and a 180 Ω resistor wired in series with a 12 V source
   b. a 150 Ω and a 180 Ω resistor wired in parallel with a 12 V source

6. **Physics in Action** A string of 35 miniature decorative lights is wired in series. If it draws 0.20 A of current when it is connected to a 120.0 V emf source, what is the resistance of each miniature bulb?

Section Review
ANSWERS

1. in the series circuit
2. a. All have equal I.
   b. 15 Ω
3. a. 5 Ω
   b. All have equal ΔV.
4. a. 2.0 Ω: 2.0 A, 4.0 V
   4.0 Ω: 2.0 A, 8.0 V
   b. 2.0 Ω: 6.0 A, 12 V
   4.0 Ω: 3.0 A, 12 V
   c. 4.0 Ω: 0.25 A, 1.0 V
   12.0 Ω: 0.25 A, 3.0 V
   d. 4.0 Ω: 1.0 A, 4.0 V
   12.0 Ω: 0.33 A, 4.0 V
5. a. 150 Ω: 0.036 A, 5.4 V
   180 Ω: 0.036 A, 6.5 V
   b. 150 Ω: 0.080 A, 12 V
   180 Ω: 0.067 A, 12 V
6. 17 Ω

(★)TEKS

p. 744: 2C, 6E
p. 745: 3C, 6E, 2C, 6E

20-3
Complex resistor combinations

Figure 20-17

Explain that the circuit breaker is shown as a switch because it contains a switch that opens when the current in the circuit becomes too large.

Q How much of the total current is in the circuit breaker when it is in series with the parallel combination of devices shown?

A *All of the total current is in the circuit breaker when it is in series.*

Teaching Tip

Point out that each device added in parallel to a circuit draws more current from the emf source.

Mathematically, because of the following equation, the more resistors that are added in parallel, the more current there will be in the main wires of the circuit.

$$I_{tot} = \frac{\Delta V}{R_1} + \frac{\Delta V}{R_2} \dots$$

If the current is too great, the main wires, plugs, and outlet connections will heat up. Excessive current can damage equipment and can even cause fires.

20-3 SECTION OBJECTIVES

- Calculate the equivalent resistance for a complex circuit involving both series and parallel portions.

- Calculate the current in and potential difference across individual elements within a complex circuit.

RESISTORS COMBINED BOTH IN PARALLEL AND IN SERIES

Series and parallel circuits are not often encountered independent of one another. Most circuits today employ both series and parallel wiring to utilize the advantages of each type.

A common example of a complex circuit is the electrical wiring typical in a home. In a home, a fuse or circuit breaker is connected in series to numerous outlets, which are wired to one another in parallel. An example of a typical household circuit is shown in **Figure 20-17**. (★)TEKS 6E

As a result of the outlets being wired in parallel, all the appliances operate independently; if one is switched off, any others remain on. Wiring the outlets in parallel ensures that an identical potential difference exists across any appliance. This way, appliance manufacturers can produce appliances that all use the same standard potential difference.

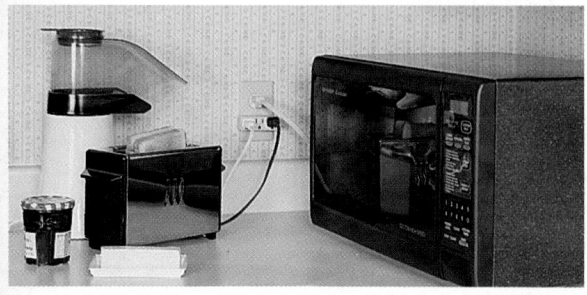

(a)

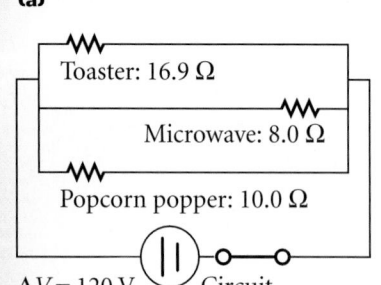

(b)

Figure 20-17

(a) When all of these devices are plugged into the same household circuit, **(b)** the result is a parallel combination of resistors in series with a circuit breaker.

To prevent excessive current, a fuse or circuit breaker must be placed in series with all of the outlets. Fuses and circuit breakers open the circuit when the current becomes too high. A fuse is a small metallic strip that melts if the current exceeds a certain value. After a fuse has melted, it must be replaced. A circuit breaker, a more modern device, triggers a switch when current reaches a certain value. The switch must be reset, rather than replaced, after the circuit overload has been removed. Both fuses and circuit breakers must be in series with the entire load to prevent excessive current from reaching any appliance. In fact, if all the devices in **Figure 20-17** were used at once, the circuit would be overloaded. The circuit breaker would interrupt the current.

Fuses and circuit breakers are carefully selected to meet the demands of a circuit. If the circuit is to carry currents as large as 30 A, an appropriate fuse or circuit breaker must be used. Because the fuse or circuit breaker is placed in series with the rest of the circuit, the current in the fuse or circuit breaker is the same as the total current in the circuit. To find this current, one must determine the equivalent resistance.

When determining the equivalent resistance for a complex circuit, you must simplify the circuit into groups of series and parallel resistors and then find the equivalent resistance for each group by using the rules for finding the equivalent resistance of series and parallel resistors.

Equivalent resistance ★TEKS 2C, 6E

PROBLEM

Determine the equivalent resistance of the complex circuit shown in Figure 20-18.

REASONING

The best approach is to divide the circuit into groups of series and parallel resistors. This way the methods presented in Sample Problems 20B and 20C can be used to calculate the equivalent resistance for each group.

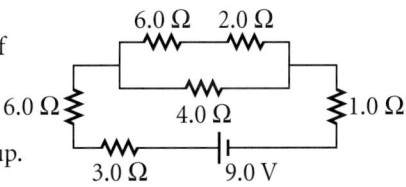

Figure 20-18

SOLUTION

1. **Redraw the circuit as a group of resistors along one side of the circuit.**

Because bends in a wire do not affect the circuit, they do not need to be represented in a schematic diagram. Redraw the circuit without the corners, keeping the arrangement of the circuit elements the same, as shown in **Figure 20-19**. For now, disregard the emf source and work only with the resistances.

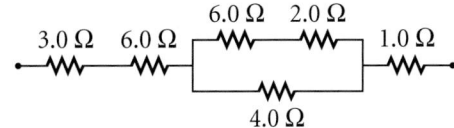

Figure 20-19

2. **Identify components in series, and calculate their equivalent resistance.**

Resistors in groups (**a**) and (**b**) in **Figure 20-20** are in series.

For group (**a**): $R_{eq} = 3.0\ \Omega + 6.0\ \Omega = 9.0\ \Omega$

For group (**b**): $R_{eq} = 6.0\ \Omega + 2.0\ \Omega = 8.0\ \Omega$

3. **Identify components in parallel, and calculate their equivalent resistance.**

Resistors in group (**c**) in **Figure 20-20** are in parallel.

For group (**c**):

$$\frac{1}{R_{eq}} = \frac{1}{8.0\ \Omega} + \frac{1}{4.0\ \Omega} = \frac{0.12}{1\ \Omega} + \frac{0.25}{1\ \Omega} = \frac{0.37}{1\ \Omega}$$

$$R_{eq} = 2.7\ \Omega$$

4. **Repeat steps 2 and 3 until the resistors in the circuit are reduced to a single equivalent resistance.**

The remainder of the resistors, group (**d**) in **Figure 20-20**, are in series.

For group (**d**): $R_{eq} = 9.0\ \Omega + 2.7\ \Omega + 1.0\ \Omega$

$$\boxed{R_{eq} = 12.7\ \Omega}$$

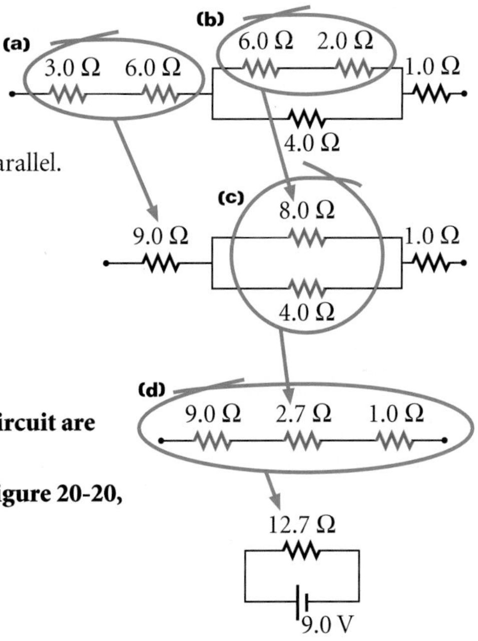

5. **Redraw the circuit as in Figure 20-20, with the equivalent resistance connected to the original emf source.**

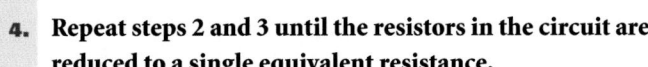

Figure 20-20

Classroom Practice

The following may be used as a teamwork exercise or for demonstration at the board or on an overhead projector.

PROBLEM

Equivalent resistance

Use the following values with the circuit in **Figure 20-21** on p. 748. What is the equivalent resistance for each circuit?

a. $R_a = 5.0\ \Omega$, $R_b = 3.0\ \Omega$, $R_c = 6.0\ \Omega$

b. $R_a = 6.0\ \Omega$, $R_b = 8.0\ \Omega$, $R_c = 2.0\ \Omega$

Answers

a. $7.0\ \Omega$

b. $7.6\ \Omega$

PRACTICE GUIDE 20C		
Solving for:		
R_{eq}	📖 **PE**	Sample, 1–2; Ch. Rvw. 23–24
	💿 **PW**	Sample, 1, 4–5
	PB	4–6
I	💿 **PW**	2–3
	PB	7–10
R	💿 **PW**	3
	PB	Sample, 1–3
P	💿 **PB**	4, 6

★TEKS

p. 746: 6E
p. 747: 2C, 6E

ANSWERS TO

Practice 20C
Equivalent resistance

1. a. 27.8 Ω
    b. 26.6 Ω
    c. 23.4 Ω
2. a. 50.9 Ω
    b. 57.6 Ω

Alternative Problem-Solving Approach

Point out that it doesn't matter in what order the operations of simplifying the circuit are done, as long as they are done so that the simpler equivalent circuits still have the same current in and potential difference across the load.

Students should be encouraged to suggest and examine alternative ways and sequences for grouping the resistors.

For example, you could first find the equivalent resistance of the 6.0 Ω and 2.0 Ω resistors shown as group (**b**) in **Figure 20-20,** then find the equivalent resistance of the 8.0 Ω and 4.0 Ω resistors shown as group (**c**). The result would be four resistors in series: 6.0 Ω, 3.0 Ω, 2.7 Ω, and 1.0 Ω. The equivalent resistance of these four resistors is 12.7 Ω.

PHYSICS **Interactive Problem-Solving Tutor**

See Module 17
"Electric Circuits" provides additional development of problem-solving skills for this chapter.

PRACTICE 20C

Equivalent resistance ⭐TEKS 2C, 6E

1. For each of the following sets of values, determine the equivalent resistance for the circuit shown in Figure 20-21.

    a. $R_a = 25.0\ \Omega$ $R_b = 3.0\ \Omega$ $R_c = 40.0\ \Omega$
    b. $R_a = 12.0\ \Omega$ $R_b = 35.0\ \Omega$ $R_c = 25.0\ \Omega$
    c. $R_a = 15.0\ \Omega$ $R_b = 28.0\ \Omega$ $R_c = 12.0\ \Omega$

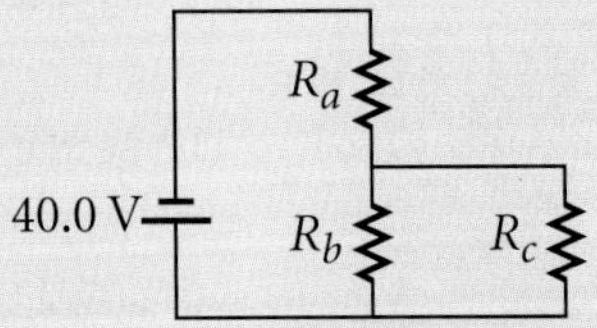

    40.0 V

    Figure 20-21

2. For each of the following sets of values, determine the equivalent resistance for the circuit shown in Figure 20-22.

    a. $R_a = 25.0\ \Omega$ $R_b = 3.0\ \Omega$ $R_c = 40.0\ \Omega$
       $R_d = 15.0\ \Omega$ $R_e = 18.0\ \Omega$

    b. $R_a = 12.0\ \Omega$ $R_b = 35.0\ \Omega$ $R_c = 25.0\ \Omega$
       $R_d = 50.0\ \Omega$ $R_e = 45.0\ \Omega$

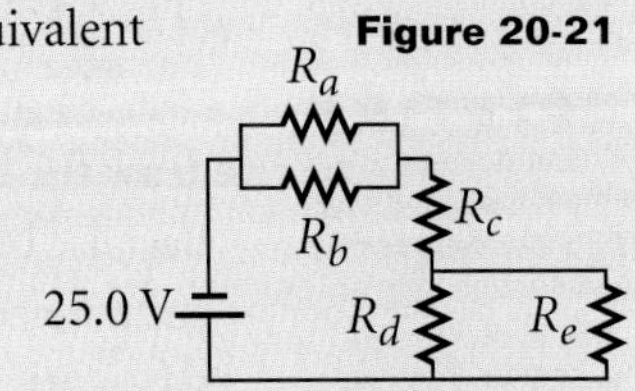

    25.0 V

    Figure 20-22

Module 20
"Electric Circuits"
provides an interactive lesson with guided problem-solving practice to teach you about many kinds of electric circuits, including complex combinations of resistors.

Work backward to find the current in and potential difference across a part of a circuit

Now that the equivalent resistance for a complex circuit has been determined, you can work backward to find the current in and potential difference across any resistor in that circuit. In the household example, substitute potential difference and equivalent resistance in $\Delta V = IR$ to find the total current in the circuit. Because the fuse or circuit breaker is in series with the load, the current in it is equal to the total current. Once this total current is determined, $\Delta V = IR$ can again be used to find the potential difference across the fuse or circuit breaker.

There is no single formula for finding the current in and potential difference across a resistor buried inside a complex circuit. Instead, $\Delta V = IR$ and the rules reviewed in **Table 20-3** must be applied to smaller pieces of the circuit until the desired values are found.

Table 20-3	Series and parallel resistors	
	Series	**Parallel**
current	same as total	add to find total
potential difference	add to find total	same as total

Current in and potential difference across a resistor ★TEKS 2C, 6E

PROBLEM

Determine the current in and potential difference across the 2.0 Ω resistor highlighted in Figure 20-23.

REASONING

First determine the total circuit current by reducing the resistors to a single equivalent resistance. Then rebuild the circuit in steps, calculating the current and potential difference for the equivalent resistance of each group until the current in and potential difference across the 2.0 Ω resistor are known.

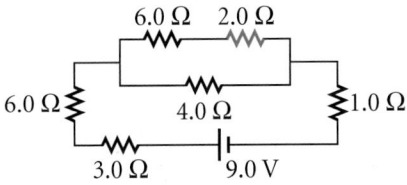

Figure 20-23

SOLUTION

1. **Determine the equivalent resistance of the circuit.**
   The equivalent resistance of the circuit is 12.7 Ω; this value is calculated in Sample Problem 20C.

2. **Calculate the total current in the circuit.**
   Substitute the potential difference and equivalent resistance in $\Delta V = IR$, and rearrange the equation to find the current delivered by the battery.

   $$I = \frac{\Delta V}{R} = \frac{9.0 \text{ V}}{12.7 \text{ Ω}} = 0.71 \text{ A}$$

3. **Determine a path from the equivalent resistance found in step 1 to the 2.0 Ω resistor.**
   Review the path taken to find the equivalent resistance in **Figure 20-24,** and work backward through this path. The equivalent resistance for the entire circuit is the same as the equivalent resistance for group (**d**). The center resistor in group (**d**) in turn is the equivalent resistance for group (**c**). The top resistor in group (**c**) is the equivalent resistance for group (**b**), and the right resistor in group (**b**) is the 2.0 Ω resistor.

4. **Follow the path determined in step 3, and calculate the current in and potential difference across each equivalent resistance. Repeat this process until the desired values are found.**

   Regroup, evaluate, and calculate.
   Replace the circuit's equivalent resistance with group

continued on next page

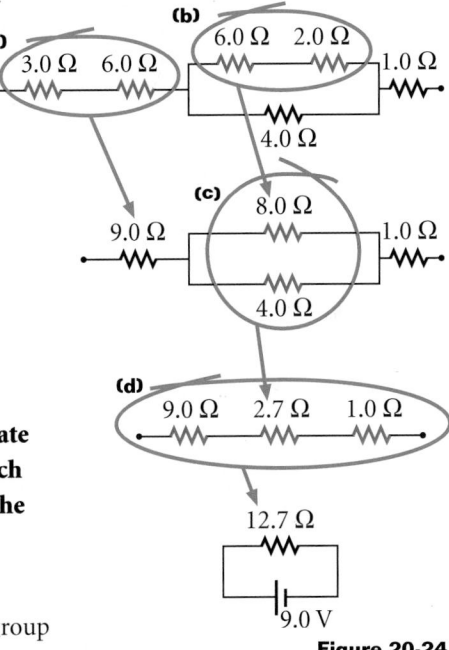

Figure 20-24

🛑 **Misconception Alert**

At first, students may believe that it is unnecessary to solve for R_{eq} first and then work back to find current in or potential difference across a particular resistor. Point out that working through these steps keeps the mathematical operations at each step simpler.

Classroom Practice

The following may be used as a teamwork exercise or for demonstration at the board or on an overhead projector.

PROBLEM

Current in and potential difference across a resistor

Use the following values with the circuit in **Figure 20-28** on p. 751. What is the current in and potential difference across each of the resistors?

$R_a = 8.0$ Ω, $R_b = 4.0$ Ω, $R_c = 6.0$ Ω, $R_d = 3.0$ Ω, $R_e = 9.0$ Ω, $R_f = 7.0$ Ω

Answers
$I_a = 0.35$ A, $\Delta V_a = 2.8$ V
$I_b = 0.35$ A, $\Delta V_b = 1.4$ V
$I_c = 0.70$ A, $\Delta V_c = 4.2$ V
$I_d = 0.80$ A, $\Delta V_d = 2.4$ V
$I_e = 0.27$ A, $\Delta V_e = 2.4$ V
$I_f = 1.05$ A, $\Delta V_f = 7.4$ V

★TEKS
p. 748: 2C, 6E
p. 749: 2C, 6E

Remind students that they can check each step by using $\Delta V = IR$ for each resistor in a set. They can also check the sum of ΔV for series circuits and the sum of I for parallel circuits.

For **Figure 20-25,** the potential difference across the 2.7 Ω resistor is 1.9 V. For the other two resistors in series in group **(d):**

$\Delta V = (0.71$ A$)(9.0$ Ω$) = 6.4$ V
$\Delta V = (0.71$ A$)(1.0$ Ω$) = 0.71$ V

The total ΔV across group **(d)** matches the terminal voltage.

1.9 V + 6.4 V + 0.71 V = 9.0 V

For **Figure 20-26,** the current across the 8.0 Ω resistor is 0.24 A. For the other resistor in group **(c):**

$$I = \frac{1.9 \text{ V}}{4.0 \text{ Ω}} = 0.48 \text{ A}$$

The total of these currents is 0.72 A, which differs from 0.71 A because of rounding.

For **Figure 20-27,** the potential difference across the 2.0 Ω resistor is 0.48 V. For the other resistor:

$\Delta V = (0.24$ A$)(6.0$ Ω$) = 1.4$ V

The total of these potential differences is 1.9 V, which was given in the previous step.

(d), as shown in **Figure 20-25.** The resistors in group **(d)** are in series; therefore, the current in each resistor is the same as the current in the equivalent resistance, which equals 0.71 A. The potential difference across the 2.7 Ω resistor in group **(d)** can be calculated using $\Delta V = IR$.

Given: $I = 0.71$ A $R = 2.7$ Ω

Unknown: $\Delta V = ?$

$\Delta V = IR = (0.71$ A$)(2.7$ Ω$) = 1.9$ V

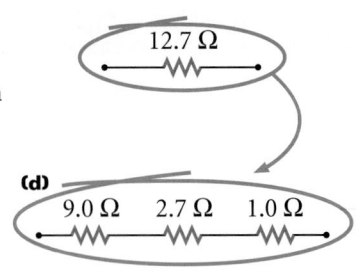

Figure 20-25

Regroup, evaluate, and calculate.

Replace the center resistor with group **(c),** as shown in **Figure 20-26.**

The resistors in group **(c)** are in parallel; therefore, the potential difference across each resistor is the same as the potential difference across the 2.7 Ω equivalent resistance, which equals 1.9 V. The current in the 8.0 Ω resistor in group **(c)** can be calculated using $\Delta V = IR$.

Given: $\Delta V = 1.9$ V $R = 8.0$ Ω

Unknown: $I = ?$

$$I = \frac{\Delta V}{R} = \frac{1.9 \text{ V}}{8.0 \text{ Ω}} = 0.24 \text{ A}$$

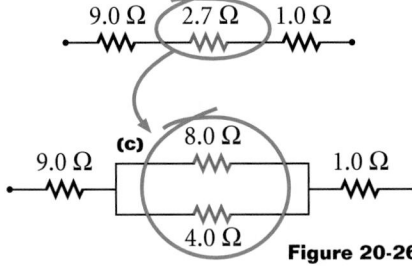

Figure 20-26

Regroup, evaluate, and calculate.

Replace the 8.0 Ω resistor with group **(b),** as shown in **Figure 20-27.**

The resistors in group **(b)** are in series; therefore, the current in each resistor is the same as the current in the 8.0 Ω equivalent resistance, which equals 0.24 A.

$I = 0.24$ A

The potential difference across the 2.0 Ω resistor can be calculated using $\Delta V = IR$.

Given: $I = 0.24$ A $R = 2.0$ Ω

Unknown: $\Delta V = ?$

$\Delta V = IR = (0.24$ A$)(2.0$ Ω$) = 0.48$ V

$\Delta V = 0.48$ V

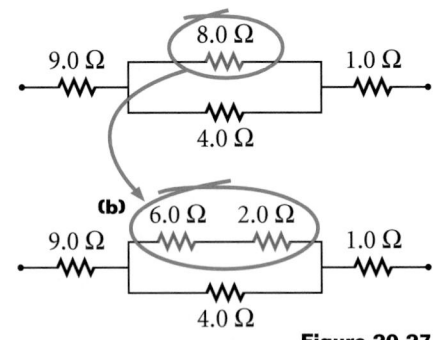

Figure 20-27

Current in and potential difference across a resistor ⊛TEKS 2C, 6E

Calculate the current in and potential difference across each of the resistors shown in the schematic diagram in **Figure 20-28**.

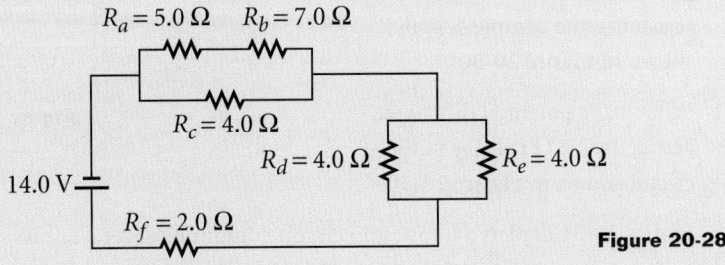

$R_a = 5.0\ \Omega$ $R_b = 7.0\ \Omega$

$R_c = 4.0\ \Omega$

$R_d = 4.0\ \Omega$ $R_e = 4.0\ \Omega$

14.0 V

$R_f = 2.0\ \Omega$

Figure 20-28

SECTION 20-3

PRACTICE GUIDE 20D

Solving for:		
I	📖 **PE**	Sample, Practice; Ch. Rvw. 25–26
	💿 **PW**	Sample, 3
	PB	Sample, 1–10
ΔV	📖 **PE**	Sample, Practice; Ch. Rvw. 25–26
	💿 **PW**	Sample, 1–3
	PB	Sample, 1–10

ANSWERS TO

Practice 20D
Current in and potential difference across a resistor
R_a: 0.50 A, 2.5 V
R_b: 0.50 A, 3.5 V
R_c: 1.5 A, 6.0 V
R_d: 1.0 A, 4.0 V
R_e: 1.0 A, 4.0 V
R_f: 2.0 A, 4.0 V

Consumer Focus    *Decorative Lights and Bulbs*

Light sets arranged in series cannot remain lit if a bulb burns out. Wiring in parallel can eliminate this problem, but each bulb must then be able to withstand 120 V. To eliminate the drawbacks of either approach, modern light sets typically contain two or three sections connected to each other in parallel, each of which contains bulbs in series.

When one bulb is removed from a modern light set, half or one-third of the lights in the set go dark because the bulbs in that section are wired in series. When a bulb *burns out*, however, all of the other bulbs in the set remain lit. How is this possible?

Modern decorative bulbs have a short loop of insulated wire called the jumper, that is wrapped

around the wires connected to the filament, as shown in **Figure 20-29**. There is no current in the insulated wire when the bulb is functioning properly. When the filament breaks, however, the current in the section is zero and the potential difference across the two wires connected to the broken filament is then 120 V. This large potential difference creates a spark across the two wires that burns the insulation off the small loop of wire. Once that occurs, the small loop closes the circuit, and the other bulbs in the section remain lit.

Because the small loop in the burnt-out bulb has very little resistance, the equivalent resistance of that portion of the light set decreases; its current increases. This increased current results in a slight increase in each bulb's brightness. As more bulbs burn out, the temperature in each bulb increases and can become a fire hazard; thus, bulbs should be replaced soon after burning out.

⊛TEKS 3C, 6E

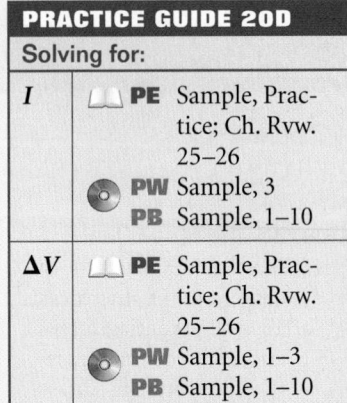

Filament

Jumper

Glass insulator

Figure 20-29

Consumer Focus

BACKGROUND

Although decorative lights are an excellent topic during classroom discussion of series and parallel circuits, many decorative light sets use the jumpers described in this feature to avoid the pitfalls of each type of circuit. In effect, the jumper functions like a switch that remains open while the filament conducts and closes to connect the wires when the filament burns out.

A more in-depth discussion of this mechanism can be found in the December 1992 edition of *The Physics Teacher*.

Section Review  (★)TEKS 2C, 6E

1. Find the equivalent resistance of the complex circuit shown in **Figure 20-30.**

2. What is the current in the 1.5 Ω resistor in the complex circuit shown in **Figure 20-30**?

3. What is the potential difference across the 1.5 Ω resistor in the circuit shown in **Figure 20-30**?

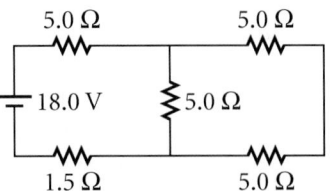

Figure 20-30

4. **Figure 20-31** depicts a household circuit containing several appliances and a circuit breaker attached to a 120 V source of potential difference.
   a. Is the current in the toaster equal to the current in the microwave?
   b. Is the potential difference across the microwave equal to the potential difference across the popcorn popper?
   c. Is the current in the circuit breaker equal to the current in all of the appliances combined?

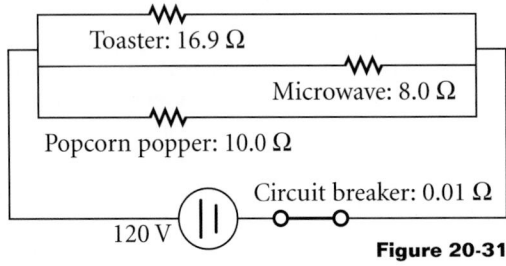

Figure 20-31

5. Determine the equivalent resistance for the household circuit shown in **Figure 20-31.**

6. Determine how much current is in the toaster in the household circuit shown in **Figure 20-31.**

7. **Physics in Action**   A certain strand of miniature lights contains 35 bulbs wired in series, with each bulb having a resistance of 15.0 Ω. What is the equivalent resistance when three such strands are connected in parallel across a potential difference of 120.0 V?

8. **Physics in Action**   What is the current in and potential difference across each of the bulbs in the strands of lights described in item 7?

9. **Physics in Action**   If one of the bulbs in one of the three strands of lights in item 7 goes out while the other bulbs remain lit, what is the current in and potential difference across each of the lit bulbs?

CHAPTER 20
Summary

KEY IDEAS

Section 20-1 Schematic diagrams and circuits

- Schematic diagrams use standardized symbols to summarize the contents of electric circuits.
- A circuit is a set of electrical components connected so that they provide one or more complete paths for the movement of charges.
- Any device that transforms nonelectrical energy into electrical energy, such as a battery or a generator, is a source of emf.
- If the internal resistance of a battery is neglected, the emf can be considered equal to the terminal voltage, the potential difference across the source's two terminals.

Section 20-2 Resistors in series or in parallel

- The equivalent resistance of a set of resistors connected in series is the sum of the individual resistances.

$$R_{eq} = R_1 + R_2 + R_3 \ldots$$
For resistors in series

- The equivalent resistance of a set of resistors connected in parallel is calculated using an inverse relationship.

$$\frac{1}{R_{eq}} = \frac{1}{R_1} + \frac{1}{R_2} + \frac{1}{R_3} \ldots$$
For resistors in parallel

Section 20-3 Complex resistor combinations

- Complex circuits can be understood by isolating segments that are in series or in parallel and simplifying them to their equivalent resistances.

Variable symbols

Quantities	Units	Conversions
I current	A amperes	= C/s = coulombs of charge per second
R resistance	Ω ohms	= V/A = volts per ampere of current
ΔV potential difference	V volts	= J/C = joules of energy per coulomb of charge

KEY TERMS

electric circuit (p. 732)

emf (p. 734)

parallel (p. 740)

schematic diagram (p. 730)

series (p. 736)

Diagram symbols

Wire or conductor	
Resistor or circuit load	—/\/\/—
Bulb or lamp	
Plug	
Battery/ direct-current emf source	
Switch	
Capacitor	—||—

Teaching Tip

Ask students to prepare a concept map for the chapter. The concept map should include most of the vocabulary terms, along with other integral terms or concepts.

★ TEKS

Review & Assess
pp. 754–759:
2A: Alt. Assess. 1
2B: Alt. Assess. 1
2C: 16–49
2D: Alt. Assess. 3–4
2F: Alt. Assess. 1
3D: Alt. Assess. 2–4
6E: 1–49, Alt. Assess. 1, 5

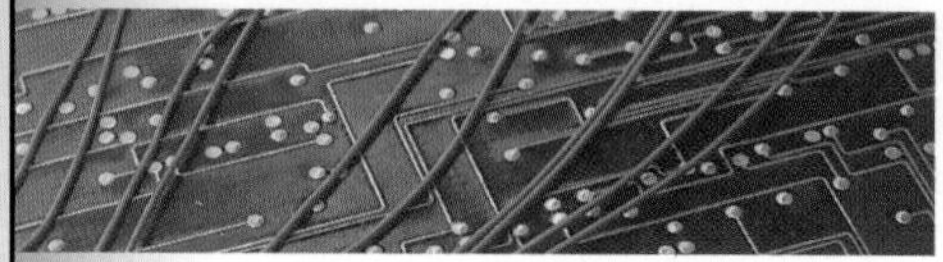

ANSWERS TO

Chapter 20
Review and Assess

1. Schematic diagrams are useful because they summarize the contents of an electric circuit.
2. Accept any schematic diagram that contains three resistors, a battery, and a switch.
3. *B*, *C*
4. 12.0 V
5. b
6. Charges move through both the emf source and the load.
7. When the circuit is open, there is no complete path for charge flow and hence no current.
8. Some of the electrical energy is dissipated by heat, and the remainder is converted into light energy.
9. A potential difference across the body from contact with a faulty wire can generate a current in the body.
10. all; *D*
11. b
12. a
13. Because the resistance is very low, the current in a short circuit is high ($I = \frac{\Delta V}{R}$). Such high currents can cause wires to overheat, causing a fire.
14. A fuse will not work in parallel because there is an alternative path for the current.

CHAPTER 20
Review and Assess

SCHEMATIC DIAGRAMS AND CIRCUITS

Review questions

1. Why are schematic diagrams useful?

2. Draw a circuit diagram for a circuit containing three 5.0 Ω resistors, a 6.0 V battery, and a switch.

3. The switch in the circuit shown in **Figure 20-32** can be set to connect to points *A*, *B*, or *C*. Which of these connections will provide a complete circuit?

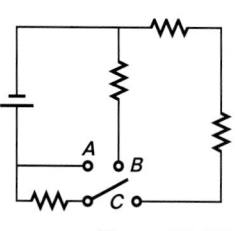

Figure 20-32

4. If the batteries in a cassette recorder provide a terminal voltage of 12.0 V, what is the potential difference across the entire recorder?

5. In a case in which the internal resistance of a battery is significant, which is greater?
- **a.** the terminal voltage
- **b.** the emf of the battery

Conceptual questions

6. Do charges move from a source of potential difference into a load or through both the source and the load?

7. Assuming that you want to create a circuit that has current in it, why should there be no openings in the circuit?

8. Suppose a 9 V battery is connected across a light bulb. In what form is the electrical energy supplied by the battery dissipated by the light bulb?

9. Why is it dangerous to turn on a light when you are in the bathtub?

10. Which of the switches in the circuit in **Figure 20-33** will complete a circuit when closed? Which will cause a short circuit?

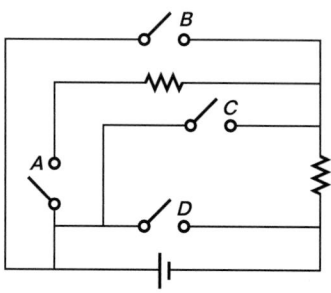

Figure 20-33

RESISTORS IN SERIES OR IN PARALLEL

Review questions

11. If four resistors in a circuit are connected in series, which of the following is the same for the resistors in the circuit?
- **a.** potential difference across the resistors
- **b.** current in the resistors

12. If four resistors in a circuit are in parallel, which of the following is the same for the resistors in the circuit?
- **a.** potential difference across the resistors
- **b.** current in the resistors

Conceptual questions

13. A short circuit is a circuit containing a path of very low resistance in parallel with some other part of the circuit. Discuss the effect of a short circuit on the current within the portion of the circuit that has very low resistance.

14. Fuses protect electrical devices by opening a circuit if the current in the circuit is too high. Would a fuse work successfully if it were connected in parallel with the device that it is supposed to protect?

15. What might be an advantage of using two identical resistors in parallel that are connected in series with another identical parallel pair, as shown in **Figure 20-34,** instead of using a single resistor?

Figure 20-34

Practice problems

16. A length of wire is cut into five equal pieces. If each piece has a resistance of 0.15 Ω, what was the resistance of the original length of wire?
(See Sample Problem 20A.)

17. A 4.0 Ω resistor, an 8.0 Ω resistor, and a 12 Ω resistor are connected in series with a 24 V battery. Determine the following:
   a. the equivalent resistance for the circuit
   b. the current in the circuit
(See Sample Problem 20A.)

18. The resistors in item 18 are connected in parallel across a 24 V battery. Determine the following:
   a. the equivalent resistance for the circuit
   b. the current in the circuit
(See Sample Problem 20B.)

19. An 18.0 Ω resistor, 9.00 Ω resistor, and 6.00 Ω resistor are connected in parallel across a 12 V battery. Determine the following:
   a. the equivalent resistance for the circuit
   b. the current in the circuit
(See Sample Problem 20B.)

COMPLEX RESISTOR COMBINATIONS

Conceptual questions

20. A technician has two resistors, each of which has the same resistance, R.
   a. How many different resistances can the technician achieve?
   b. Express the effective resistance of each possibility in terms of R.

21. The technician in item 20 finds another resistor, so now there are three resistors with the same resistance.
   a. How many different resistances can the technician achieve?
   b. Express the effective resistance of each possibility in terms of R.

22. Three identical light bulbs are connected in circuit to a battery, as shown in **Figure 20-35.** Compare the level of brightness of each bulb when all the bulbs are illuminated. What happens to the brightness of each bulb if the following changes are made to the circuit?
   a. Bulb A is removed from its socket.
   b. Bulb C is removed from its socket.
   c. A wire is connected directly between points D and E.
   d. A wire is connected directly between points D and F.

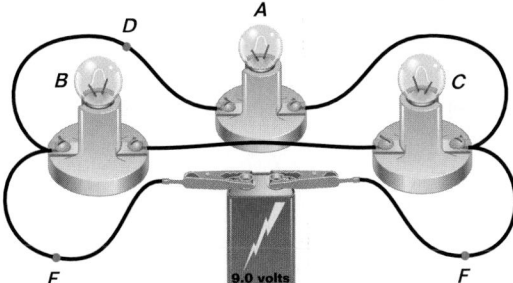

Figure 20-35

Practice problems

23. Find the equivalent resistance of the circuit shown in **Figure 20-36.**
(See Sample Problem 20C.)

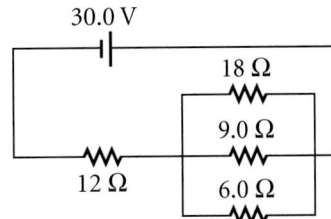

Figure 20-36

15. Because the resistors in each group are in parallel, a broken resistor does not cause a short circuit. Wiring the two groups in series increases equivalent resistance, thereby decreasing the current in the circuit.

16. 0.75 Ω

17. a. 24 Ω
   b. 1.0 A

18. a. 2.2 Ω
   b. 11 A

19. a. 2.99 Ω
   b. 4.0 A

20. a. three combinations
   b. $R, 2R, \dfrac{R}{2}$

21. a. seven combinations
   b. $R, 2R, 3R, \dfrac{R}{2}, \dfrac{R}{3}, \dfrac{2R}{3}, \dfrac{3R}{2}$

22. Bulb A is brighter than bulbs B and C. Bulbs B and C have the same brightness.
   a. Bulbs B and C stay the same.
   b. Bulb A stays the same and bulb B goes dark because the circuit is open at C.
   c. There is no change in any of the bulbs.
   d. No bulbs light because there is a short circuit across the entire wire.

23. 15 Ω

24. 13.3 Ω

25. Current: 1.1 A, 0.72 A, 1.8 A
Potential difference: 6.5 V,
6.5 V, 5.4 V

26. a. 1.7 A
b. 3.4 V
c. 5.1 V
d. 0.42 A

27. 28 V

28. 2.2 V

29. 3.8 V

30. 3.0×10^1 V

31. a. 33.0 Ω
b. 132 V
c. 4.00 A, 4.00 A

32. a. one 20 Ω resistor in series
with two parallel 50 Ω
resistors
b. Place two parallel 50 Ω
resistors in series with two
parallel 20 Ω resistors.

33. 10.0 Ω

34. 1875 Ω

---

24. Find the equivalent resistance of the circuit shown in **Figure 20-37.**
(See Sample Problem 20C.)

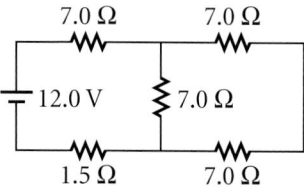

7.0 Ω   7.0 Ω

12.0 V   7.0 Ω

1.5 Ω   7.0 Ω   **Figure 20-37**

25. For the circuit shown in **Figure 20-38,** determine the current in each resistor and the potential difference across each resistor.
(See Sample Problem 20D.)

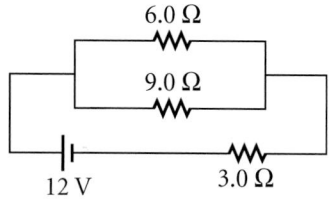

6.0 Ω

9.0 Ω

12 V   3.0 Ω   **Figure 20-38**

26. For the circuit shown in **Figure 20-39,** determine the following:

a. the current in the 2.0 Ω resistor
b. the potential difference across the 2.0 Ω resistor
c. the potential difference across the 12.0 Ω resistor
d. the current in the 12.0 Ω resistor

(See Sample Problem 20D.)

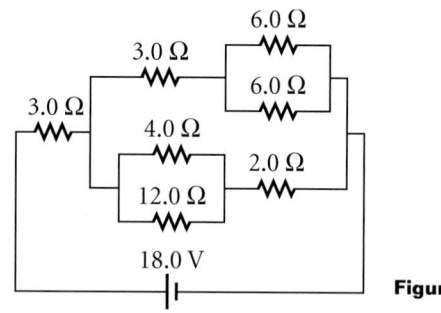

6.0 Ω
3.0 Ω
6.0 Ω
3.0 Ω
4.0 Ω
2.0 Ω
12.0 Ω
18.0 V
Figure 20-39

MIXED REVIEW

27. An 8.0 Ω resistor and a 6.0 Ω resistor are connected in series with a battery. The potential difference across the 6.0 Ω resistor is measured as 12 V. Find the potential difference across the battery.

---

28. A 9.0 Ω resistor and a 6.0 Ω resistor are connected in parallel to a battery, and the current in the 9.0 Ω resistor is found to be 0.25 A. Find the potential difference across the battery.

29. A 9.0 Ω resistor and a 6.0 Ω resistor are connected in series to a battery, and the current through the 9.0 Ω resistor is 0.25 A. What is the potential difference across the battery?

30. A 9.0 Ω resistor and a 6.0 Ω resistor are connected in series with an emf source. The potential difference across the 6.0 Ω resistor is measured with a voltmeter to be 12 V. Find the potential difference across the emf source.

31. An 18.0 Ω, 9.00 Ω, and 6.00 Ω resistor are connected in series with an emf source. The current in the 9.00 Ω resistor is measured to be 4.00 A.

a. Calculate the equivalent resistance of the three resistors in the circuit.
b. Find the potential difference across the emf source.
c. Find the current in the other resistors.

32. The stockroom has only 20 Ω and 50 Ω resistors.

a. You need a resistance of 45 Ω. How can this resistance be achieved under these circumstances?
b. Explain what you can do if you need a 35 Ω resistor.

33. The equivalent resistance of the circuit shown in **Figure 20-40** is 60.0 Ω. Use the diagram to determine the value of R.

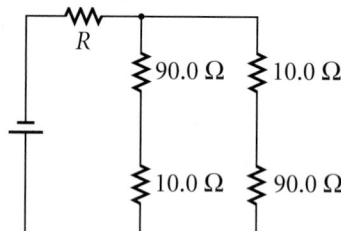

R
90.0 Ω   10.0 Ω
10.0 Ω   90.0 Ω
Figure 20-40

34. Two identical parallel-wired strings of 25 bulbs are connected to each other in series. If the equivalent resistance of the combination is 150.0 Ω when it is connected across a potential difference of 120.0 V, what is the resistance of each individual bulb?

35. Figure 20-41 depicts five resistance diagrams. Each individual resistance is 6.0 Ω.

   a. Which resistance combination has the largest equivalent resistance?

   b. Which resistance combination has the smallest equivalent resistance?

   c. Which resistance combination has an equivalent resistance of 4.0 Ω?

   d. Which resistance combination has an equivalent resistance of 9.0 Ω?

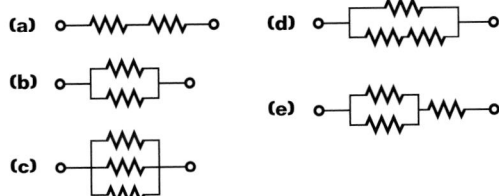

Figure 20-41

36. Three small lamps are connected to a 9.0 V battery, as shown in **Figure 20-42.**

   a. What is the equivalent resistance of this circuit?

   b. What is the total current in this circuit?

   c. What is the current in each bulb?

   d. What is the potential difference across each bulb?

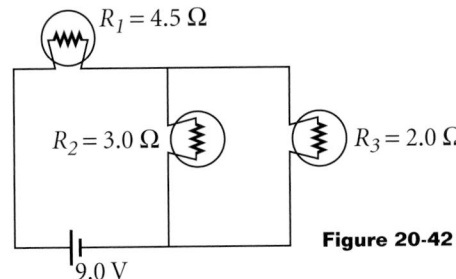

Figure 20-42

37. An 18.0 Ω resistor and a 6.0 Ω resistor are connected in series to an 18.0 V battery. Find the current in and the potential difference across each resistor.

38. A 30.0 Ω resistor is connected in parallel to a 15.0 Ω resistor. These are joined in series to a 5.00 Ω resistor and a source with a potential difference of 30.0 V.

   a. Draw a schematic diagram for this circuit.

   b. Calculate the equivalent resistance.

   c. Calculate the current in each resistor.

   d. Calculate the potential difference across each resistor.

39. A resistor with an unknown resistance is connected in parallel to a 12 Ω resistor. When both resistors are connected to an emf source of 12 V, the current across the unknown resistor is measured with an ammeter to be 3.0 A. What is the resistance of the unknown resistor?

40. The resistors described in item 37 are reconnected in parallel to the same 18.0 V battery. Find the current in each resistor and the potential difference across each resistor.

41. The equivalent resistance for the circuit shown in **Figure 20-43** drops to one-half its original value when the switch, S, is closed. Determine the value of R.

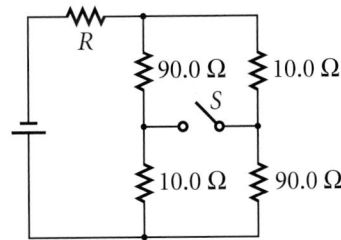

Figure 20-43

42. You can obtain only four 20.0 Ω resistors from the stockroom.

   a. How can you achieve a resistance of 50.0 Ω under these circumstances?

   b. What can you do if you need a 5.0 Ω resistor?

43. Four resistors are connected to a battery with a terminal voltage of 12.0 V, as shown in **Figure 20-44.** Determine the following:

   a. the equivalent resistance for the circuit

   b. the current in the battery

   c. the current in the 30.0 Ω resistor

   d. the power dissipated by the 50.0 Ω resistor

   e. the power dissipated by the 20.0 Ω resistor

(Hint: Remember that $P = \dfrac{(\Delta V)^2}{R} = I\Delta V$.)

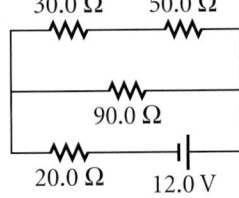

Figure 20-44

35. a. a
   b. c
   c. d
   d. e

36. a. 5.7 Ω
   b. 1.6 A
   c. 1.6 A (R_1), 0.63 A (R_2), 0.95 A (R_3)
   d. 7.2 V (R_1), 1.9 V (R_2), 1.9 V (R_3)

37. 18.0 Ω: 0.750 A, 13.5 V
   6.0 Ω: 0.750 A, 4.5 V

38. a.
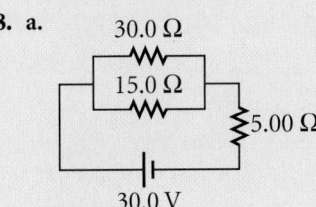

   b. 15.00 Ω
   c. 0.667 A , 1.33 A, 2.00 A
   d. 20.0 V, 20.0 V, 10.0 V

39. 4.0 Ω

40. 18.0 Ω: 1.00 A, 18.0 V
   6.0 Ω: 3.0 A, 18.0 V

41. 13.96 Ω

42. a. two resistors in series with two parallel resistors
   b. four parallel resistors

43. a. 62.4 Ω
   b. 0.192 A
   c. 0.102 A
   d. 0.520 W
   e. 0.737 W

44. 6.0 Ω (*A*), 3.0 Ω (*B*)

45. The circuit must contain three groups of resistors, each containing three resistors in parallel, that are connected to one another in series.

46. a. 14.0 Ω
   b. 2.0 A

44. Two resistors, *A* and *B*, are connected in series to a 6.0 V battery. A voltmeter connected across resistor *A* measures a potential difference of 4.0 V. When the two resistors are connected in parallel across the 6.0 V battery, the current in *B* is found to be 2.0 A. Find the resistances of *A* and *B*.

45. Draw a schematic diagram of nine 100 Ω resistors arranged in a series-parallel network so that the total resistance of the network is also 100 Ω. All nine resistors must be used.

46. For the circuit in **Figure 20-45,** find the following:
   a. the equivalent resistance of the circuit
   b. the current in the 5.0 Ω resistor

5.0 Ω   3.0 Ω   3.0 Ω

10.0 Ω   10.0 Ω   4.0 Ω

28 V

4.0 Ω   2.0 Ω   3.0 Ω   **Figure 20-45**

Technology & Learning

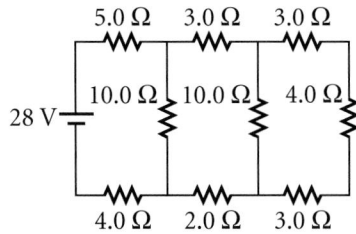

Graphing calculators

Refer to Appendix B for instructions on downloading programs for your calculator. The program "Chap20" allows you to calculate the equivalent resistance of any number of resistors in a parallel circuit.

The program "Chap20" stored on your graphing calculator makes use of the following equation for the equivalent resistance of resistors in parallel:

$$\frac{1}{R_{eq}} = \frac{1}{R_1} + \frac{1}{R_2} + \frac{1}{R_3} \cdots$$

Once the "Chap20" program is executed, your calculator will ask for number of resistors and then the resistances (R) of each of the resistors. After each resistance value entered, the calculator will enter the resistance value into the following equation:

$$Z = Z + (1/R)$$

Thus, each time a new resistance value is added, the program adds the inverse of that resistance to the total of the other inverse resistances.

   a. The variable Z does not represent the equivalent resistance of the circuit. What does Z represent?

Execute "Chap 20" on the [PRGM] menu and press [ENTER] to begin the program. When prompted with the text "Resistors," enter the number of resistors in the circuit. The calculator will then ask for the resistance of the resistors in the circuit. Enter the resistance values, pressing [ENTER] after each resistance.

When you have entered an appropriate number of values for the individual resistances, the calculator will perform the calculation and display the equivalent resistance for the parallel combination of those resistors.

Determine the equivalent resistance of the following circuits:

   b. an 8.0 Ω resistor and a 6.0 Ω resistors connected in parallel with an emf source

   c. a 4.0 Ω resistor, an 8.0 Ω resistor, and a 12 Ω resistor connected in parallel with a 12 V battery

   d. 15 resistors connected in parallel with an emf source: 1.0 Ω, 2.0 Ω, 3.0 Ω, 4.0 Ω, 5.0 Ω, 6.0 Ω, 7.0 Ω, 8.0 Ω, 9.0 Ω, 10.0 Ω, 11.0 Ω, 12.0 Ω, 13.0 Ω, 14.0 Ω, and 15.0 Ω

   e. Using the variables R and Z, write the equation for a similar program that calculates the equivalent resistance for a series circuit. Press [ENTER] to input a new value or [CLEAR] to end the program.

ANSWERS TO

Technology & Learning

   a. the inverse of the equivalent resistance

   b. 3.4 Ω

   c. 2.2 Ω

   d. 0.30 Ω

   e. Z = Z + R

47. The power supplied to the circuit shown in **Figure 20-46** is 4.00 W. Use the information in the diagram to determine the following:

   a. the equivalent resistance of the circuit

   b. the potential difference across the battery

(Hint: Remember that $P = \dfrac{(\Delta V)^2}{R}$.)

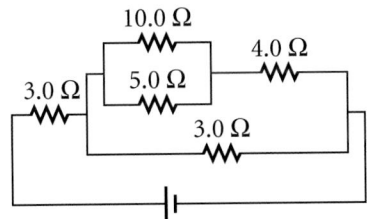

Figure 20-46

48. Your toaster oven and coffee maker each dissipate 1200 W of power. Can you operate both of these appliances at the same time if the 120 V line you use in your kitchen has a circuit breaker rated at 15 A? Explain.
(Hint: Recall that $P = I\Delta V$.)

49. An electric heater is rated at 1300 W, a toaster is rated at 1100 W, and an electric grill is rated at 1500 W. The three appliances are connected in parallel across a 120 V emf source.

   a. Find the current in each appliance.

   b. Is a 30.0 A circuit breaker sufficient in this situation? Explain.

Alternative Assessment

Performance assessment

1. How many ways can two or more batteries be connected in a circuit with a light bulb? How will the current change depending on the arrangement? First draw diagrams of the circuits you want to test. Then identify the measurements you need to make to answer the question. If your teacher approves your plan, obtain the necessary equipment and perform the experiment.

2. Research the career of an electrical engineer or technician. Prepare materials for people interested in this career field. Include information on where people in this career field work, which tools and equipment they use, and the challenges of their field. Indicate what training is typically necessary to enter the field.

Portfolio projects

3. The manager of an automotive repair shop has been contacted by two competing firms that are selling ammeters to be used in testing automobile electrical systems. One firm has published claims that its ammeter is better because it has high internal resistance. The other firm has published claims that its ammeter is better because it has low resistance. Write a report with your recommendation to the manager of the automotive repair shop. Include diagrams and calculations that explain how you reached your conclusion.

4. You and your friend want to start a business exporting small electrical appliances. You have found people willing to be your partners to distribute these appliances in Germany. Write a letter to these potential partners that describes your product line and that asks for the information you will need about the electric power, sources, consumption, and distribution in Germany.

5. Contact an electrician, builder, or contractor, and ask to see a house electrical plan. Study the diagram to identify the circuit breakers, their connections to different appliances in the home, and the limitations they impose on the circuit's design. Find out how much current, on average, is in each appliance in the house. Draw a diagram of the house, showing which circuit breakers control which appliances. Your diagram should also keep the current in each of these appliances under the performance and safety limits.

NOTE

Materials Preparation is given on pp. 728A–728B. Blank data table and sample data table are on the One-Stop Planner CD-ROM. All calculations shown use sample data.

Planning

Recommended time:

1 lab period

Classroom organization:

▸ Each lab group should have a level surface to work on that is near an electrical outlet and away from any sources of water.

▸ Each lab group should have two students.

▸ The meters and CBL procedures may be used in the same class.

▸ **Safety warnings:** Emphasize the dangers of working with electricity. For the safety of the students and of the equipment, remind students to have you check their circuits before turning on the power supply or closing the switch.

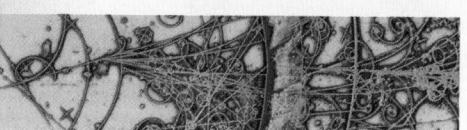
- Measure current in and potential difference across resistors in series and in parallel.
- Find the unknown resistances of two resistors.
- Calculate equivalent resistances.
- Analyze the relationships between potential difference, current, and resistance in a circuit.

MATERIALS LIST

✔ 2 resistors
✔ insulated connecting wire
✔ power supply
✔ switch

PROCEDURE

CBL AND SENSORS

✔ 2 CBL-DIN adapters
✔ CBL
✔ current and voltage probe system with amplifier box
✔ graphing calculator with link cable

METERS

✔ 2 multimeters, or 1 dc ammeter and 1 voltmeter

RESISTORS IN SERIES AND IN PARALLEL

In this lab, you will compare the circuits created by wiring two unknown resistors first in series and then in parallel. By taking measurements of the current in and the potential difference (voltage) across the resistors, and the potential difference across the whole circuit, you will find the value of the resistance of each resistor and the equivalent resistance for both resistors to compare the total current in each circuit.

SAFETY

- **Never close a circuit until it has been approved by your teacher.**
- **Never rewire or adjust any element of a closed circuit.**
- **Never work with electricity near water; be sure the floor and all work surfaces are dry.**
- **If the pointer on any kind of meter moves off scale, open the circuit immediately by opening the switch.**
- **Do not attempt this exercise with any batteries or electrical devices other than those provided by your teacher for this purpose.**
- **Use a hot mitt to handle resistors, light sources, and other equipment that may be hot. Allow all equipment to cool before storing it.**

PREPARATION

1. Determine whether you will be using the CBL and sensors procedure or the meters procedure. Read the entire lab for the appropriate procedure, and plan what steps you will take.

2. Prepare a data table in your lab notebook with six columns and three rows. In the first column, label the second row *Series* and the third row *Parallel*. In the first row, label the second through sixth columns ΔV_T, ΔV_1, I_1, ΔV_2, and I_2.

Meters procedure begins on page 762.

PROCEDURE

CBL AND SENSORS

Resistors in series

3. Construct a circuit that includes a battery, a switch, and two unequal resistors in series. ***Do not close the switch until your teacher has approved your circuit.***

4. Connect the probes to measure the current in and the potential difference across the first resistor. With the switch open, connect one current probe in series and one voltage probe in parallel with the first resistor, as shown in **Figure 20-47.**

5. Connect the CBL and graphing calculator. Connect the DIN1 lead on the amplifier box to one of the CBL-DIN adapters, and connect the adapter to the CH1 port on the CBL. Connect the DIN2 lead on the amplifier box to the other CBL-DIN adapter, and connect the adapter to the CH2 port. Connect the CURRENT1 probe to the PROBE1 port on the amplifier box, and connect the VOLTAGE1 probe to the PROBE2 port. Turn on the CBL and graphing calculator. Start the program PHYSICS on the calculator.

6. Select option *SET UP PROBES* from the MAIN MENU. Enter 2 for the number of probes. Select *MORE PROBES* from the SELECT PROBE menu. Select the *CV CURRENT* probe from the list. Enter 1 for the channel number. Select *USED STORED* from the CALIBRATION menu. Select *MORE*

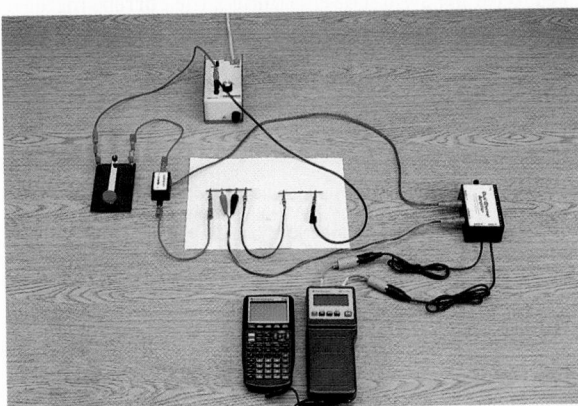

PROBES from the SELECT PROBE menu. Select the *CV VOLTAGE* probe from the list. Enter 2 for the channel number. Select *USED STORED* from the CALIBRATION menu.

7. Select the *COLLECT DATA* option from the MAIN MENU. Select the *MONITOR INPUT* option from the DATA COLLECTION menu. The calculator will begin to display values. ***Do not close the switch.***

8. When your teacher has approved your circuit, close the switch. Quickly take readings for the potential difference and current. Open the switch, and record the readings in your data table as ΔV_1 and I_1.

9. Carefully remove the probes from the first resistor. Rewire the probes to measure the potential difference across and the current in the second resistor. ***Do not close the switch.***

10. When your teacher has approved your circuit, close the switch. Measure the current in and voltage across the resistor, and open the switch. Record the information in your data table as ΔV_2 and I_2.

11. Leave the current probe in place, and carefully rewire the voltage probe to measure the potential difference across both resistors. Record this value in your data table as ΔV_T.

Resistors in parallel

12. Construct a circuit containing a power supply, a switch, and the two resistors wired in parallel. ***Do not close the switch until your teacher has approved your circuit.***

Figure 20-47

Step 6: Make sure all connections are correct and all plugs are pushed firmly in place.

Step 10: When you have recorded data for the first resistor, open the switch and connect the probes to take data for the second resistor.

Step 12: For this part of the lab, make sure the resistors are wired in parallel.

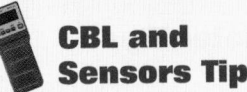
CBL and Sensors Tips

◆ Students should have the program PHYSICS on their graphing calculators. Refer to Appendix B for instructions.

Techniques to Demonstrate

If the students are using a dc power supply, show them how to adjust the voltage. Power supplies should be set at around 5.0 V for this exercise.

Explain the black (negative) and red (positive) convention, and show students how to connect the battery or power supply and the meters or probes correctly.

✔ Checkpoints

Step 9: Make sure the power supply, resistors, and probes are connected properly and are set at the proper settings. Students should be able to demonstrate that the resistors are in series. Check to make sure that students are taking both measurements for the same resistor.

Step 11: Make sure students have correctly rewired the circuit to measure voltage and current for the second resistor. Remind students to close the switch only long enough to take their readings.

Step 14: Students should be able to demonstrate that the resistors are in parallel. Check to make sure that students are taking both measurements for the same resistor.

Step 16: Make sure students have correctly rewired the circuit to measure voltage and current for the second resistor.

Meter Tips

◆ See pp. 692A–692B for instructions on setting up multimeters to measure potential difference and current.

◆ Make sure students understand how to wire current meters (in series) and voltage meters (in parallel) in a circuit.

Techniques to Demonstrate

If the students are using a dc power supply, show them how to adjust the voltage. Power supplies should be set at around 5.0 V for this exercise.

Explain the black (negative) and red (positive) convention, and show students how to connect the battery or power supply and the meters or probes correctly.

✓ Checkpoints

Step 6: Make sure the power supply, resistors, and meters are connected properly and are set at the proper settings. Students should be able to demonstrate that the resistors are in series. Check to make sure that students are taking both measurements for the same resistor.

Step 8: Make sure students have correctly rewired the circuit to measure voltage and current for the second resistor. Remind students to close the switch only long enough to take their readings.

Step 11: Students should be able to demonstrate that the resistors are in parallel. Check to make sure that students are taking both measurements for the same resistor.

---

13. With the switch open, connect the current probe in series and the voltage probe in parallel with one of the resistors. ***Do not close the switch.***

14. When your teacher has approved your circuit, close the switch. Quickly record the readings for the voltage and current in your data table. Open the switch.

15. Rewire the probes to measure the potential difference across and the current in the second resistor. ***Do not close the switch.***

PROCEDURE

METERS

Resistors in series

3. Construct a circuit that includes a battery, a switch, and two unequal resistors in series. ***Do not close the switch until your teacher has approved your circuit.***

4. With the switch open, connect the current meter in series and the voltage meter in parallel with one of the resistors. ***Do not close the switch.***

5. When your teacher has approved your circuit, close the switch. Measure the current in and the potential difference across the resistor. Record the information in your data table. Open the switch.

6. Carefully remove the meters from the first resistor. Rewire the meters to measure the potential

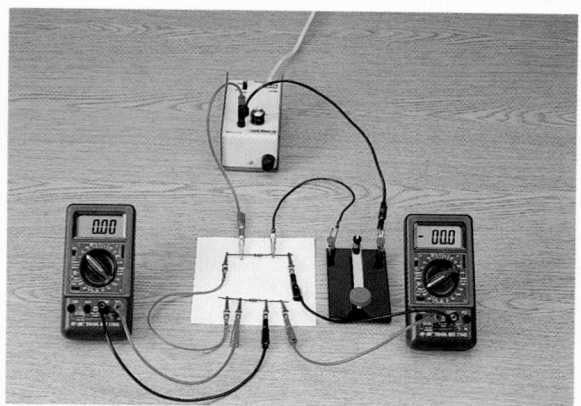

---

16. When your teacher has approved your circuit, close the switch. Measure the current in and potential difference across the resistor, and record the information in your data table. Open the switch.

17. Leave the current probe in place, and carefully rewire the voltage probe to measure the potential difference across the power supply. Record this value in your data table as ΔV_T.

Analysis and Interpretation begins on page 763.

---

difference across and the current in the second resistor. ***Do not close the switch.***

7. When your teacher has approved your circuit, close the switch. Measure the current in and potential difference across the resistor, and record the information in your data table. Open the switch.

8. Leave the current meter in place, and carefully rewire the voltage meter to measure the potential difference across both resistors. Record this value in your data table as ΔV_T.

Resistors in parallel

9. Construct a circuit containing a power supply, a switch, and the two resistors wired in parallel, as shown in **Figure 20-48**. ***Do not close the switch until your teacher has approved your circuit.***

10. With the switch open, connect the current meter in series and the voltage meter in parallel with one of the resistors. ***Do not close the switch.***

Figure 20-48

Step 4: Make sure all connections are correct. For this part of the lab, make sure the resistors are wired in series.

Step 7: When you have recorded data for the first resistor, open the switch and connect the probes to take data for the second resistor.

Step 9: For this part of the lab, make sure the resistors are wired in parallel, as shown in **Figure 20-48**.

11. When your teacher has approved your circuit, close the switch. Measure the current in and potential difference across the resistor, and record the information in your data table. Open the switch.

12. Rewire the meters to measure the potential difference across and current in the second resistor. **Do not close the switch until your teacher has approved your circuit.**

13. When your teacher has approved your circuit, close the switch. Measure the current in and potential difference across the resistor, and record the information in your data table. Open the switch.

14. Leave the current meter in place, and carefully rewire the voltage meter to measure the potential difference across the power supply. Record this value in your data table as ΔV_T.

ANALYSIS AND INTERPRETATION

Calculations and data analysis

1. Analyzing data Using your measurements for potential difference and current, compute the resistance values of R_1 and R_2 in each circuit.

2. Comparing results Compare your results from item 1 for the different circuits.

   a. Do R_1 and R_2 have the same values in each circuit?

   b. Did you expect R_1 and R_2 to have the same values? Explain. If the results are different, explain what may have caused the difference.

3. Analyzing data Compute the equivalent resistance R_{eq} using the values found in item 1 for each circuit.

Conclusions

4. Comparing results Based on your calculations in item 3, did the two resistors provide the same equivalent resistance in both circuits? If not, which combination had the greater resistance? Explain how the combination of resistors affects the total resistance in the circuit.

5. Analyzing data Compute the total current in each circuit using the calculated value for R_{eq} and the measured value for ΔV_T.

6. Comparing results Based on your calculations in item 5, do both circuits have the same total current? If not, which circuit has the greater current? Explain how the combination of resistors affects the total current in the circuit.

7. Analyzing information Compare the total current in each circuit with the current in each resistor. What is the relationship between the current in an individual resistor and the total current in the circuit?

8. Organizing conclusions For each circuit, compare the potential difference across each resistor with ΔV_T. What is the relationship?

ANSWERS TO

Analysis and Interpretation

CALCULATIONS AND DATA ANALYSIS

1. Answers must use the relationship $R = \dfrac{\Delta V}{I}$. Typical values will range from 12 Ω to 42 Ω.

2. a, b. The resistors are the same. If not, answers will vary but should include an analysis of error in the lab.

3. Make sure students use the relationships $R_{eq} = R_1 + R_2$ and $\dfrac{1}{R_{eq}} = \dfrac{1}{R_1} + \dfrac{1}{R_2}$. Answers will vary, but typical values will range from 9.3 Ω to 54 Ω.

CONCLUSIONS

4. The resistors have a higher equivalent resistance in the series circuit.

5. $I_{total} = \dfrac{\Delta V_T}{R_{eq}}$. Typical values will range from 0.11 A to 0.64 A.

6. The parallel circuit has a higher current.

7. *series:* the current in each resistor is equal to the total current in the circuit; *parallel:* the total current in the circuit is equal to the sum of the currents in each resistor.

8. *series:* ΔV_T is approximately equal to the sum of the voltage across each resistor; *parallel:* ΔV_T is equal to the voltage across each individual resistor.

CHAPTER 21 PLANNING GUIDE

Compression Guide: To shorten from 10 to 7 45-min periods (from 5 to 3½ 90-min blocks), eliminate items in magenta type.

PACING CHART	CLASSROOM RESOURCES			
	⭐ TEKS	Teacher Demonstrations	Holt Physics Transparencies	Labs (See page T52 for equipment listing for in-text labs.)
21-1 Magnets and magnetic fields 2 or 1 45-minute periods 1 or ½ 90-minute block	3A, 3C	TE *Magnetic poles*, p. 766 TE *Magnetic fields*, p. 767	T 102–103	L *Discovery Lab: Magnetism* PE *Quick Lab: Magnetic Field of a File Cabinet*, p. 768
21-2 Electromagnetism and magnetic domains 2 45-minute periods 1 90-minute block	1A, 2B, 2C, 2F, 3A, 6D, 6E	TE *Electromagnets*, p. 770 TE *Magnetic domains*, p. 772	T 104–105	PE *Quick Lab: Electromagnetism*, p. 771 PE *Magnetic Field of a Conducting Wire*, p. 786
21-3 Magnetic force 4 or 2 45-minute periods 2 or 1 90-minute blocks	2C, 3A, 3C, 6F, 8B	TE *Electromagnetic force*, p. 773 TE *Force between parallel conductors*, p. 777	T 106–109 TM 68–69	L *Invention Lab: Designing a Magnetic Spring* TL *Magnetic Field Strength*
Review and Assessment 2 45-minute periods 1 90-minute block				

Resource Key

PHYSICS (HOLT)

PE Pupil's Edition
TE Teacher's Edition

L Laboratory Experiments
TL Technology Lab Experiments
T Transparencies

🖑 **One-Stop** Planner CD-ROM **contents**

TM Transparency Masters
SR Section Review Worksheets
AA Alternative Assessment

PW Problem-Solving Workbook
PB Problem Bank
CTW Critical Thinking Worksheet

LABORATORY PLANNING: Magnetic Field of a Conducting Wire, p. 786

Materials (for each lab group):
- 1 Ω resistor, 10 W
- bare copper wire, 1 m
- power supply:
  - CENCO universal power supply, or
  - CENCO 6 V ac/dc low-voltage power supply, or
  - 20 V dc/500 mA regulated power supply, or
  - 4-output dc regulated power supply
- patch cord, black, insulated alligator clip/stacking banana plug
- patch cord, red, insulated alligator clip/stacking banana plug

- roll of self-adhesive tape, 0.5 in wide.
- tangent galvanometer kit
- wire leads with alligator clips (pkg. of 10)

Additional Equipment

CBL and Sensors Procedure
- 1-position support base and rod, 1.0 cm × 51 cm
- CBL
- CBL magnetic field probe
- CBL voltage probe

ASSIGNMENT RESOURCES

Content Mastery	Critical Thinking	Problem-Solving Practice
PE 1–5, p. 769 **SR** 21-1, *Concept Review* **PE** 1–4, p. 781	**PE** 5–6, p. 781	
PE 1–4, p. 772 **SR** 21-2, *Diagram Skills* **PE** 7–15, p. 781	**PE** 16–18, pp. 781–782	
PE 1–5, p. 779 **SR** 21-3, *Concept Review* **PE** 19–26, p. 782	**PE** 27–29, p. 782	**21A** Particle in a magnetic field: 31 items in **PE, PW,** and **PB,** see **TE** pp. 774–775 **21B** Force on a current-carrying conductor: 30 items in **PE, PW,** and **PB,** see **TE** p. 778

ASSESSMENT RESOURCES

Cumulative Review	Alternative Assessment	Traditional Assessment
SR Mixed Review, Ch. 21	**PE** 1–5, p. 785 **AA** Items for Ch. 21	Chapter 21 Test Test Generator items for Ch. 21

Scoring Rubrics for Alternative Assessment items can be found on the One-Stop Planner CD-ROM.

TECHNOLOGY RESOURCES

CNN PRESENTS
CTW Segment 21 Magnetic Attractions

PHYSICS TUTOR
Module 18 Magnetic Field of a Wire
Module 19 Magnetic Force on a Wire

PE Graphing Calculators, p. 784
(Alternative procedures for calculators without Flash-ROM technology are provided on the One-Stop Planner CD-ROM.)

The Mechanical Universe/High School Adaptation Quad VI, Magnetic Fields

internet connect

SCiLINKS NSTA
On-line Student Resources:
GO TO: www.scilinks.org
The following *sci*LINKS Internet resources can be found in the student text for this chapter.

TOPICS:
• Magnets, p. 766 (HF2211)
• Electromagnets, p. 771 (HF2212)

go.hrw.com
On-line Teacher Resources:
GO TO: go.hrw.com
KEYWORD: HF2 HOME
Visit the HRW Web site for a variety of resources related to this chapter.

Smithsonian Institution
Internet Connections
Visit www.si.edu/hrw for additional on-line resources.

CNN fyi.com
Visit www.cnnfyi.com for late-breaking news and current events stories selected just for you.

• graphing calculator with link cable
• knife switch (double pole/double throw)
• lattice rod, 1.3 cm × 30 cm
• right-angle clamp for rods up to 1.9 cm diameter
• roll of masking tape, 15 mm × 3 m
• TI Graph Link (recommended for downloading programs)
• V-jaw buret clamp

Compass Procedure
• 16 mm compass (12/pkg.)

• multimeter or voltmeter:
   basic digital multimeter, or
   student multimeter, or
   CENCO standard movement, triple range, dc voltmeter
   (0–3 V, 10 V, 15 V)
• patch cord, 30 cm, white, with banana plugs

Materials Preparation
See pages 692A–692B for instructions on using multimeters to measure voltage. Set up a sample apparatus in the laboratory for students to refer to when setting up their equipment.

Section 21-1 introduces magnets and magnetic fields and discusses magnetization.

Section 21-2 applies the right-hand rule for magnetism, explores electromagnetism and solenoids, and introduces magnetic domains.

Section 21-3 concentrates on calculations of magnetic fields and magnetic forces.

About the Illustration

Astronauts Dale A. Gardner and Joseph P. Allen IV work together to bring the *Westar VI* telecommunications satellite into the *Discovery* space shuttle's payload bay. Allen is on a mobile foot restraint, which is attached to the *Discovery*'s Remote Manipulator System. The satellite had to be recovered because a propulsion systems defect prevented it from reaching a sufficient orbital radius for telecommunications purposes.

PHYSICS Interactive Problem-Solving Tutor

See Module 18

"Magnetic Field of a Wire" provides additional development of problem-solving skills for this chapter.

See Module 19

"Magnetic Force on a Wire" provides additional development of problem-solving skills for this chapter.

CHAPTER 21

Magnetism

PHYSICS IN ACTION

Some satellites contain loops of wire called *torque coils* that can be activated by a computer or a satellite operator on Earth.

When activated, a torque coil has a current in it. As you will learn in this chapter, the current-carrying coil behaves like a magnet. Because Earth also has magnetic properties, a magnetic torque is exerted on the satellite, just as two magnets brought close together exert a force on one another. In this way, the satellite can be oriented so that its instruments point in the desired direction.

- *How does a current in the coil give the coil magnetic properties?*

- *How does the direction of the current in the wire affect the direction of the torque on the satellite?*

CONCEPT REVIEW

Force   (Section 4-1)

Force that maintains circular motion
    (Section 7-3)

Torque   (Section 8-1)

Electric fields   (Section 17-3)

Electric current   (Section 19-1)

Knowledge to Expect

✔ "Magnets attract and repel each other and certain kinds of other materials." (NRC's *National Science Education Standards*, grades K–4)

✔ "Without touching them, a magnet pulls on all things made of iron and either pushes or pulls on other magnets." (AAAS's *Benchmarks for Science Literacy*, grades 3–5)

✔ "Electric currents and magnets can exert a force on each other." (AAAS's *Benchmarks for Science Literacy*, grades 6–8)

Knowledge to Review

✔ A net force causes a change in the motion of an object. (Section 4-1)

✔ The force that maintains circular motion is always directed toward the center of the circular path. (Section 7-3)

✔ Torque is the cause of changes in rotation. Its magnitude equals the product of the force and the lever arm. (Section 8-1)

✔ Electric fields surround charged objects and exert forces on other charged objects. (Section 17-3)

✔ Electric current is the rate at which electric charges move through a cross-sectional area. (Section 19-1)

Items to Probe

✔ Electric fields: Have students relate electric-field strength to the force exerted on a charged particle in the field.

21-1
Magnets and magnetic fields

Magnetic poles

Purpose Show that all magnets have north and south poles and that there are attractive and repulsive forces between two magnets.

Materials two bar magnets, ring stand, string, various types of magnets, such as those shown in **Figure 21-1**

Procedure Use the string to suspend one bar magnet horizontally from the ring stand. Have students note that the magnet points north.

Bring the north pole of the other bar magnet near the north pole of the suspended magnet. Have students observe the reaction. Ask students to predict what will happen when you bring the south pole of the unattached magnet near the north pole of the suspended magnet. Then ask them what will happen when you bring the south pole of the unattached magnet near the south pole of the suspended magnet. Demonstrate both cases.

Perform similar demonstrations with the other types of magnets.

21-1 SECTION OBJECTIVES

- **For given situations, predict whether magnets will repel or attract each other.**

- **Describe the magnetic field around a permanent magnet.**

- **Describe the orientation of Earth's magnetic field.**

internetconnect

SCI**LINKS**

NSTA

TOPIC: Magnets
GO TO: www.scilinks.org
*sci***LINKS CODE:** HF2211

MAGNETS

Most people have had experience with different kinds of magnets, such as those shown in **Figure 21-1.** You are probably familiar with the horseshoe magnet, which can pick up iron-containing objects such as paper clips and nails, and flat magnets, such as those used to attach items to a refrigerator. In the following discussion, we will assume that the magnet has the shape of a bar. Iron objects are most strongly attracted to the ends of such a magnet. These ends are called *poles;* one is called the *north pole,* and the other is called the *south pole.* The names derive from the behavior of a magnet on Earth. If a bar magnet is suspended from its midpoint so that it can swing freely in a horizontal plane, it will rotate until its north pole points north and its south pole points south. In fact, a compass is just a magnetic needle that swings freely on a pivot.

The list of important technological applications of magnetism is very long. For instance, large electromagnets are used to pick up heavy loads. Magnets are also used in meters, motors, and loudspeakers. Magnetic tapes are routinely used in sound- and video-recording equipment, and magnetic recording material is used on computer discs. Superconducting magnets are currently being used to contain the plasmas used in controlled-nuclear-fusion research (heated to temperatures on the order of 10^8 K), and they are used to levitate modern trains. These so-called *maglev* trains are faster and provide a smoother ride than the ordinary track system because of the absence of friction between the train and the track. (★)TEKS 3C

Like poles repel each other, and unlike poles attract each other

The magnetic force between two magnets can be likened to the electric force between charged objects in that unlike poles of two magnets attract one another and like poles repel one another. Thus, the north pole of a magnet is attracted to the south pole of another magnet, and two north poles (or two south poles) brought close together repel each other. Electric charges differ from magnetic poles in that they can be isolated, while magnetic poles cannot. In fact, no matter how many times a permanent magnet is cut, each piece always has a north pole and a south pole. Thus, magnetic poles always occur in pairs.

Figure 21-1
Regardless of their shape, all magnets have both a north pole and a south pole.

Some materials can be made into permanent magnets

Just as two materials, such as rubber and wool, can become charged after they are rubbed together, an unmagnetized piece of iron can become a permanent magnet by being stroked with a permanent magnet. Magnetism can be induced by other means as well. For example, if a piece of unmagnetized iron is placed near a strong permanent magnet, the piece of iron will eventually become magnetized. The process can be reversed either by heating and cooling the iron or by hammering.

A magnetic piece of material is classified as magnetically *hard* or *soft*, depending on the extent to which it retains its magnetism. Soft magnetic materials, such as iron, are easily magnetized but also tend to lose their magnetism easily. In contrast, hard magnetic materials, such as cobalt and nickel, are difficult to magnetize, but once they are magnetized, they tend to retain their magnetism.

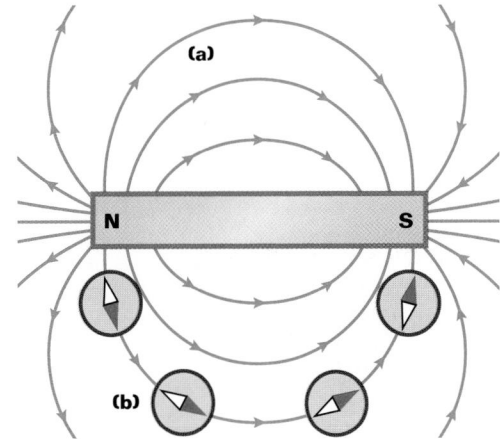

Figure 21-2
The magnetic field **(a)** of a bar magnet can be traced with a compass **(b).** Note that the north poles of the compasses point in the direction of the field lines from the magnet's north pole to its south pole.

MAGNETIC FIELDS

You know that the interaction between charged objects can be described using the concept of an electric field. A similar approach can be used to describe the **magnetic field** surrounding any magnetized material.

The direction of the magnetic field, **B**, at any location is defined as the direction in which the north pole of a compass needle points at that location. **Figure 21-2** shows how the magnetic field of a bar magnet can be traced with the aid of a compass. Note that the direction of the magnetic field is the same as the direction of the needles, and its magnitude is greatest close to the poles.

To indicate the direction of **B** when it points into or out of the page, use the conventions shown in **Table 21-1.** If **B** is directed into the page, we will use a series of blue *crosses,* representing the tails of arrows. If **B** is directed out of the page, we will use a series of blue *dots,* representing the tips of arrows.

When describing a small bar magnet as having north and south poles, it is more proper to say that it has a "north-seeking" pole and a "south-seeking" pole. This means that if such a magnet is used as a compass, the north pole of the magnet will seek, or point to, the geographic North Pole of Earth. Because unlike poles attract, we can deduce that the geographic North Pole of Earth corresponds to the magnetic south pole, and the geographic South Pole of Earth corresponds to the magnetic north pole.

The difference between true north, defined as the geographic North Pole, and north indicated by a compass varies from point to point on Earth, and the difference is referred to as *magnetic declination.* For example, along a line through South Carolina and the Great Lakes, a compass will point to true north (0° declination), whereas in Washington state it will point 25° east of true north (25° E declination).

magnetic field

a region in which a magnetic force can be detected

Table 21-1
Conventions for representing the direction of a magnetic field

In the plane of the page	↑
Into the page	
Out of the page	

Demonstration 2

Magnetic fields

Purpose Show the interaction of magnets and magnetic fields.

Materials two bar magnets, one horseshoe magnet, one blank transparency, iron filings, overhead projector

Procedure Set one of the bar magnets on the overhead projector, and lay the blank transparency over the magnet. Sprinkle the iron filings onto the transparency, and have students observe the behavior of the filings. Repeat the demonstration using the following:

a. two bar magnets about 4 cm apart, aligned with opposite poles facing each other

b. two bar magnets about 4 cm apart, aligned with like poles facing each other

c. horseshoe magnet

⬢ TEKS

p. 766: 3C

Teaching Tip

Point out that the direction of Earth's magnetic field has reversed several times during the last million years. Evidence for this is provided by basalt (an iron-containing rock) that is sometimes spewed forth by volcanic activity on the ocean floor. As the lava cools, it solidifies and retains a picture of Earth's magnetic field direction. When the basalt deposits are dated, they provide evidence for periodic reversals of Earth's magnetic field.

Quick Lab

TEACHER'S NOTES

If performing this lab in class, test the file cabinet before the lab.

Possible substitutes include iron flagpoles and iron fence posts, such as those around tennis courts.

★ TEKS

p. 768: 3A

768

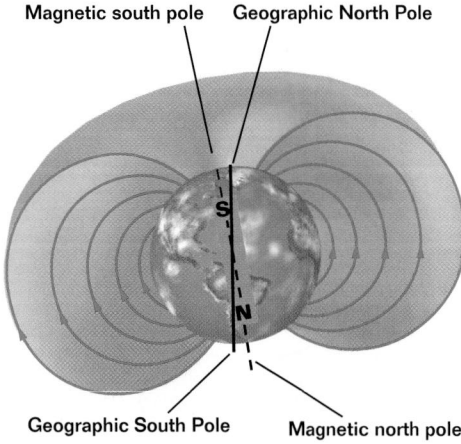

Magnetic south pole   Geographic North Pole

Geographic South Pole   Magnetic north pole

Figure 21-3
Earth's magnetic field has a configuration similar to a bar magnet's. Note that the magnetic south pole is near the geographic North Pole and that the magnetic north pole is near the geographic South Pole.

Note that the configuration of Earth's magnetic field, pictured in **Figure 21-3,** resembles the field that would be produced if a bar magnet were buried within Earth.

If a compass needle is allowed to rotate in the vertical plane as well as in the horizontal plane, the needle will be horizontal with respect to Earth's surface only near the equator. As the compass is moved northward, the needle will rotate so that it points more toward the surface of Earth. Finally, at a point just north of Hudson Bay, in Canada, the north pole of the needle will point directly downward. This site is considered to be the location of the magnetic south pole of Earth. It is approximately 1500 km from Earth's geographic North Pole. Similarly, the magnetic north pole of Earth is roughly the same distance from the geographic South Pole. Thus, it is only an approximation to say that a compass needle points toward the geographic North Pole.

Although Earth has large deposits of iron ore deep beneath its surface, the high temperatures in Earth's liquid core prevent the iron from retaining any permanent magnetization. It is considered more likely that the source of Earth's magnetic field is the movement of charges in convection currents in Earth's core. Charged ions or electrons circling in the liquid interior could produce a magnetic field. There is also evidence that the strength of a planet's magnetic field is related to the planet's rate of rotation. For example, Jupiter rotates at a faster rate than Earth, and recent space probes indicate that Jupiter's magnetic field is stronger than Earth's. Conversely, Venus rotates more slowly than Earth, and its magnetic field has been found to be weaker. Investigation into the cause of Earth's magnetism continues. ★ TEKS **3A**

Naturally occurring magnetic materials, such as magnetite, achieve their magnetism because they have been subjected to Earth's magnetic field over very long periods of time. In fact, studies have shown that a type of anaerobic bacterium that lives in swamps has a magnetized chain of magnetite as part of its internal structure. (The term *anaerobic* means that these bacteria live and grow

Quick Lab

Magnetic Field of a File Cabinet

MATERIALS LIST

✔ compass
✔ metal file cabinet

Stand in front of the file cabinet, and hold the compass face up and parallel to the ground. Now move the compass from the top of the file cabinet to the bottom. Making sure that the compass is parallel to the ground, check to see if the direction of the compass needle changes as it moves from the top of the cabinet to the bottom. If the compass needle changes direction, the file cabinet is magnetized. Can you

explain what might have caused the file cabinet to become magnetized? Remember that Earth's magnetic field has a vertical component as well as a horizontal component.

Try tracing the field around some large metal objects around your house. Can you find an object that has been magnetized by the horizontal component of Earth's magnetic field? ★ TEKS **3A**

without oxygen; in fact, oxygen is toxic to some of them.) The magnetized chain acts as a compass needle, enabling the bacteria to align with Earth's magnetic field. When they find themselves out of the mud at the bottom of the swamp, they return to their oxygen-free environment by following the magnetic-field lines of Earth. Further evidence of their magnetic-sensing ability is the discovery that the bacteria in the Northern Hemisphere have internal magnetite chains that are opposite in polarity to that of similar bacteria in the Southern Hemisphere. This is consistent with the fact that in the Northern Hemisphere Earth's magnetic field has a downward component, whereas in the Southern Hemisphere it has an upward component.

Section Review

1. For each of the cases in **Figure 21-4,** identify whether the magnets will attract or repel one another.

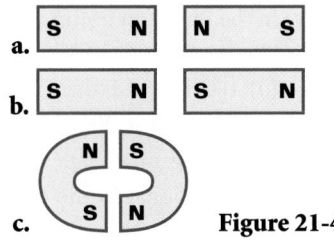

Figure 21-4

2. When you break a magnet in half, how many poles does each piece have?

3. Which of the compass-needle orientations in **Figure 21-5** might correctly describe the magnet's field at that point?

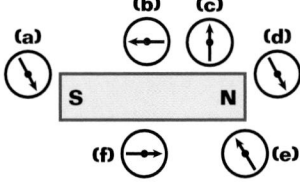

Figure 21-5

4. **Physics in Action**   Satellite ground operators use the feedback from a device called a magnetometer, which senses the direction of Earth's magnetic field, to decide which torque coil to activate. What direction will the magnetometer read for Earth's magnetic field when the satellite passes over Earth's equator?

5. **Physics in Action**   In order to protect other equipment, the body of a satellite must remain unmagnetized, even when the torque coils have been activated. Would hard or soft magnetic materials be best for building the rest of the satellite?

1. **a.** repel
   b. attract
   c. attract
2. two
3. a, b
4. parallel to Earth's surface, pointing from approximately the geographic South Pole (magnetic north pole) to approximately the geographic North Pole (magnetic south pole)
5. hard magnetic material, because it is less easily magnetized

Magnetism   **769**

Electromagnets

Purpose Show that a long, straight, current-carrying wire has a magnetic field.

Materials wire, dc power supply, small compasses, cardboard, ring stand, two clamps

Procedure Cut a small hole in the center of the cardboard. Use the clamp and ring stand to hold the cardboard parallel to the desktop. Thread the wire through the hole in the cardboard, and clamp the wire to the top of the ring stand so that it is perpendicular to the cardboard. Leave at least 10 cm of wire above and below the cardboard. Connect the wire to the dc power supply. Place the compasses on the cardboard in a circular pattern around the wire. Turn on the power supply momentarily, and have students note the deflection of the compass needles. Ask students to describe the magnetic field around the wire (*concentric circles*).

Interactive Problem-Solving Tutor

See Module 18

"Magnetic Field of a Wire" introduces the permeability of free-space and solves problems involving the magnetic field created by a current-carrying wire.

21-2
Electromagnetism and magnetic domains

21-2 SECTION OBJECTIVES

- **Describe the magnetic field produced by the current in a straight conductor and in a solenoid.**

- **Explain magnetism in terms of the domain theory of magnetism.**

Module 18
"Magnetic Field of a Wire" provides an interactive lesson with guided problem-solving practice to teach you about magnetic fields produced by current-carrying wires.

MAGNETIC FIELD OF A CURRENT-CARRYING WIRE

The experiment shown in **Figure 21-6(a)** uses iron filings to show that a current-carrying conductor produces a magnetic field. In a similar experiment, several compass needles are placed in a horizontal plane near a long vertical wire. When there is no current in the wire, all needles point in the same direction (that of Earth's magnetic field). However, when the wire carries a strong, steady current, all the needles deflect in directions tangent to concentric circles around the wire, pointing in the direction of **B,** the magnetic field due to the current. When the current is reversed, the needles reverse direction. These observations show that the direction of **B** is consistent with this rule for conventional current, known as the right-hand rule: If the wire is grasped in the right hand with the thumb in the direction of the current, as shown in **Figure 21-6(b),** the fingers will curl in the direction of **B.** (★)TEKS 3A

As shown in **Figure 21-6(a),** the lines of **B** form concentric circles about the wire. By symmetry, the magnitude of **B** is the same everywhere on a circular path centered on the wire and lying in a plane perpendicular to the wire. Experiments show that **B** is proportional to the current in the wire and inversely proportional to the distance from the wire.

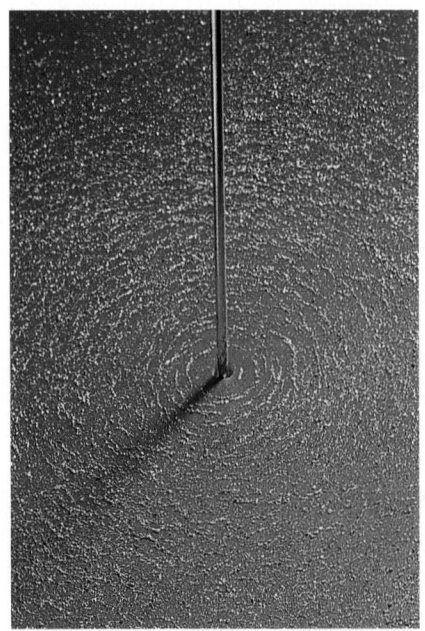

(a)

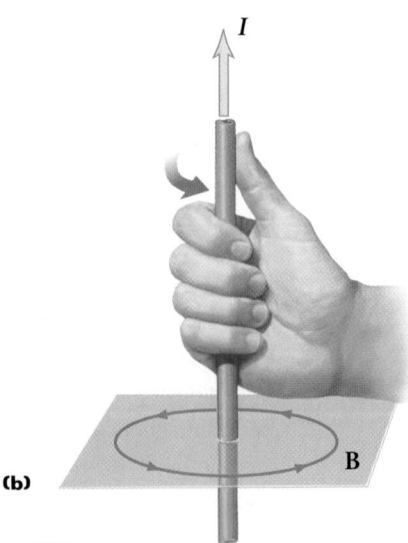

(b)

Figure 21-6
(a) When the wire carries a strong current, the iron filings show that the magnetic field due to the current forms concentric circles around the wire. **(b)** Use the right-hand rule to find the direction of this magnetic field.

MAGNETIC FIELD OF A CURRENT LOOP

The right-hand rule can also be applied to find the direction of the magnetic field of a current-carrying loop, such as the loop represented in **Figure 21-7(a).** Regardless of where on the loop you apply the right-hand rule, the field within the loop points in the same direction—upward. Note that the field lines of the current-carrying loop resemble those of a bar magnet, as shown in **Figure 21-7(b).** If a long, straight wire is bent into a coil of several closely spaced loops, as shown in **Figure 21-8,** the resulting device is called a **solenoid.**

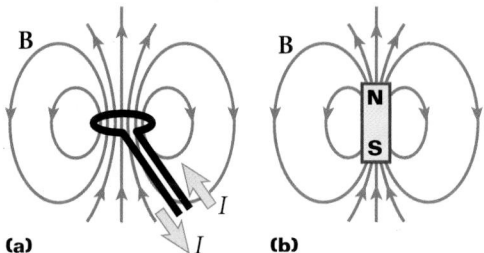

(a)

(b)

Figure 21-7
(a) The magnetic field of a current loop is similar to **(b)** that of a bar magnet.

Solenoids produce a strong magnetic field by combining several loops

A solenoid is important in many applications because it acts as a magnet when it carries a current. The magnetic field inside a solenoid increases with the current and is proportional to the number of coils per unit length. The magnetic field of a solenoid can be increased by inserting an iron rod through the center of the coil; this device is often called an *electromagnet.* The magnetic field that is induced in the rod adds to the magnetic field of the solenoid, often creating a powerful magnet.

Figure 21-8 shows the magnetic field lines of a solenoid. Note that the field lines inside the solenoid point in the same direction, are nearly parallel, are uniformly spaced, and are close together. This indicates that the field inside the solenoid is strong and nearly uniform. The field outside the solenoid is nonuniform and much weaker than the interior field.

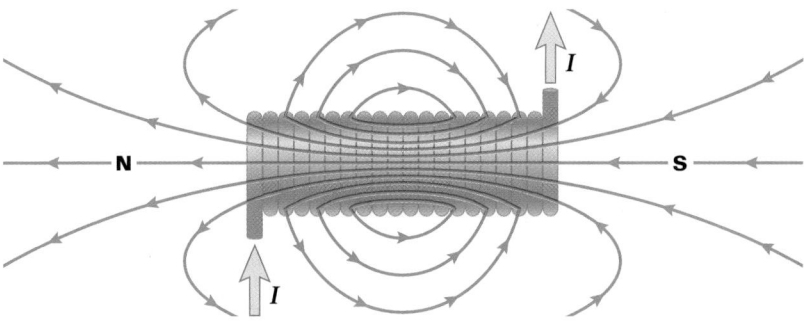

Figure 21-8
The magnetic field inside a solenoid is strong and nearly uniform. Note that the field lines resemble those of a bar magnet, so a solenoid effectively has north and south poles.

solenoid

a long, helically wound coil of insulated wire

TEKS **1A, 6D**

Quick Lab

Electromagnetism

MATERIALS LIST

✔ D-cell battery
✔ 1 m length of insulated wire
✔ large nail
✔ compass

Wind the wire around the nail as shown below. Remove the insulation from the ends of the wire, and hold these ends against the metal

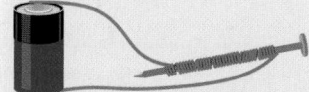

terminals of the battery. Use the compass to determine whether the nail is magnetized. Next, flip the battery so that the direction of the current is reversed. Again bring the compass toward the same part of the nail. Can you explain why the compass needle now points in a different direction?

Quick Lab

TEACHER'S NOTES

Tell students that a short circuit is being created and that the potential difference across the battery goes to zero quickly. Be sure students disconnect the battery after making their observations.

Visual Strategy

Figure 21-8

Have students practice using the right-hand rule by confirming that the direction of the magnetic field shown in **Figure 21-8** is correctly drawn at all points around the solenoid.

Q If the current in the solenoid is reversed, which end of the solenoid is the north pole?

A *the right side*

TEKS

p. 770: 3A
p. 771: 1A, 6D

Demonstration 4

Magnetic domains

Purpose Show the effects of impact on a magnetized ferromagnetic material.

Materials 2 small paper clips, bar magnet

Procedure Pick up one paper clip with the magnet. Touch the second paper clip to the bottom of the first paper clip so that both are suspended from the magnet. Remove the first paper clip from the bar magnet and have students note that the second paper clip remains suspended from the first. Have students conclude that the second paper clip has also become magnetized. Drop both paper clips onto the table top. Try to pick up one paper clip with the other, and have students note that the clips are no longer magnetic. Ask students to explain. (*The impact of the clips hitting the table caused the domains to once again return to random orientations.*)

Section Review
ANSWERS

1. concentric circles around the wire

2. The fields produced by the top and bottom of every loop all point in the same direction and are confined to a small region of space.

3. Electrons usually pair up with their spins opposite each other, and their fields cancel each other.

4. The magnetic field that is induced (by domain variation) in the rod adds to the magnetic field of the solenoid.

Figure 21-9
When a substance is unmagnetized, its domains are randomly oriented.

domain

a microscopic magnetic region composed of a group of atoms whose magnetic fields are aligned in a common direction

MAGNETIC DOMAINS

The magnetic properties of many materials are explained in terms of a model in which an electron is said to spin on its axis much like a top does. (This classical description should not be taken literally. The property of electron spin can be understood only with the methods of quantum mechanics.) The spinning electron represents a charge in motion that produces a magnetic field. In atoms containing many electrons, the electrons usually pair up with their spins opposite each other, and their fields cancel each other. That is why most substances, such as wood and plastic, are not magnets. However, in materials such as iron, cobalt, and nickel, the magnetic fields produced by the electron spins do not cancel completely. Such materials are said to be *ferromagnetic*. In ferromagnetic materials, strong coupling occurs between neighboring atoms to form large groups of atoms whose net spins are aligned; these groups are called **domains.** Domains typically range in size from about 10^{-4} cm to 0.1 cm. In an unmagnetized substance, the domains are randomly oriented, as shown in **Figure 21-9.** When an external field is applied, the orientation of the magnetic fields of each domain may change slightly to more closely align with the external magnetic field, or the domains that are already aligned with the external field may grow at the expense of the other domains. ⊛TEKS 3A

In hard magnetic materials, domain alignment persists after the external magnetic field is removed; the result is a permanent magnet. In soft magnetic materials, such as iron, once the external field is removed, the random motion of the particles in the material changes the orientation of the domains and the material returns to an unmagnetized state.

As mentioned earlier, the strength of a solenoid can be increased dramatically by the insertion of an iron rod into the coil's center. The magnetic field produced by the current in the loops causes alignment of the domains in the iron, producing a large net external field.

Section Review

1. What is the shape of the magnetic field produced by a straight current-carrying wire?

2. Why is the magnetic field inside a solenoid stronger than the magnetic field outside?

3. If electrons behave like magnets, then why aren't all atoms magnets?

4. **Physics in Action** In some satellites, torque coils are replaced by devices called torque rods. In torque rods, a ferromagnetic material is inserted inside the coil. Why does a torque rod have a stronger magnetic field than a torque coil?

CHARGED PARTICLES IN A MAGNETIC FIELD

Although experiments show that a stationary charged particle does not inter-act with a constant magnetic field, charges moving through a magnetic field do experience a magnetic force. This force has its maximum value when the charge moves perpendicular to the magnetic field, decreases in value at other angles, and becomes zero when the particle moves along the field lines. For the purposes of this book, we will limit our discussion to situations in which charges move parallel or perpendicular to the magnetic-field lines.

A charge moving through a magnetic field experiences a force

In our discussion of electric forces, the electric field at a point in space was defined as the electric force per unit charge acting on some test charge placed at that point. In a similar manner, we can describe the properties of the mag-netic field, **B,** in terms of the magnetic force exerted on a test charge at a given point. Our test object is assumed to be a positive charge, q, moving with veloc-ity **v.** It has been found experimentally that the strength of the magnetic force on the particle moving perpendicular to the field is equal to the product of the magnitude of the charge, q, the magnitude of the velocity, v, and the strength of the external magnetic field, B, as shown by the following relationship.

$$F_{magnetic} = qvB$$

This expression can be rearranged as follows:

MAGNITUDE OF A MAGNETIC FIELD

$$B = \frac{F_{magnetic}}{qv}$$

$$\text{magnetic field} = \frac{\text{magnetic force on a charged particle}}{(\text{magnitude of charge})(\text{speed of charge})}$$

If the force is in newtons, the charge is in coulombs, and the speed is in meters per second, the unit of magnetic field strength is the tesla (T). Thus, if a 1 C charge moving at 1 m/s perpendicular to a magnetic field experiences a magnetic force of 1 N, the magnitude of the magnetic field is equal to 1 T. Note that 1 C is a very large amount of charge, so most magnetic fields are much smaller than 1 T. We can express the units of the magnetic field as follows:

$$T = N/(C \bullet m/s) = N/(A \bullet m) = (V \bullet s)/m^2$$

21-3 SECTION OBJECTIVES

- **Given the force on a charge in a magnetic field, determine the strength of the magnetic field.**

- **Use the right-hand rule to find the direction of the force on a charge moving through a magnetic field.**

- **Determine the magnitude and direction of the force on a wire carrying current in a magnetic field.**

The following may be used as teamwork exercises or for demonstration at the chalkboard or on an overhead projector.

PROBLEM

Particle in a magnetic field

An electron moving north at 4.5×10^4 m/s enters a 1.0 mT magnetic field pointed upward.

a. What is the magnitude and direction of the force exerted on the electron?

b. What would the force be if the particle were a proton?

c. What would the force be if the particle were a neutron?

Answers

a. 7.2×10^{-18} N west

b. 7.2×10^{-18} N east

c. 0.0 N

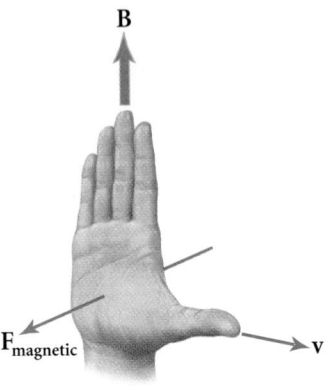

Figure 21-10
Use the right-hand rule to find the direction of the magnetic force on a positive charge.

Conventional laboratory magnets can produce magnetic fields up to about 1.5 T. Superconducting magnets that can generate magnetic fields as great as 30 T have been constructed. These values can be compared with Earth's magnetic field near its surface, which is about 50 μT (5×10^{-5} T). ★TEKS **2C**

Use the right-hand rule to find the direction of the magnetic force

Experiments show that the direction of the magnetic force is always perpendicular to both **v** and **B**. To determine the direction of the force, use the right-hand rule. As before, place your fingers in the direction of **B** with your thumb pointing in the direction of **v**, as illustrated in **Figure 21-10**. The magnetic force, **F$_{magnetic}$**, on a positive charge is directed *out* of the palm of your hand.

If the charge is negative rather than positive, the force is directed *opposite* that shown in **Figure 21-10**. That is, if q is negative, simply use the right-hand rule to find the direction of **F$_{magnetic}$** for positive q, and then reverse this direction for the negative charge. ★TEKS **3A**

The force on a moving charge due to a magnetic field is used to create pictures on a television screen. The main component of a television is the *cathode ray tube*, which is essentially a vacuum tube in which electric fields are used to form a beam of electrons. Phosphor on the television screen glows when it is struck by the electrons in the beam. Without magnetism, however, only the center of the screen would be illuminated by the beam. The direction of the beam is changed by two electromagnets, one deflecting the beam horizontally, the other deflecting the beam vertically. The direction of the beam can be changed by changing the direction of the current in each electromagnet. In this way, the beam illuminates the entire screen. ★TEKS **3C, 6F**

In a color television, three different colors of phosphor—red, green, and blue—make up the screen. Three electron beams, one for each color, scan over the screen to produce a color picture.

SAMPLE PROBLEM 21A

Particle in a magnetic field ★TEKS **2C**

PROBLEM

A proton moving east experiences a force of 8.8×10^{-19} N upward due to the Earth's magnetic field. At this location, the field has a magnitude of 5.5×10^{-5} T to the north. Find the speed of the particle.

SOLUTION

Given: $q = 1.60 \times 10^{-19}$ C $B = 5.5 \times 10^{-5}$ T

               $F_{magnetic} = 8.8 \times 10^{-19}$ N

Unknown: $v = ?$

Use the equation from page 773. Rearrange to solve for v.

$$B = \frac{F_{magnetic}}{qv}$$

$$v = \frac{F_{magnetic}}{qB}$$

$$v = \frac{8.8 \times 10^{-19} \text{ N}}{(1.60 \times 10^{-19} \text{ C})(5.5 \times 10^{-5} \text{ T})} = 1.0 \times 10^5 \text{ m/s}$$

The directions given can be used to verify the right-hand rule. Imagine standing at this location and facing north. Turn the palm of your right hand upward (the direction of the force) with your thumb pointing east (the direction of the velocity). If your palm and thumb point in these directions, your fingers point directly north in the direction of the magnetic field, as they should.

PRACTICE 21A

Particle in a magnetic field ⊛ TEKS 2C

1. A proton moves perpendicularly to a magnetic field that has a magnitude of 4.20×10^{-2} T. What is the speed of the particle if the magnitude of the magnetic force on it is 2.40×10^{-14} N?

2. A proton traveling to the right along the x-axis enters a region where there is a magnetic field of magnitude 2.5 T directed upward along the y-axis. If the proton experiences a force of 3.2×10^{-12} N, find the speed of the proton.

3. If an electron in an electron beam experiences a downward force of 2.0×10^{-14} N while traveling in a magnetic field of 8.3×10^{-2} T west, what is the direction and magnitude of the velocity?

4. A uniform 1.5 T magnetic field points north. If an electron moves vertically downward (toward the ground) with a speed of 2.5×10^7 m/s through this field, what force (magnitude and direction) will act on it?

5. A proton moves straight upward (away from the ground) through a uniform magnetic field that points from east to west and has a magnitude of 2.5 T. If the proton moves with a speed of 1.5×10^7 m/s through this field, what force (magnitude and direction) will act on it?

6. An alpha particle (the nucleus of a helium atom, carrying a charge of 3.2×10^{-19} C) moves at 5.5×10^7 m/s at a right angle to a magnetic field. If the particle experiences a force of 1.5×10^{-14} N due to the magnetic field, then what is the magnitude of the magnetic field?

PRACTICE GUIDE 21A

Solving for:	
v	📖 **PE** Sample, 1–3; Ch. Rvw. 30–31
	💿 **PW** 6–7
	PB 4–6
$F_{magnetic}$	📖 **PE** 4–5; Ch. Rvw. 35–37, 40*
	💿 **PW** Sample, 1–3
	PB 7–10
B	📖 **PE** 6; Ch. Rvw. 34, 39
	💿 **PW** 4–5
	PB Sample, 1–3

ANSWERS TO

Practice 21A
Particle in a magnetic field
1. 3.57×10^6 m/s
2. 8.0×10^6 m/s
3. 1.5×10^6 m/s north
4. 6.0×10^{-12} N west
5. 6.0×10^{-12} N south
6. 8.5×10^{-4} T

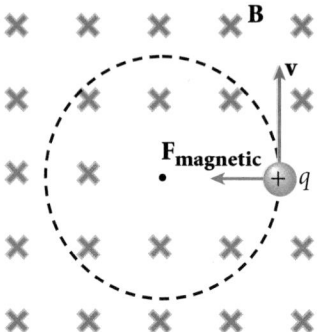

Figure 21-11

When the velocity, **v**, of a charged particle is perpendicular to a uniform magnetic field, the particle moves in a circle whose plane is perpendicular to **B**.

Module 19
"Magnetic Force on a Wire" provides an interactive lesson with guided problem-solving practice to teach you about the magnetic force on current-carrying wires that are not perpendicular to the magnetic field.

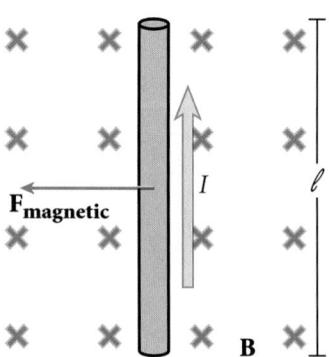

Figure 21-12

A current-carrying conductor in a magnetic field experiences a force that is perpendicular to the direction of the current.

A charge moving through a magnetic field follows a circular path

Consider a positively charged particle moving in a uniform magnetic field so that the direction of the particle's velocity is perpendicular to the field, as in **Figure 21-11.** Application of the right-hand rule for the charge q shows that the direction of the magnetic force, **F**$_{magnetic}$, at the charge's location is to the left. This causes the particle to alter its direction and to follow a curved path. Application of the right-hand rule at any point shows that the magnetic force is always directed toward the center of the circular path. Therefore, the magnetic force is, in effect, a force that maintains circular motion and changes only the direction of **v,** not its magnitude.

Now consider a charged particle traveling through a uniform magnetic field with a velocity that is neither parallel nor perpendicular to the direction of the magnetic field. In this case, the particle will follow a helical path along the direction of the magnetic field.

MAGNETIC FORCE ON A CURRENT-CARRYING CONDUCTOR

Recall that current is a collection of many charged particles in motion. If a force is exerted on a single charged particle when the particle moves through a magnetic field, it should be no surprise that a current-carrying wire also experiences a force when it is placed in a magnetic field. The resultant force on the wire is due to the sum of the individual forces on the charged particles. The force on the particles is transmitted to the bulk of the wire through collisions with the atoms making up the wire.

Consider a straight segment of wire of length ℓ carrying current, $I,$ in a uniform external magnetic field, **B,** as in **Figure 21-12.** The magnitude of the total magnetic force on the wire is given by the following relationship.

FORCE ON A CURRENT-CARRYING CONDUCTOR PERPENDICULAR TO A MAGNETIC FIELD

$$F_{magnetic} = BI\ell$$

**magnitude of magnetic force = (magnitude of magnetic field)
(current)(length of conductor within B)**

This equation can be used only when the current and the magnetic field are at right angles to each other.

The direction of the magnetic force on a wire can be obtained by using the right-hand rule. However, in this case, you must place your thumb in the direction of the current rather than in the direction of the velocity, **v.** In **Figure 21-12,** the direction of the magnetic force on the wire is to the left. When the current is either in the direction of the field or opposite the direction of the field, the magnetic force on the wire is zero.

Two parallel conducting wires exert a force on one another

Since a current in a conductor creates its own magnetic field, it is easy to understand that two current-carrying wires placed close together exert magnetic forces on each other. When the two conductors are parallel to each other, the direction of the magnetic field created by one is perpendicular to the direction of the current of the other, and vice versa. In this way, a force of $F_{magnetic} = BI\ell$ acts on each wire, where B is the magnitude of the magnetic field created by the other wire.

Consider the two long, straight, parallel wires shown in **Figure 21-13.** When the current in each is in the same direction, as shown in **Figure 21-13(a),** the two wires attract one another. You can confirm this by using the right-hand rule. Point your thumb in the direction of current in one wire, and point your fingers in the direction of the field produced by the other wire. By doing this, you find that the direction of the force (pointing out from the palm of your hand) is toward the other wire. When the currents in each wire are in opposite directions, as shown in **Figure 21-13(b),** the wires repel one another.

Loudspeakers use magnetic force to produce sound

The loudspeakers in most sound systems use a magnetic force acting on a current-carrying wire in a magnetic field to produce sound waves. One speaker design, shown in **Figure 21-14,** consists of a coil of wire, a flexible paper cone attached to the coil that acts as the speaker, and a permanent magnet. A sound signal is converted to a varying electric signal and is sent to the coil. The current causes a magnetic force to act on the coil. When the current reverses direction, the magnetic force on the coil reverses direction, and the cone accelerates in the opposite direction. This alternating force on the coil results in vibrations of the attached cone, which produce variations in the density of the air in front of it. In this way, an electric signal is converted to a sound wave that closely resembles the sound wave produced by the source.

(★) TEKS 3C, 6F, 8B

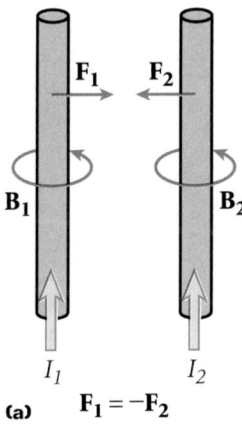

(a) $F_1 = -F_2$

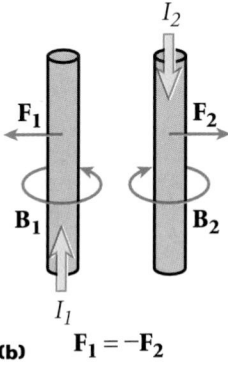

(b) $F_1 = -F_2$

Figure 21-13
Two parallel wires, each carrying a steady current, exert forces on each other. The force is **(a)** attractive if the currents have the same direction and **(b)** repulsive if the two currents have opposite directions.

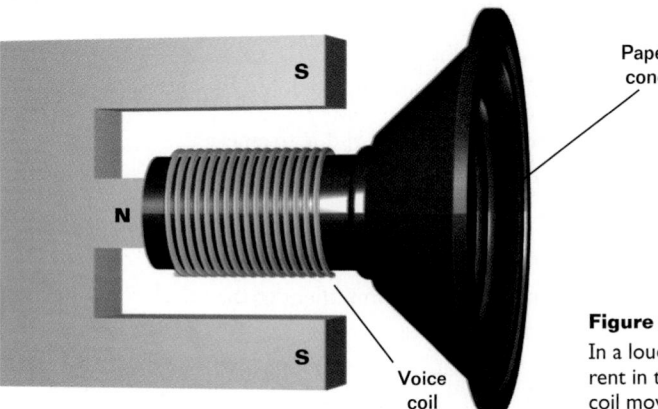

Figure 21-14
In a loudspeaker, when the direction and magnitude of the current in the coil of wire change, the paper cone attached to the coil moves, producing sound waves.

PROBLEM

A 4.5 m wire carries a current of 12.5 A from north to south. If the magnetic force on the wire due to a uniform magnetic field is 1.1×10^3 N downward, what is the magnitude and direction of the magnetic field?

Answer

2.0×10^1 T, to the west

PRACTICE GUIDE 21B		
Solving for:		
B	📖 PE	Sample, 1–3; Ch. Rvw. 32–33, 41–44
	💿 PW	5–6
	PB	3–5
F	💿 PW	9–10
	PB	8–10
ℓ	📖 PE	4
	💿 PW	7–8
	PB	Sample, 1–2
I	💿 PW	Sample, 1–4
	PB	6–7

ANSWERS TO

Practice 21B

1. 1.7×10^{-7} T in $+z$ direction
2. 0.050 T
3. 1.5 T
4. 0.59 m

SAMPLE PROBLEM 21B

Force on a current-carrying conductor ⊛TEKS 2C

PROBLEM

A wire 36 m long carries a current of 22 A from east to west. If the maximum magnetic force on the wire at this point is downward (toward Earth) and has a magnitude of 4.0×10^{-2} N, find the magnitude and direction of the magnetic field at this location.

SOLUTION

Given: $\ell = 36$ m $I = 22$ A $F_{magnetic} = 4.0 \times 10^{-2}$ N

Unknown: $B = ?$

Use the equation for the force on a current-carrying conductor perpendicular to a magnetic field, shown on page 776.

$$F_{magnetic} = BI\ell$$

Rearrange to solve for B.

$$B = \frac{F_{magnetic}}{I\ell} = \frac{4.0 \times 10^{-2}\ \text{N}}{(22\ \text{A})(36\ \text{m})} = 5.0 \times 10^{-5}\ \text{T}$$

Using the right-hand rule to find the direction of B, face north with your thumb pointing to the west (in the direction of the current) and the palm of your hand down (in the direction of the force). Your fingers point north. Thus, Earth's magnetic field is from south to north.

PRACTICE 21B

Force on a current-carrying conductor

1. A 6.0 m wire carries a current of 7.0 A toward the $+x$ direction. A magnetic force of 7.0×10^{-6} N acts on the wire in the $-y$ direction. Find the magnitude and direction of the magnetic field producing the force.

2. A wire 1.0 m long experiences a magnetic force of 0.50 N due to a perpendicular uniform magnetic field. If the wire carries a current of 10.0 A, what is the magnitude of the magnetic field?

3. The magnetic force on a straight 0.15 m segment of wire carrying a current of 4.5 A is 1.0 N. What is the magnitude of the component of the magnetic field that is perpendicular to the wire?

4. The magnetic force acting on a wire that is perpendicular to a 1.5 T uniform magnetic field is 4.4 N. If the current in the wire is 5.0 A, what is the length of the wire that is inside the magnetic field?

GALVANOMETERS

A *galvanometer* is a device used in the construction of both ammeters and voltmeters. Its operation is based on the idea that a torque acts on a current loop in the presence of a magnetic field. **Figure 21-15** shows a simplified arrangement of the main components of a galvanometer. It consists of a coil of wire wrapped around a soft iron core mounted so that it is free to pivot in the magnetic field provided by the permanent magnet. The torque experienced by the coil is proportional to the current in the coil. This means that the larger the current, the greater the torque and the more the coil will rotate before the spring tightens enough to stop the movement. Hence, the amount of deflection of the needle is proportional to the current in the coil. When there is no current in the coil, the spring returns the needle to zero. Once the instrument is properly calibrated, it can be used in conjunction with other circuit elements as an ammeter (to measure currents) or as a voltmeter (to measure potential differences).

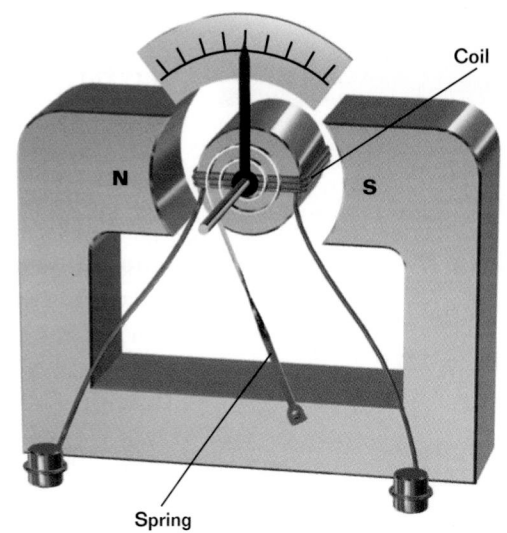

Figure 21-15
In a galvanometer, when current enters the coil, which is in a magnetic field, the magnetic force causes the coil to twist.

SECTION 21-3

Visual Strategy

Figure 21-15
Encourage students to use the right-hand rule to understand how a galvanometer works.

Q If charges move through the coil from the left terminal to the right terminal, which direction will the needle rotate?

A *clockwise*

Section Review   ⊛ TEKS 2C

1. A particle with a charge of 0.030 C experiences a magnetic force of 1.5 N while moving at right angles to a uniform magnetic field. If the speed of the charge is 620 m/s, what is the magnitude of the magnetic field the particle passes through?

2. An electron moving north encounters a uniform magnetic field. If the magnetic field points east, what is the direction of the magnetic force on the electron?

3. A straight segment of wire has a length of 25 cm and carries a current of 5.0 A. If the wire is perpendicular to a magnetic field of 0.60 T, then what is the magnitude of the magnetic force on this segment of the wire?

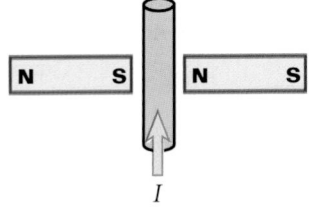

4. Two parallel wires have charges moving in the same direction. Is the force between them attractive or repulsive?

5. Find the direction of the magnetic force on the current-carrying wire in **Figure 21-16.**

Figure 21-16

Section Review
ANSWERS

1. 0.081 T
2. upward
3. 0.75 N
4. attractive
5. out of the page

⊛ TEKS

p. 778: 2C
p. 779: 2C

CHAPTER 21
Summary

Teaching Tip

Have students write an essay summarizing the various cases of current-carrying wires discussed in this chapter. Essays should include a thorough explanation of the magnetic field produced by a current-carrying wire, the magnetic force on a current-carrying wire that is in a magnetic field, and the force between two parallel current-carrying wires. Remind students to include the right-hand rule for each case.

KEY TERMS

domain (p. 772)

magnetic field (p. 767)

solenoid (p. 771)

KEY IDEAS

Section 21-1 Magnets and magnetic fields

- Like magnetic poles repel, and unlike poles attract.
- The direction of any magnetic field is defined as the direction the north pole of a magnet would point if placed in the field. The magnetic field of a magnet points from the north pole of the magnet to the south pole.
- The magnetic north pole of Earth corresponds to the geographic South Pole, and the magnetic south pole corresponds to the geographic North Pole.

Section 21-2 Electromagnetism and magnetic domains

- A magnetic field exists around any current-carrying wire; the direction of the magnetic field follows a circular path around the wire.
- The magnetic field created by a solenoid or coil is similar to the magnetic field of a permanent magnet.
- A domain is a group of atoms whose magnetic fields are aligned.

Section 21-3 Magnetic force

- The direction of the force on a positive charge moving through a magnetic field can be found using the right-hand rule. The magnitude of a magnetic field is given by the relation $B = \dfrac{F_{magnetic}}{qv}$.
- A length of wire, ℓ, in an external magnetic field undergoes a magnetic force with a magnitude of $F_{magnetic} = BI\ell$. The direction of the magnetic force on the wire can be found using the right-hand rule.
- Two parallel current-carrying wires exert on one another forces that are equal in magnitude and opposite in direction. If the currents are in the same direction, the two wires attract one another. If the currents are in opposite directions, the wires repel one another.

Diagram symbols

Magnetic field vector	
Magnetic field pointing into the page	
Magnetic field pointing out of the page	

Variable symbols

Quantities		Units		Conversions
B	magnetic field	T	tesla	$= \dfrac{N}{C \cdot m/s} = \dfrac{N}{A \cdot m}$
$F_{magnetic}$	magnetic force	N	newtons	$= \dfrac{kg \cdot m}{s^2}$
ℓ	length of conductor in field	m	meters	

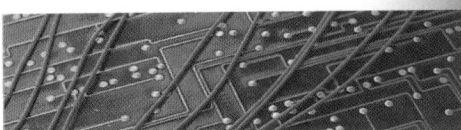

CHAPTER 21
Review and Assess

MAGNETS AND MAGNETIC FIELDS

Review questions

1. What is the minimum number of poles for a magnet?

2. When you break a magnet in half, how many poles does each piece have?

3. The north pole of a magnet is attracted to the geographic North Pole of Earth, yet like poles repel. Can you explain this?

4. Which way would a compass needle point if you were at the magnetic north pole?

Conceptual questions

5. You are an astronaut stranded on a planet with no test equipment or minerals around. The planet does not even have a magnetic field. You have two iron bars in your possession; one is magnetized, one is not. How can you determine which one is magnetized?

6. In **Figure 21-17,** two permanent magnets with holes bored through their centers are placed one over the other. Because the poles of the upper magnet are the reverse of those of the lower, the upper magnet levitates above the lower magnet. If the upper magnet were displaced slightly, either up or down, would the resulting motion be periodic? Explain. What would happen if the upper magnet were inverted?

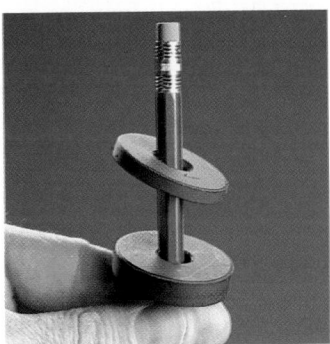

Figure 21-17

ELECTROMAGNETISM AND MAGNETIC DOMAINS

Review questions

7. What is a magnetic domain?

8. Why are iron atoms so strongly affected by magnetic fields?

9. When a magnetized steel needle is strongly heated in a Bunsen burner flame, it becomes demagnetized. Explain why.

10. What indicates that a piece of iron is magnetic, its attraction to or repulsion from another piece of iron?

11. Why does a very strong magnet attract both poles of a weak magnet?

12. A magnet attracts a piece of iron. The iron can then attract another piece of iron. Explain, on the basis of alignment of domains, what happens in each piece of iron.

13. When a small magnet is repeatedly dropped, it becomes demagnetized. Explain what happens to the magnet subatomically.

14. A conductor carrying a current is arranged so that electrons flow in one segment from east to west. If a compass is held over this segment of the wire, in what direction is the needle deflected?

15. What factors does the strength of the magnetic field of a solenoid depend on?

Conceptual questions

16. A solenoid with ends marked A and B is suspended by a thread so that the core can rotate in the horizontal plane. A current is maintained in the coil so that the electrons move clockwise when viewed from end A toward end B. How will the coil align itself in Earth's magnetic field?

17. yes, by aligning the plane of the loop perpendicular to the magnetic field

18. Yes, the north pole of the solenoid would point to Earth's geographic North Pole; No, the solenoid would oscillate back and forth as its poles continually reversed.

19. They have opposite charge.

20. The proton would go left, and the electron would go right.

21. The magnetic field of the magnet exerts a force on the moving electrons in the electron beam.

22. The proton would move up in a half circle and exit above its point of entry. The electron would move down in a half circle and exit below its point of entry.

23. Their magnetic fields point in opposite directions at the point directly between them.

24. no; Magnetic fields only exert a force on moving charges.

25. positive y direction; no; It moves at an angle between the x- and y-axes.

26. a. into the page
    b. to the right
    c. down the page

27. a. The stream moves away from the wire.
    b. The stream moves toward the wire.

28. The stream moves toward the observer.

29. For parallel wires carrying currents in opposite directions, the magnetic field exists only between the wires. By twisting the wires together, the region where the field is nonzero is very small.

30. 15 m/s

31. 2.1×10^{-3} m/s

782

---

17. Is it possible to orient a current-carrying loop of wire in a uniform magnetic field so that the loop will not tend to rotate?

18. If a solenoid were suspended by a string so that it could rotate freely, could it be used as a compass when it carried a direct current? Could it also be used if the current were alternating in direction?

MAGNETIC FORCE

Review questions

19. Two charged particles are projected into a region where there is a magnetic field perpendicular to their velocities. If the particles are deflected in opposite directions, what can you say about them?

20. Suppose an electron is chasing a proton up this page when suddenly a magnetic field pointing into the page is applied. What would happen to the particles?

21. Why does the picture on a television screen become distorted when a magnet is brought near the screen?

22. A proton moving horizontally enters a region where there is a uniform magnetic field perpendicular to the proton's velocity, as shown in **Figure 21-18**. Describe the proton's subsequent motion. How would an electron behave under the same circumstances?

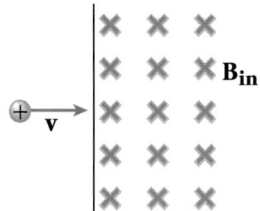

Figure 21-18

23. Explain why two parallel wires carrying currents in opposite directions repel each other.

24. Can a stationary magnetic field set a resting electron in motion? Explain.

25. At a given instant, a proton moves in the positive x direction in a region where there is a magnetic field in the negative z direction. What is the direction of the magnetic force? Does the proton continue to move along the x–axis? Explain.

---

26. Find the direction of the magnetic field for a positively charged particle moving in each situation in **Figure 21-19** if the direction of the magnetic force acting on it is as indicated.

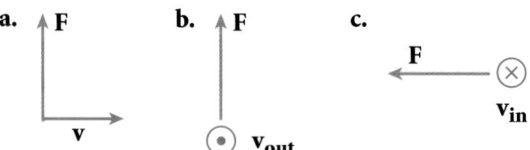

Figure 21-19

Conceptual questions

27. A stream of electrons is projected horizontally to the right. A straight conductor carrying a current is supported parallel to and above the electron stream.
    a. What is the effect on the electron stream if the current in the conductor is left to right?
    b. What is the effect if the current is reversed?

28. If the conductor in item 27 is replaced by a magnet with a downward magnetic field, what is the effect on the electron stream?

29. Two wires carrying equal but opposite currents are twisted together in the construction of a circuit. Why does this technique reduce stray magnetic fields?

Practice problems

30. A duck flying due east passes over Atlanta, where the magnetic field of the Earth is 5.0×10^{-5} T directed north. The duck has a positive charge of 4.0×10^{-8} C. If the magnetic force acting on the duck is 3.0×10^{-11} N upward, what is the duck's velocity? (See Sample Problem 21A.)

31. A proton moves eastward in the plane of Earth's magnetic equator so that its distance from the ground remains constant. What is the speed of the proton if Earth's magnetic field points north and has a magnitude of 5.0×10^{-5} T? (See Sample Problem 21A.)

32. A wire carries a 10.0 A current at an angle 90.0° from the direction of a magnetic field. If the magnitude of the magnetic force on a 5.00 m length of the wire is 15.0 N, what is the strength of the magnetic field? (See Sample Problem 21B.)

33. A thin 1.00 m long copper rod in a uniform magnetic field has a mass of 50.0 g. When the rod carries a current of 0.245 A, it floats in the magnetic field. What is the field strength of the magnetic field? (See Sample Problem 21B.)

MIXED REVIEW

34. A proton moves at 2.50×10^6 m/s horizontally at a right angle to a magnetic field.

    a. What is the strength of the magnetic field required to exactly balance the weight of the proton and keep it moving horizontally?

    b. Should the direction of the magnetic field be in a horizontal or a vertical plane?

35. Find the direction of the force on a proton moving through each magnetic field in **Figure 21-20.**

  a. 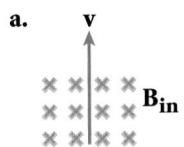   **b.**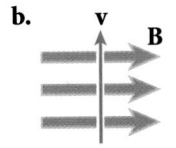

  c. 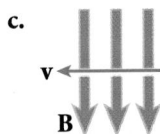   **d.**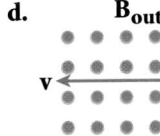

Figure 21-20

36. Find the direction of the force on an electron moving through each magnetic field in **Figure 21-20.**

37. In **Figure 21-20**, assume that in each case the velocity vector shown is replaced with a wire carrying a current in the direction of the velocity vector. Find the direction of the magnetic force acting on each wire.

38. A proton moves at a speed of 2.0×10^7 m/s at right angles to a magnetic field with a magnitude of 0.10 T. Find the magnitude of the acceleration of the proton.

39. A proton moves perpendicularly to a uniform magnetic field, **B**, with a speed of 1.0×10^7 m/s and experiences an acceleration of 2.0×10^{13} m/s^2 in the positive x direction when its velocity is in the positive z direction. Determine the magnitude and direction of the field.

40. A proton travels with a speed of 3.0×10^6 m/s at an angle of 37° west of north. A magnetic field of 0.30 T points to the north. Determine the following:

    a. the magnitude of the magnetic force on the proton

    b. the direction of the magnetic force on the proton

    c. the proton's acceleration as it moves through the magnetic field

(Hint: The magnetic force experienced by the proton in the magnetic field is proportional to the component of the proton's velocity that is perpendicular to the magnetic field.)

41. In **Figure 21-21,** a 15 cm length of conducting wire that is free to move is held in place between two thin conducting wires. All the wires are in a magnetic field. When a 5.0 A current is in the wire, as shown in the figure, the wire segment moves upward at a constant velocity. Assuming the wire slides without friction on the two vertical conductors and has a mass of 0.15 kg, find the magnitude and direction of the minimum magnetic field that is required to move the wire.

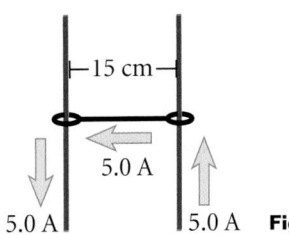

—15 cm—

5.0 A

5.0 A   5.0 A   **Figure 21-21**

42. A current, $I = 15$ A, is directed along the positive x-axis and perpendicular to a uniform magnetic field. The conductor experiences a magnetic force per unit length of 0.12 N/m in the negative y direction. Calculate the magnitude and direction of the magnetic field in the region through which the current passes.

43. A proton moves in a circular path perpendicular to a constant magnetic field so that the proton takes 1.00×10^{-6} s to complete one revolution. Determine the strength of the constant magnetic field. (Hint: The magnetic force exerted on the proton is the force that maintains circular motion, and the number of radians per time interval is the angular speed.)

32. 0.300 T

33. 2.00 T

34. a. 4.10×10^{-14} T

    b. horizontal

35. a. to the left

    b. into the page

    c. out of the page

    d. up the page

36. a. to the right

    b. out of the page

    c. into the page

    d. down the page

37. a. to the left

    b. into the page

    c. out of the page

    d. up the page

38. 1.9×10^{14} m/s^2

39. 2.1×10^{-2} T, in the negative y direction

40. a. 8.7×10^{-14} N

    b. downward

    c. 5.2×10^{13} m/s^2

41. 2.0 T, out of the page

42. 8.0×10^{-3} T, in the positive z direction

43. 6.57×10^{-2} T

44. 1.39×10^{-2} T, toward the observer

44. A singly charged positive ion that has a mass of 6.68×10^{-27} kg moves clockwise with a speed of 1.00×10^{4} m/s. The positively-charged ion moves in a circular path that has a radius of 3.00 cm. Find the direction and strength of the uniform magnetic field. (Hint: The magnetic force exerted on the positive ion is the force that maintains circular motion, and the speed given for the positive ion is its tangential speed.)

Technology & Learning

Graphing calculators

Refer to Appendix B for instructions on downloading programs for your calculator. The program "Chap21" allows you to find the magnetic field of a solenoid given the amount of current in the solenoid.

The program "Chap21" stored on your graphing calculator makes use of line-fitting techniques to find the magnetic field of a solenoid when the amount of current in the solenoid in known. Before executing the program "Chap21," you will enter two sets of data for the magnetic field strength and current in a specific solenoid. Then, when you run the program "Chap21," your graphing calculator will use this data to draw a straight line using the following equation:

$$Y_1 = aX + b$$

Next, your calculator will ask you for the value of the current (X) in the solenoid. The calculator will then find the point along the line that corresponds to that current and report the y value. This y value is the magnetic field strength (Y_1) that corresponds to the current (X) that you input. Remember that this magnetic field strength corresponds to that current only for the solenoid described by this line; new data must be entered into the calculator when you want to analyze the current and magnetic field of a different solenoid.

a. Which letter in the above equation corresponds to the slope of the line?

Clear the data lists by pressing [STAT] [4] [2nd] [L₁] [ENTER] and [STAT] [4] [2nd] [L₂] [ENTER]. Press [STAT] [1], and enter the current data into the list [L₁] and the magnetic field data into the list [L₂]. (Remember to use [2nd] [EE] to enter exponents and the [(-)] key to enter negative numbers.) Press [2nd] [QUIT] to exit the stat list editor.

Execute "Chap21" on the [PRGM] menu, and press [ENTER] to begin the program. Enter the value for the current in amperes, and press [ENTER]. Once you have entered the value for the current in the solenoid, the calculator will fit that current to the line and display the magnetic field strength of the solenoid in units of teslas. Once you have finished using the graph for one situation, press [CLEAR] to end the program. Then enter the data for the next situation and run the program again.

Determine the magnetic field strength for the solenoid with the following current and magnetic field strength/current data points:

b. a current of 2.37 A in a solenoid that has a magnetic field strength of 2.54×10^{-2} T when the current is 3.35 A and 5.20×10^{-2} T when the current is 6.90 A

c. a current of 3.54 A in a solenoid that has a magnetic field strength of 5.50×10^{-3} T when the current is 1.25 A and 1.74×10^{-2} T when the current is 3.80 A

d. Solenoid A and B carry the same current. Solenoid A produces a magnetic field of 1.5×10^{-2} T, and solenoid B produces a magnetic field of 2.5×10^{-3} T. Based on this information, which solenoid has more turns per length?

ANSWERS TO

Technology & Learning

a. a

b. 1.81×10^{-2} T

c. 1.62×10^{-2} T

d. solenoid A

45. What speed would a proton need to achieve in order to circle Earth 1000.0 km above the magnetic equator? Assume that Earth's magnetic field is everywhere perpendicular to the path of the proton and that Earth's magnetic field has an intensity of 4.00×10^{-8} T. (Hint: The magnetic force exerted on the proton is equal to the force that maintains circular motion, and the speed needed by the proton is its tangential speed. Remember that the radius of the circular orbit should also include the radius of Earth.)

46. An electron moves in a circular path perpendicular to a magnetic field that has a magnitude of 1.00×10^{-3} T. If the angular momentum of the electron as it moves around the center of the circle is 4.00×10^{-25} J•s, determine the following quantities involved in the situation:

    a. the radius of the circular path

    b. the speed of the electron

21 REVIEW & ASSESS

45. 2.82×10^7 m/s
46. a. 5.00×10^{-2} m
    b. 8.78×10^6 m/s

Alternative Assessment

Performance assessment

1. During a field investigation with your class, you find a roundish chunk of metal that attracts iron objects. Design a procedure to determine whether the object is magnetic and, if so, to locate its poles. Describe the limitations of your method. What materials would you need? How would you draw your conclusions? List all the possible results you can anticipate and the conclusions you could draw from each result.

2. Imagine you have been hired by a manufacturer interested in making kitchen magnets. The manufacturer wants you to determine how to combine several magnets to get a very strong magnet. He also wants to know what protective material to use to cover the magnets. Develop a method for measuring the strength of different magnets by recording the maximum number of paper clips they can hold under various conditions. First open a paper clip to use as a hook. Test the strength of different magnets and combinations of magnets by holding up the magnet, placing the open clip on the magnet, and hooking the rest of the paper clips so that they hang below the magnet. Examine the effect of layering different materials between the magnet and the clips. Organize your data in tables and graphs to present your conclusions.

Portfolio projects

3. Research phenomena related to one of the following topics, and prepare a report or presentation with pictures and data.

    a. How does Earth's magnetic field vary with latitude, with longitude, with the distance from Earth, and in time?

    b. How do people who rely on compasses account for these differences in Earth's magnetic field?

    c. What is the Van Allen Belt?

    d. How do solar flares occur?

    e. How do solar flares affect Earth?

4. Obtain old buzzers, bells, telephone receivers, speakers, motors from power or kitchen tools, and so on, to take apart. Identify the mechanical and electromagnetic components. Examine their connections. How do they produce magnetic fields? Work in a cooperative group to describe and organize your findings about several devices for a display entitled "Anatomy of Electromagnetic Devices."

5. Magnetic force was first described by the ancient Greeks, who mined a magnetic mineral called magnetite. Magnetite was used in early experiments on magnetic force. Research the historical development of the concept of magnetic force. Describe the work of Peregrinus, William Gilbert, Oersted, Faraday, and other scientists.

Alternative Assessment
ANSWERS

Performance assessment

1. Students' plans should be safe and logical. A compass can indicate the identity and location of the two poles.

2. Students' plans should include safe and complete plans for testing the strength of each magnet based on how many paper clips the magnet can lift.

 **Portfolio projects**

3. Earth's magnetic field is not constant through time or space; People who rely on compasses keep tables of corrections to apply, depending on their approximate location; The Van Allen belt is a cloud of charged particles around Earth; Solar flares are large outflows of charged particles that can affect Earth's magnetic field.

4. Students' answers should clearly indicate which devices are likely to produce a magnetic field (*solenoids and coils*). Students should indicate how such magnetic forces are converted into mechanical motions.

5. Peregrinus is credited with the first experiments with magnetism, using a thin piece of iron to map the magnetic field of magnetite. William Gilbert, Oersted, and Faraday studied both magnetism and electricity.

NOTE

Materials Preparation is given on pp. 764A–764B. Blank data table and sample data table are on the One-Stop Planner CD-ROM. All calculations shown use sample data.

Planning

Recommended time:
1 lab period

Classroom organization:
▶ Each lab group should have two students.

▶ The CBL and sensors procedure and the compass procedure use different methods to examine the magnetic field; be aware of these differences if using them in the same class.

▶ **Safety warnings:** Emphasize the dangers of working with electricity. Remind students to have you check their circuits before turning on the power supply or closing the switch.

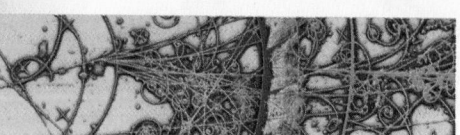

CHAPTER 21
Laboratory Exercise

✪ TEKS

pp. 786–789: 1A, 2B, 2C, 2F, 6D, 6E

OBJECTIVES

• Use a compass or magnetic field sensor to explore the existence, magnitude, and direction of the magnetic field of a current-carrying wire.

• Analyze the relationship between the magnitude of the magnetic field of a conducting wire and the current in the wire.

• Analyze the relationship between the direction of the magnetic field of a conducting wire and the direction of the current in the wire.

MATERIALS LIST

✔ 1 Ω resistor
✔ galvanometer
✔ insulated connecting wires and bare copper wire
✔ masking tape
✔ power supply
✔ switch

PROCEDURE

CBL AND SENSORS
✔ alligator clips
✔ CBL
✔ CBL magnetic field sensor
✔ CBL voltage probe
✔ graphing calculator with link cable
✔ support stand and clamp

COMPASS
✔ compass
✔ multimeter or dc ammeter

MAGNETIC FIELD OF A CONDUCTING WIRE

In this lab, you will study the magnetic field that occurs around a current-carrying wire. You will construct a circuit with a current-carrying wire and use a magnetic compass needle or a CBL and magnetic field sensor to investigate the relationship between the magnetic field and the current in the wire. You will be able to determine the magnitude and direction of the magnetic field surrounding the wire.

SAFETY

• **Never close a circuit until it has been approved by your teacher. Never rewire or adjust any element of a closed circuit. Never work with electricity near water; be sure the floor and all work surfaces are dry.**

• **If the pointer on any kind of meter moves off scale, open the circuit immediately by opening the switch.**

• **Do not attempt this exercise with any batteries, electrical devices, or magnets other than those provided by your teacher for this purpose.**

• **Wire coils may heat up rapidly during this experiment. If heating occurs, open the switch immediately and handle the equipment with a hot mitt. Allow all equipment to cool before storing it.**

PREPARATION

1. Determine whether you will be using the CBL and sensors procedure or the compass procedure. Read the entire lab for the appropriate procedure, and plan what steps you will take.

2. Prepare a data table in your lab notebook.

- **CBL and sensors** Prepare a table with four columns and thirteen rows. In the first row, label the columns *Trial, Current Direction, ΔV_R (V)*, and *$B_{Measured}$ (T)*. Label the 2nd through 13th rows *1* through *12*.

- **Compass** Prepare a table with four columns and nine rows. In the first row, label the columns *Turns, Current (A), Current direction*, and *Compass reading*. In the first column, label the second through ninth rows *One, One, Two, Two, Three, Three, Four,* and *Four*.

Compass procedure begins on page 788.

PROCEDURE

CBL AND SENSORS

Magnetic field strength

3. Set up the apparatus as shown in **Figure 21-22.** Use 1 m of copper wire to make a square loop around the coil support pins on the galvanometer apparatus. Attach alligator clips to the ends of the wire. Label one clip *A* and label the other *B.* Place the galvanometer apparatus so that you are facing the plane of the coil.

4. Use masking tape to mark a line on the stand of the galvanometer directly under the top of the coil. Make another tape line perpendicular to the first, as shown in **Figure 21-22.** The two tapes should cross in the middle of the apparatus. On the second tape, on the side away from you, mark the point 2 cm from the center. Using this point as the center point, draw a circle with a 1 cm radius.

5. Construct a circuit that contains the power supply and a 1 Ω resistor wired in series through the middle set of posts on the switch. Place the switch so that it moves from left to right. Connect the front right post of the switch to the end of the coil

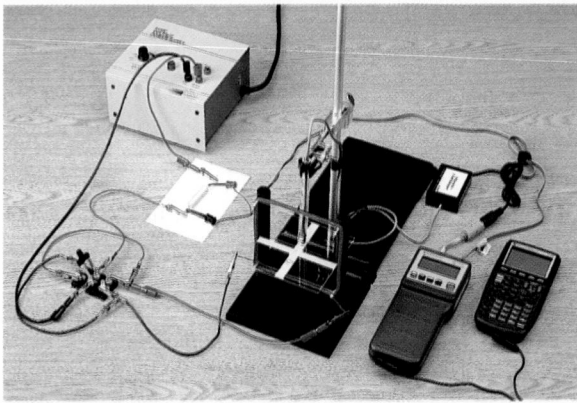

Figure 21-22
Step 3: Loop the copper wire around the support pins and attach alligator clips to the ends. Place the galvanometer with one support pin on the left and one on the right.

Step 4: Use two pieces of tape to mark perpendicular lines, and mark a circle to use as a reference for placing the sensor.

Step 5: Place the switch in front of you so that it moves from left to right. Check all connections carefully.

marked *A* and connect the rear right post of the switch to the end of the coil marked *B.* Now connect the front left post of the switch to the end of the coil marked *B* and the rear left post of the switch to the end of the coil marked *A.* **Do not close the switch or turn on the power supply until your teacher has approved your circuit.**

6. Connect the CBL to the calculator with the unit-to-unit link cable. Connect the CBL voltage probe to the CH1 port on the CBL and the magnetic field probe to the CH2 port. Connect the CBL voltage probe to measure the voltage across the resistor.

- Turn on the CBL and the graphing calculator. Start the program PHYSICS on the calculator. Select option *SET UP PROBES* from the MAIN MENU. Enter 2 for the number of probes. Select *MORE PROBES* from the SELECT PROBE menu. Select the *VOLTAGE* probe from the list. Enter 1 for the channel number.

- Select *MORE PROBES* from the SELECT PROBE menu. Select the *MAGNETIC FIELD* probe from the list. Enter 2 for the channel number. From the CALIBRATION menu, select *USE STORED.* For the MAGNETIC FIELD SETTING, select *HIGH (MTESLA).*

7. Select *OPTIONS* from the MAIN MENU. Select option *ZERO SENSOR* from the PHYSICS OPTIONS menu. Select *CHANNEL 2* to zero the magnetic field sensor.

8. Hold the magnetic field sensor vertically with the white dot facing north. The CBL will display the sensor readings in volts. When the CBL displays a constant value for the field strength, press TRIGGER on the CBL to zero the sensor.

9. Select the *COLLECT DATA* option from the MAIN MENU. Select the *MONITOR INPUT* option from the DATA COLLECTION menu. The graphing calculator will begin to display values.

CBL and Sensors Tips

◆ Students should have the program PHYSICS on their graphing calculators. Refer to Appendix B for instructions.

◆ Zeroing the magnetic field sensor in step 7 will set the sensor to read Earth's magnetic field strength as zero, so the readings taken during the lab will not include Earth's magnetic field.

Techniques to Demonstrate

Show students how to set up the wire coil on the galvanometer apparatus and how to use tape to mark the reference circle for the magnetic field sensor.

Make sure students know the correct position of the magnetic field sensor.

✔ Checkpoints

Step 5: Make sure the power supply, resistor, and switch are connected properly and are at the proper settings. Students should be able to demonstrate that all connections are properly made.

Step 8: Make sure that the magnetic field sensor is securely clamped in the right position.

Step 10: Some students may need help determining the *Current Direction*. Students should be able to demonstrate that they have recorded the correct direction.

Step 12: Make sure students reconnect the circuit properly after turning the wire loop.

Compass Tips

◆ Make sure students know how to read a compass and record the readings.

◆ See pp. 692A–692B for instructions on setting up multimeters to measure current.

◆ Make sure students understand how to wire current meters (in series) in a circuit.

Techniques to Demonstrate

Show students how to set up the wire coil on the galvanometer apparatus and how to determine the direction of the current.

✔ Checkpoints

Step 4: Make sure the power supply, resistor, current meter, and switch are connected correctly and are at the proper settings.

Step 6: Make sure students reconnect the wires properly to reverse the current.

Step 8: Make sure students have two loops of wire and that all connections are correct.

10. Set up a support stand with a buret clamp to hold the magnetic field sensor vertically. Position the magnetic sensor securely so that the white dot is facing you and the sensor is directly above the 1 cm circle marked on the tape.

11. Make sure the dial on the power supply is turned completely counterclockwise. When your teacher has approved your circuit, turn the dial on the power supply about halfway to its full value. Close the switch briefly.

12. Read the potential difference across the resistor and the strength of the magnetic field. Open the switch as soon as you have made your observations. Record ΔV_R *(V)* and $B_{Measured}$ *(T)* for *Trial 1* in your data table. Determine and record the *Current Direction (A to B or B to A)*.

13. Reverse the direction of the current by closing the switch in the opposite direction. Read and record the potential difference and the strength of the magnetic field for *Trial 2*. Open the switch as soon

as you have made your observations. Determine and record the *Current Direction (A to B or B to A)*.

14. Disconnect the alligator clips from A and B on the wire loop, and turn the wire loop 180°. Reconnect the alligator clips, making sure that clip A is on the same end of the wire it was for previous trials. Repeat steps 12 and 13 as *Trials 3 and 4*. Record all data in your data table. Always open the switch as soon as you have made your readings.

15. Return the apparatus to the same position as it was in *Trial 1*. Increase the setting on the power supply to about two-thirds of the maximum setting on the dial. Repeat the procedure in steps 10 through 14. Record all data in your data table for *Trials 5* through *8*.

16. Repeat step 15 with the power supply set to one-third of the maximum setting. Record all data in your data table as *Trials 9* through *12*.

17. Clean up your work area. Put equipment away safely so that it is ready to be used again.

Analysis and interpretation begins on page 789.

PROCEDURE

COMPASS

Magnetic field of a current-carrying wire

3. Wrap the wire once around the galvanometer. Place the large compass on the stand of the galvanometer so that the compass needle is parallel to and directly below the wire, as shown in **Figure 21-23.** Turn the galvanometer until the turn of wire is in the north-to-south plane, as indicated by the compass needle.

4. Construct a circuit that contains the power supply, a current meter, a 1 Ω resistor, and a switch, all wired in series with the galvanometer. Connect the galvanometer so that the direction of the current will be from south to north through the segment of the loop above the compass needle. ***Do not close the switch until your teacher has approved your circuit.***

5. Set the power supply to its lowest output. When your teacher has approved your circuit, close the switch

briefly. Using the power supply, adjust the current in the circuit to 1.5 A. Use the power supply to maintain this current throughout the lab. Record the current, the current direction, and the compass reading in your data table.

6. Reverse the direction of the current in the segments of the loop above the needle by reversing the wires connecting to the galvanometer.

7. Close the switch. Adjust the power supply to 1.5 A. Record your observations in your data table. Open the switch as soon as you have made your observations.

8. Remove the galvanometer from the circuit. Add a second turn of wire, and reconnect the galvanometer to the circuit so that the current direction will be south to north.

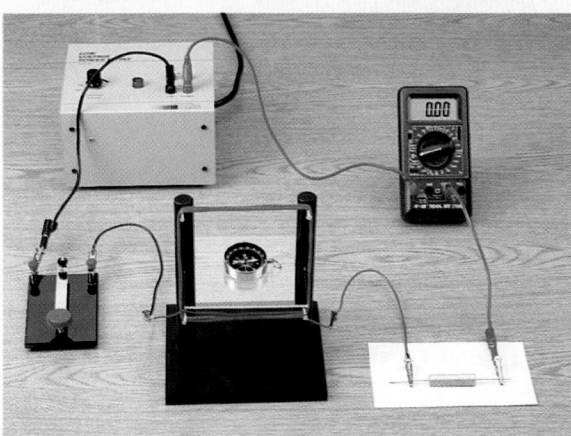

Figure 21-23
Step 3: Use the support pins on the galvanometer to wrap the wire into a loop. Adjust the apparatus so that the needle and wire are in the north-to-south plane.

9. Close the switch. Adjust the power supply to 1.5 A. Record your observations in your data table. Open the switch immediately.

10. Reverse the direction of the current through the segments of the loop above the needle by reversing the wires connecting to the galvanometer.

11. Close the switch. Adjust the power supply to 1.5 A.

12. Repeat the experiment for three turns and then four turns. For each, connect the circuit so that the direction of the current is from south to north and then north to south. Record all information.

13. Clean up your work area. Put equipment away safely so that it is ready to be used again.

ANALYSIS AND INTERPRETATION

Calculations and data analysis

1. **Analyzing data**   Use the data for each trial.
   a. **CBL and sensors**   Find the current using the equation $\Delta V = IR$.
   b. **Compass**   Find the tangent of the angle of deflection for the compass needle.

2. **Graphing data**   Use a computer, graphing calculator, or graph paper.
   a. **CBL and sensors**   Use the data from *Trials 1, 5,* and *9* to plot a graph of B_{wire} in teslas against the current in the circuit. Also plot graphs for *Trials 2, 6,* and *10; Trials 3, 7,* and *11;* and *Trials 4, 8,* and *12.*
   b. **Compass**   Plot a graph of the tangents found in item 1 against the number of turns in the wire.

Conclusions

3. **Analyzing graphs**   Use your graphs to answer the following questions.
   a. **CBL and sensors**   For each position, what is the relationship between the current in the wire loop and the magnetic field strength?
   b. **Compass**   What is the relationship between the tangent of the angle and the number of turns? Explain.

4. **Applying conclusions**   What is the relationship between the direction of current in the wire and the direction of the magnetic field? Explain.

ANSWERS TO

Analysis and Interpretation

CALCULATIONS AND DATA ANALYSIS

1. **a.** Student answers will vary. For sample data, values range from $I = 10.7$ mA to $I = 31.0$ mA.

   b. Student answers will vary. For sample data, values for $\tan\theta$ range from 0.577 to 2.747.

2. **a.** Student graphs should show that the magnetic field strength increases as the current increases.

   b. Student graphs should show that $\tan\theta$ increases as the number of turns increases.

CONCLUSIONS

3. **a.** For each position, the magnetic field strength increases as the current increases.

   b. Because $\tan\theta$ increases as the number of turns increases, students should realize that $\tan\theta$ is proportional to the magnetic field strength.

4. When the current direction changes, the direction of the magnetic field also changes.

ELECTROMAGNETIC FIELDS: CAN THEY AFFECT YOUR HEALTH?

*I*n the 1970s, many people became concerned about the electromagnetic radiation being produced by electric devices they used at work and at home. People already knew that microwaves could cook food and that exposure to ultraviolet light contributed to skin cancer. So if all these effects were possible, people reasoned, couldn't electromagnetic radiation given off by standard 60 Hz alternating current also cause harm?

Electromagnetic waves are produced when a charged particle undergoes acceleration. Accordingly, a 60 Hz alternating current in a wire produces 60 Hz electromagnetic radiation. This is fairly low on the electromagnetic spectrum, considering that AM radio waves have frequencies around 10^6 Hz, which is itself fairly low on the electromagnetic spectrum. The region of space through which electromagnetic radiation passes is an electromagnetic field. Electromagnetic waves produced by 60 Hz alternating current produce what is called an extremely low frequency electromagnetic field, or ELF.

Problems with power lines

In 1979, scientists reported that children who lived near high-voltage power-transmission lines were twice as likely to suffer from childhood leukemia as children who did not live near power lines. In 1986, another study seemed to confirm that occurrences of leukemia and other childhood cancers were linked to the presence of power lines. There were concerns that the 60 Hz electromagnetic fields from the power lines might be responsible.

Many scientists criticized these studies because the researchers did not measure the strengths of the fields the people were exposed to. Instead, they had estimated the amount of exposure from the way the power lines were arranged, the current running through the lines, and the distance of the lines from each house. Critics also pointed out that the research consisted only of epidemiological studies, which mathematically correlate the frequency of an illness to factors in the surroundings.

Partly as a result of this criticism, researchers were challenged to discover a mechanism (a specific biological change that would lead to the development of cancer) by which ELFs could affect biological systems. Soon, researchers began to report that ELFs could damage cell membranes, cause unusual expression of genes, increase the production of the hormone estrogen, and reduce the pineal gland's production of melatonin, a hormone that can limit the growth of cancerous cells.

From 1991 to 1995, scientists conducted another epidemiological study to relate the health problems of 130 000 electric utility workers to their on-the-job exposure to ELFs. The researchers made a few measurements of field strength in order to estimate the amount of exposure for each type of job. To their surprise, they did not find an increased risk of leukemia, as they had expected. Instead, they found that brain cancer occurred more than twice as often among these workers as it did among other kinds of workers. However, two other studies conducted at nearly the same time found the opposite results or no correlation at all. In these last two studies, scientists measured the ELF exposure of every worker.

A need for conclusions

By 1995, several billion dollars had been spent on research that essentially provided no conclusive results, so government and other institutions were becoming wary of pouring more money into further study. The public was also becoming increasingly uneasy over the conflicting claims about ELFs. In addition, a whole industry had built up around people's fears of ELFs. Companies were selling everything from useful devices, such as gaussmeters, which measure magnetic field strength, to questionable gadgets that promised to absorb ELFs and reradiate them in a "coherent" form that was supposedly beneficial to health.

In 1995, the American Physical Society (APS) reviewed all of the research and declared that the studies had failed to show any connection between electromagnetic fields and cancer. They added that researchers had not found any mechanism by which ELFs might cause cancer. The National Academy of Sciences (NAS) came to a similar conclusion in 1996, although it conceded that a correlation did exist between childhood leukemia and the presence of power lines. The NAS suggested that scientists should continue

to search for a reason for this correlation. In 1997, the U.S. Department of Energy announced that it would no longer fund ELF programs.

Some scientists believe funding should be continued, saying that more-refined research would reveal a firm

correlation. Others have criticized the APS, saying that the organization dismissed epidemiological studies too easily and was too focused on the lack of a confirmed mechanism. Still others were pleased that NAS supported continued research into the connection between power lines and leukemia. Scientists who defend the actions of the APS and NAS point out that research money is no longer easy to get and that it should be used in more-promising kinds of cancer research.

 **Researching the Issue**

1. Evidence indicates that the incidence of cancer is more frequent among children who live near power-transmission lines but that the cancers are not caused by electromagnetic fields. What other factors would you suggest that scientists examine? The factors do not have to involve electricity.

2. Interpret the following statement in light of the public's perception of the disagreements over ELFs: "You can prove something to be unsafe, but you can never prove something to be completely safe."

3. Find Internet sites that offer products that claim to help people avoid exposure to electromagnetic fields. List and describe products that you think would be useful and those that you think would be a waste of money. Defend your classification of each item.

4. Electromagnetic fields from household devices, especially those with electric motors, are usually stronger than fields from nearby power lines when the fields are measured in the home. How can you account for this phenomenon?

★ TEKS

pp. 790–791: 3C

CHAPTER 22 PLANNING GUIDE

Compression Guide: To shorten from 9 to 7 45-min periods (from 4½ to 3½ 90-min blocks), eliminate items in magenta type.

PACING CHART	CLASSROOM RESOURCES			
	⭐TEKS	Teacher Demonstrations	Holt Physics Transparencies	Labs (See page T52 for equipment listing for in-text labs.)
22-1 Induced current 3 or 2 45-minute periods 1½ or 1 90-minute block	1A, 2B, 2C, 2F, 3A, 3B, 3E, 5B, 6B, 6D, 6F	**TE** *Induced current*, p. 796 **TE** *Lenz's law*, p. 798 **TE** *Magnetic braking*, p. 801	**T** 110–111 **TM** 70	**L** *Discovery Lab: Electricity and Magnetism* **PE** *Electromagnetic Induction*, p. 826
22-2 Alternating current, generators, and motors 2 45-minute periods 1 90-minute block	2C, 3C, 6D, 6F	**TE** *Alternating current*, p. 803 **TE** *ac and dc*, p. 807 **TE** *Effects of alternating current*, p. 808 **TE** *Electric motor*, p. 812	**T** 112–113 **TM** 71–72	
22-3 Inductance 2 or 1 45-minute periods 1 or ½ 90-minute block	2C, 3B, 6B, 6F	**TE** *Transformers*, p. 814 **TE** *Induced potential difference*, p. 815	**T** 114–115 **TM** 73	**L** *Invention Lab: Building a Circuit Breaker*

Review and Assessment
2 45-minute periods
1 90-minute block

Resource Key

PHYSICS (HOLT)

PE Pupil's Edition
TE Teacher's Edition

L Laboratory Experiments
TL Technology Lab Experiments
T Transparencies

🖐 **One-Stop** Planner CD-ROM **contents**

TM Transparency Masters
SR Section Review Worksheets
AA Alternative Assessment

PW Problem-Solving Workbook
PB Problem Bank
CTW Critical Thinking Worksheet

LABORATORY PLANNING: Electromagnetic Induction, p. 826

Materials [for each lab group]:
- contact key
- cylindrical magnet, Alnico V
- galvanometer:
  - student galvanometer, −500 μA to +500 μA, or
  - economical galvanometer, or
  - CENCO 6 in 1 galvanometer
- 2 patch cords, red, insulated alligator clip/stacking banana plug

- power supply:
  - 1.5 V/3 V battery eliminator, or
  - 20 V dc/500 mA regulated power supply, or
  - dry cells
- primary and secondary coils, or
  - economy primary and secondary coils, or
  - CENCO primary and secondary coils
- rheostat: 10 Ω, or potentiometer

ASSIGNMENT RESOURCES

Content Mastery	Critical Thinking	Problem-Solving Practice
PE 1–6, p. 802 **SR** 22-1, *Concept Review* **PE** 1–4, p. 821	**PE** 1–2, p. 797 **PE** 5–9, p. 821	**22A** Induced emf and current: 27 items in **PE, PW,** and **PB,** see **TE** pp. 799–800
PE 1–4, p. 813 **SR** 22-2, *Concept Review* **PE** 13–18, p. 822	**PE** 19–22, p. 822	**22B** Induction in generators: 24 items in **PE, PW,** and **PB,** see **TE** pp. 805–806 **22C** rms currents and potential differences: 27 items in **PE, PW,** and **PB,** see **TE** pp. 809–810
PE 1–3, p. 819 **SR** 22-3, *Concept Review* **PE** 29–32, p. 823	**PE** 33–34, p. 823	**22D** Transformers: 27 items in **PE, PW,** and **PB,** see **TE** pp. 817–818

ASSESSMENT RESOURCES

Cumulative Review	Alternative Assessment	Traditional Assessment
SR Mixed Review, Ch. 22	**PE** 1–5, p. 825 **AA** Items for Ch. 22	Chapter 22 Test Test Generator items for Ch. 22

Scoring Rubrics for Alternative Assessment items can be found on the One-Stop Planner CD-ROM.

TECHNOLOGY RESOURCES

 PHYSICS Module 20 Induction and Transformers

 PE Graphing Calculators, p. 824
(Alternative procedures for calculators without Flash-ROM technology are provided on the One-Stop Planner CD-ROM.)

 The Mechanical Universe/High School Adaptation Quad VI, Electronmagnetic Induction Quad VI, Alternating Current

 internet connect

 On-line Student Resources:
GO TO: www.scilinks.org
The following *sci*LINKS Internet resources can be found in the student text for this chapter.

TOPICS:
• Magnetic fields, p. 795 (HF2221)
• Alternating current, p. 804 (HF2222)
• Electrical safety, p. 811 (HF2223)
• Transformers, p. 815 (HF2224)

 On-line Teacher Resources:
GO TO: go.hrw.com
KEYWORD: HF2 HOME
Visit the HRW Web site for a variety of resources related to this chapter.

 Smithsonian Institution
Internet Connections
Visit www.si.edu/hrw for additional on-line resources.

CNNfyi.com
Visit www.cnnfyi.com for late-breaking news and current events stories selected just for you.

Section 22-1 introduces induced current, discusses Lenz's law, and applies Faraday's law of induction to calculate induced emf and induced current.

Section 22-2 introduces generators and motors as devices that convert energy from one form to another and shows how to calculate the maximum emf for an electrical generator, the rms current, and the potential difference for ac circuits.

Section 22-3 examines mutual induction, discusses transformers, shows how to calculate the potential difference for a step-up or step-down transformer, and describes how self-induction occurs in a circuit.

About the Illustration

The strings of an electric guitar are magnetized by a permanent magnet located inside coils of wire beneath the strings. Two sets of coils—one at the base of the neck and the other just above the bridge—can be seen in each of the guitars in the photograph. As a string vibrates, the changing magnetic field induces in the coil an alternating current, whose frequency depends on the string's frequency.

Interactive Problem-Solving Tutor

See Module 20
"Induction and Transformers" promotes additional development of problem-solving skills.

CHAPTER 22

Induction and Alternating Current

PHYSICS IN ACTION

In an electric guitar, the vibrations of the strings are converted to an electric signal, which is then amplified outside the guitar and heard as sound from a loudspeaker.

Yet the electric guitar is not plugged directly into an electric source. Instead, it generates electric current by a process called induction. By changing the magnetic field near a coil of wire, an electric current can be induced in the coil by the vibrations of the guitar strings.

This chapter explores how induction produces and changes alternating currents.

- *Why does the generated current last only as long as the string vibrates?*

- *Why must the strings be ferromagnetic in order for an electric guitar to work?*

CONCEPT REVIEW

Resistance   (Section 19–2)

emf   (Section 20–1)

Magnetic force   (Section 21–3)

Tapping Prior Knowledge

Knowledge to Expect

✔ "Electric currents and magnets can exert a force on each other." (AAAS's *Benchmarks for Science Literacy*, grades 6–8)

✔ "Energy can be converted from one form to another." (AAAS's *Benchmarks for Science Literacy*, grades 6–8)

Knowledge to Review

✔ The definition of resistance states that the current in a circuit is proportional to the applied potential difference and inversely proportional to the resistance of the circuit. (Section 19-2)

✔ Emf increases the electrical potential energy of charges in a circuit, causing the charges to move. Thus, emf is a source of current in a circuit. (Section 20-1)

✔ A charged particle moving in a magnetic field experiences a magnetic force. When the particle moves perpendicular to the field, $F_{magnetic} = qvB$. (Section 21-3)

Items to Probe

✔ Resistance: Have students calculate current for various cases involving changing potential differences.

22-1
Induced current

22-1 SECTION OBJECTIVES

- Describe how the change in the number of magnetic field lines through a circuit loop affects the magnitude and direction of the induced current.

- Apply Lenz's law to determine the direction of an induced current.

- Calculate the induced emf and current using Faraday's law of induction.

electromagnetic induction

the production of an emf in a conducting circuit by a change in the strength, position, or orientation of an external magnetic field

MAGNETIC FIELDS AND INDUCED EMFS

Recall that in Chapter 20 you were asked if it was possible to produce an electric current using only wires and no battery. So far, all electric circuits that you have studied have used a battery or an electrical power supply to create a potential difference within a circuit. In both of these cases, an emf increases the electrical potential energy of charges in the circuit, causing them to move through the circuit and create a current. ⓧTEKS 5B

It is also possible to *induce* a current in a circuit without the use of a battery or an electrical power supply. Just as a magnetic field can be formed by a current in a circuit, a current can be formed by moving a portion of a closed electric circuit through an external magnetic field, as indicated in **Figure 22-1.** The process of inducing a current in a circuit with a changing magnetic field is called **electromagnetic induction.** ⓧTEKS 6D

Consider a circuit consisting of only a resistor in the vicinity of a magnet. There is no battery to supply a current. If neither the magnet nor the circuit is moving with respect to the other, no current will be present in the circuit. But if the circuit moves toward or away from the magnet or the magnet moves toward or away from the circuit, a current is induced. As long as there is relative motion between the two, a current can form in the circuit.

The separation of charges by the magnetic force induces an emf

It may seem strange that there can be an induced emf and a corresponding induced current without a battery or similar source of electrical energy. Recall that in the previous chapter a moving charge can be deflected by a magnetic field. This deflection can be used to explain how an emf occurs in a wire that moves through a magnetic field.

Figure 22-1
When the circuit loop crosses the lines of the magnetic field, a current is induced in the circuit, as indicated by the movement of the galvanometer needle.

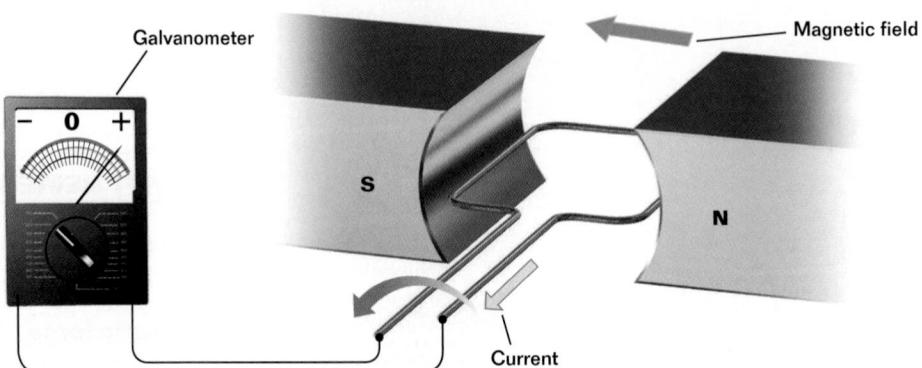

Consider a conducting wire pulled through a magnetic field, as shown on the left in **Figure 22-2.** You learned in Chapter 21 that positive charges moving with a velocity at an angle to the magnetic field will experience a magnetic force. According to the right-hand rule, this force will be directed perpendicular to both the magnetic field and the motion of the charges. For positive charges in the wire, the force is directed downward along the wire. For negative charges, the force is upward. This effect is equivalent to replacing the segment of wire and the magnetic field with a battery that has a potential difference, or emf, between its terminals, as shown on the right in **Figure 22-2.** As long as the conducting wire moves through the magnetic field, the emf will be maintained.

The polarity of the induced emf—and thus the direction of the induced current in the circuit—depends on the direction in which the wire is moved through the magnetic field. For instance, if the wire in **Figure 22-2** is moved to the right, the right-hand rule predicts that the positive charges will be pushed downward. If the wire is moved to the left, the positive charges will be pushed upward. The magnitude of the induced emf depends on the velocity with which the wire is moved through the magnetic field, the length of the wire, and the strength of the magnetic field.

The angle between a magnetic field and a circuit affects induction

To induce an emf in a wire loop, part of the loop must move through a magnetic field or the entire loop must pass into or out of the magnetic field. No emf is induced if the loop is static or the magnetic field is constant.

The magnitude of the induced emf and current depend partly on how the loop is oriented to the magnetic field, as shown in **Figure 22-3.** The induced current is largest when the plane of the loop is perpendicular to the magnetic field, as in **(a);** it is less when the plane is tilted into the field, as in **(b);** and it is zero when the plane is parallel to the field, as in **(c).**

The role of orientation on induced current can be understood in terms of the force exerted by a magnetic field on charges in the moving loop. The smaller the component of the magnetic field perpendicular to both the plane and the motion of the loop, the smaller the magnetic force on the charges in the loop. When the area of the loop is moved parallel to the magnetic field, there is no magnetic field component perpendicular to the plane of the loop and therefore no force that moves the charges through the circuit.

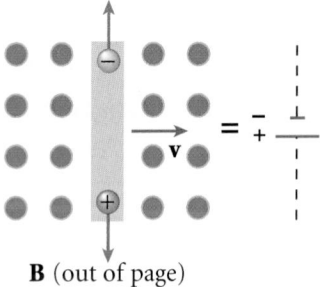

B (out of page)

Figure 22-2
The separation of positive and negative moving charges by the magnetic force creates a potential difference (emf) between the ends of the conductor.

internet**connect**

SC*i*LINKS.

NSTA

TOPIC: Magnetic fields
GO TO: www.scilinks.org
*sci*LINKS CODE: HF2221

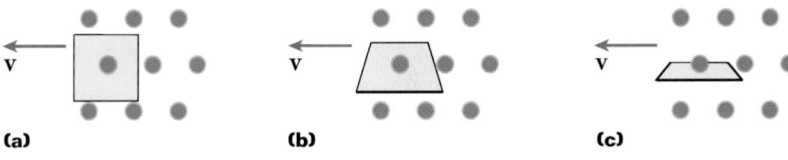

(a)   **(b)**   **(c)**

Figure 22-3
The induced emf and current are largest when the plane of the loop is perpendicular to the magnetic field **(a),** smaller when the plane of the loop is tilted into the field **(b),** and zero when the plane of the loop and the magnetic field are parallel **(c).**

Demonstration 1

Induced current

Purpose Show students an example of an induced current.

Materials flashlight bulb in holder, coil of wire, bar magnet (two may be required), connecting wires

Procedure Connect the flashlight bulb to the coil of wire with the connecting wires. Tell students to observe the demonstration with the intent of explaining the energy conversions. Move the bar magnet into and out of the coil several times in rapid succession. Have students explain the energy conversions on the board or in their notebooks. The following energy conversions should be discussed: kinetic energy is converted to electrical energy (moving magnet generates a current) and electrical energy is converted to light (the current heats the light bulb's filament).

⭐TEKS

p. 796: 3E, 6D
p. 797: 3A

Did you know?

In 1996, the space shuttle *Columbia* attempted to use a 20.7 km conducting tether to study Earth's magnetic field in space. The plan was to drag the tether through the magnetic field, inducing an emf in the tether. The magnitude of the emf would directly vary with the strength of the magnetic field. Unfortunately, the tether broke before it was fully extended, so the experiment was abandoned.

⭐TEKS 3E

Change in the number of magnetic field lines induces a current

So far, you have learned that moving a circuit loop into or out of a magnetic field can induce an emf and a current in the circuit. Changing the size of the loop or the strength of the magnetic field also will induce an emf in the circuit.

One way to predict whether a current will be induced in a given situation involves the concept of changes in magnetic field lines. For example, moving the circuit into the magnetic field causes some lines to move into the loop. Changing the size of the circuit loop or rotating the loop changes the number of field lines passing through the loop, as does changing the magnetic field's strength. **Table 22-1** summarizes these three ways of inducing a current.

CHARACTERISTICS OF INDUCED CURRENT

Suppose a bar magnet is pushed into a coil of wire. As the magnet moves into the coil, the strength of the magnetic field within the coil increases, and a current is induced in the circuit. This induced current in turn produces its own magnetic field, whose direction can be found by using the right-hand rule. If you were to apply this rule for several cases, you would notice that the induced magnetic field direction depends on the change in the applied field.

As the magnet approaches, the magnetic field passing through the coil increases in strength. The induced current in the coil must be in a direction that produces a magnetic field that opposes the increasing strength of the approaching field. The induced magnetic field is therefore in the direction opposite that of the approaching magnetic field. ⭐TEKS 6D

Table 22-1   **Ways of inducing a current in a circuit**

Description	Before	After
Circuit is moved into or out of magnetic field (either circuit or magnet moving).		
Circuit is rotated in the magnetic field (angle between area of circuit and magnetic field changes).		
Intensity of magnetic field is varied.		

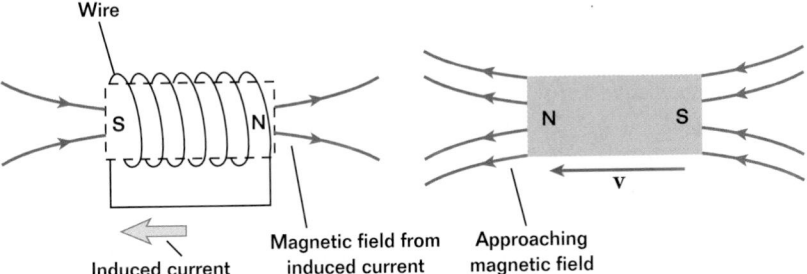

Wire

S   N

Induced current

Magnetic field from induced current

Approaching magnetic field

N   S

v

Figure 22-4

The approaching magnetic field is opposed by the induced magnetic field, which behaves like a bar magnet with the orientation shown.

The induced magnetic field is like the field of a bar magnet oriented as shown in **Figure 22-4.** Note that the coil and magnet repel each other.

If the magnet is moved away from the coil, the magnetic field passing through the coil decreases in strength. Again, the direction of current induced in the coil must be such that it produces a magnetic field that opposes the decreasing strength of the receding field. This means that the magnetic field set up by the coil must be in the same direction as the receding magnetic field.

The induced magnetic field is like the field of a bar magnet oriented as shown in **Figure 22-5.** Note that the coil and magnet attract each other.

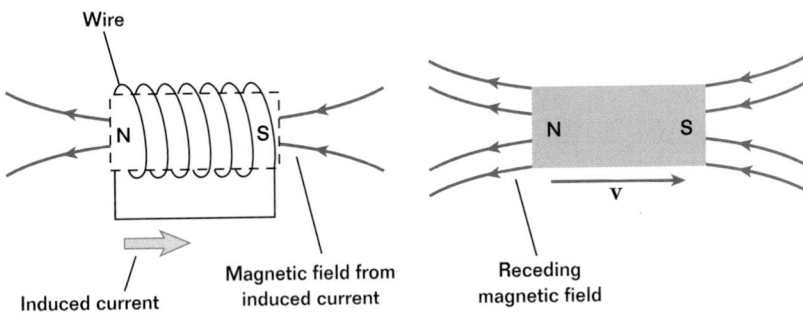

Wire

N   S

Induced current

Magnetic field from induced current

Receding magnetic field

N   S

v

Figure 22-5

The receding magnetic field is attracted by the induced magnetic field, which behaves like a bar magnet with the orientation shown.

Conceptual Challenge ⭐TEKS 3A

1. Falling magnet A bar magnet is dropped toward the floor, on which lies a large ring of conducting metal. The magnet's length—and thus the poles of the magnet—is parallel to the direction of motion. Disregarding air resistance, does the magnet during its fall toward the ring move with the same constant acceleration as a freely falling body? Explain your answer.

2. Induction in a bracelet Suppose you are wearing a bracelet that is an unbroken ring of copper. If you walk briskly into a strong magnetic field while wearing the bracelet, how would you hold your wrist with respect to the magnetic field in order to avoid inducing a current in the bracelet?

Teaching Tip

Remind students that, as seen in Chapter 21, a current creates a magnetic field whose direction can be found with the right-hand rule. If the thumb points in the direction of conventional current (moving positive charges), the right hand wraps around the wire in the direction of the magnetic field.

🛑 Misconception Alert

Some students may not distinguish between the external magnetic field that induces a current and the magnetic field that is set up by the induced current. Use the two examples discussed on this page to clarify the difference between these two magnetic fields.

ANSWERS TO

Conceptual Challenge

1. no; The magnet's acceleration is slightly smaller. The magnet induces a current in the conducting ring, and the magnetic field of this current opposes the field of the falling magnet. This induced field exerts an upward force on the magnet, reducing the magnet's net acceleration downward.

2. Upon entering and leaving the field, the plane of the bracelet must be parallel to the direction of the field.

Lenz's law

Purpose Illustrate Lenz's law experimentally.

Materials flashlight bulb in holder, coil of wire, diode, connecting wires, bar magnet (or two)

Procedure Repeat Demonstration 1, but include a diode in the circuit. Tell students that the diode allows charges to move in only one direction.

Explain to students that you will be testing Lenz's law experimentally. Point out that inserting the magnet one way (with the north pole first) will generate a current in one direction. Reversing the magnet will generate a current in the opposite direction, which the diode will stop. The light bulb will serve as an indication of the direction of current. Thus, the bulb should light up in one case but not in the other (because the diode blocks current in one direction). Perform the demonstration and verify these conclusions.

⊛ TEKS

p. 798: 3B, 6B
p. 799: 2C, 6D

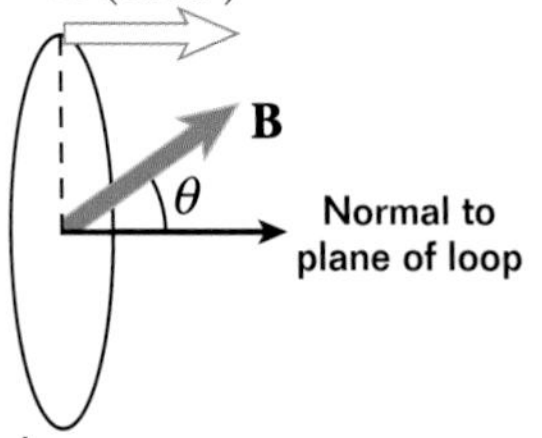

Figure 22-6
The angle θ is defined as the angle between the magnetic field and the normal to the plane of the loop. B (cos θ) equals the strength of the magnetic field perpendicular to the plane of the loop.

The rule for finding the direction of the induced current is called *Lenz's law* and is expressed as follows:

The magnetic field of the induced current opposes the change in the applied magnetic field.

Note that the field of the induced current does not oppose the applied field but rather the change in the applied field. If the applied field changes, the induced field attempts to keep the total field strength constant, according to the principle of energy conservation.

Faraday's law of induction

Because of the principle of energy conservation, Lenz's law allows you to determine the direction of an induced current in a circuit. Lenz's law does not provide information on the magnitude of the induced current or the induced emf. To calculate the magnitude of the induced emf, you must use *Faraday's law of magnetic induction*. For a single loop of a circuit, this may be expressed as follows:

$$\text{emf} = -\frac{\Delta[AB (\cos \theta)]}{\Delta t}$$

Certain features of this equation should be noted. First, a change with time of any of the three variables—applied magnetic field strength, B, circuit area, A, or angle of orientation, θ—can give rise to an induced emf. The term $B (\cos \theta)$ represents the component of the magnetic field perpendicular to the plane of the loop. The angle θ is measured between the applied magnetic field and the normal to the plane of the loop, as indicated in **Figure 22-6.**

The minus sign in front of the equation is included to indicate the polarity of the induced emf. It indicates that the induced magnetic field opposes the changing applied magnetic field according to Lenz's law.

If a circuit contains a number, N, of tightly wound loops, the average induced emf is simply N times the induced emf for a single loop. The equation thus takes the general form of Faraday's law of magnetic induction.

FARADAY'S LAW OF MAGNETIC INDUCTION

$$\text{emf} = -N\frac{\Delta[AB (\cos \theta)]}{\Delta t}$$

**average induced emf = −the number of loops in the circuit ×
the rate of change of (circuit loop area × magnetic field component
normal to the plane of loop)**

In this chapter, N is always assumed to be a whole number. ⊛TEKS **3B, 6B**

Recall that the SI unit for magnetic field strength is the tesla (T), which equals one newton per ampere-meter, or N/(A•m). When calculating induced emf, express the tesla in units of one volt-second per meter squared, or (V•s)/m².

Induced emf and current ⊛TEKS 2C, 6D

PROBLEM

A coil with 25 turns of wire is wrapped around a hollow tube with an area of 1.8 m². Each turn has the same area as the tube. A uniform magnetic field is applied at a right angle to the plane of the coil. If the field increases uniformly from 0.00 T to 0.55 T in 0.85 s, find the magnitude of the induced emf in the coil. If the resistance in the coil is 2.5 Ω, find the magnitude of the induced current in the coil.

SOLUTION

1. DEFINE

Given:

$\Delta t = 0.85$ s $A = 1.8$ m² $\theta = 0.0°$ $N = 25$ turns

$B_i = 0.00$ T $= 0.00$ V•s/m² $B_f = 0.55$ T $= 0.55$ V•s/m²

$R = 2.5$ Ω

Unknown: emf $=$? $I = $?

Diagram: Show the coil before and after the change in the magnetic field.

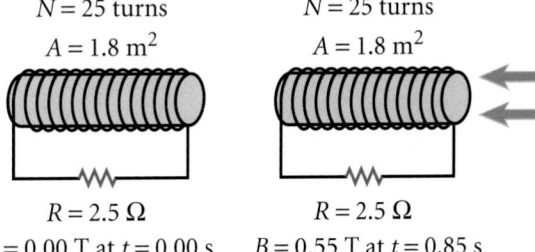

$N = 25$ turns $N = 25$ turns

$A = 1.8$ m² $A = 1.8$ m²

$R = 2.5$ Ω $R = 2.5$ Ω

$B = 0.00$ T at $t = 0.00$ s $B = 0.55$ T at $t = 0.85$ s

2. PLAN

Choose an equation(s) or situation: Use Faraday's law of magnetic induction to find the induced emf in the coil.

$$\text{emf} = -N\frac{\Delta[AB(\cos\theta)]}{\Delta t}$$

Substitute the induced emf into the definition of resistance to determine the induced current in the coil.

$$I = \frac{\text{emf}}{R}$$

Rearrange the equation(s) to isolate the unknown(s): In this example, only the magnetic field strength changes with time. The other components (the coil area and the angle between the magnetic field and the coil) remain constant.

$$\text{emf} = -NA(\cos\theta)\frac{\Delta B}{\Delta t}$$

continued on
next page

The Language of Physics

Because emf is the potential difference across a circuit (ΔV), the definition of resistance can be expressed in terms of emf, as follows:

$$I = \frac{\Delta V}{R} = \frac{\text{emf}}{R}$$

This form of the definition of resistance is used in Sample Problem 22A.

Classroom Practice

The following may be used as a teamwork exercise or for demonstration at the board or on an overhead projector.

PROBLEM

Induced emf and current

A coil with 25 turns of wire is moving in a uniform magnetic field of 1.5 T. The magnetic field is perpendicular to the plane of the coil. The coil has a cross-sectional area of 0.80 m². The coil exits the field in 1.0 s.

a. Find the induced emf.

b. Determine the induced current if the coil's resistance is 1.5 Ω.

Answers

a. 3.0×10^1 V

b. 2.0×10^1 A

Teaching Tip

Point out to students that Lenz's law is used in the fourth step of Sample Problem 22A to find the direction of the induced current.

ANSWERS TO

Practice 22A
Induced emf and current

1. 0.30 V
2. 14 A
3. 0.14 V
4. 4.83×10^{-5} T

---

3. CALCULATE **Substitute the values into the equation(s) and solve:**

$$emf = -(25)(1.8 \text{ m}^2)(\cos 0.0°)\frac{\left((0.55 - 0.00)\frac{V \cdot s}{m^2}\right)}{(0.85 \text{ s})} = -29 \text{ V}$$

$$I = \frac{-29 \text{ V}}{2.5 \text{ Ω}} = -12 \text{ A}$$

$$emf = -29 \text{ V}$$
$$I = -12 \text{ A}$$

CALCULATOR SOLUTION

Because the minimum number of significant figures for the data is two, the calculator answer, 29.11764706, should be rounded to two digits.

4. EVALUATE The induced emf, and therefore the induced current, are directed through the coil so that the magnetic field produced by the induced current opposes the change in the applied magnetic field. For the diagram shown on page 799, the induced magnetic field is directed to the right and the current that produces it is directed from left to right through the resistor.

PRACTICE 22A

Induced emf and current ⭐TEKS 2C, 6D

1. A single circular loop with a radius of 22 cm is placed in a uniform external magnetic field with a strength of 0.50 T so that the plane of the coil is perpendicular to the field. The coil is pulled steadily out of the field in 0.25 s. Find the average induced emf during this interval.

2. A coil with 205 turns of wire, a total resistance of 23 Ω, and a cross-sectional area of 0.25 m^2 is positioned with its plane perpendicular to the field of a powerful electromagnet. What average current is induced in the coil during the 0.25 s that the magnetic field drops from 1.6 T to 0.0 T?

3. A circular wire loop with a radius of 0.33 m is located in an external magnetic field of strength +0.35 T that is perpendicular to the plane of the loop. The field strength changes to −0.25 T in 1.5 s. (The plus and minus signs for a magnetic field refer to opposite directions through the coil.) Find the magnitude of the average induced emf during this interval.

4. A 505-turn circular-loop coil with a diameter of 15.5 cm is initially aligned so that its plane is perpendicular to the Earth's magnetic field. In 2.77 ms the coil is rotated 90.0° so that its plane is parallel to the Earth's magnetic field. If an average emf of 0.166 V is induced in the coil, what is the value of the Earth's magnetic field?

APPLICATIONS OF INDUCTION

In electric circuits, the need often arises for a temporary or continuously changing current to be produced. This kind of current can be generated through electromagnetic induction.

Door bells

Certain types of door bells chime when the button is briefly pushed. A hint to how these door bells work is the small electric light bulb often found behind the doorbell button. When you press the button, the light briefly goes out. This indicates that the door bell's circuit has been opened and that the current has been temporarily discontinued.

You can see the effect of opening the circuit in **Figure 22-7.** While the current is constant in the first circuit, the magnetic field in the coil of this circuit is steady. As long as this field does not change, no current is induced in the coil of the second circuit. When the current in the first circuit is interrupted by pressing the doorbell button, the magnetic field in the first coil drops rapidly, inducing a current in the second circuit and causing a magnetic field to quickly rise along the axis of the second coil. This induced magnetic field is strong enough to push the iron plunger against the chime. ⭐TEKS **6F**

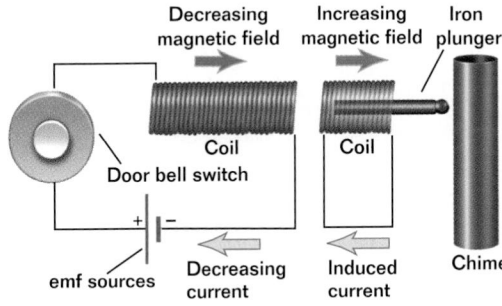

Figure 22-7
The sudden drop in current in the first circuit induces a current in the second circuit. The resulting magnetic field pushes the plunger against the door-bell chime.

Tape recorders

One common use of induced currents and emfs is found in the tape recorder. Many different types of tape recorders are made, but the same basic principles apply for all. A magnetic tape moves past a recording head and a playback head. The tape is a plastic ribbon coated with either iron oxide or chromium dioxide.

A microphone transforms a sound wave into a fluctuating electric current. This current is amplified and allowed to pass through a wire coiled around an iron ring, as shown in **Figure 22-8,** which functions as the recording head. The iron ring and wire constitute an electromagnet; the lines of the magnetic field are contained completely inside the iron except at the point where a gap is cut in the ring. The magnetic tape is passed over this gap. ⭐TEKS **6F**

Because the magnetic field does not pass as easily through the air of the gap as it does through the nearby small pieces of metal oxide in the tape, the magnetic field passes around the gap and magnetizes the metal oxide particles. As the tape moves past the slot, it becomes magnetized in a pattern that corresponds to both the frequency and the intensity of the sound signal entering the microphone.

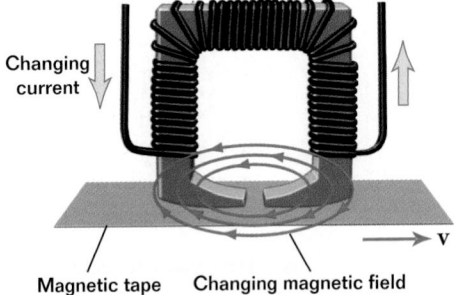

Figure 22-8
The signal fluctuations in the current induce a fluctuating magnetic field, which is recorded along the length of the tape. During playback, the changing magnetic field of the tape will induce in the coil a fluctuating current that can be converted to reproduce the original recorded sound.

Demonstration 3

Magnetic braking

Purpose Show magnetic braking as an application of induction.

Materials aluminum metal ring (approximately 8–10 cm in diameter), strong alnico bar magnet, thread, ring stand

Procedure Attach the ring with two threads to the ring stand so that the ring can swing freely. With the magnet far away, set the ring swinging and ask the class to notice any decrease in the amplitude of its swing as you count aloud the number of cycles. There should be little change during the first 10 to 20 cycles.

Restart the swinging of the aluminum ring, and place the magnet near its path. Ask students to explain the change in terms of Lenz's law. *(The metal ring crosses the magnetic field lines, and a current is induced in the ring. The direction of the current is such that its magnetic field opposes the changes in the magnetic field that induced it.)* Repeat the demonstration, telling students that magnetic braking is silent and does not wear out materials. Have students suggest some possible uses for magnetic braking. *(This principle is used in many triple-beam balances to immediately bring the scale to a specific reading.)*

⭐TEKS

p. 800: 2C, 6D
p. 801: 6F, 6F

In the recording process, the magnetic field is produced by an applied current. During the playback process, induction is used to create a current from a changing magnetic field. The sound signal is reconstructed by passing the tape over the playback head, which is an iron ring with wire wound around it. In some recorders, one head is used for both playback and recording.

When the tape moves past the playback head, the varying magnetic fields on the tape produce changing magnetic-field lines through the wire coil, inducing a current in the coil. The current corresponds to the current in the recording head that originally produced the recording on the tape. This changing electric current can be amplified and used to drive a speaker. Playback is thus an example of induction of a current by a moving magnet.

Section Review

⭐TEKS 2C, 6D

1. A circular current loop made of flexible wire is located in a magnetic field. Describe three ways an emf can be induced in the loop.

2. A spacecraft orbiting Earth has a coil of wire in it. An astronaut measures a small current in the coil, even though there is no battery connected to it and there are no magnets on the spacecraft. What is causing the current?

3. A bar magnet is positioned near a coil of wire, as shown in **Figure 22-9.** What is the direction of the current through the resistor when the magnet is moved to the left, as in (**a**)? to the right, as in (**b**)?

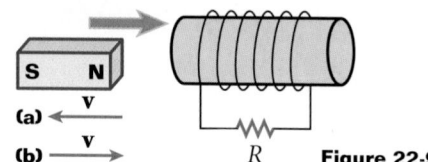

Figure 22-9

4. A 256-turn coil with a cross-sectional area of 0.0025 m² is placed in a uniform external magnetic field of strength 0.25 T so that the plane of the coil is perpendicular to the field. The coil is pulled steadily out of the field in 0.75 s. Find the average induced emf during this interval.

5. **Physics in Action** Electric guitar strings are made of ferromagnetic materials that can be magnetized. The strings lie closely over and perpendicular to a coil of wire. Inside the coil are permanent magnets that magnetize the segments of the strings overhead. Using this arrangement, explain how the vibrations of a plucked string produce an electric signal at the same frequency as the vibration of the string.

6. **Physics in Action** The magnetic field strength of a magnetized electric guitar string is 9.0×10^{-4} T. The pickup coil consists of 5200 turns of wire and has an effective area for each string of 5.4×10^{-5} m². If the string vibrates with a frequency of 440 Hz, what is the average induced emf? (Hint: Assume the magnetic field strength varies from a minimum value of 0.0 T to its maximum value during one fourth of the string's vibration cycle.)

Alternating current, generators, and motors

GENERATORS AND ALTERNATING CURRENT

In the previous section you learned that a current can be induced in a circuit either by changing the magnetic field strength or by moving the circuit loop in or out of the magnetic field. Another way to induce a current is to change the orientation of the loop with respect to the magnetic field.

This second approach to inducing a current represents a practical means of generating electrical energy. In effect, the mechanical energy used to turn the loop is converted to electrical energy. A device that does this is called an electric **generator.**

In most commercial power plants, mechanical energy is provided in the form of rotational motion. For example, in a hydroelectric plant, falling water directed against the blades of a turbine causes the turbine to turn; in a coal or natural-gas-burning plant, energy produced by burning fuel is used to convert water to steam, and this steam is directed against the turbine blades to turn the turbine.

Basically, a generator uses the turbine's rotary motion to turn a wire loop in a magnetic field. A simple generator is shown in **Figure 22-10.** As the loop rotates, the effective area of the loop changes with time, inducing an emf and a current in an external circuit connected to the ends of the loop.

A generator produces a continuously changing emf

Consider a single loop of wire that is rotated with a constant angular frequency in a uniform magnetic field. The loop can be thought of as four conducting wires similar to the conducting wire discussed on page 795 of Section 22-1. In this example, the loop is rotating counterclockwise within a magnetic field directed to the left.

Figure 22-10
In a simple generator, the rotation of conducting loops (located on the left end of the axis) through a constant magnetic field induces an alternating current in the loops.

22-2 SECTION OBJECTIVES

- **Calculate the maximum emf for an electric generator.**
- **Calculate rms current and potential difference for ac circuits.**
- **Describe how an electric motor relates to an electric generator.**

generator

a device that uses induction to convert mechanical energy to electrical energy

Misconception Alert

Some students may think that no work is required to generate electricity. Stress that this is not the case; work is required to rotate a loop in a magnetic field. In a hydroelectric power plant, this work is generated by falling water. The water loses gravitational potential energy as it does work.

Demonstration 4

Alternating current

Purpose Illustrate the effects of changing cross-sectional area on induced emf, and show how generators create ac currents.

Materials overhead projector, colored cellophane filter mounted in a frame (like a 35 mm slide)

Procedure Tell students that the light from the overhead projector represents a uniform magnetic field and the filter represents a wire loop. The amount of colored light that reaches the screen represents the induced emf and thus the current.

Slowly rotate the filter over the overhead projector, letting changing amounts of colored light reach the screen. Have students observe the changing intensity. Draw a graph of the changes in intensity over time on the board. (The graph is a sine curve.) Point out the analogy between the changing intensity and a changing current.

Misconception Alert

When trying to determine the direction of current in **Figure 22-11,** students might try using the right-hand rule for a current-carrying wire in a magnetic field instead of the right-hand rule for charges moving in a magnetic field. Explain that the latter rule must be used because charges are not initially moving along the wire; rather, they move with the wire, in a direction perpendicular to the length of the wire. Thus, when applying the right-hand rule to find the direction of induced current, the thumb points along the direction in which the wire is moving, not along the wire, and the fingers point along the direction of the magnetic field. The resulting force on the charges, and thus the current, points out of the palm of the hand. Have students apply this form of the right-hand rule for segments *a, b, c,* and *d* in each case shown in **Figure 22-11** to determine the direction of the current in each segment.

Visual Strategy

Figure 22-11

Have students compare **Figure 22-11** with the analogy introduced in Demonstration 4. After asking students the following question, plot graphs of $y = \sin x$ and $y = |\sin x|$ on the chalkboard, and use the graphs to compare the two cases.

Q The amount of colored light in Demonstration 4 varies much like the induced emf in **Figure 22-11.** How are these two cases different?

A *The emf becomes negative, while light does not.*

TOPIC: Alternating current
GO TO: www.scilinks.org
sciLINKS CODE: HF2222

Figure 22-11
For a rotating loop in a magnetic field, the induced emf is zero when the loop is perpendicular to the magnetic field, as in **(a)** and **(c)**, and is at a maximum when the loop is parallel to the field, as in **(b)** and **(d).**

When the area of the loop is perpendicular to the magnetic field lines, as shown in **Figure 22-11(a),** every segment of wire in the loop is moving parallel to the magnetic field lines. At this instant, the magnetic field does not exert force on the charges in any part of the wire, so the induced emf in each segment is therefore zero.

As the loop rotates away from this position, segments *a* and *c* cross magnetic field lines, so the magnetic force on the charges in these segments, and thus the induced emf, increases. The magnetic force on the charges in segments *b* and *d* is directed outside of the wire, so the motion of these segments does not contribute to the emf or the current. The greatest magnetic force on the charges and the greatest induced emf occur at the instant when segments *a* and *c* move perpendicularly to the magnetic field lines, as in **Figure 22-11(b).** This occurs when the plane of the loop is parallel to the field lines. ★TEKS 6D

Because segment *a* moves downward through the field while segment *c* moves upward, their emfs are in opposite directions, but both produce a counterclockwise current. As the loop continues to rotate, segments *a* and *c* cross fewer lines, and the emf decreases. When the plane of the loop is perpendicular to the magnetic field, the motion of segments *a* and *c* is again parallel to the magnetic lines and the induced emf is again zero, as shown in **Figure 22-11(c).** Segments *a* and *c* now move in directions opposite those in which they moved from their positions in **(a)** to those in **(b).** As a result, the polarity of the induced emf and the direction of the current are reversed, as shown in **Figure 22-11(d).**

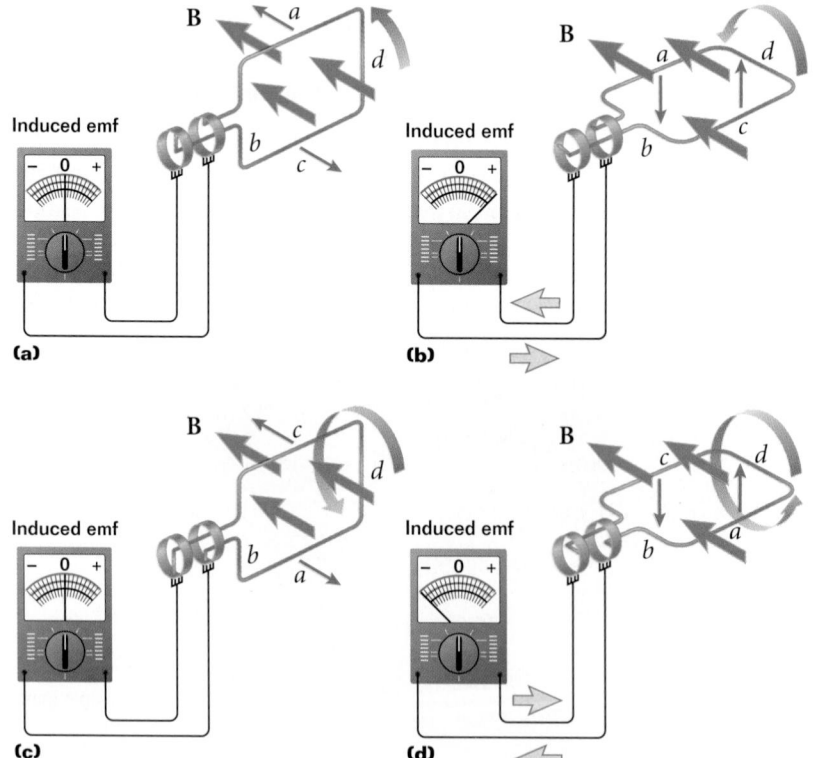

A graph of the change in emf versus time as the loop rotates is shown in **Figure 22-12.** Note the similarities between this graph and a sine curve.

The induced emf is the result of the steady change in the angle between the magnetic field lines and the normal to the loop. The following equation for the emf produced by a generator can be derived from Faraday's law of induction. The derivation is not shown here because it requires the use of calculus. In this equation, the angle of orientation, θ, has been replaced with the equivalent expression ωt, where ω is the angular frequency of rotation $(2\pi f)$.

$$\text{emf} = NAB\omega \, (\sin \omega t)$$

The equation describes the sinusoidal variation of emf with time, as graphed in **Figure 22-12.**

The emf has a maximum value when the plane of the loop is parallel to the magnetic field, that is, when $\sin (\omega t) = 1$, which occurs when $\omega t = \theta = 90°$. In this case, the expression above reduces to the following:

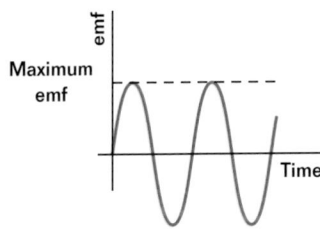

Figure 22-12
The change with time of the induced emf is depicted by a sine wave.

MAXIMUM EMF FOR A GENERATOR

$$\text{maximum emf} = NAB\omega$$

maximum induced emf = number of loops × cross-sectional area of loops × magnetic field strength × angular frequency of rotation of loops

---

SAMPLE PROBLEM 22B

Induction in generators (★ TEKS) 2C, 6D

PROBLEM

A generator consists of exactly eight turns of wire, each with an area $A = 0.095 \text{ m}^2$ and a total resistance of 12 Ω. The loop rotates in a magnetic field of 0.55 T at a constant frequency of 60.0 Hz. Find the maximum induced emf and maximum induced current in the loop.

SOLUTION

1. DEFINE **Given:** $f = 60.0 \text{ Hz}$ $A = 0.095 \text{ m}^2$ $R = 12 \text{ Ω}$
$B = 0.55 \text{ T} = 0.55 \text{ V·s/m}^2$ $N = 8$

Unknown: maximum emf = ? $I_{max} = ?$

Diagram:

$f = 60.0 \text{ Hz}$
$B = 0.55 \text{ T}$
$R = 12 \text{ Ω}$
$A = 0.095 \text{ m}^2$
$N = 8$ turns

continued on next page

 Misconception Alert

Some students may think that the boxed equation for the maximum emf of a generator holds for all cases. Be sure they understand that this equation only holds when the plane of the rotating loop and the magnetic field vectors are parallel. When the plane of the loop and the magnetic field vectors are not parallel, emf is not at a maximum, and the first equation on this page must be used.

Classroom Practice

The following may be used as a teamwork exercise or for demonstration at the board or on an overhead projector.

PROBLEM

Induction in generators

A 25 loop generator rotates with a frequency of 60.0 Hz in a 0.50 T field. The maximum emf induced is 117 V.

a. Calculate the cross-sectional area of the loops.

b. The generator is bought by another company and moved to Europe, where it is run at 50.0 Hz. Find the new maximum emf.

Answers
a. $2.5 \times 10^{-2} \text{ m}^2$
b. 98 V

(★ TEKS)

p. 804: 6D
p. 805: 2C, 6D

Teaching Tip

Point out that in Sample Problem 22B, the emf varies from a maximum value of 160 V to a minimum value of 0 V during one-fourth of a cycle ($\frac{1}{240}$ s). The current changes from a maximum value of 13 A to a minimum value of 0 A in the same time interval.

PRACTICE GUIDE 22B

Solving for:		
emf	📖 **PE**	Sample, 1–3; Ch. Rvw. 23–24
	💿 **PW**	7
	PB	4–6
I_{max}	📖 **PE**	Sample; Ch. Rvw. 24
	PB	7–8
N	📖 **PE**	4
	💿 **PW**	Sample, 1–2
	PB	9–10
ω	💿 **PW**	5–6
	PB	Sample, 1–3
B	💿 **PW**	3–4

ANSWERS TO

Practice 22B
Induction in generators

1. 87 V
2. 55 V
3. 1.7×10^{-2} V
4. 8 turns

---

2. PLAN

Choose an equation(s) or situation: Use the maximum emf equation for a generator. Use the definition of angular frequency to convert f to ω. The maximum current can be obtained from the definition for resistance.

$$\text{maximum emf} = NAB\omega \qquad \omega = 2\pi f$$

$$I_{max} = \frac{\text{maximum emf}}{R}$$

Rearrange the equation(s) to isolate the unknown(s): Substitute the angular frequency expression into the maximum emf equation.

$$\text{maximum emf} = NAB(2\pi f)$$

3. CALCULATE **Substitute the values into the equation(s) and solve:**

$$\text{maximum emf} = (8)(0.095 \text{ m}^2)(0.55 \text{ T})(2\pi)(60.0 \text{ Hz})$$

$$\boxed{\text{maximum emf} = 1.6 \times 10^2 \text{ V}}$$

$$I_{max} = \frac{1.6 \times 10^2 \text{ V}}{12 \text{ }\Omega} = 13 \text{ A}$$

$$\boxed{I_{max} = 13 \text{ A}}$$

CALCULATOR SOLUTION

Because the minimum number of significant figures for the data is two, the calculator answer, 157.5822875, should be rounded to two digits.

4. EVALUATE By expressing the units used in the calculation in terms of equivalent units (1 T = 1 (V•s/m^2) and 1 Hz = 1/s), you can see which terms cancel. In this case, m^2 and s cancel to leave the answer for emf in units of volts.

PRACTICE 22B

Induction in generators ⭐TEKS 2C, 6D

1. In a model generator, a 510-turn rectangular coil 0.082 m by 0.25 m rotates with an angular frequency of 12.8 rad/s in a uniform magnetic field of 0.65 T. What is the maximum emf induced in the coil?

2. A circular coil with a radius of $R = 0.22$ m and 17 turns is rotated in a uniform magnetic field of 1.7 T. The coil rotates at a constant frequency of 2.0 Hz. Determine the maximum value of the emf induced in the coil.

3. A square coil with an area of 0.045 m^2 consists of 120 turns of wire. The coil rotates about a vertical axis at 157 rad/s. The horizontal component of Earth's magnetic field at the location of the loop is 2.0×10^{-5} T. Calculate the maximum emf induced in the coil.

4. A maximum emf of 90.4 V is induced in a generator coil rotating with a frequency of 65 Hz. If the coil has an area of 230 cm^2 and rotates in a magnetic field of 1.2 T, how many turns are in the coil?

Alternating current changes direction at a constant frequency

The output emf of a typical generator has a sinusoidal pattern, as you can see in the graph in **Figure 22-12** on page 805. Note that the emf alternates from positive to negative. As a result, the output current from the generator changes its direction at regular intervals. This variety of current is called **alternating current,** or, more commonly, *ac.*

The rate at which the coil in an ac generator rotates determines the maximum generated emf. The frequency of the alternating current can differ from country to country. In the United States, Canada, and Central America, the frequency of rotation for commercial generators is 60 Hz. This means that the emf undergoes one full cycle of changing direction 60 times each second. In the United Kingdom, Europe, and most of Asia and Africa, 50 Hz is used. (Recall that $\omega = 2\pi f$, where f is the frequency in Hz.)

Resistors can be used in either alternating- or direct-current applications. A resistor resists the motion of charges regardless of whether they move in one continuous direction or shift direction periodically. Thus, if the definition for resistance holds for circuit elements in a dc circuit, it will also hold for the same circuit elements with alternating currents and emfs.

Effective current and potential difference are measured in ac circuits

An ac circuit consists of combinations of circuit elements and an ac generator or an ac power supply, which provides the alternating current. As shown earlier, the emf produced by a typical ac generator is sinusoidal and varies with time. The induced emf equals the instantaneous ac potential difference, which is written as Δv. The quantity for the maximum emf can be written as the maximum potential difference ΔV_{max}, and the emf produced by a generator can be expressed as follows:

$$\Delta v = \Delta V_{max}(\sin \omega t)$$

Because all circuits have some resistance, a simple ac circuit can be treated as an equivalent resistance and an ac source. In a circuit diagram, the ac source is represented by the symbol ⊘, as shown in **Figure 22-13.**

The instantaneous current that changes with the potential difference can be determined using the definition for resistance. The instantaneous current, i, is related to maximum current by the following expression:

$$i = I_{max}(\sin \omega t)$$

The rate at which electrical energy is converted to internal energy in the resistor (the power, P) has the same form as in the case of direct current. The electrical energy converted to internal energy in a resistor is proportional to the *square* of the current and is independent of the direction—or the change of direction—of the current. However, the energy produced by an alternating current with a maximum value of I_{max} is not the same as that produced by a direct current of the same value. This is because the alternating current is at its maximum value for only a very brief instant of time during a cycle.

alternating current

an electric current that changes direction at regular intervals

Did you know?

Although the light intensity from a 60 W incandescent light bulb appears to be constant, the current in the bulb fluctuates 60 times each second between −0.71 A and 0.71 A. The light appears to be steady because the fluctuations are too rapid for our eyes to perceive.

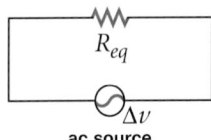

Figure 22-13
An ac circuit represented schematically consists of an ac source and an equivalent resistance.

Effects of alternating current

Purpose Illustrate the effects of alternating current.

Materials bicolored LED (available at many electronics stores), step-down transformer, resistor (100-250 Ω), 2 m of flexible wire

Procedure Connect the wire, the resistor, and the LED in series with the transformer. Explain to students that the diode is red and green, respectively, for currents with opposite directions. Turn the lights down, and show the class that the LED appears yellow. This is because the alternating polarity of the current switches the color of the LED from red to green 60 times each second.

Hold the wire about halfway between the transformer and the diode, and twirl the wire in a vertical circle so that the diode moves in a circular path. Students should see red and green bars at equally spaced intervals along the diode's path. Have students count the number of green bars they see in the circle, then have them measure the time it takes for the diode to travel 10 times around the circular path. Ask students how much time it takes for the diode to change from green to red to green $\left(\frac{1}{60} s\right)$.

rms current

the amount of direct current that dissipates as much energy in a resistor as an instantaneous alternating current does during a complete cycle

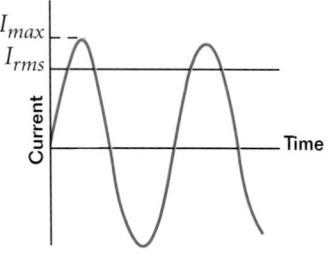

Figure 22-14
The rms current is a little more than two-thirds as large as the maximum current.

The important quantity for current in an ac circuit is the **rms current.** The rms (or *root-mean-square*) current is the same as the amount of direct current that would dissipate the same energy in a resistor as is dissipated by the instantaneous alternating current over a complete cycle.

Figure 22-14 shows a graph in which instantaneous and rms currents are compared. **Table 22-2** summarizes the notations used in this chapter for these and other ac quantities.

The equation for the power dissipated in an ac circuit has the same form as the equation for power dissipated in a dc circuit except that the dc current I is replaced by the rms current (I_{rms}).

$$P = (I_{rms})^2 R$$

This equation is identical in form to the one for direct current. However, the power dissipated in the ac circuit equals half the power dissipated in a dc circuit when the dc current equals I_{max}.

$$P = (I_{rms})^2 R = \frac{1}{2}(I_{max})^2 R$$

From this equation you may note that the rms current is related to the maximum value of the alternating current by the following equation:

$$(I_{rms})^2 = \frac{(I_{max})^2}{2}$$

$$I_{rms} = \frac{I_{max}}{\sqrt{2}} = 0.707\ I_{max}$$

This equation says that an alternating current with a maximum value of 5 A produces the same heating effect in a resistor as a direct current of $(5/\sqrt{2})$ A, or about 3.5 A.

Alternating potential differences are also best discussed in terms of *rms potential differences,* with the relationship between rms and maximum values being identical to the one for currents. The rms potential difference, ΔV_{rms}, is related to the maximum value of the potential difference, ΔV_{max}, as follows:

$$\Delta V_{rms} = \frac{\Delta V_{max}}{\sqrt{2}} = 0.707\ V_{max}$$

Table 22-2	Notation used for ac circuits	
	Potential difference	**Current**
instantaneous values	Δv	i
maximum values	ΔV_{max}	I_{max}
rms values	$\Delta V_{rms} = \dfrac{\Delta V_{max}}{\sqrt{2}}$	$I_{rms} = \dfrac{I_{max}}{\sqrt{2}}$

The ac potential difference of 120 V measured from an electric outlet is actually an rms potential difference of 120 V. A quick calculation shows that such an ac potential difference has a maximum value of about 170 V.

A resistor limits the current in an ac circuit just as it does in a dc circuit. If the definition of resistance is valid for an ac circuit, the rms potential difference across a resistor equals the rms current multiplied by the resistance. Thus, all maximum and rms values can be calculated if only one current or emf value and the circuit resistance are known.

Because ac ammeters and voltmeters measure rms values, all values of alternating current and alternating potential difference in this chapter will be given as rms values unless otherwise noted. The equations for ac circuits have the same form as those for dc circuits when rms values are used.

SAMPLE PROBLEM 22C

rms currents and potential differences ★TEKS 2C

PROBLEM

A generator with a maximum output emf of 205 V is connected to a 115 Ω resistor. Calculate the rms potential difference. Find the rms current through the resistor. Find the maximum ac current in the circuit.

SOLUTION

1. DEFINE

Given: $\Delta V_{max} = 205$ V $R = 115\ \Omega$

Unknown: $\Delta V_{rms} = ?$ $I_{rms} = ?$ $I_{max} = ?$

Diagram:

$\Delta V_{max} = 205$ V
$\Delta V_{rms} = ?$

$I_{rms} = ?$ $I_{max} = ?$
$R = 115\ \Omega$

2. PLAN

Choose an equation(s) or situation: Use the equation for the rms potential difference to find ΔV_{rms}.

$$\Delta V_{rms} = 0.707\ \Delta V_{max}$$

Rearrange the definition for resistance to calculate I_{rms}.

$$I_{rms} = \frac{\Delta V_{rms}}{R}$$

Use the equation for rms current to find I_{max}.

$$I_{rms} = 0.707\ I_{max}$$

Rearrange the equation(s) to isolate the unknown(s):
Rearrange the equation relating rms current to maximum current so that maximum current is calculated.

$$I_{max} = \frac{I_{rms}}{0.707}$$

continued on next page

SECTION 22-2

The Language of Physics
Remind students that because emf is a potential difference, maximum emf can be abbreviated as ΔV_{max}, and rms emf can be abbreviated as ΔV_{rms}.

Classroom Practice

The following may be used as a teamwork exercise or for demonstration at the board or on an overhead projector.

PROBLEM

rms currents and potential differences

A generator supplies 110 V rms to a 25 Ω circuit. What is the maximum current supplied to the circuit?

Answer
6.2 A

★TEKS

p. 809: 2C

PRACTICE GUIDE 22C

Solving for:

ΔV	📖 **PE**	Sample, 1, 4, 5a, 6; Ch. Rvw. 25–28
	💿 **PW**	Sample, 4–6
	PB	Sample, 1, 4, 5–6, 9
I	📖 **PE**	Sample, 1–2, 3a, 4, 5b; Ch. Rvw. 26–28
	💿 **PW**	Sample, 1–4, 5, 7
	PB	Sample, 1–4, 5, 7, 8–10
R	📖 **PE**	3b
	💿 **PW**	3, 5
	PB	3, 5
P	📖 **PE**	Ch. Rvw. 27
	💿 **PW**	Sample, 1
	PB	8–10

ANSWERS TO

Practice 22C
rms currents and potential differences

1. 4.8 A; 6.8 A, 170 V
2. 7.8 A
3. **a.** 7.42 A
   b. 14.8 Ω
4. 1.44 A; 2.04 A, 21.2 V
5. **a.** 1.10×10^2 V
   b. 2.1 A
6. 319 V

---

3. CALCULATE **Substitute the values into the equation(s) and solve:**

$$\Delta V_{rms} = (0.707)(205 \text{ V}) = 145 \text{ V}$$

$$I_{rms} = \frac{145 \text{ V}}{115 \text{ }\Omega} = 1.26 \text{ A}$$

$$I_{max} = \frac{1.26 \text{ A}}{0.707} = 1.78 \text{ A}$$

$$\Delta V_{rms} = 145 \text{ V}$$
$$I_{rms} = 1.26 \text{ A}$$
$$I_{max} = 1.78 \text{ A}$$

CALCULATOR SOLUTION

Because the minimum number of significant figures for the data is three, the calculator solution for the rms potential difference, 144.935, should be rounded to three digits.

4. EVALUATE The rms values for potential difference and current are a little more than two-thirds the maximum values, as expected.

PRACTICE 22C

rms currents and potential differences ⭐TEKS 2C

1. What is the rms current in a light bulb that has a resistance of 25 Ω and an rms potential difference of 120 V? What are the maximum values for current and potential difference?

2. The current in an ac circuit is measured with an ammeter. The meter gives a reading of 5.5 A. Calculate the maximum ac current.

3. A toaster is plugged into a source of alternating potential difference with an rms value of 110 V. The heating element is designed to convey a current with a maximum value of 10.5 A. Find the following:
   a. the rms current in the heating element
   b. the resistance of the heating element

4. An audio amplifier provides an alternating rms potential difference of 15.0 V. A loudspeaker connected to the amplifier has a resistance of 10.4 Ω. What is the rms current in the speaker? What are the maximum values of the current and the potential difference?

5. An ac generator has a maximum potential difference output of 155 V.
   a. Find the rms potential difference output.
   b. Find the rms current in the circuit when the generator is connected to a 53 Ω resistor.

6. The largest potential difference that can be placed across a certain capacitor at any instant is 451 V. What is the largest rms potential difference that can be placed across the capacitor without damaging it?

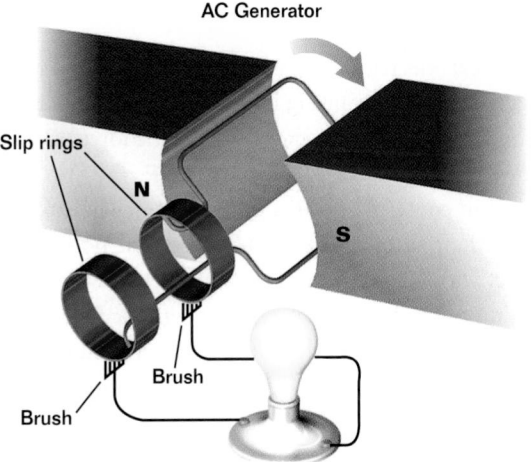

AC Generator

Slip rings

N

S

Brush

Brush

DC Generator

N

S

Brush

Commutator

Brush

Figure 22-15

A simple dc generator (shown on the right) employs the same design as an ac generator (shown on the left). A split slip ring converts alternating current to direct current.

Alternating current can be converted to direct current

The conducting loop in an ac generator must be free to rotate through the magnetic field. Yet it must also be part of an electric circuit at all times. To accomplish this, the ends of the loop are connected to conducting rings, called *slip rings,* that rotate with the loop. Connections to the external circuit are made by stationary graphite strips, called *brushes,* that make continuous contact with the slip rings. Because the current changes direction in the loop, the output current through the brushes alternates direction as well.

By varying this arrangement slightly, an ac generator can be converted to a dc generator. Note in **Figure 22-15** that the components of a dc generator are essentially the same as those of the ac generator except that the contacts to the rotating loop are made by a single split slip ring, called a *commutator.*

At the point in the loop's rotation when the current has dropped to zero and is about to change direction, each half of the commutator comes into contact with the brush that was previously in contact with the other half of the commutator. The reversed current in the loop changes directions again so that the output current has the same direction as it originally had, although it still changes from a maximum value to zero. A plot of this pulsating direct current is shown in **Figure 22-16.**

A steady direct current can be produced by using many loops and commutators distributed around the rotation axis of the dc generator. The sinusoidal currents from each loop overlap. The superposition of the currents produces a direct current output that is almost entirely free of fluctuations.

internet**connect**

SCI*LINKS*

NSTA

TOPIC: Electrical safety
GO TO: www.scilinks.org
*sci*LINKS CODE: HF2223

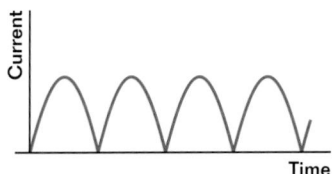

Current

Time

Figure 22-16

The output current for a dc generator is a sine wave with the negative parts of the curve made positive.

Teaching Tip

Inform students that converting from ac to dc may also be accomplished electronically. This is done, for example, by ac-to-dc power converters, such as those used for portable electronic games or CD players.

Teaching Tip

Point out that, as shown by the graph of direct current in **Figure 22-16,** the direct current produced by a generator alternates over time but does not change polarity, as does alternating current. The direct current generated by a battery, on the other hand, is steady and does not fluctuate. (See **Figure 19-5** on page 699.) Generators can produce a steady direct current similar to that produced by batteries if many loops and commutators are used.

⊛ TEKS

p. 810: 2C

Teaching Tip

Point out that generators and motors have opposite functions. Generators convert mechanical energy to electrical energy, while motors convert electrical energy to mechanical energy.

Demonstration 7

Electric motor

Purpose Illustrate the similarity between generators and motors.

Materials hand-operated generator, capacitor (1 F), 100 Ω resistor, connecting wires

Procedure Connect the capacitor to the hand-operated generator. Charge the capacitor with the generator. Do not apply a large potential difference. Release the handle, and watch the generator become a motor.

Optional: An interesting comparison can be made between the responses of the generator to different loads. Repeat the demonstration with and without the resistor attached.

⊛TEKS

p. 812: 6F
p. 813: 3C, 2C, 6D

MOTORS

Motors are devices that convert electrical energy to mechanical energy. Instead of a current being generated by a rotating loop in a magnetic field, a current is supplied to the loop by an emf source and the magnetic force on the current loop causes it to rotate (see **Figure 22-17**). ⊛TEKS **6F**

A motor is almost identical in construction to a dc generator. The coil of wire is mounted on a rotating shaft and is positioned between the poles of a magnet. Brushes make contact with a commutator, which alternates the current in the coil. This alternation of the current causes the magnetic field produced by the current to regularly reverse and thus always be repelled by the fixed magnetic field. Thus, the coil and the shaft are kept in continuous rotational motion.

A motor can perform useful mechanical work when a shaft connected to its rotating coil is attached to some external device. As the coil in the motor rotates, however, the changing normal component of the magnetic field through it induces an emf that acts to reduce the current in the coil. If this were not the case, Lenz's law would be violated. This induced emf is called the **back emf.**

The back emf increases in magnitude as the magnetic field changes at a higher rate. In other words, the faster the coil rotates, the greater the back emf becomes. The potential difference available to supply current to the motor equals the difference between the applied potential difference and the back emf. Consequently, the current in the coil is also reduced because of the presence of back emf. The faster the motor turns, the smaller the net potential difference across the motor, and the smaller the net current in the coil, becomes.

back emf

the emf induced in a motor's coil that tends to reduce the current powering the motor

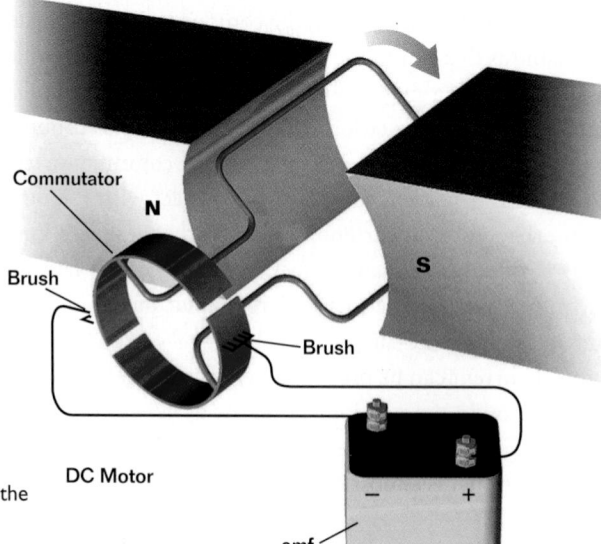

Figure 22-17

In a motor, the current in the coil interacts with the magnetic field, causing the coil and the shaft on which the coil is mounted to turn.

Consumer Focus *Avoiding Electrocution*  TEKS 3C

A person can receive an electric shock by touching a conducting or "live" wire while in contact with a lower electric potential, or ground. The ground contact might be made by touching a water pipe (which is normally at zero potential) or by standing on the floor with wet feet (because impure water is a good conductor).

Electric shock can result in fatal burns or can cause the muscles of vital organs, such as the heart, to malfunction. The degree of damage to the body depends on the magnitude of the current, the length of time it acts, and the part of the body through which it passes. A current of 100 milliamps (mA) can be fatal. If the current is larger than about 10 mA, the hand muscles contract and the person may be unable to let go of the wire.

To prevent electrocution, any wires designed to have such currents in them are wrapped in insulation, usually plastic or rubber. However, with frequent use, electric cords can fray, exposing some of the conductors. To further prevent electrocution in these and other situations in which electrical contact can be made, devices called a Ground Fault Circuit Interrupter (GFCI) and a Ground Fault Interrupter (GFI) are mounted in electrical outlets and individual appliances.

GFCIs and GFIs provide protection by comparing the current in one side of the electrical outlet socket to the current in the other socket. If there is even a 5 mA difference, the interrupter opens the circuit in a few milliseconds (thousandths of a second). If you accidentally touch a bare wire and create an alternate conducting path through you to ground, the device detects this redirection of current and you get only a small shock or tingle.

Despite these safety devices, you can still be electrocuted. Never use electrical appliances near water or with wet hands. Use a battery-powered radio near water because batteries cannot supply enough current to harm you. It is also a good idea to replace old outlets with GFCI-equipped units or to install GFI-equipped circuit breakers.

Consumer Focus

BACKGROUND

Ground Fault Interrupters make use of Faraday's law. Both the wire that leads from the wall outlet to the appliance and the wire that leads from the appliance back to the wall pass through an iron ring. Part of the iron ring is wrapped in a coil called a *sensing coil*. When the current to the appliance equals the current from the appliance, the net magnetic field through the sensing coil is zero.

If a short circuit occurs, the net magnetic field through the coil is no longer zero. Because the current in the wire is alternating, an ac potential difference is induced in the sensing coil. This induced potential difference in the coil is used to trigger a circuit breaker, stopping the current before it reaches a level that might be harmful to the person using the appliance.

Section Review   ⭐TEKS 2C, 6D

1. A loop with an area of 0.33 m^2 is rotating at 281 rad/s with its axis of rotation perpendicular to a uniform magnetic field. The magnetic field has a strength of 0.035 T. If the loop contains 37 turns of wire, what is the maximum potential difference induced in the loop?

2. A generator develops a maximum emf of 2.8 V. If the generator coil has 25 turns of wire, a cross-sectional area of 36 cm^2, and rotates with a constant frequency of 60.0 Hz, what is the strength of the magnetic field in which the coil rotates?

3. What is the purpose of a commutator in a motor? Explain what would happen if a commutator were not used in a motor.

4. **Physics in Action**   The rms current produced by a single coil of an electric guitar is 0.025 mA. How large is the maximum instantaneous current? If the coil's resistance is 4.3 kΩ, what are the rms and maximum potential differences produced by the coil?

Section Review
ANSWERS

1. 120 V
2. 8.3×10^{-2} T
3. The commutator reverses the direction of the current in the coil so that the magnetic field produced by the current always opposes the external field and the coil turns continuously. Without the commutator, the coil would turn until the coil's magnetic field was aligned with the external field. The coil would then cease to turn.
4. 0.035 mA; 0.11 V, 0.15 V

22-3
Inductance

22-3 SECTION OBJECTIVES

- **Describe how mutual induction occurs in circuits.**

- **Calculate the potential difference from a step-up or step-down transformer.**

- **Describe how self-induction occurs in an electric circuit.**

MUTUAL INDUCTANCE

The basic principle of electromagnetic induction was first demonstrated by Michael Faraday. His experimental apparatus, which resembled the arrangement shown in **Figure 22-18,** used a coil connected to a switch and a battery instead of a magnet to produce a magnetic field. This coil is called the *primary coil,* and its circuit is called the primary circuit. The magnetic field is strengthened by the magnetic properties of the iron ring around which the primary coil is wrapped. ⓣTEKS 6B

A second coil is wrapped around another part of the iron ring and is connected to a galvanometer. An emf is induced in this coil, called the *secondary coil,* when the magnetic field of the primary coil is changed. When the switch in the primary circuit is closed, the galvanometer in the secondary circuit deflects in one direction and then returns to zero. When the switch is opened, the galvanometer deflects in the opposite direction and again returns to zero. When there is a steady current in the primary circuit, the galvanometer reads zero.

The magnitude of this emf is predicted by Faraday's law of induction. However, Faraday's law can be rewritten so that the induced emf is proportional to the changing current in the primary coil. This can be done because of the direct proportionality between the magnetic field produced by a current in a coil, or solenoid, and the current itself. The form of Faraday's law in terms of changing primary current is as follows: ⓣTEKS 3B

$$\text{emf} = -N\frac{\Delta[AB(\cos\theta)]}{\Delta t} = -M\frac{\Delta I}{\Delta t}$$

mutual inductance

a measure of the ability of one circuit carrying a changing current to induce an emf in a nearby circuit

The constant, *M,* is called the **mutual inductance** of the two-coil system. The mutual inductance depends on the geometrical properties of the coils and

Figure 22-18
Faraday's electromagnetic-induction experiment used a changing current in one circuit to induce a current in another circuit.

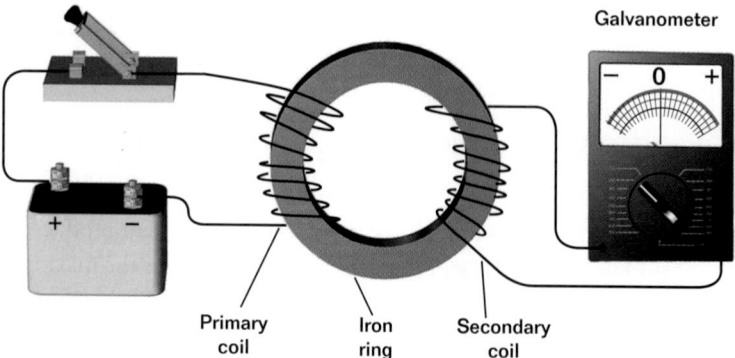

Galvanometer

Primary coil   Iron ring   Secondary coil

their orientation to each other. Because these properties are kept constant, it follows that a changing current in the secondary coil can also induce an emf in the primary circuit. The equation holds as long as the coils remain unchanged with respect to each other, so that the mutual inductance is constant.

By changing the number of turns of wire in the secondary coil, the induced emf in the secondary circuit can be changed. This arrangement is the basis of an extremely useful electrical device: the transformer.

TRANSFORMERS

It is often desirable or necessary to change a small ac potential difference to a larger one or to change a large potential difference to a smaller one. The device that makes these conversions possible is the **transformer.** (★)TEKS 6F

In its simplest form, an ac transformer consists of two coils of wire wound around a core of soft iron, like the apparatus for the Faraday experiment. The coil on the left in **Figure 22-19** has N_1 turns and is connected to the input ac potential difference source. This coil is called the primary winding, or the *primary*. The coil on the right, which is connected to a resistor R and consists of N_2 turns, is the *secondary*. As in Faraday's experiment, the iron core provides a medium for nearly all magnetic field lines passing through the two coils.

Because the strength of the magnetic field in the iron core and the cross-sectional area of the core are the same for both the primary and secondary windings, the potential differences across the two windings differ only because of the different number of turns of wire for each. The ac potential difference that gives rise to the changing magnetic field in the primary is related to that changing field by Faraday's law of induction.

$$\Delta V_1 = -N_1 \frac{\Delta(AB \cos \theta)}{\Delta t}$$

Similarly, the induced potential difference (emf) across the secondary coil is

$$\Delta V_2 = -N_2 \frac{\Delta(AB \cos \theta)}{\Delta t}$$

Taking the ratio of ΔV_1 to ΔV_2 causes all terms on the right side of both equations except for N_1 and N_2 to cancel. This result is the transformer equation.

TRANSFORMER EQUATION

$$\Delta V_2 = \frac{N_2}{N_1} \Delta V_1$$

potential difference in secondary =
$\left(\dfrac{\text{number of turns in secondary}}{\text{number of turns in primary}} \right)$ potential difference in primary

transformer

a device that changes one ac potential difference to a different ac potential difference

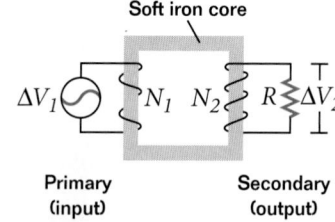

Figure 22-19
A transformer uses the alternating current in the primary circuit to induce an alternating current in the secondary circuit.

*internet*connect

SC*i*LINKS.
NSTA
TOPIC: Transformers
GO TO: www.scilinks.org
*sci*LINKS CODE: HF2224

Demonstration 9

Induced potential difference

Purpose Show that induced potential difference depends on the number of turns of wire.

Materials 9 V battery, knife switch, 30 cm (or longer) iron bar, 2 flashlight bulbs (or more) in holders, three connecting wires (each about 1 m long)

Procedure Set up the primary coil as in Demonstration 8. Set up two secondary coils, one with 25 turns of wire and one with 50 turns of wire. Close the switch, and have students compare the brightness of the two bulbs. You may wish to open and close the switch several times for repeated observations.

(★) TEKS

p. 814: 6B, 3B
p. 815: 6F

Teaching Tip

Some students may wonder whether dc transformers are possible. The dc produced by a battery would not work with a transformer. This is because a changing current is required, and a battery generates a steady direct current. However, as seen on page 811, a generator can produce a fluctuating direct current. Such a current could be used by a transformer.

Teaching Tip

Point out to students that spark plugs create a spark because the potential difference across the gap increases to 20 000 V. At such high potential differences, air is ionized and becomes a conductor.

⭐TEKS

p. 817: 6F, 2C

Another way to express this equation is to equate the ratio of the potential differences to the ratio of the number of turns.

$$\frac{\Delta V_2}{\Delta V_1} = \frac{N_2}{N_1}$$

When N_2 is greater than N_1, the secondary potential difference is greater than that of the primary, and the transformer is called a *step-up transformer*. When N_2 is less than N_1, the secondary potential difference is less than that of the primary, and the transformer is called a *step-down transformer*.

It may seem that a transformer provides something for nothing. For example, a step-up transformer can change an input potential difference from 10 V to 100 V. However, the power input at the primary must equal the power output at the secondary. An increase in potential difference at the secondary means that there must be a proportional decrease in current. If the potential difference at the secondary is 10 times that at the primary, then the current at the secondary is reduced by a factor of 10.

Real transformers are not perfectly efficient

The transformer equation assumes that there are no power losses between the transformer's primary and its secondary. Real transformers typically have efficiencies ranging from 90 percent to 99 percent. Power losses occur because of the small currents induced by changing magnetic fields in the iron core of the transformer and because of resistance in the wires of the windings.

When electric power is transmitted over large distances, it is economical to use a high potential difference and a low current. This is because the power lost to resistive heating in the transmission lines varies as I^2R. By reducing the current by a factor of 10, the power loss is reduced by a factor of 100. In practice, potential difference is stepped up to around 230 000 V at the generating station, then stepped down to 20 000 V at a regional distribution station, and finally stepped down to 120 V at the customer's utility pole. The high potential difference in long-distance transmission lines makes them especially dangerous when they are knocked down by high winds.

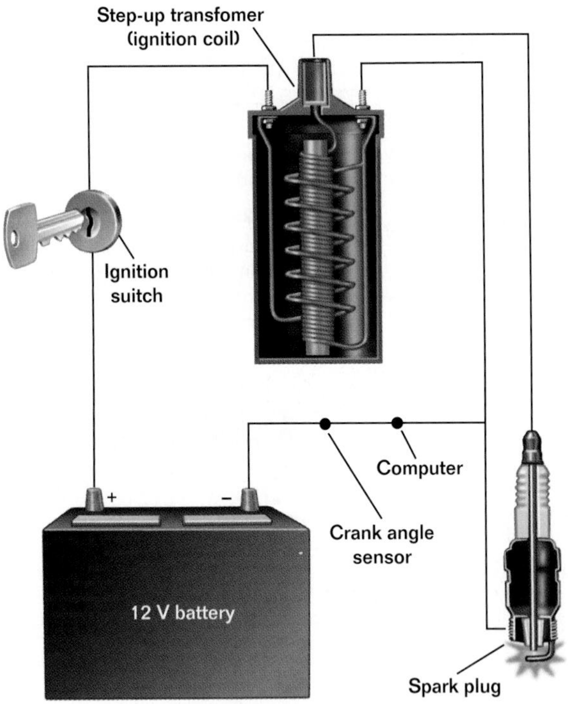

Coils in gasoline engines are transformers

An automobile ignition system uses a transformer, or *ignition coil*, to convert the car battery's 12 dc volts to a potential difference that is large enough to cause sparking between the gaps of the spark plugs. The diagram in **Figure 22-20** shows the type of ignition system that has been

Step-up transformer (ignition coil)

Ignition switch

Computer

Crank angle sensor

12 V battery

Spark plug

Figure 22-20
The transformer in an automobile engine raises the potential difference across the gap in a spark plug so that sparking occurs.

used in automobiles since 1990. In this arrangement, each cylinder has its own transformer, and a photoelectric detector called a *crank angle sensor* determines from the crankshaft's position which cylinder's contents are at maximum compression. Upon receiving this signal, the computer closes the primary circuit to the cylinder's coil, causing the current in the primary to rapidly increase and the magnetic field in the transformer to change. This, in turn, induces a potential difference of about 20 000 V across the secondary. ⊛ TEKS 6F

Module 20
"Induction and Transformers"
provides an interactive lesson with guided problem-solving practice to teach you about induction and transformers.

SAMPLE PROBLEM 22D

Transformers ⊛ TEKS 2C

PROBLEM

A step-up transformer is used on a 120 V line to provide a potential difference of 2400 V. If the primary has 75 turns, how many turns must the secondary have?

SOLUTION

1. DEFINE **Given:** $\Delta V_1 = 120 \text{ V}$ $\Delta V_2 = 2400 \text{ V}$ $N_1 = 75 \text{ turns}$

Unknown: $N_2 = ?$

Diagram:

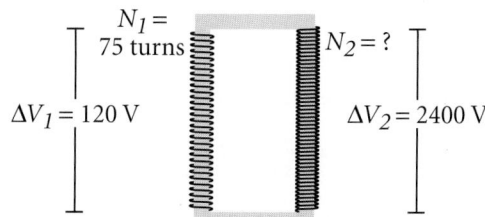

2. PLAN **Choose an equation(s) or situation:** Use the transformer equation.

$$\Delta V_2 = \frac{N_2}{N_1} \Delta V_1$$

Rearrange the equation(s) to isolate the unknown(s):

$$N_2 = \frac{\Delta V_2}{\Delta V_1} N_1$$

3. CALCULATE **Substitute the values into the equation(s) and solve:**

$$N_2 = \left(\frac{2400 \text{ V}}{120 \text{ V}}\right) 75 \text{ turns} = 1500 \text{ turns}$$

$$\boxed{N_2 = 1500 \text{ turns}}$$

4. EVALUATE The greater number of turns in the secondary accounts for the increase in the potential difference in the secondary. The step-up factor for the transformer is 20:1.

Classroom Practice

The following may be used as a teamwork exercise or for demonstration at the chalkboard or on an overhead projector.

PROBLEM

Transformers

A transformer has 75 turns on the primary and 1500 turns on the secondary.

a. If the potential difference across the primary is 120 V, what is the potential difference across the secondary?

b. If the transformer has 1625 turns on the secondary instead of 1500, what is the potential difference across the secondary?

Answers

  a. 2400 V

  b. 2600 V

Interactive Problem-Solving Tutor

See Module 20
"Induction and Transformers" provides additional development of problem-solving skills.

ANSWERS TO

Practice 22D

Transformers

1. 55 turns
2. 3.5×10^4 turns
3. 25 turns
4. 156:1
5. 2.6×10^4 V
6. 147 V

PRACTICE 22D

Transformers ★TEKS 2C, 6F

1. A step-down transformer providing electricity for a residential neighborhood has exactly 2680 turns in its primary. When the potential difference across the primary is 5850 V, the potential difference at the secondary is 120 V. How many turns are in the secondary?

2. A step-up transformer used in an automobile has a potential difference across the primary of 12 V and a potential difference across the secondary of 2.0×10^4 V. If the number of turns in the primary is 21, what is the number of turns in the secondary?

3. A step-up transformer for long-range transmission of electric power is used to create a potential difference of 119 340 V across the secondary. If the potential difference across the primary is 117 V and the number of turns in the secondary is 25 500, what is the number of turns in the primary?

4. A potential difference of 0.750 V is needed to provide a large current for arc welding. If the potential difference across the primary of a step-down transformer is 117 V, what is the ratio of the number of turns of wire on the primary to the number of turns on the secondary?

5. A television picture tube requires a high potential difference, which in older models is provided by a step-up transformer. The transformer has 12 turns in its primary and 2550 turns in its secondary. If 120 V is placed across the primary, what is the output potential difference?

6. A step-down transformer has 525 turns in its secondary and 12 500 turns in its primary. If the potential difference across the primary is 3510 V, what is the potential difference across the secondary?

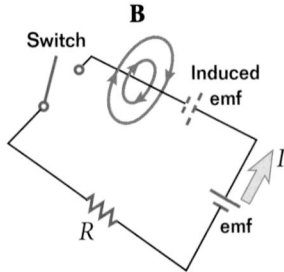

Figure 22-21

The changing magnetic field that is produced by changing current induces an emf that opposes the applied emf.

SELF-INDUCTION

Consider a circuit consisting of a switch, a resistor, and a source of emf, as shown in **Figure 22-21.** When the switch is closed, the current does not immediately change from zero to its maximum value, I_{max}. Instead, the current increases with time, and the magnetic field through the loop due to this current also increases. The increasing field induces an emf in the circuit to oppose the change in magnetic field, according to Lenz's law, and so the polarity of the induced emf must oppose the direction of the original current. The induced emf is in the direction indicated by the dashed battery. The net potential difference across the resistor is the emf of the battery minus the induced emf.

As the current increases, the *rate* of increase lessens and the induced emf decreases, as shown in **Figure 22-22.** The decrease in the induced emf results in a gradual increase in the current. Similarly, when the switch is opened, the current gradually decreases to zero. This effect is called **self-induction** because the changing magnetic field through the circuit arises from the current in the circuit itself. The induced emf in this case is called *self-induced emf.*

An example of self-induction is seen in a coil wound on a cylindrical iron core. (A practical device would have several hundred turns.) When the current is in the direction shown in **Figure 22-23(a),** a magnetic field forms inside the coil. As the current changes with time, the field through the coil changes and induces an emf in the coil.

Lenz's law indicates that this induced emf opposes the change in the current. For increasing current, the induced emf is as pictured in **Figure 22-23(b),** and for decreasing current, the induced emf is as shown in **Figure 22-23(c).**

Faraday's law of induction takes the same form for self-induction as it does for mutual induction except that the symbol for mutual inductance, M, is replaced with the symbol for *self-inductance, L.* (★)TEKS **3B**

$$emf = -N\frac{\Delta[AB(\cos\theta)]}{\Delta t} = -L\frac{\Delta I}{\Delta t}$$

The self-inductance of a coil depends on the number of turns of wire in the coil, the coil's cross-sectional area, and other geometric factors. The SI unit of inductance is the *henry* (H), which is equal to 1 volt-second per ampere.

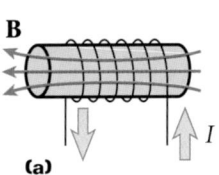

(a)

Lenz's law emf
for increasing I

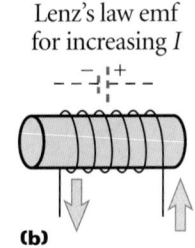

(b)

Lenz's law emf
for decreasing I

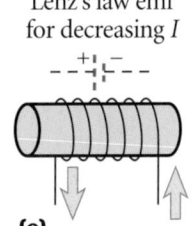

(c)

self-induction

the process by which a changing current in a circuit induces an emf in that same circuit

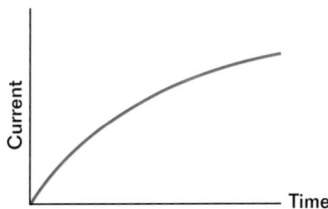

Figure 22-22
The self-induced emf reduces the rate of increase of the current in the circuit immediately after the circuit is closed.

Figure 22-23
The polarity of the self-induced emf is such that it can produce a current whose direction is opposite the change in the current in the coil.

Section Review　(★)TEKS **2C**

1. The centers of two circular loops are separated by a fixed distance. For what relative orientation of the loops will their mutual inductance be a maximum? For what orientation will it be a minimum?

2. A step-up transformer has exactly 50 turns in its primary and exactly 7000 turns in its secondary. If the potential difference across the primary is 120 V, what is the potential difference across the secondary?

3. Does the self-inductance (L) of a coil depend on the current in the coil?

The Language of Physics

The SI unit of inductance, the henry (H), was named after the American scientist Joseph Henry. Henry was a professor of mathematics and philosophy in Albany, New York. He discovered electromagnetic induction while on a one-month vacation, but he then returned to work and was not able to complete his research for publication. By the time Henry published his discovery, Faraday had already discovered the same phenomenon and published his results.

Section Review
ANSWERS

1. The planes of the two loops must be parallel for maximum mutual inductance. The planes of the loops must be perpendicular to each other for minimum mutual inductance.

2. 1.7×10^4 V

3. No, self-inductance (L) only depends on the number of turns of wire in the coil, the coil's cross-sectional area, and geometric factors.

(★)TEKS

p. 818: 2C, 6F
p. 819: 3B, 2C

CHAPTER 22
Summary

KEY TERMS

alternating current (p. 807)

back emf (p. 812)

electromagnetic induction (p. 794)

generator (p. 803)

mutual inductance (p. 814)

rms current (p. 808)

self-induction (p. 819)

transformer (p. 815)

KEY IDEAS

Section 22-1 Induced current

- Changing the magnetic field strength near a conductor induces an emf.
- The direction of an induced current in a circuit is such that its magnetic field opposes the change in the applied magnetic field.
- Induced emf can be calculated using Faraday's law of magnetic induction.

$$\text{emf} = -N\frac{\Delta[AB(\cos\theta)]}{\Delta t}$$

Section 22-2 Alternating current, generators, and motors

- Generators use induction to convert mechanical energy into electrical energy.
- Alternating current is measured in terms of rms current.
- Motors use an arrangement similar to that of generators to convert electrical energy into mechanical energy.

Section 22-3 Inductance

- Mutual inductance involves the induction of a current in one circuit by means of a changing current in a nearby circuit.
- Transformers change the potential difference of an alternating current.
- Self-induction occurs when the changing current in a circuit induces an emf in the same circuit.

Diagram symbols

Induced emf	
ac generator/ alternating current emf source	

Variable symbols

Quantities		Units	
N	number of turns		(unitless)
ΔV_{max}	maximum potential difference	V	volt
ΔV_{rms}	rms potential difference	V	volt
I_{max}	maximum current	A	ampere
I_{rms}	rms current	A	ampere
M	mutual inductance	H	$\text{henry} = \dfrac{\text{volt} \cdot \text{second}}{\text{ampere}}$
L	self-inductance	H	$\text{henry} = \dfrac{\text{volt} \cdot \text{second}}{\text{ampere}}$

INDUCED CURRENT

Review questions

1. Suppose you have two circuits. One consists of an electromagnet, a dc emf source, and a variable resistor that permits you to control the strength of the magnetic field. In the second circuit, you have a coil of wire and a galvanometer. List three ways that you can induce a current in the second circuit.

2. Explain how Lenz's law allows you to determine the direction of an induced current.

3. What four factors affect the magnitude of the induced emf in a coil of wire?

4. If you have a fixed magnetic field and a length of wire, how can you increase the induced emf across the ends of the wire?

Conceptual questions

5. Rapidly inserting the north pole of a bar magnet into a coil of wire connected to a galvanometer causes the needle of the galvanometer to deflect to the right. What will happen to the needle if you do the following?
   a. pull the magnet out of the coil
   b. let the magnet sit at rest in the coil
   c. thrust the south end of the magnet into the coil

6. Explain how Lenz's law illustrates the principle of energy conservation.

7. Does dropping a bar magnet down a long copper tube induce a current in the tube? If so, how must the magnet be oriented with respect to the tube?

8. Two bar magnets are placed side by side so that the north pole of one magnet is next to the south pole of the other magnet. If these magnets are then pushed toward a coil of wire, would you expect an emf to be induced in the coil? Explain your answer.

9. An electromagnet is placed next to a coil of wire in the arrangement shown in **Figure 22-24**. According to Lenz's law, what will be the direction of the induced current in the resistor R in the following cases?
   a. the magnetic field suddenly decreases after the switch is opened
   b. the coil is moved closer to the electromagnet

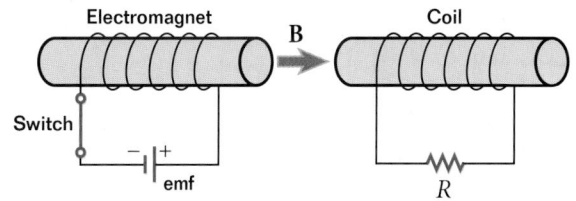

Figure 22-24

Practice problems

10. A flexible loop of conducting wire has a radius of 0.12 m and is perpendicular to a uniform magnetic field with a strength of 0.15 T, as in **Figure 22-25(a)**. The loop is grasped at opposite ends and stretched until it closes to an area of 3×10^{-3} m², as in **Figure 22-25(b)**. If it takes 0.20 s to close the loop, find the magnitude of the average emf induced in the loop during this time. (See Sample Problem 22A.)

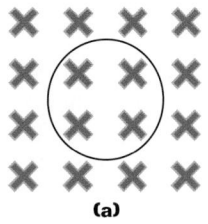   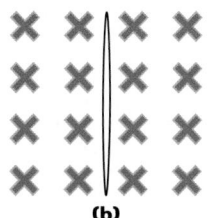

   (a)   (b)

Figure 22-25

11. A rectangular coil 0.055 m by 0.085 m is positioned so that its cross-sectional area is perpendicular to the direction of a magnetic field, B. If the coil has 75 turns and a total resistance of 8.7 Ω and the field decreases at a rate of 3.0 T/s, what is the magnitude of the induced current in the coil? (See Sample Problem 22A.)

ANSWERS TO

Chapter 22
Review and Assess

1. Place plane of coil perpendicular to magnetic field and move it into or out of field, rotate coil about axis perpendicular to magnetic field, and change strength of magnetic field with variable resistor.

2. The direction of the induced current is such that its magnetic field will oppose the change in the external magnetic field.

3. magnetic field component perpendicular to plane of coil, area of coil, time in which changes occur, number of turns of wire in coil

4. Wrap the wire into a coil that has many turns (large N) and that can be moved in and out of the B field quickly (small Δt).

5. **a.** needle deflects to the left
   b. no deflection of needle
   c. needle deflects to the left

6. By opposing changes in the external field, the induced B field prevents the system's energy from increasing or decreasing.

7. yes; The magnet's poles must be perpendicular to the tube's cross-sectional area.

8. no; Effects of one magnet cancel those of other magnet.

9. **a.** from left to right
   b. from right to left

10. 3.2×10^{-2} V

11. 0.12 A

12. −0.63 V

13. *B* field (induces emf in turning coil), wire coil (conducts induced current), slip rings (maintain contact with rest of circuit by means of conducting brushes)

14. turn the handle faster

15. *Frequency* indicates how often each second the current goes from a maximum value in one direction to a maximum value in the other direction and back.

16. Replace the slip rings with a commutator, which prevents the reversal of the current every half-cycle.

17. Battery current is constant, while dc generator current fluctuates.

18. an emf with polarity opposite that of the emf powering the motor; The coil's rotation in the *B* field induces a back emf that reduces the net potential difference across the motor.

19. The magnetic forces are greatest on charges in the sides of a loop that move perpendicular to the *B* field (that is, when the plane of the loop is parallel to the field lines).

20. The *B* field of an induced current opposes the change (due to coil rotation) in the external *B* field. Faster rotation of the coil increases this induced current and thus the opposing field.

21. Maximum values are maintained only for an instant, whereas rms values remain steady and thus are easier to measure.

22. a, b; The *B* field lines in these cases are in a plane perpendicular to the plane of the loop, so the loop crosses the field lines.

23. 3.88×10^{-2} V

24. **a.** 2.4×10^{2} V
   b. 6.9 A

12. A 52-turn coil with an area of 5.5×10^{-3} m^2 is dropped from a position where $B = 0.00$ T to a new position where $B = 0.55$ T. If the displacement occurs in 0.25 s and the area of the coil is perpendicular to the magnetic field lines, what is the resulting average emf induced in the coil? (See Sample Problem 22A.)

ALTERNATING CURRENT, GENERATORS, AND MOTORS

Review questions

13. List the essential components of an electric generator, and explain the role of each component in generating an alternating emf.

14. A student turns the handle of a small generator attached to a lamp socket containing a 15 W bulb. The bulb barely glows. What should the student do to make the bulb glow more brightly?

15. What is meant by the term *frequency* in reference to an alternating current?

16. How can an ac generator be converted to a dc generator? Explain your answer.

17. In what way is the output of a dc generator different from the output of a battery?

18. What is meant by back emf? How is it induced in an electric motor?

Conceptual questions

19. When the plane of a rotating loop of wire is parallel to the magnetic field lines, the number of lines passing through the loop is zero. Why is the current at a maximum at this point in the loop's rotation?

20. The faster the coil of loops, or *armature*, of an ac generator rotates, the harder it is to turn the armature. Use Lenz's law to explain why this happens.

21. Voltmeters and ammeters that measure ac quantities measure the rms values of emf and current, respectively. Why would this be preferred to measuring the maximum emf or current? (Hint: Think about what a meter reading would look like if ac quantities other than rms values were measured.)

22. A bar magnet is attached perpendicular to a rotating shaft. It is then placed in the center of a coil of wire. In which of the arrangements shown in **Figure 22-26** could this device be used as an electric generator? Explain your choice.

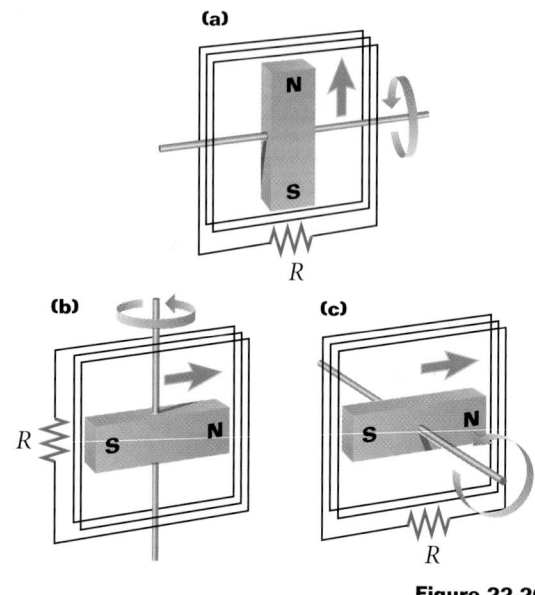

Figure 22-26

Practice problems

23. A generator can be made using the component of Earth's magnetic field that is parallel to Earth's surface. A 112-turn square wire coil with an area of 4.41×10^{-2} m^2 is mounted on a shaft so that the cross-sectional area of the coil is perpendicular to the ground. The shaft then rotates with a frequency of 25.0 Hz. The horizontal component of Earth's magnetic field at the location of the loop is 5.00×10^{-5} T. Calculate the maximum emf induced in the coil by Earth's magnetic field. (See Sample Problem 22B.)

24. An ac generator consists of 45 turns of wire with an area of 0.12 m^2. The loop rotates in a magnetic field of 0.118 T at a constant frequency of 60.0 Hz. The generator is connected across a circuit load with a total resistance of 35 Ω. Find the following:
   a. the maximum induced emf
   b. the maximum induced current
(See Sample Problem 22B.)

25. The rms potential difference across high-voltage transmission lines in Great Britain is 220 000 V. What is the maximum potential difference? (See Sample Problem 22C.)

26. The maximum potential difference across certain heavy-duty appliances is 340 V. If the total resistance of an appliance is 120 Ω, calculate the following:
    a. the rms potential difference
    b. the rms current
(See Sample Problem 22C.)

27. The maximum current that can pass through a light bulb filament is 0.909 A when its resistance is 182 Ω.
    a. What is the rms current conducted by the filament of the bulb?
    b. What is the rms potential difference across the bulb's filament?
    c. How much power does the light bulb use?
(See Sample Problem 22C.)

28. A 996 W hair dryer is designed to carry a maximum current of 11.8 A.
    a. How large is the rms current in the hair dryer?
    b. What is the rms potential difference across the hair dryer?
(See Sample Problem 22C.)

INDUCTANCE

Review questions

29. Describe how mutual induction occurs.

30. What is the difference between a step-up transformer and a step-down transformer?

31. Does a step-up transformer increase power? Explain your answer.

32. Describe how self-induction occurs.

Conceptual questions

33. In many transformers, the wire around one winding is thicker, and therefore has lower resistance, than the wire around the other winding. If the thicker wire is wrapped around the secondary winding, is the device a step-up or a step-down transformer? Explain.

34. Would a transformer work with pulsating direct current? Explain your answer.

Practice problems

35. A transformer is used to convert 120 V to 9.0 V for use in a portable CD player. If the primary, which is connected to the outlet, has 640 turns, how many turns does the secondary have? (See Sample Problem 22D.)

36. A transformer is used to convert 120 V to 6.3 V in order to power a toy electric train. If there are 210 turns in the primary, how many turns should there be in the secondary? (See Sample Problem 22D.)

MIXED REVIEW PROBLEMS

37. A student attempts to make a simple generator by passing a single loop of wire between the poles of a horseshoe magnet with a 2.5×10^{-2} T field. The area of the loop is 7.54×10^{-3} m^2 and is moved perpendicular to the magnetic field lines. In what time interval will the student have to move the loop out of the magnetic field in order to induce an emf of 1.5 V? Is this a practical generator?

38. The same student in item 37 modifies the simple generator by wrapping a much longer piece of wire around a cylinder with about one-fourth the area of the original loop (1.886×10^{-3} m^2). Again using a uniform magnetic field with a strength of 2.5×10^{-2} T, the student finds that by removing the coil perpendicular to the magnetic field lines during 0.25 s, an emf of 149 mV can be induced. How many turns of wire are wrapped around the coil?

39. A coil of 325 turns and an area of 19.5×10^{-4} m^2 is removed from a uniform magnetic field at an angle of 45° in 1.25 s. If the induced emf is 15 mV, what is the magnetic field's strength?

40. A transformer has 22 turns of wire in its primary and 88 turns in its secondary.
    a. Is this a step-up or step-down transformer?
    b. If 110 V ac is applied to the primary, what is the output potential difference?

41. The potential difference in the lines that carry electric power to homes is typically 20.0 kV. What is the ratio of the turns in the primary to the turns in the secondary of the transformer if the output potential difference is 117 V?

25. 3.1×10^5 V
26. a. 2.4×10^2 V
    b. 2.0 A
27. a. 0.643 A
    b. 117 V
    c. 75.2 W
28. a. 8.34 A
    b. 119 V
29. The changing B field produced by a changing current in one circuit induces an emf and current in a nearby circuit.
30. A step-up transformer uses the B field of an alternating current to induce an increased emf in the secondary. A step-down transformer uses the same principle to induce a smaller emf in the secondary.
31. no; The change in potential difference in a transformer is accompanied by an inverse change in the current. In an ideal transformer, power is unchanged, as expected from energy conservation.
32. Changing current in a circuit produces a changing magnetic field, which induces an opposing emf and current in the same circuit.
33. a step-down transformer; I is larger in the secondary, so wire with a lower R is needed to reduce energy dissipaton.
34. yes; Current changes continually, so its changing B field can induce an emf in a transformer's secondary.
35. 48 turns
36. 11 turns
37. 1.3×10^{-4} s; no
38. 790 turns
39. 4.2×10^{-2} T
40. a. a step-up transformer
    b. 440 V
41. 171:1

ANSWERS TO

Technology & Learning

Answers may vary slightly,
depending on viewing window
settings.

a. Y_2

b. $i = 0.238$ A; $I_{rms} = 0.177$ A

c. $i = 0.147$ A; $I_{rms} = 0.177$ A

d. $i = 0.00$ A; $I_{rms} = 0.0778$ A

e. $i = 0.0647$ A; $I_{rms} = 0.0778$ A

f. no

---

42. A bolt of lightning, such as the one shown on the left in **Figure 22-27,** behaves like a vertical wire conducting electric current. As a result, it produces a magnetic field whose strength varies with the distance from the light-ning. A 105-turn circular coil is oriented perpendicular to the magnetic field, as shown on the right in **Figure 22-27.** The coil has a radius of 0.833 m. If the magnetic field at the coil drops from 4.72×10^{-3} T to 0.00 T in 10.5 μs, what is the average emf induced in the coil?

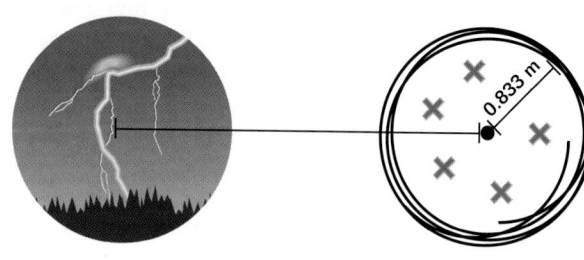

Figure 22-27

Technology & Learning

Graphing calculators

Refer to Appendix B for instructions on download-ing programs for your calculator. The program "Chap22" allows you to analyze graphs of instanta-neous and root-mean-square currents versus time in various ac circuits.

Instantaneous current and rms current, as you learned earlier in this chapter, are related to the maximum current in an ac circuit by the following equations:

$$i = I_{max} (\sin \omega t) \quad \text{and} \quad I_{rms} = \frac{I_{max}}{\sqrt{2}}$$

The program "Chap22" stored on your graphing calculator makes use of the equations for instanta-neous and rms currents. Once the "Chap22" pro-gram is executed, your calculator will ask for the frequency and maximum current. The graphing cal-culator will use the following equations to create graphs of the instantaneous current (Y_1) versus the time (X) and the rms current (Y_2) versus the time (X). The relationships in these equations are the same as those in the equations shown above.

$$Y_1 = I \sin (2\pi FX) \quad \text{and} \quad Y_2 = I/\sqrt{2}$$

a. What value in the calculator equations equals the amount of direct current that would dissipate the same energy in a resistor as is dissipated by the instantaneous ac current during a full cycle?

First make sure your calculator is in radian mode by pressing [MODE] [▼] [▼] [◄].

Execute "Chap22" on the [PRGM] menu and press [ENTER] to begin the program. Enter the values for the frequency and maximum current (shown below), pressing [ENTER] after each value.

The calculator will provide graphs of the instan-taneous current and root-mean-square current ver-sus time in various ac circuits. (If the graphs are not visible, press [WINDOW] and change the settings for the graph window, then press [GRAPH].)

Press [TRACE] and use the arrow keys to trace along the curve. The x value corresponds to the time in seconds, and the y value of the sine curve corre-sponds to the instantaneous current in amperes. The y value of the straight line gives the root-mean-square value for the current in amperes. Use the [▲] and [▼] keys to toggle between the two graphs.

Determine the instantaneous and root-mean-square values for the current in the following ac circuits:

b. an ac circuit with a maximum current of 0.250 A and a frequency of 60.0 Hz at $t = 3.27$ s

c. the same circuit at $t = 5.14$ s

d. an ac circuit with a maximum current of 0.110 A and a frequency of 110.0 Hz at $t = 2.50$ s

e. the same circuit at $t = 4.24$ s

f. Would you expect the graph of Y_2 to be above the maximum instantaneous current (Y_1)?

Press [2nd] [QUIT] to stop graphing. Press [ENTER] to input new values or [CLEAR] to end the program.

43. A generator supplies 5.0×10^3 kW of power. The output potential difference is 4500 V before it is stepped up to 510 kV. The electricity travels 410 mi (6.44×10^5 m) through a transmission line that has a resistance per unit length of 4.5×10^{-4} Ω/m.

    a. How much power is lost through transmission of the electrical energy along the line?

    b. How much power would be lost through transmission if the generator's output potential difference were not stepped up? What does this answer tell you about the role of large potential differences in power transmission?

44. The alternating potential difference of a generator is represented by the equation emf = (245 V) sin 560t, where emf is in volts and t is in seconds. Use these values to find the frequency of the potential difference and the maximum potential difference output of the source.

45. A pair of adjacent coils has a mutual inductance of 1.06 H. Determine the average emf induced in the secondary circuit when the current in the primary circuit changes from 0 A to 9.50 A in a time interval of 0.0336 s.

Alternative Assessment

Performance assessment

1. Identify the chain of electromagnetic energy transformations involved in making the blades of a ceiling fan spin. Include the fan's motor, the transformers bringing electricity to the house, and the turbines generating the electricity.

2. Two identical magnets are dropped simultaneously from the same point. One of them passes through a coil of wire in a closed circuit. Predict whether the two magnets will hit the ground at the same time. Explain your reasoning, then plan an experiment to test which of the following variables measurably affect how long each magnet takes to fall: magnetic strength, coil cross-sectional area, and the number of loops the coil has. What measurements will you make? What are the limits of precision in your measurements? If your teacher approves your plan, obtain the necessary materials and perform the experiments. Report your results to the class, describing how you made your measurements, what you concluded, and what additional questions need to be investigated.

3. What do adapters do to potential difference, current, frequency, and power? Examine the input/output information on several adapters to find out. Do they contain step-up or step-down transformers? How does the output current compare to the input? What happens to the frequency? What percentage of the energy do they transfer? What are they used for?

Portfolio projects

4. Research the debate between the proponents of alternating current and those who favored direct current in the 1880–1890s. How were Thomas Edison and George Westinghouse involved in the controversy? What advantages and disadvantages did each side claim? What uses of electricity were anticipated? What kind of current was finally generated in the Niagara Falls hydroelectric plant? Had you been in a position to fund these projects at that time, which projects would you have funded? Prepare your arguments to re-enact a meeting of businesspeople in Buffalo in 1887.

5. Research the history of telecommunication. Who invented the telegraph? Who patented it in England? Who patented it in the United States? Research the contributions of Charles Wheatstone, Joseph Henry, and Samuel Morse. How did each of these men deal with issues of fame, wealth, and credit to other people's ideas? Write a summary of your findings, and prepare a class discussion about the effect patents and copyrights have had on modern technology.

44. $f = 89$ Hz, 245 V

45. 300 V

Alternative Assessment
ANSWERS

Performance assessment

1. Students' answers will vary. Transformations include turning a generator coil in a magnetic field to induce an emf, step-up transformers, step-down transformers, and torque exerted by a B field on the coil in the fan's motor.

2. Students' plans will vary. Be sure proposed tests are safe, measure time accurately, and adjust one variable at a time.

3. Most adapters use step-down transformers. Students should note that power output is always less than input.

 Portfolio projects

4. Students' answers will vary. Edison claimed that ac was dangerous and unreliable. Westinghouse noted that dc could not be stepped up and that long-distance transmission was inefficient. Batteries can store dc for peak use, but new industries that used electricity at all hours favored ac.

5. Wheatstone (1802–1875) took out the first telegraph patent in England in 1837. In the United States, Henry (1797–1878) invented the telegraph 10 years before Morse (1791–1872), who claimed in court that he invented it by himself.

Planning

Recommended time:

1 lab period

Classroom organization:

▶ Each lab group should have two or more students.

▶ **Safety warnings:** Emphasize the dangers of working with electricity. For the safety of the students and of the equipment, remind students to have you check their circuits before turning on the power supply or closing the switch.

Techniques to Demonstrate

Show students how to set up the circuit for the second part of the lab. Make sure they know how to use the rheostat to adjust the current in the circuit.

✔ Checkpoints

Step 3: Because this lab requires no measurements, students may be unsure what to record in their notebooks. Guide students to come up with a format for recording their observations.

CHAPTER 22
Laboratory Exercise

⭐ TEKS

pp. 826–827: 1A, 2B, 2C, 2F, 6D

OBJECTIVES

- Use a galvanometer to detect an induced current.
- Determine the relationship between the magnetic field of a magnet and the current induced in a conductor.
- Determine what factors affect the direction and magnitude of an induced current.

MATERIALS LIST

✔ galvanometer

✔ insulated connecting wires

✔ momentary contact switch

✔ pair of bar magnets or a large horseshoe magnet

✔ power supply

✔ rheostat (10 Ω) or potentiometer

✔ student primary and secondary coil set with iron core

ELECTROMAGNETIC INDUCTION

In this laboratory, you will use a magnet, a conductor, and a galvanometer to explore electromagnets and the principle of self-induction.

SAFETY

- **Never close a circuit until it has been approved by your teacher. Never rewire or adjust any element of a closed circuit. Never work with electricity near water; be sure the floor and all work surfaces are dry.**

- **If the pointer on any kind of meter moves off scale, open the circuit immediately by opening the switch.**

- **Do not attempt this exercise with any batteries, electrical devices, or magnets other than those provided by your teacher for this purpose.**

PREPARATION

1. Read the entire lab, and plan what measurements you will take.

2. In your lab notebook, prepare an observation table with three wide columns. Label the columns *Sketch of setup, Experiment,* and *Observation.* For each part of the lab, you will sketch the apparatus and label the poles of the magnet, write a brief description, and record your observations.

PROCEDURE

Induction with a permanent magnet

3. Connect the ends of the smaller coil to the galvanometer. Hold the magnet still and move the coil quickly over the north pole of the magnet, as shown in **Figure 22-28.** Remove the coil quickly. Observe the galvanometer.

4. Repeat, moving the coil more slowly. Observe the galvanometer.

5. Turn the magnet over, and repeat steps 3 and 4, moving the coil over the south pole of the magnet. Observe the galvanometer.

6. Hold the coil stationary and quickly move the north pole of the magnet in and out of the coil. Repeat slowly. Turn the magnet, and repeat both quickly and slowly with the south pole. Observe the galvanometer.

Induction with an electromagnet

7. Connect the larger coil to the galvanometer. Connect the small coil in series with a switch, battery, and rheostat. Slip the smaller coil inside the larger coil, so that the arrangement resembles that shown in **Figure 22-29.** Close the switch. Adjust the rheostat so that the galvanometer reading registers on the scale. Observe the galvanometer.

8. Open the switch to interrupt the current in the small coil. Observe the galvanometer.

9. Close the switch again, and open it after a few seconds. Observe the galvanometer.

10. Adjust the rheostat to increase the current in the small coil. Close the switch, and observe the galvanometer.

11. Decrease the current in the circuit, and observe the galvanometer. Open the switch.

12. Reverse the direction of the current by reversing the battery connections. Close the switch, and observe the galvanometer.

13. Place an iron rod inside the small coil. Open and close the switch while observing the galvanometer. Record all observations in your notebook.

14. Clean up your work area. Put equipment away safely so that it is ready to be used again.

ANALYSIS AND INTERPRETATION

Calculations and data analysis

1. Organizing information Based on your observations from the first part of the lab, did the speed of the motion have any effect on the galvanometer?

2. Organizing results In the first part of the lab, did it make any difference whether the coil or the magnet moved? Explain why or why not.

Conclusions

3. Applying ideas Explain what the galvanometer readings revealed to you about the magnet and the wire coil.

4. Evaluating results Based on your observations, what conditions are required to induce a current in a wire?

5. Evaluating results Based on your observations, what factors influence the direction and magnitude of the induced current?

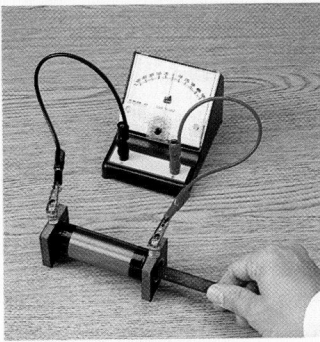

Figure 22-28
Step 3: Connect the coil to the galvanometer. Holding the magnet still, move the coil over the magnet quickly.
Step 4: Holding the magnet still, move the coil over the magnet slowly.
Step 6: Repeat the procedure, but hold the coil still while moving the magnet.

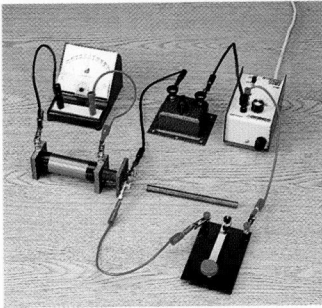

Figure 22-29
Step 7: Connect the larger coil to the galvanometer, and connect the smaller coil in series with the battery, switch, and rheostat. Place the smaller coil inside the larger coil.

Step 7: Make sure the power supply or battery, rheostat, galvanometer, coil, and switch are connected properly and set at the proper settings. Students should be able to demonstrate that all connections are properly made.

Step 12: Students should be able to demonstrate that they reversed the current in the circuit.

ANSWERS TO

Analysis and Interpretation

CALCULATIONS AND DATA ANALYSIS

1. Yes, the faster the motion, the greater the deflection of the galvanometer needle.

2. It did not matter which one moved. The important thing was the motion between them, or relative motion.

CONCLUSIONS

3. The galvanometer showed that there was a current in the wire coil. The direction of the deflection indicated the direction of the current.

4. A changing magnetic field is needed to induce a current in a circuit.

5. The direction of the current is influenced by the direction the magnet is moved and by whether the field is increasing or decreasing. The magnitude is influenced by the speed of the change.

Compression Guide: To shorten from 7 to 6 45-min periods (from 3½ to 3 90-min blocks), eliminate items in magenta type.

PACING CHART   CLASSROOM RESOURCES

PACING CHART	★TEKS	Teacher Demonstrations	Holt Physics Transparencies	Labs (See page T52 for equipment listing for in-text labs.)	
23-1 Quantization of energy 3 45-minute periods 1½ 90-minute block	2C, 3A, 3C, 3E, 5B, 8A, 8B, 9A	**TE** *A blackbody*, p. 830 **TE** *Color of thermal sources*, p. 831 **TE** *Photoelectric effect*, p. 834	**T** 116–118 **TM** 74–76	**PE** *The Photoelectric Effect*, p. 860	
23-2 Models of the atom 1 45-minute period ½ 90-minute block	1A, 3A, 3C, 3E, 8B, 9B	**TE** *Particle scattering*, p. 840 **TE** *Spectral lines*, p. 842	**T** 119–121 **TM** 77–78	**PE** *Quick Lab: Atomic Spectra*, p. 843	
23-3 Quantum mechanics 1 45-minute period ½ 90-minute block	2C, 2D, 3E, 5B, 8A, 8B, 8C, 9A, 9B	**TE** *Probability distribution*, p. 853			
Review and Assessment 2 45-minute periods 1 90-minute block					

Resource Key

PHYSICS

PE Pupil's Edition
TE Teacher's Edition

L Laboratory Experiments
TL Technology Lab Experiments
T Transparencies

One-Stop Planner CD-ROM contents

TM Transparency Masters
SR Section Review Worksheets
AA Alternative Assessment

PW Problem-Solving Workbook
PB Problem Bank
CTW Critical Thinking Worksheet

LABORATORY PLANNING: The Photoelectric Effect, p. 860

Materials (for each lab group)

- multimeter:
  - basic digital multimeter, or
  - student multimeter, or
  - general purpose multimeter, or
  - CENCO standard movement, triple range, dc voltmeter (0–3 V, 10 V, 15 V)
- black construction paper
- power supply:
  - CENCO universal power supply, or

- CENCO 6V ac/dc low-voltage power supply, or
- 20 V dc/500 mA regulated power supply, or
- 4-output dc regulated power supply
- lamp base with connecting leads
- lens or mirror support, for 4 cm lenses
- lens support, 7.5 cm
- meterstick, plain wood, 1 m long
- pair of metal meterstick supports
- patch cord, black, 60 cm, with banana clips

ASSIGNMENT RESOURCES

Content Mastery	Critical Thinking	Problem-Solving Practice
PE 1–5, p. 839 **SR** 23-1, *Concept Review* **PE** 1–8, p. 856	**PE** 1–2, p. 837 **PE** 9–13, p. 856	**23A** Quantum energy: 23 items in **PE, PW,** and **PB,** see **TE** p. 833 **23B** The photoelectric effect: 27 items in **PE, PW,** and **PB,** see **TE** p. 836
PE 1–7, p. 847 **SR** 23-2, *Concept Review* **PE** 19–25, p. 857	**PE** 1–3, p. 845 **PE** 26–28, p. 857	
PE 1–6, p. 854 **SR** 23-3, *Concept Review* **PE** 29–32, p. 857	**PE** 33–38, p. 857	**23C** De Broglie waves: 26 items in **PE, PW,** and **PB,** see **TE** p. 850

ASSESSMENT RESOURCES

Cumulative Review	Alternative Assessment	Traditional Assessment
SR Mixed Review, Ch. 23	**PE** 1–4, p. 859 **AA** Items for Ch. 23	Chapter 23 Test Test Generator items for Ch. 23

Scoring Rubrics for Alternative Assessment items can be found on the One-Stop Planner CD-ROM.

TECHNOLOGY RESOURCES

 CTW Segment 22 Atom Laser

 PE Technology and Learning, p. 858
(Alternative procedures for calculators without Flash-ROM technology are provided on the One-Stop Planner CD-ROM.)

 The Mechanical Universe/High School Adaptation Quad VII, Models of the Atom Quad VII, Wave-Particle Duality

internet connect

 On-line Student Resources:
GO TO: www.scilinks.org
The following *sci*LINKS Internet resources can be found in the student text for this chapter.

TOPICS:
- Max Planck, p. 831 (HF2231)
- Photoelectric effect, p. 834 (HF2232)
- Arthur Compton, p. 838 (HF2233)
- Early atomic theory, p. 841 (HF2234)
- Modern atomic theory, p. 844 (HF2235)

 On-line Teacher Resources:
GO TO: go.hrw.com
KEYWORD: HF2 HOME
Visit the HRW Web site for a variety of resources related to this chapter.

 Smithsonian Institution*
Internet Connections
Visit **www.si.edu/hrw** for additional on-line resources.

CNNfyi.com
Visit **www.cnnfyi.com** for late-breaking news and current events stories selected just for you.

- patch cord, red, 60 cm, with banana clips
- photoelectric effect device with amplifier and filters (red, green, blue)
- roll of self-adhesive tape, 0.5 in wide.
- set of wood blocks, 2 in. × 4 in. lumber, 4 different lengths, 10–20 cm
- xenon light bulb (4.8 V, 850 mA)

Materials Preparation

See page 2A–2B for instruction on preparation of wood blocks. For this lab, only one block may be needed, to serve as a support for the apparatus.

See page 692A–692B for instructions on using multimeters to measure current and voltage.

Use the black construction paper to make a tube between the lamp and the photoelectric effect device. This will yield better results and will also protect students' eyes during the lab.

Use the xenon light source for the blue and green filters, and a standard tungsten light for the red filter. The tungsten light will need to be placed adjacent to the red filter to give good results.

Section 23-1 introduces the quantization of energy in black-body radiation and the photo-electric effect; solves problems involving energy quanta, threshold frequency, and work function; and discusses the Compton shift as it pertains to the particle theory of light.

Section 23-2 explores Rutherford's model of the atom, introduces emission and absorption spectra, explains atomic spectra in terms of Bohr's model of hydrogen, and evaluates the strengths and weaknesses of Bohr's model.

Section 23-3 discusses the wave-particle duality of light and matter, shows how to calculate de Broglie wavelengths, introduces the uncertainty principle, and describes the quantum-mechanical picture of the atom.

About the Illustration

This photograph of the aurora borealis was taken in Denali National Park, Alaska. In the background are the Alaska Range Mountains. Auroras occur when electrically charged particles, mainly from the sun, become trapped in Earth's atmosphere over the magnetic poles and collide with other atoms, resulting in the emission of visible light.

Atomic Physics

PHYSICS IN ACTION

Colorful lights similar to these in Denali National Park, Alaska, are commonly seen in the sky in northern latitudes. This dazzling array of colors is known as the *aurora borealis,* or the *northern lights.* The northern lights, which can extend thousands of kilometers across the sky, appear as arcs, bands, or streaks of color, sometimes flickering or pulsating.

In this chapter, you will learn how classical physics has been superseded by a new theory known as *quantum mechanics.* As you will see, this scientific revolution has provided a new picture of the atom that accounts for the northern lights.

- *Why are only certain colors part of the northern lights?*

- *What do the colors of the northern lights reveal about our atmosphere?*

CONCEPT REVIEW

Standing waves (Section 12-4)

The wave model of light (Section 14-1)

Electromagnetic waves (Section 14-1)

Tapping Prior Knowledge

Knowledge to Expect

✔ "Students of all ages lack an appreciation of the very small size of particles and attribute macroscopic properties to particles." (AAAS's *Benchmarks for Science Literacy,* The Research Base)

✔ "Students learn that all matter is made up of atoms which are far too small to see directly through a microscope. Also they learn that hypotheses are valuable, even if they turn out not to be true, if they lead to fruitful investigations." (AAAS's *Benchmarks for Science Literacy,* grades 6–8)

Knowledge to Review

✔ Standing waves occur when two waves of the same frequency, amplitude, and wavelength interfere as they travel through a medium in opposite directions. (Section 12-4)

✔ Light in the wave model consists of electromagnetic waves. (Section 14-1)

✔ Electromagnetic waves are transverse waves that consist of oscillating electric and magnetic fields at right angles to each other. (Section 14-1)

Items to Probe

✔ Familiarity with atomic structure: Ask students to draw a schematic diagram of the atom.

✔ Preconceptions about the nature of light: Ask students to list arguments for and against the wave and particle theories of light.

23-1
Quantization of energy

A blackbody

Purpose Show an example of a blackbody.

Materials two closed cardboard boxes thick enough to prevent light from passing through, one with a white interior and one with a black interior; index cards with holes of different sizes

Procedure Cut a hole slightly smaller than the size of an index card in the front of each box. Have students hold the index cards up to the hole in each box one at a time, and have them look through the different index cards into each box. If the hole in the index card is not too large, they will find that even the box with the white interior appears dark. Ask them if light enters the box (*yes*) and if the light comes back out of the box (*most of it does not*). Point out that each time light hits a wall inside the box, some of the light is absorbed. If the box is reasonably insulated, the energy absorbed by the walls will cause their temperature to rise, and energy will be radiated inside the box. Tell students that both boxes are examples of blackbodies.

23-1 SECTION OBJECTIVES

- **Explain how Planck resolved the ultraviolet catastrophe in blackbody radiation.**

- **Calculate energy of quanta using Planck's equation.**

- **Solve problems involving maximum kinetic energy, work function, and threshold frequency in the photoelectric effect.**

Figure 23-1
These charcoal briquets have a bright red glow because of their high temperature.

blackbody radiation

electromagnetic radiation emitted by a blackbody, which absorbs all incoming radiation and then emits radiation based only on its temperature

BLACKBODY RADIATION

By the end of the nineteenth century, scientists thought that classical physics was nearly complete. One of the few remaining questions to be solved involved electromagnetic radiation and thermodynamics. Specifically, scientists were concerned with the glow of objects when they reach a high temperature.

All objects emit electromagnetic radiation. This radiation, which depends on the temperature and other properties of an object, typically consists of a continuous distribution of wavelengths from the infrared, visible, and ultraviolet portions of the spectrum. The distribution of the different wavelengths varies with temperature.

At low temperatures, radiation wavelengths are mainly in the infrared region. So, they cannot be seen by the human eye. As the temperature of an object increases, the wavelength given off most often by the object is in the visible region of the electromagnetic spectrum. For example, the charcoal shown in **Figure 23-1** seems to have a red glow. At even higher temperatures, the object appears to have a white glow, as in the hot tungsten filament of a light bulb.

Classical physics cannot account for blackbody radiation

One problem at the end of the 1800s was understanding the distribution of wavelengths given off by a blackbody. Most objects absorb some incoming radiation and reflect the rest. An ideal system that absorbs all incoming radiation is called a blackbody. Physicists study **blackbody radiation** by observing a hollow object with a small opening, as shown in **Figure 23-2.** The system is a good example of how a blackbody works; it traps radiation. The light given off by the opening is in equlibrium with light from the walls of the object, because the light has been given off and reabsorbed many times.

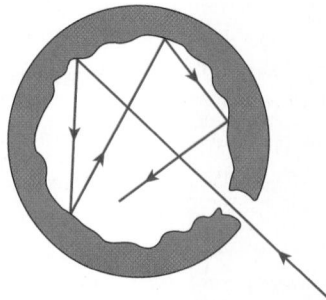

Figure 23-2
As light enters this hollow object through the small opening, part of it is reflected and part of it is absorbed on each reflection from the interior walls. After many reflections, essentially all of the incoming energy is absorbed.

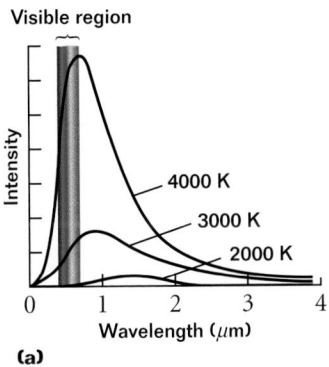

Visible region

4000 K
3000 K
2000 K

(a)

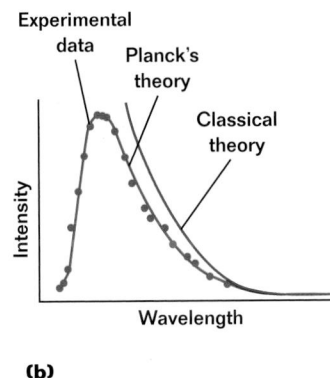

Experimental data
Planck's theory
Classical theory

Wavelength

(b)

Figure 23-3
(a) This graph shows the intensity of blackbody radiation at three different temperatures. **(b)** Classical theory's prediction for blackbody radiation (the blue curve) did not correspond to the experimental data (the red data points) at all wavelengths, while Planck's theory (the red curve) did.

Experimental data for the radiation given off by an object at three different temperatures are shown in **Figure 23-3(a).** Note that as the temperature increases, the total energy given off by the body (the area under the curve) also increases. In addition, as the temperature increases, the peak of the distribution shifts to shorter wavelengths.

Scientists could not account for these experimental results with classical physics. **Figure 23-3(b)** compares an experimental plot of the blackbody radiation spectrum (the red data points) with the theoretical picture of what this curve should look like based on classical theories (the blue curve). Classical theory predicts that as the wavelength approaches zero, the amount of energy being radiated should become infinite. This is contrary to the experimental data, which show that as the wavelength approaches zero, the amount of energy being radiated also approaches zero. This contradiction is often called the **ultraviolet catastrophe** because the disagreement occurs at the ultraviolet end of the spectrum. ⊛**TEKS** **8B**

Experimental data for blackbody radiation support the quantization of energy ⊛**TEKS** **3E**

In 1900, Max Planck (1858–1947) developed a formula for blackbody radiation that was in complete agreement with experimental data at all wavelengths. Planck's original theoretical approach is rather abstract in that it involves arguments based on entropy and thermodynamics. The arguments presented in this book are easier to visualize, and they convey the spirit and revolutionary impact of Planck's original work.

Planck proposed that blackbody radiation was produced by submicroscopic electric oscillators, which he called *resonators*. He assumed that the walls of a glowing cavity were composed of billions of these resonators, all vibrating at different frequencies. While most scientists naturally assumed that the energy of these resonators was continuous, Planck made the radical assumption that these resonators could only absorb and then give off certain discrete amounts of energy.

When he first discovered this idea, Planck was using a mathematical technique in which quantities that are known to be continuous are temporarily

ultraviolet catastrophe

contradiction between the predictions of classical physics and the experimental data for the electromagnetic radiation of a blackbody in the ultraviolet end of the spectrum

internet connect

SCI*LINKS*
NSTA
TOPIC: Max Planck
GO TO: www.scilinks.org
sci*LINKS* CODE: HF2231

Demonstration 2

Color of thermal sources

Purpose Show that different thermal sources emit different spectra.

Materials matches, candle, Bunsen burner, nail, incandescent light bulb, tongs

CAUTION *Wear goggles and an apron when using the burner.*

Procedure Ask students to rank the objects by temperature based on past experience. Tell students to carefully observe and record the colors of the different flames (match, candle, Bunsen burner) and of the hot objects (nail, light bulb) that you are about to demonstrate. Then turn on the light bulb, strike a match, light the candle, light the Bunsen burner, and hold the nail in the flame of the Bunsen burner with tongs until the nail glows. Point out that brightness does not necessarily mean a higher temperature. Have students rank the dominant color of each object by wavelengths and compare their results with their original ranking of temperatures.

⊛**TEKS**
p. 831: 8B, 3E

The Language of Physics

In the equation $E = nhf$, n represents a positive integer that can take the values 1, 2, 3, and so on. This means that the energy E of an oscillator of frequency f can be *1hf, 2hf, 3hf*, and so on, but never *0.5hf* or *2.1hf*.

Teaching Tip

Students may need to be reminded that the speed of a light wave equals frequency times wavelength ($v = f\lambda$). Ask students if the number of photons in 1 J of red light (650 nm) is greater than, equal to, or less than the number of photons in 1 J of blue light (450 nm). *(A longer wavelength corresponds to a lower frequency. Because energy is proportional to frequency, there is less energy in each quantum of red light than in each quantum of blue light. Thus, more photons of red light are contained in 1 J.)*

Teaching Tip

Point out that $1\ C \times 1\ V = 1\ J$. The charge of 1 electron is 1.60×10^{-19} C. 1 eV is the charge of 1 electron $\times$ 1 V. Therefore, $1\ eV = 1.60 \times 10^{-19}$ J.

Did you know?

Max Planck became president of the Kaiser Wilhelm Institute of Berlin in 1930. Although Planck remained in Germany during the Hitler regime, he openly protested the Nazis' treatment of his Jewish colleagues and consequently was forced to resign his presidency in 1937. Following World War II, he was reinstated as president, and the institute was renamed the Max Planck Institute in his honor.

considered to be discrete. After the calculations are made, the discrete units are taken to be infinitesimally small. Planck found that the calculations worked if he omitted this last step and considered energy to come in discrete units throughout his calculations. With this method, Planck found that the total energy of a resonator with frequency f is an integral multiple of hf, as follows:

$$E = nhf$$

In this equation, n is a positive integer called a *quantum number*, and the factor h is Planck's constant, which equals $6.626\ 075 \times 10^{-34}$ J·s. To simplify calculations, we will use the approximate value of $h = 6.63 \times 10^{-34}$ J·s in this textbook. Because the energy of each resonator comes in discrete units, it is said to be *quantized*, and the allowed energy states are called *quantum states* or *energy levels*. With the assumption that energy is quantized, Planck was able to derive the red curve shown in **Figure 23-3(b)** on page 831.

According to Planck's theory, the resonators absorb or give off energy in discrete units of light energy called *quanta* (now called *photons*) by "jumping" from one quantum state to another. As seen by the equation above, if the quantum number (n) changes by one unit, the amount of energy radiated changes by hf. For this reason, the energy of a light quantum, which corresponds to the energy difference between two adjacent levels, is given by the following equation:

ENERGY OF A LIGHT QUANTUM

$$E = hf$$

energy = Planck's constant × frequency

A resonator will radiate or absorb energy only when it changes quantum states. If a resonator remains in one quantum state, no energy is absorbed or emitted. The idea that energy comes in discrete units marked the birth of a new theory known as *quantum mechanics*. ⭐TEKS **3E**

If Planck's constant is expressed in units of J·s, the equation $E = hf$ gives the energy in joules. However, when dealing with the parts of atoms, energy is often expressed in units of the electron volt, eV. An electron volt is defined as the energy that an electron or proton gains when it is accelerated through a potential difference of 1 V. Because 1 V = 1 J/C, the relation between the electron volt and the joule is as follows:

$$1\ eV = 1.60 \times 10^{-19}\ C \cdot V = 1.60 \times 10^{-19}\ C \cdot J/C = 1.60 \times 10^{-19}\ J$$

Planck's idea that energy is quantized was so radical that most scientists, including Planck himself, did not consider the quantization of energy to be realistic. Planck thought of his assumption as a mathematical approach to be used in calculations rather than a physical explanation. Therefore, he and other scientists continued to search for a different explanation of blackbody radiation that was consistent with classical physics. ⭐TEKS **3C**

Quantum energy ⭐TEKS 2C, 8B

PROBLEM

Each photon of yellow light, the predominant color in sunlight, carries an energy of 2.5 eV. What is the frequency of this light?

SOLUTION

Given: $E = 2.5$ eV $h = 6.63 \times 10^{-34}$ J•s

Unknown: $f = ?$

First, convert the energy from electron volts to Joules.

$$E = (2.5 \text{ eV})(1.60 \times 10^{-19} \text{ J/eV}) = 4.0 \times 10^{-19} \text{ J}$$

Then, use the equation for the energy of a light quantum, given on page 832.

$$E = hf \qquad \text{or} \qquad f = \frac{E}{h}$$

So, $$f = \frac{E}{h} = \frac{4.0 \times 10^{-19} \text{ J}}{6.63 \times 10^{-34} \text{ J•s}} = 6.0 \times 10^{14} \text{ Hz}$$

$$\boxed{f = 6.0 \times 10^{14} \text{ Hz}}$$

PRACTICE 23A

Quantum energy

1. Assume that the pendulum of a grandfather clock acts as one of Planck's resonators. If it carries away an energy of 8.1×10^{-15} eV in a one quantum change, what is the frequency of the pendulum? (Note that an energy this small would not be measurable. This is why we do not notice quantum effects in the large-scale world.)

2. A vibrating mass-spring system has a frequency of 0.56 Hz. How much energy of this vibration is carried away in a one-quantum change?

3. A photon in a laboratory experiment has an energy of 5.0 eV. What is the frequency of this photon?

4. Radiation emitted from human skin reaches its peak at $\lambda = 940$ μm.
   a. What is the frequency of this radiation?
   b. What type of electromagnetic waves are these?
   c. How much energy (in electron volts) is carried by one quantum of these electromagnetic waves?

PRACTICE GUIDE 23A

Solving for:

E	📖 **PE** 2, 4*; Ch. Rvw. 15 💿 **PW** 6*, 7, 8* **PB** Sample, 1–2
f	📖 **PE** Sample, 1, 3, 4*; Ch. Rvw. 14 💿 **PW** 4*, 5*, 7, 8* **PB** 3–5
λ	💿 **PW** Sample, 1–3 **PB** 6–8

ANSWERS TO

Practice 23A
Quantum energy

1. 2.0 Hz
2. 3.7×10^{-34} J
3. 1.2×10^{15} Hz
4. a. 3.19×10^{11} Hz
   b. infrared
   c. 1.32×10^{-3} eV

⭐TEKS

p. 832: 3E, 3C
p. 833: 2C, 8B

Q Can one directly observe electrons being emitted from a metal surface?

A *No; electrons cannot be directly observed, but the effect of electrons leaving the metal surface can be detected because the metal becomes charged when it loses electrons.*

Demonstration 3

Photoelectric effect

Purpose Show that light can release electrons from a metal.

Materials electroscope, rubber rod, wool, thin rectangular piece of zinc (3 cm × 8 cm), incandescent light, ultraviolet light

Procedure Attach the zinc plate to the knob of the electroscope, charge the rubber rod with the wool, and charge the zinc plate with the rubber rod. Turn on the incandescent light, and shine it on the zinc plate. Have students note that there is no change in the separation of the leaves. Turn off the incandescent light. Turn on the ultraviolet light, and shine it on the zinc plate. Have students observe that the leaves of the electroscope collapse (discharge). Explain that this occurs because electrons are ejected from the zinc, leaving the zinc with a net positive charge.

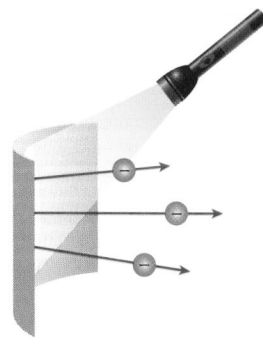

Figure 23-4
A light beam shining on this metal may eject electrons from the metal. Because this interaction involves both light and electrons, it is called the photoelectric effect.

photoelectric effect

emission of electrons from a surface that occurs when light of certain frequencies shines on the surface

internet connect

SC*i*LINKS
NSTA

TOPIC: Photoelectric effect
GO TO: www.scilinks.org
***sci*LINKS CODE:** HF2232

THE PHOTOELECTRIC EFFECT

In 1873, James Maxwell affirmed that light was a form of electromagnetic waves. Experiments by Heinrich Hertz provided experimental evidence of Maxwell's theories. However, some later experiments by Hertz could not be explained by the wave model of the nature of light. One of these was the **photoelectric effect.** The photoelectric effect can be explained in the following way. When light strikes a metal surface, the surface gives off electrons, as **Figure 23-4** illustrates. Scientists call this effect the photoelectric effect. They refer to the electrons that are emitted as *photoelectrons.* They refer to the surfaces that have the photoelectric effect as *photosensitive.* (★)TEKS **3E**

Classical physics cannot explain the photoelectric effect

The fact that light waves can eject electrons from a metal surface does not contradict the principles of classical physics. Light waves have energy, and if that energy is great enough, an electron could be stripped from its atom and have enough energy to escape the metal. However, the details of the photoelectric effect cannot be explained by classical theories. In order to see where the conflict arises, we must consider what should happen according to classical theory and then compare these predictions with experimental observations.

As was stated in Chapter 12, the energy of a wave increases as its intensity increases. Thus, according to classical physics, light waves of any frequency should have sufficient energy to eject electrons from the metal if the intensity of the light is high enough. Even at lower intensities, electrons should be ejected if light shines on the metal for a sufficient time period, because the electrons would take time to absorb the incoming energy before acquiring enough kinetic energy to escape from the metal. Furthermore, increasing the intensity of the light waves should increase the kinetic energy of the photoelectrons, and the maximum kinetic energy of any electron should be determined by the light's intensity. These classical predictions are summarized in the second column of **Table 23-1.**

Table 23-1 The photoelectric effect (★)TEKS 5B, 8A, 9A	Classical predictions	Experimental evidence
Whether electrons are ejected depends on . . .	the intensity of the light.	the frequency of the light.
The kinetic energy of ejected electrons depends on . . .	the intensity of the light.	the frequency of the light.
At low intensities, electron ejection . . .	takes time.	occurs almost instantaneously above a certain frequency.

Scientists found that *none* of these classical predictions are actually observed. No electrons are emitted if the frequency of the incoming light falls below a certain frequency, even if the intensity is very high. This frequency, known as the *threshold frequency* (f_t), varies with each metal.

If the light frequency exceeds the threshold frequency, the photoelectric effect is observed. The number of photoelectrons emitted is proportional to the light intensity, but the maximum kinetic energy of the photoelectrons is independent of the light intensity. Instead, the maximum kinetic energy of the photoelectrons increases with increasing frequency. Furthermore, electrons are emitted from the surface almost instantaneously, even at low intensities. See **Table 23-1.** (★)TEKS) **9A**

Einstein proposed that all electromagnetic waves are quantized

Albert Einstein resolved this conflict in his paper on the photoelectric effect, for which he received the Nobel Prize in 1921, by extending Planck's concept of quantization to electromagnetic waves. Einstein assumed that an electromagnetic wave can be viewed as a stream of particles called **photons.** Each photon has an energy, E, given by Planck's equation ($E = hf$). In this theory, each photon can be absorbed as a unit by an electron. When a photon's energy is transferred to an electron in a metal, the energy acquired by the electron is equal to hf. (★)TEKS) **3E**

photon

the discrete unit (or quantum) of light energy

Threshold frequency depends on the work function of the surface

In order to be ejected from a metal, an electron must escape from the metal by overcoming the force that binds it to the metal. The amount of energy the electron must have to escape the metal is known as the **work function** of the metal. The work function is equal to hf_t, where f_t is the threshold frequency for the metal. Photons with energy less than hf_t do not have enough energy to eject an electron from the metal. Because energy must be conserved, the maximum kinetic energy of the ejected photoelectrons is the difference between the photon energy and the work function of the metal. This relationship is expressed mathematically by the following equation:

work function

minimum amount of energy required for an electron to escape from a metal

---

MAXIMUM KINETIC ENERGY OF A PHOTOELECTRON

$$KE_{max} = hf - hf_t$$

maximum kinetic energy =
(Planck's constant × frequency of incoming photon) − work function

---

According to this equation, there should be a linear relationship between f and KE_{max} because h is a constant and the work function, hf_t, is constant for any given metal. Experiments have verified that this is indeed the case, as shown in **Figure 23-5,** and the slope of such a curve ($\Delta KE/\Delta f$) gives a value for h that corresponds to Planck's value. (★)TEKS) **5B, 9A**

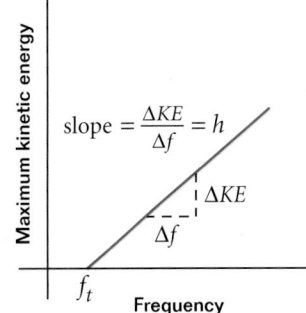

Figure 23-5
This graph shows a linear relationship between the maximum kinetic energy of emitted electrons and the frequency of incoming light. The intercept with the horizontal axis is the threshold frequency.

ANSWERS TO

Practice 23B
The photoelectric effect
1. 4.83×10^{14} Hz
2. 2.3 eV; 5.6×10^{14} Hz
3. 2.36 eV
4. lithium, cesium

SAMPLE PROBLEM 23B

The photoelectric effect ⭐TEKS 2C, 5B, 9A

PROBLEM

Light of frequency of 1.00×10^{15} Hz illuminates a sodium surface. The ejected photoelectrons are found to have a maximum kinetic energy of 1.86 eV. Find the threshold frequency for this metal.

SOLUTION

Given:
$$KE_{max} = (1.78 \text{ eV})(1.60 \times 10^{-19} \text{ J/eV})$$
$$KE_{max} = 2.85 \times 10^{-19} \text{ J} \qquad f = 1.00 \times 10^{15} \text{ Hz}$$

Unknown: $f_t = ?$

Use the expression for maximum kinetic energy, given on page 835, and solve for f_t.

$$KE_{max} = hf - hf_t \qquad f_t = \frac{hf - KE_{max}}{h}$$

$$f_t = \frac{(6.63 \times 10^{-34} \text{ J} \cdot \text{s})(1.00 \times 10^{15} \text{ Hz}) - (2.85 \times 10^{-19} \text{ J})}{6.63 \times 10^{-34} \text{ J} \cdot \text{s}}$$

$$\boxed{f_t = 5.70 \times 10^{14} \text{ Hz}}$$

PRACTICE 23B

The photoelectric effect

1. In the photoelectric effect, it is found that incident photons with energy 5.00 eV will produce electrons with a maximum kinetic energy 3.00 eV. What is the threshold frequency of this material?

2. Light of wavelength 350 nm falls on a potassium surface, and the photoelectrons have a maximum kinetic energy of 1.3 eV. What is the work function of potassium? What is the threshold frequency for potassium?

3. Calculate the work function of sodium using the information given in Sample Problem 23B.

4. Which of the following metals will exhibit the photoelectric effect when light of 7.0×10^{14} Hz frequency is shined on it?
   a. lithium, $hf_t = 2.3$ eV
   b. silver, $hf_t = 4.7$ eV
   c. cesium, $hf_t = 2.14$ eV

Photon theory accounts for observations of the photoelectric effect

The photon theory of light explains features of the photoelectric effect that cannot be understood using classical concepts. The photoelectric effect is not observed below a certain threshold frequency because the energy of the photon must be greater than or equal to the work function of the material. If the energy of each incoming photon is not equal to or greater than the work function, electrons will never be ejected from the surface, regardless of how many photons are present (how great the intensity is). Because the energy of each photon depends on the frequency of the incoming photon's waves ($E = hf$), the photoelectric effect is not observed when the waves of the incoming photon are below a certain frequency (f_t). ⭐TEKS **3A, 9A**

If the light intensity is doubled, the number of photons is doubled, and that in turn doubles the number of electrons ejected from the metal. However, the equation for the maximum kinetic energy of an electron shows that the kinetic energy depends only on the light frequency and the work function, not on the light intensity. Thus, even though there are more electrons ejected, the maximum kinetic energy of each electron does not change. ⭐TEKS **5B**

Finally, the fact that the electrons are emitted almost instantaneously is consistent with the particle theory of light, in which energy appears in small packets. Because each photon affects a single electron, there is no significant time delay between shining light on the metal and observing electrons being ejected.

Einstein's success in explaining the photoelectric effect by assuming that electromagnetic waves are quantized led scientists to realize that the quantization of energy must be considered a real description of the physical world rather than a mathematical contrivance, as most had initially supposed. The discreteness of energy had not been considered a viable possibility because the energy quantum is not detected in our everyday experiences. However, scientists began to believe that the true nature of energy is seen in the submicroscopic level of atoms and molecules, where quantum effects become important and measurable. ⭐TEKS **3C**

Conceptual Challenge

1. Photoelectric effect Even though bright red light delivers more total energy per second than dim violet light, the red light cannot eject electrons from a certain metallic surface, while the dimmer violet light can. How does Einstein's photon theory explain this observation? ⭐TEKS **3A, 9A**

2. Photographs Suppose a photograph were made of a person's face using only a few photons. According to Einstein's photon theory, would the result be simply a very faint image of the entire face? Why or why not?

3. Glowing objects As a hot object glows, the color of its glow depends on the object's temperature. As temperature increases, the color turns from red to orange to yellow and finally to white. Classical physics cannot explain this color change, while quantum mechanics can. What explanation is given by quantum mechanics?

ANSWERS TO

Conceptual Challenge

1. Although there are more photons in the red light, each photon has a lower frequency than those that make up the violet light. Energy is proportional to frequency, so the photons of red light have less energy than the photons of violet light, and thus they cannot eject electrons from the surface.

2. No; the resulting photograph would show the image only at places where photons hit the film. As more photons are used, the dots become closer together, so the image eventually appears continuous.

3. Quantum mechanics predicts that as the energy increases, the frequency also increases, and the color of the glow changes accordingly.

⭐TEKS

p. 836: 2C, 5B, 9A
p. 837: 3A, 9A, 3E, 5B, 3C, 3A, 9A

Teaching Tip

The Compton shift ($\Delta\lambda$) is given by the following equation:

$$\Delta\lambda = \lambda_f - \lambda_i = \frac{h}{mc}(1 - \cos\theta),$$

where m is the mass of the electron and θ is the angle between the directions of the scattered and incoming photons. This equation is derived by applying relativistic energy and momentum conservation to a collision between an electron and a photon. The quantity $\frac{h}{mc}$, called the Compton wavelength, has a value of 2.43×10^{-3} nm for the electron.

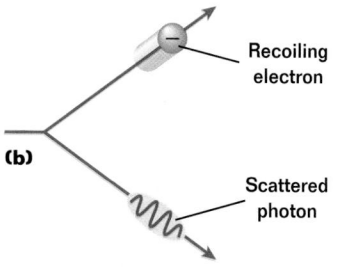

(a)

(b)

Figure 23-6
(a) When a photon collides with an electron, **(b)** the scattered photon has less energy and a longer wavelength than the incoming photon.

Compton shift

change in wavelength between incoming and scattered electromagnetic waves

internet connect

SC*i*LINKS.

NSTA

TOPIC: Arthur Compton
GO TO: www.scilinks.org
*sci*LINKS CODE: HF2233

Figure 23-7
The shading on the soundtrack varies the light intensity reaching the phototube, which varies the current sent to the speaker. This changing current reproduces the original sound waves in the speaker.

Compton shift supports the photon theory of light ⭐TEKS 3E

The American physicist Arthur Compton (1892–1962) realized that if light behaves like a particle, then a collision between an electron and a photon should be similar to a collision between two billiard balls. Photons should have momentum as well as energy; both quantities should be conserved in collisions. So, when a photon collides with an electron initially at rest, as in **Figure 23-6,** the photon transfers some of its energy and momentum to the electron. As a result, the energy and frequency of the scattered photon are lowered; its wavelength should increase. ⭐TEKS 5B

In 1923, to test this theory, Compton directed electromagnetic waves (X rays) toward a block of graphite. He found that the scattered waves had less energy and longer wavelengths than the incoming waves, just as he had predicted. This change in wavelength, known as the **Compton shift,** provides support for Einstein's photon theory of light. ⭐TEKS 8A

The amount that the wavelength shifts depends on the angle through which the photon is scattered and can range from about 10^{-6} nm to 10^{-3} nm. Note that even the largest change in wavelength is very small in relation to the wavelengths of visible light. For this reason, the Compton shift is difficult to detect using visible light, but it can be observed using electromagnetic waves with much shorter wavelengths, such as X rays. ⭐TEKS 8B

The photoelectric effect has a variety of practical applications

Many practical devices in our everyday lives depend on the photoelectric effect. Such applications often use a *phototube*, which contains a cathode made of a photosensitive material and an anode. The cathode is connected to the negative terminal of a power source and the anode is connected to the positive terminal. Thus, a potential difference exists across the cathode and the anode. When a beam of light shines on the photosensitive cathode, electrons are ejected from it. Because of the potential difference, the photoelectrons move toward the anode and a current exists. ⭐TEKS 3C

Figure 23-7 shows how the photoelectric effect is used to produce sound on movie film. The soundtrack of a movie is located along the side of the film in the form of an optical pattern of light and dark lines. A beam of light in the

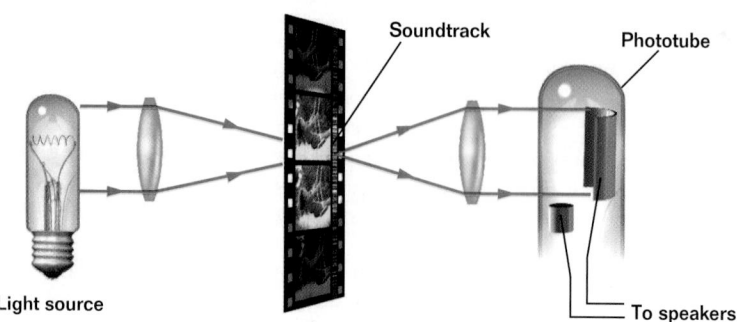

Soundtrack

Phototube

Light source

To speakers

projector is directed through the soundtrack toward a photo-tube. The variation in shading on the soundtrack varies the light intensity falling on the plate of the phototube. When the light intensity increases, the number of photoelectrons also increases because there are more photons striking the surface. Likewise, when the intensity decreases, the number of photoelectrons decreases. Hence, the current is constantly changing as the intensity changes. This changing current electrically simulates the original sound wave and reproduces it in the speaker.

Another common application of the photoelectric effect is a type of burglar alarm. In such alarms, a beam of ultraviolet light shines on a photosensitive surface in a phototube. (Ultraviolet rather than visible light is typically used to make the beam less obvious.) This beam causes the surface to emit electrons. The photoelectrons generate a current, which exists as long as the light continues to strike the surface. The current produced is amplified and used to energize an electromagnet that attracts a metal rod, as shown in **Figure 23-8(a).** (★)TEKS **3C**

If an intruder breaks the light beam, light no longer shines on the photosensitive surface and electrons are no longer ejected. This stops the current, and the electromagnet switches off. When this occurs, the spring pulls the iron rod to the right, as shown in **Figure 23-8(b).** In this position, the circuit that contains the burglar alarm is complete, and a current activates the alarm.

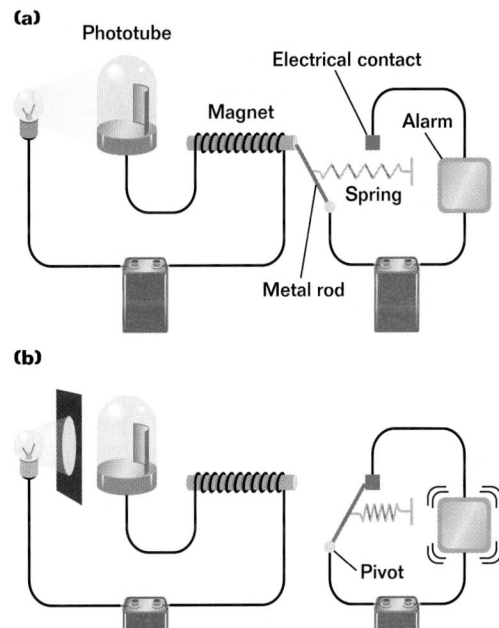

Figure 23-8
(a) When light strikes the photoelectric tube, the current in the circuit on the left energizes the magnet, keeping the burglar-alarm circuit on the right open.
(b) When the light stops shining on the tube, the alarm circuit closes, and the alarm sounds.

Section Review   (★)TEKS **2C, 3E, 5B, 9A**

1. Describe the conflict known as the ultraviolet catastrophe. How did Planck resolve this conflict? How does Planck's assumption depart from classical physics?

2. What is the energy (in eV units) carried by one photon of violet light that has a wavelength of 4.5×10^{-7} m?

3. What effect did scientists originally think that the intensity of light shining on a photosensitive surface would have on electrons ejected from that surface?

4. How does Einstein's theory that electromagnetic waves are quantized explain the fact that the frequency of light (rather than the intensity) determines whether electrons are ejected from a photosensitive surface?

5. Light with a wavelength of 1.00×10^{-7} m shines on tungsten, which has a work function of 4.6 eV. Are electrons ejected from the tungsten? If so, what is their maximum kinetic energy?

Section Review
ANSWERS

1. Classical physics predicted that radiation should increase as wavelength decreases, while experimental data showed that instead radiation increases, reaches a peak, and then decreases. Planck resolved this conflict by assuming that energy is quantized. In classical physics, energy is considered to be continuous.

2. 2.8 eV

3. A high intensity source delivers energy to the surface more rapidly than a low intensity source. Thus electrons should receive the energy necessary to escape the surface sooner using the high intensity source. However, electrons were ejected almost immediately for either source.

4. If energy depends on frequency, light at any intensity cannot eject electrons if the light's frequency is too low.

5. yes; 7.8 eV

(★)TEKS

p. 838: 3E, 5B, 8A, 8B, 3C
p. 839: 3C, 2C, 3E, 5B, 9A

Demonstration 4

Particle scattering

Purpose Demonstrate the research method used in Rutherford's gold-foil experiment.

Materials two pounds of clay, a dozen marbles, a square piece of cardboard (about 1 m × 1 m), a few books

Procedure Build a small mound with the clay, and set it on the floor. Place the cardboard over the mound, and support it with books at the corners if necessary. Explain to the class that you would like them to determine what is beneath the cardboard without looking. Then roll the marbles beneath the cardboard one at a time. Have students observe where each marble emerges. (This should be done at all locations on your side of the cardboard.)

The students should conclude that most of the space beneath the cardboard is clear because most of the marbles continue under the cardboard without being deflected. Have students try to determine the location, size, and shape of the mound of clay from the deflection of the marbles.

The frustration that the students feel when forbidden to look under the cardboard is a good analogy for what drives scientists to explore the unknown. Point out that Rutherford never got to "look beneath the cardboard" to see if his theory was correct.

23-2 SECTION OBJECTIVES

- **Explain the strengths and weaknesses of Rutherford's model of the atom.**

- **Recognize that each element has a unique emission and absorption spectrum.**

- **Explain atomic spectra using Bohr's model of the atom.**

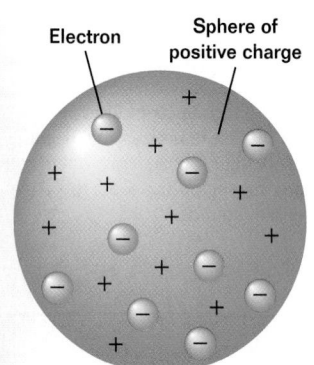

The Thomson model of the atom

Figure 23-9
In Thomson's model of the atom, electrons are embedded inside a larger region of positive charge like seeds in a watermelon.

Figure 23-10
In this experiment, positively charged alpha particles are directed at a thin metal foil. Because many particles pass through the foil and only a few are deflected, Rutherford concluded that the atom's positive charge is concentrated at the center of the atom.

EARLY MODELS OF THE ATOM

The model of the atom in the days of Newton was that of a tiny, hard, indestructible sphere. This model was a good basis for the kinetic theory of gases. However, new models had to be devised when experiments revealed the electrical nature of atoms. The discovery of the electron in 1897 prompted J. J. Thomson (1856–1940) to suggest a new model of the atom. In Thomson's model, electrons are embedded in a spherical volume of positive charge like seeds in a watermelon, as shown in **Figure 23-9.** (★)TEKS 3C, 3E

Rutherford proposed a planetary model of the atom

In 1911, Hans Geiger and Ernest Marsden, under the supervision of Ernest Rutherford (1871–1937), performed an important experiment showing that Thomson's model could not be correct. In this experiment, a beam of positively charged *alpha particles*—which consist of two protons and two neutrons—was projected against a thin metal foil, as shown in **Figure 23-10.** Most of the alpha particles passed through the foil as if it were empty space. Some of the alpha particles were deflected from their original direction through very large angles. Some particles were even deflected backward. Such deflections were completely unexpected on the basis of the Thomson model. Rutherford wrote, "It was quite the most incredible event that has ever happened to me in my life. It was almost as incredible as if you fired a 15-inch shell at a piece of tissue paper and it came back and hit you." (★)TEKS 3E

Such large deflections could not occur on the basis of Thomson's model, in which positive charge is evenly distributed throughout the atom, because the positively charged alpha particles would never come close to a positive charge concentrated enough to cause such large-angle deflections. (★)TEKS 3A

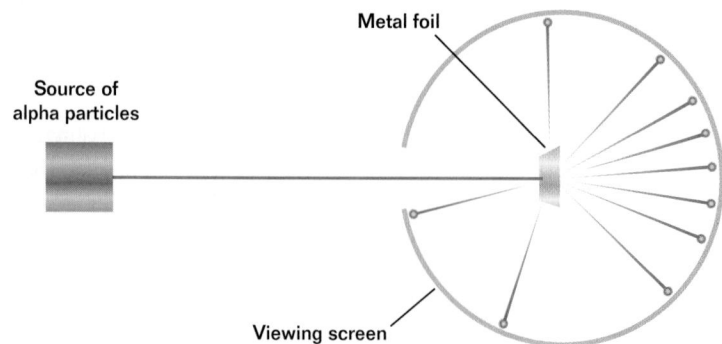

Metal foil

Source of alpha particles

Viewing screen

To explain his observations, Rutherford concluded that all of the positive charge in an atom and most of the atom's mass are found in a region that is small compared to the size of the atom. He called this concentration of positive charge and mass the *nucleus* of the atom. Any electrons in the atom were assumed to be in the relatively large volume outside the nucleus. So, according to Rutherford's theory, most alpha particles missed the nucleus entirely and passed through the foil, while only a few came close enough to the nucleus to be deflected.

Rutherford's model predicts that atoms are unstable

To explain why electrons in this outer region of the atom were not pulled into the nucleus, Rutherford viewed the electrons as moving in orbits about the nucleus, much like the planets orbit the sun, as shown in **Figure 23-11.**

However, this assumption posed a serious difficulty. If electrons orbited the nucleus, they would undergo a centripetal acceleration. According to Maxwell's theory of electromagnetism, accelerated charges should radiate electromagnetic waves, losing energy. So, the radius of an atom's orbit would steadily decrease. This would lead to an ever-increasing frequency of emitted radiation and a rapid collapse of the atom as the electron plunged into the nucleus. In fact, calculations show that according to this model, the atom would collapse in about one-billionth of a second. This difficulty with Rutherford's model led scientists to continue searching for a new model of the atom. ⊛ TEKS **3A, 3C**

ATOMIC SPECTRA

In addition to solving the problems with Rutherford's planetary model, scientists hoped that a new model of the atom would explain another mysterious fact about gases. When an evacuated glass tube is filled with a pure atomic gas and a sufficiently high voltage is applied between metal electrodes in the tube, a current is produced in the gas and the tube gives off light, as shown in **Figure 23-12.** The light's color is characteristic of the gas in the tube. This is how a neon sign works. The variety of colors seen in neon signs is the result of the light given off by different gases in the tubes.

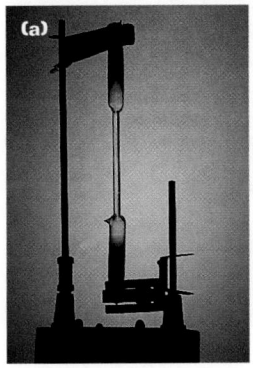
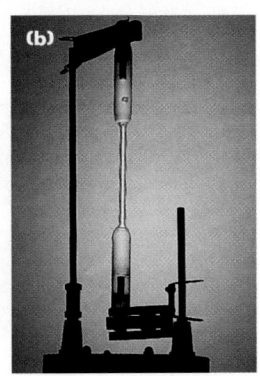
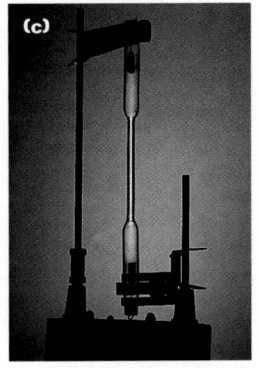

CONCEPT PREVIEW

You will learn more about the nucleus of the atom in Chapter 25.

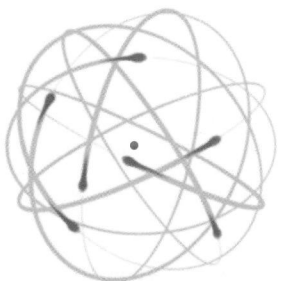

The Rutherford model

Figure 23-11

In Rutherford's model of the atom, electrons orbit the nucleus in a manner similar to planets orbiting the sun.

Figure 23-12

When a potential difference is applied across an atomic gas in a tube—here, hydrogen **(a)**, mercury **(b)**, and nitrogen **(c)**—the gas glows. The color of the glow depends on the type of gas.

Atomic Physics **841**

Key Models and Analogies

Point out that the positively charged nucleus attracts the negatively charged electrons, which end up in orbits with the right speed at the right distance, similar to satellites around Earth. But Maxwell's theory of electromagnetism predicts that charges moving on a curve (accelerating) should radiate energy.

⊛ TEKS

p. 840: 3C, 3E, 3E, 3A
p. 841: 3A, 3C

841

Spectral lines

Purpose Show line spectra of different gases.

Materials sodium and mercury vapor light sources, spectral tubes of different gases (argon, helium, nitrogen, hydrogen, etc.), power supply, diffraction gratings, dark room

Procedure Have students look at each light source through a diffraction grating and draw a diagrammatic representation of the spectra they observe. Explain that spectral lines of emission are considered to be the "fingerprints" of a gas. After they discuss why this is the case, explain that each gas emits radiation of certain characteristic wavelengths. Point out that they are only observing the spectral lines of radiation emitted in the visible spectrum. Other equipment may reveal radiation in the infrared and ultraviolet ranges or farther.

Teaching Tip

The light is first passed through slits so that the beams projected onto the screen have a well-defined, linear shape. Without any slits at all, images of the bulb itself would appear on the screen.

Did you know?

When the solar spectrum was first being studied, a set of spectral lines were found that did not correspond to any known element. A new element had been discovered. Because the Greek word for sun is *helios*, this new element was named helium. Helium was later found on Earth.

 **TEKS** 3C

emission spectrum

a unique series of spectral lines emitted by an atomic gas when a potential difference is applied across the gas

absorption spectrum

a continuous spectrum interrupted by dark lines characteristic of the medium through which the radiation has passed

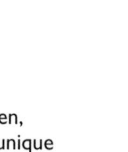 **Figure 23-14**

Each of these gases—hydrogen, mercury, and helium—has a unique emission spectrum.

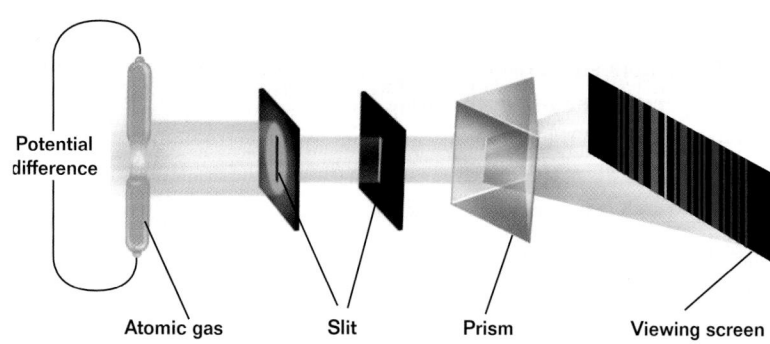

Figure 23-13

When the light from an atomic gas is passed through a prism, the dispersed light appears as a series of distinct, bright spectral lines.

Each gas has a unique emission and absorption spectrum

When the light given off (emitted) by an atomic gas is passed through a prism, as shown in **Figure 23-13,** a series of distinct bright lines is seen. Each line corresponds to a different wavelength, or color, of light. Such a series of spectral lines is commonly referred to as an **emission spectrum.**

As shown in **Figure 23-14,** the emission spectra for hydrogen, mercury, and helium are each unique. Further analysis of other substances reveals that every element has a distinct emission spectrum. In other words, the wavelengths contained in a given spectrum are characteristic of the element giving off the light. Because no two elements give off the same line spectrum, this fact represents a reliable technology for identifying elements in a mixture.

In addition to giving off light at specific wavelengths, an element can also absorb light at specific wavelengths. The spectral lines corresponding to this process form what is known as an **absorption spectrum.** An absorption spectrum can be seen by passing light containing all wavelengths through a vapor

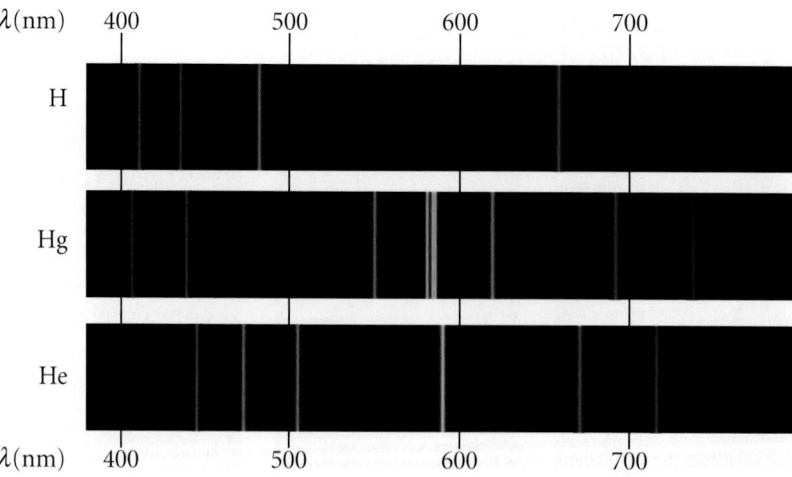

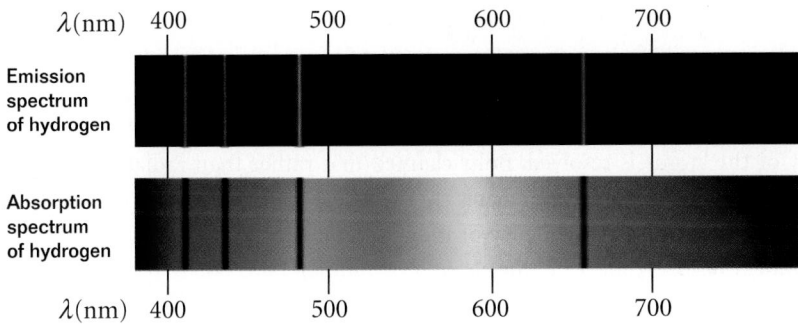

λ(nm)   400   500   600   700

Emission
spectrum
of hydrogen

Absorption
spectrum
of hydrogen

λ(nm)   400   500   600   700

Figure 23-15
Hydrogen's dark absorption lines occur at the same wavelengths as its bright emission lines.

of the element being analyzed. The absorption spectrum consists of a series of dark lines placed over the otherwise continuous spectrum.

Each line in the absorption spectrum of a given element coincides with a line in the emission spectrum of that element, as shown in **Figure 23-15** for hydrogen. In practice, more emission lines are usually seen than absorption lines. The reason for this will be discussed shortly.

The absorption spectrum of an element has many practical applications. For example, the continuous spectrum of radiation emitted by the sun must pass through the cooler gases of the solar atmosphere and then through Earth's atmosphere. The various absorption lines seen in the solar spectrum have been used to identify elements in the solar atmosphere. Scientists are able to examine the light from stars other than our sun in this fashion, but elements other than those present on Earth have never been detected.

The occurrence of atomic spectra was of great importance to scientists attempting to find a new model of the atom. Long after atomic spectra had been discovered, their cause remained unexplained. There was nothing in Rutherford's planetary model to account for the fact that each element has a unique series of spectral lines. Scientists hoped that a new model of the atom would explain why the atoms of a given element emit only certain lines and why the atoms absorb only those wavelengths that they emit. (★)TEKS **3C**

THE BOHR MODEL OF THE HYDROGEN ATOM

In 1913, the Danish physicist Niels Bohr (1885–1962) proposed a new model of the hydrogen atom that explained atomic spectra. Bohr's model of hydrogen contains some classical features and some revolutionary principles that could not be explained by classical physics. (★)TEKS **3E**

Bohr's model is similar to Rutherford's in that the electron moves in circular orbits about the nucleus. The electric force between the positively charged proton inside the nucleus and the negatively charged electron is the force that holds the electron in orbit. However, in Bohr's model, only certain orbits are stable. The electron is never found between these orbits; instead, it is said to "jump" instantly from one orbit to another without ever being between orbits.

(★)TEKS **1A, 8B**

Quick Lab

Atomic Spectra

MATERIALS LIST

✔ a diffraction grating

✔ a variety of light sources, such as:
  • a fluorescent light
  • an incandescent light
  • a clear aquarium bulb
  • a sodium-vapor street light
  • a gym light
  • a neon sign

⚡ SAFETY CAUTION

Be careful of high potential differences that may be present near some of these light sources.

Certain types of light sources produce a continuous spectrum when viewed through a diffraction grating, while others produce discrete lines. Observe a variety of different light sources through a diffraction grating, and compare your results. Try to find at least one example of a continuous spectrum and a few examples of discrete lines.

Teaching Tip

Point out that in his model, Bohr made the assumption that the radii of permitted orbits for electrons may have only discrete values ($r_1, r_2, r_3, \ldots$), which correspond to the energies ($E_1, E_2, E_3, \ldots$) of stationary states. Thus, the difference in energy between two states is $E_2 - E_1 = hf$, where f is the frequency of the corresponding spectral line.

(STOP) Misconception Alert

Students may think that electrons jump only between upper states and the lowest state. Make sure they understand that transitions between any two levels are allowed and that each transition involves the emission or absorption of radiation of the frequency that corresponds to the difference in energy. Ask students which jump would emit radiation of higher frequency, E_6 to E_5 or E_6 to E_4 (E_6 to E_4). Explain that the spectral lines we observe are produced by millions of atoms, each of which may be found at any level and may jump to any level.

Did you know?

Bohr's model assumes that the electron's angular momentum is quantized and that each value of momentum is an integral multiple of $h/2\pi$. The stable, or allowed, orbits are those that correspond to one of these values of angular momentum.

Figure 23-16

(a) When a photon is absorbed by an atom, an electron jumps to a higher energy level. **(b)** When the electron falls back to a lower energy level, the atom releases a photon.

Bohr's model further departs from classical physics by assuming that the hydrogen atom does not emit energy in the form of radiation when the electron is in any of these stable orbits. Hence, the total energy of the atom remains constant and one difficulty with the Rutherford model (the instability of the atom) is resolved. Bohr claimed that rather than radiating energy continuously, the electron radiates energy when it jumps from an outer orbit to an inner one. The frequency of the radiation emitted in the jump is *independent of the frequency of the electron's orbital motion*. The frequency of the emitted radiation is instead related to the change in the atom's energy and is found by Planck's equation ($E = E_{initial} - E_{final} = hf$).

In Bohr's model, transitions between stable orbits with different energy levels account for the discrete spectral lines

The lowest energy state in the Bohr model, which corresponds to the shortest possible radius, is often called the *ground state* of the atom, and the radius of this orbit is called the *Bohr radius*. At ordinary temperatures, most electrons are in the ground state, with the electron relatively close to the nucleus. When light of a continuous spectrum shines on the atom, only the photons whose energy (hf) matches the energy separation between two levels can be absorbed by the atom. When this occurs, an electron jumps from a lower energy state to a higher energy state, which corresponds to an orbit farther from the nucleus, as shown in **Figure 23-16(a).** This is called an *excited state*. The absorbed photons account for the dark lines in the absorption spectrum.

Once an electron is in an excited state, there is a certain probability that it will jump back to a lower energy level by emitting a photon, as shown in **Figure 23-16(b).** This process is known as *spontaneous emission*. The emitted photons are responsible for the bright lines in the emission spectrum.

In both cases, there is a correlation between the "size" of an electron's jump and the energy of the photon. For example, an electron in the fourth energy level could jump to the third level, the second level, or the ground state. Because Planck's equation gives the energy from one level to the next level, a greater jump means that more energy is emitted. Thus, jumps between different levels correspond to the various spectral lines that are observed. The jumps that correspond to the four spectral lines in the visible spectrum of

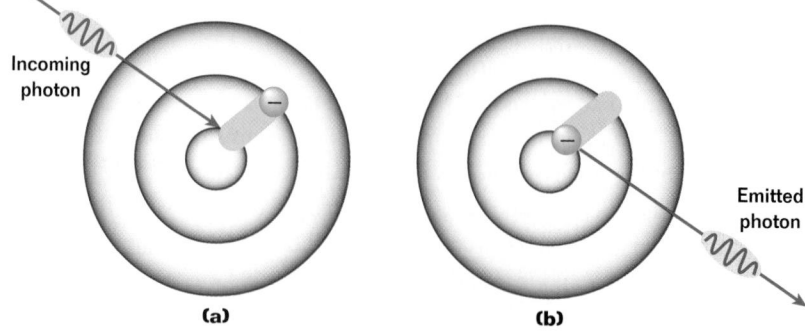

Incoming photon

Emitted photon

(a)   **(b)**

hydrogen are shown in **Figure 23-17.** Bohr's calculations successfully accounts for the wavelengths of all the spectral lines of hydrogen.

As noted earlier, fewer absorption lines than emission lines are typically observed. This is because absorption spectra are usually observed when a gas is at room temperature. Thus, most electrons are in the ground state, so all transitions observed are from a single level (E_1) to higher levels. Emission spectra, on the other hand, are seen by raising a gas to a high temperature and viewing downward transitions between any two levels. In this case, all transitions are possible, so more spectral lines are observed.

Bohr's idea of the quantum jump between energy levels provides an explanation for the aurora borealis, or northern lights. Charged particles from the sun sometimes become trapped in Earth's magnetic field and are deposited around the northern and southern magnetic poles. (Light shows in southern latitudes are called *aurora australis,* or *southern lights.*) When deposited, these charged particles from the sun collide with the electrons of the atoms in our atmosphere and transfer energy to these electrons, causing them to jump to higher energy levels. When an electron returns to its original orbit, the extra energy is released as a photon. The northern lights are the result of billions of these quantum jumps happening at the same time. (⭐TEKS) **3A**

The colors of the northern lights are determined by the type of gases in the atmosphere. The charged particles from the sun are most commonly released from Earth's magnetic field into a part of the atmosphere that contains oxygen. Oxygen releases green light; thus, green is the most common color of the northern lights. Red lights are the result of collisions with nitrogen atoms. Because each type of gas releases a unique color, the northern lights contain only a few distinct colors rather than a continuous spectrum.

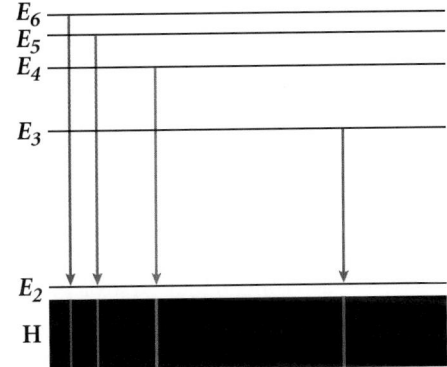

Figure 23-17

Every jump from one energy level to another corresponds to a specific spectral line. This example shows the transitions that result in the visible spectral lines of hydrogen. The lowest energy level, E_1, is not shown in this diagram.

Conceptual Challenge

1. Neon signs When a potential difference is placed across electrodes at the ends of a tube that contains neon, such as a neon sign, the neon glows. Is the light emitted by a neon sign composed of a continuous spectrum or only a few lines? Defend your answer. (⭐TEKS) **9B**

2. Energy levels If a certain atom has four possible energy levels and an electron can jump between any two energy levels of the atom, how many different spectral lines could be emitted?

3. Identifying gases Neon is not the only type of gas used in neon signs. As you have seen, a variety of gases exhibit similar effects when there is a potential difference across them. While the colors observed are sometimes different, certain gases do glow with the same color. How could you distinguish two such gases?

Tomorrow's Technology

BACKGROUND

This feature discusses the laser surface analyzer, which identifies elements by their emission spectra. This technique works because each element has a unique emission spectrum.

The laser surface analyzer will work on any form of matter: gas, liquid, or solid. The laser pulses are focused by a lens on the sample to form a laser spark. The spark light is collected by a lens and directed to a spectrograph that contains a prism or grating to spectrally disperse the light. The resulting emission spectrum is detected and displayed on the computer screen for analysis.

EXTENSION

As mentioned in the feature, other methods for testing paint for lead have been used in the past. Have students investigate one of these methods and discuss the procedure in an essay. Their essay should include a comparison between the older method and the laser surface analyzer.

Tomorrow's Technology

Laser Surface Analyzer ⓉEKS 3C

Many homes contain a silent danger to children living in them. Paints manufactured 50 years ago contain lead—a toxic metal. If children swallow flakes of this paint, they are at risk of contracting lead poisoning and of developing learning disabilities.

In the past, testing paint flakes for lead required complicated procedures available only at chemical labs or the use of instruments containing radioactive sources.

David Cremers, a scientist at Los Alamos National Laboratory, has plans to change all that with a device called a laser surface analyzer. This device uses a laser to create a tiny spark. The spark is intensely hot, around 5811 K (10 000°F). The spark vaporizes a material to produce atoms that are then excited to emit light.

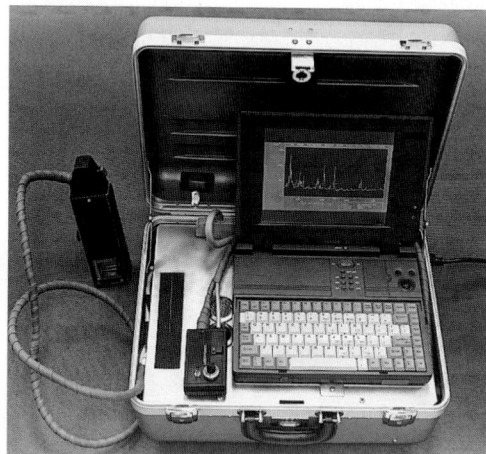

According to Cremers, "If we take this light from the spark and pass it through a prism to produce a spectrum, we'll see bright emissions in certain colors or regions of the spectrum. Because each element in the periodic table has

a unique emission spectrum, by accurately determining the wavelengths of these emissions or the colors of these emissions, we can tell you what elements the material is composed of."

Although the laser surface analyzer was originally created to test walls for toxic paint, it can be used on any surface to test for any elements. Another possible use is screening materials for recycling. A backpack model of the device has already been created for environmental workers in the field, who can use it to monitor soil for contaminants. Because the technique depends on laser light, fiber optics can be used to test samples over long distances.

Cremers hopes to develop the device's software so that the computer can automatically analyze a material and print out the results. This way, the laser surface analyzer could be operated by nearly anyone interested in checking a home for the presence of lead.

Bohr's model is incomplete

The Bohr theory of hydrogen was a tremendous success in certain areas because it explained several features of the spectra of hydrogen that had previously defied explanation. Bohr's model gave an expression for the radius of the atom and predicted the energy levels of hydrogen. This model was also successful when applied to hydrogen-like atoms, that is, atoms that contain only one electron. But while many attempts were made to extend the Bohr theory to multi-electron atoms, the results were unsuccessful.

Bohr's model of the atom also raised new questions. For example, Bohr assumed that electrons do not radiate energy when they are in a stable orbit, but his model offered no explanation for this. Another problem with Bohr's model was that it could not explain why electrons always have certain stable orbits, while other orbits are not allowed. Finally, the model seemed transitional because it followed classical physics in certain respects but radically departed from classical physics in other respects. For all of these reasons, Bohr's model was not considered to be a complete picture of the structure of the atom, and scientists continued to search for a new model that would resolve these difficulties. ⊛TEKS 3C

Section Review ⊛TEKS 3E, 9B

1. Based on the Thomson model of the atom, what did Rutherford expect to happen when he projected positively charged alpha particles against a metal foil?

2. Why did Rutherford conclude that an atom's positive charge and most of its mass are concentrated in the center of the atom?

3. What are two problems with Rutherford's model of the atom?

4. How could the atomic spectra of gases be used to identify the elements present in light from distant stars?

5. Bohr's model of the atom follows classical physics in some respects and quantum mechanics in others. Which assumptions of the Bohr model correspond to classical physics? Which correspond to quantum mechanics?

6. How does Bohr's model of the atom account for the emission and absorption spectra of an element?

7. **Physics in Action** A Norwegian scientist, Lars Vegard, determined the different wavelengths that are part of the northern lights. He found that only a few wavelengths of light, rather than a continuous spectrum, are present in the lights. How does Bohr's model of the atom account for this observation?

Section Review
ANSWERS

1. If positive charges were spread throughout the atom, as in Thomson's model, the alpha particles would be deflected slightly at most.

2. because most particles were not deflected at all, while a few were sharply deflected

3. It predicted an unstable atom, and it couldn't explain spectral lines.

4. Because spectral lines are unique to each element, an analysis of the spectra would reveal what elements are present.

5. classical physics: electron orbits the nucleus, electron radiates energy; quantum mechanics: only certain orbits are allowed, the electron radiates energy only when it jumps between levels, energy radiated depends on the difference between energy levels

6. Each line corresponds to a jump from one energy level to another.

7. Electrons can jump only from one allowed orbit to another; only lines that correspond to these jumps are found in the spectral lines.

⊛TEKS

p. 846: 3C
p. 847: 3C, 3E, 9B

Quantum mechanics

The Language of Physics

Modern physics has introduced ideas that may conflict with intuition and common sense. The dual nature of light is one example. Although the words *particle* and *wave* are familiar, students need to think about them in a new way when dealing with atomic and subatomic phenomena.

Teaching Tip

Remind students that the Compton shift (studied in Section 23-1) treats light as a particle and applies the conservation of momentum to the interaction between a photon and an electron. Explain that because the photon has no mass, the classical definition of momentum ($p = mv$) does not apply to photons. Instead, the momentum of a photon is defined as $p = \dfrac{h}{\lambda}$.

23-3 SECTION OBJECTIVES

- **Recognize the dual nature of light.**
- **Calculate the de Broglie wavelength of matter waves.**
- **Distinguish between classical ideas of measurement and Heisenberg's uncertainty principle.**
- **Describe the quantum-mechanical picture of the atom, including the electron cloud and probability waves.**

THE DUAL NATURE OF LIGHT

As discussed in Section 23-1, there is considerable evidence for the photon theory of light. In this theory, all electromagnetic waves consist of photons, particle-like pulses that have energy and momentum. On the other hand, light and other electromagnetic waves exhibit interference and diffraction effects that are considered to be wave behaviors. So which model is correct? The answer depends on the specific phenomenon being observed. ★TEKS 8B

Light is both a wave and a particle

Some experiments can be better explained or only explained by the photon concept, whereas others require a wave model. Most physicists accept both models and believe that the true nature of light is not describable in terms of a single classical picture.

For an example of how photons can be compatible with electromagnetic waves, consider radio waves at a frequency of 2.5 MHz. The energy of a photon having this frequency can be found using Planck's equation, as follows:

$$E = hf = (6.63 \times 10^{-34} \text{ J} \cdot \text{s})(2.5 \times 10^6 \text{ Hz}) = 1.7 \times 10^{-27} \text{ J}$$

From a practical viewpoint, this energy is too small to be detected as a single photon. A sensitive radio receiver might need as many as 10^{10} of these photons to produce a detectable signal. With such a large number of photons reaching the detector every second, we would not be able to detect the individual photons striking the antenna. Thus, the signal would appear as a continuous wave. ★TEKS 8A

Now consider what happens as we go to higher frequencies and hence shorter wavelengths. In the visible region, it is possible to observe both the photon and the wave characteristics of light. As we mentioned earlier, a light beam can show interference phenomena and produce photoelectrons. The interference phenomena are best explained by the wave model of light, while the photoelectrons are best explained by the particle theory of light.

At even higher frequencies and correspondingly shorter wavelengths, the momentum and energy of the photons increase. Consequently, the photon nature of light becomes very evident. In addition, as the wavelength decreases, wave effects, such as interference and diffraction, become more difficult to observe. Very indirect methods are required to detect the wave nature of very high frequency radiation, such as gamma rays. ★TEKS 8B

Thus, all forms of electromagnetic radiation can be described from two points of view. At one extreme, the electromagnetic wave description suits the overall interference pattern formed by a large number of photons. At the other extreme, the particle description is more suitable for dealing with highly energetic photons of very short wavelengths. ⭐TEKS **3A**

MATTER WAVES

In the world around us, we are accustomed to regarding things such as thrown baseballs solely as particles, and things such as sound waves solely as forms of wave motion. As already noted, this rigid distinction cannot be made with light, which has both wave and particle characteristics. In 1924, the French physicist Louis de Broglie (1892–1987) extended the wave-particle duality. In his doctoral dissertation, de Broglie proposed that all forms of matter may have both wave properties and particle properties. At that time, this was a highly revolutionary idea with no experimental support. Now, however, scientists accept the concept of matter's dual nature. ⭐TEKS **3E**

The wavelength of a photon is equal to Planck's constant (h) divided by the photon's momentum (p). De Broglie speculated that this relationship might also hold for matter waves, as follows:

WAVELENGTH OF MATTER WAVES ⭐TEKS **8A**

$$\lambda = \frac{h}{p} = \frac{h}{mv}$$

$$\text{de Broglie wavelength} = \frac{\text{Planck's constant}}{\text{momentum}}$$

As seen by this equation, the larger the momentum of an object, the smaller its wavelength. In an analogy with photons, de Broglie postulated that the frequency of a matter wave can be found with Planck's equation as illustrated below:

FREQUENCY OF MATTER WAVES ⭐TEKS **8A**

$$f = \frac{E}{h}$$

$$\text{de Broglie frequency} = \frac{\text{energy}}{\text{Planck's constant}}$$

The dual nature of matter suggested by de Broglie is quite apparent in these two equations, both of which contain particle concepts (E and mv) and wave concepts (λ and f).

Did you know?

Louis de Broglie's doctoral thesis about the wave nature of matter was so radical and speculative that his professors were uncertain about whether they should accept it. They resolved the issue by asking Einstein to read the paper. Einstein gave his approval, and de Broglie's paper was accepted. Five years after his thesis was accepted, de Broglie won the Nobel Prize for his discovery.

⭐TEKS **3E**

 Misconception Alert

The concept of matter waves is probably new and unusual to most students. Because students have no experience with matter waves, they may not understand the nature of such waves. The meaning of de Broglie's hypothesis is that we may picture the electron in two ways: as a particle in motion or as a wave occupying a certain region in space. Encourage students to discuss their mental images of electron waves to prepare them for the probability interpretation of matter waves. Also assure them that these ideas remain puzzling to many people.

⭐TEKS

p. 848: 8B, 8A, 8B
p. 849: 3A, 3E, 3E, 8A, 8A

The following may be used as teamwork exercises or for demonstration at the chalkboard or on an overhead projector.

PROBLEM

De Broglie waves

Calculate the de Broglie wavelength of an electron moving at the following speeds:

a. 8.0×10^4 m/s

b. 8.0×10^5 m/s

c. 8.0×10^6 m/s

Answer

  a. 9.1×10^{-9} m

  b. 9.1×10^{-10} m

  c. 9.1×10^{-11} m

Teaching Tip

De Broglie waves are also discussed in the "Physics on the Edge" feature in Chapter 12.

PRACTICE GUIDE 23C	
Solving for:	
m	📖 **PE** 5; Ch. Rvw. 42
	💿 **PW** 2
	PB Sample, 1–4
λ	📖 **PE** 4; Ch. Rvw. 40
	💿 **PW** Sample, 3, 5, 7*
	PB 5–8
v	📖 **PE** Sample, 1–3; Ch. Rvw. 39, 41
	💿 **PW** 1, 4, 6*
	PB 9–10

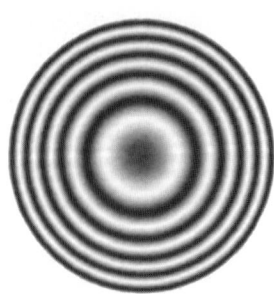

Figure 23-18

(a) Electrons show interference patterns similar to **(b)** light waves. This demonstrates that electrons sometimes behave like waves.

De Broglie's proposal that all particles also exhibit wave properties was first regarded as pure speculation. If particles such as electrons had wave properties, then under certain conditions they should exhibit interference phenomena. Three years after de Broglie's proposal, C. J. Davisson and L. Germer, of the United States, discovered that electrons can be diffracted by a single crystal of nickel. This important discovery provided the first experimental confirmation of de Broglie's theory. An example of electron diffraction compared with light diffraction is shown in **Figure 23-18.** ⊛TEKS **3E**

Electron diffraction by a crystal is possible because the de Broglie wavelength of a low-energy electron is approximately equal to the distance between atoms in a crystal. In principle, diffraction effects should be observable even for objects in our large-scale world. However, the wavelengths of material objects in our everyday world are much smaller than any possible aperture through which the object could pass, as shown below in Sample Problem 23C.

SAMPLE PROBLEM 23C

De Broglie waves ⊛TEKS **2C, 8A**

PROBLEM

With what speed would an electron with a mass of 9.109×10^{-31} kg have to move if it had a de Broglie wavelength of 7.28×10^{-11} m?

SOLUTION

Given: $m = 9.109 \times 10^{-31}$ kg $\lambda = 7.28 \times 10^{-11}$ m

              $h = 6.63 \times 10^{-34}$ J•s

Unknown: $v = ?$

Use the equation for the de Broglie wavelength, given on page 849.

$$\lambda = \frac{h}{mv} \quad \text{or} \quad v = \frac{h}{\lambda m}$$

$$v = \frac{6.63 \times 10^{-34} \text{ J•s}}{(7.28 \times 10^{-11} \text{ m})(9.109 \times 10^{-31} \text{ kg})} = 1.00 \times 10^7 \text{ m/s}$$

$$\boxed{v = 1.00 \times 10^7 \text{ m/s}}$$

De Broglie waves ⭐TEKS 2C, 8A

1. With what speed would a 50.0 g rock have to be thrown if it were to have a wavelength of 3.32×10^{-34} m?

2. If the de Broglie wavelength of an electron is equal to 5.00×10^{-7} m, how fast is the electron moving?

3. How fast would one have to throw a 0.15 kg baseball if it were to have a wavelength equal to 5.00×10^{-7} m (the same wavelength as the electron in problem 2)?

4. What is the de Broglie wavelength of a 1375 kg car traveling at 43 km/h?

5. A bacterium moving across a petri dish at 3.5 μm/s has a de Broglie wavelength of 1.9×10^{-13} m. What is the bacterium's mass?

De Broglie waves account for the allowed orbits of Bohr's model

At first, no one could explain why only some orbits were stable. Then, de Broglie saw a connection between his theory of wave character of matter and the stable orbits in the Bohr model. De Broglie assumed that an electron orbit would be stable only if it contained an integral (whole) number of electron wavelengths, as shown in **Figure 23-19.** The first orbit contains one wavelength, the second orbit contains two wavelengths, etc. ⭐TEKS 3C

De Broglie's hypothesis compares with the example of standing waves on a vibrating string of a given length, as discussed in Chapter 12. In this analogy, the circumference of the electron's orbit corresponds to the string's length. So, the condition for an electron orbit is that the circumference must contain an integral multiple of electron wavelengths. ⭐TEKS 3A

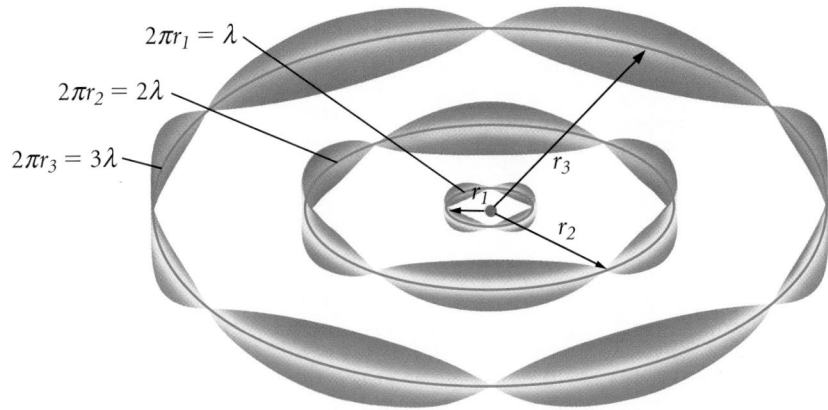

$2\pi r_1 = \lambda$

$2\pi r_2 = 2\lambda$

$2\pi r_3 = 3\lambda$

Figure 23-19
De Broglie's hypothesis that there is always an integral number of electron wavelengths around each circumference explains why only certain orbits are stable.

ANSWERS TO

Practice 23C
De Broglie waves
1. 39.9 m/s
2. 1.46×10^3 m/s
3. 8.84×10^{-27} m/s
4. 4.0×10^{-38} m
5. 1.0×10^{-15} kg

Visual Strategy

Figure 23-19
Point out that the electron has a different wavelength in each orbit. Also, just as for a satellite around Earth, the radius decreases as the speed increases.

Q The radius of the hydrogen atom at the ground state (lowest energy level) is $r_1 = 5.3 \times 10^{-11}$ m. What is the electron's wavelength in the first orbit?

A *The electron's wavelength equals the circumference of the circle of radius r_1, that is, $\lambda = 2\pi r_1 = 3.3 \times 10^{-10}$ m.*

⭐TEKS

p. 850: 3E, 2C, 8A
p. 851: 2C, 8A, 3C, 3A

uncertainty principle

states that it is impossible to simultaneously measure both the position and momentum of an object with complete certainty

THE UNCERTAINTY PRINCIPLE

In classical mechanics, there is no limitation to the accuracy of our measurements in experiments. In principle, we could always make a more precise measurement using a more finely detailed meterstick or a stronger magnifier. This unlimited precision does not hold true in quantum mechanics. The absence of such precision is not due to the limitations of our instruments or to our perturbation of the system when we make measurements. It is a fundamental limitation inherent in nature due to the wave nature of particles.

Simultaneous measurements of position and momentum cannot be completely certain

In 1927, Werner Heisenberg argued that *it is fundamentally impossible to make simultaneous measurements of a particle's position and momentum with infinite accuracy.* In fact, the more we learn about a particle's momentum, the less we know of its position, and the reverse is also true. This principle is known as Heisenberg's **uncertainty principle.** ⭐TEKS **3E**

To understand the uncertainty principle, consider the following thought experiment. Suppose you wish to measure the position and momentum of an electron as accurately as possible. You might be able to do this by viewing the electron with a powerful microscope. In order for you to see the electron and thus determine its location, at least one photon of light must bounce off the electron and pass through the microscope into your eye. This incident photon is shown moving toward the electron in **Figure 23-20(a).** When the photon strikes the electron as in **Figure 23-20(b),** it transfers some of its energy and momentum to the electron. So, in the process of attempting to locate the electron very accurately, we have caused much uncertainty in its momentum. The measurement procedure limits the accuracy to which we can determine position and momentum simultaneously. ⭐TEKS **3A**

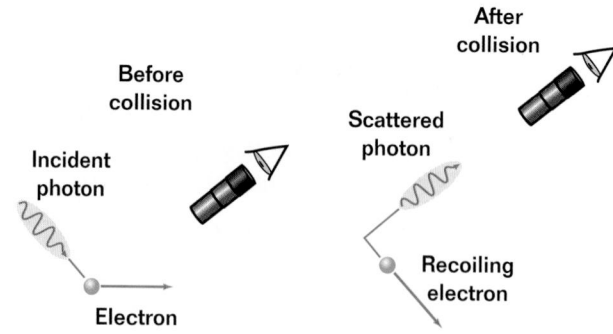

Figure 23-20

A thought experiment for viewing an electron with a powerful microscope. **(a)** The electron is viewed before colliding with the photon. **(b)** The electron recoils (is disturbed) as the result of the collision with the photon.

The mathematical form of the uncertainty principle states that the product of the uncertainties in position and momentum will always be larger than some minimum value. Arguments similar to those given here show that this minimum value is Planck's constant (h) divided by 4π.

THE ELECTRON CLOUD

In 1926, Erwin Schrödinger proposed a wave equation that described the manner in which de Broglie's matter waves change in space and time. Although this equation and its derivation are beyond the scope of this book, we will consider Schrödinger's equation qualitatively. Solving Schrödinger's equation yields a quantity called the *wave function,* represented by ψ (Greek letter *psi*). A particle is represented by a wave function, ψ, that depends on the position of the object and time. (★)TEKS **3E**

An electron's location is described by a probability wave

As discussed earlier, simultaneous measurements of position and momentum cannot be completely certain. Because the electron's location cannot be precisely determined, it is useful to discuss the *probability* of finding the electron at different locations. It turns out that the quantity $|\psi|^2$ is proportional to the probability of finding the electron at a given position. This interpretation of Schrödinger's wave function was first proposed by the German physicist Max Born in 1926. (★)TEKS **3E**

Figure 23-21 shows the probability of finding the electron at various distances from the nucleus in the ground state of hydrogen. The height of the curve at each point is proportional to the probability of finding the electron, and the x coordinate represents the electron's distance from the nucleus. Note that there is zero probability of finding the electron in the nucleus at this energy level.

The peak of this curve represents the distance from the nucleus at which the electron is most likely to be found in the ground state. Schrödinger's wave equation predicts that this distance is 5.3×10^{-11} m, which is the value of the radius of the first electron orbit in Bohr's model of hydrogen. However, as the curve indicates, there is also a probability of finding the electron at various other distances from the nucleus. In other words, the electron is not confined to a particular orbital distance from the nucleus as is assumed in the Bohr model. The electron may be found at various distances from the nucleus, but the probability of finding it at a distance corresponding to the first Bohr orbit is greater than that of finding it at any other distance. This new model of the atom is consistent with Heisenberg's uncertainty principle, which states that we cannot know the electron's location with complete certainty. Thus, the quantum-mechanical model resolves another difficulty with the Bohr model of hydrogen, which violates the uncertainty principle. (★)TEKS **3C**

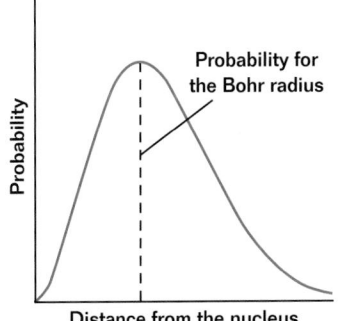

Figure 23-21
The height of this curve is proportional to the probability of finding the electron at different distances from the nucleus in the ground state of hydrogen.

Did you know?

Although Einstein was one of the founders of quantum theory, he did not believe that probability waves could be a final description of nature. His convictions in this matter led to his famous statement, "In any case, I am convinced that He [God] does not play dice."

(★)TEKS **3E**

SECTION 23-3

Demonstration 6

Probability distribution

Purpose Illustrate the meaning of the electron wave as a probability function.

Materials 100 index cards, jar, overhead projector, transparency

Procedure Write a number on each of the index cards as follows: the number *1* on 4 cards, the number *2* on 6 cards, the number *3* on 12 cards, the number *4* on 40 cards, the number *5* on 14 cards, the number *6* on 10 cards, the number *7* on 8 cards, the number *8* on 4 cards, and the number *9* on 2 cards. Place all of the cards in a jar. Draw nine concentric circles on a transparency, and label the circles 1 through 9.

Have each student pick four or five cards from the jar. Point out that each of the numbered index cards corresponds to a numbered circle on the transparency. Tell students that for every index card they pick, they should draw a dot on the corresponding circle on the transparency. A cluster of dots will form around circle 4. Tell students that this is analogous to the way electrons move in the atom. An electron could be anywhere a dot is seen, but in this example it spends most of its time on orbit 4.

(★)TEKS

p. 852: 3E, 3A
p. 853: 3E, 3E, 3C, 3E

1. Light is a wave and a particle; some experiments reveal its particle nature, and others reveal its wave nature.

2. The allowed orbits are those that contain an integral number of electron wavelengths.

3. 3.96×10^{-11} m

4. Even with perfect measuring instruments, according to Heisenberg's uncertainty principle, simultaneous measurements of position and momentum cannot be completely certain. This inherent limitation is a result of the wave nature of matter.

5. It is proportional to the probability of finding the electron at a given position.

6. It is proportional to the probability of finding the electron at various locations.

(★)TEKS

p. 854: 3C, 3C, 2C, 3E, 8A, 8B

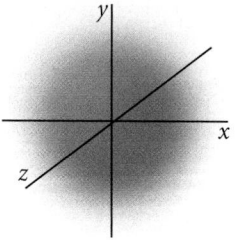

Figure 23-22
This spherical electron cloud for the ground state of hydrogen represents the probability of finding the electron at different locations. The darkest regions are the most probable, and the lightest are the least probable.

Quantum mechanics also predicts that the wave function for the hydrogen atom in the ground state is spherically symmetrical; hence, the electron can be found in a spherical region surrounding the nucleus. This is in contrast to the Bohr theory, which confines the position of the electron to points in a plane. This result is often interpreted by viewing the electron as a cloud surrounding the nucleus, called an *electron cloud,* as represented in **Figure 23-22.** The densest regions of the cloud in the figure represent the locations where the electron is most likely to be found. In this case, the densest region is a spherical shell, and the distance from any point on the shell to the nucleus is equal to the first Bohr radius. (★)TEKS **3C**

Analysis of each of the energy levels of hydrogen reveals that the most probable electron location in each case is in agreement with each of the radii predicted by the Bohr theory. The discrete energy levels that could not be explained by Bohr's theory can be derived from Schrödinger's wave equation, and the de Broglie wavelengths account for the allowed orbits that were unexplainable in Bohr's theory. Thus, the new quantum-mechanical model explains certain aspects of the structure of the atom that Bohr's model could not account for. (★)TEKS **3C**

Although probability waves and electron clouds cannot be simply visualized as Bohr's planetary model could, they offer a mathematical picture of the atom that is more complete than Bohr's model. The material presented in this chapter is only an introduction to quantum theory. Quantum mechanics has been amazingly successful in calculating the behavior of atoms and is still progressing rapidly today. While most scientists believe that quantum mechanics may be nearly the final picture of the deepest levels of nature, a few continue to search for other explanations, and debates about the implications of quantum mechanics continue.

Section Review

(★)TEKS **2C, 3E, 8A, 8B**

1. Is light considered to be a wave or a particle? Explain your answer.

2. How did de Broglie account for the fact that the electrons in Bohr's model are always found at certain distinct distances from the nucleus?

3. Calculate the de Broglie wavelength of a proton moving at 1.00×10^4 m/s.

4. In classical physics, the accuracy of measurements has always been limited by the measuring instruments used, and no instrument is perfect. How is this limitation different from that formulated by Heisenberg in the uncertainty principle?

5. What is the physical significance of the square of the Schrödinger wave function, $|\psi|^2$?

6. What does the density of an electron cloud represent?

CHAPTER 23
Summary

Chapter 23
Summary

KEY IDEAS

Section 23-1 Quantization of energy

- Blackbody radiation and the photoelectric effect contradict classical physics, but they can be explained with the assumption that energy comes in discrete units, or is quantized.
- The energy of a light quantum, or photon, depends on the frequency of the light, as follows:

$$E = hf$$

- The minimum energy required for an electron to escape from a metal depends on the threshold frequency of the metal, as follows:

$$\text{work function} = hf_t$$

- The maximum kinetic energy of photoelectrons depends on the work function and the frequency of the light shining on the metal, as follows:

$$KE_{max} = hf - hf_t$$

Section 23-2 Models of the atom

- Rutherford's scattering experiment revealed that all of an atom's positive charge and most of an atom's mass are concentrated at its center.
- Each gas has a unique emission and absorption spectrum.
- Atomic spectra are explained by Bohr's model of the atom, in which electrons move from one energy level to another when they absorb or emit photons.

Section 23-3 Quantum mechanics

- Light has both wave and particle characteristics.
- De Broglie proposed that matter has both wave and particle characteristics.
- The frequency and wavelength of de Broglie waves are found with the following expressions:

$$\lambda = \frac{h}{mv} \text{ and } f = \frac{E}{h}$$

- Simultaneous measurements of position and momentum cannot be completely certain.

Variable symbols

Quantities		Units	
E	photon energy	J	joules
		eV	electron volts
f_t	threshold frequency	Hz	hertz
hf_t	work function	eV	electron volts
KE_{max}	maximum kinetic energy	eV	electron volts

KEY TERMS

absorption spectrum (p. 842)

blackbody radiation (p. 830)

Compton shift (p. 838)

emission spectrum (p. 842)

photoelectric effect (p. 834)

photon (p. 835)

ultraviolet catastrophe (p. 831)

uncertainty principle (p. 852)

work function (p. 835)

Teaching Tip

Ask students to prepare a concept map for the chapter. The concept map should include most of the vocabulary terms, along with other integral terms or concepts.

(★) TEKS

**Review & Assess
pp. 856–859:**
2C: 14–18, 39–53, Alt.
    Assess. 1
2D: 1–4
3E: 3, 7–8, 13, 19–20,
    23–25, 30–34
5B: 4, 53
8A: 35–37, 39–44
8B: 1–3, 9, 14–18, 29,
    31, 45, 47, 49–50
8C: 3
9A: 5–7, 10–12, 43, 46,
    48, 51–52
9B: 21–22, 26–28

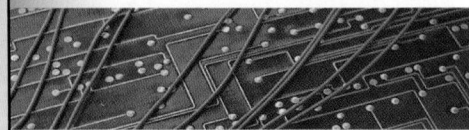

CHAPTER 23
Review and Assess

QUANTIZATION OF ENERGY

Review questions

1. Why is the term *ultraviolet catastrophe* used to describe the discrepancy between the predictions of classical physics and the experimental data for blackbody radiation?

2. What is meant by the term *quantum*?

3. What did Planck assume in order to explain the experimental data for blackbody radiation? How did Planck's assumption contradict classical physics?

4. What is the relationship between a joule and an electron volt?

5. According to classical physics, when a light illuminates a photosensitive surface, what should determine how long it took before electrons were ejected from the surface?

6. How do observations of the photoelectric effect conflict with the predictions of classical physics?

7. According to Einstein's photon theory of light, what does the frequency of light shining on a metal determine? What does the intensity of the light determine?

8. What does Compton scattering demonstrate?

Conceptual questions

9. Which has more energy, a photon of ultraviolet radiation or a photon of yellow light?

10. If the photoelectric effect is observed for one metal using light of a certain wavelength, can you conclude that the effect will also be observed for another metal under the same conditions?

11. What effect, if any, would you expect the temperature of a material to have on the ease with which electrons can be ejected from the metal in the photoelectric effect?

12. An X-ray photon is scattered by an electron. Does the frequency of this scattered photon increase, decrease, or stay the same relative to its frequency before being scattered?

13. A photon is deflected by a collision with an electron. Can the photon's wavelength ever become shorter due to the Compton shift? Explain your answer.

Practice problems

14. A quantum of electromagnetic radiation has an energy of 2.0 keV. What is its frequency?
(See Sample Problem 23A.)

15. Calculate the energy in electron volts of a photon having a wavelength in the following ranges:
 a. the microwave range, 5.00 cm
 b. the visible light range, 5.00×10^{-7} m
 c. the X-ray range, 5.00×10^{-8} m
(See Sample Problem 23A.)

16. Light of frequency 1.5×10^{15} Hz illuminates a piece of tin, and the tin emits photoelectrons of maximum kinetic energy 1.2 eV. What is the threshold frequency of the metal?
(See Sample Problem 23B.)

17. Light of wavelength 3.0×10^{-7} m shines on the metals lithium, iron, and mercury, which have work functions of 2.3 eV, 3.9 eV, and 4.5 eV, respectively.
 a. Which of these metals will exhibit the photoelectric effect?
 b. For those metals that do exhibit the photoelectric effect, what is the maximum kinetic energy of the photoelectrons?
(See Sample Problem 23B.)

18. The threshold frequency of silver is 1.14×10^{15} Hz. What is the work function of silver?
(See Sample Problem 23B.)

MODELS OF THE ATOM

Review questions

19. What did Rutherford's foil experiment reveal?

20. If Rutherford's planetary model were correct, atoms would be extremely unstable. Explain why.

21. How is an emission spectrum observed? How is an absorption spectrum observed?

22. How can the absorption spectrum of a gas be used to identify the gas?

23. What restriction does the Bohr model place on the movement of an electron in an atom?

24. How is Bohr's model of the hydrogen atom similar to Rutherford's planetary model? How are the two models different?

25. How does Bohr's model account for atomic spectra?

Conceptual questions

26. Explain why all of the wavelengths in an element's absorption spectrum are also found in that element's emission spectrum.

27. More emission lines than absorption lines are usually observed in the atomic spectra of most elements. Explain why this occurs.

28. Suppose that the electron in the hydrogen atom obeyed classical mechanics rather than quantum mechanics. Why should such a hypothetical atom emit a continuous spectrum rather than the observed line spectrum?

QUANTUM MECHANICS

Review questions

29. Name two situations in which light behaves like a wave and two situations in which light behaves like a particle.

30. What does Heisenberg's uncertainty principle claim?

31. How do de Broglie's matter waves account for the "allowed" electron orbits?

32. Describe the quantum-mechanical model of the atom. How is this model similar to Bohr's model? How are the two different?

Conceptual questions

33. Explain why it is impossible to simultaneously measure the position and momentum of a particle with infinite accuracy.

34. How does Heisenberg's uncertainty principle conflict with the Bohr model of hydrogen?

35. Discuss whether the behavior of an electron is mainly wavelike or particle-like in each of the following situations:
  a. traversing a circular orbit in a magnetic field
  b. absorbing a photon and being photoelectrically ejected from the surface of a metal
  c. forming an interference pattern

36. Use de Broglie's equation to explain why the wave properties of an electron can be observed, while those of a speeding car cannot.

37. An electron and a proton are accelerated from rest through the same potential difference. Which particle has the longer wavelength?

38. Discuss why the term *electron cloud* is used to describe the arrangement of electrons in the quantum-mechanical view of the atom.

Practice problems

39. How fast must an electron move if it is to have a de Broglie wavelenth of 5.2×10^{-11} m?
(See Sample Problem 23C.)

40. Calculate the de Broglie wavelength of a 0.15 kg baseball moving at 45 m/s.
(See Sample Problem 23C.)

41. What is the speed of a proton with a de Broglie wavelength of 4.00×10^{-14} m?
(See Sample Problem 23C.)

42. A mosquito moving at 12 m/s has a de Broglie wavelength of 5.5×10^{-30} m. What is the mass of this mosquito?
(See Sample Problem 23C.)

19. $+q$ and m are concentrated at the atom's center.

20. because electrons would radiate energy

21. by placing ΔV across a gas and passing emitted light through a prism; by passing continuous light through a gas and then through a prism

22. by comparing observed lines to known spectra of elements

23. The electron is limited to certain orbits.

24. The electron orbits the nucleus in both models. Rutherford: any orbits are possible; Bohr: only certain orbits are stable.

25. Spectral lines correspond to jumps between stable orbits.

26. In both cases, each line corresponds to a transition.

27. Absorption spectra involve upward jumps from the ground level, while emission spectra involve downward jumps to any level.

28. Classically, radiation emitted depends on the electron's orbital frequency, which constantly changes as the electron loses energy.

29. diffraction, polarization; photoelectric effect, blackbody radiation, Compton shift

30. It is impossible to simultaneously measure position and momentum with complete certainty.

31. Allowed orbits are those that contain an integral number of wavelengths.

32. The electron's location is proportional to $|\psi|^2$. Electrons can be anywhere, but the most probable locations are the Bohr radii.

Answers continued on next page.

33. This limitation results from the wave nature of matter.

34. Bohr's model predicts the electron's location exactly.

35. a. particle
    b. particle
    c. wave

36. λ is inversely proportional to m; when m is large, λ is too small to detect.

37. the electron

38. because the electron's location cannot be determined exactly

39. 1.4×10^7 m/s

40. 9.8×10^{-35} m

41. 9.91×10^6 m/s

42. 1.0×10^{-5} kg

43. a. 1.39 eV
    b. 3.35×10^{14} Hz

44. 4.0×10^{-14} m

45. red

46. 2.00 eV

ANSWERS TO

Technology & Learning

a. 2.43×10^{-7} m
b. 1.04×10^{-5} m
c. 1.32×10^{-10} m
d. 5.65×10^{-9} m

 TEKS

p. 858: 2E, 8B

MIXED REVIEW

43. Electrons are ejected from a surface with speeds ranging up to 4.6×10^5 m/s when light with a wavelength of $\lambda = 625$ nm is used.
    a. What is the work function of this surface?
    b. What is the threshold frequency for this surface?

44. What is the de Broglie wavelength of a proton traveling at 1.0×10^7 m/s?

45. A photon of a certain color of visible light has 2.8×10^{-19} J of energy. What color is the light?

46. A light source of wavelength λ illuminates a metal and ejects photoelectrons with a maximum kinetic energy of 1.00 eV. A second light source of wavelength $\frac{1}{2}\lambda$ ejects photoelectrons with a maximum kinetic energy of 4.00 eV. What is the work function of the metal?

Technology & Learning

Graphing calculators

Refer to Appendix B for instructions on downloading programs for your calculator. The program "Chap23" allows you to analyze a graph of speed versus the de Broglie wavelength for particles with a known mass and speed.

The de Broglie wavelength, as you learned earlier in this chapter, is described by the following equation:

$$\lambda = \frac{h}{mv}$$

The program "Chap23" stored on your graphing calculator makes use of the equation for the de Broglie wavelength. Once the "Chap23" program is executed, your calculator will ask for the mass and speed of the particle. The graphing calculator will use the following equation to create a graph of the de Broglie wavelength (Y_1) versus the speed (X). The relationships in this equation are the same as those in the force equation shown above.

$$Y_1 = 6.63E^-34/(MX)$$

a. If M is 1×10^{-30} and X is 1×10^3, what Ymax [WINDOW] setting would you choose?

Execute "Chap23" on the [PRGM] menu and press [ENTER] to begin the program. Enter the values for the mass and speed of the particle (shown below), pressing [ENTER] after each value. (Remember to use the [(-)] key for entering negative exponents.)

The calculator will provide a graph of the particle's speed versus its wavelength. (If the graph is not visible, press [WINDOW] and change the settings for the graph window, then press [GRAPH].)

Press [TRACE] and use the arrow keys to trace along the curve. The x value corresponds to the speed in meters per second, and the y value corresponds to the wavelength in meters.

Determine the de Broglie wavelength for the following particles:

b. an electron (m = 9.109×10^{-31} kg) with a speed of 3.00×10^3 m/s
c. the same electron with a speed of 70.0 m/s
d. a neutron (m = 1.675×10^{-27} kg) with a speed of 3.00×10^3 m/s
e. the same neutron with a speed of 70.0 m/s
f. The mass of a proton is 1.673×10^{-27} kg. If protons and electrons move at the same speed, would a proton's y value be higher or lower than an electron's?

Press [2nd] [QUIT] to stop graphing. Press [ENTER] to input a new value or [CLEAR] to end the program.

47. A 0.50 kg mass falls from a height of 3.0 m. If all of the energy of this mass could be converted to visible light of wavelength 5.0×10^{-7} m, how many photons would be produced?

48. Red light ($\lambda = 670.0$ nm) produces photoelectrons from a certain material. Green light ($\lambda = 520.0$ nm) produces photoelectrons from the same material with 1.50 times the previous maximum kinetic energy. What is the material's work function?

49. Find the de Broglie wavelength of a ball with a mass of 0.200 kg just before it strikes the Earth after it has been dropped from a building 50.0 m tall.

50. How many photons are emitted every 1.00 s by a 100.0 W sodium lamp if the wavelength of sodium light is 589.3 nm?

51. From the scattering of sunlight, Thomson found that the classical radius of the electron is 2.82×10^{-15} m. If sunlight with an intensity of 5.00×10^{2} W/m^2 falls on a disk with this radius, estimate the time required to accumulate 1.0 eV of energy. Assume that light is a classical wave and that the light striking the disk is completely absorbed. How does your estimate compare with the observation that photoelectrons are emitted within 10^{-9} s?

52. Ultraviolet light is incident normally on the surface of a substance that has a work function of 3.44 eV. The incident light has an intensity of 0.055 W/m^2, and the electrons are photoelectrically emitted with a maximum speed of 5.2×10^{5} m/s. How many electrons are emitted from a square centimeter of the surface every 1.0 s? Assume that 100.0 percent of the photons are absorbed.

53. The wave nature of electrons makes an electron microscope, which uses electrons rather than light, possible. The resolving power of any microscope is approximately equal to the wavelength used. A resolution of approximately 1.0×10^{-11} m would be required in order to "see" an atom.

    a. If electrons were used, what minimum kinetic energy of the electrons (in eV) would be required to obtain this degree of resolution?

    b. If photons were used, what minimum photon energy would be required?

Alternative Assessment

Performance assessment

1. Calculate the de Broglie wavelength for an electron, a neutron, a baseball, and your body, at speeds varying from 1.0 m/s to 3.0×10^{7} m/s. Organize your findings in a table. The distance between atoms in a crystal is approximately 10^{-10} m. Which wavelengths could produce diffraction patterns using crystal as a diffraction grating? What can you infer about the wave characteristics of large objects? Explain your conclusions.

2. A student claims that any flame that has a blue color must always be hotter than any flame that is yellow. Is this always true? Can the same inference be made when comparing the temperatures of bluish stars and reddish stars? Research the topic to prepare for a classroom debate.

Portfolio projects

3. You have been hired by a children's science museum to prepare an exhibit on solar panels and the way they work. Research the structure, size, efficiency, costs, and limitations of solar panels. Then prepare either a model or a detailed plan for the exhibit.

4. Bohr, Einstein, Planck, and Heisenberg each received the Nobel Prize for their contributions to twentieth-century physics. Their lives were also affected by the extraordinary events of World War II. Research their stories and the ways the war affected their work. What were their opinions about science and politics during and after the war? Write a report about your findings and about the opinions in your groups regarding the involvement and responsibility of scientists in politics.

47. 3.7×10^{19} photons

48. 0.80 eV

49. 1.06×10^{-34} m

50. 2.96×10^{20} photons/s

51. 1.3×10^{7} s; the estimate is much greater than the observation.

52. 8.2×10^{12} electrons/s

53. a. 1.5×10^{4} eV (15 keV)
    b. 1.2×10^{5} eV (120 keV)

Alternative Assessment
ANSWERS

Performance assessment

1. Students' tables should show that wavelength decreases as an object's mass or speed increases. Wavelengths should be near 10^{-10} m for diffraction in a crystal.

2. For a blackbody, it is true that higher frequency light will be emitted at higher temperatures, but flame color can be due to electron transitions instead.

3. Student answers will vary, but should be clearly written and include information about sources. The cells are made of two layers of semiconductor materials that constrain electrons to flow in one direction through an external circuit. The efficiency of these cells is less than 25%.

4. Student answers will vary. A good resource is *Heisenberg Probably Slept Here: The Lives, Times, and Ideas of the Great Physicists of the 20th Century,* by Richard Brennan (NY: Wiley, 1997).

Planning

Recommended time:

1 lab period

Classroom organization:

▸ You may conduct this as a demonstration for the entire class, or groups of two or more may perform the lab independently.

▸ **Safety warnings:** Emphasize the dangers of working with electricity. Remind students to have you check their circuits before turning on the power supply or closing the switch.

Techniques to Demonstrate

Give students a brief demonstration of the apparatus. Point out the parts, and explain how each item works.

✔ Checkpoints

Step 4: Make sure the voltage-adjustment knob is turned until the current meter returns to its initial value. At this point, the current has been stopped and the current meter can be adjusted to read zero.

Step 7: Repeat five times, and find the average value for the stopping voltage.

CHAPTER 23
Laboratory Exercise

★ TEKS

pp. 860–861: 1A, 2B, 2C, 2E, 2F, 5B, 9A

OBJECTIVES

• Use a phototube and photoelectric-effect device to observe and measure the photoelectric effect.

• Use the measured value for the stopping voltage to calculate the maximum kinetic energy of the emitted electrons.

MATERIALS LIST

✔ adhesive tape
✔ black construction paper
✔ insulated connecting wire
✔ lens supports
✔ light sources and power supply
✔ meterstick and meterstick supports
✔ multimeter or dc voltmeter
✔ photoelectric effect device with amplifier and filters
✔ wood block

THE PHOTOELECTRIC EFFECT

In this laboratory, you will explore the photoelectric effect.

SAFETY

• **Avoid looking directly at a light source. Looking directly at a light may cause permanent eye damage.**

• **Never close a circuit until it has been approved by your teacher. Never rewire or adjust any element of a closed circuit. Never work with electricity near water; be sure the floor and all work surfaces are dry.**

• **If the pointer on any kind of meter moves off scale, open the circuit immediately by opening the switch.**

PREPARATION

1. Read the entire lab, and plan what measurements you will take.

2. Prepare a data table with thirteen columns and four rows. In the first row, label the columns *Filter*, I_1 *(mA)*, ΔV_1 *(V)*, I_2 *(mA)*, ΔV_2 *(V)*, I_3 *(mA)*, ΔV_3 *(V)*, I_4 *(mA)*, ΔV_4 *(V)*, I_5 *(mA)*, ΔV_5 *(V)*, I_6 *(mA)*, and ΔV_{stop} *(V)*. In the first column, label the second through fourth rows *Blue*, *Green*, and *Red*.

PROCEDURE

Stopping voltage

3. Set up the apparatus as shown in **Figure 23-23.** Connect the voltage meter to measure the potential difference across the photoelectric-effect device. ***Do not turn on the power supply or light source until your teacher approves your setup.***

4. When your teacher has approved your setup, read the current meter on the device. Turn on the power supply. Turn the knob on the power supply until the current meter on the photoelectric device reads 15 mA. Set the zero point by turning the voltage-adjustment knob on the photoelectric-effect device until the applied potential difference is enough to stop the photocurrent. (The current meter will go to the initial reading when the photocurrent is zero.) When you have found this point, adjust the "Zero adjust" until the current meter on the device reads zero.

5. Turn the voltage-adjustment knob counterclockwise until the voltage meter reads zero. Record 0 V as ΔV_1 in your data table. Adjust the power supply until the current meter on the photoelectric-effect device reads about 10 mA. Record this current in your data table as I_1 (mA).

6. Slowly increase the potential difference by turning the voltage-adjustment knob clockwise. When the current is 8.0 mA, record I_2 and ΔV_2 in your data table. Continue to increase the potential difference, and record I and ΔV when the current is 6.0 mA, 4.0 mA, and 2.0 mA.

7. When the current meter reads 0.0 mA, read ΔV. Find the lowest potential difference where current is zero. This is the *stopping voltage*. Decrease potential difference; repeat to find ΔV when the current goes to zero four more times. Average all values; record current as I_6 and potential difference as ΔV_{stop} in data table.

8. Turn the light source off, and replace the blue filter with the green filter. Turn on the light source, and repeat steps 5 through 7 for the green filter. Record all data in your data table.

9. Turn the light source off, and replace the green filter with the red filter. Replace the light source with the second light source. Place the light source close to the red filter. Turn on the light source, and repeat steps 5 through 7.

10. Clean up your work area. Put equipment away safely so that it is ready to be used again.

ANALYSIS AND INTERPRETATION

Calculations and data analysis

1. **Organizing data** For each trial, calculate the maximum kinetic energy (in eV) of the emitted electrons using the equation $KE_{max} = \Delta V_{stop} \times e$, where ΔV_{stop} is the stopping voltage and $e = 1.60 \times 10^{-19}$ C.

2. **Graphing data** For each trial, graph the kinetic energy of the electron versus the wavelength of the light source.

Conclusions

3. **Interpreting graphs** Use your graph to answer the following questions.

   a. What is the relationship between the kinetic energy of the emitted electrons and the wavelength of the light?

   b. What is the wavelength of light at which the kinetic energy is zero?

4. **Evaluating data** What is the relationship between the stopping voltage and the wavelength of light?

Figure 23-23
Step 4: When the photoelectric-effect device is operating, the current meter will probably not read zero even though there is no photocurrent. Set the meter to zero by starting a photocurrent and applying a potential difference to stop it. When the current meter returns to its original setting, set the meter to zero.

Step 5: Start a photocurrent in the device.

Step 6: Adjust the potential difference to set the photocurrent to different levels.

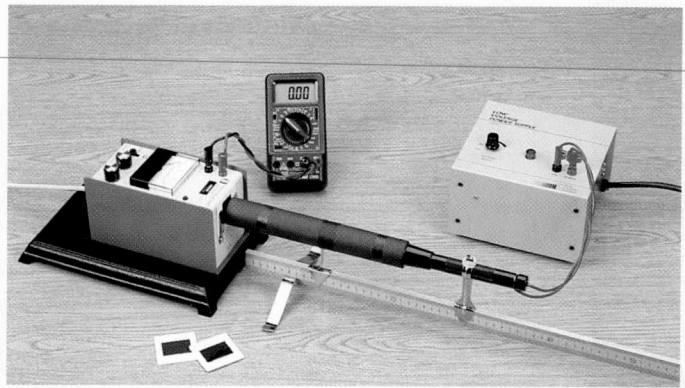

Step 9: The light source must be placed very close to the red filter for best results.

ANSWERS TO

Analysis and Interpretation

CALCULATIONS AND DATA ANALYSIS

1. blue: 2.88×10^{-20} J; green: 1.76×10^{-20} J; red: 8.0×10^{-21} J.

2. Student graphs should show a straight line pointing down and to the right.

CONCLUSIONS

3. a. The kinetic energy of the electrons decreases as the wavelength increases.

   b. Answers will vary. Students should find the answer by extending the line of the graph.

4. The stopping voltage decreases as the wavelength of light increases.

Physics and Its World *Timeline 1890–1950*

1895 – In Paris, the brothers **Auguste** and **Louis Lumière** show a motion picture to the public for the first time.

1898   Po, Ra

Marie and **Pierre Curie** are the first to isolate the radioactive elements polonium and radium.

1903 – **Wilbur** and **Orville Wright** fly the first successful heavier-than-air craft.

1905 $E_0 = mc^2$

Vol. 17 of *Annalen der Physik* contains three extraordinarily original and important papers by **Albert Einstein**. In one paper he introduces his special theory of relativity. In another he presents the quantum theory of light.

1912 – **Henrietta Leavitt** discovers the period-luminosity relation for variable stars, making them among the most accurate and useful objects for determining astronomical distances.

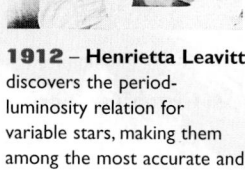

1913 $E_n = \frac{13.6}{n^2} \text{ eV}$

Niels Bohr—building on the discoveries of **Ernest Rutherford** and **J. J. Thomson,** and the quantum theories of **Max Planck** and **Albert Einstein**—develops a model of atomic structure based on energy levels that accounts for emission spectra.

1914 – World War I begins.

1922 – **James Joyce's** *Ulysses* is published.

1926

$$p = \frac{h}{\lambda}$$

Erwin Schrödinger uses the wave-particle model for light and matter to develop the theory of wave mechanics, which describes atomic systems. About the same time, **Werner Heisenberg** develops a mathematically equivalent theory called quantum mechanics, by which the probability that matter has certain properties is determined.

1929 – The New York Stock Exchange collapses, ushering in a global economic crisis known in the United States as The Great Depression.

1937 – **Pablo Picasso** paints *Guernica* in outraged response to the Nazi bombing of that town during the Spanish Civil War.

1938

$$^{1}_{0}n + ^{235}_{92}U \rightarrow ^{141}_{56}Ba + ^{92}_{36}Kr + 3\,^{1}_{0}n$$

Otto Hahn and **Fritz Strassman** achieve nuclear fission. Early the next year, **Lise Meitner** and her nephew **Otto Frisch** explain the process and introduce the term *fission* to describe the division of a nucleus into lighter nuclei.

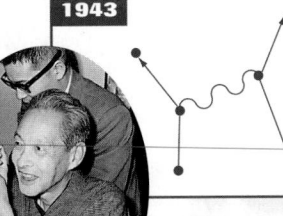

1939 – World War II begins with the Nazi invasion of Poland.

1943

Shin'ichiro Tomonaga develops quantum electrodynamics, which describes the interactions between charged particles and light at the quantum level. The theory is later independently developed by **Richard Feynman** and **Julian Schwinger**.

1948 – **Martin Luther King Jr.** graduates from Morehouse College and enters Crozer Theological Seminary where he becomes acquainted with the principles of **Mohandas Gandhi**. During the next two decades he becomes one of the most forceful and articulate voices in the US civil rights movement.

1920

1930

1940

1950

CHAPTER 24 PLANNING GUIDE

Compression Guide: To shorten the overall length of your teaching year by 7 45-minute periods or 3½ 90-minute blocks, eliminate this chapter (the items in magenta type).

PACING CHART	CLASSROOM RESOURCES			
	(★)TEKS	Teacher Demonstrations	*Holt Physics* Transparencies	Labs (See page T52 for equipment listing for in-text labs.)
24-1 Conduction in the solid state 1 45-minute period ½ 90-minute block	3A	**TE** *Orbital shapes,* p. 866	**TM** 79–81	
24-2 Semiconductor applications 3 45-minute periods 1½ 90-minute block	2C, 3A, 3C, 3E, 5B, 6E	**TE** *Hole flow,* p. 872 **TE** *n-type semiconductors,* p. 873 **TE** *p-type semiconductors,* p. 874 **TE** *p–n junction,* p. 875 **TE** *Diode current,* p. 876 **TE** *Mechanical amplification,* p. 877 **TE** *Optical amplification,* p. 878	**T** 122–125 **TM** 82–85	**PE** *Resistors and Diodes,* p. 889
24-3 Super-conductors 1 45-minute period ½ 90-minute block	2C, 3A, 3C, 3E, 8C	**TE** *Lattice imperfections,* p. 881 **TE** *Cooper pairs,* p. 882 **TE** *Meissner effect,* p. 883		
Review and Assessment 2 45-minute periods 1 90-minute block				

Resource Key

PHYSICS

PE Pupil's Edition
TE Teacher's Edition

L Laboratory Experiments
TL Technology Lab Experiments
T Transparencies

One-Stop Planner CD-ROM **contents**

TM Transparency Masters
SR Section Review Worksheets
AA Alternative Assessment

PW Problem-Solving Workbook
PB Problem Bank
CTW Critical Thinking Worksheet

LABORATORY PLANNING: Force and Acceleration, p. 158

Materials (for each lab group)
- 2 basic digital multimeters or
   2 student multimeters, or
   2 general purpose multimeters, or
   1 CENCO standard movement, triple range dc ammeter (50 mA, 500 mA, 5 A) **and** CENCO standard movement, triple range dc voltmeter (0–3, 10, 15 V)

- contact key
- decade resistance box or
   resistance substitution apparatus
- measuring resistor, 1 kΩ
- patch cord, black, insulated alligator clip/stacking banana plug
- patch cord, red, insulated alligator clip/stacking banana plug

ASSIGNMENT RESOURCES

Content Mastery	Critical Thinking	Problem-Solving Practice
PE 1–3, p. 871 **SR** 24-1, *Concept Review* **PE** 1–4, p. 886	**PE** 5–6, p. 886	
PE 1–6, p. 880 **SR** 24-2, *Concept Review* **PE** 7–17, pp. 886–888	**PE** 18–19, p. 888	
PE 1–4, p. 884 **SR** 24-3, *Concept Review* **PE** 20–22, p. 888	**PE** 23–24, p. 888	

ASSESSMENT RESOURCES

Cumulative Review	Alternative Assessment	Traditional Assessment
SR Mixed Review, Ch. 24	**PE** 1–4, p. 888 **AA** Items for Ch. 24	Chapter 24 Test Test Generator items for Ch. 24

Scoring Rubrics for Alternative Assessment items can be found on the One-Stop Planner CD-ROM.

TECHNOLOGY RESOURCES

CTW Segment 23 Portable Solar Power

PE Technology and Learning, p. 887
(Alternative procedures for calculators without Flash-ROM technology are provided on the One-Stop Planner CD-ROM.)

internet connect

On-line Student Resources:
GO TO: www.scilinks.org
The following *sci*LINKS Internet resources can be found in the student text for this chapter.

TOPICS:
• Semiconductors, p. 873 (HF2242)
• Energy levels, p. 876 (HF2241)
• Transistors, p. 878 (HF2243)
• Superconductors, p. 882 (HF2244)

On-line Teacher Resources:
GO TO: go.hrw.com
KEYWORD: HF2 HOME
Visit the HRW Web site for a variety of resources related to this chapter.

Smithsonian Institution
Internet Connections
Visit **www.si.edu/hrw** for additional on-line resources.

CNNfyi.com
Visit **www.cnnfyi.com** for late-breaking news and current events stories selected just for you.

• power supply: 4-output dc regulated power supply, or CENCO 6 V ac/dc low-voltage power supply CENCO universal power supply
• silicon diode (1N 4001)
• alternatives for more trials: germanium diode (1N34A)
  light emitting diode, dual color (forward voltages, red 2.0 V, green 2.1 V)
  light emitting diode, red (forward voltage 2.0 V)
• wire leads with alligator clips, 60 cm (pkg. of 10)

Materials Preparation
See page 692A–692B for instructions on using multimeters to measure current and voltage.

Section 24-1 discusses conductors, insulators, and semiconductors; introduces electron energy levels; identifies valence electrons; and describes the role of energy bands in conductivity.

Section 24-2 compares the roles of electrons and positively charged holes in conducting current; discusses the process of doping; examines semiconductor types, including n-type, p-type, diodes, and transistors; and explains how a transistor can be used as an amplifier.

Section 24-3 explores superconductors through the BCS theory of superconductivity and examines applications of superconductors.

About the Illustration

This photograph shows a newly imprinted silicon wafer being probed to test the integrated circuits imprinted on it. The tweezers shown in the gloved hand are used to move the wafer in and out of the testing device. Point out to students that the tweezers are coated and the worker's hand is in a rubber glove. These precautions are taken to prevent any stray particles from contaminating the wafer.

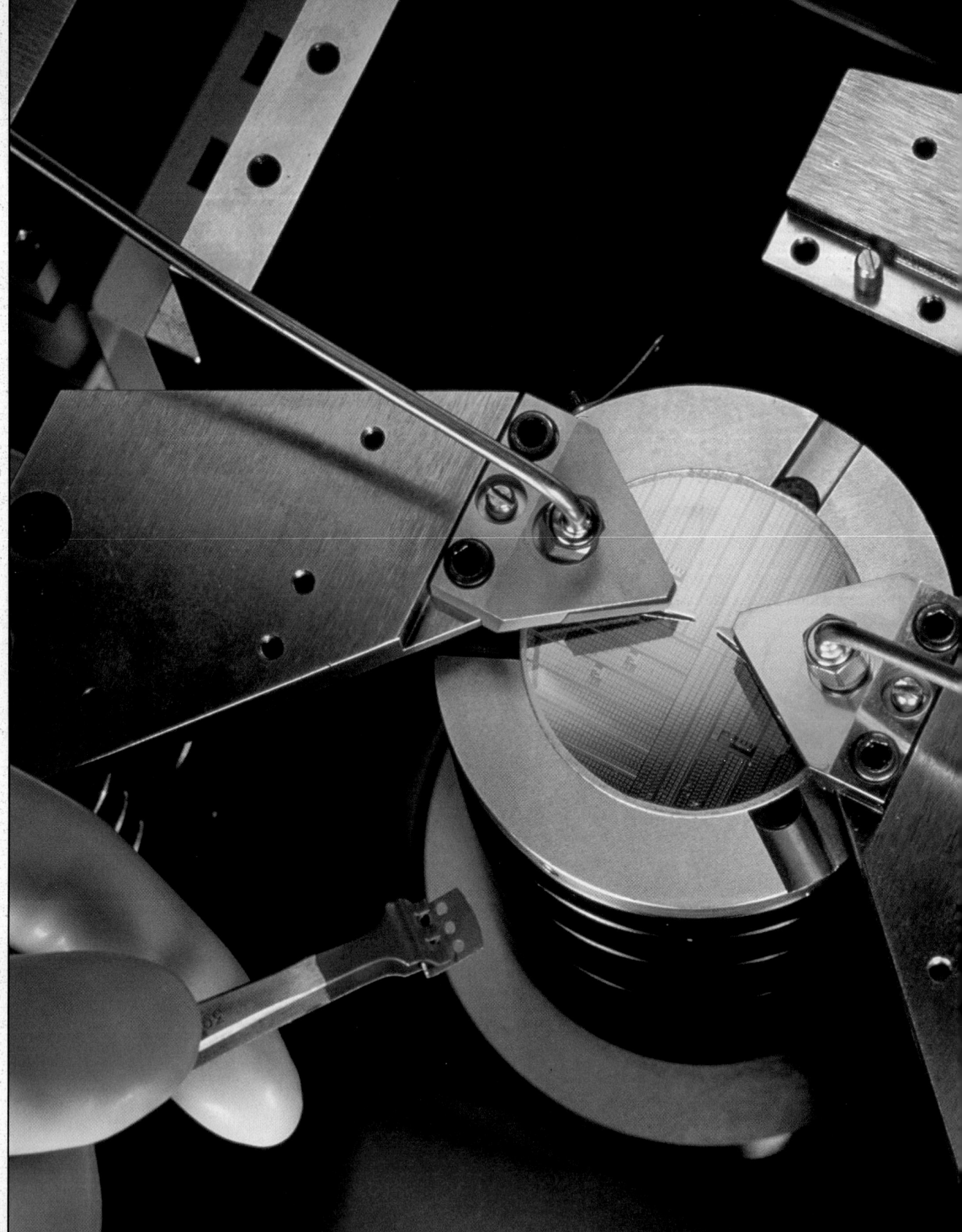

CHAPTER 24

Modern Electronics

PHYSICS IN ACTION

The wafer shown here is made of purified polycrystalline silicon derived from sand. Although silicon is usually a semiconductor, the addition of a few atoms of an impurity can alter its conductivity dramatically. Careful addition of impurities, along with precise etching, can create tiny integrated circuits with specific electrical characteristics on the wafer's surface. If no flaws are found during the testing process, shown here, the wafer will be cut up into individual chips with a diamond saw and packaged for use in electronic devices.

- *What determines whether a material is a conductor, semiconductor, or insulator?*

- *How can components made of semiconductors be used to control current?*

CONCEPT REVIEW

Work (Section 5-1)

Conductors, insulators, and semiconductors (Section 17-1)

Electric current (Section 19-1)

Tapping Prior Knowledge

Knowledge to Expect
✔ "Electrical circuits provide a means of transferring electrical energy when heat, light, sound, and chemical changes are produced." (NRC's *National Science Education Standards,* grades 5–8)

✔ "Information can be carried by many media including sound, light, and objects. In this century, the ability to code information as electric currents in wires, electromagnetic waves in space, and light in glass fibers has made communication millions of times faster than is possible by mail or sound." (AAAS's *Benchmarks for Science Literacy,* grades 6–9)

Knowledge to Review
✔ Work is the product of the magnitudes of displacement and the component of force in the direction of displacement ($W = Fd\cos\theta$). (Section 5-1)

✔ Electric current is the rate at which electric charges move through a given area $\left(I = \dfrac{\Delta q}{\Delta t}\right)$. (Section 19-1)

✔ Materials may be classified as conductors, insulators, or semiconductors, according to their ability to conduct electric charge. (Section 17-1)

Items to Probe
✔ Electric fields: Have students describe in their own words the effect of an electric field on a charged particle.

✔ Magnetic fields: Have students describe the relationship between magnetic fields and electric charges.

24-1
Conduction in the solid state

Demonstration 1

Orbital shapes

Purpose Give a visual example of the shape of electron shells.

Materials tennis ball on a string

Procedure To reinforce the idea that electrons do not orbit the nucleus as planets orbit the sun, have students look at the shapes pictured in **Figure 24-1.**

Explain that a tennis ball on a string is a limited model of an electron shell.

Then, holding one end of the string, whirl the tennis ball to demonstrate different electron "shells," such as a horizontal circle, a vertical circle, an alternating vertical circle, and an alternating horizontal circle. The alternating horizontal circle (one pass below your hand and arm and one pass above) may take some practice. These patterns can be loosely linked to shells. Additionally, the radius of the pattern can be varied to show different shell sizes.

internet connect

SciLINKS
NSTA

TOPIC: Semiconductor Industry in Texas
GO TO: www.scilinks.org
*sci***LINKS CODE:** HFX007

24-1 SECTION OBJECTIVES

- **Distinguish between conductors, insulators, and semiconductors.**

- **Identify valence electrons.**

- **Describe the role of energy bands in electrical conductivity.**

CLASSIFICATION OF SOLIDS

As you learned in Chapter 17, materials can be classified according to their ability to conduct electricity. A good *conductor* has a large number of free charge carriers that can move easily through the material, whereas an *insulator* has a small number of free charge carriers that are relatively immobile. *Semiconductors* exhibit electronic properties between those of insulators and those of conductors. There is a large variation in electrical conductivity of conductors, insulators, and semiconductors. That variation may be due to energy bands. This section will help explain those energy bands.

One of the accomplishments of *solid-state physics* is the development of a theory that uses basic physical principles to explain some of the properties of these three categories of materials.  **3A**

A more sophisticated model of atoms and solids is needed

The models that we have discussed up to this point have been of solids represented as a collection of positively charged nuclei surrounded by their associated electrons. This simple model is not very accurate. For example, it does not explain why electrons are sometimes bound and sometimes free. Further, it does not explain why the ability to conduct electricity differs between conductors and insulators as groups. The model also lacks an explanation of the differences among individual conductors and insulators themselves. Better models are needed both for the atom and for a solid. We will attempt to develop better models in this section.

Positively charged protons and neutral neutrons are contained in a small, dense nucleus at the center of an atom. The nucleus is surrounded by negatively charged electrons in a series of shells, such as those shown in **Figure 24-1.** As a whole, the atom is uncharged.

Figure 24-1
These are approximate shapes and sizes for the regions of space containing electrons of certain energy levels.

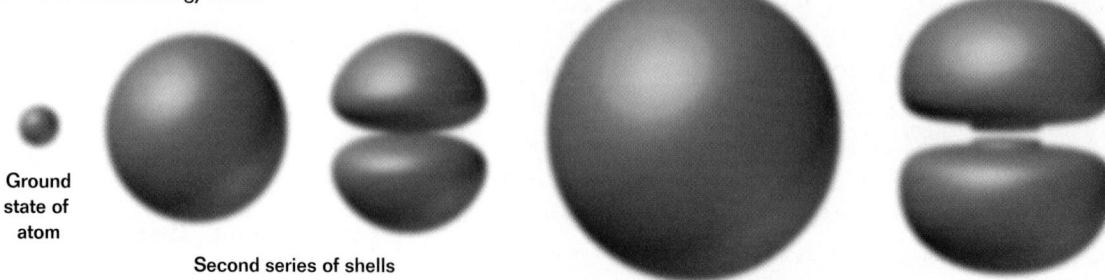

Ground state of atom

Second series of shells

Third series of shells

Valence electrons determine an atom's chemical properties

The number of electrons in an atom can range from 1 to more than 100, depending on the type of atom. The complicated arrangement of electrons around the positively charged nucleus can be simplified. To do so, use a model that groups the electrons into sets, or shells, each within a space having a specific shape, such as that of a sphere or a dumbbell (see **Figure 24-1**).

As the distance of the electron from the positively charged nucleus increases, the electrostatic force between the electron and the nucleus decreases and the electron becomes more loosely bound to the atom. This result explains in part why the valence electrons (the outer shell of the atoms) are more loosely bound than the inner shells.

Because the electrons in the outermost shell, called the **valence electrons,** are the most loosely bound, they are the ones that can interact more strongly with other atoms. As a result, the behavior of an atom's valence electrons determines the chemical properties of the atom. The inner electron shells and the nucleus can be thought of as a single point-charge surrounded by valence electrons.

Electrons occupy energy levels

As we saw in our discussion of the Bohr atom, the electrons in an atom can possess only certain amounts of energy. For this reason, the electrons are often said to occupy specific *energy levels*. Electrons in a shell sometimes form a set of closely spaced energy levels. Normally, electrons are in the lowest energy level available to them. The specific arrangement of electrons in which all are in the lowest possible energy levels of an atom is called the atom's **ground state.**

As described in Chapter 23, an atom can sometimes absorb energy from the environment. If the available energy is sufficient, one of the atom's electrons can move to a higher energy level. When this happens, the atom is said to be in an **excited state.** The electron may also absorb so much energy that it is no longer bound to the atom. The electron is then called a *free electron*.

BAND THEORY

One model that can be used to help understand why solids fall into the three categories of conductor, insulator, and semiconductor is called *band theory*. Band theory can explain the mechanisms of conduction in many solids and the large variation in the electrical conductivity of these materials.

When identical atoms are far apart, they have identical energy-level diagrams and identical wave functions. As the atoms are brought closer together, their wave functions overlap. Because no two electrons in the same system can occupy the same state, the energy level in an atom is altered by the influence of the electric field of another atom. In the case of two atoms, each energy level

CONCEPT PREVIEW

The structure within the nucleus will be studied in greater detail in Chapter 25.

valence electron

an electron in the outermost shell of an atom

ground state

the lowest energy state of a quantized system

excited state

the state of an atom that is no longer in its ground state

Teaching Tip

Valence electrons are the most loosely bound electrons for two reasons. First, the larger distance between the valence electrons and the nucleus corresponds to a weaker electric force because electric force is inversely proportional to distance squared. Second, the shielding effect of the electrons that are closer to the nucleus makes valence electrons less tightly bound than the electrons closer to the nucleus.

Key Models and Analogies

Be sure students understand the difference between the classification of materials into categories based on their ability to conduct charge (as was done in Chapter 17) and the physical characteristics that lead to the different classifications. Remind them that electrical attraction and repulsion can be explained at the atomic level by considering the transfer of electrons between objects. Then tell them that the ability of some materials to conduct charge better than others is explained at the atomic level by band theory.

⊛ TEKS

p. 866: 3A

Figure 24-2

Be sure students understand that the third case, **(c)**, is simply an extension of the second case, **(b)**, in which the energy levels are so close together that they are represented as continuous bands.

Q What causes the splitting of energy levels in each of these cases?

A *When more than one atom is present, the electric field produced by each atom affects the energy levels of other atoms, causing the splitting shown in the figure.*

Q Could splitting be observed for a single atom?

A *no*

Figure 24-2
Energy levels split when two atoms are close together **(a)**. Adding a few more nearby atoms causes further splitting **(b)**. When many atoms interact, the energy levels are so closely spaced that they can be represented as energy bands **(c)**.

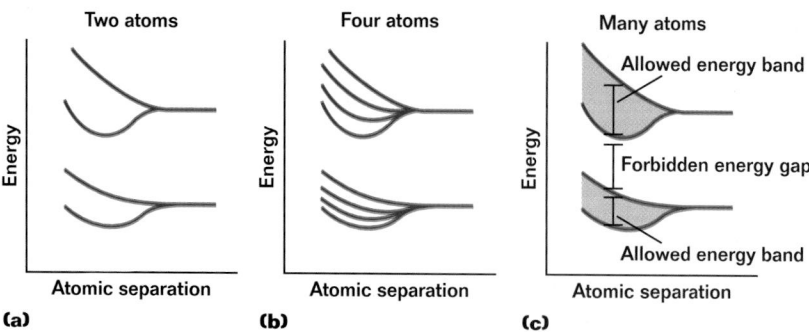

Two atoms | Four atoms | Many atoms

Energy | Energy | Energy

Atomic separation | Atomic separation | Atomic separation

(a) **(b)** **(c)**

splits into two different energy levels. **Figure 24-2** shows the splitting of two energy levels as two atoms get closer. Notice that the energy difference between two new energy levels depends on the distance between the atoms.

When more atoms are brought close together, each energy level splits into more levels. The number of splittings depends on the number of interacting particles. If there are many atoms, the energy level splits so many times and the new energy levels are so closely spaced that they may be regarded as a continuous band of energies.

When atoms are close to each other, some energy levels become energy bands

When atoms are bound together in a solid, the clearly defined energy levels of a single atom widen and blur into *energy bands*, like those shown in **Figure 24-3**.

The most important energy band is the highest band containing occupied energy levels, known as the *valence band*. In solids that are semiconductors and insulators, the valence electrons do not conduct electricity. On the other hand, in solids that are conductors, the valence electrons are able to conduct electricity.

There can be many more bands of lower energy than the valence band. These bands are completely filled. They are of little consequence in determining the electrical properties of the solid.

There are additional unoccupied bands of higher energy besides the valence band. In semiconductors and insulators, the band immediately above the valence band is called the *conduction band*. If an atom is excited, valence electrons can sometimes move to this higher energy level. Then they can contribute to the conduction of electricity, as we will see.

There is often an energy gap between bands

The range of energies lying between the valence band and the conduction band is called the **band gap,** or *energy gap,* as shown in **Figure 24-3**. An electron in an insulator or semiconductor cannot have a value for its energy that would lie within the band gap. Because these energies are not available for an electron in the insulator or semiconductor, they are often referred to collectively as "the forbidden energy gap."

The size of the band gap differs with different materials, resulting in different properties. In an insulator, the gap is so large that it is unlikely that an electron

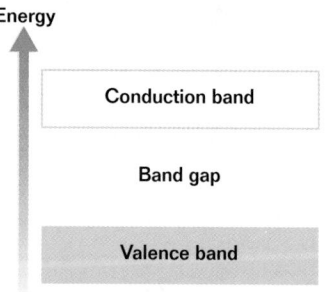

Energy

Conduction band

Band gap

Valence band

Figure 24-3
Energy levels of atoms become energy bands in solids. The valence band is the highest occupied band.

band gap

the minimum energy separation between the highest occupied state and the lowest empty state

could ever gain enough energy to move from the valence band to the conduction band. In a semiconductor, the gap is somewhat smaller, making it more likely that an electron could move to the conduction band. In a conductor, the valence band is only partially filled, so there is essentially no gap between filled levels and available unoccupied levels.

A band can be full or partially full

Electrons in a solid normally occupy the lowest available energy levels within an atom. As electrons fill the energy levels of an atom, they occupy the lowest levels first. Normally there are no electrons in higher energy levels of an atom unless all of the lower energy levels are completely filled.

In a solid, if there are more electrons than there are energy levels in the lowest energy band, then the lowest energy band is full. Because no more electrons can move into an already full band, any additional electrons must occupy energy levels in the next higher band.

In some materials, there are more energy levels in a band than there are electrons to fill them. In such cases, the band remains partially full. Whether the highest energy band is completely full or only partially full is very important in determining the electrical properties of a material. In other words, partially full energy bands have different electrical properties than energy bands which are completely full.

Electrons can move between energy levels in a solid

Recall from Chapter 23 that when an electron in an atom absorbs energy, it can be excited to a higher energy level. However, the electron can absorb only certain amounts of energy that correspond to the differences in energy between the energy levels in the atom.

A similar process can occur in a solid. Electrons that absorb energy can be excited to higher energy levels. An electron must be given sufficient energy to occupy some higher available energy level.

Figure 24-4 is a schematic diagram showing the three highest bands in a solid with an energy gap between each band. Remember that even though these bands are pictured as one solid color, they are actually made up of closely spaced energy levels for electrons. All of the energy levels in the bottom two bands are completely filled with electrons.

What transitions are possible for an electron in this solid? Because the only unfilled energy levels are in the conduction band, only the transitions represented by (a) and (b) are possible. However, transition (a) occurs much more easily than transition (b) does because transition (a) requires less energy. That is why the bands below the valence band do not appreciably affect the electrical properties of a material; the energy required for an electron in one of these low-lying bands to be excited into a higher unfilled energy level is usually too large. So, without sufficient energy, the electron will remain on a lower level and will not affect the electrical properties as much as valence electrons do. ⭐TEKS 3A

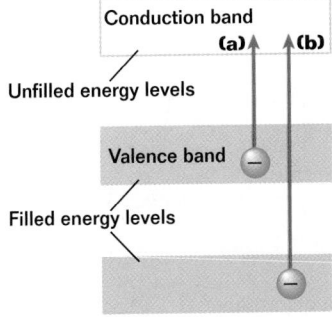

Figure 24-4
Only transitions into unfilled energy levels, such as those represented by **(a)** and **(b)**, are possible for electrons in a solid. However, transition **(b)** requires such a large amount of energy that it does not easily occur.

Teaching Tip
Point out that the forbidden energy levels lying in the energy gaps between the allowed energy levels are a consequence of quantum mechanics. As seen in Chapter 23, Bohr first formulated the idea of allowed energy levels in his model of the hydrogen atom. According to Bohr's theory, electrons in the hydrogen atom "jump" from one energy level to the next but can never be found between the allowed, or stable, levels.

Visual Strategy

Figure 24-4
Point out to students that although different transitions are possible, some are more likely than others.

Q Why is transition **(a)** more likely to occur than transition **(b)**?

A *Transition (a) is more probable because it requires less energy than transition (b).*

870

Visual Strategy

Figure 24-5

Have students visualize the band gap as a fence that electron "horses" must jump before they can wander into the open field of the conduction band.

Q Why do semiconductors conduct more easily than insulators?

A *The "fence" (the band gap) of a semiconductor is much smaller than that of an insulator, so the "horses" (the electrons) do not need to exert as much effort (energy) to cross.*

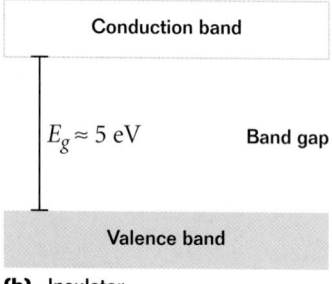

(a) Conductor

(b) Insulator

(c) Semiconductor

Figure 24-5
A conductor **(a)** has a partially filled valence band. Insulators **(b)** and semiconductors **(c)** have empty conduction bands and filled valence bands, but the band gap in a semiconductor is smaller than it is in an insulator.

CONDUCTION AND ELECTRON TRANSITIONS

So far, we have considered the band structure of a solid and the way an electron can make transitions between the energy levels within or between these bands. These transitions do not represent a physical motion of the electrons. Rather, they represent changes in the energy of electrons.

The physical motion of electrons in a solid—the conduction of electricity—depends on the arrangement of the electrons in the bands of the solid, because a moving electron moves to an unfilled final energy level.

Bound electrons need energy to escape the electrostatic forces that bind them to an individual atom. Electrons can be excited into higher energy levels by either of two important mechanisms: the application of an electric field or thermal excitation.

When an electric field is applied to a solid, the electric field does work on individual electrons, giving them enough energy to move to higher energy levels within the atom.

The absorption of thermal energy can also excite electrons in a solid. The atoms in a solid experience random vibrations (thermal energy) that can occasionally transfer enough energy to an electron to excite it to a higher energy level. At room temperature, a few electrons near the unfilled energy levels are excited to unoccupied levels. The thermal energy available for this is very small, so only the uppermost electrons can move across the band gap.

A conductor has a partially filled valence band

If the valence band overlaps the conduction band, the solid is a *conductor*. For example, **Figure 24-5(a)** shows a valence band that is also the conduction band. The band is actually partly filled. The uppermost filled energy level lies in the middle of the band. If energy is added, electrons can be excited from filled energy levels to one of many higher unfilled energy levels. ⭐TEKS **3A**

A solid conductor has many electrons that require only a small amount of energy—either from the applied field or by thermal excitation—to be moved into nearby unfilled energy levels. These electrons are then able to move freely in the conductor with only a small applied field. An important feature of a conductor is the partially filled conduction band. The rest of a conductor's band structure, including the size of the band gaps, is of little importance.

An insulator has a full valence band and a large energy gap

An *insulator* has an empty conduction band and a full valence band, as shown in **Figure 24-5(b).** It also has a very large band gap (about 5–10 eV) between the conduction band and the valence band. In fact, the band gap is so large that it is difficult for an electron to gain enough energy to be excited into the conduction band. ⭐TEKS **3A**

For example, at room temperature, thermal excitations typically give electrons about 0.025 eV, which is much smaller than the band gap. At such a temperature, very few electrons are thermally excited into the conduction band. Thus, although an insulator has many vacant states in its conduction band, there are so few electrons occupying these states that the overall electrical conductivity is very small, resulting in the high resistance typical of substances that are insulators.

A semiconductor has a full valence band and a small energy gap

Intermediate between a conductor and an insulator is a *semiconductor*. The valence band of a semiconductor, like that of an insulator, is full, as shown in **Figure 24-5(c).** However, the band gap in a semiconductor is much smaller than in an insulator (about 1 eV). In fact, the band gap in many semiconductors is small enough for electrons to be thermally excited into the conduction band rather easily. Thus, the conductivity of many semiconductors is strongly dependent on temperature.

For example, at temperatures near 0 K, most electrons are in the valence band and there is little thermal energy available for excitation. This makes semiconductors poor conductors at very low temperatures. At higher temperatures, however, a larger number of electrons can be thermally excited from the valence band to the conduction band, which has many empty energy levels. Because thermal excitation across the narrow gap is more probable at higher temperatures, the conductivity of semiconductors improves rapidly with temperature. The application of an electric field to a semiconductor also does work on electrons, increasing their energy so that they can conduct and create a current.  **3A**

Section Review
 **3A**

1. The resistance of conductors increases with increasing temperature, as described in Chapter 19. On the other hand, the resistance of a semiconductor decreases with increasing temperature. How can you explain this property of semiconductors?

2. Which of the following is likely to be a valence electron?
   a. the innermost electron of a uranium atom
   b. an electron in a calcium atom that is in the outermost energy level
   c. an electron in a bromine atom that is in the second-most outer energy level of the atom

3. Which band contains energy levels for electrons that are already free to move about within a semiconductor? Which band contains energy levels for electrons that cannot move within a semiconductor?

Teaching Tip
Ask students what factors determine whether a semiconductor acts as an insulator or a conductor (*the semiconductor's temperature and the strength of the applied electric field*).

Section Review
ANSWERS

1. At room temperature, most electrons in a semiconductor are in the valence band, so the semiconductor acts as an insulator. As temperature increases, electrons are excited to the conduction band, thereby increasing the semiconductor's conductivity.

2. b

3. conduction band; valence band

⊛ TEKS
p. 870: 3A, 3A
p. 871: 3A, 3A

Hole flow

Purpose Illustrate hole flow.
Materials nine rubber stoppers
Procedure Have 10 students stand facing the class with their right palms out. Place a stopper in the hand of every person except the student on the far right.

Beginning at the far right, have each student look to their right and place their stopper in their neighbor's palm if that person does not already have a stopper. All of the stoppers will shift one palm to the right, and the person on the far left will be without a stopper. Ask the students to consider the movement of the empty space as an electron "hole." Point out that the hole moves to the left as the stoppers (the electrons) move to the right.

24-2
Semiconductor applications

24-2 SECTION OBJECTIVES

- **Compare the roles of electrons and positively charged holes in conducting current.**

- **Describe the process of doping to create n- and p-type semiconductors.**

- **Analyze p-n junctions and their role in semiconductor devices.**

- **Explain the role of a diode as a rectifier.**

- **Explain how a transistor can be used as an amplifier.**

hole

an energy level that is not occupied by an electron in a semiconductor

Figure 24-6
An electric field can excite valence electrons into the conduction band, where they are free to move through the material. Holes in the valence band can then move in the opposite direction.

SEMICONDUCTOR DOPING

Charge carriers in a semiconductor can be negative or positive. To see why this is so, consider the valence and conduction bands of a semiconductor, as shown in **Figure 24-6.** Imagine that a few electrons are excited from the valence band to the conduction band by an electric field. The electrons in the conduction band are free to move through the material. Normally, electrons in the valence band are unable to move because all nearby energy levels are occupied. But when an electron moves from the valence band into the conduction band, it leaves a vacancy, or **hole,** in an otherwise filled valence band. Because this hole creates an empty energy level in the valence band, another valence electron from this or a nearby atom is free to move into the hole. Whenever an electron does so, a new hole is created at its former location. So, the net effect can be viewed as a hole migrating through the material in a direction opposite the motion of the electrons in the conduction band.

In a material containing only one element or compound, there are an equal number of conduction electrons and holes. Such combinations of charges are called *electron-hole pairs,* and a semiconductor that contains such pairs is called an *intrinsic semiconductor.* **Figure 24-6** is a schematic of an intrinsic semiconductor. In the presence of an electric field, the holes move in the direction of the field and the conduction electrons move opposite the field. Keep in mind that the motion of holes is always in the direction opposite the motion of the electrons.

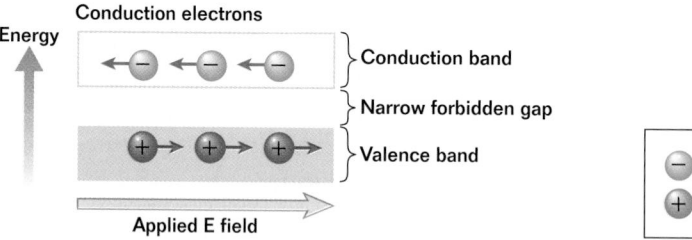

Doping adds impurities that enhance conduction

In the last section, it was explained that the concentration (the number per unit volume) of charge carriers in a semiconductor depends on temperature. Another way to change the concentration of charge carriers is to add *impurities,* atoms that are different from those of an intrinsic semiconductor.

Adding impurities is called **doping.** Only a few added impurity atoms (about one part in a million) can have a large effect on a semiconductor's resistance. The semiconductor's conductivity increases as the doping level increases. When impurities dominate conduction, the material is called an *extrinsic semiconductor.*

There are two methods for doping a semiconductor: either add impurities that have extra valence electrons or add impurities that have fewer valence electrons compared with the atoms in the intrinsic semiconductor.

Semiconductors used in commercial devices are usually doped silicon or germanium. These elements have four valence electrons. Semiconductors are doped by replacing an atom of silicon or germanium with one containing either three valence electrons or five valence electrons. Note that a doped semiconductor is electrically neutral because it is made of neutral atoms. The semiconductor atoms and the impurity atoms remain uncharged as before. The balance of positive and negative charges has not changed, but the number of charges that are free and able to move has. These charges are therefore able to participate in electrical conduction.

An n-type semiconductor has electrons as majority carriers

In **Figure 24-7(a),** an atom containing five valence electrons, such as arsenic, is added to a semiconductor. Four of the valence electrons participate in the bonding between neighboring atoms. One electron is left over. Such an impurity in effect donates an extra electron to the solid, so it is referred to as a *donor atom.* The impurity alters the band structure of the solid. Therefore, the extra electron occupies an energy level, or donor level, that lies just below the conduction band, as shown in **Figure 24-7(b).**

The energy spacings between donor levels and the bottom of the conduction band are very small (typically about 0.05 eV). So, only a small amount of thermal energy is needed to cause an electron in these donor levels to move into the conduction band. Semiconductors doped with donor atoms are called *n-type semiconductors.* The reason is that most of the charge carriers are electrons with negative charges. Positive holes that form in the donor level do not move very easily.

doping

the addition of impurity atoms to a semiconductor

internet connect

SCI**LINKS**

NSTA

TOPIC: Semiconductors
GO TO: www.scilinks.org
*sci***LINKS CODE:** HF2242

Did you know?

The average thermal energy of an electron at room temperature (22°C) is about 0.025 eV.

(★)**TEKS** **2C**

SECTION 24-2

Demonstration 3

n-type semiconductors

Purpose Illustrate an n-type semiconductor.

Materials 11 rubber stoppers

Procedure This demonstration is an extension of Demonstration 2. Have 10 students stand facing the class with both palms held out. Place a stopper in the right hand of everyone except for the student on the far left, who should have two stoppers.

Beginning with the student on the left, have each student look to their right and place one stopper in their neighbor's palm if they themselves have more stoppers than their neighbor. The extra stopper will shift to the right, and, at the end, the person on the far right will have an extra stopper. Have the students consider the movement of the extra stopper as that of an electron. The type of material demonstrated appears to be electron-rich, as in n-type semiconductors. Explain that the process of adding impurities to semiconductors is called doping.

(★)**TEKS**

p. 873: 2C

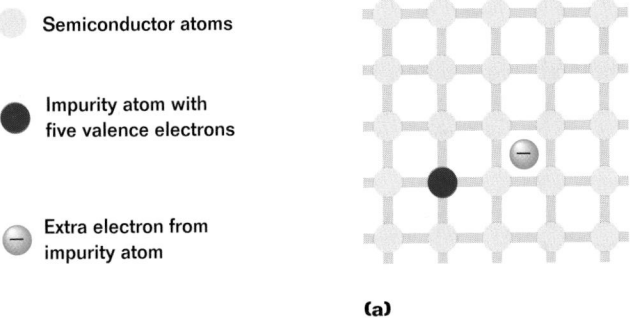

Semiconductor atoms

Impurity atom with five valence electrons

Extra electron from impurity atom

(a)

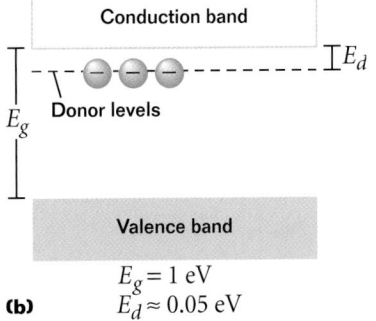

Conduction band

Donor levels

E_g

Valence band

$E_g = 1$ eV
$E_d \approx 0.05$ eV

(b)

Figure 24-7
An n-type semiconductor is doped with impurity atoms that have extra valence electrons.

p-type semiconductors

Purpose Illustrate a p-type semiconductor.

Materials 19 rubber stoppers

Procedure Repeat Demonstration 2, but this time place a stopper in the left and right hands of everyone except the student on the far right, who should have only one stopper. Beginning at the far right, have each student look to their right and place a stopper in their neighbor's palm if he or she has an empty hand. Eventually, the person on the far left will have one less stopper. Point out that the "hole" moved to the left as the stoppers moved to the right. The type of material demonstrated appears to be hole-rich, as in p-type germanium.

Tell students that approximately one atom of arsenic or gallium is added to every 100 million atoms of germanium. Have the students try to come up with a model to illustrate the size of these numbers.

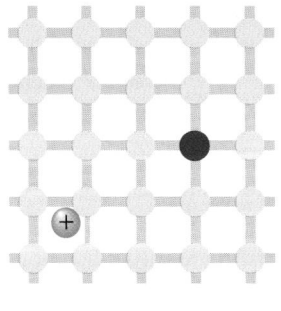

Semiconductor atoms

Impurity atom with three valence electrons

Hole, or electron deficiency in a bond

Figure 24-8
A p-type semiconductor is doped with impurity atoms having fewer valence electrons.

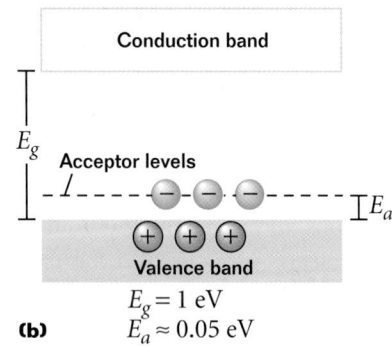

(a)   **(b)**

$E_g = 1$ eV
$E_a \approx 0.05$ eV

A p-type semiconductor has holes as majority carriers

If a semiconductor is doped with atoms containing three valence electrons, such as indium and aluminum, all three of these electrons form bonds with the neighboring atoms in the p-type semiconductor. This leaves an electron deficiency, or hole, in the fourth bond, as shown in **Figure 24-8(a).** The energy levels, or acceptor levels, of such impurities lie just above the valence band, as shown in **Figure 24-8(b).**

Electrons from the valence band have enough thermal energy at room temperature to fill these impurity levels, leaving a hole in the valence band. These holes can be filled by other nearby electrons, which leave new holes. Thus, positive charges (holes) can move throughout the material even if no electrons are in the conduction band.

Because such an impurity accepts an electron from the valence band, these impurities are referred to as *acceptors.* A semiconductor containing acceptors is known as a *p-type semiconductor* because the majority of the charge carriers are positively charged holes.

DIODES

diode

an electronic device that allows electric current to pass more easily in one direction than in the other

A **diode** is a device that has an almost infinite resistance in one direction and nearly zero resistance in the other direction. So, a diode is able to pass current in only one direction, similar to the way a one-way valve passes water into a pipeline. A diode used in this manner is called a *rectifier.* **Figure 24-9** shows the electrical circuit symbol for a diode. The arrow that is a part of the circuit symbol for the diode indicates the direction of the current in the diode under most situations.

The p-n junction is the contact between a p-type semiconductor and an n-type semiconductor

Figure 24-9
The circuit diagram symbol for a diode indicates that it allows electric current in one direction only.

Consider what happens when a p-type semiconductor is joined to an n-type semiconductor to form a *p-n junction.* A device having one p-n junction can be used as a diode in electrical circuits. It will allow current to pass through the circuit in only one direction. This is because holes and electrons on either side of the p-n junction can migrate across the junction. In doing so, they create a potential barrier (an internal electric field) that allows charge to flow one way and not the other. Let us examine this effect in more detail.

Remember that an n-type semiconductor has free electrons and a p-type semiconductor has free holes. When an n-type semiconductor and a p-type semiconductor are brought together to form a p-n junction, these free positive and negative charges move because of thermal diffusion. Electrons from the n side nearest the junction, the blue area in **Figure 24-10(a),** diffuse toward the p side. Positive ions that are fixed in the solid are left behind. Similarly, on the p side nearest the junction, electrons diffuse so that holes move to the n side, which leaves a region of fixed negative ions. When a free-moving electron and a free-moving hole meet, the charges cancel each other, leaving no charge carriers. As a result, the region where the p-type and the n-type semiconductors meet has no mobile charge carriers. For this reason, it is called a *depletion region*. The depletion region has a fixed size—the presence of the ions in it creates an internal electric field that opposes the further diffusion of electrons and holes that would cause the region to grow larger. This internal electric field is directed from right to left in **Figure 24-10(a).**

On either side of the depletion region are the p-type and n-type regions. So, the p-n junction in a diode consists of three distinct regions, as shown in **Figure 24-10(a).**

Diodes allow movement of charge in only one direction

The different regions of the diode can be understood in terms of the electric potential diagram in **Figure 24-10(b).** Because of the electric field within the depletion region, one side of the junction is held at a higher electric potential than the other side is. Positive charge carriers cannot move from left to right. Such a move requires extra energy, $q \cdot \Delta V_0$ (see Chapter 18), to overcome the internal electric field in the depletion region. For the same reason, negative charge carriers cannot move from right to left.

The extra energy needed to move negative charge carriers from right to left can be supplied by an external source of potential difference. If a large enough positive external voltage is applied to the p side of the junction, the electric potential on the left (positive) side is raised relative to the right (negative) side. Because the left side is now at a higher potential than the right side, charges will move, creating a current. When a diode is connected to an external source of potential difference in this manner, it is said to be *forward biased.* When a positive external voltage is applied to the n side of the junction, however, the potential barrier is increased even more. This increase further limits the current in the junction. Such a diode is said to be *reverse biased.*

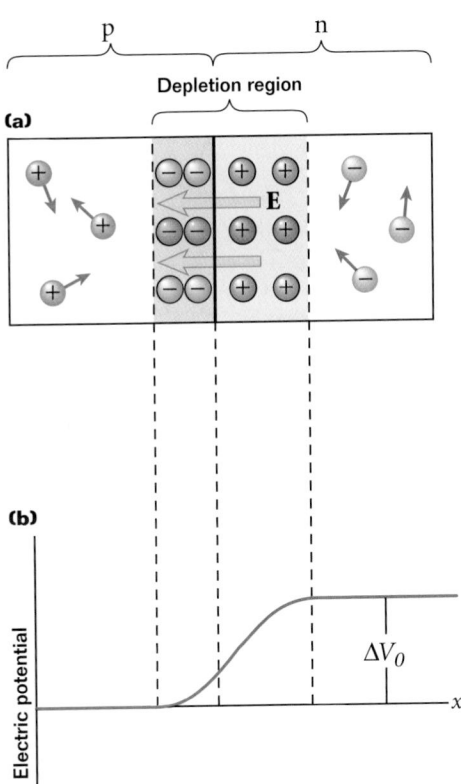

(a)

Depletion region

p   n

(b)

Electric potential

ΔV_0

x

Figure 24-10

Diffusion of charge carriers across a p-n junction sets up an electric field within the depletion region.

 **TEKS** 5B

Demonstration 5

p-n junction

Purpose Illustrate the effects of a p-n junction.

Materials LED, ohmmeter, step-down transformer, resistor (100–250 Ω), 2 m of flexible wire

Procedure Show the students that when the ohmmeter is attached in one orientation to the diode, the resistance is measurable but that when the polarity is reversed, the resistance is too high to be measured. Connect the wire, the resistor, and the LED in series with the transformer. Explain to the class that the diode lights up red if the current is directed one way and remains dark if the current is directed the opposite way. Turn the lights down, and show the class that the LED appears red because the alternating polarity of current from the outlet switches the LED on and off 60 times per second. Twirl the wire in a vertical circle with the diode at the end. This will produce red and dark bars at equally spaced intervals.

TEKS

p. 875: 5B

Diode current

Purpose Show a model explaining that diodes allow charges to move in one direction but prohibit their movement in the opposite direction.

Materials one manila folder, one sheet of paper, transparent tape, scissors

Procedure

Before class: Prepare a "paper diode" in the following manner:

First cut a 4 cm × 4 cm square out of the center of the manila folder. Then cut a 5 cm × 5 cm square of paper. Use the tape to make a hinge joining one edge of the paper with one edge of the hole in the folder. This hinge is the "top" of your diode.

In class: Demonstrate the diode as follows:

Explain that the paper diode models current through a real diode. Because the paper is larger than the hole, air (electric charge) can move in only one direction. Demonstrate this by blowing gently first from the side where the tape is and then from the other side.

The paper diode also models how the resistance of the diode (how much the flap blocks the air in the model) varies with the force of the air. Demonstrate this by again blowing from both sides at different air speeds. Have students note that at high air speed (high potential difference), the flap is almost completely out of the way (zero resistance to movement of air).

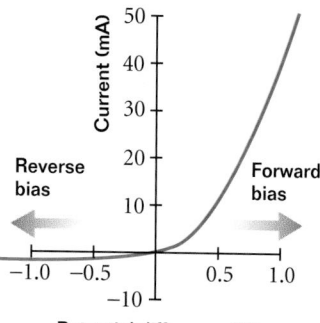

Figure 24-11

The slope at any point on this curve is equal to the reciprocal of the resistance of the diode at that voltage and current.

internetconnect

SCLINKS

NSTA

TOPIC: Energy levels
GO TO: www.scilinks.org
*sci*LINKS CODE: HF2241

Figure 24-12

The diode rectifies an alternating current **(a)** into a pulsed direct current **(b).** With the addition of other diodes, more pulses can be added to the signal **(c).** Capacitors can act as filters that smooth out the variations in current **(d).**

TEKS 6E

Figure 24-11 shows a typical plot of the current in a diode versus the potential difference across the diode. Notice that a diode does not obey Ohm's law (see Chapter 19). Ohm's law predicts that a plot of current versus potential difference will be a straight line with a slope equal to the reciprocal of the resistance. The resistance of a diode, on the other hand, is not constant. When the diode is reverse biased, the curve is nearly horizontal, which implies that the resistance is effectively infinite. (The fact that the current is slightly negative means that, in reality, a small *leakage current* still flows in the "wrong" direction in a reverse-biased diode. Because the leakage current is typically very small, we will ignore its presence.) When the diode is forward biased, the resistance varies, approaching zero for larger currents. Note that the maximum potential difference across the diode under forward bias is small (about 0.5–0.8 V).

Diodes can be used to convert alternating current to direct current

Figure 24-12 shows the effect of a diode on an applied alternating potential difference. **Figure 24-12(a)** is the current produced by the generator without the diode. **Figure 24-12(b)** is the current in the resistor with the diode in place. Only the positive part of the current remains because the diode is forward biased at these times. The diode therefore has very little resistance. When the diode is reverse biased, it has a very large resistance. In this case, the negative part of the original signal is suppressed by the diode, so there is no current in the resistor from this part of the signal. This process of converting an alternating current into a direct current is called *rectification*. When a single diode is used as a rectifier, the direct current is not constant; instead, it occurs in pulses. This current is referred to as a *pulsed direct current*.

A smoother direct current can be made from the alternating current if more than one diode and some capacitors are included in the rectifier circuit. An additional diode allows more pulses of direct current between the times of the pulses allowed by the first diode, as shown in **Figure 24-12(c).** The capacitor acts as a *filter* that smooths the output of the rectifier. Capacitors store charge when a potential difference is applied and release it when there is no potential difference. So, they keep the current in the circuit steadier, as shown in **Figure 24-12(d).**

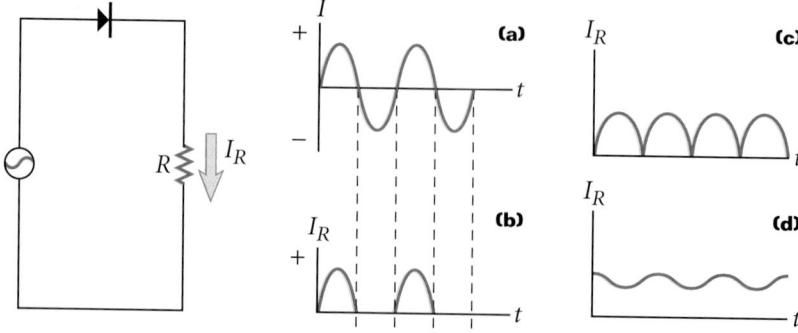

TRANSISTORS

A **transistor** is a more complicated electronic device that is used in many applications. Transistors have two p-n junctions instead of one. In this section, we will see how a transistor can be used to amplify a signal. This property of a transistor arises from the nature of the p-n junction.

There are many types of transistors, but the type we are concerned with is the *junction transistor*. The junction transistor shown in **Figure 24-13(a)** consists of a semiconducting material with a very narrow n region sandwiched between two p regions. This configuration is called a *pnp transistor*. Another configuration is the *npn transistor*, which consists of a p region sandwiched between two n regions. Because the operation of the two transistors is essentially the same, we will describe only the pnp transistor.

The structure of the pnp transistor, together with its circuit symbol, is shown in **Figure 24-13.** The arrow on the emitter lead in **Figure 24-13(b)** is necessary to distinguish a pnp transistor from an npn transistor. In an npn transistor, the arrow points in the opposite direction. The outer regions are called the *emitter* and *collector,* and the narrow central region is called the *base*. Notice that the junction transistor contains two p-n junctions: the emitter-base junction and the collector-base junction. Usually, the emitter is more heavily doped than the base, so there are more charge carriers in the emitter compared with what would be found in the base. The transistor, unlike most electrical components you have encountered, has three leads instead of two.

Setting the reference voltage is important for transistor operation

One potential difference (ΔV_{ec}) is applied between the emitter and the collector so that the emitter is at a higher electric potential than the collector, as shown at point *A* in **Figure 24-14.** The base is held at an electric potential between the emitter potential and the collector potential by the potential difference at point *B* (ΔV_{eb}). Remember from our discussion of diodes that if the

---

transistor

a device, typically containing three terminals, that can amplify a signal

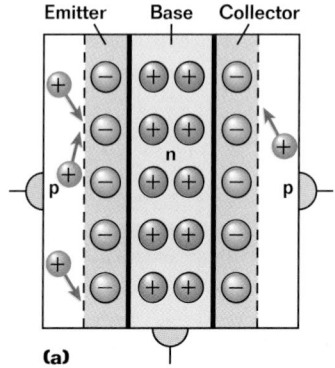

(a)

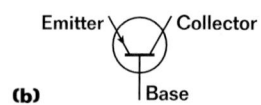

(b)

Figure 24-13

(a) A pnp junction transistor consists of an n-type semiconductor sandwiched between two p-type semiconductors. **(b)** The circuit symbol for a transistor has three leads.

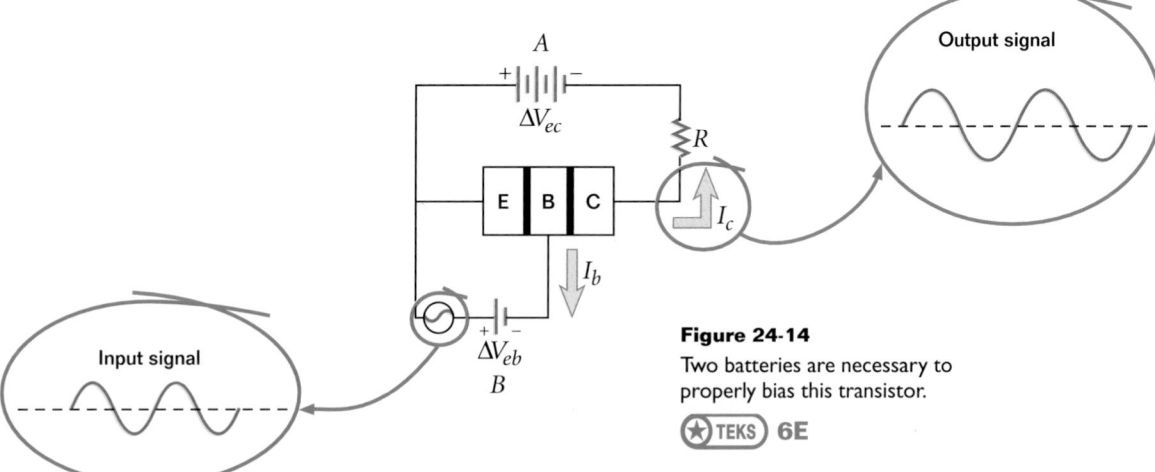

Figure 24-14

Two batteries are necessary to properly bias this transistor.

(★) TEKS 6E

SECTION 24-2

Demonstration 7

Mechanical amplification

Purpose Show a mechanical analog to amplification in transistors.

Materials meterstick, stack of books

Procedure Set up the meterstick as a class-one lever, and use a small book as the fulcrum. Place a few books at the end of the meterstick with the short moment arm. Show students that a small force at one end of the meterstick (the long lever arm) is changed into a large force at the other end.

The increased force at the long end can be explained as a type of amplification. Remind students that torque is equal to the product of the force and the lever arm. In static equilibrium, when a torque is applied at one end, there is an equal and opposite torque on the other end. Thus, $F_s d_s = F_l d_l$.

We already know that d_l is greater than d_s. To maintain the equality in the equation, F_s must be greater than F_l. In other words, the force at the long end of the meterstick has been "amplified" at the short end.

(★) TEKS

p. 876: 6E
p. 877: 6E

Optical amplification

Purpose Use light to model transistor amplification.

Materials mirror, laser

CAUTION *Avoid directing the laser or its transmitted beam toward the students.*

Procedure Shine the laser on the mirror at a glancing angle so that the reflected beam travels a considerable distance to a wall. You should see a spot of light on the wall.

Show students how a small angular movement of the mirror will produce a large sweep of the laser beam. This phenomenon can be thought of as a type of amplification.

The farther you are from the vertex, the wider the minor arc. Because the reflected laser beam is relatively far from the mirror, it will take only a small angular sweep to cause a dramatic displacement of the beam. In essence, the small movement of the mirror is "amplified" into the large sweep of the beam.

Did you know?

The discovery of the transistor by John Bardeen, Walter Brattain, and William Shockley in 1948 revolutionized the world of electronics. For this work, these three men shared a Nobel Prize in 1956.

★ TEKS 3E

internet connect

SCI*LINKS*

NSTA

TOPIC: Transistors
GO TO: www.scilinks.org
*sci*LINKS **CODE:** HF2243

p region is at a higher electric potential than the n region across a p-n junction, the p-n junction is forward biased. If we think of the transistor as two diodes back to back, we see that the emitter-base junction is forward biased and that the base-collector junction is reverse biased. This particular biasing of the two junctions in a transistor is crucial for its proper operation.

The junction transistor can be used as an amplifier

Now we can see how a transistor can amplify a signal. First consider the emitter-base junction. It is forward biased so that current enters the transistor through the emitter terminal and charge readily flows across the emitter-base junction. Because the emitter is heavily doped relative to the base, nearly all of the current consists of holes moving from the p-type emitter to the base. Few electrons move from the base to the emitter. In addition, most of the holes do not recombine with electrons in the base, because the base is very narrow.

Although only a small number of holes recombine in the base, those that do recombine limit the current that can go from the emitter to the collector through the base. This is because positive charge carriers accumulate in the base and prevent holes from flowing in.

Once the surviving holes diffuse through the base, they encounter the base-collector junction. This junction is reverse biased, so normally there is no current in this region. The base-collector barrier with its depletion region prevents electrons from migrating to the right and prevents holes from migrating to the left. In this case, however, the holes are on the n side of the junction, where the charge carriers are usually electrons. So the barrier has an opposite effect on holes migrating to the right; the holes are accelerated across the reverse-biased base-collector junction.

A small change in the properties of the base can have a great effect on the movement of charge from the emitter to the collector. One way to cause such a change is to connect the base to a second source of potential difference, labeled ΔV_{eb}, as shown at B in **Figure 24-14.** The current from this source, although small, draws some of the positive charge that would otherwise accumulate in the base. As a result, more charge can pass through the base from the emitter to the collector. The small potential difference to be amplified is placed in series with this battery. The input signal produces a small variation in the base current, resulting in a large change in collector current and therefore a large change in potential difference across the output resistor.

If the transistor is properly biased, the collector (output) current is directly proportional to the base (input) current and the transistor acts as a current amplifier. This condition may be written as follows:

$$I_c = \beta I_b$$

The quantity β (beta) is called the *current gain*. Values of the current gain typically range from 10 to 100. ★ TEKS **3A**

INTEGRATED CIRCUITS

The integrated circuit has been called the "most remarkable technology ever created by mankind." Integrated circuits form the foundation for computers, watches, cameras, automobiles, aircraft, robots, space vehicles, and all sorts of communication and switching networks.

In simplest terms, an integrated circuit is a collection of interconnected transistors, diodes, resistors, and capacitors fabricated onto a single piece of silicon, known as a *chip*. State-of-the-art chips easily contain several hundred thousand components in a very small area, as shown in **Figure 24-15.**

Integrated circuits were invented partly in an effort to achieve circuit miniaturization and partly to solve the interconnection problem spawned by the transistor. Before transistors, power and size considerations of individual components set modest limits on the number of components that could be interconnected in a given circuit. With the advent of the tiny, low-power, highly reliable transistor, design limits on the number of components disappeared and were replaced with the problem of wiring together hundreds of thousands of components. **Figure 24-16** shows a schematic diagram of one of the first types of integrated circuits, an operational amplifier.

Operational amplifiers, or *op amps,* are used extensively in electronics and computers. They are the basic element used to perform key mathematical operations, such as addition and multiplication. The numbers involved in the operation are the values of the input and output potential differences.

For example, in one configuration, a single input into an op-amp results in a single output. The magnitude of the potential difference of the output will be the magnitude of the potential difference of the input multiplied by the gain for the device.

In another configuration, several potential differences are connected as inputs to an op-amp. The result is a single output that has a potential difference with a magnitude that is the sum of the magnitudes of the individual input potential differences. ⓣ TEKS 3C

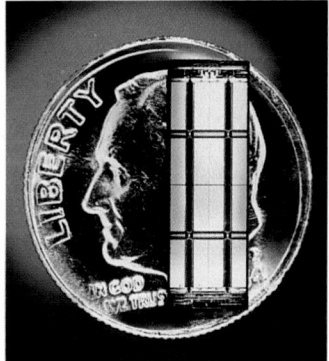

Figure 24-15
Although it is smaller than a dime, this computer memory chip can store more than 1 million digital bits of information.

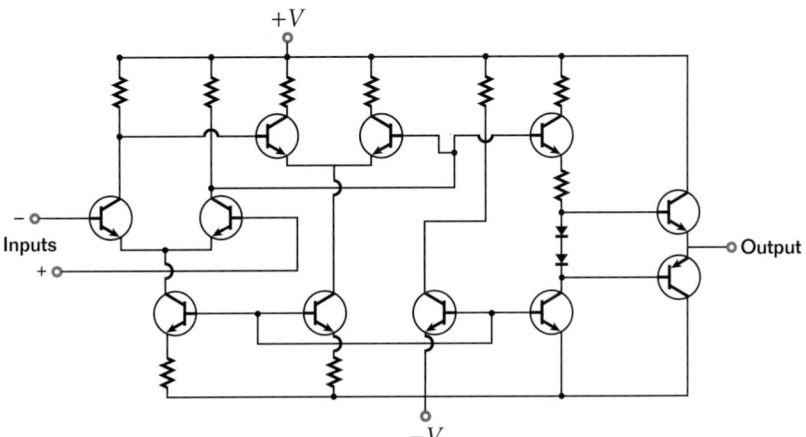

ⓣ TEKS 6E

Figure 24-16
This is a simplified schematic diagram of an operational amplifier, one of the first types of integrated circuits. Notice the many resistors, diodes, transistors, and interconnections that make up the circuit.

Teaching Tip
Pass around an etched silicon wafer so students can actually see the circuits printed on the wafer. Many wafer manufacturers will send rejects to you for this purpose.

ⓣ TEKS

p. 878: 3E, 3A
p. 879: 3C, 6E

In addition to solving the interconnection problem, integrated circuits have the advantages of miniaturization and fast response. These two features are critical for high-speed computer operation. The fast response results from the miniaturization and close packing of components. The response time of a circuit depends on the time it takes for electrical signals traveling at about 0.3 m/ns to pass from one component to another. This time is reduced by the closer packing of components. These advances, along with new parallel arrangements of processors that allow a computer to break difficult calculations into many pieces, have continued. Computers are getting smaller and faster every day. In 1997, a computer at Sandia National Laboratories became the first to perform more than 1 trillion calculations per second.

Section Review

1. Which of the following are equivalent to a conventional current from left to right?
   a. electrons traveling from left to right
   b. electrons traveling from right to left
   c. holes traveling from left to right
   d. holes traveling from right to left

2. Explain why it is not necessary for an n-type semiconductor to be negatively charged.

3. What type of fixed charge is predominant in the depletion region of a p-type semiconductor at a p-n junction?

4. A diode is connected to a source of alternating current. The current in the source fluctuates between −1.0 mA and 1.0 mA. What will be the range for the rectified current? (★)TEKS) **2C**

5. The current in the base terminal of a transistor is decreased. How will this affect the output current in the collector terminal?

6. Transistors are widely used as amplifiers in radio receivers as well as in microphones. This is because a typical radio signal from an antenna has only a few microamperes of current and a speaker often requires a current of about 0.1 A to work properly. (★)TEKS) **2C**
   a. If a transistor circuit is used to amplify a 2.5 μA signal to at least 0.1 A, what must be the minimum gain of the transistor circuit?
   b. Assume the amplifier consists of a number of transistors in series, each with a gain of 10. How many transistors are needed in this circuit? (Hint: Two such transistors in series would provide a total gain of $10 \times 10 = 100$.)

Section Review
ANSWERS

1. b, c
2. An entire neutral atom is added to an n-type semiconductor, so the semiconductor's net charge is zero. The semiconductor's conductivity increases because more electrons are able to participate in conduction.
3. negative charge
4. 0 mA to 1.0 mA
5. The output current in the collector terminal also decreases.
6. a. 4×10^4
   b. 5 transistors

24-3
Superconductors

TEMPERATURE AND CONDUCTIVITY

Recall from Chapter 19 that the resistance of many solids (other than semiconductors) increases with increasing temperature. This is because at a nonzero temperature, the atoms in a solid are always vibrating, and the higher the temperature, the larger the amplitude of the vibrations. It is more difficult for electrons to move through the solid when the atoms are moving with large amplitudes. This is similar to walking through a crowded room. It is much harder to do so when the people are in motion than when they are standing still.

If the resistance depended only on atomic vibrations, we would expect the resistance of the material that is cooled to absolute zero to go gradually to zero as the superconductivity described in Chapter 19 began to occur. Experiments have shown, however, that this does not happen. In fact, the resistances of very cold solids behave in two very different ways—either the substance suddenly begins superconducting at temperatures above absolute zero or it never superconducts, no matter how cold it gets. (★ TEKS) **3A**

Lattice imperfections cause resistance in some materials

The graph in **Figure 24-17** shows the temperature dependence of the resistance of two similar objects, one made of silver and the other made of tin. The temperature dependence of the resistance of the silver object is similar to that of a typical metal. At higher temperatures, the resistance decreases as the metal is cooled. This suggests that the amplitude of the lattice vibrations is decreasing, as expected. But at a temperature of about 10 K, the curve levels off and the resistance is constant. Cooling the metal further does not appreciably lower the resistance, even though the vibrations of the metal's atoms have been lessened.

Part of the cause of this nonzero resistance, even at absolute zero, is *lattice imperfection*. The regular, geometric pattern of the crystal, or lattice, in a solid is often flawed. A lattice imperfection occurs when some of the atoms do not line up perfectly.

Imagine you are walking through a crowded room in which the people are standing in perfect rows. It would be easy to walk through the room between two rows. Now imagine that occasionally one person stands in the middle of the aisle instead of in the row, making it harder for you to pass. This is similar to the effect of a lattice imperfection. Even in the absence of thermal vibrations, many materials exhibit a *residual resistance* due to the imperfect geometric arrangement of their atoms. (★ TEKS) **3A**

24-3 SECTION OBJECTIVES

- **Identify the cause of electrical resistance for some conductors at absolute zero.**

- **Explain the BCS theory of superconductivity.**

- **Describe some applications of superconductivity.**

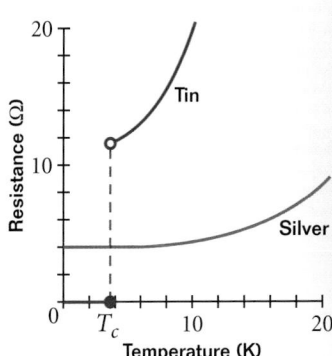

Figure 24-17
The resistance of silver exhibits the behavior of a normal metal. The resistance of tin goes to zero at temperature T_c, the temperature at which tin becomes a superconductor.

Demonstration 10

Demonstration 10

Cooper pairs

Purpose Help students visualize Cooper pairs.

Materials two tennis balls joined by a 20 cm string

CAUTION *Throw the tennis balls away from students.*

Procedure Explain that this demonstration is limited in its ability to show electron-lattice interactions but that it does show conservation of momentum (horizontally) for a pair of objects linked together, such as the electrons in a Cooper pair. Hold one of the tennis balls, whirl the free tennis ball over your head, and then release the pair into the air. The whirling pair shows horizontal conservation of momentum (disregarding gravitational effects). The momentum of an individual ball changes, but the pair acts as one object and travels in a straight line.

Did you know?

Superconductivity was discovered in 1911 by the Dutch physicist H. Kamerlingh Onnes as he and a student studied the resistivity of mercury at very low temperatures. Onnes received the Nobel Prize in physics in 1913.

(★)TEKS **3E**

internet connect

*SCI*LINKS

NSTA

TOPIC: Superconductors
GO TO: www.scilinks.org
***sci*LINKS CODE:** HF2244

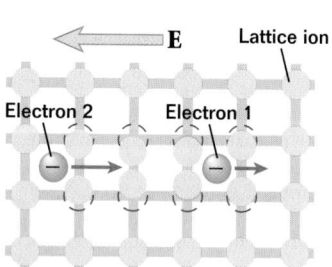

Figure 24-18
The first electron deforms the lattice, and the deformation affects the second electron. The net result is as if the two electrons were loosely bound together. Such a two-electron bound state is called a *Cooper pair.*

Figure 24-17 shows that the resistance of tin goes to zero below a certain temperature that is well above absolute zero. A solid whose resistance is zero below a certain nonzero temperature is called a *superconductor,* and the temperature at which the resistance goes to zero is the critical temperature of the superconductor, as described in Chapter 19.

BCS THEORY

Before the discovery of superconductivity, it was thought that all materials should have some nonzero resistance due to lattice vibrations and lattice imperfections, much like the behavior of the silver in **Figure 24-17.**

The first complete microscopic theory of superconductivity was not developed until 1957, almost half a century after the discovery of superconductivity. This theory is called *BCS theory* after the three scientists who first developed it: John Bardeen, Leon Cooper, and Robert Schrieffer. The crucial breakthrough of BCS theory is a new understanding of the special way that electrons traveling in pairs move through the lattice of a superconductor. According to BCS theory, electrons do suffer collisions in a superconductor, just as they do in any other material. However, the collisions do not alter the total momentum of the pair of electrons. The net effect is as if the electrons moved unimpeded through the lattice. (★)TEKS **3E**

Imagine an electron moving through a lattice, such as Electron 1 in **Figure 24-18.** There is an attractive force between the electron and the nearby positively charged atoms in the lattice. As the electron passes by, the attractive force causes the lattice atoms to be pulled toward the electron. The result is a concentration of positive charge near the electron. If a second electron is nearby, it can be attracted to this excess positive charge in the lattice before the lattice has had a chance to return to its equilibrium position.

Through the process of deforming the lattice, the first electron gives up some of its momentum. The deformed region of the lattice attracts the second electron, transferring excess momentum to the second electron. The net effect of this two-step process is a weak, delayed attractive force between the two electrons, resulting from the motion of the lattice as it is deformed by the first electron. The attractive force between these two electrons is an electron-lattice-electron interaction, where the lattice serves as the mediator of the attractive force. The two electrons travel through the lattice acting like a single particle. This particle is called a *Cooper pair.* In BCS theory, Cooper pairs are responsible for superconductivity. (★)TEKS **3A**

The reason superconductivity has been found at only low temperatures so far is that Cooper pairs are weakly bound. Random thermal motions in the lattice tend to destroy the bonds between Cooper pairs. Even at very low temperatures, Cooper pairs are constantly being formed, destroyed, and reformed in a superconducting material, usually with different pairings of electrons.

Cooper pairs maintain their momentum as they move through the lattice

Calculations of the properties of a Cooper pair have shown that this peculiar bound state of two electrons has zero total momentum in the absence of an applied electric field. When an external electric field is applied, the Cooper pairs move through the lattice under the influence of the field. However, the center of mass for every Cooper pair has exactly the same momentum. This is the crucial feature of Cooper pairs that explains superconductivity. If one electron scatters, the other electron in a pair also scatters in a way that keeps the total momentum constant. The net result is that scattering due to lattice imperfections and lattice vibrations has no net effect on Cooper pairs.

APPLICATIONS OF SUPERCONDUCTIVITY

Notice in **Table 19-3,** on page 706, that in order to be superconducting, many materials have to be cooled to very close to absolute zero. This makes the use of these superconductors for everyday applications difficult because it is expensive to cool a material to such low temperatures. In fact, early superconductors required coolant baths of either liquefied helium, which is rare and expensive, or liquid hydrogen, which is very explosive.

As shown in **Table 19-3,** a new class of materials has been found that superconducts at much higher temperatures. In particular, these superconductors can enter the superconducting state at temperatures above 77 K, which is the boiling point for liquid nitrogen. This is very important, because liquid nitrogen is inexpensive, abundant, and inert. The search continues for materials that will superconduct at room temperatures. **Figure 24-19** illustrates the dramatic increase in the highest known critical temperature for a superconductor since the discovery of this phenomenon.

Superconductors can reduce energy loss and revolutionize electronics

The property of zero resistance to direct currents would be very advantageous in low-loss electrical power transmission. A significant fraction of electrical power is converted to internal energy in normal conductors. If power transmission lines could be made to be superconducting, these losses could be eliminated, and there would be substantial savings in energy costs. (★) TEKS **3C**

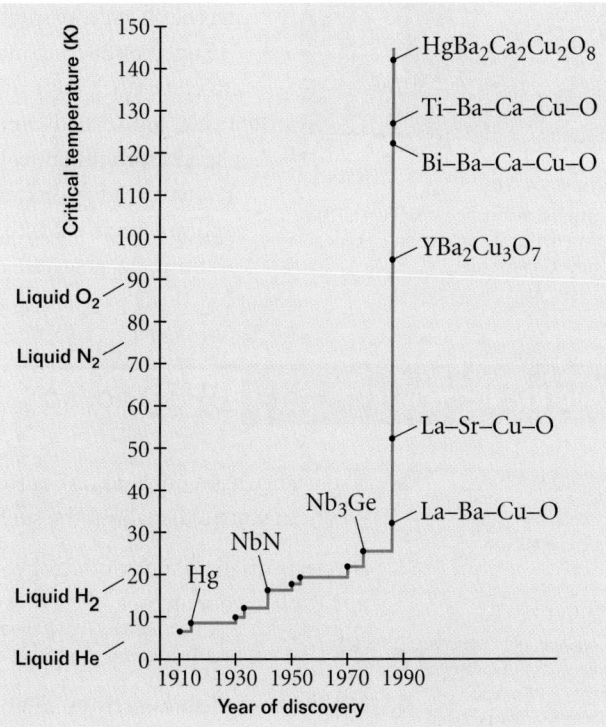

Figure 24-19
The last 20 years have yielded a dramatic rise in the temperature at which a material becomes superconducting. (★) TEKS **2C**

Demonstration 11

Meissner effect
Purpose Show the effects of superconductors.

Materials superconducting disk, small neodymium magnet, liquid nitrogen, Petri dish, plastic-foam cup, nonmagnetic tweezers, gloves

Procedure Use gloves throughout this demonstration. Carefully fill the cup with liquid nitrogen. Place the Petri dish on top of the cup, and carefully pour liquid nitrogen in the Petri dish to a depth of about $\frac{1}{2}$ cm. Wait until boiling subsides. Using nonmagnetic tweezers, place the superconducting disk in the Petri dish. Again wait until the boiling subsides. Use the nonmagnetic tweezers to carefully place a small magnet about 2 mm above the superconducting disk. When you release the magnet, it should levitate above the superconducting disk.

The magnet will levitate until the superconducting disk's temperature rises above its critical temperature.

Demonstrate a "frictionless" bearing by spinning the magnet with the tweezers.

(★) TEKS

p. 882: 3E, 3E, 3A
p. 883: 3C, 2C

High-temperature superconductors could also have a major impact on the field of electronics. For instance, the junction of two superconductors, like that of two semiconductors, has special properties. Together, the two superconductors can act as an electronic switch. Superconducting film could be used to connect computer chips, enhancing the speed of computer operation.

Superconducting rings could be used as storage devices for electrical current or energy. Because there is no resistance in a superconductor, a current introduced in a superconducting ring will continue indefinitely. The electrical energy in the current could be extracted at a later time. ⭐TEKS **3C**

Superconductors have special magnetic properties

Recall from Chapter 21 that when a conductor carries a current, a magnetic field is produced around the conductor. If a magnet is near the conductor, the magnet will be attracted to or repelled by the conductor. This is the basis of magnetic levitation, a phenomenon in which an object is suspended in the air because it is repelled by a magnet underneath. Such magnets are usually electromagnets. However, electromagnets dissipate a great deal of energy because of their electrical resistance to current. The solution to this problem is to use superconducting wires in the electromagnet.

Magnetic levitation has many potential applications in the field of transportation. In fact, a prototype for a train that runs on superconducting magnets has been constructed in Japan.

Superconducting magnets are already used in high-energy particle accelerators. Another important application of superconducting magnets is a diagnostic tool called *magnetic resonance imaging* (MRI). This technique has played a prominent role in diagnostic medicine. It uses relatively safe radio-frequency radiation rather than high-energy X rays to produce images of body sections. **Figure 24-20** shows a patient undergoing an MRI, along with the resulting images of the patient's brain. ⭐TEKS **8C**

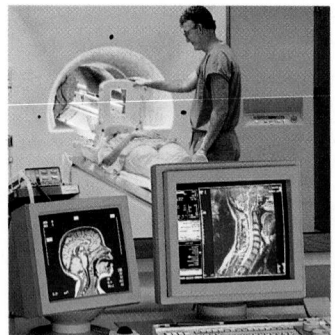

Figure 24-20
Magnetic resonance imaging (MRI) is one medical application of superconducting magnets.

Section Review   ⭐TEKS **2C, 8C**

1. A substance has a nonzero resistance even when cooled to absolute zero. What can you predict about the substance's structure?

2. Do electrons in a superconductor collide with atoms in the superconductor? Explain your answer.

3. How is a superconducting ring like a capacitor? How is it different?

4. MRI scans of the brain rely on radio-frequency waves instead of X rays. Calculate the energy of an X-ray photon with a wavelength of 1.0×10^{-10} m. Then calculate the energy carried by a radio-wave photon with a frequency of 100 MHz.

CHAPTER 24
Summary

Chapter 24
Summary

KEY IDEAS

Section 24-1 Conduction in the solid state
- Solids can be classified according to their electronic properties as conductors, insulators, or semiconductors.
- Electrons in a solid occupy sets of energy levels called bands.
- Electrons in a semiconductor can be excited from the valence band into a different band, called the conduction band. When this happens, the semiconductor conducts electric charge.

Section 24-2 Semiconductor applications
- Charge can move in a substance as electrons or as positively charged holes left by an electron-poor substance.
- The n-type semiconductor contains impurities of an atom that has five valence electrons instead of four. As a result, electrons are the majority charge carriers.
- The p-type semiconductor contains impurities of an atom that has three valence electrons instead of four. As a result, positively charged holes are the majority charge carriers.
- A diode allows current in only one direction.
- A transistor consisting of two diodes placed back to back and properly biased with batteries can be used to amplify a weak signal.

Section 24-3 Superconductors
- A superconductor is a solid whose resistance is zero below some temperature called the critical temperature.
- According to the BCS theory, electrons travel in pairs and the momentum of the pair doesn't change.

KEY TERMS

band gap (p. 868)

diode (p. 874)

doping (p. 873)

excited state (p. 867)

ground state (p. 867)

hole (p. 872)

transistor (p. 877)

valence electrons (p. 867)

Teaching Tip
Ask students to prepare a concept map for the chapter. The concept map should include most of the vocabulary terms, along with other integral terms or concepts.

Diagram symbols

electron	(−)	electric current arrow	→
hole	(+)	diode	▸⊢
electric field vector	→	transistor	(transistor symbol)

(★) TEKS

Review & Assess
pp. 886–888:
2A: 2
2B: 2
2C: 25–27
2D: 1, 3–4
2F: 2
3A: 1
3C: 3
3D: 4

CHAPTER 24
Review and Assess

ANSWERS TO

Chapter 24

Review and Assess

1. Conductors, insulators, and semiconductors differ in the size of the band gap between the conduction band and the valence band; conductivity occurs when charges are in the conduction band.

2. B, E

3. c, e, f

4. iodine

5. The excited electron, which is in the conduction band, can conduct current. A hole is left in the valence band, so positive holes moving through this band also contribute to the current.

6. Electrons in silicon's valence band readily absorb visible light and jump to the conduction band. Electrons in diamond cannot jump to the conduction band with this amount of energy, so the visible light is not absorbed.

7. to the left; to the right

8. three valence electrons

9. negative electrons

10. positive holes

11. Resistors impede current in any direction. Diodes have a small resistance to forward-biased currents but a large resistance to reverse-biased currents.

12. No, the current occurs in pulses.

VALENCE ELECTRONS AND BAND THEORY

Review questions

1. Discuss the differences in the band structures of metals, insulators, and semiconductors. How does the band structure model enable you to better understand the electrical properties of these materials?

2. Which of the energy bands in **Figure 24-21** contain unexcited valence electrons?

Band A		Band D
Band B		**Band E**
Band C		**Band F**
Semiconductor		Conductor

Figure 24-21

3. Which of the following is likely to be able to move and conduct electricity?

   a. an electron in the innermost shell of an indium atom

   b. an electron in the innermost shell of a copper atom

   c. a valence electron in a copper atom in a wire

   d. a valence electron in the semiconductor selenium

   e. an excited electron in the semiconductor selenium

   f. an electron in the conduction band of the semiconductor germanium

4. Which is likely to have the larger band gap, iodine (an insulator) or silicon (a semiconductor)?

Conceptual questions

5. When a photon is absorbed by a semiconductor, an electron-hole pair is created. Using the energy-band model, explain how this enables the semiconductor to conduct electricity.

6. The energy of visible light ranges from 1.8 to 3.2 eV. Use this fact to explain why silicon, with an energy gap of 1.1 eV, appears opaque, whereas diamond, with an energy gap of 5.5 eV, appears transparent despite the many similarities in their structures and atoms.

SEMICONDUCTORS, DIODES, AND TRANSISTORS

Review questions

7. An electric field is oriented as shown in **Figure 24-22.** In which direction will electrons in this field move? In which direction will holes move?

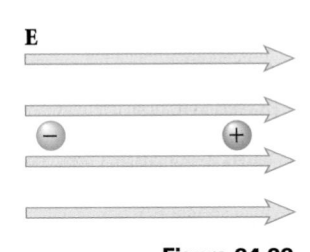

Figure 24-22

8. Which is likely to contain the most holes, a semiconductor doped with atoms with three valence electrons, four valence electrons, or five valence electrons?

9. What are the majority charge carriers in an n-type semiconductor?

10. What are the majority charge carriers in a p-type semiconductor?

11. Explain how diodes differ from resistors.

12. When a single diode is used as a rectifier, does it provide a constant current?

13. In **Figure 24-23,** the p-n diode is conducting electricity.

    a. Which way do electrons move in the circuit?

    b. Which way do holes move in the circuit?

Circuit with a conducting diode   **Figure 24-23**

14. If the circuit in **Figure 24-23** had the battery's polarity in the opposite direction, would the diode still conduct electricity?

15. Explain how transistors differ from diodes.

16. In the pnp transistor described on page 877, between which two points is the input signal connected?

17. In the pnp transistor described on page 877, between which two points is the output signal connected?

24 REVIEW & ASSESS

13. **a.** opposite the arrow
    b. in the direction of the arrow

14. no

15. Diodes, which have a single p-n junction, regulate current. Transistors, which have two p-n junctions, amplify current.

16. external circuit between the emitter and the base

17. external circuit between the emitter and the collector

Technology & Learning

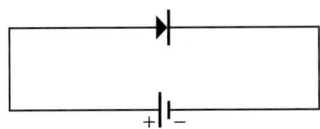

Graphing calculators

Refer to Appendix B for instructions on downloading programs for your calculator. The program "Chap24" allows you to calculate the number of free electrons per cubic centimeter of a substance.

The program "Chap24" stored on your graphing calculator makes use of the following equation for free electrons per cubic centimeter in a substance:

$$F = (D*6.02E23)/(N*M)$$

In this equation, D is the density of the substance, N is the number of atoms per free electron for that substance, and M is the atomic mass of the substance. The number 6.02×10^{23} is a constant called Avogadro's constant. You may be familiar with Avogadro's constant from your studies of the ideal gas law in chemistry.

a. Recall from your studies of conductance in Section 17-3 that the ability of a material to conduct is related to the number of free electrons in that material. If substance A has more free electrons per cubic centimeter than substance B, which substance is a better conductor?

The program "Chap24" stored on your graphing calculator makes use of the free electron equation. Once the "Chap24" program is executed, your calculator will ask for the density of the substance, the atomic mass, and the number of atoms per free electron. The graphing calculator will use these values to find the number of electrons per cubic centimeter for that substance.

Execute "Chap24" on the [PRGM] menu and press [ENTER] to begin the program. Enter the value for the mass density, then the value for the atomic mass, and finally the value for the number of atoms per free electron. Press [ENTER] after each value. Press [2nd] [EE] to enter exponents.

When you have entered the appropriate values, the calculator will perform the calculation and will display the number of electrons per cubic centimeter.

Determine the number of free electrons per cubic centimeter for the following materials:

b. copper, which has a density of 8.92 g/cm³, an average atomic mass of 63.546 u, and an average of one atom for each free electron

c. silicon, which has a density of 2.33 g/cm³, an average atomic mass of 28.086 u, and an average of 2×10^{12} atoms for each free electron

d. Which is a better insulator, copper or silicon?

Press [ENTER] to input a new value or [CLEAR] to end the program.

ANSWERS TO

Technology & Learning

a. substance A

b. 8.45×10^{22} free electrons/cm³ of copper

c. 2.50×10^{10} free electrons/cm³ of silicon

d. silicon

18. donors: P, Sb, Bi; acceptors: B, Al, In, Tl

19. more than one

20. no

21. momentum

22. no

23. No; if there is no resistance, there will be no energy dissipated by heat.

24. yes; because decreasing the I^2R loss increases the amount of energy delivered

25. a. 2.75×10^{14} Hz
   b. 1.09×10^{-6} m

26. 1.9 eV

27. 1 eV

Alternative Assessment ANSWERS

Performance assessment

1. Students should realize that the crystal is uncharged because the five electrons are balanced by the five protons in the nucleus.

2. Be sure student plans are safe and include techniques for measuring temperature, and potential difference across and current in a metal conductor. Students should graph resistance and temperature.

 **Portfolio projects**

3. Students' presentations will vary. Be certain that references to sources of information are given. About half of the periodic table's elements can be superconductors.

4. Students' presentations will vary. They may include job description, salary, and education or training needed.

Conceptual questions

18. Atoms with five valence electrons, such as arsenic, are donor atoms in a semiconductor, while those with three valence electrons, such as gallium, are acceptor atoms. Check the periodic table in Appendix E to determine what other elements might make good donors or acceptors.

19. Is the value for the current gain for an amplifying transistor likely to be one, more than one, or less than one?

SUPERCONDUCTORS

Review questions

20. If a substance is superconducting, is it likely to contain lattice imperfections?

21. What is conserved as Cooper pairs travel through a superconductor?

22. Are electrons in Cooper pairs directly attracted to each other?

Conceptual questions

23. Would a superconductor make a good heating element in a toaster oven? Explain why or why not.

24. Would a superconductor make a good material for a high-voltage transmission line? Explain why or why not.

MIXED REVIEW

25. The band gap for silicon at 300 K is 1.14 eV.
   a. Find the lowest frequency photon that will promote an electron in silicon from the valence band to the conduction band.
   b. What is the wavelength of this photon?

26. A light-emitting diode (LED) made of the semiconductor GaAsP emits red light with a wavelength of 650 nm. Determine the band gap in the semiconductor.

27. Most solar radiation has a wavelength of 10^{-6} m or less. What band gap should the material in a solar cell have in order to absorb this radiation?

Alternative Assessment

Performance assessment

1. Silicon, Si, atoms have four valence electrons. Arsenic, As, atoms used as impurities have five valence electrons. Will the resulting n-type semiconductor have a net negative charge? Draw diagrams to support or refute your claim, and present your arguments in a class discussion.

2. Design an experiment to investigate the effects of varying temperature over a range of 100°C on a metallic conductor's resistance. Review Ohm's law before planning your experiment. List the materials you will need and describe the procedure you will follow, including the measurements you will take. If your teacher approves your plan, obtain the necessary equipment and perform the experiment. What resistances do your results predict at absolute zero?

Portfolio projects

3. Research information about the physical properties, sources, and other uses of elements that can be used in superconductors. Obtain pictures of samples, if possible. Organize your findings in a brochure, poster, or computer presentation.

4. Interview someone in the electronics industry, or research careers in this industry. Find out about the types of jobs available and the education and training necessary for these jobs. Prepare a file documenting your search process, letters you wrote, notes from interviews, and other sources of information. Compile a summary report, brochure, or presentation based on your research that could be used by clients of a career guidance center.

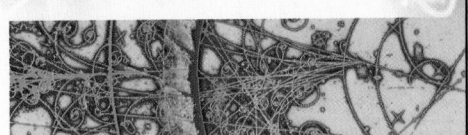
CHAPTER 24
Laboratory Exercise

RESISTORS AND DIODES

Ohm's law describes the relationship between the current and the potential difference for an ohmic resistor in a circuit. Ohmic resistors are used to regulate the current in a circuit. For a given potential difference, a higher resistance results in a smaller current, while a lower resistance results in a larger current.

Diodes are used in circuits because they have an almost infinite resistance in one direction (reverse biased) and nearly zero resistance in the other direction (forward biased). In this experiment, you will compare the performance of an ohmic resistor in a circuit with the performance of a diode in the same circuit.

SAFETY

- **Never close a circuit until it has been approved by your teacher. Never rewire or adjust any element of a closed circuit. Never work with electricity near water; be sure the floor and all work surfaces are dry.**

- **If the pointer on any kind of meter moves off scale, open the circuit immediately by opening the switch.**

- **Do not attempt this exercise with any batteries, electrical devices, or magnets other than those provided by your teacher for this purpose.**

PREPARATION

1. Read the entire lab, and plan the steps you will take.

2. Prepare a data table in your lab notebook with seven columns and eleven rows. In the first row, label the columns *Level*, I_R *(A)*, ΔV_R *(V)*, $I_{D\text{-}F}$ *(A)*, $\Delta V_{D\text{-}F}$ *(V)*, $I_{D\text{-}R}$ *(A)*, and $\Delta V_{D\text{-}R}$ *(V)*. In the first column, label the 2nd through 11th rows *1* through *10*.

PROCEDURE

Ohmic resistance

3. Construct a circuit that includes a power supply, a contact switch, a resistance box, a current meter, a resistor, and a voltage meter. ***Do not close the switch until your teacher approves your circuit.*** The current meter

---

OBJECTIVES

- **Examine the relationships between current, potential difference, and resistance in an ohmic resistor circuit.**

- **Examine the relationships between current, potential difference, and resistance in a semiconductor diode circuit.**

- **Recognize the directional character of the resistance in the diode.**

MATERIALS LIST

- ✔ **2 multimeters or dc ammeter and voltmeter**
- ✔ **diode**
- ✔ **momentary contact switch**
- ✔ **power supply**
- ✔ **resistance box**
- ✔ **resistor**
- ✔ **wire leads**

✔ Checkpoints

Step 4: Make sure the power supply, resistance box, resistor, and meters are connected correctly and set at the proper settings.

Step 5: Make sure students have correctly rewired the circuit to measure potential difference across the resistor. Remind students to close the switch only long enough to take their readings.

Step 10: Check each circuit. Students should be able to demonstrate that the diode is forward biased.

Step 12: Check each circuit. Students should be able to demonstrate that the diode is reverse biased.

should be connected in series with the power supply and the resistor. The voltage meter should be connected in parallel with the power supply. Make sure the voltage meter and the current meter are connected properly.

4. When your teacher has approved your circuit, make sure the power supply is turned to its lowest setting and the dials on the resistance box are turned all the way to the right. Turn on the power supply, and close the switch to read the voltage meter. Turn the dial on the power supply clockwise until the voltage meter reads 2.5 V. Open the switch, and turn off the power supply.

5. Remove the voltage meter from the power supply, and wire the voltage meter in parallel with the resistor. With the switch open, make sure that the meters are wired correctly. ***Do not close the switch or turn on the power supply until your teacher approves your circuit.***

6. When your teacher has approved your circuit, turn on the power supply, but do not close the switch. Slowly turn the dials on the resistance box to the left, starting with the 100 kΩ dial. Periodically close the switch just long enough to read the potential difference across the resistor. Turn the dial until the voltage meter reads 0.2 V. Record the potential difference across the resistor, ΔV_R, and the current in the resistor, I_R, in your data table under Level 1.

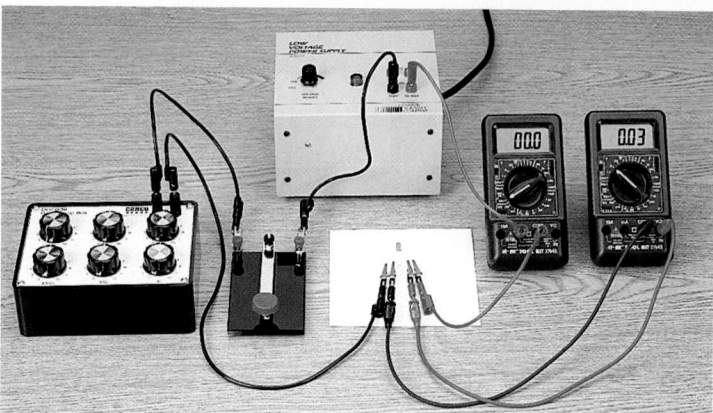

Figure 24-24

Step 3: Make sure all connections are correct.

Step 4: Use the voltage meter to measure the potential difference across the power supply. Close the switch only long enough to take your measurements.

Step 5: Rewire the circuit to measure the potential difference across the resistor.

Step 9: Replace the resistor with the diode. Connect the end of the diode with the colored bands to the black terminal of the power supply.

Step 11: Reverse the diode and repeat.

7. Repeat the procedure in step 6, turning the dials on the resistance box until the potential difference across the resistor is 0.4 V. Record ΔV_R and I_R in your data table under Level 2.

8. Repeat the procedure in step 6, increasing the potential difference across the resistor in 0.2 V steps to 2.0 V. Record the current and the potential difference for Levels 3 through 10. Open the switch.

Diode

9. Replace the resistor with the diode, as shown in **Figure 24-24.** The diode should be connected in the forward-biased position so that the end of the diode with the colored bands is connected to the wire leading to the black terminal of the power supply. Make sure that the meters are connected properly. ***Do not close the switch until your teacher approves your circuit.***

10. Repeat the procedure in steps 6 through 8 to find the current and the potential difference at each level. Adjust by steps of 0.2 V from 0.2 V to 2.0 V. Record each potential difference across the forward-biased diode, ΔV_{D-F}, and current in the forward-biased diode, I_{D-F}.

11. Remove the diode from the circuit, and reverse its connections in the circuit so that the end of the diode without the colored bands is connected to the black terminal of the power supply. **Do not close the switch until your teacher approves your circuit.**

12. Repeat the procedure in steps 6 through 8 above to find the current and the potential difference for the diode at each voltage level. Adjust the potential difference by steps of 0.2 V from 0.2 V to 2.0 V, and record the potential difference across the reverse-biased diode, ΔV_{D-R}, and the current in the reverse-biased diode, I_{D-R}, for each level.

13. Clean up your work area. Put equipment away safely so that it is ready to be used again.

ANALYSIS AND INTERPRETATION

Calculations and data analysis

1. **Organizing data**   For each level, calculate the ratio between the potential difference across the ohmic resistor and the current in the circuit.

2. **Graphing data**   Using a graphing calculator, computer, or graph paper, plot the following graphs:
   a. I_R versus ΔV_R for the ohmic resistor
   b. I_{D-F} versus ΔV_{D-F} for the forward-biased diode
   c. I_{D-R} versus ΔV_{D-R} for the reverse-biased diode

Conclusions

3. **Analyzing results**   Compare the ratios found in item 1. What is the relationship between the potential difference and the current? Is this true for all trials? Explain.

4. **Analyzing graphs**   Use your graphs to answer the following questions.
   a. What is the relationship between current and voltage in the resistance circuit?
   b. What is the relationship between current and voltage in the diode circuit?

5. **Comparing results**   Based on your results, explain the relationship between current, voltage, and resistance for the diode. Explain both the forward-biased and reverse-biased diode.

ANSWERS TO

Analysis and Interpretation

CALCULATIONS AND DATA ANALYSIS

1. Student answers will vary. For sample data, values range from 1 014 Ω to 1 111 Ω.

2. a. Student graphs should show a straight line pointing up and to the right.

   b. Student graphs should show a y-value of close to zero until a certain x-value; then the line should become nearly vertical. The value at which the line rises depends on the diode: silicon, 0.6 V; germanium, 0.4 V; green LED, 1.5 V.

   c. Student graphs should show a straight line either on the x-axis or parallel to the x-axis, with a low, constant y-value.

CONCLUSIONS

3. The ratio is equal to the resistance of the resistor. This value is constant for all trials.

4. a. $\Delta V = IR$

   b. forward biased: I is low until ΔV reaches a certain level, and then I rises rapidly; reverse biased: I is zero at any ΔV

5. forward biased: R is high until ΔV reaches a certain level, and then R falls to almost zero; reverse biased: R is infinitely high, so I is zero

What Can We Do With Nuclear Waste?

*F*or about the past 40 years, people have been arguing about what to do with radioactive waste. Since the waste is harmful to humans—as well as to the environment—deciding where to put it is a serious problem.

Protection for 10 000 years

As radioactive isotopes decay, nuclear waste emits all common forms of radioactivity–α-particles, β-particles, γ-radiation, and X rays. When this radiation penetrates living cells, it knocks electrons away from atoms, causing them to become electrically charged ions. As a result, vital biological molecules break apart or form abnormal chemical bonds with other molecules. Often, a cell can repair this damage, but if too many molecules are disrupted, the cell will die. This ionizing radiation can also damage a cell's genetic material (DNA and RNA), causing the cell to divide again and again, out of control. This condition is called cancer.

Because of these hazards, nuclear waste must be sealed and stored until the radioactive isotopes in the waste decay to the point at which radiation reaches a safe level. Some kinds of radioactive waste will require safe storage for at least 10 000 years.

Questions of disposal

Low-level waste includes materials from the nuclear medicine departments at hospitals, where radioactive isotopes are used to diagnose and treat diseases. The greatest disposal problem involves high-level waste, or HLW. Nearly all HLW consists of used fuel rods from reactors at nuclear power plants; about a third of these rods are replaced every year or two because their supply of fissionable uranium-235 becomes depleted, or spent.

When nuclear power plants in the United States began operating in 1957, engineers had planned to reprocess spent fuel to reclaim fissionable isotopes of uranium and plutonium to make new fuel rods. But people feared that the plutonium made available by reprocessing might be used to build bombs, so that plan was abandoned. Since that time, HLW has continued to accumulate at power plant sites in "temporary" storage facilities that are now nearly full. When there is no more storage space, plants will have to cease operation. Consequently, states and utility

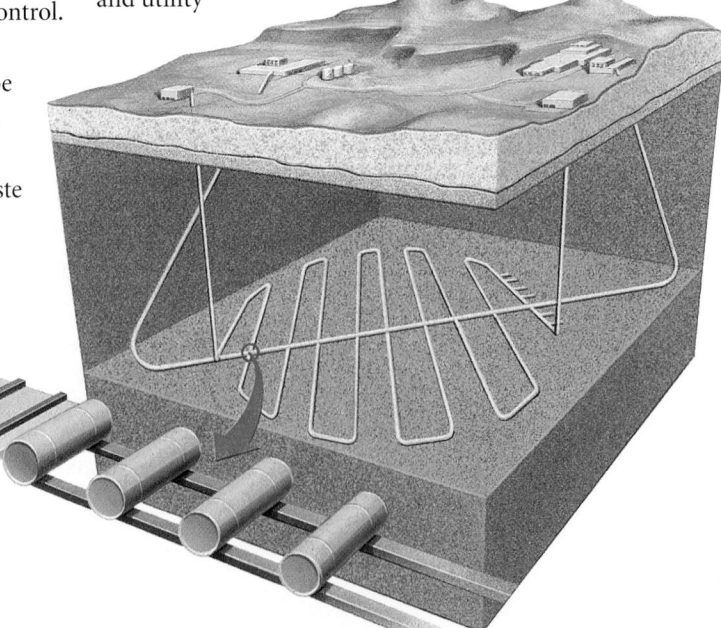

companies are demanding that the federal government honor the Nuclear Policy Act of 1982 in which the federal government agreed to provide permanent storage sites.

Disposal possibilities

Any site used for disposal of HLW must be far away from population centers and likely to remain geologically stable for thousands of years.

One possibility lies at certain deep spots of the oceans, where, some scientists claim, the seabed is geologically stable as well as devoid of life. Sealed stainless-steel canisters of waste could be packed into rocket-shaped carriers which would bury themselves deep into sediments when they hit the ocean bottom. Opponents say that the canisters have not been proven safe and that, if released, the radioactivity could kill off photosynthetic marine algae that replenish much of the world's oxygen. Proponents claim that the ocean bottom already contains many radioactive minerals and that the radioactivity from all HLWs in existence would not harm marine algae.

Scientists in the United States have considered other proposals as well, but since the Nuclear Policy Act of 1982, most of the attention has focused on the development of a disposal site beneath Yucca Mountain in Nevada. The design of this site includes sloping shafts that lead to a 570 hectare (1400 acre) storage area 300 m deep in the mountain's interior. (See cross-section on previous page.)

The U.S. Department of Energy is committed to developing Yucca Mountain. Engineers believe that it will be 2010 before the site is ready to receive waste. Until that time, there is a plan to begin moving HLW from power plants to a remote interim site in a western state. HLW would be transported to the storage site by truck or rail in sealed, steel canisters placed inside reinforced shipping casks.

Objections to Yucca Mountain

There are two main sources of opposition to the Yucca Mountain plan. One source maintains that Yucca Mountain has not been proven to be geologically secure, citing evidence that gases emitted at the lowest depth were able to reach the outside air. In addition, the group is concerned about the possibility of collisions or other accidents that might break the casks open while in transport.

The Department of Energy claims it has proven the casks safe. Several tests have been performed, including one in which a cask was loaded onto a flatbed trailer and crashed into a concrete wall at 135 km/h. In another test, a cask was struck by a locomotive at 130 km/h. No leakage occurred in these tests.

The other source of opposition to Yucca Mountain are people who maintain that the costs of overcoming legal challenges will ultimately make the plan financially infeasible. These people believe that the government should stop spending money on the Yucca Mountain project and resume the plan to reprocess waste to make new nuclear fuel.

 **Researching the Issue**

1. Learn about the policies of your city, county, or state concerning the transport and/or storage of nuclear waste. Write a report describing these policies and include your opinion of them. Give reasons for the stance you take.

2. Scientists have considered the possibility of packing waste into powerful rockets and blasting them into the sun or into solar orbit. Organize a classroom discussion of the benefits and hazards of this method.

3. Whenever public officials must decide on a location for a facility that people perceive as hazardous, they must find a way to overcome the NIMBY syndrome. What does *NIMBY* mean? How would you, as a public official, try to overcome this problem?

CHAPTER 25 PLANNING GUIDE

Compression Guide: To shorten from 11 to 8 45-min periods (from 5½ to 4 90-min blocks), eliminate items in magenta type.

PACING CHART	CLASSROOM RESOURCES			
	★ TEKS	Teacher Demonstrations	Holt Physics Transparencies	Labs (See page T52 for equipment listing for in-text labs.)
25-1 The nucleus 1 45-minute period ½ 90-minute block	2C, 3A, 3E	**TE** *Nuclear stability*, p. 899	**TM** 86–88	
25-2 Nuclear decay 2 45-minute periods 1 90-minute block	2C, 3A, 3C, 3E	**TE** *Electron beam deflection*, p. 905	**T** 126–127 **TM** 89	**PE** *Half-Life*, p. 932
25-3 Nuclear reactions 1 45-minute period ½ 90-minute block	2C, 3C, 3E		**T** 128 **TM** 90	
25-4 Particle physics 1 45-minute period ½ 90-minute block	2C, 3A, 3E		**TM** 91–96	
Review and Assessment 2 45-minute periods 1 90-minute block				

Resource Key

PHYSICS

PE Pupil's Edition
TE Teacher's Edition

L Laboratory Experiments
TL Technology Lab Experiments
T Transparencies

👆 **One-Stop Planner CD-ROM contents**

TM Transparency Masters
SR Section Review Worksheets
AA Alternative Assessment

PW Problem-Solving Workbook
PB Problem Bank
CTW Critical Thinking Worksheet

LABORATORY PLANNING: Half-Life, p. 932

Materials (for each lab group)
- basic nuclear lab station
- disposable culture dish (small)

- isogenerator set
- round-jaw symmetrical clamp, with holder
- support base with rod, 46 cm long × 0.8 cm diameter

ASSIGNMENT RESOURCES

Content Mastery	Critical Thinking	Problem-Solving Practice
PE 1–7, p. 902 **SR** 25-1, *Concept Review* **PE** 1–3, p. 927	**PE** 4–6, p. 927	**23A** Binding energy: 29 items in **PE, PW,** and **PB,** see **TE** pp. 901–902
PE 1–4, p. 912 **SR** 25-2, *Concept Review* **PE** 10–12, p. 927	**PE** 1–3, p. 909 **PE** 14–19, pp. 927–928	**25B** Nuclear decay: 36 items in **PE, PW,** and **PB,** see **TE** p. 908 **25C** Measuring nuclear decay: 35 items in **PE, PW,** and **PB,** see **TE** pp. 910–911
PE 1–6, p. 916 **SR** 25-3, *Concept Review* **PE** 13, p. 927	**PE** 20–22, p. 928	
PE 1–3, p. 923 **SR** 25-4, *Concept Review* **PE** 30–34, p. 928	**PE** 1–2, p. 920 **PE** 35–37, p. 928	

ASSESSMENT RESOURCES

Cumulative Review	Alternative Assessment	Traditional Assessment
SR Mixed Review, Ch. 25	**PE** 1–6, p. 931 **AA** Items for Ch. 25	Chapter 25 Test Test Generator items for Ch. 25

Scoring Rubrics for Alternative Assessment items can be found on the One-Stop Planner CD-ROM.

TECHNOLOGY RESOURCES

 CTW Segment 24 Wisp of Creation

 PE Technology and Learning, p. 930
(Alternative procedures for calculators without Flash-ROM technology are provided on the One-Stop Planner CD-ROM.)

 internet**connect**

On-line Student Resources:
GO TO: www.scilinks.org
The following *sci*LINKS Internet resources can be found in the student text for this chapter.

TOPICS:
• Atomic nucleus, p. 897 (HF2251)
• Radioactive decay, p. 904 (HF2252)
• Fission/fusion, p. 913 (HF2253)
• Antiparticles, p. 925 (HF2254)

 On-line Teacher Resources:
GO TO: go.hrw.com
KEYWORD: HF2 HOME
Visit the HRW Web site for a variety of resources related to this chapter.

 Smithsonian Institution*
Internet Connections
Visit **www.si.edu/hrw** for additional on-line resources.

Visit **www.cnnfyi.com** for late-breaking news and current events stories selected just for you.

Materials Preparation

The Cs-137/Ba137 isogenerator set does not present a contamination problem, because the solution has a very short half-life, no longer than an hour. However, be sure you know the cleanup and disposal procedures required by your district and local and state governments.

Section 25-1 discusses properties of the atomic nucleus; identifies mass number, atomic number, and neutron number; explains why some nuclei are unstable; and shows how to calculate binding energy.

Section 25-2 describes the three types of radioactive decay, shows how to predict the outcome of nuclear decay, defines the decay constant, and shows how to calculate the half-life of a radioactive substance.

Section 25-3 distinguishes between nuclear fission and nuclear fusion and explores nuclear chain reactions.

Section 25-4 discusses the four fundamental interactions of nature and their mediating particles, classifies the elementary particles that make up matter, and describes the standard model of the universe.

About the Illustration

The archaeologists shown in this photograph are investigating the shipwreck of La Salle's *Belle*, which was discovered in Matagorda Bay, off the Texas coast, in 1995. A cofferdam made of steel walls was built around the wreckage to simplify the excavation. The field of archaeology was revolutionized in the mid-twentieth century by the development of carbon dating, a process that has been used to date samples of wood, charcoal, bone, and shell that are 1000 to 25 000 years old. This knowledge has helped us reconstruct the history of living organisms—including humans—during this time span.

Subatomic Physics

PHYSICS IN ACTION

High-energy particles from space, called *cosmic rays*, continually create carbon-14 atoms from nitrogen in the atmosphere. Because carbon-14 atoms decay into other atoms at a specific rate, researchers can use carbon-14 to determine the age of organic materials in a process called *carbon dating*.

All living organisms have the same ratio of carbon-14 to carbon-12 atoms. When an organism dies, it no longer takes in carbon-14, and the ratio of carbon-14 to carbon-12 decreases with time as a result of the decay of carbon-14. It is therefore possible to determine the age of an organic material by measuring this ratio.

- *What makes up a carbon-14 nucleus?*
- *By what process does carbon-14 decay into other atoms?*

CONCEPT REVIEW

Electric force   (Section 17-2)

Atomic structure   (Section 23-2)

Tapping Prior Knowledge

Knowledge to Expect

✔ "The accidental discovery that minerals containing uranium darken photographic films, as light does, led to the idea of radioactivity." (AAAS's *Benchmarks for Science Literacy,* grades 6–8)

✔ "Scientific investigations sometimes result in new ideas and phenomena for study, generate new methods for investigation, or develop new technologies to improve the collection of data." (NRC's *National Science Education Standards,* grades 5–8)

Knowledge to Review

✔ The force between two charges is proportional to the magnitudes of the charges and inversely proportional to the distance between them squared:
$$F_{electric} = k_C \frac{q_1 q_2}{r^2}.$$
(Section 17-2)

✔ Atoms consist of a positive nucleus surrounded by negatively charged electrons. (Section 23-2)

internet **connect**

SCi*LINKS*
NSTA

TOPIC: Nuclear Power Plants in Texas
GO TO: www.scilinks.org
*sci*LINKS CODE: HFX003

25-1
The nucleus

25-1 SECTION OBJECTIVES

- **Identify the properties of the nucleus of an atom.**
- **Explain why some nuclei are unstable.**
- **Calculate the binding energy of various nuclei.**

PROPERTIES OF THE NUCLEUS

In Chapter 17, you learned that atoms are composed of electrons, protons, and neutrons. Except for the ordinary hydrogen nucleus, which consists of a single proton, both protons and neutrons are found in the nucleus. Together, protons and neutrons are referred to as *nucleons.*

As seen in Chapter 23, Rutherford's scattering experiment led to the conclusion that all of an atom's positive charge and most of its mass are concentrated in the nucleus. Rutherford's calculations revealed that the nucleus has a radius of no greater than about 10^{-14} m. Because such small lengths are common in nuclear physics, a convenient unit of length is the *femtometer* (fm). Sometimes called the *fermi,* this unit is equal to 10^{-15} m.

A nucleus can be specified by an atomic number and a mass number

There are a few important quantities that are used to describe the charge and mass of the nucleus. **Table 25-1** lists these quantities and the symbols commonly used to represent them. The mass number (A) represents the number of protons and neutrons—or nucleons—in the nucleus. The atomic number (Z) represents the number of protons in the nucleus, and the neutron number (N) represents the number of neutrons in the nucleus. ⊛TEKS 2C

Table 25-1	Symbols for nuclear quantities	
Symbol	**Name**	**Explanation**
A	mass number	the number of nucleons (protons and neutrons) in the nucleus
Z	atomic number	the number of protons in the nucleus
N	neutron number	the number of neutrons in the nucleus

Mass number (A)

Chemical symbol

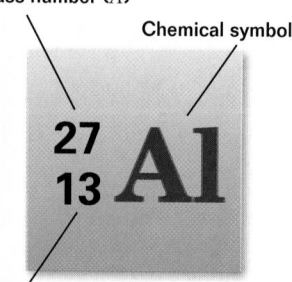

Atomic number (Z)

Figure 25-1
The chemical symbol of an element is often written with its mass number and atomic number, as shown here.

As an example, consider a nucleus of aluminum. A typical atom of aluminum has a mass number of 27 and an atomic number of 13. Therefore, it has 13 protons and 14 neutrons ($27 - 13 = 14$). A periodic table of the elements usually includes the atomic number of each element above or near the element's chemical symbol.

Often, the mass number and the atomic number of the nucleus of an atom are combined with the atom's chemical symbol, as shown in **Figure 25-1.**

The chemical symbol for aluminum is Al. The superscript refers to the mass number A (27 in the case of aluminum), and the subscript refers to the atomic number Z (13 in the case of aluminum).

An element can be identified by its atomic number, Z. Because the number of protons determines the element, the atomic number of any given element does not change. Thus, the chemical symbol, such as Al, or the name of the element, such as aluminum, can always be used to determine the atomic number. For this reason, the atomic number is sometimes omitted.

Although atomic number does not change within an element, atoms of the same element can have different mass numbers. This is because the number of neutrons in a particular element can vary. Atoms that have the same atomic number but different neutron numbers (and thus different mass numbers) are called **isotopes.** The neutron number for an isotope is found from the following relationship:

$$A = Z + N$$

This expression says that the mass number of an atom (A) equals the number of protons (Z) plus the number of neutrons (N) in the nucleus of the atom. This expression was used to find the number of neutrons in aluminum.

The natural abundance of isotopes can vary greatly. For example, $^{11}_6C$, $^{12}_6C$, $^{13}_6C$, and $^{14}_6C$ are four isotopes of carbon. The natural abundance of the $^{12}_6C$ isotope is about 98.9 percent, while that of the $^{13}_6C$ isotope is only about 1.1 percent. Some isotopes do not occur naturally but can be produced in the laboratory. Even the simplest element, hydrogen, has isotopes: ^1_1H, called hydrogen; ^2_1H, called *deuterium;* and ^3_1H, called *tritium.*

A nucleus is very dense ⭐TEKS 2C

Experiments have shown that most nuclei are approximately spherical and that the volume of a nucleus is proportional to the total number of nucleons, and thus to the mass of the nucleus. This suggests that *all nuclei have nearly the same density,* which is about 2.3×10^{17} kg/m³, 2.3×10^{14} times greater than the density of water (1.0×10^3 kg/m³). Nucleons combine to form a nucleus as though they were tightly packed spheres, as shown in **Figure 25-2.**

The unified mass unit and rest energy are used to express the mass of a nucleus

Because the mass of a nucleus is extremely small, the *unified mass unit,* u, is often used for atomic masses. This unit is sometimes referred to as the *atomic mass unit.* 1 u is defined so that 12 u is equal to the mass of one atom of carbon-12. That is, the mass of a nucleus (or atom) is measured relative to the mass of an atom of the neutral carbon-12 isotope (the nucleus plus six electrons). Based on this definition, 1 u = 1.660 5402 $\times 10^{-27}$ kg. The proton and neutron each have a mass of about 1 u, and the electron has a mass that is only a small fraction of 1 u, or about 5×10^{-4} u.

internet**connect**

SCI**LINKS**
NSTA
TOPIC: Atomic nucleus
GO TO: www.scilinks.org
*sci***LINKS CODE:** HF2251

isotopes

> atoms of an element having the same atomic number but different neutron and mass numbers

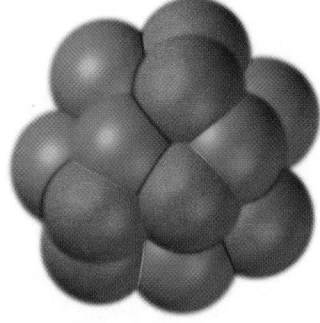

Figure 25-2
A nucleus can be visualized as a cluster of tightly packed spherical nucleons. The pink spheres represent protons, and the blue spheres represent neutrons.

🛑 Misconception Alert

To help students differentiate between *atomic mass* and *mass number,* explain that the periodic table gives the atomic number and the atomic mass (rather than the mass number) of each element. Ask students if both numbers given on the periodic table are integers. *(Atomic numbers are; atomic masses are not.)* Explain that the atomic mass reported on the periodic table is an average of many nuclei with the same atomic numbers (Z) but different mass numbers (A). Mass number, on the other hand, relates to an individual nucleus and is always an integer.

The Language of Physics

The symbol u represents the *unified mass unit,* also known as the *atomic mass unit.* Some texts use the symbol amu to represent this unit.

Teaching Tip

Have students draw schematic diagrams like the one shown in **Figure 25-2** for the four isotopes of carbon discussed on this page ($^{11}_6C$, $^{12}_6C$, $^{13}_6C$, and $^{14}_6C$). This will help students visualize the relationship between atomic number, mass number, and the structure of the nucleus.

⭐TEKS

p. 896: 2C
p. 897: 2C

The Language of Physics

As seen in Chapter 23, the electron volt, eV, is a unit of energy typically used to describe atomic processes ($1 \text{ eV} = 1.60 \times 10^{-19} \text{ J}$). Because energies at the nuclear level are much greater than those at the atomic level, units of MeV ($M = 1 \times 10^6$) are often more convenient. $1 \text{ MeV} = (1 \times 10^6)(1.60 \times 10^{-19} \text{ J}) = 1.602\ 19 \times 10^{-13} \text{ J}$.

Teaching Tip

Table 25-2 shows the masses of the proton, neutron, and electron in various units. Each unit has advantages; the choice of unit depends on the situation. For example, kg can be convenient when working with other compound SI units. When working only with atomic masses, though, u is usually more convenient. Units of MeV/c^2 are advantageous because they describe the rest energy of a particle. Have students determine the rest energy of a proton, neutron, and electron from **Table 25-2** (*938.3 MeV, 939.6 MeV, 0.5110 MeV*).

Misconception Alert

Explain to students that no particle has exactly 1 u of rest energy; the masses of the proton and neutron are only approximately equal to 1 u. By convention, $1 \text{ u} = 931.50 \text{ MeV}$. Have students compare this value to the masses of the proton and neutron in u and in kg.

Did you know?

The equivalence between mass and energy is predicted by Einstein's special theory of relativity. Another aspect of this theory is that time and length are relative, that is, they depend on an observer's frame of reference, while the speed of light is absolute.

★ TEKS   3E

strong force

the interaction that binds nucleons together in a nucleus

Alternatively, the mass of the nucleus is often expressed in terms of rest energy. As seen in the Physics on the Edge feature in Chapter 5, a particle has a certain amount of energy, called *rest energy*, associated with its mass. (Some texts distinguish between a relativistic mass, m, and a rest mass, m_0. In this book, relativistic mass is not used, and m always refers to rest mass.) The following equation expresses the relationship between mass and rest energy mathematically:

RELATIONSHIP BETWEEN REST ENERGY AND MASS

$$E_o = mc^2$$

rest energy = (mass)(speed of light)^2

This expression is often used because mass is not conserved in many nuclear processes, as we will see. Because the rest energy of a particle is given by $E_0 = mc^2$, it is convenient to express a particle's mass in terms of its energy equivalent. The equation that follows is for the rest energy of a particle with a mass of exactly 1 u.

$$E_0 = mc^2 = \frac{(1.660\ 540 \times 10^{-27} \text{ kg})(2.997\ 92 \times 10^8 \text{ m/s})^2}{1.602\ 19 \times 10^{-19} \text{ J/eV}} = 931.50 \text{ MeV}$$

Thus, the conversion of 1 u of mass into energy would produce 931.50 MeV. Nuclear physicists often express masses of small particles in terms of the unit MeV/c^2, where $1 \text{ u} = 931.50 \text{ MeV}/c^2$. ★ TEKS 2C

The masses and energy equivalent of the proton, neutron, and electron are summarized in **Table 25-2**. Notice that in order to distinguish between the mass of the proton and the mass of the neutron, you must know their masses to at least four significant figures in units of kg or u. The masses and some other properties of selected isotopes are given in Appendix F. ★ TEKS 2C

Table 25-2	Mass of the proton, neutron, and electron		
Particle	**kg**	**u**	**MeV**
proton	1.673×10^{-27}	1.007 276	938.3
neutron	1.675×10^{-27}	1.008 665	939.6
electron	9.109×10^{-31}	0.000 549	0.5110

NUCLEAR STABILITY

Given that the nucleus consists of a closely packed collection of protons and neutrons, you might be surprised that it can exist. It seems that the Coulomb repulsion between protons would cause a nucleus to fly apart. There must be some attractive force to overcome this repulsive force. This force is called the *nuclear force*, or the **strong force**.

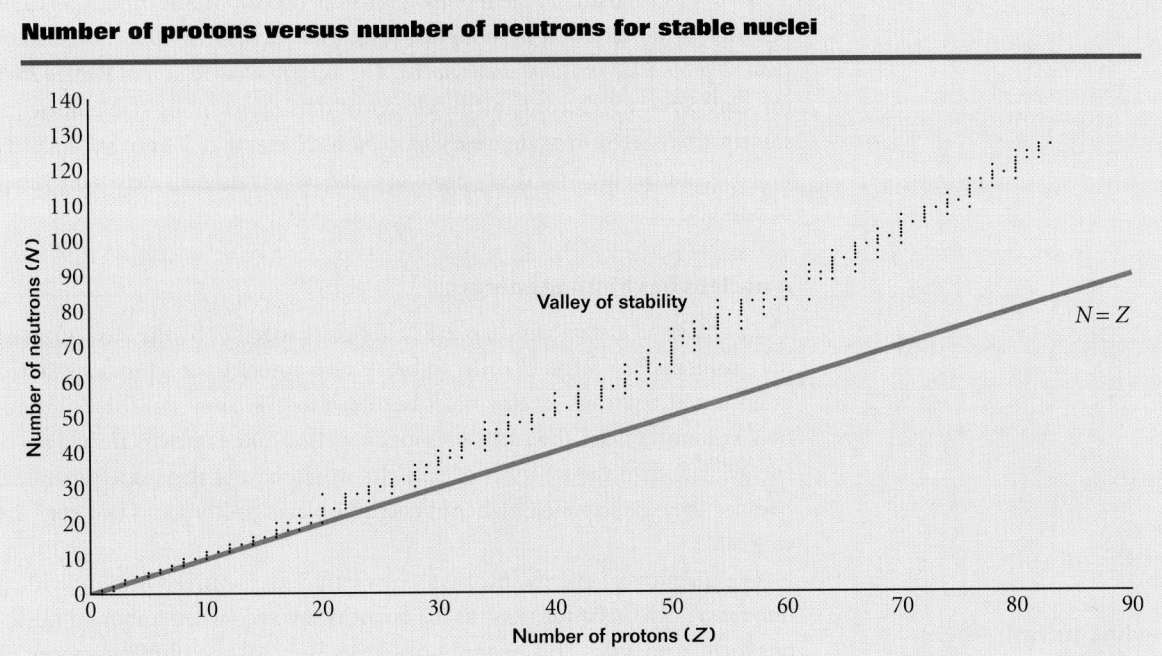

Number of protons versus number of neutrons for stable nuclei

Valley of stability

$N = Z$

Figure 25-3
Each data point in this graph represents a stable nucleus. Note that as the number of protons increases, the ratio of neutrons to protons also increases. In other words, heavy nuclei have more neutrons per proton than lighter nuclei.

⭐ TEKS 2C

The strong force has some properties that make it very much unlike other types of force. The strong force is almost completely independent of electric charge. For a given separation, the force of attraction between two protons, two neutrons, or a proton and a neutron has the same magnitude.

Another unusual property of the strong force is its very short range, only about 10^{-15} m. For longer distances, the strong force is virtually zero.

Neutrons help to stabilize a nucleus

A plot of neutron number versus atomic number (the number of protons) for stable nuclei is shown in **Figure 25-3.** The solid line in the plot shows the location of nuclei that have an equal number of protons and neutrons ($N = Z$). Notice that only light nuclei are on this line, while all heavier nuclei fall above this line. This means that heavy nuclei are stable only when they have more neutrons than protons. This can be understood in terms of the characteristics of the nuclear force (strong force). ⭐ TEKS 2C

For a nucleus to be stable, the repulsion between positively-charged protons must be balanced by the strong nuclear force's attraction between all the particles in the nucleus. The repulsive force exists between all protons in a nucleus because the electrostatic force is long range. But a proton or a neutron attracts only its nearest neighbors because of the nuclear force's short range. So, as the number of protons increases, the number of neutrons has to increase even more to add enough attractive forces to maintain stability. ⭐ TEKS 3A

Misconception Alert

Students may wonder whether elements with Z greater than 83 exist since they are unstable and do not appear in **Figure 25-3,** on page 899. Ask them to locate the last stable element on the periodic table (*Bi, or bismuth*). As seen on the periodic table, elements with atomic numbers greater than 83 (those following bismuth) do exist. After the concept of half-life has been introduced in Section 25-2, return to this question and point out that the half-lives of these unstable elements vary from fractions of seconds to thousands of years. Thus, some unstable elements exist for very long periods of time.

The Language of Physics

Make sure that students interpret the symbols and subscripts in the equations correctly. E_{bind} represents binding energy. $E_{0,unbound}$ represents the rest energy of a group of particles when they are not part of a nucleus; $E_{0,bound}$ represents the rest energy of that same group of particles when they are part of a nucleus. Ask if the total mass increases or decreases when a helium nucleus is broken down into its components of two protons and two neutrons. *(The total mass increases by the mass defect, Δm.)*

 TEKS

p. 901: 2C

binding energy

the energy involved when nucleons bind together to form a stable nucleus

For Z greater than 83, the repulsive forces between protons cannot be compensated by the addition of more neutrons. That is, *elements that contain more than 83 protons do not have stable nuclei.* The long, narrow region in **Figure 25-3** (on page 899), which contains the cluster of dots representing stable nuclei, is sometimes referred to as the *valley of stability.* Nuclei that are not stable decay into other nuclei until the decay product is one of the nuclei located in the valley of stability.

A nucleus has binding energy

The particles in a stable nucleus are held tightly together by the attractions of the strong nuclear force. In order to break such a nucleus apart into separated protons and neutrons, energy must be added to overcome this force's attraction. For most nuclei, the particles bound together in the nucleus are at a lower energy state than the same set of particles would have if they were separated. Because they are so much higher in energy, isolated protons and neutrons are very rare.

The quantity of energy involved when a nucleus is broken into individual unbound nucleons is the same as the quantity involved when unbound nucleons form a nucleus. This quantity of energy is called the **binding energy** of the nucleus. It is equal to the difference in energy between the nucleons when bound and the same nucleons when unbound. Binding energy can be calculated from the rest energies of the particles making up a nucleus by using the following equation.

$$E_{bind} = E_{0,unbound} - E_{0,bound}$$

Using the equation for rest energy on page 898, this can be written as follows:

$$E_{bind} = m_{unbound}c^2 - m_{bound}c^2 = (m_{unbound} - m_{bound})c^2$$

The mass of the nucleons when unbound minus the mass of the nucleons when bound is called the *mass defect* and is expressed as Δm. Thus, the previous equation for binding energy can be expressed as follows:

BINDING ENERGY OF A NUCLEUS

$$E_{bind} = \Delta mc^2$$

binding energy = mass defect $\times$ (speed of light)^2

Note that the total mass of a stable nucleus (m_{bound}) is always less than the sum of the masses of its individual nucleons ($m_{unbound}$). It is often useful to find the mass defect in terms of u so that it can be converted to energy as described earlier in this chapter (1 u = 931.50 MeV).

The mass of the unbound nucleus is the sum of the individual nucleon masses, and the mass of the bound nucleus is about equal to the atomic mass minus the mass of the electrons. Thus, $\Delta m = (Zm_p + Nm_n) - (\text{atomic mass} - Zm_e)$. One way to rearrange this equation is $\Delta m = (Zm_p + Zm_e) + Nm_n - (\text{atomic}$

mass). Because a hydrogen atom contains one proton and one neutron, the first term is equal to Z(atomic mass of H). Thus, the equation for mass defect can be rewritten as follows:

$$\Delta m = Z(\text{atomic mass of H}) + Nm_n - \text{atomic mass}$$

Use this equation and the atomic masses given in Appendix F to calculate mass defect when solving problems involving binding energy.

The binding energy per nucleon for light nuclei ($A < 20$) is much smaller than the binding energy per nucleon for heavier nuclei. In other words, particles in lighter nuclei are less tightly bound on average than particles in heavier nuclei. Except for the lighter nuclei, the average binding energy per nucleon is about 8 MeV. Of all nuclei, iron-58 has the greatest binding energy.

Binding energy ★TEKS 2C

PROBLEM

The nucleus of the deuterium atom, called the *deuteron,* consists of a proton and a neutron. Given that the atomic mass of deuterium is 2.014 102 u, calculate the deuteron's binding energy in MeV.

SOLUTION

1. DEFINE **Given:** $Z = 1$ atomic mass of deuterium = 2.014 102 u
$N = 1$ atomic mass of H = 1.007 825 u
$m_n = 1.008\ 665$ u

Unknown: $E_{bind} = ?$

2. PLAN **Choose an equation(s) or situation:**
First find the mass defect with the following relationship:

$$\Delta m = Z\,(\text{atomic mass of H}) + Nm_n - \text{atomic mass}$$

Then find the binding energy by converting the mass defect to rest energy.

3. CALCULATE **Substitute the values into the equation(s) and solve:**
$$\Delta m = 1(1.007\ 825\ \text{u}) + 1(1.008\ 665\ \text{u}) - 2.014\ 102\ \text{u}$$

$$\Delta m = 0.002\ 388\ \text{u}$$

$$E_{bind} = (0.002\ 388\ \text{u})\,(931.50\ \text{MeV/u})$$

$$\boxed{E_{bind} = 2.224\ \text{MeV}}$$

4. EVALUATE In order to separate a deuteron into its constituents—a proton and a neutron— it is necessary to add 2.224 MeV of energy.

Teaching Tip

Point out that the *binding energy per nucleon* is the binding energy (E_{bind}) of the entire nucleus divided by the mass number (A). Ask students to find the binding energy per nucleon of $^{56}_{26}\text{Fe}$ given $E_{bind} = 492.4$ MeV (*8.793 MeV*).

(STOP) Misconception Alert

Some students may not realize that isotopes of an element have different atomic masses. Point out that when they are solving binding-energy problems, students must find the mass number ($A = Z + N$) to determine the correct atomic mass in Appendix G.

Classroom Practice

PROBLEM

Binding energy

Calculate the binding energy of the following, referring to Appendix F for the atomic masses:

a. tritium (1 proton, 2 neutrons)

b. helium (2 protons, 2 neutrons)

Answers
a. 8.482 MeV
b. 28.297 MeV

PRACTICE GUIDE 25A	
Solving for:	
E_{bind}	📖 **PE** Sample, 1–5; Ch. Rvw. 7–9, 40, 43–44*
	💿 **PW** 1–4
	PB Sample, 1–6, 9
m	💿 **PW** 5–7
	PB 7–8, 10

ANSWERS TO

Practice 25A
Binding energy

1. 160.65 MeV; 342.06 MeV
2. 0.764 MeV
3. 7.933 MeV
4. 7.5702 MeV/nucleon
5. 0.2101 MeV; Because of the extra proton in $^{23}_{12}$Mg, the Coulomb repulsion in the Mg nucleus is greater than that in the Na nucleus. Thus, breaking the Mg nucleus into its components requires less energy; in other words, its binding energy is less.

Binding energy **2C**

1. Calculate the total binding energy of $^{20}_{10}$Ne and $^{40}_{20}$Ca. (Refer to Appendix G for this and the following problems.)

2. Determine the difference in the binding energy of $^{3}_{1}$H and $^{3}_{2}$He.

3. Calculate the binding energy of the last neutron in the $^{43}_{20}$Ca nucleus. (Hint: Compare the mass of $^{43}_{20}$Ca with the mass of $^{42}_{20}$Ca plus the mass of a neutron.)

4. Find the binding energy per nucleon of $^{238}_{92}$U in MeV.

5. Two isotopes having the same mass number are known as *isobars*. Calculate the difference in binding energy per nucleon for the isobars $^{23}_{11}$Na and $^{23}_{12}$Mg. How do you account for this difference?

Section Review
ANSWERS

1. the charge of the nucleus (the atomic number)
2. They have the same number of protons and electrons. They have a different number of neutrons and hence a different number of nucleons.
3. Z is the same for each isotope; A and N are different.
4. the strong force
5. For light nuclei, $N \approx Z$. As the number of protons (Z) increases, more neutrons per proton are needed to compensate for the Coulomb repulsion, so $N > Z$.
6. a. 805.77 MeV
   b. 1559.4 MeV
   c. 224.96 MeV
7. six; eight; six

Section Review   **2C**

1. Does the nuclear mass or the charge of the nucleus determine what element an atom is?

2. Oxygen has several isotopes. What do these isotopes have in common? How do they differ?

3. Of atomic number, mass number, and neutron number, which are the same for each isotope of an element, and which are different?

4. The protons in a nucleus repel one another with the Coulomb force. What holds these protons together?

5. Describe the relationship between the number of protons, the number of neutrons, and the stability of a nucleus.

6. Calculate the total binding energy of the following:
   a. $^{93}_{41}$Nb
   b. $^{197}_{79}$Au
   c. $^{27}_{13}$Al
   (Refer to Appendix G.)

7. **Physics in Action** How many protons are there in the nucleus $^{14}_{6}$C? How many neutrons? How many electrons are there in the neutral atom?

25-2
Nuclear decay

NUCLEAR DECAY MODES

So far, we have considered what happens when nucleons are bound together to form stable nuclei. However, not all nuclei are stable. There are about 400 stable nuclei; hundreds of others are unstable and tend to break apart into other particles. This process is called *nuclear decay*.

The nuclear decay process can be a natural event or it can be induced artificially. In either case, when a nucleus decays, radiation is emitted in the form of particles, photons, or both. The spontaneous emission of such energy, or *radiation,* is called *radioactivity*. The nucleus before decay is called the *parent nucleus,* and the nucleus remaining after decay is called the *daughter nucleus.* For example, the hands and numbers of the watch shown in **Figure 25-4** contain small amounts of radium salts. The nuclei within these salts decay, releasing light energy that causes the watch to glow in the dark. In this and all other nuclear reactions, the energy released is found by the equation $E = \Delta mc^2$.

A radioactive material can emit three types of radiation

Three types of radiation can be emitted by a nucleus as it undergoes radioactive decay: alpha (α) particles, in which the emitted particles are $^4_2$He nuclei; beta (β) particles, in which the emitted particles are either electrons or positrons (positively charged particles with a mass equal to that of the electron); and gamma (γ) rays, in which the emitted "rays" are high-energy photons. These three types of radiation are summarized in **Table 25-3.**

25-2 SECTION OBJECTIVES

- **Describe the three modes of nuclear decay.**
- **Predict the products of nuclear decay.**
- **Calculate the decay constant and the half-life of a radioactive substance.**

Figure 25-4
The radioactive decay of radium nuclei causes the hands and numbers of this watch to glow in the dark.

(★ TEKS) **3C**

Misconception Alert

Students may not understand why there are only 109 known elements but 400 stable nuclei have been identified. To clarify their understanding, have students review the meaning of isotopes (introduced in Section 25-1) and list examples.

Teaching Tip

Point out that all three types of radiation listed in **Table 25-3** are emitted from the nucleus. Have students consider possible sources for beta radiation (*protons and neutrons may break down*).

Table 25-3		Alpha, beta, and gamma radiation		
Particle	**Symbols**	**Composition**	**Charge**	**Effect on parent nucleus**
alpha	α ($^4_2$He)	2 protons, 2 neutrons	+2	mass loss; new element produced
beta	β^- ($^{0}_{-1}e$)	electron	−1	no significant change in mass; new element produced
	β^+ (^0_1e)	positron	+1	
gamma	γ	photon	0	energy loss

(★ TEKS)

p. 902: 2C, 2C
p. 903: 3C

Key Models and Analogies

Relate balancing nuclear reactions to students' knowledge of chemical equations. Point out that chemical reactions involve atomic masses and electrons, while nuclear processes involve nuclear particles. Students should realize that in both cases the procedures for balancing equations are based on the conservation of nucleon number and charge. Have students verify this in the example of the decay of uranium-238. (*The conservation of nucleon number leads to balancing the A numbers: 238 = 234 + 4; conservation of charge requires balancing the Z numbers: 92 = 90 + 2.*)

Misconception Alert

In the equation for beta decay, the symbol $_{-1}^{0}e$ suggests that the mass of the electron is zero. This may seem to contradict students' prior understanding of atomic physics. Explain that the zero means that the electron's mass is negligible compared with the other masses involved in nuclear decay. A numerical example will clarify this. Ask students to compare the mass of an electron with that of a proton by writing them with the same exponent ($m_e = 9.109 \times 10^{-31}$ kg; $m_p = 1.673 \times 10^{-27}$ kg $= 16730 \times 10^{-31}$ kg). This shows that a change in mass by beta decay is meaningful only at the fifth digit of precision.

TEKS

p. 905: 3C

---

internet connect

SC**LINKS**

NSTA

TOPIC: Radioactive decay
GO TO: www.scilinks.org
*sci*LINKS CODE: HF2252

CONCEPT PREVIEW

A positron is the antiparticle of the electron. An antiparticle is similar to a particle in most respects except that it has an opposite charge. Almost all particles have an antiparticle associated with them. Antiparticles are discussed in more detail in the feature at the end of this chapter.

The ability of radiation to pass through a material depends on the type of radiation. Alpha particles can usually be stopped by a piece of paper, beta particles can penetrate a few millimeters of aluminum, and gamma rays can penetrate several centimeters of pure lead.

Helium nuclei are emitted in alpha decay

When a nucleus undergoes alpha decay, it emits an alpha particle ($_{2}^{4}$He). Thus, the nucleus loses two protons and two neutrons. This makes the nucleus lighter and decreases its positive charge. (Because the electrons around the nuclei do not participate in nuclear reactions, they are ignored.)

For example, the nucleus of uranium-238 ($_{92}^{238}$U) can decay by alpha emission to a thorium-234 nucleus and an alpha particle. This process is written with symbols as follows:

$$_{92}^{238}\text{U} \rightarrow _{90}^{234}\text{Th} + _{2}^{4}\text{He}$$

This expression says that a parent nucleus, $_{92}^{238}$U, emits an alpha particle, $_{2}^{4}$He, and thereby changes to a daughter nucleus, $_{90}^{234}$Th (thorium-234). This nuclear reaction and all others follow the rules summarized in **Table 25-4.** These two rules can be used to determine the unknown daughter atom when a parent atom undergoes alpha decay.

Table 25-4	Rules for nuclear decay
The total of the atomic numbers on the left is the same as the total on the right because charge must be conserved.	
The total of the mass numbers on the left is the same as the total on the right because nucleon number must be conserved.	

Electrons or positrons are emitted in beta decay

When a radioactive nucleus undergoes beta decay, the nucleus emits either an electron or a positron. Thus, the atomic number is increased or decreased by one, with an opposite change in the neutron number. Because the daughter nucleus contains the same number of nucleons as the parent nucleus, the mass number does not change. Thus, beta decay does little to change the mass of a nucleus. Instead, the ratio of neutrons to protons in a nucleus is changed. This ratio affects the stability of the nucleus, as seen in Section 25-1.

A typical beta decay event involves carbon-14, as follows:

$$_{6}^{14}\text{C} \rightarrow _{7}^{14}\text{N} + _{-1}^{0}e$$

This decay produces an electron, written as $_{-1}^{0}e$. In this decay, the atomic number of the daughter nucleus is increased by 1.

Another beta decay event involves nitrogen-12, as follows:

$$^{12}_{7}\text{N} \rightarrow {}^{12}_{6}\text{C} + {}^{0}_{1}e$$

This decay produces a positron, written as $^{0}_{1}e$. In this decay, the atomic number of the daughter nucleus is decreased by 1.

The superscripts and subscripts on the carbon and nitrogen nuclei follow our usual conventions, but those on the electron and the positron may need some explanation. The -1 indicates that the electron has a charge whose magnitude is equal to that of the proton but is negative. Similarly, the 1 indicates that the positron has a charge that is equal to that of the proton in magnitude and sign. Thus, the subscript can be thought of as the charge of the particle. The 0 used for the mass number of the electron and the positron indicates that their masses are almost zero relative to those of carbon and nitrogen nuclei. Notice that both subscripts and superscripts must balance in the equations for beta decay, just as in alpha decay.

Beta decay involves the conversion of neutrons and protons

A bubble-chamber image of a positron is shown in **Figure 25-5.** The emission of electrons or positrons from a nucleus is surprising because the nucleus is made of only protons and neutrons. This apparent discrepancy can be explained by noting that in beta decay, either a neutron is transformed into a proton, creating an electron in the process, or a proton is transformed into a neutron, creating a positron in the process. These two beta decays can be written as follows:

$$^{1}_{0}n \rightarrow {}^{1}_{1}p + {}^{0}_{-1}e$$
$$^{1}_{1}p \rightarrow {}^{1}_{0}n + {}^{0}_{1}e$$

Decay events can be written in this way because other particles in the nucleus, much like the electrons around the nucleus, do not directly participate in the beta decay. The electrons and positrons involved in beta decay, on the other hand, are produced in the nuclear-decay process. Because they do not come from the shells around the nucleus, they cannot be ignored.

Neutrinos and antineutrinos are emitted in beta decay

Before we conclude our discussion of beta decay, there is one problem that must be resolved. In analyzing the experimental results of beta decay reactions, scientists noticed a disturbing fact. If carbon-14 beta decay actually occurred as described on page 904, energy, linear momentum, and angular momentum would not be conserved. In 1930, to solve this problem, Wolfgang Pauli proposed that a third particle must be missing from the equation. He reasoned that this new particle, called a *neutrino*, is necessary to conserve energy and momentum. Experimental evidence confirmed the existence of such a particle in 1956. (★ TEKS) 3C

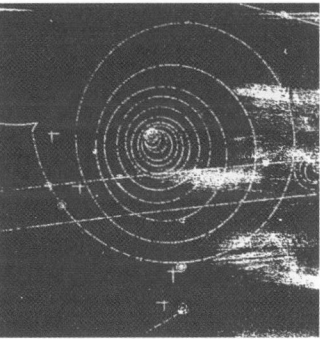

Figure 25-5
The spiral in this bubble-chamber image is the track left by a positron. This reaction took place in a magnetic field, which caused the positron to spiral as it lost energy.

Did you know?

The decay of the proton into a neutron and a positron is sometimes called *inverse-beta decay* to distinguish it from the decay of the neutron into a proton and an electron.

SECTION 25-2

Demonstration 2

Electron beam deflection

Purpose Explain the curved tracks observed in a bubble chamber.

Materials CRT, high-voltage source, horseshoe magnet

CAUTION *The deflection of the electron beam causes low-energy X rays.*

Procedure Explain to students that a CRT emits electrons when it is heated. These electrons are propelled from the anode to the cathode and beyond by a high potential difference. When the electrons hit the fluorescent screen, they produce a small amount of light. Turn off the lights in the classroom, and show the students the electron beam on the fluorescent screen. Then place the horseshoe magnet around the CRT so that the electron beam passes through the magnetic field formed at the ends of the magnet. Have students use the appropriate hand rule to predict the direction of the electron beam's deflection, then test their predictions. Reverse the polarity of the magnet so that the beam is deflected in the opposite direction.

Explain that tracks in a bubble chamber are formed by a bubbling around the ions created by the passage of charged beta or alpha particles. Point out that the direction of a particle's deflection in a magnetic field can be used to determine the particle's charge. Ask students if they could tell which tracks were created by alpha, beta, or gamma radiation, given the magnetic field. Discuss the effect of mass on curvature.

Key Models and Analogies

To help students understand the discovery of the neutrino, have them think of the C-14 decay as an explosion. If you traced the fragments' masses and velocities but a fragment got lost, the final momentum would not appear to equal the initial momentum. Similarly, by adding the directions and speeds that correspond to the tracks from N-14 and beta particles, it was revealed that a fragment of the reaction (the neutrino) appeared to be missing.

Teaching Tip

The equations for nuclear reactions apply the conservation of nucleon number and charge. Other universal conservation laws, including the conservation of momentum and the conservation of energy, must also be satisfied in nuclear processes.

Misconception Alert

Students may not understand how the antineutrino is the neutrino's antiparticle because each has zero charge. Explain that in addition to charge, antiparticles are opposite their corresponding particle in spin. (Spin is described in the Physics on the Edge feature in Chapter 8.)

The Language of Physics

Because the gamma decay follows the beta decay immediately, the two processes are sometimes expressed by a single equation:
$$^{12}_{5}B \rightarrow {}^{12}_{6}C + \gamma + {}^{0}_{-1}e + \overline{\nu}$$

Table 25-5
Properties of the neutrino

The neutrino has zero electric charge.

The neutrino has a mass much smaller than that of the electron, and in fact, its mass may be zero.

The neutrino interacts very weakly with matter and is therefore very difficult to detect.

Did you know?

The word *neutrino* means "little neutral one." It was suggested by the physicist Enrico Fermi because the neutrino had to have zero electric charge and little or no mass.

The Greek letter nu (ν) is used to represent a neutrino. When a bar is drawn above the nu ($\overline{\nu}$), the particle is an antineutrino, or the antiparticle of a neutrino. The properties of the neutrino are summarized in **Table 25-5**. Note that the neutrino has no electric charge and that its mass is very small, perhaps even zero. As a result, the neutrino is difficult to detect experimentally.

With the neutrino, we can now describe the beta decay process of carbon-14 in a form that takes energy and momentum conservation into account, as follows:

$$^{14}_{6}C \rightarrow {}^{14}_{7}N + {}^{0}_{-1}e + \overline{\nu}$$

According to this expression, carbon-14 decays into a nitrogen nucleus, releasing an electron and an antineutrino in the process.

The decay of nitrogen-12 can also be rewritten, as follows:

$$^{12}_{7}N \rightarrow {}^{12}_{6}C + {}^{0}_{1}e + \nu$$

Here we see that when $^{12}_{7}N$ decays into $^{12}_{6}C$, a positron and a neutrino are produced. To avoid confusing these two types of beta decay, keep in mind this simple rule: *In beta decay, an electron is always accompanied by an antineutrino and a positron is always accompanied by a neutrino.*

High-energy photons are emitted in gamma decay

Very often, a nucleus that undergoes radioactive decay, either alpha or beta, is left in an excited energy state. The nucleus can then undergo a gamma decay in which one or more nucleons make transitions from a higher energy level to a lower energy level. In the process, one or more photons are emitted. Such photons, or *gamma rays,* have very high energy relative to the energy of visible light. The process of nuclear de-excitation, or gamma decay, is very similar to the emission of light by an atom, in which an electron makes a transition from a state of higher energy to a state of lower energy (as discussed in Chapter 23). Note that in gamma decay energy is emitted but the parts of the nucleus are left unchanged. Thus, both the atomic number and the mass number stay the same. Nonetheless, gamma decay is still considered to be a form of nuclear decay because it involves protons or neutrons in the nucleus.

Two common reasons for a nucleus being in an excited state are alpha and beta decay. The following sequence of events represents a typical situation in which gamma decay occurs:

$$^{12}_{5}B \rightarrow {}^{12}_{6}C^{\star} + {}^{0}_{-1}e + \overline{\nu}$$

$$^{12}_{6}C^{\star} \rightarrow {}^{12}_{6}C + \gamma$$

The first step is a beta decay in which $^{12}_{5}B$ decays to $^{12}_{6}C^{\star}$. The asterisk indicates that the carbon nucleus is left in an excited state following the decay. The excited carbon nucleus then decays in the second step to the ground state by emitting a gamma ray.

NUCLEAR DECAY SERIES

If the product of a nuclear decay is stable, the decay process ends. In other cases, the decay product—the daughter nucleus—is itself unstable. The daughter nucleus then becomes the parent nucleus for an additional decay process. Such a sequence is called a *decay series*.

Figure 25-6(a) depicts the number of protons versus neutrons for all stable nuclei. A small portion of this graph is enlarged in **Figure 25-6(b),** which shows a naturally occurring decay series beginning with thorium, Th, and ending with lead, Pb. Each square in **Figure 25-6(b)** corresponds to a possible nucleus. The black dots represent stable nuclei, and the red dots represent unstable nuclei. Thus, each black dot in **Figure 25-6(b)** corresponds to a data point in the circled portion of **Figure 25-6(a).** The decay series continues until a stable nucleus is reached, in this case ^{208}Pb. Notice that there is a branch in the decay path; there are actually two ways that thorium can decay into lead.

The entire series in **Figure 25-6(b)** consists of 10 decays: 6 alpha decays and 4 beta decays. When α decay occurs, the nucleus moves down two squares and to the left two squares because it loses two protons and two neutrons. When β^- decay occurs, the nucleus moves down one square and to the right one square because it loses one neutron and gains one proton. Gamma decays are not represented because they do not alter the ratio of protons to neutrons.

The result of the decay series is to lighten the nucleus. The neutron-to-proton ratio in a parent nucleus is altered through alpha and beta decay. The proper ratio of neutrons to protons is important, as we have seen, in balancing the strong and Coulomb forces in the nucleus.

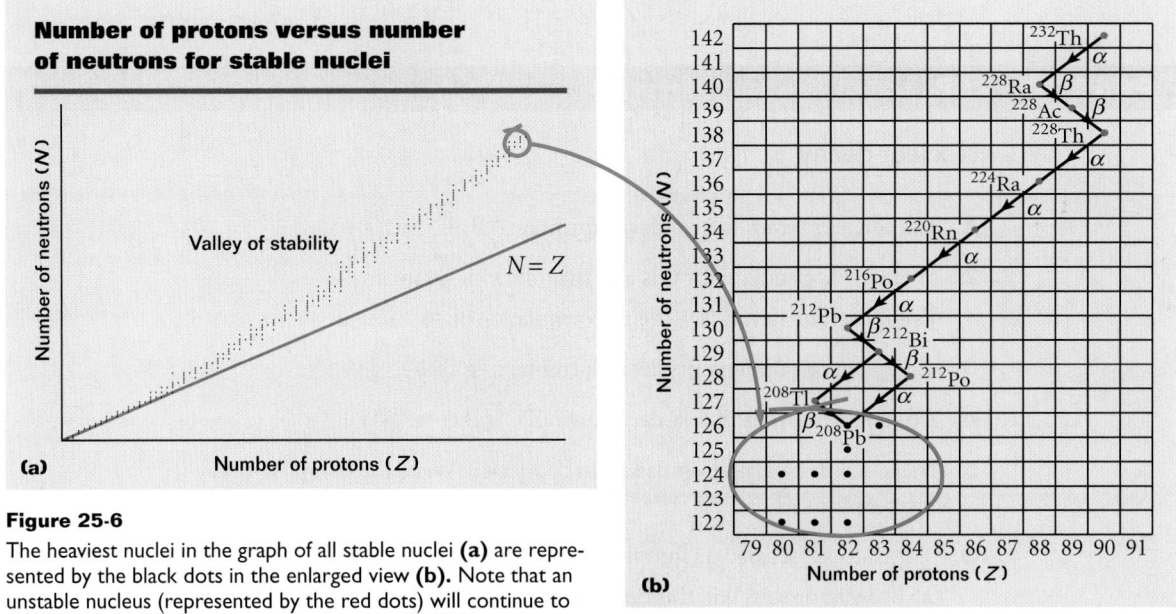

Figure 25-6
The heaviest nuclei in the graph of all stable nuclei **(a)** are represented by the black dots in the enlarged view **(b)**. Note that an unstable nucleus (represented by the red dots) will continue to decay until the daughter nucleus is stable. (★)TEKS **2C**

Visual Strategy

Figure 25-6
Be sure students understand the relationship between the two parts of this figure. Point out that the numbers shown next to each chemical symbol on the graph in **(b)** are mass numbers A ($= Z + N$). For example, with Pb, $82 + 126 = 208$.

Q Why does alpha decay move the nucleus two squares to the left and two squares down?

A *When an alpha particle is emitted, the nucleus loses two protons and two neutrons. That gives $Z - 2$ (two squares to the left) and $N - 2$ (two squares down).*

Q Write the nuclear reaction showing the decay of Ra-228 to Ac-228.

A $^{228}_{88}Ra \rightarrow {}^{228}_{89}Ac + {}^{\ 0}_{-1}e + \overline{\nu}$

Q Can you tell from the graph in **(b)** whether gamma radiation was emitted?

A *No; the graph only shows changes in mass and charge. Gamma rays have neither, so the energy released by gamma radiation is not shown on this representation.*

Classroom Practice

The following may be used as teamwork exercises or for demonstration at the chalkboard or on an overhead projector.

Nuclear decay ⭐TEKS 2C

PROBLEM

Nuclear decay

Complete the following reactions:

a. $^{228}_{88}\text{Ra} \rightarrow X + ^{0}_{-1}e + \overline{\nu}$

b. $^{220}_{86}\text{Rn} \rightarrow ^{216}_{84}\text{Po} + X$

c. $^{212}_{82}\text{Pb} \rightarrow X + ^{0}_{-1}e + \overline{\nu}$

Answers

a. $^{228}_{89}\text{Ac}$

b. $^{4}_{2}\text{He}$

c. $^{212}_{83}\text{Bi}$

PRACTICE GUIDE 25B

Solving for:	
nuclear decay	📖 **PE** Sample, 1–6; Ch. Rvw. 23–25, 41*, 46*, 47, 48–53
	💿 **PW** 1–7
	PB Sample, 1–10

ANSWERS TO

Practice 25B
Nuclear decay

1. $^{12}_{6}\text{C}$

2. $^{208}_{81}\text{Tl}$

3. $^{14}_{6}\text{C}$

4. $^{4}_{2}\text{He}$

5. $^{63}_{28}\text{Ni} \rightarrow ^{63}_{29}\text{Cu} + ^{0}_{-1}e + \overline{\nu}$

6. **a.** β^{-} decay

   b. $^{56}_{26}\text{Fe} \rightarrow ^{56}_{27}\text{Co} + ^{0}_{-1}e + \overline{\nu}$

PROBLEM

The element radium was discovered by Marie and Pierre Curie in 1898. One of the isotopes of radium, $^{226}_{88}\text{Ra}$, decays by alpha emission. What is the resulting daughter element?

SOLUTION

Given: The decay can be written symbolically as follows:

$$^{226}_{88}\text{Ra} \rightarrow X + ^{4}_{2}\text{He}$$

Unknown: the daughter element (X)

The mass numbers and atomic numbers on the two sides of the expression must be the same so that both charge and nucleon number are conserved during the course of this particular decay.

Mass number of X = 226 − 4 = 222

Atomic number of X = 88 − 2 = 86

$$^{226}_{88}\text{Ra} \rightarrow ^{222}_{86}X + ^{4}_{2}\text{He}$$

The periodic table (Appendix F) shows that the nucleus with an atomic number of 86 is radon, Rn. Thus, the process is as follows:

$$\boxed{^{226}_{88}\text{Ra} \rightarrow ^{222}_{86}\text{Rn} + ^{4}_{2}\text{He}}$$

PRACTICE 25B

Nuclear decay

1. Complete this radioactive-decay formula: $^{12}_{5}\text{B} \rightarrow ? + ^{0}_{-1}e + \overline{\nu}$

2. (Refer to Appendix F for this and the following problems.)
   Complete this radioactive-decay formula: $^{212}_{83}\text{Bi} \rightarrow ? + ^{4}_{2}\text{He}$

3. Complete this radioactive-decay formula: $? \rightarrow ^{14}_{7}\text{N} + ^{0}_{-1}e + \overline{\nu}$

4. Complete this radioactive-decay formula: $^{225}_{89}\text{Ac} \rightarrow ^{221}_{87}\text{Fr} + ?$

5. Nickel-63 decays by β^{-} emission to copper-63. Write the complete decay formula for this process.

6. The isotope $^{56}_{26}\text{Fe}$ decays into the isotope $^{56}_{27}\text{Co}$.
   a. By what process will this decay occur?
   b. Write the decay formula for this process.

MEASURING NUCLEAR DECAY

Imagine that you are studying a sample of radioactive material. You know that the atoms in the material are decaying into other types of atoms. How many of the unstable parent atoms remain after a certain amount of time?

The decay constant indicates the rate of radioactive decay

If the sample contains N radioactive parent nuclei at some instant, the number of parent nuclei that decay into daughter nuclei (ΔN) in a small time interval (Δt) is proportional to N, as follows:

$$\Delta N = -\lambda N \Delta t$$

The negative sign signifies that N decreases with time; that is, ΔN is negative. The quantity λ is called the *decay constant*. The value of λ for any isotope indicates the rate at which that isotope decays. Isotopes with a large decay constant decay quickly, and those with a small decay constant decay slowly. The number of decays per unit time, $-\Delta N/\Delta t$, is called the *decay rate*, or *activity*, of the sample. Note that the activity of a sample equals the decay constant times the number of radioactive nuclei in the sample, as follows:

$$\text{activity} = -\Delta N/\Delta t = \lambda N$$

The SI unit of activity is the *becquerel* (Bq). One becquerel is equal to 1 decay/s. The *curie* (Ci), which was the original unit of activity, is the approximate activity of 1 g of radium. One curie is equal to 3.7×10^{10} Bq.

Half-life measures how long it takes half a sample to decay

Another quantity that is useful for characterizing radioactive decay is the **half-life,** written as $T_{1/2}$. The half-life of a radioactive substance is the time it takes for half a given number of radioactive nuclei to decay. The half-life of any substance is inversely proportional to the decay constant of the substance.

half-life

the time required for half the original nuclei of a radioactive material to undergo radioactive decay

Conceptual Challenge ⓣ TEKS 3A

1. Decay series

Suppose a radioactive parent substance with a very long half-life has a daughter with a very short half-life. Describe what happens to a freshly purified sample of the parent substance.

2. Probability of decay

"The more probable the decay, the shorter the half-life." Explain this statement.

3. Decay of radium

The radioactive nucleus $^{226}_{88}\text{Ra}$ (radium-226) has a half-life of about 1.6×10^3 years. Although the solar system is approximately 5 billion years old, we still find this radium nucleus in nature. Explain how this is possible.

Classroom Practice

The following may be used as teamwork exercises or for demonstration at the chalkboard or on an overhead projector.

PROBLEM

Measuring nuclear decay

The half-life of $^{144}_{56}\text{Ba}$ is about 12 s. What is the decay constant of barium-144?

Answer

$5.8 \times 10^{-2}\,\text{s}^{-1}$

A sample of barium contains 5.0×10^9 undecayed nuclei of barium-144. How many radioactive nuclei remain after the following time periods:

a. 12 s

b. 24 s

c. 36 s

Answers

   a. $N = 2.5 \times 10^9$

   b. $N = 1.2 \times 10^9$

   c. $N = 6.2 \times 10^8$

Determine the activity in becquerels and in curies for each time period given above.

Answers

   a. 1.4×10^8 Bq; 3.9×10^{-3} Ci

   b. 7.0×10^7 Bq; 1.9×10^{-3} Ci

   c. 3.6×10^7 Bq; 9.7×10^{-4} Ci

Teaching Tip

Point out that, as seen in the Classroom Practice, the activity of a sample decreases over time because the number of radioactive nuclei (N) decreases.

Substances with large decay constants have short half-lives. The relationship between half-life and decay constant is given in the equation below. We shall not derive this equation here, but it involves the natural logarithm of 2. Because $\ln 2 = 0.693$, this factor occurs in the final equation.

HALF-LIFE

$$T_{1/2} = \frac{0.693}{\lambda}$$

$$\text{half-life} = \frac{0.693}{\text{decay constant}}$$

Consider a sample that begins with N radioactive parent nuclei. By definition, after one half-life, $\frac{1}{2}N$ radioactive parent nuclei remain; the rest have decayed into daughter nuclei. After two half-lives, half the $\frac{1}{2}N$ parent nuclei will have decayed, so $\frac{1}{4}N$ radioactive parent nuclei remain. After three half-lives, there will be $\frac{1}{8}N$ parent nuclei and $\frac{7}{8}N$ daughter nuclei, and so on.

SAMPLE PROBLEM 25C

Measuring nuclear decay ⭐TEKS 2C

PROBLEM

The half-life of the radioactive radium (^{226}Ra) nucleus is 5.0×10^{10} s. A sample contains 3.0×10^{16} nuclei. What is the decay constant for this decay? How many radium nuclei, in curies, will decay per second?

SOLUTION

1. DEFINE   **Given:** $T_{1/2} = 5.0 \times 10^{10}$ s $N = 3.0 \times 10^{16}$

Unknown: $\lambda = ?$ activity $= ?$ Ci

2. PLAN   **Choose an equation(s) or situation:**

To find the decay constant, use the equation for half-life, given on this page.

$$T_{1/2} = \frac{0.693}{\lambda}$$

The number of nuclei that decay per second is given by the equation for the activity of a sample, given on page 909:

$$\text{activity} = \lambda N$$

Rearrange the equation(s) to isolate the unknown(s):

The first equation must be rearranged to isolate the decay constant, λ.

$$\lambda = \frac{0.693}{T_{1/2}}$$

3. CALCULATE **Substitute the values into the equation(s) and solve:**

$$\lambda = \frac{0.693}{T_{1/2}} = \frac{0.693}{5.0 \times 10^{10} \text{ s}}$$

$$\boxed{\lambda = 1.4 \times 10^{-11} \text{s}^{-1}}$$

$$\text{activity} = \lambda N = (1.4 \times 10^{-11} \text{ s}^{-1})(3.0 \times 10^{16})$$

To convert the answer to Ci, divide by 3.7×10^{10} s^{-1}/Ci.

$$\text{activity} = \frac{(1.4 \times 10^{-11} \text{ s}^{-1})(3.0 \times 10^{16})}{3.7 \times 10^{10} \text{ s}^{-1}/\text{Ci}}$$

$$\boxed{\text{activity} = 1.1 \times 10^{-5} \text{ Ci}}$$

4. EVALUATE Because the half-life is on the order of 10^{10} s, the decay constant, which approximately equals 0.7 divided by the half-life, should equal a little less than 10^{-10} s^{-1}. Thus, 1.4×10^{-11} s^{-1} is a reasonable answer for the decay constant.

PRACTICE 25C

Measuring nuclear decay (★)TEKS 2C

1. The half-life of $^{214}_{84}$Po is 164 μs. A polonium-214 sample contains 2.0×10^6 nuclei. What is the decay constant for the decay? How many polonium nuclei, in curies, will decay per second?

2. The half-life of $^{214}_{83}$Bi is 19.7 min. A bismuth-214 sample contains 2.0×10^9 nuclei. What is the decay constant for the decay? How many bismuth nuclei, in curies, will decay per second?

3. The half-life of $^{131}_{53}$I is 8.07 days. Calculate the decay constant for this isotope. What is the activity in Ci for a sample that contains 2.5×10^{10} iodine-131 nuclei?

4. Suppose that you start with 1.00×10^{-3} g of a pure radioactive substance and determine 2.0 h later that only 0.25×10^{-3} g of the substance is left undecayed. What is the half-life of this substance?

5. Radon-222 ($^{222}_{86}$Rn) is a radioactive gas with a half-life of 3.82 days. A gas sample contains 4.0×10^8 radon atoms initially.

   a. Estimate how many radon atoms will remain after 12 days.
   b. Estimate how many radon nuclei will have decayed by this time.

PRACTICE GUIDE 25C

Solving for:		
λ	📖 **PE**	Sample, 1–3; Ch. Rvw. 26, 55
	💿 **PW**	5–6
	PB	6, 9
$T_{1/2}$	📖 **PE**	4; Ch. Rvw. 57
	💿 **PW**	7
	PB	Sample, 1–3
N	📖 **PE**	5; Ch. Rvw. 56*, 60–62
	💿 **PW**	1, 4
	PB	7, 10
Δt	📖 **PE**	Ch. Rvw. 27–28, 58
	💿 **PW**	Sample, 2–3
	PB	4–5
activity	📖 **PE**	Ch. Rvw. 26, 54*, 55
	PB	8

ANSWERS TO

Practice 25C
Measuring nuclear decay
1. 4.23×10^3 s^{-1}, 0.23 Ci
2. 5.86×10^{-4} s^{-1}, 3.2×10^{-5} Ci
3. 9.94×10^{-7} s^{-1}, 6.7×10^{-7} Ci
4. 1.0 h
5. a. about 5.0×10^7 atoms
   b. about 3.5×10^8 atoms

(★)TEKS

p. 910: 2C
p. 911: 2C

Misconception Alert

Some students may think that the half-life time is half of the time it takes for a sample to be used up. Ask students how many radioactive nuclei out of 600 are left after 2.0 h if $T_{1/2}$ is 1.0 h (*150*).

Visual Strategy

Figure 25-7

Be sure students interpret the graph properly.

Q Given that the half-life of C-14 is 5730 years, what years are shown on the graph points?

A *0; 5730; 11 460; 17 190*

Q If the initial sample contained 80 000 radioactive nuclei, when would it fall below 20 000 nuclei?

A *after 11 460 years (two half-lives)*

Section Review
ANSWERS

1. In α, $^4_2$He particles are emitted; in β, electrons or positrons are emitted; in γ, photons are emitted. In α and β, a new element is produced; in γ, energy is lost.

2. **a.** $^{228}_{88}$Ra
   b. $^{12}_{6}$C
   c. $^{149}_{62}$Sm

3. **a.** 1.2×10^{-10} s^{-1}
   b. 5.8×10^{9} s (180 years)

4. 17 200 years

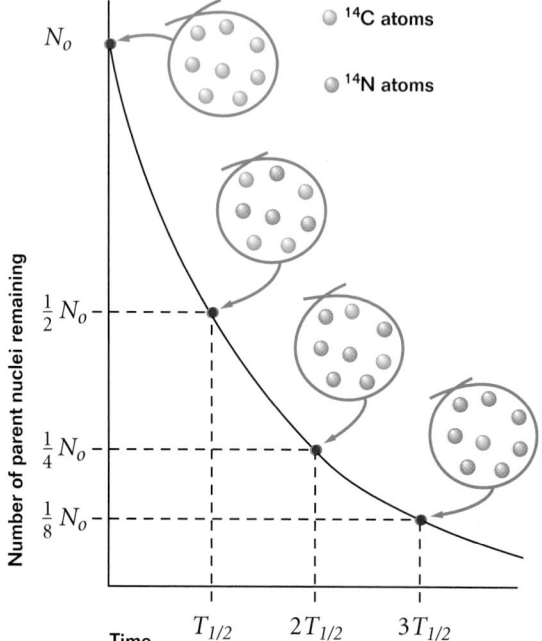

Figure 25-7

The radioactive isotope carbon-14 has a half-life of 5730 years. In each successive 5730-year period, half the remaining mass of carbon-14 decays to nitrogen-14.

★ TEKS 2C

A *decay curve* is a plot of the number of radioactive parent nuclei remaining in a sample as a function of time. A typical decay curve for a radioactive sample is shown in **Figure 25-7.** After each half-life, half the remaining parent nuclei have decayed. This is represented in the circles to the right of the decay curve. The blue spheres are the parent nuclei (carbon-14), and the red spheres are daughter nuclei (nitrogen-14). Notice that the total number of nuclei remains constant, while the number of carbon atoms continually decreases over time.

For example, the initial sample contains 8 carbon-14 atoms. After one half-life, there are 4 carbon-14 atoms and 4 nitrogen-14 atoms. By the next half-life, the number of carbon-14 atoms is reduced to 2, and the process continues. As the number of carbon-14 atoms decreases, the number of nitrogen-14 atoms increases.

Living organisms have a constant ratio of carbon-14 to carbon-12 because they continuously exchange carbon dioxide with their surroundings. When an organism dies, this ratio changes due to the decay of carbon-14. Measuring the ratio between carbon-14, which decays as shown in **Figure 25-7,** and carbon-12, which does not decay, provides an approximate date as to when the sample was alive.

Section Review

★ TEKS 2C, 3C

1. Explain the main differences between alpha, beta, and gamma decays.

2. Complete the following radioactive decay formulas:
   a. $^{232}_{90}$Th → ? + $^4_2$He
   b. $^{12}_5$B → ? + $^0_{-1}e + \overline{\nu}$
   c. ? → $^4_2$He + $^{145}_{60}$Nd

3. A radioactive sample consists of 5.3×10^5 nuclei. There is one decay every 4.2 h.
   a. What is the decay constant for the sample?
   b. What is the half-life for the sample?

4. **Physics in Action** The ^{14}C content decreases after the death of a living system with a half-life of 5730 years. If an archaeologist finds an ancient fire pit containing partially consumed firewood and if the ^{14}C content of the wood is only 12.5 percent that of an equal carbon sample from a present-day tree, what is the age of the ancient site?

25-3
Nuclear reactions

FISSION AND FUSION

Any process that involves a change in the nucleus of an atom is called a nuclear reaction. Nuclear reactions include *fission,* in which a nucleus splits into two or more nuclei, and *fusion,* in which one or more nuclei combine.

Stable nuclei can be made unstable

When a nucleus is bombarded with energetic particles, it may capture a particle, such as a neutron. As a result, the nucleus will no longer be stable and will disintegrate. For example, protons can be released when alpha particles collide with nitrogen atoms, as follows:

$$^4_2He + ^{14}_7N \rightarrow X + ^1_1H$$

According to this expression, an alpha particle (^4_2He) strikes a nitrogen nucleus ($^{14}_7N$) and produces an unknown product nucleus (X) and a proton (^1_1H). By balancing atomic numbers and mass numbers, we can conclude that the unknown product has a mass number of 17 and an atomic number of 8. Because the element with an atomic number of 8 is oxygen, the product can be written symbolically as $^{17}_8O$, and the reaction can be written as follows:

$$^4_2He + ^{14}_7N \rightarrow ^{17}_8O + ^1_1H$$

This nuclear reaction starts with two stable isotopes—helium and nitrogen—that form an unstable intermediate nucleus ($^{18}_9F$). The intermediate nucleus then disintegrates into two different stable isotopes, hydrogen and oxygen. This reaction, which was the first nuclear reaction to be observed, was detected by Rutherford in 1919. ⭐TEKS 3E

Heavy nuclei can undergo nuclear fission

Nuclear fission occurs when a heavy nucleus splits into two lighter nuclei. For fission to occur naturally, the nucleus must release energy. This means that the nucleons in the daughter nuclei must be more tightly bound and therefore have less mass than the nucleons in the parent nucleus. This decrease in mass per nucleon appears as released energy when fission occurs, often in forms such as photons or kinetic energy of the fission products. Because fission produces lighter nuclei, the binding energy per nucleon must *increase* with decreasing atomic number. **Figure 25-8** shows that this is possible only for atoms in which $A > 58$. Thus, *fission occurs naturally only for heavy atoms.*

25-3 SECTION OBJECTIVES

- **Distinguish between nuclear fission and nuclear fusion.**
- **Explain how a chain reaction is utilized by nuclear reactors.**
- **Compare fission and fusion reactors.**

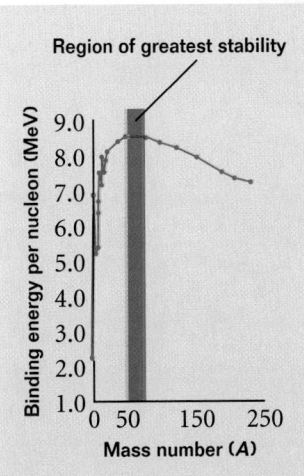

Region of greatest stability

Binding energy per nucleon (MeV) vs Mass number (A)

Figure 25-8
Light nuclei are very loosely bound. The binding energy of heavy nuclei is roughly the same for all nuclei.

⭐TEKS 2C

Visual Strategy

Figure 25-8
Point out that the graph represents empirical information about energy of all known nuclei, including isotopes of mass numbers $A = 1$ to $A \approx 230$.

Q According to the graph, what is the average binding energy per nucleon of the most stable nucleus?

A *about 8.5 MeV*

Q Suppose a nucleus of mass number $A = 24$ splits into two nuclei of equal masses. Determine the approximate binding energy per nucleon for each daughter nucleus.

A *about 7 MeV; this is less than the binding energy per nucleon of the parent, so the daughter nuclei will be less stable. This example illustrates why nuclei with $A < 58$ do not undergo fission.*

Q Repeat the last exercise with a parent nucleus of $A = 200$.

A *about 8.5 MeV; in this case, the binding energy per nucleon increases after fission.*

⭐TEKS

p. 912: 2C, 2C, 3C
p. 913: 3E, 2C

The Language of Physics

Remind students that the upper number next to the element symbol is the mass number (A) and that the lower number indicates the number of protons (Z). Also remind them that these equations show the conservation of nucleon number and charge in a nuclear reaction. Ask how many protons are found in the parents and in the daughters in the nuclear fission of U-235 ($0 + 92 = 56 + 36 + 0$). How many nucleons are found in each? ($1 + 235 = 140 + 93 + (3 \times 1)$)

Misconception Alert

Students may wonder about the meaning of "an average of about 2.5 neutrons" being emitted when ^{235}U undergoes fission, as described in the last paragraph. Point out that the products of U-235 fission may be different daughter nuclei than those shown in the diagram. Some reactions release three neutrons, while others release two. Have students find an example of each in **Figure 25-9.** (three neutrons: $^{1}_{0}n + ^{235}_{92}U \rightarrow ^{87}_{35}Br + ^{146}_{57}La + 3 ^{1}_{0}n$; two neutrons: $^{1}_{0}n + ^{235}_{92}U \rightarrow ^{144}_{55}Cs + ^{90}_{37}Rb + 2 ^{1}_{0}n$)

Teaching Tip

Point out that **Figure 25-9** is a schematic diagram. It does not represent the actual location of U-235 nuclei. There are about 2.6×10^{21} nuclei in 1 g (about 12 cm^3) of uranium. Most of these nuclei are U-238, which are more stable. Only a small portion are U-235, which will decay when hit by a neutron.

One example of this process is the fission of uranium-235. First, the nucleus is bombarded with neutrons. When the nucleus absorbs a neutron, it becomes unstable and decays. The fission of ^{235}U can be represented as follows:

$$^{1}_{0}n + ^{235}_{92}U \rightarrow ^{236}_{92}U^{\star} \rightarrow X + Y + \text{neutrons}$$

The isotope $^{236}_{92}U^{\star}$ is an intermediate state that lasts only for about 10^{-12} s before splitting into X and Y. Many combinations of X and Y are possible. In the fission of uranium, about 90 different daughter nuclei can be formed. The process also results in the production of about two or three neutrons per fission event.

A typical reaction of this type is as follows:

$$^{1}_{0}n + ^{235}_{92}U \rightarrow ^{140}_{56}Ba + ^{93}_{36}Kr + 3 ^{1}_{0}n$$

To estimate the energy released in a typical fission process, note that the binding energy per nucleon is about 7.6 MeV for heavy nuclei (those having a mass number of approximately 240) and about 8.5 MeV for nuclei of intermediate mass (see **Figure 25-8** on page 913). The amount of energy released in a fission event is the difference in these binding energies (8.5 MeV − 7.6 MeV, or about 0.9 MeV per nucleon). Assuming a total of 240 nucleons, this is about 220 MeV. This is a very large amount of energy relative to the energy released in typical chemical reactions. For example, the energy released in burning one molecule of the octane used in gasoline engines is about one hundred-millionth the energy released in a single fission event. (★)TEKS 2C

Neutrons released in fission can trigger a chain reaction

When ^{235}U undergoes fission, an average of about 2.5 neutrons are emitted per event. The released neutrons can be captured by other nuclei, making these nuclei unstable. This triggers additional fission events, which lead to the possibility of a *chain reaction,* as shown in **Figure 25-9.** Calculations show that if the chain reaction is not controlled—that is, if it does not proceed slowly—it could result in the release of an enormous amount of energy and a violent explosion. If the energy in 1 kg of ^{235}U were released, it would equal the energy released by the detonation of about 20 000 tons of TNT. This is the principle behind the first nuclear bomb, shown in **Figure 25-10,** which was essentially an uncontrolled fission reaction.

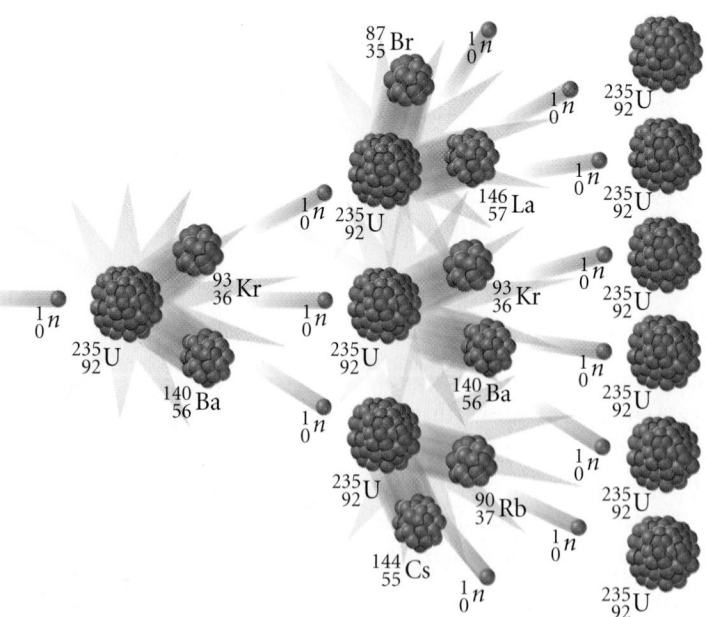

Figure 25-9

A nuclear chain reaction can be initiated by the capture of a neutron.

A *nuclear reactor* is a system designed to maintain a controlled, self-sustained chain reaction. Such a system was first achieved with uranium as the fuel in 1942 by Enrico Fermi, at the University of Chicago. Primarily, it is the uranium-235 isotope that releases energy through nuclear fission. Uranium from ore typically contains only about 0.7 percent of ^{235}U, with the remaining 99.3 percent being the ^{238}U isotope. Because uranium-238 tends to absorb neutrons instead of undergoing fission, reactor fuels must be processed to increase the proportion of ^{235}U so the reaction can sustain itself. This process is called *enrichment*. ★TEKS **3E**

At this time, all nuclear reactors operate through fission. One difficulty associated with fission reactors is the safe disposal of radioactive materials when the core is replaced. Transportation of reactor fuel and reactor wastes poses safety risks. As with all energy sources, the risks must be weighed against the benefits and the availability of the energy source. ★TEKS **3C**

Light nuclei can undergo nuclear fusion

Nuclear fusion, the opposite of nuclear fission, occurs when two light nuclei combine to form a heavier nucleus. As with fission, the product of a fusion event must have a greater binding energy than the original nuclei for energy to be released in the reaction. Because fusion reactions produce heavier nuclei, the binding energy per nucleon must increase as atomic number increases. As shown in **Figure 25-8** (page 913), this is possible only for atoms with A less than 58. Hence, *fusion occurs naturally only for light atoms.*

One example of this process is the fusion reactions that occur in stars. All stars generate energy through fusion. About 90 percent of the stars, including our sun, fuse hydrogen and possibly helium. Some other stars fuse helium or other heavier elements. The *proton-proton cycle* is a series of three nuclear-fusion reactions that are believed to be stages in the liberation of energy in our sun and other stars rich in hydrogen. In the proton-proton cycle, four protons combine to form an alpha particle and two positrons, releasing 25 MeV of energy in the process. The first two steps in this cycle are as follows:

$$^1_1\text{H} + ^1_1\text{H} \rightarrow ^2_1\text{H} + ^0_1e + \nu$$

$$^1_1\text{H} + ^2_1\text{H} \rightarrow ^3_2\text{He} + \gamma$$

This is followed by either of the following processes:

$$^1_1\text{H} + ^3_2\text{He} \rightarrow ^4_2\text{He} + ^0_1e + \nu$$

$$^3_2\text{He} + ^3_2\text{He} \rightarrow ^4_2\text{He} + ^1_1\text{H} + ^1_1\text{H}$$

The released energy is carried primarily by gamma rays, positrons, and neutrinos. These energy-liberating fusion reactions are called *thermonuclear fusion reactions*. The hydrogen (fusion) bomb, first detonated in 1952, is an example of an uncontrolled thermonuclear fusion reaction.

Figure 25-10
The first nuclear fission bomb, often called the atomic bomb, was tested in New Mexico in 1945.

Did you know?

Stars do not normally fuse elements beyond iron-58 because it is the most stable element. Fusing elements beyond iron-58 would require an input of energy rather than producing energy.

Teaching Tip

Have students use **Figure 25-8** (on page 913) to answer the following questions: If two nuclei of 90 nucleons each were to fuse, would the binding energy per nucleon in the 180-nucleon nucleus be higher or lower? *(The binding energy per nucleon would decrease from 8.2 MeV to 7.5 MeV.)* What can you infer about the stability of the product? *(The product would be less stable.)* What if two nuclei with five nucleons each were fused? *(The binding energy per nucleon would increase from 7.1 MeV to 8.5 MeV; the product would be more stable.)*

The Language of Physics

The terms *atomic bomb* and *hydrogen bomb* are misleading because terms associated with atoms and elements are used to denote *nuclear* reactions. Students should be reminded that the order of magnitude of energies at the nuclear level is a million times higher than it is at the atomic level. What has been called the *atomic bomb* since 1945 is actually a tremendous *nuclear fission* reaction. Likewise, the so-called *hydrogen bomb* is an uncontrolled *nuclear fusion* reaction in which hydrogen nuclei merge to form helium nuclei.

★TEKS
p. 914: 2C
p. 915: 3E, 3C

Teaching Tip

Have students choose a question about fusion to research and report to class. Example questions may include:

What will happen when the sun runs out of hydrogen, which currently produces helium and energy?

Why don't power plants use fusion to produce electricity?

What is true and false about *cold fusion*?

Section Review

ANSWERS

1. Both are nuclear reactions that release energy. In fission, a heavy nucleus splits; in fusion, two light nuclei combine.

2. The excess neutrons produced by each reaction are used to induce additional reactions.

3. because fusion produces more stable nuclei than fission

4. Enrichment is the process of increasing the amount of ^{235}U in reactor fuels; this is necessary because uranium from ore contains mostly ^{238}U.

5. a. $^{1}_{0}n + ^{235}_{92}U \rightarrow ^{141}_{56}Ba + ^{92}_{36}Kr + 3\ ^{1}_{0}n$

   b. three neutrons

6. Fusion reactors produce less radioactive waste, and the fuel (water) is plentiful. Difficulties include the high temperatures and densities necessary to achieve controlled fusion reactions.

916

Fusion reactors are being developed 3C

The enormous amount of energy released in fusion reactions suggests the possibility of harnessing this energy for useful purposes on Earth. Efforts are under way to create controlled thermonuclear reactions in the form of a *fusion reactor*. Because of the ready availability of its fuel source—water—controlled fusion is often called the ultimate energy source.

For example, if deuterium ($^{2}_{1}H$) were used as the fuel, 0.16 g of deuterium could be extracted from just 1 L of water at a cost of about one cent. Such rates would make the fuel costs of even an inefficient reactor almost insignificant. An additional advantage of fusion reactors is that few radioactive byproducts are formed. The proton-proton cycle shows that the end product of the fusion of hydrogen nuclei is safe, nonradioactive helium. Unfortunately, a thermonuclear reactor that can deliver a net power output for an extended time is not yet a reality. Many difficulties must be resolved before a successful device is constructed.

For example, the energy released in a gas undergoing nuclear fusion depends on the number of fusion reactions that can occur in a given amount of time. This varies with the density of the gas because collisions are more frequent in a denser gas. It also depends on the amount of time the gas is confined.

In addition, the Coulomb repulsion force between two charged nuclei must be overcome before they can fuse. The fundamental challenge is to give the nuclei enough kinetic energy to overcome this repulsive force. This can be accomplished by heating the fuel to extremely high temperatures (about 10^8 K, or about 10 times greater than the interior temperature of the sun). Such high temperatures are difficult and expensive to obtain in a laboratory or a power plant.

Section Review  3C

1. What are the similarities and differences between fission and fusion?

2. Explain how nuclear reactors utilize chain reactions.

3. Why would a fusion reactor produce less radioactive waste material than a fission reactor?

4. What is enrichment? Why is enrichment necessary when uranium is used as a reactor fuel?

5. A fission reaction leads to the formation of ^{141}Ba and ^{92}Kr when ^{235}U absorbs a neutron.

   a. How is this reaction expressed symbolically?
   b. How many neutrons are released in this reaction?

6. What are some advantages to fusion reactors (as opposed to fission reactors)? What are some difficulties in the development of a fusion reactor?

25-4
Particle physics

THE PARTICLE VIEW OF NATURE

Particle physics seeks to discover the ultimate structure of matter: *elementary particles*. Elementary particles, which are the fundamental units that compose matter, do not appear to be divisible and have neither size nor structure.

Many new particles have been produced in accelerators

Until 1932, scientists thought protons and electrons were elementary particles because they were stable. However, beginning in 1945, experiments at particle accelerators, such as the Stanford Linear Accelerator shown in **Figure 25-11,** have demonstrated that new particles are often formed in high-energy collisions between known particles. These new particles tend to be very unstable and have very short half-lives, ranging from 10^{-6}s to 10^{-23}s. So far, more than 300 new particles have been catalogued. ⭐TEKS 3E

There are four fundamental interactions in nature

The key to understanding the properties of elementary particles is to be able to describe the interactions between them. All particles in nature are subject to four fundamental interactions: *strong, electromagnetic, weak,* and *gravitational.*

The strong interaction is responsible for the binding of neutrons and protons into nuclei, as we have seen. This interaction, which represents the "glue" that holds the nucleons together, is the strongest of all the fundamental interactions. It is very short-ranged and is negligible for separations greater than about 10^{-15} m (the approximate size of a nucleus).

The electromagnetic interaction, which is about 10^{-2} times the strength of the strong interaction at nuclear distances, is responsible for the attraction of unlike charges and the repulsion of like charges. This interaction is responsible for the binding of atoms and molecules. It is a long-range interaction that decreases in strength as the inverse square of the separation between interacting particles, as described in Chapter 17.

The weak interaction is a short-range nuclear interaction that is involved in beta decay. Its strength is only about 10^{-13} times that of the strong interaction. However, because the strength of an interaction depends on the distance through which it acts, the relative strengths of two interactions differ depending on what separation distance is used. The strength of the weak interaction, for example, is sometimes cited to be as large as 10^{-6} times that of the strong interaction. Keep in mind that these relative strengths are merely estimates and they depend on the assumed separation distance.

25-4 SECTION OBJECTIVES

● **Define the four fundamental interactions of nature.**

● **Identify the elementary particles that make up matter.**

● **Describe the standard model of the universe.**

Figure 25-11
The Stanford Linear Accelerator, in California, creates high-energy particle collisions that provide evidence of new particles.

Teaching Tip

Remind students that until the end of the nineteenth century, atoms were thought to be the most fundamental and indivisible portion of matter. The discovery of the electron in 1897 suggested that the atom could actually be broken down. Higher energy reactions made it possible to break down the nucleus and discover new particles within it. It is possible that when we reach higher energy densities and measure shorter periods of time, we will be able to detect smaller and smaller particles; so far, this has produced a "subatomic zoo" with over 300 new particles. Encourage students to discuss whether there will be an end to breaking down particles into more-elementary components.

 **Misconception Alert**

Students will be familiar with electric and gravitational forces, but they may not have a clear idea of how the two compare with one another. Point out that these two forces differ in range and in order of magnitude.

⭐TEKS
p. 916: 3C, 3C
p. 917: 3E

Teaching Tip

Students may have difficulty revising their concept of force with the concepts of interactions and mediating field particles. To address this difficulty, discuss the notion of action at a distance. In the discussion, you may want to introduce the following questions: How does an electric charge "know" that there is another charge nearby? How does a compass "know" about a magnet? How does the Earth "know" that the sun (which keeps it in orbit) is nearby? Have students formulate similar questions to help them develop an intuition for interactions and mediating field particles.

Teaching Tip

A simple diagram representing the interaction between two particles is called a *Feynman diagram*. Such diagrams are named after the American scientist Richard P. Feynman (1918–1988), who developed them. **Figure 25-12** is an example of a Feynman diagram.

Teaching Tip

Make sure students understand the information contained in **Table 25-6.** Remind them that fm stands for fermi (10^{-15} m), the unit of distance that is used at the nuclear scale. Ask students the following: Which force has the shortest range? *(the weak force)* What are the mediating particles between two protons repelling each other electrostatically? *(photons)* What are the mediating particles between an apple and Earth? *(gravitons)*

918

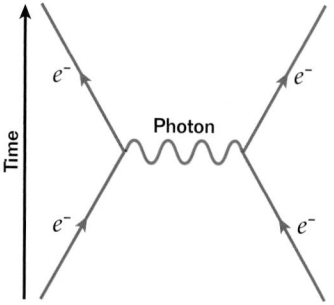

Figure 25-12

In particle physics, the electromagnetic interaction is modeled as an exchange of photons. The wavy red line represents a photon, and the blue lines represent electrons.

Did you know?

The interaction of charged particles by the exchange of photons is described by a theory called *quantum electrodynamics,* or QED.

Finally, the gravitational interaction is a long-range interaction with a strength of only about 10^{-38} times that of the strong interaction. Although this familiar interaction is what holds the planets, stars, and galaxies together, its effect on elementary particles is negligible. The gravitational interaction is the weakest of all four fundamental interactions.

A force can be thought of as mediated by an exchange of particles

Notice that in this section we have referred to a force as an *interaction*. This is because in particle physics the interaction of matter is usually described not in terms of forces but in terms of the exchange of special particles. In the case of the electromagnetic interaction, the particles are photons. Thus, it is said that the electromagnetic force is *mediated* by photons.

Figure 25-12 shows how two electrons might repel one another through the exchange of a photon. Because momentum is conserved, the electron emitting a photon changes direction slightly. As the photon is absorbed, the other electron's direction must also change. The net effect is that the two electrons change direction and move away from each other.

Likewise, the strong interaction is mediated by particles called *gluons*, the weak interaction is mediated by particles called the *W* and *Z* bosons, and the gravitational interaction is mediated by *gravitons*. All of these except gravitons have been detected. The four fundamental interactions of nature and their mediating field particles are summarized in **Table 25-6.** ⭐ TEKS 2C

Table 25-6	The fundamental interactions of nature		
Interaction (force)	Relative strength	Range of force	Mediating field particle
strong	1	≈ 1 fm	gluon
electromagnetic	10^{-2}	proportional to $1/r^2$	photon
weak	10^{-13}	$< 10^{-3}$ fm	$W^{\pm}$ and Z bosons
gravitational	10^{-38}	proportional to $1/r^2$	graviton

CLASSIFICATION OF PARTICLES

All particles other than the mediating field particles can be classified into two broad categories: *leptons* and *hadrons*. The difference between the two is whether they interact through the strong interaction. Leptons are a group of particles that participate in the weak, gravitational, and electromagnetic interactions but not in the strong interaction. Hadrons are particles that interact through all four fundamental interactions, including the strong interaction.

Leptons are thought to be elementary particles

Electrons and neutrinos are both leptons. Like all leptons, they have no measurable size or internal structure and do not seem to break down into smaller units. Because of this, *leptons appear to be truly elementary.*

The number of known leptons is small. Currently, scientists believe there are only six leptons: the electron, the muon, the tau, and a neutrino associated with each. Each of these six leptons also has an antiparticle.

Hadrons include mesons and baryons

Hadrons, the strongly interacting particles, can be further divided into two classes: *mesons* and *baryons.* Originally, mesons and baryons were classified according to their masses. Baryons were heavier than mesons, and both were heavier than leptons. However, this distinction no longer holds. Today, mesons and baryons are distinguished by their internal structure.

All mesons are unstable. Because of this, they are not constituents of normal, everyday matter. Baryons have masses equal to or greater than the proton mass. The most common examples of baryons are protons and neutrons, which are constituents of normal, everyday matter. A summary of this classification of particles is given in **Figure 25-13.**

Hadrons are thought to be made of quarks

Particle-collision experiments involving hadrons seem to involve many short-lived particles, implying that hadrons are made up of more-fundamental particles. Furthermore, there are numerous hadrons, and many of them are known to decay into other hadrons. These facts strongly suggest that hadrons, unlike leptons, cannot be truly elementary.

In 1963, Murray Gell-Mann and George Zweig independently proposed that hadrons have a more elementary substructure. According to their model, all hadrons are composed of two or three fundamental particles, which came to be called *quarks.* In the original model, there were three types of quarks, designated by the symbols *u, d,* and *s.* These were given the arbitrary names *up, down,* and *sideways* (now more commonly referred to as *strange*). Associated with each quark is an antiquark of opposite charge. (★TEKS) **3E**

The difference between mesons and baryons is due to the number of quarks that compose them. The compositions of all hadrons known when Gell-Mann and Zweig presented their models could be completely specified by three simple rules, which are summarized in **Table 25-7.**

Later evidence from collision experiments encouraged theorists to propose the existence of several more quarks, now known as *charm, top,* and *bottom.* These six quarks seem to fit together in pairs: up and down, charm and strange, and top and bottom.

All quarks have a charge associated with them. The charge of a hadron is equal to the sum of the charges of its constituent quarks and is either zero or a multiple of *e,* the fundamental unit of charge introduced in Chapter 17. This

Classification of particles

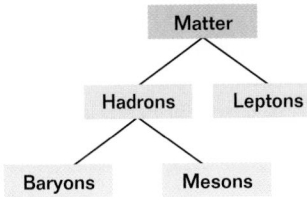

Figure 25-13
Leptons appear to be elementary, while hadrons consist of smaller particles, called quarks. As a result, hadrons can be further subdivided into baryons and mesons, based on their internal composition.

Table 25-7	**Hadrons**
Particle	**Composition**
meson	one quark and one antiquark
baryon	three quarks
antibaryon	three antiquarks

SECTION 25-4

Teaching Tip

This section contains an extensive amount of new information, including classification schemes, names, and concepts. Many of the new terms can easily be confused with each other. Students will need help making connections between the new bits of information and identifying which details are most important to remember. You may choose to use the section summary to prepare an overview of the material. Also, use a chalkboard or an overhead projector to clarify which properties or criteria are used for grouping the particles in different categories and subgroups.

(STOP) Misconception Alert

Students might not differentiate clearly between the mediating field particles and the classification of other particles. To build on their prior knowledge, suggest that they classify the particles they already know into two groups using the following system:

Building blocks or particles of matter interact with, attract, or repel other particles of matter. They may or may not be broken into smaller particles that will in turn attract and repel each other.

Mediating field particles serve as messengers of the forces between the building blocks of matter.

(★TEKS)

p. 918: 2C
p. 919: 3E

Teaching Tip

Have students revise their classification systems (see the Misconception Alert on page 919) to include more details. Ask them to place examples in each category. An example follows:

Mediating particles (field messengers): photons, bosons, gluons, gravitons

Matter particles (building blocks):

  Leptons (indivisible): electrons, muons, taus, neutrinos of each

  Hadrons (made of quarks):

    Mesons (1 quark, 1 antiquark): π^+, K^-

    Baryons (3 quarks): proton, neutron

Visual Strategy

Figure 25-14

Explain that although many of the quarks have the same charge, they have special properties called *flavor* and *color* that make them different.

Q What is the charge of a π^+ meson?

A $\left(+\frac{2}{3}e\right) + \left(+\frac{1}{3}e\right) = 1e$

ANSWERS TO

Conceptual Challenge

1. No. An antibaryon is made of three antiquarks, and a meson is made of one quark and one antiquark. Thus, a baryon, which consists of three quarks, could not be produced in this reaction.

2. Yes, although the strong interaction has a much greater effect.

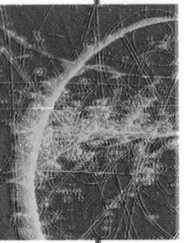

Conceptual Challenge

1. Particle-antiparticle interactions

An antibaryon interacts with a meson. Can a baryon be produced in such an interaction? Explain.

2. Strong and weak interactions

Two protons in a nucleus interact via the strong interaction. Are they also subject to the weak interaction?

★ TEKS 3A

---

implies that quarks have a very unusual property—fractional electric charge. In other words, the charge of the electron is no longer thought to be the smallest possible nonzero charge that a particle can have. The charges for all six quarks that have been discovered and their corresponding antiquarks are summarized in **Table 25-8**.

Table 25-8   Quarks and their charges

Quark	Charge	Antiquark	Charge
up (*u*)	$+\frac{2}{3}e$	$\bar{u}$	$-\frac{2}{3}e$
down (*d*)	$-\frac{1}{3}e$	$\bar{d}$	$+\frac{1}{3}e$
charm (*c*)	$+\frac{2}{3}e$	$\bar{c}$	$-\frac{2}{3}e$
strange (*s*)	$-\frac{1}{3}e$	$\bar{s}$	$+\frac{1}{3}e$
top (*t*)	$+\frac{2}{3}e$	$\bar{t}$	$-\frac{2}{3}e$
bottom (*b*)	$-\frac{1}{3}e$	$\bar{b}$	$+\frac{1}{3}e$

Figure 25-14 represents the quark compositions of several hadrons, both baryons and mesons. Just two of the quarks, *u* and *d*, are needed to construct the hadrons encountered in ordinary matter (protons and neutrons). The other quarks are needed only to construct rare forms of matter that are typically found only in high-energy situations, such as particle collisions.

The charges of the quarks that make up each hadron in **Figure 25-14** add up to zero or a multiple of *e*. For example, the proton contains three quarks (*u*, *u*, and *d*) having charges of $+\frac{2}{3}e$, $+\frac{2}{3}e$, and $-\frac{1}{3}e$. The total charge of the proton is $+e$, as you would expect. Likewise, the total charge of quarks in a neutron is zero ($+\frac{2}{3}e$, $-\frac{1}{3}e$, and $-\frac{1}{3}e$). ★ TEKS 2C

You may be wondering whether such discoveries will ever end. At present, physicists believe that six quarks and six leptons (and their antiparticles) are the fundamental particles.

Baryons   **Mesons**

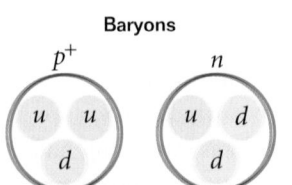
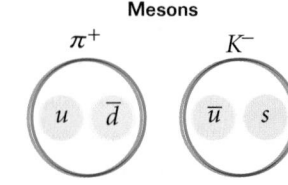

Figure 25-14

Baryons contain three quarks, while mesons contain a quark and an antiquark. The baryons represented are a proton (p^+) and a neutron (*n*). The mesons shown are a pion (π^+) and a kaon (K^-), both rare particles.

Despite many extensive efforts, no isolated quark has ever been observed. Physicists now believe that quarks are permanently confined inside ordinary particles by the strong force. This force, often called the *color force* for quarks, is analogous to the electric force between charges; like colors repel, and opposite colors attract. Of course, quarks are not really colored. Color is merely a name given to the property of quarks that allows them to attract one another and form composite particles. The attractive force between nucleons is a byproduct of the strong force between quarks.

THE STANDARD MODEL

The current model used in particle physics to understand matter is called the *standard model*. This model was developed over many years by a variety of people. Although the details of the standard model are complex, the model's essential elements can be summarized using **Figure 25-15.** The strong force, mediated by gluons, holds quarks together to form composite particles, such as protons, neutrons, and mesons. Leptons participate only in the electromagnetic, gravitational, and weak interactions. The combination of composite particles, such as protons and neutrons, with leptons, such as electrons, makes the constituents of all matter, which are atoms.

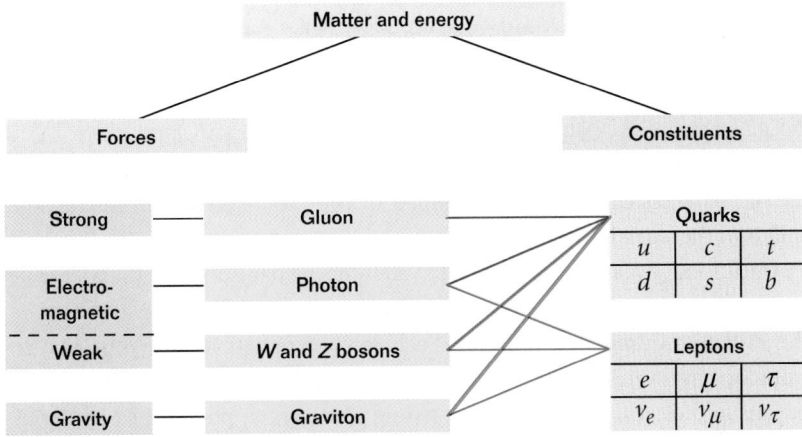

Figure 25-15
This schematic diagram summarizes the main elements of the standard model, including the fundamental forces, the mediating field particles, and the constituents of matter.

The standard model can help explain the early universe

Particle physics helps us understand not only the realm of the very small but also the origin and evolution of the universe. The current theory of the creation of the universe is called the *big bang*. According to the big bang theory, the universe erupted from a pointlike singularity about 15 billion to 20 billion years ago. In the brief instant after the big bang came such extremes in the density of matter and energy that all four fundamental interactions of physics operated in a single, unified way. The temperatures and energy present reduced everything into an undifferentiated "quark soup."

Did you know?

The word *atom* is from the Greek word *atomos,* meaning "indivisible." At one time, atoms were thought to be the indivisible constituents of matter; that is, they were regarded as elementary particles. Today quarks and leptons are considered to be elementary particles.

The Language of Physics

The Greek letter μ (mu) represents muon particles, and the Greek letter ν (nu) designates neutrino particles. The symbols ν_e, ν_μ, and ν_τ represent the neutrinos associated with the electron, the muon, and the tau, respectively.

Visual Strategy

Figure 25-15
Use the following questions to make sure students interpret the information in **Figure 25-15** properly:

Q Which forces hold quarks together?

A *gravity, the weak force, the electromagnetic force, and the strong force; the strong force is most significant.*

Q Which field particles mediate the interaction between neutrinos?

A *photons, W and Z bosons, and gravitons*

Q Which particles are held together by gluons?

A *quarks*

⭐ TEKS

p. 920: 2C, 3A

Teaching Tip

The theory of the big bang is also discussed in the Physics on the Edge feature in Chapter 13; it may be helpful for students to review this feature at this time.

Visual Strategy

Figure 25-16

Be sure students interpret the information presented in **Figure 25-16** correctly.

Q When did the strong force and the electroweak force separate?

A *at approximately 10^{-32} s*

Q Which of the colored sections shown in the figure occurred during the first second after the big bang?

A *the red section, the yellow section, the green section, and part of the blue section*

Q According to this figure, what is the present age of the universe?

A *approximately 10^{18} s*

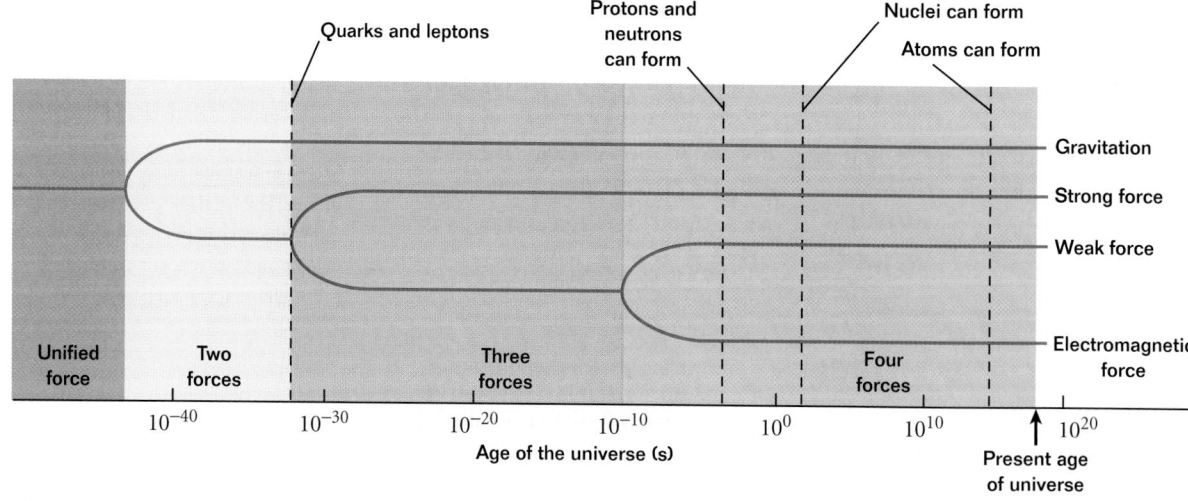

Figure 25-16
The four fundamental interactions of nature were indistinguishable during the early moments of the big bang. (★)TEKS 2C

The evolution of the four fundamental interactions from the big bang to the present is shown in **Figure 25-16.** During the first 10^{-43} s, it is presumed that the strong, electroweak (electromagnetic and weak), and gravitational interactions were joined together. From 10^{-43} s to 10^{-32} s after the big bang, gravity broke free of this unification while the strong and electroweak interactions remained as one. This was a period when particle energies were so great (greater than 10^{16} GeV) that very massive particles that are now rare, as well as quarks, leptons, and their antiparticles, existed.

Then the universe rapidly expanded and cooled, the strong and electroweak interactions parted, and the grand unification was broken. About 10^{-10} s after the big bang, as the universe continued to cool, the electroweak interaction split into the weak interaction and the electromagnetic interaction.

Until about 7×10^5 years (2×10^{13} s) after the big bang, most of the energy in the universe was in the form of radiation rather than matter. This was the era of the radiation-dominated universe. Such intense radiation prevented matter from forming even single hydrogen atoms. Matter did exist, but only in the form of ions and electrons. Electrons are strong scatterers of photons, so matter at this time was opaque to radiation. Matter continuously absorbed and re-emitted photons, thereby ensuring thermal equilibrium of radiation and matter.

By the time the universe was about 7 million years (2×10^{14} s) old, it had expanded and cooled to about 3000 K. At this temperature, protons could bind to electrons to form hydrogen atoms. Without free electrons to scatter photons, the universe suddenly became transparent. Matter and radiation no longer interacted as strongly, and each evolved separately. By this time, most of the energy in the universe was in the form of matter. Clumps of neutral matter steadily grew: first atoms, followed by molecules, gas clouds, stars, and finally galaxies. This period, referred to as the matter-dominated universe, continues to this day.

The standard model is still incomplete

While particle physicists have been exploring the realm of the very small, cosmologists have been exploring cosmic history back to the first microsecond of the big bang. Observation of the events that occur when two particles collide in an accelerator is essential to reconstructing the early moments in cosmic history. Perhaps the key to understanding the early universe is to first understand the world of elementary particles. Cosmologists and particle physicists find that they have many common goals, and they are working together to attempt to study the physical world at its most fundamental level.

Our understanding of physics at short distances is far from complete. Particle physics still faces many questions. For example, why does the photon have no mass, while the W and Z bosons do? Because of this mass difference, the electromagnetic and weak forces are quite distinct at low energies, such as those in everyday life, but they behave in similar ways at very high energies.

To account for these changes, the standard model proposes the existence of a particle called the *Higgs boson*, which exists only at the high energies at which the electromagnetic and weak forces begin to merge. The Higgs boson has not yet been found. According to the standard model, its mass should be less than 1 TeV (10^{12} eV). International efforts are under way to build a device capable of reaching energies close to 1 TeV to search for the Higgs boson.

There are still other questions that the standard model has yet to answer. Is it possible to unify the strong and electroweak theories in a logical and consistent manner? Why do quarks and leptons form three similar but distinct families? Are muons the same as electrons (apart from their different masses), or do they have other subtle differences that have not been detected? Why are some particles charged and others neutral? Why do quarks carry a fractional charge? What determines the masses of the fundamental constituents? Can isolated quarks exist? The questions go on and on. Because of the rapid advances and new discoveries in the field of particle physics, by the time you read this book, some of these questions may have been resolved, while new questions may have emerged. (★TEKS) 3C

Section Review

1. Name the four fundamental interactions and the particles that mediate each interaction.

2. What are the differences between hadrons and leptons? What are the differences between baryons and mesons?

3. Describe the main stages of the evolution of the universe according to the big bang theory.

Teaching Tip

Encourage students to formulate their own questions about the standard model and about the age of the universe.

Section Review
ANSWERS

1. The strong force is mediated by gluons, the electromagnetic force is mediated by photons, the weak force is mediated by W and Z bosons, and gravity is mediated by gravitons.

2. Hadrons consist of quarks, while leptons are thought to be indivisible. Hadrons interact with gluons (the strong interaction), but leptons do not. Mesons consist of one quark and one antiquark, and baryons consist of three quarks.

3. Up to 10^{-43} s, all forces were unified. At 10^{-43} s, gravity broke free; at 10^{-32} s, the strong force broke free; and at 10^{-10} s, the weak force broke free. From 10^{-43} s to 10^{-32} s, quarks, leptons, and their antiparticles existed. Until 7×10^5 years (2×10^{13} s), most energy was in the form of radiation. At 7 million years (2×10^{14} s), atoms began forming, and soon most energy was in the form of matter, as it is today.

(★TEKS)

p. 922: 2C
p. 923: 3C

BACKGROUND

This feature discusses the discovery of antimatter particles, the processes of pair production and annihilation, and the production of anti-hydrogen atoms in a particle accelerator. The charge of an anti-electron, also known as a positron, is equal in magnitude but opposite in sign to the charge of an electron. The positron and electron have equal masses and opposite spins. Antimatter was theoretically predicted by Dirac in the late 1920s. The anti-electron was detected experimentally in 1932, and the anti-proton and anti-neutron were detected in 1955 and 1956, respectively. Today, it is believed that every particle has a corresponding antiparticle.

Teaching Tip

Point out that charge must be conserved in pair production and annihilation. For example, in pair production, the initial photon has zero charge, and the total charge of the positron and electron is also zero. It is impossible for a photon to produce a single electron or positron because this would violate the conservation of charge.

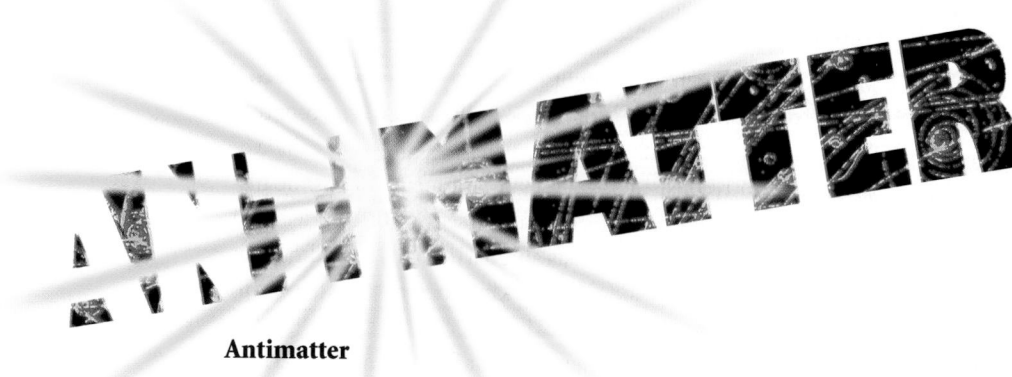

PHYSICS ON THE EDGE

Antimatter

Startling discoveries made in the twentieth century have confirmed that electrons and other particles of matter have *antiparticles*. Antiparticles have the same mass as their corresponding particle but an opposite charge.

The discovery of antiparticles

The discovery of antiparticles began in the 1920s with work by the theoretical physicist Paul Adrien Maurice Dirac (1902–1984), who developed a version of quantum mechanics that incorporated Einstein's theory of special relativity. Dirac's theory was successful in many respects, but it had one major problem: its relativistic wave equation required solutions corresponding to negative energy states. This negative set of solutions suggested the existence of something like an electron but with an opposite charge, just as the negative energy states were opposite to an electron's typical energy states. At the time, there was no experimental evidence of such antiparticles. (★)TEKS 3E

In 1932, shortly after Dirac's theory was introduced, evidence of the anti-electron was discovered by the American physicist Carl Anderson. The anti-electron, also known as the *positron,* has the same mass as the electron but is positively charged. Anderson found the positron while examining tracks created by electronlike particles in a cloud chamber placed in a magnetic field. As described in Chapter 21, such a field will cause moving particles to follow curved paths. The direction in which a particle moves depends on whether its charge is positive or negative. Anderson noted that some of the tracks had deflections typical of an electron's mass, but in the opposite direction, corresponding to a positively charged particle. (★)TEKS 3C

Pair production and annihilation

Since Anderson's initial discovery, the positron has been observed in a number of experiments. In perhaps the most common process, a gamma ray with sufficiently high energy collides with a nucleus, creating an electron-positron pair. An example of this process, known as *pair production,* is shown in **Figure 25-17.**

During pair production, the energy of the photon is completely converted into the rest energy and kinetic energy of the electron and the positron. Thus, pair

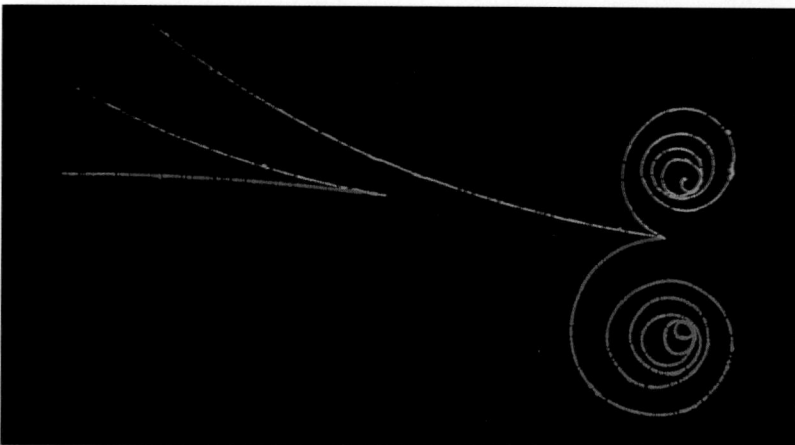

production is a striking verification of the equivalence of mass (rest energy) and other forms of energy as predicted by Einstein's special theory of relativity. (The equivalence between mass and energy is discussed in the Physics on the Edge feature at the end of Chapter 5.)

Once formed, a positron will most likely soon collide with an oppositely charged electron in a process known as *pair annihilation.* This process is the opposite of pair production—an electron-positron pair produces two photons. In the simplest example of pair annihilation, an electron and a positron initially at rest combine with each other and disappear, leaving behind two photons. Because the initial momentum of the electron-positron pair is zero, it is impossible to produce a single photon. Momentum can be conserved only if two photons moving in opposite directions, both with the same energy and magnitude of momentum, are produced.

Antimatter produced in a particle accelerator

After the positron was discovered, physicists began to search for the anti-proton and anti-neutron. However, because the proton and neutron are much more massive than the electron, a much greater amount of energy is required to produce their antiparticles. By 1955, technological advances in particle accelerators brought evidence of the anti-proton, and evidence of the anti-neutron was found a year later. (★)TEKS 3C

The discovery of other antiparticles leads to the question of whether these antiparticles can be combined to form antimatter and, if so, how that antimatter would behave. In 1995, physicists at the CERN particle accelerator in Geneva, Switzerland, succeeded in producing anti-hydrogen atoms, that is, atoms with a single anti-electron orbiting an anti-proton. Researchers observed nine anti-hydrogen atoms during a three-week period. Unfortunately, the anti-hydrogen atoms had a short lifetime—less than 37 billionths of a second—because as soon as an anti-hydrogen atom encountered ordinary matter, the two annihilated one another. Attempts to produce antimatter for greater time periods are currently under way. (★)TEKS 3E

Figure 25-17
The red and green spirals shown here are the paths of a positron and an electron moving through a magnetic field. Note that these paths have about the same shape but are opposite in direction.

internetconnect

SCiLINKS.
NSTA

TOPIC: Antiparticles
GO TO: www.scilinks.org
*sci*LINKS CODE: HF2254

CHAPTER 25
Summary

Teaching Tip

Explaining concepts in written form helps to solidify students' understanding of difficult concepts and to enforce good communication skills. Have students choose one major concept from the chapter, such as binding energy, nuclear reactions, or particle physics, and summarize the topic in an essay. Essays should explain the concept accurately and thoroughly.

internet connect

SC*LINKS*

NSTA

TOPIC: Texas Fusion Research
GO TO: www.scilinks.org
*sci***LINKS CODE:** HFX015

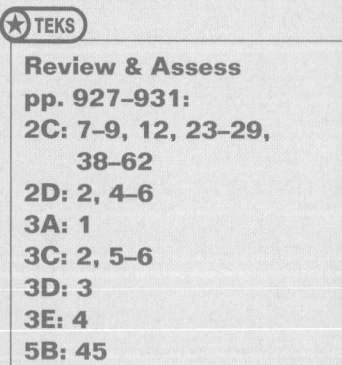

★ TEKS

Review & Assess
pp. 927–931:
2C: 7–9, 12, 23–29,
  38–62
2D: 2, 4–6
3A: 1
3C: 2, 5–6
3D: 3
3E: 4
5B: 45

KEY TERMS

binding energy (p. 900)

half-life (p. 909)

isotope (p. 897)

strong force (p. 898)

Particle symbols

Particle name	Symbol
alpha particle (helium nucleus)	$\alpha\ (^4_2\text{He})$
beta particle (electron)	$\beta^-\ (^{\ 0}_{-1}\text{e})$
beta particle (positron)	$\beta^+\ (^{\ 0}_{+1}\text{e})$
gamma ray	γ
neutron	$n\ (^1_0 n)$
proton	$p\ (^1_1 p)$
neutrino	ν
antineutrino	$\overline{\nu}$

KEY IDEAS

Section 25-1 The nucleus

- The nucleus, which consists of protons and neutrons, is the small, dense core of an atom.
- A nucleus can be characterized by a mass number, A, an atomic number, Z, and a neutron number, N.
- The binding energy of a nucleus is the difference in energy between its nucleons when bound and its nucleons when unbound.

Section 25-2 Nuclear decay

- An unstable nucleus can decay in three ways: alpha (α) decay, beta (β) decay, or gamma (γ) decay.
- The decay constant, λ, indicates the rate of radioactive decay.
- The half-life, $T_{1/2}$, is the time required for half the original nuclei of a radioactive substance to undergo radioactive decay.

Section 25-3 Nuclear reactions

- Nuclear reactions involve a change in the nucleus of an atom.
- In fission, a heavy nucleus splits into two lighter nuclei. In fusion, two light nuclei combine to form a heavier nucleus.

Section 25-4 Particle physics

- There are four fundamental interactions in nature: strong, weak, gravitational, and electromagnetic.
- The constituents of matter can be classified as leptons or hadrons, and hadrons can be further divided into mesons and baryons. Electrons and neutrinos are leptons. Protons and neutrons are baryons.
- Mesons consist of a quark-antiquark pair; baryons consist of three quarks.

Variable symbols

Quantities		Units		Conversions
m	mass	u	unified mass unit or atomic mass unit	$= 1.660\ 559 \times 10^{-27}$ kg $= 931.50$ MeV/c^2
λN	activity or decay rate	Bq Ci	becquerel curie	$= 1$ decay/s $= 3.7 \times 10^{10}$ Bq
$T_{1/2}$	half-life	s	seconds	

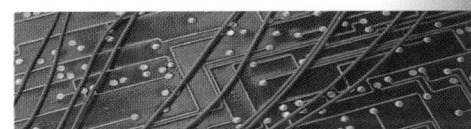

CHAPTER 25
Review and Assess

THE NUCLEUS

Review questions

1. How many protons are there in the nucleus $^{197}_{79}$Au? How many neutrons? How many electrons are there in the neutral atom?

2. What are isotopes?

3. What holds the nucleons in a nucleus together?

Conceptual questions

4. Is it possible to accurately predict an atom's mass from its atomic number? Explain.

5. What would happen if the binding energy of a nucleus was zero?

6. Why do heavier elements require more neutrons to maintain stability?

Practice problems

7. Calculate the total binding energy of $^{12}_{6}$C. (See Sample Problem 25A and refer to Appendix G.)

8. Calculate the total binding energy of tritium ($^{3}_{1}$H) and helium-3 ($^{3}_{2}$He). (See Sample Problem 25A and refer to Appendix G.)

9. Calculate the average binding energy per nucleon of $^{24}_{12}$Mg and $^{85}_{37}$Rb. (See Sample Problem 25A and refer to Appendix G.)

NUCLEAR DECAY AND REACTIONS

Review questions

10. Explain the main differences between alpha, beta, and gamma emissions.

11. What fraction of a radioactive sample has decayed after two half-lives have elapsed?

12. **Figure 25-18** shows the steps by which $^{235}_{92}$U decays to $^{207}_{82}$Pb. Draw this diagram, and enter the correct isotope symbol in each square.

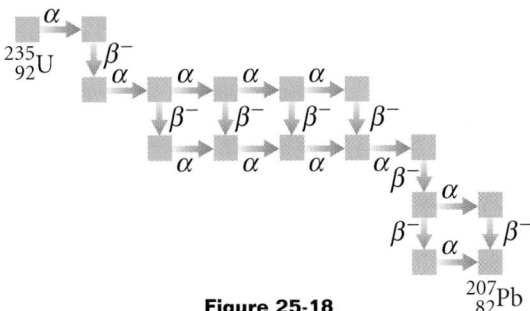

Figure 25-18

13. What factors make a fusion reaction difficult to achieve?

Conceptual questions

14. If a film is kept in a box, alpha particles from a radioactive source outside the box cannot expose the film, but beta particles can. Explain.

15. An alpha particle has twice the charge of a beta particle. Why does the beta particle deflect more when both pass between electrically charged plates, assuming they both have the same speed?

16. If a nucleus has a half-life of one year, will it be completely decayed after two years? Explain.

17. Why is carbon dating unable to provide accurate estimates of very old materials?

18. Suppose it could be shown that cosmic ray intensity was much greater thousands of years ago. How would this affect the ages we assign to ancient samples of once-living matter?

19. A free neutron undergoes beta decay with a half-life of about 15 min. Can a free proton undergo a similar decay? (Hint: Compare the masses of the proton and the neutron.)

ANSWERS TO

Chapter 25
Review and Assess

1. 79; 118; 79

2. Isotopes are atoms with the same Z but different N and A.

3. the strong force

4. No, because an atom's mass also depends on its neutrons.

5. The nucleons would spontaneously break apart.

6. because of a greater Coulomb repulsion between the protons

7. 92.163 MeV

8. 8.482 MeV; 7.718 MeV

9. 8.2609 MeV/nucleon; 8.6975 MeV/nucleon

10. See **Table 25-3** on page 903.

11. $\frac{3}{4}$

12. For each α, A decreases by four and Z decreases by two. For each β^-, A stays the same and Z increases by one.

13. the high temperatures and densities required

14. α particles cannot penetrate the box, but β particles can.

15. α has more mass and thus more momentum than β.

16. This cannot be predicted because half-life applies to a large sample of nuclei, not to a particular nucleus.

17. The amount of C-14 left undecayed would be too small to be measured accurately.

18. More C-14 would have decayed, so the samples would be older than predicted.

19. No; $m_n > m_p$, so the decay of a free proton into a neutron requires an input of energy.

20. Energy is released when heavy nuclei break down.

21. No. The three α particles have a mass of 12.007 806 u, thus $^{12}_{6}C$ is more stable than the α particles.

22. Although the Coulomb repulsion is the same in each, the extra neutron in tritium provides a stronger nuclear force, so it is easier to bring the two nuclei together.

23. $^{10}_{5}B$

24. $^{1}_{0}n$

25. a. $^{4}_{2}He$
   b. $^{4}_{2}He$

26. $4.77 \times 10^{-3} \text{ s}^{-1}$, 22 mCi

27. 560 days

28. 22 900 years old

29. 93.8%

30. Quarks make up hadrons; they have fractional charge, have never been isolated, and are attracted to one another by the strong force.

31. a. 0
   b. e
   c. e

32. a. $-e$
   b. 0

33. a. anti-proton
   b. anti-neutron

34. a. three quarks
   b. three antiquarks
   c. one quark, one antiquark
   d. one quark, one antiquark

35. The photon is a mediating field particle for the electromagnetic force, while the neutrino is a particle emitted in beta decay.

36. Yes. Hadrons consist of mesons and baryons. Thus, all mesons are hadrons, but not all hadrons are mesons; some are baryons.

37. because this discovery supported the standard model

38. 1.2×10^{-14}

928

---

20. In a fission reactor, nuclear reactions produce energy to drive a generator. How is this energy produced?

21. Is it possible for a $^{12}_{6}C$ (12.000 000 u) nucleus to spontaneously decay into three alpha particles? Explain.

22. Why is the temperature required for deuterium-tritium fusion lower than that needed for deuterium-deuterium fusion? (Hint: Consider the Coulomb repulsion and nuclear attraction for each case.)

Practice problems

23. Determine the product of the following reaction:
$$^{7}_{3}Li + ^{4}_{2}He \rightarrow ? + ^{1}_{0}n$$
(See Sample Problem 25B.)

24. A nuclear reaction of significant historical note occurred in 1932, when a beryllium target was bombarded with alpha particles. Analysis of the experiment indicated that the following reaction occurred:
$$^{4}_{2}He + ^{9}_{4}Be \rightarrow ^{12}_{6}C + X$$
What is X in this reaction?
(See Sample Problem 25B.)

25. Complete the following nuclear reactions:
   a. $? + ^{14}_{7}N \rightarrow ^{1}_{1}H + ^{17}_{8}O$
   b. $^{7}_{3}Li + ^{1}_{1}H \rightarrow ^{4}_{2}He + ?$
(See Sample Problem 25B.)

26. A radioactive sample contains 1.67×10^{11} atoms of $^{108}_{47}Ag$ (half-life = 2.42 min) at some instant. Calculate the decay constant and the activity of the sample in mCi.
(See Sample Problem 25C.)

27. How long will it take a sample of polonium-210 with a half-life of 140 days to decay to one-sixteenth its original strength?
(See Sample Problem 25C.)

28. The amount of carbon-14 ($^{14}_{6}C$) in a wooden artifact is measured to be 6.25 percent the amount in a fresh sample of wood from the same region. The half-life of carbon-14 is 5730 years. Assuming the same amount of carbon-14 was initially present in the artifact, determine the age of the artifact.
(See Sample Problem 25C.)

29. A sample of organic material is found to contain 18 g of carbon. Based on samples of pottery found at the site, investigators believe the material is about 23 000 years old. Estimate what percentage of the material's carbon-14 has decayed.
(See Sample Problem 25C.)

PARTICLE PHYSICS

Review questions

30. Describe the properties of quarks.

31. What is the electrical charge of the particles with the following quark compositions?
   a. *udd*
   b. *uud*
   c. *ud̄*

32. What is the electrical charge of the baryons with the following quark compositions?
   a. *ūūd̄*
   b. *ūd̄d̄*

33. What are each of the baryons in item 32 called?

34. How many quarks or antiquarks are there in the following particles?
   a. a baryon
   b. an antibaryon
   c. a meson
   d. an antimeson

Conceptual questions

35. Compare a neutrino with a photon.

36. Consider the statement, "All mesons are hadrons, but not all hadrons are mesons." Is this statement true? Discuss.

37. The *W* and *Z* bosons were first produced in 1983. Why was this an important discovery?

MIXED REVIEW

38. Consider the hydrogen atom to be a sphere with a radius equal to the Bohr radius, 0.53×10^{-10} m, and calculate the approximate value of the ratio of atomic density to nuclear density.

39. Certain stars are thought to collapse at the end of their lives, combining their protons and electrons to form a neutron star. Such a star could be thought of as a giant atomic nucleus. If a star with a mass equal to that of the sun (1.99×10^{30} kg) were to collapse into neutrons, what would be the radius of the star?

40. Tritium, $^3_1$H, decays to $^3_2$He by beta emission. Determine the energy released in the process.

41. Find the energy released in the alpha decay of $^{238}_{92}$U. Use the masses in the following table:

Nucleus	Mass
$^{238}_{92}$U	238.050 784 u
$^{234}_{90}$Th	234.043 593 u
$^4_2$He	4.002 602 u

42. Disregard binding energies and estimate the masses of the u and d quarks from the masses of the proton and neutron.

43. A pair of nuclei for which $Z_1 = N_2$ and $Z_2 = N_1$ are called *mirror isobars* (the atomic and neutron numbers are interchangeable). Binding energy measurements on such pairs can be used to obtain evidence of the charge independence of nuclear forces. Charge independence means that the proton-proton, proton-neutron, and neutron-neutron forces are approximately equal. Calculate the difference in binding energy for the two mirror nuclei, $^{15}_8$O (15.003 065 u) and $^{15}_7$N (15.000 108 u).

44. Find the threshold kinetic energy that the incident neutron must have to produce the following reaction: $^1_0n + {}^4_2\text{He} \rightarrow {}^2_1\text{H} + {}^3_1\text{H}$

45. A photon with an energy of 2.09 GeV creates a proton-antiproton pair in which the proton has a kinetic energy of 95 MeV. What is the kinetic energy of the antiproton?

46. Natural gold has only one stable isotope, $^{197}_{79}$Au. If gold is bombarded with slow neutrons, β^- particles are emitted.
  a. Write the appropriate reaction equation.
  b. Calculate the maximum energy of the emitted beta particles.

47. Two ways ^{235}U can undergo fission when bombarded with a neutron are described below. In each case, neutrons are also released. Find the number of neutrons released in each of the following:
  a. ^{140}Xe and ^{94}Sr released as fission fragments
  b. ^{132}Sn and ^{101}Mo released as fission fragments

48. When a $^6_3$Li nucleus is struck by a proton, an alpha particle and a product nucleus are released. What is the product nucleus?

49. Suppose $^{10}_5$B is struck by an alpha particle, releasing a proton and a product nucleus in the reaction. What is the product nucleus?

50. When ^{18}O is struck by a proton, ^{18}F and another particle are produced. What is the other particle?

51. Complete the following nuclear reactions:
  a. $^{27}_{13}\text{Al} + {}^4_2\text{He} \rightarrow ? + {}^{30}_{15}\text{P}$
  b. $^1_0n + ? \rightarrow {}^4_2\text{He} + {}^7_3\text{Li}$

52. A fission reaction that occurs when uranium-235 absorbs a neutron leads to the formation of barium-141 and krypton-92.
  a. Write the equation that describes this fission reaction.
  b. How many neutrons are released at the end of this reaction?

53. When a star has exhausted its hydrogen fuel, it may fuse other nuclear fuels, such as helium. At temperatures above 1.0×10^8 K, helium fusion can occur.
  a. Two alpha particles fuse to produce a nucleus, A, and a gamma ray. What is nucleus A?
  b. Nucleus A absorbs an alpha particle to produce a nucleus, B, and a gamma ray. What is nucleus B?

54. A piece of charcoal known to be approximately 25 000 years old contains 7.96×10^{10} C-14 atoms.
  a. Determine the number of decays per minute expected from this sample.
  b. If the radioactive background in the counter without a sample is 20.0 counts/min and we assume 100.0 percent efficiency in counting, explain why 25 000 years is close to the limit of dating with this technique.

39. 1.3×10^4 m
40. 0.6800 MeV
41. 4.275 MeV
42. 0.336 6847 u; 0.335 296 u
43. 3.53 MeV
44. 17.590 MeV
45. 120 MeV
46. a. $^1_0n + {}^{197}_{79}\text{Au} \rightarrow {}^{198}_{80}\text{Hg} + {}^{\ 0}_{-1}e + \bar{\nu}$
  b. 7.885 MeV
47. a. $2\,{}^1_0n$
  b. $3\,{}^1_0n$
48. $^3_2$He
49. $^{13}_6$C
50. ^1_0n
51. a. ^1_0n
  b. $^{10}_5$B
52. a. $^1_0n + {}^{235}\text{U} \rightarrow {}^{141}\text{Ba} + {}^{92}\text{Kr} + 3\,{}^1_0n$
  b. three neutrons
53. a. $^8_4$Be
  b. $^{12}_6$C
54. a. 18.3 decays/min
  b. The observed count rate is slightly less than the average background and thus would be difficult to measure accurately.

55. 3.81×10^{-9} s^{-1}, 2.1×10^{-10} Ci

56. 1.6×10^{16} nuclei

57. 3.8×10^{3} s

55. The half-life of radium-228 is 5.76 years. At some instant a sample contains 2.0×10^{9} nuclei. Calculate the decay constant and the activity of the sample in Ci.

56. A sample of a radioactive isotope is measured to have an activity of 240.0 mCi. If the sample has a half-life of 14 days, how many nuclei of the isotope are there at this time?

57. At some instant of time the activity of a sample of radioactive material is 5.0 μCi. If the sample contains 1.0×10^{9} radioactive nuclei, what is the half-life of the material?

58. Smoke detectors use the isotope ^{241}Am in their operation. The half-life of Am is 432 years. If the smoke detector is improperly discarded in a landfill,

Technology & Learning

Graphing calculators

Refer to Appendix B for instructions on downloading programs for your calculator. The program "Chap25" allows you to analyze a graph of time versus the amount of radioactive material remaining during nuclear decay.

The mass of a radioactive sample varies with time according to the following equation:

$$m = m_0 e^{-\lambda t}$$

The program "Chap25" stored on your graphing calculator makes use of this nuclear decay equation. In this equation m represents the mass of the sample remaining and m_0 is the original mass of the sample. Once the "Chap25" program is executed, your calculator will ask for the half-life (H) of the substance and the original mass (M) of the sample. The graphing calculator will use the following equation to create graphs of the remaining mass (Y_1) versus the time (X). The relationships in this equation are the same as those in the equations shown above.

$$Y_1 = M{*}e\wedge(-0.693{*}X/H)$$

a. What has the variable λ been replaced with in the graphing calculator equation?

Execute "Chap25" on the [PRGM] menu, and press [ENTER] to begin the program. Enter the values for the half-life of the substance and the original mass of the sample (shown below), and press [ENTER] after each value.

The calculator will provide a graph of the time versus the amount of the radioactive material remaining. (If the graph is not visible, press [WINDOW] and change the settings for the graph window, then press [GRAPH].)

Press [TRACE], and use the arrow keys to trace along the curve. The x-value corresponds to the time in years, and the y-value corresponds to the mass remaining in grams.

Determine the amount of radioactive material remaining for each of the following situations:

b. the amount of a 5.00 g sample of radium-226 with a half-life of 1600 years remaining after 250 years

c. the amount of the same sample of radium-226 remaining after 750 years

d. the amount of a 7.50 g sample of plutonium-236 with a half-life of 2.85 years remaining after 5 years

e. the amount of the same sample of plutonium-236 remaining after 25 years

f. At what x-value will the sample in item (e) have exactly half of its original mass?

Press [2nd] [QUIT] to stop graphing. Press [ENTER] to input new values or [CLEAR] to end the program.

ANSWERS TO

Technology & Learning

Answers may vary slightly, depending on viewing window settings.

a. 0.693/H

b. 4.5 g

c. 3.6 g

d. 2.2 g

e. 0.017 g

f. 2.85 years

estimate how long will it take for its activity to reduce to a relatively safe level of 0.1 percent of its original activity? (Hint: The estimation process you should use notes that the activity reduces to 50% in one half-life, to 25 % in two half-lives, and so on.)

59. It has been estimated that Earth has 9.1×10^{11} kg of natural uranium that can be economically mined. Of this total, 0.70 percent is ^{235}U. If all the world's energy needs (7.0×10^{12} J/s) were supplied by ^{235}U fission, how long would this supply last? Assume that 208 MeV of energy is released per fission event and that the mass of ^{235}U is about 3.9×10^{-25} kg.

60. If the average energy released in a fission event is 208 MeV, find the total number of fission events required to provide enough energy to keep a 100.0 W light bulb burning for 1.0 h.

61. How many atoms of ^{235}U must undergo fission to operate a 1.0×10^3 MW power plant for one day if the conversion efficiency is 30.0 percent? Assume 208 MeV released per fission event.

62. An all-electric home uses about 2.0×10^3 kW•h of electrical energy per month. How many ^{235}U atoms would be required to provide this house with its energy needs for one year? Assume 100.0 percent conversion efficiency and 208 MeV released per fission.

58. 4320 years
59. 7.8×10^{10} s (2500 years)
60. 1.1×10^{16} fission events
61. 8.7×10^{24} atoms
62. 2.6×10^{21} atoms

Alternative Assessment

Performance assessment

1. You are designing a nuclear power plant for a space station to be established on Mars. Material A is radioactive and has a half-life of two years. Material B is also radioactive and has a half-life of one year. Atoms of material B have one half the mass of atoms of material A. Discuss the benefits and drawbacks involved with each of these fuels.

2. Design a questionnaire to investigate what people in your community know about nuclear power and how they feel about it. Give the questionnaire to your classmates for their comments, and if your teacher approves, conduct a study with people in your community. Present your results in the form of a class presentation and discussion.

3. Investigate careers in nuclear medicine. Interview people who work with radiation or with isotopic tracers in a hospital. Find out what kind of patients they treat or test and the technology they use. What training is necessary for this type of career?

Portfolio projects

4. Research the lives and careers of female nuclear physicists such as Marie Curie, Lise Meitner, Ida Tacke Noddack, and Maria Goeppert-Mayer. Create a presentation about one of these scientists. The pre-sentation can be in the form of a report, poster, short video, or computer presentation.

5. Research how radioactive decay is used to date archaeological remains and fossils. What nuclear reactions are involved in the carbon-14 dating technique? What assumptions are made when the carbon-14 dating technique is used? What time scale is the carbon-14 technique suitable for? Is the carbon-14 technique appropriate to determine the age of a painting suspected to be 375 years old? Summarize your findings in a brochure or poster for visitors to a science museum.

6. "In 50 years of producing electrical power and weapons from nuclear fuel, the United States has accumulated millions of cubic meters and tens of billions of curies of radioactive waste." This statement appeared in a June 1997 special issue of *Physics Today*. Research the problem of nuclear waste in the United States. How much is there? What kinds of radioactive waste are there? Where are they produced? What are the costs and hazards associated with different techniques for disposal of radioactive waste? How do other countries deal with the problem? Choose the disposal option you think is most appropriate, and write a position paper. Include information about all options and the reasons for your choice.

Alternative Assessment
ANSWERS

Performance Assessment

1. Students should consider how long the supply will last. In two years, the rates of decay will be the same, but later, B will decrease more.

2. Students' questionnaires will vary. Questionnaires should gauge people's actual knowledge as well as their opinions.

3. Students' answers will vary. Typically, isotopic tracers use lower levels of radiation than radiation therapy.

 **Portfolio projects**

4. Student answers will vary. Curie and Noddack discovered several radioactive elements. Meitner and Goeppert-Mayer explored the phenomenon of nuclear fission. Each faced prejudice during her career.

5. ^{14}C undergoes beta decay to ^{14}N. Carbon dating can be used with organic materials only and works best for artifacts between 1000 and 25 000 years old.

6. Students' answers will vary. Be certain pros and cons are identified.

Planning

Recommended time:

1 lab period

Classroom organization:

▶ You may conduct this as a demonstration for the entire class, or groups of two or more may perform this lab independently.

▶ If students will be performing this exercise independently, be sure the isogenerator for each group is properly charged before each use.

▶ **Safety warnings:** Remind students of laboratory procedures for the use of chemicals and radioactive materials. Make sure students know the appropriate disposal methods.

Techniques to Demonstrate

Show students how to use the nuclear lab station. Point out the parts of the apparatus, and explain how each item works.

CHAPTER 25
Laboratory Exercise

(★)**TEKS**
pp. 932–933: 1A, 2B, 2C, 2E, 2F

OBJECTIVES

- Use a nuclear-lab station to measure background radiation.
- Use a nuclear lab station to measure the activity of a radioactive substance.
- Determine the half-life of a radioisotope.

MATERIALS LIST

✔ cesium-137/barium-137 isogenerator set

✔ dropper

✔ liquid soap

✔ nuclear-lab station, including counter, timer, variable high voltage, absorber set, and sources

✔ small disposable culture dish

✔ support base and rod

✔ symmetrical clamp

HALF-LIFE

The half-life of a radioactive element is the time it takes for half a given number of atoms of the element to decay. In this experiment, the activity of a radioisotope with a short half-life will be measured using a nuclear-lab station over a period of 10 s. The resulting data will be plotted on a graph, and the half-life will be calculated from points on the graph.

SAFETY

- **Do not eat or drink anything in the laboratory. Never taste chemicals or touch them with your bare hands.**

- **Tie back long hair, secure loose clothing, and remove loose jewelry to prevent their coming into contact with hazardous chemicals.**

- **Do not allow radioactive materials to come into contact with your skin, hair, clothing, or personal belongings. Although the materials used in this lab are not hazardous when used properly, radioactive materials can cause serious illness and may have permanent effects.**

- **Dispose of chemicals as instructed by your teacher. Never pour hazardous chemicals into a regular waste container. Never pour radioactive materials down the drain.**

PREPARATION

1. Read the entire lab, and plan the steps you will take.

2. Prepare a data table in your lab notebook with 2 columns and 31 rows. Label this table *Data Table*. Label the columns *Time interval (s)* and *Total count*. In the first column, label the 2nd through 31st rows in increments of 10 from *10* to *300*. Above the table, make a space to record the *Background count* (s^{-1}).

PROCEDURE

3. Turn on the nuclear-lab station. Even when no radioactive substances are near the station, counts will register. This is caused by background radiation. Take the radiation count for 10 min, and divide the total by 600 s (10 min × 60 s/min). This will give you the background count per second. Record this value in the space provided.

4. Wearing goggles, gloves, and a protective apron, set up the isogenerator vertically on a stand and clamp it securely in place, as shown in **Figure 25-19.** Carefully loosen the cap on the isogenerator, and turn the stopcock. Release 3–4 mL of the barium solution into a small culture dish. The dish now contains the radioisotope Ba-137m, a metastable isomer.

5. Add a small drop of soap to the solution so that it spreads evenly over the bottom of the dish.

6. Immediately insert the dish into the nuclear lab station, and start the counter. Read the total count at the end of each 10 s interval, and record both the time interval and the count in your data table.

7. Clean up your work area. Put equipment away safely so that it is ready to be used again. Recycle or dispose of used materials as directed by your teacher.

ANALYSIS AND INTERPRETATION

Calculations and data analysis

1. **Organizing data**   For each 10 s time interval, calculate the count per second by dividing the count by 10 s. This will give you the average count per second.

2. **Organizing data**   For each value found in item 1, calculate the actual count per second due to the source by subtracting the background count from the average count.

3. **Graphing data**   Plot a graph of the actual counts versus the total elapsed time. Use a graphing calculator, computer, or graph paper.

Conclusions

4. **Analyzing graphs**   Select a value for the actual count near the left end of the graph and read from the graph the amount of time it took for the activity to decrease to one-half that value. Record this value as the half-life.

5. **Analyzing graphs**   Repeat the procedure in step 4 three more times, each time starting at a different place on the graph. Find the average of the four values for the half-life.

6. **Analyzing graphs**   Why is using the graph a better way to find the half-life than simply using the data table?

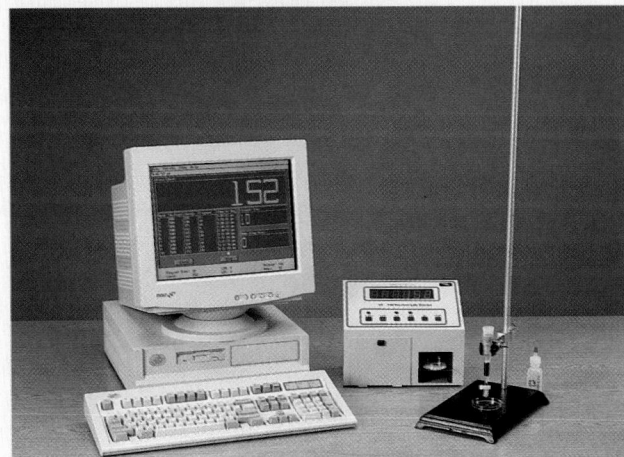

Figure 25-19

Step 3: The station will begin showing counts as soon as it is turned on. Find the background count per second so you can subtract this number from your counts to obtain an accurate result.

Step 4: Wear protective gear and work carefully when preparing the solution.

Step 6: As soon as the solution is prepared, insert the dish into the counter and begin. Record the count and time interval every 10 s until the count remains fairly constant for over a minute.

✔ **Checkpoints**

Step 3: To get more students involved in getting the background count, take the count over two 5 min intervals or three 3 min intervals, and find the average. Have different students perform different tasks during each time period.

ANSWERS TO

Analysis and Interpretation

CALCULATIONS AND DATA ANALYSIS

1. Student answers will vary. For sample data, values range from 29.6 counts/s to 55.1 counts/s.

2. For sample data, the background count is 1.24 counts/s.

3. Student graphs should show a gradual curve down and to the right.

CONCLUSIONS

4. Student answers will vary. Typical values will range from 2.08 min to 2.52 min.

5. For sample data, the average half-life value is 2.25 min.

6. Student answers will vary, but they should recognize that using the graph makes it easier to find the value at exact intervals.

Reference Section

Appendices

Appendix A
Mathematical Review

Scientific Notation

Positive exponents Many quantities that scientists deal with often have very large or very small values. For example, the speed of light is about 300 000 000 m/s, and the ink required to make the dot over an *i* in this textbook has a mass of about 0.000 000 001 kg. Obviously, it is cumbersome to work with numbers such as these. We avoid this problem by using a method based on powers of the number 10.

$$10^0 = 1$$
$$10^1 = 10$$
$$10^2 = 10 \times 10 = 100$$
$$10^3 = 10 \times 10 \times 10 = 1000$$
$$10^4 = 10 \times 10 \times 10 \times 10 = 10\ 000$$
$$10^5 = 10 \times 10 \times 10 \times 10 \times 10 = 100\ 000$$

The number of zeros determines the power to which 10 is raised, or the *exponent* of 10. For example, the speed of light, 300 000 000 m/s, can be expressed as 3×10^8 m/s. In this case, the exponent of 10 is 8.

Negative exponents For numbers less than one, we note the following:

$$10^{-1} = \frac{1}{10} = 0.1$$

$$10^{-2} = \frac{1}{10 \times 10} = 0.01$$

$$10^{-3} = \frac{1}{10 \times 10 \times 10} = 0.001$$

$$10^{-4} = \frac{1}{10 \times 10 \times 10 \times 10} = 0.0001$$

$$10^{-5} = \frac{1}{10 \times 10 \times 10 \times 10 \times 10} = 0.000\ 01$$

The value of the negative exponent equals the number of places the decimal point must be moved to be to the right of the first nonzero digit (in these cases, the digit 1). Numbers that are expressed as a number between 1 and 10 multiplied by a power of 10 are said to be in *scientific notation*. For example, 5 943 000 000 is 5.943×10^9 when expressed in scientific notation, and 0.000 083 2 is 8.32×10^{-5} when expressed in scientific notation.

Multiplication and division in scientific notation When numbers expressed in scientific notation are being multiplied, the following general rule is very useful:

$$10^n \times 10^m = 10^{(n+m)}$$

Note that n and m can be any numbers; they are not necessarily integers. For example, $10^2 \times 10^5 = 10^7$, and $10^{1/4} \times 10^{1/2} = 10^{3/4}$. The rule also applies to negative exponents. For example, $10^3 \times 10^{-8} = 10^{-5}$. When dividing numbers expressed in scientific notation, note the following:

$$\frac{10^n}{10^m} = 10^n \times 10^{-m} = 10^{(n-m)}$$

For example, $\dfrac{10^3}{10^2} = 10^{(3-2)} = 10^1$.

Fractions

The rules for multiplying, dividing, adding, and subtracting fractions are summarized in **Table A-1,** where a, b, c, and d are four numbers.

Table A-1 Basic Operations for Fractions

Operation	Rule	Example
Multiplication	$\left(\dfrac{a}{b}\right)\left(\dfrac{c}{d}\right) = \dfrac{ac}{bd}$	$\left(\dfrac{2}{3}\right)\left(\dfrac{4}{5}\right) = \dfrac{(2)(4)}{(3)(5)} = \dfrac{8}{15}$
Division	$\dfrac{\left(\dfrac{a}{b}\right)}{\left(\dfrac{c}{d}\right)} = \dfrac{ad}{bc}$	$\dfrac{\left(\dfrac{2}{3}\right)}{\left(\dfrac{4}{5}\right)} = \dfrac{(2)(5)}{(3)(4)} = \dfrac{10}{12} = \dfrac{5}{6}$
Addition and subtraction	$\dfrac{a}{b} \pm \dfrac{c}{d} = \dfrac{ad \pm bc}{bd}$	$\dfrac{2}{3} - \dfrac{4}{5} = \dfrac{(2)(5) - (3)(4)}{(3)(5)} = -\dfrac{2}{15}$

Powers

Rules of exponents When powers of a given quantity, x, are multiplied, the rule used for scientific notation applies:

$$(x^n)(x^m) = x^{(n+m)}$$

For example, $(x^2)(x^4) = x^{(2+4)} = x^6$.

When dividing the powers of a given quantity, note the following:

$$\frac{x^n}{x^m} = x^{(n-m)}$$

For example, $\dfrac{x^8}{x^2} = x^{(8-2)} = x^6$.

A power that is a fraction, such as $\frac{1}{3}$, corresponds to a root as follows:

$$x^{1/n} = \sqrt[n]{x}$$

For example, $4^{1/3} = \sqrt[3]{4} = 1.5874$. (A scientific calculator is useful for such calculations.)

Finally, any quantity, x^n, that is raised to the mth power is as follows:

$$(x^n)^m = x^{nm}$$

For example, $(x^2)^3 = x^{(2)(3)} = x^6$.

The basic rules of exponents are summarized in **Table A-2.**

Table A-2 Rules of Exponents

$x^0 = 1$	$x^1 = x$	$(x^n)(x^m) = x^{(n+m)}$
$\dfrac{x^n}{x^m} = x^{(n-m)}$	$x^{(1/n)} = \sqrt[n]{x}$	$(x^n)^m = x^{(nm)}$

Algebra

Solving for unknowns When algebraic operations are performed, the laws of arithmetic apply. Symbols such as x, y, and z are usually used to represent quantities that are not specified. Such unspecified quantities are called *unknowns.*

First, consider the following equation:

$$8x = 32$$

If we wish to solve for x, we can divide each side of the equation by the same factor without disturbing the equality. In this case, if we divide both sides by 8, we have the following:

$$\frac{8x}{8} = \frac{32}{8}$$
$$x = 4$$

Next, consider the following equation:

$$x + 2 = 8$$

In this type of expression, we can add or subtract the same quantity from each side. If we subtract 2 from each side, we get the following:

$$x + 2 - 2 = 8 - 2$$
$$x = 6$$

In general, if $x + a = b$, then $x = b - a$.

Now, consider the following equation:

$$\frac{x}{5} = 9$$

If we multiply each side by 5, we are left with x isolated on the left and a value of 45 on the right.

$$(5)\left(\frac{x}{5}\right) = (9)(5)$$

$$x = 45$$

In all cases, *whatever operation is performed on the left side of the equation must also be performed on the right side.*

Factoring

Some useful formulas for factoring an equation are given in **Table A-3.** As an example of a common factor, consider the equation $5x + 5y + 5z = 0$. This equation can be expressed as $5(x + y + z) = 0$. The expression $a^2 + 2ab + b^2$, which is an example of a perfect square, is equivalent to the expression $(a + b)^2$. For example, if $a = 2$ and $b = 3$, then $2^2 + (2)(2)(3) + 3^2 = (2 + 3)^2$, or $(4 + 12 + 9) = 5^2 = 25$. Finally, for an example of the difference of two squares, let $a = 6$ and $b = 3$. In this case, $(6^2 - 3^2) = (6 + 3)(6 - 3)$, or $(36 - 9) = (9)(3) = 27$.

Table A-3 Factoring Equations

$ax + ay + az = a(x + y + z)$	common factor
$a^2 + 2ab + b^2 = (a + b)^2$	perfect square
$a^2 - b^2 = (a + b)(a - b)$	difference of two squares

Quadratic Equations

The general form of a quadratic equation is as follows:

$$ax^2 + bx + c = 0$$

In this equation, x is the unknown quantity and a, b, and c are numerical factors known as *coefficients*. This equation has two roots, given by the following:

$$x = \frac{-b \pm \sqrt{b^2 - 4ac}}{2a}$$

If $b^2 \geq 4ac$, the value inside the square-root symbol will be positive or zero and the roots will be real. If $b^2 < 4ac$, the value inside the square-root symbol will be negative and the roots will be imaginary. In problems in this physics book, imaginary roots should not occur.

Example

Find the solutions for the equation $x^2 + 5x + 4 = 0$.

Solution

The given equation can be expressed as $(1)x^2 + (5)x + (4) = 0$. In other words, $a = 1$, $b = 5$, and $c = 4$. The two roots of this equation can be found by substituting these values into the quadratic equation, as follows:

$$x = \frac{-b \pm \sqrt{b^2 - 4ac}}{2a} = \frac{-5 \pm \sqrt{5^2 - (4)(1)(4)}}{(2)(1)} = \frac{-5 \pm \sqrt{9}}{2} = \frac{-5 \pm 3}{2}$$

The two roots are $x = \dfrac{-5 + 3}{2} = -1$ and $x = \dfrac{-5 - 3}{2} = -4$.

$$\boxed{x = -1 \text{ and } x = -4}$$

We can evaluate these answers by substituting them into the given equation and verifying that the result is zero.

$$x^2 + 5x + 4 = 0$$

For $x = -1$, $(-1)^2 + 5(-1) + 4 = 1 - 5 + 4 = 0$.

For $x = -4$, $(-4)^2 + 5(-4) + 4 = 16 - 20 + 4 = 0$.

Example

Factor the equation $2x^2 - 3x - 4 = 0$.

Solution

In this case, the given equation can be expressed as $(2)x^2 + (-3)x + (-4) = 0$. Thus, $a = 2$, $b = -3$, and $c = -4$. Substitute these values into the quadratic equation to factor the given equation.

$$x = \frac{-b \pm \sqrt{b^2 - 4ac}}{2a} = \frac{3 \pm \sqrt{(-3)^2 - (4)(2)(-4)}}{(2)(2)} = \frac{3 \pm \sqrt{41}}{4} = \frac{3 \pm 6.403}{4}$$

The two roots are $x = \dfrac{3 + 6.403}{4} = 2.351$ and $x = \dfrac{3 - 6.403}{4} = -0.851$.

$$\boxed{x = 2.351 \text{ and } x = -0.851}$$

Again, evaluate these answers by substituting them into the given equation.

$$2x^2 - 3x - 4 = 0$$

For $x = 2.351$, $2(2.351)^2 - 3(2.351) - 4 = 11.054 - 7.053 - 4 \approx 0$.

For $x = -0.851$, $2(-0.851)^2 - 3(-0.851) - 4 = 1.448 + 2.553 - 4 \approx 0$.

Linear Equations

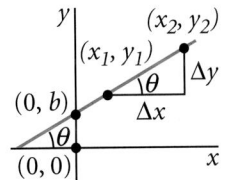

Figure A-1

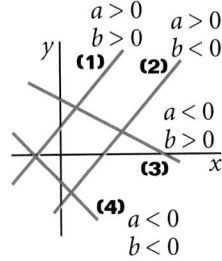

Figure A-2

A linear equation has the following general form:

$$y = ax + b$$

In this equation, a and b are constants. This equation is called linear because the graph of y versus x is a straight line, as shown in **Figure A-1.** The constant b, called the *intercept*, represents the value of y where the straight line intersects the y-axis. The constant a is equal to the *slope* of the straight line and is also equal to the tangent of the angle that the line makes with the x-axis (θ). If any two points on the straight line are specified by the coordinates (x_1, y_1) and (x_2, y_2), as in **Figure A-1,** then the slope of the straight line can be expressed as follows:

$$\text{slope} = \frac{y_2 - y_1}{x_2 - x_1} = \frac{\Delta y}{\Delta x}$$

For example, if the two points shown in **Figure A-1** are $(2, 4)$ and $(6, 9)$, then the slope of the line is as follows:

$$\text{slope} = \frac{(9 - 4)}{(6 - 2)} = \frac{5}{4}$$

Note that a and b can have either positive or negative values. If $a > 0$, the straight line has a *positive* slope, as in **Figure A-1.** If $a < 0$, the straight line has a *negative* slope. Furthermore, if $b > 0$, the y intercept is positive (above the x-axis), while if $b < 0$, the y intercept is negative (below the x-axis). **Figure A-2** gives an example of each of these four possible cases, which are summarized in **Table A-4.**

Table A-4 Linear Equations

Constants	Slope	y intercept
$a > 0, b > 0$	positive slope	positive y intercept
$a > 0, b < 0$	positive slope	negative y intercept
$a < 0, b > 0$	negative slope	positive y intercept
$a < 0, b < 0$	negative slope	negative y intercept

Solving Simultaneous Linear Equations

Consider the following equation:

$$3x + 5y = 15$$

This equation has two unknowns, x and y. Such an equation does not have a unique solution. That is, $(x = 0, y = 3)$, $(x = 5, y = 0)$, and $\left(x = 2, y = \frac{9}{5}\right)$ are all solutions to this equation.

If a problem has two unknowns, a unique solution is possible only if there are two independent equations. In general, if a problem has n unknowns, its solution requires n independent equations. There are three basic methods that can be used to solve simultaneous equations. Each of these methods is discussed below, and an example is given for each.

First method: substitution One way to solve two simultaneous equations involving two unknowns, x and y, is to solve one of the equations for one of the unknown values in terms of the other unknown value. In other words, either solve one equation for x in terms of y or solve one equation for y in terms of x. Once you have an expression for either x or y, substitute this expression into the other original equation. At this point, the equation has only one unknown quantity. This unknown can be found through algebraic manipulations and then can be used to determine the other unknown.

Example

Solve the following two simultaneous equations:

1. $5x + y = -8$
2. $2x - 2y = 4$

Solution

First solve for either x or y in one of the equations. We'll begin by solving equation 2 for x.

2. $2x - 2y = 4$
$2x = 4 + 2y$
$$x = \frac{4 + 2y}{2} = 2 + y$$

Next, we substitute this equation for x into equation 1 and solve for y.

1. $5x + y = -8$
$5(2 + y) + y = -8$
$10 + 5y + y = -8$
$6y = -18$

$$\boxed{y = -3}$$

To find x, substitute this value for y into the equation for x derived from equation 2.

$x = 2 + y = 2 + -3$

$$\boxed{x = -1}$$

There is always more than one way to solve simultaneous equations by substitution. In this example, we first solved equation 2 for x. However, we could have begun by solving equation 2 for y or equation 1 for x or y. Any of these processes would result in the same answer.

Second method: canceling one term Simultaneous equations can also be solved by multiplying both sides of one of the equations by a value that will make either the x value or the y value in that equation equal to and opposite the corresponding value in the second equation. When the two equations are added together, that unknown value drops out and only one of the unknown values remains. This unknown can be found through algebraic manipulations and then can be used to determine the other unknown.

Example
Solve the following two simultaneous equations:

1. $3x + y = -6$
2. $-4x - 2y = 6$

Solution
First, multiply each term of one of the equations by a factor that will make either the x or the y values cancel when the two equations are added together. In this case, we can multiply each term in equation 1 by the factor 2. The positive $2y$ in equation 1 will then cancel the negative $2y$ in equation 2.

1. $3x + y = -6$
$(2)(3x) + (2)(y) = -(2)(6)$
$6x + 2y = -12$

Next, add the two equations together and solve for x.

2. $-4x - 2y = 6$
$6x + 2y = -12$
$2x = -6$

$$\boxed{x = -3}$$

Then substitute this value of x into either equation to find y.
1. $3x + y = -6$
$y = -6 - 3x = -6 - (3)(-3) = -6 + 9$

$$\boxed{y = 3}$$

In this example, we multiplied both sides of equation 1 by 2 so that the y terms would cancel when the two equations were added together. As with substitution, this is only one of many possible ways to solve the equations. For example, we could have multiplied both sides of equation 2 by $\frac{3}{4}$ so that the x terms would cancel when the two equations were added together.

Third method: graphing the equations Two linear equations with two unknowns can also be solved by a graphical method. If the straight lines corresponding to the two equations are plotted in a conventional coordinate system, the intersection of the two lines represents the solution.

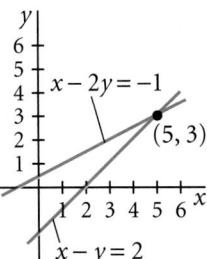

Figure A-3

Example
Solve the following two simultaneous equations:

1. $x - y = 2$
2. $x - 2y = -1$

Solution
These two equations are plotted in **Figure A-3.** To plot an equation, rewrite the equation in the form $y = ax + b$, where a is the slope and b is the y intercept. In this example, the equations can be rewritten as follows:

$$y = x - 2$$
$$y = \tfrac{1}{2}x + \tfrac{1}{2}$$

Once one point of a line is known, any other point on that line can be found with the slope of the line. For example, the slope of the first line is 1, and we know that $(0, -2)$ is a point on this line. If we choose the point $x = 2$, we have $(2, y_2)$. The coordinate y_2 can be found as follows:

$$slope = \frac{y_2 - y_1}{x_2 - x_1} = \frac{y_2 - (-2)}{2 - 0} = 1$$
$$y_2 = 0$$

Connecting the two known coordinates, $(0, -2)$ and $(2, 0)$, results in a graph of the line. The second line can be plotted with the same method.

As shown in **Figure A-3,** the intersection of the two lines has the coordinates $x = 5$, $y = 3$. This intersection represents the solution to the equations. You should check this solution using either of the analytical techniques discussed above.

Logarithms

Suppose that a quantity, x, is expressed as a power of another quantity, a.

$$x = a^y$$

The number a is called the *base number*. The *logarithm* of x with respect to the base, a, is equal to the exponent to which a must be raised in order to satisfy the expression $x = a^y$.

$$y = \log_a x$$

Conversely, the *antilogarithm* of y is the number x.

$$x = \text{antilog}_a y$$

Common and natural bases In practice, the two bases most often used are base 10, called the *common* logarithm base, and base $e = 2.718...$, called the *natural* logarithm base. When common logarithms are used, y and x are related as follows:

$$y = \log_{10} x, \text{ or } x = 10^y$$

When natural logarithms are used, the symbol ln is used to signify that the logarithm has a base of e; in other words, $\log_e x = \ln x$.

$$y = \ln x, \text{ or } x = e^y$$

For example, $\log_{10} 52 = 1.716$, so antilog$_{10}$ $1.716 = 10^{1.716} = 52$. Likewise, $\ln 52 = 3.951$, so antiln $3.951 = e^{3.951} = 52$.

Note that you can convert between base 10 and base e with the equality

$$\ln x = (2.302\ 585)\log_{10} x.$$

Some useful properties of logarithms are summarized in **Table A-5.**

Table A-5 Properties of Logarithms

Rule	Example
$\log (ab) = \log a + \log b$	$\log (2)(5) = \log 2 + \log 5$
$\log \left(\dfrac{a}{b}\right) = \log a - \log b$	$\log \frac{3}{4} = \log 3 - \log 4$
$\log (a^n) = n \log a$	$\log 7^3 = 3 \log 7$
$\ln e = 1$	
$\ln e^a = a$	$\ln e^5 = 5$
$\ln \left(\dfrac{1}{a}\right) = -\ln a$	$\ln \frac{1}{8} = -\ln 8$

Conversions Between Fractions, Decimals, and Percentages

The rules for converting numbers from fractions to decimals and percentages and from percentages to decimals are summarized in **Table A-6.**

Table A-6 Conversions

Conversion	Rule	Example
Fraction to decimal	divide numerator by denominator	$\dfrac{31}{45} = 0.69$
Fraction to percentage	convert to decimal, then multiply by 100%	$\dfrac{31}{45} = (0.69)(100\%) = 69\%$
Percentage to decimal	move decimal point two places to the left, and remove the percent sign	$69\% = 0.69$

Geometry

Table A-7 provides equations for the area and volume of several geometric shapes used throughout this text.

Table A-7 Geometrical Areas and Volumes

Geometrical shape	Useful equations
rectangle	$\text{area} = lw$
circle	$\text{area} = \pi r^2$ $\text{circumference} = 2\pi r$
triangle	$\text{area} = \frac{1}{2}bh$
sphere	$\text{surface area} = 4\pi r^2$ $\text{volume} = \frac{4}{3}\pi r^3$
cylinder	$\text{volume} = \pi r^2 l$
rectangular box	$\text{surface area} = 2(lh + lw + hw)$ $\text{volume} = lwh$

Trigonometry and the Pythagorean Theorem

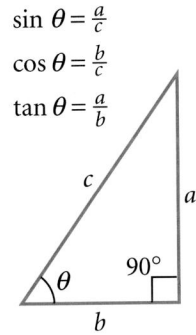

$$\sin \theta = \frac{a}{c}$$

$$\cos \theta = \frac{b}{c}$$

$$\tan \theta = \frac{a}{b}$$

Figure A-4

The portion of mathematics that is based on the special properties of a right triangle is called *trigonometry.* Many of the concepts of this branch of mathematics are of utmost importance in the study of physics. To review some of the basic concepts of trigonometry, consider the right triangle shown in **Figure A-4,** where side a is opposite the angle θ, side b is adjacent to the angle θ, and side c is the hypotenuse of the triangle (the side opposite the right angle). The most common trigonometry functions are summarized in **Table A-8,** using this figure as an example.

Table A-8 Trigonometry Functions

sine (sin)	$\sin \theta = \dfrac{\text{side opposite } \theta}{\text{hypotenuse}} = \dfrac{a}{c}$
cosine (cos)	$\cos \theta = \dfrac{\text{side adjacent to } \theta}{\text{hypotenuse}} = \dfrac{b}{c}$
tangent (tan)	$\tan \theta = \dfrac{\text{side opposite } \theta}{\text{side adjacent to } \theta} = \dfrac{a}{b}$
inverse sine ($\sin^{-1}$)	$\theta = \sin^{-1}\left(\dfrac{\text{side opposite } \theta}{\text{hypotenuse}}\right) = \sin^{-1}\left(\dfrac{a}{c}\right)$
inverse cosine ($\cos^{-1}$)	$\theta = \cos^{-1}\left(\dfrac{\text{side adjacent to } \theta}{\text{hypotenuse}}\right) = \cos^{-1}\left(\dfrac{b}{c}\right)$
inverse tangent ($\tan^{-1}$)	$\theta = \tan^{-1}\left(\dfrac{\text{side opposite } \theta}{\text{side adjacent to } \theta}\right) = \tan^{-1}\left(\dfrac{a}{b}\right)$

When $\theta = 30°$, for example, the ratio of a to c is always 0.50. In other words, $\sin 30° = 0.50$. Sine, cosine, and tangent are quantities without units because each represents the ratio of two lengths. Furthermore, note the following trigonometry identity:

$$\frac{\sin \theta}{\cos \theta} = \frac{\dfrac{\text{side opposite } \theta}{\text{hypotenuse}}}{\dfrac{\text{side adjacent to } \theta}{\text{hypotenuse}}} = \frac{\text{side opposite } \theta}{\text{side adjacent to } \theta} = \tan \theta$$

Some additional trigonometry identities are as follows:

$$\sin^2\theta + \cos^2\theta = 1$$
$$\sin \theta = \cos(90° - \theta)$$
$$\cos \theta = \sin(90° - \theta)$$

Determining an unknown side The first three functions given in **Table A-8** can be used to determine any unknown side of a right triangle when one side and one of the non-right angles are known. For example, if $\theta = 30°$ and $a = 1.0$ m, the other two sides of the triangle can be found as follows:

$$\sin \theta = \frac{a}{c}$$

$$c = \frac{a}{\sin \theta} = \frac{1.0 \text{ m}}{\sin 30°}$$

$$\boxed{c = 2.0 \text{ m}}$$

$$\tan \theta = \frac{a}{b}$$

$$b = \frac{a}{\tan \theta} = \frac{1.0 \text{ m}}{\tan 30°}$$

$$\boxed{b = 1.7 \text{ m}}$$

Determining an unknown angle In some cases, you might know the value of the sine, cosine, or tangent of an angle and need to know the value of the angle itself. The inverse sine, cosine, and tangent functions given in **Table A-8** can be used for this purpose. For example, in **Figure A-4** on page 947, suppose you know that side $a = 1.0$ m and side $c = 2.0$ m. To find the angle θ, you could use the inverse sine function, $\sin^{-1}$, as follows:

$$\theta = \sin^{-1}\left(\frac{a}{c}\right) = \sin^{-1}\left(\frac{1.0 \text{ m}}{2.0 \text{ m}}\right) = \sin^{-1}(0.50)$$

$$\boxed{\theta = 30°}$$

Pythagorean theorem Another useful equation when working with right triangles is the Pythagorean theorem. If a and b are the two legs of a right triangle and c is the hypotenuse, as in **Figure A-5,** the Pythagorean theorem can be expressed as follows:

$$c^2 = a^2 + b^2$$

In other words, the square of the hypotenuse of a right triangle equals the sum of the squares of the other two legs of the triangle. The Pythagorean theorem is useful when two sides of a right triangle are known but the third side is not. For example, if $c = 2.0$ m and $a = 1.0$ m, you could find b using the Pythagorean theorem as follows:

$$b = \sqrt{c^2 - a^2} = \sqrt{(2.0 \text{ m})^2 - (1.0 \text{ m})^2}$$

$$b = \sqrt{4.0 \text{ m}^2 - 1.0 \text{ m}^2} = \sqrt{3.0 \text{ m}^2}$$

$$\boxed{b = 1.7 \text{ m}}$$

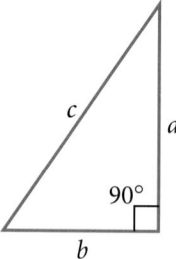

Figure A-5

Accuracy in Laboratory Calculations

Absolute error Some of the laboratory experiments in this book involve finding a value that is already known, such as free-fall acceleration. In this type of experiment, the accuracy of your measurements can be determined by comparing your results with the accepted value. The absolute value of the difference between your experimental or calculated result and the accepted value is called the *absolute error*. Thus, absolute error can be found with the following equation:

$$\text{absolute error} = |\text{experimental} - \text{accepted}|$$

Be sure not to confuse accuracy with precision. The *accuracy* of a measurement refers to how close that measurement is to the accepted value for the quantity being measured. *Precision,* on the other hand, depends on the instruments used to measure a quantity. A meterstick that includes millimeters, for example, will give a more precise result than a meterstick whose smallest unit of measure is a centimeter. Thus, a measurement of 9.61 m/s^2 for free-fall acceleration is more precise than a measurement of 9.8 m/s^2, but 9.8 m/s^2 is more accurate than 9.61 m/s^2.

Relative error Note that a measurement that has a relatively large absolute error may be more accurate than a measurement that has a smaller absolute error if the first measurement involved much larger quantities. For this reason, the percentage error, or *relative error,* is often more meaningful than the absolute error. The relative error of a measured value can be found with the following equation:

$$\text{relative error} = \frac{(\text{experimental} - \text{accepted})}{\text{accepted}}$$

In other words, the relative error is the difference between the experimental value and the accepted value divided by the accepted value. Because relative error takes the size of the quantity being measured into account, the accuracy of two different measurements can be compared by comparing their relative errors.

For example, consider two laboratory experiments in which you are determining values that are fairly well known. In the first, you determine that free-fall acceleration at Earth's surface is 10.31 m/s^2. In the second, you find that the speed of sound in air at 25°C is 355 m/s. The accepted values for these quantities are 9.81 m/s^2 and 346 m/s, respectively. Now we'll find the absolute and relative errors for each experiment.

For the first experiment, the absolute and relative errors can be calculated as follows:

$$\text{absolute error} = \left| \text{experimental} - \text{accepted} \right| = \left| 10.31 \text{ m/s}^2 - 9.81 \text{ m/s}^2 \right|$$

$$\boxed{\text{absolute error} = 0.50 \text{ m/s}^2}$$

$$\text{relative error} = \frac{(\text{experimental} - \text{accepted})}{\text{accepted}} = \frac{(10.31 \text{ m/s}^2 - 9.81 \text{ m/s}^2)}{9.81 \text{ m/s}^2}$$

$$\boxed{\text{relative error} = 0.051 = 5.1\%}$$

For the second experiment, the absolute and relative errors can be calculated as follows:

$$\text{absolute error} = \left| \text{experimental} - \text{accepted} \right| = \left| 355 \text{ m/s} - 346 \text{ m/s} \right|$$

$$\boxed{\text{absolute error} = 9 \text{ m/s}}$$

$$\text{relative error} = \frac{(\text{experimental} - \text{accepted})}{\text{accepted}} = \frac{(355 \text{ m/s} - 346 \text{ m/s})}{346 \text{ m/s}}$$

$$\boxed{\text{relative error} = 0.026 = 2.6\%}$$

Note that the *absolute* error is less in the first experiment, while the *relative* error is less in the second experiment. The absolute error is less in the first experiment because typical values for free-fall acceleration are much smaller than typical values for the speed of sound in air. The relative errors take this difference into account. Thus, comparing the relative errors shows that the speed of sound is measured with greater accuracy than is the free-fall acceleration.

Appendix B
Downloading Graphing Calculator Programs

To download required software for your graphing calculator, you will need
a Macintosh or Windows-based computer with an Internet connection,
a TI-Graph Link cable, and a TI-83 Plus or compatible calculator. Visit the
go.hrw.com Web site, and type in HF2 CALC at the Keyword prompt.

1. If your computer does not already have TI-Graph Link software
installed, click Install TI-Graph Link and follow the links for downloading and installing TI-Graph Link from the TI Web site.

2. If you will be doing the Technology and Learning exercises, click
Download Technology and Learning Programs. This will load the file
GRAPHING.ZIP onto your computer. Once the file is downloaded,
double-click the icon and the file will be extracted into a file called
Graphing.8xg. Follow the instructions for your TI-Graph Link to load
Graphing.8xg onto your TI calculator. When the file is sent to the calcu-
lator, it should expand into 25 programs (one for each chapter). These
programs should appear in the PRGM menu.

3. If you will be doing laboratory experiments and activities with the CBL
system, click Download the PHYSICS App and follow the links for downloading the PHYSICS application from the Vernier Software Web site. You
will also need to follow the instructions for your TI-Graph Link to load
the application onto your calculator. Once the PHYSICS application is
loaded onto your calculator, it should appear in the APPS menu.

4. If you need additional instructions on using the CBL system, click CBL
Made Easy for a tutorial from Vernier Software.

Troubleshooting

- Calculator and CBL instructions in the *Holt Physics* program are written
  for the TI-83 Plus, for the original CBL from TI, and for probes from
  Vernier Software. You may use other hardware, but some of the programs
  and instructions may not work exactly as described.
- If you have problems loading programs onto your calculator, you may need
  to clear programs or other data from your calculator's memory.
- Make sure that you download the correct versions of the software.
  TI-Graph Link and the PHYSICS application both have different versions
  for Windows and for Macintosh as well as different versions for different
  calculators.
- If you need additional help, both TI and Vernier Software can provide
  technical support.

Appendix C Symbols

Diagram Symbols

The entries in this table are given roughly in order of appearance in the textbook, but consistency has been maintained in later chapters as well.

Mechanics

Symbol	Meaning
→ ⇨	displacement vector, component displacement
→ ⇨	velocity vector, component velocity
→	acceleration vector
→ ⇨	force vector, component force
→	momentum vector
∠	angle marking
↰	rotational motion

Thermodynamics

Symbol	Meaning
→	energy transferred as heat
→	energy transferred as work
↻	cycle or process

Waves and Electromagnetism

Symbol	Meaning
⌄	ray (light or sound)
⊕	positive charge
⊖	negative charge
◎	electric field lines
→	electric field vector
⚡	electric current
◎	magnetic field lines
→ ✕ ●	magnetic field vector (into page, out of page)

Mathematical Symbols

Symbol	Meaning	Symbol	Meaning
Δ	(Greek *delta*) change in some quantity	≤	less than or equal to
Σ	(Greek *sigma*) sum of quantities	∝	is proportional to
θ	(Greek *theta*) any angle	≈	is approximately equal to
=	equal to	$|n|$	absolute value or magnitude of
>	greater than	sin	sine
≥	greater than or equal to	cos	cosine
<	less than	tan	tangent

Quantity Symbols Used Throughout

Symbols that are **boldfaced** refer to vector quantities that have both a magnitude and a direction. Symbols that are *italicized* refer to quantities with only a magnitude. Symbols that are neither are usually units (see Appendix D).

Symbol	Quantity
A	area
D	diameter
F, *F*	force
m	mass
M	total mass
R	radius (of a spherical body, a curved mirror, or a curved lens)
r	radius (of sphere, shell, or disk)
t	time
V	volume

Translational Mechanics Symbols Used in This Book (Chapters 2–6)

Symbols that are **boldfaced** refer to vector quantities that have both a magnitude and a direction. Symbols that are *italicized* refer to quantities with only a magnitude. Symbols that are neither are usually units (see Appendix D).

Symbol	Quantity
$\mathbf{a}$, a	acceleration
$\mathbf{d}$, d	displacement
$\mathbf{F}\Delta t$	impulse
$\mathbf{F_g}$, F_g	gravitational force (weight)
$\mathbf{F_k}$, F_k	force of kinetic friction
$\mathbf{F_n}$, F_n	normal force
$\mathbf{F_{net}}$, F_{net}	net force
$\mathbf{F_R}$, F_R	force of air resistance
$\mathbf{F_s}$, F_s	force of static friction
$\mathbf{F_{s,max}}$, $F_{s,max}$	maximum force of static friction
h	height
k	spring constant
KE	kinetic energy
KE_{trans}	translational kinetic energy
MA	mechanical advantage
ME	mechanical energy (sum of all kinetic and potential energies)
μ_k	(Greek *mu*) coefficient of kinetic friction
μ_s	(Greek *mu*) coefficient of static friction
P	power
$\mathbf{p}$, p	momentum
PE	potential energy
$PE_{elastic}$	elastic potential energy
PE_g	gravitational potential energy
r	separation between point masses
$\mathbf{v}$, v	velocity or speed
W	work
$W_{friction}$	work done by a frictional force (or work required to overcome a frictional force)
W_{net}	net work done
$\Delta\mathbf{x}$, Δx	displacement in the x direction
$\Delta\mathbf{y}$, Δy	displacement in the y direction

Rotational Mechanics Symbols Used in This Book (Chapters 7–8)

Symbols that are **boldfaced** refer to vector quantities that have both a magnitude and a direction. Symbols that are *italicized* refer to quantities with only a magnitude. Symbols that are neither are usually units (see Appendix D).

Symbol	Quantity
a_t	tangential acceleration
a_c	centripetal acceleration
α	(Greek *alpha*) angular acceleration
$d\,(\sin\theta)$	length of lever arm (for torque calculations)
$\mathbf{F_c},\ F_c$	force that maintains circular motion
I	moment of inertia
KE_{rot}	rotational kinetic energy
L	angular momentum
ℓ	length of a rotating rod
s	arc length
τ	(Greek *tau*) torque
τ_{net}	(Greek *tau*) net torque
θ	(Greek *theta*) angle of rotation
$\Delta\theta$	(Greek *delta* and *theta*) angular displacement (in radians)
v_t	tangential speed
ω	(Greek *omega*) angular speed

Fluid Dynamics and Thermodynamics Symbols Used in This Book (Chapters 9–11)

Symbols that are **boldfaced** refer to vector quantities that have both a magnitude and a direction. Symbols that are *italicized* refer to quantities with only a magnitude. Symbols that are neither are usually units (see Appendix D).

Symbol	Quantity
c_p	specific heat capacity
eff	efficiency of a simple machine, thermal efficiency of a heat engine
$\mathbf{F_B},\ F_B$	buoyant force
L	latent heat
L_f	latent heat of fusion
L_v	latent heat of vaporization
N	number of gas particles or nuclei
P	pressure
P_0	initial pressure, atmospheric pressure
P_{net}	net pressure
ρ	(Greek *rho*) mass density
Q	heat
Q_c	energy transferred as heat to or from a low-temperature (cold) substance
Q_h	energy transferred as heat to or from a high-temperature (hot) substance
Q_{net}	net amount of energy transferred as heat to or from a system
T	temperature (absolute)
T_C	temperature in degrees Celsius
T_c	temperature of a low-temperature (cool) substance
T_F	temperature in degrees Fahrenheit
T_h	temperature of a high-temperature (hot) substance
U	internal energy

Vibrations, Waves, and Optics Symbols Used in This Book (Chapters 12–16)

Symbols that are **boldfaced** refer to vector quantities that have both a magnitude and a direction. Symbols that are *italicized* refer to quantities with only a magnitude. Symbols that are neither are usually units (see Appendix D).

Symbol	Quantity
C	center of curvature for spherical mirror
d	slit separation in double-slit interference of light
$d\,(\sin\theta)$	path difference for interfering light waves
$\mathbf{F_{elastic}}$, $F_{elastic}$	spring force
F	focal point
f	focal length
f	frequency
f_n	nth harmonic frequency
h	object height
h'	image height
k	spring constant
L	length of a pendulum, vibrating string, or vibrating column of air
ℓ	path length of light wave
λ	(Greek *lambda*) wavelength
m	order number for interference fringes
M	magnification of image
n	harmonic number (sound)
n	index of refraction
p	object distance
q	image distance
T	period of a pendulum (simple harmonic motion)
θ	(Greek *theta*) angle of incidence of a beam of light (reflection)
θ	(Greek *theta*) angle of fringe separation from center of interference pattern
θ'	(Greek *theta*) angle of reflection
θ_c	(Greek *theta*) critical angle of refraction
θ_i	(Greek *theta*) angle of incidence of a beam of light (refraction)
θ_r	(Greek *theta*) angle of refraction

Electromagnetism Symbols Used in This Book (Chapters 17–22)

Symbols that are **boldfaced** refer to vector quantities that have both a magnitude and a direction. Symbols that are *italicized* refer to quantities with only a magnitude. Symbols that are neither are usually units (see Appendix D).

Symbol	Quantity
$\mathbf{B}$, B	magnetic field
C	capacitance
d	separation of plates in a capacitor
$\mathbf{E}$, E	electric field
emf	emf (potential difference) produced by a battery or electromagnetic induction
$\mathbf{F_{electric}}$, $F_{electric}$	electric force
$\mathbf{F_{magnetic}}$, $F_{magnetic}$	magnetic force
I	electric current
i	instantaneous current (ac circuit)
I_{max}	maximum current (ac circuit)
I_{rms}	root-mean-square current (ac circuit)
L	self-inductance
ℓ	length of an electrical conductor in a magnetic field
M	mutual inductance
N	number of turns in a current-carrying loop or a transformer coil
$PE_{electric}$	electrical potential energy
Q	large charge or charge on a fully charged capacitor
q	charge
R	resistance
r	separation between charges
R_{eq}	equivalent resistance
V	electric potential
ΔV	potential difference
Δv	instantaneous potential difference (ac circuit)
ΔV_{max}	maximum potential difference (ac circuit)
ΔV_{rms}	root-mean-square potential difference (ac circuit)
ω	(Greek *omega*) angular frequency

Particle and Electronic Symbols Used in This Book (Chapters 23–25)

For this part of the book, two tables are given because some symbols refer to quantities and others refer to specific particles. The symbol's context should make clear which table should be consulted.

Symbol	Quantity
A	mass number
β	(Greek *beta*) current or potential difference gain of an amplifier
E	photon energy
E_R	rest energy
f_t	threshold frequency (photoelectric effect)
hf_t	work function (photoelectric effect)
KE_{max}	maximum energy of ejected photoelectron
λ	(Greek *lambda*) decay constant
λN	decay rate (activity)
N	neutron number, number of decayed particles
n	energy quantum number
$T_{1/2}$	half-life
Z	atomic number

Symbol	Particle
α	alpha particle
$b, \bar{b}$	bottom quark, antiquark
β^+	(Greek *beta*) positron (beta particle)
β^-	(Greek *beta*) electron (beta particle)
$c, \bar{c}$	charmed quark, antiquark
$d, \bar{d}$	down quark, antiquark
$e^+, {}^{\ 0}_{+1}e$	positron
$e^-, {}^{\ 0}_{-1}e$	electron
γ	(Greek *gamma*) photon (gamma rays)
${}^4_2\text{He}$	alpha particle (helium-4 nucleus)
μ	(Greek *mu*) muon
${}^1_0 n$	neutron
${}^1_1 p$	proton
$s, \bar{s}$	strange quark, antiquark
$t, \bar{t}$	top quark, antiquark
$u, \bar{u}$	up quark, antiquark
τ	(Greek *tau*) tauon
$\nu, \bar{\nu}$	(Greek *nu*) neutrino, antineutrino
W^+, W^-	boson (weak force)
Z	boson (weak force)

Appendix D SI Units

SI Base Units Used in This Book

Symbol	Name	Quantity
A	ampere	current
K	kelvin	absolute temperature
kg	kilogram	mass
m	meter	length
s	second	time

SI Prefixes

Symbol	Name	Numerical equivalent
a	atto	10^{-18}
f	femto	10^{-15}
p	pico	10^{-12}
n	nano	10^{-9}
μ	micro	10^{-6}
m	milli	10^{-3}
c	centi	10^{-2}
d	deci	10^{-1}
k	kilo	10^{3}
M	mega	10^{6}
G	giga	10^{9}
T	tera	10^{12}
P	peta	10^{15}
E	exa	10^{18}

Other Commonly Used Units

Symbol	Name	Quantity	Conversions
atm	standard atmosphere	pressure	$1.013\,250 \times 10^5$ Pa
Btu	British thermal unit	energy	1.055×10^3 J
Cal	food calorie	energy	$= 1$ kcal $= 4.186 \times 10^3$ J
cal	calorie	energy	4.186 J
Ci	curie	decay rate or activity	3.7×10^{10} s^{-1}
°F	degree Fahrenheit	temperature	0.5556°C
ft	foot	length	0.3048 m
ft·lb	foot-pound	work and energy	1.356 J
g	gram	mass	0.001 kg
gal	gallon	volume	3.785×10^{-3} m^3
hp	horsepower	power	746 W
in	inch	length	2.54×10^{-2} m
kcal	kilocalorie	energy	4.186×10^3 J
lb	pound	force	4.45 N
mi	mile	length	1.609×10^3 m
rev	revolution	angular displacement	2π rad
°	degrees	angular displacement	$= \left(\dfrac{2\pi}{360}\right)$ rad $= 1.745 \times 10^{-2}$ rad

958

Other Units Acceptable with SI

Symbol	Name	Quantity	Conversion
Bq	becquerel	decay rate or activity	$\dfrac{1}{s}$
C	coulomb	electric charge	$1\ A{\cdot}s$
°C	degree Celsius	temperature	$1\ K$
dB	decibel	relative intensity (sound)	(unitless)
eV	electron volt	energy	$1.60 \times 10^{-19}\ J$
F	farad	capacitance	$1\dfrac{A^2{\cdot}s^4}{kg{\cdot}m^2} = 1\dfrac{C}{V}$
H	henry	inductance	$1\dfrac{kg{\cdot}m^2}{A^2{\cdot}s^2} = 1\dfrac{J}{A^2}$
h	hour	time	$3.6 \times 10^3\ s$
Hz	hertz	frequency	$\dfrac{1}{s}$
J	joule	work and energy	$1\dfrac{kg{\cdot}m^2}{s^2} = 1\ N{\cdot}m$
kW·h	kilowatt-hour	energy	$3.60 \times 10^6\ J$
L	liter	volume	$10^{-3}\ m^3$
min	minute	time	$6.0 \times 10^1\ s$
N	newton	force	$1\dfrac{kg{\cdot}m}{s^2}$
Pa	pascal	pressure	$1\dfrac{kg}{m{\cdot}s^2} = 1\dfrac{N}{m^2}$
rad	radian	angular displacement	(unitless)
T	tesla	magnetic field strength	$1\dfrac{kg}{A{\cdot}s^2} = 1\dfrac{N}{A{\cdot}m} = 1\dfrac{V{\cdot}s}{m^2}$
u	unified unit	mass (atomic masses)	$1.660\ 5402 \times 10^{-27}\ kg$
V	volt	electric potential difference	$1\dfrac{kg{\cdot}m^2}{A{\cdot}s^3} = 1\dfrac{J}{C}$
W	watt	power	$1\dfrac{kg{\cdot}m^2}{s^3} = 1\dfrac{J}{s}$
Ω	ohm	resistance	$1\dfrac{kg{\cdot}m^2}{A^2{\cdot}s^3} = 1\dfrac{V}{A}$

Appendix E Useful Tables

Fundamental Constants

Symbol	Quantity	Established value	Value used for calculations in this book
c	speed of light in a vacuum	299 792 458 m/s	3.00×10^8 m/s
e^-	elementary charge	$1.602\ 177\ 33 \times 10^{-19}$ C	1.60×10^{-19} C
e^x	base of natural logarithms	2.718 2818 28	2.72
ε_0	(Greek *epsilon*) permittivity of a vacuum	$8.854\ 187\ 817 \times 10^{-12}$ C^2/(N•m^2)	8.85×10^{-12} C^2/(N•m^2)
G	constant of universal gravitation	$6.672\ 59 \times 10^{-11}$ N•m^2/kg^2	6.673×10^{-11} N•m^2/kg^2
g	free-fall acceleration at Earth's surface	9.806 65 m/s^2	9.81 m/s^2
h	Planck's constant	$6.626\ 0755 \times 10^{-34}$ J•s	6.63×10^{-34} J•s
k_B	Boltzmann's constant (R/N_A)	$1.380\ 658 \times 10^{-23}$ J/K	1.38×10^{-23} J/K
k_C	Coulomb constant	$8.987\ 551\ 788 \times 10^9$ N•m^2/C^2	8.99×10^9 N•m^2/C^2
R	molar (universal) gas constant	8.314 510 J/(mol•K)	8.31 J/(mol•K)
π	(Greek *pi*) ratio of the circumference to the diameter of a circle	3.141 592 654	calculator value

Coefficients of Friction (Approximate Values)

	μ_s	μ_k		μ_s	μ_k
steel on steel	0.74	0.57	waxed wood on wet snow	0.14	0.1
aluminum on steel	0.61	0.47	waxed wood on dry snow	—	0.04
rubber on dry concrete	1.0	0.8	metal on metal (lubricated)	0.15	0.06
rubber on wet concrete	—	0.5	ice on ice	0.1	0.03
wood on wood	0.4	0.2	Teflon on Teflon	0.04	0.04
glass on glass	0.9	0.4	synovial joints in humans	0.01	0.003

Useful Astronomical Data

Symbol	Quantity	Value used for calculations in this book
I_E	moment of inertia of Earth	8.03×10^{37} kg•m^2
M_E	mass of Earth	5.98×10^{24} kg
R_E	radius of Earth	6.37×10^6 m
	Average Earth–moon distance	3.84×10^8 m
	Average Earth–sun distance	1.496×10^{11} m
	mass of the moon	7.36×10^{22} kg
	mass of the sun	1.99×10^{30} kg
yr	period of Earth's orbit	3.16×10^7 s

The Moment of Inertia For a Few Shapes

Shape	Moment of inertia
thin hoop about symmetry axis	MR^2
thin hoop about diameter	$\frac{1}{2}MR^2$
point mass about axis	MR^2
disk or cylinder about symmetry axis	$\frac{1}{2}MR^2$

Shape	Moment of inertia
thin rod about perpendicular axis through center	$\frac{1}{12}M\ell^2$
thin rod about perpendicular axis through end	$\frac{1}{3}M\ell^2$
solid sphere about diameter	$\frac{2}{5}MR^2$
thin spherical shell about diameter	$\frac{2}{3}MR^2$

Densities of Some Common Substances*

Substance	ρ (kg/m³)
hydrogen	0.0899
helium	0.179
steam (100°C)	0.598
air	1.29
oxygen	1.43
carbon dioxide	1.98
ethanol	0.806×10^3
ice	0.917×10^3
fresh water (4°C)	1.00×10^3
sea water (15°C)	1.025×10^3
glycerine	1.26×10^3
aluminum	2.70×10^3
iron	7.86×10^3
copper	8.92×10^3
silver	10.5×10^3
lead	11.3×10^3
mercury	13.6×10^3
gold	19.3×10^3

*All densities are measured at 0°C and 1 atm unless otherwise noted.

Specific Heat Capacities

Substance	c_p (J/kg·°C)
aluminum	8.99×10^2
copper	3.87×10^2
glass	8.37×10^2
gold	1.29×10^2
ice	2.09×10^3
iron	4.48×10^2
lead	1.28×10^2
mercury	1.38×10^2
silver	2.34×10^2
steam	2.01×10^3
water	4.186×10^3

Latent Heats of Fusion and Vaporization at Standard Pressure

Substance	Melting point (°C)	L_f (J/kg)	Boiling point (°C)	L_v (J/kg)
nitrogen	−209.97	2.55×10^4	−195.81	2.01×10^5
oxygen	−218.79	1.38×10^4	−182.97	2.13×10^5
ethyl alcohol	−114	1.04×10^5	78	8.54×10^5
water	0.00	3.33×10^5	100.00	2.26×10^6
lead	327.3	2.45×10^4	1745	8.70×10^5
aluminum	660.4	3.97×10^5	2467	1.14×10^7

Speed of Sound in Various Media

Medium	v (m/s)	Medium	v (m/s)	Medium	v (m/s)
Gases		**Liquids at 25°C**		**Solids**	
air (0°C)	331	methyl alcohol	1140	aluminum	5100
air (25°C)	346	sea water	1530	copper	3560
air (100°C)	366	water	1490	iron	5130
helium (0°C)	972			lead	1320
hydrogen (0°C)	1290			vulcanized rubber	54
oxygen (0°C)	317				

Conversion of Intensity to Decibel Level

Intensity (W/m^2)	Decibel level (dB)	Examples
1.0×10^{-12}	0	threshold of hearing
1.0×10^{-11}	10	rustling leaves
1.0×10^{-10}	20	quiet whisper
1.0×10^{-9}	30	whisper
1.0×10^{-8}	40	mosquito buzzing
1.0×10^{-7}	50	normal conversation
1.0×10^{-6}	60	air conditioning at 6 m
1.0×10^{-5}	70	vacuum cleaner
1.0×10^{-4}	80	busy traffic, alarm clock
1.0×10^{-3}	90	lawn mower
1.0×10^{-2}	100	subway, power motor
1.0×10^{-1}	110	auto horn at 1 m
1.0×10^{0}	120	threshold of pain
1.0×10^{1}	130	thunderclap, machine gun
1.0×10^{3}	150	nearby jet airplane

Indices of Refraction for Various Substances*

Solids at 20°C	n	Liquids at 20°C	n	Gases at 0°C, 1 atm	n
cubic zirconia	2.20	benzene	1.501	air	1.000 293
diamond	2.419	carbon disulfide	1.628	carbon dioxide	1.000 450
fluorite	1.434	carbon tetrachloride	1.461		
fused quartz	1.458	ethyl alcohol	1.361		
glass, crown	1.52	glycerine	1.473		
glass, flint	1.66	water	1.333		
ice (at 0°C)	1.309				
polystyrene	1.49				
sodium chloride	1.544				
zircon	1.923				

*measured with light of vacuum wavelength = 589 nm

Useful Atomic Data

Symbol	Quantity	Established value	Value used for calculations in this book
m_e	mass of electron	$9.109\ 3897 \times 10^{-31}$ kg $5.485\ 799\ 03 \times 10^{-4}$ u $0.510\ 999\ 06$ MeV	9.109×10^{-31} kg 5.49×10^{-4} u 5.110×10^{-1} MeV
m_n	mass of neutron	$1.674\ 9286 \times 10^{-27}$ kg $1.008\ 664\ 904$ u $939.565\ 63$ MeV	1.675×10^{-27} kg $1.008\ 665$ u 9.396×10^{2} MeV
m_p	mass of proton	$1.672\ 6231 \times 10^{-27}$ kg $1.007\ 276\ 470$ u $938.272\ 31$ MeV	1.673×10^{-27} kg $1.007\ 276$ u 9.383×10^{2} MeV

Appendix F
Periodic Table of the Elements

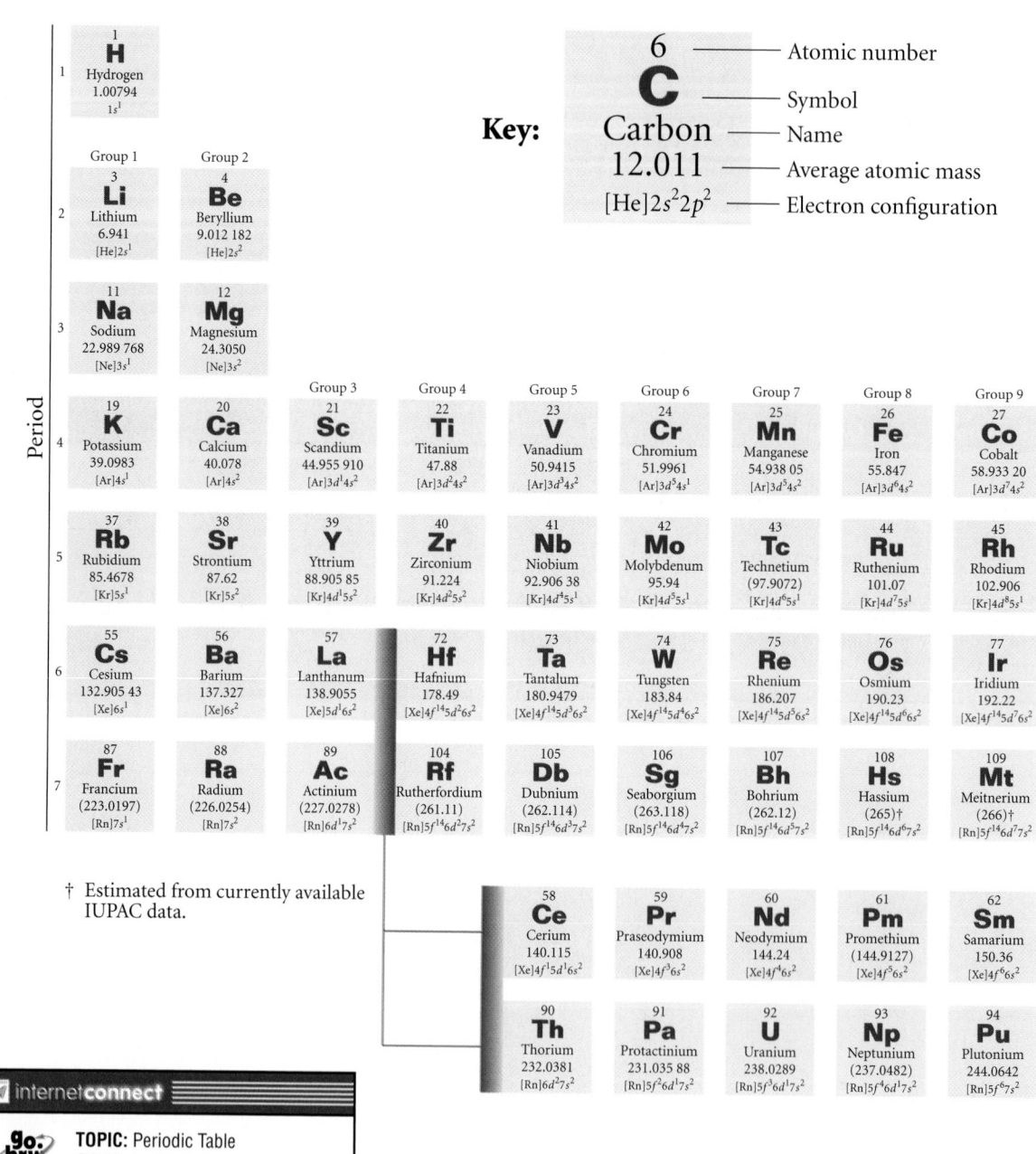

Key:

6 — Atomic number
C — Symbol
Carbon — Name
12.011 — Average atomic mass
$[He]2s^2 2p^2$ — Electron configuration

Period

Group 1	Group 2

1
H
Hydrogen
1.00794
$1s^1$

2
3 **Li** Lithium 6.941 $[He]2s^1$
4 **Be** Beryllium 9.012 182 $[He]2s^2$

3
11 **Na** Sodium 22.989 768 $[Ne]3s^1$
12 **Mg** Magnesium 24.3050 $[Ne]3s^2$

		Group 3	Group 4	Group 5	Group 6	Group 7	Group 8	Group 9

4
19 **K** Potassium 39.0983 $[Ar]4s^1$
20 **Ca** Calcium 40.078 $[Ar]4s^2$
21 **Sc** Scandium 44.955 910 $[Ar]3d^1 4s^2$
22 **Ti** Titanium 47.88 $[Ar]3d^2 4s^2$
23 **V** Vanadium 50.9415 $[Ar]3d^3 4s^2$
24 **Cr** Chromium 51.9961 $[Ar]3d^5 4s^1$
25 **Mn** Manganese 54.938 05 $[Ar]3d^5 4s^2$
26 **Fe** Iron 55.847 $[Ar]3d^6 4s^2$
27 **Co** Cobalt 58.933 20 $[Ar]3d^7 4s^2$

5
37 **Rb** Rubidium 85.4678 $[Kr]5s^1$
38 **Sr** Strontium 87.62 $[Kr]5s^2$
39 **Y** Yttrium 88.905 85 $[Kr]4d^1 5s^2$
40 **Zr** Zirconium 91.224 $[Kr]4d^2 5s^2$
41 **Nb** Niobium 92.906 38 $[Kr]4d^4 5s^1$
42 **Mo** Molybdenum 95.94 $[Kr]4d^5 5s^1$
43 **Tc** Technetium (97.9072) $[Kr]4d^6 5s^1$
44 **Ru** Ruthenium 101.07 $[Kr]4d^7 5s^1$
45 **Rh** Rhodium 102.906 $[Kr]4d^8 5s^1$

6
55 **Cs** Cesium 132.905 43 $[Xe]6s^1$
56 **Ba** Barium 137.327 $[Xe]6s^2$
57 **La** Lanthanum 138.9055 $[Xe]5d^1 6s^2$
72 **Hf** Hafnium 178.49 $[Xe]4f^{14}5d^2 6s^2$
73 **Ta** Tantalum 180.9479 $[Xe]4f^{14}5d^3 6s^2$
74 **W** Tungsten 183.84 $[Xe]4f^{14}5d^4 6s^2$
75 **Re** Rhenium 186.207 $[Xe]4f^{14}5d^5 6s^2$
76 **Os** Osmium 190.23 $[Xe]4f^{14}5d^6 6s^2$
77 **Ir** Iridium 192.22 $[Xe]4f^{14}5d^7 6s^2$

7
87 **Fr** Francium (223.0197) $[Rn]7s^1$
88 **Ra** Radium (226.0254) $[Rn]7s^2$
89 **Ac** Actinium (227.0278) $[Rn]6d^1 7s^2$
104 **Rf** Rutherfordium (261.11) $[Rn]5f^{14}6d^2 7s^2$
105 **Db** Dubnium (262.114) $[Rn]5f^{14}6d^3 7s^2$
106 **Sg** Seaborgium (263.118) $[Rn]5f^{14}6d^4 7s^2$
107 **Bh** Bohrium (262.12) $[Rn]5f^{14}6d^5 7s^2$
108 **Hs** Hassium (265)† $[Rn]5f^{14}6d^6 7s^2$
109 **Mt** Meitnerium (266)† $[Rn]5f^{14}6d^7 7s^2$

† Estimated from currently available IUPAC data.

58 **Ce** Cerium 140.115 $[Xe]4f^1 5d^1 6s^2$
59 **Pr** Praseodymium 140.908 $[Xe]4f^3 6s^2$
60 **Nd** Neodymium 144.24 $[Xe]4f^4 6s^2$
61 **Pm** Promethium (144.9127) $[Xe]4f^5 6s^2$
62 **Sm** Samarium 150.36 $[Xe]4f^6 6s^2$

90 **Th** Thorium 232.0381 $[Rn]6d^2 7s^2$
91 **Pa** Protactinium 231.035 88 $[Rn]5f^2 6d^1 7s^2$
92 **U** Uranium 238.0289 $[Rn]5f^3 6d^1 7s^2$
93 **Np** Neptunium (237.0482) $[Rn]5f^4 6d^1 7s^2$
94 **Pu** Plutonium 244.0642 $[Rn]5f^6 7s^2$

internet connect

go.hrw.com
TOPIC: Periodic Table
GO TO: go.hrw.com
KEYWORD: HF2 PERIODIC
Visit the HRW Web site for updates on the periodic table.

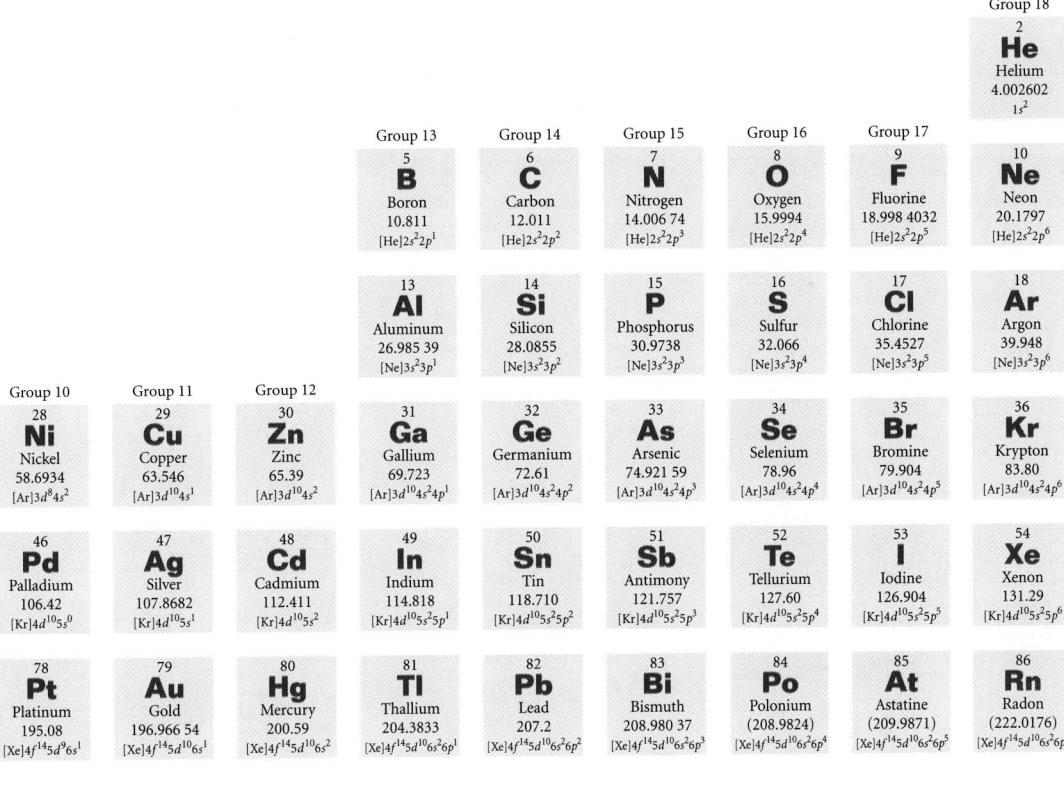

Group 18

2
He
Helium
4.002602
$1s^2$

Group 13 — 5 **B** Boron 10.811 $[He]2s^22p^1$

Group 14 — 6 **C** Carbon 12.011 $[He]2s^22p^2$

Group 15 — 7 **N** Nitrogen 14.006 74 $[He]2s^22p^3$

Group 16 — 8 **O** Oxygen 15.9994 $[He]2s^22p^4$

Group 17 — 9 **F** Fluorine 18.998 4032 $[He]2s^22p^5$

10 **Ne** Neon 20.1797 $[He]2s^22p^6$

13 **Al** Aluminum 26.985 39 $[Ne]3s^23p^1$

14 **Si** Silicon 28.0855 $[Ne]3s^23p^2$

15 **P** Phosphorus 30.9738 $[Ne]3s^23p^3$

16 **S** Sulfur 32.066 $[Ne]3s^23p^4$

17 **Cl** Chlorine 35.4527 $[Ne]3s^23p^5$

18 **Ar** Argon 39.948 $[Ne]3s^23p^6$

Group 10 — 28 **Ni** Nickel 58.6934 $[Ar]3d^84s^2$

Group 11 — 29 **Cu** Copper 63.546 $[Ar]3d^{10}4s^1$

Group 12 — 30 **Zn** Zinc 65.39 $[Ar]3d^{10}4s^2$

31 **Ga** Gallium 69.723 $[Ar]3d^{10}4s^24p^1$

32 **Ge** Germanium 72.61 $[Ar]3d^{10}4s^24p^2$

33 **As** Arsenic 74.921 59 $[Ar]3d^{10}4s^24p^3$

34 **Se** Selenium 78.96 $[Ar]3d^{10}4s^24p^4$

35 **Br** Bromine 79.904 $[Ar]3d^{10}4s^24p^5$

36 **Kr** Krypton 83.80 $[Ar]3d^{10}4s^24p^6$

46 **Pd** Palladium 106.42 $[Kr]4d^{10}5s^0$

47 **Ag** Silver 107.8682 $[Kr]4d^{10}5s^1$

48 **Cd** Cadmium 112.411 $[Kr]4d^{10}5s^2$

49 **In** Indium 114.818 $[Kr]4d^{10}5s^25p^1$

50 **Sn** Tin 118.710 $[Kr]4d^{10}5s^25p^2$

51 **Sb** Antimony 121.757 $[Kr]4d^{10}5s^25p^3$

52 **Te** Tellurium 127.60 $[Kr]4d^{10}5s^25p^4$

53 **I** Iodine 126.904 $[Kr]4d^{10}5s^25p^5$

54 **Xe** Xenon 131.29 $[Kr]4d^{10}5s^25p^6$

78 **Pt** Platinum 195.08 $[Xe]4f^{14}5d^96s^1$

79 **Au** Gold 196.966 54 $[Xe]4f^{14}5d^{10}6s^1$

80 **Hg** Mercury 200.59 $[Xe]4f^{14}5d^{10}6s^2$

81 **Tl** Thallium 204.3833 $[Xe]4f^{14}5d^{10}6s^26p^1$

82 **Pb** Lead 207.2 $[Xe]4f^{14}5d^{10}6s^26p^2$

83 **Bi** Bismuth 208.980 37 $[Xe]4f^{14}5d^{10}6s^26p^3$

84 **Po** Polonium (208.9824) $[Xe]4f^{14}5d^{10}6s^26p^4$

85 **At** Astatine (209.9871) $[Xe]4f^{14}5d^{10}6s^26p^5$

86 **Rn** Radon (222.0176) $[Xe]4f^{14}5d^{10}6s^26p^6$

63 **Eu** Europium 151.966 $[Xe]4f^76s^2$

64 **Gd** Gadolinium 157.25 $[Xe]4f^75d^16s^2$

65 **Tb** Terbium 158.925 34 $[Xe]4f^96s^2$

66 **Dy** Dysprosium 162.50 $[Xe]4f^{10}6s^2$

67 **Ho** Holmium 164.930 $[Xe]4f^{11}6s^2$

68 **Er** Erbium 167.26 $[Xe]4f^{12}6s^2$

69 **Tm** Thulium 168.934 21 $[Xe]4f^{13}6s^2$

70 **Yb** Ytterbium 173.04 $[Xe]4f^{14}6s^2$

71 **Lu** Lutetium 174.967 $[Xe]4f^{14}5d^16s^2$

95 **Am** Americium (243.0614) $[Rn]5f^77s^2$

96 **Cm** Curium (247.0703) $[Rn]5f^76d^17s^2$

97 **Bk** Berkelium (247.0703) $[Rn]5f^97s^2$

98 **Cf** Californium (251.0796) $[Rn]5f^{10}7s^2$

99 **Es** Einsteinium (252.083) $[Rn]5f^{11}7s^2$

100 **Fm** Fermium (257.0951) $[Rn]5f^{12}7s^2$

101 **Md** Mendelevium (258.10) $[Rn]5f^{13}7s^2$

102 **No** Nobelium (259.1009) $[Rn]5f^{14}7s^2$

103 **Lr** Lawrencium (262.11) $[Rn]5f^{14}6d^17s^2$

The atomic masses listed in this table reflect the precision of current measurements. (Values listed in parentheses are those of the element's most stable or most common isotope.) In calculations throughout the text, however, atomic masses have been rounded to two places to the right of the decimal.

Appendix G
Abbreviated Table of Isotopes and Atomic Masses

Z	Element	Symbol	Average atomic mass (u)	Mass number (* indicates radioactive) A	Atomic mass (u)	Percent abundance	Half-life (if radioactive) $T_{1/2}$
0	(Neutron)	n		1*	1.008 665		10.4 m
1	Hydrogen	H	1.0079	1	1.007 825	99.985	
	Deuterium	D		2	2.014 102	0.015	
	Tritium	T		3*	3.016 049		12.33 y
2	Helium	He	4.002 60	3	3.016 029	0.000 14	
				4	4.002 602	99.999 86	
				6*	6.018 886		0.81 s
3	Lithium	Li	6.941	6	6.015 121	7.5	
				7	7.016 003	92.5	
4	Beryllium	Be	9.0122	7*	7.016 928		53.3 d
				8	8.005 305		6.7×10^{-17} s
				9	9.012 174	100	
				10*	10.013 584		1.5×10^{6} y
5	Boron	B	10.81	10	10.012 936	19.9	
				11	11.009 305	80.1	
6	Carbon	C	12.011	10*	10.016 854		19.3 s
				11*	11.011 433		20.4 m
				12	12.000 000	98.9	
				13	13.003 355	1.10	
				14*	14. 003 242		5730 y
7	Nitrogen	N	14.0067	13*	13.005 738		996 m
				14	14.003 074	99.63	
				15	15.000 108	0.37	
				16*	16.006 100		7.13 s
8	Oxygen	O	15.9994	15*	15.003 065		122 s
				16	15.994 915	99.761	
				17	16.999 132	0.039	
				18	17.999 160	0.200	
				19*	19.003 577		26.9 s
9	Fluorine	F	18.998 40	18*	18.000 937		109.8 m
				19	18.998 404	100	
				20*	19.999 982		11.0 s
10	Neon	Ne	20.180	19*	19.001 880		17.2 s
				20	19.992 435	90.48	
				21	20.993 841	0.27	
				22	21.991 383	9.25	
11	Sodium	Na	22.989 87	22*	21.994 434		2.61 y
				23	22.989 767	100	
				24*	23.990 961		14.96 h
12	Magnesium	Mg	24.305	23*	22.994 124		11.3 s
				24	23.985 042	78.99	
				25	24.985 838	10.00	
				26	25.982 594	11.01	
13	Aluminum	Al	26.981 54	26*	25.986 892		7.4×10^{5} y
				27	26.981 534	100	

Z	Element	Symbol	Average atomic mass (u)	Mass number (* indicates radioactive) A	Atomic mass (u)	Percent abundance	Half-life (if radioactive) $T_{1/2}$
14	Silicon	Si	28.086	28	27.976 927	92.23	
				29	28.976 495	4.67	
				30	29.973 770	3.10	
15	Phosphorus	P	30.973 76	30*	29.978 307		2.50 m
				31	30.973 762	100	
16	Sulfur	S	32.066	32	31.972 071	95.02	
				33	32.971 459	0.75	
				34	33.967 867	4.21	
				35*	34.969 033		87.5 d
				36	35.967 081	0.02	
17	Chlorine	Cl	35.453	35	34.968 853	75.77	
				36*	35.968 307		3.0×10^5 y
				37	36.975 893	24.23	
18	Argon	Ar	39.948	36	35.967 547	0.337	
				37*	36.966 776		35.04 d
				38	37.962 732	0.063	
				39*	38.964 314		269 y
				40	39.962 384	99.600	
19	Potassium	K	39.0983	39	38.963 708	93.2581	
				40*	39.964 000	0.0117	1.28×10^9 y
				41	40.961 827	6.7302	
20	Calcium	Ca	40.08	40	39.962 591	96.941	
				41*	40.962 279		1.0×10^5 y
				42	41.958 618	0.647	
				43	42.958 767	0.135	
				44	43.955 481	2.086	
21	Scandium	Sc	44.9559	41*	40.969 250		0.596 s
				45	44.955 911	100	
22	Titanium	Ti	47.88	44*	43.959 691		49 y
				47	46.951 765	7.3	
				48	47.947 947	73.8	
23	Vanadium	V	50.9415	50*	49.947 161	0.25	1.5×10^{17} y
				51	50.943 962	99.75	
24	Chromium	Cr	51.996	48*	47.954 033		21.6 h
				52	51.940 511	83.79	
				53	52.940 652	9.50	
25	Manganese	Mn	54.938 05	54*	53.940 361		312.1 d
				55	54.938 048	100	
26	Iron	Fe	55.847	54	53.939 613	5.9	
				55*	54.938 297		2.7 y
				56	55.934 940	91.72	
27	Cobalt	Co	58.933 20	59	58.933 198	100	
				60*	59.933 820		5.27 y
28	Nickel	Ni	58.793	58	57.935 345	68.077	
				59*	58.934 350		7.5×10^4 y
				60	59.930 789	26.223	
29	Copper	Cu	63.54	63	62.929 599	69.17	
				65	64.927 791	30.83	
30	Zinc	Zn	65.39	64	63.929 144	48.6	
				66	65.926 035	27.9	
				67	66.927 129	4.1	
				68	67.924 845	18.8	

Z	Element	Symbol	Average atomic mass (u)	Mass number (* indicates radioactive) A	Atomic mass (u)	Percent abundance	Half-life (if radioactive) $T_{1/2}$
31	Gallium	Ga	69.723	69	68.925 580	60.108	
				71	70.924 703	39.892	
32	Germanium	Ge	72.61	70	69.924 250	21.23	
				72	71.922 079	27.66	
				73	72.923 462	7.73	
				74	73.921 177	35.94	
				76	75.921 402	7.44	
33	Arsenic	As	74.9216	75	74.921 594	100	
34	Selenium	Se	78.96	76	75.919 212	9.36	
				77	76.919 913	7.63	
				78	77.917 397	23.78	
				80	79.916 519	49.61	
				82*	81.916 697	8.73	1.4×10^{20} y
35	Bromine	Br	79.904	79	78.918 336	50.69	
				81	80.916 287	49.31	
36	Krypton	Kr	83.80	81*	80.916 589		2.1×10^{5} y
				82	81.913 481	11.6	
				83	82.914 136	11.4	
				84	83.911 508	57.0	
				85*	84.912 531		10.76 y
				86	85.910 615	17.3	
37	Rubidium	Rb	85.468	85	84.911 793	72.17	
				87*	86.909 186	27.83	4.75×10^{10} y
38	Strontium	Sr	87.62	86	85.909 266	9.86	
				87	86.908 883	7.00	
				88	87.905 618	82.58	
				90*	89.907 737		29.1 y
39	Yttrium	Y	88.9058	89	88.905 847	100	
40	Zirconium	Zr	91.224	90	89.904 702	51.45	
				91	90.905 643	11.22	
				92	91.905 038	17.15	
				93*	92.906 473		1.5×10^{6} y
				94	93.906 314	17.38	
41	Niobium	Nb	92.9064	93	92.906 376	100	
				94*	93.907 280		2×10^{4} y
42	Molybdenum	Mo	95.94	92	91.906 807	14.84	
				93*	92.906 811		3.5×10^{3} y
				94	93.905 085	9.25	
				95	94.905 841	15.92	
				96	95.904 678	16.68	
				97	96.906 020	9.55	
				98	97.905 407	24.13	
				100	99.907 476	9.63	
43	Technetium	Tc		97*	96.906 363		2.6×10^{6} y
				98*	97.907 215		4.2×10^{6} y
				99*	98.906 254		2.1×10^{5} y
44	Ruthenium	Ru	101.07	99	98.905 939	12.7	
				100	99.904 219	12.6	
				101	100.905 558	17.1	
				102	101.904 348	31.6	
				104	103.905 558	18.6	
45	Rhodium	Rh	102.9055	103	102.905 502	100	

Z	Element	Symbol	Average atomic mass (u)	Mass number (* indicates radioactive) A	Atomic mass (u)	Percent abundance	Half-life (if radioactive) $T_{1/2}$
46	Palladium	Pd	106.42	104	103.904 033	11.14	
				105	104.905 082	22.33	
				106	105.903 481	27.33	
				108	107.903 898	26.46	
				110	109.905 158	11.72	
47	Silver	Ag	107.868	107	106.905 091	51.84	
				109	108.904 754	48.16	
48	Cadmium	Cd	112.41	109*	108.904 984		462 d
				110	109.903 004	12.49	
				111	110.904 182	12.80	
				112	111.902 760	24.13	
				113*	112.904 401	12.22	9.3×10^{15} y
				114	113.903 359	28.73	
49	Indium	In	114.82	113	112.904 060	4.3	
				115	114.903 876	95.7	4.4×10^{14} y
50	Tin	Sn	118.71	116	115.901 743	14.53	
				117	116.902 953	7.58	
				118	117.901 605	24.22	
				119	118.903 308	8.58	
				120	119.902 197	32.59	
				121*	120.904 237		55 y
51	Antimony	Sb	121.76	121	120.903 820	57.36	
				123	122.904 215	42.64	
52	Tellurium	Te	127.60	125	124.904 429	7.12	
				126	125.903 309	18.93	
				128*	127.904 468	31.79	$> 8 \times 10^{24}$ y
				130*	129.906 228	33.87	$< 1.25 \times 10^{21}$ y
53	Iodine	I	126.9045	127	126.904 474	100	
				129*	128.904 984		1.6×10^{7} y
54	Xenon	Xe	131.29	129	128.904 779	26.4	
				131	130.905 069	21.2	
				132	131.904 141	26.9	
				134	133.905 394	10.4	
				136*	135.907 214	8.9	$> 2.36 \times 10^{21}$ y
55	Cesium	Cs	132.9054	133	132.905 436	100	
				135*	134.905 891		2×10^{6} y
				137*	136.907 078		30 y
56	Barium	Ba	137.33	133*	132.905 990		10.5 y
				137	136.905 816	11.23	
				138	137.905 236	71.70	
57	Lanthanum	La	138.905	138*	137.907 105	0.0902	1.05×10^{11} y
				139	138.906 346	99.9098	
58	Cerium	Ce	140.12	138	137.905 986	0.25	
				140	139.905 434	88.43	
				142*	141.909 241	11.13	$> 5 \times 10^{16}$ y
59	Praseodymium	Pr	140.9076	141	140.907 647	100	
60	Neodymium	Nd	144.24	142	141.907 718	27.13	
				143	142.909 809	12.18	
				144*	143.910 082	23.80	2.3×10^{15} y
				145	144.912 568	8.30	
				146	145.913 113	17.19	

Z	Element	Symbol	Average atomic mass (u)	Mass number (* indicates radioactive) A	Atomic mass (u)	Percent abundance	Half-life (if radioactive) $T_{1/2}$
61	Promethium	Pm		145*	144.912 745		17.7 y
				146*	145.914 968		5.5 y
62	Samarium	Sm	150.36	147*	146.914 894	15.0	1.06×10^{11} y
				148*	147.914 819	11.3	7×10^{15} y
				149*	148.917 180	13.8	$> 2 \times 10^{15}$ y
				150	149.917 273	7.4	
				152	151.919 728	26.7	
				154	153.922 206	22.7	
63	Europium	Eu	151.96	151	150.919 846	47.8	
				152*	151.921 740		13.5 y
				153	152.921 226	52.2	
64	Gadolinium	Gd	157.25	155	154.922 618	14.80	
				156	155.922 119	20.47	
				157	156.923 957	15.65	
				158	157.924 099	24.84	
				160	159.927 050	21.86	
65	Terbium	Tb	158.9253	159	158.925 345	100	
66	Dysprosium	Dy	162.5	161	160.926 930	18.9	
				162	161.926 796	25.5	
				163	162.928 729	24.9	
				164	163.929 172	28.2	
67	Holmium	Ho	164.9303	165	164.930 316	100	
68	Erbium	Er	167.26	166	165.930 292	33.6	
				167	166.932 047	22.95	
				168	167.932 369	27.8	
				170	169.935 462	14.9	
69	Thulium	Tm	168.9342	169	168.934 213	100	
				171*	170.936 428		1.92 y
70	Ytterbium	Yb	173.04	171	170.936 324	14.3	
				172	171.936 379	21.9	
				173	172.938 209	16.12	
				174	173.938 861	31.8	
				176	175.942 564	12.7	
71	Lutetium	Lu	174.967	175	174.940 772	97.41	
				176*	175.942 679	2.59	3.78×10^{10} y
72	Hafnium	Hf	178.49	177	176.943 218	18.606	
				178	177.943 697	27.297	
				179	178.945 813	13.029	
				180	179.946 547	35.100	
73	Tantalum	Ta	180.9479	181	180.947 993	99.988	
74	Tungsten	W	183.85	182	181.948 202	26.3	
				183	182.950 221	14.28	
				184	183.950 929	30.7	
				186	185.954 358	28.6	
75	Rhenium	Re	186.207	185	184.952 951	37.40	
				187*	186.955 746	62.60	4.4×10^{10} y
76	Osmium	Os	190.2	188	187.955 832	13.3	
				189	188.958 139	16.1	
				190	189.958 439	26.4	
				192	191.961 468	41.0	
77	Iridium	Ir	192.2	191	190.960 585	37.3	
				193	192.962 916	62.7	

Z	Element	Symbol	Average atomic mass (u)	Mass number (* indicates radioactive) A	Atomic mass (u)	Percent abundance	Half-life (if radioactive) $T_{1/2}$
78	Platinum	Pt	195.08	194	193.962 655	32.9	
				195	194.964 765	33.8	
				196	195.964 926	25.3	
79	Gold	Au	196.9665	197	196.966 543	100	
80	Mercury	Hg	200.59	198	197.966 743	9.97	
				199	198.968 253	16.87	
				200	199.968 299	23.10	
				201	200.970 276	13.10	
				202	201.970 617	29.86	
81	Thallium	Tl	204.383	203	202.972 320	29.524	
				204*	203.073 839		3.78 y
				205	204.974 400	70.476	
				208*	207.981 992		3.053 m
82	Lead	Pb	207.2	206	205.974 440	24.1	
				207	206.974 871	22.1	
				208	207.976 627	52.4	
				212*	211.991 872		10.64 h
83	Bismuth	Bi	208.9803	209	208.980 374	100	
				212*	211.991 259		60.6 m
84	Polonium	Po		209*	208.982 405		102 y
				212*	211.988 842		0.30 μs
				216*	216.001 889		0.145 s
85	Astatine	At		218*	218.008 685		1.6 s
				219*	219.011 294		0.9 m
86	Radon	Rn		220*	220.011 369		55.6 s
				222*	222.017 571		3.823 d
87	Francium	Fr		223*	223.019 733		22 m
88	Radium	Ra		224*	224.020 187		3.66 d
				226*	226.025 402		1.6×10^3 y
				228*	228.031 064		5.75 y
89	Actinium	Ac		227*	227.027 701		18.72 y
				228*	228.028 716		1.913 y
90	Thorium	Th		232*	232.038 051	100	1.40×10^{10} y
				234*	234.043 593		24.1 d
91	Protactinium	Pa		231*	231.035 880		32.760 y
				234*	234.043 300		6.7 h
92	Uranium	U		235*	235.043 924	0.720	7.04×10^8 y
				238*	238.050 784	99.2745	4.47×10^9 y
93	Neptunium	Np		236*	236.046 560		1.15×10^5 y
				237*	237.048 168		2.14×10^6 y
94	Plutonium	Pu		239*	239.052 157		2.412×10^5 y
				244*	244.064 200		8.1×10^7 y

Answers to Selected Problems

CHAPTER 1

Practice 1A, p. 14
1. 5×10^{-5} m
3. **a.** 1×10^{-8} m
   b. 1×10^{-5} mm
   c. 1×10^{-2} μm
5. 1.440×10^3 kg

1 Review and Assess, pp. 27–31
11. **a.** 2×10^2 mm
    b. 7.8×10^3 s
    c. 1.6×10^7 μg
    d. 7.5×10^4 cm
    e. 6.75×10^{-4} g
    f. 4.62×10^{-2} cm
    g. 9.7 m/s
13. 1.08×10^9 km
21. 228.8 cm
37. **a.** 22 cm; 38 cm^2
    b. 29.2 cm; 67.9 cm^2
39. 9.818×10^{-2} m
43. 1.0×10^3 kg
45. **a.** 0.677 g/cm^3
    b. 4.30×10^{16} m^2

CHAPTER 2

Practice 2A, p. 44
1. 2.0 km to the east
3. 680 m to the north
5. 0.43 h

Practice 2B, p. 49
1. 2.2 s
3. 27 s
5. **a.** 1.4 m/s
   b. 3.1 m/s

Practice 2C, p. 53
1. 21 m
3. no; the plane needs 1 km to land.
5. 24 m/s

Practice 2D, p. 55
1. 36 km/h (9.9 m/s); +0.030 km
3. −7.5 m/s; 19 m

Practice 2F, p. 64
1. **a.** 11.4 s
   b. −42.1 m/s
3. **a.** 8.0 m/s
   b. 1.63 s
5. **a.** No, the apple will reach only 1.6 m.
   b. 0.82 s

2 Review and Assess, pp. 69–75
9. 10.1 km
11. 4.22×10^1 km
13. **a.** 53.5 km/h
    b. 91.0 km
15. 0.2 km west of the flagpole
21. +8.6 m/s
23. **a.** +1.6 m/s^2
    b. +24 m/s
    c. 240 m
25. 2.2×10^2 m
27. 110 m
29. 17.5 m
31. **a.** 2.0×10^1 s
    b. No; the plane needs at least 1.2 km to land.
33. 0.99 m/s

39. 3.94 s
41. **a.** 2.55 s
    b. 2.63 s
43. 1.51 h
45. **a.** 3.0 s
    b. −24.5 m/s for each
    c. 23.6 m
47. **a.** 0.05 h
    b. 82 km
49. **a.** −24 m/s
    b. 31 m
51. **a.** 24.6 s
    b. 738 m
53. **a.** 6100 m
    b. 9 s
55. **a.** 2.33 s
    b. −32.9 m/s
57. **a.** 310 m
    b. 8.5 s
    c. 16.4 s
59. **a.** 13 s
    b. −3.1 m/s^2
    c. 12 s

CHAPTER 3

Practice 3A, p. 91
1. **a.** 23 km
   b. 17 km to the east
3. 15.7 m at 22° to the side of downfield

Practice 3B, p. 94
1. 95 km/h
3. 21 m/s, 5.7 m/s
5. 1.1×10^2 m, −53 m
7. 2.4 m/s^2, −0.77 m/s^2

Practice 3C, p. 97

1. 49 m at 7.3° to the right of down field
3. 13.0 m at 33° east of north

Practice 3D, p. 102

1. 0.66 m/s
3. 7.6 m/s

Practice 3E, p. 104

1. yes, $\Delta y = -2.3$ m
3. 2.0 s; 4.8 m
5. 17.7 m/s; 6.60 m

Practice 3F, p. 109

1. 0 m/s
3. 3.90 m/s at $(4.0 \times 10^1)°$ north of east

3 Review and Assess, pp. 113–119

9. 15.3 m at 58.4° south of east
23. 8.07 m at 42.0° south of east
25. 61.8 m at 76.0° S of E (or S of W), 25.0 m at 53.1° S of E (or S of W)
27. 2.81 km east, 1.31 km north
29. 240.0 m at 57.23° south of west
35. a. 2.77×10^5 m
    b. 284 s
37. 11 m
39. 10.8 m
41. 80 m; 210 m
47. 20 km/h south
49. a. 14.1° north of west
    b. 1.99 km/h
51. a. 23.2° upstream from straight across
    b. 8.72 m/s across the river
53. a. 41.7 m/s
    b. 3.81 s
    c. $v_{y,f} = -13.5$ m/s, $v_{x,f} = 34.2$ m/s, $v_f = 36.7$ m/s
55. 10.5 m/s

57. a. 22.2 s
    b. 2.00×10^2 s
59. 22.5 s
61. a. 32.5 m
    b. 1.78 s
63. a. 5.0×10^1 m/s
    b. 4.0×10^1 m
65. a. 57.7 km/h at 60.0° west of the vertical
    b. 28.8 km/h straight down
67. 18 m, 7.9 m
69. 6.19 m/s downfield

CHAPTER 4

Practice 4A, p. 133

1. $F_x = 60.6$ N, $F_y = 35.0$ N
3. 9.31 N at 6.48° counter-clockwise from straight down

Practice 4B, p. 138

1. 2.2 m/s^2 forward
3. 4.5 m/s^2 to the east
5. 2.1 kg

Practice 4C, p. 145

1. 0.23
3. a. 870 N, 670 N
   b. 110 N, 84 N
   c. 1000 N, 500 N
   d. 5 N, 2 N

Practice 4D, p. 147

1. 2.7 m/s^2 in the positive x direction
3. a. 0.061
   b. 3.61 m/s^2 down the ramp

4 Review and Assess, pp. 151–157

11. a. $\mathbf{F_1}$ (114 N) and $\mathbf{F_2}$ (220 N) both point right; $\mathbf{F_1}$ (114 N) points right and $\mathbf{F_2}$ (220 N) points left.
    b. 114 N to the left; 220 N first to the right, then to the left
21. 4.7×10^2 s
23. a. 6.00 s
    b. 72.0 m
    c. 63.6 m/s
25. a. 2.3 m
    b. 1.8 m
37. 0.70, 0.60
39. 0.816
41. 1.4 m/s^2 down the aisle
43. 15.9 N
45. 55 N to the right
47. 13 N down the incline
49. a. 2.00 N
    b. 6.04 N
51. a. zero
    b. 33.9 N
53. 5.0×10^1 m
55. a. 1.78 m/s^2
    b. 0.37
    c. 9.4 N
    d. 2.67 m/s
57. -5.0×10^2 N
59. 32.2 N
61. 0.144°
63. a. 2.0×10^1 m/s^2 to the right
    b. $F_1 = 4.0 \times 10^1$ N, $F_2 = 6.0 \times 10^1$ N, $F_3 = 8.0 \times 10^1$ N, all to the right
    c. m_3 by m_2: 8.0×10^1 N
       m_2 by m_3: -8.0×10^1 N
       m_2 by m_1: 1.40×10^2 N
       m_1 by m_2: -1.40×10^2 N
65. a. between the bottom block and the table
    b. 0.39
67. a. -1.2 m/s^2
    b. 0.12
    c. 45 m
69. a. 75 N
    b. 0.50

CHAPTER 5

Practice 5A, p. 170
1. 1.50×10^7 J
3. 1.6×10^3 J

Practice 5B, p. 174
1. 1.7×10^2 m/s
3. the bullet with the greater mass; 2 to 1
5. 1.6×10^3 kg

Practice 5C, p. 176
1. 7.8 m
3. 5.1 m
5. a. -1.9×10^2 J
   b. -2.8×10^2 J
   c. 7.5×10^2 J
   d. 2.8×10^2 J
   e. 7.6 m/s

Practice 5D, p. 180
1. 3.3 J
3. a. 785 J
   b. 105 J
   c. 0.00 J

Practice 5E, p. 185
1. 20.7 m/s
3. 14.1 m/s
5. 0.18 m

Practice 5F, p. 189
1. 66 kW
3. 2.61×10^8 s (8.27 years)
5. a. 7.50×10^4 J
   b. 2.50×10^4 W

5 Review and Assess, pp. 193–199
7. 53 J, −53 J
9. 47.5 J
11. a. no
    b. yes
    c. yes
19. 7.6×10^4 J

21. 2.0×10^1 m
23. a. 5400 J, 0 J; 5400 J
    b. 0 J, −5400 J; 5400 J
    c. 2700 J, −2700 J; 5400 J
25. a. $(0.5)(500.0 \text{ N/m})$
       $(4.00 \times 10^{-2} \text{ m})^2 = 0.400$ J
    b. $(\frac{1}{2})(500.0 \text{ N/m})$
       $(-3.00 \times 10^{-2} \text{ m})^2 =$
       0.225 J
    c. $(0.5)(500.0 \text{ N/m})(0 \text{ m})^2 =$
       0 J
33. 12.0 m/s
35. 17.2 s
37. a. 0.633 J
    b. 0.633 J
    c. 2.43 m/s
    d. 0.422 J, 0.211 J
39. 5.0 m
41. 9.80 m/s
43. 10.2 m
45. a. 1.2 J
    b. 5.0 m/s
    c. 6.3 J
47. a. −560 J
    b. 1.2 m
49. a. 2.25×10^4 N
    b. 1.33×10^{-4} s
51. a. 2.0 m/s
    b. 2.0×10^2 N
53. a. −100 mJ
    b. 3.1 m/s
    c. 0.82 m
55. a. 5.42 m/s
    b. 0.300
    c. −147 J
57. a. 4.4 m/s
    b. 1.5×10^5 N
59. 0.107
61. a. 66 J
    b. 2.3 m/s
    c. 66 J
    d. −16 J

CHAPTER 6

Practice 6B, p. 211
1. 3.8×10^2 N to the right
3. 16.0 kg•m/s to the south

Practice 6C, p. 213
1. 5.33 s; 53.3 m to the west
3. a. 1.22×10^4 N to the east
   b. 53.3 m to the west

Practice 6D, p. 219
1. 1.90 m/s
3. a. 12.0 m/s
   b. 9.6 m/s
   c. 11.7 m/s

Practice 6E, p. 224
1. 3.8 m/s to the south
3. 4.27 m/s to the north
5. a. 3.0 kg
   b. 5.32 m/s

Practice 6F, p. 226
1. a. 0.43 m/s to the west
   b. 17 J
3. a. 4.6 m/s to the south
   b. 3.9×10^3 J

Practice 6G, p. 229
1. a. 22.5 cm/s to the right
   b. $KE_i = 6.2 \times 10^{-4}$ J $= KE_f$
3. a. 8.0 m/s to the right
   b. $KE_i = 1.3 \times 10^2$ J $= KE_f$

6 Review and Assess, pp. 232–237
13. 5.2 kg•m/s toward home plate
15. 18 N
25. 0.037 m/s to the south
31. 1 m/s
33. 4.2 m/s
35. a. 0.81 m/s to the east
    b. 1.4×10^3 J
37. 4.0 m/s
39. 12.8 cm/s to the right
41. 42.0 m/s toward second base
43. a. 0.0 kg•m/s
    b. 1.0 kg•m/s upward
45. 23 m/s
47. 4.0×10^2 N

49. a. 2.1 m/s to the east
　　b. 4.1×10^4 J
51. 340 m/s
53. 254 s
55. a. 0.83 m/s to the right
　　b. 1.2 m/s to the left
57. a. 9.62×10^{-24} m/s upward
　　b. The velocity of Earth is so small that the Earth's movement can be disregarded.
59. 1.3×10^7 m/s, 41° below the negative *x*-axis

CHAPTER 7

Practice 7A, p. 247
1. 1.50 m
3. 37.9 m

Practice 7B, p. 248
1. 0.76 s
3. 0.33 s

Practice 7C, p. 250
1. 4.2 rad/s^2
3. a. 17 rad/s^2
　　b. 0.038 rad/s
　　c. −6.3 rad/s^2

Practice 7D, p. 252
1. 0.64 rad/s^2
3. 3.0 rad/s^2
5. 0.902 s

Practice 7E, p. 255
1. 2.2 rad/s
3. 12 rad/s

Practice 7F, p. 256
1. 1.5 m
3. 0.63 rad/s^2

Practice 7G, p. 258
1. 2.5 m/s

3. 1.5 m/s; 1.0 rad/s
5. 84 m/s^2

Practice 7H, p. 261
1. 29.6 kg
3. 18.5 kg

Practice 7I, p. 265
1. 0.692 m
3. a. 664 N
　　b. 243 N
　　c. 1×10^4 N

7 Review and Assess, pp. 269–273
5. 35 m
7. 38.2 s
9. 1.4 rad/s^2
11. a. 821 rad/s^2
　　b. 4.20×10^3 rad
21. 1.5×10^2 rad/s
23. 0.32 m
25. 7.0 m/s
37. a. 515 kg
　　b. 12.1 m/s
39. 2.50 m
41. 1.99×10^{-7} rad/s
43. a. 161 m
　　b. 5.49×10^3 N
45. 1.02 m
47. a. 0.965 m/s^2
　　b. 1.33×10^3 N
　　c. 0.0985
49. a. 4.39×10^{20} N
　　b. 1.99×10^{20} N
　　c. 3.55×10^{22} N
51. 8.3 s
53. 0.131

CHAPTER 8

Practice 8A, p. 282
1. 0.75 N•m
3. 133 N

Practice 8B, p. 288
1. 4.0×10^2 N, 592 N
3. 333 N, 567 N

Practice 8C, p. 291
1. a. −0.87 rad/s^2
　　b. −11 N•m
3. a. 2600 rad/s^2
　　b. 530 rad/s

Practice 8D, p. 294
1. 0.35 rad/s
3. 6.73 rad/s
5. 4.0×10^2 rad/s

Practice 8E, p. 297
1. 5.11 m/s, the ball wins the race
3. 1.6 s

8 Review and Assess, pp. 305–312
9. 26 N•m
11. a. 6300 N•m
　　b. 550 N
21. a. 392 N
　　b. $R_x = 339$ N, $R_y = 0$ N
23. 11 N, 1.6 N, 7.1 N
27. 1.12×10^{-2} N•m
35. 1.2 rad/s
37. 7.0 m/s
45. 220 N
47. 1800 N•m
49. 0.12 kg•m^2
51. 6.1 m
53. 57 m
55. $R = 2400$ N, $T = 1700$ N
57. a. 18 kg•m^2
　　b. −3.3 N•m
　　c. 68 rev
59. 24 m
61. −2300 N
63. (See *Teacher's Solution Manual and Answer Key.*)
65. 149 rad/s

67. a. 3.0×10^3 J, 3.5×10^3 J
  b. increased
  c. work is done as arms are lowered
69. a. 3.1 m/s^2
  b. 27 N, 9.3 N
71. 8.0 rev/s
73. 63%
75. 72%
77. a. 1320.0 N, 273 N
  b. 0.324

CHAPTER 9

Practice 9A, p. 324

1. a. 3.57×10^3 kg/m^3
  b. 6.4×10^2 kg/m^3
3. 9.4×10^3 N

Practice 9B, p. 327

1. a. 1.48×10^3 N
  b. 1.88×10^5 Pa
3. a. 1.2×10^3 Pa
  b. 6.0×10^{-2} N

Practice 9C, p. 330

1. 1.11×10^8 Pa
3. 0.20 m

Practice 9D, p. 337

1. a. 18 m/s
  b. 1.7×10^{-3} m
3. -4.4×10^{-2} Pa

Practice 9E, p. 341

1. 5.6×10^2 K
3. 1.2×10^{-7} m^3

9 Review and Assess, pp. 343–349

9. 2.1×10^3 kg/m^3
17. 6.28 N
19. a. 2.61×10^6 Pa
  b. 1.84×10^5 N
23. 2.4 m/s

29. 474 K
31. 3.4×10^{-5} m^3
33. 5.9×10^5 Pa
35. 16.5 cm^3
37. a. 1.16×10^5 Pa
  b. 52 Pa
39. 1.03×10^4 Pa
41. 4.0×10^3 m^2
43. 17 N, 31 N
45. a. 3.2×10^2 K
  b. $0.10\ V_i$
47. a. 1.0×10^3 kg/m^3
  b. 3.5×10^2 Pa
  c. 2.1×10^3 Pa
49. 17.0 cm
51. 31.6 m/s
53. 1.71×10^{-2} m
55. 1.8×10^4 Pa
57. a. 24 m
  b. 3.4×10^5 Pa
59. 0.605 m
61. 2.6×10^{10} capillaries
63. 60.9 m/s^2
65. 15 m
67. 21 Pa
69. 1.7×10^{-3} m

CHAPTER 10

Practice 10A, p. 363

1. $-89.22°$C, 183.93 K
3. $37.0°$C, $39°$C
5. $-320.5°$F, $-195.8°$C

Practice 10B, p. 370

1. 755 J
3. 0.96 J
5. 41 m/s

Practice 10C, p. 374

1. $47°$C
3. $79°$C
5. 390 J/kg•°C
7. 135 g

Practice 10D, p. 381

1. 1.29×10^5 J
3. 1.415×10^4 J
5. $76.2°$C

10 Review and Assess, pp. 387–391

9. $57.8°$C, 331 K
19. 7.9×10^4 J
31. $120°$C
33. 4.60×10^{21} J
41. a. $T_R = T_F + 459.7$, or $T_F = T_R - 459.7$
  b. $T = \frac{5}{9}\,T_R$, or $T_R = \frac{9}{5}\,T$
43. a. $-423.2°$F, 20.28 K
  b. $68.9°$F, 293.6 K
45. $T_F = \frac{9}{5}(-40°C) + 32.0 = -40°$F
47. 330 g
49. 5.7×10^3 J/min = 95 J/s

CHAPTER 11

Practice 11A, p. 405

1. a. 6.4×10^5 J
  b. -4.8×10^5 J
3. 3.3×10^2 J

Practice 11B, p. 413

1. 33 J
3. 1.00×10^4 J
5. 1.74×10^8 J

Practice 11C, p. 424

1. 0.1504
3. a. 0.247
  b. 4.9×10^4 J
5. 755 J

11 Review and Assess, pp. 431–435

11. 3.50×10^2 J
19. 2.14×10^9 J/s

21. a. 1.7×10^6 J, to the rod
  b. 3.3×10^2 J; by the rod
  c. 1.7×10^6 J; it increases
29. 0.32
39. a. $\Delta U_i = 188$ J
  b. $Q_2 = 1.400 \times 10^3$ J

CHAPTER 12

Practice 12A, p. 441
1. a. 15 N/m
  b. less stiff
3. 2.7×10^3 N/m

Practice 12B, p. 449
1. 1.4×10^2 m
3. 3.6 m

Practice 12C, p. 451
1. 2.1×10^2 N/m
3. 39.6 N/m
5. a. 1.7 s, 0.59 Hz
  b. 0.14 s, 7.1 Hz
  c. 1.6 s, 0.62 Hz

Practice 12D, p. 457
1. 0.081 m $\leq \lambda \leq 12$ m
3. 4.74×10^{14} Hz

12 Review and Assess, pp. 469–473
9. 580 N/m
19. 9.7 m
21. a. 2.000 s
  b. 9.812 m/s^2
  c. 9.798 m/s^2
37. a. a sine wave with twice the amplitude
  b. a straight line (the waves cancel each other completely)
39. y_3
41. a. 1.28 m
  b. 0.22 m
43. zero
45. a, b, d

47. 1.7 N
49. 2.0 Hz, 0.50 s, 0.30 m/s
51. 0.129 m $\leq \lambda \leq 1.73$ m
53. a. 0.15 J
  b. 18 m/s^2
55. 5.17×10^{14} Hz
57. 0.800 m/s
59. 0.319 m

CHAPTER 13

Practice 13A, p. 488
1. a. 8.0×10^{-4} W/m^2
  b. 1.6×10^{-3} W/m^2
  c. 6.4×10^{-3} W/m^2
3. 2.3×10^{-5} W
5. 4.8 m

Practice 13B, p. 499
1. 440 Hz
3. a. 82.1 Hz
  b. 115 Hz
  c. 144 Hz

13 Review and Assess, pp. 507–511
27. 70 dB
39. 443 Hz, 886 Hz, 1330 Hz
41. a. 52 cm
  b. 640 Hz, 960 Hz
  c. 350 Hz
43. a. 958 Hz
  b. 27.0 cm, 45.0 cm
45. 20 m, 2×10^{-2} m
47. a. 70.1 Hz
  b. 285
49. 750 Hz
51. $L_{closed} = 1.5\ (L_{open})$
53. 1.9×10^{-2} m

CHAPTER 14

Practice 14A, p. 523
1. 1.0×10^{-13} m

3. 85.7 m–10.1 m; The wavelengths are shorter than those of the AM radio band.
5. 5.4×10^{14} Hz

Practice 14B, p. 536
1. $p = 10.0$ cm: no image (infinite q); $p = 5.00$ cm: $q = -10.0$ cm, $M = 2.00$; virtual, upright image
3. $R = 1.00 \times 10^2$ cm; $M = 2.00$; virtual image

Practice 14C, p. 540
1. $p = 45.9$ cm; $M = 0.501$; virtual, upright image; $h = 3.39$ cm
3. $p = 43$ cm; $h = 16$ cm; $M = 0.44$; virtual, upright image
5. $q = -1.31$ cm (behind the ornament's surface); $M = 0.125$; virtual, upright image

14 Review and Assess, pp. 550–555
11. 1×10^{-6} m
13. 9.1×10^{-3} m (9.1 mm)
35. $p = 13$ cm; real, inverted; $M = -2.0$
47. inverted; $p = 6.1$ cm; $f = 2.6$ cm; real
49. $q_2 = 6.7$ cm; real; $M_1 = -0.57$, $M_2 = -0.27$; inverted
51. $p = 11.3$ cm
53. (See *Teacher's Solution Manual and Answer Key*.)
55. $R = -25.0$ cm
57. concave, $R = 48.1$ cm; $M = 2.00$; virtual
59. (See *Teacher's Solution Manual and Answer Key*.)

CHAPTER 15

Practice 15A, p. 567
1. $18.5°$
3. 1.47

Practice 15B, p. 576
1. 2.0×10^1 cm, $M = -1.00$; real, inverted image
3. -6.67 cm, $M = 0.333$; virtual, upright image

Practice 15C, p. 582
1. $42.8°$
3. $49.8°$

15 Review and Assess, pp. 587–592
11. $26°$
13. $30.0°, 19.5°, 19.5°, 30.0°$
25. 3.40; upright
37. a. $31.3°$
    b. $44.2°$
    c. $49.8°$
39. 1.31
41. 1.62; carbon disulfide
43. 7.50 cm
45. a. 6.00 cm
    b. A diverging lens cannot form an image larger than the object.
47. a. 3.01 cm
    b. 2.05 cm
49. blue: $47.8°$, red: $48.2°$
51. $48.8°$
53. 4.54 m
55. $\frac{10}{9}f$
57. a. $24.7°$
    b. It will pass through the bottom surface because $\theta_i < \theta_c$ ($\theta_c = 41.8°$).
59. 1.38
61. 58.0 m
63. a. 4.83 cm
    b. The lens must be moved 0.12 cm.
65. 1.90 cm

CHAPTER 16

Practice 16A, p. 603
1. 5.1×10^{-7} m $= 5.1 \times 10^2$ nm
3. $0.125°$

Practice 16B, p. 610
1. $0.008°, 0.02°, 0.02°$
3. 11
5. 6.62×10^3 lines/cm

16 Review and Assess, pp. 620–623
9. 460 nm
11. 160 μm
19. $3.22°$
21. a. $10.09°, 13.71°, 14.77°$
    b. $20.51°, 28.30°, 30.66°$
29. 432.0 nm
31. 1.93×10^{-3} mm $= 3\lambda$; a maximum

CHAPTER 17

Practice 17A, p. 636
1. 230 N (attractive)
3. a. 16 N
    b. attractive
    c. 2.7×10^{13} electrons
    d. 3.8×10^{13} electrons

Practice 17B, p. 639
1. 47 N, along the negative x-axis; 157 N, along the positive x-axis; 11.0×10^1 N, along the negative x-axis

Practice 17C, p. 641
1. $x = 0.64$ m
3. 5.07 m

Practice 17D, p. 647
1. 1.66×10^5 N/C, $81.1°$ above the positive x-axis
3. a. 3.2×10^{-15} N, along the negative x-axis
    b. 3.2×10^{-15} N, along the positive x-axis

17 Review and Assess, pp. 654–659
19. 45 N (attractive)
21. 1.00×10^{-7} N, $12°$ below the positive x-axis
23. $y = 0.8$ m
39. 5.7×10^3 N/C, $75°$ above the positive x-axis
41. a. 5.7×10^{-27} N, in a direction opposite $\mathbf{E}$
    b. 3.6×10^{-8} N/C
43. a. 2.0×10^7 N/C, along the positive x-axis
    b. 4.0×10^1 N
45. 5.12×10^5 N
47. 5.72×10^{13} C
49. 5.58×10^{-11} N/C, downward; 1.03×10^{-7} N/C, upward
51. $v_{electron} = 4.4 \times 10^6$ m/s; $v_{proton} = 2.4 \times 10^3$ m/s
53. 1.3×10^{-3} C
55. 5.4×10^{-14} N
57. 1.5×10^{-3} kg
59. 2.0×10^{-6} C
61. 0 N/C
63. 2.25×10^{-9} N/m
65. a. 5.27×10^{13} m/s^2
    b. 5.27×10^5 m/s
67. a. 1.3×10^4 N/C
    b. 4.2×10^6 m/s

CHAPTER 18

Practice 18A, p. 669
1. 1.46×10^{-9} m
3. 0.32 m

Practice 18B, p. 673
1. 1.4×10^{-7} V
3. 3.8×10^{4} V

Practice 18C, p. 681
1. a. 4.80×10^{-5} C
   b. 4.50×10^{-6} J
3. a. 9.00 V
   b. 5.0×10^{-12} C

18 Review and Assess
pp. 683–687
5. 1.1 m
13. −154 V
27. a. 1.3×10^{-3} C
    b. 4.2 J
29. 2.00×10^{-7} C, 3.000 m
31. 9.1 J
33. a. 90.4 V
    b. 9.04×10^{4} V/m
35. a. 4.69×10^{5} m/s
    b. 1.09×10^{4} m/s
37. 1.5×10^{5} m/s
39. −7800 V
41. a. 7.2×10^{-9} s
    b. 1.4×10^{7} m/s, 7.6×10^{3} m/s
    c. 3.1×10^{-7} s
43. 4.000×10^{-6} F
45. 2.1×10^{6} m/s
47. a. 2.8×10^{7} m/s
    b. 6.5×10^{5} m/s
49. a. 3.10×10^{-13} F
    b. 3.7×10^{-14} C
    c. 2.2×10^{-15} J
    d. 3.4×10^{-2} V
    e. 8.5×10^{-2} V

CHAPTER 19

Practice 19A, p. 695
1. 4.00×10^{2} s
3. 6.00×10^{2} s
5. a. 2.6×10^{-3} A
   b. 1.6×10^{17} electrons
   c. 5.1×10^{-3} A

Practice 19C, p. 710
1. 14 Ω
3. 22 Ω

Practice 19D, p. 712
1. a. $0.14
   b. $4.40
   c. $0.42

19 Review and Assess,
pp. 717–721
17. 2.0 s
19. 1.0 s
29. 3.4 A
41. 0.62 A; 190 Ω
43. $8.10
45. 3.0×10^{1} Ω
47. a. 32 V
    b. 0.16 V
49. 2.7 A
51. a. 720 W
    b. 8.6×10^{5} J
    c. $0.0024
53. a. 1.70×10^{9} J
    b. 5.15×10^{7} J/day, 14.3 kW•h/day
    c. $54
55. 3.2×10^{5} J
57. 13.5 h
59. 2.2×10^{-5} V

CHAPTER 20

Practice 20A, p. 739
1. a. 43.6 Ω
   b. 0.275 A

3. 1.0 V, 2.0 V, 2.5 V, 3.5 V
5. 0.5 Ω

Practice 20B, p. 744
1. 4.5 A, 2.2 A, 1.8 A, 1.3 A
3. a. 2.2 Ω
   b. 6.0 A, 3.0 A, 2.00 A

Practice 20C, p. 748
1. a. 27.8 Ω
   b. 26.6 Ω
   c. 23.4 Ω

Practice 20D, p. 751
R_a: 0.50 A, 2.5 V
R_b: 0.50 A, 3.5 V
R_c: 1.5 A, 6.0 V
R_d: 1.0 A, 4.0 V
R_e: 1.0 A, 4.0 V
R_f: 2.0 A, 4.0 V

20 Review and Assess,
pp. 754–759
11. b
17. a. 24 Ω
    b. 1.0 A
19. a. 2.99 Ω
    b. 4.0 A
23. 15 Ω
25. Current: 1.1 A, 0.72 A, 1.8 A
    Potential difference: 6.5 V, 6.5 V, 5.4 V
27. 28 V
29. 3.8 V
31. a. 33.0 Ω
    b. 132 V
    c. 4.00 A, 4.00 A
33. 10.0 Ω
35. a. a
    b. c
    c. d
    d. e
37. 18.0 Ω: 0.750 A, 13.5 V
    6.0 Ω: 0.750 A, 4.5 V
39. 4.0 Ω

41. $13.96 \ \Omega$

43. a. $62.4 \ \Omega$
   b. $0.192 \ A$
   c. $0.102 \ A$
   d. $0.520 \ W$
   e. $0.737 \ W$

45. The circuit must contain three groups of resistors, each containing three resistors in parallel, that are connected to one another in series.

47. a. $5.1 \ \Omega$
   b. $4.5 \ V$

49. a. 11 A (heater), 9.2 A (toaster), 12 A (grill)
   b. The total current is 32.2 A, so the 30.0 A circuit breaker will open the circuit if these appliances are all on.

CHAPTER 21

Practice 21A, p. 775

1. $3.57 \times 10^6 \ m/s$

3. $1.5 \times 10^6 \ m/s$ north

5. $6.0 \times 10^{-12} \ N$ south

Practice 21B, p. 778

1. $1.7 \times 10^{-7} \ T$ in $+z$ direction

3. $1.5 \ T$

21 Review and Assess, pp. 781–785

31. $2.1 \times 10^{-3} \ m/s$

33. $2.00 \ T$

35. a. to the left
   b. into the page
   c. out of the page
   d. up the page

37. a. to the left
   b. into the page
   c. out of the page
   d. up the page

39. $2.1 \times 10^{-2} \ T$, in the negative y direction

41. 2.0 T, out of the page

43. $6.57 \times 10^{-2} \ T$

45. $2.82 \times 10^7 \ m/s$

CHAPTER 22

Practice 22A, p. 800

1. $0.30 \ V$

3. $0.14 \ V$

Practice 22B, p. 806

1. $87 \ V$

3. $1.7 \times 10^{-2} \ V$

Practice 22C, p. 810

1. $4.8 \ A$; $6.8 \ A$, $170 \ V$

3. a. $7.42 \ A$
   b. $15 \ \Omega$

5. a. $1.10 \times 10^2 \ V$
   b. $2.1 \ A$

Practice 22D, p. 818

1. 55 turns

3. 25 turns

5. $2.6 \times 10^4 \ V$

22 Review and Assess, pp. 821–825

11. $\dfrac{-\Delta B}{\Delta t} = 3.0 \ T/s$

23. $3.88 \times 10^{-2} \ V$

25. $3.1 \times 10^5 \ V$

27. a. $0.643 \ A$
   b. $117 \ V$
   c. $75.2 \ W$

35. 48 turns

37. $1.3 \times 10^{-4} \ s$; no

39. $4.2 \times 10^{-2} \ T$

41. 171:1

43. a. $28 \ kW$
   b. $3.6 \times 10^5 \ kW$

45. $300 \ V$

CHAPTER 23

Practice 23A, p. 833

1. $2.0 \ Hz$

3. $1.2 \times 10^{15} \ Hz$

Practice 23B, p. 836

1. $4.83 \times 10^{14} \ Hz$

3. $2.36 \ eV$

Practice 23C, p. 851

1. $40 \ m/s$

3. $8.84 \times 10^{-27} \ m/s$

5. $1.0 \times 10^{-15} \ kg$

23 Review and Assess, pp. 856–859

15. a. $2.49 \times 10^{-5} \ eV$
   b. $2.49 \ eV$
   c. $24.9 \ eV$

17. a. lithium; iron
   b. 1.8 eV; 0.2 eV

39. $1.4 \times 10^7 \ m/s$

41. $9.91 \times 10^6 \ m/s$

43. a. $1.39 \ eV$
   b. $3.35 \times 10^{14} \ Hz$

45. red

47. 3.7×10^{19} photons

49. $1.06 \times 10^{-34} \ m$

51. $1.3 \times 10^7 \ s$; the estimate is much greater than the observation.

53. a. $1.5 \times 10^4 \ eV$ (15 keV)
   b. $1.2 \times 10^5 \ eV$ (120 keV)

CHAPTER 24

24 Review and Assess, pp. 886–888

25. a. $2.75 \times 10^{14} \ Hz$
   b. $1.09 \times 10^{-6} \ m$

27. $1 \ eV$

CHAPTER 25

Practice 25A, p. 902

1. 160.65 MeV; 342.06 MeV
3. 7.933 MeV
5. 0.2101 MeV; Because of the extra proton in $_{12}^{23}Mg$, the Coulomb repulsion in the Mg nucleus is greater than that in the Na nucleus. Thus, breaking the Mg nucleus into its components requires less energy; in other words, its binding energy is less.

Practice 25B, p. 908

1. $_6^{12}C$
3. $_6^{14}C$
5. $_{28}^{63}Ni \rightarrow _{29}^{63}Cu + _{-1}^{0}e + \bar{\nu}$

Practice 25C, p. 911

1. $4.23 \times 10^3 \, s^{-1}$, 0.23 Ci
3. $9.94 \times 10^{-7} \, s^{-1}$, 6.7×10^{-7} Ci
5. a. about 5.0×10^7 atoms
   b. about 3.5×10^8 atoms

25 Review and Assess, pp. 927–931

7. 92.163 MeV
9. 8.2609 MeV/nucleon; 8.6975 MeV/nucleon
23. $_5^{10}B$
25. a. $_2^4He$
    b. $_2^4He$
27. 560 days
29. 93.8%
39. 1.3×10^4 m
41. 4.275 MeV
43. 3.53 MeV
45. 120 MeV
47. a. $2 \, _0^1n$
    b. $3 \, _0^1n$
49. $_6^{13}C$
51. a. $_0^1n$
    b. $_5^{10}B$
53. a. $_4^8Be$
    b. $_6^{12}C$
55. $3.81 \times 10^{-9} \, s^{-1}$, 2.1×10^{10} Ci
57. 3.8×10^3 s
59. 7.8×10^{10} s (2500 years)
61. 8.7×10^{24} atoms

Credits

Illustrations

Photography

Illustrations

Abbreviations used: (t) top, (c) center, (b) bottom, (l) left, (r) right.

Photography

Holt Physics P.E. ©2002
Abbreviations used: (t) top, (c) center, (b) bottom, (l) left, (r) right.

Glossary

A

absorption spectrum a continuous spectrum interrupted by dark lines characteristic of the medium through which the radiation has passed (p. 842)

acceleration the rate of change of velocity (p. 48)

accuracy describes how close a measured value is to the true value of the quantity measured (p. 15)

action-reaction pair a pair of simultaneous equal but opposite forces resulting from the interaction of two objects (p. 139)

adiabatic process a thermodynamic process during which work is done on or by a system but no energy is transferred to or from the system by heat (p. 407)

alternating current an electric current that changes direction at regular intervals (p. 807)

amplitude the maximum displacement from equilibrium (p. 446)

angle of incidence the angle between a ray that strikes a surface and the normal to that surface at the point of contact (p. 527)

angle of reflection the angle formed by the line normal to a surface and the direction in which a reflected ray moves (p. 527)

angular acceleration the time rate of change of angular speed, expressed in radians per second per second (p. 249)

angular displacement the angle through which a point, line, or body is rotated in a specified direction and about a specified axis (p. 246)

angular momentum the product of a rotating object's moment of inertia and the angular speed about the same axis (p. 292)

angular speed the rate at which a body rotates about an axis, usually expressed in radians per second (p. 247)

antinode a point in a standing wave, halfway between two nodes, at which the largest amplitude occurs (p. 463)

average velocity total displacement divided by the time interval during which the displacement occurred (p. 43)

B

back emf the emf induced in a motor's coil that tends to reduce the current powering the motor (p. 812)

band gap the minimum energy separation between the highest occupied state and the lowest empty state (p. 868)

beat the interference of waves of slightly different frequencies traveling in the same direction, perceived as a variation in loudness (p. 502)

binding energy the energy involved when nucleons bind together to form a stable nucleus (p. 900)

blackbody radiation electromagnetic radiation emitted by a blackbody, which absorbs all incoming radiation and then emits radiation based only on its temperature (p. 830)

buoyant force a force that acts upward on an object submerged in a liquid or floating on the liquid's surface (p. 319)

C

calorimetry an experimental procedure used to measure the energy transferred from one substance to another as heat (p. 372)

capacitance the ability of a conductor to store energy in the form of electrically separated charges (p. 676)

center of mass the point at which all the mass of the body can be considered to be concentrated when analyzing translational motion (p. 283)

centripetal acceleration acceleration directed toward the center of a circular path (p. 257)

chromatic aberration the focusing of different colors of light at different distances behind a lens (p. 585)

coefficient of friction the ratio of the force of friction to the normal force acting between two objects (p. 144)

coherence the property by which two waves with identical wavelengths maintain a constant phase relationship (p. 599)

components of a vector the projections of a vector along the axes of a coordinate system (p. 92)

compression the region of a longitudinal wave in which the density and pressure are greater than normal (p. 480)

Compton shift the change in wavelength between incoming and scattered electromagnetic waves (p. 838)

concave spherical mirror an inwardly curved, mirrored surface that is a portion of a sphere and that converges incoming light rays (p. 530)

conductor material that transfers charge easily (p. 631)

constructive interference interference in which individual displacements on the same side of the equilibrium position are added together to form the resultant wave (p. 460)

contact force force that arises from the physical contact of two objects (p. 125)

controlled experiment experiment involving manipulation of only a single variable or factor (p. 9)

convex spherical mirror an outwardly curved, mirrored surface that is a portion of a sphere and that diverges incoming light rays (p. 537)

crest the highest point above the equilibrium position (p. 454)

critical angle the minimum angle of incidence for which total internal reflection occurs (p. 580)

current the rate at which electric charges move through a given area (p. 694)

cyclic process a thermodynamic process in which a system returns to the same conditions under which it started (p. 415)

D

decibel level relative intensity determined by relating the intensity of a sound wave to the intensity at the threshold of hearing (p. 490)

destructive interference interference in which individual displacements on opposite sides of the equilibrium position are added together to form the resultant wave (p. 461)

diffraction the spreading of waves into a region behind an obstruction (p. 605)

diode an electronic device that allows electric current to pass more easily in one direction than in the other (p. 874)

dispersion the process of separating polychromatic light into its component wavelengths (p. 583)

displacement the change in position of an object (p. 40)

domain a microscopic magnetic region composed of a group of atoms whose magnetic fields are aligned in a common direction (p. 772)

doping the addition of impurity atoms to a semiconductor (p. 873)

Doppler effect a frequency shift that is the result of relative motion between the source of sound waves and an observer (p. 485)

drift velocity the net velocity of a charge carrier moving in an electric field (p. 697)

E

elastic collision a collision in which the total momentum and the total kinetic energy remain constant (p. 226)

elastic potential energy the potential energy in a stretched or compressed elastic object (p. 178)

electric circuit a set of electrical components connected so that they provide one or more complete paths for the movement of charges (p. 732)

electric field a region in space around a charged object in which a stationary charged object experiences an electric force because of its charge (p. 643)

electric field lines lines that represent both the magnitude and the direction of the electric field (p. 648)

electric potential the electrical potential energy associated with a charged particle divided by the charge of the particle (p. 670)

electrical potential energy potential energy associated with an object due to its position relative to a source of electric force (p. 666)

electromagnetic induction production of an emf in a conducting circuit by a change in the strength, position, or orientation of an external magnetic field (p. 794)

electromagnetic wave a transverse wave consisting of oscillating electric and magnetic fields at right angles to each other (p. 520)

emf the energy per unit charge supplied by a source of electric current (p. 734)

emission spectrum a unique series of spectral lines emitted by an atomic gas when a potential difference is applied across the gas (p. 842)

entropy a measure of the disorder of a system (p. 426)

environment everything outside a system that can affect or be affected by the system's behavior (p. 403)

equilibrium the state in which there is no change in a body's motion (p. 134)

excited state the state of an atom that is no longer in its ground state (p. 867)

F

field force force that can exist between objects, even in the absence of physical contact between the objects (p. 125)

fluid a nonsolid state of matter in which the atoms or molecules are free to move past each other, as in a gas or a liquid (p. 318)

force the cause of an acceleration or the change in an object's velocity (p. 124)

force diagram a diagram of the objects involved in a situation and the forces exerted on the objects (p. 126)

frame of reference a coordinate system for specifying the precise location of objects in space (p. 106)

free fall the motion of an object falling with a constant acceleration (p. 60)

frequency the number of cycles or vibrations per unit of time (p. 446)

fundamental frequency the lowest frequency of vibration of a standing wave (p. 494)

generator a device that uses induction to convert mechanical energy to electrical energy (p. 803)

gravitational force the mutual force of attraction between particles of matter (p. 263)

gravitational potential energy the potential energy associated with an object due to its position relative to Earth or some other gravitational source (p. 177)

ground state the lowest-energy state of a quantized system (p. 867)

half-life the time required for half the original nuclei of a radioactive material to undergo radioactive decay (p. 909)

harmonic series a series of frequencies that includes the fundamental frequency and integral multiples of the fundamental frequency (p. 495)

heat the energy transferred between objects because of a difference in their temperatures (p. 365)

heat of fusion the energy per unit mass transferred in order to change a substance from solid to liquid or from liquid to solid at constant temperature and pressure (p. 378)

heat of vaporization the energy per unit mass transferred in order to change a substance from liquid to vapor or from vapor to liquid at constant temperature and pressure (p. 378)

hole an energy level that is not occupied by an electron in a semiconductor (p. 872)

ideal fluid a fluid that has no internal friction or viscosity and that is incompressible (p. 332)

impulse for a constant external force, the product of the force and the time over which it acts on an object (p. 210)

index of refraction the ratio of the speed of light in a vacuum to its speed in a given transparent medium (p. 564)

induction the process of charging a conductor by bringing it near another charged object and grounding the conductor (p. 632)

inertia the tendency of an object to maintain its state of motion (p. 130)

instantaneous velocity the velocity of an object at some instant (or specific point in its path) (p. 46)

insulator material that does not transfer charge easily (p. 631)

intensity the rate at which energy flows through a unit area perpendicular to the direction of wave motion (p. 487)

internal energy the energy of a substance due to the random motions of its component particles and equal to the total energy of those particles (p. 359)

isothermal process a thermodynamic process that takes place at constant temperature and in which the internal energy of a system remains unchanged (p. 407)

isotopes atoms of an element having the same atomic number but different neutron and mass numbers (p. 897)

isovolumetric process a thermodynamic process that takes place at constant volume so that no work is done on or by the system (p. 406)

kinetic energy the energy of an object due to its motion (p. 172)

kinetic friction the resistive force that opposes the relative motion of two contacting surfaces that are moving past one another (p. 143)

laser a device that produces an intense, nearly parallel beam of coherent light (p. 614)

latent heat the energy per unit mass that is transferred during a phase change of a substance (p. 379)

lens a transparent object that refracts light rays, causing them to converge or diverge to create an image (p. 568)

lever arm the perpendicular distance from the axis of rotation to a line drawn along the direction of the force (p. 279)

linear polarization the alignment of electromagnetic waves in such a way that the vibrations of the electric fields in each of the waves are parallel to each other (p. 546)

longitudinal wave a wave whose particles vibrate parallel to the direction of wave motion (p. 455)

M

magnetic field a region in which a magnetic force can be detected (p. 767)

mass density the mass per unit volume of a substance (p. 319)

mechanical energy the sum of kinetic energy and all forms of potential energy (p. 182)

mechanical wave a wave that propagates through a deformable, elastic medium (p. 452)

medium the material through which a disturbance travels (p. 452)

model a replica or description designed to show the structure or workings of an object, system, or concept (p. 6)

moment of inertia the tendency of a body rotating about a fixed axis to resist a change in rotational motion (p. 284)

momentum a vector quantity defined as the product of an object's mass and velocity (p. 208)

mutual inductance a measure of the ability of one circuit carrying a changing current to induce an emf in a nearby circuit (p. 814)

N

net external force the total force resulting from a combination of external forces on an object; sometimes called the resultant force (p. 131)

node a point in a standing wave that always undergoes complete destructive interference and therefore is stationary (p. 463)

normal force a force exerted by one object on another in a direction perpendicular to the surface of contact (p. 141)

O

order number the number assigned to interference fringes with respect to the central bright fringe (p. 602)

P

parallel describes two or more components in a circuit that are connected across common points or junctions, providing separate conducting paths for the current (p. 740)

path difference the difference in the distance traveled by two interfering light waves (p. 602)

perfectly inelastic collision a collision in which two objects stick together and move with a common velocity after colliding (p. 222)

period the time it takes to execute a complete cycle of motion (p. 446)

periodic wave a wave whose source is some form of periodic motion (p. 453)

phase change the physical change of a substance from one state (solid, liquid, or gas) to another at constant temperature and pressure (p. 377)

photoelectric effect the emission of electrons from a surface that occurs when light of certain frequencies shines on the surface (p. 834)

photon the discrete unit (or quantum) of light energy (p. 835)

pitch the perceived highness or lowness of a sound, depending on the frequency of the sound waves (p. 481)

potential difference the change in electrical potential energy associated with a charged particle divided by the charge of the particle (p. 670)

potential energy energy associated with an object due to its position (p. 174)

power the rate at which energy is transferred (p. 187)

precision the degree of exactness with which a measurement is made and stated (p. 15)

pressure the magnitude of the force on a surface per unit area (p. 325)

projectile motion the motion of objects moving in two dimensions under the influence of gravity (p. 99)

pulse wave a single nonperiodic disturbance (p. 453)

R

radian an angle whose arc length is equal to its radius, which is approximately equal to 57.3°(p. 245)

rarefaction the region of a longitudinal wave in which the density and pressure are less than normal (p. 480)

real image an image formed when rays of light actually intersect at a single point (p. 531)

reflection the turning back of an electromagnetic wave at the surface of a substance (p. 526)

refraction the bending of a wave disturbance as it passes at an angle from one medium to another (p. 562)

resistance the opposition to the flow of charge in a conductor (p. 700)

resolving power the ability of an optical instrument to separate two images that are close together (p. 612)

resonance a condition that exists when the frequency of a force applied to a system matches the natural frequency of vibration of the system (p. 492)

resultant a vector representing the sum of two or more vectors (p. 85)

rms current the amount of direct current that dissipates as much energy in a resistor as an instantaneous alternating current does during a complete cycle (p. 808)

rotational kinetic energy the energy of an object due to its rotational motion (p. 295)

rotational motion the motion of a body that spins about an axis (p. 244)

S

scalar a physical quantity that has a magnitude but no direction (p. 84)

schematic diagram a graphic representation of an electric circuit, with standardized symbols representing circuit components (p. 730)

self-induction the process by which a changing current in a circuit induces an emf in that circuit (p. 819)

series describes a circuit or portion of a circuit that provides a single conducting path without junctions (p. 736)

significant figures digits in a measurement that are known with certainty plus the first digit that is uncertain (p. 16)

simple harmonic motion vibration about an equilibrium position in which a restoring force is proportional to the displacement from equilibrium (p. 439)

solenoid a long, helically wound coil of insulated wire (p. 771)

specific heat capacity the quantity of energy needed to raise the temperature of 1 kg of a substance by 1°C at constant pressure (p. 371)

spring constant a parameter that expresses how resistant a spring is to being compressed or stretched (p. 178)

standing wave a wave pattern that results when two waves of the same frequency, wavelength, and amplitude travel in opposite directions and interfere (p. 463)

static friction the resistive force that opposes the relative motion of two contacting surfaces that are at rest with respect to one another (p. 142)

strong force the interaction that binds nucleons together in a nucleus (p. 898)

superconductor a material whose resistance is zero at or below some critical temperature, which varies with each material (p. 706)

system a set of items or interactions considered a distinct physical entity for the purpose of study (p. 7); a collection of matter within a clearly defined boundary across which no matter passes (p. 403)

T

tangential acceleration the instantaneous linear acceleration of an object directed along the tangent to the object's circular path (p. 255)

tangential speed the instantaneous linear speed of an object directed along the tangent to the object's circular path (p. 253)

temperature a measure of the average kinetic energy of the particles in a substance (p. 331)

thermal conduction the process by which energy is transferred as heat through a material between two points at different temperatures (p. 383)

thermal equilibrium the state in which two bodies in physical contact with each other have identical temperatures (p. 360)

timbre the quality of a steady musical sound that is the result of a mixture of harmonics present at different intensities (p. 500)

torque a quantity that measures the ability of a force to rotate an object around some axis (p. 279)

total internal reflection the complete reflection of light at the boundary of two transparent media; this effect occurs when the angle of incidence exceeds the critical angle (p. 580)

transformer a device that changes one ac potential difference to a different ac potential difference (p. 815)

transistor a device, typically containing three terminals, that can amplify a signal (p. 877)

transverse wave a wave whose particles vibrate perpendicular to the direction of wave motion (p. 454)

trough the lowest point below the equilibrium position (p. 454)

U

ultraviolet catastrophe contradiction between the predictions of classical physics and the experimental data for the electromagnetic radiation of a blackbody in the ultraviolet end of the spectrum (p. 831)

uncertainty principle states that it is impossible to simultaneously measure both the position and the momentum of an object with complete certainty (p. 852)

V

valence electron an electron in the outermost shell of an atom (p. 867)

vector a physical quantity that has both a magnitude and a direction (p. 84)

virtual image an image formed by light rays that only appear to intersect (p. 528)

W

wavelength the distance between two adjacent similar points of the wave, such as from crest to crest or from trough to trough (p. 454)

weight the magnitude of the force of gravity acting on an object (p. 141)

work the product of the magnitudes of the component of a force along the direction of displacement and the displacement (p. 168)

work function the minimum amount of energy required for an electron to escape from a metal (p. 835)

work–kinetic energy theorem the theorem stating that the net work done on an object is equal to the change in the kinetic energy of the object (p. 174)

Index

Page references in boldface type refer to figures. Page references preceded by the letter *t* refer to tables.

elastic potential energy, 178–180, 441

electrical potential energy: in batteries, 735; capacitors and, 679–682, 686; electric force and, 666–669; electric power and, 708–710; laboratory exercise on, 688–689

electric charge: conservation of, 629; electrons and, 630, **630**; field strength and, 644–647; fractional, 920, *t* 920; independence of, 929; induction of, 632, 796–802, 805–806, 818–820; polarization of, 632–633, **633**; properties of, 628–630, 653; quantized, 630; quarks and, 919–920; transfer of, 631–633; Van de Graaff generators and, 651–652

electric circuit(s): alternating current, 699, 807–811, 824; circuit boards and, 693, 702; closed *vs.* open, 732; complex, 746–751, 753; definition of, 732, 753; household wiring and, 746; induced, 794–802; overview, 732–735; Quick Lab on, 734; resistors in parallel in, 740–745; resistors in series in, 736–740; schematic diagrams of, 730–731, **731**; series, 736–745; short circuits, 733

electric current: alternating and direct, 699, 716, 807–811, 824, 876; conventional, 696, *t* 696, 716, 770; current gain and, 878; definition of, 694–695; diodes and, 876; drift velocity and, **696–697,** 697, 716; electrocution and, 813; graphing calculator exercise on, 824; induction of, 632, 794–800, 805–806, 818–820; laboratory exercise on, 722–725; leakage and, 876; magnetic fields and, **770,** 770–771, **771;** in parallel resistors, 741, 748–751; pulsed direct, 876, **876;** root-mean-square (rms) current, **808,** 808–810, 824; in series resistors, 702, 736–740, 748–751; sources and types of, 698–699; superconductors and, 631, **706,** 706–707, 881–885; symbols for, *t* 808; units of, 753

electric dipole, 649

electric field(s): batteries and, 675; definition of, 643; dielectric

materials and, 678, **678**; drawing, **648**; electrical potential energy and, 667; electrostatic equilibrium and, 650; examples of, **645**; field lines and, **648,** 648–649, 653; oscillating, **521,** 546, **546**; potential difference in, 670–672; strength, 643–648

electric force, 634–642, 653, 658

electric guitar(s), 792–793, 802

electric potential, 670, 672

electric power: cost of, 712; definition of, 708–709, 716; ELFs and, 790–791; energy transfer and, **708,** 708–713; units of, 711

electric shock, 813

electric vehicle(s), 690–691

electrocution, 813

electrolyte(s), 696

electromagnetic field(s), 790–791

electromagnetic induction, 794–796; laboratory exercise on, 826

electromagnetic interaction, 917–918, **918,** *t* 918, **922**

electromagnetic radiation: blackbody, 830–833, 855; Compton shift and, 838, **838**; definition of, 520; frequency range of, 456; microwaves and, 505, *t* 521, 524, 645; photoelectric effect and, *t* 834, 834–839, 860–861; spectrum of, 520–521, *t* 521; temperature and, 384; wave speed and, 522–523. *See also* light; waves.

electromagnetic wave(s), 520–525. *See also* electromagnetic radiation; waves.

electromagnetism, **5**

electromagnet(s), 707, 771, **771,** 774, 884

electron-hole pair(s), 872

electron microscope(s), 467

electron(s): in beta decay, 904–905; in the Bohr model, 843–844, **844,** 851, 853; charge, 629–630, **630**; diffraction, 850; in early atomic models, 840–841; clouds, 853–854, **854**; energy levels, **844,** 844–845, **866,** 867–869, **869**; excited state, 844; interference patterns, 850, **850**; magnetic domains and, 772; orbits, 302–303, **303**; pair production, 924–925; photoelectrons and, 834; thermal energy and, 873;

uncertainty principle and, 852–853; valence, 867, **868**

electron spin, 303, **303**

electron tunneling, 714–715

electron volt(s), 671, 832

electroporation, 705

electrostatic equilibrium, 640–641, 650–652, 660–663

electrostatic spray painting, 626–627, 629, 649, 651

elephant(s), 481

ELF(s), 790–791

emf: alternating current and, 807; back, 812; Faraday's law of induction and, 798–800; generators and, 803–806; induced, 820; magnetic fields and, 794–796; mutual inductance and, 814–815; self-induced, 819; terminal voltage and, 734, 753

emission spectra, 842, **842**

energy: alternating current and, 699, 807; bands, **868,** 868–871, 885; binding, 900–902, 913, 915; classification of, **182**; elastic potential, 178–180, 441; gap, 868–869; gravitational potential, 177–178, 444–445; heat and, 365–367, 402–408; internal, 359; mass and, 190–191; mechanical, 181–186, *t* 183, 200–205, 295–297, **444,** 666; power and, 187–189; quantization of, 831–833; relativistic kinetic, 190; rest, 191, 898, 900; rotational, 295–297, 304, **359**; symbols for, 192; temperature change, 358, 406–407; transfer by waves, 458; translational, **295, 359**; vibrational, **359**; work-kinetic energy theorem, 174–176. *See also* conservation of energy; kinetic energy; potential energy.

engine(s): Carnot, 434; efficiency of, **422,** 434; heat, 417–418, 420–424, 430; internal-combustion, 417–418, **418**; Stirling, 421; thermodynamics of, 417–418, **418**

enrichment of fuels, 915

entropy: definition of, 425–426, 430; Quick Lab on, 426; second law of thermodynamics and, 428; work and, 428–429

environment, definition of, 403

equations in physics, 21–23

598, **600,** 601; double-slit, 600, **600,** 619, 623; electrons and, 850, **850;** fringes, 599–603, **600, 601;** maxima, 602, **606;** path difference and, 601, **601;** patterns, 459, **459, 599,** 599–603; wavelength and, 598, **598;** white light and, 600, **600.** *See also* diffraction.

internal-combustion engine(s), 417–418, **418**

internal energy, 359, 406–407

Internet site(s), 7

Inupiat parka(s), 384

inverse-beta decay, 905

inverse function(s), 90–91

inverse-square law, 263, 650

iron, *t* 319, *t* 482, 901, 915

isothermal process(es), 407, **407,** *t* 411

isotope(s), 897, 909

isovolumetric process(es), 406, *t* 411

J

joule(s), 169, 173, 177, *t* 367, 667, 832

Joule, James Prescott, 169, 391, 726

joule heating, 709, 713

junction transistor(s), 877–878

K

kaon(s), **920**

Kazerooni, Homayoon, 300

kelvin(s), 331

Kelvin scale, 331, 361–362, *t* 362

Kepler, Johannes, **165**

Kilby, Jack, 880

kilocalorie(s), *t* 367

kilogram(s), **10,** *t* 11

kilowatt-hour(s), 177, 711–712

kinematics, 43

kinetic energy: conservation of mechanical energy and, *t* 183; definition of, 172–173; elastic collisions and, **227,** 227–229, *t* 230; friction and, 186, **186;** inelastic collisions and, 224–226, *t* 230, 231; photoelectrons and, 835, **835,** 855; relativistic, 190–191; rotational, 295–297; temperature and, 359, **359;** work and, 174–176

kinetic friction, 143–144

kinetic theory of gases, 329

Kolb, Greg, 421

L

laboratory exercise(s): brightness of light, 556–559; capacitance and electrical energy, 688–689; circular motion, 274–275; conservation of momentum, 238–241; converging lenses, 593–595; current and resistance, 722–725; diffraction, 624–625; electromagnetic induction, 826; electrostatics, 660–663; force and acceleration, 158–163; half life, 932–933; machines and efficiency, 313–315; magnetic field of a conducting wire, 786–789; measurement, 32–33; measuring time and motion, 76–81; pendulum and simple harmonic motion, 474–475; photoelectric effect, 860–861; resistors in series and parallel, 760–763; specific heat capacity, 392–397; speed of sound, 512–515; time and measurement, 34–37

laminar flow, 332, **332**

laser(s): active medium in, 614, **614;** applications of, 615–618; CDs and, 596–597, **606,** 615–618, **617;** coherence and, **613,** 613–615, **614;** definition of, 613; holograms and, 616, **616**

laser surface analyzer, 846

latent heat, *t* 379, 379–381

Latimer, Lewis Howard, 721

lattice imperfection(s), 881–882

Lavoisier, Antoine Laurent, 355

law of conservation of energy. *See* conservation of energy.

lead, *t* 482, 907

lead paint, 846

Leavitt, Henrietta, **862**

lens(es): aberrations, 585, **585;** characteristics of, 570–573; converging, 568–572, **571,** 575, 593–595; definition of, 568; diverging, 568, 573, 576; in eyeglasses, 577; focal length of, 569; laboratory exercise on, 593–595; magnification and, 574–575; in magnifying glasses, 572–573; in microscopes, 578; ray diagrams and, 569; in refracting telescopes, 579; sign conventions for, **574;**

symbols for, *t* 586; thin-lens equation, 573–576; types of, 568–570

Lenz's law, 798

lepton(s), 918–920

leukemia, 790–791

lever arm(s), 279–280, **280,** 299

lever(s), **298**

lift, 334, **334**

light: atomic spectra, 841–843, **842;** brightness, 525, **544, 547,** 556–559; coherent, 599, **613,** 613–615, **614;** dispersion of, 583–584; Doppler effect and, **504,** 504–505; dual nature of, 466, 848–849; monochromatic, 598; photoelectric effect and, *t* 834, 834–839, 860–861; photon theory of, 110–111, 837–838, 848; polarization of, **546,** 546–548, **547, 548;** quanta, 832; relativity and, 110–111; scattering, 548, **548;** speed of, 66–67, 110; visible, 583; white, 543–544, 600, 624–625. *See also* colors; electromagnetic radiation; lenses; refraction; waves.

light bulb(s): alternating current in, 699, 807; decorative, 751; in electric circuits, 732–733, **733;** interference and, 599; resistors in, 736; sulfur, 524

lightning, **664–665,** 675, **680**

light pipe(s), 582

linear polarization, **546,** 546–547

liquid(s), 318. *See also* fluid.

load(s), 732

lodestone, 785

Lodge, Oliver, 727

Lomonosov, Mikhail V., **355**

longitudinal wave(s), 455, 468, 481

loudspeaker(s), 777, **777**

M

machine(s), simple: efficiency of, 301; laboratory exercise on, 313–315; six types of, 298, **298;** use of, 299–301, 304

Maelzel, Johann N., 436

maglev train(s), 766

magnet(s), 766–767, **767.** *See also* electromagnet.

magnetic declination, 767

magnetic domain(s), 772

N

Fundamental Constants

Symbol	Quantity	Established Value	Value used for calculations in this book
c	speed of light in a vacuum	299 792 458 m/s	3.00×10^8 m/s
e^-	elementary charge	$1.602\ 177\ 33 \times 10^{-19}$ C	1.60×10^{-19} C
ε_0	(Greek *epsilon*) permittivity of a vacuum	$8.854\ 187\ 817 \times 10^{-12}$ C^2/(N•m^2)	8.85×10^{-12} C^2/(N•m^2)
G	constant of universal gravitation	$6.672\ 59 \times 10^{-11}$ N•m^2/kg^2	6.673×10^{-11} N•m^2/kg^2
g	free-fall acceleration at Earth's surface	$9.806\ 65$ m/s^2	9.81 m/s^2
h	Planck's constant	$6.626\ 0755 \times 10^{-34}$ J•s	6.63×10^{-34} J•s
k_B	Boltzmann's constant (R/N_A)	$1.380\ 658 \times 10^{-23}$ J/K	1.38×10^{-23} J/K
k_C	Coulomb constant	$8.987\ 551\ 788 \times 10^9$ N•m^2/C^2	8.99×10^9 N•m^2/C^2
R	molar (universal) gas constant	$8.314\ 510$ J/(mol•K)	8.31 J/(mol•K)
π	(Greek *pi*) ratio of the circumference to the diameter of a circle	$3.141\ 592\ 654$	calculator value

Useful Astronomical Data

Symbol	Quantity	Value used for calculations in this book
I_E	moment of inertia of Earth	8.03×10^{37} kg•m^2
M_E	mass of Earth	5.98×10^{24} kg
R_E	radius of Earth	6.37×10^6 m

Useful Atomic Data

Symbol	Quantity	Established Value	Value used for calculations in this book
m_e	mass of electron	$9.109\ 3897 \times 10^{-31}$ kg $5.485\ 799\ 03 \times 10^{-4}$ u $0.510\ 999\ 06$ MeV	9.109×10^{-31} kg 5.49×10^{-4} u 5.110×10^{-1} MeV
m_n	mass of neutron	$1.674\ 9286 \times 10^{-27}$ kg $1.008\ 664\ 904$ u $939.565\ 63$ MeV	1.675×10^{-27} kg $1.008\ 665$ u 9.396×10^2 MeV
m_p	mass of proton	$1.672\ 6231 \times 10^{-27}$ kg $1.007\ 276\ 470$ u $938.272\ 31$ MeV	1.673×10^{-27} kg $1.007\ 276$ u 9.383×10^2 MeV